CHILTON

GENERAL

MOTORS

SERVICE MANUAL
2004 Edition

THOMSON

DELMAR LEARNING

Australia • Canada • Mexico • Singapore • Spain • United Kingdom • United States

THOMSON

DELMAR LEARNING

Chilton
General Motors Service Manual
2004 Edition

Vice President, Technology and Trades SBU:
Alar Elken

Executive Director, Professional Business Unit:
Greg Clayton

Publisher, Professional Business Unit:
David Koontz

Channel Manager:
Beth A. Lutz

Marketing Specialist:
Brian McGrath

Production Director:
Mary Ellen Black

Production Manager:
Larry Main

Production Editor:
Elizabeth Hough

Editorial Assistant:
Kristen Shenfield

Editors:
Timothy A. Crain
Thomas A. Mellon
Richard J. Rivele
Christine L. Sheeky

NOTICE TO THE READER

Table of Contents

Model Index

EDITORIAL POLICY

Manufacturer and Model Coverage

This manual does not cover every make and model that is currently available on the market. Rather, the Chilton editorial staff makes judicious decisions as to which makes and models warrant coverage, based on which vehicles are serviced by most technicians. In general, this manual does not cover:

- Exotic vehicles (e.g. Rolls-Royce, Ferrari, Dodge Viper)
- Vehicle manufacturers with no current U.S. presence (e.g. Fiat and Peugeot)
- Vehicle models that have not sold enough units to be a factor in the repair market

Model Year Information

This manual is published toward the end of the year prior to the edition year. Every effort is made to gather current data from the Original Vehicle Manufacturers (OEMs) when they publish it. Different OEMs choose to release their new model information at different times of the year. Indeed, the same OEM can publish information early one season and late the next season. As a result, not all models are equally current when each edition of this manual is published.

Although information in this manual is based on industry sources and is as complete as possible at the time of publication, some vehicle manufacturers may make changes which cannot be included here. Information on very late models may not be available in some circumstances. While striving for total accuracy, the publisher cannot assume responsibility for any errors, changes, or omissions that may occur in the compilation of this data.

Safety Notice

Proper service and repair procedures are vital to the safe, reliable operation of all motor vehicles, as well as the personal safety of those performing the repairs. This manual outlines procedures for servicing and repairing vehicles using safe, effective methods. The procedures contain many NOTES, WARNINGS and CAUTIONS which should be followed along with standard safety procedures to reduce the possibility of personal injury or improper service which could damage the vehicle or compromise its safety.

Repair procedures, tools, parts, and technician skill and experience vary widely. It is not possible to anticipate all conceivable ways or conditions under which vehicles may be serviced, or to provide cautions for all possible hazards that may result. Standard and accepted safety precautions and equipment should be used when handling toxic or flammable fluids, and safety goggles or other protection should be used during cutting, grinding, chiseling, prying, or any other process that can cause material removal or projectiles.

Some procedures require the use of tools specially designed for a specific purpose. Before substituting another tool or procedure, you must be completely satisfied that neither your personal safety, nor the performance of the vehicle will be endangered.

LOCATING AND USING THE INFORMATION

Organization

To find where a particular model section is located, look in the Table of Contents. On the first page of each model section, main topics are listed with the page number on which they may be found. Following the main topics is an alphabetical listing of all of the procedures within the section and their page numbers.

Part Numbers

Part numbers listed in this book are not recommendations by the publisher for any product by brand name. They are references that can be used with interchanges manuals and aftermarket supplier catalogs to locate each brand supplier's discrete part number.

Special Tools

Special tools are recommended by the vehicle manufacturer to perform specific jobs. When necessary, special tools are referred to in the text by the part number of the tool manufacturer. These tools may be purchased, under the appropriate part number, from your local dealer or regional distributor, or an equivalent tool can be purchased locally from a tool supplier or parts outlet. Before substituting any tool for the one recommended, read the previous Safety Notice.

ACKNOWLEDGEMENT

Portions of materials contained herein have been reprinted with permission from General Motors Corporation, Service and Parts Operations under License Agreement #10757.

The publisher would like to express appreciation to General Motors Corporation for their assistance in producing this publication.

CHEVROLET, OLDSMOBILE AND GMC

1

Blazer • Xtreme • Trailblazer • Bravada • Envoy • Jimmy

SPECIFICATION CHARTS

ENGINE AND VEHICLE IDENTIFICATION

Engine							Model Year	
Code ①	Liters (cc)	Cu. In.	Cyl.	Fuel Sys.	Engine Type	Eng. Mfg.	Code ②	Year
S	4.2 (4200)	256	6	MFI	DOHC	CPC	Y	2000
W	4.3 (4293)	263	6	MFI	OHV	CPC	1	2001
X	4.3 (4293)	263	6	MFI	OHV	CPC	2	2002
P	5.3 (5326)	325	8	SFI	OHV	CPC	3	2003
							4	2004

CPC: Chevrolet/Pontiac/Canada

MFI: Multi-port Fuel Injection

① 8th position of VIN

② 10th position of VIN

42372-BLAZ-C01

GENERAL ENGINE SPECIFICATIONS

All measurements are given in inches.

Year	Model	Engine Displacement Liters (cc)	Engine Series (ID/VIN)	Fuel System	Net Horsepower @ rpm	Net Torque B rpm (ft. lbs.)	Bore x Stroke (in.)	Compression Ratio	Oil Pressure @ rpm
2000	Bradava	4.3 (4293)	W	MFI	①	②	4.00x3.48	9.2:1	18@2000
		4.3 (4293)	X	MFI	③	④	4.00x3.48	9.2:1	18@2000
	Envoy	4.3 (4293)	W	MFI	230@4600	285@2800	4.00x3.48	9.4:1	18@2000
		4.3 (4293)	X	MFI	③	④	4.00x3.48	9.2:1	18@2000
	Jimmy	4.3 (4293)	W	MFI	①	②	4.00x3.48	9.2:1	18@2000
		4.3 (4293)	X	MFI	③	④	4.00x3.48	9.2:1	18@2000
	Blazer	4.3 (4293)	W	MFI	190@4400	250@2800	4.00x3.48	9.4:1	18@2000
		4.3 (4293)	X	MFI	③	④	4.00x3.48	9.2:1	18@2000
2001	Bradava	4.3 (4293)	W	MFI	①	②	4.00x3.48	9.2:1	18@2000
	Envoy	4.3 (4293)	W	MFI	230@4600	285@2800	4.00x3.48	9.4:1	18@2000
	Jimmy	4.3 (4293)	W	MFI	①	②	4.00x3.48	9.2:1	18@2000
	Xtreme	4.3 (4293)	W	MFI	190@4400	250@2800	4.00x3.48	9.4:1	18@2000
	Blazer	4.3 (4293)	W	MFI	190@4400	250@2800	4.00x3.48	9.4:1	18@2000
2002	Bradava	4.2 (4200)	S	MFI	270@6000	275@3600	3.66x4.02	10.0:1	12@1200
	Envoy	4.2 (4200)	S	MFI	270@6000	275@3600	3.66x4.02	10.0:1	12@1200
	TrailBlazer	4.2 (4200)	S	MFI	270@6000	275@3600	3.66x4.02	10.0:1	12@1200
	Xtreme	4.3 (4293)	W	MFI	190@4400	250@2800	4.00x3.48	9.4:1	18@2000
	Blazer	4.3 (4293)	W	MFI	190@4400	250@2800	4.00x3.48	9.4:1	18@2000
2003-04	Bradava	4.2 (4200)	S	MFI	275@6000	275@3600	3.66x4.02	10.0:1	12@1200
	Envoy	4.2 (4200)	S	MFI	275@6000	275@3600	3.66x4.02	10.0:1	12@1200
		5.3 (5326)	P	SFI	290@5300	325@4000	3.78x3.62	9.45:1	NA
	TrailBlazer	4.2 (4200)	S	MFI	270@6000	275@3600	3.66x4.02	10.0:1	12@1200
		5.3 (5326)	P	SFI	290@5300	325@4000	3.78x3.62	9.45:1	NA
	Xtreme	4.3 (4293)	W	MFI	190@4400	250@2800	4.00x3.48	9.4:1	18@2000
	Blazer	4.3 (4293)	W	MFI	190@4400	250@2800	4.00x3.48	9.4:1	18@2000

MFI: Multi-port Fuel Injection

① 2WD: 245@2800
   4WD: 250@2800

② 2WD: 170B4400
   4WD: 180B4400

③ Below 15,000 GVWR: 180B3400
   Above 15,000 GVWR: 190B3400

④ 2WD: 180B4400
   4WD: 190B4400

42372-BLAZ-C02

GASOLINE ENGINE TUNE-UP SPECIFICATIONS

Year	Engine Displacement Liters (cc)	Engine ID/VIN	Spark Plugs Gap (in.)	Ignition Timing (deg.) MT	Ignition Timing (deg.) AT	Fuel Pump (psi)	Idle Speed (rpm) MT	Idle Speed (rpm) AT	Valve Clearance In.	Valve Clearance Ex.
2000	4.3 (4293)	W	0.060	①	①	58-64 ②	600	625	HYD	HYD
	4.3 (4293)	X	0.045	—	③	41-47	—	④	HYD	HYD
2001	4.3 (4293)	W	0.060	①	①	58-64 ②	600	625	HYD	HYD
2002	4.2 (4200)	S	0.050	—	③	NA	—	④	HYD	HYD
	4.3 (4293)	W	0.060	①	①	58-64 ②	600	625	HYD	HYD
2003-04	4.2 (4200)	S	0.050	—	③	NA	—	④	HYD	HYD
	4.3 (4293)	W	0.060	①	①	58-64 ②	600	625	HYD	HYD
	5.3 (5326)	P	0.060	①	①	55-62 ②	—	625	HYD	HYD

NOTE: The Vehicle Emission Control Information label often reflects specification changes made during production. The label figures must be used if they differ from those in this chart.

HYD: Hydraulic

① Distributorless ignition, cannot be adjusted

② With key ON and engine OFF

③ Distributorless ignition, cannot be adjusted

④ Idle speed is maintained by the PCM

42372-BLAZ-C03

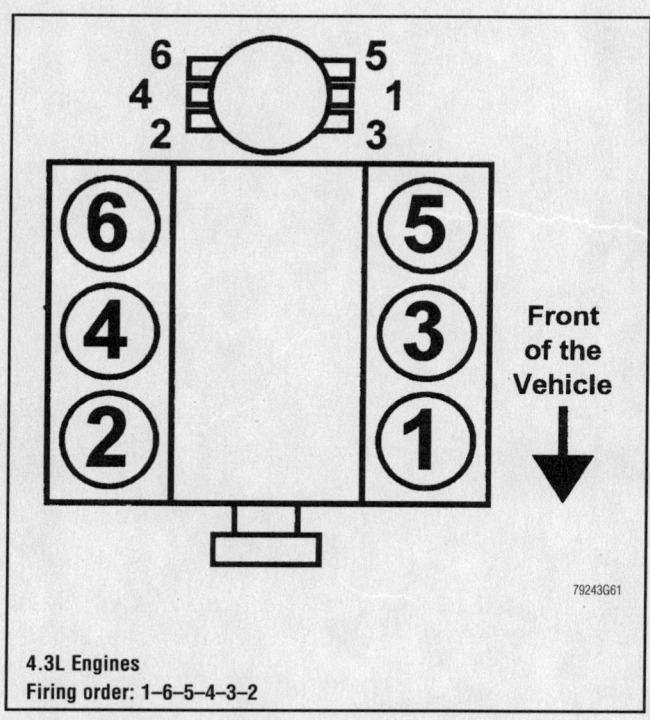

Front of the Vehicle

4.3L Engines
Firing order: 1–6–5–4–3–2

79243G61

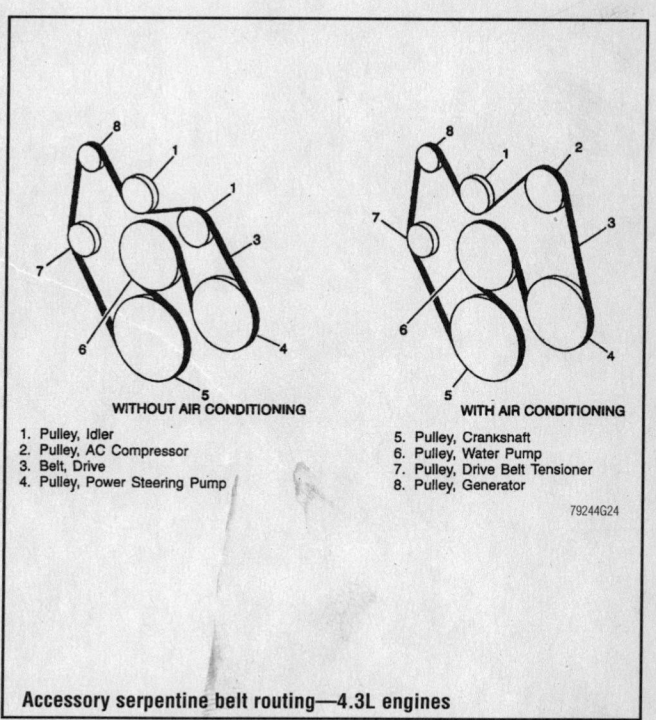

WITHOUT AIR CONDITIONING

1. Pulley, Idler
2. Pulley, AC Compressor
3. Belt, Drive
4. Pulley, Power Steering Pump

WITH AIR CONDITIONING

5. Pulley, Crankshaft
6. Pulley, Water Pump
7. Pulley, Drive Belt Tensioner
8. Pulley, Generator

79244G24

Accessory serpentine belt routing—4.3L engines

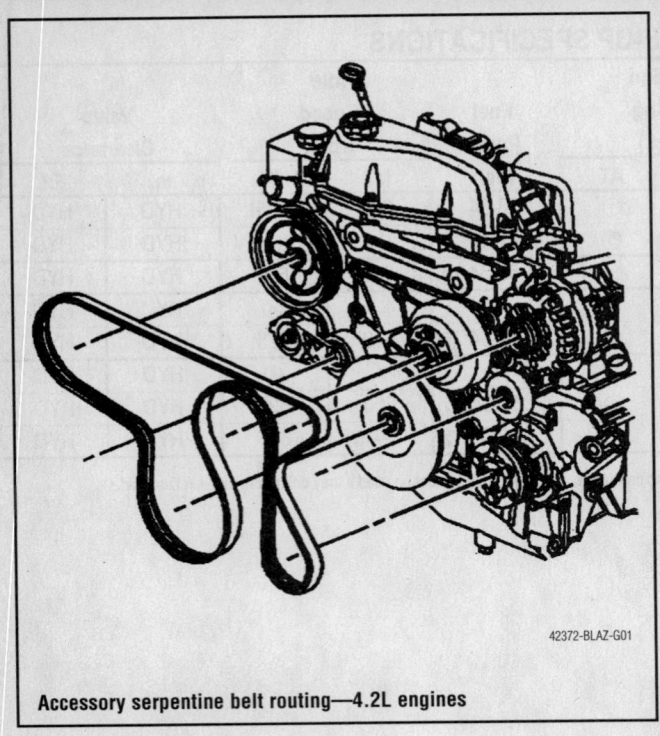

42372-BLAZ-G01

Accessory serpentine belt routing—4.2L engines

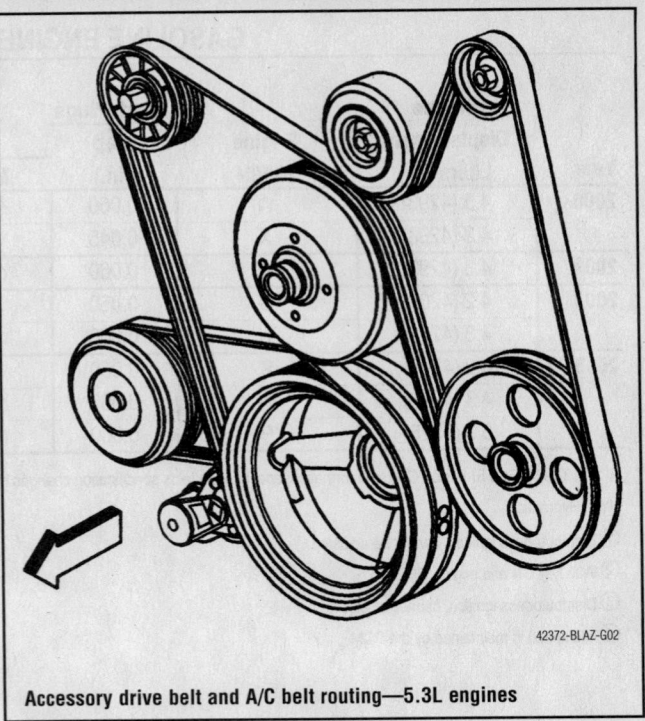

42372-BLAZ-G02

Accessory drive belt and A/C belt routing—5.3L engines

CAPACITIES

Year	Model	Engine Displacement Liters (cc)	Engine ID/VIN	Engine Oil with Filter (qts.)	Transmission (pts.) 5-Spd	Transmission (pts.) Auto.	Transfer Case (pts.)	Drive Axle Front (pts.)	Drive Axle Rear (pts.)	Fuel Tank (gal.)	Cooling System (qts.)
2000	Bravada	4.3 (4293)	W	5.0	—	11.0	2.6	2.6	3.9	18.6	11.9
		4.3 (4293)	X	5.0	—	11.0	2.6	2.6	3.9	18.6	11.9
	Envoy	4.3 (4293)	W	5.0	—	11.0	2.6	2.6	3.9	20.0	11.9
		4.3 (4293)	X	5.0	—	11.0	2.6	2.6	3.9	20.0	11.9
	Jimmy	4.3 (4293)	W	5.0	—	11.0	2.6	2.6	3.9	20.0	11.9
		4.3 (4293)	X	5.0	—	11.0	2.6	2.6	3.9	20.0	11.9
	Blazer	4.3 (4293)	W	5.0	4.4	11.0	2.6	2.6	3.9	20.0	11.9
		4.3 (4293)	X	5.0	—	11.0	2.6	2.6	3.9	20.0	11.9
2001	Bravada	4.3 (4293)	X	5.0	—	11.0	2.6	2.6	3.9	18.6	11.9
	Envoy	4.3 (4293)	X	5.0	—	11.0	2.6	2.6	3.9	20.0	11.9
	Jimmy	4.3 (4293)	W	5.0	—	11.0	2.6	2.6	3.9	20.0	11.9
	Xtreme	4.3 (4293)	W	5.0	4.4	11.0	2.6	2.6	3.9	20.0	11.9
	Blazer	4.3 (4293)	W	5.0	4.4	11.0	2.6	2.6	3.9	20.0	11.9
2002	Bravada	4.2 (4200)	S	7.0	—	10.0	2.6	2.6	4.0	18.6	13.9
	Envoy	4.2 (4200)	S	7.0	—	10.0	2.6	2.6	4.0	18.6	13.9
	TrailBlazer	4.2 (4200)	S	7.0	—	10.0	2.6	2.6	4.0	18.6	13.9
	Xtreme	4.3 (4293)	W	5.0	4.4	11.0	2.6	2.6	3.9	20.0	11.9
	Blazer	4.3 (4293)	W	5.0	4.4	11.0	2.6	2.6	3.9	20.0	11.9
2003-04	Bravada	4.2 (4200)	S	7.0	—	10.0	2.6	2.6	4.0	18.6	13.9
	Envoy	4.2 (4200)	S	7.0	—	10.0	2.6	2.6	4.0	①	13.9
		5.3 (5326)	P	6.0	—	10.0	4.0	2.6	4.0	①	14.9
	TrailBlazer	4.2 (4200)	S	7.0	—	10.0	2.6	2.6	4.0	②	13.9
		5.3 (5326)	P	6.0	—	10.0	4.0	2.6	4.0	②	14.9
	Xtreme	4.3 (4293)	W	5.0	4.4	11.0	2.6	2.6	3.9	20.0	11.9
	Blazer	4.3 (4293)	W	5.0	4.4	11.0	2.6	2.6	3.9	20.0	11.9

NOTE: All capacities are approximate. Add fluid gradually and check to be sure a proper fluid level is obtained.

① Envoy: 19.0 gal.
  Envoy XL: 25.5 gal.

② Trailblazer: 22.0 gal.

42372-BLAZ-C04

CRANKSHAFT AND CONNECTING ROD SPECIFICATIONS
All measurements are given in inches.

| Year | Engine Displacement Liters (cc) | Engine ID/VIN | Crankshaft | | | | Connecting Rod | | |
			Main Brg. Journal Dia.	Main Brg. Oil Clearance	Shaft End-play	Thrust on No.	Journal Diameter	Oil Clearance	Side Clearance
2000	4.3 (4293)	W	①	②	0.0020-0.0070	4	2.2487-2.2497	0.0013-0.0035	0.0060-0.0140
	4.3 (4293)	X	③	②	0.0020-0.0060	4	2.2487-2.2497	0.0013-0.0035	0.0060-0.0140
2001	4.3 (4293)	W	①	②	0.0020-0.0070	4	2.2487-2.2497	0.0013-0.0035	0.0060-0.0140
2002	4.2 (4200)	S	2.7567-2.7574	0.0004-0.0025	0.0044-0.0153	4	2.2337-2.2342	0.0008-0.0025	0.0019-0.0137
	4.3 (4293)	W	①	②	0.0020-0.0070	4	2.2487-2.2497	0.0013-0.0035	0.0060-0.0140
2003-04	4.2 (4200)	S	2.7567-2.7574	0.0004-0.0025	0.0044-0.0153	4	2.2337-2.2342	0.0008-0.0025	0.0019-0.0137
	4.3 (4293)	W	①	②	0.0020-0.0070	4	2.2487-2.2497	0.0013-0.0035	0.0060-0.0140
	5.3 (5326)	P	2.558	0.0008-0.0021	0.0015-0.0078	4	2.0991-2.0999	0.0009-0.0003	0.0043-0.0200

① No. 1: 2.4488-2.4495
   Nos. 2, 3: 2.4485-2.4494
   No. 4: 2.4480-2.4489

② No. 1: 0.0008-0.0020
   Nos. 2, 3: 0.0011-0.0023
   No. 4: 0.0017-0.0032

③ No. 1: 2.4484-2.4493
   Nos. 2, 3: 2.4481-2.4490
   No. 4: 2.4479-2.4488

42372-BLAZ-C05

VALVE SPECIFICATIONS

Year	Engine Displacement Liters (cc)	Engine ID/VIN	Seat Angle (deg.)	Face Angle (deg.)	Spring Test Pressure (lbs. @ in.)	Spring Installed Height (in.)	Stem-to-Guide Clearance (in.) Intake	Stem-to-Guide Clearance (in.) Exhaust	Stem Diameter (in.) Intake	Stem Diameter (in.) Exhaust
2000	4.3 (4293)	W	46	45	187-203@1.27	1.69-1.71	0.0010	0.0020	NA	NA
	4.3 (4293)	X	46	45	187-203@1.27	1.69-1.71	0.0010	0.0020	NA	NA
2001	4.3 (4293)	W	46	45	187-203@1.27	1.69-1.71	0.0010	0.0020	NA	NA
2002	4.2 (4200)	S	NA	NA	130-142@1.26	NA	0.0011-0.0025	0.0015-0.0030	NA	NA
	4.3 (4293)	W	46	45	187-203@1.27	1.69-1.71	0.0010	0.0020	NA	NA
2003-04	4.2 (4200)	S	NA	NA	130-142@1.26	NA	0.0011-0.0025	0.0015-0.0030	NA	NA
	4.3 (4293)	W	46	45	187-203@1.27	1.69-1.71	0.0010	0.0020	NA	NA
	5.3 (5326)	P	46	45	220@1.32	1.80	0.0010-0.0026	0.0010-0.0026	0.313-0.314	0.313-0.314

NA: Not Available

42372-BLAZ-C06

PISTON AND RING SPECIFICATIONS
All measurements are given in inches.

Year	Engine Displacement Liters (cc)	Engine ID/VIN	Piston Clearance	Ring Gap Top Compression	Ring Gap Bottom Compression	Ring Gap Oil Control	Ring Side Clearance Top Compression	Ring Side Clearance Bottom Compression	Ring Side Clearance Oil Control
2000	4.3 (4293)	W	0.0007-0.0017	0.010-0.030	0.018-0.026	0.065 Max.	0.0042 Max.	0.0042 Max.	0.0020-0.0070
	4.3 (4293)	X	0.0007-0.0017	0.010-0.030	0.018-0.026	0.065 Max.	0.0042 Max.	0.0042 Max.	0.0020-0.0070
2001	4.3 (4293)	W	0.0007-0.0017	0.010-0.030	0.018-0.026	0.065 Max.	0.0042 Max.	0.0042 Max.	0.0020-0.0070
2002	4.2 (4200)	S	0.0004-0.0017	0.0079-0.0157	0.0118-0.0197	0.0098-0.0299	0.0017-0.0037	0.0017-0.0037	0.0023-0.0085
	4.3 (4293)	W	0.0007-0.0017	0.010-0.030	0.018-0.026	0.065 Max.	0.0042 Max.	0.0042 Max.	0.0020-0.0070
2003-04	4.2 (4200)	S	0.0004-0.0017	0.0079-0.0157	0.0118-0.0197	0.0098-0.0299	0.0017-0.0037	0.0017-0.0037	0.0023-0.0085
	4.3 (4293)	W	0.0007-0.0017	0.010-0.030	0.018-0.026	0.065 Max.	0.0042 Max.	0.0042 Max.	0.0020-0.0070
	5.3 (5326)	P	-0.0014-0.0006	0.009-0.017	0.017-0.027	0.007-0.029	0.0015-0.0033	0.0015-0.0031	0.0005-0.0078

42372-BLAZ-C07

TORQUE SPECIFICATIONS
All readings in ft. lbs.

Year	Engine Displacement Liters (cc)	Engine ID/VIN	Cylinder Head Bolts	Main Bearing Bolts	Rod Bearing Bolts	Crankshaft Damper Bolts	Flywheel Bolts	Manifold Intake *	Manifold Exhaust	Spark Plugs	Lug Nut
2000	4.3 (4293)	W	①	77	②	74	74	③	④	11	90
	4.3 (4293)	X	①	77	②	74	74	③	④	11	90
2001	4.3 (4293)	X	①	77	②	74	74	③	④	11	90
2002	4.2 (4200)	S	⑤	⑥	⑦	⑧	⑨	12	⑩	13	92
	4.3 (4293)	X	①	77	②	74	74	③	④	11	90
2003-04	4.2 (4200)	S	⑤	⑥	⑦	⑧	⑨	12	⑩	13	103
	4.3 (4293)	X	①	77	②	74	74	③	④	11	100
	5.3 (5326)	P	⑪	⑫	⑬	⑭	⑮	⑯	⑰	11	103

* NOTE: Applies to Lower Manifold only.

① 1st pass: 22 ft. lbs.
  2nd pass:
  Short bolt: Plus 55 degrees
  Medium bolt: Plus 65 degrees
  Long bolt: Plus 75 degrees

② 20 ft. lbs. plus 70 degrees

③ Lower intake manifold:
  1st pass: 27 inch lbs.
  2nd pass: 106 inch lbs.
  Final pass: 11 ft. lbs.
  Upper manifold bolts:
  1st pass: 44 inch lbs.
  2nd pass: 88 inch lbs.

④ Tighten bolts to 12 ft. lbs.
  Retorque to 22 ft. lbs.

⑤ Cylinder head bolts (14)
  1st pass: 22 ft. lbs.
  2nd pass: Plus 155 degrees
  2 short end bolts: 15 ft. lbs.
  1 long end bolt: 13 ft. lbs.

⑥ 18 ft. lbs., plus 155 depress

⑦ 18 ft. lbs., plus 110 degrees

⑧ 111 ft. lbs., plus 180 degrees

⑨ 18 ft. lbs., plus 50 degrees

⑩ 1st pass: 18 ft. lbs.
  2nd pass: 18 ft. lbs.
  3rd pass: 18 ft. lbs.

⑪ 1st pass: 22 ft. lbs.
  2nd pass: Plus 90 degrees
  Final pass:
  Except med. bolts at front and rear: Plus 90 degrees
  Medium bolts at front and rear of head: Plus 50 degrees

⑫ Inner bolts:
  1st pass: 15 ft. lbs.
  Final pass: Plus 80 degrees
  Outer bolts:
  1st pass: 15 ft. lbs.
  Final pass: Plus 51 degrees

⑬ 15 ft. lbs. plus 75 degrees

⑭ Installation pass: 240 ft. lbs. (discard bolt)
  First pass: 37 ft. lbs. (new bolt)
  Final pass: Plus 140 degrees

⑮ 1st pass: 15 ft. lbs.
  2nd pass: 37 ft. lbs.

⑯ 1st pass: 44 inch lbs.
  2nd pass: 89 inch lbs.

⑰ 1st pass: 11 ft. lbs.
  2nd pass: 18 ft. lbs.

42372-BLAZ-C08

WHEEL ALIGNMENT

Year	Model		Caster Range (+/-Deg.)	Caster Preferred Setting (Deg.)	Camber Range (+/-Deg.)	Camber Preferred Setting (Deg.)	Toe-in (in.)	Steering Axis Inclination (Deg.)
2000	Blazer/Jimmy/Envoy	2WD	1.00	+3.00	1.00	0	0.10+/-0.05	—
		4WD	1.00	+2.00	1.00	0	0.10+/-0.05	—
	Bravada	2WD	1.00	+3.00	1.00	0	0.10+/-0.11	—
		4WD	1.00	+2.00	1.00	0	0.10+/-0.10	—
2001	Blazer/Jimmy/	2WD	1.00	①	1.00	0	0.10+/-0.05	—
	Envoy/Bravada	4WD	1.00	①	1.00	0	0.10+/-0.05	—
	Xtreme	2WD	1.00	②	1.00	0	0.10+/-0.11	—
		4WD	1.00	②	1.00	0	0.10+/-0.10	—
2002	Bravada/Envoy/	2WD	1.00	+3.50	1.00	-0.5	-0.5+/-0.5	—
	TrailBlazer	4WD	1.00	+3.50	1.00	-0.5	-0.5+/-0.5	—
	Blazer	2WD	1.00	①	1.00	0	0.10+/-0.05	—
		4WD	1.00	①	1.00	0	0.10+/-0.05	—
	Xtreme	2WD	1.00	②	1.00	0	0.10+/-0.11	—
		4WD	1.00	②	1.00	0	0.10+/-0.10	—
2003-04	Bravada/Envoy/	2WD	1.00	+3.50	1.00	-0.5	-0.5+/-0.5	—
	TrailBlazer	4WD	1.00	+3.50	1.00	-0.5	-0.5+/-0.5	—
	Blazer	2WD	1.00	①	1.00	0	0.10+/-0.05	—
		4WD	1.00	①	1.00	0	0.10+/-0.05	—
	Xtreme	2WD	1.00	②	1.00	0	0.10+/-0.11	—
		4WD	1.00	②	1.00	0	0.10+/-0.10	—

① Left Side: +2.80
   Right Side: +3.30

② Left Side: +4.70
   Right Side: +5.20

42372-BLAZ-C09

TIRE, WHEEL AND BALL JOINT SPECIFICATIONS

| Year | Model | OEM Tires | | Tire Pressures (psi) | | Wheel Size | Ball Joint Inspection |
		Standard	Optional	Front	Rear		
2000	Blazer/Jimmy/Envoy	P205/70R15	P235/70R15 P235/75R15	36	36	6-JJ	L ①
	Bravada	P235/70R15	None	33	35	7-JJ	②
2001	Blazer/Jimmy/ Envoy	P205/70R15	P235/70R15 P235/75R15	36	36	6-JJ	L ①
	Bravada	P235/70R15	None	33	35	7-JJ	②
2002	Blazer/Envoy/TrailBlazer	P205/70R15	P235/70R15 P235/75R15	36	36	6-JJ	L ①
	Bravada	P235/70R15	None	33	35	7-JJ	②
2003-04	Blazer/Envoy/TrailBlazer	P205/70R15	P235/70R15 P235/75R15	36	36	6-JJ	L ①
	Bravada	P235/70R15	None	33	35	7-JJ	②

OEM: Original Equipment Manufacturer

PSI: Pounds Per Square Inch

STD: Standard

OPT: Optional

L: Lower

U: Upper

① Do not lift truck. Inspect the boss into which the grease fitting is threaded. Replace if the boss is flush or receded below the surface of the ball joint

② Replace if any measurable movement is found.

42372-BLAZ-C10

BRAKE SPECIFICATIONS
All measurements in inches unless noted

Year	Model		Brake Disc Original Thickness	Brake Disc Minimum Thickness	Brake Disc Maximum Runout	Original Inside Diameter	Max. Wear Limit	Maximum Machine Diameter	Minimum Lining Thickness	Bracket Bolts (ft. lbs.)	Mounting Bolts (ft. lbs.)
2000	Bravada	F	1.030	0.965	0.003	—	—	—	0.030	52	①
		R	0.787	0.728	0.004	9.50	9.59	9.56	0.030	NA	—
	Envoy	F	1.030	0.965	0.003	—	—	—	0.030	52	①
		R	0.787	0.728	0.004	9.50	9.59	9.56	0.030	NA	—
	Jimmy	F	1.030	0.965	0.003	—	—	—	0.030	52	①
		R	0.787	0.728	0.004	9.50	9.59	9.56	0.030	NA	—
	Blazer	F	1.030	0.965	0.003	—	—	—	0.030	52	①
		R	0.787	0.728	0.004	9.50	9.59	9.56	0.030	NA	—
2001	Bravada	F	1.030	0.965	0.003	—	—	—	0.030	52	①
		R	0.787	0.728	0.004	9.50	9.59	9.56	0.030	NA	—
	Envoy	F	1.030	0.965	0.003	—	—	—	0.030	52	①
		R	0.787	0.728	0.004	9.50	9.59	9.56	0.030	NA	—
	Jimmy	F	1.030	0.965	0.003	—	—	—	0.030	52	①
		R	0.787	0.728	0.004	9.50	9.59	9.56	0.030	NA	—
	Blazer	F	1.030	0.965	0.003	—	—	—	0.030	52	①
		R	0.787	0.728	0.004	9.50	9.59	9.56	0.030	NA	—
	Xtreme	F	1.030	0.965	0.003	—	—	—	0.030	52	①
		R	0.787	0.728	0.004	9.50	9.59	9.56	0.030	NA	—
2002	Bravada	F	1.030	0.965	0.003	—	—	—	0.030	52	①
		R	0.787	0.728	0.004	9.50	9.59	9.56	0.030	NA	—
	Envoy	F	1.030	0.965	0.003	—	—	—	0.030	52	①
		R	0.787	0.728	0.004	9.50	9.59	9.56	0.030	NA	—
	TrailBlazer	F	1.030	0.965	0.003	—	—	—	0.030	52	①
		R	0.787	0.728	0.004	9.50	9.59	9.56	0.030	NA	—
	Blazer	F	1.030	0.965	0.003	—	—	—	0.030	52	①
		R	0.787	0.728	0.004	9.50	9.59	9.56	0.030	NA	—
	Xtreme	F	1.030	0.965	0.003	—	—	—	0.030	52	①
		R	0.787	0.728	0.004	9.50	9.59	9.56	0.030	NA	—
2003-04	Bravada	F	1.030	0.965	0.003	—	—	—	0.030	52	①
		R	0.787	0.728	0.004	9.50	9.59	9.56	0.030	NA	—
	Envoy	F	1.030	0.965	0.003	—	—	—	0.030	52	①
		R	0.787	0.728	0.004	9.50	9.59	9.56	0.030	NA	—
	TrailBlazer	F	1.030	0.965	0.003	—	—	—	0.030	52	①
		R	0.787	0.728	0.004	9.50	9.59	9.56	0.030	NA	—
	Blazer	F	1.030	0.965	0.003	—	—	—	0.030	52	①
		R	0.787	0.728	0.004	9.50	9.59	9.56	0.030	NA	—
	Xtreme	F	1.030	0.965	0.003	—	—	—	0.030	52	①
		R	0.787	0.728	0.004	9.50	9.59	9.56	0.030	NA	—

NA: Not Available

① 2WD: 38 ft. lbs.
4WD: 77 ft. lbs.

42372-BLAZ-C11

SCHEDULED MAINTENANCE INTERVALS
GM—BLAZER, JIMMY, BRAVADA, ENVOY, XTREME & TRAILBLAZER

TO BE SERVICED	TYPE OF SERVICE	7.5	15	22.5	30	37.5	45	52.5	60	67.5	75	82.5	90	97.5	105	112.5	120
Accessory drive belt	S/I								✓								✓
Air cleaner filter	R				✓				✓				✓				✓
Automatic transmission fluid	R	Every 50,000 miles															
Brake system ①	S/I	✓	✓	✓	✓	✓	✓	✓	✓	✓	✓	✓	✓	✓	✓	✓	✓
Chassis & suspension grease points	L	✓	✓	✓	✓	✓	✓	✓	✓	✓	✓	✓	✓	✓	✓	✓	✓
CV-joint boots & axle seals	S/I	✓	✓	✓	✓	✓	✓	✓	✓	✓	✓	✓		✓	✓	✓	✓
Engine coolant system ②	S/I	Every 150,000 miles															
Engine oil & filter	R	✓	✓	✓	✓	✓	✓	✓	✓	✓	✓	✓	✓	✓	✓	✓	✓
Front wheel bearings	S/I & L				✓				✓				✓				✓
Fuel filter	R				✓				✓				✓				✓
Fuel tank, cap & lines	S/I								✓								✓
PCV valve	S/I	Every 100,000 miles															
Rear/front axle fluid level	S/I	✓	✓	✓	✓	✓	✓	✓	✓	✓	✓	✓	✓	✓	✓	✓	✓
Rotate tires	S/I	✓	✓	✓	✓	✓	✓	✓	✓	✓	✓	✓	✓	✓	✓	✓	✓
Spark plug wires	S/I	Every 100,000 miles															
Spark plugs	R	Every 100,000 miles															

R: Replace   S/I: Inspect and service, if necessary   L: Lubricate

① This should be performed when the tires are removed for rotation.

② Drain, flush and refill the cooling system, inspect the system hoses, and clean the radiator and condenser.

③ 2-wheel drive models only.

FREQUENT OPERATION MAINTENANCE (SEVERE SERVICE)

If a vehicle is operated under any of the following conditions it is considered severe service:

- Towing a trailer or using a camper or car-top carrier.

- Repeated short trips of less than 5 miles in temperatures below freezing, or trips of less than 10 miles in any temperature.

- Extensive idling or low-speed driving for long distances as in heavy commercial use, such as delivery, taxi or police cars.

- Operating on rough, muddy or salt-covered roads.

- Operating on unpaved or dusty roads.

- Driving in extremely hot (over 90°) conditions.

Engine oil & filter: replace every 3000 miles or 3 months, whichever occurs first.

Chassis and suspension grease points: lubricate every 3000 miles.

Rear/front axle fluid level: inspect every 3000 miles.

Rotate the tires ever 6000 miles.

Brake system components: inspect ever 6000 miles.

Front wheel bearings (2-wheel drive only): clean, inspect and repack every 15,000 miles.

Air cleaner filter: inspect every 15,000 miles.

Automatic transmission fluid & filter: replace every 15,000 miles.

42372-BLAZ-C12

PRECAUTIONS

Before servicing any vehicle, please be sure to read all of the following precautions, which deal with personal safety, prevention of component damage, and important points to take into consideration when servicing a motor vehicle:

• Never open, service or drain the radiator or cooling system when the engine is hot; serious burns can occur from the steam and hot coolant.

• Observe all applicable safety precautions when working around fuel. Whenever servicing the fuel system, always work in a well-ventilated area. Do not allow fuel spray or vapors to come in contact with a spark, open flame, or excessive heat (a hot drop light, for example). Keep a dry chemical fire extinguisher near the work area. Always keep fuel in a container specifically designed for fuel storage; also, always properly seal fuel containers to avoid the possibility of fire or explosion. Refer to the additional fuel system precautions later in this section.

• Fuel injection systems often remain pressurized, even after the engine has been turned **OFF**. The fuel system pressure must be relieved before disconnecting any fuel lines. Failure to do so may result in fire and/or personal injury.

• Brake fluid often contains polyglycol ethers and polyglycols. Avoid contact with the eyes and wash your hands thoroughly after handling brake fluid. If you do get brake fluid in your eyes, flush your eyes with clean, running water for 15 minutes. If eye irritation persists, or if you have taken brake fluid internally, IMMEDIATELY seek medical assistance.

• The EPA warns that prolonged contact with used engine oil may cause a number of skin disorders, including cancer! You should make every effort to minimize your exposure to used engine oil. Protective gloves should be worn when changing oil. Wash your hands and any other exposed skin areas as soon as possible after exposure to used engine oil. Soap and water, or waterless hand cleaner should be used.

• All new vehicles are now equipped with an air bag system. The system must be disabled before performing service on or around system components, steering column, instrument panel components, wiring and sensors. Failure to follow safety and disabling procedures could result in accidental air bag deployment, possible personal injury and unnecessary system repairs.

• Always wear safety goggles when working with, or around, the air bag system. When carrying a non-deployed air bag, be sure the bag and trim cover are pointed away from your body. When placing a non-deployed air bag on a work surface, always face the bag and trim cover upward, away from the surface. This will reduce the motion of the module if it is accidentally deployed. Refer to the additional air bag system precautions later in this section.

• Clean, high quality brake fluid from a sealed container is essential to the safe and proper operation of the brake system. You should always buy the correct type of brake fluid for your vehicle. If the brake fluid becomes contaminated, completely flush the system with new fluid. Never reuse any brake fluid. Any brake fluid that is removed from the system should be discarded. Also, do not allow any brake fluid to come in contact with a painted surface; it will damage the paint.

• Never operate the engine without the proper amount and type of engine oil; doing so WILL result in severe engine damage.

• Timing belt maintenance is extremely important! Many models utilize an interference-type, non-freewheeling engine. If the timing belt breaks, the valves in the cylinder head may strike the pistons, causing potentially serious (also time-consuming and expensive) engine damage.Refer to the maintenance interval charts in the front of this section for the recommended replacement interval for the timing belt, and to the timing belt procedure in this section for belt replacement and inspection.

• Disconnecting the negative battery cable on some vehicles may interfere with the functions of the on-board computer system(s) and may require the computer to undergo a relearning process once the negative battery cable is reconnected.

• When servicing drum brakes, only disassemble and assemble one side at a time, leaving the remaining side intact for reference.

ENGINE REPAIR

Distributor

REMOVAL

1. Before servicing the vehicle, refer to the precautions in the beginning of this section.
2. Remove or disconnect the following:
   • Negative battery cable
   • Spark plug wires and the coil leads from the distributor
   • Electrical connector from the distributor
   • Distributor cap fasteners and the cap
3. Using a marker, matchmark the rotor-to-housing and housing-to-intake manifold positions so that they can be matched during installation.

   • Distributor hold-down bolt
   • Distributor from the engine
4. As the distributor is being removed from the engine the rotor will move in a counterclockwise direction about 42°. This will appear as slightly more than one clock position.
5. Place a second mark on the distributor to mark the position of the rotor segment. This will help to ensure the correct rotor alignment when installing the distributor.

INSTALLATION

Engine Not Disturbed

1. If installing a new distributor, place two marks on the new distributor housing in the same position as the marks on the old distributor housing.

2. Align the rotor with the second mark made on the distributor.
3. Install the distributor in the engine making sure that the mounting hole in the distributor hold-down base is aligned over the mounting hole in the intake manifold.
4. As you are installing the distributor, watch the rotor move in a clockwise direction about 42°.
5. Once the distributor is fully seated, the rotor should be aligned with the first mark made on the distributor housing. If the rotor is not aligned with the first mark made on the housing, the distributor and camshaft teeth have meshed one or more teeth out of alignment. If this is the case, remove the distributor and reinstall it so that all the marks are aligned.
6. Install or connect the following:

- Hold-down bolt and tighten the bolt to 18 ft. lbs. (25 Nm)
- Distributor cap and engage the electrical connector to the distributor
- Spark plug wires and coil leads
- Negative battery cable

Engine Disturbed

1. Remove the No. 1 cylinder spark plug. Turn the engine using a socket wrench on the large bolt on the front of the crankshaft pulley. Place a finger near the No. 1 spark plug hole and turn the crankshaft until the piston reaches Top Dead Center (TDC). As the engine approaches TDC, you will feel air being expelled by the No. 1 cylinder. If the position is not being met, turn the engine another full turn (360 degree). Once the engine position is correct, install the spark plug.

2. Align the cast arrow in the distributor housing, the driven gear roll pin and the pre-drilled indent hole in the distributor driven gear. If the driven gear is installed correctly, the dimple will be approximately 180° opposite the rotor segment when it is installed in the distributor.

➡ **Installing the distributor 180° out of alignment, or locating the rotor in the wrong holes, may cause a no start condition or can cause premature engine damage and wear.**

3. Make sure the rotor is pointing to the cap hold-down mount nearest the flat side of the housing.

4. Using a long screwdriver, align the oil pump driveshaft in the engine in the mating drive tab in the distributor.

5. Install the distributor in the engine. Make sure the spark plug towers are perpendicular to the centerline of the engine.

6. When the distributor is fully seated, the rotor segment should be aligned with the pointer cast in the distributor base. The pointer will have a "6" cast into it indicating a 6 cylinder engine. If the rotor segment is not within a few degrees of the pointer, the distributor gear may be off a tooth or more. If this is the case repeat the process until the rotor aligns with the pointer.

7. Install the cap and fasten the mounting screws.

8. Tighten the distributor mounting bolt to 18 ft. lbs. (25 Nm).

9. Engage the electrical connections and the spark plug wires.

Alternator

REMOVAL

4.2L Engine

1. Before servicing the vehicle, refer to the precautions in the beginning of this section.

2. Remove or disconnect the following:
- Negative battery cable
- Accessory belt
- Positive battery cable nut from the generator
- A/C line mounting bracket bolt at the engine lift hook
- Right engine lift hook bolts
- Engine lift hook
- Mounting bolts
- Alternator

4.3L Engine

1. Before servicing the vehicle, refer to the precautions in the beginning of this section.

2. Remove or disconnect the following:
- Negative battery cable
- Air inlet duct, if necessary
- Accessory belt
- Heater hose brace
- Wires
- Mounting bolts
- Alternator

5.3L Engine

1. Before servicing the vehicle, refer to the precautions in the beginning of this section.

2. Remove or disconnect the following:
- Negative battery cable
- Accessory belt
- Electrical connector
- Terminal stud nut, after sliding boot down
- Alternator cable
- Mounting bolts
- Alternator

INSTALLATION

4.2L Engine

1. Install or connect the following:
- Alternator and loosely install the mounting blots
- Tighten the alternator mounting bolts to 37 ft. lbs. (50 Nm)
- Positive battery cable and secure with the nut; tighten the nut to 80 inch lbs. (9 Nm)
- Engine lift hook and bolts; tighten the bolts to 37 ft. lbs. (50 Nm)
- A/C line bracket to the lift hook, then tighten the retaining bolt to 89 inch lbs. (10 Nm)
- Accessory belt
- Negative battery cable

4.3L Engine

1. Install or connect the following:
- Alternator and loosely install the mounting bolts
- Tighten the rear bolt to 37 ft. lbs. (50 Nm) and the front bolt to 18 ft. lbs. (25 Nm)
- Tighten the brace-to-alternator and brace-to-intake retainers to 18 ft. lbs. (25 Nm). Tighten the brace-to-engine stud nut to 37 ft. lbs. (50 Nm).
- Wires and the battery feed wire nut
- Heater hose bracket
- Accessory belt
- Negative battery cable

5.3L Engine

1. Install or connect the following:
- Alternator and loosely install the mounting bolts
- Tighten the bolts to 37 ft. lbs. (50 Nm)
- Alternator cable
- Terminal stud nut and tighten to 80 inch lbs. (9 Nm)
- Boot back over terminal stud
- Electrical connector
- Accessory belt
- Negative battery cable

Ignition Timing

ADJUSTMENT

The ignition timing is preset and cannot be adjusted.

Engine Assembly

REMOVAL & INSTALLATION

4.2L Engine

1. Before servicing the vehicle, refer to the precautions in the beginning of this section.

2. Drain the engine cooling system

➡ **Keep the oil drain plug removed during the engine removal and installation.**

3. Drain the engine oil. Install a suitable plug into the oil pan to prevent oil leakage during the remainder of the procedure.

4. Using the proper equipment, discharge and recover the refrigerant from the A/C system, if equipped.

5. Remove or disconnect the following:
- Hood
- Negative battery cable
- Fuel system pressure
- Air cleaner assembly
- Throttle body
- Manifold Absolute Pressure (MAP) sensor
- Windshield washer solvent container
- Air intake baffle
- Grille
- Headlight housing
- Radiator support brace
- Hood
- A/C lines from the condenser
- Transmission cooler lines from the engine, not the radiator

6. Remove the cooling fan and shroud, tilting the radiator forward, and the cooling fan and shroud rearward for clearance.
- Accessory belt
- Power steering pump bolts; position the pump aside
- Heater hoses from the heater core
- Transmission filler tube bracket nut from the Air Injector Reactor (AIR) adapter
- AIR adapter

7. Install a suitable lift hook to the AIR adapter

8. Remove or disconnect the following:
- Oxygen (O₂) sensor connector
- A/C line from the accumulator
- Front axle actuator electrical connector
- Camshaft phaser actuator valve electrical connector
- Transmission cooler lines from the clips on the right side of the engine block
- Ignition coil harness connectors
- Harness retainer from the clips
- Power brake hose from the booster
- Powertrain Control Module (PCM)
- Fuel lines from the fuel pressure regulator. Cap the lines to avoid excessive fuel leakage.
- All harnesses from the engine harness bracket
- Engine harness bracket bolt and bracket
- Starter electrical connections
- A/C pressure sensor and clutch electrical connector

- Alternator electrical connector and battery lead
- Knock Sensor (KS), Crankshaft Position (CKP) and Camshaft Position (CMP) sensor electrical connectors
- 4 ground on the left side of the block

9. Raise and safely support the vehicle.
- Left and right side driveshafts
- Propeller shaft from the front axle pinion yoke
- Engine protection shield
- Exhaust pipe from the exhaust manifold. Slide the exhaust pipe backward slightly.
- Fuel tank shield, if equipped
- Torque converter access cover and bolts

10. Place a jack on the transmission fluid pan for support.

11. Remove the transmission support.

12. Lower the transmission enough to reach the top bell housing bolts.

13. Remove the top 4 bell housing bolts, there may be 2 harness clips that will need to be removed in order to have access to 2 of the top bolts.

14. Raise the transmission.

15. Reinstall the transmission support using only 2 through bolts.

16. Remove or disconnect the following:
- Remaining bell housing bolts (11 total)
- Left and right engine lower mount nuts
- Oil level sensor electrical connector
- Oil pressure switch electrical connector

17. Carefully lower the vehicle.

18. Remove the left, then the right upper engine mount nut.

19. Install a suitable engine hoist.

20. Raise the engine out of the compartment slowly, keeping the transmission supported.

21. Remove both engine mounts for clearance.

22. Continue raising the engine out of the vehicle.

23. Place the engine on a suitable engine stand.

To install:

24. Remove the engine from the engine stand.

25. Slowly install the engine into the engine compartment, aligning the engine mounts with the brackets.

26. When the engine mounts are aligned, install the engine mounts, putting the mount up through the engine mount brackets before inserting into the chassis mount brackets.

27. Lower the engine onto the mounts and install the upper engine mounting nuts. Tighten the nuts to 51 ft. lbs. (71 Nm).

28. Remove the engine hoist.

29. Lay the radiator into the radiator support, but do not install the radiator completely.

30. Raise and safely support the vehicle.

31. Install the lower bell housing bolts, except the top four.

32. Remove the 2 through bolts secure the transmission support, then lower the transmission.

33. Install the top 4 bell housing bolts and tighten all 11 bolts to 37 ft. lbs. (50 Nm).

34. Raise the transmission.

35. Install or connect the following:
- Transmission support
- 3 torque converter bolts and tighten to 44 ft. lbs. (60 Nm)
- Torque converter bolt cover
- Fuel tank shield, if equipped
- Engine protection shield
- Propeller shaft to the front axle pinion yoke
- Exhaust pipe to the manifold and tighten the bolts to 37 ft. lbs. (50 Nm)
- Oil level switch and oil pressure sender electrical connectors
- Oil pan drain plug and tighten to 19 ft. lbs. (26 Nm)
- Lower radiator hose
- Left and right wheel driveshafts

36. Lower the vehicle.
- 4 grounds on the left side of the block
- CMP, CKP and knock sensor electrical connectors
- Alternator and starter electrical connectors and battery leads. Torque the nuts to 80 inch lbs. (9 Nm).
- Fuel lines at the fuel pressure regulator
- Engine harness bracket and bolt. Torque the bolt to 37 ft. lbs. (50 Nm).
- Front differential vent hose, to the engine harness bracket
- PCM
- Power brake hose to the booster
- Harness retainer to its original location
- Ignition coil harness connectors
- Transmission cooler lines to clips on right side of engine block
- Camshaft phaser actuator valve electrical connector
- Front axle actuator electrical connector
- A/C line at the accumulator

37. Remove the lift hook.

38. Install or connect the following:
   - AIR adapter and secure with the studs. Tighten to 18 ft. lbs. (25 Nm).
   - Transmission filler tube bracket to AIR adapter stud and secure the bracket with the nut. Torque the nut to 89 inch lbs. (10 Nm).
   - Heater hoses to the heater core
   - Power steering pump and tighten the bolts to 18 ft. lbs. (25 Nm).

39. The remainder of installation is the reverse of removal, but please note the following important steps:

40. Connect the negative battery cable

41. Check all powertrain fluid levels and add, as necessary.

42. Refill the engine crankcase.

43. Refill the engine cooling system.

44. Perform the CKP System Variation Learn Procedure, as follows:

   a. Install a suitable scan tool and check for Diagnostic Trouble Codes (DTCs). If any DTCs, other than P1336 are set, resolve those codes first, before proceeding with this procedure.

   b. With the scan tool, select the crankshaft position variation learn procedure.

   c. Observe the fuel cut-off for the 4.2L engine.

   d. The scan tool will instruct you to perform certain steps, make sure you follow all directions given by the scan tool exactly.

   e. Enable the crankshaft position system variation learn procedure.

➡ **While the learn procedure is in progress, release the throttle immediately when the engine started to decelerate. The engine control is returned to the operator and the engine responds to throttle position after the learn procedure is complete.**

   f. Slowly increase the engine speed to the RPM that you observed.

   g. Immediately release the throttle when fuel cut-out is reached.

   h. The scan tool displays: Learn Status: Learned this ignition. If the scan tool does NOT display this message and not other DTCs set, you must perform further troubleshooting.

   i. Turn the ignition **OFF** for 30 seconds after the learn procedure has been completed successfully.

45. Start and run the engine, then check for leaks.

4.3L Engine

1. Before servicing the vehicle, refer to the precautions in the beginning of this section.

2. Drain the engine cooling system

3. Drain the engine oil.

4. Remove or disconnect the following:
   - Negative battery cable
   - Fuel system pressure
   - Vacuum reservoir and/or the under-hood light from the hood, as equipped
   - Outer cowl vent grilles
   - Hood
   - Oxygen Sensor (O$_2$S) and/or wiring
   - Exhaust pipes at the manifolds and loosen the hanger at the catalytic converter. This is necessary to remove the rear catalytic converter cushion mounts for removal of the exhaust assembly.
   - Skid plate, if equipped
   - Engine-to-transmission pencil braces
   - Slave cylinder and position aside, if equipped
   - Line clamp at the bell housing
   - Wiring from the starter
   - Starter
   - Transfer case on 2000–01 models, if equipped
   - Oil filter
   - Engine mount through bolts
   - Rear engine mount crossbar, nut and washer
   - Bell housing bolts, except the upper left.
   - Battery ground (negative) cable from the engine
   - Front drive axle bolts and roll the axle downward, on 4WD vehicles
   - Air cleaner assembly
   - Upper radiator shroud
   - Fan assembly
   - Drive belt assembly
   - Water pump pulley
   - Upper radiator hose
   - Air conditioning compressor, if equipped, and position aside with the lines intact
   - Lower radiator hose
   - Oil cooler and overflow lines from the radiator, plug the openings to prevent system contamination or excessive fluid loss
   - Radiator and lower radiator shroud
   - Power steering hoses from the steering gear, then cap the openings to prevent system contamination or excessive fluid loss
   - Heater hoses from the intake manifold and the water pump
   - Wiring harness and vacuum lines from the engine
   - Throttle cables
   - Distributor cap
   - Remaining bell housing bolt
   - Fuel lines and the bracket
   - Ground strap(s) from the rear of the cylinder head
   - Front body mount bolts, on 4WD vehicles

5. Support the transmission.

6. Install a lifting device and lift the engine.

To install:

7. Install or connect the following:
   - Engine into the vehicle
   - Front body mount bolts, on 4WD vehicles
   - Ground strap(s) to the rear of the cylinder head
   - Fuel lines and the bracket
   - Upper left bell-housing bolt
   - Distributor, cap and wires
   - Throttle cables
   - Vacuum lines and wiring harness connectors
   - Heater hoses
   - Power steering hoses
   - Lower shroud and radiator
   - Oil cooler lines to the radiator and overflow hose
   - Lower radiator hose
   - Air conditioning compressor to the engine, if equipped
   - Upper radiator hose
   - Water pump pulley
   - Drive belt assembly
   - Fan assembly
   - Upper radiator shroud
   - Air cleaner assembly
   - Front drive axle, for 4WD vehicles
   - Battery ground strap to the engine block
   - Remaining bell housing bolts
   - Engine mount through-bolts. Torque them to 49 ft. lbs. (66 Nm).
   - Rear engine mount crossbar nut and washer. Tighten the nut to 33 ft. lbs. (45 Nm).
   - Oil filter
   - Starter motor
   - Flywheel cover
   - Clutch slave cylinder, if equipped
   - Pencil brace and the skid plate, as equipped
   - Catalytic converter Y-pipe assembly and hangers
   - Hood

- Outer cowl vent grilles
- Vacuum reservoir and/or the under-hood light to the hood, as equipped
- Negative battery cable

8. Check all powertrain fluid levels and add, as necessary.

9. Refill the engine crankcase.

10. Refill the engine cooling system.

11. Start and run the engine, then check for leaks.

5.3L Engine

1. Before servicing the vehicle, refer to the precautions in the beginning of this section.

2. Drain the engine cooling system

3. Drain the engine oil.

4. Remove and recover the refrigerant, if equipped with A/C.

5. Remove or disconnect the following:
- Negative battery cable
- Hood
- Radiator
- Radiator support brace
- Front axle, if 4WD
- Drive shafts
- Intake manifold
- Oil pressure sensor connector
- Oxygen (O_2) sensor connector
- Camshaft Position (CMP) sensor connector
- A/C compressor hose
- Rear auxiliary A/C compressor pipe fitting
- Rear auxiliary A/C compressor pipe nut and bolt. Tie the pipe out of the way.
- Engine Coolant Temperature (ECT) sensor
- Ground terminal bolt
- Retaining clips from the brackets
- A/C pressure switch electrical connector
- Retaining clip from the cylinder head
- Ground terminal bolts
- Starter
- Battery cable channel bolt
- Battery cable channel from the oil pan
- A/C compressor electrical connector

6. Collect all branches of the engine wiring harness, then position the harness out of the way.
- Alternator cable from the alternator
- Alternator bracket bolts, then position the bracket and alternator assembly aside
- Inlet and outlet hoses from the water outlet, using J 38185 to move the hose clamps

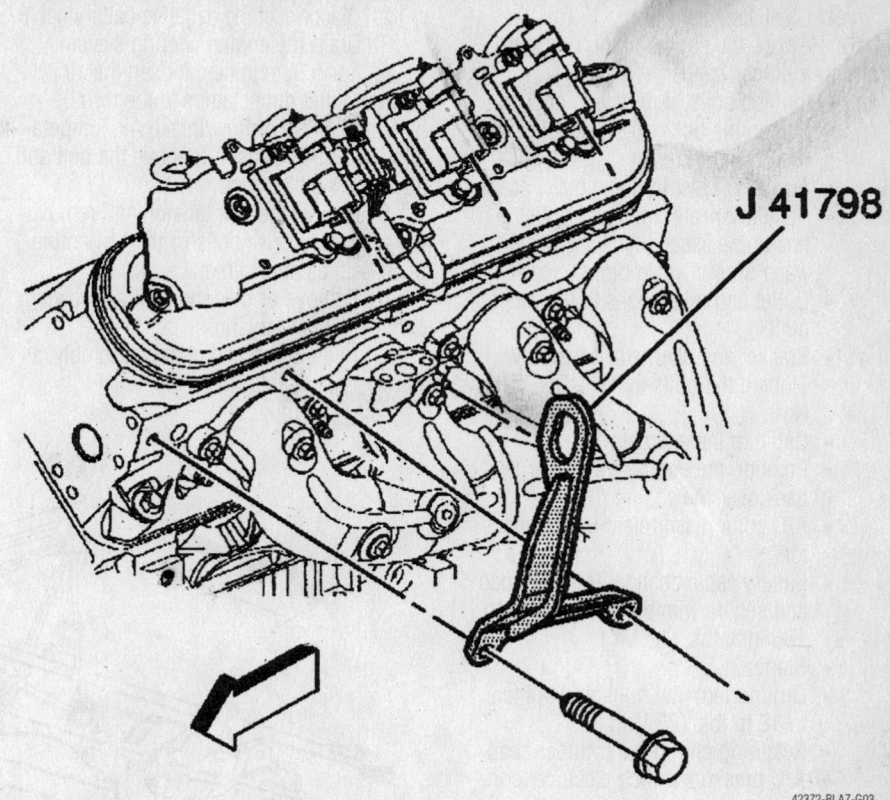

J 41798

42372-BLAZ-G03

If necessary, remove ignition coil(s) to install the engine lifting brackets—5.3L engine

- Auxiliary heater inlet and outlet hose/pipe assembly from the heater water shutoff valve pipes
- Auxiliary heater inlet and outlet hoses/pipes from the water pump, using Hose Clamp Pliers J 38185
- Remove ignition coils, if necessary, to install Engine Lifting Brackets J 41798 to the cylinder heads

7. Install Engine Lifting Brackets J 41798 to the cylinder heads. Tighten the M8 bolts to 18 ft. lbs. (25 Nm) and the M10 bolts to 37 ft. lbs. (50 Nm).
- Catalytic converter
- 3 frame engine mount bracket bolts from the right and left sides
- Torque converter bolts
- Transmission oil level dipstick tube nut and tube
- Transmission bolt and stud on the right side
- Lower transmission bolt/studs
- 3 upper transmission bolts/studs

8. Install a suitable engine hoist to the engine lifting brackets.

9. Place a floor jack under the transmission for support.

10. Separate the engine from the transmission.

11. Remove the engine from the vehicle and place on a suitable engine stand.

12. Install Converter Holding Strap J 21366 to the transmission to hold the torque converter.

To install:

13. Remove Converter Holding Strap J 21366 from the transmission.

14. Attach the engine to a hoist and remove it from the engine stand

15. Install or connect the following:
- Engine into the vehicle. Match the transmission up to the engine, then remove the floor jack.
- 3 upper transmission bolts/studs and tighten to 37 ft. lbs. (50 Nm)
- Lower transmission bolts/studs and tighten to 37 ft. lbs. (50 Nm)
- Transmission bolt and stud on the right side and tighten to 37 ft. lbs. (50 Nm)
- Transmission oil level dipstick tube and nut. Torque to 89 inch lbs. (10 Nm).
- Torque converter bolts and tighten to 44 ft. lbs. (60 Nm)
- 3 frame engine mount bracket bolts to both the right and left sides. Torque the bolts to 37 ft. lbs. (50 Nm).

- Catalytic converter

16. Remove the engine lifting brackets from the cylinder heads

- Ignition coils, if removed, and tighten the bolts to 71 inch lbs. (8 Nm)
- Auxiliary heater inlet and outlet hoses
- Auxiliary heater inlet and outlet hose/pipe assembly to the heater water shutoff valve pipes
- Outlet and inlet hoses to the water outlet
- Bracket and alternator assembly. Tighten the bolts to 37 ft. lbs. (50 Nm).
- Cable to the alternator
- Position the engine wiring harness back over the engine
- A/C compressor electrical connector
- Battery cable channel to the oil pan and secure with the bolt. Torque to 106 inch lbs. (12 Nm).
- Starter
- Ground terminal bolts and tighten to 18 ft. lbs. (25 Nm)
- Retaining clip to the cylinder head
- A/C pressure switch electrical connector
- Retaining clips to the brackets
- Ground terminal bolt and tighten to 18 ft. lbs. (25 Nm)
- ECT sensor connector
- Rear auxiliary A/C compressor pipe nut and bolt. Torque to 15 ft. lbs. (20 Nm).
- A/C compressor hose
- Oil pressure sensor connector
- O_2 sensor connector
- CMP sensor connector
- Intake manifold
- Drive shafts
- Front axle, if removed
- Radiator support brace
- Radiator

17. Recharge the A/C system
- Negative battery cable
- Hood

18. Check all powertrain fluid levels and add, as necessary.

19. Refill the engine crankcase.

20. Refill the engine cooling system.

21. Start and run the engine, then check for leaks.

Water Pump

REMOVAL & INSTALLATION

1. Before servicing the vehicle, refer to the precautions in the beginning of this section.

2. Disconnect the negative battery cable.

3. Drain the engine cooling system.

4. For 5.3L engines, loosen the air cleaner outlet duct clamps at the throttle body and Mass Airflow/Intake Air Temperature (MAP/IAT) sensor. Remove the bolt and air cleaner outlet duct.

5. Relieve the belt tension and remove the accessory drive belts or the serpentine drive belt, as applicable.

6. Remove or disconnect the following:
- Upper fan shroud
- Fan or fan and clutch assembly, as applicable
- Water pump pulley; use a suitable tool to hold the pulley while removing the bolts
- Coolant hose(s) from the water pump

➡For the hoses on some engines, removal may be easier if the hose is left attached until the pump is free from the block. Once the pump is removed from the engine, the pump may be pulled (giving a better grip and greater leverage) from the tight hose connection.

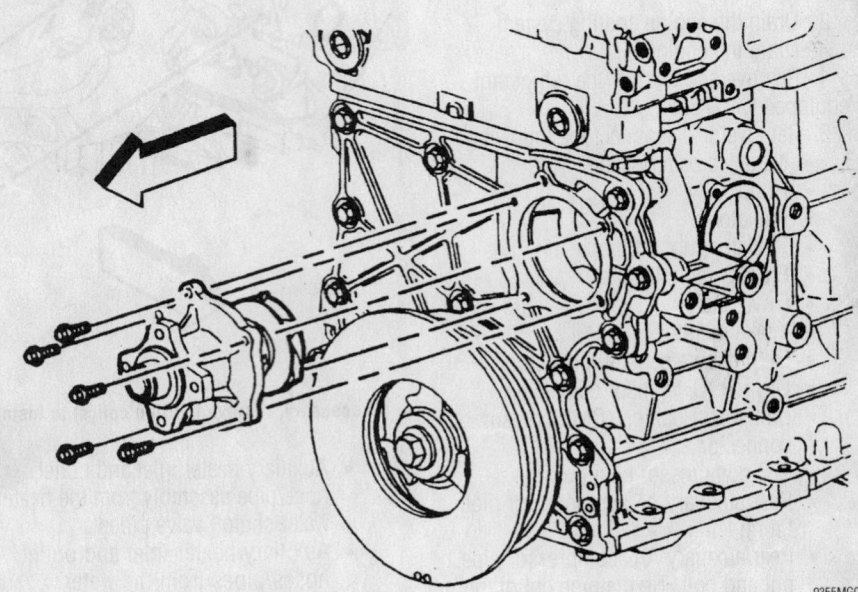

9355MG01

Exploded view of the water pump assembly mounting—4.2L engine

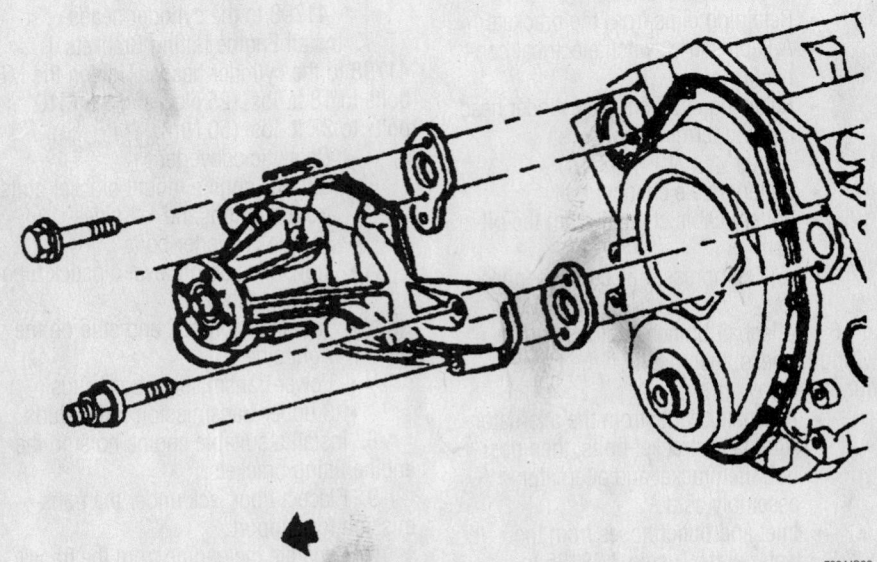

7924JG06

Exploded view of the water pump assembly mounting—4.3L engine

- Water pump retainers
- Water pump from the engine

✳✳ WARNING

Note the positions of all retainers as some engines will utilize different length fasteners in different locations and/or bolts and studs in different locations.

To install:

7. Clean the gasket mounting surfaces.

➡**The water pumps on some of the engines covered may have been installed using sealer only, no gasket, at the factory. If a gasket is supplied with the replacement part, it should be used. Otherwise, a ⅛ in. (3mm) bead of RTV sealer should be used around the sealing surface of the pump.**

8. Apply sealant to the water pump retainer threads.

9. Install or connect the following:
- Water pump using a new gasket. Tighten the water pump retainers to 89 inch lbs. (10 Nm) for 4.2L engines, or to 30 ft. lbs. (41 Nm) for 4.3L engines. For 5.3L engines, tighten the bolts to 11 ft. lbs. (15 Nm), then to 22 ft. lbs. (30 Nm).
- Coolant hose(s)
- Water pump pulley. Tighten the pulley bolts to 18 ft. lbs. (25 Nm).
- Fan or fan and clutch assembly
- Serpentine drive belt (if equipped) by positioning the belt over the pulleys and carefully allow the tensioner back into contact with the belt.
- V-belts (if equipped) and adjust the tension
- Upper fan shroud
- Negative battery cable

10. Refill the engine cooling system.

11. Run the engine and check for leaks.

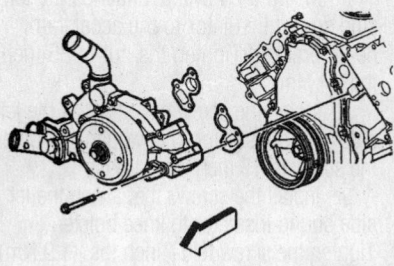

Exploded view of the water pump assembly mounting—5.3L engine

Heater Core

REMOVAL & INSTALLATION

1. Before servicing the vehicle, refer to the precautions in the beginning of this section.

2. Drain the engine cooling system.

3. Remove or disconnect the following:
- Negative battery cable
- Heater hoses from the heater core

4. Remove the instrument panel as follows:

    a. Disable the air bag system.

    b. Set the parking brake and block the wheels.

    c. Disconnect the parking brake release cable from the parking brake lever.

    d. Unfasten the screws that retain the DLC instrument panel left side sound insulator. Feed the DLC through the hole in the sound insulator.

    e. Unfasten the right side sound insulator panel screws and remove the panel.

    f. Unfasten the screws that attach the instrument panel left side sound insulator to the knee bolster and cowl panel.

    g. Unfasten the nut that attaches the left side sound insulator to the accelerator pedal bracket.

    h. Unplug the remote control door lock receiver module electrical connector.

    i. Remove the door lock receiver module from the left side sound insulator. Remove the left side sound insulator.

    j. Unfasten the screws that attach the instrument panel center sound insulator to the knee bolster, instrument panel, heater assembly and floor duct.

    k. Remove the center sound insulator.

    l. Unfasten the screws that attach the courtesy lamp to the knee bolster.

    m. Unfasten the screws that attach the knee bolster to the instrument panel.

    n. Disconnect the lap cooler duct from the knee bolster.

    o. Unplug the lighter electrical connection and remove the knee bolster.

    p. Unfasten the steering column-to-instrument panel nuts and lower the column.

    q. Unfasten the screws that attach the instrument panel accessory trim plate to the instrument panel.

    r. Remove the trim plate and unplug all necessary electrical connection.

    s. Remove the heater and/or air conditioning control assembly.

    t. Remove the radio and the storage compartment assembly (if equipped).

    u. If necessary, remove the instrument cluster.

    v. Unfasten the left and right instrument panel pivot bolts and the panel lower support bolt.

    w. Unfasten the speaker grilles retaining screws and remove the speaker grilles.

    x. Remove the windshield defroster grille using a flat-bladed prytool. Start at one end of the grille and work your way down the grille.

    y. Unfasten the 4 instrument panel upper support screws.

    z. Tag and unplug all necessary electrical connections.

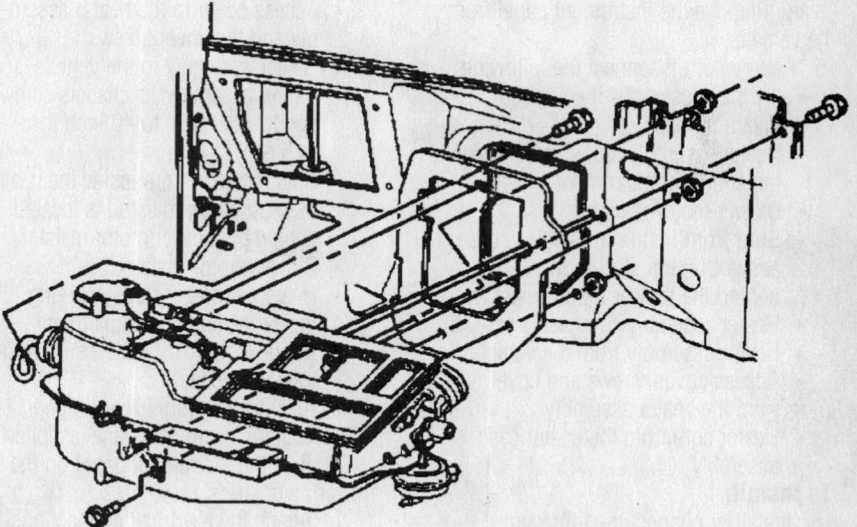

View of the heater case assembly

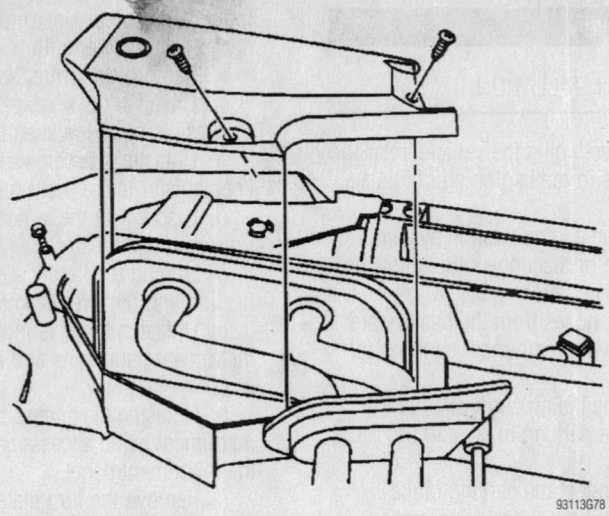

93113G78

View of the heater case cover

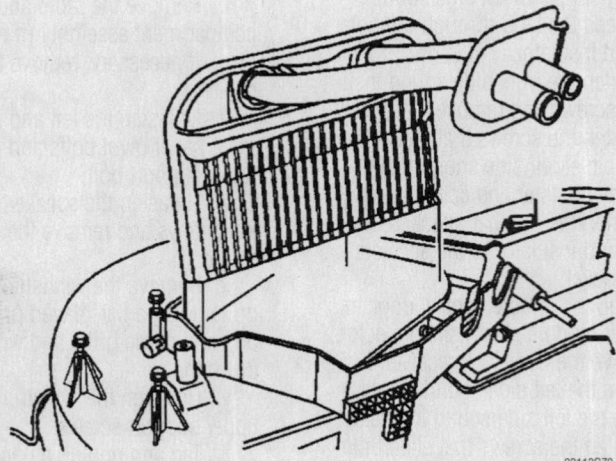

93113G79

View of the heater core

aa. Remove the instrument panel from the vehicle.

5. Remove or disconnect the following:
- Air inlet assembly, if equipped
- Vacuum hoses
- Heater assembly studs, from inside the engine compartment
- Blower motor resistor
- Stud from inside the heater case assembly; the stud is located behind the blower motor resistor
- Heater assembly-to-chassis screws
- Heater assembly from the vehicle
- Access cover screws and cover from the heater assembly
- Heater core from the heater case assembly

To install:

6. Install or connect the following:
- Heater core to the heater case assembly

- Access cover to the heater assembly and the cover screws
- Heater assembly to the vehicle
- Heater assembly-to-chassis screws and torque them to 40 inch lbs. (4.5 Nm)
- Stud, working from inside the heater case assembly;the stud is located behind the blower motor resistor
- Blower motor resistor
- Heater assembly studs, working inside the engine compartment, torque them to 17 inch lbs. (1.9 Nm)
- Vacuum hoses
- Air inlet assembly, if equipped

7. Install the instrument panel as follows:

a. Rest the instrument panel on the lower pivot studs.

b. Attach the electrical connections.

c. Install but do not tighten the 4 upper instrument panel support screws.

d. Install the left and right panel pivot bolts. Tighten the bolts to 102 inch lbs. (11.5 Nm).

e. Install the panel lower support bolt. Tighten the bolt to 102 inch lbs. (11.5 Nm).

f. Tighten the upper support screws to 17 inch lbs. (1.9 Nm).

g. Install the windshield defroster grille and the speaker grilles.

h. Install the radio and storage compartment assembly (if equipped).

i. If removed, install the instrument cluster.

j. Install the heater and/or air conditioning control assembly.

k. Attach the electrical connections to the instrument panel accessory trim plate.

l. Place the trim plate in position and install its retaining screws. Tighten the screws to 17 inch lbs. (1.9 Nm).

m. Place the steering column into position and install its retaining nuts. Tighten the nuts to 22 ft. lbs. (30 Nm).

n. Attach the lighter electrical connection and the lap cooler duct to the knee bolster.

o. Place the knee bolster into position and install its retaining screws. Tighten the Torx® head screws to 80 inch lbs. (9 Nm) and the hex head screws to 17 inch lbs. (1.9 Nm).

p. Place the courtesy lamp in position and install its screws. Tighten the screws to 17 inch lbs. (1.9 Nm).

q. Place the instrument panel center sound insulator in position. Install the screws that attach the center sound insulator to the knee bolster, instrument panel and the floor duct. Tighten the screws to 17 inch lbs. (1.9 Nm).

r. Install the screw that attaches the center sound insulator to the heater assembly. Tighten the screw to 13 inch lbs. (1.5 Nm).

s. Install the remote control door lock receiver module to the instrument panel left side sound insulator.

t. Attach the door lock receiver electrical connection.

u. Install the nut that attaches the left side sound insulator to the accelerator pedal bracket. Tighten the nut to 35 inch lbs. (4 Nm).

v. Install the screw that attaches the left side sound insulator to cowl panel. Tighten the screw to 13 inch lbs. (1.5 Nm).

w. Install the screws that attach the left side sound insulator to knee bolster. Tighten the screw to 17 inch lbs. (1.9 Nm).

x. Feed the DLC through the hole in the sound insulator, place the DLC in position and install its retaining screws.

Tighten the screws to 21 inch lbs. (2.4 Nm).

   y. Install the right side sound insulator and tighten the screws

   z. Connect the parking brake release cable to the lever.

   aa. Enable the air bag system.

8. Install the heater hoses to the heater core.

9. Refill the cooling system.

10. Connect the negative battery cable.

11. Run the engine to normal operating temperatures; then, check the climate control operation and check for leaks.

Cylinder Head

REMOVAL & INSTALLATION

4.2L Engine

1. Before servicing the vehicle, refer to the precautions in the beginning of this section.
2. Disconnect the negative battery cable.
3. Drain the engine cooling system.
4. Remove or disconnect the following:
   - Camshaft cover
   - Exhaust manifold
   - Front cover
   - Cylinder head access hole plugs
   - Timing chain tensioner shoe bolt and shoe
   - Timing chain tensioner guide bolts and guide
   - Timing chain and sprockets
5. Unfasten the cylinder head bolts by loosening them in the reverse of the torque sequence, then carefully remove the cylinder head.
6. Remove the cylinder head gasket.

To install:

7. Carefully clean and inspect the cylinder head and the gasket mounting surfaces.

➡ **The gasket surfaces on both the head and block must be clean of any foreign matter and free of nicks or heavy scratches. The cylinder bolt threads in the block and thread on the bolts must be cleaned (dirt will affect the bolt torque).**

➡ **DO NOT apply sealer to composition steel-asbestos gaskets.**

✳✳ WARNING

Make sure the number 1 cylinder is at Top Dead Center (TDC).

8. If using a steel only gasket, apply a thin and even coat of sealer to both sides of the gaskets.

9. Place a new gasket over the dowel pins with the bead or the words "This Side Up" facing upwards (as applicable), then carefully lower the cylinder head into position over the gasket and dowels.

10. Apply a coating of 12345493 or equivalent sealer to the threads of the cylinder head bolts, then thread the bolts into position until finger-tight.

11. Tighten the cylinder head bolts in sequence as follows:

   a. Tighten the long bolts (1-14), in sequence, to 30 ft. lbs. (40 Nm).

   b. Tighten the long bolts, in sequence, an additional 90 degrees.

   c. Tighten the long bolts, in sequence, an additional 60 degrees.

   d. Tighten the 2 long end bolts to 15 ft. lbs. (20 Nm).

   e. Tighten the 1 short end bolt to 13 ft. lbs. (18 Nm).

12. Install or connect the following:
   - Cylinder head access hole plugs and tighten to 44 inch lbs. (5 Nm)
   - Timing chain and sprockets
   - Front cover
   - Camshaft cover
   - Exhaust manifold
   - Negative battery cable

13. Properly refill the engine cooling system.

14. Run the engine to check for leaks.

4.3L Engine

1. Before servicing the vehicle, refer to the precautions in the beginning of this section.

2. Properly relieve the fuel system pressure, then disconnect the negative battery cable.

3. Drain the engine cooling system.
4. Remove or disconnect the following:
   - Intake manifold
   - Exhaust manifold
   - Alternator and bracket, if removing the right cylinder head
   - Cooling fan assembly, on 2000–01 models
   - Air conditioning compressor (position it aside with the refrigerant lines attached), on 2000–01 models equipped
   - Air pipe bracket and nut from the rear of the power steering pump if removing the left cylinder head on equipped 2000–01 models
   - Engine accessory bracket with power steering pump (position the pump aside with the lines attached) and brackets, if removing the left cylinder head
   - Wiring harness and clip from the rear of the cylinder head
   - Coolant sensor wire
   - Wiring from the spark plugs
   - Spark plugs, if necessary
   - Ground wires and if necessary, the fuel line bracket from the rear of the cylinder head, on 2000–01 models
   - Rocker arm cover

5. Loosen the rocker arms and remove the pushrods.

➡ **If valve train components, such as the rocker arms or pushrods, are to be reused, they must be tagged or arranged to insure installation in their original locations.**

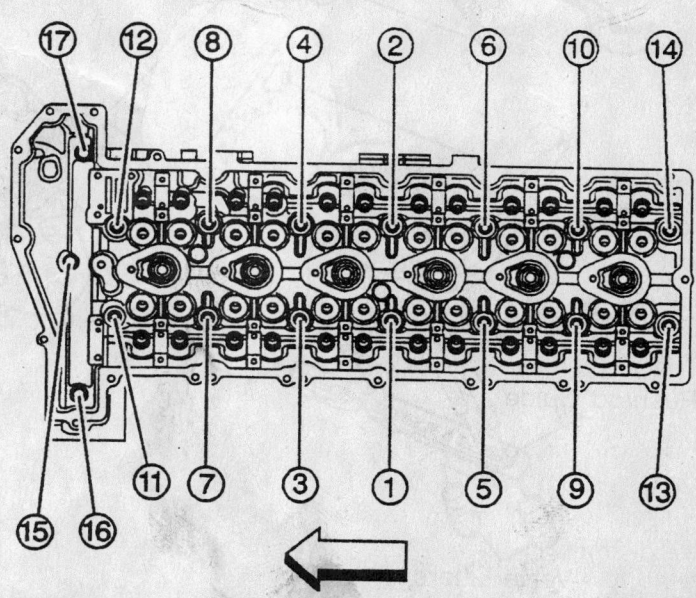

Cylinder head bolt tightening sequence—4.2L engine

9355MG02

6. Unfasten the cylinder head bolts by loosening them in the reverse of the torque sequence, then carefully remove the cylinder head.

To install:

7. Carefully clean and inspect the cylinder head and the gasket mounting surfaces.

➡ The gasket surfaces on both the head and block must be clean of any foreign matter and free of nicks or heavy scratches. The cylinder bolt threads in the block and thread on the bolts must be cleaned (dirt will affect the bolt torque).

➡ DO NOT apply sealer to composition steel-asbestos gaskets.

8. If using a steel only gasket, apply a thin and even coat of sealer to both sides of the gaskets.

9. Place a new gasket over the dowel pins with the bead or the words "This Side Up" facing upwards (as applicable), then carefully lower the cylinder head into position over the gasket and dowels.

10. Apply a coating of 12346004 or equivalent sealer to the threads of the cylinder head bolts, then thread the bolts into position until finger-tight.

11. Install the bolts in sequence to 22 ft. lbs. (30 Nm). The bolts must then be tightened again in sequence in the following order:

  a. Short length bolts: (11, 7, 3, 2, 6, 10) 55 degrees.

  b. Medium length bolts: (12, 13) 65 degrees.

  c. Long length bolts: (1, 4, 8, 5, 9) 75 degrees.

12. Install or connect the following:
- Pushrods, secure the rocker arms and adjust the valves
- Rocker arm cover
- Spark plugs, if removed
- Spark plug wires
- Attach the fuel line bracket (if removed) and ground wires to the rear of the head and tighten the bolts to 22 ft. lbs. (30 Nm) on 2000–01 models
- Air conditioning compressor and bracket, if the left cylinder head was removed

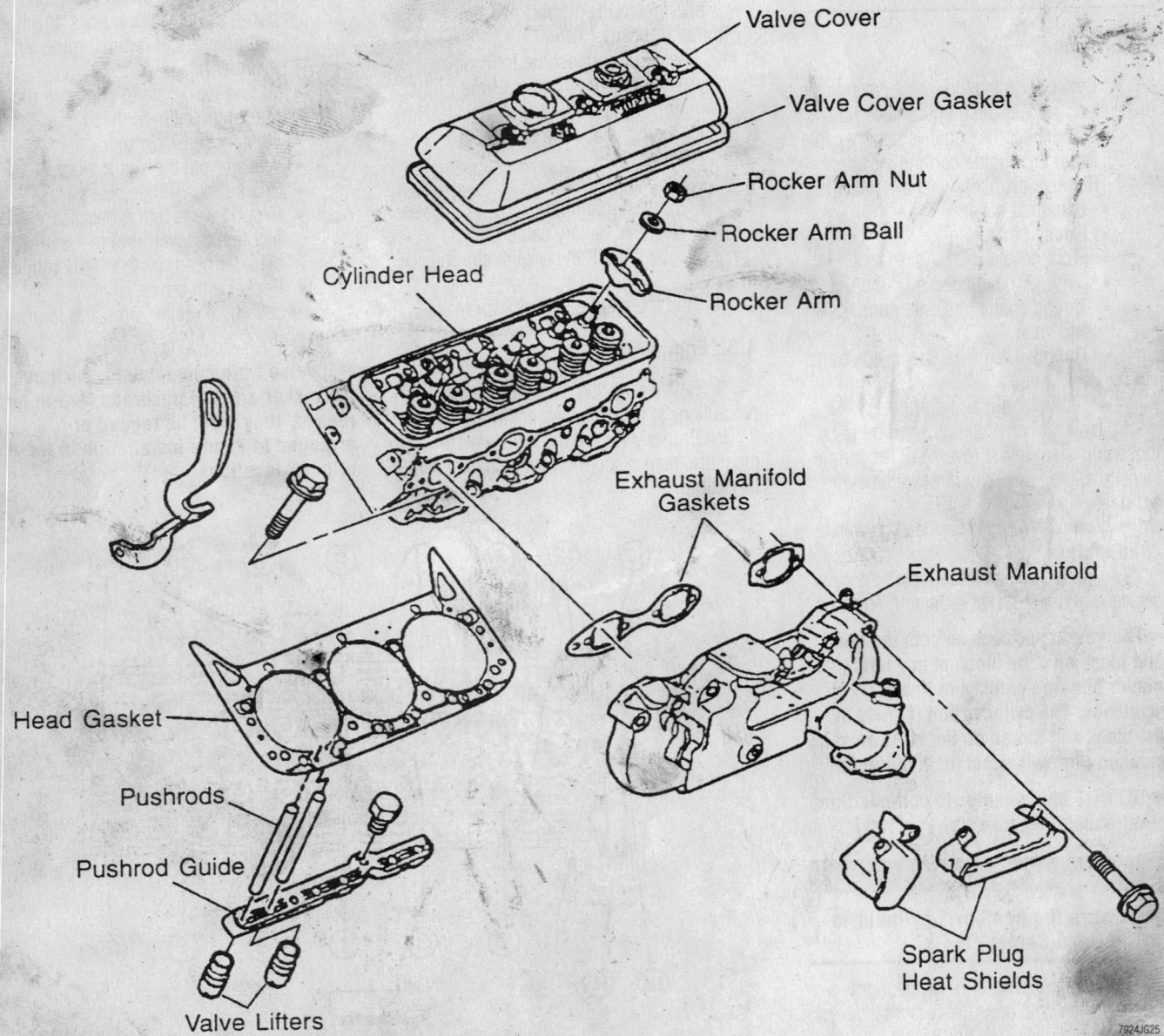

Valve Cover
Valve Cover Gasket
Rocker Arm Nut
Rocker Arm Ball
Cylinder Head
Rocker Arm
Exhaust Manifold Gaskets
Exhaust Manifold
Head Gasket
Pushrods
Pushrod Guide
Spark Plug Heat Shields
Valve Lifters

7924JG25

Cylinder head and related components—4.3L engine

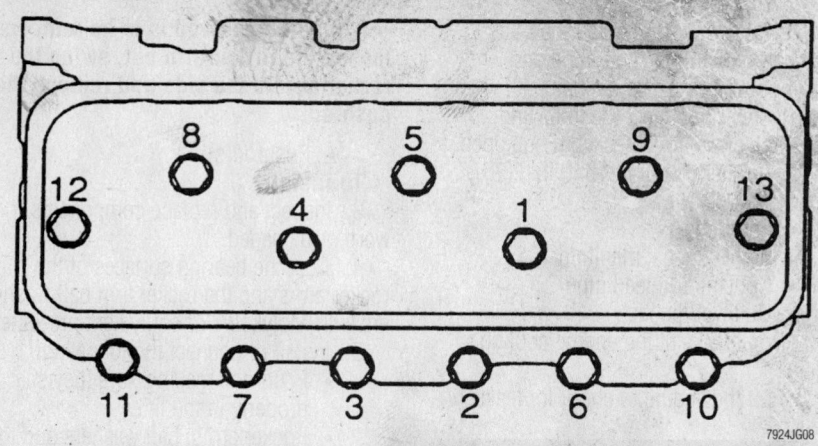

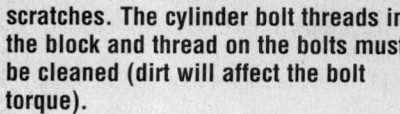

Cylinder head bolt torque sequence—4.3L engine

- Alternator and bracket, if the right cylinder head was removed
- Engine accessory bracket with power steering pump. if the left cylinder head was removed
- Air pipe bracket and nut to the rear of the power steering pump (if equipped), if the left cylinder head was removed on 2000–01 models. Tighten the nut to 30 ft. lbs. (41 Nm).
- A/C compressor, if the left cylinder head was removed
- Cooling fan assembly, if the left cylinder head was removed on 2000–01 models
- Wiring harness and clip to the rear of the cylinder head
- Coolant sensor wire
- Exhaust manifold
- Intake manifold
- Negative battery cable

13. Properly refill the engine cooling system.

14. Run the engine to check for leaks.

5.3L Engine

LEFT SIDE

1. Before servicing the vehicle, refer to the precautions in the beginning of this section.

2. Drain the engine cooling system.

3. Remove or disconnect the following:
- Negative battery cable
- Alternator bracket
- Coolant air bleed pipe
- Left exhaust manifold
- Pushrods
- Auxiliary A/C bracket bolt, if equipped
- Cylinder head bolts 1, 2 and 3. Discard the bolts
- Cylinder head
- Cylinder head gasket and discard

To install:

4. Carefully clean and inspect the cylinder head and the gasket mounting surfaces.

➡ The gasket surfaces on both the head and block must be clean of any foreign matter and free of nicks or heavy

scratches. The cylinder bolt threads in the block and thread on the bolts must be cleaned (dirt will affect the bolt torque).

➡DO NOT apply any type sealer to the cylinder head gasket, unless otherwise specified.

5. Check the cylinder head locating pins for proper installation, location (a) 0.327 in. (8.3mm), as shown.

6. Place a new gasket over the dowel pins. When installed properly, the word "FRONT" on the left side, the tab on the gasket should be left of center or closer to the front of the engine.

7. Install or connect the following:
- Cylinder head

➡You must use new cylinder head bolts during reassembly. Do NOT reuse the old head bolts.

- NEW cylinder head bolts 1, 2 and 3.

8. Tighten the cylinder head bolts in sequence as follows:

a. Tighten the M11 bolts to 22 ft. lbs. (30 Nm).

b. Tighten the M11 an additional 90 degrees.

c. Tighten M11 bolts 1–8, an additional 90 degrees and M11 bolts 9 and 10 an additional 50 degrees.

d. Tighten the M8 bolts (11–15) to 22 ft. lbs. (30 Nm). Tighten all the bolts beginning with the center bolt and working outward, alternating sides

9. Install or connect the following:
- Auxiliary A/C bracket, if equipped. Torque the bolt to 15 ft. lbs. (20 Nm).
- Pushrods
- Left exhaust manifold
- Coolant air bleed pipe
- Alternator bracket

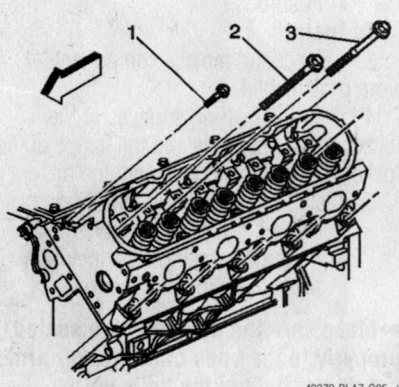

Remove and discard cylinder head bolts 1, 2 and 3—5.3L engine

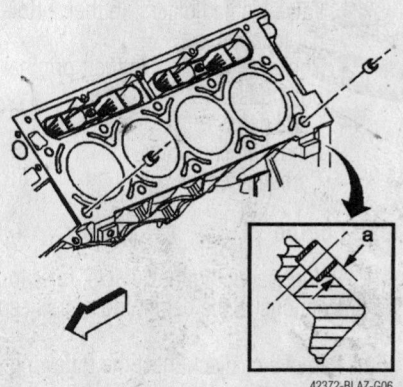

Make sure the cylinder head locating pins are properly installed, see dimension (a)

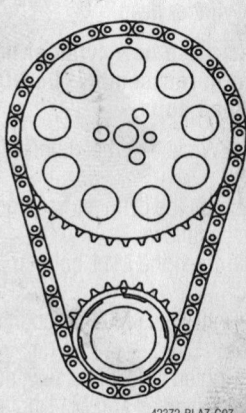

Cylinder head bolt torque sequence—5.3L engine

10. Properly refill the engine cooling system.

11. Run the engine to check for leaks.

RIGHT SIDE

1. Before servicing the vehicle, refer to the precautions in the beginning of this section.

2. Drain the engine cooling system.

3. Remove or disconnect the following:
   - Negative battery cable
   - Oil level dipstick
   - Coolant air bleed pipe
   - Right exhaust manifold
   - Pushrods
   - Auxiliary A/C bracket nut, if equipped
   - Cylinder head bolts 1, 2 and 3. Discard the bolts
   - Cylinder head
   - Cylinder head gasket and discard

To install:

4. Carefully clean and inspect the cylinder head and the gasket mounting surfaces.

➡**The gasket surfaces on both the head and block must be clean of any foreign matter and free of nicks or heavy scratches. The cylinder bolt threads in the block and thread on the bolts must be cleaned (dirt will affect the bolt torque).**

➡**DO NOT apply any type sealer to the cylinder head gasket, unless otherwise specified.**

5. Check the cylinder head locating pins for proper installation, location (a) 0.327 in. (8.3mm), as shown.

6. Place a new gasket over the dowel pins. When installed properly, the word "FRONT" on the right side, the tab on the gasket should be right of center or closer.

7. Install or connect the following:
   - Cylinder head

➡**You must use new cylinder head bolts during reassembly. Do NOT reuse the old head bolts.**

   - NEW cylinder head bolts 1, 2 and 3.

8. Tighten the cylinder head bolts in sequence as follows:
   a. Tighten the M11 bolts to 22 ft. lbs. (30 Nm).
   b. Tighten the M11 an additional 90 degrees.
   c. Tighten M11 bolts 1–8, an additional 90 degrees and M11 bolts 9 and 10 an additional 50 degrees.
   d. Tighten the M8 bolts (11–15) to 22

ft. lbs. (30 Nm). Tighten all the bolts beginning with the center bolt and working outward, alternating sides

9. Install or connect the following:
   - Auxiliary A/C bracket, if equipped. Torque the nut to 15 ft. lbs. (20 Nm).
   - Pushrods
   - Right exhaust manifold
   - Coolant air bleed pipe
   - Oil level dipstick

10. Properly refill the engine cooling system.

11. Run the engine to check for leaks.

Rocker Arms/Shafts

REMOVAL & INSTALLATION

4.2L Engine

1. Before servicing the vehicle, refer to the precautions in the beginning of this section.

2. Remove or disconnect the following:
   - Camshaft cover

➡**Make sure to place the camshaft caps in a rack to keep them in order, so they may be installed in their original locations.**

   - Camshaft cap bolts and caps
   - Camshafts

➡**If valve train components, such as the rocker arms or lash adjusters, are to be reused, they must be tagged or arranged to insure installation in their original locations.**

   - Rocker arms
   - Valve lash adjusters

To install:

3. Lubricate and fill the valve lash adjusters and the rocker arm roller with engine oil.

4. Install or connect the following
   - Valve lash adjusters, in their original locations
   - Rocker arm rollers in their original positions
   - Camshafts
   - Camshaft cap bolts
   - Camshaft cover

4.3L Engine

1. Before servicing the vehicle, refer to the precautions in the beginning of this section.

2. Remove or disconnect the following:
   - Rocker arm cover(s)
   - Rocker arm nut, rocker arm and ball washer

➡**If only the pushrod is to be removed, loosen the rocker arm nut, swing the rocker arm to the side and remove the pushrod.**

   - Pushrod(s)

To install:

3. Inspect and replace components if worn or damaged.

4. Coat the bearing surfaces of the rocker arms and the rocker arm ball washers with Molykote® or equivalent pre-lube.

5. Install or connect the following:
   - Pushrods making sure they seat properly in the lifter
   - Rocker arms, ball washers and the nuts

➡**The 4.3L engines are equipped with screw-in rocker arm studs with positive stop shoulders.**

   - Rocker arm adjusting nuts. Tighten them against the stop shoulders to 18 ft. lbs. (24 Nm). No further adjustment is necessary or possible.

6. Install the rocker arm cover(s).

7. Start and run the engine, then check for leaks and for proper ignition timing adjustment.

5.3L Engine

1. Before servicing the vehicle, refer to the precautions in the beginning of this section.

2. Remove or disconnect the following:
   - Rocker arm cover(s)

➡**If valve train components, such as the rocker arms, pushrods or pivot supports, are to be reused, they must be tagged or arranged to insure installation in their original locations.**

   - Rocker arm bolts
   - Rocker arms
   - Rocker arm pivot support
   - Pushrods

To install:

3. Inspect and replace components if worn or damaged.

4. Coat the bearing surfaces of the rocker arms, pushrods and the flange of the rocker arm bolts with clean engine oil.

5. Install or connect the following:
   - Rocker arm pivot support
   - Pushrods making sure they seat properly in the lifter sockets

➡**Make sure the pushrods are seated properly to the ends of the rocker arms, but do not tighten the bolts yet.**

   - Rocker arms and bolts

6. Rotate the crankshaft until the No. 1

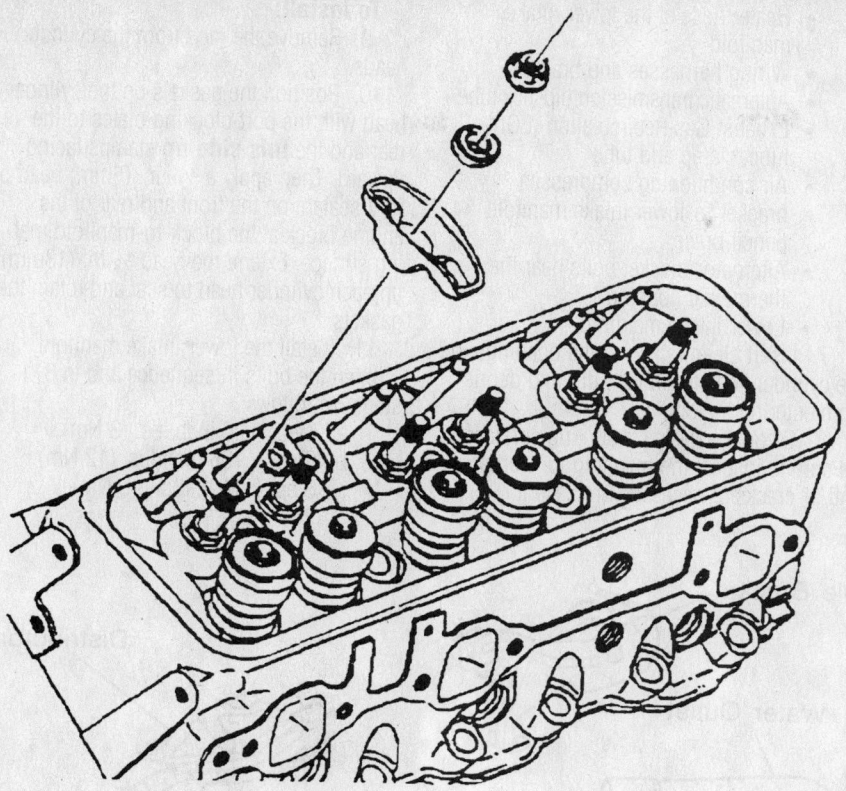

Exploded view of the rocker arm assembly

piston is at Top Dead Center (TDC) of the compressor stroke. In this position, the cylinder No. 1 rocker arms will be off lobe lift, and the crankshaft sprocket key will be at the 1:30 position.

➡**The engine firing order is: 1–8–7–2–6–5–4–3. Cylinders 1, 3, 5 and 7 are the left bank. Cylinders 2, 4, 6 and 8 are the right bank.**

7. With the engine in the No. 1 firing position, tighten the following rocker arm bolts:

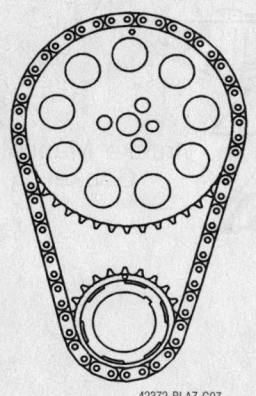

42372-BLAZ-G07

View of the crankshaft key with the No. 1 piston at TDC—5.3L engine

    a. Tighten cylinders 1, 2, 7 and 8 exhaust valve rocker arm bolts to 22 ft. lbs. (30 Nm).

    b. Tighten cylinder 1, 3, 4 and 5 intake valve rocker arm bolts to 22 ft. lbs. (30 Nm).

8. Rotate the crankshaft 360 degrees, then tighten the following rocker arm bolts:

    a. Tighten cylinders 3, 4, 5 and 6 exhaust valve rocker arm bolts to 22 ft. lbs. (30 Nm).

    b. Tighten cylinder 2, 6, 7 and 8 intake valve rocker arm bolts to 22 ft. lbs. (30 Nm).

9. Install the rocker arm cover(s).

Intake Manifold

REMOVAL & INSTALLATION

4.2L Engine

1. Before servicing the vehicle, refer to the precautions in the beginning of this section.

2. Properly relieve the fuel system pressure.

3. Disconnect the negative battery cable.

4. Drain the engine cooling system.

5. Remove or disconnect the following:

- Throttle body
- Powertrain Control Module (PCM)
- All electrical harnesses from the engine harness bracket
- Front differential vent hose from the bracket clip
- Engine harness bracket bolt and bracket
- Manifold Absolute Pressure (MAP) sensor connector
- Crankcase ventilation hose
- Brake hose from the booster
- Alternator
- Intake manifold bolts and manifold.
- Manifold gasket

To install:

6. Clean the gasket mounting surfaces. Be sure to inspect the manifold for warpage and/or cracks. If necessary, replace it.

7. Properly position a new intake manifold gasket.

8. Install or connect the following:

- Intake manifold and bolts. Torque the bolts to 16 ft. lbs. (22 Nm).
- Alternator
- Brake hose to the booster
- Crankcase ventilation hose, lubricating the inner diameter first with 12345884, or equivalent lubricant
- MAP sensor electrical connector
- Engine harness bracket. Tighten the retaining bolt to 37 ft. lbs. (50 Nm).
- Front differential vent hose to the engine harness bracket clip
- All harnesses to their original locations onto the engine harness bracket
- PCM
- Throttle body
- Negative battery cable

9. Refill the engine cooling system.

4.3L Engine

➡**If only the upper intake manifold is being removed, the fuel system pressure does not need to be released. ALWAYS release the pressure before disconnecting any fuel lines.**

1. Before servicing the vehicle, refer to the precautions in the beginning of this section.

2. Remove the engine cover, if equipped

3. Properly relieve the fuel system pressure.

4. Disconnect the negative battery cable.

5. Drain the engine cooling system.

6. Remove or disconnect the following:

- Air cleaner and air inlet duct
- Wiring harness connectors and brackets

- Throttle linkage from the upper intake manifold
- Ignition coil
- Fuel lines and bracket from the rear of the lower intake manifold
- Brake booster vacuum hose at the upper intake manifold
- Positive Crankcase Ventilation (PCV) hose at the rear of the upper intake manifold
- Vacuum hoses from both the front and rear of the upper intake
- Purge solenoid and bracket
- Upper intake manifold
- Distributor or High Voltage Switch (HVS) assembly
- Upper radiator hose at the thermostat housing

- Heater hose at the lower intake manifold
- Wiring harnesses and brackets
- Automatic transmission dipstick tube
- Exhaust Gas Recirculation (EGR) tube, clamp and tube
- Air conditioning compressor bracket-to-lower intake manifold pencil brace
- Alternator bracket bolts near the thermostat housing
- Lower intake manifold

7. Insert clean rags into the openings in the cylinder head to prevent dirt and debris from entering the engine.

8. Clean the gasket mounting surfaces. Be sure to inspect the manifold for warpage and/or cracks. If necessary, replace it.

To install:

9. Remove the rags from the cylinder heads.

10. Position the gaskets on the cylinder head with the port blocking plates to the rear and the **this side up** stamps facing upward. Then apply a ³⁄₁₆ in. (5mm) bead of RTV sealant on the front and rear of the engine block at the block-to-manifold mating surface. Extend the bead ½ in. (13mm) up each cylinder head to seal and retain the gaskets.

11. Install the lower intake manifold. Tighten the bolts in sequence and in 3 steps, as follows:

   a. Step 1: 26 inch lbs. (3 Nm).
   b. Step 2: 106 inch lbs. (12 Nm).
   c. Step 3: 11 ft. lbs. (15 Nm).

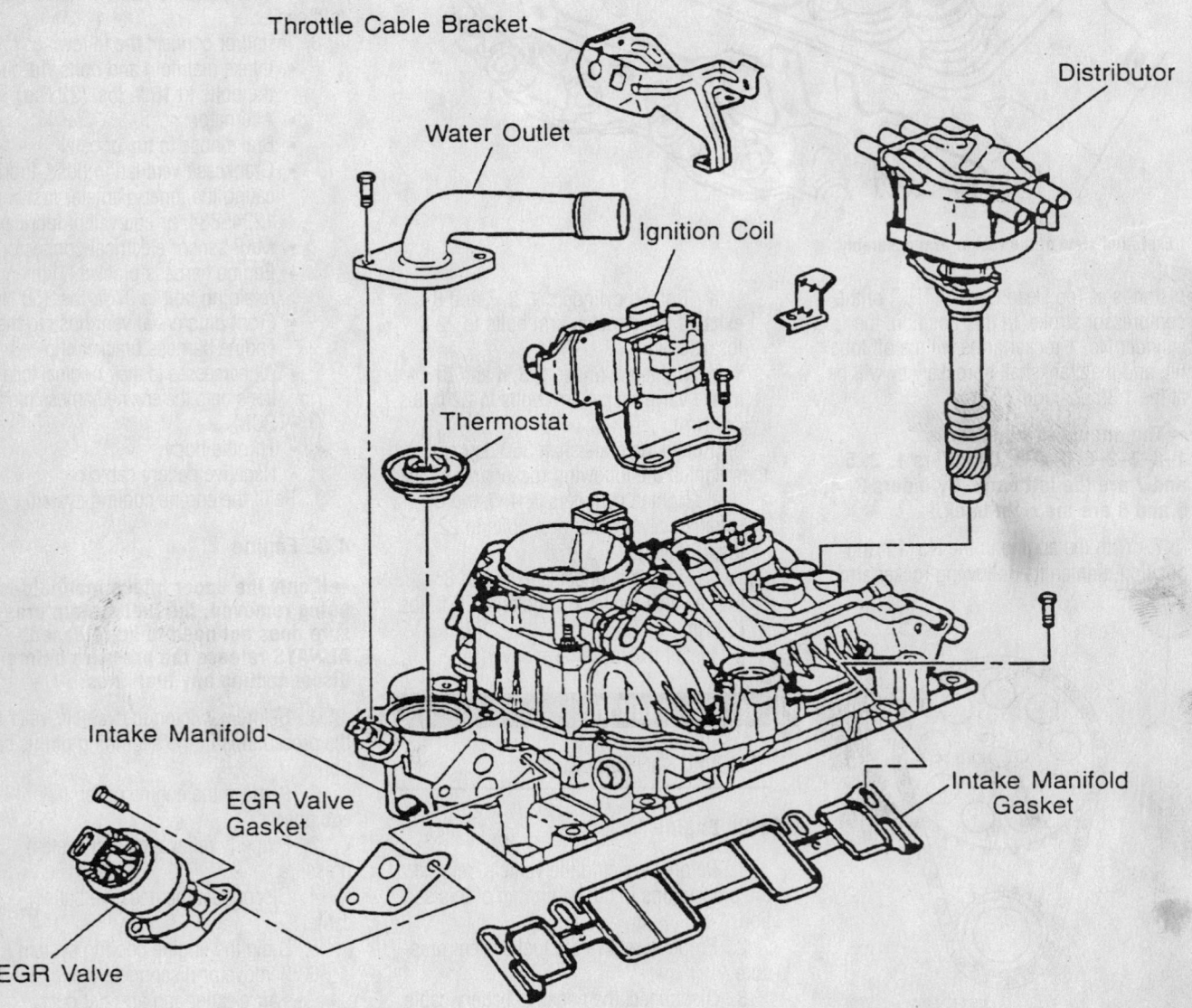

Throttle Cable Bracket

Water Outlet

Distributor

Ignition Coil

Thermostat

Intake Manifold

EGR Valve Gasket

Intake Manifold Gasket

EGR Valve

7924JG26

Intake manifold and related components

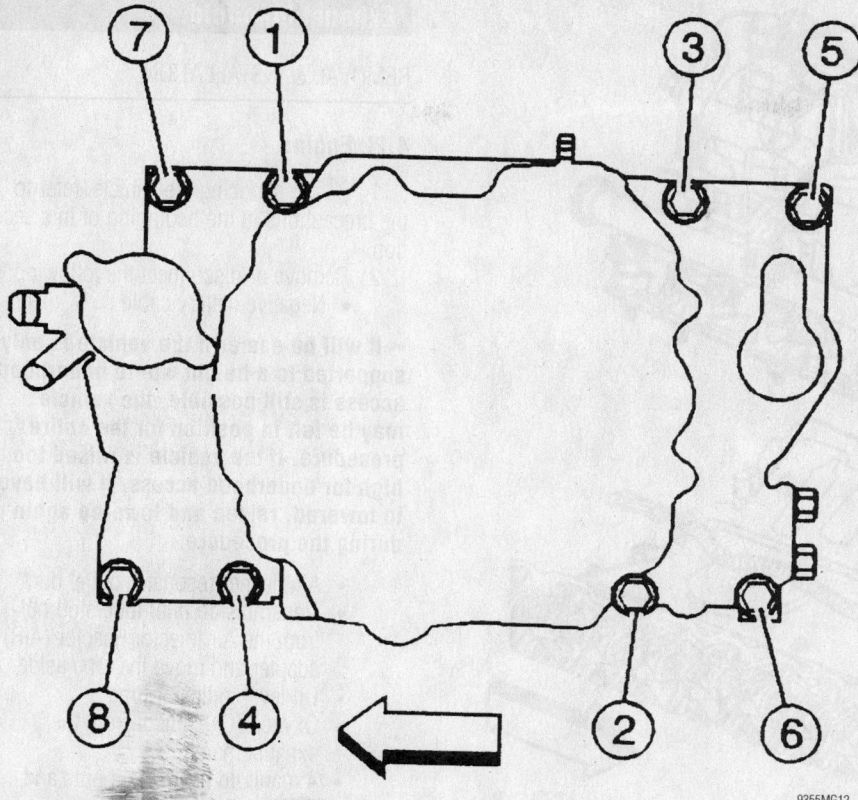

Intake manifold bolt tightening sequence—4.3L engine

9355MG12

1. Before servicing the vehicle, refer to the precautions in the beginning of this section.

2. Properly relieve the fuel system pressure.

3. Disconnect the negative battery cable.

4. Drain the engine cooling system.

5. Remove or disconnect the following:
  - Air cleaner outlet duct
  - A/C compressor pressure switch electrical connector
  - Harness clip from the cylinder head and fuel raid
  - Mass Airflow/Intake Air Temperature sensor connector

6. Disconnect the electrical connectors from the following:
  a. Main coil
  b. Electronic Throttle Control (ETC)
  c. Fuel injectors. Matchmark the connectors, pull the Connector Position Assurance (CPA) retainer up 1 click. Push the tab on the connector in, then detach the injector connector.
  - Alternator connector
  - Evaporative emission (EVAP) purge solenoid electrical connector
  - Knock Sensor (KS) electrical connector
  - Main coil
  - Fuel injector electrical connector
  - Electrical harness clips from the fuel rail
  - KS harness electrical connector from the intake manifold
  - Positive Crankcase Ventilation (PCV) valve hose and valve
  - Heater water shutoff valve actuator inlet hose from the intake manifold
  - EVAP purge solenoid vent tube
  - Vacuum brake booster hose from the rear of the intake manifold
  - Upper engine wire harness retainer nut. Position the wire harness aside.

12. Install or connect the following:
  - Alternator bracket bolt near the thermostat housing
  - EGR tube, clamp and bolt
  - Wiring harness to the lower manifold components, including the injector, EGR valve and ECT sensor
  - Air conditioning compressor bracket-to-the lower intake manifold pencil braces
  - Transmission oil dipstick tube, if necessary
  - Fuel supply and return lines to the rear of the lower intake

13. Temporarily reattach the negative battery cable, then pressurize the fuel system (by cycling the ignition without starting the engine) and check for leaks.

14. Disconnect the negative battery cable.

15. Install or connect the following:
  - Heater hose to the lower intake
  - Upper radiator hose to the thermostat housing
  - Distributor assembly and engage the wiring
  - Vacuum hoses to the upper and lower intake manifold
  - New upper intake manifold gasket, making sure the green sealing lines are facing upward
  - Upper intake manifold being careful not to pinch the fuel injector wires between the manifolds
  - Manifold retainers. Tighten them to 88 inch lbs. (10 Nm) using two passes.
  - Purge solenoid and bracket
  - Brake booster vacuum hose at the upper intake manifold
  - PCV hose to the rear of the upper intake manifold
  - Vacuum hoses to both the front and rear of the manifold assembly
  - Throttle linkage to the upper intake
  - Ignition coil
  - Wiring to the upper intake components including the TP sensor, IAC motor, MAP sensor and the IMTV
  - Plastic cover
  - Air cleaner and air inlet duct
  - Negative battery cable

16. Refill the engine cooling system.

5.3L Engine

➡The intake manifold, throttle body, fuel rail and injectors can be removed as an assembly. If you are not servicing these components individually, remove the intake manifold as a complete assembly.

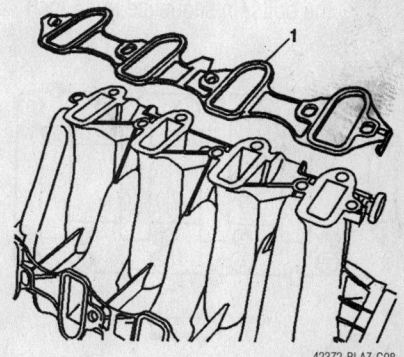

Make sure to use NEW intake manifold gaskets (1)—5.3L engine

42372-BLAZ-G08

Exploded view of the intake manifold—5.3L engine

- Intake manifold bolts
- Intake manifold and gaskets. Discard the gaskets.

To install:

7. Clean the gasket mounting surfaces. Be sure to inspect the manifold for warpage and/or cracks. If necessary, replace it.

8. Properly position a new intake manifold gasket.

9. Apply a 0.20 in. (5mm) band of a suitable threadlocking material to the intake manifold bolt threads.

10. Install or connect the following:
- Intake manifold and bolts. Torque the bolts, in sequence to 44 inch

lbs. (5 Nm), then to 89 inch lbs. (10 Nm).
- Route the electrical harness into position over the engine.
- Engine harness bracket nut and tighten to 89 inch lbs. (10 Nm)
- Vacuum brake booster hose to the rear of the intake manifold
- EVAP purge solenoid valve
- Heater water shutoff valve actuator inlet hose to the intake manifold
- PCV valve and hose
- EVAP purge solenoid, KS, MAP sensor, main coil & fuel injector electrical connectors
- Harness clips to the fuel rail
- Alternator electrical connector
- Main coil, ETC, fuel injector electrical connectors
- Electrical harness clips to the fuel rail
- A/C compressor pressure switch electrical connector
- Harness clip to the cylinder head
- Mass Airflow/Intake Air Temperature sensor connector
- Air cleaner outlet duct
- Fuel fill cap
- Negative battery cable

11. Refill the engine cooling system.

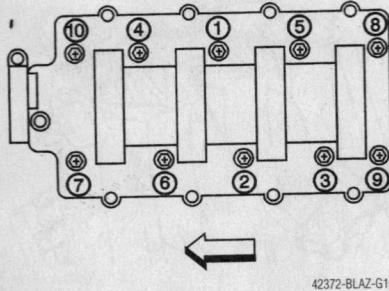

Intake manifold bolt tightening sequence—5.3L engine

Exhaust Manifold

REMOVAL & INSTALLATION

4.2L Engine

1. Before servicing the vehicle, refer to the precautions in the beginning of this section.

2. Remove or disconnect the following:
- Negative battery cable

➡ It will be easier if the vehicle is only supported to a height where underhood access is still possible, the vehicle may be left in position for the entire procedure. If the vehicle is raised too high for underhood access, it will have to lowered, raised and lowered again during the procedure.

- Air cleaner resonator outlet duct
- Transmission filler tube stud nut from the Air Injector Reactor (AIR) adapter and move the tube aside
- Oil level indicator tube
- Oxygen (O_2) sensor from the exhaust manifold
- 4 manifold heat shield nuts and shield
- Exhaust pipe bolts from the exhaust manifold
- Exhaust manifold bolts, and manifold
- Old gaskets and discard

To install:

3. Using a putty knife, clean the gasket mounting surfaces. Inspect the exhaust manifold for distortion, cracks or damage; replace if necessary.

4. Apply a threadlock such as GM 12345493 to the threads of the manifold retainers prior to installation.

5. Install or connect the following:
- Exhaust manifold to the cylinder using a new gasket, then tighten the bolts, in 3 passes, in sequence, to 18 ft. lbs. (25 Nm)
- Heat shield studs, if necessary, and tighten to 89 inch lbs. (10 Nm)
- O_2 sensor
- Exhaust manifold heat shield

➡ Apply a suitable anti-seize compound to the exhaust manifold heat shield nuts prior to installation.

- Heat shield nuts and tighten to 44 inch lbs. (5 Nm)
- Exhaust pipe to the manifold with seal and retaining nuts. Tighten the nuts to 37 ft. lbs. (50 Nm).
- Oil level indicator tube

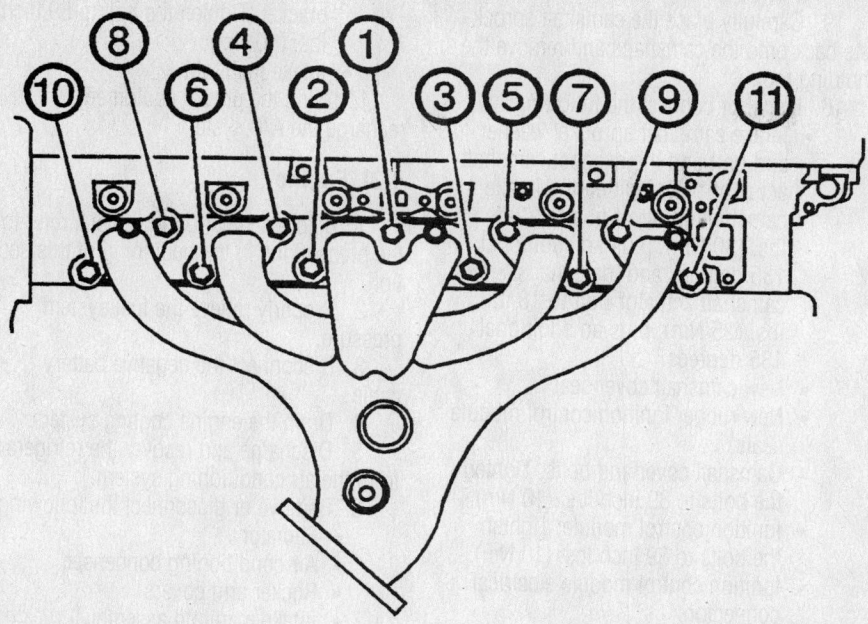

9355MG03

Exhaust manifold bolt tightening sequence—4.2L engine

- Transmission filler tube back onto the AIR adapter block stud and secure with the nut. Tighten the bracket nut to 89 inch lbs. (10 Nm).
- Air cleaner resonator outlet duct
- Negative battery cable.

4.3L Engine

1. Before servicing the vehicle, refer to the precautions in the beginning of this section.
2. Remove or disconnect the following:
   - Negative battery cable

➡ It will be easier if the vehicle is only supported to a height where underhood access is still possible, the vehicle may be left in position for the entire procedure. If the vehicle is raised too high for underhood access, it will have to lowered, raised and lowered again during the procedure.

- Exhaust pipe from the exhaust manifold. It may be necessary to remove the tires to gain access to the rear manifold bolts.
- Engine oil dipstick tube bolt, if removing the right side manifold on 2000–01 models
- Exhaust Gas Recirculation (EGR) inlet pipe from the left side manifold, if necessary.
- Engine Coolant Temperature (ECT) sensor electrical connection, on 2000–01 models

- Upper radiator support hose and nut, on 2000–01 models
- Steering intermediate shaft, if removing the left side manifold on 2000–01 models
- Wheel house extension, if removing the right side manifold on 2000–01 models
- Spark plugs wires from the plugs
- Nuts attaching the secondary air injection pipe to the manifold, on 2000–01 models
- Air injection pipe and gasket, on 2000–01 models
- Locktangs (unbend), the exhaust manifold retaining bolts, washers and tab washers
- Heat shields
- Exhaust manifold
- Old gaskets and discard

To install:

3. Using a putty knife, clean the gasket mounting surfaces. Inspect the exhaust manifold for distortion, cracks or damage; replace if necessary.
4. On 2000–01 models, apply a threadlock such as GM 12345493 to the threads of the manifold retainers prior to installation.
5. Install or connect the following:
   - Exhaust manifold to the cylinder using a new gasket, then tighten the center bolts to 11 ft. lbs. (15 Nm) and the front and rear manifold bolts to 22 ft. lbs. (30 Nm). Once the bolts are tightened, bend

the tabs on the washers back over the heads of all bolts in order to lock them in position.
- Spark plug wires to the plugs
- Fender wheelhouse extension and the tire assembly, if removed on 2000–01 models
- Secondary air injection pipe with a NEW gasket to the manifold and tighten the nuts to 18 ft. lbs. (25 Nm), if removed on 2000–01 models
- EGR inlet pipe, if removed
- ECT sensor electrical connection, if removed
- Upper radiator hose support and nut, if removed
- Steering intermediate shaft., if removed on 2000–01 models (left side manifold only)
- Engine oil dipstick tube bolt to 106 inch lbs. (12 Nm), if removed
- Exhaust pipe to the manifold
- Negative battery cable

5.3L Engine

1. Before servicing the vehicle, refer to the precautions in the beginning of this section.
2. Remove or disconnect the following:
   - Negative battery cable
   - Spark plug wires from the spark plugs. Don't disconnect the wires from the ignition coil unless necessary for clearance.
   - Exhaust manifold bolts, manifold and gasket. Discard the gasket.
   - Heat shield bolt and shield, if necessary

To install:

3. Apply a 0.2 inch (5mm) bead of threadlock GM P/N 12345493, or equivalent to the threads of the exhaust manifold bolts. Do NOT apply sealer to the first 3 threads of the bolts.
4. Install or connect the following:
   - New exhaust manifold gasket
   - Exhaust manifold
   - Exhaust manifold bolts. Tighten in two passes. First to 11 ft. lbs. (15 Nm), then to 18 ft. lbs. (25 Nm) starting with the center bolts and working outward.
5. Bend over the exposed edge of the gasket at the rear of the cylinder head using a flat punch or equivalent tool.
   - Heat shield and bolts, if removed. Torque the bolts to 80 inch lbs. (9 Nm).
   - Spark plug wires to the spark plugs
   - Negative battery cable

Camshaft and Valve Lifters

REMOVAL & INSTALLATION

4.2L Engine

1. Before servicing the vehicle, refer to the precautions in the beginning of this section.

2. Disconnect the negative battery cable.

3. Discharge and recover the refrigerant from the air conditioning system, using the proper equipment.

4. Remove or disconnect the following:
- Intake manifold
- A/C line from the oil level indicator tube
- A/C line from the accumulator
- A/C bracket bolt from the engine lift hook
- Engine lift bracket
- Ignition control module electrical connectors
- Ignition control module bolts and module

✳✳ WARNING

Be careful not to damage the clips that hold the harness housing in place.

- Engine electrical harness housing from the camshaft cover
- Fuel injection harness electrical connector
- Camshaft cover bolts and cover
- Exhaust and intake sprocket bolts

5. Install a suitable sprocket holding tool onto the cylinder head and adjust the horizontal bolts into the camshaft sprockets to maintain timing chain tension and avoid disturbing the timing chain components.

6. Carefully move the sprockets with the timing chain off of the camshafts.

➡**Make sure to place the camshaft caps in a rack to keep them in order, so they may be installed in their original locations.**

7. Remove or disconnect the following:
- Camshaft cap bolts and caps
- Camshafts

To install:

8. Coat the camshaft journals with engine oil.
- Camshafts, in their original position
- Camshaft caps, in their original locations. Tighten the bolts to 106 inch lbs. (12 Nm).

9. Carefully place the camshaft sprockets back onto the camshafts and remove the holding tool.

10. Install or connect the following:
- Intake camshaft sprocket washer and bolt and the exhaust camshaft actuator bolt. Tighten the intake camshaft sprocket bolt to 22 ft. lbs. (30 Nm), plus an additional 135 degrees and the exhaust camshaft actuator bolt to 18 ft. lbs. (25 Nm), plus an additional 135 degrees.
- New camshaft cover seal
- New rubber ignition control module seals
- Camshaft cover and bolts. Tighten the bolts to 89 inch lbs. (10 Nm).
- Ignition control module. Tighten the bolts to 89 inch lbs. (10 Nm).
- Ignition control module electrical connectors
- Fuel injector electrical connectors
- Engine electrical harness housing
- A/C line bracket to the oil level indicator tube stud and secure with the nut. Tighten the nut to 62 inch lbs. (7 Nm).
- Engine lift bracket and secure the lift hook with the bolts. Tighten the bolts to 37 ft. lbs. (50 Nm).
- A/C line bracket to the engine lift bracket. Tighten the bolt to 89 inch lbs. (10 Nm).
- Intake manifold

11. Using the proper equipment, recharge the A/C system.

4.3L Engine

1. Before servicing the vehicle, refer to the precautions in the beginning of this section.

2. Properly relieve the fuel system pressure.

3. Disconnect the negative battery cable.

4. Drain the engine cooling system.

5. Discharge and recover the refrigerant from the air conditioning system.

6. Remove or disconnect the following:
- Radiator
- Air conditioning condenser
- Rocker arm covers
- Intake manifold assembly
- Rocker arms, pushrods and lifters
- Crankshaft pulley and hub
- Engine front (timing) cover

7. Align the timing marks on the crankshaft and camshaft sprockets.
- Camshaft sprocket and timing chain
- Balance shaft drive gear, if equipped

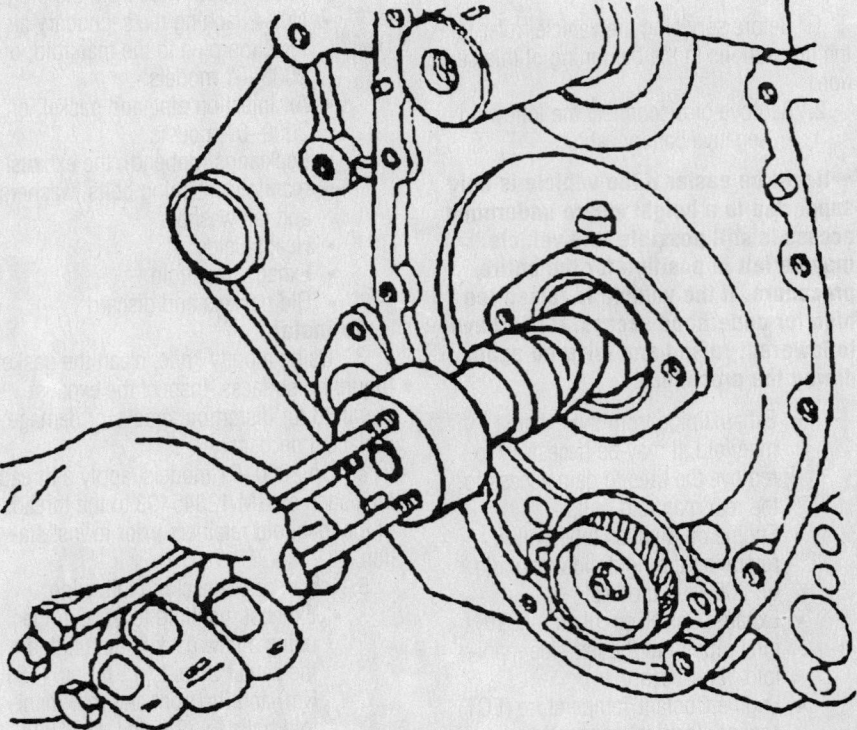

7924JG50

Thread 3 long bolts into the camshaft to use as a handle, then withdraw it from the engine

- Camshaft thrust plate
- Camshaft by installing the sprocket bolts or longer bolts the camshaft end to act as a handle; then, remove the camshaft while turning slightly from side to side, as necessary.

➡️**Take care not to damage the camshaft bearings when removing the camshaft.**

To install:

8. Lubricate the camshaft journals with clean engine oil or a suitable pre-lube.
9. Install or connect the following:
- Camshaft being extremely careful not to contact the bearings with the cam lobes
- Thrust plate. Torque the bolts to 106 inch lbs. (12 Nm).
- Balance shaft drive gear, if equipped
- Timing chain and camshaft sprocket
- Engine front (timing) cover
- Crankshaft pulley and hub
- Valve lifters, pushrods and rocker arms. Adjust the valve clearance.
- Intake manifold assembly
- Rocker arm covers
- Radiator
- Negative battery cable
10. Refill the engine cooling system.

5.3L Engine

1. Before servicing the vehicle, refer to the precautions in the beginning of this section.
2. Disconnect the negative battery cable.
3. Discharge and recover the refrigerant from the air conditioning system, using the proper equipment.
4. Remove or disconnect the following:
- Condenser
- Cylinder head and gasket
- Valve lifter guide bolts
- Valve lifters and guide

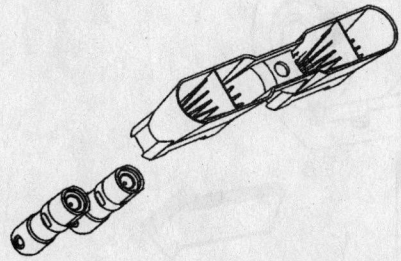

42372-BLAZ-G11

Remove the lifters from the guides, making sure to keep them in order—5.3L engine

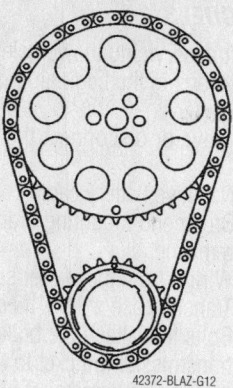

42372-BLAZ-G12

Make sure the crankshaft and camshaft timing marks are aligned

➡️**If the lifters are stuck in the bores due to built up deposits, use Valve Lifter Remover tool No. J 3049-A or equivalent to remove the lifters**

- Valve lifters from the guide

➡️**Make sure to keep the lifters in order as you are removing them. They must be installed in their original locations.**

5. Clean and inspect the lifters for damage.
- Camshaft sensor bolt and sensor
6. Rotate the crankshaft until the timing marks on the crankshaft and camshaft sprockets are aligned.
- Camshaft sprocket bolts

✳️ WARNING

Do NOT turn the crankshaft after the timing chain has been removed to avoid damaging the pistons or valves!

- Camshaft sprocket and reposition the timing chain
- Camshaft retaining bolts and retainer
- Camshaft by installing three M8-1.25 x 4.0 in. (M8-1.25 x 1.00mm) bolts in the front of the camshaft to act as a handle; then, remove the camshaft while turning slightly from side to side, as necessary. Remove the bolts from the camshaft.

➡️**Take care not to damage the camshaft bearings when removing the camshaft.**

7. Clean and inspect the camshaft and bearings.

To install:

➡️**If the camshaft must be replaced, you must also replace the lifters.**

8. Lubricate the camshaft journals with clean engine oil.
9. Install or connect the following:
- Three bolts used during removal into the bolt hold in the front of the camshaft
- Camshaft carefully into the engine block, using the bolts as a handle. Remove the bolts.
- Camshaft retainer and bolts. Make sure the retaining plate is installed with the sealing gasket facing the engine block. Tighten the bolts to 18 ft. lbs. (25 Nm).
10. Properly locate the camshaft sprocket locating pin with the cam sprocket alignment hole. The sprocket teeth and timing chain must mesh. The camshaft and crankshaft sprocket alignment marks MUST be aligned properly. Locate the camshaft sprocket alignment mark in the 6 o'clock position. It may be necessary to rotate the camshaft or crankshaft to align the marks.
- Camshaft sprocket and timing chain
- Camshaft sprocket bolts and tighten to 26 ft. lbs. (35 Nm)
- Camshaft sensor O-ring, after making sure it is not damaged and lubricating it with clean engine oil
- Camshaft sensor and bolt. Torque the bolt to 18 ft. lbs. (25 Nm).
11. Lubricate the valve lifters and engine block lifter bores with clean engine oil.
12. Install or connect the following:
- Lifters into the lifter guides. Align the area on top of the lifter with the flat area in the lifter guide bore. Push the lifter completely into the guide bore.
- Valve lifters and guide to the engine block
- Valve lifter guide bolt and tighten to 106 inch lbs. (12 Nm)
- Cylinder head and gasket
- Condenser
13. Using the proper equipment, recharge the A/C system.

Valve Lash

ADJUSTMENT

The 4.2L and 5.3L engines do not require a periodic valve lash adjustment.

The 4.3L engines are equipped with screw-in rocker arm studs with positive stop shoulders. Because the shoulders that allow the rocker arms to be tightened into proper position, no adjustments are necessary or possible. If a valvetrain problem is suspected, check that the rocker arm nuts are tightened to 18 ft. lbs. (24 Nm). When valve lash falls out of specification (valve tap is heard), replace the rocker arm, pushrod and hydraulic lifter on the offending cylinder.

Starter Motor

REMOVAL & INSTALLATION

4.2L Engine

1. Before servicing the vehicle, refer to the precautions in the beginning of this section.
2. Disconnect the negative battery cable
3. Raise and safely support the vehicle.
4. Remove the left front tire and wheel assembly.
5. Working in the left fender area, disconnect the positive battery lead from the solenoid.
6. Remove or disconnect the following:
   - Starter mount bolt and nut
   - Starter motor

To install:
7. Install or connect the following:
   - Starter motor
   - Starter mounting bolt and nut. Tighten to 37 ft. lbs. (50 Nm).
   - Positive battery cable to the starter. Tighten the nut to 80 inch lbs. (9 Nm).
   - Left front tire and wheel assembly
8. Carefully lower the vehicle, then connect the negative battery cable.

4.3L Engine

2WD MODELS

1. Before servicing the vehicle, refer to the precautions in the beginning of this section.
2. Remove or disconnect the following:
   - Negative battery cable
   - Wires from the starter solenoid
   - Starter motor mounting bolts
   - Starter motor and if equipped, the shims

To install:
3. Install or connect the following:
   - Starter motor into position
   - Starter motor inboard bolt but do not tighten it at this time
   - Starter motor shims, if equipped
   - Outboard starter motor bolt. Tighten the bolts to 32 ft. lbs. (43 Nm).
   - Wires to the solenoid
   - Negative battery cable

4WD MODELS

1. Before servicing the vehicle, refer to the precautions in the beginning of this section.
2. Remove or disconnect the following:
   - Negative battery cable
   - Brush end mounting bracket, if removed
   - Wiring from the starter solenoid
   - Transfer case shield, if equipped
   - Bolts that attach the brake pipe-to-transmission bracket to the transmission crossmember and the brackets
   - Transmission crossmember bolts, (usually three on each side)
   - Transmission mount bolts. Support the transmission assembly with a transmission jack and slide the transmission crossmember out of the way.
   - Bracket that attaches the transmission cooler lines to the flywheel housing, brace rod to the flywheel housing, and/or the lower flywheel housing as necessary
   - Starter motor mounting bolts
   - Starter and if equipped, the starter shims

To install:
3. Install or connect the following:
   - Shims in their original locations (if equipped), then place the starter motor into position
   - Starter motor bolts and tighten them to 33 ft. lbs. (45 Nm)
   - Lower flywheel cover, if removed
   - Transmission line bracket to the housing and the brace rod to the housing, if equipped
   - Crossmember and tighten the retaining bolts
   - Transfer case shield, if equipped
   - Solenoid wiring
   - Brush end bracket and tighten the nuts to 97 inch lbs. (11 Nm), if equipped
   - Negative battery cable
4. Start the vehicle to check for proper operation.

5.3L Engine

1. Before servicing the vehicle, refer to the precautions in the beginning of this section.
2. Remove or disconnect the following:
   - Negative battery cable
   - Catalytic converter
   - Engine shield bolts and shield

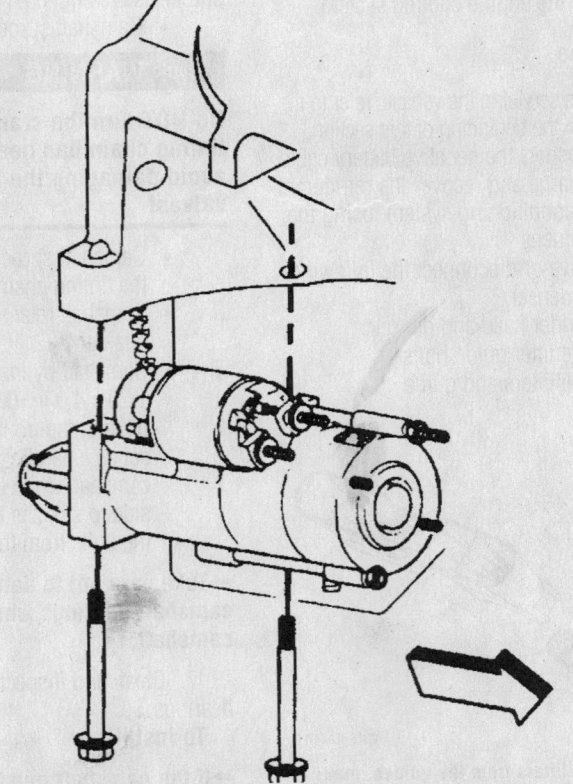

Starter mounting—4.3L engine

88452G09

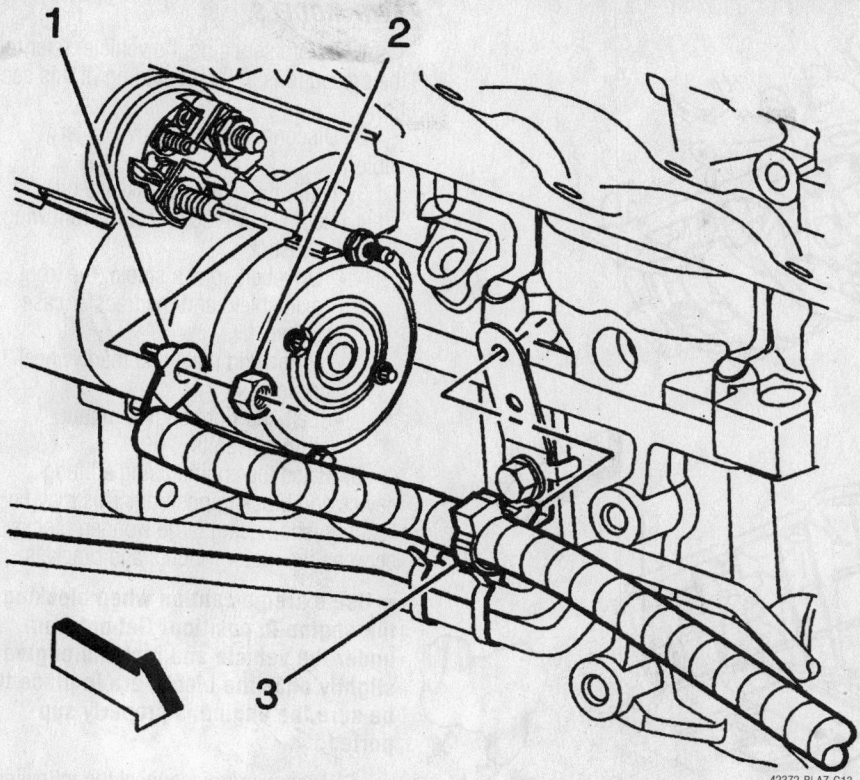

View of the starter, positive cable (1) and starter lead nut —5.3L engine

42372-BLAZ-G13

- Right transmission cover bolt
- Starter bolts
- Transmission cover and shield, after repositioning the starter

3. Position the starter down, with the terminals facing toward the front of the vehicle.

- Starter solenoid nut
- Starter lead from the solenoid stud
- Starter lead nut
- Positive cable from the starter stud
- Starter

To install:

4. Install or connect the following:

- Starter in the vehicle. Position the starter down , with the terminals facing toward the front of the vehicle.
- Positive cable to the starter stud.
- Starter lead nut and tighten to 80 inch lbs. (9 Nm)
- Starter solenoid lead to the stud
- Starter solenoid nut and tighten to 30 inch lbs. (3.4 Nm)
- Install the shield and transmission cover, after repositioning the starter

5. Slide the starter rearward.

- Starter bolts and tighten to 37 ft. lbs. (50 Nm)
- Right transmission cover bolt and tighten to 80 inch lbs. (9 Nm)

- Catalytic converter
- Negative battery cable

6. Start the vehicle to check for proper operation.

Oil Pan

REMOVAL & INSTALLATION

4.2L Engine

1. Before servicing the vehicle, refer to the precautions in the beginning of this section.

2. Disconnect the negative battery cable.

3. Remove or disconnect the following:
- A/C compressor bottom bolts and loosen the top bolts
- Oil dipstick and tube

4. Raise and safely support the vehicle.

5. Drain the engine crankcase oil.

6. Remove or disconnect the following:
- Left and right front tire and wheel assemblies
- Engine protection shield mounting bolts and shield
- Front steering gear crossmember
- Left and right driveshafts
- Front drive axle clutch fork assembly

- Prop shaft from the front axle pinion yoke
- Unclip the transmission cooler lines from the engine block
- Front differential bolts and position the differential aside
- 4 transmission bell housing-to-oil pan bolts
- Remaining oil pan bolts
- Oil pan, by placing 2 oil pan bolts in the jack screws on the oil pan and tighten evenly to release the oil pan from the engine

To install:

7. Clean the gasket mounting surfaces.

➡The alignment between the rear of the oil pan and the rear of the block is critical. When the oil pan is installed it could be inadvertently shifted front or back a small amount which could cause a transmission alignment problem. The back to the oil pan needs to be flush with the engine block.

8. Apply a 0.12 in. (3mm) bead of sealant to engine block, rather than the oil pan.

➡The oil pan MUST be installed within 10 minutes of applying the sealant to the engine block.

9. Install or connect the following:
- Oil pan, maneuvering it to clear the oil pump and screen assembly

➡After the bolts are installed, before tightening them to specifications, check the oil pan alignment. Use a straight edge on the back to the block and the oil pan transmission mounting surface.

- Oil pan bolts; tighten the side bolts to 18 ft. lbs. (25 Nm) and the end bolts to 89 inch lbs. (10 Nm)
- Transmission bell housing-to-oil pan bolts and tighten to 35 ft. lbs. (47 Nm)
- A/C compressor bottom bolts. Tighten to 37 ft. lbs. (50 Nm)
- Front differential bolts and tighten to 63 ft. lbs. (85 Nm)
- Front drive axle and clutch fork assembly
- Transmission cooler lines to block
- Prop shaft to front differential
- Steering gear crossmember
- Left and right driveshaft
- Oil pan drain plug. Tighten to 19 ft. lbs. (26 Nm)
- Engine protection shield. Tighten the bolts to 18 ft. lbs. (25 Nm)
- Left and right front wheel and tire assemblies

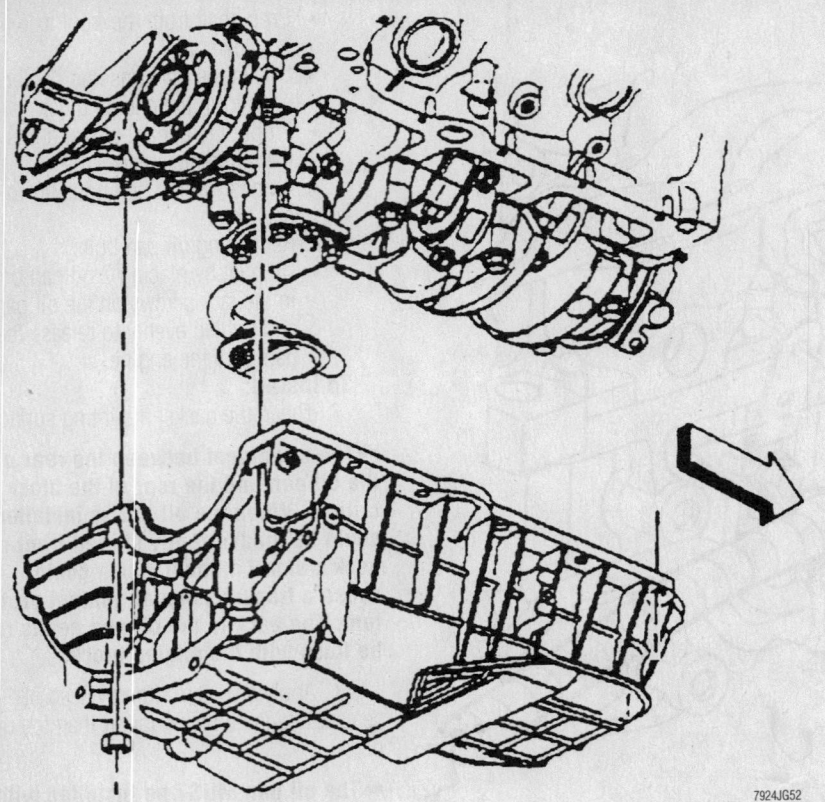

Oil pan mounting—4.3L engines

7924JG52

10. Carefully lower the vehicle.

11. Refill the crankcase with fresh oil. Start the engine, establish normal operating temperatures and check for leaks.

4.3L Engine

2WD MODELS

1. Before servicing the vehicle, refer to the precautions in the beginning of this section.

2. Remove or disconnect the following:
   • Engine
   • Oil pan retainers (nuts, studs and/or bolts) and rail reinforcements, if equipped
   • Oil pan
   • Rubber bell housing plugs and gasket

To install:

3. Clean the gasket mounting surfaces.

➡**The alignment between the rear of the oil pan and the rear of the block is critical. The oil pan must be flush or slightly forward of the rear of the block to allow for proper alignment with the transmission housing. Use a feeler gauge to measure the clearance between the 3 oil pan-to-transmission contact points. If the clearance exceeds 0.011 in. (0.3mm) at any of the 3 points, realign the oil pan.**

4. Apply sealant to the oil pan rail where it contacts the timing cover-to-block joint (front) and the crankshaft rear seal retainer-to-block joint (rear). Continue the bead of sealant about 1 in. (25mm) in both directions from each of the 4 corners.

5. Install or connect the following:
   • Rubber bell housing plugs, if equipped
   • Oil pan using a new gasket

➡**The alignment between the rear of the pan and rear of the block is critical. The two surfaces must be flush to allow for proper alignment with the transmission housing.**

6. Use a feeler gauge to check the clearance between the oil pan-to-transmission contacts. If clearance exceeds 0.011 inch (0.3mm) at any of the three contact points, readjust the pan until the clearance is within specification.

7. Once the pan is in its correct position tighten the retainers to 18 ft. lbs. (25 Nm) using the proper sequence.

8. Install the engine into the vehicle. Refill the crankcase with fresh oil. Start the engine, establish normal operating temperatures and check for leaks.

4WD MODELS

1. Before servicing the vehicle, refer to the precautions in the beginning of this section.

2. Disconnect the negative battery cable.

3. Drain the engine crankcase oil.

4. Remove or disconnect the following:
   • Dipstick
   • Drivebelt splash shield, the front axle shield and the transfer case shield
   • Front skid plate and the flywheel cover
   • Left and right engine mount through-bolts

5. Raise the engine using a lifting device and block in position. This may be accomplished using large wooden blocks between the motor mounts and brackets.

➡**Use extreme caution when blocking the engine in position. Get out from under the vehicle and rock the engine slightly once the blocks are in place to be sure the engine is properly supported.**

6. Remove or disconnect the following:
   • Oil cooler line
   • Pitman arm bolt and pitman arm
   • Idler arm bolts and idler arm
   • Front differential through-bolts
   • Front driveshaft, if necessary
   • Differential assembly by rolling it forward for clearance
   • Starter motor
   • Oil pan bolts, nuts and reinforcements
   • Oil pan and discard the gasket

To install:

7. Clean the gasket mounting surfaces.

➡**The alignment between the rear of the oil pan and the rear of the block is critical. The oil pan must be flush or slightly forward of the rear of the block to allow for proper alignment with the transmission housing. Use a feeler gauge to measure the clearance between the 3 oil pan-to-transmission contact points. If the clearance exceeds 0.011 in. (0.3mm) at any of the 3 points, realign the oil pan.**

8. Apply sealant to the oil pan rail where it contacts the timing cover-to-block joint (front) and the crankshaft rear seal retainer-to-block joint (rear). Continue the bead of sealant about 1 in. (25mm) in both directions from each of the 4 corners.

9. Install or connect the following:
   • Oil pan, using a new gasket.

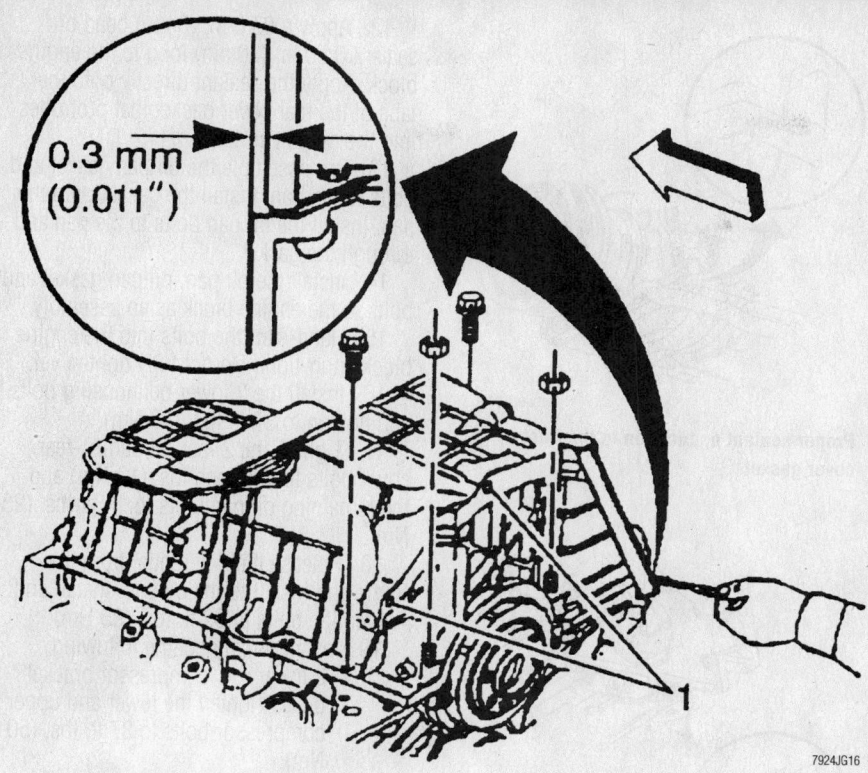

If the clearance between the 3 oil pan-to-transmission contact points exceeds 0.011 in. (0.3mm) at any of the 3 points, realign the oil pan

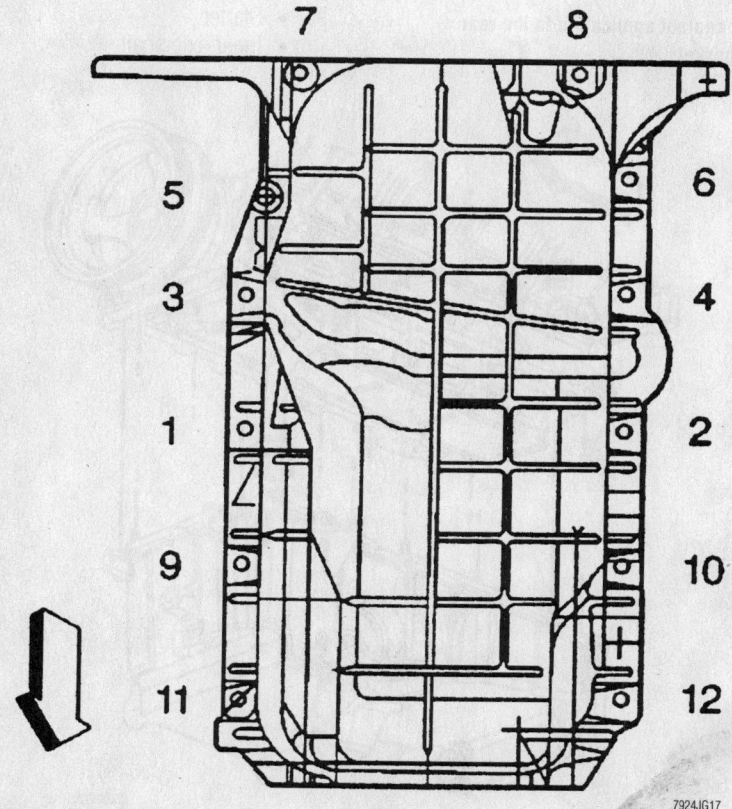

Tighten the bolts in sequence to prevent warping the sealing surface of the oil pan

Tighten the retainers, in sequence, to 18 ft. lbs. (25 Nm).
- Starter motor
- Differential by rolling it back into position
- Front driveshaft
- Front differential through-bolts
- Idler arm and secure using the retaining bolts
- Pitman arm and secure using the bolts
- Transfer case shield
- Flywheel cover
- Front skid plate
- Front axle shield
- Drive belt splash shield
- Dipstick
- Negative battery cable

10. Refill the engine crankcase.
11. Start the engine and check for leaks.

5.3L Engine

1. Before servicing the vehicle, refer to the precautions in the beginning of this section.

2. Disconnect the negative battery cable.

3. Drain the engine crankcase oil and differential oil.

4. Remove or disconnect the following:
- Oil level dipstick
- Front shock upper retaining nuts
- Tires and wheels
- Engine shield bolts and shield
- Power steering gear
- Left and right Antilock Brake System (ABS) wiring harnesses from the retainers
- Wheel Speed Sensor (WSS) electrical connectors
- Brake hose retaining bolts from the frame
- Sway bar link pins from the lower control arm on both sides

5. Place an adjustable jackstand under the lower control arm.
- Upper ball joint pinch bolt and nut
- Upper control arm from the upper ball joint

6. Lower and remove the jackstand, letting the suspension hang.
- Left driveshaft
- Right driveshaft from the intermediate shaft bearing only. Do not remove the driveshaft from the steering knuckle. Position the driveshaft aside.

7. Using wire or hooks, secure the front shock modules to the frame. Do NOT let the shocks and steering knuckle hang without being supported.

8. Matchmark the position of the propeller shaft to the front axle pinion yoke.

9. Remove or disconnect the following:

- Yoke retainer bolt and yoke retainers from the front axle pinion yoke. Wrap the bearing caps with tap to avoid losing the bearing rollers. Secure the propeller shaft to the frame.
- Transmission oil cooler lines from the retainer
- Transmission oil cooler line retaining bracket bolt and bracket
- Inner axle shaft
- Starter
- Flywheel inspection cover from the left side of the transmission
- Battery cable channel bolt from the front of the oil pan
- Battery cable channel from the oil pan
- Loosen the 2 upper A/C compressor bracket bolts
- 2 lower A/C compressor bracket bolts
- Front differential attachment bolts. Secure the front differential to the frame.
- 2 lower bellhousing bolts
- Oil pan bolts
- Oil pan by tilting the rear of the oil pan down to clear the transmission, pull the oil pan rearward past the front wire harness, then lower the oil pan clear of the vehicle

➡The oil pan gasket is reusable if it is not damaged.

10. Drill out the oil pan gasket retaining rivets, if necessary. Remove the gaskets. Discard the rivets. Inspect the gasket, if it is damaged, it must be replaced.

To install:

➡The proper alignment of the oil pan is very important. The rear bolt hold location of the oil pan provide mounting points for the transmission bellhousing. To ensure the rigidity of the powertrain and correct transmission alignment, make sure that the rear of the block and rear of the oil pan NEVER protrude beyond the engine block and transmission bellhousing plane.

➡If replacing the oil pan gasket, it is not necessary to rivet the NEW gasket to the pan.

11. Apply a 0.20 in. (5mm) bead of sealant 0.80 in. (20mm) long to the engine block. Apply the sealant directly onto the tabs of the front cover gasket that protrudes into the oil pan surface.

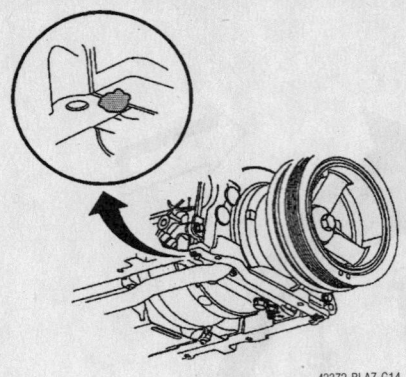

Proper sealant application to the front cover gasket

42372-BLAZ-G14

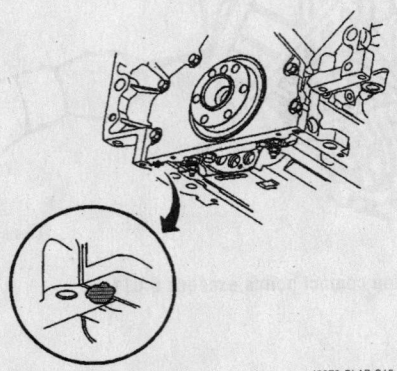

Proper sealant application to the rear cover gasket

42372-BLAZ-G15

12. Apply a 0.20 in. (5mm) bead of sealant 0.80 in. (20mm) long to the engine block. Apply the sealant directly onto the tabs of the rear cover gasket that protrudes into the oil pan surface.

13. Pre-assemble the oil pan gasket and bolts to the pan. Install the gasket onto the pan. Install the oil pan bolts to the pan and through the gasket.

14. Install the oil pan, oil pan gasket and bolts to the engine block as an assembly.

15. Hand-start the bolts into the engine block snug-tight. Do not fully tighten yet.

16. Install the 2 lower bellhousing bolts and tighten to 37 ft. lbs. (50 Nm).

17. Tighten the 2 rear oil pan-to-rear cover bolts to 106 inch lbs. (12 Nm) and the remaining oil pan bolts to 18 ft. lbs. (25 Nm).

18. Release the differential from the frame and install to the oil pan. Install and tighten the bolts to 63 ft. lbs. (85 Nm).

19. Install or connect the following:

- 2 lower A/C compressor bracket bolts. Tighten the lower and upper compressor bolts to 37 ft. lbs. (50 Nm).
- Battery cable channel to the oil pan
- Battery cable channel bolt and tighten to 106 inch lbs. (12 Nm)
- Flywheel inspection cover to the left side of the transmission
- Starter
- Inner axle shaft

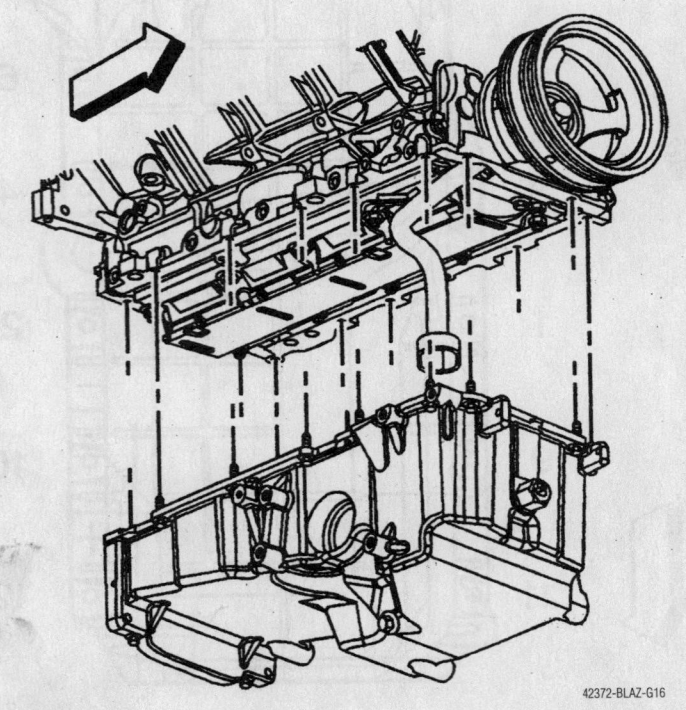

Oil pan mounting—5.3 engine

42372-BLAZ-G16

- Transmission oil cooler line retaining bracket and bolt. Torque the bolt to 80 inch lbs. (9 Nm).
- Transmission oil cooler lines to the retainer

20. Unhook the right driveshaft from the frame.

- Left and right driveshafts

21. Unsecure the shocks from the frame. Put adjustable jackstand under the lower control arm. Using the jackstand, raise the lower control arm and knuckle assembly in order to connect the upper ball joint to the upper control arm.

- Upper ball joint pinch nut and bolt and tighten to 30 ft. lbs. (40 Nm). Remove the jackstand.
- Sway bar link pins to the lower control arm on both sides
- Steering gear

22. Unsecure the prop shaft from the frame. Align the matchmarks on the prop shaft to the marks on the front axle pinion yoke.

- Propeller shaft to the front axle pinion yoke
- Yoke retainers and yoke retainer bolts to the front axle pinion yoke. Torque the bolts to 15 ft. lbs. (20 Nm).
- Brake hose retaining bolts to the frame and tighten to 18 ft. lbs. (25 Nm).
- WSS electrical connectors
- Left and right ABS wiring harnesses to the retainers
- Differential with oil
- Engine shield and bolts. Tighten the bolts to 18 ft. lbs. (25 Nm).
- Tires and wheels

23. Fill the engine with oil. Fill the power steering system with fluid.

- Upper shock nuts and tighten to 74 ft. lbs. (100 Nm).
- Oil dipstick
- Negative battery cable

Oil Pump

REMOVAL & INSTALLATION

4.2L Engine

1. Before servicing the vehicle, refer to the precautions in the beginning of this section.
2. Remove or disconnect the following:

- Engine front cover
- Oil pump cover bolts
- Oil pump cover. Mark the inner and outer gears in relation to the pump housing.

- Inner and outer pump gears
- Oil pump pressure relief valve plug
- Oil pump pressure relief valve and spring

To install:

3. Install or connect the following:

- Oil pump pressure relief valve and spring
- Oil pump pressure relief valve plug. Tighten to 10 ft. lbs. (14 Nm).
- Oil pump outer and inner gears, as marked during removal
- Oil pump cover and bolts. Tighten the bolts to 89 inch lbs. (10 Nm).
- Front cover

4.3L Engine

1. Before servicing the vehicle, refer to the precautions in the beginning of this section.
2. Remove or disconnect the following:

- Oil pan
- Oil pump and the pickup tube/shaft, if equipped

➡ **Be careful not to crack the retainer.**

To install:

3. Ensure that the pump pickup tube is tight in the pump body. If the tube should come loose, oil pressure will be lost and oil starvation will occur. If the pickup tube is loose it should be replaced.

4. If the pump has been disassembled and is being replaced or for any reason oil has been removed, it must be primed. It can either be filled with oil before installing the cover plate and oil kept within the pump during handling or the entire pump cavity can be filled with petroleum jelly.

❋❋ WARNING

If the pump is not primed, the engine could be damaged upon start up.

5. Install or connect the following:

- Oil pump by aligning the pump shaft with the distributor drive gear as necessary. Tighten oil pump/pickup tube retainer(s) to 65 ft. lbs. (90 Nm).

➡ **If the oil pump does not build up oil pressure almost immediately, remove the pan and check for a loose oil pump-to-pickup tube attachment. If necessary dismantle the pump and pack the pump cavity with petroleum jelly.**

- Oil pan

6. Refill the crankcase.
7. Disable the ignition system; crank

engine for approximately 10 seconds to aid in priming the oil pump and reducing the risk of engine damage.

❋❋ WARNING

Running the engine without measurable oil pressure will cause extensive damage.

5.3L Engine

1. Before servicing the vehicle, refer to the precautions in the beginning of this section.
2. Remove or disconnect the following:

- Oil pan
- Engine front cover
- Oil pump screen bolt and nuts
- Oil pump screen with O-ring seal
- O-ring seal from the pump screen. Discard the O-ring seal.
- Remaining crankshaft oil deflector nuts
- Crankshaft oil deflector
- Oil pump bolts
- Oil pump

➡ **Do not let any dirt or debris into the oil pump or cap end.**

- Clean and inspect the oil pump.

To install:

3. Align the splined surfaces of the crankshaft sprocket and the oil pump drive gear and install the oil pump.

4. Install or connect the following:

- Oil pump onto the crankshaft sprocket until the pump housing contacts the face of the engine block
- Oil pump bolts and tighten to 18 ft. lbs. (25 Nm)
- Crankshaft oil deflector and nuts until snug
- New oil pump screen O-ring seal into the oil pump screen, after lubricating with clean engine oil

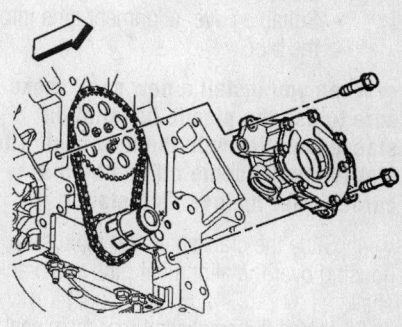

42372-BLAZ-G17

Exploded view of the oil pump mounting—5.3L engine

➡ Push the oil pump screen tube completely into the oil pump prior to tightening the bolt. Do not let the bolt pull the tube into the pump.

5. Align the oil pump screen mounting brackets with the correct crankshaft bearing cap studs.

- Oil pump screen
- Oil pump screen bolts and nuts. Tighten the bolts to 106 inch lbs. (12 Nm) and the nuts to 18 ft. lbs. (25 Nm).
- Engine front cover
- Oil pan

Rear Main Seal

REMOVAL & INSTALLATION

4.2L Engine

Please note that the transmission assembly must be removed to perform this procedure.

1. Before servicing the vehicle, refer to the precautions in the beginning of this section.

2. Remove or disconnect the following:
- Negative battery cable
- Transmission
- Flywheel
- Crankshaft rear main seal housing bolts. Install 2 bolts into the jackscrew holes to release the cover from the block
- Crankshaft and rear main seal housing
- Rear main seal from the crankshaft snout

To install:

3. Install or connect the following:
- Rear main seal, using a suitable seal installation tool, then remove the tool
- Apply a 0.12 in. (3mm) bead of 12378521, or equivalent sealant to the rear mail seal housing
- Suitable cover alignment pins into the block

➡ When you install a new seal, make sure to use the plastic installation sleeve supplies with the new seal. The sleeve should come off and be discarded after the seal is installed.

4. Slide the crankshaft rear main seal housing over the alignment pins and crankshaft.

5. Install the crankshaft rear main seal housing bolts, except the 2 in place of the guide pins.

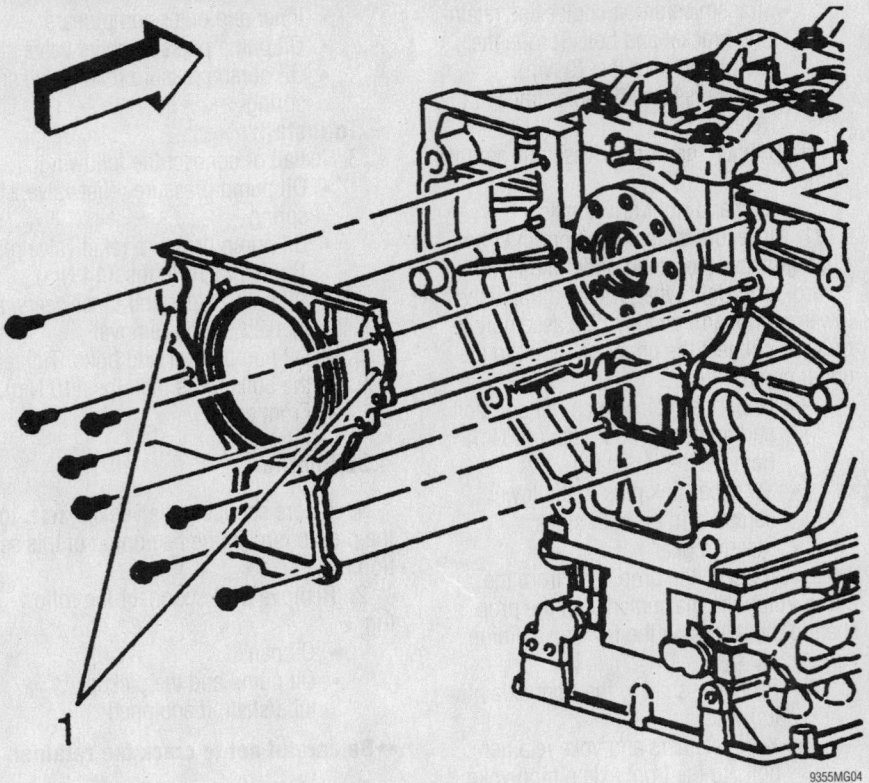

Install 2 bolts into the jackscrew holes (1) to push the cover off of the block

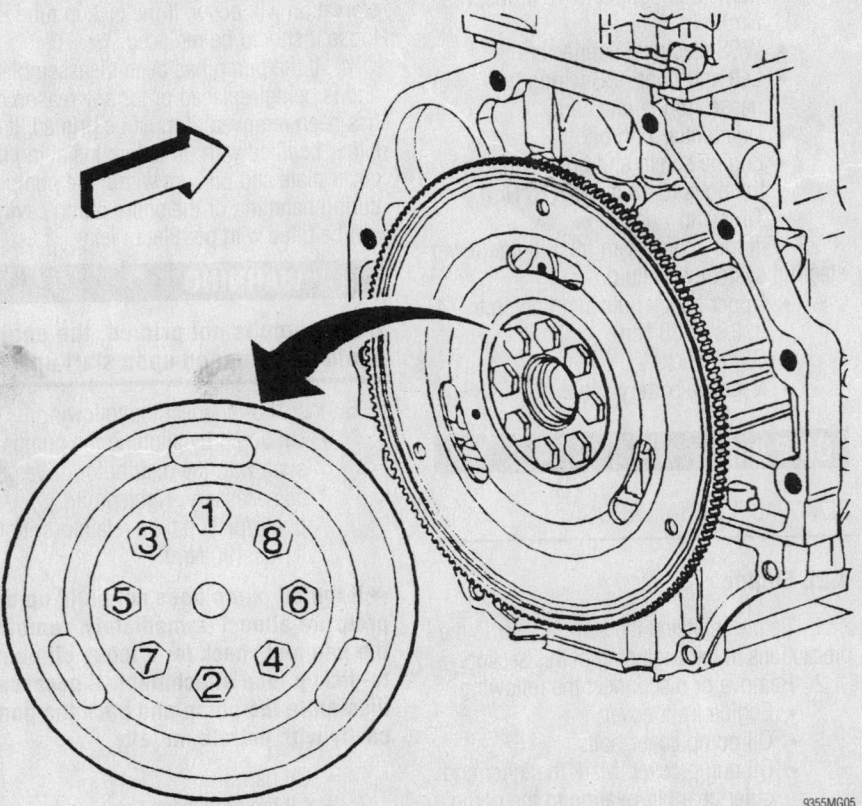

Flywheel bolt tightening sequence—4.2L engine

6. Remove the guide pins.

7. Install or connect the following:
   - Remaining 2 crankshaft rear main seal housing bolts and tighten to 89 inch lbs. (10 Nm). Wipe off any excess sealant.
   - Flywheel and secure with the mounting bolts. Tighten, in sequence, to 18 ft. lbs. (25 Nm), plus an additional 50 degrees.
   - Transmission

4.3L Engine

Please note that the transmission assembly and transfer case, if equipped, must be removed to perform this procedure.

1. Before servicing the vehicle, refer to the precautions in the beginning of this section.

2. Remove or disconnect the following:
   - Negative battery cable
   - Transfer case, if equipped
   - Transmission
   - Clutch assembly/flywheel or flexplate

3. Remove the crankshaft rear oil seal by inserting a suitable prying tool into the notches provided in the seal retainer and prying the seal out. Take care not to damage the crankshaft sealing surface.

To install:

4. Inspect the crankshaft for grit, rust or burrs and correct as necessary.

5. Clean the running surface of the crankshaft with a non-abrasive cleaner.

6. Install or connect the following:
   - New rear seal lubricated with engine oil and a seal installer
   - Flywheel and clutch or flexplate
   - Transmission
   - Transfer case, if equipped
   - Negative battery cable

7. Start the engine and verify no oil leaks.

5.3L Engine

Please note that the transmission assembly must be removed to perform this procedure.

1. Before servicing the vehicle, refer to the precautions in the beginning of this section.

2. Remove or disconnect the following:
   - Negative battery cable
   - Transmission
   - Flywheel
   - Crankshaft rear main oil seal from the rear cover

To install:

➡ **The flywheel spacer (if applicable) must be removed prior to oil seal installation. Do not lubricate the oil seal Inside Diameter (ID) or crankshaft surface. Never reuse the rear main seal. Once it is removed, it must be replaced with a new seal.**

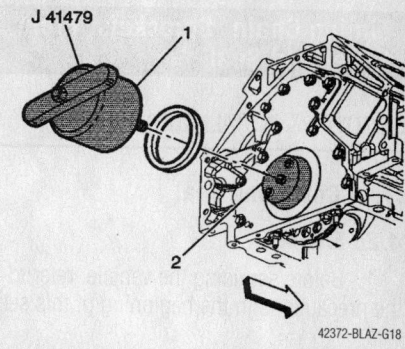

View of the rear main seal installation—5.3L engine

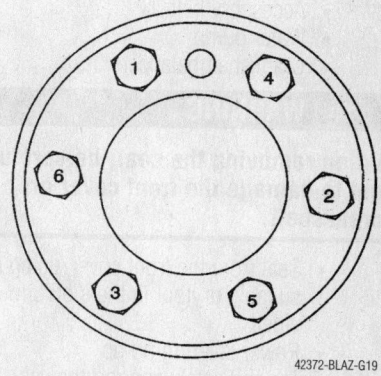

42372-BLAZ-G19

Flywheel bolt tightening sequence—5.3L engine

3. Lubricate the Outside Diameter (OD) of the rear main seal and the rear cover oil seal bore with clean engine oil. Do NOT let oil contact the seal surface or the crankshaft surface.

4. Install or connect the following:
   - Crankshaft Rear Oil Seal Installer Tool No. J 41479 tapered cone and bolts onto the rear of the crankshaft. Tighten the bolts until just snug, being careful not to overtighten.
   - Rear oil seal onto the tapered cone until the tool contacts the oil seal

5. Align the oil seal into the tool, Rotate the handle of the tool clockwise until the seal enters the rear cover and bottoms into the cover bore. Remove the tool.
   - Flywheel and secure with the mounting bolts.

6. Tighten the flywheel mounting bolts, in sequence, as follows:
   a. 1st pass: 15 ft. lbs. (20 Nm)
   b. 2nd pass: 37 ft. lbs. (50 Nm)
   c. Final pass: 74 ft. lbs. (100 Nm)
   - Transmission
   - Negative battery cable

7. Start the engine and verify no oil leaks.

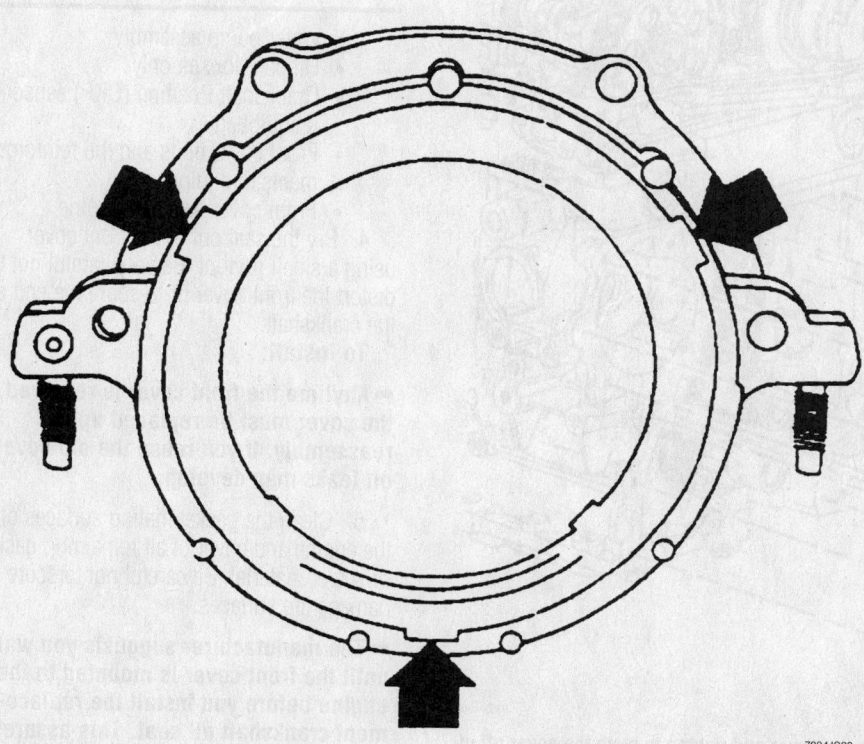

Carefully pry the rear main seal out of the retainer

Timing Chain, Sprockets, Front Cover and Seal

REMOVAL & INSTALLATION

Front Cover and Seal

4.2L ENGINE

1. Before servicing the vehicle, refer to the precautions in the beginning of this section.
2. Remove or disconnect the following:
   - Negative battery cable
   - Drain the engine cooling system.
   - Cooling fan and shroud
   - Accessory belt
   - Water pump
   - Crankshaft balancer

** WARNING

When removing the seal, be careful not to damage the front cover or crankshaft.

   - Seal from the front cover, using a suitable prytool in the slots provided
   - Power steering pump
3. Raise and safely support the vehicle.

- Oil pan, then carefully lower the vehicle
- 7mm center bolt
- Remaining front cover bolts. Place two of the front cover bolts in the jackscrew holes on the front cover and tighten the bolts evenly to release the front cover from the engine.
- 2 bolts from the front cover
- Oil pump

To install:

4. Clean the gasket mating surfaces of the engine and cover of all remaining gasket or sealer material. Be careful not to score or damage the surfaces.
5. Install or connect the following:
   - Suitable cover alignment pins, onto the engine

➡**The front cover MUST be installed within 10 minutes of applying the sealant.**

   - Apply a 0.12 in. (3mm) beat of 12378521 or equivalent sealant to the trace grooves on the back side of the engine front cover. Apply sealant on the inside 3 bolt hole bosses on the cover also.
   - Oil pump to the crankshaft splines

- Front cover and bolts, tighten the center bolt last. Tighten to 89 inch lbs. (10 Nm).
6. Remove the alignment pins and raise and safely support the vehicle. Install the oil pan, then lower the vehicle.
   - Power steering pump
   - Crankshaft balancer
   - Water pump
   - Accessory belt
   - Cooling fan and shroud
   - Negative battery cable
7. Properly refill the engine cooling system.
8. Run the engine until normal operating temperature has been reached, then check for leaks.

4.3L ENGINE

1. Before servicing the vehicle, refer to the precautions in the beginning of this section.
2. Remove or disconnect the following:
   - Negative battery cable
3. Drain the engine cooling system.
   - Crankshaft pulley and damper

** WARNING

The outer ring (weight) of the torsional damper is bonded to the hub with rubber. The damper must be removed with a puller which acts on the inner hub only. Pulling on the outer portion of the damper will break the rubber bond or destroy the tuning of the unit.

   - Water pump assembly
   - Oil pan, loosen only
   - Crankshaft Position (CKP) sensor, if equipped
   - Front cover bolts and the reinforcements, if equipped
   - Front cover from the engine
4. Pry the seal out of the front cover using a small prytool. Be very careful not to distort the front cover or to score the end of the crankshaft.
   To install:

➡**Anytime the front cover is removed, the cover must be replaced upon reassembly. If you reuse the old cover, oil leaks may develop.**

5. Clean the gasket mating surfaces of the engine and cover of all remaining gasket or sealer material. Be careful not to score or damage the surfaces.

➡**The manufacturer suggests you wait until the front cover is mounted to the engine before you install the replacement crankshaft oil seal. This assures the cover is properly supported.**

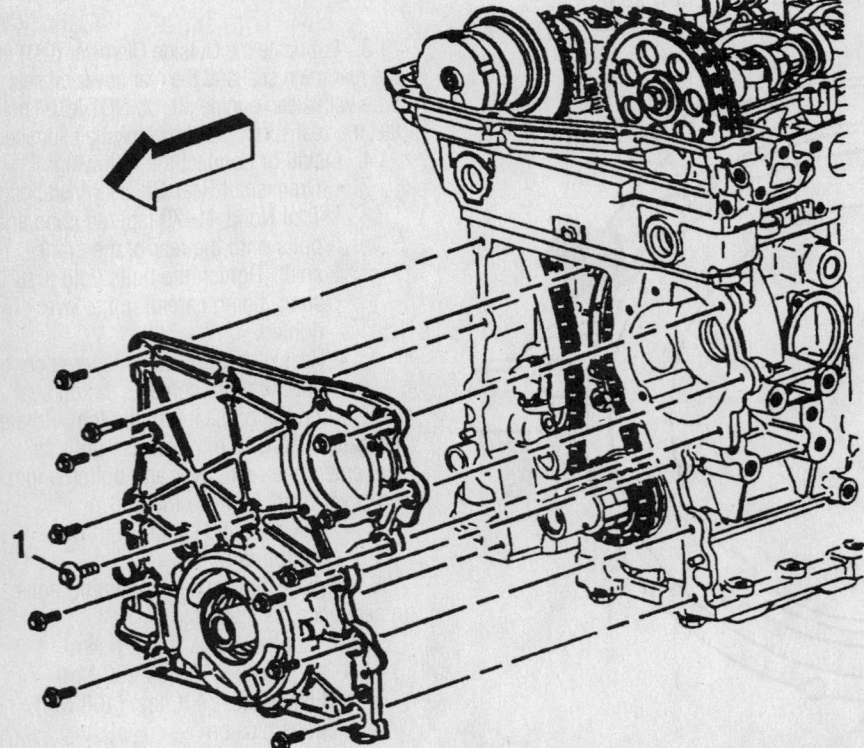

1

9355MG06

Place 2 front cover bolts in the jackscrew holes on the cover and tighten to push the cover off of the engine

6. Install or connect the following:
- New front cover gasket to the engine or cover using gasket cement to hold it in position. Lubricate the front of the oil pan seal with engine oil to aid in reassembly.
- Front cover to the engine. Take care while engaging the front of the oil pan seal with the bottom of the cover. On 2000–01 models, apply sealer 12346141 to the oil pan rail where it contacts the timing cover-to-block joint (front) and the crankshaft rear seal retainer-to-block joint (rear). Continue the bead of sealant about 1 in. (25mm) in both directions from each of the four corners.
- Front cover retaining bolts and tighten to 106 inch. lbs. (12 Nm)

7. Lightly coat the lips of the replacement crankshaft seal with clean engine oil, then position the seal with the open end facing inward the engine. Use a suitable seal installation driver to position the seal in the front cover.
- CKP sensor O-ring and the sensor, if equipped
- Tighten the Oil pan bolts

- Water pump
- Crankshaft damper and pulley
- Negative battery cable

8. Properly refill the engine cooling system.

9. Run the engine until normal operating temperature has been reached, then check for leaks.

5.3L ENGINE

1. Before servicing the vehicle, refer to the precautions in the beginning of this section.
2. Properly discharge the A/C system.
3. Drain the engine cooling system.
4. Remove or disconnect the following:
- Negative battery cable
- A/C compressor and bracket
- Water pump
- Crankshaft balancer
- Oil pan-to-front cover bolts
- Front cover bolts
- Front cover and gasket. Discard the gasket.

5. Clean and inspect the front cover.

To install:

6. Apply a 0.20 in. (5mm) bead of sealance 0.80 in. (20mm) long to the oil pan-to-engine block junction.

7. Install or connect the following:
- New front cover gasket and cover

- Front cover bolts, finger-tight
- Oil pan-to-front cover bolts, finger-tight
- Front and Rear Cover Alignment Tool No. J 41476 to the front cover. Align the tapered legs of the tool with the machined alignment surfaces on the front cover
- Crankshaft balancer bolt, finger-tight
- Oil pan-to-front cover bolts to 18 ft. lbs. (25 Nm)
- Front cover bolts to 18 ft. lbs. (25 Nm)

8. Remove the tool.

9. Install a NEW crankshaft front oil seal as follows:
   a. Remove the radiator for access.
   b. Remove the crankshaft balancer.
   c. Remove the crankshaft oil seal.
   d. Lubricate the outer edge ONLY of the NEW crankshaft oil seal with clean engine oil.

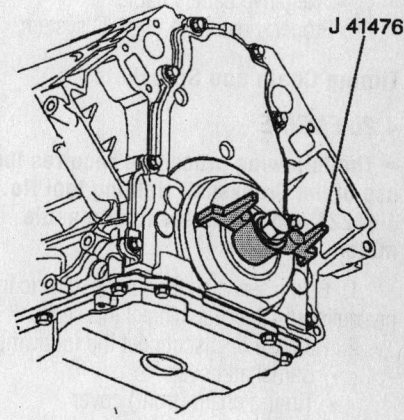

42372-BLAZ-G20

Align the tapered legs of the tool with the machined alignment surfaces on the front cover—5.3L engine

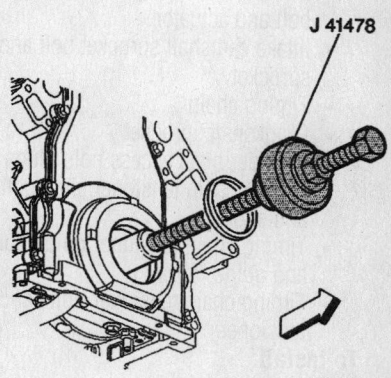

J 41478

42372-BLAZ-G21

Front cover seal installation using the proper tool—5.3L engine

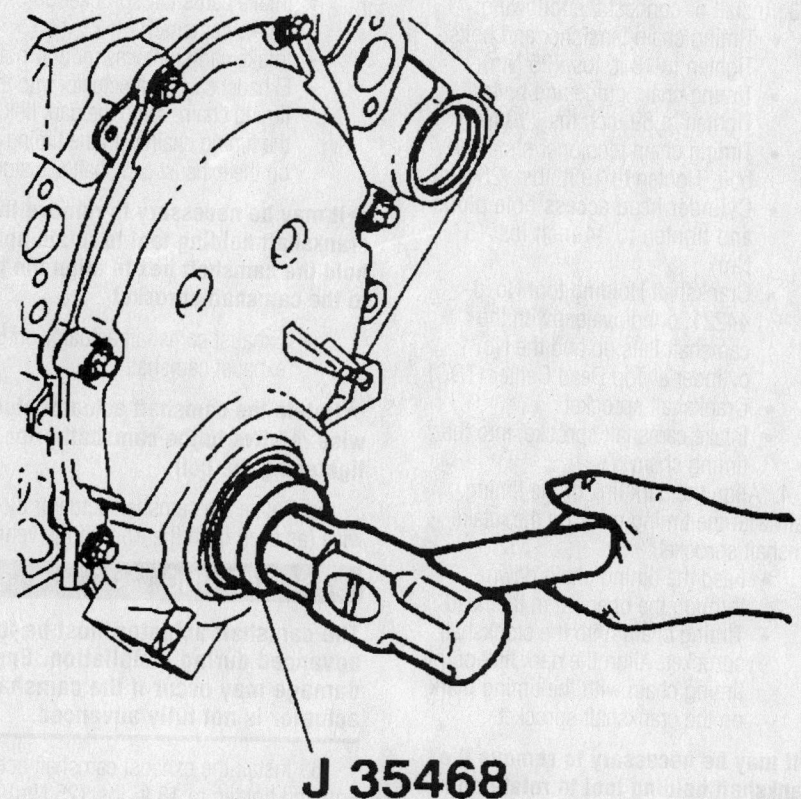

J 35468

88453GAU

Installing the crankshaft front oil seal—4.3L engine

e. Install the crankshaft front oil seal into the Crankshaft Front Seal Installation Tool No. J 41478 guide.

f. Install the J 41478 threaded rod (with nut, washer, guide and oil seal) into the end of the crankshaft.

g. Use J 41478 to install the oil seal into the cover bore. Use a wrench and hold the hex on the installer bolt. Use a second wrench to rotate the installer nut clockwise until the seal bottoms in the cover bore. Remove the tool.

h. Check the seal for proper installation. It should be installed evenly and completely into the front cover bore.

i. Install the crankshaft balancer. Tighten the bolt to 37 ft. lbs. (50 Nm), plus an additional 140 degrees using a torque angle meter.

j. Install the radiator.

10. Install or connect the following:
- Water pump
- A/C compressor and bracket
- Cooling system with coolant
- Negative battery cable

11. Properly recharge the A/C system

Timing Chain and Sprockets

4.2L ENGINE

➡The following procedure requires the use of the Crankshaft Holding tool No. J-44221 and a suitable torque angle meter.

1. Before servicing the vehicle, refer to the precautions in the beginning of this section.

2. Remove or disconnect the following:
- Camshaft cover
- Timing chain (front) cover
- Tension on the timing chain by moving the tensioner shoe in. Place a tee into the tension to hold the shoe in place.
- Top chain guide bolts and guide
- Exhaust camshaft position actuator bolt and actuator
- Intake camshaft sprocket bolt and sprocket
- Timing chain
- Crankshaft sprocket
- Cylinder head access hole plugs
- Timing chain tensioner shoe bolt and shoe
- Timing chain tensioner guide bolts and guide
- Timing chain tensioner bolts and tensioner

To install:

➡Every seventh link of the timing chain is darkened to help in aligning the timing marks.

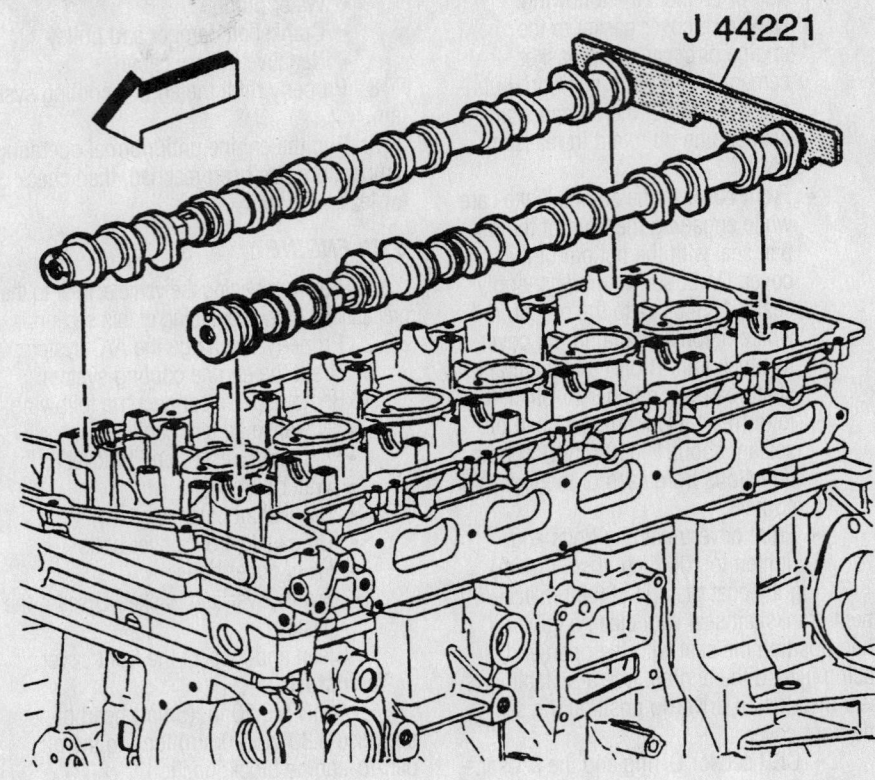

Proper installation of the crankshaft holding tool with the No. 1 cylinder at TDC

9355MG07

3. Install or connect the following:
- Timing chain tensioner and bolts. Tighten to 18 ft. lbs. (25 Nm).
- Timing chain guide and bolts. Tighten to 89 inch lbs. (10 Nm).
- Timing chain tensioner shoe and bolt. Tighten to 19 ft. lbs. (26 Nm).
- Cylinder head access hole plugs and tighten to 44 inch lbs. (5 Nm)
- Crankshaft Holding tool No. J-44221, or equivalent with the camshaft flats up and the No. 1 cylinder at Top Dead Center (TDC)
- Crankshaft sprocket
- Intake camshaft sprocket into the timing chain

4. Align the dark link of the timing chain with the timing mark on the intake camshaft sprocket.
- Feed the timing chain down through the opening in the head.
- Timing chain onto the crankshaft sprocket. Align the dark link of the timing chain with the timing mark on the crankshaft sprocket.

➡It may be necessary to remove the crankshaft holding tool to rotate and hold the camshaft hex to align the pin to the camshaft sprocket

- Intake camshaft sprocket onto the intake camshaft
- Intake camshaft washer and bolt
- Exhaust camshaft actuator into the timing chain. Align the dark link of the timing chain with the timing mark on the exhaust camshaft actuator.

➡It may be necessary to remove the crankshaft holding tool to rotate and hold the camshaft hex to align the pin to the camshaft sprocket

- Exhaust camshaft actuator onto the exhaust camshaft

➡Rotate the camshaft actuator clockwise relative to the camshaft prior to tightening the bolt.

5. Rotate the camshaft actuator clockwise (as seen from the front of the vehicle).

❈❈ WARNING

The camshaft actuator must be fully advanced during installation. Engine damage may occur if the camshaft actuator is not fully advanced.

6. Install the exhaust camshaft actuator bolt and tighten to 18 ft. lbs. (25 Nm), plus an additional 135 degrees, using a torque angle meter.

- Camshaft sprocket (along with the timing chain). If the sprocket is difficult to remove, use a plastic mallet to bump the sprocket from the camshaft.

➡**The camshaft sprocket (located by a dowel) is lightly pressed onto the camshaft and should come off easily. The chain comes off with the camshaft sprocket.**

1. If necessary use J-5825-A crankshaft sprocket removal tool to free the timing sprocket from the crankshaft.

2. If necessary, remove the crankshaft sprocket key

To install:

3. Inspect the timing chain and the timing sprockets for wear or damage, replace the damaged parts as necessary.

4. Using a putty knife, clean the gasket mounting surfaces. Using solvent, clean the oil and grease from the gasket mounting surfaces.

5. Install or connect the following:
- Crankshaft sprocket key, if removed
- Crankshaft sprocket onto the crankshaft using J-5590 crankshaft sprocket installation tool and a hammer without disturbing the position of the engine

➡**During installation, coat the thrust surfaces lightly with Molykote® or an equivalent pre-lube.**

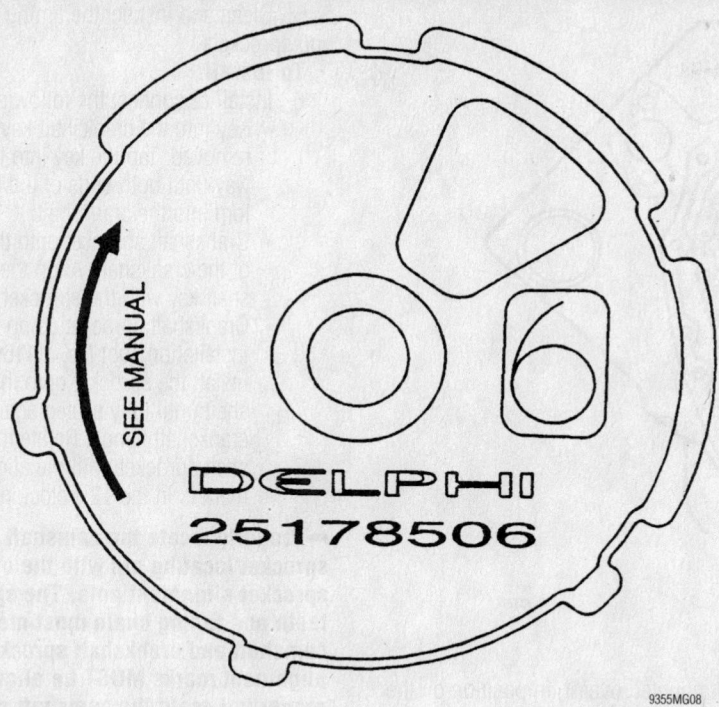

9355MG08

Rotate the camshaft actuator clockwise

7. Tighten the intake camshaft sprocket bolt to 22 ft. lbs. (30 Nm), plus an additional 135 degrees, using a torque angle meter.

8. Remove the tee from the timing chain tensioner to regain tension on the timing chain.

9. Remove the crankshaft holding tool. The dark lines on the timing chain should be aligned with the marks on the sprockets.

10. Install or connect the following:
- Top chain guide
- Suitable threadlock to the top chain guide bolt threads, then install and tighten to 89 inch lbs. (10 Nm)
- Engine front cover
- Camshaft cover

4.3L ENGINE

➡**The following procedure requires the use of the Crankshaft Sprocket Removal tool No. J-5825-A and the Crankshaft Sprocket Installation tool No. J-5590.**

1. Before servicing the vehicle, refer to the precautions in the beginning of this section.

2. Remove the timing cover from the engine.

3. Rotate the crankshaft until the No. 4 cylinder is on the Top Dead Center (TDC) of its compression stroke and the camshaft sprocket mark aligns with the mark on the crankshaft sprocket (facing each other at a point closest together in their travel) and in line with the shaft centers.

4. Remove or disconnect the following:
- Crankshaft Position (CKP) sensor reluctor ring, if equipped
- Camshaft sprocket-to-camshaft nut and/or bolts

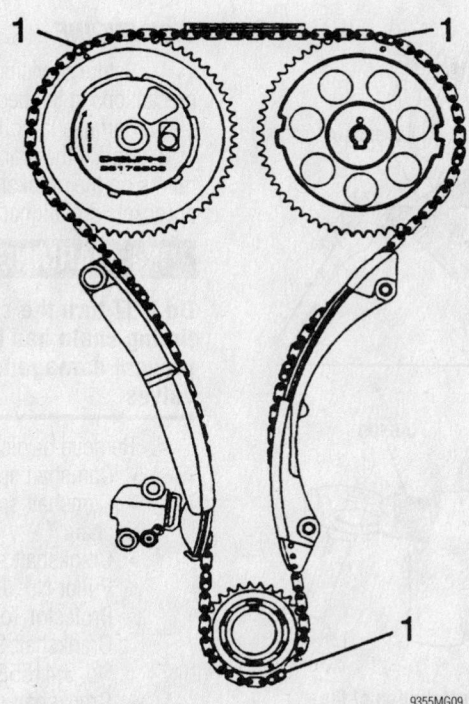

9355MG09

The dark lines on the timing chain should be aligned with the marks on the sprockets

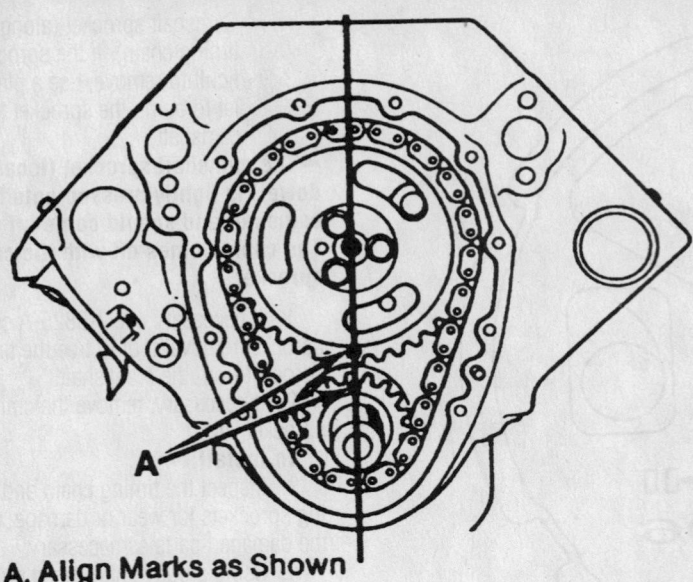

A. Align Marks as Shown

85383292

Timing mark alignment

- Timing chain over the camshaft sprocket. Arrange the camshaft sprocket in such a way that the timing marks will align between the shaft centers and the camshaft locating dowel will enter the dowel hole in the cam sprocket.
- Timing chain under the crankshaft sprocket, then place the cam sprocket, with the chain still

mounted over it, in position on the front of the camshaft
- Camshaft sprocket-to-camshaft retainers to 18 ft. lbs. (25 Nm)

6. With the timing chain installed, turn the crankshaft two complete revolutions, then check to make certain that the timing marks are in correct alignment between the shaft centers.

- CKP sensor reluctor ring, if equipped
- Timing cover

5.3L ENGINE

1. Before servicing the vehicle, refer to the precautions in the beginning of this section.
2. Remove the oil pump.
3. Rotate the crankshaft until the timing marks on the crankshaft and the camshaft sprockets are aligned.

> **✳✳ WARNING**
>
> **Do NOT turn the crankshaft after the timing chain has been removed to prevent damage to the pistons and valves.**

4. Remove or disconnect the following:
- Camshaft sprocket bolts
- Camshaft sprocket and timing chain
- Crankshaft sprocket using Pulley Puller No. J 8433, Crankshaft End Protector Tool No. J 41816-2 and Crankshaft Sprocket Removal Tool No. J 41558
- Crankshaft sprocket key, if necessary

5. Clean and inspect the timing chain and sprockets.

To install:

6. Install or connect the following:
- Key into the crankshaft keyway, if removed. Tap the key into the keyway until both ends of the key bottom into the crankshaft.
- Crankshaft sprocket onto the front of the crankshaft. Align the crankshaft key with the sprocket keyway.
- Crankshaft sprocket using Sprocket Installation Tool No. J 41665. Install the sprocket onto the crankshaft until fully seated against the crankshaft flange. Rotate the crankshaft sprocket until the alignment mark is in the 12 o'clock position.

➡ **Properly locate the camshaft sprocket locating pin with the cam sprocket alignment hole. The sprocket teeth and timing chain must mesh. The camshaft and crankshaft sprocket alignment marks MUST be aligned properly. Locate the camshaft sprocket alignment mark in the 6 o'clock posi-**

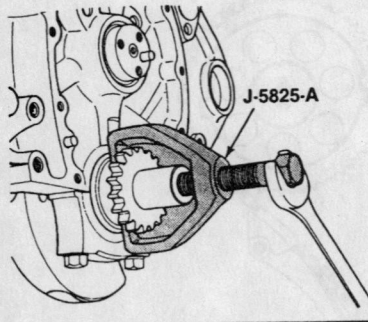

Removal (top) and installation of the crankshaft timing gear

85383293

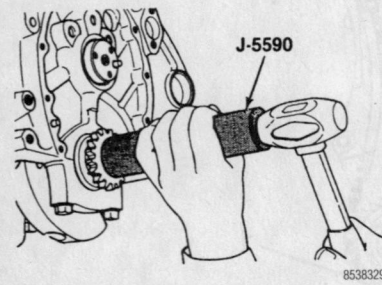

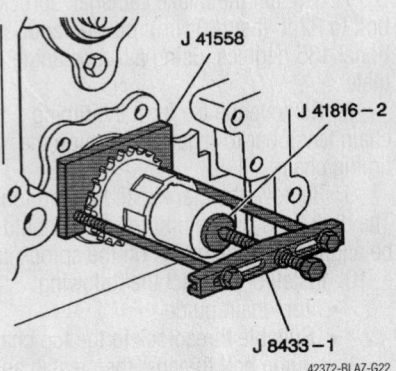

42372-BLAZ-G22

Use the proper tools to remove the crankshaft sprocket—5.3L engine

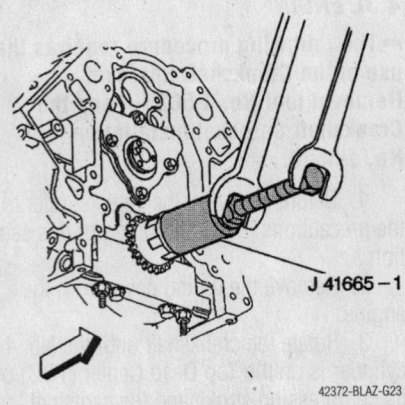

42372-BLAZ-G23

Crankshaft sprocket installation—5.3L engine

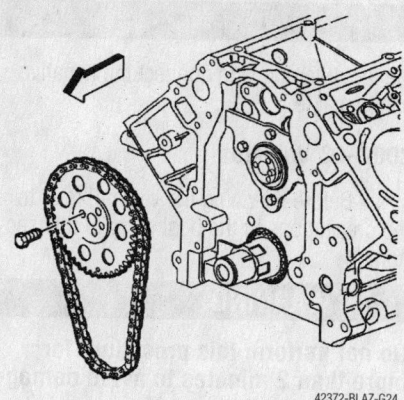

Proper alignment of the timing marks for timing chain installation—5.3L engine

tion. It may be necessary to rotate the camshaft or crankshaft to align the marks.

- Camshaft sprocket and timing chain
- Camshaft sprocket bolts and tighten to 26 ft. lbs. (35 Nm)
- Oil pump

Piston and Ring

POSITIONING

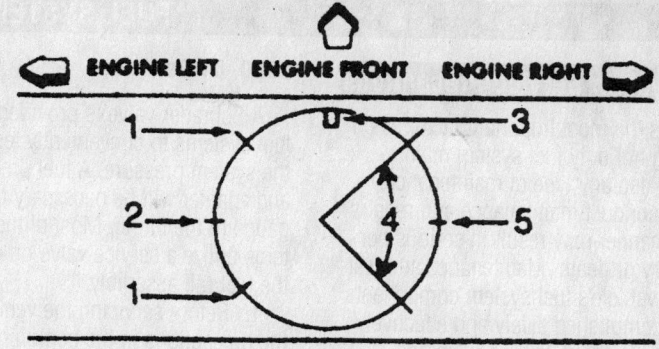

1. Oil ring rail gaps
2. 2nd Compression ring gap
3. Notch in piston
4. Oil ring spacer gap (tang in hole or slot with arc)
5. Top compression ring gap

Piston ring end-gap spacing—4.3L engine

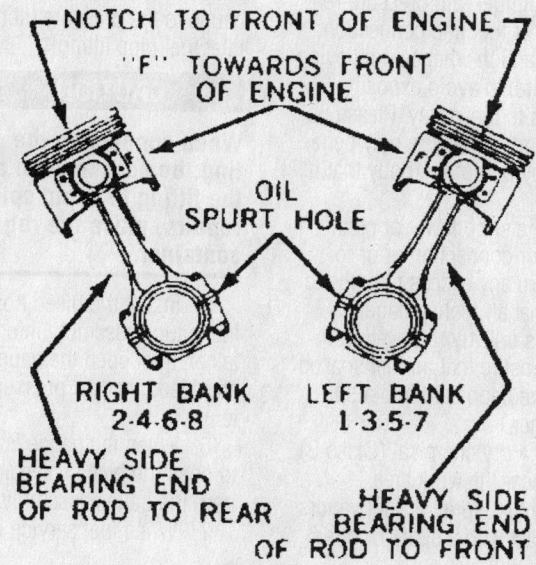

NOTCH TO FRONT OF ENGINE

"F" TOWARDS FRONT OF ENGINE

OIL SPURT HOLE

RIGHT BANK 2-4-6-8   LEFT BANK 1-3-5-7

HEAVY SIDE BEARING END OF ROD TO REAR

HEAVY SIDE BEARING END OF ROD TO FRONT

Piston and connecting rod assembly positioning—4.3L engine

Piston ring positioning—4.2L engine

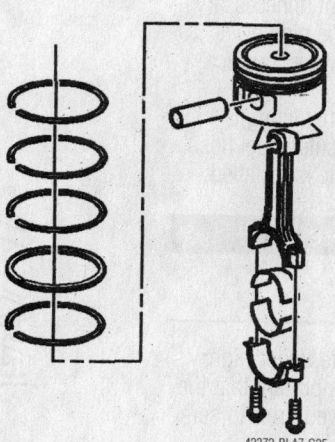

Piston ring positioning—5.3L engine

FUEL SYSTEM

Fuel System Service Precautions

Safety is the most important factor when performing not only fuel system maintenance but also any type of maintenance. Failure to conduct maintenance and repairs in a safe manner may result in serious personal injury or death. Maintenance and testing of the vehicle's fuel system components can be accomplished safely and effectively by adhering to the following rules and guidelines.

• To avoid the possibility of fire and personal injury, always disconnect the negative battery cable unless the repair or test procedure requires that battery voltage be applied.

• Always relieve the fuel system pressure prior to disconnecting any fuel system component (injector, fuel rail, pressure regulator, etc.), fitting or fuel line connection. Exercise extreme caution whenever relieving fuel system pressure, to avoid exposing skin, face and eyes to fuel spray. Please be advised that fuel under pressure may penetrate the skin or any part of the body that it contacts.

• Always place a shop towel or cloth around the fitting or connection prior to loosening to absorb any excess fuel due to spillage. Ensure that all fuel spillage (should it occur) is quickly removed from engine surfaces. Ensure that all fuel soaked cloths or towels are deposited into a suitable waste container.

• Always keep a dry chemical (Class B) fire extinguisher near the work area.

• Do not allow fuel spray or fuel vapors to come into contact with a spark or open flame.

• Always use a back-up wrench when loosening and tightening fuel line connection fittings. This will prevent unnecessary stress and torsion to fuel line piping. Always follow the proper torque specifications.

• Always replace worn fuel fitting O-rings with new. Do not substitute fuel hose or equivalent where fuel pipe is installed.

Fuel System Pressure

RELIEVING

The fuel systems operate under high fuel pressures. It is very important that the pressure be properly relieved prior to servicing the system or any of its components.

2000–01 Vehicles

A Schrader valve is provided on these fuel systems to conveniently test or release the system pressure. A fuel pressure gauge and adapter will be necessary to connect the gauge to the fitting. Most of the MFI systems utilize a service valve on one end of the fuel rail assembly.

1. Before servicing the vehicle, refer to the precautions in the beginning of this section.

2. Disconnect the negative battery cable to assure the prevention of fuel spillage if the ignition switch is accidentally turned **ON** while a fitting is still detached.

3. Loosen the fuel filler cap to release the fuel tank pressure.

4. Be sure the release valve on the fuel gauge is closed, then connect the fuel gauge to the pressure fitting located on the inlet fuel pipe fitting.

✳✳ CAUTION

When connecting the gauge to the fitting, be sure to wrap a rag around the fitting to avoid spillage. After repairs, place the rag in an approved container.

5. Install the bleed hose portion of the fuel gauge assembly into an approved container, then open the gauge release valve and bleed the fuel pressure from the system.

6. When the gauge is removed, be sure to open the bleed valve and drain all fuel from the gauge assembly.

7. When fuel service is finished, tighten the fuel filler cap and connect the negative battery cable.

2002–03 Vehicles

1. Before servicing the vehicle, refer to the precautions in the beginning of this section.

✳✳ WARNING

Do not perform this procedure for more than 2 minutes to avoid damaging the catalytic converter.

2. Loosen the fuel filler cap to release the fuel tank pressure.

3. Remove the fuel pump relay from the junction block.

4. Crank the engine, allowing it to start and stall.

5. Crank the engine for an additional 3 seconds to relieve any remaining fuel pressure.

6. Disconnect the negative battery cable to avoid repressurizing the fuel system.

7. Install the fuel pump relay in the junction block.

8. Tighten the fuel filler cap.

9. After you are finished working on the fuel system, connect the negative battery cable.

Fuel Filter

REMOVAL & INSTALLATION

1. Before servicing the vehicle, refer to the precautions in the beginning of this section.

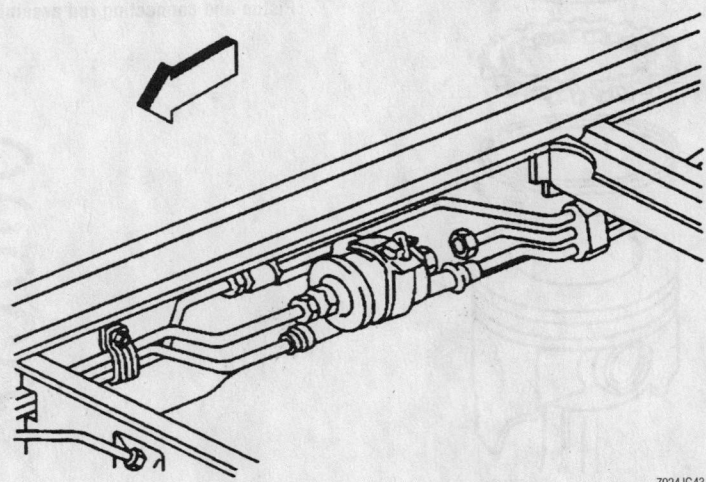

Typical fuel filter location along frame rail

7924JG43

2. Properly relieve the fuel system pressure.

3. Remove or disconnect the following:
  • Negative battery cable and fuel filler cap, if not already done
4. Raise and support the vehicle.
  • Fuel tank shield, if equipped
  • Quick connect fittings from the filter
  • Filter feed nut and the clamp bolt
  • Filter and the clamp from the vehicle

To install:

5. Install or connect the following:
  • Filter and clamp with the directional arrow facing away from the fuel tank, towards the throttle body

➡**The filter has an arrow (fuel flow direction) on the side of the case, be sure to install it correctly in the system, the with arrow facing away from the fuel tank.**

  • Tighten the fuel feed nut
  • Tighten the filter clamp assembly bolt
  • Fuel quick disconnect fittings to the filter
  • Fuel tank shield, if equipped
  • Fuel filler cap
  • Negative battery cable
6. Start the engine and check for leaks.

Fuel Pump

REMOVAL & INSTALLATION

1. Before servicing the vehicle, refer to the precautions in the beginning of this section.
2. Properly relieve the fuel system pressure.
3. Drain the fuel tank.
4. Support the fuel tank.
5. Remove or disconnect the following:
  • Negative battery cable
  • Filler neck from the tank
  • Shield from tank and tank straps
  • Fuel lines and vapor hose from pump
  • Electrical connection from fuel pump
  • Fuel tank
  • Fuel pump/sending unit assembly by turning the locking ring (located on top of the fuel tank) counterclockwise using a spanner wrench
  • Fuel pump from the fuel lever sending device

To install:

6. Install or connect the following:
  • Fuel pump in tank with new seal around opening

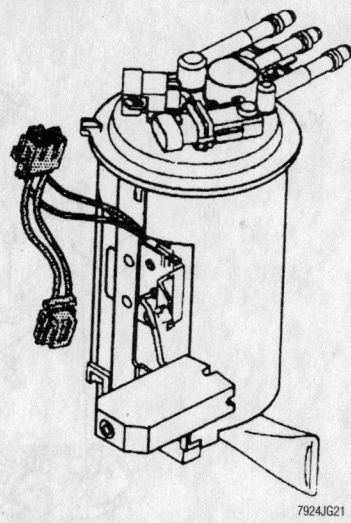
7924JG21
View of the in-tank fuel pump assembly

➡**The fuel pump strainer must be in a horizontal position when the fuel sender is installed in the tank. When installing the sender assembly, make sure that the fuel pump strainer does not block full travel of the float arm.**

  • Tank and connect fuel lines and vapor hose
  • Tank to the frame. Torque the fasteners to 33 ft. lbs. (45 nm).
  • Shield
  • Fuel filler neck and clamp
  • Negative battery cable
7. Refill the tank.
8. Run the engine and check for leaks.

Fuel Injector

REMOVAL & INSTALLATION

4.2L Engine

1. Before servicing the vehicle, refer to the precautions in the beginning of this section.
2. Relieve the fuel system pressure. Refer to the fuel system relief procedure in this section.
3. Remove or disconnect the following:
  • Negative battery cable, if not done already
  • Intake manifold

➡**Clean the fuel rail assembly with a suitable spray cleaner before proceeding. Never soak the fuel rail in a cleaning solvent.**

  • Fuel pressure regulator vacuum line
  • Fuel feed and return pipes

  • Fuel injector in-line electrical connector
  • Fuel rail attaching bolts and fuel rail
  • Fuel injector harness connector from the fuel injectors
  • Injector retaining clip
  • Injector from the fuel rail
  • Retainer clip and O-ring seals from each end of the injector and discard

To install:

➡**Each injector is calibrated. When replacing the fuel injectors, be sure to replace it with the correct injector.**

4. Lubricate the new injector O-ring seats with engine oil.
5. Install or connect the following:
  • O-rings on the injector
  • New retainer clip on the injector
6. Push the fuel injector into the fuel rail socket, making sure the connector faces outward. The retainer clip locks to a flange on the fuel rail injector socket.
  • Fuel rail assembly. Tighten the bolts to 89 inch lbs. (10 Nm).
  • Fuel feed and return lines to the rail
  • Fuel injector electrical connectors
  • Fuel pressure regulator vacuum line
  • Intake manifold
  • Negative battery cable
7. Turn the ignition **ON** for 2 seconds and then turn it **OFF** for 10 seconds. Again turn the ignition **ON** and check for leaks.

4.3L Engine

1. Before servicing the vehicle, refer to the precautions in the beginning of this section.
2. Relieve the fuel system pressure. Refer to the fuel system relief procedure in this section.
3. Remove or disconnect the following:
  • Negative battery cable
  • Fuel meter body electrical connection and the fuel feed and return hoses from the engine fuel pipes
  • Upper manifold assembly
  • Poppet nozzle out of the casting socket
  • Fuel meter body by releasing the locktabs

➡**Each injector is calibrated. When replacing the fuel injectors, be sure to replace it with the correct injector.**

  • Lower hold-down plate and nuts
4. While pulling the poppet nozzle tube downward, push with a small prytool down between the injector terminals and remove the injectors.

To install:

5. Lubricate the new injector O-ring seats with engine oil.

6. Install or connect the following:

- O-rings on the injector
- Fuel injector into the fuel meter body injector socket.
- Lower hold-down plate and nuts. Torque the nuts to 27 inch lbs. (3 Nm).
- Fuel meter body assembly into the intake manifold. Torque the fuel meter bracket retainer bolts to 88 inch. lbs. (10 Nm).

✳✳ CAUTION

To reduce the risk of fire or injury ensure that the poppet nozzles are properly seated and locked in their casting sockets

- Fuel meter body into the bracket and lock all the tabs in place
- Poppet nozzles into the casting sockets
- Electrical connections
- New O-ring seals on the fuel return and feed hoses.
- Fuel feed and return hoses and tighten the fuel pipe nuts to 22 ft. lbs. (30 Nm)
- Negative battery cable

7. Turn the ignition **ON** for 2 seconds and then turn it **OFF** for 10 seconds. Again turn the ignition **ON** and check for leaks.

8. Install the manifold plenum.

5.3L Engine

1. Before servicing the vehicle, refer to the precautions in the beginning of this section.

2. Relieve the fuel system pressure. Refer to the fuel system relief procedure in this section.

3. Remove or disconnect the following:

- Negative battery cable, if not done already
- A/C compressor pressure switch electrical connector
- Wire harness from the clip on the cylinder head
- Mass Airflow/Intake Air Temperature (MAF/IAT) sensor connector
- Alternator electrical connector
- Right side electrical connectors from the coil main electrical harness, Electronic Throttle Control (ETC) and fuel injectors.

4. To detach the injector connector: Matchmark the connectors, pull the Con-

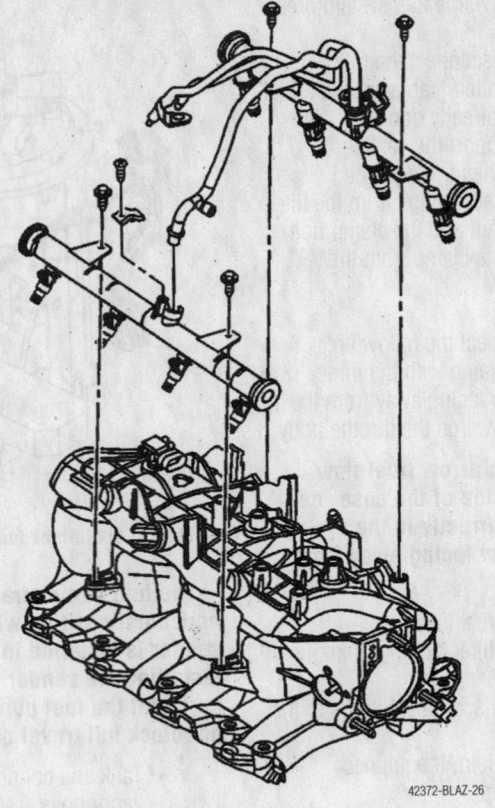

Exploded view of the fuel rail mounting—5.3L engine

nector Position Assurance (CPA) retainer up 1 click. Push the tab on the connector in, then detach the injector connector.

- Electrical harness from the clips on the ignition coil bracket
- Evaporative emission (EVAP) purge solenoid electrical connector
- Knock Sensor (KS) electrical connector
- Manifold Absolute Pressure (MAP) electrical connector
- Main coil
- Fuel injector electrical connector (right side)
- Electrical harness from the clips on the ignition coil bracket
- Upper engine wire harness retainer nut. Position the wire harness aside.
- Fuel feed and return pipes from the rail
- Fuel pressure regulator vacuum line
- Fuel rail bolts
- Fuel rail, after cleaning with a spray-type cleaner

✳✳ WARNING

Be very careful when removing the fuel rail and injectors not to damage the connector terminals or injector spray tips

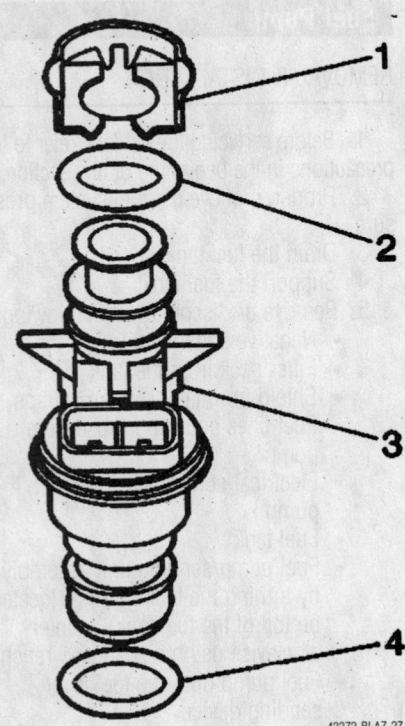

Exploded view of the fuel injector (3), retainer (1) and O-ring seals (2, 4)—5.3L engine

- Fuel injector from the fuel rail
- Fuel injector retainer clip and discard
- Fuel injector lower O-ring seals and discard

To install:

5. Install or connect the following:
   - New O-ring seals on the injectors, after lubricating with clean engine oil
   - New retainer clip on the injector
   - Fuel injector by pushing it into the fuel rail socket
   - Fuel rail
   - Apply 0.20 (5mm) band of thread-lock to the threads of the fuel rail bolts

- Fuel rail bolts and tighten to 89 inch lbs. (10 Nm)
- Fuel pressure regulator vacuum line
- Fuel feel and return pipes
- Route the upper electrical harness into position over the engine.
- Engine harness bracket nut and tighten to 89 inch lbs. (10 Nm)
- PCV valve and hose
- EVAP purge solenoid, KS, MAP sensor, main coil & fuel injector electrical connectors
- Harness to the clips on the ignition coil bracket

- Main coil, ETC, fuel injector electrical connectors
- Harness to the clips on the ignition coil bracket
- Alternator electrical connector
- MAF/IAT sensor connector
- Wire harness to the clip on the cylinder head
- A/C compressor switch electrical connector
- Air cleaner outlet duct
- Fuel fill cap
- Negative battery cable

6. Refill the engine cooling system.

DRIVE TRAIN

Manual Transmission Assembly

REMOVAL & INSTALLATION

1. Before servicing the vehicle, refer to the precautions in the beginning of this section.
2. Shift the transmission into 3rd or 4th gear position.
3. Remove or disconnect the following:
   - Negative battery cable
   - Shift lever and the if necessary, the shift housing
   - Parking brake cable for clearance
   - Propeller shaft
   - Sid plate, if equipped
   - Transfer case and shift lever, on 4WD models
   - All wiring harness that would interfere with transmission removal
   - Fuel line retainers from the rear crossmember
   - Muffler from the catalytic converter
   - Exhaust pipes from the exhaust manifold
   - Catalytic converter hanger, if necessary
   - Exhaust section
   - Bolts and nuts attaching any transmission braces to the engine and transmission
   - Hydraulic clutch quick-connect from the concentric slave cylinder following 1 of the 2 steps:

a. Use 2 small prytools at 180 degrees from each other to depress the white plastic sleeve on the quick connect to separate the clutch line from the concentric slave cylinder quick connect.

b. Use special tool J–36221 to depress the white plastic sleeve on the quick connect to separate the clutch line end from the concentric slave cylinder quick connect.

4. Remove or disconnect the following:
   - Bolts securing the clutch housing cover to the transmission, if equipped
   - Clutch plate and clutch cover, if necessary

5. Support the transmission with a suitable jack.
   - Rear crossmember from the frame rail
   - Wiring harness from the front crossmember, if equipped. Move the wiring harness away from the transmission oil pan. Lower the transmission enough to gain access to the top of the transmission.
   - Fuel line retainers or wiring harness's from the top of the transmission
   - Bolt, washer, and nut securing the wiring harness ground wires to the engine block
   - Bolts retaining the transmission to the engine. Pull the transmission straight back on the clutch hub splines.

6. Lower the transmission using the transmission jack.

To install:

Installation is the reverse of removal, but please note the following important steps.

7. Place a THIN coat of high-temperature grease on the main drive gear (input shaft) splines.

8. Secure the transmission to the floor jack and raise the transmission into position.

➡️**On some models, it may be necessary to rotate the transmission clockwise while inserting it into the clutch hub.**

9. Slowly insert the input shaft through the clutch. Rotate the output shaft slowly to engage the splines of the input shaft into

the clutch while pushing the transmission forward into place. Do not force the transmission into position, the transmission should easily fall into place once everything is properly aligned.

10. Tighten the transmission mounting bolts to 35 ft. lbs. (47 Nm).

11. Do not remove the transmission jack until the crossmembers have been installed.

12. Check the transmission fluid level and replenish as necessary.

Automatic Transmission Assembly

REMOVAL & INSTALLATION

2000–01 Vehicles

1. Before servicing the vehicle, refer to the precautions in the beginning of this section.
2. Remove or disconnect the following:
   - Negative battery cable
3. Drain the transmission fluid.
   - Driveshaft from the transmission (2WD) and transfer case, if equipped (4WD)
4. Support the transmission with a suitable transmission jack.
   - Shift cable from the transmission control lever and bracket
   - Nut and washer securing the transmission mount to the crossmember
   - Bolts and washers securing the mount to the transmission
   - Exhaust pipe from the exhaust manifold(s)
   - Bolts securing the converter pan cover to the transmission, if equipped
   - 3 bolts securing the torque converter to the flywheel

- Bolt, clip, and strap securing the three fuel lines and transmission vent hose to the transmission case
- Bolts and nut securing the transmission to the engine
- Oil filler tube and seal from the transmission
- Transmission cooler lines from the transmission. Plug the lines and the ports in the transmission.
- Wiring harness connectors from the transmission

5. Inspect for any other wiring, brackets etc. which may interfere with the removal of the transmission.

6. Since the transmission acts as a rear engine mount, properly support the rear of the engine with an underbody support or other suitable support before attempting to remove the transmission. Otherwise the rear of the engine may pitch downward and components on the rear of the engine and on the firewall may be damaged.

7. Remove the transmission from the engine by pulling the transmission rearward to disengage it from the locator dowel pins on the back of the block. Carefully lower the transmission from the vehicle. Use care that the torque converter does not fall out of the front of the transmission.

➡**Use converter holding strap tool No. J-21366, to secure the torque converter to the transmission during removal and installation procedures.**

To install:

Installation is the reverse of removal, but please note the following important steps.

8. Make sure the torque converter is fully seated in the pump drive. If not, the transmission will not fit tightly to the rear of the engine block.

9. Raise the transmission into position and remove the torque converter holding strap. Carefully slide the transmission forward until the dowel pins are engaged.

10. The torque converter should be flush with the flywheel and turn freely by hand.

11. Install the transmission–to–engine bolts. Tighten the bolts to 34 ft. lbs. (47 Nm).

12. Tighten the torque converter-to-flywheel bolts to 46 ft. lbs. (63 Nm).

13. If equipped, tighten the converter pan cover to the transmission bolts to 37 ft. lbs. (50 Nm)

14. Tighten the bolts and washers securing the transmission mount to 35 ft. lbs. (47 Nm).

15. Tighten the nut and washer securing the transmission mount to the crossmember to 38 ft. lbs. (52 Nm).

16. Refill the transmission with the proper amount and type of fluid.

17. Connect the negative battery cable. Start the vehicle and allow to warm while checking for leaks. Road test the vehicle to check for shift quality.

2002–03 Vehicles

➡**This procedure requires the use of a Converter Holding Strap tool No. J 21366 to secure the torque converter to the transmission during removal and installation.**

1. Before servicing the vehicle, refer to the precautions in the beginning of this section.

2. Remove or disconnect the following:
   - Negative, then the positive battery cables
   - Fill tube nut, located on the right side of the engine

3. Drain the transmission fluid.
   - Rear propeller shaft

4. Support the transmission with a transmission jack.
   - Nuts securing the transmission mount to the transmission support
   - Evaporative emission (EVAP) canister from its mounting bracket on the left inside of the frame to get access to the transmission support bolts. Do not disconnect the canister lines.

- Fuel tank shield
- Transmission support
- Transmission mount bolts and mount
- Front exhaust pipe assembly

5. Lower the transmission for access to the top and sides of the transmission.
   - Transfer case, if equipped
   - Range selector cable end from the transmission range selector lever ball stud and bracket
   - Transmission heat shield, transmission vent hose park/neutral position switch connector, and main connector from the transmission
   - Bolt that secures the fuel line bracket to the left side of the transmission
   - Torque converter access plug

6. Matchmark the flywheel and torque converter orientation for reassembly.
   - Flywheel-to-torque converter bolts. Be careful not to drop the bolts into the bell housing!
   - Disconnect the transmission oil cooler lines from the transmission. Plug the transmission oil cooler lines connectors in the transmission case.

7. Install a safety chain around the transmission.
   - Bolt that secures the fuel line bracket to the bell housing
   - Bolts that secure the coolant pipe to the bell housing

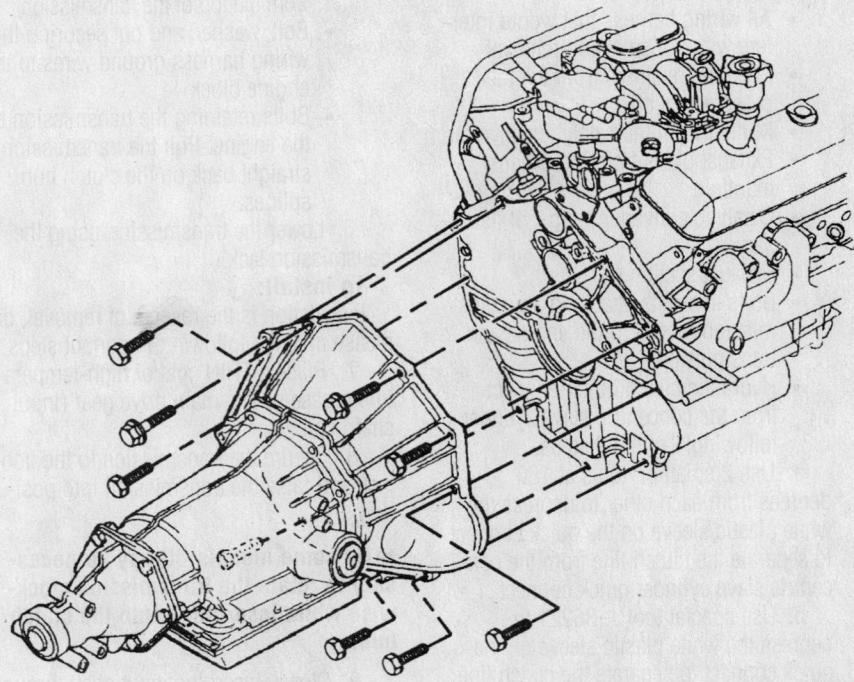

Transmission mounting on 4.3L engines

88457G34

- Remaining nuts, studs and/or bolts that secure the transmission to the engine

8. Install Converter Holding Strap tool No. J 21366 onto the transmission bell housing to hold the torque converter.

9. Pull the transmission straight back and remove it from the vehicle.

To install:

Installation is the reverse of removal, but please note the following important steps.

10. Make sure the torque converter is fully seated in the pump drive. If not, the transmission will not fit tightly to the rear of the engine block.

11. Raise the transmission into position and remove the torque converter holding strap. Carefully slide the transmission forward until the dowel pins are engaged while lining up the marks on the flywheel made during removal.

12. The torque converter should be flush with the flywheel and turn freely by hand.

13. Install the transmission-to-engine nuts, studs and or bolts. Tighten the studs and/or bolts to 37 ft. lbs. (50 Nm).

14. Tighten the bolts securing the heat shield to the transmission to 13 ft. lbs. (17 Nm).

15. Tighten the bolts and washers securing the transmission mount to 18 ft. lbs. (25 Nm).

16. Tighten the nut and washer securing the transmission mount to the transmission support to 35 ft. lbs. (46 Nm).

17. Refill the transmission with the proper amount and type of fluid.

18. Connect the negative battery cable. Start the vehicle and allow to warm while checking for leaks. Road test the vehicle to check for shift quality.

Clutch

REMOVAL & INSTALLATION

1. Before servicing the vehicle, refer to the precautions in the beginning of this section.

2. Remove or disconnect the following:
- Negative battery cable
- Transmission

3. Install a clutch alignment tool or a used transmission input shaft to support the clutch.

4. If the clutch assembly is going to be reused, mark the flywheel, clutch cover and a pressure plate lug for alignment when installing.

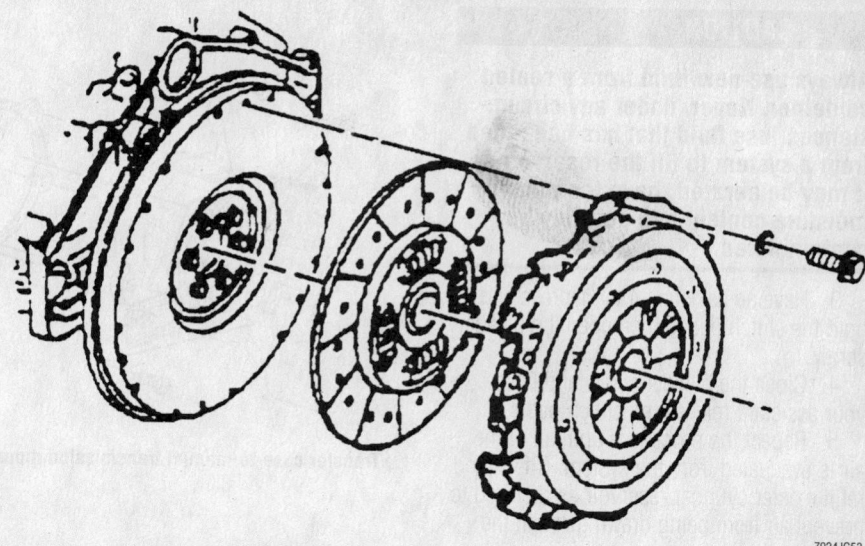

Exploded view of the clutch disc and related components

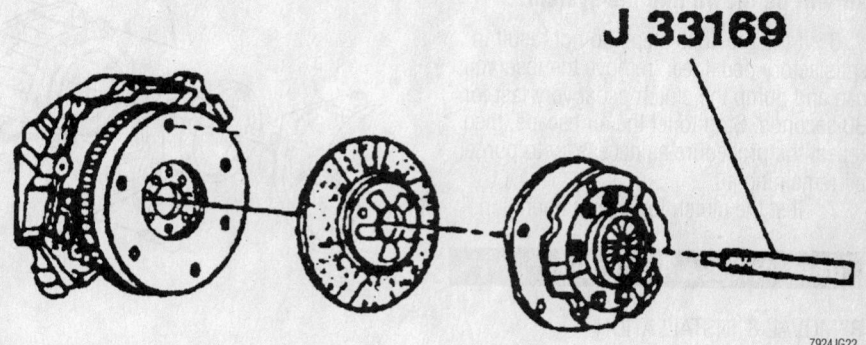

Use the clutch alignment tool to center and support the clutch disc during installation

5. Remove or disconnect the following:
- Clutch cover bolts and washers
- Clutch cover assembly and the clutch plate
- Clutch alignment tool

6. Clean all parts and inspect for damage.

To install:

7. Install or connect the following:
- Clutch alignment tool, to support the clutch
- Clutch cover by aligning the matchmarks or, if new, align the lightest part of the cover, identified by a yellow dot, with the heaviest part identified by an **X**.
- Clutch plate/clutch cover assembly to the flywheel. Tighten the bolts to 33 ft. lbs. (45 Nm) for 2.2L engines or to 29 ft. lbs. (40 Nm) for 4.3L engines.

➡**Tighten each screw 1 turn at a time to avoid warping the clutch cover.**

8. Remove the clutch alignment tool.

9. Install or connect the following:
- Transmission
- Negative battery cable

Hydraulic Clutch System

Bleeding air from the hydraulic clutch system is necessary whenever any part of the system has been disconnected or the fluid level (in the reservoir) has been allowed to fall so low, that air has been drawn into the master cylinder.

BLEEDING

1. Before servicing the vehicle, refer to the precautions in the beginning of this section.

2. Fill master cylinder reservoir with new brake fluid conforming to DOT 3 specifications.

⁂ CAUTION

Always use new fluid from a sealed container. Never, under any circumstances, use fluid that has been bled from a system to fill the reservoir as it may be aerated, have too much moisture content and possibly be contaminated.

3. Have an assistant fully depress and hold the clutch pedal, then open the bleeder screw.

4. Close the bleeder screw and have your assistant release the clutch pedal.

5. Repeat the procedure until all of the air is evacuated from the system. Check and refill master cylinder reservoir as required to prevent air from being drawn through the master cylinder.

➡ **Never release a depressed clutch pedal with the bleeder screw open or air will be drawn into the system.**

6. If the previous steps do not result in satisfactory pedal feel, remove the reservoir cap and pump the clutch pedal very fast for 30 seconds. Stop to let the air escape, then repeat the procedure as necessary to purge all remaining air.

7. Test the clutch for proper operation.

Transfer Case Assembly

REMOVAL & INSTALLATION

2000–01 Vehicles

1. Before servicing the vehicle, refer to the precautions in the beginning of this section.

2. Disconnect the negative battery cable.

3. Shift the transfer case into the **4HI** range.

4. Drain the transfer case fluid.

5. Support the transfer case.

6. Remove or disconnect the following:
- Skid plate
- Front and rear driveshafts from the transfer case. Matchmark the shafts prior to removal.
- Vacuum lines and/or the electrical connectors, as equipped
- Transfer case shift rod/cable from the case, if applicable
- Support brace-to-transfer case bolts, if applicable
- Transfer case

7. Remove all traces of old gasket material from the mating surfaces.

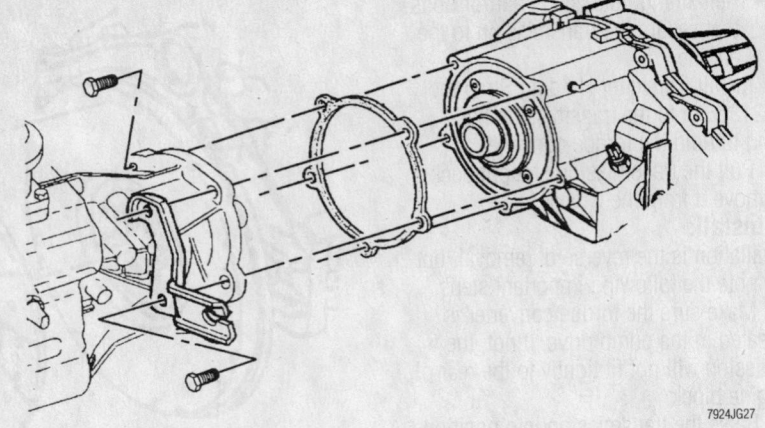

Transfer case-to-manual transmission mounting

7924JG27

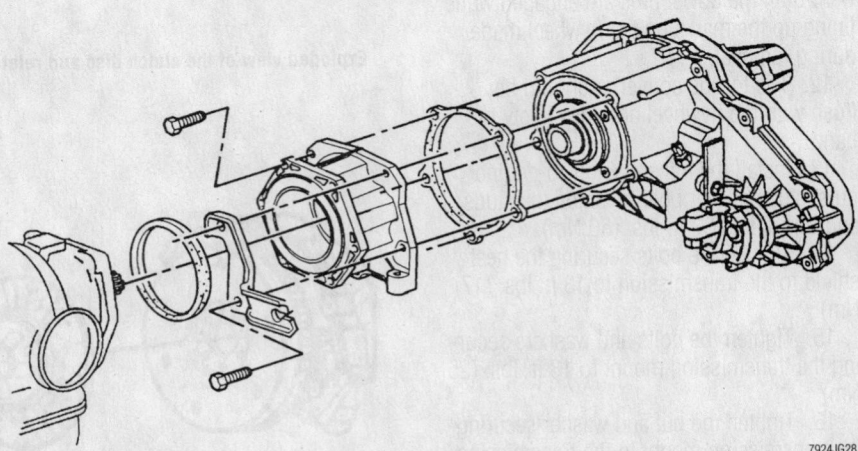

Transfer case-to-automatic transmission mounting

7924JG28

To install:

8. Install or connect the following:
- New gasket using sealer to hold it in position
- Transfer case. Torque the bolts to 33–35 ft. lbs. (45–47 Nm) on 2000–01 models.
- Support brace bolts. Torque the bolts to 35–37 ft. lbs. (47–50 Nm), if equipped.
- Shift rod to the case, if equipped
- Vacuum lines and/or electrical connections, as necessary
- Front and rear driveshafts by aligning the matchmarks

9. Refill the transfer case.

10. Install or connect the following:
- Skid plate, if equipped
- Negative battery cable

2002–03 Vehicles

1. Before servicing the vehicle, refer to the precautions in the beginning of this section.

2. Disconnect the negative battery cable.

3. Raise and support the vehicle. Drain the transfer case.

4. Remove or disconnect the following:
- Fuel tank shield mounting bolts and shield
- Front and rear propeller shaft. Matchmark the shafts prior to removal.
- Fuel lines from the retainer
- Electrical harness from the retainers on the right and left sides
- Speed sensor electrical connectors
- Motor/encoder electrical connector
- Transfer case wiring harness
- Vent hose

5. Install a transmission jack to support the transfer case.
- Transfer case mounting bolts
- Transfer case from the vehicle
- Transfer case gasket and discard if damaged

To install:

6. Install or connect the following:

➡**You must replace the transfer case gasket if it is damaged. Never use silicone sealant in place of, or with the transfer case gasket.**

- Transfer case, using a new gasket if necessary
- Transfer case mounting bolts and tighten to 35 ft. lbs. (47 Nm)

7. Remove the transmission jack.

8. The remainder of installation is the reverse of removal.

9. Refill the transfer case.

Halfshaft

REMOVAL & INSTALLATION

2000–01 Vehicles

1. Before servicing the vehicle, refer to the precautions in the beginning of this section.

2. Unlock the steering column so the steering linkage is free to move.

3. Remove or disconnect the following:
- Negative battery cable
- Front wheels

➡**Place a drift through the caliper into the edge of the rotor to keep the rotor from turning when the nut is removed.**

- Cotter pin, retainer, nut and washer.
- Brake caliper and support it with a piece of wire to avoid damaging the brake hose
- Brake rotor
- Brake line support bracket and Anti-lock Brake System (ABS) wire bracket from the upper control arm

4. Place a jackstand or jack under the lower control arm.

- Axle shaft from the hub by placing a block of wood against the outer edge of the axle (to protect the threads), then strike the block of wood sharply with a hammer. Do not remove the axle at this time.
- Tie rods from the steering knuckles
- Lower shock absorber bolts
- Upper ball joint from the steering knuckle and suspend the steering knuckle on a wire
- Skid plate, if equipped
- Halfshaft-to-axle tube bolts
- Halfshaft by moving it forward and supporting it away from the frame
- Halfshaft from the hub and bearing assembly

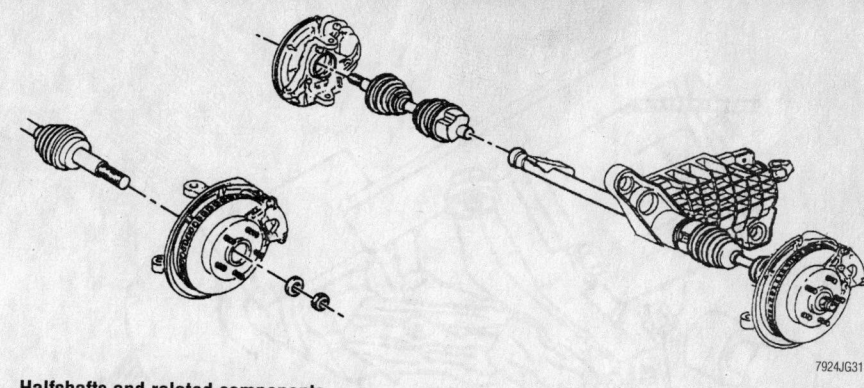

Halfshafts and related components

- Halfshaft from the differential using a block of wood and a hammer

To install:

➡**It is essential that the differential carrier and axle seals are not lubricated or damaged during installation. Prior to shaft installation, cover the shock mounting bracket, lower control arm ball stud and ALL other sharp edges with a cloth or rag to help protect the boot.**

5. Install the axle into the carrier. With both hands on the tripot housing, align the splines on the shaft with the carrier. Then center the axle into the carrier seal and push the shaft straight into the carrier until the snapring is properly seated.

➡**Be careful when supporting the lower control arm that any components are damaged with the supporting device.**

6. Raise the lower control arm using a jackstand or jack until the full weight of the arm is supported.

➡**It is necessary to slightly start the knuckle onto the axle while at the same time guiding the lower ball joint into position on the knuckle.**

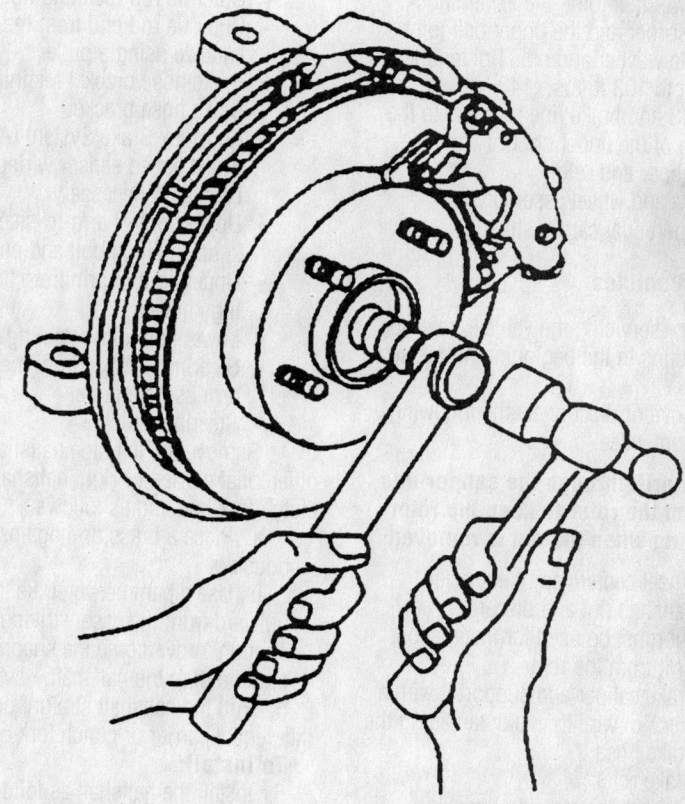

Tap the halfshaft out of the hub without damaging the threads

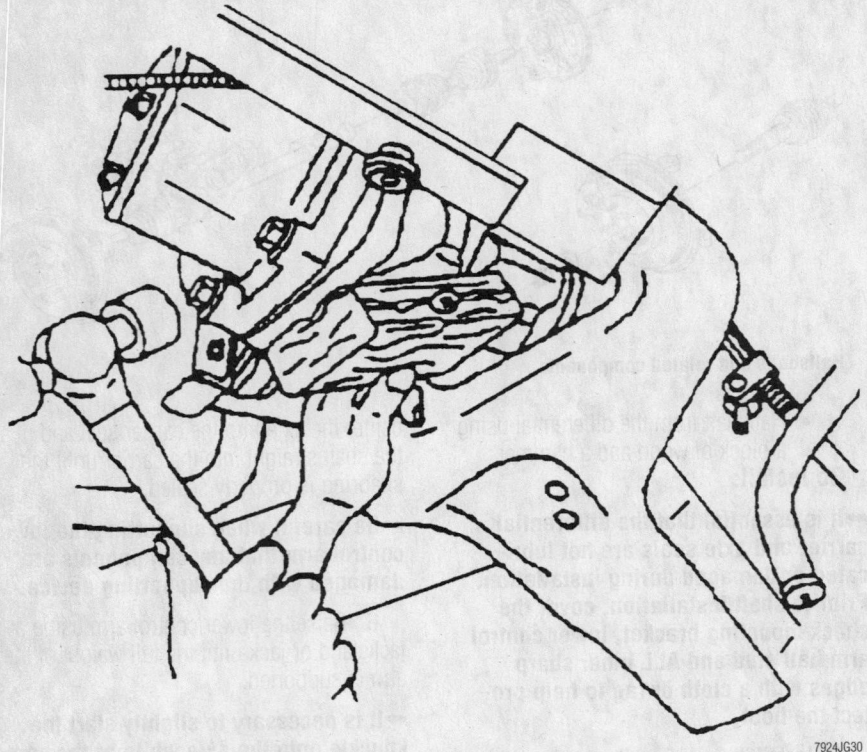

7924JG30

Using a block of wood and a mallet, disengage the halfshaft from the differential assembly

7. Install or connect the following:
- Lower ball joint, the lower shock absorber and the upper ball joint
- Axle washer and nut. Tighten the nut to 103 ft. lbs. (140 Nm).
- ABS and brake line brackets to the top of the upper control arm
- Caliper and rotor
- Tire and wheel assembly
- Differential carrier shield

2002–03 Vehicles

1. Before servicing the vehicle, refer to the precautions in the beginning of this section.
2. Remove or disconnect the following:
- Front wheel

➡**Place a drift through the caliper into the edge of the rotor to keep the rotor from turning when the nut is removed**

- Wheel center cap, if equipped
- Halfshaft nut and discard. A new nut must be used for installation.
- Drift from the rotor
- Brake caliper and support it with a piece of wire to avoid damaging the brake hose
- Brake rotor

3. To remove the steering knuckle, remove or disconnect the following:

- Wheel hub and bearing
- Outer tie rod retaining nut
- Outer tie rod end from the steering knuckle using a puller
- Brake hose bracket retaining bolts
- Brake hose bracket
- Anti-lock Brake System (ABS) wheel speed sensor wiring harness bracket, if necessary
- Upper control arm-to-steering knuckle pinch bolt and nut
- Upper control arm from the steering knuckle
- Lower ball joint retaining nut
- Steering knuckle from the control arm using a puller
- Steering knuckle

4. Remove the left side halfshaft from differential carrier, or right halfshaft from the clutch fork housing as follows:
   a. Place a brass drift against the tripot housing.
   b. Use a hammer to strike the drift outward from the case, striking hard enough to overcome the snapring tension holding the halfshaft.

5. Pull the halfshaft straight out of the differential carrier or clutch fork housing.

To install:

6. Install the halfshaft as follows:
   a. With both hands on the tripot housing, align the splines on the shaft with the differential carrier assembly (left) or clutch fork housing (right).
   b. Center the halfshaft into the differential carrier or clutch fork housing assembly seal.
   c. Firmly push the shaft straight into the differential carrier or clutch fork housing assembly until the snapring is properly seated.

7. To install the steering knuckle, install or connect the following:
- Steering knuckle to the lower control arm
- Lower ball joint retaining nut and tighten to 81 ft. lbs. (110 Nm)
- Upper control arm to the steering knuckle
- Upper control arm pinch bolt and nut and tighten to 30 ft. lbs. (40 Nm)
- ABS wheel speed sensor harness bracket
- Brake hose bracket. Tighten the bolts to 7 ft. lbs. (10 Nm).
- Outer tie rod to the steering knuckle and tighten the nut to 33 ft. lbs. (45 Nm)
- Hub and bearing

8. Install or connect the following:
- New halfshaft nut and tighten to 103 ft. lbs. (140 Nm)
- Wheel

9. Lower the vehicle. Adjust the front toe.

CV-Joints

OVERHAUL

Outer CV-Joint

1. Before servicing the vehicle, refer to the precautions in the beginning of this section.
2. Remove or disconnect the following:
- Front wheel
- Halfshaft and position it in a vise
- Large CV-joint boot clamp and discard it
- Small CV-joint boot clamp and discard it
- CV-joint boot and slide it back on the shaft
- Outer race from the halfshaft, by spreading the outer race-to-halfshaft retaining ring, using Snapring Pliers J-8059
- Retaining ring from the halfshaft and discard it
- CV-joint boot from the halfshaft and discard it, if damaged

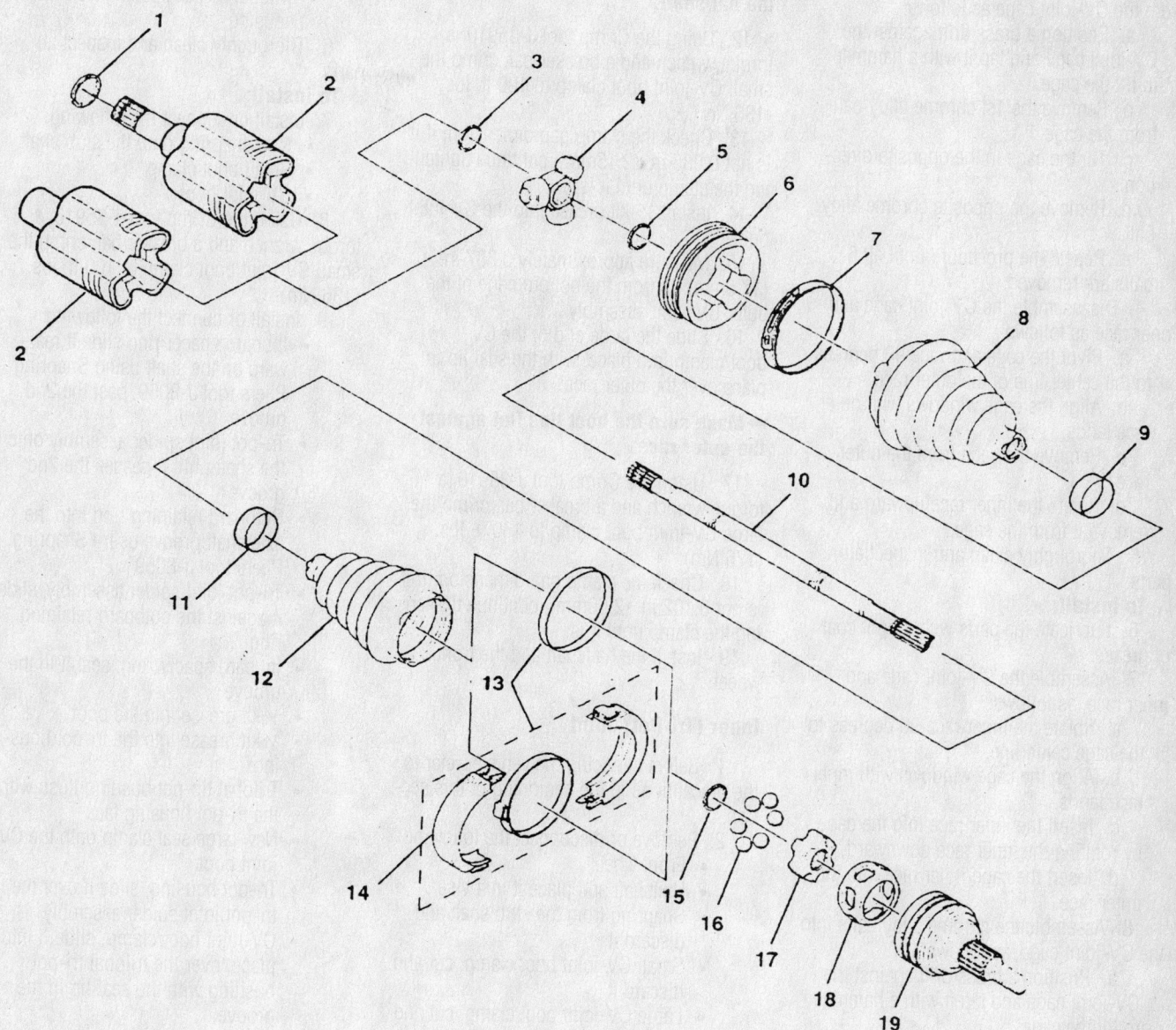

(1) Differential Shaft Ring
(2) Tripot Housing Assembly
(3) Spacer Ring
(4) Tripot Joint Spider Assembly
(5) Spacer Ring
(6) Tripot Bushing
(7) Boot Retaining Clamp
(8) Tripot Joint Boot
(9) Halfshaft Swage Ring
(10) Halfshaft Bar

(11) Halfshaft Swage Ring
(12) CV Joint Boot
(13) Swage Ring
(14) Clamp Protector
(15) Race Retaining Ring
(16) Ball
(17) CV Joint Inner Race
(18) CV Joint Cage
(19) CV Joint Outer Race

9308JG09

Exploded view of the CV-Joint Assembly

3. Disassemble the chrome alloy balls from the CV-joint cage as follows:

a. Position a brass drift against the CV-joint cage and tap it with a hammer to tilt the cage.

b. Remove the 1st chrome alloy ball from the cage.

c. Tilt the cage in the opposite direction.

d. Remove the opposite chrome alloy ball.

e. Repeat the procedure until all 6 balls are removed.

4. Disassemble the CV-joint cage and inner race as follows:

a. Pivot the cage and race 90 degrees to the center line of the outer race.

b. Align the cage windows with outer race lands.

c. Remove the cage from the outer race.

d. Rotate the inner race upward and remove it from the cage.

5. Thoroughly clean and inspect all parts.

To install:

6. Lubricate the parts with a light coat of grease.

7. Assemble the CV-joint cage and inner race, as follows:

a. Rotate the inner race 90 degrees to the cage centerline.

b. Align the cage windows with inner race lands.

c. Insert the inner race into the cage by rotating the inner race downward.

d. Insert the cage/inner race into the outer race.

8. Assemble the chrome alloy balls into the CV-joint cage, as follows:

a. Position a brass drift against the CV-joint cage and tap it with a hammer to tilt the cage.

b. Insert the 1st chrome alloy ball into the cage.

c. Tilt the cage in the opposite direction.

d. Insert the opposite chrome alloy ball.

e. Repeat the procedure until all 6 balls are inserted.

9. Install ½ kit grease into the CV-joint.

10. Install or connect the following:
- Small ring clamp on the CV boot
- New retaining ring on the halfshaft
- Large ring clamp on the CV boot
- Outer race assembly onto the halfshaft until the ring engages the halfshaft groove

11. Slide the small end of the CV-joint boot/clamp into place, with the seal lip in the halfshaft groove

➡**Make sure the boot lies flat against the halfshaft.**

12. Using the Crimp tool J-35910, a torque wrench and a breaker bar, crimp the small CV-joint boot clamp to 100 ft. lbs. (136 Nm).

13. Check the clamp gap dimension; if it is not 0.085 in. (2.15mm), continue tightening the clamp until it is.

14. Install ½ kit grease into the CV-joint boot.

15. Measure approximately 0.687 in. (17.5mm) up from the bottom edge of the outer CV-joint assembly.

16. Slide the large end of the CV boot/clamp into place, with the seal lip in place over the outer race.

➡**Make sure the boot lies flat against the outer race.**

17. Using the Crimp tool J-35910, a torque wrench and a breaker bar, crimp the large CV-joint boot clamp to 130 ft. lbs. (176 Nm).

18. Check the clamp gap dimension; if it is not 0.102 in. (2.60mm), continue tightening the clamp until it is.

19. Install the halfshaft and the front wheel.

Inner (Tri-Pot) Joint

1. Before servicing the vehicle, refer to the precautions in the beginning of this section.

2. Remove or disconnect the following:
- Front wheel
- Halfshaft and place it in a vise
- Snapring from the stub shaft and discard it
- Small CV-joint boot clamp, cut and discard it
- Large CV-joint boot clamp, cut and discard it
- CV-joint boot by sliding it away from the tri-pot joint

3. Install a Stub Shaft Removal tool J-38868-A to the stub shaft snapring groove.

4. Using a slide hammer puller, press the stub shaft from the tri-pot housing.

5. Remove or disconnect the following:
- Tri-pot housing from the tri-pot spider
- Inboard spacer ring slide it rearward on the shaft using Snapring Pliers tool J-8059
- Outboard retaining ring using Snapring Pliers tool J-8059 and discard it
- Tri-pot joint spider assembly
- Inboard spacer ring and discard it

- CV-joint boot
- Trilobal tri-pot bushing from the housing

6. Thoroughly clean and inspect all parts.

To install:

7. Install or connect the following:
- New snapring onto the stub shaft
- Small boot clamp
- CV-joint boot

8. Using the Crimp tool J-35910, a torque wrench and a breaker bar, crimp the small CV-joint boot clamp to 100 ft. lbs. (136 Nm).

9. Install or connect the following:
- Inboard spacer ring slide it rearward on the shaft using Snapring Pliers tool J-8059, past the 2nd groove
- Tri-pot joint spider assembly onto the shaft until it passes the 2nd groove
- Outboard retaining ring into the axle shaft groove using Snapring Pliers tool J-8059
- Tri-pot joint spider assembly, slide it against the outboard retaining ring
- Inboard spacer ring, seat it in the groove
- ½ kit grease into the boot
- ½ kit grease into the tri-pot housing
- Trilobal tip-pot bushing flush with the tri-pot housing face
- New large seal clamp onto the CV-joint boot
- Tri-pot housing, slide it over the tri-pot joint spider assembly
- CV-joint boot/clamp, slide it into place, over the trilobal tri-pot bushing with the seal lip in the groove

➡**Make sure the boot lies flat against the trilobal bushing.**

10. Position the CV-joint boot so it measures 4.9 in. (125mm).

11. Using the Crimp tool J-35566, latch the large CV-joint boot clamp.

12. Install the halfshaft and the front wheel.

Axle Shaft, Bearing and Seal

REMOVAL & INSTALLATION

For the Axle Shaft, Bearing and Seal, Removal and Installation, please refer to Wheel Bearing procedure located in the section.

Pinion Seal

REMOVAL & INSTALLATION

1. Before servicing the vehicle, refer to the precautions in the beginning of this section.

➡ **The following procedure requires the use of the Pinion Holding tool J-8614-10, the Pinion Flange Removal tool J-8614-1, J-8614-2, J-8614-3 and the Pinion Seal Installation tool J-23911 or J-33782.**

2. Remove or disconnect the following:
   - Driveshaft from the pinion flange. Matchmark the driveshaft prior to removal.
   - Driveshaft from the rear axle pinion flange and support the shaft up in body tunnel by wiring it to the exhaust pipe.

➡ **If the U-joint bearings are not retained by a retainer strap, use a piece of tape to hold bearings on their journals.**

3. Mark the position of the pinion stem, flange and nut for reference.

4. Use an inch lbs. torque wrench to measure the amount of torque necessary to turn the pinion, then note this measurement as it is the combined pinion bearing, seal, carrier bearing, axle bearing and seal preload.

5. Remove or disconnect the following:
   - Pinion flange nut and washer, using a Pinion Holding tool J-8614-10 and a Pinion Flange Removal tool J-8614-1, J-8614-2, J-8614-3, as applicable
   - Pinion flange
   - Pinion oil seal by driving it out of the differential with a blunt chisel; DO NOT damage the carrier

To install:

6. Examine the seal surface of pinion flange for tool marks, nicks or damage, such as a groove worn by the seal. If damaged, replace flange.

7. Examine the carrier bore and remove any burrs that might cause leaks around the O.D. of the seal.

8. Apply GM seal lubricant 1050169 to the outside diameter of the pinion flange and sealing lip of new seal.

9. Install or connect the following:
   - New pinion oil seal using a seal installer tool
   - Pinion flange and tighten nut to the same position as marked earlier.

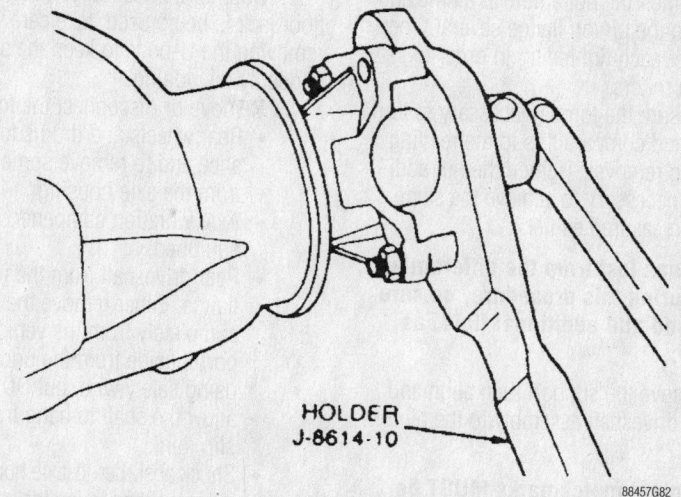

Removing the pinion nut using a pinion holding fixture tool

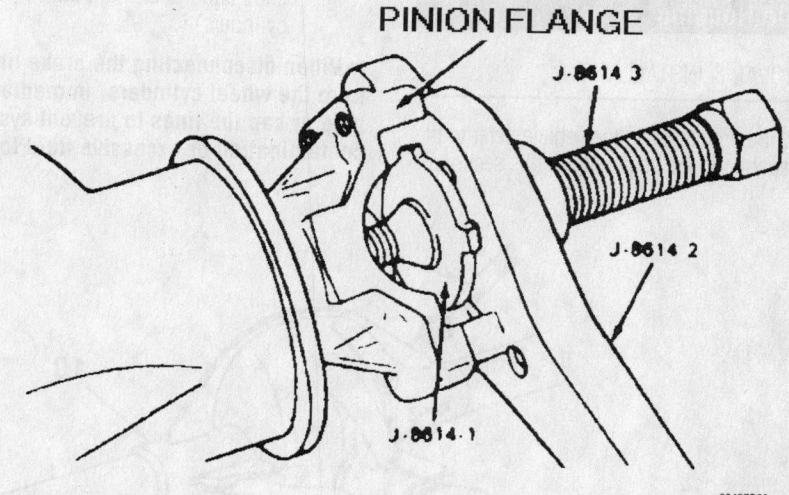

A puller and adapter should be used to withdraw the pinion from the housing

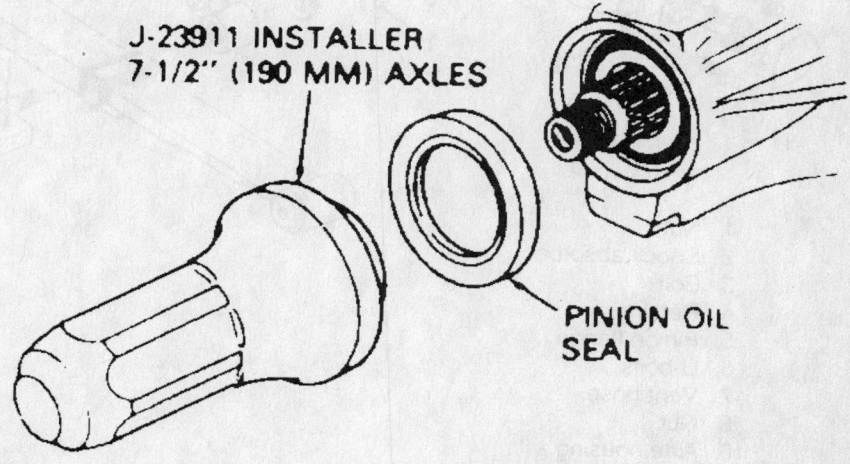

Use the appropriately sized installation tool to drive the new seal into position.

Tighten the nut a little at a time and turn the pinion flange several times after each tightening in order to set the rollers.

10. Measure the torque necessary to turn the pinion and compare this to the reading taken during removal. Tighten the nut additionally, as necessary to achieve the same preload as measured earlier.

➡**If fluid was lost from the differential housing during this procedure, be sure to check and add additional fluid, as necessary.**

11. Remove the support then align and secure the driveshaft assembly to the pinion flange.

➡**The original matchmarks MUST be aligned to assure proper shaft balance and prevent vibration.**

Axle Housing

REMOVAL & INSTALLATION

1. Before servicing the vehicle, refer to the precautions in the beginning of this section.

2. Support the rear axle housing. If a floor jack is being used, take care when removing the U-bolts to keep the axle from suddenly dislodging.

3. Remove or disconnect the following:
- Rear wheels and drums for clearance and to remove some weight from the axle housing
- Axle vibration dampener, if equipped
- Rear driveshaft from the pinion flange. Either remove the shaft completely from the vehicle or support it aside from the undercarriage using safety wire, but DO NOT allow the shaft to hang from the slip joint.
- Shock absorber-to-axle housing retainers, then swing the shock absorbers away from the axle housing
- Brake lines from the axle housing clips and the backing plates (wheel cylinders)

➡**When disconnecting the brake lines from the wheel cylinders, immediately plug or cap the lines to prevent system contamination or excessive fluid loss.**

- Speed sensor connectors at the junction block, if applicable
- Parking brake cable(s)
- Axle housing-to-spring U-bolt nuts, washers, U-bolts and the anchor plates
- Vent hose from the top of the axle housing
- Axle with the help of an assistant by moving it to clear the leaf spring

To install:

4. With the help of an assistant, carefully position the rear axle into the vehicle.

5. Install or connect the following:
- Vent hose to the axle housing

6. Be sure the housing is properly positioned on the leaf spring, then loosely install the U-bolts, anchor plates, washers and nuts.

- Tighten the U-bolt nuts in a cross pattern to 18 ft. lbs. (25 Nm) to made sure everything is evenly seated. Then tighten the nuts in steps to 74 ft. lbs. (100 Nm).
- Brakes lines secure them to the axle housing

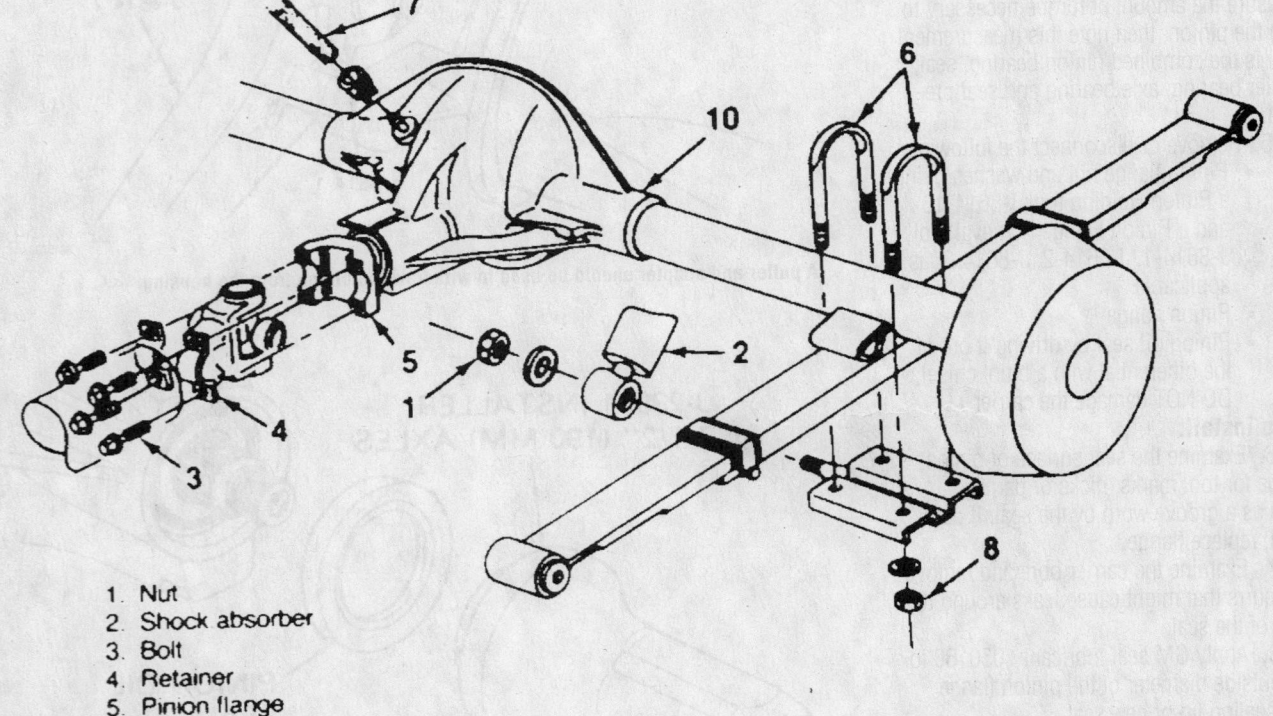

1. Nut
2. Shock absorber
3. Bolt
4. Retainer
5. Pinion flange
6. U-bolts
7. Vent hose
8. Nut
10. Axle housing

Exploded view of the rear axle mounting

88457G85

- Parking brake cable(s), if removed
- Speed sensor connectors to the junction block, if equipped
- Driveshaft assembly
- Shock absorbers to the lower

mounts, then tighten the mount nuts
- Axle vibration dampener, if equipped
- Brake drums and the tire/wheel assemblies

7. Bleed the hydraulic brake system.
8. Check the fluid level in the rear axle assembly and add, as necessary. Make sure the vehicle is level when checking and adding fluid.

STEERING AND SUSPENSION

Air Bag

✳✳ CAUTION

Some vehicles are equipped with an air bag system, also known as the Supplemental Inflatable Restraint (SIR) system. The system must be disabled before performing service on or around system components, steering column, instrument panel components, wiring and sensors. Failure to follow safety and disabling procedures could result in accidental air bag deployment, possible personal injury and unnecessary system repairs.

PRECAUTIONS

Several precautions must be observed when handling the inflator module to avoid accidental deployment and possible personal injury.
- Never carry the inflator module by the wires or connector on the underside of the module.
- When carrying a live inflator module, hold securely with both hands, and ensure that the bag and trim cover are pointed away.
- Place the inflator module on a bench or other surface with the bag and trim cover facing up.
- With the inflator module on the bench, never place anything on or close to the module, that may be thrown in the event of an accidental deployment.

DISARMING

1. Turn the steering wheel so that the vehicle's wheels are pointing straight ahead.
2. Turn the ignition switch to **LOCK**, remove the key, then disconnect the negative battery cable.
3. Remove the AIR BAG or SIR fuse from the fuse block.
4. Remove the steering column filler panel or knee bolster.
5. Unplug the Connector Position Assurance (CPA) and yellow two way connector at the base of the steering column.

6. Open the glove compartment door, lift the stop and let the door fully open.
7. Remove the Connector Position Assurance (CPA) from the passenger yellow two way connector located behind the glove box.
8. Unplug the yellow two way connector located behind the glove box.
9. Connect the negative battery cable.

➡ With the AIR BAG fuse removed, the battery cable connected and the ignition in the ON position, the AIR BAG warning lamp will be ON. This is normal and does not indicate a system malfunction.

ARMING

1. Disconnect the negative battery cable.
2. Attach the yellow two way connector located behind the glove box.
3. Install the Connector Position Assurance (CPA) to the passenger yellow two way connector located behind the glove box.
4. Close the glove compartment door.
5. Turn the ignition switch to **LOCK**, then remove the key.
6. Attach the two way connector at the base of the steering column and the Connector Position Assurance (CPA).
7. Install the steering column filler panel or knee bolster.
8. Install the AIR BAG fuse to the fuse block.
9. Connect the negative battery cable.
10. From the passenger seat, turn the ignition switch to **RUN** and make sure that the AIR BAG warning lamp flashes seven times and then shuts off. If the warning lamp does not shut off, make sure that the wiring is properly connected. If the light remains on, take the vehicle to a reputable repair facility for service.

Power Steering Gear

REMOVAL & INSTALLATION

2000–2001 Vehicles

1. Before servicing the vehicle, refer to the precautions in the beginning of this section.

2. Position a fluid catch pan under the power steering gear.
3. Remove or disconnect the following:
- Feed and return fluid hoses from the steering gear. Immediately cap or plug all openings to prevent system contamination or excessive fluid loss.
- Intermediate shaft lower coupling shield, if equipped
- Intermediate shaft-to-steering gear bolt. Matchmark the intermediate shaft-to-power steering gear and separate the shaft from the gear.
- Pitman arm from the gear pitman shaft
- Power steering gear-to-frame bolts and washers, then carefully remove the steering gear from the vehicle

To install:
4. Install or connect the following:
- Steering gear to the vehicle and secure by finger-tightening the fasteners. For some vehicles, the pitman arm must be connected to the gear while it is still removed from the vehicle or while it is partially installed and lowered for access. If necessary, align and install the pitman arm to the shaft at this time.
- Tighten the power steering gear-to-frame bolts to 55 ft. lbs. (75 Nm)
- Intermediate shaft to the power steering, then secure using the pinch bolt
- Shield over the intermediate shaft lower coupling, if equipped
- Feed and return hoses to the power steering gear
5. Bleed the power steering system.

2002–03 Vehicles

1. Before servicing the vehicle, refer to the precautions in the beginning of this section.
2. Raise and support the vehicle.
3. Position a fluid catch pan under the power steering gear.
4. Remove or disconnect the following:
- Front tire and wheel assemblies
- Outer tie rod retaining nuts

✳✳ WARNING

Do not try to separate a steering linkage joint by driving a wedge between the joint and the attached part. Doing this can cause seal damage and premature failure of the part.

- Outer tie rods from the steering knuckles using a suitable steering linkage and tie rod puller
- Lower intermediate shaft retaining bolt and shaft from the power steering gear
- Steering gear crossmember
- Feed and return fluid hoses from the steering gear. Immediately cap or plug all openings to prevent system contamination or excessive fluid loss.

5. Support the power steering gear.
- Power steering gear mounting bolts, then remove the gear from the vehicle

6. Loosen the outer tie rod jam nuts, then remove the outer tie rods from the inner tie rods and discard the jam nut.

To install:

7. Lubricate the inner tie rod threads with a suitable lubricant before installing the outer tie rod.

8. Install or connect the following:
- New jam nuts to the outer tie rods
- Outer tie rods to the inner tie rods
- Power steering gear to the vehicle. Tighten the retaining bolts to 81 ft. lbs. (110 Nm).

9. Remove the support from the power steering gear.
- Power steering hose(s) to the gear. Tighten the retaining bolt to 9 ft. lbs. (12 Nm).
- Steering gear crossmember
- Lower intermediate shaft to the power steering gear. Tighten the retaining bolt to 30 ft. lbs. (40 Nm).
- Outer tie rod ends to the steering knuckles. Tighten the retaining nuts to 33 ft. lbs. (45 Nm).
- Front tire and wheel assemblies

10. Remove the drain pan, then lower the vehicle.

11. Bleed the power steering system and adjust the front toe as necessary.

Strut/Shock Module

REMOVAL & INSTALLATION

Front—2002–03 Vehicles

➡In 2002–03, a "shock module", similar to a strut was used on these vehicles. This procedure requires the use

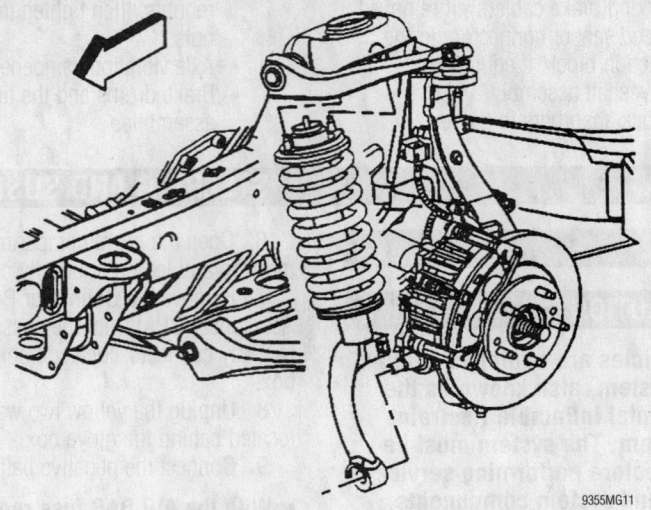

View of the shock module used on the front suspension of 2002 vehicles

of a suitable steering linkage and tie rod puller.

1. Before servicing the vehicle, refer to the precautions in the beginning of this section.
2. Remove or disconnect the following:
- Shock module upper retaining nuts
- Tire and wheel
- Shock module-to-lower control arm retaining nut
- Shock module yoke from the lower control arm using a suitable puller
- Shock module from the shock tower and lower control arm

To install:
3. Install or connect the following:
- Shock module to the shock tower and lower control arm
- Shock module yoke to the lower control arm

- Shock module upper retaining nuts and tighten to 33 ft. lbs. (45 Nm)
- Shock module-to-lower control arm retaining nut and tighten to 81 ft. lbs. (110 Nm)
- Tire and wheel

Shock Absorbers

REMOVAL & INSTALLATION

Front—2000–01 Vehicles

2WD MODELS

1. Before servicing the vehicle, refer to the precautions in the beginning of this section.

2. Remove or disconnect the following:

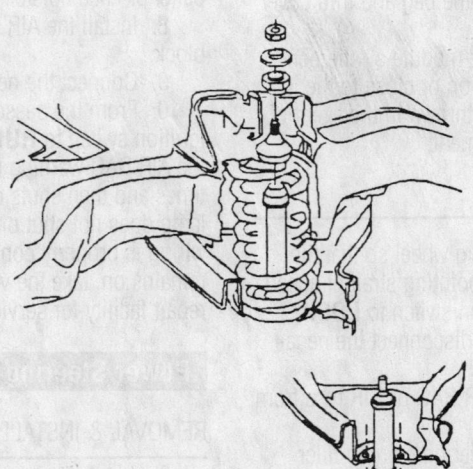

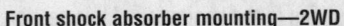

Front shock absorber mounting—2WD

- Wheel
- Mounting nut

➡ **Hold the shock absorber stem with a wrench while backing the nut off.**

- Retaining nut and grommet
- Shock absorber-to-lower control arm bolts
- Shock absorber
- Replace the parts, as necessary.

To install:

3. Fully extend the shock absorber stem, then push it up through the lower control arm and spring so that the upper stem passes through the mounting hole in the upper control arm frame bracket.

4. Install or connect the following:
- Retaining nut and grommet on the stem. Tighten the nut to 106 inch lbs. (12 Nm).
- Shock absorber-to-lower control arm bolts and tighten to 22 ft. lbs. (30 Nm).
- Wheel

4WD MODELS

1. Before servicing the vehicle, refer to the precautions in the beginning of this section.

2. Remove or disconnect the following:
- Wheel
- Lower nut/bolt and collapse the shock absorber
- Shock absorber upper nut and bolt
- Shock absorber

To install:

3. Install or connect the following:
- Shock absorber to the bracket. Tighten the nuts/bolts to 54 ft. lbs. (73 Nm).
- Wheel

Rear

1. Before servicing the vehicle, refer to the precautions in the beginning of this section.

2. Properly support the rear axle assembly.

3. Remove or disconnect the following:
- Automatic level control air lines from the shock absorber, if equipped
- Shock absorber-to-frame retainer(s) at the top of the shock
- Shock-to-axle retainer(s) at the bottom of the shock
- Shock absorber

To install:

4. Install the shock in the vehicle and loosely install the upper mounting fasteners to retain it

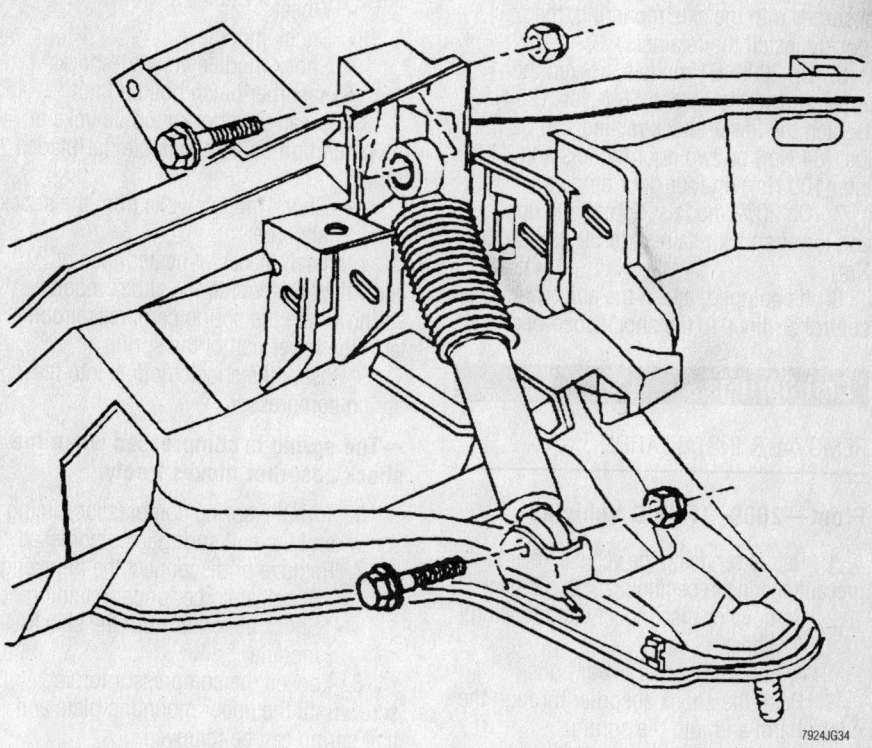

Front shock absorber mounting—4WD

7924JG34

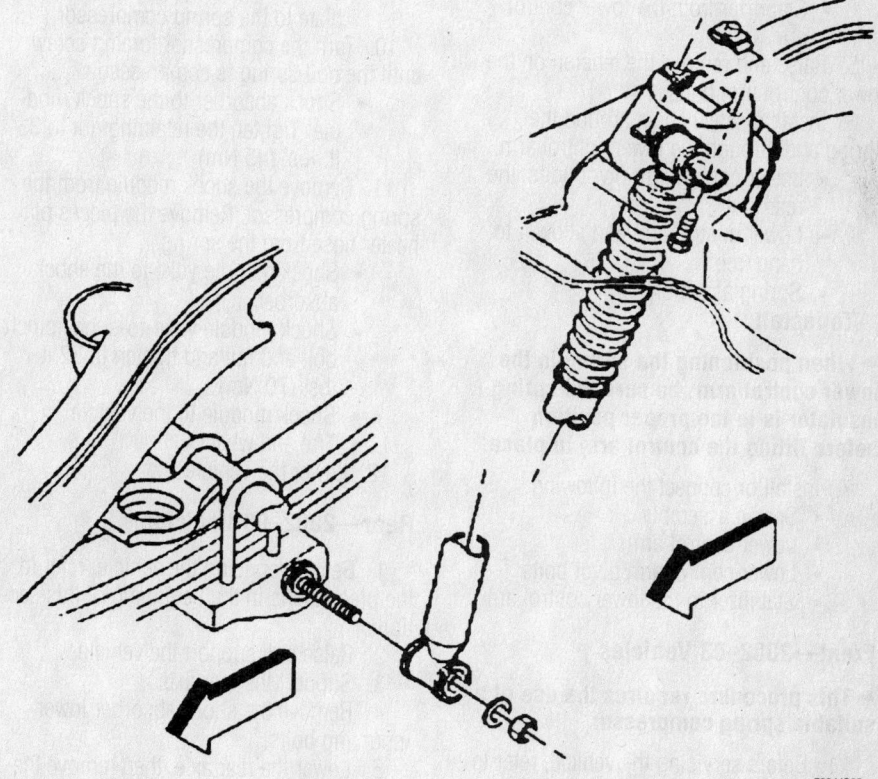

Rear shock absorber mounting

7924JG35

5. Align the lower-end of the shock absorber with the axle mounting, then loosely install the retainers.

6. On 2000–01 models, tighten the upper shock retainers to 18 ft. lbs. (25 Nm). Tighten the lower shock retainers to 62 ft. lbs. (84 Nm) on two door models and 74 ft. lbs. (100 Nm) on four door models.

7. On 2002 models, tighten the upper and lower shock retainers to 59 ft. lbs. (80 Nm).

8. If equipped, attach the automatic level control air lines to the shock absorber.

Coil Springs

REMOVAL & INSTALLATION

Front—2000–01 2WD Vehicles

1. Before servicing the vehicle, refer to the precautions in the beginning of this section.

2. Remove or disconnect the following:
- Wheel
- Shock absorber lower bolts

3. Push the shock absorber through the control arm and into the spring.

4. With the vehicle supported so the control arms hang free, install tool J-23028, onto a support and into the lower control arm bushings.
- Stabilizer bar from the control arm
- Stabilizer from the lower control arm

5. Raise and remove the tension on the lower control arm bolts.

6. Install a safety chain around the spring and through the lower control arm.
- Lower control arm pivot bolts, the rear first
- Lower control arm and allow it to hang free
- Spring assembly

To install:

➡**When positioning the spring in the lower control arm, be sure the spring insulator is in the proper position before lifting the control arm in place.**

7. Install or connect the following:
- Spring assembly
- Lower control arm
- Lower control arm pivot bolts
- Stabilizer to the lower control arm

Front—2002–03 Vehicles

➡**This procedure requires the use of a suitable spring compressor.**

1. Before servicing the vehicle, refer to the precautions in the beginning of this section.

2. Remove or disconnect the following:
- Wheel
- Shock module
- Shock module yoke-to-shock absorber pinch bolt and nut

3. Spread the shock module yoke at the pinch bolt using a suitable flat-bladed tool.
- Shock module yoke from the shock absorber

4. Install pieces of heater hose or equivalent material to the shock module spring where the spring compressor contacts the lower part of the spring.

5. Install the shock module into the spring compressor.

➡**The spring is compressed when the shock absorber moves freely.**

6. Turn the spring compressor forcing screw until the coil spring is compressed.

7. Remove or disconnect the following:
- Shock absorber upper retaining nut
- Shock absorber from the shock module

8. Loosen the compressor forcing screw until the upper mounting plate and coil spring can be removed.
- Upper mounting plate and coil spring from the spring compressor

To install:

9. Install or connect the following:
- Coil spring and upper mounting plate to the spring compressor

10. Turn the compressor forcing screw until the coil spring is compressed.
- Shock absorber to the shock module. Tighten the retaining nut to 33 ft. lbs. (45 Nm)

11. Remove the shock module from the spring compressor. Remove the pieces of heater hose from the spring..
- Shock module yoke to the shock absorber
- Shock module yoke-to-shock pinch bolt and nut and tighten to 52 ft. lbs. (70 Nm)
- Shock module to the vehicle
- Tire and wheel

12. Lower the vehicle

Rear—2002–03 Vehicles

1. Before servicing the vehicle, refer to the precautions in the beginning of this section.

2. Raise and support the vehicle.

3. Support the rear axle.

4. Remove the shock absorber lower mounting bolts.

5. Lower the rear axle, then remove the coil springs.

To install:

6. Install the coil springs, then raise the rear axle.

7. Install the shock absorber lower mounting bolts and tighten to 59 ft. lbs. (80 Nm).

8. Remove the rear axle support.

9. Lower the vehicle.

Leaf Springs

REMOVAL & INSTALLATION

Rear—2000–01 Vehicles

1. Before servicing the vehicle, refer to the precautions in the beginning of this section.

➡**The following procedure requires the use of two sets of jackstands.**

2. Support the rear axle with jackstands, support the axle and the body separately in order to relieve the load on the rear spring.

3. Remove or disconnect the following:
- Wheel
- Shock absorber
- U-bolt nuts, washers, anchor plate and bolts
- Spare tire, if equipped
- Rear exhaust hangers and lower the rear exhaust, if necessary
- Shackle-to-frame bolt, washers and nut
- Fuel tank, if necessary
- Front bracket nut, washers and bolt
- Spring
- Shackle from the spring, if necessary

To install:

4. Install or connect the following:
- Shackle to the rearward spring eye using the bolt, washers and nut, but do not fully tighten at this time
- Spring assembly
- Spring to the front bracket using the bolt, washers and nut, but do not fully tighten at this time
- Fuel tank, if removed
- Shackle-to-frame bolt, washers and nut, but do not fully tighten at this time. If used, remove the spring support.

5. Install the U-bolts, anchor plate, washers and U-bolt nuts. Torque the nuts using 2 passes of a diagonal sequence:
   a. Step 1: Torque to 18 ft. lbs. (25 Nm).
   b. Step 2: Torque to 73 ft. lbs. (100 Nm) in the sequence.

6. Position the axle to achieve an

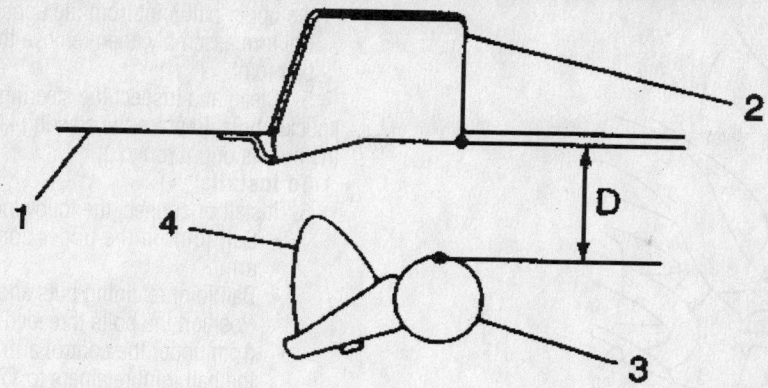

1. FRAME
2. AXLE STOP BRACKET
3. REAR AXLE (END VIEW)
4. BUMPER
5. TRIM HEIGHT 135 – 145MM
   (5.3 – 5.7 INCHES)

88268GB5

Typical rear suspension trim height measurement

approximate gap of 6.46–6.94 in. (164–176mm) between the axle housing tube and the metal surface of the rubber frame bumper bracket. Measure from the housing between the U-bolts to the metal part of the rubber bump stop on the frame.

7. While supporting the axle in this position, tighten the front and rear spring mounting fasteners to 89 ft. lbs. (122 Nm).

8. Install or connect the following:
   • Rear exhaust in position and tighten the hangers
   • Spare tire
   • Shock absorber

Torsion Bar

Instead of the coil spring used on the front suspension of 2000–01 2WD vehicles, the 4WD vehicles are equipped with a torsion bar.

REMOVAL & INSTALLATION

1. Before servicing the vehicle, refer to the precautions in the beginning of this section.

➡The following procedure requires the use of the Torsion Bar Unloader tool J-36202.

2. Remove or disconnect the following:
   • Transmission shield, if equipped
   • Torsion bar unloader tool to relax

the tension on the torsion bar adjusting arm screw; record the number of turns necessary to properly install the tool. Remove the adjusting screw and the unloader tool.
   • Lower link mount nut from one side
   • Torsion bars by disengaging them

➡Note the direction of the forward end and side of the torsion bar being removed

   • Lower link nut from the opposite side
   • Lower link mount, upper link mount nut
   • Upper link mount
   • Torsion bar from the frame
To install:
3. Install or connect the following:
   • Torsion bar and support
   • Upper link mount. Torque the nut to 48 ft. lbs. (68 Nm).
4. Place a jack under the torsion bar to release tension.
5. Install or connect the following:
   • Lower link mount bushing and nut. Torque the nut to 37 ft. lbs. (50 Nm).
   • Torsion bar unloader tool. Tighten the tool against the adjusting arm the same number turns recorded earlier and remove the tool. This loads the torsion bars.
   • Transmission shield, if removed

Upper Ball Joints

REMOVAL & INSTALLATION

2000–2001 2WD Vehicles

1. Before servicing the vehicle, refer to the precautions in the beginning of this section.

➡The following procedure requires the use of a ball joint separator tool such as J-23742 and J-9519-E ball joint remover and installer set.

2. Raise and support the front of the vehicle safely by placing stands securely under the lower control arms. Because the vehicle's weight is used to relieve spring tension on the upper control arm, the stands must be positioned between the spring seats and the lower control arm ball joints for maximum leverage.

> ✳✳ **CAUTION**

With components unbolted, the stand is holding the lower control arm in place against the coil spring. Make sure the stand is firmly positioned and cannot move, or personal injury could result.

3. Remove or disconnect the following:
   • Tire and wheel assembly
   • Brake caliper and support it from the vehicle using a coat hanger or wire. Make sure the brake line is not stretched or damaged and that the caliper's weight is not supported by the line.
   • Cotter pin and retaining nut from the upper ball joint
   • Anti-lock brake sensor wire bracket, if equipped
   • Upper ball joint from the steering knuckle using tool J-23742 and pull the steering knuckle free of the ball joint

➡After separating the steering knuckle from the upper ball joint, be sure to support the steering knuckle/hub assembly to prevent damaging the brake hose.

4. Remove the riveted upper ball joint from the upper control arm as follows:
   a. Drill a⅛in. (3mm) hole, about¼in. (6mm) deep into each rivet.
   b. Then use a½in. (13mm) drill bit, to drill off the rivet heads.
   c. Using a pin punch and the hammer, drive out the rivets in order to free

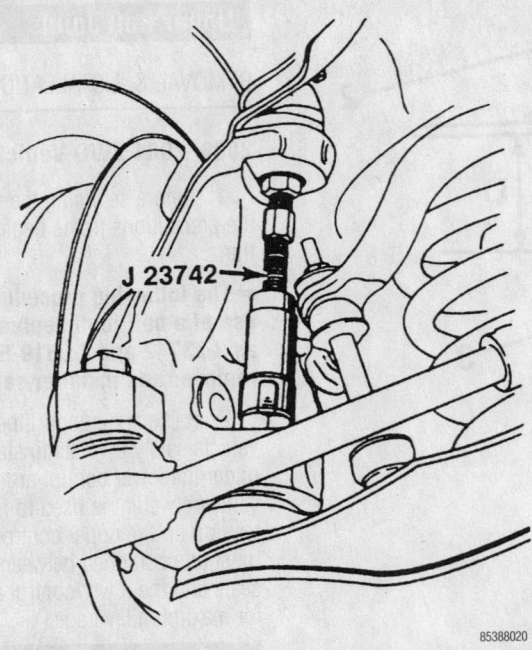

J 23742 →

85388020

Use a ball joint separator tool to drive the upper ball joint from the steering knuckle

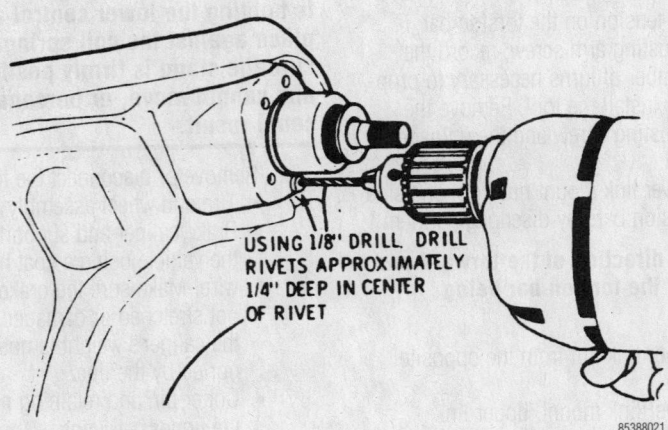

USING 1/8" DRILL. DRILL
RIVETS APPROXIMATELY
1/4" DEEP IN CENTER
OF RIVET

85388021

Drill a small guide hole into each ball joint rivet

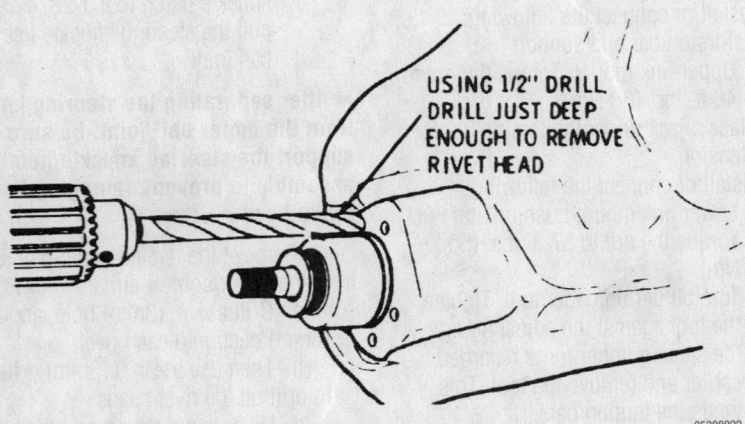

USING 1/2" DRILL
DRILL JUST DEEP
ENOUGH TO REMOVE
RIVET HEAD

85388022

Then drill off the rivet heads

the upper ball joint from the upper control arm assembly, then remove the upper ball joint.

5. Clean and inspect the steering knuckle hole. Replace the steering knuckle if the hole is out of round.

To install:

6. Install or connect the following:
- Ball joint in the upper control arm
- Ball joint retaining nuts and bolts. Position the bolts threaded upward from under the control arm. Tighten the ball joint retainers to 17 ft. lbs. (23 Nm).
- Anti-lock brake sensor wire bracket, if removed
- Ball joint to the knuckle. Make sure the joint is seated, then install the stud nut and tighten to 61 ft. lbs. (83 Nm). Insert a new cotter pin.

➡ **When installing the cotter pin, never loosen the castle nut to expose the cotter pin hole.**

- Thread the grease fitting into the ball joint. Use a grease gun to lubricate the upper ball joint until grease appears at the seal.
- Brake caliper
- Tire and wheel assembly

7. Check and adjust the front end alignment, as necessary.

2000–2001 4WD Vehicles

1. Before servicing the vehicle, refer to the precautions in the beginning of this section.

On 4WD vehicles both the upper and lower ball joints are removed in the same manner. Once the joint is separated from the steering knuckle the rivets are drilled and punched to free the joint from the control arm. Service joints are bolted into position with the retaining bolts threaded upward from beneath the control arm. In this manner, the joint is replaced in an almost identical fashion to the upper joints on 2WD vehicles.

2. Remove or disconnect the following:
- Tire and wheel assembly
- Wheel speed sensor wiring connector from the upper control arm, if removing the upper ball joint
- Cotter pin from the ball joint, then loosen the retaining nut

3. Position a suitable ball joint separator tool such as J-36607, then carefully loosen the joint in the steering knuckle. Remove the tool and the retaining nut, then separate the joint from the knuckle.

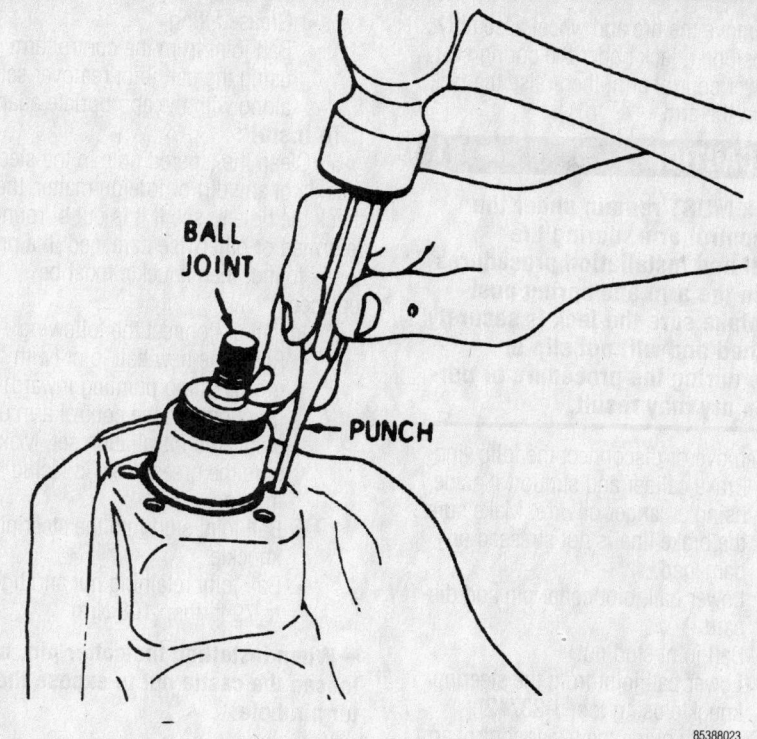

Punch the rivets out and remove the ball joint

➥**After separating the steering knuckle from the upper ball joint, be sure to support the steering knuckle/hub assembly to prevent damaging the brake hose.**

4. Remove the riveted ball joint from the control arm:

a. Drill a⅛in. (3mm) hole, about¼in. (6mm) deep into each rivet.

b. Then use a½in. (13mm) drill bit, to drill off the rivet heads.

c. Using a pin punch and the hammer, drive out the rivets in order to free the ball joint from the control arm assembly, then remove the ball joint.

To install:

5. Install or connect the following:

- Ball joint in the control arm
- Ball joint retaining nuts and bolts. Position the bolts threaded upward from under the control arm. Tighten the ball joint retainers to 17 ft. lbs. (23 Nm).
- Ball joint to the knuckle. Make sure the joint is seated, tighten the lower nut to 79 ft. lbs. (108 Nm) and the upper nut to 61 ft. lbs. (83 Nm). Install a new cotter pin.

➥**When installing the cotter pin, never loosen the castle nut to expose the cotter pin hole, but DO NOT tighten more than an additional⅛turn.**

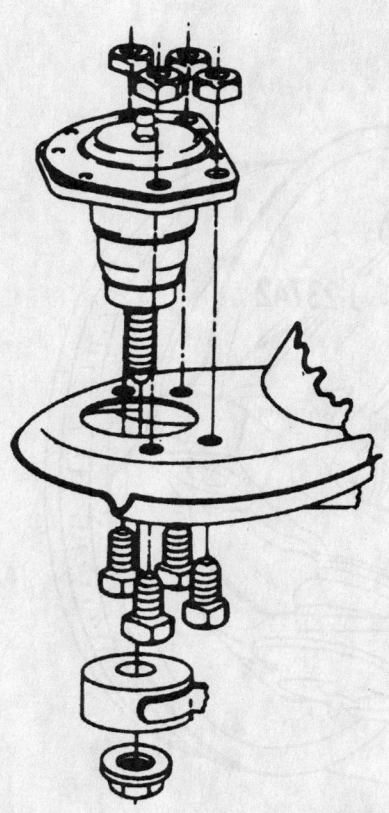

Service ball joints are bolted to the control arm

6. Use a grease gun to lubricate the upper ball joint.

- Wheel speed sensor wiring connector to the upper control arm, if the upper ball joint was removed
- Tire and wheel assembly

7. Check and adjust the front end alignment, as necessary.

2002–03 Vehicles

➥**This procedure requires the use of the following special tools: J 9519-E Lower Ball Joint Remover and Installer, J 21474-01 Control Arm Bushing Set and J 45117 Ball Joint Installation Spacer.**

1. On 4WD vehicles, remove the wheel center cap and drive axle nut.

2. Raise and support the vehicle.

3. Remove or disconnect the following:

- Tire and wheel
- Wheel hub and bearing, if necessary
- Outer tie rod retaining nut
- Out tie rod from the steering knuckle using a suitable puller
- Brake hose bracket retaining bolts and bracket

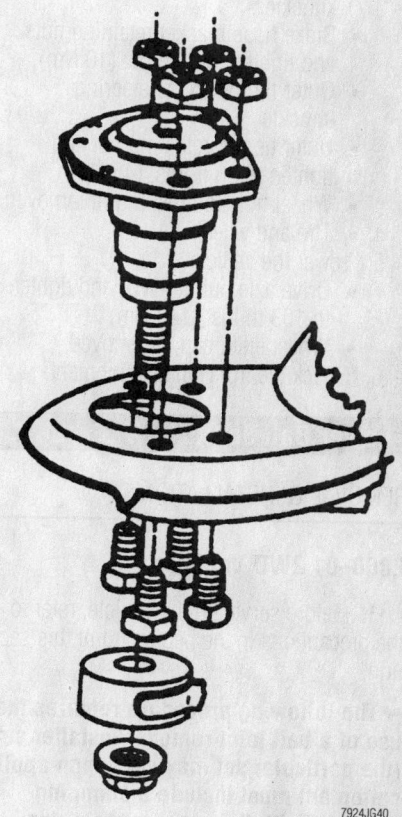

The replacement ball joint comes with nuts and bolts for installation

- Upper control arm-to-steering knuckle pinch bolt and nut
- Upper control arm from the steering knuckle
- Lower ball joint retaining nut
- Steering knuckle from the lower control arm using a suitable ball joint removal tool
- Steering knuckle from the vehicle
- Upper ball joint retaining clip
- Upper ball joint from the steering knuckle using Lower Ball Joint Removal and Installer tool No. J 9519-E

To install:

4. Install or connect the following:
   - Upper ball joint to the steering knuckle using J 9519-E, J 21474-01 and J 45117
   - Upper ball joint retaining clip
   - Steering knuckle to the lower control arm
   - Lower ball joint retaining nut and tighten to 81 ft. lbs. (110 Nm)
   - Upper control arm to the steering knuckle
   - Upper control arm pinch bolt and nut and tighten to 30 ft. lbs. (41 Nm)
   - Brake hose bracket to the steering knuckle
   - Brake hose bracket retaining nuts and tighten to 7 ft. lbs. (10 Nm)
   - Outer tie rod to the steering knuckle
   - Outer tie rod retaining nut and tighten to 33 ft. lbs. (45 Nm)
   - Wheel hub and bearing, if removed
   - Tire and wheel
5. Lower the vehicle
   - Drive axle nut, if 4WD, and tighten to 103 ft. lbs. (140 Nm)
   - Wheel enter cap, if removed
6. Check the front wheel alignment.

Lower Ball Joints

REMOVAL & INSTALLATION

2000–01 2WD Vehicles

1. Before servicing the vehicle, refer to the precautions in the beginning of this section.

➡ **The following procedure requires the use of a ball joint remover/installer set (the particular set may vary upon application but must include a clamping-type tool with the appropriately sized adapters) and a ball joint separator tool, such as J-23742.**

2. Remove the tire and wheel assembly.
3. Position a jack under the spring seat of the lower control arm, then raise the jack to support the arm.

✳✳ CAUTION

The jack MUST remain under the lower control arm, during the removal and installation procedures, to retain the arm and spring positions. Make sure the jack is securely positioned and will not slip or release during the procedure or personal injury may result.

4. Remove or disconnect the following:
   - Brake caliper and support it aside using a hanger or wire. Make sure the brake line is not stressed or damaged.
   - Lower ball joint cotter pin and discard
   - Ball joint stud nut
   - Lower ball joint from the steering knuckle using tool J-23742
5. Carefully guide the lower control arm out of the opening in the splash shield using a putty knife. Position a block of wood between the frame and upper control arm to keep the knuckle out of the way.

- Grease fitting
- Ball joint from the control arm using the ball joint remover set along with the appropriate adapters

To install:

6. Clean the tapered hole in the steering knuckle of any dirt or foreign matter, then check the hole to see if it is out of round, deformed or otherwise damaged. If a problem is found, then knuckle must be replaced.
7. Install or connect the following:
   - Press the new ball joint (with grease fitting pointing inward) until it bottoms in the control arm using a suitable installation set. Make sure the grease seal is facing inboard.
   - Ball joint stud into the steering knuckle
   - Ball joint retaining nut and tighten to 79 ft. lbs. (108 Nm)

➡ **When installing the cotter pin, never loosen the castle nut to expose the cotter pin hole.**

   - Grease fitting into the ball joint, if not already installed
8. Use a grease gun to lubricate the joint until grease appears at the seal.

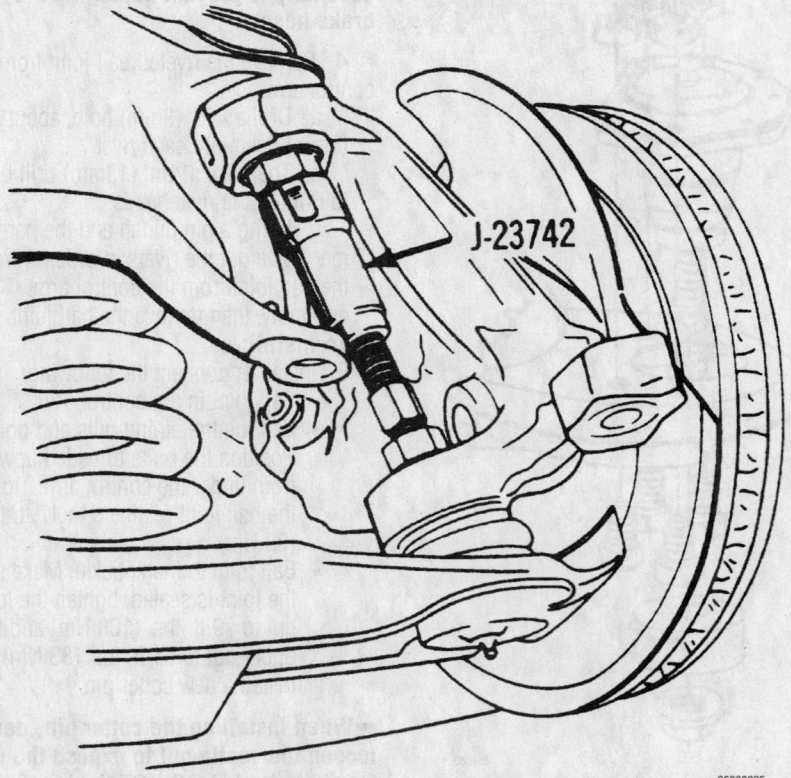

Use a ball joint separator to drive the lower joint from the knuckle

85388025

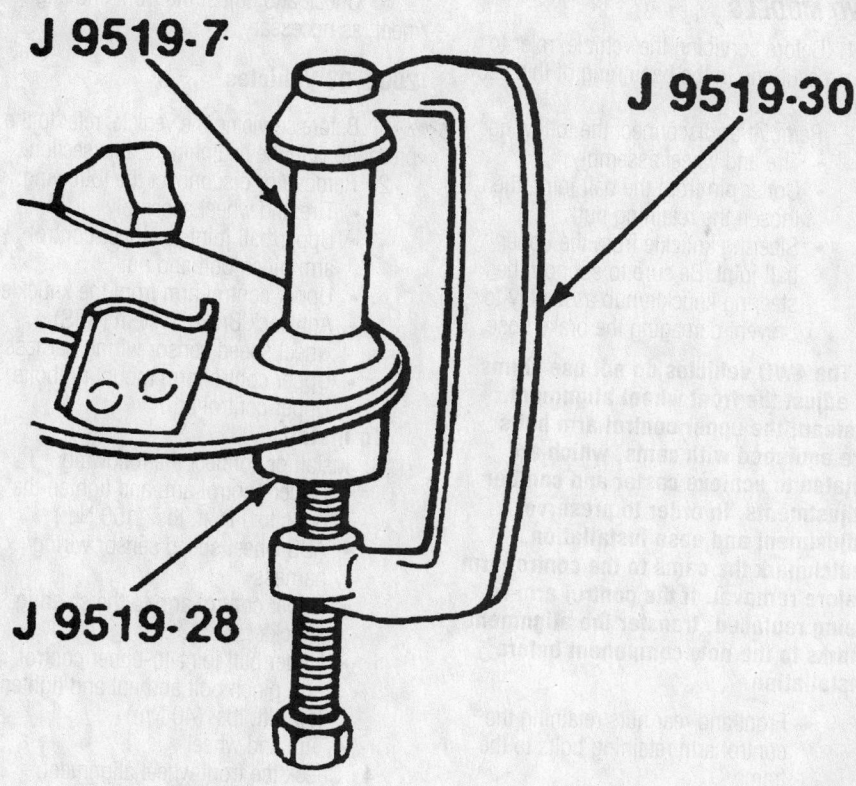

Driving the lower joint from the control arm

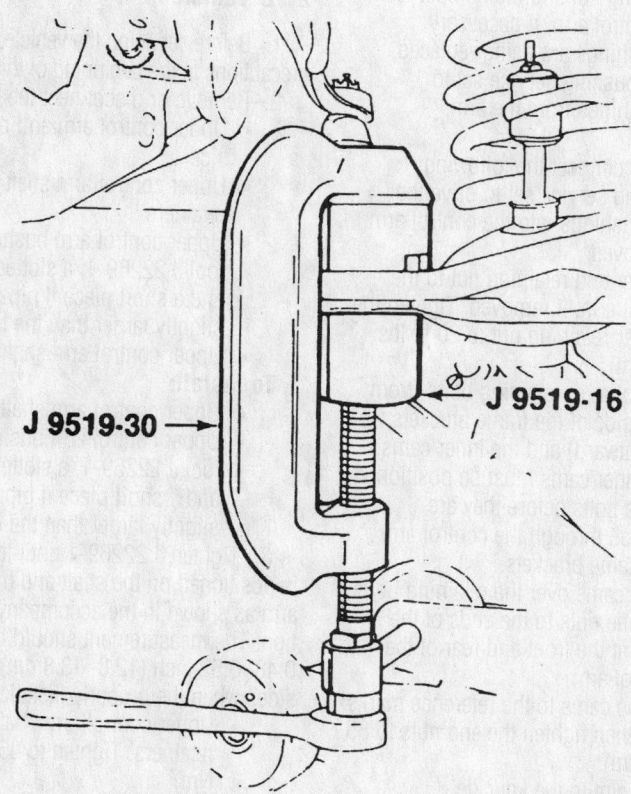

Installing a new ball joint

- Brake caliper
- Tire and wheel assembly

9. Check and adjust the front end alignment, as necessary.

2000–01 4WD Vehicles

On these vehicles both the upper and lower ball joints are removed in the same manner. Please refer to the procedure for the 2WD vehicles.

2002–03 Vehicles

➡**This procedure requires the use of the following special tools: J 9519-E Lower Ball Joint Remover and Installer, J 34874 Booster Seal Remover/Installer, J 41435 Ball Joint Installer, J 45105-1 Ball Joint Flaring Adapter and J 45105-2 Receiver.**

1. On 4WD vehicles, remove the wheel center cap and drive axle nut.
2. Raise and support the vehicle.
3. Remove or disconnect the following:
   - Tire and wheel
   - Wheel hub and bearing, if necessary
   - Outer tie rod retaining nut
   - Out tie rod from the steering knuckle using a suitable puller
   - Brake hose bracket retaining bolts and bracket
   - Upper control arm-to-steering knuckle pinch bolt and nut
   - Upper control arm from the steering knuckle
   - Lower ball joint retaining nut
   - Steering knuckle from the lower control arm using a suitable ball joint removal tool
   - Steering knuckle from the vehicle
   - Lower ball joint flange with a chisel

4. Install tools J 9519-E and J 34874 to the lower ball joint, then use those tools to remove the lower ball joint from the lower control arm.

To install:

5. Install or connect the following:
   - Lower ball joint to the lower control arm, using tools J 9519-E, J 41435 and J 45105-2
6. Remove the tools from the lower control arm.
   - Tools J 9519-E and J 45105-1 to the lower ball joint
7. Flare the lower ball joint flange with J 9519-E and J 45105-1, then remove the tools from the lower ball joint.
   - Steering knuckle to the lower control arm
   - Lower ball joint retaining nut and tighten to 81 ft. lbs. (110 Nm)

- Upper control arm to the steering knuckle
- Upper control arm pinch bolt and nut and tighten to 30 ft. lbs. (41 Nm)
- Brake hose bracket to the steering knuckle
- Brake hose bracket retaining nuts and tighten to 7 ft. lbs. (10 Nm)
- Outer tie rod to the steering knuckle
- Outer tie rod retaining nut and tighten to 33 ft. lbs. (45 Nm)
- Wheel hub and bearing, if removed
- Tire and wheel

8. Lower the vehicle
   - Drive axle nut, if 4WD, and tighten to 103 ft. lbs. (140 Nm)
   - Wheel enter cap, if removed
9. Check the front wheel alignment.

Upper Control Arm

REMOVAL & INSTALLATION

2000–01 Vehicles

2WD MODELS

1. Before servicing the vehicle, refer to the precautions in the beginning of this section.
2. Remove or disconnect the following:
   - Negative battery cable
   - Wheel
   - Wheel speed sensor harness bracket retaining bolt and nut, if equipped
   - Steering knuckle from upper control arm ball joint
   - Mounting nuts/bolts and shims

➡ **Make sure to note the location of the control arm shims prior to removal so that they may be installed in their original positions.**

   - Upper control arm

To install:
3. Install or connect the following:
   - Upper control arm

➡ **Always tighten nut on the thinner shim pack first.**

   - Mounting nuts/bolts and shims. Torque the nuts to 81–85 ft. lbs. (110–115 Nm).
   - Steering knuckle to upper control arm ball joint
   - New cotter pin

➡ **Tighten the nut to align the hole never loosen.**

   - Wheel speed sensor harness bracket retaining bolt and nut, if equipped
   - Wheel

4WD MODELS

1. Before servicing the vehicle, refer to the precautions in the beginning of this section.
2. Remove or disconnect the following:
   - Tire and wheel assembly
   - Cotter pin from the ball joint, then loosen the retaining nut
   - Steering knuckle from the upper ball joint. Be sure to support the steering knuckle/hub assembly to prevent damaging the brake hose.

➡ **The 4WD vehicles do not use shims to adjust the front wheel alignment. Instead, the upper control arm bolts are equipped with cams, which are rotated to achieve caster and camber adjustments. In order to preserve adjustment and ease installation, matchmark the cams to the control arm before removal. If the control arm is being replaced, transfer the alignment marks to the new component before installation.**

   - Front and rear nuts retaining the control arm retaining bolts to the frame
   - Outer cams from the bolts
   - Bolts and inner cams
   - Control arm from the vehicle
   - Retaining nut and the bumper from the control arm, if necessary

3. If the bushings are being replaced, use a suitable bushing service set to remove the bushings from the arm.

To install:
4. Install or connect the following:
   - Bushing service set to drive the new bushings into the control arm, if removed
   - Bumper and retaining nut to the control arm, if removed. Tighten the bumper retaining nut to 20 ft. lbs. (27 Nm).
   - Control arm, retaining bolts (from the inside of the frame brackets facing outward) and the inner cams. The inner cams must be positioned on the bolts before they are inserted through the control arm and frame brackets.
   - Outer cams over the retaining bolts, then the nuts to the ends of the bolts at the front and rear of the control arm

5. Align the cams to the reference marks made earlier, then tighten the end nuts to 85 ft. lbs. (115 Nm).
   - Ball joint to the knuckle
   - Tire and wheel assembly

6. Check and adjust the front end alignment, as necessary.

2002–03 Vehicles

1. Before servicing the vehicle, refer to the precautions in the beginning of this section.
2. Remove or disconnect the following:
   - Tire and wheel assembly
   - Upper ball joint-to-upper control arm pinch bolt and nut
   - Upper control arm from the knuckle
   - Anti-lock Brake System (ABS) wheel speed sensor wiring harness
   - Upper control arm mounting bolts
   - Upper control arm

To install:
3. Install or connect the following:
   - Upper control arm and tighten the bolts to 111 ft. lbs. (150 Nm)
   - ABS wheel speed sensor wiring harness
   - Upper control arm to the steering knuckle
   - Upper ball joint-to-upper control arm pinch bolt and nut and tighten to 30 ft. lbs. (40 Nm)
   - Tire and wheel

4. Check the front wheel alignment.

CONTROL ARM BUSHING REPLACEMENT

2WD Vehicles

1. Before servicing the vehicle, refer to the precautions in the beginning of this section.
2. Remove or disconnect the following:
   - Upper control arm and place it in a vice
   - Upper control arm shaft nuts and retainers
   - Upper control arm bushings using tool J 22269-1, a slotted washer and a short piece if pipe that is slightly larger than the bushing
   - Upper control arm shaft

To install:
   - Upper control arm shaft
   - Upper control arm bushings using tool J 22269-1, a slotted washer and a short piece if pipe that is slightly larger than the bushing

3. Tighten J 22269-1 until the bushing is positioned on the shaft and the control arm as shown in the accompanying illustration. The measurement should be 0.48–0.52 inch (12.8–13.8mm) at both sides when the properly installed.
   - Upper control arm shaft nuts and retainers. Tighten to 85 ft. lbs. (115 Nm).
   - Upper control arm

4WD Vehicles

If the bushings require replacement, refer to the control arm removal and installation procedure for bushing replacement.

Lower Control Arm

REMOVAL & INSTALLATION

2000–01 Vehicles

2WD MODELS

1. Before servicing the vehicle, refer to the precautions in the beginning of this section.
2. Remove or disconnect the following:
   - Coil Spring
   - Lower ball joint from the steering knuckle
   - Lower control arm from the vehicle

To install:

3. Install or connect the following:
   - Lower control arm
   - Lower ball joint stud into the steering knuckle
   - Ball joint-to-steering knuckle nut and tighten to specification
   - New cotter pin to the lower ball joint stud
   - Coil spring
4. Align the vehicle.

4WD MODELS

1. Before servicing the vehicle, refer to the precautions in the beginning of this section.

➡**Tools Needed: universal tie rod separator J–24319–01, torsion bar unloader J–36202, lower control arm bushing service kit J–36618 (if the control arm bushing are being replaced) and ball joint C-clamp J–9519–23. Parts Needed: whether or not the control arm or bushing are being replaced, NEW control arm retaining nut should be used once the old ones have been loosened and removed.**

2. Remove or disconnect the following:
   - Front wheels
   - 2 bolts from the front splash shield and pivot it in order to gain access to the tie rod
   - Stabilizer bar from the control arm (keeping all of the link hardware sorted for proper installation). If necessary, completely remove the bar from the vehicle for access.
   - Shock absorber
   - Inner tie rod from the relay rod using a tie rod separator
   - Outer halfshaft nut and washer
   - Bolts from the hub and bearing kit

3. Unload the torsion bar using the unloading tool J–36202. First, mark the adjuster for installation.
   - Adjustment arm. Slide the bar forward and the adapter out of the rear to remove the adjusting arm.
   - Lower ball joint cotter pin, nut and ball joint from the control arm using a ball joint separator
   - Nuts and bolts and lower control arm with the torsion bar assembly. Note the direction which the control arm retaining bolts are facing for installation purposes.

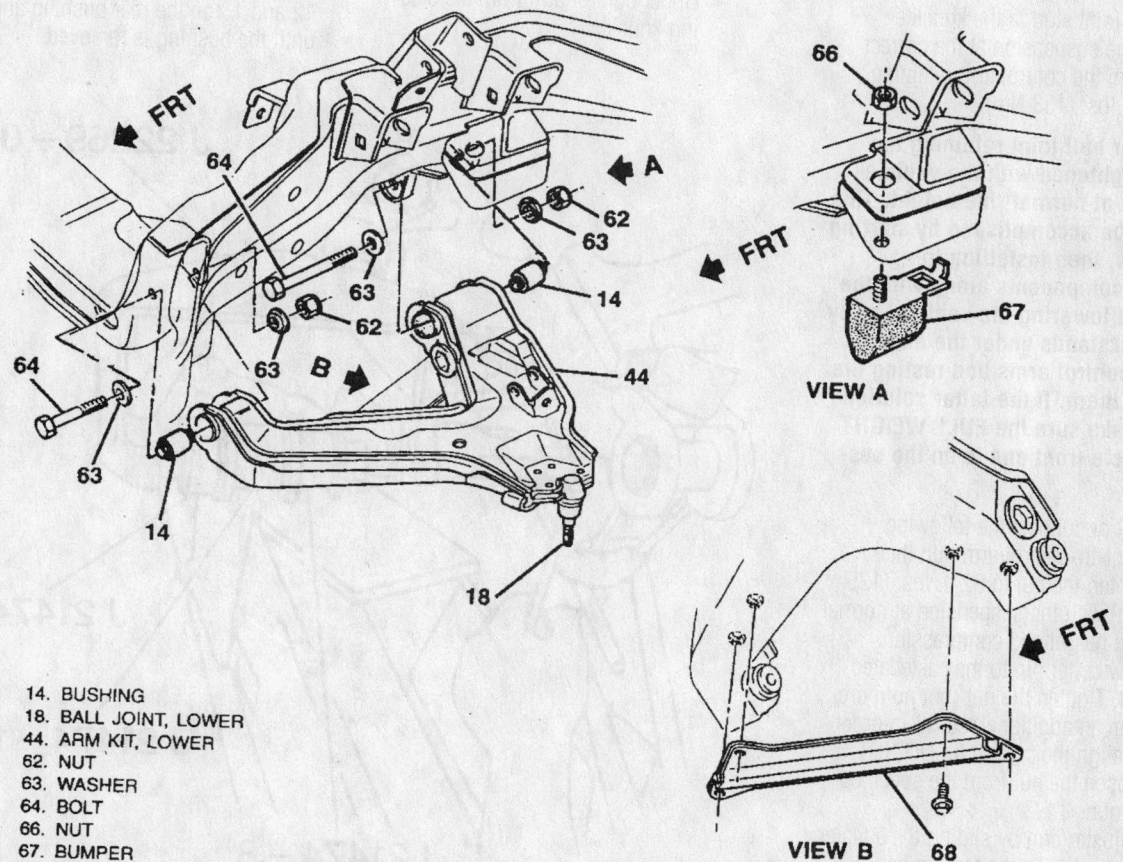

14. BUSHING
18. BALL JOINT, LOWER
44. ARM KIT, LOWER
62. NUT
63. WASHER
64. BOLT
66. NUT
67. BUMPER
68. BRACE

VIEW A

VIEW B

Exploded view of the lower control arm assembly mounting

88268GB1

To install:

4. Install or connect the following:
   - Torsion bar to the lower control arm and place the assembly into the vehicle. Position the front leg of the lower control arm into the crossmember before installing the rear leg into the frame bracket.
   - Control arm bolts (facing in the direction as noted during removal or shown in the accompanying illustration) with NEW nuts.

➡ **The control arm retainers MUST be tightened with the vehicle suspension at normal ride height. This can either be accomplished by starting the nuts now, then installing the remaining components along with the wheels and lowering the vehicle, or by moving jackstands under the ends of the lower control arms and resting the vehicle on them. If the latter solution is tried, make sure front suspension is at actual ride height compression. If you are unsure, it is best to start the nuts now and tighten them to specification once the vehicle is lowered.**

   - Ball joint stud in the knuckle
5. With the suspension at the correct height, tighten the control arm retaining nuts to 98 ft. lbs. (133 Nm).

➡ **The lower ball joint retaining nut MUST be tightened with the vehicle suspension at normal ride height. This can either be accomplished by starting the nut now, then installing the remaining components along with the wheels and lowering the vehicle, or by moving jackstands under the ends of the lower control arms and resting the vehicle on them. If the latter solution is tried, make sure the FULL WEIGHT of the vehicle front end is on the suspension.**

6. Install or connect the following:
   - Joint-to-control arm nut, then tighten the nut to 92 ft. lbs. (125 Nm) with the suspension at normal ride height and compression.
   - New cotter pin to the castellated nut. Tighten the nut (but no more than an additional⅛turn) in order to align the cotter pin. DO NOT loosen the nut from the specified torque.
   - Adjuster arm by sliding the adapter forward, over the torsion bar to install the sides of the nut. Load the torsion bar and install the adjuster bolt aligning the installation mark.

   - Drive axle through the hub and bearing assembly
   - Tighten the hub and bearing assembly retaining bolts
   - Drive axle shaft nut and washer
   - Inner tie rod end to the relay rod
   - Shock absorber
   - Stabilizer bar, if removed
   - Stabilizer link(s) to the control arm(s)
   - Splash shield
   - Front wheels
7. Recheck all fasteners for proper torque and installation before road testing.
8. Refill the differential if any fluid was lost.
9. Check and adjust the front end alignment, as necessary.

2002–03 Vehicles

1. Before servicing the vehicle, refer to the precautions in the beginning of this section.
2. Raise the vehicle.
3. Remove or disconnect the following:
   - Tire and wheel
   - Upper ball joint-to-upper control arm pinch bolt and nut
   - Upper control arm from the steering knuckle
   - Anti-lock Brake System (ABS) wheel speed sensor harness
   - Upper control arm bolt and control arm

To install:

4. Install or connect the following:
   - Upper control arm and bolts. Tighten to 111 ft. lbs. (150 Nm).
   - Connect the ABS wheel speed sensor wiring harness
   - Upper control arm to steering knuckle. Tighten the pinch bolt to 30 ft. lbs. (40 Nm).
   - Tire and wheel
5. Lower the vehicle and check the front end alignment.

CONTROL ARM BUSHING REPLACEMENT

2000–01 Vehicles

2WD MODELS

1. Before servicing the vehicle, refer to the precautions in the beginning of this section.
2. Remove lower control arm and place it in vise.
3. Install tools J 22269-01, 21474-8, 12 and 13 on the rear bushing and tighten until the bushing is removed.

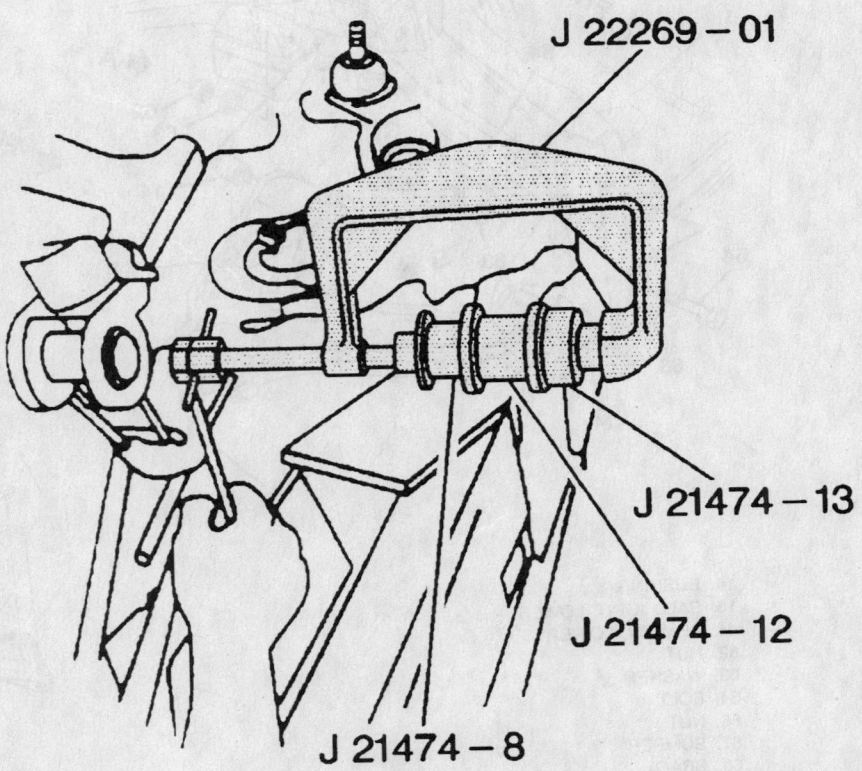

Removing the lower control arm rear bushing—all 2 wheel drive

9308JG06

9. Install the rear bushing into the control arm

10. Install tools J 22269-01, J 21474-2 and 13. Tighten until the bushing is fully seated.

11. Install the lower control arm.

4WD MODELS

1. Before servicing the vehicle, refer to the precautions in the beginning of this section.

2. Remove lower control arm and place it in vise.

3. Using bushing service set J 21474, remove the front and rear bushings.

To install:

4. Using bushing service set J 21474, install the front and rear bushings.

5. Install the lower control arm.

2002–03 Vehicles

➡This procedure requires the use of Steering Linkage and Tie Rod Puller tool No. J 24319-B and Ball Joint Remover tool No. J 43631.

1. Before servicing the vehicle, refer to the precautions in the beginning of this section.

2. Raise the vehicle.

3. Remove or disconnect the following:

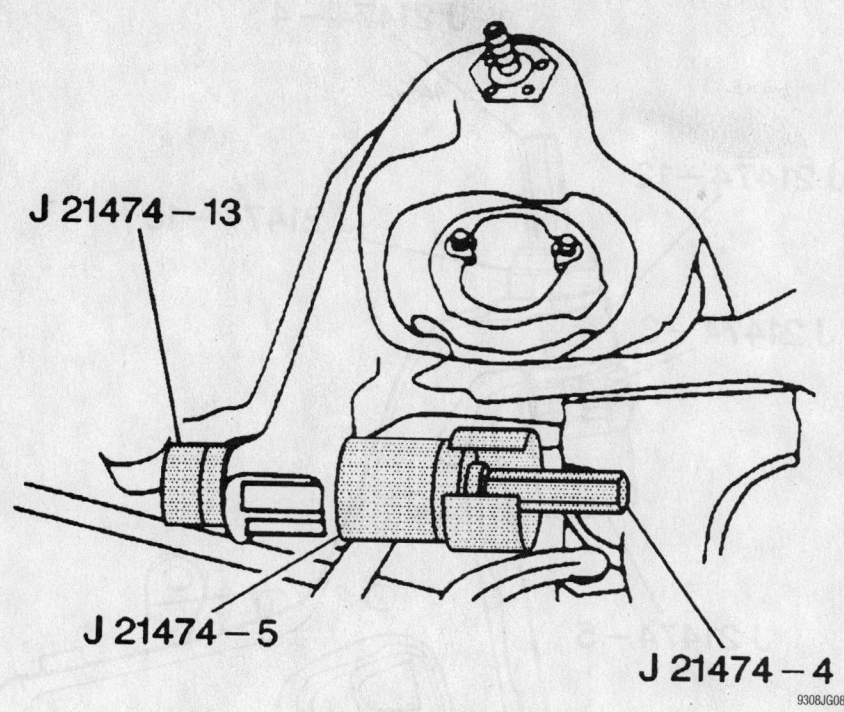

Installing the lower control arm front bushing—all 2 wheel drive

4. Using a blunt chisel, drive the front bushing flare flush with the rubber part of the bushing.

5. Place a wedge or a spacer between the bushing housing to keep the housing from bending while removing or installing the bushing.

6. Install tools J 21474-3, 4, 5 and 6 on the front bushing and tighten until the bushing is removed.

To install:

7. Install the front bushing into the control arm.

8. Install tools J 21474-4, 5 and 13. Tighten until the bushing is fully seated.

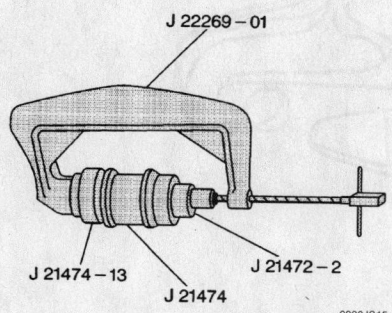

Installing the lower control arm rear bushing—all 2 wheel drive

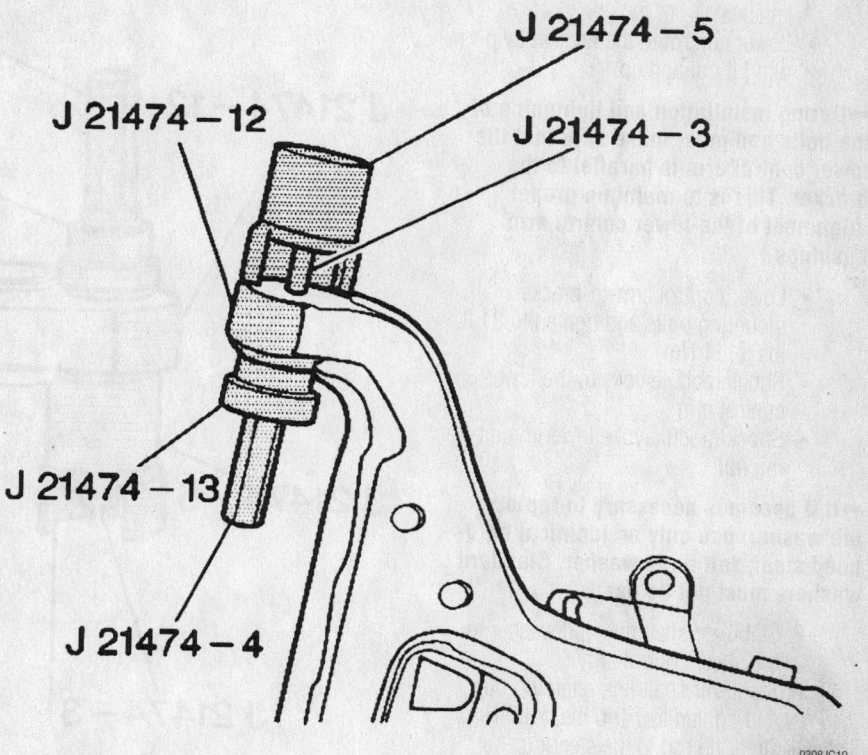

Removing the lower control arm front bushing—all 4 wheel drive

- Tire and wheel
- Outer tie rod retaining nut
- Outer tie rod from the steering knuckle using Ball Joint Removal tool No. J 43631
- Stabilizer shaft link lower nut, link and washer
- Shock module yoke lower nut and shock module using Steering Linkage and Tie Rod Puller tool No. J 24319-B
- Lower control arm-to-lower control arm bracket mounting bolts

➡ **Make sure to note the direction that the bolts are removed for installation.**

- Lower control arm-to-lower control arm bracket mounting bolts
- Lower ball joint retaining nut
- Lower ball joint from the steering knuckle using Ball Joint Removal tool No. J 43631

➡ **On 4WD vehicles, make sure not to disengage the axle shaft from the transmission.**

4. Pivot the lower control arm out and down to disengage the lower control arm from the bracket, then remove the lower control arm from the knuckle.

To install:

5. Install or connect the following:
- Lower control arm to the steering knuckle
- Lower control to the bracket by pivoting it out and up

➡ **During installation and tightening of the bolts and nuts, make sure that the lower control arm is parallel to the bracket. This is to maintain proper alignment of the lower control arm bushings.**

- Lower control arm-to-bracket mounting bolts and tighten to 81 ft. lbs. (111 Nm)
- Shock module yoke to the lower control arm
- Shock module yoke lower mounting nut

➡ **If it becomes necessary to replace the washer, use only an identical hardened steel, felt lined washer. Standard washers must not be used.**

- Stabilizer shaft link and washer to the lower control arm
- Stabilizer shaft link retaining bolt and tighten to 74 ft. lbs. (100 Nm)
- Outer tie rod to the steering knuckle. Tighten the nuts to 33 ft. lbs. (45 Nm).

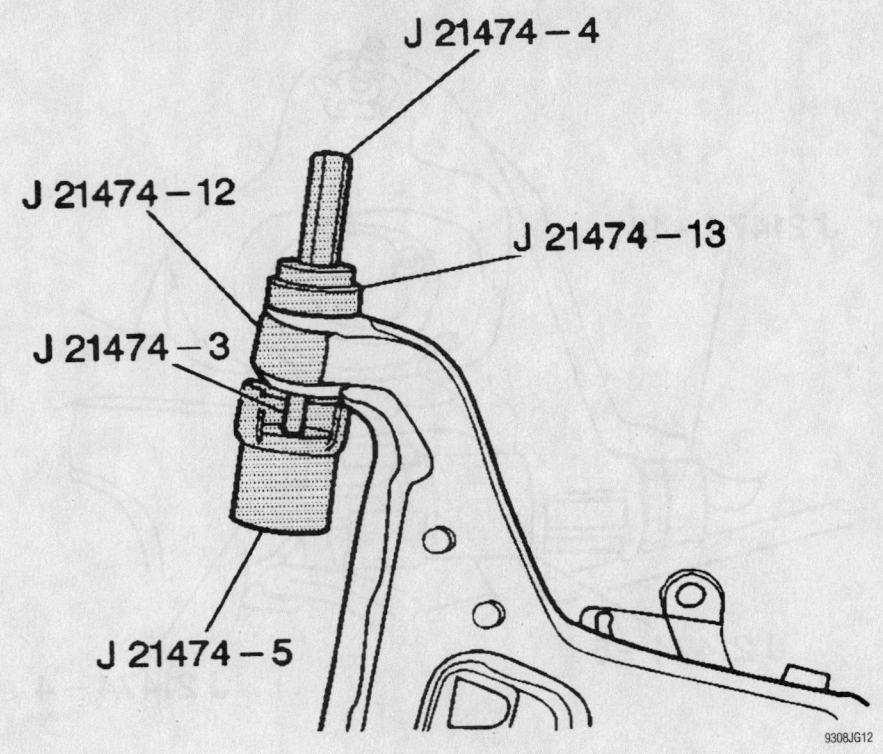

Installing the lower control arm front bushing—all 4 wheel drive

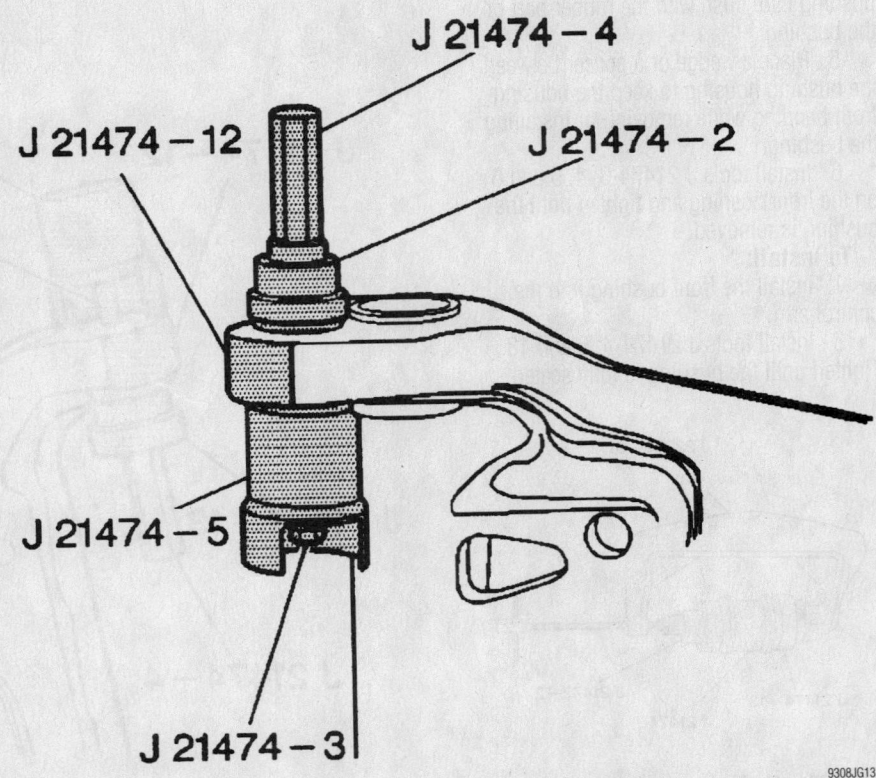

Installing the lower control arm rear bushing—all 4 wheel drive

CHEVROLET, OLDSMOBILE AND GMC
BLAZER • XTREME • TRAILBLAZER • BRAVADA • ENVOY • JIMMY **1-73**

- Tire and wheel

6. Lower the vehicle and check the front end alignment.

Wheel Bearings

ADJUSTMENT

2000 Vehicles

2WD MODELS

1. Before servicing the vehicle, refer to the precautions in the beginning of this section.

2. If equipped, remove the wheel/hub cover for access, then remove the dust cap from the hub.

3. Remove the cotter pin and loosen the spindle nut.

4. Spin the wheel forward by hand and torque the nut to 12 ft. lbs. (16 Nm) in order to fully seat the bearings and remove any burrs from the threads.

5. Back off the nut until it is just loose, then finger-tighten the nut.

6. Loosen the nut ¼–½ turn until either hole in the spindle lines up with a slot in the nut, then install a new cotter pin. This may appear to be too loose, but it is the proper adjustment.

7. Proper adjustment creates 0.001–0.005 in. (0.025–0.127mm) end-play.

4WD MODELS

The front wheel bearings on the 4-wheel drive vehicles are not adjustable. If the bearings become loose or make noise, they must be replaced.

2001–03 Vehicles

The wheel bearings on these vehicles are not adjustable. If the bearings become loose or make noise, they must be replaced.

REMOVAL & INSTALLATION

Front

2000 2WD MODELS

1. Before servicing the vehicle, refer to the precautions in the beginning of this section.

2. Remove or disconnect the following:
- Wheel
- Brake caliper with the pads without disconnecting the brake line
- Grease cap
- Cotter pin, spindle nut and washer
- Hub

※※ WARNING

Be careful not to drop the outer wheel bearing. As the hub is pulled forward, the outer wheel bearings will often fall forward and they may easily be removed at this time.

- Outer roller bearing assembly
- Inner seal by prying it out of the hub and discard it
- Inner bearing assembly

To install:

3. Clean all parts in solvent and allow to air dry, then check for excessive wear or damage. Inspect all of the parts for scoring, pitting or cracking and replace if necessary.

➡**DO NOT remove the bearing races from the hub, unless they show signs of damage.**

4. If it is necessary to remove the wheel bearing races, use the GM front bearing race removal tool J-29117 to drive the races from the hub/disc assembly. A hammer and brass drift may also be used to drive the races from the hub, but the race removal tool is quicker.

5. If the bearing races were removed, position the replacement races in the freezer for a few minutes and then install them to the hub:

a. Lightly lubricate the inside of the hub/disc assembly using wheel bearing grease.

b. Using the GM seal installation tools J-8092 and J-8850, drive the inner bearing race into the hub/disc assembly until it seats. Be sure the race is properly seated against the hub shoulder and is not cocked.

➡**When installing the bearing races, be sure to support the hub/disc assembly with GM tool J-9746-02.**

c. Using the GM seal installation tools J-8092 and J-8457, drive the outer race into the hub/disc assembly until it seats.

6. Using a high melting point wheel bearing grease, lubricate the bearings, races and spindle; be sure to place a gob of grease (inside the hub/disc assembly) between the races to provide an ample supply of lubricant.

➡**To lubricate each bearing, place a gob of grease in the palm of the hand, then scoop the bearing through the grease until it is well lubricated.**

7. Place the inner bearing in the hub, then apply a thin coating of grease to the

sealing lip and install a new inner seal, making sure the seal flange faces the bearing cup.

➡**Although a seal installation tool is preferable, a section of pipe with a smooth edge or a suitably sized socket may be used to drive the seal into position. Be sure the seal is flush with the outer surface of the hub assembly.**

8. Install or connect the following:
- Wheel hub over the spindle
- Outer bearing into the hub by hand
- Spindle washer and nut
- Brake caliper
- Wheel

9. Properly adjust the wheel bearings
- New cotter pin
- Dust cap
- Wheel cover

2000 4WD MODELS

1. Before servicing the vehicle, refer to the precautions in the beginning of this section.

2. Install Torsion Bar Unloading tool J 36202 on the torsion bar adjusting bolt and remove the bolt. To aid during installation, count the number of turns required to remove the bolt.

3. Remove the wheel.

4. Install an axle shaft boot seal protector to the Tri-pot axle joint.

5. Remove or disconnect the following:
- Cotter pin and retainer
- Castle nut and the thrust washer
- Brake caliper and support it aside using wire or a coat hanger

➡**Be sure the brake line is not stretched or damaged.**

- Brake disc from the wheel hub
- Halfshaft from the hub/bearing assembly, using a Spindle Remover tool J-28733-A to prevent damage to the shaft or hub/bearing assembly
- Hub/bearing assembly from the knuckle

6. Clean and inspect the parts for nicks, scores and/or damage, then replace them as necessary.

To install:

7. Install or connect the following:
- Hub and bearing assembly by aligning the threaded holes. Torque the bolts to 77 ft. lbs. (105 Nm).
- Tie rod end to the steering knuckle using the retaining nut
- New cotter pin
- Brake assembly
- Halfshaft nut. Tighten the nut to 180 ft. lbs. (245 Nm).

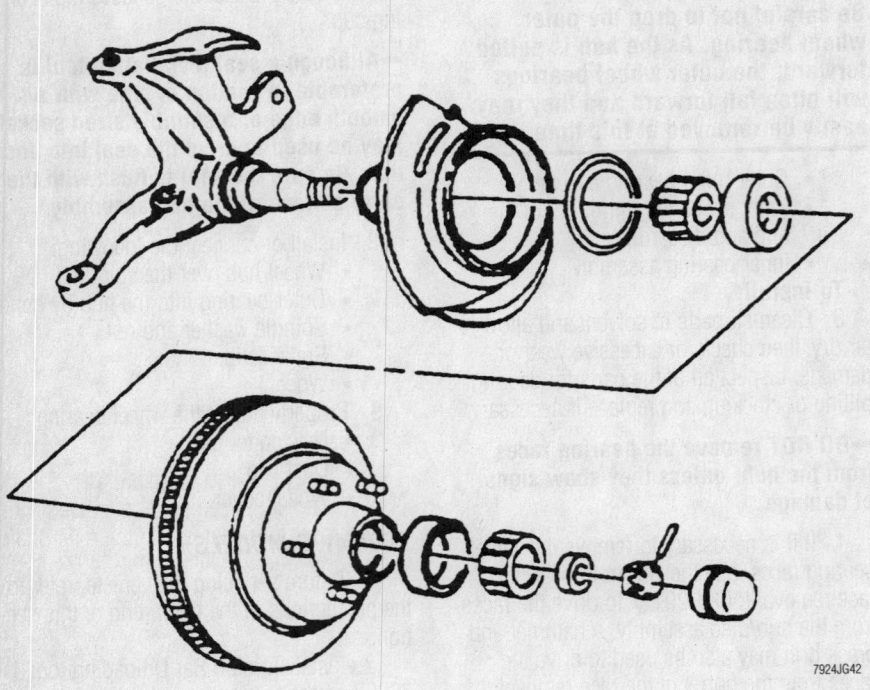

Wheel bearings, races and related components—2WD vehicles

7924JG42

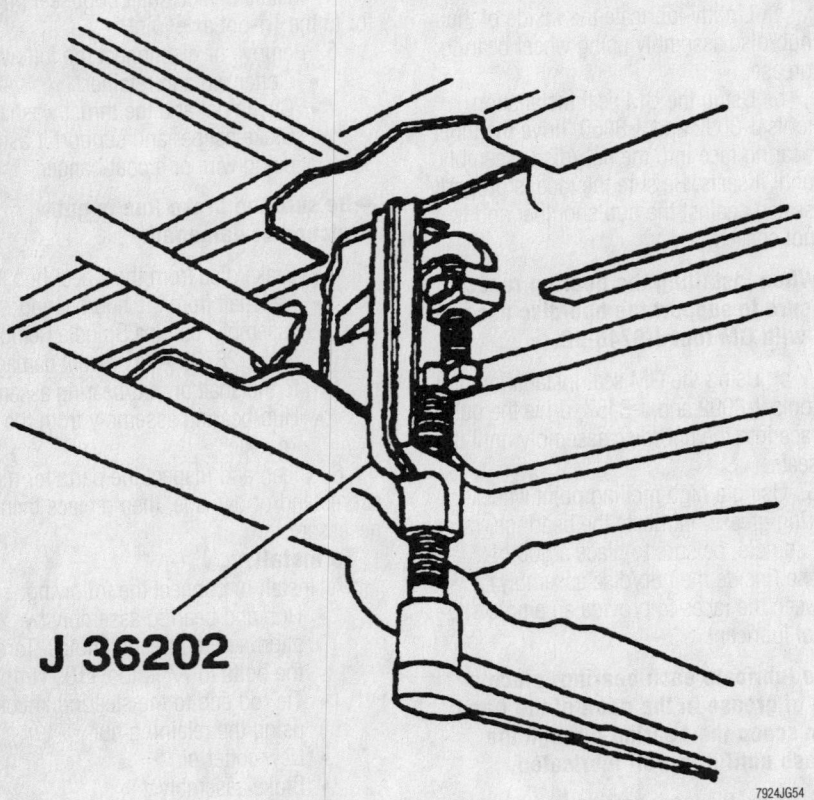

J 36202

7924JG54

Use Torsion Bar Unloading tool J 36202 to remove the adjusting bolt and unload the torsion bar

- Retainer and a new cotter pin but DO NOT back off specification in order to insert the cotter pin.
8. Remove the torsion bar unloader tool and the drive axle boot protector.
9. Install the wheel.
10. Check and/or adjust the vehicle trim height, as necessary.

2001–03 MODELS

1. Before servicing the vehicle, refer to the precautions in the beginning of this section.
2. On 4WD vehicles, remove wheel center cap, if equipped, and the drive axle nut and washer
3. Raise and support the vehicle.
4. Remove or disconnect the following:
- Tire and wheel
- Caliper, leaving the fluid lines connected
- Brake rotor
- Halfshaft from the hub and bearing on 4WD vehicles. Place a brass

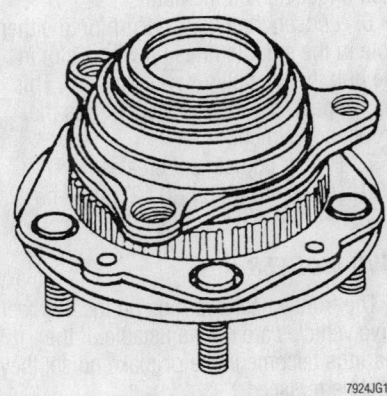

7924JG15

Hub and bearing assembly—4WD vehicles

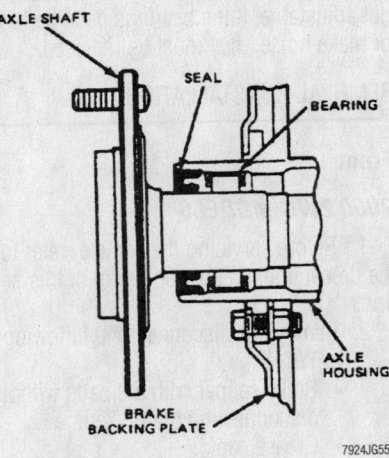

7924JG55

Cross-sectional view of the rear axle, bearing and seal assembly

drift against the outer edge of the halfshaft to protect the shaft threads. Use a hammer to sharply strike the brass drift, but to do not remove the halfshaft at this time.

- Wheel speed sensor
- Wheel hub and bearing-to-steering knuckle bolts and hub and bearing

➡ **Lay the hub and bearing on the wheel studs on the outboard side. This will avoid damaging the bearing seal.**

- Splash shield from the steering knuckle
- Seal from the hub and bearing

To install:

5. Install or connect the following:
- Wheel hub and bearing seal
- Splash shield to the steering knuckle, making sure it's properly aligned
- Hub and bearing to the steering knuckle, aligning the threaded holes
- Hub and bearing bolts and tighten to 77 ft. lbs. (105 Nm)
- Wheel speed sensor. Tighten the bolt to 13 ft. lbs. (18 Nm).
- Rotor and brake caliper
- Tire and wheel

6. Lower the vehicle

7. On 4WD vehicles, install the drive axle nut and tighten to 103 ft. lbs. (140 Nm), then install the center cap.

Rear

A new pinion shaft lockbolt should be installed whenever either of the axle shafts is removed.

The axle shaft and seal may be removed and replaced without disturbing the bearing or seal but it is highly recommended to replace the seals when removing the axle shaft.

1. Before servicing the vehicle, refer to the precautions in the beginning of this section.

2. Remove or disconnect the following:
- Rear wheels
- Brake drums

3. Using a wire brush, clean the dirt/rust from around the rear axle cover.

4. Drain the fluid.

5. Remove or disconnect the following:
- Rear pinion shaft lockbolt and the pinion shaft
- C-lock from the button end of the axle shaft by pushing the axle shaft inward
- Axle shaft from the axle housing

➡ **Be careful not to damage the oil seal.**

✲ WARNING

If equipped with an Anti-Lock Brake System (ABS), be careful not to damage the reflector ring on the axle shaft or the speed sensor bolted to the backing plate, immediately adjacent to the shaft.

6. Remove or disconnect the following:
- Oil seal by prying the it from the end of the rear axle housing

✲ WARNING

DO NOT damage the housing oil seal surface.

- Wheel bearing using the GM Slide Hammer tool J-2619, the GM Adapter tool J-2619-4 and the GM Axle Bearing Puller tool J-22813-01

To install:

7. Clean and inspect the components for excessive wear or damage and replace them, if necessary.

8. Install or connect the following:
- New or reused bearing, coated with gear lubricant, using the Axle Shaft Bearing Installer tool J-34974 to drive the bearing in until it bottoms against the seat

✲ WARNING

Be sure the bearing installer does not contact and damage the speed sensor on ABS equipped vehicles.

- New seal lubricated with gear oil using the GM Axle Shaft Seal Installer tool J-33782 to seat it in the housing until it is flush with the axle tube

➡ **Be sure the seal installer does not contact and damage the speed sensor on ABS equipped vehicles.**

- Axle shaft into the housing by engaging the splines
- C-lock retainer on the axle shaft button end

✲ WARNING

BE CAREFUL not to damage the wheel bearing seal.

- Axle shaft by pulling it outward to seat the C-lock retainer in the counterbore of the side gears
- Pinion shaft through the case and the pinions. Tighten the new lockbolt to 27 ft. lbs. (36 Nm).
- New rear axle cover gasket
- Housing cover
- Brake drums
- Wheels

9. Refill the housing.

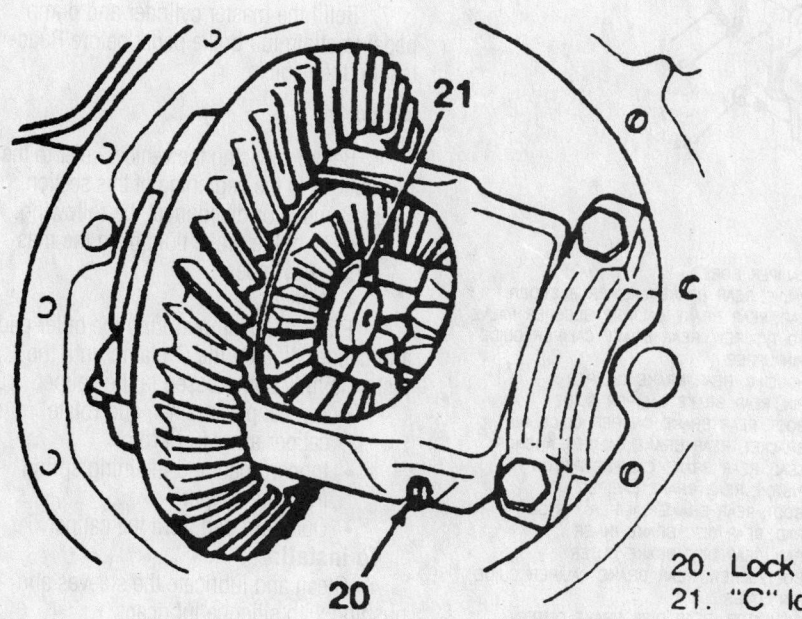

20. Lock bolt
21. "C" lock

7924JG56

Pinion shaft lockbolt and axle C-lock locations, inside the differential

BRAKES

Brake Caliper

REMOVAL & INSTALLATION

FRONT

1. Before servicing the vehicle, refer to the precautions in the beginning of this section.
2. Remove or disconnect the following:

- ⅔ of the brake fluid from the master cylinder reservoir
- Tire and wheel assembly
- Caliper fluid line, then plug
- Bolts retaining the caliper to the rotor
- Caliper from the rotor
- Disc brake pads from the caliper
- Disc brake pad retaining clips from inside the caliper

To install:

3. Clean and lubricate the sleeves and bushings with silicon grease.
4. Install or connect the followingL

- Pads in the caliper
- Caliper in position over the rotor
- Mounting bolts and tighten to 38 ft. lbs. (51 Nm)

- Fluid lines to the caliper and tighten to 33 ft. lbs. (45 Nm)
- Wheel and tire assembly

5. Refill the master cylinder to the correct level. Bleed the brake system if the fluid lines were disconnected from the caliper.

REAR

1. Before servicing the vehicle, refer to the precautions in the beginning of this section.
2. Raise and safely support the vehicle.
3. Remove or disconnect the following:

- Rear wheels
- Brake hose and cap line
- Retainers from caliper and remove caliper

To install:

4. Install or connect the following:

- Brake pads if removed
- Caliper over rotor, and onto mounts
- Retainers, and tighten to 23 ft. lbs. or (31 Nm)
- Brake hose, and tighten to 20 ft. lbs. (27 Nm)

5. Bleed brake system.
6. Install tires.
7. Refill the master cylinder and pump pedal to attain full brake pedal before Road-testing the vehicle.

Disc Brake Pads

REMOVAL & INSTALLATION

FRONT

1. Before servicing the vehicle, refer to the precautions in the beginning of this section.
2. Remove or disconnect the following:

- ⅔ of the brake fluid from the master cylinder

3. Place a C-clamp around the outer pad and caliper; tighten the C-clamp until the piston is fully compressed in the caliper.

- Brake pads
- Inboard pad and retaining spring from the caliper
- Outboard pad from the caliper
- Sleeves and bushings

To install:

4. Clean and lubricate the sleeves and bushing with silicone lubricant and install them in the caliper.
5. Clip the retaining spring onto the inboard pad and install the pad in the caliper.
6. Install or connect the following:

- Outboard pad into the caliper
- Caliper in position over the rotor and install the mounting bolts. Bend the tabs, on the outboard brake pad, over the caliper.
- Wheel and tire assemblies

7. Refill the master cylinder and pump pedal to attain full brake pedal before Road-testing the vehicle.

REAR

1. Before servicing the vehicle, refer to the precautions in the beginning of this section.
2. Remove or disconnect the following:

- ⅔ of the brake fluid from the master cylinder
- Wheels

3. Place a C-clamp around the outer pad and caliper; tighten the C-clamp until the piston is fully compressed in the caliper.

- Top caliper retainer, and rotate caliper away from rotor
- Inboard pad and retaining spring from the caliper
- Outboard pad from the caliper

To install:

4. Clean and lubricate the sleeves and bushing with silicone lubricant
5. Install or connect the following:

- Sleeves and bushings into the caliper

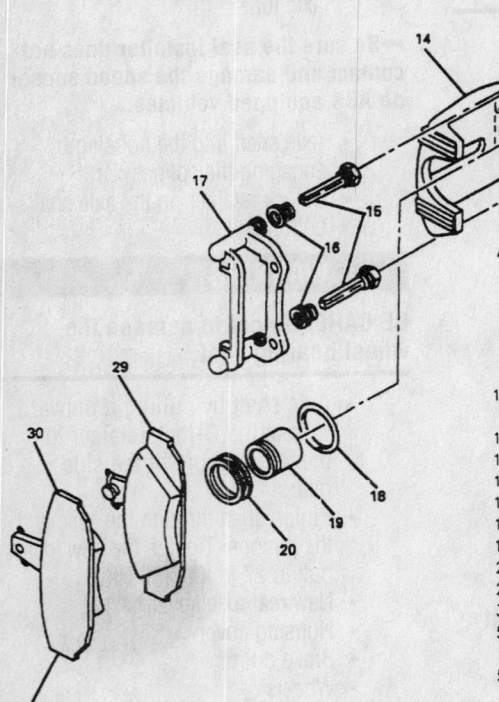

A   CALIPER BORE
6   VALVE, REAR BRAKE CALIPER BLEEDER
7   CAP, REAR BRAKE CALIPER BLEEDER VALVE
13   BOLT/SCREW, REAR BRAKE CALIPER GUIDE PIN UPPER
14   HOUSING, REAR BRAKE CALIPER
15   PIN, REAR BRAKE CALIPER GUIDE
16   BOOT, REAR BRAKE CALIPER GUIDE PIN
17   BRACKET, REAR BRAKE CALIPER ANCHOR
18   SEAL, REAR BRAKE CALIPER PISTON
19   PISTON, REAR BRAKE CALIPER
20   BOOT, REAR BRAKE CALIPER PISTON
29   PAD, REAR DISC BRAKE INNER
30   PAD, REAR DISC BRAKE OUTER
52   BOLT/SCREW, REAR BRAKE CALIPER GUIDE PIN LOWER
53   INSULATOR, REAR DISC BRAKE OUTER PAD

93026G44

Rear brake caliper—Bravada shown

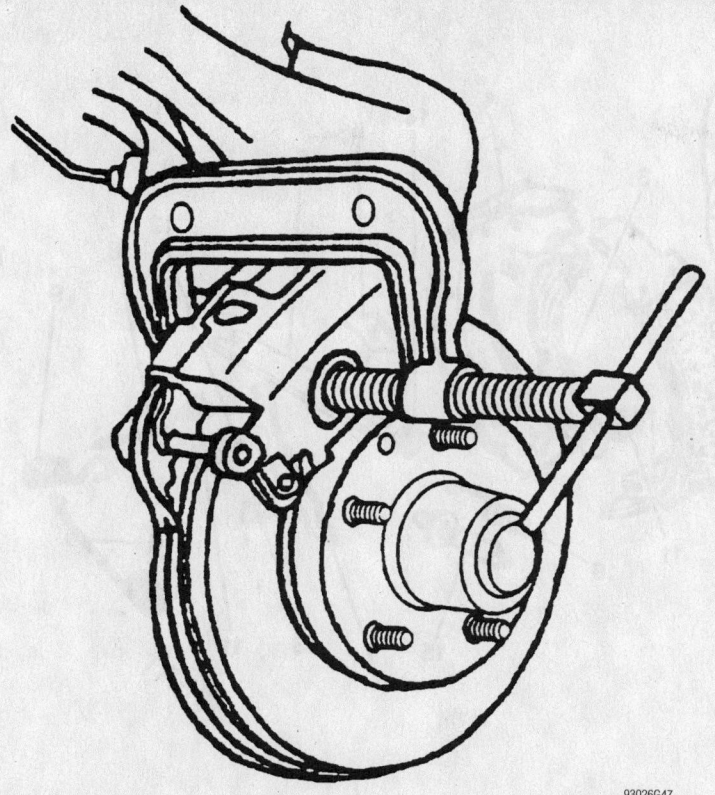

Compressing the caliper piston with a C-clamp—Blazer, Bravada, Envoy, Jimmy and Trailblazer

93026G47

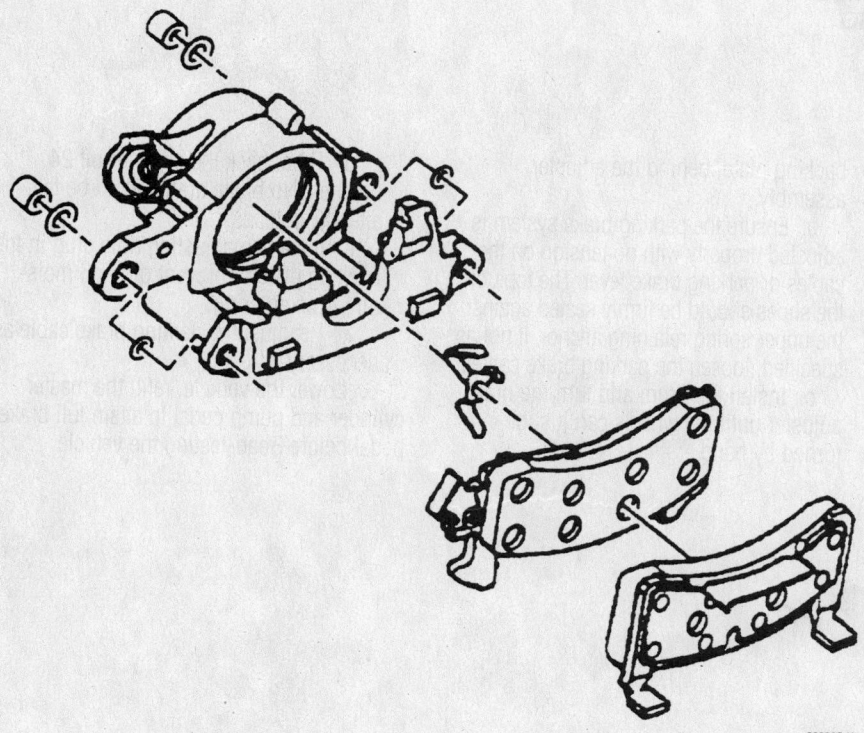

Exploded view of the disc brake assembly

93026G48

- Clip the retaining spring onto the inboard pad and install the pad in the caliper
- Outboard pad into the caliper
- Caliper in position over the rotor and install the mounting bolts
- Wheel and tire assemblies

6. Refill the master cylinder and pump pedal to attain full brake pedal before Road-testing the vehicle.

Brake Drums

REMOVAL & INSTALLATION

1. Before servicing the vehicle, refer to the precautions in the beginning of this section.
2. Remove or disconnect the following:
   - Wheel and tire assembly
   - Brake drum. If the drum will not pull of the axle, use a rubber mallet and tap it around the edge.

To install:

3. Install or connect the following:
   - Drum on the axle
   - Wheel and tire assembly

4. Refill the master cylinder and pump pedal to attain full brake pedal before road-testing the vehicle.

Brake Shoes

REMOVAL & INSTALLATION

1. Before servicing the vehicle, refer to the precautions in the beginning of this section.
2. Remove or disconnect the following:
   - Wheel and tire assembly
   - Brake drum
   - Return springs from the brake shoes
   - Remove the shoe guide
   - Hold-down springs and pins
   - Actuator lever and pivot
   - Lever return spring
   - Actuator link
   - Parking brake strut and spring
   - Parking brake lever
   - Brake shoes and the adjuster assembly

To install:

3. Lubricate the contact points on the backing plate and the adjuster with lithium grease.
4. Install or connect the following:
   - Parking brake lever, adjusting screw and spring assembly
   - Shoe assembly onto the backing plate

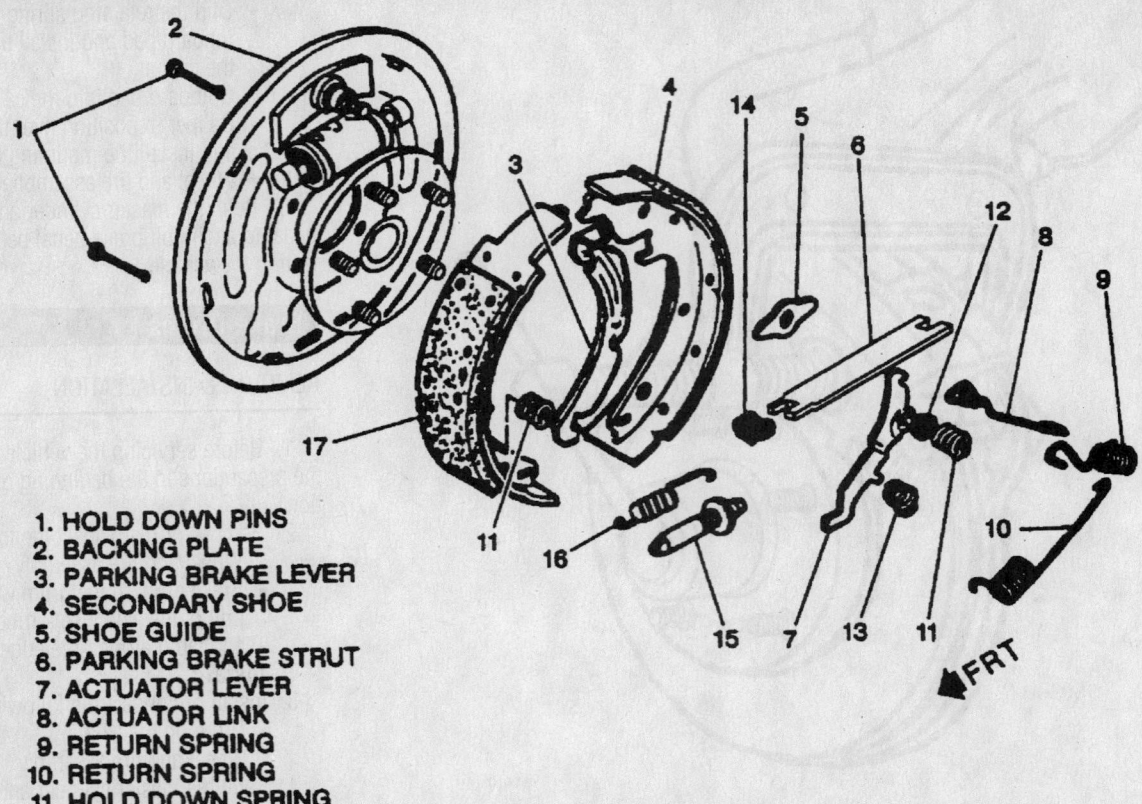

1. HOLD DOWN PINS
2. BACKING PLATE
3. PARKING BRAKE LEVER
4. SECONDARY SHOE
5. SHOE GUIDE
6. PARKING BRAKE STRUT
7. ACTUATOR LEVER
8. ACTUATOR LINK
9. RETURN SPRING
10. RETURN SPRING
11. HOLD DOWN SPRING
12. LEVER PIVOT
13. LEVER RETURN SPRING
14. STRUT SPRING
15. ADJUSTING SCREW ASSEMBLY
16. ADJUSTING SCREW SPRING
17. PRIMARY SHOE

93026G51

Exploded view of the drum brake components

- Parking brake lever, strut and strut spring
- Actuator lever and lever pivot
- Actuator link
- Lever spring, the hold-down pins and springs
- Shoe guide
- Return springs and install the brake drum in position
5. Adjust the brakes as follows:
  a. Remove the knockout area in the backing plate, behind the adjuster assembly.
  b. Ensure the parking brake system is adjusted properly with no tension on the cables or parking brake lever. The tops of the shoes should be firmly seated against the upper spring retaining anchor, if not as specified, loosen the parking brake cables.
  c. Install the drum and turn the brake adjuster until the wheels can just be turned by hand.
  d. Then, back the adjuster off 24 notches. No brake drag should be felt after 12 notches.
  e. Install an adjusting hole plug in the backing plate to prevent dirt and moisture from entering.
  f. Readjust the parking brake cable as necessary.
6. Lower the vehicle, refill the master cylinder and pump pedal to attain full brake pedal before Road-testing the vehicle.

SPECIFICATION CHARTS

ENGINE AND VEHICLE IDENTIFICATION

Engine								Model Year	
Code ①	Liters (cc)	Cu. In.	Cyl.	Fuel Sys.	Engine Type	Eng. Mfg.		Code ②	Year
E	3.4 (3350)	207	6	SFI	OHV	CPC		1	2001
								2	2002
								3	2003
								4	2004

SFI: Sequential Fuel Injection

OHV: Overhead Valves

CPC: Chevrolet/Pontiac/Canada

① 8th position of VIN

② 10th position of VIN

42372-REND-C01

GENERAL ENGINE SPECIFICATIONS

All measurements are given in inches.

Year	Model	Engine Displacement Liters (cc)	Engine Series VIN	Fuel System	Net Horsepower @ rpm	Net Torque @ rpm (ft. lbs.)	Bore x Stroke (in.)	Compression Ratio	Oil Pressure @ rpm
2001	Aztek	3.4 (3350)	E	SFI	185@5200	210@4000	3.62x3.31	9.6:1	15@1100
2002	Aztek	3.4 (3350)	E	SFI	185@5200	210@4000	3.62x3.31	9.6:1	15@1100
	Rendezvous	3.4 (3350)	E	SFI	185@5200	210@4000	3.62x3.31	9.6:1	15@1100
2003	Aztek	3.4 (3350)	E	SFI	185@5200	210@4000	3.62x3.31	9.6:1	15@1100
	Rendezvous	3.4 (3350)	E	SFI	185@5200	210@4000	3.62x3.31	9.6:1	15@1100

SFI: Sequential Fuel Injection

42372-REND-C02

ENGINE TUNE-UP SPECIFICATIONS

Year	Engine Displacement Liters (cc)	Engine ID/VIN	Spark Plug Gap (in.)	Ignition Timing (deg.)	Fuel Pump (psi)	Idle Speed (rpm)	Valve Clearance	
							Intake	Exhaust
2001	3.4 (3350)	E	0.060	①	41-47	②	HYD	HYD
2002	3.4 (3350)	E	0.060	①	41-47	②	HYD	HYD
2003	3.4 (3350)	E	0.060	①	41-47	②	HYD	HYD

NOTE: The Vehicle Emissions Control Information label often reflects specification changes made during production.

The label figures must be used if they differ from those in the chart.

HYD: Hydraulic

① Refer to underhood label for exact setting.

② Idle speed is maintained by the PCM.

42372-REND-C03

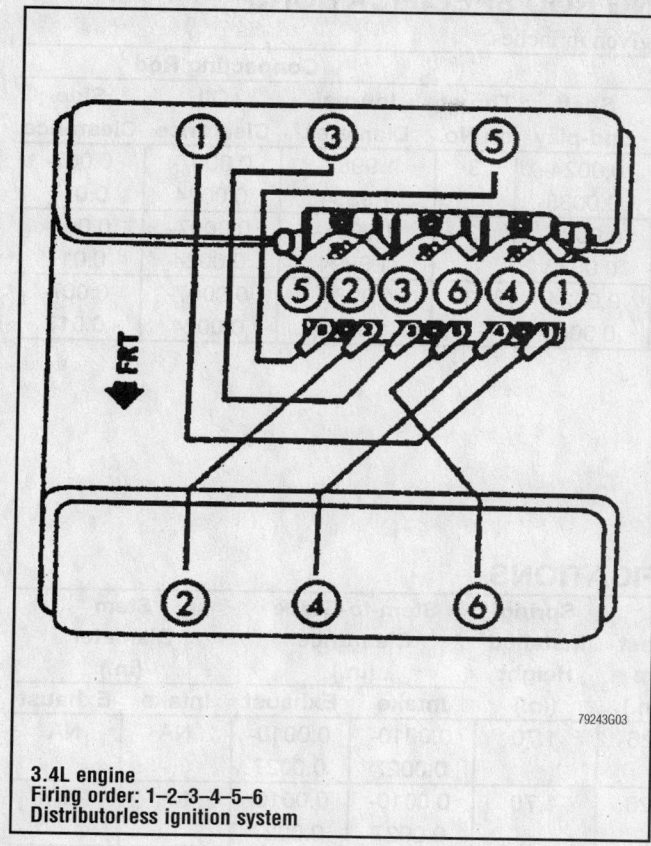

79243G03

3.4L engine
Firing order: 1–2–3–4–5–6
Distributorless ignition system

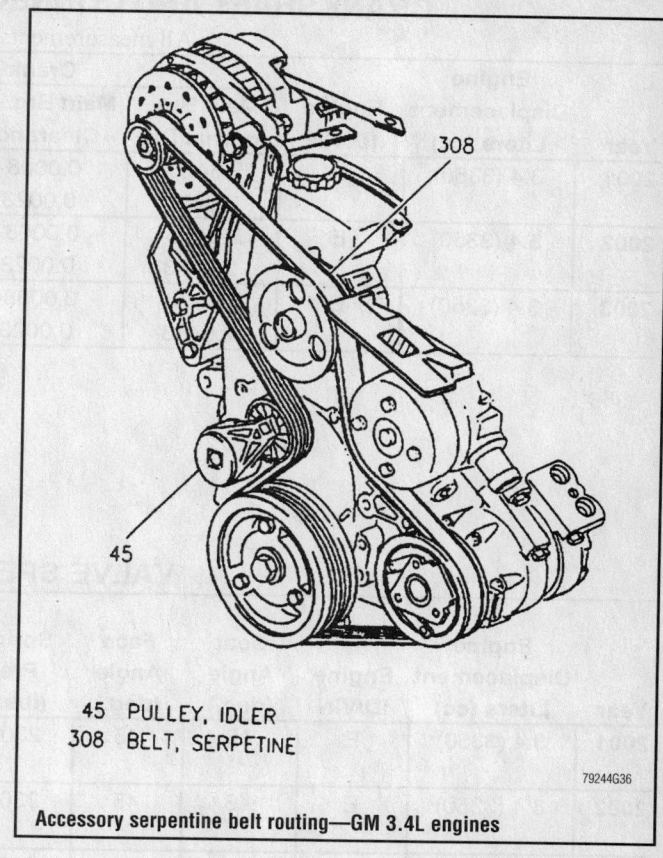

45   PULLEY, IDLER
308 BELT, SERPETINE

79244G36

Accessory serpentine belt routing—GM 3.4L engines

CAPACITIES

Year	Model	Engine Displacement Liters (cc)	Engine ID/VIN	Engine Oil with Filter (qts.)	Transmission (pts.)	Fuel Tank (gal.)	Cooling System (qts.)
2001	Aztek	3.4 (3350)	E	4.5	①	18.0	9.6
2002	Aztek	3.4 (3350)	E	4.5	②	18.5	9.6
	Rendezvous	3.4 (3350)	E	4.5	②	18.5	9.6
2003	Aztek	3.4 (3350)	E	4.5	②	18.5	9.6
	Rendezvous	3.4 (3350)	E	4.5	②	18.5	9.6

NOTE: All capacities are approximate. Add fluid gradually and check to be sure a proper fluid level is obtained.

① Drain and refill: 16 pints

   Complete overhaul: 20 pints

② Front wheel drive:

   Drain and refill: 14.8 pints

   Complete overhaul: 20 pints

   All wheel drive:

   Drain and refill: 15.6 pints

   Complete overhaul: 20.8 pints

42372-REND-C04

CRANKSHAFT AND CONNECTING ROD SPECIFICATIONS

All measurements are given in inches.

| Year | Engine Displacement Liters (cc) | Engine ID/VIN | Crankshaft | | | | Connecting Rod | | |
			Main Brg. Journal Dia.	Main Brg. Oil Clearance	Shaft End-play	Thrust on No.	Journal Diameter	Oil Clearance	Side Clearance
2001	3.4 (3350)	E	2.6473-2.6483	0.0008-0.0023	0.0024-0.0083	3	1.9987-1.9994	0.0007-0.0024	0.007-0.017
2002	3.4 (3350)	E	2.6473-2.6483	0.0008-0.0023	0.0024-0.0083	3	1.9987-1.9994	0.0007-0.0024	0.007-0.017
2003	3.4 (3350)	E	2.6473-2.6483	0.0008-0.0023	0.0024-0.0083	3	1.9987-1.9994	0.0007-0.0024	0.007-0.017

42372-REND-C05

VALVE SPECIFICATIONS

| Year | Engine Displacement Liters (cc) | Engine ID/VIN | Seat Angle (deg.) | Face Angle (deg.) | Spring Test Pressure (lbs. @ in.) | Spring Installed Height (in.) | Stem-to-Guide Clearance (in.) | | Stem Diameter (in.) | |
							Intake	Exhaust	Intake	Exhaust
2001	3.4 (3350)	E	46	45	230@1.26	1.70	0.0010-0.0027	0.0010-0.0027	NA	NA
2002	3.4 (3350)	E	46	45	230@1.26	1.70	0.0010-0.0027	0.0010-0.0027	NA	NA
2003	3.4 (3350)	E	46	45	230@1.26	1.70	0.0010-0.0027	0.0010-0.0027	NA	NA

NA: Not Available

42372-REND-C06

PISTON AND RING SPECIFICATIONS

All measurements are given in inches.

| Year | Engine Displ. Liters (cc) | Engine ID/VIN | Piston Clearance | Ring Gap | | | Ring Side Clearance | | |
				Top Compression	Bottom Compression	Oil Control	Top Compression	Bottom Compression	Oil Control
2001	3.4 (3350)	E	0.0013-0.0027	0.008-0.019	0.021-0.034	NA	0.0020-0.0034	0.0020-0.0035	NA
2002	3.4 (3350)	E	0.0013-0.0027	0.008-0.019	0.021-0.034	NA	0.0020-0.0034	0.0020-0.0035	NA
2003	3.4 (3350)	E	0.0013-0.0027	0.008-0.019	0.021-0.034	NA	0.0020-0.0034	0.0020-0.0035	NA

NA: Not available

42372-REND-C07

TORQUE SPECIFICATIONS
All readings in ft. lbs.

Year	Engine Displacement Liters (cc)	Engine ID/VIN	Cylinder Head Bolts	Main Bearing Bolts	Rod Bearing Bolts	Crankshaft Damper Bolts	Flywheel Bolts	Manifold Intake	Manifold Exhaust	Spark Plugs	Lug Nuts
2001	3.4 (3350)	E	①	②	③	④	52	⑤	12	15	100
2002	3.4 (3350)	E	①	②	③	④	52	⑤	12	15	100
2003	3.4 (3350)	E	①	②	③	④	52	⑤	12	15	100

① 37 ft. lbs. plus 90 degrees
② 37 ft. lbs. plus 77 degrees
③ 15 ft. lbs. plus 75 degrees
④ 52 ft. lbs. plus 85 degrees
⑤ 1st step: 62 inch lbs.
   2nd step: 115 inch lbs.

42372-REND-C08

BRAKE SPECIFICATIONS
All measurements in inches unless noted

Year	Model		Brake Disc Original Thickness	Brake Disc Minimum Thickness	Brake Disc Maximum Runout	Brake Drum Diameter Original Inside Diameter	Brake Drum Diameter Max. Wear Limit	Brake Drum Diameter Maximum Machine Diameter	Minimum Lining Thickness Front	Minimum Lining Thickness Rear	Brake Caliper Bracket Bolts (ft. lbs.)	Brake Caliper Mounting Bolts (ft. lbs.)
2001	Aztek	F	1.181	1.063	0.002	—	—	—	NA	—	137	26
		R	0.430	0.350	0.002	8.86	8.92	8.90	—	0.030	92	33
2002	Aztek	F	1.181	1.063	0.002	—	—	—	NA	—	137	26
		R	0.430	0.350	0.002	8.86	8.92	8.90	—	0.030	92	33
	Rendezvous	F	1.181	1.063	0.002	—	—	—	NA	—	137	26
		R	0.043	0.350	0.002	8.86	8.92	8.90	—	0.030	92	33
2003	Aztek	F	1.181	1.063	0.002	—	—	—	NA	—	137	26
		R	0.430	0.350	0.002	—	—	—	—	0.030	92	33
	Rendezvous	F	1.181	1.063	0.002	—	—	—	NA	—	137	26
		R	0.043	0.350	0.002	—	—	—	—	0.030	92	33

42372-REND-C09

TIRE, WHEEL AND BALL JOINT SPECIFICATIONS

Year	Model	OEM Tires Standard	OEM Tires Optional	Tire Pressures (psi) Front	Tire Pressures (psi) Rear	Wheel Size	Ball Joint Inspection
2001	All	P215/70R16	P215/70R16/ P235/55R17	std: 35 opt: 32	std.: 35 opt.: 32	6-JJ	U ① L: 0.090 in.
2002	All	P215/70R16	P215/70R16/ P235/55R17	std: 35 opt: 32	std.: 35 opt.: 32	6-JJ	U ① L: 0.090 in.
2003	All	P215/70R16	P215/70R16/ P235/55R17	std: 35 opt: 32	std.: 35 opt.: 32	6-JJ	U ① L: 0.090 in.

OEM: Original Equipment Manufacturer

PSI: Pounds Per Square Inch

STD: Standard

OPT: Optional

L: Lower

U: Upper

① Replace if any movement is noted or if stud can be moved by hand

42372-REND-C10

SCHEDULED MAINTENANCE INTERVALS
BUICK—RENDEZVOUS, PONTIAC—AZTEK

TO BE SERVICED	TYPE OF SERVICE	7.5	15	22.5	30	37.5	45	52.5	60	67.5	75	82.5	90	97.5	105	112	120
Accessory drive belt	I								✓								✓
Air cleaner filter	R								✓								✓
Air distributor air filter	R		✓		✓		✓		✓		✓		✓		✓		✓
Brake system	I	✓	✓	✓	✓	✓	✓	✓	✓	✓	✓	✓	✓	✓	✓	✓	✓
Engine coolant	R	Every 150,000 miles															
Engine oil & filter	S/I	✓	✓	✓	✓	✓	✓	✓	✓	✓	✓	✓	✓	✓	✓	✓	✓
Fuel tank, cap & lines	I								✓								
Transmission fluid ①	R																✓
Rotate tires	S/I	✓	✓	✓	✓	✓	✓	✓	✓	✓	✓	✓	✓	✓	✓	✓	✓
Spark plug wires	S/I	Every 100,000 miles															
Spark plugs	R	Every 100,000 miles															

R: Replace   I: Inspect   S: Service

① Change the transmission fluid every 50,000 miles if the vehicle meets any of the criteria outlined in the severe service list below.

FREQUENT OPERATION MAINTENANCE (SEVERE SERVICE)

If a vehicle is operated under any of the following conditions it is considered severe service:

- Towing a trailer or using a camper or car-top carrier.
- Repeated short trips of less than 5 miles in temperatures below freezing, or trips of less than 10 miles in any temperature.
- Extensive idling or low-speed driving for long distances as in heavy commercial use, such as delivery, taxi or police cars.
- Operating on rough, muddy or salt-covered roads.
- Operating on unpaved or dusty roads.
- Driving in extremely hot (over 90°) conditions.

Automatic transaxle fluid and filter: replace every 50,000 miles.

Tires: rotate every 6000 miles.

Brake system: inspect every 6000 miles.

Air distributor air filter: replace every 12,000 miles.

42372-REND-C11

PRECAUTIONS

Before servicing any vehicle, please be sure to read all of the following precautions, which deal with personal safety, prevention of component damage, and important points to take into consideration when servicing a motor vehicle:

• Never open, service or drain the radiator or cooling system when the engine is hot; serious burns can occur from the steam and hot coolant.

• Observe all applicable safety precautions when working around fuel. Whenever servicing the fuel system, always work in a well-ventilated area. Do not allow fuel spray or vapors to come in contact with a spark, open flame or excessive heat (a hot drop light, for example). Keep a dry chemical fire extinguisher near the work area. Always keep fuel in a container specifically designed for fuel storage; also, always properly seal fuel containers to avoid the possibility of fire or explosion. Refer to the additional fuel system precautions later in this section.

• Fuel injection systems often remain pressurized, even after the engine has been turned **OFF**. The fuel system pressure must be relieved before disconnecting any fuel lines. Failure to do so may result in fire and/or personal injury.

• Brake fluid often contains polyglycol ethers and polyglycols. Avoid contact with the eyes and wash your hands thoroughly after handling brake fluid. If you do get brake fluid in your eyes, flush your eyes with clean, running water for 15 minutes. If eye irritation persists, or if you have taken brake fluid internally, IMMEDIATELY seek medical assistance.

• The EPA warns that prolonged contact with used engine oil may cause a number of skin disorders, including cancer! You should make every effort to minimize your exposure to used engine oil. Protective gloves should be worn when changing oil. Wash your hands and any other exposed skin areas as soon as possible after exposure to used engine oil. Soap and water, or waterless hand cleaner should be used.

• All new vehicles are now equipped with an air bag system, often referred to as a Supplemental Restraint System (SRS) or Supplemental Inflatable Restraint (SIR) system. The system must be disabled before performing service on or around system components, steering column, instrument panel components, wiring and sensors. Failure to follow safety and disabling procedures could result in accidental air bag deployment, possible personal injury and unnecessary system repairs.

• Always wear safety goggles when working with, or around, the air bag system. When carrying a non-deployed air bag, be sure the bag and trim cover are pointed away from your body. When placing a non-deployed air bag on a work surface, always face the bag and trim cover upward, away from the surface. This will reduce the motion of the module if it is accidentally deployed. Refer to the additional air bag system precautions later in this section.

• Clean, high quality brake fluid from a sealed container is essential to the safe and proper operation of the brake system. You should always buy the correct type of brake fluid for your vehicle. If the brake fluid becomes contaminated, completely flush the system with new fluid. Never reuse any brake fluid. Any brake fluid that is removed from the system should be discarded. Also, do not allow any brake fluid to come in contact with a painted surface; it will damage the paint.

• Never operate the engine without the proper amount and type of engine oil; doing so WILL result in severe engine damage.

• Timing belt maintenance is extremely important! Many models utilize an interference-type, non-freewheeling engine. If the timing belt breaks, the valves in the cylinder head may strike the pistons, causing potentially serious (also time-consuming and expensive) engine damage.

• Disconnecting the negative battery cable on some vehicles may interfere with the functions of the on-board computer system(s) and may require the computer to undergo a relearning process once the negative battery cable is reconnected.

• When servicing drum brakes, only disassemble and assemble one side at a time, leaving the remaining side intact for reference.

ENGINE REPAIR

Distributor

This engine utilizes a Distributorless ignition system (DIS). There is no distributor to remove and no provision for adjustment.

Alternator

REMOVAL

1. Before servicing the vehicle, refer to the precautions in the beginning of this section.
2. Remove or disconnect the following:

• Negative battery cable
3. Rotate the engine forward.

• Alternator terminal nut, lead and electrical connector
• Serpentine belt
• Front bolts and two rear bolts
• Alternator from the bracket; position it above the drive axle
• Serpentine belt tensioner
• Bracket
• Power steering pipes from the retainer
• Fuel pressure test port cap from the injector rail

➡**Do not disconnect the power steering pipes from the pump**

• Power steering pump and reposition it to gain access to the alternator
• Alternator

INSTALLATION

1. Install or connect the following:
• Alternator
• Electrical harness to the right fender well retainer
• Power steering pump. Torque the bolts to 25 ft. lbs. (34 Nm).
• Fuel pressure test port cap to the fuel rail
• Power steering pipes to the retainer. Torque the fastener to 54 inch lbs. (6 Nm).
• Alternator bracket. Torque the bolt to 37 ft. lbs. (50 Nm).
• Serpentine belt tensioner
• Alternator to the bracket. Torque the bolts to 37 ft. lbs. (50 Nm).
• Serpentine belt
• Alternator electrical connector, lead and nut. Torque the nut to 115 inch lbs. (13 Nm).

2. Rotate the engine to its original position.

• Negative battery cable

3. Perform a charging system test and verify the proper operation of the system.

Engine Assembly

REMOVAL & INSTALLATION

1. Before servicing the vehicle, refer to the precautions in the beginning of this section.
2. Drain the cooling system.
3. Drain the engine oil.
4. Relieve the fuel system pressure.
5. Remove or disconnect the following:

- Negative battery cable
- Throttle body air inlet duct
- Cruise control cable
- Accelerator control cable
- Radiator hoses from the engine
- Heater hoses from the engine
- Engine mount struts
- Fuel lines from the fuel rail
- Engine wiring harness connectors
- Vacuum hoses
- Brake booster vacuum hose
- Automatic transaxle range selector cable
- Wiring harness grounds
- Catalytic converter three-way pipe from the right side exhaust manifold
- Rear propeller shaft, on AWD vehicles
- Front wheels
- Lower radiator baffle
- Splash shields
- Stabilizer shaft links from the lower control arms
- Tie rod ends from the steering knuckles
- Lower ball joints from the steering knuckles
- Cooler lines and bracket from the transaxle
- A/C compressor bolts and position compressor aside
- Axles from the transaxle and secure them to the steering knuckle/struts

⁂ CAUTION

Failure to remove the intermediate shaft from the steering gear may result in damage to the gear or intermediate shaft and may cause a loss of steering control.

- Intermediate shaft from the steering gear
6. Lower the vehicle until the frame is in contact with a suitable transaxle

table/engine stand (such as J 39580). Make certain that an engine stand (such as J 39580) is aligned below the engine.
- Frame bolts
7. Raise the vehicle to separate the assembly from the vehicle.
- Starter
- Flywheel-to-torque converter bolts
8. Install a suitable engine hoist
- Engine to transaxle bolts and studs
- Engine from the transaxle and place it on the engine stand

To install:
9. Install or connect the following:
- Engine to the transaxle/frame and install the bolts. Torque the bolts to 55 ft. lbs. (75 Nm).
- Engine mount-to-frame nuts. Torque the nuts to 32 ft. lbs. (43 Nm).
10. Remove the engine hoist.
- Torque converter to flywheel bolts. Torque the bolts to 47 ft. lbs. (63 Nm).
- Starter
11. Position the transaxle table with powertrain/frame under the vehicle. Lower the vehicle until the frame contacts the transaxle table.
- New frame to body bolts. Torque the front bolts to 111 ft. lbs. (150 Nm) and the rear bolts to 122 ft. lbs. (165 Nm). Remove the transaxle table.

- Intermediate shaft to the steering gear

⁂ CAUTION

When installing the intermediate shaft be certain that the shaft is seated properly before installing the pinch bolt. If the pinch bolt is inserted into the coupling before the shaft, the mating surfaces disengage. Disengagement of the two shafts may lead to a loss of steering control.

- Pinch bolt at the intermediate shaft. Torque the bolt 35 ft. lbs. (48 Nm).
- Drive axles to the transaxle
- Cooler lines and bracket to the transaxle. Torque the fasteners to 17 ft. lbs. (23 Nm).
- Lower ball joints to the steering knuckles. Torque to 40 ft. lbs. (55 Nm).
- Tie rod ends to the steering knuckles
- Stabilizer shaft links to the lower control arms. Torque the bolts 17 ft. lbs. (23 Nm).
- Inner fender splash shield. Torque the fasteners to 18 inch lbs. (2 Nm).
- Front wheels
- Catalytic converter pipe to the right

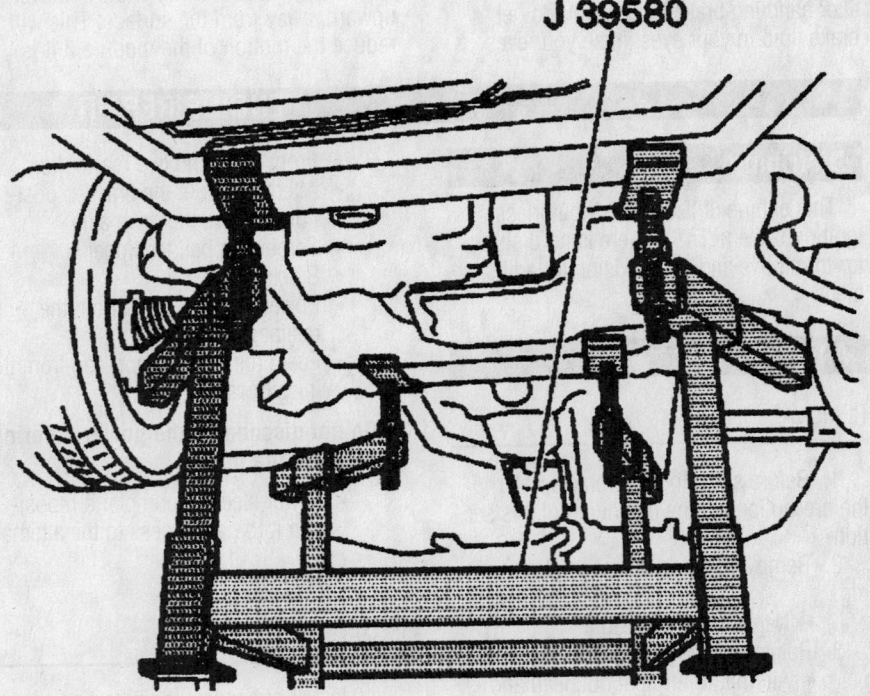

Use a engine stand to support the assembly during frame removal

9358KG02

side exhaust manifold. Torque the nuts to 25 ft. lbs. (34 Nm).
- Wiring harness grounds
- Rear propeller shaft, if removed
- Brake booster vacuum hose
- Vacuum hoses to the engine
- Range selector cable. Torque the screw to 14 ft. lbs. (20 Nm).
- Engine wiring harness connectors
- Fuel lines to the fuel rail. Torque the fasteners to 13 ft. lbs. (17 Nm).
- Throttle body brackets and cables. Torque the fasteners to 18 ft. lbs. (25 Nm).
- Engine mount strut. Torque the bolt to 35 ft. lbs. (48 Nm).
- Heater hoses
- Radiator hoses

❊❊ CAUTION

Whenever the engine has been removed from the vehicle it is necessary to install a new accelerator control cable to avoid damage or personal injury.

- New accelerator control cable
- Cruise control cable
- Throttle body air inlet duct
- Negative battery cable

12. Fill the engine with oil.
13. Fill the engine with coolant.
14. Inspect the transmission fluid level and top off, if necessary.
15. Turn the ignition to the **ON** position several times to pressurize the fuel system.
16. Start the engine and inspect for leaks, repair if necessary.
17. Check and top off the fluid levels if required.

Water Pump

REMOVAL & INSTALLATION

1. Before servicing the vehicle, refer to the precautions in the beginning of this section.

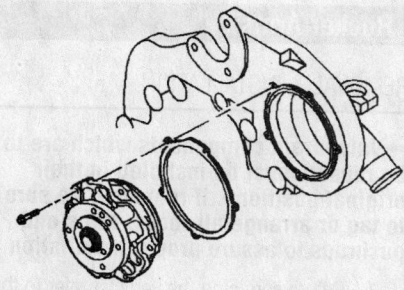

9358KG01

Exploded view of the water pump

2. Drain the coolant from the engine.
3. Remove or disconnect the following:
- Negative battery cable
- Serpentine drive belt guard
- Loosen the water pump pulley bolts
- Serpentine drive belt
- Water pump pulley
- Water pump
- Water pump gasket

To install:
4. Clean the gasket mounting surfaces.
5. Install or connect the following:
- Gasket
- Water pump. Torque the bolts to 89 inch lbs. (10 Nm).
- Water pump pulley and hand-tighten the bolts at this time
- Serpentine drive belt
- Water pump pulley bolts to 18 ft. lbs. (25 Nm)
- Serpentine drive belt guard

6. Fill the cooling system.
7. Start the engine and check for leaks, repair if necessary.
8. Road test the vehicle and verify there is no air in the cooling system.

Heater Core

REMOVAL & INSTALLATION

1. Before servicing the vehicle, refer to the precautions in the beginning of this section.
2. Drain the cooling system.
3. Remove or disconnect the following:

- Wiper module
- Brake booster vacuum hose
- Air cleaner cover and intake tube
- Throttle and cruise control cables, from the TBI bracket, also
- Transmission filler tube
- Hoses from the heater core
- Right and left instrument panel insulators
- Center floor air outlet
- Heater outlet duct
- Tie strap holding the instrument panel harness to the heater core cover
- Heater core cover
- Heater core retaining screw
- Heater core

To install:
4. Install or connect the following:
- Heater core
- Heater core retaining screw. Torque to 8 inch lbs.
- Heater core cover. Torque to 8 inch lbs.

- Heater outlet duct. Torque to 8 inch lbs.
- Tie strap holding the instrument panel harness to the heater core cover
- Center floor air outlet
- Right and left instrument panel insulators
- Hoses to the heater core
- Transmission filler tube
- Throttle and cruise control cables
- Air cleaner cover and intake tube
- Brake booster vacuum hose
- Wiper module

Cylinder Head

REMOVAL & INSTALLATION

This engine uses aluminum cylinder heads. Use care when working with light alloy parts. Valve guides are pressed in. Roller rocker arms are located on a pedestal in a slot in the cylinder head and are retained on individual threaded bolts.

The cylinder heads are retained by torque-to-yield bolts. A torque angle meter is required for proper torque during assembly. New replacement head bolts are recommended.

Before removing the cylinder head(s) from the engine and before disassembling the valve mechanism, perform a compression test and note the results. During disassembly, be sure that the valvetrain components are kept together and identified so that they can be installed in their original locations.

Left (Front) Side

1. Before servicing the vehicle, refer to the precautions in the beginning of this section.
2. Relieve the fuel system pressure using the recommended procedure.
3. Drain the cooling system.
4. Drain the oil.
5. Remove or disconnect the following:
- Negative battery cable
- Upper intake manifold
- Lower intake manifold
- Valve rocker arms and pushrods
- Exhaust crossover pipe
- Thermostat bypass pipe
- Right side engine mount strut bracket
- Oil level indicator tube
- Left side spark plug wires and spark plugs
- Left side exhaust manifold
- Left side cylinder head and gasket

To install:

6. Clean all parts well. Clean all gasket surfaces. Carefully remove all varnish soot and carbon to the bare metal. DO NOT use a motorized wire brush on any gasket surface since the soft aluminum will be damaged. If necessary, the head can be disassembled for thorough inspection and reconditioning.

7. Inspect the cylinder head for cracks. Do not attempt to weld the cylinder head. If cracked, replace it. Check the cylinder head deck, intake and exhaust manifold mating surfaces for flatness. These surfaces may be reconditioned by milling. If the surfaces are warped more than 0.005 in. (0.127mm), the surface should be milled. If more than 0.010 in. (0.251mm) of metal must be removed from the head, the head should be replaced.

8. Clean the cylinder head bolts and the bolt holes. Check the head bolts for damaged threads or stretching. New replacement head bolts are recommended.

9. Install or connect the following:
- New cylinder head gasket which is marked which side is **UP**
- Cylinder head by aligning it with the dowel pins
- New cylinder head bolts coated with a sealant (such as GM 1052080). Torque the bolts in the proper sequence (1–8) to 37 ft. lbs. (50 Nm). Using a torque angle meter turn the bolts 90 degrees in the proper sequence.
- Left side exhaust manifold. Torque the bolts to 12 ft. lbs. (16 Nm).
- Spark plugs. Torque the plugs 11 ft. lbs. (15 Nm).
- Spark plug wires
- Oil level indicator tube. Torque the fastener to 18 ft. lbs. (25 Nm).
- Right side mount strut bracket. Torque the fastener to 37 ft. lbs. (50 Nm).
- Thermostat bypass pipe. Torque it to 18 ft. lbs. (25 Nm).
- Exhaust crossover pipe. Torque the fastener to 18 ft. lbs. (25 Nm).
- Valve rocker arms and pushrods. Torque the fastener to 89 inch lbs. (10 Nm). Using a torque angle meter torque the fastener an additional 30 degrees.
- Lower intake manifold. Torque the bolts to 115 inch lbs. (13 Nm).
- Upper intake manifold. Torque the bolts to 18 ft. lbs. (25 Nm).
- Negative battery cable

10. Refill the coolant system.

11. Change the oil filter and fill the engine with clean oil.

Cylinder head bolt torque sequence—3.4L engine

12. Turn the ignition to the **ON** position several times to pressurize the fuel system. Start the engine and inspect for any leaks, repair if necessary. Check and top off the fluid levels if required.

Right (Rear) Side

1. Before servicing the vehicle, refer to the precautions in the beginning of this section.

2. Relieve the fuel system pressure using the recommended procedure.

3. Drain the coolant system.

4. Drain the oil from the engine.

5. Remove or disconnect the following:
- Negative battery cable
- Upper intake manifold
- Lower intake manifold
- Valve rocker arms and pushrods
- Exhaust crossover pipe
- Right side spark plug wires
- Right side exhaust manifold
- Right side cylinder head and gasket
- Right side spark plugs from the cylinder head

To install:

6. Clean all parts well. Clean all gasket surfaces. Carefully remove all varnish soot and carbon to the bare metal. DO NOT use a motorized wire brush on any gasket surface since the soft aluminum will be damaged. If necessary, the head can be disassembled for thorough inspection and reconditioning.

7. Inspect the cylinder head for cracks. Do not attempt to weld the cylinder head. If cracked, replace it. Check the cylinder head deck, intake and exhaust manifold mating surfaces for flatness. These surfaces may be reconditioned by milling. If the surfaces are "out of flat" by more than 0.005 inch, the surface should be milled. If more than 0.010 inch of metal must be removed from the head, the head should be replaced.

8. Clean the cylinder head bolts and the bolt holes. Check the head bolts for damaged threads or stretching. New replacement head bolts are recommended.

9. Install or connect the following:
- New cylinder head gasket
- Cylinder head on top of the gasket and make certain it is lined up properly with the dowel pins
- New cylinder head bolts coated with a sealant (such as GM 1052080). Torque the bolts in the proper sequence (1–8) to 37 ft. lbs. (50 Nm). Using a torque angle meter turn the bolts 90 degrees in the proper sequence.
- Right side exhaust manifold. Torque the bolts to 12 ft. lbs. (16 Nm).
- Spark plugs. Torque the plugs 11 ft. lbs. (15 Nm).
- Spark plug wires
- Exhaust crossover pipe. Torque the fastener to 18 ft. lbs. (25 Nm).
- Valve rocker arms and pushrods. Torque the fastener to 89 inch lbs. (10 Nm). Using a torque angle meter torque the fastener an additional 30 degrees.
- Lower intake manifold. Torque the bolts to 115 inch lbs. (13 Nm).
- Upper intake manifold. Torque the bolts to 18 ft. lbs. (25 Nm).
- Negative battery cable

10. Refill the coolant system.

11. Change the oil filter and fill the engine with clean oil.

12. Start the vehicle and verify no leaks, abnormal noises and correct engine operation.

13. Check the fluid levels and top off if necessary.

Rocker Arms

REMOVAL & INSTALLATION

➡**Valve train components which are to be reused must be installed in their original positions. If removed, be sure to tag or arrange all rocker arms and pushrods to assure proper installation.**

1. Before servicing the vehicle, refer to the precautions in the beginning of this section.

2. Remove or disconnect the following:
- Negative battery cable
- Rocker arm cover
- Rocker arm bolts
- Rocker arms

➡**Place the valve train parts in order to ensure they are installed in the proper location. Intake pushrods are yellow and measure 5.68 inches (144.18mm). Exhaust pushrods are green and measure 6.0 inches (152.51mm). When removing the pushrods, make certain they do not fall into the lifter valley.**

- Pushrods

To install:

3. Inspect and replace components if worn or damaged. Clean all old thread locking material from the pedestal bolts.

4. Coat the bearing surface of the rocker arms, pushrods and rocker arm bolts with a prelube (such as GM 1052365). Make certain to install the components in their original position.

5. Install or connect the following:
- Intake valve pushrods which are 5.68 inches (144.18mm) long
- Exhaust valve pushrods which are 6.0 inches (152.51mm) long
- Rocker arms. Torque the bolt to 14 ft. lbs. (19 Nm) plus 30 degrees.
- Rocker arm cover. Torque the bolt to 89 inch lbs. (10 Nm).
- Negative battery cable

6. Start the engine and verify the vehicle is running properly.

Intake Manifold

REMOVAL & INSTALLATION

Upper

This engine uses a 2-piece intake manifold. The upper half (often called a plenum) mounts the throttle body. The lower half of the manifold bolts to the engine and contains the fuel injectors. Please note that this engine uses a sequential multi-port fuel injection system. Injector connectors must be connected to their appropriate fuel injector assembly or engine emissions and engine performance will be seriously affected. Identify and tag for identification all wiring connectors as well as vacuum and other components as required to assure correct assembly.

1. Before servicing the vehicle, refer to the precautions in the beginning of this section.

2. Drain the engine coolant. Remove the coolant recovery bottle.

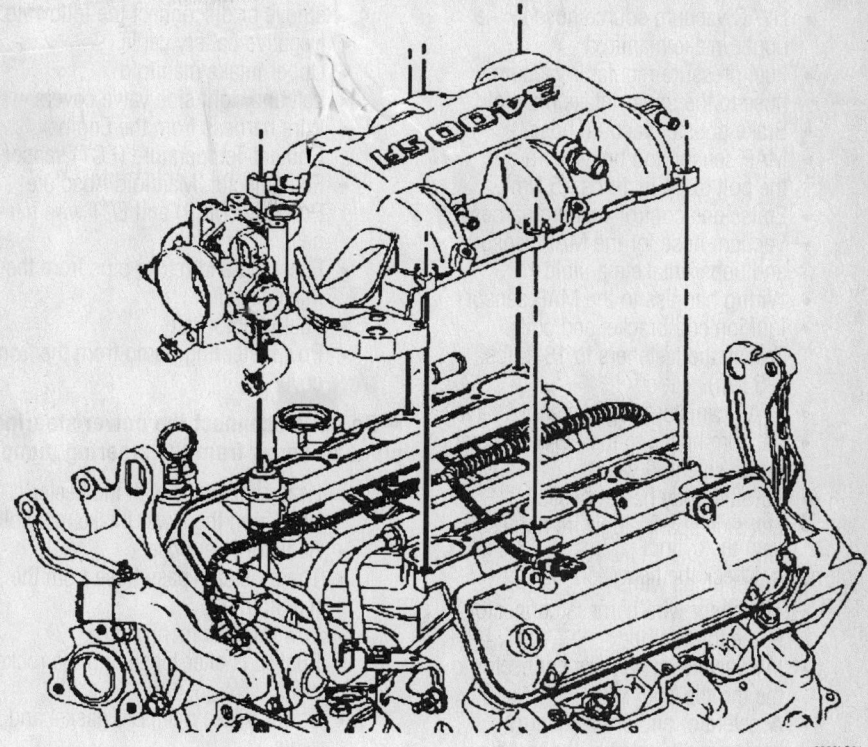

Removing the upper intake manifold

3. Relieve the fuel system pressure using the recommended procedure.

4. Remove or disconnect the following:

- Negative battery cable
- Throttle body air inlet duct
- Accelerator and cruise control cables and bracket from the throttle body
- Throttle Position (TP) sensor connector from the throttle body
- Idle Air Control (IAC) valve connector from the throttle body
- Left side spark plug wires
- Left side spark plug wire harness clip and harness
- Throttle body heater hoses
- Evaporative emissions (EVAP) canister purge solenoid valve vacuum hoses
- EVAP canister purge solenoid valve
- Ignition coil bracket and coils
- Wire harness for the Manifold Air Pressure (MAP) sensor
- Vacuum harness from the MAP sensor and upper intake manifold
- Emissions control vacuum harness
- Brake booster vacuum hose from the upper intake manifold
- Vacuum hose connection for the Heater Vent Air Conditioning (HVAC) source hose
- Vacuum hose connection for the fuel pressure regulator
- Exhaust Gas Recirculation (EGR) valve
- MAP sensor and bracket
- Alternator through-bolt bracket
- Upper intake manifold
- Upper intake manifold gasket
- Throttle body, if replacing the manifold

To install:

5. Clean all parts well. Use care in cleaning old gasket material from the machined aluminum surfaces on the plenum and manifold as sharp tools may damage sealing surfaces.

6. Clean the mating surfaces to the upper intake manifold and engine block. Remove any loose pieces of RTV sealer.

7. Install or connect the following:
- Throttle body to the upper intake manifold (if removed). Torque the bolts to 18 ft. lbs. (25 Nm).
- Upper intake manifold gasket
- Upper intake manifold
- MAP sensor and bracket. Torque the bolt to 44 inch lbs. (5 Nm).
- Upper intake manifold bolts. Torque the bolts to 18 ft. lbs. (25 Nm).
- EGR valve. Torque the fastener to 18 ft. lbs. (25 Nm).

- HVAC vacuum source hose to the upper intake manifold
- Fuel pressure regulator vacuum hose to the upper intake manifold
- Brake booster vacuum hose
- MAP sensor and bracket. Torque the bolt to 44 inch lbs. (5 Nm).
- Emissions control vacuum harness
- Vacuum hose for the MAP sensor and upper intake manifold
- Wiring harness to the MAP sensor
- Ignition coil bracket and coils. Torque the fasteners to 18 ft. lbs. (25 Nm).
- EVAP canister purge solenoid valve
- Vacuum hoses to the EVAP canister purge solenoid valve
- Throttle body heater hoses
- Left side spark plug wire harness clip
- Spark plugs wires
- TP sensor wire harness connector to the throttle body
- IAV valve wire harness connector to the throttle body
- Accelerator and cruise control cables and bracket to the throttle body. Torque the fasteners to 106 inch lbs. (12 Nm).
- Throttle body air inlet duct
- Negative battery cable

8. Fill the coolant system.
9. Fill the engine with new oil.
10. Turn the ignition to the **ON** position several times to pressurize the fuel system.
11. Start the engine and check for any leakage and repair if necessary.
12. Check and top off all fluid levels if needed.

Lower

This engine uses a 2-piece intake manifold. The upper half (often called a plenum) mounts the throttle body. The lower half of the manifold bolts to the engine and contains the fuel injectors. Please note that this engine uses a sequential multi-port fuel injection system. Injector connectors must be connected to their appropriate fuel injector assembly or engine emissions and engine performance will be seriously affected. Identify and tag for identification all wiring connectors as well as vacuum and other components as required to assure correct assembly.

1. Before servicing the vehicle, refer to the precautions in the beginning of this section.
2. Drain the engine coolant.
3. Relieve the fuel system pressure using the recommended procedure.

4. Remove or disconnect the following:
- Negative battery cable
- Upper intake manifold
- Left and right side valve covers
- Wire harness from the Engine Coolant Temperature (ECT) sensor
- Fuel injector, Manifold Absolute Pressure (MAP) and ECT wire harness
- Fuel feed and return pipe from the injector rail
- Fuel injector rail
- Power steering pump from the front cover

➡ **Do not disconnect the power steering pipes or hoses from the steering pump.**

- Heater inlet pipe with the heater hose from the lower intake manifold
- Radiator inlet hose
- Thermostat bypass hose from the manifold
- Lower intake manifold
- Pushrods after loosening the rocker arms
- Lower intake manifold gasket and seals

- ECT sensor, if replacing the manifold
- Thermostat and housing, if replacing the manifold

To install:
5. Clean the gasket mounting surfaces.
6. Inspect the intake manifold for cracks or damage, replace if necessary.
7. Install or connect the following:
- ECT sensor, if removed. Torque the sensor to 17 ft. lbs. (23 Nm).
- Thermostat and housing, if removed. Torque the fastener to 18 ft. Lbs. (25 Nm).
- Thin bead of RTV sealer (such as GM 12345739) on the ridge of the engine block where the lower intake manifold makes contact
- Lower intake manifold gaskets
- Pushrods and tighten the rocker arms. Torque the arms to 14 ft. lbs. (19 Nm) plus 30 degrees.
- Lower intake manifold. Torque the bolts to first to 62 inch lbs. (7 Nm) then to 115 inch lbs. (13 Nm) after applying a sealant (such as GM 12345382) to the threads of the bolts.

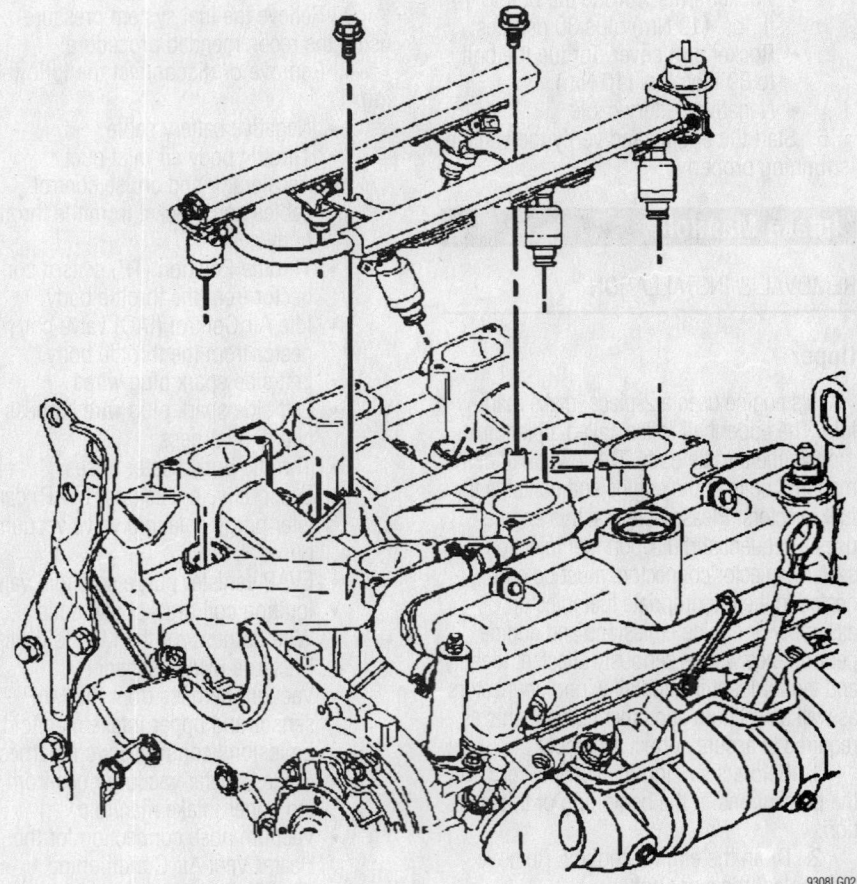

Remove the fuel injector rail

9308LG02

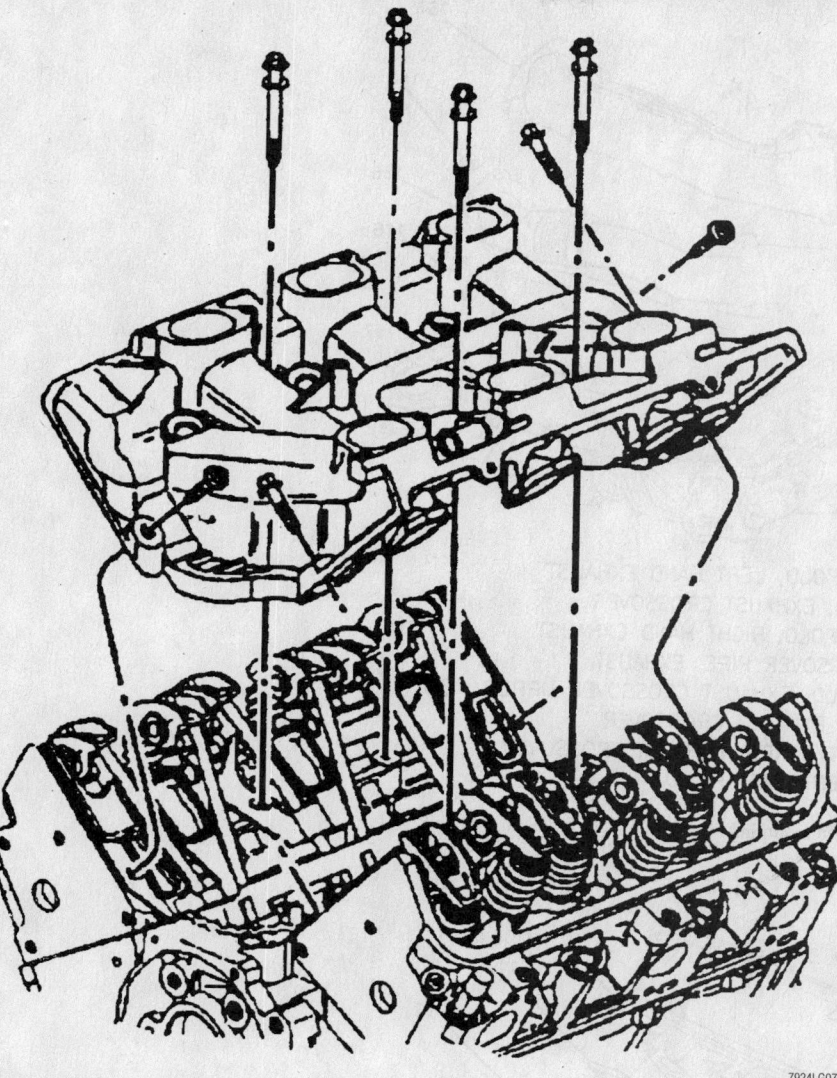

Lower intake manifold assembly

- Thermostat bypass hose to the lower intake manifold pipe. Torque the fastener to 18 ft. lbs. (25 Nm).
- Radiator inlet hose from the engine
- Power steering pump to the front cover
- Fuel injector rail. Torque the fastener to 89 inch lbs. (10 Nm).
- Fuel feed and return pipes to the injector rail. Torque the fasteners to 13 ft. lbs. (17 Nm).
- Wire harness for the fuel injector, MAP sensor and the ECT sensor
- Right side rocker arm cover. Torque the bolts to 89 inch lbs. (10 Nm).
- Left side rocker arm cover. Torque the bolts to 89 inch lbs. (10 Nm).
- Upper intake manifold. Torque the bolts to 18 ft. lbs. (25 Nm).
- Negative battery cable

8. Fill the coolant system.

9. Turn the ignition to the **ON** position several times to pressurize the fuel system.

10. Start the engine and check for any leakage and repair if necessary.

11. Check and top off all fluid levels if needed.

Exhaust Manifold

REMOVAL & INSTALLATION

The exhaust manifolds are conventional iron castings. The left and right manifolds are connected by a crossover pipe. Use care with the exhaust manifold-to-cylinder head fasteners. The cylinder heads are aluminum.

Left (Front) Side

1. Before servicing the vehicle, refer to the precautions in the beginning of this section.

2. Drain the cooling system.

3. Remove or disconnect the following:
   - Negative battery cable
   - Throttle body air inlet duct
   - Right side engine mount strut bracket
   - Radiator inlet hose
   - Thermostat bypass pipe
   - Exhaust crossover pipe heat shield
   - Exhaust crossover pipe
   - Left side exhaust manifold heat shield
   - Left side exhaust manifold and discard the gasket

To install:

4. Clean the gasket mounting surfaces.

5. Install or connect the following:
   - Left side exhaust manifold gasket
   - Left side exhaust manifold. Torque the nuts to 12 ft. lbs. (16 Nm).
   - Left side exhaust manifold heat shield. Torque the fasteners to 89 inch lbs. (10 Nm).
   - Exhaust crossover pipe. Torque the bolts to 18 ft. lbs. (25 Nm).
   - Exhaust crossover pipe heat shield. Torque the bolts to 89 inch lbs. (10 Nm).
   - Thermostat bypass pipe. Torque the fastener to 18 ft. lbs. (25 Nm).
   - Radiator inlet hose to the engine
   - Right side engine mount strut bracket. Torque the bolts to 35 ft. lbs. (48 Nm).
   - Throttle body air inlet duct
   - Negative battery cable

6. Fill the cooling system

7. Start the vehicle and check for leaks, repair if necessary.

8. Check and top off all fluid levels if necessary.

Right (Rear) Side

1. Before servicing the vehicle, refer to the precautions in the beginning of this section.

2. Remove or disconnect the following:
   - Negative battery cable
   - Throttle body air inlet duct
   - Accelerator cable bracket from the throttle body
   - Manifold Absolute Pressure (MAP) sensor
   - Exhaust Gas Recirculation (EGR) valve

3. Rotate the engine for access.
   - Ignition module
   - Ignition coils and bracket
   - Spark plug wires from the right bank spark plugs

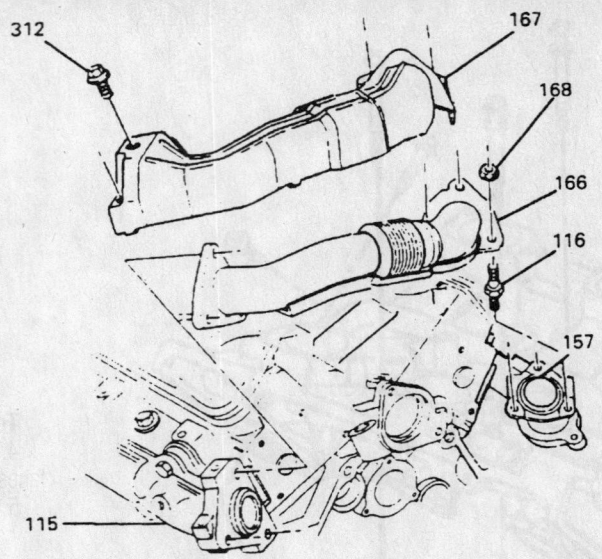

115 MANIFOLD, LEFT HAND EXHAUST
116 STUD, EXHAUST CROSSOVER
157 MANIFOLD, RIGHT HAND EXHAUST
166 CROSSOVER PIPE, EXHAUST
167 SHIELD, EXHAUST CROSSOVER UPPER HEAT
168 NUT, EXHAUST CROSSOVER
312 BOLT/SCREW, EXHAUST CROSSOVER
    UPPER HEAT SHIELD

7924LG28

Exploded view of the exhaust crossover and heat shield mounting

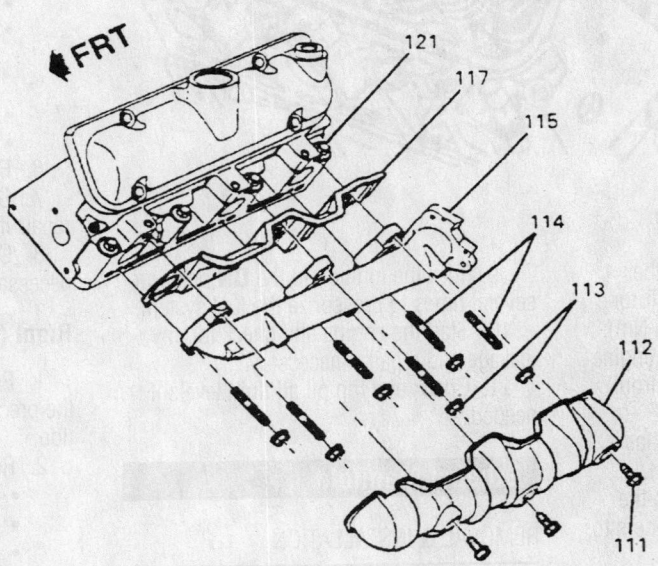

111 SCREW, LH EXHAUST MANIFOLD HEAT SHIELD
112 SHIELD LH EXHAUST MANIFOLD
113 NUT, LH EXHAUST MANIFOLD
114 STUD, LH EXHAUST MANIFOLD
115 MANIFOLD, LH EXHAUST
117 GASKET, LH EXHAUST MANIFOLD
121 HEAD, LH CYLINDER

7924LG29

Left exhaust manifold mounting

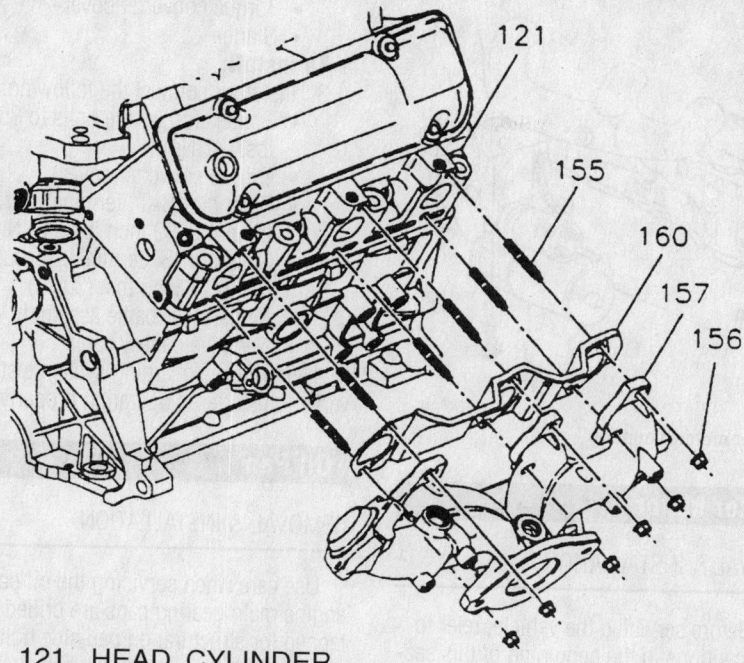

121 HEAD, CYLINDER
155 STUD, EXHAUST MANIFOLD
156 NUT, EXHAUST MANIFOLD
157 MANIFOLD RIGHT EXHAUST
160 GASKET, RIGHT EXHAUST MANIFOLD

7924LG30

Exploded view of the right exhaust manifold mounting

- Heated Oxygen (HO2) sensor electrical connector
- EVAP solenoid bracket
- Fasteners for the crossover pipe from the right bank exhaust manifold
- Catalytic converter
- Right side exhaust manifold heat shields
- Right side exhaust manifold
- Right side exhaust manifold gasket
- EGR valve pipe, if replacing the exhaust manifold
- HO2sensor, if replacing the exhaust manifold

To install:
4. Clean the gasket mounting surfaces.
5. Install or connect the following:
- HO2 sensor, if removed. Torque the sensor to 31 ft. lbs. (42 Nm).
- EGR valve pipe, if removed. Torque the fastener to 18 ft. lbs. (25 Nm).
- Exhaust manifold gasket
- Exhaust manifold. Torque the bolts to 12 ft. lbs. (16 Nm).
- Both manifold heat shields. Torque the bolts to 89 inch lbs. (10 Nm).
- Catalytic converter. Torque the fasteners to 25 ft. lbs. (34 Nm).

- Exhaust manifold crossover pipe. Torque the bolts to 18 ft. lbs. (25 Nm).
- HO2 sensor electrical connector
- EVAP solenoid bracket
- Plug wires to the spark plugs
- Ignition module
- Ignition coils and bracket. Torque the fastener to 18 ft. lbs. (25 Nm).
6. Rotate the engine to its original position.
7. Install or connect the following:
- EGR valve to the intake manifold
- MAP sensor
- Accelerator cable bracket to the throttle body. Torque the bolt to 89 inch lbs. (10 Nm).
- Throttle body air inlet duct
- Negative battery cable
8. Start the engine and inspect for leaks, repair if necessary.

Camshaft and Valve Lifters

REMOVAL & INSTALLATION

1. Before servicing the vehicle, refer to the precautions in the beginning of this section.
2. Relieve the fuel system pressure.

3. Remove or disconnect the following:
- Engine assembly

✱✱ WARNING

When removing valvetrain components they must be marked for installation in their original location. When the camshaft is being replaced, the valve lifters must also be replaced.

- Right and left side valve covers
- Upper and lower intake manifold
- Rocker arm bolts, balls, rocker arms and pushrods
- Oil splash shield
- Lifter guide bolts and the guide
- Valve lifter(s) from the bores
- Crankshaft balancer and front cover
- Timing chain and sprockets
- Oil pump driven gear bolt and gear
- Camshaft thrust plate
- Camshaft, using a large screwdriver inserted in the bolt hole to carefully rotate and pull the camshaft out of the bearings

✱✱ WARNING

Avoid damaging the camshaft bearing surfaces.

To install:
4. Coat the camshaft with Prelube.
5. Install or connect the following:
- Camshaft
- Camshaft thrust plate. Tighten the bolts to 89 inch lbs. (10 Nm).
- Oil pump driven gear. Tighten the bolt to 27 ft. lbs. (36 Nm).
- Timing chain and sprocket
- Camshaft thrust button and front cover
- Crankshaft balancer
6. Lubricate the bearing surfaces with Molykote®.

➡**Installation of a new camshaft or a wear pattern on the old valve lifter will require the replacement of the camshaft and lifters together. If camshaft replacement is not necessary, be sure to install the used valve lifters in their original position.**

7. Install or connect the following:
- Lifters in their original locations
- Lifter guide. Tighten the guide bolts to 89 inch lbs. (10 Nm).
- Oil splash shield
- Pushrods, rocker arms, balls and bolts. Tighten the nuts to 89 inch lbs. (10 Nm) plus an additional 30 degree turn.

- Lower and upper intake manifold
- Right and left side valve covers
- Engine assembly
- Negative battery cable

8. Adjust the valves, as required. Start the engine and verify no oil leaks.

Valve Lash

ADJUSTMENT

Because the rocker arm fasteners are secured and tightened, valve lash is not adjustable. If a valve train problem is suspected, check that the rocker arm pedestals bolts are tightened to specification. During initial installation the bolts are coated with thread locking compound. If they are sufficiently loosened to cause valvetrain noise, they should be removed and thoroughly cleaned. Apply thread locking compound to the rocker arm pedestal bolts. Tighten the bolts to 14 ft. lbs. (19 Nm) plus 30 degrees.

When valve lash falls out of specification (valve tap is heard) and tightening the bolts does not solve the problem, replace the rocker arm, pushrod and hydraulic lifter on the offending cylinder.

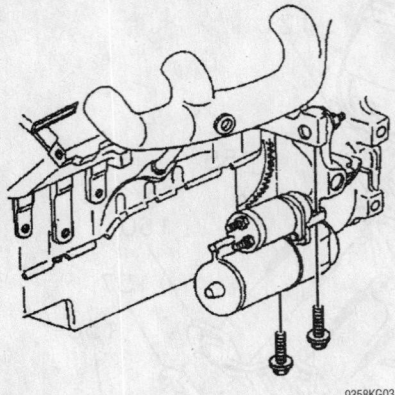

Starter motor mounting

9358KG03

Starter Motor

REMOVAL & INSTALLATION

1. Before servicing the vehicle, refer to the precautions in the beginning of this section.

2. Remove or disconnect the following:
- Negative battery cable
- Radiator air baffle assembly
- Electrical connections

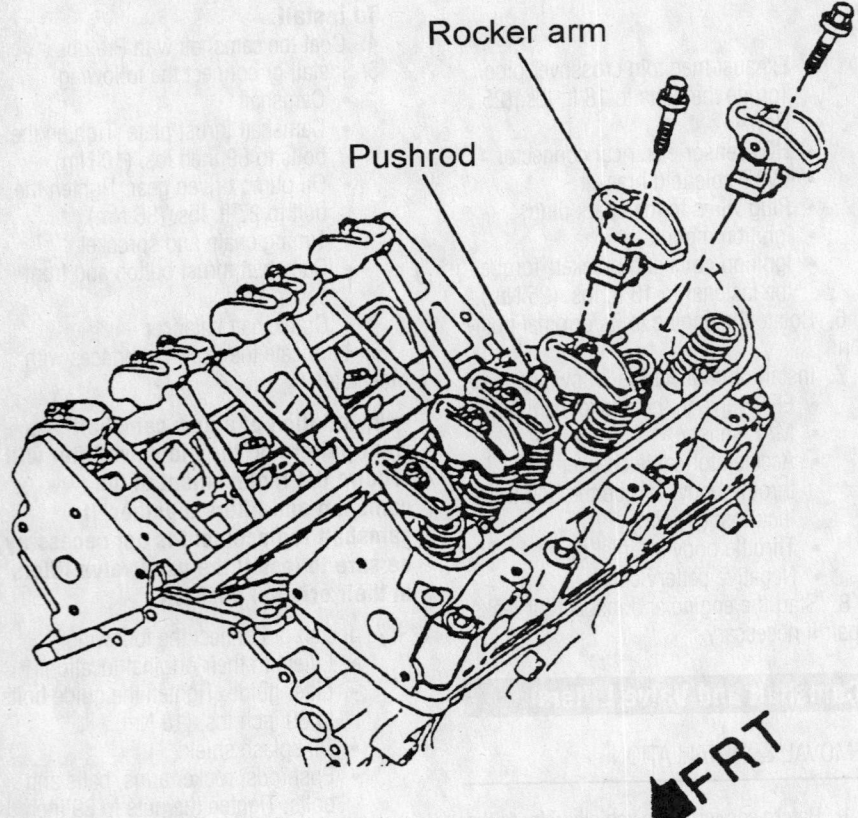

Rocker arm

Pushrod

◀FRT

7924LG33

Valve rocker arm and related components

- Torque converter cover
- Starter

To install:

3. Install or connect the following:
- Starter. Torque the bolts to 35 ft. lbs. (47 Nm).
- Torque converter cover
- Solenoid "BAT" terminal. Torque the nut to 89 inch lbs. (10 Nm).
- Solenoid "S" terminal. Torque the nut to 27 inch lbs. (3 Nm).
- Radiator air baffle assembly
- Negative battery cable

4. Perform a charging system test and verify the starter is operating properly

Oil Pan

REMOVAL & INSTALLATION

Use care when servicing the oil pan. The engine main bearing caps are drilled and tapped for structural oil pan side bolts. Do not overlook the side bolts when attempting to remove the oil pan.

1. Before servicing the vehicle, refer to the precautions in the beginning of this section.

2. Drain the engine oil.

3. Remove or disconnect the following:
- Engine mount struts
- A/C compressor and set it aside

➡It is not necessary to disconnect any of the A/C lines from the compressor.

4. Install an engine support fixture.

5. Remove or disconnect the following:
- Catalytic converter pipe from the right side exhaust manifold
- Frame bolts and make certain that an engine stand (such as J 39580) is aligned below the engine, lower the frame
- Oil level sensor wiring harness connector
- Starter
- Transaxle brace from the oil pan
- Transaxle mount lower nuts
- Engine mount lower nuts and raise the engine with the support fixture
- Engine mount and bracket from the oil pan

➡You will need to use a torque wrench adapter to remove and install the right side oil pan bolts.

- Oil pan and gasket

To install:

6. Clean the gasket mounting surfaces.

7. Apply a small amount of sealer GM 1234579 on both sides of the bearing cap.

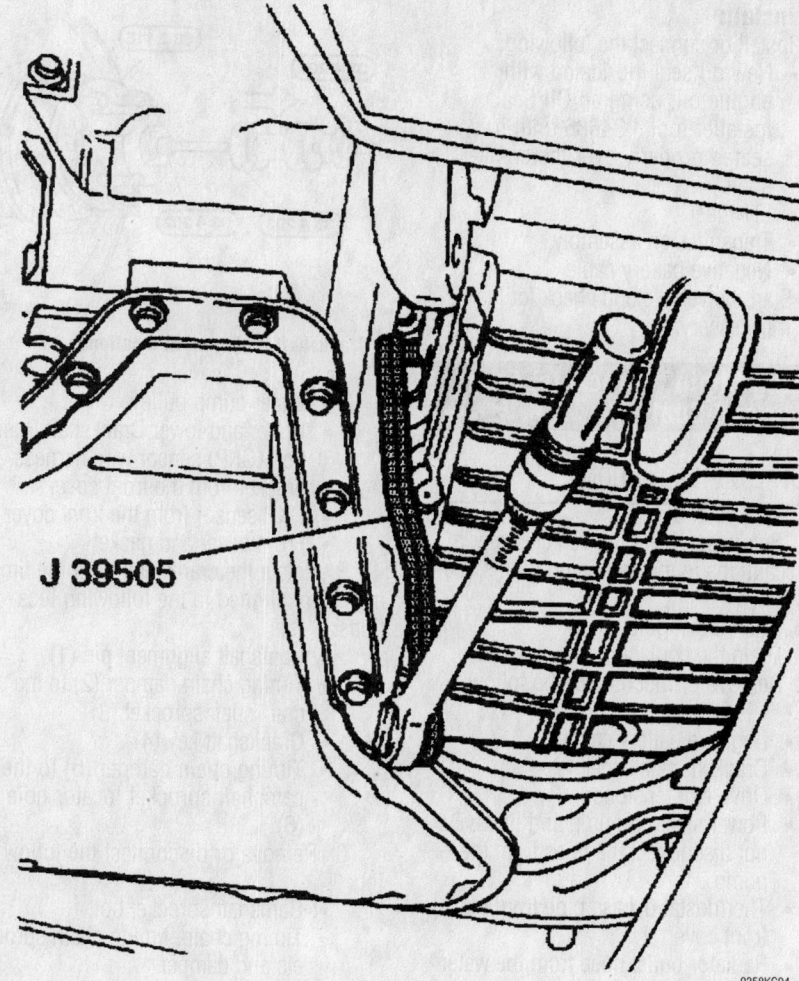

J 39505

9358KG04

Use a torque wrench adapter to tighten the oil pan side bolts

8. Install or connect the following:
   - Oil pan gasket
   - Oil pan. Tighten the bottom bolts to 18 ft. lbs. (25 Nm) and the side bolts, using a torque wrench adapter (tool J 39505), to 37 ft. lbs. (50 Nm).
   - Engine mount and bracket to the oil pan. Torque the bolt to 43 ft. lbs. (58 Nm).
   - Lower the engine into position
   - Transaxle lower nuts. Torque the nuts to 90 inch lbs. (10 Nm).
   - Transaxle brace to the oil pan. Torque the bolts to 32 ft. lbs. (43 Nm).
   - Starter. Torque the bolts to 35 ft. lbs. (47 Nm).
   - Catalytic converter pipe to the right side exhaust manifold. Torque the bolts to 26 ft. lbs. (35 Nm).
   - Frame and use new frame to body bolts. Torque the front bolts to 111 ft. lbs. (150 Nm) and the rear bolts to 122 ft. lbs. (165 Nm).
9. Remove the engine support fixture.
10. Install or connect the following:
    - A/C compressor. Torque the bolts to 37 ft. lbs. (50 Nm).
    - Engine mount struts. Torque the fasteners to 52 ft. lbs. (70 Nm).
    - Negative battery cable
11. Fill the engine with oil.
12. Start the vehicle and inspect for leaks, repair if necessary.

Oil Pump

REMOVAL & INSTALLATION

1. Before servicing the vehicle, refer to the precautions in the beginning of this section.
2. Drain the engine oil
3. Remove or disconnect the following:
   - Negative battery cable

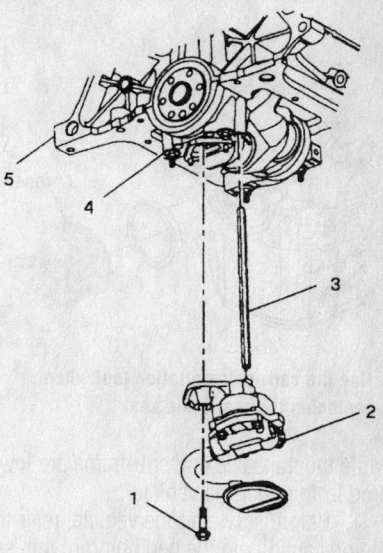

1. Oil pump bolt
2. Oil pump
3. Oil pump drive rod
4. Main bearing cap
5. Engine block

7924LG37

Exploded view of the oil pump mounting

   - Oil pan
   - Bolt attaching the oil pump to the rear crankshaft bearing cap
   - Oil pump and driveshaft
   - Crankshaft oil deflector nuts and deflector, if necessary
4. Inspect the oil pump and oil pump driveshaft.
   To install:
5. Install or connect the following:
   - Crankshaft oil deflector, if removed. Torque the nuts to 18 ft. lbs. (25 Nm).

➡**Rotate the oil pump driveshaft as necessary for proper engagement with the oil pump drive.**

   - Oil pump and driveshaft
   - Oil pump to the rear crankshaft bearing cap. Torque the bolt to 30 ft. lbs. (41 Nm).
   - Oil pan
   - Negative battery cable
6. Fill the engine with oil.
7. Start the vehicle and inspect for leaks, repair if necessary.

Rear Main Seal

REMOVAL & INSTALLATION

The transaxle assembly must be removed to perform this service. This requires special tooling to support the engine assembly

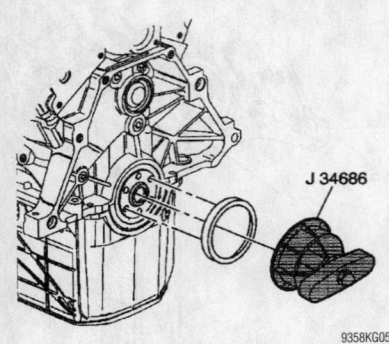

J 34686

9358KG05

Use the correct installation tool when replacing the rear main seal

while the transaxle and sub-frame are lowered from under the vehicle.

1. Before servicing the vehicle, refer to the precautions in the beginning of this section.

2. Remove or disconnect the following:

- Negative battery cable
- Transmission assembly
- Engine flywheel
- Oil seal

※※ WARNING

When removing the seal, use care so that no damage occurs to the crankshaft. Once the seal is removed, inspect the crankshaft surface for any nicks or burrs. Repair or replace crankshaft as necessary.

To install:

3. Install or connect the following:

- New oil seal lubricated with engine oil, using an Oil Seal Installer tool J 34686 until it is seated properly over the crankshaft
- Flywheel
- Transmission assembly
- Negative battery cable

4. Start the vehicle and check for leaks, repair if necessary.

Timing Chain, Sprockets, Front Cover and Seal

REMOVAL & INSTALLATION

1. Before servicing the vehicle, refer to the precautions in the beginning of this section.

2. Drain the engine oil.

3. Drain the coolant.

4. Remove or disconnect the following:

- Negative battery cable
- Crankshaft balancer
- Drive belt tensioner
- Power steering pump and lines. Do not disconnect the lines from the pump.
- Thermostat bypass pipe from the front cover
- Radiator outlet hose from the water pump

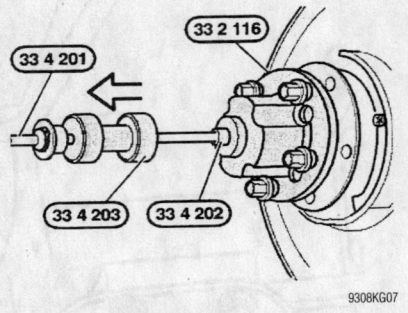

9308KG07

Crankshaft timing mark locations

- Water pump pulley
- Upper and lower Crankshaft Position (CKP) sensor wire harness bracket from the front cover
- CKP sensor from the front cover
- Front cover and gasket

5. Rotate the crankshaft until the timing marks are aligned in the following locations:

- Camshaft alignment pin (1)
- Timing chain damper (2) to the crankshaft sprocket (3)
- Crankshaft key (4)
- Timing chain damper (5) to the camshaft sprocket locator hole (6)

6. Remove or disconnect the following:

- Camshaft sprocket bolt
- Timing chain, timing chain sprockets and damper
- Front oil seal

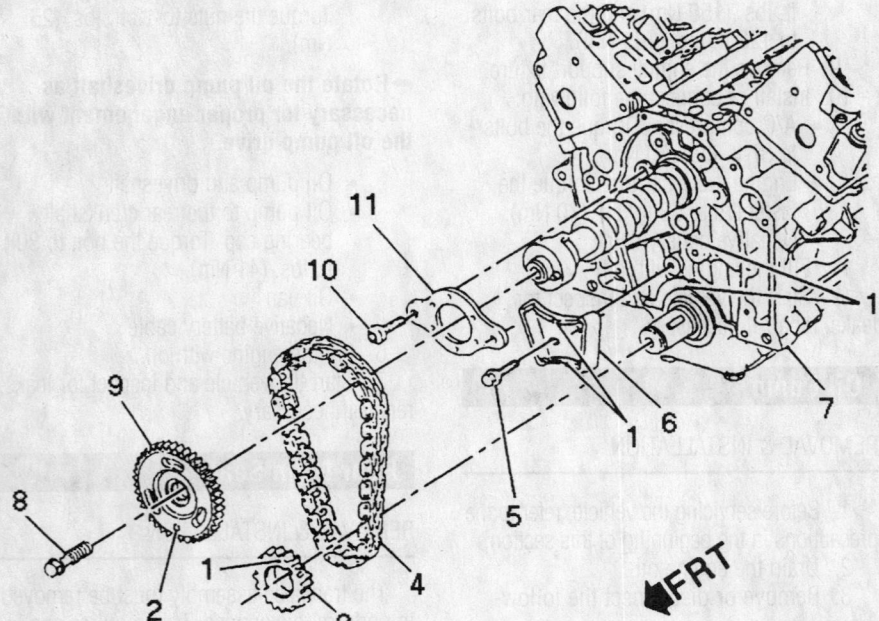

1. Timing alignment marks
2. Locator hole
3. Crankshaft sprocket
4. Timing chain
5. Timing chain dampener bolt
6. Timing chain dampener
7. Engine block
8. Camshaft sprocket bolt
9. Camshaft sprocket
10. Thrust plate bolt
11. Thrust plate

7924LG14

Exploded view of the timing chain assembly

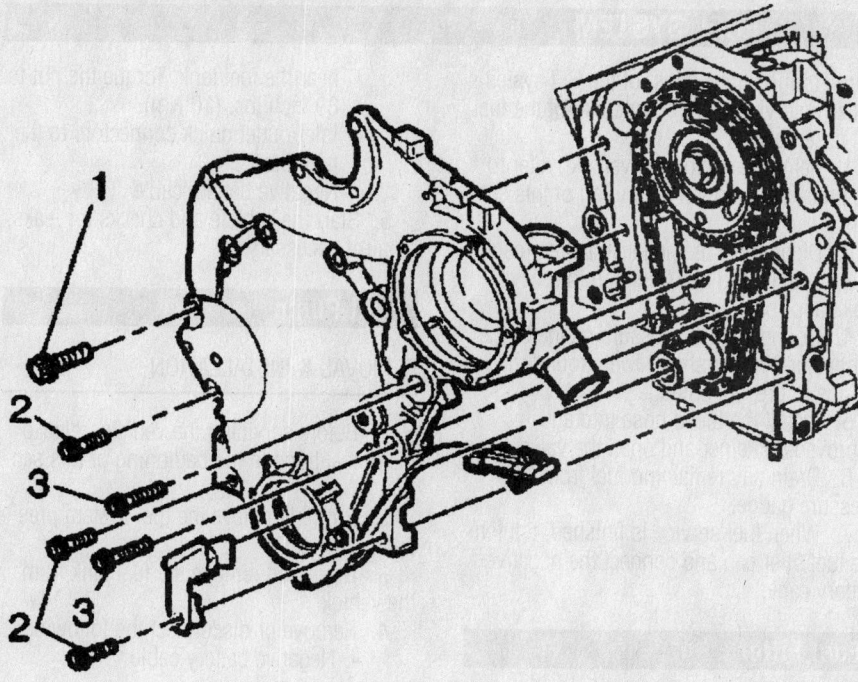

9358KG06

Timing chain front cover bolt tightening specifications

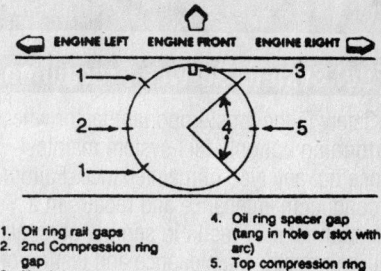

1. Oil ring rail gaps
2. 2nd Compression ring gap
3. Notch in piston
4. Oil ring spacer gap (tang in hole or slot with arc)
5. Top compression ring gap

7924AG07

Piston ring end-gap spacing—3.4L engine

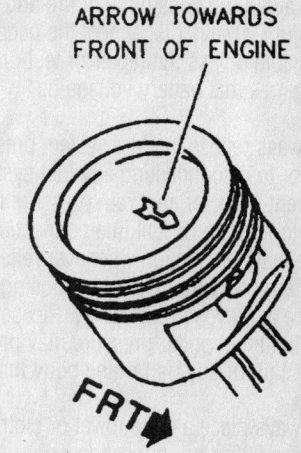

ARROW TOWARDS FRONT OF ENGINE

FRT

7922AG47

Piston positioning. Often the arrow is replaced with a notch, which must face toward the front of the engine—3.4L engine

To install:

7. Install or connect the following:
   - New front oil seal by making certain the seal is fully seated
   - Timing chain damper. Torque the bolts to 15 ft. lbs. (21 Nm).
   - Timing chain to the camshaft sprocket
   - Crankshaft sprocket
   - Timing chain to the crankshaft sprocket by making certain the chain is fully seated

8. Align the crankshaft timing mark to the bottom mark on the damper.

9. Align the timing mark on the camshaft gear center line of the locator hole with the timing mark on the top of the damper.

10. Align the dowel in the camshaft with the dowel hole in the camshaft sprocket.

11. Install or connect the following:
   - Camshaft sprocket bolt. Torque the bolt to 103 ft. lbs. (140 Nm).

12. Apply a 0.20 inch (5mm) bead of sealer to both sides of the lower tabs of the engine front cover gasket.
   - Front cover. Refer to the accompanying figure and torque bolts (2) to 15 ft. lbs. (21 Nm), bolts (3) to 41 ft. lbs. (55 Nm) and bolts (1) to 35 ft. lbs. (47 Nm).
   - Water pump to the front cover. Torque the bolts to 89 inch lbs. (10 Nm).
   - Water pump pulley. Torque the bolt to 18 ft. lbs. (25 Nm).
   - CKP sensor to the front cover
   - Upper/lower CKP wire harness brackets to the front cover
   - Radiator outlet hose to the water pump
   - Thermostat bypass pipe to the front cover
   - Power steering pump and lines
   - Drive belt tensioner
   - Crankshaft balancer
   - Negative battery cable

13. Fill the engine with oil.

14. Fill the coolant system.

15. Start the vehicle and verify that the engine is running properly.

Pistons and Ring

POSITIONING

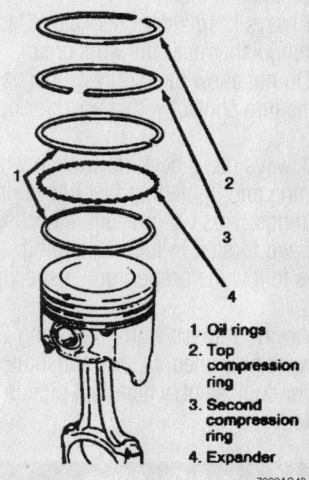

1. Oil rings
2. Top compression ring
3. Second compression ring
4. Expander

7922AG48

Piston ring positioning—3.4L engine

FUEL SYSTEM

Fuel System Service Precaution

Safety is the most important factor when performing not only fuel system maintenance but any type of maintenance. Failure to conduct maintenance and repairs in a safe manner may result in serious personal injury or death. Maintenance and testing of the vehicle's fuel system components can be accomplished safely and effectively by adhering to the following rules and guidelines.

• To avoid the possibility of fire and personal injury, always disconnect the negative battery cable unless the repair or test procedure requires that battery voltage be applied.

• Always relieve the fuel system pressure prior to disconnecting any fuel system component (injector, fuel rail, pressure regulator, etc.), fitting or fuel line connection. Exercise extreme caution whenever relieving fuel system pressure, to avoid exposing skin, face and eyes to fuel spray. Please be advised that fuel under pressure may penetrate the skin or any part of the body that it contacts.

• Always place a shop towel or cloth around the fitting or connection prior to loosening to absorb any excess fuel due to spillage. Ensure that all fuel spillage (should it occur) is quickly removed from engine surfaces. Ensure that all fuel soaked cloths or towels are deposited into a suitable waste container.

• Always keep a dry chemical (Class B) fire extinguisher near the work area.

• Do not allow fuel spray or fuel vapors to come into contact with a spark or open flame.

• Always use a back-up wrench when loosening and tightening fuel line connection fittings. This will prevent unnecessary stress and torsion to fuel line piping. Always follow the proper torque specifications.

• Always replace worn fuel fitting O-rings with new ones. Do not substitute fuel hose, or equivalent, where fuel pipe is installed.

Fuel System Pressure

RELIEVING

A Schrader valve is provided on these fuel systems to conveniently test or release the system pressure. A fuel pressure gauge and adapter will be necessary to connect the gauge to the fitting. Most of the SFI systems utilize a service valve on one end of the fuel rail assembly.

1. Before servicing the vehicle, refer to the precautions in the beginning of this section.
2. Disconnect the negative battery cable
3. Loosen the fuel filler cap to relieve tank vapor pressure.
4. Connect a fuel pressure gauge to the connector. Wrap a shop towel around the fittings to prevent spillage.
5. Install the bleed hose into an approved container and open the valve.
6. Drain any remaining fuel from the pressure gauge.
7. When fuel service is finished, tighten the fuel filler cap and connect the negative battery cable.

Fuel Filter

REMOVAL & INSTALLATION

1. Before servicing the vehicle, refer to the precautions in the beginning of this section.
2. Relieve fuel system pressure.
3. Remove or disconnect the following:
   • Negative battery cable
   • Quick connect fittings at the inlet/outlet sides of the in-pipe fuel filter
   • Fuel filter mounting bracket nut
   • Fuel filter and drain any remaining fuel

To install:
4. Install or connect the following:
   • Fuel filter to the bracket
   • Fuel filter assembly to the side rail

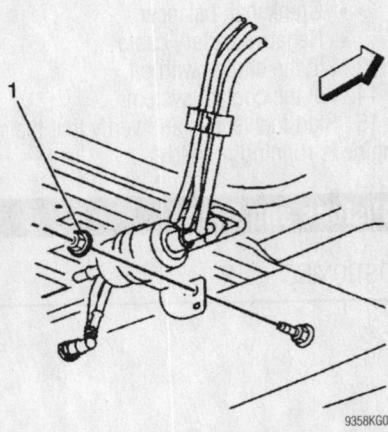

The fuel filter is located on the frame rail near the tank

9358KG08

near the fuel tank. Torque the nut to 89 inch lbs. (10 Nm).
   • Inlet/outlet quick connectors to the fuel filter
   • Negative battery cable
5. Start the vehicle and checks for leaks, repair if necessary.

Fuel Pump

REMOVAL & INSTALLATION

1. Before servicing the vehicle, refer to the precautions in the beginning of this section.
2. Properly relieve the fuel system pressure.
3. Drain and remove the fuel tank from the vehicle
4. Remove or disconnect the following:
   • Negative battery cable
   • Fuel tank
   • Fuel sender shield
   • Quick connect fittings at the fuel pump
   • Fuel Tank Pressure (FTP) electrical connector
   • Fuel pump assembly lockring

✳✳ WARNING

Do NOT pick up the fuel pump by the fuel pipes. Doing this could cause the joints to be damaged.

   • Fuel pump assembly from the fuel tank and discard the O-ring

To install:
5. Install or connect the following:
   • New O-ring on the fuel tank
   • Fuel pump into the fuel tank making certain not to fold or twist the strainer and that it does not interfere with the full travel of the float arm
   • Fuel pump locking nut
   • Quick connect fittings at the fuel pump
   • FTP sensor electrical connector
   • Fuel sender shield
   • Fuel tank and fill the tank
   • Negative battery cable
6. Prime the fuel system as follows:
   a. Turn the ignition switch **ON** for two seconds.
   b. Turn the ignition switch **OFF** for 10 seconds.
   c. Turn the ignition switch **ON** and checks for leaks. Repair if necessary.

Fuel Injector

REMOVAL & INSTALLATION

1. Before servicing the vehicle, refer to the precautions in the beginning of this section.
2. Relieve the fuel system pressure.
3. Remove or disconnect the following:
   - Negative battery cable
   - Upper intake manifold
   - Engine fuel feed pipe at the fuel rail
   - Fuel return line from the regulator
   - Main injector harness and individual injector electrical connectors
   - Electrical harness from the fuel rail
   - Coolant Temperature Sensor (CTS) electrical connector
   - Fuel rail retaining bolts
   - Fuel rail
   - Fuel injector retaining clips and injectors
   - O-rings and discard them

To install:

➡ **When replacing the fuel injector O-rings install the brown O-ring in the lower position. The lower O-ring uses a nylon collar to properly position it on the injector. Be sure to install the O-**

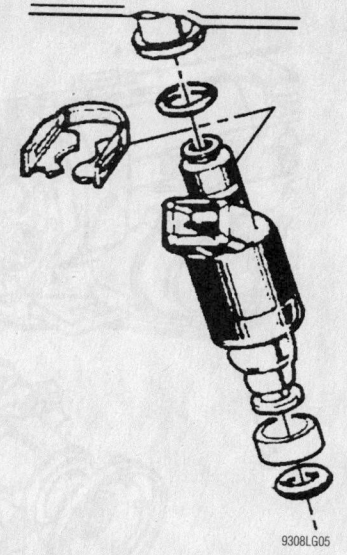

9308LG05

Fuel injector

ring backup or the sealing O-ring may move when the injector is installed to the fuel rail. If the sealing ring is not seated properly, a vacuum leak is possible thus causing driveability complaints.

4. Install or connect the following:
   - Upper O-ring to the fuel injector

   - Lower O-ring backup to the injector
   - Lower O-ring to the injector
   - Fuel injector to the fuel rail
   - Fuel rail into the intake manifold. Tilt the rail to install the injectors.
   - Fuel rail retaining bolts. Torque the bolts to 89 inch lbs. (10 Nm).
   - CTS electrical connector
   - Injector electrical harness
   - Injector electrical connectors. Make sure to push the slide locks into position.
   - Main injector harness connector
   - Fuel feed and return pipe, using new O-rings
   - Fuel feed pipe. Torque the nut to 13 ft. lbs. (17 Nm).
   - Fuel return pipe to the pressure regulator. Torque the nut to 13 ft. lbs. (17 Nm).
   - Upper intake manifold
   - Negative battery cable
5. Prime the fuel system as follows:
   a. Turn the ignition switch **ON** for two seconds.
   b. Turn the ignition switch **OFF** for 10 seconds.
   c. Turn the ignition switch **ON** and checks for leaks. Repair if necessary.

DRIVE TRAIN

Transaxle Assembly

REMOVAL & INSTALLATION

1. Install a suitable engine support fixture.
2. Remove or disconnect the following:
   - Push pins from the coolant recovery bottle. Position the bottle aside.
   - Air cleaner assembly
   - Right side engine strut
   - Transaxle range selector cable from the manual shaft
   - Transaxle range selector cable bracket
   - Wiring harness connectors from the transaxle
   - Wiring harness bracket from the side cover
   - Top four bell housing bolts
   - Propeller shaft, if equipped
   - Frame
   - Transfer case lower brace
   - Filler tube bracket retaining bolt
   - Inspection cover
   - Torque converter bolts
   - Vehicle Speed Sensor (VSS) connector

   - Right and left halfshafts from the transaxle
   - Transmission cooler lines
3. Install a transmission jack under the transmission.
   - Transfer case-to-engine mount bolts
   - Lower transaxle bolts and stud
   - Neutral safety switch connector
   - Transaxle
   - Transfer case from the transaxle

4. Flush the transaxle oil coolers, hoses and pipes.
To install:
5. Install or connect the following:
   - Transfer case to the transaxle
6. Position the flex plate alignment hole to the seven O'clock position.
7. Align the transaxle filler tube to the transmission and install the transaxle into the vehicle.

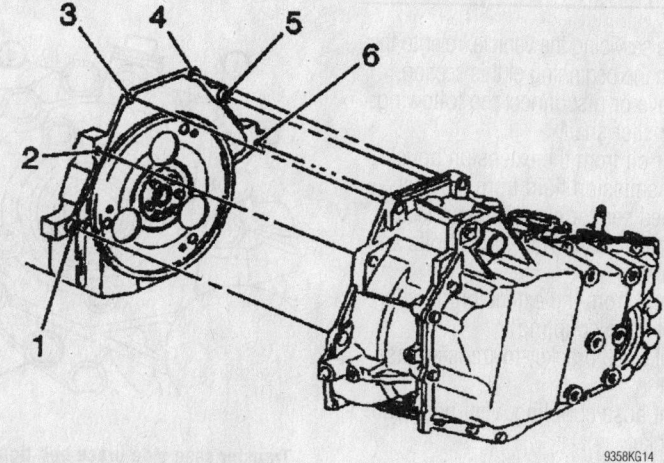

9358KG14

Transaxle mounting and bolt locations

8. Install or connect the following:
- Lower transaxle bolt and stud. Torque to 55 ft. lbs. (75 Nm).
- Wiring harness bracket to the side cover
- Neutral safety switch connector
- Transaxle brace
- Transmission cooler lines
- Right and left halfshafts
- VSS connector
- Transfer case lower brace. Torque bolts to 35 ft. lbs. (47 Nm).
- Torque converter bolts. Tighten the torque converter bolts to 46 ft. lbs. (63 Nm).
- Torque converter cover
- Filler tube bracket retaining bolt

➡ **Thoroughly clean and apply LOC-TITE® DRI-LOC 201® (GM P/N 12345493, or equivalent) to the bolt threads prior to assembly.**

- Propeller shaft, if equipped. Ensure the special washer is in place on each pair of bolts. Tighten the bolts to 24 ft. lbs. (33 Nm).
- Front wheels

9. Lower vehicle and remove the engine support fixture
- Right side engine strut
- Upper transaxle bolts and stud. Torque to 55 ft. lbs. (75 Nm).
- Wiring harness the transaxle
- Range selector cable bracket
- Range selector cable on the manual shaft
- Air cleaner assembly
- Coolant recovery bottle.

10. Check and adjust the fluid level. Inspect for fluid leaks

Transfer Case Assembly

REMOVAL & INSTALLATION

1. Before servicing the vehicle, refer to the precautions in the beginning of this section.
2. Remove or disconnect the following:
- Propeller shaft
- Gear oil from the extension housing
- Transmission fluid from the case
- Speed sensor electrical connector
- Transfer case lower brace bolts and brace
- Clamp from the extension housing vent hose coupling
- Vent hose bracket-to-transfer case bolt
- Vent hose coupling, vent hose and bracket
- Transaxle

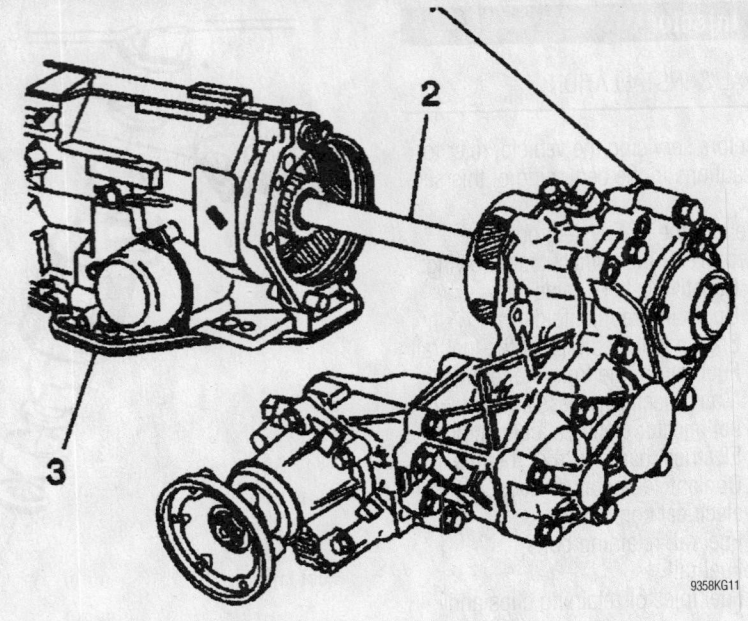

Transfer case (1), output shaft (2) and transaxle (3)

- Output shaft retaining ring
3. Rotate the transaxle 90 degrees.
- Transfer case lower brace bolt
4. Rotate the transfer case 90 degrees back to the installed position
- Transfer case side brace bolts and side brace
- Transfer case-to-transaxle bolts

❋❋ CAUTION

The transfer case weighs about 60 pounds. Be sure to lift the case properly, to avoid injury.

- Transfer case from the transaxle. Note that the output shaft withdraws from the transaxle with the transfer case.
- Output shaft

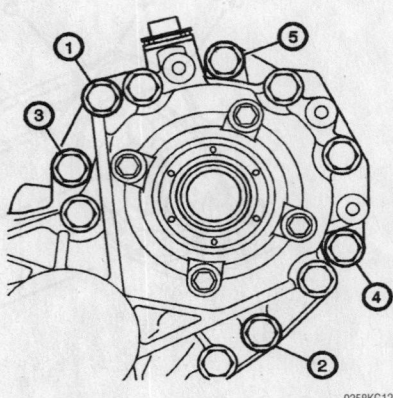

Transfer case side brace bolt tightening sequence

- Transfer case O-ring from the transfer case

To install:
5. Install or connect the following:
- O-ring seal to the transfer case
6. Rotate the transaxle so the bottom pan is facing the floor.
- Transfer case to the transaxle
7. Torque the transfer case bolts, in sequence, as follows:
   a. Bolts 1 and 2: 26 ft. lbs. (35 Nm), plus an additional 160 degrees.
   b. Bolt 3: 26 ft. lbs. (35 Nm), plus an additional 70 degrees.
   c. Bolts 4 and 5: 30 ft. lbs. (40 Nm).
- Transfer case side brace. Tighten the bolts, in sequence, to 35 ft. lbs. (47 Nm).
8. Rotate the transaxle 90 degrees.
- Transfer case lower brace-to-transaxle bolt. Torque the bolt to 35 ft. lbs. (47 Nm).
- Output shaft
- New output shaft retaining ring
- Transaxle
- Vent hose and coupling to the extension housing. Secure with the clamp.
- Vent hose bracket and bolt/stud. Tighten to 106 inch lbs. (10 Nm).
- Speed sensor electrical connector
- Propeller shaft
- Transfer case lower brace. Tighten the bolts to 35 ft. lbs. (47 Nm).
- Drain plugs and gaskets to the transfer case and extension housing. Torque the plugs to 24 ft. lbs. (32 Nm).

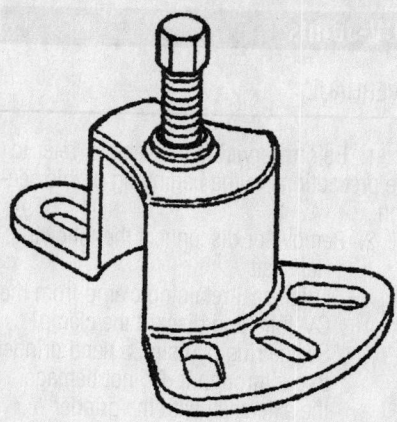

Wheel hub removal tool

9358KG10

9. Check the transaxle fluid level.
10. Remove the extension housing fill plug and fill the housing with suitable gear oil. Install the plug and tighten to 24 ft. lbs. (32 Nm).

Halfshaft

REMOVAL & INSTALLATION

These procedures requires the use of the following special tools. Slide Hammer Tool No. J 2619-01, Axle Shaft Remover Extension J 29794, Axle Shaft Puller J 33008-A and Wheel Hub Remover J 42129.

Front

1. Before servicing the vehicle, refer to the precautions in the beginning of this section.
2. Remove or disconnect the following:
   • Front wheel
   • Splash shield
   • Stabilizer shaft link
   • Wheel Speed Sensor (WSS) electrical connector
3. Insert a drift into the caliper and into the rotor to prevent the rotor from turning.

➡**Note the installed position of the 2-piece washer before removing it. The ramped sides must face each other when installing the washer.**

   • Halfshaft nut and the ramped 2-piece washer. Discard the nut.
   • Tie rod end from the steering knuckle; Do NOT loosen the tie rod jam nut
   • Lower ball joint from the steering knuckle
4. Install wheel hub removal tool (J 42129) onto the hub and secure with the lug nuts. Use the tool to disengage the halfshaft from the hub and bearing, then support the shaft.

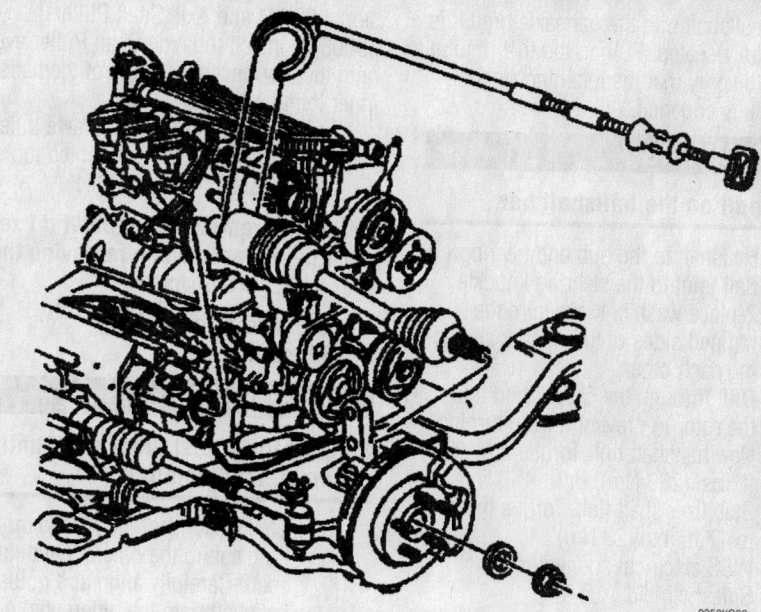

Assemble the special tools to separate the halfshaft from the transaxle

9358KG09

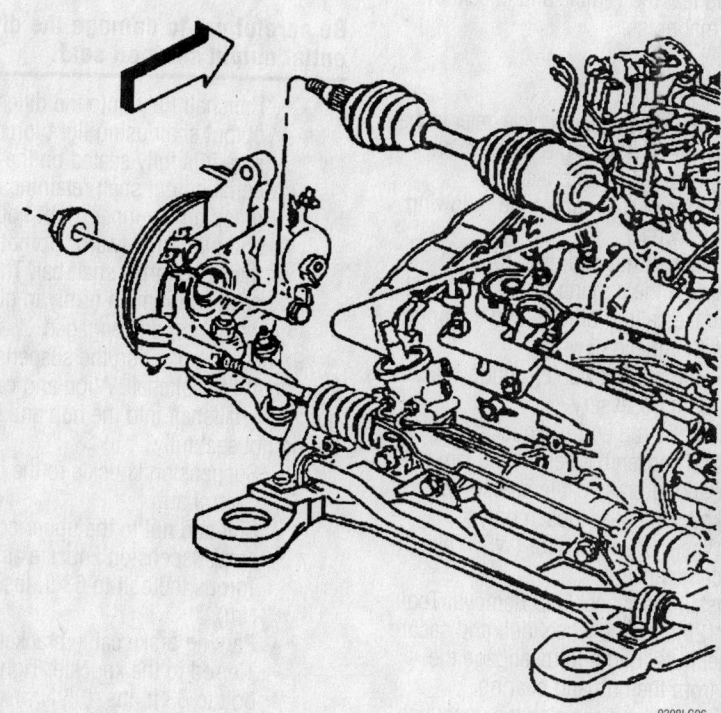

9308LG06

Install the left side halfshaft

5. Assemble the Slide Hammer Tool No. J 2619-01, Axle Shaft Remover Extension J 29794 and Axle Shaft Puller J 33008-A. Use the assembled tool to disengage the halfshaft from the transaxle.
6. Remove or disconnect the following:
   • Halfshaft from the vehicle
   • Halfshaft retaining ring and discard. On the right halfshaft, the retaining ring is on the splined shaft of the inner tripot housing. On the left halfshaft, the retaining ring is on the splined transmission output shaft.

To install:
7. Install or connect the following:
   • New halfshaft retaining ring
   • Halfshaft to the transaxle. Push the

halfshaft into the transaxle until it is fully seated. Pull on the tri-pot joint to verify that the retaining ring is fully engaged.

✳✳ WARNING

Do not pull on the halfshaft bar.

- Halfshaft to the hub and bearing.
- Ball joint to the steering knuckle
- 2-piece washer. Make sure the ramped sides of the washer are facing each other.
- Drift through the caliper and into the rotor to prevent it from turning
- New halfshaft nut. Torque it to 192 ft. lbs. (260 Nm).
- Stabilizer shaft link. Torque the nut to 17 ft. lbs. (23 Nm).
- WSS electrical connector
- Splash shield
- Front wheel

8. Check the transaxle fluid level. Check and adjust the alignment as needed.

9. Road test the vehicle and check for any abnormal noise.

Rear

1. Before servicing the vehicle, refer to the precautions in the beginning of this section.
2. Apply the parking brake.
3. Remove or disconnect the following:
   - Rear wheel
   - Halfshaft nut and discard
4. Release the parking brake.
   - Brake caliper bracket and support it with a piece of wire
   - Parking brake cable routing bracket nut, if necessary
   - Rear tie rod end-to-knuckle bolt; DO NOT loosen the tie rod end jam nut
   - Parking brake cable bracket-to-knuckle bolts; loosen only.
   - Wheel Speed Sensor (WSS) electrical connector
5. Install the Wheel Hub Removal Tool No. K 42129 onto the wheel hub and secure with wheel nuts. Begin to disengage the halfshaft from the hub and bearing.
   - Bolt and nut securing the upper control arm to the suspension knuckle
   - Halfshaft completely from the hub and bearing
6. Reposition the suspension knuckle toward the rear of the vehicle.
   - Tool from the wheel hub

➡**Support the halfshaft until it is completely removed from the vehicle.**

7. Assemble the Slide Hammer Tool No. J 2619-01, Axle Shaft Remover Exten-

sion J 29794 and Axle Shaft Puller J 33008-A. Install the Axle Shaft Puller evenly onto the rear beveled surface of the halfshaft inner joint housing.
   - Halfshaft from the rear axle differential using the assembled tools
   - Halfshaft from the vehicle

➡**The differential output shaft oil seal must be replaced when removing the rear wheel drive shaft.**

   - Halfshaft oil seal

To install:

✳✳ WARNING

Support the wheel drive shaft until it is completely installed.

8. Install or connect the following:
   - Halfshaft to the differential output shaft. Carefully align and guide the halfshaft onto the differential output shaft.

✳✳ WARNING

Be careful not to damage the differential output shaft oil seal.

   - Halfshaft fully onto the differential output shaft using light force. Make sure it is fully seated on the differential output shaft retaining ring by grasping the inner tripot housing and pulling outward. Do not pull on the wheel drive shaft bar. The halfshaft will remain firmly in place when properly engaged.

9. Begin to position the suspension knuckle to the halfshaft. Align and carefully guide the halfshaft into the hub and bearing but do not seat fully.
   - Suspension knuckle to the upper control arm
   - Bolt and nut to the upper control arm/suspension knuckle assembly. Torque the bolt to 63 ft. lbs. (85 Nm).
   - Parking brake cable bracket.
   - Tie rod to the knuckle. Tighten the bolt to 63 ft. lbs. (85 Nm).
   - Brake caliper bracket
   - WSS electrical connector
   - Parking brake cable routing bracket, if removed. Tighten the nuts to 89 inch lbs. (10 Nm).

10. Set the park brake.
   - New halfshaft nut. Tighten to 192 ft. lbs. (260 Nm).
   - Wheel

11. Lower the vehicle and release the parking brake.

CV-Joints

OVERHAUL

1. Before servicing the vehicle, refer to the precautions in the beginning of this section.
2. Remove or disconnect the following:
   - Halfshaft
   - Large seal retaining clamp from the CV-joint and discard the clamp
   - Swage ring by using a hand grinder to cut through it. Do not damage the axle shaft with the grinder
   - Halfshaft outboard seal from the CV-joint outer race and slide the seal away from the joint
   - CV-joint and boot from the halfshaft and discard the boot
3. Place a brass drift against the CV-joint cage and gently tap on it until it tilts. Remove the chrome alloy ball. Tilt the cage in the opposite direction and remove the ball. Continue to rotate the cage until all six alloy balls have been removed.
4. Pivot the cage and inner race 90 degrees to the center line of the of the outer race. Align the cage windows with the outer race lands. Lift the cage and the inner race out of the CV-joint.
5. Remove the inner race from the cage by rotating the race upward.
6. Clean the grease and contaminates with cleaning solvent from the inner/outer races; CV-joint cage and the alloy balls.
7. Remove any rust from the boot mounting area and clean the halfshaft bar.

To assemble:

8. Coat the inner and outer race grooves with grease and align the inner race with the windows of the cage.
9. Insert the inner race to the cage by rotating the race downward.
10. Insert the cage and inner race into the outer race.
11. Install the six alloy balls into the cage by tilting the cage. Repeat this process until all the balls are in place.
12. Pack the CV boot and joint with the grease supplied in the kit.
13. Install or connect the following:
   - New boot clamp onto the boot
   - CV boot on to the halfshaft bar and position the small end of the boot into the groove on the halfshaft bar. Secure the clamp to the boot with a Seal Clamp tool J 35910. Torque the clamp to 100 ft. lbs. (136 Nm).
   - Swage ring over the large diameter of the boot by pinching the ring into an oval shape

➡️**Make certain that the retaining ring side of the inner race faces the half-shaft bar before installation.**

14. Slide the joint onto the halfshaft with the retaining snapring inside of the inner race. The race is properly seated when it snaps into place. Pull on the CV-joint to verify full engagement.

15. Install the large diameter of the boot with the large swage ring in place over the outside edge of the joint outer race.

16. Clamp the boot tightly to the outer race with the large swage ring by mounting Split Plate Swage Clamp tool J 36652 in a vise.

17. Position the outboard end of the halfshaft in the bottom of the tool.

18. Align the CV boot, joint and swage ring.

19. Install the top half of the tool and align the swage ring and clamp. Install the bolts to the top of the tool and tighten snugly. Tighten each bolt an additional 180 degrees. Alternate between the bolts until both sides of the top portion of the tool touch the bottom half.

20. Loosen the bolts and remove the split plate swage clamp tool.

21. Install the halfshaft.

22. Road test the vehicle and make certain there are no abnormal noises in the front end.

23. Check and adjust the alignment if necessary.

Axle Shaft, Bearing and Seal

REMOVAL & INSTALLATION

1. Before servicing the vehicle, refer to the precautions in the beginning of this section.

2. Remove or disconnect the following:
   - Rear tire and wheel assembly
   - Rear halfshaft
   - Differential output shaft oil seal. Do not damage the differential sealing surfaces.

To install:

3. Install or connect the following:
   - Left axle shaft oil seal (1), using the J 44809

➡️**Inspect the sealing surface of the wheel drive shaft inner tri-pot housing to ensure it is free of corrosion. Use a crocus cloth in order to remove any light corrosion and clean the sealing surface with denatured alcohol, or equivalent.**

4. Lubricate the wheel drive shaft sealing surface of the oil seal with synthetic gear oil.
   - Left wheel halfshaft

- Tire and wheel assembly

5. Inspect the differential lubricant level.

Pinion Seal

REMOVAL & INSTALLATION

1. Before servicing the vehicle, refer to the precautions in the beginning of this section.

2. Drain the differential lubricant.

3. Remove or disconnect the follow-ing:
   - Torque tube assembly.
   - Drive pinion oil seal from the drive pinion housing and discard

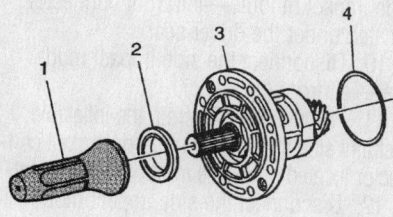

9358KG20

View of the seal installer (1), drive pinion housing shim (2), differential (3) and O-ring (4)

- Pinion housing-to-differential bolts
- Drive pinion and housing assembly
- Drive pinion housing shim from the differential
- O-ring seal from the drive pinion housing and discard

To install:

4. Lubricate the O-ring seal with synthetic gear oil.

5. Install or connect the following:
   - Oil seal to the drive pinion housing, using a seal installation tool
   - O-ring seal to the drive pinion housing
   - Drive pinion housing assembly and shim to the differential carrier
   - Pinion housing-to-differential mounting bolts. Torque the bolts to 21 ft. lbs. (28 Nm).

6. Lubricate the sealing surfaces of the drive pinion oil seal with synthetic gear oil.
   - Torque tube assembly.

7. Fill the differential with synthetic gear oil.

Differential Carrier Assembly

REMOVAL & INSTALLATION

1. Before servicing the vehicle, refer to the precautions in the beginning of this section.

2. Set the parking brake.

3. Drain the rear differential gear oil.

4. Remove or disconnect the following:
   - Right rear tire and wheel
   - Electrical connector from the clutch pump check valve
   - Right rear halfshaft
   - Front propeller shaft

5. Place an adjustable support beneath the torque tube.
   - Torque tube-to-bracket through bolt and nut; loosen only
   - Bolts from the torque tube bracket
   - Differential carrier-to-cradle mounting bolts, nuts, washers, and mounts from the differential

➡️**During the removal of the wheel drive shaft, the differential output shaft may become disengaged from the differential. If this occurs, firmly grasp and separate the output shaft from the wheel drive shaft. Align the splines on the output shaft to the differential and reposition the output shaft to the differential.**

6. While simultaneously moving the differential assembly to the right side of the vehicle, disengage the left halfshaft from the differential.
   - Rear differential and torque tube as an assembly
   - Torque tube from the differential

To install:

7. Install or connect the following:
   - Torque tube to the differential.
   - Rear differential and torque tube assembly to the suspension cradle. Simultaneously guide the left wheel drive shaft onto the differential output shaft while positioning the differential assembly to the suspension cradle.

8. Place an adjustable support under the torque tube. Ensure that the left wheel drive shaft is fully engaged to the differential output shaft.
   - Differential carrier mounts, washers, bolts, and nuts to the differential. Torque the bolts to 37 ft. lbs. (50 Nm).
   - Torque tube bracket-to-body bolts. Torque the torque tube bracket-to-body bolts to 41 ft. lbs. (55 Nm) and the torque tube-to-bracket through bolt and nut to 47 ft. lbs. (64 Nm).
   - Front propeller shaft
   - Right rear wheel halfshaft
   - Tire and wheel assembly
   - Differential drain plug and gasket.

Tighten the drain plug to 22 ft. lbs. (30 Nm).

9. Fill the axle with synthetic gear oil. Check the differential oil level to ensure it is even with, to no lower than, 0.25 in. (6mm) below the opening of the fill hole.

• Differential fill plug and gasket. Tighten the fill plug to 22 ft. lbs. (30 Nm).
• Clutch pump check valve connector

10. Remove the adjustable support from the torque tube.

11. Operate the vehicle making tight left,

then right turns in order to engage the All-Wheel-Drive (AWD) system and distribute the gear oil throughout the differential.

12. Fill the axle with synthetic gear oil. Check the differential oil level to ensure it is even with, to no lower than, 0.25 in. (6mm) below the opening of the fill hole.

STEERING AND SUSPENSION

Air Bag

✳✳ CAUTION

All models are equipped with a Supplemental Inflatable Restraint (SIR) system. Before attempting any work on or near the steering column, ALWAYS disarm the air bag to prevent a costly and possibly dangerous accidental deployment.

PRECAUTIONS

Several precautions must be observed when handling the inflator module to avoid accidental deployment and possible personal injury.

• Never carry the inflator module by the wires or connector on the underside of the module
• When carrying a live inflator module, hold securely with both hands, and ensure that the bag and trim cover are pointed away from your body
• Place the inflator module on a bench or other surface with the bag and trim cover facing up
• With the inflator module on the bench, never place anything on or close to the module that may be thrown in the event of an accidental deployment

DISARMING

1. Turn the wheels to the straight-ahead position, then turn the ignition switch to **LOCK**.

2. Remove the console accessory wiring junction block access hole cover.

3. Remove the "AIR BAG" or "SIR" fuse from the block, as applicable.

4. Remove the left-hand sound insulator, for access to the SIR wiring harness.

5. Remove the Connector Position Assurance (CPA) device, then disengage the yellow 2-way connector at the base of the steering column.

6. Remove the right hand insulator panel.

7. Remove the CPA from the inflatable restraint instrument panel module

connector located behind the RH insulator panel.

8. Disconnect the instrument panel module connector

9. If equipped with side air bags, remove the CPA from the inflatable restraint side impact module—left front connector located under the driver seat.

10. Disconnect the side impact module—left front connector.

11. Remove the CPA from the inflatable restraint side impact module—right front connector located under the front passenger seat.

12. Disconnect the side impact module—right front connector.

➥ **With the fuse removed, the AIR BAG or SIR light will illuminate if the ignition switch is turned ON at any time. This is normal and does not indicate a problem when the system is disarmed.**

To enable:

13. Be sure the ignition is in the **LOCK** position.

14. Connect the inflatable restraint side impact module—right front connector located under the front passenger seat.

15. Install the CPA to the side impact module—right front connector.

16. Connect the inflatable restraint side impact module—left front connector located under the driver seat.

17. Install the CPA to the side impact module—left front connector.

18. Engage the yellow SIR connector, then secure using the CPA device for both the drivers and passenger sides.

19. Install the sound insulator panel.

20. Install the SIR system fuse to the fuse block.

21. Turn the ignition switch to the **ON** position and verify that the AIR BAG indicator light flashes 7 times, then extinguishes. If it does not go out, troubleshoot the SIR system fault.

22. Install the instrument panel lower extension.

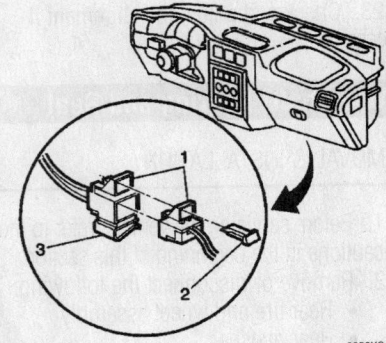

9358KG16

Passenger's side air bag connector location

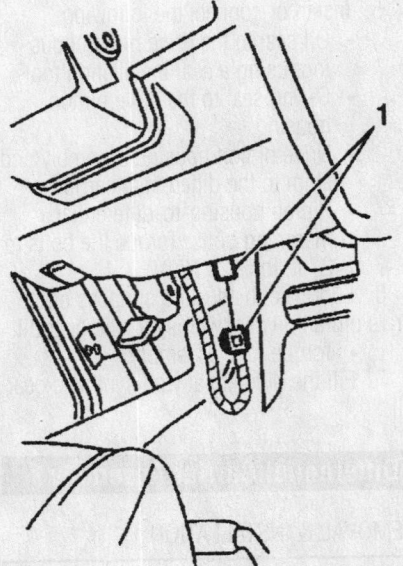

9358KG15

Driver's side air bag connector location

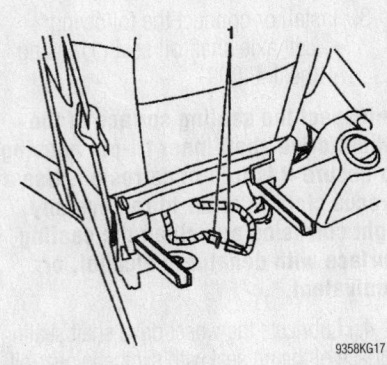

9358KG17

Side air bag connector location

Power Steering Rack and Pinion

REMOVAL & INSTALLATION

1. Before servicing the vehicle, refer to the precautions in the beginning of this section.

2. Remove or disconnect the following:

- Negative battery cable
- Left front wheel
- Stabilizer shaft
- Tie rod ends from the steering knuckle
- Intermediate shaft from the steering gear

3. Support the frame using a suitable utility stand.

- Frame rear bolts and discard them
- Power steering gear heat shield
- Cooler pipe from the power steering gear
- Pressure hose from the power steering gear
- Power steering gear through the left wheel opening

To install:

4. Install or connect the following:

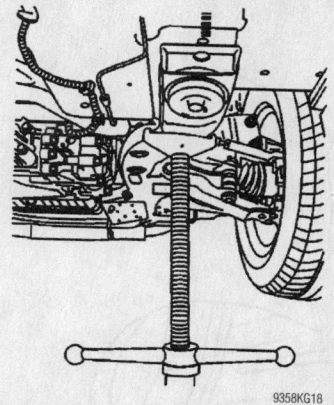

9358KG18

Use a utility stand to support the frame

- Power steering gear through the left wheel opening
- New power steering gear bolts. Torque them to 59 ft. lbs. (80 Nm).
- Pressure hose and cooler pipe to the power steering gear. Torque the fasteners to 20 ft. lbs. (27 Nm).
- Heat shield. Torque the bolts to 54 inch lbs. (6 Nm).
- Utility stand to support the frame
- New rear frame bolts. Torque them to 122 ft. lbs. (165 Nm).

5. Remove the utility stand.
6. Install or connect the following:

- Intermediate shaft to the steering gear. Torque the bolt to 35 ft. lbs. (47 Nm).
- Tie rod ends to the steering knuckle
- Stabilizer shaft. Torque the bolt to 17 ft. lbs. (23 Nm).
- Left front wheel

7. Fill and bleed the power steering system and check for leaks.

8. Road test the vehicle and adjust the toe as necessary.

Strut

REMOVAL & INSTALLATION

✷✷ CAUTION

Do not remove the top center nut from the strut assembly. This nut should only be removed when the strut assembly is out of the vehicle, mounted in a holding fixture and the coil spring is in a compressed position using the proper coil spring compressor.

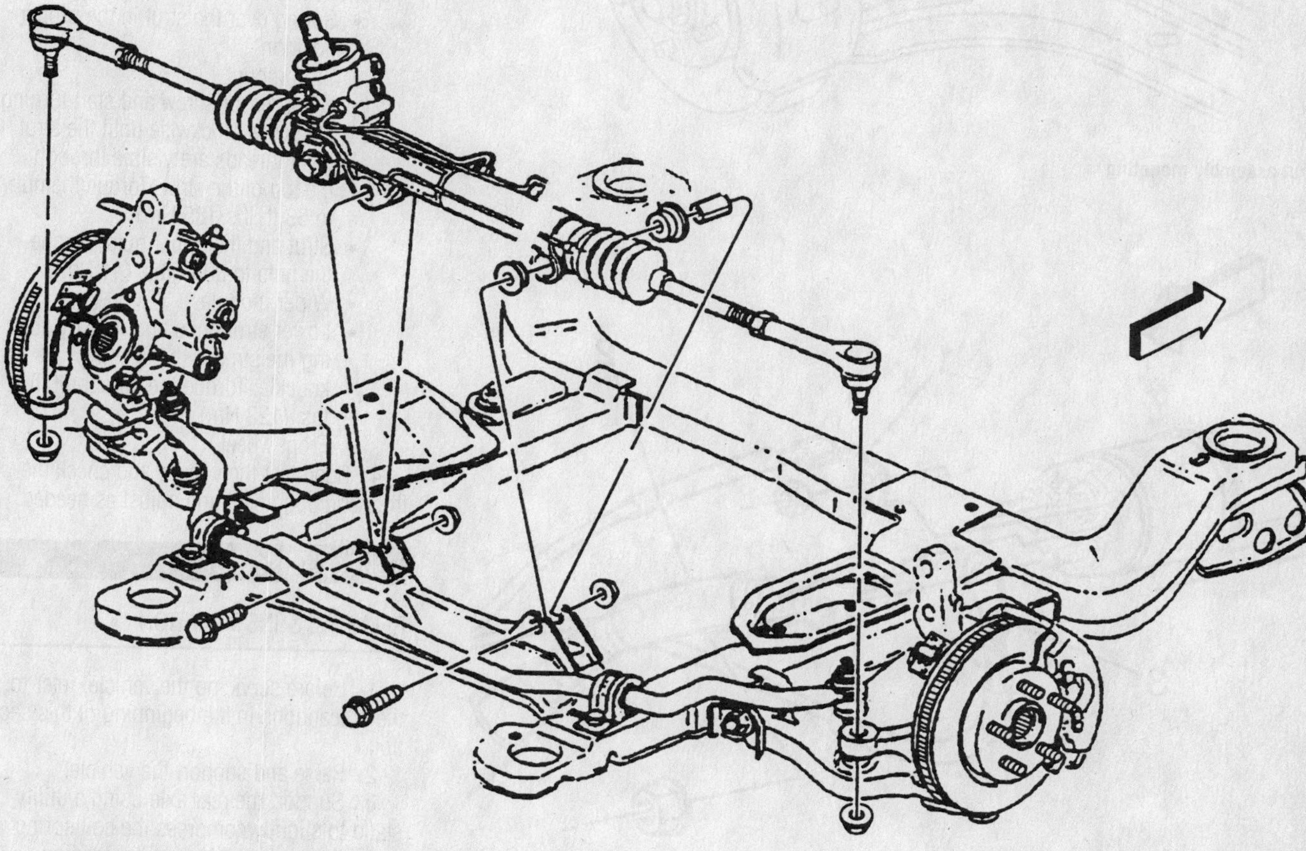

The rack and pinion steering gear is bolted to the rear of the subframe, as shown

7924LG19

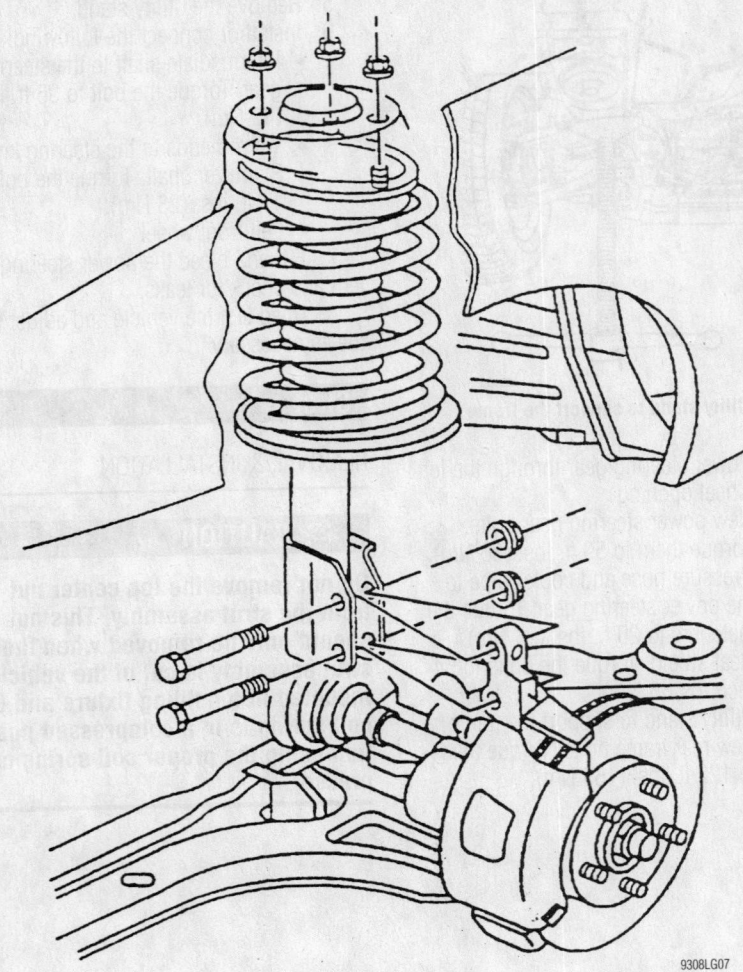

Strut assembly mounting

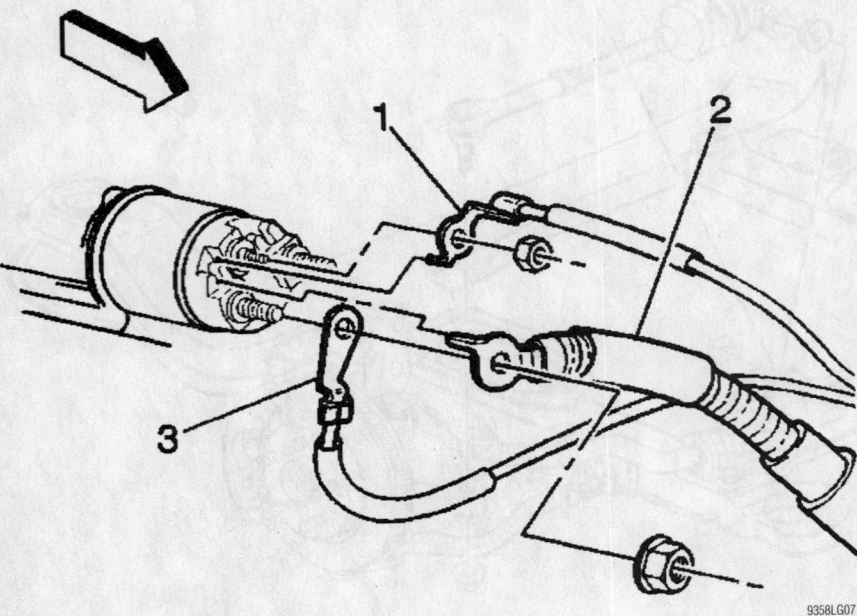

Disassembling the strut assembly

1. Before servicing the vehicle, refer to the precautions in the beginning of this section.
2. Remove or disconnect the following:
   - Wiper module
   - Three upper strut nuts

➡**Use a frame contact lift to raise the vehicle. DO NOT use a suspension contact lift.**

   - Tire and wheel
   - Lower strut bolts and nuts after marking the position of the strut to the knuckle

➡**The strut to steering knuckle position must be marked so that the camber angle will not change. If the angle is change, the wheel alignment will also be affected.**

   - Strut
   - Nut from the top of the strut by placing the assembly in a Strut Compressor tool J 34013-B and Damper Rod Clamp J 34013-20. Turn the compressor forcing screw until the spring compresses slightly
   - Strut mount
   - Spring from the strut assembly

To install:
3. Install or connect the following:
   - Spring over the strut in the proper position
   - Strut mount
   - Compressor screw and start turning the screw clockwise until the strut shaft threads are visible through the top of the strut. Torque the nut to 63 ft. lbs. (85 Nm).
   - Strut and the upper nuts. Torque the nuts to 30 ft. lbs. (41 Nm).
   - Wiper module
   - Lower strut bolts and nuts by aligning the strut to the steering knuckle. Torque the nuts to 90 ft. lbs. (123 Nm).
   - Front wheel
4. Road test the vehicle and check the front end alignment and adjust as needed.

Shock Absorber

REMOVAL & INSTALLATION

1. Before servicing the vehicle, refer to the precautions in the beginning of this section.
2. Raise and support the vehicle.
3. Support the rear axle using a utility stand to slightly compress the coil spring and relieve the shock absorber tension.

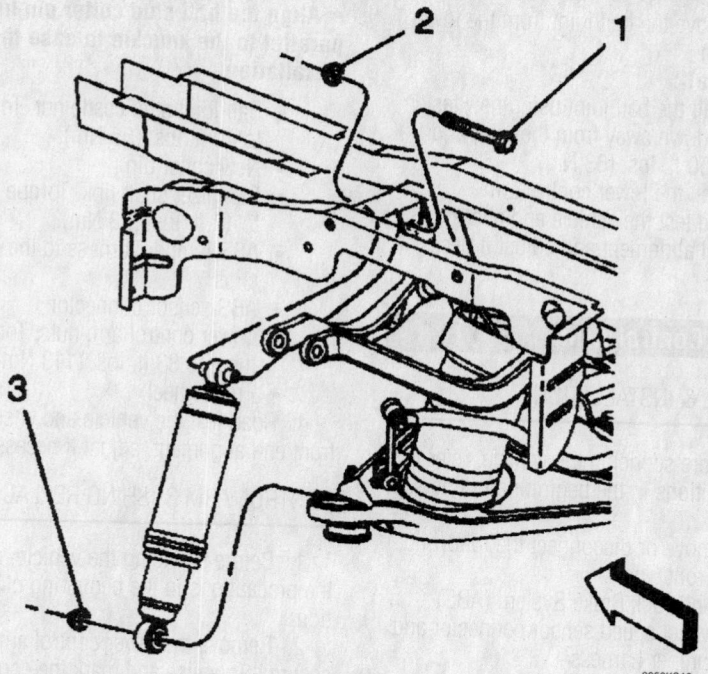

Rear shock absorber mounting—AWD vehicle shown

9358KG19

4. If the vehicle is equipped with automatic level control, disconnect the air tube connector from the shock absorber.

5. Remove or disconnect the following:
- Shock absorber upper bolt and nut
- Shock absorber lower bolt and/or nut, as applicable
- Shock absorber from the brackets by compressing it slightly

To install:

6. Inspect the shock absorber, upper and lower mounting brackets and the frame mounting hole for cracks excessive wear and burrs.

7. Install or connect the following:
- Shock absorber
- Shock absorber bolts and nuts. Torque the nuts to 63 ft. lbs. (85 Nm).
- Air tube connector to the shock absorber, if equipped with automatic level control

8. Remove the support from the rear axle.

9. Road test the vehicle.

Coil Spring

REMOVAL & INSTALLATION

Front

The service procedure for the front coil springs is covered under MacPherson Strut removal and installation.

Rear

1. Before servicing the vehicle, refer to the precautions in the beginning of this section.

2. Remove or disconnect the following:
- Brake hose bracket screw from the control arm

- Shock absorber lower bolt while using a utility stand to support the rear axle
- Tie rod from the rear axle
- Spring and insulators after lowering the rear axle

To install:

3. Install or connect the following:
- Insulators and springs on the rear axle with the paint stripe is facing rearward
- Rear axle tie rod after raising the rear axle into its proper position. Torque the nut to 92 ft. lbs. (125 Nm).
- Shock absorbers to the rear axle. Torque the upper and lower nuts to 63 ft. lbs. (85 Nm).
- Brake hose bracket to the control arm. Torque the bolt to 33 ft. lbs. (44 Nm).

4. Remove the axle supports.

Lower Ball Joint

REMOVAL & INSTALLATION

1. Before servicing the vehicle, refer to the precautions in the beginning of this section.
2. Remove the lower control arm.
3. Place the control arm in a vise.
4. Drill or grind off the ball stud rivet heads and use a punch to remove the rivets.

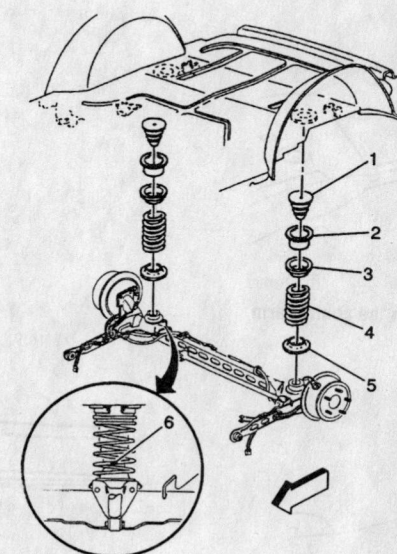

Legend
(1) Rear Suspension Jounce Bumper
(2) Rear Suspension Jounce Bumper Retainer
(3) Rear Suspension Insulator
(4) Rear Spring
(5) Rear Spring Insulator
(6) Paint Stripe

7924LG21

Exploded view of the coil spring assembly

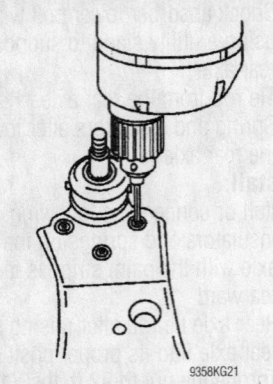

Drill off the rivet heads

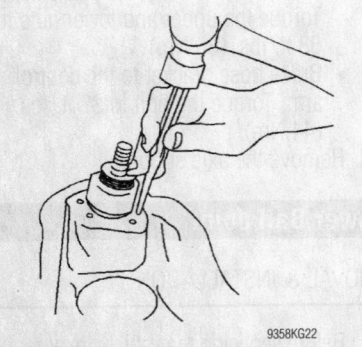

Use a punch and hammer to remove the rivets

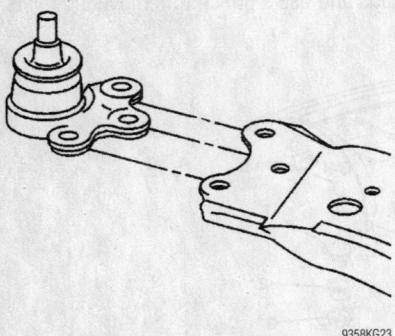

Remove the ball joint from the control arm

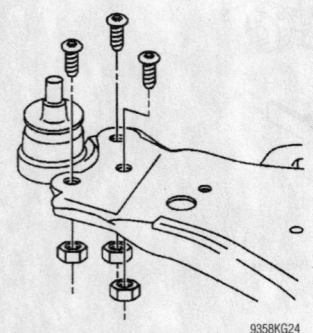

Install the ball joint fasteners facing down away from the joint

5. Remove the ball joint from the lower control arm

To install:

6. Install the ball joint using new fasteners facing down away from the joint and tighten to 50 ft. lbs. (63 Nm)

7. Install the lower control arm.

8. Road test the vehicle and check the front wheel alignment and adjust if necessary.

Lower Control Arm

REMOVAL & INSTALLATION

1. Before servicing the vehicle, refer to the precautions in the beginning of this section.

2. Remove or disconnect the following:
- Front wheel
- Anti-lock Brake System (ABS) wheel speed sensor connector and jumper harness
- Stabilizer shaft link
- Cotter pin from the ball joint stud and loosen the nut
- Ball Joint from the steering knuckle
- Lower control arm

To install:

3. Install or connect the following:
- Lower control arm
- Ball joint stud to the knuckle

➡**Align the ball stud cotter pin hole parallel to the knuckle to ease the pin installation.**

- Ball joint stud castle nut. Torque it to 40 ft. lbs. (55 Nm).
- New cotter pin
- Stabilizer shaft link. Torque the nut to 17 ft. lbs. (23 Nm).
- ABS jumper harness to the retainer clips
- ABS sensor connector
- Lower control arm nuts. Torque them to 83 ft. lbs. (113 Nm).
- Front wheel

4. Road test the vehicle and check the front end alignment, adjust if necessary.

CONTROL ARM BUSHING REPLACEMENT

1. Before servicing the vehicle, refer to the precautions in the beginning of this section.

2. Remove the lower control arm and secure it in a vise and mark the control arm along the flat edge of the bushing flange.

3. Assemble the bushing removal tool.

4. Tighten the assembly until the bushing is removed.

To install:

5. Install the bushing into the control arm by align the flat edge of the bushing to the mark in the control arm.

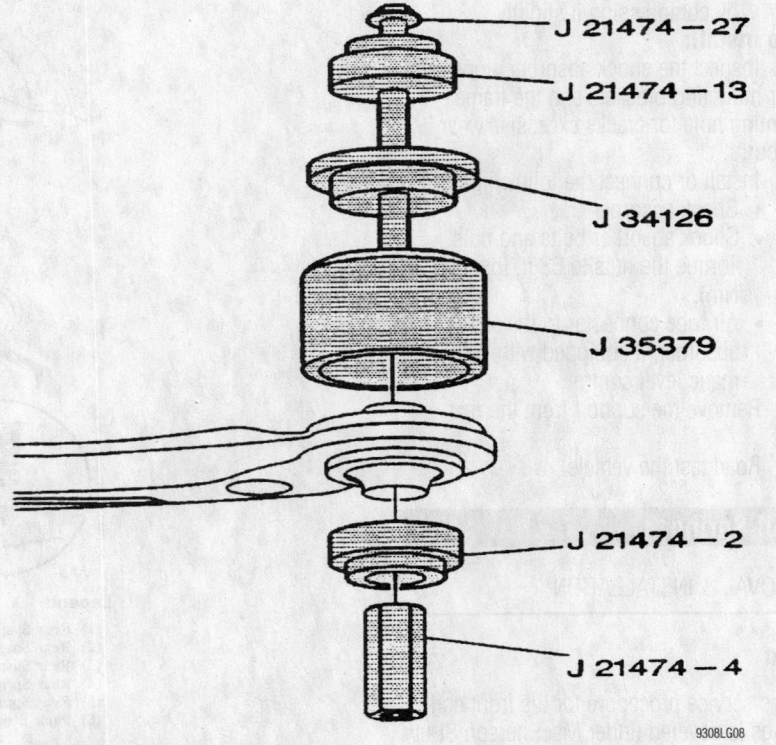

J 21474 — 27
J 21474 — 13
J 34126
J 35379
J 21474 — 2
J 21474 — 4

View of the lower control arm bushing

6. Make certain that the flat edge of the bushing is 30 degrees from the centerline of the control arm and the thin slot in the bushing is facing outboard.

7. Fully seat the bushing in the control arm.

8. Install the lower control arm.

9. Road test the vehicle and adjust the alignment, if necessary.

Wheel Bearings

ADJUSTMENT

Both front and rear wheel bearings are integral to the hub assembly and are not adjustable. If the bearings are found to be defective, the hub assembly must be replaced.

REMOVAL & INSTALLATION

Front

1. Before servicing the vehicle, refer to the precautions in the beginning of this section.

2. Remove or disconnect the following:
   - Front wheel
   - Wheel speed sensor electrical connector and the connector from the bracket
   - Brake caliper and bracket
   - Brake rotor
   - Halfshaft nut

3. Attach a front hub spindle removal tool to the wheel bearing/hub.

4. Push the halfshaft out of the wheel bearing hub assembly.

5. Remove or disconnect the following:
   - Wheel bearing/hub bolts and discard them
   - Wheel bearing/hub assembly

To install:

6. Install or connect the following:
   - Wheel bearing/hub assembly

✳✳ CAUTION

The wheel bearing/hub bolts must be replaced whenever they are loosened or removed.

   - New wheel bearing/hub bolts. Torque them to 96 ft. lbs. (130 Nm).
   - Halfshaft nut. Torque it to 118 ft. lbs. (160 Nm).
   - Brake rotor
   - Brake caliper. Torque the bolts to 63 ft. lbs. (85 Nm).

- Wheel speed sensor electrical connector to the bracket
- Wheel speed sensor electrical connector
- Front wheel

7. Road test the vehicle and check the front alignment, adjust if necessary.

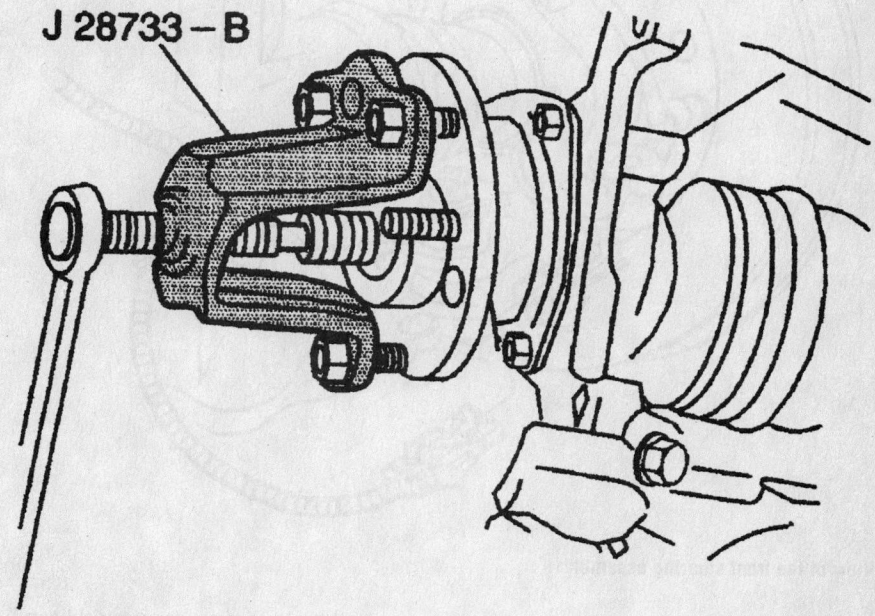

J 28733 – B

Use a pullet to separate the hub from the halfshaft

9358KG25

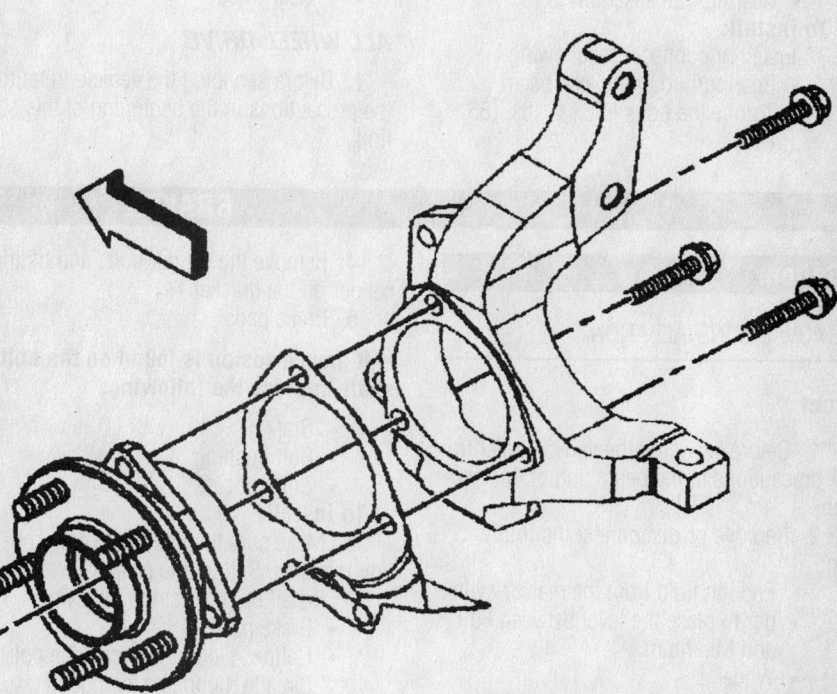

Front hub and bearing assembly mounting

9358KG26

Rear

FRONT WHEEL DRIVE

1. Before servicing the vehicle, refer to the precautions in the beginning of this section.

2. Remove or disconnect the following:

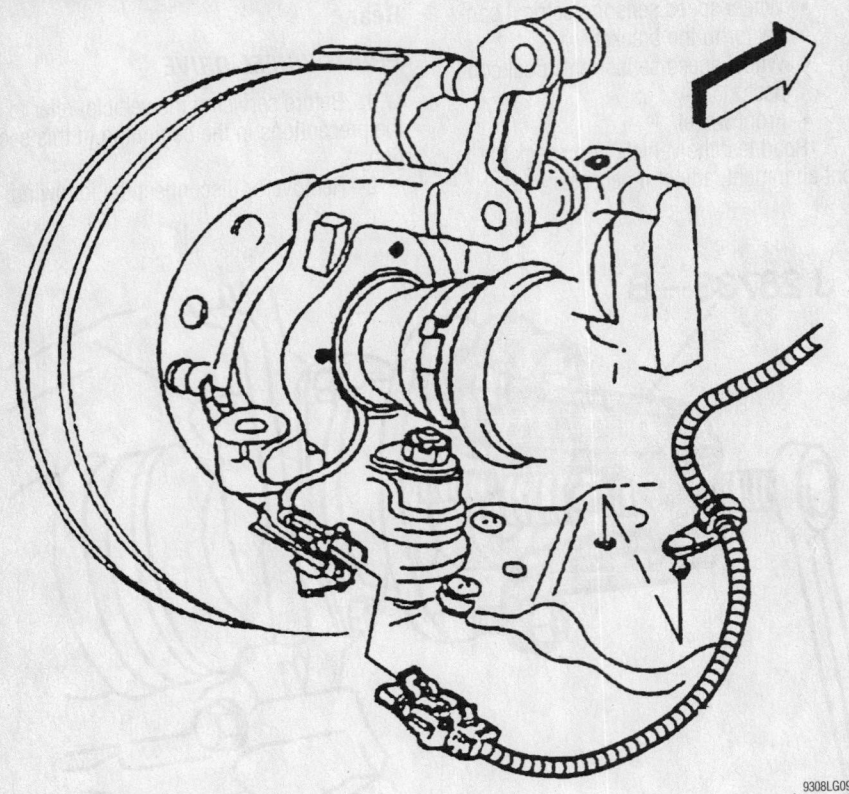

View of the front steering assembly

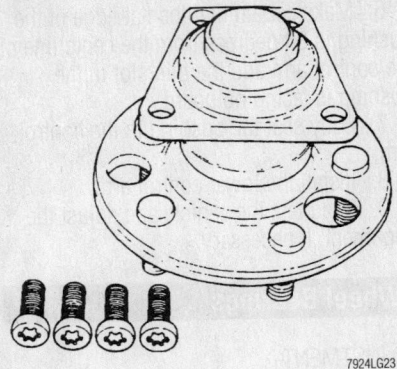

The rear wheel hub is mounted with 4 Torx® head bolts

- Rear wheel
- Brake drum
- Wheel speed sensor
- Bearing/hub assembly

To install:
3. Install or connect the following:
- Bearing/hub to the axle beam. Torque the bolts to 63 ft. lbs. (85 Nm).

- Wheel speed sensor electrical connector to the bearing/hub assembly
- Brake drum
- Rear wheel

ALL WHEEL DRIVE
1. Before servicing the vehicle, refer to the precautions in the beginning of this section.

2. Remove or disconnect the following:
- Rear wheel
- Brake caliper and support it with a piece of wire
- Brake caliper bracket
- Rotor
- Wheel speed sensor connector from the sensor and bracket
- Halfshaft from the hub and bearing
- Bearing/hub bolts and bearing/hub

To install:
3. Install or connect the following:
- Bearing/hub. Tighten the bolts to 63 ft. lbs. (85 Nm).
- Halfshaft
- Wheel speed sensor connector to the bracket and sensor
- Rotor
- Caliper bracket
- Caliper
- Tire and wheel

BRAKES

Caliper

REMOVAL & INSTALLATION

Front

1. Before servicing the vehicle, refer to the precautions in the beginning of this section.
2. Remove or disconnect the following:
- Enough fluid from the master cylinder to place the level between Full and Minimum
- Wheel
3. If the caliper is being repaired or replaced, remove the brake hose, cap the end, and discard the washers.

4. Remove the caliper bolts and lift the caliper off the bracket.
5. Brake pads

➡ If any corrosion is found on the bolt shaft, replace the following:

- Bolt
- Bolt bushing
- Bolt boot

To install:
6. Make sure that the boots are properly installed. Bottom the caliper piston.
7. Install or connect the following:
- Brake pads
- Caliper. Do not lubricate the bolt threads. Lubricate the boots. Torque the bolts to 26 ft. lbs. (35 Nm).
- Brake hose, using new washers.

Torque the bolt to 40 ft. lbs. (54 Nm).
8. If the hose was disconnected, bleed the system.
9. Install the wheel.
10. Apply approximately 175 ft. lbs. of force to the brake pedal for 10 seconds.

Rear

1. Before servicing the vehicle, refer to the precautions in the beginning of this section.
2. Remove or disconnect the following:
- Enough fluid from the master cylinder to place the level between Full and Minimum
- Wheel
3. If the caliper is being repaired or

replaced, remove the brake hose, cap the end, and discard the washers.

4. Remove the caliper bolts and lift the caliper off the bracket.

5. Brake pads

➡**If any corrosion is found on the bolt shaft, replace the following:**

- Bolt
- Bolt bushing
- Bolt boot

To install:

6. Make sure that the boots are properly installed. Bottom the caliper piston.

7. Install or connect the following:

- Brake pads
- Caliper. Do not lubricate the bolt threads. Lubricate the boots. Torque the bolts to 33 ft. lbs. (45 Nm).
- Brake hose, using new washers. Torque the bolt to 40 ft. lbs. (54 Nm).

8. If the hose was disconnected, bleed the system.

9. Install the wheel.

10. Apply approximately 175 ft. lbs. of force to the brake pedal for 10 seconds.

Brake Pads

REMOVAL & INSTALLATION

Front

1. Before servicing the vehicle, refer to the precautions in the beginning of this section.

2. Remove the wheel.

3. Remove the lower caliper bolt.

4. Rotate the caliper up.

5. Remove the pads and pad retainers.

6. Remove enough fluid from the master cylinder to place the level between Full and Minimum

7. Bottom the piston.

8. Inspect the bolt boots and piston boot for tears or deterioration. Replace as necessary.

➡**If any corrosion is found on the bolt shaft, replace the following:**

- Bolt
- Bolt bushing
- Bolt boot

❊❊ WARNING

Never attempt to polish away the corrosion.

To install:

9. Install the pad retainers and pads. Make sure that the wear indicator is at the upper edge of the inner pad.

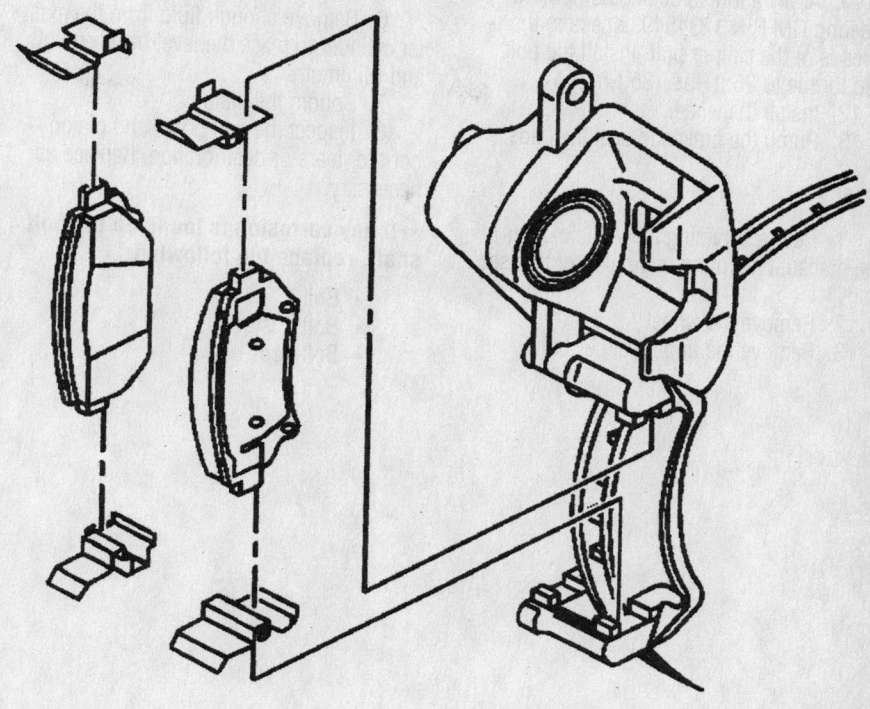

Front pads and retainers

42372-REND-G01

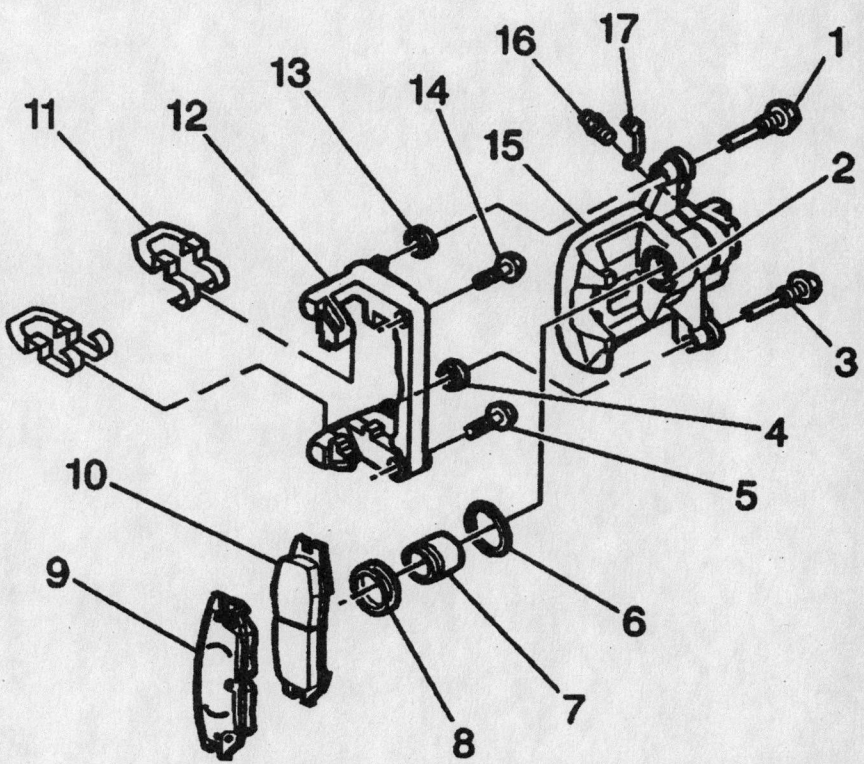

Rear disc brake parts

42372-REND-G02

10. Rotate the caliper over the pads.

11. Apply a threadlocking compound meeting GM P/N 12345493 specs to the threads of the caliper bolt. Install the bolt and torque to 26 ft. lbs. (35 Nm).

12. Install the wheel.

13. Pump the brakes to seat the pads.

Rear

1. Before servicing the vehicle, refer to the precautions in the beginning of this section.

2. Remove the wheel.

3. Remove the upper caliper bolt.

4. Rotate the caliper down.

5. Remove the pads and pad retainers.

6. Remove enough fluid from the master cylinder to place the level between Full and Minimum

7. Bottom the piston.

8. Inspect the bolt boots and piston boot for tears or deterioration. Replace as necessary.

➡ If any corrosion is found on the bolt shaft, replace the following:

- Bolt
- Bolt bushing
- Bolt boot

✳✳ WARNING

Never attempt to polish away the corrosion.

To install:

9. Install the pad retainers and pads. Make sure that the wear indicator is at the downward edge of the outer pad.

10. Rotate the caliper over the pads.

11. Install the bolt and torque to 33 ft. lbs. (45 Nm).

12. Install the wheel.

13. Pump the brakes to seat the pads.

CADILLAC

CTS

3

SPECIFICATION CHARTS

ENGINE AND VEHICLE IDENTIFICATION

			Engine						Model Year	
Code ①	Liters (cc)	Cu. In.	Cyl.	Fuel Sys.	Engine Type	Eng. Mfg.		Code ②	Year
N	3.2 (3173)	195	V6	SMFI	DOHC	Opel		3	2003
								4	2004

SMFI: Sequential Multi-port Fuel Injection

DOHC: Double Overhead Camshaft

① 8th position of VIN

② 10th position of VIN

42372-CADI-C01

GENERAL ENGINE SPECIFICATIONS

Year	Model	Engine Displacement Liters (cc)	Engine Series (ID/VIN)	Fuel System	Net Horsepower @ rpm	Net Torque @ rpm (ft. lbs.)	Bore x Stroke (in.)	Com-pression Ratio	Oil Pressure @ rpm
2003	CTS	3.2 (3173)	N	MFI	220@6000	220@3400	3.45x3.47	10.0:1	22@900

MFI: Multi-point Fuel Injection

42372-CADI-C02

ENGINE TUNE-UP SPECIFICATIONS

Year	Engine Displacement Liters (cc)	Engine ID/VIN	Spark Plug Gap (in.)	Ignition Timing (deg.) AT	Fuel Pump (psi)	Idle Speed (rpm) AT	Valve Clearance In.	Ex.
2003	3.2 (3173)	N	0.060	①	41-47	①	HYD	HYD

NOTE: The Vehicle Emission Control Information label often reflects specification changes made during production. The label figures must be used if they differ from those in this chart.

HYD: Hydraulic

① Refer to Vehicle Emission Control Information label

42372-CADI-C03

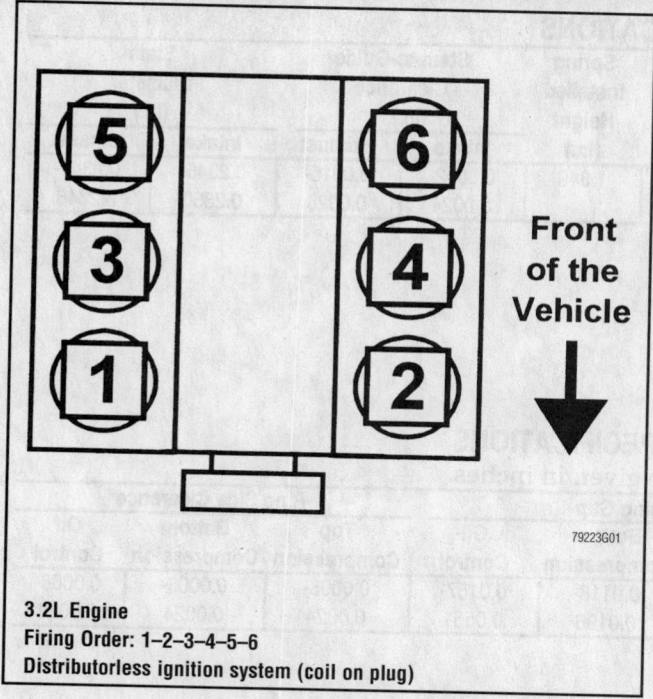

3.2L Engine
Firing Order: 1–2–3–4–5–6
Distributorless ignition system (coil on plug)

79223G01

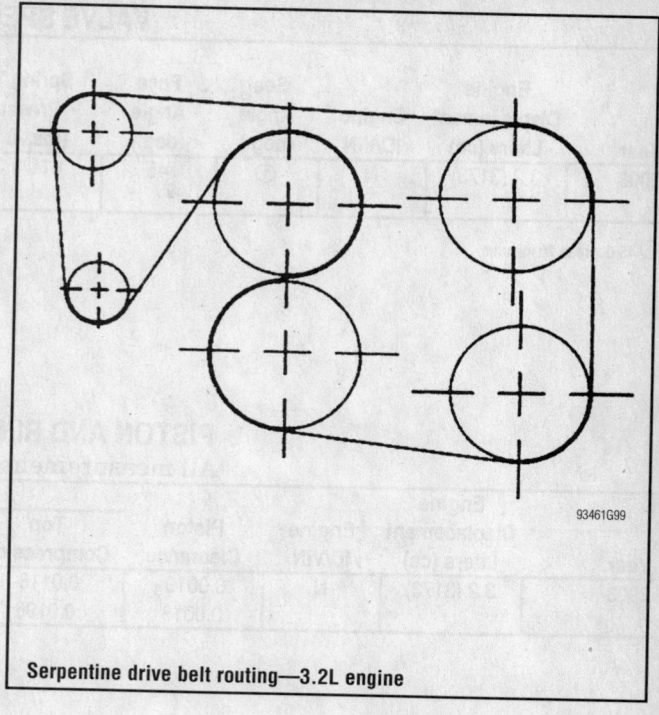

Serpentine drive belt routing—3.2L engine

93461G99

CAPACITIES

Year	Model	Engine Displacement Liters (cc)	Engine ID/VIN	Engine Oil with Filter (qts.)	Transmission (pts.) Auto.	Transmission (pts.) Man.	Drive Axle Rear (pts.)	Fuel Tank (gal.)	Cooling System (qts.)
2003	CTS	3.2 (3173)	N	5.0	18	2.6	3.5	17.0	10.4

NOTE: All capacities are approximate. Add fluid gradually and ensure a proper fluid level is obtained.

42372-CADI-C04

CRANKSHAFT AND CONNECTING ROD SPECIFICATIONS
All measurements are given in inches.

Year	Engine Displacement Liters (cc)	Engine ID/VIN	Crankshaft Main Brg. Journal Dia.	Crankshaft Main Brg. Oil Clearance	Crankshaft Shaft End-play	Crankshaft Thrust on No.	Connecting Rod Journal Diameter	Connecting Rod Oil Clearance	Connecting Rod Side Clearance
2003	3.2 (3173)	N	2.6670-2.6760	0.0006-0.0017	0.0039 0.0079	2	2.6764-2.6770	0.0008-0.0024	0.0027-0.0110

42372-CADI-C05

VALVE SPECIFICATIONS

Year	Engine Displacement Liters (cc)	Engine ID/VIN	Seat Angle (deg.)	Face Angle (deg.)	Spring Test Pressure (lbs. @ in.)	Spring Installed Height (in.)	Stem-to-Guide Clearance (in.) Intake	Exhaust	Stem Diameter (in.) Intake	Exhaust
2003	3.2 (3173)	N	①	45	61 @ 1.340	1.340	0.0012-0.0024	0.0016-0.0028	0.2345-0.2350	0.2341-0.2346

① 45 degrees 20 minutes

42372-CADI-C06

PISTON AND RING SPECIFICATIONS
All measurements are given in inches.

Year	Engine Displacement Liters (cc)	Engine ID/VIN	Piston Clearance	Ring Gap Top Compression	Bottom Compression	Oil Control	Ring Side Clearance Top Compression	Bottom Compression	Oil Control
2003	3.2 (3173)	N	0.0010-0.0018	0.0118-0.0196	0.0118-0.0196	0.0157-0.0551	0.0008-0.0024	0.0008-0.0024	0.0008-0.0024

42372-CADI-C07

TORQUE SPECIFICATIONS
All readings in ft. lbs.

Year	Engine Displacement Liters (cc)	Engine ID/VIN	Cylinder Head Bolts	Main Bearing Bolts	Rod Bearing Bolts	Crankshaft Damper Bolts	Flywheel Bolts	Manifold Intake	Exhaust	Spark Plugs	Lug Nut
2003	3.2 (3173)	N	①	②	③	15	④	15	15	18	100

① Step 1: 18 ft. lbs.
Step 2: Rotate 90 degrees
Step 3: Rotate 90 degrees
Step 4: Rotate 90 degrees
Step 5: Rotate 15 degrees

② Step 1: 37 ft. lbs.
Step 2: Rotate 60 degrees

③ Step 1: 26 ft. lbs.
Step 2: Rotate 45 degrees
Step 3: Rotate 15 degrees

④ Step 1: 48 ft. lbs.
Step 2: Rotate 30 degrees
Step 3: Rotate 15 degrees

42372-CADI-C08

BRAKE SPECIFICATIONS

All measurements in inches unless noted

| Year | Model | | Brake Disc | | | Minimum Lining Thickness | Brake Caliper | |
			Original Thickness	Minimum Thickness	Maximum Runout		Bracket Bolts (ft. lbs.)	Mounting Bolts (ft. lbs.)
2003	CTS	F	1.267	1.209	0.002	0.039	136	46
		R	1.020	0.944	0.002	0.039	88	44

F: Front

R: Rear

42372-CADI-C09

WHEEL ALIGNMENT

| Year | Model | | Caster | | Camber | | Toe-in (in.) | Steering Axis Inclination (Deg.) |
			Range (+/-Deg.)	Preferred Setting (Deg.)	Range (+/-Deg.)	Preferred Setting (Deg.)		
2003	CTS	F	1.00	+5.10	0.50	-0.50	0.02 +/- 0.02	—
		R	—	—	0.40	-1.00	0.2+/-0.2	—

42372-CADI-C10

TIRE, WHEEL AND BALL JOINT SPECIFICATIONS

| Year | Model | OEM Tires | | Tire Pressures (psi) | | Wheel Size | Ball Joint Inspection |
		Standard	Optional	Front	Rear		
2003	CTS	P225/55R16	P225/50R17	32	32	7-JJ	0.125 in. ①

OEM: Original Equipment Manufacturer

PSI: Pounds Per Square Inch

① Replace if vertical or horizontal movement exceeds specification

42372-CADI-C11

SCHEDULED MAINTENANCE INTERVALS
GM D BODY—CADILLAC CTS

TO BE SERVICED	TYPE OF SERVICE	VEHICLE MILEAGE INTERVAL (x1000)																			
		5	10	15	20	25	30	35	40	45	50	55	60	65	70	75	80	85	90	95	100
Engine oil & filter	R	✓	✓	✓	✓	✓	✓	✓	✓	✓	✓	✓	✓	✓	✓	✓	✓	✓	✓	✓	✓
Rotate tires	S/I	✓		✓			✓		✓			✓			✓			✓		✓	
Brake hoses	S/I	✓				✓		✓		✓		✓				✓		✓		✓	
Passenger Compartment Air Filter (Pollen Filter)	R			✓			✓			✓			✓			✓			✓		
Air filter element	S/I			✓			✓			✓			✓			✓				✓	
	R						✓						✓							✓	
Fuel tank, cap and lines	S/I						✓						✓								
Rear axle fluid level	S/I						✓						✓								
Automatic transaxle fluid & filter ①	R										✓										✓
Accessory drive belt(s)	S/I												✓								
Spark plugs ②	R																				✓
Ignition cables	S/I																				✓
Camshaft timing belt ③	R												✓								✓
Fuel filter	R																				✓
Engine coolant ④	R																				

R: Replace

S/I: Service or Inspect

① Automatic transaxle fluid & filter: replace at 50,000 miles (83,000 km) if the vehicle has experienced severe service usage.

② Platinum tip spark plugs: replace every 100,000 miles.

③ Replace at 60,000 miles (96,000 km) if the engine was driven without an engine coolant heater being used and where temperatures fall below -20°F (-28°C).
    Otherwise, replace the belt at 100,000 miles (160,000 km).

④ Engine coolant: replace every 150,000 miles. Use O.E. specified (DEX-COOL™) coolant only. If any silicate coolant is used, the service interval is every 30,000 miles.

FREQUENT OPERATION MAINTENANCE (SEVERE SERVICE)

If a vehicle is operated under any of the following conditions it is considered severe service:

- Extremely dusty areas.
- 50% or more of the vehicle operation is in 32°C (90°F) or higher temperatures, or constant operation in temperatures below 0°C (32°F).
- Prolonged idling (vehicle operation in stop and go traffic).
- Frequent short running periods (engine does not warm to normal operating temperatures).
- Police, taxi, delivery usage or trailer towing usage.

Oil & oil filter: change every 5000 miles

Rotate tires at 5000 miles, then every 10,000 miles.

Air filter element: service or inspect every 15,000 miles.

Camshaft timing belt: change every 60,000 miles for severe service.

42372-CADI-C12

PRECAUTIONS

Before servicing any vehicle, please be sure to read all of the following precautions, which deal with personal safety, prevention of component damage, and important points to take into consideration when servicing a motor vehicle:

• Never open, service or drain the radiator or cooling system when the engine is hot, serious burns can occur from the steam and hot coolant.

• Observe all applicable safety precautions when working around fuel. Whenever servicing the fuel system, always work in a well-ventilated area. Do not allow fuel spray or vapors to come in contact with a spark, open flame or excessive heat (a hot drop light, for example). Keep a dry chemical fire extinguisher near the work area. Always keep fuel in a container specifically designed for fuel storage; also, always properly seal fuel containers to avoid the possibility of fire or explosion. Refer to the additional fuel system precautions later in this section.

• Fuel injection systems often remain pressurized, even after the engine has been turned **OFF**. The fuel system pressure must be relieved before disconnecting any fuel lines. Failure to do so may result in fire and/or personal injury.

• Brake fluid often contains polyglycol ethers and polyglycols. Avoid contact with the eyes and wash your hands thoroughly after handling brake fluid. If you do get brake fluid in your eyes, flush your eyes with clean, running water for 15 minutes. If eye irritation persists, or if you have taken brake fluid internally, seek medical assistance IMMEDIATELY.

• The EPA warns that prolonged contact with used engine oil may cause a number of skin disorders, including cancer! You should make every effort to minimize your exposure to used engine oil. Protective gloves should be worn when changing oil. Wash your hands and any other exposed skin areas as soon as possible after exposure to used engine oil. Soap and water, or waterless hand cleaner should be used.

• All new vehicles are now equipped with an air bag system. The system must be disabled before performing service on or around system components, steering column, instrument panel components, wiring and sensors. Failure to follow safety and disabling procedures could result in accidental air bag deployment, possible personal injury and unnecessary system repairs.

• Always wear safety goggles when working with, or around, the air bag system. When carrying a non-deployed air bag, be sure the bag and trim cover are pointed away from your body. When placing a non-deployed air bag on a work surface, always face the bag and trim cover upward, away from the surface. This will reduce the motion of the module if it is accidentally deployed. Refer to the additional air bag system precautions later in this section.

• Clean, high quality brake fluid from a sealed container is essential to the safe and proper operation of the brake system. You should always buy the correct type of brake fluid for your vehicle. If the brake fluid becomes contaminated, completely flush the system with new fluid. Never reuse any brake fluid. Any brake fluid that is removed from the system should be discarded. Also, do not allow any brake fluid to come in contact with a painted surface; it will damage the paint.

• Never operate the engine without the proper amount and type of engine oil; doing so WILL result in severe engine damage.

• Timing belt maintenance is extremely important! Many models utilize an interference-type, non-freewheeling engine. If the timing belt breaks, the valves in the cylinder head may strike the pistons, causing potentially serious (also time-consuming and expensive) engine damage. Refer to the maintenance interval charts in the front of this section for the recommended replacement interval for the timing belt, and to the timing belt procedure in this section for belt replacement and inspection.

• Disconnecting the negative battery cable on some vehicles may interfere with the functions of the on-board computer system(s) and may require the computer to undergo a relearning process once the negative battery cable is reconnected.

• When servicing drum brakes, only disassemble and assemble one side at a time, leaving the remaining side intact for reference.

ENGINE REPAIR

➡**Disconnecting the negative battery cable on some vehicles may interfere with the operation of the on board computer system. The computer may undergo a relearning process once the negative battery cable is reconnected.**

Alternator

REMOVAL

1. Before servicing the vehicle, refer to the precautions in the beginning of this section.
2. Remove or disconnect the following:
   • Negative battery cable
   • Intake air resonator
   • Drive belt
   • Coolant heater, if equipped
   • Alternator cooling duct

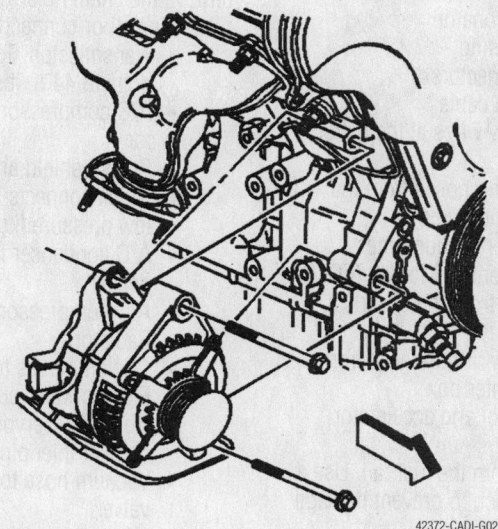

Alternator mounting

42372-CADI-G02

- Alternator electrical connectors
- Alternator

INSTALLATION

1. Install or connect the following:
   - Alternator. Tighten the bolts to 26 ft. lbs. (35 Nm).
   - Cooling duct
   - Alternator electrical connectors
   - Coolant heater, if equipped
   - Drive belt
   - Intake air resonator
   - Negative battery cable
2. Perform a charging system test and verify that the system is operating properly.

Ignition Timing

ADJUSTMENT

➡The 3.2L DOHC engine used in the CTS utilizes a Distributorless Ignition System (DIS). No ignition timing adjustment is possible.

Engine Assembly

REMOVAL & INSTALLATION

1. Before servicing the vehicle, refer to the precautions in the beginning of this section.
2. Drain the cooling system.
3. Drain the engine oil.
4. Recover the A/C refrigerant.
5. Relieve the fuel system pressure.
6. Remove or disconnect the following:
   - Battery
   - Wiper arms
   - Left and right air inlet grilles
   - Hood
   - Intake air resonator
   - Air filter housing
   - Harness connectors
   - Body ground cable
   - Power supply wires at the battery cable end
   - Power steering hoses from the power steering pump
   - Brake booster vacuum lines
   - Electronic Control Module (ECM) and harness from the electrical center
   - Relays and wiring harness from electrical center box
   - Cruise control and accelerator cables
   - Fuel lines from the fuel rail. Use a backup wrench to prevent damage to the fuel rail.
   - Coolant hose from the throttle body

   - Vacuum hose from the purge valve on the engine ventilation chamber
   - Vacuum hose from the heater control valve
   - Coolant reservoir hose from the coolant inlet pipe
   - Electric water pump with hoses attached
   - Coolant hoses from the heater core
   - Radiator
   - Refrigerant line from the A/C compressor and compressor bracket
   - Condenser line from the A/C condenser
   - A/C quick-connect hose near the low pressure service valve
   - Splash shield
   - A/C compressor electrical connector

➡Support the engine whenever the transmission is removed. The motor mounts are silicone filled and do not provide sufficient rigidity to support the engine with the transmission removed.

7. Remove the transmission from the vehicle.
8. Attach a hoist to the engine lifting eyes. Raise the engine slightly and remove the engine support fixture.
9. Remove the engine mount nuts and lift the engine out of the vehicle. Raise the engine slowly after being certain all wiring, cables and hoses have been removed.

To install:
10. Install the engine.
11. Install the engine mount. Torque the upper motor mount nuts to 30 ft. lbs. (40 Nm). Torque the lower nuts to 41 ft. lbs. (55 Nm).
12. Install an engine support fixture and remove the chain hoist.
13. Install or connect the following:
   - Transmission. Torque the mounting bolts to 44 ft. lbs. (60 Nm).
   - A/C compressor electrical connector
   - Splash shield and tighten securely
   - Quick connects for the high and low pressure fittings
   - A/C condenser line to the condenser
   - A/C compressor line
   - Radiator
   - Coolant hoses to the heater core
   - Electric water pump
   - Coolant reservoir hose to the coolant inlet pipe
   - Vacuum hose to the heater control valve
   - Vacuum hose for the purge valve on the engine ventilation chamber

   - Coolant hose to the throttle body
   - Fuel return and supply lines. Torque the lines to 11 ft. lbs. (15 Nm).
   - Cruise control and accelerator cables to the throttle body
   - Relays and wiring harness to the electrical center box
   - ECM
   - Brake booster vacuum lines
   - Power steering hoses to the steering pump. Torque the discharge hose fasteners to 21 ft. lbs. (28 Nm).
   - Power supply wires at the battery cable ends
   - Body ground cable
   - Red, white and black wiring harness connectors
   - Air filter housing
   - Intake air resonator
   - Hood
   - Left and right air inlet grilles
   - Wiper arms
   - Battery
   - Negative battery cable
14. Fill and bleed the power steering system.
15. Refill the engine oil.

➡An oil filter change is recommended.

16. Fill and bleed the coolant system.

➡When refilling the coolant system add 2 crushed engine coolant supplement sealant pellets (PN 3634621) into the reservoir.

17. Recharge the A/C system and check for leaks.
18. Start the vehicle and inspect for leaks.

Water Pump

REMOVAL & INSTALLATION

1. Before servicing the vehicle, refer to the precautions in the beginning of this section.
2. Drain the coolant.
3. Remove or disconnect the following:
   - Negative battery cable
   - Intake air resonator
   - Water pump pulley bolts, loosen only
   - Front timing belt cover
   - Water pump pulley
   - Water pump

To install:
4. Install or connect the following:
   - Water pump with a new O-ring. Torque the bolts to 18 ft. lbs. (25 Nm).

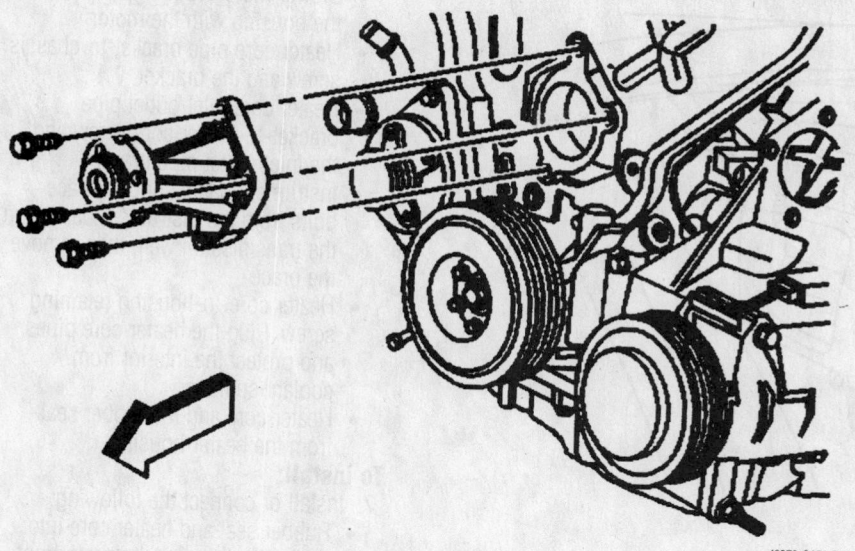

Exploded view of the water pump mounting

- Water pump pulley
- Front timing belt cover
- Intake air resonator
- Negative battery cable

5. Fill the cooling system through the reservoir tank.

➡ When refilling the cooling system, add 2 crushed engine coolant supplement sealant pellets (PN 3634621) into the coolant reservoir.

6. Start the vehicle and inspect the coolant systems for leaks.

Heater Core

REMOVAL & INSTALLATION

1. Disable the SIR system.
2. Disconnect the negative battery cable.
3. Drain the cooling system.
4. Remove the heater hose quick connects from the heater core pipes by performing the following procedure:

 a. At the passenger side, raise the air inlet screen and open the access door near the pollen filter.

 b. Unlock the quick connect collars by squeezing the tabs and carefully pulling back on the tabs to disconnect the sleeve.

 c. If green assembly marks are attached, discard them.

✳✳ WARNING

The front wheels must be maintained in the straight-ahead position and the steering column must be in the LOCK

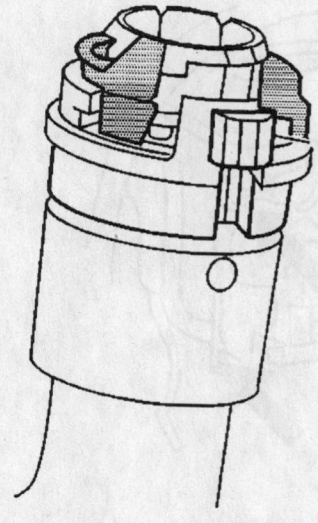

View of a heater hose connector—CTS

position. Failure to do so will cause improper alignment of some components during installation and may result to damage to the SIR coil assembly.

5. Remove the steering column by removing or disconnecting the following:

- Instrument panel driver knee bolster energy absorber and sound insulator
- Steering column electrical connector(s)
- Coupler bolt from the lower steering column connection and slightly separate the coupler to aid in the shaft removal
- Using a chisel and a hammer, rotate the forward support strap shear nut and bolt (located under the steering column) counterclockwise in order to remove them
- Steering column-to-rear support bracket bolt
- Pull the steering column straight back and through the dash panel and remove it from the vehicle.

✳✳ WARNING

Handle the steering column with care for it is very susceptible to damage. Dropping, leaning or hammering on it could cause damage to its collapsible design.

6. Remove the instrument panel carrier by removing or disconnecting the following:
- Windshield pillar moldings
- Access panel, the air deflector outlet screw, the air deflector outlet and the air outlet duct, located at the right-side of the instrument panel

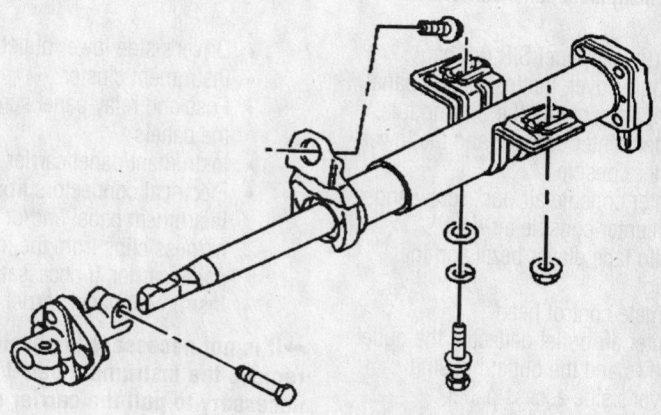

View of the steering column assembly—CTS

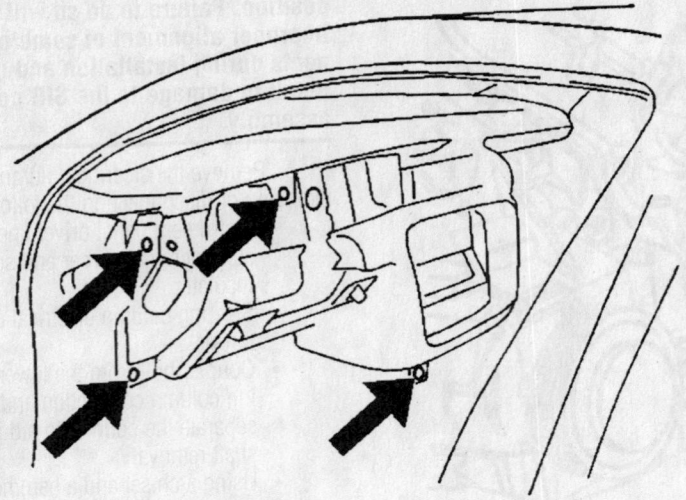

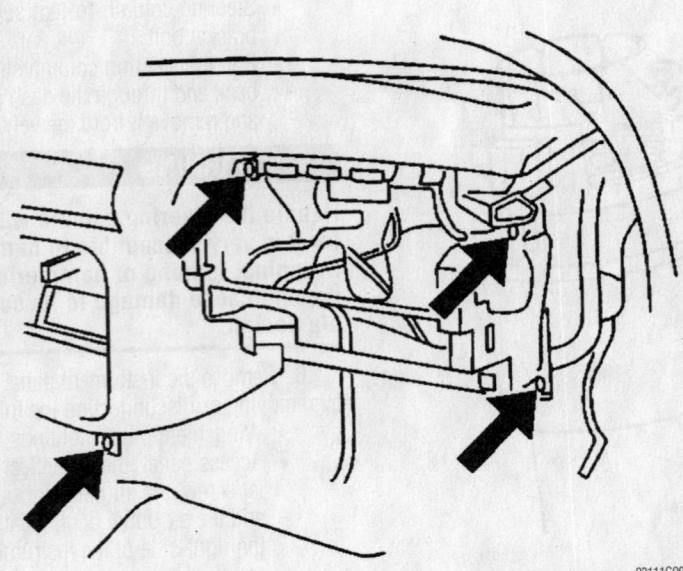

93111G09

View of the instrument panel carrier bolt locations—CTS

- Instrument panel SIR (air bag) module cover, the instrument panel compartment and the SIR module
- Upper center console and the lower center console
- Center console air duct screw and the center console air duct
- Radio tape player bezel and the radio
- Climate control head
- Center air outlet deflector, the outlet screws and the outlet housing
- Driver's side access panel
- Driver's side air outlet deflector, the outlet screw and the outlet housing

- Driver's side lower outlet duct
- Instrument cluster
- Fuse and relay panel screws and the panels
- Instrument panel carrier bolts
- Electrical connectors from the instrument panel and/or the wiring harness clips from the instrument panel carrier, if necessary
- Instrument panel carrier

➡ **It is not necessary to physically remove the instrument panel but it is necessary to pull the carrier rearward to enable the heater core to be removed from its housing.**

- Blower motor housing screws and the housing with the motor
- Heater core pipe bracket-to-chassis screw and the bracket
- Heater core inlet/outlet pipe bracket-to-heater core screw and the inlet/outlet pipe bracket
- Instrument panel support brace bolts from the instrument panel and the transmission well, then remove the brace
- Heater core-to-housing retaining screw. Plug the heater core pipes and protect the interior from coolant spills
- Heater core and the rubber seal from the heater housing

To install:

7. Install or connect the following:
- Rubber seal and heater core into the heater housing; be careful not to damage the fins
- Heater core-to-housing retainer screw
- Instrument panel support brace to the transmission well and instrument panel, then torque the bolts to 16 ft. lbs. (22 Nm)
- Heater core inlet/outlet pipe bracket to the heater core (using a new O-ring lightly coated with coolant), and torque the screw to 44 inch lbs. (5 Nm)
- Heater core pipe bracket with the screw to the chassis, be careful not to strip the screw
- Blower motor/housing assembly and torque the screws to 35 inch lbs. (4 Nm)
- Instrument panel carrier
- Instrument panel carrier by reversing the removal procedures and torque the instrument panel carrier bolts to 16 ft. lbs. (22 Nm)

8. Remove the steering column by installing or connecting the following:
- Steering column into the vehicle, through the dash panel and into the lower steering coupling
- Hand start the rear support bracket bolt, the forward support strap nut and shear bolt
- Rear support bracket bolt. Torque to 16 ft. lbs. (22 Nm)
- Forward support strap nut. Torque to 16 ft. lbs. (22 Nm)
- New forward support shear bolt. Torque to 15 ft. lbs. (21 Nm)
- Lower steering column shaft bolt and torque to 16 ft. lbs. (22 Nm)
- Steering column electrical connector(s)

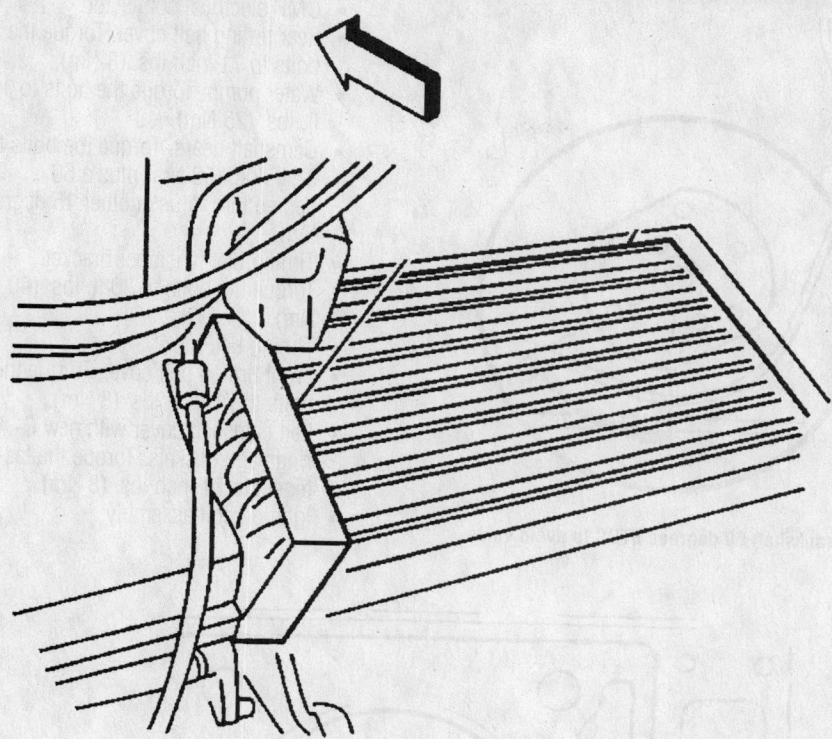

View of a heater core and seal—CTS

- Sound insulator and the instrument panel driver knee bolster energy absorber

9. Connect the heater hoses to the heater core pipes by performing the following procedure:

a. If not attached to the quick connect, discard the green assembly marker(s).

b. Push the quick connects into the pipes until they are fully seated.

c. Squeeze the locking tabs and press the retaining sleeve into the locked position.

d. At the passenger side, close the access door near the pollen filter and lower the air inlet screen.

10. Refill the cooling system by performing the following procedure:

a. Add a 50/50 mixture of water and DEX-COOL® antifreeze to the KALT/COLD mark (seam) on the surge tank.

b. Start the engine and allow it to idle for 1 min.

c. Add more coolant to the surge tank as necessary.

d. Install the radiator sure tank cap.

e. Cycle the engine, from idle to 3000 rpm, in 30 second intervals, until the engine reaches normal operating temperatures.

93111G12

➡**The cooling system will bleed itself automatically during warm-up.**

f. Turn the engine OFF and recheck the coolant level when the engine is cool

11. Connect the negative battery cable.

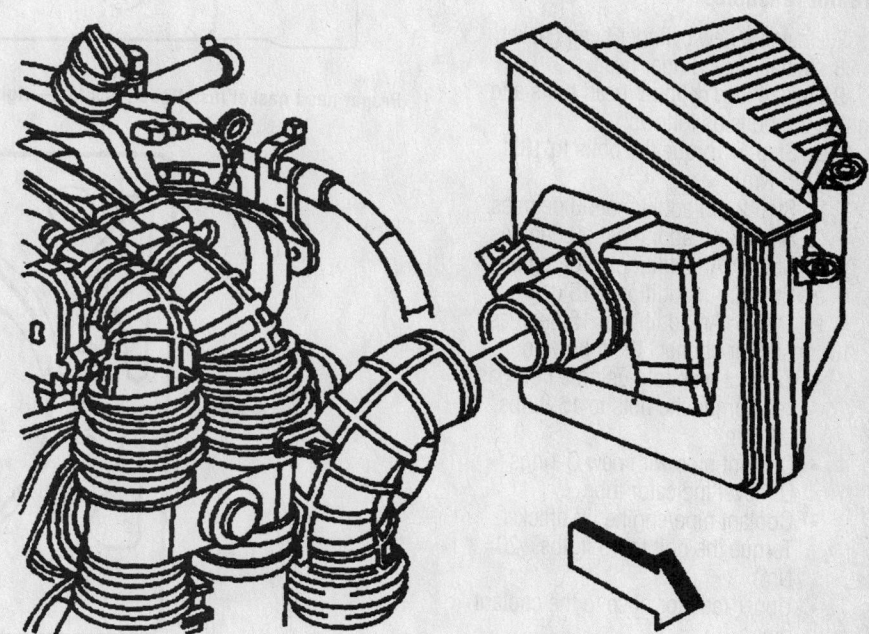

Intake air resonator

42372-CADI-G04

12. Enable the SIR system.
13. Reprogram the necessary accessories.

REMOVAL & INSTALLATION

1. Before servicing the vehicle, refer to the precautions in the beginning of this section.

2. Drain the engine coolant.

3. Relieve the fuel system pressure.

4. Remove or disconnect the following:

- Intake plenum
- Intake air resonator
- Intake manifold and spacer
- Both Engine Coolant Temperature (ECT) sensor electrical connectors
- Coolant crossover
- Ignition coil assembly
- Camshaft cover
- Front timing belt cover

5. Position the crankshaft 60 degrees Before Top Dead Center (BTDC) to avoid contact between the valves and the pistons.

6. Remove or disconnect the following:

- Timing belt
- Timing belt tensioner bracket
- Four camshaft gears
- Water pump
- Rear timing belt cover
- Camshaft Position (CMP) sensor electrical connector

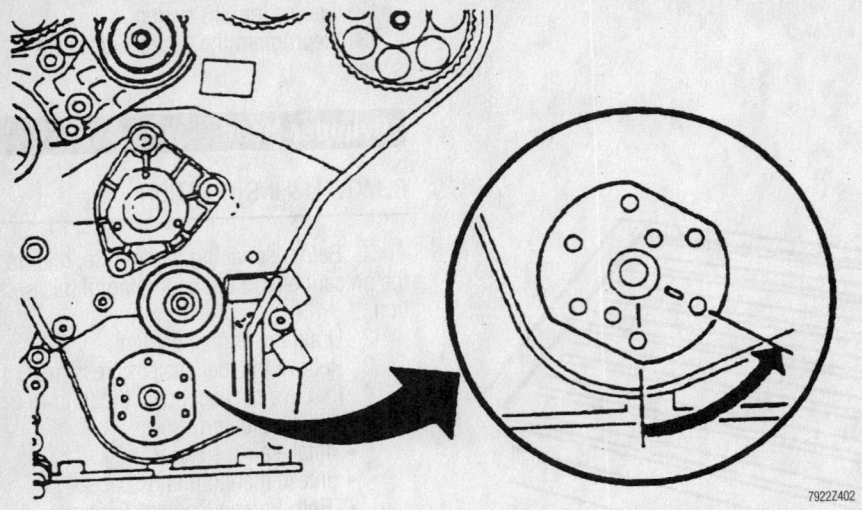

- CMP electrical connector
- Rear timing belt cover. Torque the bolts to 71 inch lbs. (8 Nm).
- Water pump. Torque the bolts to 18 ft. lbs. (25 Nm).
- Camshaft gears. Torque the bolts to 37 ft. lbs. (50 Nm) plus a 60 degree turn, plus another 15 degree turn.
- Timing belt tensioner bracket. Torque the bolts to 30 ft. lbs. (40 Nm).
- Timing belt
- Front timing belt cover. Torque the bolts to 71 inch lbs. (8 Nm).
- Left camshaft cover with new O-rings and gaskets. Torque the fasteners to 71 inch lbs. (8 Nm).
- Ignition coil assembly

Before removing the timing belt, be sure to turn the crankshaft 60 degrees BTDC to avoid valve-to-piston contact and subsequent engine damage.

- Exhaust camshaft
- Coolant pipe/engine lift bracket from the cylinder head
- Oil level indicator tube
- Upper radiator hose from the coolant pipe
- Coolant pipe
- Exhaust manifold from the cylinder head
- Cylinder head

To install:

➡ **Be sure to use new cylinder head bolts when installing the cylinder head. The old bolts have been stretched and are not reusable.**

7. Install a new cylinder head gasket.
8. Install the cylinder head.
9. Install new cylinder head bolts and tighten the bolts as follows:
   a. Step 1: Torque the bolts to 18 ft. lbs. (25 Nm).
   b. Step 2: An additional 90 degrees.
   c. Step 3: An additional 90 degrees.
   d. Step 4: An additional 90 degrees.
   e. Step 5: An additional 15 degrees.
   f. Step 6: An additional 15 degrees.
10. Install or connect the following:
    - Exhaust manifold using a new gasket. Torque the nuts to 15 ft. lbs. (20 Nm).
    - Coolant pipe with new O-rings
    - Oil level indicator tube
    - Coolant pipe/engine lift bracket. Torque the bolt to 15 ft. lbs. (20 Nm).
    - Upper radiator hose to the coolant pipe
    - Exhaust camshaft. Torque the bearing cap bolts to 71 inch lbs. (8 Nm).

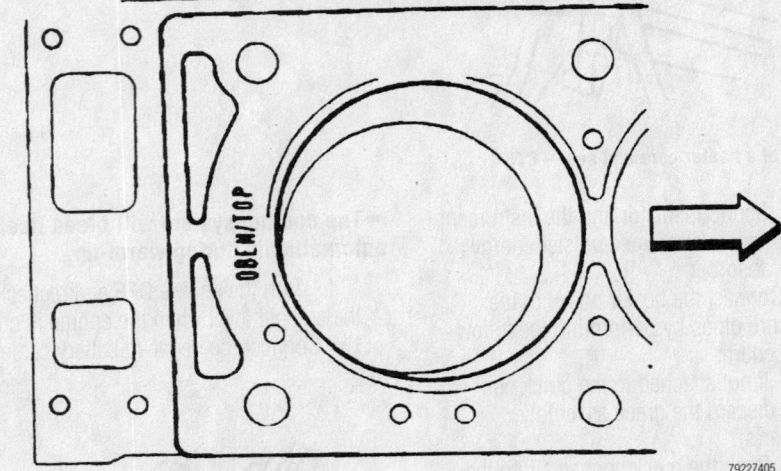

Proper head gasket installation position—right cylinder head

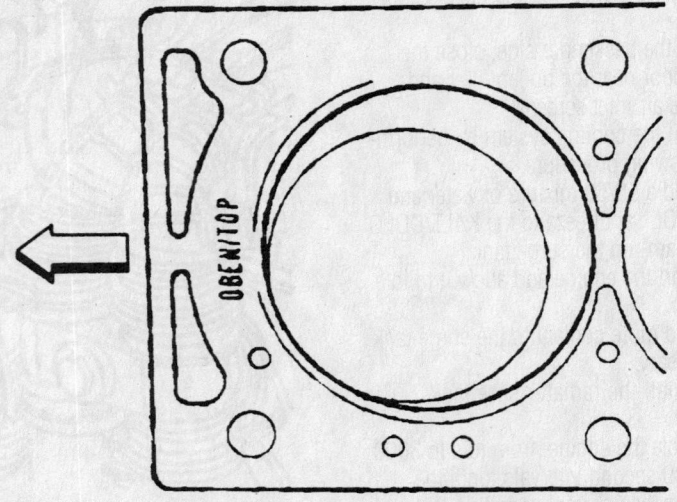

Proper head gasket installation position—left cylinder head

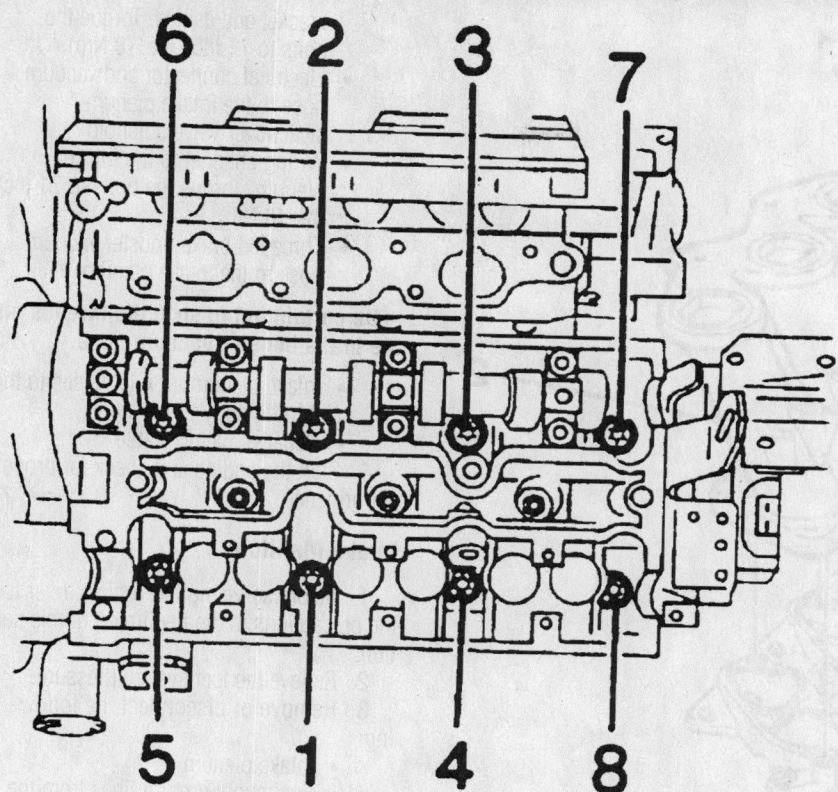

Cylinder head torque sequence—left cylinder head

7922Z403

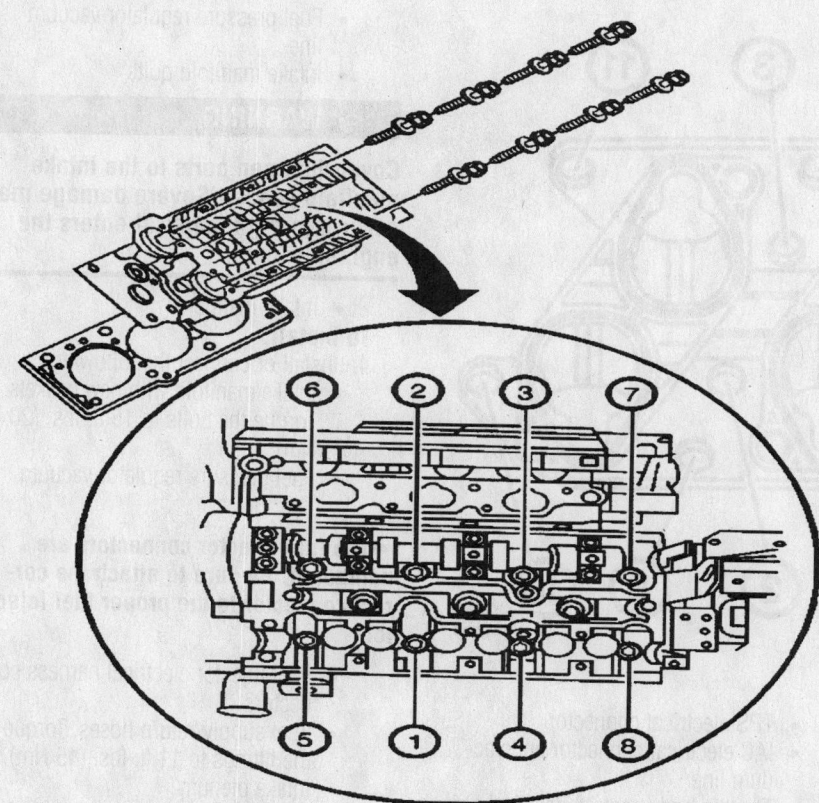

Cylinder head torque sequence—right cylinder head

9346ZGA5

- Coolant crossover. Torque the bolts to 22 ft. lbs. (30 Nm).
- Both ECT sensor electrical connectors
- Intake manifold spacer. Torque the bolts in a spiral direction, starting from the inside and working outward to 15 ft. lbs. (20 Nm).
- Intake manifold. Torque the bolts to 15 ft. lbs. (20 Nm).
- Intake air resonator. Torque the nuts to 27 inch lbs. (3 Nm).
- Intake plenum. Torque the bolts to 71 inch lbs. (8 Nm).
- Negative battery cable

➡An oil and filter change is recommended.

11. Refill the cooling system.
12. Start the engine and inspect for leakage.
13. Inspect all fluid levels and top off, if necessary.

Intake Manifold Assembly

REMOVAL & INSTALLATION

Intake Plenum

1. Before servicing the vehicle, refer to the precautions in the beginning of this section.
2. Remove or disconnect the following:
   - Negative battery cable
   - Intake plenum air inlet hoses from the throttle body
   - Brake booster vacuum hose from the intake plenum
   - Wiring harness channel from the plenum
   - Switch over valve electrical connector and vacuum line
   - Accelerator, cruise control cables and the bracket from the throttle body
   - Throttle body control electrical connector
   - Idle Air Control (IAC) inlet hose and electrical connector
   - Throttle Position Sensor (TPS) electrical connection
   - Throttle body from the intake plenum
   - Crankcase vent tube adapter and cover from the intake plenum
   - Intake plenum and O-rings

To install:
3. Install or connect the following:
   - Intake plenum with new O-rings. Torque the bolts to 71 inch lbs. (8 Nm).

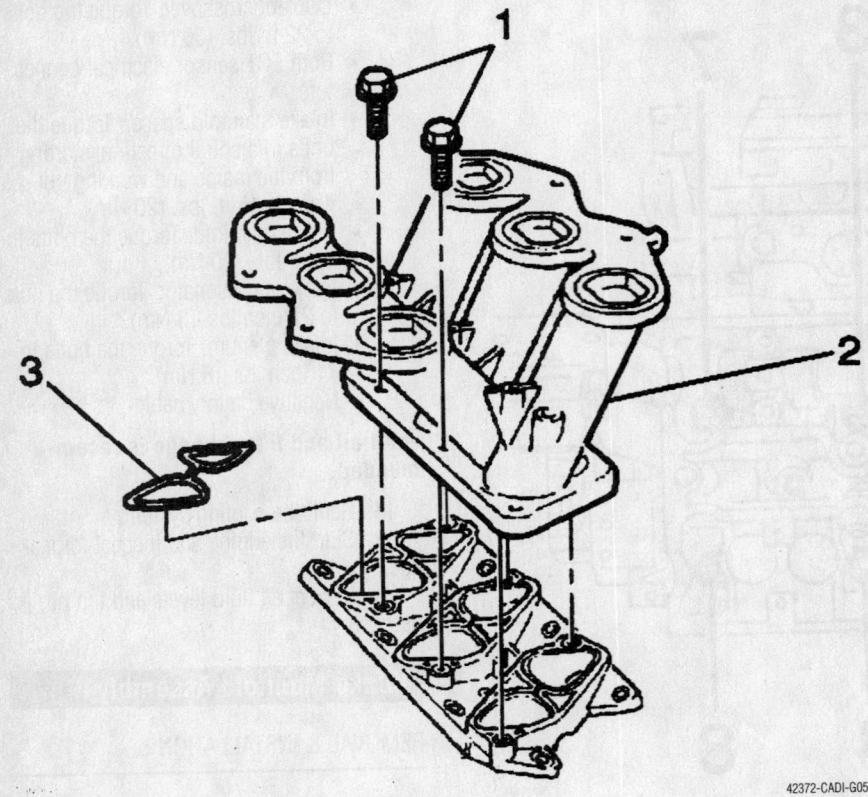

Upper intake exploded view

42372-CADI-G05

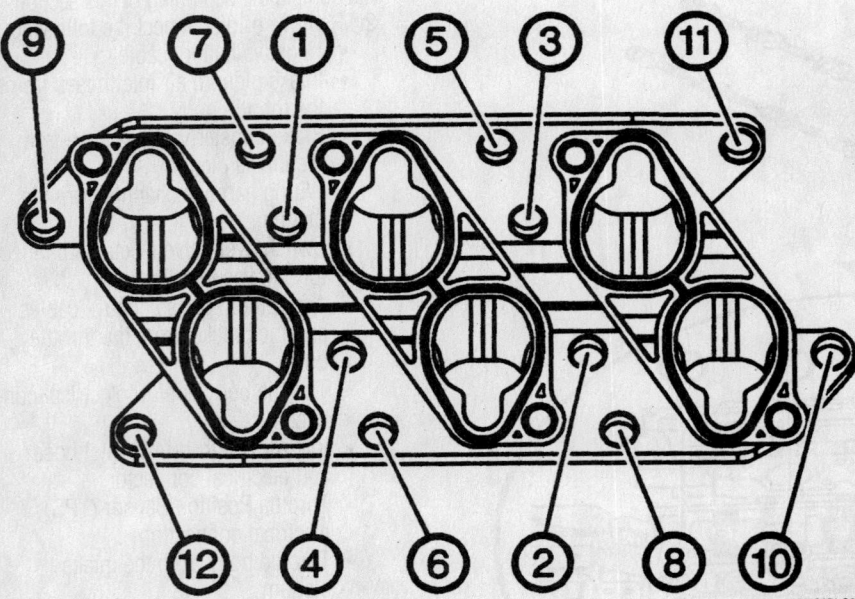

Lower intake torque sequence

42372-CADI-G06

- Crankcase vent tube adapter to the intake plenum. Torque the fastener to 71 inch lbs. (8 Nm).
- Throttle body to the intake plenum. Torque the fasteners to 106 inch lbs. (12 Nm).

- TPS electrical connector
- IAC electrical connector and vacuum line
- Throttle body control electrical connector
- Accelerator and cruise control

bracket and cables. Torque the bolts to 71 inch lbs. (8 Nm).
- Electrical connector and vacuum hose to the intake plenum switchover valve solenoid
- Wiring channel to the intake plenum. Torque the bolts to 71 inch lbs. (8 Nm).
- Threaded brake booster vacuum hose to the intake plenum

➡**Be certain not to strip the threads on the brake booster vacuum hose.**

- Intake plenum air inlet hoses to the throttle body
- Negative battery cable

4. Start the engine and check for proper performance.

Intake Manifold

1. Before servicing the vehicle, refer to the precautions in the beginning of this section.

2. Relieve the fuel system pressure.

3. Remove or disconnect the following:

- Intake plenum
- Fuel supply/return lines from the fuel rail by loosening the fittings
- Fuel injector electrical harness connector
- Fuel pressure regulator vacuum line
- Intake manifold bolts

✷✷ WARNING

Cover the open ports to the intake manifold spacer. Severe damage may occur if foreign material enters the engine.

- Intake manifold

To install:

4. Install or connect the following:

- Intake manifold with new gaskets. Torque the bolts to 15 ft. lbs. (20 Nm).
- Fuel pressure regulator vacuum line

➡**The fuel injector connectors are numbered. Be sure to attach the correct connector to the proper fuel injector.**

- Fuel injector electrical harness connectors
- Fuel supply/return hoses. Torque the fittings to 11 ft. lbs. (15 Nm).
- Intake plenum
- Negative battery cable

5. Start the vehicle and inspect for leaks.

Exhaust Manifold

REMOVAL & INSTALLATION

Left Side

1. Before servicing the vehicle, refer to the precautions in the beginning of this section.
2. Drain the engine coolant.
3. Remove or disconnect the following:

- Engine assembly
- Exhaust manifold lower/upper heat shields
- Secondary Air Injection (AIR) pipe from the exhaust manifold
- Coolant pipe and engine lift bracket from the cylinder head
- Oil level indicator tube
- Exhaust manifold

To install:

4. Install or connect the following:

- Exhaust manifold with a new gasket. Torque the nuts to 15 ft. lbs. (20 Nm).
- Oil level indicator tube. Torque the bolt to 15 ft. lbs. (20 Nm).
- Engine lifting bracket and coolant pipe. Torque the bolt to 15 ft. lbs. (20 Nm).
- Air injection pipe to the exhaust manifold. Torque the bolts to 15 ft. lbs. (20 Nm).
- Exhaust manifold upper/lower heat shields. Torque the bolts to 71 inch lbs. (8 Nm).
- Engine assembly

5. Refill the engine coolant.
6. Start the vehicle and inspect for any leaks.

Right Side

1. Before servicing the vehicle, refer to the precautions in the beginning of this section.
2. Drain the engine coolant.
3. Remove or disconnect the following:
- Transmission assembly
- Coolant intake pipe
4. Remove or disconnect the following:
- Exhaust manifold lower heat shield
- Exhaust manifold lower rear nuts
- Exhaust manifold upper rear heat shield bolt
- Catalytic converter nuts from the exhaust manifold
- Drive belt tensioner
- Exhaust manifold front two nuts
- Exhaust manifold upper heat shield bolts

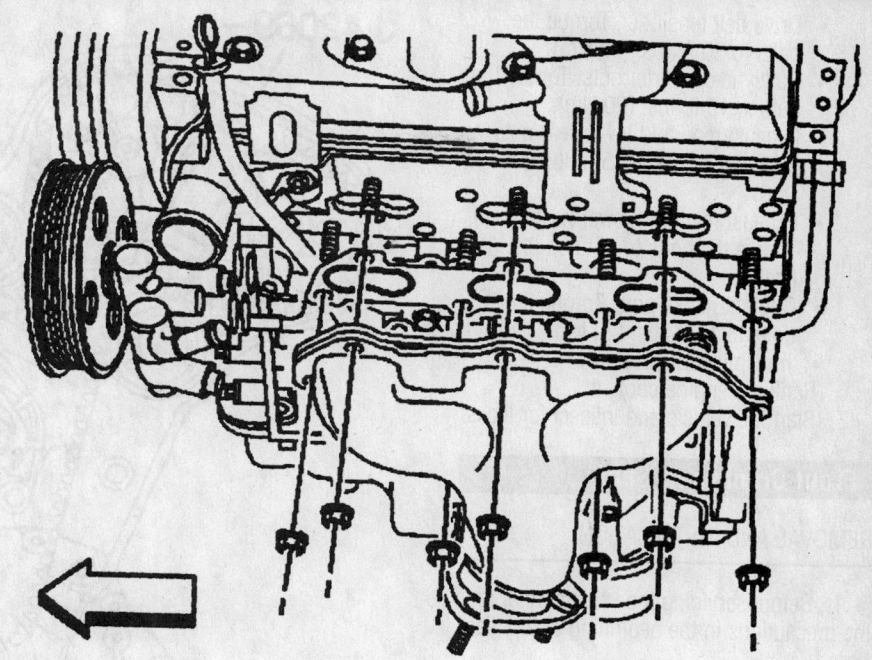

Left exhaust manifold

42372-CADI-G07

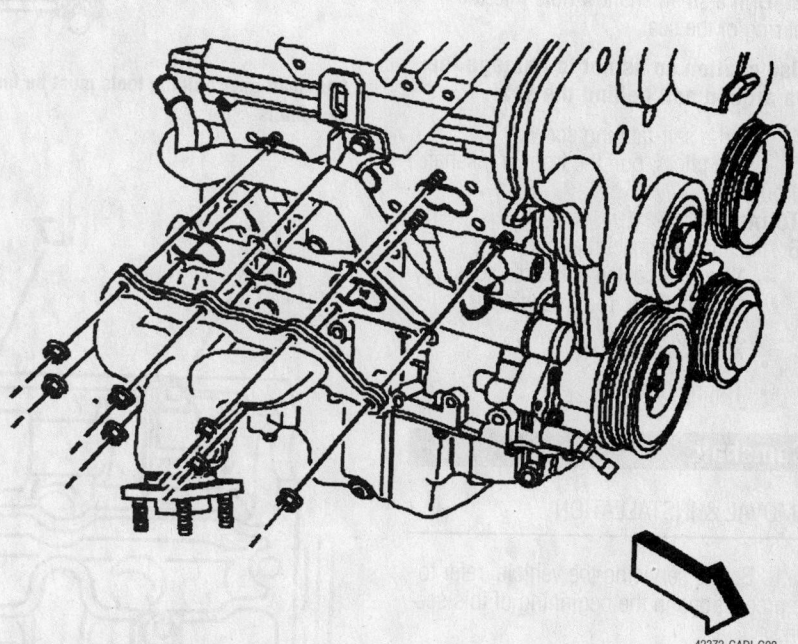

Right exhaust manifold

42372-CADI-G08

➡**It is not necessary to remove the upper heat shield.**

- Secondary Air Injection (AIR) injection pipe from the exhaust manifold
- Exhaust manifold

To install:

5. Install or connect the following:
- Exhaust manifold with a new gasket. Torque the 2 upper nuts to 15 ft. lbs. (20 Nm).

- AIR injection pipe to the exhaust manifold. Torque the bolts to 15 ft. lbs. (20 Nm).
- Upper heat shield bolts. Torque the bolts to 71 inch lbs. (8 Nm).
- Exhaust manifold upper heat shield bolts. Torque the bolts to 71 inch lbs. (8 Nm).
- Exhaust manifold lower front nuts. Torque the nuts to 15 ft. lbs. (20 Nm).

- Drive belt tensioner. Torque the bolts to 30 ft. lbs. (40 Nm).
- Catalytic converter nuts. Torque the nuts to 15 ft. lbs. (20 Nm).
- Exhaust manifold lower rear nuts. Torque the nuts to 15 ft. lbs. (20 Nm).
- Exhaust manifold lower heat shield. Torque the bolts to 71 inch lbs. (8 Nm).
- Coolant intake pipe. Torque the bolts to 15 ft. lbs. (20 Nm).
- Transmission assembly
6. Refill the engine coolant.
7. Start the vehicle and inspect for leaks.

Front Crankshaft Seal

REMOVAL AND INSTALLATION

1. Before servicing the vehicle, refer to the precautions in the beginning of this section.
2. Remove the timing belt and crankshaft gear.
3. Drill a small shallow hole into the steel ring of the seal.

➡ **Use caution so as not to damage the area around and behind the seal.**

4. Insert a self-tapping screw.
5. Using pliers, pull the front crankshaft seal out.

To install:
6. Install or connect the following:
- New seal coated with grease using Seal Installer Tool (such as J35268-A)
- Crankshaft gear
- Timing belt

Camshaft

REMOVAL & INSTALLATION

1. Before servicing the vehicle, refer to the precautions in the beginning of this section.
2. Remove or disconnect the following:
- Negative battery cable
- Intake plenum
- Intake air resonator
- Camshaft cover
- Front timing belt cover
- Timing belt
3. Rotate the crankshaft 60 degrees counterclockwise Before Top Dead Center (BTDC) to prevent valve/piston contact.
4. Use a Camshaft Locking tool to hold the camshaft gears in place and loosen the camshaft gear bolt.

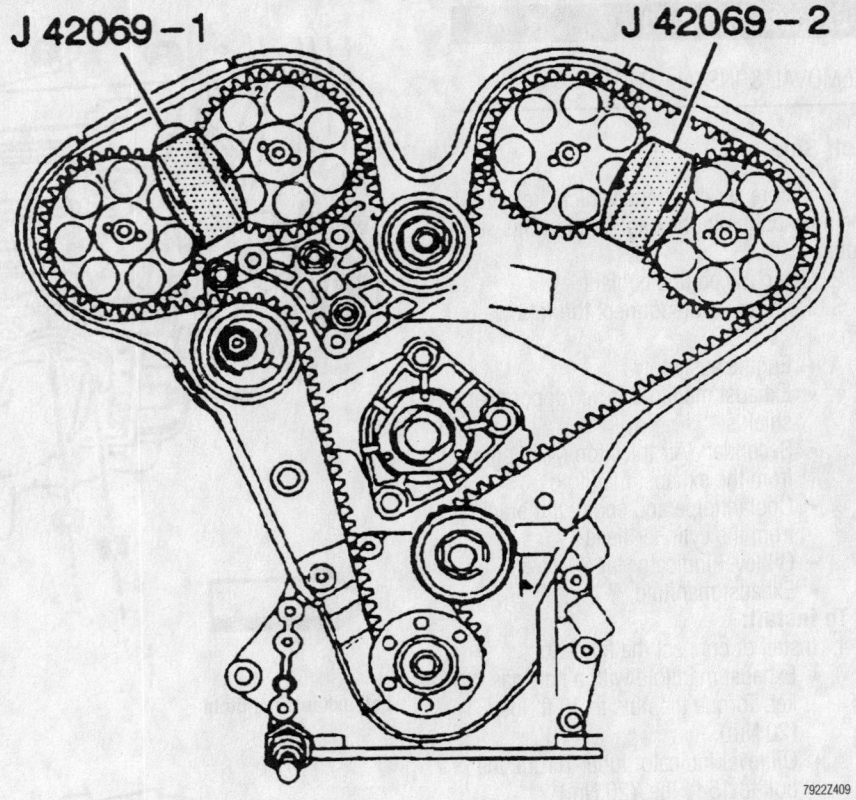

Camshaft gear holding tools must be installed before attempting to remove the gears from the camshafts

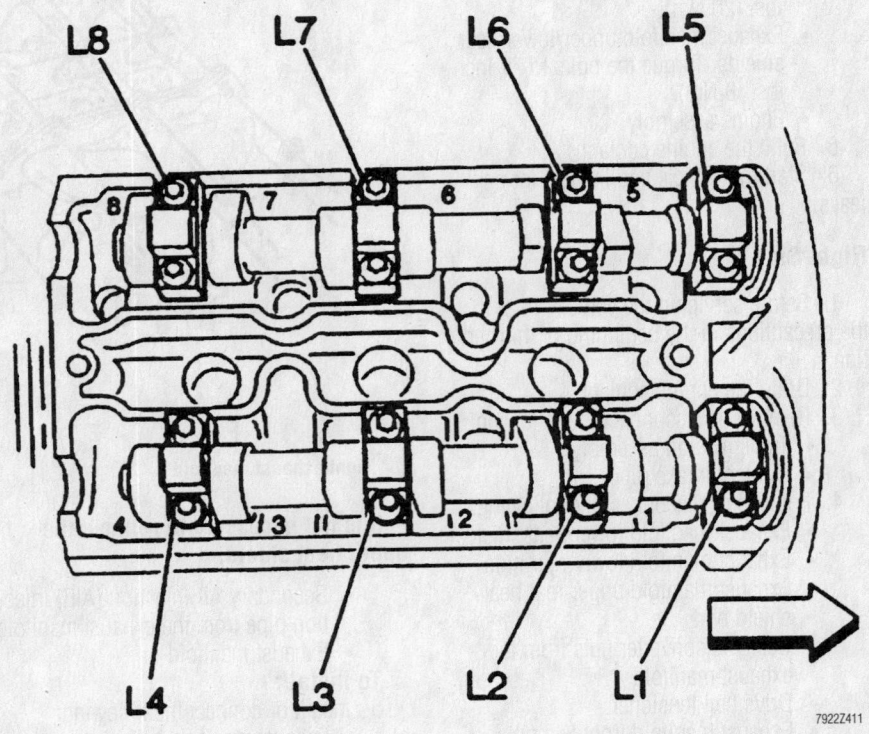

Camshaft bearing cap locations-right cylinder head

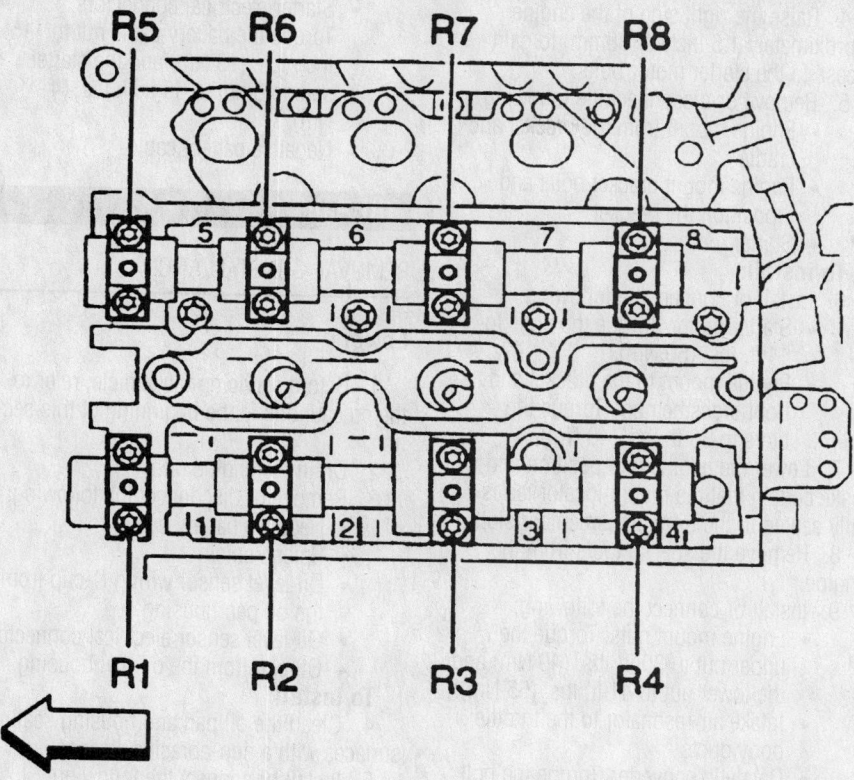

Camshaft bearing cap locations—left cylinder head

79222Z412

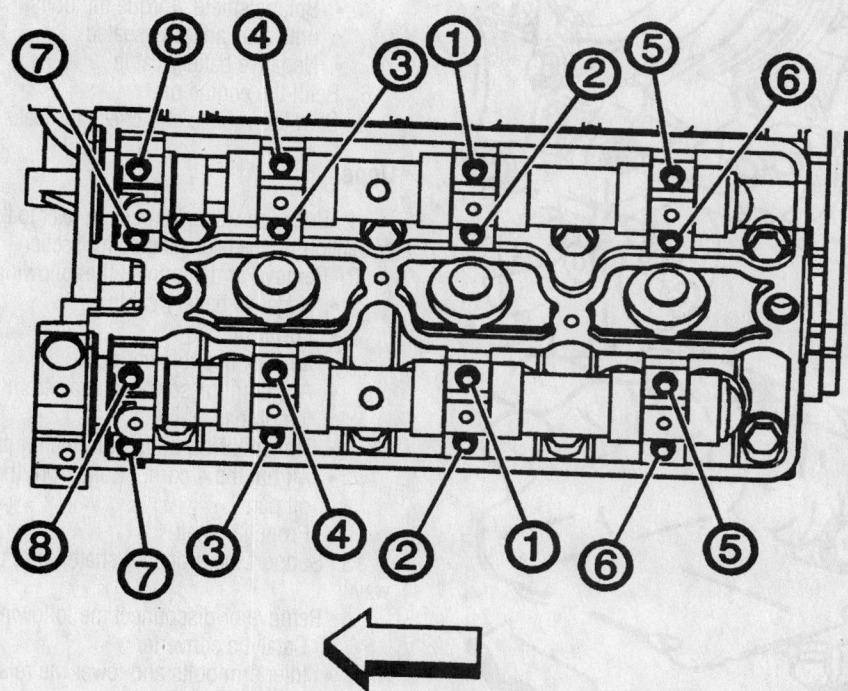

42372-CADI-G09

Camshaft bolt torque sequence

5. Remove the locking tool from the gears.
6. Remove the camshaft gear(s).

➡ **Be certain that the camshaft is not under load from the lifters.**

7. Gradually loosen the camshaft bearing cap bolts sequentially, starting in the center and working outward in a spiral. Note the identification marks on the caps.

8. Remove the camshaft from the cylinder head.

To install:

9. Lubricate the camshaft lobes, lifters and bearing journals with camshaft lubricant.

10. Install the camshaft to the cylinder head.

11. To reduce the lifter load, position the pin on the front of the camshaft as follows:

• 1 o'clock position: right exhaust camshaft
• 11 o'clock position: right intake camshaft
• 12 o'clock position: left exhaust camshaft
• 7 o'clock position: left intake camshaft

12. Place a small amount of Loctite® on the edge of the front bearing cap to ensure a good seal between the cap and surface of the cylinder head. Do not allow the sealer to get into the oil journal of the cap.

※ WARNING

The bearing caps must be installed in their original positions.

13. Install or connect the following:
• Camshaft bearing caps. Torque the bolts in sequence and in several passes to 71 inch lbs. (8 Nm).
• New camshaft seal lubricated with engine oil

➡ **Make certain that the camshaft seal is fully seated.**

• Camshaft gear. Tighten the new bolt to 37 ft. lbs. (50 Nm), plus a 60 degree turn, plus a 15 degree turn.

※ WARNING

A new camshaft gear bolt must be installed. The required tightening method will stretch the bolt making the original bolt unusable.

• Timing belt and adjust as needed
• Front timing belt cover. Torque the bolts to 71 inch lbs. (8 Nm).
• Camshaft cover. Torque the bolts to 71 inch lbs. (8 Nm).

- Intake plenum. Torque the bolts to 71 inch lbs. (8 Nm).
- Intake air resonator
- Negative battery cable

14. Start the vehicle and verify the engine is running properly.

Valve Lash

ADJUSTMENT

➡The valve lash is non-adjustable.

Starter Motor

REMOVAL & INSTALLATION

1. Before servicing the vehicle, refer to the precautions in the beginning of this section.
2. Remove or disconnect the following:
   - Negative battery cable
   - Starter electrical connectors
   - Right hand catalytic converter
   - Right side engine mount nuts
   - Intake air resonator to the throttle body ducts
3. Install an Engine Support Fixture to the right side of the engine.

4. Raise the right side of the engine approximately 1.5 inches (38mm) to gain access to the starter motor bolts.
5. Remove or disconnect the following:
   - Engine mount from the bracket and cradle
   - Engine mount bracket bolts and reposition the bracket
   - Starter motor

To install:
6. Install or connect the following:
   - Starter motor. Torque the bolts to 44 ft. lbs. (60 Nm).
   - Engine mount to the bracket and front crossmember. Torque the bolts to 30 ft. lbs. (40 Nm).
7. Lower the engine into position and make certain that the mount locator tab is fully seated in the front crossmember slot.
8. Remove the special tools from the engine.
9. Install or connect the following:
   - Engine mount nuts. Torque the upper nut to 30 ft. lbs. (40 Nm) and the lower nut to 41 ft. lbs. (55 Nm).
   - Intake air resonator to the throttle body ducts
   - Catalytic converter. Torque the bolt to 25 ft. lbs. (34 Nm) and the nuts to 18 ft. lbs. (25 Nm).

- Starter electrical connectors. Torque the battery cable nut to 115 inch lbs. (13 Nm) and the starter solenoid nut to 35 inch lbs. (4 Nm).
- Negative battery cable

Oil Pan

REMOVAL & INSTALLATION

Lower

1. Before servicing the vehicle, refer to the precautions in the beginning of this section.
2. Drain the engine oil.
3. Remove or disconnect the following:
   - Negative battery cable
   - Splash shield
   - Oil level sensor wiring C-clip from the oil pan housing
   - Oil level sensor electrical connector
   - Oil pan from the oil pan housing

To install:
4. Clean the oil pan and housing sealing surfaces with a non-abrasive cleaner.
5. Install or connect the following:
   - Oil level sensor wire connector to the oil pan housing C-clip
   - Oil pan with a new gasket. Torque the bolts to 71 inch lbs. (8 Nm).
   - Oil level sensor electrical connector
   - Splash shield. Torque the bolts until they are fully seated.
   - Negative battery cable
6. Refill the engine oil.
7. Start the vehicle and check for leaks.

Upper

1. Before servicing the vehicle, refer to the precautions in the beginning of this section.
2. Remove or disconnect the following:
   - Negative battery cable
   - Lower oil pan
   - Engine mount lower nuts
   - A/C compressor hose strap from the oil pan
   - Transmission bolts from the oil pan
   - All but the 4 corner bolts from the oil pan
   - Propeller shaft
3. Support the propeller shaft out of the way.
4. Remove or disconnect the following:
   - Catalytic converter
   - Idler arm bolts and lower the relay rod out of the way
5. Install an engine support fixture and raise the engine slightly to allow removal of the oil pan.

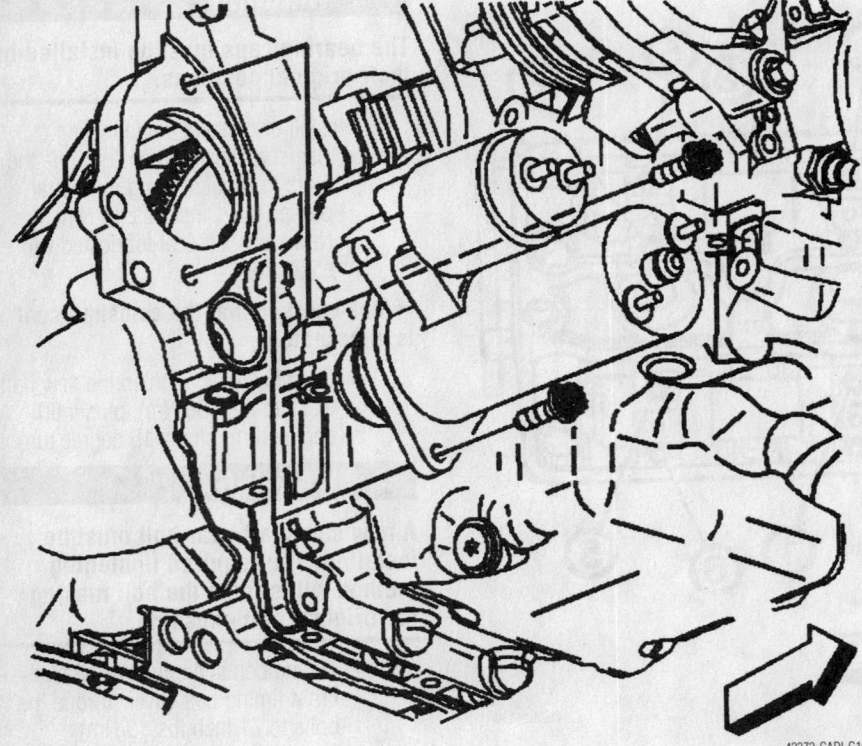

42372-CADI-G10

Remove the starter motor and bolts

6. Remove or disconnect the following:
- Remaining oil pan bolts
- Oil intake pipe
- Oil pan from the vehicle

To install:

7. Apply a bead of silicone sealer in the bottom of the upper oil pan groove.

8. Install a new rubber seal into the groove in the pan.

9. Apply a 3mm (0.12 in) bead of silicone sealant on the OUTSIDE edge of the seal, at the front of the upper oil pan and apply an identical bead on the INSIDE edge of the seal at the rear of the upper oil pan. The beads will overlap at the middle of the upper oil pan.

10. Install or connect the following:
- Upper oil pan
- Oil intake pipe. Torque the bolt to 71 inch lbs. (8 Nm).
- Four oil pan corner bolts finger tight

11. Lower the engine into place.

12. Install or connect the following:
- Remaining oil pan bolts. Torque the bolts to 11 ft. lbs. (15 Nm).
- Transmission-to-oil pan bolts. Torque the bolts to 30 ft. lbs. (40 Nm).
- Idler arm bolts. Torque the bolts to 44 ft. lbs. (60 Nm).
- Catalytic converter. Torque the nuts to 18 ft. lbs. (25 Nm).
- Propeller shaft. Torque the bolts to 70 ft. lbs. (95 Nm).
- A/C compressor hose strap. Torque the bolt to 71 inch lbs. (8 Nm).
- Engine mount lower nuts. Torque the nuts to 41 ft. lbs. (55 Nm).
- Lower oil pan

13. Refill the engine oil.

Oil Pump

REMOVAL & INSTALLATION

1. Before servicing the vehicle, refer to the precautions in the beginning of this section.

2. Drain the engine oil.

3. Drain the engine coolant.

4. Discharge and recover the A/C system.

5. Remove or disconnect the following:
- Negative battery cable
- Intake air resonator
- Front timing belt cover
- Intake plenum
- Timing belt
- Rear timing belt cover

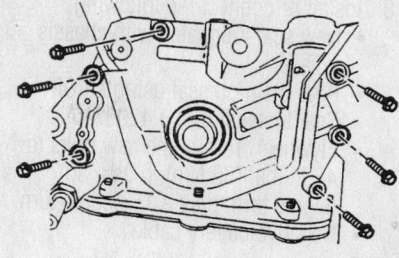

79227414

Oil pump mounting bolt locations

- A/C compressor
- A/C compressor and power steering pump bracket and position it out of the way of the oil pump housing
- Lower alternator mounting bolt and loosen the upper mounting bolt then, position the alternator aside
- Lower and upper oil pans
- Crankshaft drive gear
- Oil pump
- Front main oil seal and collar

To install:

6. Coat the pump side of the oil pump gasket with a thin layer of sealant. Do not cover or restrict the gasket openings with sealant.

7. Install or connect the following:
- Oil pump collar
- New front main oil seal
- Oil pump with a new gasket by aligning the guide pins. Torque the bolts to 9 ft. lbs. (12 Nm).
- Crankshaft gear. Torque the bolt to 184 ft. lbs. (250 Nm) plus an additional 60 degree turn.
- Upper oil pan. Torque the bolts to 11 ft. lbs. (15 Nm).
- Lower oil pan. Torque the bolts to 71 inch lbs. (8 Nm).
- Alternator. Torque the upper and lower bolts to 30 ft. lbs. (40 Nm).

8. Re-tighten the oil pump bolts to 15 ft. lbs. (20 Nm).

9. Install or connect the following:
- A/C compressor and power steering pump bracket. Torque the fasteners to 30 ft. lbs. (40 Nm).
- A/C compressor. Torque the bolts to 30 ft. lbs. (40 Nm).
- Rear timing belt cover. Torque the bolts to 71 inch lbs. (8 Nm).
- Timing belt
- Front timing belt cover. Torque the bolts to 71 inch lbs. (8 Nm).
- Intake plenum. Torque the bolts to 71 inch lbs. (8 Nm).

- Intake air resonator
- Negative battery cable

10. Fill the engine oil.

11. Fill the coolant.

12. Recharge the A/C system.

13. Start the vehicle and checks for leaks.

Rear Main Seal

REMOVAL & INSTALLATION

1. Before servicing the vehicle, refer to the precautions in the beginning of this section.

2. Remove or disconnect the following:
- Negative battery cable.
- Flywheel

3. Center punch the steel ring of the rear main seal.

4. Drill a small hole into the steel ring.

79227416

Thread a self-tapping screw into the metal part of the seal and remove the seal

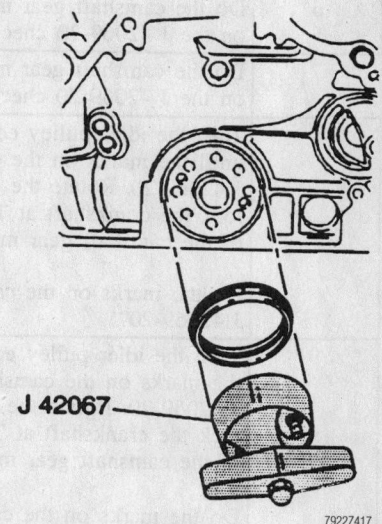

J 42067

79227417

Press the rear main seal into position using a threaded seal installer like J 42067

→Make certain not to damage any engine components on the opposite side of the seal.

5. Install a small self tapping screw.
6. Remove the rear main oil seal.

To install:

7. Clean all sealing surfaces with a non-abrasive cleaner.

8. Install or connect the following:
- New seal lubricated with chassis grease
- Rear main oil seal using an Oil Seal Installer tool J 42067
- Flywheel. Torque the new bolts to 48 ft. lbs. (65 Nm), plus a 30 degree turn, plus a 15 degree turn.
- Negative battery cable

Timing Belt

REMOVAL & INSTALLATION

1. Before servicing the vehicle, refer to the precautions in the beginning of this section.
2. Remove or disconnect the following:

TIMING BELT INSTALLATION AND ADJUSTMENT TABLE

Step	Action	Value	Yes	No
1	Install the timing belt and align marks on the belt with the marks on the camshaft gears and the crankshaft gear. Check the timing belt deflection between the idler pulley for camshafts 3 & 4 and camshaft number 4. Is the timing belt installed, the marks aligned and the timing belt deflection adjusted?	1 cm (0.4 in) maximum	Go to Step 2	—
2	Set the initial timing belt tension at the timing belt tensioner. Is the initial timing belt tension set?	—	Go to Step 3	—
3	Rotate the engine two complete revolutions and secure the crankshaft at Top Dead Center (TDC) with the J 42069-10. Has the engine been rotated and the crankshaft secured to TDC?	—	Go to Step 4	—
4	Starting with camshafts 3 and 4, check the alignment of the marks on the camshaft gears with the marks on the J 42069-20 checking gauge. Do the marks on the camshaft gears align exactly with the marks on J 42069-20?	—	Go to Step 5	Go to Step 6
5	Check the alignment of the marks on camshafts gears 1 and 2 with the marks on the J 42069-20 checking gauge. Do the marks on the camshaft gears align exactly with the marks on J 42069-20?	—	Go to Step 14	Go to Step 10
6	Do the camshaft gear marks line up to the left (BTDC) of the marks on the J 42069-20 checking gauge?	—	Go to Step 8	Go to Step 7
7	Do the camshaft gear marks line up to the right (ATDC) of the marks on the J 42069-20 checking gauge?	—	Go to Step 9	—
8	Turn the idler pulley eccentric, for camshafts 3 and 4, counterclockwise until the marks on the camshaft gear align exactly with the marks on J 42069-20. Rotate the engine two complete revolutions, lock the crankshaft at TDC with J 42069-10 and recheck the alignment of the camshaft gear marks to the marks on J 42069-20. Do the marks on the camshaft gears align exactly with the marks on J 42069-20?	—	Go to Step 5	Go to Step 6
9	Turn the idler pulley eccentric, for camshafts 3 and 4, clockwise until the marks on the camshaft gear align exactly with the marks on J 42069-20. Rotate the engine two complete revolutions, lock the crankshaft at TDC with J 42069-10 and recheck the alignment of the camshaft gear marks to the marks on J 42069-20. Do the marks on the camshaft gears align exactly with the marks on J 42069-20?	—	Go to Step 5	Go to Step 6

79225G35

Timing belt installation and adjustment table—GM 3.2L (VIN N) engine

Step	Action	Value	Yes	No
10	Do the camshaft gear marks line up to the left (BTDC) of the marks on the J 42069-20 checking gauge?	—	Go to Step 12	Go to Step 11
11	Do the camshaft gear marks line up to the right (ATDC) of the marks on the J 42069-20 checking gauge?	—	Go to Step 13	—
12	Turn the idler pulley eccentric, for camshafts 1 and 2, counterclockwise until the marks on the camshaft gear align exactly with the marks on J 42069-20. Rotate the engine two complete revolutions, lock the crankshaft at TDC with J 42069-10 and recheck the alignment of the camshaft gear marks to the marks on J 42069-20. Do the marks on the camshaft gears align exactly with the marks on J 42069-20?	—	Go to Step 14	Go to Step 10
13	Turn the idler pulley eccentric, for camshafts 1 and 2, clockwise until the marks on the camshaft gear align exactly with the marks on J 42069-20. Rotate the engine two complete revolutions, lock the crankshaft at TDC with J 42069-10 and recheck the alignment of the camshaft gear marks to the marks on J 42069-20. Do the marks on the camshaft gears align exactly with the marks on J 42069-20?	—	Go to Step 14	Go to Step 10
14	Set the final timing belt tension at the timing belt tensioner. Is the final timing belt tension set?	—	Go to Step 15	—
15	Again, rotate the engine two complete revolutions and lock the crankshaft at TDC. Do a final inspection of the camshaft gear marks' relationship to the J 42069-20 marks. The marks must align exactly. Do the marks on the camshaft gears align exactly with the marks on the J 42069-20?	—	Go to Step 16	Go to Step 2
16	Remove all checking tools and ensure all idler pulleys and the tensioner locking nut are tightened to specifications. Continue with re-assembly of the engine.	—	—	—

79225G36

Timing belt installation and adjustment table (continued)—GM 3.2L (VIN N) engine

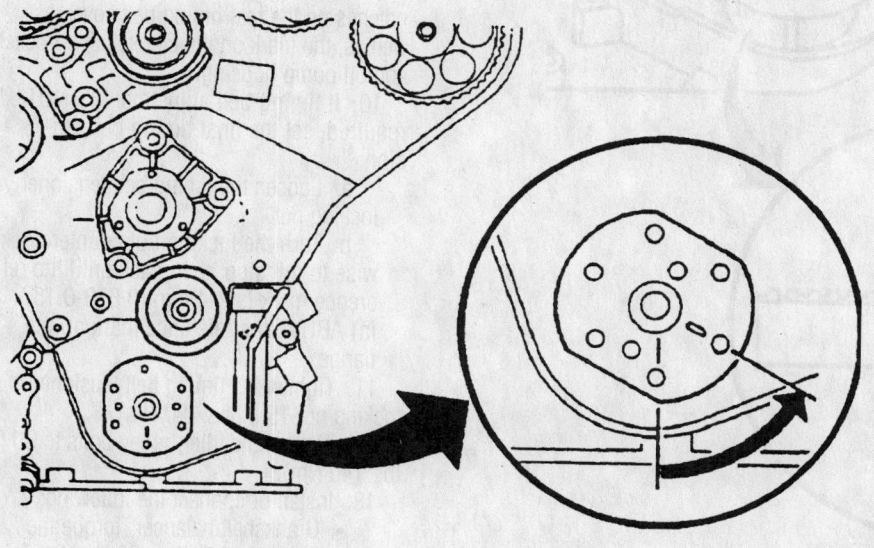

79225G30

Crankshaft alignment to 60 degrees BTDC—GM 3.2L (VIN N) engine

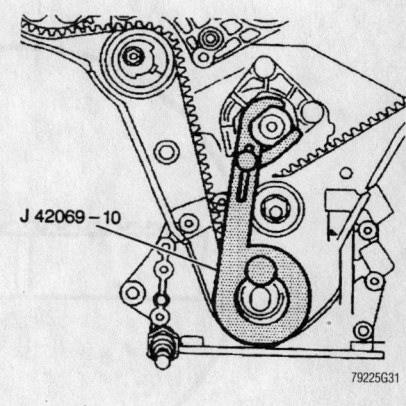

J 42069-10

79225G31

Securing the crankshaft—GM 3.2L (VIN N) engine

J 42069 – 1

J 42069 – 2

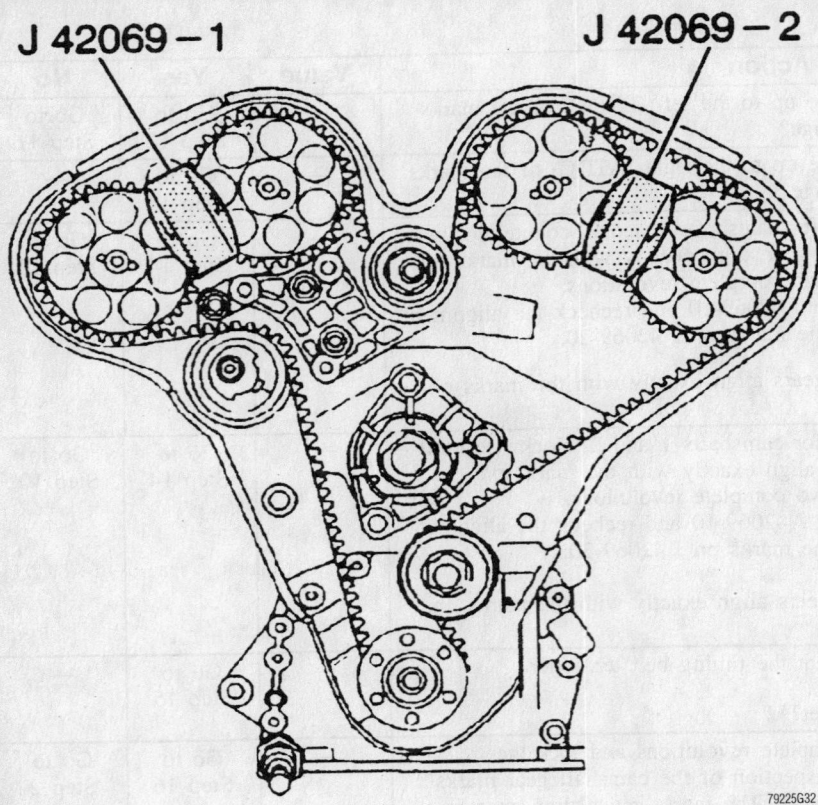

79225G32

Locking the camshaft—GM 3.2L (VIN N) engine

J 42069 – 30

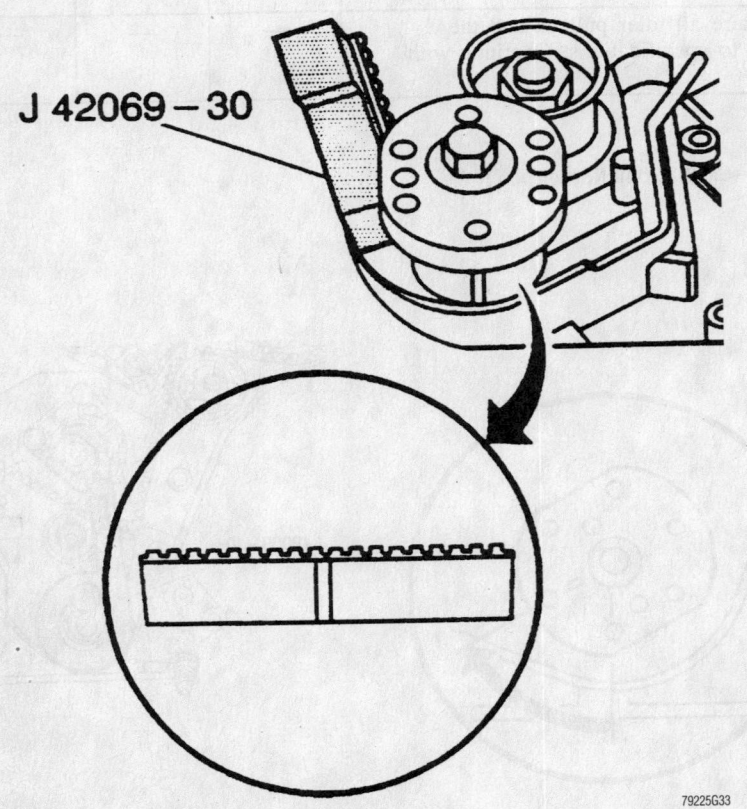

79225G33

Using the tool to pin the timing belt—GM 3.2L (VIN N) engine

- Negative battery cable
- Intake air resonator
- Intake plenum
- Splash shield
- Secondary air injection (AIR) pipe
- Accessory drive belt
- Water pump and power steering pump pulleys
- Drive belt tensioner
- Front timing belt cover
- Crankshaft balancer

3. Rotate the crankshaft clockwise to 60 degrees before top dead center (BTDC).

4. Loosen the timing belt tensioner and idler pulleys.

5. Remove the timing belt.

To install:

6. Start at the crankshaft sprocket and install the timing belt with the double dash (TDC) aligned with the marks on the oil pump and on the belt drive gear.

7. Set the initial timing belt tension as follows:

  a. Turn the tensioner nut COUNTER-CLOCKWISE to full stop, then, turn the nut back until the reference mark is 1 mm (0.003 in) over the flange.

  b. Tighten the timing belt tensioner locking nut until snug, the locking nut will be tightened to specifications after all final adjustments are made.

8. Rotate the engine in the clockwise direction two revolutions stopping at 60 degrees BTDC.

9. Inspect the alignment of the reference marks on the camshaft gears with the notches on the rear timing belt cover, as well as, the mark on the crankshaft sprocket and oil pump housing.

10. If timing belt adjustment IS NOT required, set the final timing belt tension:

  a. Loosen the timing belt tensioner locking nut.

  b. Turn the locking nut counterclockwise to full stop, then back until the reference mark is 2-4 mm (0.078-0.157 in) ABOVE the reference mark on the flange.

11. Tighten the timing belt tensioner locking nut 15 ft. lbs. (20 Nm).

12. Tighten the idler pulley bolts to 30 ft. lbs. (40 Nm).

13. Install or connect the following:

- Crankshaft balancer. Torque the bolts to 15 ft. lbs. (20 Nm).
- Front timing belt cover. Torque

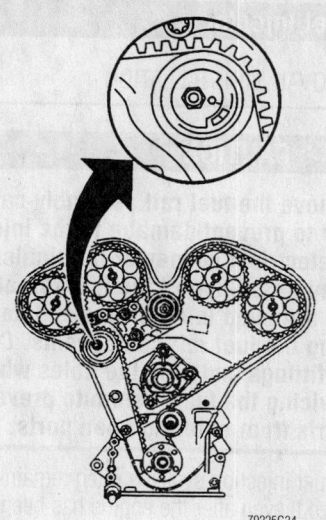

Initial timing belt tension adjustment—GM 3.2L (VIN N) engine

the bolts to 71 inch lbs. (8 Nm).
- Drive belt tensioner. Torque the bolts to 30 ft. lbs. (40 Nm).
- Water pump pulley. Torque the bolts to 71 inch lbs. (8 Nm).
- Power steering pump pulley. Torque the bolts to 15 ft. lbs. (20 Nm).
- Accessory drive belt
- AIR injection pipe
- Splash shield
- Intake plenum
- Intake air resonator

Piston and Ring

POSITIONING

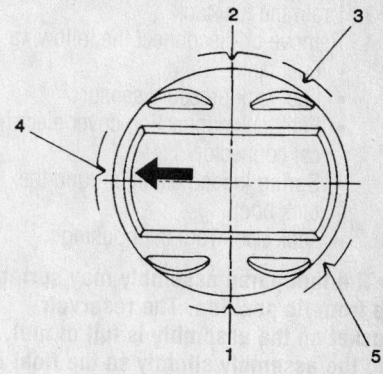

(1) 1st Compression Ring End Gap Location
(2) 2nd Compression Ring End Gap Location
(3) Oil Control Ring Upper Ring End Gap Location
(4) Oil Control Ring Spacer End Gap Location
(5) Oil Control Ring Lower Ring End Gap Location

Piston ring end-gap spacing—CTS 3.2L engine

FUEL SYSTEM

Fuel System Service Precautions

Safety is the most important factor when performing not only fuel system maintenance but any type of maintenance. Failure to conduct maintenance and repairs in a safe manner may result in serious personal injury or death. Maintenance and testing of the vehicle's fuel system components can be accomplished safely and effectively by adhering to the following rules and guidelines.

- To avoid the possibility of fire and personal injury, always disconnect the negative battery cable unless the repair or test procedure requires that battery voltage be applied.
- Always relieve the fuel system pressure prior to disconnecting any fuel system component (injector, fuel rail, pressure regulator, etc.), fitting or fuel line connection. Exercise extreme caution whenever relieving fuel system pressure, to avoid exposing skin, face and eyes to fuel spray. Please be advised that fuel under pressure may penetrate the skin or any part of the body that it contacts.
- Always place a shop towel or cloth around the fitting or connection prior to loosening to absorb any excess fuel due to spillage. Ensure that all fuel spillage (should it occur) is quickly removed from engine surfaces. Ensure that all fuel soaked cloths or towels are deposited into a suitable waste container.
- Always keep a dry chemical (Class B) fire extinguisher near the work area.

- Do not allow fuel spray or fuel vapors to come into contact with a spark or open flame.
- Always use a backup wrench when loosening and tightening fuel line connection fittings. This will prevent unnecessary stress and torsion to fuel line piping. Always follow the proper torque specifications.
- Always replace worn fuel fitting O-rings with new. Do not substitute fuel hose or equivalent, where fuel pipe is installed.

Fuel System Pressure

RELIEVING

1. Before servicing the vehicle, refer to the precautions in the beginning of this section.
2. Loosen the fuel filler cap to relieve the tank pressure.
3. Remove or disconnect the following:
   - Negative battery cable
   - Intake manifold top cover
4. Install a Fuel Pressure Gauge to the fuel pressure fitting. Wrap a shop towel around the fitting while installing the gauge.
5. Connect a bleed hose into an approved container and open the valve to bleed the system.
6. Close the valve and disconnect the gauge.
7. Drain any remaining fuel from the gauge into the approved container.

Fuel Filter

REMOVAL & INSTALLATION

1. Before servicing the vehicle, refer to the precautions in the beginning of this section.
2. Loosen the fuel filler cap to relieve pressure in the tank.
3. Relieve the fuel system pressure.
4. Remove or disconnect the following:
   - Fuel feed and return lines
   - Fuel filter from the bracket

To install:
5. Clean the bolts and fittings on both fuel lines.
6. Install or connect the following:
   - Filter in the retaining strap making certain that the flow is in the proper direction
   - Both fuel lines. Torque the fittings to 18 ft. lbs. (25 Nm).
   - Fuel filter bracket mounting bolt. Torque the bolt to 13 ft. lbs. (18 Nm).
   - Fuel filler cap
7. Crank the engine for a few seconds and check for leakage.

Fuel Pump

REMOVAL & INSTALLATION

1. Before servicing the vehicle, refer to the precautions in the beginning of this section.
2. Relieve the fuel system pressure.

3. Drain the fuel tank.
4. Remove or disconnect the following:
   - Fuel tank
   - Fuel tank pressure sensor
   - Fuel tank connection cover electrical connector
   - Spring loaded clamp around the tank boot
   - Tank boot from the housing

➡ **The fuel pump assembly may spring up from its position. The reservoir bucket on the assembly is full of fuel. Tip the assembly slightly so the float is not damaged during removal.**

   - Fuel sender locking nut from the tank using a Fuel Tank Sender Wrench (J 42219)
   - Fuel pump electrical connectors
   - Fuel tank connection cover
   - Hose from the pump
   - Fuel pump from the tank by pushing the tabs inward
   - Locking ring by pushing the tabs in and pushing the locking ring from the housing simultaneously
   - Fuel pump from the housing
   - Lip seal from the cover by sliding it downward, past the reservoir and over the float arm and discard it

5. Drain the remaining fuel from the reservoir into an approved container.

To install:

6. Install or connect the following:
   - New lip seal lubricated with engine oil
   - Fuel pump into the housing
   - Locking ring into the housing
   - New fuel inlet screen
   - Fuel pump assembly into the tank
   - Fuel line to the pump using a new clamp
   - Fuel pump electrical connectors
   - Fuel tank connection cover
   - Fuel sender locking nut. Torque the nut to 37 ft. lbs. (50 Nm).
   - Fuel tank boot to the housing
   - Fuel tank pressure sensor. Torque the fastener to 18 inch lbs. (4 Nm).
   - Air reference hose to the pressure sensor
   - Fuel feed and return lines using new clamps
   - Fuel tank pressure sensor and connection cover electrical connectors
   - Fuel tank to the vehicle. Torque the bolts to 22 ft. lbs. (30 Nm).
   - Negative battery cable

7. Refill the fuel tank.
8. Start the vehicle and inspect for leaks.

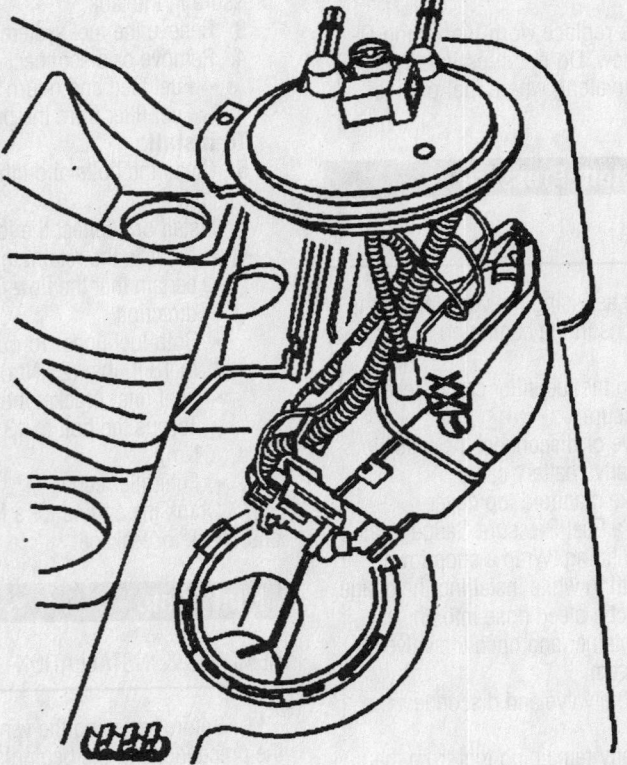

Fuel pump module

42372-CADI-G11

Fuel Injector

REMOVAL & INSTALLATION

✻✻ CAUTION

Remove the fuel rail assembly carefully to prevent damage to the injector electrical connector terminals and spray tips. Support the fuel rail after it is removed in order to avoid damaging the fuel rail components. Cap the fittings and plug the holes when servicing the fuel system to prevent debris from entering open ports.

Fuel injection systems often remain pressurized, even after the engine has been turned **OFF**. The fuel system pressure must be relieved before disconnecting any fuel lines. Failure to do so may result in fire and/or personal injury.

1. Before servicing the vehicle, refer to the precautions in the beginning of this section.
2. Relieve fuel system pressure.
3. Remove or disconnect the following:
   - Intake plenum
   - Fuel rail supply and return lines from the fuel rail
   - Fuel injector electrical connectors
   - Harness from the fuel rail
   - Fuel rail attaching bolts from the intake manifold
   - Fuel pressure regulator vacuum lines
   - Fuel rail assembly
   - Injectors from the fuel rail and discard the O-rings

To install:

4. Install or connect the following:
   - New O-rings lubricated with clean engine oil onto the fuel injectors
   - Fuel injectors with new retaining clips
   - Fuel rail to the intake manifold. Torque the bolt to 71 inch lbs. (8 Nm).
   - Fuel pressure regulator vacuum line
   - Fuel injector electrical connectors
   - Harness to the fuel rail
   - Fuel supply/return lines to the fuel rail. Torque the fasteners to 11 ft. lbs. (15 Nm).
   - Fuel line bracket. Torque the bolts to 37 ft. lbs. (50 Nm).
   - Intake plenum. Torque the bolts to 71 inch lbs. (8 Nm).
   - Negative battery cable

5. Crank the engine several times to pressurize the system.
6. Check for leaks and repair, if necessary.

DRIVE TRAIN

Transmission Assembly

REMOVAL & INSTALLATION

Automatic Transmission

1. Before servicing the vehicle, refer to the precautions in the beginning of this section.
2. Remove or disconnect the following:
   - Negative battery cable
   - Shift linkage from the transmission
   - Propeller shaft coupling bolts
   - Propeller shaft coupling from the drive flange
   - Access holes plugs from the oil pan and bell housing

➡ **Mark the flywheel to the torque converter to ease the installation procedure.**

   - Flywheel-to-torque converter bolts
   - Oil cooler inlet and outlet pipes
   - Oxygen Sensor (O₂S) connections
   - Catalytic converters
   - Transmission housing-to-oil pan bolts
3. Support the transmission.
4. Remove or disconnect the following:
   - Transmission crossmember to mount nuts
   - Transmission crossmember
   - Vent hose
   - Electrical connector from the transmission control selector switch
   - Electrical connector from the adapter case
   - Electrical connector from the main case
   - Electrical connector from the transmission speed sensor
5. Support the front of the engine.
6. Lower the transmission to gain access to the upper housing bolts.
7. Remove the upper transmission housing bolts.
8. Carefully remove the transmission from the vehicle.

To install:

✲✲ WARNING

The transmission fluid cooler and lines must be flushed out prior to installing a new or reconditioned transmission.

9. Apply thread locking compound to the transmission-to-engine mounting bolts.
10. Install or connect the following:

   - Transmission. Torque the transmission-to-engine bolts to 44 ft. lbs. (60 Nm).
   - Electrical connector to the transmission control selector switch
   - Electrical connector to the adapter case
   - Electrical connector to the main case
   - Electrical connector to the speed sensor
   - Vent hose
   - Crossmember. Torque the bolts to 33 ft. lbs. (45 Nm).
   - Crossmember-to-mount and torque the nuts to 15 ft. lbs. (20 Nm)
11. Remove the transmission support.
12. Install or connect the following:
   - Transmission housing-to-oil pan bolts and torque the bolts to 15 ft. lbs. (20 Nm)
   - Catalytic converters. Torque the bolts to 18 ft. lbs. (25 Nm).
   - O₂S connectors
   - Oil cooler inlet and outlet pipes to the center pipes
   - Torque converter to the flywheel by aligning the matchmarks. Torque the new bolts to 22 ft. lbs. (30 Nm).

➡ **Be sure the torque converter weld nuts are flush with the flywheel.**

   - Oil pan access hole plugs
   - Bell housing access hole plugs
   - Propeller shaft coupling-to-drive flange an torque the bolts to 70 ft. lbs. (95 Nm)
   - Shift linkage to the transmission
   - Negative battery cable
13. Adjust the shift lever rod as follows:
   a. Place the shift control lever in the **P** position.
   b. Loosen the adjusting bolt on the rod.
   c. Hold the selector lever on the transmission toward the rear stop to eliminate any free-play.
   d. Tighten the adjusting bolt to 71 inch lbs. (8 Nm).
14. Refill the transmission. Tighten the plug to 33 ft. lbs. (45 Nm).
15. Start the vehicle and check for leaks. Be sure the vehicle will only start in the **P** and **N** positions.

Manual Transmission

➡ **The front wheels of the vehicle must be maintained in the straight ahead position and the steering column must be in the LOCK position before disconnecting the steering column or intermediate shaft. Failure to follow these procedures will cause improper alignment of some components during installation and result in damage to the SIR coil assembly.**

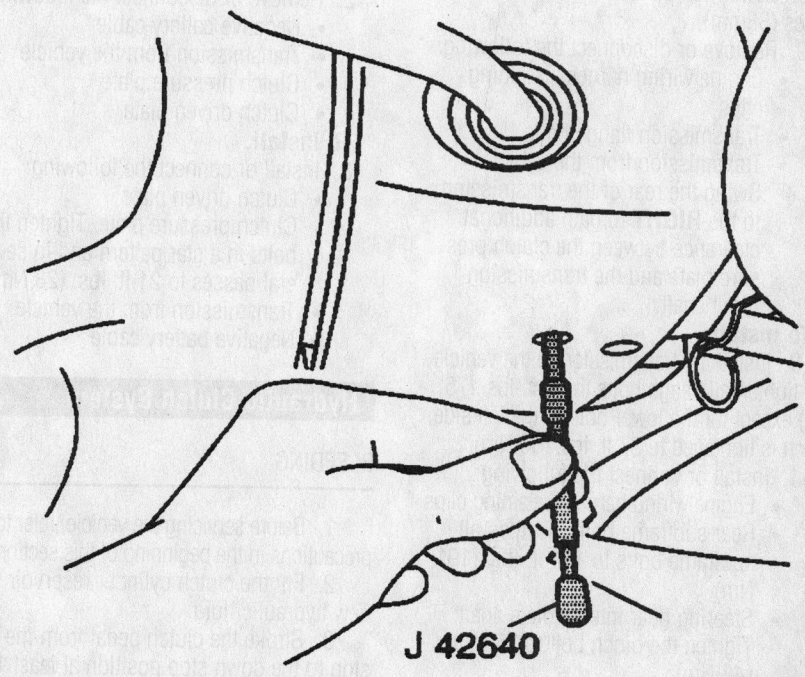

J 42640

Steering column locking tool J 42640

1. Before servicing the vehicle, refer to the precautions in the beginning of this section.

2. Turn the steering wheel so that the front wheels are pointing straight ahead.

3. Turn the ignition lock cylinder to the **LOCK** position and remove the key.

4. Lock the steering column through the access hole in the lower steering column trim cover using Steering Column Lock Tool J 42640.

5. Remove or disconnect the following:
- Negative battery cable
- Heated Oxygen (HO2S) sensor connectors
- Exhaust system
- Catalytic converter hanger bracket
- Propeller shaft
- Vehicle Speed (VSS) sensor connector
- Reverse lamp switch connector
- Hydraulic clutch slave cylinder hose connection

➥It is not necessary to plug the lower hose end or the slave cylinder fitting as they are fitted with check valves.

- Transmission mount. Support the transmission with a suitable transmission jack.
- Shift control rod
- Reaction arm
- Steering gear intermediate shaft

6. Support the subframe with a jack stand.

7. Remove the two rear bolts and loosen the front two bolts.

8. Lower the rear of the subframe 1½ inches (38mm).

9. Remove or disconnect the following:
- Engine wiring harness retaining clips
- Transmission flange bolts
- Transmission from the vehicle. Swing the rear of the transmission to the **RIGHT** to gain additional clearance between the clutch pressure plate and the transmission input shaft.

To install:

10. Install the transmission to the vehicle and tighten all flange bolts to 55 ft. lbs. (75 Nm) except for the lower bolt on the left side, which is tightened to 37 ft. lbs. (50 Nm).

11. Install or connect the following:
- Engine wiring harness retaining clips
- Rear subframe bolts. Tighten all subframe bolts to 141 ft. lbs. (191 Nm).
- Steering gear intermediate shaft. Tighten the pinch bolt to 35 ft. lbs. (48 Nm).
- Reaction arm
- Shift control rod

- Transmission mount. Tighten the bolts to 44 ft. lbs. (60 Nm).
- Hydraulic clutch slave cylinder hose connection
- Reverse lamp switch connector
- Vehicle Speed (VSS) sensor connector
- Propeller shaft. Tighten the bolts to 63 ft. lbs. (85 Nm).
- Catalytic converter hanger bracket. Tighten the bolts to 37 ft. lbs. (50 Nm).
- Exhaust system
- Heated Oxygen (HO2S) sensor connectors
- Negative battery cable

12. Remove Steering Column Lock Tool J 42640 from the steering column.

13. Bleed the clutch hydraulic system.

14. Check the transmission fluid level.

15. Road test the vehicle for proper operation.

Clutch

ADJUSTMENTS

The CTS uses a hydraulic clutch system. No adjustments are necessary.

REMOVAL & INSTALLATION

1. Before servicing the vehicle, refer to the precautions in the beginning of this section.

2. Remove or disconnect the following:
- Negative battery cable
- Transmission from the vehicle
- Clutch pressure plate
- Clutch driven plate

To install:

3. Install or connect the following:
- Clutch driven plate
- Clutch pressure plate. Tighten the bolts in a star pattern and in several passes to 21 ft. lbs. (28 Nm).
- Transmission from the vehicle
- Negative battery cable

Hydraulic Clutch System

BLEEDING

1. Before servicing the vehicle, refer to the precautions in the beginning of this section.

2. Fill the clutch cylinder reservoir with new hydraulic fluid.

3. Stroke the clutch pedal from the up stop to the down stop position at least 15 times.

4. With the pedal in the down stop

position, open the bleeder valve at the slave cylinder to release trapped air.

5. Close the bleeder valve and slowly return the clutch pedal to the up stop position.

6. Open the bleeder valve and slowly depress the clutch pedal from the up stop position to the down stop position until fluid escapes from the bleeder valve.

7. Close the bleeder valve and slowly return the clutch pedal to the up stop position.

8. Depress the clutch pedal from the up stop to the down stop position.

9. Open the bleeder valve and allow fluid with air bubbles to escape.

10. Close the bleeder valve.

11. Repeat until fluid without air bubbles escapes from the bleeder valve.

Axle Shaft, Bearing and Seal

REMOVAL & INSTALLATION

1. Before servicing the vehicle, refer to the precautions in the beginning of this section.

2. Place the gear selector in **N**.

3. Remove or disconnect the following:
- Rear wheel
- Drive axle flange bolts
- Outer end of the drive axle from the wheel bearing/hub inner flange

➥Make certain that the drive axle separator tool is properly aligned to the differential and drive axle. The tool is clearly marked "Differential Side". If installed incorrectly, damage to the Anti-lock Brake System (ABS) sensor reluctor ring is possible.

4. Remove the axle shaft.

5. Use a suitable prytool to remove the axle seal from the differential.

To install:

6. Use a Seal Installation Tool (such as J 26234) to install the new seal.

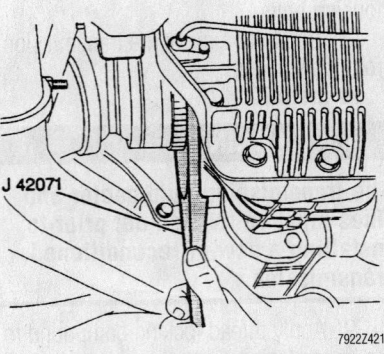

Pry the halfshaft out of the differential

7. Lubricate the splines of the halfshaft and the sealing surfaces with differential oil.

8. Install or connect the following:
- Axle shaft into the differential bore by aligning the splines

➡ **If necessary use a soft faced tool, such as a rubber mallet, to seat the halfshaft fully in the differential bore.**

- Outer end of the axle shaft into the wheel bearing/hub inner flange. Torque the bolts to 118 ft. lbs. (160 Nm).
- Rear wheel

9. Road test the vehicle.

CV-Joints

REMOVAL & INSTALLATION

Inner & outer Joint

1. Before servicing the vehicle, refer to the precautions in the beginning of this section.
2. Disconnect the negative battery cable.
3. Remove the axle shaft and place it in a vise.
4. Remove the large boot retaining clamp.
5. Remove the small boot retaining clamp.
6. Slide the boot up the shaft away from the joint.
7. With a brass drift and a hammer tap on the inner race of the cross grove.
8. Remove the CV joint retaining ring.
9. Remove the CV boot.

➡ **The joint is not serviceable, if damaged it must be replaced.**

To install:

10. Pack the CV joint with ½ of the grease provided in the service kit.
11. Install new small and large boot retaining clamps on the ends of the boot
12. Install a new boot on the halfshaft.
13. Install a new joint retaining ring onto the halfshaft.
14. Place halfshaft into a shop press and press the joint into place.
15. Slide the CV boot over the joint .
16. Using a crimping tool, a breaker bar and a torque wrench, torque the small retaining clamp to 100 ft. lbs. (140 Nm).
17. Place the remaining grease into the CV boot.
18. Using a crimping tool a breaker bar and a torque wrench, torque the large retaining clamp to 130 ft. lbs. (176 Nm).
19. Install the halfshaft in the vehicle.

Pinion Seal

REMOVAL & INSTALLATION

1. Before servicing the vehicle, refer to the precautions in the beginning of this section.
2. Place the transmission in **N**.
3. Remove or disconnect the following:
- Underbody heat shields
- Bolts from the propeller shaft center bearing bracket
- Propeller shaft from the front and rear propeller shaft couplings

- Rear propeller shaft coupling
4. Mark the position of the gear yoke, gear and nut for correct bearing pre-load.
5. Remove or disconnect the following:
- Drive pinion flange nut
- Drive pinion flange using a gear puller
- Pinion seal from the rear axle housing by using a blunt chisel to drive the seal out

To install:

6. Apply lubricant to the outer diameter of the gear yoke and the outer lip of the new seal.
7. Install or connect the following:
- New seal lubricated with chassis grease
- Drive pinion flange
- Drive pinion gear nut and hand-tighten
8. Tighten the nut to the position marked in the removal procedure. Tighten the nut an additional 0.062 inch (1.59mm).
9. Install or connect the following:
- Rear propeller shaft coupling
- Propeller shaft center bearing bracket. Torque the bolts to 15 ft. lbs. (20 Nm).
- Propeller shaft to the rear coupling. Torque the bolts to 70 ft. lbs. (95 Nm).
- Propeller shaft to the front coupling. Torque the bolts to 70 ft. lbs. (95 Nm).
- Underbody heat shields. Torque the bolts to 18 inch lbs. (2 Nm).
10. Road test the vehicle.

STEERING AND SUSPENSION

Air Bag

✷✷ CAUTION

These vehicles are equipped with an air bag system, known as a Supplemental Inflatable Restraint (SIR) system. The system must be disabled before performing service on or around system components, steering column, instrument panel components, wiring and sensors. Failure to follow safety and disabling procedures could result in accidental air bag deployment, possible personal injury and unnecessary system repairs.

PRECAUTIONS

Several precautions must be observed when handling the inflator module to avoid accidental deployment and possible personal injury.
- Never carry the inflator module by the wires or connector on the underside of the module.
- When carrying a live inflator module, hold securely with both hands, and ensure that the bag and trim cover are pointed away.
- Place the inflator module on a bench or other surface with the bag and trim cover facing up.
- With the inflator module on the bench, never place anything on or close to the module which may be thrown in the event of an accidental deployment.

DISARMING

1. Disconnect the negative battery cable.

2. Turn the steering wheel to align the wheels in the straight-ahead position.
3. Turn the ignition switch to the **LOCK** position and remove the key.
4. Remove the two TORX® screws from the back of the steering wheel.
5. Remove the inflatable restraint steering wheel module from the steering wheel.
6. Remove the Connector Position Assurance (CPA) from the inflatable restraint steering wheel module 2-way connector.
7. Disconnect the inflatable restraint steering wheel module 2-way connector
8. Remove the inflatable restraint Instrument Panel (IP) module cover.
9. Remove the CPA from the IP module connector.
10. Disconnect the IP module connector.
11. Remove the left hand outer seat track cover.

12. Disconnect the seat belt pretensioner LF connector from the connector.

13. Remove the CPA from the inflatable restraint side impact module LF connector.

14. Disconnect the side impact module LF connector.

15. Remove the right hand outer seat track cover.

16. Disconnect the seat belt pretensioner RF connector from the connector.

17. Remove the CPA from the inflatable restraint side impact module RF connector.

18. Disconnect the side impact module RF connector.

REARMING

1. Turn the steering wheel so that the wheels are pointing straight ahead.

2. Turn the ignition switch to the **LOCK** position and remove the key.

3. Connect the side impact module RF connector.

4. Install the Connector Position Assurance (CPA) from the inflatable restraint side impact module RF connector.

5. Connect the seat belt pretensioner RF connector to the connector.

6. Install the right hand outer seat track cover.

7. Connect the side impact module LF connector.

8. Install the CPA from the inflatable restraint side impact module LF connector.

9. Connect the seat belt pretensioner LF connector to the connector.

10. Install the left hand outer seat track cover.

11. Connect the IP module connector.

12. Install the CPA to the IP module connector.

13. Install the inflatable restraint Instrument Panel (IP) module cover.

14. Connect the inflatable restraint steering wheel module 2-way connector

15. Connect the CPA to the inflatable restraint steering wheel module 2-way connector.

16. Install the inflatable restraint steering wheel module to the steering wheel.

17. Install the two TORX® screws to the back of the steering wheel and tighten to 72 inch lbs. (8 Nm).

18. Reconnect the negative battery cable.

19. Wait at least 1 minute for the capacitors to recharge.

20. While staying away from the inflator modules, turn the key to the **RUN** position. The air bag warning lamp should turn ON for 3–4 seconds, then turn OFF.

Rack and Pinion Power Steering Gear

REMOVAL & INSTALLATION

➡**The front wheels of the vehicle must be maintained in the straight ahead position and the steering column must be in the LOCK position before disconnecting the steering column or intermediate shaft. Failure to follow these procedures will cause improper alignment of some components during installation and result in damage to the SIR coil assembly.**

1. Before servicing the vehicle, refer to the precautions in the beginning of this section.

2. Center the front wheels, then turn the ignition key to the **LOCK** position.

3. Lock the steering column through the access hole in the lower steering column trim cover using Steering Column Lock Tool J 42640.

4. Remove or disconnect the following:
- Front wheels
- Front air deflector
- Intermediate shaft
- Outer tie rod ends
- Variable effort steering electrical connector, if equipped
- Stabilizer shaft

- Power steering pressure and return hoses
- Power steering pressure hose retaining bolt
- Steering gear mounting bolts

5. Remove the steering gear through the left wheel opening.

To install:

6. Install or connect the following:
- Steering gear to the vehicle. Tighten the mounting bolts to 70 ft. lbs. (95 Nm).
- Power steering pressure hose retaining bolt. Tighten the bolt to 80 inch lbs. (9 Nm).
- Power steering pressure and return hoses. Use new seals and tighten the hoses to 22 ft. lbs. (30 Nm).
- Stabilizer shaft
- Variable effort steering electrical connector, if equipped
- Outer tie rod ends. Tighten the nuts to 52 ft. lbs. (70 Nm).
- Intermediate shaft. Tighten the pinch bolt to 35 ft. lbs. (47 Nm).
- Front air deflector
- Front wheels

7. Remove the Steering Column Lock Tool J 42640.

8. Bleed the power steering hydraulic system.

9. Align the front wheels as necessary.

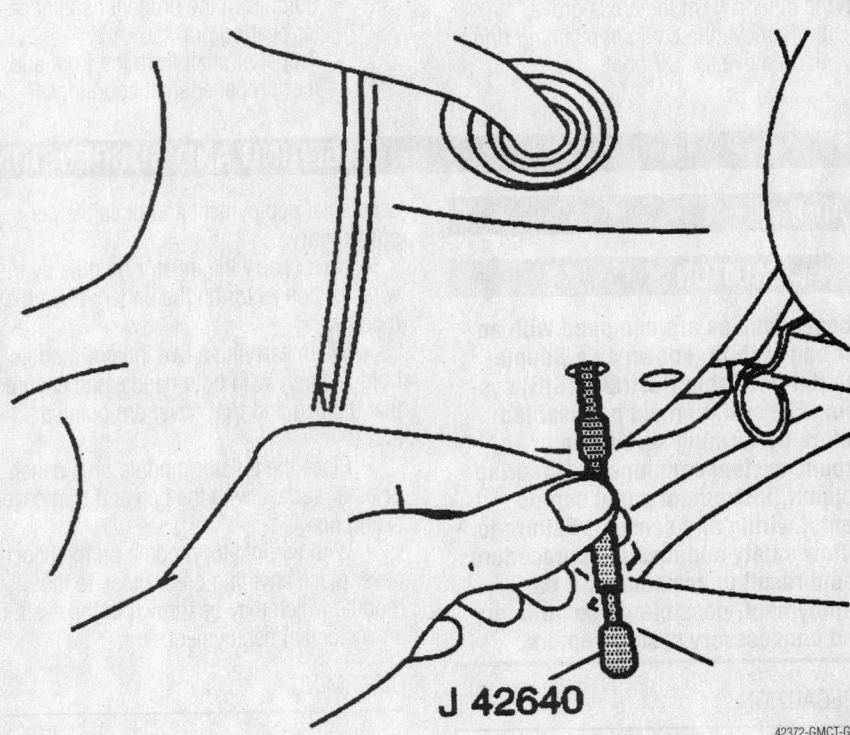

J 42640

42372-GMCT-G01

Steering column locking tool J 42640

Strut

REMOVAL & INSTALLATION

Front

1. Before servicing the vehicle, refer to the precautions in the beginning of this section.
2. Remove or disconnect the following:
   - Front wheel
   - Wheel Speed Sensor (WSS) wiring from the strut
   - Brake wear indicator wires from the strut
   - Brake flex hose clip from the strut
   - Brake caliper and bracket from the steering knuckle
   - Stabilizer shaft link
   - Lower strut mounting bolts from the steering knuckle
   - Upper support plate protective cap
   - Upper support plate
   - Strut from the wheel well

To install:
3. Install or connect the following:
   - Strut into the strut tower
   - Upper support plate and nut. Torque the nut to 41 ft. lbs. (55 Nm).
   - Cap to the upper support nut
   - Steering knuckle to the strut
   - New bolts for the steering knuckle. Torque the lower ball joint pinch bolt to 74 ft. lbs. (100 Nm).

➡**The steering knuckle-to-strut bolts will not be tightened to specifications until after camber corrections have been made.**

   - Stabilizer shaft link to the strut. Torque the nut to 48 ft. lbs. (65 Nm).

➡**Before installing the bolts for the caliper, run an M12 x 1.5 tap through the holes and coat the new bolts with a thread locking compound.**

   - Brake caliper and bracket to the steering knuckle. Torque the bracket bolts to 70 ft. lbs. (95 Nm) plus a 37 degree turn.
   - Brake flex hose to the strut and retain it with the clip
   - WSS wiring to the strut
   - Brake wear indicator wires to the strut
   - Front wheel
4. Adjust the wheel alignment.
5. Torque the steering knuckle-to-strut bolts to 52 ft. lbs. (70 Nm).
6. Road test the vehicle.

Shock Absorber

REMOVAL & INSTALLATION

Rear

1. Before servicing the vehicle, refer to the precautions in the beginning of this section.
2. Position the rear seat backs forward to allow access to the upper shock absorber mounting.
3. Remove or disconnect the following:
   - Protective cap and nut from shock tower
   - Upper mounting nut, washer and grommet
   - Lower shock mounting bolt
   - Shock absorber

To install:
4. Install or connect the following:
   - Lower rubber mounting grommet and washer
   - Shock into the tower
   - Upper shock mounting grommet, washer and nut. Torque the nut 18 ft. lbs. (25 Nm).
   - Protective cap to the shock tower
   - Lower shock mounting bolt. Torque the bolt 111 ft. lbs. (150 Nm).

➡**It may be necessary to support the lower control arm for ease of installation of the lower nut.**

5. Position the rear seats in their proper position.
6. Road test the vehicle and verify a smooth ride.

Coil Spring

REMOVAL & INSTALLATION

Front

1. Before servicing the vehicle, refer to the precautions in the beginning of this section.
2. Remove the strut from the vehicle.
3. Using a Strut Compressor Holding tool J 3289-20 place the strut in a Compressor tool J 34013-A along with an Adapter J 3413-88
4. Remove or disconnect the following:
   - Compress the spring
   - Upper bearing support nut
   - Bearing and plate assembly
5. Decompress the spring.
6. Remove or disconnect the following:
   - Upper spring support plate
   - Upper insulator
   - Strut bumper
   - Cover from the strut
   - Spring from the strut
   - Lower insulator from the strut

To install:
7. Before servicing the vehicle, refer to the precautions in the beginning of this section.
   - Lower insulator to the strut
   - Spring on the strut
   - Upper spring support plate
   - Upper insulator
   - Strut bumper
   - Cover
8. Align the spring between the upper and lower rubber isolation rings.
9. Compress the spring.
10. Install or connect the following:
    - Bearing and plate assembly
    - Upper bearing support nut. Torque the nut to 52 ft. lbs. (70 Nm).
11. Release the tension from the spring.
12. Remove the strut compressor tools from the strut.
13. Install the strut to the vehicle.
14. Road test the vehicle and verify a smooth ride and no abnormal noises from the strut and spring.

Rear

1. Before servicing the vehicle, refer to the precautions in the beginning of this section.
2. Remove or disconnect the following:
   - Retainers from the lower control arms
   - Brake pipes from the lower control arms without disconnect the brake fitting
   - Stabilizer shaft link bolts
   - Stabilizer shaft link from the lower control arms
   - Rubber exhaust insulators from the hangers. Support the exhaust system so no damage occurs to the pipes or gaskets.
   - Rear wheel speed sensor electrical connections
   - Shock lower mounting bolt after properly supporting the lower control arms
   - Support from the lower control arms and place on the rear differential
   - Rear axle cradle mounts to the body bolts
3. Lower the rear differential so that coil spring can be removed.
4. Remove or disconnect the following:
   - Rear springs with the seats on the spring
   - Seats from the spring

To install:

5. Install or connect the following:
   - Seats to the spring
   - Rear springs to the lower control arm and body. Make certain that the spring leg is aligned properly.
   - Protective shields
   - Rear differential with the cradle mounting bolts. Torque the bolts to 48 ft. lbs. (65 Nm).
   - Shock absorber. Torque the bolt to 81 ft. lbs. (110 Nm).
   - Rear wheel speed sensor electrical connections
   - Rubber exhaust insulators to the hangers
   - Stabilizer shaft link to the lower control arm. Torque the bolts to 15 ft. lbs. (20 Nm).
   - Brake pipes to the lower control arm and secure with the retainers

6. Check and/or adjust the rear toe.

Lower Ball Joint

REMOVAL & INSTALLATION

1. Before servicing the vehicle, refer to the precautions in the beginning of this section.
2. Remove or disconnect the following:
   - Front wheel
   - Lower control arm
3. Using a 0.5 inch (13mm) drill, drill out the 3 rivet heads that attach the ball stud to the lower control arm.
4. Remove the ball stud from the lower control arm.

To install:

5. Install or connect the following:
   - Ball stud to the lower control arm

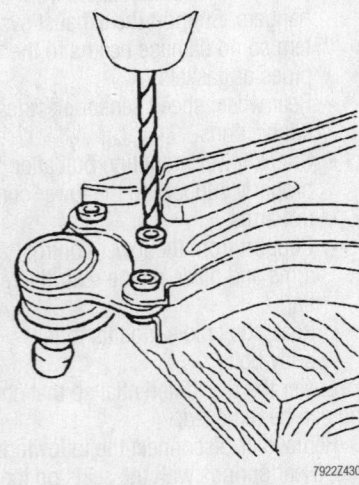

Drill out the rivets to remove the ball joint from the lower control arm

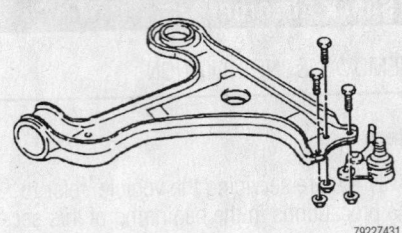

7922Z431

The replacement ball joint will be bolted to the lower control arm with the bolts supplied in the kit

   - Bolts through the upper side of the lower control arm. Torque the new bolts to 26 ft. lbs. (35 Nm).
   - Lower control arm
   - Front wheel

6. Check and/or adjust the front end alignment.

Lower Control Arm

REMOVAL & INSTALLATION

Front

1. Before servicing the vehicle, refer to the precautions in the beginning of this section.
2. Remove or disconnect the following:
   - Front wheel
   - Brake pad wear indicator rubber sleeve
   - Wheel Speed Sensor (WSS)
   - Brake hose from the strut bracket
   - Brake caliper from the steering knuckle

➡ **Do not open the hydraulic brake system.**

   - Outer tie rod from the steering knuckle
   - Stabilizer shaft nut from the shaft
   - Steering knuckle from the strut
   - Lower control arm ball stud pinch bolt
   - Steering knuckle from the lower control arm ball stud. The rotor and wheel hub will remain attached.

3. Loosen the lower control arm horizontal/vertical bolts.
4. Remove or disconnect the following:
   - Horizontal bolts
   - Vertical bolts by rotating the lower control arm from the support bracket
   - Lower control arm

To install:

5. Install or connect the following:
   - Lower control arm into the support bracket

   - Vertical bolt and hand-tighten at this time
   - Horizontal bolt by inserting it from the rear side of the vehicle. Torque all bolts to 103 ft. lbs. (140 Nm).

➡ **Be certain that the lower control arm is in the horizontal position before tightening the bolts to prevent the bushings from binding.**

   - Steering knuckle to the lower control arm ball stud. Torque the bolt to 74 ft. lbs. (100 Nm).
   - Steering knuckle to the strut with new bolts. Hand tighten the bolts at this time.
   - Stabilizer shaft link to the shaft. Torque the nut to 48 ft. lbs. (65 Nm).
   - Outer tie rod to the steering knuckle
   - New outer tie rod nut. Torque the nut 44 ft. lbs. (60 Nm).

➡ **Clean the bolt holes for the brake caliper by running a M12 x 105 tap through the holes. Use new bolts for the caliper to steering knuckle connection after applying a thread locking compound.**

   - Caliper to the steering knuckle. Torque the new bolts to 70 ft. lbs. (95 Nm) plus a 37 degree turn.
   - Brake hose to the strut bracket with a new clip
   - WSS. Torque the bolt to 71 inch lbs. (8 Nm).
   - Brake pad wear indicator sleeve to the strut
   - Front wheel

6. Adjust the wheel alignment and tighten the lower strut mounting bolts to 52 ft. lbs. (70 Nm).
7. Road test the vehicle.

Rear

1. Before servicing the vehicle, refer to the precautions in the beginning of this section.
2. Remove or disconnect the following:
   - Wheel
   - Driveshaft from the rear wheel hub flange
   - Brake pipe from the lower control arm
   - Brake caliper
   - Brake rotor
   - Parking brake cable from the actuator bracket
   - Rear hub and flange
   - Wheel bearing
   - Brake backing plate

- Exhaust system from the rubber mounts
- Rear outer tie rod from the rear control arm
- Stabilizer shaft link connection at the lower control arm
- Rear axle cradle mounting bolts

➡**Make certain that the rear differential is properly supported.**

- Lower shock mounting bolts
- Rear coil spring
- Underbody heat shield
- Propeller shaft center bearing bracket bolts
- Rear support flange bolts
- Rear support bushing bolt
- Lower control arm

To install:

3. Install or connect the following:
- Lower control arm. Torque the bolts to 74 ft. lbs. (100 Nm).
- Rear support bushing bolt. Torque the bolt to 92 ft. lbs. (125 Nm).
- Rear support flange bolts. Torque the bolts to 48 ft. lbs. (65 Nm).
- Propeller shaft center bearing bracket. Torque the bolts to 15 ft. lbs. (20 Nm).
- Underbody heat shield. Torque the bolts to 18 inch lbs. (2 Nm).
- Rear coil spring
- Rear axle cradle mount. Torque the body bolts to 48 ft. lbs. (65 Nm).
- Shock absorber. Torque the lower mount bolt to 81 ft. lbs. (110 Nm).
- Stabilizer shaft link to the lower control arm. Torque the bolt to 15 ft. lbs. (20 Nm).
- Outer tie rod. Torque the nut to 44 ft. lbs. (60 Nm).
- Exhaust system to the rubber mounts
- Brake backing plate
- Wheel bearing
- Hub and flange
- Parking brake cable and route it through the bracket in the rear lower control arm
- Brake rotor and set screw. Torque the screw to 35 inch lbs. (4 Nm).
- Brake caliper. Torque the bolts to 59 ft. lbs. (80 Nm).
- Brake pipe and clip to the lower control arm
- Driveshaft. Torque the bolts to 37 ft. lbs. (50 Nm) plus an additional 70 degree turn.
- Wheel

4. Check and/or adjust the rear toe.

CONTROL ARM BUSHING REPLACEMENT

Front

1. Before servicing the vehicle, refer to the precautions in the beginning of this section.
2. Remove the control arm from the vehicle.
3. Press the bushings from the control arm.

To install:

4. Press the new bushings into the control arm.
5. Install the control arm in the vehicle.

Rear

1. Before servicing the vehicle, refer to the precautions in the beginning of this section.

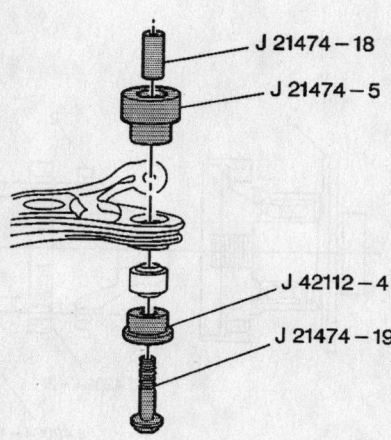

J 21474–18
J 21474–5
J 42112–4
J 21474–19

9356ZGAB

Use the following tools to install the front control arm vertical bushing

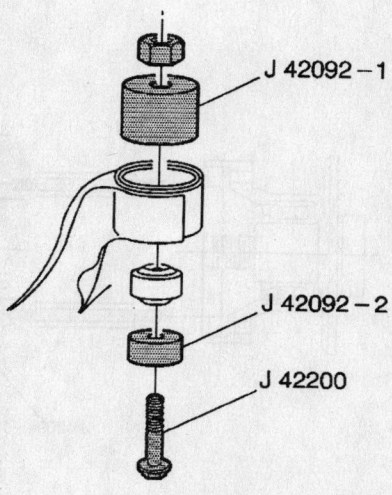

J 42092–1
J 42092–2
J 42200

9356ZGAC

Use the following tools to install the front control arm horizontal bushing

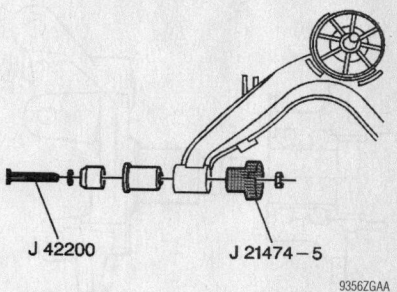

J 42200   J 21474–5

9356ZGAA

Use the following tools to remove and install the rear control arm bushing

2. Remove or disconnect the following:
- Wheel
- Rear axle lower control arm
- Collar for the inboard and outboard lower control arm bushings by cutting it off

3. Press the bushings out of the control arm.

To install:

4. Press the inboard side bushing into the control arm with the collar toward the rear differential.
5. Press the outboard side bushing into the control arm.
6. Install or connect the following:
- Lower control arm
- Wheel

Wheel Bearings

ADJUSTMENT

The wheel bearings are not adjustable.

REMOVAL & INSTALLATION

Front

1. Before servicing the vehicle, refer to the precautions in the beginning of this section.
2. Remove or disconnect the following:
- Front wheel
- Brake pad wear indicator wire from the strut bracket
- Brake hose from the strut bracket
- Caliper bolts from the steering knuckle
- Brake caliper. The brake system should remain closed.
- Brake rotor and setscrew
- Dust cap from the hub
- Hub with the bearing
- Bearing by pressing it from the hub

To install:

3. Install or connect the following:
- Bearing onto the spindle and press the outer bearing ring into the hub

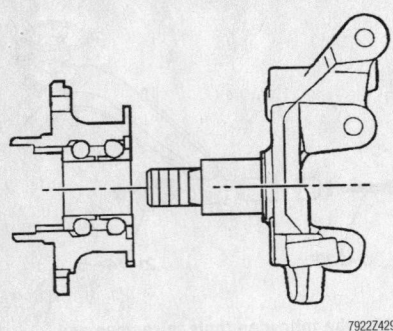

Cut-away view of the hub/wheel bearing assembly and steering knuckle

- Hub nut. Torque the nut to 236 ft. lbs. (320 Nm).
- Dust cap to the hub
- Brake rotor and setscrew. Torque the screw to 35 inch lbs. (4 Nm).
- Brake caliper to the steering knuckle after cleaning the bolt holes with an M12 x 1.5 tap and coating the new bolts with a thread locking compound. Torque the bolts to 70 ft. lbs. (95 Nm), plus a 37 degree turn.
- Brake hose to the strut bracket with a retaining clip

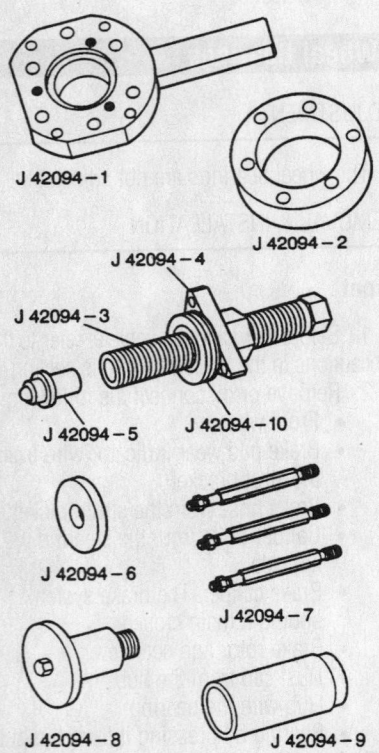

Several tools make up the J 42094 tool kit for rear wheel bearing removal and installation

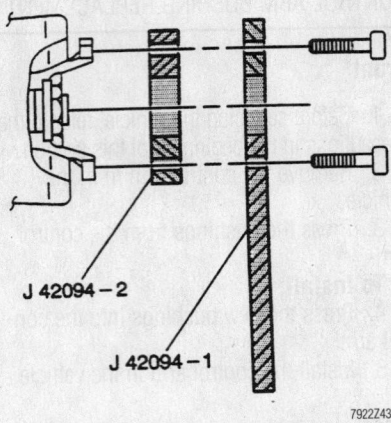

Attach the Spacer and Holding Fixture to the flange before removing the hub retaining nut

- Brake pad wear indicator wire to the strut
- Front wheel

4. Road test the vehicle.
5. Check and/or adjust the front end alignment.

Rear

➡ Several special tools are necessary to remove the rear wheel bearing from the knuckle. The tools required are as follows:

- J 36660 Torque Angle Meter
- J 42066 Holding tool
- J 42094-1 Holding Fixture
- J 42094-2 Spacer
- J 42094-3 Threaded Driver
- J 42094-4 Threaded Arbor
- J 42094-5 Ball Head

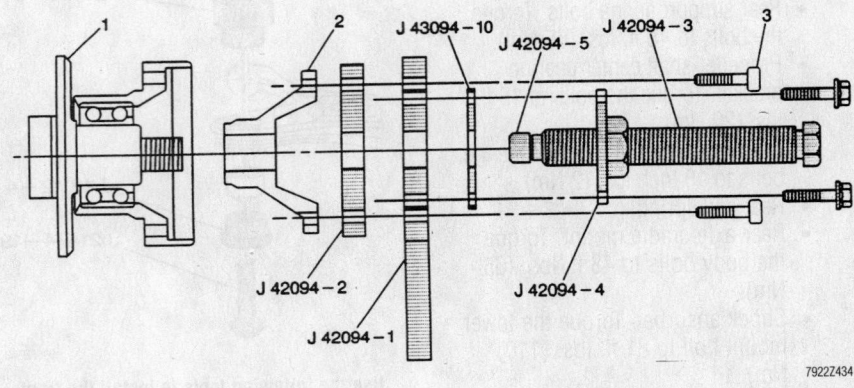

Set up the special tools as shown to remove the flange

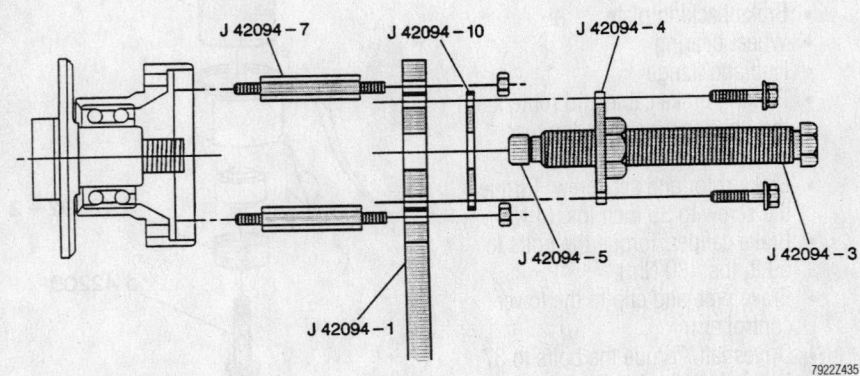

Set up the special tools as shown to remove the hub

- J 42094-6 Bearing Remover
- J 42094-7 Threaded Spacer Pin
- J 42094-8 Bearing Installer
- J 42094-9 Hub Installer
- J 42094-10 Thrust Bearing
- J 42072 Triple Hex Head (10mm) deep socket

The tool numbers mentioned are Kent-Moore tools used by GM. Equivalent tools may be available from other sources.

1. Before servicing the vehicle, refer to the precautions in the beginning of this section.

2. Remove the rear wheel.

3. Remove the outer tie rod from the knuckle.

4. Use a Holding tool J 42066 and a 0.5 inch breaker bar to counter hold the rear wheel hub.

5. Remove or disconnect the following:

- Driveshaft from the rear hub
- Brake pipe and retaining clip from the lower control arm
- Rear brake caliper and support it aside
- Set screw and brake rotor
- Three of the four brake backing plate bolts approximately 9 revolutions by installing a deep well socket
- Rear wheel hub nut by installing a Spacer J 42094-2 and Holding Fixture J 42094-1 to the rear wheel hub

6. Use a Thrust Bearing J 42094-10 as a spacer to attach a Threaded Arbor J42094-4 to the holding Fixture J 42094-7 with 3 bolts.

7. Attach a Threaded Driver into the threaded arbor.

8. Screw the bolts into the backing plate and attach the holding fixture with the stem upward.

9. Press out the rear wheel hub.

➡ **Damage to the wheel bearing seal is possible while pressing out the rear wheel hub. Inspect the seal and replace if needed.**

10. Remove the wheel bearing retaining ring.

11. Attach a Bearing Remover J 42094-6 to the end of the threaded driver.

12. Turn the driver in a clockwise manner to press out the wheel bearing.

To install:

13. Install or connect the following:

- Wheel Bearing Installer Tool J 42094-8 through the wheel bearing and on to the threaded driver
- Wheel bearing until fully seated by turning the threaded driver

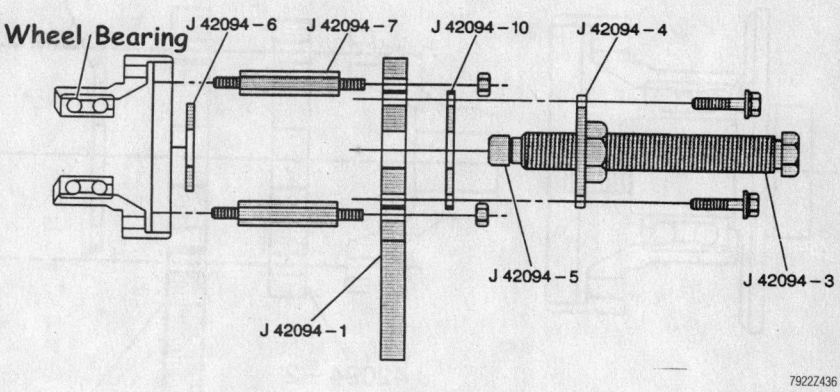

Set up the special tools as shown to remove the bearing assembly

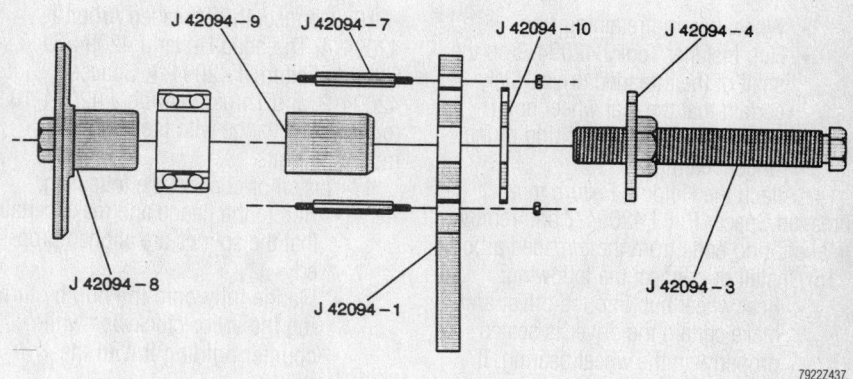

Set up the special tools as shown to install the bearing assembly

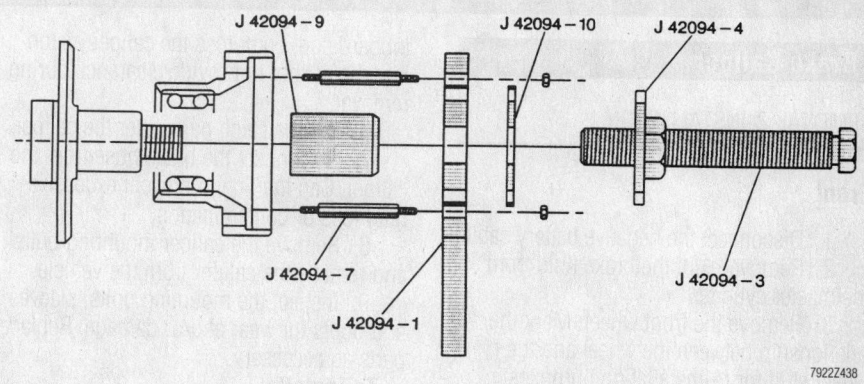

Set up the special tools as shown to pull the hub into the bearing assembly

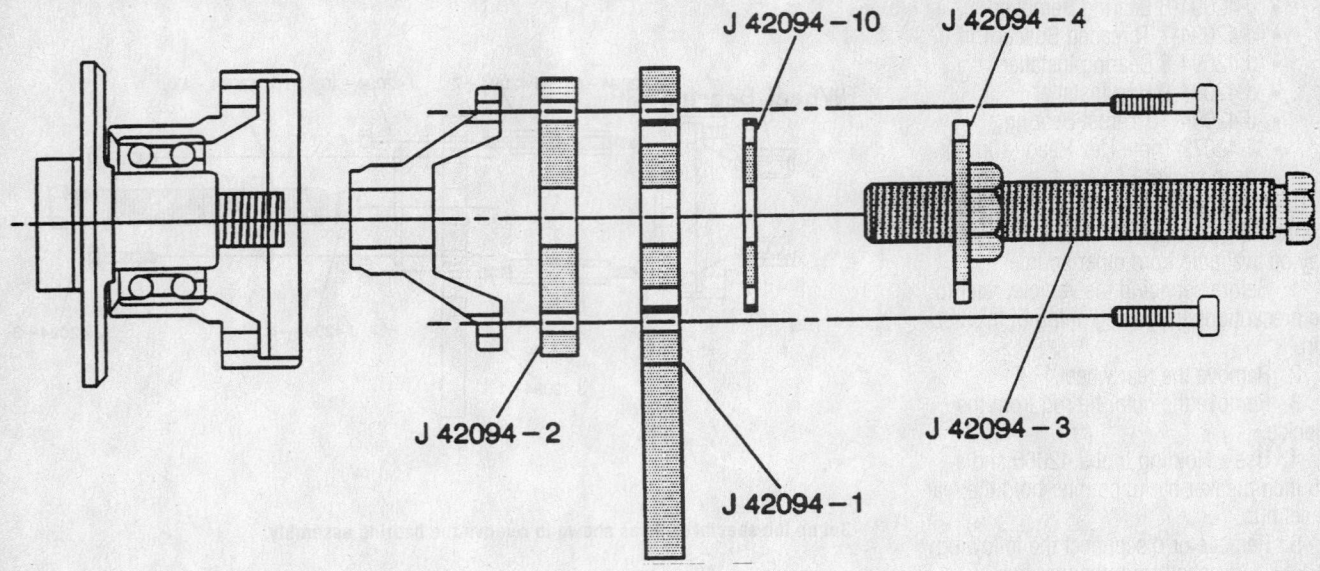

J 42094 – 10   J 42094 – 4

J 42094 – 2

J 42094 – 1

J 42094 – 3

79222439

Set up the special tools as shown to install the flange on the hub

- Wheel bearing retaining ring
- Hub Installer Tool J 42094-9 on the shaft of the threaded driver. Make certain that the rear wheel hub installer is on the inner ring of the wheel bearing

14. Attach the Holding Fixture to a Threaded Spacer Pin J 42094-7 and remove the anchoring bolts from the threaded arbor.

15. Install or connect the following:
- Rear wheel hub into the driver and make certain the driver is seated properly on the wheel bearing. If not centered properly, it may cause the hub to bind
- Wheel hub into the wheel bearing by holding the driver and turning the arbor clockwise. When fully seated, remove the tools

16. Connect the Threaded Arbor J 42094-4, Threaded Driver J 42094-3, Holding Fixture J 42094-1, Spacer J 42094-2, and Thrust Bearing J 42094-10 to the wheel flange with the halfshaft mounting bolts.

17. Install or connect the following:
- Hub to the flange and make certain that the splines are aligned properly
- Flange fully onto the hub by turning the arbor clockwise while counter holding it with the driver

18. Remove the thrust bearing, arbor and driver while leaving the holding fixture and spacer attached to the hub.

19. Install the rear wheel hub nut. Torque the nut 221 ft. lbs. (300 Nm).

20. Remove the spacer and holding fixture from the hub.

21. Install or connect the following:
- Retaining washer to the hub
- Backing plate bolts. Torque the bolts to 37 ft. lbs. (50 Nm) plus a 40 degree turn.
- Brake rotor and set screw. Torque the screw to 35 inch lbs. (4 Nm).
- Brake caliper. Torque the bolts to 59 ft. lbs. (80 Nm).
- Brake pipe to the lower control arm and secure it with a clip
- Driveshaft to the hub flange. Torque the bolts to 37 ft. lbs. (50 Nm) plus an additional 70 degree turn.
- Tie rod end. Torque the nut to 44 ft. lbs. (60 Nm).
- Rear wheel

BRAKES

Brake Caliper

REMOVAL & INSTALLATION

Front

1. Disconnect the negative battery cable.
2. Remove ⅔ of the brake fluid from the master cylinder.
3. Remove the front wheel. Mark the relationship between the wheel and the wheel stud for re-installation purposes.
4. Install 2 wheel nuts to keep the rotor in place.
5. Using a large C-clamp against the inboard pad, compress the caliper piston into the caliper to provide clearance during removal.
6. Place a catch pan under the caliper.
7. Disconnect the brake hose from the caliper. Cap the line to prevent excessive fluid loss or contamination.
8. Remove the caliper mounting bolts and remove the caliper from the vehicle.
9. Inspect the mounting bolts; sleeves and boots for wear and/or damage. Replace parts as necessary.

To install:
10. Before installing the caliper, make sure the piston is fully seated in the bore and the brake pads are properly seated.

11. Lubricate the mounting bolt shafts and inner diameter of the sleeves with silicone grease.
12. Install the caliper in the caliper mounting bracket and install the mounting bolts. Torque the mounting bolts to 38 ft. lbs. (51 Nm).
13. Connect the brake hose with the bolt and new gaskets. Torque the brake hose bolt to 33 ft. lbs. (45 Nm).
14. Refill the master cylinder and bleed the brake system.
15. Remove the 2 wheel nuts securing the rotor.
16. Install the wheel and tire assembly.
17. Connect the negative battery cable.

18. Road test the vehicle for proper brake system operation.

Rear

1. Before servicing the vehicle, refer to the precautions in the beginning of this section.
2. Remove the wheel(s).
3. Attach a hose to the bleeder screw on the caliper.
4. Open the bleeder screw.
5. Compress the pistons into the caliper housing to provide clearance.
6. Use a punch to drive the caliper retaining pins out of the caliper from the outside inward.
7. Remove or disconnect the following:
   - Brake caliper spring retainer
   - Brake pads
   - Brake caliper pipe
   - Caliper mounting bolts
   - Brake caliper

To install:

8. Install or connect the following:
   - Brake caliper. Torque the bolts to 59 ft. lbs. (80 Nm).
   - Brake caliper pipe. Torque the fitting to 12 ft. lbs. (16 Nm).
   - Brake pads
9. Install one caliper retaining pin.
10. Install the caliper spring retainer.
11. Install the second retaining pin.
12. Fill and bleed the brake system.
13. Install the wheel(s).

Disc Brake Pads

REMOVAL & INSTALLATION

Front

1. Remove ⅔ of the brake fluid from the master cylinder reservoir.
2. Remove the front wheel.
3. Remove the caliper mounting bolts.
4. Remove the caliper from the steering knuckle without disconnecting the brake hose.
5. Suspend the caliper from the coil spring with wire. Do not let the caliper hang from the brake hose.
6. Remove brake pads from anchor bracket.

To install:

7. Fully seat the caliper piston into the bore using a large C-clamp.
8. Lubricate the brake caliper mounting surfaces.
9. Install the brake pads onto the anchor bracket using new shims and clips.
10. Place the caliper onto the anchor bracket and install the mounting bolts.

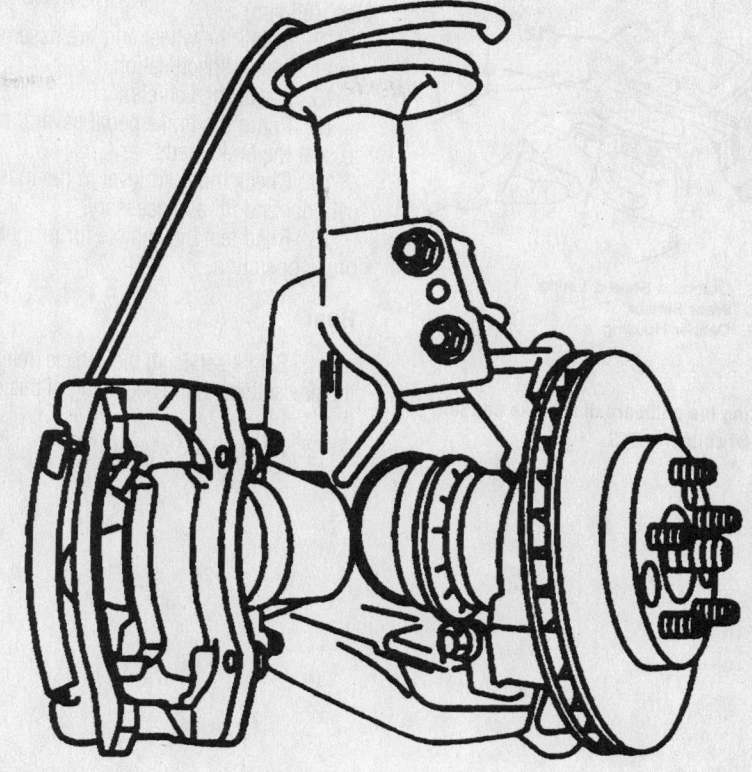

Supporting the front caliper—CTS

93006G67

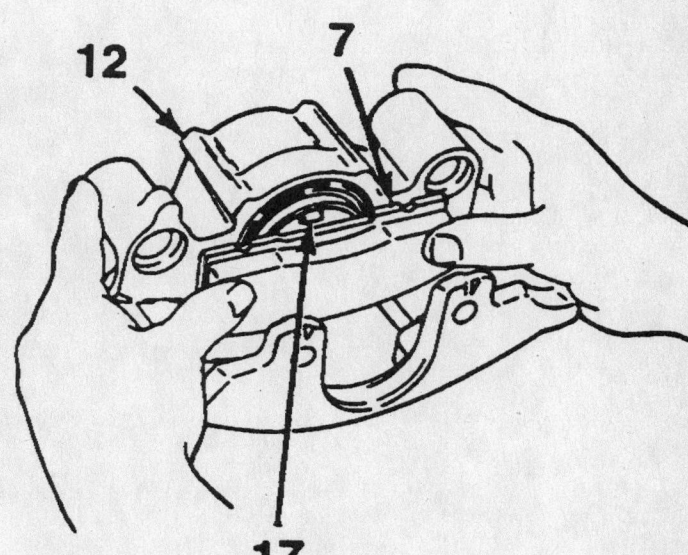

7. **Inboard Shoe & Lining**
12. **Caliper Housing**
17. **Shoe Retainer Spring**

93006G68

Installing the inboard disc brake pad into the front caliper—CTS

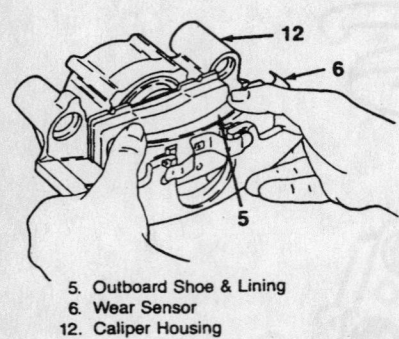

5. Outboard Shoe & Lining
6. Wear Sensor
12. Caliper Housing

93006G69

Installing the outboard disc brake pad in the front caliper—CTS

11. Torque the mounting bolts to 63 ft. lbs. (85 Nm).

12. Install the wheel and tire assembly and torque to specification.

13. Lower the vehicle.

14. Pump the brake pedal several times to seat the brake pads.

15. Check the fluid level in the master cylinder and fill as necessary.

16. Road test the vehicle for proper brake operation.

Rear

1. Before servicing the vehicle, refer to the precautions in the beginning of this section.

2. Remove the wheel(s).

3. Attach a hose to the bleeder screw on the caliper.

4. Open the bleeder screw.

5. Compress the pistons into the caliper housing to provide clearance.

6. Use a punch to drive the caliper retaining pins out of the caliper from the outside inward.

7. Remove or disconnect the following:
- Brake caliper spring retainer
- Brake pads

To install:

8. Install the brake pads.

9. Install one caliper retaining pin.

10. Install the caliper spring retainer.

11. Install the second retaining pin.

12. Fill and bleed the brake system.

13. Install the wheel(s).

CHEVROLET AND GMC

4

Astro • Safari

SPECIFICATION CHARTS

ENGINE AND VEHICLE IDENTIFICATION

Code ①	Liters (cc)	Cu. In.	Cyl.	Fuel Sys.	Engine Type	Eng. Mfg.
W	4.3 (4293)	263	6	MFI	OHV	CPC

CPC: Chevrolet/Pontiac/Canada

MFI: Multi-port Fuel Injection

① 8th position of VIN

② 10th position of VIN

Code ②	Year
Y	2000
1	2001
2	2002
3	2003
4	2004

42372-ASTR-C01

GENERAL ENGINE SPECIFICATIONS
All measurements are given in inches.

Year	Model	Engine Displacement Liters (cc)	Engine Series (ID/VIN)	Fuel System	Net Horsepower @ rpm	Net Torque @ rpm (ft. lbs.)	Bore x Stroke (in.)	Compression Ratio	Oil Pressure @ rpm
2000	Astro/Safari	4.3 (4293)	W	MFI	190@4400	250@2800	3.74x3.48	9.2:1	18@2000
2001	Astro/Safari	4.3 (4293)	W	MFI	190@4400	250@2800	3.74x3.48	9.2:1	18@2000
2002	Astro/Safari	4.3 (4293)	W	MFI	190@4400	250@2800	4.00x3.48	9.2:1	18@2000
2003-04	Astro/Safari	4.3 (4293)	W	MFI	190@4400	250@2800	4.00x3.48	9.2:1	18@2000

MFI: Multi-port Fuel Injection

42372-ASTR-C02

GASOLINE ENGINE TUNE-UP SPECIFICATIONS

Year	Engine Displacement Liters (cc)	Engine ID/VIN	Spark Plugs Gap (in.)	Ignition Timing (deg.)		Fuel Pump (psi)	Idle Speed (rpm)		Valve Clearance	
				MT	AT		MT	AT	In.	Ex.
2000	4.3 (4293)	W	0.060	①	①	58-64 ②	600	625	HYD	HYD
2001	4.3 (4293)	W	0.060	①	①	58-64 ②	600	625	HYD	HYD
2002	4.3 (4293)	W	0.060	①	①	58-64 ②	600	625	HYD	HYD
2003-04	4.3 (4293)	W	0.060	①	①	58-64 ②	600	625	HYD	HYD

NOTE: The Vehicle Emission Control Information label often reflects specification changes made during production. The label figures must be used if they differ from those in this chart.

HYD: Hydraulic

① Ignition timing is preset and cannot be adjusted

② With key ON and engine OFF

42372-ASTR-C03

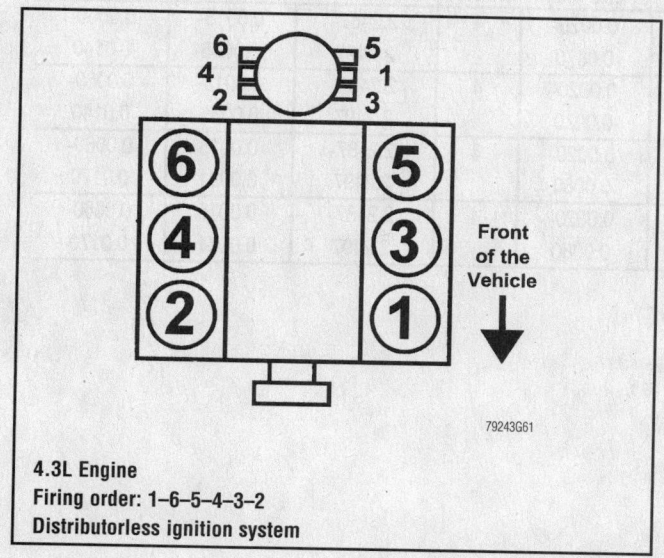

4.3L Engine
Firing order: 1–6–5–4–3–2
Distributorless ignition system

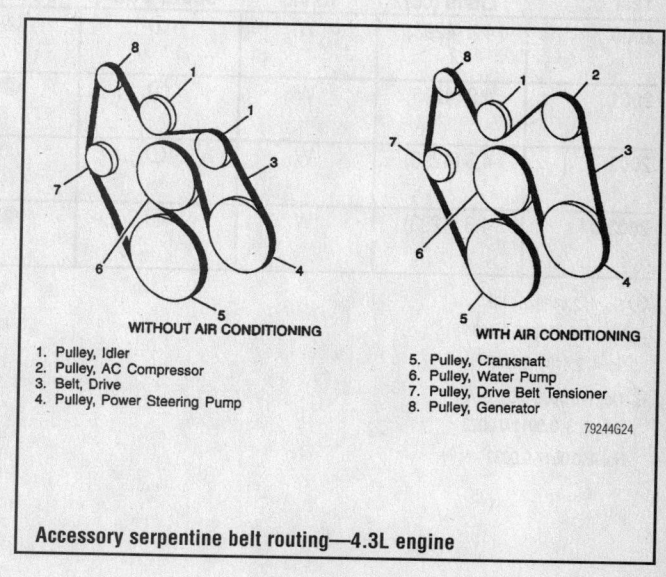

WITHOUT AIR CONDITIONING

1. Pulley, Idler
2. Pulley, AC Compressor
3. Belt, Drive
4. Pulley, Power Steering Pump

WITH AIR CONDITIONING

5. Pulley, Crankshaft
6. Pulley, Water Pump
7. Pulley, Drive Belt Tensioner
8. Pulley, Generator

79244G24

Accessory serpentine belt routing—4.3L engine

CAPACITIES

Year	Model	Engine Displacement Liters (cc)	Engine ID/VIN	Engine Oil with Filter (qts.)	Transmission (pts.) 5-Spd	Transmission (pts.) Auto.	Transfer Case (pts.)	Drive Axle Front (pts.)	Drive Axle Rear (pts.)	Fuel Tank (gal.)	Cooling System (qts.)
2000	Astro	4.3 (4293)	W	5.0	—	10.0	3.0	2.6	3.8	25.0	①
	Safari	4.3 (4293)	W	5.0	—	10.0	3.0	2.6	3.8	27.0	①
2001	Astro	4.3 (4293)	W	5.0	—	10.0	3.0	2.6	3.8	25.0	①
	Safari	4.3 (4293)	W	5.0	—	10.0	3.0	2.6	3.8	27.0	①
2002	Astro	4.3 (4293)	W	4.5	—	10.0	3.0	2.6	3.5	27.0	②
	Safari	4.3 (4293)	W	4.5	—	10.0	3.0	2.6	3.5	27.0	②
2003-04	Astro	4.3 (4293)	W	4.5	—	10.0	3.0	2.6	3.5	27.0	②
	Safari	4.3 (4293)	W	4.5	—	10.0	3.0	2.6	3.5	27.0	②

NOTE: All capacities are approximate. Add fluid gradually and check to be sure a proper fluid level is obtained.

① With rear heater: 16.5 qts.
   Without rear heater: 14.3 qts.

② With rear heater: 16.5 qts.

42372-ASTR-C04

CRANKSHAFT AND CONNECTING ROD SPECIFICATIONS
All measurements are given in inches.

Year	Engine Displacement Liters (cc)	Engine ID/VIN	Crankshaft Main Brg. Journal Dia.	Crankshaft Main Brg. Oil Clearance	Crankshaft Shaft End-play	Crankshaft Thrust on No.	Connecting Rod Journal Diameter	Connecting Rod Oil Clearance	Connecting Rod Side Clearance
2000	4.3 (4293)	W	①	②	0.0020-0.0070	4	2.2487-2.2497	0.0013-0.0035	0.0060-0.0140
2001	4.3 (4293)	W	①	②	0.0020-0.0070	4	2.2487-2.2497	0.0013-0.0035	0.0060-0.0140
2002	4.3 (4293)	W	①	②	0.0020-0.0080	4	2.2487-2.2497	0.0015-0.0031	0.0060-0.0170
2003-04	4.3 (4293)	W	①	②	0.0020-0.0080	4	2.2487-2.2497	0.0015-0.0031	0.0060-0.0170

① No. 1: 2.4488-2.4495
   Nos. 2, 3: 2.4485-2.4494
   No. 4: 2.4480-2.4489

② No. 1: 0.0008-0.0020
   Nos. 2, 3: 0.0011-0.0023
   No. 4: 0.0017-0.0032

42372-ASTR-C05

VALVE SPECIFICATIONS

Year	Engine Displacement Liters (cc)	Engine ID/VIN	Seat Angle (deg.)	Face Angle (deg.)	Spring Test Pressure (lbs. @ in.)	Spring Installed Height (in.)	Stem-to-Guide Clearance (in.)		Stem Diameter (in.)	
							Intake	Exhaust	Intake	Exhaust
2000	4.3 (4293)	W	46	45	187-203@1.27	1.69-1.71	0.0010-0.0027	0.0010-0.0027	NA	NA
2001	4.3 (4293)	W	46	45	187-203@1.27	1.69-1.71	0.0010-0.0027	0.0010-0.0027	NA	NA
2002	4.3 (4293)	W	46	45	187-203@1.27	1.67-1.70	0.0010-0.0027	0.0010-0.0027	NA	NA
2003-04	4.3 (4293)	W	46	45	187-203@1.27	1.67-1.70	0.0010-0.0027	0.0010-0.0027	NA	NA

NA: Not Available

42372-ASTR-C06

PISTON AND RING SPECIFICATIONS

All measurements are given in inches.

Year	Engine Displacement Liters (cc)	Engine ID/VIN	Piston Clearance	Ring Gap			Ring Side Clearance		
				Top Compression	Bottom Compression	Oil Control	Top Compression	Bottom Compression	Oil Control
2000	4.3 (4293)	W	0.0007-0.0017	0.010-0.030	0.018-0.026	0.065 Max.	0.0042 Max.	0.0042 Max.	0.0020-0.0070
2001	4.3 (4293)	W	0.0007-0.0024	0.010-0.016	0.015-0.023	0.010-0.029	0.0012-0.0027	0.0015-0.0031	0.0020-0.0070
2002	4.3 (4293)	W	0.0007-0.0024	0.010-0.016	0.015-0.023	0.010-0.029	0.0012-0.0027	0.0015-0.0031	0.0020-0.0070
2003-04	4.3 (4293)	W	0.0007-0.0024	0.010-0.016	0.015-0.023	0.010-0.029	0.0012-0.0027	0.0015-0.0031	0.0020-0.0070

42372-ASTR-C07

TORQUE SPECIFICATIONS
All readings in ft. lbs.

Year	Engine Displacement Liters (cc)	Engine ID/VIN	Cylinder Head Bolts	Main Bearing Bolts	Rod Bearing Bolts	Crankshaft Damper Bolts	Flywheel Bolts	Manifold Intake	Manifold Exhaust	Spark Plugs	Lug Nut
2000	4.3 (4293)	W	①	77	②	74	74	③	④	11	90
2001	4.3 (4293)	W	①	77	②	74	74	③	④	11	100
2002	4.3 (4293)	W	①	77	②	70	74	③	④	11	100
2003-04	4.3 (4293)	W	①	77	②	70	74	③	④	11	100

① 1st pass: 22 ft. lbs.
2nd pass:
Short bolts: Plus 55 degrees
Medium bolts: Plus 65 degrees
Long bolts: Plus 75 degrees

② 20 ft. lbs. plus 70 degrees

③ Lower intake manifold:
1st pass: 27 inch lbs.
2nd pass: 106 inch lbs.
Final pass: 11 ft. lbs.
Upper manifold bolts:
1st pass: 44 inch lbs.
2nd pass: 88 inch lbs.

④ Tighten bolts to 12 ft. lbs.
Retorque to 22 ft. lbs.

42372-ASTR-C08

WHEEL ALIGNMENT

Year	Model		Caster Range (+/-Deg.)	Caster Preferred Setting (Deg.)	Camber Range (+/-Deg.)	Camber Preferred Setting (Deg.)	Toe-in (in.)	Steering Axis Inclination (Deg.)
2000	RWD	Left	1.00	+3.00	1.00	+1.20	0+/-0.20	—
		Right	1.00	+3.50	1.00	+1.20	0+/-0.20	—
	AWD	Left	1.00	+3.50	1.00	0	0+/-0.20	—
		Right	1.00	+4.50	1.00	0	0+/-0.20	—
2001	RWD	Left	1.00	+3.00	1.00	+1.20	0+/-0.20	—
		Right	1.00	+3.50	1.00	+1.20	0+/-0.20	—
	AWD	Left	1.00	+3.50	1.00	0	0+/-0.20	—
		Right	1.00	+4.50	1.00	0	0+/-0.20	—
2002	RWD	Left	1.00	+3.00	1.00	+0.60	0+/-0.20	—
		Right	1.00	+3.50	1.00	+0.60	0+/-0.20	—
	AWD	Left	1.00	+3.50	1.00	0	0+/-0.20	—
		Right	1.00	+4.50	1.00	0	0+/-0.20	—
2003-04	RWD	Left	1.00	+3.00	1.00	+0.60	0+/-0.20	—
		Right	1.00	+3.50	1.00	+0.60	0+/-0.20	—
	AWD	Left	1.00	+3.50	1.00	0	0+/-0.20	—
		Right	1.00	+4.50	1.00	0	0+/-0.20	—

42372-ASTR-C09

BRAKE SPECIFICATIONS
All measurements in inches unless noted

Year	Model		Brake Disc Original Thickness	Minimum Thickness	Maximum Runout	Brake Drum Diameter Original Inside Diameter	Max. Wear Limit	Maximum Machine Diameter	Minimum Lining Thickness	Brake Caliper Bracket Bolts (ft. lbs.)	Mounting Bolts (ft. lbs.)
2000	Astro	F	①	②	0.004	—	—	—	0.030	NA	38
		R	—	—	—	9.50	9.59	9.56	0.030	NA	
	Safari	F	①	②	0.004	—	—	—	0.030	NA	38
		R	—	—	—	9.50	9.59	9.56	0.030	NA	
2001	Astro	F	①	②	0.004	—	—	—	0.030	NA	38
		R	—	—	—	9.50	9.59	9.56	0.030	NA	
	Safari	F	①	②	0.004	—	—	—	0.030	NA	38
		R	—	—	—	9.50	9.59	9.56	0.030	NA	
2002	Astro	F	①	②	0.004	—	—	—	0.030	NA	38
		R	—	—	—	9.50	9.59	9.56	0.030	NA	
	Safari	F	①	②	0.004	—	—	—	0.030	NA	38
		R	—	—	—	9.50	9.59	9.56	0.030	NA	
2003-04	Astro	F	①	②	0.004	—	—	—	0.030	NA	38
		R	—	—	—	9.50	9.59	9.56	0.030	NA	
	Safari	F	①	②	0.004	—	—	—	0.030	NA	38
		R	—	—	—	9.50	9.59	9.56	0.030	NA	

NA: Not Available

① Available with 1.040" and 1.250" rotors

② 1.040" rotors: 0.980
1.250" rotors: 1.230

42372-ASTR-C10

TIRE, WHEEL AND BALL JOINT SPECIFICATIONS

| Year | Model | OEM Tires | | Tire Pressures (psi) | | Wheel Size | Ball Joint Inspection |
		Standard	Optional	Front	Rear		
2000	Astro/Safari	P215/75R15	None	36	36	6-JJ	U: 0.125 in. L ①
2001	Astro/Safari	P215/75R15	None	36	36	6.5-JJ	U: 0.125 in. L ①
2002	Astro/Safari	P215/75R15	None	36	36	6.5-JJ	U: 0.125 in. L ①
2003-04	Astro/Safari	P215/75R15	None	36	36	6.5-JJ	U: 0.125 in. L ①

OEM: Original Equipment Manufacturer

PSI: Pounds Per Square Inch

STD: Standard

OPT: Optional

L: Lower

U: Upper

① Do not lift truck. Inspect the boss into which the grease fitting is threaded. Replace if the boss is flush or receded below the surface of the ball joint

42372-ASTR-C11

SCHEDULED MAINTENANCE INTERVALS
CHEVROLET—ASTRO & GMC—SAFARI

TO BE SERVICED	TYPE OF SERVICE	VEHICLE MILEAGE INTERVAL (x1000)															
		7.5	15	22.5	30	37.5	45	52.5	60	67.5	75	82.5	90	97.5	105	112.5	120
Accessory drive belt	S/I								✓								✓
Air cleaner filter	R			✓					✓				✓				✓
Automatic transmission fluid	R	Every 50,000 miles															
Brake system ①	S/I	✓	✓	✓	✓	✓	✓	✓	✓	✓	✓	✓	✓	✓	✓	✓	✓
Chassis & suspension grease points	L	✓	✓	✓	✓	✓	✓	✓	✓	✓	✓	✓	✓	✓	✓	✓	✓
CV-joint boots & axle seals	S/I	✓	✓	✓	✓	✓	✓	✓	✓	✓	✓	✓	✓	✓	✓	✓	✓
Engine coolant system ②	S/I	Every 150,000 miles															
Engine oil & filter	R	✓	✓	✓	✓	✓	✓	✓	✓	✓	✓	✓	✓	✓	✓	✓	✓
Front wheel bearings	S/I & L				✓				✓				✓				✓
Fuel filter	R				✓				✓				✓				✓
Fuel tank, cap & lines	S/I								✓								✓
PCV valve	S/I	Every 100,000 miles															
Rear/front axle fluid level	S/I	✓	✓	✓	✓	✓	✓	✓	✓	✓	✓	✓	✓	✓	✓	✓	✓
Rotate tires	S/I	✓	✓	✓	✓	✓	✓	✓	✓	✓	✓	✓	✓	✓	✓	✓	✓
Spark plug wires	S/I	Every 100,000 miles															
Spark plugs	R	Every 100,000 miles															

R: Replace   S/I: Inspect and service, if necessary   L: Lubricate

① This should be performed when the tires are removed for rotation.

② Drain, flush and refill the cooling system, inspect the system hoses, and clean the radiator and condenser.

FREQUENT OPERATION MAINTENANCE (SEVERE SERVICE)

If a vehicle is operated under any of the following conditions it is considered severe service:

- Towing a trailer or using a camper or car-top carrier.

- Repeated short trips of less than 5 miles in temperatures below freezing, or trips of less than 10 miles in any temperature.

- Extensive idling or low-speed driving for long distances as in heavy commercial use, such as delivery, taxi or police cars.

- Operating on rough, muddy or salt-covered roads.

- Operating on unpaved or dusty roads.

- Driving in extremely hot (over 90°) conditions.

Engine oil & filter: replace every 3000 miles or 3 months, whichever occurs first.

Chassis and suspension grease points: lubricate every 3000 miles.

Rear/front axle fluid level: inspect every 3000 miles.

Rotate the tires ever 6000 miles.

Brake system components: inspect ever 6000 miles.

Front wheel bearings (2-wheel drive only): clean, inspect and repack every 15,000 miles.

Air cleaner filter: inspect every 15,000 miles.

Automatic transmission fluid & filter: replace every 15,000 miles.

42372-ASTR-C12

PRECAUTIONS

Before servicing any vehicle, please be sure to read all of the following precautions, which deal with personal safety, prevention of component damage, and important points to take into consideration when servicing a motor vehicle:

• Never open, service or drain the radiator or cooling system when the engine is hot; serious burns can occur from the steam and hot coolant.

• Observe all applicable safety precautions when working around fuel. Whenever servicing the fuel system, always work in a well-ventilated area. Do not allow fuel spray or vapors to come in contact with a spark, open flame, or excessive heat (a hot drop light, for example). Keep a dry chemical fire extinguisher near the work area. Always keep fuel in a container specifically designed for fuel storage; also, always properly seal fuel containers to avoid the possibility of fire or explosion. Refer to the additional fuel system precautions later in this section.

• Fuel injection systems often remain pressurized, even after the engine has been turned **OFF**. The fuel system pressure must be relieved before disconnecting any fuel lines. Failure to do so may result in fire and/or personal injury.

• Brake fluid often contains polyglycol ethers and polyglycols. Avoid contact with the eyes and wash your hands thoroughly after handling brake fluid. If you do get

brake fluid in your eyes, flush your eyes with clean, running water for 15 minutes. If eye irritation persists, or if you have taken brake fluid internally, IMMEDIATELY seek medical assistance.

• The EPA warns that prolonged contact with used engine oil may cause a number of skin disorders, including cancer! You should make every effort to minimize your exposure to used engine oil. Protective gloves should be worn when changing oil. Wash your hands and any other exposed skin areas as soon as possible after exposure to used engine oil. Soap and water, or waterless hand cleaner should be used.

• All new vehicles are now equipped with an air bag system. The system must be disabled before performing service on or around system components, steering column, instrument panel components, wiring and sensors. Failure to follow safety and disabling procedures could result in accidental air bag deployment, possible personal injury and unnecessary system repairs.

• Always wear safety goggles when working with, or around, the air bag system. When carrying a non-deployed air bag, be sure the bag and trim cover are pointed away from your body. When placing a non-deployed air bag on a work surface, always face the bag and trim cover upward, away from the surface. This will reduce the motion of the module if it is accidentally

deployed. Refer to the additional air bag system precautions later in this section.

• Clean, high quality brake fluid from a sealed container is essential to the safe and proper operation of the brake system. You should always buy the correct type of brake fluid for your vehicle. If the brake fluid becomes contaminated, completely flush the system with new fluid. Never reuse any brake fluid. Any brake fluid that is removed from the system should be discarded. Also, do not allow any brake fluid to come in contact with a painted surface; it will damage the paint.

• Never operate the engine without the proper amount and type of engine oil; doing so WILL result in severe engine damage.

• Timing belt maintenance is extremely important! Many models utilize an interference-type, non-freewheeling engine. If the timing belt breaks, the valves in the cylinder head may strike the pistons, causing potentially serious (also time-consuming and expensive) engine damage.

• Disconnecting the negative battery cable on some vehicles may interfere with the functions of the on-board computer system(s) and may require the computer to undergo a relearning process once the negative battery cable is reconnected.

• When servicing drum brakes, only disassemble and assemble one side at a time, leaving the remaining side intact for reference.

ENGINE REPAIR

Distributor

REMOVAL

1. Before servicing the vehicle, refer to the precautions in the beginning of this section.

2. Remove or disconnect the following:
• Negative battery cable
• Spark plug wires and the coil leads from the distributor
• Electrical connector from the distributor
• Distributor cap fasteners and the cap

3. Using a marker, matchmark the rotor-to-housing and housing-to-intake manifold positions so that they can be matched during installation.
• Distributor hold-down bolt
• Distributor from the engine

4. As the distributor is being removed from the engine the rotor will move in a counterclockwise direction about 42°. This will appear as slightly more than one clock position.

5. Place a second mark on the distributor to mark the position of the rotor segment. This will help to ensure the correct rotor alignment when installing the distributor.

INSTALLATION

Engine Not Disturbed

1. If installing a new distributor, place two marks on the new distributor housing in the same position as the marks on the old distributor housing.

2. Align the rotor with the second mark made on the distributor.

3. Install the distributor in the engine making sure that the mounting hole in the

distributor hold-down base is aligned over the mounting hole in the intake manifold.

4. As you are installing the distributor, watch the rotor move in a clockwise direction about 42°.

5. Once the distributor is fully seated, the rotor should be aligned with the first mark made on the distributor housing. If the rotor is not aligned with the first mark made on the housing, the distributor and camshaft teeth have meshed one or more teeth out of alignment. If this is the case, remove the distributor and reinstall it so that all the marks are aligned.

6. Install or connect the following:
• Hold-down bolt and tighten the bolt to 18 ft. lbs. (25 Nm)
• Distributor cap and engage the electrical connector to the distributor
• Spark plug wires and coil leads
• Negative battery cable

Engine Disturbed

1. Remove the No. 1 cylinder spark plug. Turn the engine using a socket wrench on the large bolt on the front of the crankshaft pulley. Place a finger near the No. 1 spark plug hole and turn the crankshaft until the piston reaches Top Dead Center (TDC). As the engine approaches TDC, you will feel air being expelled by the No. 1 cylinder. If the position is not being met, turn the engine another full turn (360 degree). Once the engine position is correct, install the spark plug.

2. Align the cast arrow in the distributor housing, the driven gear roll pin and the pre-drilled indent hole in the distributor driven gear. If the driven gear is installed correctly, the dimple will be approximately 180° opposite the rotor segment when it is installed in the distributor.

➡️**Installing the distributor 180° out of alignment, or locating the rotor in the wrong holes, may cause a no start condition or can cause premature engine damage and wear.**

3. Make sure the rotor is pointing to the cap hold-down mount nearest the flat side of the housing.

4. Using a long screwdriver, align the oil pump drive shaft in the engine in the mating drive tab in the distributor.

5. Install the distributor in the engine. Make sure the spark plug towers are perpendicular to the centerline of the engine.

6. When the distributor is fully seated, the rotor segment should be aligned with the pointer cast in the distributor base. The pointer will have a "6" cast into it indicating a 6 cylinder engine. If the rotor segment is not within a few degrees of the pointer, the distributor gear may be off a tooth or more. If this is the case repeat the process until the rotor aligns with the pointer.

7. Install the cap and fasten the mounting screws.

8. Tighten the distributor mounting bolt to 18 ft. lbs. (25 Nm).

9. Engage the electrical connections and the spark plug wires.

Alternator

REMOVAL

1. Before servicing the vehicle, refer to the precautions in the beginning of this section.

2. Remove or disconnect the following:
- Negative battery cable

- Air inlet duct or air cleaner assembly, if necessary
- Accessory belt
- Heater hose brace, if necessary
- Wires
- Mounting bolts
- Alternator

INSTALLATION

Install or connect the following:
- Alternator and loosely install the mounting bolts
- Tighten the rear bolt to 37 ft. lbs. (50 Nm) and the front bolt to 18 ft. lbs. (25 Nm)
- Tighten the brace-to-alternator and brace-to-intake retainers to 18 ft. lbs. (25 Nm). Tighten the brace-to-engine stud nut to 37 ft. lbs. (50 Nm)
- Wires and the battery feed wire nut
- Heater hose bracket
- Accessory belt
- Negative battery cable

Ignition Timing

ADJUSTMENT

The ignition timing is preset and cannot be adjusted.

Engine Assembly

REMOVAL & INSTALLATION

➡️**The engine assembly is removed from the bottom of the vehicle. A special engine lifting table is necessary to perform the following procedure.**

1. Before servicing the vehicle, refer to the precautions in the beginning of this section.

2. Disconnect the negative and positive battery cables.

3. Discharge the air conditioning refrigerant.

4. Drain the coolant.

5. Drain the crankcase.

6. Remove or disconnect the following:
- Engine cover
- Battery
- Air cleaner assembly
- Throttle cable and the cruise control cable (if equipped) from the throttle body
- Air conditioning lines at the condenser and accumulator
- Radiator
- Power steering reservoir and drain the fluid
- Lines from the Hydroboost unit

- Master cylinder from the Hydroboost unit and secure it to the oil fill tube
- Steering shaft from the steering gear
- Heater hoses and vacuum lines from the engine
- Fuse box and wiring harness from the bulkhead connector

➡️**The engine/transmission assembly is removed from the bottom of the vehicle. Raise the vehicle so the rear of the vehicle is slightly higher than the front. When the frame bolts are removed, the body will be lifted away from the engine/transmission assembly.**

- Driveshaft. Matchmark it for reassembly prior to removal.
- Starter and the starter opening cover
- Torque converter bolts through the starter opening
- Shift linkage from the transmission
- Exhaust pipe from the rear of the catalytic converter
- Parking brake bracket from the frame
- Rear brake line from the Brake Pressure Modulator Valve (BPMV)

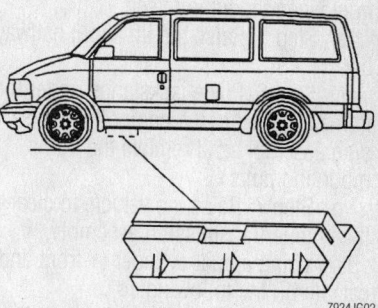

Attach the body protection pads to the pinch welds on both sides before raising the vehicle

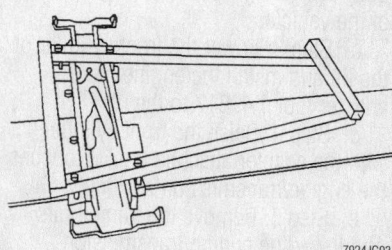

Install the engine lifting adapter to the front cylinder of the twin cylinder lift if applicable

- Front bumper and the power steering cooler from the front air deflector
- Supplemental Inflatable Restraint (SIR) connector
- Splash shields from the wheel openings
- Rear air conditioning lines at the rear crossmember, if equipped, leave the lines attached to the engine assembly
- Fuel lines at the filter and pull them through the crossmember
- Fuel tank electrical connector
- Transfer case vent tube, on all wheel drive models

7. Make sure all lines and connections are free between the engine/transmission assembly and the body.

8. If using a twin post lift (side lift), perform the following:

  a. Step 1: Lower the vehicle to the floor.

  b. Step 2: Install the body protection lift adapter tool J 41602 pads to the pinch welds on both sides of the vehicle behind the front wheels.

  c. Step 3: Position the front lifting arms of the lift under the body protection adapters.

  d. Step 4: Be sure the rear of the vehicle will be slightly higher than the front when the lift is raised.

  e. Step 5: Raise the lift about halfway up.

  f. Step 6: Place jackstands under the frame attached to the engine/transmission assembly and remove the frame mounting bolts.

  g. Step 7: Raise the vehicle to clear the engine/transmission assembly.

9. If using a dual cylinder (1 front and 1 rear) lift do the following:

  a. Step 1: Install the body protection lift adapter tool J 41602 pads to the pinch welds on both sides of the vehicle behind the front wheels.

  b. Step 2: Install stands under the body protection lift adapters and the rear of the vehicle.

  c. Step 3: Lower the front cylinder of the lift and install the engine lifting adapter tool J 41617 to the lift.

  d. Step 4: Raise the front cylinder with the adapter attached until it touches the engine/transmission assembly.

  e. Step 5: Remove the frame bolts and lower the engine/transmission assembly from the vehicle.

10. Remove the 2 right rear and 2 left front intake manifold bolts.

11. Install the Engine Lifting Bracket

tools J 41427 on the intake manifold to provide lifting points for an engine hoist. Install the hoist and raise the engine/transmission assembly.

12. Remove or disconnect the following:
- Transmission from the engine
- Engine from the frame

To install:

13. Install or connect the following:
- Engine onto the frame and the transmission to the engine. Tighten the engine mount through-bolts to 74 ft. lbs. (100 Nm).
- Engine/transmission assembly on suitable stands or on the engine lifting adapter

14. Remove the engine lifting brackets from the intake manifold and reinstall the bolts.

15. Position the engine/transmission assembly in the vehicle. Tighten the frame bolts in the following order:

  a. Step 1: Right center bolt: 114 ft. lbs. (155 Nm).

  b. Step 2: Left center bolt: 114 ft. lbs. (155 Nm).

  c. Step 3: Right front bolt: 66 ft. lbs. (90 Nm).

  d. Step 4: Left rear bolt: 66 ft. lbs. (90 Nm).

  e. Step 5: Left front bolt: 66 ft. lbs. (90 Nm).

  f. Step 6: Right rear bolt: 66 ft. lbs. (90 Nm).

16. Remove the stands or the engine lifting adapter. If the engine lifting adapter was used, raise the vehicle and remove the stands.

17. Remove the body protection adapter from the pinch welds.

18. Install or connect the following:
- Splash shields in the wheel openings
- Steering shaft to the steering gear
- Power steering cooler
- Lines to the Hydroboost unit
- Hose to the power steering reservoir
- Wiring harness to the bulkhead and fuse box
- Heater hoses and the SIR connector
- Throttle cable and the cruise control cable (if equipped)
- Radiator and air cleaner assembly
- Master cylinder to the booster
- Rear brake line to the BMPV
- Parking brake bracket to the frame
- Fuel lines and the air conditioning lines (if equipped) at the rear crossmember

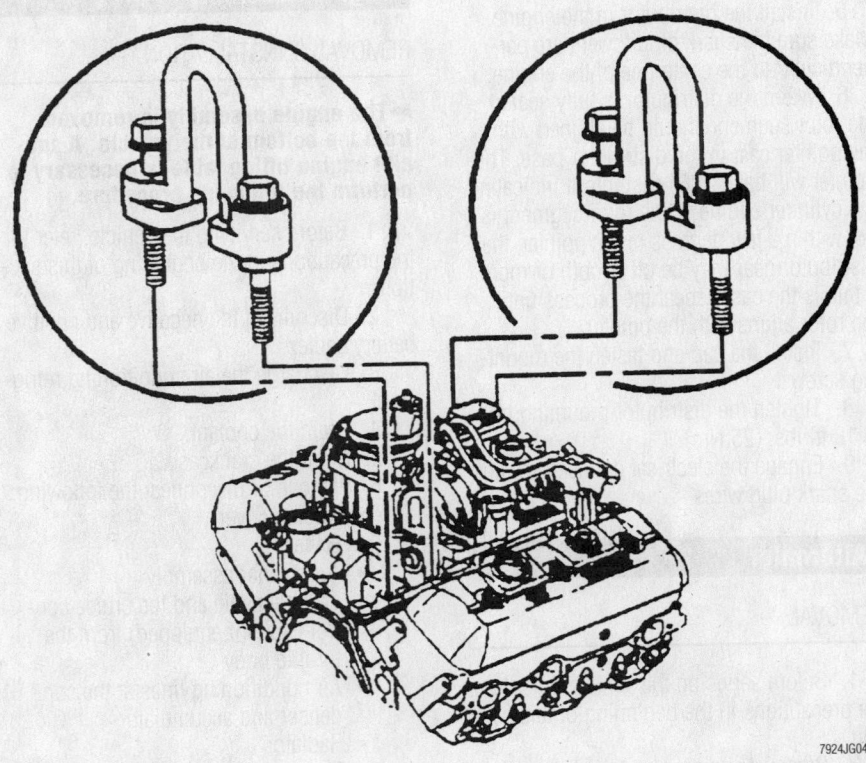

7924JG04

Install the engine lifting brackets at the right rear and left front of the intake manifold

- Transfer case vent tube, on all wheel drive models
- Front bumper and transmission shift linkage
- Torque converter bolts
- Starter and driveshaft
- Exhaust pipe
- Engine cover and battery

19. Refill the power steering, engine crankcase, brake system, cooling system and transmission.

20. Discharge the air conditioning system.

21. Bleed the brake system.

22. Start the engine and check for leaks.

Water Pump

REMOVAL & INSTALLATION

1. Before servicing the vehicle, refer to the precautions in the beginning of this section.

2. Disconnect the negative battery cable.

3. Drain the engine cooling system.

4. Remove or disconnect the following:
- Air cleaner assembly, if necessary
- Mass Air Flow (MAF) sensor
- Upper fan shroud
- Drive belt
- Fan and clutch assembly
- Water pump pulley
- Hoses from the water pump, as applicable
- Water pump

➡**On some engines, the pump retaining bolts will vary in size and thread. Be sure to note the positioning of all bolts during removal to assure proper installation.**

5. Clean gasket mounting surface.

To install:

6. Install or connect the following:
- Water pump. Torque bolts to 33 ft. lbs. (45 Nm).

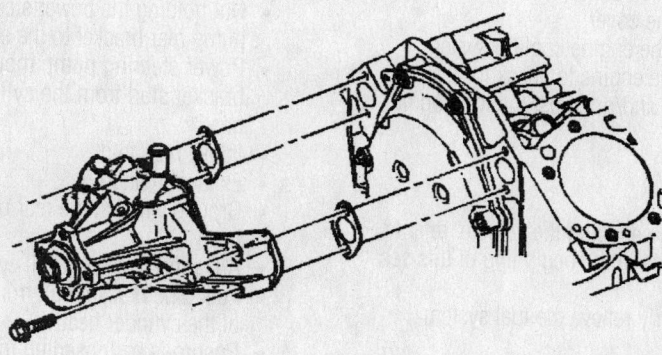

Water pump mounting—2002 shown

9358LG01

- Coolant hoses using new clamps
- Pulley and clutch assembly, as needed
- Drive belt assembly and fan shroud
- MAF sensor, if necessary
- Air cleaner assembly, if necessary
- Negative battery cable

7. Refill the cooling system

8. Run the engine and check for leaks.

Heater Core

REMOVAL & INSTALLATION

Front System

1. Before servicing the vehicle, refer to the precautions in the beginning of this section.

2. Drain the engine cooling system.

3. Remove or disconnect the following:
- Negative battery cable
- Heater hoses from the heater core
- Heater core cover-to-heater assembly screws and the cover
- Heater core-to-heater assembly strap screws and the straps
- Heater core

To install:

4. Install or connect the following:
- Heater core
- Heater core straps and the straps-to-heater assembly screws, then tighten the screws to 18 inch lbs. (2 Nm)
- Heater core cover and the cover-to-heater assembly screws, then tighten the screws to 18 inch lbs. (2 Nm)
- Heater hoses to the heater core
- Negative battery cable

5. Refill the cooling system.

6. Run the engine to normal operating temperatures; then, check the climate control operation and check for leaks.

Rear Auxiliary System

1. Before servicing the vehicle, refer to the precautions in the beginning of this section.

2. Drain the engine cooling system.

3. Remove or disconnect the following:
- Negative battery cable
- Body side front lower interior trim panel
- Clamps from the rear auxiliary heater hoses
- Heater hoses from the rear auxiliary heater case
- Screws and band clamp from the right side of the heater core
- Screw and band clamp from the left side of the heater core

➡**Place a cloth on the floor to catch any coolant that may spill from the heater core.**

- Heater core from the rear auxiliary case assembly
- Seals from the heater core

To install:

4. Install or connect the following:
- New seals to the heater core
- Heater core to the rear auxiliary case assembly
- Screws and band clamps to the right and left sides of the heater core, then tighten to 18 inch lbs. (2 Nm)
- Heater hoses to the rear auxiliary heater case
- Clamps to the rear auxiliary heater hoses
- Body side front lower interior trim panel
- Negative battery cable

5. Refill the engine cooling system.

6. Run the engine to normal operating temperatures; then, check the climate control operation and check for leaks.

Cylinder Head

REMOVAL & INSTALLATION

Right Side

1. Before servicing the vehicle, refer to the precautions in the beginning of this section.

2. Properly relieve the fuel system pressure.

3. Remove or disconnect the following:
- Engine cover
- Negative battery cable
- Engine cooling fan

4. Drain the engine cooling system.

- Rocker arm cover
- Intake manifold
- Spark plug wire support
- Exhaust manifold
- Alternator mounting bracket
- Alternator mounting bracket stud
- Wiring harness bolt/clip at the rear of the cylinder head
- Pushrods by loosening the rocker arms

➡**If valve train components, such as the rocker arms or pushrods, are to be reused, they must be tagged or arranged to insure installation in their original locations.**

- Cylinder head bolts by loosening them in the reverse of the torque sequence
- Cylinder head

To install:

5. Clean and inspect the gasket mounting surfaces.

➡**Do not apply sealer to composition steel/asbestos gaskets. If using a steel only gasket, apply a thin and even coat of sealer to both sides of the gaskets.**

6. Install or connect the following:
- New gasket over the dowel pins with the bead or the words **This Side Up** facing upwards, as applicable
- Cylinder head

7. Coat the bolts with GM sealer 1052080 , then install the bolts and tighten in sequence to 22 ft. lbs. (30 Nm). The bolts must then be tightened again in sequence in the following order:

   a. Short length bolts: (11, 7, 3, 2, 6, 10) 55 degrees.

   b. Medium length bolts: (12, 13) 65 degrees.

   c. Long length bolts: (1, 4, 8, 5, 9) 75 degrees.

8. Install or connect the following:
- Pushrods, secure the rocker arms and adjust the valves
- Exhaust manifold

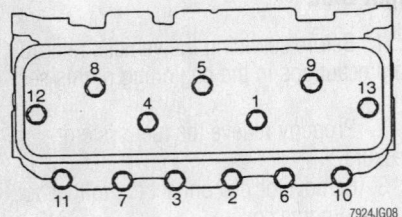

Cylinder head bolt torque sequence—4.3L engine

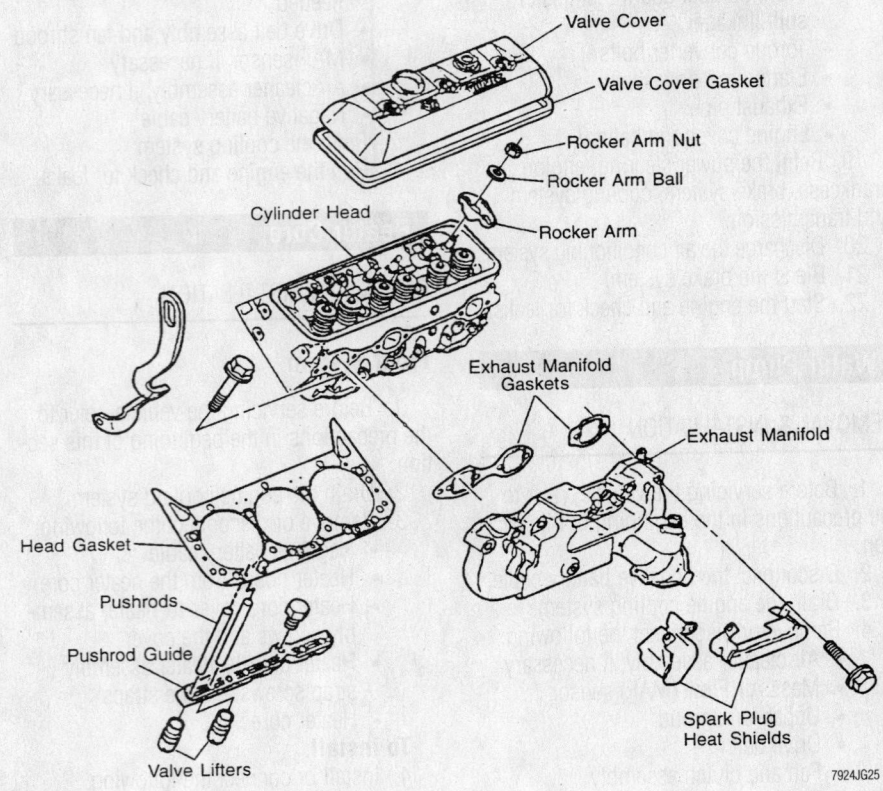

Cylinder head and related components—4.3L engine

- Spark plug wire harness and support. Tighten the bolts to 106 inch lbs. (12 Nm).
- Intake manifold
- Alternator mounting bracket stud and tighten to 15 ft. lbs. (20 Nm).
- Alternator bracket
- Bolt holding the engine wiring harness to the rear of the head. Tighten to 27 ft. lbs. (36 Nm).
- Rocker arm cover
- Negative battery cable
- Engine cover

9. Refill the engine cooling system.
10. Run the engine to check for leaks
11. Check and/or adjust the ignition timing.

Left Side

1. Before servicing the vehicle, refer to the precautions in the beginning of this section.

2. Properly relieve the fuel system pressure.

3. Properly discharge the air conditioning system, if equipped.

4. Remove or disconnect the following:
- Engine cover
- Negative battery cable
- Engine cooling fan

5. Drain the engine cooling system.
- Air conditioning compressor, if equipped

➡**The power steering pump can stay mounted on the bracket and the hoses should not be disconnected from the pump.**

- Nut holding the power steering pump rear bracket to the engine
- Power steering pump mounting bracket stud from the cylinder head
- Intake manifold
- Exhaust manifold
- Ground wires at the rear of the left head
- A/C pipe bracket nut, if equipped
- Fuel pipe bracket bolt from the rear of the cylinder head
- Pushrods by loosening the rocker arms

➡ If valve train components, such as the rocker arms or pushrods, are to be reused, they must be tagged or arranged to insure installation in their original locations.

- Cylinder head bolts by loosening them in the reverse of the torque sequence
- Cylinder head

To install:

6. Clean and inspect the gasket mounting surfaces.

➡ Do not apply sealer to composition steel/asbestos gaskets. If using a steel only gasket, apply a thin and even coat of sealer to both sides of the gaskets.

7. Install or connect the following:
- New gasket over the dowel pins with the bead or the words **This Side Up** facing upwards, as applicable
- Cylinder head

8. Coat the bolts with GM sealer 1052080 , then install the bolts and tighten in sequence to 22 ft. lbs. (30 Nm). The bolts must then be tightened again in sequence in the following order:

a. Short length bolts: (11, 7, 3, 2, 6, 10) 55 degrees.

b. Medium length bolts: (12, 13) 65 degrees.

c. Long length bolts: (1, 4, 8, 5, 9) 75 degrees.

9. Install or connect the following:
- Pushrods, secure the rocker arms and adjust the valves
- Ground wires and bolt at the rear of the head. Tighten the bolt to 26 ft. lbs. (35 Nm).
- Fuel pipe bracket and stud. Tighten the stud to 24 ft. lbs. (35 Nm).
- A/C pipe bracket and nut, and tighten to 26 ft. lbs. (35 Nm).
- Exhaust manifold
- Intake manifold
- Power steering pump mounting bracket stud. Tighten to 15 ft. lbs. (20 Nm).
- Power steering pump mounting bracket. Tighten the nut to 30 ft. lbs. (41 Nm).
- Air conditioning compressor, if equipped
- Engine cooling fan
- Negative battery cable
- Engine cover

10. Refill the engine cooling system.
11. Run the engine to check for leaks
12. Check and/or adjust the ignition timing.

Rocker Arms

REMOVAL & INSTALLATION

➡ Make sure to keep all valvetrain components in order as you remove them. They must be reinstalled in their original locations.

1. Before servicing the vehicle, refer to the precautions in the beginning of this section.

2. Remove or disconnect the following:
- Rocker arm cover(s)
- Rocker arm
- Rocker arm supports
- Pushrod(s)

To install:

3. Inspect and replace components if worn or damaged.

4. Install or connect the following:
- Pushrods making sure they seat properly in the lifter
- Rocker arm supports

5. Apply a suitable prelube to the following rocker arm contact surfaces:

a. Valve pushrod socket (1).

b. Roller pivot (2).

c. Valve stem tip (3).

6. Install the rocker arms as follows:

a. Finger-start bolt at locations 1, 2 and then 3.

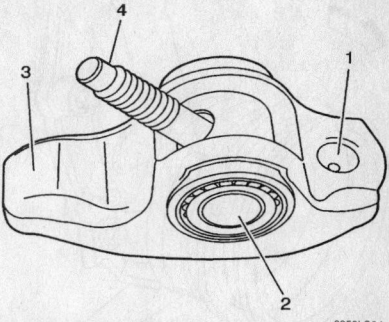

Prelube application locations on the rocker arm

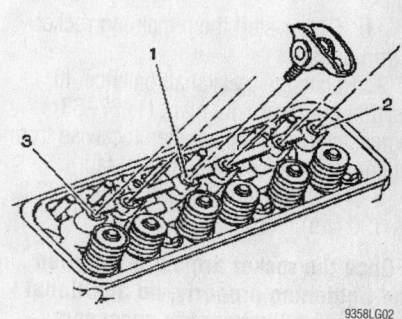

Proper rocker arm installation—2000–03 vehicles

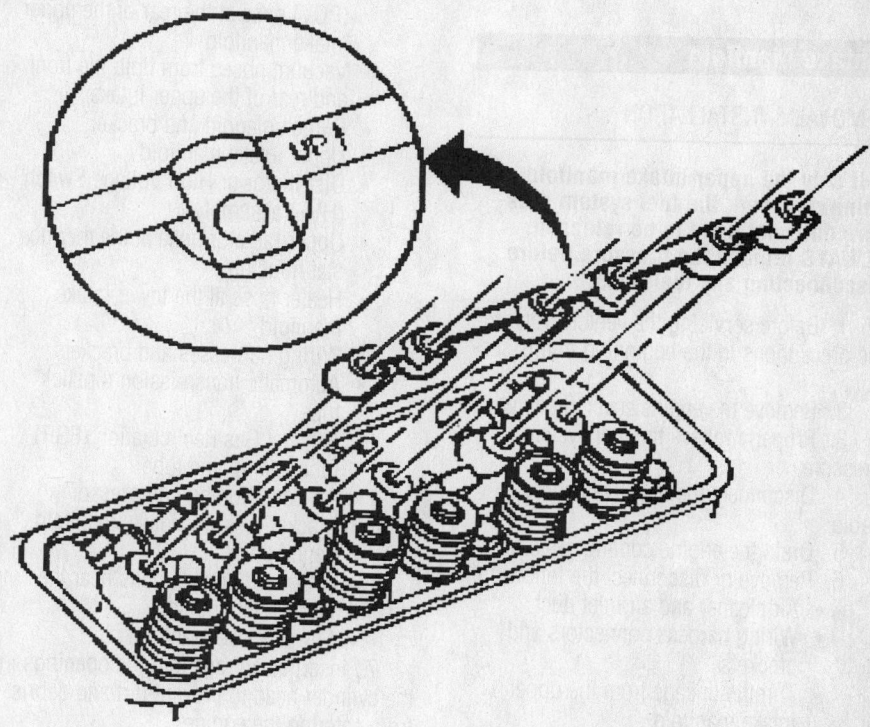

During installation, make sure the arrow on the rocker arm support is in the up position

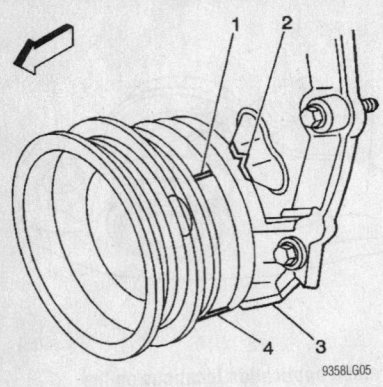

Proper crankshaft balancer alignment—
2000–03 vehicles

9358LG05

b. Finger-start the remaining rocker arm bolts.

7. Rotate the crankshaft balancer to position the alignment mark (1) 57–63 degrees clockwise or counterclockwise from the engine front cover alignment tab.

8. Tighten the rocker arm bolts to 22 ft. lbs. (30 Nm).

➡ Once the rocker arms are installed and tightening properly, no additional valve lash adjustment is necessary.

9. Install the rocker arm cover(s).

10. Start and run the engine, then check for leaks and for proper ignition timing adjustment.

Intake Manifold

REMOVAL & INSTALLATION

➡ If only the upper intake manifold is being removed, the fuel system pressure does not need to be released. ALWAYS release the pressure before disconnecting any fuel lines.

1. Before servicing the vehicle, refer to the precautions in the beginning of this section.

2. Remove the engine cover, if equipped

3. Properly relieve the fuel system pressure.

4. Disconnect the negative battery cable.

5. Drain the engine cooling system.

6. Remove or disconnect the following:
- Air cleaner and air inlet duct
- Wiring harness connectors and brackets
- Throttle linkage from the upper intake manifold
- Ignition coil
- Fuel lines and bracket from the rear of the lower intake manifold

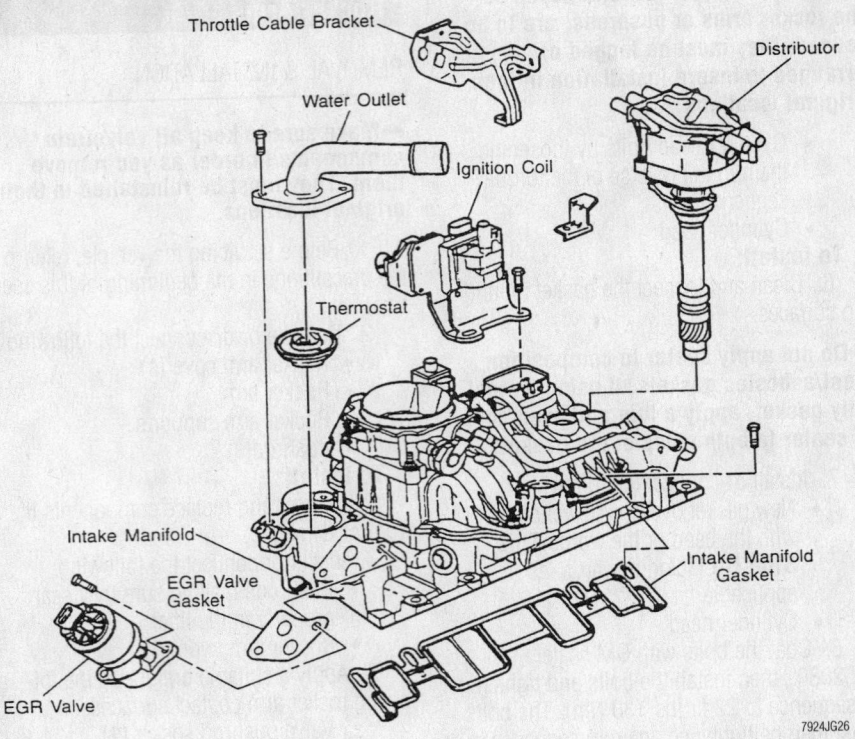

Intake manifold and related components—4.3L engine

7924JG26

- Brake booster vacuum hose at the upper intake manifold
- Positive Crankcase Ventilation (PCV) hose at the rear of the upper intake manifold
- Vacuum hoses from both the front and rear of the upper intake
- Purge solenoid and bracket
- Upper intake manifold
- Distributor or High Voltage Switch (HVS) assembly
- Upper radiator hose at the thermostat housing
- Heater hose at the lower intake manifold
- Wiring harnesses and brackets
- Automatic transmission dipstick tube
- Exhaust Gas Recirculation (EGR) tube, clamp and tube
- Air conditioning compressor bracket-to-lower intake manifold pencil brace
- Alternator bracket bolts near the thermostat housing
- Lower intake manifold

7. Insert clean rags into the openings in the cylinder head to prevent dirt and debris from entering the engine.

8. Clean the gasket mounting surfaces. Be sure to inspect the manifold for warpage and/or cracks. If necessary, replace it.

To install:

9. Remove the rags from the cylinder heads.

10. Position the gaskets on the cylinder head with the port blocking plates to the rear and the **this side up** stamps facing upward. Then apply a 3/16in. (5mm) bead of RTV sealant on the front and rear of the engine block at the block-to-manifold mating surface. Extend the bead 1/2in. (13mm) up each cylinder head to seal and retain the gaskets.

11. Install the lower intake manifold. Tighten the bolts in sequence and in 3 steps, as follows:
a. Step 1: 26 inch lbs. (3 Nm).
b. Step 2: 106 inch lbs. (12 Nm).
c. Step 3: 11 ft. lbs. (15 Nm).

12. Install or connect the following:
- Alternator bracket bolt near the thermostat housing
- EGR tube, clamp and bolt
- Wiring harness to the lower manifold components, including the injector, EGR valve and ECT sensor
- Air conditioning compressor bracket-to-the lower intake manifold pencil braces
- Transmission oil dipstick tube, if necessary
- Fuel supply and return lines to the rear of the lower intake

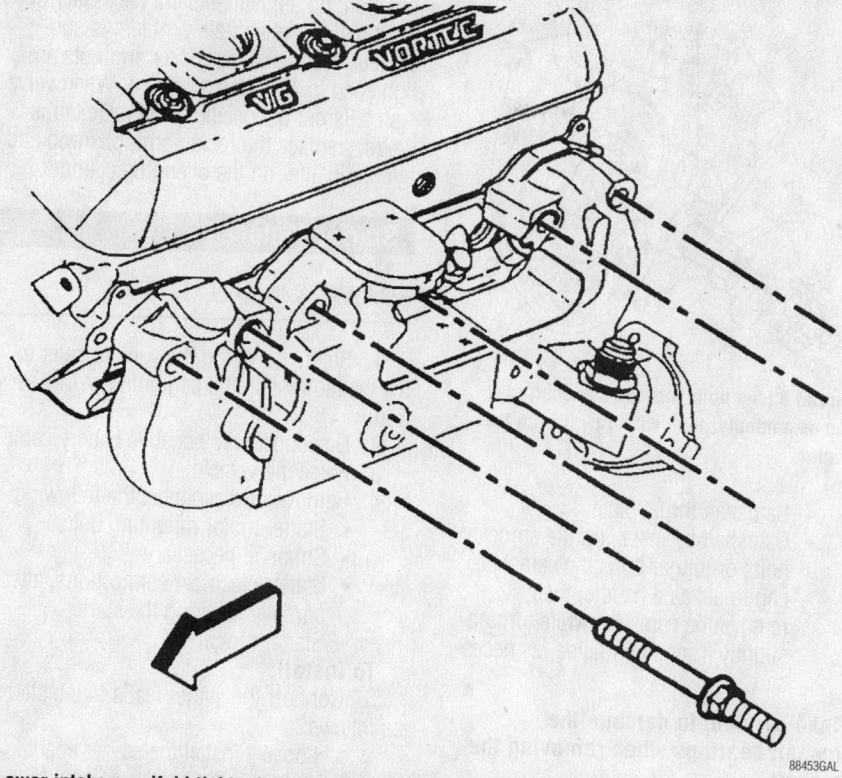

Lower intake manifold tightening sequence—4.3L engines

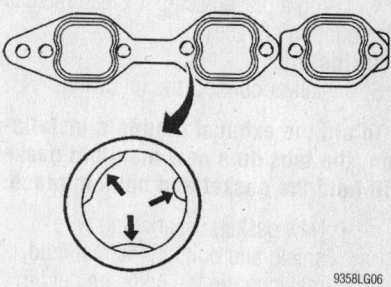

9358LG06

The tabs will help hold the gasket and bolts in place when installing the exhaust manifold

13. Temporarily reattach the negative battery cable, then pressurize the fuel system (by cycling the ignition without starting the engine) and check for leaks.
14. Disconnect the negative battery cable.
15. Install or connect the following:
   • Heater hose to the lower intake
   • Upper radiator hose to the thermostat housing
   • Distributor assembly and engage the wiring
   • Vacuum hoses to the upper and lower intake manifold
   • New upper intake manifold gasket, making sure the green sealing lines are facing upward
   • Upper intake manifold being careful not to pinch the fuel injector wires between the manifolds
   • Manifold retainers. Tighten them to 88 inch lbs. (10 Nm) using two passes.
   • Purge solenoid and bracket
   • Brake booster vacuum hose at the upper intake manifold
   • PCV hose to the rear of the upper intake manifold
   • Vacuum hoses to both the front and rear of the manifold assembly
   • Throttle linkage to the upper intake
   • Ignition coil
   • Wiring to the upper intake compo-

nents including the TP sensor, IAC motor, MAP sensor and the IMTV
   • Plastic cover
   • Air cleaner and air inlet duct
   • Negative battery cable
16. Refill the engine cooling system.

Exhaust Manifold

REMOVAL & INSTALLATION

4.3L Engines

RIGHT SIDE

1. Before servicing the vehicle, refer to the precautions in the beginning of this section.
2. Disconnect the negative battery cable
3. Raise the vehicle.
   • Remove or disconnect the following:
   • Right front tire
   • Right fender wheelhouse extension
   • Spark plug wires from the plugs
   • Catalytic converter
   • Exhaust manifold bolts through the wheelhouse
   • Exhaust manifold, with the gaskets and spark plug wire shields
4. Using a plastic scraper, clean the gasket mounting surfaces.

To install:
5. Install or connect the following:

➡ To aid the exhaust manifold installation, the tabs on a new manifold gasket will hold the gasket and bolts in place.

   • New gaskets, spark plug wire shield and bolts to the manifold, making sure the bolts are held in place by the gasket tabs.
   • Exhaust manifold. Tighten the bolts to 11 ft. lbs. (15 Nm), then to 22 ft. lbs. (30 Nm).
   • Catalytic converter
   • Spark plug wires
   • Fender wheelhouse extension
   • Front tire
6. Lower the vehicle.
7. Connect the negative battery cable
8. Start the engine and check for leaks.

LEFT SIDE

1. Before servicing the vehicle, refer to the precautions in the beginning of this section.
2. Remove or disconnect the following:
   • Negative battery cable
   • Engine cover
   • Engine Coolant Temperature (ECT) sensor
3. Raise the vehicle.
   • Left front tire
   • Left fender wheelhouse extension
   • Catalytic converter assembly
   • Spark plug wires from the spark plugs
   • Spark plug wire support bolts
   • Spark plug wire support with the wires
   • Front 4 exhaust manifold bolts through the wheelhouse
   • Rear 2 manifold bolts from the engine cover opening
   • Exhaust manifold, gasket and spark plug wire shields from underneath the vehicle

4. Using a plastic scraper, clean the gasket mounting surfaces.

To install:

5. Install or connect the following:

➡**To aid the exhaust manifold installation, the tabs on a new manifold gasket will hold the gasket and bolts in place.**

- New gaskets, spark plug wire shield and bolts to the manifold, making sure the bolts are held in place by the gasket tabs.
- Exhaust manifold. Tighten the bolts to 11 ft. lbs. (15 Nm), then to 22 ft. lbs. (30 Nm).

➡**Tighten the 2 rear bolts through the engine cover opening and the front 4 bolts through the wheelhouse.**

- Spark plug wires
- Catalytic converter
- Fender wheelhouse extension
- Front tire

6. Lower the vehicle.
- ECT sensor electrical connector
- Negative battery cable

7. Start the engine and check for leaks.

8. Once the engine has cooled sufficiently, install the engine cover to the passenger compartment.

Camshaft and Valve Lifters

REMOVAL & INSTALLATION

1. Before servicing the vehicle, refer to the precautions in the beginning of this section.

2. Properly relieve the fuel system pressure.

3. Disconnect the negative battery cable.

4. Drain the engine cooling system.

5. Discharge and recover the refrigerant from the air conditioning system.

6. Remove the engine cover and engine cooling fan.

7. Remove or disconnect the following:
- Radiator
- Air conditioning condenser
- Rocker arm covers
- Intake manifold assembly
- Rocker arms, pushrods and lifters
- Crankshaft pulley and hub
- Engine front (timing) cover

8. Align the timing marks on the crankshaft and camshaft sprockets.

9. Remove or disconnect the following:
- Camshaft sprocket and timing chain
- Balance shaft drive gear, if equipped

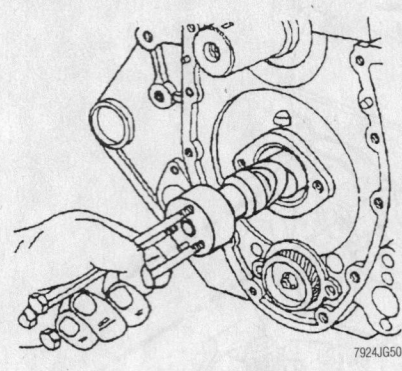

7924JG50

Thread 3 long bolts into the camshaft to use as a handle, then withdraw it from the engine

- Camshaft thrust plate
- Camshaft by installing the sprocket bolts or longer bolts the camshaft end to act as a handle; then, remove the camshaft while turning slightly from side to side, as necessary.

➡**Take care not to damage the camshaft bearings when removing the camshaft.**

To install:

10. Lubricate the camshaft journals with clean engine oil or a suitable pre-lube.

11. Install or connect the following:
- Camshaft being extremely careful not to contact the bearings with the cam lobes
- Thrust plate. Torque the bolts to 106 inch lbs. (12 Nm).
- Balance shaft drive gear, if equipped
- Timing chain and camshaft sprocket
- Engine front (timing) cover
- Crankshaft pulley and hub
- Valve lifters, pushrods and rocker arms
- Intake manifold assembly
- Rocker arm covers
- Radiator
- Negative battery cable
- Engine cooling fan and engine cover

12. Refill the engine cooling system.

Valve Lash

ADJUSTMENT

The 4.3L engines are equipped with screw-in rocker arm studs with positive stop shoulders. Because the shoulders that allow the rocker arms to be tightened into proper

position, no adjustments are necessary or possible. If a valve train problem is suspected, check that the rocker arm nuts are tightened to 18 ft. lbs. (24 Nm). When valve lash falls out of specification (valve tap is heard), replace the rocker arm, pushrod and hydraulic lifter on the offending cylinder.

Starter Motor

REMOVAL & INSTALLATION

1. Before servicing the vehicle, refer to the precautions in the beginning of this section.

2. Disconnect the negative battery cable

3. Raise the vehicle.

4. Remove or disconnect the following:
- Starter motor mounting bolts
- Shims, if necessary
- Starter electrical connections, after partially lowering the starter
- Starter motor

To install:

5. Connect the starter leads to the starter as follows:

a. Loosely install the starter enable relay lead (1) to the starter solenoid terminal. Align the terminal retaining tab to the starter solenoid.

b. Loosely install alternator output (BAT) lead (3) and the positive battery cable (2) to the solenoid terminal. Align the positive battery cable terminal retaining tab to the starter solenoid.

c. Tighten the starter enable relay lead nut to 18 inch lbs. (2 Nm) and the positive battery cable nut to 14 ft. lbs. (19 Nm).

6. Install or connect the following:
- Starter motor into position
- Starter motor mounting bolts but do not tighten it at this time
- Starter motor shims, if equipped
- Outboard starter motor bolt

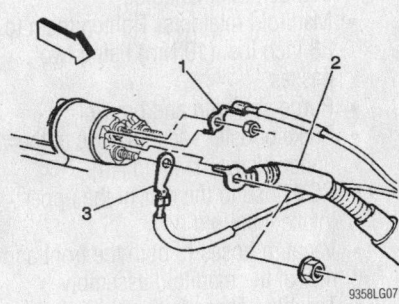

9358LG07

Starter motor wiring

Tighten the bolts to 32 ft. lbs. (43 Nm).

7. Lower the vehicle, then connect the negative battery cable

Oil Pan

REMOVAL & INSTALLATION

4.3L Engines

1. Before servicing the vehicle, refer to the precautions in the beginning of this section.
2. Remove or disconnect the following:
   - Negative battery cable
   - Oil level indicator
3. Raise the vehicle and drain the oil.
   - Oil filter
   - Bolt holding the oil cooler pipes bracket to the oil pan
   - Oil filter adapter
   - Bolt holding the bracket for the starter wiring harness and the transmission cooler pipes
   - Crankshaft Position (CKP) sensor wiring harness from the retainer
   - Starter
   - Transmission cover
   - Inner axle shaft housing support bracket to the frame nuts and the washers (AWD only)
   - Front differential carrier upper and lower mounting nuts and bolts (AWD only)
4. Lower the front differential carrier assembly only enough for removal of the oil pan.
   - Access plugs for the oil pan rear nuts
   - Transmission-to-oil pan bolts
   - Oil pan from the engine
5. Clean all sealing surfaces on the engine and the oil pan

➡**Any time the transmission and the engine oil pan are off of the engine at**

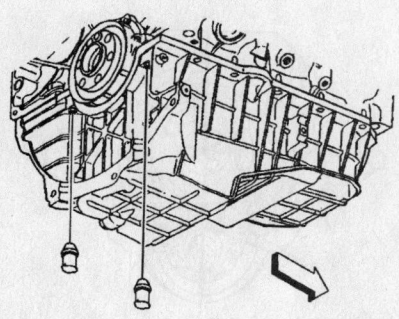

Oil pan mounting

the same time, install the transmission before the oil pan. This is to allow for the proper oil pan alignment. Failure to achieve the correct oil pan alignment can result in transmission failure.

To install:

6. Install or connect the following:
   - Oil pan to the engine
   - Transmission to oil pan bolts and nuts. Tighten the bolts to 34 ft. lbs. (47 Nm).
   - Access plugs for the oil pan rear nuts
7. Position the front differential carrier assembly.
   - Carrier upper and lower mounting bolts and nuts (AWD only)
   - Inner axle shaft housing support bracket to frame washers and nuts (AWD only)
   - Transmission cover. Tighten the bolts to 106 inch lbs. (12 Nm).
   - Starter motor
   - Bolt holding the bracket for the starter wiring harness and the transmission cooler pipes. Tighten the bolts to 89 inch lbs. (10 Nm).
   - CKP sensor wiring harness in the retainer
   - Oil filter adapter
   - Oil cooler pipes bracket bolt. Tighten the bolts to 89 inch lbs. (10 Nm).
   - Oil filter
   - Oil pan drain plug and tighten 18 ft. lbs. (25 Nm)
8. Lower the vehicle.
9. Fill the crankcase with engine oil
10. Connect the negative battery cable.

Oil Pump

REMOVAL & INSTALLATION

1. Before servicing the vehicle, refer to the precautions in the beginning of this section.
2. Remove or disconnect the following:
   - Oil pan
   - Oil pump bolt, if necessary
   - Oil pump and the pickup tube/shaft, if equipped

➡**Be careful not to crack the retainer.**

To install:

3. Ensure that the pump pickup tube is tight in the pump body. If the tube should come loose, oil pressure will be lost and oil starvation will occur. If the pickup tube is loose it should be replaced.

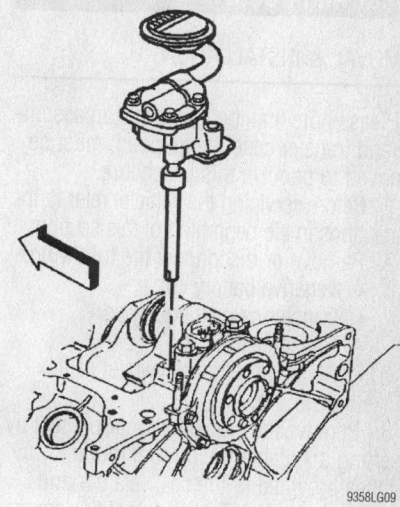

9358LG09

Exploded view of the oil pump—2002 model shown, others similar

4. If the pump has been disassembled and is being replaced or for any reason oil has been removed, it must be primed. It can either be filled with oil before installing the cover plate and oil kept within the pump during handling or the entire pump cavity can be filled with petroleum jelly.

✳✳ WARNING

If the pump is not primed, the engine could be damaged upon start up.

5. Install or connect the following:
   - Oil pump by aligning the pump shaft with the distributor drive gear as necessary. Tighten oil pump/pickup tube retainer(s) to 65 ft. lbs. (90 Nm).

➡**If the oil pump does not build up oil pressure almost immediately, remove the pan and check for a loose oil pump-to-pickup tube attachment. If necessary dismantle the pump and pack the pump cavity with petroleum jelly.**

   - Oil pan
6. Refill the crankcase.
7. Disable the ignition system; crank engine for approximately 10 seconds to aid in priming the oil pump and reducing the risk of engine damage.

✳✳ WARNING

Running the engine without measurable oil pressure will cause extensive damage.

Rear Main Seal

REMOVAL & INSTALLATION

Please note that the transmission assembly and transfer case, if equipped, must be removed to perform this procedure.

1. Before servicing the vehicle, refer to the precautions in the beginning of this section.
2. Remove or disconnect the following:
   • Negative battery cable
   • Transfer case, if equipped
   • Transmission
   • Clutch assembly/flywheel or flexplate
3. Remove the crankshaft rear oil seal by inserting a suitable prying tool into the notches provided in the seal retainer and prying the seal out. Take care not to damage the crankshaft sealing surface.

To install:
4. Inspect the crankshaft for grit, rust or burrs and correct as necessary.
5. Clean the running surface of the crankshaft with a non-abrasive cleaner.
6. Install or connect the following:
   • New rear seal lubricated with engine oil and a seal installer
   • Flywheel and clutch or flexplate
   • Transmission
   • Transfer case, if equipped
   • Negative battery cable
7. Start the engine and verify no oil leaks.

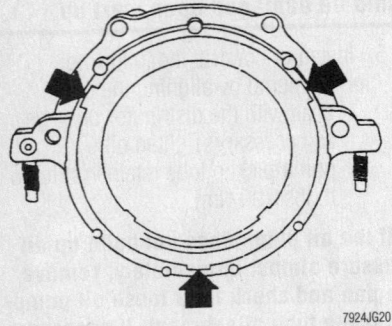

Carefully pry the rear main seal out of the retainer—4.3L engine

Timing Chain, Sprockets, Front Cover and Seal

REMOVAL & INSTALLATION

Front Cover and Seal

1. Before servicing the vehicle, refer to the precautions in the beginning of this section.

2. Drain the cooling system.
3. Remove or disconnect the following:
   • Negative battery cable
   • Oil pan
   • Crankshaft pulley and balancer
   • Water pump assembly
   • Crankshaft Position (CKP) sensor
   • Front cover bolts and the reinforcements, if equipped
   • Front cover from the engine
4. Pry the seal out of the front cover using a small prytool. Be very careful not to distort the front cover or to score the end of the crankshaft.

To install:

➡ **Anytime the front cover is removed, the cover must be replaced upon reassembly. If you reuse the old cover, oil leaks may develop.**

5. Clean the gasket mating surfaces of the engine and cover of all remaining gasket or sealer material. Be careful not to score or damage the surfaces.

➡ **The manufacturer suggests you wait until the front cover is mounted to the engine before you install the replacement crankshaft oil seal. This assures the cover is properly supported.**

6. Install or connect the following:
   • New front cover gasket to the engine or cover using gasket cement to hold it in position. Lubricate the front of the oil pan seal with engine oil to aid in reassembly.
   • Front cover to the engine. Take care while engaging the front of the oil pan seal with the bottom of the cover. Apply sealer 12346141 to the oil pan rail where it contacts the timing cover-to-block joint (front) and the crankshaft rear seal

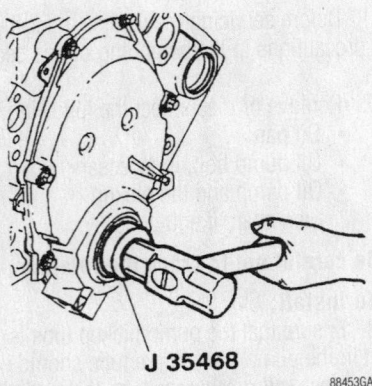

J 35468

Installing the crankshaft front oil seal—
4.3L engines

retainer-to-block joint (rear). Continue the bead of sealant about 1 inch (25mm) in both directions from each of the four corners.
   • Front cover retaining bolts and tighten to 106 inch lbs. (12 Nm)
7. Lightly coat the lips of the replacement crankshaft seal with clean engine oil, then position the seal with the open end facing inward the engine. Use a suitable seal installation driver to position the seal in the front cover.
   • Oil pan
   • CKP sensor O-ring and the sensor
   • Water pump
   • Crankshaft damper and pulley
   • Negative battery cable
8. Properly refill the engine cooling system.
9. Run the engine until normal operating temperature has been reached, then check for leaks.

Timing Chain and Sprockets

➡ **The following procedure requires the use of the Crankshaft Sprocket Removal tool No. J-5825-A and the Crankshaft Sprocket Installation tool No. J-5590.**

1. Before servicing the vehicle, refer to the precautions in the beginning of this section.
2. Remove the timing cover from the engine.
3. Rotate the crankshaft until the No. 4 cylinder is on the Top Dead Center (TDC) of its compression stroke and the camshaft sprocket mark aligns with the mark on the crankshaft sprocket (facing each other at a point closest together in their travel) and in line with the shaft centers.
4. Remove or disconnect the following:
   • Crankshaft Position (CKP) sensor reluctor ring, if equipped
   • Camshaft sprocket-to-camshaft nut and/or bolts

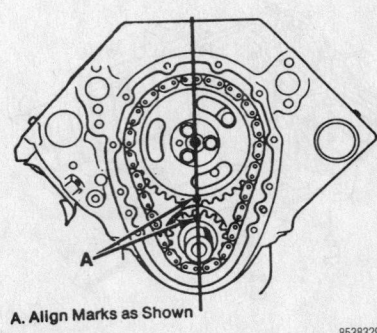

A. Align Marks as Shown

Timing mark alignment—4.3L engine

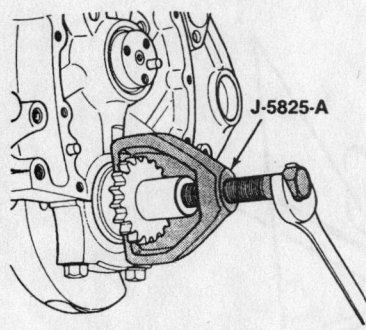

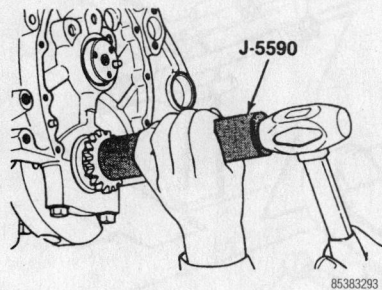

Removal (top) and installation of the crankshaft timing gear

- Camshaft sprocket (along with the timing chain). If the sprocket is difficult to remove, use a plastic mallet to bump the sprocket from the camshaft.

➡ **The camshaft sprocket (located by a dowel) is lightly pressed onto the camshaft and should come off easily. The chain comes off with the camshaft sprocket.**

5. If necessary use J-5825-A crankshaft sprocket removal tool to free the timing sprocket from the crankshaft.

6. If necessary, remove the crankshaft sprocket key.

To install:

7. Inspect the timing chain and the timing sprockets for wear or damage, replace the damaged parts as necessary.

8. Using a putty knife, clean the gasket mounting surfaces. Using solvent, clean the oil and grease from the gasket mounting surfaces.

9. Install or connect the following:
- Crankshaft sprocket key, if removed
- Crankshaft sprocket onto the crankshaft using J-5590 crankshaft sprocket installation tool and a hammer without disturbing the position of the engine

➡ **During installation, coat the thrust surfaces lightly with Molykote® or an equivalent pre-lube.**

- Timing chain over the camshaft sprocket. Arrange the camshaft sprocket in such a way that the timing marks will align between the shaft centers and the camshaft locating dowel will enter the dowel hole in the cam sprocket.
- Timing chain under the crankshaft sprocket, then place the cam sprocket, with the chain still mounted over it, in position on the front of the camshaft
- Camshaft sprocket-to-camshaft retainers to 18 ft. lbs. (25 Nm)

10. With the timing chain installed, turn the crankshaft two complete revolutions, then check to make certain that the timing marks are in correct alignment between the shaft centers.
- CKP sensor reluctor ring, if equipped
- Timing cover

Piston and Ring

POSITIONING

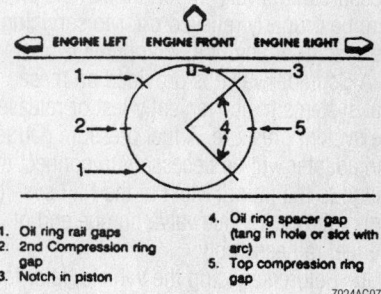

Piston ring end-gap spacing—GM 4.3L engines

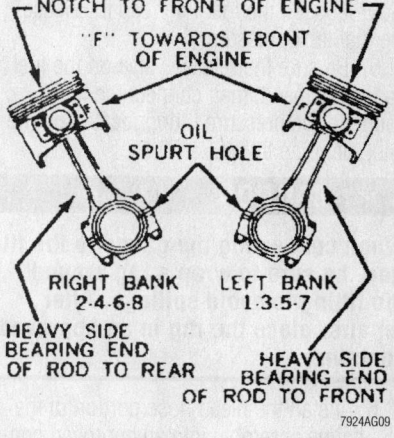

Piston and connecting rod assembly positioning—GM 4.3L engine

FUEL SYSTEM

Fuel System Service Precautions

Safety is the most important factor when performing not only fuel system maintenance but also any type of maintenance. Failure to conduct maintenance and repairs in a safe manner may result in serious personal injury or death. Maintenance and testing of the vehicle's fuel system components can be accomplished safely and effectively by adhering to the following rules and guidelines.

- To avoid the possibility of fire and personal injury, always disconnect the negative battery cable unless the repair or test procedure requires that battery voltage be applied.

- Always relieve the fuel system pressure prior to disconnecting any fuel system component (injector, fuel rail, pressure regulator, etc.), fitting or fuel line connection. Exercise extreme caution whenever relieving fuel system pressure, to avoid exposing skin, face and eyes to fuel spray. Please be advised that fuel under pressure may penetrate the skin or any part of the body that it contacts.

- Always place a shop towel or cloth around the fitting or connection prior to loosening to absorb any excess fuel due to spillage. Ensure that all fuel spillage (should it occur) is quickly removed from engine surfaces. Ensure that all fuel soaked cloths or towels are deposited into a suitable waste container.

- Always keep a dry chemical (Class B) fire extinguisher near the work area.

- Do not allow fuel spray or fuel vapors to come into contact with a spark or open flame.

- Always use a back-up wrench when loosening and tightening fuel line connection fittings. This will prevent unnecessary stress and torsion to fuel line piping. Always follow the proper torque specifications.

- Always replace worn fuel fitting O-rings with new. Do not substitute fuel hose or equivalent where fuel pipe is installed.

Fuel System Pressure

RELIEVING

The fuel systems operate under high fuel pressures. It is very important that the pressure be properly relieved prior to servicing the system or any of its components.

A Schrader valve is provided on these fuel systems to conveniently test or release the system pressure. A fuel pressure gauge and adapter will be necessary to connect the gauge to the fitting. Most of the MFI systems utilize a service valve on one end of the fuel rail assembly.

1. Before servicing the vehicle, refer to the precautions in the beginning of this section.

2. Remove the engine cover.

3. Disconnect the negative battery cable to assure the prevention of fuel spillage if the ignition switch is accidentally turned **ON** while a fitting is still detached.

4. Loosen the fuel filler cap to release the fuel tank pressure.

5. Be sure the release valve on the fuel gauge is closed, then connect the fuel gauge to the pressure fitting located on the inlet fuel pipe fitting.

❊❊ CAUTION

When connecting the gauge to the fitting, be sure to wrap a rag around the fitting to avoid spillage. After repairs, place the rag in an approved container.

6. Install the bleed hose portion of the fuel gauge assembly into an approved container, then open the gauge release valve and bleed the fuel pressure from the system.

7. When the gauge is removed, be sure to open the bleed valve and drain all fuel from the gauge assembly.

8. When fuel service is finished, tighten the fuel filler cap, connect the negative battery cable and install the engine cover.

Fuel Filter

REMOVAL & INSTALLATION

1. Before servicing the vehicle, refer to the precautions in the beginning of this section.

2. Properly relieve the fuel system pressure.

3. Remove or disconnect the following:

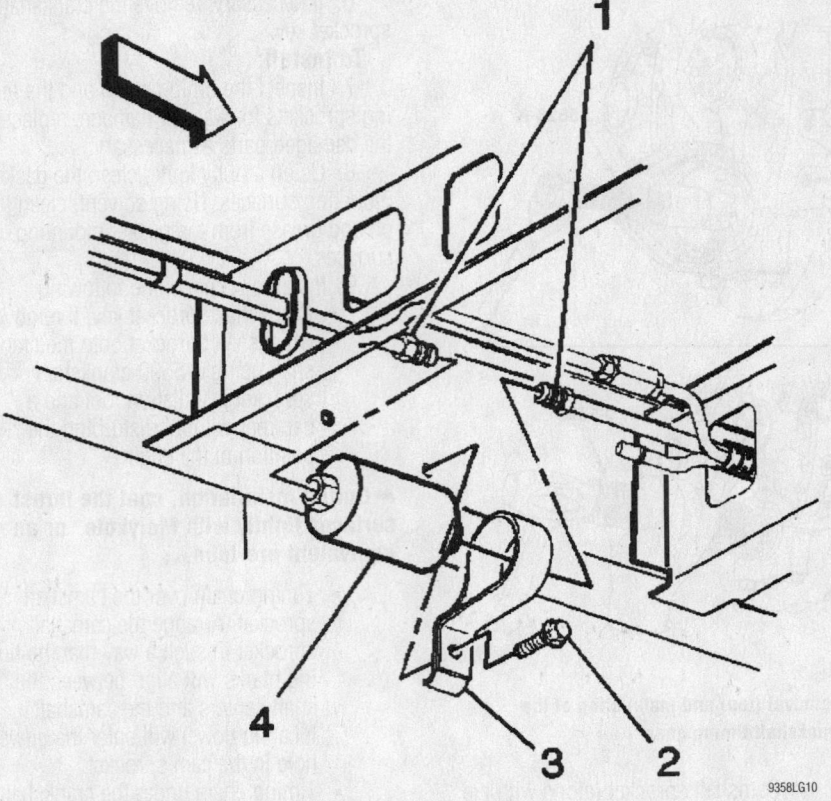

9358LG10

Exploded view of the fuel filter location along frame rail

- Negative battery cable
- Fuel filler cap
4. Raise the vehicle.
- Quick connect fittings (1) from the filter
- Filter feed nut and the clamp bolt (2)
- Filter (4) and the clamp (3) from the vehicle

To install:

5. Install or connect the following:
- Filter and clamp with the directional arrow facing away from the fuel tank, towards the throttle body

➡**The filter has an arrow (fuel flow direction) on the side of the case, be sure to install it correctly in the system, the with arrow facing away from the fuel tank.**

- Tighten the fuel feed nut
- Tighten the filter clamp assembly bolt
- Fuel quick disconnect fittings to the filter
- Fuel filler cap
- Negative battery cable
6. Start the engine and check for leaks.

Fuel Pump

REMOVAL & INSTALLATION

1. Before servicing the vehicle, refer to the precautions in the beginning of this section.

2. Properly relieve the fuel system pressure.

3. Drain the fuel tank.

4. Support the fuel tank.

5. Remove or disconnect the following:
- Negative battery cable
- Filler neck from the tank
- Shield from tank and tank straps
- Fuel lines and vapor hose from pump
- Electrical connection from fuel pump
- Fuel tank
- Fuel pump/sending unit assembly by turning the locking ring (located on top of the fuel tank) counterclockwise using a spanner wrench
- Fuel pump from the fuel lever sending device

To install:

6. Install or connect the following:

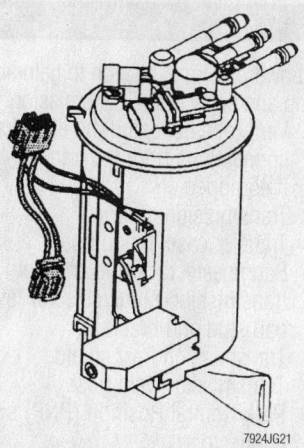

View of the in-tank fuel pump assembly

- Fuel pump in tank with new seal around opening
- Tank and connect fuel lines and vapor hose
- Tank to the frame. Torque the fasteners to 33 ft. lbs. (45 nm).
- Shield
- Fuel filler neck and clamp
- Negative battery cable
7. Refill the tank.
8. Run the engine and check for leaks.

Fuel Injector

REMOVAL & INSTALLATION

1. Before servicing the vehicle, refer to the precautions in the beginning of this section.

2. Relieve the fuel system pressure. Refer to the fuel system relief procedure in this section.

3. Remove or disconnect the following:
- Engine cover
- Negative battery cable
- Upper manifold assembly
- Fuel meter body electrical connection and the fuel feed and return hoses from the engine fuel pipes

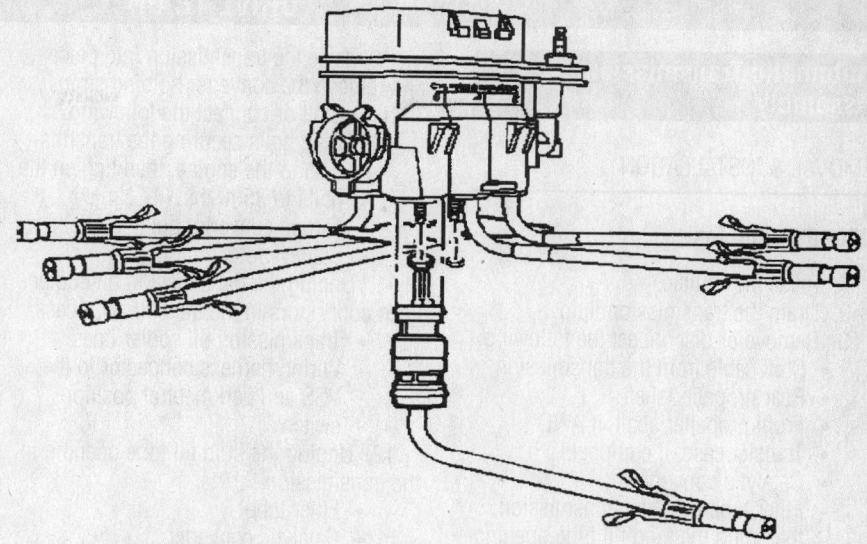

Exploded view of the fuel meter assembly, including the injectors

- Poppet nozzle out of the casting socket by squeezing the nozzle locking tabs together
- Fuel meter body by releasing the locktabs

➡ **Each injector is calibrated. When replacing the fuel injectors, be sure to replace it with the correct injector.**

- Lower hold-down plate and nuts
4. While pulling the poppet nozzle tube downward, push with a small prytool down between the injector terminals and remove the injectors.

To install:
5. Lubricate the new injector O-ring seats with engine oil.
6. Install or connect the following:
- O-rings on the injector
- Fuel injector into the fuel meter body injector socket.
- Lower hold-down plate and nuts. Torque the nuts to 27 inch lbs. (3 Nm).
- Fuel meter body assembly into the

intake manifold. Torque the fuel meter bracket retainer bolts to 88 inch. lbs. (10 Nm).

❉❉ CAUTION

To reduce the risk of fire or injury ensure that the poppet nozzles are properly seated and locked in their casting sockets

- Fuel meter body into the bracket and lock all the tabs in place
- Poppet nozzles into the casting sockets
- Electrical connections
- New o-ring seals on the fuel return and feed hoses.
- Fuel feed and return hoses and tighten the fuel pipe nuts to 22 ft. lbs. (30 Nm)
- Negative battery cable
7. Turn the ignition **ON** for 2 seconds and then turn it **OFF** for 10 seconds. Again turn the ignition **ON** and check for leaks.
8. Install the manifold plenum.

DRIVE TRAIN

Automatic Transmission Assembly

REMOVAL & INSTALLATION

2000–01 Vehicles

1. Raise the vehicle.
2. Drain the transmission fluid.
3. Remove or disconnect the following:
   - Shift cable from the transmission
   - Rear propeller shaft
   - Front propeller shaft, if AWD
   - Transfer case, if equipped
   - Catalytic converter
   - Filler tube from the transmission, then plug the fluid fill tube opening in the transmission
   - Torque converter bolts
   - Wiring harness connectors from the Vehicle Speed Sensor (VSS) and the park/neutral position switch
   - Transmission oil cooler lines, then plug the transmission oil cooler lines connectors in the transmission case
   - Starter motor
   - Nine bolts securing the transmission to the engine
4. Pull the transmission straight back from the engine.
5. Install the tool J 21366 converter holding strap, in order to keep the torque converter from sliding off of the transmission turbine shaft.
6. Remove the transmission from the vehicle.
7. Flush the transmission oil cooler and the lines.

To install:

8. Install the converter holding strap to keep the torque converter from sliding off of the transmission turbine shaft.

9. Raise the transmission into place and remove the converter holding strap.
10. Install or connect the following:
    - Nine bolts securing the transmission to the engine, then tighten the bolts to 35 ft. lbs. (47 Nm)
    - Torque converter bolts
    - Starter motor
11. Unplug the transmission oil cooler lines connectors in the transmission case.
    - Transmission oil cooler lines
    - Wiring harness connector to the VSS and park/neutral position switch
12. Unplug the fluid fill tube opening in the transmission.
    - Filler tube
    - Catalytic converter
    - Transfer case, if equipped
    - Front propeller shaft, if AWD
    - Rear propeller shaft
    - Shift cable
13. Lower the vehicle.
14. Fill the transmission with new transmission fluid.

2002–03 Vehicles

1. Disconnect the negative battery cable.
2. Raise the vehicle.
3. Drain the transmission fluid.
4. Remove the rear propeller shaft.
5. Support the transmission with a transmission jack.
6. Remove the nut (2WD) or nuts (AWD) securing the transmission mount to the transmission support.
7. Raise the transmission slightly and remove the transmission support from the vehicle.
8. Remove or disconnect the following:
   - Front exhaust pipe assembly
   - Starter motor

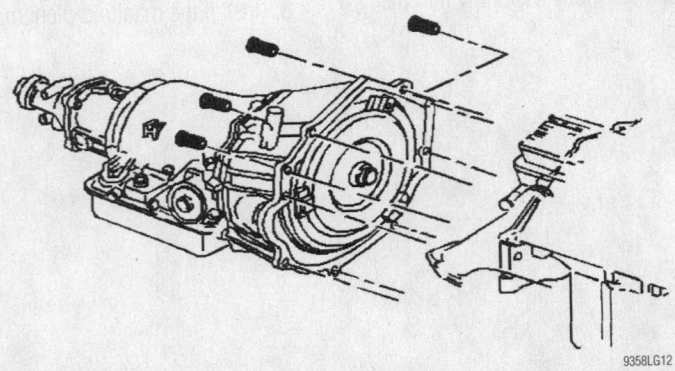

9. Lower the transmission to gain access to the top and sides of the transmission.
   - Vent tube hose and the electrical connections from the transfer case, if equipped
   - Transmission mount
   - Transfer case, if equipped
   - Range selector cable end from the transmission range selector lever ball stud and bracket
   - Transmission heat shield
   - Transmission vent hose
   - Park/Neutral Position (PNP) switch connector
   - Main electrical connector from the transmission
   - Bolt that secures the fuel line bracket to the left side of the transmission
   - Torque converter access plug
   - Flywheel-to-torque converter bolts
   - Transmission oil cooler pipes from the transmission, then plug the openings in the transmission case
   - Stud and the bolt securing the transmission to the engine
   - Six studs and one bolt securing the transmission to the engine
10. Install tool J 21366 converter holding strap onto the transmission bell housing to retain the torque converter.
11. Pull the transmission straight back.
12. Remove the transmission from the vehicle while simultaneously removing the fluid level indicator tube.

To install:

13. Install the converter holding strap onto the transmission bell housing to retain the torque converter.
14. Support the transmission with a transmission jack.
15. Raise the transmission into place while simultaneously installing the fluid indicator tube.
16. Remove the converter holding strap from the transmission.
17. Slide the transmission straight onto the locating pins while lining up the marks on the flywheel and the torque converter. The torque converter must rotate freely by hand.
18. Install or connect the following:
    - Six studs and one bolt securing the transmission to the engine. Tighten the studs and the bolt to 37 ft. lbs. (50 Nm).
    - Stud and bolt securing the transmission to the engine. Tighten the stud and the bolt to 37 ft. lbs. (50 Nm).

9358LG12

Exploded view of the automatic transmission mounting—2001 vehicle shown

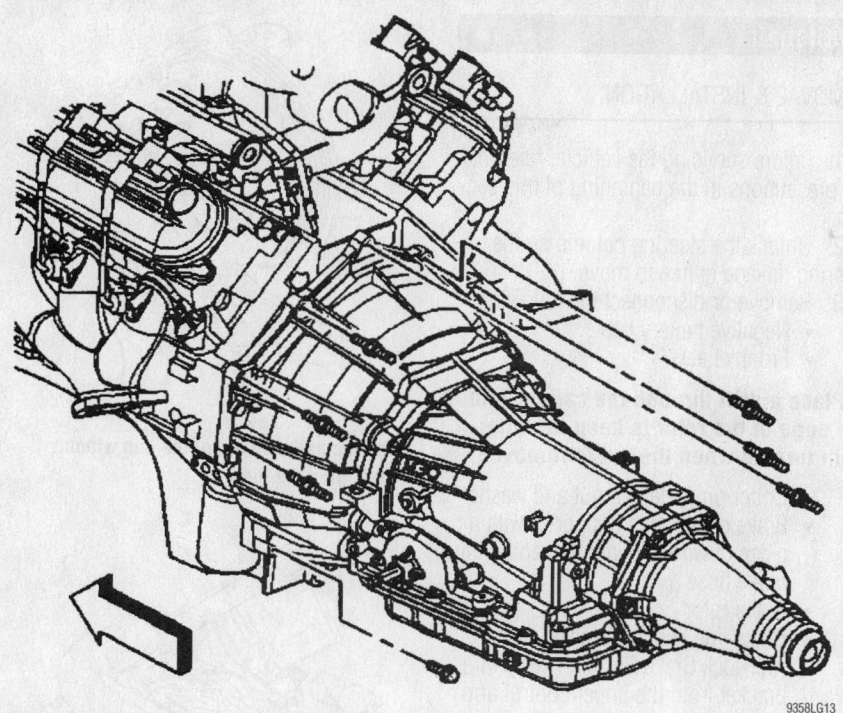

Exploded view of the automatic transmission—2002 vehicle shown

9358LG13

- Flywheel-to-torque converter bolts and tighten to 46 ft. lbs. (63 Nm)
- Torque converter access plug
- Transmission vent hose, fuel lines, and the wiring harness
- Heat shield
- Range selector cable
- Transfer case, if equipped
- Transmission mount to the transmission (2WD) or the transfer case adapter (AWD)
- Vent hose and electrical connectors to the transfer case, if equipped
- Starter motor
- Front exhaust pipe assembly
- Transmission support
- Transmission mount to transmission support nut (2WD) or nuts (AWD). Tighten the nut(s) to 29 ft. lbs. (40 Nm).
19. Remove the transmission jack.
- Rear propeller shaft.

➡**Flush the transmission oil cooler and oil cooler pipes at this time, if necessary.**

- Oil cooler pipes to the transmission
20. Lower the vehicle.
21. Connect the negative battery cable
22. Fill the transmission to the proper level with DEXRON® III transmission fluid.
23. Test the vehicle for proper operation and check for leaks.

Transfer Case Assembly

REMOVAL & INSTALLATION

2000 Vehicles

1. Raise the vehicle.
2. Remove or disconnect the following:

- Front propeller shaft
- Rear propeller shaft
- Vent hose from the transfer case
- Vehicle Speed Sensor (VSS) electrical connectors
- Electrical connectors from the transfer case motor/encoder
- Wiring harness from the transfer case.

3. Support the transfer case with a transmission jack.

- Bolt and washer securing the support brace to the transmission
- two bolts securing the support brace to the transfer case
- Two bolts and washers securing the transfer case adapter to the mount
- Six bolts securing the transfer case and bracket to the transmission adapter

4. Remove the transfer case:
   a. Start by pulling the transfer case straight back from the transmission.
   b. After the transfer case is pulled out

from the transmission turn and lower the transfer case.

5. Remove and discard the gasket.

To install:

6. Install a new gasket to the transmission. Use sealer GM P/N 12346004 in order to hold the gasket in place.

7. Raise and position the transfer case to the vehicle

8. Install the six bolts securing the transfer case and bracket to the transmission adapter. Tighten the bolts to 35 ft. lbs. (47 Nm).

9. Remove two bolts and washers securing the transfer case adapter to the mount.

10. Install or connect the following:

- Bolt securing the support brace to the transmission and tighten to 94 ft. lbs. (128 Nm)
- Two bolts securing the support brace to the transfer case and tighten to 66 ft. lbs. (90 Nm)
- Vent hose to the transfer case

11. Check the transfer case oil level.

- VSS electrical connectors
- Connectors to the transfer case motor/encoder
- Transfer case wiring
- Rear propeller shaft
- Front propeller shaft

12. Lower the vehicle.

2001–03 Vehicles

1. Raise the vehicle.
2. Remove or disconnect the following:

- Engine brace mounting bolts
- Engine brace
- Transfer case shield
- Rear propeller shaft
- Front propeller shaft
- Motor/encoder electrical connector
- Transfer case electrical wiring harness
- Transfer case vent hose

❋❋ WARNING

To perform the following, the transmission mount must be removed for access to the bottom retaining nut of the transfer case. Do not support the transmission by the oil pan.

3. Support the transmission with a suitable transmission jack.

- Transmission mount retaining nuts.

➡**Raise the transmission assembly enough to clear the rear crossmember.**

- Transmission mount bolts and mount assembly
- Transfer case bottom retaining nut

➡️The transmission mount must be replaced in it's original position to support the transmission. When reinstalling the transmission mount, it is not necessary reinstall the mounting nuts and bolts. The weight of the transmission and engine will hold the transmission mount in place until this service procedure is completed.

4. Install the transmission mount.
5. Remove the transmission jack stand from the transmission.
6. Remove the transfer case upper retaining nuts.
7. Install the transmission jack to the transfer case.
8. Remove the transfer case assembly.

✳✳ WARNING

DO NOT reuse the old gasket if it is damaged. DO NOT use any type of sealer in place of the gasket.

9. Remove the transfer case gasket.
To install:

➡️Make sure that the locating tab on the gasket is in the proper location. The locator tab should be facing up.

10. Install or connect the following:
- New transfer case gasket
- Transfer case assembly
- Transfer case retaining nuts and tighten them to 40 ft. lbs. (55 Nm)
11. Remove the transmission jack from the transfer case.
- Transmission jack to the transmission
- Transmission mount
- Transmission mount bolts and tighten them to 35 ft. lbs. (47 Nm)
12. Lower the transmission into place.
13. Remove the transmission jack stand.
14. Install or connect the following:
- Transmission mount retaining nuts and tighten to 29 ft. lbs. (40 Nm)
- Motor/encoder electrical connector
- Transfer case electrical wiring harness
- Transfer case vent hose
- Rear propeller shaft
- Front propeller shaft
- Engine brace. Tighten the mounting bolts to 37 ft. lbs. (50 Nm).
15. Inspect the transfer case fluid level.
16. Install the transfer case shield.
17. Lower the vehicle.

Halfshaft

REMOVAL & INSTALLATION

1. Before servicing the vehicle, refer to the precautions in the beginning of this section.
2. Unlock the steering column so the steering linkage is free to move.
3. Remove or disconnect the following:
- Negative battery cable
- Front wheels

➡️Place a drift through the caliper into the edge of the rotor to keep the rotor from turning when the nut is removed

- Cotter pin, retainer, nut and washer
- Brake caliper and support it with a piece of wire to avoid damaging the brake hose
- Brake rotor
- Brake line support bracket and Anti-lock Brake System (ABS) wire bracket from the upper control arm
4. Place a jackstand or jack under the lower control arm.
- Axle shaft from the hub by placing a block of wood against the outer edge of the axle (to protect the threads), then strike the block of wood sharply with a hammer. Do not remove the axle at this time.
- Tie rods from the steering knuckles
- Lower shock absorber bolts
- Upper ball joint from the steering knuckle and suspend the steering knuckle on a wire
- Skid plate, if equipped
- Halfshaft-to-axle tube bolts
- Halfshaft by moving it forward and supporting it away from the frame
- Halfshaft from the hub and bearing assembly
- Halfshaft from the differential using a block of wood and a hammer

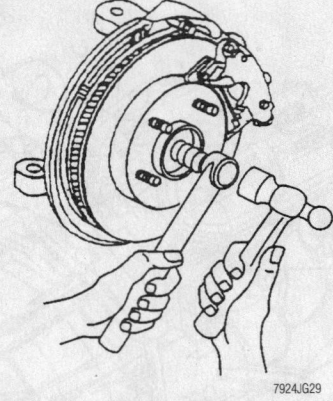

7924JG29

Tap the halfshaft out of the hub without damaging the threads

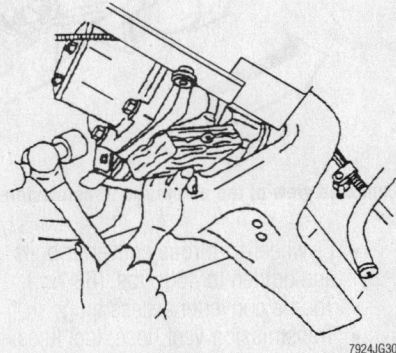

7924JG30

Using a block of wood and a mallet, disengage the halfshaft from the differential assembly

To install:

➡️It is essential that the differential carrier and axle seals are not lubricated or damaged during installation. Prior to shaft installation, cover the shock mounting bracket, lower control arm ball stud and ALL other sharp edges with a cloth or rag to help protect the boot.

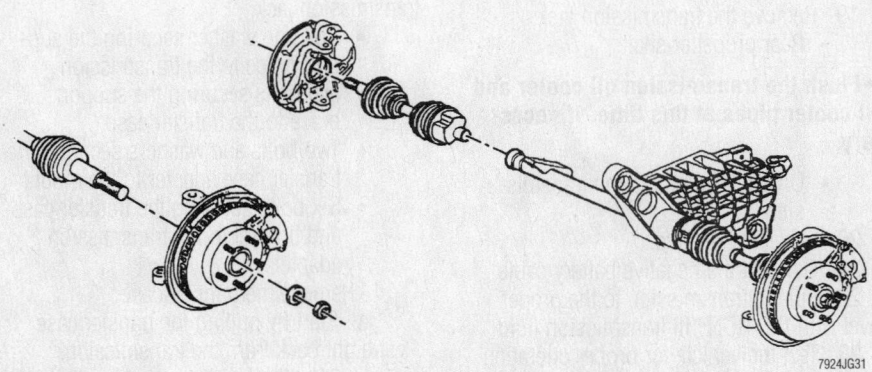

7924JG31

Halfshafts and related components

5. Install the axle into the carrier. With both hands on the tripod housing, align the splines on the shaft with the carrier. Then center the axle into the carrier seal and push the shaft straight into the carrier until the snapring is properly seated.

➡**Be careful when supporting the lower control arm that any components are damaged with the supporting device.**

6. Raise the lower control arm using a jackstand or jack until the full weight of the arm is supported.

➡**It is necessary to slightly start the knuckle onto the axle while at the same time guiding the lower ball joint into position on the knuckle.**

7. Install or connect the following:
- Lower ball joint, the lower shock absorber and the upper ball joint
- Axle washer and nut. Tighten the nut to 103 ft. lbs. (140 Nm).
- ABS and brake line brackets to the top of the upper control arm
- Caliper and rotor
- Tire and wheel assembly
- Differential carrier shield

CV-Joints

OVERHAUL

Outer CV-Joint

1. Before servicing the vehicle, refer to the precautions in the beginning of this section.

2. Remove or disconnect the following:
- Front wheel
- Halfshaft and position it in a vise
- Large CV-joint boot clamp and discard it
- Small CV-joint boot clamp and discard it
- CV-joint boot and slide it back on the shaft
- Outer race from the halfshaft, by spreading the outer race-to-half-shaft retaining ring, using Snapring Pliers J-8059
- Retaining ring from the halfshaft and discard it
- CV-joint boot from the halfshaft and discard it, if damaged

3. Disassemble the chrome alloy balls from the CV-joint cage as follows:

a. Position a brass drift against the CV-joint cage and tap it with a hammer to tilt the cage.

b. Remove the 1st chrome alloy ball from the cage.

c. Tilt the cage in the opposite direction.

d. Remove the opposite chrome alloy ball.

e. Repeat the procedure until all 6 balls are removed.

4. Disassemble the CV-joint cage and inner race as follows:

a. Pivot the cage and race 90 degrees to the center line of the outer race.

b. Align the cage windows with outer race lands.

c. Remove the cage from the outer race.

d. Rotate the inner race upward and remove it from the cage.

5. Thoroughly clean and inspect all parts.

To install:

6. Lubricate the parts with a light coat of grease.

7. Assemble the CV-joint cage and inner race, as follows:

a. Rotate the inner race 90 degrees to the cage centerline.

b. Align the cage windows with inner race lands.

c. Insert the inner race into the cage by rotating the inner race downward.

d. Insert the cage/inner race into the outer race.

8. Assemble the chrome alloy balls into the CV-joint cage, as follows:

a. Position a brass drift against the CV-joint cage and tap it with a hammer to tilt the cage.

b. Insert the 1st chrome alloy ball into the cage.

c. Tilt the cage in the opposite direction.

d. Insert the opposite chrome alloy ball.

e. Repeat the procedure until all 6 balls are inserted.

9. Install ½kit grease into the CV-joint.

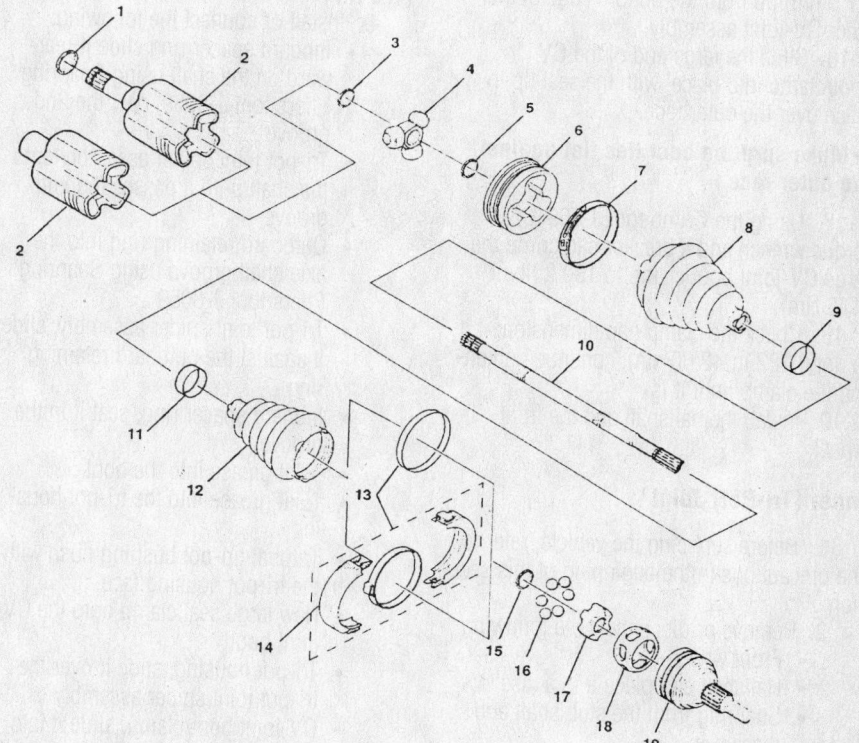

(1) Differential Shaft Ring
(2) Tripot Housing Assembly
(3) Spacer Ring
(4) Tripot Joint Spider Assembly
(5) Spacer Ring
(6) Tripot Bushing
(7) Boot Retaining Clamp
(8) Tripot Joint Boot
(9) Halfshaft Swage Ring
(10) Halfshaft Bar

(11) Halfshaft Swage Ring
(12) CV Joint Boot
(13) Swage Ring
(14) Clamp Protector
(15) Race Retaining Ring
(16) Ball
(17) CV Joint Inner Race
(18) CV Joint Cage
(19) CV Joint Outer Race

9308JG09

Exploded view of the CV-Joint Assembly

10. Install or connect the following:
- Small ring clamp on the CV boot
- New retaining ring on the halfshaft
- Large ring clamp on the CV boot
- Outer race assembly onto the half-shaft until the ring engages the halfshaft groove

11. Slide the small end of the CV-joint boot/clamp into place, with the seal lip in the halfshaft groove

➡**Make sure the boot lies flat against the halfshaft.**

12. Using the Crimp tool J-35910, a torque wrench and a breaker bar, crimp the small CV-joint boot clamp to 100 ft. lbs. (136 Nm).

13. Check the clamp gap dimension; if it is not 0.085 in. (2.15mm), continue tightening the clamp until it is.

14. Install ½ kit grease into the CV-joint boot.

15. Measure approximately 0.687 in. (17.5mm) up from the bottom edge of the outer CV-joint assembly.

16. Slide the large end of the CV boot/clamp into place, with the seal lip in place over the outer race.

➡**Make sure the boot lies flat against the outer race.**

17. Using the Crimp tool J-35910, a torque wrench and a breaker bar, crimp the large CV-joint boot clamp to 130 ft. lbs. (176 Nm).

18. Check the clamp gap dimension; if it is not 0.102 in. (2.60mm), continue tightening the clamp until it is.

19. Install the halfshaft and the front wheel.

Inner (Tri-Pot) Joint

1. Before servicing the vehicle, refer to the precautions in the beginning of this section.

2. Remove or disconnect the following:
- Front wheel
- Halfshaft and place it in a vise
- Snapring from the stub shaft and discard it
- Small CV-joint boot clamp, cut and discard it
- Large CV-joint boot clamp, cut and discard it
- CV-joint boot by sliding it away from the tri-pot joint

3. Install a Stub Shaft Removal tool J-38868-A to the stub shaft snapring groove.

4. Using a slide hammer puller, press the stub shaft from the tri-pot housing.

5. Remove or disconnect the following:

- Tri-pot housing from the tri-pot spider
- Inboard spacer ring slide it rearward on the shaft using Snapring Pliers tool J-8059
- Outboard retaining ring using Snapring Pliers tool J-8059 and discard it
- Tri-pot joint spider assembly
- Inboard spacer ring and discard it
- CV-joint boot
- Trilobal tri-pot bushing from the housing

6. Thoroughly clean and inspect all parts.

To install:

7. Install or connect the following:
- New snapring onto the stub shaft
- Small boot clamp
- CV-joint boot

8. Using the Crimp tool J-35910, a torque wrench and a breaker bar, crimp the small CV-joint boot clamp to 100 ft. lbs. (136 Nm).

9. Install or connect the following:
- Inboard spacer ring slide it rearward on the shaft using Snapring Pliers tool J-8059, past the 2nd groove
- Tri-pot joint spider assembly onto the shaft until it passes the 2nd groove
- Outboard retaining ring into the axle shaft groove using Snapring Pliers tool J-8059
- Tri-pot joint spider assembly, slide it against the outboard retaining ring
- Inboard spacer ring, seat it in the groove
- ½ kit grease into the boot
- ½ kit grease into the tri-pot housing
- Trilobal tip-pot bushing flush with the tri-pot housing face
- New large seal clamp onto the CV-joint boot
- Tri-pot housing, slide it over the tri-pot joint spider assembly
- CV-joint boot/clamp, slide it into place, over the trilobal tri-pot bushing with the seal lip in the groove

➡**Make sure the boot lies flat against the trilobal bushing.**

10. Position the CV-joint boot so it measures 4.9 in. (125mm).

11. Using the Crimp tool J-35566, latch the large CV-joint boot clamp.

12. Install the halfshaft and the front wheel.

Axle Shaft, Bearing and Seal

REMOVAL & INSTALLATION

For the Axle Shaft, Bearing and Seal, Removal and Installation, please refer to Wheel Bearing procedure located in the section.

Pinion Seal

REMOVAL & INSTALLATION

1. Before servicing the vehicle, refer to the precautions in the beginning of this section.

➡**The following procedure requires the use of the Pinion Holding tool J-8614-10, the Pinion Flange Removal tool J-8614-1, J-8614-2, J-8614-3 and the Pinion Seal Installation tool J-23911.**

2. Remove or disconnect the following:
- Driveshaft from the pinion flange. Matchmark the driveshaft prior to removal.
- Driveshaft from the rear axle pinion flange and support the shaft up in body tunnel by wiring it to the exhaust pipe.

➡**If the U-joint bearings are not retained by a retainer strap, use a piece of tape to hold bearings on their journals.**

3. Mark the position of the pinion stem, flange and nut for reference.

4. Use an inch lbs. torque wrench to measure the amount of torque necessary to turn the pinion, then note this measurement as it is the combined pinion bearing, seal, carrier bearing, axle bearing and seal preload.

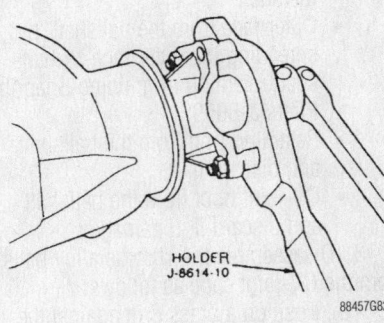

HOLDER
J-8614-10

88457G82

Removing the pinion nut using a pinion holding fixture tool

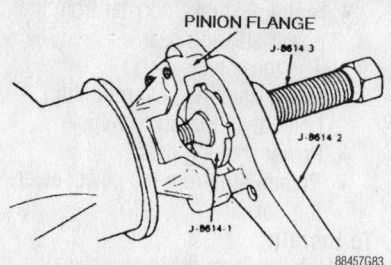

A puller and adapter should be used to withdraw the pinion from the housing

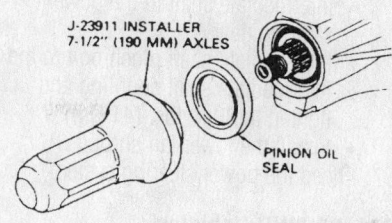

Use the appropriately sized installation tool to drive the new seal into position.

5. Remove or disconnect the following:

• Pinion flange nut and washer, using a Pinion Holding tool J-8614-10 and a Pinion Flange Removal tool J-8614-1, J-8614-2, J-8614-3, as applicable
• Pinion flange
• Pinion oil seal by driving it out of the differential with a blunt chisel; DO NOT damage the carrier

To install:

6. Examine the seal surface of pinion flange for tool marks, nicks or damage, such as a groove worn by the seal. If damaged, replace flange.

7. Examine the carrier bore and remove any burrs that might cause leaks around the O.D. of the seal.

8. Apply GM seal lubricant 1050169 to the outside diameter of the pinion flange and sealing lip of new seal.

9. Install or connect the following:

• New pinion oil seal using a seal installer tool
• Pinion flange and tighten nut to the same position as marked earlier. Tighten the nut a little at a time and turn the pinion flange several times after each tightening in order to set the rollers.

10. Measure the torque necessary to turn the pinion and compare this to the reading taken during removal. Tighten the nut additionally, as necessary to achieve the same preload as measured earlier.

➡**If fluid was lost from the differential housing during this procedure, be sure to check and add additional fluid, as necessary.**

11. Remove the support then align and secure the driveshaft assembly to the pinion flange.

➡**The original matchmarks MUST be aligned to assure proper shaft balance and prevent vibration.**

STEERING AND SUSPENSION

Air Bag

✳✳ CAUTION

Some vehicles are equipped with an air bag system, also known as the Supplemental Inflatable Restraint (SIR) system. The system must be disabled before performing service on or around system components, steering column, instrument panel components, wiring and sensors. Failure to follow safety and disabling procedures could result in accidental air bag deployment, possible personal injury and unnecessary system repairs.

PRECAUTIONS

Several precautions must be observed when handling the inflator module to avoid accidental deployment and possible personal injury.

• Never carry the inflator module by the wires or connector on the underside of the module.
• When carrying a live inflator module, hold securely with both hands, and ensure that the bag and trim cover are pointed away.
• Place the inflator module on a bench or other surface with the bag and trim cover facing up.

• With the inflator module on the bench, never place anything on or close to the module, that may be thrown in the event of an accidental deployment.

DISARMING

➡**With the AIR BAG fuse removed and the ignition switch ON, the AIR BAG warning lamp will be on. This is normal and does not indicate any system malfunction.**

1. Turn the steering wheel so that the vehicle's wheels are pointing straight ahead.
2. Turn the ignition switch to **LOCK**, remove the key, then disconnect the negative battery cable.
3. Remove the AIR BAG fuse from the fuse block.
4. Remove the steering column filler panel.
5. Disengage the Connector Position Assurance (CPA) and the yellow two way connector located at the base of the steering column.
6. Connect the negative battery cable.

ARMING

1. Disconnect the negative battery cable.
2. Turn the ignition switch to **LOCK**, then remove the key.
3. Engage the yellow SIR connector and CPA located at the base of the steering column.

4. Install the steering column filler panel.
5. Install the AIR BAG fuse to the fuse block.
6. Connect the negative battery cable.
7. Turn the ignition switch to **RUN** and make sure that the AIR BAG warning lamp flashes seven times and then shuts off. If the warning lamp does not shut off, make sure that the wiring is properly connected. If the light remains on, take the vehicle to a reputable repair facility for service.

Power Steering Gear

REMOVAL & INSTALLATION

Except 2001–03 AWD Vehicles

➡**The wheels of the vehicle must be straight ahead and the steering column in the LOCK position before disconnecting the steering column or intermediate shaft from the steering gear. Failure to do so will cause the coil assembly in the steering column to become uncentered which will cause damage to the coil assembly.**

1. Before servicing the vehicle, refer to the precautions in the beginning of this section.
2. Remove or disconnect the following:
• Lower fan shroud
• Intermediate shaft pinch bolt from the intermediate shaft coupling

- Intermediate shaft from the power steering gear

3. Place a drain pan under the steering gear.

➡ **Cap or tape the ends of the hoses and the gear fittings to prevent spillage and contamination.**

- Power brake booster outlet hose from the steering gear
- Power steering cooler pipe from the steering gear

4. Raise the vehicle.

- Pitman arm from the steering gear
- Steering gear mounting bolts and washers
- Steering gear from the vehicle

To install:

- Install or connect the following:
- Steering gear to the vehicle
- Steering gear washers and bolts. Tighten to 55 ft. lbs. (75 Nm) for 2WD vehicles, or to 105 ft. lbs. (142.5 Nm) for AWD vehicles.
- Pitman arm to the steering gear

5. Lower the vehicle.

- Power steering cooler pipe to the steering gear and tighten to 20 ft. lbs. (27 Nm)
- Power brake booster outlet hose to the steering gear and tighten the fitting to 20 ft. lbs. (27 Nm)

6. Remove the drain pan from under the vehicle.

- Intermediate shaft to the power steering gear
- Intermediate shaft pinch bolt to the intermediate shaft coupling and tighten to 30 ft. lbs. (41 Nm)
- Install the lower fan shroud

7. Bleed the power steering system

2001–03 AWD Vehicles

1. Before servicing the vehicle, refer to the precautions in the beginning of this section.

2. Raise and support the vehicle.

3. Remove or disconnect the following:

- Tire and the wheel
- Pitman arm-to connecting rod retaining nut
- Pitman arm from the connecting rod
- Stabilizer links

4. Rotate the stabilizer shaft downward.

5. Lower the vehicle.

6. Install steering column lock pin no. J 42640 into the lower steering column trim cover in order to lock the steering column.

7. Remove or disconnect the following:

- Lower fan shroud
- Intermediate shaft-to-power steering gear pinch bolt
- Intermediate shaft from the power steering gear

8. Place a drain pan under the vehicle.

- Power steering cooler hose from the power steering gear

- Hydraulic brake booster from the power steering gear

9. Raise the vehicle.

- Power steering gear mounting bolts, then lower the vehicle
- Power steering gear
- Pitman arm from the power steering gear

To install:

10. Install or connect the following:

- Pitman arm to the power steering gear
- Power steering gear to the vehicle
- Pitman arm to the connecting rod

11. Using the aid of an assistant, position the power steering gear to the frame and install the power steering mounting bolts. Tighten the power steering mounting bolts to 105 ft. lbs. (142.5 Nm.

- Power brake booster outlet hose to the power steering gear and tighten to 20 ft. lbs. (27 Nm)
- Power steering cooler hose to the power steering gear and tighten to 20 ft lbs. (27 Nm)
- Intermediate shaft to the power steering gear
- Intermediate shaft to power steering gear pinch bolt and tighten to 30 ft. lbs. (41 Nm)
- Lower fan shroud

12. Remove the lock pin from the steering column.

- Pitman arm-to-connecting rod retaining nut and tighten to 48 ft. lbs. (62 Nm)

13. Rotate the stabilizer shaft to the correct position.

- Stabilizer links
- Tire and the wheel

14. Lower the vehicle.

15. Bleed the power steering system.

Shock Absorbers

REMOVAL & INSTALLATION

Front

1. Before servicing the vehicle, refer to the precautions in the beginning of this section.

2. Support the lower control arm (front) or axle assembly (rear).

3. Remove or disconnect the following:

- Wheel
- Inner wheel well splash shield, if removing the front shock absorber
- Lower nut, washer and bolt
- Upper nut, washer and bolt

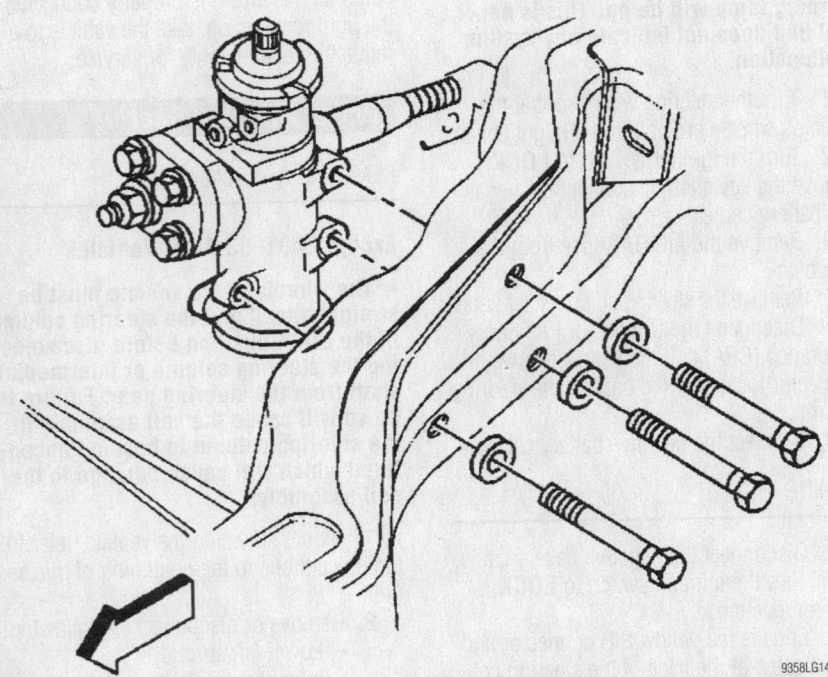

9358LG14

Exploded view of the power steering gear mounting—2002 vehicle shown, others similar

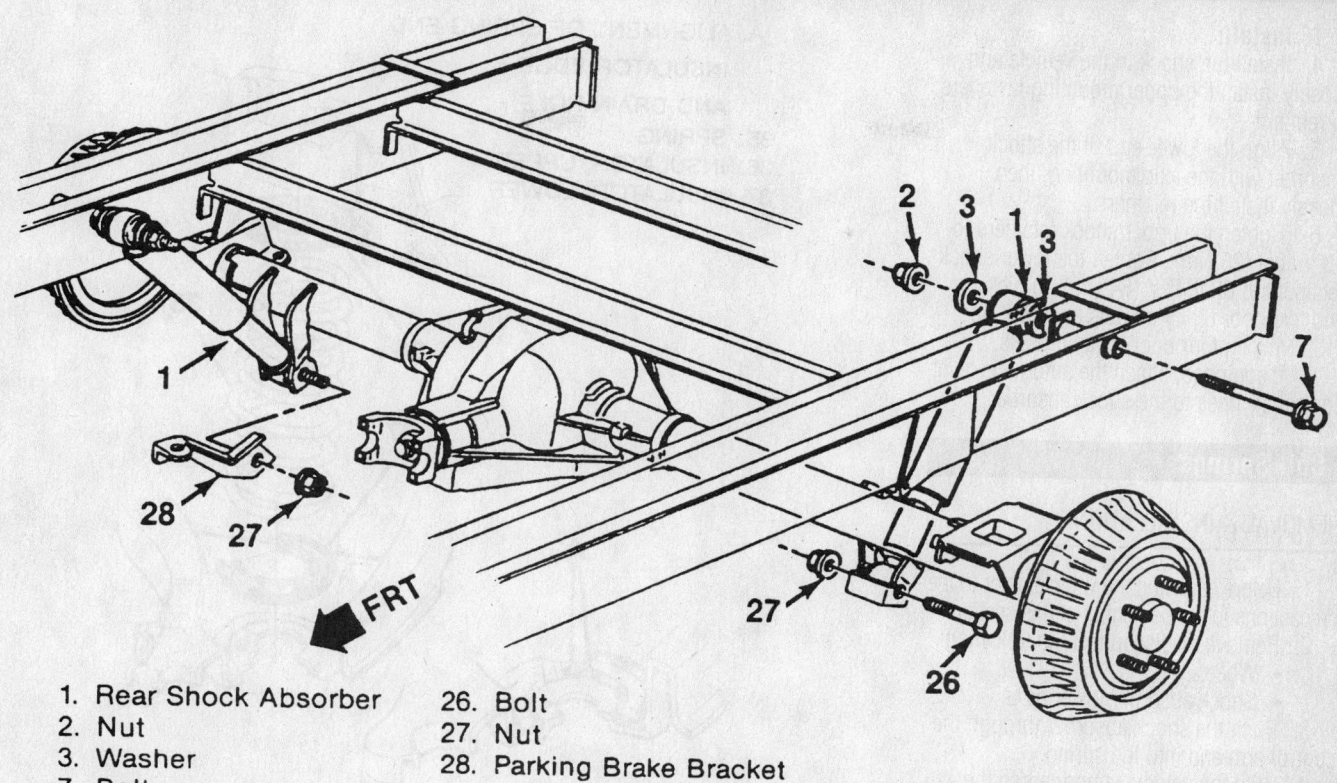

1. Rear Shock Absorber
2. Nut
3. Washer
7. Bolt
26. Bolt
27. Nut
28. Parking Brake Bracket

7924JG36

Rear shock absorber mounting

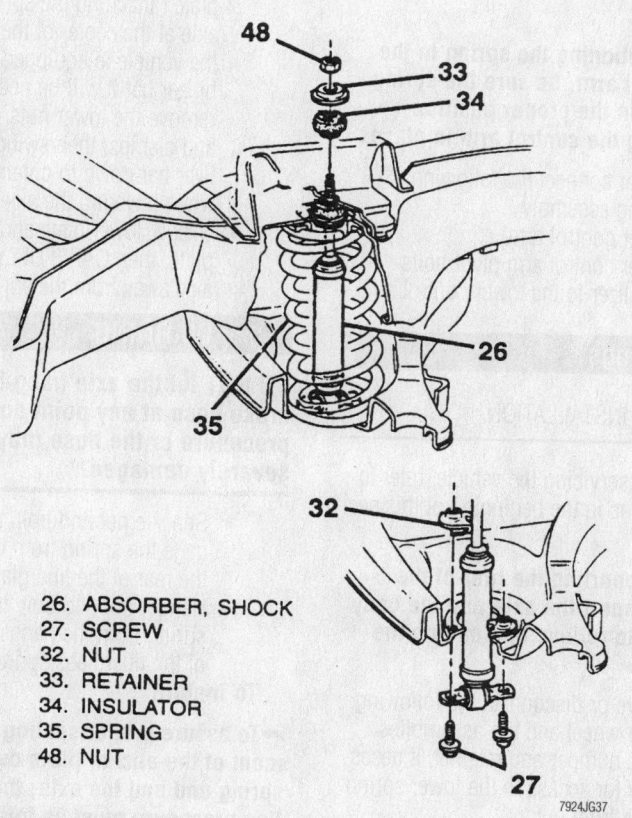

26. ABSORBER, SHOCK
27. SCREW
32. NUT
33. RETAINER
34. INSULATOR
35. SPRING
48. NUT

7924JG37

Front shock absorber mounting

➡**Compress the front shock absorber to make removal easier.**

• Shock absorber

To install:

4. Compress the front shock absorber to make installation easier.

• Shock absorber
• Lower nut. Torque the nut to 62 ft. lbs. (84 Nm).
• Upper bolt. Torque the bolts to 18 ft. lbs. (25 Nm).
• Wheel

Rear

1. Before servicing the vehicle, refer to the precautions in the beginning of this section.

2. Properly support the rear axle assembly.

3. Remove or disconnect the following:

• Automatic level control air lines from the shock absorber, if equipped
• Shock absorber-to-frame retainers at the top of the shock
• Shock-to-axle retainers at the bottom of the shock
• Shock absorber

To install:

4. Install the shock in the vehicle and loosely install the upper mounting fasteners to retain it

5. Align the lower-end of the shock absorber with the axle mounting, then loosely install the retainers.

6. Tighten the upper shock retainers to 18 ft. lbs. (25 Nm). Tighten the lower shock retainers to 62 ft. lbs. (84 Nm) on pick-up and two door utility models and 74 ft. lbs. (100 Nm) on four door utility models.

7. If equipped, attach the automatic level control air lines to the shock absorber.

Coil Springs

REMOVAL & INSTALLATION

1. Before servicing the vehicle, refer to the precautions in the beginning of this section.
2. Remove or disconnect the following:
   • Wheel
   • Shock absorber lower bolts
3. Push the shock absorber through the control arm and into the spring.
4. With the vehicle supported so the control arms hang free, install tool J-23028, onto a support and into the lower control arm bushings.
5. Remove or disconnect the following:
   • Stabilizer bar from the control arm
   • Stabilizer from the lower control arm
6. Raise and remove the tension on the lower control arm bolts.
7. Install a safety chain around the spring and through the lower control arm.
8. Remove or disconnect the following:
   • Lower control arm pivot bolts, the rear first
   • Lower control arm and allow it to hang free
   • Spring assembly

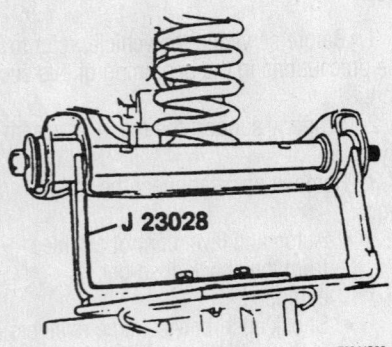

Secure tool J 23028 to a jack, then raise the jack to remove the tension on the lower control arm bolts

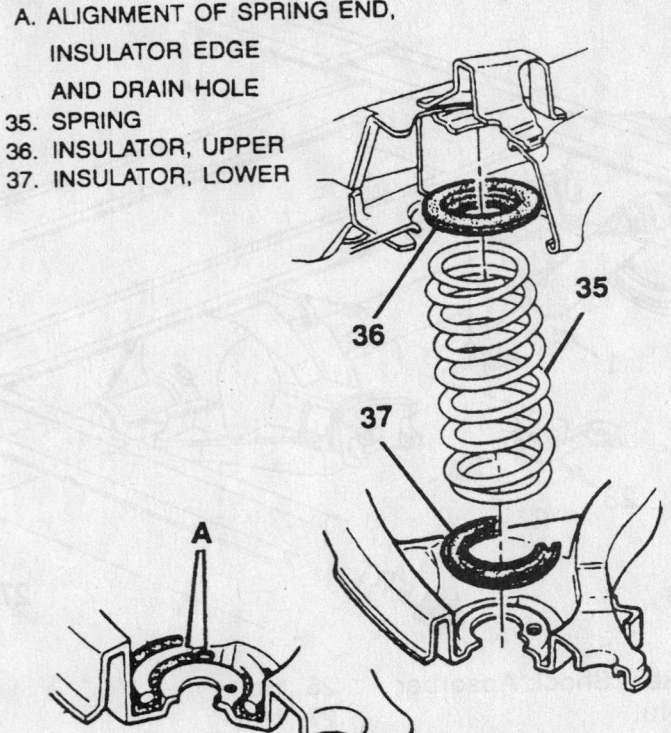

A. ALIGNMENT OF SPRING END, INSULATOR EDGE AND DRAIN HOLE
35. SPRING
36. INSULATOR, UPPER
37. INSULATOR, LOWER

Exploded view of the coil spring mounting

To install:

➡ **When positioning the spring in the lower control arm, be sure the spring insulator is in the proper position before lifting the control arm in place.**

9. Install or connect the following:
   • Spring assembly
   • Lower control arm
   • Lower control arm pivot bolts
   • Stabilizer to the lower control arm

Leaf Springs

REMOVAL & INSTALLATION

1. Before servicing the vehicle, refer to the precautions in the beginning of this section.

➡ **When supporting the rear of the vehicle, support the axle and the body separately to relieve the load on the rear spring.**

2. Remove or disconnect the following:
   • Rear wheel and tire assemblies
   • Axle bumper and retainer, if necessary for access to the lower spring plate front nut
   • Nuts securing the U-bolt and lower

plate (attaching the spring to the axle at the center of the spring). If the vehicle is equipped with a stabilizer bar it will be necessary to remove the lower nuts, washers and clamps, then swing the stabilizer bar down to obtain clearance when lowering the axle assembly.
   • U-bolt, lower plate and anchor plate, then CAREFULLY lower the axle away from the spring

❉❉ WARNING

DO NOT let the axle hang by the brake hose at any point during the procedure or the hose may be severely damaged.

   • Shackle nut and bolt, then disengage the spring from the shackle, at the rear of the fiberglass spring
   • Hanger nut and bolt, then the spring from the hanger, at the front of the fiberglass spring

To install:

➡ **To assure proper seating and attachment of the anchor plate over the spring end and the axle, the installation procedure must be followed closely.**

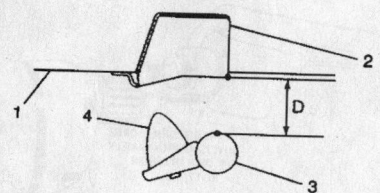

1. FRAME
2. AXLE STOP BRACKET
3. REAR AXLE (END VIEW)
4. BUMPER
5. TRIM HEIGHT 135 – 145MM
 (5.3 – 5.7 INCHES)

88268GB5

Measuring the rear suspension trim height

3. Install or connect the following:
 • Spring to the hanger, then loosely install the retaining nut and bolt
 • Spring to the shackle, then loosely install the retaining nut and bolt
4. CAREFULLY raise the axle until it contacts the spring.
5. Apply rubber lubricant to the isolator on the spring in order to aid installation of the anchor plate, then install the anchor plate to the top of the spring.
6. Install or connect the following:
 • Lower plate and U-bolt around the axle and through the anchor plate. If your van is equipped with a stabilizer bar it will be necessary to install the clamps, washers and nuts.
 • Nuts to the lower plate and U-bolts. Starting with the inner (lower plate side) nuts, gradually tighten the 4 nuts so the anchor plate moves uniformly, side-to-side, over the spring. Tighten the nuts to 41 ft. lbs. (56 Nm).

➡**After tightening the fasteners to specification, there should be no gap between the anchor plate, axle tube bracket and the lower plate. A metal-to-metal contact should exist.**

7. Raise the axle so the vehicle's weight is supported by the spring. The rear suspension height should be approximately 5.3–5.7 in. (135–145mm). With the suspension at normal ride height, tighten the shackle and hanger retainers to 74 ft. lbs. (100 Nm).
 • Axle bumper and tighten the nut to 33 ft. lbs. (45 Nm), if removed
 • Tire and wheel assemblies

Torsion Bar

Instead of the coil spring used on the front suspension of 2WD vehicles, the 4WD vehicles are equipped with a torsion bar.

REMOVAL & INSTALLATION

1. Before servicing the vehicle, refer to the precautions in the beginning of this section.
2. Remove or disconnect the following:
 • Adjustment assemblies on the torsion bar
3. Mark the adjustment bolt setting.
4. Use tool J 36202 to increase the tension on the adjustment arm.
5. Remove or disconnect the following:

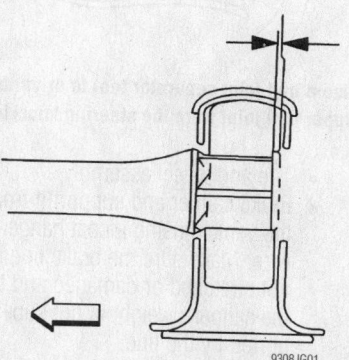

9308JG01

The rear face of the torsion bar should be within 0.04-0.10 inch (1.0-2.0 mm) of the rear face of the adjuster arm when fully installed

 • Adjustment bolt and the retaining nut, then move the tool aside
 • Torsion bar adjustment arm by sliding the bar forward
6. Slide the torsion bar partially back through the crossmember.

➡**The front end of the torsion bar are marked with a left and a right because there are different bars for both sides.**

7. Lower the front of the torsion bar down and slide it forward.
 To install:
8. Lubricate the adjuster arm and the bolt with axle grease.
9. Install or connect the following:
 • Torsion bar adjuster arm. The rear face of the torsion bar should be within 0.04-0.10 inch (1.0-2.0 mm) of the rear face of the adjuster arm when both are fully installed.
 • Adjustment retaining nut and the adjustment bolt. Using tool J 36202, increase the tension on the torsion bar
 • Retaining nut and adjustment bolt.
10. Place the adjustment bolt to the marked setting.
11. Use tool J 36202 to release the tension on the torsion bar until the load is taken up by the adjustment bolt, then remove the tool.

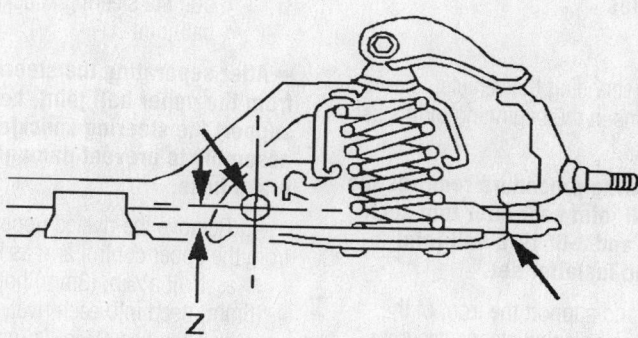

9308JG02

Z height measurement–2WD

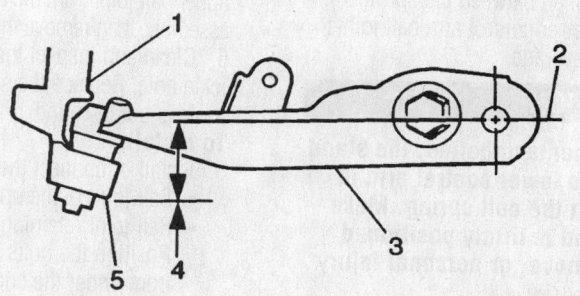

9308JG03

Z height measurement–4WD

a. Check the Z height on rear wheel drive models as follows:

b. Lift the front bumper of the vehicle up about 1.5 inch (38 mm) and then remove your hands.

c. Let the vehicle settle.

d. Repeat the previous steps twice until you have lifted the bumper 3 times.

e. Measure from the lower control arm pivot bolt center line down to the lower corner of the steering knuckle.

f. Find the average of the high measurements and the low measurements in order to determine the Z height dimension.

g. Check the Z height on rear wheel drive models as follows:

h. Lift the front bumper of the vehicle up about 1.5 inch (38 mm) and then remove your hands.

i. Let the vehicle settle.

j. Repeat the previous steps twice until you have lifted the bumper 3 times.

k. Measure from the lower control arm pivot bolt center line down to the lower corner of the steering knuckle for the Z height measurement.

Ball Joints

REMOVAL & INSTALLATION

2WD Vehicles

UPPER

1. Before servicing the vehicle, refer to the precautions in the beginning of this section.

➡The following procedure requires the use of a ball joint separator tool such as J-23742 and J-9519-E ball joint remover and installer set.

2. Raise and support the front of the vehicle safely by placing stands securely under the lower control arms. Because the vehicle's weight is used to relieve spring tension on the upper control arm, the stands must be positioned between the spring seats and the lower control arm ball joints for maximum leverage.

✳✳ CAUTION

With components unbolted, the stand is holding the lower control arm in place against the coil spring. Make sure the stand is firmly positioned and cannot move, or personal injury could result.

3. Remove or disconnect the following:

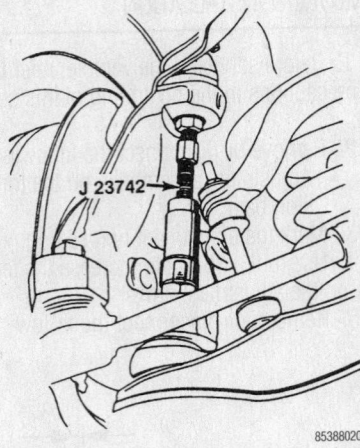

85388020

Use a ball joint separator tool to drive the upper ball joint from the steering knuckle

- Tire and wheel assembly
- Brake caliper and support it from the vehicle using a coat hanger or wire. Make sure the brake line is not stretched or damaged and that the caliper's weight is not supported by the line.
- Cotter pin and retaining nut from the upper ball joint
- Anti-lock brake sensor wire bracket, if equipped
- Upper ball joint from the steering knuckle using tool J-23742 and pull the steering knuckle free of the ball joint

➡**After separating the steering knuckle from the upper ball joint, be sure to support the steering knuckle/hub assembly to prevent damaging the brake hose.**

4. Remove the riveted upper ball joint from the upper control arm as follows:

a. Drill a⅛in. (3mm) hole, about¼in. (6mm) deep into each rivet.

b. Then use a½in. (13mm) drill bit, to drill off the rivet heads.

c. Using a pin punch and the hammer, drive out the rivets in order to free the upper ball joint from the upper control arm assembly, then remove the upper ball joint.

5. Clean and inspect the steering knuckle hole. Replace the steering knuckle if the hole is out of round.

To install:

6. Install or connect the following:

- Ball joint in the upper control arm
- Ball joint retaining nuts and bolts. Position the bolts threaded upward from under the control arm. Tighten the ball joint retainers to 17 ft. lbs. (23 Nm).

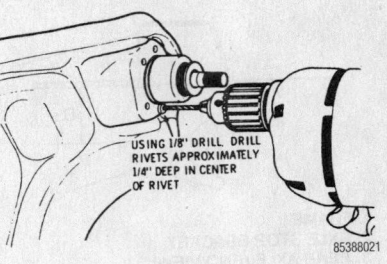

85388021

Drill a small guide hole into each ball joint rivet

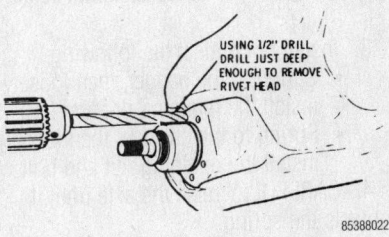

85388022

Then drill off the rivet heads

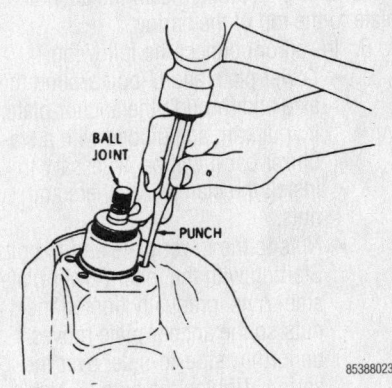

85388023

Punch the rivets out and remove the ball joint

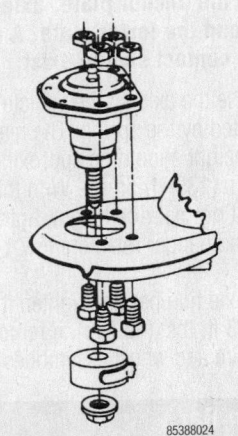

85388024

Service ball joints are bolted to the control arm

- Anti-lock brake sensor wire bracket, if removed
- Ball joint to the knuckle. Make sure the joint is seated, then install the stud nut and tighten to 61 ft. lbs. (83 Nm). Insert a new cotter pin.

➡ **When installing the cotter pin, never loosen the castle nut to expose the cotter pin hole.**

- Thread the grease fitting into the ball joint. Use a grease gun to lubricate the upper ball joint until grease appears at the seal.
- Brake caliper
- Tire and wheel assembly

7. Check and adjust the front end alignment, as necessary.

LOWER

1. Before servicing the vehicle, refer to the precautions in the beginning of this section.

➡ **The following procedure requires the use of a ball joint remover/installer set (the particular set may vary upon application but must include a clamping-type tool with the appropriately sized adapters) and a ball joint separator tool, such as J-23742.**

2. Remove the wheel.
3. Position a jack under the spring seat of the lower control arm, then raise the jack to support the arm.

✳✳ CAUTION

The jack MUST remain under the lower control arm, during the removal and installation procedures, to retain the arm and spring positions. Make sure the jack is securely positioned and will not slip or release during the procedure or personal injury may result.

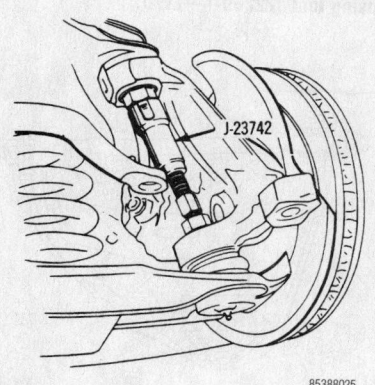

Use a ball joint separator to drive the lower joint from the knuckle

4. Remove or disconnect the following:
- Brake caliper and support it aside using a hanger or wire. Make sure the brake line is not stressed or damaged.
- Lower ball joint cotter pin and discard
- Ball joint stud nut
- Lower ball joint from the steering knuckle using tool J-23742

5. Carefully guide the lower control arm out of the opening in the splash shield using a putty knife. Position a block of wood between the frame and upper control arm to keep the knuckle out of the way.
- Grease fitting
- Ball joint from the control arm using the ball joint remover set along with the appropriate adapters

To install:

6. Clean the tapered hole in the steering knuckle of any dirt or foreign matter, then check the hole to see if it is out of round, deformed or otherwise damaged. If a problem is found, then knuckle must be replaced.

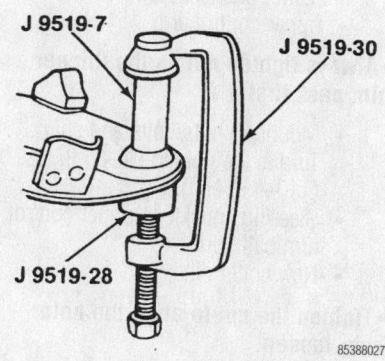

Driving the lower joint from the control arm

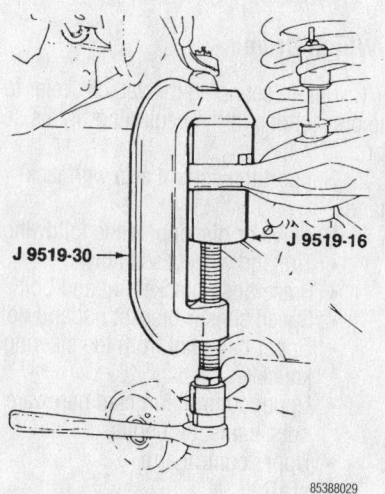

Installing a new ball joint

7. Install or connect the following:
- Press the new ball joint (with grease fitting pointing inward) until it bottoms in the control arm using a suitable installation set. Make sure the grease seal is facing inboard.
- Ball joint stud into the steering knuckle
- Ball joint retaining nut and tighten to 79 ft. lbs. (108 Nm)

➡ **When installing the cotter pin, never loosen the castle nut to expose the cotter pin hole.**

- Grease fitting into the ball joint, if not already installed

8. Use a grease gun to lubricate the joint until grease appears at the seal.
- Brake caliper
- Tire and wheel assembly

9. Check and adjust the front end alignment, as necessary.

4WD Vehicles

1. Before servicing the vehicle, refer to the precautions in the beginning of this section.

On 4WD vehicles both the upper and lower ball joints are removed in the same manner. Once the joint is separated from the steering knuckle the rivets are drilled and punched to free the joint from the control arm. Service joints are bolted into position with the retaining bolts threaded upward from beneath the control arm. In this manner, the joint is replaced in an almost identical fashion to the upper joints on 2WD vehicles.

2. Remove or disconnect the following:
- Tire and wheel assembly
- Wheel speed sensor wiring connector from the upper control arm, if removing the upper ball joint
- Cotter pin from the ball joint, then loosen the retaining nut

3. Position a suitable ball joint separator tool such as J-36607, then carefully loosen the joint in the steering knuckle. Remove the tool and the retaining nut, then separate the joint from the knuckle.

➡ **After separating the steering knuckle from the upper ball joint, be sure to support the steering knuckle/hub assembly to prevent damaging the brake hose.**

4. Remove the riveted ball joint from the control arm:
   a. Drill a ⅛in. (3mm) hole, about¼in. (6mm) deep into each rivet.

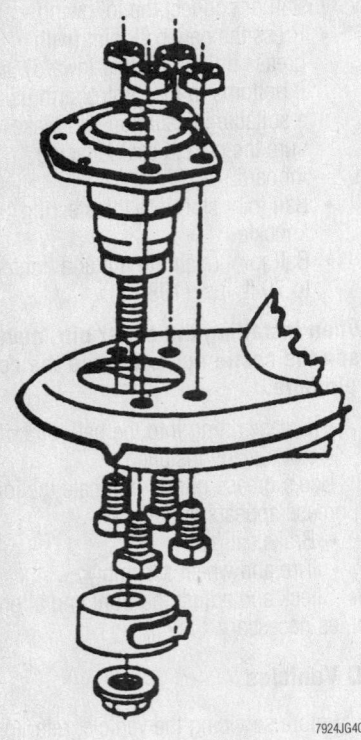

7924JG40

The replacement ball joint comes with nuts and bolts for installation

b. Then use a ½in. (13mm) drill bit, to drill off the rivet heads.

c. Using a pin punch and the hammer, drive out the rivets in order to free the ball joint from the control arm assembly, then remove the ball joint.

To install:

5. Install or connect the following:
- Ball joint in the control arm
- Ball joint retaining nuts and bolts. Position the bolts threaded upward from under the control arm. Tighten the ball joint retainers to 17 ft. lbs. (23 Nm).
- Ball joint to the knuckle. Make sure the joint is seated, tighten the lower nut to 79 ft. lbs. (108 Nm) and the upper nut to 61 ft. lbs. (83 Nm). Install a new cotter pin.

➡**When installing the cotter pin, never loosen the castle nut to expose the cotter pin hole, but DO NOT tighten more than an additional ⅛turn.**

6. Use a grease gun to lubricate the upper ball joint.
- Wheel speed sensor wiring connector to the upper control arm, if the upper ball joint was removed
- Tire and wheel assembly

7. Check and adjust the front end alignment, as necessary.

Upper Control Arm

REMOVAL & INSTALLATION

2 Wheel Drive

1. Before servicing the vehicle, refer to the precautions in the beginning of this section.
2. Remove or disconnect the following:
- Negative battery cable
- Wheel
- Wheel speed sensor harness bracket retaining bolt and nut, if equipped
- Steering knuckle from upper control arm ball joint
- Mounting nuts/bolts and shims

➡**Make sure to note the location of the control arm shims prior to removal so that they may be installed in their original positions.**

- Upper control arm

To install:

3. Install or connect the following:
- Upper control arm

➡**Always tighten nut on the thinner shim pack first.**

- Mounting nuts/bolts and shims. Torque the nuts to 81–85 ft. lbs. (110–115 Nm).
- Steering knuckle to upper control arm ball joint
- New cotter pin

➡**Tighten the nut to align the hole never loosen.**

- Wheel speed sensor harness bracket retaining bolt and nut, if equipped
- Wheel

4 Wheel Drive

1. Before servicing the vehicle, refer to the precautions in the beginning of this section.
2. Support the control arm with jackstands.
3. Remove or disconnect the following:
- Tire and wheel assembly
- Brake hose bracket nut and bolt
- Speed sensor bracket nut and bolt
- Upper ball joint from the steering knuckle
- Upper control arm cam hardware nuts, cams, and bolts
- Upper control arm

To install:

4. Install or connect the following:

- Upper control arm
- Upper control arm cam hardware bolts and the cams, making sure the bolt heads are opposed inside the bracket and that the cam lobes point down

➡**Tighten the nuts with the control arm at Z height**

- New upper control arm cam hardware nuts. Tighten the front nut first, then the rear nut to 103 ft. lbs. (140 Nm).
- Install the remaining components and check the wheel alignment

CONTROL ARM BUSHING REPLACEMENT

2 Wheel Drive

1. Before servicing the vehicle, refer to the precautions in the beginning of this section.
2. Remove or disconnect the following:
- Upper control arm and place it in a vice
- Upper control arm shaft nuts and retainers
- Upper control arm bushings using tool J 22269-1, a slotted washer and a short piece if pipe that is slightly larger than the bushing
- Upper control arm shaft

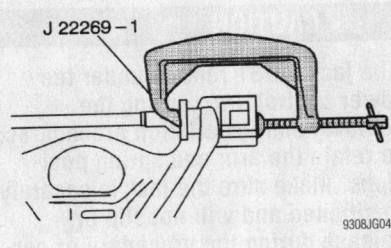

9308JG04

Removing the upper control arm bushings using tool J22269-1—2WD

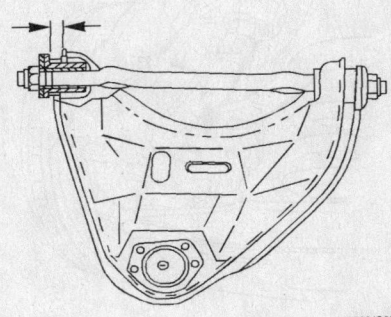

9308JG05

The bushing should be positioned as follows when correctly installed—2WD

To install:
- Upper control arm shaft
- Upper control arm bushings using tool J 22269-1, a slotted washer and a short piece if pipe that is slightly larger than the bushing
3. Tighten J 22269-1 until the bushing is positioned on the shaft and the control arm as shown in the accompanying illustration. The measurement should be 0.48–0.52 inch (12.8–13.8mm) at both sides when the properly installed.
- Upper control arm shaft nuts and retainers. Tighten to 85 ft. lbs. (115 Nm).
- Upper control arm

4 Wheel Drive

1. Before servicing the vehicle, refer to the precautions in the beginning of this section.
2. Remove or disconnect the following:
- Upper control arm
- Bushings from the arm
3. Installation is the reverse of removal

Lower Control Arm

REMOVAL & INSTALLATION

2 Wheel Drive

1. Before servicing the vehicle, refer to the precautions in the beginning of this section.
2. Remove or disconnect the following:
- Coil Spring
- Lower ball joint from the steering knuckle
- Lower control arm from the vehicle

To install:
3. Install or connect the following:
- Lower control arm
- Lower ball joint stud into the steering knuckle
- Ball joint-to-steering knuckle nut and tighten to specification
- New cotter pin to the lower ball joint stud
- Coil spring
4. Align the vehicle.

4 Wheel Drive

1. Before servicing the vehicle, refer to the precautions in the beginning of this section.

➡**Tools Needed: universal tie rod separator J–24319–01, torsion bar unloader J–36202, lower control arm bushing service kit J–36618 (if the control arm bushing are being replaced) and ball**

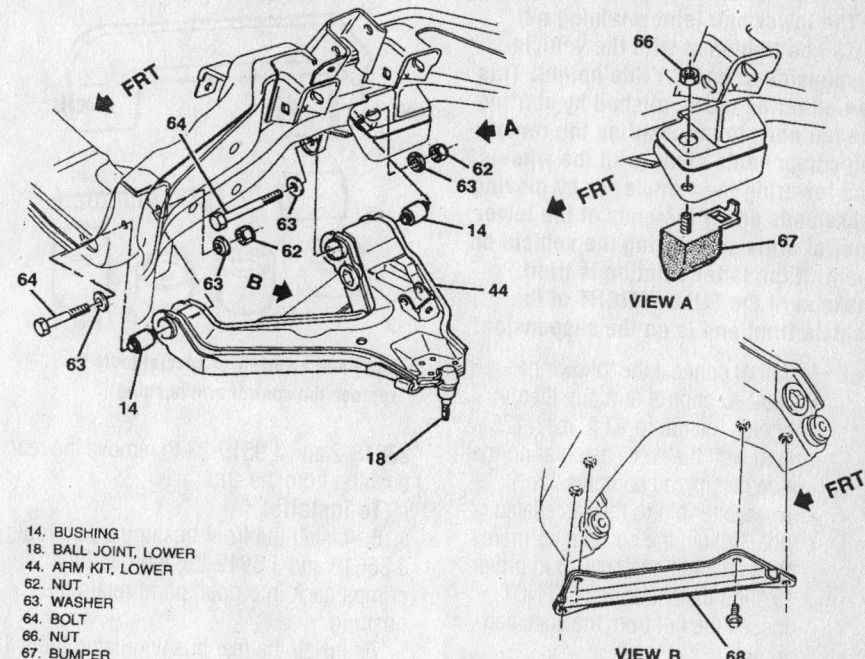

14. BUSHING
18. BALL JOINT, LOWER
44. ARM KIT, LOWER
62. NUT
63. WASHER
64. BOLT
66. NUT
67. BUMPER
68. BRACE

Exploded view of the lower control arm assembly mounting

joint C-clamp J–9519–23. Parts Needed: whether or not the control arm or bushing are being replaced, NEW control arm retaining nut should be used once the old ones have been loosened and removed.

2. Remove or disconnect the following:
- Front wheels
- 2 bolts from the front splash shield and pivot it in order to gain access to the tie rod
- Stabilizer bar from the control arm (keeping all of the link hardware sorted for proper installation). If necessary, completely remove the bar from the vehicle for access.
- Shock absorber
- Inner tie rod from the relay rod using a tie rod separator
- Outer halfshaft nut and washer
- Bolts from the hub and bearing kit
3. Unload the torsion bar using the unloading tool J–36202. First, mark the adjuster for installation.
- Adjustment arm. Slide the bar forward and the adapter out of the rear to remove the adjusting arm.
- Lower ball joint cotter pin, nut and ball joint from the control arm using a ball joint separator
- Nuts and bolts and lower control arm with the torsion bar assembly. Note the direction which the control

arm retaining bolts are facing for installation purposes.

To install:
4. Install or connect the following:
- Torsion bar to the lower control arm and place the assembly into the vehicle. Position the front leg of the lower control arm into the crossmember before installing the rear leg into the frame bracket.
- Control arm bolts (facing in the direction as noted during removal or shown in the accompanying illustration) with NEW nuts.

➡**The control arm retainers MUST be tightened with the vehicle suspension at normal ride height. This can either be accomplished by starting the nuts now, then installing the remaining components along with the wheels and lowering the vehicle, or by moving jackstands under the ends of the lower control arms and resting the vehicle on them. If the latter solution is tried, make sure front suspension is at actual ride height compression. If you are unsure, it is best to start the nuts now and tighten them to specification once the vehicle is lowered.**

- Ball joint stud in the knuckle
5. With the suspension at the correct height, tighten the control arm retaining nuts to 98 ft. lbs. (133 Nm).

➡The lower ball joint retaining nut MUST be tightened with the vehicle suspension at normal ride height. This can either be accomplished by starting the nut now, then installing the remaining components along with the wheels and lowering the vehicle, or by moving jackstands under the ends of the lower control arms and resting the vehicle on them. If the latter solution is tried, make sure the FULL WEIGHT of the vehicle front end is on the suspension.

6. Install or connect the following:
   • Joint-to-control arm nut, then tighten the nut to 92 ft. lbs. (125 Nm) with the suspension at normal ride height and compression
   • New cotter pin to the castellated nut. Tighten the nut (but no more than an additional ⅛ turn) in order to align the cotter pin. DO NOT loosen the nut from the specified torque.
   • Adjuster arm by sliding the adapter forward, over the torsion bar to install the sides of the nut. Load the torsion bar and install the adjuster bolt aligning the installation mark.
   • Drive axle through the hub and bearing assembly
   • Tighten the hub and bearing assembly retaining bolts
   • Drive axle shaft nut and washer
   • Inner tie rod end to the relay rod
   • Shock absorber
   • Stabilizer bar, if removed
   • Stabilizer link(s) to the control arm(s)
   • Splash shield
   • Front wheels
7. Recheck all fasteners for proper torque and installation before road testing.
8. Refill the differential if any fluid was lost.
9. Check and adjust the front end alignment, as necessary.

CONTROL ARM BUSHING REPLACEMENT

1. Before servicing the vehicle, refer to the precautions in the beginning of this section.
2. Remove lower control arm and place it in vise.
3. Use a punch to unbend the crimps on the front bushing.
4. Use tools J 36618-2, J 9519-23 and J 36618-1 to remove the front bushing from the arm.
5. Use tools J 36618-5, J 36618-3, J

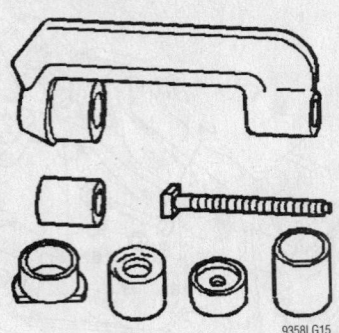

9358LG15

You need a variety of special tools to replace the control arm bushing

36618-2 and J 9519-23 to remove the rear bushing from the arm.
   To install:
6. Install the front bushing using tools J 36618 and J 5919-23 and bend the crimps back into position to retain the bushing.
7. Install the rear bushing using tools J 36618 and J 5919-23 .
8. Install the lower control arm.

Wheel Bearings

ADJUSTMENT

2WD Models

1. Before servicing the vehicle, refer to the precautions in the beginning of this section.
2. If equipped, remove the wheel/hub cover for access, then remove the dust cap from the hub.
3. Remove the cotter pin and loosen the spindle nut.
4. Spin the wheel forward by hand and torque the nut to 12 ft. lbs. (16 Nm) in order to fully seat the bearings and remove any burrs from the threads.
5. Back off the nut until it is just loose, then finger-tighten the nut.
6. Loosen the nut ¼–½ turn until either hole in the spindle lines up with a slot in the nut, then install a new cotter pin. This may appear to be too loose, but it is the proper adjustment.
7. Proper adjustment creates 0.001–0.005 in. (0.025–0.127mm) endplay.

4WD Models

The front wheel bearings on the 4-wheel drive vehicles are not adjustable. If the bearings become loose or make noise, they must be replaced.

REMOVAL & INSTALLATION

Front

2WD MODELS

1. Before servicing the vehicle, refer to the precautions in the beginning of this section.
2. Remove or disconnect the following:
   • Wheel
   • Brake caliper with the pads without disconnecting the brake line
   • Grease cap
   • Cotter pin, spindle nut and washer
   • Hub

✳✳ WARNING

Be careful not to drop the outer wheel bearing. As the hub is pulled forward, the outer wheel bearings will often fall forward and they may easily be removed at this time.

   • Outer roller bearing assembly
   • Inner seal by prying it out of the hub and discard it
   • Inner bearing assembly
   To install:
3. Clean all parts in solvent and allow to air dry, then check for excessive wear or damage. Inspect all of the parts for scoring, pitting or cracking and replace if necessary.

➡**DO NOT remove the bearing races from the hub, unless they show signs of damage.**

4. If it is necessary to remove the wheel bearing races, use the GM front bearing race removal tool J-29117 to drive the races from the hub/disc assembly. A hammer and brass drift may also be used to drive the races from the hub, but the race removal tool is quicker.
5. If the bearing races were removed, position the replacement races in the freezer for a few minutes and then install them to the hub:
   a. Lightly lubricate the inside of the hub/disc assembly using wheel bearing grease.
   b. Using the GM seal installation tools J-8092 and J-8850, drive the inner bearing race into the hub/disc assembly until it seats. Be sure the race is properly seated against the hub shoulder and is not cocked.

➡**When installing the bearing races, be sure to support the hub/disc assembly with GM tool J-9746-02.**

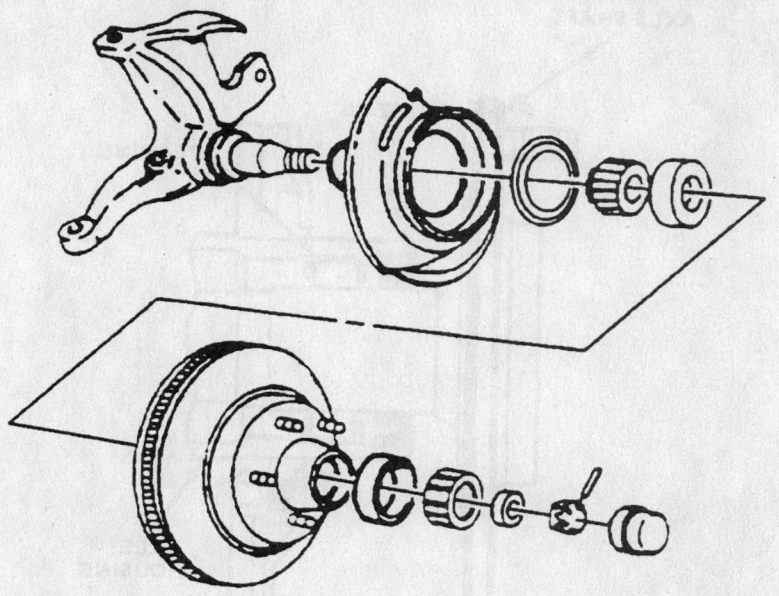

Wheel bearings, races and related components—2WD vehicles

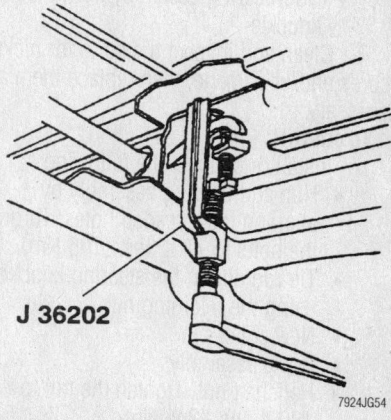

J 36202

Use Torsion Bar Unloading tool J 36202 to remove the adjusting bolt and unload the torsion bar

c. Using the GM seal installation tools J-8092 and J-8457, drive the outer race into the hub/disc assembly until it seats.

6. Using a high melting point wheel bearing grease, lubricate the bearings, races and spindle; be sure to place a gob of grease (inside the hub/disc assembly) between the races to provide an ample supply of lubricant.

➡**To lubricate each bearing, place a gob of grease in the palm of the hand, then scoop the bearing through the grease until it is well lubricated.**

7. Place the inner bearing in the hub, then apply a thin coating of grease to the sealing lip and install a new inner seal, making sure the seal flange faces the bearing cup.

➡**Although a seal installation tool is preferable, a section of pipe with a smooth edge or a suitably sized socket may be used to drive the seal into position. Be sure the seal is flush with the outer surface of the hub assembly.**

8. Install or connect the following:
  • Wheel hub over the spindle
  • Outer bearing into the hub by hand
  • Spindle washer and nut
  • Brake caliper
  • Wheel
9. Properly adjust the wheel bearings
10. Install or connect the following:
  • New cotter pin
  • Dust cap
  • Wheel cover

4WD MODELS

1. Before servicing the vehicle, refer to the precautions in the beginning of this section.

2. Install Torsion Bar Unloading tool J 36202 on the torsion bar adjusting bolt and remove the bolt. To aid during installation, count the number of turns required to remove the bolt.

3. Remove the wheel.

4. Install an axle shaft boot seal protector to the Tri-pot axle joint.

5. Remove or disconnect the following:
  • Cotter pin and retainer
  • Castle nut and the thrust washer
  • Brake caliper and support it aside using wire or a coat hanger

➡**Be sure the brake line is not stretched or damaged.**

  • Brake disc from the wheel hub
  • Halfshaft from the hub/bearing assembly, using a Spindle Remover tool J-28733-A to prevent damage to the shaft or hub/bearing assembly

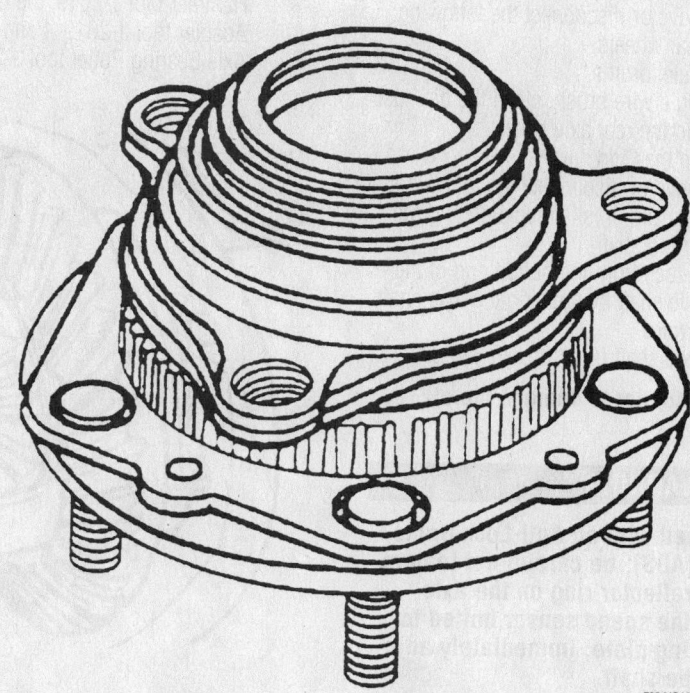

Hub and bearing assembly—4WD vehicles

- Hub/bearing assembly from the knuckle

6. Clean and inspect the parts for nicks, scores and/or damage, then replace them as necessary.

To install:

7. Install or connect the following:
- Hub and bearing assembly by aligning the threaded holes. Torque the bolts to 77 ft. lbs. (105 Nm).
- Tie rod end to the steering knuckle using the retaining nut
- New cotter pin
- Brake assembly
- Halfshaft nut. Tighten the nut to 180 ft. lbs. (245 Nm).
- Retainer and a new cotter pin but DO NOT back off specification in order to insert the cotter pin

8. Remove the torsion bar unloader tool and the drive axle boot protector.

9. Install the wheel.

10. Check and/or adjust the vehicle trim height, as necessary.

Rear

A new pinion shaft lockbolt should be installed whenever either of the axle shafts is removed.

The axle shaft and seal may be removed and replaced without disturbing the bearing or seal but it is highly recommended to replace the seals when removing the axle shaft.

1. Before servicing the vehicle, refer to the precautions in the beginning of this section.

2. Remove or disconnect the following:
- Rear wheels
- Brake drums

3. Using a wire brush, clean the dirt/rust from around the rear axle cover.

4. Drain the fluid.

5. Remove or disconnect the following:
- Rear pinion shaft lockbolt and the pinion shaft
- C-lock from the button end of the axle shaft by pushing the axle shaft inward
- Axle shaft from the axle housing

➡Be careful not to damage the oil seal.

❋❋ WARNING

If equipped with an Anti-Lock Brake System (ABS), be careful not to damage the reflector ring on the axle shaft or the speed sensor bolted to the backing plate, immediately adjacent to the shaft.

6. Remove or disconnect the following:

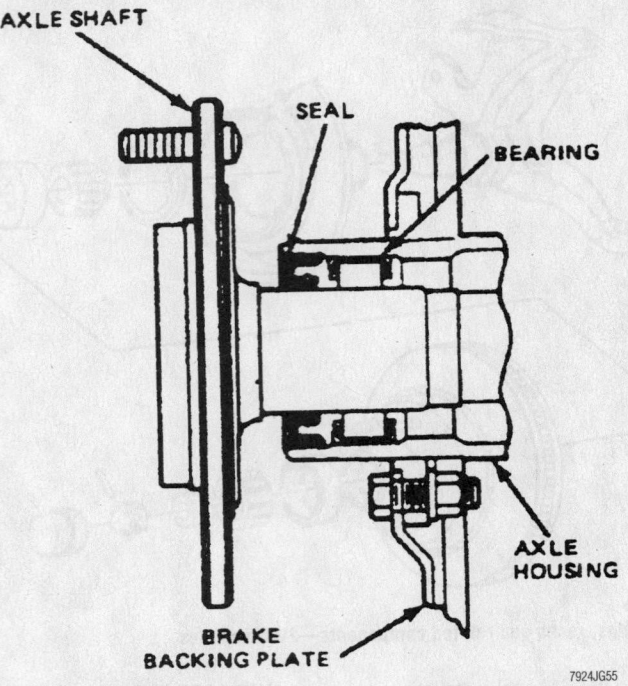

Cross-sectional view of the rear axle, bearing and seal assembly

- Oil seal by prying the it from the end of the rear axle housing

❋❋ WARNING

DO NOT damage the housing oil seal surface.

- Wheel bearing using the GM Slide Hammer tool J-2619, the GM Adapter tool J-2619-4 and the GM Axle Bearing Puller tool J-22813-01

To install:

7. Clean and inspect the components for excessive wear or damage and replace them, if necessary.

8. Install or connect the following:
- New or reused bearing, coated with gear lubricant, using the Axle Shaft Bearing Installer tool J-34974 to drive the bearing in until it bottoms against the seat

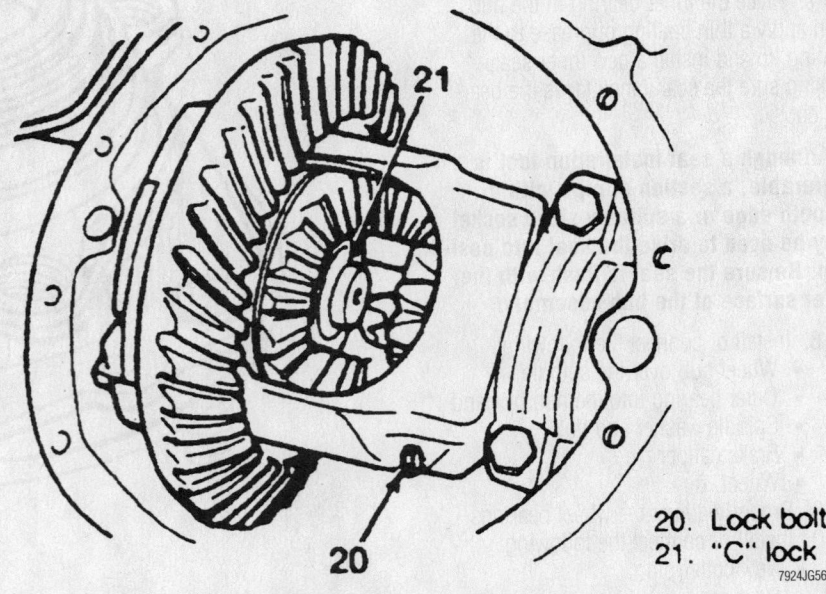

20. Lock bolt
21. "C" lock

Pinion shaft lockbolt and axle C-lock locations, inside the differential

✳✳ WARNING

Be sure the bearing installer does not contact and damage the speed sensor on ABS equipped vehicles.

- New seal lubricated with gear oil using the GM Axle Shaft Seal Installer tool J-33782 to seat it in the housing until it is flush with the axle tube

➡**Be sure the seal installer does not contact and damage the speed sensor on ABS equipped vehicles.**

- Axle shaft into the housing by engaging the splines
- C-lock retainer on the axle shaft button end

✳✳ WARNING

BE CAREFUL not to damage the wheel bearing seal.

- Axle shaft by pulling it outward to seat the C-lock retainer in the counterbore of the side gears
- Pinion shaft through the case and the pinions. Tighten the new lockbolt to 27 ft. lbs. (36 Nm).
- New rear axle cover gasket
- Housing cover
- Brake drums
- Wheels
9. Refill the housing.

BRAKES

Brake Caliper

REMOVAL & INSTALLATION

Front

1. Before servicing the vehicle, refer to the precautions in the beginning of this section.
2. Remove or disconnect the following:
   - ⅔ of the brake fluid from the master cylinder reservoir
   - Tire and wheel assembly
   - Brake fluid line and plug it

- Bolts retaining the caliper to the rotor
- Caliper from the rotor
- Disc brake pads from the caliper
- Disc brake pad retaining clips from inside the caliper

To install:

3. Clean and lubricate the sleeves and bushings with silicon grease.
4. Install or connect the following:
   - Sleeves and bushings
   - Pads in the caliper
   - Caliper in position over the rotor
   - Mounting bolts and tighten the bolts to 38 ft. lbs. (51 Nm) for

single piston calipers; 85 ft. lbs. (115 Nm) for dual piston calipers
- Fluid lines to the caliper, if disconnected, and tighten to 33 ft. lbs. (45 Nm)
- Wheel and tire assembly
5. Refill the master cylinder to the correct level. Bleed the brake system if the fluid lines were disconnected from the caliper.

Rear

1. Before servicing the vehicle, refer to the precautions in the beginning of this section.

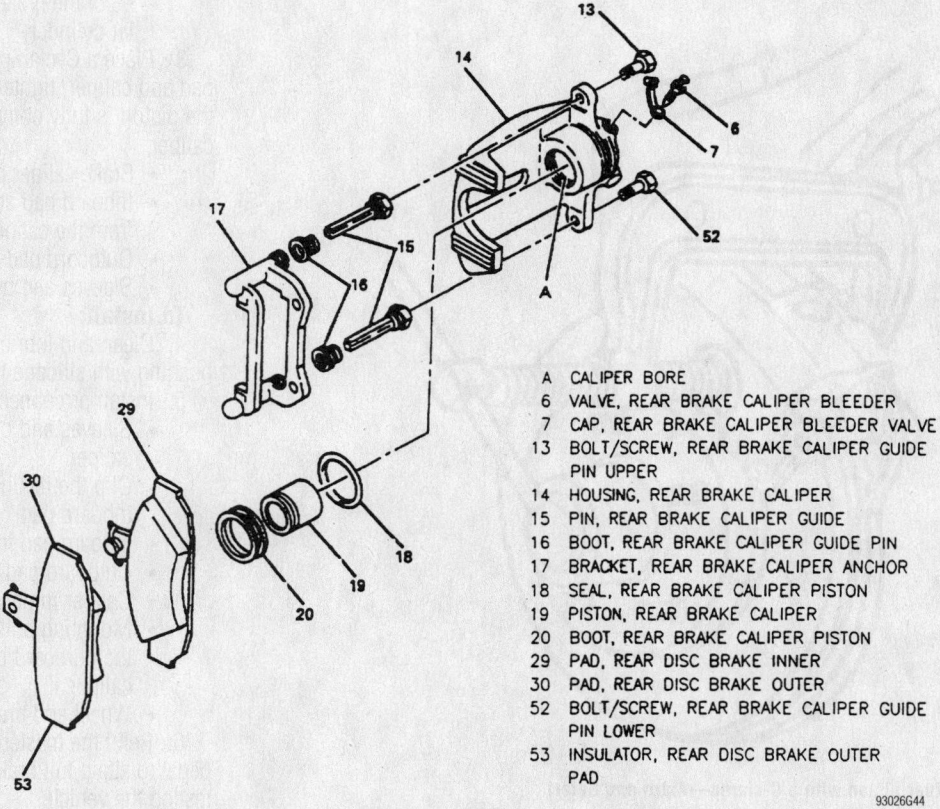

A   CALIPER BORE
6   VALVE, REAR BRAKE CALIPER BLEEDER
7   CAP, REAR BRAKE CALIPER BLEEDER VALVE
13 BOLT/SCREW, REAR BRAKE CALIPER GUIDE PIN UPPER
14 HOUSING, REAR BRAKE CALIPER
15 PIN, REAR BRAKE CALIPER GUIDE
16 BOOT, REAR BRAKE CALIPER GUIDE PIN
17 BRACKET, REAR BRAKE CALIPER ANCHOR
18 SEAL, REAR BRAKE CALIPER PISTON
19 PISTON, REAR BRAKE CALIPER
20 BOOT, REAR BRAKE CALIPER PISTON
29 PAD, REAR DISC BRAKE INNER
30 PAD, REAR DISC BRAKE OUTER
52 BOLT/SCREW, REAR BRAKE CALIPER GUIDE PIN LOWER
53 INSULATOR, REAR DISC BRAKE OUTER PAD

93026G44

Rear brake caliper

2. Remove or disconnect the following:
- Rear wheels
- Brake hose and cap line
- Retainers from caliper and remove caliper

To install:

3. Install or connect the following:
- Brake pads if removed
- Caliper over rotor, and onto mounts
- Retainers, and tighten to 23 ft. lbs. or (31 Nm)
- Brake hose, and tighten to 20 ft. lbs. (27 Nm)

4. Bleed brake system.
- Tires and wheels

5. Refill the master cylinder and pump pedal to attain full brake pedal before Road-testing the vehicle.

Disc Brake Pads

REMOVAL & INSTALLATION

Front

1. Before servicing the vehicle, refer to the precautions in the beginning of this section.

2. Remove or disconnect the following:

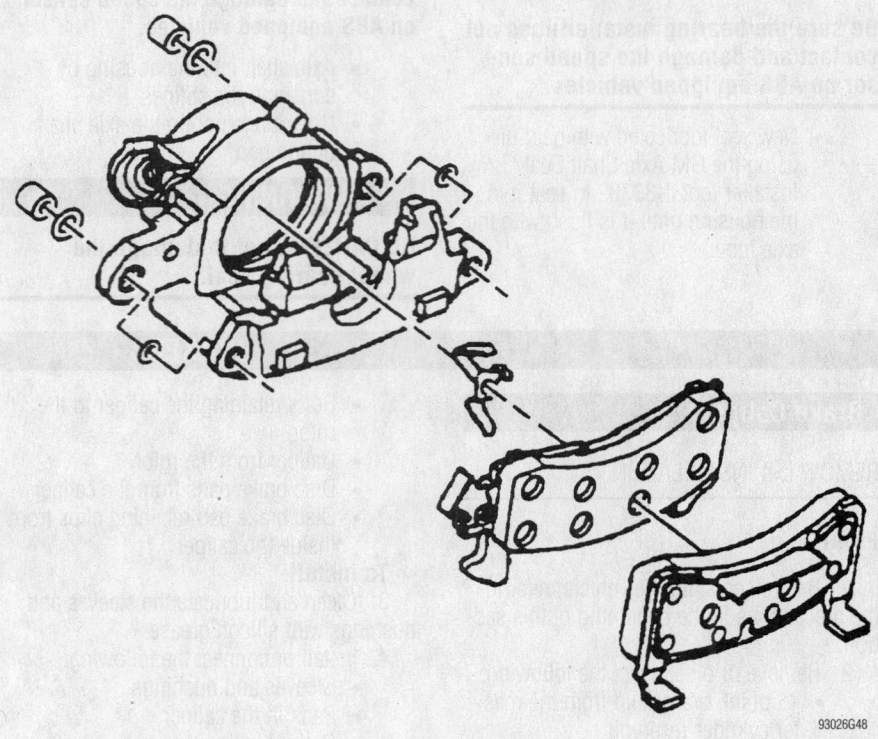

Exploded view of the disc brake assembly—Astro and Safari

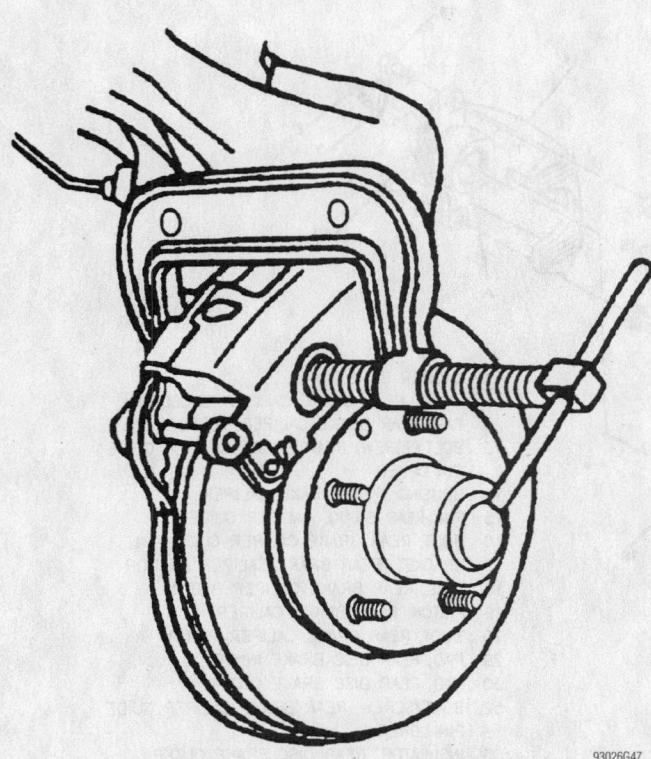

Compressing the caliper piston with a C-clamp—Astro and Safari

- ⅔ of the brake fluid from the master cylinder

3. Place a C-clamp around the outer pad and caliper; tighten the C-clamp until the piston is fully compressed in the caliper.
- Brake caliper pads
- Inboard pad and retaining spring from the caliper
- Outboard pad from the caliper
- Sleeves and bushings

To install:

4. Clean and lubricate the sleeves and bushing with silicone lubricant.

5. Install or connect the following:
- Sleeves and bushings in the caliper
- Clip the retaining spring onto the inboard pad
- Inboard pad in the caliper
- Outboard pad into the caliper
- Caliper in position over the rotor
- Mounting bolts. Bend the tabs, on the outboard brake pad, over the caliper.
- Wheel and tire assemblies

6. Refill the master cylinder and pump pedal to attain full brake pedal before Road-testing the vehicle.

Rear

1. Before servicing the vehicle, refer to the precautions in the beginning of this section.
2. Remove or disconnect the following:
   - ⅔ of the brake fluid from the master cylinder
   - Wheels and tires
3. Place a C-clamp around the outer pad and caliper; tighten the C-clamp until the piston is fully compressed in the caliper. Remove top caliper retainer, and rotate caliper away from rotor.
   - Inboard pad and retaining spring from the caliper
   - Outboard pad from the caliper

To install:
4. Clean and lubricate the sleeves and bushing with silicone lubricant.
5. Install or connect the following:
   - Sleeves and bushings in the caliper
   - Clip the retaining spring onto the inboard pad
   - Inboard pad in the caliper
   - Outboard pad into the caliper

- Caliper in position over the rotor
- Mounting bolts
- Wheel and tire assemblies
6. Refill the master cylinder and pump pedal to attain full brake pedal before Road-testing the vehicle.

Brake Drums

REMOVAL & INSTALLATION

1. Before servicing the vehicle, refer to the precautions in the beginning of this section.
2. Remove or disconnect the following:
   - Wheel and tire assembly
   - Brake drum. If the drum will not pull of the axle, use a rubber mallet and tap it around the edge.

To install:
3. Install or connect the following:
   - Drum on the axle
   - Wheel and tire assembly
4. Refill the master cylinder and pump pedal to attain full brake pedal before road-testing the vehicle.

Brake Shoes

REMOVAL & INSTALLATION

1. Before servicing the vehicle, refer to the precautions in the beginning of this section.
2. Remove or disconnect the following:
   - Wheel and tire assembly
   - Brake drum
   - Return springs from the brake shoes
   - Shoe guide
   - Hold-down springs and pins
   - Actuator lever and pivot
   - Lever return spring
   - Actuator link
   - Parking brake strut and spring
   - Parking brake lever
   - Brake shoes and the adjuster assembly

To install:
3. Lubricate the contact points on the backing plate and the adjuster with lithium grease.
4. Install or connect the following:

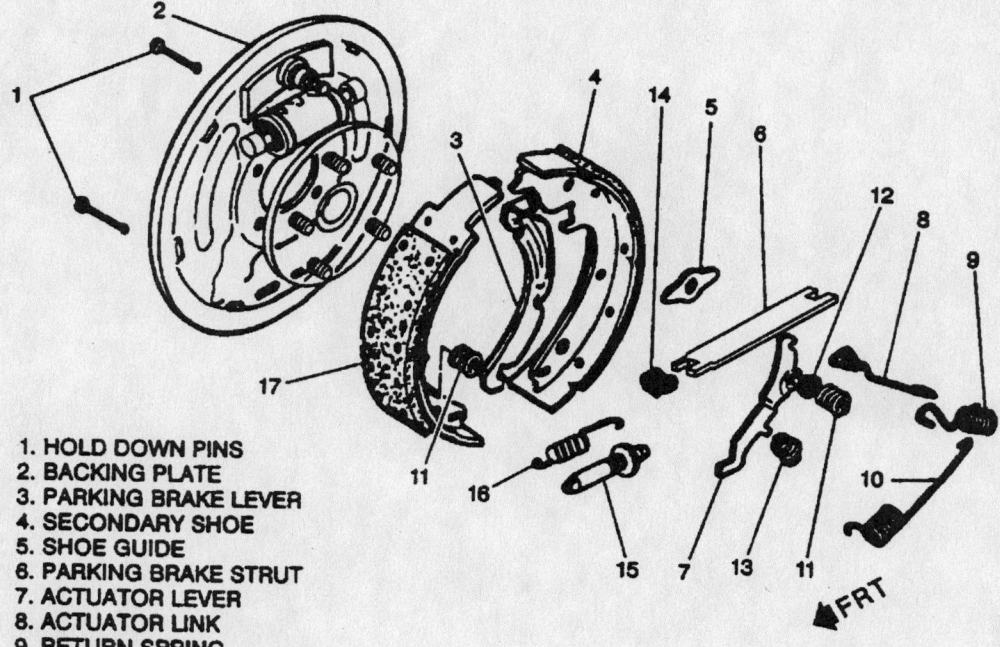

1. HOLD DOWN PINS
2. BACKING PLATE
3. PARKING BRAKE LEVER
4. SECONDARY SHOE
5. SHOE GUIDE
6. PARKING BRAKE STRUT
7. ACTUATOR LEVER
8. ACTUATOR LINK
9. RETURN SPRING
10. RETURN SPRING
11. HOLD DOWN SPRING
12. LEVER PIVOT
13. LEVER RETURN SPRING
14. STRUT SPRING
15. ADJUSTING SCREW ASSEMBLY
16. ADJUSTING SCREW SPRING
17. PRIMARY SHOE

93026G51

Exploded view of the drum brake components—Astro & Safari

- Parking brake lever, adjusting screw and spring assembly
- Shoe assembly onto the backing plate
- Parking brake lever, strut and strut spring
- Actuator lever and lever pivot
- Actuator link
- Lever spring, the hold-down pins and springs
- Shoe guide
- Return springs and install the brake drum in position

5. Adjust the brakes as follows:

a. Remove the knockout area in the backing plate, behind the adjuster assembly.

b. Ensure the parking brake system is adjusted properly with no tension on the cables or parking brake lever. The tops of the shoes should be firmly seated against the upper spring retaining anchor, if not as specified, loosen the parking brake cables.

c. Install the drum and turn the brake adjuster until the wheels can just be turned by hand.

d. Then, back the adjuster off 24 notches. No brake drag should be felt after 12 notches.

e. Install an adjusting hole plug in the backing plate to prevent dirt and moisture from entering.

f. Readjust the parking brake cable as necessary.

6. Install the wheel and tire assemblies.

7. Refill the master cylinder and pump pedal to attain full brake pedal before Road-testing the vehicle.

CHEVROLET AND GMC

Sierra • Silverado • Express • Savana

5

SPECIFICATION CHARTS

ENGINE AND VEHICLE IDENTIFICATION

Engine						
Code ①	Liters (cc)	Cu. In.	Cyl.	Fuel Sys.	Engine Type	Eng. Mfg.
W	4.3 (4293)	263	6	MFI	OHV	CPC
X	4.3 (4293)	263	6	MFI	OHV	CPC
V	4.8 (4802)	293	8	MFI	OHV	CPC
M	5.0 (4999)	305	8	MFI	OHV	CPC
T	5.3 (5327)	325	8	MFI	OHV	CPC
Z	5.3 (5327)	325	8	MFI	OHV	CPC
R	5.7 (5735)	350	8	MFI	OHV	CPC
U	6.0 (5966)	364	8	MFI	OHV	CPC
N	6.0 (5966)	364	8	MFI	OHV	CPC
F	6.5 (6473)	395	8	DSL	OHV	CPC
1	6.6 (6599)	364	8	DSL	OHV	CPC
J	7.4 (7440)	454	8	MFI	OHV	CPC
G	8.1 (8128)	496	8	MFI	OHV	CPC

Model Year	
Code ②	Year
Y	2000
1	2001
2	2002
3	2003
4	2004

CPC: Chevrolet/Pontiac/Canada

DSL: Diesel

MFI: Multi-port Fuel Injection

① 8th position of VIN

② 10th position of VIN

42372-SILV-C01

GENERAL ENGINE SPECIFICATIONS
All measurements are given in inches.

Year	Model	Engine Displacement Liters (cc)	Engine Series (ID/VIN)	Fuel System	Net Horsepower @ rpm	Net Torque @ rpm (ft. lbs.)	Bore x Stroke (in.)	Compression Ratio	Oil Pressure @ rpm
2000	C1500	4.3 (4293)	W	MFI	230@4600	285@2800	3.74x3.48	9.4:1	18@2000
		5.0 (4999)	M	MFI	230@4600	285@2800	3.74x3.48	9.4:1	18@2000
		5.7 (5735)	R	MFI	255@4600	335@2800	4.00x3.48	9.4:1	18@2000
	C2500	4.3 (4293)	W	MFI	230@4600	285@2800	3.74x3.48	9.4:1	18@2000
		5.0 (4999)	M	MFI	230@4600	285@2800	3.74x3.48	9.4:1	18@2000
		5.7 (5735)	R	MFI	255@4600	335@2800	4.00x3.48	9.4:1	18@2000
		6.5 (6374)	F	DSL	195@3400	430@1800	4.05x3.80	21.5:1	30-43@2000
	C3500	6.5 (6374)	F	DSL	195@3400	430@1800	4.05x3.80	21.5:1	30-43@2000
		7.4 (7440)	J	MFI	290@4200	410@3200	4.25x4.00	9.0:1	40@2000
	G1500	4.3 (4293)	W	MFI	230@4600	285@2800	3.74x3.48	9.4:1	18@2000
		5.0 (4999)	M	MFI	230@4600	285@2800	3.74x3.48	9.4:1	18@2000
		5.7 (5735)	R	MFI	255@4600	335@2800	4.00x3.48	9.4:1	18@2000
	G2500	5.0 (4999)	M	MFI	230@4600	285@2800	3.74x3.48	9.4:1	18@2000
		5.7 (5735)	R	MFI	255@4600	335@2800	4.00x3.48	9.4:1	18@2000
	G3500	6.5 (6374)	F	DSL	195@3400	430@1800	4.05x3.80	21.5:1	30-43@2000
		7.4 (7440)	J	MFI	290@4200	410@3200	4.25x4.00	9.0:1	40@2000
	K1500	4.3 (4293)	W	MFI	230@4600	285@2800	3.74x3.48	9.4:1	18@2000
		5.0 (4999)	M	MFI	230@4600	285@2800	3.74x3.48	9.4:1	18@2000
		5.7 (5735)	R	MFI	255@4600	335@2800	4.00x3.48	9.4:1	18@2000
		6.5 (6374)	F	DSL	195@3400	430@1800	4.05x3.80	21.5:1	30-43@2000
	K2500	5.0 (4999)	M	MFI	230@4600	285@2800	3.74x3.48	9.4:1	18@2000
		5.7 (5735)	R	MFI	255@4600	335@2800	4.00x3.48	9.4:1	18@2000
		6.5 (6374)	F	DSL	195@3400	430@1800	4.05x3.80	21.5:1	30-43@2000
	K3500	5.7 (5735)	R	MFI	255@4600	335@2800	4.00x3.48	9.4:1	18@2000
		6.5 (6374)	F	DSL	195@3400	430@1800	4.05x3.80	20:01	30-43@2000
		7.4 (7440)	J	MFI	290@4200	410@3200	4.25x4.00	9.0:1	40@2000
	Sierra	4.3 (4293)	W	MFI	200@4600	260@2800	4.00x3.48	9.2:1	18@2000
		4.8 (4802)	V	MFI	255@5200	285@4000	3.78x3.27	9.5:1	18@2000
		5.3 (5327)	T	MFI	270@5000	315@4000	3.78x3.62	9.5:1	18@2000
		6.0 (5966)	U	MFI	300@4800	355@4000	4.00x3.62	9.4:1	18@2000
		6.5 (6374)	F	DSL	195@3400	430@1800	4.05x3.80	20:01	30-43@2000
	Silverado	4.3 (4293)	W	MFI	200@4600	260@2800	4.00x3.48	9.2:1	18@2000
		4.8 (4802)	V	MFI	255@5200	285@4000	3.78x3.27	9.5:1	18@2000
		5.3 (5327)	T	MFI	270@5000	315@4000	3.78x3.62	9.5:1	18@2000
		6.0 (5966)	U	MFI	300@4800	355@4000	4.00x3.62	9.4:1	18@2000
		6.5 (6374)	F	DSL	195@3400	430@1800	4.05x3.80	20:01	30-43@2000
2001	G1500	4.3 (4293)	W	MFI	200@4600	260@2800	4.00x3.48	9.2:1	18@2000
		5.0 (4999)	M	MFI	220@4600	280@2800	3.74x3.48	9.4:1	18@2000
		5.7 (5735)	R	MFI	255@4600	330@2800	4.00x3.48	9.4:1	18@2000
	G2500	4.3 (4293)	W	MFI	200@4600	260@2800	4.00x3.48	9.2:1	18@2000
		5.0 (4999)	M	MFI	230@4600	285@2800	3.74x3.48	9.4:1	18@2000
		5.7 (5735)	R	MFI	255@4600	335@2800	4.00x3.48	9.4:1	18@2000
		6.5 (6374)	F	DSL	195@3400	430@1800	4.05x3.80	21.5:1	30-43@2000
	G3500	5.7 (5735)	R	MFI	255@4600	335@2800	4.00x3.48	9.4:1	18@2000
		6.5 (6374)	F	DSL	195@3400	430@1800	4.05x3.80	21.5:1	30-43@2000
		8.1 (8128)	G	MFI	340@4200	455@3200	4.25x4.37	9.1:1	10@2000

GENERAL ENGINE SPECIFICATIONS

All measurements are given in inches.

Year	Model	Engine Displacement Liters (cc)	Engine Series (ID/VIN)	Fuel System	Net Horsepower @ rpm	Net Torque @ rpm (ft. lbs.)	Bore x Stroke (in.)	Compression Ratio	Oil Pressure @ rpm
2001 (cont.)	Sierra	4.3 (4293)	W	MFI	200@4600	260@2800	4.00x3.48	9.2:1	18@2000
		4.8 (4802)	V	MFI	270@5200	285@4000	3.78x3.27	9.5:1	18@2000
		5.3 (5327)	T	MFI	285@4000	360@4000	3.78x3.62	9.5:1	18@2000
		6.0 (5967)	U	MFI	300@4400	360@4000	4.00x3.62	9.4:1	18@2000
		6.6 (6599)	1	DSL	300@3000	520@1800	4.00x3.90	17.5:1	57@3250
		8.1 (8128)	G	MFI	340@4200	455@3200	4.25x4.37	9.1:1	10@2000
	Silverado	4.3 (4293)	W	MFI	200@4600	260@2800	4.00x3.48	9.2:1	18@2000
		4.8 (4802)	V	MFI	270@5200	285@4000	3.78x3.27	9.5:1	18@2000
		5.3 (5327)	T	MFI	285@4000	360@4000	3.78x3.62	9.5:1	18@2000
		6.0 (5967)	U	MFI	300@4400	360@4000	4.00x3.62	9.4:1	18@2000
		6.6 (6599)	1	DSL	300@3000	520@1800	4.00x3.90	17.5:1	57@3250
		8.1 (8128)	G	MFI	340@4200	455@3200	4.25x4.37	9.1:1	10@2000
2002	G1500	4.3 (4293)	W	MFI	200@4600	260@2800	4.00x3.48	9.2:1	18@2000
		5.0 (4999)	M	MFI	220@4600	280@2800	3.74x3.48	9.4:1	18@2000
		5.7 (5735)	R	MFI	255@4600	330@2800	4.00x3.48	9.4:1	18@2000
	G2500	4.3 (4293)	W	MFI	200@4600	260@2800	4.00x3.48	9.2:1	18@2000
		5.0 (4999)	M	MFI	230@4600	285@2800	3.74x3.48	9.4:1	18@2000
		5.7 (5735)	R	MFI	255@4600	335@2800	4.00x3.48	9.4:1	18@2000
		6.5 (6374)	F	DSL	195@3400	430@1800	4.05x3.80	21.5:1	30-43@2000
	G3500	5.7 (5735)	R	MFI	255@4600	335@2800	4.00x3.48	9.4:1	18@2000
		6.5 (6374)	F	DSL	195@3400	430@1800	4.05x3.80	21.5:1	30-43@2000
		8.1 (8128)	G	MFI	340@4200	455@3200	4.25x4.37	9.1:1	10@2000
	Sierra	4.3 (4293)	W	MFI	200@4600	260@2800	4.00x3.48	9.2:1	18@2000
		4.8 (4802)	V	MFI	270@5200	285@4000	3.78x3.27	9.5:1	18@2000
		5.3 (5327)	T	MFI	285@4000	360@4000	3.78x3.62	9.5:1	18@2000
		6.0 (5967)	U	MFI	300@4400	360@4000	4.00x3.62	9.4:1	18@2000
		6.6 (6599)	1	DSL	300@3000	520@1800	4.00x3.90	17.5:1	57@3250
		8.1 (8128)	G	MFI	340@4200	455@3200	4.25x4.37	9.1:1	10@2000
	Silverado	4.3 (4293)	W	MFI	200@4600	260@2800	4.00x3.48	9.2:1	18@2000
		4.8 (4802)	V	MFI	270@5200	285@4000	3.78x3.27	9.5:1	18@2000
		5.3 (5327)	T	MFI	285@4000	360@4000	3.78x3.62	9.5:1	18@2000
		6.0 (5967)	U	MFI	300@4400	360@4000	4.00x3.62	9.4:1	18@2000
		6.6 (6599)	1	DSL	300@3000	520@1800	4.00x3.90	17.5:1	57@3250
		8.1 (8128)	G	MFI	340@4200	455@3200	4.25x4.37	9.1:1	10@2000
2003-04	Express	4.3 (4293)	W	MFI	200@4600	260@2800	4.00x3.48	9.2:1	18@2000
		4.8 (4802)	V	MFI	270@5200	285@4000	3.78x3.27	9.5:1	18@2000
		5.3 (5327)	T	MFI	285@4000	360@4000	3.78x3.62	9.5:1	18@2000
		6.0 (5967)	U	MFI	300@4400	360@4000	4.00x3.62	9.4:1	18@2000
	Savana	4.3 (4293)	W	MFI	200@4600	260@2800	4.00x3.48	9.2:1	18@2000
		4.8 (4802)	V	MFI	270@5200	285@4000	3.78x3.27	9.5:1	18@2000
		5.3 (5327)	T	MFI	285@4000	360@4000	3.78x3.62	9.5:1	18@2000
		6.0 (5967)	U	MFI	300@4400	360@4000	4.00x3.62	9.4:1	18@2000
	Sierra	4.3 (4293)	W	MFI	200@4600	260@2800	4.00x3.48	9.2:1	18@2000
		4.8 (4802)	V	MFI	270@5200	285@4000	3.78x3.27	9.5:1	18@2000
		5.3 (5327)	T	MFI	285@4000	360@4000	3.78x3.62	9.5:1	18@2000
		6.0 (5967)	U	MFI	300@4400	360@4000	4.00x3.62	9.4:1	18@2000
		6.6 (6599)	1	DSL	300@3000	520@1800	4.00x3.90	17.5:1	57@3250
		8.1 (8128)	G	MFI	340@4200	455@3200	4.25x4.37	9.1:1	10@2000

GENERAL ENGINE SPECIFICATIONS
All measurements are given in inches.

Year	Model	Engine Displacement Liters (cc)	Engine Series (ID/VIN)	Fuel System	Net Horsepower @ rpm	Net Torque @ rpm (ft. lbs.)	Bore x Stroke (in.)	Compression Ratio	Oil Pressure @ rpm
2003-04 (cont.)	Silverado	4.3 (4293)	X	MFI	200@4600	260@2800	4.00x3.48	9.2:1	18@2000
		4.8 (4802)	V	MFI	270@5200	285@4000	3.78x3.27	9.5:1	18@2000
		5.3 (5327)	T	MFI	285@4000	360@4000	3.78x3.62	9.5:1	18@2000
		6.0 (5967)	U	MFI	300@4400	360@4000	4.00x3.62	9.4:1	18@2000
		6.6 (6599)	1	DSL	300@3000	520@1800	4.00x3.90	17.5:1	57@3250
		8.1 (8128)	G	MFI	340@4200	455@3200	4.25x4.37	9.1:1	10@2000

DSL: Diesel

MFI: Multi-port Fuel Injection

① Below 15,000 GVWR: 180@3400
    Above 15,000 GVWR: 190@3400

② 2WD: 180@4400
    4WD: 190@4400

③ 2WD: 235@2800
    4WD: 240@2800

④ Below 8500 GVWR: 275@1700
    Above 8500 GVWR: 290@1700

42372-SILV-C04

GASOLINE ENGINE TUNE-UP SPECIFICATIONS

Year	Engine Displacement Liters (cc)	Engine ID/VIN	Spark Plugs Gap (in.)	Ignition Timing (deg.) MT	Ignition Timing (deg.) AT	Fuel Pump (psi)	Idle Speed (rpm) MT	Idle Speed (rpm) AT	Valve Clearance In.	Valve Clearance Ex.
2000	4.3 (4293)	W	0.060	①	①	58-64 ②	600	625	HYD	HYD
	4.8 (4802)	V	0.060	①	①	55-62 ②	③	③	HYD	HYD
	5.0 (4999)	M	0.060	①	①	60-66 ②	650 ④	550	HYD	HYD
	5.3 (5327)	T	0.060	①	①	55-62 ②	③	③	HYD	HYD
	5.7 (5735)	R	0.060	①	①	60-66 ②	660 ⑥	525	HYD	HYD
	6.0 (5966)	U	0.060	①	①	55-62 ②	③	③	HYD	HYD
	7.4 (7440)	J	0.060	①	①	60-66 ②	750 ⑤	675 ⑤	HYD	HYD
2001	4.3 (4293)	W	0.060	①	①	58-64 ②	600	625	HYD	HYD
	4.8 (4802)	V	0.060	①	①	55-62 ②	③	③	HYD	HYD
	5.0 (4999)	M	0.060	①	①	60-66 ②	650 ④	550	HYD	HYD
	5.3 (5327)	T	0.060	①	①	55-62 ②	③	③	HYD	HYD
	5.7 (5735)	R	0.060	①	①	60-66 ②	660 ⑤	525	HYD	HYD
	6.0 (5966)	U	0.060	①	①	55-62 ②	③	③	HYD	HYD
	8.1 (8128)	G	0.060	①	①	55-62 ②	③	③	HYD	HYD
2002	4.3 (4293)	W	0.060	①	①	58-64 ②	600	625	HYD	HYD
	4.8 (4802)	V	0.060	①	①	55-62 ②	③	③	HYD	HYD
	5.0 (4999)	M	0.060	①	①	60-66 ②	650 ④	550	HYD	HYD
	5.3 (5327)	T	0.060	①	①	55-62 ②	③	③	HYD	HYD
	5.7 (5735)	R	0.060	①	①	60-66 ②	660 ⑤	525	HYD	HYD
	6.0 (5966)	U	0.060	①	①	55-62 ②	③	③	HYD	HYD
	8.1 (8128)	G	0.060	①	①	55-62 ②	③	③	HYD	HYD
2003-04	4.3 (4293)	X	0.060	①	①	58-64 ②	600	625	HYD	HYD
	4.8 (4802)	V	0.060	①	①	55-62 ②	③	③	HYD	HYD
	5.3 (5327)	T	0.060	①	①	55-62 ②	③	③	HYD	HYD
	5.3 (5327)	Z	0.060	①	①	55-62 ②	③	③	HYD	HYD
	6.0 (5966)	U	0.060	①	①	55-62 ②	③	③	HYD	HYD
	6.0 (5966)	N	0.060	①	①	55-62 ②	③	③	HYD	HYD
	8.1 (8128)	G	0.060	①	①	55-62 ②	③	③	HYD	HYD

NOTE: The Vehicle Emission Control Information label often reflects specification changes made during production. The label figures must be used if they differ from those in this chart.

HYD: Hydraulic

① Ignition timing is preset and cannot be adjusted
② With key ON and engine OFF
③ Idle speed is maintained by the Powertrain Control Module (PCM)
④ Under 8500 GVW
⑤ Over 8500 GVW

42372-SILV-C05

DIESEL ENGINE TUNE-UP SPECIFICATIONS

Year	Engine Displacement cu. in. (cc)	Engine ID/VIN	Valve Clearance Intake (in.)	Valve Clearance Exhaust (in.)	Intake Valve Opens (deg.)	Injection Pump Setting (deg.)	Injection Nozzle Pressure (psi) New	Injection Nozzle Pressure (psi) Used	Idle Speed (rpm)	Cranking Compression Pressure (psi)
2000	6.5 (6473)	F	HYD	HYD	①	①	1800	1700	①	380-400
2001	6.5 (6473)	F	HYD	HYD	①	①	1800	1700	①	380-400
	6.6 (6599)	1	HYD	HYD	①	①	1800	1700	①	380-400
2002	6.5 (6473)	F	HYD	HYD	①	①	1800	1700	①	380-400
	6.6 (6599)	1	HYD	HYD	①	①	1800	1700	①	380-400
2003-04	6.6 (6599)	1	HYD	HYD	①	①	1800	1700	①	380-400

NOTE: The Vehicle Emission Control Information label often reflects specification changes made during production. The label figures must be used if they differ from those in this chart.

HYD: Hydraulic

NA: Not Available

① Refer to Vehicle Emission Control Information label

42372-SILV-C06

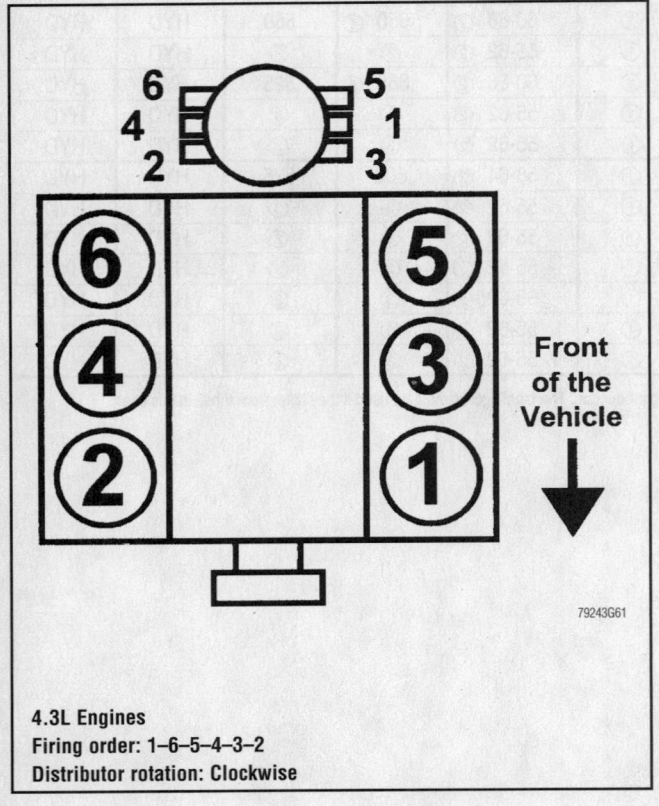

4.3L Engines
Firing order: 1–6–5–4–3–2
Distributor rotation: Clockwise

79243G61

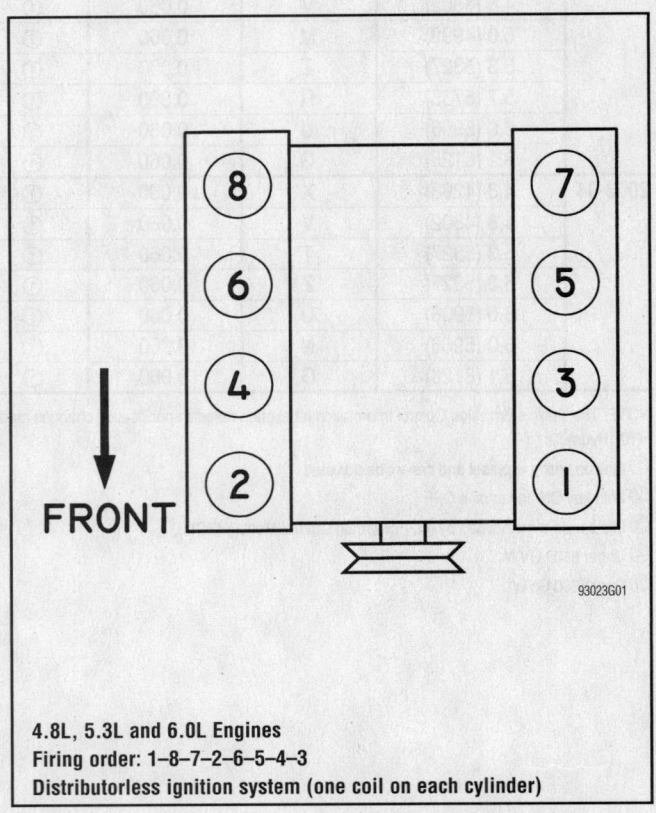

4.8L, 5.3L and 6.0L Engines
Firing order: 1–8–7–2–6–5–4–3
Distributorless ignition system (one coil on each cylinder)

93023G01

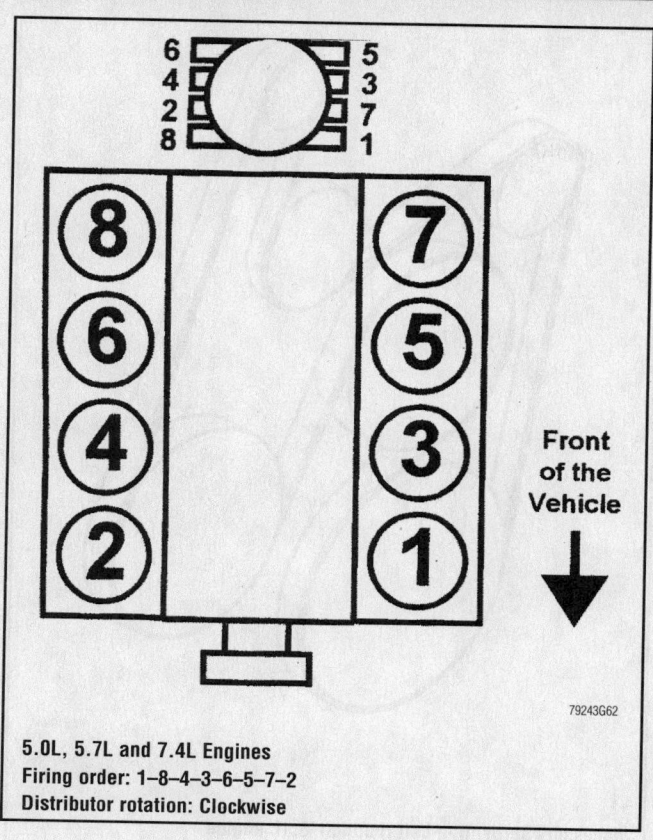

5.0L, 5.7L and 7.4L Engines
Firing order: 1–8–4–3–6–5–7–2
Distributor rotation: Clockwise

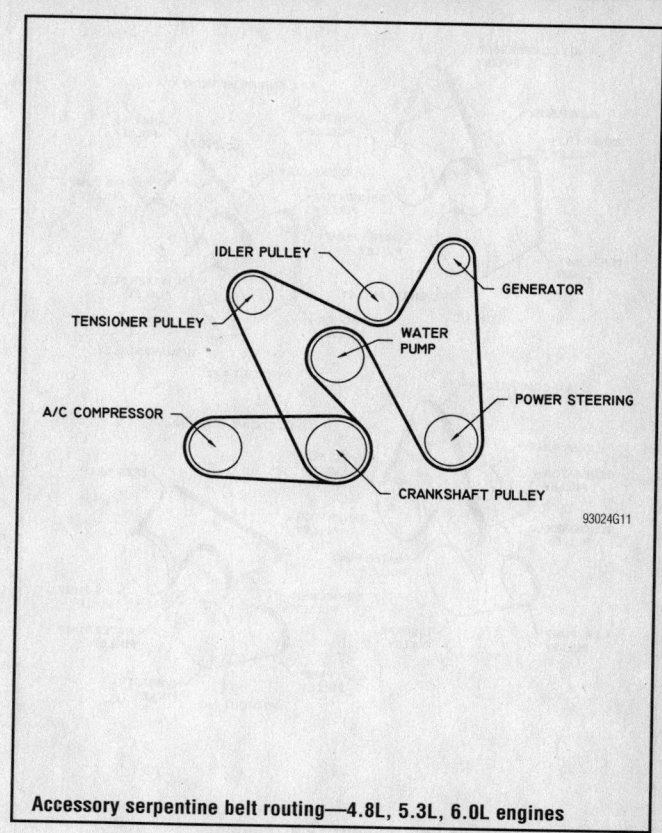

Accessory serpentine belt routing—4.8L, 5.3L, 6.0L engines

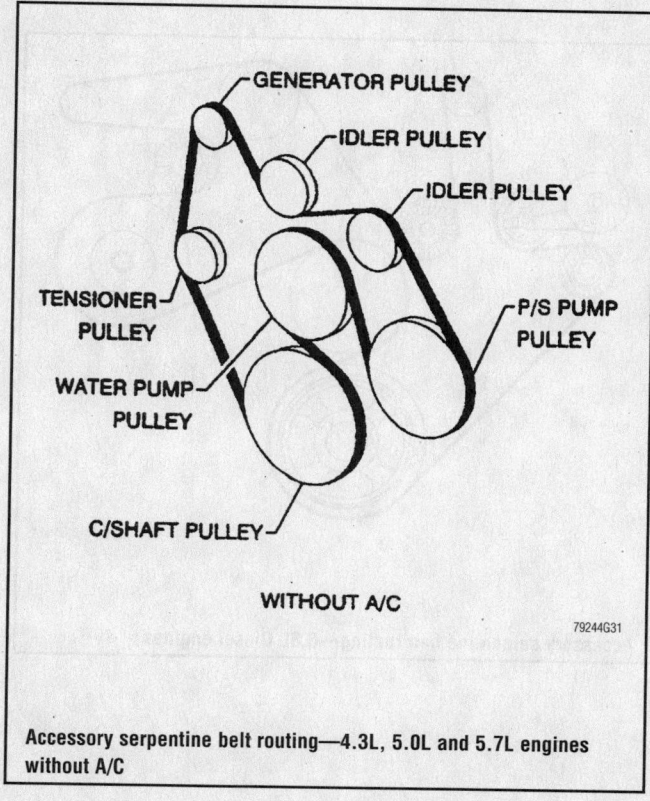

Accessory serpentine belt routing—4.3L, 5.0L and 5.7L engines without A/C

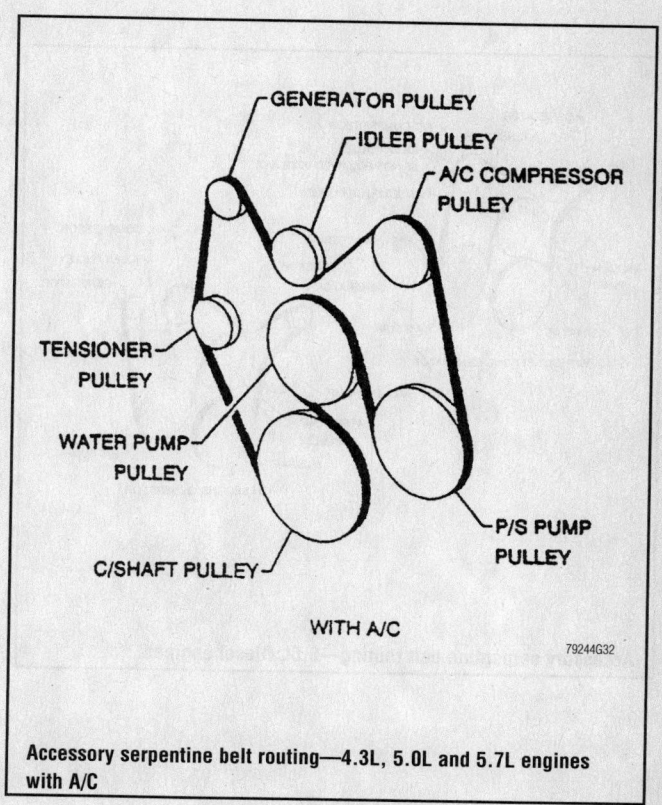

Accessory serpentine belt routing—4.3L, 5.0L and 5.7L engines with A/C

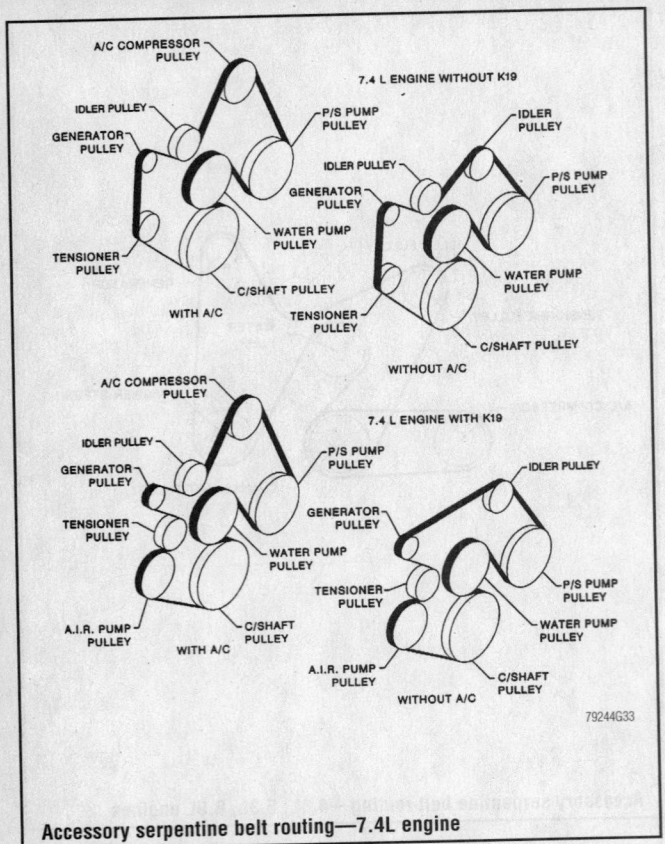

Accessory serpentine belt routing—7.4L engine

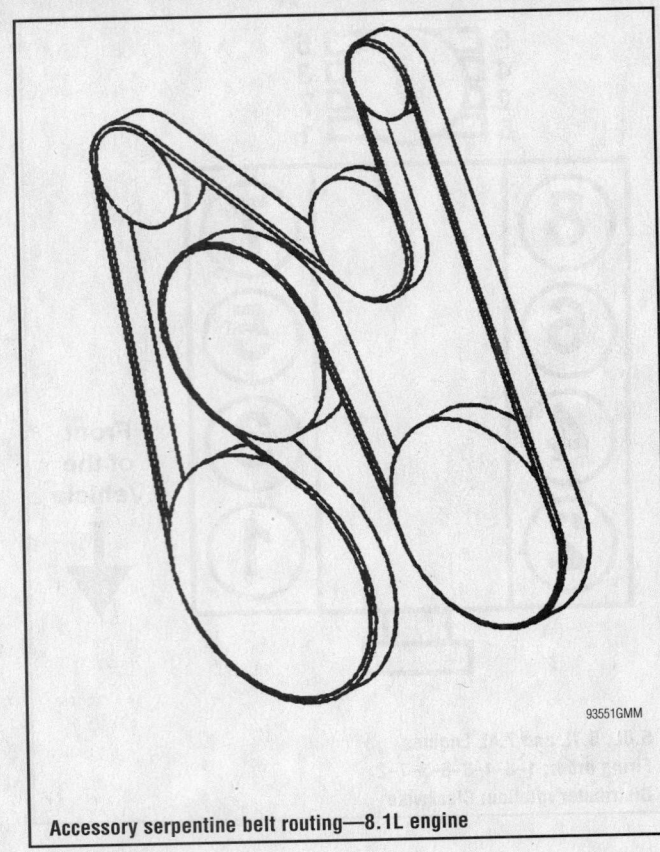

Accessory serpentine belt routing—8.1L engine

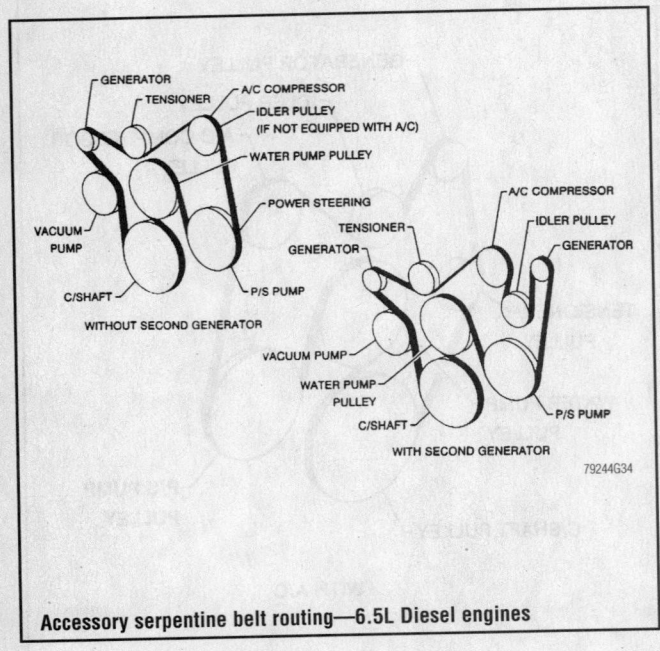

Accessory serpentine belt routing—6.5L Diesel engines

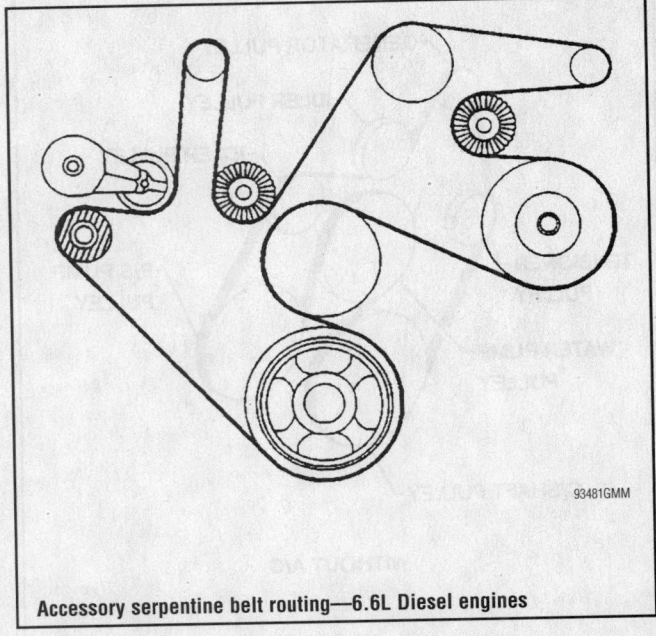

Accessory serpentine belt routing—6.6L Diesel engines

CAPACITIES

Year	Model	Engine Displacement Liters (cc)	Engine ID/VIN	Engine Oil with Filter (qts.)	Transmission (pts.) 5-Spd	Transmission (pts.) Auto.	Transfer Case (pts.)	Drive Axle Front (pts.)	Drive Axle Rear (pts.)	Fuel Tank (gal.)	Cooling System (qts.)
2000	C1500	5.0 (4999)	M	5.0	①	②	—	—	③	④	18.0
	C1500	5.7 (5735)	R	5.0	①	②	—	—	③	④	18.0
	C1500	4.3 (4293)	W	5.0	①	②	—	—	③	④	13.0
	C2500	6.5 (6473)	F	7.0	①	②	—	—	③	④	27.5
	C2500	5.0 (4999)	M	5.0	①	②	—	—	③	④	18.0
	C2500	5.7 (5735)	R	5.0	①	②	—	—	③	④	18.0
	C2500	6.5 (6473)	S	7.0	①	②	—	—	③	④	27.5
	C2500	4.3 (4293)	W	5.0	①	②	—	—	③	④	13.0
	C3500	6.5 (6473)	F	7.0	①	②	—	—	③	④	27.5
	C3500	7.4 (7440)	J	6.0	①	②	—	—	③	④	25.0 ⑤
	G/P1500	5.0 (4999)	M	5.0	①	②	—	—	③	22.0 ⑥	17.0 ⑦
	G/P1500	5.7 (5735)	R	5.0	①	②	—	—	③	⑥	18.0 ⑦
	G/P1500	4.3 (4293)	W	5.0	①	②	—	—	③	22.0 ⑥	13.0 ⑦
	G/P2500	5.0 (4999)	M	5.0	①	②	—	—	③	22.0 ⑥	17.0 ⑦
	G/P2500	5.7 (5735)	R	5.0	①	②	—	—	③	⑥	18.0 ⑦
	G/P3500	6.5 (6473)	F	7.0	①	②	—	—	③	22.0 ⑥	27.5 ⑦
	G/P3500	7.4 (7440)	J	6.0	①	②	—	—	③	⑧	24.5 ⑦
	K1500	6.5 (6473)	F	7.0	①	②	⑨	⑩	③	④	27.5
	K1500	5.0 (4999)	M	5.0	①	②	⑨	⑩	③	④	18.0
	K1500	5.7 (5735)	R	5.0	①	②	⑨	⑩	③	④	18.0
	K1500	4.3 (4293)	W	5.0	①	②	⑨	⑩	③	④	11.0
	K2500	6.5 (6473)	F	7.0	①	②	⑨	⑩	③	④	27.5
	K2500	5.0 (4999)	M	5.0	①	②	⑨	⑩	③	④	18.0
	K2500	5.7 (5735)	R	5.0	①	②	⑨	⑩	③	④	18.0
	K3500	6.5 (6473)	F	7.0	①	②	⑨	⑩	③	④	27.5
	K3500	7.4 (7440)	J	6.0	①	②	⑨	⑩	③	④	25.0
	Sierra	4.3 (4293)	W	5.0	8.0	10.0	4.8	3.5	③	⑪	14.5
	Sierra	4.8 (4802)	V	6.0	8.0	10.0	4.8	3.5	③	⑪	14.5
	Sierra	5.3 (5327)	T	6.0	8.0	10.0	4.8	3.5	③	⑪	14.5
	Sierra	6.0 (5966)	U	6.0	8.0	10.0	4.8	3.5	③	⑪	14.5
	Sierra	6.5 (6374)	F	6.0	8.0	10.0	4.8	3.5	③	⑪	14.5
	Silverado	4.3 (4293)	W	5.0	8.0	10.0	4.8	3.5	③	⑪	14.5
	Silverado	4.8 (4802)	V	6.0	8.0	10.0	4.8	3.5	③	⑪	14.5
	Silverado	5.3 (5327)	T	6.0	8.0	10.0	4.8	3.5	③	⑪	14.5
	Silverado	6.0 (5966)	U	6.0	8.0	10.0	4.8	3.5	③	⑪	14.5
	Silverado	6.5 (6374)	F	6.0	8.0	10.0	4.8	3.5	③	⑪	14.5
2001	G/P1500	4.3 (4293)	W	5.0	①	②	—	—	③	22.0 ⑥	13.0 ⑦
	G/P1500	5.0 (4999)	M	5.0	①	②	—	—	③	22.0 ⑥	17.0 ⑦
	G/P1500	5.7 (5735)	R	5.0	①	②	—	—	③	⑥	18.0 ⑦
	G/P2500	4.3 (4293)	W	5.0	①	②	—	—	③	22.0 ⑥	13.0 ⑦
	G/P2500	5.0 (4999)	M	5.0	①	②	—	—	③	22.0 ⑥	17.0 ⑦
	G/P2500	5.7 (5735)	R	5.0	①	②	—	—	③	⑥	18.0 ⑦
	G/P2500	6.5 (6473)	F	7.0	①	②	—	—	③	22.0 ⑥	27.5 ⑦
	G/P3500	6.5 (6473)	F	7.0	①	②	—	—	③	22.0 ⑥	27.5 ⑦
	G/P3500	8.1 (8128)	G	6.5	⑫	⑬	4.8	⑭	⑮	⑪	㉑

42372-SILV-C07

CAPACITIES

Year	Model	Engine Displacement Liters (cc)	Engine ID/VIN	Engine Oil with Filter (qts.)	Transmission (pts.) 5-Spd	Auto.	Transfer Case (pts.)	Drive Axle Front (pts.)	Rear (pts.)	Fuel Tank (gal.)	Cooling System (qts.)
2001 (cont.)	Sierra	4.3 (4293)	W	4.5	[12]	[13]	4.8	[14]	[15]	[11]	[16]
	Sierra	4.8 (4802)	V	6.0	[12]	[13]	4.8	[14]	[15]	[11]	[17]
	Sierra	5.3 (5327)	T	6.0	[12]	[13]	4.8	[14]	[15]	[11]	[18]
	Sierra	6.0 (5966)	U	6.0	[12]	[13]	4.8	[14]	[15]	[11]	[19]
	Sierra	6.6 (6599)	1	8.0	[12]	[13]	4.8	[14]	[15]	[11]	[20]
	Sierra	8.1 (8128)	G	6.5	[12]	[13]	4.8	[14]	[15]	[11]	[21]
	Silverado	4.3 (4293)	W	4.5	[12]	[13]	4.8	[14]	[15]	[11]	[16]
	Silverado	4.8 (4802)	V	6.0	[12]	[13]	4.8	[14]	[15]	[11]	[17]
	Silverado	5.3 (5327)	T	6.0	[12]	[13]	4.8	[14]	[15]	[11]	[18]
	Silverado	6.0 (5966)	U	6.0	[12]	[13]	4.8	[14]	[15]	[11]	[19]
	Silverado	6.6 (6599)	1	8.0	[12]	[13]	4.8	[14]	[15]	[11]	[20]
	Silverado	8.1 (8128)	G	6.5	[12]	[13]	4.8	[14]	[15]	[11]	[21]
2002	G/P1500	4.3 (4293)	W	5.0	[1]	[2]	—	—	[3]	22.0 [6]	13.0 [7]
	G/P1500	5.0 (4999)	M	5.0	[1]	[2]	—	—	[3]	22.0 [6]	17.0 [7]
	G/P1500	5.7 (5735)	R	5.0	[1]	[2]	—	—	[3]	[6]	18.0 [7]
	G/P2500	4.3 (4293)	W	5.0	[1]	[2]	—	—	[3]	22.0 [6]	13.0 [7]
	G/P2500	5.0 (4999)	M	5.0	[1]	[2]	—	—	[3]	22.0 [6]	17.0 [7]
	G/P2500	5.7 (5735)	R	5.0	[1]	[2]	—	—	[3]	[6]	18.0 [7]
	G/P2500	6.5 (6473)	F	7.0	[1]	[2]	—	—	[3]	22.0 [6]	27.5 [7]
	G/P3500	6.5 (6473)	F	7.0	[1]	[2]	—	—	[3]	22.0 [6]	27.5 [7]
	G/P3500	8.1 (8128)	G	6.5	[12]	[13]	4.8	[14]	[15]	[11]	[21]
	Sierra	4.3 (4293)	W	4.5	[12]	[13]	4.8	[14]	[15]	[11]	[16]
	Sierra	4.8 (4802)	V	6.0	[12]	[13]	4.8	[14]	[15]	[11]	[17]
	Sierra	5.3 (5327)	T	6.0	[12]	[13]	4.8	[14]	[15]	[11]	[18]
	Sierra	6.0 (5966)	U	6.0	[12]	[13]	4.8	[14]	[15]	[11]	[19]
	Sierra	6.6 (6599)	1	8.0	[12]	[13]	4.8	[14]	[15]	[11]	[20]
	Sierra	8.1 (8128)	G	6.5	[12]	[13]	4.8	[14]	[15]	[11]	[21]
	Silverado	4.3 (4293)	W	4.5	[12]	[13]	4.8	[14]	[15]	[11]	[16]
	Silverado	4.8 (4802)	V	6.0	[12]	[13]	4.8	[14]	[15]	[11]	[17]
	Silverado	5.3 (5327)	T	6.0	[12]	[13]	4.8	[14]	[15]	[11]	[18]
	Silverado	6.0 (5966)	U	6.0	[12]	[13]	4.8	[14]	[15]	[11]	[19]
	Silverado	6.6 (6599)	1	8.0	[12]	[13]	4.8	[14]	[15]	[11]	[20]
	Silverado	8.1 (8128)	G	6.5	[12]	[13]	4.8	[14]	[15]	[11]	[21]
2003-04	Express	4.3 (4293)	X	5.0	[1]	[2]	—	—	[3]	22.0 [6]	13.0 [7]
	Express	4.8 (4802)	V	5.0	[1]	[2]	—	—	[3]	22.0 [6]	17.0 [7]
	Express	5.3 (5327)	T	5.0	[1]	[2]	—	—	[3]	[6]	18.0 [7]
	Express	6.0 (5966)	U	5.0	[1]	[2]	—	—	[3]	22.0 [6]	13.0 [7]
	Savana	4.3 (4293)	X	5.0	[1]	[2]	—	—	[3]	22.0 [6]	17.0 [7]
	Savana	4.8 (4802)	V	5.0	[1]	[2]	—	—	[3]	[6]	18.0 [7]
	Savana	5.3 (5327)	T	7.0	[1]	[2]	—	—	[3]	22.0 [6]	27.5 [7]
	Savana	6.0 (5966)	U	7.0	[1]	[2]	—	—	[3]	22.0 [6]	27.5 [7]
	Sierra	4.3 (4293)	X	4.5	[12]	[13]	4.8	[14]	[15]	[11]	[16]
	Sierra	4.8 (4802)	V	6.0	[12]	[13]	4.8	[14]	[15]	[11]	[17]
	Sierra	5.3 (5327)	T	6.0	[12]	[13]	4.8	[14]	[15]	[11]	[18]
	Sierra	6.0 (5966)	U	6.0	[12]	[13]	4.8	[14]	[15]	[11]	[19]
	Sierra	6.6 (6599)	1	8.0	[12]	[13]	4.8	[14]	[15]	[11]	[20]
	Sierra	8.1 (8128)	G	6.5	[12]	[13]	4.8	[14]	[15]	[11]	[21]

42372-SILV-C08

CAPACITIES

Year	Model	Engine Displacement Liters (cc)	Engine ID/VIN	Engine Oil with Filter (qts.)	Transmission (pts.)		Transfer Case (pts.)	Drive Axle		Fuel Tank (gal.)	Cooling System (qts.)
					5-Spd	Auto.		Front (pts.)	Rear (pts.)		
2003-04 (cont.)	Silverado	4.3 (4293)	X	4.5	⑫	⑬	4.8	⑭	⑮	⑪	⑯
	Silverado	4.8 (4802)	V	6.0	⑫	⑬	4.8	⑭	⑮	⑪	⑰
	Silverado	5.3 (5327)	T	6.0	⑫	⑬	4.8	⑭	⑮	⑪	⑱
	Silverado	6.0 (5966)	U	6.0	⑫	⑬	4.8	⑭	⑮	⑪	⑲
	Silverado	6.6 (6599)	1	8.0	⑫	⑬	4.8	⑭	⑮	⑪	⑳
	Silverado	8.1 (8128)	G	6.5	⑫	⑬	4.8	⑭	⑮	⑪	㉑

NOTE: All capacities are approximate. Add fluid gradually and check to be sure a proper fluid level is obtained.

① New Venture gear 4500: 8.0 pts.
New Venture gear 5LM60: 4.4 pts.

② 4L60E trans.: 10.0 pts.
4L80E trans.: 14.5 pts.

③ 8.5 in. ring gear: 4.2 pts.
9.5 in. ring gear: 6.5 pts.
9.75 in. ring gear: 6.0 pts.
10.5 in. ring gear: 6.5 pts.

④ Std. available with 25 and 34 gallon tanks
Chassis cab available with 22, 30 and 34 gallon tanks

⑤ 3500HD: 28.5 qts. capacity

⑥ Optional 31 and 40 gallon tanks

⑦ Add three qts. with rear heater

⑧ Available with a variety of fuel tanks

⑨ NV241 and NV243: 4.5 pts.
4401 and 4470: 6.6 pts.

⑩ K2 models: 1.75 qts.
K3 models: 2.25 qts.

⑪ Short bed: 26 gals.
Long bed: 34 gals.

⑫ New Venture Gear 3500: 4.8 pts.
New Venture Gear 4500: 8.0 pts.
6-speed M/T: 12.6 pts.

⑬ 4L60-E: 10 pts.
4L80-E: 15.4 pts.
Allison: 14.8 pts.

⑭ 8.25 in ring gear: 3.5 pts.
9.25 ring gear: 3.7 pts.

⑮ 8.6 in. ring gear: 4.8 pts.
9.5 & 10.5 in. ring gear: 5.5 pts.
11.5 in. ring gear: 7.7 pts.

⑯ With A/T: 12.6 qts.
With M/T: 12.9 qts.

⑰ With A/T: 13.4 qts.
With M/T: 13.7 qts.
With A/T and front A/C: 14.4 qts.
With A/T and front and rear A/C: 15.8 qts.

⑱ With A/T: 13.4 qts.
With A/T and optional A/C: 14.9 qts.
With A/T and front A/C: 14.4 qts.
With A/T and front and rear A/C: 15.8 qts.

⑲ With A/T: 14.8 qts.
With A/T and optional oil cooler: 14.4 qts.
With M/T: 15.2 qts.
With M/T and optional oil cooler: 15.8 qts.

⑳ With A/T: 20.3 qts.
With M/T: 20.7 qts.

㉑ With A/T: 20.7 qts.
With M/T: 21.1 qts.

42372-SILV-C09

VALVE SPECIFICATIONS

Year	Engine Displacement Liters (cc)	Engine ID/VIN	Seat Angle (deg.)	Face Angle (deg.)	Spring Test Pressure (lbs. @ in.)	Spring Installed Height (in.)	Stem-to-Guide Clearance (in.)		Stem Diameter (in.)	
							Intake	Exhaust	Intake	Exhaust
2000	4.3 (4293)	W	46	45	187-203@1.27	1.67-1.70	0.0010-0.0027	0.0010-0.0027	NA	NA
	4.8 (4802)	V	46	45	220@1.32	1.80	0.0010-0.0037	0.0010-0.0037	0.3132-0.3110	0.3132-0.3110
	5.0 (4999)	M	46	45	187-203@1.27	1.69-1.71	0.0010-0.0027	0.0010-0.0027	NA	NA
	5.3 (5327)	T	46	45	220@1.32	1.80	0.0010-0.0037	0.0010-0.0037	0.3132-0.3110	0.3132-0.3110
	5.7 (5735)	R	46	45	187-203@1.27	1.69-1.71	0.0010-0.0027	0.0010-0.0027	NA	NA
	6.0 (5966)	U	46	45	220@1.32	1.80	0.0010-0.0037	0.0010-0.0037	0.3132-0.3110	0.3132-0.3110
	6.5 (6473)	F	46	45	230@1.40	1.80	0.0010-0.0027	0.0010-0.0027	NA	NA
	7.4 (7440)	J	46	45	238-262@1.34	1.83	0.0010-0.0029 ①	0.0012-0.0031 ①	NA	NA
2001	4.3 (4293)	W	46	45	187-203@1.27	1.67-1.70	0.0010-0.0027	0.0010-0.0027	NA	NA
	4.8 (4802)	V	46	45	220@1.32	1.80	0.0010-0.0026	0.0010-0.0026	0.3132-0.3140	0.3132-0.3140
	5.0 (4999)	M	46	45	187-203@1.27	1.69-1.71	0.0010-0.0027	0.0010-0.0027	NA	NA
	5.3 (5327)	T	46	45	220@1.32	1.80	0.0010-0.0026	0.0010-0.0026	0.3132-0.3140	0.3132-0.3140
	5.7 (5735)	R	46	45	187-203@1.27	1.69-1.71	0.0010-0.0027	0.0010-0.0027	NA	NA
	6.0 (5966)	U	46	45	220@1.32	1.80	0.0010-0.0026	0.0010-0.0026	0.3132-0.3140	0.3132-0.3140
	6.5 (6473)	F	46	45	230@1.40	1.80	0.0010-0.0027	0.0010-0.0027	NA	NA
	6.6 (6599)	1	45	45	NA	1.61	0.0012-0.0025	0.0015-0.0028	0.2737-0.2744	0.2734-0.2741
	8.1 (8128)	G	46	45	216-236@1.34	1.81-1.84	0.0010-0.0029	0.0012-0.0031	0.3715-0.3722	0.3713-0.3720
2002	4.3 (4293)	W	46	45	187-203@1.27	1.67-1.70	0.0010-0.0027	0.0010-0.0027	NA	NA
	4.8 (4802)	V	46	45	220@1.32	1.80	0.0010-0.0026	0.0010-0.0026	0.3132-0.3140	0.3132-0.3140
	5.0 (4999)	M	46	45	187-203@1.27	1.69-1.71	0.0010-0.0027	0.0010-0.0027	NA	NA
	5.3 (5327)	T	46	45	220@1.32	1.80	0.0010-0.0026	0.0010-0.0026	0.3132-0.3140	0.3132-0.3140
	5.7 (5735)	R	46	45	187-203@1.27	1.69-1.71	0.0010-0.0027	0.0010-0.0027	NA	NA

42372-SILV-C10

VALVE SPECIFICATIONS

Year	Engine Displacement Liters (cc)	Engine ID/VIN	Seat Angle (deg.)	Face Angle (deg.)	Spring Test Pressure (lbs. @ in.)	Spring Installed Height (in.)	Stem-to-Guide Clearance (in.)		Stem Diameter (in.)	
							Intake	Exhaust	Intake	Exhaust
2002 (cont.)	6.0 (5966)	U	46	45	220@1.32	1.80	0.0010-0.0026	0.0010-0.0026	0.3132-0.3140	0.3132-0.3140
	6.5 (6473)	F	46	45	230@1.40	1.80	0.0010-0.0027	0.0010-0.0027	NA	NA
	6.6 (6599)	1	45	45	NA	1.61	0.0012-0.0025	0.0015-0.0028	0.2737-0.2744	0.2734-0.2741
	8.1 (8128)	G	46	45	216-236@1.34	1.81-1.84	0.0010-0.0029	0.0012-0.0031	0.3715-0.3722	0.3713-0.3720
2003-04	4.3 (4293)	X	46	45	187-203@1.27	1.67-1.70	0.0010-0.0027	0.0010-0.0027	NA	NA
	4.8 (4802)	V	46	45	220@1.32	1.80	0.0010-0.0026	0.0010-0.0026	0.3132-0.3140	0.3132-0.3140
	5.3 (5327)	T	46	45	220@1.32	1.80	0.0010-0.0026	0.0010-0.0026	0.3132-0.3140	0.3132-0.3140
	5.3 (5327)	Z	46	45	220@1.32	1.80	0.0010-0.0026	0.0010-0.0026	0.3130-0.3140	0.3130-0.3140
	6.0 (5966)	U	46	45	220@1.32	1.80	0.0010-0.0026	0.0010-0.0026	0.3132-0.3140	0.3132-0.3140
	6.0 (5966)	N	46	45	220@1.32	1.80	0.0010-0.0026	0.0010-0.0026	0.3130-0.3140	0.3130-0.3140
	6.6 (6599)	1	45	45	NA	1.61	0.0012-0.0025	0.0015-0.0028	0.2737-0.2744	0.2734-0.2741
	8.1 (8128)	G	46	45	216-236@1.34	1.81-1.84	0.0010-0.0029	0.0012-0.0031	0.3715-0.3722	0.3713-0.3720

NA: Not Available

① Service limit:
  Intake: 0.0037 MAX
  Exhaust: 0.0049 MAX

42372-SILV-C11

TORQUE SPECIFICATIONS
All readings in ft. lbs.

Year	Engine Displacement Liters (cc)	Engine ID/VIN	Cylinder Head Bolts	Main Bearing Bolts	Rod Bearing Bolts	Crankshaft Damper Bolts	Flywheel Bolts	Manifold Intake *	Manifold Exhaust	Spark Plugs	Lug Nut
2000	4.3 (4293)	W	①	77	②	74	74	③	④	11	90
	4.8 (4802)	V	⑤	⑥	⑦	⑧	⑨	⑩	⑪	12	140
	5.0 (4999)	M	⑤	⑮	⑫	74	74	⑯	④	15	⑮
	5.3 (5327)	T	⑤	⑥	⑦	⑧	⑨	⑩	⑪	12	140
	5.7 (5735)	R	⑫	⑬	⑭	74	74	⑫	④	15	⑮
	6.0 (5966)	U	⑤	⑥	⑦	⑧	⑨	⑩	⑪	12	140
	6.5 (6473)	F	⑫	⑰	48	200	65	31	26	—	⑮
	7.4 (7440)	J	85	100	45	110	67	30	22	15	⑮
2001	4.3 (4293)	W	①	77	②	70	74	③	④	11	⑱
	4.8 (4802)	V	⑤	⑥	⑦	⑧	⑨	⑩	⑪	12	⑱
	5.0 (4999)	M	⑤	⑮	⑫	74	74	⑯	④	15	⑱
	5.3 (5327)	T	⑤	⑥	⑦	⑧	⑨	⑩	⑪	12	⑱
	5.7 (5735)	R	⑫	⑬	⑭	74	74	⑯	④	15	⑱
	6.0 (5966)	U	⑤	⑥	⑦	⑧	⑨	⑩	⑪	12	⑱
	6.5 (6473)	F	⑫	⑰	48	200	65	31	26	—	⑱
	6.6 (6374)	1	⑲	⑳	㉑	260	㉒	15	28	—	⑱
	8.1 (8128)	G	㉓	㉔	㉔	189	㉕	㉖	㉗	15	⑱
2002	4.3 (4293)	W	①	77	②	70	74	③	④	11	⑱
	4.8 (4802)	V	⑤	⑥	⑦	⑧	⑨	⑩	⑪	12	⑱
	5.0 (4999)	M	⑤	⑮	⑫	74	74	⑯	④	15	⑱
	5.3 (5327)	T	⑤	⑥	⑦	⑧	⑨	⑩	⑪	12	⑱
	5.7 (5735)	R	⑫	⑬	⑭	74	74	⑯	④	15	⑱
	6.0 (5966)	U	⑤	⑥	⑦	⑧	⑨	⑩	⑪	12	⑱
	6.5 (6473)	F	⑫	⑰	48	200	65	31	26	—	⑱
	6.6 (6374)	1	⑲	⑳	㉑	260	㉒	15	28	—	⑱
	8.1 (8128)	G	㉓	㉔	㉔	189	㉕	㉖	㉗	15	⑱

42372-SILV-C12

TORQUE SPECIFICATIONS
All readings in ft. lbs.

Year	Engine Displacement Liters (cc)	Engine ID/VIN	Cylinder Head Bolts	Main Bearing Bolts	Rod Bearing Bolts	Crankshaft Damper Bolts	Flywheel Bolts	Manifold Intake *	Manifold Exhaust	Spark Plugs	Lug Nut
2003-04	4.3 (4293)	X	①	77	②	70	74	③	④	11	⑱
	4.8 (4802)	V	⑤	⑥	⑦	⑧	⑨	⑩	⑪	12	⑱
	5.3 (5327)	T	⑤	⑥	⑦	⑧	⑨	⑩	⑪	12	⑱
	5.3 (5327)	Z	⑤	⑥	⑦	⑧	⑨	⑩	⑪	12	⑱
	6.0 (5966)	U	⑤	⑥	⑦	⑧	⑨	⑩	⑪	12	⑱
	6.0 (5966)	N	⑤	⑥	⑦	⑧	⑨	⑩	⑪	12	⑱
	6.6 (6374)	1	⑲	⑳	㉑	260	㉒	15	28	—	⑱
	8.1 (8128)	G	㉓	㉔	㉔	189	㉕	㉖	㉗	15	⑱

* NOTE: Applies to Lower Manifold only.

① 1st pass: 22 ft. lbs.
2nd pass:
Short bolt: Plus 55 degrees
Medium bolt: Plus 65 degrees
Long bolt: Plus 75 degrees

② 20 ft. lbs. plus 70 degrees

③ Lower intake manifold:
1st pass: 27 in. lbs.
2nd pass: 106 in. lbs.
Final pass: 11 ft. lbs.
Upper manifold bolts:
1st pass: 44 in. lbs.
2nd pass: 88 in. lbs.

④ Tighten bolts to 12 ft. lbs.
Retorque to 22 ft. lbs.

⑤ Step 1: 22 ft. lbs.
Step 2: 90 degrees
Step 3: 90 degrees,
(except medium length bolts at front and rear)
Step 4: Tighten medium length bolts,
at front and rear an additional 50 degrees

⑥ Inner bolts;
Step 1: 15 ft. lbs.

⑦ Step 1: 15 ft. lbs.
Step 2: 60 degrees

⑧ Use a new bolt
Step 1: 37 ft. lbs.
Step 2: 140 degrees

⑨ Step 1: 15 ft. lbs.
Step 2: 37 ft. lbs.
Step 3: 74 ft. lbs.

⑩ Step 1: 44 in. lbs.
Step 2: 89 in. lbs.

⑪ Nuts: 39 ft. lbs.
Stud: 16 ft. lbs.

⑫ Step 1: 20 ft. lbs.
Step 2: 50 ft. lbs.
Step 3: 50 ft. lbs.

⑬ Outer bolts on caps 2-4: 67 ft. lbs.
All others: 74 ft. lbs.

⑭ Tighten all bolts to 20 ft. lbs.
Retorque to 50 ft. lbs.

⑮ All 5 & 6 stud single rear wheels: 110 ft. lbs.
All 8 stud single rear wheels: 120 ft. lbs.
All 8 stud dual rear wheels: 140 ft. lbs.
All 10 stud dual wheels: 175 ft. lbs.

⑯ Step 1: 22 ft. lbs.
Step 2:
Short bolt: Plus 55 degrees
Medium bolt: Plus 65 degrees
Long bolt: Plus 75 degrees

⑰ Outer bolts: 100 ft. lbs.
Inner bolts: 111 ft. lbs.

⑱ Single wheels: 140 ft. lbs.
Dual rear wheels: 175 ft. lbs.

⑲ M8 bolts: 18 ft. lbs.
M12 bolts
1st pass: 37 ft. lbs.
2nd pass: 59 ft. lbs.
3rd pass: Plus 90 degrees
4th pass: Plus 75 degrees

⑳ 1st pass: 72 ft. lbs.
2nd pass: 97 ft. lbs.
3rd pass: Plus 30 degrees

㉑ 1st pass: 47 ft. lbs.
2nd pass: Plus 30 degrees
3rd pass: Plus 30 degrees

㉒ 1st pass: 58 ft. lbs.
2nd pass: Plus 60 degrees
3rd pass: Plus 60 degrees

㉓ 1st pass: 22 ft. lbs.
2nd pass: 22 ft. lbs.,
plus 120 degrees
Final pass:
Short bolt: Plus 60 degrees
Medium bolt: Plus 45 degrees
Long bolt: Plus 30 degrees

㉔ 22 ft. lbs., plus 90 degrees

㉕ 1st pass: 30 ft. lbs.
2nd pass: 59 ft. lbs.
3rd pass: 70 ft. lbs.

㉖ 1st & 2nd pass: 44 inch lbs.
3rd pass: 89 inch lbs.
4th pass: 106 inch lbs.

㉗ Center bolt: 26 ft. lbs.
Nut: 12 ft. lbs.
Stud: 15 ft. lbs.

42372-SILV-C13

CRANKSHAFT AND CONNECTING ROD SPECIFICATIONS
All measurements are given in inches.

Year	Engine Displacement Liters (cc)	Engine ID/VIN	Crankshaft Main Brg. Journal Dia.	Crankshaft Main Brg. Oil Clearance	Crankshaft Shaft End-play	Thrust on No.	Connecting Rod Journal Diameter	Connecting Rod Oil Clearance	Connecting Rod Side Clearance
2000	4.3 (4293)	W	①	②	0.0020-0.0070	4	2.2487-2.2497	0.0013-0.0035	0.0060-0.0140
	4.8 (4802)	V	2.5580-2.5593	0.0007-0.0021	0.0015-0.0078	5	2.0990-2.1000	0.0006-0.0030	0.0043-0.0200
	5.0 (4999)	M	③	④	0.0020-0.0080	5	2.0978-2.0998	0.0013-0.0035	0.0060-0.0140
	5.3 (5327)	T	2.5580-2.5593	0.0007-0.0021	0.0015-0.0078	5	2.0990-2.1000	0.0006-0.0030	0.0043-0.0200
	5.7 (5735)	R	③	④	0.0020-0.0080	5	2.0978-2.0998	0.0013-0.0035	0.0060-0.0140
	6.0 (5966)	U	2.5580-2.5593	0.0007-0.0021	0.0015-0.0078	5	2.0990-2.1000	0.0006-0.0030	0.0043-0.0200
	6.5 (6473)	F	⑤	⑥	0.0039-0.0100	3	⑦	0.0018-0.0039	0.0067-0.0248
	7.4 (7440)	J	2.7482-2.7489	⑧	0.0050-0.0110	5	2.1990-2.1996	0.0011-0.0029	0.0013-0.0230
2001	4.3 (4293)	W	①	②	0.0020-0.0070	4	2.2487-2.2497	0.0013-0.0035	0.0060-0.0140
	4.8 (4802)	V	2.5580-2.5593	0.0007-0.0021	0.0015-0.0078	5	2.0990-2.1000	0.0006-0.0030	0.0043-0.0200
	5.0 (4999)	M	③	④	0.0020-0.0080	5	2.0978-2.0998	0.0013-0.0035	0.0060-0.0140
	5.3 (5327)	T	2.5580-2.5593	0.0007-0.0021	0.0015-0.0078	5	2.0990-2.1000	0.0006-0.0030	0.0043-0.0200
	5.7 (5735)	R	③	④	0.0020-0.0080	5	2.0978-2.0998	0.0013-0.0035	0.0060-0.0140
	6.0 (5966)	U	2.5580-2.5593	0.0007-0.0021	0.0015-0.0078	5	2.0990-2.1000	0.0006-0.0030	0.0043-0.0200
	6.5 (6473)	F	⑤	⑥	0.0039-0.0100	3	⑦	0.0018-0.0039	0.0067-0.0248
	6.6 (6599)	1	3.1459-3.1466	0.0015-0.0028	0.0016-0.0081	NA	2.4789-2.4795	0.0014-0.0030	0.0122-0.0193
	8.1 (8128)	G	2.7482-2.7489	⑨	0.0050-0.0110	NA	2.1990-2.1996	0.0008-0.0025	0.0151-0.0270
2002	4.3 (4293)	W	①	②	0.0020-0.0070	4	2.2487-2.2497	0.0013-0.0035	0.0060-0.0140
	4.8 (4802)	V	2.5580-2.5593	0.0007-0.0021	0.0015-0.0078	5	2.0990-2.1000	0.0006-0.0030	0.0043-0.0200
	5.0 (4999)	M	③	④	0.0020-0.0080	5	2.0978-2.0998	0.0013-0.0035	0.0060-0.0140
	5.3 (5327)	T	2.5580-2.5593	0.0007-0.0021	0.0015-0.0078	5	2.0990-2.1000	0.0006-0.0030	0.0043-0.0200
	5.7 (5735)	R	③	④	0.0020-0.0080	5	2.0978-2.0998	0.0013-0.0035	0.0060-0.0140
	6.0 (5966)	U	2.5580-2.5593	0.0007-0.0021	0.0015-0.0078	5	2.0990-2.1000	0.0006-0.0030	0.0043-0.0200

42372-SILV-C14

CRANKSHAFT AND CONNECTING ROD SPECIFICATIONS
All measurements are given in inches.

Year	Engine Displacement Liters (cc)	Engine ID/VIN	Crankshaft				Connecting Rod		
			Main Brg. Journal Dia.	Main Brg. Oil Clearance	Shaft End-play	Thrust on No.	Journal Diameter	Oil Clearance	Side Clearance
2002 (cont.)	6.5 (6473)	F	⑤	⑥	0.0039-0.0100	3	⑦	0.0018-0.0039	0.0067-0.0248
	6.6 (6599)	1	3.1459-3.1466	0.0015-0.0028	0.0016-0.0081	NA	2.4789-2.4795	0.0014-0.0030	0.0122-0.0193
	8.1 (8128)	G	2.7482-2.7489	⑨	0.0050-0.0110	NA	2.1990-2.1996	0.0008-0.0025	0.0151-0.0270
2003-04	4.3 (4293)	Z	①	②	0.0020-0.0070	4	2.2487-2.2497	0.0013-0.0035	0.0060-0.0140
	4.8 (4802)	V	2.5580-2.5593	0.0007-0.0021	0.0015-0.0078	5	2.0990-2.1000	0.0006-0.0030	0.0043-0.0200
	5.3 (5327)	T	2.5580-2.5593	0.0007-0.0021	0.0015-0.0078	5	2.0990-2.1000	0.0006-0.0030	0.0043-0.0200
	5.3 (5327)	Z	2.5580-2.5590	0.0008-0.0021	0.0015-0.0078	5	2.0991-2.0999	0.0006-0.0030	0.0043-0.2000
	6.0 (5966)	U	2.5580-2.5593	0.0007-0.0021	0.0015-0.0078	5	2.0990-2.1000	0.0006-0.0030	0.0043-0.0200
	6.0 (5966)	N	2.5580-2.5590	0.0008-0.0021	0.0015-0.0078	5	2.0991-2.0999	0.0006-0.0030	0.0043-0.2000
	6.6 (6599)	1	3.1459-3.1466	0.0015-0.0028	0.0016-0.0081	NA	2.4789-2.4795	0.0014-0.0030	0.0122-0.0193
	8.1 (8128)	G	2.7482-2.7489	⑨	0.0050-0.0110	NA	2.1990-2.1996	0.0008-0.0025	0.0151-0.0270

NA - Not Available

① No. 1: 2.4488 in.-2.4495 in.
Nos. 2, 3: 2.4485 in.-2.4494 in.
No. 4: 2.4480 in.-2.4489 in.

② No. 1: 0.0008 in.-0.0020 in.
Nos. 2, 3: 0.0011 in.-0.0023 in.
No. 4: 0.0017 in.-0.0032 in.

③ No. 1: 2.4484 in.-2.4493 in.
Nos. 2, 3, 4: 2.4481 in.-2.4490 in.
No. 5: 2.4479 in.-2.4488 in.

④ No. 1: 0.0007 in.-0.0021 in.
Nos. 2, 3, 4: 0.0009 in.-0.0024 in.
No. 5: 0.0010 in.-0.0027 in.

⑤ No. 1, 2, 3, 4: 2.9517 in.-2.9520 in. (Blue)
2.9520 in.-2.9524 in. (Orange/Red)
2.9524 in.-2.9527 in. (White)
No. 5: 2.9515 in.-2.9518 in. (Blue)
2.9518 in.-2.9522 in. (Orange/Red)
2.9522 in.-2.9525 in. (White)

⑥ No. 1, 2, 3, 4: 0.0018 in.-0.0033 in.
No. 5: 0.0022 in.-0.0037 in.

⑦ 2.399 in.-2.400 in. (Green)
2.400 in.-2.401 in. (Yellow)

⑧ No. 1: 0.0017 in.-0.0030 in.
No. 2, 3, 4: 0.0011 in.-0.0024 in.
No. 5: 0.0025 in.-0.0038 in.

⑨ No. 1, 2, 3, 4: 0.0008-0.0020 in.
No. 5: 0.0014-0.0026 in.

42372-SILV-C15

PISTON AND RING SPECIFICATIONS

All measurements are given in inches.

Year	Engine Displacement Liters (cc)	Engine ID/VIN	Piston Clearance	Ring Gap			Ring Side Clearance		
				Top Compression	Bottom Compression	Oil Control	Top Compression	Bottom Compression	Oil Control
2000	4.3 (4293)	W	0.0007-0.0017	0.010-0.030	0.018-0.026	0.065 Max.	0.0042 Max.	0.0042 Max.	0.0020-0.0070
	4.8 (4802)	V	0.0010-0.0024	0.009-0.015	0.017-0.025	0.007-0.027	0.0016-0.0033	0.0016-0.0031	0.0004-0.0087
	5.0 (4999)	M	0.0007-0.0021	0.010-0.020	0.018-0.026	0.010-0.030	0.0012-0.0032	0.0012-0.0032	0.0020-0.0070
	5.3 (5327)	T	0.0010-0.0024	0.009-0.015	0.017-0.025	0.007-0.027	0.0016-0.0033	0.0016-0.0031	0.0004-0.0087
	5.7 (5735)	R	0.0007-0.0021	0.010-0.020	0.018-0.026	0.010-0.030	0.0012-0.0032	0.0012-0.0032	0.0020-0.0070
	6.0 (5966)	U	0.0010-0.0024	0.009-0.015	0.017-0.025	0.007-0.027	0.0016-0.0033	0.0016-0.0031	0.0004-0.0087
	6.5 (6473)	F	①	0.010-0.020	0.030-0.039	0.010-0.020	0.0015-0.0031	0.0015-0.0031	0.0016-0.0035
	7.4 (7440)	J	0.0018-0.0030	0.010-0.0180	0.016-0.0240	0.010-0.030	0.0012-0.0029	0.0012-0.0029	0.0050-0.0065
2001	4.3 (4293)	W	0.0007-0.0017	0.010-0.030	0.018-0.026	0.065 Max.	0.0042 Max.	0.0042 Max.	0.0020-0.0070
	4.8 (4802)	V	0.0010-0.0024	0.009-0.015	0.017-0.025	0.007-0.027	0.0016-0.0033	0.0016-0.0031	0.0004-0.0087
	5.0 (4999)	M	0.0007-0.0021	0.010-0.020	0.018-0.026	0.010-0.030	0.0012-0.0032	0.0012-0.0032	0.0020-0.0070
	5.3 (5327)	T	0.0010-0.0024	0.009-0.015	0.017-0.025	0.007-0.027	0.0016-0.0033	0.0016-0.0031	0.0004-0.0087
	5.7 (5735)	R	0.0007-0.0021	0.010-0.020	0.018-0.026	0.010-0.030	0.0012-0.0032	0.0012-0.0032	0.0020-0.0070
	6.0 (5966)	U	0.0010-0.0024	0.009-0.015	0.017-0.025	0.007-0.027	0.0016-0.0033	0.0016-0.0031	0.0004-0.0087
	6.5 (6473)	F	①	0.010-0.020	0.030-0.039	0.010-0.020	0.0015-0.0031	0.0015-0.0031	0.0016-0.0035
	6.6 (6599)	1	0.0002-0.0007	0.012-0.018	0.020-0.026	0.006-0.014	0.0030-0.0067	0.0004-0.0012	0.0004-0.0012
	8.1 (8128)	G	②	0.012-0.018	0.017-0.025	0.010-0.030	0.0012-0.0029	0.0012-0.0029	0.002-0.008
2002	4.3 (4293)	W	0.0007-0.0017	0.010-0.030	0.018-0.026	0.065 Max.	0.0042 Max.	0.0042 Max.	0.0020-0.0070
	4.8 (4802)	V	0.0010-0.0024	0.009-0.015	0.017-0.025	0.007-0.027	0.0016-0.0033	0.0016-0.0031	0.0004-0.0087
	5.0 (4999)	M	0.0007-0.0021	0.010-0.020	0.018-0.026	0.010-0.030	0.0012-0.0032	0.0012-0.0032	0.0020-0.0070
	5.3 (5327)	T	0.0010-0.0024	0.009-0.015	0.017-0.025	0.007-0.027	0.0016-0.0033	0.0016-0.0031	0.0004-0.0087
	5.7 (5735)	R	0.0007-0.0021	0.010-0.020	0.018-0.026	0.010-0.030	0.0012-0.0032	0.0012-0.0032	0.0020-0.0070
	6.0 (5966)	U	0.0010-0.0024	0.009-0.015	0.017-0.025	0.007-0.027	0.0016-0.0033	0.0016-0.0031	0.0004-0.0087

42372-SILV-C16

PISTON AND RING SPECIFICATIONS
All measurements are given in inches.

Year	Engine Displacement Liters (cc)	Engine ID/VIN	Piston Clearance	Ring Gap			Ring Side Clearance		
				Top Compression	Bottom Compression	Oil Control	Top Compression	Bottom Compression	Oil Control
2002 (cont.)	6.5 (6473)	F	①	0.010-0.020	0.030-0.039	0.010-0.020	0.0015-0.0031	0.0015-0.0031	0.0016-0.0035
	6.6 (6599)	1	0.0002-0.0007	0.012-0.018	0.020-0.026	0.006-0.014	0.0030-0.0067	0.0004-0.0012	0.0004-0.0012
	8.1 (8128)	G	②	0.012-0.018	0.017-0.025	0.010-0.030	0.0012-0.0029	0.0012-0.0029	0.002-0.008
2003-04	4.3 (4293)	Z	0.0007-0.0017	0.010-0.030	0.018-0.026	0.065 Max.	0.0042 Max.	0.0042 Max.	0.0020-0.0070
	4.8 (4802)	V	0.0010-0.0024	0.009-0.015	0.017-0.025	0.007-0.027	0.0016-0.0033	0.0016-0.0031	0.0004-0.0087
	5.3 (5327)	T	0.0010-0.0024	0.009-0.015	0.017-0.025	0.007-0.027	0.0016-0.0033	0.0016-0.0031	0.0004-0.0087
	5.3 (5327)	Z	-0.0014 0.0006	0.009-0.002	0.017-0.027	0.007-0.029	0.0016-0.0033	0.0016-0.0031	0.0005-0.0087
	6.0 (5966)	U	0.0010-0.0024	0.009-0.015	0.017-0.025	0.007-0.027	0.0016-0.0033	0.0016-0.0031	0.0004-0.0087
	6.0 (5966)	N	-0.0009 0.0012	0.012-0.020	0.020-0.030	0.012-0.034	0.0014-0.0031	0.0013-0.0030	0.0005-0.0008
	6.6 (6599)	1	0.0002-0.0007	0.012-0.018	0.020-0.026	0.006-0.014	0.0030-0.0067	0.0004-0.0012	0.0004-0.0012
	8.1 (8128)	G	②	0.012-0.018	0.017-0.025	0.010-0.030	0.0012-0.0029	0.0012-0.0029	0.002-0.008

① 1-6: 0.0037-0.0047 in.
　7-8: 0.0042-0.0052 in.

② Interference fit (coated piston)

42372-SILV-C17

BRAKE SPECIFICATIONS
All measurements in inches unless noted

Year	Model		Brake Disc — Original Thickness	Brake Disc — Minimum Thickness	Brake Disc — Maximum Runout	Brake Drum — Original Inside Diameter	Brake Drum — Max. Wear Limit	Brake Drum — Maximum Machine Diameter	Minimum Lining Thickness	Brake Caliper — Bracket Bolts (ft. lbs.)	Brake Caliper — Mounting Bolts (ft. lbs.)
2000	C1500	F	1.250	1.230	0.004	—	—	—	0.030	NA	—
		R	—	—	—	①	②	③	0.030	NA	38
	C2500	F	1.500	1.480	0.004	—	—	—	0.030	NA	—
		R	—	—	—	①	②	③	0.030	NA	38
	C3500	F	1.500	1.480	0.004	—	—	—	0.030	NA	—
		R	—	—	—	①	②	③	0.030	NA	38
	G/P1500	F	④	⑤	0.004	—	—	—	0.030	NA	—
		R	—	—	—	①	②	③	0.030	NA	38
	G/P2500	F	④	⑤	0.004	—	—	—	0.030	NA	—
		R	—	—	—	①	②	③	0.030	NA	38
	G/P3500	F	④	⑤	0.004	—	—	—	0.030	NA	—
		R	—	—	—	①	②	③	0.030	NA	38
	K1500	F	1.500	1.480	0.004	—	—	—	0.030	NA	—
		R	—	—	—	①	②	③	0.030	NA	38
	K2500	F	1.500	1.480	0.004	—	—	—	0.030	NA	—
		R	—	—	—	①	②	③	0.030	NA	38
	K3500	F	1.500	1.480	0.004	—	—	—	0.030	NA	—
		R	—	—	—	①	②	③	0.030	NA	80
	Sierra	F	⑥	⑦	0.005	—	—	—	0.030	NA	⑩
		R	⑧	⑨	0.005	—	—	—	0.030	NA	80
	Silverado	F	⑥	⑦	0.005	—	—	—	0.030	NA	⑩
		R	⑧	⑨	0.005	—	—	—	0.030	NA	38
2001	G/P1500	F	④	⑤	0.004	—	—	—	0.030	NA	—
		R	—	—	—	①	②	③	0.030	NA	38
	G/P2500	F	④	⑤	0.004	—	—	—	0.030	NA	—
		R	—	—	—	①	②	③	0.030	NA	38
	G/P3500	F	④	⑤	0.004	—	—	—	0.030	NA	—
		R	—	—	—	①	②	③	0.030	⑪	80
	Sierra	F	⑥	⑬	0.005	—	—	—	0.030	⑫	⑩
		R	⑧	⑭	0.005	—	—	—	0.030	⑪	80
	Silverado	F	⑥	⑬	0.005	—	—	—	0.030	⑫	⑩
		R	⑧	⑭	0.005	—	—	—	0.030	NA	38
2002	G/P1500	F	④	⑤	0.004	—	—	—	0.030	NA	—
		R	—	—	—	①	②	③	0.030	NA	38
	G/P2500	F	④	⑤	0.004	—	—	—	0.030	NA	—
		R	—	—	—	①	②	③	0.030	NA	38
	G/P3500	F	④	⑤	0.004	—	—	—	0.030	NA	—
		R	—	—	—	①	②	③	0.030	⑪	80
	Sierra	F	⑥	⑬	0.005	—	—	—	0.030	⑫	⑩
		R	⑧	⑭	0.005	—	—	—	0.030	⑪	80
	Silverado	F	⑥	⑬	0.005	—	—	—	0.030	⑫	⑩
		R	⑧	⑭	0.005	—	—	—	0.030	⑪	⑩

42372-SILV-C18

BRAKE SPECIFICATIONS
All measurements in inches unless noted

Year	Model		Brake Disc Original Thickness	Brake Disc Minimum Thickness	Brake Disc Maximum Runout	Brake Drum Diameter Original Inside Diameter	Brake Drum Diameter Max. Wear Limit	Brake Drum Diameter Maximum Machine Diameter	Minimum Lining Thickness	Brake Caliper Bracket Bolts (ft. lbs.)	Brake Caliper Mounting Bolts (ft. lbs.)
2003-04	Express	F	④	⑤	0.004	—	—	—	0.030	NA	38
		R	—	—	—	①	②	③	0.030	NA	—
	Savana	F	④	⑤	0.004	—	—	—	0.030	NA	38
		R	—	—	—	①	②	③	0.030	NA	—
	Sierra	F	⑥	⑬	0.005	—	—	—	0.030	⑪	80
		R	⑧	⑭	0.005	—	—	—	0.030	⑫	⑩
	Silverado	F	⑥	⑬	0.005	—	—	—	0.030	⑪	80
		R	⑧	⑭	0.005	—	—	—	0.030	⑫	⑩

NA: Not Available

① Available with 1 in., 11.15 in. and 13 in. drums

② 10 in. drum: 10.05
11.15 in. drum: 11.24
1 in. drum: 13.09

③ 1 in. drum: 10.09
11.15 in. drum: 11.21
1 in. drum: 13.06

④ Available with 1.280 in. and 1.540 in. discs

⑤ 1.2 in. disc: 1.230
1.5 in. disc: 1.480

⑥ Vacuum: 1.14 in.
Hydraulic: 1.50 in.

⑦ Vacuum: 1.08 in.
Hydraulic: 1.44 in.

⑧ Vacuum: 0.787 in.
Hydraulic: 1.14 in.

⑨ Vacuum: 0.728 in.
Hydraulic: 1.08 in.

⑩ 15 series: 31 ft. lbs.
25/35 series: 80 ft. lbs.

⑪ 15 series: 129 ft. lbs.
25/35 series: 221 ft. lbs.

⑫ Vacuum: 148 ft. lbs.
Hydraulic (9000 lbs.): 122 ft. lbs.
Hydraulic (12,000 lbs.): 221 ft. lbs.

⑬ Vacuum: 1.10 in.
Hydraulic: 1.46 in.

⑭ Vacuum: 0.748 in.
Hydraulic: 1.10 in.

42372-SILV-C19

WHEEL ALIGNMENT

Year	Model		Caster Range (+/-Deg.)	Caster Preferred Setting (Deg.)	Camber Range (+/-Deg.)	Camber Preferred Setting (Deg.)	Toe-in (Deg.)	Steering Axis Inclination (Deg.)
2000	C1500/2500	2wd/4wd	1.00	+3.75	0.50	+0.50	0.24+/-0.20	—
	C3500	2wd/4wd	Not Adj.	Not Adj.	0.50	+1.25	0.12+/-0.12	—
	K Series below 8600 gvw	2wd/4wd	1.00	+3.00	1.00	+0.65	0.24+/-0.20	—
	K series 8600 gvw +	2wd/4wd	1.00	+3.00	1.00	+0.50	0.24+/-0.20	—
	Silverado 15 Series	2wd	1.00	L +3.75 R +4.00	0.50	+0.25	0.10+/-0.20	—
	Silverado 25 Series	2wd	1.00	L +4.50 R +4.75	0.50	+0.25	0.10+/-0.20	—
	Silverado 15 Series	4wd	1.00	+4.25	0.50	+0.25	0.10+/-0.20	—
	Silverado 25 Series	4wd	1.00	+4.25	0.50	+0.25	0.10+/-0.20	—
2001	Silverado C15 Series Reg. & Ext. Cab	2wd/4wd	1.00	L +3.75 R +4.00	1.00	+0.25	0.10+/-0.20	—
	Silverado C15 Series Crew Cab	2wd/4wd	1.00	L +4.50 R +4.75	1.00	+0.25	0.10+/-0.20	—
	Silverado K15 Series Reg. & Ext. Cab	2wd/4wd	1.00	L +3.40 R +4.25	1.00	+0.25	0.10+/-0.20	—
	Silverado K15 Series Crew Cab	2wd/4wd	1.00	L +4.20 R +4.75	1.00	+0.25	0.10+/-0.20	—
	C25 LD	2wd/4wd	1.00	L +4.50 R +4.75	1.00	+0.25	0.10+/-0.20	—
	K25 LD	2wd/4wd	1.00	L +4.25 R +4.75	1.00	+0.25	0.10+/-0.20	—
	C25 HD	2wd/4wd	1.00	L +4.25 R +4.75	1.00	+0.25	0.10+/-0.20	—
	K25 HD	2wd/4wd	1.00	L +4.00 R +4.75	1.00	+0.25	0.10+/-0.20	—
	C35 HD	2wd/4wd	1.00	L +4.25 R +4.75	1.00	+0.25	0.10+/-0.20	—
	K35 HD	2wd/4wd	1.00	L +4.00 R +4.75	1.00	+0.25	0.10+/-0.20	—
2002	Silverado C15 Series Reg. & Ext. Cab	2wd/4wd	1.00	L +3.75 R +4.00	1.00	+0.25	0.10+/-0.20	—
	Silverado C15 Series Crew Cab	2wd/4wd	1.00	L +4.50 R +4.75	1.00	+0.25	0.10+/-0.20	—
	Silverado K15 Series Reg. & Ext. Cab	2wd/4wd	1.00	L +3.40 R +4.25	1.00	+0.25	0.10+/-0.20	—
	Silverado K15 Series Crew Cab	2wd/4wd	1.00	L +4.20 R +4.75	1.00	+0.25	0.10+/-0.20	—
	C25 LD	2wd/4wd	1.00	L +4.50 R +4.75	1.00	+0.25	0.10+/-0.20	—
	K25 LD	2wd/4wd	1.00	L +4.25 R +4.75	1.00	+0.25	0.10+/-0.20	—

42372-SILV-C20

WHEEL ALIGNMENT

Year	Model		Caster Range (+/-Deg.)	Caster Preferred Setting (Deg.)	Camber Range (+/-Deg.)	Camber Preferred Setting (Deg.)	Toe-in (Deg.)	Steering Axis Inclination (Deg.)
2002 (cont.)	C25 HD	2wd/4wd	1.00	L +4.25 R +4.75	1.00	+0.25	0.10+/-0.20	—
	K25 HD	2wd/4wd	1.00	L +4.00 R +4.75	1.00	+0.25	0.10+/-0.20	—
	C35 HD	2wd/4wd	1.00	L +4.25 R +4.75	1.00	+0.25	0.10+/-0.20	—
	K35 HD	2wd/4wd	1.00	L +4.00 R +4.75	1.00	+0.25	0.10+/-0.20	—
2003-04	Silverado C15 Series Reg. & Ext. Cab	2wd/4wd	1.00	L +3.75 R +4.00	1.00	+0.25	0.10+/-0.20	—
	Silverado C15 Series Crew Cab	2wd/4wd	1.00	L +4.50 R +4.75	1.00	+0.25	0.10+/-0.20	—
	Silverado K15 Series Reg. & Ext. Cab	2wd/4wd	1.00	L +3.40 R +4.25	1.00	+0.25	0.10+/-0.20	—
	Silverado K15 Series Crew Cab	2wd/4wd	1.00	L +4.20 R +4.75	1.00	+0.25	0.10+/-0.20	—
	C25 LD	2wd/4wd	1.00	L +4.50 R +4.75	1.00	+0.25	0.10+/-0.20	—
	K25 LD	2wd/4wd	1.00	L +4.25 R +4.75	1.00	+0.25	0.10+/-0.20	—
	C25 HD	2wd/4wd	1.00	L +4.25 R +4.75	1.00	+0.25	0.10+/-0.20	—
	K25 HD	2wd/4wd	1.00	L +4.00 R +4.75	1.00	+0.25	0.10+/-0.20	—
	C35 HD	2wd/4wd	1.00	L +4.25 R +4.75	1.00	+0.25	0.10+/-0.20	—
	K35 HD	2wd/4wd	1.00	L +4.00 R +4.75	1.00	+0.25	0.10+/-0.20	—

42372-SILV-C21

TIRE, WHEEL AND BALL JOINT SPECIFICATIONS

| Year | Model | OEM Tires | | Tire Pressures (psi) | | Wheel Size | Ball Joint Inspection |
		Standard	Optional	Front	Rear		
2000	1500 PU 2wd	P235/75R15	None	36	36	6-JJ	L ①
	1500 PU 4wd	P245/75R16	None	36	36	7-JJ	L ①
	2500 PU	LT225/75R16D	LT245/75R16C	36	36	7-JJ	L ①
			LT245/75R16E	36	36		
	3500 PU SRW	LT245/75R16E	None	36	36	7-JJ	0.125 in.②
	3500 PU DRW	LT225/75R16D	LT215/85R16D	36	36	7-JJ	0.125 in.②
2001	1500 PU 2wd	P235/75R15	None	36	36	6-JJ	L ①
	1500 PU 4wd	P245/75R16	None	36	36	7-JJ	L ①
	2500 PU	LT225/75R16D	LT245/75R16C	36	36	7-JJ	L ①
			LT245/75R16E	36	36		
	3500 PU SRW	LT245/75R16E	None	36	36	7-JJ	0.125 in.②
	3500 PU DRW	LT225/75R16D	LT215/85R16D	36	36	7-JJ	0.125 in.②
2002	1500 PU 2wd	P235/75R15	None	36	36	6-JJ	L ①
	1500 PU 4wd	P245/75R16	None	36	36	7-JJ	L ①
	2500 PU	LT225/75R16D	LT245/75R16C	36	36	7-JJ	L ①
			LT245/75R16E	36	36		
	3500 PU SRW	LT245/75R16E	None	36	36	7-JJ	0.125 in.②
	3500 PU DRW	LT225/75R16D	LT215/85R16D	36	36	7-JJ	0.125 in.②
2003-04	1500 PU 2wd	P235/75R15	None	36	36	6-JJ	L ①
	1500 PU 4wd	P245/75R16	None	36	36	7-JJ	L ①
	2500 PU	LT225/75R16D	LT245/75R16C	36	36	7-JJ	L ①
			LT245/75R16E	36	36		
	3500 PU SRW	LT245/75R16E	None	36	36	7-JJ	0.125 in.②
	3500 PU DRW	LT225/75R16D	LT215/85R16D	36	36	7-JJ	0.125 in.②

OEM: Original Equipment Manufacturer

PSI: Pounds Per Square Inch

STD: Standard

OPT: Optional

L: Lower

U: Upper

① Do not lift truck. Inspect the boss into which the grease fitting is threaded. Replace if the boss is flush or receded below the surface of the ball joint

② Applies to both upper and lower

42372-SILV-C22

SCHEDULED MAINTENANCE INTERVALS
GENERAL MOTORS 2000 CHEVROLET—SILVERADO & GMC—SIERRA PICK-UP—GASOLINE

TO BE SERVICED	TYPE OF SERVICE	7.5	15	22.5	30	37.5	45	52.5	60	67.5	75	82.5	90	97.5	105	112.5	120
		VEHICLE MILEAGE INTERVAL (x1000)															
Accessory drive belt	S/I								✓								✓
Automatic transmission fluid ①	R	Every 50,000 miles															
Brake system	S/I	✓	✓	✓	✓	✓	✓	✓	✓	✓	✓	✓	✓	✓	✓	✓	✓
Chassis & suspension grease points	L	✓	✓	✓	✓	✓	✓	✓	✓	✓	✓	✓	✓	✓	✓	✓	✓
Cooling fan operation	S/I		✓		✓		✓		✓		✓		✓		✓		✓
CV-joint boots & axle seals	S/I	✓	✓	✓	✓	✓	✓	✓	✓	✓	✓	✓	✓	✓	✓	✓	✓
EGR system	S/I								✓								✓
Engine coolant	R	Every 150,000 miles															
Engine oil & filter	R	✓	✓	✓	✓	✓	✓	✓	✓	✓	✓	✓	✓	✓	✓	✓	✓
EVAP system	S/I								✓								✓
Front wheel bearings ②	S/I & L			✓					✓				✓				✓
Fuel filter	R								✓								✓
Fuel system	S/I								✓								✓
Rear/front axle fluid level	S/I	✓	✓	✓	✓	✓	✓	✓	✓	✓	✓	✓	✓	✓	✓	✓	✓
Rotate tires	S/I	✓	✓	✓	✓	✓	✓	✓	✓	✓	✓	✓	✓	✓	✓	✓	✓
Shields & underhood insulation ①	S/I		✓		✓		✓		✓		✓		✓				✓
Spark plugs	R	Every 100,000 miles															
Spark plug wires	S/I	Every 100,000 miles															

R: Replace   S/I: Inspect and service, if necessary   L: Lubricate

① Vehicles with a GVWR or 8500 lbs. or more only.

② 2-wheel drive models only.

FREQUENT OPERATION MAINTENANCE (SEVERE SERVICE)

If a vehicle is operated under any of the following conditions it is considered severe service:

- Towing a trailer or using a camper or car-top carrier.
- Repeated short trips of less than 5 miles in temperatures below freezing, or trips of less than 10 miles in any temperature.
- Extensive idling or low-speed driving for long distances as in heavy commercial use, such as delivery, taxi or police cars.
- Operating on rough, muddy or salt-covered roads.
- Operating on unpaved or dusty roads.
- Driving in extremely hot (over 90°) conditions.

Engine oil & filter: replace every 3000 miles or 3 months, whichever occurs first.

Chassis and suspension grease points: lubricate every 3000 miles.

Rear/front axle fluid level: inspect every 3000 miles.

Rotate the tires ever 6000 miles.

Brake system components: inspect ever 6000 miles.

Front wheel bearings (2-wheel drive only): clean, inspect and repack every 15,000 miles.

Shields & underhood insulation (vehicles w/GVWR over 8500 lbs. only): inspect every 15,000 miles.

Cooling fan system hoses & connections: inspect every 15,000 miles.

Fuel filter: replace every 30,000 miles.

Air cleaner filter: inspect every 45,000 miles.

Automatic transmission fluid & filter: replace every 50,000 miles.

Accessory drive belt: inspect every 60,000 miles.

Fuel system tank, cap and lines: inspect every 60,000 miles.

EVAP system: inspect every 60,000 miles.

EGR system: inspect every 60,000 miles.

PCV system: inspect every 100,000 miles.

Engine cooling system components: inspect and clean every 150,000 miles.

42372-SILV-C23

SCHEDULED MAINTENANCE INTERVALS
GENERAL MOTORS 2001-03 CHEVROLET—SILVERADO & GMC—SIERRA

TO BE SERVICED	TYPE OF SERVICE	VEHICLE MILEAGE INTERVAL (x1000)														
		7.5	15	22.5	30	37.5	45	52.5	60	67.5	75	82.5	90	97.5	100	150
Engine oil & filter	R	✓	✓	✓	✓	✓	✓	✓	✓	✓	✓	✓	✓	✓		
Chassis lubrication	S/I	✓	✓	✓	✓	✓	✓	✓	✓	✓	✓	✓	✓	✓		
Oil Life Monitor	S/I	✓	✓	✓	✓	✓	✓	✓	✓	✓	✓	✓	✓	✓		
Front/Rear Axle Fluid	S/I ①	✓	✓	✓	✓	✓	✓	✓	✓	✓	✓	✓	✓	✓		
CV joints & axle seals	S/I	✓	✓	✓	✓	✓	✓	✓	✓	✓	✓	✓	✓	✓		
Rotate tires	S/I	✓	✓	✓	✓	✓	✓	✓	✓	✓	✓	✓	✓	✓		
Passenger Compartment Air Filter	R		✓		✓		✓		✓		✓		✓			
Underhood Sound Shield	S/I	✓		✓			✓			✓		✓		✓		
Fuel filter	R				✓				✓				✓			
Automatic transmission fluid & filter	R ②														✓	
Engine accessory drive belt	S/I								✓							
Fuel Tank, cap and lines	S/I								✓							
EGR System	S/I								✓							
EVAP System	S/I								✓							
Spark plugs	R														✓	
Spark Plug Wires	S/I														✓	
PCV Valve	S/I														✓	
Coolant	R															✓
Air cleaner filter	R				✓				✓				✓			

R: Replace   S/I: Service or Inspect

① If the vehicle is used for continuous trailer towing, change the fluid in the rear axle after the first 500 miles, then, every 7,500 miles

② Vehicles over 8,600 lbs. GVWR: every 50,000 miles

FREQUENT OPERATION MAINTENANCE (SEVERE SERVICE)

If a vehicle is operated under any of the following conditions it is considered severe service:

- Extremely dusty areas.
- 50% or more of the vehicle operation is in 32°C (90°F) or higher temperatures, or constant operation in temperatures below 0°C (32°F).
- Prolonged idling (vehicle operation in stop and go traffic.
- Frequent short running periods (engine does not warm to normal operating temperatures).
- Police, taxi, delivery usage or trailer towing usage.

Oil & oil filter change: change every 3000 miles

Lubricate chassis every 3000 miles

Drive axle: check every 3000 miles

Rotate tires every 6000 miles

Air cleaner filter: change every 24,000 miles

42372-SILV-C24

SCHEDULED MAINTENANCE INTERVALS
GENERAL MOTORS CHEVROLET—SIERRA & GMC—SILVERADO—DIESEL

TO BE SERVICED	TYPE OF SERVICE	5	8	10	15	20	23	25	30	35	38	40	45	50	53	55	60	65	68	70	75	80	83	85	90	95	98
Air intake system	S/I			✓		✓			✓			✓		✓			✓			✓		✓			✓		
Automatic transmission fluid ①	R	Every 50,000 miles																									
Brake system	S/I	✓	✓		✓		✓		✓		✓		✓		✓		✓		✓		✓		✓		✓		✓
Chassis & suspension grease points	L	✓		✓	✓	✓		✓	✓	✓		✓	✓	✓		✓	✓	✓		✓	✓	✓		✓	✓	✓	
Cooling fan operation	S/I			✓		✓			✓			✓		✓			✓			✓		✓			✓		
Crankcase depression regular valve system hoses	S/I															✓											
CV-joint boots & axle seals	S/I	✓		✓	✓	✓		✓	✓	✓		✓	✓	✓		✓	✓	✓		✓	✓	✓		✓	✓	✓	
EGR system ②	S/I															✓									✓		
Engine coolant	R	Every 150,000 miles																									
Engine cooling system hoses & radiator	S/I & C	Initially at 100,000 miles, then every 50,000 miles																									
Engine oil & filter	R	✓		✓	✓	✓		✓	✓	✓		✓	✓	✓		✓	✓	✓		✓	✓	✓		✓	✓	✓	
Front wheel bearings ③	S/I & L								✓							✓									✓		
Fuel filter	R								✓							✓									✓		
Rear/front axle fluid level	S/I	✓		✓	✓	✓		✓	✓	✓		✓	✓	✓		✓	✓	✓		✓	✓	✓		✓	✓	✓	
Rotate tires	S/I	✓	✓		✓		✓		✓		✓		✓		✓		✓		✓		✓		✓		✓		✓
Shields & underhood insulation ①	S/I			✓		✓			✓			✓		✓			✓			✓		✓			✓		

R: Replace   S/I: Inspect and service, if necessary   L: Lubricate   C: Clean
① Vehicles with a GVWR of 8500 lbs or more only.
② If equipped.
③ 2-wheel drive models only.

FREQUENT OPERATION MAINTENANCE (SEVERE SERVICE)

If a vehicle is operated under any of the following conditions it is considered severe service:
- Towing a trailer or using a camper or car-top carrier.
- Repeated short trips of less than 5 miles in temperatures below freezing, or trips of less than 10 miles in any temperature.
- Extensive idling or low-speed driving for long distances as in heavy commercial use, such as delivery, taxi or police cars.
- Operating on rough, muddy or salt-covered roads.
- Operating on unpaved or dusty roads.
- Driving in extremely hot (over 90°) conditions.

Engine oil & filter: replace every 2500 miles.

Chassis and suspension grease points: lubricate every 2500 miles

Rear/front axle fluid level: inspect initially at 5000 miles, then every 2500 miles thereafter.

Rotate tires: every 7500 miles.

Air cleaner filter: inspect every 15,000 miles.

Front wheel bearings (2-wheel drive only): clean, inspect and repack every 15,000 miles.

42372-SILV-C25

SCHEDULED MAINTENANCE INTERVALS
GENERAL MOTORS—G SERIES VAN, EXPRESS & SAVANA—GASOLINE

TO BE SERVICED	TYPE OF SERVICE	7.5	15	22.5	30	37.5	45	52.5	60	67.5	75	82.5	90	97.5	105	112.5	120
Accessory drive belt	S/I								✓								✓
Air cleaner filter	R				✓				✓				✓				✓
Automatic transmission fluid ①	R	Every 50,000 miles														
Chassis & suspension grease points	L	✓	✓	✓	✓	✓	✓	✓	✓	✓	✓	✓	✓	✓	✓	✓	✓
CV-joint boots & axle seals	S/I	✓	✓	✓	✓	✓	✓	✓	✓	✓	✓	✓	✓	✓	✓	✓	✓
EGR system	S/I								✓								✓
Engine coolant	R	Every 150,000 miles														
Engine oil & filter	R	✓	✓	✓	✓	✓	✓	✓	✓	✓	✓	✓	✓	✓	✓	✓	✓
EVAP system	S/I								✓								✓
Front wheel bearings	S/I & L				✓								✓				✓
Fuel filter	R				✓								✓				✓
Fuel system	S/I								✓								✓
PCV system	S/I	Every 100,000 miles														
Rear axle fluid level	S/I	✓	✓	✓	✓	✓	✓	✓	✓	✓	✓	✓	✓	✓	✓	✓	✓
Rotate tires	S/I	✓	✓	✓	✓	✓	✓	✓	✓	✓	✓	✓	✓	✓	✓	✓	✓
Shields & underhood insulation ①	S/I		✓		✓		✓		✓		✓		✓		✓		✓
Spark plugs	R	Every 100,000 miles														
Spark plug wires	S/I	Every 100,000 miles														

R: Replace   S/I: Inspect and service, if necessary   L: Lubricate

① Vehicles with a GVWR or 8500 lbs. or more only.

FREQUENT OPERATION MAINTENANCE (SEVERE SERVICE)

If a vehicle is operated under any of the following conditions it is considered severe service:

- Towing a trailer or using a camper or car-top carrier.
- Repeated short trips of less than 5 miles in temperatures below freezing, or trips of less than 10 miles in any temperature.
- Extensive idling or low-speed driving for long distances as in heavy commercial use, such as delivery, taxi or police cars.
- Operating on rough, muddy or salt-covered roads.
- Operating on unpaved or dusty roads.
- Driving in extremely hot (over 90°) conditions.

Engine oil & filter: replace every 3000 miles or 3 months, whichever occurs first.

Chassis and suspension grease points: lubricate every 3000 miles.

Rear/front axle fluid level: inspect every 3000 miles.

Rotate the tires ever 6000 miles.

Brake system components: inspect ever 6000 miles.

Front wheel bearings (2-wheel drive only): clean, inspect and repack every 15,000 miles.

Shields & underhood insulation (vehicles w/GVWR over 8500 lbs. Only): inspect every 15,000 miles

Cooling fan system hoses & connections: inspect every 15,000 miles.

Fuel filter: replace every 30,000 miles.

Air cleaner filter: inspect every 45,000 miles.

Automatic transmission fluid & filter: replace every 50,000 miles.

Accessory drive belt: inspect every 60,000 miles.

Fuel system tank, cap and lines: inspect every 60,000 miles.

EVAP system: inspect every 60,000 miles.

EGR system: inspect every 60,000 miles.

PCV system: inspect every 100,000 miles.

Engine cooling system components: inspect and clean every 150,000 miles.

42372-SILV-C26

SCHEDULED MAINTENANCE INTERVALS
GENERAL MOTORS—G SERIES VAN, EXPRESS & SAVANA—DIESEL

TO BE SERVICED	TYPE OF SERVICE	5	10	15	20	25	30	35	40	45	50	55	60	65	70	75	80	85	90	95	100	105	110	115	120
																									VEHICLE MILEAGE INTERVAL (x1000)
Air cleaner filter	R						✓						✓						✓						✓
Air intake system	S/I		✓		✓		✓		✓		✓		✓		✓		✓		✓		✓		✓		✓
Automatic transmission fluid ①	R										✓										✓				
Chassis & suspension grease points	L	✓	✓	✓	✓	✓	✓	✓	✓	✓	✓	✓	✓	✓	✓	✓	✓	✓	✓	✓	✓	✓	✓	✓	✓
Cooling fan, ducts & hoses	S/I												✓												✓
Crankcase depression regulator valve system hoses	S/I												✓												✓
CV-joint boots & axle seals	S/I	✓	✓	✓	✓	✓	✓	✓	✓	✓	✓	✓	✓	✓	✓	✓	✓	✓	✓	✓	✓	✓	✓	✓	✓
EGR system ②	S/I												✓												
Engine coolant	R	Every 100,000 miles																							
Engine cooling system hoses & radiator	S/I & C	Initially at 100,000 miles, then every 50,000 miles																							
Engine oil & filter ③	R	✓	✓	✓	✓	✓	✓	✓	✓	✓	✓	✓	✓	✓	✓	✓	✓	✓	✓	✓	✓	✓	✓	✓	✓
Front wheel bearings	S/I & L						✓						✓						✓						✓
Fuel filter	R												✓												✓
Rear axle fluid level	S/I	✓	✓	✓	✓	✓	✓	✓	✓	✓	✓	✓	✓	✓	✓	✓	✓	✓	✓	✓	✓	✓	✓	✓	✓
Rotate tires	S/I	✓	✓	✓	✓	✓	✓	✓	✓	✓	✓	✓	✓	✓	✓	✓	✓	✓	✓	✓	✓	✓	✓	✓	✓
Shields & underhood insulation	S/I						✓						✓						✓						✓

R: Replace   S/I: Inspect and service, if necessary   L: Lubricate   C: Clean

① For vehicles with a GVWR of 8500 lbs. or more.

② If equipped.

③ Perform at the mileage specified or every 3 months, whichever occurs first.

FREQUENT OPERATION MAINTENANCE (SEVERE SERVICE)

If a vehicle is operated under any of the following conditions it is considered severe service:

- Towing a trailer or using a camper or car-top carrier.
- Repeated short trips of less than 5 miles in temperatures below freezing, or trips of less than 10 miles in any temperature.
- Extensive idling or low-speed driving for long distances as in heavy commercial use, such as delivery, taxi or police cars.
- Operating on rough, muddy or salt-covered roads.
- Operating on unpaved or dusty roads.
- Driving in extremely hot (over 90°) conditions.

Engine oil & filter: replace every 2500 miles.

Chassis and suspension grease points: lubricate every 2500 miles.

Rear axle fluid level: inspect every 2500 miles.

CV-joint boots and axle seals: inspect for leakage every 2500 miles.

Rotate tires: every 7500 miles.

Air cleaner filter: inspect every 15,000 miles.

Front wheel bearings (2-wheel drive only): clean, inspect and repack every 15,000 miles.

Automatic transmission fluid & filter: replace every 50,000 miles.

42372-SILV-C27

PRECAUTIONS

Before servicing any vehicle, please be sure to read all of the following precautions, which deal with personal safety, prevention of component damage, and important points to take into consideration when servicing a motor vehicle:

• Never open, service or drain the radiator or cooling system when the engine is hot; serious burns can occur from the steam and hot coolant.

• Observe all applicable safety precautions when working around fuel. Whenever servicing the fuel system, always work in a well-ventilated area. Do not allow fuel spray or vapors to come in contact with a spark, open flame, or excessive heat (a hot drop light, for example) Keep a dry chemical fire extinguisher near the work area. Always keep fuel in a container specifically designed for fuel storage; also, always properly seal fuel containers to avoid the possibility of fire or explosion. Refer to the additional fuel system precautions later in this section.

• Fuel injection systems often remain pressurized, even after the engine has been turned **OFF**. The fuel system pressure must be relieved before disconnecting any fuel lines. Failure to do so may result in fire and/or personal injury.

• Brake fluid often contains polyglycol ethers and polyglycols. Avoid contact with the eyes and wash your hands thoroughly after handling brake fluid. If you do get brake fluid in your eyes, flush your eyes with clean, running water for 15 minutes. If eye irritation persists, or if you have taken brake fluid internally, IMMEDIATELY seek medical assistance.

• The EPA warns that prolonged contact with used engine oil may cause a number of skin disorders, including cancer! You should make every effort to minimize your exposure to used engine oil. Protective gloves should be worn when changing oil. Wash your hands and any other exposed skin areas as soon as possible after exposure to used engine oil. Soap and water, or waterless hand cleaner should be used.

• All new vehicles are now equipped with an air bag system. The system must be disabled before performing service on or around system components, steering column, instrument panel components, wiring and sensors. Failure to follow safety and disabling procedures could result in accidental air bag deployment, possible personal injury and unnecessary system repairs.

• Always wear safety goggles when working with, or around, the air bag system. When carrying a non-deployed air bag, be sure the bag and trim cover are pointed away from your body. When placing a non-deployed air bag on a work surface, always face the bag and trim cover upward, away from the surface. This will reduce the motion of the module if it is accidentally deployed. Refer to the additional air bag system precautions later in this section.

• Clean, high quality brake fluid from a sealed container is essential to the safe and proper operation of the brake system. You should always buy the correct type of brake fluid for your vehicle. If the brake fluid becomes contaminated, completely flush the system with new fluid. Never reuse any brake fluid. Any brake fluid that is removed from the system should be discarded. Also, do not allow any brake fluid to come in contact with a painted surface; it will damage the paint.

• Never operate the engine without the proper amount and type of engine oil; doing so WILL result in severe engine damage.

• Timing belt maintenance is extremely important! Many models utilize an interference-type, non-freewheeling engine. If the timing belt breaks, the valves in the cylinder head may strike the pistons, causing potentially serious (also time-consuming and expensive) engine damage. Refer to the maintenance interval charts in the front of this section for the recommended replacement interval for the timing belt, and to the timing belt procedure in this section for belt replacement and inspection.

• Disconnecting the negative battery cable on some vehicles may interfere with the functions of the on-board computer system(s) and may require the computer to undergo a relearning process once the negative battery cable is reconnected.

• When servicing drum brakes, only disassemble and assemble one side at a time, leaving the remaining side intact for reference.

GASOLINE ENGINE REPAIR

Distributor

REMOVAL

4.3L, 5.0L, 5.7L and 7.4L Engines

1. Before servicing the vehicle, refer to the precautions in the beginning of this section.
2. Remove or disconnect the following:
   • Negative battery cable
   • Spark plug wires and the coil leads from the distributor
   • Electrical connector at the base of the distributor
   • Distributor cap
3. Matchmark the rotor-to-housing and housing-to-engine block positions so that they can be matched during installation.

   • Distributor hold-down bolt
   • Distributor from the engine

4.8L, 5.3L and 6.0L Engines

➡ If the Malfunction Indicator Lamp turns on, and a DTC code P1345 sets after installing the distributor, this indicates an incorrectly installed distributor. Engine damage or distributor damage may occur.

1. Before servicing the vehicle, refer to the precautions in the beginning of this section.
2. Turn **OFF** the ignition switch.
3. Remove or disconnect the following:
   • Spark plug wires from the distributor cap
   • Electrical connector from the base of the distributor
   • Two screws that hold the distributor cap to the housing. Discard the screws.
   • Distributor cap from the housing
4. Use a grease pencil in order to note the position of the rotor in relation to the distributor housing.
5. Mark the distributor housing and the intake manifold with a grease pencil.
6. Remove or disconnect the following:
   • Mounting clamp hold-down bolt
   • Distributor
7. As the distributor is being removed from the engine, watch the rotor move in a counterclockwise direction about 42 degrees. This will appear as slightly more than the 1 o'clock position. Note the position of the rotor segment. Place a second mark on the base of the distributor. This will aid in achieving proper rotor alignment during the distributor installation.

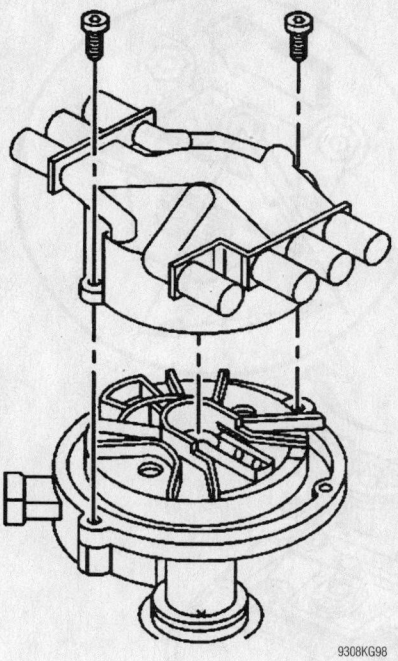

Distributor cap—4.8L, 5.3L and 6.0L engines

INSTALLATION

4.3L, 5.0L, 5.7L and 7.4L Engines

TIMING NOT DISTURBED

1. Install or connect the following:
   - Distributor, aligning the match-marks are properly alignment
   - Distributor hold-down bolt
   - Distributor cap
   - Electrical connector at the base of the distributor
   - Spark plug wires and coil leads
   - Negative battery cable

TIMING DISTURBED

1. Remove the No. 1 cylinder spark plug. Turn the engine using a socket wrench on the large bolt on the front of the crankshaft pulley. Place a finger near the No. 1 spark plug hole and turn the crankshaft until the piston reaches Top Dead Center (TDC). As the engine approaches TDC, you will feel air being expelled through the No. 1 cylinder spark plug hole. The timing mark on the crankshaft pulley should now be aligned with the **0** mark on the timing scale. If the position is not being met, turn the engine another full turn (360 degrees) Once the engines position is correct, install the spark plug.

➡ **Before installation, position the rotor so it points to the No. 2 terminal on the cap. As the distributor is lowered into the engine, the rotor will rotate clockwise and stop at the No. 1 terminal. This is the desired position.**

2. Turn the rotor so that it will point to the No. 1 terminal of the distributor cap when it is fully seated in the engine.
3. Install or connect the following:
   - Distributor. It may be necessary to turn the rotor a little in either direction, in order to engage the gears.

➡ **If the distributor will not seat completely in the engine, remove the distributor and align the groove on the top of the oil pump drive shaft with a long screwdriver to match the tab on the bottom of the distributor shaft. Reinstall the distributor.**

4. Tap the starter a few times to ensure that the oil pump shaft is mated to the distributor shaft.
5. Bring the engine to TDC again and check that the rotor is pointed toward the No. 1 terminal of the cap. If the marks are all aligned.
6. Install or connect the following:
   - Hold-down bolt and tighten
   - Cap and fasten the mounting screws
   - Electrical connections and the spark plug wires

4.8L, 5.3L and 6.0L Engines

TIMING NOT DISTURBED

1. If installing a new distributor assembly, place two marks on the new distributor housing in the same location as the two marks on the original housing. Remove the new distributor cap, if necessary. Align the rotor with mark made at location 2.
2. Guide the distributor into the engine. Align the hole in the distributor hold-down base over the mounting hole in the intake manifold.
3. As the distributor is being installed, observe the rotor moving in a clockwise direction about 42 degrees. Once the distributor is completely seated, the rotor segment should be aligned with the mark on the distributor base in location number 1. If the rotor segment is not aligned with the number 1 mark, the driven gear teeth and the camshaft have meshed one or more teeth out of alignment.
4. Install or connect:
   - Distributor mounting clamp bolt and tighten to 18 ft. lbs. (25 Nm)

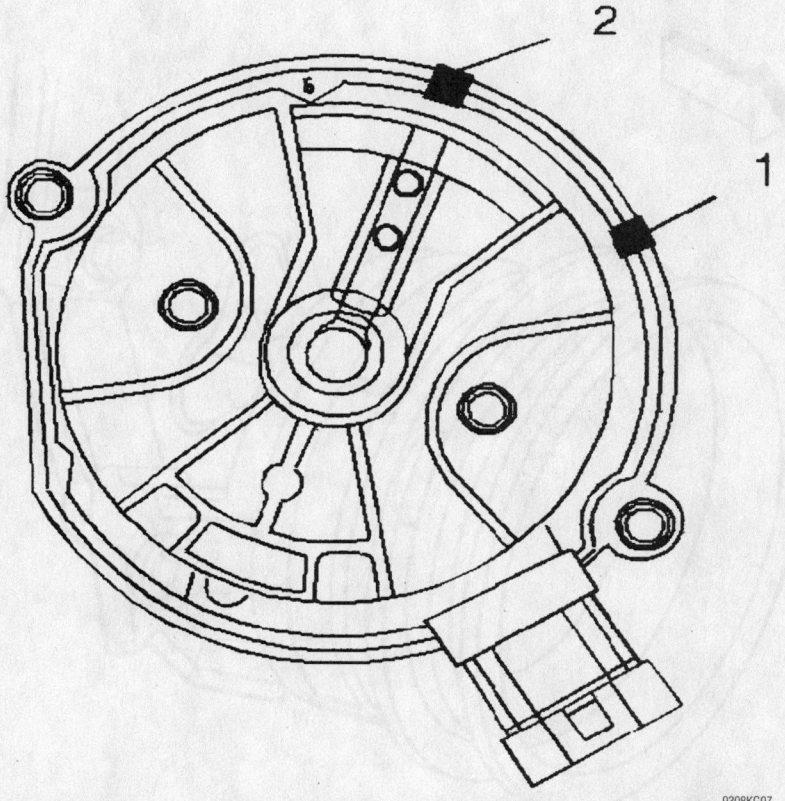

Distributor rotor starting point (1) and 42 degrees counterclockwise (2)—4.8L, 5.3L and 6.0L engines

5. Distributor cap with NEW distributor cap screws. Tighten to 21 inch lbs. (2.4 Nm).

- Electrical connector
- Spark plug wires
- Ignition coil wire

➡ **If the Malfunction Indicator lamp is turned on after installing the distributor, and a DTC P1345 is found, the distributor has been installed incorrectly.**

TIMING DISTURBED

1. Rotate the number 1 cylinder to TDC of the compression stroke. The engine front cover has 2 alignment tabs and the crankshaft balancer has 2 alignment marks (spaced 90 degrees apart) which are used for positioning number 1 piston at top dead center (TDC). With the piston on the compression stroke and at top dead center, the crankshaft balancer alignment mark must align with the engine front cover tab and the crankshaft balancer alignment mark must align with the engine front cover tab.

2. Align the white paint mark on the bottom stem of the distributor with the pre-drilled indent hole in the bottom of the gear. If the driven gear is installed incorrectly, the dimple will be approximately 180 degrees opposite of the rotor segment when it is installed in the distributor.

The OBD II ignition system distributor driven gear and rotor may be installed in multiple positions. In order to avoid mistakes, mark the distributor on the following components in order to ensure the same mounting position upon reassembly:

- The distributor driven gear
- The distributor shaft
- The rotor holes

Installing the driven gear 180 degrees out of alignment, or locating the rotor in the wrong holes, will cause a no-start condition. Premature engine wear or damage may result.

3. Using a long screwdriver, align the oil pump drive shaft to the drive tab of the distributor. Guide the distributor into the engine. Ensure that the spark plug towers are perpendicular to the centerline of the engine.

Once the distributor is fully seated, the rotor segment should be aligned with the pointer cast into the distributor base.

This pointer may have a 6 cast into it, indicating that the distributor is to be used on a 6 cylinder engine or a 8 cast into it, indicating that the distributor is to be used on a 8 cylinder engine.

If the rotor segment does not come within a few degrees of the pointer, the gear

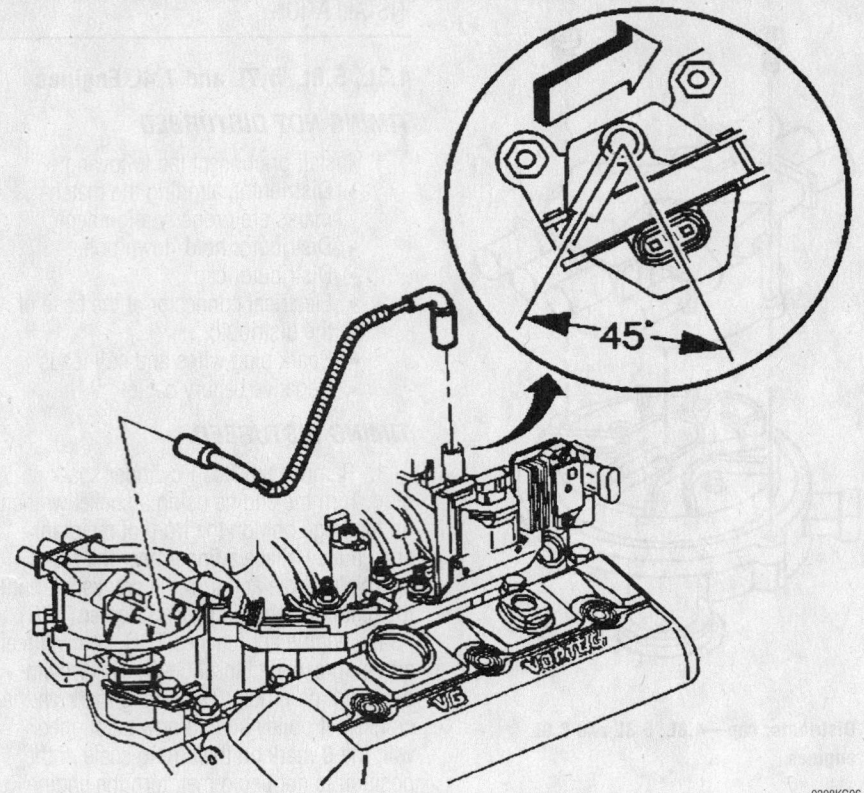

Distributor electrical connection—4.8L, 5.3L and 6.0L engines

9308KG96

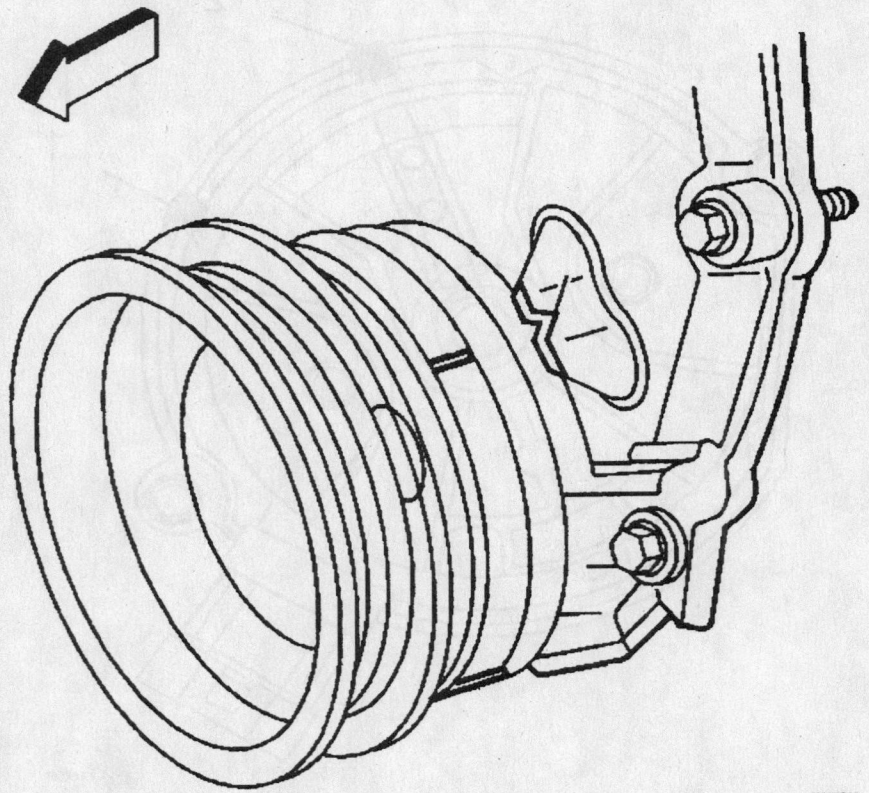

Engine at TDC compression—4.8L, 5.3L and 6.0L engines

9308KG95

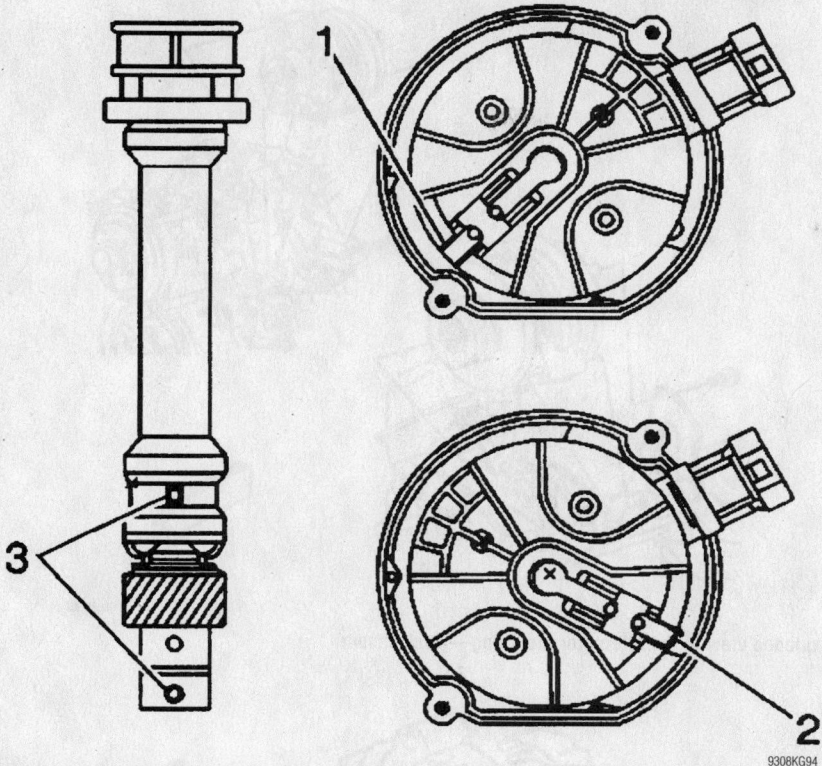

Distributor alignment. 1 is the starting point; 2 is installed; 3 are the shaft alignment marks—4.8L, 5.3L and 6.0L engines

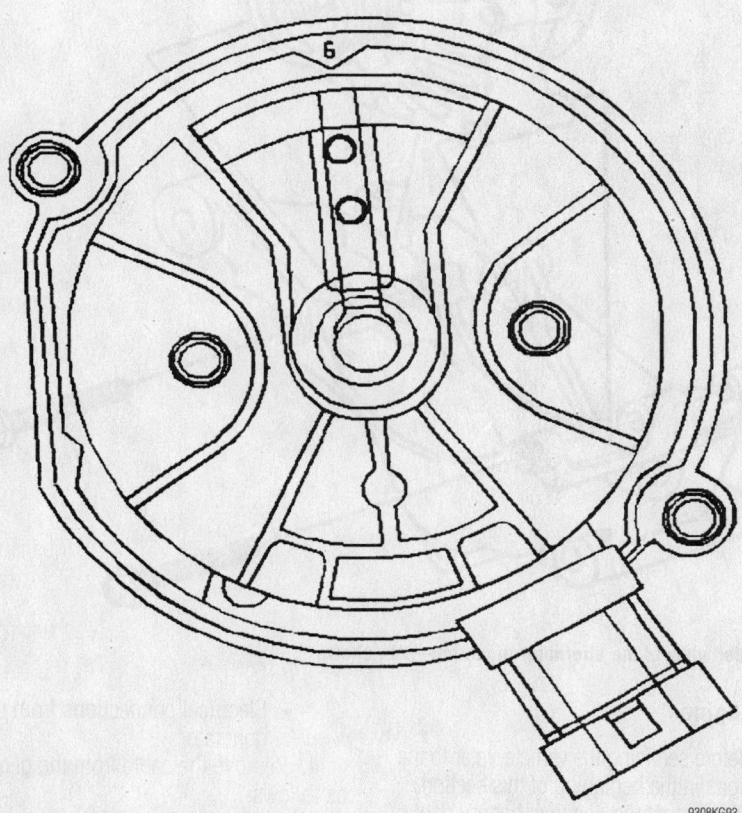

Distributor fully seated—4.8L, 5.3L and 6.0L

mesh between the distributor and the camshaft may be off a tooth or more.

If this is the case, repeat the procedure again in order to achieve proper alignment.

➡**Use the correct fastener in the correct location. Replacement fasteners must be the correct part number for that application. Fasteners requiring replacement or fasteners requiring the use of thread locking compound or sealant are identified in the service procedure. Do not use paints, lubricants, or corrosion inhibitors on fasteners or fastener joint surfaces unless specified. These coatings affect fastener torque and joint clamping force and may damage the fastener. Use the correct tightening sequence and specifications when installing fasteners in order to avoid damage to parts and systems.**

4. Install or connect the following:
   • Distributor mounting clamp bolt and tighten to 18 ft. lbs. (25 Nm)
   • Distributor cap with NEW distributor cap screws. Tighten to 21 inch lbs. (2.4 Nm).
   • Electrical connector
   • Spark plug wires
   • Ignition coil wire.

➡**If the Malfunction Indicator lamp is turned on after installing the distributor, and a DTC P1345 is found, the distributor has been installed incorrectly.**

Alternator

REMOVAL & INSTALLATION

4.3L, 5.0L, 5.7L and 7.4L Engines

SILVERADO & SIERRA

1. Before servicing the vehicle, refer to the precautions in the beginning of this section.
2. Remove or disconnect the following:
   • Negative battery cable
   • Wires
   • Accessory belt(s)
   • Mounting bracket, if necessary
   • Alternator

To install:
3. Install or connect the following:
   • Alternator
   • Mounting bracket. Torque the bolts to 18 ft lbs. (25 Nm).
   • Mounting bolts. Torque the right bolt to 18 ft lbs. (25 Nm) and left bolt to 37 ft lbs. (50 Nm).
   • Accessory belt(s)

- Wires. Torque the battery feed wire to 71 inch lbs. (8 Nm).
- Negative battery cable

EXPRESS AND SAVANA

1. Before servicing the vehicle, refer to the precautions in the beginning of this section.
2. Remove or disconnect the following:
   - Negative battery cable
   - Coolant reservoir
   - Air cleaner
   - Upper fan shroud
   - Accessory belt(s)
   - Heater hose pipe
   - Oilfill tube from bracket
   - Support bracket
   - Mounting bracket
   - Mounting bolts
   - Alternator
   - Wires

To install:

3. Install or connect the following:
   - Wires. Torque the battery feed wire to 15 ft lbs. (20 Nm).
   - Alternator
   - Mounting bolts. Torque the front bolt to 37 ft lbs. (50 Nm) and the rear bolt to 18 ft lbs. (25 Nm).
   - Oilfill tube support bracket
   - Oilfill tube to bracket
   - Accessory belt(s)
   - Upper fan shroud
   - Air cleaner
   - Coolant reservoir
   - Negative battery cable

4.8L, 5.3L and 6.0L Engines

1. Before servicing the vehicle, refer to the precautions in the beginning of this section.
2. Disconnect the negative battery cable.
3. Remove or disconnect the following:
   - Accessory drive belt
   - Engine sight shield, if necessary
   - Electrical connections from the generator
   - Mounting bolts
   - Generator

To install:

4. Install the generator.
5. Install or connect the following:
   - Generator mounting bolts. Tighten the bolts to 37 ft. lbs. (50 Nm).
   - Electrical connections to the generator. Tighten the B+ nut to 13 ft. lbs. (18 Nm).
   - Engine sight shield, if removed
   - Accessory drive belt
6. Connect the negative battery cable. Tighten to bolt to 13 ft. lbs. (17 Nm).

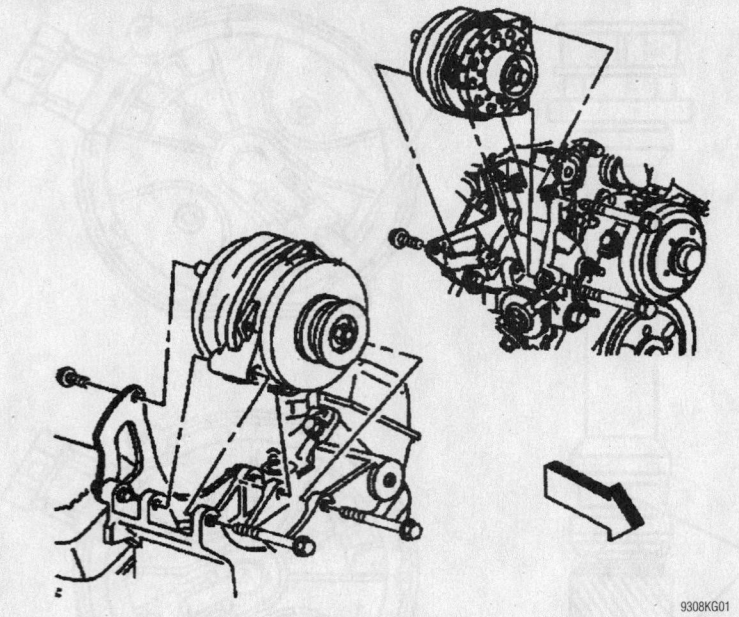

Exploded view of the alternator mounting—Truck shown

9308KG01

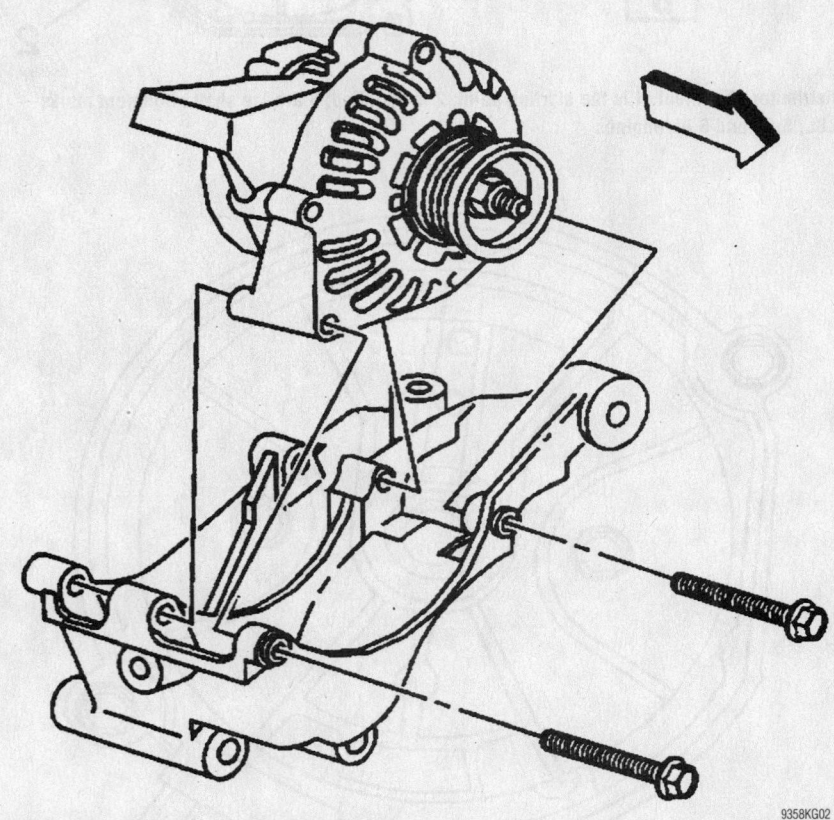

Exploded view of the alternator mounting—Van shown

9358KG02

8.1L Engine

1. Before servicing the vehicle, refer to the precautions in the beginning of this section.
2. Disconnect the negative battery cable.
3. Remove or disconnect the following:
   - Electrical connections from the generator
4. Remove the cable from the generator as follows:
   a. Slide the boot down, to reveal the terminal stud.

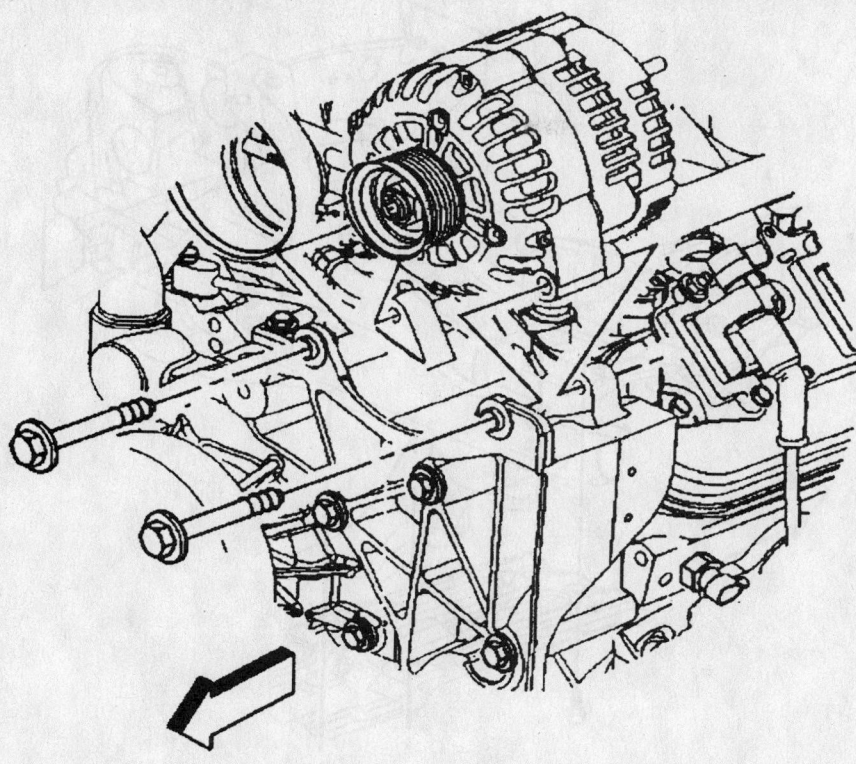

Alternator mounting—4.8L, 5.3L & 6.0L engines

9308KG99

b. Unfasten the cable nut from the stud, then remove the generator cable.

- Accessory drive belt
- Mounting bolts
- Generator
- Mounting bolts securing the generator to the brace and bracket
- Generator

To install:

5. Install or connect the following:
- Generator
- Generator mounting bolts. Tighten the bolts to 37 ft. lbs. (50 Nm).
- Accessory drive belt

6. Connect the generator cable, secure with the nut and tighten to 80 inch lbs. (9 Nm). Slide the boot back over the terminal stud.
- Electrical connections to the generator

7. Connect the negative battery cable.

Ignition Timing

ADJUSTMENT

Always refer to the Vehicle Emissions Control Information label in the engine compartment for base ignition timing specification and adjustment procedures.

Engine Assembly

REMOVAL & INSTALLATION

4.3L (Exc. Silverado and Sierra), 5.0L and 5.7L Engines

1. Before servicing the vehicle, refer to the precautions in the beginning of this section.
2. Drain the cooling system.
3. Drain the engine oil.
4. Remove or disconnect the following:
- Negative battery cable
- Hood
- Air cleaner
- Accessory drive belt
- Fan
- Water pump pulley
- Radiator and shroud
- Heater hoses at the engine
- Accelerator, cruise control and detent linkage if used
- Air conditioning compressor, if used, and lay aside
- Power steering pump, if used, and lay aside
- Wiring from the engine
- Fuel line
- Vacuum lines from the intake manifold

- Exhaust pipes from the manifold
- Strut rods at the engine mountings, if used
- Flywheel or torque converter cover
- Wiring along the oil pan rail
- Starter
- Wire for the fuel gauge
- Converter-to-flex plate bolts, if equipped with automatic transmission

5. Support the transmission
- Bell housing to engine bolts
- Rear engine mounting to frame bolts and the front through bolts and the engine

To install:

6. Lower the engine.
7. Install or connect the following:
- Engine mounting bolts. Torque the rear engine mounting to frame bolts or nuts to 45 ft. lbs. (54 Nm), the front through-bolts to 70 ft. lbs. (97 Nm) and the front nuts to 50 ft. lbs. (67 Nm).
- Bell housing to engine bolts and torque to 35 ft. lbs. (47 Nm)

8. Remove the transmission support.
- Converter-to-flex plate bolts and tighten to 35 ft. lbs. (47 Nm)
- Fuel gauge wiring
- Starter
- Flywheel or torque converter cover
- Strut rods at the engine mountings, if used
- Exhaust pipes at the manifold
- Vacuum lines to the intake manifold
- Fuel line
- Engine wiring harness
- Power steering pump, if used
- Air conditioning compressor, if used
- Accelerator, cruise control and detent linkage
- Heater hoses
- Radiator and shroud
- Accessory drive belts
- Hood
- Negative battery cable

9. Refill coolant and engine oil.

7.4L Engine

1. Before servicing the vehicle, refer to the precautions in the beginning of this section.
2. Drain the cooling system.
3. Remove or disconnect the following:
- Hood
- Negative battery cable
- Air cleaner
- Radiator and fan shroud
- Engine wiring

- Accelerator, cruise control linkage
- Fuel supply lines
- Vacuum wires
- Air conditioning compressor, if used, and lay aside
- Power steering pump and position it out of the way. It's not necessary to disconnect the fluid lines.
- Exhaust pipes from the manifold
- Starter
- Torque converter cover
- Converter-to-flexplate bolts

4. Support the transmission
- Bellhousing-to-engine bolts
- Rear engine mounting-to-frame bolts and the front through bolts
- Engine

To install:

5. Lower the engine into the vehicle.
6. Install or connect the following:
- Engine mounting bolts. Torque the rear engine mounting-to-frame bolts or nuts to 45 ft. lbs. (54 Nm), the front through bolts to 70 ft. lbs. (97 Nm) and the front nuts to 50 ft. lbs. (67 Nm).
- Bellhousing-to-engine bolts. Torque the bolts to 35 ft. lbs. (47 Nm).

7. Remove transmission support.
- Converter-to-flexplate bolts and torque them to 35 ft. lbs. (47 Nm)
- Fuel gauge wiring
- Starter
- Torque converter cover
- Exhaust pipes at the manifold
- Power steering pump
- Air conditioning compressor
- Vacuum hoses
- Fuel supply line
- Accelerator, cruise control linkage
- Engine wiring
- Radiator and fan shroud
- Air cleaner
- Hood
- Negative battery cable

8. Refill the coolant.

4.3L Silverado and Sierra

1. Before servicing the vehicle, refer to the precautions in the beginning of this section.

2. Remove or disconnect the following:
- Battery negative cable
- Coolant
- A/C refrigerant, if equipped
- Oil pan skid plate
- Engine shield
- Starter.
- Transmission cover
- Bolt holding the bracket for the

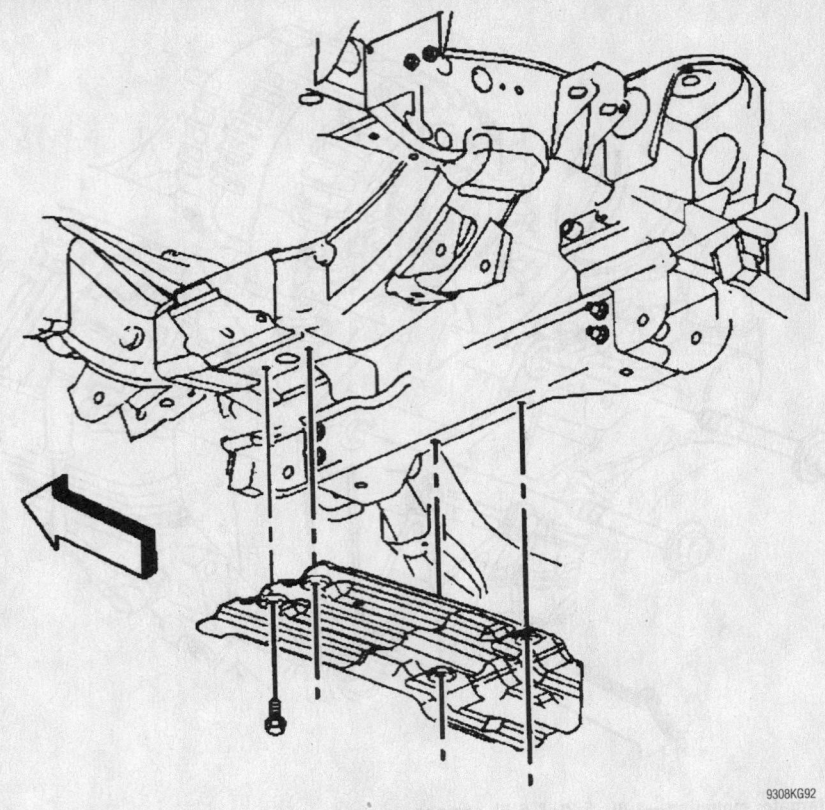

Engine shield removal—4.8L, 5.3L and 6.0L engines

9308KG92

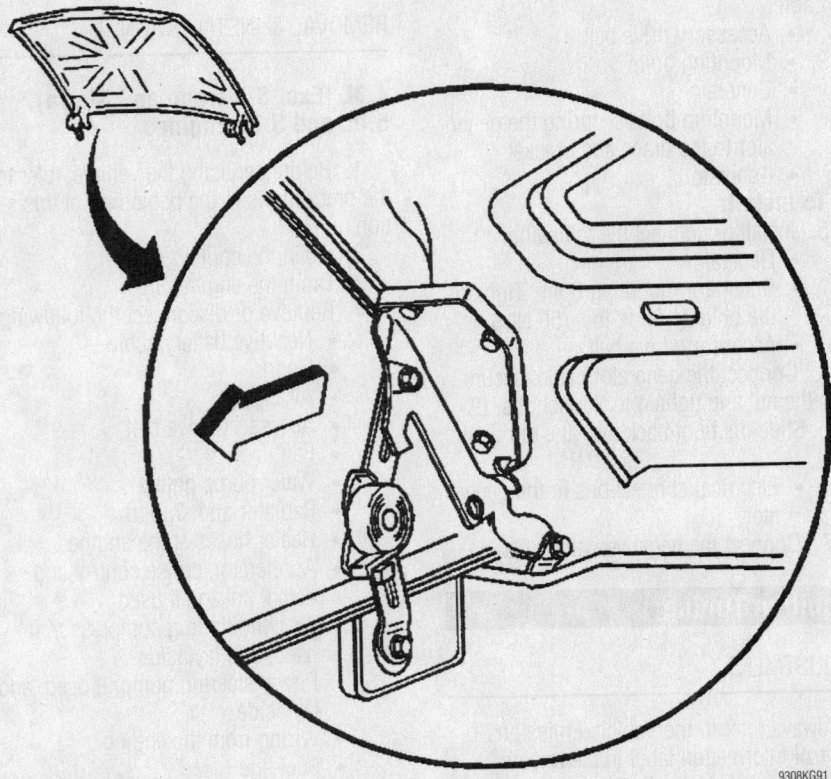

Hood in the service position—4.8L, 5.3L and 6.0L engines

9308KG91

starter cables and transmission cooler lines, if equipped
- Nuts at the catalytic converter pipe
- Exhaust pipes from the exhaust manifolds
- Bolts holding the brackets to the oil pan for both battery cables
- Crankshaft Position (CKP) sensor electrical connector and remove the harness from the retainer
- Low oil level sensor electrical connector and remove the wire harness from the retainer
- Bolt holding the battery negative cable and ground cable to the engine
- Torque converter-to-flywheel bolts, if equipped, through the starter opening
- Engine to transmission bolts

3. Move the hood hinge bolts to hold the hood in the service position.

4. Remove or disconnect the following:
- Positive Crankcase Ventilation (PCV) hose from the air cleaner outlet duct
- Air cleaner outlet duct from the throttle body and the air cleaner assembly
- Fan shroud
- Drive belt
- Engine cooling fan
- Radiator inlet and outlet hoses from the engine

✳✳ CAUTION

In order to avoid possible injury or vehicle damage, always replace the accelerator control cable with a NEW cable whenever you remove the engine from the vehicle.

✳✳ WARNING

In order to avoid cruise control cable damage, position the cable out of the way while you remove or install the engine.

- Accelerator control cable
- Cruise control cable from the throttle body and the bracket on the throttle body and intake manifold, if equipped
- Engine wiring harness and clip from the accelerator control cable bracket
- Accelerator control cable bracket from the throttle body
- A/C hoses from the compressor and the accumulator, if equipped

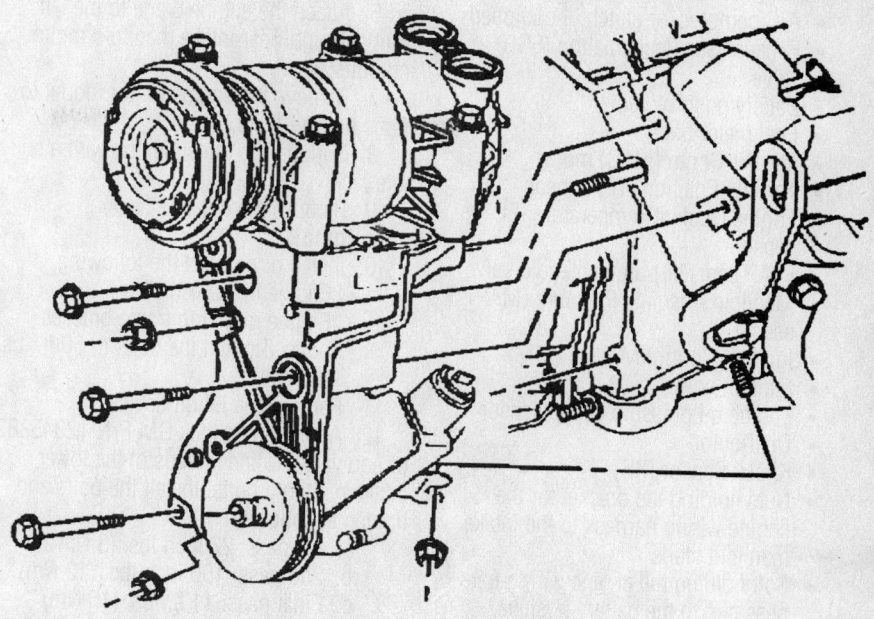

Power steering mount bracket removal—4.8L, 5.3L and 6.0L engines

9308KG90

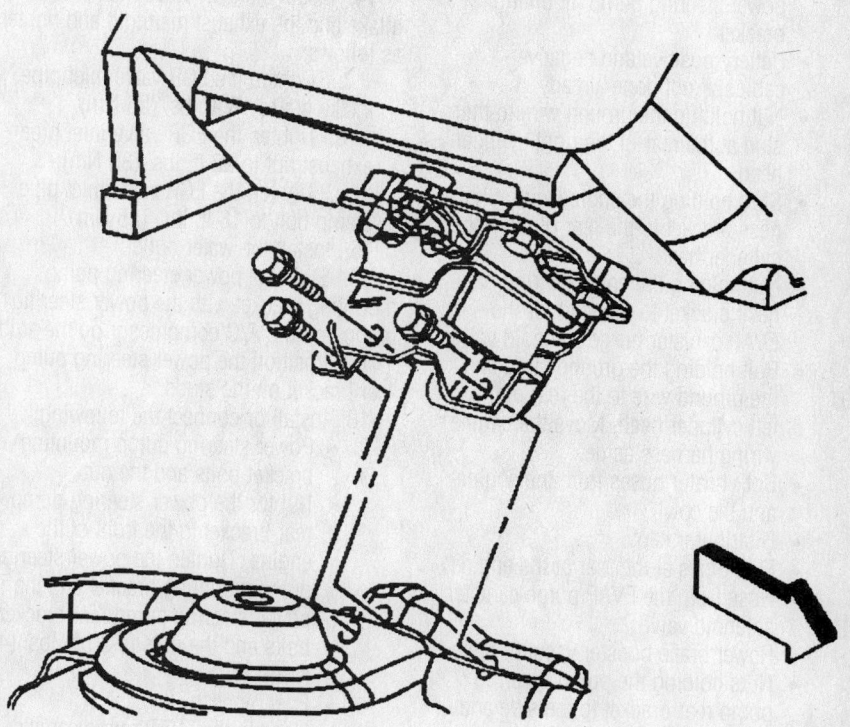

Engine mount disconnect—4.8L, 5.3L and 6.0L engines

9308KG89

- Secondary Air Injection (AIR) crossover pipe from the AIR pipe assemblies

➡**Remove the AIR pipes before engine removal. The AIR pipes can break or damage easily causing erratic engine operation.**

- AIR pipe assemblies from the left exhaust manifold, if equipped
- AIR pipe assembly from the AIR pump
- AIR pipe assembly from the right exhaust manifold, if equipped.
- A/C pressure switch, if equipped

- A/C compressor clutch, if equipped
- Exhaust Gas Recirculation (EGR) valve
- Battery positive cable
- Fuel meter body
- Idle Air Control (IAC) motor
- Throttle Position (TP) sensor
- Engine Coolant Temperature (ECT) sensor
- EVAP canister purge solenoid valve
- Manifold Absolute Pressure (MAP) sensor
- Ignition Control Module (ICM)
- Ignition coil
- Engine oil pressure gauge sensor
- Distributor
- Knock Sensor (KS)
- Nuts holding the bracket for the engine wiring harness to the intake manifold studs
- Bolt holding the engine wiring harness clip to the battery positive cable junction block bracket
- Bracket for the battery positive cable junction block from the power steering pump mounting bracket
- Battery positive and negative cables, if not done already
- Nut holding the ground wire to the stud at the rear of the right cylinder head
- Stud holding the engine wiring harness bracket to the rear of the right cylinder head
- Nut holding the engine wiring harness bracket to the stud for the EVAP canister purge solenoid valve
- Bolt holding the ground strap and the ground wire to the rear of the left cylinder head. Move the engine wiring harness aside.
- Both heater hoses from the engine and the cowl
- Distributor cap
- Fuel pipes at the rear of the engine
- Hose from the EVAP purge canister solenoid valve
- Power brake booster vacuum hose
- Nuts holding the power steering pump rear bracket to the side and front of the engine
- Three bolts and the nut holding the power steering pump mounting bracket to the engine

5. With the power steering pump and the A/C compressor still attached, slide the power steering pump mount bracket off the stud and set aside.
- Water outlet
- EGR valve inlet pipe from the intake and exhaust manifold

6. Attach the engine crane to the left front and right rear intake manifold mounting bolts.

7. Remove the engine motor mount to frame bracket bolts.

8. Support the transmission with a suitable jack.

9. Remove the engine.

To install:

10. Install or connect the following:
- Engine in the vehicle
- Engine mount to frame bracket bolts. Tighten the bolts to 50 ft. lbs. (65 Nm).

11. Remove the lifting device.

12. Apply thread lock GM P/N 12345382 or equivalent to the threads of the lower intake manifold bolts. Install the bolts and tighten as follows:
  a. 1st pass: 27 inch lbs. (3 Nm)
  b. 2nd pass: 106 inch lbs. (12 Nm)
  c. Final pass: 11 ft. lbs. (15 Nm)

13. Loosely install one transmission to engine bolt. Remove the support jack from under the transmission.

14. Install the EGR valve inlet pipe to the intake and the exhaust manifold and tighten as follows:
  a. Tighten the EGR valve inlet pipe intake nut to 18 ft. lbs. (25 Nm).
  b. Tighten the EGR valve inlet pipe exhaust nut to 22 ft. lbs. (30 Nm).
  c. Tighten the EGR valve inlet pipe clamp bolt to 18 ft. lbs. (25 Nm).

15. Install the water outlet.

16. Slide the power steering pump mounting bracket with the power steering pump and the A/C compressor on the stud.

17. Position the power steering pump rear bracket on the studs.

18. Install or connect the following:
- Power steering pump mounting bracket bolts and the nut
- Nut for the power steering pump rear bracket to the front of the engine. Tighten the power steering pump mounting bracket and the power steering pump rear bracket bolts and the nuts to 30 ft. lbs. (41 Nm).
- Fuel pipes
- Hose to the EVAP purge canister solenoid valve
- Vacuum brake booster hose to engine and the vacuum brake booster
- Distributor cap
- Both heater hoses to the engine and the cowl
- AIR pipe assembly with new gaskets to the right exhaust manifold, if equipped

- AIR pipe nuts and bracket bolt. Tighten the nuts to 18 ft. lbs. (25 Nm) and the bolt to 89 inch lbs. (10 Nm).
- AIR pipe assembly to the AIR pump
- AIR pipe assembly with new gaskets to the left exhaust manifold, if equipped. Install the AIR pipe nuts and bracket bolt. Tighten the to 18 ft. lbs. (25 Nm) and the bolt to 89 inch lbs. (10 Nm).
- AIR crossover pipe to the AIR pipe assemblies
- Engine wiring harness
- A/C pressure switch
- A/C compressor clutch
- EGR valve
- Generator battery positive cable
- Fuel meter body assembly
- IAC motor
- TP sensor
- ECT sensor
- EVAP canister purge solenoid valve
- MAP sensor
- ICM
- Ignition coil
- Engine oil pressure gauge sensor
- Distributor
- KS

19. Install the bolt holding the ground strap and the ground wire to the rear of the left cylinder head. Tighten the ground strap and ground wire bolt to 12 ft. lbs. (16 Nm).

20. Position the engine wiring harness bracket on the EVAP purge canister solenoid valve stud and install the nut.

21. Install the stud holding the wire harness bracket to the rear of the right cylinder head. Tighten the nut on the EVAP solenoid to 80 inch lbs. (9 Nm). Tighten the stud at rear of the cylinder head to 18 ft. lbs. (25 Nm).

22. Install the nut holding the ground wire on the stud at the rear of the right cylinder head. Tighten the ground wire nut to 12 ft. lbs. (16 Nm).

23. Position the battery positive and negative cables. Do not connect the negative battery cable to the battery.

24. Install or connect the following:
- Battery positive cable junction block bracket and bolt to the power steering pump mounting bracket. Tighten the bolt to 18 ft. lbs. (25 Nm).
- Bolt holding the engine wiring harness bracket to battery positive cable junction block bracket
- Engine wiring harness bracket on the intake manifold studs. Tighten the nuts to 106 inch lbs. (12 Nm) and the bolt to 80 inch lbs. (9 Nm).

- A/C hoses
- Accelerator control cable bracket. Tighten the nuts to 80 inch lbs. (9 Nm).
- Engine wire harness and clip to the accelerator control cable bracket

✳✳ CAUTION

In order to avoid possible injury or vehicle damage, always replace the accelerator control cable with a NEW cable whenever you remove the engine from the vehicle.

- NEW accelerator control cable
- Cruise control cable
- Radiator inlet and outlet hoses
- Engine cooling fan
- Drive belt
- Upper and lower radiator shroud
- Air cleaner outlet duct to the throttle body and the air cleaner assembly
- PCV hose to the air inlet duct

25. Move the hood hinge bolts from the service position to the normal operating position.
26. Raise the vehicle.
27. Install or connect the following:
- Remaining transmission to engine bolts except for the one where the transmission cover mounts
- Torque converter to flywheel
- Transmission cover and bolts. Tighten the transmission cover to oil pan bolt to 106 inch lbs. (12 Nm). Tighten the transmission cover to transmission bolt to 34 ft. lbs. (47 Nm).
- Bolt holding the bracket for the starter cables and the transmission cooler pipe and tighten to 80 inch lbs. (9 Nm)
- Positive and negative battery cable brackets-to-oil pan bolts and tighten to 106 inch lbs. (12 Nm).
- Bolt for the battery negative cable and ground wire to the front of the engine. Tighten the battery negative cable and ground wire bolt to 18 ft. lbs. (25 Nm).
- CKP sensor and install the harness in the retainer
- Low oil level sensor and install the wire harness in the retainer
- Exhaust pipe to the exhaust manifolds and tighten the nuts at the catalytic converter flange
- Starter motor
- Oil pan skid plate and tighten bolt to 15 ft. lbs. (20 Nm)

- Engine shield
28. Lower the vehicle.
- Battery negative cable
- Engine oil
- Coolant
29. Recharge the A/C system.

4.8L, 5.3L and 6.0L Engines

✳✳ CAUTION

Before servicing any electrical component, the ignition key must be in the OFF or LOCK position and all electrical loads must be OFF, unless instructed otherwise in these procedures.

1. Before servicing the vehicle, refer to the precautions in the beginning of this section.
2. Remove or disconnect the following:
- Negative battery cable
- Coolant
- A/C refrigerant
3. Raise the hood to the servicing position. Move the hood hinge bolt to hold the hood in the servicing position.
- Upper and the lower radiator hoses from the engine
- Air cleaner duct from the engine
- A/C condenser mounting bolts

- Radiator support from the vehicle
- A/C compressor
- Coolant hose from the throttle body
- Heater hoses from the engine and the cowl
- Engine sight shield from the intake manifold
- Accelerator control cable mounting bracket from the intake manifold

✳✳ CAUTION

In order to avoid possible injury or vehicle damage, always replace the accelerator control cable with a NEW cable whenever you remove the engine from the vehicle. In order to avoid cruise control cable damage, position the cable out of the way while you remove or install the engine.

- Accelerator control cable and the cruise control cable, if equipped, from the throttle shaft
4. Open the large electrical harness retainer. Remove one 10 mm nut in order to release the engine harness from the intake manifold.
5. Disconnect the electrical connectors from the following:

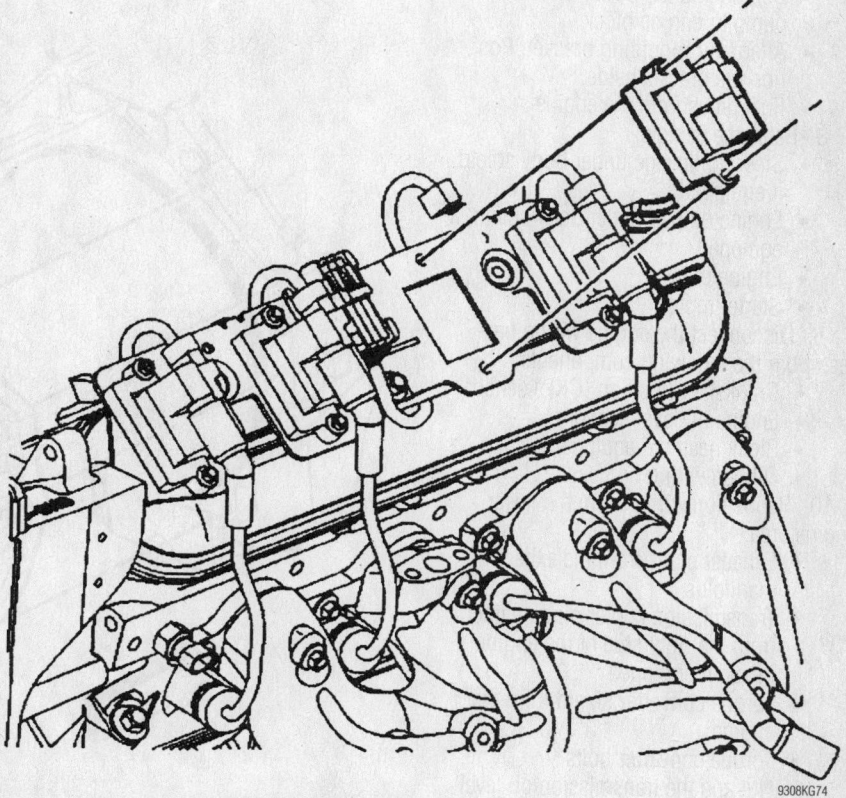

Ignition coil removal—4.8L, 5.3L and 6.0L engines

- Eight injectors
- Idle Air Control (IAC) motor
- Throttle Position (TP) sensor
- Evaporative Emissions (EVAP) canister purge solenoid
- Manifold Absolute Pressure (MAP) sensor
- Camshaft Position (CMP) sensor
- Ground splice at the rear of the right side of the block
- Ground splice and the ground strap at the rear of the left side of the block
- Coolant Temperature (CTS) sensor
- Oil pressure sensor/switch
- Electrical connector from intake and disconnect from harness
- Junction block bracket from alternator bracket

6. Set the electrical harness aside.
7. Remove or disconnect the following:
   - EVAP canister purge solenoid vent tube from the solenoid by squeezing the retainer, then release the tube from the solenoid
   - Battery negative cable from the engine block
   - Drive belt
   - Bolts holding the alternator mounting bracket to the cylinder head and block
   - Bolt behind the power steering pump to engine block
   - Alternator mounting bracket. Position the bracket aside.
   - Fuel pipes from the engine
8. Raise the vehicle.
   - Steering linkage under body shield, if equipped
   - Engine oil pan under body shield, if equipped
   - Engine oil
   - Starter motor
9. Disconnect the engine wiring harness from the following components:
   - Crankshaft Position (CKP) sensor
   - Engine oil level sensor
   - Block heater, if equipped
   - Oil pan wiring harness
10. Reposition wiring from the lower engine area.
   - Exhaust pipes from the exhaust manifolds
   - Transmission cooler pipe retainer from the right side of the engine block, if equipped
   - Torque converter shield from the engine
   - Torque converter bolts
   - Nut and the transmission oil level indicator tube from the bellhousing stud

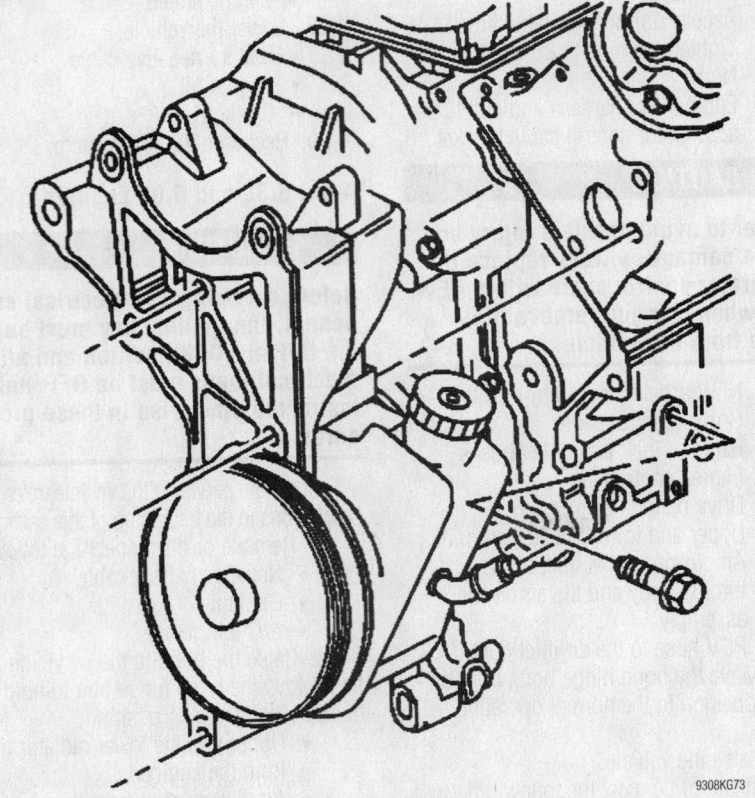

Power steering pump removal—4.8L, 5.3L and 6.0L engines

9308KG73

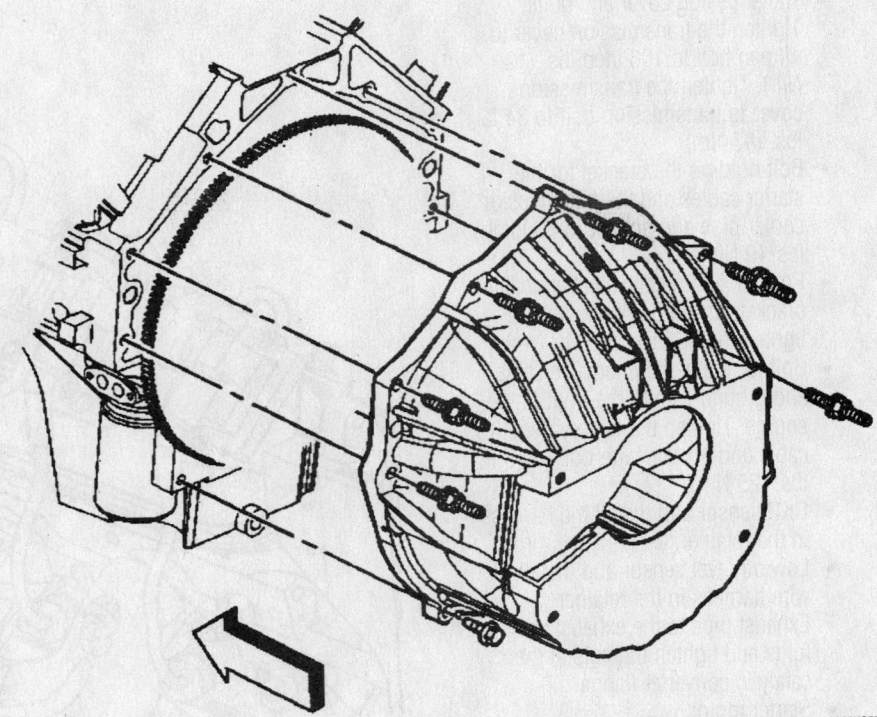

Bellhousing bolt removal—4.8L, 5.3L and 6.0L engines

9308KG75

- Lower bellhousing studs from the engine
11. Lower the vehicle.
   - Remaining bellhousing bolts
   - Engine electrical harness aside
   - Ignition coil(s)
12. Install an engine crane.
13. Install a floor jack or stands to transmission for support.
14. Remove the engine mount bolts.

➠**Use care while moving the engine assembly in order to avoid breaking the MAP sensor locating tabs. Broken MAP sensor tabs may result in decreased engine performance.**

15. Remove the engine from the vehicle.
To install:
16. Install or connect the following:
   - Engine to the vehicle
   - Engine mount bolts
   - Upper bellhousing bolts
17. Remove transmission support apparatus.
18. Remove the lifting device.
19. Remove the lift brackets from both cylinder heads.
20. Install the ignition coil(s) and the spark plug wire(s).
21. Route the engine wiring harness to the lower right hand side of the engine.
22. Raise the vehicle.
23. Install or connect the following:
   - Remaining bellhousing bolts
   - Torque converter bolts
   - Torque converter shield
   - Transmission oil level indicator tube and nut to bellhousing stud
   - A/C compressor
   - Transmission cooler pipe retainer to right side of engine block
   - Engine exhaust pipes to the exhaust manifolds
24. Reroute wiring to lower engine area and install bolt to oil pan.
25. Connect electrical connectors to the CKP sensor, the engine oil level sensor and the block heater, if equipped.
26. Install or connect the following:
   - Starter motor
   - Engine oil pan under body shield, if equipped
   - Steering linkage under body shield
27. Lower the vehicle.
   - Fuel pipes to the engine
   - Alternator mounting bracket to the cylinder head using the nuts and the bolts. Tighten the bolts to 37 ft. lbs. (50 Nm).
   - Bolt at the rear of the power steering pump to the engine block and tighten to 37 ft. lbs. (50 Nm)

   - Alternator
   - Drive belt
   - Battery negative cable to the engine block
   - EVAP canister purge solenoid to the intake manifold
28. Route the engine harness over the top of the engine. Attach the connectors for following components:
   - Eight injectors
   - IAC motor
   - TP sensor
   - EVAP canister purge solenoid.
   - MAP sensor
   - CMP sensor
   - Ground splice at the rear of the right side of engine block
   - Ground splice and the ground strap at the rear of the left side of engine block
   - CTS sensor
29. Install or connect the following:
   - Nut to the engine wiring harness bracket and tighten to 89 inch lbs. (10 Nm)

✳✳ CAUTION

In order to avoid possible injury or vehicle damage, always replace the accelerator control cable with a NEW cable whenever you remove the engine from the vehicle. In order to avoid cruise control cable damage, position the cable out of the way while you remove or install the engine.

   - NEW accelerator control cable
   - Cruise control cable, if equipped, to the throttle shaft
   - Bolts for the accelerator control cable mounting bracket and tighten to 89 inch lbs. (10 Nm)
   - Engine sight shield to the intake manifold
   - Heater hoses to the cowl and the engine
   - Coolant hose to the throttle body
   - Radiator support in the vehicle
   - A/C condenser mounting bolts
   - Air cleaner duct
   - Lower radiator hoses to the engine
30. Lower the hood.
31. Fill the engine with oil.
32. Fill the engine with coolant.
33. Connect the negative battery cable.

8.1L Engine

1. Before servicing the vehicle, refer to the precautions in the beginning of this section.
2. Raise the hood to the servicing posi-

tion. Move the hood hinge bolt to hold the hood in the servicing position.
3. Release the fuel system pressure.
4. Remove or disconnect the following:
   - Negative, then positive battery cables
   - Coolant
   - A/C refrigerant
   - Engine oil cooler lines from the engine block
   - Transmission-to-engine bolts
   - Clutch pressure plate bolts, if equipped
   - Torque converter bolts, if equipped
   - Catalytic converter
   - Exhaust manifold pipe
   - Hoses from power steering pump, then plug the lines and ports
   - Starter motor
5. Raise the vehicle.
   - Engine electrical harness and tie aside
   - Alternator
   - Ground cable bolt from engine block
   - Exhaust Gas Recirculation (EGR) valve adapter
   - Vacuum lines (tag before removal)
   - Throttle Actuator Control (TAC) module electrical connector
6. Install Engine Lift Brackets part No. J 36857, or equivalent, to the rear of the right cylinder head and the front of the left cylinder head.
7. Install the attaching bolt and washer. Use part No. 9428217 with 1560963. Tighten the bolts to 30 ft. lbs. (40 Nm).
8. Remove or disconnect the following:
   - Engine mount heat shield bolt and shields
   - Engine mount-to-engine mount bracket bolts

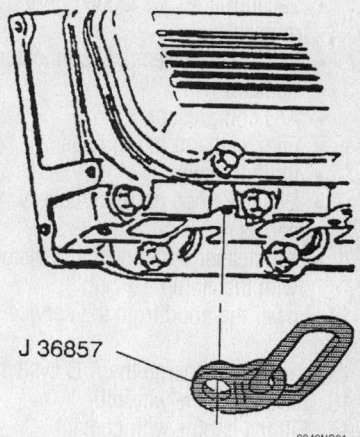

J 36857

Install suitable lift brackets to the rear of the right head and the front of the left head

- Engine from the vehicle, using a suitable lifting device. Place on a suitable stand.
- A/C compressor/power steering pump bracket from the cylinder head
- Lift brackets from the cylinder head

To install:

9. Install Engine Lift Brackets part No. J 36857, or equivalent, to the rear of the right cylinder head and the front of the left cylinder head.

10. Install the attaching bolt and washer. Use part No. 9428217 with 1560963. Tighten the bolts to 30 ft. lbs. (40 Nm).

11. Install or connect the following:
- A/C compressor/power steering mounting bracket. Tighten the bolts and nut to 37 ft. lbs. (50 Nm).
- Alternator bracket
- Engine into the vehicle
- Engine mount-to-engine mount bracket bolts
- Engine mount heat shield and bolts

12. Remove the lift hooks from the cylinder heads, then raise the vehicle.
- Engine oil cooler lines
- Transmission-to-engine bolts
- Clutch pressure plate bolts, if equipped
- Torque converter bolts, if equipped
- Catalytic converter
- Exhaust manifold pipe
- Hoses to the power steering pump
- Starter motor

13. Lower the vehicle.
- Engine electrical harness. Make sure the harness is properly routed.
- Alternator
- Ground cable bolt to engine block and tighten to 12 ft. lbs. (16 Nm)
- EGR valve adapter
- Vacuum lines, as tagged during removal
- TAC module electrical connector
- Radiator
- A/C compressor
- Fuel feed and return lines
- Ignition coils
- Positive, then negative battery cables
- Air cleaner outlet duct and secure with the clamp

14. Lower the hood from the service position.

15. Properly recharge the A/C system.

16. Fill the engine with oil.

17. Fill the engine with coolant.

18. Perform the Crankshaft Position (CKP) sensor variation learn procedure:
    a. Install a suitable scan tool and check for Diagnostic Trouble Codes

(DTCs). If any DTCs, other than P1336 are set, resolve those codes first, before proceeding with this procedure.
    b. With the scan tool, select the crankshaft position variation learn procedure.
    c. Observe the fuel cut-off for the 8.1L engine.
    d. The scan tool will instruct you to perform certain steps, make sure you follow all directions given by the scan tool exactly.
    e. Enable the crankshaft position system variation learn procedure.

➡ **While the learn procedure is in progress, release the throttle immediately when the engine started to decelerate. The engine control is returned to the operator and the engine responds to throttle position after the learn procedure is complete.**

    f. Slowly increase the engine speed to the RPM that you observed.
    g. Immediately release the throttle when fuel cut-out is reached.
    h. The scan tool displays: Learn Status: Learned this ignition. If the scan tool does NOT display this message and not other DTCs set, you must perform further troubleshooting.
    i. Turn the ignition **OFF** for 30 seconds after the learn procedure has been completed successfully.

19. Start and run the engine, then check for leaks.

Water Pump

REMOVAL & INSTALLATION

4.3L, 5.0L, 5.7L and 7.4L Engines

1. Before servicing the vehicle, refer to the precautions in the beginning of this section.

2. Drain the radiator.

3. Remove or disconnect the following:
- Fan shroud
- Negative battery cable
- Drive belt(s)
- Alternator and other accessories, if necessary
- Fan, fan clutch and pulley
- Accessory brackets that might interfere with water pump removal
- Lower radiator hose from the water pump inlet
- Heater hose from the nipple on the pump

➡ **On the 7.4L engine, remove the bypass hose.**

- Water pump assembly away from the timing cover

To install:

4. Clean all old gasket material from the timing chain cover.

5. Install or connect the following:
- Pump assembly with a new gasket. Torque the bolts to 30 ft. lbs. (41 Nm).
- Hose between the water pump inlet and the pump
- Heater hose and the bypass hose (7.4L only)
- Fan, fan clutch and pulley
- Alternator and other accessories, if necessary
- Drive belt(s)
- Upper radiator shroud

6. Refill the cooling system.

7. Connect the battery.

4.8L, 5.3L and 6.0L Engines

1. Remove or disconnect the following:
- Air inlet duct
- Coolant
- Inlet radiator hose from the water pump

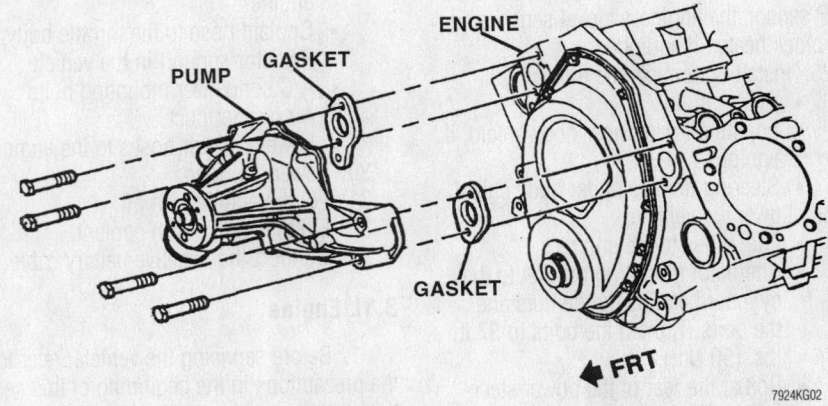

Exploded view of the water pump mounting—4.3L engine

7924KG02

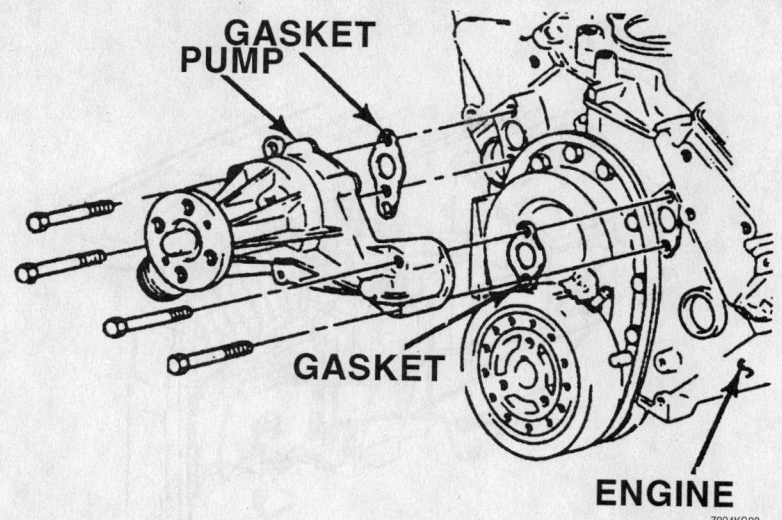

Exploded view of the water pump mounting—5.0L and 5.7L engines

7924KG03

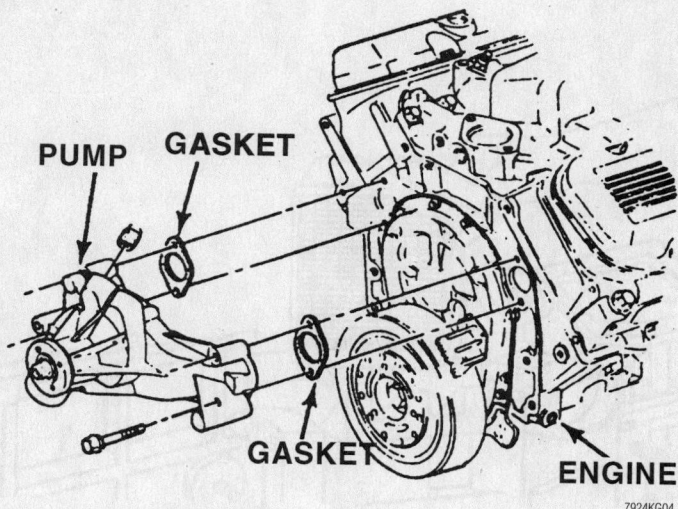

Exploded view of the water pump mounting—7.4L engine

7924KG04

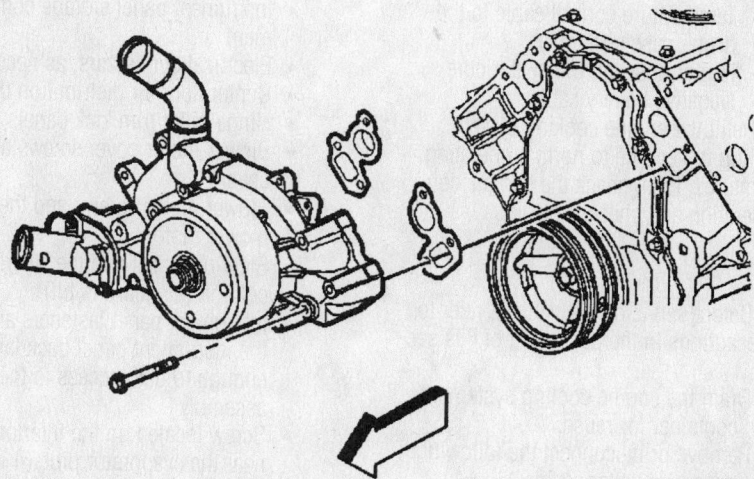

Exploded view of the water pump assembly—4.8L, 5.3L and 6.0L engines

9302KG01

- Upper fan shroud
- Cooling fan and clutch assembly
- Drive belt
- Radiator outlet hose from the coolant pump
- Surge tank hose
- Heater hose
- Water pump

To install:

➡**DO NOT use cooling system seal tabs (or similar compounds) unless otherwise instructed. The use of cooling system seal tabs (or similar compounds) may restrict coolant flow through the passages of the cooling system or the engine components. Restricted coolant flow may cause engine overheating and/or damage to the cooling system or the engine components/assembly.**

2. Install or connect the following:
   - Water pump. Install the water pump bolts. Tighten the water pump bolts first pass to 11 ft. lbs. (15 Nm); tighten the bolts final pass to 22 ft. lbs. (30 Nm).
   - Water pump drive belt pulley and bolts (if applicable). Tighten the pulley bolts first pass to 89 inch lbs. (10 Nm); tighten the bolts final pass to 18 ft. lbs. (25 Nm).
   - Surge tank hose
   - Heater hose
   - Outlet radiator hose to the coolant pump
   - Drive belt
   - Cooling fan and clutch assembly
   - Upper fan shroud
   - Inlet radiator hose to the water pump
   - Air inlet duct
   - Coolant

8.1L Engines

1. Before servicing the vehicle, refer to the precautions in the beginning of this section.
2. Remove or disconnect the following:
   - Coolant
   - Drive belt
   - Fan clutch
   - Outlet hose clamp and hose
3. Reposition the bypass hose clamps at the water pump and water crossover
   - Bypass hose
   - Water pump bolt and pump. Discard the water pump gaskets.

To install:

4. Install or connect the following:
   - New water pump gaskets.
   - Water pump and bolts. Tighten the

water pump bolts 37 ft. lbs. (50 Nm).
- Bypass hose and clamps
- Outlet hose and clamp
- Fan clutch
- Drive belt
- Surge tank hose
- Heater hose
- Outlet radiator hose to the coolant pump
- Drive belt
- Cooling fan and clutch assembly
- Upper fan shroud
- Inlet radiator hose to the water pump
- Air inlet duct
- Coolant

Heater Core

REMOVAL & INSTALLATION

Silverado

1. Before servicing the vehicle, refer to the precautions in the beginning of this section.
2. Drain the engine cooling system into a clean container for reuse.
3. Remove or disconnect the following:
   - Negative battery cable
   - Heater hoses from the heater core
   - Temperature control cable from the heater case assembly
   - Disconnect the mode control cable from the heater case assembly
   - Instrument panel carrier to provide access to the heater case assembly
   - Electrical connectors that may interfere with the heater case assembly removal
   - Heater case assembly-to-chassis screws/nuts and the assembly. Place the heater case assembly on a bench.
   - Heater core cover screws
   - Heater core from the heater case

To install:

4. Install or connect the following:
   - Heater core to the heater case
   - Heater core cover screws and tighten to 14 inch lbs. (1.6 Nm)
   - Heater case assembly and the assembly-to-chassis screws, then, tighten the screws to 35 inch lbs. (4 Nm) and the nuts to 80 inch lbs. (9 Nm)
   - Electrical connectors, as necessary
   - Instrument panel carrier
   - Mode control cable to the heater case assembly

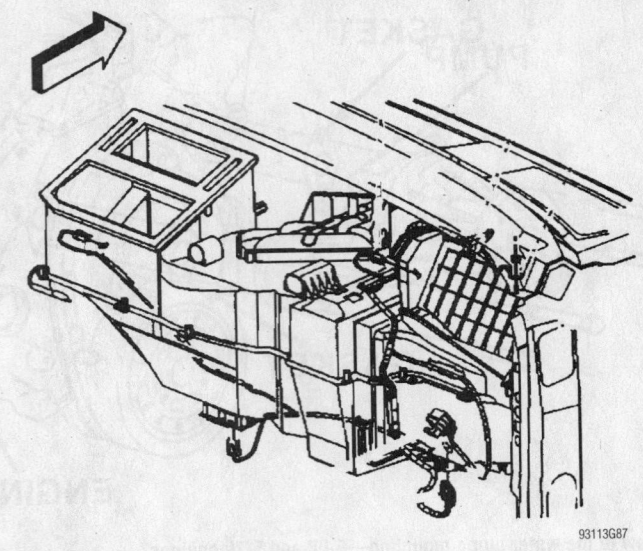

View of the heater case assembly—Silverado

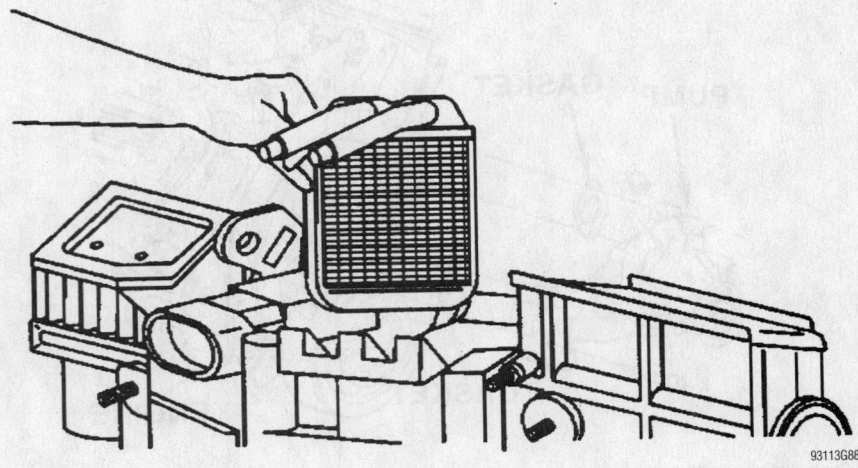

View of the heater core—Silverado

   - Temperature control cable to the heater case assembly
   - Heater hoses to the heater core
   - Negative battery cable.
5. Refill the engine cooking system.
6. Run the engine to normal operating temperatures; then, check the climate control operation and check for leaks.

Sierra

1. Before servicing the vehicle, refer to the precautions in the beginning of this section.
2. Drain the engine cooling system into a clean container for reuse.
3. Remove or disconnect the following:
   - Negative battery cable
   - Heater hoses from the heater core

   - Instrument panel storage compartment
   - Electrical connectors, as necessary
   - Center floor air distribution duct
   - Hinge pillar trim kick panels
   - Blower motor cover screws and the cover
   - Blower motor screws and the blower motor
   - Steering wheel and the steering column (standard & tilt)
   - Instrument panel fasteners and pull the instrument panel back far enough to gain access to the heater assembly
   - Screw located on the interior side near the evaporator pipe (if equipped), while holding the heater assembly against the firewall

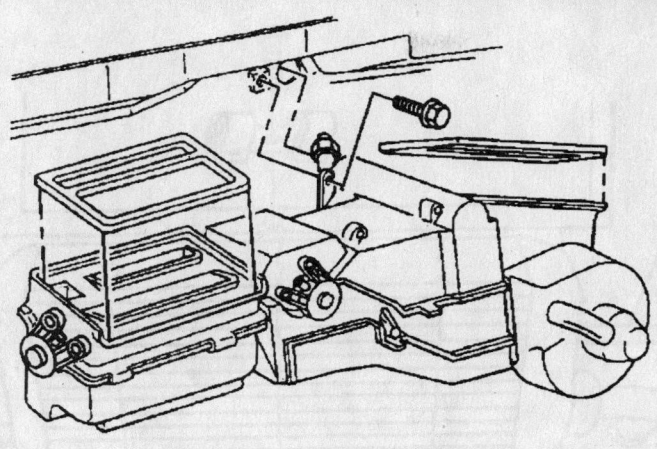

View of the front heater assembly—Sierra

93113G80

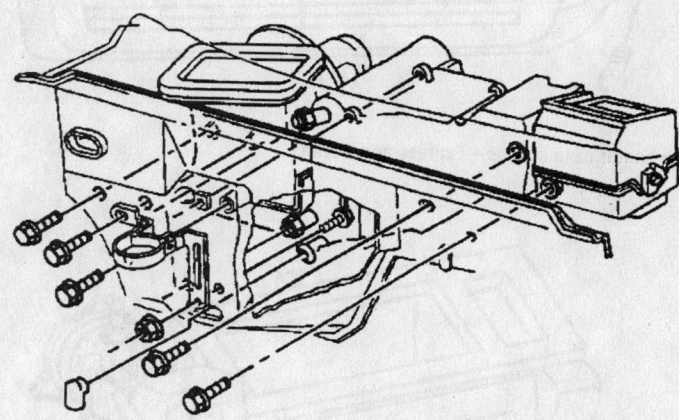

Location of the front heater assembly-to-chassis fasteners—Sierra

93113G81

- 4 heater assembly-to-chassis screws and the 2 heater assembly-to-chassis nuts, working in the engine compartment.

➡**Removal of the heater assembly may require the help of an assistant.**

- 7 heater cover-to-heater assembly screws and the cover
- Heater core from the heater assembly

To install:

4. Install or connect the following:
   - Heater core to the heater assembly
   - Heater cover and the 7 heater cover-to-heater assembly screws

➡**Installation of the heater assembly may require the help of an assistant.**

- 4 heater assembly-to-chassis screws and the 2 heater assembly-to-chassis nuts. Torque the screws to 17 inch lbs. (1.9 Nm) and the nuts to 25 inch lbs. (2.8 Nm).
- Screw located on the interior side near the evaporator pipe (if

equipped), while holding the heater assembly against the firewall. Torque the screw to 97 inch lbs. (11 Nm).
- Instrument panel and the instrument panel fasteners
- Steering column (standard & tilt) and the steering wheel
- Blower motor and the blower motor screws
- Blower motor the cover and the cover screws
- Hinge pillar trim kick panels
- Center floor air distribution duct
- Electrical connectors
- Instrument panel storage compartment
- Heater hoses to the heater core

5. Refill the engine cooling system.
6. Connect the negative battery cable.
7. Run the engine to normal operating temperatures; then, check the climate control operation and check for leaks.

Express and Savana

FRONT HEATER

1. Before servicing the vehicle, refer to the precautions in the beginning of this section.
2. Drain the engine cooling system into a clean container for reuse.
3. Discharge and recover the air conditioning system refrigerant.
4. Remove or disconnect the following:
   - Negative battery cable

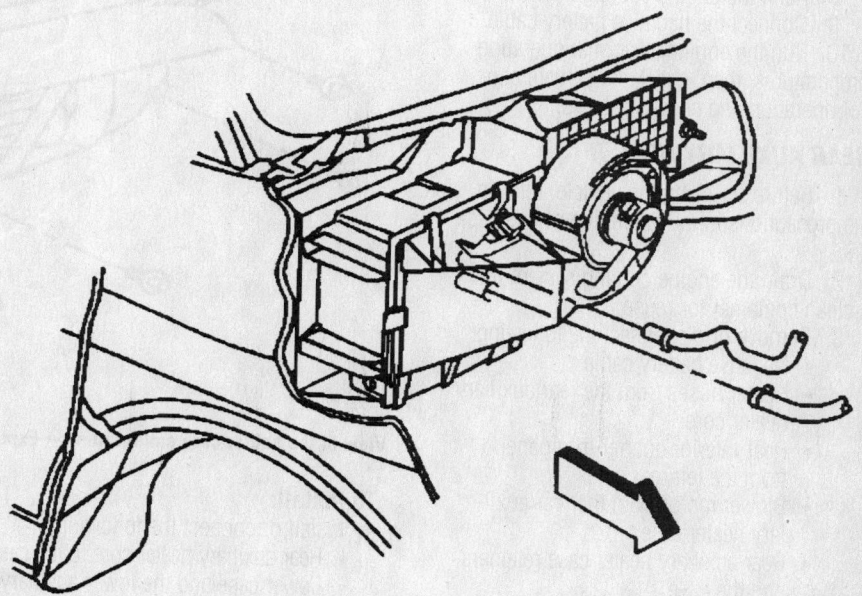

93113G83

Underhood view of the heater assembly—Express and Savana

- Heater hoses from the heater core
- Surge tank (diesel) or the coolant recovery reservoir (except diesel).
- Positive battery cable, the battery hold-down and the battery
- Refrigeration lines from the air conditioning accumulator and discard the gaskets
- Air conditioning accumulator
- Lower right kick panel and the knee bolster, from the right side
- Lower outer floor air outlet duct
- Heater case screws

5. Carefully open the heater core access door.
- Heater core-to-heater case retainers and the heater core

To install:

6. Install or connect the following:
- Heater core and the heater core-to-heater case retainers. Carefully, close the heater core access door.
- Heater case screws
- Lower outer floor air outlet duct
- Knee bolster and the lower right kick panel
- Air conditioning accumulator.
- Refrigeration lines to the air conditioning accumulator, using new gaskets
- Battery, the battery hold-down and the positive battery cable
- Surge tank (diesel) or the coolant recovery reservoir (except diesel)
- Heater hoses to the heater core

7. Evacuate and charge the air conditioning system.
8. Refill the engine cooling system.
9. Connect the negative battery cable.
10. Run the engine to normal operating temperatures; then, check the climate control operation and check for leaks.

REAR AUXILIARY HEATER

1. Before servicing the vehicle, refer to the precautions in the beginning of this section.
2. Drain the engine cooling system into a clean container for reuse.
3. Remove or disconnect the following:
- Negative battery cable
- Heater hoses from the rear auxiliary heater core
- Rear interior quarter trim panel from the left rear side
- Blower motor from the rear auxiliary heater case
- Rear auxiliary heater case retainers and the case
- Lower auxiliary heater case retainers and the lower case
- Rear heater core from the case

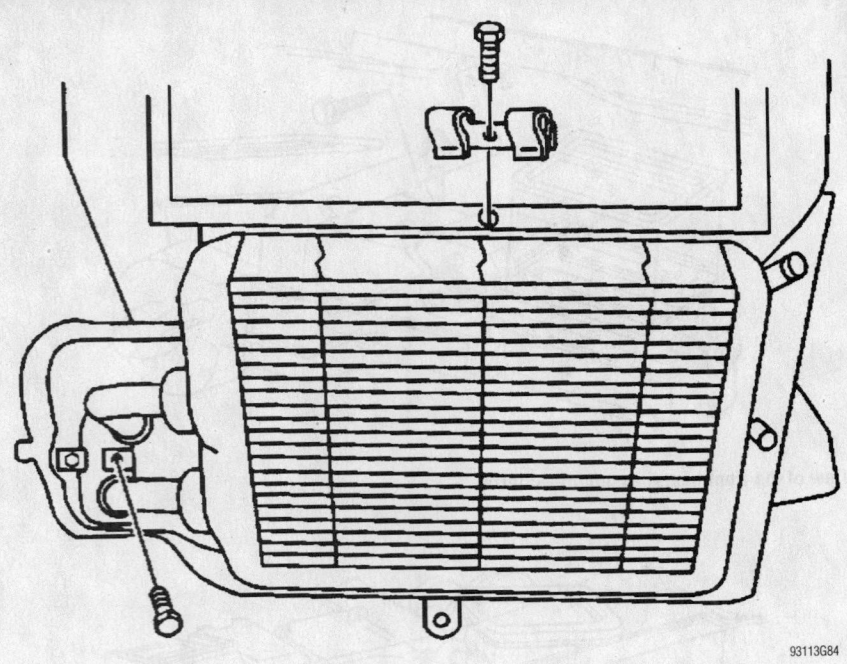

View of the heater case screws—Express and Savana

93113G84

View of the heater core and retainers—Express and Savana

93113G85

To install:
4. Install or connect the following:
- Rear auxiliary heater core to the case
- Lower case and the lower auxiliary heater case retainers
- Rear auxiliary heater case retainers and the case

- Blower motor to the rear auxiliary heater case
- Rear interior quarter trim panel
- Heater hoses to the rear auxiliary heater core

5. Refill the engine cooling system.
6. Connect the negative battery cable.

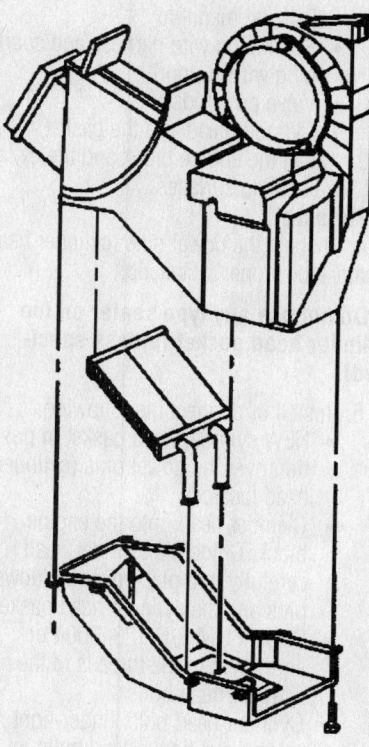

View of the rear auxiliary heater core, case and retainers— Express and Savana

93113G86

Cylinder Head

REMOVAL & INSTALLATION

4.3L Engine—Van Models

1. Before servicing the vehicle, refer to the precautions in the beginning of this section.
2. Drain the coolant.
3. Remove or disconnect the following:

- Negative battery cable
- Engine cover, if equipped
- Intake manifold
- Exhaust manifold
- Air pipe at the rear of the right cylinder head, if applicable
- Alternator mounting bolt at the right cylinder head
- Alternator, if necessary
- Power steering pump and brackets from the left cylinder head and lay aside
- Air conditioner compressor, and lay aside
- Spark plug wires at their brackets
- Ground strap from the right side and the coolant sensor wire from the left head

- Cylinder cover
- Spark plugs
- Pushrods. Identify the pushrods so that they can be installed in their original positions.
- Cylinder head bolts in the reverse order of the tightening sequence
- Cylinder head and gasket

To install:
4. Clean all gasket mating surfaces.
5. Install or connect the following:
- New gasket

➡**Be sure the gasket has the word HEAD up.**

- Cylinder head

➡**Coat a steel gasket on both sides with sealer. If a composition gasket is used, do not use sealer.**

6. Clean the cylinder head bolts, apply sealer to the threads, and hand-tighten.
7. Install the cylinder head bolts in sequence to 22 ft. lbs. (30 Nm) The bolts must, then be tightened again in sequence in the following order:
   a. Step 1: Short length bolt: (11, 7, 3, 2, 6, 10) 55 degrees.
   b. Step 2: Medium length bolt: (12, 13) 65 degrees.
   c. Step 3: Long length bolts: (1, 4, 8, 5, 9) 75 degrees.
8. Install or connect the following
- Pushrods (adjust the rocker arms, if necessary)
- Spark plugs
- Rocker arm cover
- Air conditioner compressor
- Power steering pump and brackets
- Alternator or the alternator mounting bolt at the cylinder head
- Air pipe at the rear of the head if removed
- Exhaust manifold
- Intake manifold
- Engine cover, if removed

9. Refill the engine with coolant.
10. Connect the negative battery cable.

4.3L Engine—Silverado and Sierra

LEFT SIDE

1. Before servicing the vehicle, refer to the precautions in the beginning of this section.
2. Remove or disconnect the following:
- Battery negative cable
- Coolant
- Cooling fan assembly
- Power steering pump mounting bracket
- Power steering pump mounting bracket stud from the cylinder head
- Lower intake manifold
- Exhaust manifold
- Spark plug wire harness and the spark plug wire support
- Valve pushrods
- Ground strap and ground wire bolt from the rear of the cylinder head
- Engine Coolant Temperature (ECT) sensor (if applicable)
- ECT gauge sensor (if applicable)
- Spark plugs
- Spark plug wire support
- Cylinder head bolts
- Cylinder head and the gasket

➡**Clean all dirt, debris, and coolant from the engine block cylinder head bolt holes. Failure to remove all foreign material may result in damaged threads, improperly tightened fasteners or damage to components.**

3. Clean the cylinder head bolts and the engine block bolt holes.
To install:
4. Inspect the dowel pins (cylinder head locator) for proper installation.

➡**Do not use any type sealer on the cylinder head gasket (unless specified).**

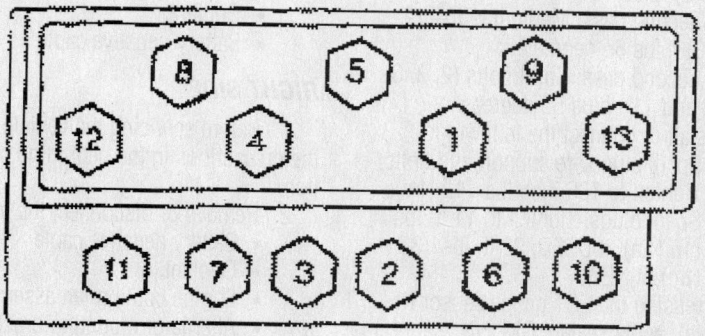

Cylinder head bolt tightening sequence—4.3L engine

7924KG06

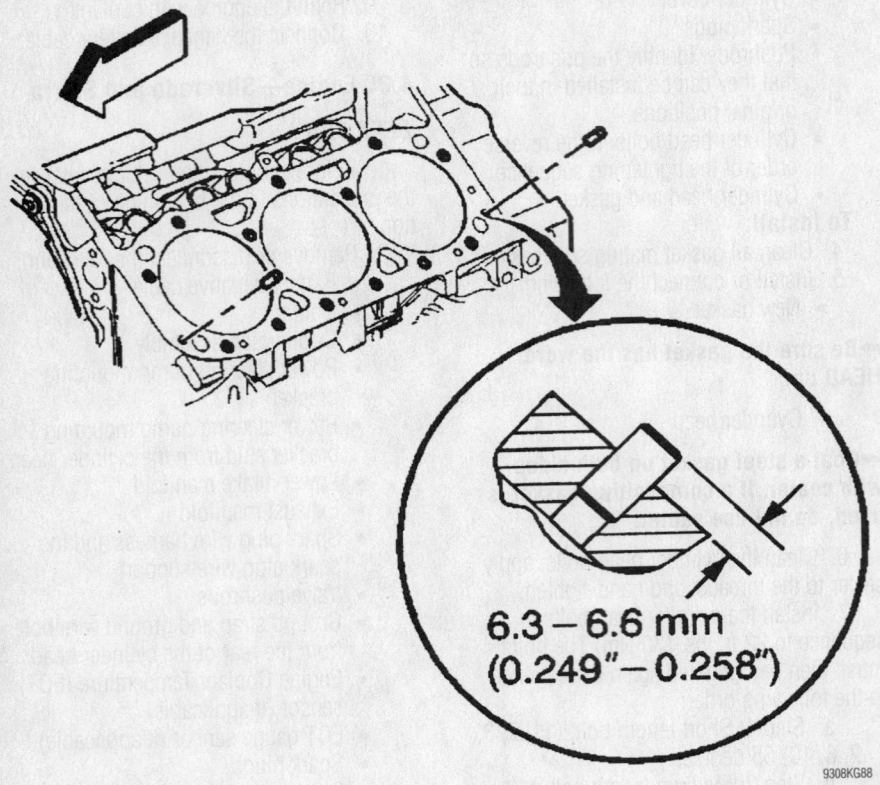

6.3 – 6.6 mm
(0.249" – 0.258")

9308KG88

Dowel pin installation—4.3L engine

5. Install or connect the following:
- NEW cylinder head gasket in position over the dowel pins (cylinder head locator)
- Cylinder head onto the engine block. Guide the cylinder head carefully into place over the dowel pins and the cylinder head gasket.
- Sealant GM P/N 12346004, or equivalent, to the threads of the cylinder head bolts
- Cylinder head bolts finger-tight

6. Tighten the cylinder head bolts in sequence:
a. First pass: 22 ft. lbs. (30 Nm).
b. Second pass: Long bolts (1, 4, 5, 8, and 9)–plus 75 degrees.
c. Second pass: Medium bolts (12 and 13)–plus 65 degrees.
d. Second pass: Short bolts (2, 3, 6, 7, 10, and 11)–plus 55 degrees.

7. Install or connect the following:
- Spark plug wire support and bolts. Tighten to 106 inch lbs. (12 Nm).
- Spark plugs. Tighten to 11 ft. lbs. (15 Nm), if USED; 22 ft. lbs. (30 Nm), if NEW.

8. If reusing the ECT gauge sensor (if applicable), apply sealant GM P/N 12346004 or equivalent to the threads of the ECT gauge sensor. Install the ECT gauge

sensor (if applicable). Tighten the sensor to 15 ft. lbs. (20 Nm).

9. Install or connect the following:
- Ground strap and the ground wire bolt. Tighten the bolt to 12 ft. lbs. (16 Nm).
- Valve pushrods
- Lower intake manifold
- Exhaust manifold
- Stud for the power steering pump mounting bracket to the cylinder head. Tighten the power steering pump mounting bracket stud to 15 ft. lbs. (20 Nm).
- Power steering pump mounting bracket
- Engine cooling fan assembly
- Coolant
- Battery negative cable

RIGHT SIDE

1. Before servicing the vehicle, refer to the precautions in the beginning of this section.

2. Remove or disconnect the following:
- Battery negative cable
- Coolant
- Engine cooling fan assembly
- Alternator mounting bracket
- Alternator mounting bracket stud from the cylinder head

- Lower intake manifold
- Exhaust manifold
- Spark plug wire harness and spark plug wire support
- Valve pushrods
- Cylinder head and the gasket

3. Clean the engine block and the cylinder head sealing surfaces.

To install:

4. Inspect the dowel pins (cylinder head locator) for proper installation.

➡**Do not use any type sealer on the cylinder head gasket (unless specified).**

5. Install or connect the following:
- NEW cylinder head gasket in position over the dowel pins (cylinder head locator)
- Cylinder head onto the engine block. Guide the cylinder head carefully into place over the dowel pins and the cylinder head gasket.
- Sealant GM P/N 12346004 or equivalent to the threads of the cylinder head bolts
- Cylinder head bolts finger-tight

6. Tighten the cylinder head bolts in sequence:
a. First pass: 22 ft. lbs. (30 Nm).
b. Second pass: Long bolts (1, 4, 5, 8, and 9)–plus 75 degrees.
c. Second pass: Medium bolts (12 and 13)–plus 65 degrees.
d. Second pass: Short bolts (2, 3, 6, 7, 10, and 11)–plus 55 degrees.

7. Install or connect the following:
- Spark plug wire support and bolts. Tighten only the rear support bolt to 106 inch lbs. (12 Nm).

➡**The front spark plug wire support bolt is used to fasten the oil level indicator tube, and will be installed within the oil level indicator tube installation procedure.**

- Front spark plug wire support bolt
- Spark plugs. Tighten to 11 ft. lbs. (15 Nm), if USED; 22 ft. lbs. (30 Nm), if NEW.
- Valve pushrods
- Lower intake manifold
- Spark plug wire harness and wire support. Tighten to 106 inch lbs. (12 Nm).
- Exhaust manifold
- Stud for the alternator mounting bracket. Tighten the alternator mounting bracket stud to 15 ft. lbs. (20 Nm).
- Alternator mounting bracket
- Engine cooling fan assembly

- Coolant
- Battery negative cable

4.8L, 5.3L and 6.0L Engines

RIGHT SIDE

✳✳ CAUTION

Before servicing any electrical component, the ignition key must be in the OFF or LOCK position and all electrical loads must be OFF, unless instructed otherwise in these procedures.

1. Before servicing the vehicle, refer to the precautions in the beginning of this section.
2. Remove or disconnect the following:
   - Negative battery cable
   - Intake manifold
   - Push rods
   - Exhaust manifold(s)
   - Alternator
   - Alternator mounting bracket-to-cylinder head bolts
   - Bolt behind the power steering pump
   - Alternator mounting bracket and set it aside
   - Bolt holding the oil level indicator tube to the right side cylinder head
   - Oil level indicator tube
   - Cylinder head(s) from the engine
   - Spark plugs

➡ **The M11 cylinder head bolts are NOT reusable. Install NEW M11 cylinder head bolts during reassembly.**

   - Cylinder head bolts

➡ **After removal, place the cylinder head on two wood blocks to prevent damage.**

3. Remove the gasket. Discard the gasket. Discard the M11 cylinder head bolts.
   To install:

➡ **Do not use any type sealant on the cylinder head gasket (unless specified). The cylinder head gaskets must be installed in the proper direction and position.**

4. Clean the engine block cylinder head bolt holes (if required). Thread repair tool J 42385-107 may be used to clean the threads of old threadlocking material.
5. Spray cleaner GM P/N 12346139, P/N 12377981, or equivalent into the hole.
6. Clean the cylinder head bolt holes with compressed air.
7. Check the cylinder head locating pins for proper installation.

➡ **When properly installed, the tab on the right cylinder head gasket will be located right of center or closer to the front of the engine.**

8. Install or connect the following:
   - NEW right cylinder head gasket onto the locating pins
   - Cylinder head onto the locating pins and the gasket
   - NEW M11 cylinder head bolts. Apply a 0.20 in. (5mm) band of threadlock GM P/N 12345382 or equivalent to the threads of the M8 cylinder head bolts.
   - M8 cylinder head bolts.

9. Tighten the cylinder head bolts as follows:
   a. M11 bolts 1st pass: in sequence to 22 ft. lbs. (30 Nm).
   b. M11 bolts 2nd pass: in sequence plus 90 degrees.
   c. M11 bolts (1,2,3,4,5,6,7,8): plus 90 degrees.
   d. M11 bolts (9 and 10): plus 50 degrees.
   e. M8 cylinder head bolts (11,12,13,14,15) to 22 ft. lbs. (30 Nm). Begin with the center bolt (11) and alternating side-to-side, work outward tightening all of the bolts.

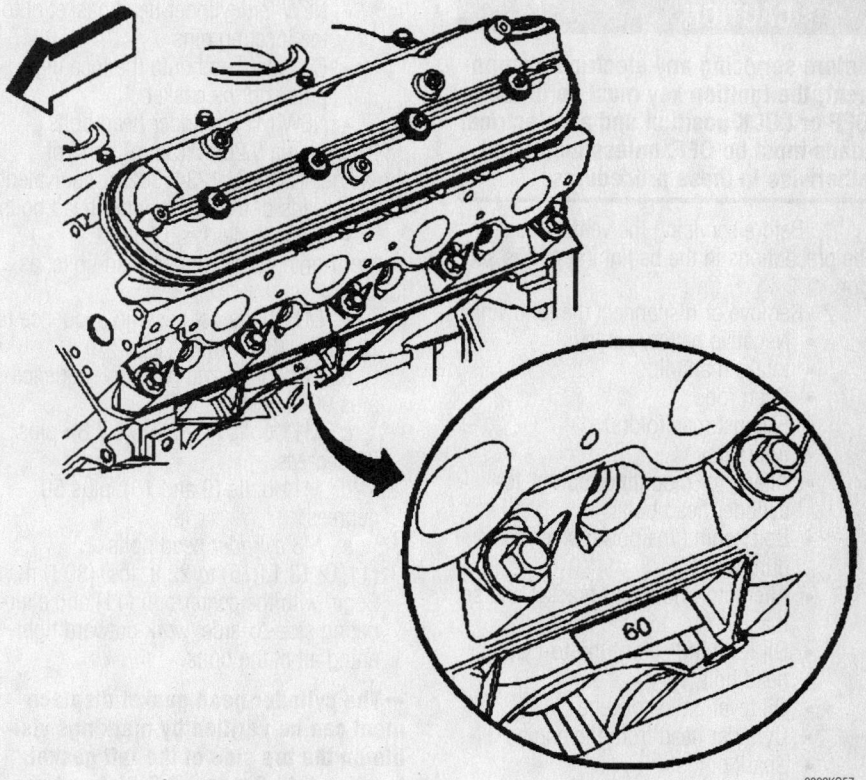

Locating tab—4.8L, 5.3L and 6.0L engines

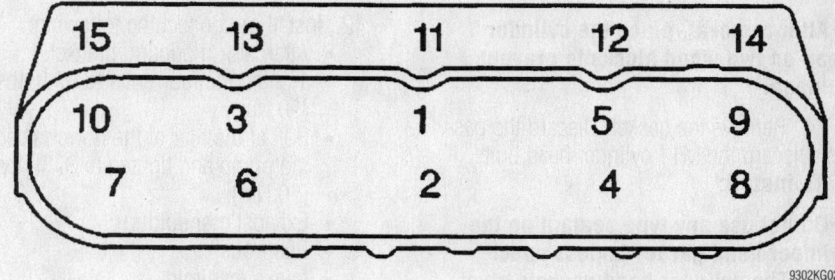

Cylinder head bolt tightening sequence—4.8L, 5.3L and 6.0L engines

➡ The cylinder head gasket displacement can be verified by markings visible on the underside of the right gasket locating tab. Some 4.8/5.3L head gaskets may have 53 stamped onto the locating tab. Some 6.0L head gaskets may have 60 stamped onto the locating tab.

10. Install or connect the following:
- Alternator
- Exhaust manifold(s)
- Pushrods
- Intake manifold
- Negative battery cable

LEFT SIDE

✳✳ CAUTION

Before servicing any electrical component, the ignition key must be in the OFF or LOCK position and all electrical loads must be OFF, unless instructed otherwise in these procedures.

1. Before servicing the vehicle, refer to the precautions in the beginning of this section.
2. Remove or disconnect the following:
- Negative battery cable
- Intake manifold
- Push rods
- Exhaust manifold(s)
- Alternator
- Alternator mounting bracket-to-cylinder head bolts
- Bolt behind the power steering pump
- Alternator mounting bracket and set it aside
- Oil level indicator tube-to-cylinder head bolt
- Oil level indicator tube
- Cylinder head from the engine
- Spark plugs

➡ The M11 cylinder head bolts are NOT reusable. Install NEW M11 cylinder head bolts during assembly.

3. Remove the cylinder head bolts.

➡ After removal, place the cylinder head on two wood blocks to prevent damage.

4. Remove the gasket. Discard the gasket. Discard the M11 cylinder head bolts.
To install:

➡ Do not use any type sealant on the cylinder head gasket (unless specified). The cylinder head gaskets must be installed in the proper direction and position.

5. Clean the engine block cylinder head bolt holes (if required). Thread repair tool J 42385-107 may be used to clean the threads of old threadlocking material.
6. Spray cleaner GM P/N 12346139, P/N 12377981, or equivalent into the hole.
7. Clean the cylinder head bolt holes with compressed air.
8. Check the cylinder head locating pins for proper installation.

➡ When properly installed, the tab on the left cylinder head gasket will be located left of center or closer to the front of the engine.

9. Install or connect the following:
- NEW left cylinder head gasket onto the locating pins
- Cylinder head onto the locating pins and the gasket
- NEW M11 cylinder head bolts.
10. Apply a 0.20 in. (5mm) band of threadlock GM P/N 12345382 or equivalent to the threads of the M8 cylinder head bolts.
- M8 cylinder head bolts
11. Tighten the cylinder head bolts, as follows:
a. M11 bolts 1st pass: in sequence to 22 ft. lbs. (30 Nm).
b. M11 bolts 2nd pass: in sequence plus 90 degrees.
c. M11 bolts (1,2,3,4,5,6,7,8): plus 90 degrees.
d. M11 bolts (9 and 10): plus 50 degrees.
e. M8 cylinder head bolts (11,12,13,14,15) to 22 ft. lbs. (30 Nm). Begin with the center bolt (11) and alternating side-to-side, work outward tightening all of the bolts.

➡ The cylinder head gasket displacement can be verified by markings visible on the top side of the left gasket locating tab. Some 4.8/5.3L head gaskets may have 53 stamped onto the locating tab. Some 6.0L head gaskets may have 60 stamped onto the locating tab.

12. Install or connect the following:
- Alternator mounting bracket. Tighten the four bolts to 37 ft. lbs. (50 Nm).
- Bolt at the rear of the power steering pump and tighten to 37 ft. lbs. (50 Nm).
- Exhaust manifold(s)
- Pushrods
- Intake manifold
- Negative battery cable

5.0L and 5.7L Engines

1. Before servicing the vehicle, refer to the precautions in the beginning of this section.
2. Drain the coolant.
3. Remove or disconnect the following:
- Negative battery cable
- Engine cover
- Coolant recovery reservoir, if applicable
- Intake manifold
- Exhaust manifolds and position them out of the way
- Ground strap at the rear of the right AIR pipe, if equipped
4. For vans with A/C, remove the A/C compressor and the forward mounting bracket and lay the compressor aside. Do not disconnect any of the refrigerant lines.
- Exhaust Gas Recirculation (EGR) inlet tube
5. On the right side cylinder head, disconnect the fuel pipe, spark plug wires and wiring harness bracket.
- Nut and stud attaching the main accessory bracket to the cylinder head

➡ You may have to loosen the remaining bolts and studs in order to remove the head.

- Coolant sensor wire
- Spark plug wire bracket
- Cylinder head covers
- Spark plugs
- Pushrods. Identify the pushrods so that they can be installed in their original positions.
- Cylinder head bolts in the reverse order of the tightening sequence
- Heads
To install:
6. Inspect the cylinder head and block mating surfaces. Clean all old gasket material.
7. Install or connect the following:
- Cylinder heads using new gaskets. Install the gaskets with the word **HEAD** up.

➡ Coat a steel gasket on both sides with sealer. If a composition gasket is used, do not use sealer.

8. Clean the bolts, apply sealer to the threads, and hand-tighten.
9. Install the cylinder head bolts in sequence to 22 ft. lbs. (30 Nm). The bolts must be tightened once, then be tightened again in sequence in the following order:
a. Step 1: Short length bolt: (3, 4, 7, 8, 11, 12, 15, 16), plus 55 degrees.

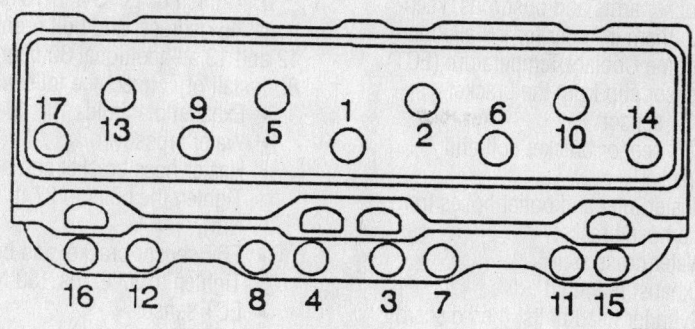

Cylinder head bolt tightening sequence—5.0L and 5.7L engines

b. Step 2: Medium length bolt: (14, 17), plus 65 degrees.

c. Step 3: Long length bolts: (1, 2, 5, 6, 9, 10, 13), plus 75 degrees.

10. Install or connect the following:
- Pushrods in their original positions
- Cylinder head covers
- Spark plugs
- Coolant sensor wire
- Spark plug wire bracket
- Main accessory bracket to the cylinder head
- EGR vent tube
- Fuel pipe
- Spark plug wires
- Wiring harness bracket
- A/C compressor and forward mounting bracket
- Ground strap to the rear of the right AIR pipe
- Exhaust manifolds
- Intake manifold
- Coolant recovery reservoir, if removed
- Engine cover
- Negative battery cable

11. Refill the engine with coolant.

7.4L Engines

1. Before servicing the vehicle, refer to the precautions in the beginning of this section.

2. Drain the cooling system.

3. Remove or disconnect the following:
- Negative battery cable
- Engine cover
- Intake manifold
- Exhaust manifolds
- Alternator and bracket
- Air pump, if equipped
- Air conditioning compressor and the forward mounting bracket. Do not disconnect any of the refrigerant lines.
- Rocker arm cover
- Spark plugs

- Air pipes at the rear of the head, if equipped
- Ground strap at the rear of the head
- Temperature sensor wire
- Pushrods. Identify the pushrods so that they can be installed in their original positions.
- Cylinder head bolts and the heads

To install:

➡**The cylinder head should be cleaned and inspected for warpage or damage before installation.**

4. Thoroughly clean the mating surfaces of the head and block. Clean the bolt holes thoroughly.

➡**Coat a steel gasket on both sides with sealer. If a composition gasket is used, do not use sealer.**

5. Install or connect the following:
- New gaskets, with the word **HEAD** up
- Cylinder heads

6. Apply sealer to the threads of the head bolts, then install and tighten in sequence, in 3 stages, as follows:
  a. Step 1: 30 ft. lbs. (40 Nm).
  b. Step 2: 60 ft. lbs. (80 Nm).
  c. Step 3: 85 ft. lbs. (115 Nm).
- Intake and exhaust manifolds

- Pushrods
- Rocker arms
- Temperature sensor wire
- Ground strap at the rear of the head
- AIR pipes at the rear of the head
- Spark plugs
- Rocker arm cover
- Air conditioning compressor and forward mounting bracket
- AIR pump
- Alternator
- Engine cover

7. Connect the battery cable and refill the cooling system.

8.1L Engine

LEFT SIDE

1. Before servicing the vehicle, refer to the precautions in the beginning of this section.

2. Drain the cooling system.

3. Remove or disconnect the following:
- Negative battery cable
- Intake manifold
- Valve cover
- Rocker arms and pushrods, keeping them in order for installation
- Engine harness ground bolts

4. Reposition the engine harness grounds and ground straps from the cylinder head.
- Water crossover
- Exhaust manifold
- Cylinder head bolts, then discard

➡**The cylinder head bolts must be replaced for installation.**

- Cylinder head. Place the head on 2 wood blocks to protect the sealing surfaces while it is removed.

To install:

➡**The cylinder head should be cleaned and inspected for warpage or damage before installation.**

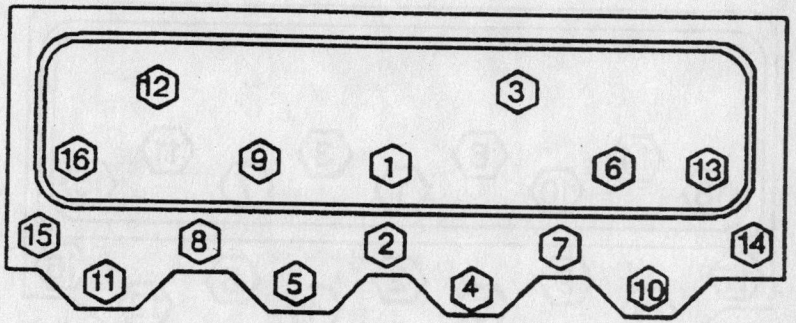

Cylinder head bolt tightening sequence—7.4L engines

5. Thoroughly clean the mating surfaces of the head and block. Clean the bolt holes thoroughly.

➡**If a composition gasket is used, do not use sealer.**

6. Align the cylinder head gasket locating marks to face up. Make sure that the gasket tabs are located of the No. 1 and 2 cylinder for proper installation.

7. Install or connect the following:
- New cylinder head gasket
- Cylinder head
- Sealer to the threads of new cylinder head bolts, if not pre-applied

8. Tighten the head bolts, in sequence, in 3 stages, as follows:
  a. Step 1: 30 ft. lbs. (40 Nm).
  b. Step 2: Additional 120 degrees using a torque angle meter.
  c. Step 3: Torque bolt numbers. 1, 2, 3, 6, 7, 8, 9, 10, 11, 14, 16 and 17 an additional 60 degrees.
  d. Tighten bolts 15 and 18 an additional 45 degrees, and bolt numbers 4, 5, 12 and 13 an additional 30 degrees.

9. Install or connect the following:
- Exhaust manifold
- Water crossover
- Engine harness grounds and ground strap
- Rocker arms and pushrods
- Valve cover
- Intake manifold

10. Connect the battery cable and refill the cooling system.

RIGHT SIDE

1. Before servicing the vehicle, refer to the precautions in the beginning of this section.

2. Drain the cooling system.

3. Remove or disconnect the following:
- Negative battery cable
- Intake manifold
- Valve cover

- Rocker arms and pushrods, keeping them in order for installation
- Engine Coolant Temperature (ECT) sensor clip from the bracket
- ECT sensor
- ECT sensor bracket bolt and bracket
- Heater inlet and outlet hoses from the hose bracket
- Water crossover
- Exhaust manifold
- Cylinder head bolts, then discard

➡**The cylinder head bolts must be replaced for installation.**

- Cylinder head. Place the head on 2 wood blocks to protect the sealing surfaces while it is removed.

To install:

➡**The cylinder head should be cleaned and inspected for warpage or damage before installation.**

4. Thoroughly clean the mating surfaces of the head and block. Clean the bolt holes thoroughly.

➡**If a composition gasket is used, do not use sealer.**

5. Align the cylinder head gasket locating marks to face up. Make sure that the gasket tabs are located of the no. 1 and 2 cylinder for proper installation.

6. Install or connect the following:
- New cylinder head gasket
- Cylinder head
- Sealer to the threads of new cylinder head bolts, if not pre-applied

7. Tighten the head bolts, in sequence, in 3 stages, as follows:
  a. Step 1: 30 ft. lbs. (40 Nm).
  b. Step 2: Additional 120 degrees using a torque angle meter.
  c. Step 3: Torque bolt numbers. 1, 2, 3, 6, 7, 8, 9, 10, 11, 14, 16 and 17 an additional 60 degrees.

  d. Tighten bolts 15 and 18 an additional 45 degrees, and bolt numbers 4, 5, 12 and 13 an additional 30 degrees.

8. Install or connect the following:
- Exhaust manifold
- Water crossover
- Heater hose bracket and bolts. Tighten the bolts to 37 ft. lbs. (50 Nm).
- ECT sensor bracket and bolt. Tighten to 37 ft. lbs. (50 Nm).
- ECT sensor
- ECT sensor clip
- Rocker arms and pushrods
- Valve cover
- Intake manifold

9. Connect the battery cable and refill the cooling system.

Rocker Arms

REMOVAL & INSTALLATION

4.3L, 5.0L and 5.7L Engines

1. Before servicing the vehicle, refer to the precautions in the beginning of this section.

2. Remove or disconnect the following:
- Engine cover
- Cylinder head cover
- Rocker arm nut. If you are only replacing the pushrod, back the nut off until you can swing the rocker out of the way.
- Rocker arms and balls as a unit

➡**Always remove each set of rocker arms (1 set per cylinder) as a unit.**

- Pushrods and pushrod guides

To install:

3. Install or connect the following:
- Pushrods and their guides. Be sure that they seat properly in each lifter.

4. Position a set of rocker arms (for 1 cylinder) in the proper location.

➡**Install the rocker arms for each cylinder only when the lifters are off the cam lobe and both valves are closed.**

5. Coat the replacement rocker arm with Molykote® or its equivalent, and the rocker arm and pivot with SAE 90 gear oil, and install the pivots.
- Nuts and tighten alternately
- Engine cover

4.8L, 5.3L and 6.0L Engines

1. Before servicing the vehicle, refer to the precautions in the beginning of this section.

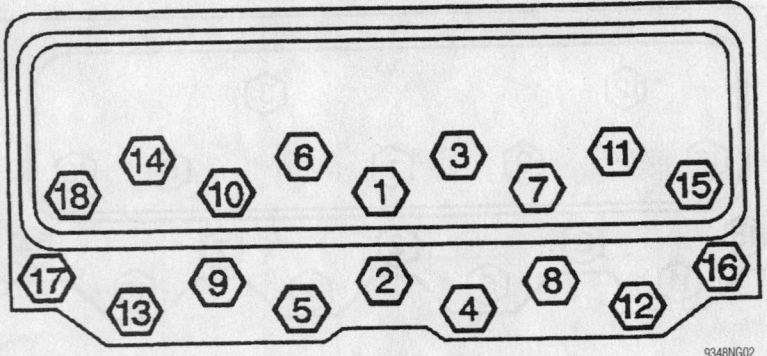

Cylinder head bolt tightening sequence—8.1L engine

9348NG02

→Do not remove the ignition coils from the valve rocker arm cover unless required. Do not remove the oil fill tube from the cover unless service is required. If the oil fill tube has been removed from the cover, install a NEW tube during assembly.

On the right side:

2. Remove or disconnect the following:
   - Ignition coil bracket bolts from the rocker arm cover, if required
   - Ignition coil and bracket assembly from the cover
   - Valve rocker arm cover bolts
   - Valve rocker arm cover
   - Gasket from the cover. Discard the gasket. The bolt grommets may be reused if not damaged.
   - Oil fill cap from the oil fill tube
   - Oil fill tube, if required. Discard the oil fill tube.

On the left side:

→Do not remove the Positive Crankcase Ventilation (PCV) valve grommet from the cover unless service is required.

3. Remove or disconnect the following:
   - Ignition coil bracket bolts from the rocker arm cover (if required)
   - Ignition coil and bracket assembly from the cover
   - Valve rocker arm cover bolts
   - Valve rocker arm cover
   - Gasket from the cover. Discard the gasket. The bolt grommets may be reused if not damaged.
   - Valve rocker arm bolts
   - Valve rocker arms
   - Valve rocker arm pivot support
   - Pushrods

To install:

→Valve lash is built in. No valve adjustment is required.

4. Lubricate the valve rocker arms and pushrods with clean engine oil.

5. Lubricate the flange of the valve rocker arm bolts with clean engine oil.

6. Lubricate the flange or washer surface of the bolt that will contact the valve rocker arm.

7. Install or connect the following:
   - Valve rocker arm pivot support

→Make sure that the pushrods seat properly to the valve lifter sockets.

   - Pushrods

→Make sure that the pushrods seat properly to the ends of the rocker arms.

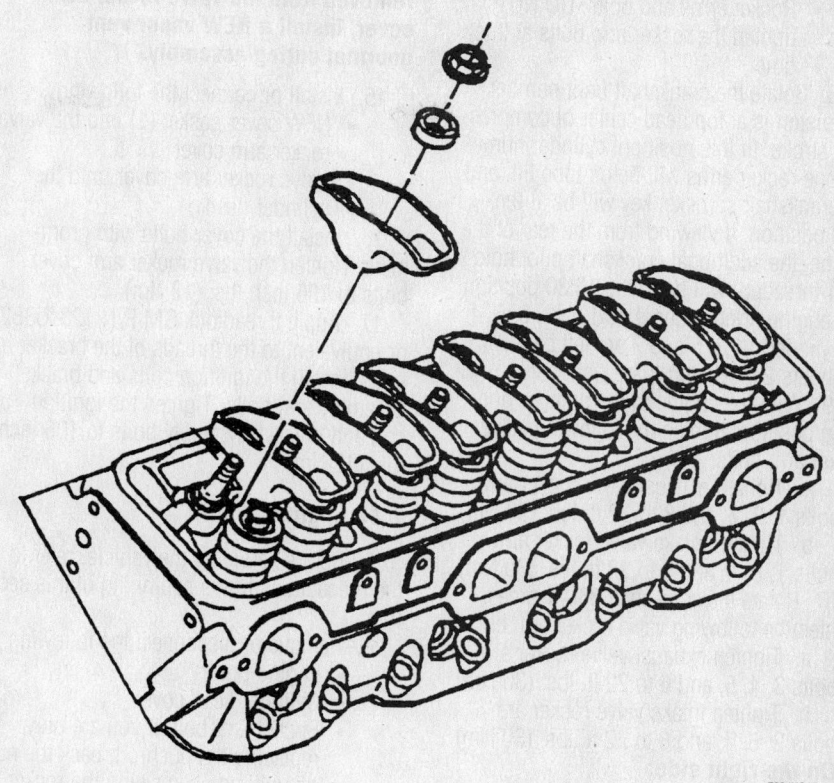

Exploded view to the rocker arm and related components—4.3L, 5.0L and 5.7L engines

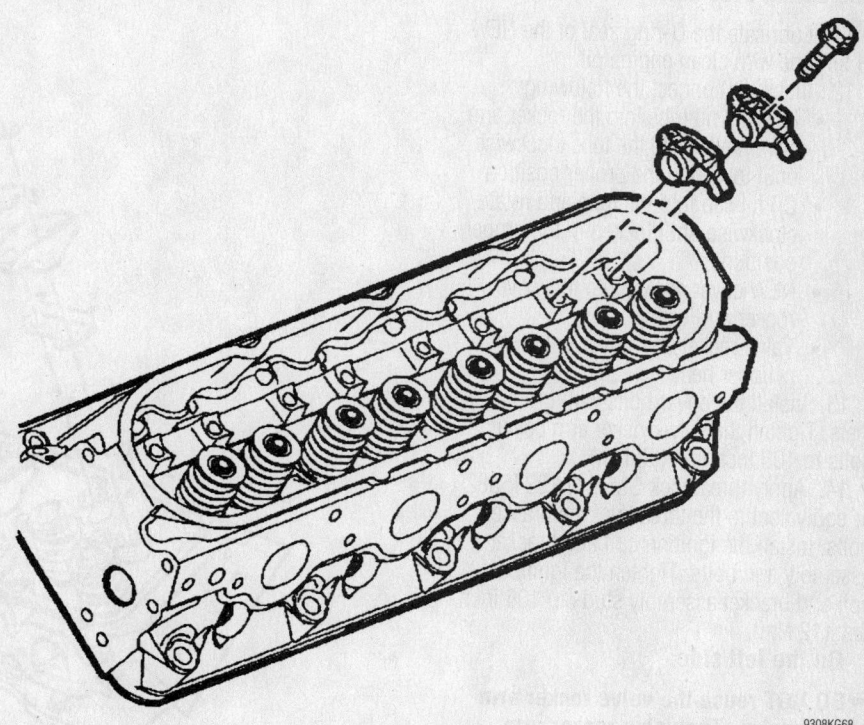

Rocker arm removal—4.8L, 5.3L and 6.0L engines

- Rocker arms and bolts. DO NOT tighten the rocker arm bolts at this time

8. Rotate the crankshaft until number one piston is at top dead center of compression stroke. In this position, cylinder number one rocker arms will be off lobe lift, and the crankshaft sprocket key will be at the 1:30 position. If viewing from the rear of the engine, the additional crankshaft pilot hole (non-threaded) will be in the 10:30 position. The engine firing order is 1, 8, 7, 2, 6, 5, 4, 3. Cylinders 1, 3, 5 and 7 are left bank. Cylinders 2, 4, 6, and 8 are right bank.

9. With the engine in the number one firing position, tighten the following valve rocker arm bolts:

    a. Tighten exhaust valve rocker arm bolts 1, 2, 7, and 8 to 22 ft. lbs. (30 Nm).

    b. Tighten intake valve rocker arm bolts 1, 3, 4, and 5 to 22 ft. lbs. (30 Nm).

10. Rotate the crankshaft 360 degrees. Tighten the following valve rocker arm bolts:

    a. Tighten exhaust valve rocker arm bolts 3, 4, 5, and 6 to 22 ft. lbs. (30 Nm).

    b. Tighten intake valve rocker arm bolts 2, 6, 7, and 8 to 22 ft. lbs. (30 Nm).

On the right side:

➡**The valve rocker arm cover bolt grommets may be reused. If the oil fill tube has been removed from the valve rocker arm cover, install a NEW oil fill tube during assembly.**

11. Lubricate the O-ring seal of the NEW oil fill tube with clean engine oil.

12. Install or connect the following:
- NEW oil fill tube into the rocker arm cover and rotate the tube clockwise until locked in the proper position
- Oil fill cap into the tube and rotate clockwise until locked in the proper position
- NEW cover gasket into the valve rocker arm cover
- Valve rocker arm cover onto the cylinder head

13. Install the cover bolts with grommets. Tighten the valve rocker arm cover bolts to 106 inch lbs. (12 Nm).

14. Apply threadlock GM P/N 12345382 or equivalent to the threads of the bracket bolts. Install the ignition coil and bracket assembly and bolts. Tighten the ignition coil and bracket assembly studs to 106 inch lbs. (12 Nm).

On the left side:

➡**DO NOT reuse the valve rocker arm cover gasket. The valve rocker arm cover bolt grommets may be reused. If the vapor vent grommet has been**

removed from the valve rocker arm cover, install a NEW vapor vent gourmet during assembly.

15. Install or connect the following:
- NEW cover gasket (1) into the valve rocker arm cover
- Valve rocker arm cover onto the cylinder head

16. Install the cover bolts with grommets. Tighten the valve rocker arm cover bolts to 106 inch lbs. (12 Nm).

17. Apply threadlock GM P/N 12345382 or equivalent to the threads of the bracket bolts. Install the ignition coils and bracket assembly and bolts. Tighten the ignition coil and bracket assembly bolts to 106 inch lbs. (12 Nm).

7.4L Engines

1. Before servicing the vehicle, refer to the precautions in the beginning of this section.

2. Remove or disconnect the following:
- Engine cover
- Cylinder head cover
- Rocker arm bolt. If you are only replacing the pushrod, back the nut off until you can swing the rocker out of the way.
- Rocker arms and balls as a unit

➡Always remove each set of rocker arms (1 set per cylinder) as a unit.

- Pushrods and pushrod guides

To install:

3. Install or connect the following:
- Pushrods and their guides, Be sure that they seat properly in each lifter.

4. Position a set of rocker arms (for 1 cylinder) in the proper location.

➡**Install the rocker arms for each cylinder only when the lifters are off the cam lobe and both valves are closed.**

5. Coat the replacement rocker arm with Molykote® or its equivalent, and the rocker arm and pivot with SAE 90 gear oil, and install the pivots.
- Rocker arm bolts. Torque the bolts to 45 ft. lbs. (61 Nm).
- Engine cover

8.1L Engine

➡**Always make sure to keep all removed valve train components in order for reassembly. They must be installed in the same position from which they were removed.**

1. Before servicing the vehicle, refer to the precautions in the beginning of this section.

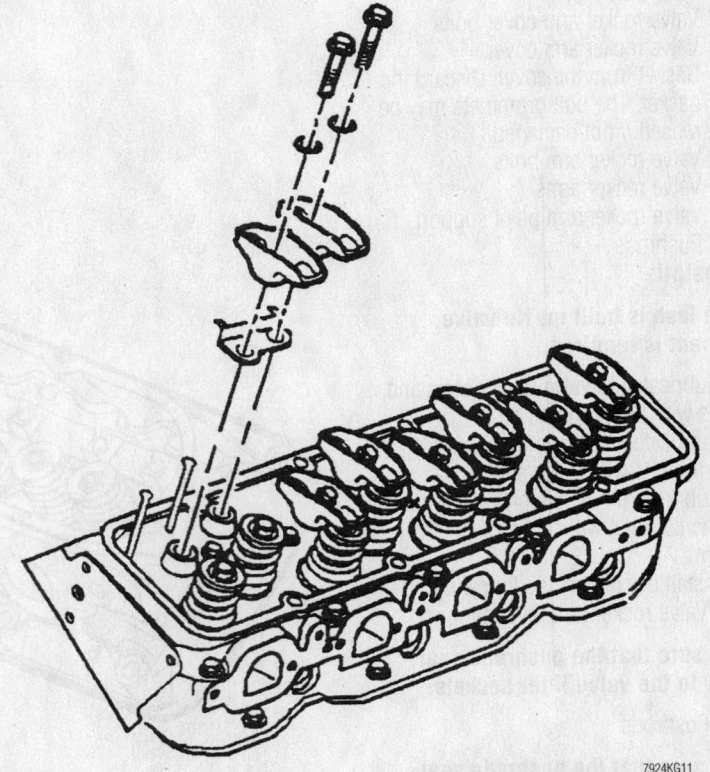

Exploded view of the rocker arms and related components—7.4L engines

7924KG11

2. Remove or disconnect the following:
- Valve (rocker arm) cover
- Rocker arm nuts, balls and rocker arms

➡ **The intake pushrods are shorter than the exhaust pushrods.**

- Pushrods
- Rocker arm guides and pushrod guides

3. Clean and inspect all components for damage.

To install:

4. Apply a suitable sealer to the rocker arm stud-to-cylinder head threads.

5. Install or connect the following:
- Pushrod guides and rocker arm studs. Tighten to 37 ft. lbs. (50 Nm).
- Pushrods

6. Coat the rocker arm and ball bearing surfaces with a suitable prelube.
- Rocker arms, balls and nuts. Tighten the nuts slowly to 18 ft. lbs. (25 Nm) while guiding the tips of the rocker arms over the tips of the valves.
- Valve (rocker arm) cover

Intake Manifold

REMOVAL & INSTALLATION

4.3L Engine

1. Before servicing the vehicle, refer to the precautions in the beginning of this section.

2. Relieve the fuel system pressure

3. Remove or disconnect the following:
- Negative battery cable
- Air intake duct
- Wiring harness connectors and brackets from the manifold
- Throttle linkage and bracket from the upper manifold
- Cruise control cable, if equipped
- Fuel lines at the rear of the lower intake manifold
- Brake booster vacuum hose from the upper intake manifold
- Ignition coil and bracket
- Purge solenoid and bracket
- Studs and intake manifold attaching bolts, mark for reassembly
- Upper intake manifold
- Distributor housing and rotor, mark for reassembly
- Upper radiator hose from the thermostat housing
- Heater hoses and the bypass hose from the lower intake manifold

- Exhaust Gas Recirculation (EGR) valve
- Transmission dipstick tube, if equipped
- Positive Crankcase Ventilation (PCV) valve and hoses
- Air conditioning compressor and bracket. Without disconnecting, position aside
- Alternator bracket and bolt next to the thermostat housing, if needed
- Lower intake manifold mounting bolts and the lower manifold

To install:

4. Clean all gasket mating surfaces thoroughly.

5. Position the new gaskets on the cylinder heads with the port blocking plates at the rear and the words **THIS SIDE UP** facing up.

6. Apply a ³⁄₁₆ inch (5mm) bead of RTV to the front and rear sealing surfaces on the engine block. Extend the bead ½ inch (13mm) up each cylinder head to retain the gasket.

7. Carefully position the lower intake manifold onto the engine.

8. Apply GM 1052080 or equivalent sealer to the lower intake manifold bolts

9. Torque the bolts using 3 steps in the sequence shown:
   a. Step 1: 24 inch lbs. (3 Nm).
   b. Step 2: 108 inch lbs. (12 Nm).
   c. Step 3: 11 ft. lbs. (15 Nm).

10. Install or connect the following:
- Alternator bracket and bolts near the thermostat housing, if removed
- Air conditioning compressor
- PCV valve and hose
- Transmission dipstick tube, if equipped
- EGR valve
- Upper radiator and bypass hose to the thermostat housing
- Distributor.

11. Position the upper intake manifold gasket on the lower manifold.

✳✳ WARNING

Be careful not to pinch the injector tubes between the upper and lower manifolds.

- Upper intake manifold. Torque the bolts and studs to 88 inch lbs. (10 Nm).
- Purge control bracket and valve
- Ignition coil
- Brake booster vacuum
- Fuel lines
- Accelerator cable
- Cruise control cable, if equipped
- Wiring harness brackets and connections
- Air intake duct
- Negative battery cable

12. Refill and bleed the cooling system.

13. Pressurize the fuel system and check for leaks.

4.8L, 5.3L and 6.0L Engines

➡ **The intake manifold, throttle body, fuel injection rail, and fuel injectors may be removed as an assembly. If not servicing the individual components, remove the manifold as a complete assembly.**

1. Before servicing the vehicle, refer to the precautions in the beginning of this section.

2. Remove or disconnect the following:
- Positive Crankcase Ventilation (PCV) hose and valve
- Manifold Absolute Pressure (MAP) sensor, if required
- Engine coolant air bleed clamp and hose from the throttle body
- Accelerator control cable bracket and bolts, if required

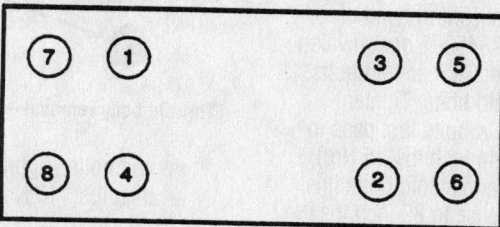

◀ FRT

INTAKE SEQUENCE

7924KG14

Lower intake manifold bolt tightening sequence—4.3L engines

- EVAP solenoid, bolt, and isolator
- Intake manifold bolts
- Intake manifold with gaskets
- Intake manifold-to-cylinder head gaskets from the manifold. Discard the intake manifold gaskets.
- Fuel rail with injectors
- Throttle body and gasket

3. Clean the intake manifold in solvent.

4. Dry the intake manifold with compressed air.

5. Inspect the throttle body and wire harness studs and stud inserts for looseness or damaged threads.

6. Inspect the fuel rail bolt inserts for looseness or damaged threads.

7. Inspect the intake manifold vacuum passages for debris or restrictions.

8. Inspect for damaged or broken vacuum fittings, damaged MAP sensor mounting bore, or broken MAP sensor retaining tabs.

9. Inspect the composite intake manifold assembly for cracks or other damage.

10. Inspect the areas between the intake runners. Inspect all the gasket sealing surfaces for damage.

11. Inspect the fuel injector bores for excessive scoring or damage. Inspect the intake manifold cylinder head deck for warpage.

12. Locate a straight edge across the intake manifold cylinder head deck surface. Position the straight edge across a minimum of two runner port openings.

13. Insert a feeler gauge between the intake manifold and the straight edge. A intake manifold with warpage in excess of 0.118 in. (3mm) over a 7.87 in. (200mm) area is warped and should be replaced.

To install:

14. Install or connect the following:
- MAP sensor
- EVAP solenoid, bolt, and isolator. Tighten the bolt to 89 inch lbs. (10 Nm).
- NEW intake manifold-to-cylinder head gaskets
- Intake manifold

15. Apply a 0.20 in. (5mm) band of threadlock GM P/N 12345382 or equivalent to the threads of the intake manifold bolts.
- Intake manifold bolts. Tighten intake manifold bolts first pass in sequence to 44 inch lbs. (5 Nm). Tighten intake manifold bolts final pass in sequence to 89 inch lbs. (10 Nm).
- PCV valve and hose
- Coolant air bleed hose and clamp onto the throttle body
- Accelerator control cable bracket

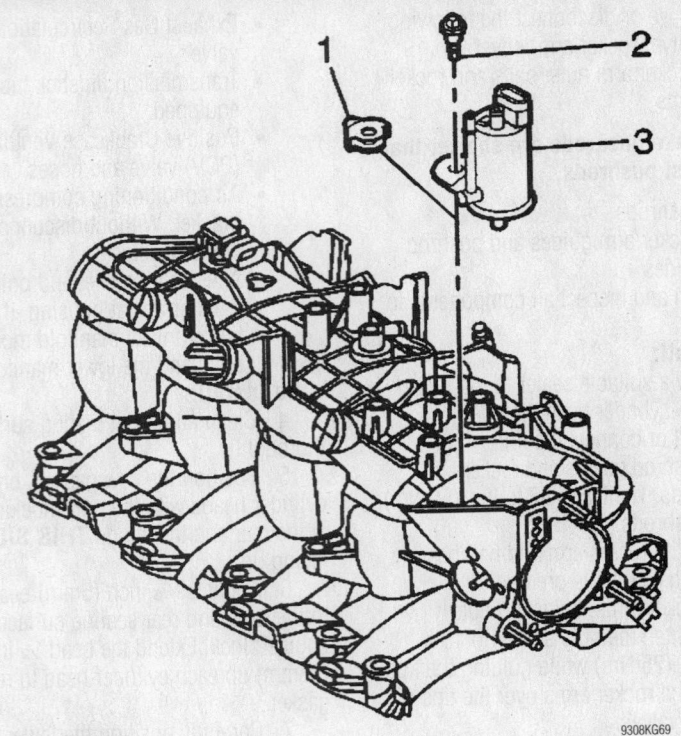

EVAP solenoid removal—4.8L, 5.3L and 6.0L engines

9308KG69

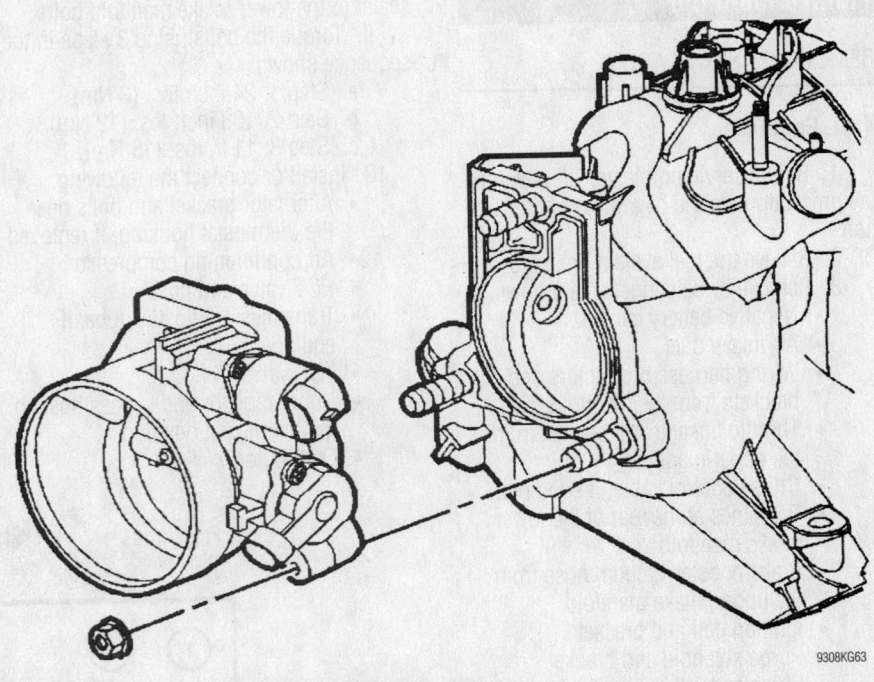

Throttle body removal—4.8L, 5.3L and 6.0L engines

9308KG63

and bolts. Tighten the bolts to 89 inch lbs. (10 Nm).

5.0L, 5.7L and 7.4L Engines

1. Before servicing the vehicle, refer to the precautions in the beginning of this section.

2. Remove or disconnect the following:
- Negative battery cable
- Engine cover
- Air cleaner intake duct
- Coolant reservoir
- Wiring harness connectors and brackets

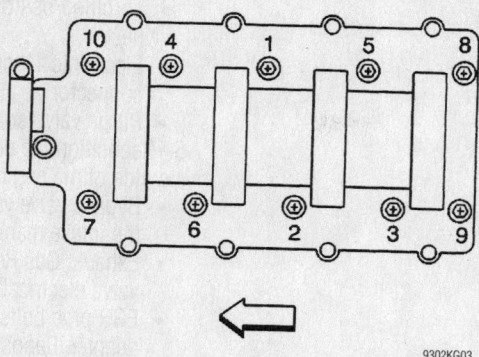

Lower intake manifold bolt tightening sequence—4.8L, 5.3L and 6.0L engines

- Throttle linkage and bracket from the upper intake manifold
- Cruise control cable, if equipped
- Fuel lines and the bracket from the rear of the intake manifold
- Positive Crankcase Ventilation (PCV) valve and hoses
- Ignition coil and bracket
- Purge solenoid and bracket

➡ **Note the location of the manifold bolts and studs before removal for reassembly in their original positions.**

- Intake manifold bolts and studs
- Upper intake manifold

3. Clean the old gasket residue from both mating surfaces.
- Distributor
- Upper radiator hose from the thermostat housing
- Heater hose from the lower intake manifold
- Coolant bypass hose
- Exhaust Gas Recirculation (EGR) valve
- Fuel pressure and return lines from the lower intake manifold
- Wiring harnesses and brackets from the lower manifold
- Left side valve cover
- Transmission oil level indicator and tube, if equipped
- EGR tube, clamp and bolt
- Air conditioning compressor and bracket, but do not disconnect the lines

4. Loosen the compressor mounting bracket and slide it forward, but do not remove it.
- Power brake vacuum tube
- Lower intake manifold bolts and lower intake manifold

To install:
5. Clean all gasket surfaces completely.
6. Install the intake manifold gaskets with the port blocking plates facing the rear.

Factory gaskets should have the words **This Side Up** visible.
7. Apply gasket sealer to the front and rear sealing surfaces of the engine block. Extend the sealer approximately ½ inch (13mm) onto the heads.
8. Install the lower intake manifold.
9. Apply sealer to the lower intake manifold bolts prior to installation.
10. On the 5.0L and 5.7L engines, install the bolts and torque in sequence as follows:
   a. Step 1: 71 inch lbs. (8 Nm).
   b. Step 2: 106 inch lbs. (12 Nm).
   c. Step 3: 11 ft. lbs. (15 Nm).
11. On the 7.4L engine, torque the bolts to 30 ft. lbs. (40 Nm) in the sequence shown.
12. Install or connect the following:
   - Power brake vacuum tube
   - PCV valve and hose
   - EGR tube, clamp and bolt
   - Transmission oil level indicator and tube, if equipped
   - Left side valve cover
   - Wiring harnesses and brackets to the lower manifold
   - Fuel pressure and return lines to the lower intake manifold
   - EGR valve
   - Coolant bypass hose
   - Heater hose to the lower intake manifold
   - Upper radiator hose to the thermostat housing
   - Air conditioning compressor and bracket
   - Distributor
   - Upper intake manifold gasket
   - Upper intake manifold

✲✲ WARNING

When installing the upper intake manifold be careful not to pinch the injector wires between the upper and lower intake manifolds.

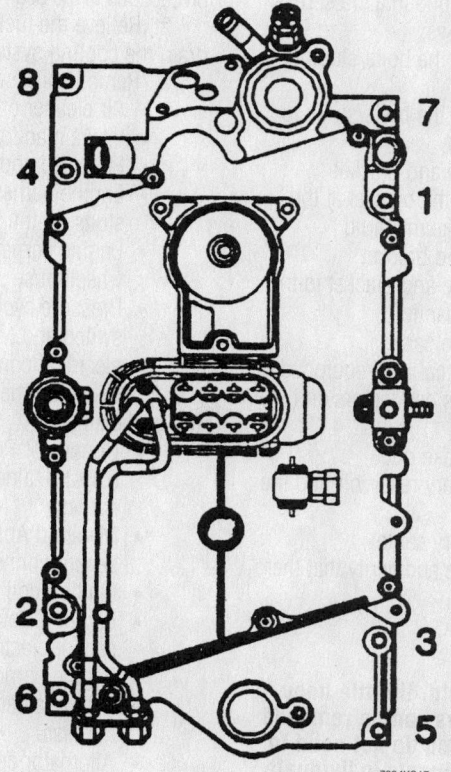

Lower intake manifold bolt tightening sequence—5.0L and 5.7L engines

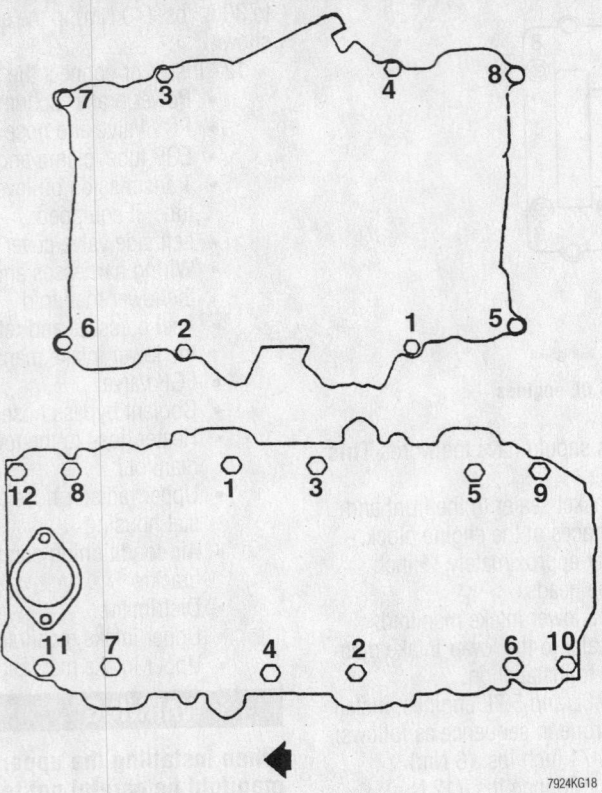

Upper and lower intake manifold bolt tightening sequence—7.4L engines

- Upper intake manifold mounting bolts/studs, torque in a crisscross pattern as follows:
  a. Step 1: Torque the bolts/studs to 44 inch lbs. (5 Nm).
  b. Step 2: Torque the bolts/studs to 83 inch lbs. (10 Nm).
  - Purge solenoid and bracket
  - Fuel lines and the bracket at the rear of the intake manifold
  - Ignition coil and bracket
  - Throttle linkage and bracket to the upper intake manifold
  - Throttle linkage cable
  - Cruise control cable, if equipped
  - Wiring harness connectors and brackets
  - Air cleaner intake duct
  - Coolant recovery reservoir and the engine cover
  - Negative battery cable

13. Start the vehicle and verify that there are no leaks.

8.1L Engine

➥The intake manifold, throttle body, fuel rail and injectors can be removed as an assembly. If you do not need to service these components individually, remove the manifold as a complete assembly.

1. Before servicing the vehicle, refer to the precautions in the beginning of this section.
2. Relieve the fuel system pressure and drain the cooling system.
3. Remove or disconnect the following:
   - Air cleaner outlet duct
   - Intake manifold sight shield
   - Fuel feed and return pipes
   - Engine harness clips from the studs on the front of the dash
   - Engine harness clip from the wheelhouse splash shield
   - Pressure cycling switch, surge tank switch and Mass Air Flow (MAF) electrical connectors
4. Reposition the engine harness to the top of the engine
   - Connector Position Assurance (CPA) retainer from the ignition coil harness
   - Manifold Absolute Pressure (MAP) sensor connector
   - Ignition coil connector(s)
   - Engine Coolant Temperature (ECT) sensor electrical connector
   - Engine harness bolt and studs
   - CPA retainer from the ignition coil harness
   - Alternator connector
   - Injector harness connector
   - Ignition coil harness connector

- Throttle Position (TP) sensor connector
- Electronic Throttle Control (ETC) connector
- Purge valve solenoid connector
5. Reposition the engine harness to the drivers side of the engine compartment.
   - Bypass valve vacuum hose from the intake manifold
   - Exhaust Gas Recirculation (EGR) valve electrical connector
   - EGR pipe bolts from the EGR adapter. Reposition the EGR pipe
   - EGR valve pipe gasket and discard
   - Secondary Air Injection (AIR) pipe nut from the fuel rail stud, if equipped
   - AIR pipe bolts from the exhaust manifold
   - AIR pipe from the AIR pump pipe
   - AIR pipe gasket and discard
   - AIR pipe nut from the fuel rail stud
   - AIR pipe bolts from the exhaust manifold
   - AIR pipe from the AIR pump pipe
   - AIR pipe gasket
6. Reposition the AIR pump hose clamp, then remove the air pump hose from the pump pipe.
   - AIR pump pipe bolt from the cylinder head
   - AIR pump pipe
   - Fuel pressure regulator vacuum hose
   - Fuel rail studs and fuel rail, ONLY if replacing the manifold
   - Intake manifold bolts

✴✴ WARNING

Do NOT try to remove the intake manifold by prying under the sealing surfaces.

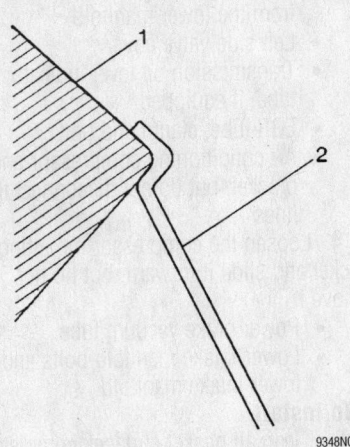

Make sure that the splash shield snap fits between the cylinder heads

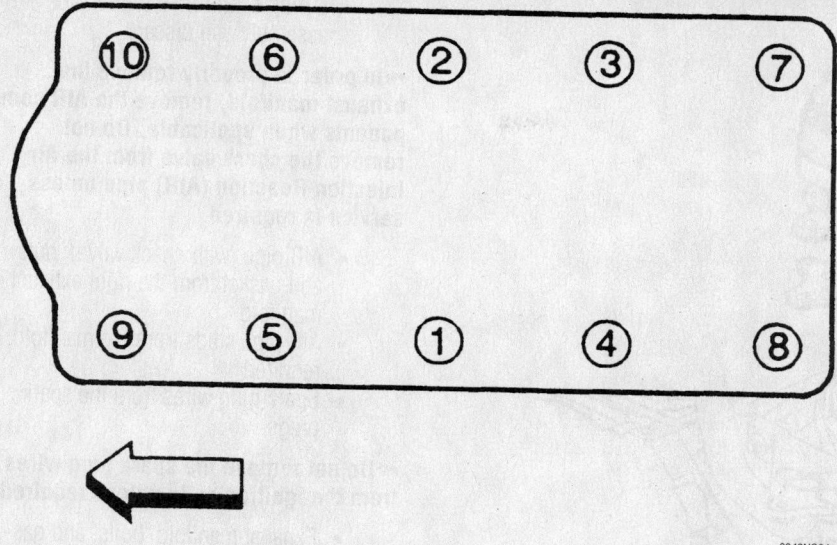

Intake manifold bolt tightening sequence—8.1L engine

9348NG04

- Intake manifold
- Intake manifold side gaskets and end seals and discard

➡**The splash shield is reusable and secured using a snap-in fit. Do not distort the shield during removal.**

- Splash shield

To install:

7. Clean all gasket surfaces completely.
8. Install or connect the following:
- Splash shield. Make sure the shield fits properly between the cylinder head.

➡**Make sure the manifold gasket tabs align with the hole in the head gasket.**

- New intake manifold end seals
- New intake manifold side gaskets onto the heads. Make sure the stamped THIS SIDE UP is showing.
- Intake manifold to the block
- Apply a suitable thread locking material to at least 8 threads of the intake manifold bolts

9. Install the intake manifold bolts and tighten, in the sequence shown, in 4 passes:
a. 1st pass: 44 inch lbs. (5 Nm).
b. 2nd pass: 44 inch lbs. (5 Nm). Check the manifold joints for shifting and fix as necessary.
c. 3rd pass: 89 inch lbs. (10 Nm).
d. 4th pass: 106 inch lbs. (12 Nm).
10. Install the remaining components in the reverse order of the removal procedure.
11. Fill the cooling system, then connect the negative battery cable
12. Start the vehicle and verify that there are no leaks.

Exhaust Manifold

REMOVAL & INSTALLATION

4.3L Engines

1. Before servicing the vehicle, refer to the precautions in the beginning of this section.
2. Remove or disconnect the following:
- Negative battery cable
- Engine cover, if equipped
- Exhaust Gas Recirculation (EGR) valve inlet pipe (left side manifold), if necessary
- Exhaust pipe from the exhaust manifold
- Spark plug wires from the plugs and the retaining clips
- Heat shields
3. If removing the left side manifold:
a. Remove the power steering/alternator rear bracket, if needed.
b. Check for sufficient clearance between the manifold and the intermediate steering shaft. On some models it will be necessary to disconnect the intermediate shaft from the steering gear in order to reposition the shaft for clearance.
4. If removing the right side manifold:
a. Remove air conditioning compressor and bracket, then position the assembly aside, if necessary. Do not disconnect the lines or allow them to become kinked or otherwise damaged.
b. Remove the spark plugs, dipstick tube and wiring, if necessary.
5. Unbend the exhaust manifold bolt lock tangs.

6. Remove or disconnect the following:
- Exhaust manifold retaining bolts, washers and tab washers
- Exhaust manifold
- Old gaskets and discard

To install:

7. Clean the gasket mounting surfaces.
8. Inspect the exhaust manifold for distortion, cracks or damage; replace if necessary.
9. Install or connect the following:
- Exhaust manifold to the cylinder using a new gasket. Torque the exhaust manifold-to-cylinder head bolts to 26 ft. lbs. (36 Nm) on the center exhaust tube and to 20 ft. lbs. (28 Nm) on the front and rear exhaust tubes.

➡**Once the bolts are tightened, bend the tabs on the washers back over the heads of all bolts in order to lock them in position.**

10. On the right side install:
- Spark plugs
- Dipstick tube
- Wiring
- Air conditioning compressor and bracket assembly, if unbolted
11. If the left manifold was removed install:
- Intermediate shaft to the steering gear, if unbolted
- Power steering/alternator rear bracket
- Air cleaner along with the heat stove pipe and cold air intake pipe
- Spark plug wires to the retainer clips and plugs
- Exhaust pipe to the manifold
- Engine cover, on van models
- Negative battery cable

4.8L, 5.3L and 6.0L Engines

RIGHT SIDE

➡**Do not remove the Exhaust Gas Recirculation (EGR) valve from the pipe assembly unless service is required.**

1. Before servicing the vehicle, refer to the precautions in the beginning of this section.
2. Remove or disconnect the following:
3. EGR valve, gasket, and bolts
- EGR pipe bolt from the intake manifold
- EGR pipe bolts and gasket from the exhaust manifold
- EGR pipe bolts from the cylinder head
- EGR pipe assembly

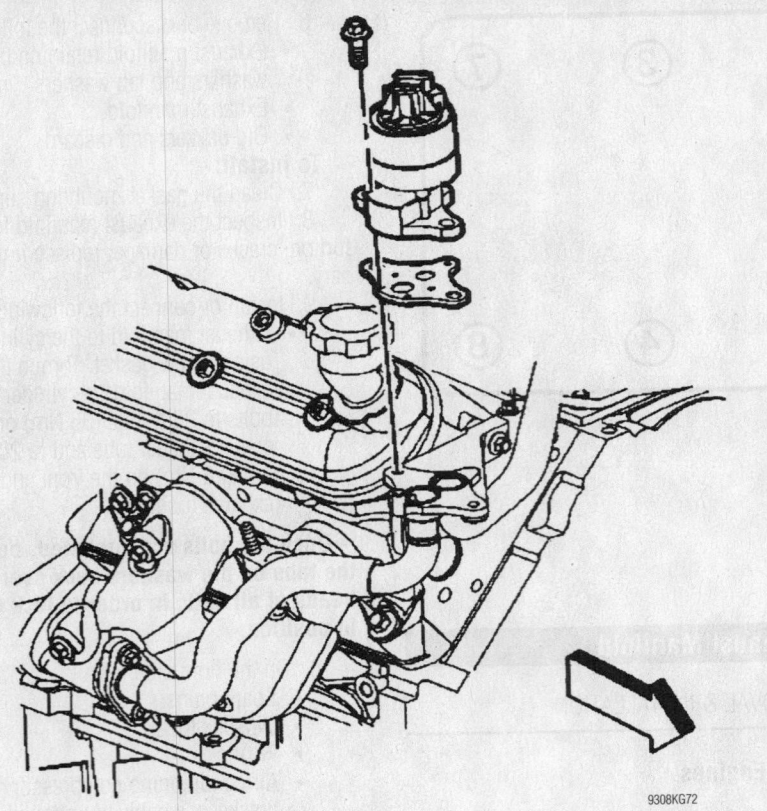

EGR valve removal—4.8L, 5.3L and 6.0L engines

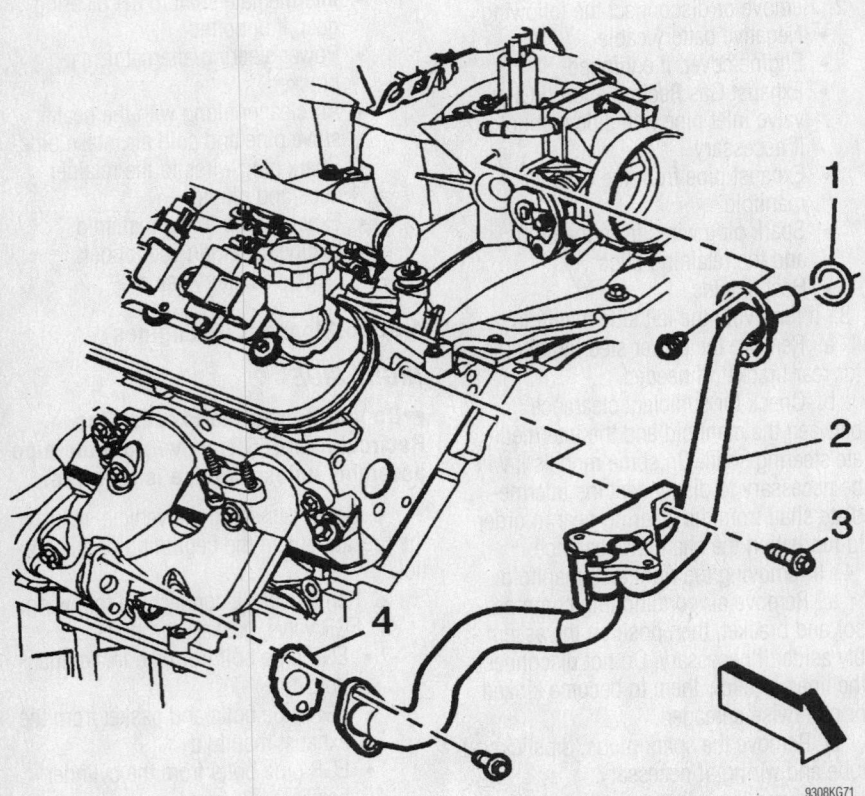

EGR pipe removal—4.8L, 5.3L and 6.0L engines

- O-ring seal from the EGR pipe assembly and discard

➡ **In order to properly remove the exhaust manifold, remove the AIR components when applicable. Do not remove the check valve from the Air Injection Reaction (AIR) pipe unless service is required.**

- AIR pipe (with check valve), nuts and gasket from the right exhaust manifold.
- AIR pipe studs from the manifold, if required
- Spark plug wires from the spark plugs

➡ **Do not remove the spark plug wires from the ignition coils unless required.**

- Exhaust manifold, bolts, and gasket. Discard the gasket.
- Heat shield and bolts from the manifold, if required

To install:

➡ **Do not reuse the exhaust manifold-to-cylinder head gaskets. Upon installation of the exhaust manifold, install a NEW gasket. A improperly installed gasket or leaking exhaust system may effect On-Board Diagnostics (OBD) II system performance.**

4. Clean the exhaust manifold and heat shield in solvent. Dry the exhaust manifold with compressed air.

5. Use a straight edge and a feeler gauge and measure the exhaust manifold cylinder head deck for warpage. An exhaust manifold deck with warpage in excess of 0.01 in. (0.25mm) within the two front or two rear runners or 0.02 in. (0.5mm) overall, may cause an exhaust leak and may effect OBD II system performance. Exhaust manifolds not within specifications must be replaced.

➡ **Do not reuse EGR valve and pipe gaskets or seals during assembly. Install NEW gaskets and O-ring seal.**

6. Apply a 0.2 in. (5mm) wide band of threadlock GM P/N 12345493 or equivalent to the threads of the exhaust manifold bolts.

7. Install the exhaust manifold gasket and exhaust manifold

8. Install the exhaust manifold bolts and tighten, beginning with the center two bolts. Alternate from side-to-side, and work toward the outside bolts.

a. Tighten the exhaust manifold bolts first pass to 11 ft. lbs. (15 Nm).

b. Tighten the exhaust manifold bolts final pass to 22 ft. lbs. (30 Nm). Using a

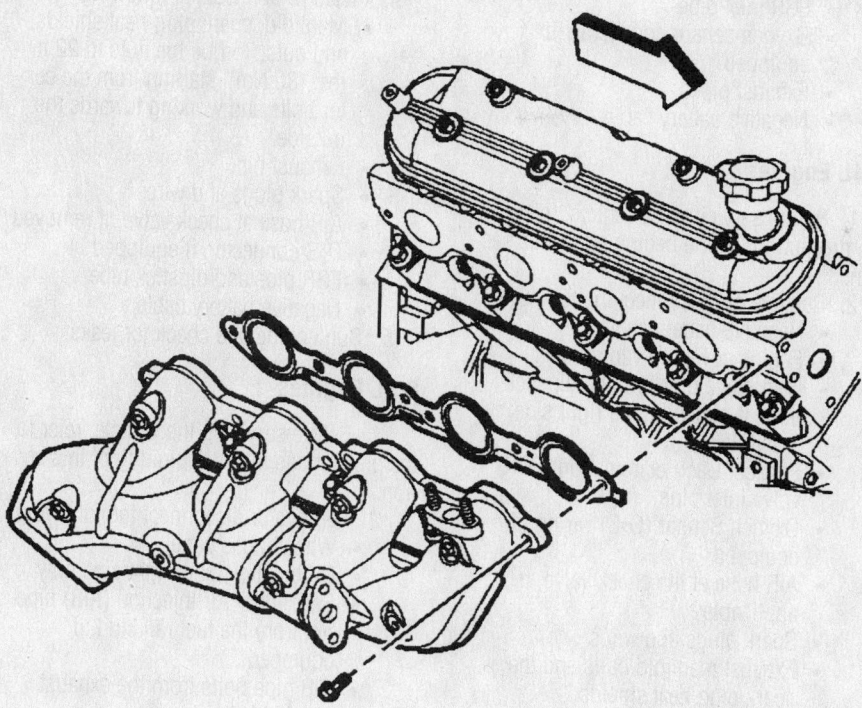

Right exhaust manifold removal—4.8L, 5.3L and 6.0L

9308KG70

flat punch, bend over the exposed edge of the exhaust manifold gasket at the front of the right cylinder head.

9. Install or connect the following:
- Heat shield and bolts and tighten to 80 inch lbs. (9 Nm)
- AIR pipe studs (if required) and tighten to 45 inch lbs. (5 Nm)
- AIR pipe (with check valve), NEW gasket and nuts (if required). Tighten the AIR pipe to exhaust manifold nuts to 18 ft. lbs. (25 Nm).
- AIR hose assembly and clamps
- AIR pipe bracket-to-cylinder head bolt and tighten to 37 ft. lbs. (50 Nm)

10. Apply a light coating of clean engine oil to a NEW O-ring seal and install the seal onto the EGR pipe. Insert the EGR pipe into the intake manifold.

11. Start the EGR pipe to intake manifold bolt. Do not tighten the bolt at this time. Install the EGR pipe to cylinder head bolts. Do not tighten the bolts at this time.

12. Install a NEW EGR pipe exhaust manifold gasket and bolts:

   a. Tighten the EGR pipe to intake manifold bolt to 89 inch lbs. (10 Nm).

   b. Tighten the EGR pipe to cylinder head bolts to 37 ft. lbs. (50 Nm).

   c. Tighten the EGR pipe to exhaust manifold bolts to 22 ft. lbs. (30 Nm).

13. Install the EGR valve, a NEW gasket, and bolts. Tighten the EGR valve bolts a first pass to 89 inch lbs. (10 Nm). Tighten the EGR valve bolts a second pass to 18 ft. lbs. (25 Nm).

LEFT SIDE

➡ **In order to properly remove the exhaust manifold, remove the AIR components when applicable.**

1. Before servicing the vehicle, refer to the precautions in the beginning of this section.

2. Remove or disconnect the following:
- AIR center pipe bolt
- AIR hose clamps and remove the hose assembly

➡ **Do not remove the check valve from the AIR pipe unless service is required.**

- AIR pipe (with check valve), nuts and gasket from the left exhaust manifold
- AIR pipe studs from the manifold (if required)
- Spark plug wires from the spark plugs. Do not remove the spark plug wires from the ignition coils unless required.

3. Exhaust manifold, bolts, and gasket. Discard the gasket.

4. Heat shield and bolts from the manifold, if required

➡ **Do not reuse the exhaust manifold-to-cylinder head gaskets. Upon installation of the exhaust manifold, install a NEW gasket. An improperly installed gasket or leaking exhaust system may effect On-Board Diagnostics (OBD) II system performance.**

5. Clean the exhaust manifold and heat shield in solvent. Dry the exhaust manifold with compressed air.

6. Use a straight edge and a feeler gauge and measure the exhaust manifold cylinder head deck for warpage. An exhaust manifold deck with warpage in excess of 0.01 in. (0.25mm) within the two front or two rear runners or 0.02 in. (0.5mm) overall, may cause an exhaust leak and may effect OBD II system performance. Exhaust manifolds not within specifications must be replaced.

To install:

➡ **Do not apply sealant to the first three threads of the bolt.**

7. Apply a 0.2 in. (5mm) wide band of threadlock GM P/N 12345493 or equivalent to the threads of the exhaust manifold bolts. Install the exhaust manifold and NEW exhaust manifold gasket.

8. Install the exhaust manifold bolts and tighten, beginning with the center two bolts. Alternate from side-to-side, and work toward the outside bolts.

   a. Tighten the exhaust manifold bolts a first pass to 11 ft. lbs. (15 Nm).

   b. Tighten the exhaust manifold bolts a final pass to 18 ft. lbs. (25 Nm).

9. Using a flat punch, bend over the exposed edge of the exhaust manifold gasket at the rear of the left cylinder head.

10. Install the heat shield and bolts and tighten to 80 inch lbs. (9 Nm).

11. Install the Air Injection Reaction (AIR) pipe studs (if required). Tighten the studs to 45 inch lbs. (5 Nm).

12. Install the AIR pipe (with check valve), NEW gasket and nuts (if required). Tighten the AIR pipe to exhaust manifold nuts to 18 ft. lbs. (25 Nm).

5.0L and 5.7L Engines

1. Before servicing the vehicle, refer to the precautions in the beginning of this section.

2. Remove or disconnect the following:
- Negative battery cable
- Engine cover
- Air cleaner, if needed
- Exhaust pipe at the manifold
- Oxygen Sensor (O_2S) wiring, if equipped

- AIR hose at the check valve
- Exhaust Gas Recirculation (EGR) valve, inlet pipe
- Heat stove pipe and the dipstick tube bracket, if working on the right side of the engine
- Power steering pump rear bracket at the manifold, if removing the left side manifold
- Loosen the alternator and remove the lower bracket, if necessary
- Air conditioner compressor rear bracket and the diverter valve and bracket. if needed

➡**On models with air conditioning, it may be necessary to remove the compressor, do not disconnect the compressor lines.**

- Manifold bolts and the manifold(s) Some models have lock tabs on the front and rear manifold bolts which must be removed before removing the bolts.

To install:

3. Clean gasket surfaces, and inspect manifold for cracks replace as necessary.

4. Install the manifold and torque it in the following steps:
   a. Step 1: 15 ft. lbs. (20 Nm).
   b. Step 2: 22 ft. lbs. (30 Nm).

5. Install or connect the following:
- Alternator, if removed
- Air conditioning compressor, if removed
- Diverter, if removed
- Power steering brackets, if removed
- Dipstick tube on right side

- EGR inlet pipe
- Oxygen sensor connector, if equipped
- Exhaust pipes
- Negative battery cable

7.4L Engines

1. Before servicing the vehicle, refer to the precautions in the beginning of this section.

2. Remove or disconnect the following:
- Negative battery cable
- Engine cover, on van models
- Heat stove pipe and the dipstick tube, if removing the right side manifold
- Exhaust Gas Recirculation (EGR) valve inlet pipe
- Oxygen Sensor (O2S) wiring, if equipped
- AIR hose at the check valve, if applicable
- Spark plugs and wires
- Exhaust manifold bolts and the spark plug heat shields

➡**Leave the front nut (left manifold) or rear nut (right manifold) in place for support.**

- Heat shield bolts from the engine mount and bell housing
- Heat shield, if equipped
- Exhaust pipe at the manifold
- Exhaust manifold

To install:

3. Clean the mating surfaces and the retainer threads.

4. Install or connect the following:
- Manifold, spark plug heat shields and nuts. Torque the nuts to 22 ft. lbs. (30 Nm), starting from the center bolts and working towards the outside.
- Exhaust pipe
- Spark plugs and wires
- AIR hose at check valve, if removed
- O2S connector, if equipped
- EGR pipe and dipstick tube
- Negative battery cable

5. Run engine and check for leaks.

8.1L Engines

1. Before servicing the vehicle, refer to the precautions in the beginning of this section.

2. Remove or disconnect the following:
- Wheelhouse panel
- Oil dipstick tube, right side only
- Secondary Air Injection (AIR) pipe nut from the fuel rail stud, if equipped
- AIR pipe bolts from the exhaust manifold
- AIR pipe from the AIR pump pipe
- AIR pipe gasket and discard
- Spark plug wires
- Spark plugs

3. If removing the right exhaust manifold, perform the following:
   a. Remove the Exhaust Gas Recirculation (EGR) pipe nuts from the manifold.
   b. Remove the EGR pipe bracket bolt.
   c. Remove and discard the EGR pipe gasket.
   d. Reposition the EGR pipe.

4. Raise and support the vehicle.

5. Remove or disconnect the following:
- Exhaust manifold heat shield bolts and shield

6. Lower the vehicle.
- Exhaust manifold bolt and nuts
- Exhaust manifold
- Exhaust manifold gasket and discard

To install:

7. Clean the mating surfaces and the retainer threads.

8. Install or connect the following:
- New exhaust manifold gasket
- Exhaust manifold
- Exhaust manifold bolt and nuts. Tighten the bolt to 26 ft. lbs. (35 Nm) and the nuts to 12 ft. lbs. (16 Nm).

9. Raise the vehicle.
- Heat shield. Tighten the retaining bolts and nuts to 18 ft. lbs. (25 Nm).

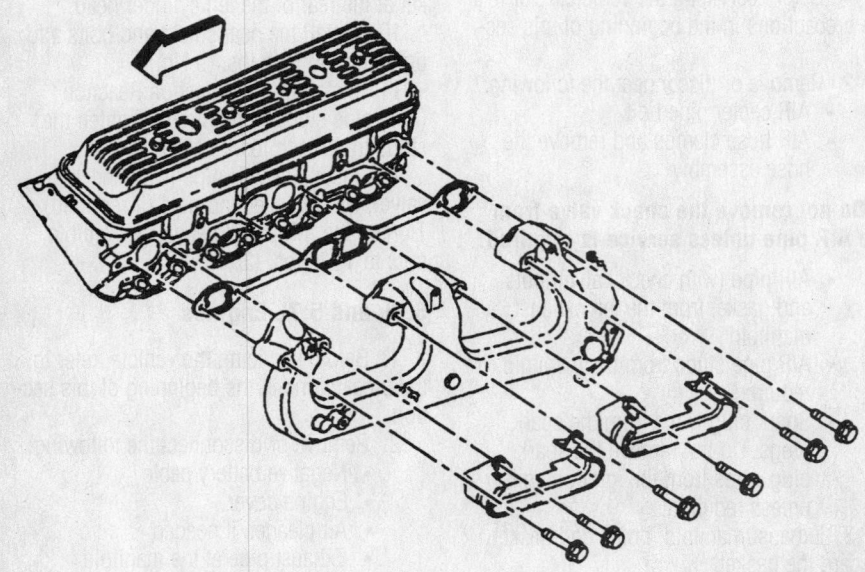

7924KG19

Exploded view of the left exhaust manifold, the right side is similar—5.0L and 5.7L engines

10. Lower the vehicle.
- EGR pipe, right side only. Tighten the pipe bracket bolt to 37 ft. lbs. (50 Nm) and the nuts to 22 ft. lbs. (30 Nm).
- Spark plugs and plug wires
- Air pipe, using a new gasket and reversing the removal procedure
- Wheel house panel

Camshaft and Valve Lifters

REMOVAL & INSTALLATION

4.3L Engines

1. Before servicing the vehicle, refer to the precautions in the beginning of this section.
2. Properly relieve the fuel system pressure.
3. Drain the engine cooling system.
4. Remove or disconnect the following:
- Negative battery cable
- Radiator
- Cooling fan
- Water pump
- Rocker arm covers from the engine
- Intake manifold assembly
- Rocker arms, pushrods and lifters
- Crankshaft pulley and hub
- Engine front cover
5. Align the timing marks on the crankshaft and camshaft sprockets.
- Camshaft sprocket and timing chain
- Balance shaft drive gear, if equipped
- Camshaft thrust plate

➡**Install the sprocket bolts or longer bolts of the same thread into the end of the camshaft as a handle.**

- Camshaft

To install:

6. Lubricate the camshaft journals with clean engine oil or a suitable pre-lube.
7. Install or connect the following:
- Camshaft
- Camshaft thrust plate
- Balance shaft drive gear, if equipped
- Timing chain and camshaft sprocket
- Engine front cover
- Crankshaft pulley and hub
- Valve lifters
- Pushrods and rocker arms, properly adjust the valve clearance
- Intake manifold assembly
- Rocker arm covers to the engine

- Radiator to the vehicle
- Negative battery cable
8. Refill the engine cooling system.

4.8L, 5.3L and 6.0L Engines

1. Before servicing the vehicle, refer to the precautions in the beginning of this section.
2. Raise the hood to the servicing position and secure it. Move the hood hinge bolt to hold the hood in the servicing position.
3. Remove or disconnect the following:
- Battery negative cable
- Coolant
- Upper and lower radiator hoses from the engine
- Air cleaner duct from the engine
- A/C condenser mounting bolts, if equipped
- Radiator support and radiator
- Engine cooling fan
- Drive belt
- A/C drive belt, if equipped
- Engine sight shield
- Electrical wiring harness from the thermostat housing
- Water pump
4. Raise the vehicle.
- Starter motor

- Right side closeout cover and bolt
- Crankshaft balancer
- Engine oil pan
- Engine front cover
- Cylinder heads from the engine
- Valve lifters from the engine
5. Align the timing marks on the camshaft and crankshaft sprockets. Make sure that the number 1 piston is in the firing position.
- Camshaft sprocket
- Camshaft sensor bolt and sensor
- Camshaft retainer bolts and retainer

➡**All camshaft journals are the same diameter, so care must be used in removing or installing the camshaft to avoid damage to the camshaft bearings.**

6. Install the three M8-1.25 x 100 mm bolts in the camshaft front bolt holes. Using the bolts as a handle, carefully rotate and pull the camshaft out of the engine block. Remove the bolts from the front of the camshaft.
7. Clean and inspect all sealing surfaces.

To install:

➡**If camshaft replacement is required, the valve lifters must also be replaced.**

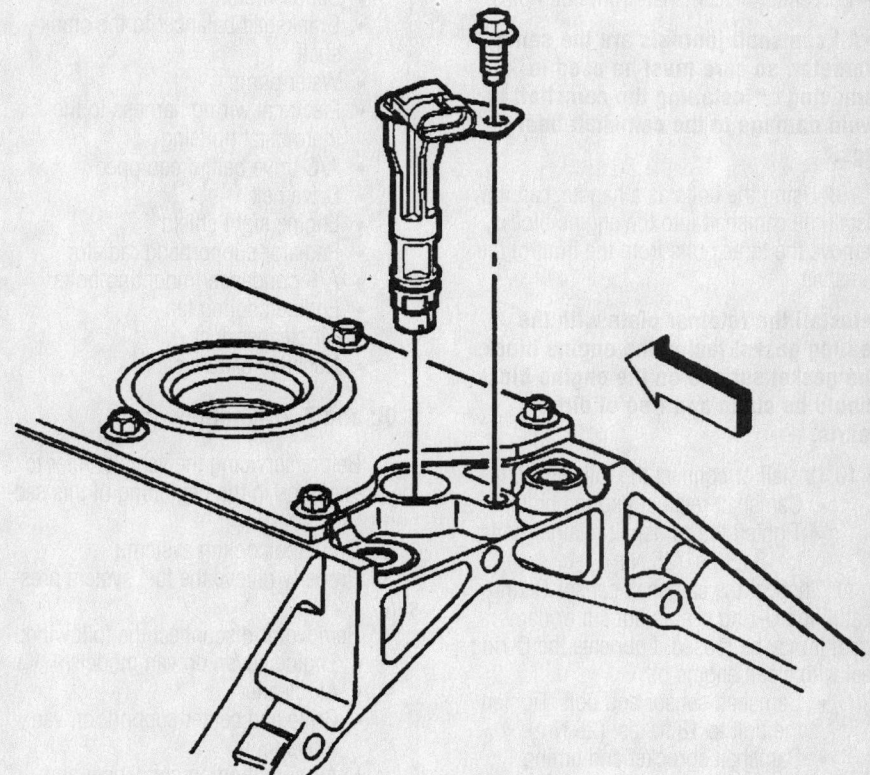

Camshaft sensor removal—4.8L, 5.3L and 6.0L engines

9308KG66

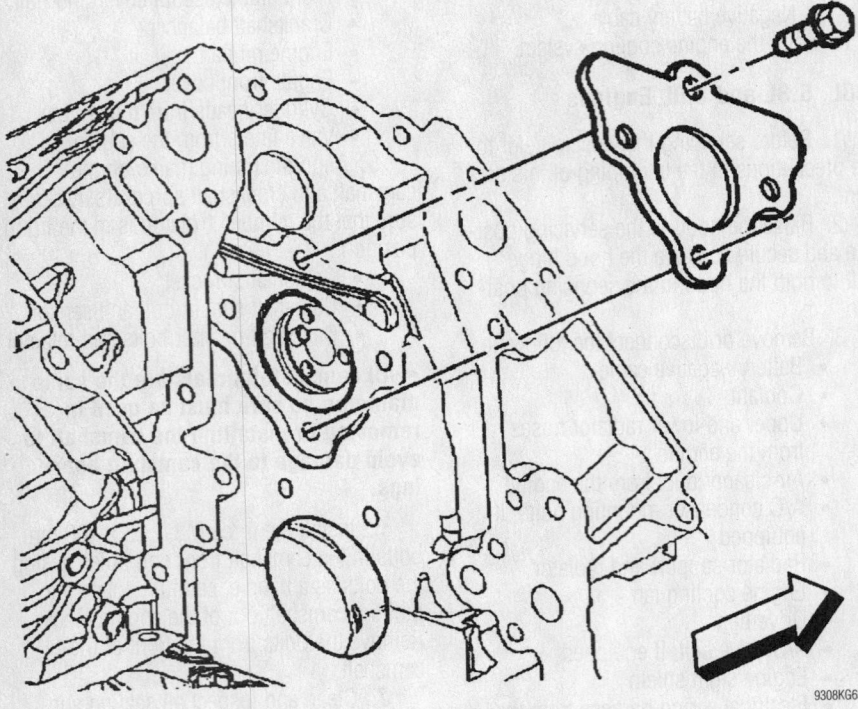

Camshaft retainer removal—4.8L, 5.3L and 6.0L engines

9308KG65

8. Lubricate the camshaft journals and the bearings with clean engine oil. Install three M8-1.25 x 100 mm (M8-1.25 x 4.0 in) bolts into the camshaft front bolt holes.

➡**All camshaft journals are the same diameter, so care must be used in removing or installing the camshaft to avoid damage to the camshaft bearings.**

9. Using the bolts as a handle, carefully install the camshaft into the engine block. Remove the three bolts from the front of the camshaft.

➡**Install the retainer plate with the sealing gasket facing the engine block. The gasket surface on the engine block should be clean and free of dirt or debris.**

10. Install or connect the following:
   • Camshaft retainer and the bolts. Tighten the camshaft retainer bolts to 18 ft. lbs. (25 Nm).

11. Inspect the camshaft sensor O-ring seal. If the O-ring seal is not cut or damaged, it may be reused. Lubricate the O-ring seal with clean engine oil.
   • Camshaft sensor and bolt. Tighten the bolt to 18 ft. lbs. (25 Nm).
   • Camshaft sprocket and timing chain
   • Valve lifters
   • Cylinder heads
   • Engine front cover to the engine
   • Oil pan
   • Right side closeout cover
   • Starter motor
   • Crankshaft balancer to the crankshaft
   • Water pump
   • Electrical wiring harness to the thermostat housing
   • A/C drive belt, if equipped
   • Drive belt
   • Engine sight shield
   • Radiator support and radiator
   • A/C condenser mounting bolts
   • Engine cooling fan
   • Air cleaner duct
   • Negative battery cable

5.0L and 5.7L Engines

1. Before servicing the vehicle, refer to the precautions in the beginning of this section.
2. Drain the cooling system.
3. Properly relieve the fuel system pressure.
4. Remove or disconnect the following:
   • Engine cover, on van models
   • Air cleaner
   • Grille and center support, on van models
   • Air conditioning condenser and swing the condenser forward from its mounting, if equipped
   • Fan, the shroud and the radiator
   • Valve covers
   • Water pump assembly
5. Align the timing marks and remove the torsional damper.
   • Timing chain cover
   • Electrical and vacuum connections at the intake manifold
   • Distributor assembly, mark the distributor rotor-to-housing location
   • Intake manifold, pushrods and hydraulic lifters
   • Camshaft sprocket bolts
   • Camshaft sprocket and timing chain
   • Crankshaft sprocket, as required
   • Front engine mount through-bolts and raise the engine to gain sufficient clearance for camshaft removal, as required
6. Install 2 or 3 ⁵⁄₁₆–18 bolts 4–5 in. (102–127mm) long into the camshaft threaded holes.
   • Camshaft

➡**Inspect the shaft for signs of excessive wear or damage.**

To install:

➡**Liberally coat camshaft and bearing with heavy engine oil or engine assembly lubricant.**

7. Install or connect the following:
   • Camshaft, align the timing marks on the camshaft and crankshaft gears
   • Engine mount through-bolts
   • Camshaft sprocket and chain. Torque the bolts to 18 ft. lbs. (25 Nm).
   • Hydraulic lifters and pushrods
   • Distributor assembly
   • Timing chain cover
   • Torsional damper
   • Water pump
   • Valve covers
   • Fan, the shroud and radiator
   • Air conditioning condenser, if equipped
   • Grille and center support, on van models
   • Air cleaner
   • Engine cover, on van models
   • Negative battery cable
8. Refill the cooling system.

7.4L Engines

1. Before servicing the vehicle, refer to the precautions in the beginning of this section.
2. Drain the cooling system.

3. Properly relieve the fuel system pressure.

4. Properly discharge the air conditioning system.

5. Remove or disconnect the following:
- Negative battery cable
- Engine cover, on van models
- Air cleaner assembly
- Grille and center support section, as required
- Air conditioning compressor, condenser and auxiliary fan, if equipped
- Fan, the shroud
- Radiator
- Accessory belt, as required
- Alternator, as required
- Valve covers
- Hoses from the water pump
- Water pump

6. Align the timing marks at Top Dead Center (TDC).
- Harmonic balancer and pulley
- Engine front cover

7. Mark the distributor rotor-to-housing location.
- Distributor assembly
- Intake manifold assembly
- Lifters, pushrods, and rocker arms

8. Rotate the camshaft so the timing marks align.
- Camshaft sprocket bolts
- Camshaft sprocket and timing
- Engine mount through-bolts

9. Install 2 or 3 ⁵⁄₁₆–18 bolts in the holes in the front of the camshaft and carefully pull the camshaft from the block.

To install:

10. Liberally coat camshaft and bearing with heavy engine oil or engine assembly lubricant

11. Align the timing marks on the camshaft sprocket and crankshaft gears.

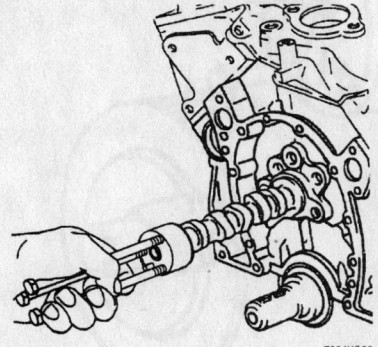

Install 2 or 3 long bolts into the camshaft to use as a handle for easy removal or installation—7.4L engine shown, other engines similar

7924KG20

12. Install or connect the following:
- Camshaft
- Camshaft sprocket and chain. Torque the bolts to 25 ft. lbs. (34 Nm)
- Engine mount bolts
- Lifters and pushrods and adjust the valves
- Intake manifold
- Distributor using the locating marks made during removal
- Engine front cover
- Harmonic balancer and pulley
- Water pump
- Hoses at the water pump
- Valve covers
- Alternator, if removed
- Accessory belt, if removed
- Fan shroud and radiator
- Air conditioning condenser and compressor
- Grille and center support, if removed
- Air cleaner assembly
- Negative battery cable

13. Fill the cooling system with the proper type and quantity of antifreeze.

8.1L Engines

1. Before servicing the vehicle, refer to the precautions in the beginning of this section.

2. Properly discharge the air conditioning system.

3. Remove or disconnect the following:
- Grille
- A/C condenser
- Intake manifold
- Rocker arms and pushrods
- Valve lifter guide retainer bolts and retainer
- Valve lifter guides, keeping them in proper order for reassembly
- Valve lifters

➡ **If any lifters are stuck in their bores, use a suitable valve lifter to remove them.**

- Timing chain and sprocket
- Camshaft retaining bolts
- Camshaft retainer

❊❊ **WARNING**

All of the cam journals are the same size so be very careful when removing and installing the camshaft that you do not damage the bearings.

4. Install three 8-1.25 x 100mm bolts in the holes in the front of the camshaft and carefully pull the camshaft from the block.

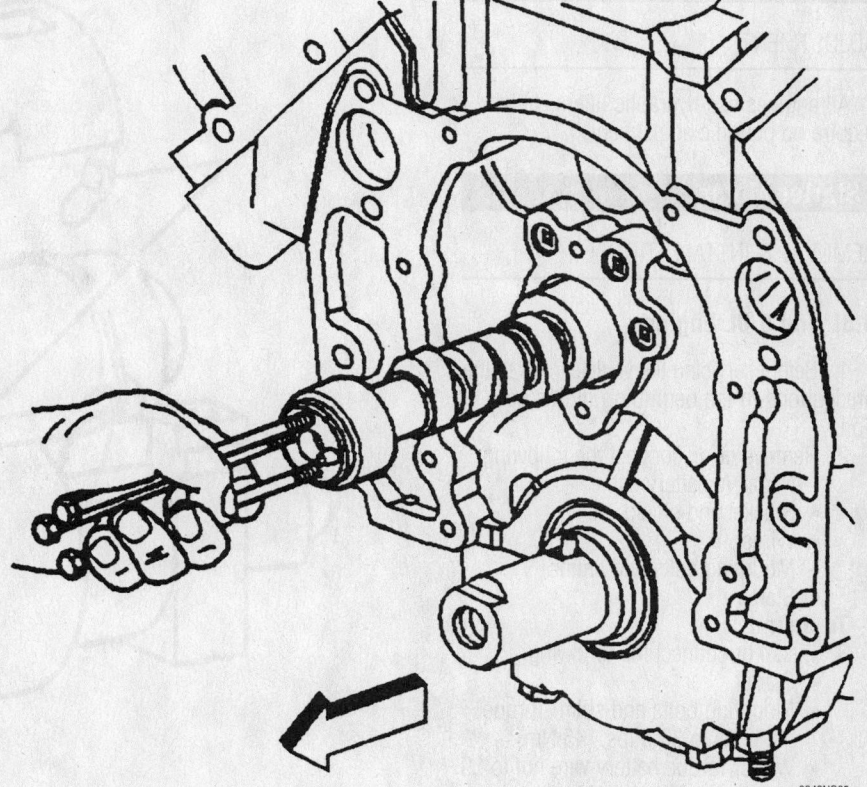

Use the 3 bolts as a handle to carefully remove and install the camshaft—8.1L engine shown

9348NG05

5. Remove the bolts from the front of the camshaft.

6. Clean and inspect the camshaft for damage.

To install:

7. Liberally coat camshaft and bearings with heavy engine oil or engine assembly lubricant.

8. Install the camshaft, using the 3 bolts threaded into the camshaft bolt holes as a handle, then remove the bolts.

9. Install or connect the following:
- Camshaft retainer and bolts. Tighten to 106 inch lbs. (12 Nm).

➡**If a new camshaft is installed, you MUST install new valve lifters.**

- Timing chain and sprocket
- Valve lifters
- Valve lifter guides over the flats on the lifters. Make sure the rollers of the lifters are properly aligned with the cam lobes.
- Valve lifter guide retainer. Tighten the bolts to 18 ft. lbs. (25 Nm).
- Rocker arms and pushrods
- Intake manifold
- A/C condenser
- Grille

10. Recharge the A/C system.

Valve Lash

ADJUSTMENT

All engines use hydraulic lifters, which require no periodic adjustment.

Starter Motor

REMOVAL & INSTALLATION

4.3L and 5.0L Engines

1. Before servicing the vehicle, refer to the precautions in the beginning of this section.

2. Remove or disconnect the following:
- Negative battery cable
- Bracket and shield
- Wires
- Mounting bolts and shims
- Starter

To install:

3. Install or connect the following:
- Starter
- Mounting bolts and shim. Torque the bolts to 33 ft lbs. (45 Nm).
- Wires. Torque battery wire nut to 89 inch lbs. (10 Nm) and ignition nut to 18 inch lbs. (2 Nm).

- Bracket and shield. Torque the nuts to 53 inch lbs. (6 Nm).
- Negative battery cable

5.7L and 7.4L Engines

1. Before servicing the vehicle, refer to the precautions in the beginning of this section.

2. Remove or disconnect the following:
- Negative battery cable
- Mounting bolts and shims
- Wires
- Heat shield
- Starter

To install:

3. Install or connect the following:
- Starter
- Wires. Torque battery wire nut to 89 inch lbs. (10 Nm), and ignition nut to 18 inch lbs. (2 Nm).
- Heat shield. Torque the bolts to 53 inch lbs. (6 Nm) and the nuts to 35 inch lbs. (3 Nm).
- Mounting bolts and shim. Torque the bolts to 33 ft lbs. (45 Nm).
- Negative battery cable

4.8L, 5.3L and 6.0L Engines

1. Before servicing the vehicle, refer to the precautions in the beginning of this section.

2. Disconnect the negative battery cable.

3. Raise and support the vehicle.

4. Remove or disconnect the following:
- Protective shields (as necessary)
- Starter solenoid shield
- Starter-to-transmission close out cover bolt
- Engine oil level sensor connection
- Front axle mounting bracket through bolt nut

5. Reposition the front axle mounting bracket through bolt until the bolt tip is flush with the support bushing. Do not remove the bolt.
- Mounting bolts from the engine block

6. Slide the starter forward until the starter clears the transmission.
- Starter transmission close out cover
- Positive battery cable and wiring harness from the starter
- Starter

To install:

7. Install or connect the following:
- Starter
- Positive battery cable. Tighten the nut to 12 ft. lbs. (16 Nm).
- Starter transmission close out cover

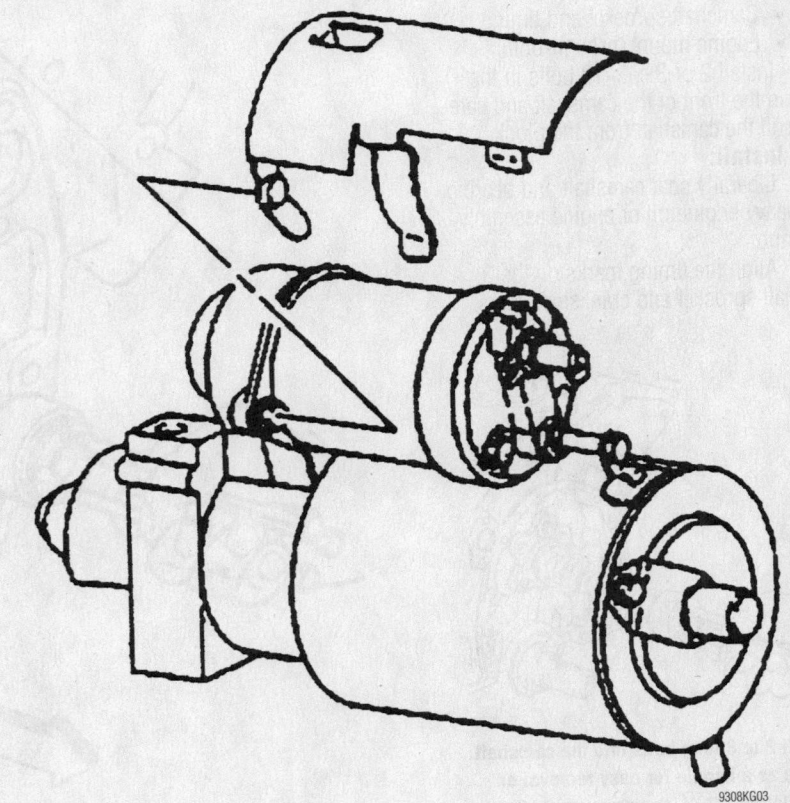

Exploded view of the starter motor.

9308KG03

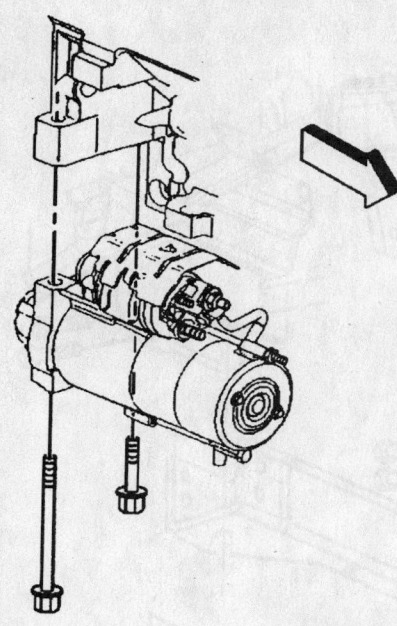

Starter removal—4.8L, 5.3L and 6.0L engines

- Mounting bolts to the engine block and tighten to 37 ft. lbs. (50 Nm)

8. Reposition the front axle mounting bracket through bolt until the bolt is fully seated.

- Front axle mounting bracket through bolt nut and tighten to 67 ft. lbs. (90 Nm)
- Oil level sensor connection
- Starter-to-transmission close out cover bolt
- Starter solenoid shield
- Protective shields (as necessary)

9. Remove the safety stands.
10. Lower the vehicle.
11. Connect the negative battery cable.

8.1L Engine

1. Before servicing the vehicle, refer to the precautions in the beginning of this section.
2. Remove or disconnect the following:

- Negative battery cable
- Positive battery cable nut
- Positive cable from the solenoid
- Engine harness ground nut and ground from the solenoid
- Mounting bolts and starter
- Heat shield bolts, nut and shield, if necessary

To install:
3. Install or connect the following:
- Heat shield, bolts and nut if

removed. Tighten the bolts to 35 inch lbs. (3 Nm) and the nut to 44 inch lbs. (5 Nm).
- Starter and bolts. Tighten to 37 ft. lbs. (50 Nm).
- Ground wire and nut. Tighten to 30 inch lbs. (3.4 Nm).
- Positive cable and nut. Tighten to 80 inch lbs. (9 Nm).
- Negative battery cable

Oil Pan

REMOVAL & INSTALLATION

4.3L Engines

1. Before servicing the vehicle, refer to the precautions in the beginning of this section.
2. Drain the engine oil.
3. Remove or disconnect the following:

- Negative battery cable
- Exhaust crossover pipe
- Torque converter cover, if equipped with automatic transmission
- Cooler lines from guides and the oil filter adapter
- Strut rods at the flywheel/flexplate cover, if equipped

- Strut rod at the front engine mounts, if equipped
- Starter assembly
- Front drive axle tube nuts and lower axle bushing bolts
- Oil pan bolts/nuts and reinforcements
- Oil pan and gaskets

To install:
4. Thoroughly clean all gasket surfaces
5. Install or connect the following:

- New gasket
- Oil pan and new gaskets
- Oil pan bolts, nuts and reinforcements. Torque bolts to 18 ft. lbs. (25 Nm).
- Front drive axle tube nuts and lower axle bushing bolts
- Starter
- Strut rod brackets at the front engine mounts
- Strut rods at the flywheel/flexplate cover
- Cooler lines into guides and oil filter adapter with new filter
- Torque converter cover, if equipped with automatic transmission
- Exhaust crossover pipe
- Negative battery cable

6. Refill the engine with oil.

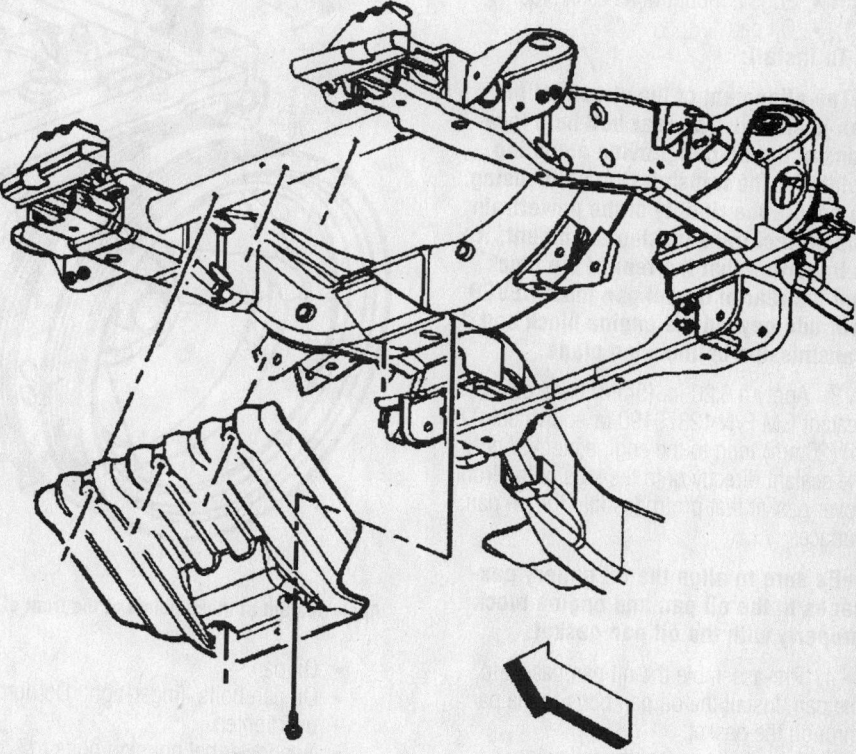

Oil pan shield—4.8L, 5.3L and 6.0L engines

4.8L, 5.3L and 6.0L Engines

➡ The original oil pan gasket is retained and aligned to the oil pan by rivets. When installing a new gasket, it is not necessary to install new rivets. DO NOT reuse the oil pan gasket. When installing the oil pan, install a NEW oil pan gasket.

1. Before servicing the vehicle, refer to the precautions in the beginning of this section.

2. Remove or disconnect the following:
- Negative battery cable
- Front differential if equipped with four wheel drive
- Under body shield from the vehicle
- Oil pan shield
- Cross brace if equipped
- Engine oil and filter
- Transmission-to-oil pan bolts
- Oil level sensor electrical connector
- Two front wiring harness retainer bolts
- Engine wiring harness retainer bolts from the engine oil pan
- Engine oil cooler pipe-to-oil pan bolt
- Transmission oil cooler pipe retainer and the bolt from the oil pan
- Closeout covers and bolts (one each side of engine)
- Engine mount bolts each side
- Oil pan

To install:

➡ The alignment of the structural oil pan is critical. The rear bolt hole locations of the oil pan provide mounting points for the transmission bellhousing. To ensure the rigidity of the powertrain and correct transmission alignment, it is important that the rear of the block and the rear of the oil pan must NEVER protrude beyond the engine block and transmission bellhousing plane.

3. Apply a 0.20 in. (5mm) bead of sealant GM P/N 12378190 or equivalent 0.8 in. (20mm) long to the engine block. Apply the sealant directly onto the tabs of the front cover gasket that protrudes into the oil pan surface.

➡ Be sure to align the oil gallery passages in the oil pan and engine block properly with the oil pan gasket.

4. Pre-assemble the oil pan gasket to the pan. Install the oil pan bolts to the pan through the gasket.

5. Install or connect the following:
- Oil pan gasket

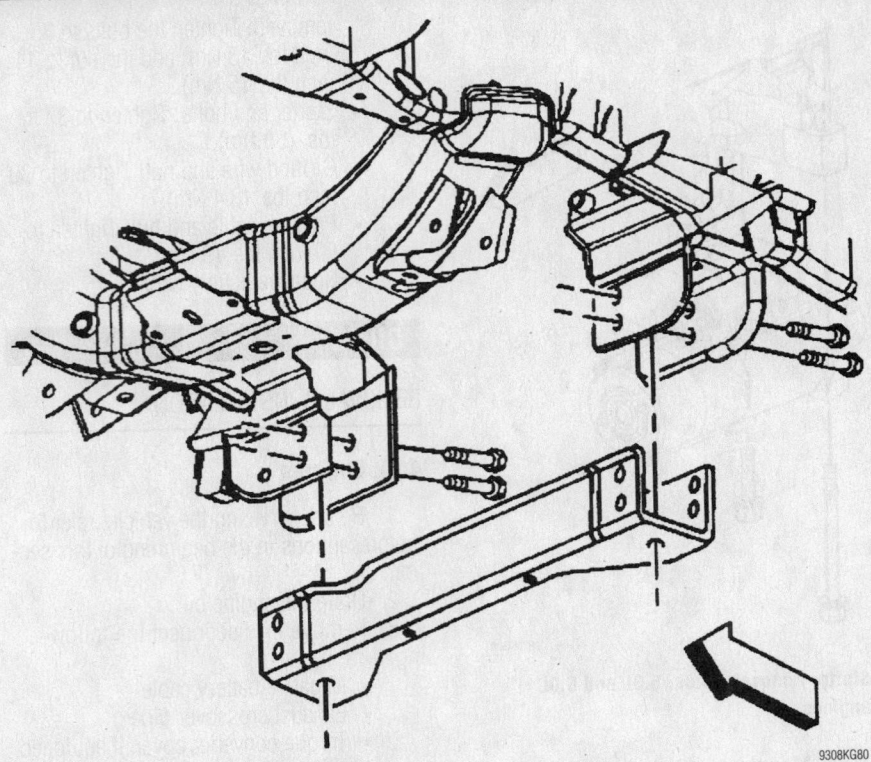

Cross brace—4.8L, 5.3L and 6.0L engines

9308KG80

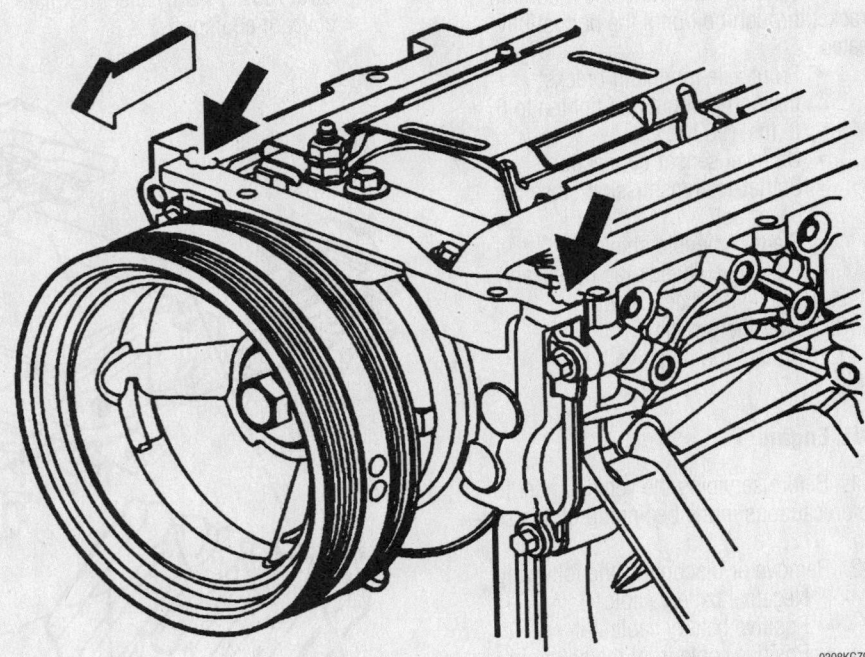

Apply sealant at these points at the front of the block—4.8L, 5.3L and 6.0L engines

9308KG79

- Oil pan
- Oil pan bolts, finger-tight. Do not overtighten.
- Two lower bellhousing bolts to position the oil pan correctly

6. Snug the lower bellhousing bolt finger-tight. Do not overtighten. Tighten the oil pan-to-block and oil pan-to-oil pan front cover bolts to 18 ft. lbs. (25 Nm). Tighten the oil pan-to-rear cover bolts to 106 inch

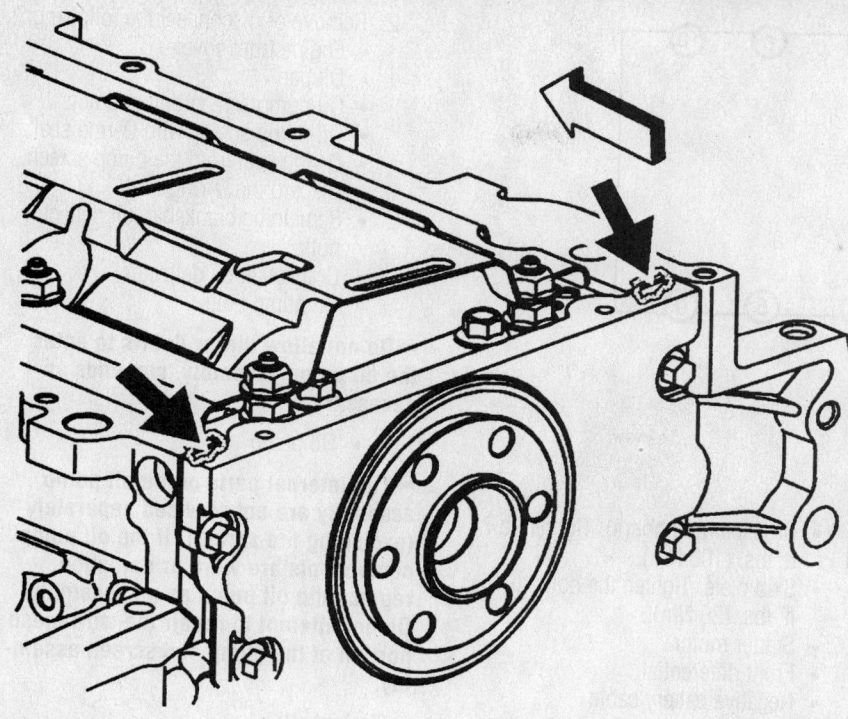

Apply sealant at these points at the rear of the block—4.8L, 5.3L and 6.0L engines

9308KG78

lbs. (12 Nm). Tighten the bellhousing bolts to 37 ft. lbs. (50 Nm).

- Transmission oil cooler pipe retainer and the bolt to the oil pan
- Engine oil cooler pipe-to-oil pan bolt and tighten to 89 inch lbs. (10 Nm)
- Engine wiring harness retainer bolts to the engine oil pan
- Oil level sensor electrical connector
- Transmission-to-oil pan bolts and tighten to 41 ft. lbs. (55 Nm)
- Front differential, if equipped with four wheel drive
- Underbody shield

7. Lower the vehicle. Fill the engine with oil and install the engine oil filter.

8. Connect the negative battery cable.

5.0L and 5.7L Engines

1. Before servicing the vehicle, refer to the precautions in the beginning of this section.

2. Drain the engine oil.

3. Remove or disconnect the following:
- Negative battery cable
- Underbody protector shields
- Transmission and engine oil lines from guides
- Front driveshaft, if needed
- Front drive axles, if needed
- Exhaust crossover pipe

- Flywheel/flexplate or torque converter cover
- Oil filter and adapter
- Strut rods at the front engine mounting, if equipped
- Oil pan bolts, nuts and reinforcements
- Oil pan and gaskets

To install:

4. Thoroughly clean all gasket surfaces.

5. Install or connect the following:
- New gasket
- Oil pan
- Oil pan bolts, nuts and reinforcements. Torque the bolts to 18 ft. lbs. (25 Nm).
- Strut rods at the front engine mounting
- Oil filter and adapter
- Torque converter or flywheel/flexplate cover
- Exhaust crossover pipe
- Front drive axles and driveshaft, if removed
- Transmission and engine oil lines to guides
- Underbody protectors
- Negative battery cable

6. Fill the crankcase with oil.

7.4L Engines

➡**Removal of the transmission may be necessary on vans.**

1. Before servicing the vehicle, refer to the precautions in the beginning of this section.

2. Drain the engine oil.

3. Remove or disconnect the following:
- Negative battery cable
- Fan shroud
- Air cleaner
- Distributor cap
- Underbody protectors, if needed
- Front driveshaft and front drive axles, if needed
- Starter, if equipped with manual transmission
- Torque converter or clutch housing cover
- Oil filter and adapter
- Oil pressure line from the side of the block

4. Support the engine
- Engine mount through-bolts
- Oil pan bolts
- Oil pan and discard the gaskets

To install:

5. Clean all sealing surfaces.

6. Apply RTV gasket material to the front and rear corners of the gaskets.

7. Install or connect the following:
- New gaskets, coat the gaskets with adhesive sealer and position them on the block
- Rear pan seal in the pan with the seal ends mating with the gaskets.
- Front seal on the bottom of the front cover, pressing the locating tabs into the holes in the cover.
- Oil pan.
- Pan bolts, clips and reinforcements. Torque the bolts to 18 ft. lbs. (25 Nm).

8. Lower the engine onto the mounts.
- Engine mount through-bolts
- Oil pressure line
- Oil filter
- Starter, if removed
- Torque converter or clutch housing cover
- Front drive axles and driveshaft, if removed
- Underbody protectors, if removed
- Distributor cap
- Air cleaner
- Fan shroud
- Negative battery cable

9. Fill the crankcase with oil.

8.1L Engine

➡**Removal of the transmission may be necessary on vans.**

1. Before servicing the vehicle, refer to the precautions in the beginning of this section.

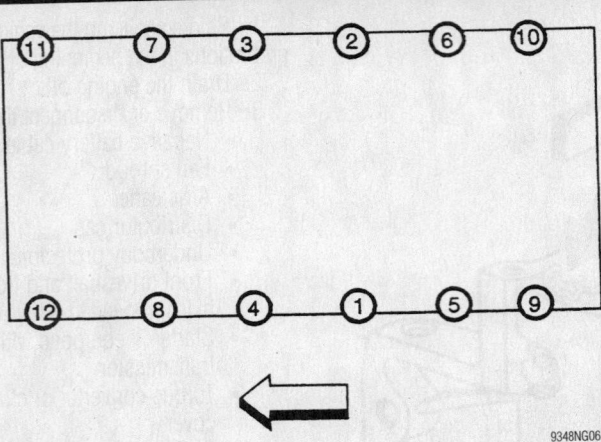

9348NG06

Oil pan bolt tightening sequence—8.1L engine

2. Disconnect the negative battery cable and drain the engine oil.

3. Remove or disconnect the following:
- Front differential, if equipped with 4WD
- Starter motor
- Oil pan skid plate bolts and plate
- Crossbar bolt(s) and crossbar
- Oil level dipstick
- Oil level sensor electrical connector
- Engine harness clip from the oil pan
- Battery cable channel bolt
- Battery cable channel and reposition
- Oil pan bolts, oil pan and gasket

➡You can reuse the oil pan gasket, if it is not damaged

To install:

➡You must install the oil pan within 5 minutes of applying the sealer.

4. Before servicing the vehicle, refer to the precautions in the beginning of this section.

5. Apply sealant to the sides of the front and rear crankshaft bearing caps on the left and right sides.

6. Install or connect the following:
- Oil pan gasket into the oil pan groove
- Oil pan and bolts

7. Tighten the oil pan bolts, in sequence, as follows:
  a. 1st pass: 89 inch lbs. (10 Nm).
  b. 2nd pass: 18 ft. lbs. (25 Nm).

8. Install or connect the following:
- Battery cable channel and bolt. Tighten to 80 inch lbs. (9 Nm).
- Oil level sensor and tighten to 15 ft. lbs. (20 Nm)
- Engine harness clip
- Oil level sensor connector
- Oil level dipstick

- Crossbar and bolt(s). Tighten to 74 ft. lbs. (100 Nm).
- Skid plate. Tighten the bolts to 15 ft. lbs. (20 Nm).
- Starter motor
- Front differential
- Negative battery cable

9. Fill the crankcase with oil.

Oil Pump

REMOVAL & INSTALLATION

4.8L, 5.3L and 6.0L Engines

1. Before servicing the vehicle, refer to the precautions in the beginning of this section.

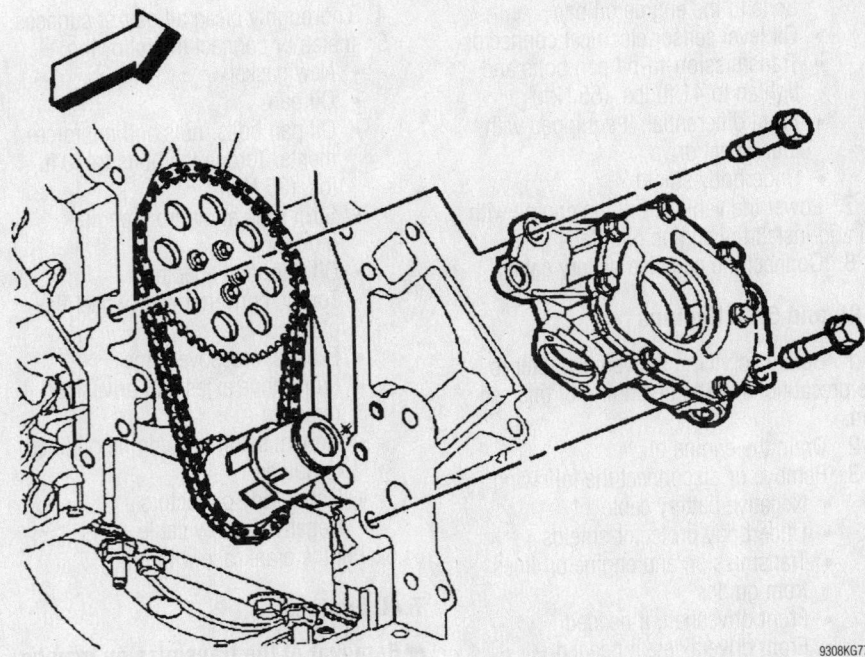

Oil pump removal—4.8L, 5.3L and 6.0L engines

9308KG77

2. Remove or disconnect the following:
- Engine front cover
- Oil pan
- Oil pump screen bolt and nuts
- Oil pump screen with O-ring seal.
- O-ring seal from the pump screen. Discard the O-ring seal.
- Remaining crankshaft oil deflector nuts.
- Crankshaft oil deflector
- Oil pump bolts

➡Do not allow dirt or debris to enter the oil pump assembly, cap ends as necessary.

- Oil pump

➡The internal parts of the oil pump assembly are not serviced separately (excluding the spring). If the oil pump components are worn or damaged, replace the oil pump as an assembly. Do not attempt to repair the wire mesh portion of the pump and screen assembly.

To install:

➡Inspect the oil pump and engine block oil gallery passages. These surfaces must be clear and free of debris or restrictions.

3. Align the splined surfaces of the crankshaft sprocket and the oil pump drive gear and install the oil pump. Install the oil pump onto the crankshaft sprocket until the

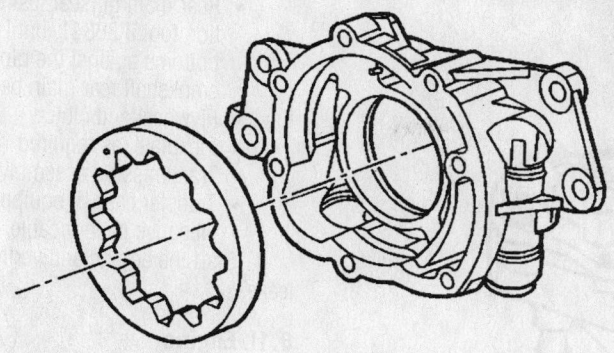

Oil pump disassembly—4.8L, 5.3L and 6.0L engines

9308KG64

pump housing contacts the face of the engine block.

4. Install or connect the following:
- Oil pump bolts. Tighten the oil pump bolts to 18 ft. lbs. (25 Nm).
- Crankshaft oil deflector

➡**Lubricate a NEW oil pump screen O-ring seal with clean engine oil.**

- NEW O-ring seal onto the oil pump screen

➡**Push the oil pump screen tube completely into the oil pump prior to tightening the bolt. Do not allow the bolt to pull the tube into the pump.**

5. Align the oil pump screen mounting brackets with the correct crankshaft bearing cap studs.

6. Install or connect the following:
- Oil pump screen
- Oil pump screen bolt and the deflector nuts. Tighten the bolt to 106 inch lbs. (12 Nm) and the nuts to 18 ft. lbs. (25 Nm).
- Oil pan
- Engine front cover

4.3L, 5.0L, 5.7L and 7.4L Engines

1. Before servicing the vehicle, refer to the precautions in the beginning of this section.

2. Remove or disconnect the following:
- Oil pan
- Oil pump attaching bolt, if equipped
- Pick-up tube nut/bolt
- Pump along with the pick-up tube and shaft, as necessary

3. Clean all sealing surfaces

To install:

4. Ensure that the pump pick-up tube is tight in the pump body. If the tube should come loose, oil pressure will be lost and oil starvation will occur. If the pick-up tube is loose it should be replaced.

5. If the pump has been disassembled and is being replaced or for any reason oil has been removed, it must be primed. It can either be filled with oil before installing the cover plate and oil kept within the pump during handling or the entire pump cavity can be filled with petroleum jelly.

➡**If the pump is not primed, the engine could be damaged before it receives adequate lubrication when the engine is started.**

6. Install or connect the following:
- Pump, aligning the pump shaft with the oil pump drive gear as necessary. Torque oil pump/pick-up tube retainer(s) to 65 ft. lbs. (90 Nm).
- Oil pan

7. Refill the engine crankcase
8. Disable the ignition system; crank engine for approximately 10 seconds to aid in priming the oil pump and reducing the risk of engine damage.

➡**If the oil pump does not build up oil pressure almost immediately, remove the pan and check for a loose oil pump-to-pick-up tube attachment. If necessary dismantle the pump and pack the pump cavity with petroleum jelly. Running the engine without measurable oil pressure will cause extensive damage.**

8.1L Engine

1. Before servicing the vehicle, refer to the precautions in the beginning of this section.

2. Remove or disconnect the following:
- Oil pan
- Oil pump screen bolt
- Oil pump, retainer and driveshaft. Discard the driveshaft retainer
- Crankshaft oil deflector nuts
- Crankshaft oil deflector
- Oil pump bolts
- Oil pump

3. Clean and inspect the oil pump

To install:

4. Install the crankshaft oil deflector. Tighten the nuts to 37 ft. lbs. (50 Nm).

➡**Always replace the retainer between the oil pump and the shaft, when installing the oil pump. During assembly, install a new oil pump driveshaft retainer. To ease installation, slightly heat the retainer to above room temperature.**

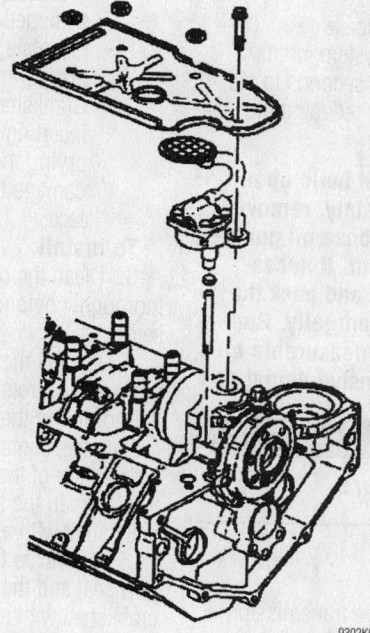

Exploded view of the oil pump mounting—4.3L, 5.0L, 5.7L and 7.4L engines

9302KG04

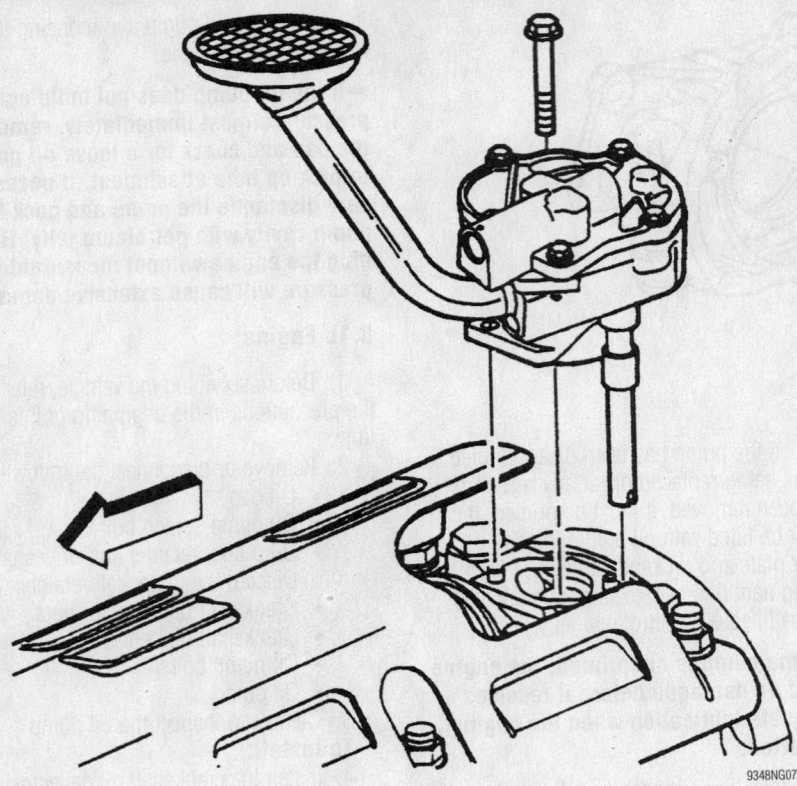

Oil pump removal—8.1L engine

5. Assemble the oil pump, driveshaft and a new retainer.

6. Install or connect the following:
- Oil pump, positioning it on the locating pins
- Oil pump bolt and tighten to 56 ft. lbs. (75 Nm)
- Oil pan

7. Refill the engine crankcase

8. Disable the ignition system; crank engine for approximately 10 seconds to aid in priming the oil pump and reducing the risk of engine damage.

➡ **If the oil pump does not build up oil pressure almost immediately, remove the pan and check for a loose oil pump-to-pick-up tube attachment. If necessary dismantle the pump and pack the pump cavity with petroleum jelly. Running the engine without measurable oil pressure will cause extensive damage.**

Rear Main Seal

REMOVAL & INSTALLATION

Except 8.1L Engine

Please note that the entire transmission assembly and flywheel/flexplate must be removed to perform this procedure.

1. Before servicing the vehicle, refer to the precautions in the beginning of this section.

2. Remove or disconnect the following:
- Negative battery cable
- Transfer case, if equipped
- Transmission assembly
- Clutch assembly and flywheel, if equipped with manual transmission
- Flexplate, if equipped with automatic transmission
- Crankshaft rear main oil seal by inserting a suitable prying tool and prying the seal out. Take care not to damage the crankshaft sealing surface.

To install:

3. Clean the oil seal bore in the block thoroughly before installation of the new seal.

4. Inspect the crankshaft for grit, rust or burrs and correct as necessary. Also inspect the portion of the crankshaft where the oil seal makes contact, for wear due to the rubbing action of the oil seal.

5. Clean the seal running surface of the crankshaft with a non-abrasive cleaner.

6. Lubricate the inner diameter of the new seal and the outer diameter of the crankshaft with engine oil.

7. Install or connect the following:

- Rear main oil seal, using installation tool J 38841, until the tool bottoms against the block and crankshaft rear main bearing cap.
- Flywheel and clutch
- Flexplate, as required
- Transmission assembly
- Transfer case, if equipped
- Negative battery cable

8. Start the engine and verify no oil leaks.

8.1L Engine

Please note that the entire transmission assembly and flywheel/flexplate must be removed to perform this procedure. This procedure requires the use of the following tools: Crankshaft Rear Seal Puller tool No. J 43320 and Crankshaft Rear Seal Installer tool No. J 42849.

1. Before servicing the vehicle, refer to the precautions in the beginning of this section.

2. Remove or disconnect the following:
- Negative battery cable
- Transfer case, if equipped
- Transmission assembly
- Clutch assembly and flywheel, if equipped with manual transmission
- Flexplate, if equipped with automatic transmission

3. Install the guide pins from the Crankshaft Rear Sear Puller into the crankshaft.

4. Install the Rear Seal Puller over the guide pins.

5. Using a drill, insert 8 of the self-drilling sheet metal screws into the rear crankshaft seal, using a crisscross pattern as shown. The self tapping screws are included with the Crankshaft Rear Seal Puller.

6. Thread the center bolt of the Crankshaft Rear Seal Puller into the crankshaft to remove the seal.

7. Remove the guide pins from the crankshaft.

To install:

8. Make sure there is no dirt, rust or loose burrs on the crankshaft.

9. Apply a light coating of engine oil to the crankshaft sealing surface. Do NOT get oil on the sealing surface of the engine block.

10. Install the new rear main seal onto the Crankshaft Rear Seal Installation Tool.

11. Position the Rear Seal Installation Tool against the crankshaft. Thread the attaching screws into the tapped holes in the crankshaft.

12. Use a screwdriver to tighten the

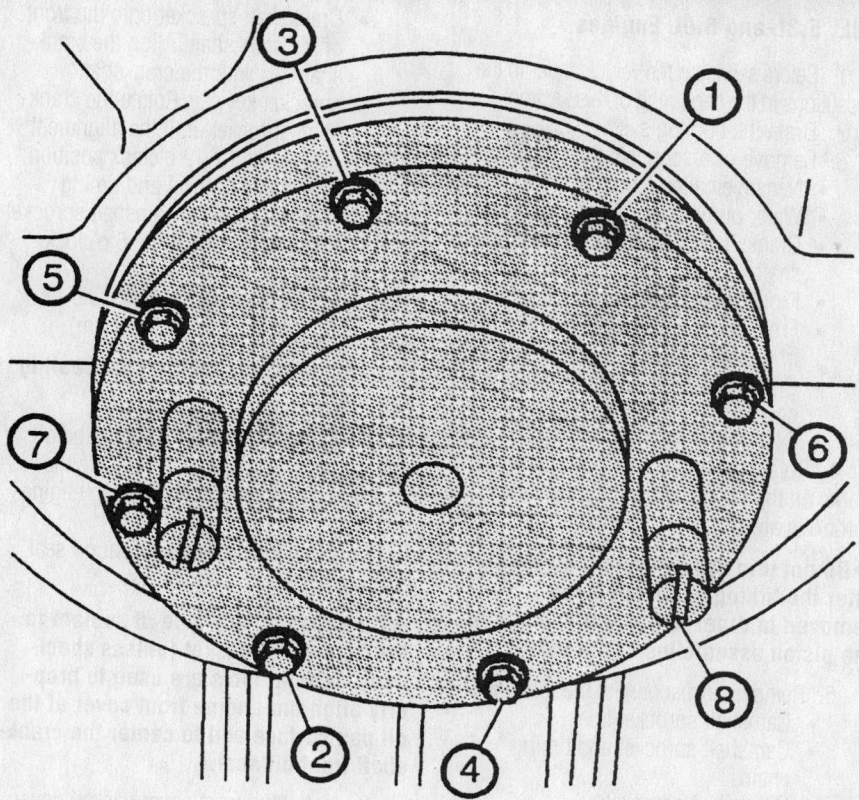

You must drill the screws into the rear main seal using a crisscross pattern—8.1L engine

9348NG08

screws securely to make sure the seal is squarely installed against the crankshaft.

13. Rotate the center nut until the installation tool bottoms, then remove the seal installation tool.

14. Install or connect the following:
- Flexplate, if equipped with automatic transmission
- Clutch assembly and flywheel, if equipped with manual transmission
- Transmission assembly
- Transfer case, if equipped
- Negative battery cable

Timing Chain, Sprockets, Front Cover and Seal

The manufacturer recommends that the front cover oil seal be replaced whenever the cover is removed.

REMOVAL & INSTALLATION

4.3L, 5.0L, 5.7L and 7.4L Engines

1. Before servicing the vehicle, refer to the precautions in the beginning of this section.
2. Drain the cooling system.
3. Remove or disconnect the following:
- Negative battery cable
- Fan shroud assembly

- Belts, pulleys and water pump assembly
- Crankshaft pulley and damper
- Oil pan-to-front cover bolts

➡ If equipped with a composite front cover, it must be replaced with a new one. Reusing the front cover may result in oil leaks.

- Screws holding the timing chain cover to the block
- Cover and gaskets.

4. Use a suitable tool to pry the old seal out of the front face of the cover.

5. Rotate the crankshaft until the timing marks on the camshaft and crankshaft sprockets are in proper alignment.

6. Remove or disconnect the following:
- Camshaft sprocket-to-camshaft nut and/or bolts
- Camshaft sprocket (along with the timing chain), if the sprocket is difficult to remove, use a plastic mallet to bump the sprocket from the camshaft.

➡ The camshaft sprocket (located by a dowel) is lightly pressed onto the camshaft and should come off easily. The chain comes off with the camshaft sprocket.

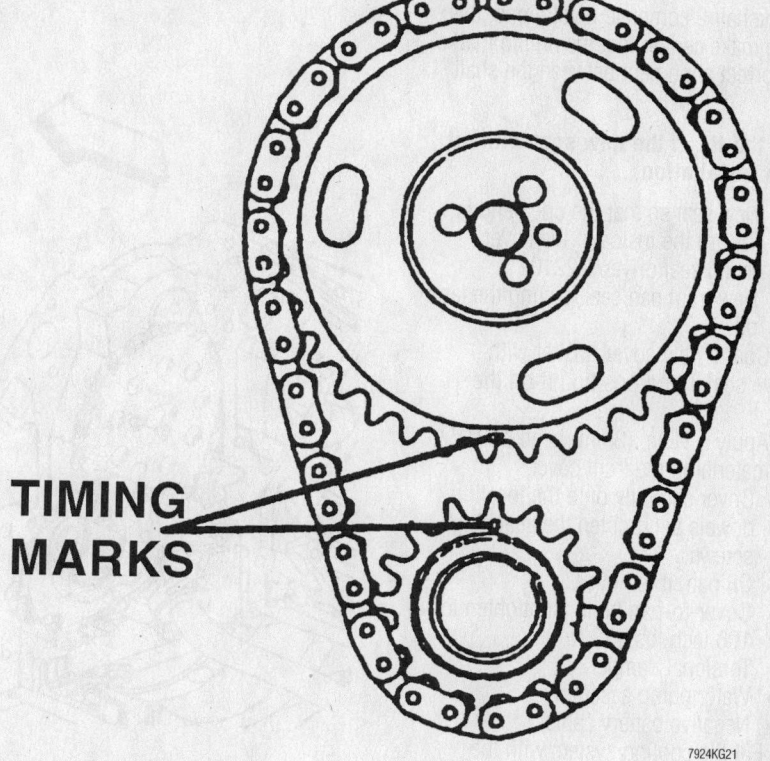

7924KG21

Timing mark alignment for timing chain removal and installation

7. If necessary use J-5825-A, or equivalent, crankshaft sprocket removal tool to free the timing sprocket from the crankshaft.

To install:

8. Inspect the timing chain and the timing sprockets for wear or damage, replace the damaged parts as necessary.

9. Clean the gasket mounting surfaces of all remaining traces of old gasket.

➡**During installation, coat the thrust surfaces lightly with Molykote® or equivalent pre-lube.**

10. Install or connect the following:
- Crankshaft sprocket onto the crankshaft, use tool J-5590, crankshaft sprocket installation tool, and a hammer, without disturbing the position of the engine.
- Timing chain, arrange the camshaft sprocket in such a way that the timing marks will align between the shaft centers and the camshaft locating dowel will enter the dowel hole in the cam sprocket.
- Cam sprocket, with the chain mounted under it in position on the front of the camshaft. Torque the camshaft sprocket-to-camshaft retainer bolts to 106 inch lbs. (12 Nm).

11. With the timing chain installed, turn the crankshaft 2 complete revolutions, then check to make certain that the timing marks are in correct alignment between the shaft centers.

➡**Coat the lip of the new seal with oil prior to installation.**

- New seal so that the open end is toward the inside of the cover, using seal driver J-22102
- New front pan seal, cutting the tabs off.

12. Coat a new cover gasket with adhesive sealer and position it on the block.

13. Apply a ⅛ in. (3mm) bead of RTV gasket material to the front cover.
- Cover carefully onto the locating dowels and tighten the attaching screws
- Oil pan, if removed
- Cover-to-pan bolts and tighten to 106 inch lbs. (12 Nm)
- Torsional damper
- Water pump assembly
- Negative battery cable

14. Fill the cooling system with the proper type and quantity of antifreeze.

4.8L, 5.3L and 6.0L Engines

1. Before servicing the vehicle, refer to the precautions in the beginning of this section.

2. Drain the cooling system.

3. Remove or disconnect the following:
- Negative battery cable
- Water pump
- Crankshaft balancer from the crankshaft
- Front cover bolts
- Front cover and gasket. Discard the front cover gasket.
- Crankshaft front oil seal from the cover
- Oil pump

4. Rotate the crankshaft until the timing marks on the crankshaft and the camshaft sprockets are aligned.

➡**Do not turn the crankshaft assembly after the timing chain has been removed in order to prevent damage to the piston assemblies or the valves.**

5. Remove or disconnect the following:
- Camshaft sprocket bolts
- Camshaft sprocket and timing chain
- Crankshaft sprocket
- Crankshaft sprocket key

To install:

6. Install or connect the following:
- Key into the crankshaft keyway
- Crankshaft sprocket onto the front of the crankshaft. Align the crankshaft key with the crankshaft sprocket keyway. Rotate the crankshaft sprocket until the alignment mark is in the 12 o'clock position.
- Camshaft sprocket and timing chain. Locate the camshaft sprocket alignment mark in the 6 o'clock position.
- Camshaft sprocket bolts and tighten to 26 ft. lbs. (35 Nm)

➡**Do not lubricate the oil seal sealing surface.**

7. Lubricate the outer edge of the oil seal with clean engine oil. Lubricate the front cover oil seal bore with clean engine oil.

8. Install the crankshaft front oil seal with an installer.

➡**Do not apply any type of sealant to the front cover gasket (unless specified). Special tools are used to properly align the engine front cover at the oil pan surface and to center the crankshaft front oil seal.**

9. Install the front cover gasket, cover, and bolts onto the engine. Tighten the cover bolts finger-tight. Do not overtighten.

10. Start the J41480 tool-to-front cover bolts. Don't tighten the bolts yet.

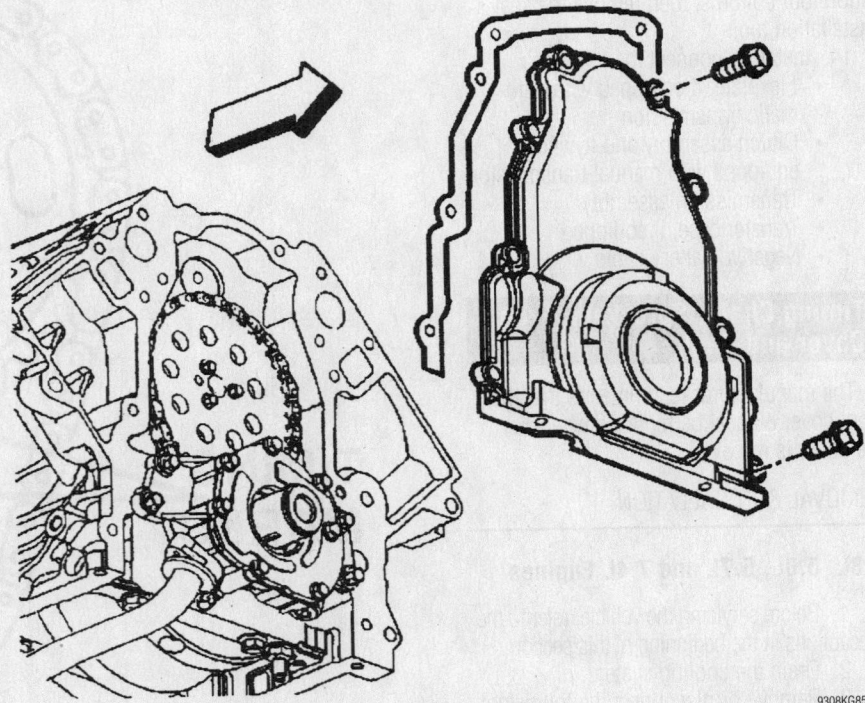

Front cover and gasket—4.8L, 5.3L and 6.0L engines

9308KG85

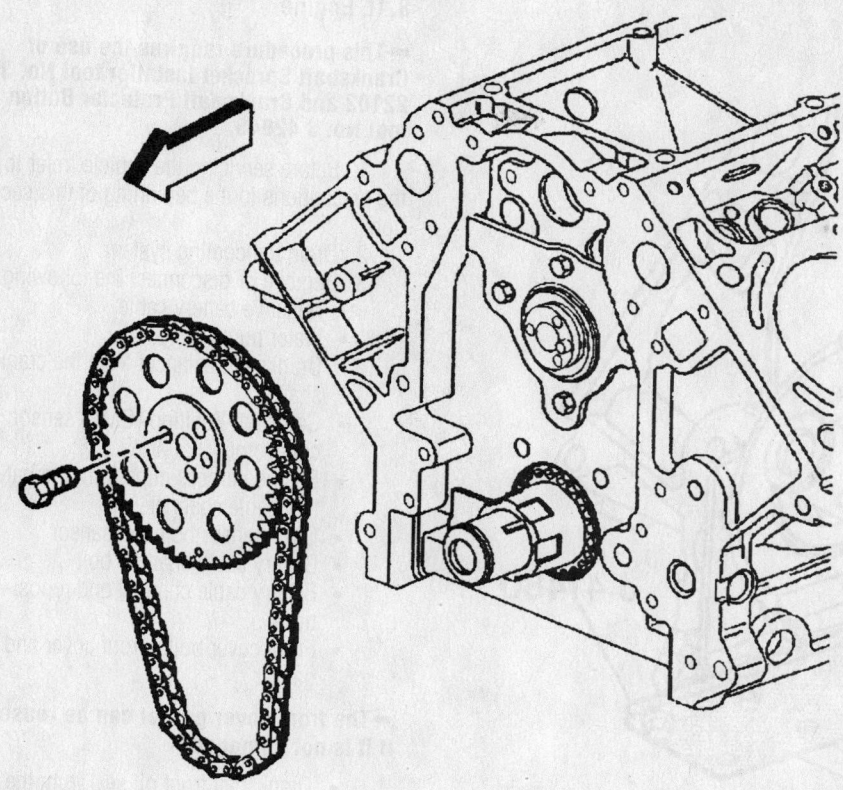

Sprocket and chain removal—4.8L, 5.3L and 6.0L engines

9308KG76

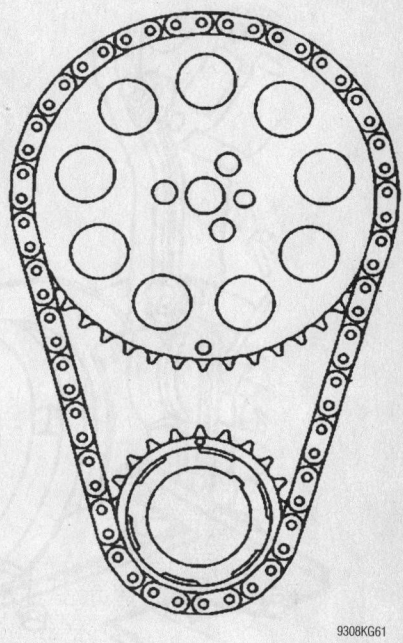

9308KG61

Timing mark alignment—4.8L, 5.3L and 6.0L engines

➡Align the tapered legs of the tool with the machined alignment surfaces on the front cover.

11. Install tool J41476 . Install the crankshaft balancer bolt. Tighten the crankshaft balancer bolt by hand until snug. Do not overtighten. Tighten the J41480 bolts and front cover bolts to 18 ft. lbs. (25 Nm).

12. Remove the tools.

13. Place a straight edge across the engine block and front cover oil pan sealing surfaces. Avoid contact with the portion of the gasket that protrudes into the oil pan surface. Insert a feeler gauge between the front cover and the straight edge tool. The cover must be flush with the oil pan surface or no more than 0.02 in. (0.5mm) below flush. If the front cover-to-engine block oil pan surface alignment is not within specifications, repeat the cover alignment procedure. If the correct front cover-to-engine block alignment cannot be obtained, replace the front cover.

14. Install the crankshaft balancer bolt. Tighten the crankshaft balancer bolt by hand until snug. Do not overtighten the bolt.

15. Snug the oil pan-to-cover bolts in order to position the cover at the pan rail.

16. Tighten the oil pan-to-front cover bolts to 18 ft. lbs. (25 Nm).

17. Tighten the front cover bolts to 18 ft. lbs. (25 Nm).

18. Install the water pump.

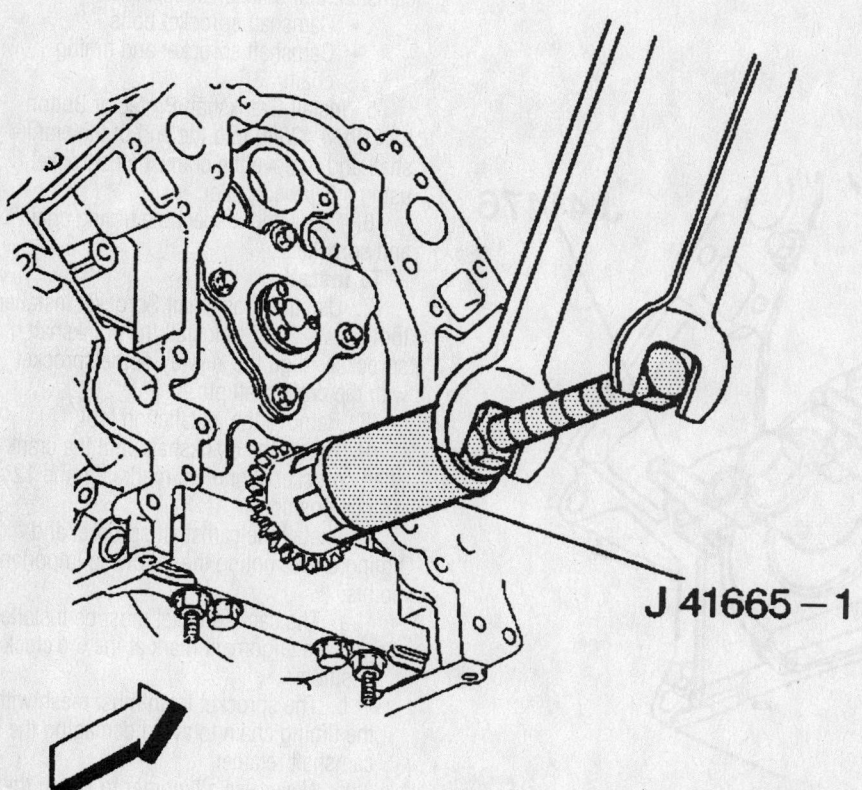

J 41665 — 1

9308KG62

Crankshaft sprocket installation—4.8L, 5.3L and 6.0L engines

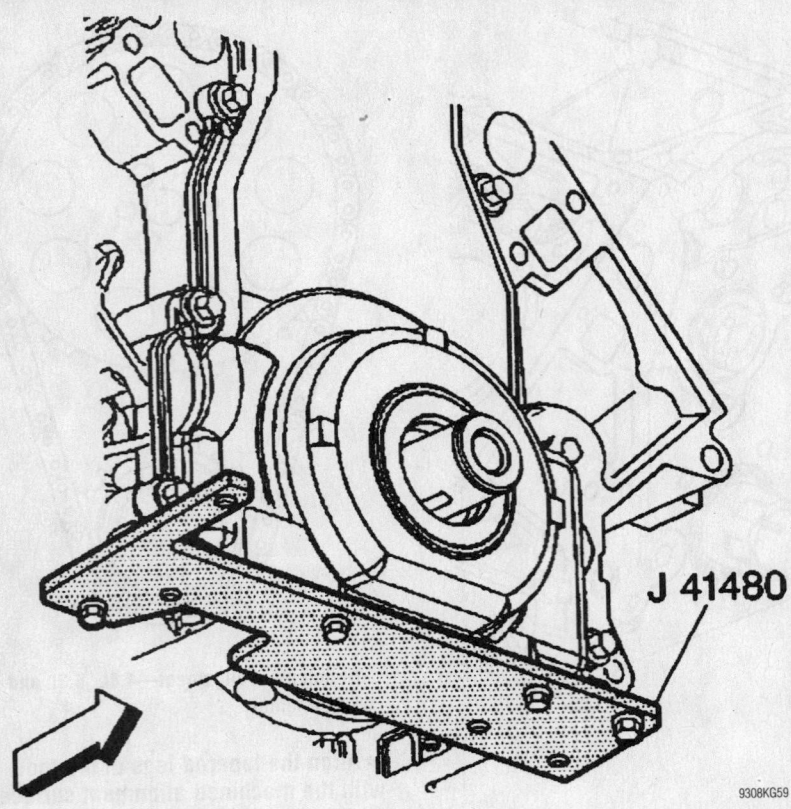

J 41480

J41480 installation—4.8L, 5.3L and 6.0L engines

9308KG59

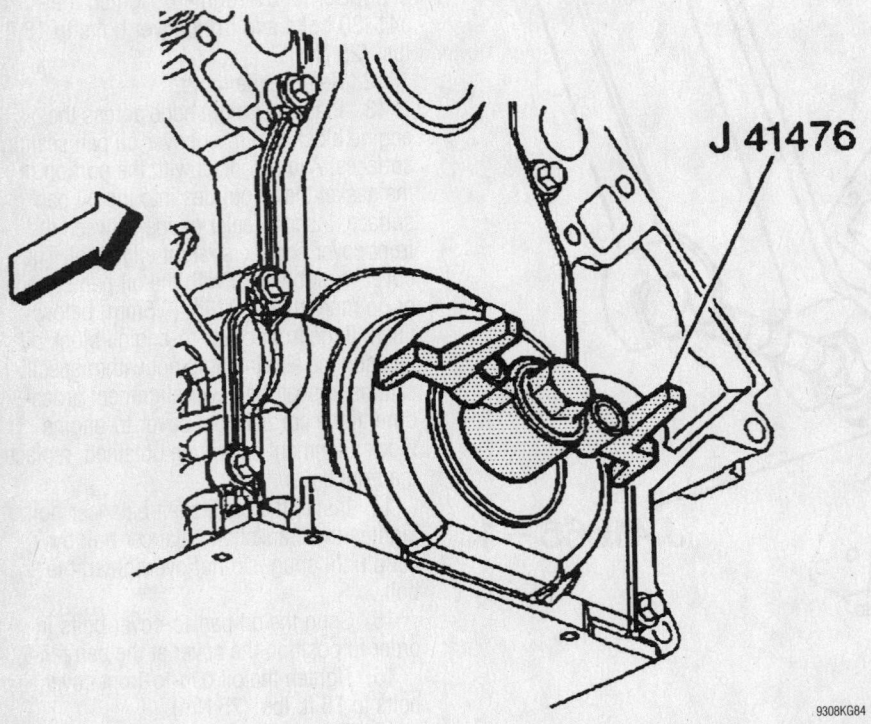

J 41476

Seal alignment tool installation—4.8L, 5.3L and 6.0L engines

9308KG84

8.1L Engine

➡**This procedure requires the use of Crankshaft Sprocket Installer tool No. J 22102 and Crankshaft Protector Button tool No. J 42846.**

1. Before servicing the vehicle, refer to the precautions in the beginning of this section.
2. Drain the cooling system.
3. Remove or disconnect the following:
   - Negative battery cable
   - Water pump
   - Crankshaft balancer from the crankshaft
   - Camshaft Position (CMP) sensor connector
   - Engine harness clips from the battery cable channel
   - CMP sensor bolt and sensor
   - Battery cable channel bolt
   - Battery cable channel and reposition
   - Front cover bolts, front cover and gasket

➡**The front cover gasket can be reused if it is not damaged.**

   - Crankshaft front oil seal from the front cover
4. Align the timing marks on the camshaft and crankshaft sprockets.
   - Camshaft sprocket bolts
   - Camshaft sprocket and timing chain
5. Install Crankshaft Protector Button tool No. J 42846 into the end of the crankshaft and remove the crankshaft sprocket using a 3-jawed puller.
6. Clean and inspect the timing chain and sprockets.

To install:

7. Use the Crankshaft Sprocket Installer tool No. J 22102 to install the crankshaft sprocket. Align the keyway of the sprocket with the crankshaft pin.
8. Remove the installation tool.
9. Rotate the crankshaft until the crankshaft sprocket alignment mark is in the 12 o'clock position.
10. Install the camshaft sprocket and timing chain, noting the following important points:
    a. The cam sprocket must be installed with the alignment mark at the 6 o'clock position.
    b. The sprocket teeth must mesh with the timing chain to avoid damaging the camshaft retainer.
    c. Never use a hammer to install the sprocket onto the camshaft.
11. Make sure the crankshaft sprocket is

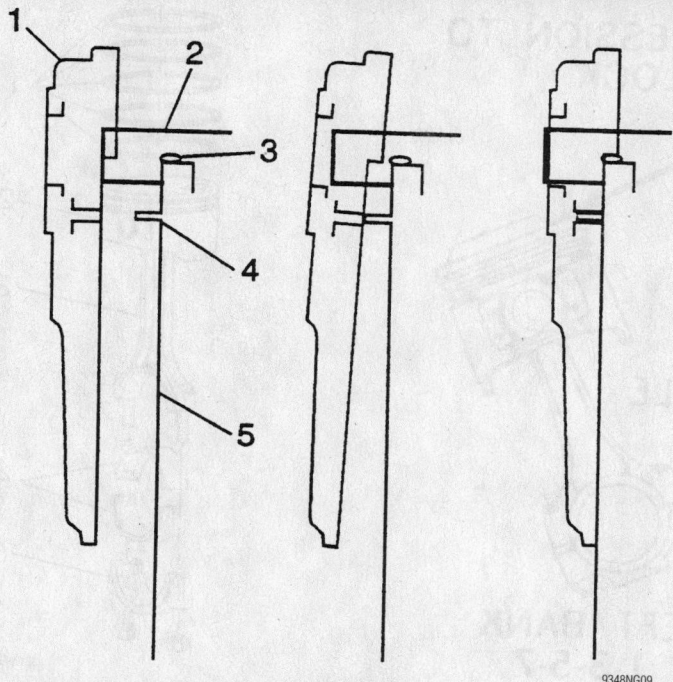

Proper front cover installation sequence—8.1L engine

9348NG09

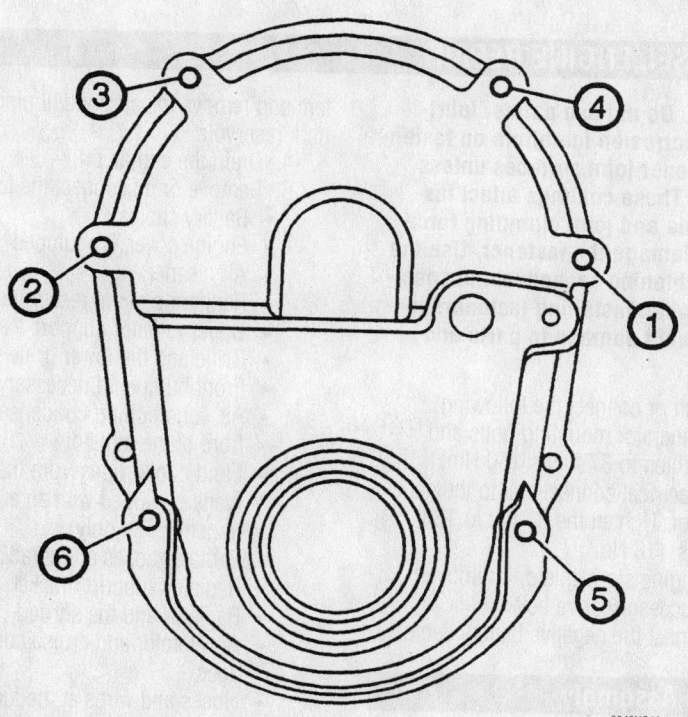

Engine front cover bolt tightening sequence—8.1L engine

9348NG10

alignment at the 12 o'clock position and the cam sprocket is at the 6 o'clock position.

12. Install the camshaft sprocket bolts and tighten, in two passes, to 22 ft. lbs. (30 Nm)

13. Use clean engine oil to lubricate the sealing surfaces of the front oil seal.

14. Install or connect the following:
- New seal into the front cover, using a suitable seal installation tool

➡**The front cover must be installed while the sealant is still wet to the touch.**

- Sealant to the 2 places on the engine block where the front cover meets the oil pan
- Front cover gasket into the cover

15. Install the front cover, referring to the accompanying figure and using the following steps only:

a. Hold the front cover (1) up to the crankshaft (2).

b. Lift the cover (1) while sliding the cover over the crankshaft (2).

c. Slide the front cover toward the engine block (5) while keeping the cover raised.

d. Lower the cover down over the dowel pin (4), allowing the front cover to rest on the sealant (3).

16. Install the front cover bolts and tighten, in sequence, as follows:

a. 1st pass: 53 inch lbs. (6 Nm)

b. 2nd pass: 106 inch lbs. (12 Nm)

17. Install or connect the following:
- Battery cable channel and bolt. Tighten to 80 inch lbs. (9 Nm).
- CMP sensor. Inspect the O-ring first, replace if necessary and coat with oil before installation
- CMP sensor bolt to 106 inch lbs. (12 Nm)
- Engine harness clips to the battery cable channel
- CMP sensor electrical connector
- Crankshaft balancer
- Water pump
- Negative battery cable.

18. Fill the cooling system with the proper type and quantity of antifreeze.

Piston and Ring

POSITIONING

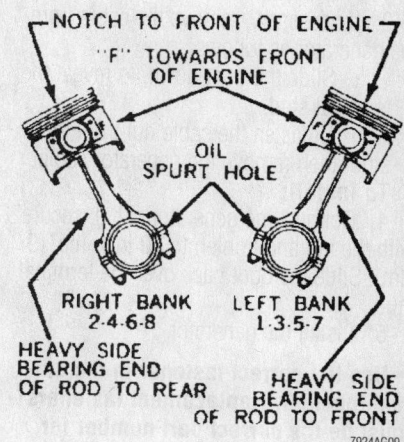

7924AG09

Piston and connecting rod assembly positioning—4.3L, 4.8L, 5.0L, 5.3L, 5.7L and 6.0L engines

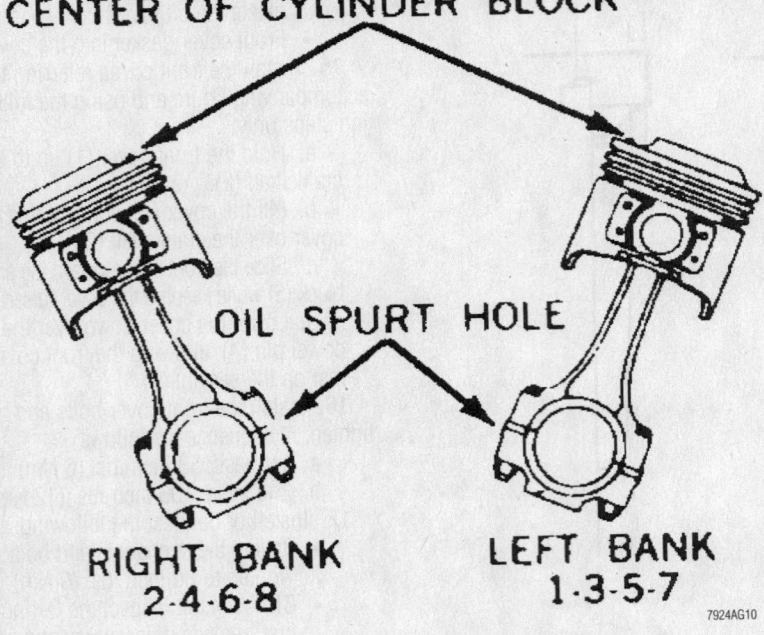

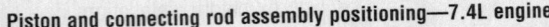

Piston and connecting rod assembly positioning—7.4L engine

7924AG10

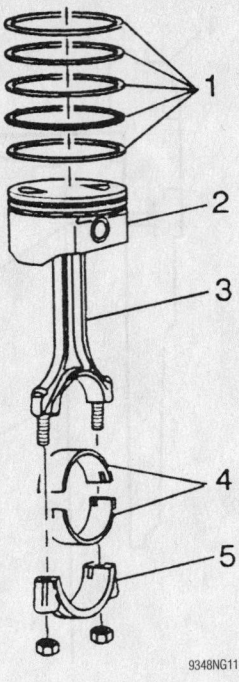

9348NG11

Piston rings (1), piston (2), connecting rod
(3) and related components—8.1L engine

DIESEL ENGINE REPAIR

Alternator

REMOVAL & INSTALLATION

1. Disconnect the negative battery cable.
2. Remove or disconnect the following:
   - Accessory drive belt
   - Engine sight shield, if necessary
   - Electrical connections from the generator
   - Mounting bolts
   - Generator
3. If necessary, remove the cable from the generator as follows:
   a. Slide the boot down, to reveal the terminal stud.
   b. Unfasten the cable nut from the stud, then remove the generator cable.

To install:

4. Connect the generator cable, secure with the nut and tighten to 80 inch lbs. (9 Nm). Slide the boot back over the terminal stud.
5. Install the generator.

➡**Use the correct fastener in the correct location. Replacement fasteners must be the correct part number for that application. Fasteners requiring replacement or fasteners requiring the use of thread locking compound or sealant are identified in the service**

procedure. Do not use paints, lubricants, or corrosion inhibitors on fasteners or fastener joint surfaces unless specified. These coatings affect fastener torque and joint clamping force and may damage the fastener. Use the correct tightening sequence and specifications when installing fasteners in order to avoid damage to parts and systems.

6. Install or connect the following:
   - Generator mounting bolts and tighten to 37 ft. lbs. (50 Nm)
   - Electrical connections to the generator. Tighten the B+ nut to 13 ft. lbs. (18 Nm).
   - Engine sight shield, if removed
   - Accessory drive belt
7. Connect the negative battery cable.

Engine Assembly

REMOVAL & INSTALLATION

6.5L Engine

1. Before servicing the vehicle, refer to the precautions in the beginning of this section.
2. Drain the cooling system.
3. Discharge the air conditioning sys-

tem and remove the air conditioning vacuum reservoir.

4. Drain the engine oil.
5. Remove or disconnect the following:
   - Battery cables
   - Engine cover, if equipped
   - Air cleaner
   - Radiator coolant reservoir bottle
   - Upper radiator support
   - Grille and the lower grille valance
   - Front bumper, if necessary
   - Air conditioning condenser from in front of the radiator
   - Fluid cooler lines from the radiator (vans equipped with an automatic transmission only)
   - Radiator hoses at the radiator
   - Radiator support bracket
   - Radiator and the shroud
   - Accelerator and cruise control linkages
   - Hoses and wires at the fuel unit
   - Fuel supply unit and cap the lines
   - Intake manifold
   - Turbocharger assembly, if equipped
   - Lower intake manifold, if equipped
   - Exhaust manifolds, if necessary
   - Engine wiring harness from the firewall connection
   - Power steering pump, it's not necessary to disconnect the hoses; just move aside

- Heater hoses at the engine
- Thermostat housing, if necessary
- Oil filler and automatic transmission tubes
- Cruise control servo, servo bracket and transducer, if equipped
- Exhaust pipes at the manifolds
- Driveshaft and plug the end of the transmission
- Transmission shift linkage and the speedometer cable
- Fuel line from the fuel tank and pump
- Transmission mounting bolts

6. Support the transmission and engine.

7. Install lifting hooks J-41427 as follows:

a. Remove the 2 right rear lower intake manifold retainers and install lifting hook J-41427 (the one marked "right"). Tighten the bolts to 11 ft. lbs. (15 Nm).

b. Remove the air conditioning compressor and the accessory drive bracket.

c. If equipped, disconnect the EGR tube and the 2 left lower bolts from the intake manifold.

d. Install the lifting hook J-41427 (the one marked "left") and tighten the bolts to 11 ft. lbs. (15 Nm).

8. Remove or disconnect the following:
- Engine mount bracket-to-frame bolts
- Engine mount through-bolts

9. Raise the engine slightly and remove the engine mounts, support the engine with wood between the oil pan and the crossmember.

10. Remove the manual transmission and clutch as follows:

a. Remove the clutch housing rear bolts.

b. Remove the bolts attaching the clutch housing to the engine and remove the transmission and clutch as a unit.

➡ **Support the transmission as the last bolt is being removed to prevent damaging the clutch.**

c. Remove the starter and clutch housing rear cover.

d. Loosen the clutch mounting bolts a little at a time to prevent distorting the disc until spring pressure is released. Remove all of the bolts, the clutch disc and the pressure plate.

11. Remove the automatic transmission as follows:

a. Lower the engine and support it on blocks.

b. Remove the starter and converter housing cover.

c. Remove the flexplate-to-converter attaching bolts.

d. Support the transmission on blocks.

e. Disconnect the detent cable on the Turbo Hydra-Matic.

f. Remove the transmission-to-engine mounting bolts.

12. Attach an engine crane to the engine.

13. Remove the blocks from the engine only and glide the engine away from the transmission.

To install:

14. Install the manual transmission and clutch as follows:

a. Install the clutch disc and the pressure plate. Tighten the clutch mounting bolts a little at a time to prevent distorting the disc.

b. Install the starter and clutch housing rear cover.

c. Install the bolts attaching the clutch housing to the engine and install the transmission and clutch as a unit. Tighten the bolts to specification.

d. Install the clutch housing rear bolts.

15. Install the automatic transmission as follows:

a. Position the transmission.

b. Install the transmission-to-engine mounting bolts.

c. Connect the throttle linkage and detent cable.

d. Install the flexplate-to-converter attaching bolts. Torque the bolts to 65 ft lbs. (90 Nm).

e. Install the starter and converter housing cover.

16. Install or connect the following:
- Engine mount through-bolts and tighten to 50 ft lbs. (68 Nm)
- Engine mount bracket-to-frame bolts. Torque the bolts to 44 ft lbs. (59 Nm).
- Clutch cross-shaft
- Transmission mounting bolts and tighten to 75 ft lbs. (100 Nm)
- Transmission shift linkage and the speedometer cable
- Driveshaft

17. Remove the lifting hooks.
- Compressor
- EGR valve tube
- Intake manifold retaining bolts
- Condenser
- Hood latch support
- Lower fan shroud and filler panel
- Transmission dipstick tube and the accelerator cable at the tube

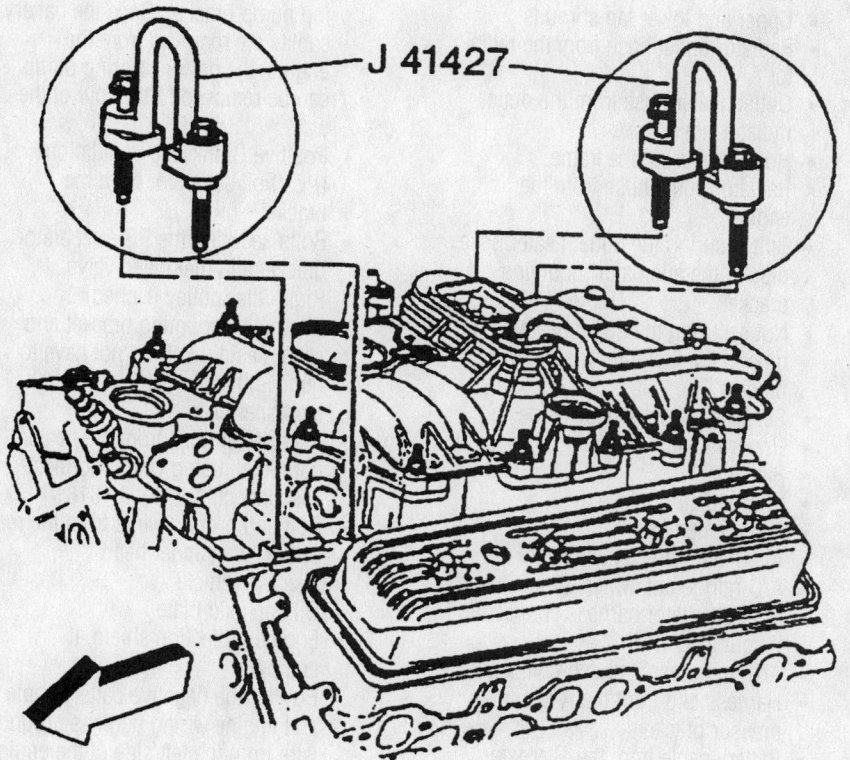

7924KG01

For engine removal and installation, universal lift brackets should be installed in place of the proper intake manifold bolts

- Coolant hose at the intake manifold and the PCV valve
- Cruise control servo, bracket and transducer
- Oil filler pipe and automatic transmission filler pipe
- Engine dipstick tube
- Thermostat housing, if removed
- Heater hoses at the engine
- Engine wiring harness to the firewall connection
- Radiator and the shroud
- Radiator support bracket
- Exhaust manifolds if removed
- Lower intake manifold if removed
- Intake manifold
- Fuel supply unit
- Lines to the fuel supply unit
- Accelerator and cruise control linkages
- Windshield wiper jar and bracket
- Air conditioning condenser
- Air conditioning vacuum reservoir, if equipped
- Fluid cooler lines at the radiator, if equipped with an automatic transmission
- Radiator coolant reservoir bottle
- Hoses at the radiator
- Upper radiator support
- Grille and the lower grille valance
- Air cleaner
- Air stove pipe
- Engine cover
- Battery cables

18. Refill the cooling system.
19. Recharge the air conditioning system.

6.6L Engine

1. Before servicing the vehicle, refer to the precautions in the beginning of this section.

➡ **In order to remove the engine, the vehicle must be on a lift. the front tires also need to be removed. You will have to support the vehicle by it's frame for tire removal.**

2. Drain the cooling system.
3. Discharge and recover the air conditioning system.
4. Drain the engine oil.
5. Raise the hood to the servicing position. Move the hood hinge bolt to hold the hood in the servicing position.
6. Disconnect the battery cables.
7. Remove the upper intake manifold sight shield as follows:
  a. Remove the retaining bolt in the front of the shield.

  b. Lift up on the front of the shield, then left the shield off the rear bracket.

8. Remove or disconnect the following:

➡ **After you remove the duct, cover the turbocharger openings and ducts with tape to prevent foreign objects from entering.**

- Air cleaner outlet duct from the air cleaner and turbocharger.
- Mass Air Flow (MAF) switch connector
- A/C pressure cycling switch connector
- Surge tank switch
- Engine wire harness clip from the accumulator
- Engine wire harness clips from the wheelhouse inner panel and engine bracket
- Air cleaner assembly and bracket
- Surge tank

9. Raise the vehicle.
- Front tires and wheels
- Both front fender wheelhouse inner panels

10. Lower the vehicle.
- Charged air cooler pipes and hoses from the engine and charged air cooler
- Radiator inlet hose form the radiator and engine
- Upper and lower fan shrouds
- Radiator outlet hose from the radiator
- Outlet heater hose from the outlet radiator hose
- Hose clips from the frame
- Radiator outlet hose from the engine
- Bolt securing the outlet heater hose pipe to the alternator mounting bracket
- Nut securing the outlet heater hose pipe to the fuel filter mounting bracket
- Secure the heater hose aside
- Upper radiator support
- Radiator
- Charged air cooler
- A/C condenser
- Alternator harness connector
- A/C refrigerant switch connector
- Dual alternator harness connector, if equipped
- A/C compressor clutch connector
- Harness clip from the A/C compressor bracket
- Battery cable from the alternator and auxiliary alternator, if equipped
- Battery cable harness clips from the bracket

- Bolt securing the battery cable junction block from the power steering pump
- Move and secure the battery cables aside
- Both fuel injection control module harness connectors, by flipping the latch up
- Engine wire harness from the retainer
- Fuel lines at the engine
- Remove the nut and the fuel line bracket from the upper valve rocker arm cover stud
- Fuel lines aside
- Power supply cable from the glow plug relay
- Drive belt
- Suction hose from the accumulator. You can leave the compressor end on the compressor.
- A/C compressor bolts, then move the compressor, with the hoses attached, to the right side of the engine compartment
- Wiring harness to the left side of the engine and tie aside
- Bolts holding the power steering pump front bracket to the pump and A/C compressor mounting bracket
- A/C compressor and power steering pump bracket. Once the battery cables are removed from the engine, the power steering pump can be removed further out of the way
- Positive Crankcase Ventilation (PCV) oil separator from the bracket
- Bolts securing the PCV separator bracket and fuel bleed valve
- Right idler pulley (ribbed)
- Alternator mounting bracket and secure aside. You do not have to remove the alternator or the belt tensioner
- Inlet heater hose from the heater core inlet, using Quick Connect-Disconnect tool No. J 43181
- Bolt and ground wires from the rear of the left cylinder head

11. Raise the vehicle
- Oil pan skid plate
- Engine protection shield, if equipped
- Bolt for the negative battery cable and engine wiring harness ground wire from the left side of the engine
- Bolts holding the battery cable channel retainer to the lower crankcase

- Engine coolant heater cord
- Starter motor
- Nut securing the battery cable bracket to the right side of the lower crankcase
- Bolt holding the auxiliary negative battery cable and the engine wiring harness ground wires to the right side of the engine
- Position the battery cables aside
- Exhaust pipe-to-exhaust outlet clamp
- Lower oil pan, if 4WD

12. If equipped with an automatic transmission, matchmark the installed position of the flywheel and torque converter.

- Torque converter bolts through the starter opening
- Transmission oil line clip nut if equipped with A/T
- Nuts securing the transmission fluid fill tube bracket, if equipped with A/T
- Transmission-to-engine stud and bolts. Note the location of the studs and any brackets attached to the studs

13. Lower the vehicle to work through the wheel opening

- Engine mount-to-frame bracket bolts

14. Lower the vehicle.

15. Install Engine Lifting Bracket tool No. J 36857 to the rear of the left cylinder head with a suitable bolt.

16. Install Engine Lifting Bracket tool No. J 36857 to the front of the right cylinder head with a suitable bolt.

17. Install a suitable lifting device. The engine will have to be angled to remove it. Use a load positioning sling to help in angling the engine.

18. Raise the vehicle off the engine mounts.

19. Remove the left and right engine mount frame brackets.

20. Remove the engine assembly from the vehicle.

21. Secure the engine on an engine stand by removing the following components:

- Flywheel/flexplate
- Rear main seal
- Exhaust outlet
- Oil pan
- Flywheel housing

To install:

22. Install Engine Lifting Bracket tool No. J 36857 to the rear of the left cylinder head with a suitable bolt.

23. Install Engine Lifting Bracket tool No. J 36857 to the front of the right cylinder head with a suitable bolt.

24. Install a suitable lifting device. The engine will have to be angled to install it. Use a load positioning sling to help in angling the engine.

25. Install or connect the following:

- Engine in the vehicle

- 2 transmission-to-engine bolts, loosely
- Left and right side engine mount frame brackets and tighten to 55 ft. lbs. (75 Nm)
- Engine mount-to-frame bracket bolts and tighten to 50 ft. lbs. (65 Nm)

26. Remove the lifting brackets from the cylinder heads.

- Transmission-to-engine bolts/studs and tighten to 37 ft. lbs. (50 Nm)
- Torque converter bolts and tighten to 44 ft. lbs. (60 Nm)

27. Install the remaining components in the reverse of the removal procedure, noting the following important specifications and steps:

- Transmission fluid tube bracket nuts: 13 ft. lbs. (18 Nm)
- Transmission oil cooler line clip bolt: 80 inch lbs. (9 Nm)
- Exhaust pipe clamp: 30 ft. lbs. (40 Nm)
- Ground wire bolt: 25 ft. lbs. (34 Nm)
- Battery cable bracket bolts: 106 inch lbs. (12 Nm)
- Cable and ground wire bolts: 25 ft. lbs. (34 Nm)

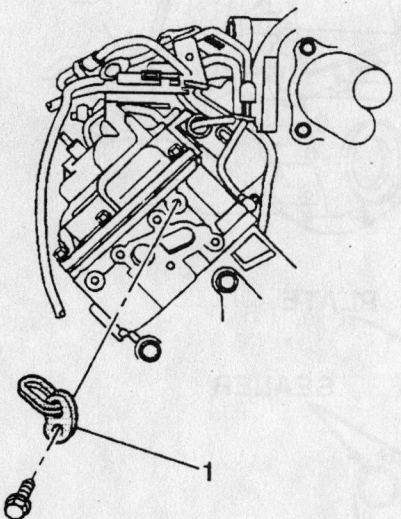

9348NG12

Install the engine lifting bracket to the rear of the left cylinder head

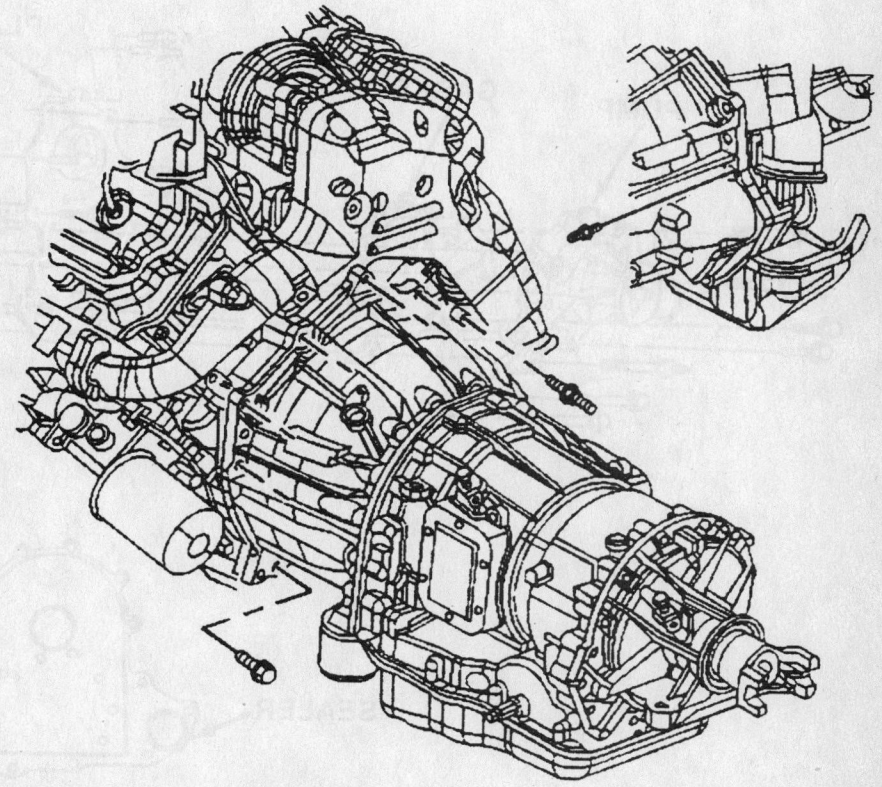

9348NG13

Transmission-to-engine mounting—6.6L engine with A/T

- Oil pan skid plate bolts: 15 ft. lbs. (20 Nm)
- Engine protection shield bolts: 15 ft. lbs. (20 Nm)
- Alternator bracket bolt: 37 ft. lbs. (50 Nm)
- Idler pulley bolt: 32 ft. lbs. (43 Nm)
- A/C compressor bolts: 37 ft. lbs. (50 Nm)

28. Refill the crankcase and the cooling system.

29. Recharge the air conditioning system.

Water Pump

REMOVAL & INSTALLATION

6.5L Engine

1. Before servicing the vehicle, refer to the precautions in the beginning of this section.
2. Drain the engine coolant.
3. Remove or disconnect the following:
   - Negative battery cables
   - Fan and fan shroud
   - Air conditioning hose bracket and/or the oil filler tube, as required

- Accessory drive belt(s)
- Vacuum pump mounting bracket nuts/bolt
- Vacuum pump and bracket
- Power steering pump and bracket
- Coolant hoses from the pump
- Water pump plate retaining bolts
- Pump and plate assembly from the engine

➡ **Remove the bolt on the rear of the water pump plate.**

- Separate the pump and gasket from the plate

To install:

4. Install or connect the following:
   - Water pump and a new gasket to the plate. Torque the retaining bolt (at the rear of the plate) to 20 ft. lbs. (28 Nm).

5. Be sure the block mating surface and the plate flanges are free of oil. Apply an anaerobic sealer GM part 1052357 or equivalent.

➡ **The sealer must be wet to the touch when the bolts are tightened.**

- Water pump and plate assembly. Torque the bolts to 20 ft. lbs. (28 Nm).

- Coolant hoses to the pump assembly
- Power steering pump and bracket
- Vacuum pump and bracket, along with the bolt holding the pump and alternator
- Fan and pulley
- Accessory drive belt(s)
- Oil filler tube and/or air conditioning hose bracket nuts, if removed
- Fan shroud
- Batteries

6. Refill the radiator.

6.6L Engine

1. Before servicing the vehicle, refer to the precautions in the beginning of this section.
2. Remove the left front fender wheelhouse inner panel.
3. Drain the coolant.
4. Remove or disconnect the following:
   - Thermostat housing crossover
   - Fan clutch
   - Crankshaft balancer
   - Water pump outlet pipe-to-water pump nuts
   - Engine wiring harness retainer front the inner stud
   - Water pump bolts, noting their

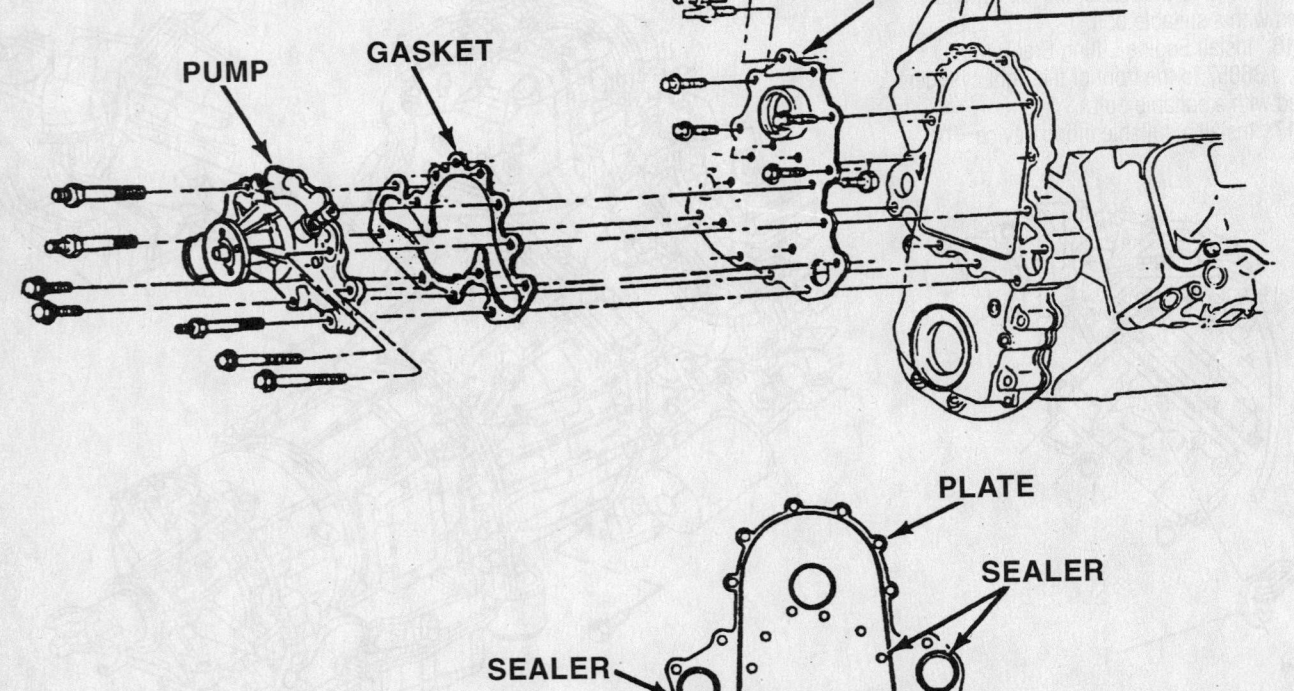

Exploded view of the water pump assembly and related components—6.5L diesel engines

7924KG05

locations as they are different lengths
  • Water pump and gasket

To install:

5. Lubricate the water pump O-ring with engine oil.

6. Install or connect the following:
  • Water pump
  • Water pump bolts and tighten to 15 ft. lbs. (20 Nm)
  • Water pump-to-water pump outlet gasket
  • Engine wiring harness retainer on the water pump outlet pipe inner stud
  • Water pump-to-water pump outlet pipe nuts and tighten to 15 ft. lbs. (20 Nm)
  • Thermostat housing crossover
  • Crankshaft balancer
  • Fan clutch

7. Fill the cooling system and install the left front fender wheelhouse inner panel.

Heater Core

REMOVAL & INSTALLATION

Please refer to Heater Core under Gasoline Engine Repair for this procedure.

Glow Plugs

REMOVAL & INSTALLATION

6.5L Engine

1. Before servicing the vehicle, refer to the precautions in the beginning of this section.

2. Remove or disconnect the following:
  • Negative battery cables
  • Glow plug lead wires
  • Plugs
  • Right front tire
  • Inner splash shield from the fender well
  • Lead wire from the plug at the No. 2 cylinder and the lead wires from plugs in the Nos. 4 and 6 cylinders at the harness connectors
  • Heat shroud for the plug in the No. 4 and 6 cylinder. Slide the shrouds back just far enough to allow access so you can unplug the wires.
  • Glow plugs in cylinders No. 2, 4 and 6.

3. Reach up under the vehicle and disconnect the lead wire at No. 8. Remove the glow plug.

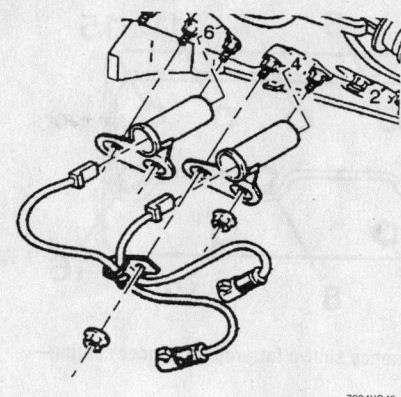

7924KG42

Exploded view of the heat shrouds and glow plug wiring—6.5L diesel engine

➡ **You may find that removing the exhaust pipe down might make this a bit easier when working on Nos. 6 and 8.**

To install:

4. Install or connect the following:
  • Glow plugs and tighten to 16 ft. lbs. (22 Nm)
  • Heat shrouds and electrical connection
  • Exhaust pipe, if removed
  • Splash shields, if removed
  • Negative battery cable

6.6L Engine

BANK 1 OR 2

1. Before servicing the vehicle, refer to the precautions in the beginning of this section.

2. Remove or disconnect the following:
  • Negative battery cables
  • Front tire
  • Inner splash shield from the fender well
  • Air cleaner outlet duct
  • Electrical nuts from the glow plug(s)
  • Harness from the glow plug(s)

➡ **On vehicles with Federal emissions systems, there is a buss bar connecting the glow plugs on each bank of the engine.**

  • Glow plug(s)

To install:

3. Install or connect the following:
  • Glow plug and tighten to 13 ft. lbs. (18 Nm)
  • Buss bar and wiring
  • Glow plug electrical nut and tighten to 13 inch lbs. (1.5 Nm)
  • Air cleaner outlet duct

  • Splash shield to the fender well
  • Negative battery cables

Cylinder Head

REMOVAL & INSTALLATION

6.5L Engine

1. Before servicing the vehicle, refer to the precautions in the beginning of this section.

2. Relieve the fuel system pressure.

3. Drain the coolant system.

4. Discharge the air conditioning system.

5. Remove or disconnect the following:
  • Negative battery cables
  • Intake manifold
  • Fan upper shroud
  • Compressor assembly, if equipped
  • Turbocharger, if equipped
  • Exhaust manifold
  • Valve cover
  • Rocker arm assemblies and pushrods

➡ **Mark all components so they may be returned to their original location.**

  • Air cleaner resonator and bracket
  • Transmission and oil dipstick tube; remove the oil fill tube from the coolant crossover pipe
  • Heater, radiator and bypass hoses
  • Alternator upper bracket
  • Alternator
  • Power steering pump
  • Vacuum pump
  • Fuel bleeder valve at the coolant crossover pipe
  • Fuel return crossover line clamp bolts from both cylinder heads
  • Wire connector from the sensor in the coolant crossover pipe
  • Electrical connection and brackets from cylinder head
  • Coolant crossover pipe/thermostat assembly
  • Head bolts and the cylinder heads

To install:

6. Clean the mating surfaces of the heads and block thoroughly.

7. Clean the head bolts thoroughly. Coat the threads of the head bolts with sealing compound GM part 1052080 or equivalent, before installation.

8. Install a new gaskets, and the cylinder head and bolts. Torque the bolts, in sequence, as follows:
  a. Step 1: 20 ft. lbs. (25 Nm)
  b. Step 2: 50 ft. lbs. (65 Nm)

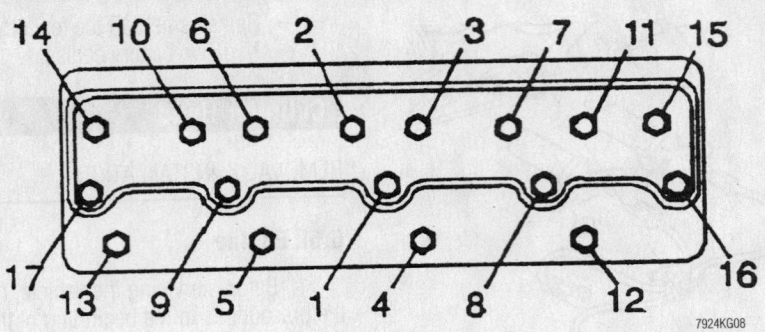

Tighten the cylinder head bolts according to the sequence shown for proper cylinder sealing—6.5L diesel engines

c. Step 3: An additional 90 degrees (¼ turn).
9. Install or connect the following:
   • Coolant crossover pipe and thermostat
   • Fuel valve
   • Bypass hose
   • Upper radiator hose
   • Heater hoses at the head
   • Transmission and oil dipstick tube
   • Air cleaner resonator and bracket
   • Pushrods, hardened ends facing up
   • Rocker arm assemblies
10. Adjust the valves.
   • Valve cover
   • Alternator and upper bracket
   • Exhaust manifolds. Torque bolts to 22 ft. lbs. (30 Nm).
   • Upper fan shroud
   • Intake manifold
   • Turbocharger, if equipped
   • Vacuum pump, if equipped
   • Air conditioning compressor, if equipped
   • Engine electrical connection
   • Negative battery cables
11. Refill the cooling system with the proper type and quantity of antifreeze.
12. Evacuate and recharge the air conditioning system.

6.6L Engine

1. Before servicing the vehicle, refer to the precautions in the beginning of this section.
2. Relieve the fuel system pressure.
3. Drain the coolant system.
4. Remove or disconnect the following:
   • Negative battery cables
   • Left or right front splash shield from the fender well, as applicable
   • Turbocharger
   • Turbocharger charged air cooler inlet duct
   • Thermostat housing crossover
   • Left or right intake manifold, as necessary

   • Upper left or right valve cover
   • Fuel rail assembly
   • Left or right exhaust manifold
   • Bolt and ground straps from the rear of the cylinder head
   • Lower left or right valve cover
   • Rocker arm shaft assembly
   • Glow plugs

   • Fuel injector return pipe eye bolts and washers
   • Fuel injector return pipe assembly
   • Fuel injector bracket bolts
   • Fuel injectors with the brackets, using a suitable removal tool
   • Injector bracket pins
   • Cylinder head bolts, in the proper sequence
   • Cylinder head and gasket. Discard the gasket

To install:
5. Clean the mating surfaces of the heads and block thoroughly.
6. Position a new left or right side head gasket on the block. Note that the left and right side gaskets are NOT interchangeable.

➡ **The cylinder head bolts on these vehicles are pre-coated with an application of a molybdenum disulfide for thread lubrication. Do not remove the coating or add any additional lubrication.**

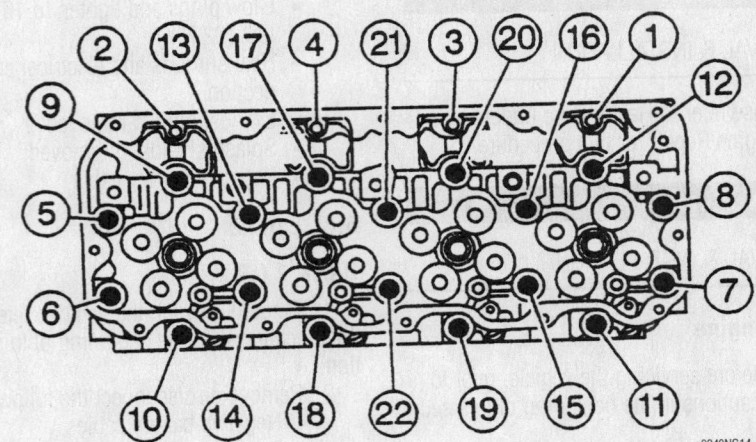

Cylinder head bolt loosening sequence—6.6L diesel engine

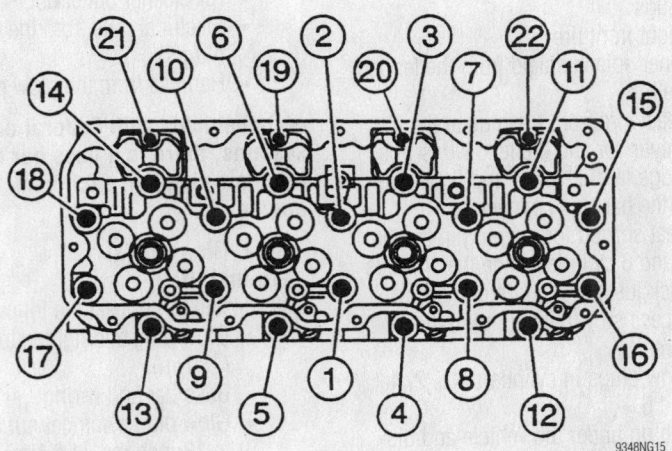

Cylinder head bolt tightening sequence—6.6L diesel engines

7. Install the cylinder head and bolts.

8. Tighten the cylinder head bolts, in sequence, as follows:

a. Step 1: M12 bolts to 37 ft. lbs. (50 Nm).

b. Step 2: M12 bolts to 59 ft. lbs. (80 Nm).

c. Step 3: Tighten the M12 bolts an additional 150 degrees using a torque angle meter.

d. Step 4: M8 bolts to 18 ft. lbs. (25 Nm).

9. Install or connect the following:

- New O-ring onto the fuel injectors after coating with clean engine oil
- New copper washer into the fuel injector bore in the cylinder head
- Fuel injector bracket pin

➡**If you are reusing the old injectors, clean the carbon from the tips, but do not use a wire brush.**

- Fuel injector bracket bolt and tighten to 37 ft. lbs. (50 Nm)
- Fuel injector return pipe assembly
- Fuel injector return pipe-to-injector eye bolts and washers. Tighten to 11 ft. lbs. (15 Nm).
- Fuel return pipe-to-cylinder head eye bolts and washers. Tighten to 11 ft. lbs. (15 Nm).
- Bolt and ground straps to the rear of the cylinder head. Tighten to 18 ft. lbs. (25 Nm).
- Valve rocker shaft assembly
- Lower and upper valve covers
- Glow plugs
- Exhaust manifold
- Fuel rail assembly
- Intake manifold
- Thermostat housing crossover
- Turbocharger charged air cooler duct
- Clamp and hose to the charged air cooler. Tighten to 53 inch lbs. (6 Nm).
- Turbocharger
- Fender splash shield
- Negative battery cables

10. Refill the cooling system with the proper type and quantity of antifreeze.

11. Evacuate and recharge the air conditioning system.

Starter Motor

REMOVAL & INSTALLATION

1. Before servicing the vehicle, refer to the precautions in the beginning of this section.

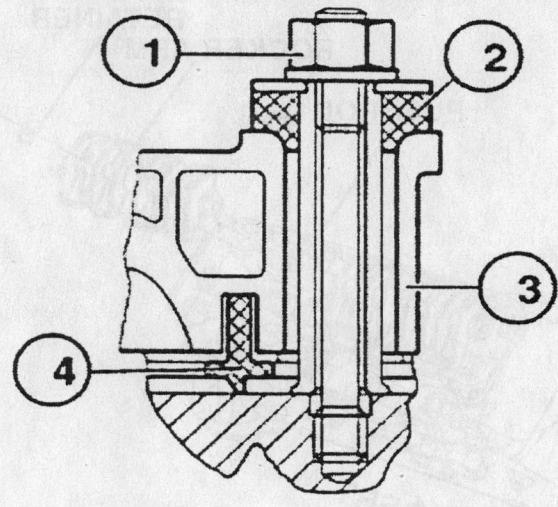

(1) Nut

(2) Decoupling element

(3) Intake air manifold

(4) Seal

9308KG04

Exploded view of the starter motor—6.5L engine shown

2. Remove or disconnect the following:

- Negative battery cables
- Right front wheel and fender splash shield, on 6.6L engines
- Mounting bolts/nuts and shim, if used
- Starter
- Wires
- Heat shield and bracket

To install:

3. Install or connect the following:

- Heat shield and bracket. Torque the bolts to 13 ft lbs. (17 Nm).
- Wires. Torque battery wire nut to 89 inch lbs. (10 Nm), and ignition nut to 18 inch lbs. (2 Nm) on 6.5L engines. On 6.6L engines, tighten the solenoid nut to 30 inch lbs. (3.4 Nm) and the positive battery cable nut to 80 inch lbs. (9 Nm).
- Starter
- Mounting bolts/nuts and shim, if used. Torque the bolts to 33 ft lbs. (45 Nm) and the nut to 75 inch lbs. (8.5 Nm) for 6.5L engines. For 6.6L engines tighten the starter bolts to 58 ft. lbs. (78 Nm).
- Right front fender splash shield and wheel, on 6.6L engines
- Negative battery cables

Rocker Arms/Shaft

REMOVAL & INSTALLATION

6.5L Engine

1. Before servicing the vehicle, refer to the precautions in the beginning of this section.

2. Remove or disconnect the following:

- Engine cover

➡**Rotate the engine until the mark on the crankshaft balancer is at the 2 o'clock position. Rotate the crankshaft counterclockwise 3½ in. (88mm) aligning the crankshaft balancer mark with the first lower water pump bolt, at about the 12:30 position. This will ensure that no valves are close to a piston crown**

- Cylinder head cover
- Rocker shaft assembly

➡**The rocker assemblies are mounted on 2 short rocker shafts per cylinder head, with each shaft operating 4 rockers. 2 bolts secure each rocker shaft assembly, Mark the shafts so they can be installed in their original locations.**

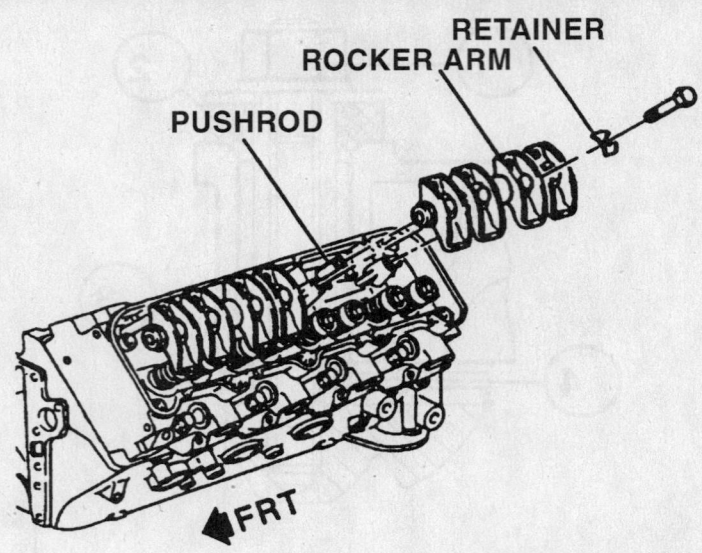

RETAINER
ROCKER ARM
PUSHROD
FRT

Rocker shaft assembly and related components—diesel engines

7924KG22

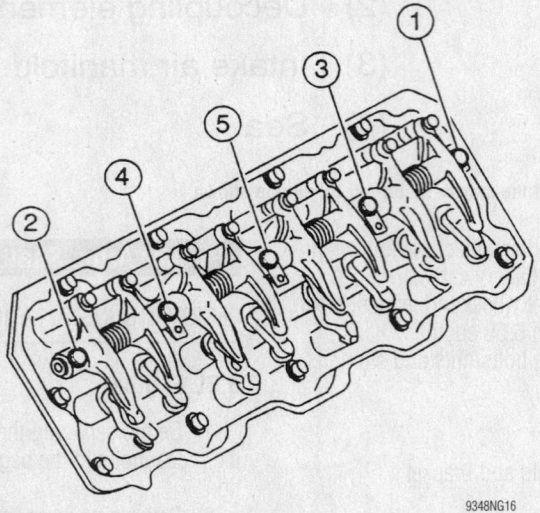

Rocker arm shaft bolt loosening sequence—6.6L engine

9348NG16

- Pushrods. The pushrods MUST be installed in the original direction! A paint stripe usually identifies the upper end of each rod, but if you can't see it, be sure to mark each rod yourself.

3. Insert a small prybar into the end of the rocker shaft bore and break off the end of the nylon retainers. Pull off the retainers with pliers, then slide off the rockers.

To install:

4. Be sure first that the rocker arms and springs go back on the shafts in the exact order in which they were removed. It's a good idea to coat them with engine oil.

5. Center the rockers on the corresponding holes in the shaft

6. Install or connect the following:
- New plastic retainers using a ½ in. (13mm) drift
- Pushrods with there marked ends up
- Rocker shaft assemblies and be sure that the ball ends of the pushrods seat themselves in the rockers

7. Rotate the engine clockwise until the mark on the torsional damper aligns with the **0** on the timing tab. Rotate the engine counterclockwise 3 ½ in. (88mm) measured at the damper. You can estimate this by checking that the mark on the damper is now aligned with the FIRST lower water pump bolt. BE CAREFUL! This ensures that the piston is away from the valves.
- Rocker shaft bolts and tighten to 40 ft. lbs. (55 Nm)
- Cylinder head cover
- Engine cover

6.6L Engine

1. Before servicing the vehicle, refer to the precautions in the beginning of this section.

2. Remove the lower valve (rocker arm) covers

3. Loosen the valve clearance lock nuts on each rocker arm

4. Loosen the valve clearance adjusting screw on each rocker arm to relieve tension on the valve train

➡ **The rocker arm bolts retain the rocker arms on the shaft. Do not remove the bolts from the rocker arm shaft brackets.**

5. Loosen the rocker arm shaft bolts in the proper sequence, leaving the bolts in the rocker arm shaft brackets.

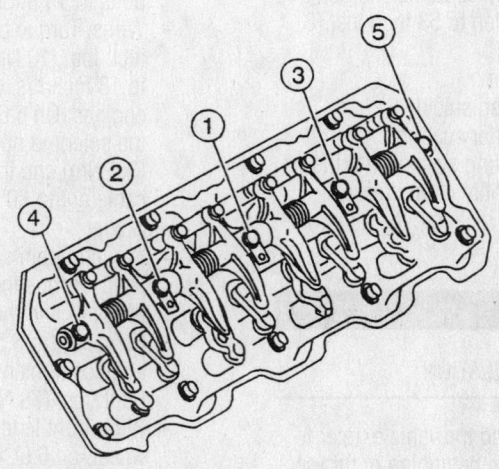

Rocker arm shaft tightening sequence—6.6L engines

9348NG17

6. Remove or disconnect the following:
- Rocker arm shaft assemblies from the cylinder head
- Valve bridge pins
- Valve bridges
- Valve push rods

7. Clean all parts in a suitable solvent. Disassemble the rocker arm shaft as necessary.

To install:

8. Lubricate the rocker arm shaft and the inside of the rocker arms with engine oil.

9. If disassembled, install or connect the following:
- Rocker arm bracket on one end of the rocker arm shaft with the bolt
- Rocker arm intake, spring, exhaust and the bracket with bolt. Continue in the same sequence to the last bracket.
- Push the bracket to compress the springs and then install the bolt

10. Lubricate the top of the valves, the valve bridge stem, the valve bridge and the valve bridge pins.

11. Install or connect the following:
- Valve bridge pins
- Valve bridges
- Pushrods. Make sure it is fully installed by gently pulling up on it. You should feel resistance from the pushrod trying to lift the valve lifter

12. Use clean engine oil to lubricate the rocker arm shaft bolt threads, tops of the push rods, rocker arms and rocker arm shaft.
- Rocker arm shaft assembly to the cylinder head
- Rocker arm shaft assembly bolts and tighten, in the proper sequence to 30 ft. lbs. (40 Nm)

13. Adjust the valve clearance, as follows:

a. Remove the fan clutch.

b. Remove both upper valve covers.

c. Rotate the engine in the normal direction and place the No. 1 piston at Top Dead Center (TDC) of the compression stroke. The No. 1 cylinder is at the right side front. While turning the engine, watch the intake valve to open and close. Align the mark on the crankshaft balancer with the pointer on the engine.

d. Loosen the valve clearance adjusting screws for the valve being adjusted.

e. Insert the feeler gauge between the tip of the rocker arm and the valve bridge.

f. Adjust the intake and the exhaust valve clearance to 0.012 in. (0.3mm) with

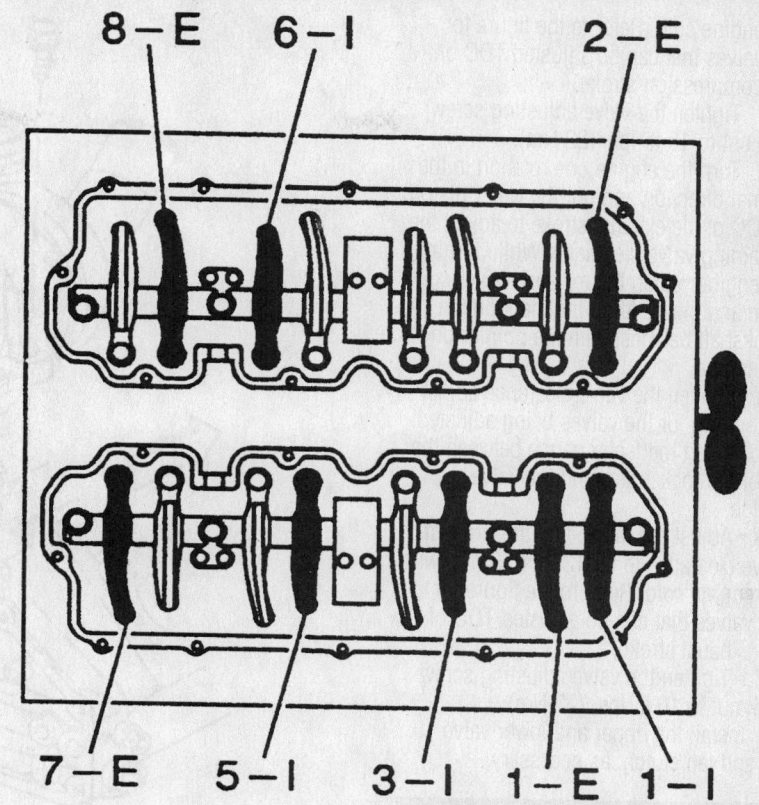

Location of the valves that are adjusted at TDC of the compression stroke—6.6L engine

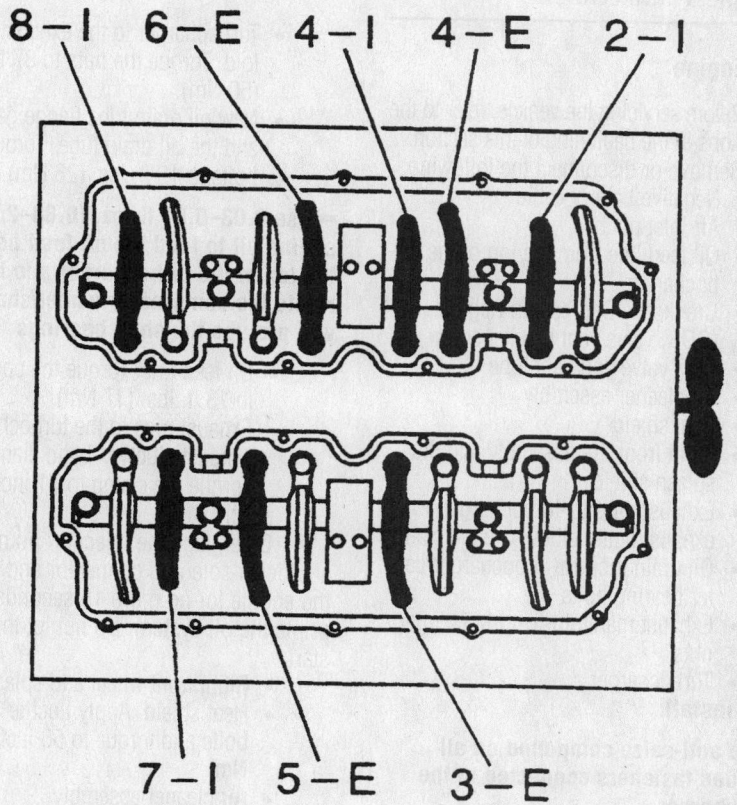

Location of the valves that are adjusted at TDC of the exhaust stroke—6.6L engine

the engine cold. Refer to the figure for the valves that can be adjusted TDC of the compression stroke.

g. Tighten the valve adjusting screw lock nut to 16 ft. lbs. (22 Nm).

h. Turn the engine one rotation in the normal direction and put the No. 1 piston at TDC of the exhaust stroke to adjust the remaining valve clearance. While turning the engine, watch the exhaust valve to open and close. Align the mark on the crankshaft balancer with the pointer on the engine.

i. Loosen the valve clearance adjusting screws for the valves being adjusted.

j. Insert the feeler gauge between the tip of the rocker arm and the valve bridge.

k. Adjust the intake and the exhaust valve clearance to 0.012 in. (0.3mm) with the engine cold. Refer to the figure for the valves that can be adjusted TDC of the exhaust stroke.

l. Tighten the valve adjusting screw lock nut to 16 ft. lbs. (22 Nm).

14. Install the upper and lower valve cover and fan clutch, as necessary.

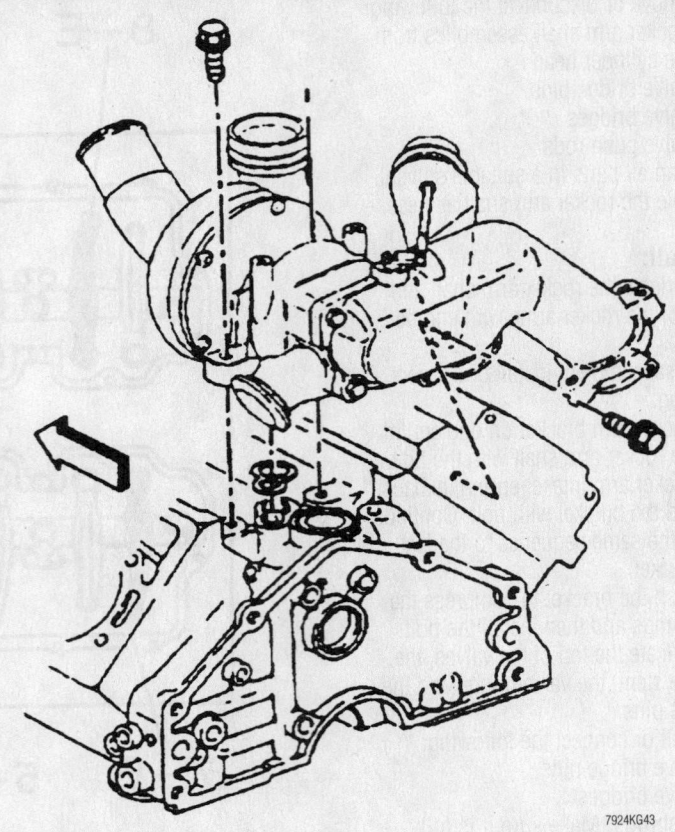

Turbocharger mounting—6.5L diesel engine

7924KG43

Turbocharger

REMOVAL & INSTALLATION

6.5L Engine

1. Before servicing the vehicle, refer to the precautions in the beginning of this section.
2. Remove or disconnect the following:
   - Negative battery cable
   - Air inlet duct
   - Oil feed line from the top of the turbocharger
   - Crankcase Depression Regulator (CDR) valve vent bracket screw
   - CDR valve and vent tube
   - Air cleaner assembly
   - Heat shield
   - Right front tire assembly and the splash shield
   - Exhaust pipe-to-turbocharger exhaust outlet elbow V-band clamp
   - Oil drain tube-to-turbocharger center bearing bolts
   - Exhaust manifold-to-turbocharger nuts
   - Turbocharger

   To install:

   ➡**Use anti-seize compound on all threaded fasteners connected to the turbocharger**

   3. Install or connect the following:

   - Turbocharger to the exhaust manifold. Torque the nuts to 37 ft. lbs. (50 Nm).
   - New oil drain tube flange gasket and the oil drain tube. Torque the bolts to 19 ft. lbs. (26 Nm).

   ➡**Use 0.03–0.07 fl. oz. (0.88–2ml) of engine oil to feed the oil feed hole at the top of the turbocharger and hand rotate the compressor wheel/shaft. This will prelube the shaft bearings**

   - Oil feed line. Torque the connection to 13 ft. lbs. (17 Nm).
   - Exhaust pipe to the turbocharger exhaust elbow V-band clamp. Torque the clamp to 71 inch. lbs. (8 Nm).

   4. Disengage the injection pump fuel shutdown solenoid connector and crank the engine for no more 15 seconds to prime the oil system. Do not let the engine start.

   - Right front wheel and splash shield
   - Heat shield. Apply Loctite® to the bolts and torque to 56 inch lbs. (6 Nm)
   - Air cleaner assembly
   - Turbocharger compressor outlet
   - CDR valve, tube and bracket
   - Air intake duct

   ➡**Operate the engine at idle for at least 3 minutes after installing the turbocharger**

6.6L Engine

1. Disconnect the negative battery cables.
2. Open the hood and move the hinge bolts to the service position.
3. Raise the vehicle.
4. Drain the coolant.
5. Remove or disconnect the following:
   - Exhaust pipe-to-exhaust outlet clamp. Move the clamp onto the exhaust pipe
   - Transmission fluid fill tube-to-bell housing nuts if equipped with an A/T. Position the tube to the right side of the vehicle; it does not need to be removed from the transmission
   - 3 nuts and left exhaust heat shield from the front of the lower dash panel
   - Left exhaust pipe heat shield bolts

6. Position the left exhaust pipe heat shield to access the left exhaust pipe-to-manifold bolts. Do not remove the heat shield from the vehicle at this time.

➡**Do not bend the exhaust pipe at the expansion area.**

- Left, then the right exhaust pipe-to-exhaust manifold bolts
- Gaskets and discard
- Lower bolt for the exhaust outlet shield

7. Lower the vehicle.
- Upper intake manifold sight shield front retaining bolt
- Sight shield
- Air cleaner outlet duct from the air cleaner and turbocharger. Cover the openings to prevent debris from entering
- Charged air cooler outlet duct-to-intake hose clamps (loosen only)
- Hose from the charged air cooler duct-to-intake manifold tube
- A/C compressor clutch electrical connector
- A/C cut-out switch connector
- Drive belt
- A/C compressor mounting bolts; position the compressor aside with the lines attached
- Turbocharger inlet coolant hose from the bypass valve
- Turbocharger outlet coolant hose from the turbocharger
- Crankcase hose from the left valve cover and position aside
- Wire connector from the intake heater
- Intake air heater relay, if equipped
- Heat shield-to-turbocharger bolts and heat shield
- Remaining 2 bolts from the exhaust outlet heat shield
- Exhaust outlet heat shield
- 4 bolts and 2 nuts from the exhaust outlet. You do not have to remove the outlet for turbocharger removal

8. Move the exhaust outlet to one side in order to access the right exhaust pipe-to-turbocharger bolts.
- Exhaust outlet gasket and discard
- Right exhaust pipe-to-turbocharger bolts
- Right exhaust pipe and gasket

9. Move the exhaust outlet to one side for access to the left pipe.
- Left exhaust pipe heat shield
- Left exhaust pipe-to-turbocharger bolts
- Left exhaust pipe and gasket
- Turbocharger oil supply hose eye bolt and washers. Move the hose aside
- Turbocharger oil drain pipe nuts from the flywheel housing

- Turbocharger mounting bolts
- Turbocharger with the oil drain pipe

10. If replacing the turbocharger, remove the oil drain pipe and coolant hose.

To install:

11. Thoroughly clean the gasket surfaces.

12. Install or connect the following:
- Turbocharger oil drain pipe and new gasket. Tighten the bolts to 16 ft. lbs. (21 Nm).
- Turbocharger inlet coolant hose
- Turbocharger oil supply hose to the engine block
- Turbocharger oil supply hose eye bolt and washers and tighten to 31 ft. lbs. (42 Nm)
- Turbocharger lower heat shield
- Turbocharger. Tighten the 3 mounting bolts to 80 ft. lbs. (108 Nm).
- New gasket for oil drain pipe
- Oil drain pipe nuts and tighten to 15 ft. lbs. (20 Nm)

13. If installing a new turbocharger, pour 4–5 oz. of clean engine oil into the turbocharger supply hose opening, while rotating the impeller.
- Oil supply hose, using new washers. Tighten the eye bolt to 31 ft. lbs. (42 Nm).

14. Install the remaining components in the reverse order of removal, noting the following important points:
- When installing the exhaust pipe, use new gaskets and align the tabs and make sure the proper pipe flange is towards the turbocharger, as they are different. Tighten the exhaust pipe-to-turbocharger bolts to 39 ft. lbs. (53 Nm).
- Tighten the turbocharger heat shield bolts to 80 inch lbs. (9 Nm).
- Tighten the A/C compressor bolts to 37 ft. lbs. (50 Nm).
- Tighten the exhaust pipe clamp to 30 ft. lbs. (40 Nm).

15. Fill the cooling system and connect the negative battery cables.

➡**Operate the engine at idle for at least 3 minutes after installing the turbocharger**

Intake Manifold

REMOVAL & INSTALLATION

6.5L Engine

1. Before servicing the vehicle, refer to the precautions in the beginning of this section.

2. Recover air conditioning system, and reposition air conditioning lines.

3. Remove or disconnect the following:
- Negative battery cable
- Air cleaner assembly
- Fuel lines, and electrical connections
- Engine and transmission oil level tubes

❊❊ WARNING

Do not remove the center intake and side intakes as an assembly. Damage to the center intake and turbocharger may occur.

- Center intake assembly, and glow plug relay
- Side intake bolts and fuel retaining clips
- Side intakes

4. Clean all gaskets surface.

To install:

5. Install or connect the following:
- Side intakes and new gaskets. Torque the bolts to 31 ft. lbs. (42 Nm).
- Fuel lines retaining clips and electrical connection
- Center intake with new gaskets. Torque the bolts to 17 ft. lbs. (23 Nm).
- Engine oil and transmission oil level tubes
- Glow plug relay
- Air conditioning lines
- Air cleaner assembly
- Negative battery cable

6. Recharge air conditioning system.

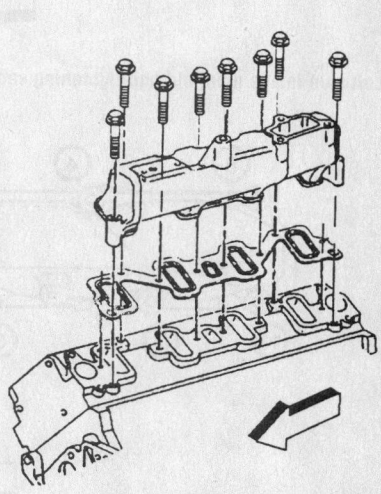

Exploded view of the side intake manifold mounting—6.5L engines

7924KG40

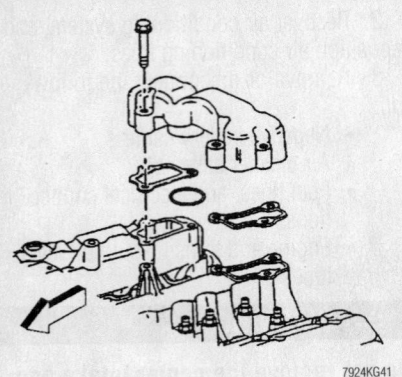

Center intake manifold mounting—6.5L engines

7924KG41

6.6L Engines

➡**This procedure is for replacement of the left or right intake manifold.**

1. Before servicing the vehicle, refer to the precautions in the beginning of this section.
2. Drain the cooling system.
3. Remove or disconnect the following:
   - Batteries cables
   - Turbocharger

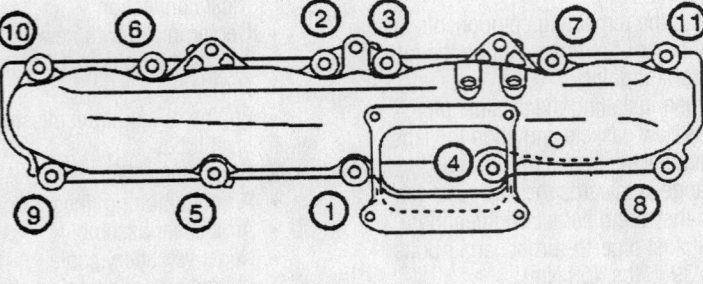

Left side intake manifold bolt tightening sequence—6.6L engine

9348NG20

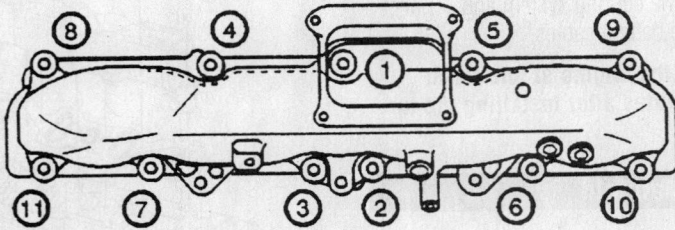

9348NG21

Right side intake manifold bolt tightening sequence—6.6L engine

- Fuel junction block
- Left or right fuel rail
- Intake manifold tube
- 9 bolts and 2 nuts from the intake manifold. A bolt is located in the manifold opening.

➡**The intake manifold uses sealer. If necessary, pry at the area by the common rail bolt holes and be careful to avoid damaging the sealing surfaces.**

- Intake manifold from the head. Cover the head openings to prevent debris from entering.

4. Clean all gaskets surface.

To install:

5. Install or connect the following:
   - A ⅛ in. (2–3mm) wide to 1⁄16 in (0.5–1.5mm) high bead of sealant to the sealing surface of the intake manifold

➡**The left and right side manifolds are NOT interchangeable.**

- Intake manifold
- Bolts and nuts. Tighten to 15 ft. lbs. (20 Nm), in sequence.
- Intake manifold tube

- Fuel rail
- Fuel junction block
- Turbocharger
- Negative battery cables

6. Fill cooling system.

Exhaust Manifold

REMOVAL & INSTALLATION

6.5L Engine

1. Before servicing the vehicle, refer to the precautions in the beginning of this section.
2. Remove or disconnect the following:
   - Batteries cables
   - Exhaust pipe from the manifold flange
   - Engine oil and transmission oil fill tubes
   - Engine cover and disconnect the glow plug wires
   - Glow plugs
   - Turbocharger assembly, as required
   - Air conditioner compressor rear bracket, as required
   - Manifold bolts and the manifold

To install:

3. Install or connect the following:
   - Exhaust manifold. Torque the bolts to 26 ft. lbs. (35 Nm).
   - Exhaust pipe
   - Glow plugs and electrical connection
   - Engine and transmission oil fill tubes
   - Air conditioning compressor bracket
   - Negative battery cable

6.6L Engine

LEFT SIDE

1. Before servicing the vehicle, refer to the precautions in the beginning of this section.
2. Raise the vehicle.
3. Remove or disconnect the following:
   - Bolts securing the left exhaust pipe heat shield and move the heat shield aside
   - Left exhaust pipe-to-manifold bolts
   - Left front wheel
   - Left front fender splash shield
   - Charge air cooler duct
   - Exhaust manifold heat shield bolts and shield
   - 2 nuts and 6 bolts with the plain washer and bell view washer from the left manifold
   - Exhaust manifold by removing it from the rear, then the front studs

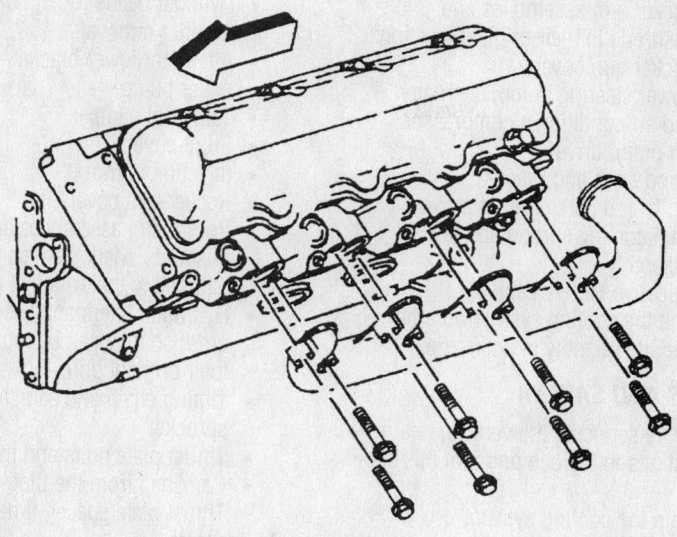

7924KG44

Exploded view of the left exhaust manifold mounting—6.5L diesel engines

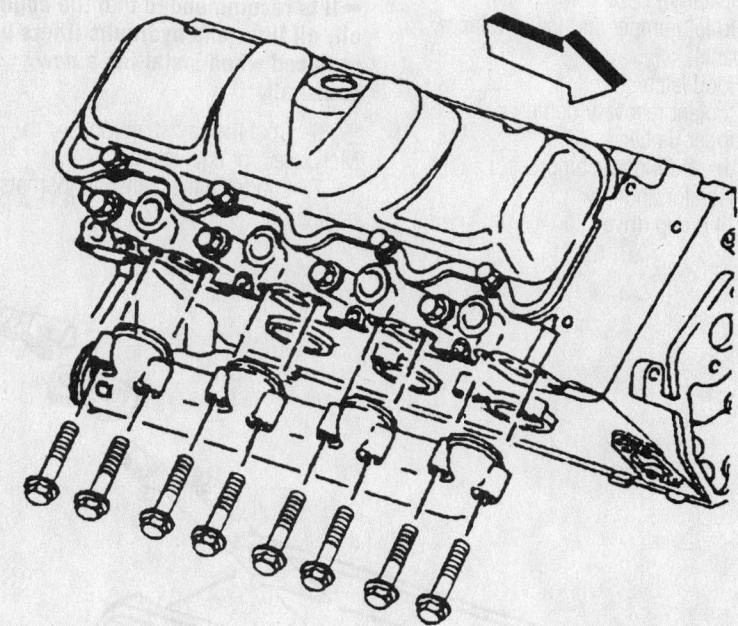

7924KG45

Exploded view of the right exhaust manifold mounting—6.5L diesel engines

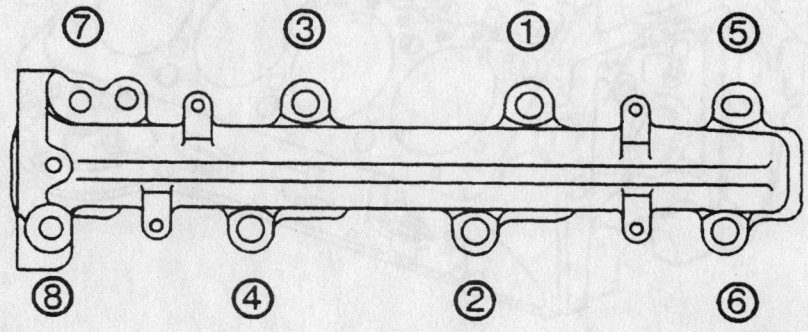

9348NG22

Left and right side exhaust manifold bolt torque sequence—6.6L engine

and sliding it out the bottom, past the oil filter
- Exhaust manifold gasket and discard

To install:

4. Installation is the reverse of the removal procedure. Tighten the retainers as follows:

a. Exhaust manifold nuts and bolts, in sequence, in 2 passes: 25 ft. lbs. (34 Nm).

b. Heat shield bolts: 71 inch lbs. (8 Nm).

c. Exhaust pipe-to-manifold bolts: 39 ft. lbs. (59 Nm).

RIGHT SIDE

1. Before servicing the vehicle, refer to the precautions in the beginning of this section.

2. Raise the vehicle.

3. Remove or disconnect the following:
- Right front wheel
- Right front fender splash shield
- Exhaust manifold heat shield bolts and shield
- Right exhaust pipe-to-manifold bolts
- 2 nuts and 6 bolts with the plain washer and bell view washer from the left manifold
- Exhaust manifold by removing it from the rear, then the front studs and sliding it out the bottom, past the oil filter
- Bolt for the oil level dipstick tube, to remove the gasket
- Exhaust manifold gasket and discard

To install:

4. Installation is the reverse of the removal procedure. Tighten the retainers as follows:

a. Oil level dipstick tube: 15 ft. lbs. (20 Nm).

b. Exhaust manifold nuts and bolts, in sequence, in 2 passes: 25 ft. lbs. (34 Nm).

c. Heat shield bolts: 71 inch lbs. (8 Nm).

d. Exhaust pipe-to-manifold bolts: 39 ft. lbs. (59 Nm).

Camshaft and Valve Lifters

REMOVAL & INSTALLATION

6.5L Engine

SILVERADO AND SIERRA

1. Before servicing the vehicle, refer to the precautions in the beginning of this section.

2. Drain the cooling system.
3. Discharge the air conditioning system.
4. Relieve the fuel system pressure.
5. Remove or disconnect the following:
- Battery cables
- Radiator, condenser, shroud and fan assembly
- Grille and parking light assembly
- Hood latch and brace assembly
- Oil pump drive
- Power steering pump and position aside
- Alternator
- Air conditioner compressor and position aside
- Rocker arm covers
- Rocker arm assemblies and pushrods. Mark them so they can be returned to their original position.
- Cylinder heads
- Hydraulic lifters; keep them in order so they can be returned to their original bore
- Front cover
- Timing chain and camshaft sprocket
- Injector pump
- Front engine mounting through-bolts
- Air conditioner condenser mounting bolts and lift the condenser out
- Thrust plate bolts and thrust plate
- Camshaft from the block
- Thrust plate spacer, if necessary

To install:
6. Install the spacer with the I.D. chamfer toward the camshaft.

➡ **It is recommended that the engine oil, oil filter and hydraulic lifters be replaced when installing a new camshaft.**

7. Coat the camshaft lobes with Molykote® or equivalent.
8. Lubricate the camshaft journals with engine oil.
9. Install or connect the following:
- Camshaft carefully into the block
- Thrust plate and bolts. Tighten to 17 ft. lbs. (23 Nm).
- Engine mount through-bolts
- Timing chain and sprockets, align the timing marks
- Air conditioner condenser, if equipped
- Injector pump
- Front cover
- Cylinder head
- Hydraulic lifters in the same bore as they were removed

- Rocker arm assemblies and pushrods in their original locations
- Rocker arm covers
- Power steering pump, alternator and air conditioner compressor
- Oil pump drive
- Hood latch and brace
- Grille and parking light assembly
- Radiator, the shroud and fan assembly
- Negative battery cables
10. Refill the cooling system with the proper type and quantity of antifreeze.

EXPRESS AND SAVANA

1. Before servicing the vehicle, refer to the precautions in the beginning of this section.
2. Drain the cooling system.
3. Relieve the fuel system pressure.
4. Remove or disconnect the following:
- Battery cables
- Headlight bezels
- Grille, bumper and lower valance panel
- Hood latch
- Coolant recovery bottle
- Upper tie bar
- Air conditioner compressor
- Radiator and fan
- Oil pump drive

- Cylinder heads to gain clearance for lifter removal
- Alternator lower bracket
- Water pump
- Torsional damper
- Front cover
- Injection pump
- Rocker arm covers
- Rocker arm assemblies and pushrods. Mark them so they can be returned to their original position.
- Hydraulic lifters and keep them in order so they can be returned to their original bore.
- Timing chain and camshaft sprocket
- Thrust plate bolts and thrust plate
- Camshaft from the block
- Thrust plate spacer, if necessary

To install:
5. Install the spacer with the I.D. chamfer toward the camshaft.

➡ **It is recommended that the engine oil, oil filter and hydraulic lifters be replaced when installing a new camshaft.**

6. Coat the camshaft lobes with Molykote®, or equivalent.
7. Lubricate the camshaft journals with engine oil.

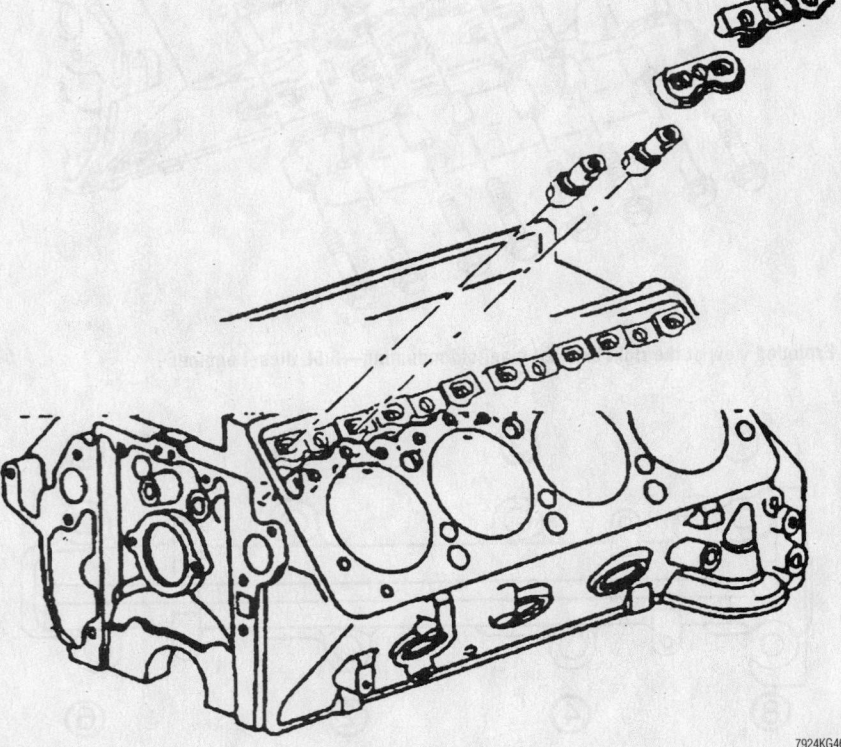

Exploded view of the lifter, guide plate and clamp—6.5L diesel engines

7924KG46

8. Install or connect the following:
  - Camshaft carefully into the block
  - Thrust plate and bolts. Torque the bolts to 17 ft. lbs. (23 Nm).
  - Timing chain and sprockets, align the timing marks
  - Hydraulic lifters in the same bore as they were removed
  - Rocker arm assemblies and pushrods in their original locations
  - Rocker arm covers
  - Fuel pump
  - Front cover
  - Torsional damper and water pump
  - Alternator lower bracket
  - Cylinder heads
  - Oil pump drive
  - Radiator and fan
  - Air conditioner compressor
  - Upper tie bar
  - Coolant recovery bottle
  - Hood latch
  - Grille, bumper and lower valence panel
  - Headlight bezels
  - Battery cables
9. Refill the cooling system.
10. Evacuate and charge the air conditioner system.

6.6L Engines

➡This procedure requires the use of the following special tools: Flywheel Holding Tool No. J 44643, Magnetic Base J 26900-13 and Dial Indicator J 26900-12.

1. Before servicing the vehicle, refer to the precautions in the beginning of this section.
2. Properly discharge the A/C system.
3. Remove or disconnect the following:
  - Both cylinder heads
  - Valve lifter guide hold-down bracket bolts
  - Valve lifter guide hold-down brackets
  - Valve lifter guides
  - Valve lifters
  - Charged air cooler
  - A/C condenser
  - Starter
4. Install the Flywheel Holding Tool No. J 44643 in the starter opening. Make sure the tool is flush to the flywheel opening. The holding tool will be used to remove the crankshaft balancer bolt and camshaft drive gear bolt.
  - Engine front cover
  - Oil pump driven gear nut and gear

➡The crankshaft reluctor and oil pump drive gear are timed together at the factory. Do NOT remove the reluctor from the oil pump drive gear.

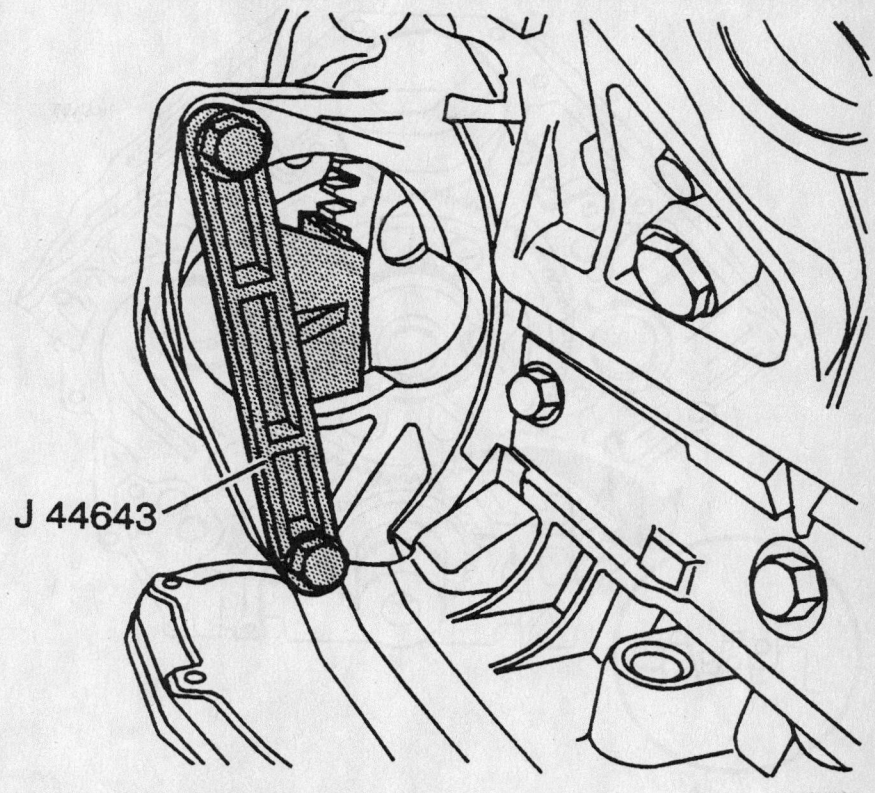

J 44643

9348NG25

Proper installation of the flywheel holding tool in the starter opening

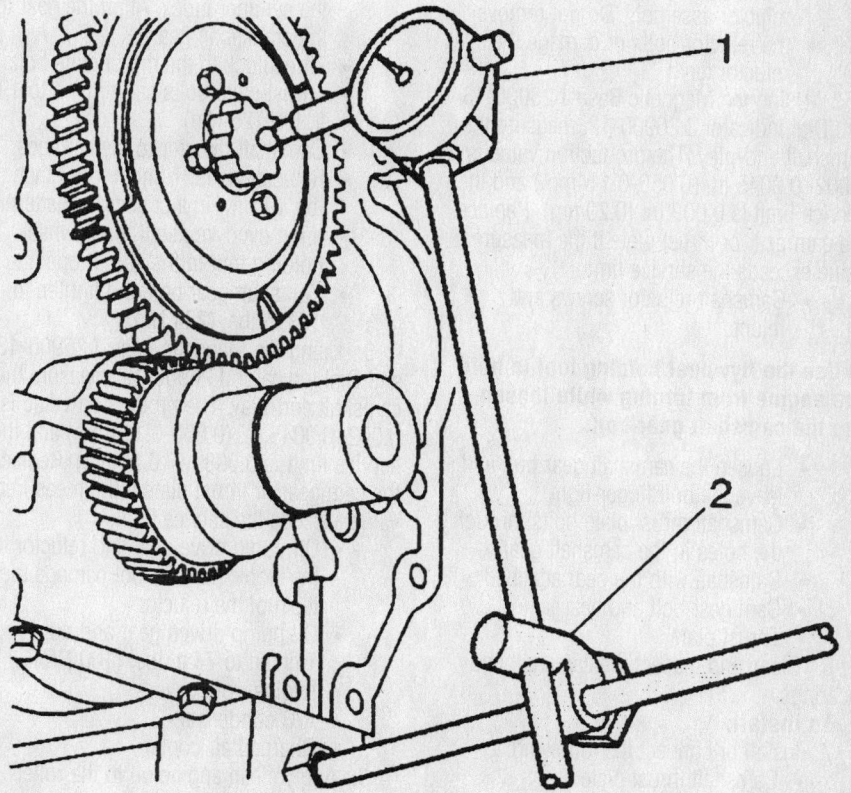

1

2

9348NG23

Use the dial indicator (1) and magnetic base (2) to measure the camshaft end-play

9348NG24

Camshaft and crankshaft gear alignment—6.5L engine

- Oil pump drive gear and crankshaft reluctor assembly. Do not remove the reluctor bolts or damage the reluctor teeth

5. Using the Magnetic Base J 26900-13 and Dial Indicator J 26900-12, measure the camshaft end-play. The production value is 0.002–0.0045 in. (0.050-0.114mm) and the service limit is 0.008 in. (0.20mm). Replace the cam gear or thrust plate if the measured value exceeds the service limit.

- Camshaft reluctor screws and reluctor

➡**Use the flywheel holding tool to hold the engine from turning while loosening the camshaft gear bolt.**

- Loosen the camshaft gear bolt and leave the bolt finger-tight
- Camshaft thrust plate bolts through the holes in the camshaft gear
- Camshaft with the gear attached
- Cam gear bolt and gear
- Thrust plate

6. Clean and inspect the camshaft and bearings.

To install:

7. Install or connect the following:
- Camshaft thrust plate
- Camshaft driven gear
- Driven gear bolt (finger-tight)

- Camshaft and gear assembly into the cylinder block. Align the gear to the crankshaft gear
- Threadlock to the thrust plate bolts
- Thrust plate bolts and tighten to 19 ft. lbs. (26 Nm)
- Camshaft reluctor to the cam gear
- Reluctor bolts. Tighten to 80 inch lbs. (9 Nm) in a crisscross pattern.
- If removed, reinstall the flywheel holding tool in the starter opening
- Camshaft gear bolt and tighten to 173 ft. lbs. (234 Nm)

8. Using the Magnetic Base J 26900-13 and Dial Indicator J 26900-12, measure the camshaft end-play. The production value is 0.002–0.0045 in. (0.050-0.114mm) and the service limit is 0.008 in. (0.20mm). Replace the cam gear or thrust plate if the measured value exceeds the service limit.

- Oil pump drive gear and reluctor to the crankshaft. Do not damage the teeth of the reluctor.
- Oil pump driven gear and nut. Tighten to 74 ft. lbs. (100 Nm).
- Engine front cover
- A/C condenser
- Charged air cooler

9. Apply clean engine oil to the roller and outside of the lifters.
- Valve lifters

- Valve lifter guides
- Valve lifter guide hold-down brackets. Make sure that both tabs of the bracket are in the holes of the valve lifter guides.
- Valve lifter guide hold-down bracket bolts. Tighten to 97 inch lbs. (11 Nm).

Valve Lash

ADJUSTMENT

All engines use hydraulic lifters, which require no periodic adjustment.

Oil Pan

REMOVAL & INSTALLATION

6.5L Engine

SILVERADO AND SIERRA

1. Before servicing the vehicle, refer to the precautions in the beginning of this section.
2. Drain the engine oil.
3. Remove or disconnect the following:
- Battery cables
- Oil dipstick
- Flywheel/flexplate cover
- Oil cooler line guides
- Front driveshaft
- Front axle, if needed
- Exhaust pipes from the manifolds
- Front engine mount through-bolts
- Oil pan bolts and the oil pan
- Oil pan rear seal

To install:

4. Clean all sealing surfaces.
5. Apply a ³⁄₁₆ in. (5mm) bead of RTV sealant to the oil pan sealing surface, inboard of the bolt holes. The sealant must be wet to the touch when the oil pan is to be installed.
6. Install or connect the following:
- Oil pan rear seal
- Oil pan to the engine. Torque all bolts except the rear 2 bolts to 84 inch lbs. (9.4 Nm). Tighten the rear bolts to 17 ft. lbs. (23 Nm).
- Engine mounting through-bolt and nut
- Front axles and front driveshaft, if removed
- Oil cooler lines in guides
- Oil dipstick
- Exhaust pipes to the manifolds
- Flywheel/flexplate cover
- Battery cables

7. Refill with the proper grade and quantity of oil.

EXPRESS AND SAVANA

1. Before servicing the vehicle, refer to the precautions in the beginning of this section.

2. Drain the engine oil.

3. Remove or disconnect the following:
- Battery cables
- Engine cover
- Engine oil dipstick
- Transmission flywheel/flexplate cover
- Oil cooler lines at the block
- Starter
- Transmission cooler lines, battery cables and attaching clamps from the oil pan
- Oil pan bolts
- Oil pan and oil pan rear seal

To install:

4. Clean all sealing surfaces

5. Apply a ³⁄₁₆ in. (5mm) bead of RTV sealant to the oil pan sealing surface, inboard of the bolt holes. The sealant must be wet to the touch when the oil pan is to be installed.

6. Install or connect the following:
- Oil pan rear seal
- Oil pan to the engine and the retaining bolts.
- Starter
- Transmission cooler lines, battery cables and attaching clamps to the oil pan
- Engine oil cooler lines
- Transmission flywheel/flexplate cover
- Engine oil dipstick tube
- Engine cover
- Battery cables

7. Refill engine with oil.

6.6L Engine

LOWER OIL PAN

1. Before servicing the vehicle, refer to the precautions in the beginning of this section.

2. Drain the engine oil.

3. Remove or disconnect the following:
- Oil pan skid plate (2WD vehicles)
- Crossbar
- Oil level sensor connector
- Lower oil pan bolts and nuts
- Lower oil pan from the lower crankcase
- Lower oil pan

To install:

4. Clean all sealing surfaces

5. Apply a ⅛ in. (2mm) bead of sealant to the oil pan sealing surface.

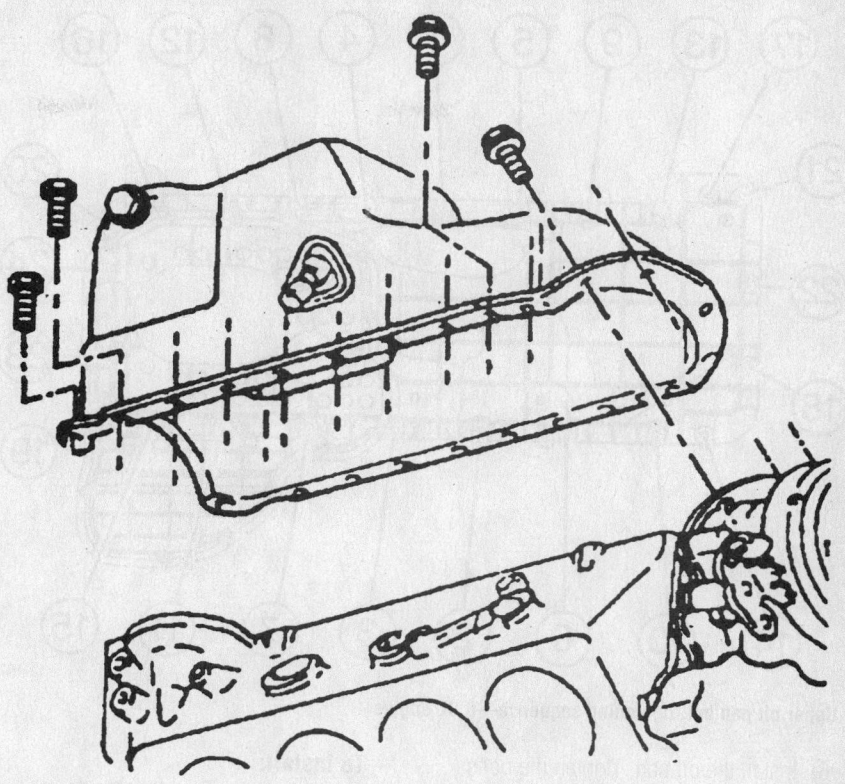

Exploded view of the oil pan mounting—6.5L diesel engines

7924KG47

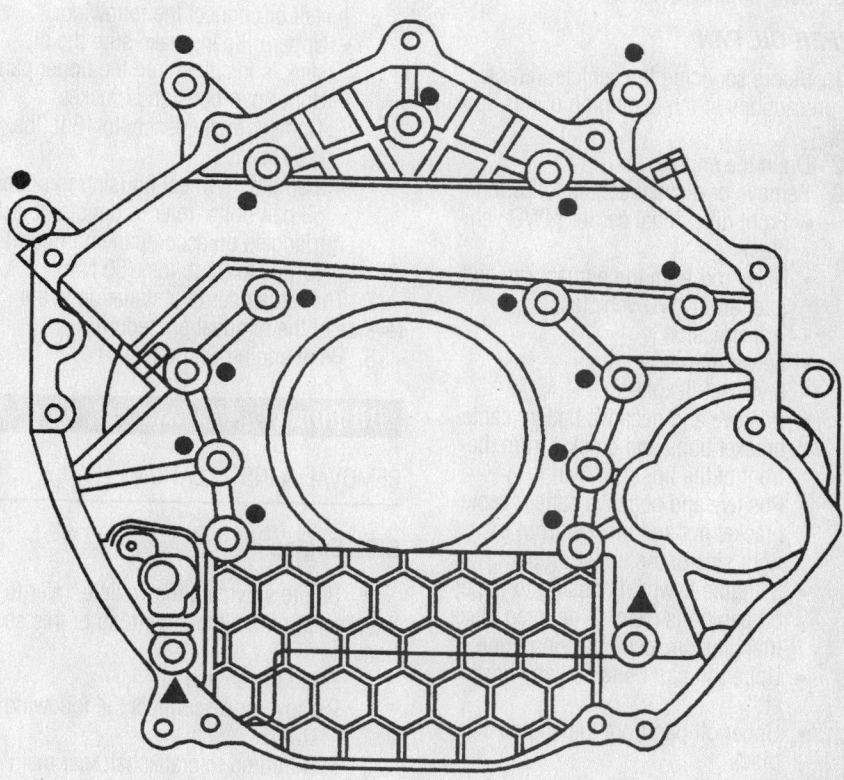

Remove only the flywheel housing-to-upper oil pan bolts designated with a black triangle—6.6L engine

9348NG26

Upper oil pan bolt tightening sequence—6.6L engine

6. Install the oil pan. Tighten the bolts and nuts to 89 inch lbs. (10 Nm)

7. The remainder of installation is the reverse of the removal procedure.

8. Refill engine with oil.

UPPER OIL PAN

1. Before servicing the vehicle, refer to the precautions in the beginning of this section.

2. Drain the engine oil.

3. Remove or disconnect the following:
- Front differential carrier (4WD vehicles)
- Relay rod from the pitman arm and idler arm (2WD vehicles)
- Transmission
- Lower oil pan
- Flywheel/flexplate
- Positive and negative battery cable bracket bolts and bracket from the front of the upper oil pan
- Positive and negative battery cable bracket nut and bracket from the right side of the upper oil pan
- 2 engine flywheel housing to upper oil pan bolts (refer to denoted black triangles on accompanying figure)
- Upper oil pan bolts and any brackets
- Upper oil pan from the engine block
- Upper oil pan. The oil dipstick tube needs to be removed while lowering the upper oil pan.

To install:

4. Clean all sealing surfaces

5. Apply a ⅛ in. (2mm) bead of sealant to the oil pan and flywheel sealing surfaces.

6. Install or connect the following:
- Upper oil pan; make sure the dipstick is installed into the upper pan
- Upper pan bolts and brackets. Tighten, in sequence, to 15 ft. lbs. (20 Nm).
- 2 engine flywheel housing to upper oil pan bolts (refer to denoted black triangles on accompanying figure). Torque to 37 ft. lbs. (50 Nm).

7. The remainder of installation is the reverse of the removal procedure.

8. Refill engine with oil.

Oil Pump

REMOVAL & INSTALLATION

6.5L Engine

1. Before servicing the vehicle, refer to the precautions in the beginning of this section.

2. Drain the engine oil

3. Remove or disconnect the following:
- Oil pan
- Oil pump to crankshaft rear main bearing attaching bolt
- Oil pump and hex drive

To install:

4. Inspect the oil pan pick up tube and screen for damage and the hex drive for cracks.

5. Install or connect the following:
- Oil pump and extension shaft to the engine. Align the extension shaft hex with the drive hex, the oil pump should push easily into place.
- Oil pump bolt and tighten to 65 ft. lbs. (90 Nm)
- Oil pan

6. Refill the crankcase with oil.

6.6L Engine

1. Before servicing the vehicle, refer to the precautions in the beginning of this section.

2. Drain the engine oil

3. Remove or disconnect the following:
- Engine flywheel housing (2WD vehicles)
- Engine front cover
- Upper oil pan
- Oil pump pipe and screen and gasket

4. Block the crankshaft from turning with a wooden dowel.
- Oil pump driven gear nut
- Oil pump driven gear

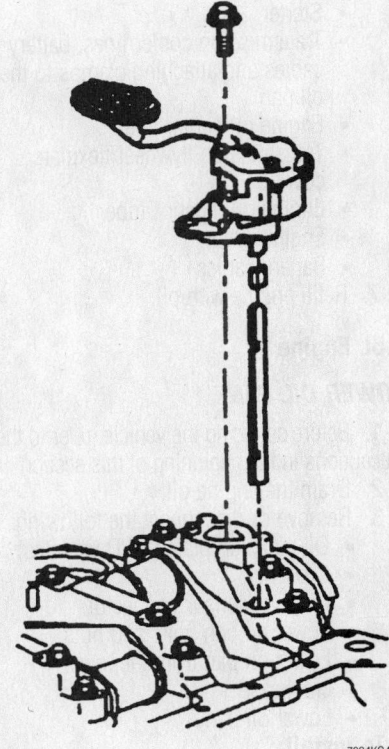

Exploded view of the oil pump mounting— 6.5L diesel engines

➡**The crankshaft reluctor and oil pump drive gear are timed together at the factory. Do NOT remove the reluctor from the oil pump drive gear or damage the reluctor teeth.**

- Oil pump drive gear and crankshaft reluctor assembly using a brass drift and tapping as close to the center of the reluctor assembly
- 3 hex head and 1 Allen head bolt
- Oil pump
- Oil pump O-ring seal
- Oil pump gear cover bolts and cover

5. Measure the clearance between the gear teeth and oil pump housing using a feeler gauge. The production clearance is 0.0049–0.0087 in. (0.125–0.221mm) and the service limit is 0.0087 in. (0.221mm). Replace the pump if the clearance exceeds the service limit.

6. Use a feeler gauge and a straight-edge to measure the clearance between the side of the gear and the cover. The production clearance is 0.0025–0.0043 in. (0.064–0.109mm) and the service limit is 0.0043 in. (0.109mm). Replace the pump if the clearance exceeds the service limit.

7. Calculate the driven gear shaft-to-bushing clearance:

  a. Measure the driven gear shaft outside diameter. The production specification is 0.7853–0.7858 in. (19.947–19.960mm) and the service limit is 0.7819 in. (19.86mm).

  b. Measure the driven gear bushing inside diameter. The production value is 0.7874 in. (20mm).

  c. Calculate the driven gear shaft-to-bushing clearance. The service limit is 0.0055 in. (0.14mm).

  d. Replace the pump if the clearance exceeds the service limit.

To install:

8. Install or connect the following:
- Oil pump gear cover and bolts. Tighten to 15 ft. lbs. (20 Nm).
- New O-ring seal for the oil pump
- Oil pump and bolts. Tighten to 15 ft. lbs. (20 Nm).

9. Check the oil pump drive gear for wear and replace the gear pin if necessary.
- Oil pump drive gear and reluctor
- Oil pump driven gear and nut. Block the crankshaft from moving, then tighten to 74 ft. lbs. (100 Nm).
- Oil pump pipe and screen gasket to the oil pump (4WD vehicle)
- Oil pump pipe and screen (4WD vehicle)
- Oil pump pipe and screen bolts and

nuts (4WD vehicle). Tighten to 18 ft. lbs. (25 Nm).
- Engine front cover
- Engine flywheel housing (2WD vehicle)
- Upper oil pan

10. Refill the crankcase with oil.

Rear Main Seal

REMOVAL & INSTALLATION

Please note that the entire transmission assembly must be removed before performing this procedure. Before a new seal is installed, the Crankcase Depression Regulator (CDR) and crankcase ventilation system should be cleaned and inspected. In addition, use care removing the flywheel. Some models use a heavy, dual mass flywheel that must be handled with care.

1. Before servicing the vehicle, refer to the precautions in the beginning of this section.

2. Remove or disconnect the following:
- Negative battery cables
- Transfer case, if equipped
- Transmission assembly
- Clutch assembly and flywheel, if equipped with manual transmission
- Flexplate, if equipped with automatic transmission
- Crankshaft rear main oil seal by inserting a suitable crankshaft seal removal tool and prying the seal out

To install:

3. Clean the oil seal bore in the block thoroughly before installation of the new seal.

4. Inspect the crankshaft for grit, rust or burrs and correct as necessary. Also inspect the portion of the crankshaft where the oil seal makes contact, for wear due to the rubbing action of the oil seal.

➡**Because of rear crankshaft wear or grooving, the new oil seal should be seated in a new location. The J 39084 installation tool will control the seal positioning. This will provide a new surface on the crankshaft for the seal to ride on.**

5. Clean the running surface of the crankshaft with a non-abrasive cleaner.

6. Lubricate the inner diameter of the new seal and the outer diameter of the crankshaft with engine oil.

7. Install or connect the following:
- Rear main oil seal using a crankshaft rear oil seal installation tool
- Flywheel.

- Transmission assembly
- Transfer case, if equipped
- Negative battery cables

8. Start the engine and verify no oil leaks.

Timing Chain, Sprockets, Front Cover and Seal

REMOVAL & INSTALLATION

6.5L Engine

1. Before servicing the vehicle, refer to the precautions in the beginning of this section.

2. Drain the cooling system.

3. Remove or disconnect the following:
- Negative battery cables
- Water pump and pulleys

4. Rotate the crankshaft to align the marks on the torsional damper with the **0** mark on the timing tab.

5. Scribe a mark aligning the injection pump flange and the front cover, if not already marked.

➡**The outer ring (weight) of the torsional damper is bonded to the hub with rubber. The damper must be removed with a puller that acts on the inner hub only. Pulling on the outer portion of the damper will break the rubber bond or destroy the tuning of the unit.**

6. Remove or disconnect the following:
- Crankshaft pulley and torsional damper
- Front cover-to-oil pan bolts (4)
- 2 fuel return line clips
- Injection pump gear
- Injection pump retaining nuts from the front cover
- Crankshaft sensor
- Baffle
- Cover bolts remaining and the front cover
- Injection pump gear

7. Align the camshaft timing gear marks
- Bolt and washer attaching the camshaft gear
- Camshaft sprocket with the timing chain
- Crankshaft sprocket.

To install:

8. Install or connect the following:
- Cam sprocket, timing chain and crankshaft sprocket as a unit, aligning the timing marks on the sprockets.

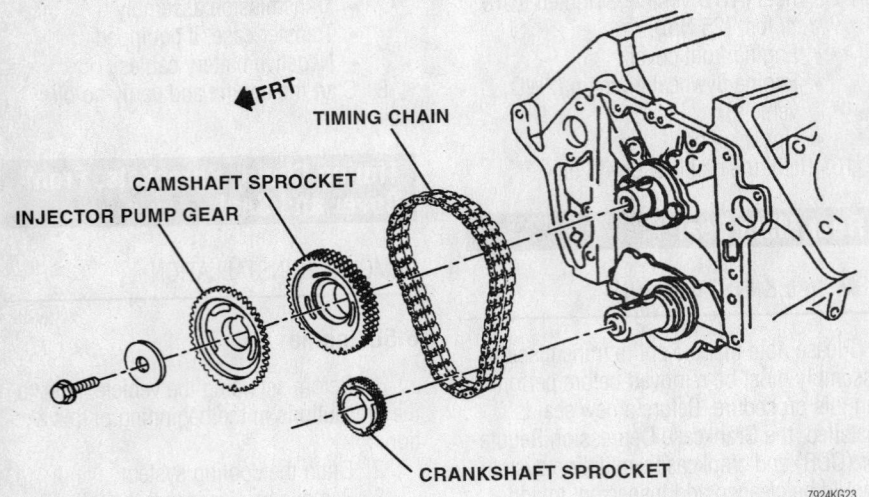

Timing chain and related components—6.5L diesel engines

9. Rotate the crankshaft to align the injection pump and camshaft gears.
- Injection pump gear

10. If the front cover oil seal is to be replaced, it can now be pried out of the cover with a suitable prying tool. Press the new seal into the cover evenly.

11. Clean both sealing surfaces until all traces of old sealer are gone. Apply a ³⁄₃₂ in. (2mm) bead of GM sealant 1052357 or equivalent to the sealing surface. Apply a ³⁄₁₆ in. (5mm) bead of RTV type sealer to the bottom portion of the front cover which attaches to the oil pan.
- Front cover
- Baffle
- Injection pump. Torque the nuts to 31 ft. lbs. (42 Nm), making sure the scribe marks on the pump and front cover are aligned.
- Injection pump driven gear. Torque the injection pump gear bolts to 17 ft. lbs. (23 Nm), making sure the marks on the cam gear and pump are aligned.

➡**Verify that there is a minimum clearance of 0.040 in. (1.0mm) between the injection pump gear and baffle or noise may be result.**

- Fuel line clips
- Front cover-to-oil bolts
- Torsional damper
- Crankshaft pulley. Torque the bolts to 80 inch lbs. (9 Nm).
- Oil pan bolts. Torque the bolts to 106 inch lbs. (12 Nm).
- Water pump
- Pulley assembly
- Negative battery cables

12. Refill the cooling system with the proper type and quantity of antifreeze.

13. Inspect the engine for leaks.

Timing Gears, Front Cover and Seal

REMOVAL & INSTALLATION

6.6L Engine

➡**The 6.6L engine uses gears in place of a timing chain. For removal and installation, please see the Camshaft and Lifters procedure. This procedure** covers the removal of the front cover and seal.

1. Before servicing the vehicle, refer to the precautions in the beginning of this section.

2. Remove the upper intake manifold sight shield as follows:

   a. Remove the retaining bolt in the front of the shield.

   b. Lift up on the front of the shield, then lift the shield off the rear bracket.

3. Drain the cooling system.

4. Remove or disconnect the following:
- Negative battery cables
- Right front wheel
- Right front fender splash shield
- Upper fan shroud
- Fan clutch
- Drive belt
- Oil dipstick tube
- Thermostat housing crossover
- Crankshaft balancer
- Crankshaft front oil seal
- Water pump
- Camshaft sensor electrical connector
- Camshaft sensor bolt and sensor
- Crankshaft Position (CKP) sensor connector, bolt and sensor
- CKP sensor spacer bolts and spacer
- 5 bolts securing the upper oil pan to the front cover
- Bracket bolts and the bracket for the turbocharger outlet coolant pipe
- Engine front cover bolts

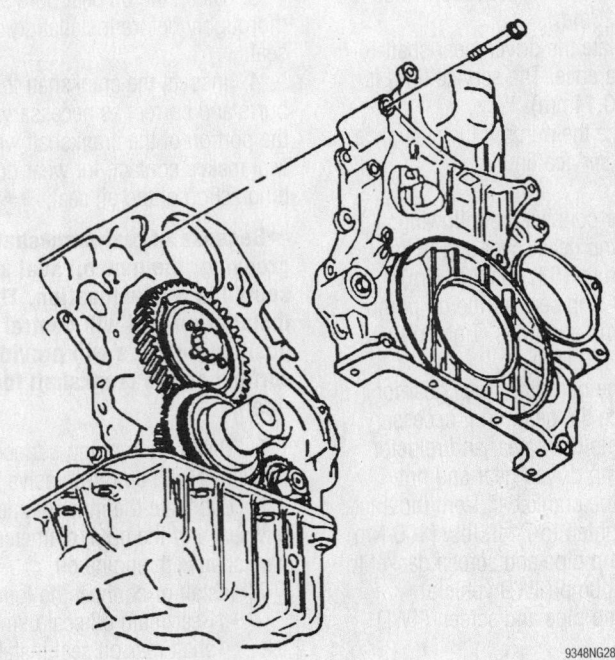

Engine front cover—6.6L engine

- Use a suitable seal cutter to separate the front cover from the cylinder block and upper oil pan

➡**Do not bend the turbocharger outlet pipe.**

- O-ring from the front cover
- Oil pressure relief valve from the front cover

To install:

5. Clean and inspect all sealing surfaces.

6. Install or connect the following:

- Oil pressure relief valve with a new O-ring. Tighten to 30 ft. lbs. (41 Nm).
- Apply a ⅛ in. (2–3mm) wide to ¹⁄₁₆ in. (0.5–1.5mm) high bead of sealant to the front cover sealing surfaces to the engine block and oil pan.
- New front cover O-ring after lubricating it with engine oil
- Front cover and bolts. Tighten to 15 ft. lbs. (20 Nm).
- Upper oil pan-to-front cover bolts. Tighten to 15 ft. lbs. (20 Nm).
- Turbocharger coolant outlet pipe bracket and bolts. Tighten to 15 ft. lbs. (20 Nm).
- Camshaft sensor and bolt. Tighten to 80 inch lbs. (9 Nm).
- Camshaft sensor connector

➡**The CKP sensor spacers are machined with different timing positions. If you have to replace a spacer, make sure it has the same part number.**

- CKP sensor spacer and spacer bolts. Tighten to 89 inch lbs. (10 Nm).
- CKP sensor and bolt. Tighten to 89 inch lbs. (10 Nm).
- Water pump
- Crankshaft front oil seal
- Crankshaft balancer
- Thermostat housing crossover
- Oil fill tube
- Drive belt
- Upper fan shroud
- Right front fender splash shield and wheel
- Negative battery cables

7. Refill the cooling system with the proper type and quantity of antifreeze.

8. Inspect the engine for leaks.

Piston and Ring

POSITIONING

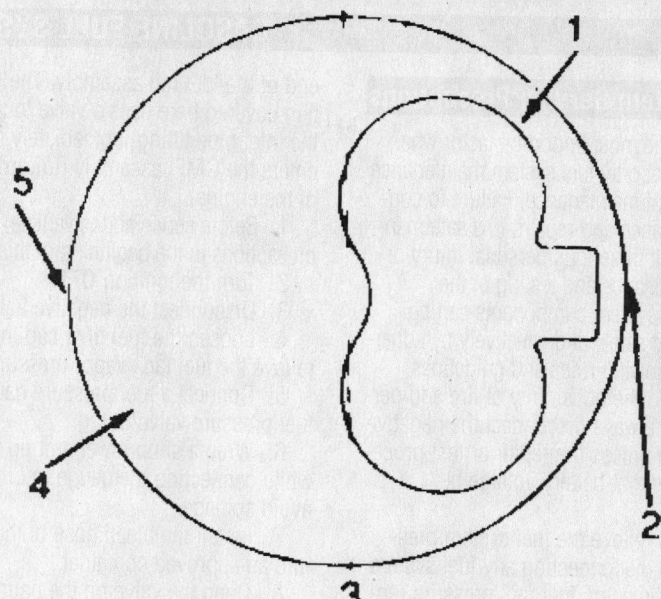

1. Oil control ring expander gap
2. Second compression ring gap
3. Centerline of piston pin
4. Oil control ring gap
5. Top compression ring gap

7924AG11

Piston ring end-gap spacing —6.5L diesel engines

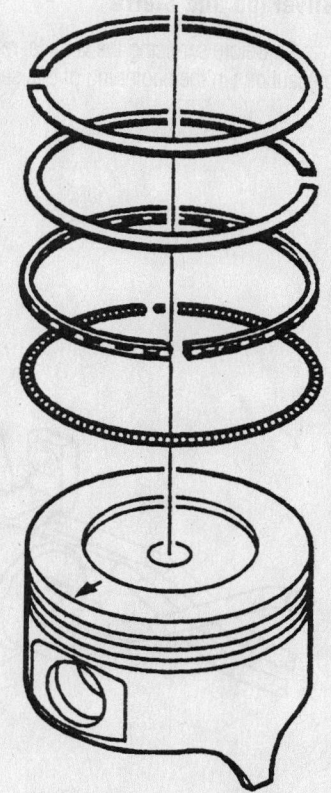

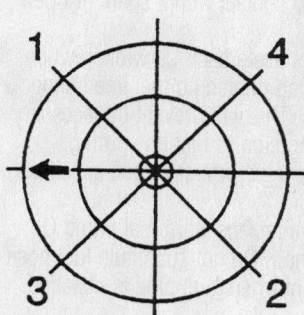

9348NG29

Piston ring positioning—6.6L diesel engines

GASOLINE FUEL SYSTEM

Fuel System Service Precautions

Safety is the most important factor when performing not only fuel system maintenance but any type of maintenance. Failure to conduct maintenance and repairs in a safe manner may result in serious personal injury or death. Maintenance and testing of the vehicle's fuel system components can be accomplished safely and effectively by adhering to the following rules and guidelines.

• To avoid the possibility of fire and personal injury, always disconnect the negative battery cable unless the repair or test procedure requires that battery voltage be applied.

• Always relieve the fuel system pressure prior to disconnecting any fuel system component (injector, fuel rail, pressure regulator, etc.), fitting or fuel line connection. Exercise extreme caution whenever relieving fuel system pressure, to avoid exposing skin, face and eyes to fuel spray. Please be advised that fuel under pressure may penetrate the skin or any part of the body that it contacts.

• Always place a shop towel or cloth around the fitting or connection prior to loosening to absorb any excess fuel due to spillage. Ensure that all fuel spillage (should it occur) is quickly removed from engine surfaces. Ensure that all fuel soaked cloths or towels are deposited into a suitable waste container.

• Always keep a dry chemical (Class B) fire extinguisher near the work area.

• Do not allow fuel spray or fuel vapors to come into contact with a spark or open flame.

• Always use a back-up wrench when loosening and tightening fuel line connection fittings. This will prevent unnecessary stress and torsion to fuel line piping. Always follow the proper torque specifications.

• Always replace worn fuel fitting O-rings with new. Do not substitute fuel hose or equivalent where fuel pipe is installed.

Fuel System Pressure

RELIEVING

A Schrader valve is provided on these fuel systems, in order to conveniently test or release the system pressure. A fuel pressure gauge and adapter will be necessary to connect the gauge to the fitting. Most of the MFI systems utilize a service valve on one end of the fuel rail assembly. The CMFI system covered here uses a valve located on the inlet pipe fitting, immediately before it enters the CMFI assembly (towards the rear of the engine)

1. Before servicing the vehicle, refer to the precautions in the beginning of this section.
2. Turn the ignition **OFF**.
3. Disconnect the negative battery cable.
4. Loosen the fuel filler cap in order to relieve the fuel tank vapor pressure.
5. Connect a fuel pressure gauge to the fuel pressure valve/fitting.
6. Wrap a shop towel around the fitting while connecting the gauge in order to avoid spillage.
7. Install the bleed hose of the gauge into an approved container.
8. Open the valve on the gauge to bleed the system pressure.

The fuel connections are now safe for servicing. Drain any fuel remaining in the gauge into an approved container.

Fuel Filter

REMOVAL & INSTALLATION

Silverado and Sierra

1. Before servicing the vehicle, refer to the precautions in the beginning of this section.

2. Disconnect the negative battery cable.
3. Relieve the fuel system pressure.
4. Raise the vehicle.
5. Clean all the fuel filter connections and the surrounding areas before disconnecting the fuel pipes in order to avoid possible contamination of the fuel system.
6. Disconnect the threaded fittings from the fuel filter.
7. Cap the fuel pipes in order to prevent possible fuel system contamination.
8. Slide the fuel filter from the bracket.
9. Inspect the fuel pipe O-rings for cuts, nicks, swelling, or distortion. Replace the O-rings if necessary.

To install:
10. Slide the fuel filter into the bracket. Remove the caps from the fuel pipes.
11. Connect the threaded fittings to the fuel filter. Tighten the fittings to 18 ft. lbs. (25 Nm).
12. Lower the vehicle.
13. Tighten the fuel filler cap.
14. Connect the negative battery cable.
15. Turn the ignition **ON** for 2 seconds.
16. Turn the ignition **OFF** for 10 seconds.
17. Turn the ignition **ON**.
18. Inspect for fuel leaks.

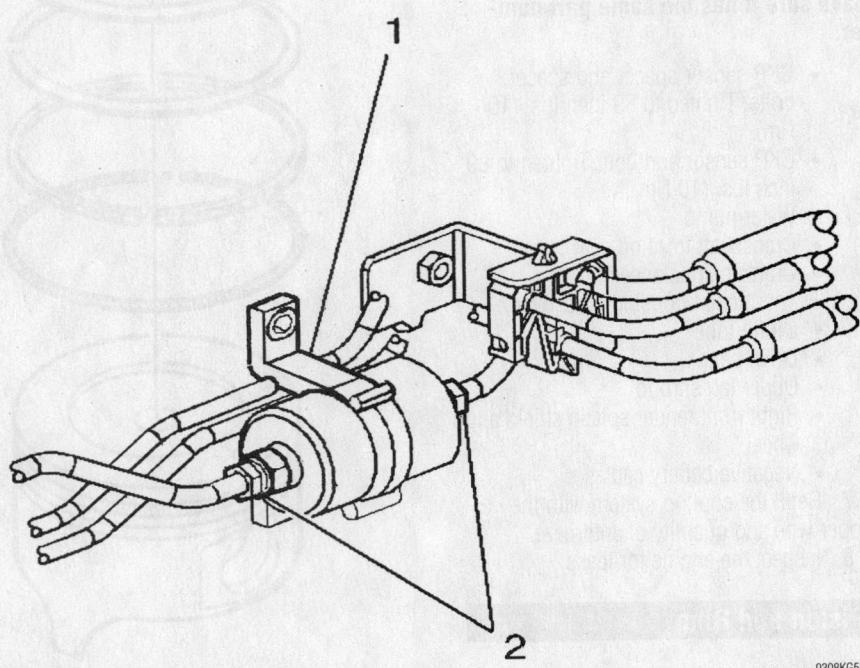

Fuel filter—4.8L, 5.3L and 6.0L engines

9308KG55

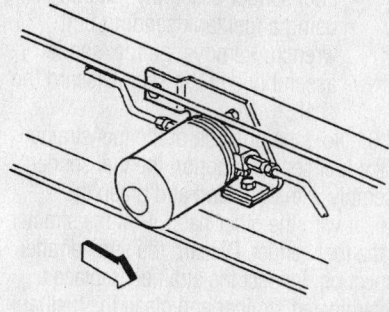

7924KG24

The spin-on fuel filter is serviced in the same manner as a inline oil filter

Express and Savana

The fuel filter is normally located along the frame rail of the vehicle. On some vehicles however, it may have been relocated to the engine compartment. When in doubt, trace a fuel line from the engine backwards or from the tank forward in order to locate the filter.

Some vehicles utilize a spin-on fuel filter located on the frame rail. This filter can be turned counterclockwise after the fuel pressure is relieved.

1. Before servicing the vehicle, refer to the precautions in the beginning of this section.
2. Properly relieve the fuel system pressure.
3. Remove or disconnect the following:
   • Negative battery cable

• Fuel line connections from the filter or unscrew the filter in the case of the spin-on type
• In line filters, remove the bolt from the filter mounting clamp, then remove the clamp and filter assembly. Separate the filter from the clamp.

To install:

→ The inline filter has an arrow (fuel flow direction) on the side of the case, be sure to install it correctly in the system, with the arrow facing away from the fuel tank.

4. Install or connect the following:
   • In line filters place filter in clamp
   • Install filter and clamp
   • Spin-on filters, lubricate the gasket before installation. Then tighten the filter an additional ¾ of a turn from the point when the gasket touches the filter adapter. Always check for leaks after a new filter is installed.

Fuel Pump

REMOVAL & INSTALLATION

2000 Vehicles—Except Silverado And Sierra

1. Before servicing the vehicle, refer to the precautions in the beginning of this section.

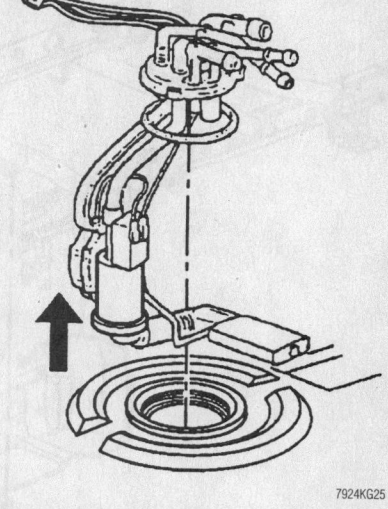

7924KG25

Lift the fuel pump assembly out of the tank after removing the locking ring

2. Properly relieve the fuel system pressure.
3. Drain the fuel from the vehicle.
4. Remove or disconnect the following:

   • Negative battery cable
   • Fuel tank from the vehicle
   • Locking ring (located on top of the fuel tank) counterclockwise
   • Lift the fuel pump assembly out of the tank

To install:
5. Install or connect the following:
   • Fuel pump and secure with locking ring
   • Fuel tank to vehicle and refill
   • Negative battery cable
6. Run engine and check for fuel leaks.

Silverado And Sierra & 2001–03 Vehicles

1. Before servicing the vehicle, refer to the precautions in the beginning of this section.
2. Remove or disconnect the following:

   • Negative battery cable
3. Relieve the fuel system pressure.
4. Drain the fuel tank.
5. Remove or disconnect the following:
   • Fuel tank

✳✳ WARNING

Do not handle the fuel sender assembly by the fuel pipes. The amount of leverage generated by handling the fuel pipes could damage the joints.

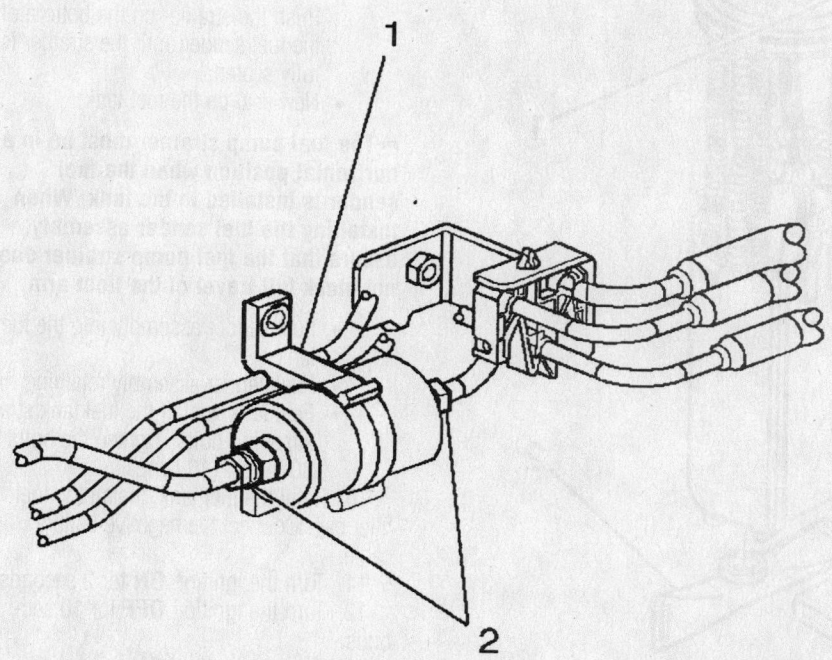

Typical in-line fuel filter mounting location

9302KG05

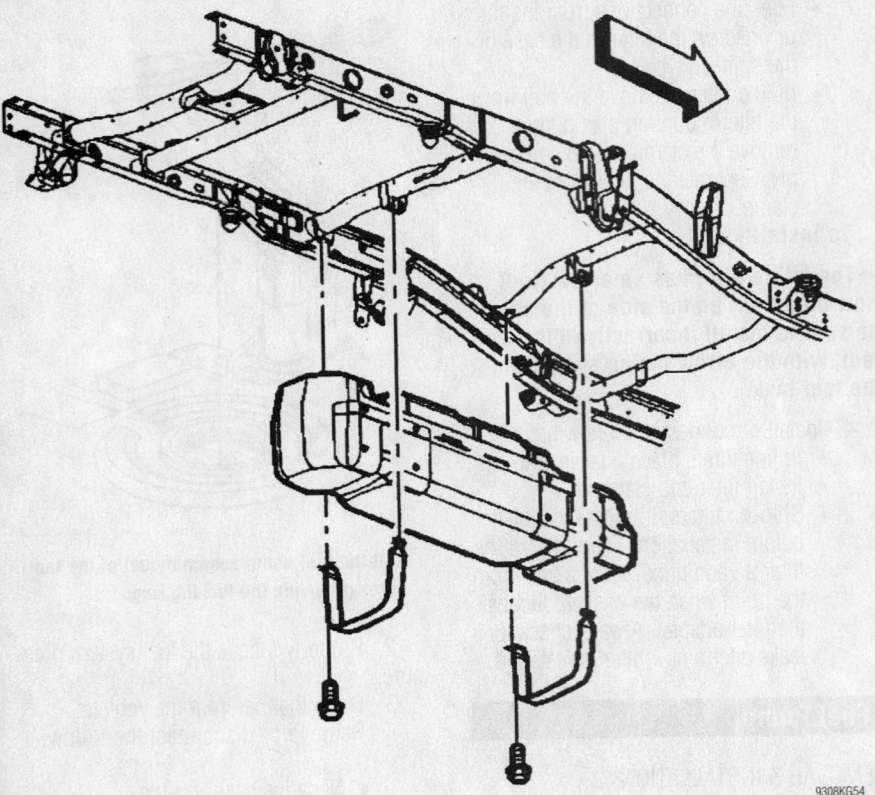

Fuel tank—Silverado shown

9308KG54

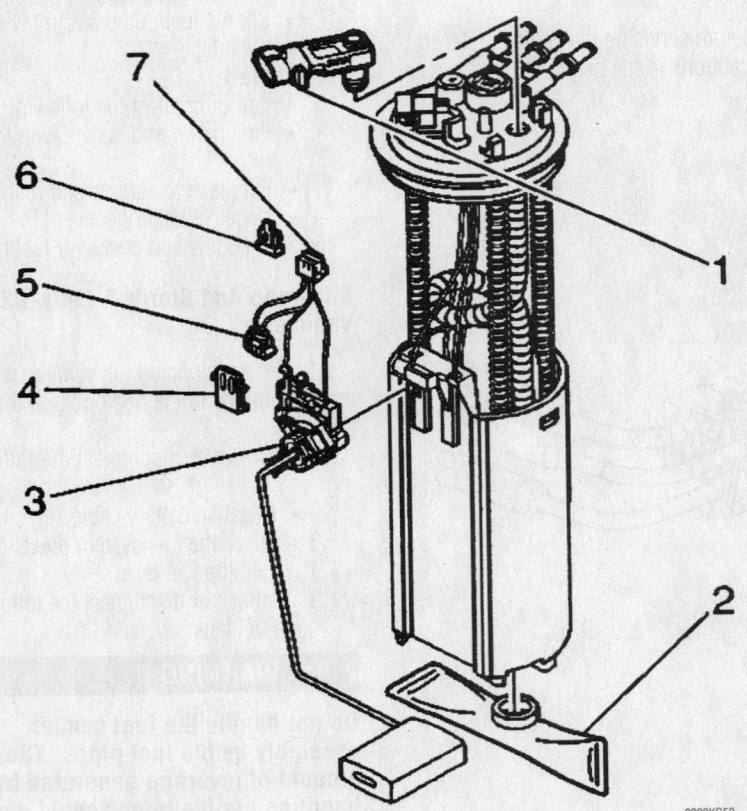

Fuel pump/sender assembly—4.8L, 5.3L and 6.0L engines

9308KG53

- Fuel sender assembly retaining ring using a fuel tank sending unit wrench. Remove the fuel sender assembly and the seal. Discard the seal.

6. Note the position of the fuel strainer on the fuel sender. Support the fuel sender assembly with one hand and grasp the strainer with the other hand. Pull the strainer off the fuel sender. Discard the strainer after inspection. Inspect the strainer. Replace a contaminated strainer and clean the fuel tank.

- Fuel pump electrical connector
- Electrical connector retaining clip from the fuel level sensor
- Sensor electrical connector from under the fuel sender cover
- Fuel level sensor retaining clip

7. Squeeze the locking tangs and remove the fuel level sensor.

8. Remove the fuel pressure sensor.

To install:

9. Install or connect the following:
- Fuel pressure sensor
- Fuel level sensor
- Sensor retaining clip
- Electrical connector to the fuel level sensor
- Electrical connector retaining clip to the fuel level sensor
- Fuel pump electrical connector

➡**Always install a new fuel strainer when replacing the fuel tank fuel pump module.**

- New fuel strainer in the same position as noted during disassembly. Push the strainer on the bottom of the fuel sender until the strainer is fully seated.
- New seal on the fuel tank

➡**The fuel pump strainer must be in a horizontal position when the fuel sender is installed in the tank. When installing the fuel sender assembly, assure that the fuel pump strainer does not block full travel of the float arm.**

- Fuel sender assembly into the fuel tank
- Fuel sender assembly retaining ring
- Fuel tank. Install the fuel tank strap attaching bolts. Tighten the bolts to 30 ft. lbs. (40 Nm).

10. Refill the fuel tank. Install the fuel filler cap. Connect the negative battery cable.

11. Turn the ignition **ON** for 2 seconds.

12. Turn the ignition **OFF** for 10 seconds.

13. Turn the ignition **ON**.

14. Inspect for fuel leaks.

Fuel Injector

REMOVAL & INSTALLATION

4.3L, 5.0L, and 5.7L Engines

1. Before servicing the vehicle, refer to the precautions in the beginning of this section.

➡ Use care when removing the injectors to prevent damage to the electrical connector pins on the injector and the nozzle. The fuel injector is serviced as a complete assembly only. Since the injector is an electrical component, it should not be immersed in any type of cleaner.

2. Relieve the fuel system pressure.
3. Remove or disconnect the following:
   - Negative battery cable
   - Intake manifold plenum
   - Fuel rail assembly
   - Wiring harness
   - Injector clip and discard it
   - Injector O-ring seals from both ends of the injector. Save the O-ring backups for use on reassembly

To install:

4. Install or connect the following:
   - O-ring backups before installing the O-rings
5. Lubricate the new injector O-rings with clean engine oil
   - O-ring to injector assembly
   - Fuel injector into the fuel rail injector socket with the electrical connectors facing outward

- New injector retaining clips on the injector fuel rail assembly by sliding the clip into the injector groove as it snaps onto the fuel rail
- Fuel rail assembly
- Manifold plenum
- Wiring harness
- Negative battery cable

4.8L, 5.3L and 6.0L Engines

1. Relieve the fuel system pressure.
2. Remove or disconnect the following:
   - Negative battery cable
   - Engine sight shield bolts and bracket
   - Accelerator control and cruise control cables from the cable bracket and throttle body
   - Upper engine wire harness retainer nut
   - Evaporative Emission (EVAP) purge valve harness connector
3. Position the upper engine wire harness aside
4. Tag the injector connectors for identification, then pull the top part of the injector connector up. Do not pull the top part of the connector past the top of the white portion.
5. Push the tab on the lower side of the injector connector to release the connect from the injection. Perform these steps on each injector connector.
6. Remove or disconnect the following:
   - Fuel feed and return pipes from the fuel rail
   - Fuel pressure regulator vacuum line

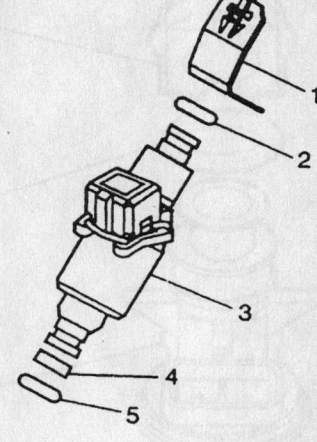

1. Clip - SFI Fuel Injector Retainer
2. O-ring - SFI Fuel Injector Upper
3. Injector Asm - SFI Fuel
4. O-ring - Backup
5. O-ring - SFI Fuel Injector Lower

9308KG06

Exploded view of the fuel injector assembly—MFI systems

- Crossover tube-to-right fuel rail retainer screw
- Fuel rail attaching bolts and fuel rail

➡ Use care in removing the fuel injectors in order to prevent damage to the electrical connector pins on the injector and to prevent damage to the nozzle. Service the fuel injector as a complete assembly only. The fuel injector is an electrical component. DO NOT immerse the fuel injector in any type of cleaner.

- Injector retainer clip. Insert the fork of a fuel injector assembly removal tool behind the injector connector between the fuel rail pod and the 3 protruding retaining clip ledges. Use a prying motion while inserting the tool in order to force the injector out of the fuel rail pod.
- Injector retainer clip
- Injector O-ring seals from both ends of the injector. Discard the O-ring seals.

To install:

➡ When ordering new fuel injectors, be sure to order the correct injector for the application being serviced. The fuel injector assembly is stamped with a part number identification.

7. Lubricate the new injector O-ring seals with clean engine oil.
8. Install or connect the following:

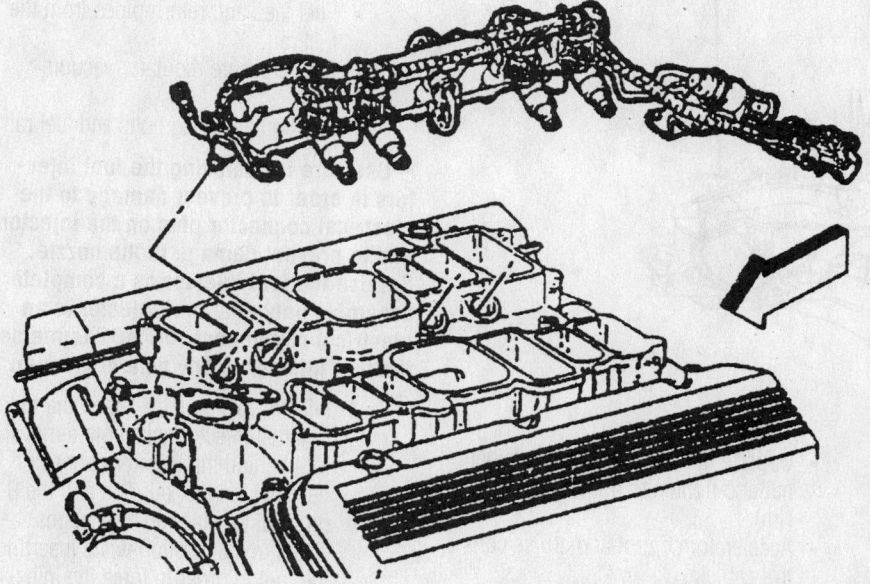

9308KG05

Exploded view of the fuel rail assembly—MFI systems

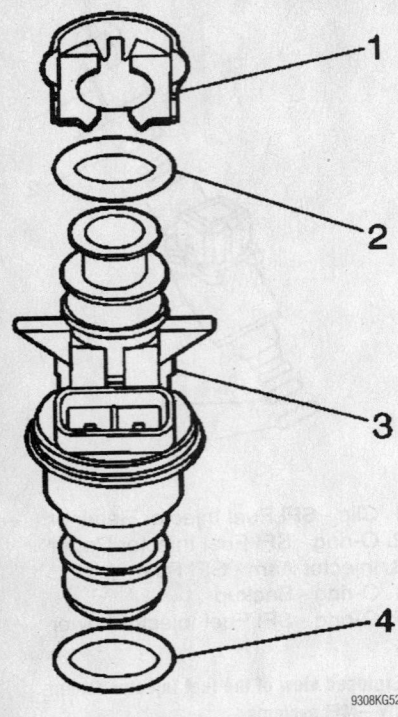

Fuel injector—4.8L, 5.3L and 6.0L

9308KG52

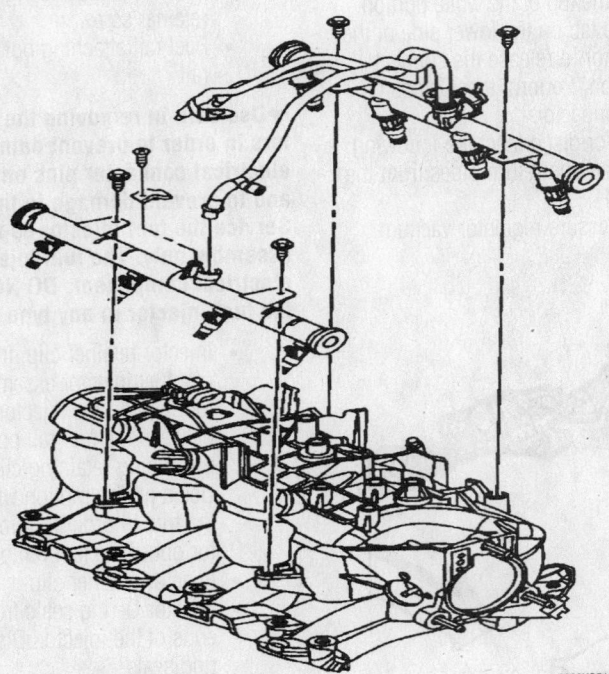

Fuel rail assembly—4.8L, 5.3L and 6.0L engines

9308KG51

- New injector O-ring seals on the injector
- New retainer clip on the injector

9. Push the fuel injector into the fuel rail injector socket with the electrical connector facing outward. The retainer clip locks on to a flange on the fuel rail injector socket.

10. Remove the crossover tube-to-right fuel rail retainer, then remove the crossover tube.

11. Replace the crossover tube O-ring with a new, lubricated one.

12. Install or connect the following:
- Crossover tube and loosely install the retainer
- Fuel rail to the intake manifold
- Apply a 0.020 in. (5mm) band of threadlock to the fuel rail retaining bolts
- Fuel rail bolts and tighten to 89 inch lbs. (10 Nm)
- Crossover pipe retainer and tighten to 34 inch lbs. (3.8 Nm)
- Fuel pressure regulator vacuum line
- Fuel feed and return pipes
- Fuel injector electrical connectors, as tagged. Rotate the injectors as necessary to avoid stretching the wire harness.
- Upper engine wire harness
- EVAP purge solenoid electrical connector

- Upper engine wire harness retainer nut and tighten to 49 inch lbs. (5.5 Nm)
- Accelerator control and cruise control cables
- Engine sight shield mounting bracket and bolts

13. Tighten the fuel cap.
14. Connect the negative battery cable.
15. Turn the ignition **ON** for 2 seconds.
16. Turn the ignition **OFF** for 10 seconds.
17. Turn the ignition **ON**.
18. Inspect for fuel leaks.
19. Install the engine sight shield. Tighten the engine sight shield bolts to 89 inch lbs. (10 Nm)

8.1L Engine

1. Relieve the fuel system pressure.
2. Remove or disconnect the following:
- Negative battery cable
- Engine sight shield nuts and bracket
- Alternator harness connector
- Evaporative Emission (EVAP) purge valve harness connector
- Throttle Position (TP) sensor electrical connector
- Electronic Throttle Control (ETC) electrical connector
- Upper engine wire harness bracket studs, and position the harness aside
- Secondary Air Injection (AIR) crossover pipe, if equipped

3. Tag the injector connectors for identification, then pull the top part of the injector connector up. Do not pull the top part of the connector past the top of the white portion.

4. Push the tab on the lower side of the injector connector to release the connect from the injection. Perform these steps on each injector connector.

5. Remove or disconnect the following:
- Fuel feed and return pipes from the fuel rail
- Fuel pressure regulator vacuum line
- Fuel rail attaching bolts and fuel rail

➡**Use care in removing the fuel injectors in order to prevent damage to the electrical connector pins on the injector and to prevent damage to the nozzle. Service the fuel injector as a complete assembly only. The fuel injector is an electrical component. DO NOT immerse the fuel injector in any type of cleaner.**

- Injector retainer clip. Insert the fork of a fuel injector assembly removal tool behind the injector connector between the fuel rail pod and the 3 protruding retaining clip ledges. Use a prying motion while inserting the tool in order to force the injector out of the fuel rail pod.

- Injector retainer clip
- Injector from the fuel rail pod
- Injector O-ring seals from both ends of the injector. Discard the O-ring seals.

To install:

→ **When ordering new fuel injectors, be sure to order the correct injector for the application being serviced. The fuel injector assembly is stamped with a part number identification.**

6. Lubricate the new injector O-ring seals with clean engine oil.
7. Install or connect the following:
   - New injector O-ring seals on the injector
   - New retainer clip on the injector
8. Push the fuel injector into the fuel rail

injector socket with the electrical connector facing outward. The retainer clip locks on to a flange on the fuel rail injector socket.

9. Install or connect the following:
   - Fuel rail to the intake manifold
   - Apply a 0.020 (5mm) band of threadlock to the fuel rail retaining bolts
   - Fuel rail bolts and tighten to 106 inch lbs. (12 Nm)
   - Fuel pressure regulator vacuum line
   - Fuel feed and return pipes
   - Fuel injector electrical connectors, as tagged. Rotate the injectors as necessary to avoid stretching the wire harness
   - AIR crossover pipe, if equipped
   - Upper engine wire harness bracket

   - Retainer studs to the upper engine wire harness and tighten the nut to 89 inch lbs. (10 Nm)
   - Alternator electrical connector
   - EVAP purge solenoid electrical connector
   - TP and ETC sensor connectors
   - Engine sight shield mounting bracket and bolts
10. Tighten the fuel cap.
11. Connect the negative battery cable.
12. Turn the ignition **ON** for 2 seconds.
13. Turn the ignition **OFF** for 10 seconds.
14. Turn the ignition **ON**.
15. Inspect for fuel leaks.
16. Install the engine sight shield. Tighten the engine sight shield bolts to 89 inch lbs. (10 Nm).

DIESEL FUEL SYSTEM

Fuel System Pressure

RELIEVING

Fuel system pressure can be released by wrapping a fuel fitting in a heavy shop towel and slightly loosening the fitting. NEVER perform this with any source of ignition nearby!

Fuel System Air

BLEEDING

1. Before servicing the vehicle, refer to the precautions in the beginning of this section.
2. Open the air bleed valve on the fuel manager/filter.
3. Connect a hose to the air bleed valve and place the other of the hose in a suitable container.

✳✳ CAUTION

The diesel/water mixture is flammable and may be hot. To avoid personal injury or property damage, do not allow the diesel/water mixture to come in contact with skin, open flame or a hot engine. Do not overfill the container holding the fuel mixture as heat from a warm engine or any another heat source may cause the fuel to expand and leak from the container that may lead to a fire.

4. Remove the F/SOL fuse from the fuse panel.
5. Crank the engine in short intervals of

10-to-15 seconds until clear fuel is observed at the air bleed hose (wait for 1 minute between cranking intervals)
6. Remove the hose and close the air bleed valve.
7. Install the F/SOL fuse and start the vehicle. Allow the vehicle to run at idle for 5 minutes.
8. Check for fuel leaks, and clear any Diagnostic Trouble Code's (DTC's)

Idle Speed

ADJUSTMENT

Idle speed and injection timing is controlled by the PCM. There is no provision for adjustment.

Fuel Filter

REMOVAL & INSTALLATION

6.5L Engine

1. Before servicing the vehicle, refer to the precautions in the beginning of this section.
2. Turn the ignition **OFF**. Remove the fuel tank cap to release any pressure or vacuum in the tank.

→ **It is not necessary to drain all the fuel from the header in order to change the element since the fuel will remain in the header's cavity.**

3. Remove or disconnect the following:
   - Open the air bleed valve to relieve residual pressure

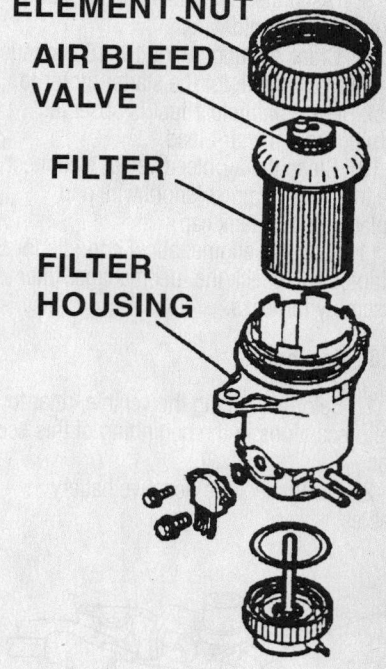

ELEMENT NUT
AIR BLEED VALVE
FILTER
FILTER HOUSING

7924KG26

Exploded view of the fuel filter assembly—6.5L diesel engines

- Element nut, turning it by hand to the left. If necessary, a strap wrench may be used to loosen the nut.
- Element by lifting straight up and out of the header assembly

To install:

4. Be sure the mating surface between the element assembly and the header assembly is clean.
5. Install or connect the following:

- New element by aligning the widest key slot located under the element assembly cap with the widest key in the header assembly.

6. Carefully push the element downward until the mating surfaces make contact.

- Element nut and tighten securely by hand

7. If not already done, open the air bleed valve on top of the fuel manager/filter assembly, then connect a length of hose placing the other end in a suitable container.

❊❊ CAUTION

Be extremely cautious when handling diesel fuel. Do not expose the fuel to sparks or open flames. Also, be cautious as the fuel coming out of the drain hose could be hot.

8. Disconnect the fuel injection pump shutdown solenoid wire.

9. Crank the engine for 10–15 seconds, then wait 1 minute for the starter motor to cool. Repeat until clear fuel is observed coming from the air bleed.

10. Close the air bleed valve, reconnect the injection pump solenoid wire and replace the fuel tank cap.

11. Start the engine, allow it to idle for 5 minutes and check the fuel manager/filter assembly for leaks.

6.6L Engine

1. Before servicing the vehicle, refer to the precautions in the beginning of this section.

2. Disconnect the negative battery cables.

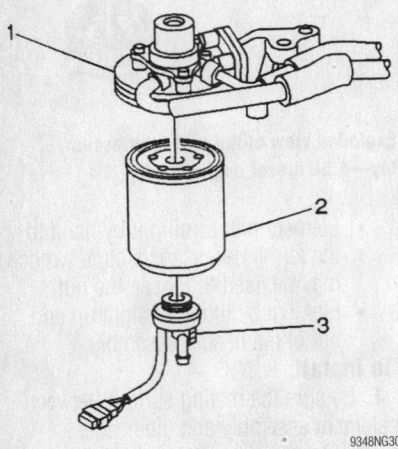

9348NG30

Exploded view of the fuel filter (2)—6.6L engine

3. Drain the fuel from the fuel filter as follows:

   a. Install a hose on the water drain on the water-in-fuel sensor.

   b. Place the other end of the hose into an approved container.

   c. Drain as much fuel as possible from the fuel filter housing.

   d. Tighten the water drain on the water-in-fuel sensor.

4. Remove or disconnect the following:

- Water-in-fuel sensor harness connector
- Fuel filter from the fuel filter/heater element housing
- Water-in-fuel sensor from the fuel filter

To install:

5. Install or connect the following:

- Water-in-fuel sensor in the fuel filter

➡**Check the fuel filter/heater element housing and the filter for a dislocated filter seal or foreign debris. Contamination on the filter/heater housing may cause leakage at the fuel filter. Coat the seal with clean engine oil.**

- Fuel filter on the fuel filter/heater element housing
- Water-in-fuel harness connector
- Negative battery cables

6. Prime the fuel system:

   a. Pump the primer located on top of the fuel filter 30 times or until stiff.

   b. Try to start and run the engine. If the engine does not start, repeat the previous step.

   c. Allow the engine to run for 5 minutes at idle.

   d. Check for fuel leaks and clear all Diagnostic Trouble Codes (DTCs).

Diesel Injection Pump

All vehicles are equipped with an electronically controlled pump. The electronic pump is driven by gears and rotates at the same speed as the camshaft. An electronic stepper motor used to control injection timing and a fuel solenoid driver used to control the fuel injection solenoid on the electronic model.

REMOVAL & INSTALLATION

6.5L Engine

1. Before servicing the vehicle, refer to the precautions in the beginning of this section.

2. Relieve the fuel system pressure.

3. Remove or disconnect the following:

- Negative battery cables
- Intake manifold
- Fuel injection and inlet lines
- Cables wires, and hoses at the injection pump
- Fuel return line at the top of the injection pump
- Fuel feed line at the injection pump, if necessary
- Oil filler tube grommet

➡**Do not engage the starter in order to rotate the engine with the injection pump removed. The pump driven gear could jam in the front housing resulting in a sheared crankshaft or camshaft gear key and possible valvetrain damage.**

4. Scribe or paint a matchmark on the front cover and the injection pump flange.

5. Rotate the crankshaft by hand and remove the injection pump driven gear bolts, accessing the bolts through the oil filler neck hole.

6. Remove the injection pump-to-front cover attaching nuts. Remove the pump. Be sure to cap all open lines and nozzles in order to prevent system contamination and damage.

To install:

7. Align the locating pin on the pump hub with the slot in the injection pump driven gear (the SLOT not the hole in the gear) At the same time, align the timing marks.

8. Attach the injection pump to the front cover. Torque the nuts to 30 ft. lbs. (40 Nm) checking the timing marks before tightening.

9. Install or connect the following:

- Driven gear-to-injection pump bolts and tighten to 20 ft. lbs. (27 Nm)
- Grommet and oil fill tube
- Air conditioning bracket. If applicable
- Fuel feed line. Torque to 20 ft. lbs. (27 Nm).
- Fuel return line to the pump, if removed
- Cables, wires and hoses previously removed
- Injector lines
- Intake manifold
- Negative battery cables

10. Start the engine and check for leaks.

6.6L Engine

1. Before servicing the vehicle, refer to the precautions in the beginning of this section.

2. Drain the cooling system

3. Remove or disconnect the following:

- Negative battery cables
- Air intake duct. Cover the end to prevent dirt from entering
- Upper fan shroud
- Fan blade assembly
- Drive belt
- Bolt holding the positive battery cable junction box and bracket and position aside
- A/C compressor and power steering pump and position aside with the lines attached
- Oil dipstick tube
- A/C and power steering pump bracket
- Alternator
- Thermostat housing bracket, wiring and fuel test port and 2 nuts
- Positive Crankcase Ventilation (PCV) catch tank from the PCV bracket and the bolt below holding the lower line, then position aside
- Alternator bracket
- Turbo cooling hose return line clamp and hose
- Upper radiator hose at the outlet pipe. Remove the bracket and support the bracket at the valve cover and swing out of the way
- Bolt holding the wiring support bracket at the thermostat housing

4. Move the main wiring harness by disconnecting the following:

  a. The fuel pressure regulator connector on the fuel injection pump

  b. Fuel injection control module connectors

5. Flip the wire harness and harness tray towards the back and position aside.

6. Remove or disconnect the following:

- Heater pipe bolt and temperature sensor wire from the thermostat housing
- Air intake pipe
- Water crossover assembly
- Hose from the turbo water feed line

➡**Cap all open fuel connections to prevent contaminants from entering.**

- High pressure fuel lines and support pipe and hose at the fuel injection pump and junction block
- Fuel return hose from the fuel injection pump
- Y-junction banjo fitting from the junction block
- Bolts securing the fuel injection pump at the front cover and block

➡**When removing the pump, be careful not to damage any of the mating surfaces.**

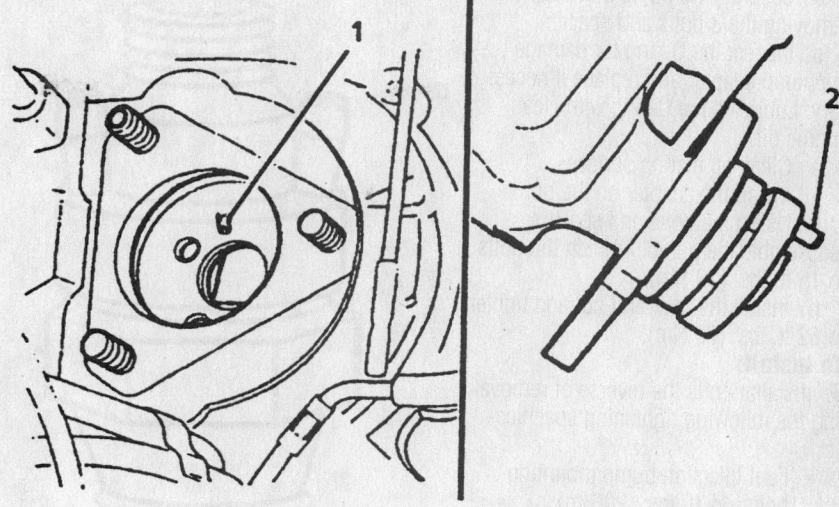

1   SLOT IN DRIVEN GEAR

2   PUMP HUB

7924KG27

Align the pin on the pump hub with the slot in the driven gear, NOT into the hole in the gear—diesel engines

9348NG31

Diesel fuel injection pump—6.6L engine

- Fuel injection pump from the block using 2 prytools to work the pump from the block toward the rear of the engine, keeping the pump straight.

7. Prepare the fuel pump as follows:

  a. Hold the fuel pump by the drive gear in a vice with copper jaw liners.

  b. Loosen the gear nut until the nut is even with the end of the gear shaft.

c. Separate the pump and adapter by removing the 3 bolts and spacers.

d. Inspect the O-ring for damage on the pump adapter and replace if necessary. Lubricate the O-ring with clean engine oil.

e. Clean all mating surfaces.

f. Install the adapter on the pump

g. Using the bolts and spacers, reassemble the pump. Tighten the bolts to 15 ft. lbs. (20 Nm).

h. Install the gear and nut and tighten to 52 ft. lbs. (70 Nm).

To install:

8. Installation is the reverse of removal, noting the following tightening specifications:

- Fuel injection pump mounting bolts: 15 ft. lbs. (20 Nm)
- Y-junction banjo fitting: 11 ft. lbs. (15 Nm)
- High pressure fuel lines: 40 ft. lbs. (54 Nm)
- Heater pipe bracket and water crossover bolts: 15 ft. lbs. (20 Nm)
- Upper radiator hose mounting bolts: 89 inch lbs. (10 Nm)
- A/C and power steering pump bracket bolts: 34 ft. lbs. (46 Nm)

9. Refill the cooling system, then start the engine and check for leaks.

Fuel Injectors

REMOVAL & INSTALLATION

6.5L Engine

1. Before servicing the vehicle, refer to the precautions in the beginning of this section.

➡ **Special tool J–29873, or its equivalent, an injection nozzle socket, will be necessary for this procedure.**

2. Remove or disconnect the following:
- Batteries.
- Fuel line clip(s)
- Fuel return hose
- Fuel injection lines

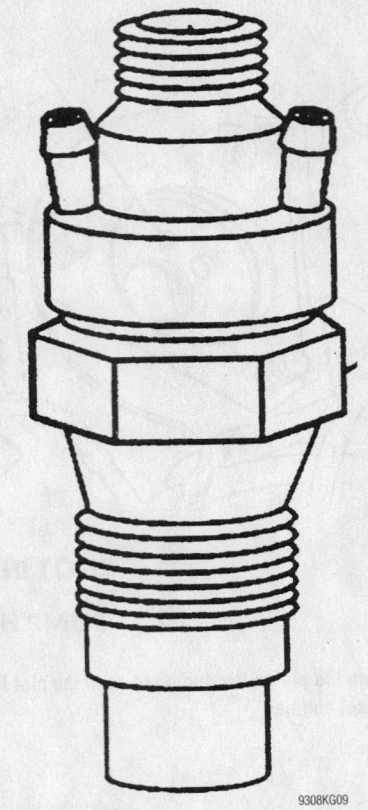

9308KG09

View of the fuel injector nozzle.

3. Using GM special tool J–29873, remove the injector. Always remove the injector by turning the 30mm hex portion of the injector; turning the round portion will damage the injector. Always cap the injector and fuel lines when disconnected, to prevent contamination.

To install:

4. Always install the injector by turning the 30mm hex portion of the injector; turning the round portion will damage the injector.

5. Install or connect the following:
- Injector with a new gasket. Torque to 50 ft. lbs. (70 Nm).
- Injection line. Torque the nut to 20 ft. lbs. (25 Nm).
- Fuel return hose
- Fuel line clip(s)
- Batteries

6.6L Engine

The following procedure is for the left or right bank fuel injector replacement.

1. Before servicing the vehicle, refer to the precautions in the beginning of this section.

➡ **Special tool J 44639, or its equivalent, an injector removal tool, will be necessary for this procedure.**

2. Remove or disconnect the following:
- Negative battery cables
- Lower valve cover
- Spill line from the injectors
- Retainer bolt from the injector bracket

3. Install the fuel injector removal tool onto the injector retainer bracket.

➡ **Check to see which side of the banjo washers have the largest hole.**

4. Install a wrench on the fuel injector removal tool and pry away from the injector.

5. Remove the faulty injectors.

6. Remove the copper compression washer from the injection hole if the washer did not come off with the injector.

To install:

7. If the injector sleeve is pulled from the cylinder head when removing the injector, the injector sleeve must be installed as follows:

a. Set the new injector sleeve gaskets to the injector sleeve.

b. Apply threadlock to the lower sealing area of the injector sleeve.

c. Use a large brass drift to drive the injector sleeve into the cylinder head until fully seated.

8. Replace the copper compression washer. Assembly grease may be needed to hold the washer in place.

9. Install or connect the following:
- New injectors
- Retainer bolt on the injector bracket. Tighten to 36 ft. lbs. (46 Nm).
- Spill line and tighten the bolts to 108 inch lbs. (12 Nm)
- Lower valve cover
- Negative battery cables

DRIVE TRAIN

Transmission Assembly

REMOVAL & INSTALLATION

Manual Transmission

SILVERADO AND SIERRA—NV3500

1. Shift the transmission into 3rd or 4th speed gear.
2. Remove or disconnect the following:

- Shift lever
- Shift tower
- Transmission oil
- If equipped with a transfer case, remove the front propeller shaft
- Rear propeller shaft.

3. If equipped with a transfer case, remove or disconnect the following:

- Two transfer case shields
- Manual transfer case shift linkage
- Bolt securing the left side support brace to the transmission
- Bolt and stud securing the left side support brace to the transfer case
- Bolt securing the right side support brace to the transmission

- Bolt securing the right side support brace to the transfer case

4. Using tool J42371, push back on the white plastic sleeve on the quick connect in order to separate the hydraulic clutch line from the concentric slave cylinder quick connect.
5. Disconnect the wiring harness and connectors from the vehicle speed sensor, backup lamp switch, and transmission harness retainers.
6. If equipped with a 4.3L engine, remove the two bolts securing the clutch housing cover. Remove the transmission rear mount. Support the transmission with a transmission jack.
7. Remove or disconnect the following:

- Bolts securing the bottom right side of the transmission to the engine
- Stud securing the right side of the transmission to the engine
- Bolt and six studs securing the transmission to the engine

8. Pull the transmission straight back on the clutch hub splines. Do not let the transmission hang from the clutch plate and the clutch cover.

- Transmission from the vehicle
- Clutch plate and the clutch cover from the engine flywheel, if required

To install:

9. Install the clutch plate and the clutch cover to the engine flywheel if removed.
10. Ensure the transmission is positioned in the 3rd or 4th speed gear. Rotate the transmission clockwise onto the clutch hub splines. Install the bolt and the studs securing the transmission to the engine. Tighten the bolts to 37 ft. lbs. (50 Nm).
11. Install or connect the following:

- Stud securing the right side of the transmission to the engine and tighten to 37 ft. lbs. (50 Nm)
- Bolts securing the bottom right side of the transmission to the engine and tighten to 37 ft. lbs. (50 Nm)
- Clutch housing cover using the two bolts (4.3L engine). Tighten the bolts to 10 ft. lbs. (14 Nm).
- Transmission rear mount
- Clutch line to the concentric slave cylinder

12. If equipped with a transfer case, install or connect the following:

- Right side support brace-to-transmission bolt and tighten to 37 ft. lbs. (50 Nm)
- Right side support brace-to-transfer case bolts and tighten to 37 ft. lbs. (50 Nm)
- Left side support brace-to-transfer case bolt(s) and stud and tighten to 37 ft. lbs. (50 Nm)
- Left side support brace-to-transmission bolts and tighten to 37 ft. lbs. (50 Nm)
- Manual transfer case shift linkage.
- Two transfer case shields
- Front propeller shaft

13. Install or connect the following:

- The rear propeller shaft.
- The shift tower.
- Transmission with transmission fluid
- Shift lever

SILVERADO AND SIERRA—NV4500

1. Shift the transmission into 3rd or 4th speed gear.
2. Remove or disconnect the following:

- The shift lever
- The shift tower
- The transmission oil
- Front propeller shaft, 4WD only

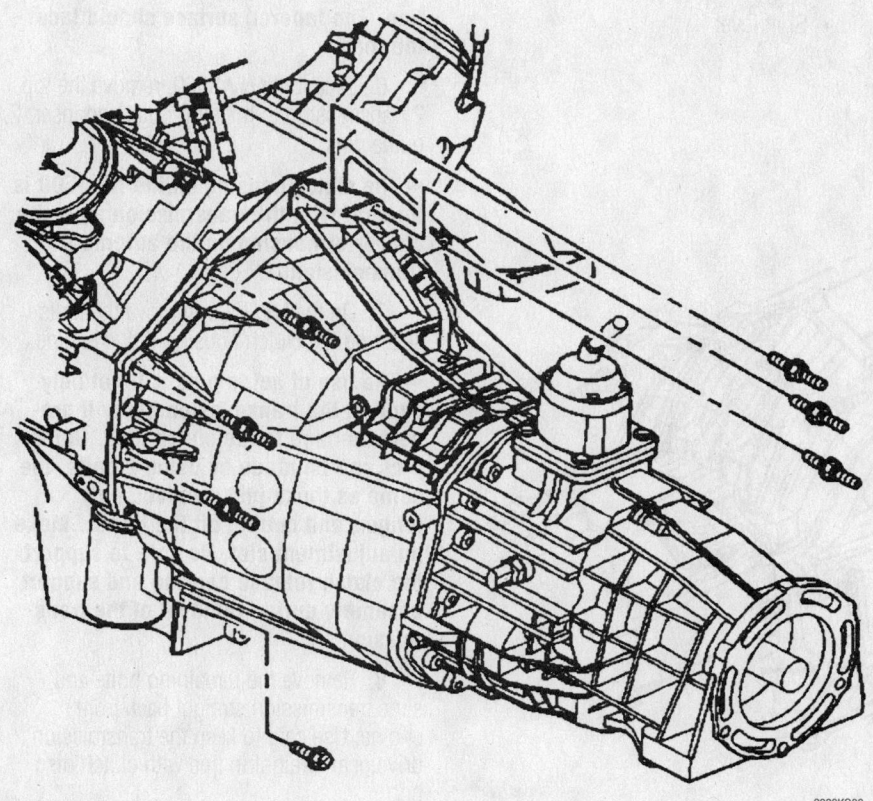

9308KG39

NV3500 removal—Silverado

- Rear propeller shaft
- Two transfer case shields
- Manual transfer case shift linkage, if equipped
- Two bolts securing the right side support bracket to the transmission

3. Using tool J42371, push back on the white plastic sleeve on the quick connect in order to separate the hydraulic clutch line from the concentric slave cylinder quick connect.

- Vehicle Speed Sensor (VSS) connector and harness
- Backup lamp switch connector and harness
- Transmission harness retainers
- Clutch housing cover-to-transmission bolts (4)
- Left and right side transmission-to-engine cover bolts
- Transmission rear mount. Support the transmission with a transmission jack.
- Bolts and studs securing the transmission to the engine.

4. Pull the transmission straight back on the clutch hub splines. Do not let the transmission hang from the clutch plate and the clutch cover. Remove the transmission from the vehicle.

5. Remove the clutch plate and the clutch cover from the engine flywheel if required.

To install:

6. Install or connect the following:
- Clutch plate and the clutch cover to the engine flywheel, if removed

7. Ensure the transmission is positioned in the 3rd or 4th speed gear. Rotate the transmission clockwise onto the clutch hub splines. Install the bolt and the studs securing the transmission to the engine. Tighten the bolts to 37 ft. lbs. (50 Nm).

8. Install or connect the following:
- Right and left side transmission to engine cover bolts and tighten to 10 ft. lbs. (14 Nm)
- Clutch cover-to-transmission bolts and tighten to 10 ft. lbs. (14 Nm)
- Transmission rear mount
- Clutch line to the slave cylinder
- Right side support bracket-to-transmission bolts (2). Tighten to 37 ft. lbs. (50 Nm).
- Manual transfer case shift linkage, if removed
- Two transfer case shields, if equipped
- Front propeller shaft, if equipped
- Rear propeller shaft
- Shift tower
- Transmission with transmission fluid
- Shift lever

EXPRESS AND SAVANA

1. Before servicing the vehicle, refer to the precautions in the beginning of this section.

2. Drain the transmission.

3. Remove or disconnect the following:

- Negative battery cable
- Shifter boot and lever
- Exhaust pipes
- Parking brake cables
- Driveshaft, matchmark for reassembly
- Transfer case, if equipped
- Transmission to engine braces (vehicles equipped with diesel engines have only 1 brace)
- Wiring harness at the transmission

4. Support the transmission with a transmission jack.
- Nut securing the transmission mount to the crossmember

5. Position a transmission jack or equivalent, under the transmission for support.
- Crossmember. Visually inspect to see if other equipment, brackets or lines, must be removed to permit removal of transmission.

➥ **Mark position of crossmember when removing to prevent incorrect installation. The tapered surface should face the rear.**

6. Except the NV 3500, remove the top 2 transmission to housing bolts and insert 2 guide pins.

➥ **The clutch housing on the NV 3500 is integral with the transmission as is the converter housing on the automatic transmissions.**

7. On the NV 3500, remove the bolts securing the clutch housing to the engine.

➥ **The use of guide pins will not only support the transmission but will prevent damage to the clutch disc. Guide pins can be made by using 2 bolts, the same as those just removed only longer, and cutting off the heads. Make an adjustment slot. Be sure to support the clutch release bearing and support assembly during removal of the transmission.**

8. Remove the remaining bolts and slide transmission straight back from engine. Use care to keep the transmission drive gear straight in line with clutch disc hub.

9. Remove or disconnect the following:

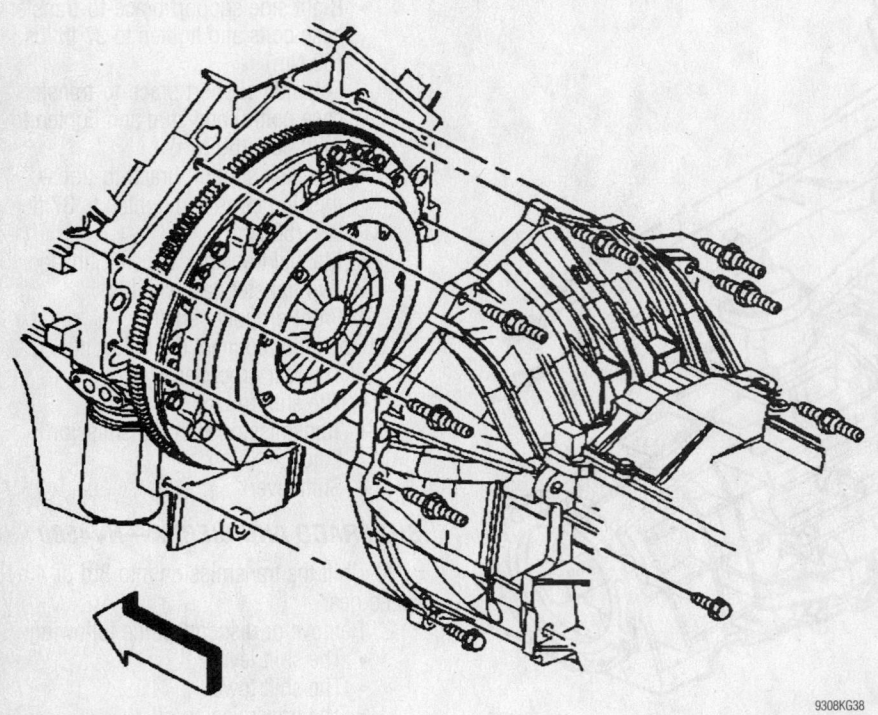

9308KG38

NV4500 removal—Silverado

- Wiring, clips, tubes and brackets etc., which would interfere with the removal of the transmission

➡**Ensure that the engine is supported with a jack stand before detaching the transmission from the engine.**

- Transmission from the engine

10. Carefully lower the transmission using the transmission jack.

To install:

11. Check the area behind the torque converter for leaks. Replace the front seal, if required.

12. It is good practice to examine the area around the rear crankshaft seal, checking for leaks. If necessary, remove the flywheel and replace the seal.

13. Inspect the flywheel ring gear teeth. If damaged, replace the flywheel.

14. Perform the following steps:

a. Place the transmission in high gear. Lightly coat the input shaft splines with high temperature grease.

b. Raise the transmission into position.

c. On transmissions with a separate clutch housing, install the guide pins in the top 2 bolt holes if they have been removed.

d. Roll the transmission forward and engage the clutch splines. Keep pushing the transmission forward until it mates with the engine.

e. On transmissions with a separate clutch housing, remove the guide pins and install the bolts, tighten the bolts to 23 ft. lbs. (31 Nm).

f. On the NV 3500, install the transmission-to-engine bolts. Tighten the bolts to 35 ft. lbs. (47 Nm)

15. When satisfied that the transmission is properly seated, install and tighten the transmission-to-engine bolts and/or studs to 34 ft. lbs. (47 Nm).

16. Install or connect the following:

- Wiring, clips, tubes and brackets etc
- Shifter cable
- Starter
- Exhaust pipes
- Transmission crossmember. Torque the bolts to 56 ft. lbs. (77 Nm).
- Transmission mount on the transmission. Torque the bolts to 35 ft. lbs. (47 Nm).
- Nut and washer that secure the transmission mount to the crossmember. Torque the nut to 38 ft. lbs. (52 Nm).
- Transmission to engine brace(s) Torque the bolts to 41 ft. lbs. (55

Nm) for gasoline engines and to 51 ft. lbs. (70 Nm) for diesel engines.

- Transfer case, if equipped

17. Remove the transmission jack and engine support stands.

- Driveshaft
- Shifter lever and boot
- Negative battery cable

18. Refill the transmission with fluid.

19. Road test the vehicle and test for proper operation. Check for leaks.

Automatic Transmission

SILVERADO AND SIERRA—4L60E

1. Before servicing the vehicle, refer to the precautions in the beginning of this section.

2. Remove or disconnect the following:

- Transmission fluid
- Transmission oil level indicator tube and seal from the transmission

➡**Plug the oil level indicator tube opening in the transmission.**

- Shift cable end from the transmission shift lever ball stud
- Front propeller shaft, if 4WD
- Rear propeller shaft.

3. Plug the transmission oil cooler line connectors in the transmission case.

4. Remove or disconnect the following:

- Starter motor

5. Support the transmission with a transmission jack.

6. Remove or disconnect the following:

- Torque converter access plug
- Flywheel-to-torque converter bolts
- Transmission rear mount-to-transmission bolts and nut
- Heat shield-to-transmission bolts
- Transmission vent hose from the transmission
- Fuel lines from the transmission
- Wiring harness from the transmission
- Transmission-to-engine stud and bolt
- Six studs and bolt securing the transmission to the engine.

7. Install tool J21366 onto the transmission bell housing to retain the torque converter. Pull the transmission straight back.

8. The transmission from the vehicle

9. Flush the transmission oil cooler and cooling lines.

To install:

10. Install Tool J21366 onto the transmission bell housing to retain the torque converter.

11. Support the transmission with a transmission jack.

12. Raise the transmission into place and remove the tool from the transmission.

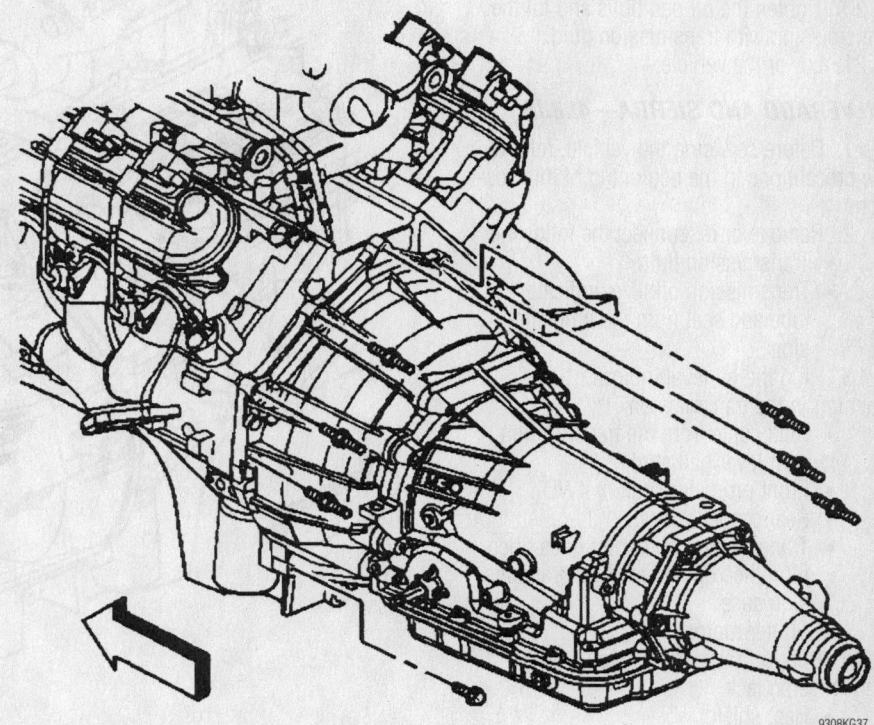

4L60E removal—Silverado

9308KG37

13. Slide the transmission straight onto the locating pins while lining up the marks on the flywheel and the torque converter. The torque converter must be flush onto the flywheel and rotate freely by hand.

14. Install or connect the following:
- Studs and bolt securing the transmission to the engine. Tighten to 37 ft. lbs. (50 Nm).
- Flywheel to torque converter bolts
- Torque converter access plug
- Transmission vent hose to the transmission
- Fuel lines to the transmission
- Wiring harness to the transmission.
- Heat shield-to-transmission bolts and tighten to 13 ft. lbs. (17 Nm)
- Transmission rear mount-to-transmission bolt and nut and tighten to 18 ft. lbs. (25 Nm)

15. Remove the transmission jack from the transmission.

16. Unplug the transmission oil cooler line connectors in the transmission case.

17. Install or connect the following:
- Transmission oil cooler lines
- Front propeller shaft, if equipped
- Rear propeller shaft
- Shift cable end to the transmission shift lever ball stud

18. Unplug the oil level indicator tube opening in the transmission.

19. Install the transmission oil level indicator tube and seal to the transmission.

20. Tighten the oil pan bolts and fill the transmission with transmission fluid.

21. Lower the vehicle.

SILVERADO AND SIERRA—4L80E

1. Before servicing the vehicle, refer to the precautions in the beginning of this section.

2. Remove or disconnect the following:
- Transmission fluid
- Transmission oil level indicator tube and seal from the transmission
3. Plug the oil level indicator tube opening in the transmission.
- Shift cable from the transmission shift lever ball stud
- Front propeller shaft, if 4WD
- Rear propeller shaft.
- Transmission oil cooler lines, then plug thee openings in the transmission case
- Starter motor
4. Support the transmission with a transmission jack.
- Heat shield
- Transmission vent hose
- Fuel lines from the transmission

- Wiring harness from the transmission
- Transmission brace-to-engine bracket and transmission nut and bolt
- Torque converter cover
- Flywheel to torque converter bolts
- Transmission rear mount
- Stud and bolt on the right side securing the transmission to the engine
- Remaining six studs and the bolt securing the transmission to the engine

5. Install Tool J21366 onto the transmission bell housing to retain the torque converter.

6. Pull the transmission straight back. Remove the transmission from the vehicle.

7. Flush the transmission oil cooler and cooling lines when you remove the transmission.

To install:

8. Install Tool J21366 onto the transmission bell housing to retain the torque converter.

9. Support the transmission with a transmission jack.

10. Raise the transmission into place and remove the tool from the transmission.

11. Slide the transmission straight onto the locating pins while lining up the marks on the flywheel and the torque converter. The torque converter must be flush onto the flywheel and rotate freely by hand.

12. Install or connect the following:
- Six studs and bolt securing the transmission to the engine. Tighten to 37 ft. lbs. (50 Nm).
- Stud and bolt on the right side securing the transmission to the engine. Tighten to 37 ft. lbs. (50 Nm).
- Flywheel-to-torque converter bolts and tighten to 44 ft. lbs. (60 Nm).
- Torque converter cover-to-engine bolts and tighten to 37 ft. lbs. (50 Nm)
- Torque converter cover-to-transmission stud and bolt and tighten to 24 ft. lbs. (33 Nm).
- Transmission vent hose
- Fuel lines
- Wiring harness
- Heat shield. Tighten the bolts to 13 ft. lbs. (17 Nm).
- Transmission rear mount-to-transmission nuts and bolt. Tighten to 18 ft. lbs. (25 Nm).
- Transmission brace. Tighten the bolts and nut to 37 ft. lbs. (50 Nm).

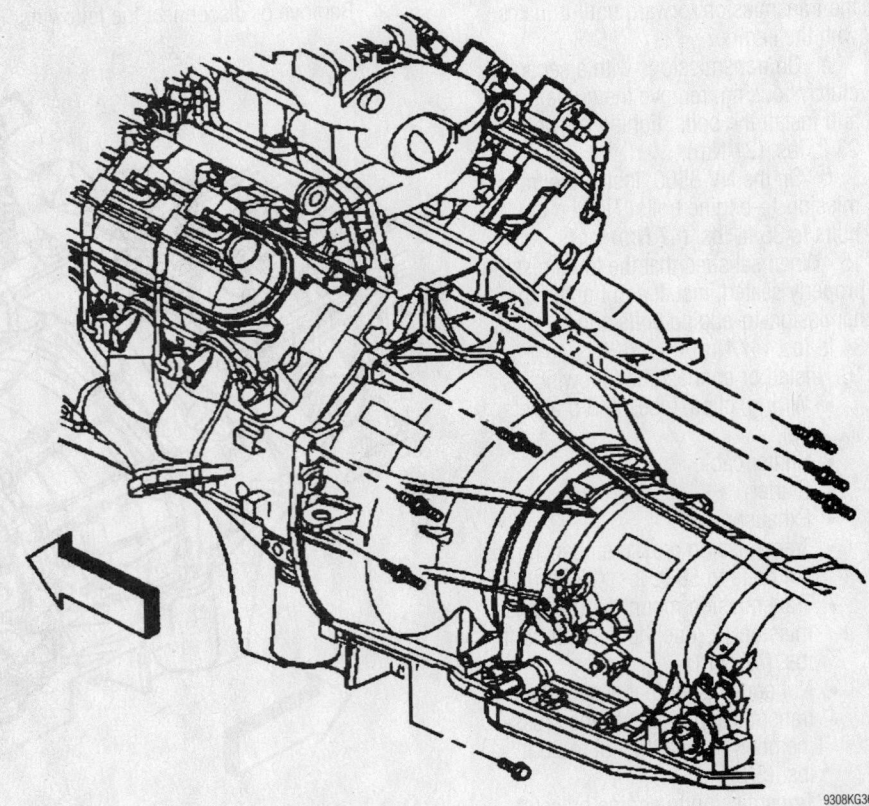

4L80E removal—Silverado

9308KG36

13. Remove the transmission jack from the transmission.
 • Starter motor
14. Unplug the transmission oil cooler line connectors in the transmission case.
15. Connect the transmission oil cooler lines to the transmission.
16. Install or connect the following:
 • Rear propeller shaft
 • Front propeller shaft, if 4WD
 • Shift cable end to the transmission shift lever ball stud
17. Unplug the oil level indicator tube opening in the transmission.
18. Install the transmission oil level indicator tube and seal to the transmission.
19. Tighten the oil pan bolts and fill the transmission with transmission fluid.
20. Lower the vehicle.

EXPRESS AND SAVANA

1. Before servicing the vehicle, refer to the precautions in the beginning of this section.
2. Drain the transmission.
3. Remove or disconnect the following:
 • Negative battery cable
 • Shift cable, control lever and bracket
 • Exhaust pipes
 • Parking brake cables
 • Driveshaft, matchmark for reassembly
 • Transfer case, if equipped
 • Transmission to engine braces (vehicles equipped with diesel engines have only 1 brace)
 • Wiring harness at the transmission
4. Support the transmission with a transmission jack.
 • Nut securing the transmission mount to the crossmember
5. Position a transmission jack or equivalent, under the transmission for support.
 • Crossmember. Visually inspect to see if other equipment, brackets or lines, must be removed to permit removal of transmission.

➡**Mark position of crossmember when removing to prevent incorrect installation. The tapered surface should face the rear.**

6. Perform the following:
 a. Remove the torque converter inspection cover.
 b. Mark the alignment of the torque converter to the flexplate.
 c. Remove the torque converter to flexplate bolts. Remove the dipstick tube and seal from the transmission. Plug the opening to avoid contamination.

 d. Disconnect both transmission lines at the transmission and plug them to avoid contamination and leakage.
 e. Position a J 21366 converter holding strap onto the transmission/torque converter to keep the torque converter from sliding off of the transmission turbine shaft.

➡**The use of guide pins will not only support the transmission but will prevent damage to the clutch disc. Guide pins can be made by using 2 bolts, the same as those just removed only longer, and cutting off the heads. Make an adjustment slot.**

7. Remove the remaining bolts and slide transmission straight back from engine. Use care to keep the transmission drive gear straight in line with clutch disc hub.
8. Remove or disconnect the following:
 • Wiring, clips, tubes and brackets etc., which would interfere with the removal of the transmission

➡**Ensure that the engine is supported with a jack stand before detaching the transmission from the engine.**

 • Transmission from the engine
9. Carefully lower the transmission using the transmission jack.
To install:
10. Check the area behind the torque converter for leaks. Replace the front seal, if required.
11. It is good practice to examine the area around the rear crankshaft seal, checking for leaks. If necessary, remove the flexplate and replace the seal.
12. Inspect the flexplate ring gear teeth. If damaged, replace the flexplate.
13. Perform the following steps:
 a. With tool J 21366 or equivalent, torque converter holding strap in place, raise the transmission into position with a transmission jack.
 b. Remove the torque converter holding strap and slide the transmission into place. Slide the transmission straight onto the locating pins while lining up the marks on the flexplate and the torque converter. Be sure the transmission is fully seated against the rear of the engine block and the locating pins are completely engaged.

✳✳ WARNING

DO NOT attempt to draw the transmission to the block with the mounting bolts. If the transmission is not

properly seated, the bolts will break the transmission case.

➡**The torque converter must be flush with the flexplate and rotate freely by hand.**

14. When satisfied that the transmission is properly seated, install and tighten the transmission-to-engine bolts and/or studs to 34 ft. lbs. (47 Nm).
15. Perform the following steps:
 a. Install the dipstick tube and seal.
 b. Check the alignment marks on the torque converter and flexplate to be sure that they are properly aligned. Install the torque converter bolts. Finger-tighten the bolts to ensure proper converter seating. When the converter is properly seated, tighten the bolts to 46 ft. lbs. (63 Nm).
 c. Install the torque converter cover. Tighten the retaining bolts to 24 ft. lbs. (33 Nm) on the 4.3L engines or 89 inch lbs. (10 Nm) on the V8 engines.
16. Install or connect the following:
 • Wiring, clips, tubes and brackets etc
 • Transmission cooling lines
 • Shifter cable
 • Starter
 • Exhaust pipes
 • Transmission crossmember. Torque the bolts to 56 ft. lbs. (77 Nm).
 • Transmission mount on the transmission. Torque the bolts to 35 ft. lbs. (47 Nm).
 • Nut and washer that secure the transmission mount to the crossmember. Torque the nut to 38 ft. lbs. (52 Nm).
 • Transmission to engine brace(s) Torque the bolts to 41 ft. lbs. (55 Nm) for gasoline engines and to 51 ft. lbs. (70 Nm) for diesel engines.
 • Transfer case, if equipped
17. Remove the transmission jack and engine support stands.
 • Driveshaft
 • Shifter lever and boot, on manual transmissions
 • Negative battery cable
18. Refill the transmission with fluid.
19. Road test the vehicle and test for proper operation. Check for leaks.

Clutch

ADJUSTMENTS

The hydraulic clutch system requires no periodic adjustment.

REMOVAL & INSTALLATION

Silverado and Sierra

1. Before servicing the vehicle, refer to the precautions in the beginning of this section.

2. Remove or disconnect the following:
- Transmission
- Quick disconnect from the actuator cylinder

3. Install a clutch alignment tool.

4. Mark the flywheel and a clutch pressure plate lug for the installation alignment.

5. Remove the pressure plate bolts and the washers.

6. Secure the clutch pressure plate and the clutch driven plate to the flywheel.

7. Remove the clutch alignment tool.

To install:

8. Install the bolts and the washers securing the clutch pressure plate and the clutch driven plate to the flywheel.

9. Install the clutch alignment tool.

10. Align the marks made during removal or, if new align the lightest part of the clutch pressure plate identified by a yellow dot, to the heaviest part of the flywheel, identified by an "X". Tighten the clutch pressure plate to the flywheel bolts to 30 ft. lbs. (41 Nm).

11. Remove the clutch alignment tool.

12. Install the transmission.

13. Install the quick disconnect to the concentric slave cylinder.

Express and Savana

1. Before servicing the vehicle, refer to the precautions in the beginning of this section.

2. Remove or disconnect the following:
- Negative battery cable
- Slave cylinder
- Transmission assembly

3. Install the clutch removal tool and support the clutch assembly.

➡**Before removing the clutch from the flywheel, matchmark the flywheel, clutch cover and 1 pressure plate lug, so these parts may be assembled in their same relative positions and retain the factory balance.**

4. Loosen the clutch plate retaining bolts slowly and evenly one at a time until all pressure is released from the pressure plate assembly.

5. Remove the clutch, pressure plate

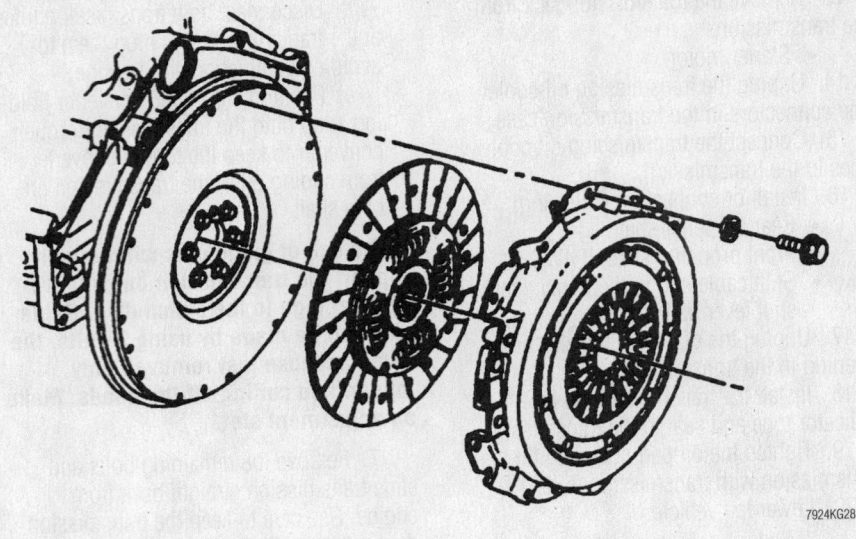

7924KG28

Exploded view of the typical clutch assembly

and removal tool. Check the flywheel for damage, repair or replace, as required.

6. Check the clutch assembly and flywheel for signs of wear, scoring, overheating, etc. If the clutch plate, flywheel or pressure plate is oil-soaked, inspect the engine rear main seal and the transmission input shaft seal and correct leakage as required. Replace any damaged parts.

To install:

7. Assemble the pressure plate and disc assembly, as required.

➡**The manufacturer recommends that new pressure plate bolts and washers be used.**

8. Turn the flywheel until the previously applied mark is at the bottom.

9. Install or connect the following:
- Clutch disc, pressure plate and cover using a suitable clutch aligning tool

10. Turn the clutch until the matchmark on the clutch cover aligns with the mark on the flywheel.
- Attaching bolts and tighten in a crossing pattern until the spring pressure is taken up. Torque the bolts to 29 ft. lbs. (34 Nm).

11. Remove the aligning tool.

12. Apply high temperature grease to transmission input shaft.

13. Install or connect the following:
- Transmission. Torque the bell housing bolts to 35 ft. lbs. (47 Nm).
- Negative battery cable

14. Bleed the hydraulic system.

Hydraulic Clutch System

BLEEDING

Bleeding air from the hydraulic clutch system is necessary whenever any part of the system has been disconnected or the fluid level (in the reservoir) has been allowed to fall so low, that air has been drawn into the master cylinder.

1. Fill master cylinder reservoir with new brake fluid conforming to DOT 3 specifications.

2. Have an assistant fully depress and hold the clutch pedal, then open the bleeder screw.

3. Close the bleeder screw and have your assistant release the clutch pedal.

4. Repeat the procedure until all of the air is evacuated from the system. Check and refill master cylinder reservoir as required to prevent air from being drawn through the master cylinder.

➡**Never release a depressed clutch pedal with the bleeder screw open or air will be drawn into the system.**

5. Test the clutch for proper operation.

Transfer Case Assembly

REMOVAL & INSTALLATION

Silverado and Sierra

1. Before servicing the vehicle, refer to the precautions in the beginning of this section.

2. Remove or disconnect the following:

- Transfer case shields
- Front propeller shaft
- Rear propeller shaft
- Shift rod from the transfer case
- Vent hose from the transfer case
- Vehicle Speed Sensor (VSS) electrical connectors
- All necessary wiring harnesses from the transfer case

3. Support the transfer case with a transmission jack.

4. If equipped with a NV3500 manual transmission, remove or disconnect the following:

- Bolt securing the left side support brace to the transmission
- Bolt and stud securing the left side support brace to the transfer case
- Two bolts securing the right side support brace to the transmission and transfer case

5. Remove or disconnect the following:

- Six nuts securing the transfer case and bracket to the transmission or transmission adapter, as applicable
- Transfer case
- Gasket, then discard

To install:

6. Install a new gasket to the transmission. Use Teflon pipe sealant GM P/N 12346004 in order to hold the gasket in place.

7. Raise and position the transfer case to the vehicle.

8. Install or connect the following:

- Six nuts securing the transfer case and bracket to the transmission adapter or transmission. Tighten to 37 ft. lbs. (50 Nm).

9. If equipped with a manual transmission, install or connect the following:

- Bolt securing the left side support brace to the transmission and tighten to 37 ft. lbs. (50 Nm)
- Bolt and stud securing the left side support brace to the transfer case and tighten to 37 ft. lbs. (50 Nm)
- Two bolts securing the right side support brace to the transmission and transfer case and tighten to 37 ft. lbs. (50 Nm)

10. Install or connect the following:

- Vent hose to the transfer case

11. Check the transfer case oil level.

12. VSS electrical connectors

- Wiring harness to the transfer case
- Shift rod to the transfer case
- Front and rear propeller shafts
- Transfer case shields

13. Lower the vehicle.

Express and Savana

1. Before servicing the vehicle, refer to the precautions in the beginning of this section.

2. Drain transfer case of lubricant.

3. Remove or disconnect the following:

- Negative battery cable
- Skid plate, if equipped
- Vent hose clamp at the transfer case
- Front driveshaft at the transfer case and support it aside
- Rear driveshaft and support it aside
- Electrical connections from the transfer case
- Transfer case shift linkage

4. Support the transfer case with a transmission jack.

- Transmission-to-transfer case bolts and spring washers
- Transfer case assembly and gasket

5. Carefully lower the transfer case.

To install:

6. Carefully raise the transfer case into position.

7. Install or connect the following:

- New gasket to the transmission, using gasket sealer to hold it in place
- Transfer case onto the transmission or transmission adapter. Torque the bolts to 33 ft. lbs. (45 Nm).
- Electrical harness connectors to the transfer case connections
- Transfer case shift linkage and make the proper adjustments
- Front and rear driveshafts

8. Refill the transfer case with DEXRON®IIE automatic transmission fluid.

- Skid plate, if equipped
- Negative battery cable

9. Test drive for proper operation.

Halfshaft

REMOVAL & INSTALLATION

Silverado and Sierra

1. Before servicing the vehicle, refer to the precautions in the beginning of this section.

2. Remove or disconnect the following:

- Wheels

3. Insert a drift or a large screwdriver through the brake caliper into one of the brake rotor vanes in order to prevent the drive axle wheel drive shaft from turning.

4. Remove or disconnect the following:

- Nut and the washer from the hub

➥ **Do not reuse the hub nut. A new nut must be used when installing the wheel drive shaft.**

- Bolts (6) securing the wheel drive shaft inboard flange to the output shaft flange
- Drift from the rotor
- Stabilizer shaft link from the lower control arm

5. Wrap shop towels around both the inner and the outer wheel drive shaft boots in order to avoid damage to the boots during removal and installation.

6. Pull the wheel drive shaft through the lower control arm opening.

To install:

7. Wrap shop towels around both the inner and the outer wheel drive shaft boots in order to avoid damage to the boots during removal and installation.

➥ **Clean the steering knuckle and the wheel drive shaft splines and threads. These areas must be dry and free of grease, dirt, and contamination.**

8. Insert the wheel drive shaft splined shank into the knuckle hub.

➥ **Use only a genuine GM front wheel drive shaft nut. Installation of anything but an OEM front wheel drive shaft nut could cause damage to the vehicle.**

9. Install or connect the following:

- Washer and the new hub nut to the wheel driveshaft. Do not tighten.
- The wheel drive shaft inboard flange to the output shaft flange using the inboard flange bolts

10. Insert a drift or a large screwdriver through the brake caliper into 1 of the brake rotor vanes in order to prevent the wheel drive shaft from turning. Tighten the inboard flange bolts to 58 ft. lbs. (78 Nm). Tighten the hub nut to 155 ft. lbs. (210 Nm).

11. Remove the drift from the rotor.

12. Install the stabilizer shaft link.

13. Install the wheel and tire assembly.

Express and Savana

1. Before servicing the vehicle, refer to the precautions in the beginning of this section.

2. Remove or disconnect the following:

- Front wheel and tire assembly
- Skid plate, as required
- Drive axle hub nut and washer
- Brake line and wheel speed sensor support bracket from the upper control arm
- Left outer tie rod attaching nut and cotter pin
- Tie rod from the steering knuckle

3. Position the tie rod aside and push steering linkage to the opposite side of the vehicle.

- Lower shock attaching nut and bolt; position the shock aside
- Left stabilizer bar bracket and bushing at the frame
- Stabilizer bar bolt, spacer and bushings at the lower control arm

4. Take pressure off the upper control arm by placing a support below the lower control arm between the spring seat and the ball joint.

- Upper ball joint cotter pin and loosen (do not remove) the upper ball joint attaching nut. Separate the ball joint stud from the steering knuckle. Remove the attaching nut.

➡**Cover the shock mounting bracket and lower ball joint stud with a towel to prevent the axle boot from tearing during removal and installation.**

5. Separate the axle shaft from the hub and rotor using tool J-28733 or equivalent.

- Axle shaft inner flange bolts and shaft

To install:

6. Lubricate the axle and hub splines with an approved high temperature wheel bearing grease.

7. Install or connect the following:

- Axle shaft in the hub
- Inboard CV-joint-to-flange bolts and tighten to 60 ft. lbs. (80 Nm)
- Upper ball joint to steering knuckle. Torque the stud nut to 61 ft. lbs. (83 Nm).
- New cotter pin through the upper ball joint stud and nut, lubricate the ball joint as required.
- Left stabilizer bar bracket and bushing at the frame
- Stabilizer bar bolt, spacer and bushings at the lower control arm
- Lower shock in the mount bracket and the attaching nut and bolt
- Left tie rod end at the steering knuckle. Torque the nut to 35 ft. lbs. (47 Nm).
- New cotter pin through the tie rod stud and nut
- Brake line bracket to the control arm, ensuring the line and/or hose is not twisted or kinked
- Skid plate, as required
- Axle hub washer and nut. Insert a drift through the rotor vanes to keep the axle from turning. Torque the hub nut to 180 ft. lbs. (245 Nm)
- Wheel and tire assembly

CV-Joints

OVERHAUL

Silverado and Sierra

INNER JOINT

1. Before servicing the vehicle, refer to the precautions in the beginning of this section.

➡**With removal of the halfshaft for any reason, the transmission sealing surface (the tripot male/female shank of the halfshaft) should be inspected for corrosion. If corrosion is evident, the surface should be cleaned with 320 grit cloth or equivalent. Transmission fluid may be used to clean off any remaining debris. The surface should be wiped dry and the halfshaft reinstalled free of any buildup.**

2. Before servicing the vehicle, refer to the precautions in the beginning of this section.

3. Use a hand grinder in order to cut through the swage ring.

4. Remove the tripot housing from the halfshaft. Wipe the grease off of the tripot assembly roller bearings and the tripot

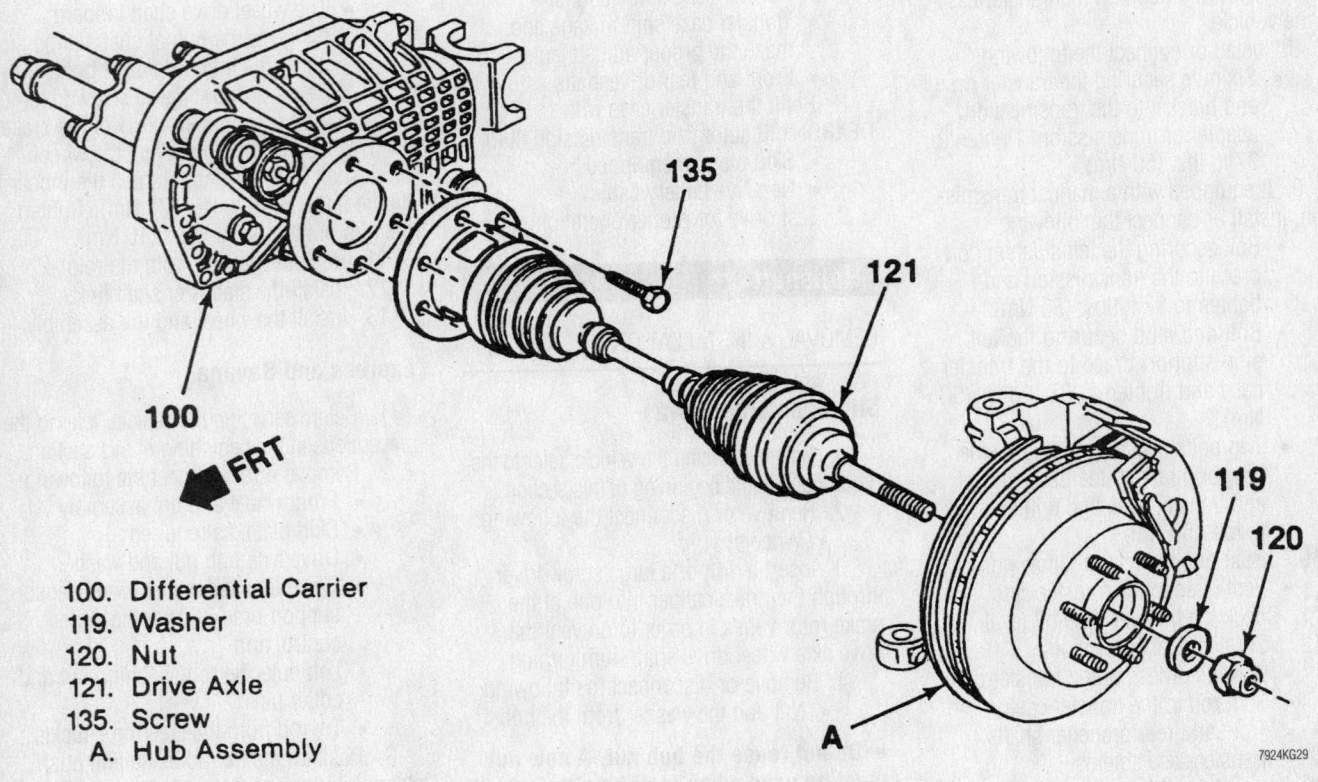

100. Differential Carrier
119. Washer
120. Nut
121. Drive Axle
135. Screw
A. Hub Assembly

7924KG29

The halfshaft is mounted to the flange on the differential and through the hub assembly—4-wheel drive models

housing. Thoroughly degrease the tripot housing. Allow the tripot housing to dry prior to assembly.

➡ **Handle the tripot spider assembly with care. Tripot balls and needle rollers may separate from the spider trunnion if the tripot balls and needle rollers are not handled carefully.**

5. Use side cutters to cut away the small boot clamp.

6. Compress the tripot boot up the half-shaft away from the tripot spider assembly toward the outboard (CV joint assembly) end of the halfshaft.

7. Spread the spider spacer ring with tool J8059, or equivalent.

8. Remove the following items from the halfshaft bar:
   a. The spacer ring.
   b. The spider assembly.
   c. The tripot boot.

9. Clean the halfshaft bar. Use a wire brush in order to remove any rust in the boot mounting area (grooves).

10. Inspect the needle rollers, needle bearings, and trunnion. Check the tripot housing for unusual wear, cracks, or other damage. Replace any damaged parts.

To assemble:

11. Place the new small boot clamp onto the small end of the joint boot.

12. Compress the joint boot and small boot clamp onto the halfshaft bar.

13. Position the small end of the joint boot into the joint boot groove on the half-shaft bar.

14. Secure the small boot clamp with tool J35910, or equivalent, a breaker bar, and a torque wrench. Tighten the small boot clamp (1) to 100 ft. lbs. (136 Nm).

15. Check the gap dimension on the clamp ear. Continue tightening until the gap dimension is reached.

➡ **Assemble the CV joint with the convolute retainer in the correct position, as illustrated.**

16. Install the convolute retainer over the inboard joint boot, being sure to capture three convolutions.

17. Install the tripot spider assembly onto the halfshaft bar with the counterbore towards the end of the halfshaft bar.

18. Install the spacer ring in the groove at the end of the halfshaft bar.

19. Push the spider assembly back toward the end of the halfshaft bar until the spacer ring is covered by the spider assembly counterbore.

20. Pack the tripot boot and the tripot housing with the grease supplied in the kit. The amount of grease supplied in this kit has been pre-measured for this application.

21. Reassemble the tripot housing and the tripot boot using the following procedure:
   a. Pinch the swage ring slightly by

hand in order to distort it into an oval shape.
   b. Slide the distorted swage ring over the large diameter of the boot.
   c. Place the tripot housing over the spider assembly.
   d. Install the boot onto the tripot housing.
   e. Align the tripot boot with the swage ring in place, over the flat area on the tripot housing.

22. Mount tool J36652 in a vise. Install the bottom half of the split-plate swage clamp. For K15 models, use tool J36652-98. For K25 models, use tool J36652-1.

23. Check the inboard stroke position. Use measurement A for the K15 models. Use measurement B for the K25 models.

24. Position the inboard end (tripot end) of the halfshaft assembly in tool J36652. Install the top half of the proper size tool on the lower half of the tool. For K15 models, use tool J36652-98. For K25 models, use tool J36652-1.

25. Align the swage ring and the swage ring clamp. Insert the bolts. Hand tighten the bolts in tool J36652 until the bolts are snug.

26. Align the following during this procedure:
   a. The tripot boot.
   b. The housing.
   c. The swage ring. Tighten each bolt 180 degrees at a time. Alternate between the bolts until both sides of the top half of J36652 touch the bottom half of the tool.

27. Loosen the bolts and remove the halfshaft assembly from J36652.

28. Remove the convolute retainer from the boot.

OUTER JOINT

1. Before servicing the vehicle, refer to the precautions in the beginning of this section.

2. Place protective covers over the vise jaws. Place the halfshaft in the vise.

3. Use a hand grinder to cut through the swage ring. Use side cutters to cut off the small boot clamp.

4. Slide the boot down the halfshaft bar and away from the CV-joint outer race. Wipe all grease away from the face of the CV joint.

5. Find the halfshaft bar retaining snap ring, which is located in the inner race.

6. Spread the snapring ears apart.

7. Pull the CV joint and the CV joint boot from the halfshaft bar. Discard the old CV joint boot.

8. Place a brass drift against the CV

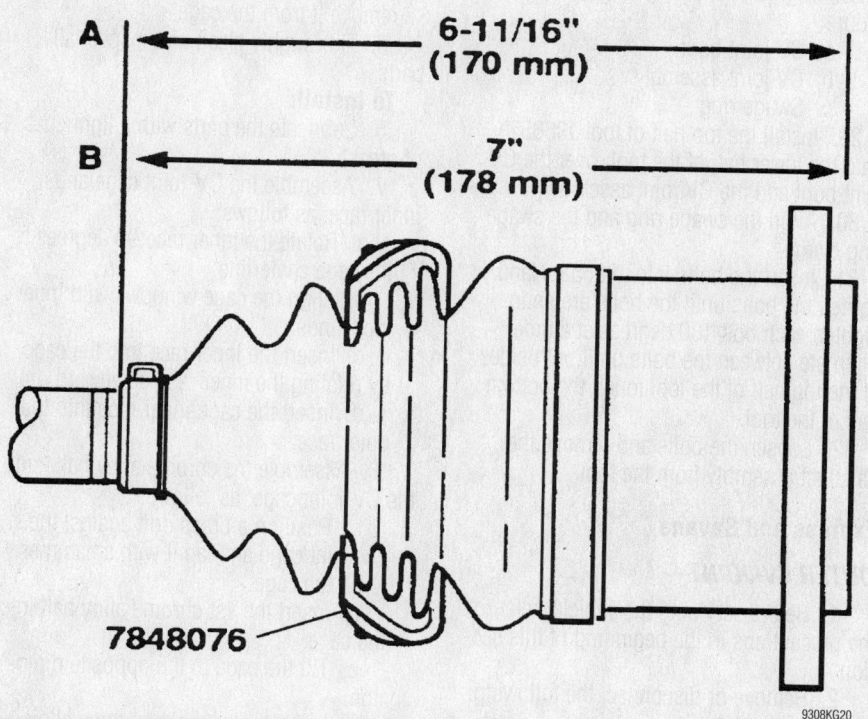

A ← 6-11/16" (170 mm) →

B ← 7" (178 mm) →

7848076

9308KG20

Assembled joint measurement—15 Series Silverado

joint cage. Tap gently on the brass drift with a hammer in order to tilt the cage.

9. Remove the first chrome alloy ball when the CV joint cage tilts. Tilt the CV joint cage (1) in the opposite direction to remove the opposing chrome alloy ball. Repeat this process to remove all six of the balls.

10. Pivot the CV joint cage and the inner race 90 degrees to the center line of the outer race. At the same time, align the cage windows with the lands of the outer race. Lift out the cage and the inner race.

11. Remove the inner race from the cage by rotating the inner race upward. Clean the following items thoroughly with cleaning solvent. Remove all traces of old grease and any contaminates.

    a. Inner and outer race assemblies.
    b. CV joint cage.
    c. Chrome alloy balls.

12. Dry all the parts. Check the CV joint assembly for unusual wear, cracks, or other damage. Replace any damaged parts. Clean the halfshaft bar. Use a wire brush to remove any rust in the boot mounting area (grooves).

To assemble:

13. Inspect all of the parts for unusual wear, cracks, or other damage. Replace the CV joint assembly if necessary. Put a light coat of the recommended grease on the inner and the outer race grooves.

14. Hold the inner race at 90 degrees to the centerline of the cage. Align the lands of the inner race with the windows of the cage. Insert the inner race into the cage by rotating the inner race downward.

15. Insert the cage and inner race into the outer race.

16. Place a brass drift against the CV joint cage. Tap gently on the brass drift with a hammer in order to tilt the cage. Install the first chrome alloy ball when the CV joint cage tilts. Tilt the CV joint cage in the opposite direction to install the opposing chrome alloy ball. Repeat this process in order to install all six of the balls.

17. Pack the CV joint boot and the CV joint assembly with the grease supplied in the kit. The amount of grease supplied in this kit has been pre-measured for this application.

18. Place the new small boot clamp onto the CV joint boot.

19. Slide the CV joint boot onto the halfshaft bar.

20. Position the small end of the CV joint boot into the joint boot groove on the halfshaft bar.

21. Secure the small boot clamp, a breaker bar, and a torque wrench. Tighten the small clamp (1) to 100 ft. lbs. (136 Nm).

22. Check the gap dimension on the clamp ear. Continue tightening until the gap dimension is reached.

23. Pinch the new swage ring slightly by hand to distort it into an oval shape. Slide the distorted swage ring over the large diameter of the boot.

➡ Be sure that the retaining ring side of the CV joint inner race faces the halfshaft bar (3) before installation.

24. Slide the CV joint onto the halfshaft bar. The retaining snap ring inside of the inner race engages in the halfshaft bar groove with a click when the CV joint is in the proper position.

25. Pull on the CV joint to verify engagement.

26. Slide the large diameter of the CV joint boot with the large swage ring in place, over the outside edge of the CV joint outer race.

27. Clamp the CV joint boot tightly to the CV joint outer race with the large swage ring, using the following procedure:

    a. Mount tool J36652 in a vise.
    b. Install the bottom half of the split-plate swage clamp. For K15 models, use tool J36652-98.
    c. For K25 models, use tool J36652-1.
    d. Position the CV joint end (outboard end) of the halfshaft assembly in the bottom half of tool J36652.

28. Align the following during this procedure:

    a. CV joint boot.
    b. CV joint assembly.
    c. Swage ring.

29. Install the top half of tool J36652 onto the lower half of the tool, over the CV joint boot and the CV joint assembly.

30. Align the swage ring and the swage ring clamp.

31. Insert the bolts into J36652. Hand tighten the bolts until the bolts are snug. Tighten each bolt 180 degrees at a time. Alternate between the bolts until both sides of the top half of the tool touch the bottom half of the tool.

32. Loosen the bolts and remove the halfshaft assembly from the tool.

Express and Savana

OUTER CV-JOINT

1. Before servicing the vehicle, refer to the precautions in the beginning of this section.

2. Remove or disconnect the following:
    • Front wheel
    • Halfshaft and position it in a vise

    • Large CV-joint boot clamp and discard it
    • Small CV-joint boot clamp and discard it
    • CV-joint boot and slide it back on the shaft
    • Outer race from the halfshaft by spreading the outer race-to-halfshaft retaining ring using Snapring Pliers J-8059
    • Retaining ring from the halfshaft and discard it
    • CV-joint boot from the halfshaft and discard it if damaged

3. Disassemble the chrome alloy balls from the CV-joint cage as follows:

    a. Position a brass drift against the CV-joint cage and tap it with a hammer to tilt the cage.
    b. Remove the 1st chrome alloy ball from the cage.
    c. Tilt the cage in the opposite direction.
    d. Remove the opposite chrome alloy ball.
    e. Repeat the procedure until all 6 balls are removed.

4. Disassemble the CV-joint cage and inner race as follows:

    a. Pivot the cage and race 90 degrees to the center line of the outer race.
    b. Align the cage windows with outer race lands.
    c. Remove the cage from the outer race.
    d. Rotate the inner race upward and remove it from the cage.

5. Thoroughly clean and inspect all parts.

To install:

6. Lubricate the parts with a light coat of grease.

7. Assemble the CV-joint cage and inner race, as follows:

    a. Rotate the inner race 90 degrees to the cage centerline.
    b. Align the cage windows with inner race lands.
    c. Insert the inner race into the cage by rotating the inner race downward.
    d. Insert the cage/inner race into the outer race.

8. Assemble the chrome alloy balls into the CV-joint cage, as follows:

    a. Position a brass drift against the CV-joint cage and tap it with a hammer to tilt the cage.
    b. Insert the 1st chrome alloy ball into the cage.
    c. Tilt the cage in the opposite direction.
    d. Insert the opposite chrome alloy ball.

e. Repeat the procedure until all 6 balls are inserted.

9. Install ½ of the kit grease into the CV-joint.

10. Install or connect the following:
- Small ring clamp on the CV boot
- New retaining ring on the halfshaft
- Large ring clamp on the CV boot
- Outer race assembly onto the half-shaft until the ring engages the halfshaft groove

11. Slide the small end of the CV-joint boot/clamp into place, with the seal lip in the halfshaft groove

➡ **Make sure the boot lies flat against the halfshaft.**

12. Using the Crimp tool J-35910, a torque wrench and a breaker bar, crimp the small CV-joint boot clamp to 100 ft. lbs. (136 Nm).

13. Check the clamp gap dimension; if it is not 0.085 in. (2.15mm), continue tightening the clamp until it is.

14. Install ½ of the kit grease into the CV-joint boot.

15. Slide the large end of the CV boot/clamp into place, with the seal lip in place over the outer race.

➡ **Make sure the boot lies flat against the outer race.**

16. Using the Crimp tool J-35910, a torque wrench and a breaker bar, crimp the large CV-joint boot clamp to 130 ft. lbs. (176 Nm).

17. Check the clamp gap dimension; if it is not 0.102 in. (2.60mm), continue tightening the clamp until it is.

18. Install the halfshaft and the front wheel.

INNER (TRI-POT) JOINT

1. Before servicing the vehicle, refer to the precautions in the beginning of this section.

2. Remove or disconnect the following:
- Front wheel
- Halfshaft and place it in a vise
- Snapring from the stub shaft and discard it
- Small CV-joint boot clamp, cut and discard it
- Large CV-joint boot clamp, cut and discard it
- CV-joint boot by sliding it away from the tri-pot joint

3. Install a Stub Shaft Removal tool J-38868-A to the stub shaft snapring groove.

4. Using a slide hammer puller, press the stub shaft from the tri-pot housing.

5. Remove or disconnect the following:
- Tri-pot housing from the tri-pot spider
- Inboard spacer ring, slide it rear-ward on the shaft using Snapring Pliers tool J-8059
- Outboard retaining ring using Snapring Pliers tool J-8059 and discard it
- Tri-pot joint spider assembly
- Inboard spacer ring and discard it
- CV-joint boot
- Trilobal tri-pot bushing from the housing

6. Thoroughly clean and inspect all parts.

To install:

7. Install or connect the following:
- New snapring onto the stub shaft
- Small boot clamp
- CV-joint boot

8. Using the Crimp tool J-35910, a torque wrench and a breaker bar, crimp the small CV-joint boot clamp to 100 ft. lbs. (136 Nm).

9. Install or connect the following:
- Inboard spacer ring slide it rear-ward on the shaft using Snapring Pliers tool J-8059, past the 2nd groove

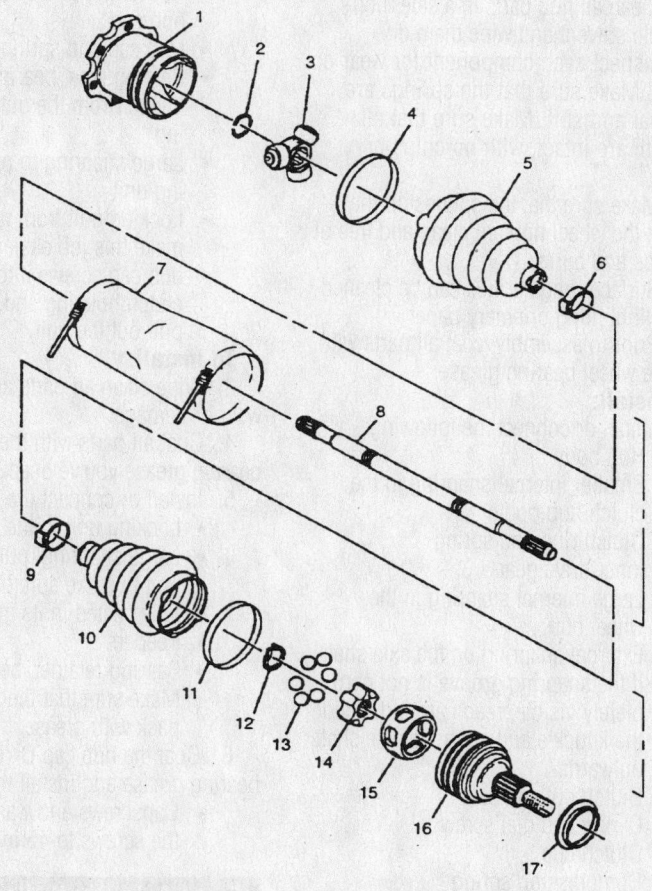

Legend

(1) Tripot Housing Assembly
(2) Spacer Ring
(3) Tripot Joint Spider Assembly
(4) Swage Ring
(5) Tripot Joint Seal
(6) Small Seal Retaining Clamp
(7) Drive Axle Seal Cover (Optional)
(8) Drive Axle Shaft

(9) CV Joint Seal
(10) Race Retaining Ring
(11) Ball
(12) CV Joint Inner Race
(13) CV Joint Cage
(14) CV Joint Outer Race
(15) Deflector Ring

9308KG10

Exploded view of the CV-Joint assembly

- Tri-pot joint spider assembly onto the shaft until it passes the 2nd groove
- Outboard retaining ring into the axle shaft groove using Snapring Pliers tool J-8059
- Tri-pot joint spider assembly, slide it against the outboard retaining ring
- Inboard spacer ring, seat it in the groove
- ½ of the kit grease into the boot
- ½ of the kit grease into the tri-pot housing
- Trilobal tip-pot bushing flush with the tri-pot housing face
- New large seal clamp onto the CV-joint boot
- Tri-pot housing, slide it over the tri-pot joint spider assembly
- CV-joint boot/clamp, slide it into place, over the trilobal tri-pot bushing with the seal lip in the groove

➡**Make sure the boot lies flat against the trilobal bushing.**

10. Using the Crimp tool J-35910, a torque wrench and a breaker bar, crimp the large CV-joint boot clamp to 130 ft. lbs. (176 Nm)

11. Check the clamp gap dimension; if it is not 0.102 in. (2.60mm), continue tightening the clamp until it is.

12. Install the halfshaft and the front wheel.

Manual Locking Hubs

The engagement and disengagement of the hubs is a manual operation which must be performed at each hub assembly. The hubs should be placed FULLY in either Lock or Free position or damage will result.

✳✳ WARNING

Do not place the transfer case in either 4-wheel mode unless the hubs are in the Lock position!

Locking hubs should be run in the Lock position periodically for a few miles to assure proper differential lubrication.

REMOVAL & INSTALLATION

1. Before servicing the vehicle, refer to the precautions in the beginning of this section.

2. Remove or disconnect the following:
   - Wheels
3. Lock the hubs.

- Outer retaining plate
- Allen head bolts
- Plate, O-ring, and knob assembly
- External snapring from the axle shaft
- Compression spring
- Clutch cup
- O-ring and dial screw
- Clutch nut and seal
- Large internal snapring from the wheel hub
- Inner drive gear
- Clutch ring and spring
- Smaller internal snapring from the clutch hub body
- Hub body

4. Clean all hub parts in a safe, non-flammable solvent and wipe them dry.

5. Inspect each component for wear or damage. Make sure that the springs are functional and stiff. Make sure that all gear teeth are intact, with no chips or burrs.

6. Make sure that the splines on the inside of the wheel hub are clean and free of dirt, chips and burrs.

7. Surface irregularities can be cleaned up with light filing or emery paper.

8. Prior to assembly, coat all parts with the same wheel bearing grease.

To install:

9. Install or connect the following:
   - Hub body
   - Smaller internal snapring in the clutch hub body
   - Clutch ring and spring
   - Inner drive gear
   - Large internal snapring in the wheel hub
   - External snapring on the axle shaft. If the snapring groove is not completely visible, reach around, inside the knuckle and push the axle shaft outwards.
   - Clutch nut and seal
   - O-ring and dial screw
   - Clutch cup
   - Compression spring

10. Place the hub dial in the Lock position.

11. Coat the hub dial assembly O-ring with wheel bearing grease and position the hub dial and retainer on the hub.
   - Allen head bolts. Make sure that you used any washers that were there originally. Torque to 45 inch lbs. (5 Nm).

12. Rotate the hub dial to the free position and turn the wheel hubs to make sure that the axle is free.
   - Wheels

Automatic Locking Hubs

REMOVAL & INSTALLATION

The following procedure covers removal & installation only, for the hub assembly. The hub should be disassembled ONLY if overhaul is necessary. In that event, an overhaul kit will be required. Follow the instructions in the overhaul kit to rebuild the hub.

1. Before servicing the vehicle, refer to the precautions in the beginning of this section.

2. Remove or disconnect the following:
   - Capscrews and washer from the hub cap
   - Hub cap and spring
   - Bearing race, bearing and retainer
   - Keeper from the outer clutch housing
   - Large snapring to release the locking unit
   - Locking unit from the hub. You can make this job easier by threading 2 hub cap screws into the outer clutch housing and hold these to pull out the unit.

To install:

3. Wipe clean all parts and check for wear or damage.

4. Coat all parts with the same wheel bearing grease you've used on the bearings.

5. Install or connect the following:
   - Locking unit in the hub
   - Large snapring, pull outward on the unit to make sure the snapring is fully seated in its groove
   - Keepers
   - Bearing retainer, bearing and race. Make sure that the bearing is fully pack with grease.

6. Coat the hub cap O-ring with wheel bearing grease and install the hub cap.
   - Capscrews and washers. Tighten the screws to 45 inch lbs. (5 Nm).

Front Axle Shaft, Bearing and Seal

REMOVAL & INSTALLATION

Silverado and Sierra

1. Before servicing the vehicle, refer to the precautions in the beginning of this section.

2. Remove or disconnect the following:
   - Halfshaft assembly
   - Front axle fluid

- Electrical connectors
- Axle shaft (output shaft) tube nuts from the bracket
- Bracket bolts from the frame. Do not remove the bracket. The bolts are removed in order to provide clearance.
- Axle shaft (output shaft) bolts from the carrier

➡ **Keep the open end of the tube up.**

- Axle shaft (output shaft) tube from the carrier. Ensure the spring is not lost during removal. In a vise, hold the axle shaft (output shaft) tube by the mounting flange.

3. Remove the following components:
   a. The shift shaft.
   b. The damper spring.
   c. The fork.
   d. The clip assembly.
4. Remove or disconnect the following:
   - Sleeve
   - Gear
   - Thrust washer
   - Axle shaft (output shaft). Tap out the axle shaft (output shaft) with a soft mallet.
   - Deflector by prying it out with a suitable tool
   - Seal by prying it out with a suitable tool
   - Bearing, using a slide hammer
5. Clean the parts in suitable solvent. Clean the gasket surfaces on the axle shaft (output shaft) tube and carrier housing.

To install:

6. Install or connect the following:
   - New bearing into the axle shaft tube using a driver. Apply axle lubricant to the bearing.
   - New seal using a driver. Coat the seal lips with axle lubricant.
   - Deflector
   - Axle shaft
   - Thrust washer. Use grease in order to hold the thrush washer in place. Ensure the tabs on the thrush washer align with the slot in the axle shaft tube.
   - Gear
   - Sleeve
   - Shift shaft
   - Damper spring
   - Fork
   - Clip assembly
7. Apply sealant GM P/N 12345739 or the equivalent to the carrier sealing surfaces.
8. Install or connect the following:
   - Spring into the carrier case
   - Axle shaft tube to the carrier

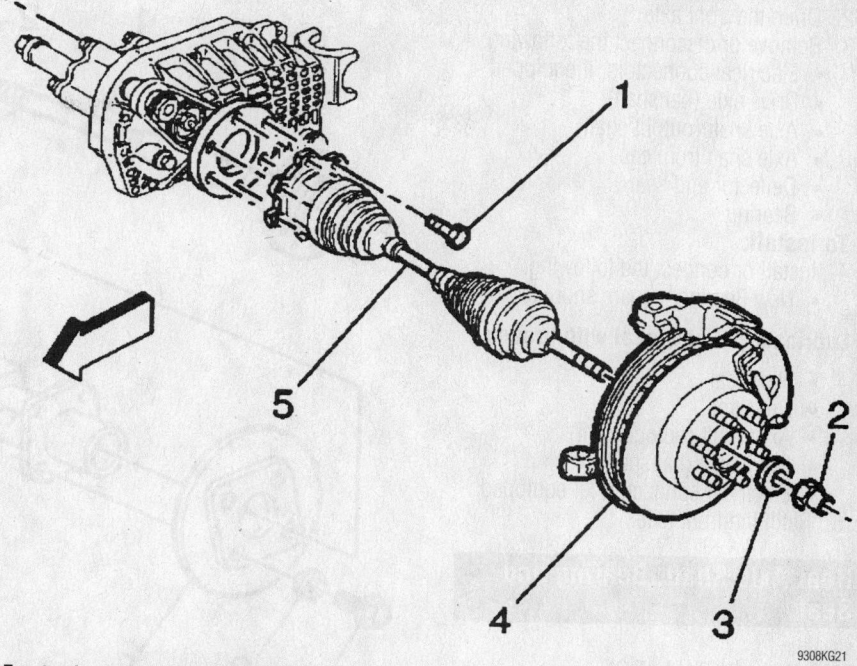

Front axle shaft removal—15 Series Silverado

9308KG21

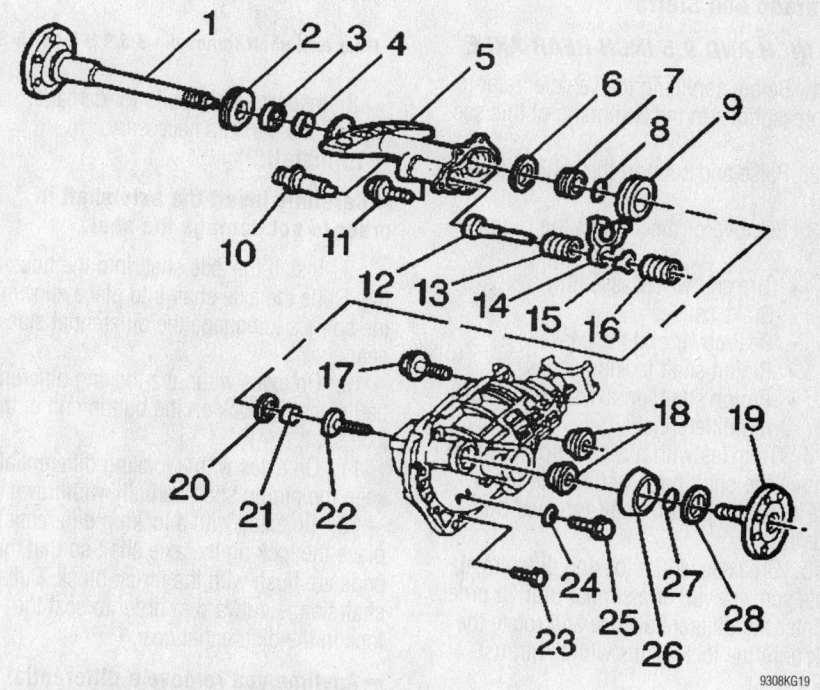

Front axle shaft exploded view—15 Series Silverado

9308KG19

- Axle shaft (output shaft) bolts and tighten to 30 ft. lbs. (40 Nm).
- Bracket bolts to the frame and tighten to 67 ft. lbs. (90 Nm).
- Axle shaft (output shaft) tube nuts to the bracket. Tighten the nuts to 75 ft. lbs. (100 Nm).
- Halfshaft assembly
- Electrical connectors

9. Fill the front differential with lubricant until the level is 0.5 in. (12mm) below the fill plug.

Express and Savana

1. Before servicing the vehicle, refer to the precautions in the beginning of this section.

2. Drain the front axle.
3. Remove or disconnect the following:
   - Electrical connectors, if equipped
   - Drive axle (halfshaft)
   - Axle shaft (output shaft)
   - Axle shaft from case
   - Deflector and seal
   - Bearing

To install:
4. Install or connect the following:
   - New Bearing, square shoulder in

➡ **Lubricate the new seal with grease.**

   - New seal
   - Deflector
   - Axle shaft (output shaft)
   - Drive axle (halfshaft)
   - Electrical connectors, if equipped
5. Refill the front axle.

Rear Axle Shaft, Bearing and Seal

REMOVAL & INSTALLATION

Silverado and Sierra

8.5 INCH AND 9.5 INCH REAR AXLE

1. Before servicing the vehicle, refer to the precautions in the beginning of this section.
2. Raise and support the vehicle on a hoist.
3. Remove or disconnect the following:
   - Tire and wheel assembly
   - Brake caliper
   - Rear cover and the gasket
   - Pinion shaft locking screw.
   - Pinion shaft, on axles without locking differential
4. On axles with a locking differential, remove the shaft part way. Rotate the case until the pinion shaft touches the housing.
5. On axles with a locking differential, use a screwdriver, or a similar tool, in order to enter the differential case and rotate the lock until the lock aligns with the thrust block.
6. Push the flange of the axle shaft toward the differential. Remove the lock from the button end of the axle shaft.

➡ **When removing the axle shaft, do not rotate the shaft. Rotating the shaft will misalign the gears. Misaligning the gears will make the assembly difficult.**

7. Remove the axle shaft from the housing.

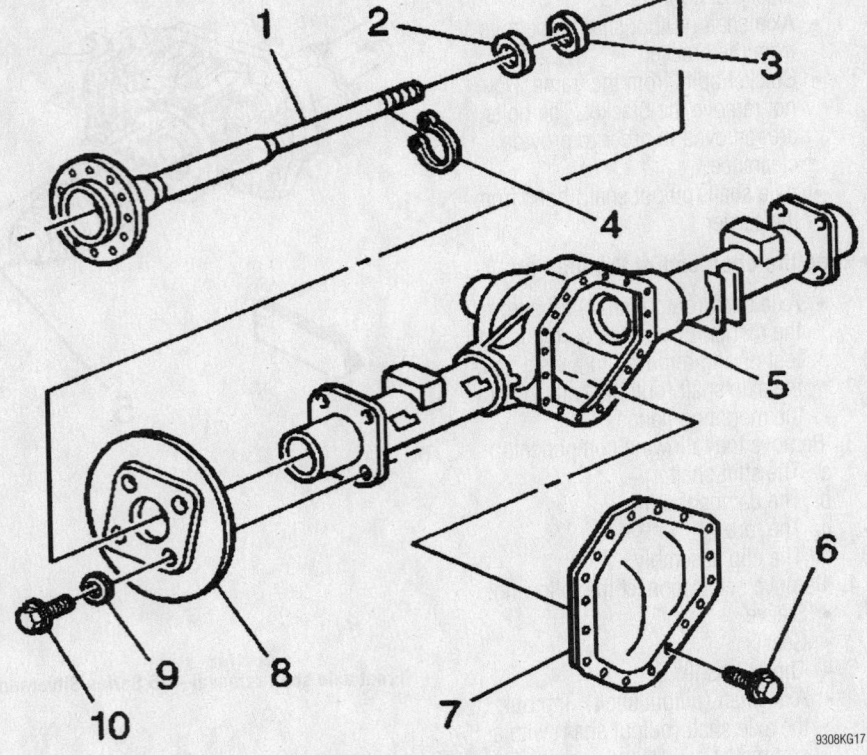

Rear axle shaft removal—8.5/9.5 inch 15 Series Silverado

9308KG17

8. Inspect all the parts for damage. Replace the parts as necessary.

To install:

➡ **Carefully insert the axle shaft in order to not damage the seal.**

9. Install the axle shaft into the housing. Slide the axle shaft into place allowing the splines to engage the differential side gear.
10. On axles without a locking differential, place the lock on the button end of the axle shaft.
11. On axles with a locking differential, keep the pinion shaft partially withdrawn.
12. On axles with a locking differential, place the lock on the axle shaft so that the ends are flush with the thrust block. Pull the shaft flange outward in order to seat the lock in the differential gear.

➡ **Anytime you remove a differential pinion shaft locking screw, coat the screw threads with Loctite® 242 before reinstalling the screws. The screw has an adhesive coating in order to prevent the screw from loosening in the case. Removing the screw removes the adhesive on the screw.**

13. Align the hole in the pinion shaft with the screw hole in the differential case.
14. Install or connect the following:

   - Pinion flange locking screw and tighten to 25 ft. lbs. (34 Nm).
   - Rear cover and the gasket
   - Brake caliper
   - Tire and wheel assembly
15. Fill the rear axle.
16. Remove the supports and lower the vehicle.

10.5 INCH REAR AXLE

1. Before servicing the vehicle, refer to the precautions in the beginning of this section.
2. Remove or disconnect the following:
   - Tire and wheel
   - Brake caliper
   - Brake rotor
   - Flange bolts
3. Lightly rap the axle shaft with a soft-faced hammer in order to loosen the shaft. Grip the rib on the axle shaft flange with a locking pliers. Twist the axle shaft flange in order to start the axle shaft removal. Remove the axle shaft from the tube.
4. Remove the gasket.
5. Clean the axle shaft flange and the outside face of the hub assembly. Inspect all the parts. Replace the parts as necessary.
To install:
6. Install or connect the following:
   - Gasket onto the axle shaft
   - Gasket and axle shaft into the tube.

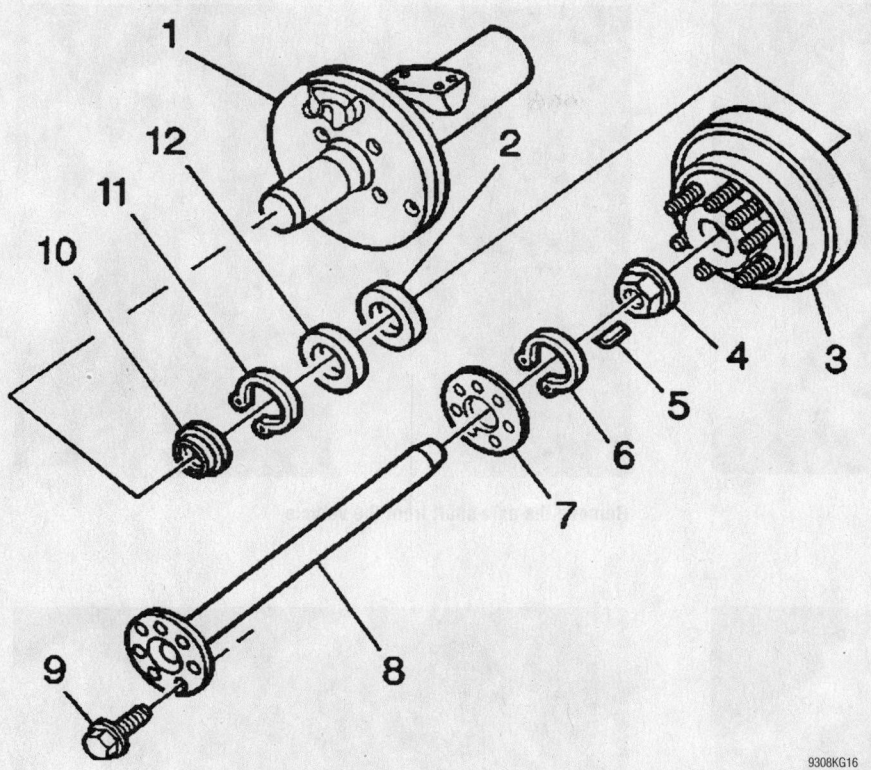

Rear axle shaft removal—10.5 inch 25 Series Silverado

Ensure the shaft splines mesh into the differential side gear. Align the holes in the axle flange and the gasket with the holes in the hub.

- Axle flange bolts and tighten to 110 ft. lbs. (150 Nm).
- Rotor
- Caliper
- Wheel and tire

Express and Savana

SEMI-FLOATING NON-LOCKING DIFFERENTIALS

1. Before servicing the vehicle, refer to the precautions in the beginning of this section.
2. Remove the wheels and brake drums.
3. Remove the differential cover
4. Turn the differential until you can reach the differential pinion shaft lockscrew. Remove the lockscrew and the pinion shaft.
5. Push in on the axle end. Remove the C-lock from the inner (button) end of the shaft.
6. Remove the shaft, being careful of the oil seal.
7. You can pry the oil seal out of the housing by placing the inner end of the axle shaft behind the steel case of the seal, then prying it out carefully.

8. A puller or a slide hammer is required to remove the bearing from the housing.

To install:

9. Pack the new or reused bearing with wheel bearing grease and lubricate the cavity between the seal lips with the same grease.
10. The bearing has to be driven into the housing. Don't use a drift, you might cock the bearing in its bore. Use a piece of pipe or a large socket instead. Drive only on the outer bearing race. In a similar manner, drive the seal in flush with the end of the tube.
11. Slide the shaft into place, turning it slowly until the splines are engaged with the differential. Be careful of the oil seal.
12. Install the C-lock on the inner axle end. Pull the shaft out so that the C-lock seats in the counterbore of the differential side gear.
13. Position the differential pinion shaft through the case and the pinion gears, aligning the lockscrew hole. Install the lockscrew.
14. Install the cover with a new gasket and tighten the bolts evenly in a criss-cross pattern.
15. Fill the axle with lubricant.
16. Replace the brake drums and wheels.

SEMI-FLOATING LOCKING DIFFERENTIALS

This axle uses a thrust block on the differential pinion shaft.

1. Before servicing the vehicle, refer to the precautions in the beginning of this section.
2. Remove the wheels and brake drums.
3. Clean off the differential cover area, loosen the cover to drain the lubricant, and remove the cover.
4. Rotate the differential case so that you can remove the lockscrew and support the pinion shaft so it can't fall into the housing. Remove the differential pinion shaft lockscrew.
5. Carefully pull the pinion shaft partway out and rotate the differential case until the shaft touches the housing at the top.
6. Use a screwdriver to position the C-lock with its open end directly inward. You can't push in the axle shaft till you do this.
7. Push the axle shaft in and remove the C-lock.
8. Remove the shaft, being careful of the oil seal.
9. You can pry the oil seal out of the housing by placing the inner end of the axle shaft behind the steel case of the seal, then prying it out carefully.
10. A puller or a slide hammer is required to remove the bearing from the housing.

To install:

11. Pack the new or reused bearing with wheel bearing grease and lubricate the cavity between the seal lips with the same grease.
12. The bearing has to be driven into the housing. Don't use a drift, you might cock the bearing in its bore. Use a piece of pipe or a large socket instead. Drive only on the outer bearing race. In a similar manner, drive the seal in flush with the end of the tube.
13. Slide the shaft into place, turning it slowly until the splines are engaged with the differential. Be careful of the oil seal.
14. Keep the pinion shaft partway out of the differential case while installing the C-lock on the axle shaft. Put the C-lock on the axle shaft and carefully pull out on the axle shaft until the C-lock is clear of the thrust block.
15. Position the differential pinion shaft through the case and the pinion gears, aligning the lockscrew hole. Install the lockscrew.
16. Install the cover with a new gasket and tighten the bolts evenly in a criss-cross pattern.

Remove the differential pinion shaft lockscrew

Remove the axle shaft from the vehicle

Remove the pinion shaft

Use a puller to remove the oil seal

Remove the C-lock from the inner (button) end of the shaft

Install the oil seal using a seal installer

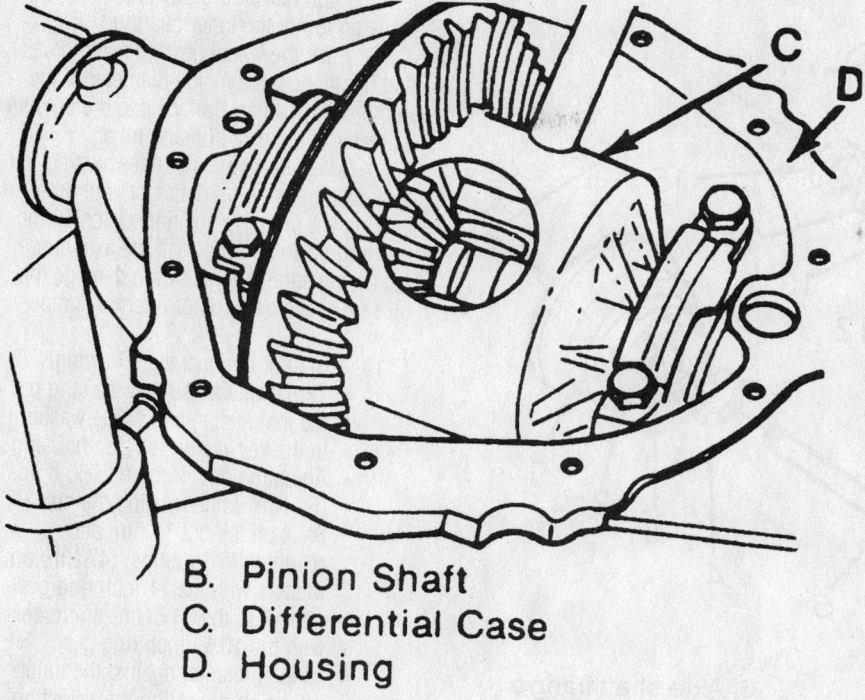

B. **Pinion Shaft**
C. **Differential Case**
D. **Housing**

Positioning the case for the best clearance—semi-floating axle w/locking differential

84907357

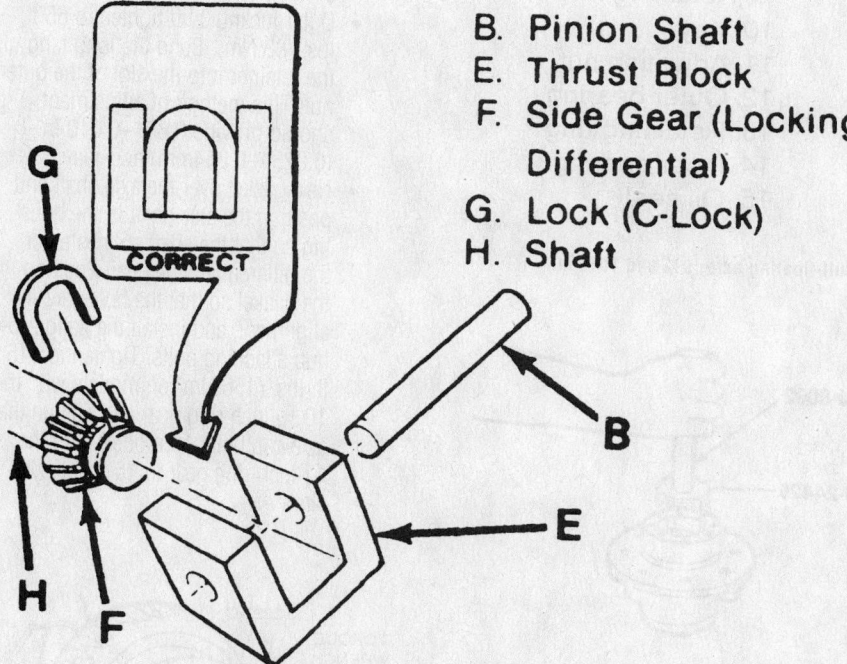

B. **Pinion Shaft**
E. **Thrust Block**
F. **Side Gear (Locking Differential)**
G. **Lock (C-Lock)**
H. **Shaft**

84907358

Aligning the lock—semi-floating axle w/locking differential

17. Fill the axle with lubricant.
18. Replace the brake drums and wheels.

FULL-FLOATING AXLES

The procedures are the same for locking and non-locking axles.

The best way to remove the bearings from the wheel hub is with an arbor press. Use of a press reduces the chances of damaging the bearing races, cocking the bearing in its bore, or scoring the hub walls. A local machine shop is probably equipped with the tools to remove and install bearings and seals. However, if one is not available, the hammer and drift method outlined can be used.

1. Before servicing the vehicle, refer to the precautions in the beginning of this section.
2. Support the axles on jackstands.
3. Remove the wheels.
4. Remove the bolts and lock washers that attach the axle shaft flange to the hub.
5. Rap on the flange with a soft faced hammer to loosen the shaft. Grip the rib on the end of the flange with a pair of locking pliers and twist to start shaft removal. Remove the shaft from the axle tube.
6. The hub and drum assembly must be removed to remove the bearings and oil seals. You will need a large socket to remove and later adjust the bearing adjustment nut. There are also special tools available.
7. Remove or disconnect the following:
   - Tang of the locknut retainer from the slot or slat of the locknut
   - Locknut from the housing tube
   - Tang of the retainer from the slot or flat of the adjusting nut
   - Retainer from the housing tub
   - Adjusting nut from the housing tube
   - Thrust washer from the housing tube
   - Hub and drum straight off the axle housing
   - Oil seal and discard.
8. Use a hammer and a long drift to knock the inner bearing, cup, and oil seal from the hub assembly.
   - Outer bearing snapring with a pair of pliers. It may be necessary to tap the bearing outer race away from the retaining ring slightly by tapping on the ring to remove the ring.
   - Outer bearing from the hub with a hammer and drift

To install:
9. Place the outer bearing into the hub. The larger outside diameter of the bearing should face the outer end of the hub. Drive the bearing into the hub using a washer that will cover both the inner and outer races of the bearing. Place a socket on top of this washer, then drive the bearing into place with a series of light taps. If available, an arbor press should be used for this job.
10. Drive the bearing past the snapring groove, and install the snapring. Then, turning the hub assembly over, drive the bearing back against the snapring. Protect the bearing by placing a washer on top of it. You can use the thrust washer that fits between

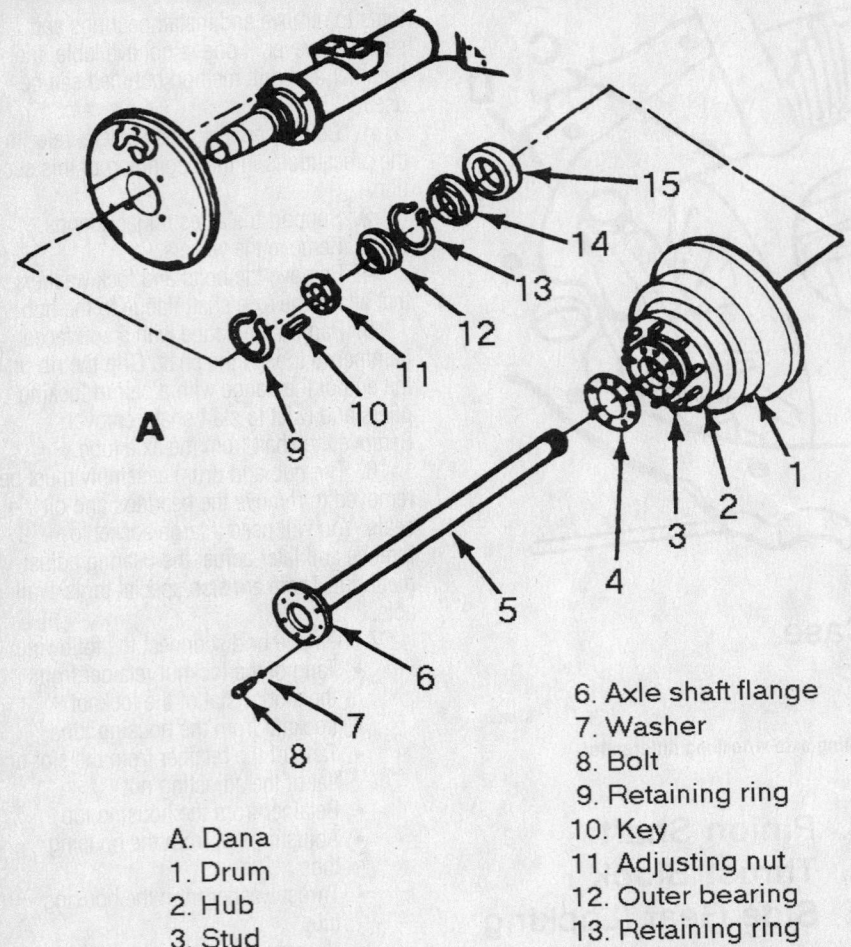

A. Dana
1. Drum
2. Hub
3. Stud
4. Gasket
5. Shaft
6. Axle shaft flange
7. Washer
8. Bolt
9. Retaining ring
10. Key
11. Adjusting nut
12. Outer bearing
13. Retaining ring
14. Inner bearing
15. Oil seal

84907359

Exploded view of the axle, hub and drum assembly—full-floating axle, 9¾ and 10½ in.

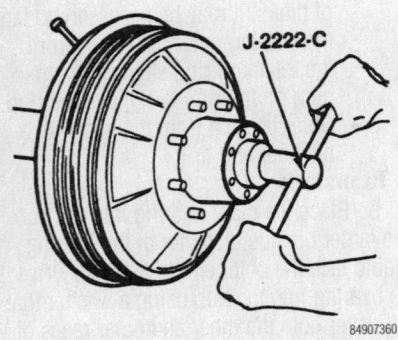

84907360

Removing the bearing adjusting nut—full-floating axle, 9¾ and 10½ in.

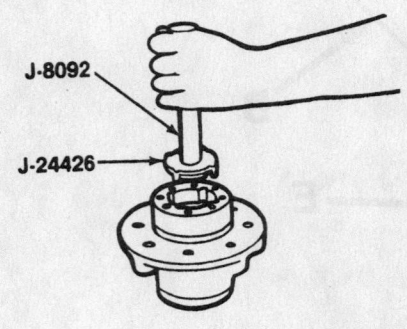

84907361

Removing the bearing outer cup—full-floating axle, 9¾ and 10½ in.

it in the hub bore. Carefully press it into place on top of the inner bearing.

13. Pack the wheel bearings with grease, and lightly coat the inside diameter of the hub bearing contact surface and the outside diameter of the axle housing tube.

14. Make sure that the inner bearing, oil seal, axle housing oil deflector, and outer bearing are properly positioned. Install the hub and drum assembly on the axle housing, being careful so as not to damage the oil seal or dislocate other internal components.

15. Install or connect the following:
- Thrust washer so that the tang on the inside diameter of the washer is in the keyway on the axle housing
- Adjusting nut. Tighten to 50 ft. lbs. (68 Nm) while rotating the hub. Back off the nut ¼ turn and retighten to 35 ft. lbs. (47 Nm) on models with the 11 inch ring gear and 13 ft. lbs. (17 Nm) on models with the 10 ½ inch ring gear.
- Tanged retainer against the inner adjusting nut. Align the adjusting nut so that the short tang of the retainer will engage the nearest slot on the adjusting nut.
- Outer locknut and tighten to 65 ft. lbs. (88 Nm). Bend the long tang of the retainer into the slot of the outer nut. This method of adjustment should provide 0.001–0.010 in. (0.0254–0.254mm) end-play.
- New gasket over the axle shaft and position the axle shaft in the housing so that the shaft splines enter the differential side gear. Position the gasket so that the holes are in alignment, and install the flange-to-hub attaching bolts. Tighten to 115 ft. lbs. (156 Nm) on models with the 10 ½ inch ring gear and tighten the axle cap bolts on models with the 11 inch ring gear to 15 ft. lbs. (20 Nm).

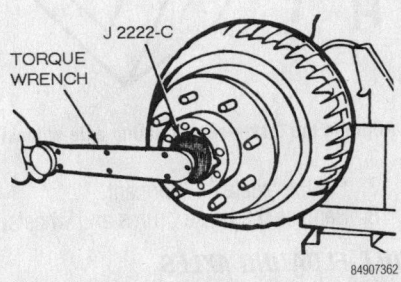

84907362

Tightening the adjusting nut—full-floating axle, 9¾ and 10½ in.

the bearing and the adjusting nut for the job.

11. Place the inner bearing into the hub. The thick edge should be toward the shoulder in the hub. Press the bearing into the hub until it seats against the shoulder,

using a washer and socket as outlined earlier. Make certain that the bearing is not cocked and that it is fully seated on the shoulder.

12. Pack the cavity between the oil seal lips with wheel bearing grease, and position

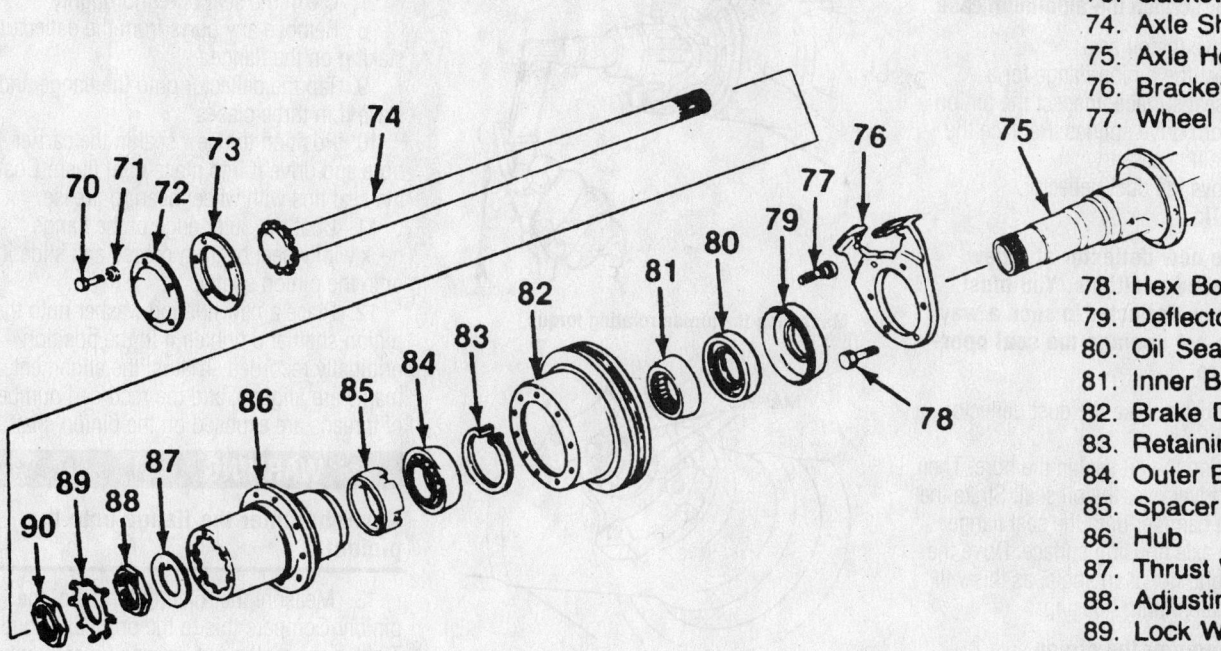

73. Gasket
74. Axle Shaft
75. Axle Housing
76. Bracket
77. Wheel Stud

78. Hex Bolt
79. Deflector
80. Oil Seal
81. Inner Bearing
82. Brake Disc
83. Retaining Ring
84. Outer Bearing
85. Spacer
86. Hub
87. Thrust Washer
88. Adjusting Nut
89. Lock Washer
90. Nut

84907363

Exploded view of the axle and hub assembly—full-floating axle, 12 in.

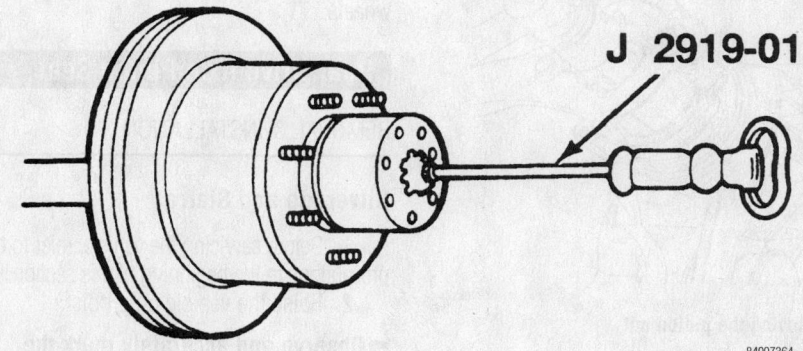

J 2919-01

84907364

Removing the axle shaft—full-floating axle, 12 in.

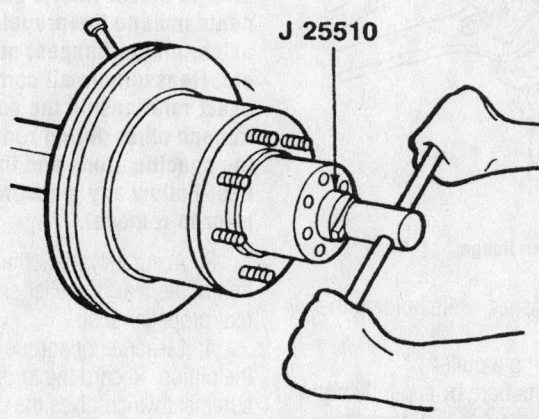

J 25510

84907365

Removing the wheel bearing nut—full-floating axle, 12 in.

➡ To prevent lubricant from leaking through the flange holes, apply a non-hardening sealer to the bolt threads. Use the sealer sparingly.

• Wheels

Front Drive Axle Pinion Seal

REMOVAL & INSTALLATION

Silverado and Sierra

1. Before servicing the vehicle, refer to the precautions in the beginning of this section.
2. Raise the vehicle on a hoist.
3. Remove the propeller shaft from the axle.
4. Tie the propeller shaft to a frame rail or the crossmember.
5. Measure the torque required in order to rotate the pinion. Record the torque value for reassembly.
6. Scribe a line on the pinion stem, the pinion nut and the companion flange. Record the number of exposed threads on the pinion stem.
7. Remove the nut.
8. Position tool J8614-01 on the flange so that the 4 notches on the tool face the flange.
9. Remove the flange. Use the special nut and the forcing screw.

➡Carefully pry the seal from the bore. Do not distort or scratch the aluminum case.

10. Remove the oil seal.

11. Inspect the pinion flange for a smooth oil seal surface. Inspect the pinion flange for worn drive splines. Replace the pinion flange if necessary.

12. Remove the dust deflector.

To install:

➡Stake the new deflector at 3 new equally spaced positions. You must stake the new deflector in such a way that you do not damage the seal operating surface.

13. Install and stake the dust deflector on the flange.

14. Position the oil seal in the bore. Then place the a driver over the oil seal. Strike the driver with a hammer until the seal flange seats on the axle housing surface. Drive the seal in straight, not at an angle, as this will damage the aluminum housing.

➡Do not hammer the pinion flange/yoke onto the pinion shaft. Pinion components may be damaged if the pinion flange/yoke is hammered onto the pinion shaft.

15. Install the flange onto the pinion using tool J8614-01. Place the washer and a new nut on the pinion threads. Tighten the nut to the original scribed position using the scribe marks and the exposed threads as reference.

16. Measure the rotating torque of the pinion. Compare the measurement with the rotating torque recorded earlier. Tighten the pinion nut by small increments until the torque required in order to rotate the pinion is 3 inch lbs. (0.35 Nm) greater than the original torque.

17. Install the propeller shaft.

18. Lower the vehicle.

Express and Savana

1. Before servicing the vehicle, refer to the precautions in the beginning of this section.

2. Raise and support the front end on jackstands.

3. Matchmark and disconnect the front driveshaft at the carrier.

4. Remove or disconnect the following:
- Wheels
- Calipers and wire them up, out of the way

5. Position an inch pound torque wrench on the pinion nut. Measure the torque needed to rotate the pinion one full revolution. Record the figure.

6. Matchmark the pinion flange, shaft and nut. Count and record the number of exposed threads on the pinion shaft.

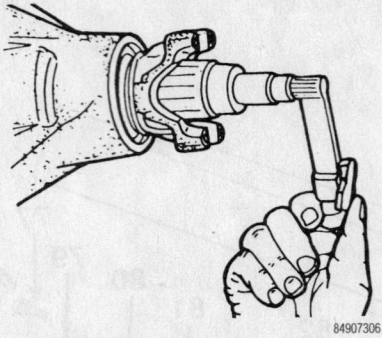

Measuring the pinion rotating torque

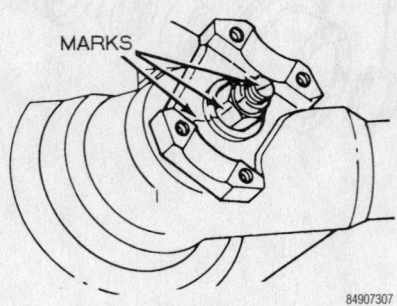

View of the scribed marks

Removing the pinion nut

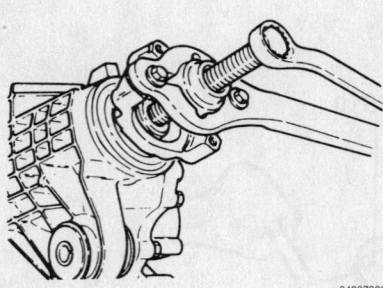

Removing the pinion flange

- Nut and washer, while holding the flange
- Flange, using a puller
- Seal from its bore by carefully prying it. Be careful to avoid scratching the seal bore.
- Deflector from the flange

To install:

7. Clean the seal bore thoroughly.

8. Remove any burrs from the deflector staking on the flange.

9. Tap the deflector onto the flange and stake it in three places.

10. Position the new seal in the carrier bore and drive it into place until flush. Coat the seal lips with wheel bearing grease.

11. Coat the outer edge of the flange neck with wheel bearing grease and slide it onto the pinion shaft.

12. Place a new nut and washer onto the pinion shaft and tighten it to the position originally recorded. That is, the alignment marks are aligned, and the recorded number of threads are exposed on the pinion shaft.

❊❊ WARNING

Never hammer the flange onto the pinion!

13. Measure the rotating torque of the pinion. Compare this to the original torque. Tighten the pinion nut, in small increments, until the rotating torque is 3 inch lbs. (0.35 Nm) GREATER than the original torque.

14. Install the driveshaft.

15. Install the calipers and install the wheels.

Rear Drive Axle Pinion Seal

REMOVAL & INSTALLATION

Silverado and Sierra

1. Before servicing the vehicle, refer to the precautions in the beginning of this section.

2. Raise the vehicle on a hoist.

➡Observe and accurately mark the positions of all driveline components relative to the propeller shaft and axles prior to disassembly. These components include the propeller shaft, drive axles, pinion flanges, output shafts, etc. Reassemble all components in the exact relationship the components had to each other during removal. Follow the specifications and the torque values. Follow any measurements made prior to removal.

3. Accurately mark the installed position of the rear propeller shaft. Remove the rear propeller shaft.

4. Measure the torque required to turn the pinion. Record the torque number measurement which gives the combined pinion bearing, seal, carrier bearing, axle bearing and seal preload.

5. Make and accurate alignment mark on the pinion flange. Record the number of exposed threads on the pinion stem.

6. Remove the pinion flange nut and the washer. Use a container in order to catch any lubricant.

➡**Use care not to damage any of the machined surfaces.**

7. Remove the pinion flange.

➡**The pinion flange has an oil seal that is part of the pinion flange assembly. The pinion flange must be inspected to ensure that the seal is not damaged.**

8. Pry the oil seal from the bore.

9. Thoroughly clean any foreign material from the contact area. Replace any parts as necessary.

To install:

10. Lubricate the cavity between the lips of the oil seal with wheel bearing lubricant.

11. Install the oil seal into the bore using a driver.

➡**Do not hammer the pinion flange onto the pinion stem.**

12. Install the pinion flange. Use the alignment marks in the installation of the pinion flange.

13. Install the washer and a new nut. Tighten the nut on the pinion stem as close as possible to the alignment marks without going past the marks. Use the alignment marks and the thread count as a reference. Tighten the nut a little at a time. Turn the pinion flange several times after each tightening in order to seat the rollers.

➡**If the recorded preload torque value was less than 3 ft. lbs. (4 Nm), reset the torque specification to 3–5 ft. lbs. (4–7 Nm).**

14. Measure the torque required to rotate the pinion flange. Compare the measured torque with the recorded value. Continue tightening the pinion nut and measuring the torque until you achieve the recorded value.

15. Align the propeller shaft with the alignment marks. Connect the propeller shaft.

16. Install the retainers and the bolts. Tighten the bolts to 15 ft. lbs. (20 Nm).

17. Fill the rear axle.

18. Lower the vehicle.

Express and Savana

SEMI-FLOATING AXLES

1. Before servicing the vehicle, refer to the precautions in the beginning of this section.

2. Raise and support the truck on jack-
stands. It would help to have the front end slightly higher than the rear to avoid fluid loss.

3. Matchmark and remove the driveshaft.

4. Release the parking brake.

5. Remove the rear wheels. Rotate the rear wheels by hand to make sure that there is absolutely no brake drag. If there is brake drag, remove the drums.

6. Using a torque wrench on the pinion nut, record the force needed to rotate the pinion.

7. Matchmark the pinion shaft, nut and flange. Count the number of exposed threads on the pinion shaft

8. Instal a holding tool on the pinion. A very large adjustable wrench will do, or, if one is not available, set the parking brake as tightly as possible.

9. Remove the pinion nut.

10. Slide the flange off of the pinion. A puller may be necessary.

11. Centerpunch the oil seal to distort it and pry it out of the bore. Be careful to avoid scratching the bore.

To install:

12. Pack the cavity between the lips of the seal with lithium-based chassis lube.

13. Position the seal in the bore and carefully drive it into place. A seal installer is VERY helpful in doing this.

14. Pack the cavity between the end of the pinion splines and the pinion flange with Permatex No.2 sealer, or equivalent non-hardening sealer.

15. Place the flange on the pinion and push it on as far as it will go.

16. Install the pinion washer and nut on the shaft and force the pinion into place by turning the nut.

✳✳ WARNING

Never hammer the flange into place!

17. Tighten the nut until the exact number of threads previously noted appear and the matchmarks align.

18. Measure the rotating torque of the pinion under the same circumstances as before. Compare the two readings. As necessary, tighten the pinion nut in VERY small increments until the torque necessary to rotate the pinion is 3 inch lbs. (0.35 Nm) higher than the originally recorded torque.

19. Install the driveshaft.

FULL-FLOATING AXLES

1. Before servicing the vehicle, refer to the precautions in the beginning of this section.

2. Raise and support the truck on jack-
stands. It would help to have the front end slightly higher than the rear to avoid fluid loss.

3. Matchmark and remove the driveshaft.

4. Matchmark the pinion shaft, nut and flange. Count the number of exposed threads on the pinion shaft.

5. Install a holding tool on the pinion. A very large adjustable wrench will do, or, if one is not available, set the parking brake as tightly as possible.

6. Remove the pinion nut.

7. Slide the flange off of the pinion. A puller may be necessary.

8. Centerpunch the oil seal to distort it and pry it out of the bore. Be careful to avoid scratching the bore.

To install:

9. Pack the cavity between the lips of the seal with lithium-based chassis lube.

10. Position the seal in the bore and carefully drive it into place. A seal installer is VERY helpful in doing this.

14. Place the flange on the pinion and push it on as far as it will go.

✳✳ WARNING

Never hammer the flange into place!

12. Install the pinion washer and nut on the shaft and force the pinion into place by turning the nut.

13. On models with the 11 inch ring gear, tighten the nut to 440–500 ft. lbs. (596–678 Nm)

14. On models with the 10 ½ inch ring gear Tighten the nut until the exact number of threads previously noted appear and the matchmarks align.

15. Install the driveshaft.

Front Drive Axle Differential Carrier

REMOVAL & INSTALLATION

Silverado and Sierra

1. Before servicing the vehicle, refer to the precautions in the beginning of this section.

2. Remove or disconnect the following:

- Front axle fluid
- Front propeller shaft
- Left and the right drive axle wheel drive shaft
- Axle tube nuts from the bracket
- Wiring at the axle
- Vent hose at the axle
- Carrier assembly lower mounting bolt and the nut

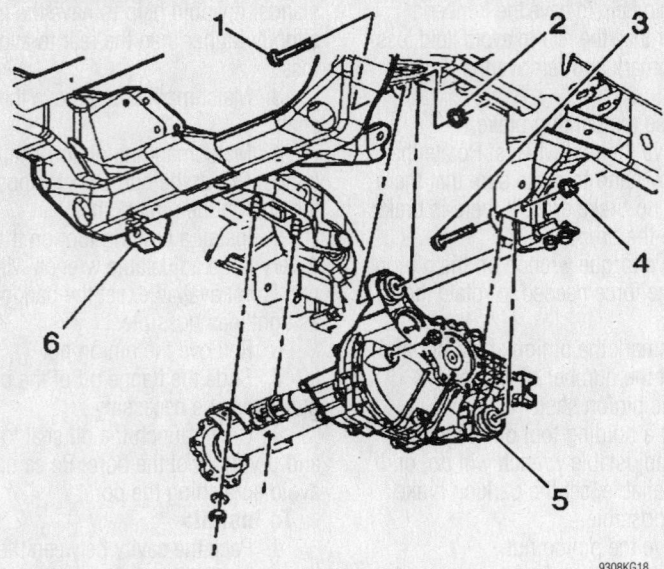

Front differential carrier removal—15 Series Silverado

- Idler arm from the relay rod
- Pitman arm from the relay rod

3. Attach a transmission jack to the carrier assembly.

- Upper carrier assembly mounting bolt and the nut
- Carrier assembly from the vehicle

To install:

4. Install or connect the following:

- Carrier assembly in the vehicle
- Carrier assembly upper mounting bolt and nut
- Lower carrier assembly mounting bolt and nut. Tighten to 75 ft. lbs. (100 Nm).
- Pitman arm to the relay rod
- Idler arm to the relay rod
- Axle tube nuts to the bracket. Tighten the bolts to 75 ft. lbs. (100 Nm).
- Vent hose
- Wiring to the axle
- Left and the right drive axle wheel drive shaft
- Front propeller shaft to the pinion flange

5. Fill the front differential with lubricant.

Express and Savana

1. Before servicing the vehicle, refer to the precautions in the beginning of this section.

2. Remove or disconnect the following:

- Wheels
- Skid plate
- Fluid

3. Matchmark and remove the front driveshaft.

- Right axle shaft at the tube flange
- Left axle shaft at the carrier flange

➡**Wire both axle shafts out of the way.**

- Connectors at the indicator switch and actuator
- Carrier vent hose
- Axle tube-to-frame bolts, washers and nuts
- Lower carrier mounting bolt
- Right side inner tie rod end at the relay rod

➡**Depending on the model, it may be necessary to remove the engine oil filter.**

4. Support the carrier on a floor jack
- Upper carrier mounting bolt
5. Lower the carrier assembly from the truck.

To install:

6. Raise the carrier into position.
7. Install or connect the following:

- Upper and lower carrier mounting bolt, washers and nut. Tighten the bolts to 80 ft. lbs. (110 Nm).
- Oil filter
- Tie rod end. Tighten the nut to 35 ft. lbs. (47 Nm).
- Axle tube-to-frame bolts, washers and nuts. Tighten the nuts to 75 ft. lbs. (100 Nm) for 15 and 25 series; 107 ft. lbs. (145 Nm) for 35 series.
- Vent hose
- Wiring
- Axle shafts at the flanges. Tighten the bolts to 59 ft. lbs. (80 Nm).
- Driveshaft. Tighten the bolts to 15 ft. lbs. (20 Nm).

- Gear oil
- Wheels

8. Add any engine oil lost when the filter was removed.

Rear Drive Axle Housing

REMOVAL & INSTALLATION

Silverado and Sierra

1. Before servicing the vehicle, refer to the precautions in the beginning of this section.

2. Remove or disconnect the following:

- Axle lubricant
- Propeller shaft
- Wheel assemblies
- Parking brake cable
- Brake calipers
- Shock absorbers from the axle brackets
- Vent hose from the rear axle vent fitting
- Nuts and the washers from the U-bolts
- U-bolts, the spring plates and the spacers form the axle assembly

3. Lower the axle assembly.

To install:

4. Place the rear axle assembly under the vehicle. Align the rear axle assembly with the springs. Connect the spacers, the spring plates and the U-bolts to the rear axle. Raise the rear axle assembly into position.

5. Install or connect the following:

- Washers and nuts to the U-bolts. Tighten the nuts to 59 ft. lbs. (80 Nm) first, then to 89 ft. lbs. (120 Nm).
- Vent hose to the rear axle vent fitting
- Shock absorbers to the rear axle
- Brake calipers
- Parking brake cable
- Wheel assemblies
- Propeller shaft

6. Fill the rear axle.
7. Bleed the brake system.
8. Remove the supports and lower the vehicle.

Express and Savana

1. Before servicing the vehicle, refer to the precautions in the beginning of this section.

2. Drain the lubricant from the axle housing

3. Remove or disconnect the following:
- Driveshaft

- Wheel, the brake drum or hub and the drum assembly
- Parking brake cable from the lever and at the brake flange plate
- Hydraulic brake lines from the connectors
- Shock absorbers from the axle brackets
- Vent hose from the axle vent fitting, if equipped
- Height sensing and brake proportional valve linkage, if equipped

- Stabilizer shaft, if equipped
4. Support the axle assembly with a jack.
- U-bolts
- Spring plates and spacers
- Axle assembly

To install:
5. Raise the axle assembly into position.
6. Install or connect the following:
- U-bolts
- Spring plates and spacers
- Nuts and washers on the U-bolts. Torque to 81 ft. lbs. (110 Nm).

- Stabilizer shaft, if equipped
- Height sensing and brake proportional valve linkage, if equipped
- Vent hose at the axle vent fitting, if equipped
- Shock absorbers at the axle brackets
- Hydraulic brake lines
- Parking brake cable
- Wheels
- Driveshaft
7. Fill the axle housing.

STEERING AND SUSPENSION

Air Bag

✳✳ CAUTION

Some vehicles are equipped with an air bag system. The system must be disabled before performing service on or around system components, steering column, instrument panel components, wiring and sensors. Failure to follow safety and disabling procedures could result in accidental air bag deployment, possible personal injury and unnecessary system repairs.

PRECAUTIONS

Several precautions must be observed when handling the inflator module to avoid accidental deployment and possible personal injury.

- Never carry the inflator module by the wires or connector on the underside of the module
- When carrying a live inflator module, hold securely with both hands, and ensure that the bag and trim cover are pointed away
- Place the inflator module on a bench or other surface with the bag and trim cover facing up
- With the inflator module on the bench, never place anything on or close to the module that may be thrown in the event of an accidental deployment

DISARMING

1. Turn the front wheels to the straight-ahead position.
2. Turn the ignition switch to the **LOCK** position and remove the key.

➡ **If the key is in the RUN position when the Air Bag fuse is removed or open**

(blown), the Air Bag warning lamp in the dash will light up. This is normal operation, not a sign of a malfunction.

3. Remove the Air Bag fuse from the fuse panel.
4. Remove the drivers side knee bolster and unplug the yellow 2-pin connector at

the base of the steering column to disarm the driver's side Air Bag. Remove the passenger side knee bolster and unplug the yellow 2-pin connector to disable the passenger's side Air Bag.

5. Reverse the procedure to arm the Air Bag restraint system.

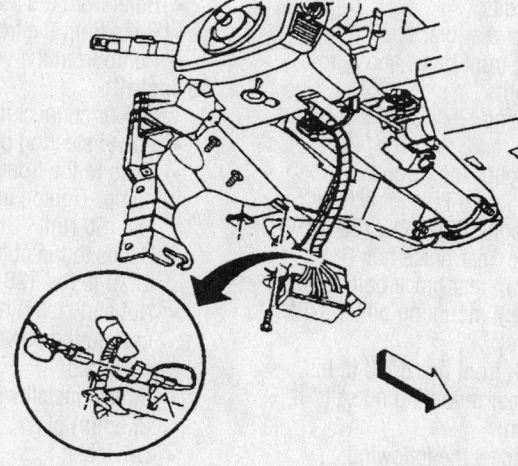

Typical air bag connector location—driver's side

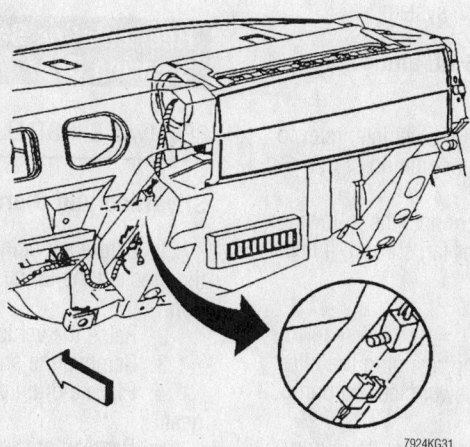

Typical air bag connector location—passenger's side

Power Steering Pump

REMOVAL & INSTALLATION

4.3L, 5.0L, 5.7L, 6.5L and 7.4L Engines

1. Before servicing the vehicle, refer to the precautions in the beginning of this section.
2. Remove or disconnect the following:
   • Hoses at the pump.

➡When the hoses are disconnected, secure the ends in a raised position to prevent leakage. Cap the ends of the hoses and pump fittings to prevent the entrance of dirt.

   • Pump drive belt, by loosening the tensioner
   • Pulley with a puller such as J–29785–A
3. Remove the following fasteners:
   • 4.3L, 5.0L and 5.7L engines: front mounting bolts
   • 7.4L engine: rear brace
   • 6.5L diesel: front brace and rear mounting nuts
4. Lift out the pump.

To install:
5. Observe the following torques:
   • 4.3L, 5.0L, and 5.7L engine front mounting bolts: 37 ft. lbs. (50 Nm)
   • 7.4L engine, rear brace nut: 61 ft. lbs. (82 Nm); rear brace bolt: 24 ft. lbs. (32 Nm); mounting bolts: 37 ft. lbs. (50 Nm)
   • 6.5L diesel, front brace: 30 ft. lbs. (40 Nm); rear mounting nuts: 17 ft. lbs. (23 Nm)
6. Install or connect the following:
   • Pulley with J–25033–B
   • Drive belt
   • Hoses
7. Fill and bleed the system.

4.8L, 5.3L, 6.0L, 6.6L and 8.1L Engines

1. Before servicing the vehicle, refer to the precautions in the beginning of this section.
2. Remove or disconnect the following:
   • Upper radiator fan shroud, if necessary
   • Drive belt
   • Pulley
   • Nut and clamp retaining the filler neck to the power steering pump, if equipped
3. Place a drain pan under the pump
   • Hoses from the pump

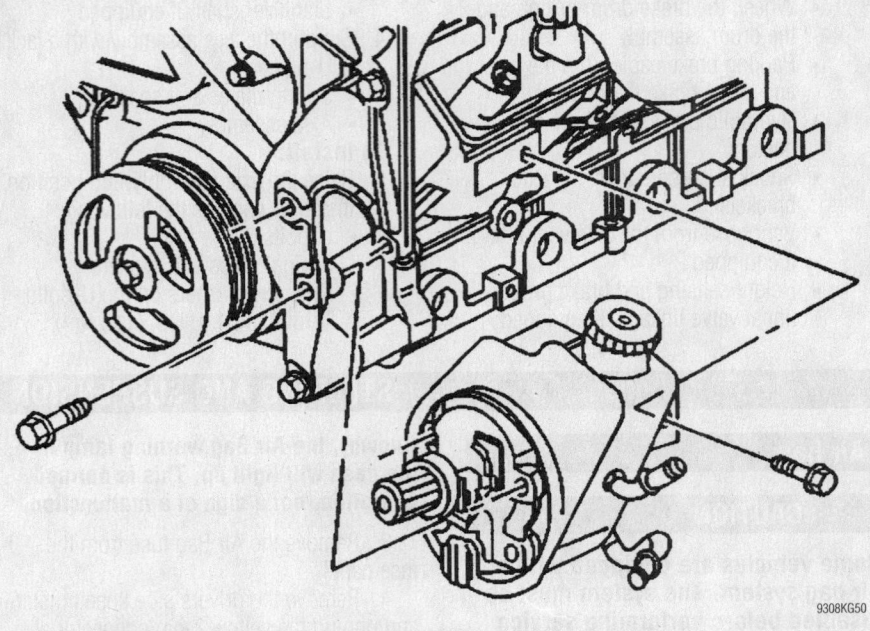

Power steering pump—4.8L, 5.3L and 6.0L engines shown

9308KG50

   • Bolts from the rear of the pump
   • Bolts from the front of the pump
   • Pump from the vehicle

To install:
4. Install or connect the following:
   • Power steering pump
   • Bolts to the front and the rear of the pump. Tighten the bolts to 37 ft. lbs. (50 Nm)
   • Hoses to the pump. Tighten the nut to 20 ft. lbs. (28 Nm)
   • Nut and clamp retaining the filler neck to the power steering pump, if equipped
   • Pulley; install with 0.020 in. (0.5mm) play
   • Drive belt
   • Upper radiator shroud
5. Fill and bleed the power steering system.

Recirculating Ball Power Steering Gear

REMOVAL & INSTALLATION

Silverado and Sierra

1. Before servicing the vehicle, refer to the precautions in the beginning of this section.
2. Raise the vehicle.
3. Remove the shield.
4. Place a drain pan below the steering gear.
5. Remove or disconnect the following:
   • Hoses from the steering gear

   • Intermediate shaft from the steering gear
   • Pitman arm from the relay rod
   • Steering gear frame bolts and the steering gear

To install:
6. Place the steering gear in position.
7. Install or connect the following:
   • Steering gear to the frame bolts and tighten to 100 ft. lbs. (135 Nm)
   • Pitman arm
   • Intermediate shaft
8. Remove the plugs and the caps from the steering gear and the hoses.
   • Hoses to the steering gear. Tighten the hose connection to 20 ft. lbs. (28 Nm).
   • Shield
9. Fill and bleed the system.
10. Lower the vehicle.

Express and Savana

These vehicles use a conventional power steering gear with a recirculating ball system. All tubes, hoses and fittings should be inspected for leakage at regular intervals. Fittings must be tight. Be sure the clips, clamps and supporting tubes and hoses are in place and properly secured. Inspect the hoses with the wheels in the straight-ahead position. Then, turn the wheels fully to the left and right while observing the movement of the hoses. Correct any hose contact with other parts of the vehicle that could cause chafing or wear. Power steering hoses and pipes should not be twisted, kinked or tightly bent. The hoses should have suffi-

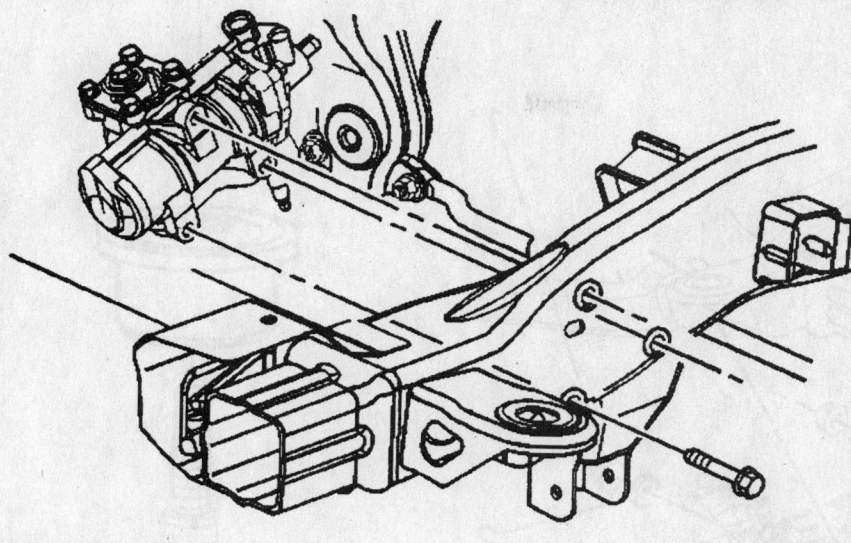

Recirculating ball gear—Silverado

9308KG48

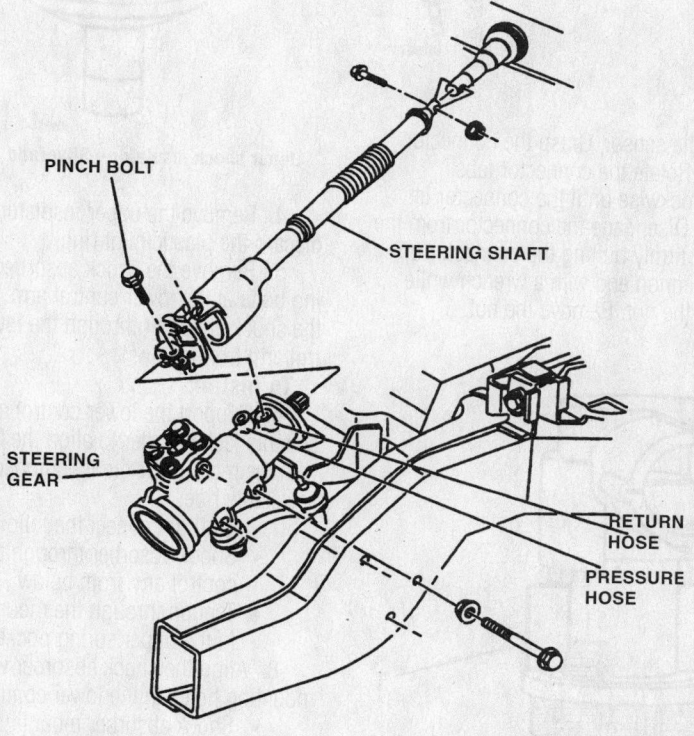

PINCH BOLT

STEERING SHAFT

STEERING GEAR

RETURN HOSE

PRESSURE HOSE

7924KG32

3 long bolts attach the power steering gear to the driver's side frame rail

cient natural curvature in the routing to absorb movement and hose shortening during vehicle operation.

1. Before servicing the vehicle, refer to the precautions in the beginning of this section.

2. Set the front wheels in the straight-ahead position.

3. Place a drain pan under the steering gear and disconnect the fluid lines. Cap the openings to protect the system from contamination.

4. Remove or disconnect the following:
- Negative battery cable
- Adapter and shield from the gear and flexible coupling
- Flexible coupling clamp and steering box input shaft, matchmark for reassembly
- Flexible coupling pinch bolt
- Pitman arm from the pitman shaft, matchmark for reassembly
- Pitman shaft nut and lockwasher
- Pitman arm from the shaft using the proper puller
- Steering gear-to-frame bolts
- Gear assembly

To install:

5. Install or connect the following:
- Steering gear in position, guiding the input shaft into the flexible coupling. Align the flat in the coupling with the flat on the input shaft.
- Steering gear-to-frame bolts and tighten to 100 ft. lbs. (135 Nm)
- Flexible coupling pinch bolt. Torque to 22 ft. lbs. (30 Nm).

➡ **Check that the relationship of the flexible coupling to the flange is ¼–¾ in. (6–19mm) of flat.**

- Pitman arm onto the Pitman shaft, lining up the marks made at removal. Torque the nut to 215 ft. lbs. (285 Nm).
- Adapter and shield
- Fluid lines and refill the reservoir with the proper power steering fluid

6. Properly bleed the system and verify no leaks.

7. Road test the vehicle for proper steering system operation.

Rack & Pinion Steering Gear

REMOVAL & INSTALLATION

Silverado and Sierra

1. Before servicing the vehicle, refer to the precautions in the beginning of this section.

2. Remove or disconnect the following:
- Wheel assemblies
- Engine shield, if equipped
- Stabilizer shaft
- Power steering high and low pressure lines
- Coupler clamp bolt from the intermediate shaft
- Intermediate shaft from the rack and pinion assembly
- Rack and pinion assembly mounting nuts, washers and bolts
- Rack and pinion assembly from the vehicle

To install:

3. Install or connect the following:
- Rack and pinion assembly into the vehicle
- Rack and pinion assembly mount-

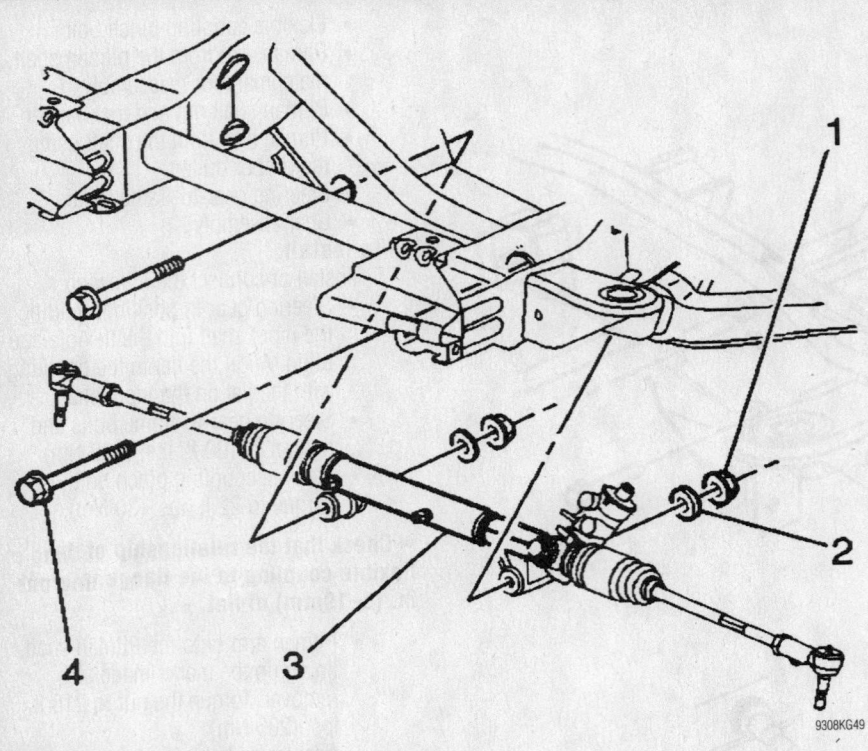

Rack and pinion gear—Silverado

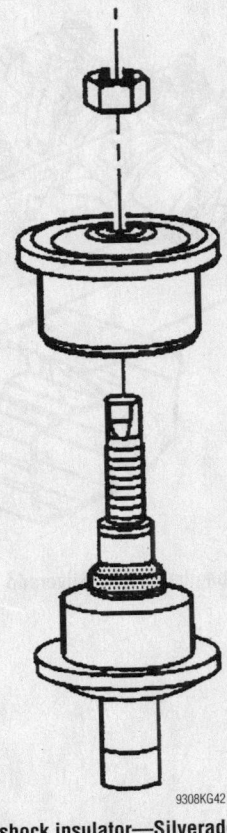

Upper shock insulator—Silverado

ing bolts, washers and nuts. Tighten the nuts to 136 ft. lbs. (185 Nm).

- Intermediate shaft to the rack and pinion assembly
- Coupler clamp bolt to the intermediate shaft. Tighten the bolt to 33 ft. lbs. (45 Nm).
- Low pressure hose
- High pressure hose. Tighten the hoses to 20 ft. lbs. (27 Nm).
- Engine protection shield, if equipped
- Stabilizer shaft
- Wheels

4. Lower the vehicle.
5. Fill and bleed the power steering system.

Shock Absorber

REMOVAL & INSTALLATION

Silverado and Sierra

2WD FRONT

1. Before servicing the vehicle, refer to the precautions in the beginning of this section.
2. Raise and support the vehicle.
3. If equipped with selectable ride, disconnect the Real Time Damping (RTD) link

rod from the sensor. Grasp the connector lock tabs. Rotate the connector tabs counter-clockwise until the connector is unlocked. Disengage the connector from the tennon by firmly pulling the connector up. Hold the tennon end with a wrench while removing the nut. Remove the nut.

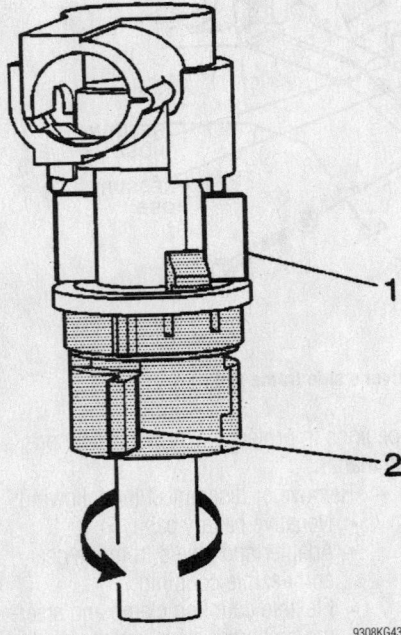

RTD connector—Silverado

4. Remove the upper insulator. Do not discard the plastic pilot ring.
5. Remove the shock absorber mounting bolts at the lower control arm. Remove the shock absorber through the lower control arm from below.

To install:

6. Support the lower control arm with a suitable jack in order to align the tennon with the mounting hole if equipped with selectable ride.
7. Install or connect the following:

- Shock absorber through the lower control arm from below
- Tennon through the mounting hole in the upper spring pocket

8. Align the shock absorber with the mounting holes in the lower control arm.

- Shock absorber mounting bolts to the lower control arm. Tighten to 18 ft. lbs. (25 Nm).

➡The upper insulators are substantially larger that the lower insulators. The upper insulator must be installed above the shock mounting bracket on the frame. The plastic pilot ring will assist the alignment of the isolators.

- Upper insulator to the shock absorber

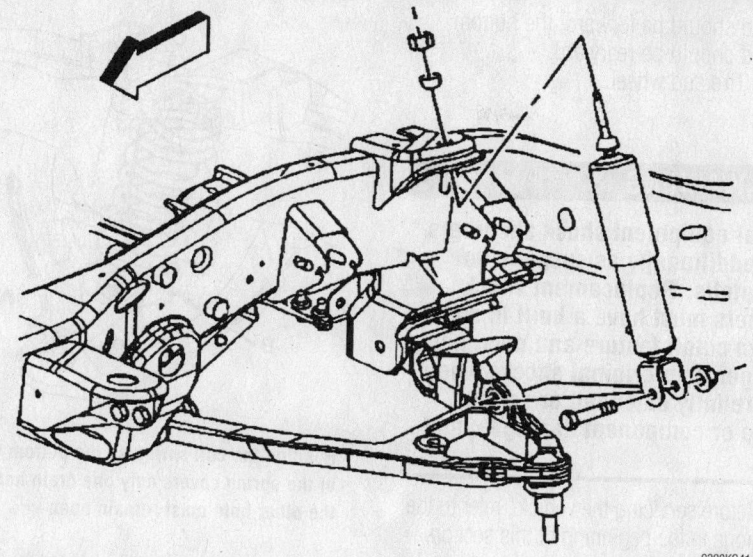

Shock absorber removal—4WD Silverado

9308KG41

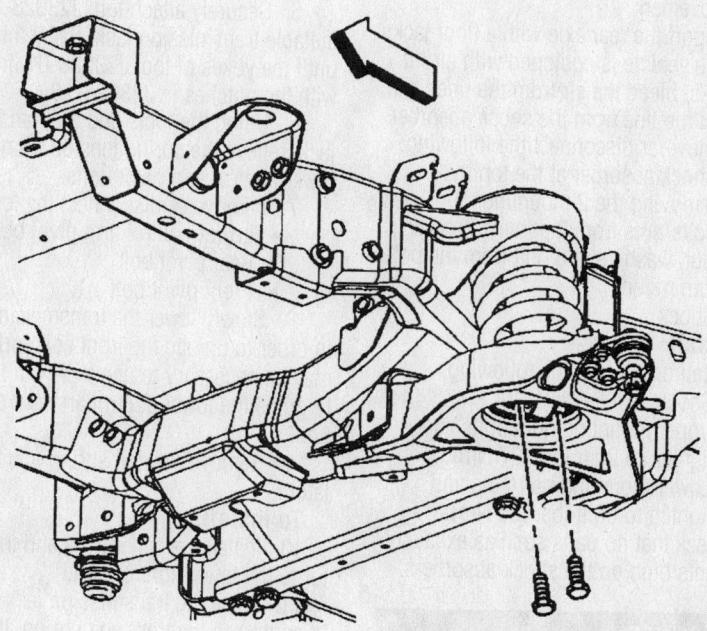

Shock absorber removal—2WD Silverado

9308KG40

- Nut to the tennon end. Do not tighten the nut.
- RTD link rod to the sensor (if equipped).

9. Remove the safety stands.

10. Lower the vehicle. Hold the tennon end with a wrench while torquing the nut. Tighten the nut to 15 ft. lbs. (20 Nm).

11. Connect the electrical connector using the following procedure:

a. Verify that the connector is unlocked.

b. Align the connector so that the tabs are perpendicular to the wrench flats on the tennon end.

c. Engage the connector to the tennon by firmly pushing the connector down.

d. Grasp the connector lock tabs. Rotate the connector counter clockwise.

12. The connector is locked into place when you hear an audible snap and the tabs are aligned.

4WD FRONT

1. Before servicing the vehicle, refer to the precautions in the beginning of this section.

2. Raise and support the vehicle.

3. Remove or disconnect the following:

- Real Time Damping (RTD) link rod from the sensor, if equipped
- Electrical connector, if equipped with selectable ride. Grasp the connector lock tabs. Rotate the connector tabs counter clockwise until the connector is unlocked. Disengage the connector from the tennon by firmly pulling the connector up. Hold the tennon end with a wrench while removing the nut. Remove the nut.
- Upper insulator. Do not discard the plastic pilot ring.
- Shock absorber mounting bolt at the lower control arm

➡ **The lower shock mounting bushing is serviceable by driving the bushing out with the appropriate tool.**

- Shock absorber

To install:

4. Install the shock absorber. Insert the stem through the hole in the shock bracket on the frame. Align the shock absorber with the mounting holes in the lower control arm.

5. Install or connect the following:

- Shock absorber through bolt to the lower control arm
- Shock absorber through bolt nut and tighten to 59 ft. lbs. (80 Nm)

➡ **The upper insulators are substantially larger that the lower insulators. The upper insulator must be installed above the shock mounting bracket on the frame. The plastic pilot ring will assist the alignment of the isolators.**

- Upper insulator to the shock absorber
- Nut to the tennon end. Do not tighten the nut
- RTD link rod to the sensor, if equipped

6. Remove the safety stands. Lower the vehicle. Hold the tennon end with a wrench while torquing the nut. Tighten the nut to 15 ft. lbs. (20 Nm).

7. Connect the electrical connector using the following procedure if equipped with selectable ride.

a. Verify that the connector is unlocked.

b. Align the connector so that the tabs (1) are perpendicular to the wrench flats on the tennon end.

c. Engage the connector to the tennon by firmly pushing the connector down.

d. Grasp the connector lock tabs (1,

2). Rotate the connector counter clockwise. The connector is locked into place when you hear an audible snap and the tabs are aligned.

REAR

1. Before servicing the vehicle, refer to the precautions in the beginning of this section.
2. Raise and support the vehicle.
3. Remove or disconnect the following:
   • Electrical connector, if equipped with selectable ride
   • Upper shock absorber nut and bolt
   • Lower shock absorber nut and bolt
   • Shock absorber

To install:
4. Installation is the reverse of removal. Tighten the nuts to 70 ft. lbs. (95 Nm).
5. Connect the electrical connector if equipped with Selectable Ride. Remove the safety stands. Lower the vehicle.

Express and Savana

FRONT

> ※※ **WARNING**
>
> **The front shock absorbers are multifunctional. They not only aid in a smooth ride, they serve as the suspension stop when the suspension is fully extended. When replacing front shocks, a shock of equivalent length and strength must be used. Use of a shock that does not comply may result in suspension over travel and component failure.**

1. Before servicing the vehicle, refer to the precautions in the beginning of this section.
2. Support the front of the vehicle safely under the lower control arms.
3. Remove or disconnect the following:
   • Tire and wheel assembly
   • Upper and lower shock absorber retaining fastener(s)

➡ **Vehicles equipped with quad shocks have a spacer between them.**

   • Shock absorber

To install:
4. Install or connect the following:
   • Shock absorber and fastener(s)
5. On 2-wheel drive vehicles torque the upper bolt to 12 ft. lbs. (16 Nm) and the lower bolts to 24 ft. lbs. (33 Nm). On 4-wheel drive vehicles, torque the nuts to 66 ft. lbs. (90 Nm). Be sure the bolts are inserted in the proper direction. The upper

bolt head should be forward; the bottom bolt head should be rearward.
   • Tire and wheel

REAR

> ※※ **WARNING**
>
> **Original equipment shock absorbers serve additionally as suspension drop cutoffs. Replacement shock absorbers must have a built in suspension cutoff feature and must not be longer than original shocks when they are fully extended or serious vehicle or component damage could result.**

1. Before servicing the vehicle, refer to the precautions in the beginning of this section.
2. The vehicle's weight should rest on correctly placed safety stands located under the frame. Chock the front wheels to prevent vehicle movement.
3. Support the rear axle with a floor jack.
4. If the vehicle is equipped with air lift type shocks, bleed the air from the lines and disconnect the line from the shock absorber.
5. Remove or disconnect the following:
   • Shock absorber at the top by removing the 2 mounting bolts/nuts from the frame bracket
   • Nut, washers and bolt from the bottom mount
   • Shock

To install:
6. Install or connect the following:
   • Shock
   • Upper mounting nuts/bolts and tighten to 20 ft lbs. (27 Nm).
   • Lower mounting bolt/nuts and tighten to 60 ft lbs. (80 Nm).
7. Check that no parts such as exhaust components bind on the shock absorbers.

Coil Springs

REMOVAL & INSTALLATION

Silverado and Sierra

FRONT

1. Raise and support the vehicle.
2. Remove or disconnect the following:
   • Engine protection shield
   • Frame cross bar (25 series only)
   • Tire and wheel assembly
   • Shock absorber
   • Front stabilizer shaft link
3. Install tool J23028-15 using the outboard locating tab (15 Series), or, the inboard locating tab (25 Series).

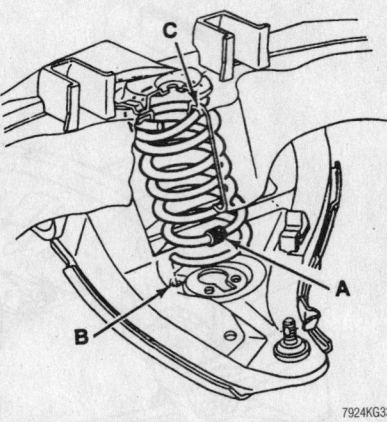

7924KG33

Position the coil spring so the bottom end of the spring covers only one drain hole—the other hole must remain open

4. Attach the retaining hook to the control arm. Tighten the wing nut until free-play is eliminated.
5. Securely attach tool J23028-01 to a suitable transmission jack. Raise the jack until the yokes of tool J23028-01 line up with the notches in J23028-15.
6. Using the tools and the transmission jack, relieve the spring tension from the lower control arm pivot bolts.
7. Remove or disconnect the following:
   • Lower control arm pivot bolt nuts
   • Rear pivot bolt
   • Front pivot bolt
8. Slowly lower the transmission jack in order to unload the front coil spring. It may be necessary to use a pry bar in order to guide the lower control arm out of position.
9. Remove the coil spring and the insulator.

To install:
10. Install the coil spring and the insulator to the lower control arm.
11. Raise the transmission jack in order to compress the front coil spring. It may be necessary to use a pry bar in order to guide the lower control arm into position.
12. Install or connect the following:
   • Front pivot bolt
   • Rear pivot bolt
   • Lower control arm pivot nuts. Tighten the pivot bolt nuts to 107 ft. lbs. (145 Nm).
13. Lower the jack. Remove the tool from the control arm.
   • Front stabilizer shaft link
   • Shock absorber
   • Tire and wheel assembly
   • Frame cross bar (25 series only). Tighten the nuts to 74 ft. lbs. (100 Nm).

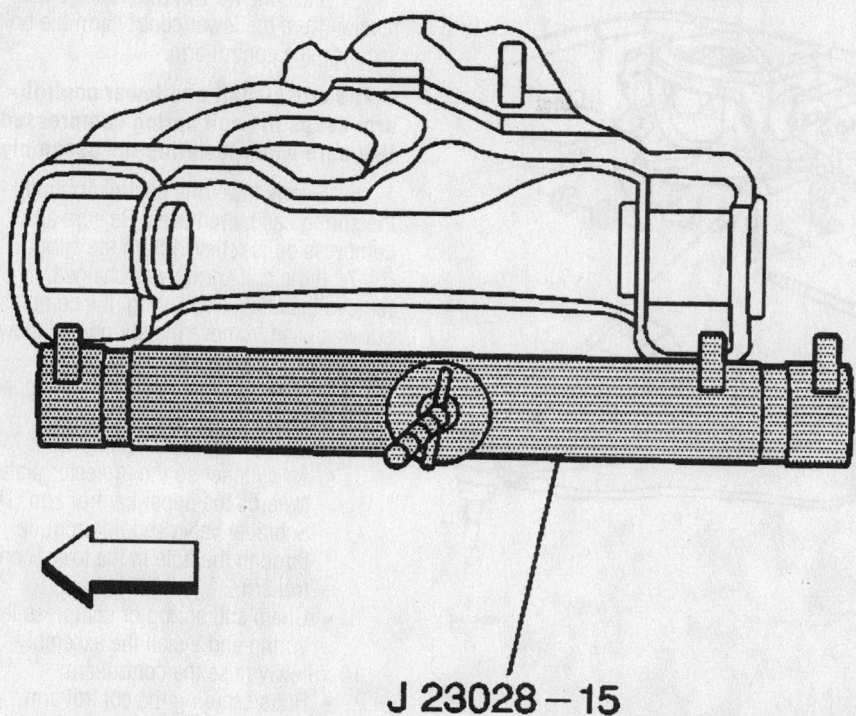

J 23028—15

Installing J23028-15 on the 25 Series

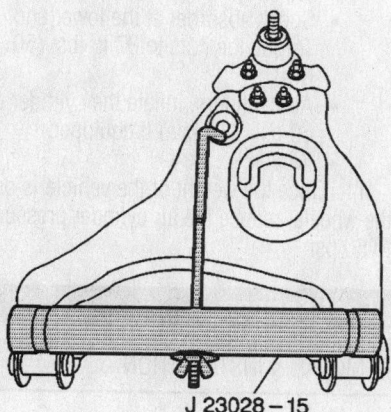

J 23028—15

9308KG46

Retaining hook installation

14. Install the engine protection shield.
15. Remove the safety stands. Lower the vehicle.

REAR

1. Raise and support the vehicle.
2. Disconnect the Real Time Damping (RTD) sensor, if equipped.
3. Remove the lower shock absorber nuts and bolt from the rear axle.
4. Lower the rear axle until the springs are fully unloaded.
5. Remove the spring and the upper and lower insulators.

To install:

6. Position the spring and the upper and lower insulators.
7. Install the rear spring to the rear axle.
8. Raise the rear axle. Install the lower shock absorber nuts to the rear axle.
9. Connect the RTD sensor, if equipped.
10. Remove the rear axle support. Lower the vehicle.

Express and Savana

1. Before servicing the vehicle, refer to the precautions in the beginning of this section.
2. Support the vehicle safely under the frame rails. The control arms should hang freely.
3. Remove or disconnect the following:
   • Wheel
   • Shock absorber lower end mounting nut/bolt
   • Stabilizer bar from the lower control arm
4. Support the lower control arm and install a spring compressor on the spring or chain the spring to the control arm as a safety precaution.

9308KG47

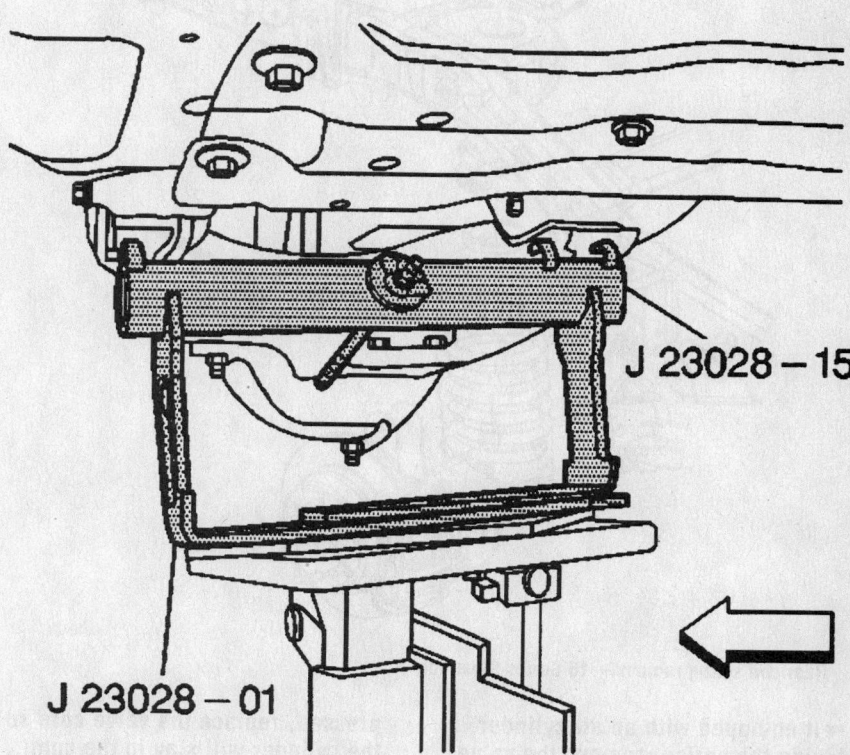

J 23028—15

J 23028—01

Tool attached to a jack

9308KG45

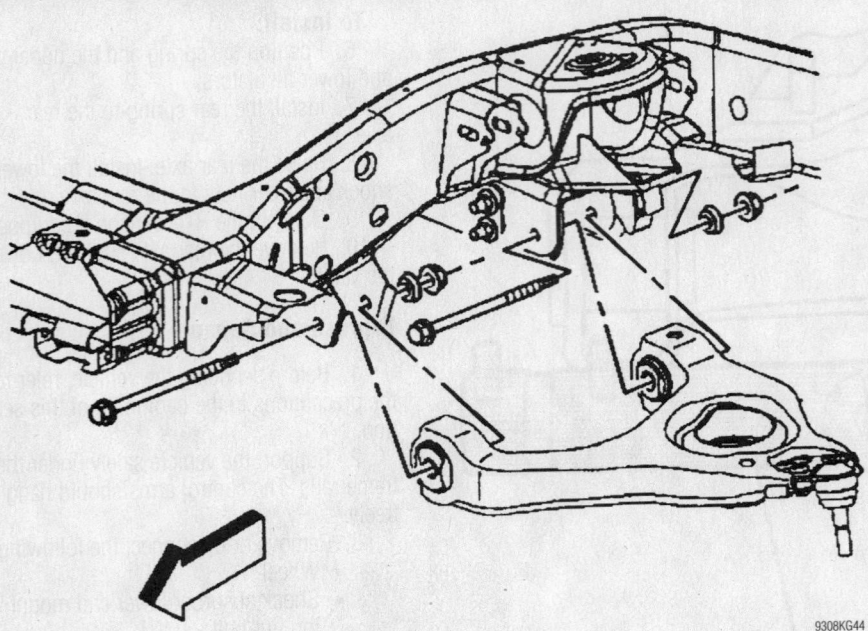

Lower control arm removal—Silverado

9308KG44

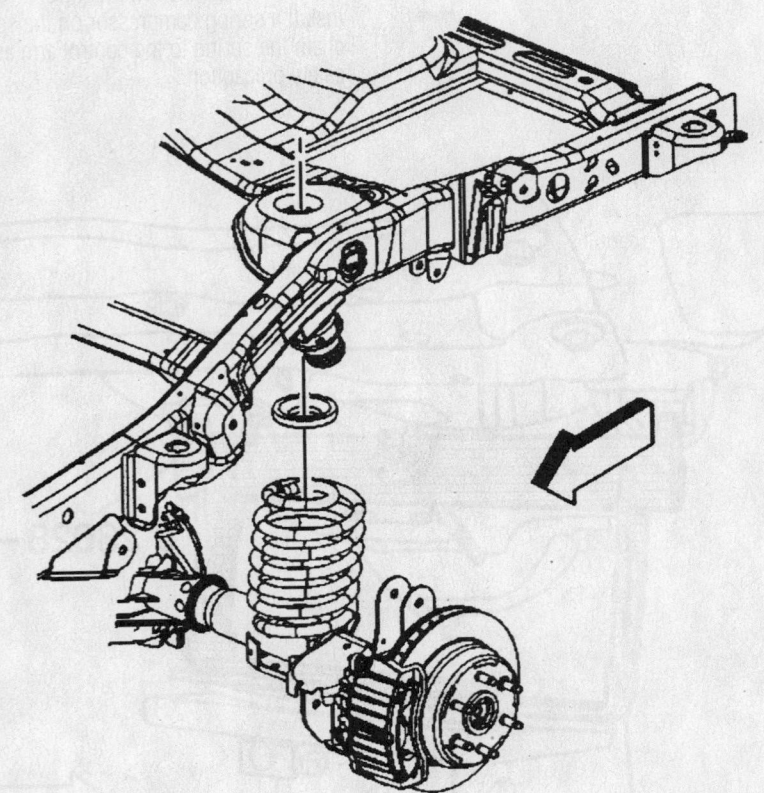

Rear coil spring removal—15 Series Silverado

9308KG24

➡If equipped with an air cylinder inside the spring, remove the valve core from the cylinder and expel the air by compressing the cylinder with a prybar. With the cylinder com- pressed, replace the valve core so the cylinder will stay in the com- pressed position. Push the cylinder as far as possible towards the top of the spring.

5. Raise the front end to remove the tension from the lower control arm the bolts securing the control arm.

➡**The cross-shaft and lower control arm keeps the coil spring compressed. Use care when lowering the assembly.**

6. Slowly lower the control arm until the spring can be removed. Be sure all compression is relieved from the spring.

7. If the coil spring was chained, remove the chain and spring. If a compressor was used, remove the spring and slowly release the compressor.

8. Remove the air cylinder, if equipped.

To install:

9. Install or connect the following:
- Air cylinder so the protector plate is towards the upper control arm. The Schrader valve should protrude through the hole in the lower control arm.
- Chain and spring or compress the spring and install the assembly

10. Slowly raise the control arm.
- Bolts securing the control arm. Tighten to 115 ft. lbs. (155 Nm).
- Stabilizer bar to the lower control arm. Torque the nuts to 24 ft. lbs. (34 Nm).
- Shock absorber at the lower end. Torque the nuts to 37 ft. lbs. (50 Nm).
- Air cylinders, inflate the cylinder to 60 psi (414 kpa) if equipped
- Wheel

11. Once the weight of the vehicle is on the wheels, reduce the air cylinder pressure to 50 psi.

Leaf Springs

REMOVAL & INSTALLATION

Silverado and Sierra

1. Before servicing the vehicle, refer to the precautions in the beginning of this section.

2. Raise and support the vehicle.

3. Support the rear axle independently in order to relieve the tension on the leaf springs.

4. Remove or disconnect the following:
- Real Time Damping (RTD) sensors, if equipped
- Trailer hitch if equipped
- Fuel tank for left side applications
- U-bolt nuts and U-bolts
- Spring spacer and anchor plate
- Shackle to the frame bracket nut and the bolt

- Front spring bracket bolt
- Leaf spring assembly from the vehicle
- Shackle from the spring

To install:

5. Loosely assemble the spring shackle bracket to the frame. Install the shackle bolt. Install the shackle nut.

6. Install the leaf spring assembly to the vehicle.

7. Loosely assemble the spring to the front hanger bracket.

8. Install or connect the following:

- Front spring hanger bracket bolt
- Front spring hanger bracket nut
- Shackle to the spring bolt
- Shackle to the spring nut

➡**Do not reuse the U-bolts.**

- Spring spacer
- U-bolts
- Anchor plate
- U-bolt nuts

9. Observe the following torques:

- 14mm U-bolt nuts: 59 ft. lbs. (80 Nm)
- 16mm U-bolt nuts: 89 ft. lbs. (120 Nm)
- Front hanger bracket nut: 92 ft. lbs. (125 Nm)
- Shackle-to-frame nut: 70 ft. lbs. (95 Nm)
- Shackle-to-spring nut: 70 ft. lbs. (95 Nm)

10. Install the fuel tank for left side applications.

11. Install the trailer hitch if equipped.

12. Connect the RTD sensors, if equipped

13. Remove the rear axle support.

14. Remove the safety stands. Lower the vehicle.

Express and Savana

1. Before servicing the vehicle, refer to the precautions in the beginning of this section.

2. Raise the vehicle and support it so that there is no tension on the leaf spring assembly.

3. Remove or disconnect the following:

- U-bolt nuts, plates, and spacer(s)
- Anchor plate
- Spring-to-shackle retaining bolts. (Do not remove these bolts)
- Bolts which attach the shackle to the rear bracket
- Bolt which attaches the spring to the front bracket
- Spring from the vehicle

4. Inspect the spring and replace any damaged components.

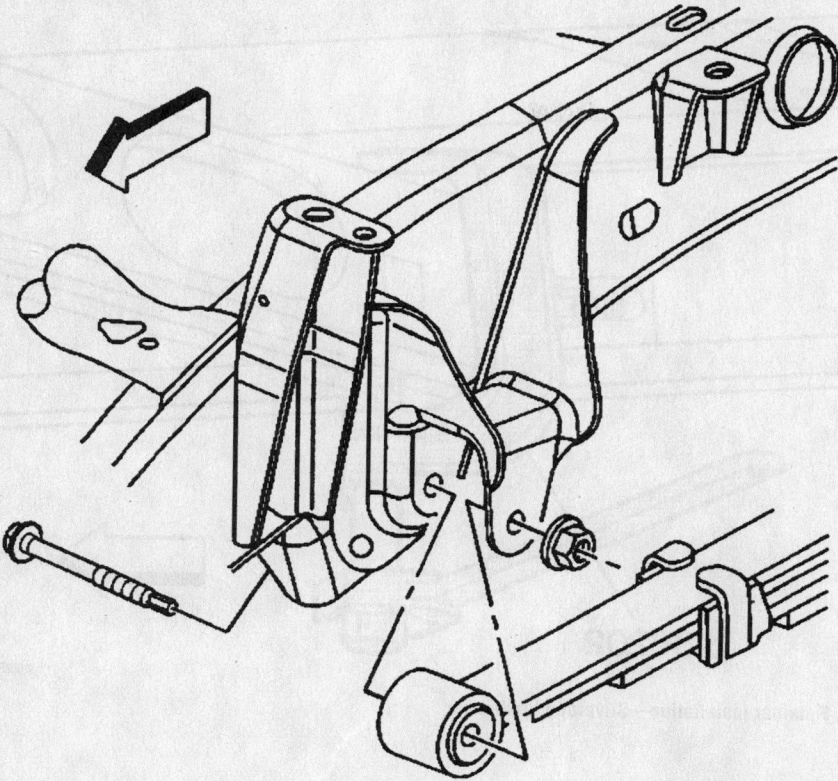

Rear leaf spring front shackle—Silverado

9308KG23

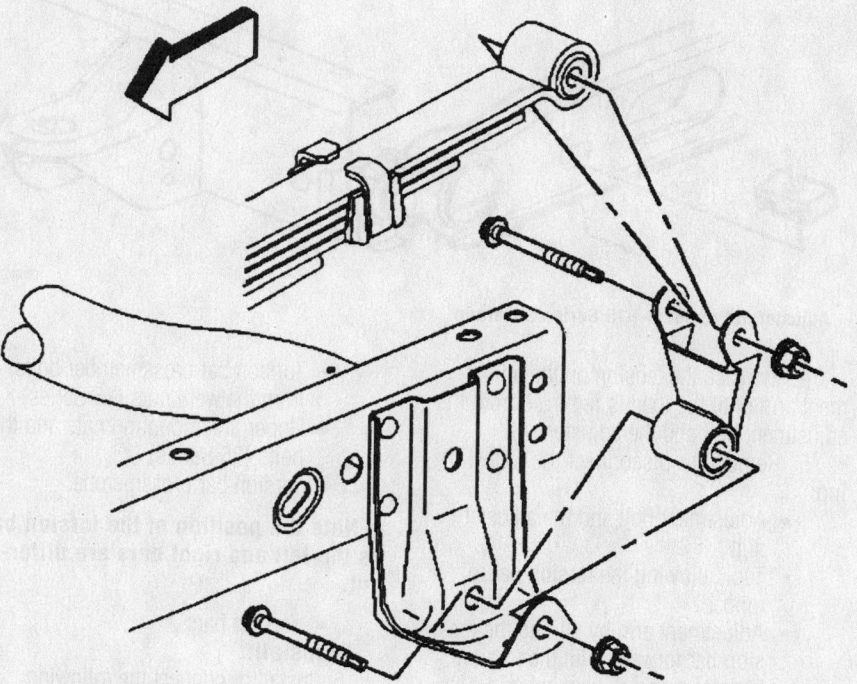

Rear leaf spring rear shackle—Silverado

9308KG22

To install:

➡If the spring bushings are defective, use the following procedures for replacement. On bushings that are staked in place, the stakes must first be straightened. Using a press or vise, remove the bushing and install the new one. When a new, previously staked bushing is installed, stake it in 3 equally spaced locations.

5. Place the spring assembly onto the axle housing. Position the front and rear of the spring at the brackets. Raise the axle with a floor jack as necessary to make the alignments.

6. Install or connect the following:
- Front and rear brackets bolts loosely
- Spacers and spring plate
- New u-bolts, washers and nuts
- Anchor plate
- U-bolt nuts. Torque them in a diagonal sequence, to 17 ft. lbs. (23 Nm) When the spring is evenly seated, tighten the nuts to 81 ft. lbs. (110 Nm).

7. Check that the hanger and shackle bolts are properly installed. All bolt heads should be inboard. Don't tighten them yet.

8. Using the floor jack, raise the axle until the distance between the bottom of the rebound bumper and its contact point on the axle is 182mm plus or minus 6mm.

9. When the spring is properly positioned, tighten all the hanger and shackle nuts to 70 ft. lbs. (95 Nm).

10. Tighten the following:
- Leaf spring-to-shackle nuts 15/25/35 series: 70 ft. lbs. (95 Nm)
- Leaf spring-to-shackle nuts C3HD series: 157 ft. lbs. (213 Nm)
- Shackle-to-bracket nuts C3HD series: 157 ft. lbs. (213 Nm)

Torsion Bars

REMOVAL & INSTALLATION

Silverado and Sierra

➡This procedure requires the removal of both torsion bars.

1. Before servicing the vehicle, refer to the precautions in the beginning of this section.
2. Raise and support the vehicle.
3. Mark the adjustment bolt setting. Install tool J36202 to the adjustment arm and the crossmember.

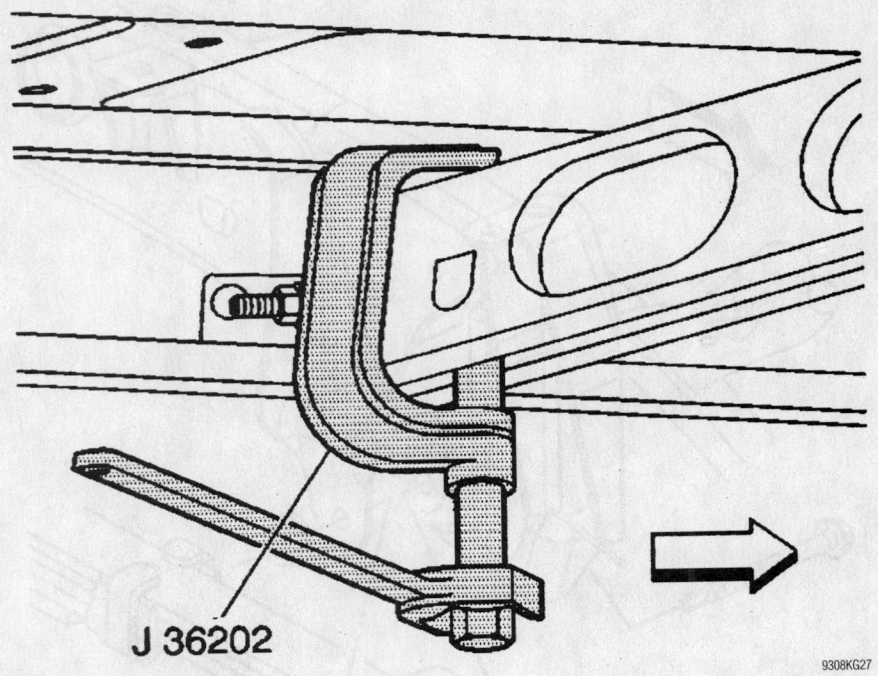

J 36202

Retainer installation—Silverado torsion bar

9308KG27

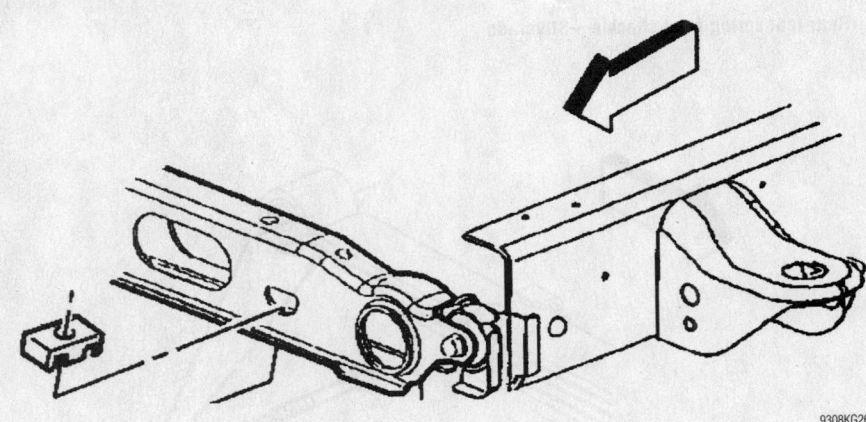

Adjuster nut removal—15 Series Silverado

9308KG26

4. Increase the tension on the adjustment arm until the load is removed from the adjustment bolt and the adjuster nut.

5. Remove or disconnect the following:

- Adjustment bolt and the adjuster nut
- Tool, allowing the torsion bar to unload.
- Adjustment arm by sliding the torsion bar forward until the torsion bar clears the adjustment arm. Use your hand to support the adjustment arm as the adjustment arm releases from the torsion bar.

- Torsion bar crossmember bolts from the weld nuts (15 Series)
- Upper link mounting nuts and the bolts (25 Series)
- Torsion bar crossmember

➡Note the position of the torsion bars as the left and right bars are different.

- Torsion bars

To install:

6. Install or connect the following:
- Torsion bars
- Torsion bar crossmember
- Torsion bar crossmember bolts to

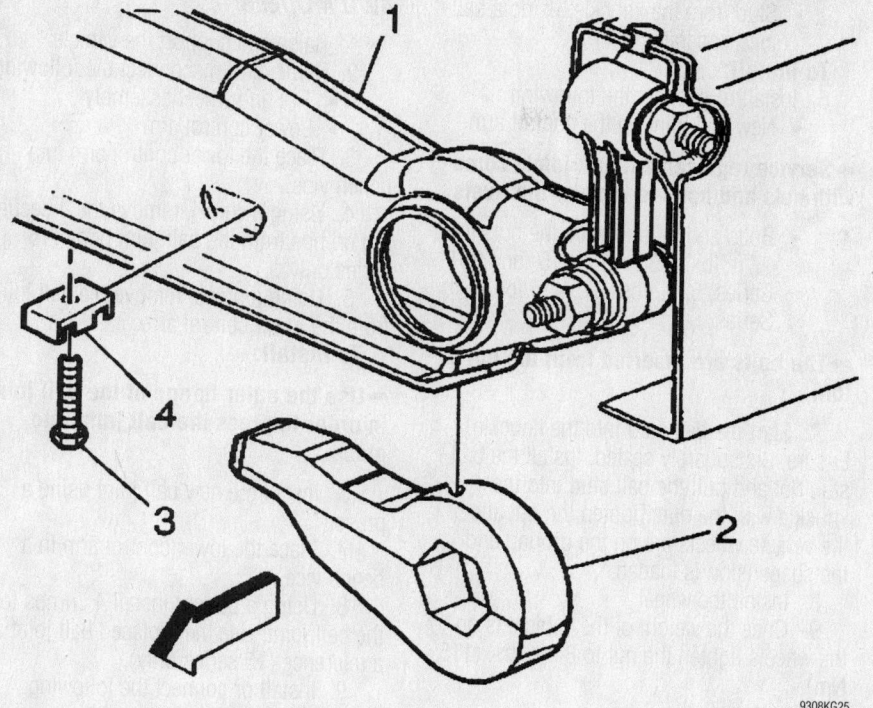

Adjuster bolt removal—15 Series Silverado

the weld nuts (15 Series). Tighten the bolt to 70 ft. lbs. (95 Nm)
- Upper link mounting nuts and the bolts (25 Series). Tighten the nut to 70 ft. lbs. (95 Nm)

7. While supporting the adjustment arm, slide the torsion bar rearward until the torsion bar fully engages the adjustment arm. Install tool J36202 to the adjustment arm and the crossmember. Increase the tension on the adjustment arm in order to load the torsion bar.
- Adjustment bolt and the adjuster nut

8. Remove the tool, releasing the tension on the torsion bar until the load is taken up by the adjustment bolt.

9. Remove the safety stands.

10. Lower the vehicle.

11. Measure the ride height.

12. Turn the adjustment bolt clockwise to increase the ride height and counterclockwise to decrease it.

Express and Savana

➡ Special tool J–36202, or its equivalent, is necessary for this procedure.

1. Before servicing the vehicle, refer to the precautions in the beginning of this section.

2. Remove or disconnect the following:
- Wheels

3. Support the lower control arm with a floor jack.

4. Matchmark both torsion bar adjustment bolt positions.

5. Using tool J–36202, increase the tension on the adjusting arm.

6. Remove or disconnect the following:
- Adjustment bolt and retaining plate

7. Move the tool aside, and slide the torsion bars forward.
- Adjusting arms
- Torsion bar support crossmember and slide the support crossmember rearwards
- Torsion bars, matchmark the position. They are not interchangeable.
- Support crossmember
- Retainer, spacer and bushing from the support crossmember

To install:

8. Install or connect the following:
- Retainer, spacer and bushing to the support crossmember
- Support assembly on the frame, out of the way
- Torsion bars, sliding them forward until they are supported. Align the marks made when removed.

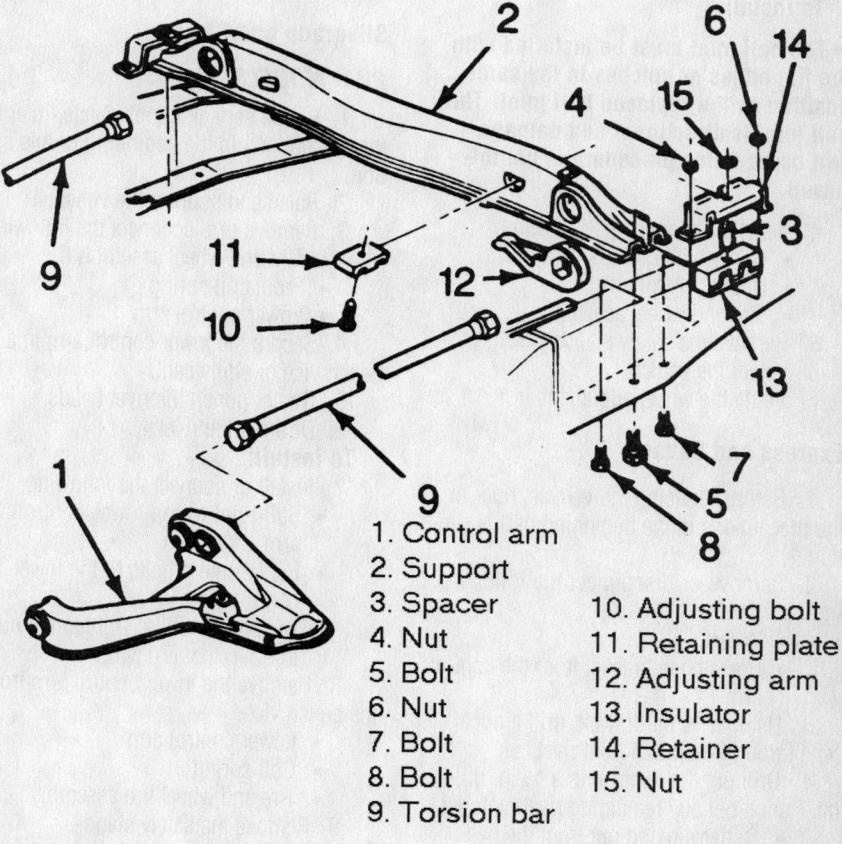

1. Control arm
2. Support
3. Spacer
4. Nut
5. Bolt
6. Nut
7. Bolt
8. Bolt
9. Torsion bar
10. Adjusting bolt
11. Retaining plate
12. Adjusting arm
13. Insulator
14. Retainer
15. Nut

Installing the torsion bar—K-Series

- Support crossmember into position. Torque the center nut to 18 ft. lbs. (24 Nm), the edge nuts to 46 ft. lbs. (62 Nm).
- Adjuster retaining plate and bolt on each torsion bar

9. Using tool J–36202, increase tension on both torsion bars.

10. Set the adjustment bolt to the marked position.

11. Release the tension on the torsion bar until the load is taken up by the adjustment bolt.

12. Install both wheels

Upper Ball Joint

REMOVAL & INSTALLATION

Silverado and Sierra

1. Before servicing the vehicle, refer to the precautions in the beginning of this section.

2. Raise and support the vehicle.

3. Remove or disconnect the following:
- Tire and wheel assembly
- Upper control arm
- Upper ball joint, using a press

To install:

➡**The ball joint must be installed with the flat edges or notches in the same position as the replaced ball joint. The ball joint is directional and damage will occur if this procedure is not followed.**

4. Install or connect the following:
- Upper ball joint, using a press
- Upper control arm
- Tire and wheel assembly

5. Remove the safety stands.

6. Lower the vehicle.

7. Verify the wheel alignment.

Express and Savana

1. Before servicing the vehicle, refer to the precautions in the beginning of this section.

2. Remove or disconnect the following:
- Wheel
- Brake hose bracket from the control arm

3. Using a ⅛ in. drill bit, drill a pilot hole through each ball joint rivet.

4. Drill out the rivets with a ½ in. drill bit. Punch out any remaining rivet material.
- Cotter pin and nut from the ball stud

5. Support the lower control arm.

- Stud from the knuckle, using a ball joint separator

To install:

6. Install or connect the following:
- New ball joint on the control arm

➡**Service replacement ball joints come with nuts and bolts to replace the rivets.**

- Bolts and nuts. Torque the nuts to 17 ft. lbs. (23 Nm) for 15- and 25-Series, 52 ft. lbs. (70 Nm) for 35-Series.

➡**The bolts are inserted from the bottom.**

7. Start the ball stud into the knuckle. Ensure it is squarely seated. Install the ball stud nut and pull the ball stud into the knuckle with the nut. Tighten the nut after the vehicle wheels are on the ground and the suspension is loaded.

8. Install the wheel.

9. Once the weight of the vehicle is on the wheels tighten the nut to 84 ft. lbs. (115 Nm).

Lower Ball Joint

REMOVAL & INSTALLATION

Silverado and Sierra

2WD MODELS

1. Before servicing the vehicle, refer to the precautions in the beginning of this section.

2. Raise and support the vehicle.

3. Remove or disconnect the following:
- Tire and wheel assembly
- Front coil spring
- Lower control arm

4. Secure the lower control arm in a bench vice or equivalent.

5. Center punch the rivet heads.

6. Drill out the rivets.

To install:

7. Install or connect the following:
- Ball joint to the lower control arm
- Replacement bolts to the lower control arm
- Nuts to the bolts. Tighten the nuts to 52 ft. lbs. (70 Nm).

8. Remove the lower control arm from the bench vice.
- Lower control arm
- Coil spring
- Tire and wheel tire assembly

9. Remove the safety stands.

10. Lower the vehicle.

11. Verify the wheel alignment.

4WD MODELS

1. Raise and support the vehicle.

2. Remove or disconnect the following:
- Tire and wheel assembly
- Lower control arm

3. Place the lower control arm in a bench vice.

4. Using a chisel, remove the 4 securing crimps from the ball joint body (15 series only).

5. Using a press, remove the ball joint from the lower control arm.

To install:

➡**Use the outer flange of the ball joint in order to press the ball joint into place.**

6. Install the new ball joint using a press.

7. Place the lower control arm in a bench vice.

8. Using a punch, install 4 crimps to the ball joint. Use the replaced ball joint as a reference (15 series only).

9. Install or connect the following:
- Lower control arm
- Tire and wheel assembly

10. Remove the safety stands.

11. Lower the vehicle.

12. Verify the wheel alignment.

Express and Savana

2-WHEEL DRIVE MODELS

1. Before servicing the vehicle, refer to the precautions in the beginning of this section.

2. Place jack under lower control arm, then raise the jack slightly.

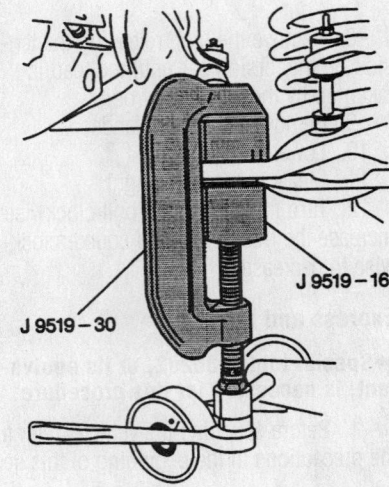

Installing the lower ball joint into the lower control arm—2-wheel drive

7924KG34

3. Remove or disconnect the following:
- Tire and wheel assembly
- Brake caliper and position it to the side
- Cotter pin and the lower ball joint retaining nut. Using the proper tool separate the ball joint from its mounting. Support the knuckle assembly so its weight will not damage the brake hose.
- Ball joint out of the lower control arm, using tool J-9519-30-D or equivalent.

To install:

4. Start the new ball joint into the control arm. Position the bleed vent in the rubber boot facing inward.

5. Install or connect the following:
- Ball joint into the control arm until fully seated
- Lower ball joint stud into the steering knuckle
- Brake caliper, if removed
- Ball stud nut. Torque the nut to 90 ft. lbs. (122 Nm) plus additional necessary to align the cotter pin hole. Do not exceed 130 ft. lbs. (175 Nm) and NEVER back the nut off to align the holes with the pin.
- New lube fitting and lubricate the new joint
- Tire and wheel

4-WHEEL DRIVE MODELS

1. Before servicing the vehicle, refer to the precautions in the beginning of this section.

2. Remove or disconnect the following:
- Wheel
- Splash shield from the knuckle
- Inner tie rod end from the relay rod using a ball joint separator
- Hub nut and washer. Insert a long drift or dowel through the vanes in the brake rotor to hold the rotor in place.
- Axle shaft inner flange bolts

3. Using a puller, force the outer end of the axle shaft out of the hub.
- Axle shaft
- Cotter pin and nut from the ball stud

4. Support the lower control arm.

5. Matchmark both torsion bar adjustment bolt positions.

6. Using tool J-36202 or equivalent, increase the tension on the adjusting arm.

7. Remove or disconnect the following:
- Adjustment bolt and retaining plate

8. Move the tool aside and slide the torsion bars forward.
- Ball joint from the knuckle, using a screw-type forcing tool

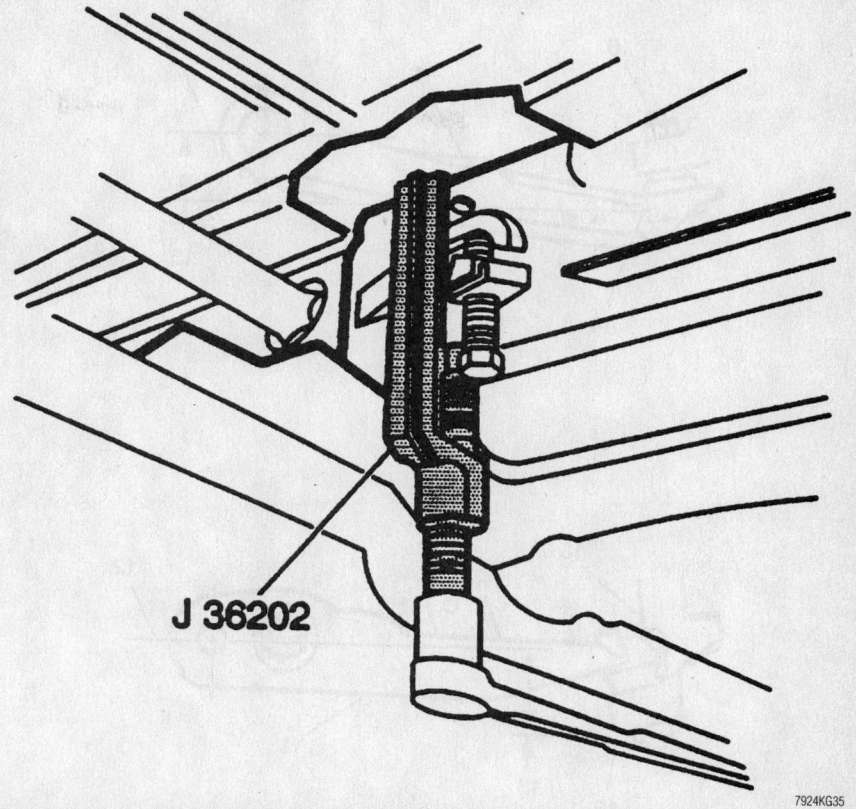

J 36202

7924KG35

A special tool is available for removing or installing the torsion bar adjusting bolt—4-wheel drive

- Lower control arm
- Lower ball joint out of control arm with tool J-9519-E or equivalent ball joint press

To install:

9. Install or connect the following:
- New ball joint into the control arm with tool J-9519-E or equivalent
- Lower control arm

10. Using tool J-36202 or equivalent, increase tension on both torsion bars.
- Adjustment retainer plate and bolt on both torsion bars

11. Set the adjustment bolt to the marked position.

12. Release the tension on the torsion bar until the load is take up by the adjustment bolt and remove the tool.
- Shaft in the hub and the washer and hub nut. Leave the drift in the rotor vanes and torque the hub nut to 175 ft. lbs. (238 Nm).
- Flange bolts. Tighten them to 59 ft. lbs. (80 Nm), remove the drift.
- Inner tie rod end at the steering relay rod. Torque the nut to 35 ft. lbs. (48 Nm).
- Splash shield
- Wheel

13. Once the weight of the vehicle is on the wheels follow these steps:

a. Lift the front bumper about 1 ½ in. (38mm) and let it drop.

b. Repeat this procedure 2–3 more times.

c. Draw a line on the side of the lower control arm from the centerline of the control arm pivot shaft, dead level to the outer end of the control arm.

d. Measure the distance between the lowest corner of the steering knuckle and the line on the control arm, record the figure.

e. Push down about 1½ in. (38mm) on the front bumper and let it return. Repeat the procedure 2–3 more times.

f. Re-measure the distance at the control arm.

g. Determine the average of the 2 measurements. This is the "Z" height measurement. The "Z" height should be as specified in the chart.

h. If the figure is correct, tighten the control arm pivot nuts to 94 ft. lbs. (128 Nm).

i. If the figure is not correct, tighten the pivot bolts to 94 ft. lbs. (128 Nm) and have the front end alignment corrected.

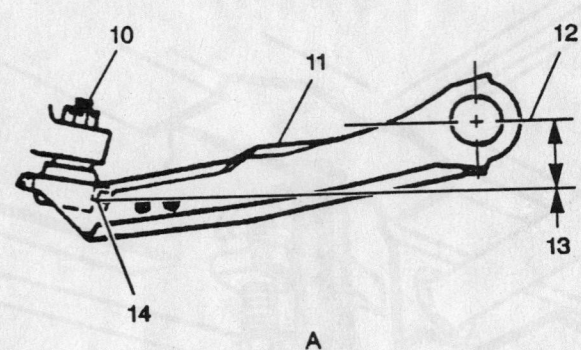

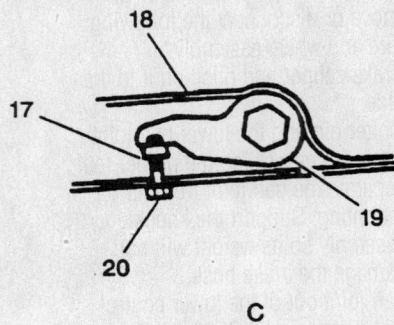

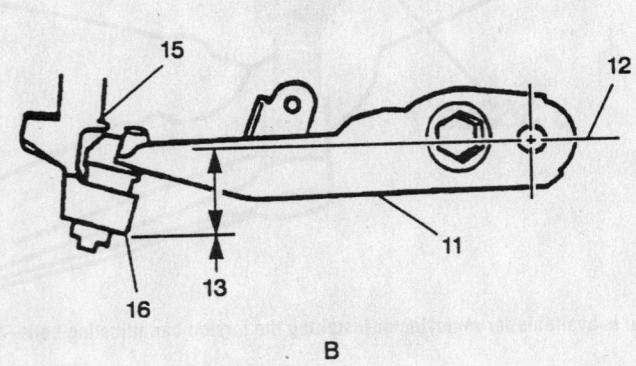

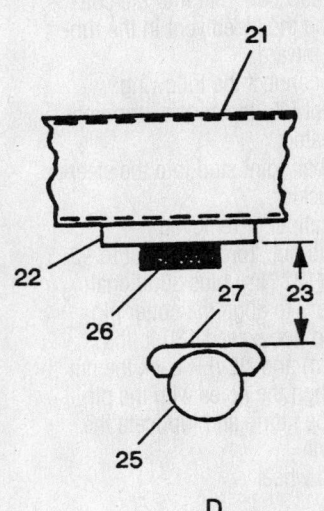

A. "C" MODEL
B. "K" MODEL
C. "K" MODEL TORSION BAR ADJUSTER
D. "CK" MODEL REAR SUSPENSION
10. LOWER BALL JOINT
11. LOWER CONTROL ARM
12. PIVOT BOLT CENTER LINE
13. "Z" HEIGHT
    C 1,2,3 95.0 ± 6.0mm
    K 1,2   157.0 ± 6.0mm
    K 3   145.0 ± 6.0mm
14. LOWER BALL JOINT EXTRUSION

15. STEERING KNUCKLE
16. STEERING KNUCKLE LOWER CORNER
17. NUT
18. TORSION BAR SUPPORT ASM.
19. TORSION BAR ADJUSTMENT ARM
20. BOLT – ONE TURN EQUALS 6mm HEIGHT CHANGE
21. FRAME
22. BOTTOM SURFACE OF JOUNCE BRACKET
23. "D" HEIGHT
25. REAR AXLE
26. JOUNCE BUMPER
27. AXLE JOUNCE PAD

7924KG36

Use these specifications and diagrams to determine if the vehicle ride height is correct

Upper Control Arm

REMOVAL & INSTALLATION

Silverado and Sierra

1. Before servicing the vehicle, refer to the precautions in the beginning of this section.
2. Raise and support the vehicle.
3. Remove or disconnect the following:

- Tire and wheel assembly
- Real Time Damping (RTD) link rod from the sensor, if equipped
- Retaining bolt for the brake hose and the wheel speed sensor brackets
- Halfshaft
- Nut at the upper ball joint. Discard the nut
- Upper control arm from the steering knuckle
- Upper control arm nuts and the adjustment cams

- Upper control arm bolts
- Upper control arm

To install:

4. Install or connect the following:

- Upper control arm
- Upper control arm bolts
- Upper control arm nuts and the adjustment cams. Tighten the nuts to 140 ft. lbs. (190 Nm)
- Upper control arm to the steering knuckle
- Halfshaft

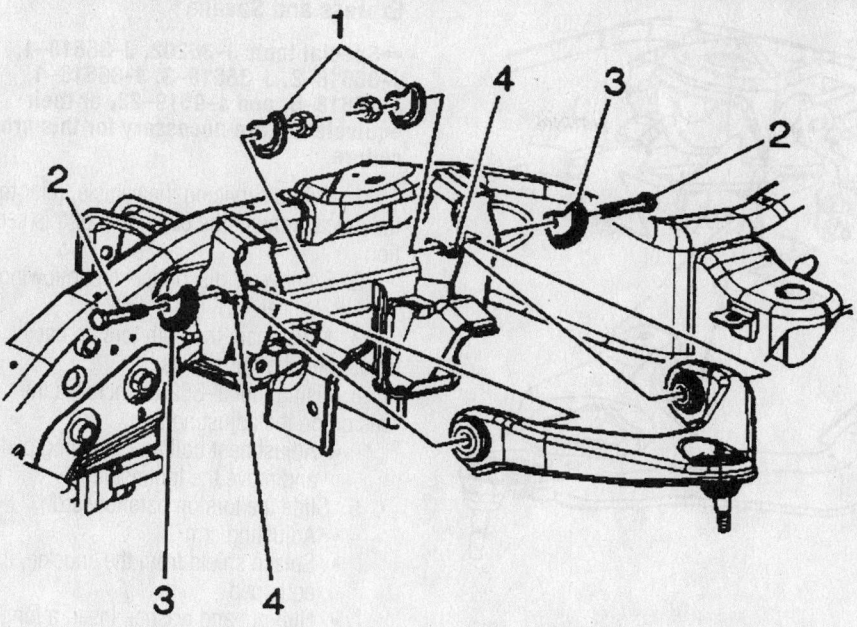

Upper control arm—Silverado

9308KG31

- New nut to the upper ball joint stud. Tighten the nut to 37 ft. lbs. (50 Nm).
- Retaining bolts for the brake hose and wheel speed sensor brackets. Tighten the bolts to 80 inch lbs. (9 Nm).
- RTD link rod to the sensor, if equipped
- Tire and wheel assembly

5. Remove the safety stands.

6. Lower the vehicle. Verify the wheel alignment.

Express and Savana

1. Before servicing the vehicle, refer to the precautions in the beginning of this section.

2. Support the lower control arm with a floor jack.

3. Remove or disconnect the following:

- Wheel
- Brake hose bracket from the control arm
- Air cleaner extension, if necessary
- Brake hose bracket retainer and wire the hose aside
- Cotter pin from the upper control arm ball stud and loosen the stud nut until the bottom surface of the nut is slightly below the end of the stud
- Ball joint from the knuckle
- Control arm to the frame brackets
- Shims and spacers

To install:

4. Install or connect the following:

- Control arm in position
- Shims and spacers, bolts and new nuts. Both bolt heads **must** be inboard of the control arm brackets. Tighten the nuts finger-tight for now.

➡ **Do not tighten the bolts yet. The bolts must be torqued with the truck at its proper ride height.**

- Ball joint to the knuckle. Torque the nut to 84 ft. lbs. (115 Nm).
- Cotter pin. Never back off the nut to install the cotter pin. Always advance it. Never advance it more than 1/6 turn.

5. Lower the truck. Once the weight of the truck is on the wheels. Torque the control arm pivot nuts to 140 ft. lbs. (190 Nm).

CONTROL ARM BUSHING REPLACEMENT

1. The control arm bushings are removed and installed using a press.

Lower Control Arm and Bushing

REMOVAL & INSTALLATION

Silverado and Sierra

2WD MODELS

1. Before servicing the vehicle, refer to the precautions in the beginning of this section.

2. Raise and support the vehicle.

3. Remove or disconnect the following:

- Tire and wheel assembly
- Real Time Damping (RTD) link rod from the sensor, if equipped
- Shock absorber
- Front stabilizer shaft link
- Front coil spring
- Lower control arm nuts and the washers
- Lower control arm bolts
- Lower ball joint stud nut
- Lower ball joint stud from the steering knuckle
- Lower control arm

To install:

4. Install or connect the following:

- Lower control arm
- Ball joint stud to the steering knuckle
- Lower ball joint stud nut. Tighten the lower ball joint stud nut to 74 ft. lbs. (100 Nm)
- Front coil spring
- Lower control arm bolt
- Lower control arm nuts and the washers. Tighten the Nuts to 107 ft. lbs. (145 Nm)
- Front stabilizer shaft link.
- Shock absorber
- Tire and wheel assembly

5. Remove the safety stands. Lower the vehicle. Verify the wheel alignment.

4WD

1. Before servicing the vehicle, refer to the precautions in the beginning of this section.

2. Raise and support the vehicle.

3. Remove or disconnect the following:

- Tire and wheel assembly
- Real Time Damping (RTD) link rod from the sensor, if equipped
- Stabilizer shaft links from the lower control arm
- Shock absorber nut and the bolt
- Torsion bars
- Halfshaft
- Lower ball joint stud nut
- Lower ball joint stud from the steering knuckle
- Lower control arm nuts and the washers
- Lower control arm bolts
- Lower control arm

To install:

- Lower control arm
- Lower control arm bolts
- Washers with the shoulder facing the arm

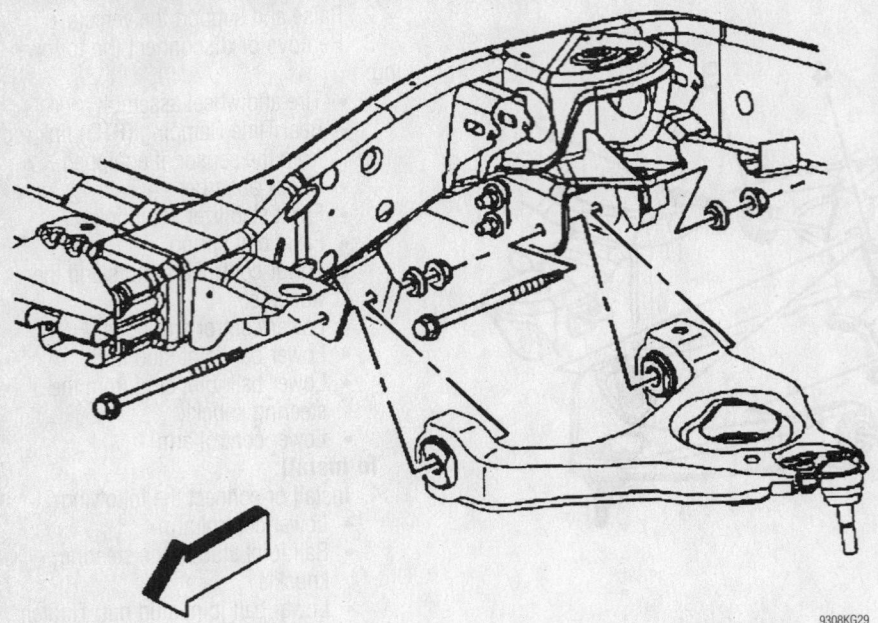

2WD lower control arm—15 Series Silverado

9308KG29

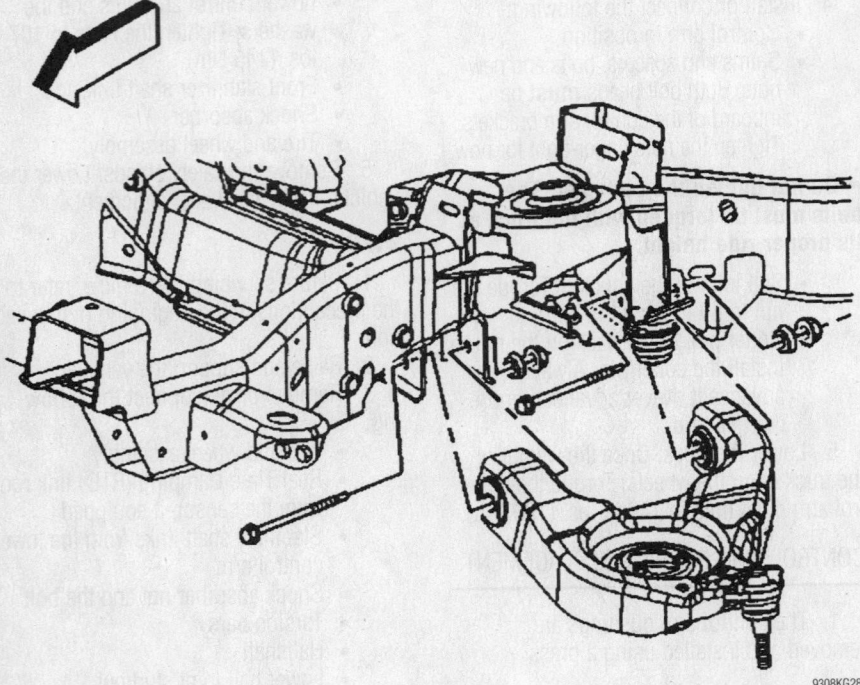

2WD lower control arm—25 Series Silverado

9308KG28

- Nuts and tighten to 107 ft. lbs. (145 Nm)
- Halfshaft
- Lower ball joint stud to the steering knuckle. Install the nut to the ball joint stud. Tighten the nut to 74 ft. lbs. (100 Nm).
- Torsion bars

- Shock absorber through nut and bolt
- Stabilizer shaft links to the lower control arm
- RTD link rod to the sensor (if equipped)
- Tire and wheel assembly

4. Remove the safety stands. Lower the vehicle. Verify the wheel alignment.

Express and Savana

➡ Special tools J–36202, J–36618–1, J–36618–2, J–36618–3, J–36618–4, J–36618–5, and J–9519–23, or their equivalents, are necessary for this procedure.

1. Before servicing the vehicle, refer to the precautions in the beginning of this section.
2. Remove or disconnect the following:
- Wheel
3. Matchmark the both torsion bar adjustment bolt positions.
4. Using tool J–36202, increase the tension on the adjusting arm.
- Adjustment bolt and retaining plate, and move the tool aside
5. Slide the torsion bars forward.
- Adjusting arm
- Splash shield from the knuckle, if equipped
- Hub nut and washer. Insert a long drift or dowel through the vanes in the brake rotor to hold the rotor in place.
- Axle shaft from the hub
- Brake caliper and wire it aside
- Rotor
- Shock absorber from control arm
- Inner tie rod end from the relay rod
6. Support the lower control arm with a floor jack.
- Stabilizer bar from the control arm
- Cotter pin from the lower ball stud and loosen the nut
- Ball joint from control arm
- Control arm-to-frame bracket bolts, nuts and washers
- Lower control arm and torsion bar as a unit
7. Separate the control arm and torsion bar.

To install:
8. Install or connect the following:
- Control arm assembly into position. Insert the front leg of the control arm into the crossmember first, then the rear leg into the frame bracket.
- Mounting bolts, front one first. The bolts **must** be installed with the front bolt head heads towards the front of the truck and the rear bolt head towards the rear of the truck!

➡ Do not tighten the bolts yet. The bolts must be torqued with the truck at its proper ride height.

- Ball joint into the knuckle. Torque the nut to 94 ft lbs. (128 Nm).
- Adjuster arm

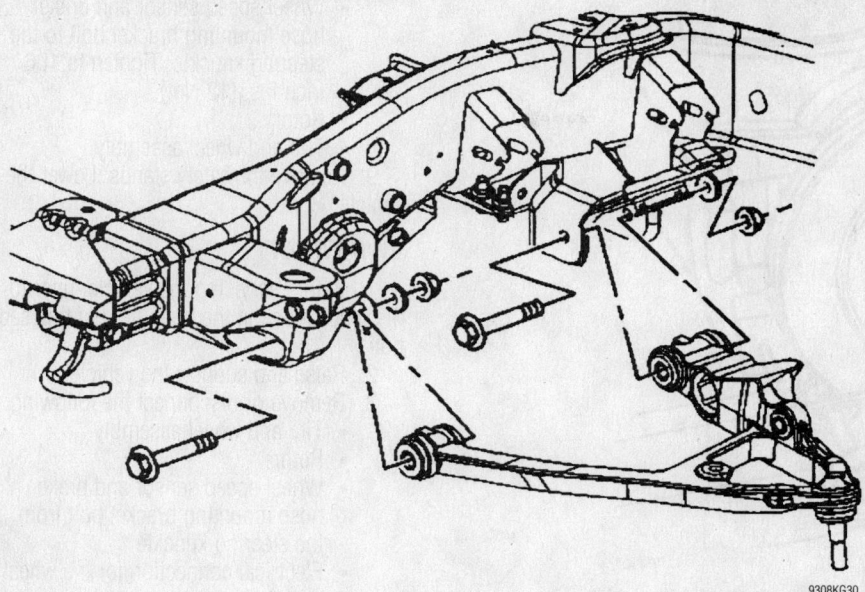

4WD lower control arm—15 Series Silverado

9. Using tool J–36202, increase tension on both torsion bars.
- Adjustment retainer plate and bolt on both torsion bars and set to marked positions. Release the tension on the torsion bar until the load is taken up by the adjustment bolt.
- Wheel

10. Lower the truck. Once the weight of the truck is on the wheels. Torque the bolts to 121 ft. lbs. (165 Nm).

CONTROL ARM BUSHING REPLACEMENT

Front Bushing

➡**On 15 and 25 Series, the bushings are not replaceable. If they are damaged, the control arm will have to be replaced.**

1. Before servicing the vehicle, refer to the precautions in the beginning of this section.
2. Remove or disconnect the following:
- Wheel
- Lower control arm
- Unbend the crimps with a punch on the front bushing

3. Press out the bushings with tools J–36618–2, J–9519–23, J–36618–4 and 36618–1.
To install:
4. Lubricate the outer case of the bushing.
5. Install or connect the following:
- Bushing into control arm

6. Press in the bushings with tools J–36618–2, J–9519–23, J–36618–4 and 36618–1 until the bushing is seated in.
7. After bushing is installed crimp it in place.
- Control arm and mounting bolts. Torque the front nut first then the rear to 140 ft lbs. (190 Nm).
- Wheel

Rear Bushing

➡**On 15 and 25 Series, the bushings are not replaceable. If they are damaged, the control arm will have to be replaced.**

1. Before servicing the vehicle, refer to the precautions in the beginning of this section.
2. Remove or disconnect the following:
- Wheel
- Lower control arm

3. Press out the bushings with tools. J–36618–5, J–9519–23, J–36618–3 and J–36618–2. There are no crimps.
To install:
4. Lubricate the outer case of the bushing.
5. Install or connect the following:
- Bushing into control arm

6. Press in the bushings with tools J–36618–5, J–9519–23, J–36618–3 and J–36618–2. There are no crimps.
- Control arm and mounting bolts. Torque the front nut first then the rear to 140 ft lbs. (190 Nm).
- Wheel

Wheel Bearings

ADJUSTMENT

➡**The front wheel bearings on 2-wheel drive vehicles (exc. Silverado and 2000–03 vehicles) are adjustable.**

1. Before servicing the vehicle, refer to the precautions in the beginning of this section.
2. Remove the dust cap, cotter pin.
3. Loosen the spindle nut.
4. Spin the wheel hub by hand and tighten the nut until it is just snug—12 ft. lbs. (16 Nm). Back off the nut until it is loose, then tighten it finger-tight. Loosen the nut until either hole in the spindle lines up with a slot in the nut and insert a new cotter pin. There should be 0.001–0.008 in. (0.025–0.200mm) end-play. This can be measured with a dial indicator, if you wish.
5. Replace the dust cap, wheel and tire.

REMOVAL & INSTALLATION

Silverado and Sierra

2WD—FRONT

1. Before servicing the vehicle, refer to the precautions in the beginning of this section.
2. Raise and support the vehicle.
3. Remove or disconnect the following:
- Tire and wheel assembly
- Rotor
- Wheel speed sensor and brake hose mounting bracket bolt from the steering knuckle
- Electrical connection for the wheel speed sensor
- Hub and bearing assembly mounting bolts
- Hub and bearing assembly
- O-ring seal from the steering knuckle bore (25 Series)

4. Clean and inspect the O-ring seal (25 Series).
To install:
5. Clean all corrosion or contaminates from the steering knuckle bore and the hub and bearing assembly.
6. Install the O-ring to the steering knuckle (25 Series).
7. Lubricate the steering knuckle bore with wheel bearing grease or the equivalent.
8. Install or connect the following:
- Hub and bearing assembly
- Hub and bearing assembly mounting bolts Tighten to 133 ft. lbs. (180 Nm).
- Electrical connection for the wheel speed sensor

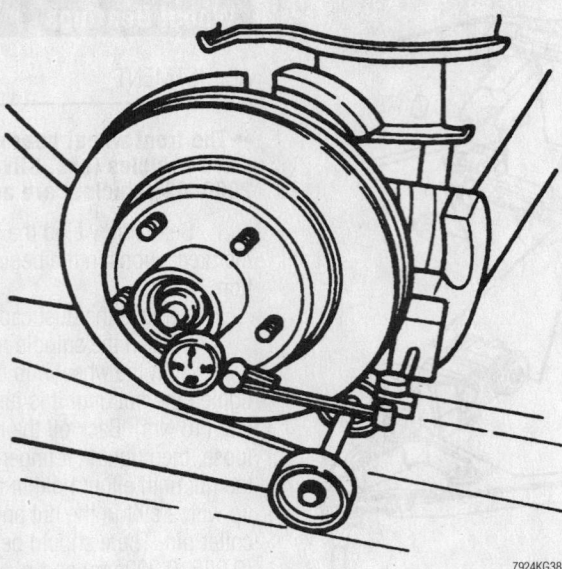

Use a dial indicator to measure the wheel bearing end-play—2-wheel drive vehicles

7924KG38

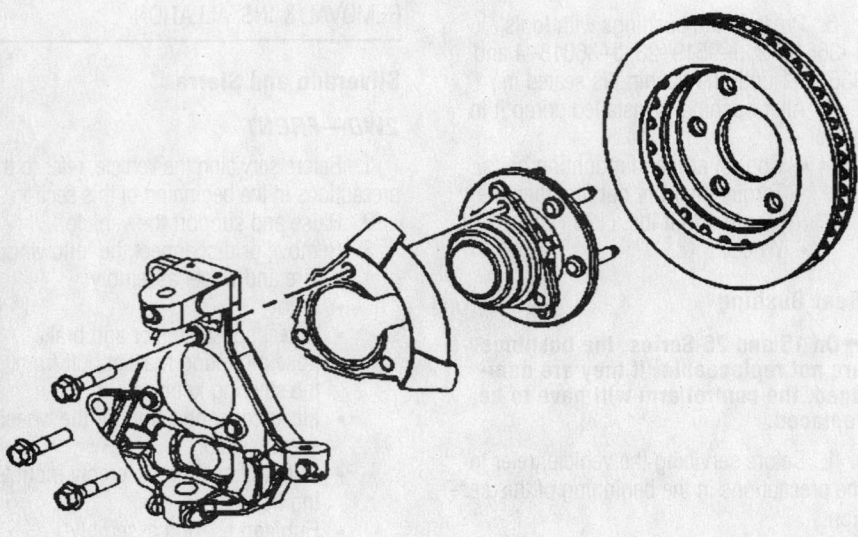

2WD front hub—15 Series Silverado

9308KG33

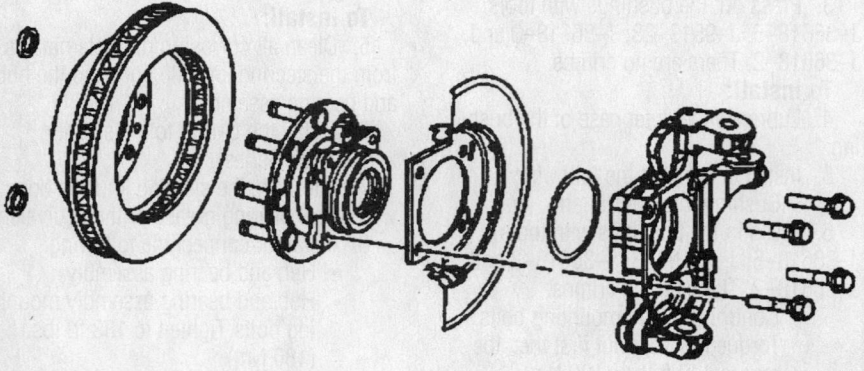

2WD front hub—25 Series Silverado

9308KG32

- Wheel speed sensor and brake hose mounting bracket bolt to the steering knuckle. Tighten to 106 inch lbs. (12 Nm).
- Rotor
- Tire and wheel assembly

9. Remove the safety stands. Lower the vehicle.

4WD FRONT

1. Before servicing the vehicle, refer to the precautions in the beginning of this section.

2. Raise and support the vehicle.

3. Remove or disconnect the following:
- Tire and wheel assembly
- Rotor
- Wheel speed sensor and brake hose mounting bracket bolt from the steering knuckle
- Electrical connection for the wheel speed sensor
- Front drive halfshaft assembly
- Hub and bearing assembly mounting bolts
- Hub and bearing assembly
- O-ring seal from the steering knuckle bore (25 Series)

4. Clean and inspect the O-ring seal (25 Series).

To install:

5. Clean all corrosion or contaminates from the steering knuckle bore and the hub and bearing assembly.

6. Install the O-ring to the steering knuckle (25 Series).

7. Lubricate the steering knuckle bore with wheel bearing grease or the equivalent.

8. Install or connect the following:
- Hub and bearing assembly
- Hub and bearing assembly mounting bolts. Tighten the bolts to 133 ft. lbs. (180 Nm).
- Front drive halfshaft assembly
- Electrical connection for the wheel speed sensor
- Wheel speed sensor and brake hose mounting bracket bolt to the steering knuckle. Tighten to 106 inch lbs. (12 Nm).
- Rotor
- Tire and wheel assembly.

Express and Savana

2WD—FRONT

1. Before servicing the vehicle, refer to the precautions in the beginning of this section.

2. Remove or disconnect the following:
- Wheel
- Caliper and wire it out of the way

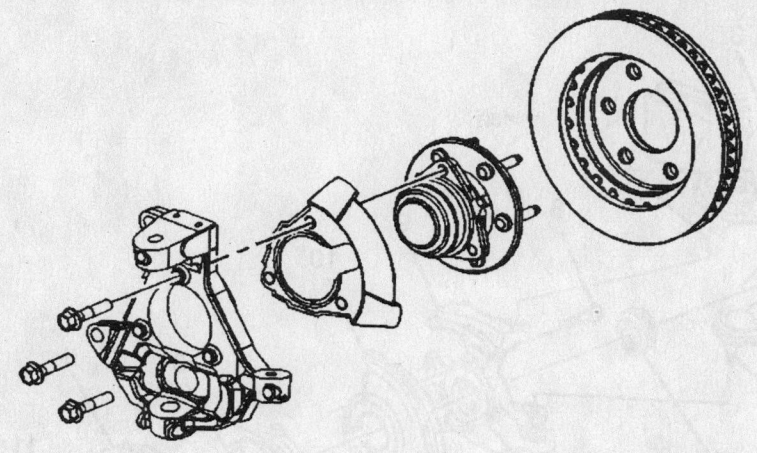

4WD front hub—15 Series Silverado

9308KG35

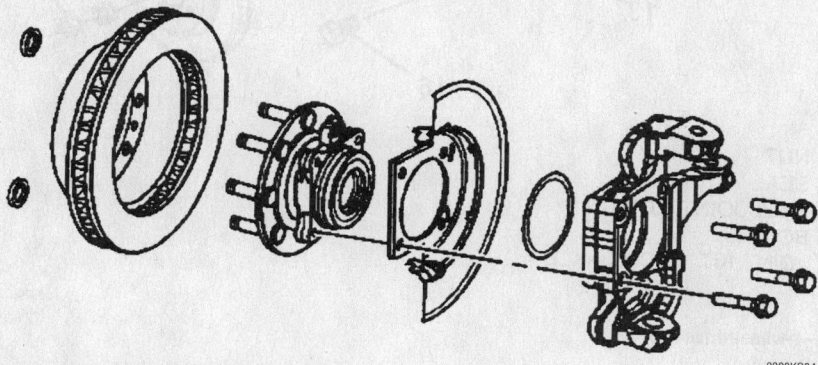

4WD front hub—25 Series Silverado

9308KG34

- Grease cap
- Cotter pin, spindle nut, and washer
- Hub.

✼✼ WARNING

Do not drop the wheel bearings.

- Outer roller bearing assembly from the hub

3. The inner bearing assembly will remain in the hub and may be removed after prying out the inner seal. Discard the seal.

4. Clean all parts in a non-flammable solvent and let them air dry. Never spin-dry a bearing with compressed air! Check for excessive wear and damage.

5. If necessary for replacement, remove the bearing races from the hub using a hammer and drift. They are driven out from the inside out.

To install:

6. Install or connect the following:
- New bearing races, if required. When installing new races, ensure that they are not cocked and that they are fully seated against the hub shoulder.

7. Pack both wheel bearings using high melting point wheel bearing grease for disc brakes.

- Inner bearing in the hub and a new inner seal, making sure that the seal flange faces the bearing race.

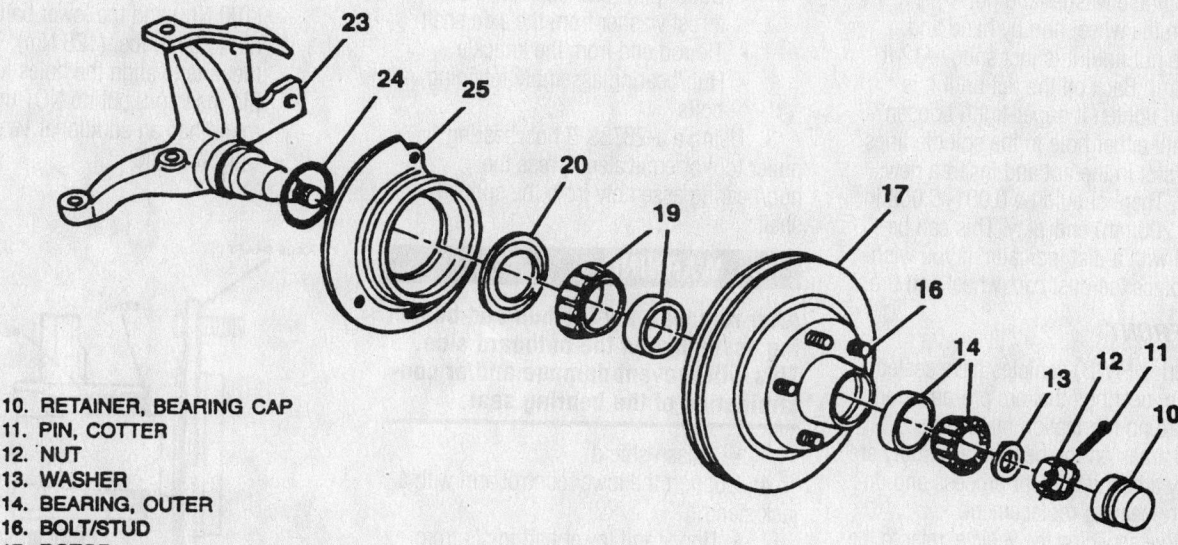

10. RETAINER, BEARING CAP
11. PIN, COTTER
12. NUT
13. WASHER
14. BEARING, OUTER
16. BOLT/STUD
17. ROTOR
19. BEARING, INNER
20. SEAL
23. KNUCKLE
24. GASKET
25. SHIELD

7924KG53

Exploded view of the front wheel bearing and related components—2-wheel drive models

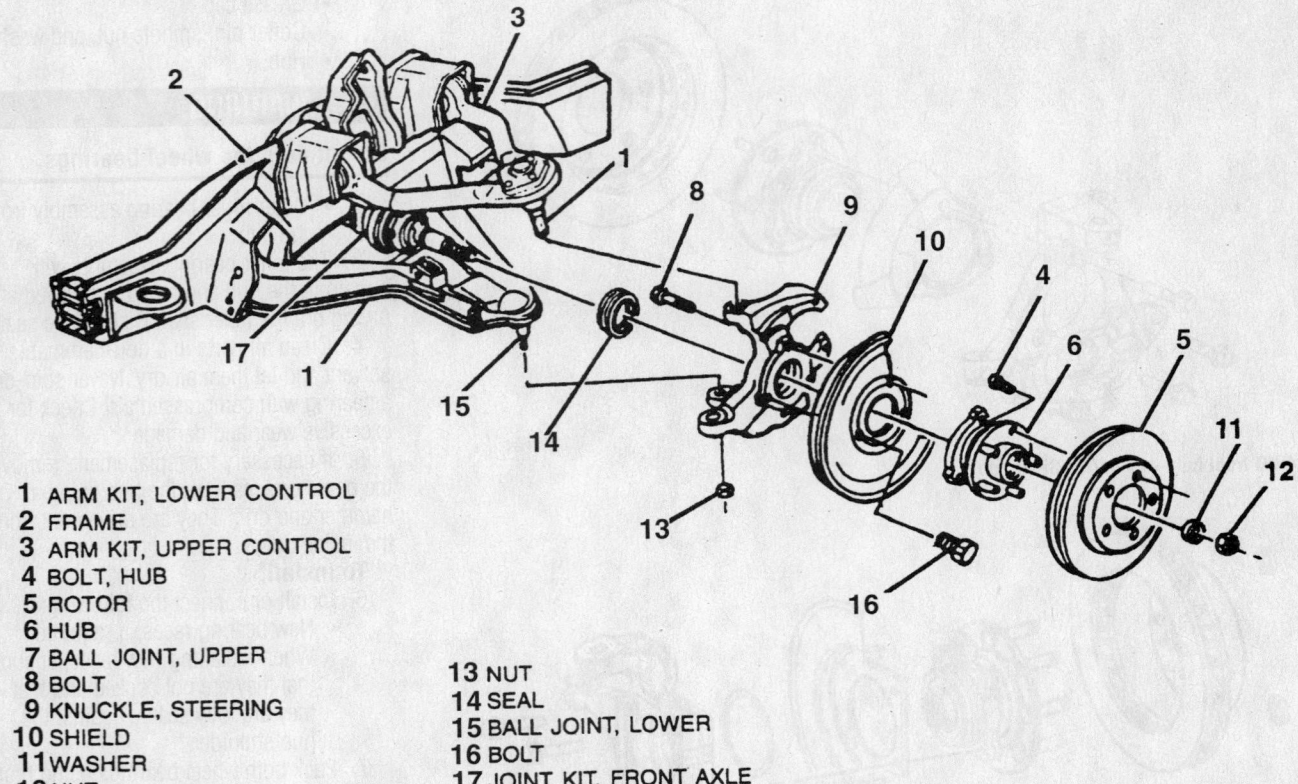

1 ARM KIT, LOWER CONTROL
2 FRAME
3 ARM KIT, UPPER CONTROL
4 BOLT, HUB
5 ROTOR
6 HUB
7 BALL JOINT, UPPER
8 BOLT
9 KNUCKLE, STEERING
10 SHIELD
11 WASHER
12 NUT
13 NUT
14 SEAL
15 BALL JOINT, LOWER
16 BOLT
17 JOINT KIT, FRONT AXLE

7924KG37

Exploded view of the front hub and knuckle assembly—4-wheel drive

- Wheel hub over the spindle
- Outer bearing into the hub
- Spindle washer and nut

8. Spin the wheel hub by hand and tighten the nut until it is just snug—12 ft. lbs. (16 Nm). Back off the nut until it is loose, then tighten it finger-tight. Loosen the nut until either hole in the spindle lines up with a slot in the nut and insert a new cotter pin. There should be 0.001–0.008 in. (0.025–0.200mm) end-play. This can be measured with a dial indicator, if you wish.

9. Replace the dust cap, wheel and tire.

4WD—FRONT

"K" Series (4WD) vehicles have sealed front wheel bearings that are pre-adjusted and require no lubrication maintenance. Darkened areas on the bearing assembly are caused by a heat treatment process and do not require bearing replacement.

1. Before servicing the vehicle, refer to the precautions in the beginning of this section.

2. Remove or disconnect the following:
- Wheel

➡Wrap shop towels around the CV-Joint boots to protect them from damage during this procedure.

- Brake caliper and support it aside
- Brake rotor
- Cotter pin, retainer, castle nut, and thrust washer from the axle shaft
- Tie rod end from the knuckle
- Hub/bearing assembly retaining bolts

3. Using a J-28733-B hub/bearing puller tool or equivalent, press the hub/bearing assembly from the splined shaft.

✲✲ WARNING

After removal, lay the hub and bearing assembly on the outboard side. This will prevent damage and/or contamination of the bearing seal.

- Splash shield

4. Support the lower control arm with a jack stand.
- Upper and lower ball joints from the steering knuckle
- Steering knuckle
- Seal from the steering knuckle

To install:

5. Install or connect the following:
- New seal in the steering knuckle, using a J 36605 seal installer

- Steering knuckle on the ball joints and the retaining nuts. Torque the upper ball joint nut to 74 ft. lbs. (100 Nm) and the lower ball joint nut to 94 ft. lbs. (128 Nm). Tighten the nuts to align the holes for cotter pin insertion, but do NOT tighten more than an additional 1/6 turn.
- Splash shield

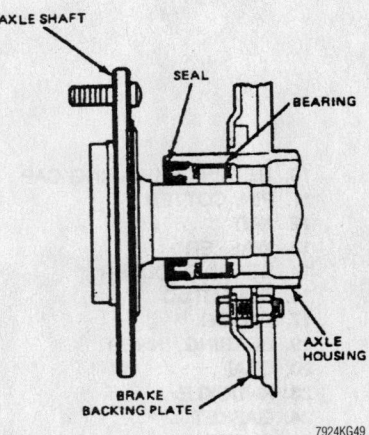

7924KG49

Cutaway view of the rear axle shaft and bearing assembly

- Hub/bearing assembly over the splined shaft, making sure the splines line up correctly. Torque the bolts to 133 ft. lbs. (180 Nm).
- Tie rod end at the steering knuckle
- Thrust washer and axle nut, Torque the nut to 165 ft. lbs. (225 Nm).
- Retainer and cotter pin
- Rotor and caliper

6. Remove the shop towels from the CV-Joint boot.

- Tire and wheel assemblies

7. Check and adjust the front end alignment and road test the vehicle.

REAR

1. Before servicing the vehicle, refer to the precautions in the beginning of this section.

A new pinion shaft lockbolt should be installed whenever either of the axle shafts is removed.

→Axle shaft seal removal and installation uses the following special tools: the GM Axle Shaft Seal Installer tool No. J-33782 (seal driver) or equivalent and the Axle Shaft Bearing Installer tool No. J-34974 (bearing driver) or equivalent.

2. Place a catch pan under the differential, then remove the drain plug (if equipped) or rear axle cover and drain the fluid (discard the old fluid)

3. Remove or disconnect the following:

- Rear wheel assemblies
- Brake drums

4. Using a wire brush, clean the dirt/rust from around the rear axle cover.

- Rear pinion shaft lockbolt and the pinion shaft, at the differential
- C-lock from the button end of the axle shaft, push the axle shafts inward.
- Axle shaft from the axle housing. Be careful not to damage the oil seal.

❊❊ WARNING

On vehicles equipped with an Anti-Lock Brake System (ABS) be careful not to damage the reluctor ring on the axle shaft or the speed sensor bolted to the backing plate, immediately adjacent to the shaft.

5. Clean the gasket mounting surfaces.

→It is recommended, when the axle shaft is removed, to replace the oil seal.

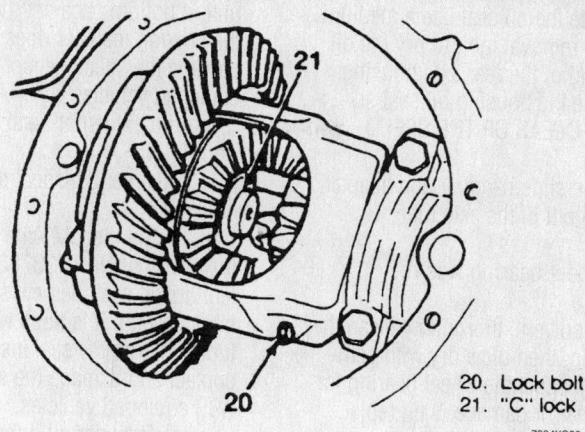

20. Lock bolt
21. "C" lock

7924KG50

Remove the lockbolt and pinion shaft, then push in the axle shaft and remove the C-lock

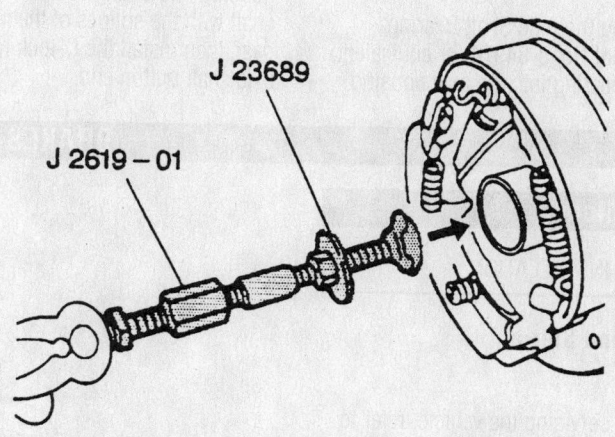

J 23689

J 2619 – 01

7924KG51

Using a slide hammer and adapters, remove the axle bearing and seal

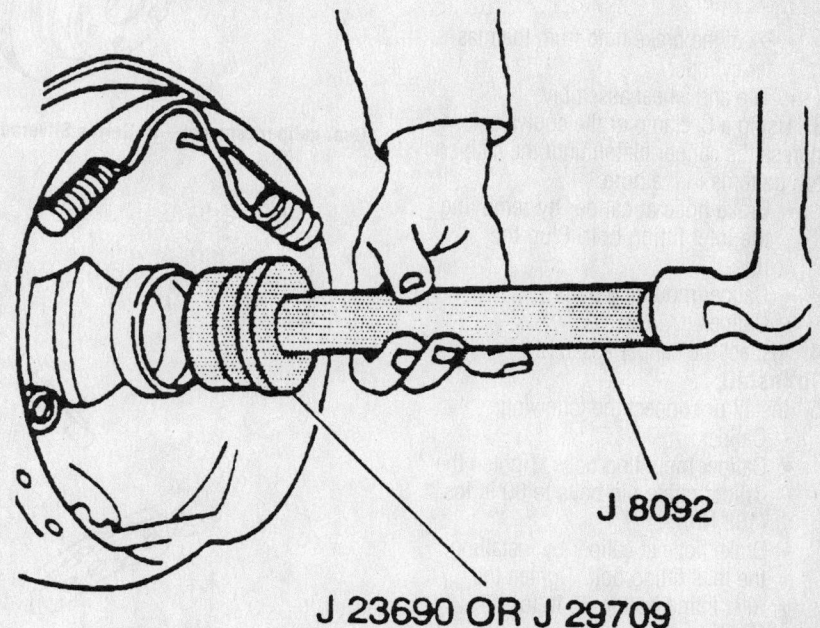

J 8092

J 23690 OR J 29709

7924KG52

Axle and seal installation using a bearing driver

6. To replace the oil seal use a medium prybar or a seal removal tool, to pry the oil seal from the end of the rear axle housing. DO NOT damage the housing oil seal surface. And STAY CLEAR OF THE SPEED SENSOR.

7. Using the slide hammer and adapter, pull the bearing out of the axle tube.

To install:

8. If the wheel bearing was removed:

   a. Using solvent, thoroughly clean the wheel bearing, then blow dry with compressed air. Inspect the wheel bearing for excessive wear or damage, then replace it, if necessary.

   b. With a new or the reused bearing, thoroughly coat the bearing with gear lubricant.

   c. Using the Axle Shaft Bearing Installer tool No. J-34974, or equivalent, drive the bearing into the axle housing until it bottoms against the seat. Be sure the bearing installer does not contact and damage the speed sensor on ABS equipped vehicles.

9. If the axle shaft seal was removed:

   a. Clean and inspect the axle tube housing.

   b. Using the GM Axle Shaft Seal Installer tool No. J-33782, or an equivalent driver, seat the new seal into the housing until it is flush with the axle tube. Be sure the seal installer does not contact and damage the speed sensor on ABS equipped vehicles.

   c. Using gear oil, lubricate the new seal lips.

10. Slide the axle shaft into the rear axle housing and engage the splines of the axle shaft with the splines of the rear axle side gear, then install the C-lock retainer on the axle shaft button end.

※※ WARNING

BE CAREFUL not to damage the wheel bearing seal with the splines on the axle shaft.

11. After the C-lock is installed, pull the axle shaft outward to seat the C-lock retainer in the counterbore of the side gears.

12. Install or connect the following:
   - Pinion shaft through the case and the pinions
   - New pinion shaft lockbolt. Torque to 25 ft. lbs. (34 Nm).
   - Housing cover use a new rear axle cover gasket
   - Brake drums
   - Tire and wheel assemblies

13. Refill the housing. REMEMBER that the vehicle must be completely level, meaning that if the rear is still raised and supported, the front should also be raised.

BRAKES

Brake Caliper

REMOVAL & INSTALLATION

Silverado and Sierra

FRONT

1. Before servicing the vehicle, refer to the precautions in the beginning of this section.

2. Remove or disconnect the following:
   - ⅔ of the brake fluid from the master cylinder
   - Tire and wheel assembly

3. Using a C-clamp or the equivalent, compress the caliper piston until the caliper piston bottoms in the bore.
   - Brake hose at caliper by removing the inlet fitting bolt. Plug the line.
   - Caliper mounting bolts
   - Caliper

4. Inspect the caliper assembly.

To install:

5. Install or connect the following:
   - Caliper
   - Caliper mounting bolts. Tighten the caliper guide pin bolts to 80 ft. lbs. (108 Nm).
   - Brake hose at caliper by installing the inlet fitting bolt. Tighten the inlet fitting bolt to 33 ft. lbs. (45 Nm).

6. Bleed the brakes.
   - Tire and wheel assembly

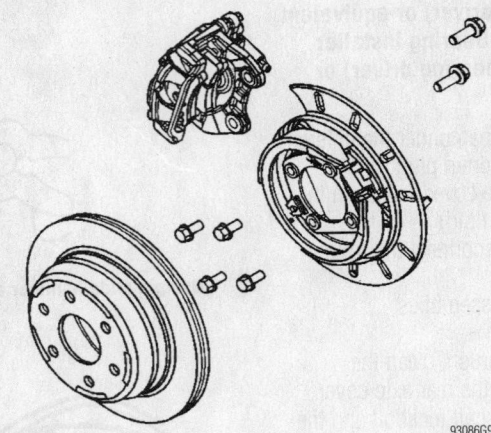

Rear caliper removal—15 Series Silverado/Sierra

93086G96

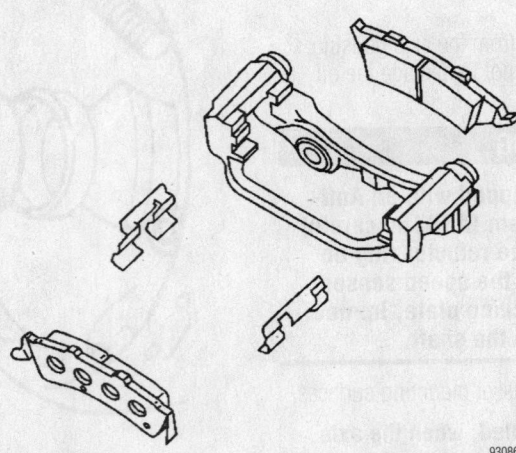

Rear pad removal—15 Series Silverado/Sierra

93086G95

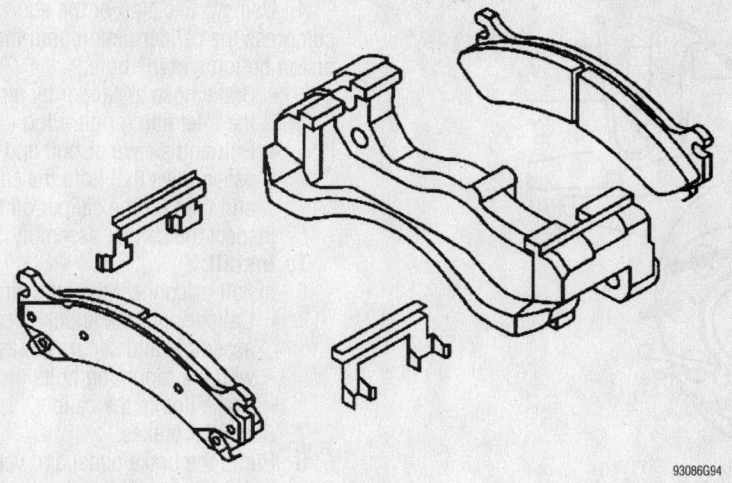

Rear pad removal—25 Series Silverado

93086G94

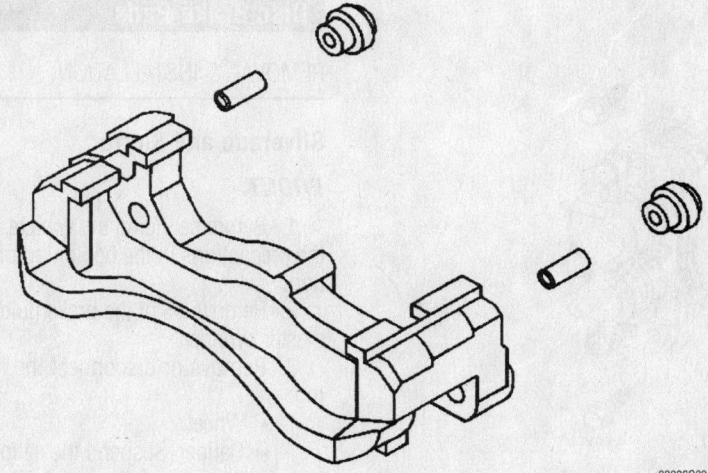

Sleeve and bushing removal—15 Series Silverado/Sierra

93086G93

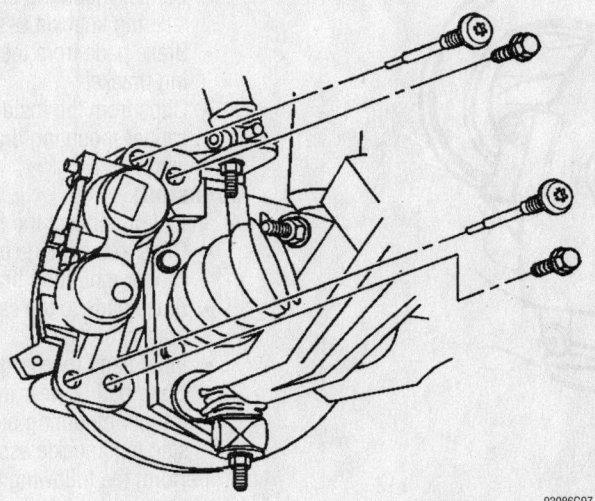

Front caliper removal—Silverado/Sierra

93086G97

REAR

1. Before servicing the vehicle, refer to the precautions in the beginning of this section.

2. Remove or disconnect the following:
   - ⅔ of the brake fluid from the master cylinder
   - Tire and wheel assembly

3. Using a C-clamp or the equivalent, compress the caliper piston until the caliper piston bottoms in the bore.
   - Brake hose at caliper by removing the inlet fitting bolt. Plug the line.
   - Caliper mounting bolts
   - Caliper

4. Inspect the caliper assembly.

To install:

5. Install or connect the following:
   - Caliper

6. Perform the following procedure before installing the caliper guide pin bolts (15 Series only).
   a. Remove all traces of the original adhesive patch.
   b. Clean the threads of the bolt with brake parts cleaner or the equivalent and allow to dry.
   c. Apply Red Loctite® #272 to the threads of the bolt.

7. Install or connect the following:
   - Caliper mounting bolts. Tighten the caliper guide pin bolts to 31 ft. lbs. (42 Nm) on the 15 series; 80 ft. lbs. (108 Nm) on the 25 series.
   - Brake hose at the caliper by installing the inlet fitting bolt. Tighten the bolt to 33 ft. lbs. (45 Nm).

8. Bleed the brakes.
   - Tire and wheel assembly

9. Refill the brake master cylinder to the proper level with fresh brake fluid.

Express and Savana

➡There are 2 caliper designs and they can be identified by the method used to secure the assembly to the spindle bracket. The Delco caliper is secured by a bolt and sleeve combination. The Bendix caliper assembly is secured by a slider, spring and bolt.

1. Before servicing the vehicle, refer to the precautions in the beginning of this section.

2. Before servicing the vehicle, refer to the precautions in the beginning of this section.

3. Remove or disconnect the following:
   - ⅔ of the brake fluid from the master cylinder
   - Tire and wheel assembly

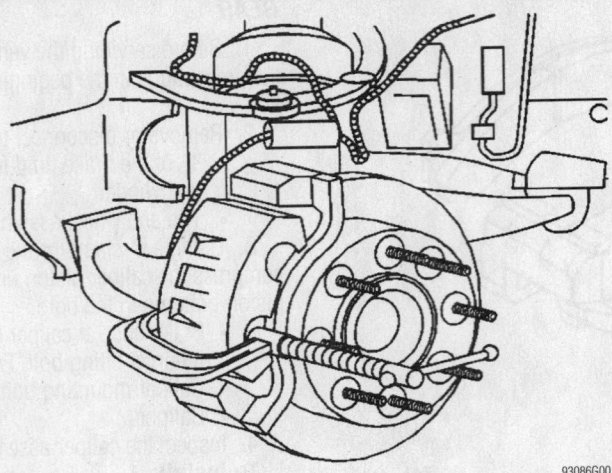

Compressing the rear caliper piston—15 Series Silverado/Sierra

93086G00

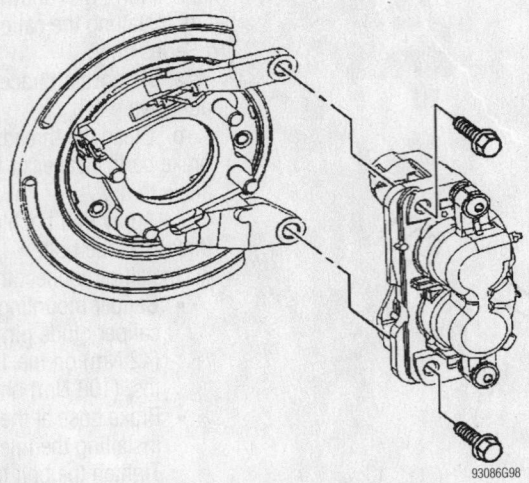

Rear caliper removal—25 Series Silverado

93086G98

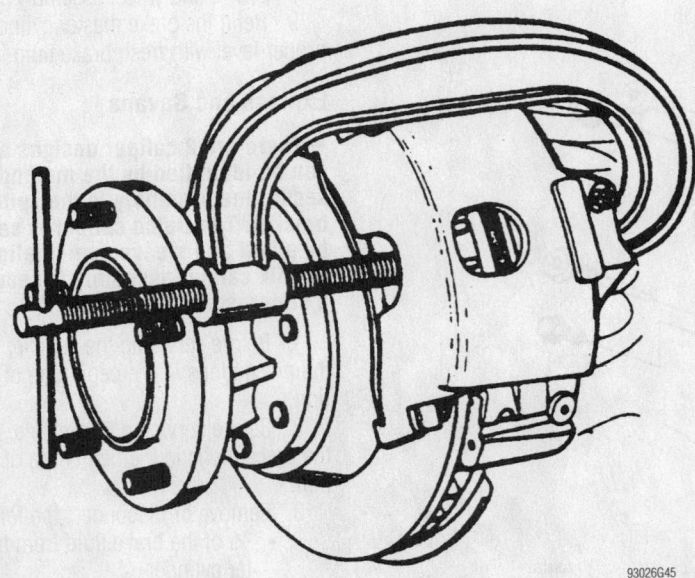

Compressing the caliper piston—Express & Savana

93026G45

4. Using a C-clamp or the equivalent, compress the caliper piston until the caliper piston bottoms in the bore.
  • Brake hose at caliper by removing the inlet fitting bolt. Plug the line.
  • Bolt and sleeve or bolt and slider assemblies that hold the caliper and then lift the caliper off the rotor

5. Inspect the caliper assembly.

To install:

6. Install or connect the following:
  • Caliper onto the knuckle/rotor assembly and secure the assembly with the mounting bolts or sliders
  • Brake line to the caliper

7. Bleed the brakes.

8. Pump the brake pedal and verify there is minimal brake pedal travel.

9. Check the brake fluid level. Install the tire and wheel assembly.

Disc Brake Pads

REMOVAL & INSTALLATION

Silverado and Sierra

FRONT

1. Before servicing the vehicle, refer to the precautions in the beginning of this section.

2. Remove ⅔ of the brake fluid from the master cylinder.

3. Remove or disconnect the following:
  • Wheel
  • Caliper. Suspend the caliper from the frame with mechanic's wire. Do not allow the caliper to hang from the brake hose.
  • Caliper mounting bracket bolts
  • Caliper mounting bracket from the steering knuckle assembly
  • Brake pads from the caliper mounting bracket
  • Clips from the inside ends of the caliper mounting bracket and discard

To install:

4. Install or connect the following:
  • Clips to the inside ends of the caliper mounting bracket
  • Brake pads to the caliper mounting bracket
  • Inner pad (1 wear indicator)
  • Outer pad (2 wear indicators)
  • Caliper mounting bracket to the steering knuckle assembly

5. Perform the following procedure before installing the caliper mounting bracket bolts:

a. Remove all traces of the original adhesive patch.

b. Clean the threads of the bolt with brake parts cleaner or the equivalent and allow to dry.

c. Apply red Loctite®272 to the threads of the bolt.

6. Install or connect the following:
- Caliper mounting bracket bolts to the steering knuckle. Tighten to 129 ft. lbs. (175 Nm) on the 15 series; 221 ft. lbs. (300 Nm) on the 25 series
- Caliper
- Tire and wheel assembly

7. Refill the master cylinder to the proper level with fresh brake fluid. Pump the brake pedal slowly and firmly in order to seat the brake pads. Burnish the brakes as needed.

REAR

1. Before servicing the vehicle, refer to the precautions in the beginning of this section.
2. Remove or disconnect the following:
- ⅔ of the brake fluid from the master cylinder
- Tire and wheel assembly
- Caliper. Suspend the caliper from the frame with mechanic's wire. Do not allow the caliper to hang from the brake hose.
- Caliper mounting bracket bolts from the backing plate
- Brake pads from the caliper mounting bracket
- Clips from the inside ends of the caliper mounting bracket and discard

To install:

3. Install or connect the following:
- Clips to the inside ends of the caliper mounting bracket
- Brake pads to the caliper mounting bracket
- Inner pad (1 wear indicator)
- Outer pad (2 wear indicators)
- Clips to the inside ends of the caliper mounting bracket
- Caliper mounting bracket to the backing plate assembly (15 series).

4. Install the caliper mounting bracket to the backing plate assembly (25 series). Perform the following procedure before installing the caliper mounting bracket bolts.

a. Remove all traces of the original adhesive patch.

b. Clean the threads of the bolt with brake parts cleaner or the equivalent and allow to dry.

c. Apply red Loctite®272 to the threads of the bolt.

➡Use the correct fastener in the correct location. Replacement fasteners must be the correct part number for that application. Fasteners requiring replacement or fasteners requiring the use of thread locking compound or sealant are identified in the service procedure. Do not use paints, lubricants, or corrosion inhibitors on fasteners or fastener joint surfaces unless specified. These coatings affect fastener torque and joint clamping force and may damage the fastener. Use the correct tightening sequence and specifications when installing fasteners in order to avoid damage to parts and systems.

5. Install or connect the following:
- Caliper mounting bracket bolts to the steering knuckle. Tighten to 148 ft. lbs. (200 Nm) on the 15 series; 122 ft. lbs. (165 Nm) on the 25 series.
- Caliper
- Tire and wheel assembly

6. Refill the master cylinder to the proper level with fresh brake fluid. Pump the brake pedal slowly and firmly in order to seat the brake pads. Burnish the brakes as needed.

Express and Savana

DELCO TYPE

1. Before servicing the vehicle, refer to the precautions in the beginning of this section.
2. Remove or disconnect the following:
- ⅔ of the brake fluid from the master cylinder
- Tire and wheel assembly

3. Compress the brake piston back into its bore using a C-clamp.
- 2 bolts holding the caliper and then lift the caliper off the disc
- Inboard and outboard pads
- Pad support spring from the piston, if equipped

To install:

4. Thoroughly inspect, clean and lubricate all caliper slide points, bolts and hardware.

5. Install or connect the following:
- Retainer spring on the inner pad and insert the assembly into the center cavity of the piston

6. Push down on the inner pad until it lays flat against the caliper. It is important to push the piston all the way into the caliper if new linings are installed or the caliper will not fit over the rotor.

- Outboard pad with the ears of the pad over the caliper ears and the tab at the bottom engaged in the caliper cutout
- Caliper over the brake disc and align the holes in the caliper with those of the mounting bracket
- Mounting bracket bolts through the sleeves in the inboard caliper ears and through the mounting bracket, making sure the ends of the bolts pass under the retaining ears on the inboard pad
- Mounting bolts to 38 ft. lbs. (51 Nm). After both calipers are mounted pump the brake pedal to seat the pad against the rotor. Use a pair of locking pliers to bend over the upper ears of the outer pad so it isn't loose.
- Wheels

7. Add fluid to the master cylinder reservoirs so they are ¼ in. (6.35mm) from the top.

8. Test the brake pedal by pumping it to obtain a hard pedal. Check the fluid level again and add fluid as necessary. Do not move the vehicle until a pedal is obtained.

BENDIX TYPE

1. Before servicing the vehicle, refer to the precautions in the beginning of this section.
2. Remove or disconnect the following:
- ⅔ of the brake fluid from the master cylinder
- Tire and wheel assembly

3. Compress the brake piston back into its bore using a C-clamp.
- Bolt at the caliper slider. Use a brass drift pin to remove the slider and spring.
- Caliper up and forward from the bottom and lift it off the caliper support. Tie the caliper out of the way with a piece of wire. Be careful not to damage the brake line.
- Inner shoe from the caliper support. Discard the inner shoe clip.
- Outer shoe from the caliper

To install:

4. Thoroughly clean, inspect and lubricate the caliper, slider and spring with silicone.

5. Install or connect the following:
- New inboard shoe clip on the shoe
- Lower end of the inboard shoe into the groove provided in the support. Slide the upper end of the shoe into position. Be sure the clip remains in position.

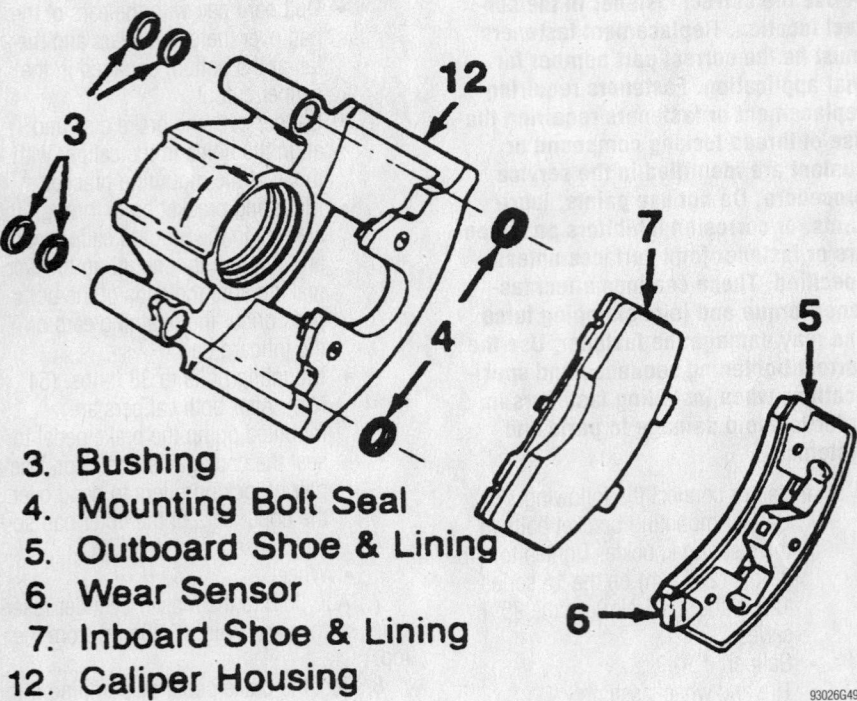

3. **Bushing**
4. **Mounting Bolt Seal**
5. **Outboard Shoe & Lining**
6. **Wear Sensor**
7. **Inboard Shoe & Lining**
12. **Caliper Housing**

93026G49

Replacing the disc brake pads—Delco type

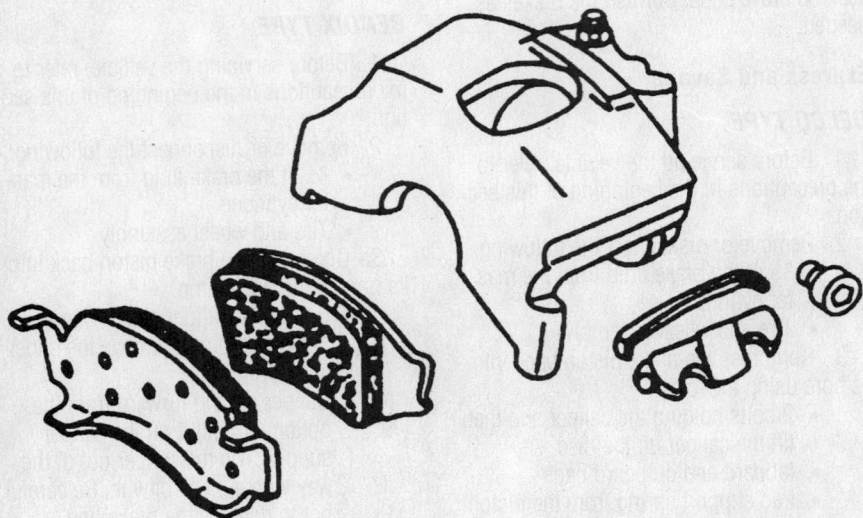

93026G50

Replacing the disc brake pads—Bendix type

- Outboard shoe in the caliper, with the ears at the top of the shoe over the caliper ears and the tab at the bottom of the shoe engaged in the caliper cutout. If assembly is difficult, a C-clamp may be used. Be careful not to damage the lining.
- Caliper over the brake disc, top edge first. Rotate the caliper downward onto the support.

- Spring over the caliper support key, install the assembly between the support and lower caliper groove. Tap into place until the key retaining screw can be installed.
- Screw and torque to 15 ft. lbs. (20 Nm). The boss must fit fully into the circular cutout in the key.

6. Install the wheel and add brake fluid as necessary.

Brake Drums

REMOVAL & INSTALLATION

With Semi-Floating Axles

1. Before servicing the vehicle, refer to the precautions in the beginning of this section.
2. Raise and support the vehicle safely.
3. Mark the relationship of the wheel to the hub and remove the wheel.
4. Mark the relationship of the drum to the hub and pull the drum from the brake assembly. If the brake drums have been scored from worn linings, the brake adjuster must be backed off so the brake shoes will retract from the drum. The adjuster can be backed off by inserting a brake adjusting tool through the access hole provided. In some cases the access hole is provided in the brake drum. A metal cover plate is over the hole. This may be removed by using a hammer and chisel.

To install:

5. Align the mark on the drum to mark on hub and install drum
6. Align the mark on the wheel to mark on drum and install wheel
7. Adjust brake lining as needed. Pump brakes

With Full Floating Axles

To remove the drums from full floating rear axles, the axle shaft will have to be removed. Full-floating rear axles can be identified by a bearing housing that protrudes through the center of the wheel.

1. Before servicing the vehicle, refer to the precautions in the beginning of this section.
2. Remove or disconnect the following:
   - Wheel
   - Axle shaft
   - Retaining ring, key and adjusting nut
   - Hub and drum

To install:

3. Install or connect the following:
   - Hub and drum to the tube
   - Adjusting nut
   - Key and retaining ring
   - Axle shaft and wheel

Brake Shoes

REMOVAL & INSTALLATION

Leading/Trailing Brakes

1. Before servicing the vehicle, refer to the precautions in the beginning of this section.

2. Remove or disconnect the following:
- Tire and wheel assembly
- Brake drums

3. Raise the lever arm of the actuator until the upper end is clear of the slot in the adjuster screw.
- Actuator off the adjuster pin
- Actuator spring from the shoe
- Hold-down spring assemblies and pins

4. Pull the bottom ends of the shoes apart and lift the lower return spring over the anchor plate. Allow the shoe ends to come together and remove the spring.
- Shoe assembly, along with the upper return spring and the adjusting screw assembly
- Upper return spring and the adjusting screw assembly from the shoes
- Retaining ring, pin, spring washer, and parking brake lever

To install:

5. Clean adjuster wheel and the backing plates with a suitable cleaner. Lubricate the backing plate contact points, levers and adjuster with a suitable lubricant.

6. Assemble the parking lever, spring washer (concave side facing the brake lever), pin, and retaining ring onto the rearward shoe.

7. Install or connect the following:
- Adjuster pin in the forward shoe with the pin projecting 0.276 in. (7mm) from the side of the shoe web where the adjuster actuator is installed.
- Upper return spring, with the brake shoes resting on a flat surface (the shoe with the parking lever to the rear of the vehicle)
- Adjuster screw assembly with the spring clip facing the backing plate
- Shoes in position on the backing plate. Do not place the lower shoe webs under the anchor plate.
- Lower return spring, spread the bottom of the shoes and position the shoe against the backing plate
- Hold-down pins and spring assemblies
- Adjuster actuator over the end of the adjuster pin so the top leg engages the notch in the adjuster screw
- Actuator spring, being careful not to over-stretch it more than 3.27 in. (83mm)
- Parking brake cable to the lever

8. Adjust the parking brake if the shoes will not totally retract.

- Drum, tire and wheel assembly

9. Adjust the rear brakes and lower the vehicle.

Duo-Servo Brakes

1. Before servicing the vehicle, refer to the precautions in the beginning of this section.

2. Remove or disconnect the following:
- Tire and wheel assembly
- Brake drums
- Shoe return springs, using a brake tool
- Shoe guide
- Hold-down springs and pins
- Actuator lever and pivot
- Lever return spring
- Actuator link, parking brake strut, spring retaining ring
- Parking brake lever and washer
- Shoe assemblies
- Adjuster screw and spring from the shoe assembly

To install:

3. Use a brake cleaning fluid to remove dirt from the brake drum. Check the drums for scoring, cracks and for out-of-round; service the drums as necessary.

4. Check the wheel cylinders by carefully pulling the lower edges of the wheel

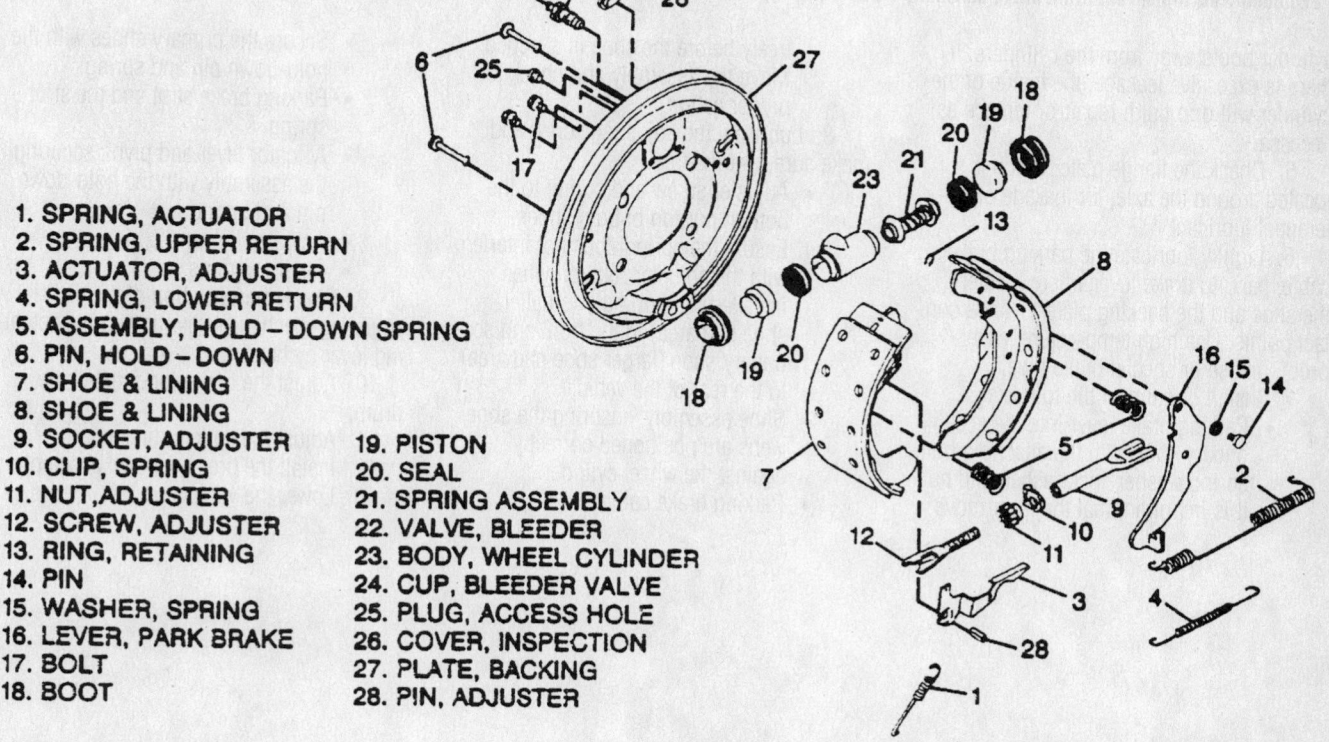

1. SPRING, ACTUATOR
2. SPRING, UPPER RETURN
3. ACTUATOR, ADJUSTER
4. SPRING, LOWER RETURN
5. ASSEMBLY, HOLD–DOWN SPRING
6. PIN, HOLD–DOWN
7. SHOE & LINING
8. SHOE & LINING
9. SOCKET, ADJUSTER
10. CLIP, SPRING
11. NUT, ADJUSTER
12. SCREW, ADJUSTER
13. RING, RETAINING
14. PIN
15. WASHER, SPRING
16. LEVER, PARK BRAKE
17. BOLT
18. BOOT
19. PISTON
20. SEAL
21. SPRING ASSEMBLY
22. VALVE, BLEEDER
23. BODY, WHEEL CYLINDER
24. CUP, BLEEDER VALVE
25. PLUG, ACCESS HOLE
26. COVER, INSPECTION
27. PLATE, BACKING
28. PIN, ADJUSTER

Exploded view of the rear drum brake assembly—Leading/Trailing type

93026G52

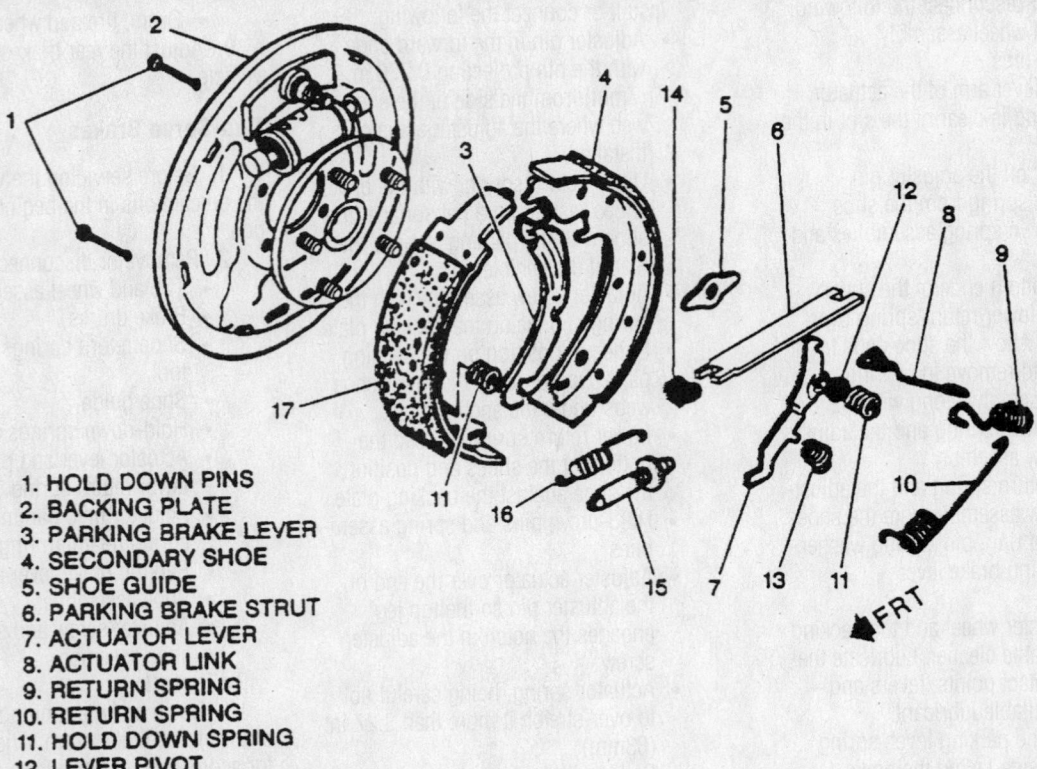

1. HOLD DOWN PINS
2. BACKING PLATE
3. PARKING BRAKE LEVER
4. SECONDARY SHOE
5. SHOE GUIDE
6. PARKING BRAKE STRUT
7. ACTUATOR LEVER
8. ACTUATOR LINK
9. RETURN SPRING
10. RETURN SPRING
11. HOLD DOWN SPRING
12. LEVER PIVOT
13. LEVER RETURN SPRING
14. STRUT SPRING
15. ADJUSTING SCREW ASSEMBLY
16. ADJUSTING SCREW SPRING
17. PRIMARY SHOE

93026G53

Exploded view of the rear drum brake assembly—Duo-Servo type

cylinder boots away from the cylinders. If there is excessive leakage, the inside of the cylinder will drip fluid; repair or replace as necessary.

5. Check the flange plate, which is located around the axle, for leakage of differential lubricant.

6. Lightly lubricate the parking brake cable, parking brake lever where it enters the shoe and the backing plate-to-shoe contact points. Use high temperature, waterproof, grease or special brake lube.

7. Install or connect the following:
• Parking brake lever into the secondary shoe with the attaching bolt, spring washer, lockwasher, and nut. It is important that the lever move

freely before the shoe is attached. Move the assembly and check for proper action.

8. Lubricate the adjusting screw and make sure it works freely.
• Adjuster screw and spring to the bottom portion of both shoes. Ensure the spring does not interfere with the adjuster rotation when installed. The primary (smaller shoe pad area) to the front and secondary shoe (larger shoe pad area) to the rear of the vehicle.
• Shoe assembly, ensuring the shoe webs are positioned correctly against the wheel cylinder
• Parking brake cable

• Secure the primary shoes with the hold-down pin and spring
• Parking brake strut and the strut spring
• Actuator lever and pivot, securing the assembly with the hold-down pin and spring
• Actuator link and spring
• Return springs

9. Check the operation of the self-adjusting mechanism by moving the actuating lever by hand.

10. Adjust the brakes and install the drum.

11. Adjust the parking brake.

12. Install the tire and wheel assembly.

13. Lower the vehicle.

SPECIFICATION CHARTS

ENGINE AND VEHICLE IDENTIFICATION

Engine							Model Year	
Code ①	Liters (cc)	Cu. In.	Cyl.	Fuel Sys.	Engine Type	Eng. Mfg.	Code ②	Year
V	4.8 (4802)	293	8	MFI	OHV	CPC	Y	2000
T	5.3 (5327)	325	8	MFI	OHV	CPC	1	2001
R	5.7 (5735)	350	8	MFI	OHV	CPC	2	2002
U	6.0 (5966)	364	8	MFI	OHV	CPC	3	2003
N	6.0 (5966)	364	8	MFI	OHV	CPC	4	2004
S	6.5 (6473)	395	8	DSL	OHV	CPC		
J	7.4 (7440)	454	8	MFI	OHV	CPC		
G	8.1 (8128)	496	8	MFI	OHV	CPC		

CPC: Chevrolet/Pontiac/Canada

DSL: Diesel

MFI: Multi-port Fuel Injection

① 8th position of VIN

② 10th position of VIN

42372-YUKO-C01

GENERAL ENGINE SPECIFICATIONS
All measurements are given in inches.

Year	Model	Engine Displacement Liters (cc)	Engine Series (ID/VIN)	Fuel System	Net Horsepower @ rpm	Net Torque @ rpm (ft. lbs.)	Bore x Stroke (in.)	Compression Ratio	Oil Pressure @ rpm
2000	Denali	5.7 (5735)	R	MFI	255@4600	335@2800	4.00x3.48	9.4:1	18@2000
	Denali	6.5 (6374)	S	DSL	180@3400	360@1700	4.06x3.82	21.5:1	30-43@2000
	Escalade	5.7 (5735)	R	MFI	255@4600	335@2800	4.00x3.48	9.4:1	18@2000
	Escalade	6.5 (6374)	S	DSL	180@3400	360@1700	4.06x3.82	21.5:1	30-43@2000
	Suburban	5.7 (5735)	R	MFI	255@4600	335@2800	4.00x3.48	9.4:1	18@2000
	Suburban	7.4 (7440)	J	MFI	290@4200	410@3200	4.25x4.00	9.0:1	40@2000
	Suburban	6.5 (6374)	S	DSL	180@3400	360@1700	4.06x3.82	21.5:1	30-43@2000
	Tahoe/Yukon	5.7 (5735)	R	MFI	255@4600	335@2800	4.00x3.48	9.4:1	18@2000
2001	Denali	6.0 (5967)	U	MFI	300@4400	360@4000	4.00x3.62	9.4:1	18@2000
	Denali XL	6.0 (5967)	U	MFI	300@4400	360@4000	4.00x3.62	9.4:1	18@2000
	Suburban	5.3 (5327)	T	MFI	285@4000	360@4000	3.78x3.62	9.5:1	18@2000
	Suburban	6.0 (5967)	U	MFI	300@4400	360@4000	4.00x3.62	9.4:1	18@2000
	Suburban	8.1 (8128)	G	MFI	340@4200	455@3200	4.25x4.37	9.1:1	10@2000
	Tahoe	4.8 (4802)	V	MFI	270@5200	285@4000	3.78x3.27	9.5:1	18@2000
	Tahoe	5.3 (5327)	T	MFI	285@4000	360@4000	3.78x3.62	9.5:1	18@2000
	Yukon	4.8 (4802)	V	MFI	270@5200	285@4000	3.78x3.27	9.5:1	18@2000
	Yukon	5.3 (5327)	T	MFI	285@4000	360@4000	3.78x3.62	9.5:1	18@2000
	Yukon XL	5.3 (5327)	T	MFI	285@4000	360@4000	3.78x3.62	9.5:1	18@2000
	Yukon XL	6.0 (5967)	U	MFI	300@4400	360@4000	4.00x3.62	9.4:1	18@2000
	Yukon XL	8.1 (8128)	G	MFI	340@4200	455@3200	4.25x4.37	9.1:1	10@2000
2002	Denali	6.0 (5967)	U	MFI	300@4400	360@4000	4.00x3.62	9.4:1	18@2000
	Denali XL	6.0 (5967)	U	MFI	300@4400	360@4000	4.00x3.62	9.4:1	18@2000
	Escalade	5.3 (5327)	T	MFI	285@4000	360@4000	3.78x3.62	9.5:1	18@2000
	Escalade	6.0 (5967)	U	MFI	300@4400	360@4000	4.00x3.62	9.4:1	18@2000
	Suburban	5.3 (5327)	T	MFI	285@4000	360@4000	3.78x3.62	9.5:1	18@2000
	Suburban	6.0 (5967)	U	MFI	300@4400	360@4000	4.00x3.62	9.4:1	18@2000
	Suburban	8.1 (8128)	G	MFI	340@4200	455@3200	4.25x4.37	9.1:1	10@2000
	Tahoe	4.8 (4802)	V	MFI	270@5200	285@4000	3.78x3.27	9.5:1	18@2000
	Tahoe	5.3 (5327)	T	MFI	285@4000	360@4000	3.78x3.62	9.5:1	18@2000
	Yukon	4.8 (4802)	V	MFI	270@5200	285@4000	3.78x3.27	9.5:1	18@2000
	Yukon	5.3 (5327)	T	MFI	285@4000	360@4000	3.78x3.62	9.5:1	18@2000
	Yukon XL	5.3 (5327)	T	MFI	285@4000	360@4000	3.78x3.62	9.5:1	18@2000
	Yukon XL	6.0 (5967)	U	MFI	300@4400	360@4000	4.00x3.62	9.4:1	18@2000
	Yukon XL	8.1 (8128)	G	MFI	340@4200	455@3200	4.25x4.37	9.1:1	10@2000

42372-YUKO-C02

GENERAL ENGINE SPECIFICATIONS

All measurements are given in inches.

Year	Model	Engine Displacement Liters (cc)	Engine Series (ID/VIN)	Fuel System	Net Horsepower @ rpm	Net Torque @ rpm (ft. lbs.)	Bore x Stroke (in.)	Compression Ratio	Oil Pressure @ rpm
2003-04	Denali	6.0 (5967)	N	MFI	300@4400	360@4000	4.00x3.62	9.4:1	18@2000
	Denali XL	6.0 (5967)	N	MFI	300@4400	360@4000	4.00x3.62	9.4:1	18@2000
	Escalade	5.3 (5327)	T	MFI	285@4000	360@4000	3.78x3.62	9.5:1	18@2000
	Escalade	6.0 (5967)	N	MFI	300@4400	360@4000	4.00x3.62	9.4:1	18@2000
	Suburban	5.3 (5327)	T	MFI	285@4000	360@4000	3.78x3.62	9.5:1	18@2000
	Suburban	6.0 (5967)	N	MFI	300@4400	360@4000	4.00x3.62	9.4:1	18@2000
	Suburban	8.1 (8128)	G	MFI	340@4200	455@3200	4.25x4.37	9.1:1	10@2000
	Tahoe	4.8 (4802)	V	MFI	270@5200	285@4000	3.78x3.27	9.5:1	18@2000
	Tahoe	5.3 (5327)	T	MFI	285@4000	360@4000	3.78x3.62	9.5:1	18@2000
	Yukon	4.8 (4802)	V	MFI	270@5200	285@4000	3.78x3.27	9.5:1	18@2000
	Yukon	5.3 (5327)	T	MFI	285@4000	360@4000	3.78x3.62	9.5:1	18@2000
	Yukon XL	5.3 (5327)	T	MFI	285@4000	360@4000	3.78x3.62	9.5:1	18@2000
	Yukon XL	6.0 (5967)	N	MFI	300@4400	360@4000	4.00x3.62	9.4:1	18@2000
	Yukon XL	8.1 (8128)	G	MFI	340@4200	455@3200	4.25x4.37	9.1:1	10@2000

DSL: Diesel

MFI: Multi-port Fuel Injection

42372-YUKO-C03

GASOLINE ENGINE TUNE-UP SPECIFICATIONS

Year	Engine Displacement Liters (cc)	Engine ID/VIN	Spark Plugs Gap (in.)	Ignition Timing (deg.) MT	Ignition Timing (deg.) AT	Fuel Pump (psi)	Idle Speed (rpm) MT	Idle Speed (rpm) AT	Valve Clearance In.	Valve Clearance Ex.
2000	5.7 (5735)	R	0.060	—	①	60-66 ②	—	525	HYD	HYD
	7.4 (7440)	J	0.060	—	①	60-66 ②	—	675 ③	HYD	HYD
2001	4.8 (4802)	V	0.060	—	①	55-62 ②	—	④	HYD	HYD
	5.3 (5327)	T	0.060	—	①	55-62 ②	—	④	HYD	HYD
	6.0 (5966)	U	0.060	—	①	55-62 ②	—	④	HYD	HYD
	8.1 (8128)	G	0.060	—	①	55-62 ②	—	④	HYD	HYD
2002	4.8 (4802)	V	0.060	—	①	55-62 ②	—	④	HYD	HYD
	5.3 (5327)	T	0.060	—	①	55-62 ②	—	④	HYD	HYD
	6.0 (5966)	U	0.060	—	①	55-62 ②	—	④	HYD	HYD
	8.1 (8128)	G	0.060	—	①	55-62 ②	—	④	HYD	HYD
2003-04	4.8 (4802)	V	0.060	—	①	55-62 ②	—	④	HYD	HYD
	5.3 (5327)	T	0.060	—	①	55-62 ②	—	④	HYD	HYD
	6.0 (5966)	N	0.060	—	①	55-62 ②	—	④	HYD	HYD
	8.1 (8128)	G	0.060	—	①	55-62 ②	—	④	HYD	HYD

NOTE: The Vehicle Emission Control Information label often reflects specification changes made during production. The label figures must be used if they differ from those in this chart.

HYD: Hydraulic

① Ignition timing is preset and cannot be adjusted

② With key ON and engine OFF

③ Over 8500 GVW

④ Idle speed is maintained by the Powertrain Control Module (PCM)

42372-YUKO-C04

DIESEL ENGINE TUNE-UP SPECIFICATIONS

Year	Engine Displacement cu. in. (cc)	Engine ID/VIN	Valve Clearance		Intake Valve Opens (deg.)	Injection Pump Setting (deg.)	Injection Nozzle Pressure (psi)		Idle Speed (rpm)	Cranking Compression Pressure (psi)
			Intake (in.)	Exhaust (in.)			New	Used		
					①	①	1800	1700	①	380-400
2000	6.5 (6473)	S	HYD	HYD						

NOTE: The Vehicle Emission Control Information label often reflects specification changes made during production. The label figures must be used if they differ from those in this chart.

HYD: Hydraulic

NA: Not Available

① Refer to Vehicle Emission Control Information label

42372-YUKO-C05

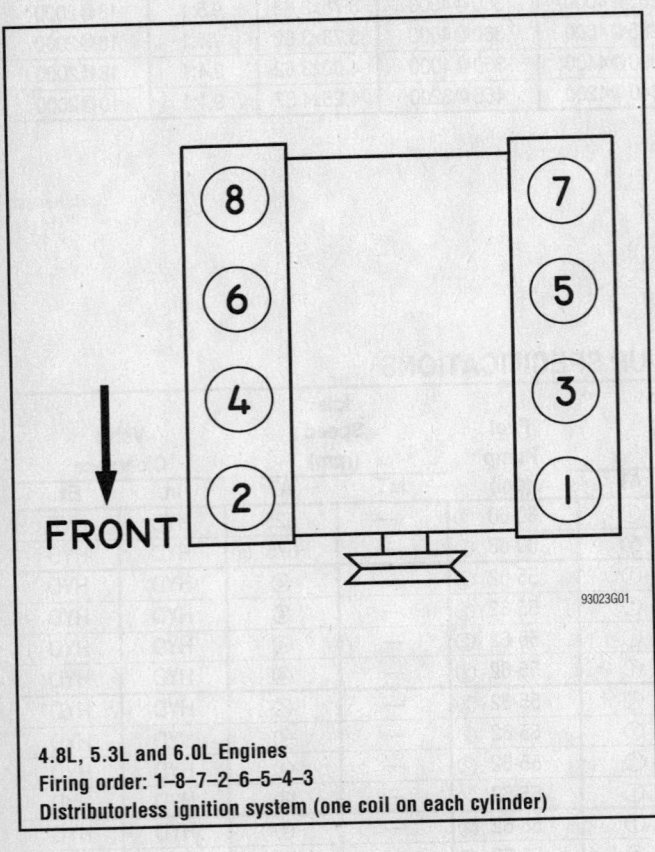

4.8L, 5.3L and 6.0L Engines
Firing order: 1–8–7–2–6–5–4–3
Distributorless ignition system (one coil on each cylinder)

93023G01

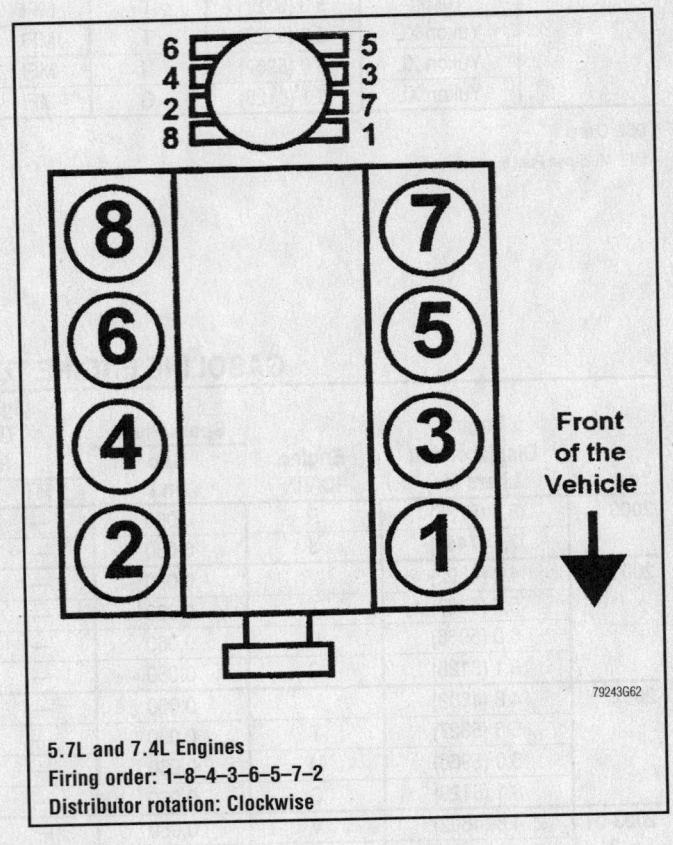

5.7L and 7.4L Engines
Firing order: 1–8–4–3–6–5–7–2
Distributor rotation: Clockwise

Front of the Vehicle

79243G62

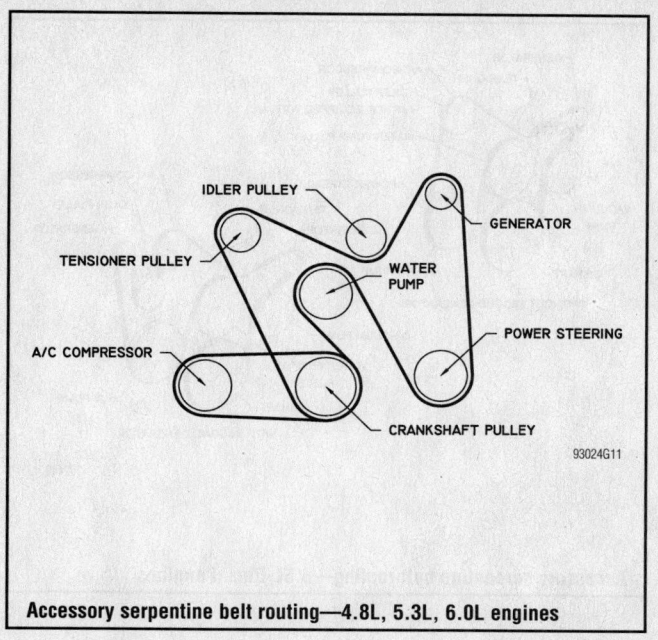

Accessory serpentine belt routing—4.8L, 5.3L, 6.0L engines

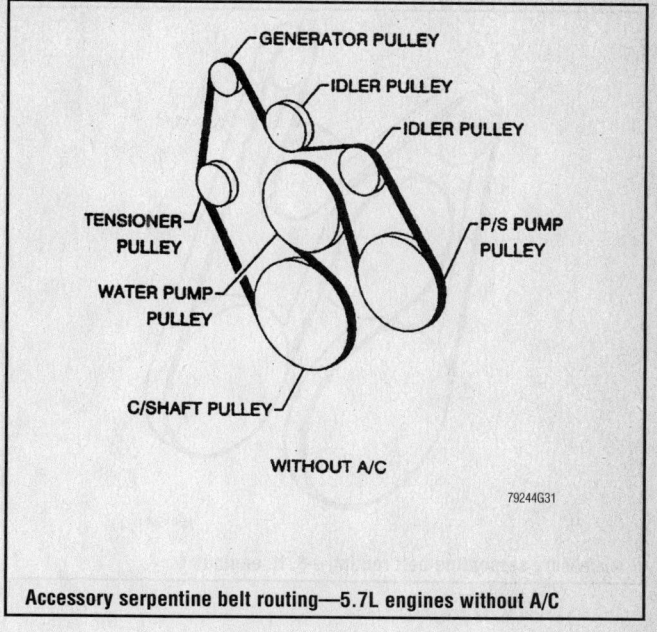

Accessory serpentine belt routing—5.7L engines without A/C

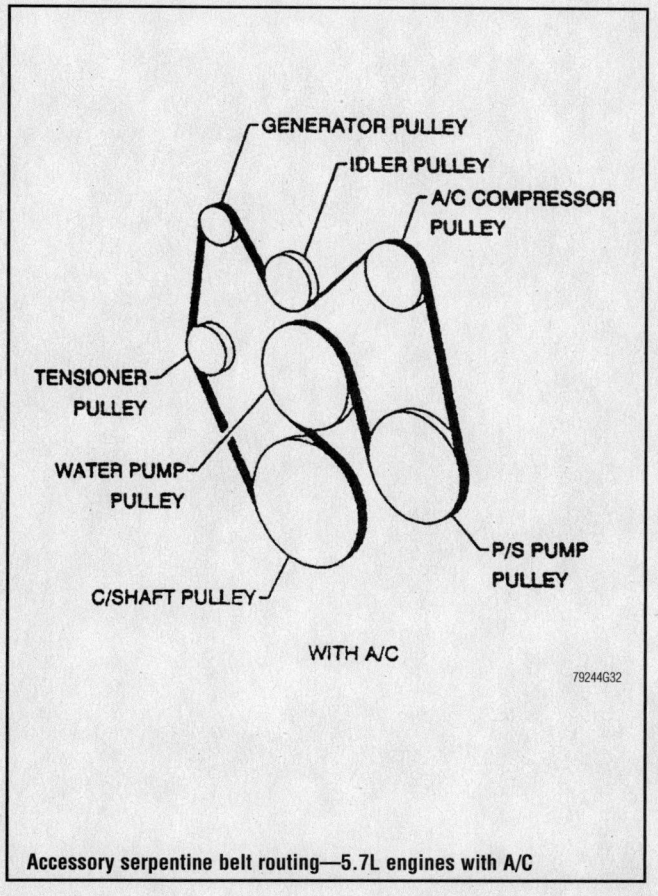

Accessory serpentine belt routing—5.7L engines with A/C

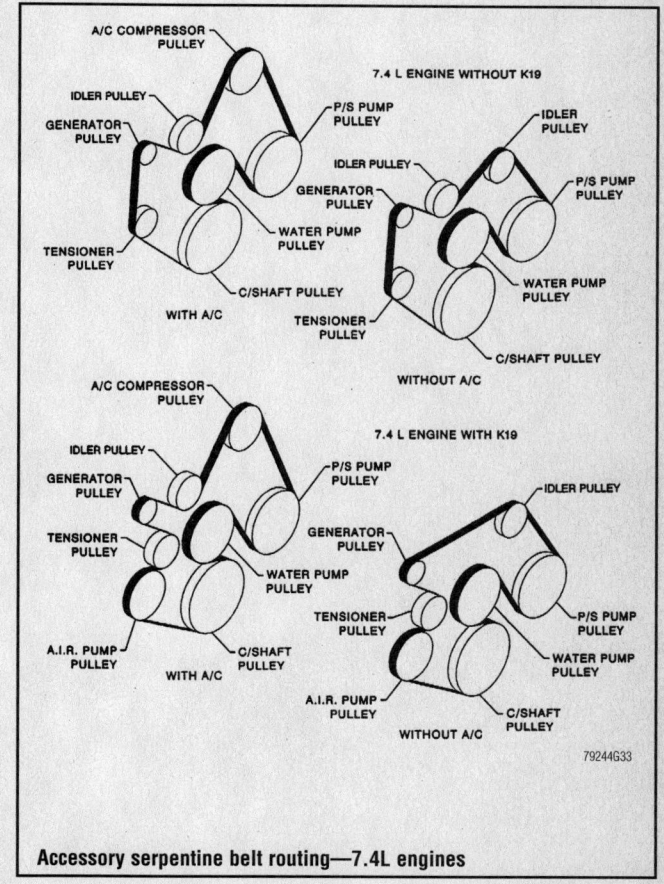

Accessory serpentine belt routing—7.4L engines

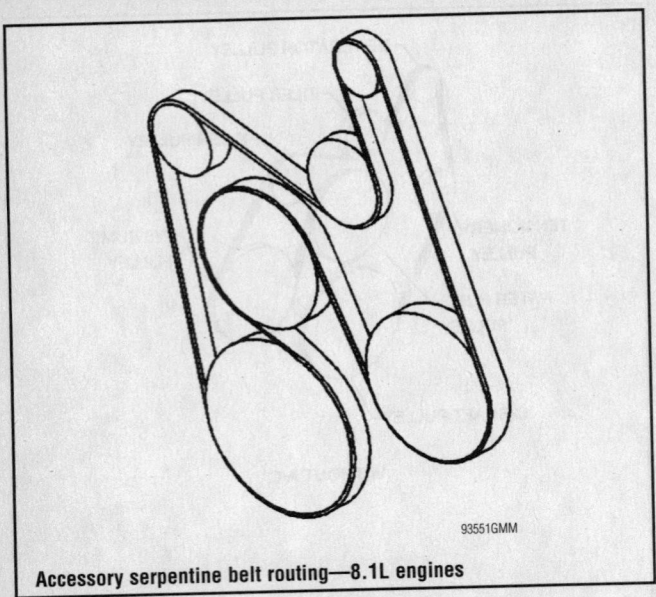

93551GMM

Accessory serpentine belt routing—8.1L engines

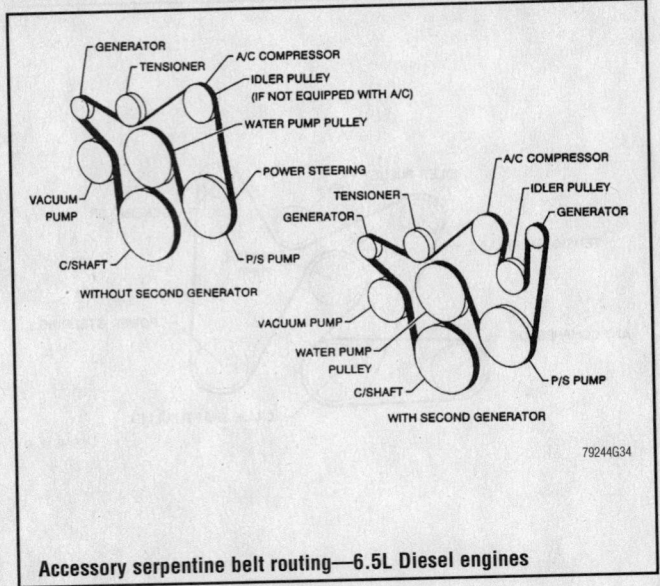

79244G34

Accessory serpentine belt routing—6.5L Diesel engines

CAPACITIES

Year	Model	Engine Displacement Liters (cc)	Engine ID/VIN	Engine Oil with Filter (qts.)	Transmission (pts.) 5-Spd	Transmission (pts.) Auto.	Transfer Case (pts.)	Drive Axle Front (pts.)	Drive Axle Rear (pts.)	Fuel Tank (gal.)	Cooling System (qts.)
2000	Denali	5.7 (5735)	R	5.0	—	①	②	③	④	⑤	18.0
	Denali	6.5 (6473)	S	7.0	—	①	—	—	④	⑤	23.8
	Escalade	5.7 (5735)	R	5.0	—	①	②	③	④	⑤	18.0
	Escalade	6.5 (6473)	S	7.0	—	①	—	—	④	⑤	23.8
	Suburban	7.4 (7440)	J	6.0	—	①	②	③	④	25.0 ⑥	24.5
	Suburban	5.7 (5735)	R	5.0	—	①	—	—	④	⑤	18.0
	Tahoe	5.7 (5735)	R	5.0	—	①	②	③	④	⑤	18.0
	Tahoe	6.5 (6473)	S	7.0	—	①	—	—	④	⑤	23.8
	Yukon	5.7 (5735)	R	5.0	—	①	②	③	④	⑤	18.0
	Yukon	6.5 (6473)	S	7.0	—	①	—	—	④	⑤	23.8
2001	Denali	6.0 (5967)	U	6.0	—	⑦	②	⑧	⑨	26.0	14.8 ⑪
	Denali XL	6.0 (5967)	U	6.0	—	⑦	②	⑧	⑨	⑩	14.8 ⑪
	Suburban	5.3 (5327)	T	6.0	—	⑦	②	⑧	⑨	⑩	13.4 ⑫
	Suburban	6.0 (5967)	U	6.0	—	⑦	②	⑧	⑨	⑩	15.8 ⑬
	Suburban	8.1 (8128)	G	6.5	—	⑦	②	⑧	⑨	⑩	20.7
	Tahoe	4.8 (4802)	V	6.0	—	⑦	②	⑧	⑨	26.0	13.4 ⑫
	Tahoe	5.3 (5327)	T	6.0	—	⑦	②	⑧	⑨	26.0	13.4 ⑫
	Yukon	4.8 (4802)	V	6.0	—	⑦	②	⑧	⑨	26.0	13.4 ⑫
	Yukon	5.3 (5327)	T	6.0	—	⑦	②	⑧	⑨	26.0	13.4 ⑫
	Yukon XL	5.3 (5327)	T	6.0	—	⑦	②	⑧	⑨	⑩	13.4 ⑫
	Yukon XL	6.0 (5967)	U	6.0	—	⑦	②	⑧	⑨	⑩	15.8 ⑬
	Yukon XL	8.1 (8128)	G	6.5	—	⑦	②	⑧	⑨	⑩	20.7
2002	Denali	6.0 (5967)	U	6.0	—	⑦	②	⑧	⑨	26.0	14.8 ⑪
	Denali XL	6.0 (5967)	U	6.0	—	⑦	②	⑧	⑨	⑩	14.8 ⑪
	Escalade	5.3 (5327)	T	6.0	—	⑦	②	⑧	⑨	⑩	13.4 ⑫
	Escalade	6.0 (5967)	U	6.0	—	⑦	②	⑧	⑨	⑩	15.8 ⑬
	Suburban	5.3 (5327)	T	6.0	—	⑦	②	⑧	⑨	⑩	13.4 ⑫
	Suburban	6.0 (5967)	U	6.0	—	⑦	②	⑧	⑨	⑩	15.8 ⑬
	Suburban	8.1 (8128)	G	6.5	—	⑦	②	⑧	⑨	⑩	20.7
	Tahoe	4.8 (4802)	V	6.0	—	⑦	②	⑧	⑨	26.0	13.4 ⑫
	Tahoe	5.3 (5327)	T	6.0	—	⑦	②	⑧	⑨	26.0	13.4 ⑫
	Yukon	4.8 (4802)	V	6.0	—	⑦	②	⑧	⑨	26.0	13.4 ⑫
	Yukon	5.3 (5327)	T	6.0	—	⑦	②	⑧	⑨	26.0	13.4 ⑫
	Yukon XL	5.3 (5327)	T	6.0	—	⑦	②	⑧	⑨	⑩	13.4 ⑫
	Yukon XL	6.0 (5967)	U	6.0	—	⑦	②	⑧	⑨	⑩	15.8 ⑬
	Yukon XL	8.1 (8128)	G	6.5	—	⑦	②	⑧	⑨	⑩	20.7

42372-YUKO-C06

CAPACITIES

Year	Model	Engine Displacement Liters (cc)	Engine ID/VIN	Engine Oil with Filter (qts.)	Transmission (pts.)		Transfer Case (pts.)	Drive Axle		Fuel Tank (gal.)	Cooling System (qts.)
					5-Spd	Auto.		Front (pts.)	Rear (pts.)		
2003-04	Denali	6.0 (5967)	N	6.0	—	⑦	②	⑧	⑨	26.0	14.8 ⑪
	Denali XL	6.0 (5967)	N	6.0	—	⑦	②	⑧	⑨	⑩	14.8 ⑪
	Escalade	5.3 (5327)	T	6.0	—	⑦	②	⑧	⑨	⑩	13.4 ⑫
	Escalade	6.0 (5967)	N	6.0	—	⑦	②	⑧	⑨	⑩	15.8 ⑬
	Suburban	5.3 (5327)	T	6.0	—	⑦	②	⑧	⑨	⑩	13.4 ⑫
	Suburban	6.0 (5967)	N	6.0	—	⑦	②	⑧	⑨	⑩	15.8 ⑬
	Suburban	8.1 (8128)	G	6.5	—	⑦	②	⑧	⑨	⑩	20.7
	Tahoe	4.8 (4802)	V	6.0	—	⑦	②	⑧	⑨	26.0	13.4 ⑫
	Tahoe	5.3 (5327)	T	6.0	—	⑦	②	⑧	⑨	26.0	13.4 ⑫
	Yukon	4.8 (4802)	V	6.0	—	⑦	②	⑧	⑨	26.0	13.4 ⑫
	Yukon	5.3 (5327)	T	6.0	—	⑦	②	⑧	⑨	⑩	13.4 ⑫
	Yukon XL	5.3 (5327)	T	6.0	—	⑦	②	⑧	⑨	⑩	15.8 ⑬
	Yukon XL	6.0 (5967)	N	6.0	—	⑦	②	⑧	⑨	⑩	20.7
	Yukon XL	8.1 (8128)	G	6.5	—	⑦	②	⑧	⑨	⑩	20.7

NOTE: All capacities are approximate. Add fluid gradually and check to be sure a proper fluid level is obtained.

① 4L60E trans.: 10.0 pts.
　4L80E trans.: 14.5 pts.

② NV241 and NV243: 4.5 pts.
　4401 and 4470: 6.6 pts.

③ K2 models: 3.5 pts.
　K3 models: 4.5 pts.

④ 8.5 in. ring gear: 4.2 pts.
　9.5 in. ring gear: 6.5 pts.
　9.75 in. ring gear: 6.0 pts.
　10.5 in. ring gear: 6.5 pts.

⑤ Std. available with 25 and 34 gallon tanks
　Chassis cab available with 22, 30 and 34 gallon tanks

⑥ Optional 31 and 40 gallon tanks

⑦ 4L60E trans.: 10.0 pts.
　4L80E trans.: 15.4 pts.

⑧ 8.25 in ring gear: 3.5 pts.
　9.25 ring gear: 3.7 pts.

⑨ 8.6 in. ring gear: 4.8 pts.
　9.5 & 10.5 in. ring gear: 5.5 pts.
　11.5 in. ring gear: 7.7 pts.

⑩ 1500 XL: 32.5 gallon tank
　2500 XL: 38.5 gallon tank

⑪ With optional oil cooler: 15.4 qts.

⑫ With optional A/C: 14.9 qts.
　With front A/C: 14.4 qts.
　With front and rear A/C: 15.8 qts.

⑬ With optional oil cooler: 15.4 qts.

42372-YUKO-C07

VALVE SPECIFICATIONS

Year	Engine Displacement Liters (cc)	Engine ID/VIN	Seat Angle (deg.)	Face Angle (deg.)	Spring Test Pressure (lbs. @ in.)	Spring Installed Height (in.)	Stem-to-Guide Clearance (in.)		Stem Diameter (in.)	
							Intake	Exhaust	Intake	Exhaust
2000	5.7 (5735)	R	46	45	187-203@1.27	1.69-1.71	0.0010-0.0027	0.0010-0.0027	NA	NA
	6.5 (6473)	S	46	45	230@1.40	1.80	0.0010-0.0027	0.0010-0.0027	NA	NA
	7.4 (7440)	J	46	45	238-262@1.34	1.83	0.0010-0.0029 ①	0.0012-0.0031 ①	NA	NA
2001	4.8 (4802)	V	46	45	220@1.32	1.80	0.0010-0.0026	0.0010-0.0026	0.3132-0.3140	0.3132-0.3140
	5.3 (5327)	T	46	45	220@1.32	1.80	0.0010-0.0026	0.0010-0.0026	0.3132-0.3140	0.3132-0.3140
	6.0 (5966)	U	46	45	220@1.32	1.80	0.0010-0.0026	0.0010-0.0026	0.3132-0.3140	0.3132-0.3140
	8.1 (8128)	G	46	45	216-236@1.34	1.81-1.84	0.0010-0.0029	0.0012-0.0031	0.3715-0.3722	0.3713-0.3720
2002	4.8 (4802)	V	46	45	220@1.32	1.80	0.0010-0.0026	0.0010-0.0026	0.3132-0.3140	0.3132-0.3140
	5.3 (5327)	T	46	45	220@1.32	1.80	0.0010-0.0026	0.0010-0.0026	0.3132-0.3140	0.3132-0.3140
	6.0 (5966)	U	46	45	220@1.32	1.80	0.0010-0.0026	0.0010-0.0026	0.3132-0.3140	0.3132-0.3140
	8.1 (8128)	G	46	45	216-236@1.34	1.81-1.84	0.0010-0.0029	0.0012-0.0031	0.3715-0.3722	0.3713-0.3720
2003-04	4.8 (4802)	V	46	45	220@1.32	1.80	0.0010-0.0026	0.0010-0.0026	0.3132-0.3140	0.3132-0.3140
	5.3 (5327)	T	46	45	220@1.32	1.80	0.0010-0.0026	0.0010-0.0026	0.3132-0.3140	0.3132-0.3140
	6.0 (5966)	N	46	45	220@1.32	1.80	0.0010-0.0026	0.0010-0.0026	0.3132-0.3140	0.3132-0.3140
	8.1 (8128)	G	46	45	216-236@1.34	1.81-1.84	0.0010-0.0029	0.0012-0.0031	0.3715-0.3722	0.3713-0.3720

NA: Not Available

① Service limit:
  Intake: 0.0037 MAX
  Exhaust: 0.0049 MAX

42372-YUKO-C08

CRANKSHAFT AND CONNECTING ROD SPECIFICATIONS

All measurements are given in inches.

Year	Engine Displacement Liters (cc)	Engine ID/VIN	Crankshaft				Connecting Rod		
			Main Brg. Journal Dia.	Main Brg. Oil Clearance	Shaft End-play	Thrust on No.	Journal Diameter	Oil Clearance	Side Clearance
2000	5.7 (5735)	R	①	②	0.0020-0.0080	5	2.0978-2.0998	0.0013-0.0035	0.0060-0.0140
	6.5 (6473)	S	③	④	0.0039-0.0100	3	⑤	0.0018-0.0039	0.0067-0.0248
	7.4 (7440)	J	2.7482-2.7489	⑥	0.0050-0.0110	5	2.1990-2.1996	0.0011-0.0029	0.0013-0.0230
2001	4.8 (4802)	V	2.5580-2.5593	0.0007-0.0021	0.0015-0.0078	5	2.0990-2.1000	0.0006-0.0030	0.0043-0.0200
	5.3 (5327)	T	2.5580-2.5593	0.0007-0.0021	0.0015-0.0078	5	2.0990-2.1000	0.0006-0.0030	0.0043-0.0200
	6.0 (5966)	U	2.5580-2.5593	0.0007-0.0021	0.0015-0.0078	5	2.0990-2.1000	0.0006-0.0030	0.0043-0.0200
	8.1 (8128)	G	2.7482-2.7489	⑦	0.0050-0.0110	NA	2.1990-2.1996	0.0008-0.0025	0.0151-0.0270
2002	4.8 (4802)	V	2.5580-2.5593	0.0007-0.0021	0.0015-0.0078	5	2.0990-2.1000	0.0006-0.0030	0.0043-0.0200
	5.3 (5327)	T	2.5580-2.5593	0.0007-0.0021	0.0015-0.0078	5	2.0990-2.1000	0.0006-0.0030	0.0043-0.0200
	6.0 (5966)	U	2.5580-2.5593	0.0007-0.0021	0.0015-0.0078	5	2.0990-2.1000	0.0006-0.0030	0.0043-0.0200
	8.1 (8128)	G	2.7482-2.7489	⑦	0.0050-0.0110	NA	2.1990-2.1996	0.0008-0.0025	0.0151-0.0270
2003-04	4.8 (4802)	V	2.5580-2.5593	0.0007-0.0021	0.0015-0.0078	5	2.0990-2.1000	0.0006-0.0030	0.0043-0.0200
	5.3 (5327)	T	2.5580-2.5593	0.0007-0.0021	0.0015-0.0078	5	2.0990-2.1000	0.0006-0.0030	0.0043-0.0200
	6.0 (5966)	N	2.5580-2.5593	0.0007-0.0021	0.0015-0.0078	5	2.0990-2.1000	0.0006-0.0030	0.0043-0.0200
	8.1 (8128)	G	2.7482-2.7489	⑦	0.0050-0.0110	NA	2.1990-2.1996	0.0008-0.0025	0.0151-0.0270

NA: Not Available

① No. 1: 2.4484 in.-2.4493 in.
  Nos. 2, 3, 4: 2.4481 in.-2.4490 in.
  No. 5: 2.4479 in.-2.4488 in.

② No. 1: 0.0007 in.-0.0021 in.
  Nos. 2, 3, 4: 0.0009 in.-0.0024 in.
  No. 5: 0.0010 in.-0.0027 in.

③ No. 1, 2, 3, 4: 2.9517 in.-2.9520 in. (Blue)
  2.9520 in.-2.9524 in. (Orange/Red)
  2.9524 in.-2.9527 in. (White)
  No. 5: 2.9515 in.-2.9518 in. (Blue)
  2.9518 in.-2.9522 in. (Orange/Red)
  2.9522 in.-2.9525 in. (White)

④ No. 1, 2, 3, 4: 0.0018 in.-0.0033 in.
  No. 5: 0.0022 in.-0.0037 in.

⑤ 2.399 in.-2.400 in. (Green)
  2.400 in.-2.401 in. (Yellow)

⑥ No. 1: 0.0017 in.-0.0030 in.
  No. 2, 3, 4: 0.0011 in.-0.0024 in.
  No. 5: 0.0025 in.-0.0038 in.

⑦ No. 1, 2, 3, 4: 0.0008-0.0020 in.
  No. 5: 0.0014-0.0026 in.

42372-YUKO-C09

PISTON AND RING SPECIFICATIONS
All measurements are given in inches.

Year	Engine Displacement Liters (cc)	Engine ID/VIN	Piston Clearance	Ring Gap			Ring Side Clearance		
				Top Compression	Bottom Compression	Oil Control	Top Compression	Bottom Compression	Oil Control
1999	5.7 (5735)	R	0.0007-0.0021	0.010-0.020	0.018-0.026	0.010-0.030	0.0012-0.0032	0.0012-0.0032	0.0020-0.0070
	6.5 (6473)	S	①	0.010-0.020	0.030-0.039	0.010-0.020	0.0015-0.0031	0.0015-0.0031	0.0016-0.0035
	7.4 (7440)	J	0.0018-0.0030	0.010-0.0180	0.016-0.0240	0.010-0.030	0.0012-0.0029	0.0012-0.0029	0.0050-0.0065
2000	5.7 (5735)	R	0.0007-0.0021	0.010-0.020	0.018-0.026	0.010-0.030	0.0012-0.0032	0.0012-0.0032	0.0020-0.0070
	6.5 (6473)	S	①	0.010-0.020	0.030-0.039	0.010-0.020	0.0015-0.0031	0.0015-0.0031	0.0016-0.0035
	7.4 (7440)	J	0.0018-0.0030	0.010-0.0180	0.016-0.0240	0.010-0.030	0.0012-0.0029	0.0012-0.0029	0.0050-0.0065
2001	4.8 (4802)	V	0.0010-0.0024	0.009-0.015	0.017-0.025	0.007-0.027	0.0016-0.0033	0.0016-0.0031	0.0004-0.0087
	5.3 (5327)	T	0.0010-0.0024	0.009-0.015	0.017-0.025	0.007-0.027	0.0016-0.0033	0.0016-0.0031	0.0004-0.0087
	6.0 (5966)	U	0.0010-0.0024	0.009-0.015	0.017-0.025	0.007-0.027	0.0016-0.0033	0.0016-0.0031	0.0004-0.0087
	8.1 (8128)	G	②	0.012-0.018	0.017-0.025	0.010-0.030	0.0012-0.0029	0.0012-0.0029	0.002-0.008
2002-03	4.8 (4802)	V	0.0010-0.0024	0.009-0.015	0.017-0.025	0.007-0.027	0.0016-0.0033	0.0016-0.0031	0.0004-0.0087
	5.3 (5327)	T	0.0010-0.0024	0.009-0.015	0.017-0.025	0.007-0.027	0.0016-0.0033	0.0016-0.0031	0.0004-0.0087
	6.0 (5966)	N	0.0010-0.0024	0.009-0.015	0.017-0.025	0.007-0.027	0.0016-0.0033	0.0016-0.0031	0.0004-0.0087
	8.1 (8128)	G	②	0.012-0.018	0.017-0.025	0.010-0.030	0.0012-0.0029	0.0012-0.0029	0.002-0.008

① 1-6: 0.0037-0.0047 in.
7-8: 0.0042-0.0052 in.

② Interference fit (coated piston)

42372-YUKO-C10

TORQUE SPECIFICATIONS
All readings in ft. lbs.

Year	Engine Displacement Liters (cc)	Engine ID/VIN	Cylinder Head Bolts	Main Bearing Bolts	Rod Bearing Bolts	Crankshaft Damper Bolts	Flywheel Bolts	Manifold Intake *	Manifold Exhaust	Spark Plugs	Lug Nut
2000	5.7 (5735)	R	①	②	③	74	74	④	⑤	15	⑥
	6.5 (6473)	S	①	⑦	48	200	65	31	26	—	⑥
	7.4 (7440)	J	85	100	45	110	67	30	22	15	⑥
2001	4.8 (4802)	V	⑧	⑨	⑩	⑪	⑫	⑬	⑭	12	140
	5.3 (5327)	T	⑧	⑨	⑩	⑪	⑫	⑬	⑭	12	140
	6.0 (5966)	U	⑧	⑨	⑩	⑪	⑫	⑬	⑭	12	140
	8.1 (8128)	G	⑮	⑯	⑯	189	⑰	⑱	⑲	15	140
2002	4.8 (4802)	V	⑧	⑨	⑩	⑪	⑫	⑬	⑭	12	140
	5.3 (5327)	T	⑧	⑨	⑩	⑪	⑫	⑬	⑭	12	140
	6.0 (5966)	U	⑧	⑨	⑩	⑪	⑫	⑬	⑭	12	140
	8.1 (8128)	G	⑮	⑯	⑯	189	⑰	⑱	⑲	15	140
2003-04	4.8 (4802)	V	⑧	⑨	⑩	⑪	⑫	⑬	⑭	12	140
	5.3 (5327)	T	⑧	⑨	⑩	⑪	⑫	⑬	⑭	12	140
	6.0 (5966)	N	⑧	⑨	⑩	⑪	⑫	⑬	⑭	12	140
	8.1 (8128)	G	⑮	⑯	⑯	189	⑰	⑱	⑲	15	140

* NOTE: Applies to Lower Manifold only.

① Outer bolts on caps 2-4: 67 ft. lbs.
All others: 74 ft. lbs.

② Step 1: 22 ft. lbs.
Step 2:
Short bolt: Plus 55 degrees
Medium bolt: Plus 65 degrees
Long bolt: Plus 75 degrees

③ Tighten bolts to 12 ft. lbs.
Retorque to 22 ft. lbs.

④ All 5 & 6 stud single rear wheels: 110 ft. lbs.
All 8 stud single rear wheels: 120 ft. lbs.

⑤ Step 1: 20 ft. lbs.
Step 2: 50 ft. lbs.
Step 3: 50 ft. lbs.
Step 4: Plus 90-100 degrees

⑥ Outer bolts: 100 ft. lbs.
Inner bolts: 111 ft. lbs.

⑦ Tighten all bolts to 20 ft. lbs.
Retorque to 50 ft. lbs.

⑧ Step 1: 22 ft. lbs.
Step 2: 90 degrees
Step 3: 90 degrees,
(except medium length bolts at front and rear)
Step 4: Tighten medium length bolts,
at front and rear an additional 50 degrees

⑨ Inner bolts;
Step 1: 15 ft. lbs.

⑩ Step 1: 15 ft. lbs.
Step 2: 60 degrees

⑪ Use a new bolt
Step 1: 37 ft. lbs.
Step 2: 140 degrees

⑫ Step 1: 15 ft. lbs.
Step 2: 37 ft. lbs.
Step 3: 74 ft. lbs.

⑬ Step 1: 44 in. lbs.
Step 2: 89 in. lbs.

⑭ Nuts: 39 ft. lbs.
Stud: 16 ft. lbs.

⑯ 1st pass: 22 ft. lbs.
2nd pass: 22 ft. lbs.,
plus 120 degrees
Final pass:
Short bolt: Plus 60 degrees
Medium bolt: Plus 45 degrees
Long bolt: Plus 30 degrees

⑮ 22 ft. lbs., plus 90 degrees

⑰ 1st pass: 30 ft. lbs.
2nd pass: 59 ft. lbs.
3rd pass: 70 ft. lbs.

⑱ 1st & 2nd pass: 44 inch lbs.
3rd pass: 89 inch lbs.
4th pass: 106 inch lbs.

⑲ Center bolt: 26 ft. lbs.
Nut: 12 ft. lbs.
Stud: 15 ft. lbs.

42372-YUKO-C11

WHEEL ALIGNMENT

Year	Model	Caster Range (+/-Deg.)	Caster Preferred Setting (Deg.)	Camber Range (+/-Deg.)	Camber Preferred Setting (Deg.)	Toe-in (in.)	Steering Axis Inclination (Deg.)
2000	2WD	1.00	+3.75	0.50	+0.50	0.24+/-0.20	—
	2WD HD	not adjustable	not adjustable	0.50	+1.25	0.12+/-0.12	—
	4WD	1.00	+3.00	0.50	+0.65	0.24+/-0.20	—
	4WD HD	1.00	+3.00	0.50	+0.50	0.24+/-0.20	—
2001	2WD	1.00	①	0.50	+0.25	0.10+/-0.20	—
	2WD HD	1.00	②	0.50	+0.25	0.10+/-0.12	—
	4WD	1.00	③	0.50	+0.25	0.10+/-0.20	—
	4WD HD	1.00	②	0.50	+0.25	0.10+/-0.20	—
2002	2WD	1.00	①	0.50	+0.25	0.10+/-0.20	—
	2WD HD	1.00	②	0.50	+0.25	0.10+/-0.20	—
	4WD	1.00	③	0.50	+0.25	0.10+/-0.12	—
	4WD HD	1.00	②	0.50	+0.25	0.10+/-0.20	—
2003-04	2WD	1.00	①	0.50	+0.25	0.10+/-0.20	—
	2WD HD	1.00	②	0.50	+0.25	0.10+/-0.12	—
	4WD	1.00	③	0.50	+0.25	0.10+/-0.20	—
	4WD HD	1.00	②	0.50	+0.25	0.10+/-0.20	—

① Left side: 3.90
   Right side: 4.70
② Left side: 4.50
   Right side: 4.75
③ Left side: 3.50
   Right side: 4.50

42372-YUKO-C12

TIRE, WHEEL AND BALL JOINT SPECIFICATIONS

Year	Model	OEM Tires Standard	OEM Tires Optional	Tire Pressures (psi) Front	Tire Pressures (psi) Rear	Wheel Size	Ball Joint Inspection
2000	Denali	P265/70R16	None	36	36	7-JJ	U ② L: 0.090 in.
	Escalade	P265/70R16	None	36	36	7-JJ	U ② L: 0.090 in.
	Tahoe/Yukon, 2wd	P235/75R15	None	36	36	6.5-JJ	U ① L: 0.090 in.
	Tahoe/Yukon, 4wd	P245/75R16	P265/75R16	36	36	7-JJ	②
	1500 Suburban 2wd	P235/75R15XL	LT245/75R16E	36	36	7-JJ	L ③
	1500 Suburban 4wd	P245/75R16C	LT245/75R16E	36	36	6.5-JJ	L ③
	2500 Suburban	LT245/75R16E	None	36	36	6.5-JJ	U: 0.125 in. L ③
2001	Denali	P265/70R16	None	36	36	7-JJ	U ② L: 0.090 in.
	Tahoe/Yukon, 2wd	P235/75R15	None	36	36	6.5-JJ	U ① L: 0.090 in.
	Tahoe/Yukon, 4wd	P245/75R16	P265/75R16	36	36	7-JJ	②
	1500 Suburban 2wd	P235/75R15XL	LT245/75R16E	36	36	7-JJ	L ③
	1500 Suburban 4wd	P245/75R16C	LT245/75R16E	36	36	6.5-JJ	L ③
	2500 Suburban	LT245/75R16E	None	36	36	6.5-JJ	U: 0.125 in. L ③
2002	Denali	P265/70R16	None	36	36	7-JJ	U ② L: 0.090 in.
	Escalade	P265/70R16	None	36	36	7-JJ	U ② L: 0.090 in.
	Tahoe/Yukon, 2wd	P235/75R15	None	36	36	6.5-JJ	U ① L: 0.090 in.
	Tahoe/Yukon, 4wd	P245/75R16	P265/75R16	36	36	7-JJ	②
	1500 Suburban 2wd	P235/75R15XL	LT245/75R16E	36	36	7-JJ	L ③
	1500 Suburban 4wd	P245/75R16C	LT245/75R16E	36	36	6.5-JJ	L ③
	2500 Suburban	LT245/75R16E	None	36	36	6.5-JJ	U: 0.125 in. L ③
2003-04	Denali	P265/70R16	None	36	36	7-JJ	U ② L: 0.090 in.
	Escalade	P265/70R16	None	36	36	7-JJ	U ② L: 0.090 in.
	Tahoe/Yukon, 2wd	P235/75R15	None	36	36	6.5-JJ	U ① L: 0.090 in.
	Tahoe/Yukon, 4wd	P245/75R16	P265/75R16	36	36	7-JJ	②
	1500 Suburban 2wd	P235/75R15XL	LT245/75R16E	36	36	7-JJ	L ③
	1500 Suburban 4wd	P245/75R16C	LT245/75R16E	36	36	6.5-JJ	L ③
	2500 Suburban	LT245/75R16E	None	36	36	6.5-JJ	U: 0.125 in. L ③

OEM: Original Equipment Manufacturer

PSI: Pounds Per Square Inch

STD: Standard

OPT: Optional

L: Lower

U: Upper

① Replace if any movement is noted or if stud can be moved by hand

② Ball joint is adjustable, refer to manual for procedure

③ Do not lift truck. Inspect the boss into which the grease fitting is threaded. Replace if the boss is flush or receded below the surface of the ball joint

42372-YUKO-C13

BRAKE SPECIFICATIONS
All measurements in inches unless noted

Year	Model		Brake Disc Original Thickness	Brake Disc Minimum Thickness	Maximum Runout	Brake Drum Diameter Original Inside Diameter	Brake Drum Diameter Max. Wear Limit	Brake Drum Diameter Maximum Machine Diameter	Minimum Lining Thickness	Brake Caliper Bracket Bolts (ft. lbs.)	Brake Caliper Mounting Bolts (ft. lbs.)
2000	Denali	F	1.500	1.480	0.004	—	—	—	0.030	NA	38
		R	—	—	—	①	②	③	0.030	NA	—
	Envoy	F	1.030	0.965	0.003	—	—	—	0.030	52	④
		R	0.787	0.728	0.004	9.50	9.59	9.56	0.030	NA	—
	Escalade	F	1.500	1.480	0.004	—	—	—	0.030	NA	38
		R	—	—	—	①	②	③	0.030	NA	—
	Suburban	F	1.500	1.480	0.004	—	—	—	0.030	NA	38
		R	—	—	—	①	②	③	0.030	NA	—
	Tahoe	F	1.500	1.480	0.004	—	—	—	0.030	NA	38
		R	—	—	—	①	②	③	0.030	NA	—
	Yukon	F	1.500	1.480	0.004	—	—	—	0.030	NA	38
		R	—	—	—	①	②	③	0.030	NA	—
2001	Denali	F	⑤	⑥	0.005	—	—	—	0.030	⑦	80
		R	⑧	⑨	0.005	—	—	—	0.030	⑩	⑪
	Denali XL	F	⑤	⑥	0.005	—	—	—	0.030	⑦	80
		R	⑧	⑨	0.005	—	—	—	0.030	⑩	⑪
	Suburban	F	⑤	⑥	0.005	—	—	—	0.030	⑦	80
		R	⑧	⑨	0.005	—	—	—	0.030	⑩	⑪
	Tahoe	F	⑤	⑥	0.005	—	—	—	0.030	⑦	80
		R	⑧	⑨	0.005	—	—	—	0.030	⑩	⑪
	Yukon	F	⑤	⑥	0.005	—	—	—	0.030	⑦	80
		R	⑧	⑨	0.005	—	—	—	0.030	⑩	⑪
	Yukon XL	F	⑤	⑥	0.005	—	—	—	0.030	⑦	80
		R	⑧	⑨	0.005	—	—	—	0.030	⑩	⑪
2002	Denali	F	⑤	⑥	0.005	—	—	—	0.030	⑦	80
		R	⑧	⑨	0.005	—	—	—	0.030	⑩	⑪
	Denali XL	F	⑤	⑥	0.005	—	—	—	0.030	⑦	80
		R	⑧	⑨	0.005	—	—	—	0.030	⑩	⑪
	Escalade	F	⑤	⑥	0.005	—	—	—	0.030	⑦	80
		R	⑧	⑨	0.005	—	—	—	0.030	⑩	⑪
	Suburban	F	⑤	⑥	0.005	—	—	—	0.030	⑦	80
		R	⑧	⑨	0.005	—	—	—	0.030	⑩	⑪
	Tahoe	F	⑤	⑥	0.005	—	—	—	0.030	⑦	80
		R	⑧	⑨	0.005	—	—	—	0.030	⑩	⑪
	Yukon	F	⑤	⑥	0.005	—	—	—	0.030	⑦	80
		R	⑧	⑨	0.005	—	—	—	0.030	⑩	⑪
	Yukon XL	F	⑤	⑥	0.005	—	—	—	0.030	⑦	80
		R	⑧	⑨	0.005	—	—	—	0.030	⑩	⑪

42372-YUKO-C14

BRAKE SPECIFICATIONS
All measurements in inches unless noted

Year	Model		Brake Disc Original Thickness	Brake Disc Minimum Thickness	Brake Disc Maximum Runout	Brake Drum Diameter Original Inside Diameter	Brake Drum Diameter Max. Wear Limit	Brake Drum Diameter Maximum Machine Diameter	Minimum Lining Thickness	Brake Caliper Bracket Bolts (ft. lbs.)	Brake Caliper Mounting Bolts (ft. lbs.)
2003-04	Denali	F	⑤	⑥	0.005	—	—	—	0.030	⑦	80
		R	⑧	⑨	0.005	—	—	—	0.030	⑩	⑪
	Denali XL	F	⑤	⑥	0.005	—	—	—	0.030	⑦	80
		R	⑧	⑨	0.005	—	—	—	0.030	⑩	⑪
	Escalade	F	⑤	⑥	0.005	—	—	—	0.030	⑦	80
		R	⑧	⑨	0.005	—	—	—	0.030	⑩	⑪
	Suburban	F	⑤	⑥	0.005	—	—	—	0.030	⑦	80
		R	⑧	⑨	0.005	—	—	—	0.030	⑩	⑪
	Tahoe	F	⑤	⑥	0.005	—	—	—	0.030	⑦	80
		R	⑧	⑨	0.005	—	—	—	0.030	⑩	⑪
	Yukon	F	⑤	⑥	0.005	—	—	—	0.030	⑦	80
		R	⑧	⑨	0.005	—	—	—	0.030	⑩	⑪
	Yukon XL	F	⑤	⑥	0.005	—	—	—	0.030	⑦	80
		R	⑧	⑨	0.005	—	—	—	0.030	⑩	⑪

NA: Not Available

① Available with 1 in., 11.15 in. and 13 in. drums

② 10 in. drum: 10.05
   11.15 in. drum: 11.24
   1 in. drum: 13.09

③ 1 in. drum: 10.09
   11.15 in. drum: 11.21
   1 in. drum: 13.06

④ 2WD: 38 ft. lbs.
   4WD: 77 ft. lbs.

⑤ Vacuum: 1.14 in.
   Hydraulic: 1.50 in.

⑥ Vacuum: 1.10 in.
   Hydraulic: 1.46 in.

⑦ 15 series: 129 ft. lbs.
   25 series: 221 ft. lbs.

⑧ Vacuum: 0.787 in.
   Hydraulic: 1.14 in.

⑨ Vacuum: 0.748 in.
   Hydraulic: 1.10 in.

⑩ Vacuum: 148 ft. lbs.
   Hydraulic (9000 lbs.): 122 ft. lbs.
   Hydraulic (12,000 lbs.): 221 ft. lbs.

⑪ 15 series: 31 ft. lbs.
   25 series: 80 ft. lbs.

42372-YUKO-C15

SCHEDULED MAINTENANCE INTERVALS
DENALI, DENALI XL, ESCALADE, SUBURBAN, TAHOE, YUKON & YUKON XL—GASOLINE

TO BE SERVICED	TYPE OF SERVICE	VEHICLE MILEAGE INTERVAL (x1000)															
		7.5	15	22.5	30	37.5	45	52.5	60	67.5	75	82.5	90	97.5	105	112.5	120
Accessory drive belt	S/I								✓								✓
Automatic transmission fluid ①	R	Every 50,000 miles															
Brake system	S/I	✓	✓	✓	✓	✓	✓	✓	✓	✓	✓	✓	✓	✓	✓	✓	✓
Chassis & suspension grease points	L	✓	✓	✓	✓	✓	✓	✓	✓	✓	✓	✓	✓	✓	✓	✓	✓
Cooling fan operation	S/I		✓		✓		✓		✓		✓		✓		✓		
CV-joint boots & axle seals	S/I	✓	✓	✓	✓	✓	✓	✓	✓	✓	✓	✓	✓	✓	✓	✓	✓
EGR system	S/I								✓								✓
Engine coolant	R	Every 150,000 miles															
Engine oil & filter	R	✓	✓	✓	✓	✓	✓	✓	✓	✓	✓	✓	✓	✓	✓	✓	✓
EVAP system	S/I								✓								✓
Front wheel bearings ①	S/I & L				✓				✓				✓				✓
Fuel filter	R								✓								✓
Fuel system	S/I								✓								✓
Rear/front axle fluid level	S/I	✓	✓	✓	✓	✓	✓	✓	✓	✓	✓	✓	✓	✓	✓	✓	✓
Rotate tires	S/I	✓	✓	✓	✓	✓	✓	✓	✓	✓	✓	✓	✓	✓	✓	✓	✓
Shields & underhood insulation ②	S/I		✓		✓		✓		✓		✓		✓		✓		✓
Spark plugs	R	Every 100,000 miles															
Spark plug wires	S/I	Every 100,000 miles															

R: Replace   S/I: Inspect and service, if necessary   L: Lubricate
① 2-wheel drive models only.
② Vehicles with a GVWR or 8500 lbs. or more only.

FREQUENT OPERATION MAINTENANCE (SEVERE SERVICE)
If a vehicle is operated under any of the following conditions it is considered severe service:
- Towing a trailer or using a camper or car-top carrier.
- Repeated short trips of less than 5 miles in temperatures below freezing, or trips of less than 10 miles in any temperature.
- Extensive idling or low-speed driving for long distances as in heavy commercial use, such as delivery, taxi or police cars.
- Operating on rough, muddy or salt-covered roads.
- Operating on unpaved or dusty roads.
- Driving in extremely hot (over 90°) conditions.

Engine oil & filter: replace every 3000 miles or 3 months, whichever occurs first.
Chassis and suspension grease points: lubricate every 3000 miles.
Rear/front axle fluid level: inspect every 3000 miles.
Rotate the tires ever 6000 miles.
Brake system components: inspect ever 6000 miles.
Front wheel bearings (2-wheel drive only): clean, inspect and repack every 15,000 miles.
Shields & underhood insulation (vehicles w/GVWR over 8500 lbs. only): inspect every 15,000 miles.
Cooling fan system hoses & connections: inspect every 15,000 miles.
Fuel filter: replace every 30,000 miles.
Air cleaner filter: inspect every 45,000 miles.
Automatic transmission fluid & filter: replace every 50,000 miles.
Accessory drive belt: inspect every 60,000 miles.
Fuel system tank, cap and lines: inspect every 60,000 miles.
EVAP system: inspect every 60,000 miles.
EGR system: inspect every 60,000 miles.
PCV system: inspect every 100,000 miles.
Engine cooling system components: inspect and clean every 150,000 miles.

42372-YUKO-C16

SCHEDULED MAINTENANCE INTERVALS
DENALI, DENALI XL, SUBURBAN, TAHOE, YUKON & YUKON XL—DIESEL

TO BE SERVICED	TYPE OF SERVICE	5	8	10	15	20	23	25	30	35	38	40	45	50	53	55	60	65	68	70	75	80	83	85	90	95	98
Air intake system	S/I			✓		✓			✓			✓		✓			✓			✓	✓				✓		
Automatic transmission fluid ①	R	Every 50,000 miles																									
Brake system	S/I	✓	✓		✓		✓		✓		✓		✓		✓		✓		✓		✓		✓		✓		✓
Chassis & suspension grease points	L	✓		✓	✓	✓		✓	✓	✓		✓	✓			✓	✓	✓		✓	✓	✓		✓	✓	✓	
Cooling fan operation	S/I			✓		✓			✓			✓		✓			✓			✓		✓			✓		
Crankcase depression regular valve system hoses	S/I																✓										
CV-joint boots & axle seals	S/I	✓		✓	✓	✓		✓	✓	✓		✓	✓	✓		✓	✓	✓		✓	✓	✓		✓	✓	✓	
																	✓								✓		
EGR system ②	S/I																										
Engine coolant	R	Every 150,000 miles																									
Engine cooling system hoses & radiator	S/I & C	Initially at 100,000 miles, then every 50,000 miles																									
Engine oil & filter	R	✓		✓	✓	✓		✓	✓	✓		✓	✓	✓		✓	✓			✓	✓	✓		✓	✓	✓	
Front wheel bearings ③	S/I & L								✓								✓								✓		
Fuel filter	R								✓								✓										
Rear/front axle fluid level	S/I	✓		✓	✓	✓		✓	✓	✓		✓	✓	✓		✓	✓	✓		✓	✓	✓		✓	✓	✓	✓
Rotate tires	S/I	✓	✓		✓		✓		✓		✓		✓		✓		✓		✓		✓		✓		✓		✓
Shields & underhood insulation ①	S/I				✓		✓		✓			✓		✓			✓			✓		✓			✓		

R: Replace   S/I: Inspect and service, if necessary   L: Lubricate   C: Clean
① Vehicles with a GVWR of 8500 lbs or more only.
② If equipped.
③ 2-wheel drive models only.

FREQUENT OPERATION MAINTENANCE (SEVERE SERVICE)

If a vehicle is operated under any of the following conditions it is considered severe service:

- Towing a trailer or using a camper or car-top carrier.
- Repeated short trips of less than 5 miles in temperatures below freezing, or trips of less than 10 miles in any temperature.
- Extensive idling or low-speed driving for long distances as in heavy commercial use, such as delivery, taxi or police cars.
- Operating on rough, muddy or salt-covered roads.
- Operating on unpaved or dusty roads.
- Driving in extremely hot (over 90°) conditions.

Engine oil & filter: replace every 2500 miles.

Chassis and suspension grease points: lubricate every 2500 miles

Rear/front axle fluid level: inspect initially at 5000 miles, then every 2500 miles thereafter.

Rotate tires: every 7500 miles.

Air cleaner filter: inspect every 15,000 miles.

Front wheel bearings (2-wheel drive only): clean, inspect and repack every 15,000 miles.

42372-YUKO-C17

PRECAUTIONS

Before servicing any vehicle, please be sure to read all of the following precautions, which deal with personal safety, prevention of component damage, and important points to take into consideration when servicing a motor vehicle:

• Never open, service or drain the radiator or cooling system when the engine is hot; serious burns can occur from the steam and hot coolant.

• Observe all applicable safety precautions when working around fuel. Whenever servicing the fuel system, always work in a well-ventilated area. Do not allow fuel spray or vapors to come in contact with a spark, open flame, or excessive heat (a hot drop light, for example) Keep a dry chemical fire extinguisher near the work area. Always keep fuel in a container specifically designed for fuel storage; also, always properly seal fuel containers to avoid the possibility of fire or explosion. Refer to the additional fuel system precautions later in this section.

• Fuel injection systems often remain pressurized, even after the engine has been turned **OFF**. The fuel system pressure must be relieved before disconnecting any fuel lines. Failure to do so may result in fire and/or personal injury.

• Brake fluid often contains polyglycol ethers and polyglycols. Avoid contact with the eyes and wash your hands thoroughly after handling brake fluid. If you do get brake fluid in your eyes, flush your eyes with clean, running water for 15 minutes. If

eye irritation persists, or if you have taken brake fluid internally, IMMEDIATELY seek medical assistance.

• The EPA warns that prolonged contact with used engine oil may cause a number of skin disorders, including cancer! You should make every effort to minimize your exposure to used engine oil. Protective gloves should be worn when changing oil. Wash your hands and any other exposed skin areas as soon as possible after exposure to used engine oil. Soap and water, or waterless hand cleaner should be used.

• All new vehicles are now equipped with an air bag system. The system must be disabled before performing service on or around system components, steering column, instrument panel components, wiring and sensors. Failure to follow safety and disabling procedures could result in accidental air bag deployment, possible personal injury and unnecessary system repairs.

• Always wear safety goggles when working with, or around, the air bag system. When carrying a non-deployed air bag, be sure the bag and trim cover are pointed away from your body. When placing a non-deployed air bag on a work surface, always face the bag and trim cover upward, away from the surface. This will reduce the motion of the module if it is accidentally deployed. Refer to the additional air bag system precautions later in this section.

• Clean, high quality brake fluid from a

sealed container is essential to the safe and proper operation of the brake system. You should always buy the correct type of brake fluid for your vehicle. If the brake fluid becomes contaminated, completely flush the system with new fluid. Never reuse any brake fluid. Any brake fluid that is removed from the system should be discarded. Also, do not allow any brake fluid to come in contact with a painted surface; it will damage the paint.

• Never operate the engine without the proper amount and type of engine oil; doing so WILL result in severe engine damage.

• Timing belt maintenance is extremely important! Many models utilize an interference-type, non-freewheeling engine. If the timing belt breaks, the valves in the cylinder head may strike the pistons, causing potentially serious (also time-consuming and expensive) engine damage. Refer to the maintenance interval charts in the front of this section for the recommended replacement interval for the timing belt, and to the timing belt procedure in this section for belt replacement and inspection.

• Disconnecting the negative battery cable on some vehicles may interfere with the functions of the on-board computer system(s) and may require the computer to undergo a relearning process once the negative battery cable is reconnected.

• When servicing drum brakes, only disassemble and assemble one side at a time, leaving the remaining side intact for reference.

GASOLINE ENGINE REPAIR

Distributor

REMOVAL

5.7L and 7.4L Engines

1. Before servicing the vehicle, refer to the precautions in the beginning of this section.
2. Remove or disconnect the following:
   • Negative battery cable
   • Spark plug wires and the coil leads from the distributor
   • Electrical connector at the base of the distributor
   • Distributor cap
3. Matchmark the rotor-to-housing and housing-to-engine block positions so that they can be matched during installation.

   • Distributor hold-down bolt
   • Distributor from the engine

4.8L, 5.3L and 6.0L Engines

➡If the Malfunction Indicator Lamp turns on, and a DTC code P1345 sets after installing the distributor, this indicates an incorrectly installed distributor. Engine damage or distributor damage may occur.

1. Before servicing the vehicle, refer to the precautions in the beginning of this section.
2. Turn OFF the ignition switch.
3. Remove or disconnect the following:
   • Spark plug wires from the distributor cap
   • Electrical connector from the base of the distributor
   • Two screws that hold the distributor

cap to the housing. Discard the screws.
   • Distributor cap from the housing
4. Use a grease pencil in order to note the position of the rotor in relation to the distributor housing.
5. Mark the distributor housing and the intake manifold with a grease pencil.
6. Remove or disconnect the following:
   • Mounting clamp hold-down bolt
   • Distributor
7. As the distributor is being removed from the engine, watch the rotor move in a counterclockwise direction about 42 degrees. This will appear as slightly more than the 1 o'clock position. Note the position of the rotor segment. Place a second mark on the base of the distributor. This will aid in achieving proper rotor alignment during the distributor installation.

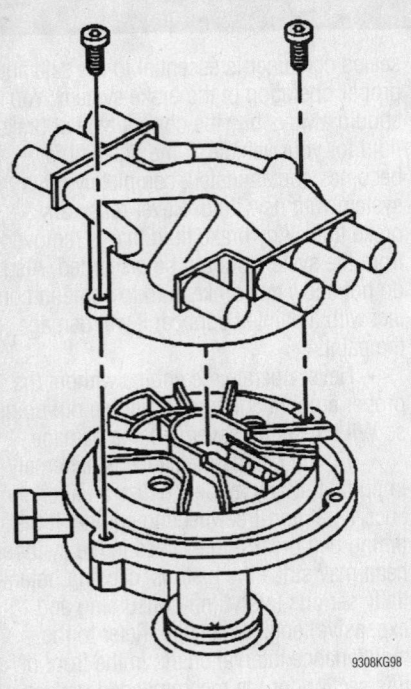

Distributor cap—4.8L, 5.3L and 6.0L engines

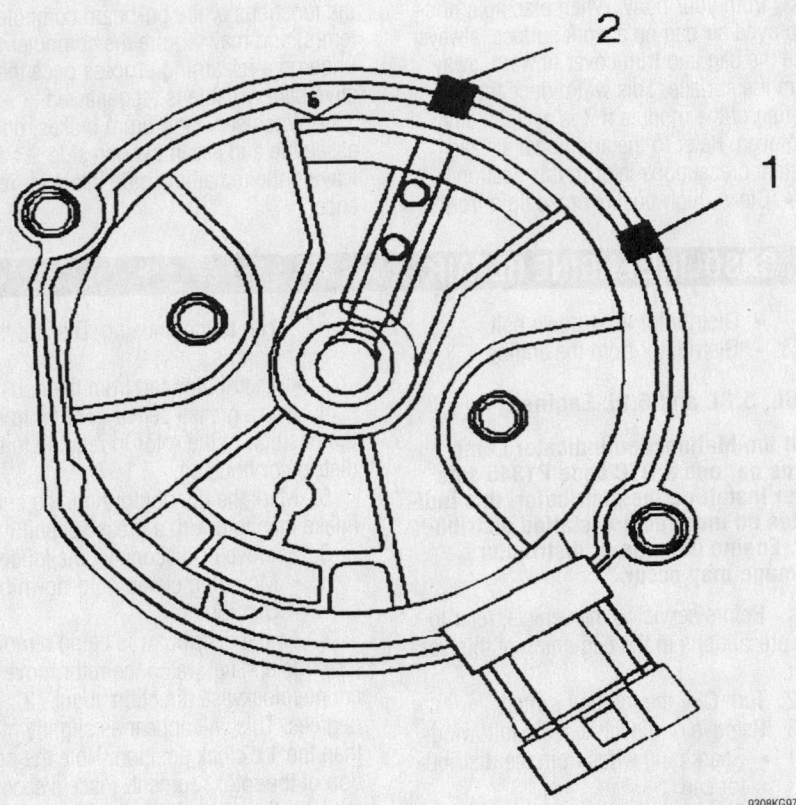

Distributor rotor starting point (1) and 42 degrees counterclockwise (2)—4.8L, 5.3L and 6.0L engines

INSTALLATION

5.7L and 7.4L Engines

TIMING NOT DISTURBED

1. Install or connect the following:
   - Distributor, aligning the match-marks properly
   - Distributor hold-down bolt
   - Distributor cap
   - Electrical connector at the base of the distributor
   - Spark plug wires and coil leads
   - Negative battery cable

TIMING DISTURBED

1. Remove the No. 1 cylinder spark plug. Turn the engine using a socket wrench on the large bolt on the front of the crankshaft pulley. Place a finger near the No. 1 spark plug hole and turn the crankshaft until the piston reaches TDC. As the engine approaches TDC, you will feel air being expelled through the No. 1 cylinder spark plug hole. The timing mark on the crankshaft pulley should now be aligned with the **0** mark on the timing scale. If the position is not being met, turn the engine another full turn (360 degrees) Once the engines position is correct, install the spark plug.

➡Before installation, position the rotor so it points to the No. 2 terminal on the cap. As the distributor is lowered into the engine, the rotor will rotate clockwise and stop at the No. 1 terminal. This is the desired position.

2. Turn the rotor so that it will point to the No. 1 terminal of the distributor cap when it is fully seated in the engine.

3. Install the distributor. It may be necessary to turn the rotor a little in either direction, in order to engage the gears.

➡If the distributor will not seat completely in the engine, remove the distributor and align the groove on the top of the oil pump drive shaft with a long screwdriver to match the tab on the bottom of the distributor shaft. Reinstall the distributor.

4. Tap the starter a few times to ensure that the oil pump shaft is mated to the distributor shaft.

5. Bring the engine to TDC again and check that the rotor is pointed toward the No. 1 terminal of the cap. If the marks are all aligned.

6. Install or connect the following:
   - Hold-down bolt and tighten
   - Cap and fasten the mounting screws
   - Electrical connections and the spark plug wires

4.8L, 5.3L and 6.0L Engines

TIMING NOT DISTURBED

1. If installing a new distributor assembly, place two marks on the new distributor housing in the same location as the two marks on the original housing. Remove the new distributor cap, if necessary. Align the rotor with mark made at location 2.

2. Guide the distributor into the engine. Align the hole in the distributor hold-down base over the mounting hole in the intake manifold.

3. As the distributor is being installed, observe the rotor moving in a clockwise direction about 42 degrees. Once the distributor is completely seated, the rotor segment should be aligned with the mark on the distributor base in location number 1. If the rotor segment is not aligned with the number 1 mark, the driven gear teeth and the camshaft have meshed one or more teeth out of alignment.

4. Install or connect the following:

- Distributor mounting clamp bolt and tighten to 18 ft. lbs. (25 Nm)
- Distributor cap. Install two NEW distributor cap screws and tighten to 21 inch lbs. (2.4 Nm).
- Electrical connector to the distributor
- Spark plug wires to the distributor cap
- Ignition coil wire

➡ **If the Malfunction Indicator lamp is turned on after installing the distributor, and a DTC P1345 is found, the distributor has been installed incorrectly.**

TIMING DISTURBED

1. Rotate the number 1 cylinder to TDC of the compression stroke. The engine front cover has 2 alignment tabs and the crankshaft balancer has 2 alignment marks (spaced 90 degrees apart) which are used for positioning number 1 piston at top dead center (TDC). With the piston on the compression stroke and at top dead center, the crankshaft balancer alignment mark must align with the engine front cover tab and the crankshaft balancer alignment mark must align with the engine front cover tab.

2. Align the white paint mark on the bottom stem of the distributor with the predrilled indent hole in the bottom of the gear. If the driven gear is installed incorrectly, the dimple will be approximately 180 degrees opposite of the rotor segment when it is installed in the distributor.

The OBD II ignition system distributor driven gear and rotor may be installed in multiple positions. In order to avoid mistakes, mark the distributor on the following components in order to ensure the same mounting position upon reassembly:

- The distributor driven gear
- The distributor shaft
- The rotor holes

Installing the driven gear 180 degrees out of alignment, or locating the rotor in the wrong holes, will cause a no-start condition. Premature engine wear or damage may result.

3. Using a long screwdriver, align the oil pump drive shaft to the drive tab of the distributor. Guide the distributor into the engine. Ensure that the spark plug towers are perpendicular to the centerline of the engine.

Once the distributor is fully seated, the rotor segment should be aligned with the pointer cast into the distributor base.

This pointer may have a 6 cast into it, indicating that the distributor is to be used on a 6 cylinder engine or a 8 cast into it,

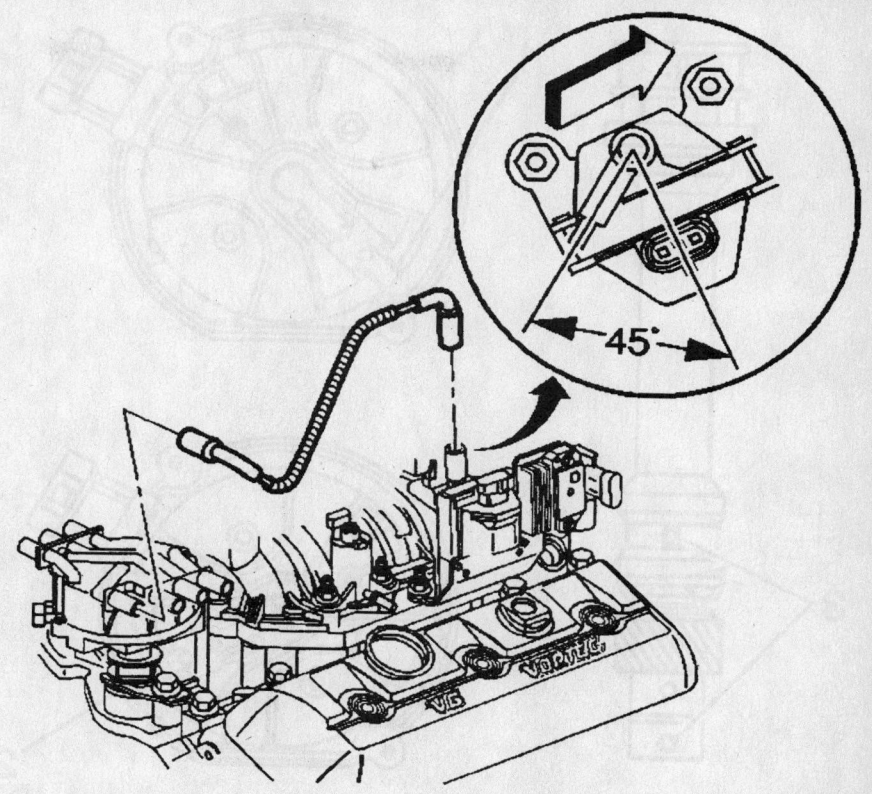

Distributor electrical connection—4.8L, 5.3L and 6.0L engines

9308KG96

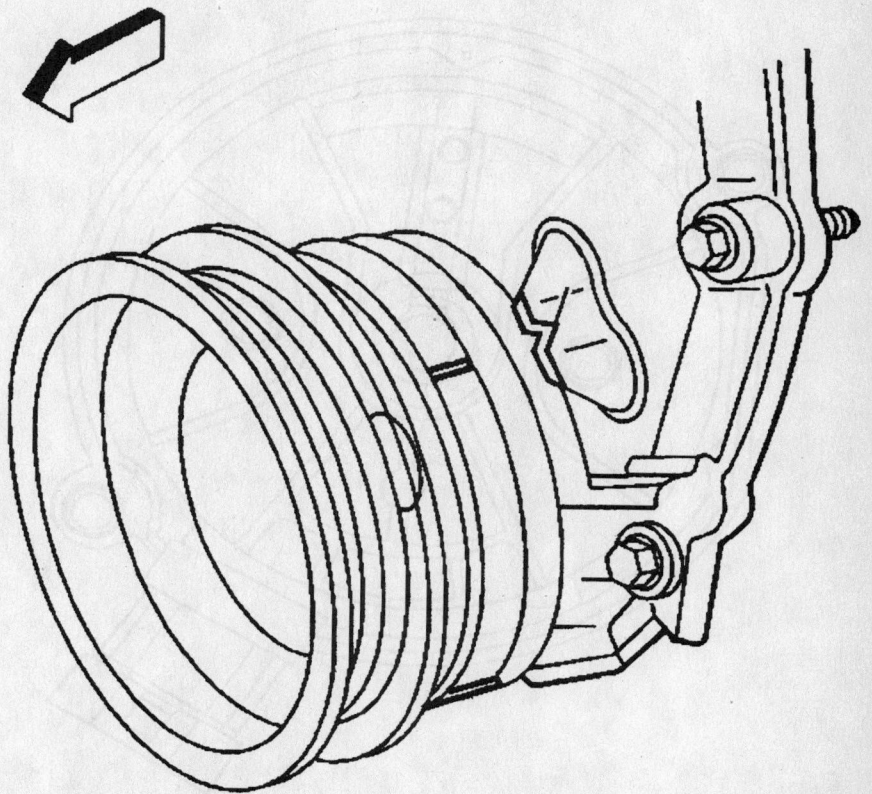

Engine at TDC compression—4.8L, 5.3L and 6.0L engines

9308KG95

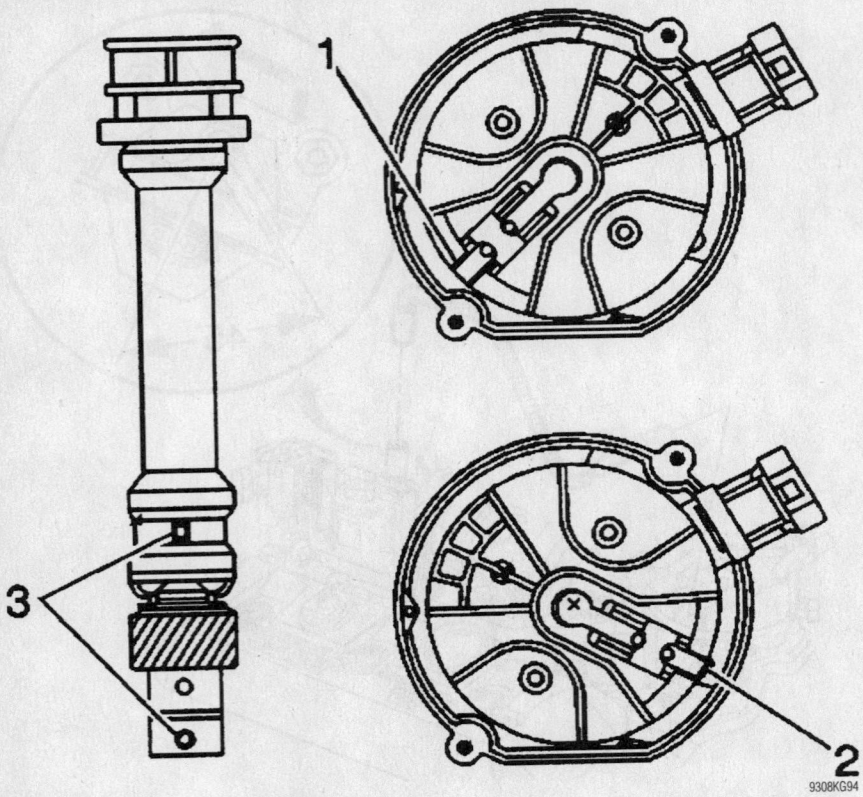

Distributor alignment. 1 is the starting point; 2 is installed; 3 are the shaft alignment marks—4.8L, 5.3L and 6.0L engines

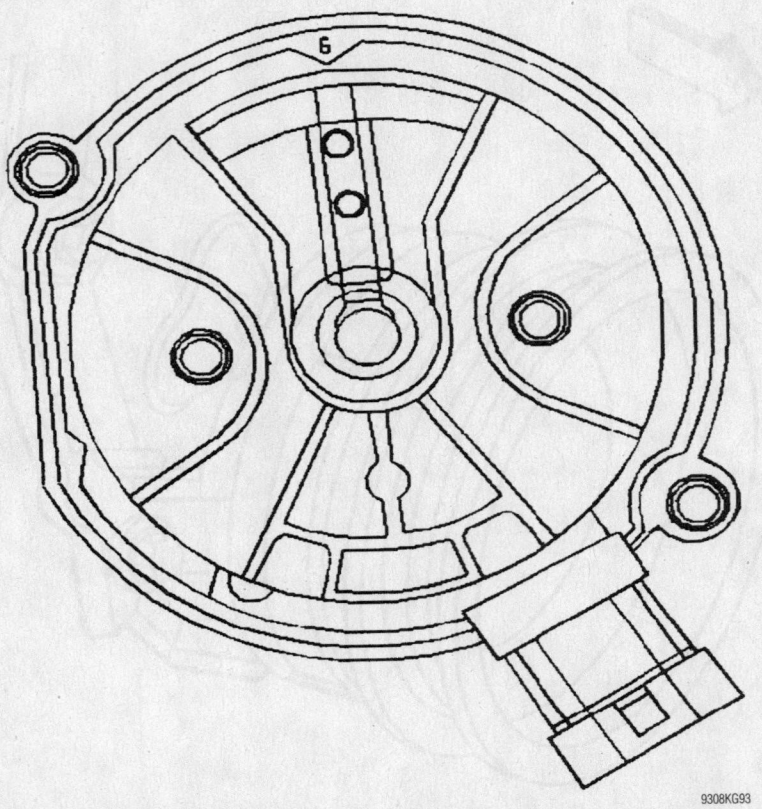

Distributor fully seated—4.8L, 5.3L and 6.0L engines

indicating that the distributor is to be used on a 8 cylinder engine.

If the rotor segment does not come within a few degrees of the pointer, the gear mesh between the distributor and the camshaft may be off a tooth or more.

If this is the case, repeat the procedure again in order to achieve proper alignment.

4. Install or connect the following:
- Distributor mounting clamp bolt. Tighten to 18 ft. lbs. (25 Nm).
- Distributor cap. Install two NEW distributor cap screws and tighten to 21 inch lbs. (2.4 Nm).
- Distributor electrical connector
- Spark plug wires to the distributor cap
- Ignition coil wire

➡ If the Malfunction Indicator lamp is turned on after installing the distributor, and a DTC P1345 is found, the distributor has been installed incorrectly.

Alternator

REMOVAL & INSTALLATION

5.7L and 7.4L Engines

1. Before servicing the vehicle, refer to the precautions in the beginning of this section.
2. Remove or disconnect the following:
- Negative battery cable
- Wires
- Accessory belt(s)
- Mounting bracket, if necessary
- Alternator

To install:

3. Install or connect the following:
- Alternator
- Mounting bracket. Torque bolts to 18 ft. lbs. (25 Nm).
- Mounting bolts. Torque the right mounting bolt to 18 ft. lbs. (25 Nm) and left bolt to 37 ft. lbs. (50 Nm).
- Accessory belt(s)
- Wires. Torque the battery feed wire to 71 inch lbs. (8 Nm).
- Negative battery cable

4.8L, 5.3L and 6.0L Engines

1. Disconnect the negative battery cable.
2. Remove or disconnect the following:
- Accessory drive belt
- Engine sight shield, if necessary
- Electrical connections from the generator
- Mounting bolts
- Generator

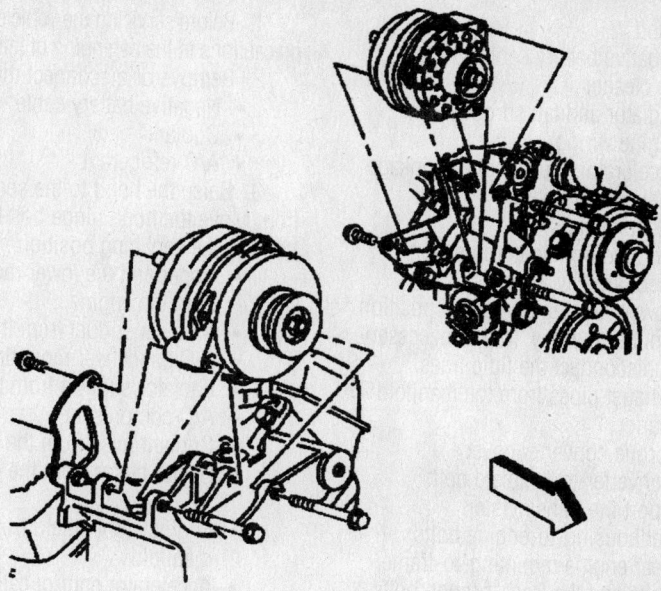

Exploded view of the alternator mounting

9308KG01

To install:

3. Install or connect the following:
   - Generator
   - Generator mounting bolts and tighten to 37 ft. lbs. (50 Nm).
   - Electrical connections to the generator. Tighten the B+ nut to 13 ft. lbs. (18 Nm).

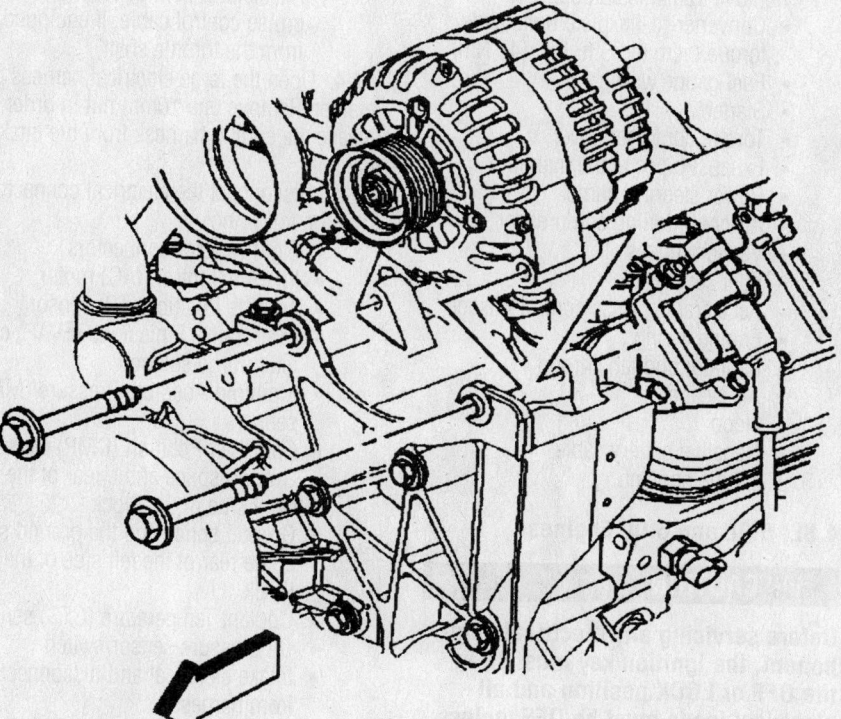

Alternator mounting—4.8L, 5.3L and 6.0L engines

9308KG99

   - Engine sight shield, if removed
   - Accessory drive belt
   - Negative battery cable

8.1L Engine

1. Disconnect the negative battery cable.
2. Remove or disconnect the following:

   - Electrical connections from the generator
3. Remove the cable from the generator as follows:
   a. Slide the boot down, to reveal the terminal stud.
   b. Unfasten the cable nut from the stud, then remove the generator cable.
4. Remove or disconnect the following:
   - Accessory drive belt
   - Mounting bolts
   - Generator
   - Mounting bolts securing the generator to the brace and bracket
   - Generator

To install:

5. Install or connect the following:
   - Generator
   - Generator mounting bolts and tighten to 37 ft. lbs. (50 Nm)
   - Accessory drive belt
   - Generator cable, secure with the nut and tighten to 80 inch lbs. (9 Nm)
   - Boot back over the terminal stud.
   - Electrical connections to the generator
   - Negative battery cable

Ignition Timing

ADJUSTMENT

Always refer to the Vehicle Emissions Control Information label in the engine compartment for base ignition timing specification and adjustment procedures.

Engine Assembly

REMOVAL & INSTALLATION

5.7L Engines

1. Before servicing the vehicle, refer to the precautions in the beginning of this section.
2. Drain the cooling system.
3. Drain the engine oil.
4. Remove or disconnect the following:
   - Negative battery cable
   - Hood
   - Air cleaner
   - Accessory drive belt
   - Fan
   - Water pump pulley
   - Radiator and shroud
   - Heater hoses at the engine
   - Accelerator, cruise control and detent linkage if used
   - Air conditioning compressor, if used, and lay aside

- Power steering pump, if used, and lay aside
- Wiring from the engine
- Fuel line
- Vacuum lines from the intake manifold
- Exhaust pipes from the manifold
- Strut rods at the engine mountings, if used
- Flywheel or torque converter cover
- Wiring along the oil pan rail
- Starter
- Wire for the fuel gauge
- Converter-to-flex plate bolts, if equipped with automatic transmission

5. Support the transmission
- Bell housing to engine bolts
- Rear engine mounting to frame bolts and the front through bolts and the engine

To install:

6. Lower the engine.
7. Install or connect the following:
- Engine mounting bolts. Torque the rear engine mounting to frame bolts or nuts to 45 ft. lbs. (54 Nm), the front through-bolts to 70 ft. lbs. (97 Nm) and the front nuts to 50 ft. lbs. (67 Nm)
- Bell housing to engine bolts and torque to 35 ft. lbs. (47 Nm)

8. Remove the transmission support.
- Converter-to-flex bolts and torque to 35 ft. lbs. (47 Nm)
- Fuel gauge wiring
- Starter
- Flywheel or torque converter cover
- Strut rods at the engine mountings, if used
- Exhaust pipes at the manifold
- Vacuum lines to the intake manifold
- Fuel line
- Engine wiring harness
- Power steering pump, if used
- Air conditioning compressor, if used
- Accelerator, cruise control and detent linkage
- Heater hoses
- Radiator and shroud
- Accessory drive belts
- Hood
- Negative battery cable

9. Refill coolant and engine oil.

7.4L Engine

1. Before servicing the vehicle, refer to the precautions in the beginning of this section.
2. Drain the cooling system.
3. Remove or disconnect the following:

- Hood
- Negative battery cable
- Air cleaner
- Radiator and fan shroud
- Engine wiring
- Accelerator, cruise control linkage
- Fuel supply lines
- Vacuum wires
- Air conditioning compressor, if used, and lay aside
- Power steering pump and position it out of the way. It's not necessary to disconnect the fluid lines.
- Exhaust pipes from the manifold
- Starter
- Torque converter cover
- Converter-to-flexplate bolts

4. Support the transmission
- Bellhousing-to-engine bolts
- Rear engine mounting-to-frame bolts and the front through bolts
- Engine

To install:

5. Lower the engine.
6. Install or connect the following:
- Engine mounting bolts. Torque the rear engine mounting-to-frame bolts or nuts to 45 ft. lbs. (54 Nm), the front through bolts to 70 ft. lbs. (97 Nm) and the front nuts to 50 ft. lbs. (67 Nm)
- Bellhousing-to-engine bolts. Torque the bolts to 35 ft. lbs. (47 Nm)

7. Remove transmission support.
- Converter-to-flexplate bolts and torque them to 35 ft. lbs. (47 Nm)
- Fuel gauge wiring
- Starter
- Torque converter cover
- Exhaust pipes at the manifold
- Power steering pump
- Air conditioning compressor
- Vacuum hoses
- Fuel supply line
- Accelerator, cruise control linkage
- Engine wiring
- Radiator and fan shroud
- Air cleaner
- Hood
- Negative battery cable

8. Refill the coolant.

4.8L, 5.3L and 6.0L Engines

✳✳ CAUTION

Before servicing any electrical component, the ignition key must be in the OFF or LOCK position and all electrical loads must be OFF, unless instructed otherwise in these procedures.

1. Before servicing the vehicle, refer to the precautions in the beginning of this section.
2. Remove or disconnect the following:
- Negative battery cable
- Coolant
- A/C refrigerant

3. Raise the hood to the servicing position. Move the hood hinge bolt to hold the hood in the servicing position.
- Upper and the lower radiator hoses from the engine
- Air cleaner duct from the engine
- A/C condenser mounting bolts
- Radiator support from the vehicle
- A/C compressor
- Coolant hose from the throttle body
- Heater hoses from the engine and the cowl
- Engine sight shield from the intake manifold
- Accelerator control cable mounting bracket from the intake manifold

✳✳ CAUTION

In order to avoid possible injury or vehicle damage, always replace the accelerator control cable with a NEW cable whenever you remove the engine from the vehicle. In order to avoid cruise control cable damage, position the cable out of the way while you remove or install the engine.

- Accelerator control cable and the cruise control cable, if equipped, from the throttle shaft

4. Open the large electrical harness retainer. Remove one 10mm nut in order to release the engine harness from the intake manifold.
5. Disconnect the electrical connectors from the following:
- Eight injector connectors
- Idle Air Control (IAC) motor
- Throttle Position (TP) sensor
- Evaporative Emissions (EVAP) canister purge solenoid
- Manifold Absolute Pressure (MAP) sensor
- Camshaft Position (CMP) sensor
- Ground splice at the rear of the right side of the block
- Ground splice and the ground strap at the rear of the left side of the block
- Coolant Temperature (CTS) sensor
- Oil pressure sensor/switch
- Intake electrical and disconnect from harness
- Junction block bracket from alternator bracket

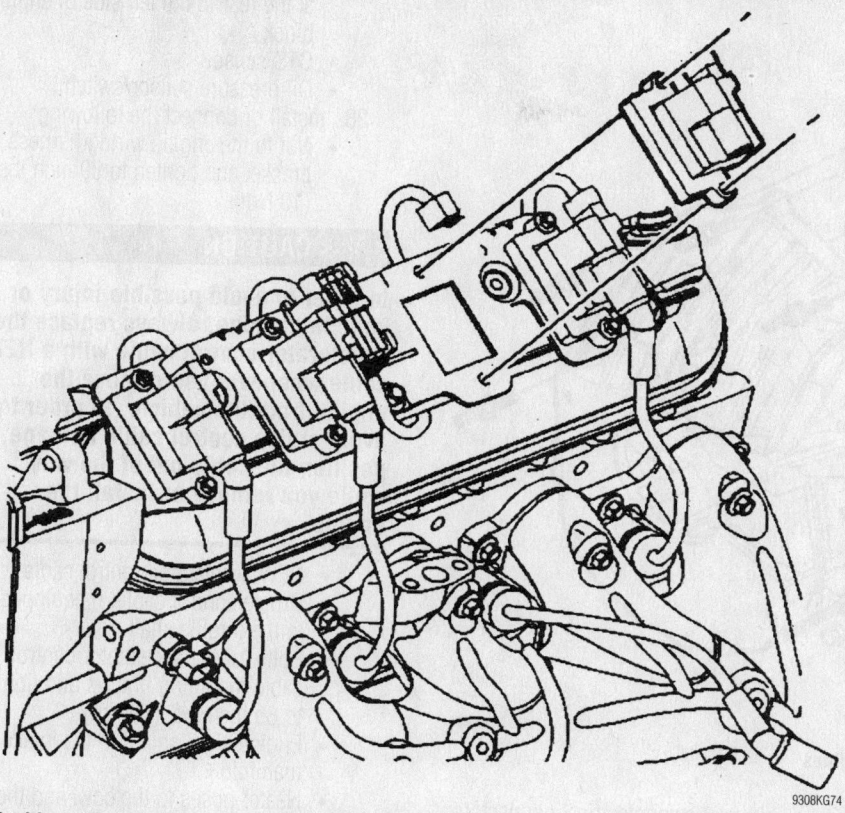

Ignition coil removal—4.8L, 5.3L and 6.0L engines

9308KG74

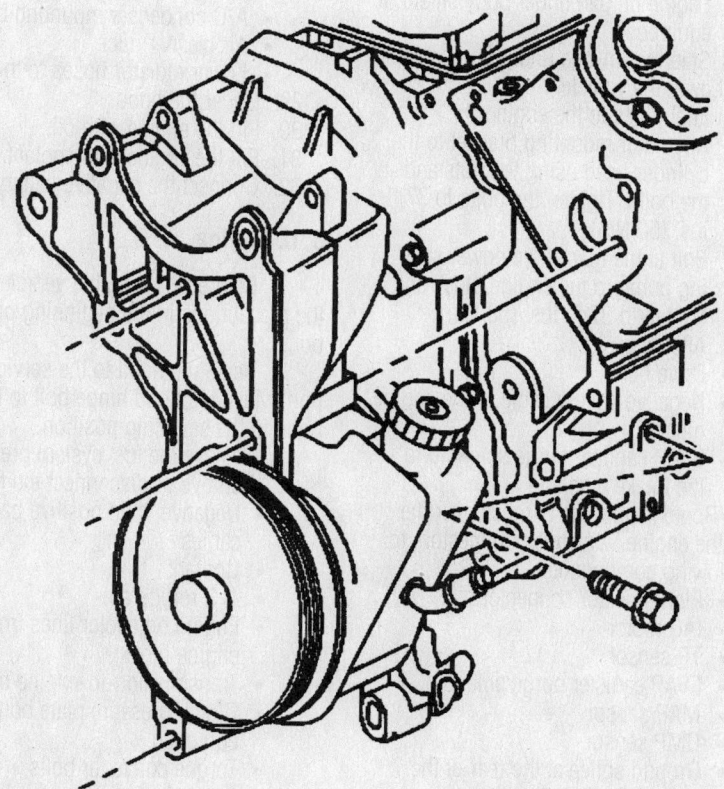

Power steering pump removal—4.8L, 5.3L and 6.0L engines

9308KG73

6. Set the electrical harness aside.
7. Remove or disconnect the following:
- EVAP canister purge solenoid vent tube from the solenoid by squeezing the retainer, then release the tube from the solenoid
- Battery negative cable from the engine block
- Drive belt
- Bolts holding the alternator mounting bracket to the cylinder head and block
- Bolt behind the power steering pump to engine block
- Alternator mounting bracket. Position the bracket aside.
- Fuel pipes from the engine
8. Raise the vehicle.
- Steering linkage under body shield, if equipped
- Engine oil pan under body shield, if equipped
- Engine oil
- Starter motor
9. Disconnect the engine wiring harness from the following components:
- Crankshaft Position (CKP) sensor
- Engine oil level sensor
- Block heater, if equipped
- Wiring harness from the oil pan
- Reposition wiring from lower engine area
10. Remove or disconnect the following:
- Exhaust pipes from the exhaust manifolds
- Transmission cooler pipe retainer from the right side of the engine block, if equipped
- Torque converter shield from the engine
- Torque converter bolts
- Nut and the transmission oil level indicator tube from the bellhousing stud
- Lower bellhousing studs from the engine
11. Lower the vehicle.
- Remaining bellhousing bolts
- Engine electrical harness; position aside
- Ignition coil(s)
12. Install an engine crane.
13. Install a floor jack or stands to transmission for support.
14. Remove the engine mount bolts.

➡**Use care while moving the engine assembly in order to avoid breaking the MAP sensor locating tabs. Broken MAP sensor tabs may result in decreased engine performance.**

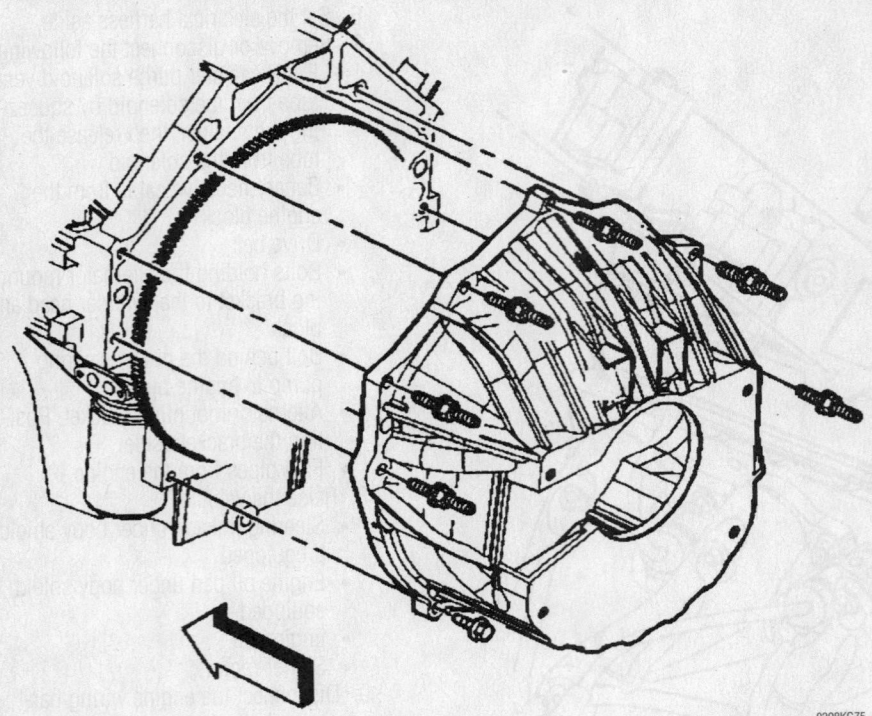

9308KG75

Bellhousing bolt removal—4.8L, 5.3L and 6.0L engines

15. Remove the engine from the vehicle.

To install:

16. Install or connect the following:
    - Engine to the vehicle
    - Engine mount bolts
    - Upper bellhousing bolts

17. Remove transmission support apparatus.

18. Remove the lifting device.

19. Remove the lift brackets from both cylinder heads.

20. Install the ignition coil(s) and the spark plug wire(s).

21. Route the engine wiring harness to the lower right hand side of the engine.

22. Raise the vehicle.

23. Install or connect the following:
    - Remaining bellhousing bolts
    - Torque converter bolts
    - Torque converter shield
    - Transmission oil level indicator tube and nut to bellhousing stud
    - A/C compressor
    - Transmission cooler pipe retainer to right side of engine block
    - Engine exhaust pipes to the exhaust manifolds

24. Reroute wiring to lower engine area and install bolt to oil pan.

25. Install or connect the following:
    - CKP sensor electrical connector
    - Engine oil level sensor and the block heater electrical connectors, if equipped.
    - Starter motor
    - Engine oil pan under body shield, if equipped
    - Steering linkage under body shield

26. Lower the vehicle.
    - Fuel pipes to the engine
    - Alternator mounting bracket to the cylinder head using the nuts and the bolts. Tighten the bolts to 37 ft. lbs. (50 Nm).
    - Bolt at the rear of the power steering pump to the engine block and tighten to 37 ft. lbs. (50 Nm).
    - Alternator
    - Drive belt
    - Negative battery cable to the engine block
    - EVAP canister purge solenoid to the intake manifold

27. Route the engine harness over the top of the engine. Attach the connectors to the following components:
    - Eight injector connectors
    - IAC motor
    - TP sensor
    - EVAP canister purge solenoid.
    - MAP sensor
    - CMP sensor
    - Ground splice at the rear of the right side of engine block
    - Ground splice and the ground strap at the rear of the left side of engine block
    - CTS sensor
    - Oil pressure sensor/switch

28. Install or connect the following:
    - Nut to the engine wiring harness bracket and tighten to 89 inch lbs. (10 Nm)

> **❊❊ CAUTION**
>
> In order to avoid possible injury or vehicle damage, always replace the accelerator control cable with a NEW cable whenever you remove the engine from the vehicle. In order to avoid cruise control cable damage, position the cable out of the way while you remove or install the engine.

- NEW accelerator control cable
- Cruise control cable, if equipped, to the throttle shaft
- Bolts for the accelerator control cable mounting bracket and tighten to 89 inch lbs. (10 Nm)
- Engine sight shield to the intake manifold
- Heater hoses to the cowl and the engine
- Coolant hose to the throttle body
- Radiator support in the vehicle
- A/C condenser mounting bolts
- Air cleaner duct
- Lower radiator hoses to the engine

29. Lower the hood.

30. Fill the engine with oil.

31. Fill the engine with coolant.

32. Connect the negative battery cable.

8.1L Engine

1. Before servicing the vehicle, refer to the precautions in the beginning of this section.

2. Raise the hood to the servicing position. Move the hood hinge bolt to hold the hood in the servicing position.

3. Release the fuel system pressure.

4. Remove or disconnect the following:
    - Negative, then positive battery cables
    - Coolant
    - A/C refrigerant
    - Engine oil cooler lines from the engine block
    - Transmission-to-engine bolts
    - Clutch pressure plate bolts, if equipped
    - Torque converter bolts, if equipped
    - Catalytic converter
    - Exhaust manifold pipe

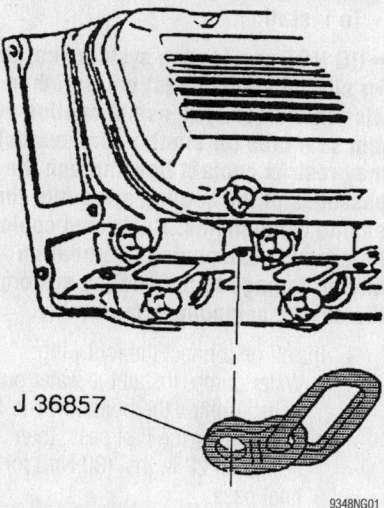

J 36857

9348NG01

Install suitable lift brackets to the rear of the right head and the front of the left head

- Hoses from power steering pump, then plug the lines and ports
- Starter motor
5. Raise the vehicle.
- Engine electrical harness and tie aside
- Alternator
- Ground cable bolt from engine block
- Exhaust Gas Recirculation (EGR) valve adapter
- Vacuum lines (tag before removal)
- Throttle Actuator Control (TAC) module electrical connector
6. Install Engine Lift Brackets part No. J 36857, or equivalent, to the rear of the right cylinder head and the front of the left cylinder head.
7. Install the attaching bolt and washer. Use part No. 9428217 with 1560963. Tighten the bolts to 30 ft. lbs. (40 Nm).
8. Remove or disconnect the following:
- Engine mount heat shield bolt and shields
- Engine mount-to-engine mount bracket bolts
- Engine from the vehicle, using a suitable lifting device. Place on a suitable stand.
- A/C compressor/power steering pump bracket from the cylinder head
- Lift brackets from the cylinder head
To install:
9. Install Engine Lift Brackets part No. J 36857, or equivalent, to the rear of the right cylinder head and the front of the left cylinder head.
10. Install the attaching bolt and washer. Use part No. 9428217 with 1560963. Tighten the bolts to 30 ft. lbs. (40 Nm).

11. Install or connect the following:
- A/C compressor/power steering mounting bracket. Tighten the bolts and nut to 37 ft. lbs. (50 Nm)
- Alternator bracket
- Engine into the vehicle
- Engine mount-to-engine mount bracket bolts
- Engine mount heat shield and bolts
12. Remove the lift hooks from the cylinder heads, then raise the vehicle.
- Engine oil cooler lines
- Transmission-to-engine bolts
- Clutch pressure plate bolts, if equipped
- Torque converter bolts, if equipped
- Catalytic converter
- Exhaust manifold pipe
- Hoses to the power steering pump
- Starter motor
13. Lower the vehicle.
- Engine electrical harness. Make sure the harness is properly routed.
- Alternator
- Ground cable bolt to engine block. Tighten to 12 ft. lbs. (16 Nm).
- EGR valve adapter
- Vacuum lines, as tagged during removal
- TAC module electrical connector
- Radiator
- A/C compressor
- Fuel feed and return lines
- Ignition coils
- Positive, then negative battery cables
- Air cleaner outlet duct and secure with the clamp
14. Lower the hood from the service position.
15. Properly recharge the A/C system.
16. Fill the engine with oil.
17. Fill the engine with coolant.
18. Perform the Crankshaft Position (CKP) sensor variation learn procedure:
  a. Install a suitable scan tool and check for Diagnostic Trouble Codes (DTCs). If any DTCs, other than P1336 are set, resolve those codes first, before proceeding with this procedure.
  b. With the scan tool, select the crankshaft position variation learn procedure.
  c. Observe the fuel cut-off for the 8.1L engine.
  d. The scan tool will instruct you to perform certain steps, make sure you follow all directions given by the scan tool exactly.
  e. Enable the crankshaft position system variation learn procedure.

➡ While the learn procedure is in progress, release the throttle immediately when the engine started to decelerate. The engine control is returned to the operator and the engine responds to throttle position after the learn procedure is complete.

  f. Slowly increase the engine speed to the RPM that you observed.
  g. Immediately release the throttle when fuel cut-out is reached.
  h. The scan tool displays: Learn Status: Learned this ignition. If the scan tool does NOT display this message and not other DTCs set, you must perform further troubleshooting.
  i. Turn the ignition **OFF** for 30 seconds after the learn procedure has been completed successfully.
19. Start and run the engine, then check for leaks.

Water Pump

REMOVAL & INSTALLATION

5.7L and 7.4L Engines

1. Before servicing the vehicle, refer to the precautions in the beginning of this section.
2. Drain the radiator.
3. Remove or disconnect the following:
- Fan shroud
- Negative battery cable
- Drive belt(s)
- Alternator and other accessories, if necessary
- Fan, fan clutch and pulley
- Accessory brackets that might interfere with water pump removal
- Lower radiator hose from the water pump inlet
- Heater hose from the nipple on the pump
- Bypass hose, 7.4L engine only
- Water pump assembly away from the timing cover
To install:
4. Clean all old gasket material from the timing chain cover and water pump.
5. Install or connect the following:
- Pump assembly with a new gasket. Torque the bolts to 30 ft. lbs. (41 Nm)
- Hose between the water pump inlet and the pump
- Heater hose and the bypass hose (7.4L only)
- Fan, fan clutch and pulley
- Alternator and other accessories, if necessary

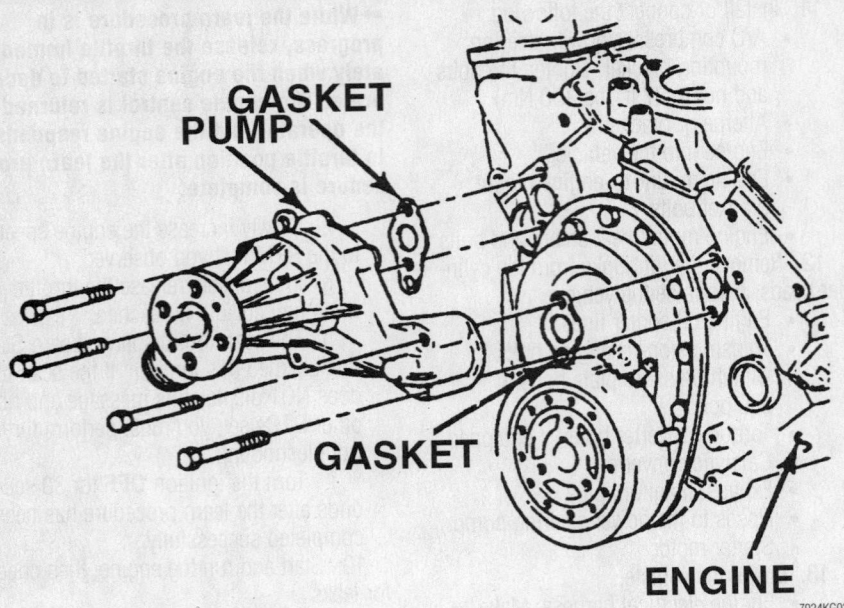

Exploded view of the water pump mounting—5.7L engines

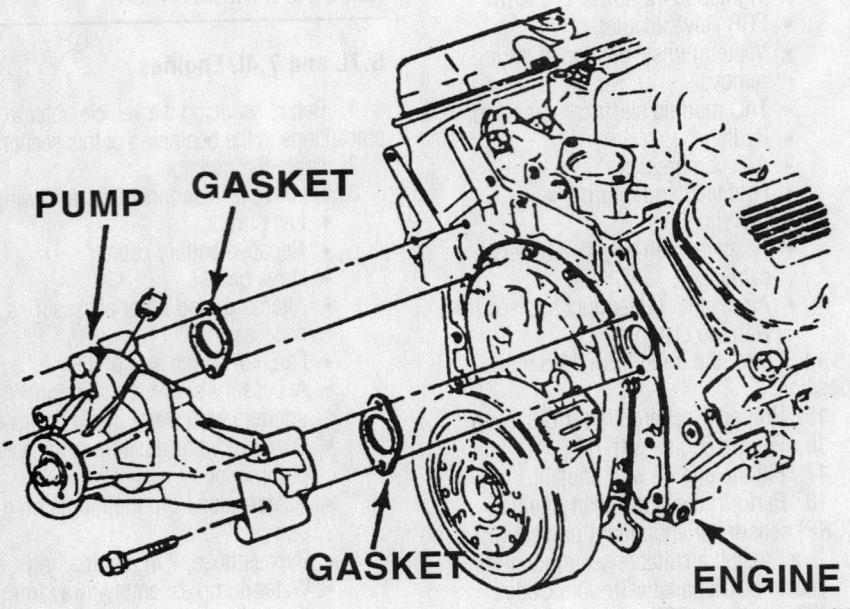

Exploded view of the water pump mounting—7.4L engine

- Drive belt(s)
- Upper radiator shroud
6. Refill the cooling system.
7. Connect the battery.

4.8L, 5.3L and 6.0L Engines

1. Before servicing the vehicle, refer to the precautions in the beginning of this section.

2. Remove or disconnect the following:
- Air inlet duct

- Coolant
- Inlet radiator hose from the water pump
- Upper fan shroud
- Cooling fan and clutch assembly
- Drive belt
- Radiator outlet hose from the coolant pump
- Surge tank hose
- Heater hose
- Water pump

To install:

➡DO NOT use cooling system seal tabs (or similar compounds) unless otherwise instructed. The use of cooling system seal tabs (or similar compounds) may restrict coolant flow through the passages of the cooling system or the engine components. Restricted coolant flow may cause engine overheating and/or damage to the cooling system or the engine components/assembly.

3. Install or connect the following:
- Water pump. Install the water pump bolts. Tighten the bolts to 11 ft. lbs. (15 Nm) for the first pass; then tighten to 22 ft. lbs. (30 Nm) for the final pass.
- Water pump drive belt pulley and bolts (if applicable). Tighten the bolts to 89 inch lbs. (10 Nm) for the first pass; then tighten to 18 ft. lbs. (25 Nm) for the final pass.
- Surge tank hose
- Heater hose
- Outlet radiator hose to the coolant pump
- Drive belt
- Cooling fan and clutch assembly
- Upper fan shroud
- Inlet radiator hose to the water pump
- Air inlet duct
- Coolant

8.1L Engines

1. Before servicing the vehicle, refer to the precautions in the beginning of this section.

2. Remove or disconnect the following:
- Coolant
- Drive belt
- Fan clutch
- Outlet hose clamp and hose

3. Reposition the bypass hose clamps at the water pump and water crossover.
- Bypass hose
- Water pump bolt and pump. Discard the water pump gaskets.

To install:

4. Install or connect the following:
- New water pump gaskets
- Water pump and bolts. Tighten the water pump bolts 37 ft. lbs. (50 Nm).
- Bypass hose and clamps
- Outlet hose and clamp
- Fan clutch
- Drive belt
- Surge tank hose
- Heater hose

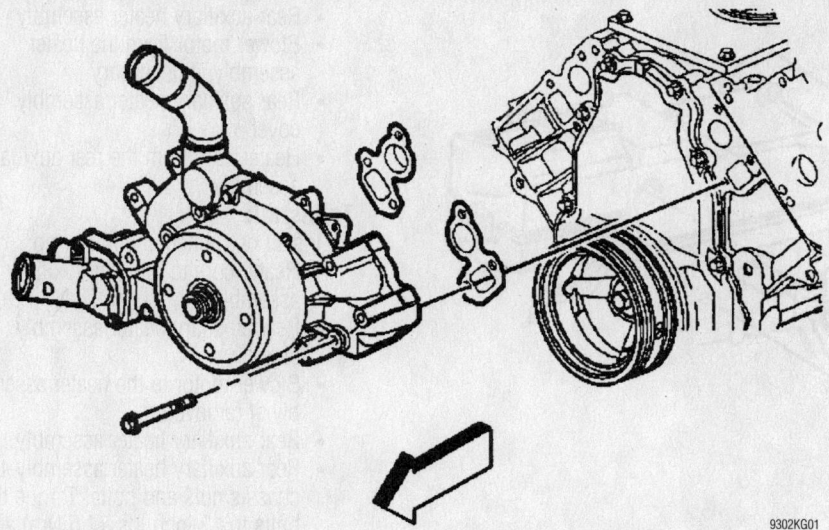

9302KG01

Exploded view of the water pump assembly—4.8L, 5.3L and 6.0L engines

- Outlet radiator hose to the coolant pump
- Drive belt
- Cooling fan and clutch assembly
- Upper fan shroud
- Inlet radiator hose to the water pump
- Air inlet duct
- Coolant

Heater Core

REMOVAL & INSTALLATION

Front Heater

1. Before servicing the vehicle, refer to the precautions in the beginning of this section.
2. Drain the engine cooling system into a clean container for reuse.
3. Remove or disconnect the following:
   - Negative battery cable
   - Heater hoses from the heater core
   - Instrument panel storage compartment
   - Electrical connectors, as necessary, that may be in the way
   - Center floor air distribution duct
   - Hinge pillar trim kick panels
   - Blower motor cover screws and the cover
   - Blower motor screws and the blower motor
   - Steering wheel and the steering column
   - Instrument panel fasteners and pull the instrument panel back far enough to gain access to the heater assembly

- Screw located on the interior side near the evaporator pipe, if equipped, while holding the heater assembly against the firewall
- 4 heater assembly-to-chassis screws and the 2 heater assembly-to-chassis nuts, working in the engine compartment

➡ **Removal of the heater assembly may require the help of an assistant.**

- 7 heater cover-to-heater assembly screws and the cover
- Heater core from the heater assembly

To install:

4. Install or connect the following:
   - Heater core to the heater assembly
   - Heater cover and the 7 heater cover-to-heater assembly screws

➡ **Installation of the heater assembly may require the help of an assistant.**

- 4 heater assembly-to-chassis screws and the 2 heater assembly-to-chassis nuts. Torque the screws to 17 inch lbs. (1.9 Nm) and the nuts to 25 inch lbs. (2.8 Nm).
- Screw located on the interior side near the evaporator pipe, if equipped. Torque the screw to 97 inch lbs. (11 Nm).
- Instrument panel and the instrument panel fasteners
- Steering column and the steering wheel
- Blower motor and the blower motor screws
- Blower motor the cover and the cover screws
- Hinge pillar trim kick panels
- Center floor air distribution duct
- Electrical connectors that were disconnected
- Instrument panel storage compartment
- Heater hoses to the heater core
- Negative battery cable

5. Refill the engine cooling system.
6. Run the engine to normal operating temperatures; then, check the climate control operation and check for leaks.

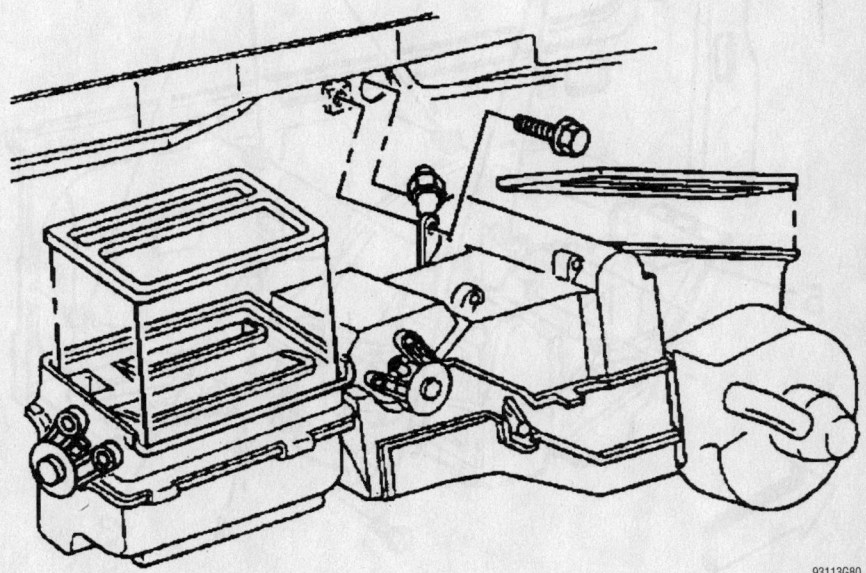

93113G80

View of the front heater assembly

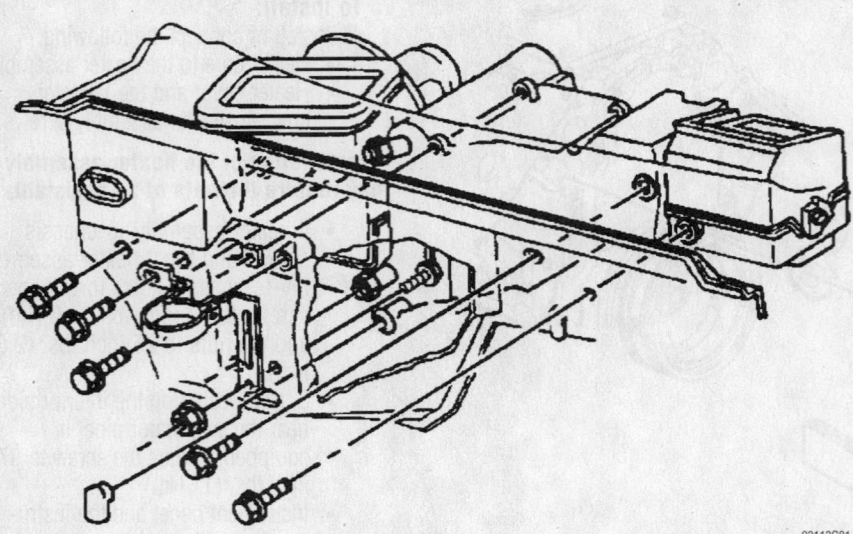

93113G81

Location of the front heater assembly-to-chassis fasteners

Rear Auxiliary Heater

1. Before servicing the vehicle, refer to the precautions in the beginning of this section.
2. Drain the engine cooling system into a clean container for reuse.
3. Remove or disconnect the following:
   - Negative battery cable
   - Rear quarter trim panel, as necessary
   - Right rear quarter trim panel
   - Right rear wheelhouse
   - Heater hoses from the rear auxiliary heater core
   - Electrical connectors, as necessary
   - Drain valve
   - Rear auxiliary heater assembly-to-chassis nuts and bolts

- Rear auxiliary heater assembly
- Blower motor from the heater assembly, if necessary
- Rear auxiliary heater assembly cover
- Heater core from the rear auxiliary assembly

To install:

4. Install or connect the following:
   - Heater core to the rear auxiliary assembly
   - Rear auxiliary heater assembly cover
   - Blower motor to the heater assembly, if removed
   - Rear auxiliary heater assembly
   - Rear auxiliary heater assembly-to-chassis nuts and bolts. Torque the bolts to 13 inch lbs. (1.5 Nm) and the nuts to 89 inch lbs. (10 Nm).
   - Drain valve
   - Electrical connectors, as necessary
   - Heater hoses to the rear auxiliary heater core
   - Right rear wheelhouse
   - Right rear quarter trim panel
   - Rear quarter trim panel, as necessary
5. Refill the engine cooling system.
6. Connect the negative battery cable.
7. Run the engine to normal operating temperatures; then, check the climate control operation and check for leaks.

Cylinder Head

REMOVAL & INSTALLATION

4.8L, 5.3L and 6.0L Engines

RIGHT SIDE

☀ CAUTION

Before servicing any electrical component, the ignition key must be in the OFF or LOCK position and all electrical loads must be OFF, unless instructed otherwise in these procedures.

1. Before servicing the vehicle, refer to the precautions in the beginning of this section.
2. Remove or disconnect the following:
   - Negative battery cable
   - Intake manifold
   - Push rods
   - Exhaust manifold(s)
   - Alternator
   - Three bolts holding the alternator mounting bracket to the cylinder head

93113G82

View of the rear auxiliary heater assembly—Suburban shown

- Bolt behind the power steering pump
- Alternator mounting bracket and set it aside
- Bolt holding the oil level indicator tube to the right side cylinder head
- Oil level indicator tube
- Spark plugs from the cylinder head

➡**The M11 cylinder head bolts are NOT reusable. Install NEW M11 cylinder head bolts during assembly.**

- Cylinder head bolts
- Cylinder head(s) from the engine

➡**After removal, place the cylinder head on two wood blocks to prevent damage.**

3. Remove and discard the gasket. Discard the M11 cylinder head bolts.

To install:

➡**Do not use any type of sealant on the cylinder head gasket (unless specified). The cylinder head gaskets must be installed in the proper direction and position.**

4. Clean the engine block cylinder head bolt holes (if required). Thread repair tool J 42385-107 may be used to clean the threads of old threadlocking material.

5. Spray cleaner GM P/N 12346139, P/N 12377981, or equivalent into the hole.

6. Clean the cylinder head bolt holes with compressed air.

7. Check the cylinder head locating pins for proper installation.

➡**When properly installed, the tab on the right cylinder head gasket will be located right of center or closer to the front of the engine.**

8. Install or connect the following:
- NEW right cylinder head gasket onto the locating pins
- Cylinder head onto the locating pins and the gasket
- NEW M11 cylinder head bolts. Apply a 0.20 in. (5mm) band of threadlock GM P/N 12345382 or equivalent to the threads of the M8 cylinder head bolts.
- M8 cylinder head bolts.

9. Tighten the cylinder head bolts, as follows:
   a. M11 bolts, first pass: in sequence to 22 ft. lbs. (30 Nm).
   b. M11 bolts, second pass: in sequence + 90 degrees.
   c. M11 bolts (1,2,3,4,5,6,7,8): + 90 degrees.
   d. M11 bolts (9 and 10): + 50 degrees.

   e. M8 bolts (11,12,13,14,15): to 22 ft. lbs. (30 Nm). Begin with the center bolt (11) and alternating side-to-side, work outward tightening all of the bolts.

➡**The cylinder head gasket displacement can be verified by markings visible on the underside of the right gasket locating tab. Some 4.8 and5.3L head gaskets may have 53 stamped onto the locating tab. Some 6.0L head gaskets may have 60 stamped onto the locating tab.**

- Install or connect the following:
- Alternator
- Exhaust manifold(s)

- Pushrods
- Intake manifold
- Negative battery cable

LEFT SIDE

1. Before servicing the vehicle, refer to the precautions in the beginning of this section.

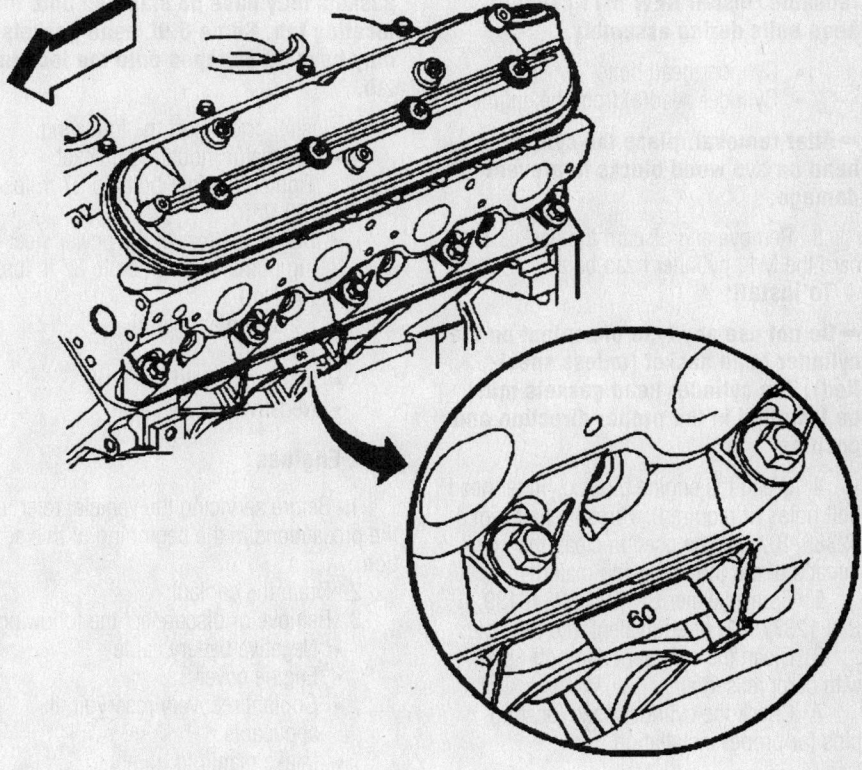

Locating tab—4.8L, 5.3L and 6.0L engines

9308KG57

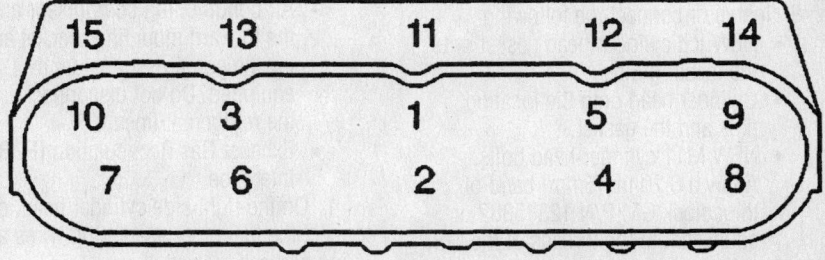

Cylinder head bolt tightening sequence—4.8L, 5.3L and 6.0L engines

9302KG02

2. Remove or disconnect the following:
- Negative battery cable
- Intake manifold
- Pushrods
- Exhaust manifold(s)
- Alternator
- Three bolts holding the alternator mounting bracket to the cylinder head
- The bolt behind the power steering pump
- Alternator mounting bracket and set it aside
- Bolt holding the oil level indicator tube to the right side cylinder head
- Oil level indicator tube
- Spark plugs from the cylinder head

➡ **The M11 cylinder head bolts are NOT reusable. Install NEW M11 cylinder head bolts during assembly.**

- Cylinder head bolts
- Cylinder head(s) from the engine

➡ **After removal, place the cylinder head on two wood blocks to prevent damage.**

3. Remove and discard the gasket. Discard the M11 cylinder head bolts.

To install:

➡ **Do not use any type of sealant on the cylinder head gasket (unless specified). The cylinder head gaskets must be installed in the proper direction and position.**

4. Clean the engine block cylinder head bolt holes (if required). Thread repair tool J 42385-107 may be used to clean the threads of old threadlocking material.
5. Spray cleaner GM P/N 12346139, P/N 12377981, or equivalent into the hole.
6. Clean the cylinder head bolt holes with compressed air.
7. Check the cylinder head locating pins for proper installation.

➡ **When properly installed, the tab on the left cylinder head gasket will be located left of center or closer to the front of the engine.**

8. Install or connect the following:
- NEW left cylinder head gasket onto the locating pins
- Cylinder head onto the locating pins and the gasket
- NEW M11 cylinder head bolts. Apply a 0.20 in. (5mm) band of threadlock GM P/N 12345382 or equivalent to the threads of the M8 cylinder head bolts.
- M8 cylinder head bolts

9. Tighten the cylinder head bolts, as follows:
  a. M11 bolts, first pass: in sequence to 22 ft. lbs. (30 Nm).
  b. M11 bolts, second pass: in sequence + 90 degrees.
  c. M11 bolts (1,2,3,4,5,6,7,8): + 90 degrees.
  d. M11 bolts (9 and 10): + 50 degrees.
  e. M8 bolts (11,12,13,14,15): to 22 ft. lbs. (30 Nm). Begin with the center bolt (11) and alternating side-to-side, work outward tightening all of the bolts.

➡ **The cylinder head gasket displacement can be verified by markings visible on the top side of the left gasket locating tab. Some 4.8 and 5.3L head gaskets may have 53 stamped onto the locating tab. Some 6.0L head gaskets may have 60 stamped onto the locating tab.**

10. Install or connect the following:
- Alternator mounting bracket. Tighten the four bolts to 37 ft. lbs. (50 Nm).
- Bolt at the rear of the power steering pump and tighten to 37 ft. lbs. (50 Nm).
- Exhaust manifold(s)
- Pushrods
- Intake manifold
- Negative battery cable

5.7L Engines

1. Before servicing the vehicle, refer to the precautions in the beginning of this section.
2. Drain the coolant.
3. Remove or disconnect the following:
- Negative battery cable
- Engine cover
- Coolant recovery reservoir, if applicable
- Intake manifold
- Exhaust manifolds and position them out of the way
- Ground strap at the rear of the right AIR pipe, if equipped
- Air conditioning compressor and the forward mounting bracket and lay the compressor aside, if equipped. Do not disconnect any of the refrigerant lines.
- Exhaust Gas Recirculation (EGR) inlet tube

4. On the right side cylinder head, disconnect the fuel pipe, spark plug wires and wiring harness bracket.
- Nut and stud attaching the main

accessory bracket to the cylinder head.

➡ **You may have to loosen the remaining bolts and studs in order to remove the head.**

- Coolant sensor wire
- Spark plug wire bracket
- Cylinder head covers
- Spark plugs
- Pushrods, Identify the pushrods so that they can be installed in their original positions.
- Cylinder head bolts in the reverse order of the tightening sequence
- Cylinder head(s)

To install:

5. Inspect the cylinder head and block mating surfaces. Clean all old gasket material.
6. Install or connect the following:
- Cylinder heads using new gaskets. Install the gaskets with the word **HEAD** up.

➡ **Coat a steel gasket on both sides with sealer. If a composition gasket is used, do not use sealer.**

7. Clean the bolts, apply sealer to the threads, and hand-tighten.
8. Install the cylinder head bolts and tighten, in sequence, to 22 ft. lbs. (30 Nm). The bolts must be tightened once, then be tightened again in sequence in the following order:
  a. Step 1: Short bolts (3, 4, 7, 8, 11, 12, 15, 16), plus 55 degrees.
  b. Step 2: Medium bolts (14, 17), plus 65 degrees.
  c. Step 3: Long bolts (1, 2, 5, 6, 9, 10, 13), plus 75 degrees.
9. Install or connect the following:
- Pushrods, in their original positions
- Cylinder head covers
- Spark plugs
- Coolant sensor wire
- Spark plug wire bracket
- Main accessory bracket to the cylinder head
- EGR vent tube
- Fuel pipe
- Spark plug wires
- Wiring harness bracket
- Air conditioning compressor and forward mounting bracket, if equipped
- Ground strap to the rear of the right AIR pipe
- Exhaust manifolds
- Intake manifold
- Coolant recovery reservoir, if removed

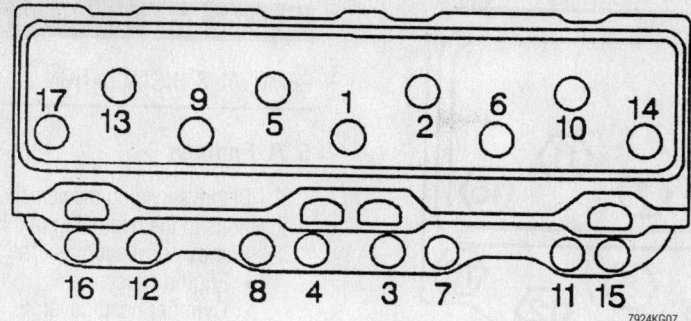

Cylinder head bolt tightening sequence—5.7L engines

- Engine cover.
- Negative battery cable
10. Refill the engine with coolant.

7.4L Engines

1. Before servicing the vehicle, refer to the precautions in the beginning of this section.
2. Drain the cooling system.
3. Remove or disconnect the following:
- Negative battery cable
- Engine cover
- Intake manifold
- Exhaust manifolds
- Alternator and bracket
- Air pump, if equipped
- Air conditioning compressor and the forward mounting bracket. Do not disconnect any of the refrigerant lines.
- Rocker arm cover
- Spark plugs
- Air pipes at the rear of the head, if equipped
- Ground strap at the rear of the head
- Temperature sensor wire
- Pushrods. Identify the pushrods so that they can be installed in their original positions.
- Cylinder head bolts and the heads

To install:

➡The cylinder head should be cleaned and inspected for warpage or damage before installation.

4. Thoroughly clean the mating surfaces of the head and block. Clean the bolt holes thoroughly.

➡Coat a steel gasket on both sides with sealer. If a composition gasket is used, do not use sealer.

5. Install or connect the following:
- New gaskets, with the word **HEAD** up
- Cylinder heads
- Cylinder head bolts, apply sealer to the threads, and hand-tighten.

6. Tighten the head bolts a little at a time, in the proper sequence, in 3 stages:
a. Step 1: Torque to 30 ft. lbs. (40 Nm).
b. Step 2: Torque to 60 ft. lbs. (80 Nm).
c. Step 3: Torque to 85 ft. lbs. (115 Nm).
- Intake and exhaust manifolds
- Pushrods
- Rocker arms
- Temperature sensor wire
- Ground strap at the rear of the head
- AIR pipes at the rear of the head
- Spark plugs
- Rocker arm cover
- Air conditioning compressor and the forward mounting bracket
- AIR pump
- Alternator
- Engine cover
7. Connect the battery cable and refill the cooling system.

8.1L Engine

LEFT SIDE

1. Before servicing the vehicle, refer to the precautions in the beginning of this section.
2. Drain the cooling system.
3. Remove or disconnect the following:

- Negative battery cable
- Intake manifold
- Valve cover
- Rocker arms and pushrods, keeping them in order for installation
- Engine harness ground bolts
4. Reposition the engine harness grounds and ground straps from the cylinder head.
- Water crossover
- Exhaust manifold
- Cylinder head bolts, then discard

➡The cylinder head bolts must be replaced for installation.

- Cylinder head. Place the head on 2 wood blocks to protect the sealing surfaces while it is removed.

To install:

➡The cylinder head should be cleaned and inspected for warpage or damage before installation.

5. Thoroughly clean the mating surfaces of the head and block. Clean the bolt holes thoroughly.

➡If a composition gasket is used, do not use sealer.

6. Align the cylinder head gasket locating marks to face up. Make sure that the gasket tabs are located of the no. 1 and 2 cylinder for proper installation.
7. Install or connect the following:
- New cylinder head gasket
- Cylinder head
- Sealer to the threads of new cylinder head bolts, if not pre-applied
- Cylinder head bolts and hand-tighten
8. Tighten the head bolts a little at a time, in the proper sequence, in 3 stages:
a. Step 1: Torque to 30 ft. lbs. (40 Nm).
b. Step 2: Tighten an additional 120 degrees using a torque angle meter.

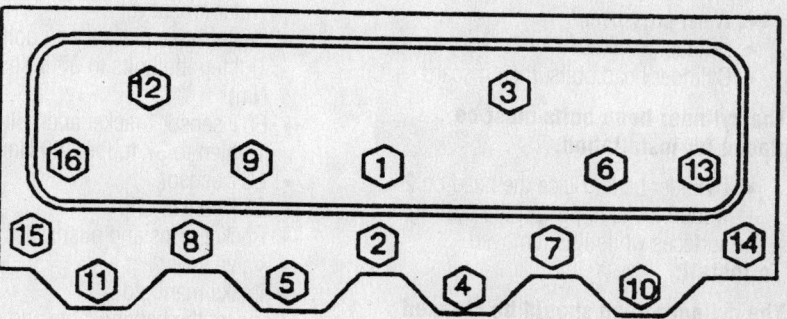

Cylinder head bolt tightening sequence—7.4L engines

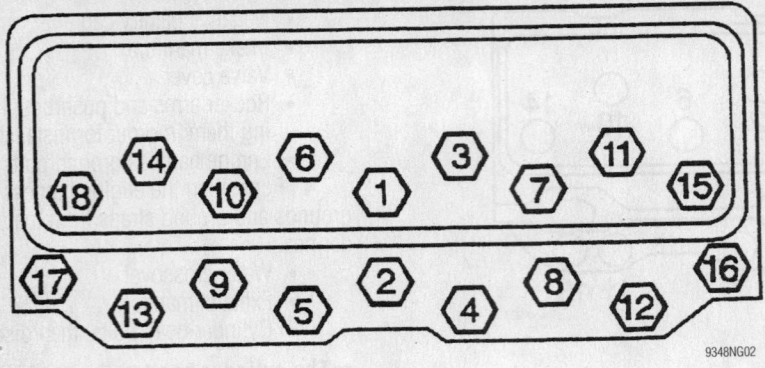

9348NG02

Cylinder head bolt tightening sequence—8.1L engine

c. Step 3: Torque bolt numbers. 1, 2, 3, 6, 7, 8, 9, 10, 11, 14, 16 and 17 an additional 60 degrees. Tighten bolts 15 and 18 an additional 45 degrees, and bolt numbers 4, 5, 12 and 13 an additional 30 degrees.

9. Install or connect the following:
- Exhaust manifold
- Water crossover
- Engine harness grounds and ground strap
- Rocker arms and pushrods
- Valve cover
- Intake manifold

10. Connect the battery cable and refill the cooling system.

RIGHT SIDE

1. Before servicing the vehicle, refer to the precautions in the beginning of this section.
2. Drain the cooling system.
3. Remove or disconnect the following:
- Negative battery cable
- Intake manifold
- Valve cover
- Rocker arms and pushrods, keeping them in order for installation
- Engine Coolant Temperature (ECT) sensor clip from the bracket
- ECT sensor
- ECT sensor bracket bolt and bracket
- Heater inlet and outlet hoses from the hose bracket
- Water crossover
- Exhaust manifold
- Cylinder head bolts, then discard

➡ **The cylinder head bolts must be replaced for installation.**

- Cylinder head. Place the head on 2 wood blocks to protect the sealing surfaces while it is removed.

To install:

➡ **The cylinder head should be cleaned and inspected for warpage or damage before installation.**

4. Thoroughly clean the mating surfaces of the head and block. Clean the bolt holes thoroughly.

➡ **If a composition gasket is used, do not use sealer.**

5. Align the cylinder head gasket locating marks to face up. Make sure that the gasket tabs are located of the no. 1 and 2 cylinder for proper installation.
6. Install or connect the following:
- New cylinder head gasket
- Cylinder head
- Sealer to the threads of new cylinder head bolts, if not pre-applied
- Cylinder head bolts and hand-tighten

7. Tighten the head bolts a little at a time, in the proper sequence, in 3 stages:
a. Step 1: Torque to 30 ft. lbs. (40 Nm).
b. Step 2: Tighten an additional 120 degrees using a torque angle meter.
c. Step 3: Torque bolt numbers. 1, 2, 3, 6, 7, 8, 9, 10, 11, 14, 16 and 17 an additional 60 degrees. Tighten bolts 15 and 18 an additional 45 degrees, and bolt numbers 4, 5, 12 and 13 an additional 30 degrees.
8. Install or connect the following:
- Exhaust manifold
- Water crossover
- Heater hose bracket and bolts. Tighten the bolts to 37 ft. lbs. (50 Nm).
- ECT sensor bracket and bolt. Tighten to 37 ft. lbs. (50 Nm).
- ECT sensor
- ECT sensor clip
- Rocker arms and pushrods
- Valve cover
- Intake manifold

9. Connect the battery cable and refill the cooling system.

Rocker Arms

REMOVAL & INSTALLATION

5.7L Engines

1. Before servicing the vehicle, refer to the precautions in the beginning of this section.
2. Remove or disconnect the following:
- Engine cover
- Cylinder head cover.
- Rocker arm nut. If you are only replacing the pushrod, back the nut off until you can swing the rocker out of the way.
- Rocker arms and balls as a unit

➡ **Always remove each set of rocker arms (1 set per cylinder) as a unit.**

- Pushrods and pushrod guides

To install:
3. Install or connect the following:
- Pushrods and their guides. Be sure that they seat properly in each lifter.
4. Position a set of rocker arms (for 1 cylinder) in the proper location.

➡ **Install the rocker arms for each cylinder only when the lifters are off the cam lobe and both valves are closed.**

5. Coat the replacement rocker arm with Molykote® or its equivalent, and the rocker arm and pivot with SAE 90 gear oil, and install the pivots.
6. Install or connect the following:
- Nuts and tighten alternately
- Engine cover

4.8L, 5.3L and 6.0L Engines

➡ **Do not remove the ignition coils from the valve rocker arm cover unless required. Do not remove the oil fill tube from the cover unless service is required. If the oil fill tube has been removed from the cover, install a NEW tube during assembly.**

On the right side:
1. Remove or disconnect the following:
- Ignition coil bracket bolts from the rocker arm cover (if required)
- Ignition coil and bracket assembly from the cover
- Valve rocker arm cover bolts
- Valve rocker arm cover
- Gasket from the cover. Discard the gasket. The bolt grommets may be reused if not damaged.
- Oil fill cap from the oil fill tube
- Oil fill tube (if required). Discard the oil fill tube.

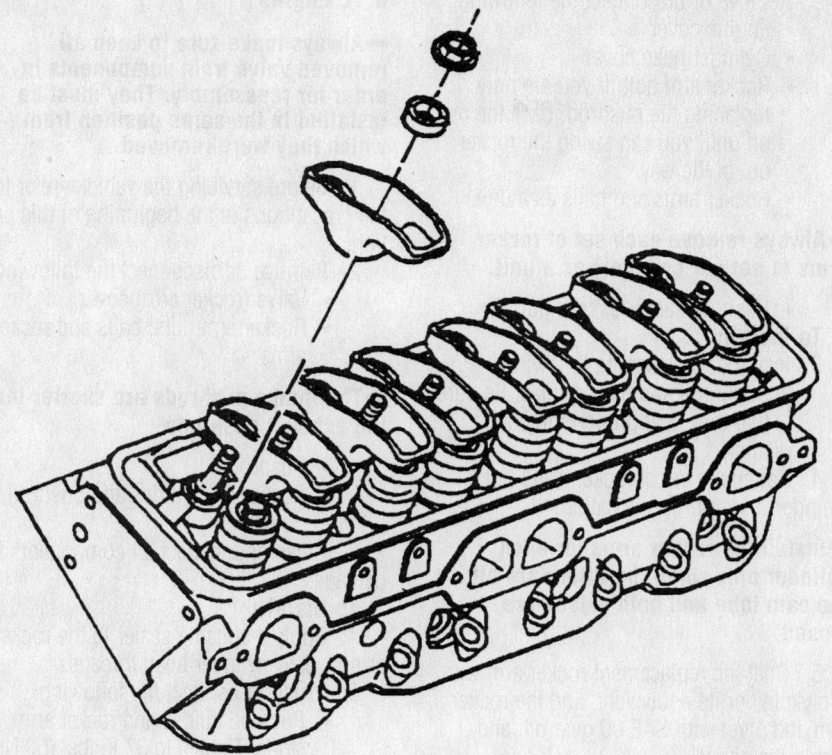

Exploded view to the rocker arm and related components—5.7L engines

7924KG10

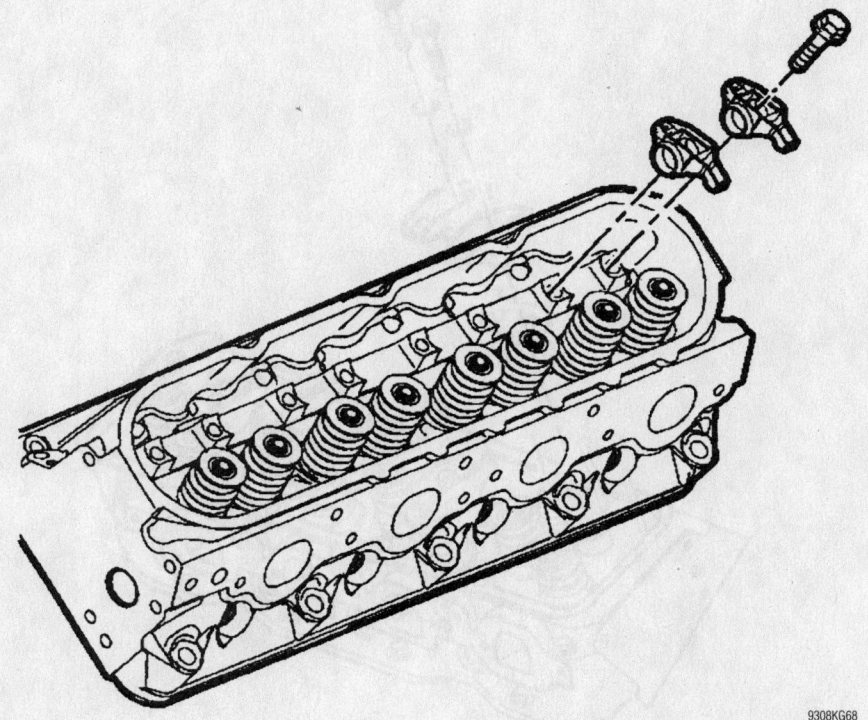

Rocker arm removal—4.8L, 5.3L and 6.0L engines

9308KG68

On the left side:

➡ Do not remove the Positive Crankcase Ventilation (PCV) valve grommet from the cover unless service is required.

2. Remove or disconnect the following:
- Ignition coil bracket bolts from the rocker arm cover (if required)
- Ignition coil and bracket assembly from the cover
- Valve rocker arm cover bolts
- Valve rocker arm cover
- Gasket from the cover. Discard the gasket. The bolt grommets may be reused if not damaged.
- Valve rocker arm bolts
- Valve rocker arms
- Valve rocker arm pivot support
- Pushrods

To install:

➡ Valve lash is built in. No valve adjustment is required.

3. Lubricate the valve rocker arms and pushrods with clean engine oil.
4. Lubricate the flange of the valve rocker arm bolts with clean engine oil.
5. Lubricate the flange or washer surface of the bolt that will contact the valve rocker arm.
6. Install or connect the following:
- Valve rocker arm pivot support

➡ Make sure that the pushrods seat properly to the valve lifter sockets.

- Pushrods

➡ Make sure that the pushrods seat properly to the ends of the rocker arms.

- Rocker arms and bolts. DO NOT tighten the rocker arm bolts at this time

7. Rotate the crankshaft until number one piston is at top dead center of compression stroke. In this position, cylinder number one rocker arms will be off lobe lift, and the crankshaft sprocket key will be at the 1:30 position. If viewing from the rear of the engine, the additional crankshaft pilot hole (non-threaded) will be in the 10:30 position. The engine firing order is 1, 8, 7, 2, 6, 5, 4, 3. Cylinders 1, 3, 5 and 7 are left bank. Cylinders 2, 4, 6, and 8 are right bank.

8. With the engine in the number one firing position, tighten the following valve rocker arm bolts:
   a. Tighten exhaust valve rocker arm bolts 1, 2, 7, and 8 to 22 ft. lbs. (30 Nm)
   b. Tighten intake valve rocker arm bolts 1, 3, 4, and 5 to 22 ft. lbs. (30 Nm)
9. Rotate the crankshaft 360 degrees.

Tighten the following valve rocker arm bolts:

  a. Tighten exhaust valve rocker arm bolts 3, 4, 5, and 6 to 22 ft. lbs. (30 Nm)

  b. Tighten intake valve rocker arm bolts 2, 6, 7, and 8 to 22 ft. lbs. (30 Nm)

On the right side:

➡**The valve rocker arm cover bolt grommets may be reused. If the oil fill tube has been removed from the valve rocker arm cover, install a NEW oil fill tube during assembly.**

10. Lubricate the O-ring seal of the NEW oil fill tube with clean engine oil.

11. Install or connect the following:
  - NEW oil fill tube into the rocker arm cover and rotate the tube clockwise until locked in the proper position
  - Oil fill cap into the tube and rotate clockwise until locked in the proper position
  - NEW cover gasket into the valve rocker arm cover
  - Valve rocker arm cover onto the cylinder head
  - Cover bolts with grommets and tighten to 106 inch lbs. (12 Nm)

12. Apply threadlock GM P/N 12345382 or equivalent to the threads of the bracket bolts.
  - Ignition coil and bracket assembly and bolts. Tighten to 106 inch lbs. (12 Nm).

On the left side:

➡**DO NOT reuse the valve rocker arm cover gasket. The valve rocker arm cover bolt grommets may be reused. If the vapor vent grommet has been removed from the valve rocker arm cover, install a NEW vapor vent gourmet during assembly.**

13. Install or connect the following:
  - NEW cover gasket (1) into the valve rocker arm cover
  - Valve rocker arm cover onto the cylinder head
  - Cover bolts with grommets and tighten to 106 inch lbs. (12 Nm)

14. Apply threadlock GM P/N 12345382 or equivalent to the threads of the bracket bolts.
  - Ignition coils and bracket assembly and bolts. Tighten the bolts to 106 inch lbs. (12 Nm).

7.4L Engines

1. Before servicing the vehicle, refer to the precautions in the beginning of this section.

2. Remove or disconnect the following:
  - Engine cover
  - Cylinder head cover
  - Rocker arm bolt. If you are only replacing the pushrod, back the nut off until you can swing the rocker out of the way.
  - Rocker arms and balls as a unit

➡**Always remove each set of rocker arms (1 set per cylinder) as a unit.**

  - Pushrods and pushrod guides

To install:

3. Install or connect the following:
  - Pushrods and their guides, Be sure that they seat properly in each lifter.

4. Position a set of rocker arms (for 1 cylinder) in the proper location.

➡**Install the rocker arms for each cylinder only when the lifters are off the cam lobe and both valves are closed.**

5. Coat the replacement rocker arm with Molykote® or its equivalent, and the rocker arm and pivot with SAE 90 gear oil, and install the pivots.
  - Rocker arm bolts and tighten to 45 ft. lbs. (61 Nm)
  - Engine cover

8.1L Engine

➡**Always make sure to keep all removed valve train components in order for reassembly. They must be installed in the same position from which they were removed.**

1. Before servicing the vehicle, refer to the precautions in the beginning of this section.

2. Remove or disconnect the following:
  - Valve (rocker arm) cover
  - Rocker arm nuts, balls and rocker arms

➡**The intake pushrods are shorter than the exhaust pushrods.**

  - Pushrods
  - Rocker arm guides and pushrod guides

3. Clean and inspect all components for damage.

To install:

4. Apply a suitable sealer to the rocker arm stud-to-cylinder head threads.

5. Install or connect the following:
  - Pushrod guides and rocker arm studs. Tighten to 37 ft. lbs. (50 Nm).
  - Pushrods

6. Coat the rocker arm and ball bearing surfaces with a suitable prelube.

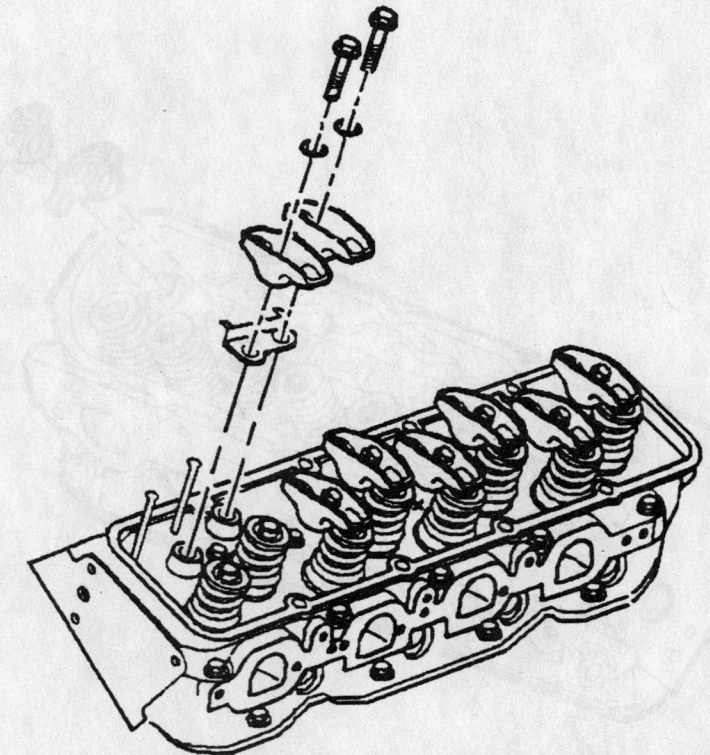

Exploded view of the rocker arms and related components—7.4L engines

7924KG11

- Rocker arms, balls and nuts. Tighten the nuts slowly to 18 ft. lbs. (25 Nm) while guiding the tips of the rocker arms over the tips of the valves.
- Valve (rocker arm) cover

Intake Manifold

REMOVAL & INSTALLATION

4.8L, 5.3L and 6.0L Engines

➡The intake manifold, throttle body, fuel injection rail, and fuel injectors may be removed as an assembly. If not servicing the individual components, remove the manifold as a complete assembly.

1. Before servicing the vehicle, refer to the precautions in the beginning of this section.
2. Remove or disconnect the following:
   - Positive Crankcase Ventilation (PCV) hose and valve
   - Manifold Absolute Pressure (MAP) sensor, if required
   - Engine coolant air bleed clamp and hose from the throttle body
   - Accelerator control cable bracket and bolts, if required
   - EVAP solenoid, bolt, and isolator
   - Intake manifold bolts
   - Intake manifold with gaskets
   - Intake manifold-to-cylinder head gaskets from the manifold. Discard the intake manifold gaskets.
   - Fuel rail with injectors
   - Throttle body and gasket
3. Clean the intake manifold in solvent.
4. Dry the intake manifold with compressed air.
5. Inspect all components for damage, and replace the necessary parts.
6. Locate a straight edge across the intake manifold cylinder head deck surface. Position the straight edge across a minimum of two runner port openings.
7. Insert a feeler gauge between the intake manifold and the straight edge. A intake manifold with warpage in excess of 0.118 in. (3mm) over a 7.87 in. (200mm) area is warped and should be replaced.

To install:

8. Install or connect the following:
   - MAP sensor
   - EVAP solenoid, bolt, and isolator. Tighten the bolt to 89 inch lbs. (10 Nm).
   - NEW intake manifold-to-cylinder head gaskets

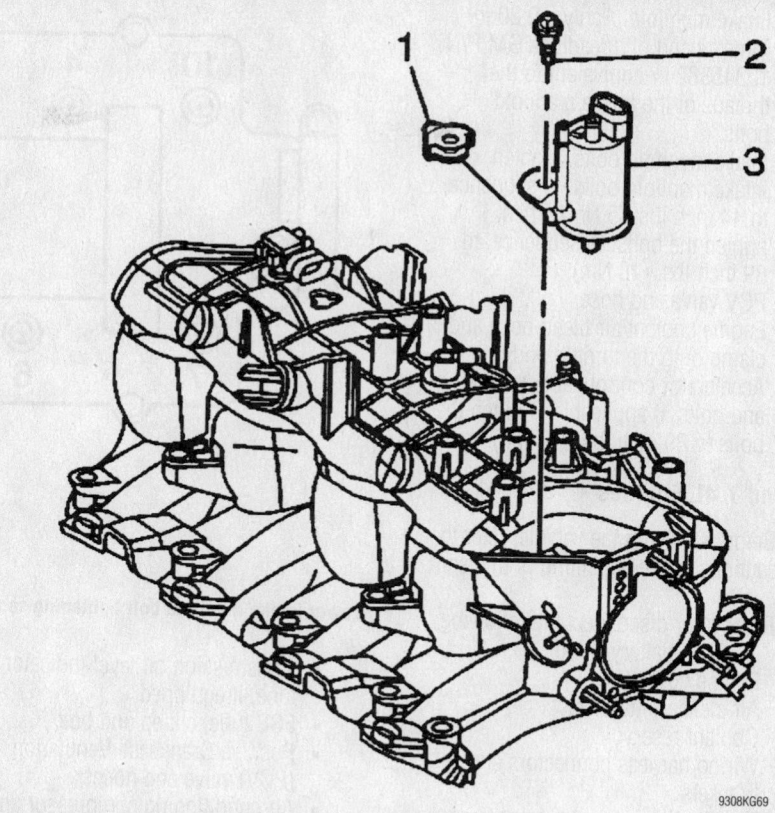

EVAP solenoid removal—4.8L, 5.3L and 6.0L engines

9308KG69

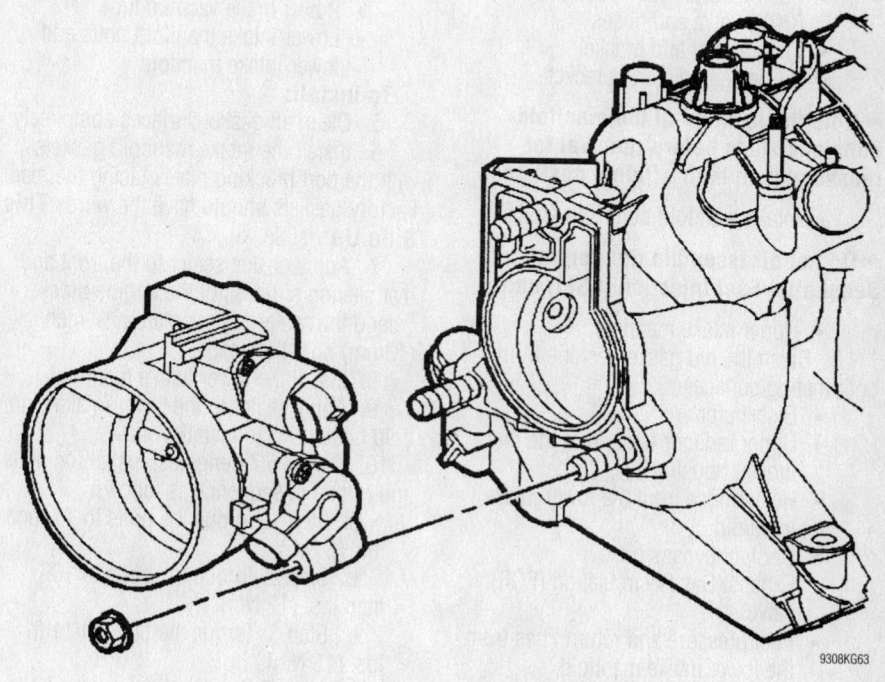

Throttle body removal—4.8L, 5.3L and 6.0L engines

9308KG63

- Intake manifold. Apply a 0.20 in. (5mm) band of threadlock GM P/N 12345382 or equivalent to the threads of the intake manifold bolts.
- Intake manifold bolts. Tighten intake manifold bolts, in sequence, to 44 inch lbs. (5 Nm). Then, tighten the bolts, in sequence, to 89 inch lbs. (10 Nm).
- PCV valve and hose
- Engine coolant air bleed hose and clamp onto the throttle body
- Accelerator control cable bracket and bolts, if applicable. Tighten the bolts to 89 inch lbs. (10 Nm).

5.7L and 7.4L Engines

1. Before servicing the vehicle, refer to the precautions in the beginning of this section.
2. Remove or disconnect the following:
- Negative battery cable
- Engine cover
- Air cleaner intake duct
- Coolant reservoir
- Wiring harness connectors and brackets
- Throttle linkage and bracket from the upper intake manifold
- Cruise control cable, if equipped
- Fuel lines and the bracket from the rear of the intake manifold
- Positive Crankcase Ventilation (PCV) valve and hoses
- Ignition coil and bracket
- Purge solenoid and bracket

➡**Note the location of the manifold bolts and studs before removal for reassembly in their original positions.**

- Intake manifold bolts and studs

➡**Do not disassemble the Central Sequential Fuel Injection (CSFI) unit.**

- Upper intake manifold
3. Clean the old gasket residue from both mating surfaces.
- Distributor
- Upper radiator hose from the thermostat housing
- Heater hose from the lower intake manifold
- Coolant bypass hose
- Exhaust Gas Recirculation (EGR) valve
- Fuel pressure and return lines from the lower intake manifold
- Wiring harnesses and brackets from the lower manifold
- Left side valve cover

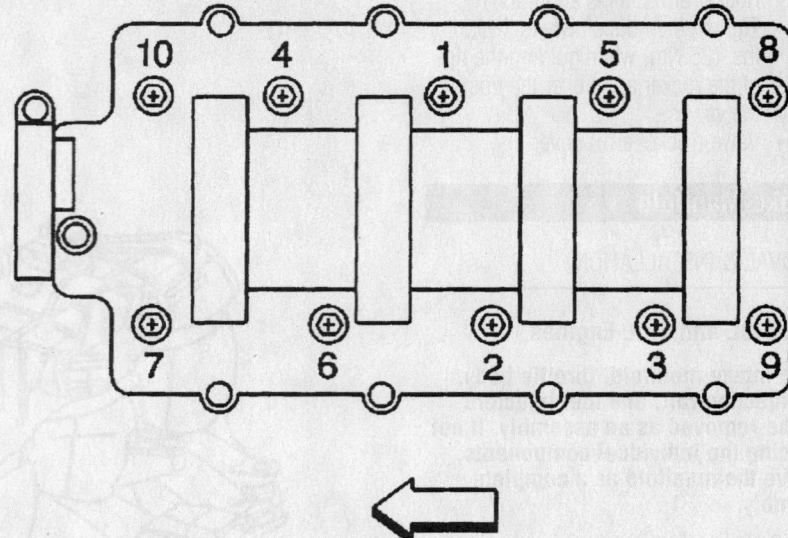

Lower intake manifold bolt tightening sequence—4.8L, 5.3L and 6.0L engines

9302KG03

- Transmission oil level indicator and tube, if equipped
- EGR tube, clamp and bolt
- Positive Crankcase Ventilation (PCV) valve and hoses
- Air conditioning compressor and bracket, but do not disconnect the lines
4. Loosen the compressor mounting bracket and slide it forward, but do not remove it.
- Power brake vacuum tube
- Lower intake manifold bolts and lower intake manifold

To install:
5. Clean all gasket surfaces completely.
6. Install the intake manifold gaskets with the port blocking plates facing the rear. Factory gaskets should have the words **This Side Up** visible.
7. Apply gasket sealer to the front and rear sealing surfaces of the engine block. Extend the sealer approximately ½ inch (13mm) onto the heads.
8. Install the lower intake manifold.
9. Apply sealer to the lower intake manifold bolts prior to installation.
10. On the 5.7L engines, install the bolts and torque in sequence as follows:
   a. Step 1: Torque the bolts to 71 inch lbs. (8 Nm).
   b. Step 2: Torque the bolts to 106 inch lbs. (12 Nm).
   c. Step 3: Torque the bolts to 11 ft. lbs. (15 Nm).
11. On the 7.4L engine, torque the bolts to 30 ft. lbs. (40 Nm) in the sequence shown.

12. Install or connect the following:
- Power brake vacuum tube
- PCV valve and hose
- EGR tube, clamp and bolt
- Transmission oil level indicator and tube, if equipped
- Left side valve cover
- Wiring harnesses and brackets to the lower manifold
- Fuel pressure and return lines to the lower intake manifold
- EGR valve
- Coolant bypass hose
- Heater hose to the lower intake manifold
- Upper radiator hose to the thermostat housing
- Air conditioning compressor and bracket
- Distributor
- Upper intake manifold gasket
- Upper intake manifold

⁑ WARNING

When installing the upper intake manifold be careful not to pinch the injector wires between the upper and lower intake manifolds.

- Upper intake manifold mounting bolts/studs, torque in a crisscross pattern as follows:
   a. Step 1: Torque to 44 inch lbs. (5 Nm).
   b. Step 2: Torque to 83 inch lbs. (10 Nm).
- Purge solenoid and bracket
- PCV hose

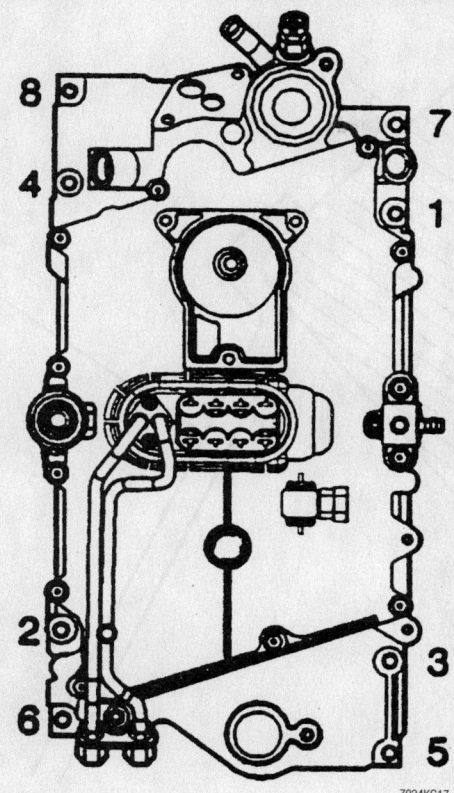

Lower intake manifold bolt tightening sequence—5.7L engines

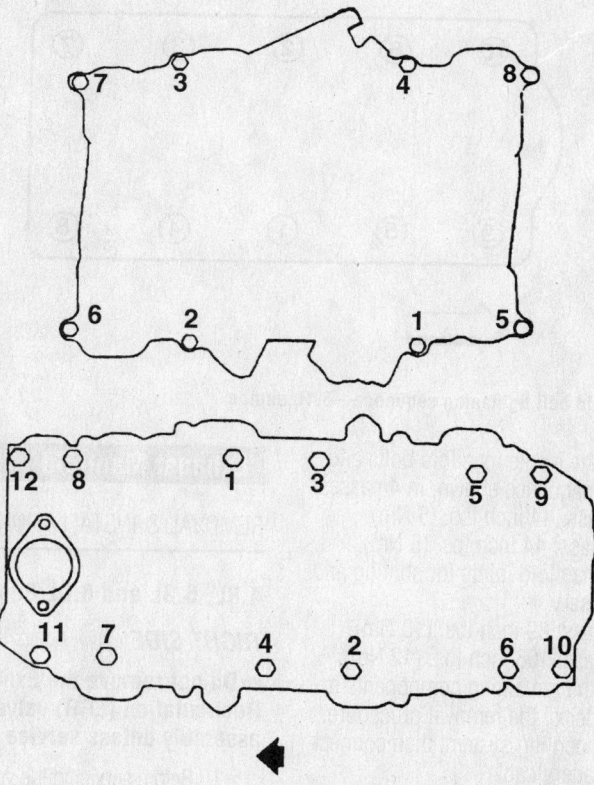

Upper and lower intake manifold bolt tightening sequence—7.4L engines

- Fuel lines and the bracket at the rear of the intake manifold
- Ignition coil and bracket
- Throttle linkage and bracket to the upper intake manifold
- Throttle linkage cable
- Cruise control cable, if equipped
- Wiring harness connectors and brackets
- Air cleaner intake duct
- Coolant recovery reservoir and the engine cover
- Negative battery cable

13. Start the vehicle and verify that there are no leaks.

8.1L Engine

→The intake manifold, throttle body, fuel rail and injectors can be removed as an assembly. If you do not need to service these components individually, remove the manifold as a complete assembly.

1. Before servicing the vehicle, refer to the precautions in the beginning of this section.

2. Relieve the fuel system pressure and drain the cooling system.

3. Remove or disconnect the following:

- Air cleaner outlet duct
- Intake manifold sight shield
- Fuel feed and return pipes
- Engine harness clips from the studs on the front of the dash
- Engine harness clip from the wheelhouse splash shield
- Pressure cycling switch, surge tank switch and Mass Air Flow (MAF) electrical connectors

4. Reposition the engine harness to the top of the engine

- Connector Position Assurance (CPA) retainer from the ignition coil harness
- Manifold Absolute Pressure (MAP) sensor electrical connector
- Ignition coil electrical connector
- Engine Coolant Temperature (ECT) sensor electrical connector
- Engine harness bolt and studs
- CPA retainer from the ignition coil harness
- Alternator, injector harness and ignition coil harness connectors
- Throttle Position (TP) sensor connector
- Electronic Throttle Control (ETC) and purge valve solenoid connectors

5. Reposition the engine harness to the drivers side of the engine compartment.

- Bypass valve vacuum hose from the intake manifold
- Exhaust Gas Recirculation (EGR) valve electrical connector
- EGR pipe bolts from the EGR adapter. Reposition the EGR pipe
- EGR valve pipe gasket and discard
- Secondary Air Injection (AIR) pipe nut from the fuel rail stud, if equipped
- AIR pipe bolts from the exhaust manifold
- AIR pipe from the AIR pump pipe
- AIR pipe gasket and discard
- AIR pipe nut from the fuel rail stud
- AIR pipe bolts from the exhaust manifold
- AIR pipe from the AIR pump pipe
- AIR pipe gasket

6. Reposition the AIR pump hose clamp, then remove the air pump hose from the pump pipe.

- AIR pump pipe bolt from the cylinder head
- AIR pump pipe
- Fuel pressure regulator vacuum hose
- Fuel rail studs and fuel rail, ONLY if replacing the manifold
- Intake manifold bolts

✵✵ WARNING

Do NOT try to remove the intake manifold by prying under the sealing surfaces.

- Intake manifold
- Intake manifold side gaskets and end seals and discard

➡**The splash shield is reusable and secured using a snap-in fit. Do not distort the shield during removal.**

- Splash shield

To install:

7. Clean all gasket surfaces completely.
8. Install or connect the following:
- Splash shield. Make sure the shield fits properly between the cylinder head.

➡**Make sure the manifold gasket tabs align with the hole in the head gasket.**

- New intake manifold end seals
- New intake manifold side gaskets onto the heads. Make sure the stamped THIS SIDE UP is showing.
- Intake manifold to the block
- Apply a suitable thread locking material to at least 8 threads of the intake manifold bolts

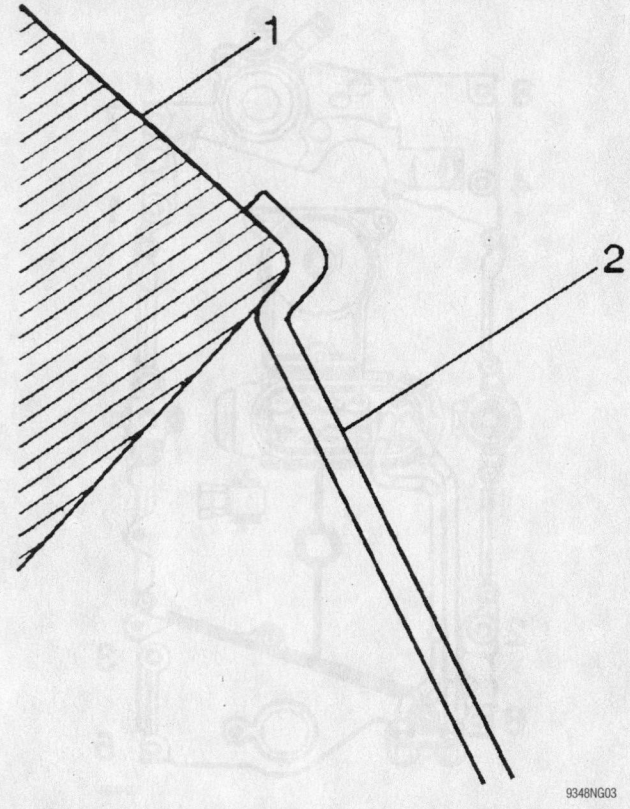

Make sure that the splash shield snap fits between the cylinder heads

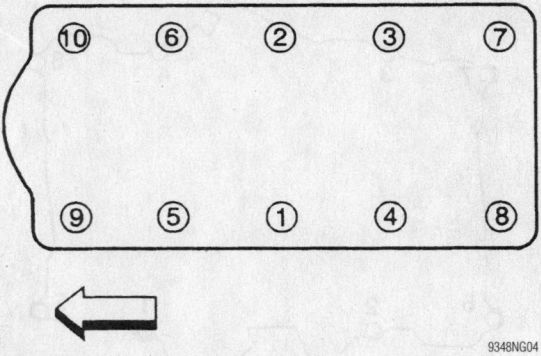

Intake manifold bolt tightening sequence—8.1L engine

9. Install the intake manifold bolts and tighten, in the sequence shown, in 4 passes:
   a. 1st pass: 44 inch lbs. (5 Nm).
   b. 2nd pass: 44 inch lbs. (5 Nm). Check the manifold joints for shifting and fix as necessary.
   c. 3rd pass: 89 inch lbs. (10 Nm).
   d. 4th pass: 106 inch lbs. (12 Nm).
10. Install the remaining components in the reverse order of the removal procedure.
11. Fill the cooling system, then connect the negative battery cable
12. Start the vehicle and verify that there are no leaks.

Exhaust Manifold

REMOVAL & INSTALLATION

4.8L, 5.3L and 6.0L Engines

RIGHT SIDE

➡**Do not remove the Exhaust Gas Recirculation (EGR) valve from the pipe assembly unless service is required.**

1. Before servicing the vehicle, refer to the precautions in the beginning of this section.

2. Remove or disconnect the following:
- EGR valve, gasket, and bolts
- EGR pipe bolt from the intake man-ifold
- EGR pipe bolts and gasket from the exhaust manifold
- EGR pipe bolts from the cylinder head
- EGR pipe assembly. With mild force, pull the EGR pipe from the intake manifold.
- O-ring seal from the EGR pipe assembly. Discard the exhaust manifold gasket and O-ring seal.

➡**In order to properly remove the exhaust manifold, remove the AIR components when applicable. Do not remove the check valve from the Air Injection Reaction (AIR) pipe unless service is required.**

- AIR pipe (with check valve), nuts and gasket from the right exhaust manifold.
- AIR pipe studs from the manifold, if required
- Spark plug wires from the spark plugs. Do not remove the spark plug wires from the ignition coils unless required.
- Exhaust manifold, bolts, and gas-ket. Discard the gasket.
- Heat shield and bolts from the manifold, if required

To install:

➡**Do not reuse the exhaust manifold-to-cylinder head gaskets. Upon instal-lation of the exhaust manifold, install a NEW gasket. A improperly installed gasket or leaking exhaust system may effect On-Board Diagnostics (OBD) II system performance.**

3. Clean the exhaust manifold and heat shield in solvent. Dry the exhaust manifold with compressed air. Inspect the compo-nents for restrictions or damage and replace as necessary.

4. Use a straight edge and a feeler gauge and measure the exhaust manifold cylinder head deck for warpage. An exhaust manifold deck with warpage in excess of 0.01 in. (0.25mm) within the two front or two rear runners or 0.02 in. (0.5mm) overall, may cause an exhaust leak and may effect OBD II system performance. Exhaust manifolds not within specifications must be replaced.

➡**Do not reuse Exhaust Gas Recircula-tion (EGR) valve and pipe gaskets or seals during assembly. Install NEW gaskets and O-ring seal.**

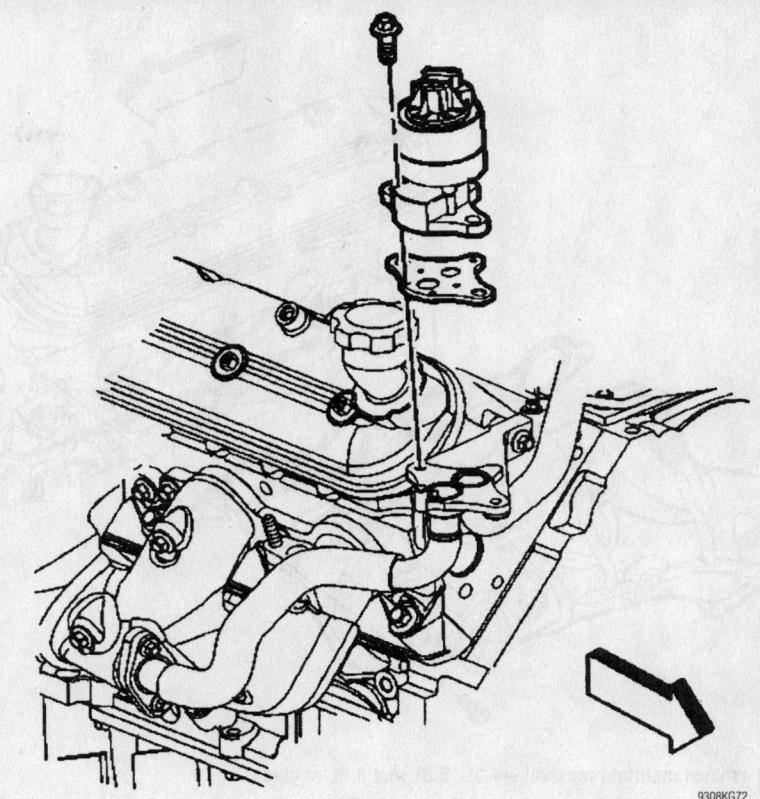

9308KG72

EGR valve removal—4.8L, 5.3L and 6.0L engines

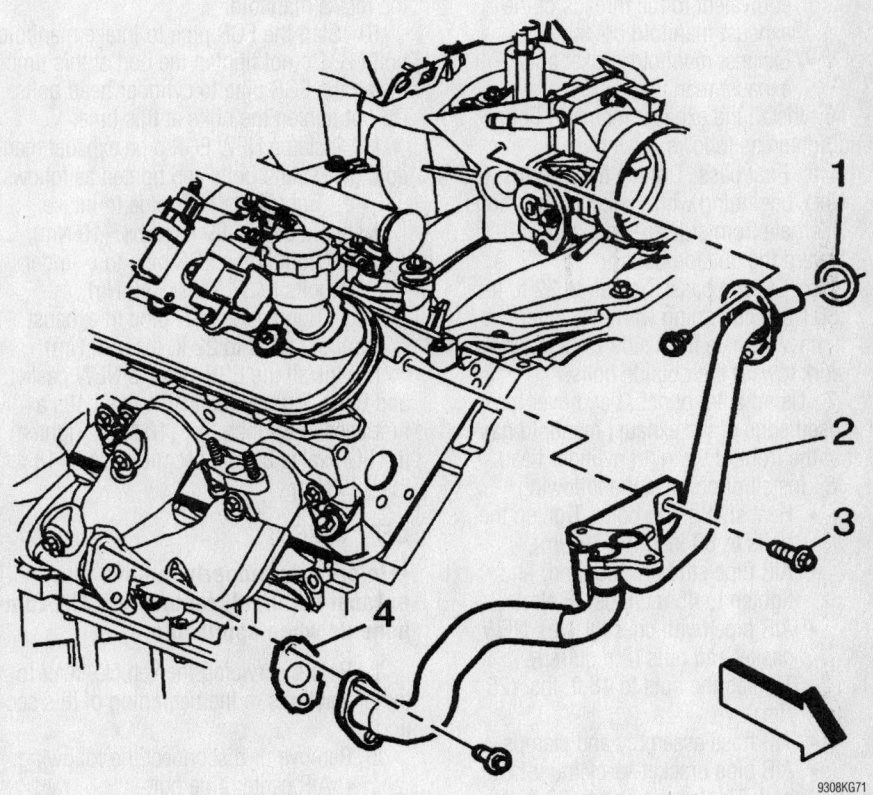

9308KG71

EGR pipe removal—4.8L, 5.3L and 6.0L engines

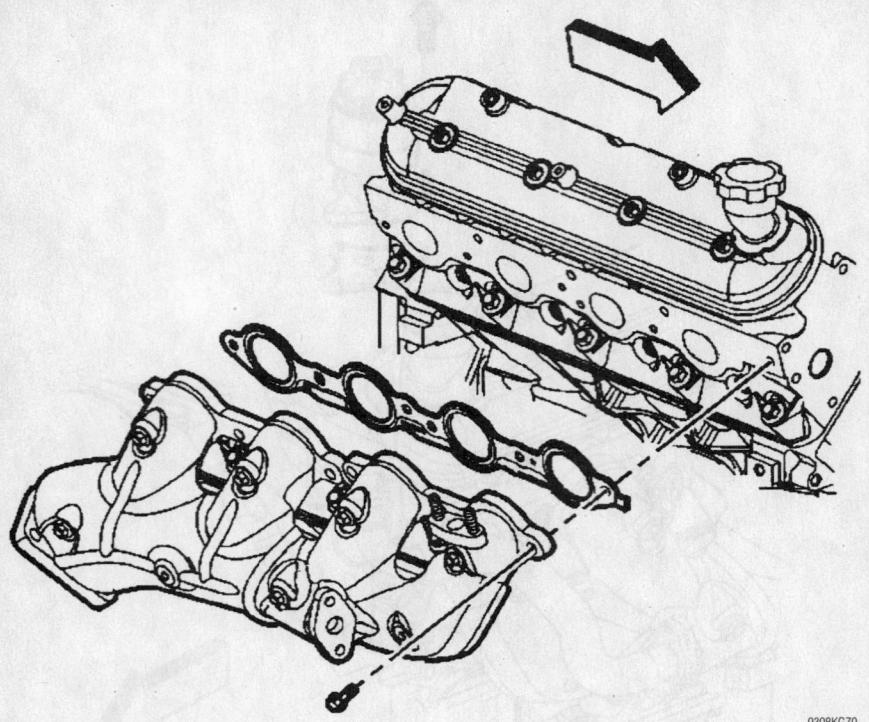

9308KG70

Right exhaust manifold removal—4.8L, 5.3L and 6.0L engines

5. Install or connect the following:
- A 0.2 in. (5mm) wide band of threadlock GM P/N 12345493 or equivalent to the threads of the exhaust manifold bolts.
- Exhaust manifold gasket and exhaust manifold

6. Install the exhaust manifold bolts and tighten as follows:

a. First pass: Tighten to 11 ft. lbs. (15 Nm), beginning with the center two bolts. Alternate from side-to-side, and work toward the outside bolts.

b. Second pass: Tighten to 22 ft. lbs. (30 Nm), beginning with the center two bolts. Alternate from side-to-side, and work toward the outside bolts.

7. Using a flat punch, bend over the exposed edge of the exhaust manifold gasket at the front of the right cylinder head.

8. Install or connect the following:
- Heat shield and bolts. Tighten the bolts to 80 inch lbs. (9 Nm).
- AIR pipe studs (if required) and tighten to 45 inch lbs. (5 Nm).
- AIR pipe (with check valve), NEW gasket and nuts (if required). Tighten the nuts to 18 ft. lbs. (25 Nm).
- AIR hose assembly and clamps.
- AIR pipe bracket-to-cylinder head bolt. Tighten the bolt to 37 ft. lbs. (50 Nm).

9. Apply a light coating of clean engine oil to a NEW O-ring seal and install the seal onto the EGR pipe. Insert the EGR pipe into the intake manifold.

10. Start the EGR pipe to intake manifold bolt (1). Do not tighten the bolt at this time. Install the EGR pipe to cylinder head bolts. Do not tighten the bolts at this time.

11. Install a NEW EGR pipe exhaust manifold gasket and bolts and tighten as follows:

a. Tighten the EGR pipe to intake manifold bolt to 89 inch lbs. (10 Nm).

b. Tighten the EGR pipe to cylinder head bolts to 37 ft. lbs. (50 Nm).

c. Tighten the EGR pipe to exhaust manifold bolts to 22 ft. lbs. (30 Nm).

12. Install the EGR valve, a NEW gasket, and bolts. Tighten the EGR valve bolts a first pass to 89 inch lbs. (10 Nm). Tighten the EGR valve bolts a second pass to 18 ft. lbs. (25 Nm).

LEFT SIDE

➡️**In order to properly remove the exhaust manifold, remove the AIR components when applicable.**

1. Before servicing the vehicle, refer to the precautions in the beginning of this section.

2. Remove or disconnect the following:
- AIR center pipe bolt
- AIR hose clamps and remove the hose assembly

➡️**Do not remove the check valve from the AIR pipe unless service is required.**

- AIR pipe (with check valve), nuts and gasket from the left exhaust manifold
- AIR pipe studs from the manifold (if required)
- Spark plug wires from the spark plugs. Do not remove the spark plug wires from the ignition coils unless required.
- Exhaust manifold, bolts, and gasket. Discard the gasket.
- Heat shield and bolts from the manifold, if required

➡️**Do not reuse the exhaust manifold-to-cylinder head gaskets. Upon installation of the exhaust manifold, install a NEW gasket. An improperly installed gasket or leaking exhaust system may effect On-Board Diagnostics (OBD) II system performance.**

3. Clean the exhaust manifold and heat shield in solvent. Dry the exhaust manifold with compressed air. Inspect the components for restrictions or damage and replace as necessary.

4. Use a straight edge and a feeler gauge and measure the exhaust manifold cylinder head deck for warpage. An exhaust manifold deck with warpage in excess of 0.01 in. (0.25mm) within the two front or two rear runners or 0.02 in. (0.5mm) overall, may cause an exhaust leak and may effect OBD II system performance. Exhaust manifolds not within specifications must be replaced.

To install:

➡️**Do not apply sealant to the first three threads of the bolt.**

5. Apply a 0.2 in. (5mm) wide band of threadlock GM P/N 12345493 or equivalent to the threads of the exhaust manifold bolts. Install the exhaust manifold and NEW exhaust manifold gasket.

6. Install the exhaust manifold bolts and tighten as follows:

a. First pass: Tighten to 11 ft. lbs. (15 Nm), beginning with the center two bolts. Alternate from side-to-side, and work toward the outside bolts.

b. Second pass: Tighten to 22 ft. lbs. (30 Nm), beginning with the center two bolts. Alternate from side-to-side, and work toward the outside bolts.

7. Using a flat punch, bend over the exposed edge of the exhaust manifold gasket at the front of the right cylinder head.

8. Install or connect the following:

- Heat shield and bolts. Tighten the bolts to 80 inch lbs. (9 Nm).
- AIR pipe studs (if required) and tighten to 45 inch lbs. (5 Nm)
- AIR pipe (with check valve), NEW gasket and nuts (if required). Tighten the nuts to 18 ft. lbs. (25 Nm).

5.7L Engines

1. Before servicing the vehicle, refer to the precautions in the beginning of this section.
2. Remove or disconnect the following:
   - Negative battery cable
   - Engine cover
   - Air cleaner, if needed
   - Exhaust pipe at the manifold
   - Oxygen Sensor (O2S) wiring, if equipped
   - AIR hose at the check valve
   - Exhaust Gas Recirculation (EGR) valve, inlet pipe
   - Heat stove pipe and the dipstick tube bracket, if working on the right side of the engine
   - Power steering pump rear bracket at the manifold, if removing the left side manifold
   - Loosen the alternator and remove the lower bracket, if necessary
   - Air conditioner compressor rear bracket and the diverter valve and bracket. if needed

➡**On models with air conditioning, it may be necessary to remove the compressor, do not disconnect the compressor lines.**

   - Manifold bolts and the manifold(s) Some models have lock tabs on the front and rear manifold bolts which must be removed before removing the bolts.

To install:

3. Clean gasket surfaces, and inspect manifold for cracks replace as necessary.
4. Install the manifold, then torque the bolts in 2 steps, as follows:
   a. Step 1: Torque to 15 ft. lbs. (20 Nm).
   b. Step 2: Torque to 22 ft. lbs. (30 Nm).
5. Install or connect the following:
   - Alternator, if removed
   - Air conditioning compressor, if removed
   - Diverter, if removed
   - Power steering brackets, if removed
   - Dipstick tube on right side
   - EGR inlet pipe

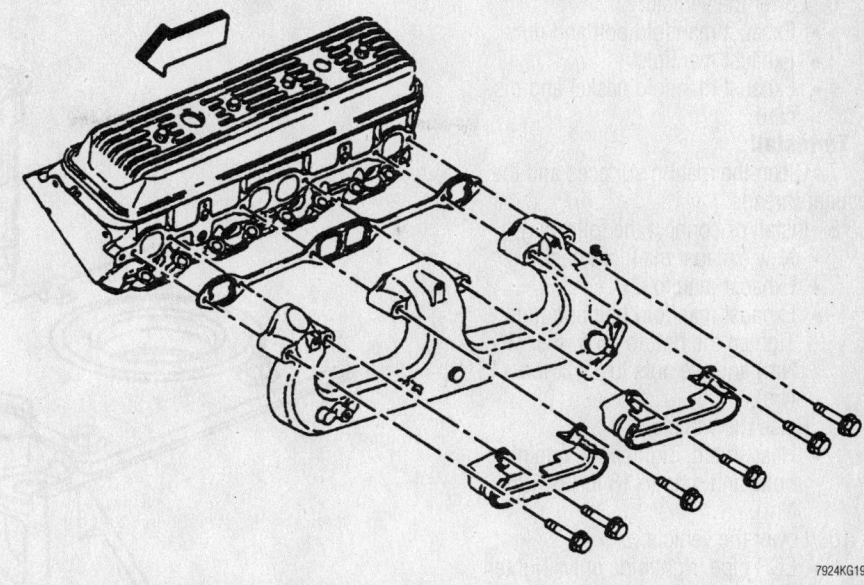

Exploded view of the left exhaust manifold, the right side is similar—5.7L engines

7924KG19

   - Oxygen sensor connector, if equipped
   - Exhaust pipes
   - Negative battery cable

7.4L Engines

1. Before servicing the vehicle, refer to the precautions in the beginning of this section.
2. Remove or disconnect the following:
   - Negative battery cable
   - Heat stove pipe and the dipstick tube, if removing the right side manifold
   - Exhaust Gas Recirculation (EGR) valve inlet pipe
   - Oxygen Sensor (O2S) wiring, if equipped
   - AIR hose at the check valve, if applicable
   - Spark plugs and wires
   - Exhaust manifold bolts and the spark plug heat shields

➡**Leave the front nut (left manifold) or rear nut (right manifold) in place for support.**

   - Heat shield bolts from the engine mount and bell housing
   - Heat shield, if equipped
   - Exhaust pipe at the manifold
   - Exhaust manifold

To install:

3. Clean the mating surfaces and the retainer threads.
4. Install or connect the following:
   - Manifold, spark plug heat shields and nuts. Torque the nuts to 22 ft.

lbs. (30 Nm), starting from the center bolts and working outside.
   - Exhaust pipe
   - Spark plugs and wires
   - AIR hose at check valve, if removed
   - O2S connector, if equipped
   - EGR pipe and dipstick tube
   - Negative battery cable
5. Run engine and check for leaks.

8.1L Engines

1. Before servicing the vehicle, refer to the precautions in the beginning of this section.
2. Remove or disconnect the following:
   - Wheelhouse panel
   - Oil dipstick tube, right side only
   - Secondary Air Injection (AIR) pipe nut from the fuel rail stud, if equipped
   - AIR pipe bolts from the exhaust manifold
   - AIR pipe from the AIR pump pipe
   - AIR pipe gasket and discard
   - Spark plug wires
   - Spark plugs
3. If removing the right exhaust manifold, perform the following:
   a. Remove the Exhaust Gas Recirculation (EGR) pipe nuts from the manifold.
   b. Remove the EGR pipe bracket bolt.
   c. Remove and discard the EGR pipe gasket.
   d. Reposition the EGR pipe.
4. Raise and support the vehicle.
5. Remove or disconnect the following:
   - Exhaust manifold heat shield bolts and shield

6. Lower the vehicle.
   - Exhaust manifold bolt and nuts
   - Exhaust manifold
   - Exhaust manifold gasket and discard

To install:

7. Clean the mating surfaces and the retainer threads.

8. Install or connect the following:
   - New exhaust manifold gasket
   - Exhaust manifold
   - Exhaust manifold bolt and nuts. Tighten the bolt to 26 ft. lbs. (35 Nm) and the nuts to 12 ft. lbs. (16 Nm).

9. Raise the vehicle.
   - Heat shield. Tighten the retaining bolts and nuts to 18 ft. lbs. (25 Nm).

10. Lower the vehicle.
   - EGR pipe, right side only. Tighten the pipe bracket bolt to 37 ft. lbs. (50 Nm) and the nuts to 22 ft. lbs. (30 Nm).
   - Spark plugs and plug wires
   - Air pipe, using a new gasket and reversing the removal procedure
   - Wheel house panel

Camshaft and Valve Lifters

REMOVAL & INSTALLATION

4.8L, 5.3L and 6.0L Engines

1. Before servicing the vehicle, refer to the precautions in the beginning of this section.

2. Raise the hood to the servicing position and secure it. Move the hood hinge bolt to hold the hood in the servicing position.

3. Remove or disconnect the following:
   - Battery negative cable
   - Coolant
   - Upper and lower radiator hoses from the engine
   - Air cleaner duct from the engine
   - A/C condenser mounting bolts, if equipped
   - Radiator support and radiator form vehicle
   - Engine cooling fan
   - Drive belt
   - A/C drive belt, if equipped
   - Engine sight shield
   - Electrical wiring harness from the thermostat housing
   - Water pump

4. Raise the vehicle.
   - Starter motor

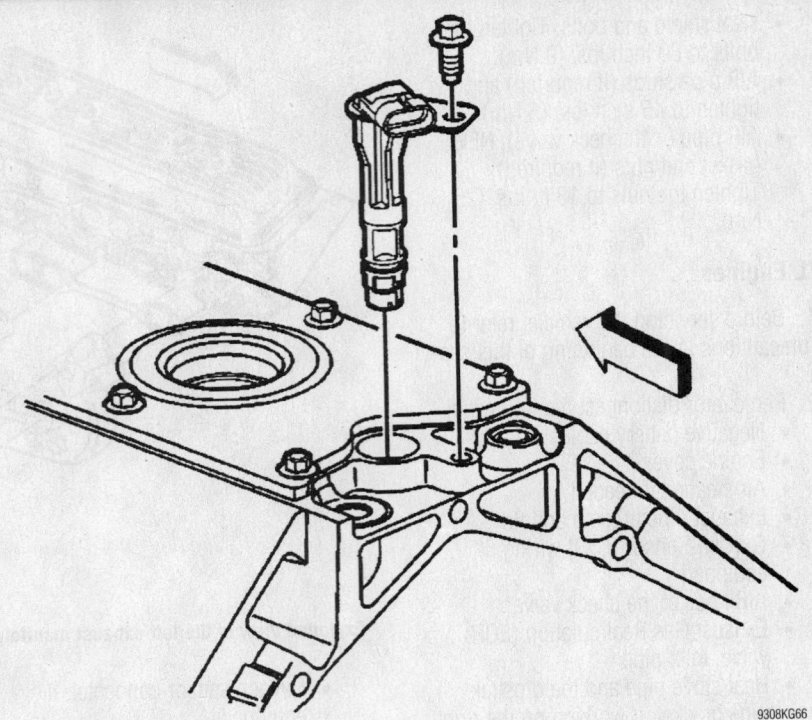

Camshaft sensor removal—4.8L, 5.3L and 6.0L engines

   - Right side closeout cover and bolt
   - Crankshaft balancer
   - Engine oil pan
   - Engine front cover
   - Cylinder heads from the engine
   - Valve lifters from the engine

5. Align the timing marks on the camshaft and crankshaft sprockets. Make sure that the number 1 piston is in the firing position.
   - Camshaft sprocket
   - Camshaft sensor bolt and the sensor

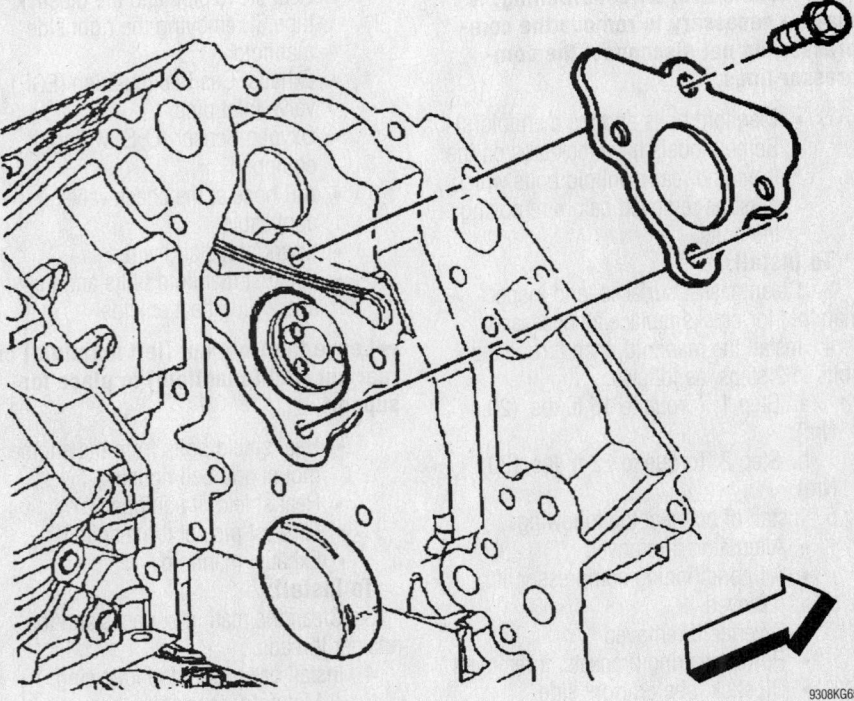

Camshaft retainer removal—4.8L, 5.3L and 6.0L engines

- Camshaft retainer bolts and the retainer

➡**All camshaft journals are the same diameter, so care must be used in removing or installing the camshaft to avoid damage to the camshaft bearings.**

6. Install the three M8-1.25 x 100 mm bolts in the camshaft front bolt holes. Using the bolts as a handle, carefully rotate and pull the camshaft out of the engine block. Remove the bolts from the front of the camshaft.

7. Clean and inspect all sealing surfaces.

To install:

➡**If camshaft replacement is required, the valve lifters must also be replaced.**

8. Lubricate the camshaft journals and the bearings with clean engine oil. Install three M8-1.25 x 100 mm (M8-1.25 x 4.0 in) bolts into the camshaft front bolt holes.

➡**All camshaft journals are the same diameter, so care must be used in removing or installing the camshaft to avoid damage to the camshaft bearings.**

9. Using the bolts as a handle, carefully install the camshaft into the engine block. Remove the three bolts from the front of the camshaft.

➡**Install the retainer plate with the sealing gasket facing the engine block. The gasket surface on the engine block should be clean and free of dirt or debris.**

10. Install or connect the following:
- Camshaft retainer and the bolts. Tighten the bolts to 18 ft. lbs. (25 Nm).

11. Inspect the camshaft sensor O-ring seal. If the O-ring seal is not cut or damaged, it may be reused. Lubricate the O-ring seal with clean engine oil.
- Camshaft sensor and bolt and tighten to 18 ft. lbs. (25 Nm)
- Camshaft sprocket and timing chain
- Valve lifters
- Cylinder heads
- Engine front cover to the engine
- Oil pan
- Right side closeout cover
- Starter motor
- Crankshaft balancer to the crankshaft
- Water pump

- Electrical wiring harness to the thermostat housing
- A/C drive belt, if equipped
- Drive belt
- Engine sight shield
- Radiator support and radiator
- A/C condenser mounting bolts
- Engine cooling fan
- Air cleaner duct
- Battery negative cable to the battery

5.7L Engines

1. Before servicing the vehicle, refer to the precautions in the beginning of this section.
2. Drain the cooling system.
3. Properly relieve the fuel system pressure.
4. Remove or disconnect the following:
- Air cleaner
- Air conditioning condenser and swing the condenser forward from its mounting, if equipped.
- Fan, the shroud and the radiator
- Valve covers
- Water pump assembly

5. Align the timing marks and remove the torsional damper.
- Timing chain cover
- Electrical and vacuum connections at the intake manifold
- Distributor assembly, mark the distributor rotor-to-housing location
- Intake manifold, pushrods and hydraulic lifters
- Camshaft sprocket bolts
- Camshaft sprocket and timing chain
- Crankshaft sprocket, as required
- Front engine mount through-bolts and raise the engine to gain sufficient clearance for camshaft removal, as required

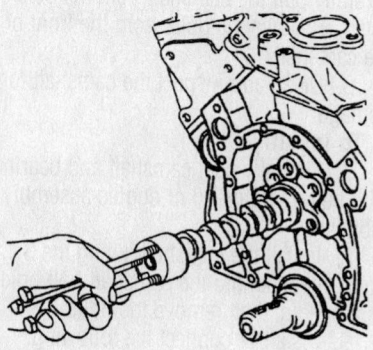
7924KG20

Install 2 or 3 long bolts into the camshaft to use as a handle for easy removal or installation—7.4L engine shown, other engines similar

6. Install 2 or 3 5/16–18 bolts 4–5 in. (102–127mm) long into the camshaft threaded holes.
- Camshaft

➡**Inspect the camshaft for signs of excessive wear or damage.**

To install:

➡**Liberally coat camshaft and bearing with heavy engine oil or engine assembly lubricant.**

7. Install or connect the following:
- Camshaft, align the timing marks on the camshaft and crankshaft gears
- Engine mount through-bolts
- Camshaft sprocket and chain. Torque the bolts to 18 ft. lbs. (25 Nm).
- Hydraulic lifters and pushrods
- Distributor assembly
- Timing chain cover
- Torsional damper
- Water pump
- Valve covers
- Fan, the shroud and radiator
- Air conditioning condenser, if equipped
- Air cleaner
- Negative battery cable

8. Refill the cooling system.

7.4L Engines

1. Before servicing the vehicle, refer to the precautions in the beginning of this section.
2. Drain the cooling system.
3. Properly relieve the fuel system pressure.
4. Properly discharge the air conditioning system.
5. Remove or disconnect the following:
- Negative battery cable
- Air cleaner assembly
- Grille and center support section, as required
- Air conditioning compressor, condenser and auxiliary fan, if equipped
- Fan, the shroud
- Radiator
- Accessory belt, as required
- Alternator, as required
- Valve covers
- Hoses from the water pump
- Water pump

6. Align the timing marks at TDC.
- Harmonic balancer and pulley
- Engine front cover

7. Mark the distributor rotor-to-housing location.

- Distributor assembly
- Intake manifold assembly
- Lifters, pushrods, and rocker arms

8. Rotate the camshaft so the timing marks align.
- Camshaft sprocket bolts
- Camshaft sprocket and timing
- Engine mount through-bolts

9. Install 2 or 3 ⁵⁄₁₆–18 bolts in the holes in the front of the camshaft and carefully pull the camshaft from the block.

To install:

10. Liberally coat camshaft and bearing with heavy engine oil or engine assembly lubricant

11. Align the timing marks on the camshaft sprocket and crankshaft gears.

12. Install or connect the following:
- Camshaft
- Camshaft sprocket and chain. Torque the bolts to 25 ft. lbs. (34 Nm).
- Engine mount bolts
- Lifters and pushrods and adjust the valves
- Intake manifold
- Distributor using the locating marks made during removal
- Engine front cover
- Harmonic balancer and pulley
- Water pump
- Hoses at the water pump
- Valve covers
- Alternator, if removed
- Accessory belt, if removed
- Fan shroud and radiator
- Air conditioning condenser and compressor
- Grille and center support, if removed
- Air cleaner assembly
- Negative battery cable

13. Fill the cooling system with the proper type and quantity of antifreeze.

8.1L Engines

1. Before servicing the vehicle, refer to the precautions in the beginning of this section.

2. Properly discharge the air conditioning system.

3. Remove or disconnect the following:
- Grille
- A/C condenser
- Intake manifold
- Rocker arms and pushrods
- Valve lifter guide retainer bolts and retainer
- Valve lifter guides, keeping them in proper order for reassembly
- Valve lifters

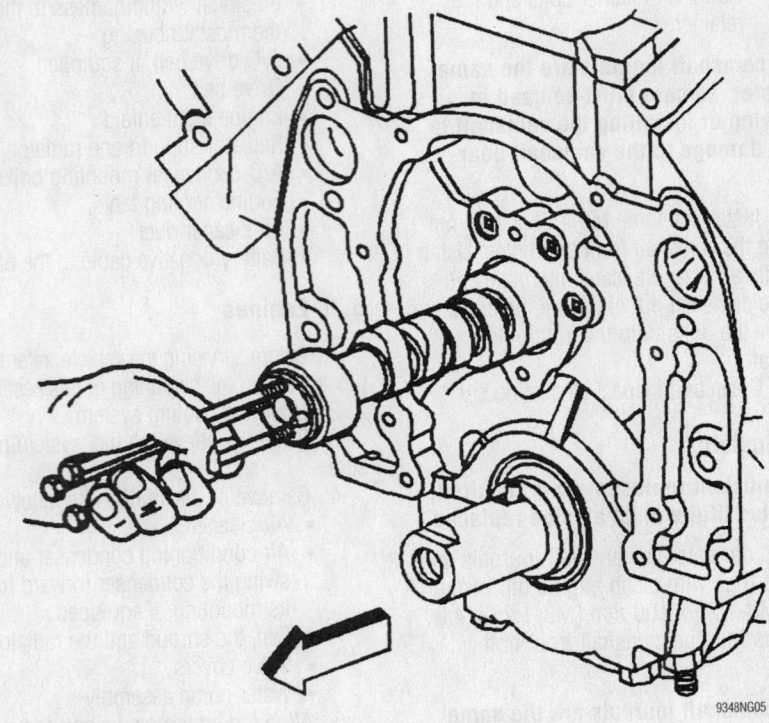

Use the 3 bolts as a handle to carefully remove and install the camshaft

➡ If any lifters are stuck in their bores, use a suitable valve lifter to remove them.

- Timing chain and sprocket
- Camshaft retaining bolts
- Camshaft retainer

❋❋ WARNING

All of the cam journals are the same size so be very careful when removing and installing the camshaft that you do not damage the bearings.

4. Install three 8-1.25 x 100mm bolts in the holes in the front of the camshaft and carefully pull the camshaft from the block.

5. Remove the bolts from the front of the camshaft.

6. Clean and inspect the camshaft for damage.

To install:

7. Liberally coat camshaft and bearings with heavy engine oil or engine assembly lubricant.

8. Install the camshaft, using the 3 bolts threaded into the camshaft bolt holes as a handle, then remove the bolts.

9. Install or connect the following:
- Camshaft retainer and bolts. Tighten to 106 inch lbs. (12 Nm).

➡ If a new camshaft is installed, you MUST install new valve lifters.

- Timing chain and sprocket
- Valve lifters
- Valve lifter guides over the flats on the lifters. Make sure the rollers of the lifters are properly aligned with the cam lobes.
- Valve lifter guide retainer and tighten the bolts to 18 ft. lbs. (25 Nm)
- Rocker arms and pushrods
- Intake manifold
- A/C condenser
- Grille

10. Recharge the A/C system.

Valve Lash

ADJUSTMENT

All engines use hydraulic lifters, which require no periodic adjustment.

Starter Motor

REMOVAL & INSTALLATION

5.7L and 7.4L Engines

1. Before servicing the vehicle, refer to the precautions in the beginning of this section.

2. Remove or disconnect the following:

- Negative battery cable
- Mounting bolts and shims
- Wires
- Heat shield
- Starter

To install:
3. Install or connect the following:
- Starter
- Wires. Torque battery wire nut to 89 inch lbs. (10 Nm), and ignition nut to 18 inch lbs. (2 Nm)
- Heat shield. Torque the bolts to 53 inch lbs. (6 Nm) and the nuts to 35 inch lbs. (3 Nm)
- Mounting bolts and shim. Torque the bolts to 33 ft. lbs. (45 Nm)
- Negative battery cable

4.8L, 5.3L and 6.0L Engines

✳✳ CAUTION

Before servicing any electrical component, the ignition key must be in the OFF or LOCK position and all electrical loads must be OFF, unless instructed otherwise in these procedures.

1. Before servicing the vehicle, refer to the precautions in the beginning of this section.
2. Disconnect the negative battery cable.
3. Raise and support the vehicle.
4. Remove or disconnect the following:

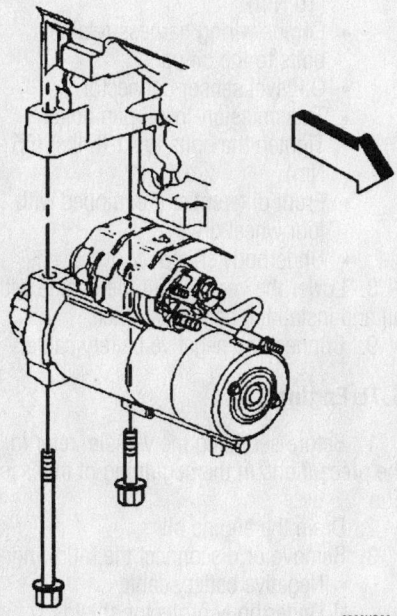

Starter removal—4.8L, 5.3L and 6.0L engines

- Protective shields, as necessary
- Starter solenoid shield
- Starter-to-transmission close out cover bolt
- Engine oil level sensor connection
- Front axle mounting bracket through bolt nut

5. Reposition the front axle mounting bracket through bolt until the bolt tip is flush with the support bushing. Do not remove the bolt.
- Mounting bolts from the engine block. Slide the starter forward until the starter clears the transmission.
- Starter transmission close out cover
- Positive battery cable and wiring harness from the starter
- Starter from the vehicle

To install:
6. Install or connect the following:
- Starter
- Positive battery cable to the starter. Tighten the nut to 12 ft. lbs. (16 Nm).
- Starter transmission close out cover
- Mounting bolts to the engine block and tighten to 37 ft. lbs. (50 Nm)

7. Reposition the front axle mounting bracket through bolt until the bolt is fully seated.
- Front axle mounting bracket through bolt nut and tighten to 67 ft. lbs. (90 Nm)
- Engine oil level sensor connection
- Starter-to-transmission close out cover bolt
- Starter solenoid shield
- Protective shields as necessary

8. Remove the safety stands.
9. Lower the vehicle.
10. Connect the negative battery cable.

8.1L Engine

1. Before servicing the vehicle, refer to the precautions in the beginning of this section.
2. Remove or disconnect the following:
- Negative battery cable
- Positive battery cable nut
- Positive cable from the solenoid
- Engine harness ground nut and ground from the solenoid
- Mounting bolts and starter
- Heat shield bolts, nut and shield, if necessary

To install:
3. Install or connect the following:
- Heat shield, bolts and nut if removed. Tighten the bolts to 35

inch lbs. (3 Nm) and the nut to 44 inch lbs. (5 Nm).
- Starter and bolts. Tighten to 37 ft. lbs. (50 Nm).
- Ground wire and nut. Tighten to 30 inch lbs. (3.4 Nm).
- Positive cable and nut. Tighten to 80 inch lbs. (9 Nm).
- Negative battery cable

Oil Pan

REMOVAL & INSTALLATION

4.8L, 5.3L and 6.0L Engines

➡**The original oil pan gasket is retained and aligned to the oil pan by rivets. When installing a new gasket, it is not necessary to install new rivets. DO NOT reuse the oil pan gasket. When installing the oil pan, install a NEW oil pan gasket.**

1. Before servicing the vehicle, refer to the precautions in the beginning of this section.
2. Remove or disconnect the following:
- Negative battery cable
- Front differential, if equipped with four wheel drive
- Underbody shield from the vehicle
- Oil pan shield
- Cross brace if equipped
- Engine oil and filter
- Transmission to oil pan bolts
- Oil level sensor electrical connector
- Two front wiring harness retainer bolts
- Engine wiring harness retainer bolts from the engine oil pan
- Engine oil cooler pipe to oil pan bolt
- Transmission oil cooler pipe retainer and the bolt from the oil pan
- Closeout covers and bolts (one each side of engine)
- Engine mount bolts each side
- Oil pan

To install:

➡**The alignment of the structural oil pan is critical. The rear bolt hole locations of the oil pan provide mounting points for the transmission bellhousing. To ensure the rigidity of the powertrain and correct transmission alignment, it is important that the rear of the block and the rear of the oil pan must NEVER protrude beyond the engine block and transmission bellhousing plane.**

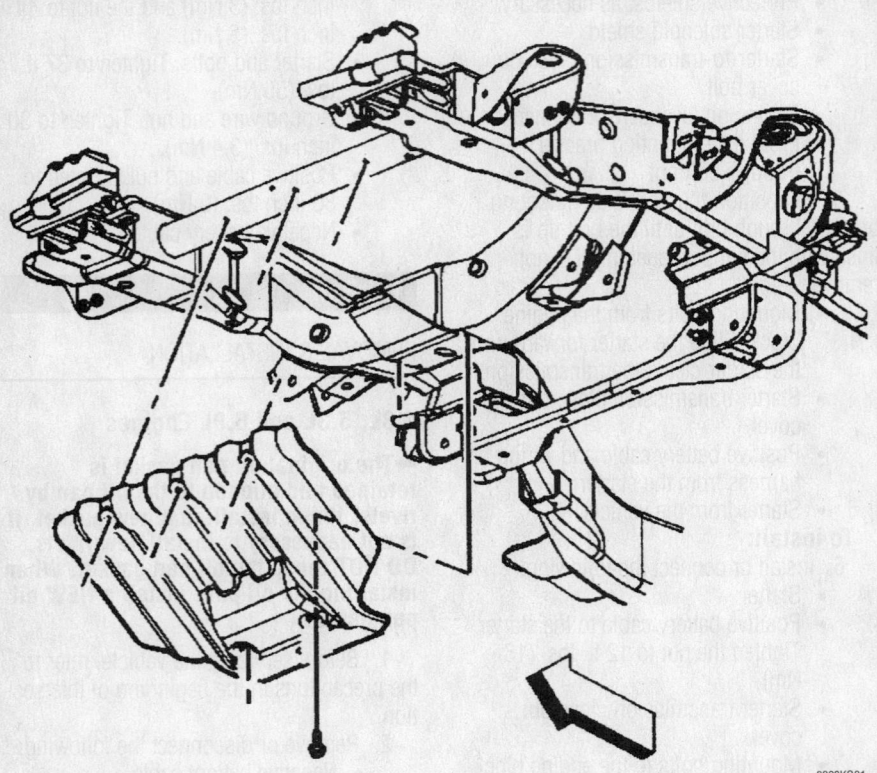

Oil pan shield—4.8L, 5.3L and 6.0L engines

9308KG81

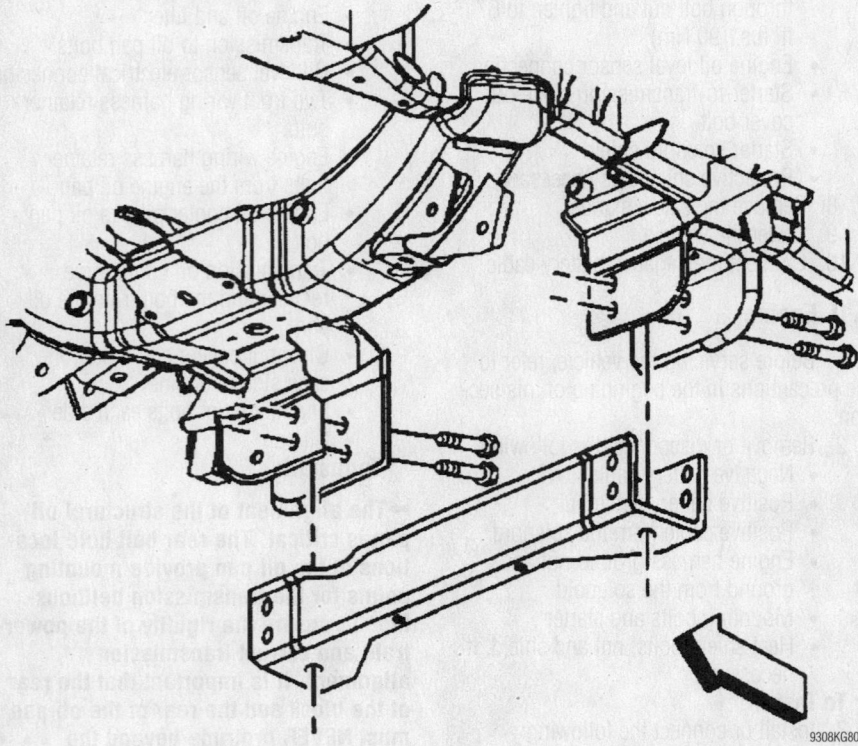

Cross brace—4.8L, 5.3L and 6.0L engines

9308KG80

3. Apply a 0.20 in. (5mm) bead of sealant GM P/N 12378190 or equivalent 0.8 in. (20mm) long to the engine block. Apply the sealant directly onto the tabs of the front cover gasket that protrudes into the oil pan surface.

4. Apply a 0.20 in. (5mm) bead of sealant GM P/N 12378190 or equivalent 0.8 in. (20mm) long to the engine block. Apply the sealant directly onto the tabs of the rear cover gasket that protrudes into the oil pan surface.

➡ **Be sure to align the oil gallery passages in the oil pan and engine block properly with the oil pan gasket.**

5. Pre-assemble the oil pan gasket to the pan. Install the oil pan bolts to the pan through the gasket.

6. Install or connect the following:
- Oil pan, gasket and bolts to the engine block. Snug the oil pan bolts finger-tight. Do not over-tighten.
- Two lower bellhousing bolts to position the oil pan correctly

7. Snug the lower bellhousing bolt finger-tight. Do not overtighten. Tighten the oil pan-to-block and oil pan-to-oil pan front cover bolts to 18 ft. lbs. (25 Nm). Tighten the oil pan-to-rear cover bolts to 106 inch lbs. (12 Nm). Tighten the bellhousing bolts to 37 ft. lbs. (50 Nm).
- Transmission oil cooler pipe retainer and the bolt to the oil pan
- Engine oil cooler pipe to oil pan bolt. Tighten the nut to 89 inch lbs. (10 Nm).
- Engine wiring harness retainer bolts to the oil pan
- Oil level sensor connector
- Transmission-to-oil pan bolts. Tighten the bolts to 41 ft. lbs. (55 Nm).
- Front differential if equipped with four wheel drive
- Underbody shield

8. Lower the vehicle. Fill the engine with oil and install the engine oil filter.

9. Connect the negative battery cable.

5.7L Engines

1. Before servicing the vehicle, refer to the precautions in the beginning of this section.

2. Drain the engine oil.

3. Remove or disconnect the following:
- Negative battery cable
- Under body protector shields
- Transmission and engine oil lines from guides

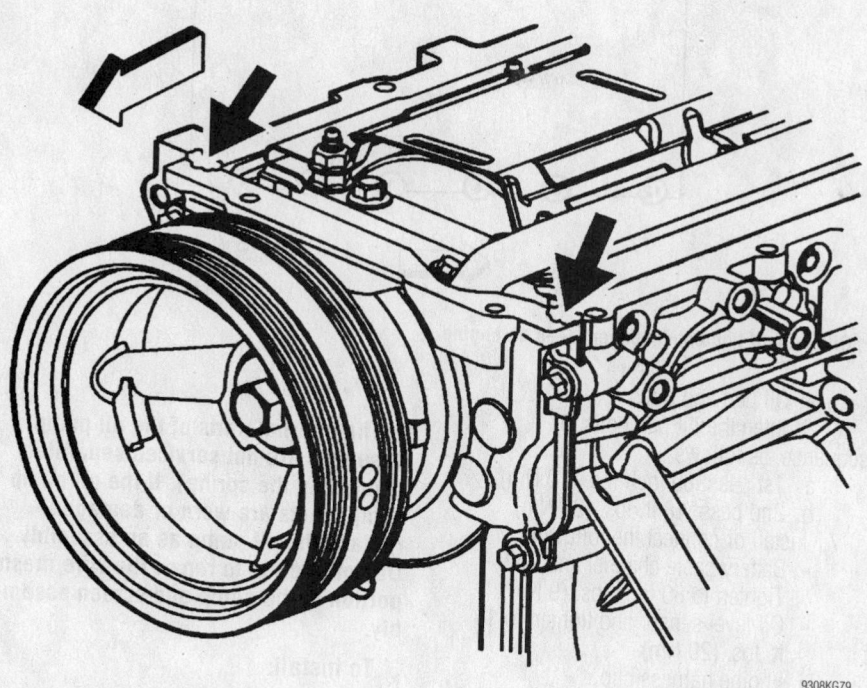

Apply sealant at these points at the front of the block—4.8L, 5.3L and 6.0L engines

9308KG79

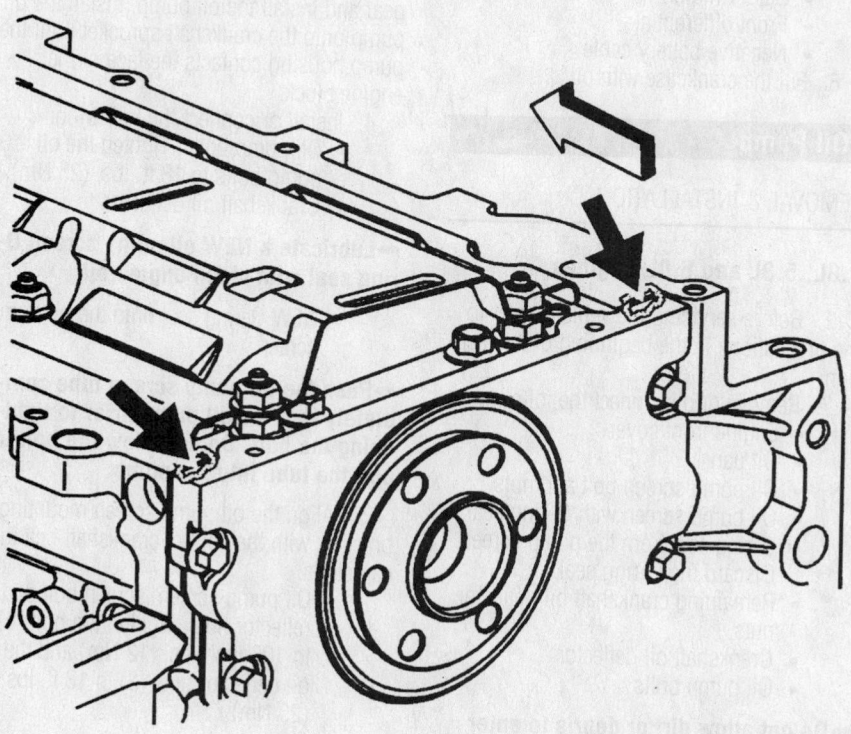

Apply sealant at these points at the rear of the block—4.8L, 5.3L and 6.0L engines

9308KG78

- Front driveshaft, if needed
- Front drive axles, if needed
- Exhaust crossover pipe
- Flywheel/flexplate or torque converter cover
- Oil filter and adapter
- Strut rods at the front engine mounting, if equipped
- Oil pan bolts, nuts and reinforcements
- Oil pan and gaskets

To install:

4. Thoroughly clean all gasket surfaces.
5. Install or connect the following:
   - New gasket
   - Oil pan and new gaskets
   - Oil pan bolts, nuts and reinforcements. Torque the bolts to 18 ft. lbs. (25 Nm).
   - Strut rods at the front engine mounting
   - Oil filter and adapter
   - Torque converter or flywheel/flexplate cover
   - Exhaust crossover pipe
   - Front drive axles and driveshaft, if removed
   - Transmission and engine oil lines to guides
   - Under body protectors
   - Negative battery cable
6. Fill the crankcase with oil.

7.4L Engines

1. Before servicing the vehicle, refer to the precautions in the beginning of this section.
2. Drain the engine oil.
3. Remove or disconnect the following:
   - Negative battery cable
   - Fan shroud
   - Air cleaner
   - Distributor cap
   - Underbody protectors, if needed
   - Front driveshaft and front drive axles, if needed
   - Starter, if equipped with manual transmission
   - Torque converter or clutch housing cover
   - Oil filter and adapter
   - Oil pressure line from the side of the block
4. Support the engine
   - Engine mount through-bolts
   - Oil pan bolts
   - Oil pan and discard the gaskets

To install:

5. Clean all sealing surfaces.
6. Apply RTV gasket material to the front and rear corners of the gaskets.

7. Install or connect the following:
- New gaskets, coat the gaskets with adhesive sealer and position them on the block
- Rear pan seal in the pan with the seal ends mating with the gaskets
- Front seal on the bottom of the front cover, pressing the locating tabs into the holes in the cover
- Oil pan
- Pan bolts, clips and reinforcements. Torque the bolts to 18 ft. lbs. (25 Nm).

8. Lower the engine onto the mounts.
- Engine mount through-bolts
- Oil pressure line
- Oil filter
- Starter, if removed
- Torque converter or clutch housing cover
- Front drive axles and driveshaft, if removed
- Underbody protectors, if removed
- Distributor cap
- Air cleaner
- Fan shroud
- Negative battery cable

9. Fill the crankcase with oil.

8.1L Engine

1. Before servicing the vehicle, refer to the precautions in the beginning of this section.

2. Disconnect the negative battery cable and drain the engine oil.

3. Remove or disconnect the following:
- Front differential, if equipped with 4WD
- Starter motor
- Oil pan skid plate bolts and plate
- Crossbar bolt(s) and crossbar
- Oil level dipstick
- Oil level sensor electrical connector
- Engine harness clip from the oil pan
- Battery cable channel bolt
- Battery cable channel and reposition
- Oil pan bolts, oil pan and gasket

➡You can reuse the oil pan gasket, if it is not damaged

To install:

➡You must install the oil pan within 5 minutes of applying the sealer.

4. Apply sealant to the sides of the front and rear crankshaft bearing caps on the left and right sides.

5. Install or connect the following:
- Oil pan gasket into the oil pan groove

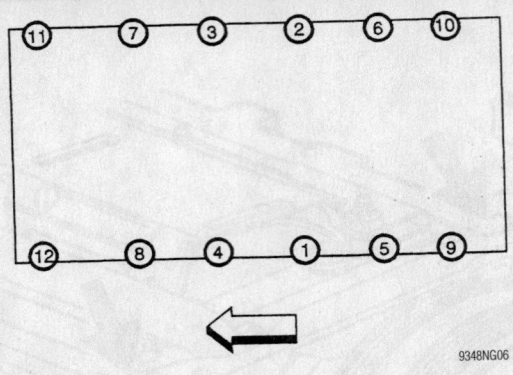

Oil pan bolt tightening sequence—8.1L engine

9348NG06

- Oil pan and bolts

6. Tighten the oil pan bolts, in sequence, as follows:
  a. 1st pass: 89 inch lbs. (10 Nm).
  b. 2nd pass: 18 ft. lbs. (25 Nm).

7. Install or connect the following:
- Battery cable channel and bolt. Tighten to 80 inch lbs. (9 Nm).
- Oil level sensor and tighten to 15 ft. lbs. (20 Nm)
- Engine harness clip
- Oil level sensor connector
- Oil level dipstick
- Crossbar and bolt(s). Tighten to 74 ft. lbs. (100 Nm).
- Skid plate. Tighten the bolts to 15 ft. lbs. (20 Nm).
- Starter motor
- Front differential
- Negative battery cable

8. Fill the crankcase with oil.

Oil Pump

REMOVAL & INSTALLATION

4.8L, 5.3L and 6.0L Engines

1. Before servicing the vehicle, refer to the precautions in the beginning of this section.

2. Remove or disconnect the following:
- Engine front cover
- Oil pan
- Oil pump screen bolt and nuts
- Oil pump screen with O-ring seal.
- O-ring seal from the pump screen. Discard the O-ring seal.
- Remaining crankshaft oil deflector nuts
- Crankshaft oil deflector
- Oil pump bolts

➡Do not allow dirt or debris to enter the oil pump assembly, cap ends as necessary.

- Oil pump

➡The internal parts of the oil pump assembly are not serviced separately (excluding the spring). If the oil pump components are worn or damaged, replace the oil pump as an assembly. Do not attempt to repair the wire mesh portion of the pump and screen assembly.

To install:

➡Inspect the oil pump and engine block oil gallery passages. These surfaces must be clear and free of debris or restrictions.

3. Align the splined surfaces of the crankshaft sprocket and the oil pump drive gear and install the oil pump. Install the oil pump onto the crankshaft sprocket until the pump housing contacts the face of the engine block.

4. Install or connect the following:
- Oil pump bolts. Tighten the oil pump bolts to 18 ft. lbs. (25 Nm).
- Crankshaft oil deflector.

➡Lubricate a NEW oil pump screen O-ring seal with clean engine oil.

- NEW O-ring seal onto the oil pump screen

➡Push the oil pump screen tube completely into the oil pump prior to tightening the bolt. Do not allow the bolt to pull the tube into the pump.

5. Align the oil pump screen mounting brackets with the correct crankshaft bearing cap studs.
- Oil pump screen, screen bolt and reflector nuts. Tighten the bolt (4) to 106 inch lbs. (12 Nm) and the oil deflector nuts (2) to 18 ft. lbs. (25 Nm).
- Oil pan
- Engine front cover

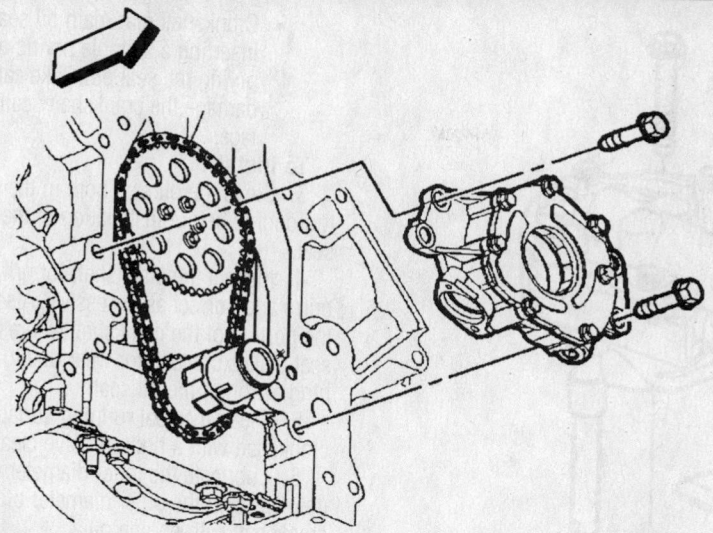

Oil pump removal—4.8L, 5.3L and 6.0L engines

9308KG77

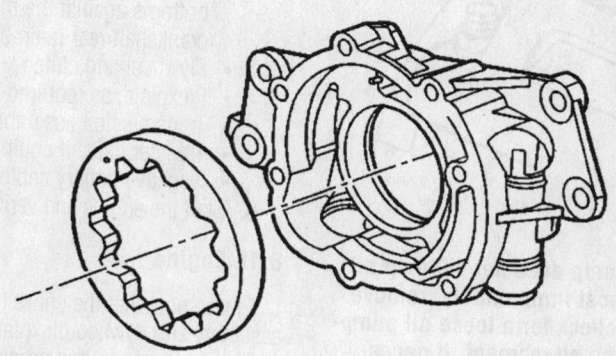

Oil pump disassembly—4.8L, 5.3L and 6.0L engines

9308KG64

5.7L and 7.4L Engines

1. Before servicing the vehicle, refer to the precautions in the beginning of this section.

2. Remove or disconnect the following:

- Oil pan
- Oil pump attaching bolt, if equipped
- Pick-up tube nut/bolt
- Pump along with the pick-up tube and shaft, as necessary

3. Clean all sealing surfaces

To install:

4. Ensure that the pump pick-up tube is tight in the pump body. If the tube should come loose, oil pressure will be lost and oil starvation will occur. If the pick-up tube is loose it should be replaced.

5. If the pump has been disassembled and is being replaced or for any reason oil has been removed, it must be primed. It can either be filled with oil before installing the

cover plate and oil kept within the pump during handling or the entire pump cavity can be filled with petroleum jelly.

➡**If the pump is not primed, the engine could be damaged before it receives adequate lubrication when the engine is started.**

6. Install or connect the following:

- Pump, aligning the pump shaft with the oil pump drive gear as necessary. Torque oil pump/pick-up tube retainer(s) to 65 ft. lbs. (90 Nm).
- Oil pan

7. Refill the engine crankcase

8. Disable the ignition system; crank engine for approximately 10 seconds to aid in priming the oil pump and reducing the risk of engine damage.

➡**If the oil pump does not build up oil pressure almost immediately, remove the pan and check for a loose oil pump-to-pick-up tube attachment. If necessary dismantle the pump and pack the pump cavity with petroleum jelly. Running the engine without measurable oil pressure will cause extensive damage.**

8.1L Engine

1. Before servicing the vehicle, refer to the precautions in the beginning of this section.

2. Remove or disconnect the following:

- Oil pan
- Oil pump screen bolt

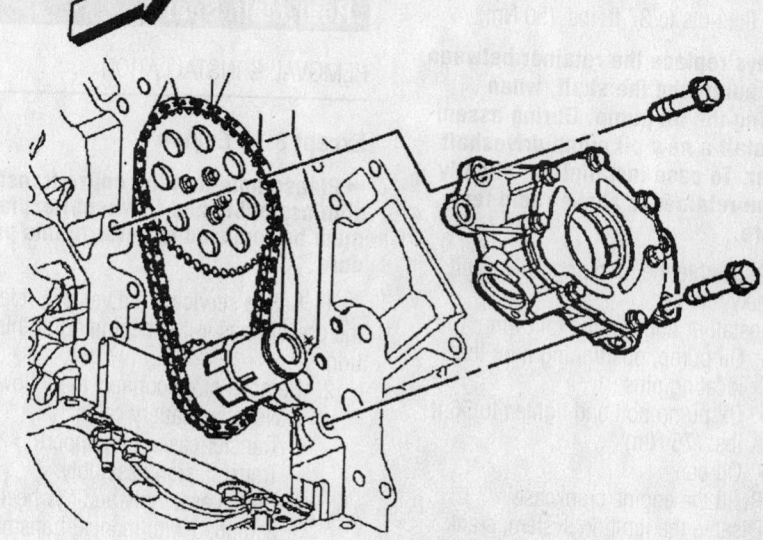

Exploded view of the oil pump mounting—4.8L, 5.3L and 6.0L engines

9302KG04

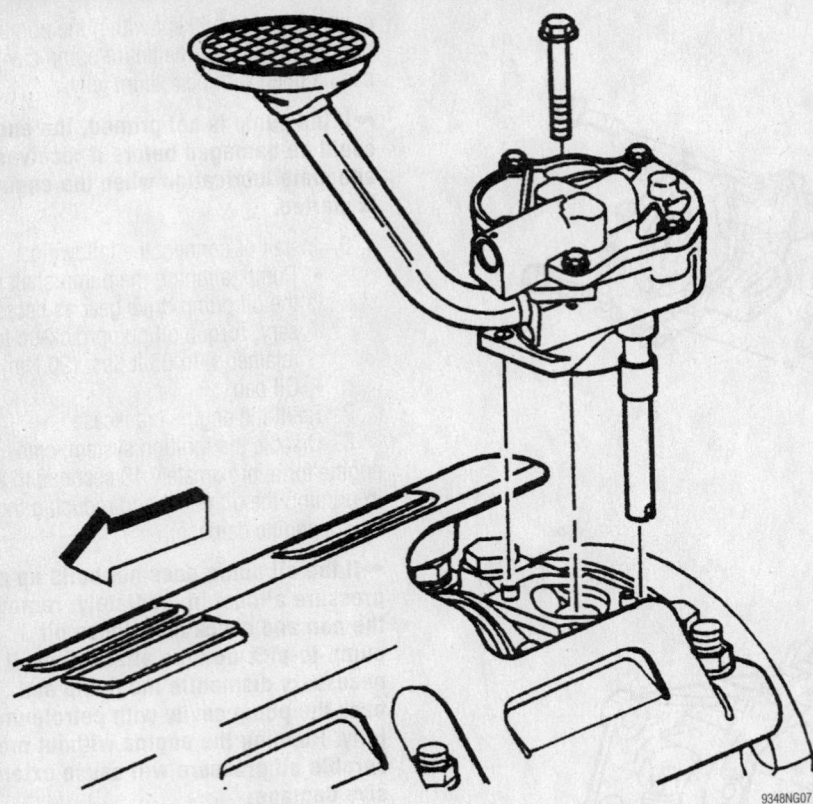

9348NG07

Oil pump removal—8.1L engine

- Oil pump, retainer and driveshaft. Discard the driveshaft retainer.
- Crankshaft oil deflector nuts
- Crankshaft oil deflector
- Oil pump bolts
- Oil pump

3. Clean and inspect the oil pump

To install:

4. Install the crankshaft oil deflector. Tighten the nuts to 37 ft. lbs. (50 Nm).

➡**Always replace the retainer between the oil pump and the shaft, when installing the oil pump. During assembly, install a new oil pump driveshaft retainer. To ease installation, slightly heat the retainer to above room temperature.**

5. Assemble the oil pump, driveshaft and a new retainer.

6. Install or connect the following:
   - Oil pump, positioning it on the locating pins
   - Oil pump bolt and tighten to 56 ft. lbs. (75 Nm)
   - Oil pan

7. Refill the engine crankcase

8. Disable the ignition system; crank engine for approximately 10 seconds to aid in priming the oil pump and reducing the risk of engine damage.

➡**If the oil pump does not build up oil pressure almost immediately, remove the pan and check for a loose oil pump-to-pick-up tube attachment. If necessary dismantle the pump and pack the pump cavity with petroleum jelly. Running the engine without measurable oil pressure will cause extensive damage.**

Rear Main Seal

REMOVAL & INSTALLATION

Except 8.1L Engine

➡**Please note that the entire transmission assembly and flywheel/flexplate must be removed to perform this procedure.**

1. Before servicing the vehicle, refer to the precautions in the beginning of this section.

2. Remove or disconnect the following:
   - Negative battery cable
   - Transfer case, if equipped
   - Transmission assembly
   - Clutch assembly and flywheel, if equipped with manual transmission
   - Flexplate, if equipped with automatic transmission

- Crankshaft rear main oil seal by inserting a suitable prying tool and prying the seal out. Take care not to damage the crankshaft sealing surface.

To install:

3. Clean the oil seal bore in the block thoroughly before installation of the new seal.

4. Inspect the crankshaft for grit, rust or burrs and correct as necessary. Also inspect the portion of the crankshaft where the oil seal makes contact, for wear due to the rubbing action of the oil seal.

5. Clean the seal running surface of the crankshaft with a non-abrasive cleaner.

6. Lubricate the inner diameter of the new seal and the outer diameter of the crankshaft with engine oil.

7. Install or connect the following:
   - Rear main oil seal, using installation tool J 38841, until the tool bottoms against the block and crankshaft rear main bearing cap.
   - Flywheel and clutch
   - Flexplate, as required
   - Transmission assembly
   - Transfer case, if equipped
   - Negative battery cable

8. Start the engine and verify no oil leaks.

8.1L Engine

Please note that the entire transmission assembly and flywheel/flexplate must be removed to perform this procedure. This procedure requires the use of the following tools: Crankshaft Rear Seal Puller tool No. J 43320 and Crankshaft Rear Seal Installer tool No. J 42849.

1. Before servicing the vehicle, refer to the precautions in the beginning of this section.

2. Remove or disconnect the following:
   - Negative battery cable
   - Transfer case, if equipped
   - Transmission assembly
   - Clutch assembly and flywheel, if equipped with manual transmission
   - Flexplate, if equipped with automatic transmission

3. Install the guide pins from the Crankshaft Rear Sear Puller into the crankshaft.

4. Install the Rear Seal Puller over the guide pins.

5. Using a drill, insert 8 of the self-drilling sheet metal screws into the rear crankshaft seal, using a crisscross pattern as shown. The self tapping screws are included with the Crankshaft Rear Seal Puller.

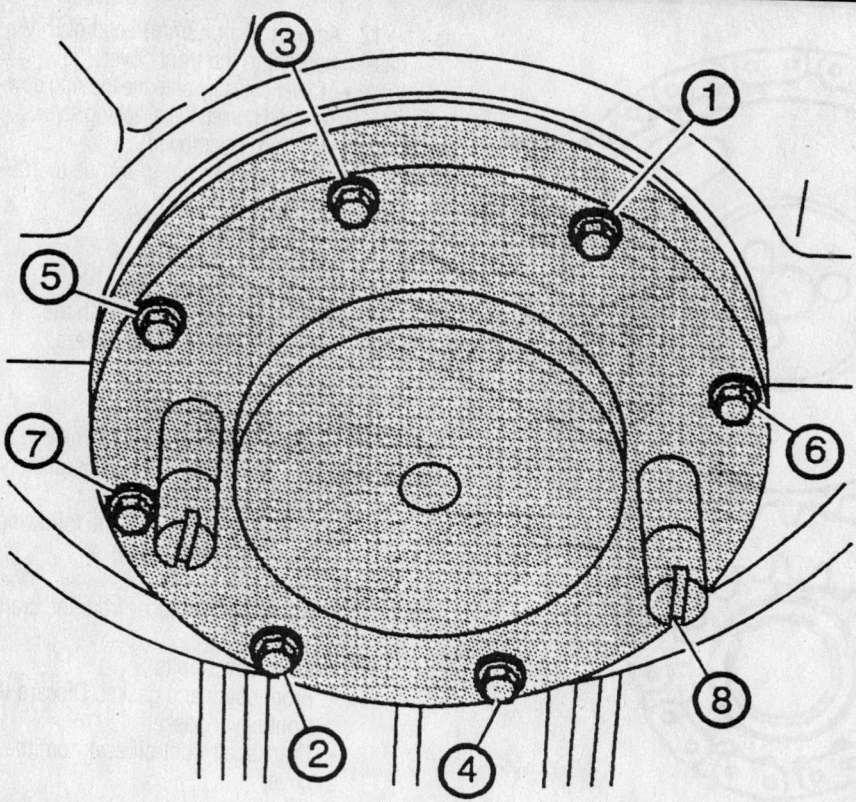

You must drill the screws into the rear main seal using a crisscross pattern—8.1L engine

9348NG08

6. Thread the center bolt of the Crankshaft Rear Seal Puller into the crankshaft to remove the seal.

7. Remove the guide pins from the crankshaft.

To install:

8. Make sure there is no dirt, rust or loose burrs on the crankshaft.

9. Apply a light coating of engine oil to the crankshaft sealing surface. Do NOT get oil on the sealing surface of the engine block.

10. Install the new rear main seal onto the Crankshaft Rear Seal Installation Tool.

11. Position the Rear Seal Installation Tool against the crankshaft. Thread the attaching screws into the tapped holes in the crankshaft.

12. Use a screwdriver to tighten the screws securely to make sure the seal is squarely installed against the crankshaft.

13. Rotate the center nut until the installation tool bottoms, then remove the seal installation tool.

14. Install or connect the following:
- Flexplate, if equipped with automatic transmission
- Clutch assembly and flywheel, if equipped with manual transmission
- Transmission assembly
- Transfer case, if equipped
- Negative battery cable

Timing Chain, Sprockets, Front Cover and Seal

➡**The manufacturer recommends that the front cover oil seal be replaced whenever the cover is removed.**

REMOVAL & INSTALLATION

5.7L and 7.4L Engines

1. Before servicing the vehicle, refer to the precautions in the beginning of this section.

2. Drain the cooling system.

3. Remove or disconnect the following:
- Negative battery cable
- Fan shroud assembly
- Belts, pulleys and water pump assembly
- Crankshaft pulley and damper
- Oil pan-to-front cover bolts

➡**If equipped with a composite front cover, it must be replaced with a new one. Reusing the front cover may result in oil leaks.**

- Screws holding the timing chain cover to the block
- Cover and gaskets

4. Use a suitable tool to pry the old seal out of the front face of the cover.

5. Rotate the crankshaft until the timing marks on the camshaft and crankshaft sprockets are in proper alignment.
- Camshaft sprocket-to-camshaft nut and/or bolts
- Camshaft sprocket (along with the timing chain). If the sprocket is difficult to remove, use a plastic mallet to bump the sprocket from the camshaft.

➡**The camshaft sprocket (located by a dowel) is lightly pressed onto the camshaft and should come off easily. The chain comes off with the camshaft sprocket.**

6. If necessary use J-5825-A, or equivalent, crankshaft sprocket removal tool to free the timing sprocket from the crankshaft.

To install:

7. Inspect the timing chain and the timing sprockets for wear or damage, replace the damaged parts as necessary.

8. Clean the gasket mounting surfaces of all remaining traces of old gasket.

➡**During installation, coat the thrust surfaces lightly with Molykote® or equivalent pre-lube.**

9. Install or connect the following:
- Crankshaft sprocket onto the crankshaft, use tool J-5590, or equivalent, crankshaft sprocket installation tool, and a hammer, without disturbing the position of the engine.
- Timing chain, arrange the camshaft sprocket in such a way that the timing marks will align between the shaft centers and the camshaft locating dowel will enter the dowel hole in the cam sprocket.
- Cam sprocket, with the chain mounted under it in position on the front of the camshaft. Torque the camshaft sprocket-to-camshaft retainer bolts to 106 inch lbs. (12 Nm).

10. With the timing chain installed, turn the crankshaft 2 complete revolutions, then check to make certain that the timing marks are in correct alignment between the shaft centers.

➡**Coat the lip of the new seal with oil prior to installation.**

- New seal so that the open end is toward the inside of the cover, Using seal driver J-22102, or equivalent.
- New front pan seal, cutting the tabs off

11. Coat a new cover gasket with adhesive sealer and position it on the block.

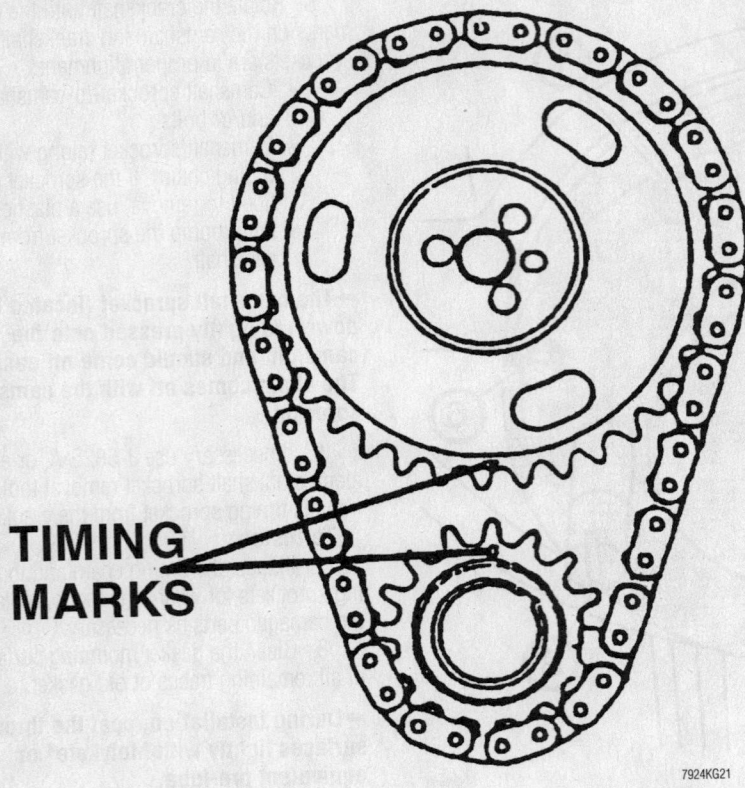

TIMING MARKS

7924KG21

Timing mark alignment for timing chain removal and installation—gasoline engines

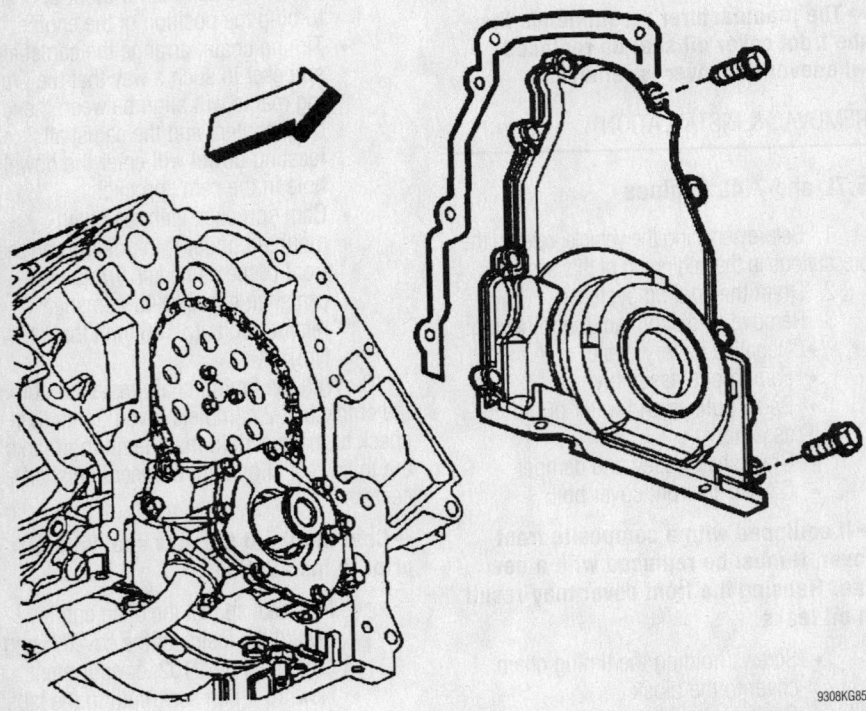

9308KG85

Front cover and gasket—4.8L, 5.3L and 6.0L engines

12. Apply a ⅛ in. (3mm) bead of RTV gasket material to the front cover.
- Cover carefully onto the locating dowels and tighten the attaching screws
- Oil pan, if removed
- Cover-to-pan bolts, Torque to 106 inch lbs. (12 Nm).
- Torsional damper
- Water pump assembly
- Negative battery cable

13. Fill the cooling system with the proper type and quantity of antifreeze.

4.8L, 5.3L and 6.0L Engine

1. Before servicing the vehicle, refer to the precautions in the beginning of this section.
2. Drain the cooling system.
3. Remove or disconnect the following:
- Negative battery cable
- Water pump
- Crankshaft balancer from the crankshaft
- Front cover bolts
- Front cover and gasket. Discard the front cover gasket.
- Crankshaft front oil seal from the cover
- Oil pump

4. Rotate the crankshaft until the timing marks on the crankshaft and the camshaft sprockets are aligned.

➡ **Do not turn the crankshaft assembly after the timing chain has been removed in order to prevent damage to the piston assemblies or the valves.**

- Camshaft sprocket bolts
- Camshaft sprocket and timing chain
- Crankshaft sprocket
- Crankshaft sprocket key

To install:

5. Install or connect the following:
- Key into the crankshaft keyway
- Crankshaft sprocket onto the front of the crankshaft. Align the crankshaft key with the crankshaft sprocket keyway. Rotate the crankshaft sprocket until the alignment mark is in the 12 o'clock position.
- Camshaft sprocket and timing chain. Locate the camshaft sprocket alignment mark in the 6 o'clock position.
- Camshaft sprocket bolts. Tighten the bolts to 26 ft. lbs. (35 Nm).

➡ **Do not lubricate the oil seal sealing surface.**

6. Lubricate the outer edge of the oil seal with clean engine oil. Lubricate the front cover oil seal bore with clean engine oil.

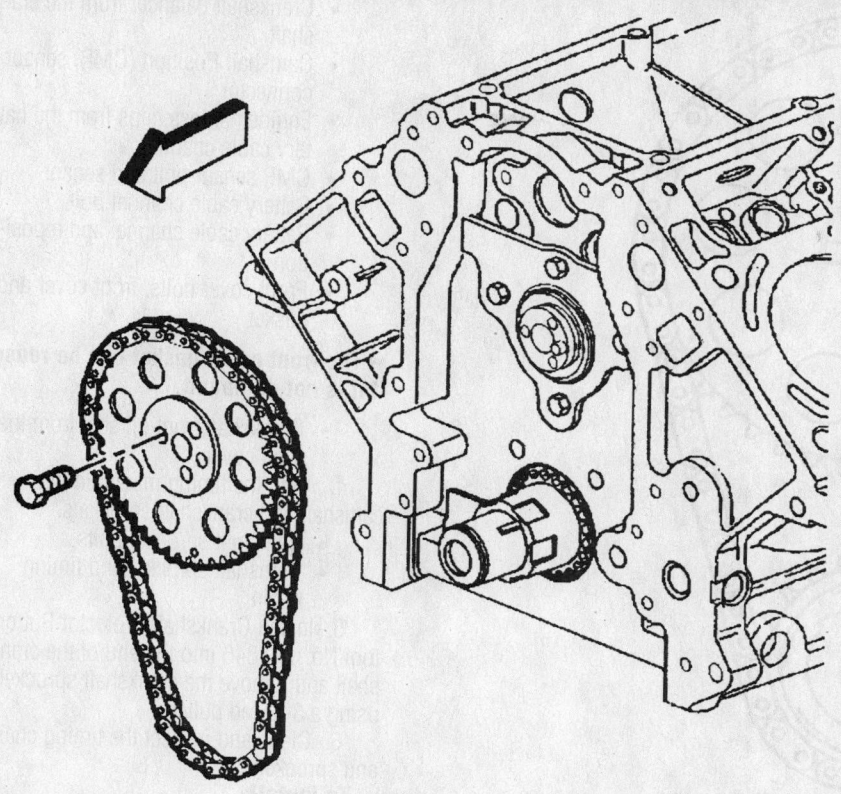

Sprocket and chain removal—4.8L, 5.3L and 6.0L engines

9308KG76

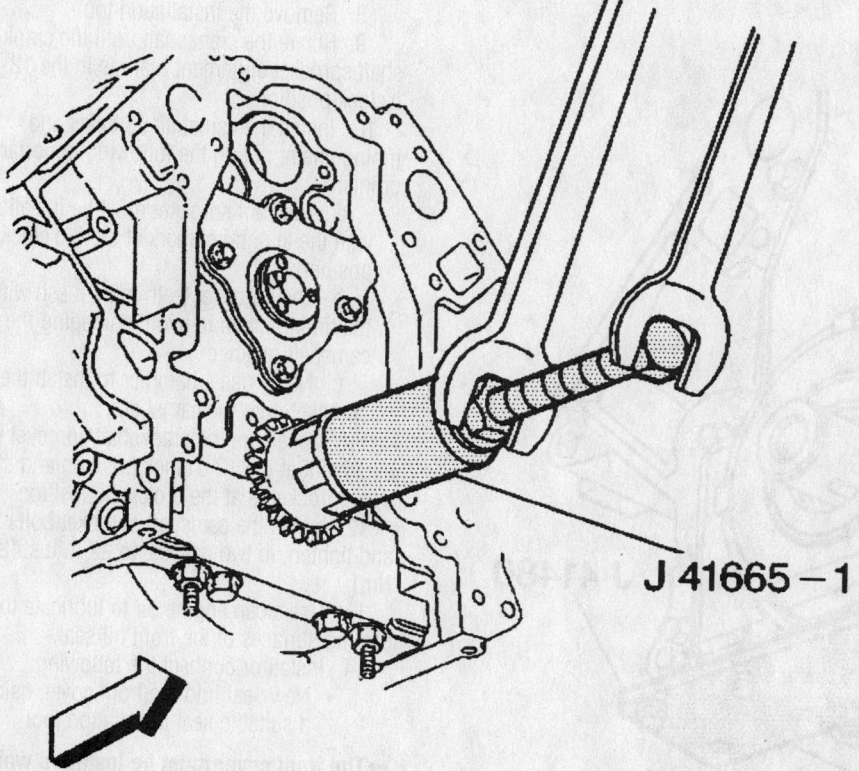

J 41665 — 1

Crankshaft sprocket installation—4.8L, 5.3L and 6.0L engines

9308KG62

7. Install the crankshaft front oil seal with an installer.

➡Do not apply any type sealant to the front cover gasket (unless specified). Special tools are used to properly align the engine front cover at the oil pan surface and to center the crankshaft front oil seal.

8. Install the front cover gasket, cover, and bolts onto the engine. Tighten the cover bolts finger-tight. Do not overtighten.

9. Start the J41480 tool-to-front cover bolts. Don't tighten the bolts yet.

➡Align the tapered legs of the tool with the machined alignment surfaces on the front cover.

10. Install tool J41476 . Install the crankshaft balancer bolt. Tighten the crankshaft balancer bolt by hand until snug. Do not overtighten. Tighten the J41480 bolts and front cover bolts to 18 ft. lbs. (25 Nm).

11. Remove the tools.

12. Place a straight edge across the engine block and front cover oil pan sealing surfaces. Avoid contact with the portion of the gasket that protrudes into the oil pan surface. Insert a feeler gauge between the front cover and the straight edge tool. The cover must be flush with the oil pan surface or no more than 0.02 in. (0.5mm) below flush. If the front cover-to-engine block oil pan surface alignment is not within specifications, repeat the cover alignment procedure. If the correct front cover-to-engine block alignment cannot be obtained, replace the front cover.

13. Install the crankshaft balancer bolt. Tighten the crankshaft balancer bolt by hand until snug. Do not overtighten the bolt.

14. Snug the oil pan-to-cover bolts in order to position the cover at the pan rail.

15. Tighten the oil pan-to-front cover bolts to 18 ft. lbs. (25 Nm).

16. Tighten the front cover bolts to 18 ft. lbs. (25 Nm).

17. Install the water pump.

8.1L Engine

➡This procedure requires the use of Crankshaft Sprocket Installer tool No. J 22102 and Crankshaft Protector Button tool No. J 42846.

1. Before servicing the vehicle, refer to the precautions in the beginning of this section.

2. Drain the cooling system.

3. Remove or disconnect the following:
• Negative battery cable
• Water pump

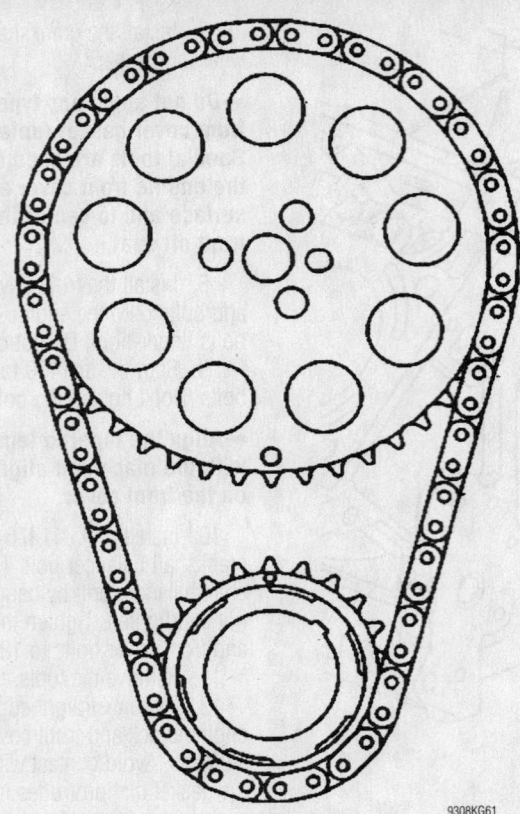

Timing mark alignment—4.8L, 5.3L and 6.0L engines

9308KG61

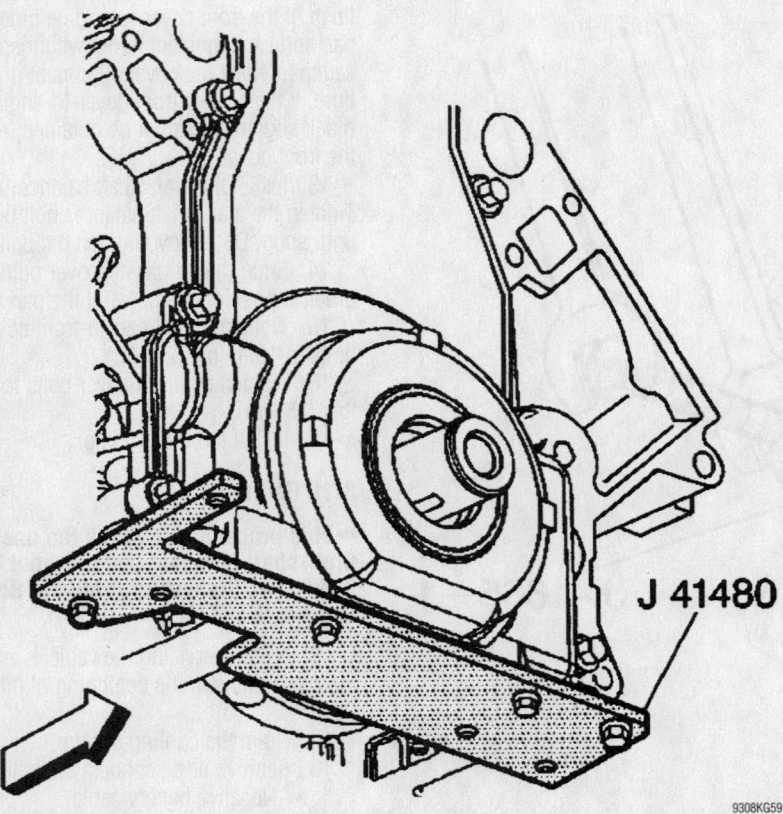

J 41480

J41480 installation—4.8L, 5.3L and 6.0L engines

9308KG59

- Crankshaft balancer from the crankshaft
- Camshaft Position (CMP) sensor connector
- Engine harness clips from the battery cable channel
- CMP sensor bolt and sensor
- Battery cable channel bolt
- Battery cable channel and reposition
- Front cover bolts, front cover and gasket

➡ **The front cover gasket can be reused if it is not damaged.**

- Crankshaft front oil seal from the front cover

4. Align the timing marks on the camshaft and crankshaft sprockets.

- Camshaft sprocket bolts
- Camshaft sprocket and timing chain

5. Install Crankshaft Protector Button tool No. J 42846 into the end of the crankshaft and remove the crankshaft sprocket using a 3-jawed puller.

6. Clean and inspect the timing chain and sprockets.

To install:

7. Use the Crankshaft Sprocket Installer tool No. J 22102 to install the crankshaft sprocket. Align the keyway of the sprocket with the crankshaft pin.

8. Remove the installation tool.

9. Rotate the crankshaft until the crankshaft sprocket alignment mark is in the 12 o'clock position.

10. Install the camshaft sprocket and timing chain, noting the following important points:

a. The cam sprocket must be installed with the alignment mark at the 6 o'clock position.

b. The sprocket teeth must mesh with the timing chain to avoid damaging the camshaft retainer.

c. Never use a hammer to install the sprocket onto the camshaft.

11. Make sure the crankshaft sprocket is alignment at the 12 o'clock position and the cam sprocket is at the 6 o'clock position.

12. Install the camshaft sprocket bolts and tighten, in two passes, to 22 ft. lbs. (30 Nm)

13. Use clean engine oil to lubricate the sealing surfaces of the front oil seal.

14. Install or connect the following:

- New seal into the front cover, using a suitable seal installation tool

➡ **The front cover must be installed while the sealant is still wet to the touch.**

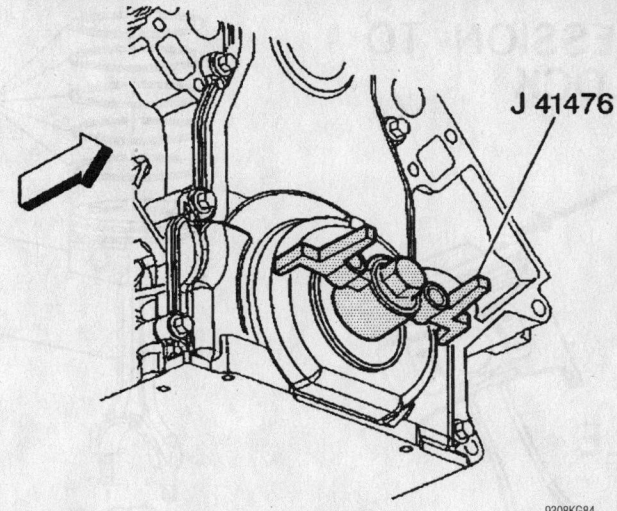

Seal alignment tool installation—4.8L, 5.3L and 6.0L engines

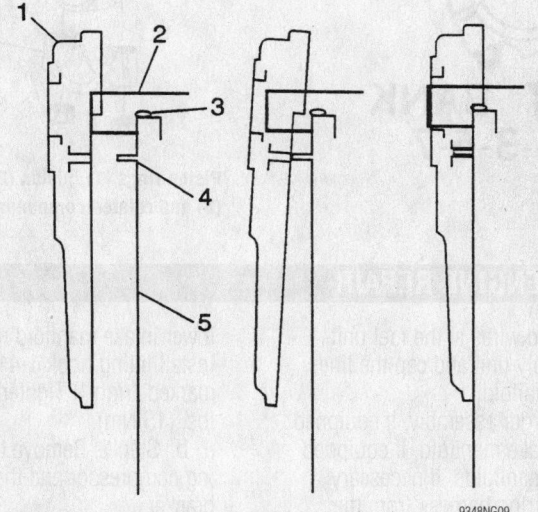

Proper front cover installation sequence—8.1L engine

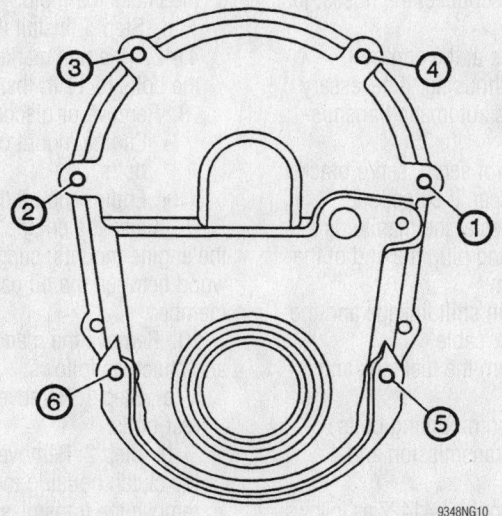

Engine front cover bolt tightening sequence—8.1L engine

- Sealant to the 2 places on the engine block where the front cover meets the oil pan
- Front cover gasket into the cover

15. Install the front cover, referring to the accompanying figure and using the following steps only:

   a. Hold the front cover (1) up to the crankshaft (2).

   b. Lift the cover (1) while sliding the cover over the crankshaft (2).

   c. Slide the front cover toward the engine block (5) while keeping the cover raised.

   d. Lower the cover down over the dowel pin (4), allowing the front cover to rest on the sealant (3).

16. Install the front cover bolts and tighten, in sequence, as follows:

   a. 1st pass: 53 inch lbs. (6 Nm).

   b. 2nd pass: 106 inch lbs. (12 Nm).

17. Install or connect the following:

- Battery cable channel and bolt. Tighten to 80 inch lbs. (9 Nm).
- CMP sensor. Inspect the O-ring first, replace if necessary and coat with oil before installation.
- CMP sensor bolt to 106 inch lbs. (12 Nm).
- Engine harness clips to the battery cable channel
- CMP sensor electrical connector
- Crankshaft balancer
- Water pump
- Negative battery cable.

18. Fill the cooling system with the proper type and quantity of antifreeze.

Piston and Ring

POSITIONING

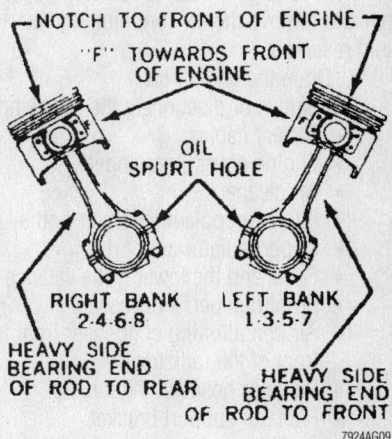

Piston and connecting rod assembly positioning —4.8L, 5.3L, 5.7L and 6.0L engines

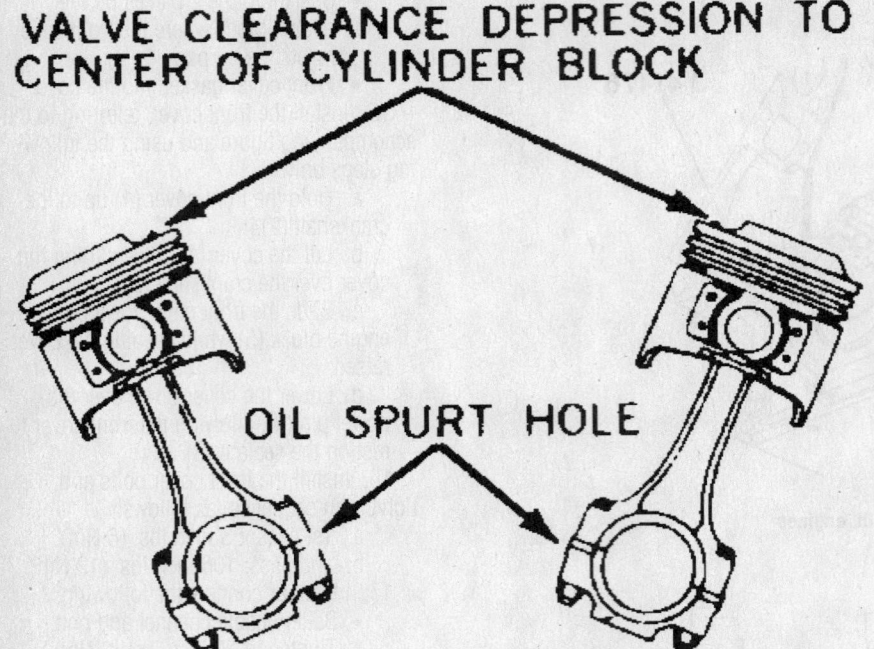

VALVE CLEARANCE DEPRESSION TO CENTER OF CYLINDER BLOCK

OIL SPURT HOLE

RIGHT BANK 2-4-6-8

LEFT BANK 1-3-5-7

7924AG10

Piston and connecting rod assembly positioning —7.4L engine

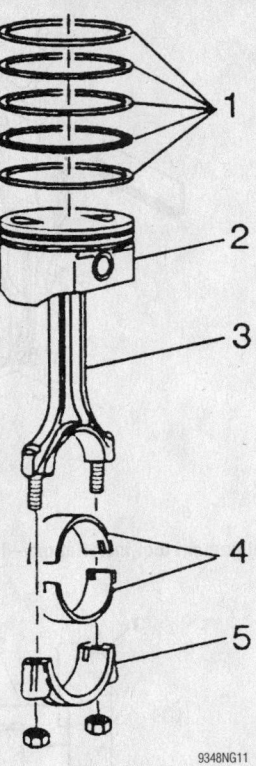

9348NG11

Piston rings (1), piston (2), connecting rod (3) and related components—8.1L engine

DIESEL ENGINE REPAIR

Engine Assembly

REMOVAL & INSTALLATION

6.5L Engine

1. Before servicing the vehicle, refer to the precautions in the beginning of this section.
2. Drain the cooling system.
3. Discharge the air conditioning system and remove the air conditioning vacuum reservoir.
4. Drain the engine oil.
5. Remove or disconnect the following:
   - Battery cables
   - Engine cover, if equipped
   - Air cleaner
   - Radiator coolant reservoir bottle
   - Upper radiator support
   - Grille and the lower grille valance
   - Front bumper, if necessary
   - Air conditioning condenser from in front of the radiator
   - Radiator hoses at the radiator
   - Radiator support bracket
   - Radiator and the shroud
   - Accelerator and cruise control linkages

- Hoses and wires at the fuel unit
- Fuel supply unit and cap the lines
- Intake manifold
- Turbocharger assembly, if equipped
- Lower intake manifold, if equipped
- Exhaust manifolds, if necessary
- Engine wiring harness from the firewall connection
- Power steering pump, it's not necessary to disconnect the hoses; just move aside
- Heater hoses at the engine
- Thermostat housing, if necessary
- Oil filler and automatic transmission tubes
- Cruise control servo, servo bracket and transducer, if equipped
- Exhaust pipes at the manifolds
- Driveshaft and plug the end of the transmission
- Transmission shift linkage and the speedometer cable
- Fuel line from the fuel tank and pump
- Transmission mounting bolts

6. Support the transmission and engine.
7. Install lifting hooks J-41427 as follows:
   a. Step 1: Remove the 2 right rear

lower intake manifold retainers and install lifting hook J-41427 (the one marked "right") Tighten the bolts to 11 ft. lbs. (15 Nm).
   b. Step 2: Remove the air conditioning compressor and the accessory drive bracket.
   c. Step 3: If equipped, disconnect the EGR tube and the 2 left lower bolts from the intake manifold.
   d. Step 4: Install the lifting hook J-41427 (the one marked "left") and tighten the bolts to 11 ft. lbs. (15 Nm).

8. Remove or disconnect the following:
   - Engine mount bracket-to-frame bolts
   - Engine mount through-bolts

9. Raise the engine slightly and remove the engine mounts, support the engine with wood between the oil pan and the crossmember.
10. Remove the manual transmission and clutch as follows:
   a. Step 1: Remove the clutch housing rear bolts.
   b. Step 2: Remove the bolts attaching the clutch housing to the engine and remove the transmission and clutch as a unit.

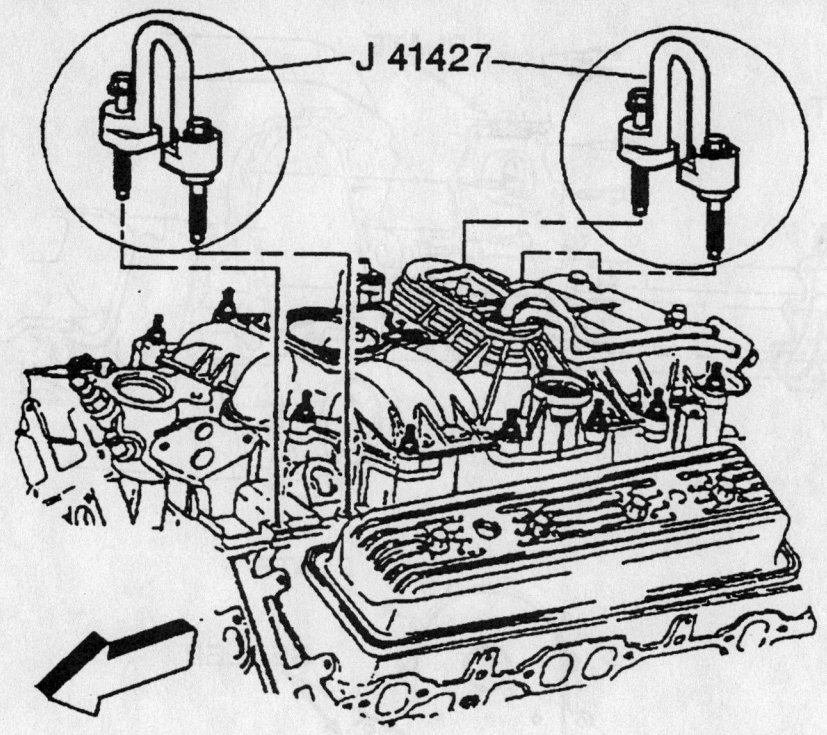

J 41427

7924KG01

For engine removal and installation, universal lift brackets should be installed in place of the proper intake manifold bolts

→**Support the transmission as the last bolt is being removed to prevent damaging the clutch.**

    c. Step 3: Remove the starter and clutch housing rear cover.

    d. Step 4: Loosen the clutch mounting bolts a little at a time to prevent distorting the disc until spring pressure is released. Remove all of the bolts, the clutch disc and the pressure plate.

11. Remove the automatic transmission as follows:

    a. Step 1: Lower the engine and support it on blocks.

    b. Step 2: Remove the starter and converter housing cover.

    c. Step 3: Remove the flexplate-to-converter attaching bolts.

    d. Step 4: Support the transmission on blocks.

    e. Step 5: Disconnect the detent cable on the Turbo Hydra-Matic.

    f. Step 6: Remove the transmission-to-engine mounting bolts.

12. Attach an engine crane to the engine.

    g. Step 7: Remove the blocks from the engine only and glide the engine away from the transmission.

To install:

13. Install the manual transmission and clutch as follows:

    a. Step 1: Install the clutch disc and the pressure plate. Tighten the clutch mounting bolts a little at a time to prevent distorting the disc.

    b. Step 2: Install the starter and clutch housing rear cover.

    c. Step 3: Install the bolts attaching the clutch housing to the engine and install the transmission and clutch as a unit. Tighten the bolts to specification.

    d. Step 4: Install the clutch housing rear bolts.

14. Install the automatic transmission as follows:

    a. Step 1: Position the transmission.

    b. Step 2: Install the transmission-to-engine mounting bolts.

    c. Step 3: Connect the throttle linkage and detent cable.

    d. Step 4: Install the flexplate-to-converter attaching bolts. Torque the bolts to 65 ft. lbs. (90 Nm).

    e. Step 5: Install the starter and converter housing cover.

15. Install or connect the following:

- Engine mount through-bolts. Torque the bolts to 50 ft. lbs. (68 Nm).
- Engine mount bracket-to-frame bolts. Torque the bolts to 44 ft. lbs. (59 Nm).
- Clutch cross-shaft

- Transmission mounting bolts. Torque the bolts to 75 ft. lbs. (100 Nm).
- Transmission shift linkage and the speedometer cable
- Driveshaft

16. Remove the lifting hooks.

- Compressor
- EGR valve tube
- Intake manifold retaining bolts
- Condenser
- Hood latch support
- Lower fan shroud and filler panel
- Transmission dipstick tube and the accelerator cable at the tube
- Coolant hose
- Cruise control servo, bracket and transducer
- Oil filler pipe and automatic transmission filler pipe
- Engine dipstick tube
- Thermostat housing, if removed
- Heater hoses at the engine
- Engine wiring harness to the firewall connection
- Radiator and the shroud
- Radiator support bracket
- Exhaust manifolds if removed
- Lower intake manifold if removed
- Intake manifold
- Fuel supply unit
- Lines to the fuel supply unit
- Accelerator and cruise control linkages
- Windshield wiper jar and bracket
- Air conditioning condenser
- Air conditioning vacuum reservoir, if equipped
- Fluid cooler lines at the radiator, if equipped with an automatic transmission
- Radiator coolant reservoir bottle
- Hoses at the radiator
- Upper radiator support
- Grille and the lower grille valance
- Air cleaner
- Air stove pipe
- Engine cover
- Battery cables

17. Refill the cooling system.

18. Recharge the air conditioning system.

Water Pump

REMOVAL & INSTALLATION

6.5L Engine

1. Before servicing the vehicle, refer to the precautions in the beginning of this section.

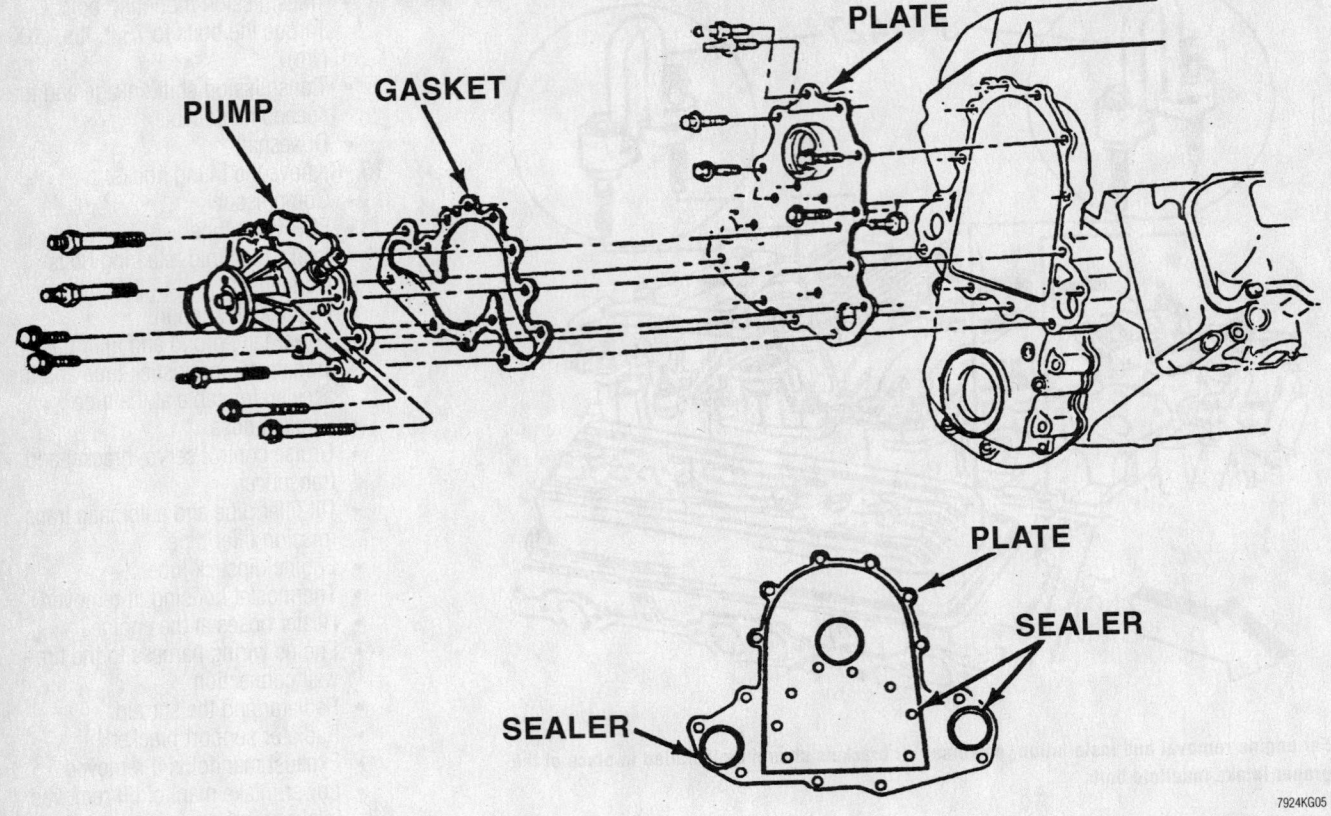

Exploded view of the water pump assembly and related components—6.5L diesel engines

2. Drain the engine coolant.
3. Remove or disconnect the following:
- Negative battery cables
- Fan and fan shroud
- Air conditioning hose bracket and/or the oil filler tube, as required
- Accessory drive belt(s)
- Vacuum pump mounting bracket nuts/bolt
- Vacuum pump and bracket
- Power steering pump and bracket
- Coolant hoses from the pump
- Water pump plate retaining bolts
- Pump and plate assembly from the engine

➡**Remove the bolt on the rear of the water pump plate.**

- Separate the pump and gasket from the plate

To install:

4. Install or connect the following:
- Water pump and a new gasket to the plate. Torque the retaining bolt (at the rear of the plate) to 20 ft. lbs. (28 Nm).
5. Be sure the block mating surface and the plate flanges are free of oil. Apply an anaerobic sealer GM part 1052357 or equivalent.

➡**The sealer must be wet to the touch when the bolts are tightened.**

- Water pump and plate assembly. Torque the bolts to 20 ft. lbs. (28 Nm).
- Coolant hoses to the pump assembly
- Power steering pump and bracket
- Vacuum pump and bracket, along with the bolt holding the pump and alternator
- Fan and pulley
- Accessory drive belt(s)
- Oil filler tube and/or air conditioning hose bracket nuts. If removed
- Fan shroud
- Batteries
6. Refill the radiator.

Heater Core

REMOVAL & INSTALLATION

For Heater Core Removal & Installation, please refer to the procedure under Gasoline Engine Repair.

Glow Plugs

REMOVAL & INSTALLATION

6.5L Engine

1. Before servicing the vehicle, refer to the precautions in the beginning of this section.
2. Remove or disconnect the following:
- Negative battery cables
- Glow plug lead wires
- Plugs
- Right front tire
- Inner splash shield from the fender well
- Lead wire from the plug at the No. 2 cylinder and the lead wires from plugs in the Nos. 4 and 6 cylinders at the harness connectors
- Heat shroud for the plug in the No. 4 and 6 cylinder. Slide the shrouds back just far enough to allow access so you can unplug the wires.
- Glow plugs in cylinders No. 2, 4 and 6.
3. Reach up under the vehicle and dis-

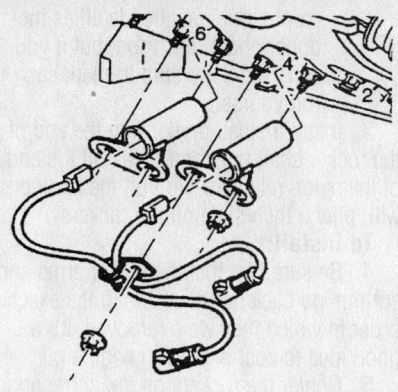

Exploded view of the heat shrouds and glow plug wiring—6.5L diesel engine

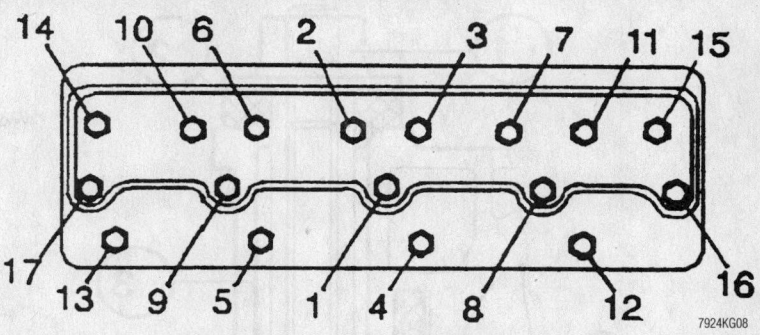

Tighten the cylinder head bolts according to the sequence shown for proper cylinder sealing—6.5L diesel engines

connect the lead wire at No. 8. Remove the glow plug.

➡ **You may find that removing the exhaust pipe down mite make this a bit easier when working on Nos. 6 and 8.**

To install:
4. Install or connect the following:
- Glow plugs and tighten to 16 ft. lbs. (22 Nm)
- Heat shrouds and electrical connection
- Exhaust pipe, if removed
- Splash shields, if removed
- Negative battery cable

Cylinder Head

REMOVAL & INSTALLATION

6.5L Engine

1. Before servicing the vehicle, refer to the precautions in the beginning of this section.
2. Relieve the fuel system pressure.
3. Drain the coolant system.
4. Discharge the air conditioning system.
5. Remove or disconnect the following:
- Negative battery cables
- Intake manifold
- Fan upper shroud
- Compressor assembly, if equipped
- Turbocharger, if equipped
- Exhaust manifold
- Valve cover
- Rocker arm assemblies and pushrods

➡ **Mark all components so they may be returned to their original location.**

- Air cleaner resonator and bracket
- Transmission and oil dipstick tube;

remove the oil fill tube from the coolant crossover pipe
- Heater, radiator and bypass hoses
- Alternator upper bracket
- Alternator
- Power steering pump
- Vacuum pump
- Fuel bleeder valve at the coolant crossover pipe
- Fuel return crossover line clamp bolts from both cylinder heads
- Wire connector from the sensor in the coolant crossover pipe
- Electrical connection and brackets from cylinder head
- Coolant crossover pipe/thermostat assembly
- Head bolts and the cylinder heads

To install:
6. Clean the mating surfaces of the heads and block thoroughly.
7. Clean the head bolts thoroughly. Coat the threads of the head bolts with sealing compound GM part 1052080 or equivalent, before installation.
8. Install or connect the following:
- New gasket
- Cylinder head and bolts. Torque the bolts in the follows:
  a. Step 1: Torque the bolts to 20 ft. lbs. (25 Nm).
  b. Step 2: Torque the bolts to 50 ft. lbs. (65 Nm).
  c. Step 3: Then an additional 90 degrees (¼ turn).
- Coolant crossover pipe and thermostat
- Fuel valve
- Bypass hose
- Upper radiator hose
- Heater hoses at the head
- Transmission and oil dipstick tube
- Air cleaner resonator and bracket
- Pushrods, hardened ends facing up
- Rocker arm assemblies

9. Adjust the valves.
- Valve cover
- Alternator and upper bracket
- Exhaust manifolds. Torque bolts to 22 ft. lbs. (30 Nm).
- Upper fan shroud
- Intake manifold
- Turbocharger, if equipped
- Vacuum pump, if equipped
- Air conditioning compressor, if equipped
- Engine electrical connection
- Negative battery cables

10. Refill the cooling system with the proper type and quantity of antifreeze.
11. Evacuate and recharge the air conditioning system.

Starter Motor

REMOVAL & INSTALLATION

1. Before servicing the vehicle, refer to the precautions in the beginning of this section.
2. Remove or disconnect the following:
- Negative battery cables
- Mounting bolts/nuts and shim, if used
- Starter
- Wires
- Heat shield and bracket

To install:
3. Install or connect the following:
- Heat shield and bracket. Torque the bolts to 13 ft. lbs. (17 Nm).
- Wires. Torque battery wire nut to 89 inch lbs. (10 Nm), and ignition nut to 18 inch lbs. (2 Nm).
- Starter
- Mounting bolts/nuts and shim, if used. Torque the bolts to 33 ft. lbs. (45 Nm) and the nut to 75 inch lbs. (8.5 Nm).
- Negative battery cables

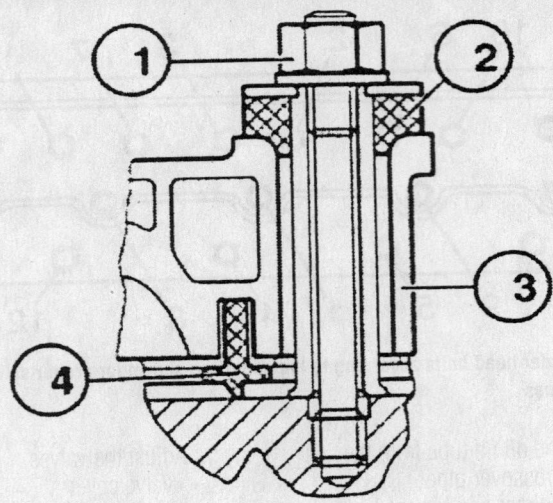

(1)   Nut

(2)   Decoupling element

(3)   Intake air manifold

(4)   Seal

9308KG04

Exploded view of the starter motor—6.5L engine shown

Rocker Arms/Shaft

REMOVAL & INSTALLATION

6.5L Engine

1. Before servicing the vehicle, refer to the precautions in the beginning of this section.

2. Remove or disconnect the following:

➡**Rotate the engine until the mark on the crankshaft balancer is at the 2 o'clock position. Rotate the crankshaft counterclockwise 3½ in. (88mm) aligning the crankshaft balancer mark with the first lower water pump bolt, at about the 12:30 position. This will ensure that no valves are close to a piston crown.**

- Cylinder head cover
- Rocker shaft assembly

➡**The rocker assemblies are mounted on 2 short rocker shafts per cylinder head, with each shaft operating 4 rockers. 2 bolts secure each rocker shaft assembly, Mark the shafts so they can be installed in their original locations.**

- Pushrods. The pushrods MUST be installed in the original direction! A

paint stripe usually identifies the upper end of each rod, but if you can't see it, be sure to mark each rod yourself.

3. Insert a small prybar into the end of the rocker shaft bore and break off the end of the nylon retainers. Pull off the retainers with pliers, then slide off the rockers.

To install:

4. Be sure first that the rocker arms and springs go back on the shafts in the exact order in which they were removed. It's a good idea to coat them with engine oil.

5. Center the rockers on the corresponding holes in the shaft

6. Install or connect the following:

- New plastic retainers using a ½ in. (13mm) drift
- Pushrods with there marked ends up
- Rocker shaft assemblies and be sure that the ball ends of the pushrods seat themselves in the rockers

7. Rotate the engine clockwise until the mark on the torsional damper aligns with the **0** on the timing tab. Rotate the engine counterclockwise 3 ½ in. (88mm) measured at the damper. You can estimate this by checking that the mark on the damper is

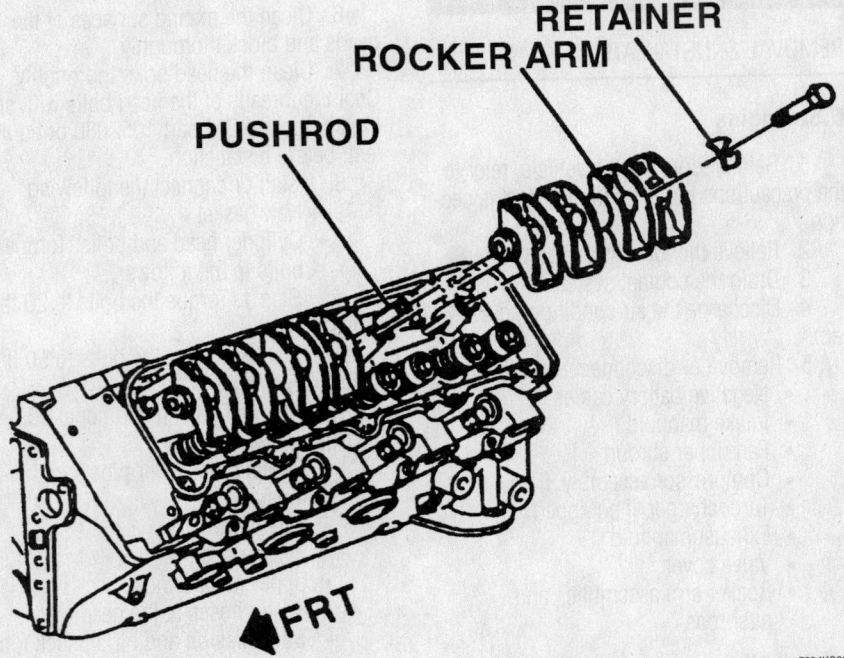

7924KG22

Rocker shaft assembly and related components—diesel engines

now aligned with the FIRST lower water pump bolt. BE CAREFUL! This ensures that the piston is away from the valves.

- Rocker shaft bolts. Torque them to 40 ft. lbs. (55 Nm).
- Cylinder head cover

Turbocharger

REMOVAL & INSTALLATION

6.5L Engine

1. Before servicing the vehicle, refer to the precautions in the beginning of this section.
2. Remove or disconnect the following:
   - Negative battery cable
   - Air inlet duct
   - Oil feed line from the top of the turbocharger
   - Crankcase Depression Regulator (CDR) valve vent bracket screw
   - CDR valve and vent tube
   - Air cleaner assembly
   - Heat shield
   - Right front tire assembly and the splash shield
   - Exhaust pipe-to-turbocharger exhaust outlet elbow V-band clamp

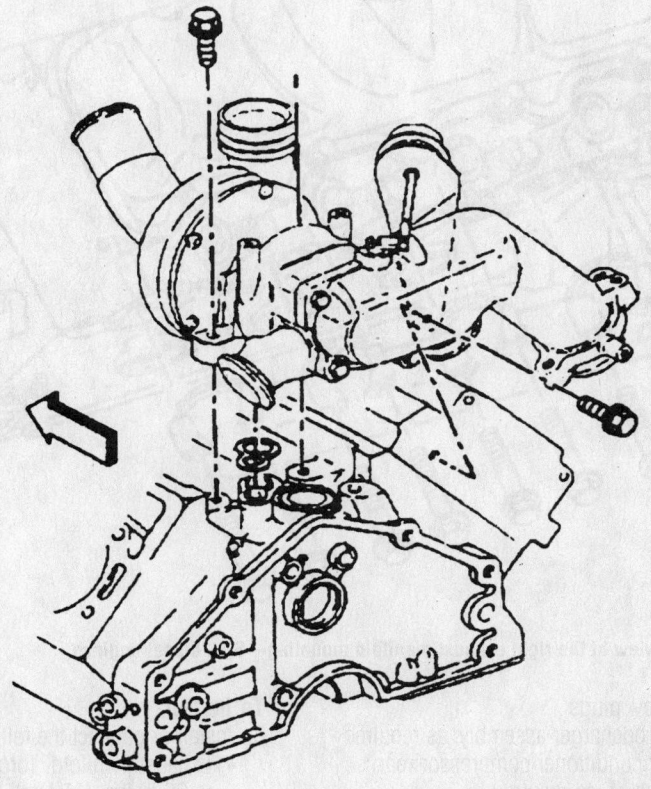

Turbocharger mounting—diesel engines

7924KG43

- Oil drain tube-to-turbocharger center bearing bolts
- Exhaust manifold-to-turbocharger nuts
- Turbocharger

To install:

➡Use anti-seize compound on all threaded fasteners connected to the turbocharger

3. Install or connect the following:
   - Turbocharger to the exhaust manifold. Torque the nuts to 37 ft. lbs. (50 Nm).
   - New oil drain tube flange gasket and the oil drain tube. Torque the bolts to 19 ft. lbs. (26 Nm).

➡Use 0.03–0.07 fl. oz. (0.88–2ml) of engine oil to feed the oil feed hole at the top of the turbocharger and hand rotate the compressor wheel/shaft. This will pre-lube the shaft bearings

   - Oil feed line. Torque the connection to 13 ft. lbs. (17 Nm).
   - Exhaust pipe to the turbocharger exhaust elbow V-band clamp. Torque the clamp to 71 inch. lbs. (8 Nm).
4. Disengage the injection pump fuel

shutdown solenoid connector and crank the engine for no more 15 seconds to prime the oil system. Do not let the engine start.

   - Right front wheel and splash shield
   - Heat shield. Apply Loctite® or equivalent to the bolts. Torque to 56 inch. lbs. (6 Nm).
   - Air cleaner assembly
   - Turbocharger compressor outlet
   - CDR valve, tube and bracket
   - Air intake duct

➡Operate the engine at idle for at least 3 minutes after installing the turbocharger

Intake Manifold

REMOVAL & INSTALLATION

6.5L Engine

1. Before servicing the vehicle, refer to the precautions in the beginning of this section.
2. Recover air conditioning system, and reposition air conditioning lines.
3. Remove or disconnect the following:
   - Negative battery cable
   - Air cleaner assembly
   - Fuel lines, and electrical connections
   - Engine and transmission oil level tubes

✽✽ WARNING

Do not remove the center intake and side intakes as an assembly. Damage to the center intake and turbocharger may occur.

   - Center intake assembly, and glow plug relay
   - Side intake bolts and fuel retaining clips
   - Side intakes
4. Clean all gaskets surface.

To install:
5. Install or connect the following:
   - Side intakes and new gaskets. Torque the bolts to 31 ft. lbs. (42 Nm).
   - Fuel lines retaining clips and electrical connection
   - Center intake with new gaskets. Torque the bolts to 17 ft. lbs. (23 Nm).
   - Engine oil and transmission oil level tubes
   - Glow plug relay

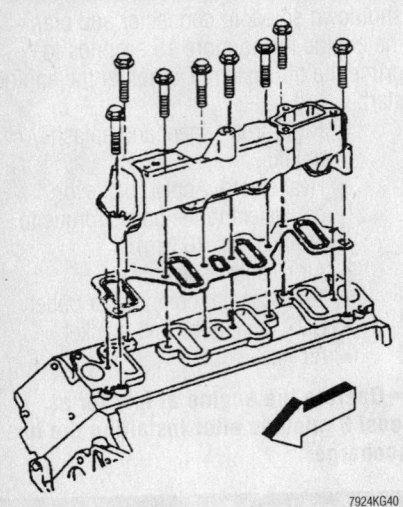

Exploded view of the side intake manifold mounting—6.5L engines

7924KG40

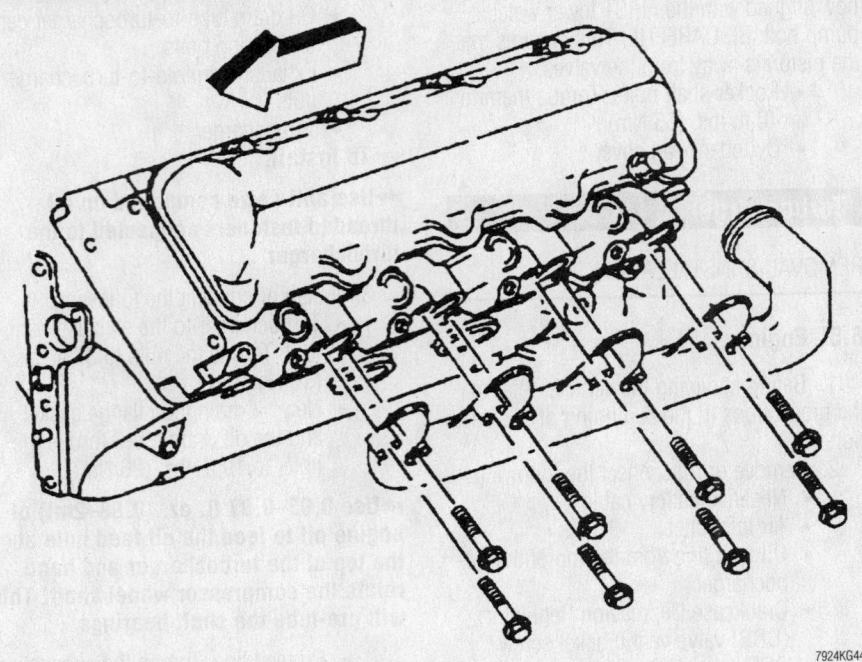

7924KG44

Exploded view of the left exhaust manifold mounting—6.5L diesel engines

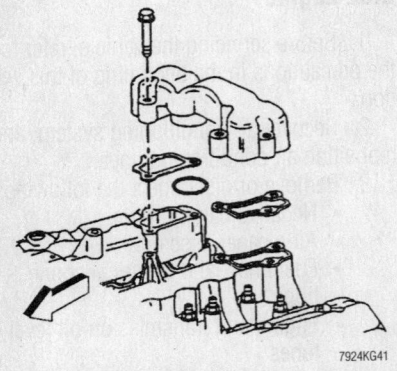

7924KG41

Center intake manifold mounting—6.5L engines

- Air conditioning lines
- Air cleaner assembly
- Negative battery cable
6. Recharge air conditioning system.

Exhaust Manifold

REMOVAL & INSTALLATION

6.5L Engine

1. Before servicing the vehicle, refer to the precautions in the beginning of this section.
2. Remove or disconnect the following:
   - Batteries cables
   - Exhaust pipe from the manifold flange
   - Engine oil and transmission oil fill tubes
   - Engine cover and disconnect the glow plug wires

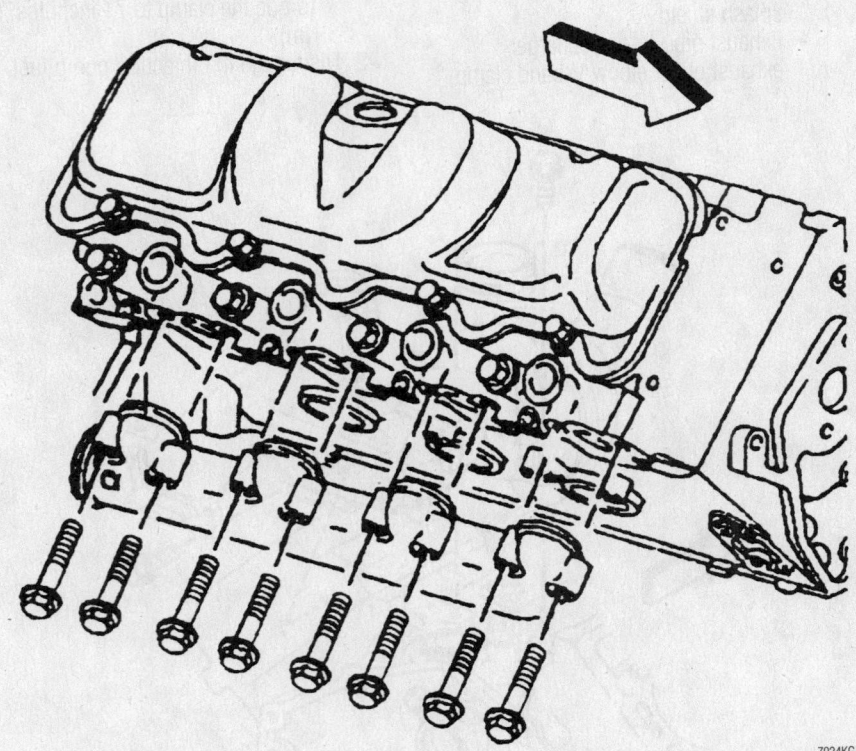

7924KG45

Exploded view of the right exhaust manifold mounting—6.5L diesel engines

- Glow plugs
- Turbocharger assembly, as required
- Air conditioner compressor rear bracket, as required
- Manifold bolts and the manifold

To install:
3. Install or connect the following:
   - Exhaust manifold. Torque the bolts to 26 ft. lbs. (35 Nm).
   - Exhaust pipe

- Glow plugs and electrical connection
- Engine and transmission oil fill tubes
- Air conditioning compressor bracket
- Negative battery cable

Camshaft and Valve Lifters

REMOVAL & INSTALLATION

6.5L Engine

1. Before servicing the vehicle, refer to the precautions in the beginning of this section.
2. Drain the cooling system.
3. Discharge the air conditioning system.
4. Relieve the fuel system pressure.
5. Remove or disconnect the following:
- Battery cables
- Radiator, condenser, shroud and fan assembly
- Grille and parking light assembly
- Hood latch and brace assembly
- Oil pump drive
- Power steering pump and position aside
- Alternator, air conditioner compressor and position aside
- Rocker arm covers
- Rocker arm assemblies and pushrods. Mark them so they can be returned to their original position.
- Cylinder heads
- Hydraulic lifters and keep them in order so they can be returned to their original bore
- Front cover
- Timing chain and camshaft sprocket
- Injector pump
- Front engine mounting through-bolts
- Air conditioner condenser mounting bolts and lift the condenser out
- Thrust plate bolts and thrust plate
- Camshaft from the block
- Thrust plate spacer, if necessary

To install:

6. Install the spacer with the ID chamfer toward the camshaft.

➡️**It is recommended that the engine oil, oil filter and hydraulic lifters be replaced when installing a new camshaft.**

7. Coat the camshaft lobes with Molykote® or equivalent.

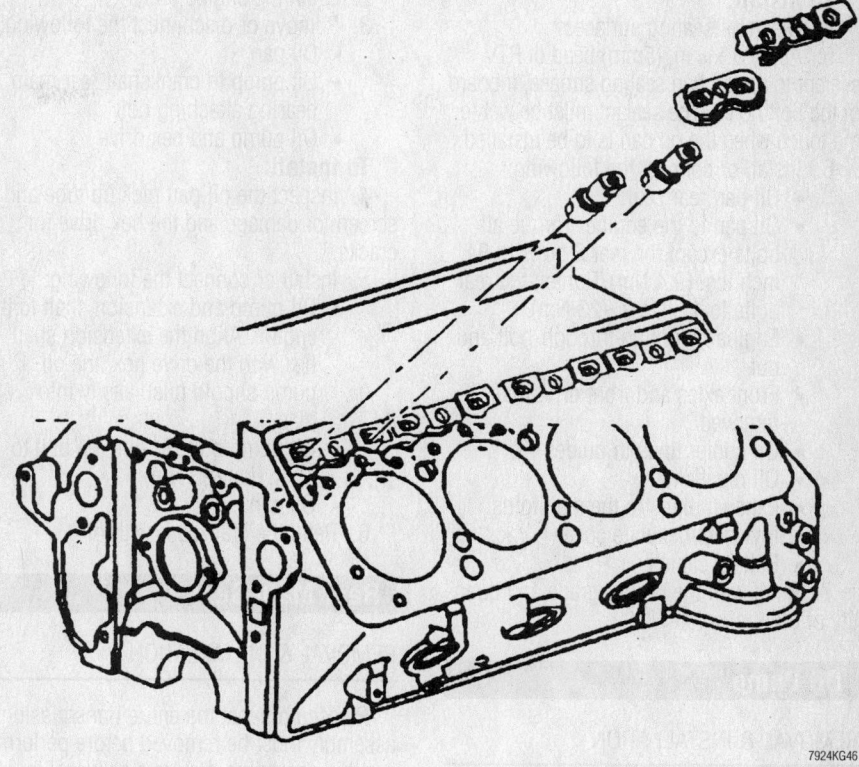

Exploded view of the lifter, guide plate and clamp—6.5L diesel engines

7924KG46

8. Lubricate the camshaft journals with engine oil.
9. Install or connect the following:
- Camshaft carefully into the block
- Thrust plate and bolts. Tighten to 17 ft. lbs. (23 Nm).
- Engine mount through-bolts
- Timing chain and sprockets, align the timing marks
- Air conditioner condenser, if equipped
- Injector pump
- Front cover
- Cylinder head
- Hydraulic lifters in the same bore as they were removed
- Rocker arm assemblies and pushrods in their original locations
- Rocker arm covers
- Power steering pump, alternator and air conditioner compressor
- Oil pump drive
- Hood latch and brace
- Grille and parking light assembly
- Radiator, the shroud and fan assembly
- Negative battery cables
10. Refill the cooling system with the proper type and quantity of antifreeze.

Valve Lash

ADJUSTMENT

All engines use hydraulic lifters, which require no periodic adjustment.

Oil Pan

REMOVAL & INSTALLATION

6.5L Engine

1. Before servicing the vehicle, refer to the precautions in the beginning of this section.
2. Drain the engine oil.
3. Remove or disconnect the following:

- Battery cables
- Oil dipstick
- Flywheel/flexplate cover
- Oil cooler line guides
- Front driveshaft
- Front axle, if needed
- Exhaust pipes from the manifolds
- Front engine mount through-bolts
- Oil pan bolts and the oil pan
- Oil pan rear seal

To install:

4. Clean all sealing surfaces.

5. Apply a ³⁄₁₆ in. (5mm) bead of RTV sealant to the oil pan sealing surface, inboard of the bolt holes. The sealant must be wet to the touch when the oil pan is to be installed.

6. Install or connect the following:
- Oil pan rear seal
- Oil pan to the engine. Torque all bolts except the rear 2 bolts to 84 inch lbs. (9.4 Nm) Tighten the rear bolts to 17 ft. lbs. (23 Nm).
- Engine mounting through-bolt and nut
- Front axles and front driveshaft, if removed
- Oil cooler lines in guides
- Oil dipstick
- Exhaust pipes to the manifolds
- Flywheel/flexplate cover
- Battery cables

7. Refill with the proper grade and quantity of oil.

Oil Pump

REMOVAL & INSTALLATION

6.5L Engine

1. Before servicing the vehicle, refer to the precautions in the beginning of this section.

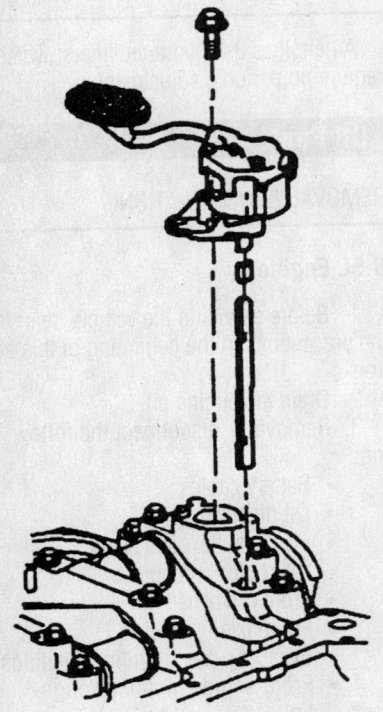

Exploded view of the oil pump mounting—
6.5L diesel engines

2. Drain the engine oil.
3. Remove or disconnect the following:
- Oil pan
- Oil pump to crankshaft rear main bearing attaching bolt
- Oil pump and hex drive

To install:

4. Inspect the oil pan pick up tube and screen for damage and the hex drive for cracks.

5. Install or connect the following:
- Oil pump and extension shaft to the engine. Align the extension shaft hex with the drive hex, the oil pump should push easily into place.
- Oil pump bolt. Torque the bolt to 65 ft. lbs. (90 Nm).
- Oil pan

6. Refill the crankcase with oil.

Rear Main Seal

REMOVAL & INSTALLATION

Please note that the entire transmission assembly must be removed before performing this procedure. Before a new seal is installed, the Crankcase Depression Regulator (CDR) and crankcase ventilation system should be cleaned and inspected. In addition, use care removing the flywheel. Some models use a heavy, dual mass flywheel that must be handled with care.

1. Before servicing the vehicle, refer to the precautions in the beginning of this section.

2. Remove or disconnect the following:
- Negative battery cables
- Transfer case, if equipped
- Transmission assembly
- Clutch assembly and flywheel, if equipped with manual transmission
- Flexplate, if equipped with automatic transmission
- Crankshaft rear main oil seal by inserting a suitable crankshaft seal removal tool and prying the seal out

To install:

3. Clean the oil seal bore in the block thoroughly before installation of the new seal.

4. Inspect the crankshaft for grit, rust or burrs and correct as necessary. Also inspect the portion of the crankshaft where the oil seal makes contact, for wear due to the rubbing action of the oil seal.

➡**Because of rear crankshaft wear or grooving, the new oil seal should be seated in a new location. The J 39084**

installation tool will control the seal positioning. This will provide a new surface on the crankshaft for the seal to ride on.

5. Clean the running surface of the crankshaft with a non-abrasive cleaner.

6. Lubricate the inner diameter of the new seal and the outer diameter of the crankshaft with engine oil.

7. Install or connect the following:
- Rear main oil seal using a crankshaft rear oil seal installation tool
- Flywheel.
- Transmission assembly
- Transfer case, if equipped
- Negative battery cables

8. Start the engine and verify no oil leaks.

Timing Chain, Sprockets, Front Cover and Seal

REMOVAL & INSTALLATION

6.5L Engine

1. Before servicing the vehicle, refer to the precautions in the beginning of this section.

2. Drain the cooling system.

3. Remove or disconnect the following:
- Negative battery cables
- Water pump and pulleys

4. Rotate the crankshaft to align the marks on the torsional damper with the **0** mark on the timing tab.

5. Scribe a mark aligning the injection pump flange and the front cover, if not already marked.

➡**The outer ring (weight) of the torsional damper is bonded to the hub with rubber. The damper must be removed with a puller that acts on the inner hub only. Pulling on the outer portion of the damper will break the rubber bond or destroy the tuning of the unit.**

6. Remove or disconnect the following:
- Crankshaft pulley and torsional damper
- Front cover-to-oil pan bolts (4)
- 2 fuel return line clips
- Injection pump gear
- Injection pump retaining nuts from the front cover
- Crankshaft sensor
- Baffle
- Cover bolts remaining and the front cover
- Injection pump gear

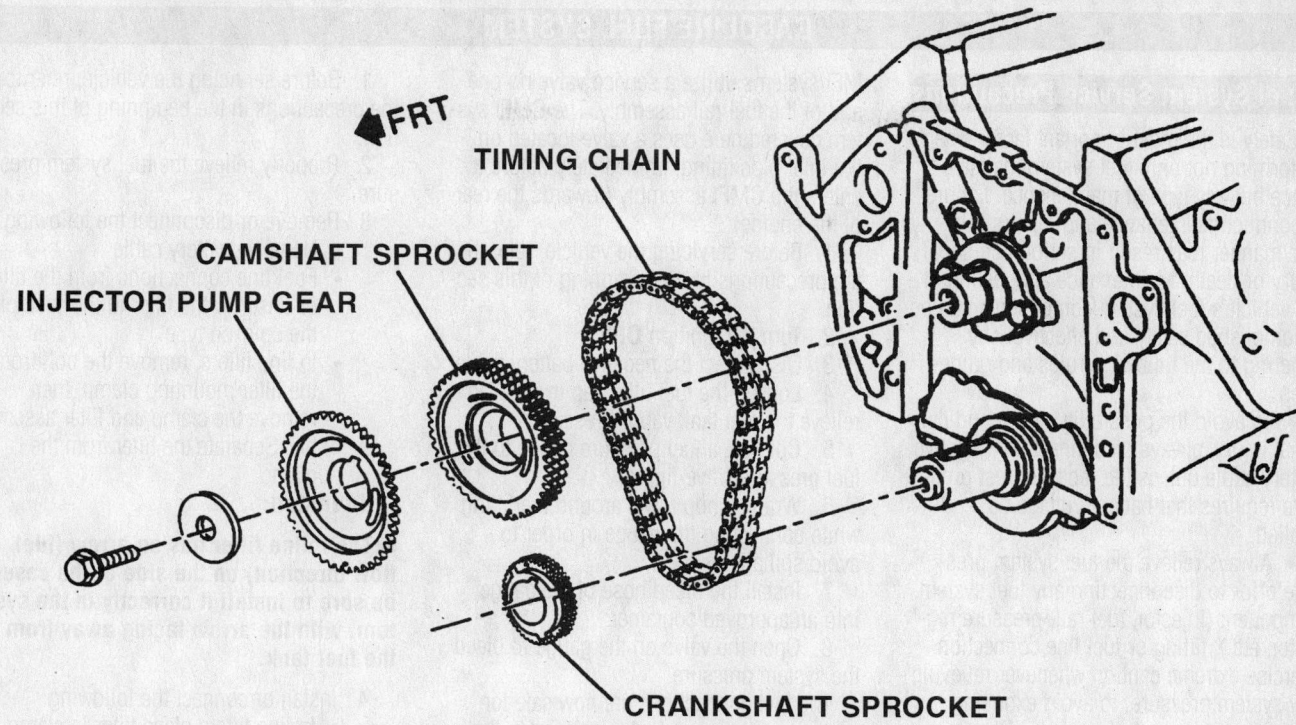

◄ FRT

TIMING CHAIN

CAMSHAFT SPROCKET

INJECTOR PUMP GEAR

CRANKSHAFT SPROCKET

7924KG23

Timing chain and related components—6.5L diesel engines

7. Align the camshaft timing gear marks
8. Remove the bolt and washer attaching the camshaft gear.
   - Camshaft sprocket with the timing chain
   - Crankshaft sprocket

To install:

9. Install or connect the following:
   - Cam sprocket, timing chain and crankshaft sprocket as a unit, aligning the timing marks on the sprockets

10. Rotate the crankshaft to align the injection pump and camshaft gears.
   - Injection pump gear

11. If the front cover oil seal is to be replaced, it can now be pried out of the cover with a suitable prying tool. Press the new seal into the cover evenly.

12. Clean both sealing surfaces until all traces of old sealer are gone. Apply a ³⁄₃₂ in. (2mm) bead of GM sealant 1052357 or equivalent to the sealing surface. Apply a ³⁄₁₆ in. (5mm) bead of RTV type sealer to the bottom portion of the front cover which attaches to the oil pan. Install the front cover.

13. Install or connect the following:
   - Baffle
   - Injection pump. Torque the nuts to 31 ft. lbs. (42 Nm), making sure the scribe marks on the pump and front cover are aligned.
   - Injection pump driven gear. Torque

the injection pump gear bolts to 17 ft. lbs. (23 Nm), making sure the marks on the cam gear and pump are aligned.

➡ **Verify that there is a minimum clearance of 0.040 in. (1.0mm) between the injection pump gear and baffle or noise may be result.**

   - Fuel line clips
   - Front cover-to-oil bolts
   - Torsional damper
   - Crankshaft pulley. Torque the bolts to 80 inch lbs. (9 Nm).

   - Oil pan bolts. Torque the bolts to 106 inch lbs. (12 Nm).
   - Water pump
   - Pulley assembly
   - Negative battery cables

14. Refill the cooling system with the proper type and quantity of antifreeze.

15. Inspect the engine for leaks.

Piston and Ring

POSITIONING

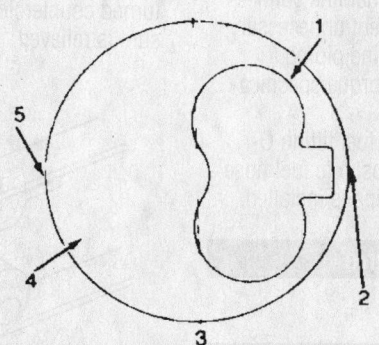

1 Oil control ring expander gap	3 Centerline of piston pin
2 Second compression ring gap	4 Oil control ring gap
	5 Top compression ring gap

7924AG11

Piston ring end-gap spacing—6.5L diesel engines

GASOLINE FUEL SYSTEM

Fuel System Service Precautions

Safety is the most important factor when performing not only fuel system maintenance but any type of maintenance. Failure to conduct maintenance and repairs in a safe manner may result in serious personal injury or death. Maintenance and testing of the vehicle's fuel system components can be accomplished safely and effectively by adhering to the following rules and guidelines.

• To avoid the possibility of fire and personal injury, always disconnect the negative battery cable unless the repair or test procedure requires that battery voltage be applied.

• Always relieve the fuel system pressure prior to disconnecting any fuel system component (injector, fuel rail, pressure regulator, etc.), fitting or fuel line connection. Exercise extreme caution whenever relieving fuel system pressure, to avoid exposing skin, face and eyes to fuel spray. Please be advised that fuel under pressure may penetrate the skin or any part of the body that it contacts.

• Always place a shop towel or cloth around the fitting or connection prior to loosening to absorb any excess fuel due to spillage. Ensure that all fuel spillage (should it occur) is quickly removed from engine surfaces. Ensure that all fuel soaked cloths or towels are deposited into a suitable waste container.

• Always keep a dry chemical (Class B) fire extinguisher near the work area.

• Do not allow fuel spray or fuel vapors to come into contact with a spark or open flame.

• Always use a back-up wrench when loosening and tightening fuel line connection fittings. This will prevent unnecessary stress and torsion to fuel line piping. Always follow the proper torque specifications.

• Always replace worn fuel fitting O-rings with new. Do not substitute fuel hose or equivalent where fuel pipe is installed.

Fuel System Pressure

RELIEVING

A Schrader valve is provided on these fuel systems, in order to conveniently test or release the system pressure. A fuel pressure gauge and adapter will be necessary to connect the gauge to the fitting. Most of the MFI systems utilize a service valve on one end of the fuel rail assembly. The CMFI system covered here uses a valve located on the inlet pipe fitting, immediately before it enters the CMFI assembly (towards the rear of the engine)

1. Before servicing the vehicle, refer to the precautions in the beginning of this section.
2. Turn the ignition **OFF**.
3. Disconnect the negative battery cable.
4. Loosen the fuel filler cap in order to relieve the fuel tank vapor pressure.
5. Connect a fuel pressure gauge to the fuel pressure valve/fitting.
6. Wrap a shop towel around the fitting while connecting the gauge in order to avoid spillage.
7. Install the bleed hose of the gauge into an approved container.
8. Open the valve on the gauge to bleed the system pressure.

The fuel connections are now safe for servicing. Drain any fuel remaining in the gauge into an approved container.

Fuel Filter

REMOVAL & INSTALLATION

2000 Vehicles

The fuel filter is normally located along the frame rail of the vehicle. On some vehicles however, it may have been relocated to the engine compartment. When in doubt, trace a fuel line from the engine backwards or from the tank forward in order to locate the filter.

Some vehicles utilize a spin-on fuel filter located on the frame rail. This filter can be turned counterclockwise after the fuel pressure is relieved.

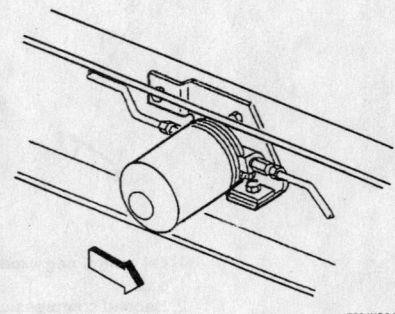

7924KG24

The spin-on fuel filter is serviced in the same manner as a spin-on oil filter

1. Before servicing the vehicle, refer to the precautions in the beginning of this section.
2. Properly relieve the fuel system pressure.
3. Remove or disconnect the following:
   • Negative battery cable
   • Fuel line connections from the filter or unscrew the filter in the case of the spin-on type
   • In line filters, remove the bolt from the filter mounting clamp, then remove the clamp and filter assembly. Separate the filter from the clamp.

To install:

➡**The inline filter has an arrow (fuel flow direction) on the side of the case, be sure to install it correctly in the system, with the arrow facing away from the fuel tank.**

4. Install or connect the following:
   • In line filters place filter in clamp
   • Install filter and clamp
   • Spin-on filters, lubricate the gasket before installation. Then tighten the filter an additional ¾ of a turn from the point when the gasket touches the filter adapter. Always check for leaks after a new filter is installed.

2001–03 Vehicles

1. Before servicing the vehicle, refer to the precautions in the beginning of this section.
2. Disconnect the negative battery cable.
3. Relieve the fuel system pressure.
4. Raise the vehicle.
5. Clean all the fuel filter connections and the surrounding areas before disconnecting the fuel pipes in order to avoid possible contamination of the fuel system.
6. Disconnect the threaded fittings from the fuel filter.
7. Cap the fuel pipes in order to prevent possible fuel system contamination.
8. Slide the fuel filter from the bracket.
9. Inspect the fuel pipe O-rings for cuts, nicks, swelling, or distortion. Replace the O-rings if necessary.

To install:

10. Slide the fuel filter into the bracket. Remove the caps from the fuel pipes.
11. Connect the threaded fittings to the fuel filter. Tighten the fittings to 18 ft. lbs. (25 Nm).

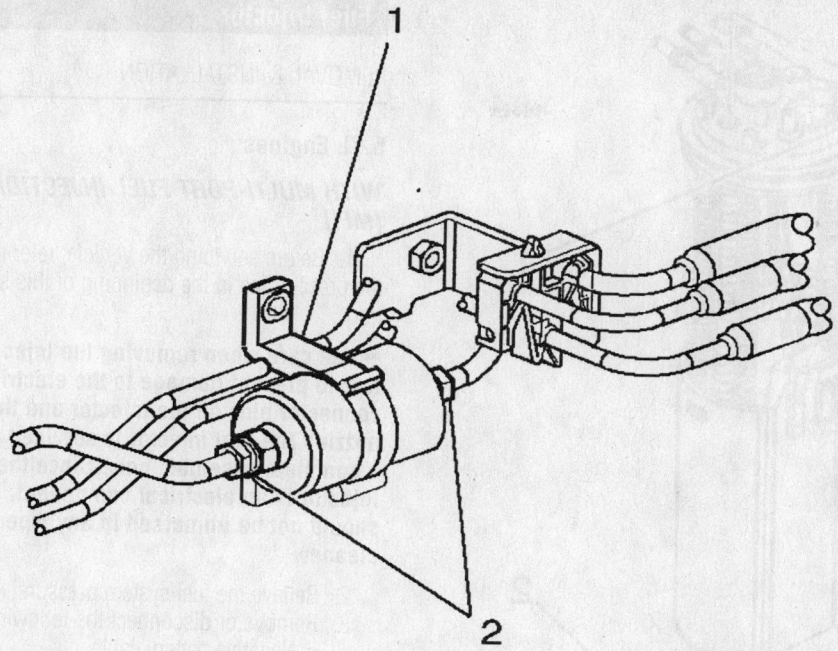

Typical in-line fuel filter mounting location

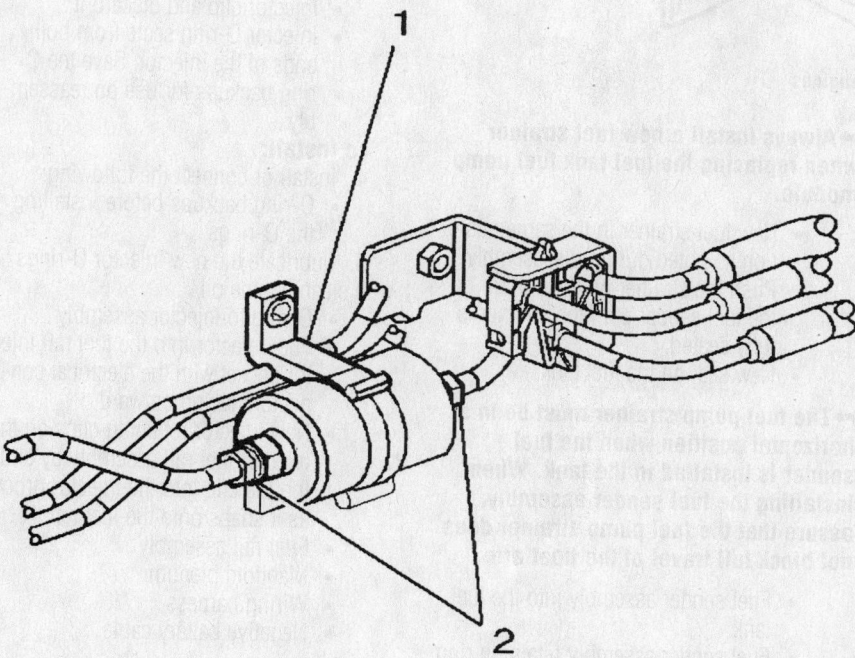

Fuel filter—4.8L, 5.3L and 6.0L engines

12. Lower the vehicle.
13. Tighten the fuel filler cap.
14. Connect the negative battery cable.
15. Turn the ignition **ON** for 2 seconds.
16. Turn the ignition **OFF** for 10 seconds.
17. Turn the ignition **ON**.
18. Inspect for fuel leaks.

Fuel Pump

REMOVAL & INSTALLATION

2000 Vehicles

1. Before servicing the vehicle, refer to the precautions in the beginning of this section.

2. Properly relieve the fuel system pressure.
3. Drain the fuel from the vehicle.
4. Remove or disconnect the following:
   • Negative battery cable
   • Fuel tank from the vehicle
   • Locking ring (located on top of the fuel tank) counterclockwise
   • Lift the fuel pump assembly out of the tank

To install:

5. Install or connect the following:
   • Fuel pump and secure with locking ring
   • Fuel tank to vehicle and refill
   • Negative battery cable
6. Run engine and check for fuel leaks.

2001–03 Vehicles

1. Before servicing the vehicle, refer to the precautions in the beginning of this section.
2. Remove or disconnect the following:
   • Negative battery cable
3. Relieve the fuel system pressure.
4. Drain the fuel tank.
5. Remove or disconnect the following:
   • Fuel tank

✳✳ WARNING

Do not handle the fuel sender assembly by the fuel pipes. The amount of leverage generated by handling the fuel pipes could damage the joints.

   • Fuel sender assembly retaining ring using a fuel tank sending unit wrench. Remove the fuel sender

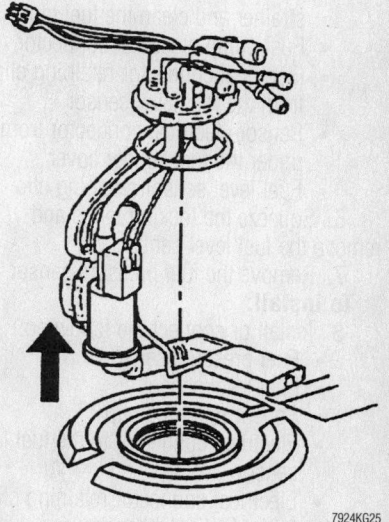

Lift the fuel pump assembly out of the tank after removing the locking ring

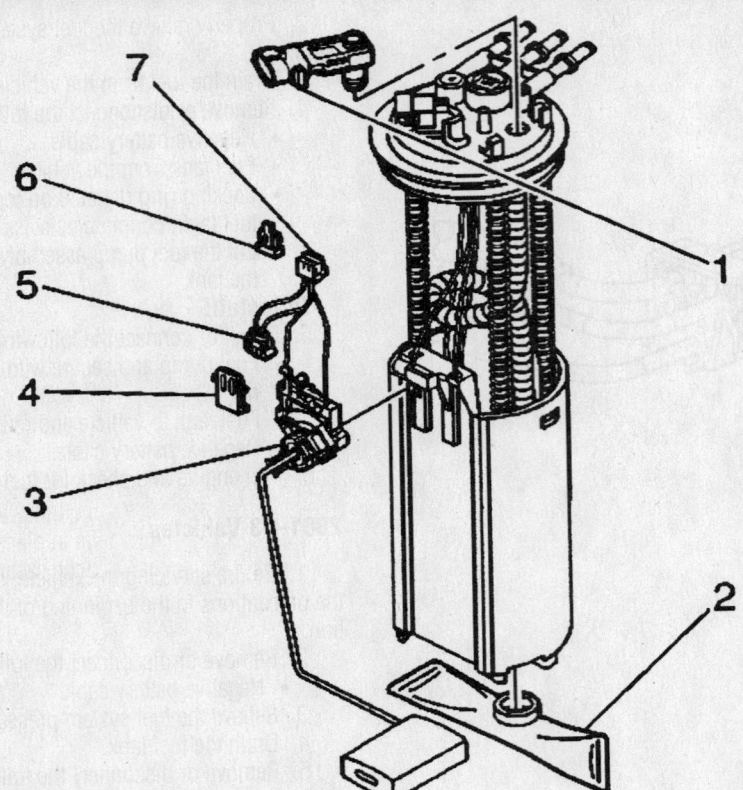

Fuel pump/sender assembly—4.8L, 5.3L and 6.0L engines

assembly and the seal. Discard the seal.
- Note the position of the fuel strainer on the fuel sender. Support the fuel sender assembly with one hand and grasp the strainer with the other hand. Pull the strainer off the fuel sender. Discard the strainer after inspection. Inspect the strainer. Replace a contaminated strainer and clean the fuel tank.
- Fuel pump electrical connector
- Electrical connector retaining clip from the fuel level sensor
- Sensor electrical connector from under the fuel sender cover
- Fuel level sensor retaining clip

6. Squeeze the locking tangs and remove the fuel level sensor.

7. Remove the fuel pressure sensor.

To install:

8. Install or connect the following:
- Fuel pressure sensor
- Fuel level sensor
- Sensor retaining clip
- Electrical connector to the fuel level sensor
- Electrical connector retaining clip to the fuel level sensor
- Fuel pump electrical connector

➡**Always install a new fuel strainer when replacing the fuel tank fuel pump module.**

- New fuel strainer in the same position as noted during disassembly. Push the strainer on the bottom of the fuel sender until the strainer is fully seated.
- New seal on the fuel tank

➡**The fuel pump strainer must be in a horizontal position when the fuel sender is installed in the tank. When installing the fuel sender assembly, assure that the fuel pump strainer does not block full travel of the float arm.**

- Fuel sender assembly into the fuel tank
- Fuel sender assembly retaining ring
- Fuel tank. Install the fuel tank strap attaching bolts. Tighten the bolts to 30 ft. lbs. (40 Nm).

9. Refill the fuel tank. Install the fuel filler cap. Connect the negative battery cable.

10. Turn the ignition **ON** for 2 seconds.

11. Turn the ignition **OFF** for 10 seconds.

12. Turn the ignition **ON**.

13. Inspect for fuel leaks.

Fuel Injector

REMOVAL & INSTALLATION

5.7L Engines

WITH MULTI-PORT FUEL INJECTION (MFI)

1. Before servicing the vehicle, refer to the precautions in the beginning of this section.

➡**Use care when removing the injectors to prevent damage to the electrical connector pins on the injector and the nozzle. The fuel injector is serviced as a complete assembly only. Since the injector is an electrical component, it should not be immersed in any type of cleaner.**

2. Relieve the fuel system pressure.
3. Remove or disconnect the following:
- Negative battery cable
- Intake manifold plenum
- Fuel rail assembly
- Wiring harness
- Injector clip and discard it
- Injector O-ring seals from both ends of the injector. Save the O-ring backups for use on reassembly.

To install:
4. Install or connect the following:
- O-ring backups before installing the O-rings

5. Lubricate the new injector O-rings with clean engine oil
- O-ring to injector assembly
- Fuel injector into the fuel rail injector socket with the electrical connectors facing outward
- New injector retaining clips on the injector fuel rail assembly by sliding the clip into the injector groove as it snaps onto the fuel rail
- Fuel rail assembly
- Manifold plenum
- Wiring harness
- Negative battery cable

WITH CENTRAL SEQUENTIAL FUEL INJECTION (CSFI)

1. Before servicing the vehicle, refer to the precautions in the beginning of this section.
2. Relieve the fuel system pressure.
3. Remove or disconnect the following:
- Negative battery cable
- Electrical connection
- Fuel feed and return hoses from the engine fuel pipes

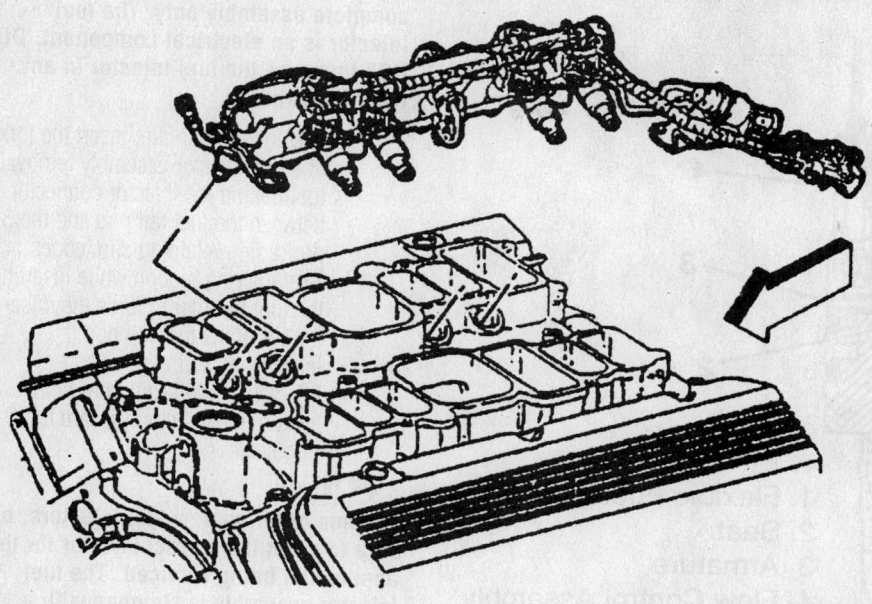

Exploded view of the fuel rail assembly—MFI systems

- Upper manifold assembly
- Poppet nozzle out of the casting socket
- Fuel meter body by releasing the locktabs

➡ **Each injector is calibrated. When replacing the fuel injectors, be sure to replace it with the correct injector.**

- Lower hold-down plate and nuts

4. While pulling the poppet nozzle tube downward, push with a small prytool down between the injector terminals and remove the injectors.

To install:

5. Lubricate the new injector O-ring seats with engine oil.

6. Install or connect the following:

- O-rings on the injector
- Fuel injector into the fuel meter body injector socket
- Lower hold-down plate and nuts, Torque the nuts to 27 inch lbs. (3 Nm).
- Fuel meter body assembly into the intake manifold. Torque the fuel meter bracket retainer bolts to 88 inch. lbs. (10 Nm).

✳✳ CAUTION

To reduce the risk of fire or injury ensure that the poppet nozzles are properly seated and locked in their casting sockets

- Fuel meter body into the bracket and lock all the tabs in place
- Poppet nozzles into the casting sockets
- Electrical connections
- New O-ring seals on the fuel return and feed hoses
- Fuel feed and return hoses. Torque the fuel pipe nuts to 22 ft. lbs. (30 Nm).
- Negative battery cable

7. Turn the ignition **ON** for 2 seconds and then turn it **OFF** for 10 seconds. Again turn the ignition **ON** and check for leaks.

- Manifold plenum

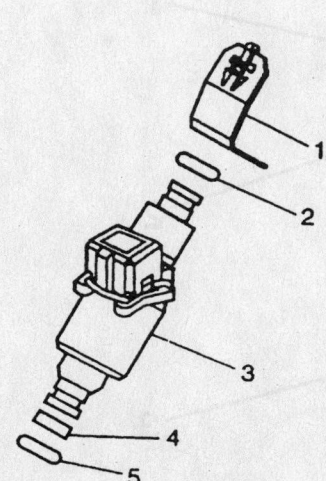

1. Clip - SFI Fuel Injector Retainer
2. O-ring - SFI Fuel Injector Upper
3. Injector Asm - SFI Fuel
4. O-ring - Backup
5. O-ring - SFI Fuel Injector Lower

Exploded view of the fuel injector assembly—MFI systems

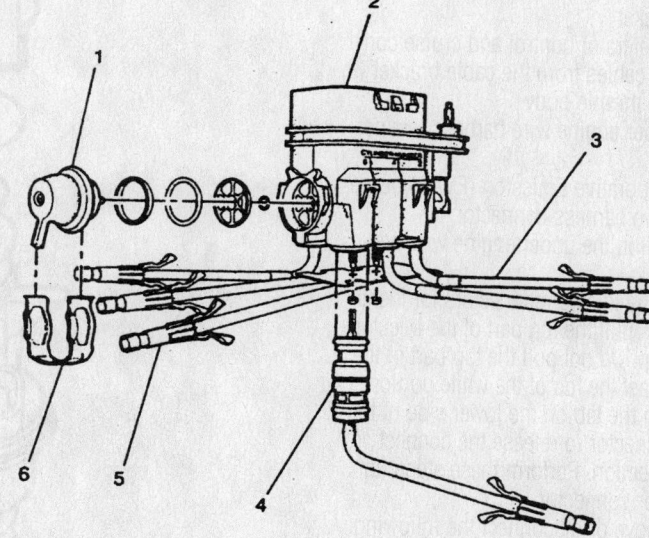

1. Regulator Assembly
2. Fuel Meter Body
3. Flexible Fuel Line
4. Injector Assembly
5. Poppet Nozzle
6. Regulator Retainer

Exploded view of the CSFI fuel meter body assembly

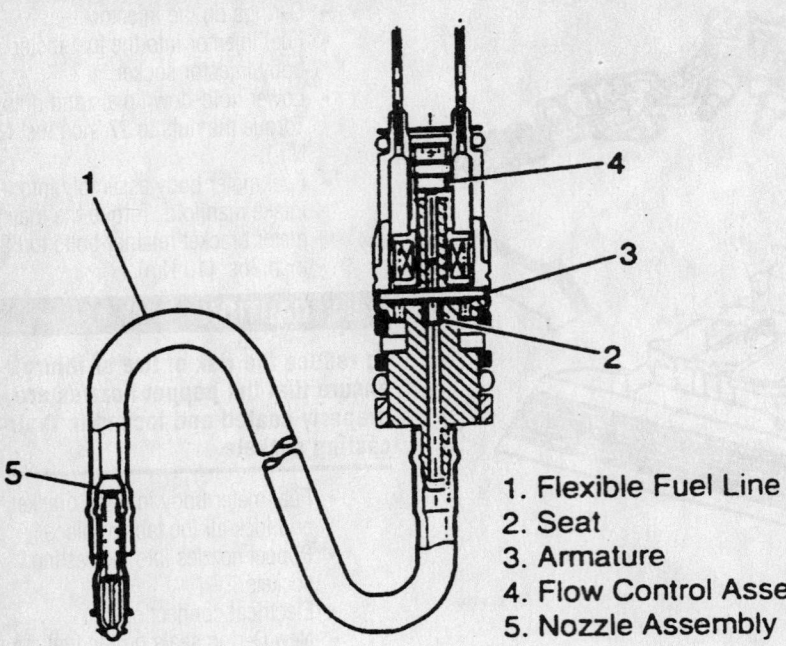

1. Flexible Fuel Line
2. Seat
3. Armature
4. Flow Control Assembly
5. Nozzle Assembly

9308KG08

Exploded view of the fuel injector assembly–CSFI systems

4.8L, 5.3L and 6.0L Engines

1. Before servicing the vehicle, refer to the precautions in the beginning of this section.

2. Relieve the fuel system pressure.

3. Remove or disconnect the following:
- Negative battery cable
- Engine sight shield bolts and bracket
- Accelerator control and cruise control cables from the cable bracket and throttle body
- Upper engine wire harness retainer nut
- Evaporative Emission (EVAP) purge valve harness connector

4. Position the upper engine wire harness aside

5. Tag the injector connectors for identification, then pull the top part of the injector connector up. Do not pull the top part of the connector past the top of the white portion.

6. Push the tab on the lower side of the injector connector to release the connect from the injection. Perform these steps on each injector connector.

7. Remove or disconnect the following:
- Fuel feed and return pipes from the fuel rail
- Fuel pressure regulator vacuum line
- Crossover tube-to-right fuel rail retainer screw
- Fuel rail attaching bolts and fuel rail

➡Use care in removing the fuel injectors in order to prevent damage to the electrical connector pins on the injector and to prevent damage to the nozzle. Service the fuel injector as a complete assembly only. The fuel injector is an electrical component. DO NOT immerse the fuel injector in any type of cleaner.

- Injector retainer clip. Insert the fork of a fuel injector assembly removal tool behind the injector connector between the fuel rail pod and the 3 protruding retaining clip ledges. Use a prying motion while inserting the tool in order to force the injector out of the fuel rail pod.
- Injector retainer clip
- Injector O-ring seals from both ends of the injector. Discard the O-ring seals.

To install:

➡When ordering new fuel injectors, be sure to order the correct injector for the application being serviced. The fuel injector assembly is stamped with a part number identification.

8. Lubricate the new injector O-ring seals with clean engine oil.

9. Install or connect the following:
- New injector O-ring seals on the injector
- New retainer clip on the injector

10. Push the fuel injector into the fuel rail injector socket with the electrical con-

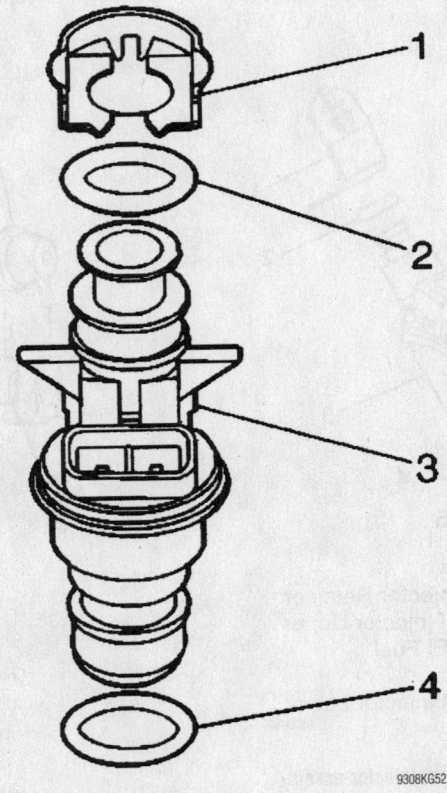

Fuel injector—4.8L, 5.3L and 6.0L engines

9308KG52

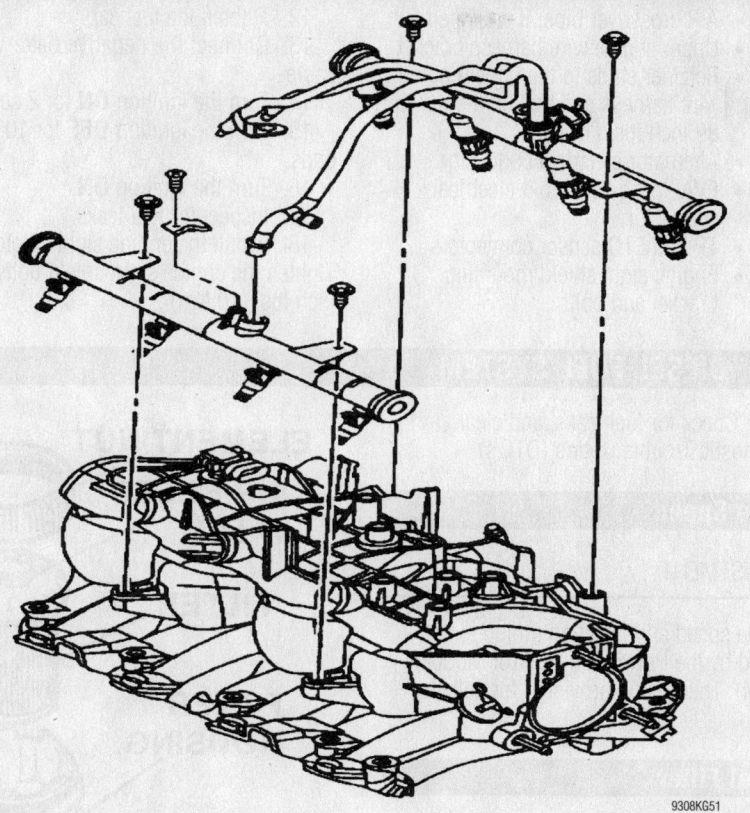

Fuel rail assembly—4.8L, 5.3L and 6.0L engines

nector facing outward. The retainer clip locks on to a flange on the fuel rail injector socket.

11. Remove the crossover tube-to-right fuel rail retainer, then remove the crossover tube.

12. Replace the crossover tube O-ring with a new, lubricated one.

13. Install or connect the following:
- Crossover tube and loosely install the retainer
- Fuel rail to the intake manifold
- Apply a 0.020 (5mm) band of threadlock to the fuel rail retaining bolts, then install and tighten to 89 inch lbs. (10 Nm). Tighten the crossover pipe retainer to 34 inch lbs. (3.8 Nm).
- Fuel pressure regulator vacuum line
- Fuel feed and return pipes
- Fuel injector electrical connectors, as tagged during removal. Rotate the injectors as necessary to avoid stretching the wire harness.
- Upper engine wire harness
- EVAP purge solenoid electrical connector
- Upper engine wire harness retainer nut and tighten to 49 inch lbs. (5.5 Nm)

- Accelerator control and cruise control cables to the bracket and throttle body
- Engine sight shield mounting bracket and bolts

14. Tighten the fuel cap.

15. Connect the negative battery cable.

16. Turn the ignition **ON** for 2 seconds.

17. Turn the ignition **OFF** for 10 seconds.

18. Turn the ignition **ON**.

19. Inspect for fuel leaks.

20. Install the engine sight shield. Tighten the engine sight shield bolts to 89 inch lbs. (10 Nm)

8.1L Engine

1. Before servicing the vehicle, refer to the precautions in the beginning of this section.

2. Relieve the fuel system pressure.

3. Remove or disconnect the following:
- Negative battery cable
- Engine sight shield nuts and bracket
- Alternator harness connector
- Evaporative Emission (EVAP) purge valve harness connector
- Throttle Position (TP) sensor electrical connector

- Electronic Throttle Control (ETC) electrical connector

4. Upper engine wire harness bracket studs, and position the harness aside

5. Secondary Air Injection (AIR) crossover pipe, if equipped.

6. Tag the injector connectors for identification, then pull the top part of the injector connector up. Do not pull the top part of the connector past the top of the white portion.

7. Push the tab on the lower side of the injector connector to release the connect from the injection. Perform these steps on each injector connector.

8. Remove or disconnect the following:
- Fuel feed and return pipes from the fuel rail
- Fuel pressure regulator vacuum line
- Fuel rail attaching bolts and fuel rail

➡**Use care in removing the fuel injectors in order to prevent damage to the electrical connector pins on the injector and to prevent damage to the nozzle. Service the fuel injector as a complete assembly only. The fuel injector is an electrical component. DO NOT immerse the fuel injector in any type of cleaner.**

- Injector retainer clip. Insert the fork of a fuel injector assembly removal tool behind the injector connector between the fuel rail pod and the 3 protruding retaining clip ledges. Use a prying motion while inserting the tool in order to force the injector out of the fuel rail pod.
- Injector retainer clip
- Injector from the fuel rail pod
- Injector O-ring seals from both ends of the injector. Discard the O-ring seals.

To install:

➡**When ordering new fuel injectors, be sure to order the correct injector for the application being serviced. The fuel injector assembly is stamped with a part number identification.**

9. Lubricate the new injector O-ring seals with clean engine oil.

10. Install or connect the following:
- New injector O-ring seals on the injector
- New retainer clip on the injector

11. Push the fuel injector into the fuel rail injector socket with the electrical connector facing outward. The retainer clip locks on to a flange on the fuel rail injector socket.

- Fuel rail to the intake manifold

- Apply a 0.020 (5mm) band of threadlock to the fuel rail retaining bolts, then install and tighten to 106 inch lbs. (12 Nm)
- Fuel pressure regulator vacuum line
- Fuel feed and return pipes
- Fuel injector electrical connectors, as tagged during removal. Rotate the injectors as necessary to avoid stretching the wire harness.

- AIR crossover pipe, if equipped
- Upper engine wire harness bracket
- Retainer studs to the upper engine wire harness and tighten the nut to 89 inch lbs. (10 Nm)
- Alternator electrical connector
- EVAP purge solenoid electrical connector
- TP and ETC sensor connectors
- Engine sight shield mounting bracket and bolts

12. Tighten the fuel cap.
13. Connect the negative battery cable.
14. Turn the ignition **ON** for 2 seconds.
15. Turn the ignition **OFF** for 10 seconds.
16. Turn the ignition **ON**.
17. Inspect for fuel leaks.
18. Install the engine sight shield. Tighten the engine sight shield bolts to 89 inch lbs. (10 Nm)

DIESEL FUEL SYSTEM

Fuel System Pressure

RELIEVING

Fuel system pressure can be released by wrapping a fuel fitting in a heavy shop towel and slightly loosening the fitting. NEVER perform this with any source of ignition nearby!

Fuel System Air

BLEEDING

1. Before servicing the vehicle, refer to the precautions in the beginning of this section.
2. Open the air bleed valve on the fuel manager/filter.
3. Connect a hose to the air bleed valve and place the other of the hose in a suitable container.

✳✳ CAUTION

The diesel/water mixture is flammable and may be hot. To avoid personal injury or property damage, do not allow the diesel/water mixture to come in contact with skin, open flame or a hot engine. Do not overfill the container holding the fuel mixture as heat from a warm engine or any another heat source may cause the fuel to expand and leak from the container that may lead to a fire.

4. Remove the F/SOL fuse from the fuse panel.
5. Crank the engine in short intervals of 10-to-15 seconds until clear fuel is observed at the air bleed hose (wait for 1 minute between cranking intervals)
6. Remove the hose and close the air bleed valve.
7. Install the F/SOL fuse and start the vehicle. Allow the vehicle to run at idle for 5 minutes.

8. Check for fuel leaks, and clear any Diagnostic Trouble Code's (DTC's)

Idle Speed

ADJUSTMENT

Idle speed and injection timing is controlled by the Powertrain Control Module (PCM). There is no provision for adjustment.

Fuel Filter

REMOVAL & INSTALLATION

6.5L Engine

1. Before servicing the vehicle, refer to the precautions in the beginning of this section.
2. Turn the ignition **OFF**. Remove the fuel tank cap to release any pressure or vacuum in the tank.

➡It is not necessary to drain all the fuel from the header in order to change the element since the fuel will remain in the header's cavity.

3. Open the air bleed valve to relieve residual pressure.
4. Remove or disconnect the following:
   - Element nut, turning it by hand to the left. If necessary, a strap wrench may be used to loosen the nut.
   - Element by lifting straight up and out of the header assembly
   To install:
5. Be sure the mating surface between the element assembly and the header assembly is clean.
6. Install or connect the following:
   - New element by aligning the widest key slot located under the element assembly cap with the widest key in the header assembly.

ELEMENT NUT
AIR BLEED VALVE
FILTER
FILTER HOUSING

7924KG26

Exploded view of the fuel filter assembly—6.5L diesel engines

7. Carefully push the element downward until the mating surfaces make contact.
   - Element nut and tighten securely by hand
8. If not already done, open the air bleed valve on top of the fuel manager/filter assembly, then connect a length of hose placing the other end in a suitable container.

➡Be extremely cautious when handling diesel fuel. Do not expose the fuel to sparks or open flames. Also, be cautious as the fuel coming out of the drain hose could be hot.

9. Disconnect the fuel injection pump shutdown solenoid wire.
10. Crank the engine for 10–15 seconds, then wait 1 minute for the starter motor to

cool. Repeat until clear fuel is observed coming from the air bleed.

11. Close the air bleed valve, reconnect the injection pump solenoid wire and replace the fuel tank cap.

12. Start the engine, allow it to idle for 5 minutes and check the fuel manager/filter assembly for leaks.

Diesel Injection Pump

All vehicles are equipped with an electronically controlled pump. The electronic pump is driven by gears and rotates at the same speed as the camshaft. An electronic stepper motor used to control injection timing and a fuel solenoid driver used to control the fuel injection solenoid on the electronic model.

REMOVAL & INSTALLATION

6.5L Engine

1. Before servicing the vehicle, refer to the precautions in the beginning of this section.
2. Relieve the fuel system pressure.
3. Remove or disconnect the following:
   • Negative battery cables
   • Intake manifold
   • Fuel injection and inlet lines
   • Cables wires, and hoses at the injection pump
   • Fuel return line at the top of the injection pump
   • Fuel feed line at the injection pump, if necessary
   • Oil filler tube grommet

➡ Do not engage the starter in order to rotate the engine with the injection pump removed. The pump driven gear could jam in the front housing resulting in a sheared crankshaft or camshaft gear key and possible valvetrain damage.

4. Scribe or paint a matchmark on the front cover and the injection pump flange.
5. Rotate the crankshaft by hand and remove the injection pump driven gear bolts, accessing the bolts through the oil filler neck hole.
6. Remove the injection pump-to-front cover attaching nuts. Remove the pump. Be sure to cap all open lines and nozzles in order to prevent system contamination and damage.

To install:

7. Align the locating pin on the pump hub with the slot in the injection pump driven gear (the SLOT not the hole in the gear) At the same time, align the timing marks.
8. Attach the injection pump to the front cover. Torque the nuts to 30 ft. lbs. (40 Nm) checking the timing marks before tightening.
9. Install or connect the following:
   • Driven gear-to-injection pump bolts. Torque the bolts to 20 ft. lbs. (25 Nm).
   • Grommet and oil fill tube
   • Air conditioning bracket. If applicable
   • Fuel feed line. Torque to 20 ft. lbs. (25 Nm).
   • Fuel return line to the pump, if removed
   • Cables, wires and hoses previously removed
   • Injector lines
   • Intake manifold
   • Negative battery cables
10. Start the engine and check for leaks.

Fuel Injectors

REMOVAL & INSTALLATION

6.5L Engine

1. Before servicing the vehicle, refer to the precautions in the beginning of this section.

➡ Special tool J–29873, or its equivalent, an injection nozzle socket, will be necessary for this procedure.

2. Remove or disconnect the following:
   • Batteries.
   • Fuel line clip

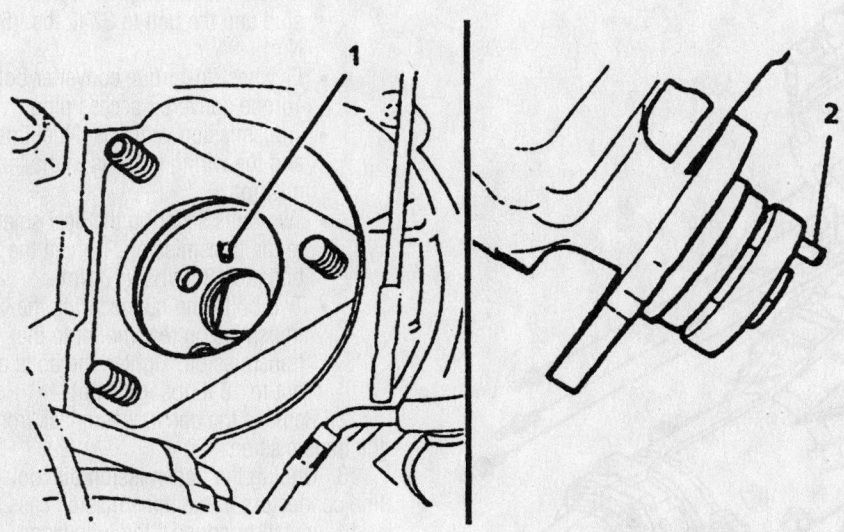

1   SLOT IN DRIVEN GEAR

2   PUMP HUB

7924KG27

Align the pin on the pump hub with the slot in the driven gear, NOT into the hole in the gear—diesel engines

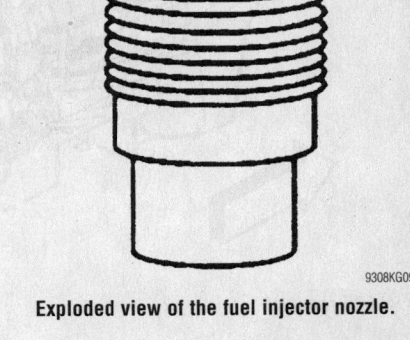

9308KG09

Exploded view of the fuel injector nozzle.

- Fuel return hose
- Fuel injection lines

3. Using GM special tool J–29873, remove the injector. Always remove the injector by turning the 30mm hex portion of the injector; turning the round portion will damage the injector. Always cap the injector

and fuel lines when disconnected, to prevent contamination.

To install:

4. Always install the injector by turning the 30mm hex portion of the injector; turning the round portion will damage the injector.

5. Install or connect the following:

- Injector with a new gasket. Torque to 50 ft. lbs. (70 Nm).
- Injection line. Torque the nut to 20 ft. lbs. (25 Nm).
- Fuel return hose
- Fuel line clips
- Batteries

DRIVE TRAIN

Transmission Assembly

REMOVAL & INSTALLATION

Automatic Transmission

4L60E

1. Before servicing the vehicle, refer to the precautions in the beginning of this section.

2. Remove or disconnect the following:
- Transmission fluid
- Transmission oil level indicator tube and seal from the transmission
- Plug the oil level indicator tube opening in the transmission.

3. Remove or disconnect the following:
- Shift cable end from the transmission shift lever ball stud
- Front propeller shaft, if equipped with a transfer case

- Rear propeller shaft.

4. Plug the transmission oil cooler line connectors in the transmission case.
- Starter motor

5. Support the transmission with a transmission jack.
- Torque converter access plug
- Flywheel-to-torque converter bolts
- Two bolts and nut securing the transmission rear mount to the transmission
- Two bolts securing the heat shield to the transmission
- Transmission vent hose, fuel lines, and the wiring harness from the transmission
- Stud and the bolt securing the transmission to the engine
- Six studs and bolt securing the transmission to the engine. Install tool J21366 onto the transmission bell housing to retain the torque

converter. Pull the transmission straight back.
- Transmission from the vehicle

6. Flush the transmission oil cooler and cooling lines when you remove the transmission.

To install:

7. Install Tool J21366 onto the transmission bell housing to retain the torque converter.

8. Support the transmission with a transmission jack.

9. Raise the transmission into place and remove the tool from the transmission.

10. Slide the transmission straight onto the locating pins while lining up the marks on the flywheel and the torque converter. The torque converter must be flush onto the flywheel and rotate freely by hand.

11. Install or connect the following:
- Six studs and one bolt securing the transmission to the engine. Tighten the studs and the bolt to 37 ft. lbs. (50 Nm).
- Stud and bolt securing the transmission to the engine. Tighten the stud and the bolt to 37 ft. lbs. (50 Nm).
- Flywheel-to-torque converter bolts
- Torque converter access plug
- Transmission vent hose, fuel lines, and the wiring harness to the transmission
- Two bolts securing the heat shield to the transmission. Tighten the bolts to 13 ft. lbs. (17 Nm).
- Two bolts and nut securing the transmission rear mount to the transmission. Tighten the bolts and nut to 18 ft. lbs. (25 Nm).

12. Remove the transmission jack from the transmission.

13. Unplug the transmission oil cooler line connectors in the transmission case.

14. Install or connect the following:
- Transmission oil cooler lines to the transmission
- Front propeller shaft, if equipped with a transfer case
- Rear propeller shaft
- Shift cable end to the transmission shift lever ball stud

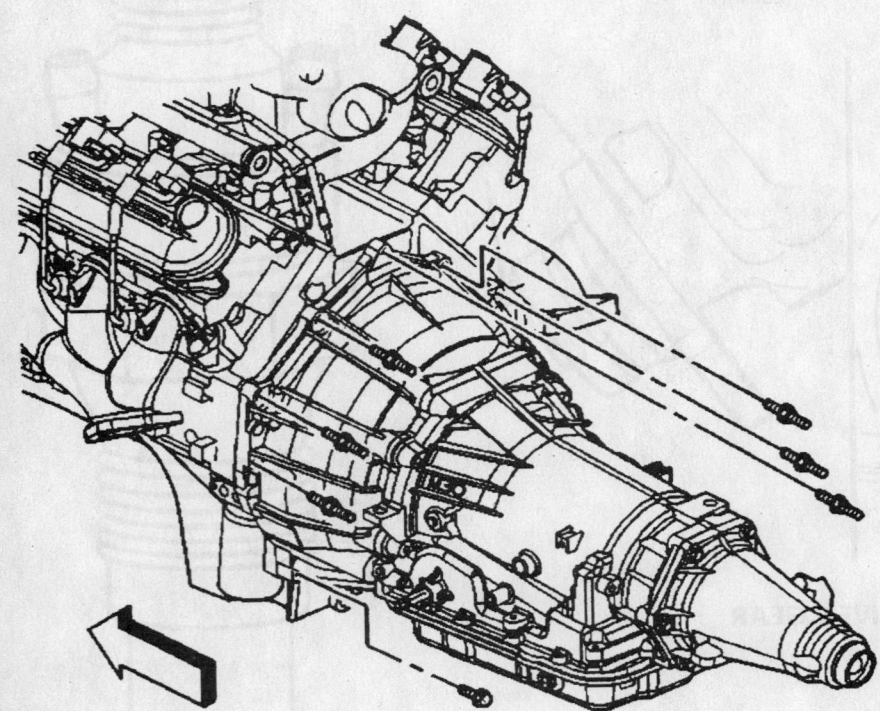

9308KG37

4L60E removal

15. Unplug the oil level indicator tube opening in the transmission.

16. Install the transmission oil level indicator tube and seal to the transmission.

17. Tighten the oil pan bolts and fill the transmission with transmission fluid.

18. Lower the vehicle.

4L80E

1. Before servicing the vehicle, refer to the precautions in the beginning of this section.

2. Remove or disconnect the following:
   • Transmission fluid
   • Transmission oil level indicator tube and seal from the transmission
3. Plug the oil level indicator tube opening in the transmission.

4. Remove or disconnect the following:
   • Shift cable from the transmission shift lever ball stud
   • Propeller shaft(s), as necessary
   • Transmission oil cooler lines from the transmission
5. Plug the transmission oil cooler line connectors in the transmission case.
   • Starter motor
6. Support the transmission with a transmission jack.
   • Two bolts securing the heat shield to the transmission
   • Transmission vent hose, fuel lines, and the wiring harness from the transmission
   • Nut and bolt securing the transmission brace to the engine bracket and transmission
   • Two bolts securing the torque converter cover to the engine
   • Four bolts securing the torque converter cover to the transmission
   • Six flywheel-to-torque converter bolts
   • Two bolts and nut securing the transmission rear mount to the transmission
   • Stud and the bolt on the right side securing the transmission to the engine
   • Remaining six studs and the one bolt securing the transmission to the engine
7. Install Tool J21366 onto the transmission bell housing to retain the torque converter

8. Pull the transmission straight back. Remove the transmission from the vehicle.

9. Flush the transmission oil cooler and cooling lines when you remove the transmission.

To install:

10. Install Tool J21366 onto the transmission bell housing to retain the torque converter

11. Support the transmission with a transmission jack.

12. Raise the transmission into place and remove the tool from the transmission.

13. Slide the transmission straight onto the locating pins while lining up the marks on the flywheel and the torque converter. The torque converter must be flush onto the flywheel and rotate freely by hand.

14. Install or connect the following:
   • Six studs and one bolt securing the transmission to the engine. Tighten the studs and the bolt to 37 ft. lbs. (50 Nm).
   • Stud and bolt on the right side securing the transmission to the engine. Tighten the stud and the bolt to 37 ft. lbs. (50 Nm).
   • Six flywheel-to-torque converter bolts. Tighten the bolts to 44 ft. lbs. (60 Nm).
   • Two bolts securing the torque converter cover to the engine. Tighten the bolt to 37 ft. lbs. (50 Nm).
   • Four bolts securing the torque converter cover to the transmission.

Tighten the stud and the bolt to 24 ft. lbs. (33 Nm).
   • Transmission vent hose, fuel lines, and the wiring harness to the transmission
   • Two bolts securing the heat shield to the transmission. Tighten the bolt to 13 ft. lbs. (17 Nm).
   • Two bolts and nut securing the transmission rear mount to the transmission. Tighten the bolts and nut to 18 ft. lbs. (25 Nm).
   • Flywheel-to-torque converter bolts
   • Nut and bolt securing the transmission brace to the engine bracket and transmission. Tighten the bolts and nut to 37 ft. lbs. (50 Nm).

15. Remove the transmission jack from the transmission.
   • Starter motor
16. Unplug the transmission oil cooler line connectors in the transmission case.

17. Connect the transmission oil cooler lines to the transmission.

18. Install or connect the following:
   • Transfer case
   • Rear propeller shaft
   • Shift cable end to the transmission shift lever ball stud

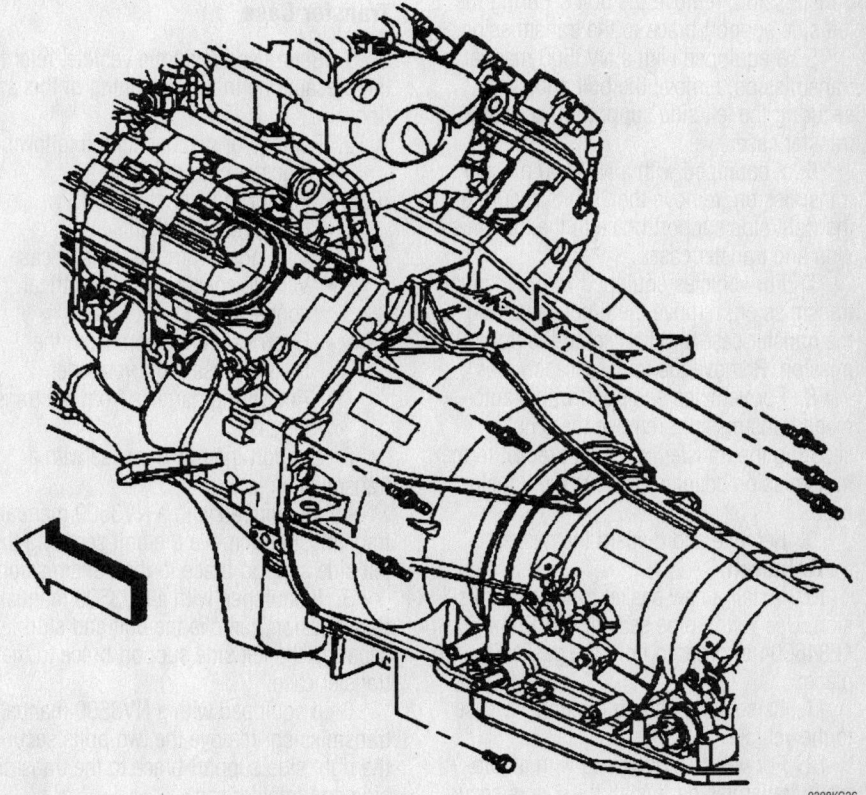

4L80E removal

9308KG36

19. Unplug the oil level indicator tube opening in the transmission.

20. Install the transmission oil level indicator tube and seal to the transmission.

21. Tighten the oil pan bolts and fill the transmission with transmission fluid.

22. Lower the vehicle.

Transfer Case Assembly

REMOVAL & INSTALLATION

NVG261-NP2 2-Speed Manual Transfer Case

1. Before servicing the vehicle, refer to the precautions in the beginning of this section.

2. Remove or disconnect the following:
   • Transfer case shields
   • Front propeller shaft
   • Rear propeller shaft
   • Shift rod from the transfer case
   • Vent hose from the transfer case
   • Vehicle speed sensor electrical connectors
   • Any wiring harness from the transfer case

3. Support the transfer case with a transmission jack.

4. If equipped with a NV3500 manual transmission, remove the bolt securing the left side support brace to the transmission.

5. If equipped with a NV3500 manual transmission, remove the bolt and stud securing the left side support brace to the transfer case.

6. If equipped with a NV3500 manual transmission, remove the 2 bolts securing the right side support brace to the transmission and transfer case.

7. For vehicles equipped with a manual transmission, remove the 6 nuts securing the transfer case and bracket to the transmission. Remove the transfer case.

8. For vehicles equipped with a automatic transmission, remove the 6 nuts securing the transfer case and bracket to the transmission adapter. Remove the transfer case.

9. Remove and discard the gasket.

To install:

10. Install a new gasket to the transmission. Use Teflon pipe sealant GM P/N 12346004 in order to hold the gasket in place.

11. Raise and position the transfer case to the vehicle

12. For vehicles equipped with a automatic transmission, install the 6 nuts securing the transfer case and bracket to the

transmission adapter. Tighten the nuts to 37 ft. lbs. (50 Nm).

13. For vehicles equipped with a manual transmission, install the 6 nuts securing the transfer case and bracket to the transmission. Tighten the nuts to 37 ft. lbs. (50 Nm).

14. If equipped with a NVG 261manual transmission, install the bolt securing the left side support brace to the transmission. Tighten the bolt to 37 ft. lbs. (50 Nm).

15. If equipped with a NVG 261 manual transmission, Install the bolt and stud securing the left side support brace to the transfer case. Tighten the bolt and stud to 37 ft. lbs. (50 Nm).

16. If equipped with a NVG 261manual transmission, install the 2 bolts securing the right side support brace to the transmission and transfer case. Tighten the bolts to 37 ft. lbs. (50 Nm).

17. Install or connect the following:
   • Vent hose to the transfer case

18. Check the transfer case oil level.
   • Speed sensor electrical connectors
   • Wiring harness to the transfer case
   • Shift rod to the transfer case
   • Rear propeller shaft
   • Front propeller shaft
   • Transfer case shields

19. Lower the vehicle.

NVG246-NP8 2-Speed Automatic Transfer Case

1. Before servicing the vehicle, refer to the precautions in the beginning of this section.

2. Remove or disconnect the following:
   • Transfer case shields
   • Front propeller shaft
   • Rear propeller shaft
   • Vent hose from the transfer case
   • Vehicle speed sensor electrical connectors
   • Electrical connectors from the transfer case motor/encoder
   • Any wiring harness from the transfer case

3. Support the transfer case with a transmission jack.

4. If equipped with a NV3500 manual transmission, remove the bolt securing the left side support brace to the transmission.

5. If equipped with a NV3500 manual transmission, remove the bolt and stud securing the left side support brace to the transfer case.

6. If equipped with a NV3500 manual transmission, remove the two bolts securing the right side support brace to the transmission and transfer case.

7. For vehicles equipped with a manual

transmission, remove the six nuts securing the transfer case and bracket to the transmission. Remove the transfer case.

8. For vehicles equipped with a automatic transmission, remove the six nuts securing the transfer case and bracket to the transmission adapter. Remove the transfer case.

9. Remove and discard the gasket.

To install:

10. Install a new gasket to the transmission. Use Teflon Pipe Sealant GM P/N 12346004 in order to hold the gasket in place.

11. Raise and position the transfer case to the vehicle

12. For vehicles equipped with a automatic transmission, install the six nuts securing the transfer case and bracket to the transmission adapter. Tighten the nuts to 37 ft. lbs. (50 Nm).

13. For vehicles equipped with a manual transmission, install the six nuts securing the transfer case and bracket to the transmission. Tighten the nuts to 37 ft. lbs. (50 Nm).

14. If equipped with a NV3500 manual transmission, install the bolt securing the left side support brace to the transmission. Tighten the bolt to 37 ft. lbs. (50 Nm).

15. If equipped with a NV3500 manual transmission, Install the bolt and stud securing the left side support brace to the transfer case. Tighten the bolt and stud to 37 ft. lbs. (50 Nm).

16. If equipped with a NV3500 manual transmission, install the two bolts securing the right side support brace to the transmission and transfer case. Tighten the bolts to 37 ft. lbs. (50 Nm).

17. Install or connect the following:
   • Vent hose to the transfer case
   • Oil
   • Speed sensor electrical connectors
   • Electrical connectors to the transfer case motor/encoder
   • Any wiring harness to the transfer case
   • Rear propeller shaft
   • Front propeller shaft
   • Transfer case shields

Halfshaft

REMOVAL & INSTALLATION

1. Before servicing the vehicle, refer to the precautions in the beginning of this section.

2. Remove or disconnect the following:
   • Wheels

3. Insert a drift or a large screwdriver through the brake caliper into one of the brake rotor vanes in order to prevent the drive axle wheel drive shaft from turning.
- Nut and the washer from the hub. Do not reuse the nut. A new nut must be used when installing the wheel drive shaft.
- The 6 bolts securing the wheel drive shaft inboard flange to the output shaft flange
- The drift from the rotor
- The stabilizer shaft link from the lower control arm

4. Wrap shop towels around both the inner and the outer wheel drive shaft boots in order to avoid damage to the boots during removal and installation.

5. Pull the wheel drive shaft through the lower control arm opening.

To install:

6. Wrap shop towels around both the inner and the outer wheel drive shaft boots in order to avoid damage to the boots during removal and installation.

➡**Clean the steering knuckle and the wheel drive shaft splines and threads. These areas must be dry and free of grease, dirt, and contamination.**

7. Insert the wheel drive shaft splined shank into the knuckle hub.

➡**Use only a genuine GM front wheel drive shaft nut. Installation of anything but an OEM front wheel drive shaft nut could cause damage to the vehicle.**

8. Install or connect the following:

9. Install the washer and new hub nut to the wheel drive shaft. Do not tighten.

10. Install the wheel drive shaft inboard flange to the output shaft flange using the inboard flange bolts

11. Insert a drift or a large screwdriver through the brake caliper into 1 of the brake rotor vanes in order to prevent the wheel drive shaft from turning. Tighten the inboard flange bolts to 58 ft. lbs. (78 Nm). Tighten the hub nut to 155 ft. lbs. (210 Nm).

12. Remove the drift from the rotor.

13. Install the stabilizer shaft link.

14. Install the wheel and tire assembly.

CV-Joints

OVERHAUL

Inner Joint

➡**With removal of the halfshaft for any reason, the transmission sealing surface (the tripot male/female shank of the halfshaft) should be inspected for corrosion. If corrosion is evident, the surface should be cleaned with 320 grit cloth or equivalent. Transmission fluid may be used to clean off any remaining debris. The surface should be wiped dry and the halfshaft reinstalled free of any buildup.**

1. Before servicing the vehicle, refer to the precautions in the beginning of this section.

2. Use a hand grinder in order to cut through the swage ring.

3. Remove the tripot housing from the halfshaft. Wipe the grease off of the tripot assembly roller bearings and the tripot housing. Thoroughly degrease the tripot housing. Allow the tripot housing to dry prior to assembly.

➡**Handle the tripot spider assembly with care. Tripot balls and needle rollers may separate from the spider trunnion if the tripot balls and needle rollers are not handled carefully.**

4. Before servicing the vehicle, refer to the precautions in the beginning of this section.

5. Use side cutters to cut away the small boot clamp.

6. Compress the tripot boot up the halfshaft away from the tripot spider assembly toward the outboard (CV joint assembly) end of the halfshaft.

7. Spread the spider spacer ring with tool J8059, or equivalent.

8. Remove the following items from the halfshaft bar:
   a. The spacer ring.
   b. The spider assembly.
   c. The tripot boot.

9. Clean the halfshaft bar. Use a wire brush in order to remove any rust in the boot mounting area (grooves).

10. Inspect the needle rollers, needle bearings, and trunnion. Check the tripot housing for unusual wear, cracks, or other damage. Replace any damaged parts.

To assemble:

11. Place the new small boot clamp onto the small end of the joint boot.

12. Compress the joint boot and small boot clamp onto the halfshaft bar.

13. Position the small end of the joint boot into the joint boot groove on the halfshaft bar.

14. Secure the small boot clamp with tool J35910, or equivalent, a breaker bar, and a torque wrench. Tighten the small boot clamp (1) to 100 ft. lbs. (136 Nm).

15. Check the gap dimension on the clamp ear. Continue tightening until the gap dimension is reached.

➡**Assemble the CV joint with the convolute retainer in the correct position, as illustrated.**

16. Install the convolute retainer over the inboard joint boot, being sure to capture three convolutions.

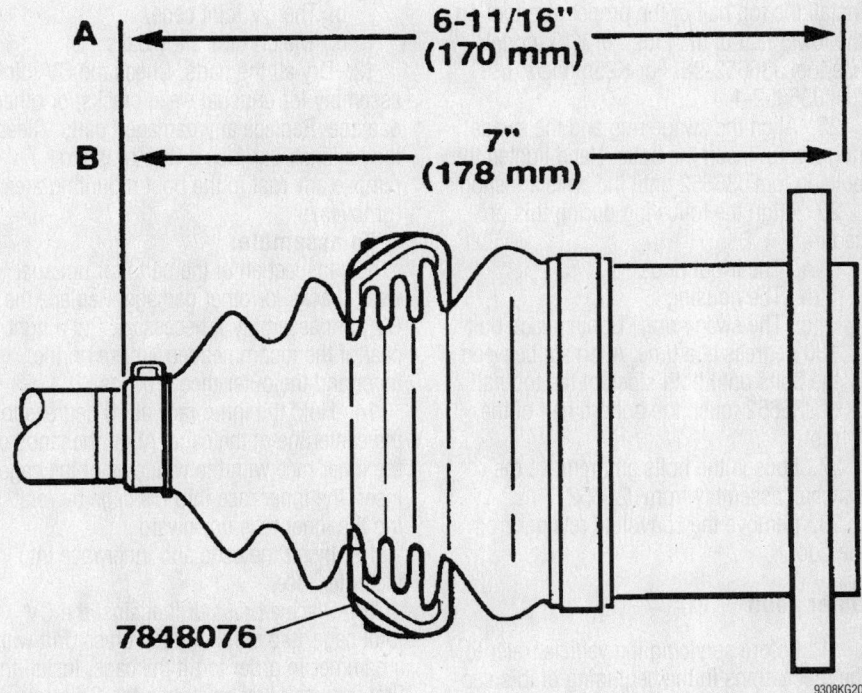

A 6-11/16" (170 mm)
B 7" (178 mm)
7848076

Assembled joint measurement

9308KG20

17. Install the tripot spider assembly onto the halfshaft bar with the counterbore towards the end of the halfshaft bar.

18. Install the spacer ring in the groove at the end of the halfshaft bar.

19. Push the spider assembly back toward the end of the halfshaft bar until the spacer ring is covered by the spider assembly counterbore.

20. Pack the tripot boot and the tripot housing with the grease supplied in the kit. The amount of grease supplied in this kit has been pre-measured for this application.

21. Reassemble the tripot housing and the tripot boot using the following procedure:

    a. Pinch the swage ring slightly by hand in order to distort it into an oval shape.

    b. Slide the distorted swage ring over the large diameter of the boot.

    c. Place the tripot housing over the spider assembly.

    d. Install the boot onto the tripot housing.

    e. Align the tripot boot with the swage ring in place, over the flat area on the tripot housing.

22. Mount tool J36652 in a vise. Install the bottom half of the split-plate swage clamp. For K15 models, use tool J36652-98. For K25 models, use tool J36652-1.

23. Check the inboard stroke position. Use measurement A for the K15 models. Use measurement B for the K25 models.

24. Position the inboard end (tripot end) of the halfshaft assembly in tool J36652. Install the top half of the proper size tool on the lower half of the tool. For K15 models, use tool J36652-98. For K25models, use tool J36652-1.

25. Align the swage ring and the swage ring clamp. Insert the bolts. Hand tighten the bolts in tool J36652 until the bolts are snug.

26. Align the following during this procedure:

    a. The tripot boot.

    b. The housing.

    c. The swage ring. Tighten each bolt 180 degrees at a time. Alternate between the bolts until both sides of the top half of J36652 touch the bottom half of the tool.

27. Loosen the bolts and remove the halfshaft assembly from J36652.

28. Remove the convolute retainer from the boot.

Outer Joint

1. Before servicing the vehicle, refer to the precautions in the beginning of this section.

2. Place protective covers over the vise jaws. Place the halfshaft in the vise.

3. Use a hand grinder to cut through the swage ring. Use side cutters to cut off the small boot clamp.

4. Slide the boot down the halfshaft bar and away from the CV-joint outer race. Wipe all grease away from the face of the CV joint.

5. Find the halfshaft bar retaining snap ring, which is located in the inner race.

6. Spread the snapring ears apart.

7. Pull the CV joint and the CV joint boot from the halfshaft bar. Discard the old CV joint boot.

8. Place a brass drift against the CV joint cage. Tap gently on the brass drift with a hammer in order to tilt the cage.

9. Remove the first chrome alloy ball when the CV joint cage tilts. Tilt the CV joint cage (1) in the opposite direction to remove the opposing chrome alloy ball. Repeat this process to remove all six of the balls.

10. Pivot the CV joint cage and the inner race 90 degrees to the center line of the outer race. At the same time, align the cage windows with the lands of the outer race. Lift out the cage and the inner race.

11. Remove the inner race from the cage by rotating the inner race upward. Clean the following items thoroughly with cleaning solvent. Remove all traces of old grease and any contaminates.

    a. The inner and outer race assemblies.

    b. The CV joint cage.

    c. The chrome alloy balls.

12. Dry all the parts. Check the CV joint assembly for unusual wear, cracks, or other damage. Replace any damaged parts. Clean the halfshaft bar. Use a wire brush to remove any rust in the boot mounting area (grooves).

To assemble:

13. Inspect all of the parts for unusual wear, cracks, or other damage. Replace the CV joint assembly if necessary. Put a light coat of the recommended grease on the inner and the outer race grooves.

14. Hold the inner race at 90 degrees to the centerline of the cage. Align the lands of the inner race with the windows of the cage. Insert the inner race into the cage by rotating the inner race downward.

15. Insert the cage and inner race into the outer race.

16. Place a brass drift against the CV joint cage. Tap gently on the brass drift with a hammer in order to tilt the cage. Install the first chrome alloy ball when the CV joint cage tilts. Tilt the CV joint cage in the oppo-

site direction to install the opposing chrome alloy ball. Repeat this process in order to install all six of the balls.

17. Pack the CV joint boot and the CV joint assembly with the grease supplied in the kit. The amount of grease supplied in this kit has been pre-measured for this application.

18. Place the new small boot clamp onto the CV joint boot.

19. Slide the CV joint boot onto the halfshaft bar.

20. Position the small end of the CV joint boot into the joint boot groove on the halfshaft bar.

21. Secure the small boot clamp, a breaker bar, and a torque wrench. Tighten the small clamp (1) to 100 ft. lbs. (136 Nm).

22. Check the gap dimension on the clamp ear. Continue tightening until the gap dimension is reached.

23. Pinch the new swage ring slightly by hand to distort it into an oval shape. Slide the distorted swage ring over the large diameter of the boot.

➡ **Be sure that the retaining ring side of the CV joint inner race faces the halfshaft bar (3) before installation.**

24. Slide the CV joint onto the halfshaft bar. The retaining snap ring inside of the inner race engages in the halfshaft bar groove with a click when the CV joint is in the proper position.

25. Pull on the CV joint to verify engagement.

26. Slide the large diameter of the CV joint boot with the large swage ring in place, over the outside edge of the CV joint outer race.

27. Clamp the CV joint boot tightly to the CV joint outer race with the large swage ring, using the following procedure:

    a. Mount tool J36652 in a vise.

    b. Install the bottom half of the split-plate swage clamp. For K15 models, use tool J36652-98.

    c. For K25 models, use tool J36652-1.

    d. Position the CV joint end (outboard end) of the halfshaft assembly in the bottom half of tool J36652.

28. Align the following during this procedure:

    a. The CV joint boot.

    b. The CV joint assembly.

    c. The swage ring.

29. Install the top half of tool J36652 onto the lower half of the tool, over the CV joint boot and the CV joint assembly.

30. Align the swage ring and the swage ring clamp.

31. Insert the bolts into J36652. Hand tighten the bolts until the bolts are snug. Tighten each bolt 180 degrees at a time. Alternate between the bolts until both sides of the top half of the tool touch the bottom half of the tool.

32. Loosen the bolts and remove the halfshaft assembly from the tool.

Manual Locking Hubs

The engagement and disengagement of the hubs is a manual operation which must be performed at each hub assembly. The hubs should be placed FULLY in either Lock or Free position or damage will result.

❄❄ WARNING

Do not place the transfer case in either 4-wheel mode unless the hubs are in the Lock position!

Locking hubs should be run in the Lock position periodically for a few miles to assure proper differential lubrication.

REMOVAL & INSTALLATION

1. Before servicing the vehicle, refer to the precautions in the beginning of this section.
2. Remove or disconnect the following:
   • Wheels
3. Lock the hubs. Remove the outer retaining plate, Allen head bolts and take off the plate, O-ring, and knob assembly.
   • External snapring from the axle shaft
   • Compression spring
   • Clutch cup
   • O-ring and dial screw
   • Clutch nut and seal
   • Large internal snapring from the wheel hub
   • Inner drive gear
   • Clutch ring and spring
   • Smaller internal snapring from the clutch hub body
   • Hub body
4. Clean all hub parts in a safe, nonflammable solvent and wipe them dry.
5. Inspect each component for wear or damage. Make sure that the springs are functional and stiff. Make sure that all gear teeth are intact, with no chips or burrs.
6. Make sure that the splines on the inside of the wheel hub are clean and free of dirt, chips and burrs.
7. Surface irregularities can be cleaned up with light filing or emery paper.
8. Prior to assembly, coat all parts with the same wheel bearing grease.

To install:
9. Install or connect the following:
   • Hub body
   • Smaller internal snapring in the clutch hub body
   • Clutch ring and spring
   • Inner drive gear
   • Large internal snapring in the wheel hub
   • External snapring on the axle shaft. If the snapring groove is not completely visible, reach around, inside the knuckle and push the axle shaft outwards.
   • Clutch nut and seal
   • O-ring and dial screw
   • Clutch cup
   • Compression spring
10. Place the hub dial in the Lock position.
11. Coat the hub dial assembly O-ring with wheel bearing grease and position the hub dial and retainer on the hub.
   • Allen head bolts. Make sure that you used any washers that were there originally. Torque these bolts to 45 inch lbs. (5 Nm)
12. Rotate the hub dial to the free position and turn the wheel hubs to make sure that the axle is free.
   • Wheels

Automatic Locking Hubs

REMOVAL & INSTALLATION

The following procedure covers removal & installation only, for the hub assembly. The hub should be disassembled ONLY if overhaul is necessary. In that event, an overhaul kit will be required. Follow the instructions in the overhaul kit to rebuild the hub.

1. Before servicing the vehicle, refer to the precautions in the beginning of this section.
2. Remove or disconnect the following:
   • Capscrews and washer from the hub cap
   • Hub cap and spring
   • Bearing race, bearing and retainer
   • Keeper from the outer clutch housing
   • Large snapring to release the locking unit
   • Locking unit from the hub. You can make this job easier by threading 2 hub cap screws into the outer clutch housing and hold these to pull out the unit.

To install:
3. Wipe clean all parts and check for wear or damage.

4. Coat all parts with the same wheel bearing grease you've used on the bearings.
5. Install or connect the following:
   • Locking unit in the hub
   • Large snapring, pull outward on the unit to make sure the snapring is fully seated in its groove
   • Keepers
   • Bearing retainer, bearing and race. Make sure that the bearing is fully pack with grease.
6. Coat the hub cap O-ring with wheel bearing grease and install the hub cap.
   • Capscrews and washers. Tighten the screws to 45 inch lbs. (5 Nm).

Front Axle Shaft, Bearing and Seal

REMOVAL & INSTALLATION

1. Before servicing the vehicle, refer to the precautions in the beginning of this section.
2. Remove or disconnect the following:
   • Halfshaft assembly
   • Front axle fluid
   • Electrical connectors
   • Axle shaft (output shaft) tube nuts from the bracket
   • Bracket bolts from the frame. Do not remove the bracket. The bolts are removed in order to provide clearance.
   • Axle shaft (output shaft) bolts from the carrier

➡ **Keep the open end of the tube up.**

   • Axle shaft (output shaft) tube from the carrier. Ensure the spring is not lost during removal. In a vise, hold the axle shaft (output shaft) tube by the mounting flange.
3. Remove the following components:
   a. The shift shaft.
   b. The damper spring.
   c. The fork.
   d. The clip assembly.
4. Remove or disconnect the following:
   • Sleeve
   • Gear
   • Thrust washer
   • Axle shaft (output shaft). Tap out the axle shaft (output shaft) with a soft mallet.
   • Deflector. Pry out the deflector with a suitable prytool.
   • Seal. Pry out the seal with a suitable prytool.
   • Bearing, using a slide hammer.
5. Clean the parts in suitable solvent.

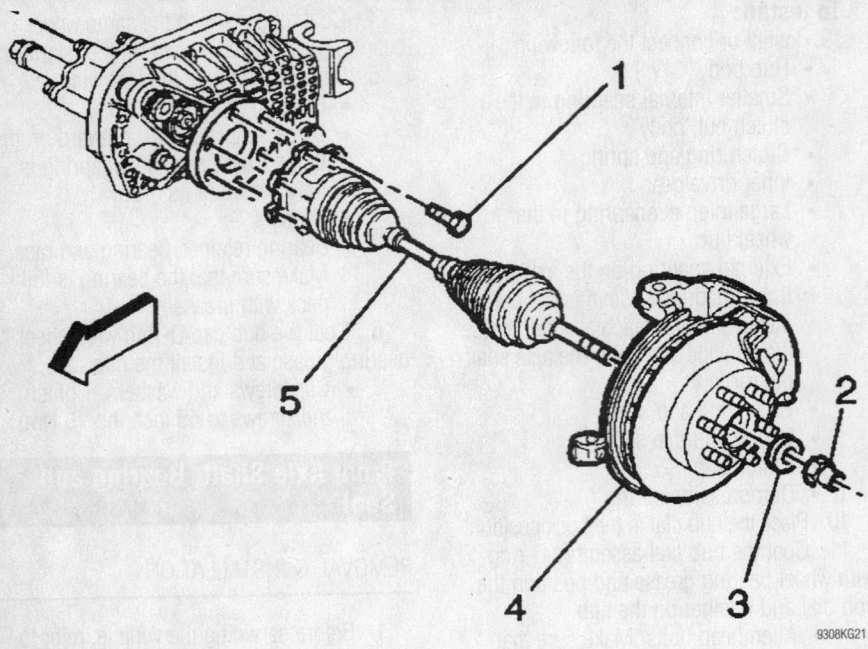

Front axle shaft removal

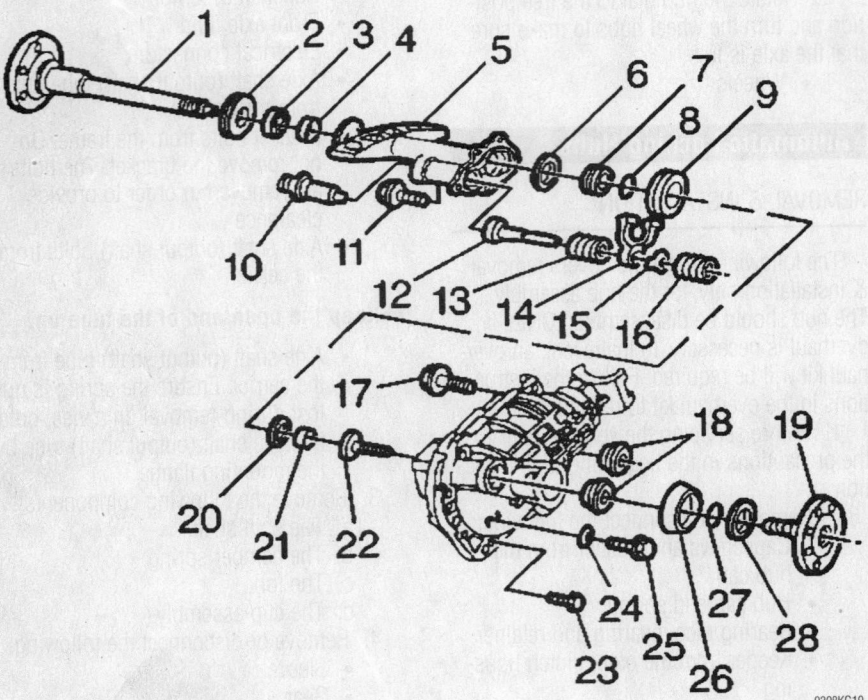

Front axle shaft exploded view

Clean the gasket surfaces on the axle shaft (output shaft) tube and carrier housing.

To install:

6. Install or connect the following:
- New bearing into the axle shaft tube using a driver. Apply axle lubricant to the bearing.

- New seal using a driver. Coat the seal lips with axle lubricant.
- Deflector
- Axle shaft
- Thrust washer. Use grease in order to hold the thrush washer in place. Ensure the tabs on the thrush

washer align with the slot in the axle shaft tube.
- Gear
- Sleeve
- Shift shaft
- Damper spring
- Fork
- Clip assembly

7. Apply sealant GM P/N 12345739 or the equivalent to the carrier sealing surfaces.
- Spring into the carrier case
- Axle shaft tube to the carrier
- Axle shaft (output shaft) bolts. Tighten the bolts to 30 ft. lbs. (40 Nm).
- Bracket bolts to the frame. Tighten the bolts to 67 ft. lbs. (90 Nm).
- Axle shaft (output shaft) tube nuts to the bracket. Tighten the nuts to 75 ft. lbs. (100 Nm).
- Halfshaft assembly
- Electrical connectors

8. Fill the front differential with lubricant until the level is 0.5 in. (12mm) below the fill plug.

Rear Axle Shaft, Bearing and Seal

REMOVAL & INSTALLATION

8.5 Inch and 9.5 Inch Rear Axle

1. Before servicing the vehicle, refer to the precautions in the beginning of this section.
2. Raise and support the vehicle on a hoist.
3. Remove or disconnect the following:
- Tire and wheel assembly
- Brake caliper
- Rear cover and the gasket
- Pinion shaft locking screw
- Pinion shaft, on vehicles without a locking differential

4. On axles with a locking differential, remove the shaft part way. Rotate the case until the pinion shaft touches the housing.
5. On axles with a locking differential, use a screwdriver, or a similar tool, in order to enter the differential case and rotate the lock until the lock aligns with the thrust block.
6. Push the flange of the axle shaft toward the differential. Remove the lock from the button end of the axle shaft.

➡**When removing the axle shaft, do not rotate the shaft. Rotating the shaft will misalign the gears. Misaligning the gears will make the assembly difficult.**

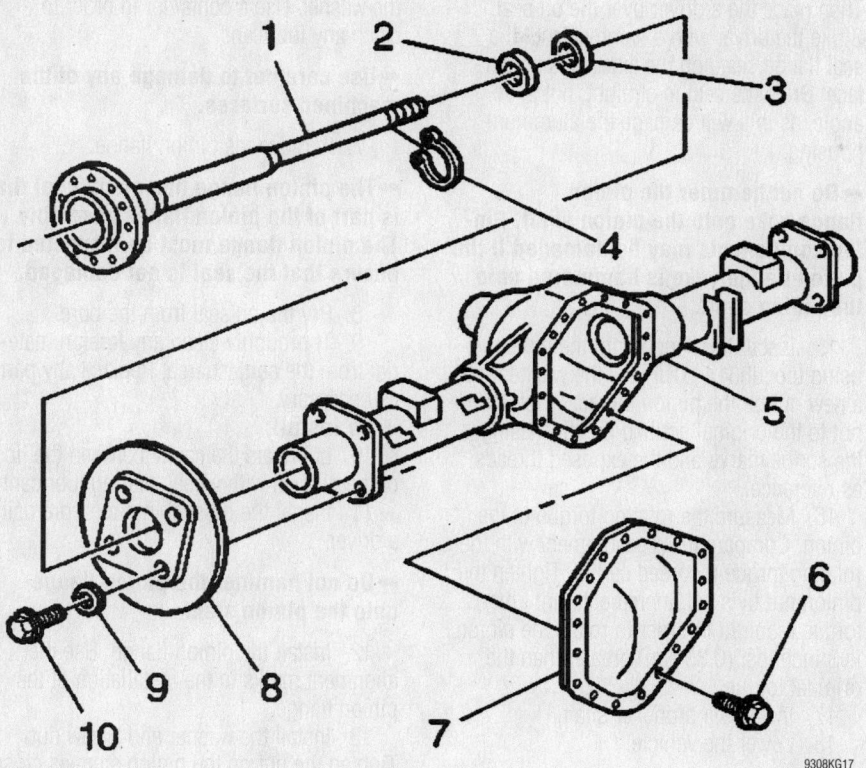

Rear axle shaft removal—8.5/9.5 inch

7. Remove the axle shaft from the housing.

8. Inspect all the parts for damage. Replace the parts as necessary.

To install:

➡ **Carefully insert the axle shaft in order to not damage the seal.**

9. Install the axle shaft into the housing. Slide the axle shaft into place allowing the splines to engage the differential side gear.

10. On axles without a locking differential, place the lock on the button end of the axle shaft.

11. On axles with a locking differential, keep the pinion shaft partially withdrawn.

12. On axles with a locking differential, place the lock on the axle shaft so that the ends are flush with the thrust block. Pull the shaft flange outward in order to seat the lock in the differential gear.

➡ **Anytime you remove a differential pinion shaft locking screw, coat the screw threads with LOCTITE® 242 before reinstalling the screws. The screw has an adhesive coating in order to prevent the screw from loosening in the case. Removing the screw removes the adhesive on the screw.**

13. Align the hole in the pinion shaft with the screw hole in the differential case.

14. Install or connect the following:
- Pinion flange locking screw. Tighten the screw to 25 ft. lbs. (34 Nm).
- Rear cover and the gasket
- Brake caliper
- Tire and wheel assembly

15. Fill the rear axle.

16. Remove the supports and lower the vehicle.

10.5 Inch Rear Axle

1. Before servicing the vehicle, refer to the precautions in the beginning of this section.

2. Remove or disconnect the following:
- Tire and wheel
- Brake caliper
- Brake rotor
- Flange bolts

3. Lightly rap the axle shaft with a soft-faced hammer in order to loosen the shaft. Grip the rib on the axle shaft flange with a locking pliers. Twist the axle shaft flange in order to start the axle shaft removal. Remove the axle shaft from the tube.

4. Remove the gasket.

5. Clean the axle shaft flange and the

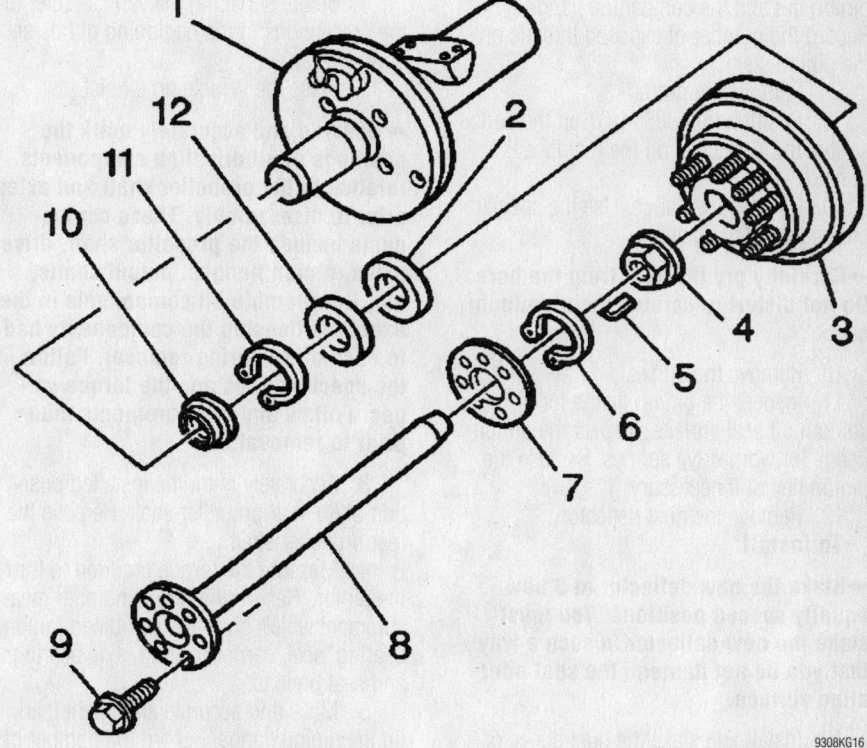

Rear axle shaft removal—10.5 inch 25 Series

outside face of the hub assembly. Inspect all the parts. Replace the parts as necessary.

To install:

6. Install the gasket onto the axle shaft.

7. Install the gasket and the axle shaft into the tube. Ensure the shaft splines mesh into the differential side gear. Align the holes in the axle flange and the gasket with the holes in the hub.

8. Install or connect the following:
   - Axle flange bolts and tighten to 110 ft. lbs. (150 Nm)
   - Brake rotor
   - Brake caliper
   - Wheel and tire

Front Drive Axle Pinion Seal

REMOVAL & INSTALLATION

1. Before servicing the vehicle, refer to the precautions in the beginning of this section.

2. Raise the vehicle on a hoist.

3. Remove the propeller shaft from the axle.

4. Tie the propeller shaft to a frame rail or the crossmember.

5. Measure the torque required in order to rotate the pinion. Record the torque value for reassembly.

6. Scribe a line on the pinion stem, the pinion nut and the companion flange. Record the number of exposed threads on the pinion stem.

7. Remove the nut.

8. Position tool J8614-01 on the flange so that the 4 notches on the tool face the flange.

9. Remove the flange. Use the special nut and the forcing screw.

➥**Carefully pry the seal from the bore. Do not distort or scratch the aluminum case.**

10. Remove the oil seal.

11. Inspect the pinion flange for a smooth oil seal surface. Inspect the pinion flange for worn drive splines. Replace the pinion flange if necessary.

12. Remove the dust deflector.

To install:

➥**Stake the new deflector at 3 new equally spaced positions. You must stake the new deflector in such a way that you do not damage the seal operating surface.**

13. Install and stake the dust deflector on the flange.

14. Position the oil seal in the bore.

Then place the a driver over the oil seal. Strike the driver with a hammer until the seal flange seats on the axle housing surface. Drive the seal in straight, not at an angle, as this will damage the aluminum housing.

➥**Do not hammer the pinion flange/yoke onto the pinion shaft. Pinion components may be damaged if the pinion flange/yoke is hammered onto the pinion shaft.**

15. Install the flange onto the pinion using tool J8614-01. Place the washer and a new nut on the pinion threads. Tighten the nut to the original scribed position using the scribe marks and the exposed threads as reference.

16. Measure the rotating torque of the pinion. Compare the measurement with the rotating torque recorded earlier. Tighten the pinion nut by small increments until the torque required in order to rotate the pinion is 3 inch lbs. (0.35 Nm) greater than the original torque.

17. Install the propeller shaft.

18. Lower the vehicle.

Rear Drive Axle Pinion Seal

REMOVAL & INSTALLATION

1. Before servicing the vehicle, refer to the precautions in the beginning of this section.

2. Raise the vehicle on a hoist.

➥**Observe and accurately mark the positions of all driveline components relative to the propeller shaft and axles prior to disassembly. These components include the propeller shaft, drive axles, pinion flanges, output shafts, etc. Reassemble all components in the exact relationship the components had to each other during removal. Follow the specifications and the torque values. Follow any measurements made prior to removal.**

3. Accurately mark the installed position of the rear propeller shaft. Remove the rear propeller shaft.

4. Measure the torque required to turn the pinion. Record the torque number measurement which gives the combined pinion bearing, seal, carrier bearing, axle bearing and seal preload.

5. Make and accurate alignment mark on the pinion flange. Record the number of exposed threads on the pinion stem.

6. Remove the pinion flange nut and

the washer. Use a container in order to catch any lubricant.

➥**Use care not to damage any of the machined surfaces.**

7. Remove the pinion flange.

➥**The pinion flange has an oil seal that is part of the pinion flange assembly. The pinion flange must be inspected to ensure that the seal is not damaged.**

8. Pry the oil seal from the bore.

9. Thoroughly clean any foreign material from the contact area. Replace any parts as necessary.

To install:

10. Lubricate the cavity between the lips of the oil seal with wheel bearing lubricant.

11. Install the oil seal into the bore using a driver.

➥**Do not hammer the pinion flange onto the pinion stem.**

12. Install the pinion flange. Use the alignment marks in the installation of the pinion flange.

13. Install the washer and a new nut. Tighten the nut on the pinion stem as close as possible to the alignment marks without going past the marks. Use the alignment marks and the thread count as a reference. Tighten the nut a little at a time. Turn the pinion flange several times after each tightening in order to seat the rollers.

➥**If the recorded preload torque value was less than 3 ft. lbs. (4 Nm), reset the torque specification to 4–7 ft. lbs. (3–4 Nm).**

14. Measure the torque required to rotate the pinion flange. Compare the measured torque with the recorded value. Continue tightening the pinion nut and measuring the torque until you achieve the recorded value.

15. Align the propeller shaft with the alignment marks. Connect the propeller shaft.

16. Install the retainers and the bolts. Tighten the bolts to 15 ft. lbs. (20 Nm).

17. Fill the rear axle.

18. Lower the vehicle.

Front Drive Axle Differential Carrier

REMOVAL & INSTALLATION

1. Before servicing the vehicle, refer to the precautions in the beginning of this section.

2. Remove or disconnect the following:

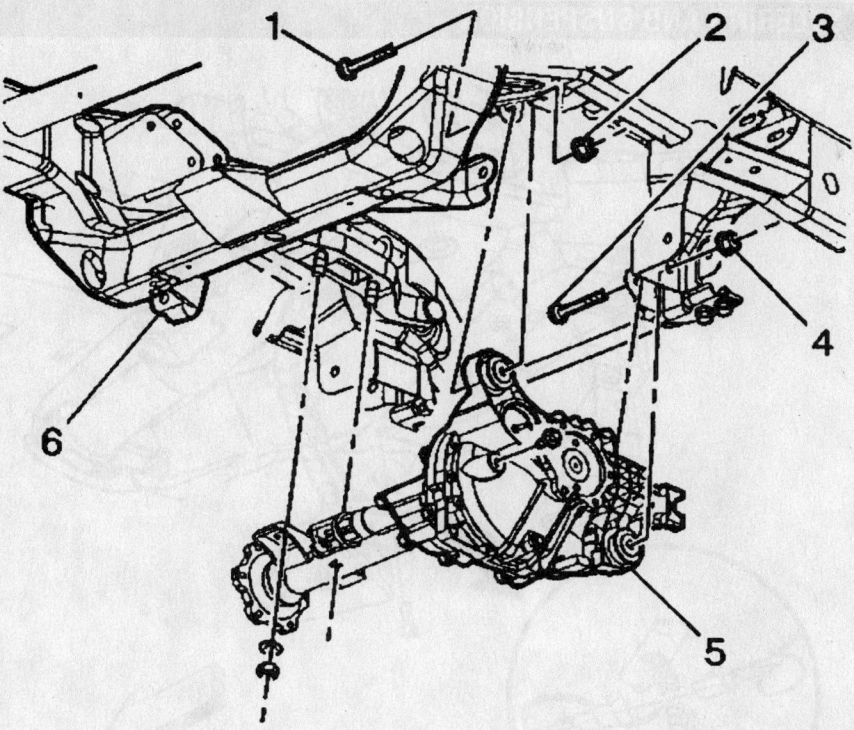

9308KG18

Front differential carrier removal

- Front axle fluid
- Front propeller shaft
- Left and the right drive axle wheel drive shaft
- Axle tube nuts from the bracket
- Wiring at the axle
- Vent hose at the axle
- Carrier assembly lower mounting bolt and the nut
- Idler arm from the relay rod
- Pitman arm from the relay rod

3. Attach a transmission jack to the carrier assembly.
- Remove the upper carrier assembly mounting bolt and the nut.
- Remove the carrier assembly from the vehicle.

To install:
4. Install or connect the following:

- Carrier assembly in the vehicle
- Carrier assembly upper mounting bolt and the nut
- Lower carrier assembly mounting bolt and the nut. Tighten the bolts to 75 ft. lbs. (100 Nm).
- Pitman arm to the relay rod
- Idler arm to the relay rod
- Axle tube nuts to the bracket. Tighten the bolts to 75 ft. lbs. (100 Nm).
- Vent hose
- Wiring to the axle
- Left and the right drive axle wheel drive shaft
- Front propeller shaft to the pinion flange

5. Fill the front differential with lubricant.

Rear Drive Axle Housing

REMOVAL & INSTALLATION

1. Before servicing the vehicle, refer to the precautions in the beginning of this section.
2. Remove or disconnect the following:

- Axle lubricant
- Propeller shaft
- Wheel assemblies
- Parking brake cable
- Brake calipers
- Shock absorbers from the axle brackets
- Vent hose from the rear axle vent fitting
- Nuts and the washers from the U-bolts
- U-bolts, the spring plates and the spacers form the axle assembly

3. Lower the axle assembly.

To install:
4. Place the rear axle assembly under the vehicle. Align the rear axle assembly with the springs. Connect the spacers, the spring plates and the U-bolts to the rear axle. Raise the rear axle assembly into position.
5. Install or connect the following:

- Washers and nuts to the U-bolts. Tighten the nuts to 59 ft. lbs. (80 Nm). first, then to 89 ft. lbs. (120 Nm).
- Vent hose to the rear axle vent fitting
- Shock absorbers to the rear axle
- Brake calipers
- Parking brake cable
- Wheel assemblies
- Propeller shaft

6. Fill the rear axle.
7. Bleed the brake system.
8. Remove the supports and lower the vehicle.

STEERING AND SUSPENSION

Air Bag

✳✳ CAUTION

Some vehicles are equipped with an air bag system. The system must be disabled before performing service on or around system components, steering column, instrument panel components, wiring and sensors. Failure to follow safety and disabling procedures could result in accidental air bag deployment, possible personal injury and unnecessary system repairs.

PRECAUTIONS

Several precautions must be observed when handling the inflator module to avoid accidental deployment and possible personal injury.

• Never carry the inflator module by the wires or connector on the underside of the module

• When carrying a live inflator module, hold securely with both hands, and ensure that the bag and trim cover are pointed away

• Place the inflator module on a bench or other surface with the bag and trim cover facing up

• With the inflator module on the bench, never place anything on or close to the module that may be thrown in the event of an accidental deployment

DISARMING

1. Turn the front wheels to the straight-ahead position.

2. Turn the ignition switch to the **LOCK** position and remove the key.

➡ **If the key is in the RUN position when the Air Bag fuse is removed or open (blown), the Air Bag warning lamp in the dash will light up. This is normal operation, not a sign of a malfunction.**

3. Remove the Air Bag fuse from the fuse panel.

4. Remove the drivers side knee bolster and unplug the yellow 2-pin connector at the base of the steering column to disarm the driver's side Air Bag. Remove the passenger side knee bolster and unplug the yellow 2-pin connector to disable the passenger's side Air Bag.

5. Reverse the procedure to arm the Air Bag restraint system.

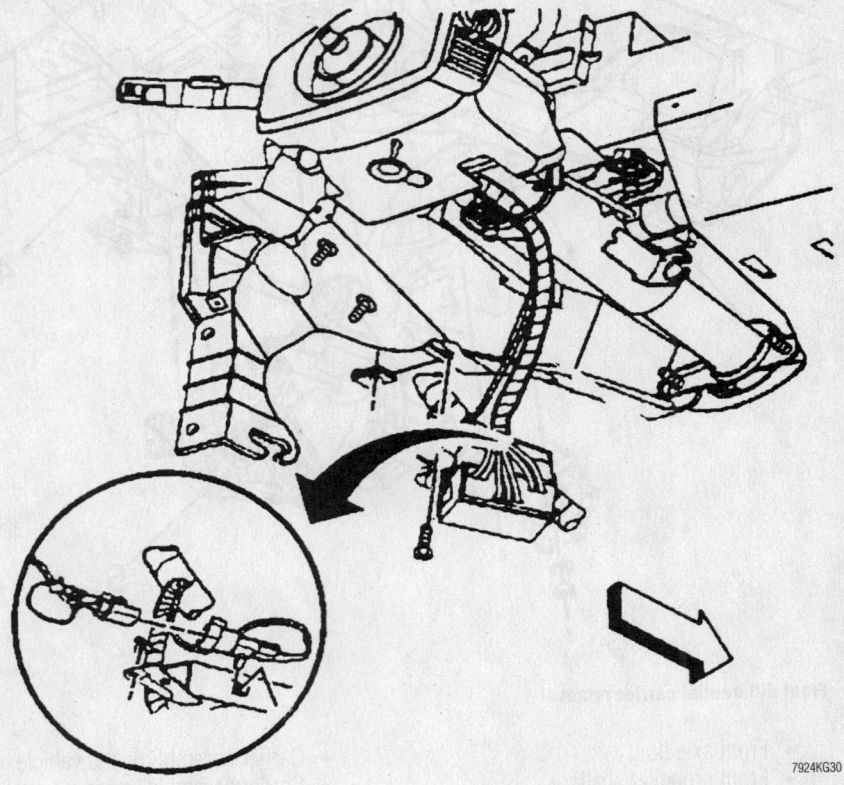

7924KG30

Typical air bag connector location—driver's side

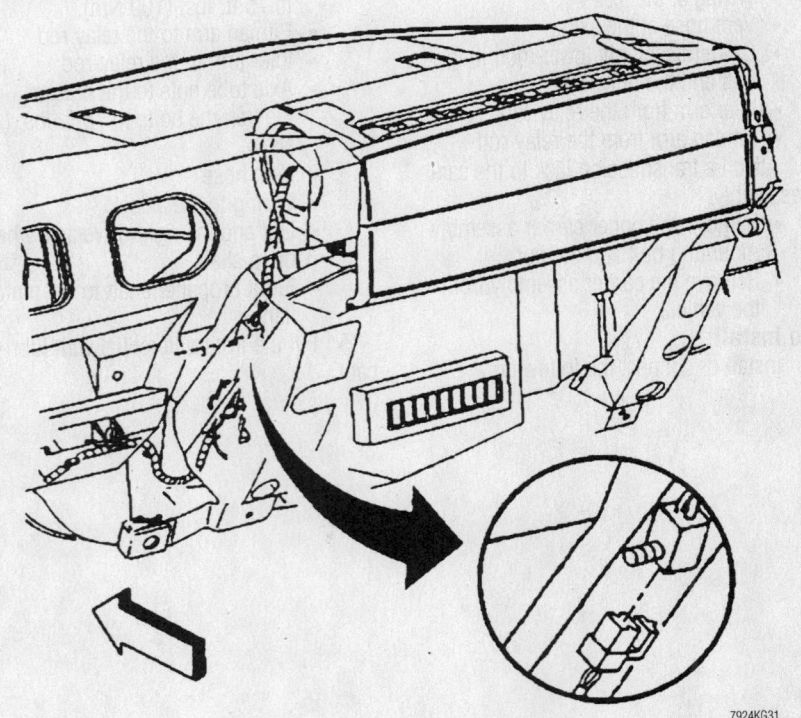

7924KG31

Typical air bag connector location—passenger's side

Power Steering Pump

REMOVAL & INSTALLATION

5.7L, 6.5L and 7.4L Engines

1. Before servicing the vehicle, refer to the precautions in the beginning of this section.

2. Disconnect the hoses at the pump. When the hoses are disconnected, secure the ends in a raised position to prevent leakage. Cap the ends of the hoses to prevent the entrance of dirt.

3. Cap the pump fittings.

4. Loosen the belt tensioner.

5. Remove or disconnect the following:
   - Pump drive belt
   - Pulley with a puller such as J–29785–A
   - Front mounting bolts, on the 5.7L engines
   - Rear brace, on the 7.4L engine
   - Front brace and rear mounting nuts, on the 6.5L diesel engine

6. Lift out the pump.

To install:

7. Observe the following torques:
   - 5.7L engines: front mounting bolts to 37 ft. lbs. (50 Nm)
   - 7.4L engine: rear brace nut: 61 ft. lbs. (82 Nm); rear brace bolt: 24 ft. lbs. (32 Nm); mounting bolts: 37 ft. lbs. (50 Nm)
   - 6.5L diesel: front brace to 30 ft. lbs. (40 Nm); rear mounting nuts to 17 ft. lbs. (23 Nm).
   - Pulley with J–25033–B
   - Drive belt
   - Hoses

8. Fill and bleed the system.

4.8L, 5.3L, 6.0L, and 8.1L Engines

1. Before servicing the vehicle, refer to the precautions in the beginning of this section.

2. Remove or disconnect the following:
   - Upper radiator fan shroud, if necessary
   - Drive belt
   - Pulley
   - Nut and clamp retaining the filler neck to the power steering pump, if equipped

3. Place a drain pan under the pump. Remove the hoses from the pump.
   - Bolts from the rear of the pump
   - Bolts from the front of the pump
   - Pump from the vehicle

To install:

4. Install or connect the following:
   - Power steering pump

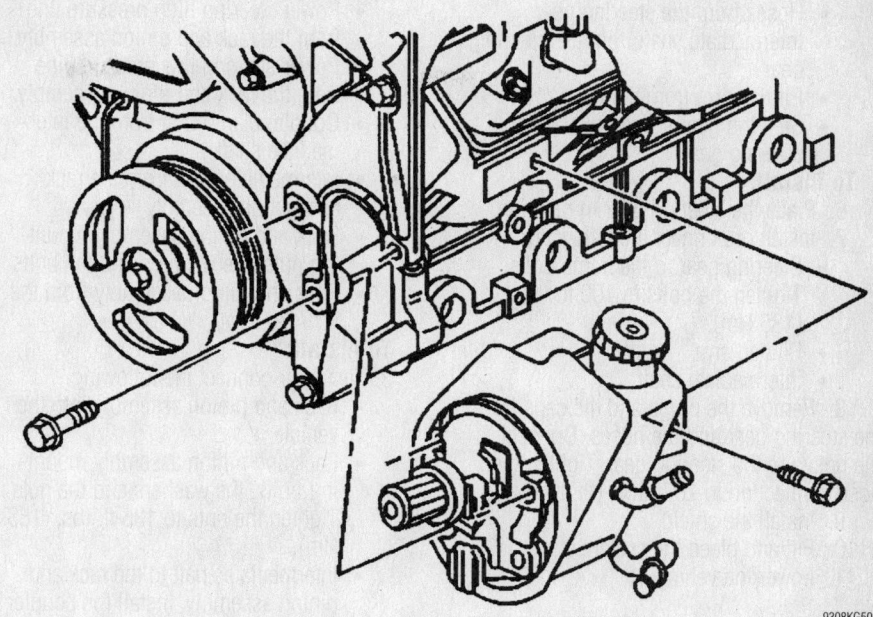

Power steering pump—4.8L, 5.3L and 6.0L engines shown

9308KG50

   - Bolts to the front and the rear of the pump. Tighten the bolts to 37 ft. lbs. (50 Nm).
   - Hoses to the pump. Tighten the nut to 20 ft. lbs. (28 Nm).
   - Nut and clamp retaining the filler neck to the power steering pump, if equipped
   - Pulley. Install the pulley with 0.020 in. (0.5mm) play.
   - Drive belt
   - Upper radiator shroud

5. Fill and bleed the power steering system.

Recirculating Ball Power Steering Gear

REMOVAL & INSTALLATION

1. Before servicing the vehicle, refer to the precautions in the beginning of this section.

2. Raise the vehicle.

3. Remove the shield.

4. Place a drain pan below the steering gear.

5. Remove or disconnect the following:

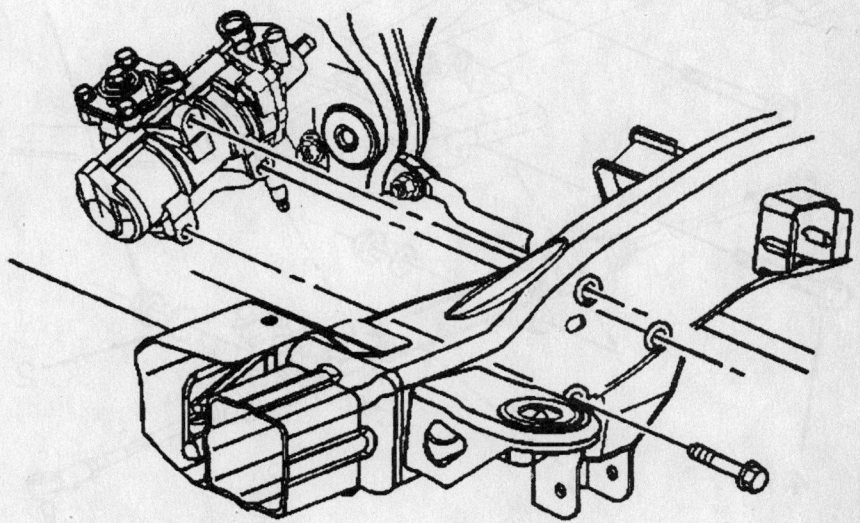

Recirculating ball gear

9308KG48

- Hoses from the steering gear
- Intermediate shaft from the steering gear
- Pitman arm from the relay rod
- Steering gear frame bolts and the steering gear

To install:

6. Place the steering gear in position.
7. Install or connect the following:
    - Steering gear to the frame bolts. Tighten the bolts to 100 ft. lbs. (135 Nm).
    - Pitman arm
    - Intermediate shaft
8. Remove the plugs and the caps from the steering gear and the hoses. Connect the hoses to the steering gear. Tighten the hose connection to 20 ft. lbs. (28 Nm).
9. Install the shield.
10. Fill and bleed the system.
11. Lower the vehicle.

Rack & Pinion Steering Gear

REMOVAL & INSTALLATION

1. Before servicing the vehicle, refer to the precautions in the beginning of this section.
2. Remove or disconnect the following:
    - Wheel assemblies
    - Engine shield, if equipped
    - Stabilizer shaft

- Power steering high pressure line from the rack and pinion assembly
- Power steering low pressure line from the rack and pinion assembly
- Coupler clamp bolt from the intermediate shaft
- Intermediate shaft from the rack and pinion assembly
- Rack and pinion assembly mounting nuts, the washers and the bolts
- Rack and pinion assembly from the vehicle

To install:

3. Install or connect the following:
    - Rack and pinion assembly into the vehicle
    - Rack and pinion assembly mounting bolts, the washers and the nuts. Tighten the nuts to 136 ft. lbs. (185 Nm).
    - Intermediate shaft to the rack and pinion assembly. Install the coupler clamp bolt to the intermediate shaft. Tighten the coupler clamp bolt to 33 ft. lbs. (45 Nm).
    - Power steering low pressure hose to the rack and pinion assembly
    - Power steering high pressure hose to the rack and pinion assembly. Tighten the hoses to 20 ft. lbs. (27 Nm).
    - Engine protection shield, if equipped

- Stabilizer shaft
- Wheels
4. Lower the vehicle.
5. Fill and bleed the power steering system.

Shock Absorber

REMOVAL & INSTALLATION

2WD Front

1. Before servicing the vehicle, refer to the precautions in the beginning of this section.
2. Raise and support the vehicle.
3. If equipped with selectable ride, disconnect the Real Time Damping (RTD) link rod from the sensor. Grasp the connector lock tabs. Rotate the connector tabs counter clockwise until the connector is unlocked. Disengage the connector from the tennon by firmly pulling the connector up. Hold the tennon end with a wrench while removing the nut. Remove the nut.
4. Remove the upper insulator. Do not discard the plastic pilot ring.
5. Remove the shock absorber mounting bolts at the lower control arm. Remove the shock absorber through the lower control arm from below.

To install:

6. Support the lower control arm with a suitable jack in order to align the tennon with the mounting hole if equipped with selectable ride.

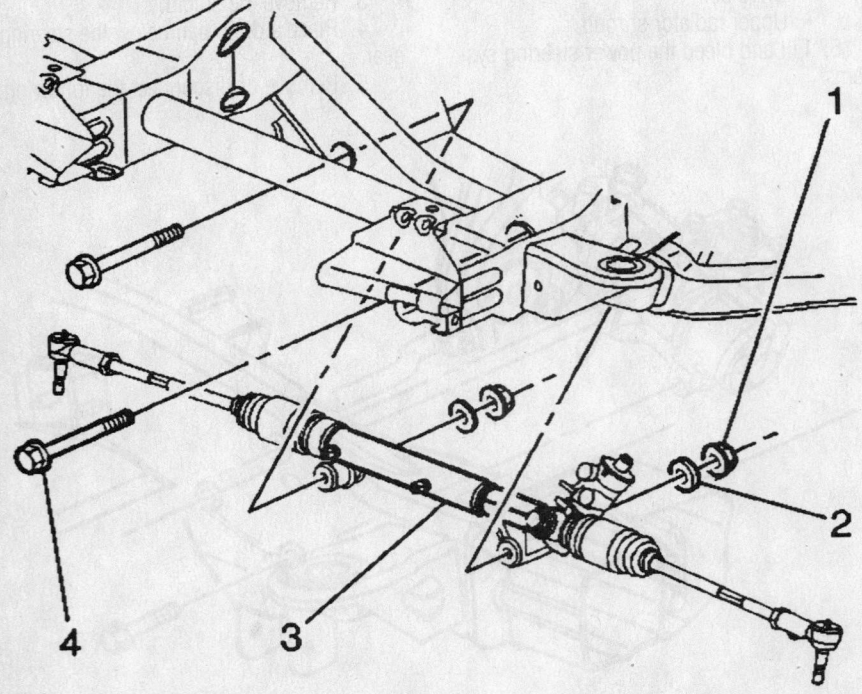

Rack and pinion gear

9308KG49

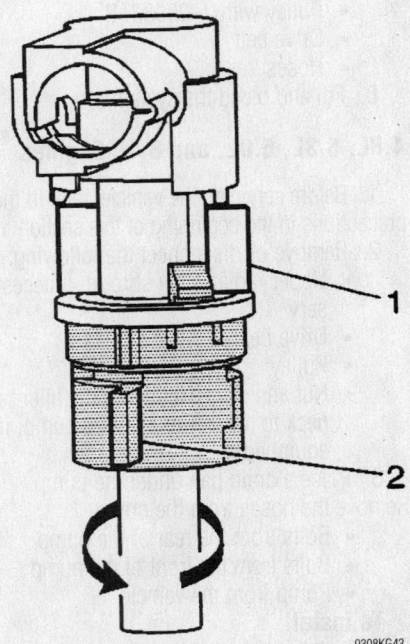

RTD connector

9308KG43

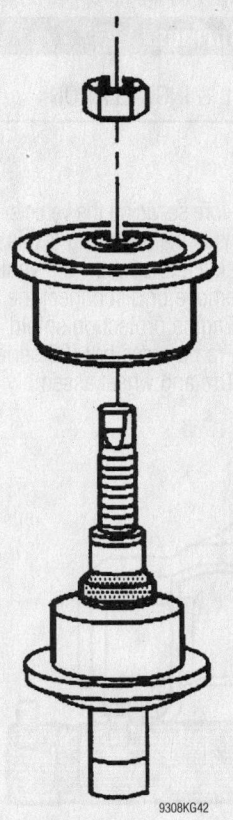

Upper shock insulator

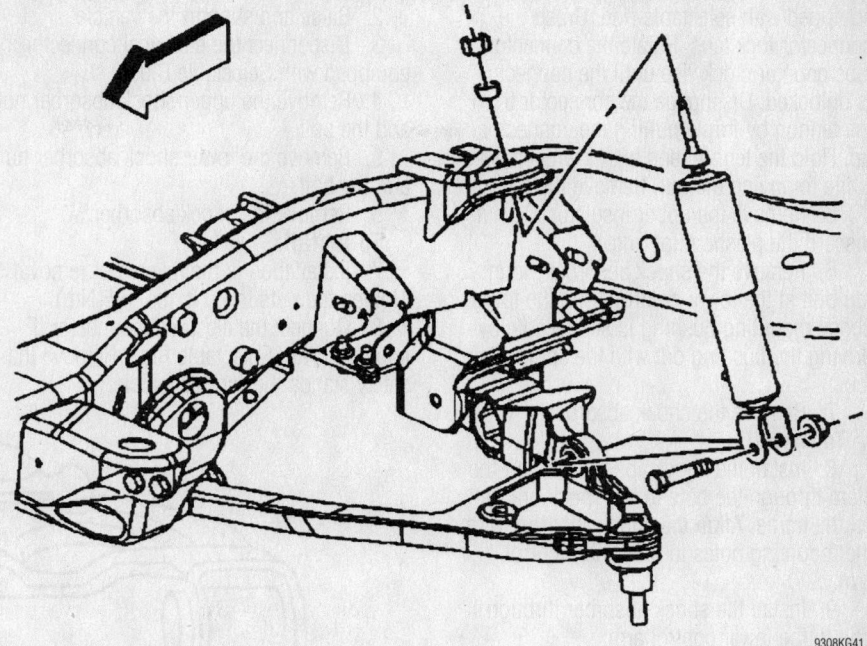

Shock absorber removal—4WD vehicles

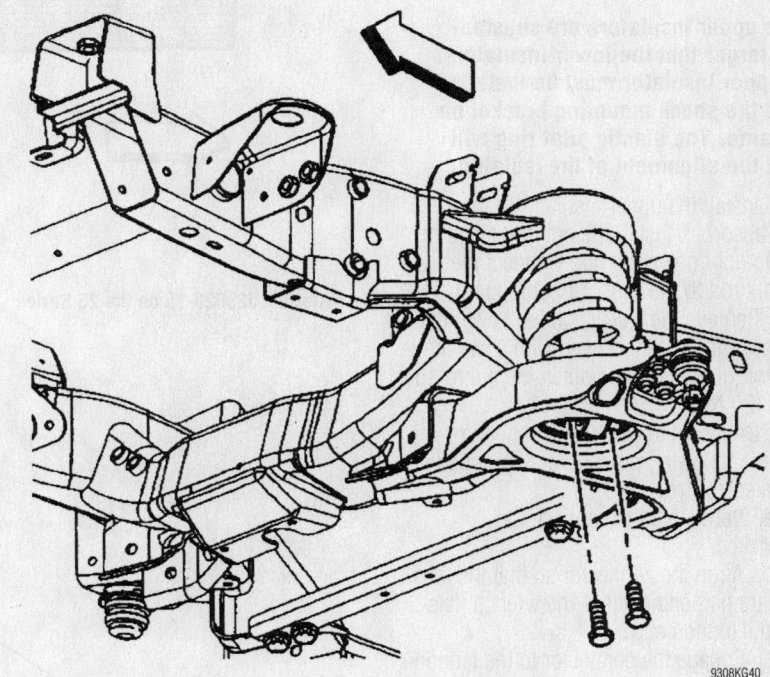

Shock absorber removal—2WD vehicles

7. Install the shock absorber through the lower control arm from below. Insert the tennon through the mounting hole in the upper spring pocket. Align the shock absorber with the mounting holes in the lower control arm.

8. Install the shock absorber mounting bolts to the lower control arm. Tighten the bolts to 18 ft. lbs. (25 Nm).

➡ **The upper insulators are substantially larger that the lower insulators. The upper insulator must be installed above the shock mounting bracket on the frame. The plastic pilot ring will assist the alignment of the isolators.**

9. Install the upper insulator to the shock absorber. Install the nut to the tennon end. Do not tighten the nut.

10. Connect the RTD link rod to the sensor (if equipped).

11. Remove the safety stands.

12. Lower the vehicle. Hold the tennon end with a wrench while torquing the nut. Tighten the nut to 15 ft. lbs. (20 Nm).

13. Connect the electrical connector using the following procedure:

   a. Verify that the connector is unlocked.

   b. Align the connector so that the tabs are perpendicular to the wrench flats on the tennon end.

   c. Engage the connector to the tennon by firmly pushing the connector down.

   d. Grasp the connector lock tabs. Rotate the connector counter clockwise.

14. The connector is locked into place when you hear an audible snap and the tabs are aligned.

4WD Front

1. Before servicing the vehicle, refer to the precautions in the beginning of this section.

2. Raise and support the vehicle.

3. Disconnect the (RTD) link rod from the sensor (if equipped).

4. Disconnect the electrical connector if

equipped with selectable ride. Grasp the connector lock tabs. Rotate the connector tabs counter clockwise until the connector is unlocked. Disengage the connector from the tennon by firmly pulling the connector up. Hold the tennon end with a wrench while removing the nut. Remove the nut.

5. Remove the upper insulator. Do not discard the plastic pilot ring.

6. Remove the shock absorber mounting bolt at the lower control arm. The lower shock mounting bushing is serviceable by driving the bushing out with the appropriate tool.

7. Remove the shock absorber.

To install:

8. Install the shock absorber. Insert the stem through the hole in the shock bracket on the frame. Align the shock absorber with the mounting holes in the lower control arm.

9. Install the shock absorber through bolt to the lower control arm.

10. Install the shock absorber through bolt nut. Tighten the nut to 59 ft. lbs. (80 Nm).

➡The upper insulators are substantially larger that the lower insulators. The upper insulator must be installed above the shock mounting bracket on the frame. The plastic pilot ring will assist the alignment of the isolators.

11. Install the upper insulator to the shock absorber. Install the nut to the tennon end. Do not tighten the nut. Connect the RTD link rod to the sensor (if equipped).

12. Remove the safety stands. Lower the vehicle. Hold the tennon end with a wrench while torquing the nut. Tighten the nut to 15 ft. lbs. (20 Nm).

13. Connect the electrical connector using the following procedure if equipped with selectable ride.

  a. Verify that the connector is unlocked.

  b. Align the connector so that the tabs (1) are perpendicular to the wrench flats on the tennon end.

  c. Engage the connector to the tennon by firmly pushing the connector down.

  d. Grasp the connector lock tabs (1, 2). Rotate the connector counter clockwise. The connector is locked into place when you hear an audible snap and the tabs are aligned.

Rear

1. Before servicing the vehicle, refer to the precautions in the beginning of this section.

2. Raise and support the vehicle.

3. Disconnect the electrical connector if equipped with Selectable Ride.

4. Remove the upper shock absorber nut and the bolt.

5. Remove the lower shock absorber nut and the bolt.

6. Remove the shock absorber.

To install:

7. Installation is the reverse of removal. Tighten the nuts to 70 ft. lbs. (95 Nm).

8. Connect the electrical connector if equipped with Selectable Ride. Remove the safety stands. Lower the vehicle.

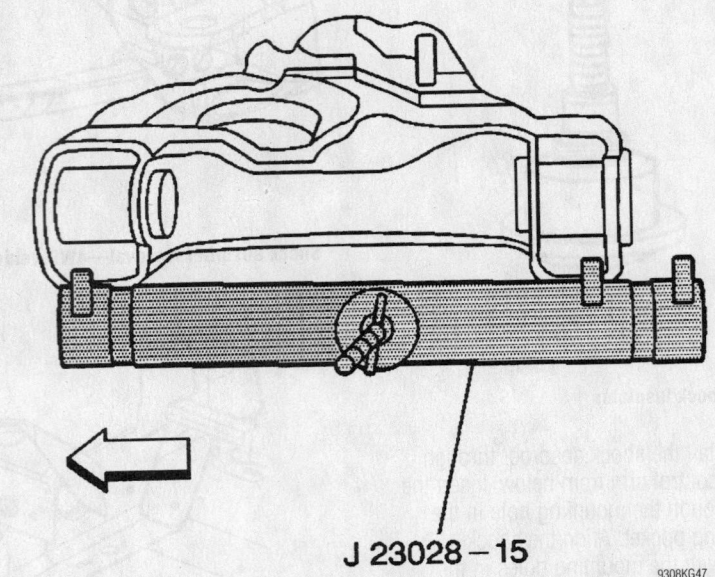

J 23028 – 15

9308KG47

Installing J23028-15 on the 25 Series

Coil Springs

REMOVAL & INSTALLATION

Front

1. Before servicing the vehicle, refer to the precautions in the beginning of this section.

2. Raise and support the vehicle.

3. Remove or disconnect the following:
  • Engine protection shield
  • Frame cross bar (25 series only)
  • Tire and wheel assembly

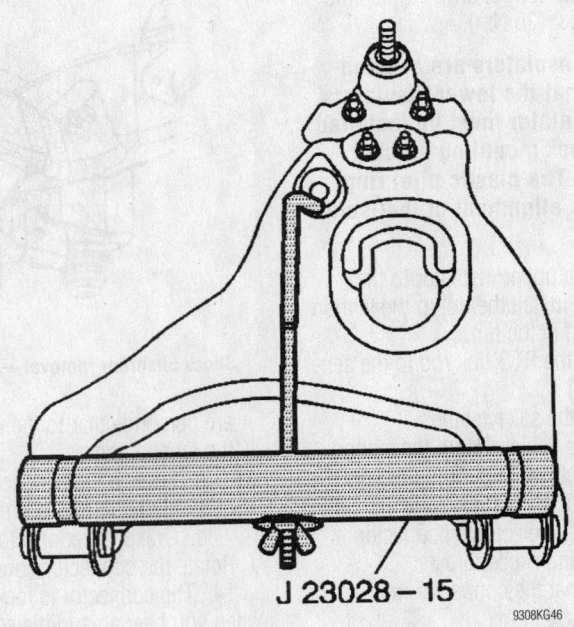

J 23028 – 15

9308KG46

Retaining hook installation

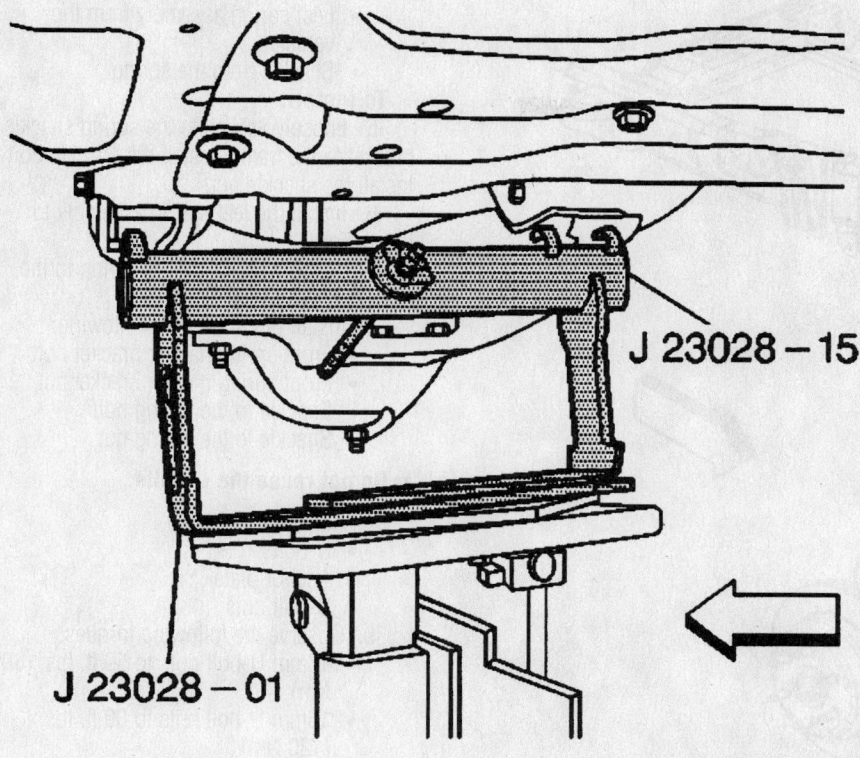

J 23028 – 15

J 23028 – 01

9308KG45

Tool attached to a jack

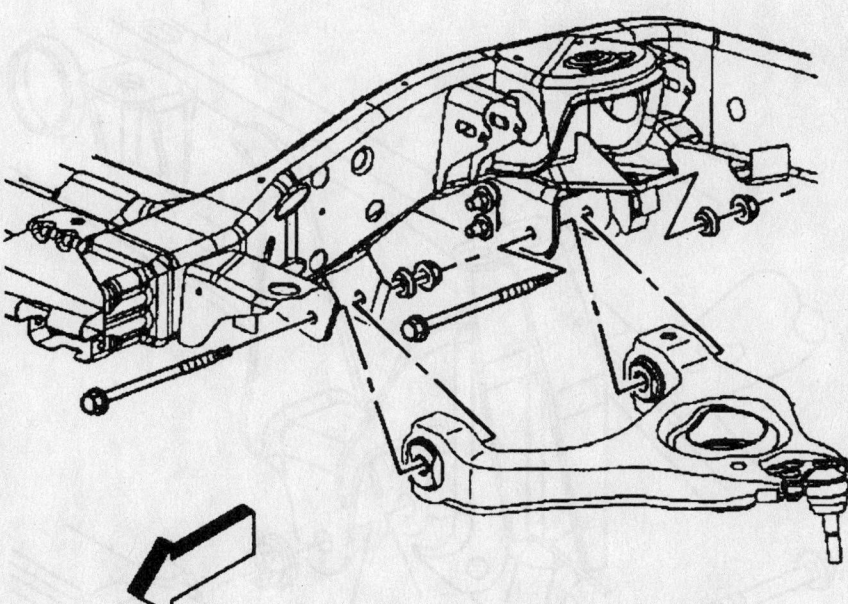

9308KG44

Lower control arm removal

- Shock absorber
- Front stabilizer shaft link

4. Install tool J23028-15 using the outboard locating tab (15 Series), or, the inboard locating tab (25 Series).

5. Attach the retaining hook to the control arm. Tighten the wing nut until you eliminate any free play.

6. Securely attach tool J23028-01 to a suitable transmission jack. Raise the jack until the yokes of tool J23028-01 line up with the notches in J23028-15.

7. Using the tools and the transmission jack, relieve the spring tension from the lower control arm pivot bolts.

8. Remove or disconnect the following:
- Lower control arm pivot bolt nuts (15 Series)
- Rear pivot bolt
- Front pivot bolt
- Lower control arm pivot bolt nuts (25 Series)
- Rear pivot bolt
- Front pivot bolt

9. Slowly lower the transmission jack in order to unload the front coil spring. It may be necessary to use a pry bar in order to guide the lower control arm out of position.
- Coil spring and the insulator

To install:

10. Install or connect the following:
- Coil spring and the insulator to the lower control arm

11. Raise the transmission jack in order to compress the front coil spring. It may be necessary to use a pry bar in order to guide the lower control arm into position.
- Front pivot bolt (15 Series)
- Rear pivot bolt
- Lower control arm pivot nuts. Tighten the pivot bolt nuts to 107 ft. lbs. (145 Nm).
- Front pivot bolt (25 Series)
- Rear pivot bolt
- Lower control arm pivot nuts. Tighten the pivot bolt nuts to 107 ft. lbs. (145 Nm).

12. Lower the jack. Remove the tool from the control arm.
- Front stabilizer shaft link
- Shock absorber
- Tire and wheel assembly
- Frame cross bar (25 series only). Tighten the nuts to 74 ft. lbs. (100 Nm).
- Engine protection shield

13. Remove the safety stands. Lower the vehicle.

Rear

1. Before servicing the vehicle, refer to the precautions in the beginning of this section.

2. Raise and support the vehicle.

3. Disconnect the Real Time Damping (RTD) sensor, if equipped.

4. Remove the lower shock absorber nuts and bolt from the rear axle.

5. Lower the rear axle until the springs are fully unloaded.

6. Remove the spring and the upper and lower insulators.

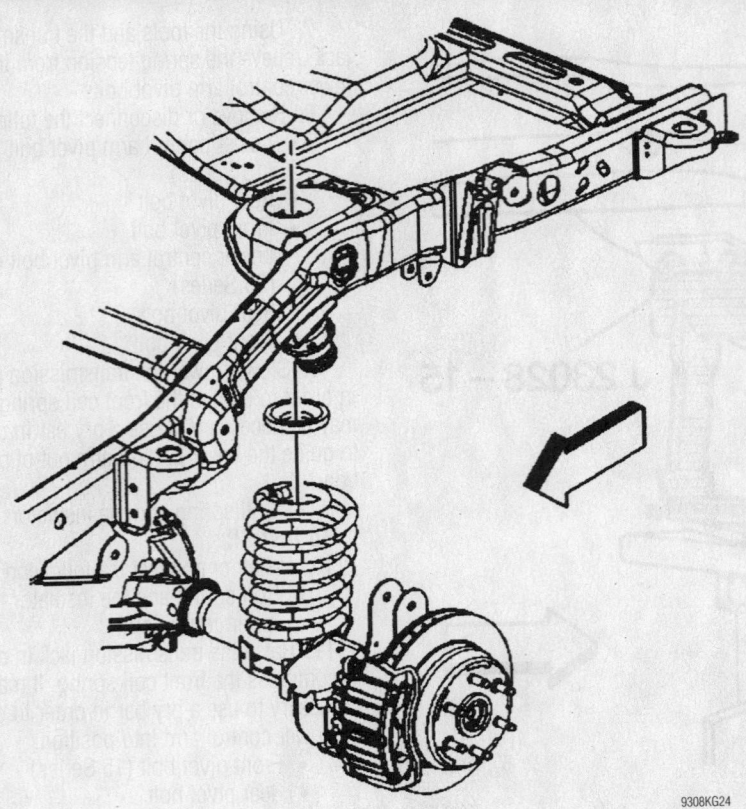

Rear coil spring removal—15 Series

To install:

7. Position the spring and the upper and lower insulators.

8. Install the rear spring to the rear axle.

9. Raise the rear axle. Install the lower shock absorber nuts to the rear axle.

10. Connect the RTD sensor, if equipped.

11. Remove the rear axle support. Lower the vehicle.

Leaf Springs

REMOVAL & INSTALLATION

1. Before servicing the vehicle, refer to the precautions in the beginning of this section.

2. Raise and support the vehicle.

3. Support the rear axle independently in order to relieve the tension on the leaf springs.

4. Remove or disconnect the following:
- Real Time Damping (RTD) sensors, if equipped
- Trailer hitch if equipped
- Fuel tank for left side applications
- U-bolt nuts and U-bolts
- Spring spacer and anchor plate
- Shackle to the frame bracket nut and the bolt

- Front spring bracket bolt
- Leaf spring assembly from the vehicle
- Shackle from the spring

To install:

5. Loosely assemble the spring shackle bracket to the frame. Install the shackle bolt. Install the shackle nut.

6. Install the leaf spring assembly to the vehicle.

7. Loosely assemble the spring to the front hanger bracket.

8. Install or connect the following:
- Front spring hanger bracket bolt
- Front spring hanger bracket nut
- Shackle to the spring bolt
- Shackle to the spring nut

➡**Do not reuse the U-bolts.**

- Spring spacer
- U-bolts
- Anchor plate
- U-bolt nuts

9. Observe the following torques:
- 14mm U-bolt nuts to 59 ft. lbs. (80 Nm)
- 16mm U-bolt nuts to 89 ft. lbs. (120 Nm)
- Front hanger bracket nut to 92 ft. lbs. (125 Nm)

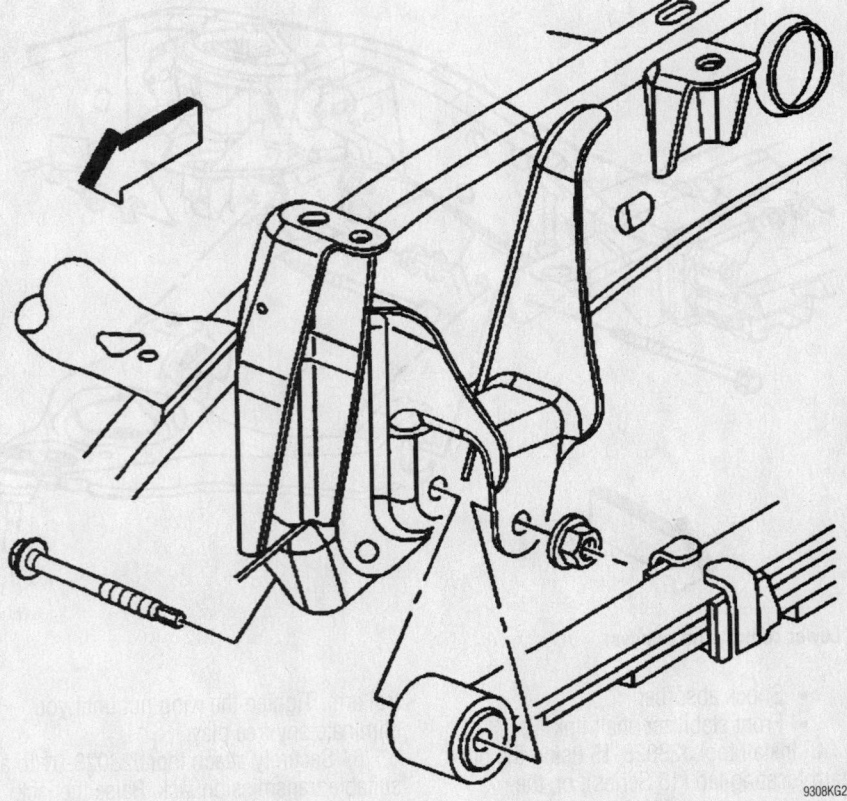

Rear leaf spring front shackle

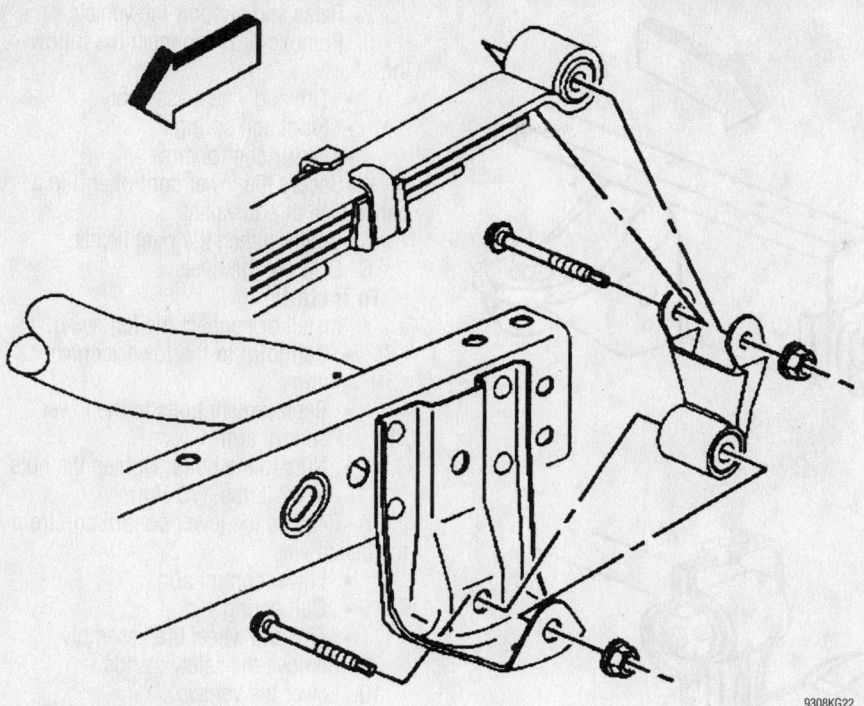

9308KG22

Rear leaf spring rear shackle

- Shackle to the frame nut to 70 ft. lbs. (95 Nm)
- Shackle to the spring nut to 70 ft. lbs. (95 Nm)
10. Install or connect the following:
    - Fuel tank for left side applications
    - Trailer hitch if equipped
    - RTD sensors (if equipped)
11. Remove the rear axle support.
12. Remove the safety stands. Lower the vehicle.

Torsion Bars

REMOVAL & INSTALLATION

1. Before servicing the vehicle, refer to the precautions in the beginning of this section.

➡ **This procedure requires the removal of both torsion bars.**

2. Raise and support the vehicle.
3. Mark the adjustment bolt setting. Install tool J36202 to the adjustment arm and the crossmember.
4. Increase the tension on the adjustment arm until the load is removed from the adjustment bolt and the adjuster nut.
5. Remove or disconnect the following:
    - Adjustment bolt and the adjuster nut
    - Tool, allowing the torsion bar to unload

- Adjustment arm by sliding the torsion bar forward until the torsion bar clears the adjustment arm. Use your hand to support the adjustment arm as the adjustment arm releases from the torsion bar.
- Torsion bar crossmember bolts from the weld nuts (15 Series)

- Upper link mounting nuts and the bolts (25 Series)
- Torsion bar crossmember

➡ **Note the position of the torsion bars as the left and right bars are different.**

- Torsion bars

To install:
6. Install or connect the following:
    - Torsion bars
    - Torsion bar crossmember
    - Torsion bar crossmember bolts to the weld nuts (15 Series). Tighten the bolt to 70 ft. lbs. (95 Nm)
    - Upper link mounting nuts and the bolts (25 Series). Tighten the nut to 70 ft. lbs. (95 Nm)
7. While supporting the adjustment arm, slide the torsion bar rearward until the torsion bar fully engages the adjustment arm. Install tool J36202 to the adjustment arm and the crossmember. Increase the tension on the adjustment arm in order to load the torsion bar.
    - Adjustment bolt and the adjuster nut
8. Remove the tool, releasing the tension on the torsion bar until the load is taken up by the adjustment bolt.
9. Remove the safety stands.
10. Lower the vehicle.
11. Measure the ride height.
12. Turn the adjustment bolt clockwise to increase the ride height and counterclockwise to decrease it.

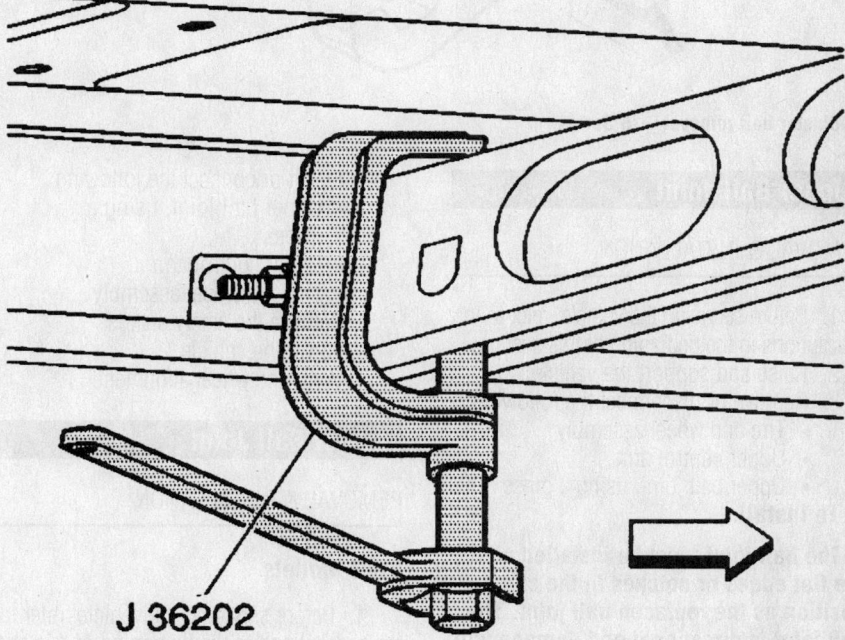

J 36202

Retainer installation—torsion bar

9308KG27

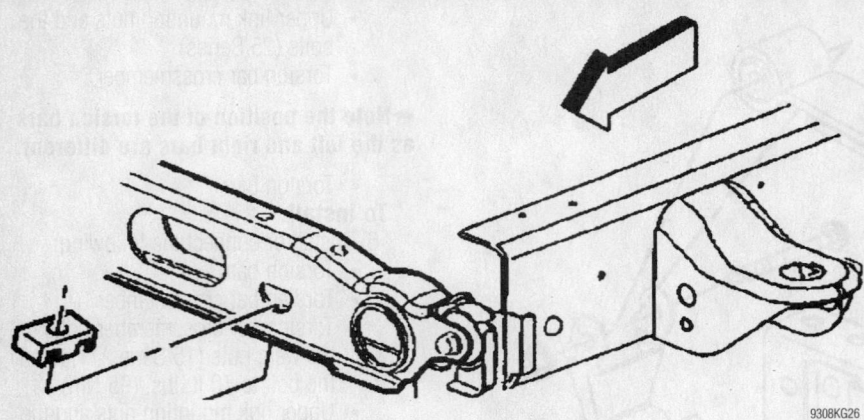

Adjuster nut removal—15 Series

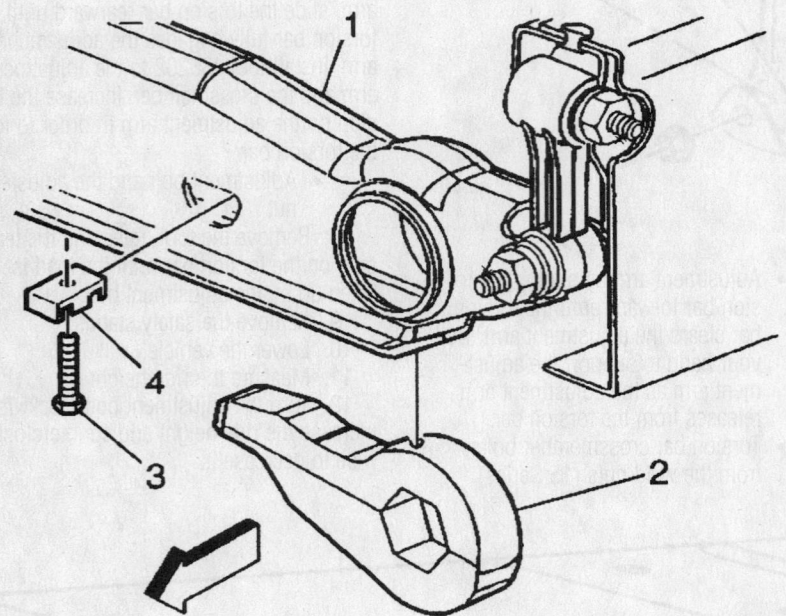

Adjuster bolt removal—15 Series

Upper Ball Joint

REMOVAL & INSTALLATION

1. Before servicing the vehicle, refer to the precautions in the beginning of this section.
2. Raise and support the vehicle.
3. Remove or disconnect the following:
   • Tire and wheel assembly
   • Upper control arm
   • Upper ball joint, using a press

To install:

➡**The ball joint must be installed with the flat edges or notches in the same position as the replaced ball joint. The ball joint is directional and damage will occur if this procedure is not followed.**

4. Install or connect the following:
   • Upper ball joint, using a press
   • Upper control arm
   • Tire and wheel assembly
5. Remove the safety stands.
6. Lower the vehicle.
7. Verify the wheel alignment.

Lower Ball Joint

REMOVAL & INSTALLATION

2WD Models

1. Before servicing the vehicle, refer to the precautions in the beginning of this section.

2. Raise and support the vehicle.
3. Remove or disconnect the following:
   • Tire and wheel assembly
   • Front coil spring
   • Lower control arm
4. Secure the lower control arm in a bench vice or equivalent.
5. Center punch the rivet heads.
6. Drill out the rivets.

To install:

7. Install or connect the following:
   • Ball joint to the lower control arm
   • Replacement bolts to the lower control arm
   • Nuts to the bolts. Tighten the nuts to 52 ft. lbs. (70 Nm).
8. Remove the lower control arm from the bench vice.
   • Lower control arm
   • Coil spring
   • Tire and wheel tire assembly
9. Remove the safety stands.
10. Lower the vehicle.
11. Verify the wheel alignment.

4WD Models

1. Before servicing the vehicle, refer to the precautions in the beginning of this section.
2. Raise and support the vehicle.
3. Remove or disconnect the following:
   • Tire and wheel assembly
   • Lower control arm
4. Place the lower control arm in a bench vice.
5. Using a chisel, remove the 4 securing crimps from the ball joint body (15 series only).
6. Using a press, remove the ball joint from the lower control arm.

To install:

➡**Use the outer flange of the ball joint in order to press the ball joint into place.**

7. Install the new ball joint using a press.
8. Place the lower control arm in a bench vice.
9. Using a punch, install 4 crimps to the ball joint. Use the replaced ball joint as a reference (15 series only).
10. Install or connect the following:
    • Lower control arm
    • Tire and wheel assembly
11. Remove the safety stands.
12. Lower the vehicle.
13. Verify the wheel alignment.

Upper Control Arm

REMOVAL & INSTALLATION

1. Before servicing the vehicle, refer to the precautions in the beginning of this section.
2. Raise and support the vehicle.
3. Remove or disconnect the following:
   - Tire and wheel assembly
   - Real Time Damping (RTD) link rod from the sensor, if equipped
   - Retaining bolt for the brake hose and the wheel speed sensor brackets
   - Halfshaft
   - Nut at the upper ball joint. Discard the nut.
   - Upper control arm from the steering knuckle
   - Upper control arm nuts and the adjustment cams (15 Series RWD, 4WD, and 25 Series RWD)
   - Upper control arm bolts (15 Series RWD, 4WD, and 25 Series RWD)
   - Upper control arm nuts and the adjustment cams (25 Series 4WD)
   - Upper control arm bolts (25 Series 4WD)
   - Upper control arm

To install:
4. Install or connect the following:
   - Upper control arm
   - Upper control arm bolts (25 Series 4WD)
   - Upper control arm nuts and the adjustment cams (25 Series 4WD). Tighten the nuts to 140 ft. lbs. (190 Nm).

- Upper control arm bolts (15 Series RWD, 4WD, and 25 Series RWD)
- Upper control arm nuts and the adjustment cams (2) (15 Series RWD, 4WD, and 25 Series RWD). Tighten the nuts to 140 ft. lbs. (190 Nm).
- Upper control arm to the steering knuckle
- Halfshaft
- New nut to the upper ball joint stud. Tighten the nut to 37 ft. lbs. (50 Nm).
- Retaining bolts for the brake hose and wheel speed sensor brackets. Tighten the bolts to 80 inch lbs. (9 Nm).
- RTD link rod to the sensor, if equipped
- Tire and wheel assembly
5. Remove the safety stands.
6. Lower the vehicle. Verify the wheel alignment.

CONTROL ARM BUSHING REPLACEMENT

1. The control arm bushings are removed and installed using a press.

Lower Control Arm and Bushing

REMOVAL & INSTALLATION

2WD Models

1. Before servicing the vehicle, refer to the precautions in the beginning of this section.

2. Raise and support the vehicle.
3. Remove or disconnect the following:
   - Tire and wheel assembly
   - Real Time Damping (RTD) link rod from the sensor, if equipped
   - Shock absorber
   - Front stabilizer shaft link
   - Front coil spring
   - Lower control arm nuts and the washers or bolts, as applicable
   - Lower control arm nuts and washers and/or bolts, as applicable
   - Lower ball joint stud nut
   - Lower ball joint stud from the steering knuckle
   - Lower control arm

To install:
4. Install or connect the following:
   - Lower control arm
   - Ball joint stud to the steering knuckle
   - Lower ball joint stud nut. Tighten the lower ball joint stud nut to 74 ft. lbs. (100 Nm).
   - Front coil spring
   - Lower control arm bolts (15 Series)
   - Lower control arm nuts and the washers (15 Series). Tighten the lower control arm nuts to 107 ft. lbs. (145 Nm).
   - Lower control arm bolt (25 Series)
   - Lower control arm nuts and the washers (25 Series). Tighten the lower control arm nuts to 107 ft. lbs. (145 Nm).
   - Front stabilizer shaft link.
   - Shock absorber
   - RTD sensor, if equipped
   - Tire and wheel assembly
5. Remove the safety stands. Lower the vehicle. Verify the wheel alignment.

4WD Models

1. Before servicing the vehicle, refer to the precautions in the beginning of this section.
2. Raise and support the vehicle.
3. Remove or disconnect the following:
   - Tire and wheel assembly
   - Real Time Damping (RTD) link rod from the sensor, if equipped
   - Stabilizer shaft links from the lower control arm
   - Shock absorber nut and the bolt
   - Torsion bars
   - Halfshaft
   - Lower ball joint stud nut
   - Lower ball joint stud from the steering knuckle
   - Lower control arm nuts and the washers (15 Series)

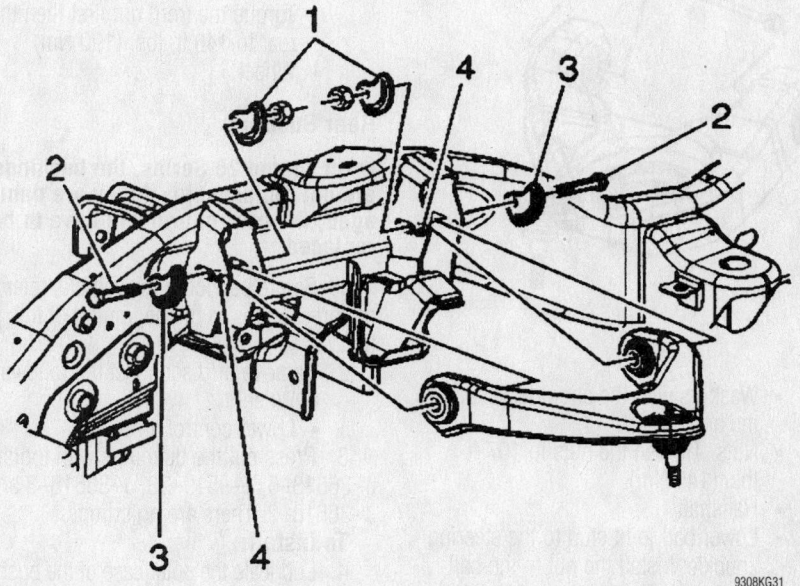

9308KG31

Upper control arm

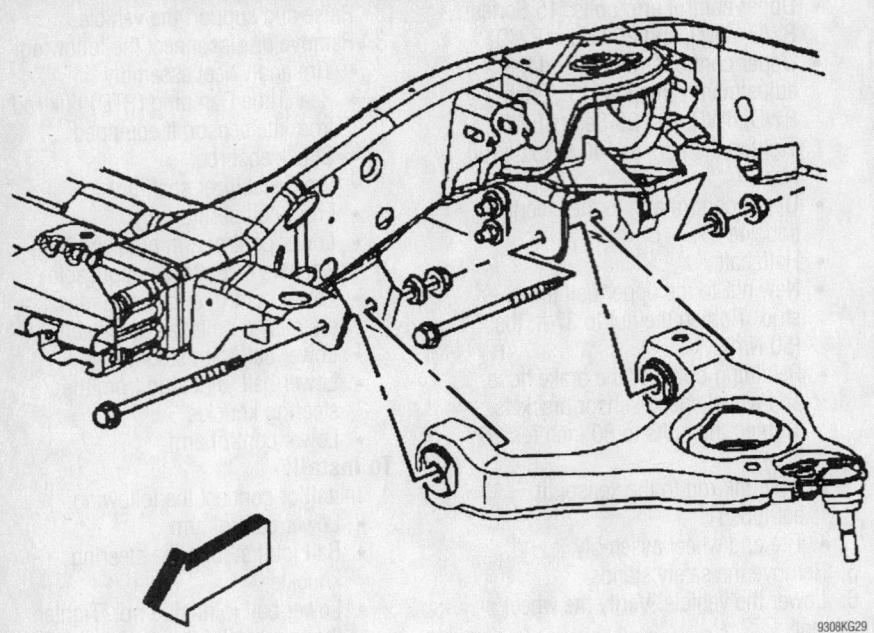

2WD lower control arm—15 Series

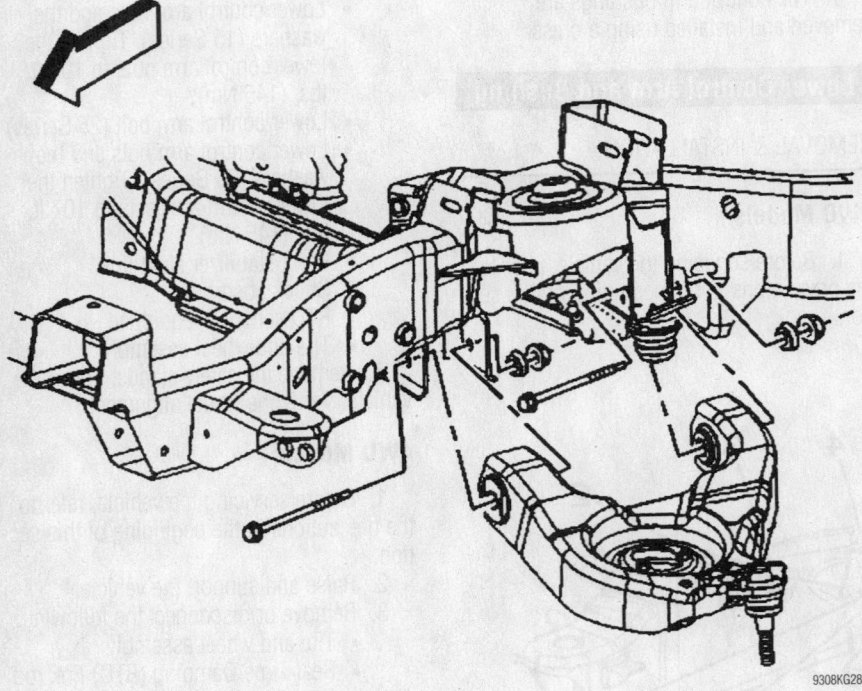

2WD lower control arm

- Lower control arm bolts
- Lower control arm nuts and the washers (25 Series)
- Lower control arm bolts
- Lower control arm

To install:
- Lower control arm
- Lower control arm bolts (15 Series)

- Washers with the shoulder facing the arm
- Nuts. Tighten the nuts to 107 ft. lbs. (145 Nm).
- Halfshaft
- Lower ball joint stud to the steering knuckle. Install the nut to the ball joint stud. Tighten the nut to 74 ft. lbs. (100 Nm).

- Torsion bars
- Shock absorber through nut and bolt
- Stabilizer shaft links to the lower control arm
- RTD link rod to the sensor, if equipped
- Tire and wheel assembly

4. Remove the safety stands. Lower the vehicle. Verify the wheel alignment.

CONTROL ARM BUSHING REPLACEMENT

Front Bushing

1. Before servicing the vehicle, refer to the precautions in the beginning of this section.

2. On 15 and 25 Series, the bushings are not replaceable. If they are damaged, the control arm will have to be replaced.

3. Remove or disconnect the following:
- Wheel
- Lower control arm
- Unbend the crimps with a punch on the front bushing

4. Press out the bushings with tools J–36618–2, J–9519–23, J–36618–4 and 36618–1.

To install:

5. Lubricate the outer case of the bushing.

6. Install or connect the following:
- Bushing into control arm

7. Press in the bushings with tools J–36618–2, J–9519–23, J–36618–4 and 36618–1 until the bushing is seated in.

8. After bushing is installed crimp it in place.
- Control arm and mounting bolts. Torque the front nut first then the rear to 140 ft. lbs. (190 Nm)
- Wheel

Rear Bushing

➡ **On 15 and 25 Series, the bushings are not replaceable. If they are damaged, the control arm will have to be replaced.**

1. Before servicing the vehicle, refer to the precautions in the beginning of this section.

2. Remove or disconnect the following:
- Wheel
- Lower control arm

3. Press out the bushings with tools. J–36618–5, J–9519–23, J–36618–3 and J–36618–2. There are no crimps.

To install:

4. Lubricate the outer case of the bushing.

5. Install or connect the following:

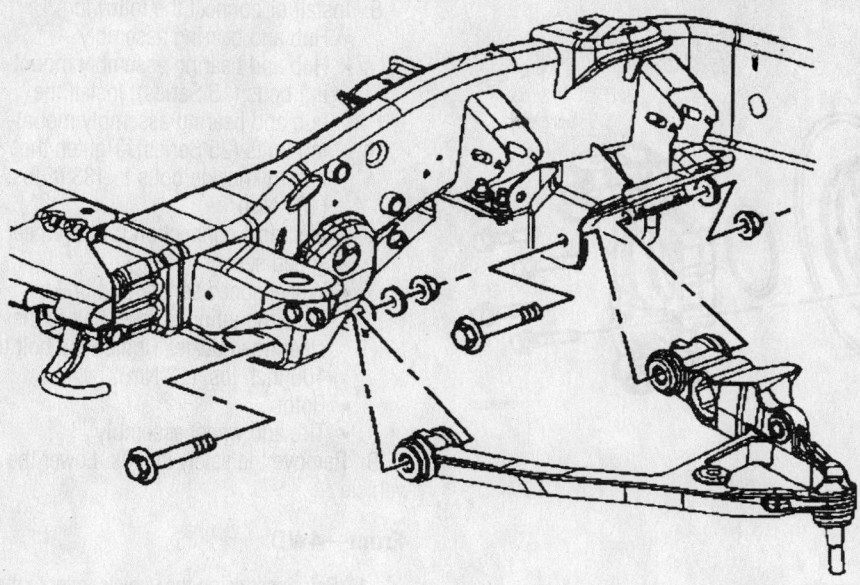

4WD lower control arm—15 Series

- Bushing into control arm
6. Press in the bushings with tools J–36618–5, J–9519–23, J–36618–3 and J–36618–2. There are no crimps.
- Control arm and mounting bolts. Torque the front nut first then the rear to 140 ft. lbs. (190 Nm).
- Wheel

Wheel Bearings

ADJUSTMENT

➡ **The front wheel bearings on 2-wheel drive vehicles (exc. 2000–01 vehicles) are adjustable.**

1. Before servicing the vehicle, refer to the precautions in the beginning of this section.
2. Remove the dust cap, cotter pin.
3. Loosen the spindle nut.
4. Spin the wheel hub by hand and tighten the nut until it is just snug—12 ft. lbs. (16 Nm) Back off the nut until it is loose, then tighten it finger-tight. Loosen the nut until either hole in the spindle lines up with a slot in the nut and insert a new cotter pin. There should be 0.001–0.008 in. (0.025–0.200mm) end-play. This can be measured with a dial indicator, if you wish.
5. Replace the dust cap, wheel and tire.

REMOVAL & INSTALLATION

Front—2WD

1. Before servicing the vehicle, refer to the precautions in the beginning of this section.
2. Raise and support the vehicle.

3. Remove or disconnect the following:
- Tire and wheel assembly
- Rotor
- Wheel speed sensor and brake hose mounting bracket bolt from the steering knuckle
- Electrical connection for the wheel speed sensor
- Hub and bearing assembly mounting bolts
- Hub and bearing assembly
- O-ring seal from the steering knuckle bore (25 Series)
4. Clean and inspect the O-ring seal (25 Series).

To install:
5. Clean all corrosion or contaminates from the steering knuckle bore and the hub and bearing assembly.
6. Install the O-ring to the steering knuckle (25 Series).
7. Lubricate the steering knuckle bore with wheel bearing grease or the equivalent.

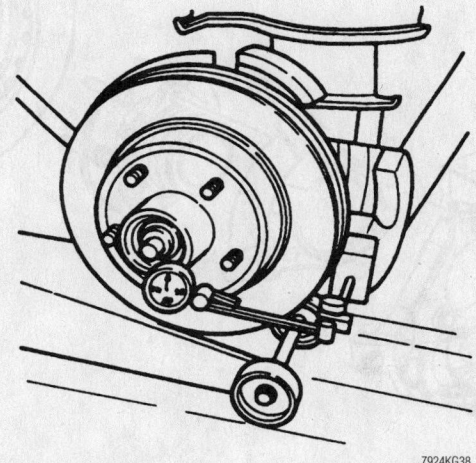

Use a dial indicator to measure the wheel bearing end-play—2-wheel drive vehicles

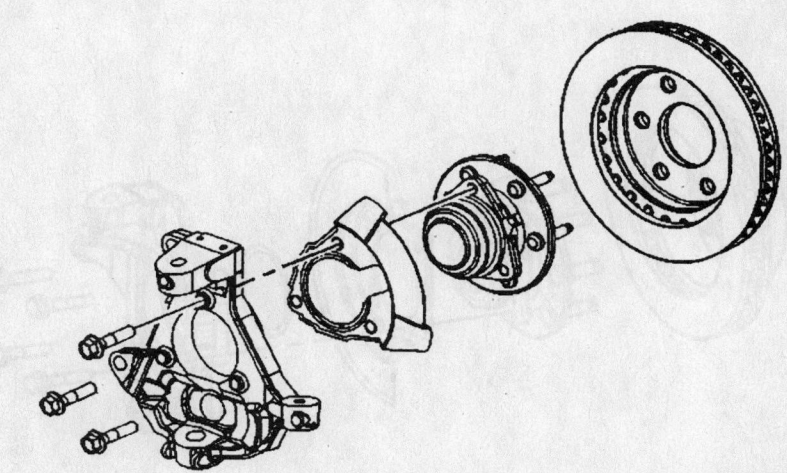

2WD front hub—15 Series

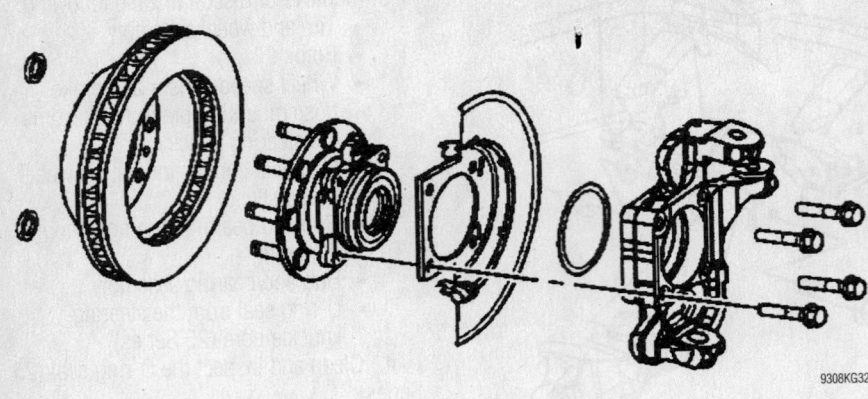

2WD front hub—25 Series

9308KG32

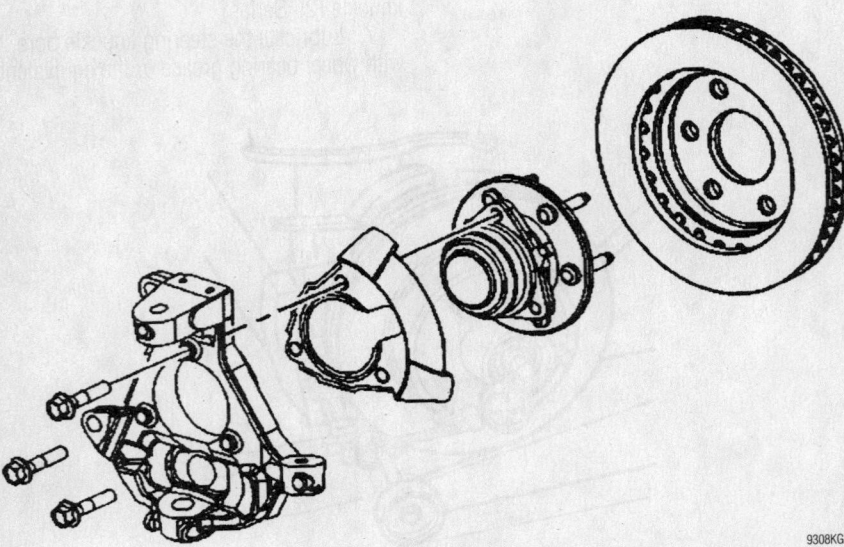

4WD front hub—15 Series

9308KG35

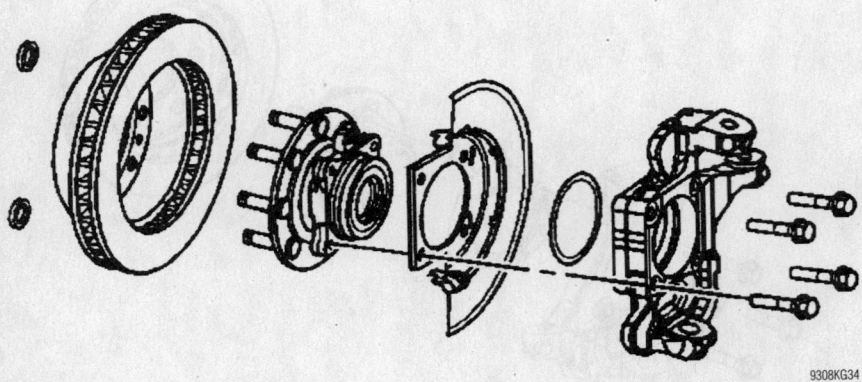

4WD front hub—25 Series

9308KG34

8. Install or connect the following:
- Hub and bearing assembly
- Hub and bearing assembly mounting bolts (15 Series). Install the hub and bearing assembly mounting bolts (25 Series). Tighten the hub to knuckle bolts to 133 ft. lbs. (180 Nm).
- Electrical connection for the wheel speed sensor
- Wheel speed sensor and brake hose mounting bracket bolt to the steering knuckle. Tighten the bolt to 106 inch lbs. (12 Nm).
- Rotor
- Tire and wheel assembly

9. Remove the safety stands. Lower the vehicle .

Front—4WD

1. Before servicing the vehicle, refer to the precautions in the beginning of this section.
2. Raise and support the vehicle.
3. Remove or disconnect the following:
- Tire and wheel assembly
- Rotor
- Wheel speed sensor and brake hose mounting bracket bolt from the steering knuckle
- Electrical connection for the wheel speed sensor
- Front drive halfshaft assembly
- Hub and bearing assembly mounting bolts

4. Remove the hub and bearing assembly.
5. Remove the O-ring seal from the steering knuckle bore (25 Series).
6. Clean and inspect the O-ring seal (25 Series).

To install:

7. Clean all corrosion or contaminates from the steering knuckle bore and the hub and bearing assembly.
8. Install the O-ring to the steering knuckle (25 Series).
9. Lubricate the steering knuckle bore with wheel bearing grease or the equivalent.
10. Install or connect the following:
- Hub and bearing assembly
- Hub and bearing assembly mounting bolts Tighten the hub to knuckle bolts to 133 ft. lbs. (180 Nm)
- Front drive halfshaft assembly
- Electrical connection for the wheel speed sensor
- Wheel speed sensor and brake hose mounting bracket bolt to the steering knuckle. Tighten the bolt to 106 inch lbs. (12 Nm).
- Rotor
- Tire and wheel assembly

BRAKES

Brake Caliper

REMOVAL & INSTALLATION

➡There are 2 caliper designs and they can be identified by the method used to secure the assembly to the spindle bracket. The Delco caliper is secured by a bolt and sleeve combination. The Bendix caliper assembly is secured by a slider, spring and bolt.

1. Before servicing the vehicle, refer to the precautions in the beginning of this section.
2. Remove or disconnect the following:
   - ⅔ of the brake fluid from the master cylinder
   - Front wheels and tires
3. Position a C-clamp around the outside pad and caliper; tighten the C-clamp until the caliper piston bottoms in its bore.

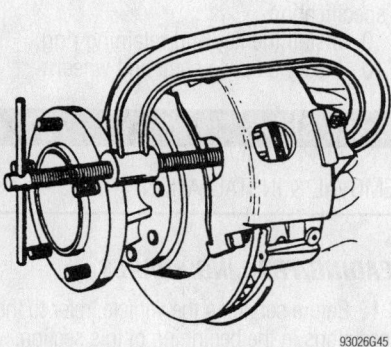

Compressing the caliper piston

- Brake hose from the caliper by removing the inlet fitting
- Bolt and sleeve or bolt and slider assemblies that hold the caliper and then lift the caliper off the rotor
- Inboard and outboard pad

To install:
4. Install or connect the following:
   - Pads onto the caliper
   - Caliper onto the knuckle/rotor assembly and secure the assembly with the mounting bolts or sliders
   - Brake line to the caliper
5. Bleed the brakes
6. Pump the brake pedal and verify there is minimal brake pedal travel.
7. Check the brake fluid level. Install the tire and wheel assembly.
8. Lower the vehicle.

Disc Brake Pads

REMOVAL & INSTALLATION

DELCO TYPE

1. Before servicing the vehicle, refer to the precautions in the beginning of this section.
   - ⅔ of the fluid from the brake master cylinder
   - Wheels
2. Compress the brake piston back into its bore using a C-clamp.
   - 2 bolts holding the caliper and then lift the caliper off the disc
   - Inboard and outboard shoe
   - Pad support spring from the piston, if equipped

To install:
3. Thoroughly inspect, clean and lubricate all caliper slide points, bolts and hardware.
4. Install or connect the following:
   - Retainer spring on the inner pad and insert the assembly into the center cavity of the piston
5. Push down on the inner pad until it lays flat against the caliper. It is important to push the piston all the way into the caliper if new linings are installed or the caliper will not fit over the rotor.
   - Outboard pad with the ears of the pad over the caliper ears and the tab at the bottom engaged in the caliper cutout
   - Caliper over the brake disc and align the holes in the caliper with those of the mounting bracket
   - Mounting bracket bolts through the sleeves in the inboard caliper ears and through the mounting bracket, making sure the ends of the bolts pass under the retaining ears on the inboard pad
6. Tighten the mounting bolts to 38 ft. lbs. (51 Nm). After both calipers are mounted pump the brake pedal to seat the pad against the rotor. Use a pair of locking pliers to bend over the upper ears of the outer pad so it isn't loose.
   - Wheels
7. Add fluid to the master cylinder reservoirs so they are ¼ in. (6.35mm) from the top.
8. Test the brake pedal by pumping it to obtain a hard pedal. Check the fluid level again and add fluid as necessary. Do not move the vehicle until a pedal is obtained.

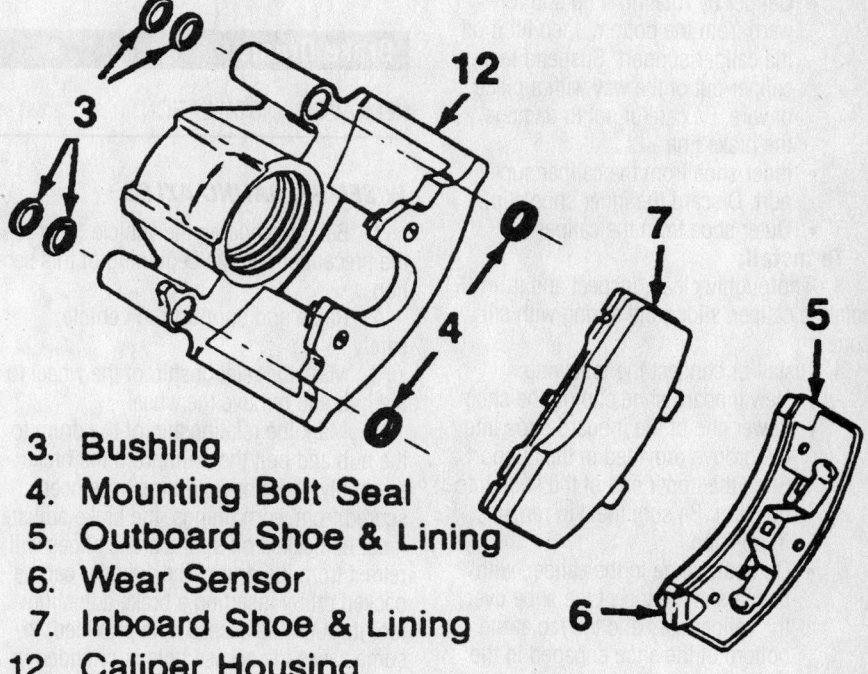

3. **Bushing**
4. **Mounting Bolt Seal**
5. **Outboard Shoe & Lining**
6. **Wear Sensor**
7. **Inboard Shoe & Lining**
12. **Caliper Housing**

Replacing the disc brake pads—Delco type

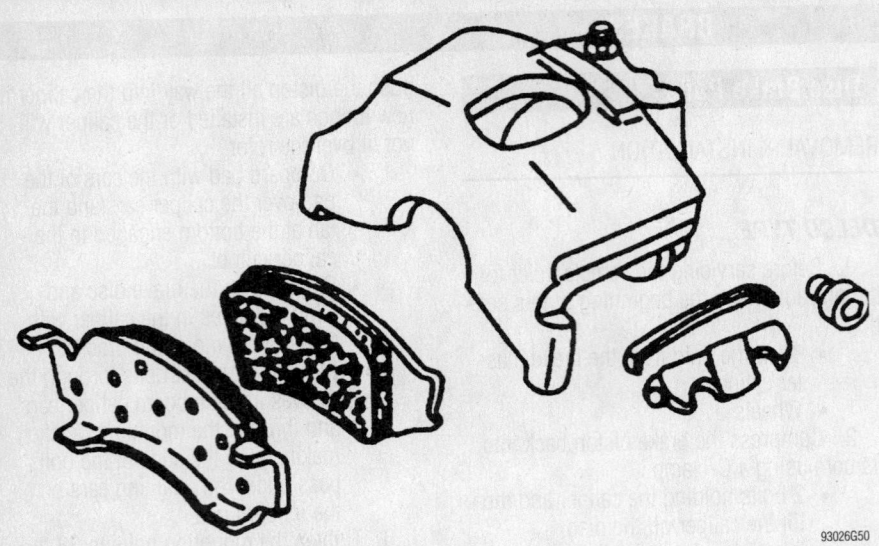

93026G50

Replacing the disc brake pads—Bendix type

BENDIX TYPE

1. Before servicing the vehicle, refer to the precautions in the beginning of this section.
   - ⅔ of the fluid from the brake master cylinder
   - Wheels
2. Compress the brake piston back into its bore using a C-clamp.
   - Bolt at the caliper slider. Use a brass drift pin to remove the slider and spring.
   - Caliper by rotating it up and forward from the bottom, then lift it off the caliper support. Suspend the caliper out of the way with a piece of wire. Be careful not to damage the brake line.
   - Inner shoe from the caliper support. Discard the inner shoe clip.
   - Outer shoe from the caliper

To install:

3. Thoroughly clean, inspect and lubricate the caliper, slider and spring with silicone.
4. Install or connect the following:
   - New inboard shoe clip on the shoe
   - Lower end of the inboard shoe into the groove provided in the support. Slide the upper end of the shoe into position. Be sure the clip remains in position.
   - Outboard shoe in the caliper, with the ears at the top of the shoe over the caliper ears and the tab at the bottom of the shoe engaged in the caliper cutout. If assembly is difficult, a C-clamp may be used. Be careful not to damage the lining.

- Caliper over the brake disc, top edge first. Rotate the caliper downward onto the support.
- Spring over the caliper support key, install the assembly between the support and lower caliper groove. Tap into place until the key retaining screw can be installed.
- Screw and torque to 15 ft. lbs. (20 Nm). The boss must fit fully into the circular cutout in the key.
- Wheel and add brake fluid as necessary.

Brake Drums

REMOVAL & INSTALLATION

W/SEMI-FLOATING AXLES

1. Before servicing the vehicle, refer to the precautions in the beginning of this section.
2. Raise and support the vehicle safely.
3. Mark the relationship of the wheel to the hub and remove the wheel.
4. Mark the relationship of the drum to the hub and pull the drum from the brake assembly. If the brake drums have been scored from worn linings, the brake adjuster must be backed off so the brake shoes will retract from the drum. The adjuster can be backed off by inserting a brake adjusting tool through the access hole provided. In some cases the access hole is provided in the brake drum. A metal cover plate is over the hole. This may be removed by using a hammer and chisel.

To install:

5. Align the mark on the drum to mark on hub and install drum
6. Align the mark on the wheel to mark on drum and install wheel
7. Adjust brake lining as needed. Pump brakes

W/FULL FLOATING AXLES

To remove the drums from full floating rear axles, the axle shaft will have to be removed. Full-floating rear axles can be identified by a bearing housing that protrudes through the center of the wheel.

1. Before servicing the vehicle, refer to the precautions in the beginning of this section.
2. Raise and support the vehicle safely.
3. Remove the wheel.
4. Remove the axle shaft.
5. Remove the retaining ring, key and adjusting nut.
6. Remove the hub and drum.

To install:

7. Install the hub and drum to the tube.
8. Install the adjusting nut and torque to specification.
9. Install the key and retaining ring.
10. Install the axle shaft and wheel.

Brake Shoes

REMOVAL & INSTALLATION

LEADING/TRAILING BRAKES

1. Before servicing the vehicle, refer to the precautions in the beginning of this section.
2. Remove or disconnect the following:
   - Tire and wheel assembly
   - Brake drums
   - Raise the lever arm of the actuator until the upper end is clear of the slot in the adjuster screw.
   - Actuator off the adjuster pin by sliding it
   - Actuator spring from the shoe
   - Hold-down spring assemblies and pins
   - Bottom ends of the shoes apart and lift the lower return spring over the anchor plate. Allow the shoe ends to come together and remove the spring.
   - Shoe assembly, along with the upper return spring and the adjusting screw assembly
   - Upper return spring and the adjusting screw assembly from the shoes
   - Retaining ring, pin, spring washer, and parking brake lever

To install:

3. Clean adjuster wheel and the backing plates with a suitable cleaner. Lubricate the backing plate contact points, levers and adjuster with a suitable lubricant.

4. Assemble the parking lever, spring washer (concave side facing the brake lever), pin, and retaining ring onto the rearward shoe.

5. Install or connect the following:

- Adjuster pin in the forward shoe with the pin projecting 0.276 in. (7mm) from the side of the shoe web where the adjuster actuator is installed
- Upper return spring, with the brake shoes resting on a flat surface (the shoe with the parking lever to the rear of the vehicle)
- Adjuster screw assembly with the spring clip facing the backing plate
- Shoes in position on the backing plate. Do not place the lower shoe webs under the anchor plate.
- Lower return spring, spread the bottom of the shoes and position the shoe against the backing plate
- Hold-down pins and spring assemblies
- Adjuster actuator over the end of

the adjuster pin so the top leg engages the notch in the adjuster screw
- Actuator spring, being careful not to over-stretch it more than 3.27 in. (83mm)
- Parking brake cable to the lever

6. Adjust the parking brake if the shoes will not totally retract.

7. Install the drum, tire and wheel assembly. Adjust the rear brakes and lower the vehicle.

DUO-SERVO BRAKES

1. Before servicing the vehicle, refer to the precautions in the beginning of this section.

2. Remove or disconnect the following:

- Tire and wheel assembly
- Brake drums
- Shoe return springs, using a brake tool
- Shoe guide
- Hold-down springs and pins
- Actuator lever and pivot
- Lever return spring
- Actuator link, parking brake strut, spring retaining ring
- Parking brake lever and washer
- Shoe assemblies

- Adjuster screw and spring from the shoe assembly

To install:

3. Use a brake cleaning fluid to remove dirt from the brake drum. Check the drums for scoring, cracks and for out-of-round; service the drums as necessary.

4. Check the wheel cylinders by carefully pulling the lower edges of the wheel cylinder boots away from the cylinders. If there is excessive leakage, the inside of the cylinder will drip fluid; repair or replace as necessary.

5. Check the flange plate, which is located around the axle, for leakage of differential lubricant.

6. Lightly lubricate the parking brake cable, parking brake lever where it enters the shoe and the backing plate-to-shoe contact points. Use high temperature, waterproof, grease or special brake lube.

7. Install or connect the following:

- Parking brake lever into the secondary shoe with the attaching bolt, spring washer, lockwasher, and nut. It is important that the lever move freely before the shoe is attached. Move the assembly and check for proper action.

8. Lubricate the adjusting screw and make sure it works freely.

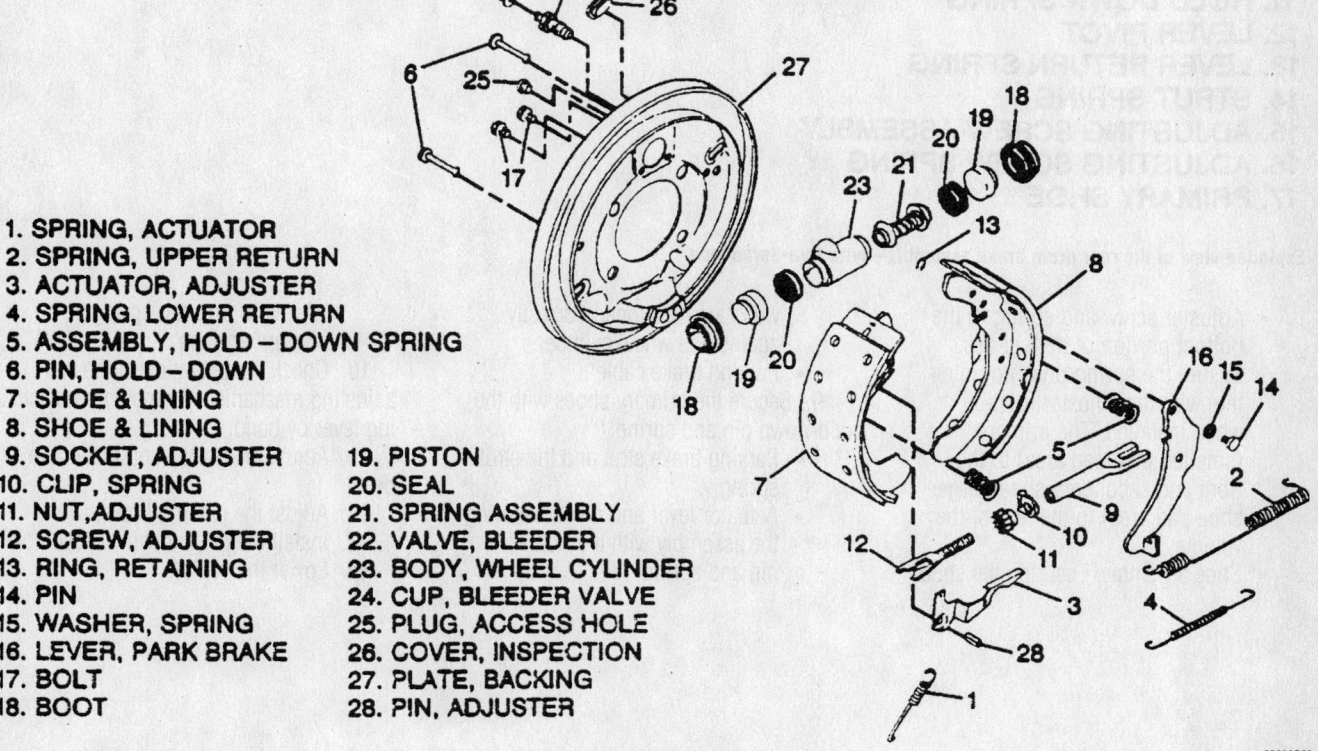

1. SPRING, ACTUATOR
2. SPRING, UPPER RETURN
3. ACTUATOR, ADJUSTER
4. SPRING, LOWER RETURN
5. ASSEMBLY, HOLD–DOWN SPRING
6. PIN, HOLD–DOWN
7. SHOE & LINING
8. SHOE & LINING
9. SOCKET, ADJUSTER
10. CLIP, SPRING
11. NUT, ADJUSTER
12. SCREW, ADJUSTER
13. RING, RETAINING
14. PIN
15. WASHER, SPRING
16. LEVER, PARK BRAKE
17. BOLT
18. BOOT
19. PISTON
20. SEAL
21. SPRING ASSEMBLY
22. VALVE, BLEEDER
23. BODY, WHEEL CYLINDER
24. CUP, BLEEDER VALVE
25. PLUG, ACCESS HOLE
26. COVER, INSPECTION
27. PLATE, BACKING
28. PIN, ADJUSTER

Exploded view of the rear drum brake assembly—with Leading/Trailing type

93026G52

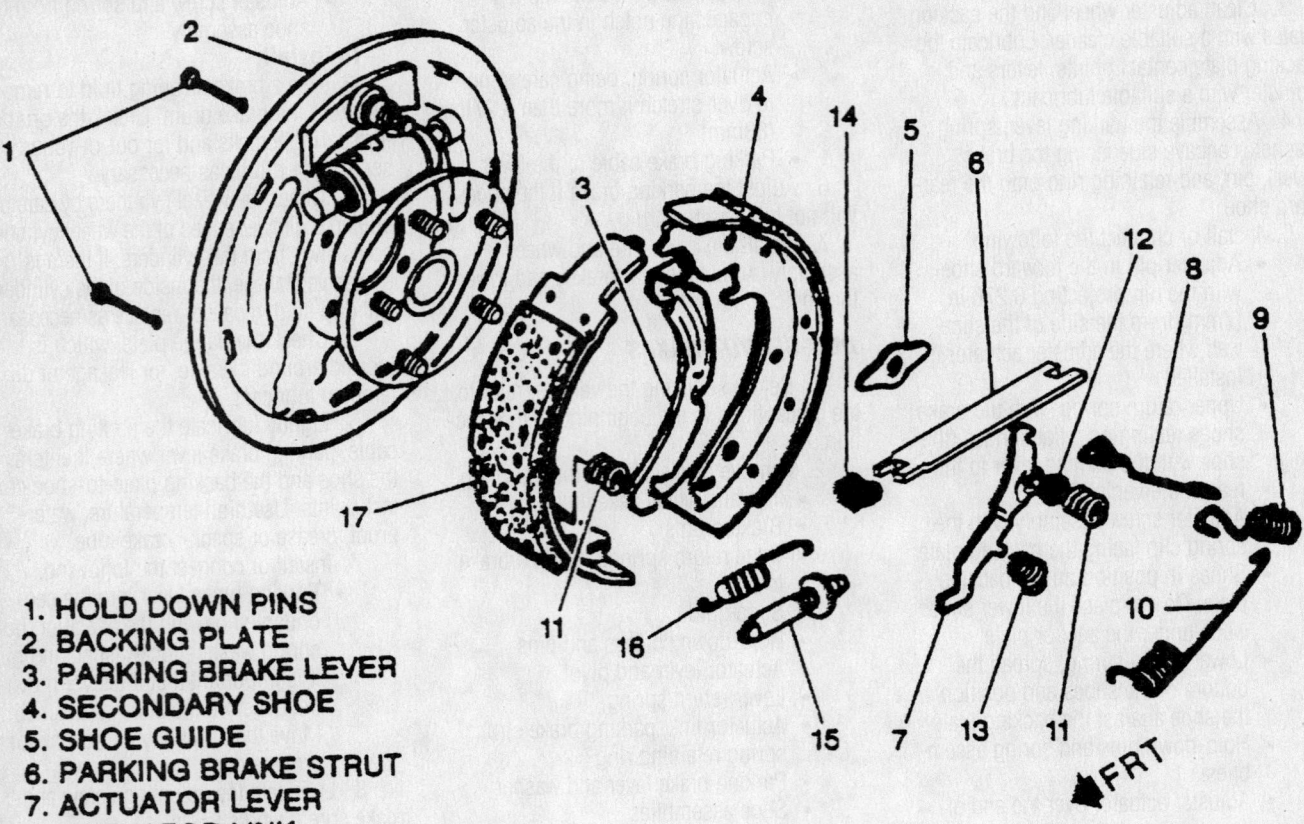

1. HOLD DOWN PINS
2. BACKING PLATE
3. PARKING BRAKE LEVER
4. SECONDARY SHOE
5. SHOE GUIDE
6. PARKING BRAKE STRUT
7. ACTUATOR LEVER
8. ACTUATOR LINK
9. RETURN SPRING
10. RETURN SPRING
11. HOLD DOWN SPRING
12. LEVER PIVOT
13. LEVER RETURN SPRING
14. STRUT SPRING
15. ADJUSTING SCREW ASSEMBLY
16. ADJUSTING SCREW SPRING
17. PRIMARY SHOE

93026G53

Exploded view of the rear drum brake assembly—with Duo-Servo type

- Adjuster screw and spring to the bottom portion of both shoes. Ensure the spring does not interfere with the adjuster rotation when installed. The primary (smaller shoe pad area) to the front and secondary shoe (larger shoe pad area) to the rear of the vehicle.
- Shoe assembly. Ensuring the shoe webs are positioned correctly against the wheel cylinder.
- Parking brake cable

9. Secure the primary shoes with the hold-down pin and spring.
- Parking brake strut and the strut spring
- Actuator lever and pivot, securing the assembly with the hold-down pin and spring

- Actuator link and spring
- Return springs

10. Check the operation of the self-adjusting mechanism by moving the actuating lever by hand.
11. Adjust the brakes and install the drum.
12. Adjust the parking brake.
13. Install the tire and wheel assembly.
14. Lower the vehicle.

SPECIFICATION CHARTS

ENGINE AND VEHICLE IDENTIFICATION

		Engine						Model Year	
Code ①	Liters (cc)	Cu. In.	Cyl.	Fuel Sys.	Engine Type	Eng. Mfg.	Code ②		Year
4	2.2 (2189)	134	4	MFI	OHV	CPC	Y		2000
W	4.3 (4293)	263	6	MFI	OHV	CPC	1		2001
							2		2002
							3		2003
							4		2004

CPC: Chevrolet/Pontiac/Canada

MFI: Multi-port Fuel Injection

① 8th position of VIN

② 10th position of VIN

42372-S10P-C01

GENERAL ENGINE SPECIFICATIONS
All measurements are given in inches.

Year	Model	Engine Displacement Liters (cc)	Engine Series (ID/VIN)	Fuel System	Net Horsepower @ rpm	Net Torque @ rpm (ft. lbs.)	Bore x Stroke (in.)	Compression Ratio	Oil Pressure @ rpm
2000	S10	2.2 (2189)	4	MFI	120@5000	140@3600	3.50x3.46	9.0:1	56@3000
		4.3 (4293)	W	MFI	230@4600	285@2800	4.00x3.48	9.4:1	18@2000
	Sonoma	2.2 (2189)	4	MFI	118@5200	130@2800	3.50x3.46	9.0:1	56@3000
		4.3 (4293)	W	MFI	①	②	4.00x3.48	9.2:1	18@2000
2001	S10	2.2 (2189)	4	MFI	120@5000	140@3600	3.50x3.46	9.0:1	56@3000
		4.3 (4293)	W	MFI	230@4600	285@2800	4.00x3.48	9.4:1	18@2000
	Sonoma	2.2 (2189)	4	MFI	118@5200	130@2800	3.50x3.46	9.0:1	56@3000
		4.3 (4293)	W	MFI	①	②	4.00x3.48	9.2:1	18@2000
2002	S10	2.2 (2189)	4	MFI	120@5000	140@3600	3.50x3.46	9.0:1	56@3000
		4.3 (4293)	W	MFI	230@4600	285@2800	4.00x3.48	9.4:1	18@2000
	Sonoma	2.2 (2189)	4	MFI	118@5200	130@2800	3.50x3.46	9.0:1	56@3000
		4.3 (4293)	W	MFI	①	②	4.00x3.48	9.2:1	18@2000
2003-04	S10	2.2 (2189)	4	MFI	120@5000	140@3600	3.50x3.46	9.0:1	56@3000
		4.3 (4293)	W	MFI	230@4600	285@2800	4.00x3.48	9.4:1	18@2000
	Sonoma	2.2 (2189)	4	MFI	118@5200	130@2800	3.50x3.46	9.0:1	56@3000
		4.3 (4293)	W	MFI	①	②	4.00x3.48	9.2:1	18@2000

MFI: Multi-port Fuel Injection

① Below 15,000 GVWR: 180@3400
　Above 15,000 GVWR: 190@3400

② 2WD: 180@4400
　4WD: 190@4400

③ 2WD: 245@2800
　4WD: 250@2800

④ 2WD: 170@4400
　4WD: 180@4400

42372-S10P-C02

GASOLINE ENGINE TUNE-UP SPECIFICATIONS

Year	Engine Displacement Liters (cc)	Engine ID/VIN	Spark Plugs Gap (in.)	Ignition Timing (deg.)		Fuel Pump (psi)	Idle Speed (rpm)		Valve Clearance	
				MT	AT		MT	AT	In.	Ex.
2000	2.2 (2189)	4	0.060	①	①	41-47	②	②	HYD	HYD
	4.3 (4293)	W	0.060	①	①	58-64 ③	600	625	HYD	HYD
2001	2.2 (2189)	4	0.040	①	①	41-47	②	②	HYD	HYD
	4.3 (4293)	W	0.060	①	①	58-64 ③	600	625	HYD	HYD
2002	2.2 (2189)	4	0.040	①	①	41-47	②	②	HYD	HYD
	4.3 (4293)	W	0.060	①	①	58-64 ③	600	625	HYD	HYD
2003-04	2.2 (2189)	4	0.040	①	①	41-47	②	②	HYD	HYD
	4.3 (4293)	W	0.060	①	①	58-64 ③	600	625	HYD	HYD

NOTE: The Vehicle Emission Control Information label often reflects specification changes made during production. The label figures must be used if they differ from those in this chart.

HYD: Hydraulic

① Ignition timing is preset and cannot be adjusted

② Idle speed is maintained by the PCM

③ With key ON and engine OFF

42372-S10P-C03

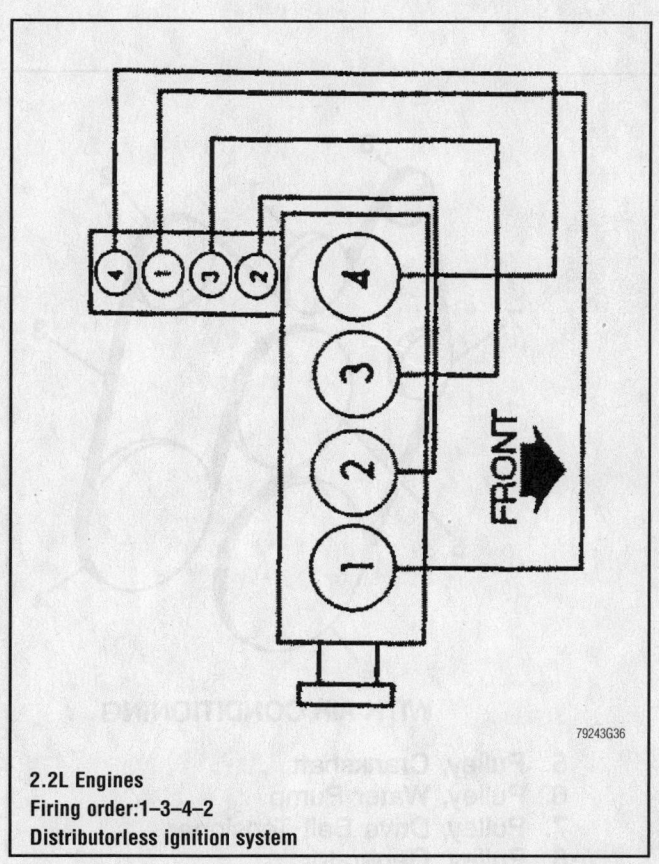

2.2L Engines
Firing order:1–3–4–2
Distributorless ignition system

79243G36

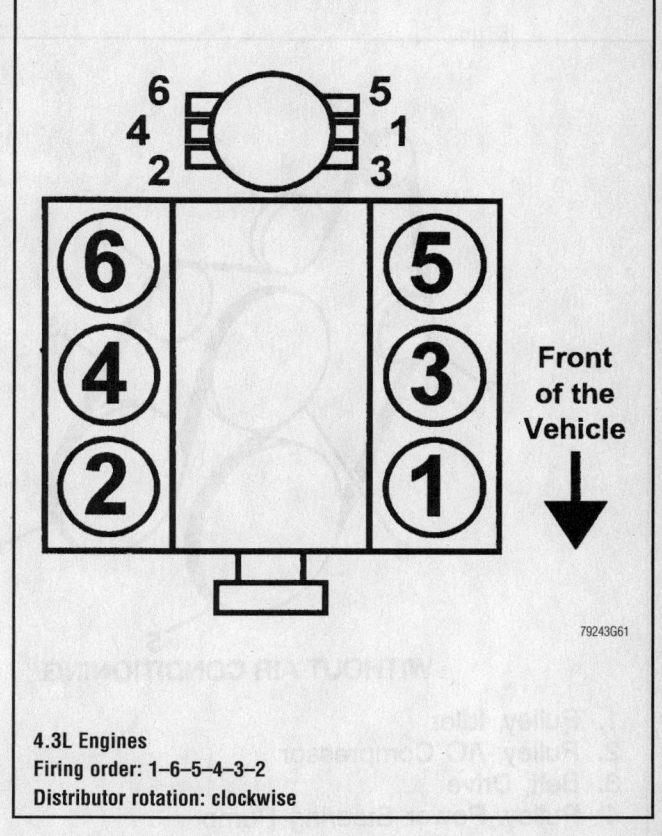

4.3L Engines
Firing order: 1–6–5–4–3–2
Distributor rotation: clockwise

79243G61

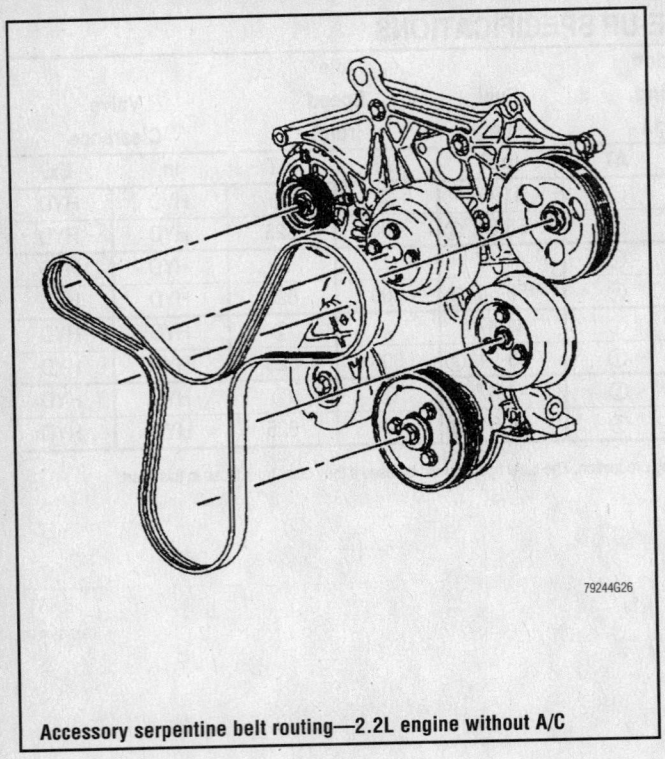

Accessory serpentine belt routing—2.2L engine without A/C

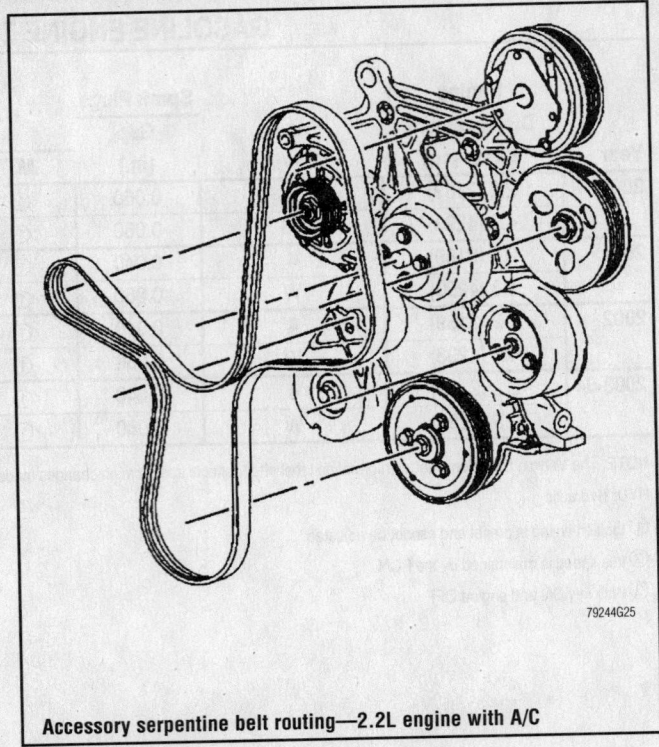

Accessory serpentine belt routing—2.2L engine with A/C

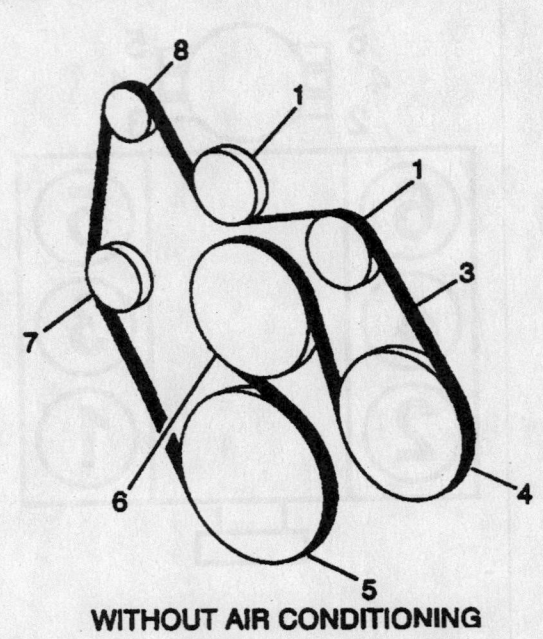

WITHOUT AIR CONDITIONING

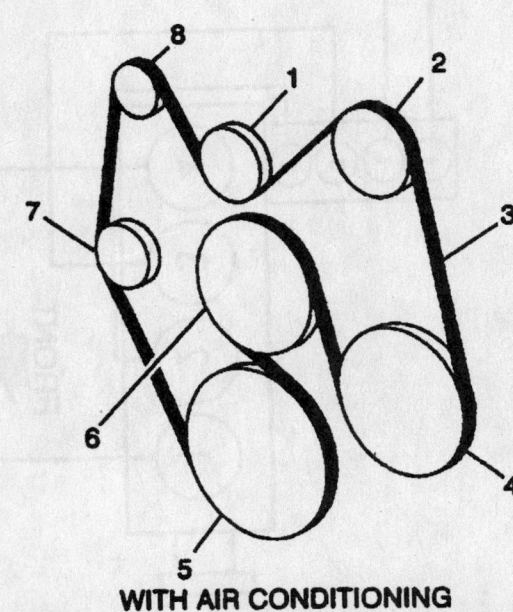

WITH AIR CONDITIONING

1. Pulley, Idler
2. Pulley, AC Compressor
3. Belt, Drive
4. Pulley, Power Steering Pump

5. Pulley, Crankshaft
6. Pulley, Water Pump
7. Pulley, Drive Belt Tensioner
8. Pulley, Generator

Accessory serpentine belt routing—4.3L engine

CAPACITIES

Year	Model	Engine Displacement Liters (cc)	Engine ID/VIN	Engine Oil with Filter (qts.)	Transmission (pts.)		Transfer Case (pts.)	Drive Axle		Fuel Tank (gal.)	Cooling System (qts.)
					5-Spd	Auto.		Front (pts.)	Rear (pts.)		
2000	S10	2.2 (2189)	4	4.0	4.4	11.0	—	—	3.9	19.0	11.5
		4.3 (4293)	W	5.0	4.4	11.0	2.6	2.6	3.9	19.0	11.9
	Sonoma	2.2 (2189)	4	4.0	4.4	11.0	—	—	3.9	19.0	11.5
		4.3 (4293)	W	5.0	4.4	11.0	2.6	2.6	3.9	19.0	11.9
2001	S10	2.2 (2189)	4	4.0	4.4	11.0	—	—	3.9	18.0	11.5
		4.3 (4293)	W	5.0	4.4	11.0	2.6	2.6	3.9	18.0②	11.9
	Sonoma	2.2 (2189)	4	4.0	4.4	11.0	—	—	3.9	18.0	11.5
		4.3 (4293)	W	5.0	4.4	11.0	2.6	2.6	3.9	18.0	11.9
2002	S10	2.2 (2189)	4	4.5	③	10.0	—	—	④	⑤	9.9
		4.3 (4293)	W	4.5	③	10.0	2.6	2.6	④	⑤	14.0
	Sonoma	2.2 (2189)	4	4.5	③	10.0	—	—	④	⑤	9.9
		4.3 (4293)	W	4.5	③	10.0	2.6	2.6	④	⑤	14.0
2003-04	S10	2.2 (2189)	4	4.5	③	10.0	—	—	④	⑤	9.9
		4.3 (4293)	W	4.5	③	10.0	2.6	2.6	④	⑤	14.0
	Sonoma	2.2 (2189)	4	4.5	③	10.0	—	—	④	⑤	9.9
		4.3 (4293)	W	4.5	③	10.0	2.6	2.6	④	⑤	14.0

NOTE: All capacities are approximate. Add fluid gradually and check to be sure a proper fluid level is obtained.

① Available with 20 gallon tank

② 4-dr. 4wd Crew Cab: 17.5

③ NV1500: 5.8
   NV3500: 4.4

④ 7.6 inch: 3.6
   8.6 inch: 4.0

⑤ 2-dr & crew: 19.0
   4-dr: 18.0
   Extended: 18.5

42372-S10P-C04

CRANKSHAFT AND CONNECTING ROD SPECIFICATIONS

All measurements are given in inches.

Year	Engine Displacement Liters (cc)	Engine ID/VIN	Crankshaft				Connecting Rod		
			Main Brg. Journal Dia.	Main Brg. Oil Clearance	Shaft End-play	Thrust on No.	Journal Diameter	Oil Clearance	Side Clearance
2000	2.2 (2189)	4	2.4945-2.4954	0.0006-0.0019	0.0020-0.0070	4	1.9983-1.9994	0.0010-0.0031	0.0039-0.0149
	4.3 (4293)	W	①	②	0.0020-0.0070	4	2.2487-2.2497	0.0013-0.0035	0.0060-0.0140
2001	2.2 (2189)	4	2.4945-2.4954	0.0006-0.0019	0.0020-0.0070	4	1.9983-1.9994	0.0010-0.0031	0.0039-0.0149
	4.3 (4293)	W	①	②	0.0020-0.0070	4	2.2487-2.2497	0.0013-0.0035	0.0060-0.0140
2002	2.2 (2189)	4	2.4945-2.4954	0.0006-0.0019	0.0020-0.0070	4	1.9983-1.9994	0.0010-0.0031	0.0039-0.0149
	4.3 (4293)	W	①	②	0.0020-0.0070	4	2.2487-2.2497	0.0013-0.0035	0.0060-0.0140
2003-04	2.2 (2189)	4	2.4945-2.4954	0.0006-0.0019	0.0020-0.0070	4	1.9983-1.9994	0.0010-0.0031	0.0039-0.0149
	4.3 (4293)	W	①	②	0.0020-0.0070	4	2.2487-2.2497	0.0013-0.0035	0.0060-0.0140

① No. 1: 2.4488-2.4495
   Nos. 2, 3: 2.4485-2.4494
   No. 4: 2.4480-2.4489

② No. 1: 0.0008-0.0020
   Nos. 2, 3: 0.0011-0.0023
   No. 4: 0.0017-0.0032

42372-S10P-C05

VALVE SPECIFICATIONS

Year	Engine Displacement Liters (cc)	Engine ID/VIN	Seat Angle (deg.)	Face Angle (deg.)	Spring Test Pressure (lbs. @ in.)	Spring Installed Height (in.)	Stem-to-Guide Clearance (in.)		Stem Diameter (in.)	
							Intake	Exhaust	Intake	Exhaust
2000	2.2 (2189)	4	46	45	228@1.28	1.71	0.0010-0.0020	0.0010-0.0030	NA	NA
	4.3 (4293)	W	46	45	187-203@1.27	1.69-1.71	0.0010	0.0020	NA	NA
2001	2.2 (2189)	4	46	45	201-215@1.18	1.71	0.0007-0.0020	0.0014-0.0030	NA	NA
	4.3 (4293)	W	46	45	187-203@1.27	1.69-1.71	0.0010-0.0027	0.0010-0.0027	NA	NA
2002	2.2 (2189)	4	46	45	201-215@1.18	1.71	0.0007-0.0020	0.0014-0.0030	NA	NA
	4.3 (4293)	W	46	45	187-203@1.27	1.69-1.71	0.0010-0.0027	0.0010-0.0027	NA	NA
2003-04	2.2 (2189)	4	46	45	201-215@1.18	1.71	0.0007-0.0020	0.0014-0.0030	NA	NA
	4.3 (4293)	W	46	45	187-203@1.27	1.69-1.71	0.0010-0.0027	0.0010-0.0027	NA	NA

NA: Not Available

42372-S10P-C06

PISTON AND RING SPECIFICATIONS

All measurements are given in inches.

Year	Engine Displacement Liters (cc)	Engine ID/VIN	Piston Clearance	Ring Gap			Ring Side Clearance		
				Top Compression	Bottom Compression	Oil Control	Top Compression	Bottom Compression	Oil Control
2000	2.2 (2189)	4	0.0007-0.0017	0.010-0.020	0.010-0.020	0.010-0.03	0.0019-0.0027	0.0019-0.0027	0.0019-0.0082
	4.3 (4293)	W	0.0007-0.0017	0.010-0.030	0.018-0.026	0.065 Max.	0.0042 Max.	0.0042 Max.	0.0020-0.0070
2001	2.2 (2189)	4	0.0007-0.0017	0.010-0.020	0.012-0.018	0.010-0.03	0.0020-0.0035	0.0008-0.0031	0.0005-0.0087
	4.3 (4293)	W	0.0007-0.0024	0.010-0.016	0.015-0.023	0.010-0.029	0.0012-0.0027	0.0015-0.0031	0.0020-0.0070
2002	2.2 (2189)	4	0.0007-0.0017	0.010-0.020	0.012-0.018	0.010-0.03	0.0020-0.0035	0.0008-0.0031	0.0005-0.0087
	4.3 (4293)	W	0.0007-0.0024	0.010-0.016	0.015-0.023	0.010-0.029	0.0012-0.0027	0.0015-0.0031	0.0020-0.0070
2003-04	2.2 (2189)	4	0.0007-0.0017	0.010-0.020	0.012-0.018	0.010-0.03	0.0020-0.0035	0.0008-0.0031	0.0005-0.0087
	4.3 (4293)	W	0.0007-0.0024	0.010-0.016	0.015-0.023	0.010-0.029	0.0012-0.0027	0.0015-0.0031	0.0020-0.0070

42372-S10P-C07

TORQUE SPECIFICATIONS
All readings in ft. lbs.

Year	Engine Displacement Liters (cc)	Engine ID/VIN	Cylinder Head Bolts	Main Bearing Bolts	Rod Bearing Bolts	Crankshaft Damper Bolts	Flywheel Bolts	Manifold Intake *	Manifold Exhaust	Spark Plugs	Lug Nut
2000	2.2 (2189)	4	①	70	38	77	55	②	10	11	100
	4.3 (4293)	W	③	77	④	74	74	⑤	⑥	11	100
2001	2.2 (2189)	4	①	70	38	77	55	17	10	11	100
	4.3 (4293)	W	③	77	④	74	74	⑤	⑥	11	100
2002	2.2 (2189)	4	①	70	38	77	55	17	10	11	100
	4.3 (4293)	W	③	77	④	74	74	⑤	⑥	11	100
2003-04	2.2 (2189)	4	①	70	38	77	55	17	10	11	100
	4.3 (4293)	W	③	77	④	74	74	⑤	⑥	11	100

* NOTE: Applies to Lower Manifold only.

① Short bolts: 43 ft. lbs. plus 90 degrees
   Long bolts: 46 ft. lbs. plus 90 degrees

② Lower intake manifold nuts: 24 ft. lbs.
   Lower intake manifold studs: 22 ft. lbs.
   Upper intake manifold bolts: 22 ft. lbs.

③ 1st pass: 22 ft. lbs.
   2nd pass:
   Short bolt: Plus 55 degrees
   Medium bolt: Plus 65 degrees
   Long bolt: Plus 75 degrees

④ 20 ft. lbs. plus 70 degrees

⑤ Lower intake manifold:
   1st pass: 27 inch lbs.
   2nd pass: 106 inch lbs.
   Final pass: 11 ft. lbs.
   Upper manifold bolts:
   1st pass: 44 inch lbs.
   2nd pass: 88 inch lbs.

⑥ Tighten bolts to 12 ft. lbs.
   Retorque to 22 ft. lbs.

42372-S10P-C08

WHEEL ALIGNMENT

Year	Model		Caster Range (+/-Deg.)	Caster Preferred Setting (Deg.)	Camber Range (+/-Deg.)	Camber Preferred Setting (Deg.)	Toe-in (in.)	Steering Axis Inclination (Deg.)
2000	exc. ZR2/Z85	Left	1.0	+3.0	1.0	0	0.10+/-0.10	—
	ZM6/G51	Right	1.0	+3.0	1.0	0	0.10+/-0.10	—
	ZR2/Z85	Left	1.0	+2.0	1.0	0	0.10+/-0.10	—
	ZM6/G51	Right	1.0	+2.0	1.0	0	0.10+/-0.10	—
2001	Exc. ZQ8/Z87	Left	1.0	2.8	1.0	0	0.10+/-0.10	—
		Right	1.0	3.3	—	—	—	—
	ZQ8/Z87	Left	1.0	+4.7	1.0	0	0.10+/-0.10	—
		Right	1.0	+5.2	—	—	—	—
2002	Exc. ZQ8/Z87	Left	1.0	2.8	1.0	0	0.10+/-0.10	—
		Right	1.0	3.3	—	—	—	—
	ZQ8/Z87	Left	1.0	+4.7	1.0	0	0.10+/-0.10	—
		Right	1.0	+5.2	—	—	—	—
2003-04	Exc. ZQ8/Z87	Left	1.0	2.8	1.0	0	0.10+/-0.10	—
		Right	1.0	3.3	—	—	—	—
	ZQ8/Z87	Left	1.0	+4.7	1.0	0	0.10+/-0.10	—
		Right	1.0	+5.2	—	—	—	—

42372-S10P-C09

BRAKE SPECIFICATIONS
All measurements in inches unless noted

Year	Model		Brake Disc			Brake Drum Diameter			Minimum Lining Thickness	Brake Caliper	
			Original Thickness	Minimum Thickness	Maximum Runout	Original Inside Diameter	Max. Wear Limit	Maximum Machine Diameter		Bracket Bolts (ft. lbs.)	Mounting Bolts (ft. lbs.)
2000	S10	F	1.030	0.965	0.003	—	—	—	0.030	52	①
		R	0.787	0.728	0.004	9.50	9.59	9.56	0.030	—	—
	Sonoma	F	1.030	0.965	0.003	—	—	—	0.030	52	38
		R	—	—	—	9.50	9.59	9.56	0.030	—	—
2001	S10	F	1.030	0.965	0.003	—	—	—	0.030	52	①
		R	0.787	0.728	0.004	9.50	9.59	9.56	0.030	—	—
	Sonoma	F	1.030	0.965	0.003	—	—	—	0.030	52	38
		R	—	—	—	9.50	9.59	9.56	0.030	—	—
2002	S10	F	1.140	1.130	0.002	—	—	—	0.030	②	③
		R	0.787	0.735	0.002	9.50	9.59	9.56	0.030	②	23
	Sonoma	F	1.140	1.130	0.002	—	—	—	0.030	②	③
		R	0.787	0.735	0.002	9.50	9.59	9.56	0.030	②	23
2003-04	S10	F	1.140	1.130	0.002	—	—	—	0.030	②	③
		R	0.787	0.735	0.002	9.50	9.59	9.56	0.030	②	23
	Sonoma	F	1.140	1.130	0.002	—	—	—	0.030	②	③
		R	0.787	0.735	0.002	9.50	9.59	9.56	0.030	②	23

NA: Not Available

① 2WD: 38 ft. lbs.
  4WD: 77 ft. lbs.

② Dual piston caliper-to-knuckle: 133 ft. lbs

③ Single piston: 38 ft. lbs.
  Dual piston 85 ft. lbs.

42372-S10P-C10

TIRE, WHEEL AND BALL JOINT SPECIFICATIONS

Year	Model	OEM Tires		Tire Pressures (psi)		Wheel Size	Ball Joint Inspection
		Standard	Optional	Front	Rear		
2000	S-10 2wd, base	P205/70R15	P235/70R15	36	36	6-JJ	U: 0.125 in. L ①
	S-10 2wd, Sport	P215/65R15	None	36	36	6-JJ	U: 0.125 in. L ①
	S-10 4wd, Reg. Cab, w/117.9 WB	P235/70R15	P235/75R15	36	36	6-JJ	U: 0.125 in. L ①
	S-10 4wd, all others	P235/75R15	None	36	36	6-JJ	U: 0.125 in. L ①
2001	S-10 2wd, base	P205/70R15	P235/70R15	36	36	6-JJ	U: 0.125 in. L ①
	S-10 2wd, Sport	P215/65R15	None	36	36	6-JJ	U: 0.125 in. L ①
	S-10 4wd, Reg. Cab, w/117.9 WB	P235/70R15	P235/75R15	36	36	6-JJ	U: 0.125 in. L ①
	S-10 4wd, all others	P235/75R15	None	36	36	6-JJ	U: 0.125 in. L ①
2002	S-10 2wd, base	P205/70R15	P235/70R15	36	36	6-JJ	U: 0.125 in. L ①
	S-10 2wd, Sport	P215/65R15	None	36	36	6-JJ	U: 0.125 in. L ①
	S-10 4wd, Reg. Cab, w/117.9 WB	P235/70R15	P235/75R15	36	36	6-JJ	U: 0.125 in. L ①
	S-10 4wd, all others	P235/75R15	None	36	36	6-JJ	U: 0.125 in. L ①
2003-04	S-10 2wd, base	P205/70R15	P235/70R15	36	36	6-JJ	U: 0.125 in. L ①
	S-10 2wd, Sport	P215/65R15	None	36	36	6-JJ	U: 0.125 in. L ①
	S-10 4wd, Reg. Cab, w/117.9 WB	P235/70R15	P235/75R15	36	36	6-JJ	U: 0.125 in. L ①
	S-10 4wd, all others	P235/75R15	None	36	36	6-JJ	U: 0.125 in. L ①

OEM: Original Equipment Manufacturer

PSI: Pounds Per Square Inch

STD: Standard

OPT: Optional

L: Lower

U: Upper

① Do not lift truck. Inspect the boss into which the grease fitting is threaded. Replace if the boss is flush or receded below the surface of the ball joint

42372-S10P-C11

SCHEDULED MAINTENANCE INTERVALS
GENERAL MOTORS—S-SERIES PICK-UP & SONOMA

TO BE SERVICED	TYPE OF SERVICE	VEHICLE MILEAGE INTERVAL (x1000)															
		7.5	15	22.5	30	37.5	45	52.5	60	67.5	75	82.5	90	97.5	105	112.5	120
Accessory drive belt	S/I								✓								
Air cleaner filter	R			✓					✓				✓				✓
Automatic transmission fluid	R	Every 50,000 miles															
Brake system ①	S/I	✓	✓	✓	✓	✓	✓	✓	✓	✓	✓	✓	✓	✓	✓	✓	✓
Chassis & suspension grease points	L	✓	✓	✓	✓	✓	✓	✓	✓	✓	✓	✓	✓	✓	✓	✓	✓
CV-joint boots & axle seals	S/I	✓	✓	✓	✓	✓	✓	✓	✓	✓	✓	✓	✓	✓	✓	✓	✓
Engine coolant system ②	S/I	Every 150,000 miles															
Engine oil & filter	R	✓	✓	✓	✓	✓	✓	✓	✓	✓	✓	✓	✓	✓	✓	✓	✓
Front wheel bearings	S/I & L				✓				✓				✓				✓
Fuel filter	R				✓				✓				✓				✓
Fuel tank, cap & lines	S/I								✓								✓
PCV valve	S/I	Every 100,000 miles															
Rear/front axle fluid level	S/I	✓	✓	✓	✓	✓	✓	✓	✓	✓	✓	✓	✓	✓	✓	✓	✓
Rotate tires	S/I	✓	✓	✓	✓	✓	✓	✓	✓	✓	✓	✓	✓	✓	✓	✓	✓
Spark plug wires	S/I	Every 100,000 miles															
Spark plugs	R	Every 100,000 miles															

R: Replace   S/I: Inspect and service, if necessary   L: Lubricate

① This should be performed when the tires are removed for rotation.

② Drain, flush and refill the cooling system, inspect the system hoses, and clean the radiator and condenser.

③ 2-wheel drive models only.

FREQUENT OPERATION MAINTENANCE (SEVERE SERVICE)

If a vehicle is operated under any of the following conditions it is considered severe service:

- Towing a trailer or using a camper or car-top carrier.
- Repeated short trips of less than 5 miles in temperatures below freezing, or trips of less than 10 miles in any temperature.
- Extensive idling or low-speed driving for long distances as in heavy commercial use, such as delivery, taxi or police cars.
- Operating on rough, muddy or salt-covered roads.
- Operating on unpaved or dusty roads.
- Driving in extremely hot (over 90°) conditions.

Engine oil & filter: replace every 3000 miles or 3 months, whichever occurs first.

Chassis and suspension grease points: lubricate every 3000 miles.

Rear/front axle fluid level: inspect every 3000 miles.

Rotate the tires ever 6000 miles.

Brake system components: inspect ever 6000 miles.

Front wheel bearings (2-wheel drive only): clean, inspect and repack every 15,000 miles.

Air cleaner filter: inspect every 15,000 miles.

Automatic transmission fluid & filter: replace every 15,000 miles.

42372-S10P-C12

PRECAUTIONS

Before servicing any vehicle, please be sure to read all of the following precautions, which deal with personal safety, prevention of component damage, and important points to take into consideration when servicing a motor vehicle:

• Never open, service or drain the radiator or cooling system when the engine is hot; serious burns can occur from the steam and hot coolant.

• Observe all applicable safety precautions when working around fuel. Whenever servicing the fuel system, always work in a well-ventilated area. Do not allow fuel spray or vapors to come in contact with a spark, open flame, or excessive heat (a hot drop light, for example). Keep a dry chemical fire extinguisher near the work area. Always keep fuel in a container specifically designed for fuel storage; also, always properly seal fuel containers to avoid the possibility of fire or explosion. Refer to the additional fuel system precautions later in this section.

• Fuel injection systems often remain pressurized, even after the engine has been turned **OFF**. The fuel system pressure must be relieved before disconnecting any fuel lines. Failure to do so may result in fire and/or personal injury.

• Brake fluid often contains polyglycol ethers and polyglycols. Avoid contact with the eyes and wash your hands thoroughly after handling brake fluid. If you do get brake fluid in your eyes, flush your eyes with clean, running water for 15 minutes. If

eye irritation persists, or if you have taken brake fluid internally, IMMEDIATELY seek medical assistance.

• The EPA warns that prolonged contact with used engine oil may cause a number of skin disorders, including cancer! You should make every effort to minimize your exposure to used engine oil. Protective gloves should be worn when changing oil. Wash your hands and any other exposed skin areas as soon as possible after exposure to used engine oil. Soap and water, or waterless hand cleaner should be used.

• All new vehicles are now equipped with an air bag system. The system must be disabled before performing service on or around system components, steering column, instrument panel components, wiring and sensors. Failure to follow safety and disabling procedures could result in accidental air bag deployment, possible personal injury and unnecessary system repairs.

• Always wear safety goggles when working with, or around, the air bag system. When carrying a non-deployed air bag, be sure the bag and trim cover are pointed away from your body. When placing a non-deployed air bag on a work surface, always face the bag and trim cover upward, away from the surface. This will reduce the motion of the module if it is accidentally deployed. Refer to the additional air bag system precautions later in this section.

• Clean, high quality brake fluid from a

sealed container is essential to the safe and proper operation of the brake system. You should always buy the correct type of brake fluid for your vehicle. If the brake fluid becomes contaminated, completely flush the system with new fluid. Never reuse any brake fluid. Any brake fluid that is removed from the system should be discarded. Also, do not allow any brake fluid to come in contact with a painted surface; it will damage the paint.

• Never operate the engine without the proper amount and type of engine oil; doing so WILL result in severe engine damage.

• Timing belt maintenance is extremely important! Many models utilize an interference-type, non-freewheeling engine. If the timing belt breaks, the valves in the cylinder head may strike the pistons, causing potentially serious (also time-consuming and expensive) engine damage. Refer to the maintenance interval charts in the front of this section for the recommended replacement interval for the timing belt, and to the timing belt procedure in this section for belt replacement and inspection.

• Disconnecting the negative battery cable on some vehicles may interfere with the functions of the on-board computer system(s) and may require the computer to undergo a relearning process once the negative battery cable is reconnected.

• When servicing drum brakes, only disassemble and assemble one side at a time, leaving the remaining side intact for reference.

ENGINE REPAIR

Alternator

REMOVAL

2.2L Engine

1. Before servicing the vehicle, refer to the precautions in the beginning of this section.
2. Remove or disconnect the following:
• Negative battery cable
• Passenger side wheel assembly
• Alternator brace-to-block bolt, the brace-to-intake nut and the brace-to-engine stud
• Alternator wiring
• Accessory belt
• Mounting bolts
• Alternator

4.3L Engine

1. Before servicing the vehicle, refer to the precautions in the beginning of this section.
2. Remove or disconnect the following:
• Negative battery cable
• Air inlet duct, if necessary
• Accessory belt
• Heater hose brace
• Wires
• Mounting bolts
• Alternator

INSTALLATION

2.2L Engine

Install or connect the following:
• Alternator

• Mounting bolts. Torque the left bolt to 22 ft. lbs. (30 Nm) and the right bolt to 32 ft. lbs. (43 Nm).
• Wires. Torque the battery feed wire nut to 71 inch lbs. (8 Nm).
• Alternator brace. Torque the nuts and bolts to 22 ft. lbs. (30 Nm).
• Accessory belt
• Negative battery cable

4.3L Engine

Install or connect the following:
• Alternator and loosely install the mounting bolts
• Tighten the rear bolt to 37 ft. lbs. (50 Nm) and the front bolt to 18 ft. lbs. (25 Nm)
• Tighten the brace-to-alternator and brace-to-intake retainers to 18 ft.

lbs. (25 Nm). Tighten the brace-to-engine stud nut to 37 ft. lbs. (50 Nm).

- Wires and the battery feed wire nut
- Heater hose bracket
- Accessory belt
- Negative battery cable

Ignition Timing

ADJUSTMENT

The ignition timing is preset and cannot be adjusted.

Engine Assembly

REMOVAL & INSTALLATION

2.2L Engine

➡ In certain cases on some models the A/C system will have to be evacuated because the compressor may need to be removed from the vehicle to allow clearance for engine removal. On other models you maybe able to set the compressor and lines to one side and still have enough clearance to remove the engine. In this case the system does not have to be evacuated because the lines do not have to be disconnected from the compressor. To check if your system has to be evacuated, unplug the electrical connectors from the compressor, then unbolt the compressor assembly. Unfasten any brackets holding the refrigerant lines and try to set the components aside so that you will have enough clearance for engine removal. If there is not enough clearance for engine removal you must recover the refrigerant from the A/C system with an approved recovery station before attempting to remove the engine from your vehicle. DO NOT attempt this without the proper equipment. R-134a should NOT be mixed with R-12 refrigerant and, depending on your local laws, attempting to service this system could be illegal.

1. Disconnect the negative battery cable and properly relieve the fuel system pressure.
2. Drain the engine cooling system and the engine oil into separate drain pans.
3. Remove or disconnect the following:
   - Hood
   - Oxygen Sensor (O_2S) electrical connection
   - Exhaust pipe from the manifold

➡ On some models it may also be necessary to disconnect the catalytic converter from the exhaust pipe.

- Braces from the engine and the transmission, if equipped
- Starter motor
- Transmission and separate it from the engine or, if necessary, remove it from the vehicle
- Alternator rear brace by unfastening the bolt and nuts
- Ground straps from the engine block
- Drive belt
- A/C compressor and bracket. If possible, set the compressor and bracket to one side without disconnecting the lines.
- Hoses and transmission coolant lines engaged to the radiator
- Radiator
- Power steering pump and cap the power steering lines to avoid contamination
- Heater hoses from the heater core
- 12 volt supply from the mega fuse, if necessary
- All electrical connections and wiring harnesses
- All vacuum lines
- Throttle cable, and if equipped the cruise control cable
- Exhaust Gas Recirculation (EGR) pipe and the EGR valve
- Fuel lines

4. Install a suitable lifting device to the engine.
5. Remove the engine mount bolts and carefully lift the engine from the vehicle. Pause several times while lifting the engine to make sure no wires or hoses have become snagged.

To install:

6. Carefully lower the engine into the vehicle and install the engine mount bolts. Remove the engine lifting device.
7. Install or connect the following:
   - Fuel lines
   - 12 volt supply to the mega fuse, if removed
   - All vacuum lines, electrical connections and wiring harnesses
   - EGR valve and pipe, if removed
   - Throttle and if equipped, the cruise control cable
   - Heater hoses to the heater core
   - Power steering pump and attach the lines
   - A/C compressor
   - Radiator, all hoses and fluid cooler lines
   - Water pump, if removed
   - Drive belt
   - Ground strap to the engine
   - Alternator rear brace and tighten the bolt and nuts, if removed

- Transmission to the engine
- Starter motor, if removed
- Braces to the engine and the transmission, if equipped
- Exhaust pipe to the manifold
- Catalytic converter to the exhaust pipe, if removed
- O_2S electrical connection
- Battery
- Hood

8. Check all powertrain fluid levels and add, as necessary. Be sure to properly fill the engine crankcase with clean engine oil.
9. Connect the battery cables and properly fill the engine cooling system.
10. Start and run the engine, then check for leaks.

4.3L Engine

1. Before servicing the vehicle, refer to the precautions in the beginning of this section.
2. Drain the engine cooling system
3. Drain the engine oil.
4. Remove or disconnect the following:
   - Negative battery cable
   - Fuel system pressure
   - Vacuum reservoir and/or the underhood light from the hood, as equipped
   - Outer cowl vent grilles
   - Hood
   - Oxygen Sensor (O_2S) and/or wiring
   - Exhaust pipes at the manifolds and loosen the hanger at the catalytic converter. This is necessary to remove the rear catalytic converter cushion mounts for removal of the exhaust assembly.
   - Skid plate, if equipped
   - Engine-to-transmission pencil braces
   - Slave cylinder and position aside, if equipped
   - Line clamp at the bell housing
   - Wiring from the starter
   - Starter
   - Transfer case
   - Oil filter
   - Engine mount through bolts
   - Rear engine mount crossbar, nut and washer
   - Bell housing bolts, except the upper left.
   - Battery ground (negative) cable from the engine
   - Front drive axle bolts and roll the axle downward, on 4WD vehicles
   - Air cleaner assembly
   - Upper radiator shroud
   - Fan assembly

- Drive belt assembly
- Water pump pulley
- Upper radiator hose
- Air conditioning compressor, if equipped, and position aside with the lines intact
- Lower radiator hose
- Oil cooler and overflow lines from the radiator, plug the openings to prevent system contamination or excessive fluid loss
- Radiator and lower radiator shroud
- Power steering hoses from the steering gear, then cap the openings to prevent system contamination or excessive fluid loss
- Heater hoses from the intake manifold and the water pump
- Wiring harness and vacuum lines from the engine
- Throttle cables
- Remaining bell housing bolt
- Fuel lines and the bracket
- Ground strap(s) from the rear of the cylinder head
- Front body mount bolts, on 4WD vehicles

5. Support the transmission.
6. Install a lifting device and lift the engine.

To install:

7. Install or connect the following:
- Engine into the vehicle
- Front body mount bolts, on 4WD vehicles
- Ground strap(s) to the rear of the cylinder head
- Fuel lines and the bracket
- Upper left bell-housing bolt
- Throttle cables
- Vacuum lines and wiring harness connectors
- Heater hoses
- Power steering hoses
- Lower shroud and radiator
- Oil cooler lines to the radiator and overflow hose
- Lower radiator hose
- Air conditioning compressor to the engine, if equipped
- Upper radiator hose
- Water pump pulley
- Drive belt assembly
- Fan assembly
- Upper radiator shroud
- Air cleaner assembly
- Front drive axle, for 4WD vehicles
- Battery ground strap to the engine block
- Remaining bell housing bolts
- Engine mount through-bolts. Torque them to 49 ft. lbs. (66 Nm).

- Rear engine mount crossbar nut and washer. Tighten the nut to 33 ft. lbs. (45 Nm).
- Oil filter
- Starter motor
- Flywheel cover
- Clutch slave cylinder, if equipped
- Pencil brace and the skid plate, as equipped
- Catalytic converter Y-pipe assembly and hangers
- Hood
- Outer cowl vent grilles
- Vacuum reservoir and/or the under-hood light to the hood, as equipped
- Negative battery cable

8. Check all powertrain fluid levels and add, as necessary.
9. Refill the engine crankcase.
10. Refill the engine cooling system.
11. Start and run the engine, then check for leaks.

Water Pump

REMOVAL & INSTALLATION

1. Before servicing the vehicle, refer to the precautions in the beginning of this section.
2. Disconnect the negative battery cable.

3. Drain the engine cooling system.
4. Relieve the belt tension and remove the accessory drive belts or the serpentine drive belt, as applicable.
5. Remove or disconnect the following:
- Upper fan shroud
- Fan or fan and clutch assembly, as applicable
- Water pump pulley
- Coolant hose(s) from the water pump

➡For the hoses on some engines, removal may be easier if the hose is left attached until the pump is free from the block. Once the pump is removed from the engine, the pump may be pulled (giving a better grip and greater leverage) from the tight hose connection.

- Water pump retainers
- Water pump from the engine

❊❊ WARNING

Note the positions of all retainers as some engines will utilize different length fasteners in different locations and/or bolts and studs in different locations.

To install:
6. Clean the gasket mounting surfaces.

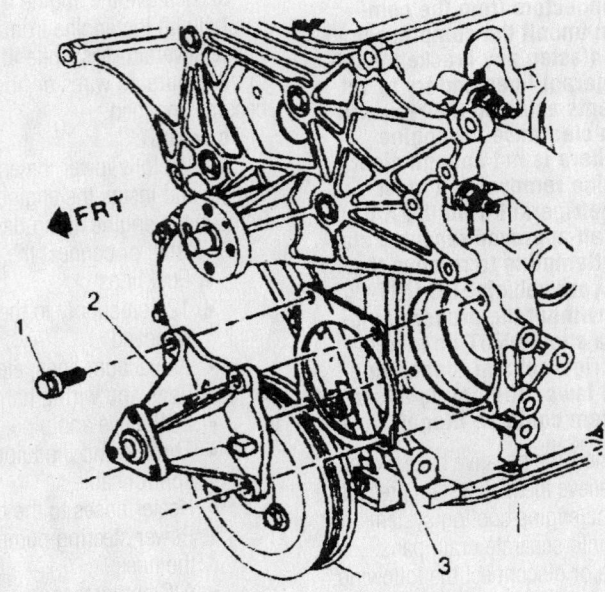

1. BOLT
2. PUMP, COOLANT
3. GASKET

7924JG05

Exploded view of the water pump mounting—2.2L engine

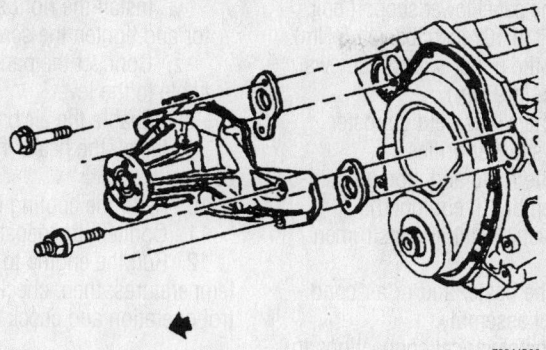

Exploded view of the water pump assembly mounting—4.3L engine

7924JG06

➡**The water pumps on some of the earlier engines covered may have been installed using sealer only, no gasket, at the factory. If a gasket is supplied with the replacement part, it should be used. Otherwise, a ⅛ in. (3mm) bead of RTV sealer should be used around the sealing surface of the pump.**

7. Apply sealant to the water pump retainer threads.

8. Install or connect the following:
- Water pump using a new gasket. Tighten the water pump retainers to 18 ft. lbs. (25 Nm) for 2.2L engine or to 30 ft. lbs. (41 Nm) for 4.3L engine.
- Coolant hose(s)
- Water pump pulley
- Fan or fan and clutch assembly
- Serpentine drive belt (if equipped) by positioning the belt over the pulleys and carefully allow the tensioner back into contact with the belt
- V-belts (if equipped) and adjust the tension
- Upper fan shroud
- Negative battery cable

9. Refill the engine cooling system.

10. Run the engine and check for leaks.

Heater Core

REMOVAL & INSTALLATION

1. Before servicing the vehicle, refer to the precautions in the beginning of this section.

2. Disconnect the negative battery cable.

3. Drain the engine cooling system.

4. Remove or disconnect the following:
- Heater hoses from the heater core

5. Remove the instrument panel as follows:

a. Disable the air bag system.

b. Set the parking brake and block the wheels.

c. Disconnect the parking brake release cable from the parking brake lever.

d. Unfasten the screws that retain the DLC instrument panel left side sound insulator. Feed the DLC through the hole in the sound insulator.

e. Unfasten the right side sound insulator panel screws and remove the panel.

f. Unfasten the screws that attach the instrument panel left side sound insulator to the knee bolster and cowl panel.

g. Unfasten the nut that attaches the left side sound insulator to the accelerator pedal bracket.

h. Unplug the remote control door lock receiver module electrical connector.

i. Remove the door lock receiver module from the left side sound insulator. Remove the left side sound insulator.

j. Unfasten the screws that attach the instrument panel center sound insulator to the knee bolster, instrument panel, heater assembly and floor duct.

k. Remove the center sound insulator.

l. Unfasten the screws that attach the courtesy lamp to the knee bolster.

m. Unfasten the screws that attach the knee bolster to the instrument panel.

n. Disconnect the lap cooler duct from the knee bolster.

o. Unplug the lighter electrical connection and remove the knee bolster.

p. Unfasten the steering column-to-instrument panel nuts and lower the column.

q. Unfasten the screws that attach the instrument panel accessory trim plate to the instrument panel.

r. Remove the trim plate and unplug all necessary electrical connection.

s. Remove the heater and/or air conditioning control assembly.

t. Remove the radio and the storage compartment assembly (if equipped).

u. If necessary, remove the instrument cluster.

v. Unfasten the left and right instrument panel pivot bolts and the panel lower support bolt.

w. Unfasten the speaker grilles retaining screws and remove the speaker grilles.

x. Remove the windshield defroster grille using a flat-bladed prytool. Start at one end of the grille and work your way down the grille.

y. Unfasten the 4 instrument panel upper support screws.

z. Tag and unplug all necessary electrical connections.

aa. Remove the instrument panel from the vehicle.

6. Remove or disconnect the following:
- Air inlet assembly, if equipped
- Vacuum hoses
- Heater assembly studs, from inside the engine compartment
- Blower motor resistor
- From inside the heater case assembly, the stud; the stud is located behind the blower motor resistor
- Heater assembly-to-chassis screws

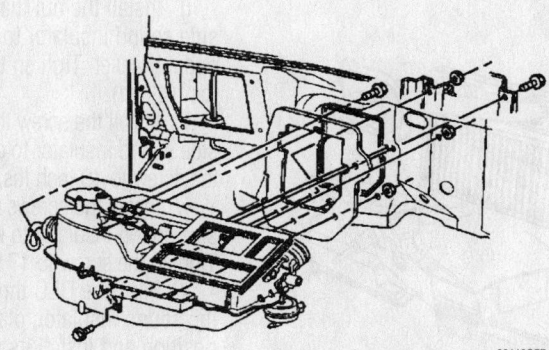

93113G77

View of the heater case assembly

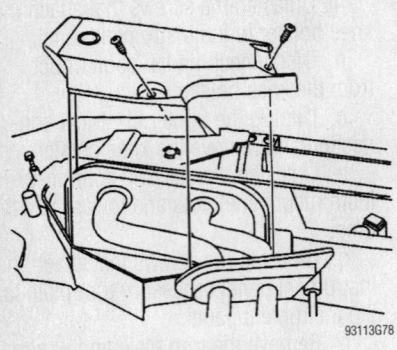

View of the heater case cover

- Heater assembly from the vehicle
- Access cover screws and cover from the heater assembly
- Heater core from the heater case assembly

To install:

7. Install or connect the following:
- Heater core to the heater case assembly
- Access cover to the heater assembly and the cover screws
- Heater assembly to the vehicle
- Heater assembly-to-chassis screws and torque them to 40 inch lbs. (4.5 Nm)
- The stud, working inside the heater case assembly; the stud is located behind the blower motor resistor
- Blower motor resistor
- Heater assembly studs, working inside the engine compartment, and torque them to 17 inch lbs. (1.9 Nm)
- Vacuum hose
- Air inlet assembly, if equipped

8. Install the instrument panel as follows:
a. Rest the instrument panel on the lower pivot studs.
b. Attach the electrical connections.
c. Install but do not tighten the 4 upper instrument panel support screws.
d. Install the left and right panel pivot bolts. Tighten the bolts to 102 inch lbs. (11.5 Nm).

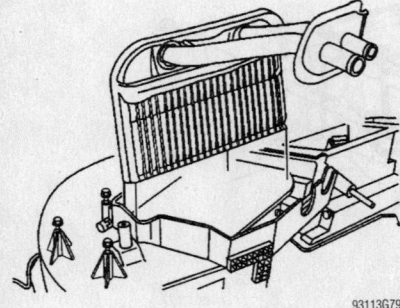

View of the heater core

e. Install the panel lower support bolt. Tighten the bolt to 102 inch lbs. (11.5 Nm).
f. Tighten the upper support screws to 17 inch lbs. (1.9 Nm).
g. Install the windshield defroster grille and the speaker grilles.
h. Install the radio and storage compartment assembly (if equipped).
i. If removed, install the instrument cluster.
j. Install the heater and/or air conditioning control assembly.
k. attach the electrical connections to the instrument panel accessory trim plate.
l. Place the trim plate in position and install its retaining screws. Tighten the screws to 17 inch lbs. (1.9 Nm).
m. Place the steering column into position and install its retaining nuts. Tighten the nuts to 22 ft. lbs. (30 Nm).
n. Attach the lighter electrical connection and the lap cooler duct to the knee bolster.
o. Place the knee bolster into position and install its retaining screws. Tighten the Torx® head screws to 80 inch lbs. (9 Nm) and the hex head screws to 17 inch lbs. (1.9 Nm).
p. Place the courtesy lamp in position and install its screws. Tighten the screws to 17 inch lbs. (1.9 Nm).
q. Place the instrument panel center sound insulator in position. Install the screws that attach the center sound insulator to the knee bolster, instrument panel and the floor duct. Tighten the screws to 17 inch lbs. (1.9 Nm).
r. Install the screw that attaches center sound insulator to the heater assembly. Tighten the screw to 13 inch lbs. (1.5 Nm).
s. Install the remote control door lock receiver module to the instrument panel left side sound insulator.
t. Attach the door lock receiver electrical connection.
u. Install the nut that attaches the left side sound insulator to the accelerator pedal bracket. Tighten the nut to 35 inch lbs. (4 Nm).
v. Install the screw that attaches the left side sound insulator to cowl panel. Tighten the screw to 13 inch lbs. (1.5 Nm).
w. Install the screws that attach the left side sound insulator to knee bolster. Tighten the screw to 17 inch lbs. (1.9 Nm).
x. Feed the DLC through the hole in the sound insulator, place the DLC in position and install its retaining screws. Tighten the screws to 21 inch lbs. (2.4 Nm).

y. Install the right side sound insulator and tighten the screws
z. Connect the parking brake release cable to the lever.
aa. Enable the air bag system.
9. Install the heater hoses to the heater core.
10. Refill the cooling system.
11. Connect the negative battery cable.
12. Run the engine to normal operating temperatures; then, check the climate control operation and check for leaks.

Cylinder Head

REMOVAL & INSTALLATION

2.2L Engine

1. Before servicing the vehicle, refer to the precautions in the beginning of this section.
2. Relieve the fuel system pressure.
3. Disconnect the negative battery cable.
4. Drain the engine cooling system.
5. Remove or disconnect the following:
- Air duct from the air inlet
- Upper radiator hose and upper fan shroud
- Radiator assembly
- Lower fan shroud
- Fan assembly
- Drive belt assembly
- Water pump pulley
- Heater hose from the intake manifold and the thermostat housing
- Thermostat housing
- Alternator support brace and the alternator wiring
- Air conditioning compressor with brackets, if equipped, move it aside without disconnecting the lines
- Accessory bracket along with the alternator and power steering pump still attached. Be careful not to damage the steering pump lines.
- Throttle cable and cable support linkage
- Heater hose from the water pump
- Oil fill tube
- Exhaust pipe
- Oxygen Sensor (O$_2$S)
- Exhaust manifold
- Electrical wiring and the vacuum hoses from the upper intake manifold
- Upper intake manifold
- Wiring from the lower intake manifold
- Fuel lines and the spark plug wires
- Lower intake manifold
- Rocker arm cover

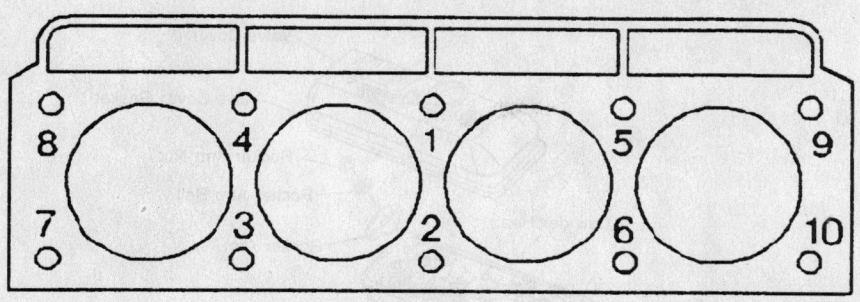

Cylinder head bolt torque sequence—2.2L engine

7924JG07

- Cylinder head bolt threads coated with sealer 1052080. Tighten the bolts within 15 minutes of sealer application, in sequence, to 46 ft. lbs. (63 Nm) for long bolts and to 43 ft. lbs. (58 Nm) for short bolts; then, tighten all bolts an additional 90 degree turn using a torque angle meter.
- Engine lift bracket
- Rocker arms and pushrods
- Rocker arm cover
- Lower intake manifold
- Spark plug wires and the fuel lines
- Lower intake manifold and wiring
- Upper intake manifold
- Vacuum hoses and electrical wiring to the upper intake
- Oil fill tube assembly

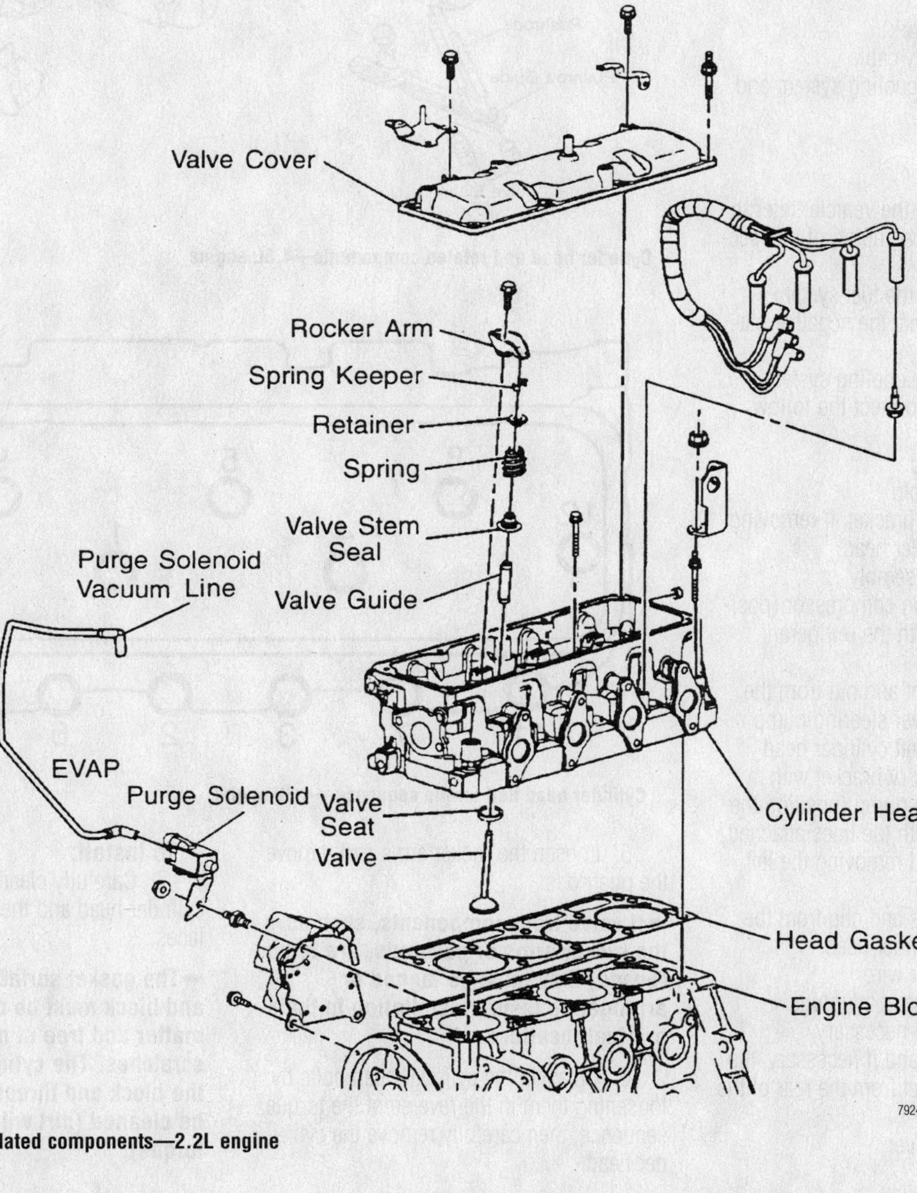

- Rocker arms and pushrods
- Engine lift bracket from the rear of the engine
- Cylinder head bolts and studs
- Cylinder head from the engine

To install:

6. Clean and inspect the gasket mounting surfaces.
7. Install or connect the following:
- Cylinder head using a new gasket

Valve Cover

Rocker Arm

Spring Keeper

Retainer

Spring

Valve Stem Seal

Valve Guide

Purge Solenoid Vacuum Line

EVAP

Purge Solenoid

Valve Seat

Valve

Cylinder Head

Head Gasket

Engine Block

7924JG23

Cylinder head and related components—2.2L engine

- Exhaust manifold
- Exhaust pipe and O_2S
- Heater hose to the water pump
- Throttle cable support and throttle cable
- Accessory support bracket and components
- Air conditioning compressor, if equipped
- Power steering support brace
- Alternator support brace and wiring
- Thermostat housing and the heater hose
- Water pump pulley and drive belt assembly
- Fan assembly
- Radiator and the lower fan shroud
- Upper fan shroud and upper radiator hose
- Air inlet ductwork
- Negative battery cable

8. Refill the engine cooling system and check for leaks.

4.3L Engine

1. Before servicing the vehicle, refer to the precautions in the beginning of this section.

2. Properly relieve the fuel system pressure, then disconnect the negative battery cable.

3. Drain the engine cooling system.

4. Remove or disconnect the following:

- Intake manifold
- Exhaust manifold
- Alternator and bracket, if removing the right cylinder head
- Cooling fan assembly
- Air conditioning compressor (position it aside with the refrigerant lines attached)
- Air pipe bracket and nut from the rear of the power steering pump if removing the left cylinder head
- Engine accessory bracket with power steering pump (position the pump aside with the lines attached) and brackets, if removing the left cylinder head
- Wiring harness and clip from the rear of the cylinder head
- Coolant sensor wire
- Wiring from the spark plugs
- Spark plugs, if necessary
- Ground wires and if necessary, the fuel line bracket from the rear of the cylinder head
- Rocker arm cover

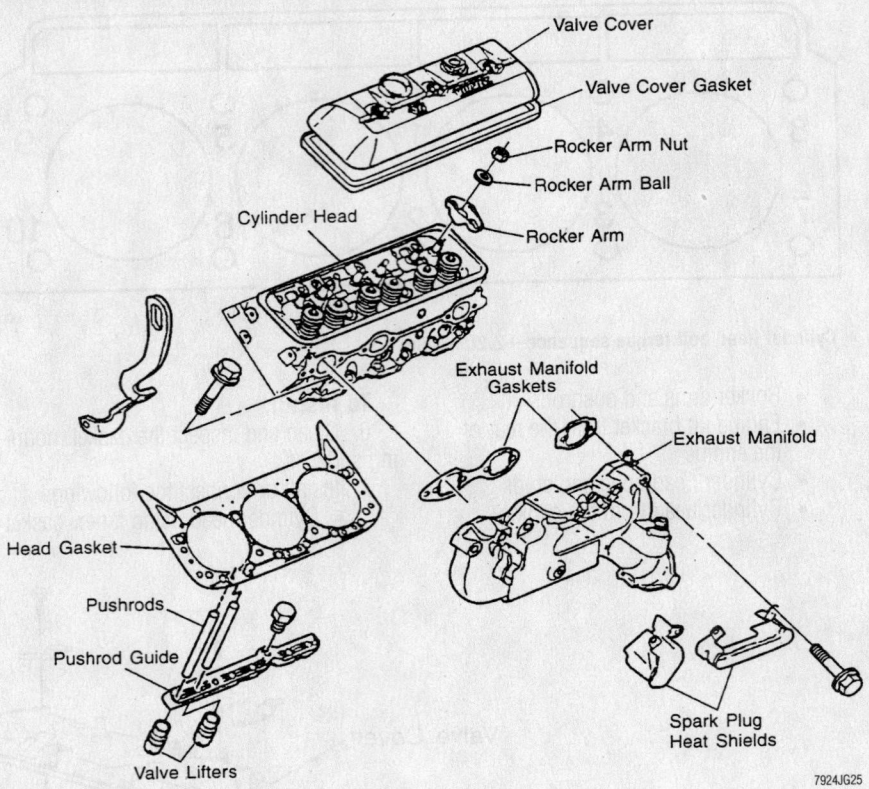

Cylinder head and related components—4.3L engine

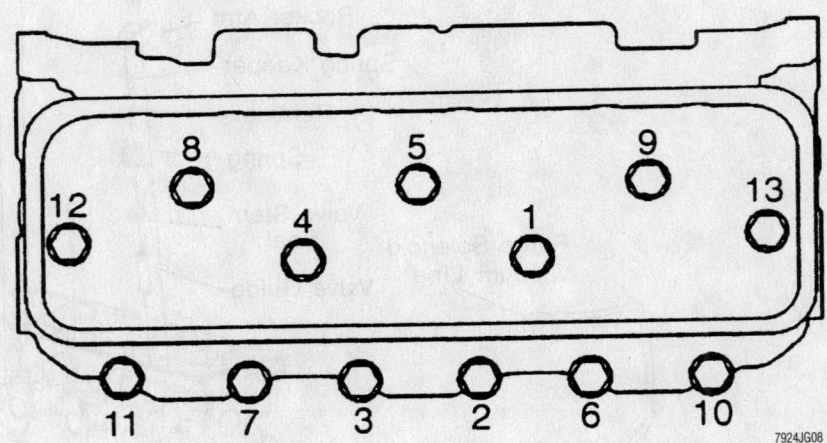

Cylinder head bolt torque sequence—4.3L engine

5. Loosen the rocker arms and remove the pushrods.

➡ If valve train components, such as the rocker arms or pushrods, are to be reused, they must be tagged or arranged to insure installation in their original locations.

6. Unfasten the cylinder head bolts by loosening them in the reverse of the torque sequence, then carefully remove the cylinder head.

To install:

7. Carefully clean and inspect the cylinder head and the gasket mounting surfaces.

➡ The gasket surfaces on both the head and block must be clean of any foreign matter and free of nicks or heavy scratches. The cylinder bolt threads in the block and thread on the bolts must be cleaned (dirt will affect the bolt torque).

➡**DO NOT apply sealer to composition steel-asbestos gaskets.**

8. If using a steel only gasket, apply a thin and even coat of sealer to both sides of the gaskets.

9. Place a new gasket over the dowel pins with the bead or the words "This Side Up" facing upwards (as applicable), then carefully lower the cylinder head into position over the gasket and dowels.

10. Apply a coating of 12346004 or equivalent sealer to the threads of the cylinder head bolts, then thread the bolts into position until finger-tight.

11. Install the bolts in sequence to 22 ft. lbs. (30 Nm). The bolts must then be tightened again in sequence in the following order:

    a. Short length bolts: (11, 7, 3, 2, 6, 10) 55 degrees.

    b. Medium length bolts: (12, 13) 65 degrees.

    c. Long length bolts: (1, 4, 8, 5, 9) 75 degrees.

12. Install or connect the following:
- Pushrods, secure the rocker arms and adjust the valves
- Rocker arm cover
- Spark plugs, if removed
- Spark plug wires
- Attach the fuel line bracket (if removed) and ground wires to the rear of the head and tighten the bolts to 22 ft. lbs. (30 Nm)
- Air conditioning compressor and bracket, if the left cylinder head was removed
- Alternator and bracket, if the right cylinder head was removed
- Engine accessory bracket with power steering pump. if the left cylinder head was removed
- Air pipe bracket and nut to the rear of the power steering pump (if equipped), if the left cylinder head was removed. Tighten the nut to 30 ft. lbs. (41 Nm).
- A/C compressor, if the left cylinder head was removed
- Cooling fan assembly, if the left cylinder head was removed
- Wiring harness and clip to the rear of the cylinder head
- Coolant sensor wire
- Exhaust manifold
- Intake manifold
- Negative battery cable

13. Properly refill the engine cooling system.

14. Run the engine to check for leaks.

Rocker Arms

REMOVAL & INSTALLATION

2.2L Engine

1. Before servicing the vehicle, refer to the precautions in the beginning of this section.

2. Remove or disconnect the following:
- Rocker arm cover
- Rocker arm retaining nut, arm and ball
- Pushrod, if necessary

➡**Valvetrain components, being reused, must be installed in their original positions. If removed, be sure to tag or arrange all rocker arms and pushrods to assure proper installation.**

To install:

3. Inspect the rocker arms, balls and pushrods for damage or wear and replace as necessary.

4. Check the rocker arms, balls and their mating surfaces. Be sure the surfaces are smooth and free from scoring or other damage.

5. Check the rocker arm areas that contact the valve stems and the sockets that contact the pushrods, be sure these areas are smooth and free of both damage and wear.

6. Be sure the pushrods are not bent which can be determined by rolling them on a flat surface. Check the ends of the pushrods for scoring or roughness

7. Inspect the rocker arm bolts for thread damage. Check the rocker arm bolts in the shoulder area for contact damage with the rocker arm.

8. Install or connect the following:
- Pushrods making sure they are seated within the lifters, if removed
- New rocker arms and balls by coating the friction surfaces using Dri-Slide Molykote® or equivalent pre-lube

✳✳ WARNING

When tightening a rocker arm retainer, be sure the lifter for that valve is resting on the base circle of the camshaft and not on the lobe, otherwise the valve train can be damaged. Do not over-tighten the retainers.

- Rocker arms and ball. Tighten the nuts to 22 ft. lbs. (30 Nm).
- Rocker arm cover

9. Start and run the engine to check for leaks.

4.3L Engines

1. Before servicing the vehicle, refer to the precautions in the beginning of this section.

2. Remove or disconnect the following:
- Rocker arm cover(s)
- Rocker arm nut, rocker arm and ball washer

➡**If only the pushrod is to be removed, loosen the rocker arm nut, swing the rocker arm to the side and remove the pushrod.**

- Pushrod(s)

To install:

3. Inspect and replace components if worn or damaged.

4. Coat the bearing surfaces of the

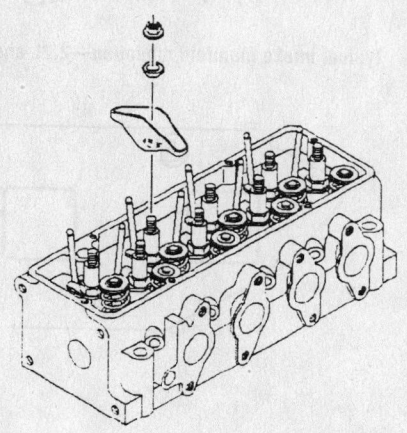

Exploded view of the rocker arm assembly—2.2L engine

7924JG45

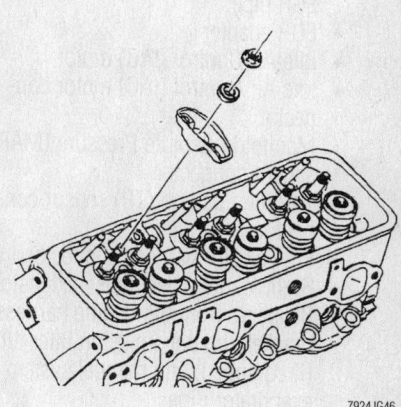

Exploded view of the rocker arm assembly—4.3L engine

7924JG46

rocker arms and the rocker arm ball washers with Molykote® or equivalent pre-lube.

5. Install or connect the following:
- Pushrods making sure they seat properly in the lifter
- Rocker arms, ball washers and the nuts

➡ **The 4.3L engines are equipped with screw-in rocker arm studs with positive stop shoulders.**

- Rocker arm adjusting nuts. Tighten them against the stop shoulders to 18 ft. lbs. (24 Nm). No further adjustment is necessary or possible.

6. Install the rocker arm cover(s).

7. Start and run the engine, then check for leaks and for proper ignition timing adjustment.

Intake Manifold

REMOVAL & INSTALLATION

2.2L Engine

1. Before servicing the vehicle, refer to the precautions in the beginning of this section.

2. Remove or disconnect the following:
- Negative battery cable and remove the air cleaner resonator.
- Three vacuum hoses from the throttle body
- Throttle cable support bracket and the throttle body assembly
- Upper fan shroud and disconnect the vacuum brake booster hose, if necessary
- Exhaust Gas Recirculation (EGR) pipe-to-manifold bolts and the EGR pipe-to-EGR adapter bolt, and the EGR pipe.
- EGR adapter
- Idle Air Control (IAC) motor
- Idle Air Control (IAC) motor connector
- Manifold Absolute Pressure (MAP) sensor connector
- Throttle Position (TP) sensor connector
- Fuel njector harness connector
- Right fender wheelhouse extension
- Retainers from the engine harness bracket, the transmission filler tube (if equipped) and the fuel system evaporator pipe
- Fuel pipes from the fuel rail
- Accelerator cable and if equipped, the cruise control cable

- Spark plug wires from the plugs
- Spark plug wire harness retainer from the heater hose pipe and set aside the harness
- Alternator rear brace by accessing the retaining nuts and bolts through the wheelhouse, if necessary
- Engine wiring harness bracket located at the rear of the cylinder head (if necessary), by unfastening the bracket-to-valve cover and bracket-to-cylinder head retainers, then slide the bracket off the bolt at the rear of the cylinder head
- Intake manifold bolts
- Fuel rail bracket
- Intake manifold and gasket

To install:

3. Carefully remove all traces of gasket material from the mating surfaces. Check the EGR passage to be sure it is free of excessive carbon deposits and clean, as necessary.

4. Install or connect the following:
- Lower intake manifold using a new gasket, then tighten the retaining

bolts to 17 ft. lbs. (24 Nm) using the sequence illustrated
- Engine wiring harness bracket, if removed. Tighten the bracket-to-valve cover bolts to 88 inch lbs. (10 Nm) and the bracket-to-cylinder head bolt to 18 ft. lbs. (25 Nm).
- Generator rear brace, if removed. Tighten the nuts and bolts to 18 ft. lbs. (25 Nm).
- Spark plug wire harness and retainer and attach the spark plug wires to the plugs
- Throttle body assembly, if removed and the throttle cable support bracket
- Three vacuum lines to the throttle body
- Accelerator cable and if equipped, the cruise control cable
- Fuel lines
- Retainers to the engine harness bracket, the transmission filler tube (if equipped) and the fuel system evaporator pipe
- IAC motor connector
- MAP sensor connector

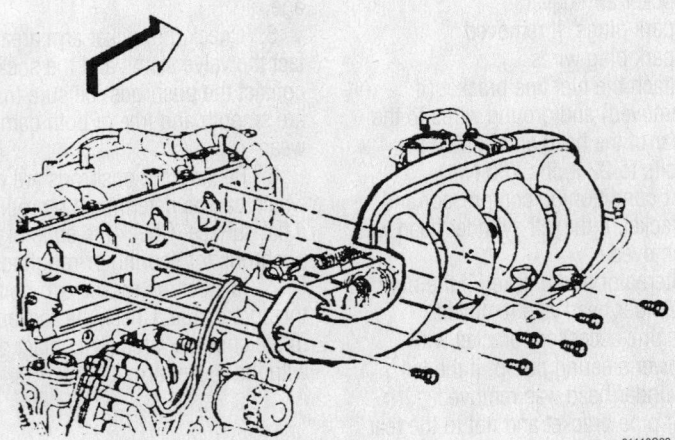

Typical intake manifold mounting—2.2L engine shown

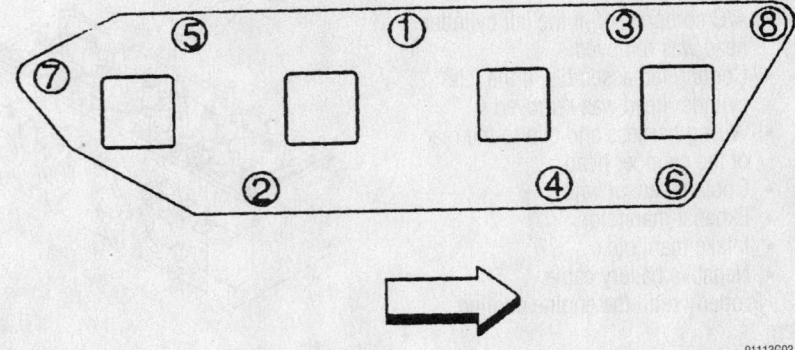

Intake manifold bolt tightening sequence—2.2L engine

- TP sensor connector
- Fuel injector harness connector
- Wheelhouse extension
- EGR adapter and tighten the retainers to 97 inch lbs. (11 Nm)
- EGR pipe to the EGR adapter and tighten the bolt to 18 ft. lbs. (25 Nm)
- EGR pipe-to-intake manifold bolts and tighten the bolts to 89 inch lbs. (10 Nm)
- Upper fan shroud and the brake booster hose
- Air cleaner resonator and connect the negative battery cable.

5. Start the engine and check for leaks.

4.3L Engines

➡**If only the upper intake manifold is being removed, the fuel system pressure does not need to be released. ALWAYS release the pressure before disconnecting any fuel lines.**

1. Before servicing the vehicle, refer to the precautions in the beginning of this section.

2. Remove the engine cover, if equipped

3. Properly relieve the fuel system pressure.

4. Drain the engine cooling system.

5. Remove or disconnect the following:
- Negative battery cable
- Air cleaner and air inlet duct
- Wiring harness connectors and brackets
- Throttle linkage from the upper intake manifold
- Ignition coil
- Fuel lines and bracket from the rear of the lower intake manifold
- Brake booster vacuum hose at the upper intake manifold
- Positive Crankcase Ventilation (PCV) hose at the rear of the upper intake manifold
- Vacuum hoses from both the front and rear of the upper intake
- Purge solenoid and bracket
- Upper intake manifold
- High Voltage Switch (HVS) assembly
- Upper radiator hose at the thermostat housing
- Heater hose at the lower intake manifold
- Wiring harnesses and brackets.
- Automatic transmission dipstick tube
- Exhaust Gas Recirculation (EGR) tube, clamp and tube

- Air conditioning compressor bracket-to-lower intake manifold pencil brace
- Alternator bracket bolts near the thermostat housing
- Lower intake manifold

6. Insert clean rags into the openings in the cylinder head to prevent dirt and debris from entering the engine.

7. Clean the gasket mounting surfaces. Be sure to inspect the manifold for warpage and/or cracks. If necessary, replace it.

To install:

8. Remove the rags from the cylinder heads.

9. Position the gaskets on the cylinder head with the port blocking plates to the rear and the **this side up** stamps facing upward. Then apply a ³⁄₁₆ in. (5mm) bead of RTV sealant on the front and rear of the engine block at the block-to-manifold mating surface. Extend the bead ½ in. (13mm) up each cylinder head to seal and retain the gaskets.

10. Install the lower intake manifold. Tighten the bolts in sequence and in 3 steps, as follows:
- a. Step 1: 26 inch lbs. (3 Nm).
- b. Step 2: 106 inch lbs. (12 Nm).
- c. Step 3: 11 ft. lbs. (15 Nm).

11. Install or connect the following:
- Alternator bracket bolt near the thermostat housing

- EGR tube, clamp and bolt
- Wiring harness to the lower manifold components, including the injector, EGR valve and ECT sensor
- Air conditioning compressor bracket-to-the lower intake manifold pencil braces
- Transmission oil dipstick tube, if necessary
- Fuel supply and return lines to the rear of the lower intake

12. Temporarily reattach the negative battery cable, then pressurize the fuel system (by cycling the ignition without starting the engine) and check for leaks.

13. Disconnect the negative battery cable.

14. Install or connect the following:
- Heater hose to the lower intake
- Upper radiator hose to the thermostat housing
- Vacuum hoses to the upper and lower intake manifold
- New upper intake manifold gasket, making sure the green sealing lines are facing upward
- Upper intake manifold being careful not to pinch the fuel injector wires between the manifolds
- Manifold retainers. Tighten them to 88 inch lbs. (10 Nm) using two passes.

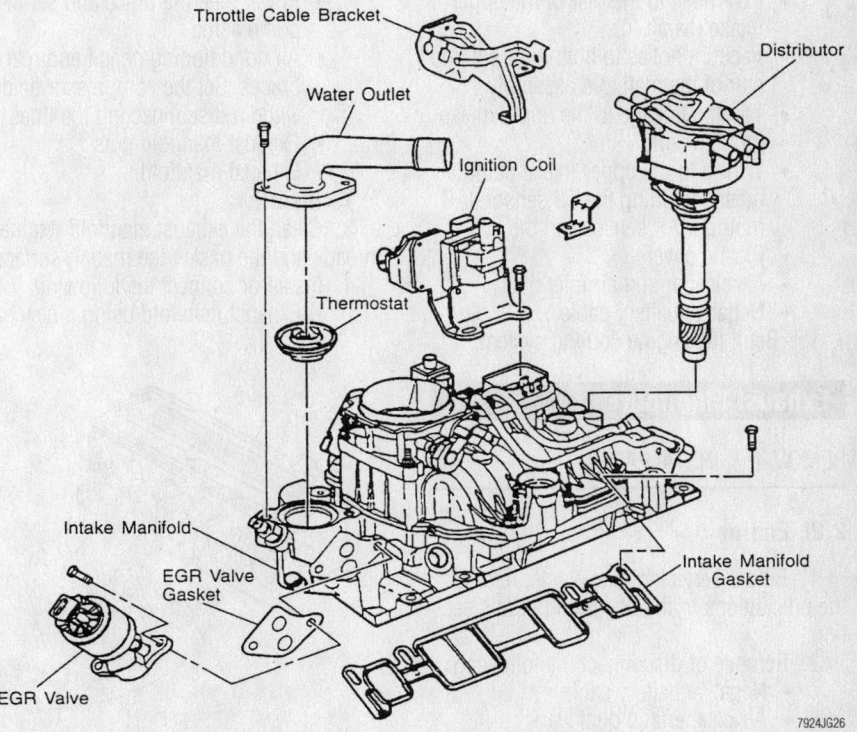

Intake manifold and related components—4.3L engine

7924JG26

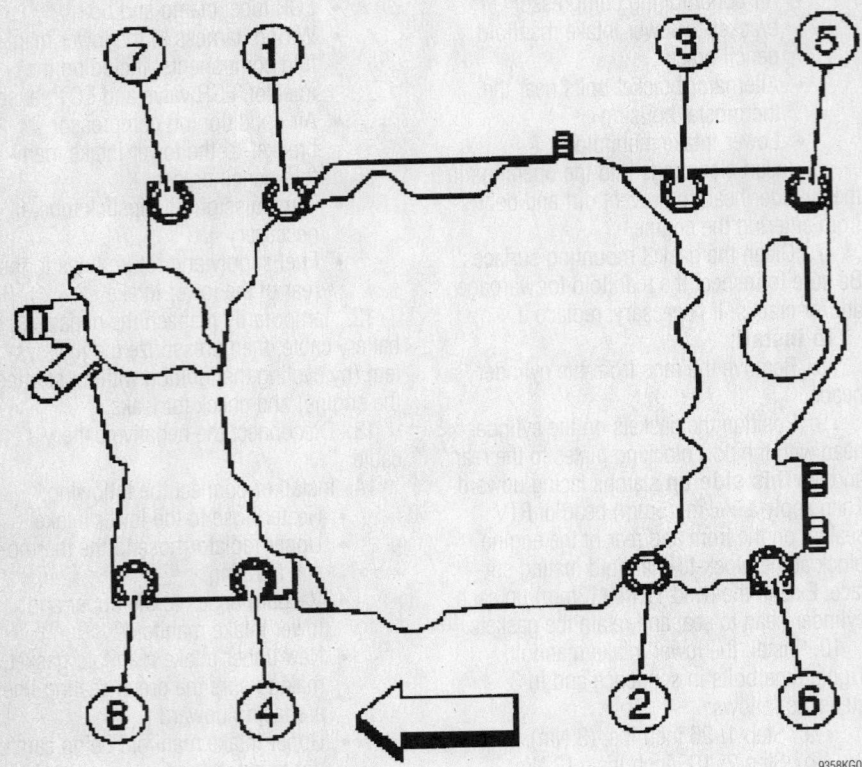

Lower intake manifold tightening sequence—4.3L engines

- Purge solenoid and bracket
- Brake booster vacuum hose at the upper intake manifold.
- PCV hose to the rear of the upper intake manifold
- Vacuum hoses to both the front and rear of the manifold assembly
- Throttle linkage to the upper intake
- Ignition coil
- Wiring to the upper intake components including the TP sensor, IAC motor, MAP sensor and the IMTV.
- Plastic cover
- Air cleaner and air inlet duct
- Negative battery cable

15. Refill the engine cooling system.

Exhaust Manifold

REMOVAL & INSTALLATION

2.2L Engine

1. Before servicing the vehicle, refer to the precautions in the beginning of this section.
2. Remove or disconnect the following:
   - Negative battery cable
   - Air cleaner and duct work
   - Oxygen Sensor (O2S) from the manifold, if replacing it

- Drive belt
- Oil fill tube assembly
- Heater hose brace
- Power steering brace and set the pump aside
- Air conditioning pencil and rear braces. Set the compressor aside without disconnecting the lines
- Exhaust manifold nuts
- Exhaust manifold

To install:

3. Clean the exhaust manifold retainer threads and the gasket the mating surfaces.
4. Install or connect the following:
   - Exhaust manifold using a new gas-

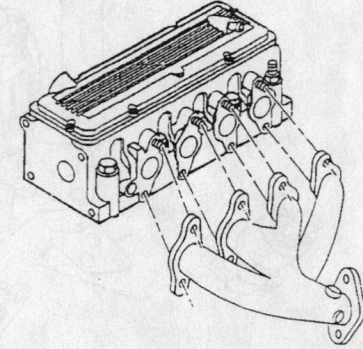

Exhaust manifold mounting—2.2L engine

ket. Torque the nuts to 115 inch lbs. (13 Nm).
- Exhaust pipe to the manifold
- Air conditioning pencil and rear braces
- Heater hose
- Power steering and heater hose braces
- Oil fill tube assembly
- O2S. Torque the sensor to 31 ft. lbs. (42 Nm), if necessary.
- Drive belt
- Air cleaner and duct work
- Negative battery cable

4.3L Engines

1. Before servicing the vehicle, refer to the precautions in the beginning of this section.
2. Remove or disconnect the following:
   - Negative battery cable

➡It will be easier if the vehicle is only supported to a height where underhood access is still possible, the vehicle may be left in position for the entire procedure. If the vehicle is raised too high for underhood access, it will have to lowered, raised and lowered again during the procedure.

- Exhaust pipe from the exhaust manifold. It may be necessary to remove the tires to gain access to the rear manifold bolts.
- Engine oil dipstick tube bolt, if removing the right side manifold
- Exhaust Gas Recirculation (EGR) inlet pipe from the left side manifold, if necessary
- Engine Coolant Temperature (ECT) sensor electrical connection
- Upper radiator support hose and nut
- Steering intermediate shaft, if removing the left side manifold
- Wheel house extension, if removing the right side manifold
- Spark plugs wires from the plugs
- Nuts attaching the secondary air injection pipe to the manifold
- Air injection pipe and gasket
- Locktangs (unbend), the exhaust manifold retaining bolts, washers and tab washers
- Heat shields
- Exhaust manifold
- Old gaskets and discard

To install:

3. Using a putty knife, clean the gasket mounting surfaces. Inspect the exhaust manifold for distortion, cracks or damage; replace if necessary.

4. Apply a threadlock such as GM 12345493 to the threads of the manifold retainers prior to installation.

5. Install or connect the following:
- Exhaust manifold to the cylinder using a new gasket, then tighten the center bolts to 11 ft. lbs. (15 Nm) and the front and rear manifold bolts to 22 ft. lbs. (30 Nm). Once the bolts are tightened, bend the tabs on the washers back over the heads of all bolts in order to lock them in position.
- Spark plug wires to the plugs
- Fender wheelhouse extension and the tire assembly, if removed
- Secondary air injection pipe with a NEW gasket to the manifold and tighten the nuts to 18 ft. lbs. (25 Nm), if removed
- EGR inlet pipe, if removed
- ECT sensor electrical connection, if removed
- Upper radiator hose support and nut, if removed
- Steering intermediate shaft, if removed (left side manifold only)
- Engine oil dipstick tube bolt to 106 inch lbs. (12 Nm), if removed
- Exhaust pipe to the manifold
- Negative battery cable

Camshaft and Valve Lifters

REMOVAL & INSTALLATION

2.2L Engine

1. Before servicing the vehicle, refer to the precautions in the beginning of this section.

2. Properly relieve the fuel system pressure.

3. Disconnect the negative battery cable.

4. Drain the engine cooling system and the engine oil.

5. Remove or disconnect the following:
- Radiator
- Rocker arm cover
- Cylinder head
- Anti-rotation bracket bolts and brackets
- Valve lifters
- Oil pump drive retaining bolt and the drive by lifting and twisting
- Camshaft Position (CMP) sensor, if equipped
- Crankshaft pulley and hub
- Drive belt idler pulley
- Timing cover from the engine

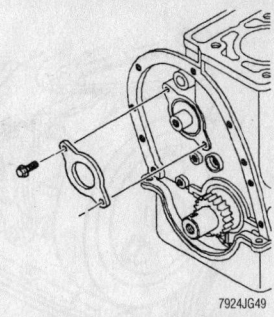

7924JG49

Remove the camshaft thrust plate and withdraw the camshaft from the engine—2.2L engine

- Timing chain and camshaft sprocket
- Camshaft thrust plate
- Camshaft by pulling it straight out of the engine, while turning it slightly as it is withdrawn and taking care not to damage the bearings.

To install:

6. Inspect the camshaft, journals and lobes for wear and replace, if necessary.

7. If removed, use the camshaft bearing tool to install a new set of bearings.

8. Coat the camshaft lobes and journals with a high viscosity oil with zinc such as GM 12345501.

9. Install or connect the following:
- Camshaft by turning it slightly from side-to-side as it is inserted
- Thrust plate. Torque the bolts to 106 inch lbs. (12 Nm).
- Timing chain and camshaft sprocket
- Timing cover
- Serpentine drive belt idler pulley
- Crankshaft pulley and hub
- Oil pump drive by inserting while twisting. Torque the fasteners to 18 ft. lbs. (25 Nm).
- Valve lifters and the anti-rotation brackets
- Cylinder head. Torque the bolts to 46 ft. lbs. (62 Nm) plus an additional 90 degrees turn.
- Rocker arm cover
- Radiator
- Negative battery cable

10. Refill the engine cooling system.

4.3L Engines

1. Before servicing the vehicle, refer to the precautions in the beginning of this section.

2. Properly relieve the fuel system pressure.

3. Disconnect the negative battery cable.

4. Drain the engine cooling system.

5. Discharge and recover the refrigerant from the air conditioning system.

6. Remove or disconnect the following:
- Radiator
- Air conditioning condenser
- Rocker arm covers
- Intake manifold assembly
- Rocker arms, pushrods and lifters
- Crankshaft pulley and hub
- Engine front (timing) cover

7. Align the timing marks on the crankshaft and camshaft sprockets.

8. Remove or disconnect the following:
- Camshaft sprocket and timing chain
- Balance shaft drive gear, if equipped
- Camshaft thrust plate
- Camshaft by installing the sprocket bolts or longer bolts the camshaft end to act as a handle; then, remove the camshaft while turning slightly from side to side, as necessary.

➡**Take care not to damage the camshaft bearings when removing the camshaft.**

To install:

9. Lubricate the camshaft journals with clean engine oil or a suitable pre-lube.

10. Install or connect the following:
- Camshaft being extremely careful not to contact the bearings with the cam lobes
- Thrust plate. Torque the bolts to 106 inch lbs. (12 Nm).
- Balance shaft drive gear, if equipped
- Timing chain and camshaft sprocket
- Engine front (timing) cover
- Crankshaft pulley and hub

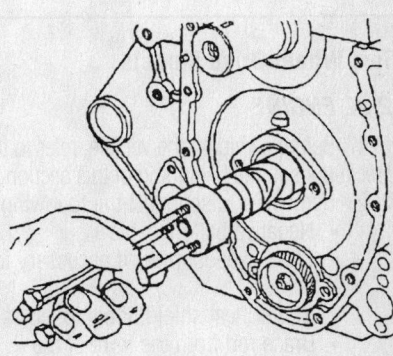

7924JG50

Thread 3 long bolts into the camshaft to use as a handle, then withdraw it from the engine

- Valve lifters, pushrods and rocker arms. Adjust the valve clearance.
- Intake manifold assembly
- Rocker arm covers
- Radiator
- Negative battery cable

11. Refill the engine cooling system.

Valve Lash

ADJUSTMENT

2.2L Engine

Because the rocker arm fasteners are secured and tightened, valve lash is not adjustable on the 2.2L engine. If a valvetrain problem is suspected, check that the rocker arm nuts are tightened to 22 ft. lbs. (30 Nm). Be very careful not to over-tighten the rocker arm nuts. ONLY tighten the nuts when the hydraulic lifter is resting on the base circle of the camshaft and not when it is held upward on the lobe. When valve lash falls out of specification (valve tap is heard), replace the rocker arm, pushrod and hydraulic lifter on the offending cylinder.

4.3L Engines

The 4.3L engines are equipped with screw-in rocker arm studs with positive stop shoulders. Because the shoulders that allow the rocker arms to be tightened into proper position, no adjustments are necessary or possible. If a valvetrain problem is suspected, check that the rocker arm nuts are tightened to 18 ft. lbs. (24 Nm). When valve lash falls out of specification (valve tap is heard), replace the rocker arm, pushrod and hydraulic lifter on the offending cylinder.

Starter Motor

REMOVAL & INSTALLATION

Two Wheel Drive Models

2.2L ENGINE

1. Before servicing the vehicle, refer to the precautions in the beginning of this section.
2. Remove or disconnect the following:
   - Negative battery cable
   - Front exhaust pipe, if necessary for access
   - Starter heat shield, if equipped
   - Brace rod from the front of the engine and the bell housing
   - Drivers side wheel to access the starter motor wires and the starter motor attaching bracket-to-engine

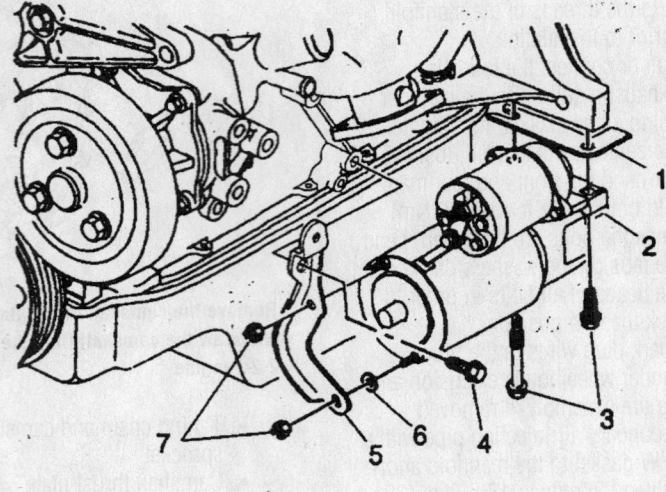

1. SHIM
2. STARTER ASSEMBLY
3. BOLT, 43 N·m (32 LBS. FT.)
4. BOLT, 43 N·m (32 LBS. FT.)
5. BRACKET, STARTER MOTOR
6. WASHER
7. NUT, 11 N·m (97 LBS. IN.)

88452G08

Starter motor and related components—2.2L engine

bolt through the opening in the wheel well
- Wires from the starter solenoid
- Attaching bracket-to-engine mount bolt
- Starter-to-engine block bolts. When removing the last bolt, be sure to support the starter to keep it from falling and possibly injuring you.
- Starter and shims (if equipped) from the vehicle
- Bracket from the starter assembly, if equipped

To install:
3. Install or connect the following:
   - Bracket to the starter, if removed. Tighten the bracket nuts to 97 inch lbs. (11 Nm).
   - Starter and shims (if equipped) into position in the vehicle and thread one of the retaining bolts to hold it in position.
   - Bracket-to-engine mount bolt (loosely), if equipped
   - Starter mounting bolt, then tighten all mounting fasteners to 32 ft. lbs. (43 Nm)
   - Wiring to the solenoid
   - Brace rod and tighten the retainers
   - Front exhaust pipe and tighten the fasteners, if removed

- Starter heat shield, if equipped
- Driver's side wheel, if removed
- Negative battery cable

4.3L ENGINE

1. Before servicing the vehicle, refer to the precautions in the beginning of this section.
2. Remove or disconnect the following:
   - Negative battery cable
   - Wires from the starter solenoid
   - Starter motor mounting bolts
   - Starter motor and if equipped, the shims

To install:
3. Install or connect the following:
   - Starter motor into position
   - Starter motor inboard bolt but do not tighten it at this time
   - Starter motor shims, if equipped
   - Outboard starter motor bolt. Tighten the bolts to 32 ft. lbs. (43 Nm).
   - Wires to the solenoid
   - Negative battery cable

Four Wheel Drive Models

1. Before servicing the vehicle, refer to the precautions in the beginning of this section.

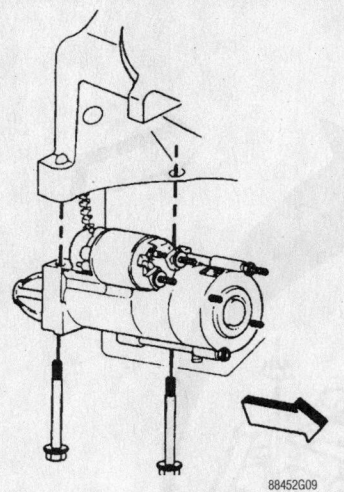

The starter motor on 4.3L engines is retained by two long bolts

2. Remove or disconnect the following:

- Negative battery cable

➥In some cases it may be easier to access the starter motor bolts if you raise the vehicle and remove the wheel assembly.

- Wheel assembly, if necessary
- Engine mounts
- Transmission mount and support the transmission assembly
- Starter-to-engine bolts and support the starter

3. Rotate the starter as necessary for access, then tag and disconnect the solenoid wiring.

4. Carefully lower the starter and shims (if equipped) from the vehicle. Note the location of any shims for installation purposes.

5. If necessary, remove the shield from the starter assembly.

To install:

6. Install or connect the following:

- Starter into position in the vehicle along with any shims (making sure they are in their original positions), then tighten the mounting bolts to 32 ft. lbs. (43 Nm).
- Shield to the starter assembly and tighten the retaining nuts to 106 inch lbs. (12 Nm), if removed
- Wiring to the solenoid
- Transmission mount and remove the supports
- Secure the engine mounts, then remove the lifting device
- Wheel assembly, if removed

7. Connect the negative battery cable.

Oil Pan

REMOVAL & INSTALLATION

2.2L Engine

1. Before servicing the vehicle, refer to the precautions in the beginning of this section.

2. Drain the engine oil.

3. Remove or disconnect the following:

- Engine
- Clutch pressure plate and disc, if equipped
- Flywheel
- Oil pan retainers and the pan

To install:

4. Clean the gasket mating surfaces

5. Install or connect the following:

- New gasket and seal onto the oil pan using a thin bead of sealant at either side of the seal
- Oil pan. Torque the bolts to 89 inch. lbs. (10 Nm).
- Flywheel
- Pressure plate and disc, if equipped
- Engine

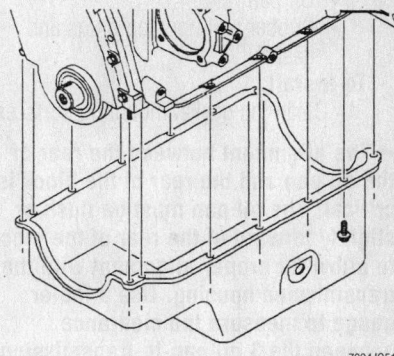

Oil pan mounting—2.2L engine

4.3L Engines

2WD MODELS

1. Before servicing the vehicle, refer to the precautions in the beginning of this section.

2. Drain the engine oil.

3. Remove or disconnect the following:

- Engine
- Oil level sensor, if equipped, and discard
- Oil pan retainers (nuts, studs and/or bolts) and rail reinforcements, if equipped

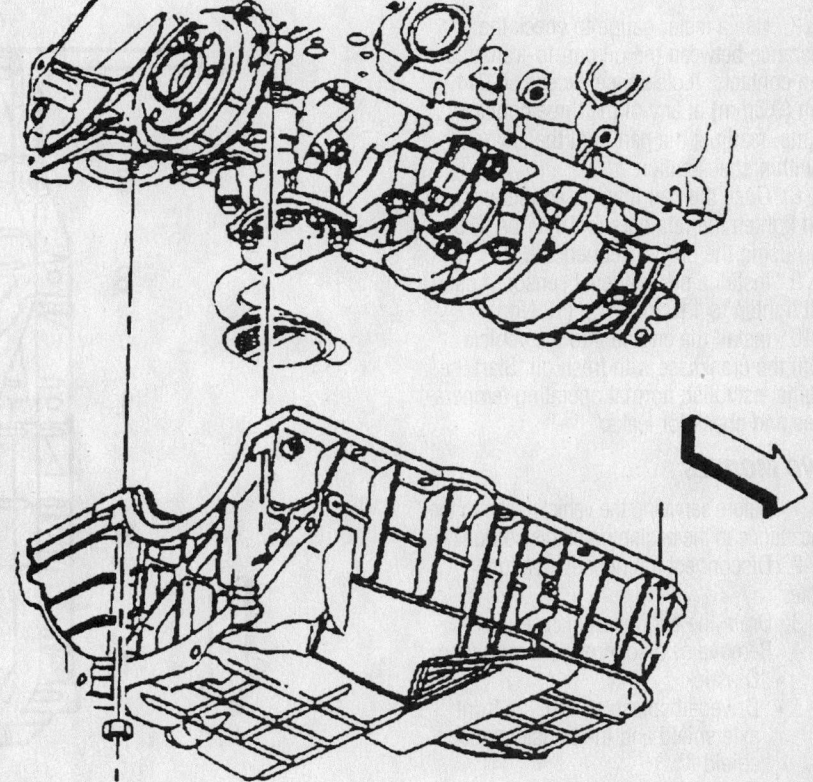

Oil pan mounting—4.3L engines

- Oil pan
- Rubber bell housing plugs and gasket

To install:

4. Clean the gasket mounting surfaces.

➡The alignment between the rear of the oil pan and the rear of the block is critical. The oil pan must be flush or slightly forward of the rear of the block to allow for proper alignment with the transmission housing. Use a feeler gauge to measure the clearance between the 3 oil pan-to-transmission contact points. If the clearance exceeds 0.011 in. (0.3mm) at any of the 3 points, realign the oil pan.

5. Apply sealant to the oil pan rail where it contacts the timing cover-to-block joint (front) and the crankshaft rear seal retainer-to-block joint (rear). Continue the bead of sealant about 1 in. (25mm) in both directions from each of the 4 corners.

6. Install or connect the following:
- Rubber bell housing plugs, if equipped
- Oil pan using a new gasket

➡The alignment between the rear of the pan and rear of the block is critical. The two surfaces must be flush to allow for proper alignment with the transmission housing.

7. Use a feeler gauge to check the clearance between the oil pan-to-transmission contacts. If clearance exceeds 0.011 inch (0.3mm) at any of the three contact points, readjust the pan until the clearance is within specification.

8. Once the pan is in its correct position tighten the retainers to 18 ft. lbs. (25 Nm) using the proper sequence.

9. Install a new oil level sensor, if used and tighten to 115 inch lbs. (13 Nm).

10. Install the engine into the vehicle. Refill the crankcase with fresh oil. Start the engine, establish normal operating temperatures and check for leaks.

4WD MODELS

1. Before servicing the vehicle, refer to the precautions in the beginning of this section.

2. Disconnect the negative battery cable.

3. Drain the engine crankcase oil.

4. Remove or disconnect the following:
- Dipstick
- Drivebelt splash shield, the front axle shield and the transfer case shield
- Front skid plate and the flywheel cover

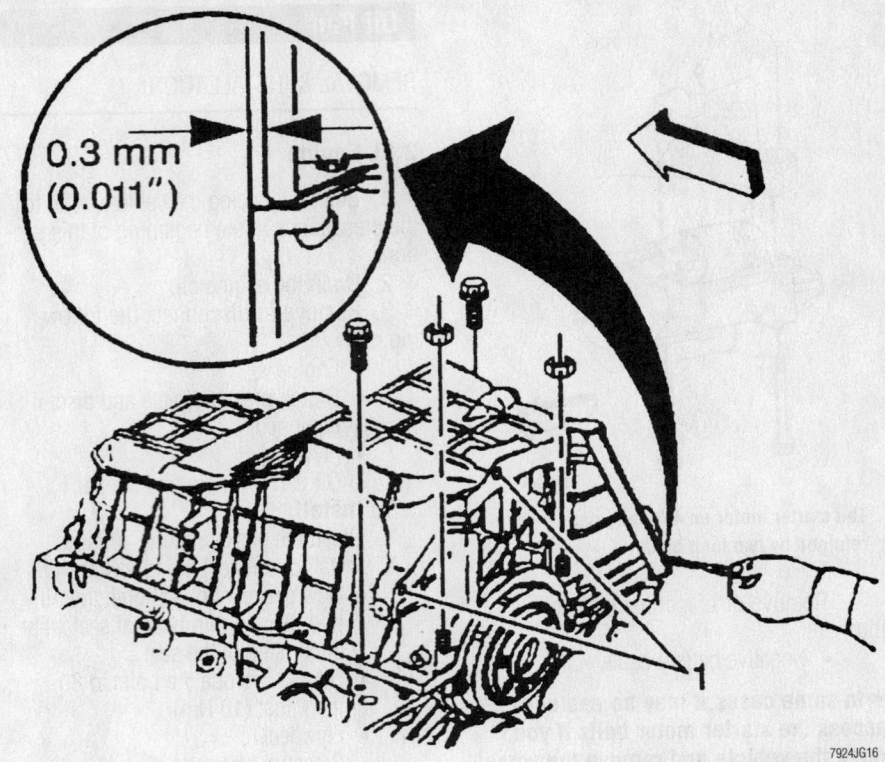

If the clearance between the 3 oil pan-to-transmission contact points exceeds 0.011 in. (0.3mm) at any of the 3 points, realign the oil pan—4.3L engine

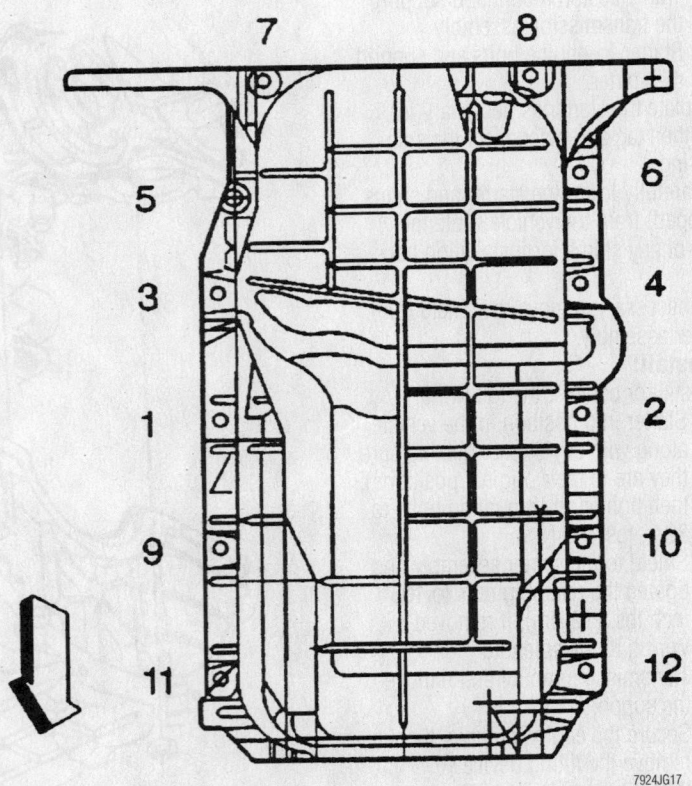

Tighten the bolts in sequence to prevent warping the sealing surface of the oil pan—4.3L vehicles

- Left and right engine mount through-bolts

5. Raise the engine using a lifting device and block in position. This may be accomplished using large wooden blocks between the motor mounts and brackets.

➡Use extreme caution when blocking the engine in position. Get out from under the vehicle and rock the engine slightly once the blocks are in place to be sure the engine is properly supported.

6. Remove or disconnect the following:
- Oil cooler line
- Pitman arm bolt and pitman arm
- Idler arm bolts and idler arm
- Front differential through-bolts
- Front driveshaft, if necessary
- Differential assembly by rolling it forward for clearance
- Starter motor
- Oil pan bolts, nuts and reinforcements
- Oil pan and discard the gasket

To install:

7. Clean the gasket mounting surfaces.

➡The alignment between the rear of the oil pan and the rear of the block is critical. The oil pan must be flush or slightly forward of the rear of the block to allow for proper alignment with the transmission housing. Use a feeler gauge to measure the clearance between the 3 oil pan-to-transmission contact points. If the clearance exceeds 0.011 in. (0.3mm) at any of the 3 points, realign the oil pan.

8. Apply sealant to the oil pan rail where it contacts the timing cover-to-block joint (front) and the crankshaft rear seal retainer-to-block joint (rear). Continue the bead of sealant about 1 in. (25mm) in both directions from each of the 4 corners.

9. Install or connect the following:
- Oil pan, using a new gasket. Tighten the retainers, in sequence, to 18 ft. lbs. (25 Nm).
- Starter motor
- Differential by rolling it back into position
- Front driveshaft
- Front differential through-bolts
- Idler arm and secure using the retaining bolts
- Pitman arm and secure using the bolts
- Transfer case shield
- Flywheel cover
- Front skid plate
- Front axle shield

- Drive belt splash shield
- Dipstick
- Negative battery cable

10. Refill the engine crankcase.
11. Start the engine and check for leaks.

Oil Pump

REMOVAL & INSTALLATION

1. Before servicing the vehicle, refer to the precautions in the beginning of this section.

2. Remove or disconnect the following:
- Oil pan
- Oil pump and the pickup tube/shaft, if equipped
- Extension shaft and retainer from the pump, if necessary for the 2.2L engine

➡Be careful not to crack the retainer.

To install:

3. For the 2.2L engine, if the extension shaft was removed, heat the extension shaft retainer in hot water, then install the shaft and retainer to the oil pump. Be sure the retainer does not crack during installation.

4. Ensure that the pump pickup tube is tight in the pump body. If the tube should come loose, oil pressure will be lost and oil starvation will occur. If the pickup tube is loose it should be replaced.

5. If the pump has been disassembled

and is being replaced or for any reason oil has been removed, it must be primed. It can either be filled with oil before installing the cover plate and oil kept within the pump during handling or the entire pump cavity can be filled with petroleum jelly.

✳✳ WARNING

If the pump is not primed, the engine could be damaged upon start up.

6. Install or connect the following:
- Oil pump. Tighten oil pump/pickup tube retainer(s) to 32 ft. lbs. (44 Nm) for the 2.2L engine or to 65 ft. lbs. (90 Nm), for the 4.3L engine.

➡If the oil pump does not build up oil pressure almost immediately, remove the pan and check for a loose oil pump-to-pickup tube attachment. If necessary dismantle the pump and pack the pump cavity with petroleum jelly.

- Oil pan
7. Refill the crankcase.
8. Disable the ignition system; crank engine for approximately 10 seconds to aid in priming the oil pump and reducing the risk of engine damage.

✳✳ WARNING

Running the engine without measurable oil pressure will cause extensive damage.

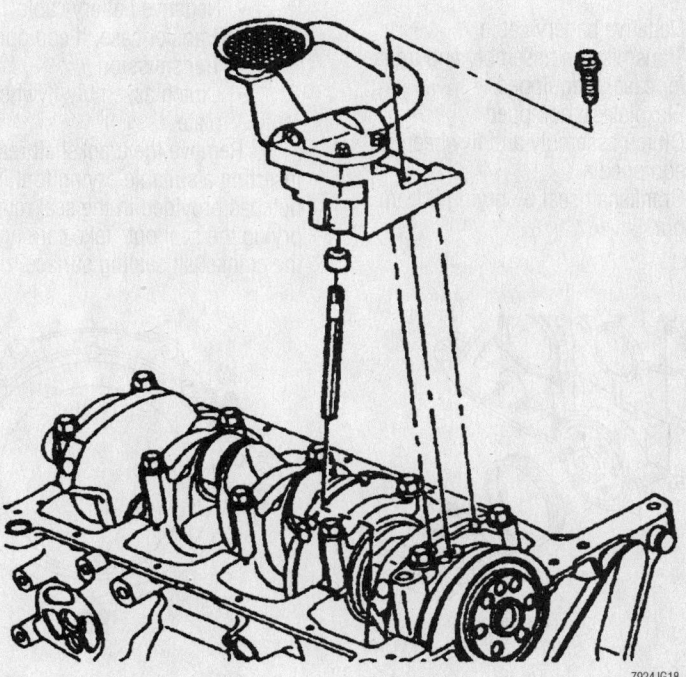

Exploded view of the oil pump mounting—2.2L engine

7924JG18

Rear Main Seal

REMOVAL & INSTALLATION

2.2L Engines

Please note that the transmission assembly and transfer case, if equipped, must be removed to perform this procedure.

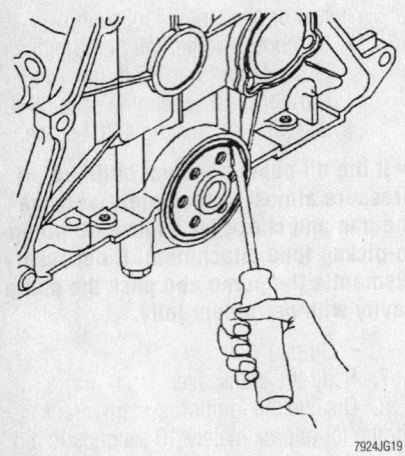

Carefully pry the rear main oil seal out of its bore—2.2L engine

1. Before servicing the vehicle, refer to the precautions in the beginning of this section.
2. Remove or disconnect the following:

- Negative battery cable
- Transmission assembly and transfer case, if equipped
- Flexplate, if equipped
- Clutch assembly and flywheel, if equipped
- Crankshaft seal by prying it from out

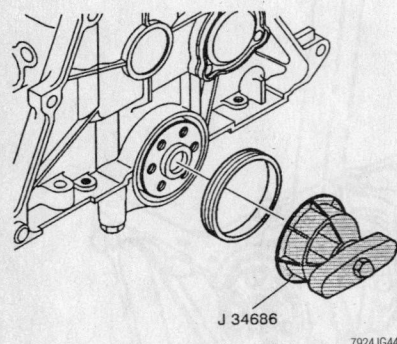

Rear main oil seal installation using tool J-34686—2.2L engine

➡ **Be careful not to damage the crankshaft seal surface with the prying tool.**

To install:

3. Install the new rear seal by lubricating it with engine oil and using a seal tool J-34686.
4. Slide the seal over the mandrel until the dust lip bottoms squarely against the tool collar.
5. Align the dowel pin of the tool with the dowel pinhole in the crankshaft and attach the tool to crankshaft.
6. Tighten the T-handle of the tool to push the seal into the bore. Continue until the tool collar is flush against the block.
7. Loosen the T-handle completely. Remove the attaching screws and tool. Check to be sure the seal is seated squarely in the bore.
8. Install or connect the following:

- Flywheel/clutch assembly or flexplate
- Transmission assembly and transfer case, if equipped
- Negative battery cable

9. Start the engine and check for leaks.

4.3L Engines

Please note that the transmission assembly and transfer case, if equipped, must be removed to perform this procedure.

1. Before servicing the vehicle, refer to the precautions in the beginning of this section.
2. Remove or disconnect the following:

- Negative battery cable
- Transfer case, if equipped
- Transmission
- Clutch assembly/flywheel or flexplate

3. Remove the crankshaft rear oil seal by inserting a suitable prying tool into the notches provided in the seal retainer and prying the seal out. Take care not to damage the crankshaft sealing surface.

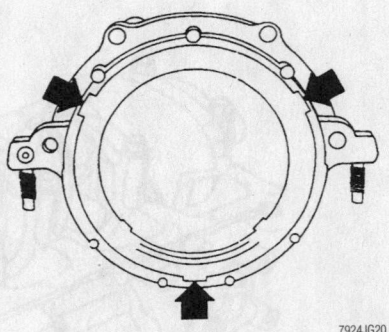

Carefully pry the rear main seal out of the retainer—4.3L engine

To install:

4. Inspect the crankshaft for grit, rust or burrs and correct as necessary.
5. Clean the running surface of the crankshaft with a non-abrasive cleaner.
6. Install or connect the following:

- New rear seal lubricated with engine oil and a seal installer
- Flywheel and clutch or flexplate
- Transmission
- Transfer case, if equipped
- Negative battery cable

7. Start the engine and verify no oil leaks.

Timing Chain, Sprockets, Front Cover and Seal

REMOVAL & INSTALLATION

Front Cover and Seal

2.2L ENGINES

1. Before servicing the vehicle, refer to the precautions in the beginning of this section.
2. Remove or disconnect the following:

- Negative battery cable
- Drive belt
- Cooling fan assembly and pulley
- Crankshaft pulley and hub
- Belt tensioner/idler pulley assembly
- Front oil pan-to-front cover nuts or studs
- Starter
- Alternator and brackets from the engine, then position them aside
- Oil pan bolts, loosen but do not remove
- Crankcase (timing) front cover bolts and the cover. Make sure all bolts are removed and be careful not to force and damage the cover.
- Old crankshaft seal from the cover using a suitable prytool. Be very careful not to distort the front cover or to score the end of the crankshaft.

To install:

3. Carefully remove all traces of gasket or sealant from the mating surfaces.
4. Lubricate the lips of a new seal with clean engine oil, then use a seal centering tool (such as J-35468) to install the seal to the front cover. Leave the tool in position in the seal until the cover is installed.
5. Apply a ³⁄₈ in. (10mm) wide by ⁵⁄₁₆ (5mm) thick bead of RTV sealer to the oil pan at the front crankcase cover sealing surface. Then apply a ¼ in. (6mm) by ⅛ in. (3mm) thick bead of RTV to the

crankcase front cover at the block sealing surface.

6. Install or connect the following:
- Crankcase front cover to the engine using the seal tool to assure it is properly centered and prevent damage to the hub. Tighten the cover retaining bolts to 97 inch lbs. (11 Nm), then remove the seal centering tool.
- Oil pan bolts
- Starter
- Alternator with brackets
- Belt tensioner/idler pulley assembly
- Belt assembly
- Crankshaft pulley and hub
- Cooling fan assembly and pulley
- Negative battery cable

4.3L ENGINES

1. Before servicing the vehicle, refer to the precautions in the beginning of this section.
2. Remove or disconnect the following:
- Negative battery cable
- Drain the engine cooling system.
- Crankshaft pulley and damper

✳✳ WARNING

The outer ring (weight) of the torsional damper is bonded to the hub with rubber. The damper must be removed with a puller that acts on the inner hub only. Pulling on the outer portion of the damper will break the rubber bond or destroy the tuning of the unit.

- Water pump assembly
- Crankshaft Position (CKP) sensor

➡**Depending upon the year of your truck, you may just need to loosen the oil pan bolts, or you may need to remove it completely.**

- Oil pan or loosen bolts, as applicable
- Crankshaft Position (CKP) sensor, if equipped
- Front cover bolts and the reinforcements, if equipped
- Front cover from the engine

3. Pry the seal out of the front cover using a small prytool. Be very careful not to distort the front cover or to score the end of the crankshaft.

To install:

➡**Anytime the front cover is removed, the cover must be replaced upon reassembly. If you reuse the old cover, oil leaks may develop.**

4. Clean the gasket mating surfaces of the engine and cover of all remaining gasket or sealer material. Be careful not to score or damage the surfaces.

➡**The manufacturer suggests you wait until the front cover is mounted to the engine before you install the replacement crankshaft oil seal. This assures the cover is properly supported.**

5. Install or connect the following:
- New front cover gasket to the engine or cover using gasket cement to hold it in position. Lubricate the front of the oil pan seal with engine oil to aid in reassembly.
- Front cover to the engine. Take care while engaging the front of the oil pan seal with the bottom of the cover. Apply sealer 12346141 to the oil pan rail where it contacts the timing cover-to-block joint (front) and the crankshaft rear seal retainer-to-block joint (rear). Continue the bead of sealant about 1 in. (25mm) in both directions from each of the four corners.

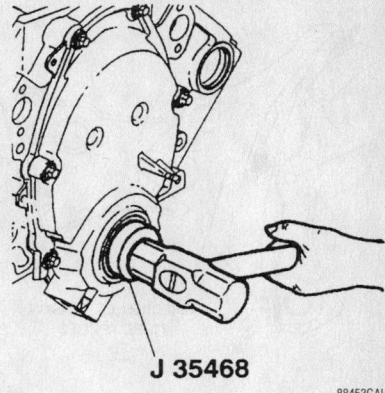

J 35468

88453GAU

Installing the crankshaft front oil seal— 4.3L engines

- Front cover retaining bolts and tighten to 106 inch. lbs. (12 Nm)

6. Lightly coat the lips of the replacement crankshaft seal with clean engine oil, then position the seal with the open end facing inward the engine. Use a suitable seal installation driver to position the seal in the front cover.

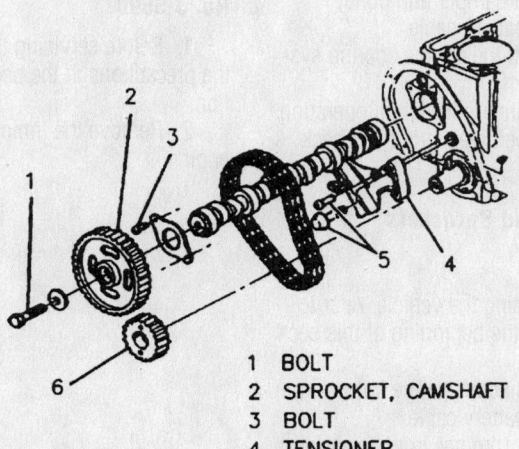

1	BOLT
2	SPROCKET, CAMSHAFT
3	BOLT
4	TENSIONER
5	BOLTS
6	SPOCKET, CRANKSHAFT

A ALIGN TABS ON TENSIONER WITH MARKS ON CAMSHAFT & CRANKSHAFT SPROCKETS.

85383289

Timing chain, sprocket and camshaft mounting—2.2L engine

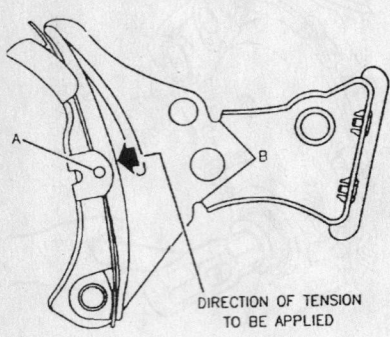

A INSERT PIN AFTER TENSION HAS BEEN APPLIED
B TABS, USED FOR CAMSHAFT AND CRANKSHAFT
ALIGNMENT

85383290

Locking the timing chain tensioner into position for chain installation—2.2L engine

- CKP sensor O-ring and the sensor, if equipped
- Oil pan, if removed
- Tighten the oil pan bolts
- Water pump
- Crankshaft damper and pulley
- Negative battery cable

7. Properly refill the engine cooling system.

8. Run the engine until normal operating temperature has been reached, then check for leaks.

Timing Chain and Sprockets

2.2L ENGINES

1. Before servicing the vehicle, refer to the precautions in the beginning of this section.

2. Remove or disconnect the following:
- Negative battery cable
- Crankcase (timing) front cover from the engine

3. Turn the crankshaft until the timing marks on the sprockets are in alignment. The marks should also be in alignment with the tabs on the tensioner.
- Tensioner retaining bolts
- Camshaft sprocket retaining bolts
- Camshaft sprocket and timing chain at the same time
- Tensioner assembly
- Crankshaft Position (CKP) sensor, if equipped
- Crankshaft sprocket using J-22888-20, if equipped

To install:

4. Install or connect the following:
- Crankshaft sprocket using a suitable installer such as J-5590, if

removed. Make sure the sprocket is fully seated against the crankshaft.

5. Compress the tensioner spring and insert a cotter pin or nail in the hole provided to hold the tensioner in position.
- Tensioner retaining bolts
- Camshaft sprocket in the timing chain, position the chain under the crankshaft sprocket and the camshaft sprocket to the camshaft

6. Verify that the timing marks are all properly aligned, then loosely install the camshaft sprocket bolt.
- CKP sensor, if equipped
- Tighten the tensioner bolts to 18 ft. lbs. (24 Nm), then tighten the camshaft sprocket bolt to 96 ft. lbs. (130 Nm)

7. Remove the cotter pin or nail holding the tensioner in position off the chain.
- Timing cover to the engine

4.3L ENGINES

➡**The following procedure requires the use of the Crankshaft Sprocket Removal tool No. J-5825-A and the Crankshaft Sprocket Installation tool No. J-5590.**

1. Before servicing the vehicle, refer to the precautions in the beginning of this section.

2. Remove the timing cover from the engine.

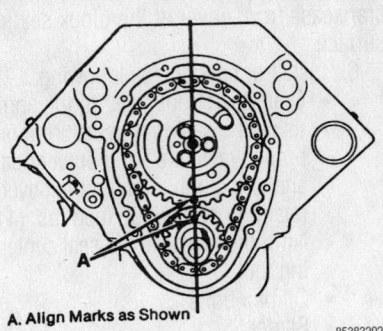

A. Align Marks as Shown

85383292

Timing mark alignment—4.3L engine

3. If equipped, remove the crankshaft reluctor ring.

4. Rotate the crankshaft until the No. 4 cylinder is on the Top Dead Center (TDC) of its compression stroke and the camshaft sprocket mark aligns with the mark on the crankshaft sprocket (facing each other at a point closest together in their travel) and in line with the shaft centers.

5. Remove or disconnect the following:
- Crankshaft Position (CKP) sensor reluctor ring, if equipped
- Camshaft sprocket-to-camshaft nut and/or bolts
- Camshaft sprocket (along with the timing chain). If the sprocket is difficult to remove, use a plastic mallet to bump the sprocket from the camshaft

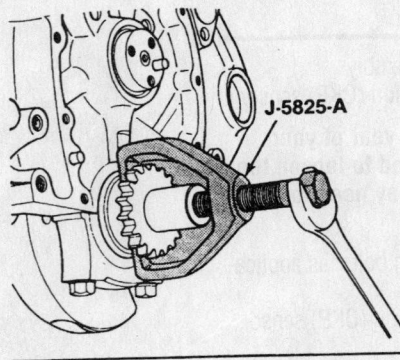

J-5825-A

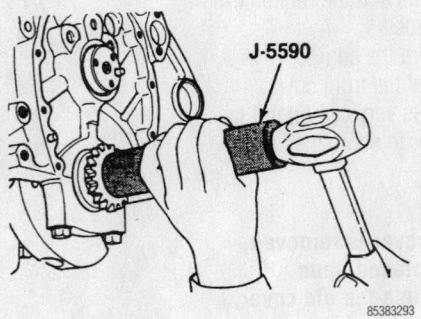

J-5590

85383293

Removal (top) and installation of the crankshaft timing gear

➥The camshaft sprocket (located by a dowel) is lightly pressed onto the camshaft and should come off easily. The chain comes off with the camshaft sprocket.

6. If necessary use J-5825-A crankshaft sprocket removal tool to free the timing sprocket from the crankshaft.

7. If necessary, remove the crankshaft sprocket key.

To install:

8. Inspect the timing chain and the timing sprockets for wear or damage, replace the damaged parts as necessary.

9. Using a putty knife, clean the gasket mounting surfaces. Using solvent, clean the oil and grease from the gasket mounting surfaces.

10. Install or connect the following:
   - Crankshaft sprocket key, if removed
   - Crankshaft sprocket onto the crankshaft using J-5590 crankshaft sprocket installation tool and a hammer without disturbing the position of the engine

➥During installation, coat the thrust surfaces lightly with Molykote, or an equivalent pre-lube.

   - Timing chain over the camshaft sprocket. Arrange the camshaft sprocket in such a way that the timing marks will align between the shaft centers and the camshaft locating dowel will enter the dowel hole in the cam sprocket.
   - Timing chain under the crankshaft sprocket, then place the cam sprocket, with the chain still mounted over it, in position on the front of the camshaft
   - Camshaft sprocket-to-camshaft retainers to 18 ft. lbs. (25 Nm)

11. With the timing chain installed, turn the crankshaft two complete revolutions, then check to make certain that the timing marks are in correct alignment between the shaft centers.
   - CKP sensor reluctor ring, if equipped
   - Timing cover

Piston and Ring

POSITIONING

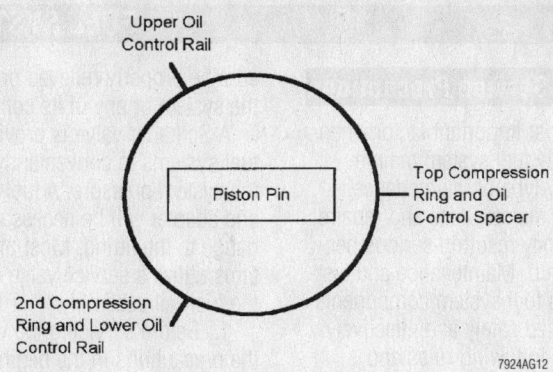

Piston ring end-gap spacing—2.2L engine

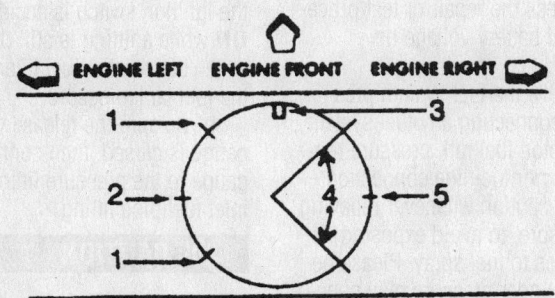

1. Oil ring rail gaps
2. 2nd Compression ring gap
3. Notch in piston
4. Oil ring spacer gap (tang in hole or slot with arc)
5. Top compression ring gap

Piston ring end-gap spacing—4.3L engines

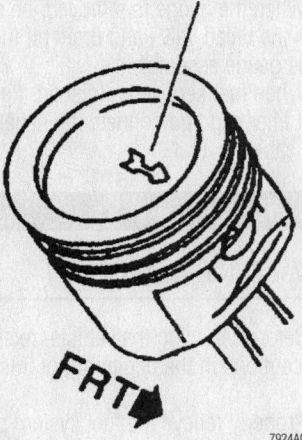

ARROW TOWARDS FRONT OF ENGINE

FRT

Piston/connecting rod-to-engine positioning—2.2L engines

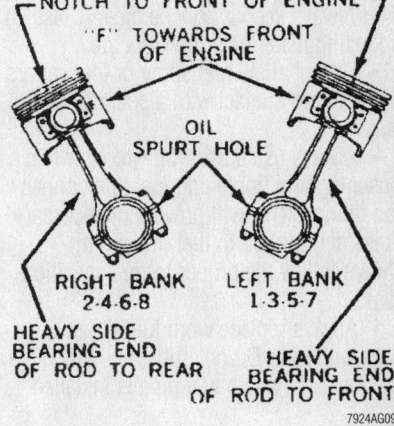

NOTCH TO FRONT OF ENGINE
"F" TOWARDS FRONT OF ENGINE
OIL SPURT HOLE
RIGHT BANK 2-4-6-8
LEFT BANK 1-3-5-7
HEAVY SIDE BEARING END OF ROD TO REAR
HEAVY SIDE BEARING END OF ROD TO FRONT

Piston and connecting rod assembly positioning—4.3L engine

FUEL SYSTEM

Fuel System Service Precautions

Safety is the most important factor when performing not only fuel system maintenance but also any type of maintenance. Failure to conduct maintenance and repairs in a safe manner may result in serious personal injury or death. Maintenance and testing of the vehicle's fuel system components can be accomplished safely and effectively by adhering to the following rules and guidelines.

• To avoid the possibility of fire and personal injury, always disconnect the negative battery cable unless the repair or test procedure requires that battery voltage be applied.

• Always relieve the fuel system pressure prior to disconnecting any fuel system component (injector, fuel rail, pressure regulator, etc.), fitting or fuel line connection. Exercise extreme caution whenever relieving fuel system pressure, to avoid exposing skin, face and eyes to fuel spray. Please be advised that fuel under pressure may penetrate the skin or any part of the body that it contacts.

• Always place a shop towel or cloth around the fitting or connection prior to loosening to absorb any excess fuel due to spillage. Ensure that all fuel spillage (should it occur) is quickly removed from engine surfaces. Ensure that all fuel soaked cloths or towels are deposited into a suitable waste container.

• Always keep a dry chemical (Class B) fire extinguisher near the work area.

• Do not allow fuel spray or fuel vapors to come into contact with a spark or open flame.

• Always use a back-up wrench when loosening and tightening fuel line connection fittings. This will prevent unnecessary stress and torsion to fuel line piping. Always follow the proper torque specifications.

• Always replace worn fuel fitting O-rings with new. Do not substitute fuel hose or equivalent where fuel pipe is installed.

Fuel System Pressure

RELIEVING

Multi-Port Fuel Injection and Central Port Injection Systems

The fuel systems operate under high fuel pressures. It is very important that the pressure be properly relieved prior to servicing the system or any of its components.

A Schrader valve is provided on these fuel systems to conveniently test or release the system pressure. A fuel pressure gauge and adapter will be necessary to connect the gauge to the fitting. Most of the MFI systems utilize a service valve on one end of the fuel rail assembly.

1. Before servicing the vehicle, refer to the precautions in the beginning of this section.

2. Disconnect the negative battery cable to assure the prevention of fuel spillage if the ignition switch is accidentally turned **ON** while a fitting is still detached.

3. Loosen the fuel filler cap to release the fuel tank pressure.

4. Be sure the release valve on the fuel gauge is closed, then connect the fuel gauge to the pressure fitting located on the inlet fuel pipe fitting.

✳✳ CAUTION

When connecting the gauge to the fitting, be sure to wrap a rag around the fitting to avoid spillage. After repairs, place the rag in an approved container.

5. Install the bleed hose portion of the fuel gauge assembly into an approved container, then open the gauge release valve and bleed the fuel pressure from the system.

6. When the gauge is removed, be sure to open the bleed valve and drain all fuel from the gauge assembly.

7. When fuel service is finished, tighten the fuel filler cap and connect the negative battery cable.

Fuel Filter

REMOVAL & INSTALLATION

1. Before servicing the vehicle, refer to the precautions in the beginning of this section.

2. Properly relieve the fuel system pressure.

3. Remove or disconnect the following:

• Negative battery cable
• Fuel filler cap
• Quick connect fittings from the filter
• Filter feed nut and the clamp bolt

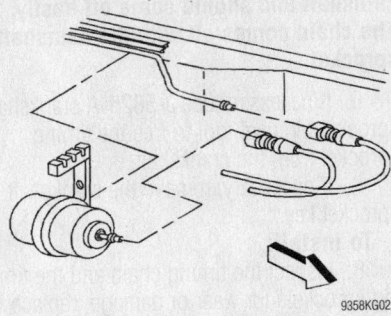

9358KG02

Typical fuel filter location along frame rail

• Filter and the clamp from the vehicle

To install:

4. Install or connect the following:

• Filter and clamp with the directional arrow facing away from the fuel tank, towards the throttle body

➡**The filter has an arrow (fuel flow direction) on the side of the case, be sure to install it correctly in the system, the with arrow facing away from the fuel tank.**

• Fuel feed nut
• Filter clamp assembly bolt
• Fuel quick disconnect fittings to the filter
• Fuel filler cap
• Negative battery cable

5. Start the engine and check for leaks.

Fuel Pump

REMOVAL & INSTALLATION

1. Before servicing the vehicle, refer to the precautions in the beginning of this section.

2. Properly relieve the fuel system pressure.

3. Disconnect the negative battery cable.

4. Lower the spare tire.

5. Remove or disconnect the following:

• Rear tail lamp assemblies
• Frame-to-pickup box bolts
• Wiring harness ground wire from the frame
• License plate lamp
• Fuel filler neck-to-pickup box ground wire and screws
• Pickup box
• Fuel sender electrical connectors
• Fuel sender and Evaporative Emission (EVAP) pipes

✳✳ WARNING

The fuel sender assembly may spring up from the fuel tank. When removing the sender from the fuel tank, keep in mind that the reservoir bucket is full of fuel, so you must tip the sender slightly during removal to avoid damaging the float. Discard the fuel sender O-ring and replace with a new on during installation.

6. While holding the module fuel sender down, remove the snapring from the designated slots (1) found on the retainer.

To install:

7. Install a new O-ring on the fuel sender to the tank.

8. Align the tab on the front of the sender with the slot on the front of the retainer snapring.

9. Slowly apply pressure to the top of the spring-loaded sender until it aligns flush with the retainer on the tank.

➡ **Make sure that the snapring is properly and fully seated in the tab slots.**

10. Install or connect the following:
- Snapring into the proper slots
- Fuel and EVAP pipes
- Electrical connectors
- Negative battery cable

11. Check for fuel leaks as follows:
- a. Turn the ignition **ON** for 2 seconds.
- b. Turn the ignition **OFF** for 10 seconds.
- c. Turn the ignition **ON**.
- d. Check for leaks.

12. Install or connect the following:
- Pickup box to the truck
- Filler neck-to-pickup box screws and ground wire
- License plate lamp
- Wiring harness ground wire to the frame
- Frame-to-pickup box bolts and tighten to 52 ft. lbs. (70 Nm)
- Rear tail lamp assemblies

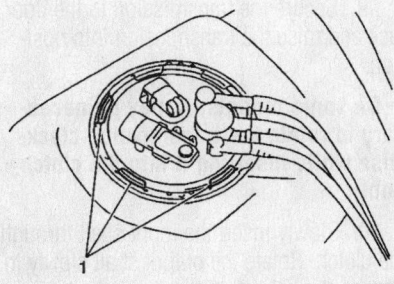

Hold the fuel sender down, and remove the snapring from the designated slots (1) found on the retainer

Fuel Injector

REMOVAL & INSTALLATION

2.2L Engines

1. Before servicing the vehicle, refer to the precautions in the beginning of this section.

2. Relieve the fuel system pressure.

3. Remove or disconnect the following:
- Negative battery cable
- Intake manifold, if necessary
- Fuel injector electrical connections by pushing in the wire connector clip and gently pulling on the connector
- Fuel feed inlet pipe from the rail

➡ **Use a back-up wrench on the fuel rail return fitting to prevent it from turning.**

- Fuel return pipe from the fuel pressure regulator
- Fuel pressure regulator
- Fuel rail attaching bolts and lift the fuel rail assembly from the cylinder head
- Fuel rail by moving the rail towards the front of the engine
- Fuel injector retaining clip
- Fuel injector

➡ **Because each injector is calibrated for a specific flow rate, make sure you only replace fuel injectors using an IDENTICAL part number to the old injectors.**

To install:

➡ **When installing the injector care should be taken not to tear or misalign O-rings.**

4. Lubricate the injector O-ring seals with clean engine oil and install them injector.

5. Install or connect the following:
- Upper O-ring, lower back-up O-ring and lower O-ring
- Fuel injector to the fuel rail
- Fuel injector retaining clip
- Fuel rail and insert it into the cylinder head
- Fuel rail retaining bolts and tighten to 18 ft. lbs. (25 Nm)
- Fuel pressure regulator

➡ **Use a back-up wrench on the fuel rail return fitting to prevent it from turning.**

- Return pipe to the fuel pressure regulator. Tighten the fuel pipe nut to 22 ft. lbs. (30 Nm).
- Fuel feed inlet pipe to rail

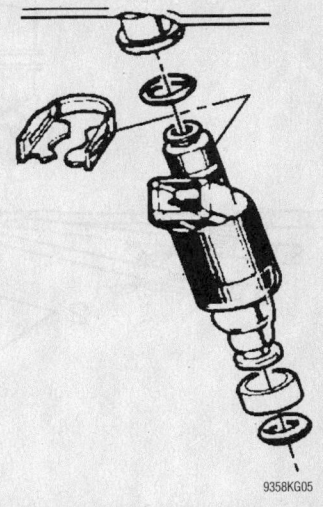

9358KG05

Exploded view of a typical fuel injector—2.2L engine

➡ **Rotate the fuel injectors as necessary to avoid stretching the wire harness.**

- Injector electrical connections
- Intake manifold, if removed
- Negative battery cable

6. Inspect for leaks as follows:
- a. Turn the switch to the **ON** position for 2 seconds.
- b. Turn the ignition switch **OFF** for 10 seconds.
- c. Turn the ignition switch to the **ON** position and check for leaks.

4.3L Engines

1. Before servicing the vehicle, refer to the precautions in the beginning of this section.

2. Relieve the fuel system pressure. Refer to the fuel system relief procedure in this section.

3. Relieve the fuel system pressure.

4. Remove or disconnect the following:
- Negative battery cable
- Fuel meter body electrical connection and the fuel feed and return hoses from the engine fuel pipes
- Upper manifold assembly
- Poppet nozzle out of the casting socket
- Fuel meter body by releasing the locktabs

➡ **Each injector is calibrated. When replacing the fuel injectors, be sure to replace it with the correct injector.**

- Lower hold-down plate and nuts

5. While pulling the poppet nozzle tube downward, push with a small prytool down

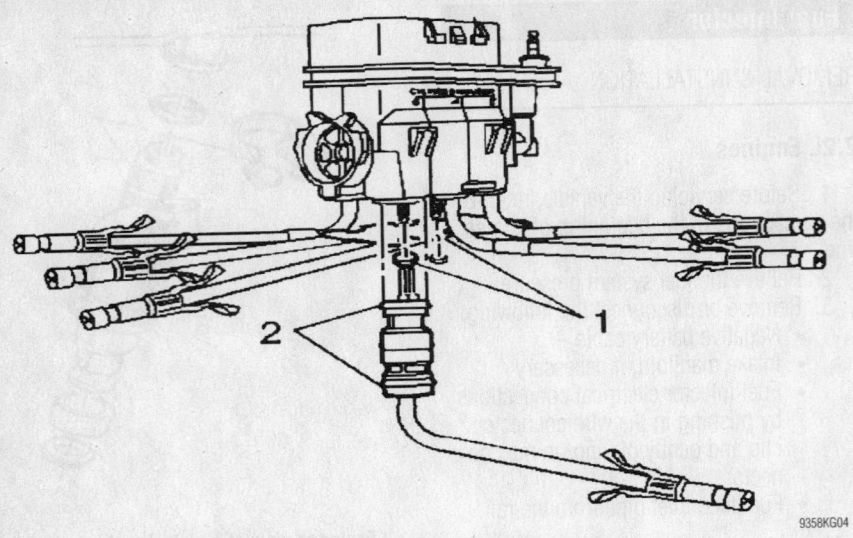

9358KG04

Exploded view of the fuel meter assembly, including the injectors

between the injector terminals and remove the injectors.

To install:

6. Lubricate the new injector O-ring seats with engine oil.

7. Install or connect the following:

- O-rings on the injector
- Fuel injector into the fuel meter body injector socket.
- Lower hold-down plate and nuts. Torque the nuts to 27 inch lbs. (3 Nm).

- Fuel meter body assembly into the intake manifold. Torque the fuel meter bracket retainer bolts to 88 inch. lbs. (10 Nm).

✳✳ CAUTION

To reduce the risk of fire or injury ensure that the poppet nozzles are properly seated and locked in their casting sockets

- Fuel meter body into the bracket and lock all the tabs in place
- Poppet nozzles into the casting sockets
- Electrical connections
- New o-ring seals on the fuel return and feed hoses
- Fuel feed and return hoses and tighten the fuel pipe nuts to 22 ft. lbs. (30 Nm)
- Negative battery cable

8. Turn the ignition **ON** for 2 seconds and then turn it **OFF** for 10 seconds. Again turn the ignition **ON** and check for leaks.

9. Install the manifold plenum.

DRIVE TRAIN

Manual Transmission Assembly

REMOVAL & INSTALLATION

1. Before servicing the vehicle, refer to the precautions in the beginning of this section.

2. Shift the transmission into 3rd or 4th gear position.

3. Remove or disconnect the following:

- Negative battery cable
- Shift lever and the if necessary, the shift housing
- Parking brake cable for clearance
- Propeller shaft
- Sid plate, if equipped
- Transfer case and shift lever, on 4WD models
- All wiring harness that would interfere with transmission removal
- Fuel line retainers from the rear crossmember
- Muffler from the catalytic converter
- Exhaust pipes from the exhaust manifold
- Catalytic converter hanger, if necessary
- Exhaust section
- Bolts and nuts attaching any transmission braces to the engine and transmission

4. Disconnect the hydraulic clutch quick-connect from the concentric slave cylinder following 1 of the 2 steps:

a. Use 2 small prytools at 180 degrees from each other to depress the white plastic sleeve on the quick connect to separate the clutch line from the concentric slave cylinder quick connect.

b. Use special tool J–36221 to depress the white plastic sleeve on the quick connect to separate the clutch line end from the concentric slave cylinder quick connect.

5. Remove or disconnect the following:

- Bolts securing the clutch housing cover to the transmission, if equipped
- Clutch plate and clutch cover, if necessary

6. Support the transmission with a suitable jack.

- Rear crossmember from the frame rail
- Wiring harness from the front crossmember, if equipped. Move the wiring harness away from the transmission oil pan. Lower the transmission enough to gain access to the top of the transmission.
- Fuel line retainers or wiring har-

ness's from the top of the transmission

- Bolt, washer, and nut securing the wiring harness ground wires to the engine block
- Bolts retaining the transmission to the engine. Pull the transmission straight back on the clutch hub splines.

7. Lower the transmission using the transmission jack.

To install:

Installation is the reverse of removal, but please note the following important steps.

8. Place a THIN coat of high-temperature grease on the main drive gear (input shaft) splines.

9. Secure the transmission to the floor jack and raise the transmission into position.

➡ **On some models, it may be necessary to rotate the transmission clockwise while inserting it into the clutch hub.**

10. Slowly insert the input shaft through the clutch. Rotate the output shaft slowly to engage the splines of the input shaft into the clutch while pushing the transmission forward into place. Do not force the transmission into position, the transmission

should easily fall into place once everything is properly aligned.

11. Tighten the transmission mounting bolts to 35 ft. lbs. (47 Nm).

12. Do not remove the transmission jack until the crossmembers have been installed.

13. Check the transmission fluid level and replenish as necessary.

Automatic Transmission Assembly

REMOVAL & INSTALLATION

1. Before servicing the vehicle, refer to the precautions in the beginning of this section.

2. Remove or disconnect the following:
   - Negative battery cable
3. Drain the transmission fluid.
   - Driveshaft from the transmission (2WD) and transfer case, if equipped (4WD)
4. Support the transmission with a suitable transmission jack.
   - Shift cable from the transmission control lever and bracket
   - Nut and washer securing the transmission mount to the crossmember
   - Bolts and washers securing the mount to the transmission
   - Exhaust pipe from the exhaust manifold(s)
   - Bolts securing the converter pan cover to the transmission, if equipped
   - 3 bolts securing the torque converter to the flywheel
   - Bolt, clip, and strap securing the three fuel lines and transmission vent hose to the transmission case
   - Bolts and nut securing the transmission to the engine
   - Oil filler tube and seal from the transmission
   - Transmission cooler lines from the transmission. Plug the lines and the ports in the transmission.
   - Wiring harness connectors from the transmission.
5. Inspect for any other wiring, brackets etc. which may interfere with the removal of the transmission.

6. Since the transmission acts as a rear engine mount, properly support the rear of the engine with an underbody support or other suitable support before attempting to remove the transmission. Otherwise the rear of the engine may pitch downward and components on the rear of the engine and on the firewall may be damaged.

7. Remove the transmission from the engine by pulling the transmission rearward to disengage it from the locator dowel pins on the back of the block. Carefully lower the transmission from the vehicle. Use care that the torque converter does not fall out of the front of the transmission.

➡Use converter holding strap tool No. J-21366, to secure the torque converter to the transmission during removal and installation procedures.

To install:

Installation is the reverse of removal, but please note the following important steps.

8. Make sure the torque converter is fully seated in the pump drive. If not, the transmission will not fit tightly to the rear of the engine block.

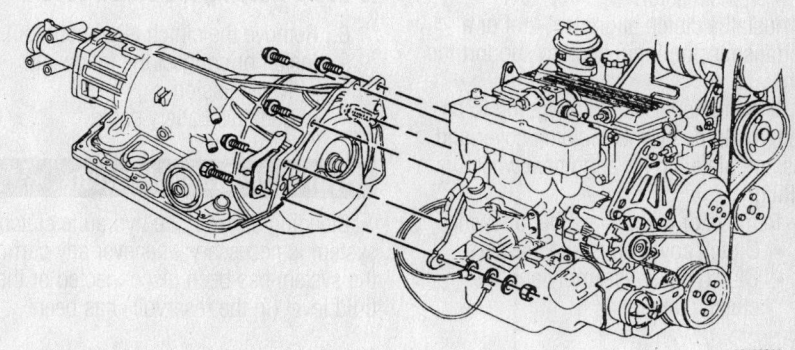

Transmission mounting on 2.2L engines

88457G33

9. Raise the transmission into position and remove the torque converter holding strap and carefully. Slide the transmission forward until the dowel pins are engaged.

10. The torque converter should be flush with the flywheel and turn freely by hand.

11. Install the transmission–to–engine bolts. Tighten the bolts to 34 ft. lbs. (47 Nm).

12. Tighten the torque converter-to-flywheel bolts to 46 ft. lbs. (63 Nm).

13. If equipped, tighten the converter pan cover to the transmission bolts to 37 ft. lbs. (50 Nm)

14. Tighten the bolts and washers securing the transmission mount to 35 ft. lbs. (47 Nm).

15. Tighten the nut and washer securing

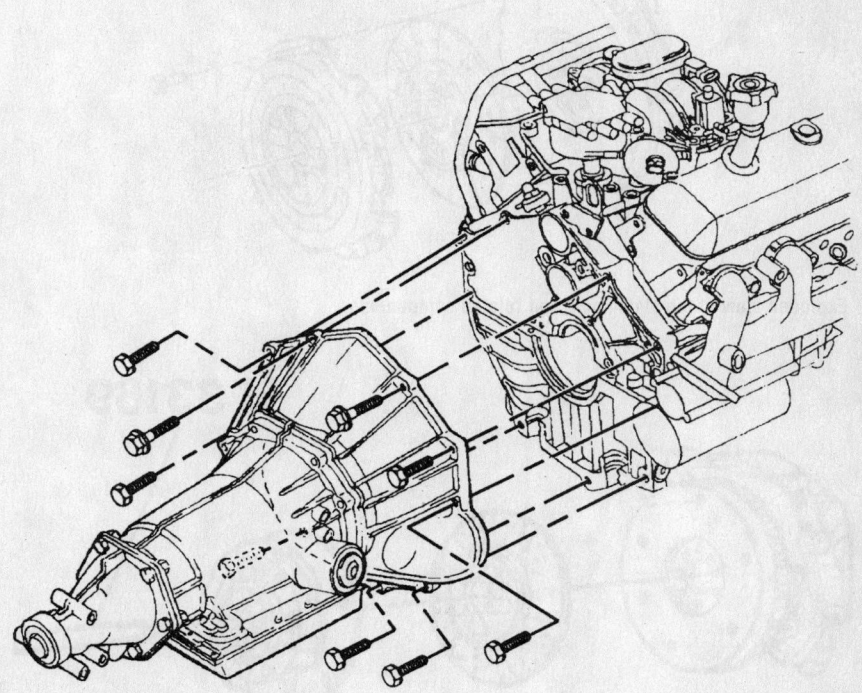

Transmission mounting on 4.3L engines

88457G34

the transmission mount to the crossmember to 38 ft. lbs. (52 Nm).

16. Refill the transmission with the proper amount and type of fluid.

17. Connect the negative battery cable. Start the vehicle and allow to warm while checking for leaks. Road test the vehicle to check for shift quality.

Clutch

REMOVAL & INSTALLATION

1. Before servicing the vehicle, refer to the precautions in the beginning of this section.

2. Remove or disconnect the following:
   - Negative battery cable
   - Transmission

3. Install a clutch alignment tool or a used transmission input shaft to support the clutch.

4. If the clutch assembly is going to be reused, mark the flywheel, clutch cover and a pressure plate lug for alignment when installing.

5. Remove or disconnect the following:
   - Clutch cover bolts and washers
   - Clutch cover assembly and the clutch plate

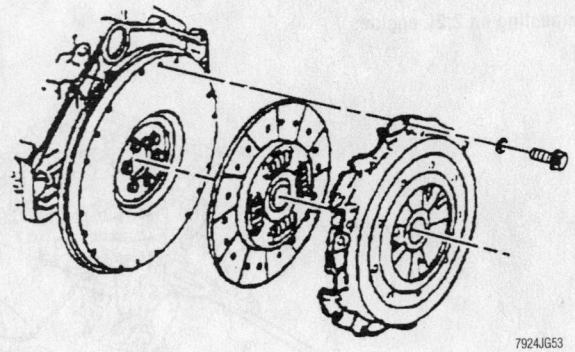

Exploded view of the clutch disc and related components

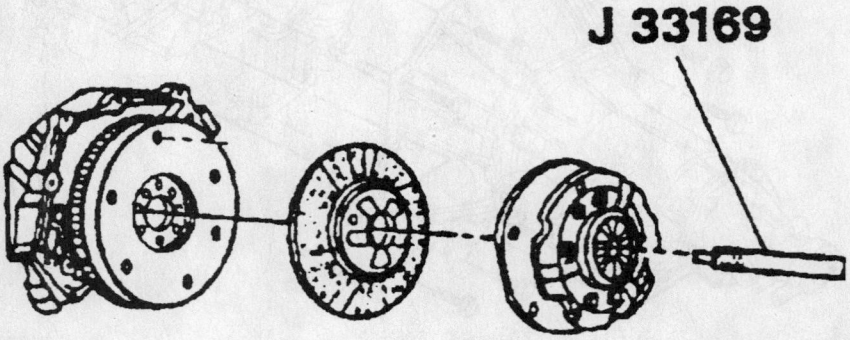

Use the clutch alignment tool to center and support the clutch disc during installation

- Clutch alignment tool

6. Clean all parts and inspect for damage.

To install:

7. Install or connect the following:
   - Clutch alignment tool, to support the clutch
   - Clutch cover by aligning the matchmarks or, if new, align the lightest part of the cover, identified by a yellow dot, with the heaviest part identified by an **X**.
   - Clutch plate/clutch cover assembly to the flywheel. Tighten the bolts to 33 ft. lbs. (45 Nm) for 2.2L engines or to 29 ft. lbs. (40 Nm) for 4.3L engines.

➡ **Tighten each screw 1 turn at a time to avoid warping the clutch cover.**

8. Remove the clutch alignment tool.
9. Install or connect the following:
   - Transmission
   - Negative battery cable

Hydraulic Clutch System

Bleeding air from the hydraulic clutch system is necessary whenever any part of the system has been disconnected or the fluid level (in the reservoir) has been allowed to fall so low, that air has been drawn into the master cylinder.

BLEEDING

1. Before servicing the vehicle, refer to the precautions in the beginning of this section.

2. Fill master cylinder reservoir with new brake fluid conforming to DOT 3 specifications.

✳✳ CAUTION

Always use new fluid from a sealed container. Never, under any circumstances, use fluid that has been bled from a system to fill the reservoir as it may be aerated, have too much moisture content and possibly be contaminated.

3. Have an assistant fully depress and hold the clutch pedal, then open the bleeder screw.

4. Close the bleeder screw and have your assistant release the clutch pedal.

5. Repeat the procedure until all of the air is evacuated from the system. Check and refill master cylinder reservoir as required to prevent air from being drawn through the master cylinder.

➡ **Never release a depressed clutch pedal with the bleeder screw open or air will be drawn into the system.**

6. If the previous steps do not result in satisfactory pedal feel, remove the reservoir cap and pump the clutch pedal very fast for 30 seconds. Stop to let the air escape, then repeat the procedure as necessary to purge all remaining air.

7. Test the clutch for proper operation.

Transfer Case Assembly

REMOVAL & INSTALLATION

1. Before servicing the vehicle, refer to the precautions in the beginning of this section.

2. Disconnect the negative battery cable.

3. Shift the transfer case into the **4HI** range.

4. Drain the transfer case fluid.

5. Support the transfer case.

6. Remove or disconnect the following:
   - Skid plate
   - Front and rear driveshafts from the transfer case. Matchmark the shafts prior to removal.

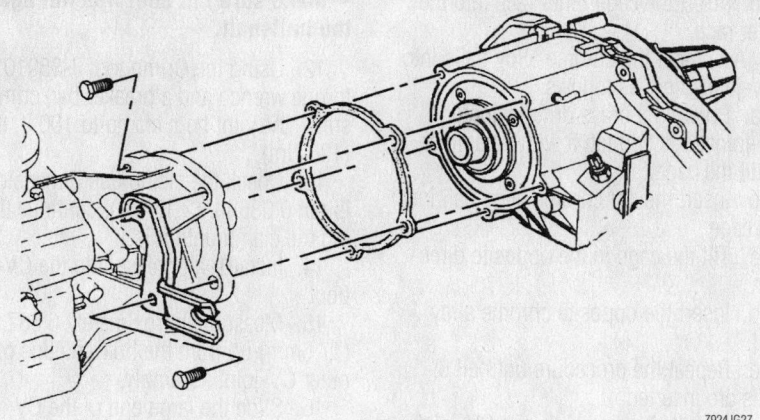

Transfer case-to-manual transmission mounting—Typical

7924JG27

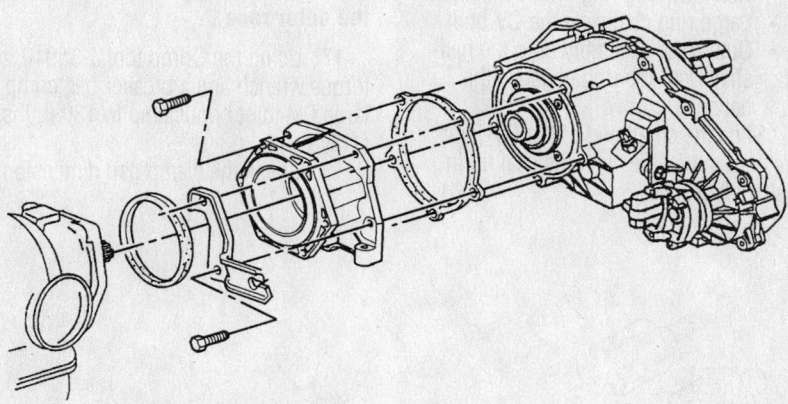

Transfer case-to-automatic transmission mounting—Typical

7924JG28

- Vacuum lines and/or the electrical connectors, as equipped
- Transfer case shift rod/cable from the case, if applicable
- Support brace-to-transfer case bolts, if applicable
- Transfer case

7. Remove all traces of old gasket material from the mating surfaces.

To install:

8. Install or connect the following:
- New gasket using sealer to hold it in position
- Transfer case. Torque the bolts to 33–35 ft. lbs. (45–47 Nm).
- Support brace bolts. Torque the bolts to 35–37 ft. lbs. (47–50 Nm), if equipped.
- Shift rod to the case, if equipped
- Vacuum lines and/or electrical connections, as necessary
- Front and rear driveshafts by aligning the matchmarks

9. Refill the transfer case.
- Skid plate, if equipped
- Negative battery cable

Halfshaft

REMOVAL & INSTALLATION

1. Before servicing the vehicle, refer to the precautions in the beginning of this section.

2. Unlock the steering column so the steering linkage is free to move.

3. Remove or disconnect the following:
- Negative battery cable
- Front wheels
- Skid plate
- 6 bolts and the flange
- Snapring
- Spindle washer

✳✳ CAUTION

The coil spring is under extreme pressure. Make sure the control arm is firmly supported with a hydraulic jack before removing the lower ball joint nut. After the lower ball joint nut has been removed, lower the hydraulic jack slowly to relieve coil spring pressure. If this precaution is not observed, serious bodily injury may result.

4. Support the lower control arm with a hydraulic jack.

5. Remove the cotter pin and nut attaching the ball joint to the lower control arm.

6. Remove the nuts and bolts connecting the strut to the steering knuckle. Separate the steering knuckle from the strut and lower control arm.

7. Slowly lower the hydraulic jack until coil spring pressure is relieved.

8. Remove the outer CV-joint from the steering knuckle.

9. If removing the right side halfshaft, place tool J 37780 or equivalent, between the front axle housing and the inner CV-joint. Gently tap the inner CV-joint away and out of the front axle housing.

10. If removing the left side halfshaft, scribe a reference mark on the left inner axle shaft flange and the inner CV-joint flange to ensure correct installation. Remove the three bolts and three nuts and separate the inner CV-joint from the left inner axle shaft.

11. Remove the halfshaft from the vehicle.

To install:

12. If installing the right halfshaft, install the inner CV-joint into the axle housing, making sure the snapring seats in the differential side gear.

13. If installing the left halfshaft, install the left inner axle shaft flange to the inner CV-joint flange, aligning the reference marks made during removal. Install the three bolts and three nuts and tighten to 41 ft. lbs. (55 Nm).

14. Install the outer CV-joint into the steering knuckle.

15. Support the lower control arm with the hydraulic jack.

16. Attach the steering knuckle and lower ball joint to the lower control arm. Tighten the strut bolts and nuts to 65 ft. lbs. (90 Nm). Tighten the ball joint nut to 63 ft. lbs. (85 Nm) and install a new cotter pin.

17. Remove the hydraulic jack from the lower control arm.

18. Install the spindle washer and snapring to the end of the halfshaft.

19. Apply sealer, GM part no. 1052366, or equivalent, to the flange. Install the flange. Torque the flange bolts to 35 ft. lbs. (48 Nm).

20. If equipped, install the locking hub.

21. Install the front wheel.

22. Install the skid plate, if equipped.

23. Lower the vehicle.

CV-Joints

OVERHAUL

Outer CV-Joint

1. Before servicing the vehicle, refer to the precautions in the beginning of this section.
2. Remove or disconnect the following:
   - Front wheel
   - Halfshaft and position it in a vise
   - Large CV-joint boot clamp and discard it
   - Small CV-joint boot clamp and discard it
   - CV-joint boot and slide it back on the shaft
   - Outer race from the halfshaft, by spreading the outer race-to-half-shaft retaining ring, using Snapring Pliers J-8059
   - Retaining ring from the halfshaft and discard it
   - CV-joint boot from the halfshaft and discard it, if damaged
3. Disassemble the chrome alloy balls from the CV-joint cage as follows:
   a. Position a brass drift against the CV-joint cage and tap it with a hammer to tilt the cage.
   b. Remove the 1st chrome alloy ball from the cage.
   c. Tilt the cage in the opposite direction.
   d. Remove the opposite chrome alloy ball.
   e. Repeat the procedure until all 6 balls are removed.
4. Disassemble the CV-joint cage and inner race as follows:
   a. Pivot the cage and race 90 degrees to the center line of the outer race.
   b. Align the cage windows with outer race lands.
   c. Remove the cage from the outer race.
   d. Rotate the inner race upward and remove it from the cage.
5. Thoroughly clean and inspect all parts.

To install:

6. Lubricate the parts with a light coat of grease.
7. Assemble the CV-joint cage and inner race, as follows:
   a. Rotate the inner race 90 degrees to the cage centerline.
   b. Align the cage windows with inner race lands.
   c. Insert the inner race into the cage by rotating the inner race downward.

d. Insert the cage/inner race into the outer race.
8. Assemble the chrome alloy balls into the CV-joint cage, as follows:
   a. Position a brass drift against the CV-joint cage and tap it with a hammer to tilt the cage.
   b. Insert the 1st chrome alloy ball into the cage.
   c. Tilt the cage in the opposite direction.
   d. Insert the opposite chrome alloy ball.
   e. Repeat the procedure until all 6 balls are inserted.
9. Install ½ kit grease into the CV-joint.
10. Install or connect the following:
    - Small ring clamp on the CV boot
    - New retaining ring on the halfshaft
    - Large ring clamp on the CV boot
    - Outer race assembly onto the halfshaft until the ring engages the halfshaft groove
11. Slide the small end of the CV-joint boot/clamp into place, with the seal lip in the halfshaft groove

➡ **Make sure the boot lies flat against the halfshaft.**

12. Using the Crimp tool J-35910, a torque wrench and a breaker bar, crimp the small CV-joint boot clamp to 100 ft. lbs. (136 Nm).
13. Check the clamp gap dimension; if it is not 0.085 in. (2.15mm), continue tightening the clamp until it is.
14. Install ½ kit grease into the CV-joint boot.
15. Measure approximately 0.687 in. (17.5mm) up from the bottom edge of the outer CV-joint assembly.
16. Slide the large end of the CV boot/clamp into place, with the seal lip in place over the outer race.

➡ **Make sure the boot lies flat against the outer race.**

17. Using the Crimp tool J-35910, a torque wrench and a breaker bar, crimp the large CV-joint boot clamp to 130 ft. lbs. (176 Nm).
18. Check the clamp gap dimension; if it

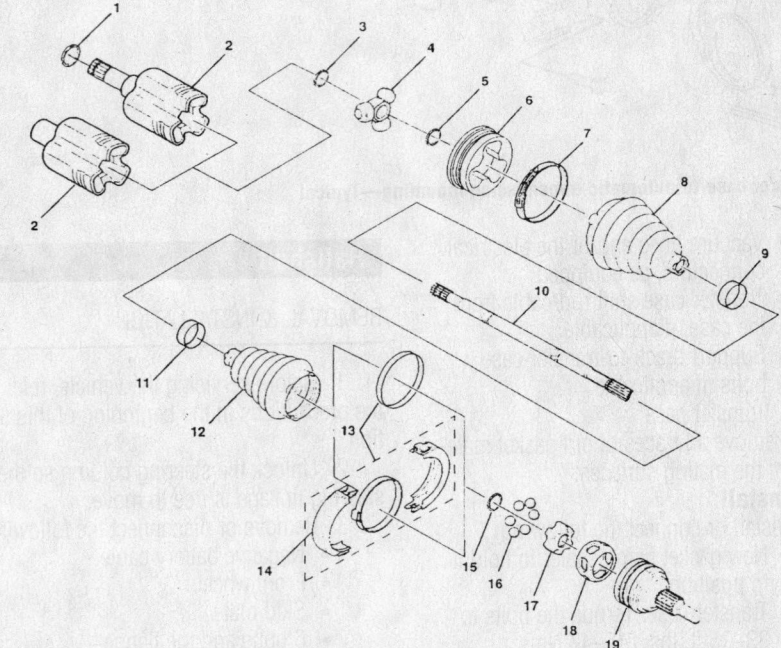

(1) Differential Shaft Ring	(11) Halfshaft Swage Ring
(2) Tripot Housing Assembly	(12) CV Joint Boot
(3) Spacer Ring	(13) Swage Ring
(4) Tripot Joint Spider Assembly	(14) Clamp Protector
(5) Spacer Ring	(15) Race Retaining Ring
(6) Tripot Bushing	(16) Ball
(7) Boot Retaining Clamp	(17) CV Joint Inner Race
(8) Tripot Joint Boot	(18) CV Joint Cage
(9) Halfshaft Swage Ring	(19) CV Joint Outer Race
(10) Halfshaft Bar	

9308JG09

Exploded view of the CV-Joint assembly

is not 0.102 in. (2.60mm), continue tightening the clamp until it is.

19. Install the halfshaft and the front wheel.

Inner (Tri-Pot) Joint

1. Before servicing the vehicle, refer to the precautions in the beginning of this section.

2. Remove or disconnect the following:
- Front wheel
- Halfshaft and place it in a vise
- Snapring from the stub shaft and discard it
- Small CV-joint boot clamp, cut and discard it
- Large CV-joint boot clamp, cut and discard it
- CV-joint boot by sliding it away from the tri-pot joint

3. Install a Stub Shaft Removal tool J-38868-A to the stub shaft snapring groove.

4. Using a slide hammer puller, press the stub shaft from the tri-pot housing.

5. Remove or disconnect the following:
- Tri-pot housing from the tri-pot spider
- Inboard spacer ring slide it rearward on the shaft using Snapring Pliers tool J-8059
- Outboard retaining ring using Snapring Pliers tool J-8059 and discard it
- Tri-pot joint spider assembly
- Inboard spacer ring and discard it
- CV-joint boot
- Trilobal tri-pot bushing from the housing

6. Thoroughly clean and inspect all parts.

To install:

7. Install or connect the following:
- New snapring onto the stub shaft
- Small boot clamp
- CV-joint boot

8. Using the Crimp tool J-35910, a torque wrench and a breaker bar, crimp the small CV-joint boot clamp to 100 ft. lbs. (136 Nm).

9. Install or connect the following:
- Inboard spacer ring slide it rearward on the shaft using Snapring Pliers tool J-8059, past the 2nd groove
- Tri-pot joint spider assembly onto the shaft until it passes the 2nd groove
- Outboard retaining ring into the axle shaft groove using Snapring Pliers tool J-8059
- Tri-pot joint spider assembly, slide

it against the outboard retaining ring
- Inboard spacer ring, seat it in the groove
- ½kit grease into the boot
- ½kit grease into the tri-pot housing
- Trilobal tip-pot bushing flush with the tri-pot housing face
- New large seal clamp onto the CV-joint boot
- Tri-pot housing, slide it over the tri-pot joint spider assembly
- CV-joint boot/clamp, slide it into place, over the trilobal tri-pot bushing with the seal lip in the groove

➡**Make sure the boot lies flat against the trilobal bushing.**

10. Position the CV-joint boot so it measures 4.9 in. (125mm).

11. Using the Crimp tool J-35566, latch the large CV-joint boot clamp.

12. Install the halfshaft and the front wheel.

Axle Shaft, Bearing and Seal

REMOVAL & INSTALLATION

For the Axle Shaft, Bearing and Seal, Removal and Installation, please refer to Wheel Bearing procedure located in the section.

Pinion Seal

REMOVAL & INSTALLATION

1. Before servicing the vehicle, refer to the precautions in the beginning of this section.

➡**The following procedure requires the use of the Pinion Holding tool J-8614-10, the Pinion Flange Removal tool J-8614-1, J-8614-2, J-8614-3 and the Pinion Seal Installation tool J-23911.**

2. Remove or disconnect the following:
- Driveshaft from the pinion flange. Matchmark the driveshaft prior to removal.
- Driveshaft from the rear axle pinion flange and support the shaft up in body tunnel by wiring it to the exhaust pipe.

➡**If the U-joint bearings are not retained by a retainer strap, use a piece of tape to hold bearings on their journals.**

3. Mark the position of the pinion stem, flange and nut for reference.

4. Use an inch lbs. torque wrench to measure the amount of torque necessary to turn the pinion, then note this measurement as it is the combined pinion bearing, seal, carrier bearing, axle bearing and seal preload.

5. Remove or disconnect the following:

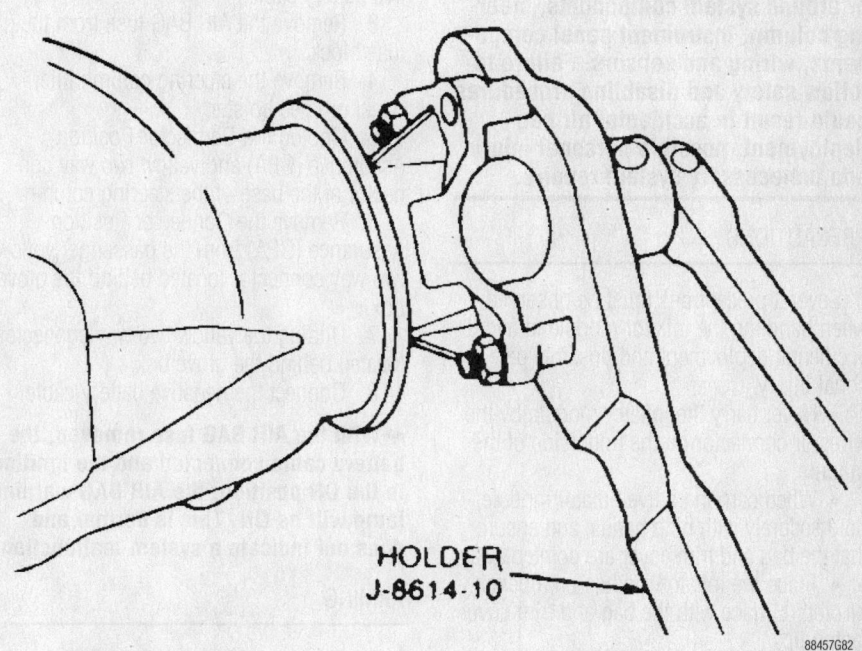

HOLDER
J-8614-10

88457G82

Removing the pinion nut using a pinion holding fixture tool

- Pinion flange nut and washer, using a Pinion Holding tool J-8614-10 and a Pinion Flange Removal tool J-8614-1, J-8614-2, J-8614-3, as applicable
- Pinion flange
- Pinion oil seal by driving it out of the differential with a blunt chisel; DO NOT damage the carrier

To install:

6. Examine the seal surface of pinion flange for tool marks, nicks or damage, such as a groove worn by the seal. If damaged, replace flange.

7. Examine the carrier bore and remove any burrs that might cause leaks around the O.D. of the seal.

8. Apply GM seal lubricant 1050169 to the outside diameter of the pinion flange and sealing lip of new seal.

9. Install or connect the following:
- New pinion oil seal using a seal installer tool
- Pinion flange and tighten nut to the

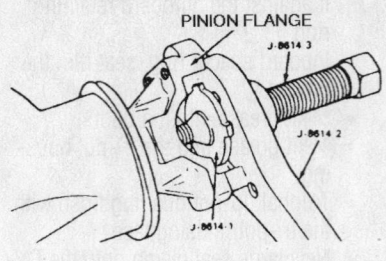

A puller and adapter should be used to withdraw the pinion from the housing

same position as marked earlier. Tighten the nut a little at a time and turn the pinion flange several times after each tightening in order to set the rollers.

10. Measure the torque necessary to turn the pinion and compare this to the reading taken during removal. Tighten the nut additionally, as necessary to achieve the same preload as measured earlier.

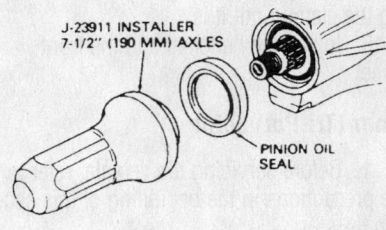

Use the appropriately sized installation tool to drive the new seal into position.

➡If fluid was lost from the differential housing during this procedure, be sure to check and add additional fluid, as necessary.

11. Remove the support then align and secure the driveshaft assembly to the pinion flange.

➡The original matchmarks MUST be aligned to assure proper shaft balance and prevent vibration.

STEERING AND SUSPENSION

Air Bag

✳✳ CAUTION

Some vehicles are equipped with an air bag system, also known as the Supplemental Inflatable Restraint (SIR) system. The system must be disabled before performing service on or around system components, steering column, instrument panel components, wiring and sensors. Failure to follow safety and disabling procedures could result in accidental air bag deployment, possible personal injury and unnecessary system repairs.

PRECAUTIONS

Several precautions must be observed when handling the inflator module to avoid accidental deployment and possible personal injury.
- Never carry the inflator module by the wires or connector on the underside of the module.
- When carrying a live inflator module, hold securely with both hands, and ensure that the bag and trim cover are pointed away.
- Place the inflator module on a bench or other surface with the bag and trim cover facing up.
- With the inflator module on the bench, never place anything on or close to the

module, that may be thrown in the event of an accidental deployment.

DISARMING

1. Turn the steering wheel so that the vehicle's wheels are pointing straight ahead.
2. Turn the ignition switch to **LOCK**, remove the key, then disconnect the negative battery cable.
3. Remove the AIR BAG fuse from the fuse block.
4. Remove the steering column filler panel or knee bolster.
5. Unplug the Connector Position Assurance (CPA) and yellow two way connector at the base of the steering column.
6. Remove the Connector Position Assurance (CPA) from the passenger yellow two way connector located behind the glove box.
7. Unplug the yellow two way connector located behind the glove box.
8. Connect the negative battery cable.

➡With the AIR BAG fuse removed, the battery cable connected and the ignition in the ON position, the AIR BAG warning lamp will be ON. This is normal and does not indicate a system malfunction.

ARMING

1. Disconnect the negative battery cable.

2. Attach the yellow two way connector located behind the glove box.
3. Install the Connector Position Assurance (CPA) to the passenger yellow two way connector located behind the glove box.
4. Turn the ignition switch to **LOCK**, then remove the key.
5. Attach the two way connector at the base of the steering column and the Connector Position Assurance (CPA).
6. Install the steering column filler panel or knee bolster.
7. Install the AIR BAG fuse to the fuse block.
8. Connect the negative battery cable.
9. From the passenger seat, turn the ignition switch to **RUN** and make sure that the AIR BAG warning lamp flashes seven times and then shuts off. If the warning lamp does not shut off, make sure that the wiring is properly connected. If the light remains on, take the vehicle to a reputable repair facility for service.

Power Steering Gear

REMOVAL & INSTALLATION

Two Wheel Drive

1. Before servicing the vehicle, refer to the precautions in the beginning of this section.
2. Position a fluid catch pan under the power steering gear.

✳✳ WARNING

Do NOT rotate the steering shaft after the steering column has been removed.

3. Lock the steering column through the access hole in the steering column lower trim cover using steering column anti-rotation pin J 42640.

4. Remove or disconnect the following:
- Air cleaner assembly
- Intermediate shaft from the steering gear
- Feed and return fluid hoses from the steering gear. Immediately cap or plug all openings to prevent system contamination or excessive fluid loss.
- Intermediate shaft lower coupling shield, if equipped
- Lower intermediate shaft coupling bolt
- Matchmark the lower intermediate shaft coupling and the steering shaft
- Lower intermediate shaft coupling from the steering shaft
- Pitman arm from the gear pitman shaft
- Power steering gear-to-frame bolts and washers, then carefully remove the steering gear from the vehicle

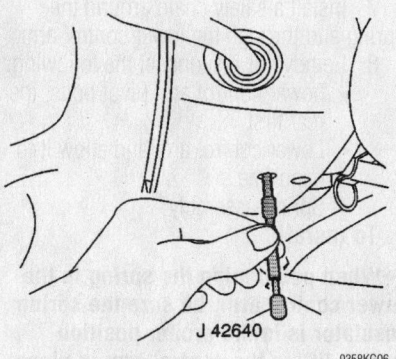

J 42640

9358KG06

Use J 42640 to lock the steering column on 2000–03 vehicles

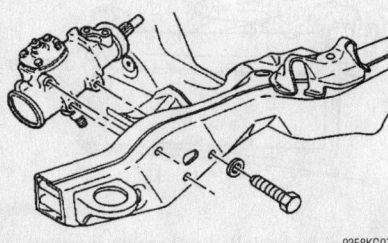

9358KG07

Power steering gear mounting

To install:

5. Install or connect the following:
- Steering gear to the vehicle and tighten the bolts to 55 ft. lbs. (75 Nm)
- Pitman arm
- Intermediate shaft to the power steering, making sure the matchmarks line up. Tighten the bolt to 26 ft. lbs. (35 Nm).
- Pressure and return hoses to the power steering gear. Tighten the pressure hose to 18 ft. lbs. (25 Nm) for 2.2L engines and 22 ft. lbs. (30 Nm) for 4.3L engines. Tighten the return hose to 18 ft. lbs. (25 Nm).
- Shield over the intermediate shaft lower coupling, if equipped
- Air cleaner assembly

6. Remove the steering column lock pin
7. Bleed the power steering system

Four Wheel Drive

This procedure requires the use of the following special tools: J 24319-B Steering Linkage and Tie Rod Puller, J 42640 Steering Column Anti-Rotation Pin, J 29193 Steering Linkage Installer (12mm), J 29194 Steering Linkage Installer (14mm).

1. Before servicing the vehicle, refer to the precautions in the beginning of this section.

2. Position a fluid catch pan under the power steering gear.

✳✳ WARNING

Do NOT rotate the steering shaft after the steering column has been removed.

3. On 2000–03 vehicles, lock the steering column through the access hole in the steering column lower trim cover using steering column anti-rotation pin J 42640.

4. Remove or disconnect the following:
- Air cleaner assembly
- Intermediate shaft lower coupling shield, if equipped
- Wiring harness clip from the power steering return hose at the power steering gear
- Feed and return fluid hoses from the steering gear. Immediately cap or plug all openings to prevent system contamination or excessive fluid loss.
- Lower intermediate shaft coupling bolt
- Matchmark the lower intermediate

shaft coupling and the steering shaft
- Lower intermediate shaft coupling from the steering shaft

5. Raise the vehicle.
- Steering linkage shield
- Differential carrier shield mounting bolts
- Differential carrier shield
- Pitman arm ball stud cotter pin and nut at the relay rod
- Pitman arm from relay rod using a suitable puller
- Steering gear mounting bolts and the washers from the frame
- Steering gear
- Pitman arm

To install:

6. Install or connect the following:
- Pitman arm
- Steering gear
- Power steering gear to the frame washers and mounting bolts. Tighten the bolts to 55 ft. lbs. (75 Nm).
- Relay rod to the pitman arm ball stud. Ensure the seal is on the stud

7. Seat the taper using a J 29193 J 29194 and tighten the tool to 48 ft. lbs. (62 Nm).

8. Remove the special tool from the pitman arm ball stud.

9. Install or connect the following:
- New nut and cotter pin to the pitman arm ball stud at the relay rod and tighten the pitman arm ball stud nut at the relay rod to 61 ft. lbs. (83 Nm)
- Differential carrier shield
- Differential carrier shield mounting bolts
- Steering linkage shield

10. Lower the vehicle.
- Intermediate shaft to the power steering, making sure the matchmarks line up. Tighten the bolt to 26 ft. lbs. (35 Nm).
- Pressure and return hoses to the power steering gear. Tighten the pressure hose to 22 ft. lbs. (30 Nm) and the return hose to 18 ft. lbs. (25 Nm).
- Wiring harness clip to the power steering return hose at the power steering gear
- Shield over the intermediate shaft lower coupling, if equipped
- Air cleaner assembly

11. Remove the steering column lock pin
12. Bleed the power steering system

Shock Absorbers

REMOVAL & INSTALLATION

Front

2WD MODELS

1. Before servicing the vehicle, refer to the precautions in the beginning of this section.
2. Remove or disconnect the following:
   • Wheel
   • Mounting nut

➡ **Hold the shock absorber stem with a wrench while backing the nut off.**

   • Retaining nut and grommet
   • Shock absorber-to-lower control arm bolts
   • Shock absorber
   • Replace the parts, as necessary.

To install:

3. Fully extend the shock absorber stem, then push it up through the lower control arm and spring so that the upper stem passes through the mounting hole in the upper control arm frame bracket.

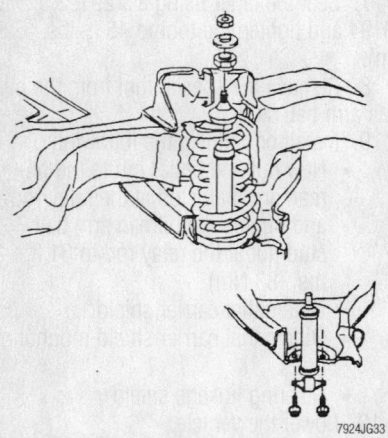

7924JG33

Front shock absorber mounting—2WD

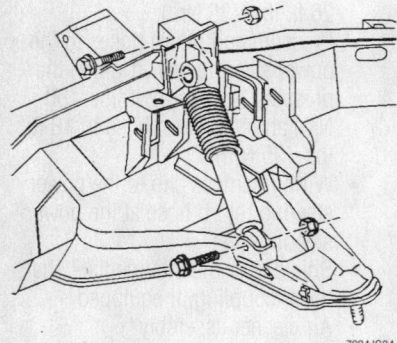

7924JG34

Front shock absorber mounting—4WD vehicles

4. Install or connect the following:
   • Retaining nut and grommet on the stem. Tighten the nut to 106 inch lbs. (12 Nm).
   • Shock absorber-to-lower control arm bolts and tighten to 22 ft. lbs. (30 Nm)
   • Wheel

4WD MODELS

1. Before servicing the vehicle, refer to the precautions in the beginning of this section.
2. Remove or disconnect the following:
   • Wheel
   • Lower nut/bolt and collapse the shock absorber
   • Shock absorber upper nut and bolt
   • Shock absorber

To install:

3. Install or connect the following:
   • Shock absorber to the bracket. Tighten the nuts/bolts to 54 ft. lbs. (73 Nm).
   • Wheel

Rear

1. Before servicing the vehicle, refer to the precautions in the beginning of this section.
2. Properly support the rear axle assembly.
3. Remove or disconnect the following:
   • Automatic level control air lines from the shock absorber, if equipped
   • Shock absorber-to-frame retainers at the top of the shock
   • Shock-to-axle retainers at the bottom of the shock
   • Shock absorber

To install:

4. Install the shock in the vehicle and loosely install the upper mounting fasteners to retain it.

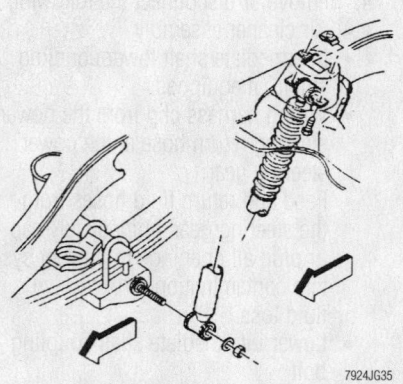

7924JG35

Rear shock absorber mounting

5. Align the lower-end of the shock absorber with the axle mounting, then loosely install the retainers.
6. Tighten the upper shock retainers to 18 ft. lbs. (25 Nm). Tighten the lower shock retainers to 62 ft. lbs. (84 Nm).
7. If equipped, attach the automatic level control air lines to the shock absorber.

Coil Springs

REMOVAL & INSTALLATION

2000 Vehicles

1. Before servicing the vehicle, refer to the precautions in the beginning of this section.
2. Remove or disconnect the following:
   • Wheel
   • Shock absorber lower bolts
3. Push the shock absorber through the control arm and into the spring.
4. With the vehicle supported so the control arms hang free, install Coil Spring Remover and Installer tool J-23028, or equivalent onto a support and into the lower control arm bushings.
5. Remove or disconnect the following:
   • Stabilizer bar from the control arm
   • Stabilizer from the lower control arm
6. Raise and remove the tension on the lower control arm bolts.
7. Install a safety chain around the spring and through the lower control arm.
8. Remove or disconnect the following:
   • Lower control arm pivot bolts, the rear first
   • Lower control arm and allow it to hang free
   • Spring assembly

To install:

➡ **When positioning the spring in the lower control arm, be sure the spring insulator is in the proper position before lifting the control arm in place.**

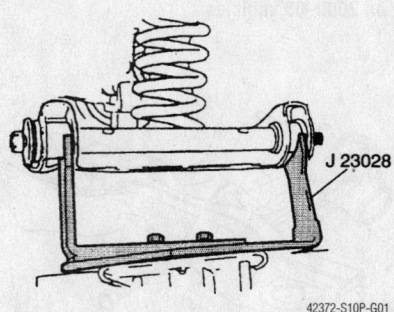

42372-S10P-G01

View of the installed coil spring removal tool—2001–03 vehicles

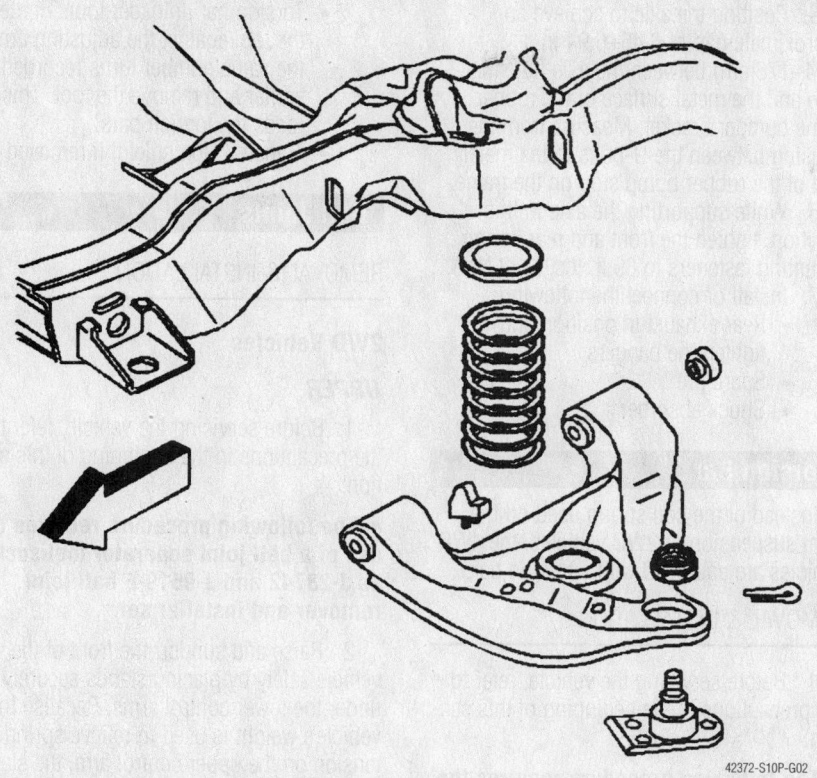

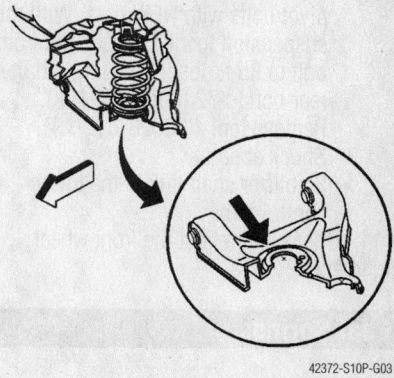

42372-S10P-G03

The coil spring must cover all or part of one inspection drain hole. The other hole must be partly or completely uncovered

42372-S10P-G02

Exploded view of the coil spring removal—2001–03 vehicles

9. Install or connect the following:
   • Spring assembly
   • Lower control arm
   • Lower control arm pivot bolts
   • Stabilizer to the lower control arm

2001–03 Vehicles

1. Before servicing the vehicle, refer to the precautions in the beginning of this section.

2. Remove or disconnect the following:
   • Wheel
   • Stabilizer shaft link from the lower control arm
   • Shock absorber

3. Secure Coil Spring Remover & Installer tool J 23028-01, or equivalent to the end of a suitable jack. Cradle the lower control arms using the tool. Raise the jack to relieve tension on the lower control arm pivot bolts.

4. Turn the steering wheel to one side, to allow the steering linkage to clear the lower control arm front pivot bolt.

5. Remove or disconnect the following:
   • Lower control arm rear and front pivot bolts and nuts

6. Lower tool J 23028-01 slowly to relieve tension from the coil spring.
   • Front coil spring and insulators. While removing the coil spring, do

not apply any force to the lower control arm and/or ball joint.

To install:

7. Install or connect the following:
   • Front coil spring and insulators on the lower control arm

➡ Make sure that the coil spring covers all or part of one inspection drain hole. The other hole must be partly or completely uncovered. Rotate the coil spring as necessary.

8. Support the control arm using tool J 23028-01. Position the coil spring and insulator in the upper spring seat on the frame.

9. Raise the lower control arm using tool J 23028-01.

10. Install or connect the following:
   • Lower control arm to the frame

➡ You must install the bolts in the direction shown to keep proper steering linkage clearance.

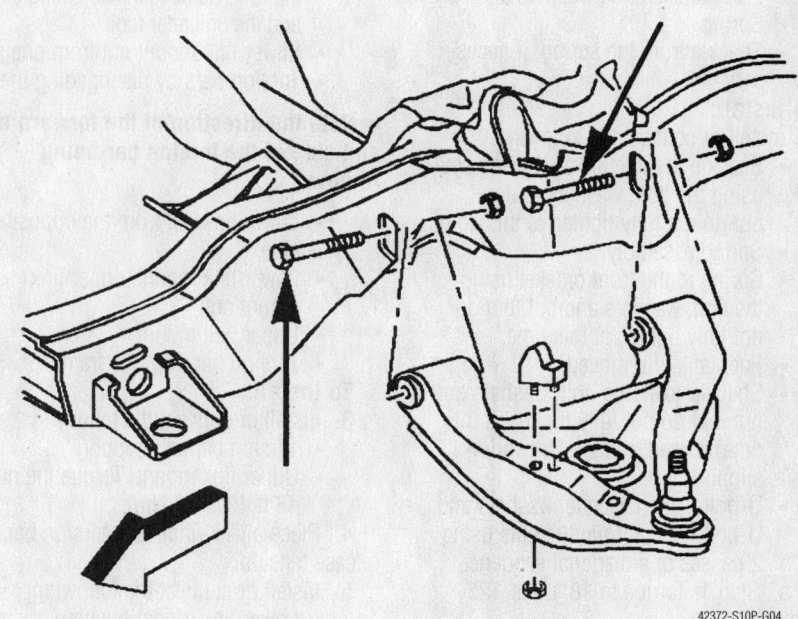

42372-S10P-G04

For steering linkage clearance, the control arm pivot bolts must be installed as shown—2001–03 vehicles

- Lower control arm front and rear pivot bolts with NEW nuts. With the suspension loaded, torque the front bolt to 85 ft. lbs. (115 Nm) and the rear bolt to 72 ft. lbs. (98 Nm). Remove tool J 23028-01.
- Shock absorber
- Stabilizer shaft link to the lower control arm

11. Check and adjust the front wheel alignment.

Leaf Springs

REMOVAL & INSTALLATION

1. Before servicing the vehicle, refer to the precautions in the beginning of this section.

➡The following procedure requires the use of two sets of jackstands.

2. Support the rear axle with jackstands, support the axle and the body separately in order to relieve the load on the rear spring.
3. Remove or disconnect the following:
- Wheel
- Shock absorber
- U-bolt nuts, washers, anchor plate and bolts
- Spare tire, if equipped
- Rear exhaust hangers and lower the rear exhaust, if necessary
- Shackle-to-frame bolt, washers and nut
- Fuel tank, if necessary
- Front bracket nut, washers and bolt
- Spring
- Shackle from the spring, if necessary

To install:
4. Install or connect the following:
- Shackle to the rearward spring eye using the bolt, washers and nut, but do not fully tighten at this time.
- Spring assembly
- Spring to the front bracket using the bolt, washers and nut, but do not fully tighten at this time.
- Fuel tank, if removed
- Shackle-to-frame bolt, washers and nut, but do not fully tighten at this time. If used, remove the spring support.
- U-bolts, anchor plate, washers and U-bolt nuts. Torque the nuts using 2 passes of a diagonal sequence:
  a. Step 1: Torque to 18 ft. lbs. (25 Nm).
  b. Step 2: Torque to 73 ft. lbs. (100 Nm) in the sequence.

5. Position the axle to achieve an approximate gap of 6.46–6.94 in. (164–176mm) between the axle housing tube and the metal surface of the rubber frame bumper bracket. Measure from the housing between the U-bolts to the metal part of the rubber bump stop on the frame.
6. While supporting the axle in this position, tighten the front and rear spring mounting fasteners to 89 ft. lbs. (122 Nm).
7. Install or connect the following:
- Rear exhaust in position and tighten the hangers
- Spare tire
- Shock absorber

Torsion Bar

Instead of the coil spring used on the front suspension of 2WD vehicles, the 4WD vehicles are equipped with a torsion bar.

REMOVAL & INSTALLATION

1. Before servicing the vehicle, refer to the precautions in the beginning of this section.

➡The following procedure requires the use of the Torsion Bar Unloader tool J-36202.

2. Remove or disconnect the following:
- Transmission shield, if equipped
- Torsion bar unloader tool to relax the tension on the torsion bar adjusting arm screw; record the number of turns necessary to properly install the tool. Remove the adjusting screw and the unloader tool.
- Lower link mount nut from one side
- Torsion bars by disengaging them

➡Note the direction of the forward end and side of the torsion bar being removed

- Lower link nut from the opposite side
- Lower link mount, upper link mount nut
- Upper link mount
- Torsion bar from the frame

To install:
3. Install or connect the following:
- Torsion bar and support
- Upper link mount. Torque the nut to 48 ft. lbs. (68 Nm).
4. Place a jack under the torsion bar to release tension.
5. Install or connect the following:
- Lower link mount bushing and nut. Torque the nut to 37 ft. lbs. (50 Nm).

- Torsion bar unloader tool. Tighten the tool against the adjusting arm the same number turns recorded earlier and remove the tool. This loads the torsion bars.
- Transmission shield, if removed

Ball Joints

REMOVAL & INSTALLATION

2WD Vehicles

UPPER

1. Before servicing the vehicle, refer to the precautions in the beginning of this section.

➡The following procedure requires the use of a ball joint separator tool such as J-23742 and J-9519-E ball joint remover and installer set.

2. Raise and support the front of the vehicle safely by placing stands securely under the lower control arms. Because the vehicle's weight is used to relieve spring tension on the upper control arm, the stands must be positioned between the spring seats and the lower control arm ball joints for maximum leverage.

✳✳ CAUTION

With components unbolted, the stand is holding the lower control arm in place against the coil spring. Make sure the stand is firmly positioned and cannot move, or personal injury could result.

3. Remove or disconnect the following:
- Tire and wheel assembly
- Brake caliper and support it from the vehicle using a coat hanger or wire. Make sure the brake line is not stretched or damaged and that the caliper's weight is not supported by the line.
- Cotter pin and retaining nut from the upper ball joint
- Anti-lock brake sensor wire bracket, if equipped
- Upper ball joint from the steering knuckle using tool J-23742 and pull the steering knuckle free of the ball joint

➡After separating the steering knuckle from the upper ball joint, be sure to support the steering knuckle/hub assembly to prevent damaging the brake hose.

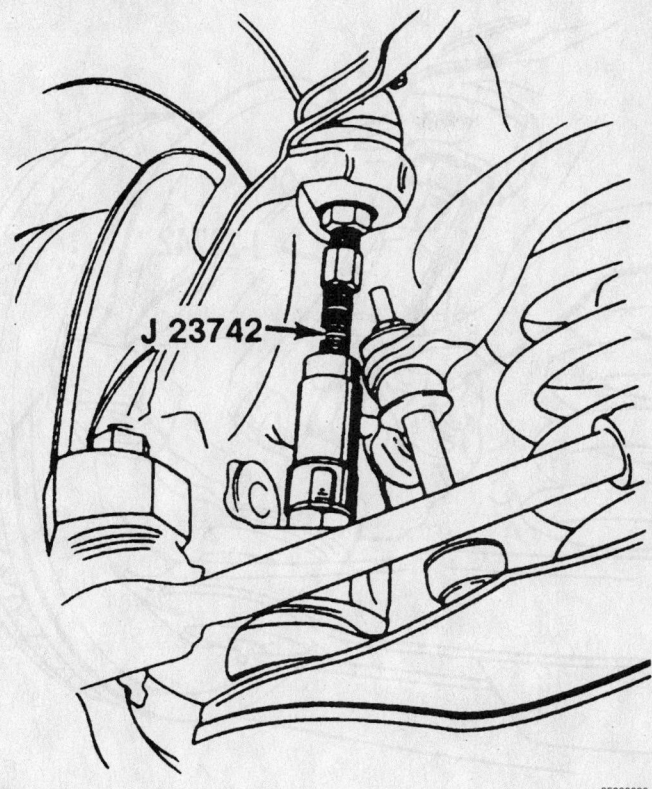

Use a ball joint separator tool to drive the upper ball joint from the steering knuckle

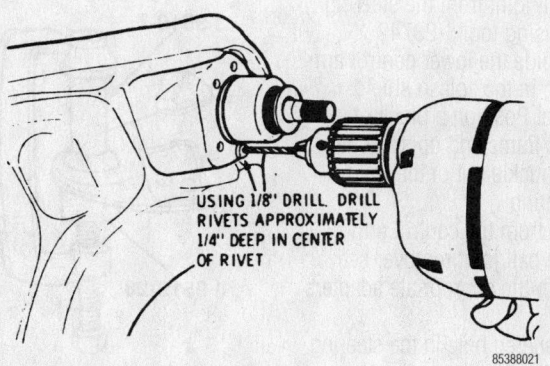

Drill a small guide hole into each ball joint rivet

USING 1/8" DRILL, DRILL RIVETS APPROXIMATELY 1/4" DEEP IN CENTER OF RIVET

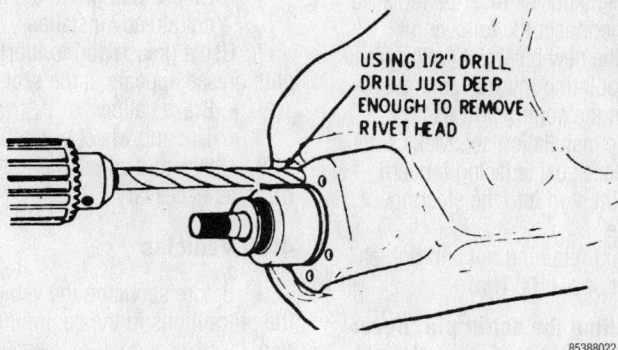

USING 1/2" DRILL DRILL JUST DEEP ENOUGH TO REMOVE RIVET HEAD

Then drill off the rivet heads

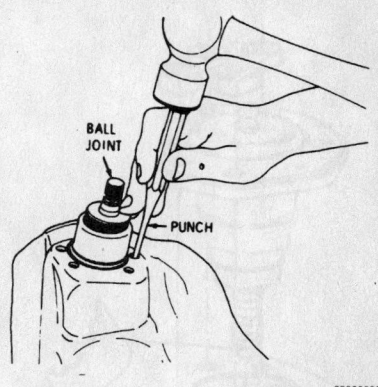

BALL JOINT

PUNCH

Punch the rivets out and remove the ball joint

4. Remove the riveted upper ball joint from the upper control arm as follows:

   a. Drill a⅛in. (3mm) hole, about¼in. (6mm) deep into each rivet.

   b. Then use a½in. (13mm) drill bit, to drill off the rivet heads.

   c. Using a pin punch and the hammer, drive out the rivets in order to free the upper ball joint from the upper control arm assembly, then remove the upper ball joint.

5. Clean and inspect the steering knuckle hole. Replace the steering knuckle if the hole is out of round.

To install:

6. Install or connect the following:
- Ball joint in the upper control arm
- Ball joint retaining nuts and bolts. Position the bolts threaded upward from under the control arm. Tighten the ball joint retainers to 17 ft. lbs. (23 Nm).
- Anti-lock brake sensor wire bracket, if removed
- Ball joint to the knuckle. Make sure the joint is seated, then install the stud nut and tighten to 61 ft. lbs. (83 Nm). Insert a new cotter pin.

➡**When installing the cotter pin, never loosen the castle nut to expose the cotter pin hole.**

- Thread the grease fitting into the ball joint. Use a grease gun to lubricate the upper ball joint until grease appears at the seal.
- Brake caliper
- Tire and wheel assembly

7. Check and adjust the front end alignment, as necessary.

LOWER

1. Before servicing the vehicle, refer to the precautions in the beginning of this section.

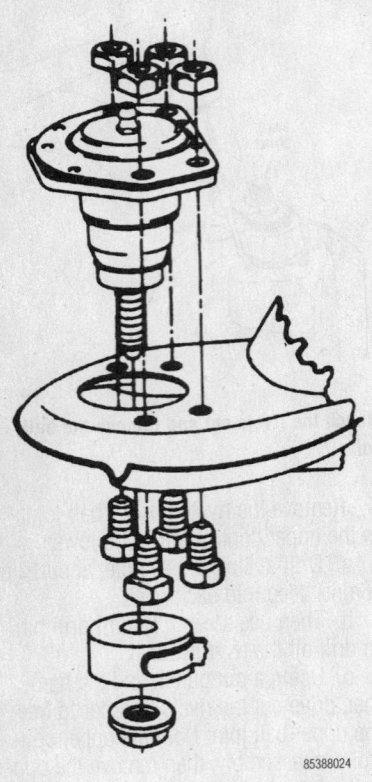

Service ball joints are bolted to the control arm

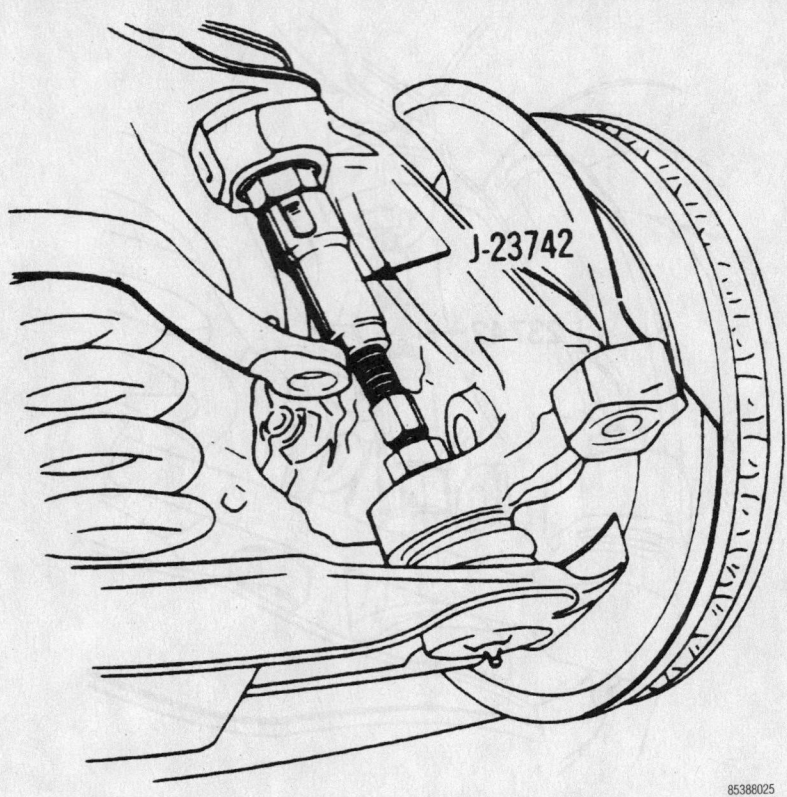

Use a ball joint separator to drive the lower joint from the knuckle

➥The following procedure requires the use of a ball joint remover/installer set (the particular set may vary upon application but must include a clamping-type tool with the appropriately sized adapters) and a ball joint separator tool, such as J-23742.

  • Tire and wheel assembly
2. Position a jack under the spring seat of the lower control arm, then raise the jack to support the arm.

✹✹ CAUTION

The jack MUST remain under the lower control arm, during the removal and installation procedures, to retain the arm and spring positions. Make sure the jack is securely positioned and will not slip or release during the procedure or personal injury may result.

3. Remove or disconnect the following:
  • Brake caliper and support it aside using a hanger or wire. Make sure the brake line is not stressed or damaged.
  • Lower ball joint cotter pin and discard
  • Ball joint stud nut

  • Lower ball joint from the steering knuckle using tool J-23742
4. Carefully guide the lower control arm out of the opening in the splash shield using a putty knife. Position a block of wood between the frame and upper control arm to keep the knuckle out of the way.
  • Grease fitting
  • Ball joint from the control arm using the ball joint remover set along with the appropriate adapters

To install:
5. Clean the tapered hole in the steering knuckle of any dirt or foreign matter, then check the hole to see if it is out of round, deformed or otherwise damaged. If a problem is found, then knuckle must be replaced.
6. Install or connect the following:
  • Press the new ball joint (with grease fitting pointing inward) until it bottoms in the control arm using a suitable installation set. Make sure the grease seal is facing inboard.
  • Ball joint stud into the steering knuckle
  • Ball joint retaining nut and tighten to 79 ft. lbs. (108 Nm)

➥When installing the cotter pin, never loosen the castle nut to expose the cotter pin hole.

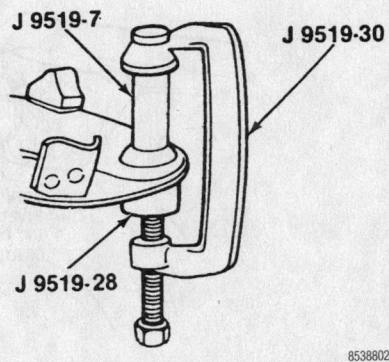

Driving the lower joint from the control

  • Grease fitting into the ball joint, if not already installed
7. Use a grease gun to lubricate the joint until grease appears at the seal.
  • Brake caliper
  • Tire and wheel assembly
8. Check and adjust the front end alignment, as necessary.

4WD Vehicles

1. Before servicing the vehicle, refer to the precautions in the beginning of this section.
  On 4WD vehicles both the upper and

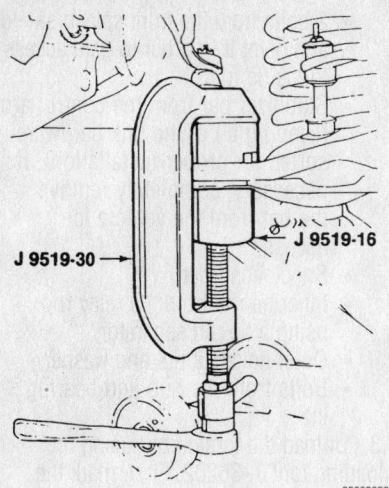

J 9519-30 → ← J 9519-16

85388029

Installing a new ball joint

lower ball joints are removed in the same manner. Once the joint is separated from the steering knuckle the rivets are drilled and punched to free the joint from the control arm. Service joints are bolted into position with the retaining bolts threaded upward from beneath the control arm. In this manner, the joint is replaced in an almost identical fashion to the upper joints on 2WD vehicles.

2. Remove or disconnect the following:
   • Tire and wheel assembly
   • Wheel speed sensor wiring connector from the upper control arm, if removing the upper ball joint
   • Cotter pin from the ball joint, then loosen the retaining nut

3. Position a suitable ball joint separator tool such as J-36607, then carefully loosen the joint in the steering knuckle. Remove the tool and the retaining nut, then separate the joint from the knuckle.

➡**After separating the steering knuckle from the upper ball joint, be sure to support the steering knuckle/hub assembly to prevent damaging the brake hose.**

4. Remove the riveted ball joint from the control arm:

   a. Drill a ⅛ in. (3mm) hole, about ¼ in. (6mm) deep into each rivet.

   b. Then use a ½ in. (13mm) drill bit, to drill off the rivet heads.

   c. Using a pin punch and the hammer, drive out the rivets in order to free the ball joint from the control arm assembly, then remove the ball joint.

To install:

5. Install or connect the following:
   • Ball joint in the control arm
   • Ball joint retaining nuts and bolts. Position the bolts threaded upward

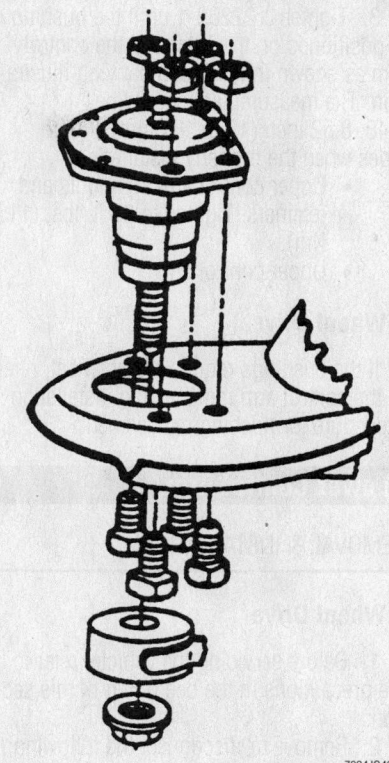

7924JG40

The replacement ball joint comes with nuts and bolts for installation

from under the control arm. Tighten the ball joint retainers to 17 ft. lbs. (23 Nm).
   • Ball joint to the knuckle. Make sure the joint is seated, tighten the lower nut to 79 ft. lbs. (108 Nm) and the upper nut to 61 ft. lbs. (83 Nm). Install a new cotter pin.

➡**When installing the cotter pin, never loosen the castle nut to expose the cotter pin hole, but DO NOT tighten more than an additional ⅙ turn.**

6. Use a grease gun to lubricate the upper ball joint.
   • Wheel speed sensor wiring connector to the upper control arm, if the upper ball joint was removed
   • Tire and wheel assembly

7. Check and adjust the front end alignment, as necessary.

Upper Control Arm

REMOVAL & INSTALLATION

2 Wheel Drive

1. Before servicing the vehicle, refer to the precautions in the beginning of this section.

2. Remove or disconnect the following:
   • Negative battery cable
   • Wheel
   • Wheel speed sensor harness bracket retaining bolt and nut, if equipped
   • Steering knuckle from upper control arm ball joint
   • Mounting nuts/bolts and shims

➡**Make sure to note the location of the control arm shims prior to removal so that they may be installed in their original positions.**

   • Upper control arm

To install:

3. Install or connect the following:
   • Upper control arm

➡**Always tighten nut on the thinner shim pack first.**

   • Mounting nuts/bolts and shims. Torque the nuts to 81–85 ft. lbs. (110–115 Nm).
   • Steering knuckle to upper control arm ball joint
   • New cotter pin

➡**Tighten the nut to align the hole never loosen.**

   • Wheel speed sensor harness bracket retaining bolt and nut, if equipped
   • Wheel

4 Wheel Drive

1. Before servicing the vehicle, refer to the precautions in the beginning of this section.

2. Remove or disconnect the following:
   • Tire and wheel assembly
   • Cotter pin from the ball joint, then loosen the retaining nut
   • Steering knuckle from the upper ball joint. Be sure to support the steering knuckle/hub assembly to prevent damaging the brake hose.

➡**The 4WD vehicles do not use shims to adjust the front wheel alignment. Instead, the upper control arm bolts are equipped with cams, which are rotated to achieve caster and camber adjustments. In order to preserve adjustment and ease installation, matchmark the cams to the control arm before removal. If the control arm is being replaced, transfer the alignment marks to the new component before installation.**

- Front and rear nuts retaining the control arm retaining bolts to the frame
- Outer cams from the bolts
- Bolts and inner cams
- Control arm from the vehicle
- Retaining nut and the bumper from the control arm, if necessary

3. If the bushings are being replaced, use a suitable bushing service set to remove the bushings from the arm.

To install:

4. Install or connect the following:
- Bushing service set to drive the new bushings into the control arm, if removed
- Bumper and retaining nut to the control arm, if removed. Tighten the bumper retaining nut to 20 ft. lbs. (27 Nm).
- Control arm, retaining bolts (from the inside of the frame brackets facing outward) and the inner cams. The inner cams must be positioned on the bolts before they are inserted through the control arm and frame brackets.
- Outer cams over the retaining bolts, then the nuts to the ends of the bolts at the front and rear of the control arm

5. Align the cams to the reference marks made earlier, then tighten the end nuts to 85 ft. lbs. (115 Nm).
- Ball joint to the knuckle
- Tire and wheel assembly

6. Check and adjust the front end alignment, as necessary.

CONTROL ARM BUSHING REPLACEMENT

2 Wheel Drive

1. Before servicing the vehicle, refer to the precautions in the beginning of this section.
2. Remove or disconnect the following:
- Upper control arm and place it in a vice
- Upper control arm shaft nuts and retainers
- Upper control arm bushings using tool J 22269-1, a slotted washer and a short piece if pipe that is slightly larger than the bushing
- Upper control arm shaft

To install:
- Upper control arm shaft
- Upper control arm bushings using tool J 22269-1, a slotted washer and a short piece if pipe that is slightly larger than the bushing

3. Tighten J 22269-1 until the bushing is positioned on the shaft and the control arm as shown in the accompanying illustration. The measurement should be 0.48–0.52 inch (12.8–13.8mm) at both sides when the properly installed.
- Upper control arm shaft nuts and retainers. Tighten to 85 ft. lbs. (115 Nm).
- Upper control arm

4 Wheel Drive

If the bushings require replacement, refer to the control arm removal and installation procedure for bushing replacement.

Lower Control Arm

REMOVAL & INSTALLATION

2 Wheel Drive

1. Before servicing the vehicle, refer to the precautions in the beginning of this section.
2. Remove or disconnect the following:
- Coil spring
- Lower ball joint from the steering knuckle
- Lower control arm from the vehicle

To install:
3. Install or connect the following:
- Lower control arm
- Lower ball joint stud into the steering knuckle
- Ball joint-to-steering knuckle nut and tighten to specification
- New cotter pin to the lower ball joint stud
- Coil spring
4. Align the vehicle.

4 Wheel Drive

1. Before servicing the vehicle, refer to the precautions in the beginning of this section.

➡Tools Needed: universal tie rod separator J–24319–01, torsion bar unloader J–36202, lower control arm bushing service kit J–36618 (if the control arm bushing are being replaced) and ball joint C-clamp J–9519–23. Whether or not the control arm or bushing are being replaced, NEW control arm retaining nut should be used once the old ones have been loosened and removed.

2. Remove or disconnect the following:
- Front wheels

- 2 bolts from the front splash shield and pivot it in order to gain access to the tie rod
- Stabilizer bar from the control arm (keeping all of the link hardware sorted for proper installation). If necessary, completely remove the bar from the vehicle for access.
- Shock absorber
- Inner tie rod from the relay rod using a tie rod separator
- Outer halfshaft nut and washer
- Bolts from the hub and bearing kit

3. Unload the torsion bar using the unloading tool J–36202. First, mark the adjuster for installation.
- Adjustment arm. Slide the bar forward and the adapter out of the rear to remove the adjusting arm.
- Lower ball joint cotter pin, nut and ball joint from the control arm using a ball joint separator
- Nuts and bolts and lower control arm with the torsion bar assembly. Note the direction which the control arm retaining bolts are facing for installation purposes.

To install:
4. Install or connect the following:
- Torsion bar to the lower control arm and place the assembly into the vehicle. Position the front leg of the lower control arm into the crossmember before installing the rear leg into the frame bracket.
- Control arm bolts (facing in the direction as noted during removal or shown in the accompanying illustration) with NEW nuts.

➡The control arm retainers MUST be tightened with the vehicle suspension at normal ride height. This can either be accomplished by starting the nuts now, then installing the remaining components along with the wheels and lowering the vehicle, or by moving jackstands under the ends of the lower control arms and resting the vehicle on them. If the latter solution is tried, make sure front suspension is at actual ride height compression. If you are unsure, it is best to start the nuts now and tighten them to specification once the vehicle is lowered.

- Ball joint stud in the knuckle
5. With the suspension at the correct height, tighten the control arm retaining nuts to 98 ft. lbs. (133 Nm).

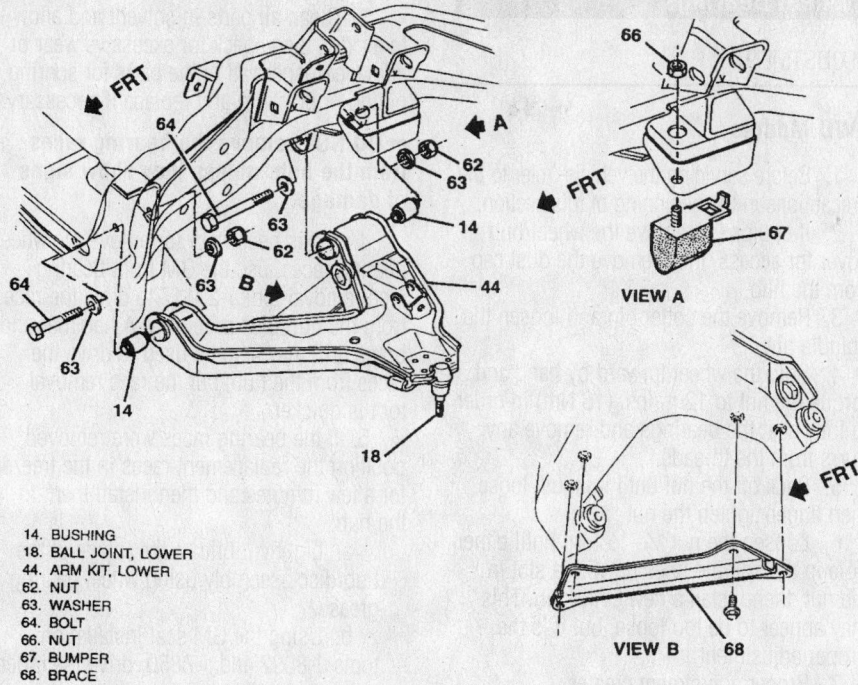

14. BUSHING
18. BALL JOINT, LOWER
44. ARM KIT, LOWER
62. NUT
63. WASHER
64. BOLT
66. NUT
67. BUMPER
68. BRACE

Exploded view of the lower control arm assembly mounting

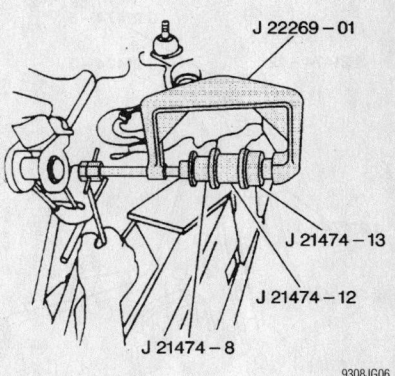

Removing the lower control arm rear bushing—all 2 wheel drive

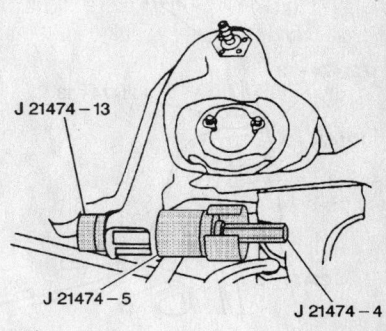

Installing the lower control arm front bushing—all 2 wheel drive

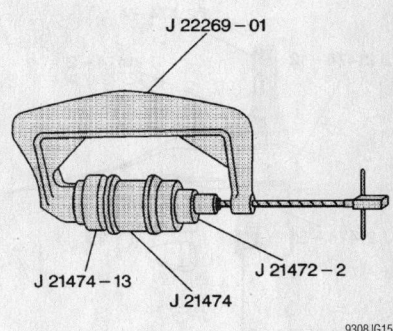

Installing the lower control arm rear bushing—all 2 wheel drive

→The lower ball joint retaining nut MUST be tightened with the vehicle suspension at normal ride height. This can either be accomplished by starting the nut now, then installing the remaining components along with the wheels and lowering the vehicle, or by moving jackstands under the ends of the lower control arms and resting the vehicle on them. If the latter solution is tried, make sure the FULL WEIGHT of the vehicle front end is on the suspension.

6. Install or connect the following:
• Joint-to-control arm nut, then tighten the nut to 92 ft. lbs. (125 Nm) with the suspension at normal ride height and compression
• New cotter pin to the castellated nut. Tighten the nut (but no more than an additional ⅛ turn) in order to align the cotter pin. DO NOT loosen the nut from the specified torque.
• Adjuster arm by sliding the adapter forward, over the torsion bar to install the sides of the nut. Load the torsion bar and install the adjuster bolt aligning the installation mark.
• Drive axle through the hub and bearing assembly
• Tighten the hub and bearing assembly retaining bolts
• Drive axle shaft nut and washer

• Inner tie rod end to the relay rod
• Shock absorber
• Stabilizer bar, if removed
• Stabilizer link(s) to the control arm(s)
• Splash shield
• Front wheels

7. Recheck all fasteners for proper torque and installation before road testing.
8. Refill the differential if any fluid was lost.
9. Check and adjust the front end alignment, as necessary.

CONTROL ARM BUSHING REPLACEMENT

2 Wheel Drive

1. Before servicing the vehicle, refer to the precautions in the beginning of this section.
2. Remove lower control arm and place it in vise.
3. Install tools J 22269-01, 21474-8, 12 and 13 on the rear bushing and tighten until the bushing is removed.
4. Using a blunt chisel, drive the front bushing flare flush with the rubber part of the bushing.
5. Place a wedge or a spacer between the bushing housing to keep the housing from bending while removing or installing the bushing.
6. Install tools J 21474-3, 4, 5 and 6 on the front bushing and tighten until the bushing is removed.

To install:
7. Install the front bushing into the control arm.
8. Install tools J 21474-4, 5 and 13. Tighten until the bushing is fully seated.
9. Install the rear bushing into the control arm
10. Install tools J 22269-01, J 21474-2 and 13. Tighten until the bushing is fully seated.
11. Install the lower control arm.

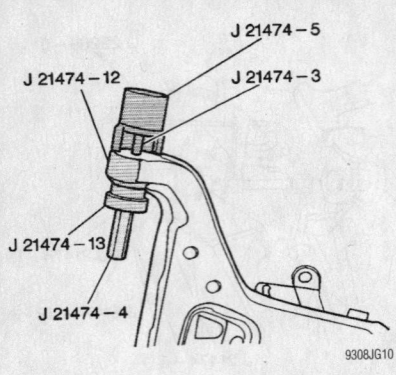

Removing the lower control arm front bushing—4 wheel drive

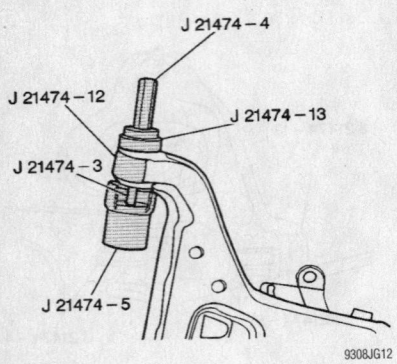

Installing the lower control arm front bushing—4 wheel drive

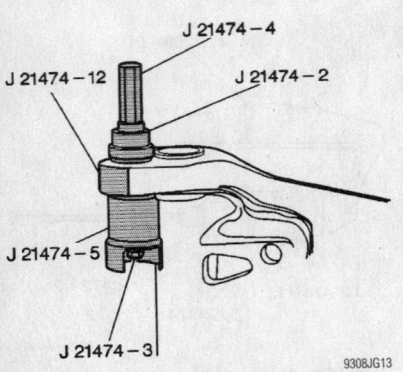

Installing the lower control arm rear bushing—4 wheel drive

4 Wheel Drive

1. Before servicing the vehicle, refer to the precautions in the beginning of this section.

2. Remove lower control arm and place it in vise.

3. Using bushing service set J 21474, remove the front and rear bushings.

To install:

4. Using bushing service set J 21474, install the front and rear bushings.

5. Install the lower control arm.

Wheel Bearings

ADJUSTMENT

2WD Models

1. Before servicing the vehicle, refer to the precautions in the beginning of this section.

2. If equipped, remove the wheel/hub cover for access, then remove the dust cap from the hub.

3. Remove the cotter pin and loosen the spindle nut.

4. Spin the wheel forward by hand and torque the nut to 12 ft. lbs. (16 Nm) in order to fully seat the bearings and remove any burrs from the threads.

5. Back off the nut until it is just loose, then finger-tighten the nut.

6. Loosen the nut ¼–½ turn until either hole in the spindle lines up with a slot in the nut, then install a new cotter pin. This may appear to be too loose, but it is the proper adjustment. Proper adjustment creates 0.001–0.005 in. (0.025–0.127mm) end-play.

4WD Models

The front wheel bearings on the 4-wheel drive vehicles are not adjustable. If the bearings become loose or make noise, they must be replaced.

REMOVAL & INSTALLATION

Front

2WD MODELS

1. Before servicing the vehicle, refer to the precautions in the beginning of this section.

2. Remove or disconnect the following:
   - Wheel
   - Brake caliper with the pads without disconnecting the brake line
   - Grease cap
   - Cotter pin, spindle nut and washer
   - Hub

✳✳ WARNING

Be careful not to drop the outer wheel bearing. As the hub is pulled forward, the outer wheel bearings will often fall forward and they may easily be removed at this time.

   - Outer roller bearing assembly
   - Inner seal by prying it out of the hub and discard it
   - Inner bearing assembly

To install:

3. Clean all parts in solvent and allow to air dry, then check for excessive wear or damage. Inspect all of the parts for scoring, pitting or cracking and replace if necessary.

➡**DO NOT remove the bearing races from the hub, unless they show signs of damage.**

4. If it is necessary to remove the wheel bearing races, use the GM front bearing race removal tool J-29117 to drive the races from the hub/disc assembly. A hammer and brass drift may also be used to drive the races from the hub, but the race removal tool is quicker.

5. If the bearing races were removed, position the replacement races in the freezer for a few minutes and then install them to the hub:

   a. Lightly lubricate the inside of the hub/disc assembly using wheel bearing grease.

   b. Using the GM seal installation tools J-8092 and J-8850, drive the inner bearing race into the hub/disc assembly until it seats. Be sure the race is properly seated against the hub shoulder and is not cocked.

➡**When installing the bearing races, be sure to support the hub/disc assembly with GM tool J-9746-02.**

   c. Using the GM seal installation tools J-8092 and J-8457, drive the outer race into the hub/disc assembly until it seats.

6. Using a high melting point wheel bearing grease, lubricate the bearings, races and spindle; be sure to place a gob of grease (inside the hub/disc assembly) between the races to provide an ample supply of lubricant.

➡**To lubricate each bearing, place a gob of grease in the palm of the hand, then scoop the bearing through the grease until it is well lubricated.**

7. Place the inner bearing in the hub, then apply a thin coating of grease to the sealing lip and install a new inner seal, making sure the seal flange faces the bearing cup.

➡**Although a seal installation tool is preferable, a section of pipe with a smooth edge or a suitably sized socket may be used to drive the seal into position. Be sure the seal is flush with the outer surface of the hub assembly.**

8. Install or connect the following:
   - Wheel hub over the spindle

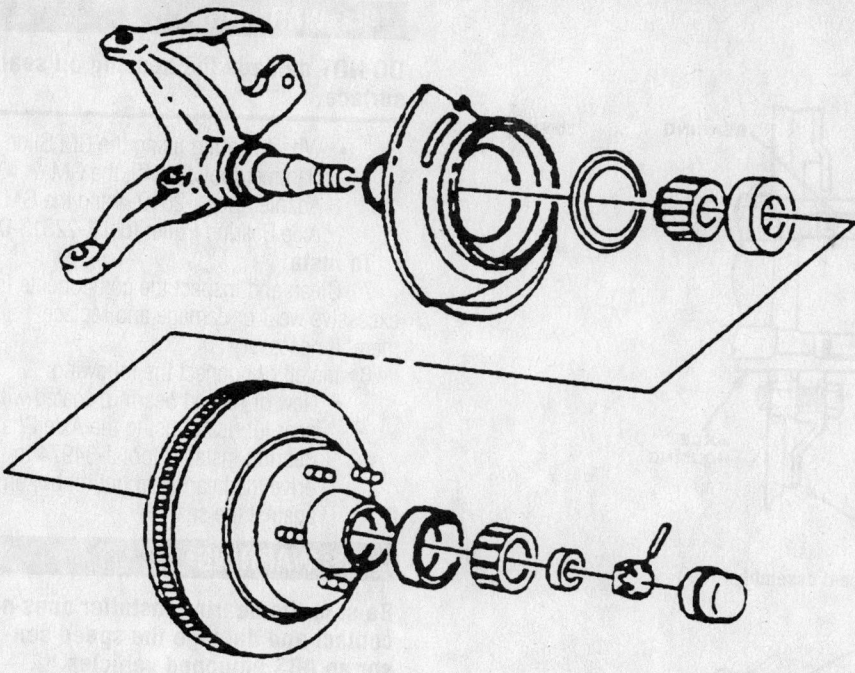

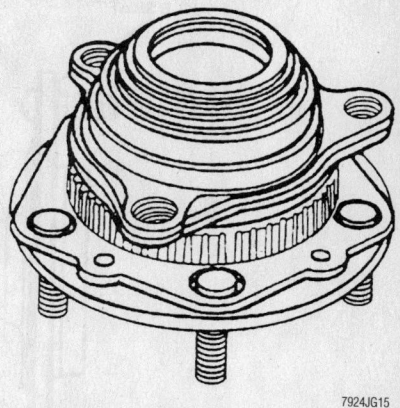

Hub and bearing assembly—4WD vehicles

Wheel bearings, races and related components—2WD vehicles

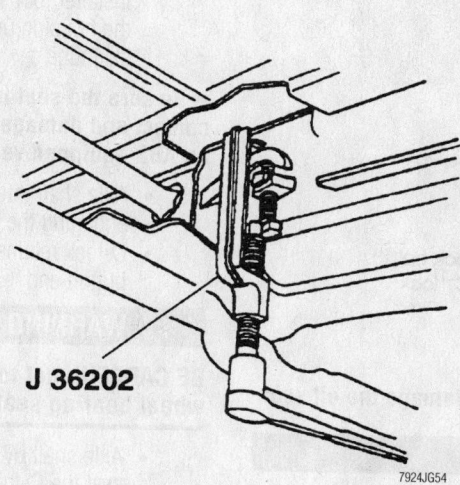

J 36202

Use Torsion Bar Unloading tool J 36202 to remove the adjusting bolt and unload the torsion bar

- Outer bearing into the hub by hand
- Spindle washer and nut
- Brake caliper
- Wheel
9. Properly adjust the wheel bearings
10. Install or connect the following:
  - New cotter pin
  - Dust cap
  - Wheel cover

4WD MODELS

1. Before servicing the vehicle, refer to the precautions in the beginning of this section.

2. Install Torsion Bar Unloading tool J 36202 on the torsion bar adjusting bolt and remove the bolt. To aid during installation, count the number of turns required to remove the bolt.

3. Remove the wheel.

4. Install an axle shaft boot seal protector to the Tri-pot axle joint.

5. Remove or disconnect the following:

- Cotter pin and retainer
- Castle nut and the thrust washer
- Brake caliper and support it aside using wire or a coat hanger

➡ **Be sure the brake line is not stretched or damaged.**

- Brake disc from the wheel hub
- Halfshaft from the hub/bearing assembly, using a Spindle Remover tool J-28733-A to prevent damage to the shaft or hub/bearing assembly
- Hub/bearing assembly from the knuckle

6. Clean and inspect the parts for nicks, scores and/or damage, then replace them as necessary.

To install:

7. Install or connect the following:

- Hub and bearing assembly by aligning the threaded holes. Torque the bolts to 77 ft. lbs. (105 Nm).
- Tie rod end to the steering knuckle using the retaining nut
- New cotter pin
- Brake assembly
- Halfshaft nut. Tighten the nut to 180 ft. lbs. (245 Nm).
- Retainer and a new cotter pin but DO NOT back off specification in order to insert the cotter pin.

8. Remove the torsion bar unloader tool and the drive axle boot protector.

9. Install the wheel.

10. Check and/or adjust the vehicle trim height, as necessary.

Rear

A new pinion shaft lockbolt should be installed whenever either of the axle shafts is removed.

The axle shaft and seal may be removed and replaced without disturbing the bearing or seal but it is highly recommended to replace the seals when removing the axle shaft.

1. Before servicing the vehicle, refer to the precautions in the beginning of this section.

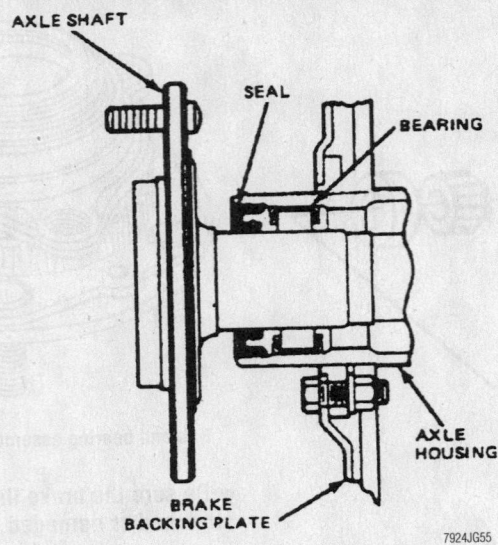

Cross-sectional view of the rear axle, bearing and seal assembly

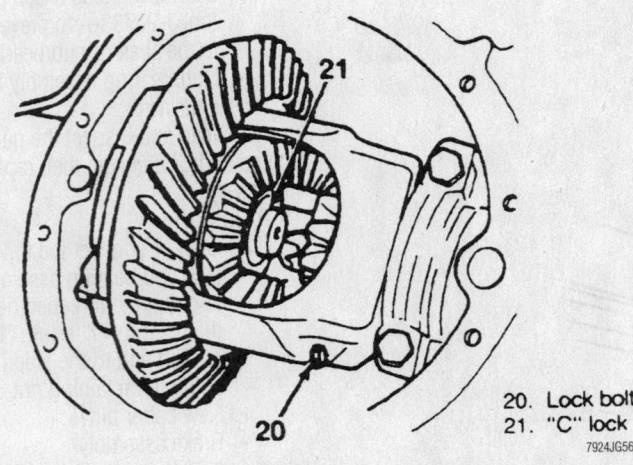

20. Lock bolt
21. "C" lock

Pinion shaft lockbolt and axle C-lock locations, inside the differential

2. Remove or disconnect the following:
   - Rear wheels
   - Brake drums
3. Using a wire brush, clean the dirt/rust from around the rear axle cover.
4. Drain the fluid.
5. Remove or disconnect the following:
   - Rear pinion shaft lockbolt and the pinion shaft
   - C-lock from the button end of the axle shaft by pushing the axle shaft inward
   - Axle shaft from the axle housing

➡ Be careful not to damage the oil seal.

※ WARNING

If equipped with an Anti-Lock Brake System (ABS), be careful not to damage the reflector ring on the axle shaft or the speed sensor bolted to the backing plate, immediately adjacent to the shaft.

6. Remove or disconnect the following:
   - Oil seal by prying the it from the end of the rear axle housing

※ WARNING

DO NOT damage the housing oil seal surface.

   - Wheel bearing using the GM Slide Hammer tool J-2619, the GM Adapter tool J-2619-4 and the GM Axle Bearing Puller tool J-22813-01

To instal

7. Clean and inspect the components for excessive wear or damage and replace them, if necessary.
8. Install or connect the following:
   - New or reused bearing, coated with gear lubricant, using the Axle Shaft Bearing Installer tool J-34974 to drive the bearing in until it bottoms against the seat

※ WARNING

Be sure the bearing installer does not contact and damage the speed sensor on ABS equipped vehicles.

   - New seal lubricated with gear oil using the GM Axle Shaft Seal Installer tool J-33782 to seat it in the housing until it is flush with the axle tube

➡ **Be sure the seal installer does not contact and damage the speed sensor on ABS equipped vehicles.**

   - Axle shaft into the housing by engaging the splines
   - C-lock retainer on the axle shaft button end

※ WARNING

BE CAREFUL not to damage the wheel bearing seal.

   - Axle shaft by pulling it outward to seat the C-lock retainer in the counterbore of the side gears
   - Pinion shaft through the case and the pinions. Tighten the new lockbolt to 27 ft. lbs. (36 Nm).
   - New rear axle cover gasket
   - Housing cover
   - Brake drums
   - Wheels
9. Refill the housing.

BRAKES

Brake Caliper

REMOVAL & INSTALLATION

Front

1. Before servicing the vehicle, refer to the precautions in the beginning of this section.
2. Remove or disconnect the following:
   - ⅔ of the brake fluid from the master cylinder reservoir
   - Tire and wheel assembly
   - Brake caliper fluid line, then plug it
   - Bolts retaining the caliper to the rotor
   - Caliper from the rotor
   - Disc brake pads from the caliper
   - Brake pad retaining clips from inside the caliper

To install:

3. Clean and lubricate the sleeves and bushings with silicon grease.
4. Install or connect the following:
   - Pads in the caliper
   - Caliper in position over the rotor
   - Mounting bolts. Tighten to 38 ft. lbs. (51 Nm) for single piston calipers; 85 ft. lbs. (115 Nm) for dual piston calipers.
   - Fluid lines to the caliper, if disconnected, and tighten to 33 ft. lbs. (45 Nm)
   - Wheel and tire assembly
5. Refill the master cylinder to the correct level. Bleed the brake system if the fluid lines were disconnected from the caliper.

Rear

1. Before servicing the vehicle, refer to the precautions in the beginning of this section.
2. Remove or disconnect the following:
   - Rear tires and wheels
   - Brake hose; and cap the line
   - Retainers from caliper and remove caliper
   - Brake pads, if necessary

To install:

3. Install or connect the following:
   - Brake pads, if removed
   - Caliper over rotor, and onto mounts
   - Retainers, and tighten to 23 ft. lbs. or (31 Nm)
   - Brake hose, and tighten to 20 ft. lbs. (27 Nm)

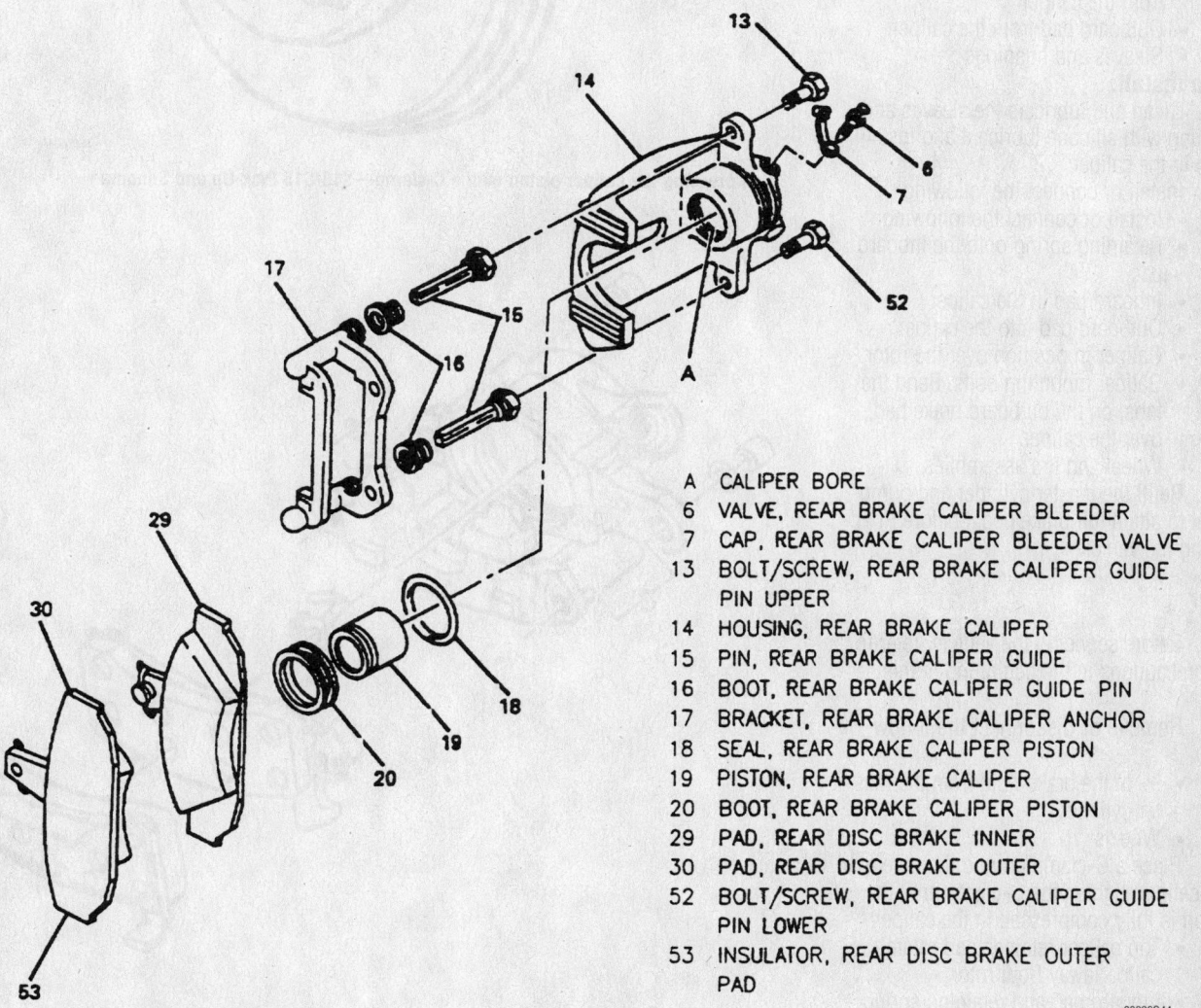

A   CALIPER BORE
6   VALVE, REAR BRAKE CALIPER BLEEDER
7   CAP, REAR BRAKE CALIPER BLEEDER VALVE
13   BOLT/SCREW, REAR BRAKE CALIPER GUIDE PIN UPPER
14   HOUSING, REAR BRAKE CALIPER
15   PIN, REAR BRAKE CALIPER GUIDE
16   BOOT, REAR BRAKE CALIPER GUIDE PIN
17   BRACKET, REAR BRAKE CALIPER ANCHOR
18   SEAL, REAR BRAKE CALIPER PISTON
19   PISTON, REAR BRAKE CALIPER
20   BOOT, REAR BRAKE CALIPER PISTON
29   PAD, REAR DISC BRAKE INNER
30   PAD, REAR DISC BRAKE OUTER
52   BOLT/SCREW, REAR BRAKE CALIPER GUIDE PIN LOWER
53   INSULATOR, REAR DISC BRAKE OUTER PAD

93026G44

Rear brake caliper

- Rear tires and wheels

4. Refill the master cylinder to the correct level. Bleed the brake system if the fluid lines were disconnected from the caliper.

Disc Brake Pads

REMOVAL & INSTALLATION

Front

1. Before servicing the vehicle, refer to the precautions in the beginning of this section.
2. Remove or disconnect the following:
   - ⅔ of the brake fluid from the master cylinder
3. Place a C-clamp around the outer pad and caliper; tighten the C-clamp until the piston is fully compressed in the caliper.
   - Remove top caliper retainer, and rotate caliper away from rotor
   - Inboard pad and retaining spring from the caliper
   - Outboard pad from the caliper
   - Sleeves and bushings

To install:

4. Clean and lubricate the sleeves and bushing with silicone lubricant and install them in the caliper.
5. Install or connect the following:
   - Install or connect the following:
   - Retaining spring onto the inboard pad
   - Inboard pad in the caliper
   - Outboard pad into the caliper
   - Caliper in position over the rotor
   - Caliper mounting bolts. Bend the tabs, on the outboard brake pad, over the caliper.
   - Wheel and tire assemblies

6. Refill the master cylinder and pump pedal to attain full brake pedal before Road-testing the vehicle.

Rear

1. Before servicing the vehicle, refer to the precautions in the beginning of this section.
2. Remove or disconnect the following:
   - ⅔ of the brake fluid from the master cylinder
   - Wheels
3. Place a C-clamp around the outer pad and caliper; tighten the C-clamp until the piston is fully compressed in the caliper.
   - Top caliper retainer, and rotate caliper away from rotor
   - Inboard pad and retaining spring from the caliper
   - Outboard pad from the caliper

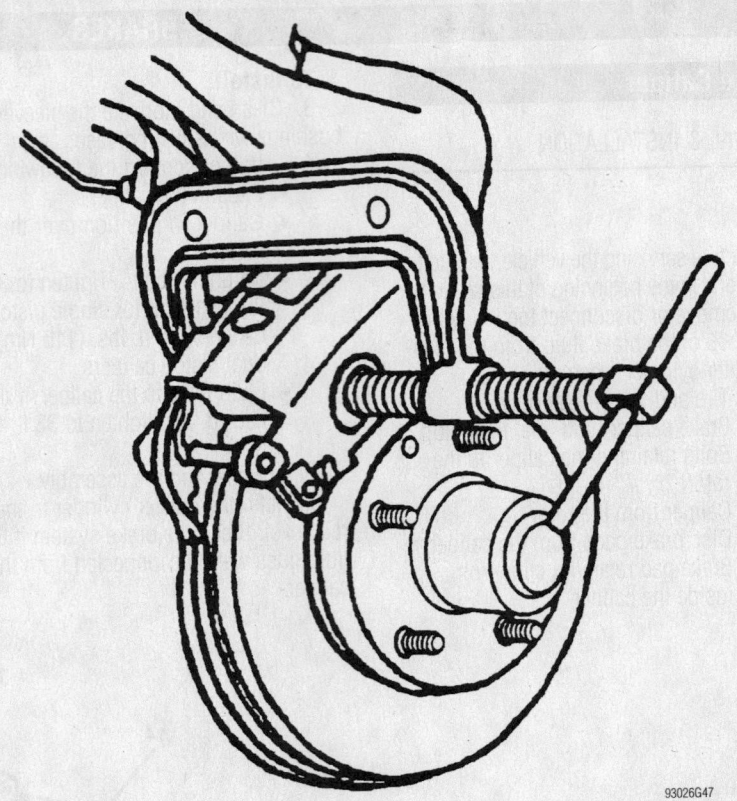

93026G47

Compressing the caliper piston with a C-clamp—S10/S15 Pick-Up and Sonoma

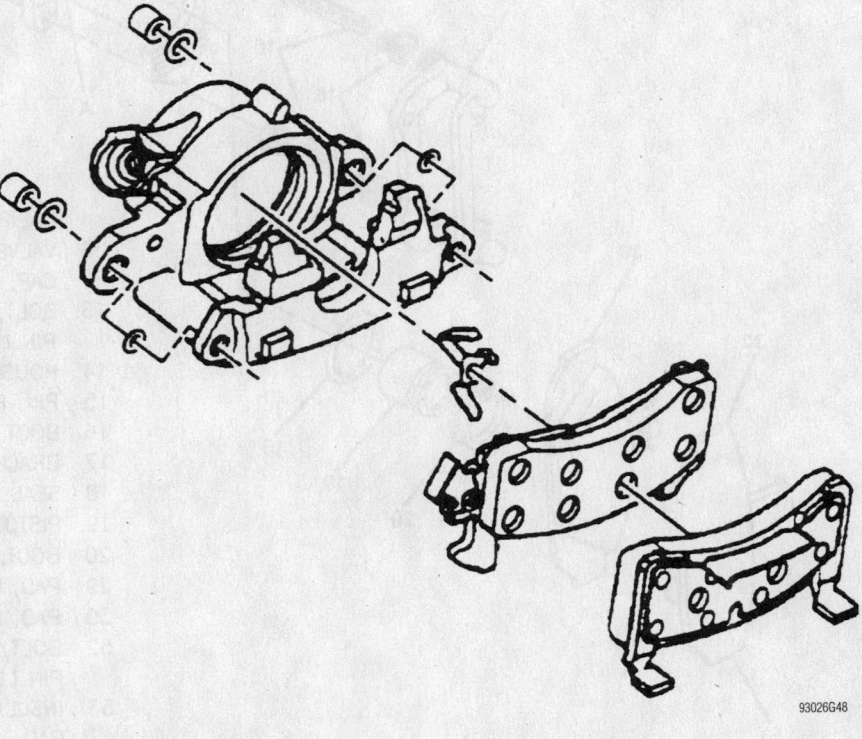

93026G48

Exploded view of the disc brake assembly—S10/S15 Pick-Up and Sonoma

To install:

4. Clean and lubricate the sleeves and bushing with silicone lubricant and install them in the caliper.

5. Install or connect the following:
- Retaining spring onto the inboard pad
- Inboard pad in the caliper
- Outboard pad into the caliper
- Caliper in position over the rotor
- Caliper mounting bolts
- Wheel and tire assemblies

6. Refill the master cylinder and pump pedal to attain full brake pedal before Road-testing the vehicle.

Brake Drums

REMOVAL & INSTALLATION

1. Before servicing the vehicle, refer to the precautions in the beginning of this section.

2. Remove or disconnect the following:
- Wheel and tire assembly
- Brake drum. If the drum will not pull of the axle, use a rubber mallet and tap it around the edge.

To install:

3. Install or connect the following:
- Drum on the axle
- Wheel and tire assembly

4. Refill the master cylinder and pump pedal to attain full brake pedal before road-testing the vehicle.

Brake Shoes

REMOVAL & INSTALLATION

1. Before servicing the vehicle, refer to the precautions in the beginning of this section.

2. Remove or disconnect the following:
- Wheel and tire assembly

- Brake drum
- Return springs from the brake shoes
- Shoe guide
- Hold-down springs and pins
- Actuator lever and pivot
- Lever return spring
- Actuator link
- Parking brake strut and spring
- Parking brake lever
- Brake shoes and the adjuster assembly

To install:

3. Lubricate the contact points on the backing plate and the adjuster with lithium grease.

4. Install or connect the following:
- Parking brake lever, adjusting screw and spring assembly
- Shoe assembly onto the backing plate
- Parking brake lever, strut and strut spring

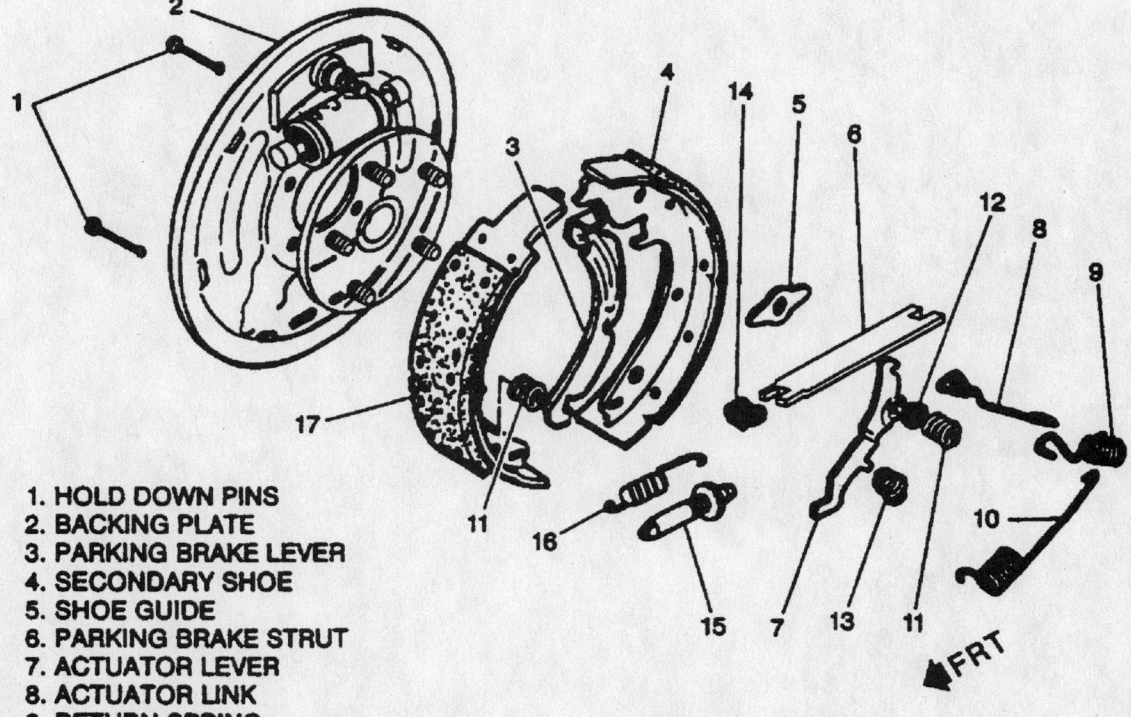

1. HOLD DOWN PINS
2. BACKING PLATE
3. PARKING BRAKE LEVER
4. SECONDARY SHOE
5. SHOE GUIDE
6. PARKING BRAKE STRUT
7. ACTUATOR LEVER
8. ACTUATOR LINK
9. RETURN SPRING
10. RETURN SPRING
11. HOLD DOWN SPRING
12. LEVER PIVOT
13. LEVER RETURN SPRING
14. STRUT SPRING
15. ADJUSTING SCREW ASSEMBLY
16. ADJUSTING SCREW SPRING
17. PRIMARY SHOE

93026G51

Exploded view of the drum brake components—S10/S15 Pick-Up & Sonoma

- Actuator lever and lever pivot
- Actuator link
- Lever spring, the hold-down pins and springs
- Shoe guide
- Return springs
- Brake drum in position

5. Adjust the brakes as follows:
   a. Remove the knockout area in the backing plate, behind the adjuster assembly.

b. Ensure the parking brake system is adjusted properly with no tension on the cables or parking brake lever. The tops of the shoes should be firmly seated against the upper spring retaining anchor, if not as specified, loosen the parking brake cables.

c. Install the drum and turn the brake adjuster until the wheels can just be turned by hand.

d. Then, back the adjuster off 24 notches. No brake drag should be felt after 12 notches.

e. Install an adjusting hole plug in the backing plate to prevent dirt and moisture from entering.

f. Readjust the parking brake cable as necessary.

6. Install the wheel and tire assemblies.

7. Refill the master cylinder and pump pedal to attain full brake pedal before Road-testing the vehicle.

SPECIFICATION CHARTS

ENGINE AND VEHICLE IDENTIFICATION

		Engine						Model Year	
Code ①	Liters (cc)	Cu. In.	Cyl.	Fuel Sys.	Engine Type	Eng. Mfg.	Code ②		Year
6	1.0 (993)	(61)	3	MFI	SOHC	Suzuki	X		1999
8	1.8 (1803)	(110)	4	MFI	DOHC	Toyota	Y		2000
9	1.3 (1300)	(79)	4	MFI	DOHC	Suzuki	1		2001
							2		2002
							3		2003
							4		2004

MFI: Multi-point Fuel Injection
DOHC: Dual Overhead Camshaft
SOHC: Single Overhead Camshaft

① 8th position of VIN
② 10th position of VIN

23728C01

GENERAL ENGINE SPECIFICATIONS

Year	Model	Engine Displacement Liters (cc)	Engine Series (ID/VIN)	Fuel System	Net Horsepower @ rpm	Net Torque @ rpm (ft. lbs.)	Bore x Stroke (in.)	Compression Ratio	Oil Pressure @ rpm
2000	Metro	1.0 (993)	6	MFI	55@5700	58@3300	2.91x3.03	9.5:1	54@3000
	Metro	1.3 (1300)	9	MFI	70@5500	74@3500	2.91x3.03	9.5:1	54@3000
	Prizm	1.8 (1803)	8	MFI	115@5200	117@2800	3.20x3.40	9.5:1	36-71@3000
2001	Metro	1.0 (993)	6	MFI	55@5700	58@3300	2.91x3.03	9.5:1	54@3000
	Metro	1.3 (1300)	9	MFI	70@5500	74@3500	2.91x3.03	9.5:1	54@3000
	Prizm	1.8 (1803)	8	MFI	115@5200	117@2800	3.20x3.40	9.5:1	36-71@3000
2002	Prizm	1.8 (1803)	8	MFI	115@5200	117@2800	3.20x3.40	9.5:1	36-71@3000

MFI: Multi-point Fuel Injection

23728C02

TUNE-UP SPECIFICATIONS

Year	Engine Displacement Liters (cc)	Engine ID/VIN	Spark Plug Gap (in.)	Ignition Timing (deg.)		Fuel Pump (psi)	Idle Speed (rpm)		Valve Clearance	
				MT	AT		MT	AT	Intake	Exhaust
2000	1.0 (993)	6	0.041	5B ①	5B ①	23-30	800	850	HYD	HYD
	1.3 (1300)	9	0.041	5B ①	5B ①	23-30	800	850	HYD	HYD
	1.8 (1803)	8	0.031	10B ②	10B ②	31-37	700-750	700-750	0.0060-0.0100	0.0100-0.0140
2001	1.0 (993)	6	0.041	5B ①	5B ①	23-30	800	850	HYD	HYD
	1.3 (1300)	9	0.041	5B ①	5B ①	23-30	800	850	HYD	HYD
	1.8 (1803)	8	0.031	10B ②	10B ②	31-37	700-750	700-750	0.0060-0.0100	0.0100-0.0140
2002	1.8 (1803)	8	0.031	10B ②	10B ②	31-37	700-750	700-750	0.0060-0.0100	0.0100-0.0140

NOTE: The Vehicle Emission Control Information label often reflects specification changes made during production. The label figures must be used if they differ from those in this chart.

B: Before top dead center

HYD: Hydraulic

① Connect a fused jumper from Duty Check cavity 4 to cavity 5 for fixed timing (DLC connector located at left strut tower)

② Insert jumper wire between terminals in DLC connector E1 and TE1

23728C03

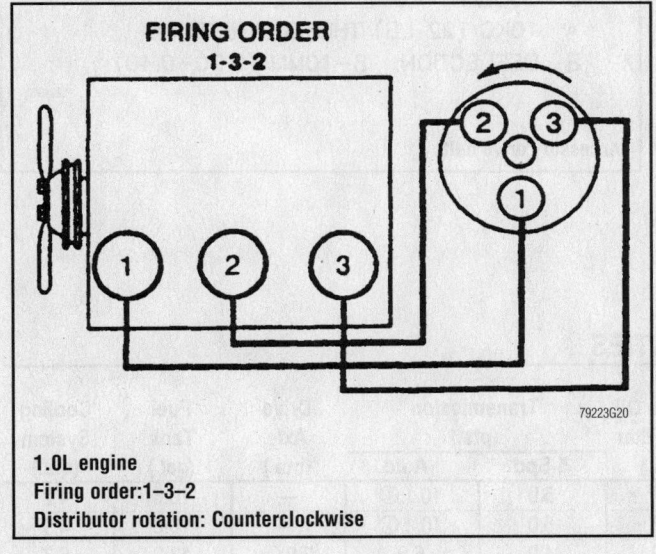

FIRING ORDER 1-3-2

1.0L engine
Firing order:1–3–2
Distributor rotation: Counterclockwise

79223G20

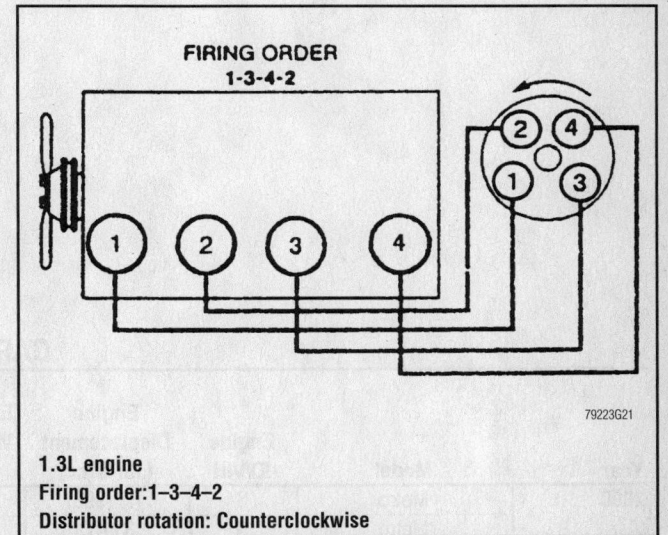

FIRING ORDER 1-3-4-2

1.3L engine
Firing order:1–3–4–2
Distributor rotation: Counterclockwise

79223G21

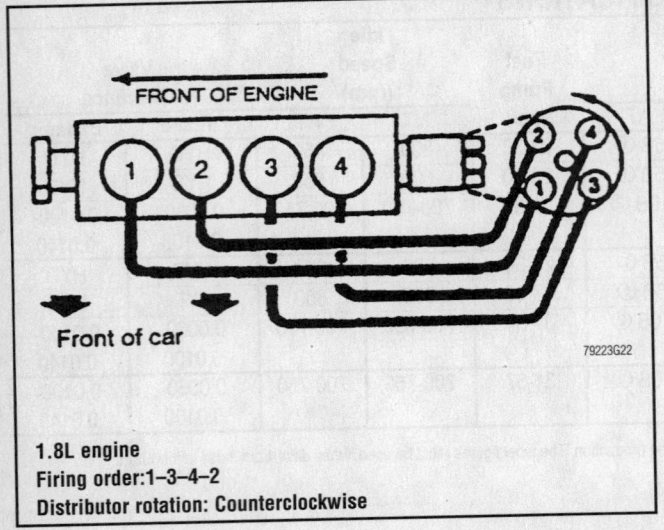

1.8L engine
Firing order:1–3–4–2
Distributor rotation: Counterclockwise

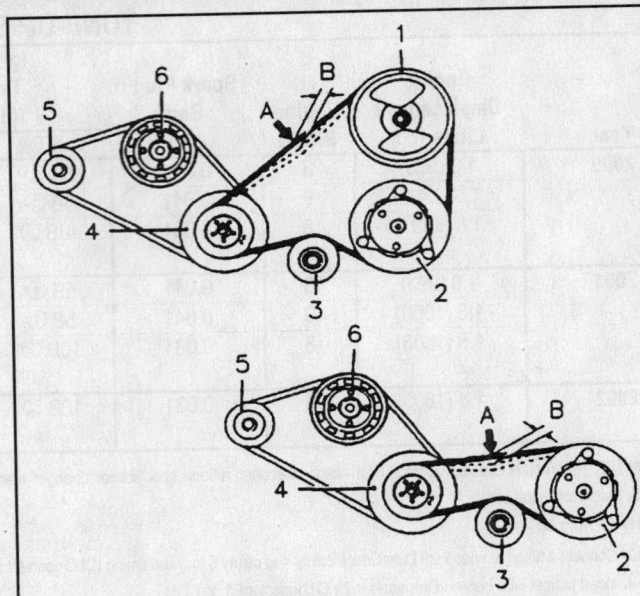

1 POWER STEERING PUMP PULLEY
2 COMPRESSOR CLUTCH PULLEY
3 TENSIONER PULLEY
4 CRANKSHIFT PULLEY
5 GENERATOR PULLY
6 COOLANT PUMP PULLEY
A 10KG (22 LB) THUMB PRESSURE
B DEFLECTION: 8–10MM (0.30–0.40)

Accessory drive belts

CAPACITIES

Year	Model	Engine ID/VIN	Engine Displacement Liters (cc)	Engine Oil With Filter (qts.)	Transmission (pts.)		Drive Axle (pts.)	Fuel Tank (gal.)	Cooling System (qts.)
					5-Spd	Auto.			
2000	Metro	6	1.0 (993)	3.7	5.0	10.1 ①	—	10.6	4.2
	Metro	9	1.3 (1300)	3.7	5.0	10.1 ①	—	10.6	4.9
	Prizm	8	1.8 (1803)	3.9	4.0	6.6	3.0 ②	13.2	6.7
2001	Metro	6	1.0 (993)	3.7	5.0	10.1 ①	—	10.6	4.2
	Metro	9	1.3 (1300)	3.7	5.0	10.1 ①	—	10.6	4.9
	Prizm	8	1.8 (1803)	3.9	4.0	6.6	3.0 ②	13.2	6.7
2002	Prizm	8	1.8 (1803)	3.9	4.0	6.6	3.0 ②	13.2	6.7

NOTE: All capacities are approximate. Add fluid gradually and ensure a proper fluid level is obtained.
① Automatic transmission: Specification is after complete overhaul. Drain and fill will be less
② 3 speed automatic only

CRANKSHAFT AND CONNECTING ROD SPECIFICATIONS
All measurements are given in inches.

Year	Engine Displacement Liters (cc)	Engine ID/VIN	Crankshaft				Connecting Rod		
			Main Brg. Journal Dia.	Main Brg. Oil Clearance	Shaft End-play	Thrust on No.	Journal Diameter	Oil Clearance	Side Clearance
2000	1.0 (993)	6	①	0.00008-0.00015	0.0044-0.0122	NA	1.6529-1.6535	0.0012-0.0019	0.0039-0.0078
	1.3 (1300)	9	①	0.00008-0.00015	0.0044-0.0122	NA	1.6529-1.6535	0.0012-0.0019	0.0039-0.0078
	1.8 (1803)	8	②	0.0006-0.0013	0.0006-0.0087	3	1.8893-1.8898	0.0008-0.0019	0.0050-0.0150
2001	1.0 (993)	6	①	0.00008-0.00015	0.0044-0.0122	NA	1.6529-1.6535	0.0012-0.0019	0.0039-0.0078
	1.3 (1300)	9	①	0.00008-0.00015	0.0044-0.0122	NA	1.6529-1.6535	0.0012-0.0019	0.0039-0.0078
	1.8 (1803)	8	②	0.0006-0.0013	0.0006-0.0087	3	1.8893-1.8898	0.0008-0.0019	0.0050-0.0150
2002	1.8 (1803)	8	②	0.0006-0.0013	0.0006-0.0087	3	1.8893-1.8898	0.0008-0.0019	0.0050-0.0150

NA: Not Available

① 1 Stamping: 1.7714-1.7716
  2 Stamping: 1.7712-1.7714
  3 Stamping: 1.7710-1.7712

② 0 Stamping: 1.8895-1.8898
  1 Stamping: 1.8893-1.8895
  2 Stamping: 1.8891-1.8893

23728C05

VALVE SPECIFICATIONS

Year	Engine Displacement Liters (cc)	Engine ID/VIN	Seat Angle (deg.)	Face Angle (deg.)	Spring Test Pressure (lbs. @ in.)	Spring Installed Height (in.)	Stem-to-Guide Clearance (in.)		Stem Diameter (in.)	
							Intake	Exhaust	Intake	Exhaust
2000	1.0 (993)	6	45	45	46.1-51.8@1.28	1.28	0.0008-0.0022	0.0018-0.0028	0.2148-0.2157	0.2142-0.2148
	1.3 (1300)	9	45	45	54.7-64.3@1.63	1.63	0.0008-0.0019	0.0014-0.0025	0.2742-0.2748	0.2737-0.2742
	1.8 (1803)	8	45	45.5	37.3@1.25	1.25	0.0010-0.0024	0.0012-0.0026	0.2350-0.2356	0.2348-0.2354
2001	1.0 (993)	6	45	45	46.1-51.8@1.28	1.28	0.0008-0.0022	0.0018-0.0028	0.2148-0.2157	0.2142-0.2148
	1.3 (1300)	9	45	45	54.7-64.3@1.63	1.63	0.0008-0.0019	0.0014-0.0025	0.2742-0.2748	0.2737-0.2742
	1.8 (1803)	8	45	45.5	37.3@1.25	1.25	0.0010-0.0024	0.0012-0.0026	0.2350-0.2356	0.2348-0.2354
2002	1.8 (1803)	8	45	45.5	37.3@1.25	1.25	0.0010-0.0024	0.0012-0.0026	0.2350-0.2356	0.2348-0.2354

23728C06

PISTON AND RING SPECIFICATIONS
All measurements are given in inches.

Year	Engine Displacement Liters (cc)	Engine ID/VIN	Piston Clearance	Ring Gap			Ring Side Clearance		
				Top Compression	Bottom Compression	Oil Control	Top Compression	Bottom Compression	Oil Control
2000	1.0 (993)	6	0.0008-0.0015	0.0079-0.0118	0.0079-0.0118	0.0079-0.0275	0.0012-0.0027	0.0008-0.0023	—
	1.3 (1300)	9	0.0008-0.0015	0.0079-0.0118	0.0079-0.0012	0.0079-0.0275	0.0014-0.0027	0.0008-0.0023	—
	1.8 (1803)	8	0.0033-0.0041	0.0098-0.0138	0.0138-0.0197	0.0039-0.0157	0.0018-0.0033	0.0012-0.0028	—
2001	1.0 (993)	6	0.0008-0.0015	0.0079-0.0118	0.0079-0.0118	0.0079-0.0275	0.0012-0.0027	0.0008-0.0023	—
	1.3 (1300)	9	0.0008-0.0015	0.0079-0.0118	0.0079-0.0012	0.0079-0.0275	0.0014-0.0027	0.0008-0.0023	—
	1.8 (1803)	8	0.0033-0.0041	0.0098-0.0138	0.0138-0.0197	0.0039-0.0157	0.0018-0.0033	0.0012-0.0028	—
2002	1.8 (1803)	8	0.0033-0.0041	0.0098-0.0138	0.0138-0.0197	0.0039-0.0157	0.0018-0.0033	0.0012-0.0028	—

23728C07

TORQUE SPECIFICATIONS
All readings in ft. lbs.

Year	Engine Displacement Liters (cc)	Engine ID/VIN	Cylinder Head Bolts	Main Bearing Bolts	Rod Bearing Bolts	Crankshaft Damper Bolt	Flywheel Bolts	Manifold		Spark Plugs	Lug Nuts
								Intake	Exhaust		
2000	1.0 (993)	6	54	40	26	96 ①	55	17	17	21	44
	1.3 (1300)	9	49	40	26	94 ①	55	17	17	21	44
	1.8 (1803)	8	②	③	④	105 ①	⑤	13	36	21	76
2001	1.0 (993)	6	54	40	26	96 ①	55	17	17	21	44
	1.3 (1300)	9	49	40	26	94 ①	55	17	17	21	44
	1.8 (1803)	8	②	③	④	105 ①	⑤	13	36	21	76
2002	1.8 (1803)	8	②	③	④	105 ①	⑤	13	36	21	76

① Crankshaft timing belt sprocket

② Step 1: 18 ft. lbs.
   Step 2: 36 ft. lbs.
   Step 3: 36 ft. lbs. Plus 90 degrees

③ Step 1: 16 ft. lbs.
   Step 2: 32 ft. lbs.
   Step 3: 32 ft. lbs. Plus 90 degrees

④ 15 ft. lbs. plus 90 degrees

⑤ Manual transaxle: 36 ft. lbs. Plus 90 degrees
   Automatic transaxle: 61 ft. lbs.

23728C08

WHEEL ALIGNMENT

Year	Model		Caster Range (+/-Deg.)	Caster Preferred Setting (Deg.)	Camber Range (+/-Deg.)	Camber Preferred Setting (Deg.)	Toe-in (in.)	Steering Axis Inclination (Deg.)
2000	Metro	F	2.00	+3.00 ①	1.00	+0.50	0.08 +/- 0.08	—
		R	—	—	1.00	0 ①	0.23 +/- 0.07	—
	Prizm	F	0.75	+1.32 ①	0.75	-0.18	0.05 +/- 0.10	—
		R	—	—	0.75	-0.92 ①	0.20 +/- 0.10	—
2001	Metro	F	2.00	+3.00 ①	1.00	+0.50	0.08 +/- 0.08	—
		R	—	—	1.00	0 ①	0.23 +/- 0.07	—
	Prizm	F	0.75	+1.32 ①	0.75	-0.18	0.05 +/- 0.10	—
		R	—	—	0.75	-0.92 ①	0.20 +/- 0.10	—
2002	Prizm	F	0.75	+1.32 ①	0.75	-0.18	0.05 +/- 0.10	—
		R	—	—	0.75	-0.92 ①	0.20 +/- 0.10	—

① Not adjustable, for reference only

23728C09

TIRE, WHEEL AND BALL JOINT SPECIFICATIONS

Year	Model	OEM Tires Standard	OEM Tires Optional	Tire Pressures (psi) Front	Tire Pressures (psi) Rear	Wheel Size	Ball Joint Inspection
2000	Metro	P155/80R13	None	32	32	4.5-B	①
	Prizm	P175/65R14	P185/65R14	30	30	5.5J	①
	Prizm Lsi	P185/65R14	None	30	30	5.5J	①
2001	Metro	P155/80R13	None	32	32	4.5-B	①
	Prizm	P175/65R14	P185/65R14	30	30	5.5J	①
	Prizm Lsi	P185/65R14	None	30	30	5.5J	①
2002	Prizm	P175/65R14	P185/65R14	30	30	5.5J	①
	Prizm Lsi	P185/65R14	None	30	30	5.5J	①

OEM: Original Equipment Manufacturer
PSI: Pounds Per Square Inch
STD: Standard
OPT: Optional
① Replace if any measurable movement is found.

23728C10

BRAKE SPECIFICATIONS
All measurements in inches unless noted

Year	Model	Brake Disc Original Thickness	Brake Disc Minimum Thickness	Brake Disc Maximum Runout	Brake Drum Diameter Original Inside Diameter	Brake Drum Diameter Max. Wear Limit	Brake Drum Diameter Maximum Machine Diameter	Minimum Lining Thickness ① Front	Minimum Lining Thickness ① Rear	Brake Caliper Bracket Bolts (ft. lbs.)	Brake Caliper Mounting Bolts (ft. lbs.)
2000	Metro	0.670	0.590	0.004	②	②	②	0.236	0.111	29-43	22
	Prizm	0.866	0.787	0.003	7.87	7.91	7.91	0.390	0.390	65	25
2001	Metro	0.670	0.590	0.004	②	②	②	0.236	0.111	29-43	22
	Prizm	0.866	0.787	0.003	7.87	7.91	7.91	0.390	0.390	65	25
2002	Prizm	0.866	0.787	0.003	7.87	7.91	7.91	0.390	0.390	65	25

① Minimum lining thickness includes pad/shoe backing
② 2 door: 7.09
4 door: 7.87
③ Front: 18 ft. lbs.
Rear: 14 ft. lbs.

23728C11

SCHEDULED MAINTENANCE INTERVALS
GEO/CHEVROLET—METRO

TO BE SERVICED	TYPE OF SERVICE	7.5	15	22.5	30	37.5	45	52.5	60	67.5	75	82.5	90	97.5
Engine oil & filter	R	✓	✓	✓	✓	✓	✓	✓	✓	✓	✓	✓	✓	✓
Chassis lubrication	S/I	✓	✓	✓	✓	✓	✓	✓	✓	✓	✓	✓	✓	✓
Locking front hubs	S/I	✓	✓	✓	✓	✓	✓	✓	✓	✓	✓	✓	✓	✓
Lubricate parking brake cable guides, underbody contact points & linkage	S/I	✓	✓	✓	✓	✓	✓	✓	✓	✓	✓	✓	✓	✓
Rotate tires	S/I	✓	✓	✓	✓	✓	✓	✓	✓	✓	✓	✓	✓	✓
Brake system	S/I	✓		✓		✓		✓		✓		✓		✓
Exhaust system	S/I	✓		✓		✓		✓		✓		✓		✓
Fuel tank, cap & lines	S/I		✓		✓		✓		✓		✓		✓	
Air cleaner filter	R				✓				✓				✓	
Engine coolant	R				✓				✓				✓	
Fuel filter	R				✓				✓				✓	
Manual transaxle oil	R				✓				✓				✓	
Spark plugs	R				✓				✓				✓	
Accessory drive belt(s)	S/I				✓				✓				✓	
Cooling system	S/I				✓				✓				✓	
Automatic transaxle fluid & filter	S/I				✓				✓				✓	
EGR system	S/I				✓				✓					
Engine timing	S/I								✓					
Ignition cables	S/I								✓				✓	
PCV system	S/I				✓				✓					
Brake fluid	R													
PCV valve ①	R													
Timing belt	R								✓					
Throttle body unit mount bolt torque ②	S/I													

R: Replace   S/I: Service or Inspect

① PCV valve: replace at 50,000 miles.

② Torque to 16 ft. lbs. (22 Nm) at 6000 miles.

FREQUENT OPERATION MAINTENANCE (SEVERE SERVICE)

If a vehicle is operated under any of the following conditions it is considered severe service:

- Extremely dusty areas.
- 50% or more of the vehicle operation is in 32°C (90°F) or higher temperatures, or constant operation in temperatures below 0°C (32°F).
- Prolonged idling (vehicle operation in stop and go traffic).
- Frequent short running periods (engine does not warm to normal operating temperatures).
- Police, taxi, delivery usage or trailer towing usage.

Oil & oil filter: change every 3000 miles.

Chassis lubrication: lubricate every 6000 miles.

Throttle body mount bolt torque: torque to 16 ft. lbs. (22 Nm) at 6000 miles.

Air cleaner filter: service or inspect every 15,000 miles.

Rotate tires: rotate at 6000 miles and then every 15,000 miles thereafter.

Automatic transaxle fluid & filter: replace every 50,000 miles.

23728C12

SCHEDULED MAINTENANCE INTERVALS
GEO/CHEVROLET—PRIZM

TO BE SERVICED	TYPE OF SERVICE	VEHICLE MILEAGE INTERVAL (x1000)												
		7.5	15	22.5	30	37.5	45	52.5	60	67.5	75	82.5	90	97.5
Engine oil & filter	R	✓	✓	✓	✓	✓	✓	✓	✓	✓	✓	✓	✓	✓
Lubricate parking brake cable guides, underbody contact points & linkage	S/I	✓	✓	✓	✓	✓	✓	✓	✓	✓	✓	✓	✓	✓
Rotate tires	S/I	✓	✓	✓	✓	✓	✓	✓	✓	✓	✓	✓	✓	✓
Brake system	S/I	✓		✓		✓		✓		✓		✓		✓
Engine idle speed	S/I	✓		✓		✓		✓		✓		✓		✓
Exhaust system	S/I	✓		✓		✓		✓		✓		✓		✓
Chassis lubrication	S/I		✓		✓		✓		✓		✓		✓	
Fuel tank, cap & lines	S/I				✓				✓				✓	
Valve clearance	S/I								✓					
Air cleaner filter	R				✓				✓				✓	
Engine coolant	R				✓				✓				✓	
Fuel tank cap gasket	R				✓				✓				✓	
Manual transaxle oil	R				✓				✓				✓	
Spark plugs	R				✓				✓				✓	
Accessory drive belt(s)	S/I								✓		✓		✓	✓
Cooling system	S/I						✓				✓			
Automatic transaxle fluid & filter	S/I				✓				✓				✓	
EGR system	S/I				✓				✓				✓	
Ignition cables	S/I								✓					
PCV system	S/I				✓				✓				✓	
Timing belt	R								✓					
EVAP canister	S/I								✓					
Throttle body unit mount bolt torque ①	S/I													

R: Replace   S/I: Service or Inspect

① Torque to 16 ft. lbs. (22 Nm) at 6000 miles.

FREQUENT OPERATION MAINTENANCE (SEVERE SERVICE)
If a vehicle is operated under any of the following conditions it is considered severe service:
- Extremely dusty areas.
- 50% or more of the vehicle operation is in 32°C (90°F) or higher temperatures, or constant operation in temperatures below 0°C (32°F).
- Prolonged idling (vehicle operation in stop and go traffic).
- Frequent short running periods (engine does not warm to normal operating temperatures).
- Police, taxi, delivery usage or trailer towing usage.

Oil & oil filter: change every 3000 miles.

Chassis lubrication: lubricate every 6000 miles.

Throttle body mount bolt torque: torque to 16 ft. lbs. (22 Nm) at 6000 miles.

Air cleaner filter: service or inspect every 15,000 miles.

Differential fluid: replace every 15,000 miles.

Rotate tires: rotate at 6000 miles and then every 15,000 miles thereafter.

Automatic transaxle fluid & filter replace every 15,000 miles.

23728C13

PRECAUTIONS

Before servicing any vehicle, please be sure to read all of the following precautions, which deal with personal safety, prevention of component damage, and important points to take into consideration when servicing a motor vehicle:

• Never open, service or drain the radiator or cooling system when the engine is hot; serious burns can occur from the steam and hot coolant.

• Observe all applicable safety precautions when working around fuel. Whenever servicing the fuel system, always work in a well-ventilated area. Do not allow fuel spray or vapors to come in contact with a spark, open flame or excessive heat (a hot drop light, for example). Keep a dry chemical fire extinguisher near the work area. Always keep fuel in a container specifically designed for fuel storage; also, always properly seal fuel containers to avoid the possibility of fire or explosion. Refer to the additional fuel system precautions later in this section.

• Fuel injection systems often remain pressurized, even after the engine has been turned **OFF**. The fuel system pressure must be relieved before disconnecting any fuel lines. Failure to do so may result in fire and/or personal injury.

• Brake fluid often contains polyglycol ethers and polyglycols. Avoid contact with the eyes and wash your hands thoroughly after handling brake fluid. If you do get brake fluid in your eyes, flush your eyes with clean, running water for 15 minutes. If eye irritation persists, or if you have taken brake fluid internally, IMMEDIATELY seek medical assistance.

• The EPA warns that prolonged contact with used engine oil may cause a number of skin disorders, including cancer! You should make every effort to minimize your exposure to used engine oil. Protective gloves should be worn when changing oil. Wash your hands and any other exposed skin areas as soon as possible after exposure to used engine oil. Soap and water, or waterless hand cleaner should be used.

• All new vehicles are now equipped with an air bag system. The system must be disabled before performing service on or around system components, steering column, instrument panel components, wiring and sensors. Failure to follow safety and disabling procedures could result in accidental air bag deployment, possible personal injury and unnecessary system repairs.

• Always wear safety goggles when working with, or around, the air bag system. When carrying a non-deployed air bag, be sure the bag and trim cover are pointed away from your body. When placing a non-deployed air bag on a work surface, always face the bag and trim cover upward, away from the surface. This will reduce the motion of the module if it is accidentally deployed. Refer to the additional air bag system precautions later in this section.

• Clean, high quality brake fluid from a sealed container is essential to the safe and proper operation of the brake system. You should always buy the correct type of brake fluid for your vehicle. If the brake fluid becomes contaminated, completely flush the system with new fluid. Never reuse any brake fluid. Any brake fluid that is removed from the system should be discarded. Also, do not allow any brake fluid to come in contact with a painted surface; it will damage the paint.

• Never operate the engine without the proper amount and type of engine oil; doing so WILL result in severe engine damage.

• Timing belt maintenance is extremely important! Many models utilize an interference-type, non-freewheeling engine. If the timing belt breaks, the valves in the cylinder head may strike the pistons, causing potentially serious (also time-consuming and expensive) engine damage. Refer to the maintenance interval charts in the front of this section for the recommended replacement interval for the timing belt, and to the timing belt procedure in this section for belt replacement and inspection.

• Disconnecting the negative battery cable on some vehicles may interfere with the functions of the on-board computer system(s) and may require the computer to undergo a relearning process once the negative battery cable is reconnected.

• When servicing drum brakes, only disassemble and assemble one side at a time, leaving the remaining side intact for reference.

ENGINE REPAIR

Distributor

REMOVAL

1. Before servicing the vehicle, refer to the precautions in the beginning of this section.
2. Remove or disconnect the following:
   • Negative battery cable
   • Distributor electrical connector
   • Spark plug wires from the distributor cap
   • Distributor cap
3. Mark the position of the distributor rotor in relation to the distributor body and mark the position of the distributor body in relation to the cylinder head.
4. Remove the distributor hold-down bolts.
5. Remove the distributor from the engine.

INSTALLATION

Timing Not Disturbed

1. Align the reference marks made during removal. Position the distributor carefully and be sure the drive gear engages properly within the slot.
2. Install or connect the following:
   • Hold-down bolts but do not tighten
   • Distributor electrical connector
   • Spark plug wires
   • Distributor cap
   • Negative battery cable
3. Check and/or adjust the ignition timing.
4. Tighten the distributor hold-down bolts to 11 ft. lbs. (15 Nm).

Timing Disturbed

1. Crank the engine until the number 1 cylinder is at Top Dead Center (TDC) on the compression stroke. This can be accomplished by the following:
   a. Remove the FI fuse from the fuse and relay box.
   b. Remove the number 1 spark plug
   c. Install a spark plug port adapter and a compression gauge into the spark plug cavity.
   d. Crank the engine and observe the compression gauge. The TDC of the cylinder's compression stroke occurs when the compression reading is the highest.
   e. Align the timing mark on the crankshaft pulley to the zero position on the timing indicator.
   f. Position the rotor to the number 1 cylinder position.
   g. Install the distributor.
   h. Remove the spark plug port adapter and the compression gauge.

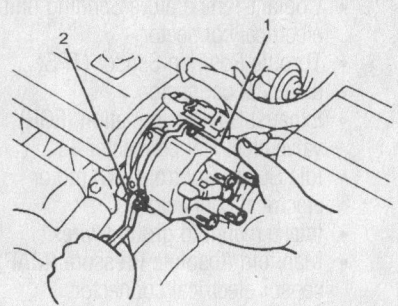

1. Distributor
2. Distributor mounting bolts

79222Z701

Be sure to tighten the distributor mounting bolts to prevent the timing from changing—Prizm

   i. Install the number 1 spark plug and tighten to 21 ft. lbs. (28 Nm).
   j. Install the FI fuse into the fuse and relay box. Set the ignition timing.
  • Negative battery cable
2. Check and/or adjust ignition timing.
3. Tighten the distributor hold-down bolts to 11 ft. lbs. (15 Nm).

Alternator

REMOVAL

Metro

1. Before servicing the vehicle, refer to the precautions in the beginning of this section.
2. Remove or disconnect the follow-ing:
  • Negative battery cable
  • Air cleaner assembly
  • Alternator electrical connectors
  • Accessory drive belt
  • Upper and lower mounting bolts
  • Alternator

Prizm

1. Before servicing the vehicle, refer to the precautions in the beginning of this section.
2. Remove or disconnect the following:
  • Negative battery cable
  • Accessory drive belt
  • Alternator electrical connectors
  • Alternator

INSTALLATION

Metro

1. Install or connect the following:
  • Alternator and the bolts, do not tighten the bolts at this time

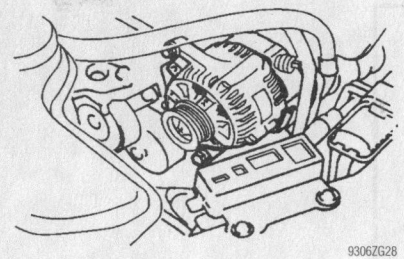

Exploded view of the alternator—Metro and Prizm models

2. Install the alternator belt and adjust the tension until a deflection of 0.24–0.31 inch (6–8mm) deflection with 22 lbs. (10 kg) of pressure exerted when installing a used belt. On a new belt the belt deflection should be 0.20–0.27 inch (5–7mm) with 22 lbs. (10 kg) of pressure exerted.
  • Alternator electrical connectors, then torque the alternator bolts to 17 ft. lbs. (23 Nm)
  • Air cleaner assembly
  • Negative battery cable

Prizm

1. Install the alternator and hand-tighten the bolts.
2. Install the alternator belt and adjust the tension until a deflection of 0.24–0.31 inch (6–8mm) deflection with 22 lbs. (10 kg) of pressure exerted when installing a used belt. On a new belt the belt deflection should be 0.20–0.27 inch (5–7mm) with 22 lbs. (10 kg) of pressure exerted.
3. Install or connect the following:
  • Upper alternator bolt to 18 ft. lbs. (25 Nm)
  • Lower alternator bolt to 40 ft. lbs. (54 Nm)
  • Alternator electrical connectors
  • Accessory drive belt
  • Negative battery cable

Ignition Timing

ADJUSTMENT

➡**The ignition timing on the Prizm is not adjustable.**

Metro

1. Before servicing the vehicle, refer to the precautions in the beginning of this section.
2. Run the engine until it reaches normal operating temperature. Stop the engine, but keep the ignition switch in the **ON** position for approximately 5 seconds. Start the engine again.

3. Run the engine at 2000 RPM for approximately 5 minutes. After 5 minutes, allow it to run at idle speed.
4. Be sure all accessories are turned **OFF**.
5. Fully engage the parking brake.
6. Be sure the shift lever is placed in the **NEUTRAL** or **PARK** depending on transaxle type.
7. Connect a tachometer to the negative terminal of the ignition coil. Connect a timing light to the No. 1 spark plug wire. Check the engine idle speed and adjust, if needed.
8. Refer to the vehicle emission control information label (located under the hood) for ignition timing specifications.
9. Remove the cap from the diagnostic check connector, located next to the Powertrain Control Module (PCM) behind the glove box. Insert a fused jumper wire between appropriate the terminals (illustrated.
10. On 4 terminal connectors, hold the connector with the locking tab at the top and jump the lower 2 terminals (**C** and **D**).
11. On 6 terminal connectors, hold the connector with the locking tab at the top and jump the 2 terminals at the lower left (**D** and **E**).
12. Loosen the distributor hold-down bolt and rotate the distributor until the correct timing marks are aligned.
13. Tighten the distributor hold-down bolt to 11 ft. lbs. (15 Nm) and recheck the timing. Be sure the timing advances according to engine speed.
14. Remove the diagnostic check jumper. Remove the timing light and tachometer.

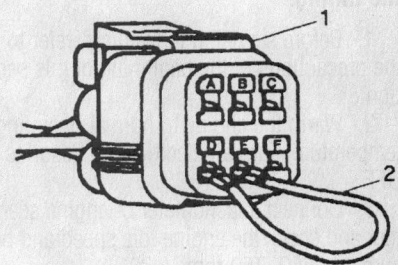

A. Blank (cavity 1)
B. Diagnostic request terminal (cavity 2)
C. Diagnostic output terminal (cavity 3)
D. Ground terminal (cavity 4)
E. Test switch terminal (cavity 5)
F. Duty check terminal (cavity 6)
1. Duty check DLC
2. Jumper

79222Z702

Jumping terminals in the 6 terminal DLC—Metro models

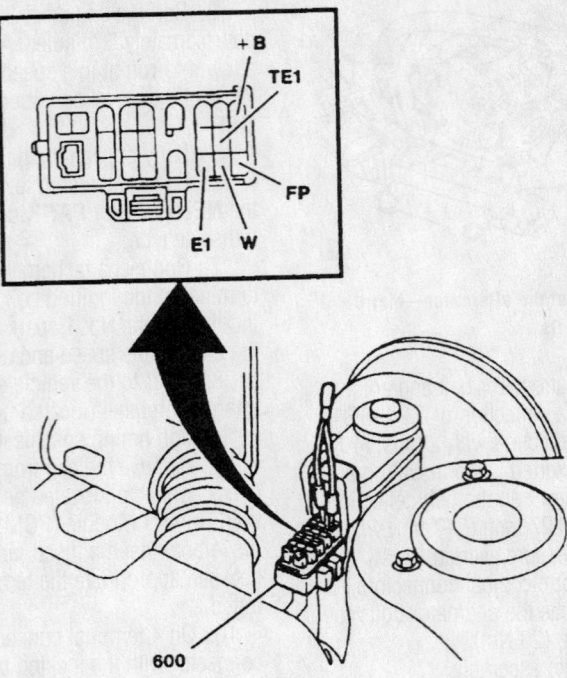

+B   SYSTEM VOLTAGE
E1   GROUND TERMINAL
FP   FUEL PUMP TERMINAL
TE1   DIAGNOSTIC REQUEST TERMINAL
W   MALFUNCTION INDICATOR LAMP (MIL) TERMINAL
600   DATA LINK CONNECTOR (DLC) (UNDER HOOD)

7922Z703

DLC pin locations—Prizm models

Prizm

➡ If the ignition timing is out of specification, a possible engine electrical failure may have occurred. The following procedure may be used to check the timing:

1. Before servicing the vehicle, refer to the precautions in the beginning of this section.

2. Warm the engine to normal operating temperature. Turn all electrical accessories OFF.

3. Connect a tachometer or engine scan tool and check the engine idle speed and be sure it is 650–750 rpm.

4. Connect a timing light. Remove the cap on the Data Link Connector (DLC). Using a fused jumper wire, connect terminals **E1** and **TE1**.

5. Start the engine and check timing. With the jumper wire connected, the timing should be at 10 degrees Before Top Dead Center (BTDC).

6. Remove the jumper wire from the DLC and reinstall the cap.

7. Shut the engine **OFF** and disconnect all test equipment.

Engine Assembly

REMOVAL & INSTALLATION

Metro

1.0L ENGINE

1. Before servicing the vehicle, refer to the precautions in the beginning of this section.

2. Properly relieve the fuel system pressure.

3. Drain the cooling system.

4. Drain the engine oil.

5. Drain the transmission fluid, if equipped with an automatic transmission.

6. Remove or disconnect the following:
- Hood after scribing the hood hinge to the hood
- Air cleaner
- Coil wire from the distributor
- Radiator and cooling fan assembly
- Evaporative emissions (EVAP) purge solenoid electrical connector
- Distributor electrical connector
- Engine Coolant Temperature (ECT) sensor electrical connector
- Coolant Temperature Sending unit electrical connector
- Throttle Position Sensor (TPS) electrical connector
- Exhaust Gas Recirculation (EGR) valve electrical connector
- Idle Speed Control (ISC) motor electrical connector
- Intake manifold ground wires
- Manifold Absolute Pressure (MAP) sensor electrical connector
- Heated Oxygen (HO2S) sensor electrical connector
- Throttle Body Injection (TBI) unit electrical connector
- Early Fuel Evaporation (EFE) heater electrical connector, if equipped
- Oil pressure switch electrical connector
- Alternator connectors
- Starter solenoid electrical connector
- A/C compressor clutch coil electrical connector, if equipped
- EVAP canister vacuum hose
- MAP sensor vacuum hose
- Brake booster hose
- Accelerator cable from the throttle body
- Clutch cable, if equipped
- Gear select cable for automatic transmissions
- Speedometer cable
- Heater inlet and outlet hoses
- Fuel feed and return hoses

7. Support the engine.
- Exhaust pipe from the manifold
- Gearshift control shaft and extension rod from the transmission
- Backup lamp switch electrical connector
- Crankshaft Position (CKP) sensor electrical connector
- Both halfshafts

➡ It is not necessary to disconnect the halfshaft from the knuckle.

- A/C compressor without disconnecting the hoses
- Power steering hoses from the pump, if equipped

8. Attach a lifting device to the engine.
- Left side transmission mount
- Right side and rear engine mount
- Engine/transmission assembly

To install:

9. Install or connect the following:
- Engine/transmission assembly
- Left transmission mount and torque the nut to 41 ft. lbs. (55 Nm)
- Right and rear engine mounts and torque the bolts to 41 ft. lbs. (55 Nm)

10. Remove the engine lifting device and properly support the engine assembly.
- Exhaust pipe to the manifold and torque the bolts to 33 ft. lbs. (45 Nm)
- Gearshift control shaft and extension
- Backup lamp switch
- Halfshafts
- Power steering hoses to the pump
- A/C compressor and torque the bolts to 37 ft. lbs. (50 Nm)
- Speedometer cable
- Gear select cable, if equipped
- Clutch cable, if equipped
- Accelerator cable
- Heater inlet/outlet hoses
- Engine vacuum lines
- Alternator and torque the bolts to 17 ft. lbs. (23 Nm)
- Fuel return and feed lines
- Distributor electrical connector
- Starter solenoid electrical connector
- CKP sensor electrical connector
- MAP sensor electrical connector
- EVAP purge solenoid electrical connector
- ECT sensor connectors
- Coolant temperature sending unit electrical connector
- TPS electrical connector
- EGR valve electrical connector
- ISC motor electrical connector
- Fuel injector connectors
- HO$_2$S sensor connector
- Intake manifold ground wires
- EFE heater electrical connector
- Oil pressure switch electrical connector
- Alternator electrical connector
- A/C Compressor clutch coil electrical connector, if equipped
- Starter solenoid electrical connector
- MAP sensor electrical connector
- EVAP canister vacuum hose
- Map sensor vacuum hose
- Brake booster vacuum hose
- Accelerator cable
- Clutch cable, if equipped
- Speedometer cable
- Radiator and cooling fan and torque the bolts to 89 inch lbs. (10 Nm)
- Air cleaner assembly
- Hood and torque the bolts to 20 ft. lbs. (27 Nm)
- Negative battery cable

11. Adjust the clutch pedal free-play, gear select cable and accelerator cable play, if equipped.

12. Fill the engine with clean oil.
13. Fill the cooling system.
14. Fill the transmission.
15. Fill the power steering reservoir.
16. Start the engine and check for leaks, repair if necessary.

1.3L ENGINE

1. Before servicing the vehicle, refer to the precautions in the beginning of this section.
2. Properly relieve the fuel system pressure.
3. Drain the cooling system.
4. Evacuate and recover the A/C system.
5. Drain the engine oil.
6. Drain the transmission fluid.
7. Remove or disconnect the following:
- Positive and negative battery cables
- Hood
- Radiator and cooling fan assembly
- Air cleaner assembly
- Engine Coolant Temperature (ECT) sensor electrical connector
- Throttle Position (TPS) sensor electrical connector
- Evaporative emissions (EVAP) purge solenoid electrical connector
- Camshaft Position (CMP) sensor electrical connector
- Crankshaft Position (CKP) sensor electrical connector
- Idle Speed Control (ISC) motor electrical connector
- Heated Oxygen (HO$_2$S) sensor electrical connector
- Fuel injector electrical connectors
- Manifold Absolute Pressure (MAP) sensor electrical connector
- Power steering pump pressure switch electrical connector, if equipped
- Oil pressure switch electrical connector
- A/C compressor clutch electrical connector, if equipped
- Park/Neutral Position (PNP) switch electrical connector
- Ignition coil connectors
- Alternator electrical connector
- Starter solenoid electrical connector
- Backup lamp switch electrical connector
- Vehicle Speed (VSS) sensor electrical connector
- Direct clutch and second brake solenoid electrical connector
- Brake booster hose
- Accelerator cable
- Clutch cable, if equipped

- Shift select cable, if equipped
- Speedometer cable
- Heater inlet and outlet hoses
- Fuel feed and return hoses

8. Install and engine support fixture.
- Right side engine mount
- Left transmission mount
- Rear engine mount
- Exhaust pipe from the manifold
- Gearshift control shaft and extension rod, if equipped
- Both halfshafts
- Power steering pump, if equipped
- A/C compressor, if equipped

9. Place an engine table under the vehicle.
10. Remove the engine support fixture.
11. Raise the vehicle from the engine/transmission assembly.

To install:

12. Place the engine/transmission assembly on the table, under the vehicle.
13. Lower the vehicle onto the engine/transmission.
14. Install or connect the following:
- Left transmission mount and torque the bolts to 41 ft. lbs. (55 Nm)
- Right side engine mount and torque the bolts to 41 ft. lbs. (55 Nm)
- Rear engine mount and torque the bolts to 41 ft. lbs. (55 Nm)
- Power steering pump and torque the bolts to 25 ft. lbs. (35 Nm)
- A/C compressor and torque the bolts to 18 ft. lbs. (25 Nm)
- Both halfshafts
- Gearshift control shaft and extension rod
- Exhaust pipe to the manifold and torque the bolts to 37 ft. lbs. (50 Nm)

15. Remove the engine support fixture.
- Fuel feed and return hoses
- Heater inlet and outlet hoses
- Accelerator cable
- Clutch cable, if equipped
- Shift select cable, if equipped
- Speedometer cable
- Brake booster hose
- ECT sensor electrical connector
- TPS sensor electrical connector
- EVAP purge solenoid electrical connector
- CMP sensor electrical connector
- CKP sensor electrical connector
- ISC motor electrical connector
- HO$_2$S sensor electrical connector
- Fuel injector electrical connectors
- MAP sensor electrical connector
- Power steering pump pressure switch electrical connector, if equipped

- Oil pressure switch electrical connector
- A/C compressor clutch electrical connector, if equipped
- PNP switch electrical connector
- Ignition coil connectors
- Alternator electrical connector
- Starter solenoid electrical connector
- Backup lamp switch electrical connector
- VSS sensor electrical connector
- Direct clutch and second brake solenoid electrical connector
- Radiator and cooling fan assembly and torque the bolts to 89 inch lbs. (10 Nm)
- Air cleaner assembly
- Hood
- Positive and negative battery cables
16. Fill the cooling system.
17. Fill the engine oil.
18. Evacuate and recharge the A/C system.
19. Fill the transmission fluid.
20. Start the engine and check for proper operation.

Prizm

1. Before servicing the vehicle, refer to the precautions in the beginning of this section.
2. Properly relieve the fuel system pressure.
3. Drain the cooling system.
4. Drain the engine oil.
5. Drain the transmission fluid.
6. Evacuate the A/C system, if equipped.
7. Remove or disconnect the following:
- Hood after scribing the hood hinge to the hood
- Ignition coil electrical connectors
- Accessory drive belt
- A/C compressor, if equipped
- Alternator
- Crankshaft pulley, if equipped with manual transmission
- Drive belt tensioner, if equipped with manual transmission
- Accelerator cable
- Intake Air Temperature (IAT) sensor electrical connector
- Air cleaner assembly
- Upper radiator hose
- Windshield washer reservoir
- Oxygen (O_2S) sensor
- Engine wiring harness and place aside
- Lower radiator hose from the thermostat housing

- Splash shields
- Both halfshafts
- Starter motor electrical connectors
- Starter motor
- Torque converter bolts, if equipped with an automatic transmission
- Power steering pump, do not disconnect the hoses
- Exhaust pipe from the manifold
- Transmission-to-engine bolts
- Manifold Absolute Pressure (MAP) sensor hose
- Brake booster hose
- MAP sensor
- Data Link Connector (DLC) and A/C pressure switch wiring harnesses
- Ground wires from the intake manifold and fenders
- Heater inlet and outlet hoses
- Fuel feed and return hoses
- Shift select cable
8. Attach an engine hoist to the engine.
9. Remove or disconnect the following:
- Front and rear transmission mounts
- Right engine mount
- Engine assembly

To install:
10. Install or connect the following:
- Engine assembly
- Engine-to-transmission bolts and torque them to 47 ft. lbs. (64 Nm)
- Right side engine mount and torque the bolts to 40 ft. lbs. (54 Nm)
- Front and rear transmission mounts and torque the front mount bolts to 47 ft. lbs. (64 Nm) and the through-bolt to 64 ft. lbs. (87 Nm). Torque the rear transmission nuts to 42 ft. lbs. (52 Nm) and the through-bolt to 64 ft. lbs. (87 Nm).
11. Remove the engine hoist and support the engine.
- Both halfshafts
- Front exhaust pipe and torque the bolts to 46 ft. lbs. (62 Nm)
- O_2S sensor and torque the nuts to 32 ft. lbs. (40 Nm)
- Power steering pump and torque the bolts to 27 ft. lbs. (37 Nm)
- Torque converter bolts and tighten them to 26 ft. lbs. (35 Nm)
- Starter motor and torque the bolts to 22 ft. lbs. (30 Nm)
- Engine splash shields and torque the bolts to 89 inch lbs. (10 Nm)
- Lower radiator hose
- Engine wiring harness
- Alternator and torque the bolts to 17 ft. lbs. (23 Nm)
- Windshield washer reservoir and

torque the bolts to 89 inch lbs. (10 Nm)
- Upper radiator hose
- Air cleaner assembly and torque the bolts to 89 inch lbs. (10 Nm)
- IAT sensor electrical connector
- Accelerator cable to the throttle lever
- Accessory drive belt
- A/C compressor and torque the bolts to 18 ft. lbs. (25 Nm)
- Alternator and torque the upper bolt to 18 ft. lbs. (25 Nm) and the lower bolt to 40 ft. lbs. (54 Nm)
- Drive belt tensioner (if equipped with manual transmission) and torque the bolt to 51 ft. lbs. (69 Nm) and the nut to 21 ft. lbs. (29 Nm)
- Crankshaft pulley, if equipped with manual transmission, and toque the bolt to 105 ft. lbs. (142 Nm)
- Ignition coil electrical connectors
- Shift cable
- Fuel return hose
- Fuel feed pipe to the fuel rail, using new gaskets and torque the bolt to 22 ft. lbs. (29 Nm)
- Heater inlet and outlet hoses
- Engine ground wires to the intake manifold and fender
- Radiator and cooling fan and torque the bolts to 9 ft. lbs. (13 Nm)
- Hood and torque the bolts to 20 ft. lbs. (27 Nm)
- Negative battery cable
12. Adjust the clutch pedal free-play, gear select cable and accelerator cable play, if equipped.
13. Recharge the A/C system.
14. Fill the engine with clean oil.
15. Fill the cooling system.
16. Fill the transmission.
17. Fill the power steering reservoir.
18. Start the engine and check for leaks, repair if necessary.

Water Pump

REMOVAL & INSTALLATION

Metro

1. Before servicing the vehicle, refer to the precautions in the beginning of this section.
2. Drain the cooling system.
3. Remove or disconnect the following:
- Negative battery cable
- Air cleaner assembly

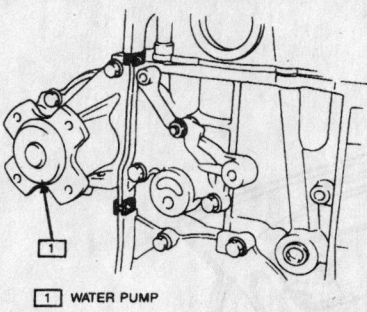

1 WATER PUMP

7922Z704

To ensure a tight seal, be sure gasket surfaces are properly prepared—Metro

- A/C compressor suction pipe bracket, if equipped
- Water pump pulley bolts, loosen but do not remove the bolts
- Right side lower splash shield
- A/C compressor drive belt, if equipped
- Lower alternator cover plate
- Water pump/alternator drive belt
- Crankshaft pulley
- Water pump pulley
- Timing belt
- Oil level indicator tube
- Upper alternator bracket from the water pump
- Water pump bolts and nuts
- Water pump

To install:

4. Install or connect the following:
- Water pump with a new gasket and torque the bolts to 115 inch lbs. (13 Nm)
- New rubber seals
- Upper alternator adjusting bracket and torque the bolt to 17 ft. lbs. (23 Nm)
- Oil level indicator tube
- Timing belt
- Crankshaft pulley and torque the bolts to 12 ft. lbs. (16 Nm)
- Water pump/alternator drive belt
- Lower alternator cover plate and torque the bolts to 89 inch lbs. (10 Nm)
- A/C compressor drive belt, if equipped
- Right side lower splash shield
- Water pump pulley and torque the bolts to 18 ft. lbs. (24 Nm)

5. Adjust the water pump drive belt tension and torque the alternator adjustment bolt to 17 ft. lbs. (23 Nm).
- A/C compressor suction pipe bracket, if equipped
- Air cleaner assembly
- Negative battery cable

6. Refill the cooling system.
7. Start the engine and check for leaks, repair if necessary.

Prizm

1. Before servicing the vehicle, refer to the precautions in the beginning of this section.
2. Drain the engine coolant.
3. Remove or disconnect the following:
- Negative battery cable
- Accessory drive belt
- Right side lower splash shield
- Water pump bolts
- Water pump

To install:

4. Install or connect the following:
- Water pump with a new O-ring and torque the bolts to 8 ft. lbs. (11 Nm)
- Right side lower splash shield
- Accessory drive belt
- Negative battery cable

5. Fill the cooling system.
6. Start the vehicle and check for leaks, repair if necessary.

Heater Core

REMOVAL & INSTALLATION

Metro

1. Disconnect the negative battery cable.
2. Properly drain the cooling system into a clean container for reuse.
3. Remove the instrument panel.
4. Remove or disconnect the following:
- Heater hoses from the heater core tubes at the firewall
- Temperature and mode control cables from the heater housing
- Air duct from the heater housing to the blower motor assembly (on models not equipped with A/C); otherwise, detach the heater housing from the evaporator assembly
- 2 bolts and nuts and remove the heater case assembly
- Heater case halves
- Heater core from the heater case assembly

To install:

5. Install or connect the following:
- Heater core into the heater case assembly
- Assemble the heater case assembly
- Heater assembly and install the 2 bolts and nuts
- Heater case assembly to the evaporator case assembly, if equipped with A/C

- Mode control and the temperature cables to the heater housing
- Heater hoses to the heater core

6. Install the instrument panel.
7. Refill the cooling system.
8. Connect the negative battery cable.
9. Operate the engine to normal operating temperatures; then, check for leaks and the heater operation.

Prizm

1. Disconnect the negative battery cable.
2. Drain the cooling system into a clean container for reuse.
3. Discharge and recover the air conditioning system refrigerant.
4. Remove the heater hoses from the heater core.
5. Remove the instrument panel by removing or disconnecting the following:
- Steering wheel-disarm air bag system if equipped
- Right and left front pillar garnish trim
- Floor console bin
- Engine hood release lever
- Lower finish No. 1 trim panel
- Steering column cover
- Center cluster finish panel
- Cluster finish panel
- Radio
- Stereo opening cover or center differential control switch if so equipped
- Combination meter assembly
- Lower finish No. 2 panel with glove compartment door
- Heater control assembly
- Lower center finish panel

➡ **The defroster nozzle has a boss on the reverse side for clamping onto the clip on the body side. When removing, pull upward at an angle.**

- No. 1 and No. 2 side defroster nozzles
- Safety pad assembly from the vehicle
- Instrument panel reinforcement
- Blower motor electrical connector and resistor electrical connector
- Air conditioning compressor control module electrical connector, if equipped
- Cruise control servo
- Refrigerant lines from the evaporator core; then, plug the openings to prevent contamination
- Ventilation ducts from the heater/air conditioning housing
- 6 heater/air conditioning housing

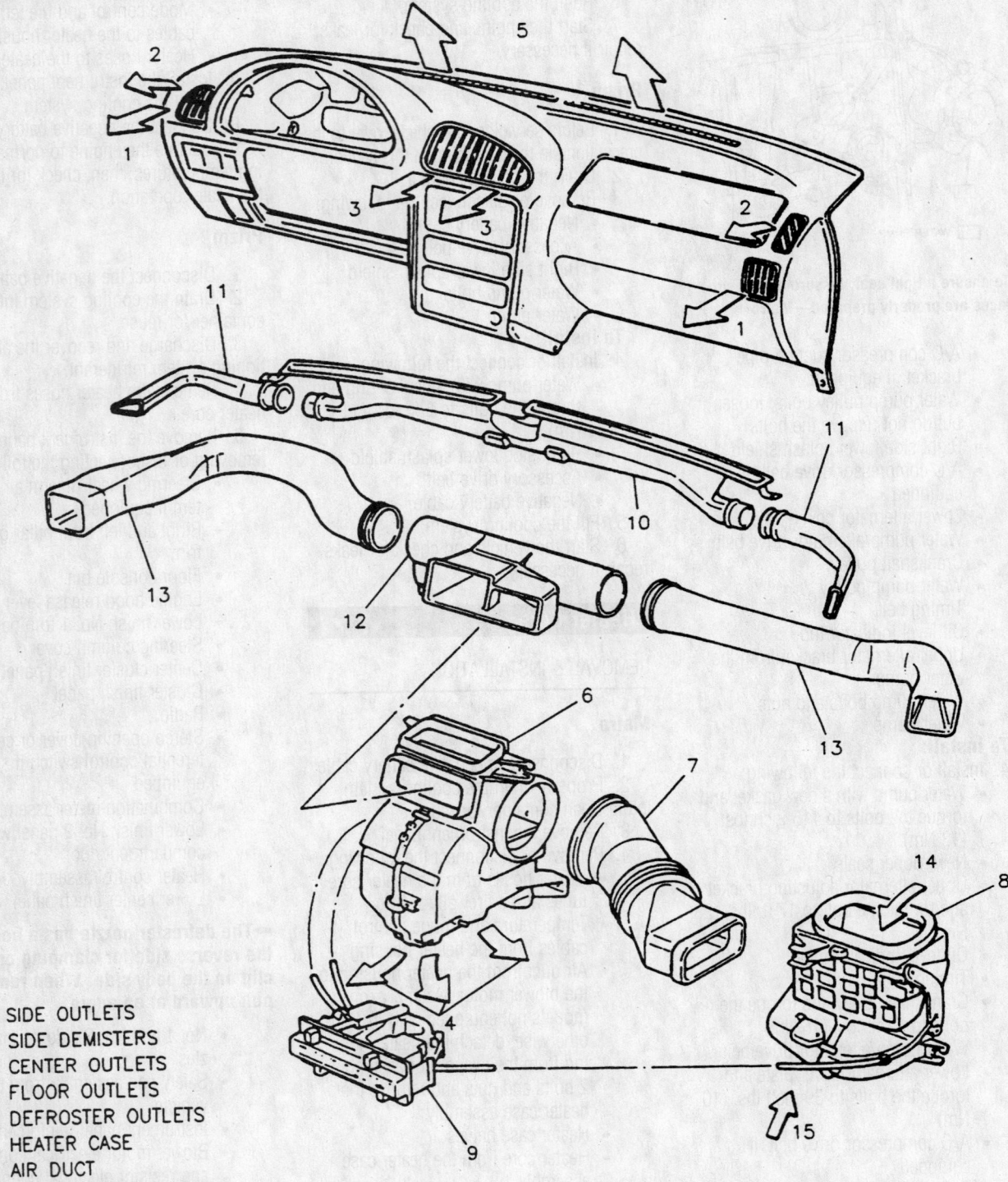

1 SIDE OUTLETS
2 SIDE DEMISTERS
3 CENTER OUTLETS
4 FLOOR OUTLETS
5 DEFROSTER OUTLETS
6 HEATER CASE
7 AIR DUCT
8 BLOWER MOTOR CASE
9 HEATER CONTROL UNIT
10 DEFROSTER DUCT
11 SIDE DEMISTER DUCTS
12 CENTER VENT DUCT
13 SIDE VENT DUCTS
14 OUTSIDE AIR
15 RECIRCULATED AIR

93111GC3

Exploded view of the instrument panel, heater housing, ventilation ducts and related components—Metro

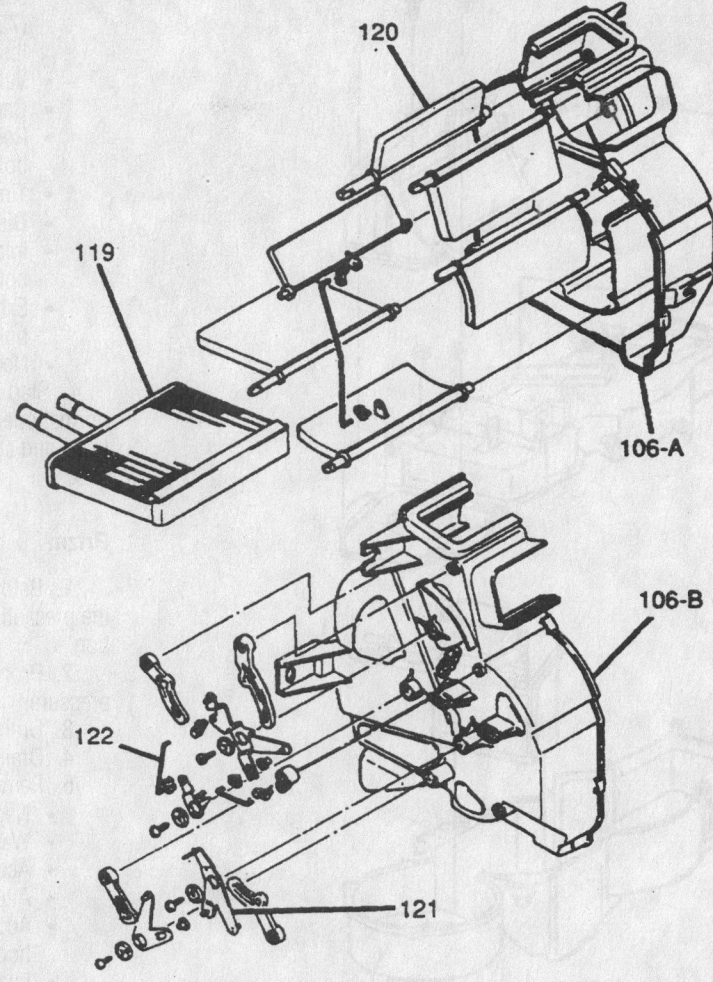

106-A HEATER CASE—RIGHT HALF
106-B HEATER CASE—LEFT HALF
119 HEATER CORE
120 CONTROL DOOR (DAMPER)
121 CONTROL LEVEL LINKAGE
122 CONTROL SHAFT

93111GC4

Exploded view of the heater core, heater case and related components—Metro

- assembly-to-chassis nuts and the assembly
- Heater core-to-heater/air conditioning housing clamp screws and the clamps
- Heater core

To install:
6. Install or connect the following:
- Heater core
- Heater core-to-heater/air conditioning housing clamps and the clamp screws
- Heater/air conditioning housing assembly and the 6 assembly-to-chassis nuts

- Ventilation ducts to the heater/air conditioning housing
- Refrigerant lines to the evaporator core
- Cruise control servo, if equipped
- Air conditioning compressor control module electrical connector
- Blower motor electrical connector and resistor electrical connector
- Instrument panel reinforcement
- Safety pad assembly in the vehicle
- No. 1 and No. 2 side defroster nozzles
- Lower center finish panel
- Heater control assembly

- Lower finish No. 2 panel with glove compartment door
- Combination meter assembly
- Stereo opening cover or center differential control switch if so equipped
- Radio
- Cluster finish panel
- Center cluster finish panel
- Steering column covers
- No. 1 lower finish panel
- Engine hood release lever
- Floor console box
- Right and left front pillar garnish trim

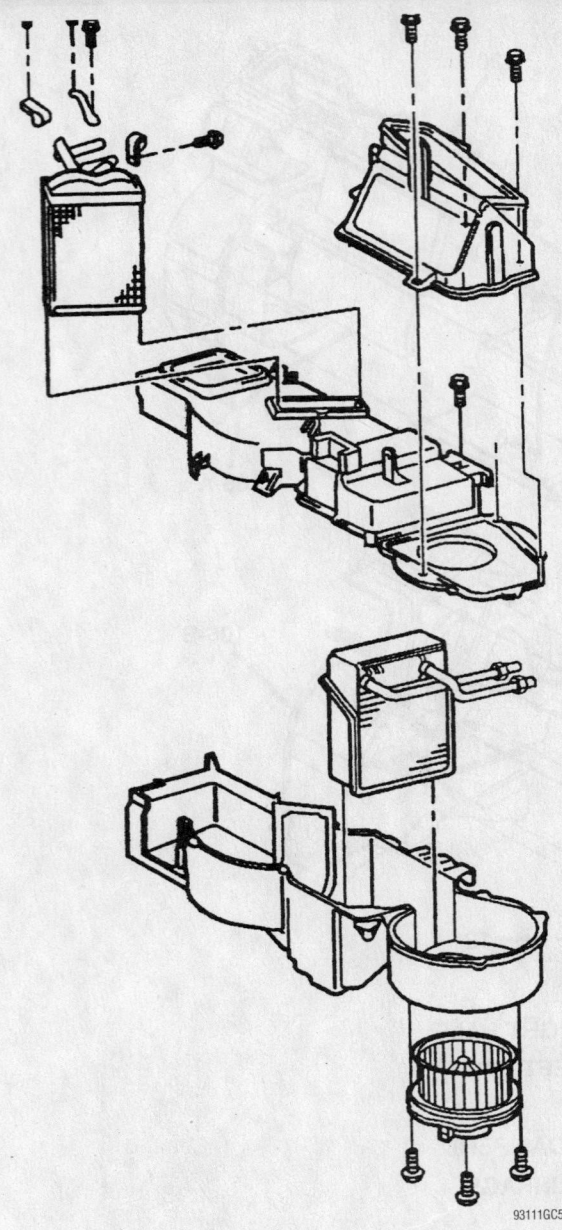

Exploded view of the heater core, the evaporator core, the heater/air conditioning housing and related components—Prizm

93111GC5

- Steering wheel
- Heater hoses to the heater core

7. Evacuate, charge and leak test the air conditioning system refrigerant.

8. Refill the cooling system.

9. Connect the negative battery cable.

Cylinder Head

REMOVAL & INSTALLATION

Metro

1. Before servicing the vehicle, refer to the precautions in the beginning of this section.

2. Remove or disconnect the following:
- Negative battery cable
- Intake manifold
- Exhaust manifold
- Rocker arm cover
- Timing belt and tensioner
- Camshaft
- Valve lifters
- Distributor
- Cylinder head bolts
- Cylinder head

To install:

3. Install or connect the following:
- Cylinder head with a new gasket and torque the bolts in 3 even stages, in sequence, to 54 ft. lbs. (73 Nm) on 1.0L engines or 49 ft. lbs. (68 Nm) on 1.3L engines
- Valve lifters
- Camshaft
- Rocker arm cover and torque the bolts to 44 inch lbs. (5 Nm)
- Timing belt and tensioner
- Distributor
- Intake manifold and torque the bolts to 17 ft. lbs. (23 Nm)
- Exhaust manifold and torque the bolts to 17 ft. lbs. (23 Nm)
- Negative battery cable

4. Start the engine and allow it to reach normal operating temperature. Check for leaks and adjust the ignition timing, if necessary.

Prizm

1. Before servicing the vehicle, refer to the precautions in the beginning of this section.

2. Properly relieve the fuel system pressure.

3. Drain the engine coolant.

4. Drain the engine oil.

5. Remove or disconnect the following:
- Negative battery cable
- Washer tank and pump
- Accessory drive belt
- Alternator
- Accelerator cable from the throttle body
- Intake Air Temperature (IAT) sensor electrical connector
- Air cleaner hose
- Throttle Position (TPS) sensor electrical connector
- Idle Air Control (IAC) valve electrical connector
- Manifold Absolute Pressure (MAP) sensor electrical connector
- Coolant bypass hoses
- Fuel injector electrical connectors

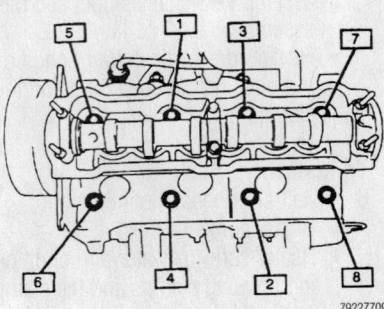

79222709

Cylinder head torque sequence—1.0L engine

CAPACITIES

Year	Model	Engine Displacement Liters (cc)	Engine ID/VIN	Engine Oil with Filter (qts.)	Transmission (pts.)		Transfer Case (pts.)	Drive Axle		Fuel Tank (gal.)	Cooling System (qts.)
					5-Spd	Auto.		Front (pts.)	Rear (pts.)		
2000	Tracker	1.6 (1590)	6	4.75	3.2	10.6	3.6	2.4	4.6	11.0	5.5
	Tracker	2.0 (1997)	C	5.90	3.2	10.6	3.6	2.4	4.6	11.0	6.5
2001	Tracker	2.0 (1997)	C	5.90	3.2	①	3.6	2.4	4.6	②	6.9
	Tracker	2.5 (2494)	1	6.0	5.5	①	3.6	2.1	4.6	②	8.5
2002	Tracker	2.0 (1997)	C	5.90	3.2	①	3.6	2.4	4.6	②	6.9
	Tracker	2.5 (2494)	1	6.0	5.5	①	3.6	2.1	4.6	②	8.5
2003	Tracker	2.0 (1997)	C	5.90	3.2	①	3.6	2.4	4.6	17.0	6.9
	Tracker	2.5 (2494)	1	6.0	5.5	①	3.6	2.1	4.6	17.0	8.5

Note: All capacities are approximate. Add fluid gradually and check to be sure a proper fluid level is obtained.

① 2WD: 6.0 pts
  4WD: 5.2 pts.

② 2dr: 14.8
  4dr: 17.4

42372-TRAK-C04

CRANKSHAFT AND CONNECTING ROD SPECIFICATIONS
All measurements are given in inches.

Year	Engine Displacement Liters (cc)	Engine ID/VIN	Crankshaft				Connecting Rod		
			Main Brg. Journal Dia.	Main Brg. Oil Clearance	Shaft End-play	Thrust on No.	Journal Diameter	Oil Clearance	Side Clearance
2000	1.6 (1590)	6	2.0465-2.0472	0.0006-0.0023	0.0044-0.0149	3	1.7316-1.7322	0.0008-0.0031	NA
	2.0 (1997)	C	2.2828-2.2834	0.0008-0.0023	0.0039-0.0165	3	1.9678-1.9685	0.0016-0.0031	NA
2001	2.5 (2494)	1	2.5583-2.5590	0.0008-0.0023	0.0044-0.0149	2	1.9678-1.9685	0.0016-0.0031	NA
	2.0 (1997)	C	2.2828-2.2834	0.0008-0.0023	0.0039-0.0165	3	1.9678-1.9685	0.0016-0.0031	NA
2002	2.5 (2494)	1	2.5583-2.5590	0.0008-0.0023	0.0044-0.0149	2	1.9678-1.9685	0.0016-0.0031	NA
	2.0 (1997)	C	2.2828-2.2834	0.0008-0.0023	0.0039-0.0165	3	1.9678-1.9685	0.0016-0.0031	NA
2003	2.5 (2494)	1	2.5583-2.5590	0.0008-0.0023	0.0044-0.0149	2	1.9678-1.9685	0.0016-0.0031	NA
	2.0 (1997)	C	2.2828-2.2834	0.0008-0.0023	0.0039-0.0165	3	1.9678-1.9685	0.0016-0.0031	NA

NA: Not Available

42372-TRAK5C05

VALVE SPECIFICATIONS

Year	Engine Displacement Liters (cc)	Engine ID/VIN	Seat Angle (deg.)	Face Angle (deg.)	Spring Test Pressure (lbs. @ in.)	Spring Installed Height (in.)	Stem-to-Guide Clearance (in.)		Stem Diameter (in.)	
							Intake	Exhaust	Intake	Exhaust
2000	1.6 (1590)	6	45	45	23.6-27.5@ 1.24	1.24	0.0008-0.0027	0.0018-0.0035	0.2152-0.2157	0.2142-0.2148
	2.0 (1997)	C	45	45	①	②	0.0008-0.0027	0.0018-0.0035	0.2348-0.2354	0.2339-0.2344
2001	2.5 (2494)	1	45	45	①	1.250	0.0008-0.0027	0.0018-0.0035	0.2348-0.2354	0.2339-0.2344
	2.0 (1997)	C	45	45	①	②	0.0008-0.0027	0.0018-0.0035	0.2348-0.2354	0.2339-0.2344
2002	2.5 (2494)	1	45	45	①	1.250	0.0008-0.0027	0.0018-0.0035	0.2348-0.2354	0.2339-0.2344
	2.0 (1997)	C	45	45	①	②	0.0008-0.0027	0.0018-0.0035	0.2348-0.2354	0.2339-0.2344
2003	2.5 (2494)	1	45	45	①	1.250	0.0008-0.0027	0.0018-0.0035	0.2348-0.2354	0.2339-0.2344
	2.0 (1997)	C	45	45	①	②	0.0008-0.0027	0.0018-0.0035	0.2348-0.2354	0.2339-0.2344

① Inner: 13.6-17.4@1.08
  Outer: 30.4-39.2@1.25

② Inner spring: 1.08
  Outer spring: 1.25

42372-TRAK-C06

PISTON AND RING SPECIFICATIONS
All measurements are given in inches.

Year	Engine Displacement Liters (cc)	Engine ID/VIN	Piston Clearance	Ring Gap			Ring Side Clearance		
				Top Compression	Bottom Compression	Oil Control	Top Compression	Bottom Compression	Oil Control
2000	1.6 (1590)	6	0.0008-0.0015	0.0079-0.0275	0.0138-0.0275	0.0039-0.0669	0.0012-0.0027	0.0008-0.0023	NA
	2.0 (1997)	C	0.0008-0.0015	0.0079-0.0276	0.0138-0.0276	0.0079-0.0709	0.0012-0.0027	0.0008-0.0023	NA
2001	2.5 (2494)	1	0.0008-0.0015	0.0079-0.0276	0.0138-0.0276	0.0079-0.0709	0.0012-0.0027	0.0008-0.0023	NA
	2.0 (1997)	C	0.0008-0.0015	0.0079-0.0276	0.0138-0.0276	0.0079-0.0709	0.0012-0.0027	0.0008-0.0023	NA
2002	2.5 (2494)	1	0.0008-0.0015	0.0079-0.0276	0.0138-0.0276	0.0079-0.0709	0.0012-0.0027	0.0008-0.0023	NA
	2.0 (1997)	C	0.0008-0.0015	0.0079-0.0276	0.0138-0.0276	0.0079-0.0709	0.0012-0.0027	0.0008-0.0023	NA
2003	2.5 (2494)	1	0.0008-0.0015	0.0079-0.0276	0.0138-0.0276	0.0079-0.0709	0.0012-0.0027	0.0008-0.0023	NA
	2.0 (1997)	C	0.0008-0.0015	0.0079-0.0276	0.0138-0.0276	0.0079-0.0709	0.0012-0.0027	0.0008-0.0023	NA

NA: Not Available

42372-TRAK-C07

TORQUE SPECIFICATIONS

All readings in ft. lbs.

Year	Engine Displacement Liters (cc)	Engine ID/VIN	Cylinder Head Bolts	Main Bearing Bolts	Rod Bearing Bolts	Crankshaft Damper Bolts	Flywheel Bolts	Manifold		Spark Plugs	Lug Nut
								Intake	Exhaust		
2000	1.6 (1590)	6	①	39 ②	25.5	94 ③	58	17	17	21	70
	2.0 (1997)	C	④	⑤	33	109	51	17	17	18	70
2001	2.5 (2494)	1	④	⑤	33	109	51	17	22	18	70
	2.0 (1997)	C	④	⑤	33	109	51	17	17	18	70
2002	2.5 (2494)	1	④	⑤	33	109	51	17	22	18	70
	2.0 (1997)	C	④	⑤	33	109	51	17	17	18	70
2003	2.5 (2494)	1	④	⑤	33	109	51	17	22	18	70
	2.0 (1997)	C	④	⑤	33	109	51	17	17	18	70

① Step 1: 26 ft. lbs.

Step 2: 41 ft. lbs.

Step 3: Loosen in reverse order to 0 ft. lbs.

Step 4: 26 ft. lbs.

Step 5: 52 ft. lbs.

② Use multiple passes to arrive at final torque.

③ Value shown is for crankshaft timing belt sprocket

④ Step 1: 38 ft. lbs.

Step 2: 61 ft. lbs.

Step 3: Loosen in reverse order to 0 ft. lbs.

Step 4: 38 ft. lbs.

Step 5: 76 ft. lbs.

Step 6: Tighten 6mm bolt to 8 ft. lbs.

⑤ 10mm: 43.5 ft. lbs.

8mm: 19.5 ft. lbs.

42372-TRAK-C08

WHEEL ALIGNMENT

Year	Model	Caster		Camber		Toe-in (in.)	Steering Axis Inclination (Deg.)
		Range (+/-Deg.)	Preferred Setting (Deg.)	Range (+/-Deg.)	Preferred Setting (Deg.)		
2000	Tracker	1.00	+2.67	1.00	0	0+/-0.16	—
2001	Tracker	1.00	+2.67	1.00	0	0+/-0.16	—
2002	Tracker	1.00	+2.67	1.00	0	0+/-0.16	—
2003	Tracker	1.00	+2.67	1.00	0	0+/-0.16	—

42372-TRAK-C09

TIRE, WHEEL AND BALL JOINT SPECIFICATIONS

Year	Model	OEM Tires Standard	OEM Tires Optional	Tire Pressures (psi) Front	Tire Pressures (psi) Rear	Wheel Size	Ball Joint Inspection
2000	Tracker 2wd	P195/75R15	None	23	23	5.5-JJ	①
	Tracker 4wd	P205/75R15	None	23	23	5.5-JJ	①
2001	Tracker 2wd	P195/75R15	None	23	23	5.5-JJ	①
	Tracker 4wd	P205/75R15	None	23	23	5.5-JJ	①
2002	Tracker 2wd	P195/75R15	None	23	23	5.5-JJ	①
	Tracker 4wd	P205/75R15	None	23	23	5.5-JJ	①
2003	Tracker 2wd	P195/75R15	None	23	23	5.5-JJ	①
	Tracker 4wd	P205/75R15	None	23	23	5.5-JJ	①

OEM: Original Equipment Manufacturer

PSI: Pounds Per Square Inch

STD: Standard

OPT: Optional

① Replace if any measurable movement is found.

42372-TRAK-C10

BRAKE SPECIFICATIONS
All measurements in inches unless noted

Year	Model	Brake Disc Original Thickness	Brake Disc Minimum Thickness	Brake Disc Maximum Runout	Brake Drum Diameter Original Inside Diameter	Brake Drum Diameter Max. Wear Limit	Brake Drum Diameter Maximum Machine Diameter	Minimum Lining Thickness Front	Minimum Lining Thickness Rear	Brake Caliper Bracket Bolts (ft. lbs.)	Brake Caliper Mounting Bolts (ft. lbs.)
2000	Tracker ①	0.670	0.590	0.006	8.66	8.74	8.74	0.08	0.04	61.5	19-21
	Tracker ②	0.670	0.590	0.006	8.66	8.74	8.74	0.08	0.04	61.5	19-21
2001	Tracker ①	0.670	0.590	0.006	8.66	8.74	8.74	0.08	0.04	61.5	19-21
	Tracker ②	0.670	0.590	0.006	8.66	8.74	8.74	0.08	0.04	61.5	19-21
2002	Tracker ①	0.670	0.590	0.006	8.66	8.74	8.74	0.08	0.04	61.5	19-21
	Tracker ②	0.670	0.590	0.006	8.66	8.74	8.74	0.08	0.04	61.5	19-21
2003	Tracker ①	0.670	0.590	0.006	8.66	8.74	8.74	0.08	0.04	61.5	19-21
	Tracker ②	0.670	0.590	0.006	8.66	8.74	8.74	0.08	0.04	61.5	19-21

① 2-door model

② 4-door model

42372-TRAK-C11

SCHEDULED MAINTENANCE INTERVALS
CHEVROLET—TRACKER

TO BE SERVICED	TYPE OF SERVICE	VEHICLE MILEAGE INTERVAL (x1000)												
		7.5	15	22.5	30	37.5	45	52.5	60	67.5	75	82.5	90	97.5
Engine oil & filter	R	✓	✓	✓	✓	✓	✓	✓	✓	✓	✓	✓	✓	
Automatic transmission fluid ①	S/I	✓	✓	✓	✓	✓	✓	✓	✓	✓	✓	✓	✓	✓
Manual transmission oil ②	S/I	✓	✓	✓	✓	✓	✓	✓	✓	✓	✓	✓	✓	✓
Steering system	S/I	✓	✓	✓	✓	✓	✓	✓	✓	✓	✓	✓	✓	✓
Transfer & differential oil ②	S/I	✓	✓	✓	✓	✓	✓	✓	✓	✓	✓	✓	✓	✓
Wheel discs & free wheeling hubs	S/I	✓	✓	✓	✓	✓	✓	✓	✓	✓	✓	✓	✓	✓
Suspension system	S/I	✓	✓	✓	✓	✓	✓	✓	✓	✓	✓	✓	✓	✓
Brake discs & pads (front)	S/I		✓		✓		✓		✓		✓		✓	
Brake drums & shoes (rear)	S/I		✓		✓		✓		✓		✓		✓	
Brake fluid ③	S/I		✓		✓		✓		✓		✓		✓	
Brake hoses & pipes	S/I		✓		✓		✓		✓		✓		✓	
Brake pedal	S/I		✓		✓		✓		✓		✓		✓	
Brake lever & cable	S/I		✓		✓		✓		✓		✓		✓	
Clutch	S/I		✓		✓		✓		✓		✓		✓	
Idle speed	S/I		✓		✓		✓		✓		✓		✓	
Propeller shafts	S/I		✓		✓		✓		✓		✓		✓	
Valve lash (clearance)	S/I		✓		✓		✓		✓		✓		✓	
Wheel bearings	S/I		✓		✓		✓		✓		✓		✓	
Air cleaner filter element	R				✓				✓				✓	
Engine coolant	R				✓				✓				✓	
Fuel filter	R				✓				✓				✓	
Spark plugs	R				✓				✓				✓	
Cooling system hoses	S/I				✓				✓				✓	
Drive belt(s)	S/I				✓				✓				✓	
Exhaust pipes & mountings	S/I				✓				✓				✓	
Fuel lines & connections	S/I				✓				✓				✓	
Camshaft timing belt	R								✓				✓	
Distributor cap & rotor	S/I								✓					
Emission-related hoses	S/I								✓					
Oxygen sensor	S/I											✓		
EVAP canister	R	every 100,000 miles												
PCV valve	R							✓						

42372-TRAK-C12

SCHEDULED MAINTENANCE INTERVALS
CHEVROLET—TRACKER

TO BE SERVICED	TYPE OF SERVICE	VEHICLE MILEAGE INTERVAL (x1000)												
		7.5	15	22.5	30	37.5	45	52.5	60	67.5	75	82.5	90	97.5
EGR system	S/I							✓						
Fuel Injectors	S/I	every 100,000 miles												
TWC converter	S/I	every 100,000 miles												

R: Replace   S/I: Service or Inspect

① Replace at 100,000 miles.

② Replace oil every 30,000 miles.

③ Replace every 60,000 miles.

FREQUENT OPERATION MAINTENANCE (SEVERE SERVICE)

If a vehicle is operated under any of the following conditions it is considered severe service:

- Extremely dusty areas.

- 50% or more of the vehicle operation is in 32°C (90°F) or higher temperatures, or constant operation in temperatures below 0°C (32°F).

- Prolonged idling (vehicle operation in stop and go traffic).

- Frequent short running periods (engine does not warm to normal operating temperatures).

- Police, taxi, delivery usage or trailer towing usage.

Oil & oil filter: replace every 3000 miles.

Air cleaner filter element: service or inspect every 3000 miles & replace every 15,000 miles.

Steering wheel free play, gear box oil & linkage: service or inspect every 3000 miles.

Brake & nuts on chassis: tighten every 6000 miles.

Brake discs & pads (front): service or inspect every 6000 miles.

Brake drums & shoes (rear): service or inspect every 6000 miles.

Exhaust pipes & mountings: tighten every 6000 miles.

Propeller shafts: service or inspect every 6000 miles.

Automatic transmission fluid & filter: replace every 15,000 miles.

Distributor cap & ignition wires: service or inspect every 15,000 miles.

Drive belt(s): service or inspect every 15,000 miles.

Manual transmission oil: replace every 15,000 miles.

Transfer & differential oil: replace every 15,000 miles.

42372-TRAK-C13

PRECAUTIONS

Before servicing any vehicle, please be sure to read all of the following precautions, which deal with personal safety, prevention of component damage and important points to take into consideration when servicing a motor vehicle:

• Never open, service or drain the radiator or cooling system when the engine is hot; serious burns can occur from the steam and hot coolant.

• Observe all applicable safety precautions when working around fuel. Whenever servicing the fuel system, always work in a well-ventilated area. Do not allow fuel spray or vapors to come in contact with a spark, open flame, or excessive heat (a hot drop light, for example). Keep a dry chemical fire extinguisher near the work area. Always keep fuel in a container specifically designed for fuel storage; also, always properly seal fuel containers to avoid the possibility of fire or explosion. Refer to the additional fuel system precautions later in this section.

• Fuel injection systems often remain pressurized, even after the engine has been turned **OFF**. The fuel system pressure must be relieved before disconnecting any fuel lines. Failure to do so may result in fire and/or personal injury.

• Brake fluid often contains polyglycol ethers and polyglycols. Avoid contact with the eyes and wash your hands thoroughly after handling brake fluid. If you do get brake fluid in your eyes, flush your eyes with clean, running water for 15 minutes. If

eye irritation persists, or if you have taken brake fluid internally, IMMEDIATELY seek medical assistance.

• The EPA warns that prolonged contact with used engine oil may cause a number of skin disorders, including cancer. You should make every effort to minimize your exposure to used engine oil. Protective gloves should be worn when changing oil. Wash your hands and any other exposed skin areas as soon as possible after exposure to used engine oil. Soap and water, or waterless hand cleaner should be used.

• All new vehicles are now equipped with an air bag system, often referred to as a Supplemental Restraint System (SRS) or Supplemental Inflatable Restraint (SIR) system. The system must be disabled before performing service on or around system components, steering column, instrument panel components, wiring and sensors. Failure to follow safety and disabling procedures could result in accidental air bag deployment, possible personal injury, and unnecessary system repairs.

• Always wear safety goggles when working with, or around, the air bag system. When carrying a non-deployed air bag, be sure the bag and trim cover are pointed away from your body. When placing a non-deployed air bag on a work surface, always face the bag and trim cover upward, away from the surface. This will reduce the motion of the module if it is accidentally deployed. Refer to the additional air bag system precautions later in this section.

• Clean, high quality brake fluid from a sealed container is essential to the safe and proper operation of the brake system. You should always buy the correct type of brake fluid for your vehicle. If the brake fluid becomes contaminated, completely flush the system with new fluid. Never reuse any brake fluid. Any brake fluid that is removed from the system should be discarded. Also, do not allow any brake fluid to come in contact with a painted surface; it will damage the paint.

• Never operate the engine without the proper amount and type of engine oil; doing so WILL result in severe engine damage.

• Timing belt maintenance is extremely important. Many models utilize an interference-type, non-freewheeling engine. If the timing belt breaks, the valves in the cylinder head may strike the pistons, causing potentially serious (also time-consuming and expensive) engine damage. Refer to the maintenance interval charts in the front of this section for the recommended replacement interval for the timing belt, and to the timing belt procedure in this section for belt replacement and inspection.

• Disconnecting the negative battery cable on some vehicles may interfere with the functions of the on-board computer system(s) and may require the computer to undergo a relearning process once the negative battery cable is reconnected.

• When servicing drum brakes, only disassemble and assemble one side at a time, leaving the remaining side intact for reference.

ENGINE REPAIR

➡**Disconnecting the negative battery cable on some vehicles may interfere with the functions of the on board computer system. The computer may undergo a relearning process once the negative battery cable is reconnected.**

Distributor

REMOVAL

All engines are equipped with a Distributorless Ignition System (DIS).

Alternator

REMOVAL

1. Before servicing the vehicle, refer to the precautions in the beginning of this section.

2. Remove or disconnect the following:

• Negative battery cable
• Evaporative Emission (EVAP) canister
• Accessory drive belt
• Alternator harness connectors
• Alternator mounting bracket
• Alternator

INSTALLATION

Install or connect the following:
• Alternator
• Alternator mounting bracket. Tighten the bolts to 20 ft. lbs. (27 Nm).
• Alternator harness connectors
• Accessory drive belt. Tighten the alternator bolts to 24 ft. lbs. (33 Nm).

• EVAP canister
• Negative battery cable

Ignition Timing

ADJUSTMENT

1.6L Engine

This engine is equipped with a Distributorless Ignition System (DIS). All timing functions are controlled by the Powertrain Control Module (PCM). No adjustment is possible.

2.0L and 2.5L Engines

➡**The 2.0L and 2.5L engines use a Camshaft Position (CMP) sensor that is rotated to set base timing.**

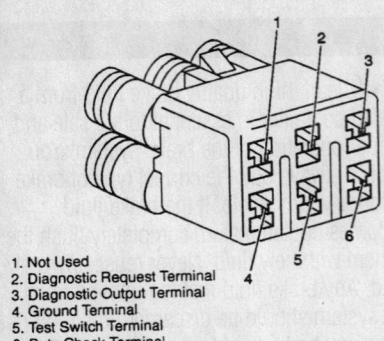

1. Not Used
2. Diagnostic Request Terminal
3. Diagnostic Output Terminal
4. Ground Terminal
5. Test Switch Terminal
6. Duty Check Terminal

7924HG01

Duty Check Data Link Connector terminal identification for ignition timing

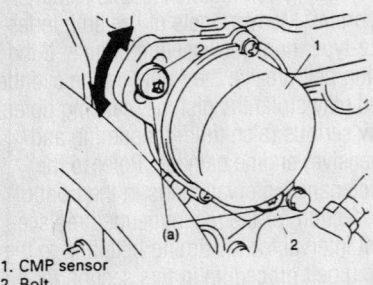

1. CMP sensor
2. Bolt

7924HG82

Camshaft position sensor

➡ **Check and adjust the ignition timing with the engine at normal operating temperature, all electrical accessories OFF and transmission in P, N for automatic transmission or neutral for manual transmission.**

1. Before servicing the vehicle, refer to the precautions in the beginning of this section.

2. With the engine **OFF**, connect a jumper wire between terminals **4** and **5** of the Data Link Connector (DLC) for Tracker or between terminals **D** and **E** of the DLC for all others.

3. Connect a timing light to the No. 1 spark plug wire and start the engine.

4. Ignition timing at idle should be 4–6 degrees Before Top Dead Center (BTDC).

5. Adjust the timing as necessary, then turn the engine **OFF**.

6. Remove the jumper wire from the DLC and remove the timing light.

Engine Assembly

REMOVAL & INSTALLATION

1.6L Engine

1. Before servicing the vehicle, refer to the precautions in the beginning of this section.

2. Relieve the fuel system pressure.
3. Drain the cooling system and engine oil.
4. Remove or disconnect the following:
- Negative battery cable
- Hood
- Strut tower bar, if equipped
- Cooling fan and shroud
- Heater hoses
- Radiator hoses
- Bypass hose
- Radiator
- Air intake tube
- Accelerator cable
- Transmission cable, if equipped
- Positive Crankcase Ventilation (PCV) valve and hose
- Exhaust Gas Recirculation (EGR) valve and temperature sensor
- EGR vacuum valve connector
- EGR bypass valve connector
- Idle Air Control (IAC) valve hoses and connector
- Fuel lines
- Main engine control wiring harness connectors at the firewall
- Throttle Position (TP) sensor connector
- Heated Oxygen (HO2S) sensor connectors
- Engine Coolant Temperature (ECT) sensor connector
- Temperature gauge sender connector
- A/C temperature switch, if equipped
- Injector harness connectors
- Alternator wiring connectors
- Manifold Absolute Pressure (MAP) sensor connector and vacuum line
- Brake booster vacuum line
- Evaporative Emission (EVAP) canister and hoses
- Distributor, if equipped
- Front skidplate, if equipped
- Power steering hoses
- A/C compressor, if equipped
- Flywheel access cover
- Torque converter, if equipped
- Clutch cable, if equipped
- Transmission cooler lines, if equipped
- Exhaust front pipe
- Starter motor
- Transmission flange fasteners and support the transmission
- Left and right engine mounts
- Engine

5. Install or connect the following:
- Engine. Tighten the mount bolts to 40 ft. lbs. (54 Nm).
- Transmission flange fasteners. Tighten them to 62 ft. lbs. (85 Nm).

- Starter motor
- Exhaust front pipe
- Transmission cooler lines, if equipped
- Clutch cable, if equipped
- Torque converter, if equipped. Tighten the bolts to 40 ft. lbs. (54 Nm).
- Flywheel access cover
- A/C compressor, if equipped
- Power steering hoses
- Front skidplate, if equipped
- Distributor, if equipped
- EVAP canister and hoses
- Brake booster vacuum line
- MAP sensor connector and vacuum line
- Alternator wiring connectors
- Injector harness connectors
- A/C temperature switch, if equipped
- Temperature gauge sender connector
- ECT sensor connector
- HO2S sensor connectors
- TP sensor connector
- Main engine control wiring harness connectors at the firewall
- Fuel lines
- IAC valve hoses and connector
- EGR bypass valve connector
- EGR vacuum valve connector
- EGR valve and temperature sensor
- PCV valve and hose
- Transmission cable, if equipped
- Accelerator cable
- Air intake tube
- Radiator
- Bypass hose
- Radiator hoses
- Heater hoses
- Cooling fan and shroud
- Strut tower bar, if equipped
- Hood
- Negative battery cable

6. Fill the crankcase to the correct level.
7. Fill the cooling system.
8. Start the engine and check for leaks.

2.0L and 2.5L Engines

1. Before servicing the vehicle, refer to the precautions in the beginning of this section.

2. Relieve the fuel system pressure.
3. Drain the cooling system.
4. Drain the engine oil.
5. Remove or disconnect the following:
- Negative battery cable
- Hood
- Heater hoses
- Radiator hoses

- Cooling fan and shroud
- Radiator overflow tank
- Radiator
- Accelerator cable
- Transmission cable, if equipped
- Strut tower bar, if equipped
- Air intake assembly
- Engine oil dipstick tube
- Transmission oil dipstick tube, if equipped
- Ignition coil covers
- Ignition coil connectors
- Injector connectors
- Camshaft Position (CMP) sensor connector
- Crankshaft Position (CKP) sensor connector
- Throttle Position (TP) sensor connector
- Mass Air Flow (MAF) sensor connector
- Idle Air Control (IAC) valve
- Intake manifold ground cable
- Evaporative Emission (EVAP) canister purge valve
- Exhaust Gas Recirculation (EGR) valve connector
- Heated Oxygen (HO2S) sensor connectors
- Engine Coolant Temperature (ECT) sensor connector
- Alternator wiring connectors
- Oil pressure gauge sender connector
- Power Steering Pressure (PSP) switch connector
- Alternator bracket ground cable
- Brake booster vacuum line
- Tank pressure control vacuum valve hose
- Fuel lines
- EVAP canister
- Power steering pump
- A/C compressor
- Steering shaft lower assembly
- Front differential housing, if equipped
- Exhaust front pipe and bracket
- Transmission oil cooler lines, if equipped
- Transmission stiffener brackets, if equipped
- Flywheel access cover
- Torque converter, if equipped
- Starter motor
- Transmission flange fasteners and support the transmission
- Left and right engine mounts
- Engine

To install:
6. Install or connect the following:
- Engine

- Left and right engine mounts. Tighten the nuts to 36 ft. lbs. (50 Nm).
- Transmission flange fasteners. Tighten them to 58 ft. lbs. (80 Nm).
- Starter motor
- Torque converter. Tighten the bolts to 47 ft. lbs. (65 Nm).
- Flywheel access cover
- Transmission stiffener brackets. Tighten the bolts to 36 ft. lbs. (50 Nm).
- Transmission oil cooler lines, if equipped
- Exhaust front pipe and bracket
- Front differential housing, if equipped
- Steering shaft lower assembly
- A/C compressor
- Power steering pump
- EVAP canister
- Fuel lines
- Tank pressure control vacuum valve hose
- Brake booster vacuum line
- Alternator bracket ground cable
- PSP switch connector
- Oil pressure gauge sender connector
- Alternator wiring connectors
- ECT sensor connector
- HO2S sensor connectors
- EGR valve connector
- EVAP canister purge valve
- Intake manifold ground cable
- IAC valve
- MAF sensor connector
- TP sensor connector
- CKP sensor connector
- CMP sensor connector
- Injector connectors
- Ignition coil connectors
- Ignition coil covers
- Transmission oil dipstick tube, if equipped
- Engine oil dipstick tube
- Air intake assembly
- Strut tower bar, if equipped
- Transmission cable, if equipped
- Accelerator cable
- Radiator
- Radiator overflow tank
- Cooling fan and shroud
- Radiator hoses
- Heater hoses
- Hood
- Negative battery cable
7. Fill the crankcase to the correct level.
8. Fill the cooling system.
9. Start the engine and check for leaks.

Water Pump

REMOVAL & INSTALLATION

1.6L Engines

1. Before servicing the vehicle, refer to the precautions in the beginning of this section.
2. Drain the cooling system.
3. Remove or disconnect the following:
- Negative battery cable
- Accessory drive belts
- Cooling fan and shroud
- Front cover
- Timing belt. Refer to the Timing Belt unit repair section.
- Oil dipstick tube
- Alternator bracket
- Timing belt tensioner
- Water pump

To install:
4. Install or connect the following:
- Water pump with a new gasket. Tighten the bolts to 106 inch lbs. (12 Nm).
- Timing belt tensioner
- Alternator bracket
- Oil dipstick tube. Tighten the bolt to 97 inch lbs. (11 Nm).
- Timing belt. Refer to the Timing Belt unit repair section.
- Front cover
- Cooling fan and shroud
- Accessory drive belts
- Negative battery cable
5. Fill the cooling system.
6. Start the engine and check for leaks.

2.0L Engines

1. Before servicing the vehicle, refer to the precautions in the beginning of this section.
2. Drain the cooling system.
3. Remove or disconnect the following:

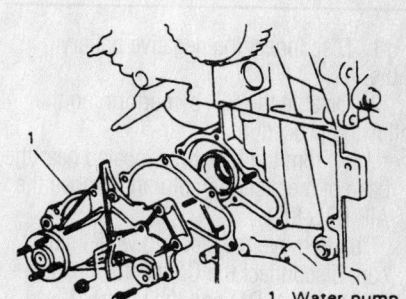

1. Water pump

7924HG04

Exploded view of the water pump mounting—1.6L engines

- Negative battery cable
- Radiator hose at the thermostat housing
- Heater outlet pipe bolt
- Alternator belt
- Water pump

To install:

→**Use new water pump bolts for assembly.**

4. Install or connect the following:
- Water pump with a new O-ring seal. Tighten the bolts to 19 ft. lbs. (27 Nm).
- Alternator belt
- Heater outlet pipe bolt
- Radiator hose at the thermostat housing
- Negative battery cable

5. Fill the cooling system.
6. Start the engine and check for leaks.

2.5L Engine

1. Before servicing the vehicle, refer to the precautions in the beginning of this section.
2. Drain the cooling system.
3. Remove or disconnect the following:

- Negative battery cable
- Accessory drive belts
- Front cover
- Water pump

To install:

4. Install or connect the following:
- Water pump with a new O-ring seal. Tighten the bolts to 19 ft. lbs. (27 Nm).
- Front cover
- Accessory drive belts
- Negative battery cable

5. Fill the cooling system.
6. Start the engine and check for leaks.

Heater Core

REMOVAL & INSTALLATION

1. Disconnect the negative battery cable.
2. Disable the SIR by performing the following procedure:
    a. From the fuse box, located near the base of the steering column, remove the AIR BAG fuse.
    b. Remove the steering wheel side cap, disconnect the Connector Positive Assurance (CPA) and the yellow 2-way driver's inflator module connectors.
    c. Pull the instrument panel compartment out by pushing the right-side and left-side stoppers (located on both sides) inward.
    d. Disconnect the CPA and the yellow 4-way passenger's inflator module connectors.

→**With the AIR BAG fuse removed and the ignition switch turned ON, the AIR BAG warning light will be ON; this is normal operation and does not indicate a SIR system malfunction.**

3. Drain the cooling system into a clean container for reuse.
4. Remove the instrument panel as follows:
    a. Remove the center console.
    b. Remove the lower steering column cover by loosening the mounting screws.
    c. Remove the glove box.
    d. Detach the wiring harness connectors from the heater unit and the blower motor assembly.
    e. Detach the wiring harness connectors from the ignition switch, contact coil and combination switch.
    f. Open the hood.
    g. Remove the steering column shaft joint bolt, then separate the steering column shaft from the lower steering shaft.
    h. Loosen all of the steering column-to-firewall and instrument panel brace bolts.
    i. If equipped, remove the shift (key) interlock cable screw. Disconnect the cable from the ignition switch.
    j. Remove the steering column from the vehicle.

✳✳ WARNING

Do not rest the steering column assembly on the steering wheel with the air bag module facing downward and the column vertical—personal injury may be the result.

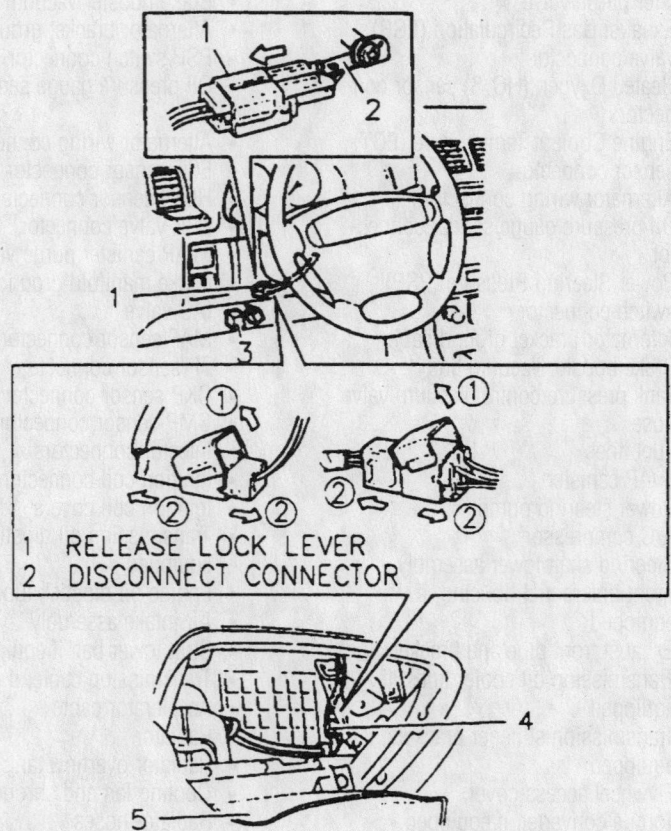

```
1  RELEASE LOCK LEVER
2  DISCONNECT CONNECTOR
```

```
1  YELLOW 2-WAY SIR CONNECTOR (DRIVER)
2  CONNECTOR POSITION ASSURANCE (CPA)
3  AIR BAG FUSE
4  YELLOW 4-WAY SIR CONNECTOR (PASSENGER)
5  GLOVE BOX
```

93113G90

Disabling the air bag system

k. Disconnect the speedometer cable from the speedometer, then remove the instrument cluster.

l. Remove the hood latch handle.

m. Remove the radio, the heater control panel and the heater control cables from the instrument panel.

n. Disconnect and label all wiring harness connectors from the instrument panel.

o. Remove the instrument panel mounting screws and bolts. Remove the side cover plates and the instrument panel mounting fasteners from the side of the assembly. Then, remove the upper cover plates and loosen the remaining mounting fasteners

p. Have an assistant help you carefully lift the instrument panel up and out of the vehicle. When separating the instrument panel from the firewall, ensure that all of the cables, wires and hoses are disconnected form the instrument panel.

5. Remove the 2 bolts and the right-side instrument panel center support.

6. If equipped with air conditioning, remove the evaporator.

7. Remove the 2 screws securing the SIR harness clip on the Sensing and Diagnostic Module (SDM) bracket.

8. Disconnect the SDM electrical connector.

9. Remove the 4 screws and the SDM bracket from the vehicle.

10. Remove the speedometer cable and antenna cable (if equipped) from the heater case.

11. Remove the floor duct from the heater case.

12. If equipped with air conditioning, disconnect the electrical jumper harness for the air conditioning amplifier.

13. Remove the relay bracket screws and the relay bracket.

14. From the engine compartment, remove the 2 heater assembly-to-chassis nuts and the 2 bolts.

15. Remove the heater case from the vehicle.

16. Remove the dampers and linkages from the heater case.

17. Remove the heater core bracket screw and the bracket.

18. Remove the heater core from the heater case.

To install:

19. Install the heater core to the heater case.

20. Install the heater core bracket and the bracket screw.

21. Install the dampers and linkages to the heater case.

22. Install the heater case to the vehicle.

23. In the engine compartment, install the 2 heater assembly-to-chassis nuts and the 2 bolts. Torque the nuts/bolts to 89 inch lbs. (10 Nm).

24. Install the relay bracket and the relay bracket screws.

25. If equipped with air conditioning, connect the electrical jumper harness for the air conditioning amplifier.

26. Install the floor duct to the heater case.

27. Install the speedometer cable and antenna cable (if equipped) to the heater case.

28. Install the SDM bracket and the 4 screws to the vehicle. Torque the screws to 49 inch lbs. (5.5 Nm).

29. Connect the SDM electrical connector.

30. Install the 2 screws securing the SIR harness clip on the Sensing and Diagnostic Module (SDM) bracket. Torque the screws to 49 inch lbs. (5.5 Nm)

31. If equipped with air conditioning, install the evaporator.

32. Install the 2 bolts and the right-side instrument panel center support.

33. Install the instrument panel as follows:

a. Have an assistant help you position the instrument panel in the vehicle. When installing the instrument panel on the firewall, ensure that all of the cables, wires and hoses are routed properly.

b. Install and tighten the instrument panel mounting screws and bolts.

c. Reattach all wiring harness connectors to the instrument panel.

d. Install the radio, the heater control panel and the heater control cables. Be sure to adjust the heater control cables.

100	HEATER CONTROL UNIT
106	BLOWER MOTOR CASE-TO-HEATER CASE DUCT
107	HEATER CASE
114	TEMPERATURE CONTROL CABLE
115	MODE CONTROL CABLE
116	FRESH/RECIRC CONTROL CABLE
117	HEATER CORE
118	DAMPERS

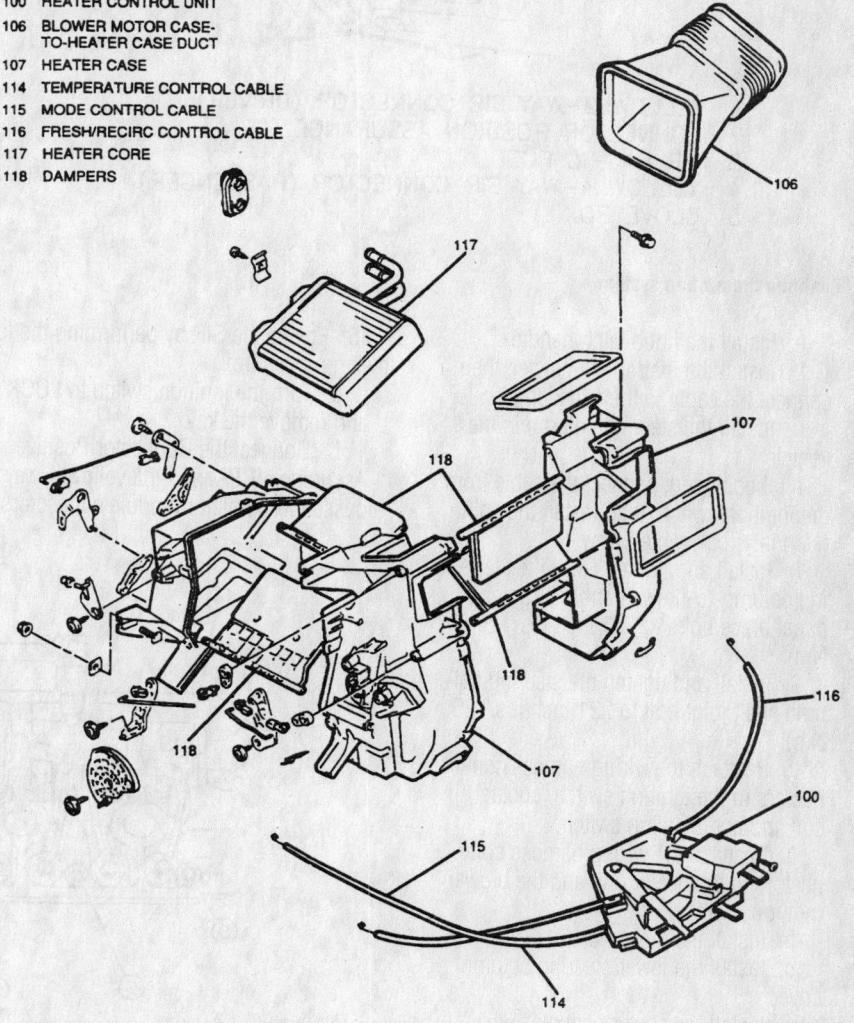

Exploded view of the heater case assembly and related components

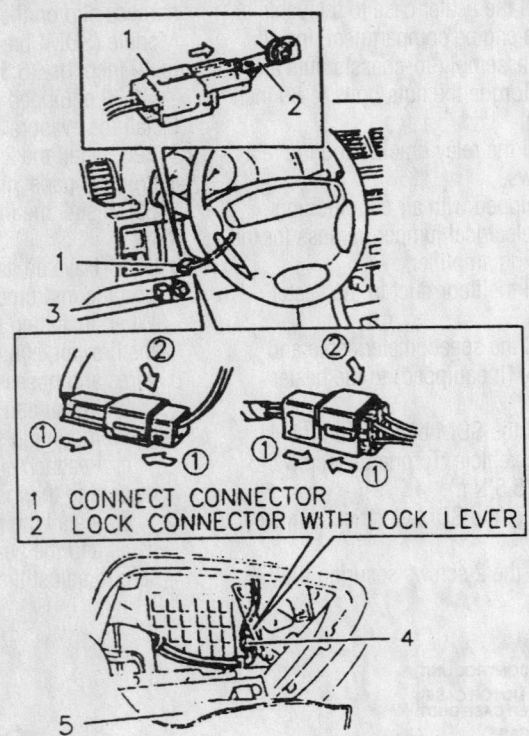

1 YELLOW 2-WAY SIR CONNECTOR (DRIVER)
2 CONNECTOR POSITION ASSURANCE (CPA)
3 AIR BAG-IG FUSE
4 YELLOW 4-WAY SIR CONNECTOR (PASSENGER)
5 GLOVE BOX

93113G91

Enabling the air bag system

e. Install the hood latch handle.

f. Install the instrument cluster, then connect the cable to the speedometer.

g. Install the steering column in the vehicle.

h. If equipped, connect the cable from the ignition switch, then install the shift (key) interlock cable screw.

i. Install and tighten all of the steering column-to-firewall and instrument panel brace bolts to 221 inch lbs. (25 Nm).

j. Install and tighten the steering column shaft joint bolt to 221 inch lbs. (25 Nm).

k. Reattach the wiring harness connectors to the ignition switch, contact coil and combination switch.

l. Reattach the wiring harness connectors to the heater unit and the blower motor assembly.

m. Install the glove box.

n. Install the lower steering column cover.

o. Install the center console.

34. Refill the cooling system.

35. Enable the SIR by performing the following procedure:

a. Turn the ignition switch to LOCK and remove the key.

b. Connect the Connector Positive Assurance (CPA) and the yellow 4-way passenger's inflator module connectors.

c. Close the instrument panel compartment.

d. Connect the Connector Positive Assurance (CPA) and the yellow 2-way driver Inflator module connectors and install the steering wheel side cap.

e. At the fuse box, located near the base of the steering column, install the AIR BAG fuse.

36. Connect the negative battery cable.

37. Run the engine to normal operating temperatures; then, check the climate control operation and check for leaks.

Cylinder Head

REMOVAL & INSTALLATION

1.6L Engine

1. Before servicing the vehicle, refer to the precautions in the beginning of this section.

2. Relieve the fuel system pressure.

3. Drain the cooling system.

4. Remove or disconnect the following:
- Negative battery cable
- Accessory drive belts
- Air intake pipe
- Fuel lines
- Upper radiator hose
- Coolant bypass hose
- Alternator bracket
- Intake manifold brackets
- Intake manifold
- Heated Oxygen (HO_2S) sensor connectors
- Exhaust manifold heat shield
- Exhaust manifold bracket
- Exhaust front pipe
- Exhaust manifold

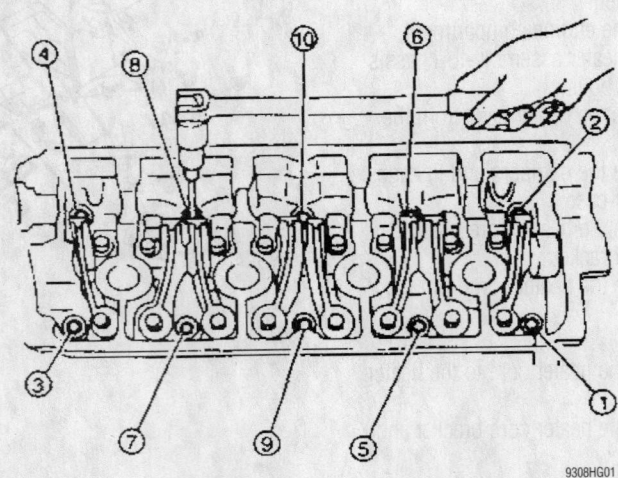

9308HG01

Cylinder head loosening sequence—1.6L engine

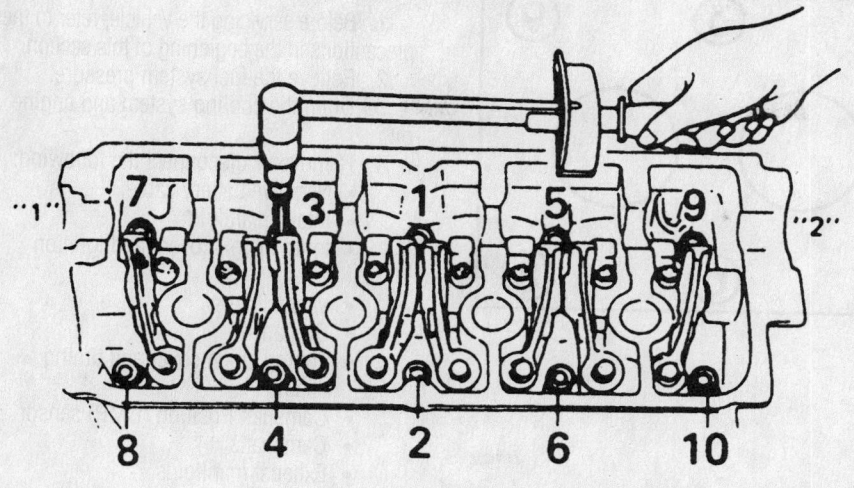

"1": Camshaft pulley side
"2": Distributor side

7924HG08

Cylinder head torque sequence—1.6L engine

- A/C compressor
- Power steering pump
- Front cover
- Timing belt. Refer to the Timing Belt unit repair section.
- Camshaft timing sprocket
- Rear timing belt cover
- Distributor and spark plug wires, if equipped
- Ignition coils and wiring, if equipped with Distributorless Ignition System (DIS)
- Valve cover
- Cylinder head. Loosen the bolts in the sequence shown.

To install:

5. Install the cylinder head with a new gasket.

6. Tighten the bolts in sequence as follows:

    a. Step 1: 26 ft. lbs. (35 Nm)

    b. Step 2: 41 ft. lbs. (55 Nm)

    c. Step 3: Loosen all bolts to 0 ft. lbs. (0 Nm)

    d. Step 4: 26 ft. lbs. (35 Nm)

    e. Step 5: 52 ft. lbs. (70 Nm)

7. Install or connect the following:

- Valve cover
- Ignition coils and wiring, if equipped with DIS
- Distributor and spark plug wires, if equipped
- Rear timing belt cover
- Camshaft timing sprocket
- Timing belt
- Front cover
- Power steering pump
- A/C compressor

- Exhaust manifold
- Exhaust front pipe
- Exhaust manifold bracket
- Exhaust manifold heat shield
- HO2S sensor connectors
- Intake manifold
- Intake manifold brackets
- Alternator bracket
- Coolant bypass hose
- Upper radiator hose
- Fuel lines
- Air intake pipe
- Accessory drive belts
- Negative battery cable

8. Fill the cooling system.

9. Start the engine and check for leaks.

2.0L Engines

1. Before servicing the vehicle, refer to the precautions in the beginning of this section.

2. Relieve the fuel system pressure.

3. Drain the cooling system.

4. Drain the engine oil.

5. Remove or disconnect the following:

- Negative battery cable
- Strut tower brace
- Air intake tube
- Exhaust Gas Recirculation (EGR) valve connector
- Idle Air Control (IAC) valve connector
- Throttle Position (TP) sensor connector
- Evaporative Emission (EVAP) canister purge valve connector and hose
- Intake manifold ground cable
- Heated Oxygen (HO2S) sensor connectors
- Camshaft Position (CMP) sensor connector
- Engine Coolant Temperature (ECT) sensor connector
- Fuel injector connectors
- Ignition coils
- Accelerator cable
- Transmission cable, if equipped
- Brake booster vacuum hose
- Radiator hose
- Bypass hose
- Heater hose
- Fuel lines
- Intake manifold bracket
- Water pipe
- Valve cover
- Accessory drive belts
- Oil pan
- Front cover
- Timing chains
- Camshafts
- Exhaust front pipe
- Exhaust manifold bracket
- Cylinder head. Loosen the bolts in the sequence shown.

1. Crankshaft pulley side
2. Flywheel side
3. Bolt (M6)

7924HG09

Cylinder head loosening sequence—2.0L engines

6 MM BOLT

Cylinder head torque sequence—2.0L engines

To install:

6. Install the cylinder head with a new gasket.

7. Tighten the bolts in sequence as follows:

   a. Step 1: 38 ft. lbs. (52 Nm)

   b. Step 2: 61 ft. lbs. (84 Nm)

   c. Step 3: Loosen all bolts to 0 ft. lbs. (0 Nm)

   d. Step 4: 38 ft. lbs. (52 Nm)

   e. Step 5: 76 ft. lbs. (105 Nm)

   f. Step 6: 6mm bolt to 96 inch lbs. (8 Nm)

8. Install or connect the following:

- Exhaust manifold bracket
- Exhaust front pipe
- Camshafts
- Timing chains
- Front cover
- Oil pan
- Accessory drive belts
- Valve cover
- Water pipe
- Intake manifold bracket
- Fuel lines
- Heater hose
- Bypass hose
- Radiator hose
- Brake booster vacuum hose
- Transmission cable, if equipped
- Accelerator cable
- Ignition coils
- Fuel injector connectors
- ECT sensor connector
- CMP sensor connector
- HO2S sensor connectors
- Intake manifold ground cable
- EVAP canister purge valve connector and hose
- TP sensor connector
- IAC valve connector

- EGR valve connector
- Air intake tube
- Strut tower brace
- Negative battery cable

9. Fill the crankcase to the correct level.

10. Fill the cooling system.

11. Start the engine and check for leaks.

2.5L Engine

1. Before servicing the vehicle, refer to the precautions in the beginning of this section.

2. Relieve the fuel system pressure.

3. Drain the cooling system and engine oil.

4. Remove or disconnect the following:

- Negative battery cable
- Intake manifold
- Ignition coil covers and ignition coils
- Valve covers
- Oil pan
- Timing chain cover and timing chains
- Camshaft Position (CMP) sensor
- Camshafts
- Exhaust manifolds
- Water outlet caps
- Cylinder heads. Loosen the bolts in the sequence shown.

To install:

5. Install the cylinder heads with new gaskets. Tighten the bolts in sequence as follows:

   a. Step 1: 38 ft. lbs. (52 Nm)

RH bank

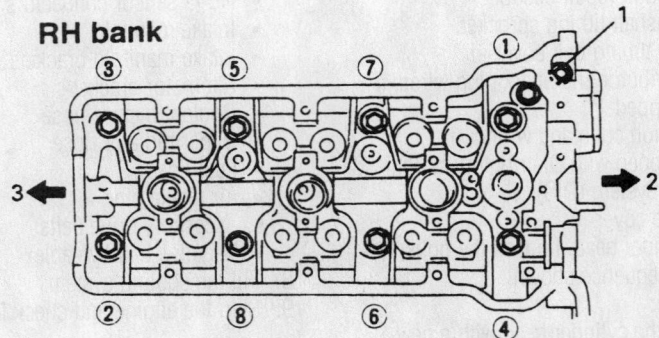

LH bank

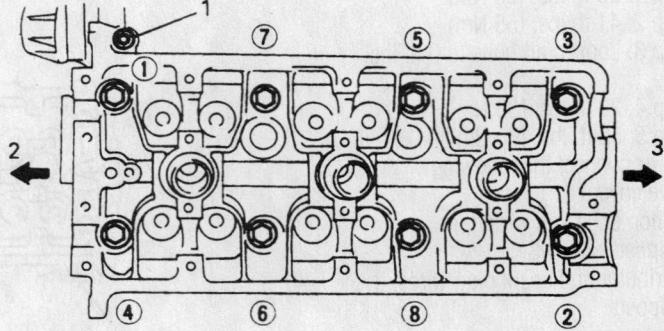

1. Hex hole bolt
2. Timing chain side
3. Flywheel side

Cylinder head loosening sequence—2.5L engine

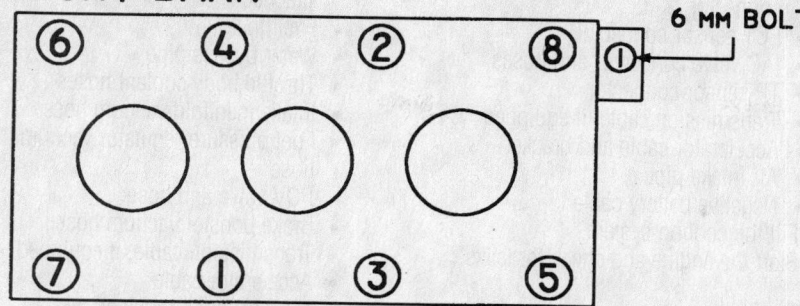

RIGHT BANK

6 MM BOLT

LEFT BANK

6 MM BOLT

Cylinder head torque sequence—2.5L engine

9308HG08

b. Step 2: 61 ft. lbs. (84 Nm)
c. Step 3: Loosen all bolts to 0 ft. lbs. (0 Nm)
d. Step 4: 38 ft. lbs. (52 Nm)
e. Step 5: 76 ft. lbs. (105 Nm)
f. Step 6: 6mm bolt to 96 inch lbs. (8 Nm)
6. Install or connect the following:
• Water outlet caps
• Exhaust manifolds
• Camshafts
• CMP sensor
• Timing chain cover and timing chains
• Oil pan
• Valve covers
• Ignition coil covers and ignition coils
• Intake manifold
• Negative battery cable
7. Fill the crankcase to the correct level
8. Fill the cooling system.
9. Start the engine and check for leaks.

Rocker Arms/Shafts

REMOVAL & INSTALLATION

1.6L Engine

1. Before servicing the vehicle, refer to the precautions in the beginning of this section.

2. Drain the cooling system.
3. Remove or disconnect the following:
• Negative battery cable
• Accessory drive belts
• Cooling fan and shroud
• Radiator and hoses
• Distributor, if equipped
• Camshaft Position (CMP) sensor, if equipped
• Valve cover
• Front cover
• Timing belt. Refer to the Timing Belt unit repair section.
• Camshaft
• Rocker arm shaft plug
• Rear timing belt cover

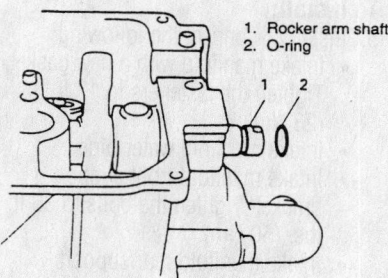

1. Rocker arm shaft
2. O-ring

9308HG02

Rocker arm shaft and O-ring—1.6L engine

4. Loosen the rocker arm locknuts and back the valve adjusters off until all rocker arms move freely with no tension.
➤**Keep all valvetrain components in order for assembly.**
5. Remove or disconnect the following:
• Intake rocker arms and clips
• Rocker arm shaft bolts
6. Push the rocker arm shaft towards the rear of the cylinder head and remove the rocker arm shaft O-ring.
7. Remove the exhaust rocker arms and springs by pulling the rocker arm shaft out of the front of the cylinder head.
To install:
8. Insert the rocker arm shaft into the front of the cylinder head, while installing the exhaust rocker arms and springs in their original positions.
9. Push the end of the rocker arm shaft out of the rear of the cylinder head and install a new O-ring.
10. Install or connect the following:
• Rocker arm shaft bolts. Tighten them to 96 inch lbs. (8 Nm).
• Intake rocker arms and clips in their original positions
• Rear timing belt cover
• Rocker arm shaft plug. Tighten it to 24 ft. lbs. (33 Nm).
• Camshaft
• Timing belt and adjust the valve clearance
• Valve cover
• Front cover
• Distributor, if equipped
• CMP sensor, if equipped
• Radiator and hoses
• Cooling fan and shroud
• Accessory drive belts
• Negative battery cable
11. Fill the cooling system.
12. Start the engine and check for leaks.

2.0L and 2.5L Engines

The 2.0L and 2.5L engines do not utilize rocker arms or rocker arm shafts.

Intake Manifold

REMOVAL & INSTALLATION

1.6L Engine

1. Before servicing the vehicle, refer to the precautions in the beginning of this section.
2. Relieve the fuel system pressure.
3. Drain the cooling system.
4. Remove or disconnect the following:

- Negative battery cable
- Air intake pipe
- Accelerator cable and bracket
- Transmission cable, if equipped
- Throttle Position (TP) sensor connector
- Idle Air Control (IAC) valve connector and hoses
- Engine Coolant Temperature (ECT) sensor connector
- Coolant temperature gauge sender connector
- A/C coolant temperature switch connector, if equipped
- Evaporative Emission (EVAP) canister purge valve connector and vacuum line
- Exhaust Gas Recirculation (EGR) temperature sensor connector
- EGR vacuum valve connector
- EGR bypass valve connector
- EGR vacuum lines
- Fuel injector connectors
- Intake manifold ground cable
- Transmission vacuum line, if equipped
- Brake booster vacuum line
- Manifold Absolute Pressure (MAP) sensor vacuum line
- Fuel lines
- Upper radiator hose
- Coolant bypass hose
- Alternator bracket
- Intake manifold brackets
- Intake manifold

To install:

5. Install or connect the following:
   - Intake manifold with a new gasket. Tighten the nuts to 17 ft. lbs. (23 Nm).
   - Intake manifold brackets. Tighten the fasteners to 36 ft. lbs. (50 Nm).
   - Alternator bracket. Tighten the fasteners to 36 ft. lbs. (50 Nm).
   - Coolant bypass hose
   - Upper radiator hose
   - Fuel lines
   - MAP sensor vacuum line
   - Brake booster vacuum line
   - Transmission vacuum line, if equipped
   - Intake manifold ground cable
   - Fuel injector connectors
   - EGR vacuum lines
   - EGR bypass valve connector
   - EGR vacuum valve connector
   - EGR temperature sensor connector
   - EVAP canister purge valve connector and vacuum line
   - A/C coolant temperature switch connector, if equipped

- Coolant temperature gauge sender connector
- ECT sensor connector
- IAC valve connector and hoses
- TP sensor connector
- Transmission cable, if equipped
- Accelerator cable and bracket
- Air intake pipe
- Negative battery cable

6. Fill the cooling system.
7. Start the engine and check for leaks.

2.0L Engines

1. Before servicing the vehicle, refer to the precautions in the beginning of this section.
2. Relieve the fuel system pressure.
3. Drain the cooling system.
4. Remove or disconnect the following:
   - Negative battery cable
   - Air intake tube
   - Exhaust Gas Recirculation (EGR) valve connector
   - Idle Air Control (IAC) valve connector
   - Throttle Position (TP) sensor connector
   - Evaporative Emissions (EVAP) canister purge valve connector and hose
   - Intake manifold ground cable
   - Manifold Absolute Pressure (MAP) sensor connector
   - Accelerator cable
   - Transmission cable, if equipped
   - Brake booster vacuum hose
   - Positive Crankcase Ventilation (PCV) valve and hose
   - Fuel pressure regulator vacuum hose
   - Intake manifold vacuum hose
   - Throttle body coolant hoses
   - Water bypass pipe
   - Fuel lines
   - Fuel supply manifold with injectors attached
   - Intake manifold support brackets
   - Intake manifold water pipe
   - Intake manifold

To install:

5. Install or connect the following:
   - Intake manifold with a new gasket. Tighten the fasteners to 17 ft. lbs. (23 Nm).
   - Intake manifold water pipe
   - Intake manifold front support bracket. Tighten the bolts to 36 ft. lbs. (50 Nm).
   - Intake manifold rear support bracket. Tighten the bolts to 18 ft. lbs. (25 Nm).

- Fuel supply manifold with injectors attached
- Fuel lines
- Water bypass pipe
- Throttle body coolant hoses
- Intake manifold vacuum hose
- Fuel pressure regulator vacuum hose
- PCV valve and hose
- Brake booster vacuum hose
- Transmission cable, if equipped
- Accelerator cable
- MAP sensor connector
- Intake manifold ground cable
- EVAP canister purge valve connector and hose
- TP sensor connector
- IAC valve connector
- EGR valve connector
- Air intake tube
- Negative battery cable

6. Fill the cooling system.
7. Start the engine and check for leaks.

2.5L Engine

1. Before servicing the vehicle, refer to the precautions in the beginning of this section.
2. Relieve the fuel system pressure.
3. Drain the cooling system.
4. Remove or disconnect the following:
   - Negative battery cable
   - Strut tower bar
   - Intake Air Temperature (IAT) sensor connector
   - Surge tank cover
   - Air intake assembly
   - Accelerator cable
   - Transmission cable, if equipped
   - Throttle body coolant hoses
   - Fuel injector connectors
   - Throttle Position (TP) sensor connector
   - Mass Air Flow (MAF) sensor connector
   - Idle Air Control (IAC) valve connector
   - Intake manifold ground cables
   - Brake booster vacuum hose
   - Evaporative Emissions (EVAP) canister purge valve connector and hoses
   - Exhaust Gas Recirculation (EGR) valve connector
   - Positive Crankcase Ventilation (PCV) valve and hose
   - Heater hoses
   - EGR pipe
   - Fuel lines
   - Throttle body and intake collector
   - Intake manifold

To install:

5. Install or connect the following:
   - Intake manifold with new gaskets. Tighten the fasteners to 16 ft. lbs. (23 Nm).
   - Throttle body and intake collector with new gaskets. Tighten the fasteners to 102 inch lbs. (12 Nm).
   - Fuel lines
   - EGR pipe
   - Heater hoses
   - PCV valve and hose
   - EGR valve connector
   - EVAP canister purge valve connector and hoses
   - Brake booster vacuum hose
   - Intake manifold ground cables
   - IAC valve connector
   - MAF sensor connector
   - TP sensor connector
   - Fuel injector connectors
   - Throttle body coolant hoses
   - Transmission cable, if equipped
   - Accelerator cable
   - Air intake assembly
   - Surge tank cover
   - IAT sensor connector
   - Strut tower bar
   - Negative battery cable
6. Fill the cooling system.
7. Start the engine and check for leaks.

Exhaust Manifold

REMOVAL & INSTALLATION

1.6L and 2.0L Engines

1. Before servicing the vehicle, refer to the precautions in the beginning of this section.
2. Remove or disconnect the following:
   - Negative battery cable
   - Strut tower bar, if equipped
   - Air intake assembly and bracket
   - Heated Oxygen (HO$_2$S) sensor connector
   - Exhaust front pipe
   - Exhaust manifold heat shield
   - Exhaust manifold bracket, if equipped
   - Exhaust manifold

To install:

3. Install or connect the following:
   - Exhaust manifold with a new gasket. Tighten the fasteners to 13–20 ft. lbs. (18–28 Nm).
   - Exhaust manifold bracket, if equipped. Tighten the bolts to 36–43 ft. lbs. (50–60 Nm).
   - Exhaust manifold heat shield

   - Exhaust front pipe. Tighten the fasteners to 29–43 ft. lbs. (40–60 Nm).
   - HO$_2$S sensor connector
   - Air intake assembly and bracket
   - Strut tower bar, if equipped. Tighten the fasteners to 66 ft. lbs. (90 Nm).
   - Negative battery cable
4. Start the engine and check for leaks.

2.5L Engine

1. Before servicing the vehicle, refer to the precautions in the beginning of this section.
2. Remove or disconnect the following:
   - Negative battery cable
   - Strut tower bar
   - Air intake assembly
   - Heated Oxygen (HO$_2$S) sensor connectors
   - Oil dipstick tube
   - Exhaust Gas Recirculation (EGR) pipe
   - Exhaust manifold heat shields
   - Evaporative Emissions (EVAP) canister
   - Front driveshaft, if equipped
   - Exhaust front pipe
   - Exhaust manifold brace
   - Exhaust manifolds

To install:

3. Install or connect the following:
   - Exhaust manifolds with new gaskets. Tighten the nuts to 21 ft. lbs. (30 Nm).
   - Exhaust manifold brace
   - Exhaust front pipe. Tighten the fasteners to 37 ft. lbs. (50 Nm).
   - Front driveshaft, if equipped
   - EVAP canister
   - Exhaust manifold heat shields
   - EGR pipe
   - Oil dipstick tube
   - HO$_2$S sensor connectors
   - Air intake assembly
   - Strut tower bar
   - Negative battery cable
4. Start the engine and check for leaks.

Front Crankshaft Seal

REMOVAL & INSTALLATION

1.6L Engine

1. Before servicing the vehicle, refer to the precautions in the beginning of this section.
2. Drain the cooling system.
3. Remove or disconnect the following:
   - Negative battery cable

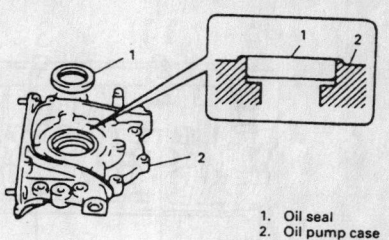

1. Oil seal
2. Oil pump case

7924HG13

Install the new oil pump seal flush with the oil pump housing—1.6L engine

   - Accessory drive belts
   - Cooling fan and shroud
   - Water pump pulley
   - Crankshaft pulley
   - Front cover
   - Timing belt. Refer to the Timing Belt unit repair section.
   - Crankshaft timing sprocket
   - Front crankshaft seal

To install:

4. Install or connect the following:
   - Front crankshaft seal flush with the oil pump housing
   - Crankshaft timing sprocket. Tighten the bolt to 94 ft. lbs. (128 Nm).
   - Timing belt
   - Front cover
   - Crankshaft pulley. Tighten the bolts to 12 ft. lbs. (16 Nm).
   - Water pump pulley
   - Cooling fan and shroud
   - Accessory drive belts
   - Negative battery cable
5. Start the engine and check for leaks.

Camshaft and Valve Lifters

REMOVAL & INSTALLATION

1.6L Engine

1. Before servicing the vehicle, refer to the precautions in the beginning of this section.
2. Drain the cooling system.
3. Remove or disconnect the following:
   - Negative battery cable
   - Radiator
   - Accessory drive belts
   - Crankshaft pulley
   - Front cover
   - Timing belt. Refer to the Timing Belt unit repair section.
   - Camshaft sprocket
   - Valve cover
   - Distributor and case, if equipped
   - Camshaft Position (CMP) sensor and case, if equipped

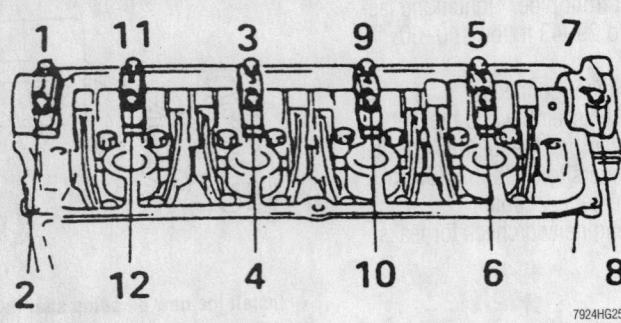

Camshaft housing torque sequence—1.6L engine

7924HG25

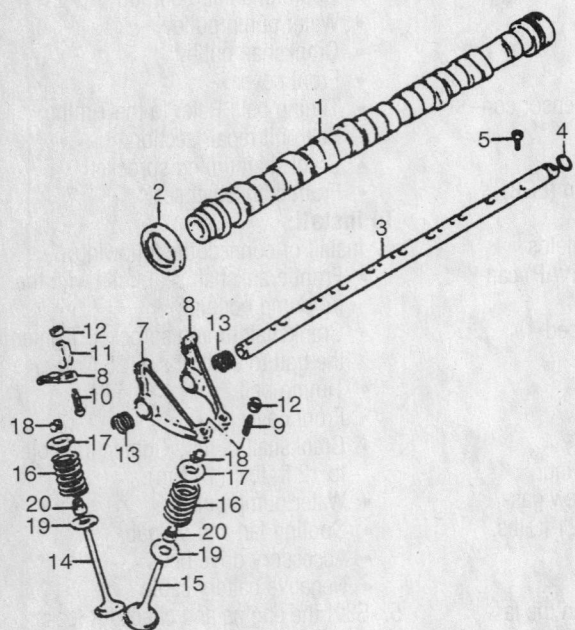

1. Camshaft
2. Camshaft oil seal
3. Rocker arm shaft
4. O ring
5. Rocker shaft bolt
6. Rocker arm (IN)
7. Rocker arm No. 1 (EX)
8. Rocker arm No. 2 (EX)
9. Valve adjusting screw
10. Valve adjusting screw
11. Clip
12. Lock nut
13. Rocker arm spring
14. Intake valve
15. Exhaust valve
16. Valve spring
17. Valve spring retainer
18. Valve cotter
19. Valve spring seat
20. Valve stem seal

7924HG24

Exploded view of the valve train components—1.6L engine

4. Loosen the rocker arm locknuts and back the valve adjusters off until all rocker arms move freely with no tension.
5. Remove or disconnect the following:
   - Camshaft bearing caps. Loosen the bolts in reverse of the tightening sequence.
   - Camshaft

To install:
6. Install or connect the following:
   - Camshaft
   - Camshaft bearing caps. Tighten the bolts in sequence to 96 inch lbs. (11 Nm).
   - CMP sensor and case, if equipped
   - Distributor and case, if equipped
   - Camshaft sprocket. Tighten the bolt to 44 ft. lbs. (60 Nm).
   - Timing belt and adjust the valve clearance
   - Valve cover

- Front cover
- Crankshaft pulley. Tighten the bolts to 12 ft. lbs. (16 Nm).
- Accessory drive belts
- Radiator
- Negative battery cable

7. Fill the cooling system.
8. Start the engine and check for leaks.

2.0L Engines

1. Before servicing the vehicle, refer to the precautions in the beginning of this section.
2. Drain the engine oil.
3. Drain the cooling system.
4. Remove or disconnect the following:
   - Negative battery cable
   - Oil pan
   - Valve cover
   - Accessory drive belts
   - Crankshaft pulley
   - Front cover
   - Secondary timing chain
   - Camshaft Position (CMP) sensor

➡**Keep all valvetrain components in order for installation.**

   - Camshaft bearing caps. Loosen the bolts in several steps in reverse of the tightening sequence.
   - Camshafts
   - Hydraulic lash adjusters

To install:
5. Install or connect the following:
   - Hydraulic lash adjusters in their original positions
   - Camshafts
   - Camshaft bearing caps in their original positions. Tighten the bolts in several steps in sequence to 96 inch lbs. (11 Nm).
   - CMP sensor
   - Secondary timing chain
   - Front cover
   - Crankshaft pulley. Tighten the bolt to 109 ft. lbs. (148 Nm).
   - Accessory drive belts
   - Valve cover
   - Oil pan
   - Negative battery cable

6. Fill the crankcase to the correct level.
7. Fill the cooling system.
8. Start the engine and check for leaks.

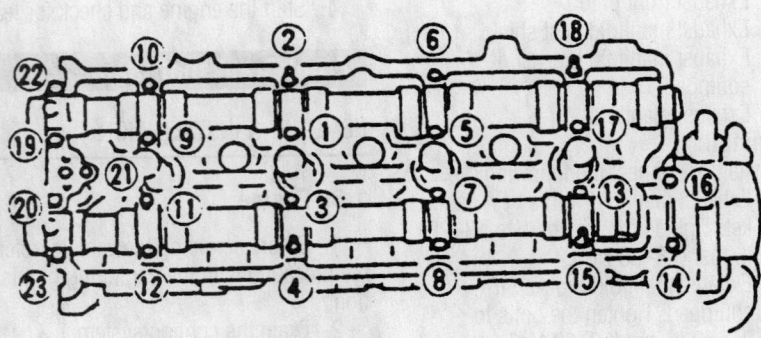

Camshaft housing torque sequence—2.0L engines

7924HG26

✻✻ WARNING

Wait ½ hour after installing the lash adjusters and camshafts before cranking or starting the engine to allow the lash adjusters to bleed down. Operating the engine before this time period may result in interference between the valves and pistons.

2.5L Engine

1. Before servicing the vehicle, refer to the precautions in the beginning of this section.
2. Drain the engine oil.
3. Drain the cooling system.
4. Remove or disconnect the following:

- Negative battery cable
- Intake manifold
- Oil pan
- Accessory drive belts
- Water pump pulley
- Crankshaft pulley
- Timing chain cover

5. Align the timing marks as shown.

✻✻ WARNING

Do not allow the crankshaft or camshafts to rotate once the timing chains have been removed. Valve or piston damage could result.

6. Remove or disconnect the following:
- Left bank secondary timing chain
- Primary timing chain
- Valve covers

➡**Keep all valvetrain components in order for assembly.**

- Right bank camshaft bearing caps. Loosen the bolts in several steps and in the sequence shown.
- Right bank secondary timing chain, exhaust and intake camshafts as an assembly
- Camshaft Position (CMP) sensor
- Left bank camshaft bearing caps. Loosen the bolts in several steps and in the sequence shown.
- Left bank camshafts
- Hydraulic lash adjusters

To install:

7. Install or connect the following:
- Hydraulic lash adjusters in their original positions
- Left bank camshafts
- Left bank camshaft bearing caps. Tighten the bolts in several steps and in reverse of the loosening sequence to 102 inch lbs. (12 Nm).
- CMP sensor
- Right bank secondary timing chain, exhaust and intake camshafts as an assembly
- Right bank camshaft bearing caps. Tighten the bolts in several steps

and in reverse of the loosening sequence to 102 inch lbs. (12 Nm).

✻✻ WARNING

Wait ½ hour after installing the lash adjusters and camshafts before cranking or starting the engine to allow the lash adjusters to bleed down. Operating the engine before this time period may result in interference between the valves and pistons.

- Valve covers. Tighten the bolts to 90 inch lbs. (10.5 Nm).
- Primary timing chain
- Left bank secondary timing chain
- Timing chain cover
- Crankshaft pulley. Tighten the bolt to 109 ft. lbs. (148 Nm).
- Water pump pulley
- Accessory drive belts
- Oil pan
- Intake manifold
- Negative battery cable

8. Fill the crankcase to the correct level.
9. Fill the cooling system.
10. Start the engine and check for leaks.

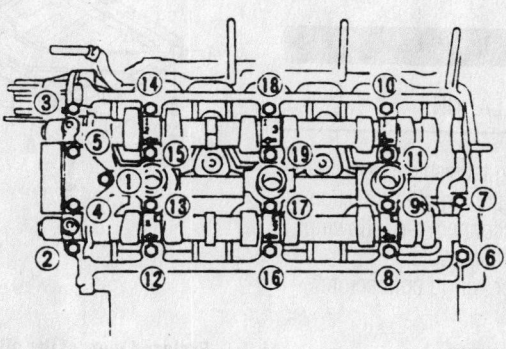

Left bank camshaft housing loosening sequence—2.5L engine

9302HG03

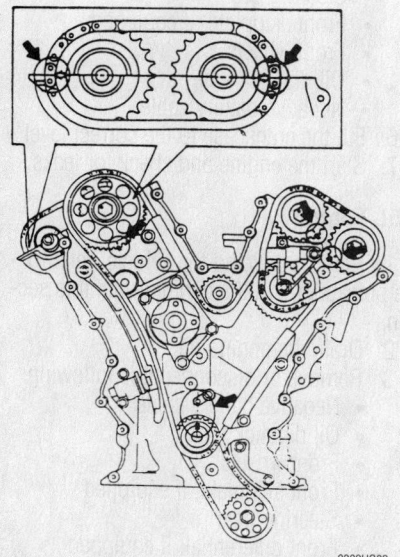

Timing mark alignment—2.5L engine

9302HG02

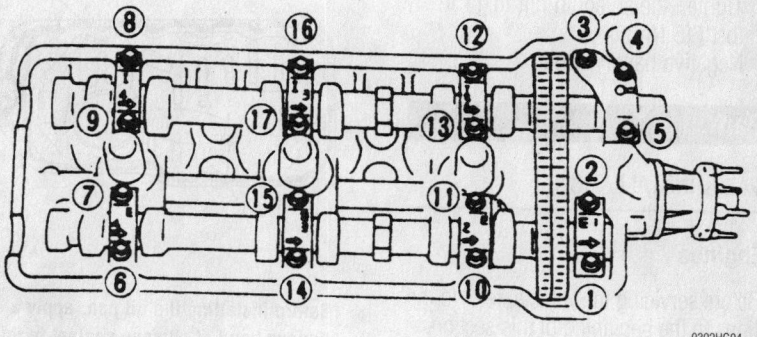

Right bank camshaft housing loosening sequence—2.5L engine

9302HG04

Valve Lash

ADJUSTMENT

1.6L Engines

➡ **Measure valve clearance with the engine cold.**

1. Before servicing the vehicle, refer to the precautions in the beginning of this section.
2. Remove the valve cover.
3. Set the engine to Top Dead Center (TDC) of the compression stroke for the cylinder to be adjusted.
4. Check the valve clearance. The valve clearance specifications are as follows:
   - Intake valves: 0.005–0.007 inches (0.13–0.17mm)
   - Exhaust valves: 0.005–0.007 inches (0.13–0.17mm)
5. After adjustment, tighten the locknuts to 11–14 ft. lbs. (15–19 Nm).
6. Repeat for each valve to be adjusted.

2.0L and 2.5L Engines

2.0L and 2.5L engines utilize automatic hydraulic lash adjusters to maintain proper valve lash at all times. Periodic valve lash inspection and adjustment is not necessary or possible.

Starter Motor

REMOVAL & INSTALLATION

1. Before servicing the vehicle, refer to the precautions in the beginning of this section.
2. Remove or disconnect the following:
   - Negative battery cable
   - Starter motor wiring connectors
   - Starter motor

To install:
3. Install or connect the following:
   - Starter motor. Tighten the bolts to 22 ft. lbs. (30 Nm).
   - Starter motor wiring connectors. Tighten the solenoid nut to 11 ft. lbs. (15 Nm).
   - Negative battery cable

Oil Pan

REMOVAL & INSTALLATION

1.6L Engines

1. Before servicing the vehicle, refer to the precautions in the beginning of this section.
2. Drain the engine oil.

3. Remove or disconnect the following:
   - Negative battery cable
   - Front skidplate, if equipped
   - Front differential, if equipped
   - Crankshaft Position (CKP) sensor
   - Left transmission stiffener bracket, if equipped
   - Flywheel access panel
   - Oil pan and oil pump pickup tube

To install:
4. Apply a bead of silicone sealant to the oil pan flange.
5. Install or connect the following:
   - New oil pump pickup tube O-ring seal
   - Oil pan and oil pump pickup tube. Tighten the fasteners to 97 inch lbs. (11 Nm).
   - Flywheel access panel
   - Left transmission stiffener bracket, if equipped
   - CKP sensor
   - Front differential, if equipped
   - Front skidplate, if equipped. Tighten the bolts to 40 ft. lbs. (55 Nm).

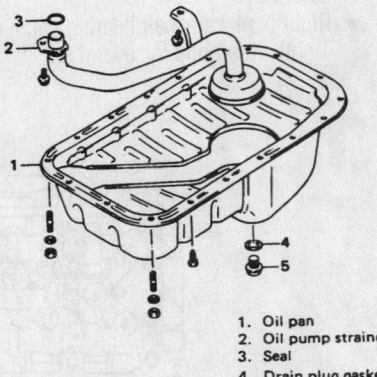

1. Oil pan
2. Oil pump strainer
3. Seal
4. Drain plug gasket
5. Drain plug

7924HG11

Exploded view of the oil pan and pump pickup mounting—1.6L engine

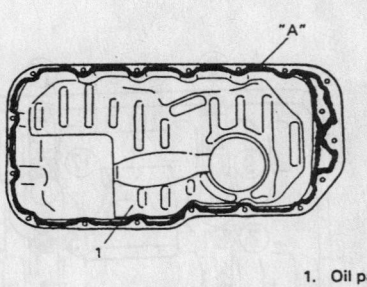

7924HG79

Before installing the oil pan, apply a continuous bead of silicone sealant to the oil pan mating flange—all engines

   - Negative battery cable
6. Fill the crankcase to the correct level.
7. Start the engine and check for leaks.

2.0L Engines

1. Before servicing the vehicle, refer to the precautions in the beginning of this section.
2. Drain the engine oil.
3. Remove or disconnect the following:
   - Negative battery cable
   - Oil dipstick tube
   - Front wheels
   - Front skidplate, if equipped
   - Steering gear
   - Front differential, if equipped
   - Left transmission stiffener bracket, if equipped
   - Flywheel access panel
   - Exhaust front pipe
   - Left and right motor mounts and raise the engine about 1 inch (25mm) for clearance
   - Oil pan and oil pump pickup tube

To install:
4. Apply a bead of silicone sealant to the oil pan flange. Install new oil pump pickup tube O-ring seals.
5. Install or connect the following:
   - Oil pan and oil pump pickup tube. Tighten the fasteners to 97 inch lbs. (11 Nm).
   - Left and right engine mounts. Tighten the nuts to 36 ft. lbs. (50 Nm).
   - Exhaust front pipe
   - Flywheel access panel
   - Left transmission stiffener bracket, if equipped
   - Front differential, if equipped
   - Steering gear
   - Front skidplate, if equipped
   - Front wheels
   - Oil dipstick tube
   - Negative battery cable
6. Fill the crankcase to the correct level.
7. Start the engine and check for leaks.

2.5L Engine

1. Before servicing the vehicle, refer to the precautions in the beginning of this section.
2. Drain the engine oil.
3. Remove or disconnect the following:
   - Negative battery cable
   - Oil dipstick tube
   - Front wheels
   - Front skidplate, if equipped
   - Steering gear
   - Front differential, if equipped
   - Lower oil pan

1. O-ring

9302HG05

Lower crankcase O-ring seal

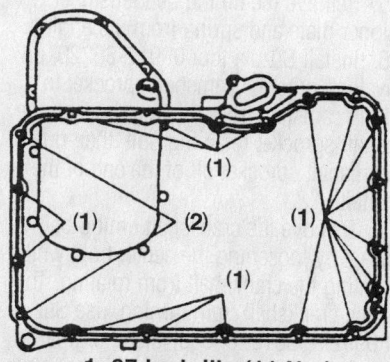

1: 97 Inch ilb. (11 Nm)
2: 20 Ft. lbs. (27 Nm)

9308HG03

Upper oil pan bolt torque values—2.5L engine

- Oil pickup tube bracket
- Radiator outlet pipe
- Upper oil pan and oil pickup tube

To install:

4. Install a new O-ring to the lower crankcase.

5. Apply a bead of silicone sealant to the upper oil pan flange.

6. Install or connect the following:
   - New oil pump pickup tube O-ring seals
   - Upper oil pan and oil pump pickup tube and tighten the fasteners as shown
   - Radiator outlet pipe
   - Oil pickup tube bracket
   - Lower oil pan. Tighten the bolts to 97 inch lbs. (11 Nm).
   - Front differential, if equipped
   - Steering gear
   - Front skidplate, if equipped
   - Front wheels
   - Oil dipstick tube
   - Negative battery cable

7. Fill the crankcase to the correct level.

8. Start the engine and check for leaks.

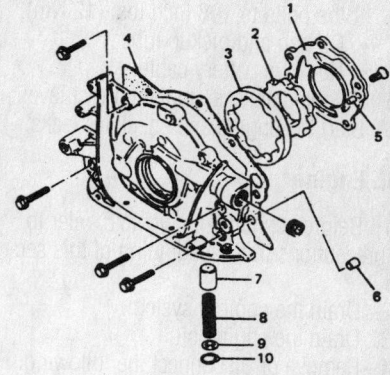

1. Rotor plate
2. Inner rotor
3. Outer rotor
4. Gasket
5. Pin
6. Pin
7. Relief valve
8. Spring
9. Retainer
10. Retainer ring

7924HG12

Exploded view of the oil pump housing—1.6L engines

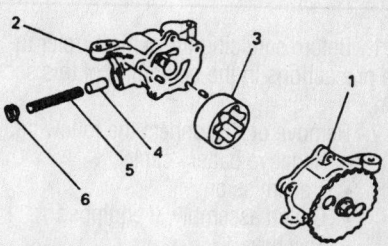

1. Oil pump case No.1
2. Oil pump case No.2
3. Outer rotor
4. Relief valve
5. Relief spring
6. Retainer

7924HG15

Exploded view of oil pump—2.0L and 2.5L engines

Oil Pump

REMOVAL & INSTALLATION

1.6L Engines

1. Before servicing the vehicle, refer to the precautions in the beginning of this section.
2. Drain the engine oil.
3. Drain the cooling system.
4. Remove or disconnect the following:
   - Negative battery cable
   - Accessory drive belts
   - Crankshaft pulley
   - Front cover
   - Timing belt. Refer to the Timing Belt unit repair section.
   - Alternator and bracket
   - A/C compressor and bracket, if equipped
   - Oil pan and oil pump pickup tube
   - Crankshaft timing sprocket
   - Oil pump

➡ The oil pump bolts are different lengths. Note their location for assembly.

To install:

5. Install or connect the following:
   - Oil pump with a new gasket. Tighten the bolts to 97 inch lbs. (11 Nm).
   - Crankshaft timing sprocket. Tighten the bolt to 94 ft. lbs. (130 Nm).
   - Oil pan and oil pump pickup tube
   - A/C compressor and bracket, if equipped
   - Alternator and bracket
   - Timing belt
   - Front cover
   - Crankshaft pulley. Tighten the bolts to 12 ft. lbs. (16 Nm).
   - Accessory drive belts
   - Negative battery cable

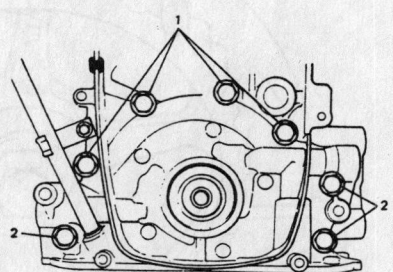

1. No. 1 bolts (short)
2. No. 2 bolts (long)

7924HG14

Oil pump housing short (1) and long bolt (2) locations—1.6L engines

6. Fill the crankcase to the correct level.
7. Start the engine and check for leaks.

2.0L Engines

1. Before servicing the vehicle, refer to the precautions in the beginning of this section.
2. Drain the engine oil.
3. Remove or disconnect the following:
   - Negative battery cable
   - Oil pan and pickup tube
   - Oil pump sprocket cover
   - Oil pump

✳✳ WARNING

Do not remove the sprocket from the oil pump. Damage to the oil pump center shaft and abnormal pump operation may result.

To install:

4. Install or connect the following:
   - Oil pump. Tighten the bolts to 20 ft. lbs. (27 Nm).

- Oil pump sprocket cover. Tighten the bolts to 108 inch lbs. (12 Nm).
- Oil pan and pickup tube
- Negative battery cable
5. Fill the crankcase to the correct level.
6. Start the engine and check for leaks.

2.5L Engine

1. Before servicing the vehicle, refer to the precautions in the beginning of this section.
2. Drain the cooling system.
3. Drain the engine oil.
4. Remove or disconnect the following:
- Negative battery cable
- Accessory drive belts
- Intake manifold
- Oil pan and oil pickup tube
- Front cover
- Oil pump chain guide
- Oil pump

✳✳ WARNING

Do not remove the sprocket from the oil pump. Damage to the oil pump center shaft and abnormal pump operation may result.

To install:
5. Install or connect the following:
- Oil pump. Tighten the bolts to 20 ft. lbs. (27 Nm).
- Oil pump chain guide. Tighten the bolts to 97 inch lbs. (11 Nm).
- Front cover
- Oil pan and oil pickup tube
- Intake manifold
- Accessory drive belts
- Negative battery cable
6. Fill the crankcase to the correct level.
7. Fill the cooling system.
8. Start the engine and check for leaks.

Rear Main Seal

REMOVAL & INSTALLATION

1. Before servicing the vehicle, refer to the precautions in the beginning of this section.
2. Remove or disconnect the following:
- Negative battery cable
- Transmission
- Clutch assembly, if equipped
- Flywheel
- Rear main seal
To install:
3. Install or connect the following:
- Rear main seal flush with the cylinder block

- Flywheel. Tighten the bolts in a crossing pattern to 58 ft. lbs. (79 Nm) for 1.6L engines or to 51 ft. lbs. (69 Nm) for all other engines.
- Clutch assembly, if equipped
- Transmission
- Negative battery cable

Timing Belt & Cover

REMOVAL & INSTALLATION

1.6L 16-Valve Engine

The 1.6L 16-valve engine is known as an interference motor, because it is fabricated with such close tolerances between the pistons and valves that, if the timing belt is incorrectly positioned, jumps teeth on one of the sprockets or breaks, the valve and pistons will come into contact. This can cause severe internal engine damage

➡**Do not rotate the crankshaft counterclockwise or attempt to rotate the crankshaft by turning the camshaft sprocket.**

1. Remove the timing belt cover.
2. If the timing belt is not already marked with a directional arrow, use white paint, a grease pencil or correction fluid to do so.
3. Rotate the crankshaft clockwise until the timing mark on the camshaft sprocket and the V mark on the timing belt inside cover are aligned, and the punch mark on the crankshaft sprocket is aligned with the mark on the engine.

✳✳ WARNING

Do not rotate the crankshaft or camshaft once the timing belt is removed, because the valves and pistons can come into contact, which may cause internal engine damage.

4. Disconnect one end of the tensioner spring. Loosen the timing belt tensioner bolt and stud, then, using your finger, press the tensioner plate up and remove the timing belt from the crankshaft and camshaft sprockets.
5. Remove the timing belt tensioner, tensioner plate and spring from the engine.
6. Install Suzuki tool 09917-68220, or equivalent, onto the camshaft sprocket to hold the camshaft from rotating. Loosen the camshaft sprocket retaining bolt, then pull the camshaft sprocket off of the end of the camshaft.
7. Remove the crankshaft timing belt sprocket by loosening the center bolt, while preventing the crankshaft from rotating. To hold the crankshaft from turning, use Suzuki tool 09927-56010, or equivalent, or a large prybar inserted in the transmission housing slot and the flywheel teeth. Pull the sprocket off of the end of the crankshaft. Be sure to retain the crankshaft sprocket key and belt guide for assembly.
8. If necessary, remove the timing belt inside cover from the cylinder head.
To install:
9. If necessary, install the timing belt inside cover.
10. Slide the timing belt guide on the crankshaft so that the concave side faces

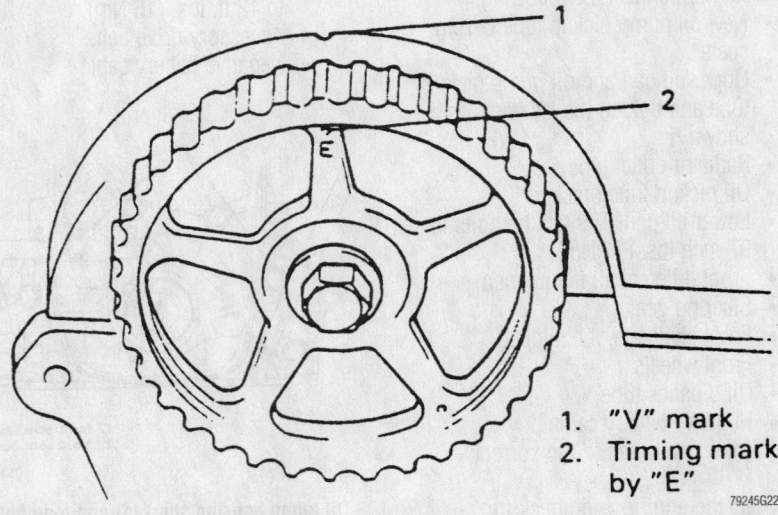

1. "V" mark
2. Timing mark by "E"

79245G22

Camshaft timing marks —1.6L 16-valve engine

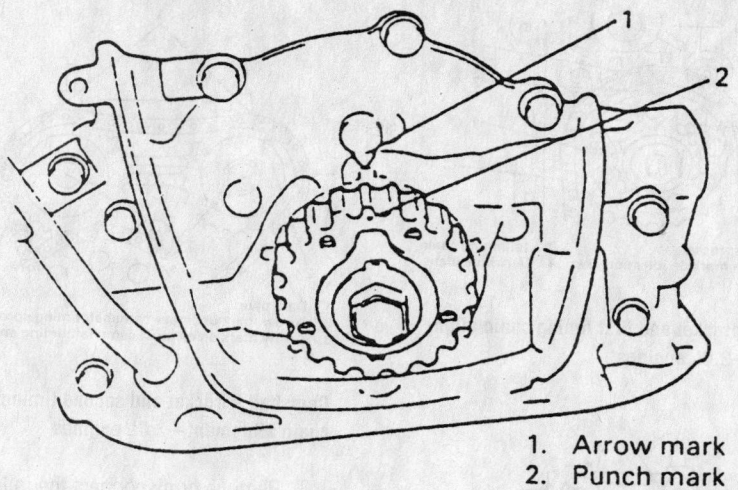

1. Arrow mark
2. Punch mark

79245G23

Align the punch mark with the arrow for proper timing belt installation —1.6L 16-valve engine

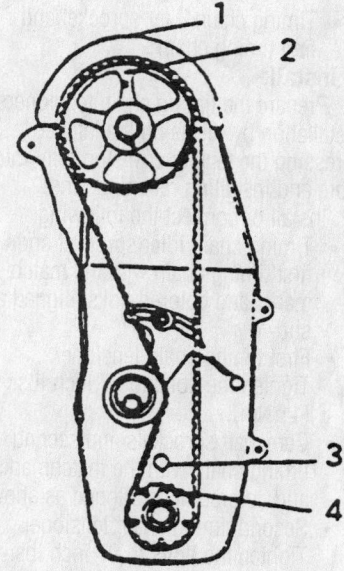

1. "V" mark on cylinder head cover
2. Timing mark by "E" on camshaft timing belt pulley
3. Arrow mark on oil pump case
4. Punch mark on crankshaft timing belt pulley

79245G47

Rotate the crankshaft clockwise until the camshaft and crankshaft timing marks are aligned — 1.6L 16-valve engine

the oil pump, then install the sprocket key in the groove in the crankshaft.

11. Slide the pulley onto the crankshaft, and install the center retaining bolt. Tighten the center bolt to 80 ft. lbs. (110 Nm). To hold the crankshaft from turning, use Suzuki tool 09927-56010, or equivalent, or a large prybar inserted in the transmission housing slot and the flywheel teeth.

12. Install the timing belt camshaft sprocket, ensuring that the slot in the sprocket engages the camshaft (pulley) pin; this ensures that the sprocket is properly positioned on the end of the camshaft.

Secure the camshaft with the holding tool used during removal, then tighten the sprocket bolt to 44 ft. lbs. (60 Nm).

13. Assemble the timing belt tensioner plate and the tensioner, making sure that the lug of the tensioner plate engages the tensioner.

✳✳ WARNING

If any binding is felt when adjusting the timing belt tension by turning the crankshaft, STOP turning the engine, because the pistons may be hitting the valves.

14. Install the timing belt tensioner, tensioner plate and spring on the engine. Tighten the mounting bolt and stud only finger-tight at this time. Ensure that when the tensioner is moved in a counterclockwise direction, the tensioner moves in the same direction. If the tensioner does not move, remove it and the tensioner plate to reassemble them properly.

15. Loosen all rocker arm valve lash locknuts and adjusting screws. This will permit movement of the camshaft without any rocker arm associated drag, which is essential for proper timing belt tensioning. If the camshaft does not rotate freely (free of rocker arm drag), the belt will not be properly tensioned.

16. Rotate the camshaft sprocket clockwise until the timing mark on the sprocket and the V mark on the timing belt inside cover are aligned.

17. Using a wrench, or socket and breaker bar, on the crankshaft sprocket center bolt, turn the crankshaft clockwise until the punch mark on the sprocket is aligned with the arrow mark on the oil pump.

18. With the camshaft and crankshaft marks properly aligned, push the tensioner up with your finger and install the timing belt on the 2 sprockets, ensuring that the drive side of the belt is free of all slack. Release your finger from the tensioner. Be sure to install the timing belt so that the directional arrow is pointing in the appropriate direction.

➡️ **In this position, the No. 4 cylinder is at Top Dead Center (TDC) on the compression stroke.**

19. Rotate the crankshaft clockwise 2 full revolutions, then tighten the tensioner stud to 97 inch lbs. (11 Nm). Then, tighten the tensioner bolt to 18 ft. lbs. (24 Nm).

20. Ensure that all 4 timing marks are still aligned as before; if they are not, remove the timing belt, and install and tension it again.

21. Install the timing belt cover and all related components.

Timing Chain, Sprockets, Front Cover and Seal

REMOVAL & INSTALLATION

2.0L Engines

1. Before servicing the vehicle, refer to the precautions in the beginning of this section.

2. Drain the cooling system.

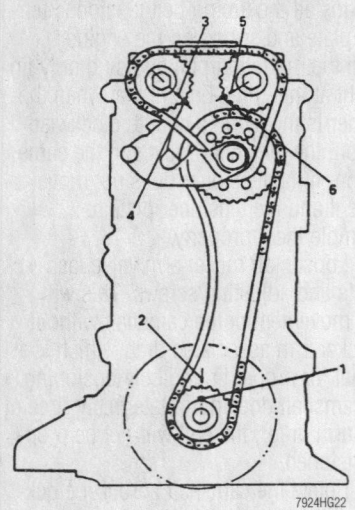

Timing mark alignment—2.0L engines

3. Drain the engine oil.
4. Remove or disconnect the following:
- Negative battery cable
- Oil pan and pickup tube
- Valve cover
- Bypass pipe and hose
- Accessory drive belts
- Cooling fan and shroud
- Water pump pulley
- Alternator belt tensioner and idler pulleys
- Upper radiator hose
- A/C compressor and bracket, if equipped
- Crankshaft pulley
- Front crankshaft seal
- Front cover

5. Rotate the crankshaft to align the timing marks as shown.

✳✳ WARNING

Do not allow the crankshaft or camshafts to rotate once the timing chains have been removed. Valve or piston damage could result.

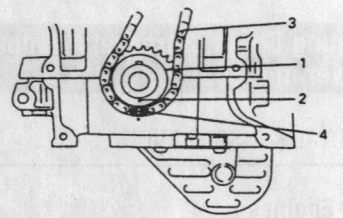

1. Crankshaft timing sprocket   3. 1st timing chain
2. Match mark   4. Yellow plate

Crankshaft and first timing chain alignment—2.0L engines

1. Idler sprocket   3. 1st timing chain
2. Match mark on idler sprocket 4. Dark blue plate

Idler sprocket and first timing chain alignment—2.0L engines

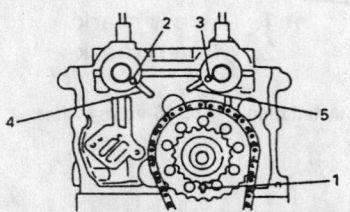

1. Arrow mark on idler sprocket
2. Knock pin of intake camshaft
3. Knock pin of exhaust camsaft
4. Timing mark of intake side
5. Timing mark of exhaust side

Idler sprocket and camshaft alignment—2.0L engines

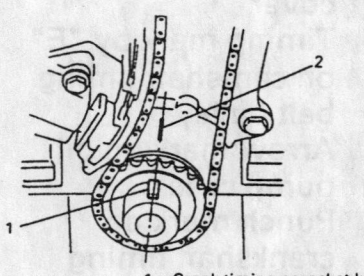

1. Crank timing sprocket key
2. Timing mark

Crankshaft sprocket alignment—2.0L engines

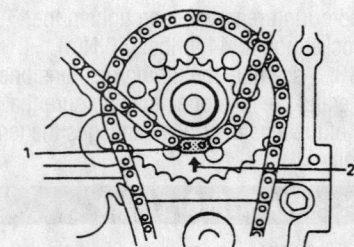

1. Yellow plate
2. Match mark of 2nd timing chain (Arrow mark)

Idler sprocket and second timing chain alignment—2.0L engines

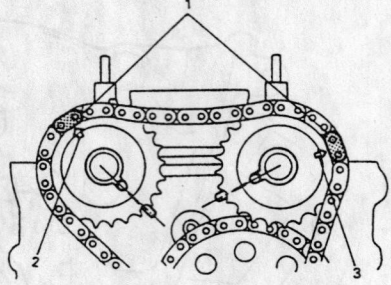

1. Dark blue
2. Arrow mark on intake camshaft timing sprocket
3. Arrow mark on exhaust camshaft timing sprocket

Camshaft sprocket and second timing chain alignment—2.0L engines

6. Remove or disconnect the following:
- Second timing chain tensioner
- Camshaft sprockets and second timing chain
- First timing chain tensioner
- Timing chain idler sprocket and first timing chain

To install:
7. Prepare the timing chain tensioners for installation by releasing the latches, compressing the tensioner piston fully into the bore and installing retaining pins.
8. Install or connect the following:
- Timing chain idler sprocket and first timing chain with the match-marks and colored links aligned as shown
- First timing chain tensioner. Tighten the bolts to 97 inch lbs. (11 Nm).
- Camshaft sprockets and second timing chain with the matchmarks and colored links aligned as shown
- Second timing chain tensioner. Tighten the bolts to 97 inch lbs. (11 Nm) and the nut to 33 ft. lbs. (45 Nm).

9. Tighten the camshaft sprocket bolts to 59 ft. lbs. (80 Nm).
10. Remove the timing chain tensioner retaining pins.
11. Rotate the crankshaft two complete turns and check that the timing marks align.
12. Install or connect the following:
- Front cover. Apply sealant as shown.
- Front crankshaft seal
- Crankshaft pulley. Tighten the bolt to 109 ft. lbs. (130 Nm).
- A/C compressor and bracket, if equipped
- Upper radiator hose
- Alternator belt tensioner and idler pulleys

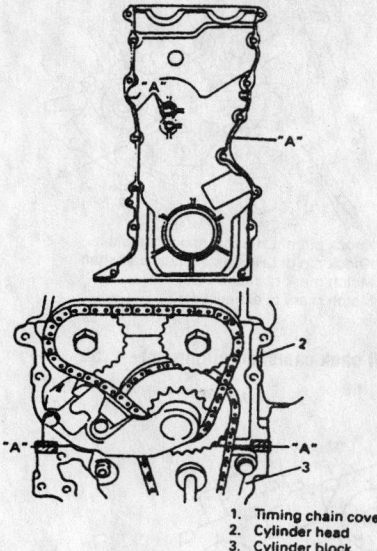

1. Timing chain cover
2. Cylinder head
3. Cylinder block

7924HG10

Prior to installing the timing chain cover on the engine block and cylinder head, apply silicone sealant to the cover as indicated (areas marked A)—2.0L and 2.5L engines

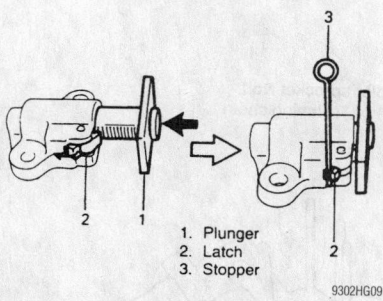

1. Plunger
2. Latch
3. Stopper

9302HG09

Preparing the No. 1 timing chain tensioner adjuster for installation—2.0L and 2.5L engines

- Water pump pulley
- Cooling fan and shroud
- Accessory drive belts
- Bypass pipe and hose
- Valve cover
- Oil pan and pickup tube
- Negative battery cable

13. Fill the crankcase to the correct level.
14. Fill the cooling system.
15. Start the engine and check for leaks.

2.5L Engine

1. Before servicing the vehicle, refer to the precautions in the beginning of this section.
2. Drain the cooling system.
3. Drain the engine oil.

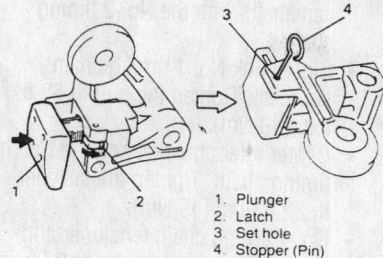

1. Plunger
2. Latch
3. Set hole
4. Stopper (Pin)

9302HG13

No. 2 timing chain tensioner—left bank tensioner shown—2.0L and 2.5L engines

4. Remove or disconnect the following:
- Negative battery cable
- Intake manifold
- Ignition coils
- Valve covers
- Accessory drive belts
- Cooling fan and shroud
- Water pump pulley
- Radiator
- Power steering pump and brackets
- Oil pan and pickup tube
- Crankshaft pulley
- Front crankshaft seal
- Crankshaft Position (CKP) sensor
- Front cover
5. Rotate the crankshaft so that the timing marks are aligned as shown.

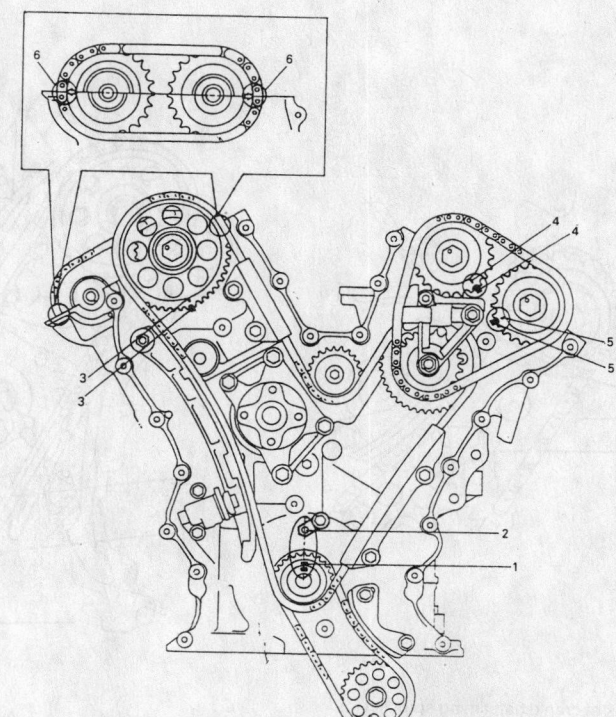

Timing chain alignment marks—2.5L engine

9302HG07

6. Remove or disconnect the following:
- Left bank No. 2 timing chain tensioner
- Left bank intake and exhaust camshaft sprockets with the No. 2 timing chain
- No. 1 timing chain guides
- No. 1 timing chain tensioner
- Center idler sprocket and the No. 1 timing chain
- Right bank No. 1 timing chain sprocket
7. The right bank No. 2 timing chain is removed with the intake and exhaust camshafts.

To install:

8. Prepare the timing chain tensioners for installation by releasing the latches, compressing the tensioner piston fully into the bore and installing retaining pins.

9. Align the timing chain sprocket matchmarks and colored chain links as shown during assembly.

10. Install or connect the following:

1. Crankshaft Keyway
2. Oil Jet
3. Right Bank No. 1 Chain Marks
4. Left Bank Intake Cam Marks
5. Left Bank Exhaust Cam Marks
6. Right Bank Intake and Exhaust Cam Marks

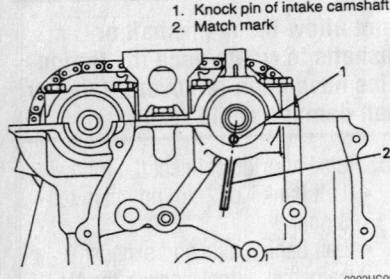

1. Knock pin of intake camshaft
2. Match mark

9302HG08

Right bank camshaft timing marks—2.5L engine

- Right bank intake and exhaust camshafts with the No. 2 timing chain
- Right bank No. 1 timing chain sprocket. Tighten the bolt to 58 ft. lbs. (80 Nm).
- Center idler sprocket and the No. 1 timing chain. Tighten the fastener to 32 ft. lbs. (45 Nm).
- No. 1 timing chain tensioner and guides. Tighten the bolts to 97 inch lbs. (11 Nm).
- Left bank intake and exhaust camshaft sprockets with the No. 2

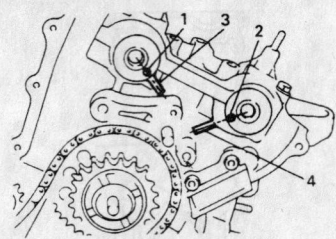

1. Knock pin of LH bank intake camshaft
2. Knock pin of LH bank exhaust camshaft
3. Match mark of intake side
4. Match mark of exhaust side

9302HG11

Left bank camshaft alignment—2.5L engine

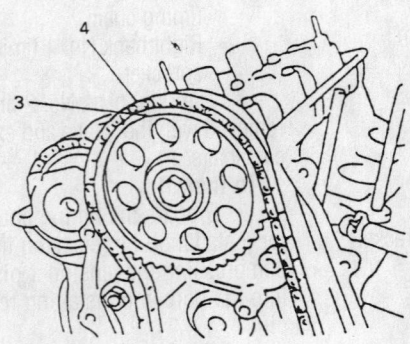

3. Match mark of RH bank 1st timing chain sprocket
4. Silver plate (LH) of 1st timing chain

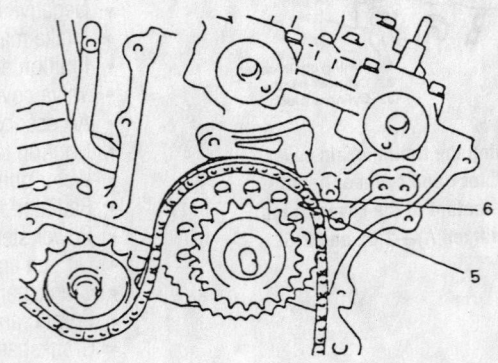

5. Match mark of idler sprocket No.2
6. Silver plate (RH) of 1st timing chain

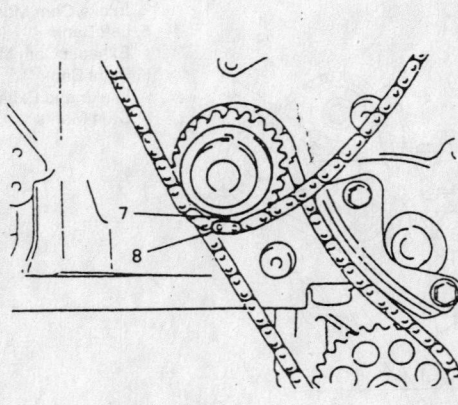

7. Match mark of crankshaft timing sprocket
8. Gold or Yellow plate of 1st timing chain

No. 1 timing chain alignment—2.5L engine

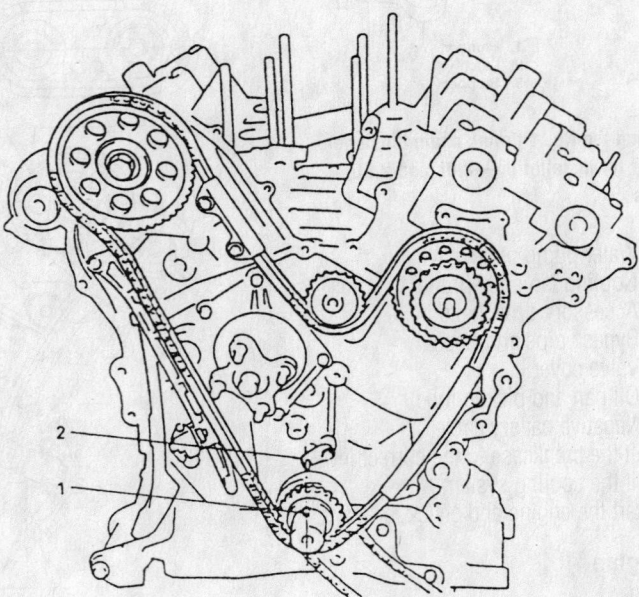

1. Crank timing pulley key
2. Oil jet

9302HG10

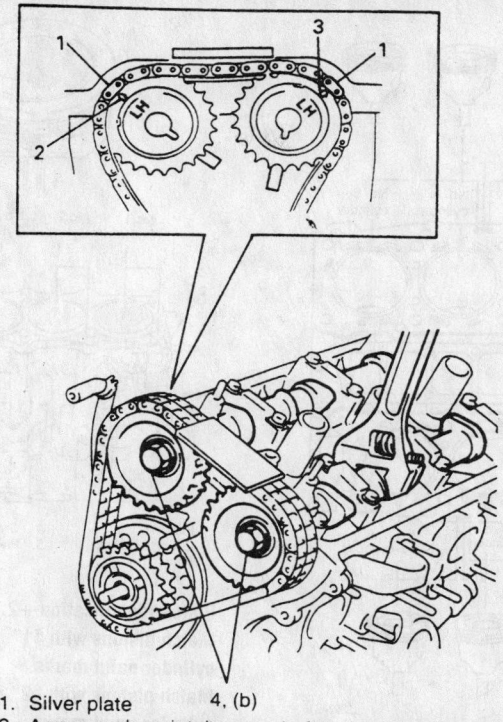

1. Silver plate 4, (b)
2. Arrow mark on intake camshaft timing sprocket
3. Arrow mark on exhaust camshaft timing sprocket
4. Sprocket bolt

9302HG12

Align the left bank No. 2 chain silver links—2.5L engine

timing chain. Tighten the bolts to 57 ft. lbs. (80 Nm).

• Left bank No. 2 timing chain tensioner. Tighten the bolts to 97 inch lbs. (11 Nm).

11. Remove the retaining pins from the timing chain tensioners.

12. Rotate the crankshaft two complete turns and check that the timing marks align.

13. Install or connect the following:
• Front cover. Tighten the bolts to 97 inch lbs. (11 Nm).
• CKP sensor
• Front crankshaft seal
• Crankshaft pulley. Tighten the bolt to 109 ft. lbs. (148 Nm).
• Oil pan and pickup tube
• Power steering pump and brackets
• Radiator
• Water pump pulley
• Cooling fan and shroud
• Accessory drive belts
• Valve covers
• Ignition coils
• Intake manifold
• Negative battery cable

14. Fill the crankcase to the correct level.
15. Fill the cooling system.
16. Start the engine and check for leaks.

Piston and Ring

POSITIONING

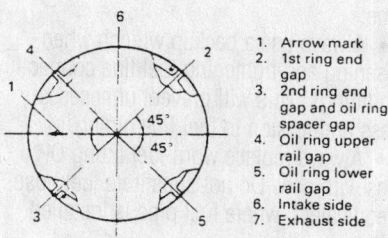

1. Arrow mark
2. 1st ring end gap
3. 2nd ring end gap and oil ring spacer gap
4. Oil ring upper rail gap
5. Oil ring lower rail gap
6. Intake side
7. Exhaust side

7924AG67

Piston ring end-gap spacing—All engines

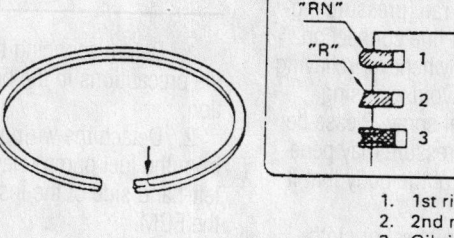

1. 1st ring
2. 2nd ring
3. Oil ring

7924AG68

Compression ring identification marks—All engines

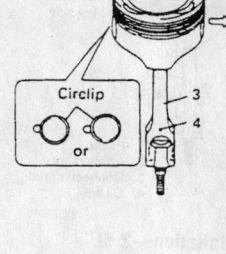

1. Piston
2. Arrow mark
3. Connecting rod
4. Oil hole

The oil hole should come on intake side

7924AG69

Piston and connecting rod positioning— 1.6L engine

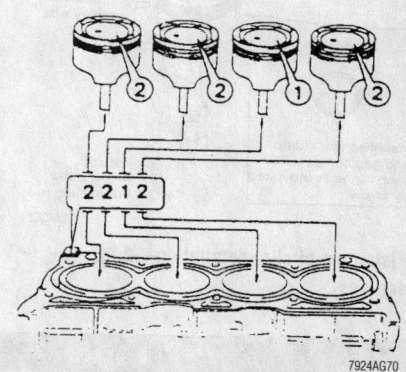

7924AG70

Piston installation—1.6L engine

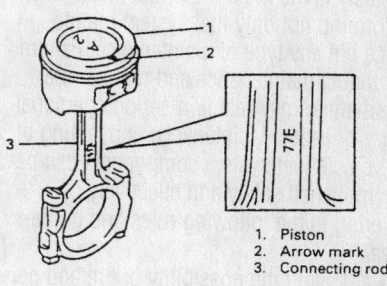

1. Piston
2. Arrow mark
3. Connecting rod

7924AG61

Piston and connecting rod positioning— 2.0L and 2.5L engines

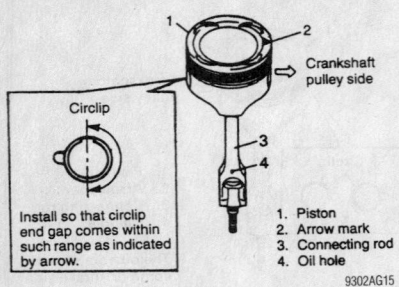

Install so that circlip end gap comes within such range as indicated by arrow.

1. Piston
2. Arrow mark
3. Connecting rod
4. Oil hole

9302AG15

Piston pin circlip installation—2.5L engine

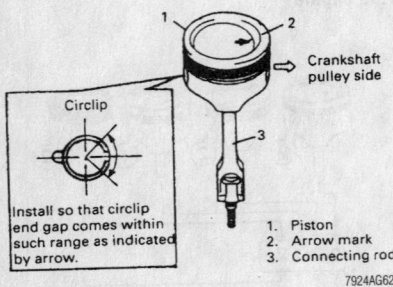

Install so that circlip end gap comes within such range as indicated by arrow.

1. Piston
2. Arrow mark
3. Connecting rod

7924AG62

Piston pin circlip installation—2.0L engines

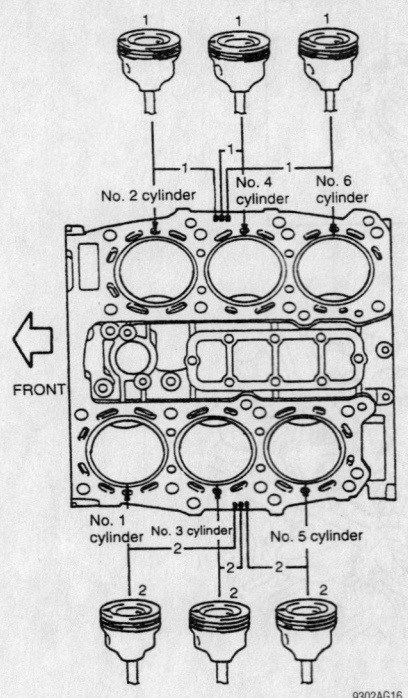

No. 2 cylinder   No. 4 cylinder   No. 6 cylinder

FRONT

No. 1 cylinder   No. 3 cylinder   No. 5 cylinder

9302AG16

Piston identification—2.5L engine

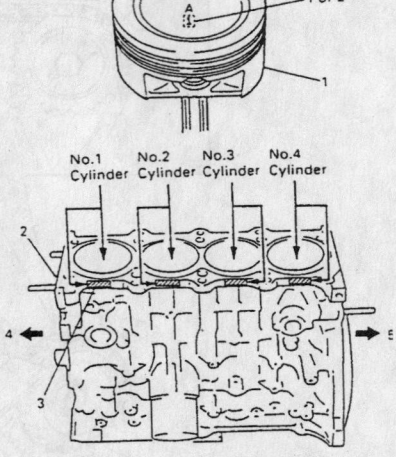

No.1 Cylinder   No.2 Cylinder   No.3 Cylinder   No.4 Cylinder

1. Piston
2. Cylinder block
3. Paint
4. Crank shaft pulley side
5. Flywheel side

7924AG65

Piston identification—2.0L engines
Match pistons with "1" indicators to red cylinder paint marks
Match pistons with "2" indicators to blue cylinder paint marks

FUEL SYSTEM

Fuel System Service Precautions

Safety is the most important factor when performing not only fuel system maintenance but any type of maintenance. Failure to conduct maintenance and repairs in a safe manner may result in serious personal injury or death. Maintenance and testing of the vehicle fuel system components can be accomplished safely and effectively by adhering to the following rules and guidelines.

• To avoid the possibility of fire and personal injury, always disconnect the negative battery cable unless the repair or test procedure requires that battery voltage be applied.

• Always relieve the fuel system pressure prior to disconnecting any fuel system component (injector, fuel rail, pressure regulator, etc.), fitting or fuel line connection. Exercise extreme caution whenever relieving fuel system pressure to avoid exposing skin, face and eyes to fuel spray. Please be advised that fuel under pressure may penetrate the skin or any part of the body that it contacts.

• Always place a shop towel or cloth around the fitting or connection prior to loosening to absorb any excess fuel due to spillage. Ensure that all fuel spillage (should it occur) is quickly removed from

engine surfaces. Ensure that all fuel soaked cloths or towels are deposited into a suitable waste container.

• Always keep a dry chemical (Class B) fire extinguisher near the work area.

• Do not allow fuel spray or fuel vapors to come into contact with a spark or open flame.

• Always use a backup wrench when loosening and tightening fuel line connection fittings. This will prevent unnecessary stress and torsion to fuel line piping.

• Always replace worn fuel fitting O-rings with new. Do not substitute fuel hose or equivalent, where fuel pipe is installed.

Fuel System Pressure

RELIEVING

1. Before servicing the vehicle, refer to the precautions in the beginning of this section.

2. Detach the wiring harness connector from the fuel pump relay, located under the left-hand side of the instrument panel near the ECM.

3. Start the engine and run it until it stops from lack of fuel. Crank the engine 2–3 times for a 3 second period. The fuel lines should now be depressurized.

4. After servicing, reattach the wiring harness connector to the fuel pump relay.

Fuel Filter

REMOVAL & INSTALLATION

1. Before servicing the vehicle, refer to the precautions in the beginning of this section.

2. Relieve fuel system pressure.

3. Remove or disconnect the following:
• Negative battery cable
• Fuel lines from the fuel filter
• Fuel filter

To install:

4. Install or connect the following:
• Fuel filter and tighten the bracket bolt. Note the fuel flow directional arrow.
• Fuel lines to the fuel filter.
• Negative battery cable

5. Start the engine and inspect the fuel filter connections for leaks.

Fuel Pump

REMOVAL & INSTALLATION

1. Before servicing the vehicle, refer to the precautions in the beginning of this section.

2. Relieve the fuel system pressure.
3. Remove or disconnect the following:

- Negative battery cable
- Fuel filler hose and vent hose
- Fuel tank inlet valve and drain the fuel tank
- Fuel filter inlet hose
- Evaporative Emissions (EVAP) vapor hose
- Fuel return line
- Fuel tank skidplate
- Fuel pump connector
- Fuel tank pressure sensor connector
- Fuel tank
- Fuel pump module

To install:

4. Install or connect the following:

- Fuel pump module with a new seal. Tighten the bolts to 44 inch lbs. (5 Nm).
- Fuel tank. Tighten the strap bolts to 37 ft. lbs. (50 Nm).
- Fuel tank pressure sensor connector
- Fuel pump connector
- Fuel tank skidplate
- Fuel return line
- EVAP vapor hose
- Fuel filter inlet hose
- Fuel tank inlet valve and drain the fuel tank
- Fuel filler hose and vent hose
- Negative battery cable

5. Fill the fuel tank.
6. Start the engine and check for leaks.

Fuel Injector

REMOVAL & INSTALLATION

1.6L and 2.0L Engines

1. Before servicing the vehicle, refer to the precautions in the beginning of this section.
2. Relieve the fuel system pressure.
3. Remove or disconnect the following:

- Negative battery cable
- Front intake manifold bracket, if equipped
- Positive Crankcase Ventilation (PCV) valve and hose
- Fuel injector harness connectors
- Fuel line bracket
- Fuel supply manifold
- Fuel injectors

To install:

4. Install or connect the following:

- Fuel injectors with new O-ring seals
- Fuel supply manifold. Tighten the bolts to 17 ft. lbs. (23 Nm).
- Fuel line bracket
- Fuel injector harness connectors
- PCV valve and hose
- Front intake manifold bracket, if equipped
- Negative battery cable

5. Start the engine and check for leaks.

2.5L Engine

1. Before servicing the vehicle, refer to the precautions in the beginning of this section.
2. Relieve the fuel system pressure.
3. Remove or disconnect the following:

- Negative battery cable
- Air intake tube
- Throttle body intake collector
- Fuel lines
- Fuel pressure regulator vacuum line
- Fuel injector harness connectors
- Fuel supply manifold connect pipe
- Fuel supply manifolds
- Fuel injectors

To install:

4. Install or connect the following:

- Fuel injectors with new O-ring seals
- Fuel supply manifolds. Tighten the bolts to 17 ft. lbs. (23 Nm).
- Fuel supply manifold connect pipe. Tighten the bolts to 22 ft. lbs. (30 Nm).
- Fuel injector harness connectors
- Fuel pressure regulator vacuum line
- Fuel lines
- Throttle body intake collector
- Air intake tube
- Negative battery cable

5. Start the engine and check for leaks.

DRIVE TRAIN

Transmission Assembly

REMOVAL & INSTALLATION

Manual Transmission

1. Before servicing the vehicle, refer to the precautions in the beginning of this section.
2. Drain the transmission fluid.
3. Drain the transfer case fluid, if equipped.
4. Remove or disconnect the following:

- Negative battery cable
- Shift lever boots
- Gear shift lever
- Transfer case shift lever, if equipped
- 4WD switch connector, if equipped
- Reverse light switch connector
- Starter motor
- Front driveshaft, if equipped
- Rear driveshaft
- Speedometer cable, if equipped

- Vehicle Speed (VSS) sensor, if equipped
- Clutch slave cylinder or cable
- Flywheel access cover
- Transmission flange bolts and nuts
- Transmission braces, if equipped

5. Support the transmission with a jack and remove the transmission mount and crossmember.
6. Place a wooden block at the rear of the cylinder head as shown to support the engine when the transmission is removed.
7. Lower the transmission away from the vehicle.

To install:

8. Install or connect the following:

- Transmission. Tighten the flange fasteners to 62–72 ft. lbs. (85–98 Nm).
- Transmission mount and crossmember. Tighten the fasteners to 29–43 ft. lbs. (40–60 Nm).
- Transmission braces, if equipped

Tighten the bolts to 62–72 ft. lbs. (85–98 Nm).

- Flywheel access cover
- Clutch slave cylinder or cable
- VSS sensor, if equipped

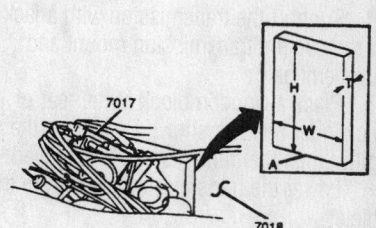

A	WOOD BLOCK
H	200 mm (8.0")
T	45 mm (1.8")
W	100–150 mm (4.0–6.0")
7017	DISTRIBUTOR CAP
7018	BULKHEAD

7924HG37

Support the engine with a wooden block between the cylinder head and the firewall—All models

- Speedometer cable, if equipped
- Rear driveshaft
- Front driveshaft, if equipped
- Starter motor
- Reverse light switch connector
- 4WD switch connector, if equipped
- Transfer case shift lever, if equipped
- Gear shift lever
- Shift lever boots
- Negative battery cable

9. Fill the transmission to the correct level.

10. Fill the transfer case, if equipped.

Automatic Transmission

1. Before servicing the vehicle, refer to the precautions in the beginning of this section.

2. Drain the transfer case oil, if equipped.

3. Remove or disconnect the following:
- Negative battery cable
- Center console and transfer case shift lever, if equipped
- Transmission dipstick tube
- Transmission wiring harness connectors
- Starter motor
- Front driveshaft, if equipped
- Rear driveshaft
- Gear select cable and bracket
- Throttle Valve (TV) cable, if equipped
- Exhaust front pipe
- Transmission oil cooler lines
- Transmission brace
- Flywheel access cover
- Torque converter
- Speedometer cable, if equipped
- Vehicle Speed (VSS) sensor connector, if equipped
- Transmission flange bolts and nuts
- Transmission braces, if equipped

4. Support the transmission with a jack and remove the transmission mount and crossmember.

5. Place a wooden block at the rear of the cylinder head as shown to support the engine when the transmission is removed.

6. Lower the transmission away from the vehicle.

To install:

7. Install or connect the following:
- Transmission. Tighten the flange fasteners to 62–72 ft. lbs. (85–98 Nm).
- Transmission mount and crossmember. Tighten the fasteners to 29–43 ft. lbs. (40–60 Nm).
- Transmission braces, if equipped.

Tighten the bolts to 62–72 ft. lbs. (85–98 Nm).
- VSS sensor connector, if equipped
- Speedometer cable, if equipped
- Torque converter. Tighten the bolts to 47 ft. lbs. (65 Nm).
- Flywheel access cover
- Transmission brace
- Transmission oil cooler lines
- Exhaust front pipe
- TV cable, if equipped
- Gear select cable and bracket
- Rear driveshaft
- Front driveshaft, if equipped
- Starter motor
- Transmission wiring harness connectors
- Transmission dipstick tube
- Center console and transfer case shift lever, if equipped
- Negative battery cable

8. Fill the transmission to the correct level.

9. Fill the transfer case, if equipped.

Clutch

ADJUSTMENTS

These vehicles are equipped with a hydraulic clutch system. No adjustment is necessary.

REMOVAL & INSTALLATION

1. Before servicing the vehicle, refer to the precautions in the beginning of this section.

2. Remove the transmission.

3. Loosen the pressure plate mounting bolts in a 2-step crisscross sequence until the spring tension is relieved.

4. Remove the pressure plate and the clutch disc.

To install:

5. Using a clutch alignment tool, assemble the clutch disc and pressure plate onto the flywheel.

6. Tighten the pressure plate bolts in multiple passes to 17 ft. lbs. (23 Nm).

7. Install the transmission.

8. Check for proper clutch operation.

Hydraulic Clutch System

BLEEDING

1. Before servicing the vehicle, refer to the precautions in the beginning of this section.

2. Fill the master cylinder reservoir to

the MAX line with clean brake fluid and keep it at least half full throughout the bleeding procedure.

3. From beneath the vehicle, remove the bleeder plug cap, then attach a clear vinyl tube to the slave cylinder bleeder plug. Insert the open end of the hose into a container.

4. Have an assistant depress the clutch pedal. Open the bleeder after the pedal is depressed.

5. Close the bleeder before releasing the clutch pedal.

6. Repeat until all air bubbles are gone from the hydraulic fluid.

7. Install the bleeder plug cap.

8. Fill the clutch master cylinder fluid reservoir to the specified full level.

Transfer Case Assembly

REMOVAL & INSTALLATION

1. Before servicing the vehicle, refer to the precautions in the beginning of this section.

2. Drain the transfer case oil.

3. Remove or disconnect the following:
- Negative battery cable
- Distributor or Camshaft Position (CMP) sensor, if equipped
- Center console
- Transmission shift lever and case, if equipped with a manual transmission
- Transfer case shift lever
- Front and rear driveshafts
- Exhaust center pipe
- Speedometer cable or Vehicle Speed (VSS) sensor, as equipped
- Vent hose
- 4WD switch connector

4. Support the transmission with a jack and remove the transmission mount and crossmember.

5. Place a wooden block at the rear of the cylinder head as shown to support the engine when the transfer case is removed.

6. Lower the transfer case away from the vehicle.

To install:

7. Install or connect the following:
- Transfer case. Tighten the bolts to 30 ft. lbs. (41 Nm).
- 4WD switch connector
- Vent hose
- Speedometer cable or VSS sensor, as equipped
- Exhaust center pipe
- Front and rear driveshafts. Tighten the bolts to 36 ft. lbs. (50 Nm).

- Transfer case shift lever
- Transmission shift lever and case, if equipped with a manual transmission
- Center console
- Distributor or CMP sensor, if equipped
- Negative battery cable

8. Fill the transfer case.

Halfshaft

REMOVAL & INSTALLATION

Left

1. Before servicing the vehicle, refer to the precautions in the beginning of this section.

2. Remove or disconnect the following:

- Front wheel
- Hub drive flange or locking hub, as equipped
- Snapring
- Thrust washer
- Halfshaft flange fasteners
- Halfshaft

To install:

3. Install or connect the following:

- Halfshaft. Tighten the flange bolts to 37 ft. lbs. (50 Nm).
- Thrust washer
- Snapring
- Locking hub, if equipped. Tighten the bolts to 24 ft. lbs. (33 Nm).

- Hub drive flange, if equipped. Tighten the bolts to 35 ft. lbs. (48 Nm).
- Front wheel

Right

1. Before servicing the vehicle, refer to the precautions in the beginning of this section.

2. Remove or disconnect the following:

- Front wheel
- Hub drive flange or locking hub, as equipped
- Snapring
- Thrust washer
- Brake caliper
- Wheel speed sensor, if equipped
- Brake rotor
- Stabilizer bar link
- Outer tie rod end
- Lower ball joint
- Strut bracket bolts
- Steering knuckle and wheel hub

3. Pry the inboard joint out of the differential and remove the halfshaft.

To install:

4. Insert the inboard joint into the differential until the circlip is felt to seat.

5. Install or connect the following:

- Steering knuckle and wheel hub
- Strut bracket bolts. Tighten them to 70 ft. lbs. (95 Nm).
- Lower ball joint. Tighten the nut to 40 ft. lbs. (55 Nm).
- Outer tie rod end. Tighten the nut to 35 ft. lbs. (48 Nm).

- Stabilizer bar link. Tighten the nut to 21 ft. lbs. (29 Nm).
- Brake rotor
- Wheel speed sensor, if equipped
- Brake caliper
- Thrust washer
- Snapring
- Locking hub, if equipped. Tighten the bolts to 24 ft. lbs. (33 Nm).
- Hub drive flange, if equipped. Tighten the bolts to 35 ft. lbs. (48 Nm).
- Front wheel

6. Check the wheel alignment and adjust as necessary.

CV-Joints

OVERHAUL

Outer CV-Joint

The outer CV-joint is serviced with the axle shaft as an assembly. The outer CV-joint boot can be serviced by removing the inner CV-joint.

Inner CV-Joint

1. Before servicing the vehicle, refer to the precautions in the beginning of this section.

2. Remove or disconnect the following:

- Halfshaft from the vehicle
- Grease boot clamps
- Outer race snapring
- Outer race
- Shaft snapring
- Inner race, cage and balls

To install:

3. Install or connect the following:

- Inner race, cage and balls
- Shaft snapring
- Outer race
- Outer race snapring

4. Fill the outer race and the grease boot with CV-joint grease and tighten the boot clamps.

5. Install the axle halfshaft.

Manual Locking Hubs

REMOVAL & INSTALLATION

1. Before servicing the vehicle, refer to the precautions in the beginning of this section.

2. Set the selector knob to the **FREE** position.

3. Remove or disconnect the following:

1. Drive shaft oil seal
2. Double off-set joint (DOJ)
3. Joint circlip
4. DOJ boot
5. Ball joint boot
6. Ball joint assembly (RH side)
7. Drive shaft assembly (LH side)
8. Left drive shaft
9. Drive shaft bearing circlip
10. Drive shaft bearing

7924HG31

Exploded view of the left- and right-hand halfshaft assemblies

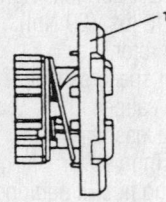

1. Cover

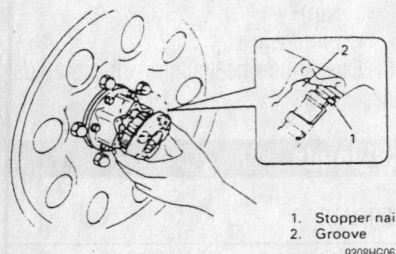

1. Stopper nail
2. Groove

9308HG06

Manual hub alignment—All models

- Hub cover assembly
- Hub body assembly

To install:

4. Install the hub body. Tighten the bolts to 18 ft. lbs. (25 Nm).

5. Align the hub cover stopper nail with the groove in the hub body and install the hub cover. Tighten the bolts to 90 inch lbs. (10 Nm).

6. Check for proper hub operation.

Automatic Locking Hubs

REMOVAL & INSTALLATION

1. Before servicing the vehicle, refer to the precautions in the beginning of this section.

2. Unlock the hub by setting the transfer case in the 2H position and driving backwards at least 6.5 feet (2 meters).

3. Remove or disconnect the following:

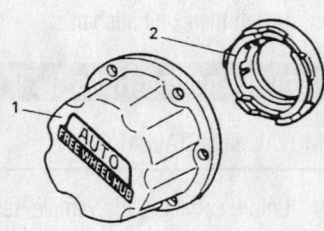

1. Free wheeling hub sub assembly
2. Free wheeling hub brake assembly

9308HG04

Automatic hub—All models

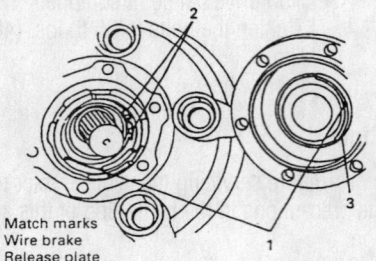

1. Match marks
2. Wire brake
3. Release plate

9308HG05

Automatic hub matchmarks—All models

- Hub sub assembly
- Hub brake assembly

To install:

4. Align the brake assembly key with the slot in the spindle and install the brake assembly.

5. Align the matchmark on the sub assembly with the mark on the brake assembly and install the sub assembly. Tighten the hub bolts to 24 ft. lbs. (33 Nm).

6. Check for proper hub operation.

Spindle Bearings

REMOVAL, PACKING & INSTALLATION

1. Before servicing the vehicle, refer to the precautions in the beginning of this section.

2. Support the control arm with a stand or floor jack.

3. Remove or disconnect the following:

- Front wheel
- Locking hub or drive flange, as equipped
- Brake caliper and rotor
- Wheel hub and bearing assembly
- Outer tie rod end
- Lower ball joint
- Strut bracket bolts
- Wheel spindle and steering knuckle assembly
- Inner oil seal
- Spindle bearing

To install:

4. Fill the recess in the wheel spindle with lithium grease.

5. Coat the spindle bearing and wheel spindle mating surfaces with sealant.

6. Press or drive the spindle bearing into the wheel spindle.

7. Install or connect the following:

- Inner oil seal
- Wheel spindle and steering knuckle assembly
- Strut bracket bolts
- Lower ball joint

- Outer tie rod end
- Wheel hub and bearing assembly
- Brake caliper and rotor
- Locking hub or drive flange, as equipped
- Front wheel

Axle Shaft, Bearing and Seal

REMOVAL & INSTALLATION

1. Before servicing the vehicle, refer to the precautions in the beginning of this section.

2. Loosen the parking brake cable for clearance.

3. Remove or disconnect the following:

- Rear wheel
- Brake drum
- Wheel speed sensor, if equipped
- Bearing retainer nuts
- Axle shaft and bearing
- Axle shaft inner oil seal

4. If equipped with ABS, grind a flat spot on the wheel speed sensor tone ring, then split the ring with a chisel.

5. Grind flat spots on the bearing retainer and split it with a chisel.

6. Press the wheel bearing off the axle shaft.

7. Remove the bearing retainer and the outer oil seal.

To install:

8. Install or connect the following:

- Outer oil seal to the bearing retainer
- Bearing retainer to the axle shaft
- Bearing and retainer ring pressed onto the axle shaft
- Wheel speed sensor tone ring pressed onto the axle shaft, if equipped
- Axle shaft inner oil seal
- Axle shaft and bearing
- Bearing retainer nuts. Tighten them to 17 ft. lbs. (23 Nm).
- Wheel speed sensor, if equipped
- Brake drum
- Rear wheel

9. Fill the rear differential to the correct level.

Pinion Seal

REMOVAL & INSTALLATION

1. Before servicing the vehicle, refer to the precautions in the beginning of this section.

2. Remove or disconnect the following:

- Driveshaft
- Wheels
- Brake calipers and pads or brake drum

➡ **The brake calipers and pads or brake drum must be removed so that there is no additional drag when measuring pinion bearing preload.**

3. Use an inch lb. torque wrench and measure and record the amount of torque required to maintain pinion rotation through several revolutions.

4. Remove or disconnect the following:
- Pinion flange
- Pinion seal
- Pinion bearing
- Collapsible spacer

To install:

➡ **Use a new collapsible spacer and flange nut for assembly.**

5. Install or connect the following:
- Collapsible spacer
- Pinion bearing
- Pinion seal
- Pinion flange

6. Rotate the pinion flange occasionally while tightening the flange nut to make sure the pinion bearings seat correctly.

7. Take frequent bearing preload torque readings. Tighten the flange nut to achieve the preload torque readings originally recorded.

✵✵ CAUTION

Never loosen the pinion nut to reduce bearing preload. If it is necessary to

reduce bearing preload, install a new collapsible spacer and pinion nut.

8. Install or connect the following:
- Driveshaft
- Brake calipers and pads or brake drum
- Wheels

9. Fill the differential with gear lubricant and check for leaks.

Axle Housing Assembly

REMOVAL & INSTALLATION

1. Before servicing the vehicle, refer to the precautions in the beginning of this section.

2. Drain the gear oil.

3. Support the vehicle at the frame with a hoist or jackstands.

4. Support the rear axle with a floor jack.

5. Remove or disconnect the following:
- Rear wheels
- Rear brake drums
- Rear axle shafts
- Load sensing proportioning valve linkage, if equipped
- Brake fluid hose
- Brake backing plates
- Wheel speed sensor connector, if equipped
- Axle vent tube
- Rear driveshaft
- Differential carrier assembly

- Shock absorber lower bolts
- Coil springs
- Upper rods
- Lower rods
- Lateral rod
- Axle housing

To install:

6. Install or connect the following:
- Axle housing
- Upper rods
- Lower rods
- Coil springs
- Lateral rod
- Shock absorber lower bolts
- Differential carrier assembly. Tighten the nuts to 40 ft. lbs. (55 Nm).
- Rear driveshaft
- Axle vent tube
- Wheel speed sensor connector, if equipped
- Brake backing plates
- Brake fluid hose
- Load sensing proportioning valve linkage, if equipped
- Rear axle shafts
- Rear brake drums
- Rear wheels

7. Fill the rear axle to the correct level.

8. Lower the vehicle so that the rear suspension is at curb height.

9. Tighten the upper, lower and lateral rod fasteners to 65 ft. lbs. (90 Nm).

10. Tighten the lower shock absorber fasteners to 62 ft. lbs. (85 Nm).

STEERING AND SUSPENSION

Air Bag

✵✵ CAUTION

Some vehicles are equipped with an air bag system. The system must be disarmed before performing service on, or around, system components, the steering column, instrument panel components, wiring and sensors. Failure to follow the safety precautions and the disarming procedure could result in accidental air bag deployment, possible injury and unnecessary system repairs.

PRECAUTIONS

Several precautions must be observed when handling the inflator module to avoid

accidental deployment and possible personal injury.

• Never carry the inflator module by the wires or connector on the underside of the module.

• When carrying a live inflator module, hold securely with both hands and ensure that the bag/trim cover are pointed away.

• Place the inflator module on a bench or other surface with the bag and trim cover facing up.

• With the inflator module on the bench, never place anything on or close to the module which may be thrown in the event of an accidental deployment.

• Never use air bag component parts from another vehicle.

• If there is a chance of electrical shock to any of the air bag components, remove the air bag module before servicing the vehicle.

DISARMING

1. Before servicing the vehicle, refer to the precautions in the beginning of this section.

2. Remove or disconnect the following:
- Negative battery cable
- AIR BAG fuse
- Driver air bag connector
- Glove box
- Passenger air bag connector

ARMING

When repairs are complete, install or connect the following:
- Passenger air bag connector
- Glove box
- Driver air bag connector
- AIR BAG fuse
- Negative battery cable

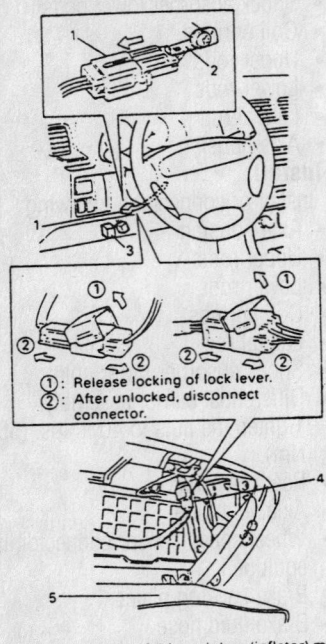

1. Yellow connector of driver air bag (inflator) module
2. Connector stay
3. Air bag fuse box
4. Yellow connector of passenger air bag (inflator) module
5. Glove box

7924HG36

Air bag component location and identification—All models

Recirculating Ball Steering Gear

REMOVAL & INSTALLATION

1. Before servicing the vehicle, refer to the precautions in the beginning of this section.
2. Remove or disconnect the following:
   - Skidplate, if equipped
   - Coolant overflow tank
   - Intermediate shaft pinch bolt
   - Power steering hoses
   - Pitman arm center link joint
   - Steering gearbox
   - Pitman arm

To install:

3. Install or connect the following:
   - Pitman arm. Tighten the nut to 102 ft. lbs. (140 Nm).
   - Steering gearbox. Tighten the bolts to 62 ft. lbs. (85 Nm).
   - Pitman arm center link joint. Tighten the nut to 37 ft. lbs. (50 Nm).
   - Power steering hoses
   - Intermediate shaft pinch bolt. Tighten it to 18 ft. lbs. (25 Nm).
   - Coolant overflow tank

- Skidplate, if equipped
4. Fill the power steering system.
5. Start the engine and check for leaks.
6. Check the wheel alignment and adjust as necessary.

Power Rack and Pinion Steering Gear

REMOVAL & INSTALLATION

1. Before servicing the vehicle, refer to the precautions in the beginning of this section.
2. Remove or disconnect the following:
   - Power steering hoses
   - Intermediate shaft pinch bolt
   - Front wheels
   - Outer tie rod ends
   - Steering gear

To install:

3. Install or connect the following:
   - Steering gear. Tighten the bolts to 40 ft. lbs. (55 Nm).
   - Outer tie rod ends. Tighten the nuts to 32 ft. lbs. (43 Nm).
   - Front wheels
   - Intermediate shaft pinch bolt. Tighten it to 18 ft. lbs. (25 Nm).
   - Power steering hoses
4. Fill the power steering system.
5. Start the engine and check for leaks.
6. Check the wheel alignment and adjust as necessary.

Strut

REMOVAL & INSTALLATION

1. Before servicing the vehicle, refer to the precautions in the beginning of this section.
2. Support the control arm with a stand or floor jack.
3. Remove or disconnect the following:
   - Front wheel
   - Brake hose bracket
   - Strut bracket bolts
   - Upper strut mount nuts
   - Strut

To install:

4. Install or connect the following:
   - Strut. Tighten the upper mount nuts to 40 ft. lbs. (55 Nm) and the bracket bolts to 70 ft. lbs. (95 Nm).
   - Brake hose bracket
   - Front wheel
5. Check the wheel alignment and adjust as necessary.

Shock Absorber

REMOVAL & INSTALLATION

1. Before servicing the vehicle, refer to the precautions in the beginning of this section.
2. Support the rear axle housing with a hydraulic jack or stand.

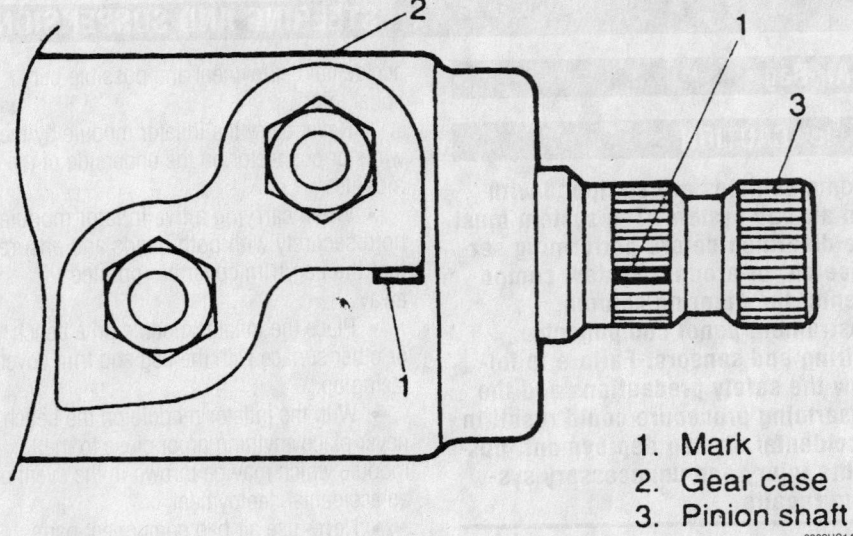

1. Mark
2. Gear case
3. Pinion shaft

9302HG14

Steering gear centering marks

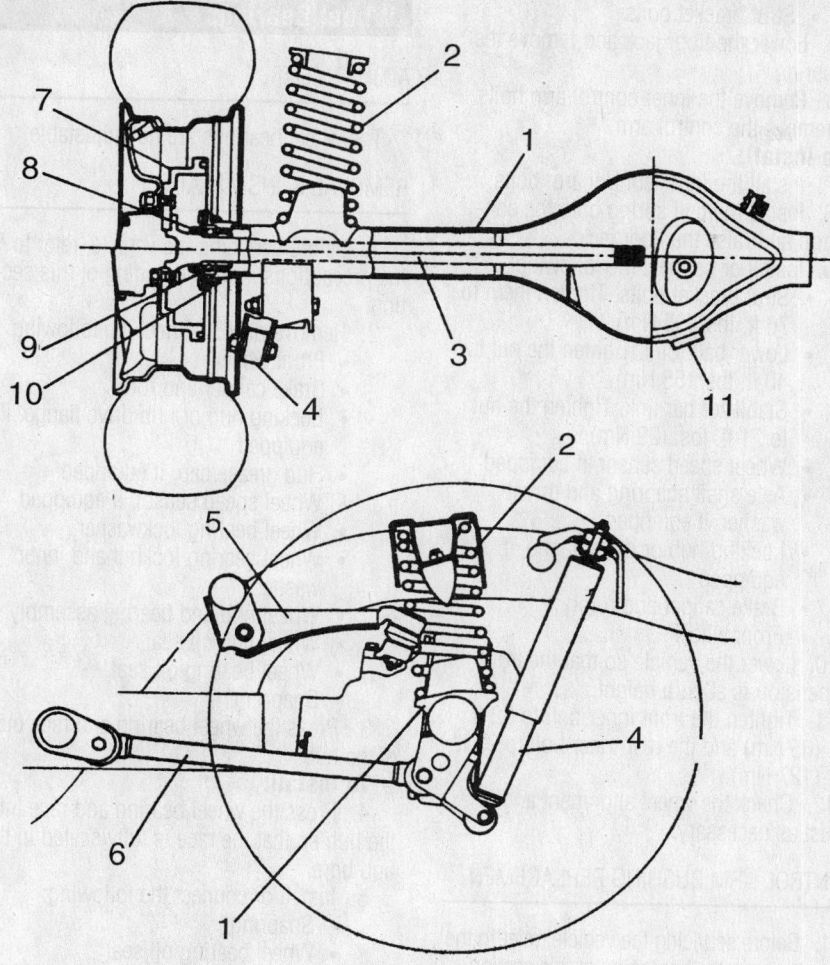

1. **Rear axle housing**
2. **Coil spring**
3. **Axle shaft**
4. **Shock absorber**
5. **Upper arm**
6. **Trailing rod**
7. **Brake drum**
8. **Wheel bearing retainer**
9. **Rear wheel bearing**
10. **Brake back plate**
11. **Oil drain plug**

7924HG32

Rear suspension component identification

3. Remove or disconnect the following:
   - Shock absorber upper locknut and retaining nut
   - Lower shock absorber mounting nut and bolt
   - Rear shock absorber

To install:

4. Install or connect the following:
   - Rear shock absorber

   - Lower mounting nut and bolt
   - Upper retaining nut and locknut

5. Torque the upper mounting nuts to 21 ft. lbs. (22–35 Nm) and the lower mounting nut/bolt to 62 ft. lbs. (85 Nm) for 1999–01 models; 74 ft. lbs. (100 Nm) for 2002 models.

6. Remove the jack or stand from the rear axle assembly.

Coil Spring

REMOVAL & INSTALLATION

Front

1. Before servicing the vehicle, refer to the precautions in the beginning of this section.

2. Support the vehicle at the frame with a hoist or jackstand.

3. Support the control arm with a floor jack.

4. Remove or disconnect the following:
   - Front wheel
   - Brake caliper and rotor
   - Locking hub or drive flange, if equipped
   - Axle shaft snapring and thrust washer, if equipped
   - Wheel speed sensor, if equipped
   - Stabilizer bar link
   - Lower ball joint
   - Strut bracket bolts

5. Lower the floor jack and remove the coil spring.

To install:

➡ **The bottom of the spring has a larger diameter than the top.**

6. Install the coil spring onto the control arm and raise the floor jack.

7. Install or connect the following:
   - Strut bracket bolts. Tighten them to 70 ft. lbs. (95 Nm).
   - Lower ball joint. Tighten the nut to 40 ft. lbs. (55 Nm).
   - Stabilizer bar link. Tighten the nut to 21 ft. lbs. (29 Nm).
   - Wheel speed sensor, if equipped
   - Axle shaft snapring and thrust washer, if equipped
   - Locking hub or drive flange, if equipped
   - Brake caliper and rotor
   - Front wheel

8. Check the wheel alignment and adjust as necessary.

Rear

1. Before servicing the vehicle, refer to the precautions in the beginning of this section.

2. Support the vehicle at the frame with a hoist or jackstand.

3. Support the rear axle housing with a floor jack.

4. Remove or disconnect the following:
   - Rear wheels
   - Parking brake cable hanger
   - Shock absorber lower mounting bolts

- Wheel speed sensor harness clamps, if equipped
- Brake pipe E-ring
- Axle vent hose

5. Lower the floor jack and remove the coil springs.

To install:

6. Install the coil springs onto the axle spring seats and raise the floor jack.

7. Install or connect the following:
- Axle vent hose
- Brake pipe E-ring
- Wheel speed sensor harness clamps, if equipped
- Shock absorber lower mounting bolts. Tighten them to 62 ft. lbs. (85 Nm) for 1999–01 models; 74 ft. lbs. (100 Nm) for 2002 models.
- Parking brake cable hanger
- Rear wheels

Lower Ball Joint

REMOVAL & INSTALLATION

The lower ball joint is serviced with the lower control arm as an assembly.

Lower Control Arm

REMOVAL & INSTALLATION

1. Before servicing the vehicle, refer to the precautions in the beginning of this section.

2. Support the vehicle at the frame with a hoist or jackstand.

3. Support the control arm with a floor jack.

4. Remove or disconnect the following:
- Front wheel
- Brake caliper and rotor
- Locking hub or drive flange, if equipped
- Axle shaft snapring and thrust washer, if equipped
- Wheel speed sensor, if equipped
- Stabilizer bar link
- Lower ball joint

- Strut bracket bolts

5. Lower the floor jack and remove the coil spring.

6. Remove the inner control arm bolts and remove the control arm.

To install:

7. Install the inner control arm bolts.

8. Install the coil spring onto the control arm and raise the floor jack.

9. Install or connect the following:
- Strut bracket bolts. Tighten them to 70 ft. lbs. (95 Nm).
- Lower ball joint. Tighten the nut to 40 ft. lbs. (55 Nm).
- Stabilizer bar link. Tighten the nut to 21 ft. lbs. (29 Nm).
- Wheel speed sensor, if equipped
- Axle shaft snapring and thrust washer, if equipped
- Locking hub or drive flange, if equipped
- Brake caliper and rotor
- Front wheel

10. Lower the vehicle so that the front suspension is at curb height.

11. Tighten the front inner bolt to 62 ft. lbs. (85 Nm) and the rear inner bolt to 92 ft. lbs. (127 Nm).

12. Check the wheel alignment and adjust as necessary.

CONTROL ARM BUSHING REPLACEMENT

1. Before servicing the vehicle, refer to the precautions in the beginning of this section.

2. Remove the control arm from the vehicle.

3. Remove the control arm bushings with a hydraulic press.

To install:

4. Lubricate the control arm bushings with liquid soap.

5. Press the bushings into the control arm until the bushing flange contacts the housing edge of the control arm.

6. Install the control arm to the vehicle.

7. Check the wheel alignment and adjust as necessary.

Wheel Bearings

ADJUSTMENT

The wheel bearings are not adjustable.

REMOVAL & INSTALLATION

1. Before servicing the vehicle, refer to the precautions in the beginning of this section.

2. Remove or disconnect the following:
- Front wheel
- Brake caliper and rotor
- Locking hub or hub drive flange, if equipped
- Hub grease cap, if equipped
- Wheel speed sensor, if equipped
- Wheel bearing lockwasher
- Wheel bearing locknut and inner washer
- Wheel hub and bearing assembly
- Wheel hub oil seal
- Wheel bearing oil seal
- Snapring

3. Press the wheel bearing and race out of the hub.

To install:

4. Press the wheel bearing and race into the hub so that the race is fully seated in the hub bore.

5. Install or connect the following:
- Snapring
- Wheel bearing oil seal
- Wheel hub oil seal
- Wheel hub and bearing assembly
- Wheel bearing locknut and inner washer. Tighten the nut to 157 ft. lbs. (216 Nm).
- Wheel bearing lockwasher. Tighten the retaining screws to 13 inch lbs. (1.5 Nm).
- Wheel speed sensor, if equipped
- Hub grease cap, if equipped
- Locking hub or hub drive flange, if equipped
- Brake caliper and rotor
- Front wheel

BRAKES

Brake Caliper

REMOVAL & INSTALLATION

1. Raise and safely support the vehicle.
2. Remove the wheels.
3. Disconnect and plug the brake line.
4. Remove the caliper mounting bolts (guide pins) and remove the caliper from the vehicle.

To install:

5. Install the caliper on the vehicle. Tighten the mounting bolts to 20 ft. lbs. (27 Nm).
6. Connect the hydraulic brake line, using 2 new washers. Torque the union bolt to 17 ft. lbs. (23 Nm).
7. Replace the front wheels.
8. Lower the vehicle.
9. Fill the brake reservoir and bleed the hydraulic brake system.

Disc Brake Pads

REMOVAL & INSTALLATION

1. Siphon about ⅔ of the fluid out of the master cylinder.
2. Raise and safely support the vehicle.
3. Remove the wheels.
4. Remove the brake caliper mounting bolts and remove the caliper from the mounting bracket.
5. Support the caliper with a wire.
6. Using a large pair of plies or a C-clamp compress the caliper piston back into the bore.
7. Remove the disc brake pads and any shims from the caliper mounting bracket.

To install:

8. Install the brake pads and any shims removed from the caliper mounting bracket.

9. Install the caliper on the mounting bracket and install the mounting bolts. Tighten the mounting bolts to 20 ft. lbs. (27 Nm).
10. Install the front wheels and lower the vehicle.

✳✳ CAUTION

Do not attempt to drive the vehicle until after the following step is performed.

11. Depress the brake pedal repeatedly until a firm pedal is obtained. Do not attempt to drive the vehicle unless a firm pedal is obtained.
12. Check the fluid level in the master cylinder. Add fresh brake fluid, as necessary.
13. Road-test the vehicle.

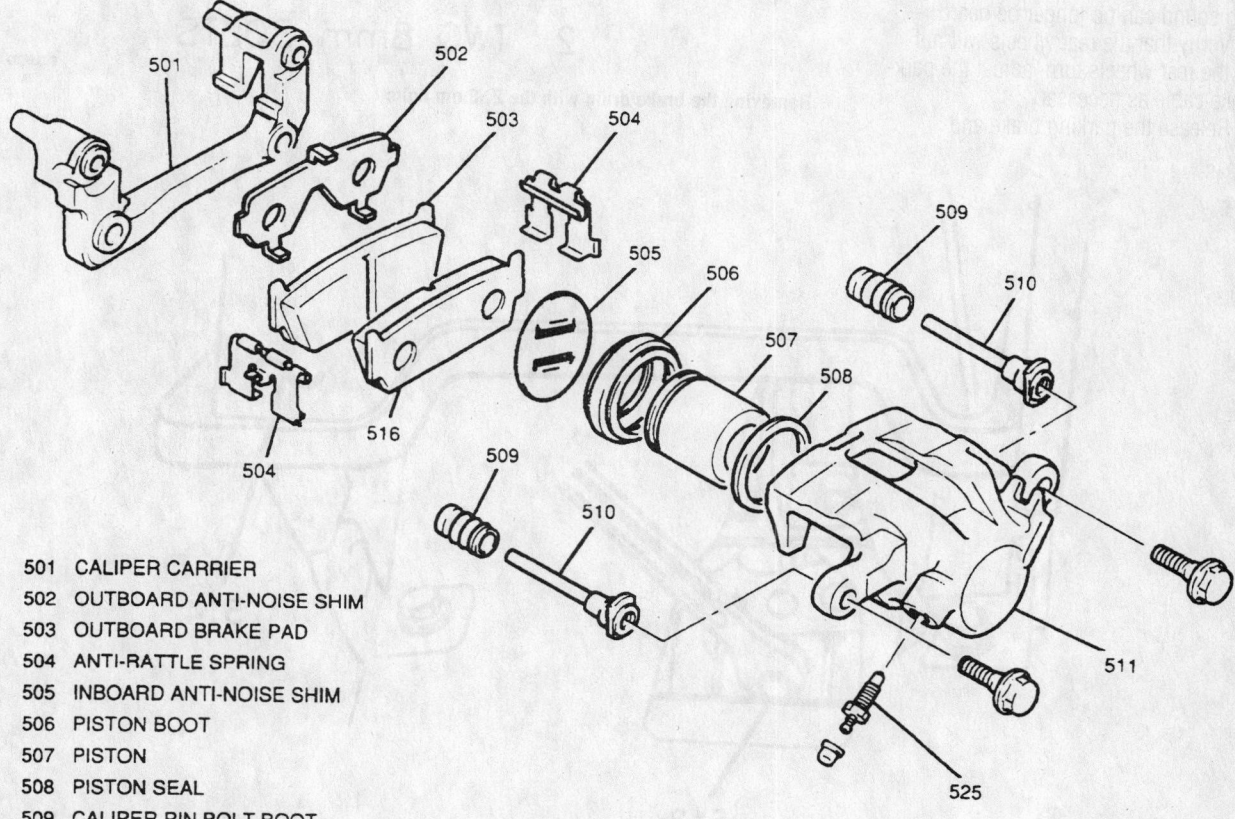

501 CALIPER CARRIER
502 OUTBOARD ANTI-NOISE SHIM
503 OUTBOARD BRAKE PAD
504 ANTI-RATTLE SPRING
505 INBOARD ANTI-NOISE SHIM
506 PISTON BOOT
507 PISTON
508 PISTON SEAL
509 CALIPER PIN BOLT BOOT
510 CALIPER PIN BOLT
511 CALIPER
516 INBOARD BRAKE PAD
525 BLEEDER VALVE

Front disc brake components

93026G37

Brake Drums

REMOVAL & INSTALLATION

1. Raise and safely support the vehicle.

2. Remove the rear wheel(s).

3. Release the parking brake.

4. Remove the parking brake lever cover screws and loosen the brake cable locking nut.

5. Install 2, 8mm bolts into the brake drum holes and uniformly tighten each bolt. Tighten each bolt until the brake drum is removed from the vehicle. If there is difficulty in removing the drum, insert a small tool through the hole in the rear of the backing plate, and hold the automatic adjusting lever away from the adjuster. Using another narrow, flat tool at the same time, reduce the brake shoe adjuster by turning the adjusting wheel.

To install:

6. Install the brake drum and pull the parking brake lever all the way up until a clicking sound can no longer be heard.

7. Verify that the rear wheels will not turn. If the rear wheels turn, adjust the parking brake cable as necessary.

8. Release the parking brake and

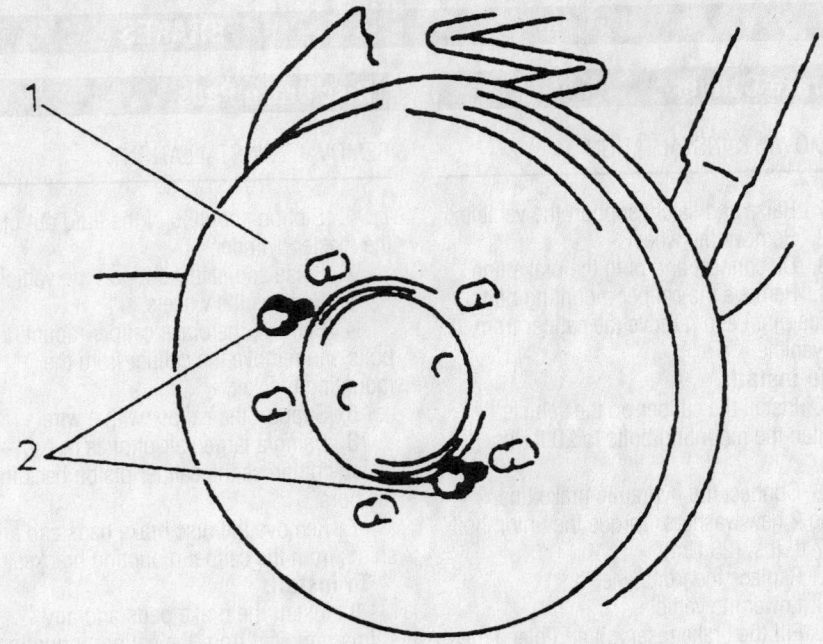

1 DRUM
2 TWO 8mm BOLTS

93026G39

Removing the brake drum with the 2, 8mm bolts

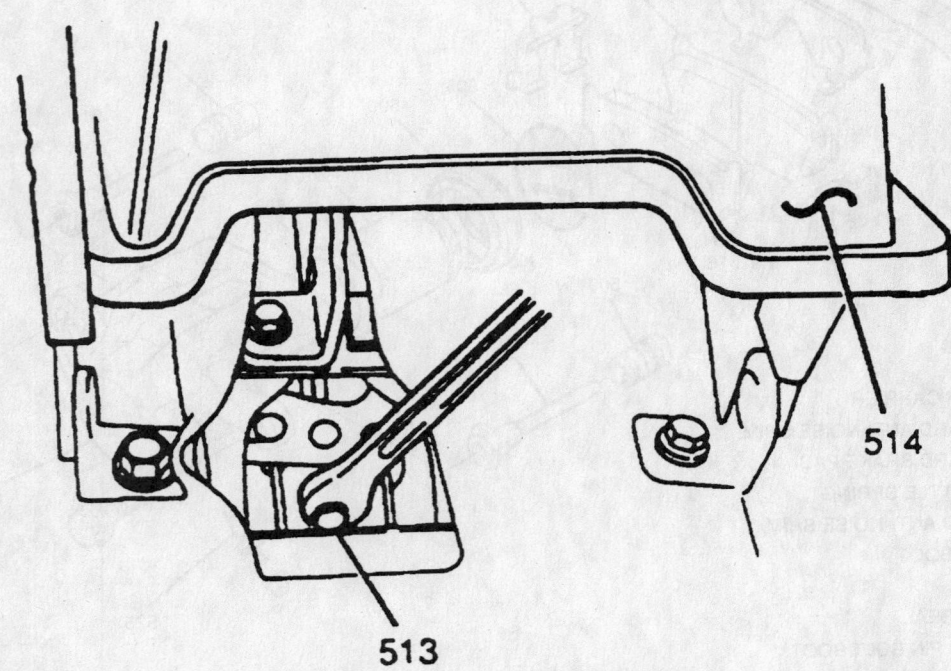

513 PARKING BRAKE CABLE LOCKNUT
514 PARKING BRAKE LEVER COVER

93026G38

Reducing the adjuster to remove the brake drum

Brake Shoes

REMOVAL & INSTALLATION

1. Raise and safely support the vehicle.
2. Remove the rear wheel(s).
3. Remove the brake drum.
4. Using a suitable tool, remove the brake shoe return spring.
5. Using a brake spring hold-down tool, disengage the hold-down spring and retainers from the front shoe. Remove the hold-down retainer pinch
6. Disconnect the anchor spring from the front shoe and remove the front shoe.
7. Remove the anchor spring from the rear shoe. Using a brake spring hold-down tool, disengage the hold-down spring and retainers from the rear shoe. Remove the hold-down pinch
8. Disengage the parking brake lever from the parking brake cable and remove the rear shoe.
9. Remove the C-washer, the automatic adjuster lever and spring, the C-washer, and the parking brake lever from the rear shoe.
10. Thoroughly clean the backing plate and brake hardware with brake cleaning solvent. Apply high temperature grease to the backing plate shoe contact points, anchor plate and shoe contact points, adjusting bolt, and adjuster and brake shoe contact points.

To install:
11. Reinstall the automatic adjuster lever and the parking brake lever to the rear shoe using new C-washers.
12. Connect the parking brake lever to the parking brake cable. Set the adjuster and spring to the rear shoe.
13. Set the rear brake shoe in place, install the hold-down pin and install the hold-down spring and retainers. Make sure that the shoe is inserted in the wheel cylinder and that the other end is in the anchor plate.
14. Install the anchor spring to the rear shoe.
15. Install the front shoe to the other end of the anchor spring and set the front shoe in place. Make sure that the front shoe engages the wheel cylinder, adjuster mechanism and spring, and the anchor plate.
16. Reinstall the front brake shoe hold-down pin and secure with the hold-down spring and retainers using a suitable tool.
17. Install the return spring.
18. Install the brake drum and pull the parking brake lever all the way up until a clicking sound can no longer be heard.
19. Verify that the rear wheels will not turn. If the rear wheels turn, adjust the parking brake cable as necessary.
20. Release the parking brake and

1 WHEEL CYLINDER
2 ADJUSTER
3 SHOE RETURN LOWER SPRING
4 BRAKE SHOES
5 SHOE HOLD DOWN SPRING
6 ADJUSTER SPRING
7 PAWL LEVER
8 SHOE RETURN UPPER SPRING
9 BACKING PLATE
10 SHOE HOLD DOWN PIN

93026G40

Exploded view of the rear brake components

remove the brake drum. Measure the diameter of the brake shoes. Outer diameter should be as follows:
- 2 door models: 8.638 (0.0012 inches (219 (0.3mm)
- 4 door models: 9.980 (0.0079 inches (253.5 (0.2mm)
9. If the brake shoe clearance is not correct, adjust the brake shoes until the clearance is correct.
10. Reinstall the brake drum, replace the wheel(s), and safely lower the vehicle.
11. Adjust the parking brake and install the cover with the 2 screws.
12. Road-test the vehicle for proper brake operation.

remove the brake drum. Measure the diameter of the brake shoes. Brake diameter should be as follows:

- 2 door models: 8.638 (0.0012 inches (219 (0.3mm)

- 4 door models: 9.980 (0.0079 inches (253.5 (0.2mm)

21. If the brake shoe clearance is not correct, adjust the brake shoes until the clearance is correct.

22. Reinstall the brake drum, replace the wheel(s), and safely lower the vehicle.

23. Road-test the vehicle for proper brake operation.

SPECIFICATION CHARTS

ENGINE AND VEHICLE IDENTIFICATION

		Engine						Model Year	
Code ①	Liters (cc)	Cu. In.	Cyl.	Fuel Sys.	Engine Type	Eng. Mfg.		Code ②	Year
1ZZ-FE	1.8 (1794)	109	4	EFI	DOHC	Toyota		3	2003
2ZZ-GE	1.8 (1796)	109.5	4	EFI	DOHC	Toyota		4	2004

EFI: Electronic Fuel Injection

DOHC: Double Overhead Camshaft

① 8th digit of VIN

② 10th digit of VIN

42372-VIBE-C01

GENERAL ENGINE SPECIFICATIONS

Year	Model	Engine Displacement Liters (cc)	Engine Series (ID/VIN)	Fuel System	Net Horsepower @ rpm	Net Torque @ rpm (ft. lbs.)	Bore x Stroke (in.)	Com-pression Ratio	Oil Pressure @ idle
2003	Vibe	1.8 (1794)	1ZZ-FE	EFI	①	②	3.11x3.60	10.0:1	4.2
	Vibe	1.8 (1796)	2ZZ-GE	EFI	180@7600	130@6800	3.23x3.35	11.5:1	2.8
2004	Vibe	1.8 (1794)	1ZZ-FE	EFI	①	②	3.11x3.60	10.0:1	4.2
	Vibe	1.8 (1796)	2ZZ-GE	EFI	180@7600	130@6800	3.23x3.35	11.5:1	2.8

EFI: Electronic Fuel Injection

① 2WD models: 130@6000
   4WD models: 123@6000

② 2WD models: 126@4200
   4WD models: 119@4200

42372-VIBE-C02

ENGINE TUNE-UP SPECIFICATIONS

Year	Engine Displacement Liters (cc)	Engine ID/VIN	Spark Plug Gap (in.)	Ignition Timing (deg.)	Fuel Pump (psi)	Idle Speed (rpm) MT	Idle Speed (rpm) AT	Valve Clearance In.	Valve Clearance Ex.
2003	1.8 (1794)	1ZZ-FE	0.043	①	44-50	650-750	650-750	0.0059-0.0098	0.0098-0.0138
	1.8 (1796)	2ZZ-GE	0.043	②	44-50	750-850	700-800	0.0031-0.0071	0.0087-0.0126
2004	1.8 (1794)	1ZZ-FE	0.043	①	44-50	650-750	650-750	0.0059-0.0098	0.0098-0.0138
	1.8 (1796)	2ZZ-GE	0.043	②	44-50	750-850	700-800	0.0031-0.0071	0.0087-0.0126

Note: The Vehicle Emission Control Information label often reflects specification changes made during production. The label figures must be used if they differ from those in this chart.

① With terminal TC and CG of DLC3 connected: 8-12 degrees BTDC
   With terminal TC and CG of DLC3 disconnected: 10-18 degrees BTDC

② With terminal TC and CG of DLC3 connected: 8-12 degrees BTDC
   With terminal TC and CG of DLC3 disconnected:
   A/T: 10-18 degrees BTDC
   M/T: 4-12 degrees BTDC

42372-VIBE-C03

CAPACITIES

Year	Model	Engine Displacement Liters (cc)	Engine ID/VIN	Engine Oil with Filter	Transmission (pts.)			Drive Axle		Fuel Tank (gal.)	Cooling System (qts.)
					5-Spd	6-Spd	Auto.	Front (pts.)	Rear (pts.)		
2003	Vibe	1.8 (1794)	1ZZ-FE	3.9	4.0	4.0	5.2	3.0	—	13.2	①
	Vibe	1.8 (1796)	2ZZ-GE	4.7	4.8	4.8	6.1	②	—	13.2	6.0
2004	Vibe	1.8 (1794)	1ZZ-FE	3.9	4.0	4.0	5.2	3.0	—	13.2	①
	Vibe	1.8 (1796)	2ZZ-GE	4.7	4.8	4.8	6.1	②	—	13.2	6.0

Note: All capacities are approximate. Add fluid gradually and check to be sure a proper fluid level is obtained.

① M/T with Nippodenso radiator: 5.6
   A/T with Nippodenso radiator: 6.2
   M/T with Harrison radiator: 6.3
   A/T with Harrison radiator: 6.2

② Included in transaxle capacity

42372-VIBE-C04

VALVE SPECIFICATIONS

Year	Engine Displacement Liters (cc)	Engine ID/VIN	Seat Angle (deg.)	Face Angle (deg.)	Spring Test Pressure (lbs. @ in.)	Spring Installed Height (in.)	Stem-to-Guide Clearance (in.)		Stem Diameter (in.)	
							Intake	Exhaust	Intake	Exhaust
2003	1.8 (1794)	1ZZ-FE	45	44.5	31.3-34.8 @ 1.252	1.323	0.0010-0.0024	0.0012-0.0026	0.2154-0.2159	0.2152-0.2158
	1.8 (1796)	2ZZ-GE	45	44.5	①	1.516	0.0010-0.0023	0.0012-0.0025	0.2145-0.2156	0.2144-0.2154
2004	1.8 (1794)	1ZZ-FE	45	44.5	31.3-34.8 @ 1.252	1.323	0.0010-0.0024	0.0012-0.0026	0.2154-0.2159	0.2152-0.2158
	1.8 (1796)	2ZZ-GE	45	44.5	①	1.516	0.0010-0.0023	0.0012-0.0025	0.2145-0.2156	0.2144-0.2154

① Intake: 49.6-55.5 @ 1.516
   Exhaust: 47.6-52.6 @ 1.516

42372-VIBE-C05

TORQUE SPECIFICATIONS
All readings in ft. lbs.

Year	Engine Displacement Liters (cc)	Engine ID/VIN	Cylinder Head Bolts	Main Bearing Bolts	Rod Bearing Bolts	Crankshaft Damper Bolts	Flywheel Bolts	Manifold		Spark Plugs	Lug Nuts
								Intake	Exhaust		
2003	1.8 (1794)	1ZZ-FE	①	②	③	102	①	22	27	18	76
	1.8 (1796)	2ZZ-GE	④	⑤	⑥	87	①	⑦	37	13	76
2004	1.8 (1794)	1ZZ-FE	①	②	③	102	①	22	27	18	76
	1.8 (1796)	2ZZ-GE	④	⑤	⑥	87	①	⑦	37	13	76

① Step 1: 36 ft. lbs.
   Step 2: 90 degree turn

② 12 pointed bolts:
   Step 1: 33 ft. lbs.
   Step 2: 90 degree turn
   Hex head bolts: 14 ft. lbs.

③ Step 1: 15 ft. lbs.
   Step 2: 90 degree turn

④ Step 1: 26 ft. lbs.
   Step 2: 180 degree turn

⑤ 12 pointed bolts:
   Step 1: 16 ft. lbs.
   Step 2: 32 ft. lbs.
   Step 3: 45 degree turn
   Step 4: 45 degree turn
   Hex head bolts: 13 ft. lbs.

⑥ Step 1: 22 ft. lbs.
   Step 2: 90 degree turn

⑦ Bolt A: 25 ft. lbs.
   Bolt B: 34 ft. lbs.

42372-VIBE-C06

PISTON AND RING SPECIFICATIONS
All measurements are given in inches.

Year	Engine Displacement Liters (cc)	Engine ID/VIN	Piston Clearance	Ring Gap			Ring Side Clearance		
				Top Compression	Bottom Compression	Oil Control	Top Compression	Bottom Compression	Oil Control
2003	1.8 (1762)	1ZZ-FE	0.0026-0.0035	0.0098-0.0138	0.0138-0.0197	0.0059-0.0197	0.0008-0.0028	0.0012-0.0028	0.0012-0.0043
	1.8 (1796)	2ZZ-GE	0.0003-0.0015	0.0098-0.0138	0.0138-0.0197	NA	0.0009-0.0028	0.0012-0.0028	NA
2004	1.8 (1762)	1ZZ-FE	0.0026-0.0035	0.0098-0.0138	0.0138-0.0197	0.0059-0.0197	0.0008-0.0028	0.0012-0.0028	0.0012-0.0043
	1.8 (1796)	2ZZ-GE	0.0003-0.0015	0.0098-0.0138	0.0138-0.0197	NA	0.0009-0.0028	0.0012-0.0028	NA

NA - Not available

42372-VIBE-C07

CRANKSHAFT AND CONNECTING ROD SPECIFICATIONS
All measurements are given in inches.

Year	Engine Displacement Liters (cc)	Engine ID/VIN	Crankshaft				Connecting Rod		
			Main Brg. Journal Dia.	Main Brg. Oil Clearance	Shaft End-play	Thrust on No.	Journal Diameter	Oil Clearance	Side Clearance
2003	1.8 (1762)	1ZZ-FE	1.8893-1.8898	0.0006-0.0013	0.0008-0.0087	3	1.7320-1.7323	0.0011-0.0024	0.0063-0.0135
	1.8 (1796)	2ZZ-GE	1.8893-1.8898	0.0006-0.0013	0.0016-0.0094	3	1.7713-1.7717	0.0011-0.0020	0.0063-0.0135
2004	1.8 (1762)	1ZZ-FE	1.8893-1.8898	0.0006-0.0013	0.0008-0.0087	3	1.7320-1.7323	0.0011-0.0024	0.0063-0.0135
	1.8 (1796)	2ZZ-GE	1.8893-1.8898	0.0006-0.0013	0.0016-0.0094	3	1.7713-1.7717	0.0011-0.0020	0.0063-0.0135

42372-VIBE-C08

WHEEL ALIGNMENT

Year	Model		Caster		Camber		Toe-in (in.)	Steering Axis Inclination (Deg.)
			Range (+/-Deg.)	Preferred Setting (Deg.)	Range (+/-Deg.)	Preferred Setting (Deg.)		
2003	Vibe - 2WD	F	0.75	+2.78	0.75	-0.77	0+/-0.08	12.47+/-0.75
		R	—	—	0.50	-1.45	0.11+/-0.11	—
	Vibe - 4WD	F	0.75	+2.77	0.75	-0.48	0+/-0.08	12.22+/-0.75
		R	—	—	0.75	-0.73	0.08+/-0.08	—
2004	Vibe - 2WD	F	0.75	+2.78	0.75	-0.77	0+/-0.08	12.47+/-0.75
		R	—	—	0.50	-1.45	0.11+/-0.11	—
	Vibe - 4WD	F	0.75	+2.77	0.75	-0.48	0+/-0.08	12.22+/-0.75
		R	—	—	0.75	-0.73	0.08+/-0.08	—

42372-VIBE-C09

TIRE, WHEEL AND BALL JOINT SPECIFICATIONS

Year	Model	OEM Tires		Tire Pressures (psi)		Wheel Size	Ball Joint Inspection
		Standard	Optional	Front	Rear		
2003	Vibe	205/55R16	—	33	33	6.5-JJ	9-26 in. ①
2004	Vibe	205/55R16	—	33	33	6.5-JJ	9-26 in. ①

OEM: Original Equipment Manufacturer

PSI: Pounds Per Square Inch

STD: Standard

OPT: Optional

① Torque required in inch lbs. to rotate ball joint when removed from the knuckle

42372-VIBE-C10

BRAKE SPECIFICATIONS
All measurements in inches unless noted

Year	Model		Brake Disc			Brake Drum Diameter			Minimum Lining Thickness	Brake Caliper	
			Original Thickness	Minimum Thickness	Maximum Runout	Original Inside Diameter	Max. Wear Limit	Maximum Machine Diameter		Bracket Bolts (ft. lbs.)	Mounting Bolts (ft. lbs.)
2003	Vibe	F	0.984	0.906	0.0020	—	—	—	0.039	25	79
		R	0.354	0.295	0.0059	9.00	—	9.04	0.039	—	34
2004	Vibe	F	0.984	0.906	0.0020	—	—	—	0.039	25	79
		R	0.354	0.295	0.0059	9.00	—	9.04	0.039	—	34

F: Front

R: Rear

42372-VIBE-C11

SCHEDULED MAINTENANCE INTERVALS
PONTIAC—VIBE

TO BE SERVICED	TYPE OF SERVICE	__VEHICLE MILEAGE INTERVAL (x1000)												
		7.5	15	22.5	30	37.5	45	52.5	60	67.5	75	82.5	90	97.5
Engine oil & filter	R	✓	✓	✓	✓	✓	✓	✓	✓	✓	✓	✓	✓	✓
Drive belts	S/I								✓	✓	✓	✓	✓	✓
Automatic transaxle fluid & filter	S/I		✓		✓		✓		✓		✓		✓	
Ball joints & dust covers	S/I		✓		✓		✓		✓		✓		✓	
Bolts & nuts on body & chassis	S/I		✓		✓		✓		✓		✓		✓	
Brake line pipes & hoses	S/I		✓		✓		✓		✓		✓		✓	
Brake linings & drums	S/I		✓		✓		✓		✓		✓		✓	
Brake pads & discs (front & rear if equipped)	S/I		✓		✓		✓		✓		✓		✓	
Differential oil	S/I		✓		✓		✓		✓		✓		✓	
Drive shaft boots (except Supra)	S/I		✓		✓		✓		✓		✓		✓	
Manual transaxle oil	S/I		✓		✓		✓		✓		✓		✓	
Steering gear housing oil	S/I		✓		✓		✓		✓		✓		✓	
Steering linkage	S/I		✓		✓		✓		✓				✓	
Air filter	R				✓				✓				✓	
Spark plugs	R				✓				✓					
Spark plugs (platinum tip)	R								✓					
Exhaust system	S/I				✓				✓				✓	
Fuel lines & connections	S/I				✓				✓				✓	
Valve clearance	S/I				✓				✓					
Engine coolant	R						✓				✓			
Fuel tank cap gasket	R								✓					
Charcoal canister	S/I								✓					

R: Replace   S/I: Service or Inspect

FREQUENT OPERATION MAINTENANCE (SEVERE SERVICE)

If a vehicle is operated under any of the following conditions it is considered severe service:

- Extremely dusty areas.
- 50% or more of the vehicle operation is in 32°C (90°F) or higher temperatures, or constant operation in temperatures below 0°C (32°F).
- Prolonged idling (vehicle operation in stop and go traffic).
- Frequent short running periods (engine does not warm to normal operating temperatures).
- Police, taxi, delivery usage or trailer towing usage.

Oil & oil filter: change every 6000 miles.

Bolts & nuts on chassis & body: tighten every 7500 miles.

Ball joints & dust covers: service or inspect every 12,000 miles.

Brake linings & drums: service or inspect ever 12,000 miles.

Brake pads & discs (front & rear if equipped): service or inspect every 12,000 miles.

Drive shaft boots & except Supra): service or inspect every 12,000 miles.

Steering linkage: service or inspect every 12,000 miles.

Air filter: service or inspect every 15,000 miles.

Exhaust system: service or inspect every 15,000 miles.

Timing belt: replace every 60,000 miles.

42372-VIBE-C12

PRECAUTIONS

Before servicing any vehicle, please be sure to read all of the following precautions, which deal with personal safety, prevention of component damage, and important points to take into consideration when servicing a motor vehicle:

• Never open, service or drain the radiator or cooling system when the engine is hot; serious burns can occur from the steam and hot coolant.

• Observe all applicable safety precautions when working around fuel. Whenever servicing the fuel system, always work in a well-ventilated area. Do not allow fuel spray or vapors to come in contact with a spark, open flame or excessive heat (a hot drop light, for example). Keep a dry chemical fire extinguisher near the work area. Always keep fuel in a container specifically designed for fuel storage; also, always properly seal fuel containers to avoid the possibility of fire or explosion. Refer to the additional fuel system precautions later in this section.

• Fuel injection systems often remain pressurized, even after the engine has been turned **OFF**. The fuel system pressure must be relieved before disconnecting any fuel lines. Failure to do so may result in fire and/or personal injury.

• Brake fluid often contains polyglycol ethers and polyglycols. Avoid contact with the eyes and wash your hands thoroughly after handling brake fluid. If you do get brake fluid in your eyes, flush your eyes with clean, running water for 15 minutes. If eye irritation persists, or if you have taken brake fluid internally, IMMEDIATELY seek medical assistance.

• The EPA warns that prolonged contact with used engine oil may cause a number of skin disorders, including cancer! You should make every effort to minimize your exposure to used engine oil. Protective gloves should be worn when changing oil. Wash your hands and any other exposed skin areas as soon as possible after exposure to used engine oil. Soap and water, or waterless hand cleaner should be used.

• All new vehicles are now equipped with an air bag system. The system must be disabled before performing service on or around system components, steering column, instrument panel components, wiring and sensors. Failure to follow safety and disabling procedures could result in accidental air bag deployment, possible personal injury and unnecessary system repairs.

• Always wear safety goggles when working with, or around, the air bag system. When carrying a non-deployed air bag, be sure the bag and trim cover are pointed away from your body. When placing a non-deployed air bag on a work surface, always face the bag and trim cover upward, away from the surface. This will reduce the motion of the module if it is accidentally deployed. Refer to the additional air bag system precautions later in this section.

• Clean, high quality brake fluid from a sealed container is essential to the safe and proper operation of the brake system. You should always buy the correct type of brake fluid for your vehicle. If the brake fluid becomes contaminated, completely flush the system with new fluid. Never reuse any brake fluid. Any brake fluid that is removed from the system should be discarded. Also, do not allow any brake fluid to come in contact with a painted surface; it will damage the paint.

• Never operate the engine without the proper amount and type of engine oil; doing so WILL result in severe engine damage.

• Timing belt maintenance is extremely important! Many models utilize an interference-type, non-freewheeling engine. If the timing belt breaks, the valves in the cylinder head may strike the pistons, causing potentially serious (also time-consuming and expensive) engine damage.

• Disconnecting the negative battery cable on some vehicles may interfere with the functions of the on-board computer system(s) and may require the computer to undergo a relearning process once the negative battery cable is reconnected.

• When servicing drum brakes, only disassemble and assemble one side at a time, leaving the remaining side intact for reference.

• Only an MVAC-trained, EPA-certified automotive technician should service the air conditioning system or its components.

ENGINE REPAIR

Alternator

REMOVAL & INSTALLATION

1. Before servicing the vehicle, refer to the precautions in the beginning of this section.
2. Remove or disconnect the following:
   • Negative battery cable
   • Drive belt
   • Wire clamp from the clip on the end frame
   • Rubber clamp and nut
   • Alternator wiring and connector
   • Alternator

To install:

3. Install or connect the following:
   • Alternator. Torque the 12mm bolt to 18 ft. lbs. (25 Nm) and the 14mm bolt to 39 ft. lbs. (54 Nm).
   • Alternator connector and wiring

   • Rubber clamp and nut
   • Wire clamp
   • Drive belt
   • Negative battery cable

Ignition Timing

ADJUSTMENT

➡ **The timing on engines equipped with Distributorless Ignition Systems (DIS) is not adjustable.**

Engine Assembly

REMOVAL & INSTALLATION

1. Before servicing the vehicle, refer to the precautions in the beginning of this section.
2. Relieve the fuel system pressure.

3. Drain the cooling system.
4. Drain the engine oil.
5. Drain the transaxle fluid and transfer fluid, if equipped.
6. Remove or disconnect the following:
   • Negative battery cable. Wait at least 90 seconds before proceeding.
   • Battery
   • Hood
   • Undercovers
   • Radiator inlet and outlet hoses
   • Radiator hose outlet
   • Oil cooler inlet and outlet tubes
   • Upper radiator support and radiator, if equipped with A/C
   • Battery
   • Air cleaner assembly
   • Fuel pipe clamp
   • Fuel tube sub-assembly
   • Accelerator control cable
   • Cruise control actuator, if equipped
   • Union-to-connector tube hose

- Heater inlet and outlet hoses
- Transmission shift cable(s)
- Clutch release cylinder, on manual transaxle
- Glove compartment door
- Engine relay block cover
- 3 connectors from the relay block
- 2 ground cables
- Engine wire from the Engine Control Module (ECM) and junction block
- Engine wire from the cabin
- Drive belt
- Compressor and magnetic clutch, if equipped with A/C. Unbolt and position aside, DO NOT disconnect the lines.
- Vane pump oil reservoir from the bracket
- Return tube
- Right side front door scuff plate
- Right side cowl side trim plate
- Right side rear door scuff plate, AWD
- Lower right side center pillar garnish, AWD
- Right front seat, AWD
- Column hole cover silencer sheet
- Steering intermediate shaft
- Front floor panel brace, FWD
- Center exhaust pipe, AWD
- Propeller shaft with center bearing shaft, AWD
- Front exhaust pipe
- Front hub nuts
- Tie rod ends from the steering knuckles

7. Separate the front stabilizer links and lower control arm ball joints
- Front halfshafts

8. Remove the engine from the vehicle, as follows:
   a. Set the engine lifter.
   b. Remove the bolts and nuts, then remove the engine mounting insulator.
   c. Remove the through bolt and nut, then detach the engine mounting insulator from the vehicle.
   d. Remove the 6 bolts as shown.

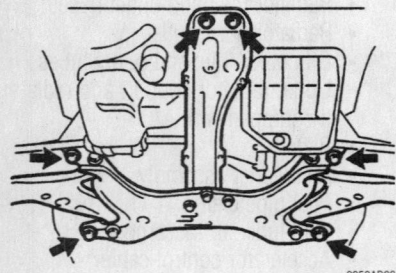

Remove the 6 bolts, as indicated by arrows

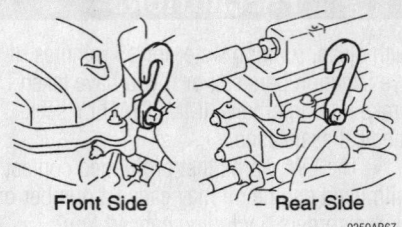

Install the engine hangers—1ZZ-FE shown, 2ZZ-GE similar

   e. Use a suitable tool to suspend the engine assembly, as shown in the figure.
   f. No. 1 engine hanger: P/N 12281-15040 (1ZZ-FE), 12281-88600 (2ZZ-GE).
   g. No. 2 engine hanger: P/N 12281-22021 (1ZZ-FE), 12281-88600 (2ZZ-GE).
   h. Bolt: P/N 91512-B1016.
   i. Torque the bolts to 28 ft. lbs. (38 Nm).

✳✳ WARNING

Do not try to suspend the engine by hooking the chain to any other part.

   j. Attach an engine chain hoist to the hangers.
   k. Using the chain block and sling device, suspend the engine.
   l. Remove the engine and transaxle assembly from the vehicle.

9. Remove or disconnect the following components, as necessary:
- Vane pump
- Steering gear
- Crossmember
- Manifold stay
- Oxygen (O_2) sensor
- Exhaust manifold
- Starter
- Transaxle
- Transfer case
- Clutch
- Flywheel
- Alternator
- Ignition coil
- Fuel delivery pipe
- Intake manifold
- Oil level gauge
- Water inlet and bypass pipes
- Thermostat
- Oil pressure switch
- Crankshaft Position (CKP) sensor
- Knock Sensor (KS)
- Drive belt tensioner
- Engine mounts and brackets
- Coolant Temperature Sensor (CTS)

To install:

10. Install any removed components to the engine and transaxle assembly.

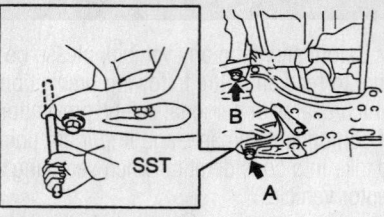

Insert the SST to the positioning holes of the right handle crossmember and on the right handle of the vehicle. Temporarily tighten bolt A, then bolt B

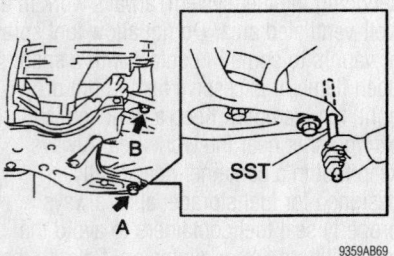

Insert the SST to the positioning holes of the left handle crossmember and on the left handle of the vehicle. Temporarily tighten bolt A, then bolt B

11. To install the engine:
   a. Place the engine and transaxle on an engine lifter.
   b. Install the engine with the transaxle to the vehicle.
   c. Temporarily install the crossmember and 6 bolts.
   d. Install the left engine mounting insulator. Tighten the bolts to 59 ft. lbs. (80 Nm).
   e. Install the right engine mounting insulator. Tighten the bolts to 38 ft. lbs. (52 Nm).
   f. Insert SST 09670-00010 to the positioning holes of the right handle crossmember and on the right handle of the vehicle. Temporarily tighten bolt A, then bolt B.
   g. Insert SST 09670-00010 to the

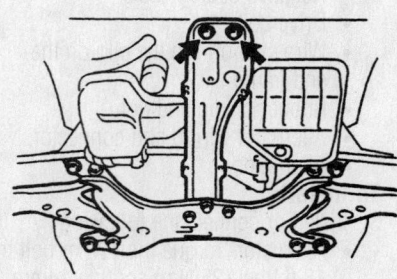

Tighten the 2 crossmember bolts, indicated by arrows

positioning holes of the left handle crossmember and on the left handle of the vehicle. Temporarily tighten bolt A, then bolt B.

h. Insert the SST to the positioning holes on the right-handle crossmember and right handle. Tighten bolt A to 116 ft. lbs. (157 Nm) and bolt B to 83 ft. lbs. (113 Nm).

i. Insert the SST to the positioning holes on the left-handle crossmember and left handle. Tighten bolt A to 116 ft. lbs. (157 Nm) and bolt B to 83 ft. lbs. (113 Nm).

j. Tighten the 2 crossmember bolts, shown in the figure, to 29 ft. lbs. (39 Nm).

12. Installation of the remaining components is the reverse of the removal procedure.

13. Make sure all fluid levels are accurate, then start the engine check for leaks.

Water Pump

REMOVAL & INSTALLATION

1ZZ-FE Engine

1. Before servicing the vehicle, refer to the precautions in the beginning of this section.
2. Drain the cooling system.
3. Remove or disconnect the following:
   - Negative battery cable
   - Right-hand engine under cover
   - Drive belt

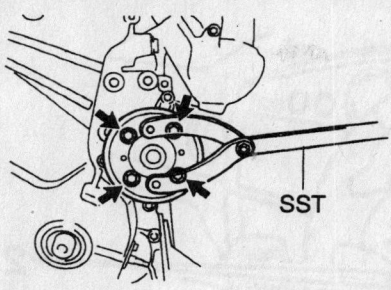

View of the special tool needed to remove and install the water pump pulley

   - Alternator
   - Water pump

To install:
4. Install or connect the following:
   - Water pump. Torque bolts marked **A** (short) to 80 inch lbs. (9 Nm) and bolts marked **B** (long) to 96 inch lbs. (11 Nm).
   - Alternator
   - Drive belt
   - Right engine under cover
   - Negative battery cable
5. Fill the cooling system to the proper level.
6. Start the vehicle, check for leaks and repair if necessary.

2ZZ-GE Engine

1. Before servicing the vehicle, refer to the precautions in the beginning of this section.
2. Drain the cooling system.
3. Remove or disconnect the following:

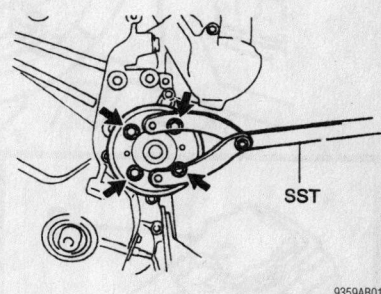

Water pump mounting and bolt locations—2ZZ-GE engine

   - Negative battery cable
   - Right-hand engine under cover
   - Drive belt
   - Alternator
   - Water pump pulley, using SST 09960-10010
   - Water pump and O-ring

To install:
4. Install or connect the following:
   - Water pump with new O-ring. Torque the bolts to 80 inch lbs. (9 Nm).
   - Water pump pulley, using SST 09960-10010. Torque the bolts to 11 ft. lbs. (15 Nm).
   - Alternator
   - Drive belt
   - Right engine under cover
   - Negative battery cable
5. Fill the cooling system to the proper level.
6. Start the vehicle, check for leaks and repair if necessary.

Heater Core

REMOVAL & INSTALLATION

1. Before servicing the vehicle, refer to the precautions in the beginning of this section.
2. Drain the cooling system.
3. Discharge and recover the A/C system refrigerant using approved equipment.
4. Remove or disconnect the following:
   - Negative battery cable

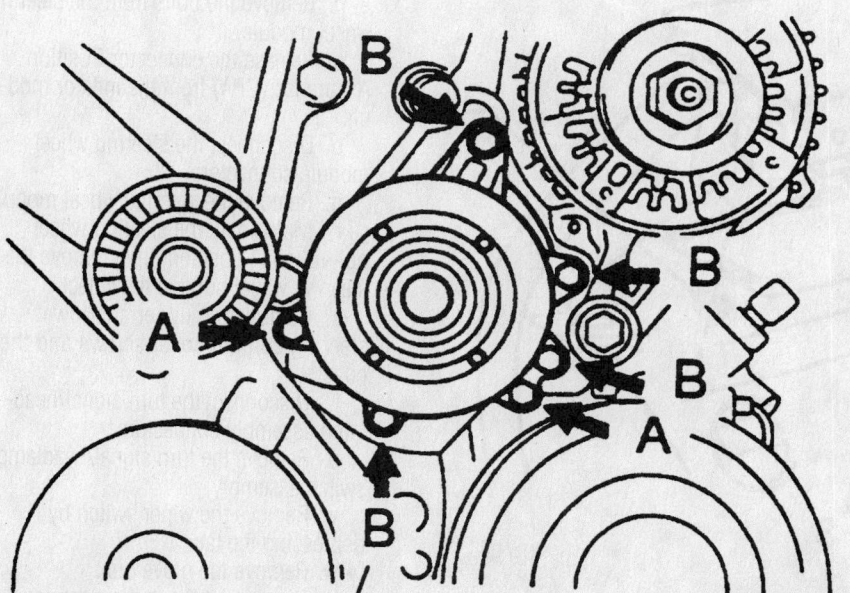

Water pump bolt identification—1.8L (1ZZ-FE) engine

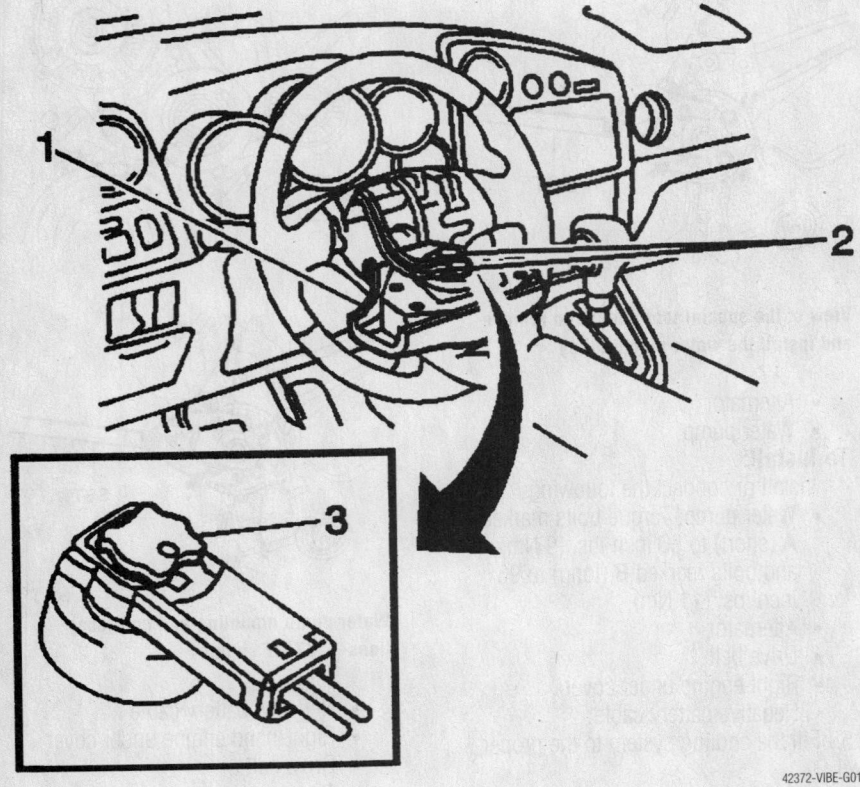

Exploded view of the CPA assembly

42372-VIBE-G01

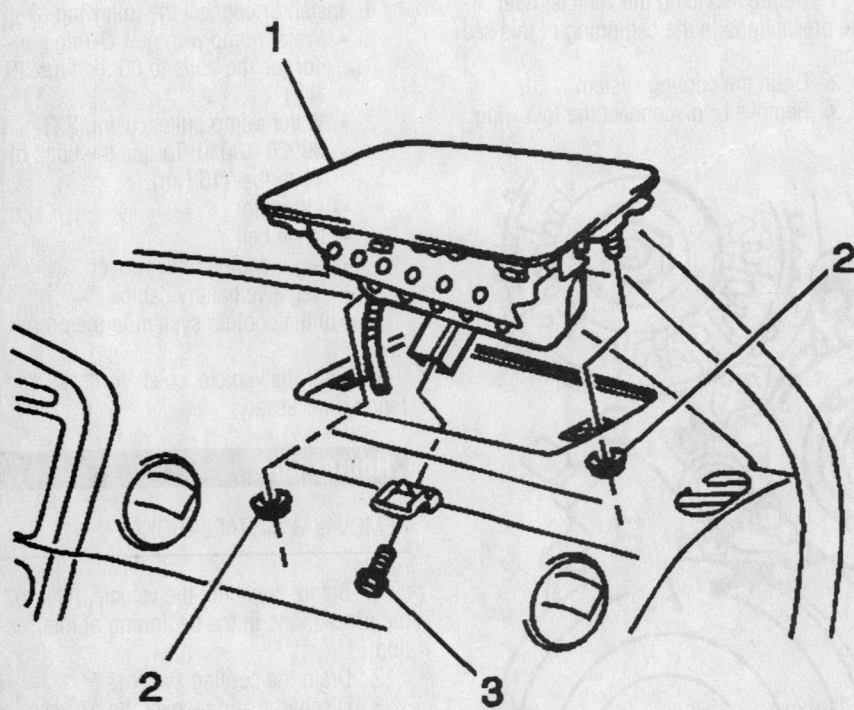

Remove the instrument panel module connectors and the passenger air bag assembly

42372-VIBE-G02

- Heater hoses from the core
- Evaporator inlet and outlet tubes from the evaporator and cap the lines to avoid system contamination

5. Remove the instrument panel as follows:

a. Disable the air bag system.

b. Using a taped flat–bladed tool, carefully pry the retaining clips attaching the center trim plate to the instrument panel.

c. Disconnect the A/C switch, hazard switch; rear defogger switch and passenger seat belt indicator switch electrical connections.

d. Remove the radio retaining screws, clamp from the radio bracket, slide the radio forward to disconnect the power and antenna connections. Remove the radio.

e. Remove the A/C switch and screw.

f. Remove the hazard switch.

g. Remove the rear defogger switch.

h. Remove the manual transmission shift knob.

i. Using a taped flat–bladed tool, carefully pry the retaining clips attaching the front floor console trim plate to the floor console assembly.

j. Disconnect the 2 cigar lighter connectors.

k. Disconnect the accessory power receptacle connectors.

l. Remove the cigar lighters and power receptacle.

m. Place both wheels in the straight ahead position.

n. Remove the bolts from the steering wheel module.

o. Release the connector Position Assurance (CPA) from the inflator module.

p. Disconnect the steering wheel module connectors.

q. Remove the steering wheel module.

r. Matchmark the steering wheel nut–to–shaft position, then remove the steering wheel nut and the wheel.

s. Remove the upper and lower steering column cover screws and the covers.

t. Disconnect the turn signal/headlamp assembly connectors.

u. Remove the turn signal/headlamp switch assembly

v. Remove the wiper switch by depressing the tab.

w. Remove the glove box.

x. Disconnect the instrument panel connector.

y. Remove the instrument panel module connectors and the passenger air bag assembly.

z. Remove the cluster trim plate by disengaging the clips.

aa. Remove the cluster screw and disengage the 2 lower clips.

bb. Disconnect the cluster electrical connectors and remove the cluster.

cc. Remove the windshield garnish moldings.

dd. Using a taped flat–bladed tool, carefully pry the retaining clips attaching the instrument panel left trim plate to the instrument panel.

ee. Disconnect the power mirror and dimmer switch connectors.

ff. Remove the power mirror and dimmer switches.

gg. Disconnect any remaining electrical connections.

hh. Remove the upper instrument panel screws and the panel by pulling towards the rear to disengage the tabs.

ii. Disconnect the steering wheel coil connector.

jj. Release the 3 claws and remove the coil assembly.

kk. If the vehicle is equipped with an automatic transmission, insert the key into the cylinder, turn to the ACC position, push in the release butt, disconnect the park lock cable, remove the key from the cylinder and lock the steering wheel.

ll. Move the silencer pad from the column.

mm. Matchmark the steering shaft coupling to the shaft.

nn. Loosen the upper bolt on the coupling.

oo. Remove the lower bolt from the coupling.

pp. Move the coupling onto the column shaft.

qq. Disconnect the wiring harness clamps from the column.

rr. Remove the 3 bolts and the column.

ss. Remove the body hinge trim panels.

tt. Remove the sill plates.

uu. Remove the front floor console storage door.

vv. Remove the screws attaching the console to the instrument panel, pull the console rewards and up and remove the front floor console.

ww. Remove the HVAC retaining screw; disconnect the electrical connectors and module control, temperature control and A/C cables. Remove the unit.

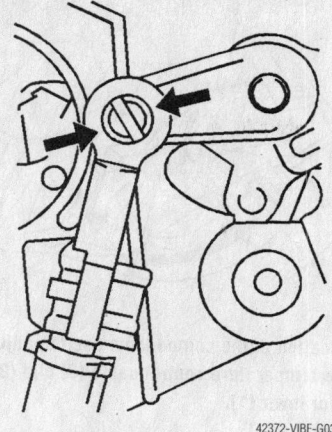

Push in the clip and disconnect the cable from the manual selector shifter assembly

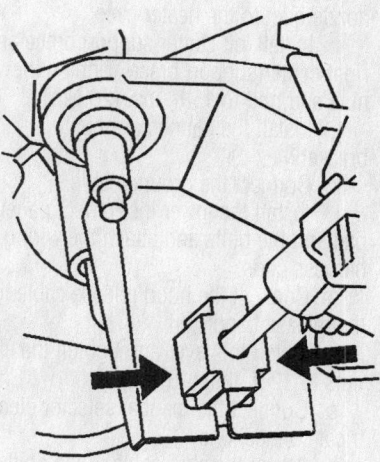

Using a suitable prytool, disconnect the park lock cable from the bracket

xx. Push in the clip and disconnect the cable from the manual selector shifter assembly.

yy. Using a suitable prytool, disconnect the park lock cable from the bracket.

zz. Disconnect the shift select cable from the manual selector lever.

aaa. Using a suitable prytool, disconnect the shift select cable from the shift lever plate.

bbb. Disconnect the electrical connectors and the wire harness clip.

ccc. Remove the nuts from the selector and remove the selector.

ddd. Disconnect the hood release cable from the release handle.

eee. Remove the 8 bolts, 4 push retainers and the wire harness clamps from the lower instrument panel and remove the panel.

fff. Disengage the wiring harness

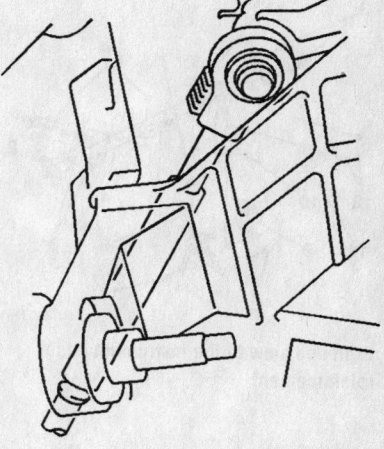

Disconnect the shift select cable from the manual selector lever

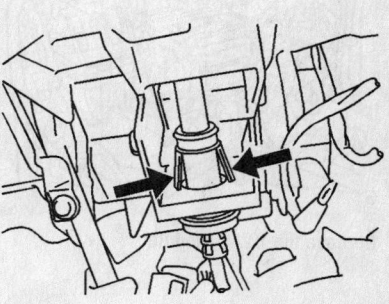

Using a suitable prytool, disconnect the shift select cable from the shift lever plate

clips, remove the bolts retaining the lower instrument panel pad and the pad.

ggg. Remove the ground cable.

hhh. Remove the connector housing bracket from the right instrument panel center support brace.

iii. Remove the left brace nut, right brace nut, left brace bolt, right brace bolt, left center support brace and right center support brace.

jjj. Remove the windshield defroster nozzle duct from the heater case.

kkk. Remove the 5 bolts and the nuts from the instrument panel reinforcement at the hinge pillars.

lll. Remove the instrument panel reinforcement.

mmm. Disconnect the blower motor connector.

nnn. Disconnect the rear ducts from the HVAC module.

ooo. Remove the HVAC module.

ppp. Remove the 12 bolts from the core case.

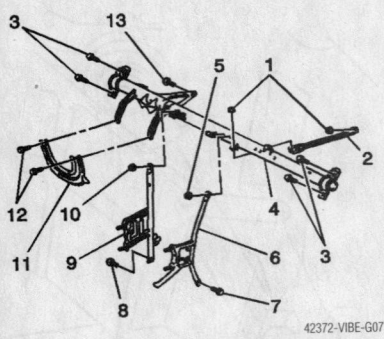

Exploded view of the instrument panel reinforcement

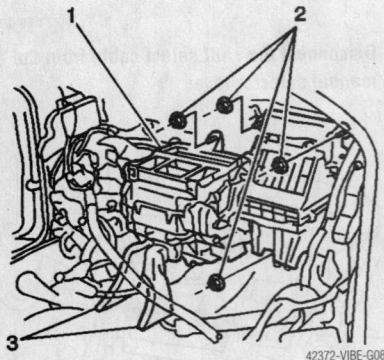

Remove the HVAC module

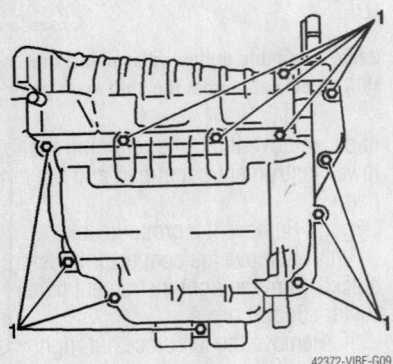

Remove the 12 bolts from the core case to access the heater core

qqq. Remove the heater core.

To install:

a. Install the heater core.

b. Install the 12 bolts from the core case and tighten to 89 inch lbs. (10 Nm).

c. Install the HVAC module.

d. Connect the rear ducts to the HVAC module.

e. Connect the blower motor connector.

f. Install the instrument panel reinforcement.

g. Install the 5 bolts and the nuts to

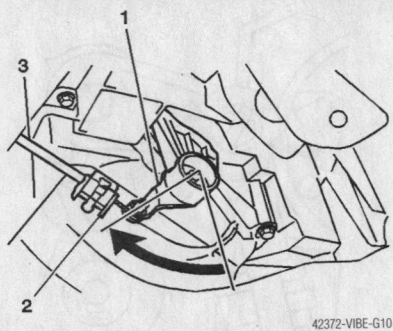

Location of the components used to adjust the temperature control cable (3) clip (2), door lever (1).

the instrument panel reinforcement at the hinge pillars. Tighten to 21 ft. lbs. (28 Nm).

h. Install the windshield defroster nozzle duct to the heater case.

i. Install left center support brace and right center support brace. Tighten the nuts and bolt to 15 ft. lbs. (20 Nm).

j. Install the connector housing bracket.

k. Connect the ground cable.

l. Install the lower instrument panel pad and the bolts and attach the wiring harness clips.

m. Connect the hood release cable to the release handle.

n. Install the lever and tighten the nuts to 12 ft. lbs. (18 Nm).

o. Connect the manual selector electrical connections.

p. Attach the shift cable to the shift lever plate.

q. Connect the shift select cable to the selector lever.

r. Install the park lock cable to the shift lever plate.

s. Connect the park lock cable to the manual selector lever.

t. Connect the electrical connectors and module control, temperature control and A/C cables. Install the HVAC unit.

u. Adjust the temperature control cable by setting the temperature control dial to coldest. Hold the door lever fully rearwards, clockwise. Attach the cable to the control clip.

v. Adjust the Mode linkage by setting the dial to defrost. Hold the door lever fully rearwards, clockwise. Attach the cable to the control clip.

w. Install the front floor console and tighten the screws.

x. Install the front floor console storage door.

y. Install the sill plates.

z. Install the column.

aa. Install the 3 bolts and tighten the lower bolt to 16 ft. Lbs. (21 Nm) and the 2 upper bolts to 16 ft. Lbs. (21 Nm).

bb. Align the matchmarks made prior to removal.

cc. Lower the coupling onto the shaft. Install the bolts and tighten to 26 ft. lbs. (35 Nm).

dd. Connect the wiring harness clamps to the column.

ee. Move the silencer pad to the column.

ff. If the vehicle is equipped with an automatic transmission, insert the key into the cylinder, turn to the ACC position, insert the park lock cable making sure the release button engages. Make sure the key will not rotate to the lock position unless the shifter is in the park position, remove the key from the cylinder and lock the steering wheel.

gg. Install the body hinge trim panels.

hh. Make sure the turn signal switch is in the neutral position.

ii. If installing a new coil, remove the lock pin.

jj. Install the coil making sure the 3 claws engage.

kk. While holding the coil casing, turn the coil center casing counterclockwise until the coil reaches its stop.

ll. Turn the coil center casing clockwise 2 ½ turns.

mm. Align the center casing with the arrow on the outer casing.

nn. Connect the coil electrical connector.

oo. Install the upper instrument panel and screws.

pp. Connect the electrical connections.

qq. Install the power mirror and dimmer switches.

rr. Connect the power mirror and dimmer switch connectors.

ss. Install the instrument panel left trim plate to the instrument panel.

tt. Install the windshield garnish moldings.

uu. Connect the cluster electrical connectors and install the cluster.

vv. Engage the cluster lower clips and install the screw.

ww. Install the cluster trim plate.

xx. Install the passenger air bag assembly and the instrument panel module connectors.

yy. Connect the instrument panel connector.

zz. Install the glove box.

aaa. Install the wiper switch.

bbb. Install the turn signal/headlamp switch assembly

ccc. Connect the turn signal/headlamp assembly connectors.

ddd. Install the upper and lower steering column covers and screws.

eee. Install the steering wheel and nut aligning the matchmarks made prior to removal and tighten the nut to 37 ft. lbs. (50 Nm).

fff. Connect the steering wheel module connectors.

ggg. Install the CPA to the inflator module.

hhh. Install the steering wheel module and tighten the retainers 78 inch lbs. (9 Nm)

iii. Install cigar lighters and power receptacle.

jjj. Connect the accessory power receptacle connectors.

kkk. Connect the 2 cigar lighter connectors.

lll. Install the front floor console trim plate to the floor console assembly.

mmm. Install the manual transmission shift knob.

nnn. Install the rear defogger switch.

ooo. Install the hazard switch.

ppp. Install the A/C switch and screw.

qqq. Install the radio.

rrr. Connect the A/C switch, hazard switch, rear defogger switch and passenger seat belt indicator switch electrical connections.

sss. Install the center trim plate to the instrument panel.

ttt. Connect the evaporator inlet and outlet tubes.

uuu. Connect the heater hoses to the core.

vvv. Connect the negative battery cable.

www. Recharge the A/C system and fill the cooling system.

Cylinder Head

REMOVAL & INSTALLATION

1. Before servicing the vehicle, refer to the precautions in the beginning of this section.
2. Drain the cooling system.
3. Remove or disconnect the following:
   - Right side engine under cover
   - Right front wheel and tire
   - Cylinder head cover
   - Air cleaner assembly with hose
   - Accelerator control cable

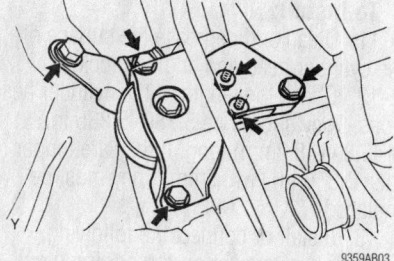

With the engine supported, remove the right side engine mount—1ZZ-FE engine shown, 2ZZ-GE similar

   - Wire harness clamp and suction hose assembly, 2ZZ-GE engine only
   - Water bypass hoses
   - Fuel pipe clamp
   - Fuel tube sub-assembly
   - Union-to-connector tube hose
   - Radiator and heater inlet hoses
   - Drive belt
4. Separate the vane pipe assembly, but do not disconnect the hose, 1ZZ-FE engine.
   - Alternator bracket, 2ZZ-GE
   - Alternator
5. Separate the compressor and magnetic clutch, on 2ZZ-GE engines with air conditioning.
   - Front exhaust pipe assembly
   - Power steering pump reservoir and position it aside, 1ZZ-FE engine
6. Place a jack with a wooden block under the vehicle for support, then remove the 4 bolts and 2 nuts and remove the right side engine mount.
7. Remove the engine wire, on 1ZZ-FE engines as follows:
   a. Remove the 5 clamps from the brackets.
   b. Detach the connectors.
   c. Remove the ignition coil connectors.
   d. Bolt and nut holding the engine wire.
8. Remove or disconnect the following:
   - Ignition coil assembly
   - Positive Crankcase Ventilation (PCV) hoses
   - Valve (cylinder head) cover sub-assembly
9. Set the No. 1 cylinder to Top Dead Center (TDC) of the compressor stroke as follows:
   a. Turn the crankshaft pulley, and align its groove with the "0" timing mark of the timing chain cover.
   b. Make sure the point marks of the camshaft timing sprockets and Variable Valve Timing (VVT) timing sprockets are in a straight line as shown. If not, turn the crankshaft 1 complete revolution (360°) and align the marks.

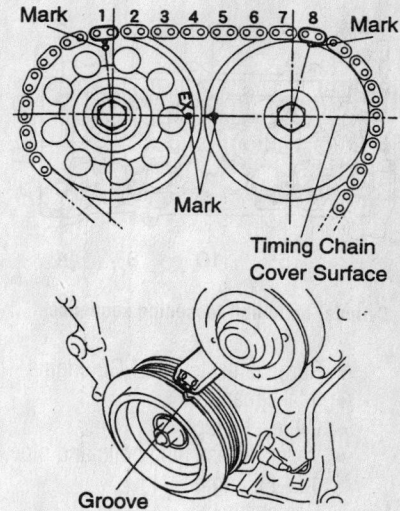

Proper timing mark alignment for TDC

10. Remove or disconnect the following:
    - Crankshaft pulley, using SST 09960-10010
    - Belt tensioner
    - Exhaust manifold stay and head insulator, 2ZZ-GE engine
    - Water pump pulley and pump
    - Transverse engine mounting bracket
    - Crankshaft Position (CKP) sensor
    - No. 1 chain tensioner assembly, making sure not to revolve the crankshaft without the tensioner
    - Timing chain or belt cover
    - Timing gear cover oil seal
    - CKP sensor plate No. 1
    - Timing chain tensioner slipper
    - Timing chain vibration damper No. 1

➡**In case you turn the camshafts with the timing chain removed, turn the crankshaft ¼ turn for the valve to avoid contact with the pistons.**

   - Timing chain sub-assembly. Remove the chain with the crankshaft gear, using screwdrivers as shown.

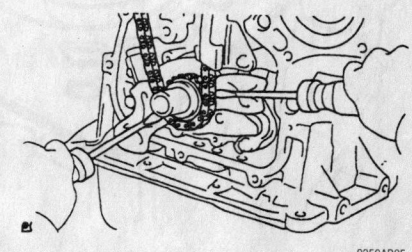

Remove the timing chain with the crankshaft gear

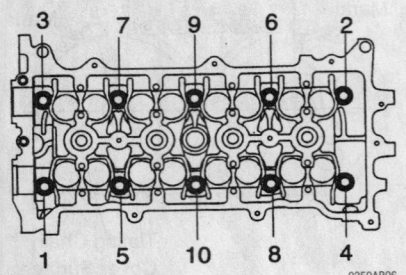

Cylinder head bolt loosening sequence

- Surge tank stay, 2ZZ-GE engine
- Intake manifold
- Oil level gauge
- Water bypass pipe bolts and pipe, 1ZZ-FE engine
- Camshafts
- Camshaft timing oil control valve, 1ZZ-FE engine
- Manifold stay, 1ZZ-FE engine
- Cylinder head bolts in sequence. To prevent damage to the cylinder head, loosen each bolt about ¼ of a turn during each pass until the bolts are loose.
- Cylinder head

To install:

11. Clean and degrease the surface of the cylinder head and engine block.

12. Check the length of the cylinder head bolts. They should be 5.780–5.835 in. (146.8–148.2mm) long. If they are longer than 5.846 in. (148.5mm), they must be replaced.

13. Install or connect the following:
- New gasket on the engine block with the Lot No. stamp facing up.
- Cylinder head
- Apply a light coat of oil to cylinder head bolt threads and tighten in sequence, in several passes, to 36 ft. lbs. (49 Nm) for 1ZZ-FE engines or to 26 ft. lbs. (35 Nm) for 2ZZ-GE engines.
- Tighten each head bolt, in sequence, an additional 90 degree turn for 1ZZ-FE engines, or 180 degree turn for 2ZZ-GE engines.
- Manifold stay, 1ZZ-FE engine. Tighten the bolts to 36 ft. lbs. (49 Nm).
- Camshaft timing oil control valve,

on 1ZZ-FE engines, and tighten to 80 inch lbs. (9 Nm)
- Camshaft
- Water by-pass pipe, on 1ZZ-FE engines, and tighten to 80 inch lbs. (9 Nm)
- Oil level gauge
- Intake manifold
- Surge tank stay, 2ZZ-GE engine. Tighten to 18 ft. lbs. (24 Nm).
- Timing chain
- Timing chain vibration damper. Tighten the bolts to 80 inch lbs. (9 m).
- Timing chain tensioner slipper and tighten the bolt to 14 ft. lbs. (19 Nm).
- Crankshaft position sensor plate, with the "F" mark facing forward.
- Timing gear cover oil seal
- Timing cover. For 1ZZ-FE engine, tighten the "A" bolts to 10 ft. lbs. (13 Nm), the "B" bolts to 14 ft. lbs. (19 Nm) and the stud bolt to 84 inch lbs. (9.5 Nm), using a Torx® wrench. For 2ZZ-GE engines, tighten the M8 bolts to 15 ft. lbs.

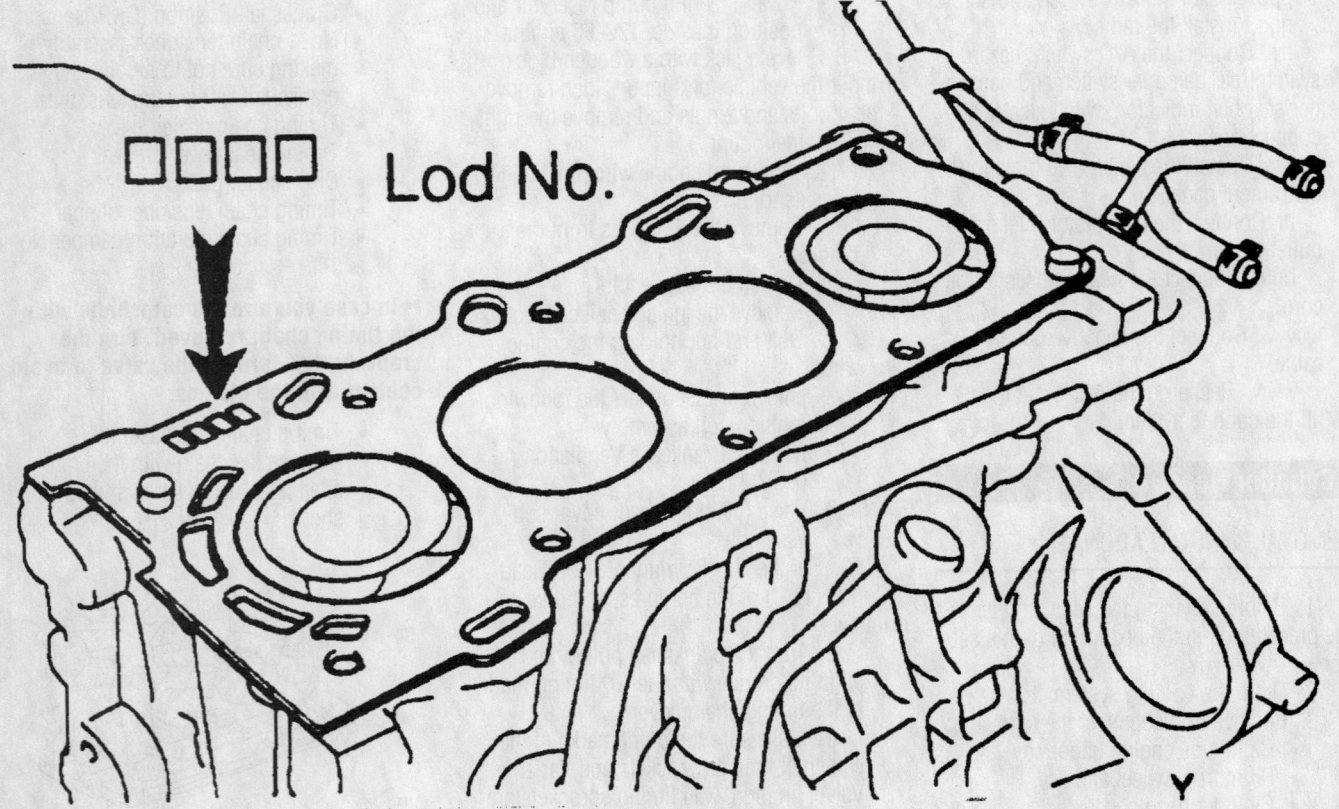

Lod No.

Position the head gasket correctly on the cylinder head—1.8L (1ZZ-FE) engine

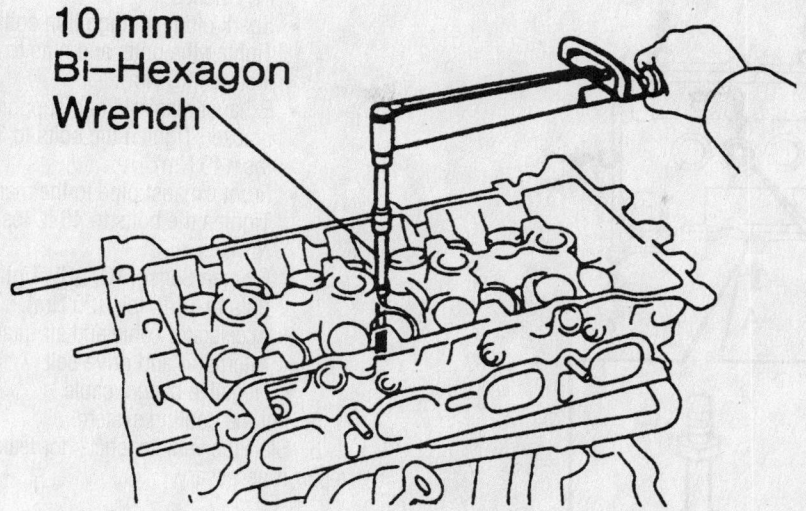

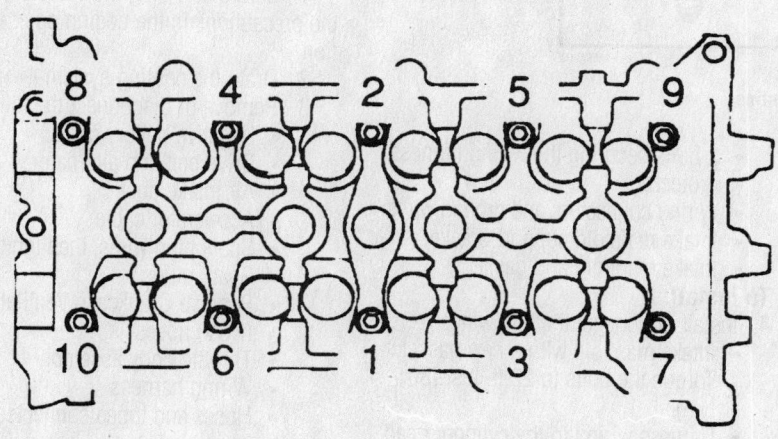

Cylinder head bolt tightening sequence—1ZZ-FE and 2ZZ-GE engines

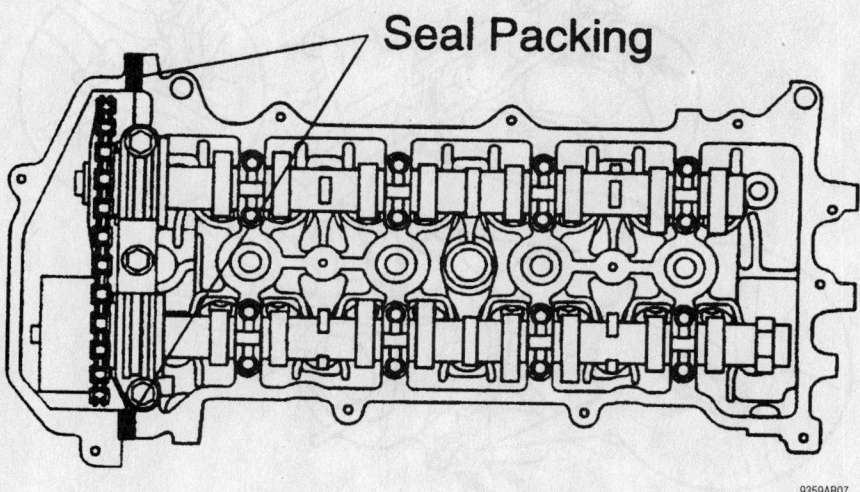

Seal packing installation locations

(21 m), the M6 bolts to 8 ft. lbs. (11 Nm) and the stud bolt to 84 inch lbs. (9.5 Nm).

➡ **When installing the tensioner, make sure to set the hook again if the hook releases the plunger.**

- Timing chain tensioner. Torque the nuts to 80 inch lbs. (9 Nm).
- CKP sensor and tighten the bolts to 80 inch lbs. (9 Nm)
- Transverse engine mounting bracket. Tighten the bolts to 35 ft. lbs. (47 Nm).
- Water pump and pulley
- Exhaust manifold stay and heat insulator, 2ZZ-GE engine
- Belt tensioner. Tighten the nut to 21 ft. lbs. (29 Nm) and the bolt to 51 ft. lbs. (69 Nm) on 1ZZ-FE engines or to 74 ft. lbs. (100 Nm) on 2ZZ-GE engines.

14. Install the crankshaft pulley, as follows:

   a. Align the pulley set key with the key groove of the pulley and slide on the pulley.

   b. Use SST 09960-11010 to install the bolt and tighten to 102 ft. lbs. (138 Nm) for 1ZZ-FE engine or to 87 ft. lbs. (118 Nm) on 2ZZ-GE engines.

   c. Turn the crankshaft counterclockwise and disconnect the plunger knock pin from the hook.

   d. Turn the crankshaft clockwise and check that the slipper is pushed by the plunger. If the plunger does not spring out, press the slipper into the chain tensioner with a screwdriver so that the hook is released from the knock pin and the plunger springs out.

15. Install or connect the following:

- Cylinder head sub-assembly cover. Install seal packing into the locations shown and install within 3 minutes. Tighten the "A" bolts to 8 ft. lbs. (11 Nm) and the "B" bolts to 80 inch lbs. (9 Nm) for 1ZZ-FE engines. For 2ZZ-GE engines, tighten the bolts to 7 ft. lbs. (10 Nm).
- Ignition coil assembly. Torque the bolts to 80 inch lbs. (9 Nm).
- Engine wire and tighten to 80 inch lbs. (9 Nm), 1ZZ-FE
- Right side engine mount. Tighten to 38 ft. lbs. (52 Nm).
- Front exhaust pipe
- Vane pump, 1ZZ-FE
- Compressor and magnetic clutch, 2ZZ-GE
- Alternator bracket, 2ZZ-GE engine
- Alternator

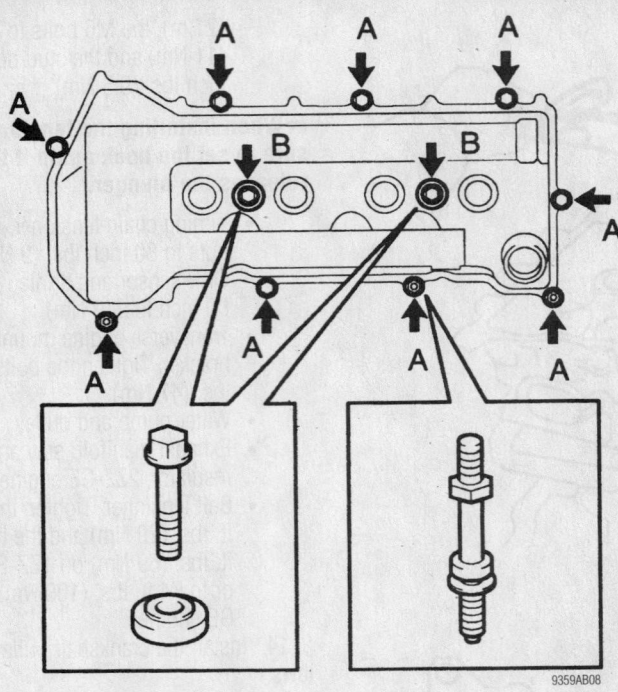

Cylinder head (valve) cover bolt locations—1ZZ-FE engine

- Suction hose and wire harness clamp, 2ZZ-GE engine
- Air cleaner and hose
- Main cylinder head cover and tighten to 62 inch lbs. (7 Nm)
- Right front wheel and tire. Tighten the lug nuts to 76 ft. lbs. (103 Nm).

16. Fill the cooling system to the proper level.

17. Start the vehicle, check for leaks and repair if necessary.

Intake Manifold

REMOVAL & INSTALLATION

1ZZ-FE Engine

1. Before servicing the vehicle, refer to the precautions in the beginning of this section.

2. Drain the cooling system.

3. Remove or disconnect the following:
   - Negative battery cable
   - Drive belt and alternator
   - Air intake duct
   - Accelerator cable
   - Exhaust pipe from the manifold.
   - Exhaust manifold support bracket
   - Spark plug wires, then ignition coils
   - Spark plugs
   - Positive Crankcase Ventilation (PCV) hoses
   - Throttle body assembly

- 2 bolts securing the wiring harness protector
- Wiring connectors and ground wires
- Intake manifold support bracket
- Intake manifold and gasket

To install:

4. Install or connect the following:
   - Intake manifold with a new gasket. Torque the bolts to 22 ft. lbs. (30 Nm).
   - Harness wiring to the cylinder head and harness protector

- Fuel injectors, throttle body and the PCV hoses
- Spark plugs and ignition coils. Tighten the bolts and nuts to 80 inch lbs. (9 Nm).
- Exhaust manifold and support bracket. Tighten the bolts to 37 ft. lbs. (49 Nm).
- Front exhaust pipe to the manifold. Tighten the bolts to 46 ft. lbs. (62 Nm).
- Oxygen Sensor ($O_{2>S}S$). Tighten the nuts to 14 ft. lbs. (20 Nm).
- Accelerator cable and air intake duct
- Alternator and drive belt
- Negative battery cable

5. Fill the cooling system.

6. Start the vehicle, check for leaks and repair if necessary.

2ZZ-GE Engine

1. Before servicing the vehicle, refer to the precautions in the beginning of this section.

2. Drain the cooling system.

3. Remove or disconnect the following:
   - Negative battery cable
   - Drive belt and alternator
   - Air intake duct
   - Accelerator cable
   - Spark plug wires, then ignition coils
   - Spark plugs
   - Positive Crankcase Ventilation (PCV) hoses
   - Throttle body assembly
   - Wiring harness
   - Hoses and tubes connected to the head

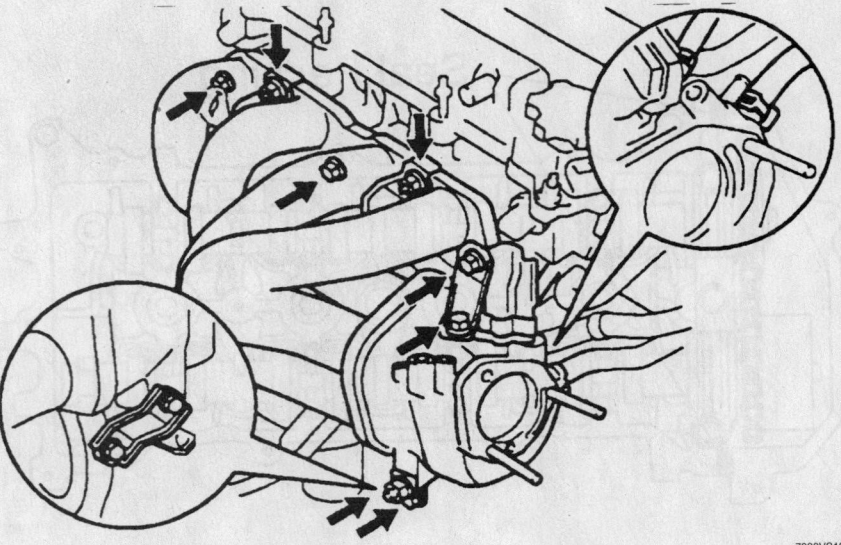

Intake manifold mounting fastener locations—1.8L (1ZZ-FE) engine

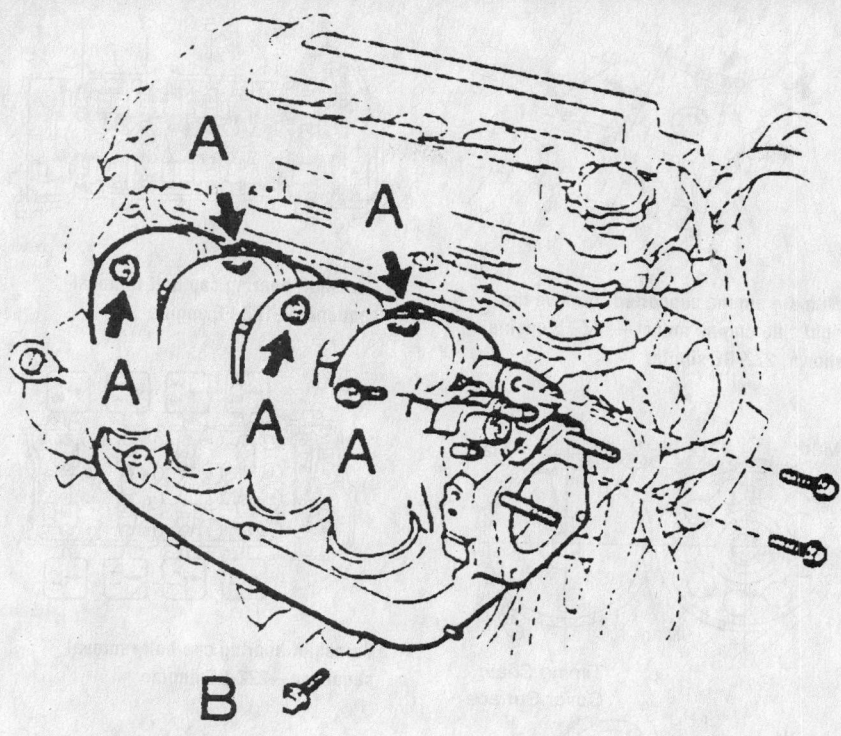

Intake manifold bolt installation—2ZZ-GE engine

9307WG93

- Intake manifold support bracket
- Intake manifold and gasket

To install:

4. Install or connect the following:
  - Intake manifold with a new gasket. Tighten bolts A to 25 ft. lbs. (34 Nm) and bolt B to 34 ft. lbs. (46 Nm).
  - Harness wiring to the cylinder head and harness protector
  - Fuel injectors, throttle body and the PCV hoses
  - Spark plugs and ignition coils. Tighten the bolts and nuts to 80 inch lbs. (9 Nm).
  - Oxygen Sensor (O$_2$S). Tighten the nuts to 14 ft. lbs. (20 Nm).
  - Accelerator cable and air intake duct
  - Alternator and drive belt
  - Negative battery cable
5. Fill the cooling system.
6. Start the vehicle, check for leaks and repair if necessary.

Exhaust Manifold

REMOVAL & INSTALLATION

1ZZ-FE Engine

1. Before servicing the vehicle, refer to the precautions in the beginning of this section.

2. Drain the cooling system.
3. Remove or disconnect the following:
  - Negative battery cable
  - Drive belt and alternator
  - Air intake duct

- Accelerator cable
- Exhaust pipe from the manifold
- Exhaust manifold support bracket
- Heat insulator from the dash panel
- Upper heat insulator
- Exhaust manifold and gasket
- If necessary, the lower heat insulator from the exhaust manifold.

To install:

4. Install or connect the following:
  - Lower heat insulator on the exhaust manifold. Tighten the bolts to 108 inch lbs. (12 Nm).
  - Exhaust manifold using a new gasket. Tighten the nuts, in several passes, to 27 ft. lbs. (37 Nm).
  - Upper heat insulator. Tighten the bolts to 108 inch lbs. (12 Nm).
  - Heat insulator on the dash panel
  - Exhaust manifold support bracket. Tighten the bolts, in an alternating pattern, to 37 ft. lbs. (49 Nm).
  - Front exhaust pipe to the manifold. Tighten the bolts to 46 ft. lbs. (62 Nm).
  - Oxygen Sensor (O$_2$S). Tighten the nuts to 14 ft. lbs. (20 Nm).
  - Accelerator cable and air intake duct
  - Alternator and drive belt
  - Negative battery cable
5. Fill the cooling system.
6. Start the vehicle, check for leaks and repair if necessary.

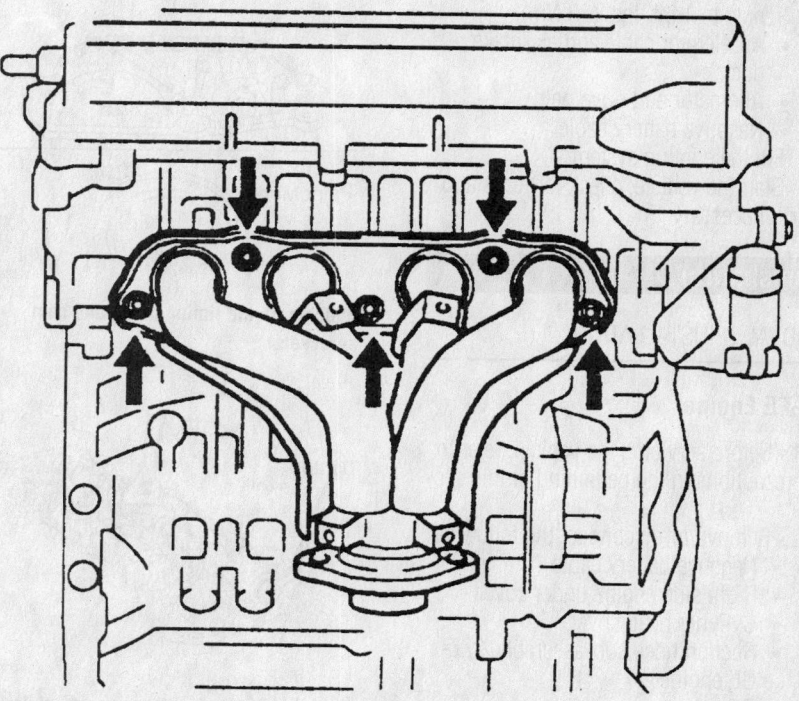

Exhaust manifold mounting nut locations—1.8L (1ZZ-FE) engine

7923VG22

2ZZ-GE Engine

1. Before servicing the vehicle, refer to the precautions in the beginning of this section.
2. Drain the cooling system.
3. Remove or disconnect the following:
   - Negative battery cable
   - Drive belt and alternator
   - Air intake duct
   - Accelerator cable
   - Exhaust pipe from the manifold
   - Exhaust manifold support bracket
   - Heat insulator from the dash panel
   - Upper heat insulator
   - Exhaust manifold and gasket
   - If necessary, the lower heat insulator from the exhaust manifold

To install:

4. Install or connect the following:
   - Lower heat insulator on the exhaust manifold. Tighten the bolts to 15 ft. lbs. (20 Nm).
   - Exhaust manifold using a new gasket. Tighten the nuts, in several passes to 37 ft. lbs. (50 Nm).
   - Upper heat insulator. Tighten the bolts to 15 ft. lbs. (20 Nm).
   - Heat insulator on the dash panel.
   - Exhaust manifold support bracket. Tighten the bolts to 37 ft. lbs. (49 Nm).
   - Front exhaust pipe to the manifold. Tighten the bolts to 46 ft. lbs. (62 Nm).
   - Oxygen Sensor (O_2S). Tighten the nuts to 14 ft. lbs. (20 Nm).
   - Accelerator cable and air intake duct.
   - Alternator and drive belt.
   - Negative battery cable
5. Fill the cooling system.
6. Start the vehicle, check for leaks and repair if necessary.

Camshaft(s)

REMOVAL & INSTALLATION

1ZZ-FE Engine

1. Before servicing the vehicle, refer to the precautions in the beginning of this section.
2. Remove or disconnect the following:
   - Negative battery cable
   - Right side engine under cover
   - Cylinder head cover
   - Suction hose sub-assembly, 2ZZ-GE engine
   - Drive belt
   - Power steering pump reservoir

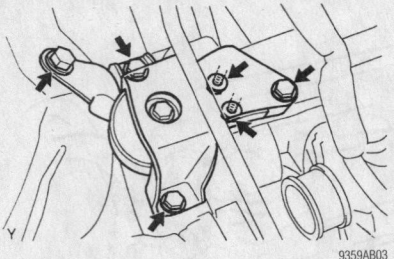

With the engine supported, remove the right side engine mount—1ZZ-FE engine shown, 2ZZ-GE similar

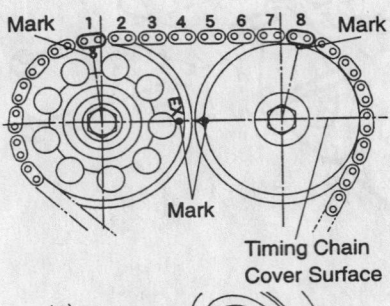

Proper timing mark alignment for TDC

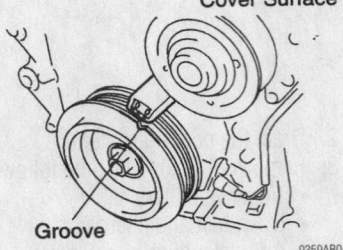

Matchmark the timing chain and cam sprockets

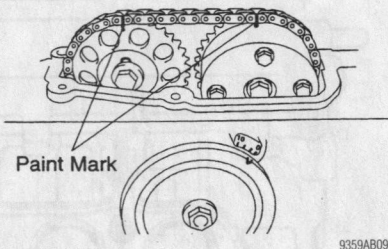

Hold the camshaft with a wrench while removing the set bolt

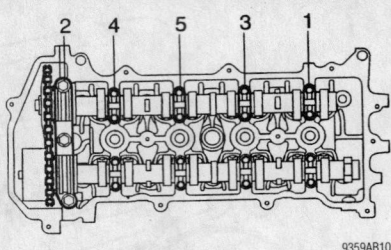

Camshaft bearing cap bolt removal sequence—1ZZ-FE engine

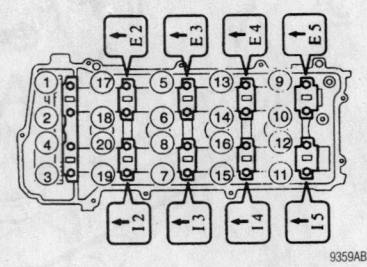

Camshaft bearing cap bolt removal sequence—2ZZ-GE engine

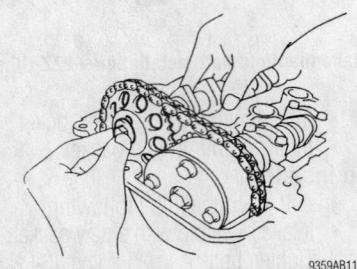

Carefully remove the cam and timing gear

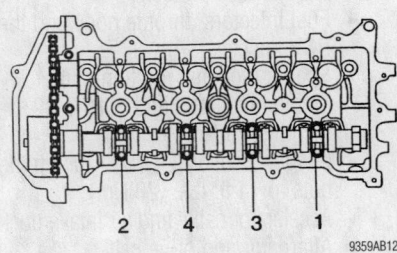

Camshaft bearing cap bolt removal sequence—1ZZ-FE engine

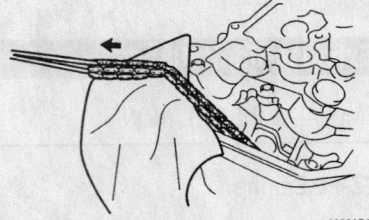

Secure the timing chain with string to prevent it from slipping down into the timing chain cover

and position it aside, 1ZZ-FE engine

3. Place a jack with a wooden block under the vehicle for support, then remove the 4 bolts and 2 nuts and remove the right side engine mount.

4. Remove the engine wire, on 1ZZ-FE engines:

   a. Remove the 5 clamps from the brackets.

b. Detach the connectors.

c. Remove the ignition coil connectors.

d. Bolt and nut holding the engine wire.

5. Remove or disconnect the following:
   • Ignition coil assembly
   • Positive Crankcase Ventilation (PCV) hoses from the valve cover
   • Valve (cylinder head) cover sub-assembly

6. Set the No. 1 cylinder to Top Dead Center (TDC) of the compressor stroke as follows:

   a. Turn the crankshaft pulley, and align its groove with the "0" timing mark of the timing chain cover.

   b. Make sure the point marks of the camshaft timing sprockets and VVT timing sprockets are in a straight line as shown. If not, turn the crankshaft 1 com-

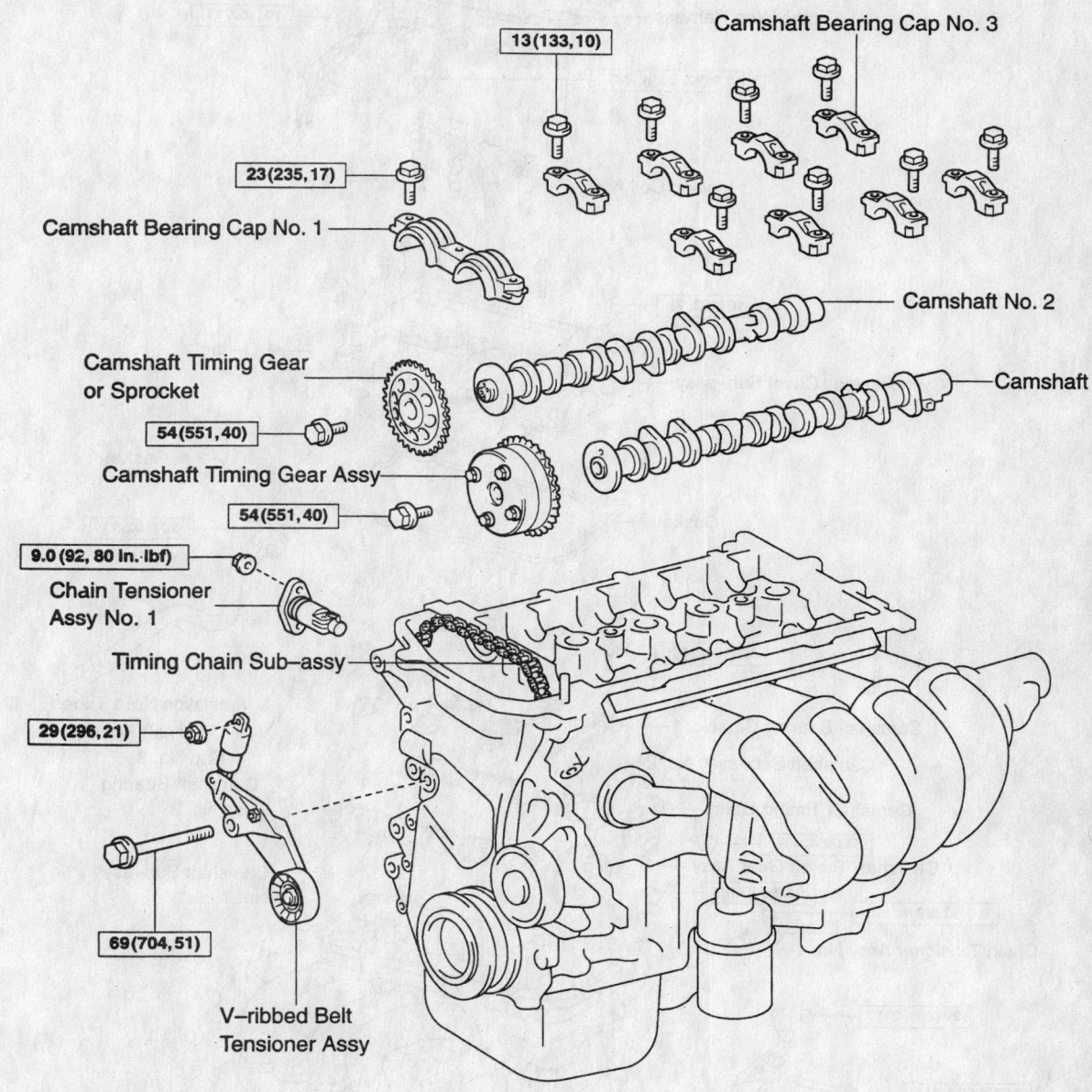

13(133,10)

Camshaft Bearing Cap No. 3

23(235,17)

Camshaft Bearing Cap No. 1

Camshaft No. 2

Camshaft Timing Gear or Sprocket

54(551,40)

Camshaft Timing Gear Assy

Camshaft

54(551,40)

9.0 (92, 80 in.·lbf)

Chain Tensioner Assy No. 1

Timing Chain Sub–assy

29(296,21)

69(704,51)

V–ribbed Belt Tensioner Assy

N·m (kgf·cm, ft·lbf) : Specified torque

9359AB21

Exploded view of the camshafts and related components—1ZZ-FE engine

plete revolution (360°) and align the marks.

7. Remove the drive belt tensioner.

✸✸ WARNING

Do not turn the crankshaft without the tensioner installed.

8. Make sure the No. 1 cylinder is at TDC of the compression stroke.

9. Matchmark the timing chain and camshaft sprockets

10. Remove the 2 nuts and chain tensioner.

11. Hold the camshafts with a wrench and loosen the camshaft set bolt.

12. Using several passes, gradually remove the bearing cap bolts from the No. 2 camshaft, in the proper sequence.

13. Remove the camshaft and timing gear as shown.

14. Using several passes, gradually remove the bearing cap bolts from the other camshaft, in the proper sequence.

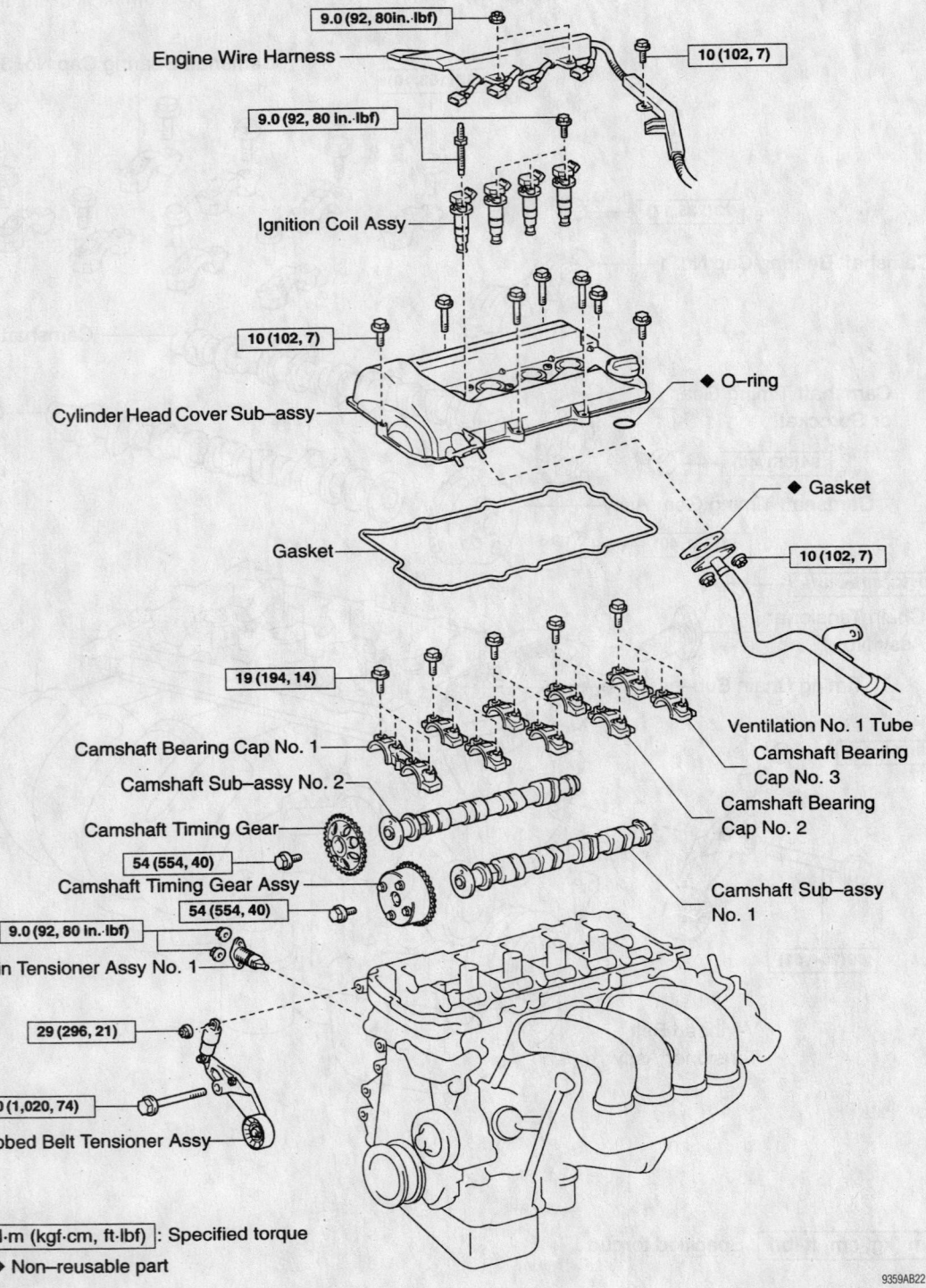

9.0 (92, 80 in.·lbf)
Engine Wire Harness
10 (102, 7)
9.0 (92, 80 in.·lbf)
Ignition Coil Assy
10 (102, 7)
Cylinder Head Cover Sub–assy
◆ O–ring
◆ Gasket
10 (102, 7)
Gasket
19 (194, 14)
Ventilation No. 1 Tube
Camshaft Bearing Cap No. 1
Camshaft Bearing Cap No. 3
Camshaft Sub–assy No. 2
Camshaft Bearing Cap No. 2
Camshaft Timing Gear
54 (554, 40)
Camshaft Timing Gear Assy
54 (554, 40)
Camshaft Sub–assy No. 1
9.0 (92, 80 in.·lbf)
Chain Tensioner Assy No. 1
29 (296, 21)
100 (1,020, 74)
V–ribbed Belt Tensioner Assy

N·m (kgf·cm, ft·lbf) : Specified torque

◆ Non–reusable part

9359AB22

Exploded view of the camshafts and related components—2ZZ-GE engine

15. Remove the camshaft while holding the timing chain.

✳✳ WARNING

Do not let anything drop down into the timing chain cover while the camshafts are removed.

16. Tie the timing chain with a string as shown, to prevent it from dropping down into the timing chain cover.

To install:

17. Position the camshaft on the cylinder head, then install the timing chain on the cam timing gear, with the painted links aligned with the marks on the timing gear.

18. Check the front marks and numbers and torque the camshaft cap bolts, in sequence, to 10 ft. lbs. (13 Nm) for 1ZZ-FE engine, or to 14 ft. lbs. (19 Nm) for 2ZZ-GE engines.

19. Put camshaft No. 2 on the cylinder head, with the painted links of the chain aligned with the mark on the timing gear.

20. Tighten the camshaft gear set bolt temporarily.

21. Check the front marks and numbers and torque the camshaft cap bolts, in sequence, to 10 ft. lbs. (13 Nm). Install the No. 1 bearing cap and tighten to 17 ft. lbs. (23 Nm).

22. Hold the camshaft secure with a wrench and tighten the set bolt to 40 ft. lbs. (54 Nm). Be careful not the damage the lifters.

23. Check to be sure the matchmarks on the timing chain and cam sprockets, and the alignment of the pulley groove with the timing mark on the cover are still aligned.

24. Install the chain tensioner:

   a. Make sure the O-ring is clean, then set the hook as shown.

   b. Oil the tensioner, then install and tighten to 80 inch lbs. (9 Nm).

➡ **When installing the tensioner, set the hook again if the hook releases the plunger.**

   c. Turn the crankshaft counterclockwise, and disconnect the plunger knock pin from the hook.

   d. Turn the crankshaft clockwise and check that the slipper is pushed by the plunger. If the plunger does not spring out, press the slipper into the chain tensioner with a screwdriver so that the hook is released from the knock pin and the plunger springs out.

25. Check the valve clearance and make adjustments as needed.

26. Install or connect the following:

- Belt tensioner. Tighten the nut to 21 ft. lbs. (29 Nm) and the bolt to 51 ft. lbs. (69 Nm).
- Cylinder head sub-assembly cover. Install seal packing into the locations shown and install within 3 minutes. Tighten the "A" bolts to 8 ft. lbs. (11 Nm) and the "B" bolts to 80 inch lbs. (9 Nm) for 1ZZ-FE

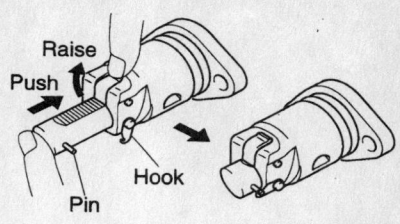

Set the timing chain tensioner hook properly

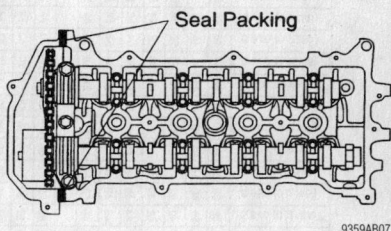

Seal packing installation locations

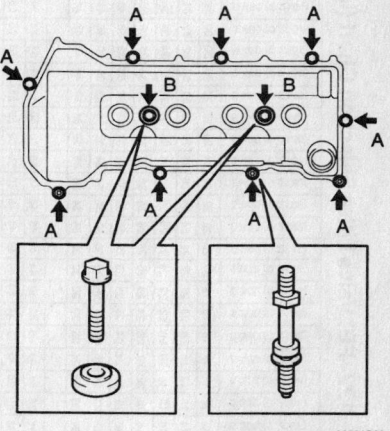

Cylinder head (valve) cover bolt locations—1ZZ-FE engine

engine and to 7 ft. lbs. (10 Nm) for 2ZZ-GE engines.

- Ignition coil assembly. Torque the bolts to 80 inch lbs. (9 Nm).
- Engine wire and tighten to 80 inch lbs. (9 Nm)
- Right side engine mount. Tighten to 38 ft. lbs. (52 Nm).
- Cylinder head (valve) cover
- Negative battery cable

Valve Lash

ADJUSTMENT

1ZZ-FE Engine

➡ **Adjust the valve clearance when the engine is cold.**

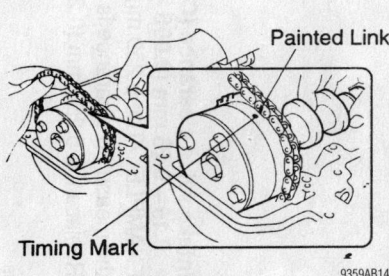

Make sure the alignment marks on the timing chain and camshaft gear match up

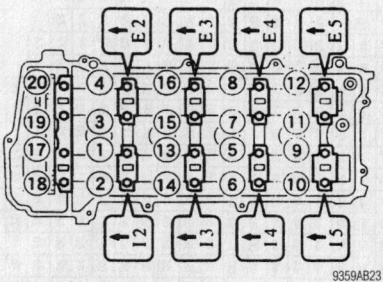

Camshaft cap bolt tightening sequence—2ZZ-GE engine

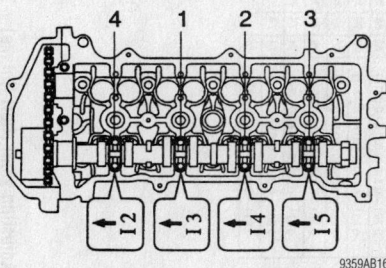

Camshaft cap bolt tightening sequence—1ZZ-FE engine

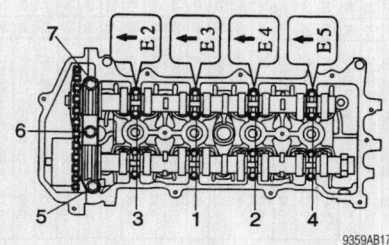

Camshaft cap bolt tightening sequence—1ZZ-FE

1ZZ-FE: Valve Lifter Selection Chart (Intake)

9307W670

(Large triangular Valve Lifter Selection Chart matrix. The vertical axis lists "Measured clearance mm (in.)" ranges; the horizontal axis lists "Installed lifter thickness mm (in.)" values. Cells contain new lifter part numbers.)

Measured clearance mm (in.):

- 0.000 – 0.030 (0.0000 – 0.0012)
- 0.031 – 0.050 (0.0012 – 0.0020)
- 0.051 – 0.070 (0.0020 – 0.0028)
- 0.071 – 0.090 (0.0028 – 0.0035)
- 0.091 – 0.110 (0.0036 – 0.0043)
- 0.111 – 0.130 (0.0044 – 0.0051)
- 0.131 – 0.150 (0.0052 – 0.0059)
- 0.150 – 0.250 (0.0059 – 0.0098)
- 0.251 – 0.270 (0.0099 – 0.0105)
- 0.271 – 0.290 (0.0107 – 0.0114)
- 0.291 – 0.310 (0.0115 – 0.0122)
- 0.311 – 0.330 (0.0122 – 0.0130)
- 0.331 – 0.350 (0.0130 – 0.0138)
- 0.351 – 0.370 (0.0138 – 0.0146)
- 0.371 – 0.390 (0.0146 – 0.0154)
- 0.391 – 0.410 (0.0154 – 0.0161)
- 0.411 – 0.430 (0.0162 – 0.0169)
- 0.431 – 0.450 (0.0170 – 0.0177)
- 0.451 – 0.470 (0.0178 – 0.0185)
- 0.471 – 0.490 (0.0185 – 0.0193)
- 0.491 – 0.510 (0.0193 – 0.0201)
- 0.511 – 0.530 (0.0201 – 0.0209)
- 0.531 – 0.550 (0.0209 – 0.0217)
- 0.551 – 0.570 (0.0217 – 0.0224)
- 0.571 – 0.590 (0.0225 – 0.0232)
- 0.591 – 0.610 (0.0233 – 0.0240)
- 0.611 – 0.630 (0.0241 – 0.0248)
- 0.631 – 0.650 (0.0248 – 0.0256)
- 0.651 – 0.670 (0.0256 – 0.0264)
- 0.671 – 0.690 (0.0264 – 0.0272)
- 0.691 – 0.710 (0.0272 – 0.0280)
- 0.711 – 0.730 (0.0280 – 0.0287)
- 0.731 – 0.750 (0.0288 – 0.0295)
- 0.751 – 0.770 (0.0296 – 0.0303)
- 0.771 – 0.790 (0.0304 – 0.0311)
- 0.791 – 0.810 (0.0311 – 0.0319)
- 0.811 – 0.830 (0.0319 – 0.0327)
- 0.831 – 0.850 (0.0327 – 0.0335)
- 0.851 – 0.870 (0.0335 – 0.0343)
- 0.871 – 0.890 (0.0343 – 0.0350)
- 0.891 – 0.910 (0.0351 – 0.0358)
- 0.911 – 0.930 (0.0359 – 0.0366)

New lifter thickness   mm (in.)

Lifter No.	Thickness	Lifter No.	Thickness	Lifter No.	Thickness
06	5.060 (0.1992)	30	5.300 (0.2087)	54	5.540 (0.2181)
08	5.080 (0.2000)	32	5.320 (0.2094)	56	5.560 (0.2189)
10	5.100 (0.2008)	34	5.340 (0.2102)	58	5.580 (0.2197)
12	5.120 (0.2016)	36	5.360 (0.2110)	60	5.600 (0.2205)
14	5.140 (0.2024)	38	5.380 (0.2118)	62	5.620 (0.2213)
16	5.160 (0.2031)	40	5.400 (0.2126)	64	5.640 (0.2220)
18	5.180 (0.2039)	42	5.420 (0.2134)	66	5.660 (0.2228)
20	5.200 (0.2047)	44	5.440 (0.2142)	68	5.680 (0.2236)
22	5.220 (0.2055)	46	5.460 (0.2150)	70	5.700 (0.2244)
24	5.240 (0.2063)	48	5.480 (0.2157)	72	5.720 (0.2252)
26	5.260 (0.2071)	50	5.500 (0.2165)	74	5.740 (0.2260)
28	5.280 (0.2079)	52	5.520 (0.2173)		

Intake valve clearance (Cold):
0.15 – 0.25 mm (0.006 – 0.010 in.)
EXAMPLE: The 5.250 mm (0.2067 in.) lifter is installed, and the measured clearance is 0.400 mm (0.0157 in.).
Replace the 5.250 mm (0.2067 in.) lifter with a new No. 48 lifter.

Adjusting shim chart (intake)—1ZZ-FE engine

1ZZ-FE: Valve Lifter Selection Chart (Exhaust)

New lifter thickness mm (in.)

Lifter No.	Thickness	Lifter No.	Thickness	Lifter No.	Thickness
06	5.060 (0.1992)	30	5.300 (0.2087)	54	5.540 (0.2181)
08	5.080 (0.2000)	32	5.320 (0.2094)	56	5.560 (0.2189)
10	5.100 (0.2008)	34	5.340 (0.2102)	58	5.580 (0.2197)
12	5.120 (0.2016)	36	5.360 (0.2110)	60	5.600 (0.2205)
14	5.140 (0.2024)	38	5.380 (0.2118)	62	5.620 (0.2213)
16	5.160 (0.2031)	40	5.400 (0.2126)	64	5.640 (0.2220)
18	5.180 (0.2039)	42	5.420 (0.2134)	66	5.660 (0.2228)
20	5.200 (0.2047)	44	5.440 (0.2142)	68	5.680 (0.2236)
22	5.220 (0.2055)	46	5.460 (0.2150)	70	5.700 (0.2244)
24	5.240 (0.2063)	48	5.480 (0.2157)	72	5.720 (0.2252)
26	5.260 (0.2071)	50	5.500 (0.2165)	74	5.740 (0.2260)
28	5.280 (0.2079)	52	5.520 (0.2173)		

Exhaust valve clearance (Cold):
0.25 – 0.35 mm (0.010 – 0.014 in.)

EXAMPLE: The 5.340 mm (0.2102 in.) lifter is installed, and the measured clearance is 0.440 mm (0.0173 in.).
Replace the 5.340 mm (0.2102 in.) lifter with a new No. 48 lifter.

Adjusting shim chart (exhaust)—1ZZ-FE engine

Installed lifter thickness mm (in.) — column headings across the chart:

5.060 (0.1992), 5.080 (0.2008), 5.100 (0.2008), 5.120 (0.2016), 5.140 (0.2024), 5.160 (0.2031), 5.180 (0.2039), 5.200 (0.2047), 5.220 (0.2055), 5.240 (0.2063), 5.260 (0.2071), 5.280 (0.2079), 5.300 (0.2087), 5.320 (0.2094), 5.340 (0.2102), 5.360 (0.2110), 5.380 (0.2118), 5.400 (0.2126), 5.420 (0.2134), 5.440 (0.2142), 5.460 (0.2150), 5.480 (0.2157), 5.500 (0.2165), 5.520 (0.2173), 5.540 (0.2181), 5.560 (0.2189), 5.580 (0.2197), 5.600 (0.2205), 5.620 (0.2213), 5.640 (0.2220), 5.660 (0.2228), 5.680 (0.2236), 5.700 (0.2244), 5.720 (0.2252), 5.740 (0.2260)

Measured clearance mm (in.) — row headings down the chart:

Measured clearance mm (in.)
0.000 – 0.030 (0.0000 – 0.0012)
0.031 – 0.050 (0.0012 – 0.0020)
0.051 – 0.070 (0.0020 – 0.0028)
0.071 – 0.090 (0.0028 – 0.0035)
0.091 – 0.110 (0.0036 – 0.0043)
0.111 – 0.130 (0.0044 – 0.0051)
0.131 – 0.150 (0.0052 – 0.0059)
0.151 – 0.170 (0.0059 – 0.0067)
0.171 – 0.190 (0.0067 – 0.0075)
0.191 – 0.210 (0.0075 – 0.0083)
0.211 – 0.230 (0.0083 – 0.0091)
0.231 – 0.249 (0.0091 – 0.0098)
0.250 – 0.350 (0.0098 – 0.0138)
0.351 – 0.370 (0.0138 – 0.0146)
0.371 – 0.390 (0.0146 – 0.0154)
0.391 – 0.410 (0.0154 – 0.0161)
0.411 – 0.430 (0.0162 – 0.0169)
0.431 – 0.450 (0.0170 – 0.0177)
0.451 – 0.470 (0.0178 – 0.0185)
0.471 – 0.490 (0.0185 – 0.0193)
0.491 – 0.510 (0.0193 – 0.0201)
0.511 – 0.530 (0.0201 – 0.0209)
0.531 – 0.550 (0.0209 – 0.0217)
0.551 – 0.570 (0.0217 – 0.0225)
0.571 – 0.590 (0.0225 – 0.0232)
0.591 – 0.610 (0.0233 – 0.0240)
0.611 – 0.630 (0.0241 – 0.0248)
0.631 – 0.650 (0.0248 – 0.0256)
0.651 – 0.670 (0.0256 – 0.0264)
0.671 – 0.690 (0.0264 – 0.0272)
0.691 – 0.710 (0.0272 – 0.0280)
0.711 – 0.730 (0.0280 – 0.0287)
0.731 – 0.750 (0.0288 – 0.0295)
0.751 – 0.770 (0.0296 – 0.0303)
0.771 – 0.790 (0.0304 – 0.0311)
0.791 – 0.810 (0.0311 – 0.0319)
0.811 – 0.830 (0.0319 – 0.0327)
0.831 – 0.850 (0.0327 – 0.0335)
0.851 – 0.870 (0.0335 – 0.0343)
0.871 – 0.890 (0.0343 – 0.0350)
0.891 – 0.910 (0.0351 – 0.0358)
0.911 – 0.930 (0.0359 – 0.0366)
0.931 – 0.950 (0.0367 – 0.0374)
0.951 – 0.970 (0.0374 – 0.0382)
0.971 – 0.990 (0.0382 – 0.0390)
0.991 – 1.010 (0.0390 – 0.0398)
1.011 – 1.030 (0.0398 – 0.0406)

9307WG71

9359AB26

Adjusting shim chart (intake)—2ZZ-GE engine

Intake valve clearance (Cold):
0.08 – 0.18 mm (0.0031 – 0.0071 in.)

EXAMPLE: The 2.200 mm (0.0826 in.) shim is installed, and the measured clearance is 0.400 mm (0.0157 in.).

Replace the 2.600 mm (0.1024 in.) shim with a new No. 60 shim.

New Shim thickness   mm (in.)

Shim No.	Thickness	Shim No.	Thickness	Shim No.	Thickness
00	2.000(0.0787)	28	2.280(0.0898)	56	2.560(0.1008)
02	2.020(0.0795)	30	2.300(0.0906)	58	2.580(0.1016)
04	2.040(0.0803)	32	2.320(0.0913)	60	2.600(0.1024)
06	2.060(0.0811)	34	2.340(0.0921)	62	2.620(0.1031)
08	2.080(0.0819)	36	2.360(0.0929)	64	2.640(0.1039)
10	2.100(0.0827)	38	2.380(0.0937)	66	2.660(0.1047)
12	2.120(0.0835)	40	2.400(0.0945)	68	2.680(0.1055)
14	2.140(0.0843)	42	2.420(0.0953)	70	2.700(0.1063)
16	2.160(0.0850)	44	2.440(0.0961)	72	2.720(0.1071)
18	2.180(0.0858)	46	2.460(0.0969)	74	2.740(0.1079)
20	2.200(0.0866)	48	2.480(0.0976)	76	2.760(0.1087)
22	2.220(0.0874)	50	2.500(0.0984)	78	2.780(0.1094)
24	2.240(0.0882)	52	2.520(0.0992)	80	2.800(0.1102)
26	2.260(0.0890)	54	2.540(0.1000)		

Adjusting shim chart

Installed shim thickness mm(in.) (top axis): 2.000(0.0787), 2.020(0.0795), 2.040(0.0803), 2.060(0.0811), 2.080(0.0819), 2.100(0.0827), 2.120(0.0835), 2.140(0.0843), 2.160(0.0850), 2.180(0.0858), 2.200(0.0866), 2.210(0.0870), 2.220(0.0874), 2.230(0.0878), 2.240(0.0882), 2.250(0.0886), 2.260(0.0890), 2.270(0.0894), 2.280(0.0898), 2.290(0.0902), 2.300(0.0906), 2.310(0.0909), 2.320(0.0913), 2.330(0.0917), 2.340(0.0921), 2.340(0.0921), 2.350(0.0925), 2.360(0.0929), 2.370(0.0933), 2.380(0.0937), 2.390(0.0941), 2.400(0.0945), 2.410(0.0949), 2.420(0.0953), 2.430(0.0957), 2.440(0.0961), 2.450(0.0965), 2.460(0.0969), 2.470(0.0972), 2.480(0.0976), 2.490(0.0980), 2.500(0.0984), 2.510(0.0988), 2.520(0.0992), 2.530(0.0996), 2.540(0.1000), 2.550(0.1004), 2.560(0.1008), 2.580(0.1016), 2.600(0.1024), 2.620(0.1031), 2.640(0.1039), 2.660(0.1047), 2.680(0.1055), 2.700(0.1063), 2.720(0.1071), 2.740(0.1079), 2.760(0.1087), 2.780(0.1094), 2.800(0.1102)

Measure clearance mm(in.) (side axis):

Measure clearance mm	Measure clearance (in.)
0.000 – 0.030	0.0000 – 0.0012
0.031 – 0.050	0.0012 – 0.0020
0.051 – 0.070	0.0020 – 0.0028
0.071 – 0.090	0.0028 – 0.0035
0.091 – 0.099	0.0036 – 0.0039
0.100 – 0.160	0.0039 – 0.0063
0.161 – 0.180	0.0063 – 0.0071
0.181 – 0.200	0.0071 – 0.0079
0.201 – 0.220	0.0079 – 0.0087
0.221 – 0.240	0.0087 – 0.0094
0.241 – 0.260	0.0095 – 0.0102
0.261 – 0.280	0.0103 – 0.0110
0.281 – 0.300	0.0111 – 0.0118
0.301 – 0.320	0.0119 – 0.0126
0.321 – 0.340	0.0126 – 0.0134
0.341 – 0.360	0.0134 – 0.0142
0.361 – 0.380	0.0142 – 0.0150
0.381 – 0.400	0.0150 – 0.0157
0.401 – 0.420	0.0158 – 0.0165
0.421 – 0.440	0.0166 – 0.0173
0.441 – 0.460	0.0174 – 0.0181
0.461 – 0.480	0.0181 – 0.0189
0.481 – 0.500	0.0189 – 0.0197
0.501 – 0.520	0.0197 – 0.0205
0.521 – 0.540	0.0205 – 0.0213
0.541 – 0.560	0.0213 – 0.0220
0.561 – 0.580	0.0221 – 0.0228
0.581 – 0.600	0.0229 – 0.0236
0.601 – 0.620	0.0237 – 0.0244
0.621 – 0.640	0.0244 – 0.0252
0.641 – 0.660	0.0252 – 0.0260
0.661 – 0.680	0.0260 – 0.0268

Exhaust valve clearance (Cold):
0.22 – 0.32 mm (0.0087 – 0.0126 in.)
EXAMPLE: The 2.200 mm (0.0862 in.) shim is installed, and
the measured clearance is 0.500 mm (0.0197 in.).
Replace the 2.540 mm (0.1000 in.) shim with a new No. 54 shim.

Adjusting shim chart (exhaust)—2ZZ-GE engine

New Shim thickness   mm (in.)

Shim No.	Thickness	Shim No.	Thickness	Shim No.	Thickness
00	2.000(0.0787)	28	2.280(0.0898)	56	2.560(0.1008)
02	2.020(0.0795)	30	2.300(0.0906)	58	2.580(0.1016)
04	2.040(0.0803)	32	2.320(0.0913)	60	2.600(0.1024)
06	2.060(0.0811)	34	2.340(0.0921)	62	2.620(0.1031)
08	2.080(0.0819)	36	2.360(0.0929)	64	2.640(0.1039)
10	2.100(0.0827)	38	2.380(0.0937)	66	2.660(0.1047)
12	2.120(0.0835)	40	2.400(0.0945)	68	2.680(0.1055)
14	2.140(0.0843)	42	2.420(0.0953)	70	2.700(0.1063)
16	2.160(0.0850)	44	2.440(0.0961)	72	2.720(0.1071)
18	2.180(0.0858)	46	2.460(0.0969)	74	2.740(0.1079)
20	2.200(0.0866)	48	2.480(0.0976)	76	2.760(0.1087)
22	2.220(0.0874)	50	2.500(0.0984)	78	2.780(0.1094)
24	2.240(0.0882)	52	2.520(0.0992)	80	2.800(0.1102)
26	2.260(0.0890)	54	2.540(0.1000)		

9359AB27

Adjusting shim chart matrix — Installed shim thickness mm(in.) (columns, 2.000/0.0787 through 2.800/0.1102) versus Measure clearance mm(in.) (rows, 0.000–0.030 through 0.801–0.820). Body cells contain new shim numbers (00–80).

1. Before servicing the vehicle, refer to the precautions in the beginning of this section.

2. Remove or disconnect the following:
- Negative battery cable.
- Cylinder head covers
- Engine wire
- Ignition coil
- Positive Crankcase Ventilation (PCV) hoses
- Cylinder head cover sub-assembly

3. Set the No. 1 cylinder to Top Dead Center (TDC) of the compressor stroke as follows:

a. Turn the crankshaft pulley, and align its groove with the "0" timing mark of the timing chain cover.

b. Make sure the point marks of the camshaft timing sprockets and VVT timing sprockets are in a straight line as shown. If not, turn the crankshaft 1 complete revolution (360°) and align the marks.

4. Check the valve clearance of the first set of the valves shown:

a. Use a feeler gauge to measure the clearance between the valve lifter and

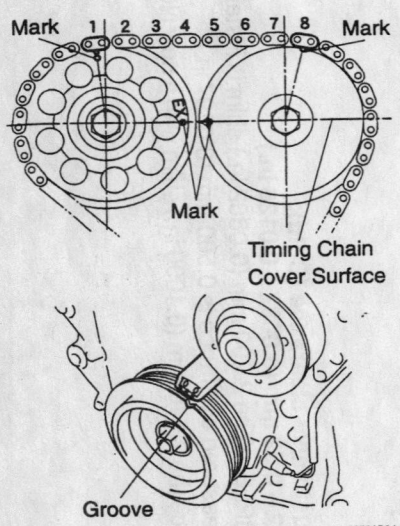

Proper timing mark alignment for TDC—1ZZ-FE and 2ZZ-GE engines

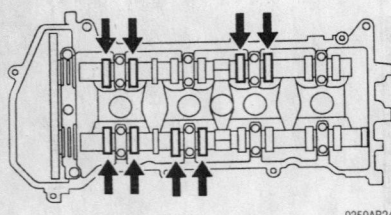

Check the clearance of the 1st set of valves—1ZZ-FE engine

camshaft. The clearance of the intake valves should be 0.0059–0.0098 in. (0.15–0.25mm). The clearance of the exhaust valves should be 0.0098–0.0138 in. (0.25–0.35mm).

b. Note the out-of-specification valve clearance measurements. You will need them later to determine the required replacement valve lifter.

c. Turn the crankshaft 1 revolution (360°) to set the No. 4 cylinder to TDC.

5. Check the valve clearance of the second set of the valves shown:

a. Use a feeler gauge to measure the clearance between the valve lifter and camshaft. The clearance of the intake valves should be 0.0059–0.0098 in. (0.15–0.25mm). The clearance of the exhaust valves should be 0.0098–0.0138 in. (0.25–0.35mm).

b. Note the out-of-specification valve clearance measurements. You will need them later to determine the required replacement valve lifter.

6. Remove or disconnect the following:
- Drive belt
- Right side engine mount
- Drive belt tensioner

❈❈ WARNING

DO NOT turn the crankshaft while the tensioner is removed!

7. Set the No. 1 cylinder to TDC of the compression stroke.
- Camshafts
- Valve lifters.

8. Use a micrometer to measure the thickness of the used lifter. Calculate the thickness of a new lifter. so the valve clearance comes within the specified value:

a. A: Thickness of new lifter.

b. B: Thickness of used lifter.

c. C: Measured valve clearance.

d. Intake valve clearance: A = B + (C−0.0079 in. (0.20mm).

e. Exhaust valve clearance: A = B + (C−0.0118 in. (0.30mm).

f. Select a new lifter with a thickness as close as possible to the calculated

values. Lifters come in 35 sizes in increments of 0.0008 in. (0.020mm) from 0.1992–0.2260 in (5.060–5.740mm).

9. Install or connect the following:
- Camshafts
- Drive belt tensioner
- Right hand engine mount
- Cylinder head (valve) cover sub-assembly
- Ignition coil
- Engine wire
- Cylinder head (valve) cover
- Negative battery cable

2ZZ-GE Engine

➡**Adjust the valve clearance when the engine is cold.**

1. Before servicing the vehicle, refer to the precautions in the beginning of this section.

2. Remove or disconnect the following:
- Negative battery cable.
- Right side engine under cover
- Cylinder head cover
- Ignition coil assembly
- Wire harness clamp
- Suction hose sub-assembly
- Cylinder head cover sub-assembly
- Drive belt
- Right side engine mount

3. Set the No. 1 cylinder to Top Dead Center (TDC) of the compressor stroke as follows:

a. Turn the crankshaft pulley, and

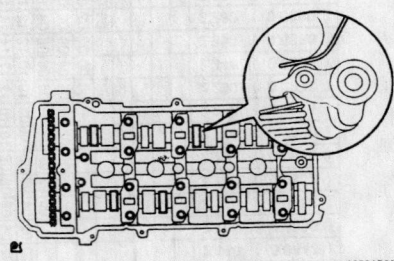

Check the clearance of the 1st set of valves–2ZZ-GE engine

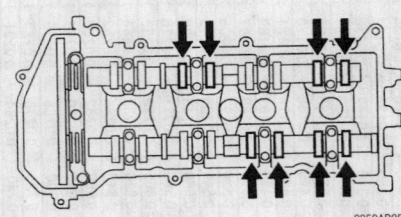

Check the clearance of the 2nd set of valves–1ZZ-FE engine

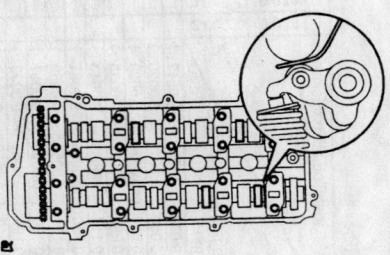

Check the clearance of the 2nd set of valves–2ZZ-GE engine

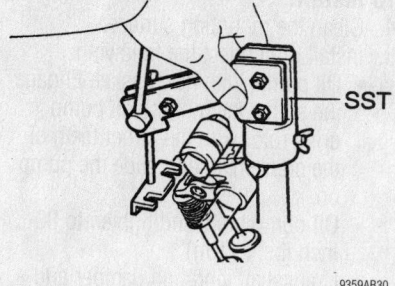

Insert the special tool into the plug tube—2ZZ-GE

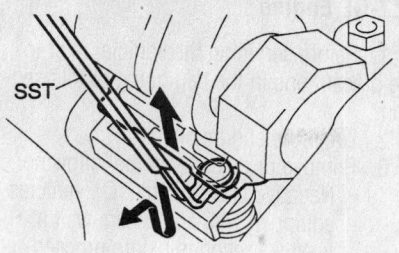

Setting the tool from the right side, makes shim removal easier—2ZZ-GE

align its groove with the "0" timing mark of the timing chain cover.

b. Make sure the point marks of the camshaft timing sprockets and VVT timing sprockets are in a straight line as shown. If not, turn the crankshaft 1 complete revolution (360°) and align the marks.

4. Check the valve clearance of the first set of the valves shown:

a. Use a feeler gauge to measure the clearance between the valve lifter and camshaft. The clearance of the intake valves should be 0.0031–0.0071 in. (0.08–0.18mm). The clearance of the exhaust valves should be 0.0087–0.0126 in. (0.22–0.32mm).

b. Note the out-of-specification valve clearance measurements. You will need them later to determine the required replacement valve lifter.

c. Turn the crankshaft 1 revolution (360°) to set the No. 4 cylinder to TDC.

5. Check the valve clearance of the second set of the valves shown:

a. Use a feeler gauge to measure the clearance between the valve lifter and camshaft. The clearance of the intake valves should be 0.0031–0.0071 in.

(0.08–0.18mm). The clearance of the exhaust valves should be 0.0087–0.0126 in. (0.22–0.32mm).

b. Note the out-of-specification valve clearance measurements. You will need them later to determine the required replacement valve lifter.

6. To adjust the intake valve clearance:

a. Set the SST. Turn the crankshaft so the related rocker arm, where the valve clearance is adjusted, is fully pushed down.

➡ **Remove he spark plug and take off the compression.**

b. Insert SST 09248-77010 into the plug tube. The tool cannot be inserted unless the set screw is loosened.

c. Operate the lever so that the SST's seat surface comes to contact with the valve retainer and lock them with the set screw. Clearance between the valve retainer and SST's set surface is not allowed. Be careful not to make clearance when inserting the SST, since clearance may unlock the keeper.

d. Lock the set screw on the tube side of the SST.

e. Rotate the crankshaft so that the camshaft is position as shown. During rotation, pay attention to the direction, to

prevent the nose of the camshaft from interfering with the SST's shaft. Do not rotate the crankshaft excessively.

f. Lift the rocker arm to make room and remove the adjusting shim using SST 09248-77010.

7. Determine the size of the replaced shim according to the chart or the following formula:

a. Use a dial indicator to measure the thickness of the removed shim.

b. Calculate the thickness of a new shim so that the valve clearance comes within the specified value.

c. A: Thickness of new shim.

d. B: Thickness of used shim.

e. C: Measured valve clearance.

f. Intake: A = B + (C—0.005 in. [0.13mm])

g. Exhaust: A = B + (C—0.011 in. [0.27mm])

h. Select a new shim with a thickness as close as possible to the calculated values. Shims come in 41 sizes in increments of 0.0008 in. (0.020mm) from 0.0787–0.1102 in (2.0–2.8mm).

8. Lift the rocker arm to make room, then install the adjusting shim using the SST. To remove the tool from the shim, push down on the rocker arm.

9. Turn the crankshaft so the related rocker arm, where the valve clearance is adjusted, is fully pushed down.

10. Loosen the 2 set-screws, then remove the SST.

11. Install all components in the reverse of the removal procedure.

Starter

REMOVAL & INSTALLATION

1. Before servicing the vehicle, refer to the precautions in the beginning of this section.

2. Remove or disconnect the following:
- Negative battery cable
- Right side engine undercover

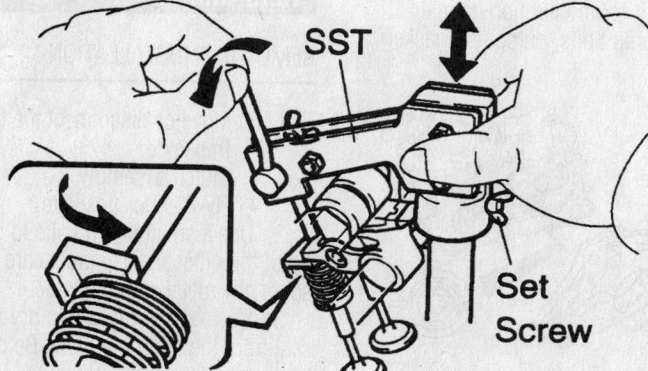

Operate the lever so that the SST's seat surface comes to contact with the valve retainer and lock them with the set screw

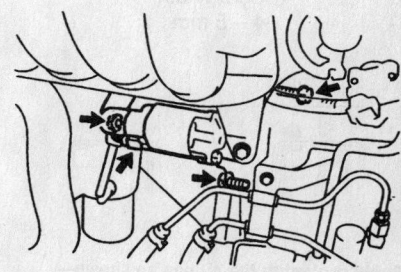

Starter mounting—Vibe

- Starter wiring
- Starter

3. Installation is the reverse of removal. Torque the bolts to 27 ft. lbs. (37 Nm) and the nut to 7 ft. lbs. (10 Nm).

Oil Pan

REMOVAL & INSTALLATION

1ZZ-FE Engine

1. Before servicing the vehicle, refer to the precautions in the beginning of this section.
2. Drain the engine oil.
3. Remove or disconnect the following:

- Negative battery cable
- Undercovers
- Front exhaust pipe
- Oil pan mounting bolts and nuts
- Oil pan, cutting off the applied sealer.

To install:

4. Remove any old sealant from the oil pan flange and thoroughly clean the sealing surface.
5. Install or connect the following:

- Oil pan. Tighten the bolts and nuts in several passes to 80 inch lbs. (9 Nm).
- Front exhaust pipe
- Negative battery cable
- Undercovers

6. Fill the engine with clean oil.
7. Start the vehicle, check for leaks and repair if necessary.

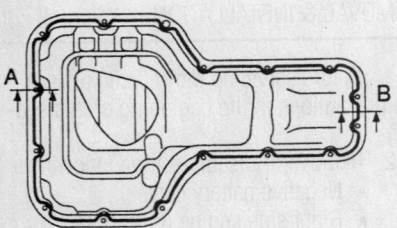

Seal Width
4 – 5 mm

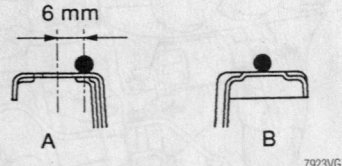

Apply sealant to the oil pan as shown—1.8L (1ZZ-FE) engine

7923VG72

2ZZ-GE Engine

1. Before servicing the vehicle, refer to the precautions in the beginning of this section.
2. Drain the engine oil.
3. Remove or disconnect the following:

- Negative battery cable. On vehicles equipped with an air bag, wait at least 90 seconds before proceeding.
- Undercovers
- Front exhaust pipe
- Oil pan mounting bolts and nuts
- Oil pan, cutting off the applied sealer

To install:

4. Remove any old sealant from the oil pan flange and thoroughly clean the sealing surface.
5. Install or connect the following:

- Oil pan. Tighten the bolts and nuts in several passes to 80 inch lbs. (9 Nm).
- Front exhaust pipe
- Negative battery cable
- Undercovers

6. Fill the engine with clean oil.
7. Start the vehicle, check for leaks and repair if necessary.

Oil Pump

REMOVAL & INSTALLATION

1ZZ-FE Engine

1. Before servicing the vehicle, refer to the precautions in the beginning of this section.
2. Drain the engine oil.
3. Remove or disconnect the following:

- Negative battery cable
- Timing chain and crankshaft sprocket
- Timing chain vibration damper
- Oil pump bolts, pump and gasket

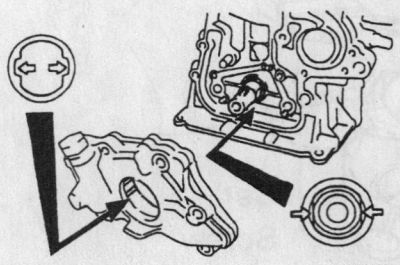

Oil pump mounting—1ZZ-FE and 2ZZ-GE engines

9359AB34

To install:

4. Clean the mounting surface.
5. Install or connect the following:

- Oil pump, with new gasket. Engage the spline teeth of the oil pump drive rotor with the larger teeth of the crankshaft, and slide the pump on.
- Oil pump bolts and tighten to 80 inch lbs. (9 Nm)
- Crankshaft vibration damper and tighten to 80 inch lbs. (9 Nm)
- Crankshaft sprocket and timing chain
- Negative battery cable

6. Fill the engine with clean oil.
7. Start the vehicle, check for leaks and repair if necessary.

2ZZ-GE Engine

1. Before servicing the vehicle, refer to the precautions in the beginning of this section.
2. Drain the engine oil..
3. Remove or disconnect the following:

- Negative battery cable
- Timing chain and crankshaft sprocket
- Oil pump and gasket

To install:

4. Clean the mounting surface.
5. Install or connect the following:

- Oil pump, with new gasket. Engage the spline teeth of the oil pump drive rotor with the larger teeth of the crankshaft, and slide the pump on.
- Oil pump bolts and tighten to 80 inch lbs. (9 Nm)
- Crankshaft sprocket and timing chain
- Negative battery cable

6. Fill the engine with clean oil.
7. Start the vehicle, check for leaks and repair if necessary.

Rear Main Seal

REMOVAL & INSTALLATION

1. Remove or disconnect the following:

- Transaxle
- Clutch assembly
- Flywheel or flexplate

2. Use a small sharp knife to cut off the lip of the oil seal. Take great care not to score any metal with the knife.
3. Use a small prytool to pry the old seal from the retaining plate. Be careful not to damage the plate. Protect the tip of the tool with tape and pad the fulcrum point with cloth.

4. Inspect the crankshaft and seal lip contact surfaces for any sign of damage.

To install:

5. Apply a light coat of multi-purpose grease to the lip of a new oil seal. Loosely fit the seal into place by hand, making sure it is not crooked.

6. Use a seal driver of the correct size to install the seal. Tap it into place until the surface of the seal is flush with the edge of the housing.

Timing Chain, Sprockets, Front Cover and seal

REMOVAL & INSTALLATION

1. Before servicing the vehicle, refer to the precautions in the beginning of this section.

2. Drain the cooling system.

3. Remove or disconnect the following:
- Right side engine under cover
- Right front wheel and tire
- Cylinder head cover
- Wire harness clamp and suction hose assembly, 2ZZ-GE engine
- Drive belt

4. Separate the vane pipe assembly, but do not disconnect the hose, 1ZZ-FE engine.
- Alternator bracket, 2ZZ-GE
- Alternator
- Power steering pump reservoir and position it aside, 1ZZ-FE engine

5. Place a jack with a wooden block under the vehicle for support, then remove the 4 bolts and 2 nuts and remove the right side engine mount.

6. Remove the engine wire as follows, on 1ZZ-FE engines:

a. Remove the 5 clamps from the brackets.

b. Detach the connectors.

c. Remove the ignition coil connectors.

d. Bolt and nut holding the engine wire.

7. Remove the engine wire as follows, on 2ZZ-GE engines:

a. Detach the ignition coil, oil control valve and Crankshaft Position Sensor (CKP) sensor electrical connectors.

b. Bolt and nut for the engine ground, then position the engine wire aside

8. Remove or disconnect the following:
- Ignition coil assembly
- Positive Crankcase Ventilation (PCV) hoses from the cylinder head cover, if necessary
- Cylinder head (valve) cover sub-assembly

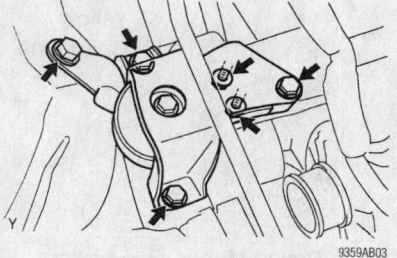

With the engine supported, remove the right side engine mount—1ZZ-FE engine shown, 2ZZ-GE similar

9. Set the No. 1 cylinder to Top Dead Center (TDC) of the compressor stroke as follows:

a. Turn the crankshaft pulley, and align its groove with the "0" timing mark of the timing chain cover.

b. Make sure the point marks of the camshaft timing sprockets and VVT timing sprockets are in a straight line as shown. If not, turn the crankshaft 1 complete revolution (360°) and align the marks.
- Crankshaft pulley, using SST 09960-10010
- Belt tensioner
- Water pump pulley, if equipped, and pump
- Transverse engine mounting bracket
- Crankshaft Position (CKP) sensor
- No. 1 chain tensioner assembly, making sure not to revolve the crankshaft without the tensioner
- Timing chain cover. The cover is retained with 11 bolts and nuts and a Torx® stud bolt. Pry the cover between the cylinder head and block to remove it.
- Timing gear cover oil seal
- CKP sensor plate No. 1
- Timing chain tensioner slipper

➡**In case you turn the camshafts with the timing chain removed, turn the crankshaft ¼ turn for the valve to avoid contact with the pistons.**

- Timing chain sub-assembly. Remove the chain with the crankshaft gear, using screwdrivers as shown.

To install:

10. Set the No. 1 cylinder to TDC of the compression stroke:

a. Turn the hexagonal wrench head part of the camshafts, and align the point marks of the cam sprockets.

b. Using the crankshaft pulley bolt,

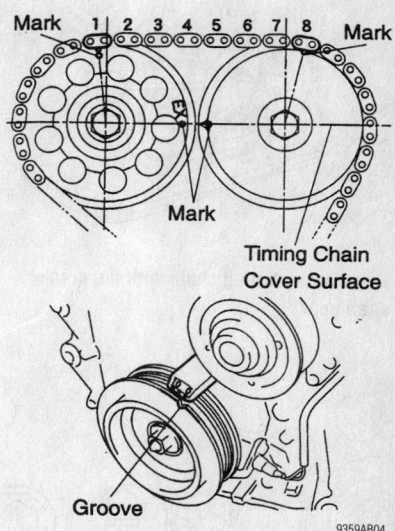

Proper timing mark alignment for TDC

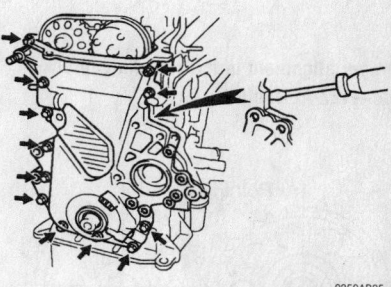

Timing chain cover mounting—1ZZ-FE engine shown, 2ZZ-GE similar

turn the crankshaft and position the crankshaft set key upward.

11. Install or connect the following:
- Timing chain on the crank sprocket with the yellow link aligned with the mark on the crank sprocket. There are 3 yellow links on the timing chain.
- Crankshaft sprocket, using SST 09223-22010
- Timing chain on the camshaft sprockets with the yellow links aligned with the marks on the cam sprockets
- Timing chain tensioner slipper and tighten the bolt to 14 ft. lbs. (19 Nm)
- Crankshaft position sensor plate, with the "F" mark facing forward
- Timing gear cover oil seal
- Timing cover. For 1ZZ-FE engine, tighten the "A" bolts to 10 ft. lbs. (13 Nm), the "B" bolts to 14 ft. lbs. (19 Nm) and the stud bolt to 84 inch lbs. (9.5 Nm), using a Torx® wrench. For 2ZZ-GE engines,

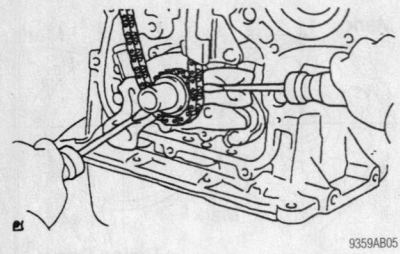

Remove the timing chain with the crankshaft gear

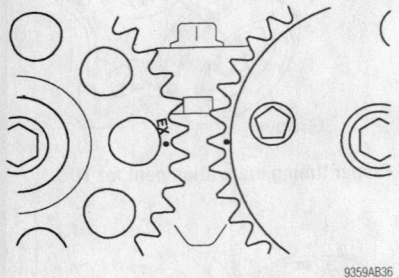

Proper alignment of the camshaft sprockets—1ZZ-FE engine

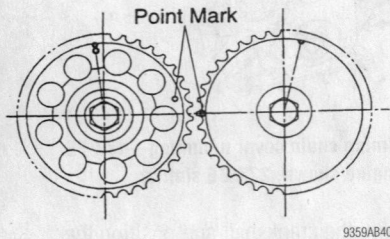

Proper alignment of the camshaft sprockets—2ZZ-GE engine

tighten the M8 bolts to 15 ft. lbs. (21 m), the M6 bolts to 8 ft. lbs. (11 Nm) and the stud bolt to 84 inch lbs. (9.5 Nm).

➡When installing the tensioner, make sure to set the hook again if the hook releases the plunger.

- Timing chain tensioner. Torque the nuts to 80 inch lbs. (9 Nm).
- CKP sensor and tighten the bolts to 80 inch lbs. (9 Nm)
- Transverse engine mounting bracket. Tighten the bolts to 35 ft. lbs. (47 Nm).
- Water pump and pulley
- Drive belt tensioner. Tighten the nut to 21 ft. lbs. (29 Nm) and the bolt to 51 ft. lbs. (69 Nm) on 1ZZ-

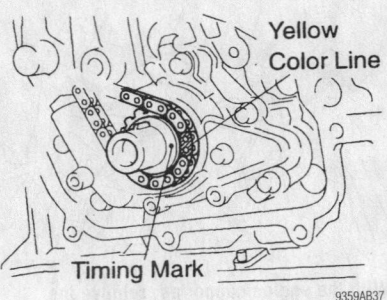

Make sure the yellow link is aligned with the crankshaft sprocket timing mark—1ZZ-FE and 2ZZ-GE engines

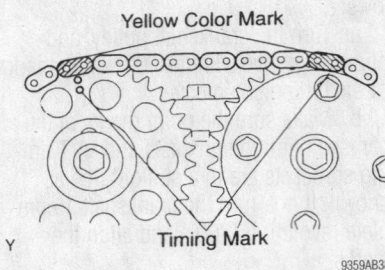

The yellow links of the timing chain must align with the camshaft sprocket timing marks—1ZZ-FE and 2ZZ-GE engines

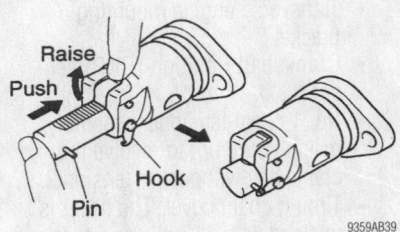

Timing chain tensioner—1ZZ-FE engine

FE engines or to 74 ft. lbs. (100 Nm) on 2ZZ-GE engines.
12. Install the crankshaft pulley, as follows:
    a. Align the pulley set key with the key groove of the pulley and slide on the pulley.
    b. Use SST 09960-11010 to install the bolt and tighten to 102 ft. lbs. (138 Nm) for 1ZZ-FE engine or to 87 ft. lbs. (118 Nm) on 2ZZ-GE engines.
    c. Turn the crankshaft counterclockwise and disconnect the plunger knock pin from the hook.
    d. Turn the crankshaft clockwise and check that the slipper is pushed by the plunger. If the plunger does not spring out, press the slipper into the chain tensioner with a screwdriver so that the

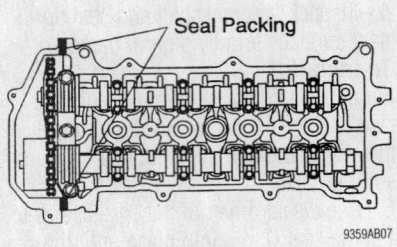

Seal packing installation locations

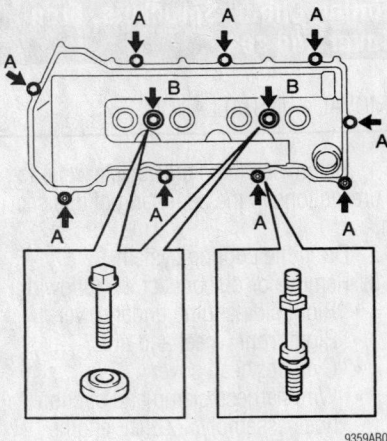

Cylinder head (valve) cover bolt locations—1ZZ-FE engine

hook is released from the knock pin and the plunder springs out.

- Cylinder head sub-assembly cover. Install seal packing into the locations shown and install within 3 minutes. Tighten the "A" bolts to 8 ft. lbs. (11 Nm) and the "B" bolts to 80 inch lbs. (9 Nm) for 1ZZ-FE engines. For 2ZZ-GE engines, tighten the bolts to 7 ft. lbs. (10 Nm).
- Ignition coil assembly. Torque the bolts to 80 inch lbs. (9 Nm).
- Engine wire and tighten to 80 inch lbs. (9 Nm)
- Right side engine mount. Tighten to 38 ft. lbs. (52 Nm).
- Alternator bracket, 2ZZ-GE engine
- Alternator
- Vane pump, 1ZZ-FE
- Main cylinder head cover and tighten to 62 inch lbs. (7 Nm)
- Right front wheel and tire. Tighten the lug nuts to 76 ft. lbs. (103 Nm).

13. Fill the cooling system to the proper level.
14. Start the vehicle, check for leaks and repair if necessary.

Piston and Ring Positioning

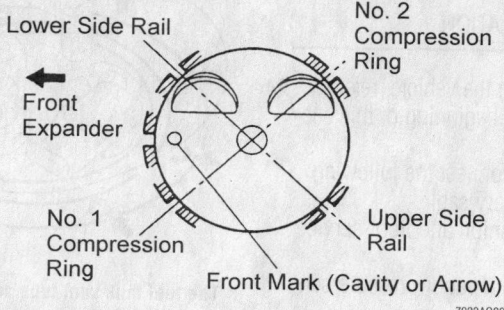

Piston ring end-gap spacing —1ZZ-FE and 2ZZ-GE engines

7923AG90

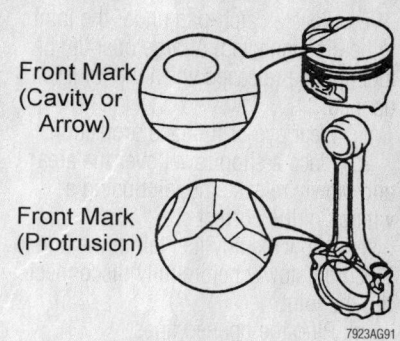

7923AG91

Piston-to-connecting rod assembly —1ZZ-FE and 2ZZ-GE engines

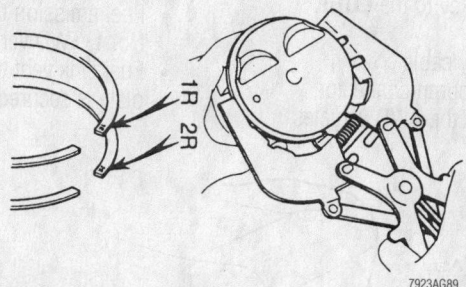

7923AG89

Piston ring identification mark locations—1ZZ-FE and 2ZZ-GE engines

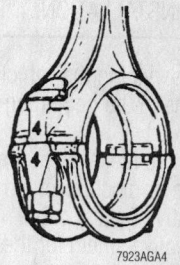

7923AGA4

Before removing the caps from the connecting rods, be sure to matchmark them as shown

FUEL SYSTEM

Fuel System Service Precautions

Safety is the most important factor when performing not only fuel system maintenance, but any type of maintenance. Failure to conduct maintenance and repairs in a safe manner may result in serious personal injury or death. Work on a vehicle's fuel system components can be accomplished safely and effectively by adhering to the following rules and guidelines.

• To avoid the possibility of fire and personal injury, always disconnect the negative battery cable unless the repair or test procedure requires that battery voltage by applied.

• Always relieve the fuel system pressure prior to disconnecting any fuel system component (injector, fuel rail, pressure regulator, etc.) fitting or fuel line connection. Exercise extreme caution whenever relieving fuel system pressure, to avoid exposing skin, face and eyes to fuel spray. Please be advised that fuel under pressure may penetrate the skin or any part of the body that it contacts.

• Always place a shop towel or rag around the fitting or connection prior to loosening to absorb any excess fuel due to spillage. Ensure that all fuel spillage is quickly remove from engine surfaces. Ensure that all fuel-soaked cloths or towels are deposited into a flame-proof waste container with a lid.

• Always keep a dry chemical (Class B) fire extinguisher near the work area.

• Do not allow fuel spray or fuel vapors to come into contact with a light bulb, spark or open flame.

• Always use a second wrench when loosening or tightening fuel line connections fittings. This will prevent unnecessary stress and torsion to fuel piping. Always follow the proper torque specifications.

• Always replace worn fuel fitting O-rings with new ones. Do not substitute fuel hose where rigid pipe is installed.

Fuel System Pressure

RELIEVING

✳✳ CAUTION

Failure to relieve fuel pressure before repairs or disassembly can cause serious personal injury and/or property damage. Fuel pressure is maintained within the fuel lines, even if the engine is OFF or has not been run in a period of time. This pressure must be safely relieved before any fuel-bearing line or component is loosened or removed. On vehicles equipped with inflatable restraints or air bag systems, wait at least 90 seconds after disconnecting the battery cable before performing any other work. The back-up power will keep the restraint system energized for a period of time after the battery is disconnected.

1. Before servicing the vehicle, refer to the precautions in the beginning of this section.

2. Perform the following:

a. Remove the rear seat cushion.

b. Remove the rear floor service hole cover.

c. Disconnect the fuel pump connector.

d. Start and run the engine, until it stalls.

e. Turn the ignition key to the **LOCK** position.

f. Disconnect the negative battery cable.

g. Connect the fuel pump connector.

h. Install the service hole cover and rear seat cushion.

i. Place a catch-pan under the joint to be disconnected. A large quantity of fuel may be released when the joint is opened.

j. Wear eye or full face protection.

k. Place a shop towel over the area and slowly release the joint using a wrench of the correct size.

l. Allow the any fuel left in the line to bleed off slowly before fully disconnecting the joint.

m. Plug the opened lines.

Fuel Filter

REMOVAL & INSTALLATION

1. Before servicing the vehicle, refer to the precautions in the beginning of this section.
2. Relieve the fuel system pressure.
3. Remove or disconnect the following:
   - Negative battery cable
   - Protective shield for the fuel filter
   - Air cleaner hose and cap, if necessary
   - Charcoal canister, if necessary
   - Slowly loosen the lower flare nut fitting until all the pressure is relieved
   - Banjo fitting and 2 metal gaskets. Discard the gaskets.
   - Fuel line with the flared nut from the filter
   - Filter from the mounting bracket

To install:
4. Install or connect the following:
   - New fuel filter
   - Banjo fitting with a new metal gasket on each side and install the union bolt. Bolt: 22 ft. lbs. (30 Nm).
   - Flare nut to the lower connection. Nut: 22 ft. lbs. (30 Nm).
   - Charcoal canister
   - Air cleaner hose and cap
   - Protective shield
   - Negative battery cable

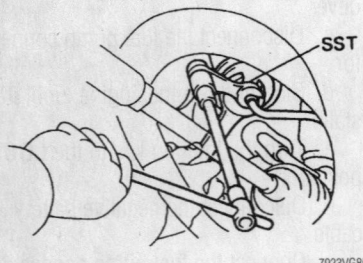

7923VG85

A line wrench with an extension may be needed to loosen the inlet line at the filter

Fuel Pump

REMOVAL & INSTALLATION

1. Before servicing the vehicle, refer to the precautions in the beginning of this section.
2. Remove or disconnect the following:
   - Negative battery cable
   - Rear seat cushion and floor service hole cover
   - Fuel pump and vapor pressure sensor connectors
   - Start and run the engine, until it stalls
3. Turn the ignition key to the **LOCK** position.
   - Negative battery cable
4. Connect the fuel pump connector.
   - Fuel tank protector, AWD vehicles

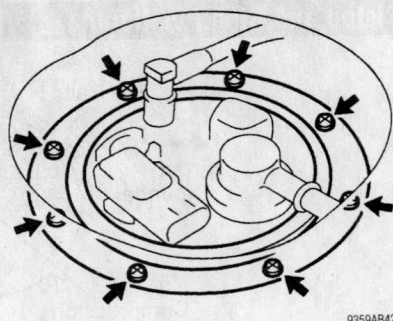

9359AB42

The fuel tank vent tube set plate is secured with 8 bolts on FWD vehicles

- Fuel tank main tube sub-assembly
- Fuel emission tube sub-assembly No. 1, FWD vehicles
- Fuel tank vent tube set plate. The plate is secured with 8 bolts on

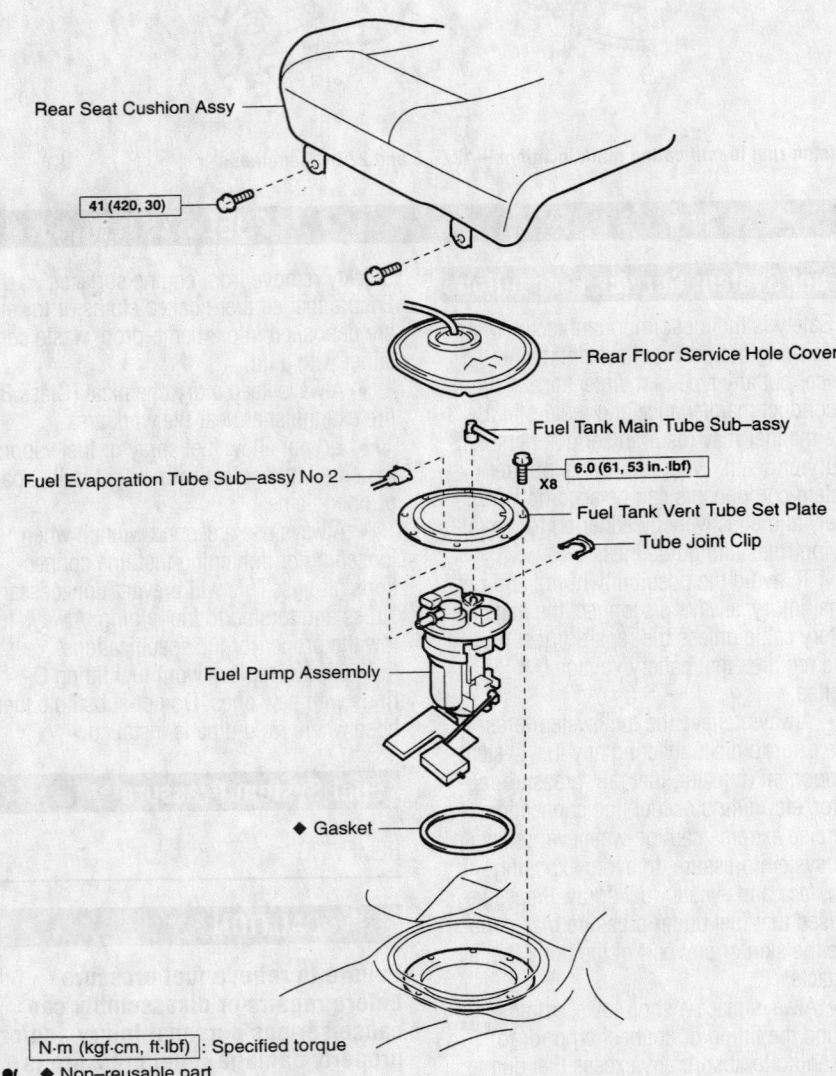

Rear Seat Cushion Assy

41 (420, 30)

Rear Floor Service Hole Cover

Fuel Tank Main Tube Sub-assy

Fuel Evaporation Tube Sub-assy No 2

6.0 (61, 53 in.·lbf) X8

Fuel Tank Vent Tube Set Plate

Tube Joint Clip

Fuel Pump Assembly

◆ Gasket

N·m (kgf·cm, ft·lbf) : Specified torque
◆ Non-reusable part

9359AB41

Exploded view of the fuel pump mounting—FWD shown, AWD similar

FWD vehicles, or 5 bolts on AWD vehicles.
- Fuel pump assembly, being careful not to damage the filter or bend the arm of the fuel sender gauge
- Fuel suction tube set gasket
- Fuel suction support No. 2
- Fuel pump rubber cushion
- Fuel sender gauge assembly. Unplug the connector, then use a

screwdriver to unlock the gauge and slide it to remove.
- Fuel section plate sub-assembly
- Vapor pressure sensor
- Fuel pump harness
- Fuel pump
- Fuel pump filter
- Fuel pressure regulator and O-ring

To install:
5. Install or connect the following:

- New regulator O-ring and regulator
- Fuel pump filter
- Fuel pump
- Vapor pressure sensor
- Fuel suction tube set gasket
- Fuel pump assembly
- Fuel tank vent tube set plate. Tighten the bolts to 53 inch lbs. (6 Nm).

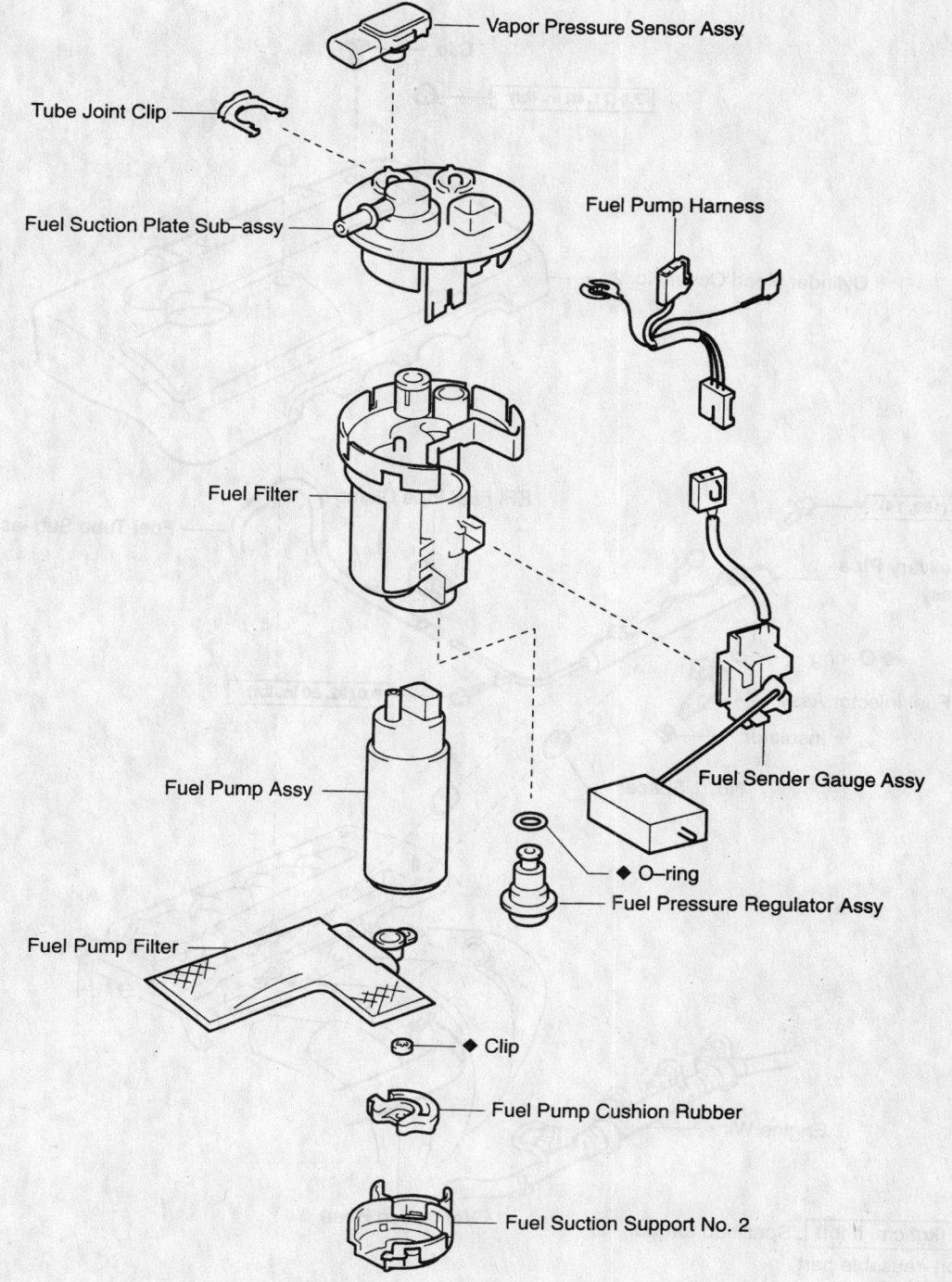

Fuel pump assembly components—FWD vehicles shown, AWD similar

9359AB43

- Connect the fuel emission tube sub-assembly
- Fuel tank main tube sub-assembly
- Fuel tank protector No. 2, AWD vehicles
- Negative battery cable. Check for fuel leaks.
- Floor service hole cover. Use butyl tape to seal the cover.
- Rear seat cushion

Fuel Injectors

REMOVAL & INSTALLATION

1ZZ-FE Engine

1. Before servicing the vehicle, refer to the precautions in the beginning of this section.

2. Properly relieve the fuel system pressure.

3. Remove or disconnect the following:

- Negative battery cable.
- No. 2 cylinder head cover
- Positive Crankcase Ventilation (PCV) hose
- Engine wire, unplugging the injector connectors and clamps
- Fuel pipe clamp
- Fuel line/tube sub-assembly

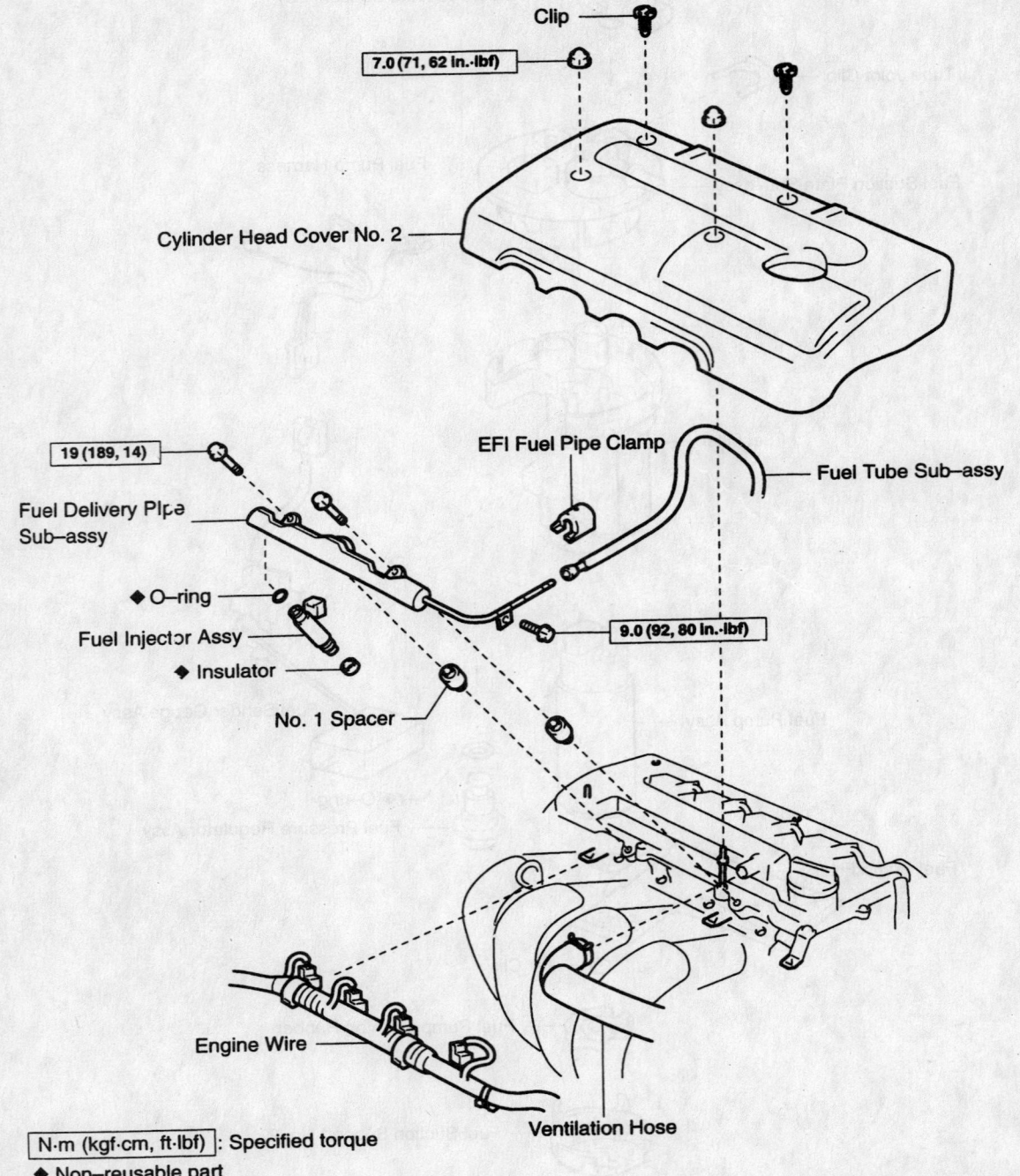

Fuel injector removal and installation—1ZZ-FE engine

9359AB44

✲✲ WARNING

Be careful not to drop the fuel injectors when removing the delivery pipe.

- Fuel delivery pipe sub-assembly with the injectors attached
- Delivery pipe and injectors
- Spacers from the head
- Injectors from the delivery pipe
- O-ring and grommet from each injector

To install:

4. Install or connect the following:

- New grommets
- New O-rings coated with light machine oil
- Injectors on the delivery pipe

➡ **Coat the contact point on the pipe with light machine oil and twist the injectors into place. The connector should face outward.**

- Spacers

➡ **Coat the seats in the head where the injectors contact, with light machine oil.**

- Delivery pipe and injectors

5. Loosely install the hold-down bolts and check that the injectors rotate smoothly. If they don't, the probable cause is incorrect O-ring installation. Torque the delivery pipe hold-down bolts to 14 ft. lbs. (19 Nm) and the fuel pipe bolt to 80 inch lbs. (9 Nm).

- Engine wire, attaching the injector connectors and clamps
- Fuel line/tube sub-assembly
- PCV hose
- No. 2 cylinder head (valve) cover

2ZZ-GE ENGINE

1. Before servicing the vehicle, refer to the precautions in the beginning of this section.

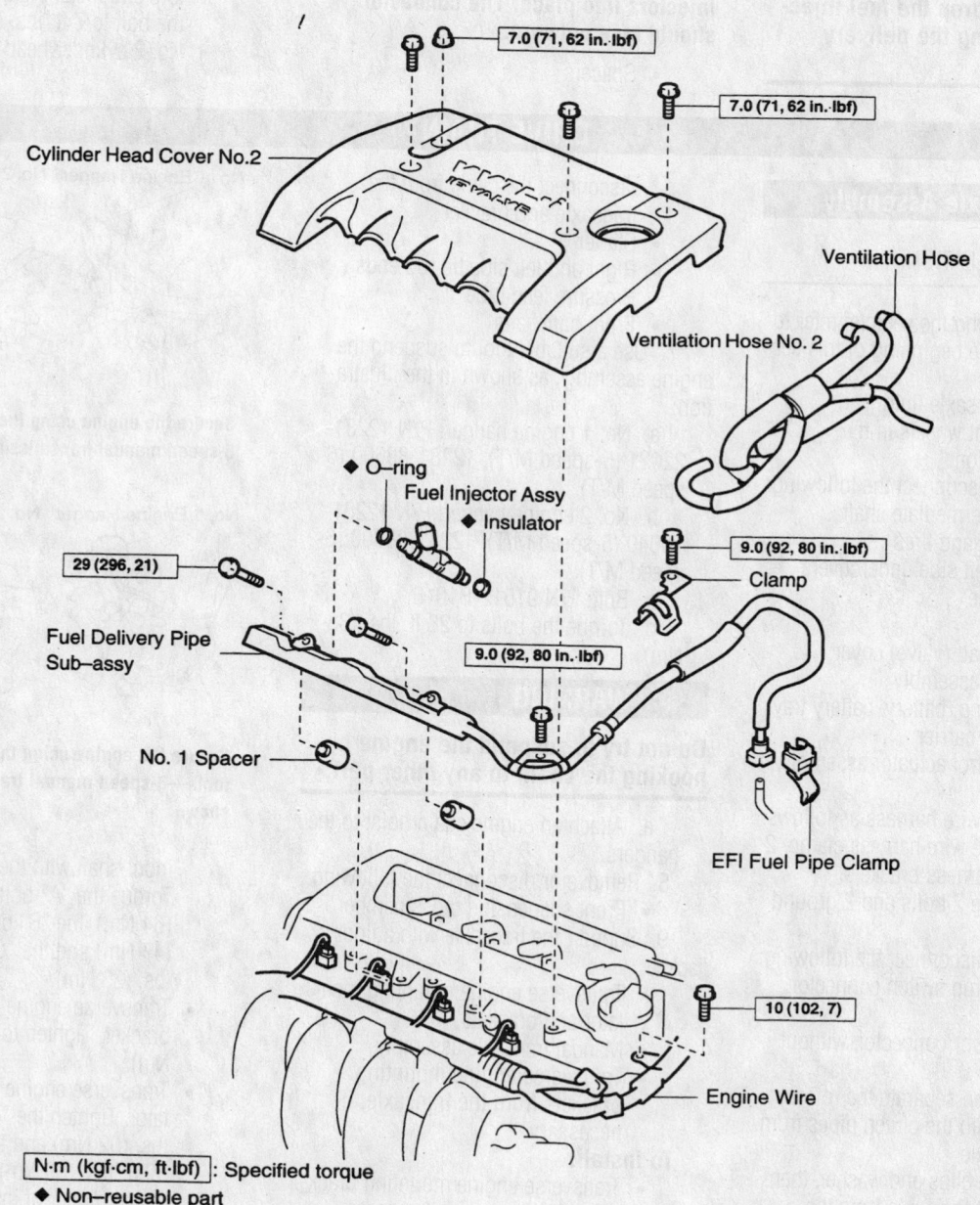

Cylinder Head Cover No.2

7.0 (71, 62 in. lbf)
7.0 (71, 62 in. lbf)

Ventilation Hose
Ventilation Hose No. 2

◆ O-ring
Fuel Injector Assy
◆ Insulator

9.0 (92, 80 in. lbf)
Clamp

29 (296, 21)

Fuel Delivery Pipe Sub-assy

9.0 (92, 80 in. lbf)

No. 1 Spacer

EFI Fuel Pipe Clamp

10 (102, 7)

Engine Wire

N·m (kgf·cm, ft·lbf): Specified torque
◆ Non-reusable part

Fuel injector removal and installation—2ZZ-GE engine

9359AB45

2. Properly relieve the fuel system pressure.

3. Remove or disconnect the following:

- Negative battery cable.
- No. 2 cylinder head cover
- Positive Crankcase Ventilation (PCV) hose
- Engine wire, by removing the bolt, then unplugging the injector and Camshaft Position (CMP) sensor connectors
- Fuel pipe clamp

✳✳ WARNING

Be careful not to drop the fuel injectors when removing the delivery pipe.

- Fuel delivery pipe sub-assembly with the injectors attached
- Delivery pipe and injectors
- Spacers from the head
- Injectors from the delivery pipe
- O-ring and grommet from each injector

To install:

4. Install or connect the following:

- New grommets
- New O-rings coated with light machine oil
- Injectors on the delivery pipe

➡**Coat the contact point on the pipe with light machine oil and twist the injectors into place. The connector should face outward.**

- Spacers

➡**Coat the seats in the head where the injectors contact, with light machine oil.**

- Delivery pipe and injectors

5. Loosely install the hold-down bolts and check that the injectors rotate smoothly. If they don't, the probable cause is incorrect O-ring installation. Torque the delivery pipe hold-down bolts to 14 ft. lbs. (19 Nm) and the fuel pipe bolt to 80 inch lbs. (9 Nm).

- Fuel line/tube sub-assembly
- PCV hose
- Engine wire, by connecting the CMP sensor and injector connectors and installing the bolt. Tighten the bolt to 7 ft. lbs. (10 Nm).
- No. 2 cylinder head (valve) cover

DRIVE TRAIN

Manual Transaxle Assembly

REMOVAL & INSTALLATION

1. Before servicing the vehicle, refer to the precautions in the beginning of this section.

2. Drain the transaxle fluid.

3. Place the front wheels in the straight-ahead position.

4. Remove or disconnect the following:

- Steering intermediate shaft
- Front wheel and tires
- Right and left side undercovers
- Exhaust pipe
- Hood
- Cylinder head (valve) cover
- Air cleaner assembly
- Battery clamp, battery, battery tray and battery carrier
- Cruise control actuator assembly, if equipped

5. Remove the wire harness as follows:

a. Remove the wire harness clamp, 2 bolts and wire harness brackets.

b. Remove the 2 bolts and 2 ground cables.

6. Remove or disconnect the following:

- Back-up lamp switch connector, with ABS
- Speed sensor connector, without ABS
- 5 bolts, then separate the release cylinder with the clutch pipes from the transaxle
- Shift cable clips and washer, then disconnect the cable from the transaxle and bracket
- Select cable clips and washer, then

disconnect the cable from the transaxle and bracket
- Starter
- Right and left side tie rod ends
- Pressure feed tube
- Front halfshafts

7. Use a suitable tool to suspend the engine assembly, as shown in the illustration:

a. No. 1 engine hanger: P/N 12281-22021 (5-speed M/T), 12281-88600 (6-speed M/T).

b. No. 2 engine hanger: P/N 12281-15040 (5-speed M/T), 12281-88600 (6-speed M/T).

c. Bolt: P/N 91512-B1016.

d. Torque the bolts to 28 ft. lbs. (38 Nm).

✳✳ WARNING

Do not try to suspend the engine by hooking the chain to any other part.

e. Attach an engine chain hoist to the hangers.

8. Remove or disconnect the following:

- Front suspension crossmember

9. Support the transaxle with a floor jack.

- Transverse engine mounting insulator and brackets
- Manual transaxle assembly
- Transverse engine mounting brackets from the transaxle, if necessary

To install:

- Transverse engine mounting brackets to the transaxle, if necessary
- Manual transaxle, by aligning the

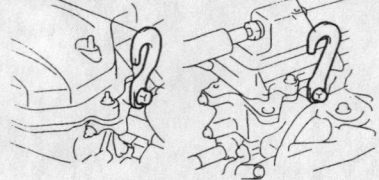

No. 1 Engine Hanger   No. 2 Engine Hanger

9359AB46

Secure the engine using the proper tools—5-speed manual transmission shown

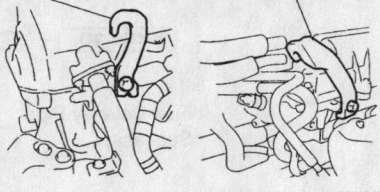

No. 1 Engine Hanger   No. 2 Engine Hanger

9359AB49

Secure the engine using the proper tools—6-speed manual transmission shown

input shaft with the clutch disc. Torque the "A" bolts to 47 ft. lbs. (64 Nm), the "B" bolts to 35 ft. lbs. (47 Nm) and the "C" bolts to 17 ft. lbs. (23 Nm).

- Transverse engine mounting bracket. Tighten to 38 ft. lbs. (52 Nm).
- Transverse engine mounting insulator. Tighten the "A" bolts to 38 ft. lbs. (52 Nm) and the "B" bolts to 59 ft. lbs. (80 Nm).

10. The remainder of installation is the reverse of the removal procedure, noting the following specifications:

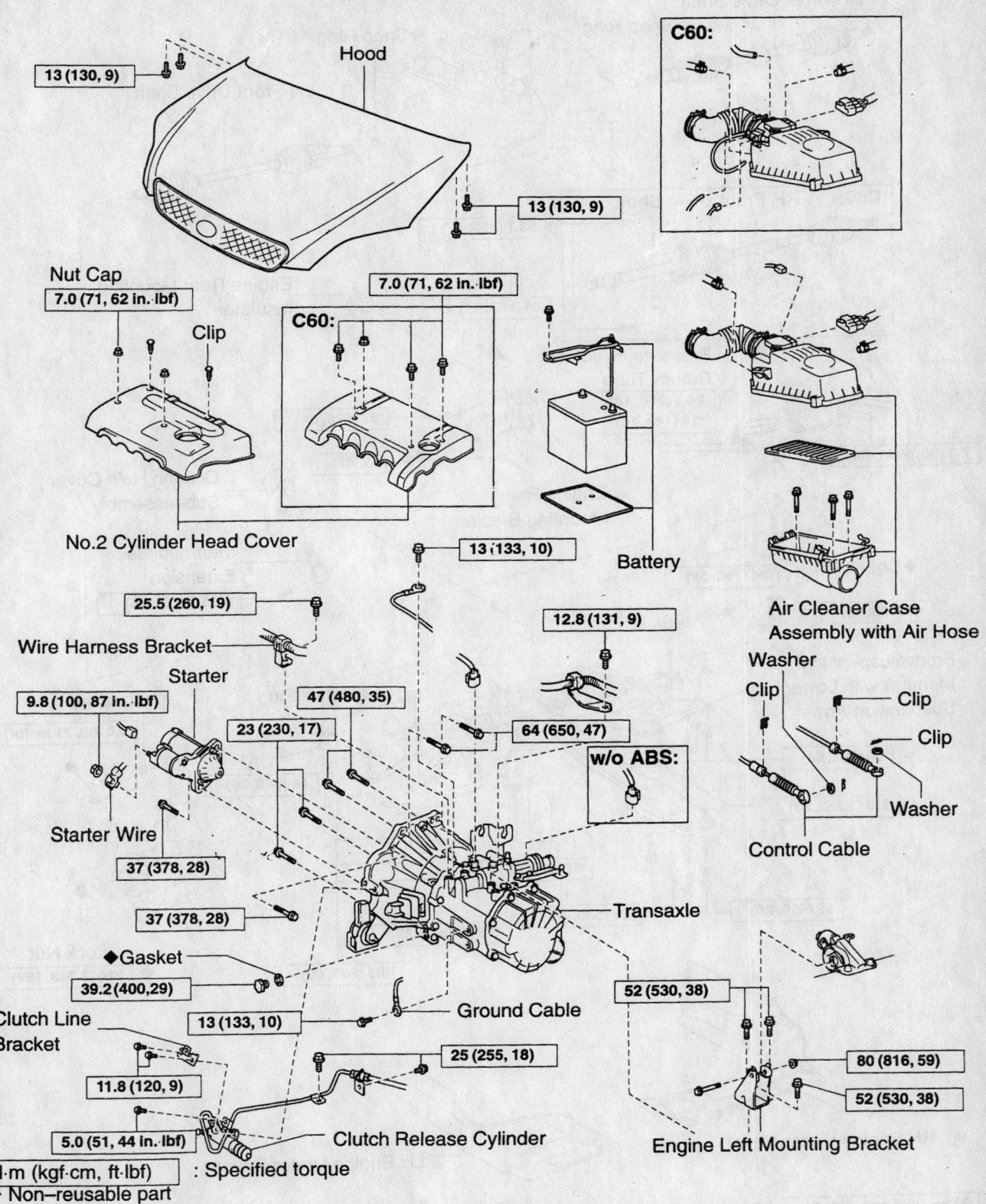

13 (130, 9)

Hood

C60:

13 (130, 9)

Nut Cap
7.0 (71, 62 in.·lbf)

7.0 (71, 62 in.·lbf)

Clip

C60:

No.2 Cylinder Head Cover

13 (133, 10)

Battery

Air Cleaner Case
Assembly with Air Hose

25.5 (260, 19)

Wire Harness Bracket

12.8 (131, 9)

Washer

Clip

Clip

Starter

Clip

9.8 (100, 87 in.·lbf)

47 (480, 35)

23 (230, 17)

64 (650, 47)

Washer

Starter Wire

w/o ABS:

Control Cable

37 (378, 28)

37 (378, 28)

Transaxle

◆Gasket

39.2 (400,29)

52 (530, 38)

Clutch Line
Bracket

13 (133, 10)

Ground Cable

11.8 (120, 9)

25 (255, 18)

80 (816, 59)

52 (530, 38)

5.0 (51, 44 in.·lbf)

Clutch Release Cylinder

Engine Left Mounting Bracket

N·m (kgf·cm, ft·lbf) : Specified torque
◆ Non–reusable part

Exploded view of the manual transaxle (1 of 2)

9359AB47

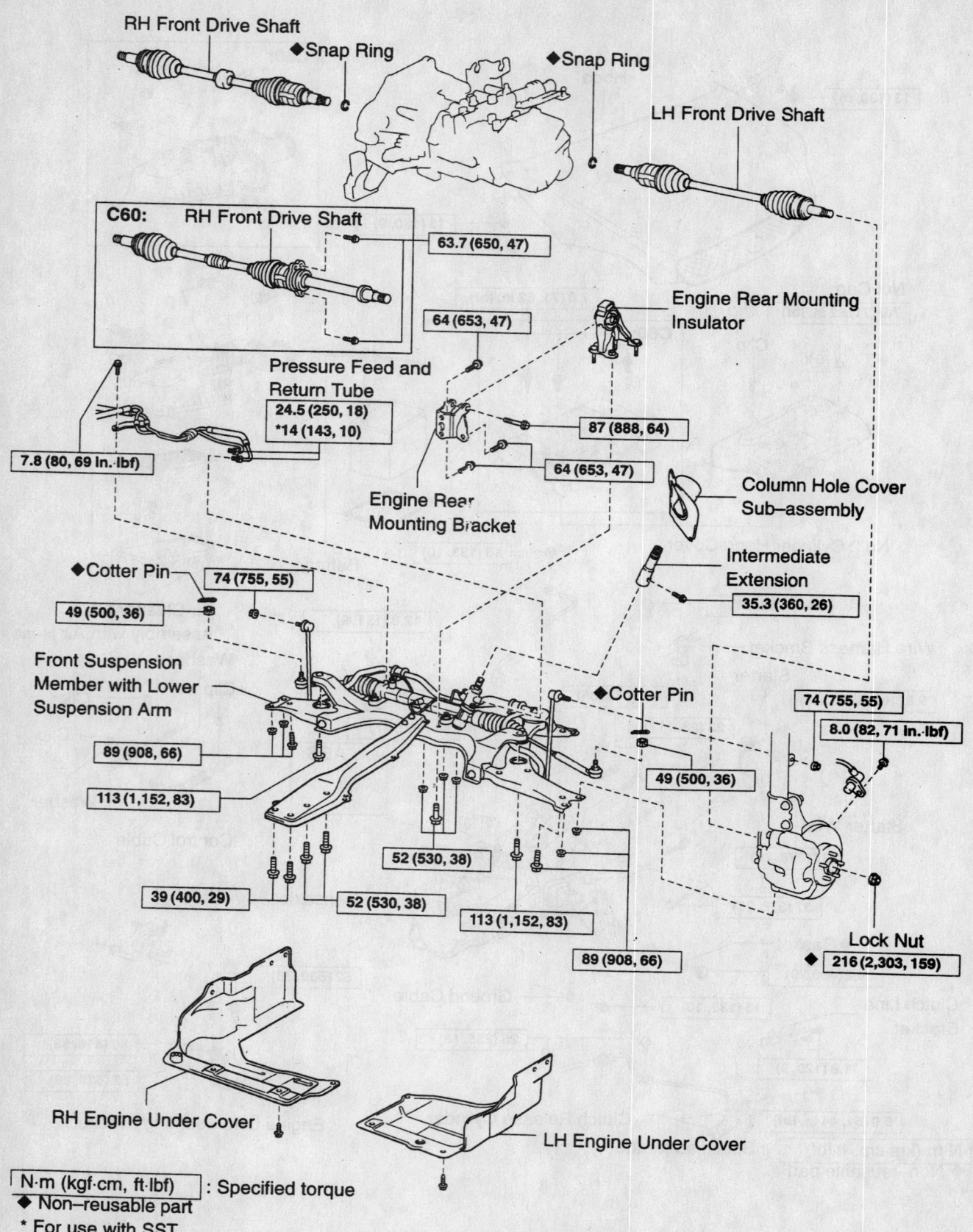

RH Front Drive Shaft

◆Snap Ring

◆Snap Ring

LH Front Drive Shaft

C60: RH Front Drive Shaft

63.7 (650, 47)

64 (653, 47)

Engine Rear Mounting Insulator

Pressure Feed and Return Tube

24.5 (250, 18)
*14 (143, 10)

7.8 (80, 69 in.·lbf)

87 (888, 64)

64 (653, 47)

Engine Rear Mounting Bracket

Column Hole Cover Sub–assembly

◆Cotter Pin

74 (755, 55)

49 (500, 36)

Intermediate Extension

35.3 (360, 26)

Front Suspension Member with Lower Suspension Arm

◆Cotter Pin

74 (755, 55)

8.0 (82, 71 in.·lbf)

89 (908, 66)

113 (1,152, 83)

49 (500, 36)

52 (530, 38)

39 (400, 29)

52 (530, 38)

113 (1,152, 83)

89 (908, 66)

Lock Nut
◆ 216 (2,303, 159)

RH Engine Under Cover

LH Engine Under Cover

N·m (kgf·cm, ft·lbf) : Specified torque
◆ Non–reusable part
* For use with SST

9359AB48

Exploded view of the manual transaxle (2 of 2)

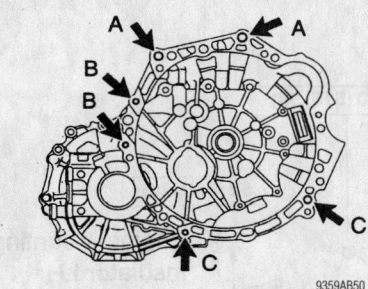

Manual transaxle bolt installation locations

9359AB50

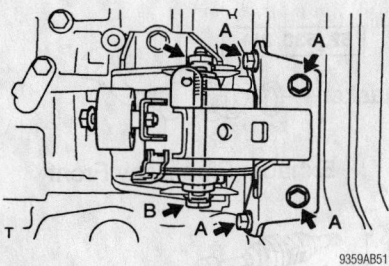

Transverse engine mounting insulator bolt locations

9359AB51

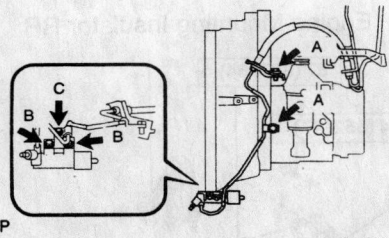

Clutch release cylinder bolt locations

9359AB52

a. Starter mounting bolts: 27 ft. lbs. (37 Nm).

b. Clutch release cylinder bolts: "A" bolts 19 ft. lbs. (25 Nm), "B" bolts 9 ft. lbs. (12 Nm) and "C" bolts 44 inch lbs. (5 Nm).

c. Battery carrier bolts: 10 ft. lbs. (13 Nm).

d. Battery clamp bolt: 44 inch lbs. (5 Nm).

e. Battery clamp nut: 31 inch lbs. (3.5 Nm).

f. Cylinder head cover bolts: 62 inch lbs. (7 Nm).

g. Hood bolts: 10 ft. lbs. (13 Nm).

h. Wheel lug nuts: 76 ft. lbs. (103 Nm).

11. Fill the transaxle fluid to the proper level.

12. Start the vehicle, check for leaks and repair if necessary.

Automatic Transaxle

REMOVAL & INSTALLATION

FWD—A246E & U240E Transaxles

1. Before servicing the vehicle, refer to the precautions in the beginning of this section.

2. Drain the transaxle fluid.

3. Remove or disconnect the following:
- Negative battery cable
- Hood
- No. 2 cylinder head cover
- Battery and battery carrier
- Air cleaner assembly with hose
- Floor shift cable transmission control shift
- Transmission control cable support
- No. 1 transmission control cable bracket
- Wiring harness and brackets
- Transmission wire connector
- Park/neutral position switch connector, with Anti-lock Brake System (ABS)
- Speedometer sensor connector, without ABS
- Transmission revolution sensor connectors, if equipped
- Transmission fluid filler tube
- No. 1 oil cooler inlet and outlet tubes
- Foot rest
- Floor carpet
- Oxygen (O_2) sensor connector

4. Suspend the engine as follows:

a. Disconnect the 2 Positive Crankcase Ventilation (PCV) hoses.

b. Install the No. 1 and No. 2 engine hangers in the correct direction.

c. No. 1 engine hanger: P/N 12281-22021 (A246E) or 12281-88600 (U240E).

d. No. 2 engine hanger: P/N 12281-15040 (A246E) or 12281-88600 (U240E)

e. Bolt: P/N 91512-B1016.

f. Torque the bolt to 28 ft. lbs. (38 Nm).

g. Attach an engine chain hoist to the engine hangers.
- Front wheels
- Right and left engine undercovers
- Front floor panel brace, U240E transaxle
- Front exhaust pipe
- Front halfshafts
- Automatic transmission case protector
- Starter

5. Support the transaxle with a floor jack

- Left side transverse engine mounting insulator and bracket
- Right side front and rear engine mount insulators
- 4 bolts, dynamic damper and member sub-assembly
- Front and rear right side transverse engine mounting brackets
- Flywheel housing undercover
- Automatic transaxle. Turn the crankshaft for access to the 6 bolts while holding the crankshaft pulley bolt with a wrench.
- Torque converter clutch

6. Installation is the reverse of the removal procedure, noting the following specifications:

a. Automatic transaxle: Bolt "A" to 47 ft. lbs. (64 Nm), bolt "B" to 34 ft. lbs. (47 Nm) and bolt "C" to 17 ft. lbs. (23 Nm).

b. Torque converter bolts: 20 ft. lbs. (28 Nm).

c. Front and rear right transverse engine mounting bracket bolts: 47 ft. lbs. (64 Nm).

d. Member sub-assembly center bolts: "A" bolts to 29 ft. lbs. (39 Nm) and "B" bolts to 38 ft. lbs. (52 Nm).

e. Right rear engine mounting insulator-to-engine mounting bracket bolt: 64 ft. lbs. (87 Nm).

f. Right rear engine mount insulator nuts and bolt: 38 ft. lbs. (52 Nm).

g. Left side engine mounting bracket-to-transaxle bolts: 38 ft. lbs. (52 Nm).

h. Left side engine mounting insulator bolts and nut: Bolt "A" to 38 ft. lbs. (52 Nm), Bolt "B" and Nut "B" to 59 ft. lbs. (80 Nm).

i. Front right engine mount insulator-to-mounting bracket bolt and nut: 38 ft. lbs. (52 Nm).

j. Starter bolts: 29 ft. lbs. (39 Nm).

k. Automatic transmission case protector bolts: 14 ft. lbs. (18 Nm).

l. Wheel lug nuts: 76 ft. lbs. (103 Nm).

m. Oil cooler clamp bolts: 49 inch lbs. (5.5 Nm).

n. Oil cooler inlet and outlet tubes: 25 ft. lbs. (34 Nm).

o. Wire harness bracket bolt: 9 ft. lbs. (13 Nm).

p. Transmission control cable bracket bolts: 9 ft. lbs. (12 Nm).

q. Transmission control cable support: 9 ft. lbs. (12 Nm).

r. Battery carrier: 10 ft. lbs. (13 Nm).

s. Air cleaner assembly: 62 inch lbs. (7 Nm).

t. Cylinder head cover bolts: 62 inch lbs. (7 Nm).

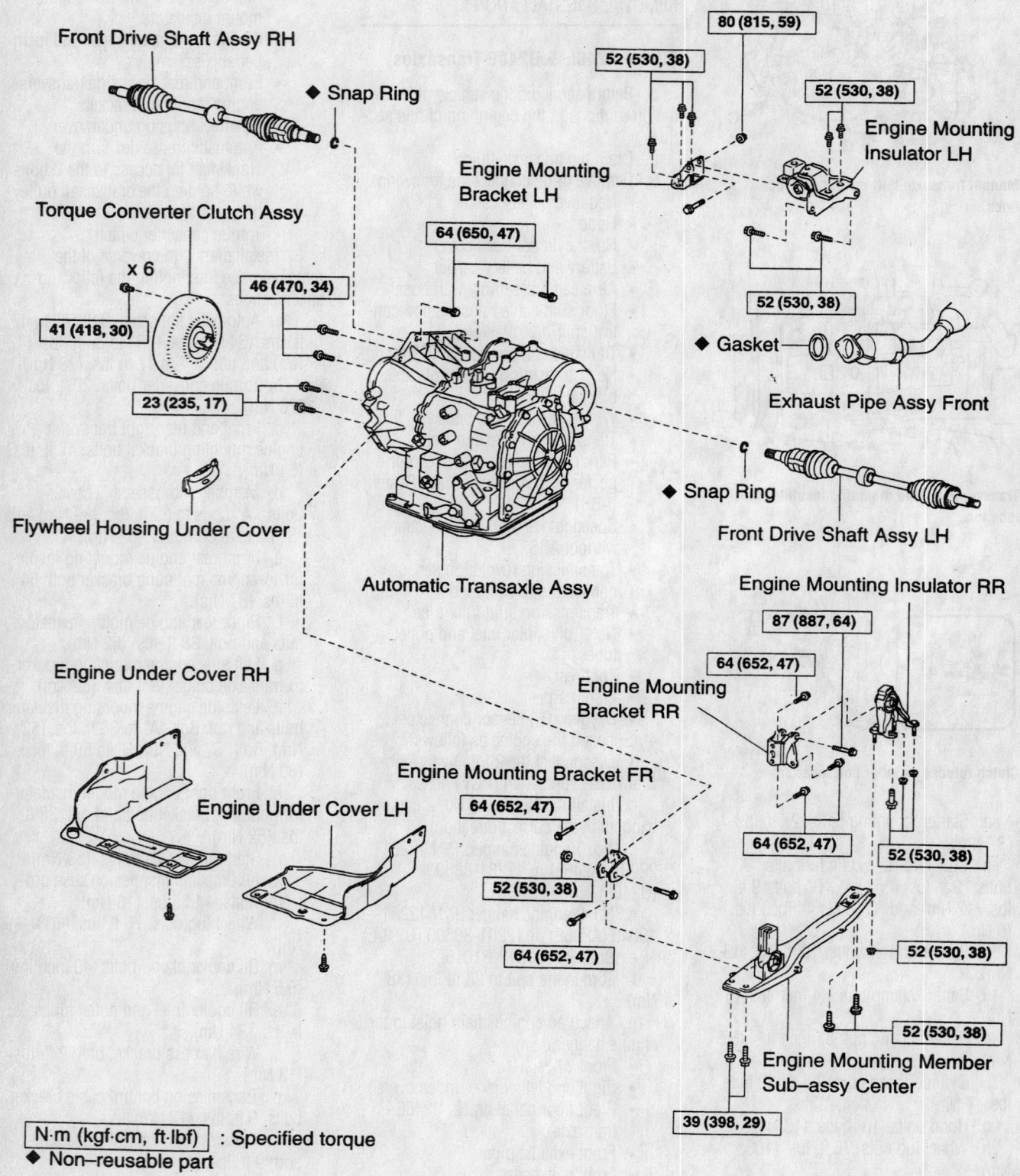

Front Drive Shaft Assy RH

◆ Snap Ring

Torque Converter Clutch Assy

x 6

41 (418, 30)

46 (470, 34)

23 (235, 17)

Flywheel Housing Under Cover

64 (650, 47)

Engine Mounting
Bracket LH

80 (815, 59)

52 (530, 38)

52 (530, 38)

Engine Mounting
Insulator LH

52 (530, 38)

◆ Gasket

Exhaust Pipe Assy Front

Automatic Transaxle Assy

◆ Snap Ring

Front Drive Shaft Assy LH

Engine Mounting Insulator RR

87 (887, 64)

64 (652, 47)

Engine Mounting
Bracket RR

64 (652, 47)

52 (530, 38)

Engine Under Cover RH

Engine Under Cover LH

Engine Mounting Bracket FR

64 (652, 47)

52 (530, 38)

64 (652, 47)

52 (530, 38)

52 (530, 38)

Engine Mounting Member
Sub–assy Center

39 (398, 29)

N·m (kgf·cm, ft·lbf) : Specified torque
◆ Non–reusable part

9359AB55

Automatic transaxle and related components—U240E transaxle shown, A246E similar

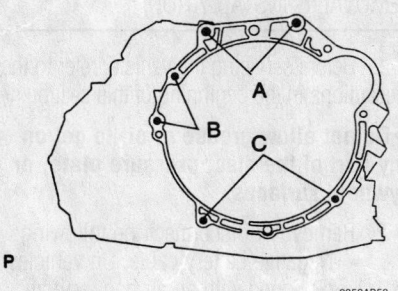

Automatic transaxle bolt locations

9359AB53

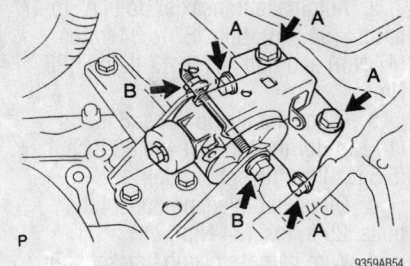

Left side engine mount insulator bolt and nut locations

9359AB54

u. Hood bolts: 10 ft. lbs. (13 Nm).

7. Fill the transaxle fluid to the proper level.

8. Start the vehicle, check for leaks and repair if necessary.

AWD—U341F Transaxle

1. Before servicing the vehicle, refer to the precautions in the beginning of this section.

2. Drain the transaxle fluid.

3. Remove or disconnect the following:

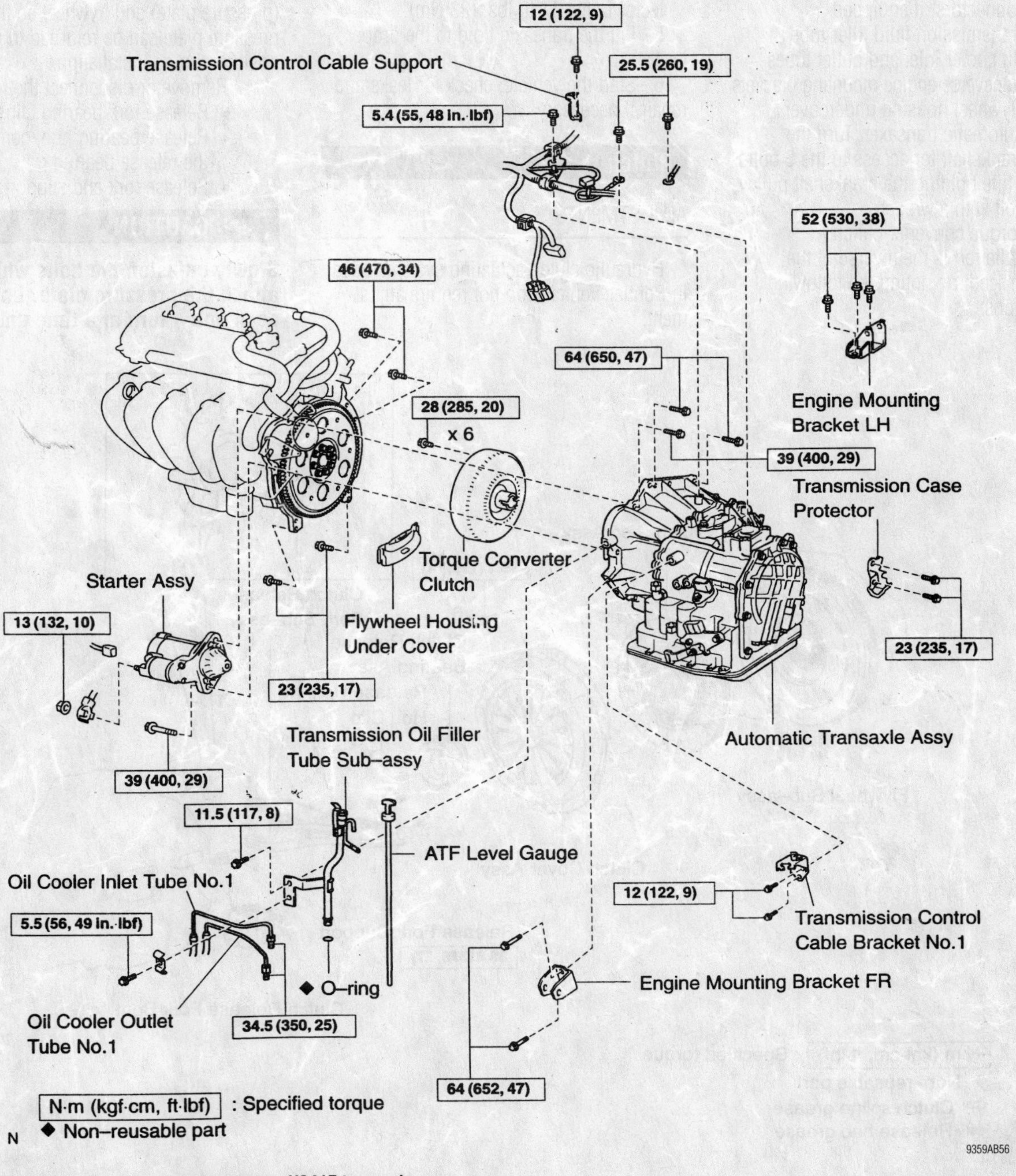

Transmission Control Cable Support

12 (122, 9)

25.5 (260, 19)

5.4 (55, 48 in. lbf)

52 (530, 38)

64 (650, 47)

Engine Mounting Bracket LH

46 (470, 34)

28 (285, 20) x 6

39 (400, 29)

Transmission Case Protector

Torque Converter Clutch

Starter Assy

Flywheel Housing Under Cover

13 (132, 10)

Automatic Transaxle Assy

23 (235, 17)

23 (235, 17)

39 (400, 29)

Transmission Oil Filler Tube Sub–assy

11.5 (117, 8)

ATF Level Gauge

Oil Cooler Inlet Tube No.1

12 (122, 9)

Transmission Control Cable Bracket No.1

5.5 (56, 49 in. lbf)

◆ O–ring

Oil Cooler Outlet Tube No.1

34.5 (350, 25)

Engine Mounting Bracket FR

64 (652, 47)

N·m (kgf·cm, ft·lbf) : Specified torque

N   ◆ Non–reusable part

9359AB56

Automatic transaxle and related components—U341F transaxle

- Negative battery cable
- Engine and transaxle assembly
- Transfer case
- Automatic transmission case protector
- Front left side halfshaft
- Transmission control cable support and bracket
- Wire harness clamp bracket, bolts and 2 wire harnesses
- Transmission wire connector
- Park/neutral position switch connector
- Transmission revolution sensor connectors, if equipped
- Transmission fluid filler tube
- Oil cooler inlet and outlet tubes
- Transverse engine mounting brackets
- Flywheel housing undercover
- Automatic transaxle. Turn the crankshaft for access to the 6 bolts while holding the crankshaft pulley bolt with a wrench.
- Torque converter clutch

4. Installation is the reverse of the removal procedure, noting the following specifications:

a. Automatic transaxle: Bolt "A" to 47 ft. lbs. (64 Nm), bolt "B" to 34 ft. lbs. (47 Nm) and bolt "C" to 17 ft. lbs. (23 Nm).

b. Oil cooler clamp bolts: 8 ft. lbs. (11 Nm) for the top bolt and 49 inch lbs. (5.5 Nm) for the bottom bolt

c. Oil cooler inlet and outlet tube bolts: 25 ft. lbs. (34 Nm).

d. Wire harness clamp bracket bolt: 48 inch lbs. (5 Nm).

e. Transmission control cable bracket and support bolts: 9 ft. lbs. (12 Nm).

f. Automatic transmission case protector bolts: 17 ft. lbs. (23 Nm).

5. Fill the transaxle fluid to the proper level.

6. Start the vehicle, check for leaks and repair if necessary.

Clutch

ADJUSTMENTS

Hydraulic clutch actuating systems used in Pontiac vehicles do not require adjustment.

REMOVAL & INSTALLATION

1. Before servicing the vehicle, refer to the precautions in the beginning of this section.

➡ **Do not allow grease or oil to get on any part of the disc, pressure plate, or flywheel surfaces.**

2. Remove or disconnect the following:
   - Negative battery cable. On vehicles equipped with an air bag, wait at least 90 seconds before proceeding
   - Transaxle assembly

3. Make matchmarks on the clutch cover (pressure plate) and flywheel so that the pressure plate can be returned to its original position during installation.

4. Remove or disconnect the following:
   - Release fork bearing clips
   - Release bearing hub, complete with the release bearing
   - Release fork and support

✳✳ CAUTION

Slowly unfasten the bolts which attach the pressure plate. Loosen each bolt 1 turn at a time until the

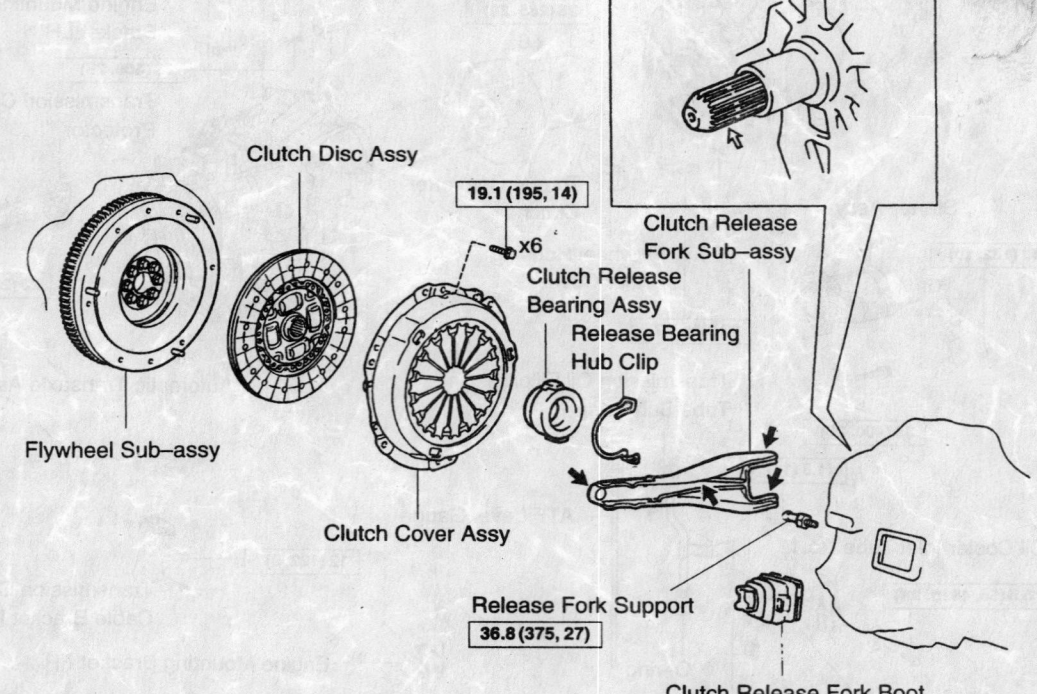

N·m (kgf·cm, ft·lbf) : Specified torque
◆ Non–reusable part
⇐ Clutch spline grease
⬅ Release hub grease

Exploded view of the clutch components

9359AB57

spring tension is released. **If the bolts are released improperly the clutch assembly could fly apart, causing possible injury.**

• Pressure plate from the clutch cover/spring assembly

5. Inspect the disc, pressure plate and flywheel for damage and wear using a caliper to measure depth and width and a dial indicator to measure runout.

a. The minimum clutch disc rivet head depth is 0.012 in. (0.3mm).

b. The maximum clutch disc runout is 0.031 in. (0.8mm).

c. The maximum pressure plate spring depth is 0.024 in. (0.6mm).

d. The maximum pressure plate spring width is 0.197 in. (5.0mm).

e. The maximum flywheel runout is 0.004 in. (0.1mm).

6. Replace or machine parts as necessary.

To install:

7. When reassembling, apply a thin coating of multipurpose grease to the release bearing hub and release fork contact points. Also, pack the groove inside the clutch hub with multipurpose grease and lubricate the pivot points of the release fork.

8. Install or connect the following:
• Clutch disc and pressure plate. The bolts should be tightened in 2 or 3 steps, gradually and evenly. Final bolt torque is 14 ft. lbs. (19 Nm).
• Release bearing, fork and boot
• Transaxle assembly
• Negative battery cable

Hydraulic Clutch System

BLEEDING

➡**If any maintenance on the clutch system was performed or the system is suspected of containing air, bleed the system. Use care; brake fluid will remove the paint from any surface. If the brake fluid spills onto any painted surface, wash it off immediately with soap and water.**

1. Before servicing the vehicle, refer to the precautions in the beginning of this section.

2. Fill the clutch reservoir with brake fluid. Check the reservoir level frequently and add fluid as needed.

3. Connect one end of a vinyl tube to the bleeder plug on the slave cylinder and submerge the other end into a clear container half-filled with brake fluid.

4. Slowly pump the clutch pedal several times.

5. Have an assistant hold the clutch pedal down and loosen the bleeder plug until fluid and/or air starts to run out of the bleeder plug. Close the bleeder plug while the pedal is held to the floor.

➡**Do not allow the pedal to rise back-up while the bleeder is still open. If this happens, it will allow air to re-enter the slave cylinder and cause the clutch system not to work properly.**

6. Repeat Steps 2 and 3 until all the air bubbles are removed from the system.

7. Tighten the bleeder plug when all the air is gone.

8. Refill the master cylinder to the proper level as required.

9. Check the system for leaks.

Transfer Case

REMOVAL & INSTALLATION

1. Before servicing the vehicle, refer to the precautions in the beginning of this section.

2. Drain the transfer case fluid.

3. Remove or disconnect the following:
• Negative battery cable. Due to the air bag system, wait at least 90 seconds before proceeding
• Engine and transaxle assembly
• Separate vane pump
• Steering gear
• Crossmember
• Manifold stay
• Oxygen (O$_2$) sensor
• Exhaust manifold heat shield

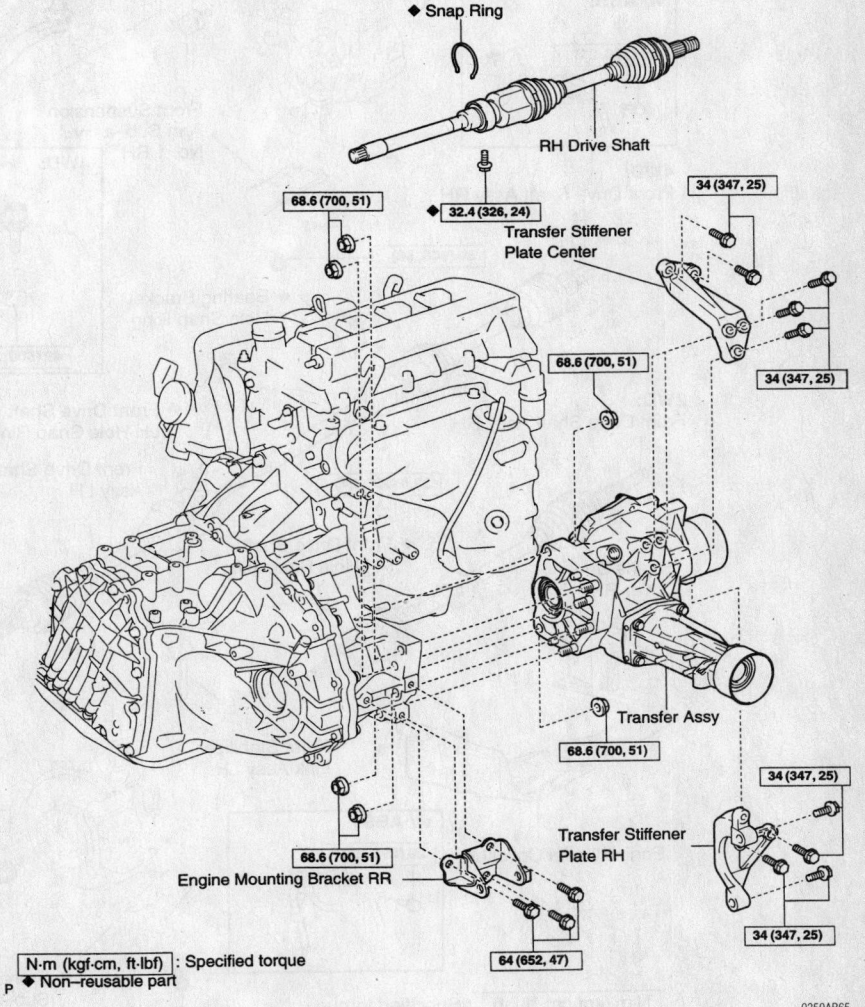

Exploded view of the transfer case mounting

- Exhaust manifold
- Starter
- Right side halfshaft
- Transverse engine mounting bracket
- Center and right side transfer stiffener plates

✳✳ WARNING

When removing the transfer case, DO NOT touch the oil seal.

- Transfer case bolts, and transfer assembly, using a mallet to dislodge it from the transaxle
4. Installation is the reverse of the removal procedure, noting the following specifications:

   a. Transfer case stiffener case bolts: 25 ft. lbs. (34 Nm).

   b. Engine mounting bracket bolts: 47 ft. lbs. (64 Nm).

5. Add fluid to the transfer case, and check for leaks.

Halfshaft

REMOVAL & INSTALLATION

➡ **The hub bearing could be damaged if subjected to the full weight of the vehicle, such as if the vehicle is moved** without the halfshafts. If it is absolutely necessary to place the full vehicle weight on the hub bearing, first support the bearing with SST No. 09608–16041.

1. Before servicing the vehicle, refer to the precautions in the beginning of this section.
2. Drain the transaxle fluid.
3. Remove or disconnect the following:
   - Negative battery cable. Due to the air bag system, wait at least 90 seconds before proceeding.
   - Both front wheels
   - Cotter pin, locknut cap, and the hub nut

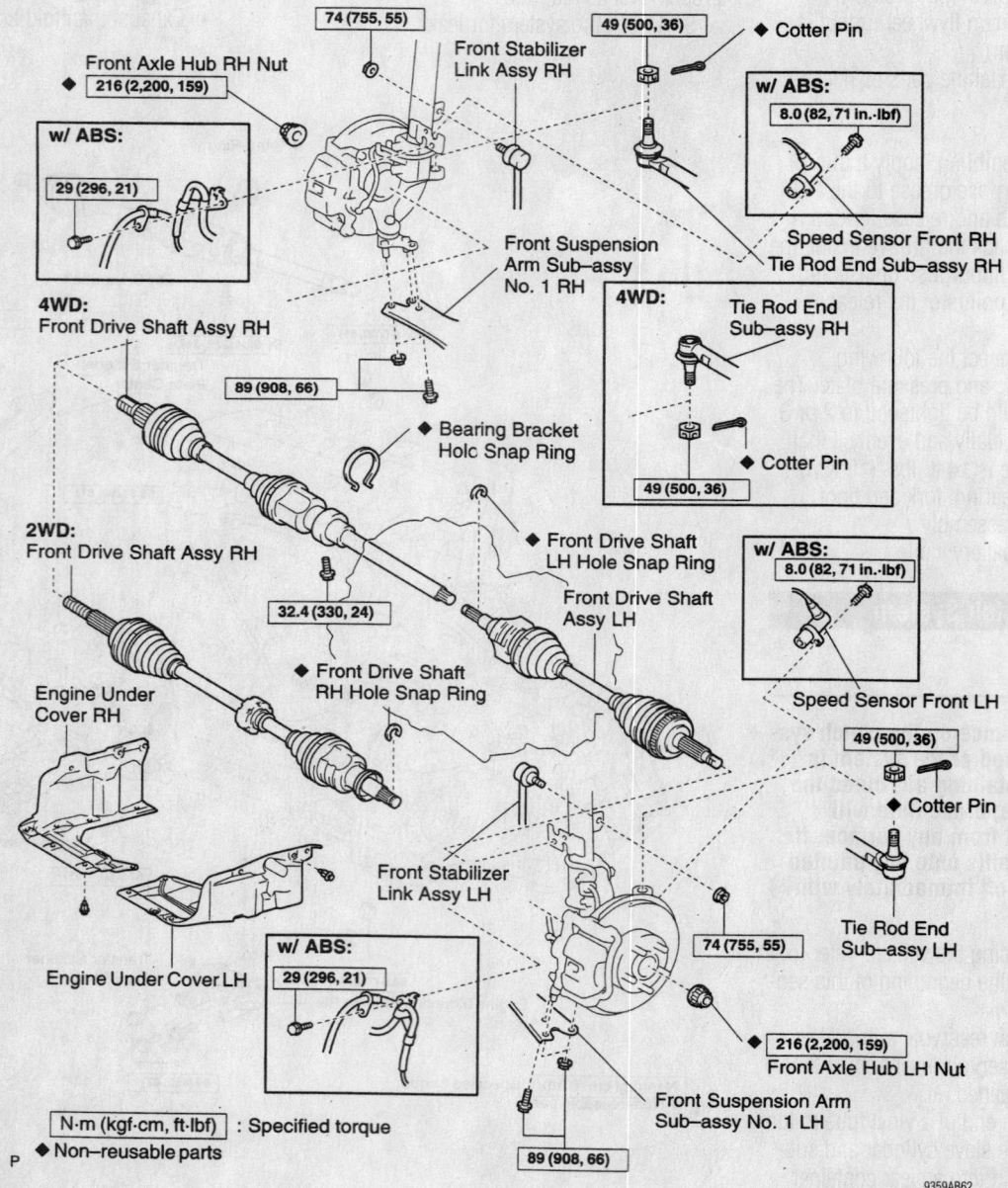

Front Axle Hub RH Nut — 216 (2,200, 159)

w/ ABS: 29 (296, 21)

4WD: Front Drive Shaft Assy RH

2WD: Front Drive Shaft Assy RH

Engine Under Cover RH

Engine Under Cover LH

74 (755, 55)

Front Stabilizer Link Assy RH

89 (908, 66)

Bearing Bracket Hold Snap Ring

32.4 (330, 24)

Front Drive Shaft RH Hole Snap Ring

Front Suspension Arm Sub–assy No. 1 RH

Front Drive Shaft LH Hole Snap Ring

Front Drive Shaft Assy LH

Front Stabilizer Link Assy LH

w/ ABS: 29 (296, 21)

49 (500, 36)

Cotter Pin

w/ ABS: 8.0 (82, 71 in.·lbf)

Speed Sensor Front RH
Tie Rod End Sub–assy RH

4WD: Tie Rod End Sub–assy RH

49 (500, 36)

Cotter Pin

w/ ABS: 8.0 (82, 71 in.·lbf)

Speed Sensor Front LH

49 (500, 36)

Cotter Pin

Tie Rod End Sub–assy LH

74 (755, 55)

216 (2,200, 159)
Front Axle Hub LH Nut

Front Suspension Arm Sub–assy No. 1 LH

89 (908, 66)

N·m (kgf·cm, ft·lbf) : Specified torque
◆ Non–reusable parts

9359AB62

Halfshafts and related components

- Undercovers
- Speed sensors
- Tie rod ball joint from the steering knuckle
- Stabilizer bar link from the lower suspension arm
- Lower ball joint from the lower suspension arm
- Halfshaft from the knuckle

➡**Be careful not to damage the inner oil seal or the ABS sensor rotor on the halfshaft.**

4. To remove the left side halfshaft, separate the halfshaft from the transaxle.

5. To remove the right side halfshaft perform the following steps:

- Remove the 2 bolts of the center bearing bracket
- Pull the halfshaft out together with the center bearing case and the center halfshaft.
- Remove the center shaft with the right-hand halfshaft from the transaxle through the bearing bracket.

➡**Do not damage the oil seal lip.**

To install:

6. Install or connect the following:

- Snapring opening side facing downward, on the oiled inboard joint tulip
- Left side halfshaft into the transaxle
- Right side halfshaft, with the bearing case and center shaft, into the transaxle
- Center bearing case (right side).

7. After installing either halfshaft, check that there is 0.08–0.12 in. (2–3mm) of axial play. Check that the halfshaft is making contact with the pinion shaft and that the halfshaft cannot be pulled out.

8. Install or connect the following:

- Halfshaft into the knuckle
- Lower suspension arm to the lower ball joint. Torque the bolt and nuts to 66 ft. lbs. (89 Nm).
- Tie rod end to the steering knuckle. Tighten the nut to 36 ft. lbs. (49 Nm).
- Stabilizer bar link to the lower suspension arm. Torque the nuts to 55 ft. lbs. (74 Nm).
- Front wheels
- Hub nut and washer and tighten to 159 ft. lbs. (216 Nm).
- Negative battery cable
- Locknut cap and a new cotter pin.
- Speed sensors
- Undercover

9. Fill the transaxle fluid to the proper level

10. Start the vehicle, check for leaks and repair if necessary.

CV-Joints

OVERHAUL

1. Before servicing the vehicle, refer to the precautions in the beginning of this section.

2. Remove or disconnect the following:

- Inboard joint boot clips
- Inboard joint tulip from the driveshaft
- Snapring
- Using a brass rod and hammer, the tri-pot joint off the driveshaft without hitting the joint roller

- Inboard joint boot
- Clamp and driveshaft damper
- Clamps and the outboard drive boot. DO NOT disassemble the outboard joint.

To assemble:

3. Install or connect the following:

➡**Before installing the boot, wrap the spline end of the shaft with masking tape to prevent damage to the boot.**

- Driveshaft damper with a new clamp
- Temporarily, the inboard boot with new clamp to the drive joint

➡**The inboard boot and clamp are larger than those of the outboard boot.**

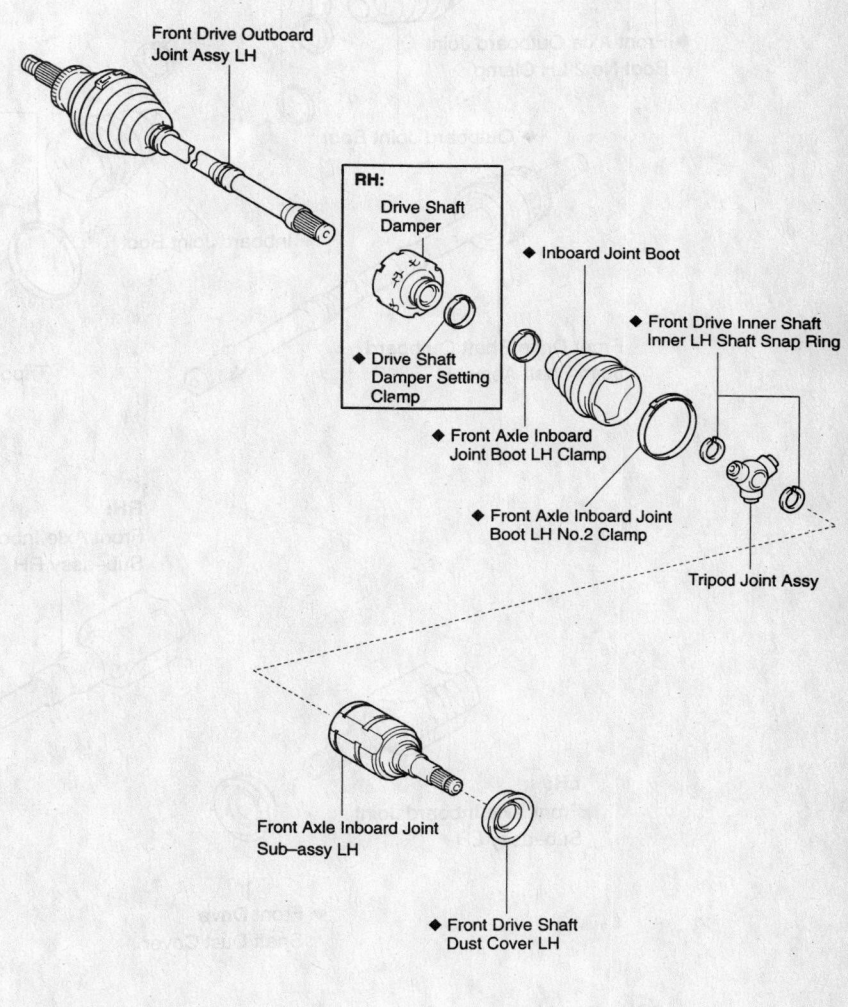

2WD:

Front Drive Outboard Joint Assy LH

RH:
Drive Shaft Damper

◆ Drive Shaft Damper Setting Clamp

◆ Inboard Joint Boot

◆ Front Axle Inboard Joint Boot LH Clamp

◆ Front Axle Inboard Joint Boot LH No.2 Clamp

◆ Front Drive Inner Shaft Inner LH Shaft Snap Ring

Tripod Joint Assy

Front Axle Inboard Joint Sub–assy LH

◆ Front Drive Shaft Dust Cover LH

◆ Non–reusable parts

P

9359AB63

Exploded view of the CV-joint—FWD vehicles

- The tri-pot onto the driveshaft with a brass rod and hammer without hitting the joint roller
- The snapring

4. Pack the outboard tulip joint and the outboard boot with about 0.26–0.33 lbs. ounces of grease that was supplied with boot kit.

5. Install or connect the following:

- Boot onto the outboard joint

6. Pack the inboard tulip joint and boot with ½ lb. of grease that was supplied with the boot kit.

- Inboard tulip joint onto the driveshaft

- Boot onto the driveshaft

7. Before checking the standard length, bend the band and lock it. Make sure that the boot is not stretched or squashed when the driveshaft is at standard length. Standard driveshaft length: LH: 540.2 mm (21.268 in.); RH: 857.4 mm (33.756 in.)

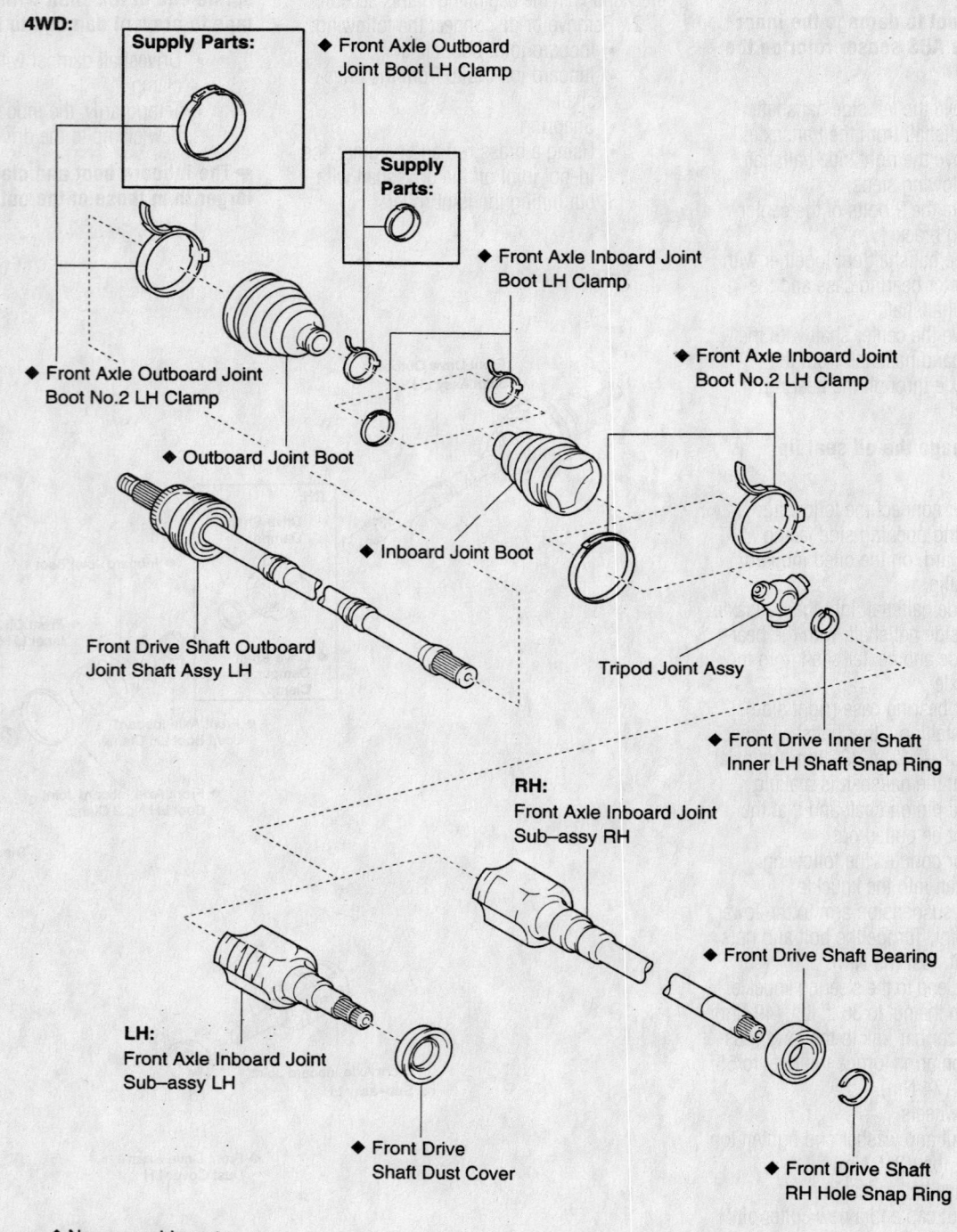

4WD:

Supply Parts:

◆ Front Axle Outboard Joint Boot LH Clamp

Supply Parts:

◆ Front Axle Inboard Joint Boot LH Clamp

◆ Front Axle Inboard Joint Boot No.2 LH Clamp

◆ Front Axle Outboard Joint Boot No.2 LH Clamp

◆ Outboard Joint Boot

◆ Inboard Joint Boot

Front Drive Shaft Outboard Joint Shaft Assy LH

Tripod Joint Assy

◆ Front Drive Inner Shaft Inner LH Shaft Snap Ring

RH: Front Axle Inboard Joint Sub–assy RH

◆ Front Drive Shaft Bearing

LH: Front Axle Inboard Joint Sub–assy LH

◆ Front Drive Shaft Dust Cover

◆ Front Drive Shaft RH Hole Snap Ring

◆ Non–reusable parts

P

9359AB64

Exploded view of the CV-joint—AWD vehicles

STEERING AND SUSPENSION

Air Bag

PRECAUTIONS

Several precautions must be observed when handling the inflator module to avoid accidental deployment and possible personal injury.

• Never carry the inflator module by the wires or connector on the underside of the module.

• When carrying a live inflator module, hold securely with both hands, and ensure that the bag and trim cover are pointed away.

• Place the inflator module on a bench or other surface with the bag and trim cover facing up.

• With the inflator module on the bench, never place anything on or close to the module that may be thrown in the event of an accidental deployment.

DISARMING

To avoid personal injury when working on vehicles equipped with an air bag, the negative battery cable must be disconnected and at least 90 seconds must elapse before working on the system. Failure to do so may result in deployment of the air bag.

REARMING

After vehicle service is completed, reattach the battery cables (positive cable first!) to rearm the air bag system.

Rack and Pinion Steering Gear

REMOVAL & INSTALLATION

1. Before servicing the vehicle, refer to the precautions in the beginning of this section.
2. Position the front wheels straight ahead.
3. Remove or disconnect the following:
   • Negative battery cable. Because these vehicles are equipped with air bags, wait at least 90 seconds before proceeding.
   • Horn button
   • Steering wheel
   • Front wheels
   • Left and right engine undercovers
   • Left and right tie rod ends
   • Column hose cover silencer sheet
   • Steering intermediate shaft
   • Pressure feed and return tubes

1ZZ–FE:

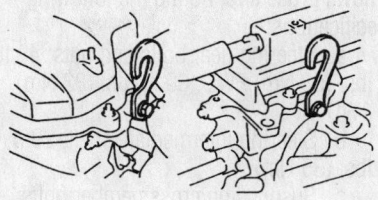

2ZZ–GE:

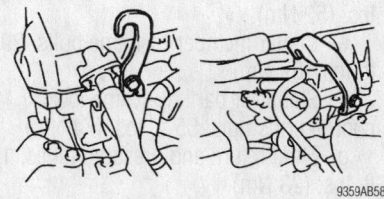

9359AB58

Proper installation of engine hangers

• Left and right side front stabilizer links
• Right and left front lower control arms from the ball joints
• Hood
• No. 2 cylinder head (valve) cover

4. Install an engine support and tension it to support the engine without raising it.

   a. No. 1 engine hanger: P/N 12281-22021 1ZZ-FE, 12281-88600 2ZZ-GE.

   b. No. 2 engine hanger: P/N 12281-15040 1ZZ-FE, 12281-88600 2ZZ-GE.

   c. Bolt: P/N 91512-B1016.

   d. Torque the bolts to 28 ft. lbs. (38 Nm).

✳✳ WARNING

Do not try to suspend the engine by hooking the chain to any other part.

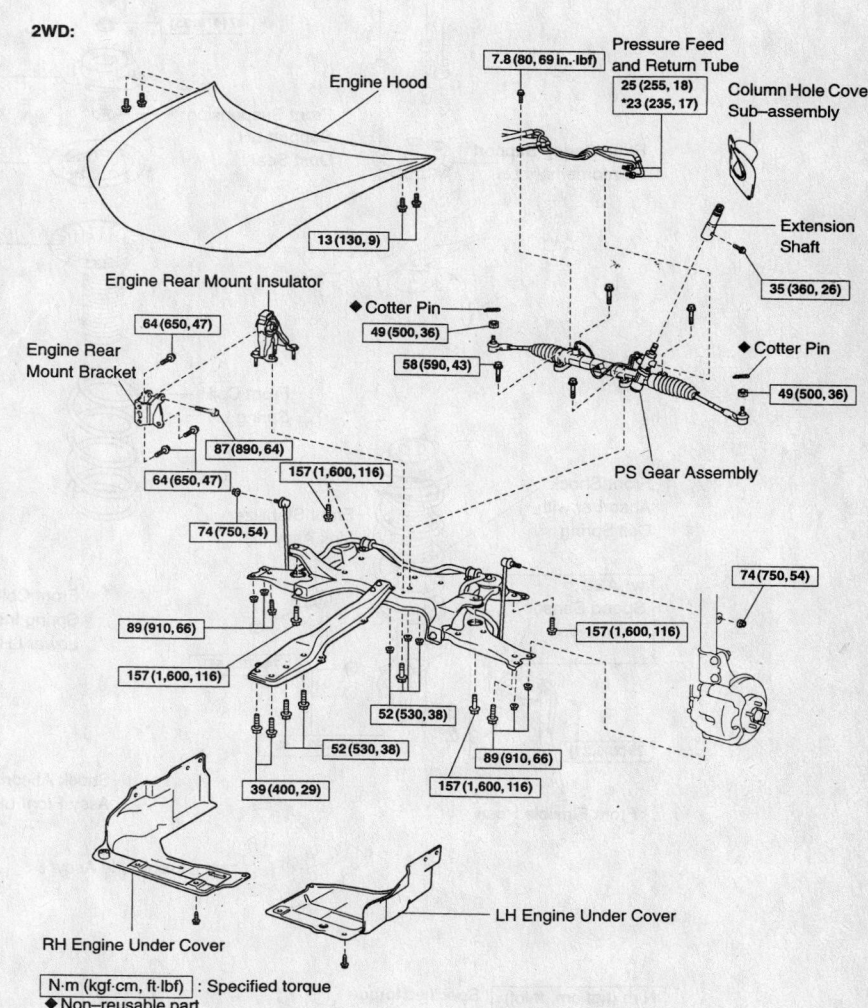

Exploded view of a typical power rack and pinion steering gear unit—FWD shown

9359AB59

e. Attach an engine chain hoist to the hangers.

✳✳ WARNING

The engine hoist is now in place and under tension. Use care when repositioning the vehicle and make necessary adjustments to the engine support.

5. Remove or disconnect the following:
- Bolt and nuts holding in the middle of the crossmember and support the crossmember with a jack
- Bolts from the outer side of the suspension crossmember
- Suspension crossmember with the steering gear assembly
- Steering intermediate shaft, after matchmarking it

- Rack and pinion steering gear from the crossmember

6. Installation is the reverse of the removal procedure, noting the following specifications:
a. Steering gear bolts and nuts: 43 ft. lbs. (58 Nm) FWD, 60 ft. lbs. (82 Nm) AWD.
b. Steering intermediate shaft: 26 ft. lbs. (35 Nm).
c. Suspension crossmember bolts: 116 ft. lbs. (157 Nm).
d. Engine mount insulator bolts: 38 ft. lbs. (52 Nm).
e. Center member-to-frame bolts: 29 ft. lbs. (39 Nm).
f. Stabilizer bar link-to-the lower control arms nuts: 55 ft. lbs. (74 Nm).
g. Fluid return and pressure tubes: 17 ft. lbs. (23 Nm).

h. Tie rod ends: 36 ft. lbs. (49 Nm).
i. Wheel lug nuts: 76 ft. lbs. (103 Nm).
7. Check and top off the power steering fluid.
8. Check and adjust the alignment, if needed.

Strut and Coil Spring

REMOVAL & INSTALLATION

Front

1. Before servicing the vehicle, refer to the precautions in the beginning of this section.
2. Remove or disconnect the following:
- Negative battery cable. Because of the air bag system, wait at least 90 seconds before proceeding

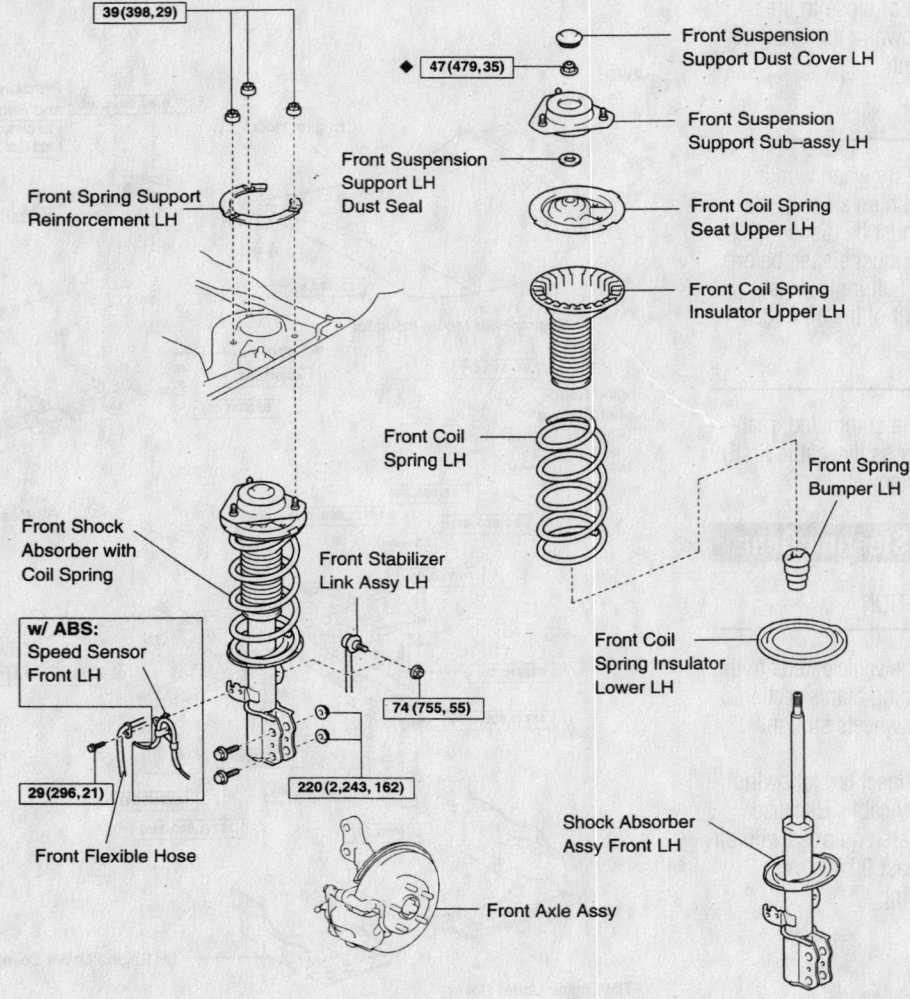

N·m (kgf·cm, ft·lbf): Specified torque
◆ Non-reusable part

Common coil spring and strut component assembly

9359AB60

Do not support the weight of the vehicle on the suspension arm; the arm will deform under its weight.

- Wheel
- Stabilizer link from the strut
- Bolt, and disconnect the brake hose from the strut
- With ABS brakes, speed sensor wiring harness from the strut
- Lower strut bolts and nuts
- Upper strut nuts
- Strut from the steering knuckle
- Strut

3. To disassemble the strut:
- Install a bolt and 2 nuts to the bracket at the lower portion of the strut shell and secure it in a vise
- Compress the coil spring
- Dust cover and hold the spring seat so that it will not turn
- Nut on the top of the strut
- Suspension support, bearing, dust seal, spring seat, spring, insulators and bumper

To install:

4. To assemble the strut:
- Install the spring bumper to piston

5. Using a spring compressor, compress the spring.
- Coil spring to the strut. Fit the lower end of the coil spring into the gap of the lower seat.
- Spring seat with the insulator
- Dust seal on the spring seat
- Suspension support and tighten 35 ft. lbs. (47 Nm). After the nut has been tighten, release the compressor tool tension.

6. Pack multipurpose grease into the suspension support.
- Dust cover.

➡**Do not use an impact wrench to tighten the nut. Also, check that the**

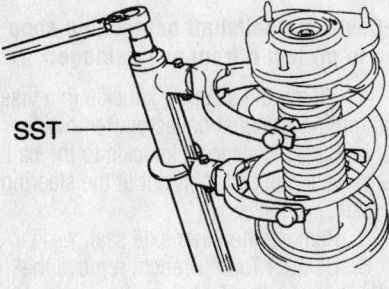

SST

9359AB61

Proper method of supporting the strut in a vise

bearing fits into the recess in the suspension support.

- Strut
- Nuts holding the strut to the strut tower. Tighten the nuts to 29 ft. lbs. (39 Nm).
- 2 lower strut bolts and nuts. Tighten to 162 ft. lbs. (220 Nm).
- Brake line to the steering knuckle. Tighten the line bolt to 21 ft. lbs. (29 Nm).
- Secure the wiring harness, if equipped with ABS
- Stabilizer link. Tighten the nut to 55 ft. lbs. (74 Nm).
- Wheel. Tighten the lug nuts to 76 ft. lbs. (103 Nm).
- Negative battery cable

7. Check and adjust the alignment, if needed.

Rear

1. Before servicing the vehicle, refer to the precautions in the beginning of this section.
2. Remove or disconnect the following:
- Negative battery cable. Because of the air bag system, wait at least 90 seconds before proceeding.
- Rear wheel
- Rear deck board, luggage compartment tray and any trim necessary to access the strut towers
- Shock absorber head cover

3. On AWD vehicles, separate the rear stabilizer link.
4. For FWD vehicles:
a. Support the axle beam with a jack.
b. Remove the strut tower nuts and bolt.
c. Remove the lower strut nut, cushion retainer and strut .
5. For AWD vehicles:
a. Support the rear control arm.
b. Remove the bolt and nut from the rear control arm.
c. Remove the strut tower nuts.
d. Remove the 3 rear control arm bolts.
e. Press the rear control arm down to the outside of the vehicle, then remove the strut.
6. To disassemble the strut:
a. Place the strut assembly in a pipe vise or strut vise.

Do not attempt to clamp the strut assembly in a flat jaw vise as this will result in damage to the strut tube.

b. Compress the spring until the upper suspension support is free of any spring tension. Do not over-compress the spring.
c. Hold the upper support, then remove the nut on the end of the shock piston rod.
d. Remove the support, coil spring, insulator, and bumper.
7. Inspect the strut as follows:
a. Check the shock absorber by moving the piston shaft through its full range of travel. It should move smoothly and evenly throughout its entire travel without any trace of binding or notching.
b. Use a small straightedge to check the piston shaft for any bending or deformation.
c. Inspect the spring for any sign of deterioration or cracking. The waterproof coating on the coils should be intact to prevent rusting.

To install:

➡**Never reuse a self-locking nut. Always replace self-locking nuts and cotter pins as applicable.**

8. Assemble the strut as follows:
a. Loosely assemble all components onto the strut assembly. Be sure the spring end aligns with the hollow in the lower seat.
b. Align the upper suspension support with the piston rod and install the support.
c. Align the suspension support with the strut lower bracket. This assures the spring will be properly seated top and bottom.
d. Compress the spring to expose the strut piston rod threads.
e. Install a new strut piston nut and tighten to 41 ft. lbs. (56 Nm).
f. Remove the spring compressor. Be sure the paint mark on the upper support faces the outside of the strut.
9. Install or connect the following:
- Strut on the vehicle. Tighten the strut-to-strut tower nuts to 59 ft. lbs. (80 Nm).
- Strut to the axle carrier and install the nut and cushion retainer/bolt snug. Do not fully tighten at this time.
- Strut head cover
- Rear control arm (AWD). Tighten the bolts to 48 ft. lbs. (65 Nm).
- Rear stabilizer link (AWD)
- Trunk tray, deckboard and any other trim pieces removed
- Wheel

10. With the vehicle's weight on the sus-

pension, tighten the bolt holding the strut to the axle carrier to 59 ft. lbs. (80 Nm) for FWD vehicles, or 103 ft. lbs. (140 Nm) for AWD vehicles.

- Negative battery cable

11. Check and adjust the rear wheel alignment.

Lower Ball Joint

REMOVAL & INSTALLATION

1. Before servicing the vehicle, refer to the precautions in the beginning of this section.
2. Remove or disconnect the following:
   - Negative battery cable. Wait at least 90 seconds before proceeding.
   - Front wheel
3. Depress the brake pedal and loosen the hub nut
   - ABS speed sensor, if equipped
   - Cotter pin and nut from the tie rod end. Using a tie rod end removal tool, separate the tie rod end from the steering knuckle.
   - Lower control arm ball joint, using a suitable puller
   - Separate the front halfshaft
   - Lower ball joint cotter pin and castle nut
   - Lower ball joint from the steering knuckle using a puller

To install:
4. Install or connect the following:
   - Lower ball joint to the lower arm. Tighten the castle nut to 76 ft. lbs. (103 Nm).
   - New cotter pin
   - Front halfshaft
   - Lower control arm
   - Tie rod end to the knuckle
   - ABS speed sensor
   - Hub nut
   - Wheel
   - Negative battery cable

SST

Removing the ball joint from the knuckle

9359AB71

5. Check and adjust the alignment, if needed.

Upper Control Arm

REMOVAL & INSTALLATION

Rear—AWD Only

1. Before servicing the vehicle, refer to the precautions in the beginning of this section.
2. Remove or disconnect the following:
   - Negative battery cable. Wait at least 90 seconds before proceeding
   - Rear wheel
   - Exhaust pipe
   - Propeller shaft with center bearing shaft
   - Rear stabilizer links
   - Rear hub nuts
   - Rear brake drum
   - Speed sensor
   - Front brake shoe
   - Parking brake shoe strut set
   - Rear brake shoe
   - Parking brake cables
   - Rear brake hoses
   - Separate the rear suspension arms
   - Separate the upper control arm
   - Rear drive axle assembly
   - Rear strut nut and bolt
   - Rear strut
   - Rear suspension arm
   - Rear suspension member
   - Upper control arm assembly. Matchmark the camber adjust cams and rear suspension member prior to removal.
3. Installation is the reverse of the removal procedure.

Lower Control Arm

REMOVAL & INSTALLATION

1. Before servicing the vehicle, refer to the precautions in the beginning of this section.
2. Remove or disconnect the following:
   - Negative battery cable. Wait at least 90 seconds before proceeding..
   - Front wheel
   - Stabilizer link
   - Bolt and nuts and separate the lower control arm from the lower ball joint
   - Bolts and nuts, then separate the steering gear. Loosen the bolt, since the nut cannot be rotated, then suspend the steering gear.

3. Support the engine, using the engine lifting hooks and the procedure under Engine Removal & Installation.
   - Crossmember
   - Lower control arm from the crossmember
4. Installation is the reverse of the removal procedure.

Wheel Bearings

REMOVAL & INSTALLATION

Front

1. Before servicing the vehicle, refer to the precautions in the beginning of this section.
2. Remove or disconnect the following:
   - Negative battery cable. On vehicles equipped with an air bag, wait at least 90 seconds before proceeding.
   - Wheels
   - Hub nut
   - Front stabilizer link
   - Anti-lock Brake System (ABS) speed sensor
   - Brake caliper
   - Rotor
   - Tie rod end from the steering knuckle
   - Lower control arm ball joint
   - Front halfshaft from the hub, using a mallet to tap it out. Be careful not to damage the boot or speed sensor.
3. Loosen the nuts on the lower side of the strut assembly. Do not remove at this time.
   - Lower ball joint using a puller
   - Tie rod end from the steering knuckle
   - Steering knuckle from the lower control arm
   - Knuckle from the strut assembly
   - Hub

➡ **Cover the halfshaft boot with a shop rag to protect it from any damage.**

4. Clamp the steering knuckle in a vise and remove the dust deflector. Remove the nut holding the steering knuckle to the ball joint. Press the ball joint out of the steering knuckle.
5. Remove the inner axle seal.
6. Using a Torx® wrench, remove the bolts securing the dust cover.
7. Using hub puller, remove the hub and backing plate from the steering knuckle.

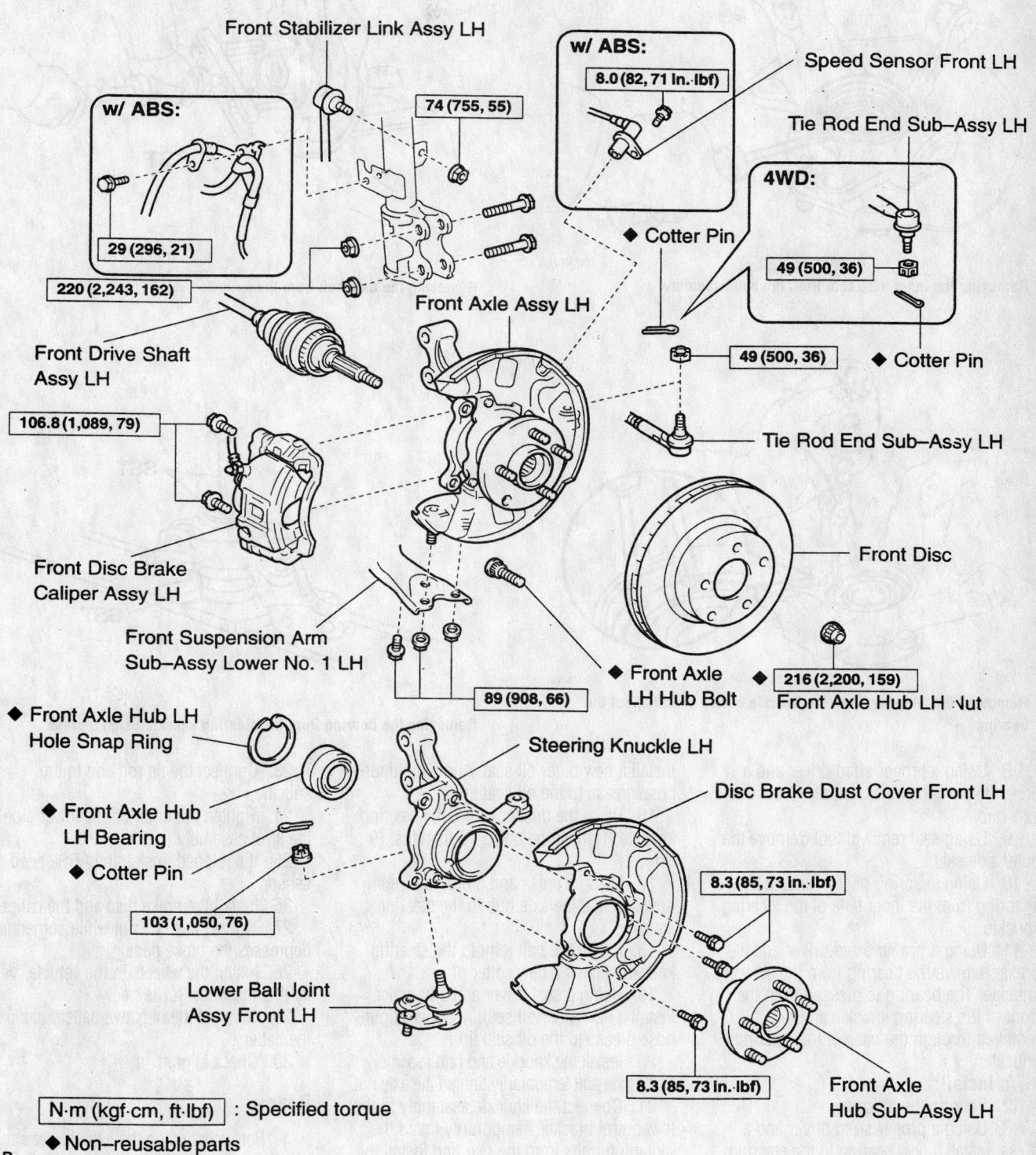

Front Stabilizer Link Assy LH

74 (755, 55)

w/ ABS:

8.0 (82, 71 In.·lbf)

Speed Sensor Front LH

Tie Rod End Sub–Assy LH

w/ ABS:

29 (296, 21)

220 (2,243, 162)

4WD:

49 (500, 36)

◆ Cotter Pin

Front Axle Assy LH

49 (500, 36)

◆ Cotter Pin

Front Drive Shaft Assy LH

Tie Rod End Sub–Assy LH

106.8 (1,089, 79)

Front Disc Brake Caliper Assy LH

Front Disc

Front Suspension Arm Sub–Assy Lower No. 1 LH

89 (908, 66)

◆ Front Axle LH Hub Bolt

216 (2,200, 159)

Front Axle Hub LH Nut

◆ Front Axle Hub LH Hole Snap Ring

Steering Knuckle LH

Disc Brake Dust Cover Front LH

◆ Front Axle Hub LH Bearing

◆ Cotter Pin

8.3 (85, 73 in.·lbf)

103 (1,050, 76)

Lower Ball Joint Assy Front LH

8.3 (85, 73 in.·lbf)

Front Axle Hub Sub–Assy LH

N·m (kgf·cm, ft·lbf) : Specified torque

◆ Non–reusable parts

P

9359AB72

Exploded view of the front hub and bearing, and related components

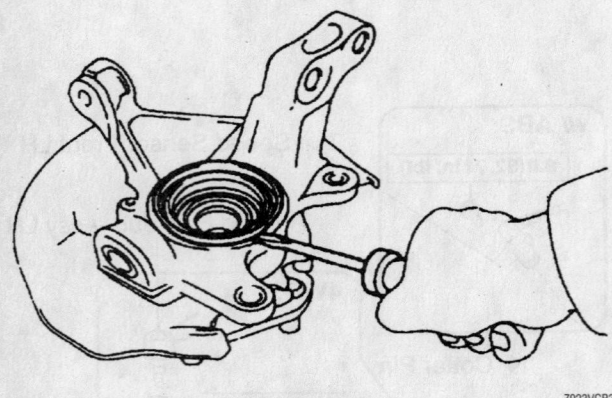

Removing the inner axle seal from the hub assembly

7923VGB3

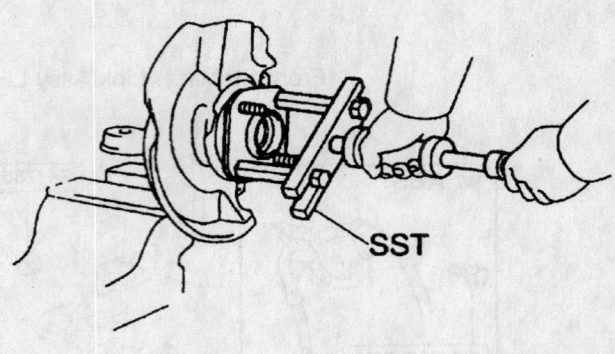

Removing the axle hub from the knuckle

7923VGB4

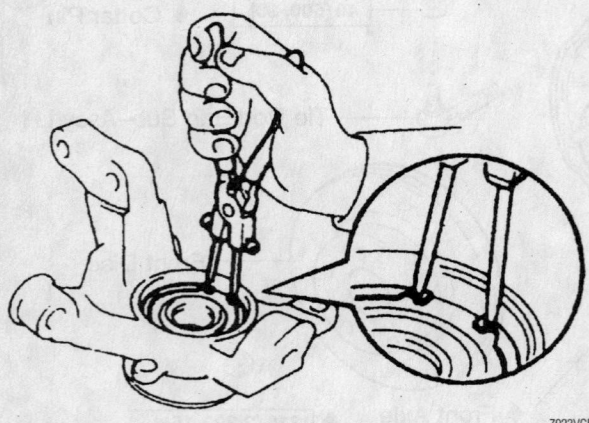

Removing the snapring from the knuckle before pressing out the bearing

7923VGB5

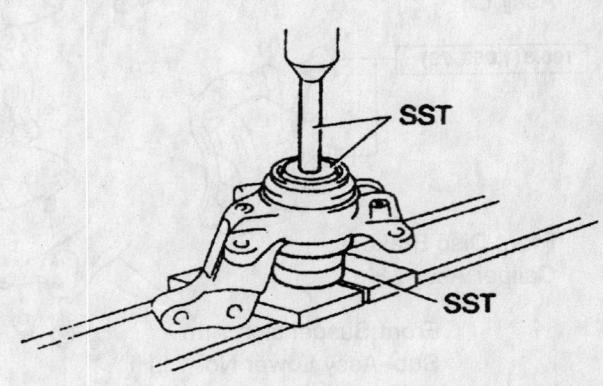

Removing the bearing from the steering knuckle using a press

7923VGB6

8. Using a proper sized driver and a press, remove the inner hub race from the axle hub.

9. Using seal removal tool, remove the outer axle seal.

10. Using snapring pliers, remove the snapring from the inner side of the steering knuckle.

11. Using a proper sized driver and a press, remove the bearing from the steering knuckle. The bearing is pressed from the front of the steering knuckle and is removed through the back of the steering knuckle.

To install:

12. Perform the following:

13. Using a proper sized driver and a press, install a new bearing to the steering knuckle.

14. Install the snapring to the steering knuckle using snapring pliers.

15. Using a seal driver and a hammer,

install a new outer oil seal. Apply multipurpose grease to the oil seal lip.

16. Place the dust cover on the steering knuckle. Tighten the bolts: 78 inch lbs. (9 Nm).

17. Using a press and a proper sized driver, install the axle hub to the steering knuckle.

18. Attach the ball joint to the steering knuckle. Install a new cotter pin.

19. Using a seal driver and a hammer, install a new inner oil seal. Apply multipurpose grease to the oil seal lip.

20. Install the knuckle and hub assembly to the axle and temporarily tighten the axle nut.

21. Connect the knuckle assembly to the lower strut bracket. Temporarily insert the mounting bolts from the rear and install the nuts making sure the matchmarks made earlier are in alignment.

22. Connect the lower ball joint to lower arm.

23. Connect the tie rod end to the knuckle.

24. Tighten the bolts on the lower side of the strut assembly.

25. If equipped, install the ABS speed sensor.

26. Install the brake disc and the caliper.

27. Tighten the axle nut while someone depresses the brake pedal.

28. Install the wheels to the vehicle. Verify that the wheel turns freely.

29. Connect the negative battery cable to the battery.

30. Check alignment.

Rear

1. Before servicing the vehicle, refer to the precautions in the beginning of this section.

2. Remove or disconnect the following:

- Negative battery cable. On vehicles

Disc Rear Brake Type:

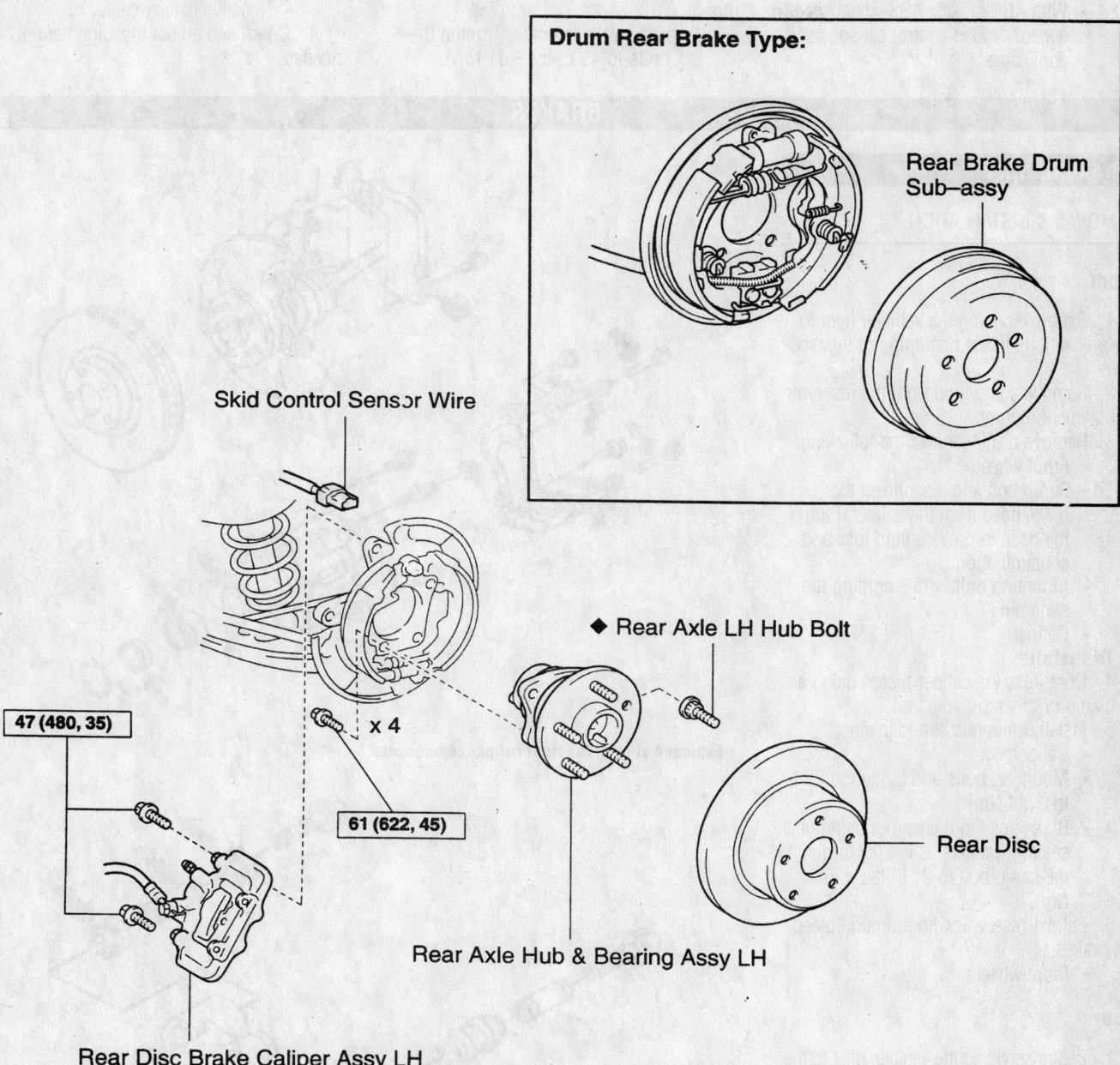

Drum Rear Brake Type:

Rear Brake Drum Sub-assy

Skid Control Sensor Wire

◆ Rear Axle LH Hub Bolt

47 (480, 35)

x 4

61 (622, 45)

Rear Disc

Rear Axle Hub & Bearing Assy LH

Rear Disc Brake Caliper Assy LH

N·m (kgf·cm, ft·lbf) : Specified torque

◆ Non–reusable part

9359AB73

Exploded view of the hub and wheel bearing assembly

equipped with an air bag, wait at least 90 seconds before proceeding.
- Wheel
- Brake drum or rotor
- With ABS brakes, ABS wheel speed sensor or skid control sensor, as applicable

- 4 hub retaining bolts
- Hub

To install:

3. Install or connect the following:

- Hub to the knuckle. Tighten the bolts to 45 ft. lbs. (61 Nm).

- ABS wheel speed or skid control sensor, if equipped
- Brake drum or rotor
- Wheel
- Negative battery cable

4. Check and adjust the alignment, if needed.

BRAKES

Brake Caliper

REMOVAL & INSTALLATION

Front

1. Before servicing the vehicle, refer to the precautions in the beginning of this section.

2. Remove some fluid from the reservoir with a suction pump.

3. Remove or disconnect the following:

- Front wheels
- Banjo bolt and disconnect the brake hose from the caliper. Plug the hose to prevent fluid loss and contamination.
- Mounting bolts while holding the slide pin
- Caliper

To Install:

4. Compress the caliper piston using a C–clamp or other suitable tool.

5. Install or connect the following:

- Caliper
- Mounting bolts and tighten to 25 ft. lbs. (34 Nm)
- Brake hose to the caliper using new sealing washers. Carefully torque the banjo bolt to 21 ft. lbs. (29 Nm).

6. Fill the reservoir with fluid and bleed the brakes.

- Front wheels

Rear

1. Before servicing the vehicle, refer to the precautions in the beginning of this section.

2. Remove some fluid from the reservoir with a suction pump.

3. Remove or disconnect the following:

- Rear wheels
- Clip and both anti-rattle springs
- Two pad guide pins
- Pads with the shims
- Banjo bolt and disconnect the brake hose from the caliper. Plug the hose to prevent fluid loss and contamination.
- 2 caliper mounting bolts and the caliper from its mounting bracket

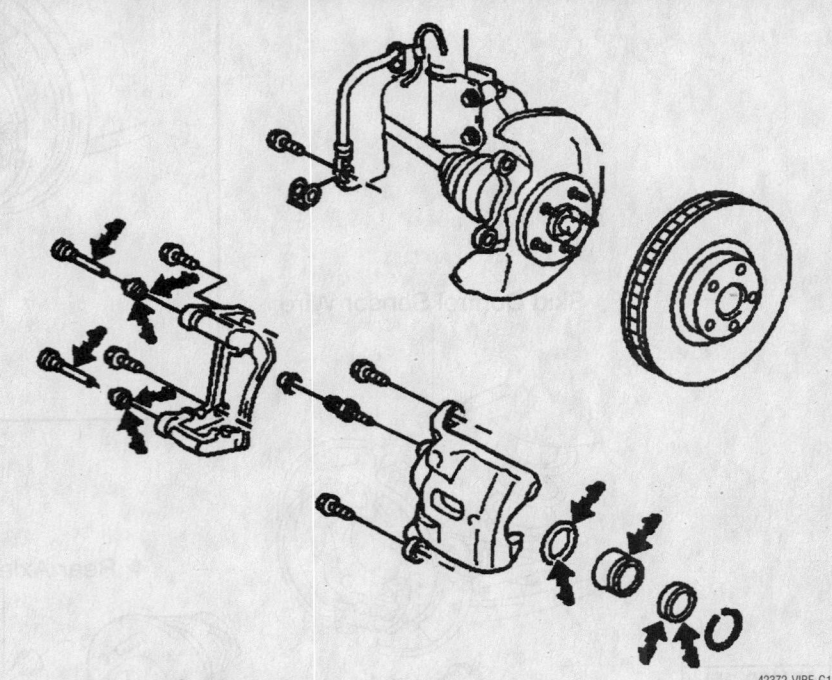

Exploded view of the front caliper components

42372-VIBE-G11

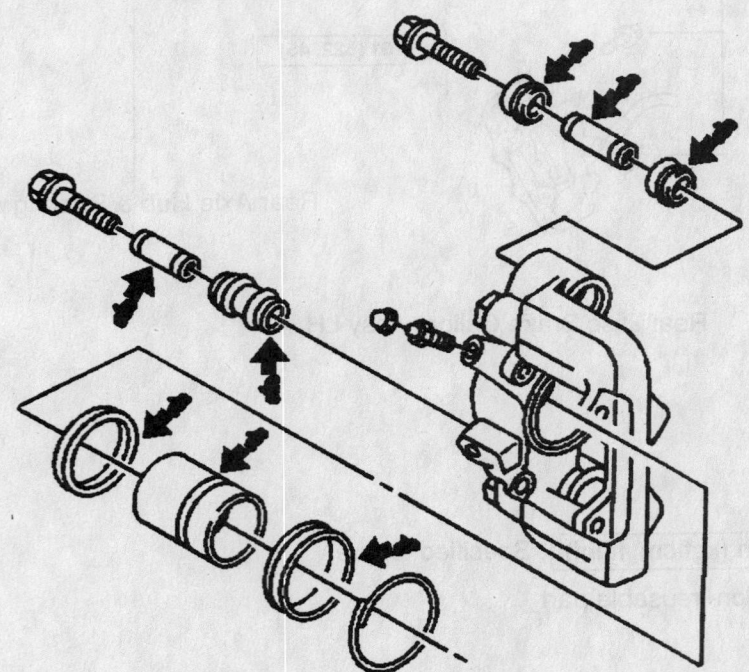

Exploded view of the rear caliper components

42372-VIBE-G12

To Install:

4. Compress the caliper piston using a C–clamp or other suitable tool.

5. Install or connect the following:

- Caliper. Tighten the caliper bolts to 34 ft. lbs. (46 Nm).
- Brake hose with new sealing washers. Tighten the banjo bolt to 21 ft. lbs. (29 Nm).
- New anti-squeal shims, apply disc brake grease to the inside of the shim before installation
- Inner pad with the wear indicator facing upwards
- Outer pad
- Two pad guide pins
- Anti-rattle springs and the clip

6. Fill the reservoir with fluid and bleed the brake system. Adjust the parking brake if necessary.

- Rear wheels

Disc Brake Pads

REMOVAL & INSTALLATION

Front

1. Before servicing the vehicle, refer to the precautions in the beginning of this section.

2. Remove some fluid from the reservoir with a suction pump.

3. Remove or disconnect the following:

- Front wheels
- Mounting bolts while holding the slide pin
- Caliper
- Pads and shims
- Both anti-squeal shims from the pads
- Wear indicator plates from the pads

To Install:

4. Compress the caliper piston using a C–clamp or other suitable tool.

5. Install or connect the following:

42372-VIBE-G13

Exploded view of the front pads and related components

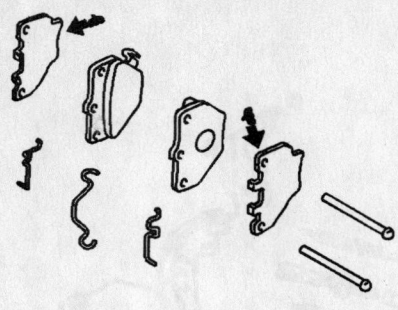

42372-VIBE-G14

Exploded view of the front pads and related components

- Wear indicator plates
- Both anti-squeal shims
- Pads and shims
- Caliper
- Mounting bolts and tighten to 25 ft. lbs. (34 Nm)

6. Fill the reservoir with fluid and bleed the brakes, if necessary.

- Front wheels

Rear

1. Before servicing the vehicle, refer to the precautions in the beginning of this section.

2. Remove some fluid from the reservoir with a suction pump.

3. Remove or disconnect the following:

- Rear wheels
- Clip and both anti-rattle springs
- Two pad guide pins
- Pads with the shims

To Install:

4. Compress the caliper piston using a C–clamp or other suitable tool.

5. Install or connect the following:

- New anti-squeal shims, apply disc brake grease to the inside of the shim before installation
- Inner pad with the wear indicator facing upwards
- Outer pad
- Two pad guide pins
- Anti-rattle springs and the clip

6. Fill the reservoir with fluid and bleed the brake system. Adjust the parking brake if necessary.

- Rear wheels

Brake Drums

REMOVAL & INSTALLATION

1. Before servicing the vehicle, refer to the precautions in the beginning of this section.

2. Remove or disconnect the following:

- Wheel
- Brake drum. If the drum will not pull of the axle, back off the automatic adjuster by turning the adjusting wheel.

To install:

3. Install or connect the following:

- Drum on the axle
- Wheel

4. Refill the master cylinder and pump pedal to attain full brake pedal before road-testing the vehicle.

Brake Shoes

REMOVAL & INSTALLATION

1. Before servicing the vehicle, refer to the precautions in the beginning of this section.

2. Remove or disconnect the following:

- Wheel
- Brake drum. If the drum will not pull of the axle, back off the automatic adjuster by turning the adjusting wheel.
- Upper side tension spring with the spacer
- Anchor side spring using needle nosed pliers
- Hold-down springs and pins from the front shoe
- Upper side return spring from the front shoe
- Front shoe
- Left hand parking brake shoe strut set from the front shoe
- Automatic adjuster lever from the front shoe
- Hold-down springs and pins from the rear shoe
- Upper side return spring from the rear shoe
- Parking brake cable from the rear shoe using needle nosed pliers
- Rear brake shoe
- C–washer using a suitable pry tool from the shoe
- Parking brake lever from the shoe

To install:

3. Lubricate the contact points on the backing plate and the adjuster with lithium grease.

4. Install or connect the following:

- Parking brake lever, and attach using a new C–washer
- Parking brake lever to the shoe lever

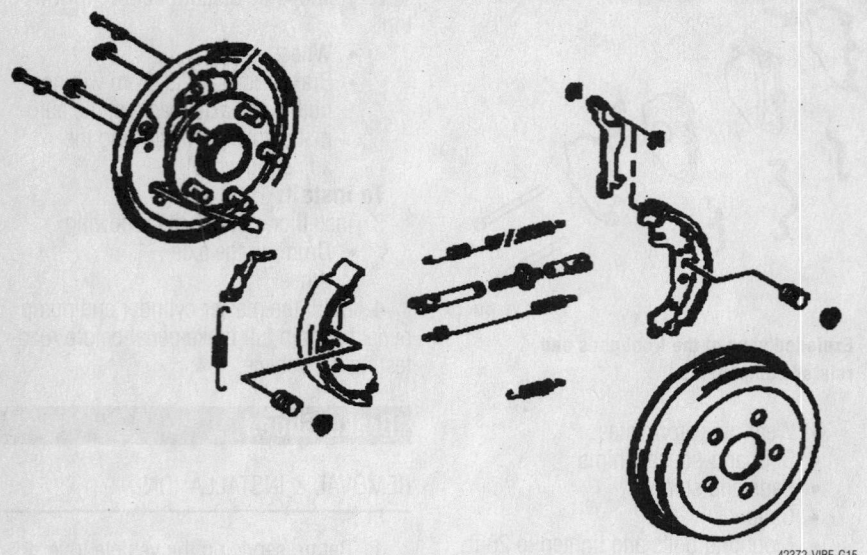

42372-VIBE-G15

Exploded view of the drum brake assembly—2WD shown—AWD similar

- Upper return spring to the rear shoe
- Shoe assembly onto the backing plate
- Hold-down springs and pins to the rear shoe
- Automatic adjuster lever

5. Apply lithium grease to the adjuster bolt.

- Left hand parking brake shoe strut set

- Upper side return spring to the front shoe
- Hold-down springs and pins to the front shoe
- Anchor side spring using needle nosed pliers
- Upper side tension spring with the spacer

6. Adjust the rear brakes as follows:
   a. Temporarily install the drum and hub nuts.

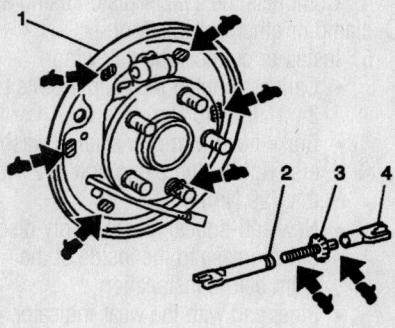

42372-VIBE-G16

Lubricate the contact points on the backing plate and the adjuster with lithium

b. Remove the hole plug from the backing plate.

c. Turn the adjuster to expand the shoe until the drum locks.

d. Back off the adjuster eight notches using a suitable adjustment tool.

e. Install the hole plug into the backing plate to prevent dirt and moisture from entering.

f. Readjust the parking brake cable as necessary.

7. Install the wheels.

8. Refill the master cylinder and pump pedal to attain full brake pedal before Road-testing the vehicle.

SPECIFICATION CHARTS

ENGINE AND VEHICLE IDENTIFICATION

Engine							Model Year	
Code ①	Liters (cc)	Cu. In.	Cyl.	Fuel Sys.	Engine Type	Eng. Mfg.	Code ②	Year
7	1.9 (1901)	116	4	MFI	DOHC	Saturn	Y	2000
8	1.9 (1901)	116	4	MFI	SOHC	Saturn	1	2001
F	2.2 (2199)	134	4	SFI	DOHC	Saturn	2	2002
R	3.0 (3000)	183	6	SFI	DOHC	Saturn	3	2003
							4	2004

MFI: Multi-point Fuel Injection

SFI: Sequential Fuel Injection

DOHC: Double Overhead Camshafts

SOHC: Single Overhead Camshaft

① 8th digit of VIN

② 10th digit of VIN

42372-SCAR-C01

GENERAL ENGINE SPECIFICATIONS

Year	Model	Engine Displacement Liters (cc)	Engine ID/VIN	Fuel System Type	Net Horsepower @ rpm	Net Torque @ rpm (ft. lbs.)	Bore x Stroke (in.)	Compression Ratio	Oil Pressure @ rpm
2000	Sedan	1.9 (1901)	7	MFI	124@5600	122@4800	3.23x3.54	9.5:1	29@2000
	Sedan	1.9 (1901)	8	MFI	100@5000	114@2400	3.23x3.54	9.3:1	36@2000
	Wagon	1.9 (1901)	7	MFI	124@5600	122@4800	3.23x3.54	9.5:1	29@2000
	Wagon	1.9 (1901)	8	MFI	100@5000	114@2400	3.23x3.54	9.3:1	36@2000
	Sedan	2.2 (1901)	F	SFI	137@5800	147@4400	3.38x3.5	9.5:1	50-80@1000
	Wagon	2.2 (1901)	F	SFI	137@5800	147@4400	3.38x3.50	9.5:1	50-80@1000
	Sedan	3.0 (3000)	R	SFI	182@6000	184@3600	3.38x3.50	10.0:1	22@1000
	Wagon	3.0 (3000)	R	SFI	182@6000	184@3600	3.38x3.50	10.0:1	22@1000
2001	Sedan	1.9 (1901)	7	MFI	124@5600	122@4800	3.23x3.54	9.5:1	29@2000
	Sedan	1.9 (1901)	8	MFI	100@5000	114@2400	3.23x3.54	9.3:1	36@2000
	Wagon	1.9 (1901)	7	MFI	124@5600	122@4800	3.23x3.54	9.5:1	29@2000
	Wagon	1.9 (1901)	8	MFI	100@5000	114@2400	3.23x3.54	9.3:1	36@2000
	Sedan	2.2 (1901)	F	SFI	137@5800	147@4400	3.38x3.5	9.5:1	50-80@1000
	Wagon	2.2 (1901)	F	SFI	137@5800	147@4400	3.38x3.50	9.5:1	50-80@1000
	Sedan	3.0 (3000)	R	SFI	182@6000	184@3600	3.38x3.50	10.0:1	22@1000
	Wagon	3.0 (3000)	R	SFI	182@6000	184@3600	3.38x3.50	10.0:1	22@1000
2002	Sedan	1.9 (1901)	7	MFI	124@5600	122@4800	3.23x3.54	9.5:1	29@2000
	Sedan	1.9 (1901)	8	MFI	100@5000	114@2400	3.23x3.54	9.3:1	36@2000
	Wagon	1.9 (1901)	7	MFI	124@5600	122@4800	3.23x3.54	9.5:1	29@2000
	Wagon	1.9 (1901)	8	MFI	100@5000	114@2400	3.23x3.54	9.3:1	36@2000
	Sedan	2.2 (1901)	F	SFI	137@5800	147@4400	3.38x3.5	9.5:1	50-80@1000
	Wagon	2.2 (1901)	F	SFI	137@5800	147@4400	3.38x3.50	9.5:1	50-80@1000
	Sedan	3.0 (3000)	R	SFI	182@6000	184@3600	3.38x3.50	10.0:1	22@1000
	Wagon	3.0 (3000)	R	SFI	182@6000	184@3600	3.38x3.50	10.0:1	22@1000
2003-04	Sedan	1.9 (1901)	7	MFI	124@5600	122@4800	3.23x3.54	9.5:1	29@2000
	Sedan	1.9 (1901)	8	MFI	100@5000	114@2400	3.23x3.54	9.3:1	36@2000
	Wagon	1.9 (1901)	7	MFI	124@5600	122@4800	3.23x3.54	9.5:1	29@2000
	Wagon	1.9 (1901)	8	MFI	100@5000	114@2400	3.23x3.54	9.3:1	36@2000
	Sedan	2.2 (1901)	F	SFI	137@5800	147@4400	3.38x3.5	9.5:1	50-80@1000
	Wagon	2.2 (1901)	F	SFI	137@5800	147@4400	3.38x3.50	9.5:1	50-80@1000
	Sedan	3.0 (3000)	R	SFI	182@6000	184@3600	3.38x3.50	10.0:1	22@1000
	Wagon	3.0 (3000)	R	SFI	182@6000	184@3600	3.38x3.50	10.0:1	22@1000
	ION	2.2 (1901)	F	SFI	137@5800	147@4400	3.38x3.50	9.5:1	50-80@1000

MFI: Multi-port Fuel Injection

SFI: Sequential Fuel Injection

42372-SCAR-C02

ENGINE TUNE-UP SPECIFICATIONS

Year	Engine Displacement Liters (cc)	Engine ID/VIN	Spark Plug Gap (in.)	Ignition Timing (deg.) MT	Ignition Timing (deg.) AT	Fuel Pump (psi) ①	Idle Speed (rpm) MT ②	Idle Speed (rpm) AT ②	Valve Clearance In.	Valve Clearance Ex.
2000	1.9 (1901)	7	0.040	③	③	40-55	850	750	HYD	HYD
	1.9 (1901)	8	0.040	③	③	40-55	750	650	HYD	HYD
	2.2 (2199)	F	0.045	③	③	55-65	④	④	HYD	HYD
	3.0 (3000)	R	0.043	③	③	39-49	④	④	HYD	HYD
2001	1.9 (1901)	7	0.040	③	③	40-55	850	750	HYD	HYD
	1.9 (1901)	8	0.040	③	③	40-55	750	650	HYD	HYD
	2.2 (2199)	F	0.045	③	③	55-65	④	④	HYD	HYD
	3.0 (3000)	R	0.043	③	③	39-49	④	④	HYD	HYD
2002	1.9 (1901)	7	0.040	③	③	40-55	850	750	HYD	HYD
	1.9 (1901)	8	0.040	③	③	40-55	750	650	HYD	HYD
	2.2 (2199)	F	0.045	③	③	55-65	④	④	HYD	HYD
	3.0 (3000)	R	0.043	③	③	39-49	④	④	HYD	HYD
2003-04	1.9 (1901)	7	0.040	③	③	40-55	850	750	HYD	HYD
	1.9 (1901)	8	0.040	③	③	40-55	750	650	HYD	HYD
	2.2 (2199)	F	0.045	③	③	55-65	④	④	HYD	HYD
	3.0 (3000)	R	0.043	③	③	39-49	④	④	HYD	HYD

NOTE: The Vehicle Emission Control Information label often reflects specification changes made during production. The label figures must be used if they differ from those in this chart.

HYD: Hydraulic

① Pressure measured at idle

② Idle speed measured with manual transmission in Neutral; automatic transmission in D (drive)

③ Engines equipped with Distributorless Ignition System (DIS). Ignition timing is not adjustable

④ Refer to the Vehicle Emission Control Information label

42372-SCAR-C03

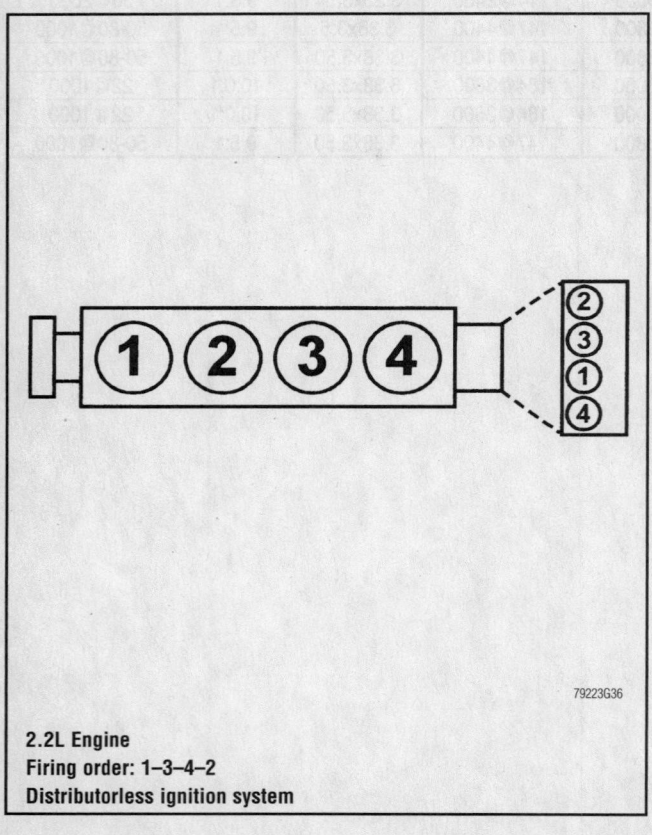

2.2L Engine
Firing order: 1–3–4–2
Distributorless ignition system

79223G36

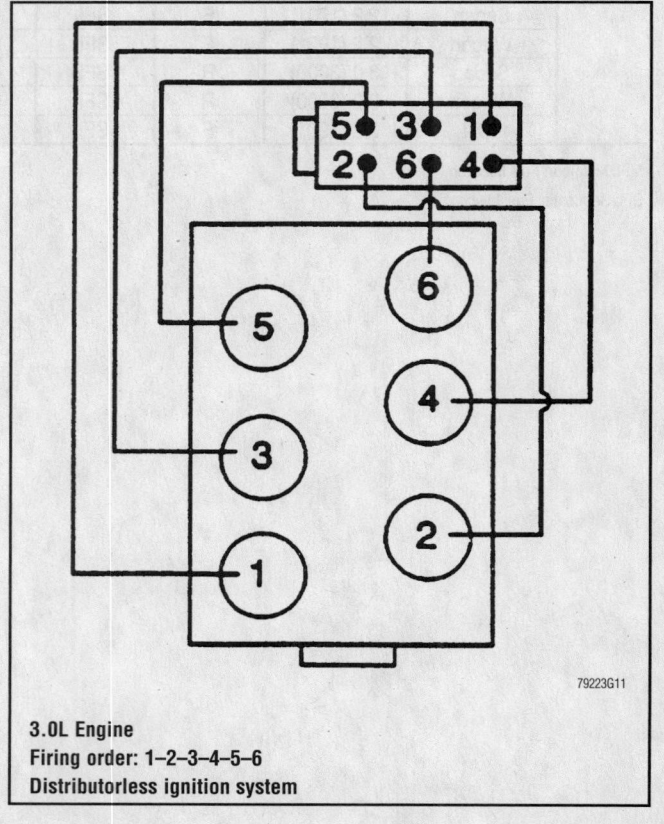

3.0L Engine
Firing order: 1–2–3–4–5–6
Distributorless ignition system

79223G11

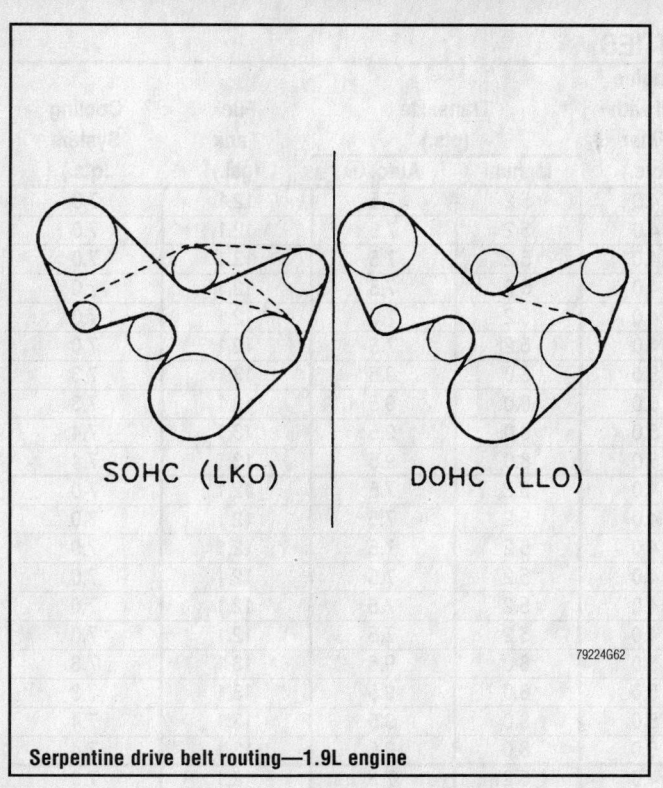

Serpentine drive belt routing—1.9L engine

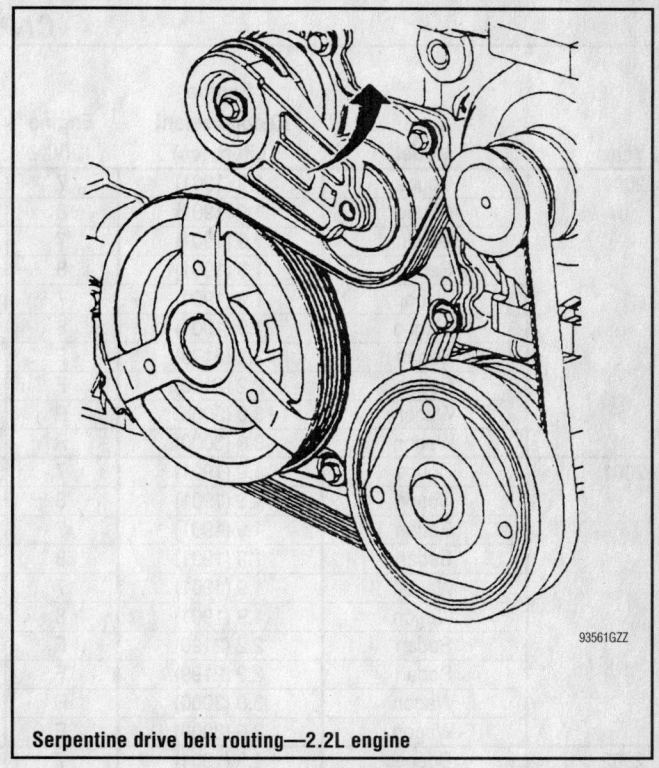

Serpentine drive belt routing—2.2L engine

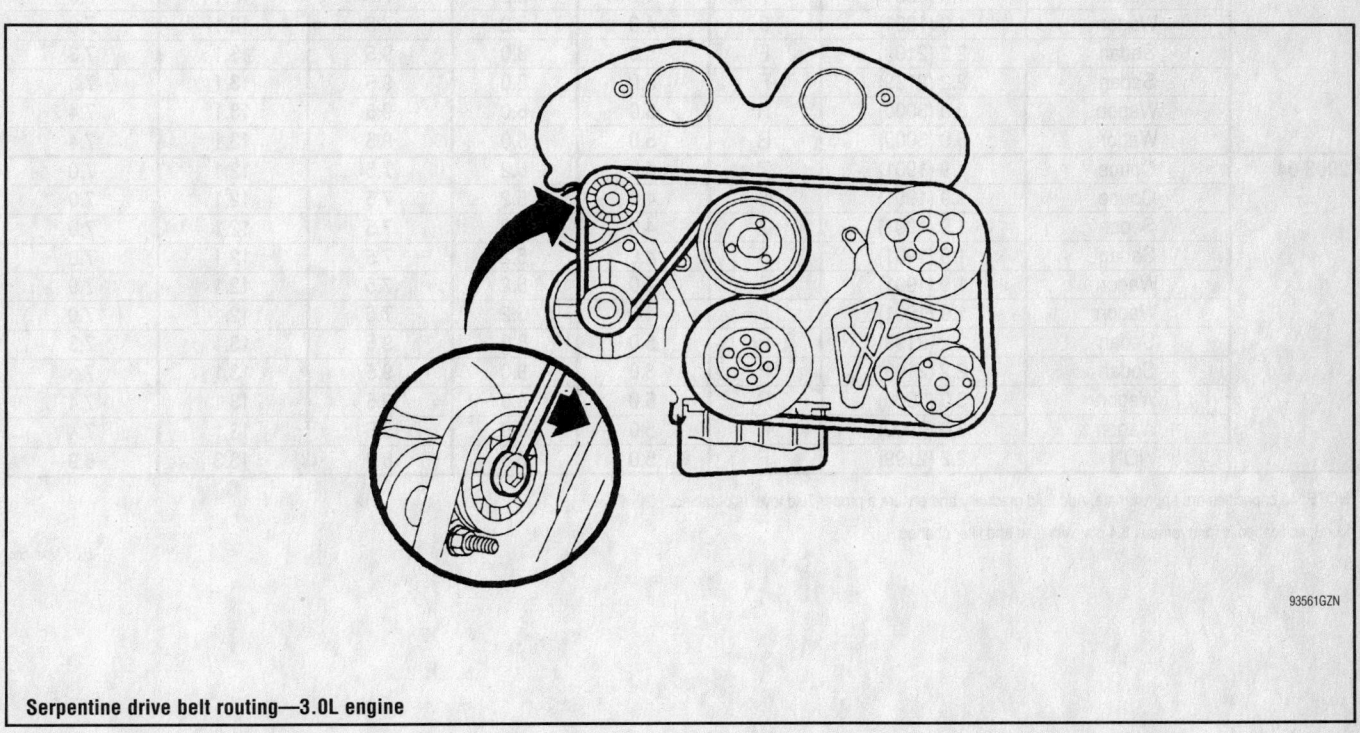

Serpentine drive belt routing—3.0L engine

CAPACITIES

Year	Model	Engine Displacement Liters (cc)	Engine ID/VIN	Engine Oil with Filter (qts.)	Transaxle (pts.) Manual	Transaxle (pts.) Auto. ①	Fuel Tank (gal.)	Cooling System (qts.)
2000	Coupe	1.9 (1901)	7	4.0	5.2	7.5	12.1	7.0
	Coupe	1.9 (1901)	8	4.0	5.2	7.5	12.1	7.0
	Sedan	1.9 (1901)	7	4.0	5.2	7.5	12.1	7.0
	Sedan	1.9 (1901)	8	4.0	5.2	7.5	12.1	7.0
	Wagon	1.9 (1901)	7	4.0	5.2	7.5	12.1	7.0
	Wagon	1.9 (1901)	8	4.0	5.2	7.5	12.1	7.0
	Sedan	2.2 (2199)	F	5.0	8.0	9.5	13.1	7.3
	Sedan	2.2 (2199)	F	5.0	8.0	9.5	13.1	7.3
	Wagon	3.0 (3000)	R	5.0	8.0	9.5	13.1	7.4
	Wagon	3.0 (3000)	R	5.0	8.0	9.5	13.1	7.4
2001	Coupe	1.9 (1901)	7	4.0	5.2	7.5	12.1	7.0
	Coupe	1.9 (1901)	8	4.0	5.2	7.5	12.1	7.0
	Sedan	1.9 (1901)	7	4.0	5.2	7.5	12.1	7.0
	Sedan	1.9 (1901)	8	4.0	5.2	7.5	12.1	7.0
	Wagon	1.9 (1901)	7	4.0	5.2	7.5	12.1	7.0
	Wagon	1.9 (1901)	8	4.0	5.2	7.5	12.1	7.0
	Sedan	2.2 (2199)	F	5.0	8.0	9.5	13.1	7.3
	Sedan	2.2 (2199)	F	5.0	8.0	9.5	13.1	7.3
	Wagon	3.0 (3000)	R	5.0	8.0	9.5	13.1	7.4
	Wagon	3.0 (3000)	R	5.0	8.0	9.5	13.1	7.4
2002	Coupe	1.9 (1901)	7	4.0	5.2	7.5	12.1	7.0
	Coupe	1.9 (1901)	8	4.0	5.2	7.5	12.1	7.0
	Sedan	1.9 (1901)	7	4.0	5.2	7.5	12.1	7.0
	Sedan	1.9 (1901)	8	4.0	5.2	7.5	12.1	7.0
	Wagon	1.9 (1901)	7	4.0	5.2	7.5	12.1	7.0
	Wagon	1.9 (1901)	8	4.0	5.2	7.5	12.1	7.0
	Sedan	2.2 (2199)	F	5.0	8.0	9.5	13.1	7.3
	Sedan	2.2 (2199)	F	5.0	8.0	9.5	13.1	7.3
	Wagon	3.0 (3000)	R	5.0	8.0	9.5	13.1	7.4
	Wagon	3.0 (3000)	R	5.0	8.0	9.5	13.1	7.4
2003-04	Coupe	1.9 (1901)	7	4.0	5.2	7.5	12.1	7.0
	Coupe	1.9 (1901)	8	4.0	5.2	7.5	12.1	7.0
	Sedan	1.9 (1901)	7	4.0	5.2	7.5	12.1	7.0
	Sedan	1.9 (1901)	8	4.0	5.2	7.5	12.1	7.0
	Wagon	1.9 (1901)	7	4.0	5.2	7.5	12.1	7.0
	Wagon	1.9 (1901)	8	4.0	5.2	7.5	12.1	7.0
	Sedan	2.2 (2199)	F	5.0	8.0	9.5	13.1	7.3
	Sedan	2.2 (2199)	F	5.0	8.0	9.5	13.1	7.3
	Wagon	3.0 (3000)	R	5.0	8.0	9.5	13.1	7.4
	Wagon	3.0 (3000)	R	5.0	8.0	9.5	13.1	7.4
	ION	2.2 (2199)	F	5.0	5.6	8.4	13.3	6.9

NOTE: All capacities are approximate. Add fluid gradually and ensure a proper fluid level is obtained.

① Specification is for overhaul. 8.4 pts. with fluid and filter change

42372-SCAR-C04

VALVE SPECIFICATIONS

Year	Engine Displacement Liters (cc)	Engine ID/VIN	Seat Angle (deg.)	Face Angle (deg.)	Spring Test Pressure (lbs. @ in.)	Spring Free-Length (in.)	Stem-to-Guide Clearance (in.)		Stem Diameter (in.)	
							Intake	Exhaust	Intake	Exhaust
2000	1.9 (1901)	7	44.5-45.4	45-45.5	163-180@ 0.984	1.5600	0.0010- 0.0025	0.0015- 0.0032	0.2736- 0.2740	0.2729- 0.2736
	1.9 (1901)	8	44.5-45.4	45-45.25	202-211@ 1.280	1.8898- 1.9134	0.0010- 0.0025	0.0015- 0.0032	0.2736- 0.2741	0.2736- 0.2740
	2.2 (2199)	F	44.5-45.4	45-45.5	① ②	1.6100	0.0012 0.0022	0.0016 0.0026	0.2344 0.2355	0.2341 0.2347
	3.0 (3000)	R	45	45	56.6@1.338	NA	0.0012 0.0022	0.0016 0.0026	0.2344 0.2350	0.2341 0.2346
2001	1.9 (1901)	7	44.5-45.4	45-45.5	163-180@ 0.984	1.5600	0.0010- 0.0025	0.0015- 0.0032	0.2736- 0.2740	0.2729- 0.2736
	1.9 (1901)	8	44.5-45.4	45-45.25	202-211@ 1.280	1.8898- 1.9134	0.0010- 0.0025	0.0015- 0.0032	0.2736- 0.2741	0.2736- 0.2740
	2.2 (2199)	F	44.5-45.4	45-45.5	① ②	1.6100	0.0012 0.0022	0.0016 0.0026	0.2344 0.2355	0.2341 0.2347
	3.0 (3000)	R	45	45	56.6@1.338	NA	0.0012 0.0022	0.0016 0.0026	0.2344 0.2350	0.2341 0.2346
2002	1.9 (1901)	7	44.5-45.4	45-45.5	163-180@ 0.984	1.5600	0.0010- 0.0025	0.0015- 0.0032	0.2736- 0.2740	0.2729- 0.2736
	1.9 (1901)	8	44.5-45.4	45-45.25	202-211@ 1.280	1.8898- 1.9134	0.0010- 0.0025	0.0015- 0.0032	0.2736- 0.2741	0.2736- 0.2740
	2.2 (2199)	F	44.5-45.4	45-45.5	① ②	1.6100	0.0012 0.0022	0.0020 0.0026	0.2344 0.2355	0.2337 0.2343
	3.0 (3000)	R	45	45	56.6@1.338	NA	0.0012 0.0022	0.0016 0.0026	0.2344 0.2350	0.2341 0.2346
2003-04	1.9 (1901)	7	44.5-45.4	45-45.5	163-180@ 0.984	1.5600	0.0010- 0.0025	0.0015- 0.0032	0.2736- 0.2740	0.2729- 0.2736
	1.9 (1901)	8	44.5-45.4	45-45.25	202-211@ 1.280	1.8898- 1.9134	0.0010- 0.0025	0.0015- 0.0032	0.2736- 0.2741	0.2736- 0.2740
	2.2 (2199)	F	44.5-45.4	45-45.5	① ②	1.6100	0.0012 0.0022	0.0020 0.0026	0.2344 0.2355	0.2337 0.2343
	3.0 (3000)	R	45	45	56.6@1.338	NA	0.0012 0.0022	0.0016 0.0026	0.2344 0.2350	0.2341 0.2346

NA: Not available

① Valve spring load closed: 245-271 N

② Valve spring load open: 525-575 N

42372-SCAR-C05

CRANKSHAFT AND CONNECTING ROD SPECIFICATIONS

All measurements are given in inches.

Year	Engine Displacement Liters (cc)	Engine ID/VIN	Crankshaft				Connecting Rod		
			Main Brg. Journal Dia.	Main Brg. Oil Clearance	Shaft End-play	Thrust on No.	Journal Diameter	Oil Clearance	Side Clearance
2000	1.9 (1901)	7	2.2438-2.2444	0.0002-0.0020	0.0020-0.0079	3	1.9761-1.9767	0.0001-0.0021	0.0065-0.0171
	1.9 (1901)	8	2.2438-2.2444	0.0002-0.0020	0.0020-0.0079	3	1.9761-1.9767	0.0001-0.0021	0.0065-0.0171
	2.2 (2199)	F	2.2045-2.2050	0.0012-0.0026	0.0012-0.0150	3	1.9291-1.9297	0.0001-0.0021	0.0028-0.0146
	3.0 (3000)	R	2.6763-2.6766	0.0060-0.0017	0.0004-0.0300	3	1.927-1.9280	0.0001-0.0021	0.0027-0.0110
2001	1.9 (1901)	7	2.2438-2.2444	0.0002-0.0020	0.0020-0.0079	3	1.9761-1.9767	0.0001-0.0021	0.0065-0.0171
	1.9 (1901)	8	2.2438-2.2444	0.0002-0.0020	0.0020-0.0079	3	1.9761-1.9767	0.0001-0.0021	0.0065-0.0171
	2.2 (2199)	F	2.2045-2.2050	0.0012-0.0026	0.0012-0.0150	3	1.9291-1.9297	0.0001-0.0021	0.0028-0.0146
	3.0 (3000)	R	2.6763-2.6766	0.0060-0.0017	0.0004-0.0300	3	1.927-1.9280	0.0001-0.0021	0.0027-0.0110
2002	1.9 (1901)	7	2.2438-2.2444	0.0002-0.0020	0.0020-0.0079	3	1.9761-1.9767	0.0001-0.0021	0.0065-0.0171
	1.9 (1901)	8	2.2438-2.2444	0.0002-0.0020	0.0020-0.0079	3	1.9761-1.9767	0.0001-0.0021	0.0065-0.0171
	2.2 (2199)	F	2.2045-2.2050	0.0012-0.0026	0.0012-0.0150	3	1.9291-1.9297	0.0001-0.0021	0.0028-0.0146
	3.0 (3000)	R	2.6763-2.6766	0.0060-0.0017	0.0004-0.0300	3	1.927-1.9280	0.0001-0.0021	0.0027-0.0110
2003-04	1.9 (1901)	7	2.2438-2.2444	0.0002-0.0020	0.0020-0.0079	3	1.9761-1.9767	0.0001-0.0021	0.0065-0.0171
	1.9 (1901)	8	2.2438-2.2444	0.0002-0.0020	0.0020-0.0079	3	1.9761-1.9767	0.0001-0.0021	0.0065-0.0171
	2.2 (2199)	F	2.2045-2.2050	0.0012-0.0026	0.0012-0.0150	3	1.9291-1.9297	0.0001-0.0021	0.0028-0.0146
	3.0 (3000)	R	2.6763-2.6766	0.0060-0.0017	0.0004-0.0300	3	1.927-1.9280	0.0001-0.0021	0.0027-0.0110

NA: Not available

42372-SCAR-C06

PISTON AND RING SPECIFICATIONS
All measurements are given in inches.

Year	Engine Displacement Liters (cc)	Engine ID/VIN	Piston Clearance	Ring Gap			Ring Side Clearance		
				Top Compression	Bottom Compression	Oil Control	Top Compression	Bottom Compression	Oil Control
2000	1.9 (1901)	7	①	0.0071-0.0130	0.0138-0.0216	0.0039-0.0197	0.0016-0.0032	0.0012-0.0031	SNUG
	1.9 (1901)	8	①	0.0071-0.0130	0.0138-0.0216	0.0039-0.0197	0.0016-0.0032	0.0012-0.0031	SNUG
	2.2 (2199)	F	0.0004-0.0016	0.008-0.016	0.0014 0.0022	0.0010 0.0030	0.0028-0.0146	0.0005-0.0024	SNUG
	3.0 (3000)	R	0.0010-0.0018	0.0008-0.0015	0.0118 0.0196	0.0157 0.0551	0.0027-0.0110	0.0005-0.0024	SNUG
2001	1.9 (1901)	7	①	0.0071-0.0130	0.0138-0.0216	0.0039-0.0197	0.0016-0.0032	0.0012-0.0031	SNUG
	1.9 (1901)	8	①	0.0071-0.0130	0.0138-0.0216	0.0039-0.0197	0.0016-0.0032	0.0012-0.0031	SNUG
	2.2 (2199)	F	0.0004-0.0016	0.008-0.016	0.0014 0.0022	0.0010 0.0030	0.0028-0.0146	0.0005-0.0024	SNUG
	3.0 (3000)	R	0.0010-0.0018	0.0008-0.0015	0.0118 0.0196	0.0157 0.0551	0.0027-0.0110	0.0005-0.0024	SNUG
2002	1.9 (1901)	7	①	0.0071-0.0130	0.0138-0.0216	0.0039-0.0197	0.0016-0.0032	0.0012-0.0031	SNUG
	1.9 (1901)	8	①	0.0071-0.0130	0.0138-0.0216	0.0039-0.0197	0.0016-0.0032	0.0012-0.0031	SNUG
	2.2 (2199)	F	0.0004-0.0016	0.008-0.016	0.0014 0.0022	0.0010 0.0030	0.0028-0.0146	0.0005-0.0024	SNUG
	3.0 (3000)	R	0.0010-0.0018	0.0008-0.0015	0.0118 0.0196	0.0157 0.0551	0.0027-0.0110	0.0005-0.0024	SNUG
2003-04	1.9 (1901)	7	①	0.0071-0.0130	0.0138-0.0216	0.0039-0.0197	0.0016-0.0032	0.0012-0.0031	SNUG
	1.9 (1901)	8	①	0.0071-0.0130	0.0138-0.0216	0.0039-0.0197	0.0016-0.0032	0.0012-0.0031	SNUG
	2.2 (2199)	F	0.0004-0.0016	0.008-0.016	0.0014 0.0022	0.0010 0.0030	0.0028-0.0146	0.0005-0.0024	SNUG
	3.0 (3000)	R	0.0010-0.0018	0.0008-0.0015	0.0118 0.0196	0.0157 0.0551	0.0027-0.0110	0.0005-0.0024	SNUG

NA: Not available

① Piston No. 2 and 3: 0.0002-0.0017
Piston No. 1, 4: 0.0003-0.0021

42372-SCAR-C07

TORQUE SPECIFICATIONS
All readings in ft. lbs.

Year	Engine Displacement Liters (cc)	Engine ID/VIN	Cylinder Head Bolts	Main Bearing Bolts	Rod Bearing Bolts	Crankshaft Damper Bolts	Flywheel Bolts	Manifold		Spark Plugs	Lug Nuts
								Intake	Exhaust		
2000	1.9 (1901)	7	①	37	33	159	59②	22③	13③	20	103
	1.9 (1901)	8	①	37	33	159	59②	22③	16③	20	103
	2.2 (2199	F	④	⑤	18	⑥	59②	⑦	13	15	138
	3.0 (3000)	R	⑧	⑨	26	15	59②	15	15	18	139
2001	1.9 (1901)	7	①	37	33	159	59②	22③	13③	20	103
	1.9 (1901)	8	①	37	33	159	59②	22③	16③	20	103
	2.2 (2199	F	④	⑤	18	⑥	59②	⑦	13	15	138
	3.0 (3000)	R	⑧	⑨	26	15	59②	15	15	18	139
2002	1.9 (1901)	7	①	37	33	159	59②	22③	13③	20	103
	1.9 (1901)	8	①	37	33	159	59②	22③	16③	20	103
	2.2 (2199	F	④	⑤	18	⑥	②	⑦	13	15	138
	3.0 (3000)	R	⑧	⑨	26	15	②	15	15	18	139
2003-04	1.9 (1901)	7	①	37	33	159	59②	22③	13③	20	103
	1.9 (1901)	8	①	37	33	159	59②	22③	16③	20	103
	2.2 (2199	F	④	⑤	18	⑥	②	⑦	13	15	138
	3.0 (3000)	R	⑧	⑨	26	15	②	15	15	18	139

① Step 1: 22 ft. lbs.
   Step 2: 33 ft. lbs.
   Step 3: 90 degrees
② Flexplate specification: 39 ft. lbs. Plus 25 degrees
③ Studs: 106 inch lbs.
④ Except ION models.
   Step 1: 22 ft. lbs.
   Step 2: 155 degrees
   Step 3: Front 4 bolts to 15 ft. lbs.
   ION Models:
   Step 1: 22 ft. lbs.
   Step 2: 155 degrees
   Step 3: Front 4 bolts to 26 ft. lbs.
⑤ 15 ft. lbs. Plus 70 degrees Plus 20 degrees
⑥ 89 inch lbs.
⑦ 74 ft. lbs. Plus 75 degrees

⑧ Step 1: 22 ft. lbs.
   Step 2: plus 90 degrees
   Step 3: plus 90 degrees
   Step 4: plus 90 degrees
   Step 5: plus 15 degrees
⑨ Step 1: 37 ft. lbs.
   Step 2: plus 60 degrees
   Step 3: plus 15 degrees

42372-SCAR-C08

WHEEL ALIGNMENT

Year	Model		Caster Range (+/-Deg.)	Caster Preferred Setting (Deg.)	Camber Range (+/-Deg.)	Camber Preferred Setting (Deg.)	Toe-in (in.)	Steering Axis Inclination (Deg.)
2000	S Series	F	0.60	+1.70	0.70	-0.50	0.20 +/- 0.10	—
		R	—	—	0.70	-0.70	0.20 +/- 0.10	—
	L Series	F	1.00	+3.70	0.50	-1.00	0.20 +/- 0.15	—
		R	—	—	0.60	-1.00	0.15 +/- 0.07	—
2001	L Series	F	1.00	+3.70	0.50	-1.00	0.20 +/- 0.15	—
		R	—	—	0.60	-1.00	0.15 +/- 0.07	—
	S Series	F	0.60	+1.70	0.70	-0.50	0.20 +/- 0.10	—
		R	—	—	0.70	-0.70	0.20 +/- 0.10	—
2002	L Series	F	1.00	+3.70	0.50	-1.00	0.20 +/- 0.15	—
		R	—	—	0.60	-1.00	0.15 +/- 0.07	—
	S Series	F	0.60	+1.70	0.70	-0.50	0.20 +/- 0.10	—
		R	—	—	0.70	-0.70	0.20 +/- 0.10	—
2003-04	L Series	F	1.00	+3.70	0.50	-1.00	0.20 +/- 0.15	—
		R	—	—	0.60	-1.00	0.15 +/- 0.07	—
	S Series	F	0.60	+1.70	0.70	-0.50	0.20 +/- 0.10	—
		R	—	—	0.70	-0.70	0.20 +/- 0.10	—
	ION	F	0.75	3.25	0.75	1.00	0.10 +/- 0.20	—
		R	—	—	0.75	-1.40	0.06 +/- 0.35	—

42372-SCAR-C09

TIRE, WHEEL AND BALL JOINT SPECIFICATIONS

| Year | Model | OEM Tires | | Tire Pressures (psi) | | Wheel | Ball Joint |
		Standard	Optional	Front	Rear	Size	Inspection
2000	LS, LS1	P195/65R15	None	30	26	6J	NS
	LS2	P205/65R15	None	30	30	6J	NS
	LW1	P195/65R15	None	30	26	6J	NS
	LW2	P205/65R15	None	30	30	6J	NS
	SC1, SL, SL1	P185/65R14	None	30	26	6J	NS
	SC2	P195/60R15	None	30	26	6J	NS
	SL2, SW2	P185/65R15	None	30	26	6J	NS
2001	LS, LS1	P195/65R15	None	30	26	6J	NS
	LS2	P205/65R15	None	30	30	6J	NS
	LW1	P195/65R15	None	30	26	6J	NS
	LW2	P205/65R15	None	30	30	6J	NS
	SC1, SL, SL1	P185/65R14	None	30	26	6J	NS
	SC2	P195/60R15	None	30	26	6J	NS
	SL2, SW2	P185/65R15	None	30	26	6J	NS
2002	LS, LS1	P195/65R15	None	30	26	6J	NS
	LS2	P205/65R15	None	30	30	6J	NS
	LW1	P195/65R15	None	30	26	6J	NS
	LW2	P205/65R15	None	30	30	6J	NS
	SC1, SL, SL1	P185/65R14	None	30	26	6J	NS
	SC2	P195/60R15	None	30	26	6J	NS
	SL2, SW2	P185/65R15	None	30	26	6J	NS
2003-04	LS, LS1	P195/65R15	None	30	26	6J	NS
	LS2	P205/65R15	None	30	30	6J	NS
	LW1	P195/65R15	None	30	26	6J	NS
	LW2	P205/65R15	None	30	30	6J	NS
	SC1, SL, SL1	P185/65R14	None	30	26	6J	NS
	SC2	P195/60R15	None	30	26	6J	NS
	SL2, SW2	P185/65R15	None	30	26	6J	NS
	ION	①	①	①	①	①	①

OEM: Original Equipment Manufacturer

PSI: Pounds Per Square Inch

STD: Standard

OPT: Optional

NS: Not specified by manufacturer

①: For tire size and information, check the label located on the inside of the glove compartment door.

42372-SCAR-C10

BRAKE SPECIFICATIONS
All measurements in inches unless noted

Year	Model		Brake Disc Original Thickness	Brake Disc Minimum Thickness	Brake Disc Maximum Runout	Brake Drum Diameter Original Inside Diameter	Brake Drum Diameter Max. Wear Limit	Brake Drum Diameter Maximum Machine Diameter	Minimum Lining Thickness	Brake Caliper Bracket Bolt (ft. lbs.)	Brake Caliper Mounting Bolt (ft. lbs.)
2000	Coupe ①	F	0.710	0.633	0.0024	—	—	—	0.080	81	27
		R	0.430	0.370	0.0024	7.87	7.93	7.91	0.040	63	27
	Sedan ①	F	0.710	0.633	0.0024	—	—	—	0.080	81	27
		R	0.430	0.370	0.0024	7.87	7.93	7.91	0.040	63	27
	Wagon ①	F	0.710	0.633	0.0024	—	—	—	0.080	81	27
		R	0.430	0.370	0.0024	7.87	7.93	7.91	0.040	63	27
	Coupe ②	F	0.980	0.900	0.001	—	—	—	0.080	70	22
		R	0.390	0.350	0.001	9.05	9.09	9.08	0.080	59	27
	Sedan ②	F	0.980	0.900	0.001	—	—	—	0.080	70	22
		R	0.390	0.350	0.001	9.05	9.09	9.08	0.080	59	27
	Wagon ②	F	0.980	0.900	0.001	—	—	—	0.080	70	22
		R	0.390	0.350	0.001	9.05	9.09	9.08	0.080	59	27
2001	Coupe ①	F	0.710	0.633	0.0024	—	—	—	0.080	81	27
		R	0.430	0.370	0.0024	7.87	7.93	7.91	0.040	63	27
	Sedan ①	F	0.710	0.633	0.0024	—	—	—	0.080	81	27
		R	0.430	0.370	0.0024	7.87	7.93	7.91	0.040	63	27
	Wagon ①	F	0.710	0.633	0.0024	—	—	—	0.080	81	27
		R	0.430	0.370	0.0024	7.87	7.93	7.91	0.040	63	27
	Coupe ②	F	0.980	0.900	0.001	—	—	—	0.080	70	22
		R	0.390	0.350	0.001	9.05	9.09	9.08	0.080	59	27
	Sedan ②	F	0.980	0.900	0.001	—	—	—	0.080	70	22
		R	0.390	0.350	0.001	9.05	9.09	9.08	0.080	59	27
	Wagon ②	F	0.980	0.900	0.001	—	—	—	0.080	70	22
		R	0.390	0.350	0.001	9.05	9.09	9.08	0.080	59	27
2002	Coupe ①	F	0.710	0.633	0.0024	—	—	—	0.080	81	27
		R	0.430	0.370	0.0024	7.87	7.93	7.91	0.040	63	27
	Sedan ①	F	0.710	0.633	0.0024	—	—	—	0.080	81	27
		R	0.430	0.370	0.0024	7.87	7.93	7.91	0.040	63	27
	Wagon ①	F	0.710	0.633	0.0024	—	—	—	0.080	81	27
		R	0.430	0.370	0.0024	7.87	7.93	7.91	0.040	63	27
	Coupe ②	F	0.980	0.900	0.001	—	—	—	0.080	70	22
		R	0.390	0.350	0.001	9.05	9.09	9.08	0.080	59	27
	Sedan ②	F	0.980	0.900	0.001	—	—	—	0.080	70	22
		R	0.390	0.350	0.001	9.05	9.09	9.08	0.080	59	27
	Wagon ②	F	0.980	0.900	0.001	—	—	—	0.080	70	22
		R	0.390	0.350	0.001	9.05	9.09	9.08	0.080	59	27
2003-04	Coupe ①	F	0.710	0.633	0.0024	—	—	—	0.080	81	27
		R	0.430	0.370	0.0024	7.87	7.93	7.91	0.040	63	27
	Sedan ①	F	0.710	0.633	0.0024	—	—	—	0.080	81	27
		R	0.430	0.370	0.0024	7.87	7.93	7.91	0.040	63	27
	Wagon ①	F	0.710	0.633	0.0024	—	—	—	0.080	81	27
		R	0.430	0.370	0.0024	7.87	7.93	7.91	0.040	63	27
	Coupe ②	F	0.980	0.900	0.001	—	—	—	0.080	70	22
		R	0.390	0.350	0.001	9.05	9.09	9.08	0.080	59	27
	Sedan ②	F	0.980	0.900	0.001	—	—	—	0.080	70	22
		R	0.390	0.350	0.001	9.05	9.09	9.08	0.080	59	27
	Wagon ②	F	0.980	0.900	0.001	—	—	—	0.080	70	22
		R	0.390	0.350	0.001	9.05	9.09	9.08	0.080	59	27
	ION	F	0.933	0.896	0.001	—	—	—	0.039	85	25
		R	—	—	—	9.06	9.094	9.075	0.020	—	—

NA: Not Available
F: Front
R: Rear
① S series
② L series

42372-SCAR-C11

SCHEDULED MAINTENANCE INTERVALS

SATURN—ION, L & S SERIES

TO BE SERVICED	TYPE OF SERVICE	VEHICLE MILEAGE INTERVAL (x1000)												
		3	6	9	12	15	18	21	24	27	30	33	36	39
Engine oil & filter	R		✓		✓		✓		✓		✓		✓	
Lubricate chassis, suspension and steering linkage	S/I		✓		✓		✓		✓		✓		✓	
Lubricate transaxle shift linkage and parking brake cable guides	S/I		✓		✓		✓		✓		✓		✓	
Lubricate underbody contact points & linkage	S/I		✓		✓		✓		✓		✓		✓	
Driveshaft boots, suspension bushings & ball joint seals	S/I		✓				✓		✓		✓		✓	
Exhaust system & throttle linkage	S/I		✓		✓		✓		✓		✓		✓	
Rotate tires	S/I		✓		✓		✓				✓			
Brake hoses & brake lining	S/I		✓				✓				✓			
Accessory drive belt(s)	S/I						✓						✓	
Engine coolant level, hoses & clamps	S/I						✓						✓	
Air filter element	R										✓			
Engine coolant	R												✓	
Manual transaxle oil	R		✓											
Spark plugs ①	R										✓			
Automatic transaxle fluid & filter	S/I										✓			
Ignition cables & fuel systems	S/I										✓			
Vacuum line/hose	S/I										✓			
Fuel filter ②	R													

S/I: Service or Inspect

R: Replace

① Platinum tip spark plugs: replace every 100,000 miles

② Replace every 60,000 miles

FREQUENT OPERATION MAINTENANCE (SEVERE SERVICE)

If a vehicle is operated under any of the following conditions it is considered severe service:

- Extremely dusty areas

- 50% or more of the vehicle operation is in 32°C (90°F) or higher temperatures, or constant operation in temperatures below 0°C (32°F)

#NAME?

- Frequent short running periods (engine does not warm to normal operating temperatures)

- Police, taxi, delivery usage or trailer towing usage

Engine oil & oil filter: change every 3000 miles

42372-SCAR-C12

PRECAUTIONS

Before servicing any vehicle, please be sure to read all of the following precautions, which deal with personal safety, prevention of component damage, and important points to take into consideration when servicing a motor vehicle:

• Never open, service or drain the radiator or cooling system when the engine is hot; serious burns can occur from the steam and hot coolant.

• Observe all applicable safety precautions when working around fuel. Whenever servicing the fuel system, always work in a well-ventilated area. Do not allow fuel spray or vapors to come in contact with a spark, open flame or excessive heat (a hot drop light, for example). Keep a dry chemical fire extinguisher near the work area. Always keep fuel in a container specifically designed for fuel storage; also, always properly seal fuel containers to avoid the possibility of fire or explosion. Refer to the additional fuel system precautions later in this section.

• Fuel injection systems often remain pressurized, even after the engine has been turned **OFF**. The fuel system pressure must be relieved before disconnecting any fuel lines. Failure to do so may result in fire and/or personal injury.

• Brake fluid often contains polyglycol ethers and polyglycols. Avoid contact with the eyes and wash your hands thoroughly after handling brake fluid. If you do get brake fluid in your eyes, flush your eyes with clean, running water for 15 minutes. If eye irritation persists, or if you have taken brake fluid internally, IMMEDIATELY seek medical assistance.

• The EPA warns that prolonged contact with used engine oil may cause a number of skin disorders, including cancer! You should make every effort to minimize your exposure to used engine oil. Protective gloves should be worn when changing oil. Wash your hands and any other exposed skin areas as soon as possible after exposure to used engine oil. Soap and water, or waterless hand cleaner should be used.

• All new vehicles are now equipped with an air bag system. The system must be disabled before performing service on or around system components, steering column, instrument panel components, wiring and sensors. Failure to follow safety and disabling procedures could result in accidental air bag deployment, possible personal injury and unnecessary system repairs.

• Always wear safety goggles when working with, or around, the air bag system. When carrying a non-deployed air bag, be sure the bag and trim cover are pointed away from your body. When placing a non-deployed air bag on a work surface, always face the bag and trim cover upward, away from the surface. This will reduce the motion of the module if it is accidentally deployed. Refer to the additional air bag system precautions later in this section.

• Clean, high quality brake fluid from a sealed container is essential to the safe and proper operation of the brake system. You should always buy the correct type of brake fluid for your vehicle. If the brake fluid becomes contaminated, completely flush the system with new fluid. Never reuse any brake fluid. Any brake fluid that is removed from the system should be discarded. Also, do not allow any brake fluid to come in contact with a painted surface; it will damage the paint.

• Never operate the engine without the proper amount and type of engine oil; doing so WILL result in severe engine damage.

• Timing belt maintenance is extremely important! Many models utilize an interference-type, non-freewheeling engine. If the timing belt breaks, the valves in the cylinder head may strike the pistons, causing potentially serious (also time-consuming and expensive) engine damage. Refer to the maintenance interval charts in the front of this section for the recommended replacement interval for the timing belt, and to the timing belt procedure in this section for belt replacement and inspection.

• Disconnecting the negative battery cable on some vehicles may interfere with the functions of the on-board computer system(s) and may require the computer to undergo a relearning process once the negative battery cable is reconnected.

• When servicing drum brakes, only disassemble and assemble one side at a time, leaving the remaining side intact for reference.

• Only an MVAC-trained, EPA-certified automotive technician should service the air conditioning system or its components.

ENGINE REPAIR

Ignition Timing

ADJUSTMENT

The engines covered in this section utilize a Distributorless Ignition System (DIS), no adjustment is possible.

Alternator

REMOVAL

1.9L Engines

1. Before servicing the vehicle, refer to the precautions in the beginning of this section.
2. Remove or disconnect the following:

• Negative battery cable
• Accessory drive belt
• Right front wheel
• Right front wheel well splash shield
• Alternator electrical connectors

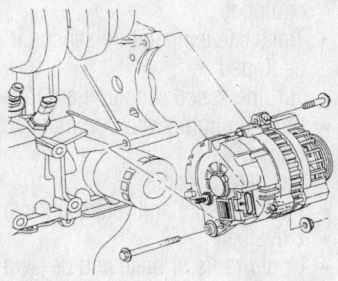

9306ZG39

Exploded view of the alternator—1.9L engines

• Alternator bolts
• Alternator through the wheel well opening

2.2L Engine

1. Before servicing the vehicle, refer to the precautions in the beginning of this section.
2. Remove or disconnect the following:
• Negative battery cable
• Throttle body air duct
• Accessory drive belt
• Alternator electrical connectors
• Alternator bolts
• Alternator

3.0L Engine

1. Before servicing the vehicle, refer to the precautions in the beginning of this section.

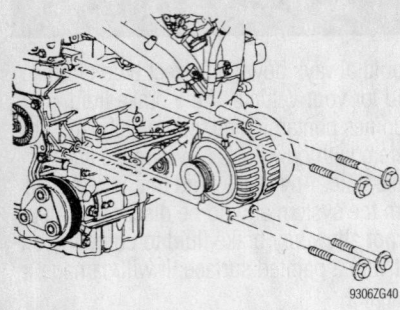

9306ZG40

Exploded view of the alternator—2.2L engine

2. Remove or disconnect the following:
- Negative battery cable
- Accessory drive belt and tensioner
- Upper alternator bolts
- Alternator lower bolts, turn the steering wheel to the right to access the bolt
- Alternator electrical connectors
- Alternator

INSTALLATION

1.9L Engines

➡Make certain the alternator shield is in place before installation.

1. Install or connect the following:
- Alternator to the cylinder block bracket and torque the bolts to 24 ft. lbs. (32 Nm)
- Alternator electrical connectors
- Splash shield
- Right front wheel
- Accessory drive belt
- Negative battery cable

2.2L Engine

Install or connect the following:
- Alternator and torque the bolts to 15 ft. lbs. (20 Nm)
- Alternator electrical connectors
- Accessory drive belt
- Throttle body air duct
- Negative battery cable

3.0L Engine

Install or connect the following:
- Alternator electrical connectors
- Alternator and torque the bolts to 30 ft. lbs. (40 Nm)
- Drive belt tensioner and torque the bolts to 30 ft lbs. (40 Nm)
- Accessory drive belt
- Negative battery cable

Engine Assembly

REMOVAL & INSTALLATION

1.9L Engines

➡The manufacturer recommends that the engine and transaxle be removed as a complete unit. Disconnect the cradle and lower the entire assembly instead of lifting the assembly out of the vehicle. Both the Single Over Head Camshaft (SOHC) and Dual Over Head Camshaft (DOHC) engines are removed or installed in the same manner. We have found that it is often possible, however, to remove the engine or transaxle alone on some models with automatic transaxles by raising it up and out of the engine compartment.

1. Before servicing the vehicle, refer to the precautions in the beginning of this section.
2. Properly disable the Supplemental Inflatable Restraint (SIR) system.
3. Properly relieve the fuel system pressure.
4. Drain the engine coolant.
5. Drain the engine oil.
6. Remove or disconnect the following:
- Both battery cables
- Air cleaner and intake duct assembly
- Engine Coolant Temperature (ECT) sensor electrical connector
- Oxygen Sensor (O_2S) and clip
- Idle Air Control (IAC) valve
- Ignition coil connectors
- Throttle Position Sensor (TPS)
- Manifold Absolute Pressure (MAP) sensor
- Exhaust Gas Recirculation (EGR) solenoid
- Brake booster hose
- Ground connectors at the rear of the cylinder block
- Fuel injector connectors
- Neutral Safety Selector (NSS) switch, if equipped
- Valve body actuator connector, if equipped
- Transaxle temperature sensor, if equipped
- Turbine speed sensor, if equipped
- Back-up light switch on manual transaxles
- Accelerator cable
- Fuel pressure and return lines
- Drive belt
- Upper radiator hose and de-aeration hose
- A/C compressor without removing the hoses
- Transaxle cooler lines, if equipped
- Automatic transaxle shifter cable, if equipped
- Manual transaxle shifter cables
- Hydraulic clutch actuator, if equipped
- Tie the radiator, condenser and fan module to the crossbar
- Front wheels
- Splash shields
- Brake caliper brackets and secure the calipers to the shock tower
- Struts from the knuckles
- Lower radiator hose
- Heater inlet and outlet hoses
- Steering shaft
- Power steering pressure switch electrical connector, if equipped
- Front exhaust pipe from the exhaust manifold
- Powertrain stiffening bracket
- Flywheel cover
- Torque converter bolts, if equipped
- Starter motor electrical connectors
- Alternator electrical connector
- Oil pressure sensor electrical connector
- Knock Sensor (KS) electrical connector
- Crankshaft Position (CKP) sensor electrical connector
- Electronic Variable Orifice (EVO) solenoid electrical connector, if equipped
- Vehicle Speed Sensor (VSS) electrical connector
- Evaporative Emissions (EVAP) purge solenoid electrical connector
- Powertrain Control Module (PCM) electrical connector
- Antilock Brake System (ABS) connectors, if equipped
- Brake lines from the rear of the cradle
- Electrical harness from the engine and position it out of the way

7. Place a block of wood between the torque strut and cradle.
8. Remove the 3 right side upper engine torque axis-to-front cover nuts and the 2 mount-to-midrail bracket nuts, allowing the powertrain to rest on the block of wood.
9. Properly support the engine/transaxle assembly.
10. Remove or disconnect the following:
- 2 right side front engine mount torque strut bracket-to-cradle nuts
- 4 cradle attaching bolts
- Powertrain/cradle assembly from the vehicle
- Spark plug wires from the ignition module

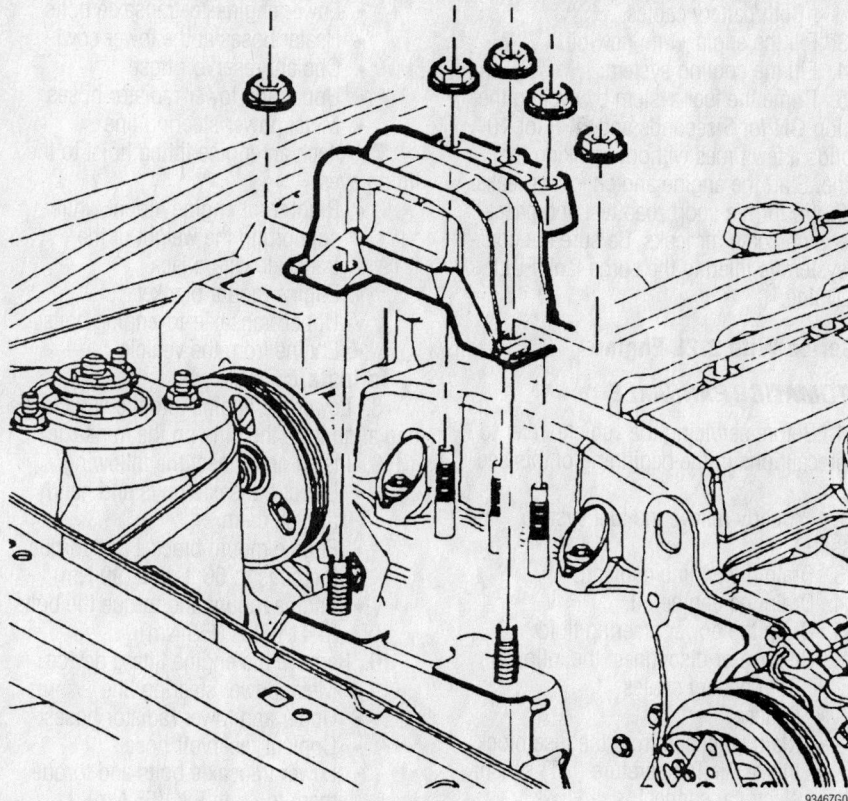

Right hand upper engine torque axis—1.9L engines

9346ZG04

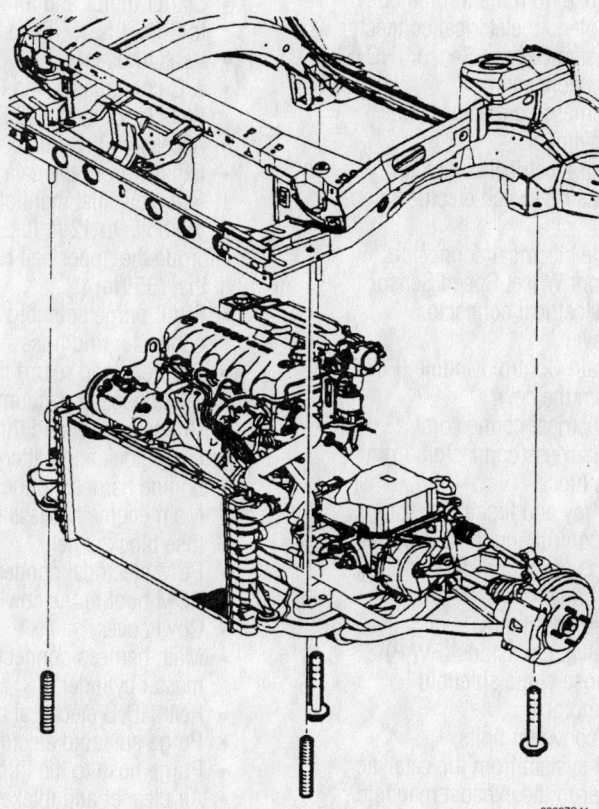

Exploded view of the cradle to body bolts—1.9L engines

9306ZG41

- Power steering pump and bracket, if equipped
11. Install an engine lifting device to the service support brackets.
- Front mount assembly
- Intake manifold support brace, DOHC engines only
- 3 axle shaft bracket support bolts and allow the bracket to rotate rearward
- Transaxle attaching bolts
- Engine from the transaxle

To install:

12. Install or connect the following:
- Engine to the transaxle and torque the lower transaxle bolts to 96 ft. lbs. (130 Nm) and the upper transaxle bolts to 66 ft. lbs. (90 Nm)
- Front engine mount and torque the bolts to 41 ft. lbs. (55 Nm)
- Engine mount torque strut bracket and torque the bolts to 52 ft. lbs. (70 Nm)
- Engine strut-to-engine bracket and torque the bolts to 52 ft. lbs. (70 Nm)
- Axle shaft bracket and torque the bolts to 41 ft. lbs. (55 Nm)
- Starter bracket and torque the bolts to 80 inch lbs. (9 Nm)
- Engine/transaxle assembly to the chassis and torque the cradle-to-body bolts to 151 ft. lbs. (205 Nm)
- Brake lines to the cradle
- Steering shaft and torque the bolts to 35 ft. lbs. (47 Nm)
- Starter motor electrical connectors
- Alternator electrical connectors
- Oil pressure sensor electrical connector
- KS sensor electrical connector
- CKP sensor electrical connector
- EVAP canister purge valve electrical connector
- Wiring harness PCM ground
- Wiring harness to the transaxle/engine block and torque the fasteners to 18 ft. lbs. (25 Nm)
- EVO solenoid electrical connector, if equipped
- VSS electrical connector
- ABS wheel sensor electrical connectors, if equipped
- Front exhaust pipe to the manifold and torque the bolts to 23 ft. lbs. (31 Nm)
- Exhaust pipe-to-stiffening bracket and torque the bolts to 35 ft. lbs. (47 Nm)
- Exhaust pipe-to-support bracket

and torque the bolts to 23 ft. lbs. (31 Nm)
- Heater hoses
- Lower radiator hose
- Steering knuckle to the strut and torque the bolts to 148 ft. lbs. (200 Nm)
- Brake caliper and torque the bolts to 81 ft. lbs. (110 Nm)
- Hydraulic clutch slave cylinder and torque the fasteners to 19 ft. lbs. (25 Nm), if equipped
- Shift cables
- Transaxle oil cooler lines, if equipped
- A/C compressor bracket and torque the bolts to 22 ft. lbs. (30 Nm)
- A/C compressor and torque the front bolt to 40 ft. lbs. (54 Nm) and the rear bolt to 22 ft. lbs. (30 Nm)
- Accessory drive belt
- Engine mount-to-midrail bracket and torque the bolts to 37 ft. lbs. (50 Nm)
- Engine mount to the front cover and torque the bolts to 37 ft. lbs. (50 Nm)
- Torque converter-to-flexplate bolts and torque them to 52 ft. lbs. (70 Nm)
- Dust cover and torque the fasteners to 89 inch lbs. (10 Nm)
- Splash shields
- Both front wheels
- ECT sensor electrical connector
- O2S electrical connector
- IAC valve electrical connector
- Fuel injector connectors
- TPS electrical connectors
- MAP sensor electrical connector
- EGR valve electrical connector
- A/C compressor electrical connector, if equipped
- Brake booster vacuum hose
- Ground connections
- NSS switch connectors
- PNP switch connectors
- Valve body actuator connector
- Back-up light switch connector, if equipped
- Temperature sensor electrical connector
- Accelerator cable
- Fuel feed and return lines
- Upper radiator and de-aeration hoses

➡Check the upper cooling module grommets for misalignment. The cooling module must be able to move freely.

- Air cleaner and intake duct assembly

- Both battery cables
13. Fill the engine with new oil.
14. Fill the cooling system.
15. Prime the fuel system by cycling the ignition **ON** for 5 seconds and **OFF** for 10 seconds a few times without cranking the engine. Start the engine and check for leaks.
16. Perform a short road test and check the engine again for leaks. Be sure the cooling system is filled to the surge tank FULL COLD line.

L–Series With 2.2L Engine

AUTOMATIC TRANSAXLE

1. Before servicing the vehicle, refer to the precautions in the beginning of this section.
2. Properly relieve the fuel system pressure.
3. Drain the engine coolant.
4. Drain the engine oil.
5. Drain the power steering fluid.
6. Remove or disconnect the following:
- Both battery cables
- Battery
- Main wire feed from the fuse block
- Intake Air Temperature (IAT) sensor electrical connector
- Air cleaner and intake duct assembly
- Purge hose from the throttle body
- Purge solenoid electrical connector
- Rear Heated Oxygen Sensor (HO2S) electrical connector
- Main harness connector by the master cylinder
- Alternator electrical connector
- A/C pressure switch electrical connector
- Transaxle electrical connectors
- Right front Wheel Speed Sensor (WSS) electrical connector
- Cowl cover
- Powertrain Control Module (PCM) boot from the cowl
- PCM electrical connectors
- Engine harness connectors from the fuse block
- Battery tray and fuse block tray
- Cruise control and throttle cables
- Brake assist vacuum line from the throttle body
- Fuel feed and return
- Evaporative Emissions (EVAP) purge hose at the solenoid
- Starter motor
- Torque converter bolts
- Exhaust system from the catalytic converter to the exhaust manifold
- A/C compressor without removing the hoses

- Lower engine-to-transaxle bolts
- Heater hoses at the lower cowl
- Coolant reservoir hose
- Upper and lower radiator hoses
- Metal power steering line
7. Attach an engine lifting hoist to the lifting eyes.
- Right front engine mount while supporting the weight of the transaxle with a jack
- Engine mount bracket
- Upper transaxle-to-engine bolts
- Engine from the vehicle

To install:
8. Lower the engine into the vehicle and align it on the pins on the transaxle.
9. Install or connect the following:
- Upper transaxle bolts and hand-tighten them
- Engine mount bracket and torque the bolts to 66 ft. lbs. (90 Nm)
- Engine mount and torque the bolts to 41 ft. lbs. (55 Nm)
10. Remove the engine lifting device.
- Metal power steering line
- Upper and lower radiator hoses
- Coolant reservoir hose
- Lower transaxle bolts and torque them to 48 ft. lbs. (65 Nm)
- Torque converter bolts and torque them to 33 ft. lbs. (45 Nm)
- Starter motor and torque the bolts to 37 ft. lbs. (50 Nm)
- Heater hoses
- A/C compressor and torque the bolts to 18 ft. lbs. (25 Nm)
- Accessory drive belt
- Exhaust from the catalytic converter to the exhaust manifold and torque the bolts to 12 ft. lbs. (16 Nm)
11. Torque the upper bell housing bolts to 48 ft. lbs. (65 Nm).
- EVAP purge solenoid electrical connector and hose
- Fuel feed and return lines
- Brake booster vacuum hose
- Cruise control and throttle cables
- Fuse block and battery trays
- Engine harness to the fuse block
- Main engine harness under the fuse block panel
- PCM electrical connector
- PCM boot to the cowl
- Cowl cover
- Main harness connector near the master cylinder
- Rear HO2S electrical connector
- Purge solenoid electrical connector
- Purge hose to the throttle body
- Air cleaner and intake duct assembly
- IAT sensor electrical connector

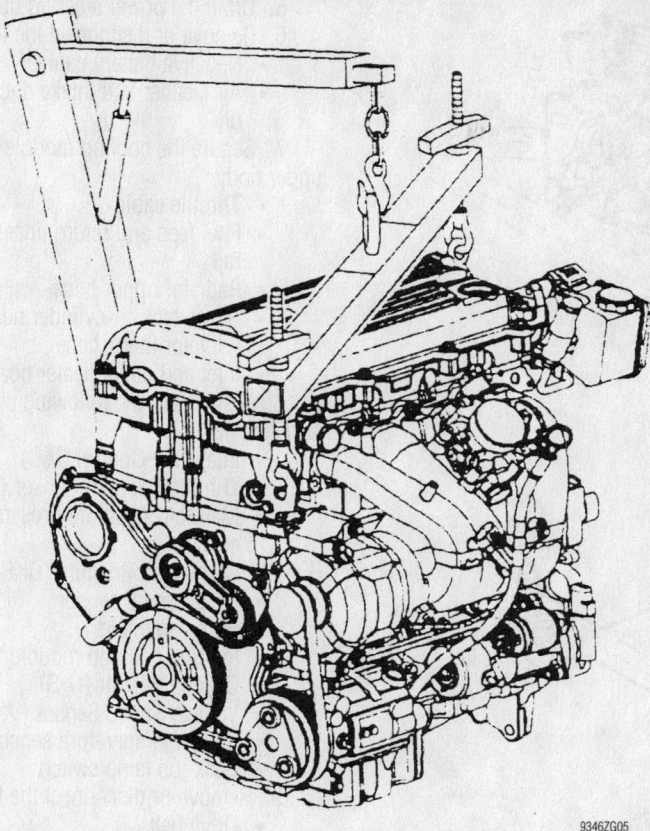

9346ZG05

Remove the engine with the lifting hoist—L–Series With 2.2L engine

- Ground wire to the left fender well
- Main wire feed to the fuse block
- Battery and both cables

12. Fill the engine with coolant.

13. Fill the engine with new oil.

14. Fill the power steering fluid.

15. Prime the fuel system by cycling the ignition **ON** for 5 seconds and **OFF** for 10 seconds a few times without cranking the engine.

16. Start the engine, check for leaks, and repair if necessary.

MANUAL TRANSAXLE

1. Before servicing the vehicle, refer to the precautions in the beginning of this section.

2. Properly relieve the fuel system pressure.

3. Drain the engine coolant.

4. Drain the engine oil.

5. Drain the power steering fluid.

6. Remove or disconnect the following:
- Both battery cables
- Steering gear-to-intermediate shaft pinch bolt
- Main wire feed to the fuse block
- Ground wire on the left fender well
- Intake Air Temperature (IAT) sensor electrical connector
- Air cleaner and intake duct assembly
- Purge hose from the throttle body
- Purge solenoid electrical connector
- Rear Heated Oxygen Sensor (HO$_2$S) electrical connector
- Back-up switch electrical connector
- Front dash cover
- Powertrain Control Module (PCM) boot from the front of the dash
- Main harness connector by the master cylinder
- PCM electrical connectors
- Engine harness connectors from the fuse block
- Battery, tray and fuse block tray
- Cruise control and throttle cables
- Brake booster vacuum hose
- Control assembly rod-to-control shaft lever pinch bolt
- Retaining clips connecting the control shaft lever to the transaxle
- Control shaft lever assembly
- Heater hoses at the lower cowl
- Upper and lower radiator hoses
- Fuel feed and return lines

7. Secure the radiator to the upper radiator support.
- Both front wheels
- Right front splash shield
- Left front wheel liner push pin from the frame

8. Install wood blocks between the transaxle case and frame and the crank pulley and frame.
- Right side engine mount
- Left side transaxle mount
- Exhaust system from the catalytic converter to the exhaust manifold
- A/C compressor without removing the hoses
- Tie rods from the steering knuckles
- Stabilizer bar links from the struts
- Lower ball joints from the steering knuckles
- Suspension support assemblies
- Suspension support cage nuts from the body

9. Properly support the powertrain and frame assembly.
- Frame-to-body attachment fasteners
- Powertrain/frame assembly from the vehicle

To install:

10. Raise the powertrain into the vehicle.

11. Install or connect the following:
- Powertrain/frame assembly
- New frame-to-body bolts do not tighten at this time
- Suspension supports with new bolts and torque the suspension support and frame bolts to 74 ft. lbs. (100 Nm) plus 45 degrees
- Ball joint to the steering knuckle and torque the nuts to 74 ft. lbs. (100 Nm)
- Stabilizer links to the struts and torque the fasteners to 50 ft. lbs. (65 Nm)
- Tie rods to the steering knuckles and torque the new nuts to 45 ft. lbs. (65 Nm)
- Exhaust system to the exhaust manifold and torque the bolts to 12 ft. lbs. (16 Nm)
- A/C compressor and torque the bolts to 18 ft. lbs. (25 Nm)
- Rear HO$_2$S electrical connector
- Right side engine mount and torque the bolts to 41 ft. lbs. (55 Nm)
- Left side transaxle mount and torque the bolts to 41 ft. lbs. (55 Nm)

12. Remove the wood blocks.
- Left front wheel liner push pins
- Right front splash shield
- Both front wheels
- Fuel feed and return lines
- Upper and lower radiator hoses
- Heater hoses

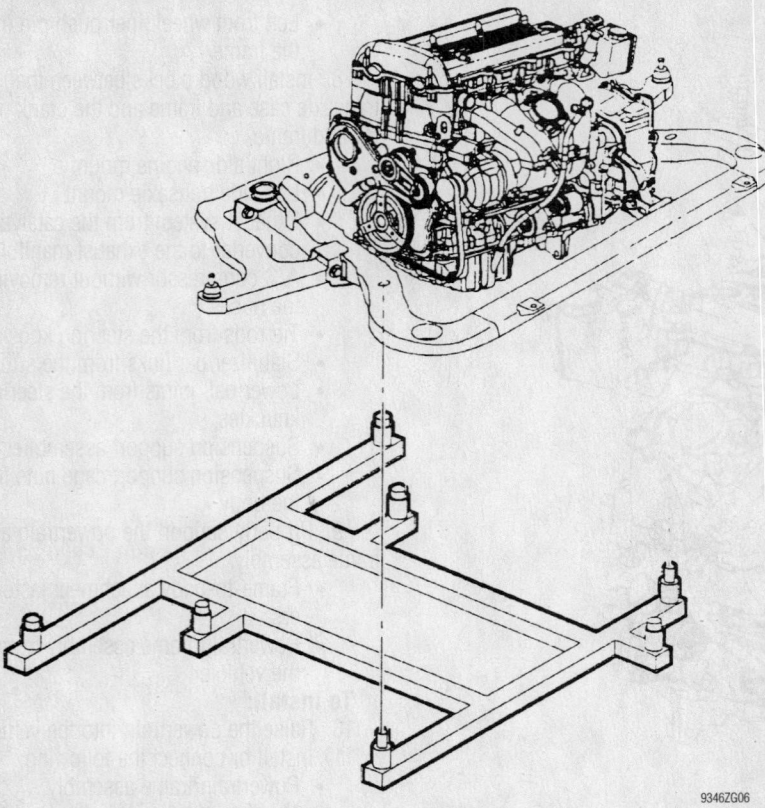

Powertrain/frame assembly with lifting table—L–Series With 2.2L engine

9346ZG06

- Brake booster vacuum hose
- Shift control rod

13. Adjust the shift control rod as follows:

a. Install the shift rod into the shift linkage assembly.

b. Rotate the transaxle shift control shaft clockwise and depress the spring loaded transaxle shift linkage lock pin on the case.

c. Pull up the shift lever boot.

d. Move the shift lever to the 5 (reverse gate) and use a ⅜ inch punch to hold it in place.

e. Tighten the pinch bolt to 9 ft. lbs. (12 Nm) plus 180 degrees

14. Install or connect the following:

- Cruise control and throttle cables
- Fuse block and battery trays
- Fuse block
- Engine harness connector to the bottom of the fuse block
- Fuse block
- PCM electrical connector
- PCM boot to the cowl
- Cowl cover
- EVAP purge solenoid electrical connector and hose
- Main harness connector near the master cylinder

- Transaxle back-up switch electrical connector
- Purge hose to the throttle body
- Air cleaner and intake duct assembly
- IAT sensor electrical connector
- Ground wire to the left fender well
- Main wire feed to the fuse block
- Battery and both cables
- Steering gear-to-intermediate shaft and torque the pinch bolt to 22 ft. lbs. (30 Nm)

15. Fill the engine with coolant.

16. Fill the engine with new oil.

17. Fill the power steering fluid.

18. Prime the fuel system by cycling the ignition **ON** for 5 seconds and **OFF** for 10 seconds a few times without cranking the engine.

19. Start the engine, check for leaks, and repair if necessary.

ION With 2.2L Engine

1. Before servicing the vehicle, refer to the precautions in the beginning of this section.

2. Properly relieve the fuel system pressure.

3. Drain the engine coolant.

4. Drain the engine oil.

5. Drain the power steering fluid.

6. Remove or disconnect the following:
- Negative battery cable
- Air cleaner and intake duct assembly

7. Secure the cooling module to the upper body.
- Throttle cable
- Fuel feed and return lines from the rail
- Radiator upper hose
- Surge tank-to-cylinder head hose
- Radiator lower hose
- Inlet and outlet heater hoses

8. Disconnect the following electrical connections:
- Intake Air Control (IAC)
- Throttle Position Sensor (TPS)
- Manifold Absolute Pressure (MAP) sensor
- Crankshaft Position (CKP) sensor
- Oil pressure sensor
- Purge solenoid
- Ignition coil and module
- Oxygen Sensor (O_2S)
- Vehicle Speed Sensor (VSS)
- Engine temperature sensor
- Back–up lamp switch

9. Remove or disconnect the following:
- Drive belt
- A/C compressor bolts and set the compressor aside without disconnecting the lines
- Starter and alternator wiring
- Front exhaust pipe from the manifold
- Transmission harness connectors
- Transmission shift cable

10. Use suitable blocks of wood to support the powertrain assembly between the assembly and the frame.

11. Support the engine with a jack using a piece of wood between the jack and the engine to avoid damage.
- Engine mount, nuts and bolts.

12. Remove the side transmission mount on models with a manual transmission as follows:

a. Remove the under hood electrical tray.

b. Disconnect the wiring harness from the tray bracket.

c. Lift the electrical center up and swing it back out of the way.

d. Remove the mount-to-transmission bolts and mount-to-rail bolts.

e. Remove the mount.

13. Remove the side transmission mount on models with a automatic 5-speed transmission as follows:

a. Support the transmission with a jack.

b. Remove the front and rear transmission mount through bolts.

c. Remove the under hood electrical center cover.

d. Disconnect the engine control module harness connector.

e. Disconnect the positive battery cable from the under hood electrical center.

f. Make sure the surge tank hose is disconnected.

g. Remove the electrical center tray.

h. Disconnect the harness from the tray bracket.

i. Lift the electrical center up and swing it back out of the way.

j. Remove the mount-to-transmission bolts and mount-to-rail bolts.

k. Remove the mount.

14. Remove the side transmission mount on models with a automatic 5-speed continuously variable transmission as follows:

a. Remove the engine cradle.

b. Remove the under hood electrical center cover.

c. Disconnect the engine control module harness connector.

d. Disconnect the positive battery cable from the under hood electrical center.

e. Make sure the surge tank hose is disconnected.

f. Remove the electrical center tray.

g. Disconnect the harness from the tray bracket.

h. Lift the electrical center up and swing it back out of the way.

i. Install an engine support fixture. If not already removed, remove the engine mount.

j. Disconnect the upper transmission cooler line from the radiator.

k. Remove the mount-to-transmission bolts

➡**Make sure the A/C clutch does not contact the inner metal rail of the vehicle.**

l. Lower and tilt the powertrain assembly down approximately 4 inches.

m. Remove the mount-to-rail bolts.

n. Remove the mount.

15. Remove or disconnect the following:
- Stabilizer link from the stabilizer bar
- Outer tie rod ends from knuckles

✳✳ CAUTION

To prevent air bag system deployment, do not attempt to rotate the steering shaft.

- Intermediate shaft from the steering gear
- Lower control arms from the knuckles
- Half shafts from the knuckles

16. Remove the front frame as follows:

a. Mark the frame-to-body position.

b. Lower the vehicle until it is 3 feet off the ground so that a lift table may be placed in position. As necessary, use blocks of wood between the frame and table to support the assembly.

c. Slowly remove the front frame bolts, the partially unscrew the bolts until 1 ½ of the bolt shank is exposed.

d. Slowly lower the table to the floor

17. Attach the engine hooks to a hoist.

18. Remove or disconnect the following:
- Starter
- Torque converter-to-flywheel bolts, if equipped
- Engine-to-transmission bolts and the engine from the transmission.

To install:

19. Install or connect the following:
- Engine to the transmission and tighten the bolts to 55 ft. lbs. (75 Nm)
- Torque converter-to-flywheel bolts, if equipped and tighten the bolts to 46 ft. lbs. (62 Nm)
- Starter

20. Install the front frame as follows:

a. Slowly raise the table to the into position and align the frame to the marks made during removal.

b. Hand tighten the frame bolts

c. Tighten the frame bolts to 74 ft. lbs. (100 Nm) plus an additional torque of 180 degrees using a torque angle meter.

d. Remove the lift table.

21. Install or connect the following:
- Half shafts to the knuckles
- Lower control arms to the knuckles

✳✳ CAUTION

To prevent air bag system deployment, do not attempt to rotate the steering shaft.

- Intermediate shaft to the steering gear
- Outer tie rod ends to knuckles
- Stabilizer link to the stabilizer bar

22. Install the side transmission mount on models with a manual transmission as follows:

a. Place the mount in position and tighten the mount-to-rail bolts to 25 ft. lbs. (34 Nm).

23. Use a jack to align the mount when it is in position.

a. Install and hand tighten the mount-to-transmission bolts. Do not pry the transmission or mount to align the holes. Hand start the bolts in the following sequence; rear bolt, middle and then front bolt. Using the same sequence, tighten the bolts to 33 ft. lbs. (45 Nm).

b. Place the electrical center in position.

c. Connect the wiring harness to the tray bracket.

d. Install the under hood electrical tray.

24. Install the side transmission mount on models with a automatic 5-speed transmission as follows:

a. Place the mount in position and tighten the mount-to-rail bolts to 25 ft. lbs. (34 Nm).

25. Use a jack to align the mount when it is in position.

a. Install and hand tighten the mount-to-transmission bolts. Do not pry the transmission or mount to align the holes. Hand start the bolts in the following sequence; rear bolt, middle and then front bolt. Using the same sequence, tighten the bolts to 33 ft. lbs. (45 Nm).

b. Place the electrical center in position.

c. Connect the wiring harness to the tray bracket.

d. Install the under hood electrical tray.

e. Connect the positive battery cable to the under hood electrical center.

f. Connect the engine control module harness connector.

g. Install the under hood electrical center cover.

h. Install the front and rear transmission mount through bolts. Tighten to 74 ft. lbs. (100 Nm).

26. Install the side transmission mount on models with a automatic 5-speed continuously variable transmission as follows:

a. Place the mount in position and tighten the mount-to-rail bolts to 25 ft. lbs. (34 Nm).

27. Use a jack raise the engine back into position and to align the mount when it is in position.

a. Install and hand tighten the mount-to-transmission bolts. Do not pry the transmission or mount to align the holes. Hand start the bolts in the following sequence; rear bolt, middle and then front bolt. Using the same sequence, tighten the bolts to 33 ft. lbs. (45 Nm).

b. If installed, install the engine

mount and remove the engine support fixture.

   c. Connect the upper transmission cooler line to the radiator.

   d. Place the electrical center in position.

   e. Connect the wiring harness to the tray bracket.

   f. Install the under hood electrical tray.

   g. Connect the positive battery cable to the under hood electrical center.

   h. Connect the engine control module harness connector.

   i. Install the under hood electrical center cover.

   j. Install the front and rear transmission mount through bolts. Tighten to 74 ft. lbs. (100 Nm).

➡ **The engine mount bolts must be hand started. Do not pry on the engine mount to align the holes.**

28. Install or connect the following:
- Engine mount nuts and bolts. Tighten the nuts to 74 ft. lbs. (100 Nm) and the bolts to 37 ft. lbs. (50 Nm).

29. Remove the blocks of wood used to support the powertrain assembly between the assembly and the frame.
- Transmission shift cable
- Transmission harness connectors
- Front exhaust pipe to the manifold
- Starter and alternator wiring
- A/C compressor
- Drive belt

30. Connect the following electrical connections:
- Back–up lamp switch
- Engine temperature sensor
- VSS
- O_2S
- Ignition coil and module
- Purge solenoid
- Oil pressure sensor
- CKP
- MAP sensor
- TPS
- IAC

31. Install or connect the following:
- Inlet and outlet heater hoses
- Radiator lower hose
- Surge tank-to-cylinder head hose
- Radiator upper hose
- Fuel feed and return lines to the rail
- Throttle cable
- Cooling module
- Air cleaner and intake duct assembly
- Negative battery cable

32. Fill the engine with coolant.

33. Fill the engine with new oil.

34. Prime the fuel system by cycling the ignition **ON** for 5 seconds and **OFF** for 10 seconds a few times without cranking the engine.

35. Start the engine, check for leaks, and repair if necessary.

3.0L Engine

1. Before servicing the vehicle, refer to the precautions in the beginning of this section.

2. Properly relieve the fuel system pressure.

3. Drain the engine coolant.

4. Drain the engine oil.

5. Drain the power steering fluid.

6. Remove or disconnect the following:
- Both battery cables
- Air cleaner assembly
- Mass Air Flow (MAF) sensor electrical connector
- Battery ground cable from the fuse block
- Transaxle Control Module (TCM) main connector from under the cowl cover
- TCM inline connector by the brake master cylinder
- A/C pressure connector
- Black engine harness connector from under the fuse block panel
- Lower weather pack connector from inside the fuse block and secure the engine harness to the engine
- Fuse block
- Battery tray
- Evaporative Emissions (EVAP) purge connector
- Right front speed sensor connector
- Front Oxygen Sensor (O_2S) from the down pipe
- Transaxle ground electrical connector
- Transaxle main electrical connector
- Transaxle shift control electrical connector
- Rear connector on the Engine Control Module (ECM)
- Brake booster vacuum hose
- Fuel lines
- EVAP purge hose and solenoid
- Starter motor
- Torque converter bolts
- Exhaust system from the catalytic converter to the exhaust manifold
- Transaxle nose bracket
- A/C compressor without removing the hoses
- Heater hoses at the lower cowl
- Lower engine-to-transaxle bolts

- Upper and lower radiator hoses
- Coolant reservoir hose
- Metal power steering line
- Power steering reservoir from the radiator support

7. Attach an engine lifting device to the engine lifting eyes.
- Right front engine mount and bracket
- Upper transaxle bolts
- Engine from the vehicle

To install:

8. Lower the engine into the vehicle and align it on the pins on the transaxle.

9. Install or connect the following:
- Upper transaxle bolts and hand-tighten them
- Engine mount bracket and torque the bolt to 30 ft. lbs. (40 Nm)
- Engine mount and torque the upper bolt to 30 ft. lbs. (40 Nm) and the lower bolt to 41 ft. lbs. (55 Nm)
- Transaxle nose bracket and torque the bolts to 30 ft. lbs. (40 Nm)

10. Remove the engine lifting device.
- Metal power steering line
- Power steering reservoir to the radiator support
- Upper and lower radiator hoses
- Coolant reservoir hose
- Lower transaxle bolts and torque them to 48 ft. lbs. (65 Nm)
- Torque converter and torque them to 48 ft. lbs. (65 Nm)
- Starter motor and torque the bolts to 30 ft. lbs. (40 Nm)
- Heater hoses at the lower cowl
- A/C compressor and torque the bolts to 30 ft. lbs. (40 Nm)
- Exhaust system to the exhaust manifold and torque the nuts to 15 ft. lbs. (20 Nm)
- EVAP purge solenoid and hose
- Fuel feed and return lines
- Brake booster vacuum hose
- ECM rear connector
- Transaxle shift control electrical connector
- Transaxle ground connector
- Transaxle main connector
- Right front speed sensor electrical connector
- Front O_2S electrical connector
- Right front speed sensor
- EVAP purge solenoid electrical connector

11. Torque the upper bell housing bolts to 48 ft. lbs. (65 Nm).
- Battery tray
- Fuse block and route the engine harness to the block
- Lower weather pack connector

- Main wire feed to the fuse block
- A/C pressure connector
- TCM inline connector
- TCM main connector under the cowl cover and secure it to the engine
- Positive main feed cable at the fuse block
- Battery ground cable at the wheel housing
- Battery and both cables
- Air cleaner and intake duct assembly
- MAF sensor electrical connector

12. Fill the engine with coolant.

13. Fill the engine with new oil.

14. Fill the power steering fluid.

15. Prime the fuel system by cycling the ignition **ON** for 5 seconds and **OFF** for 10 seconds a few times without cranking the engine.

16. Start the engine, check for leaks, and repair if necessary.

Heater Core

REMOVAL & INSTALLATION

S–Series

1. Before servicing the vehicle, refer to the precautions in the beginning of this section.

2. Disconnect the negative battery cable.

3. Drain the cooling system into a clean container for reuse.

4. Raise and safely support the vehicle.

5. Heater hoses from the heater core

➡**Carefully, use compressed air to blow the remainder of the coolant from the heater core to prevent spillage of coolant in the interior when removing the heater core.**

6. Remove or disconnect the following
- Lower trim panel extensions (located at both sides of the console-to-center instrument panel), by pulling outward at the dual lock locations; then, rotate the panels outward to disengage the hinges at the console
- Lower instrument panel closeout panel (located on the right side of the heater/air conditioning housing), by pulling it out at the top edge and rotating the top downward
- Lower heater duct-to-heater/air conditioning housing 2 screws and

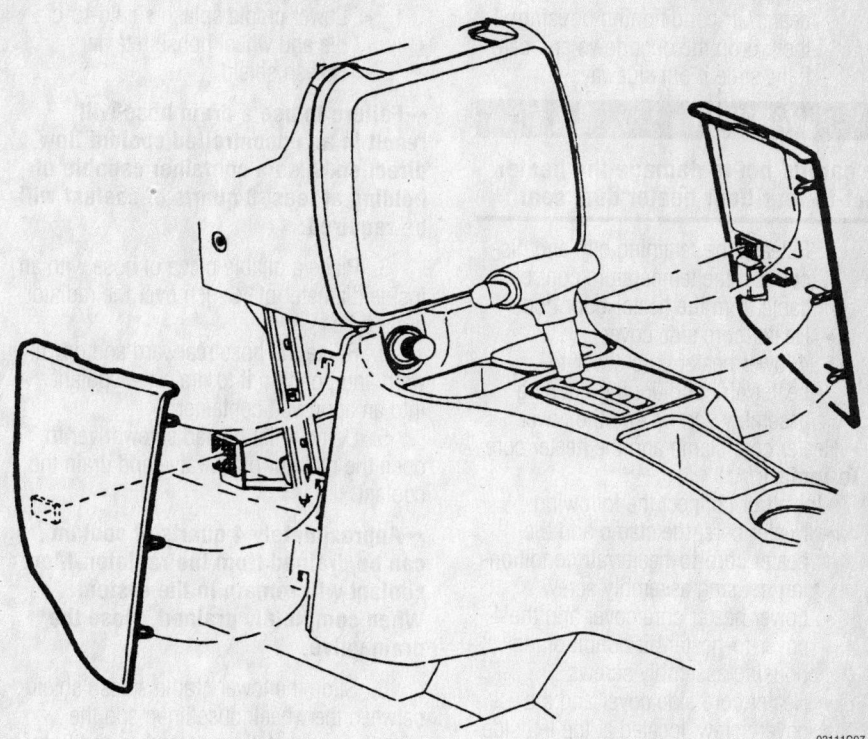

View of the lower trim panel extensions—S–Series

93111G97

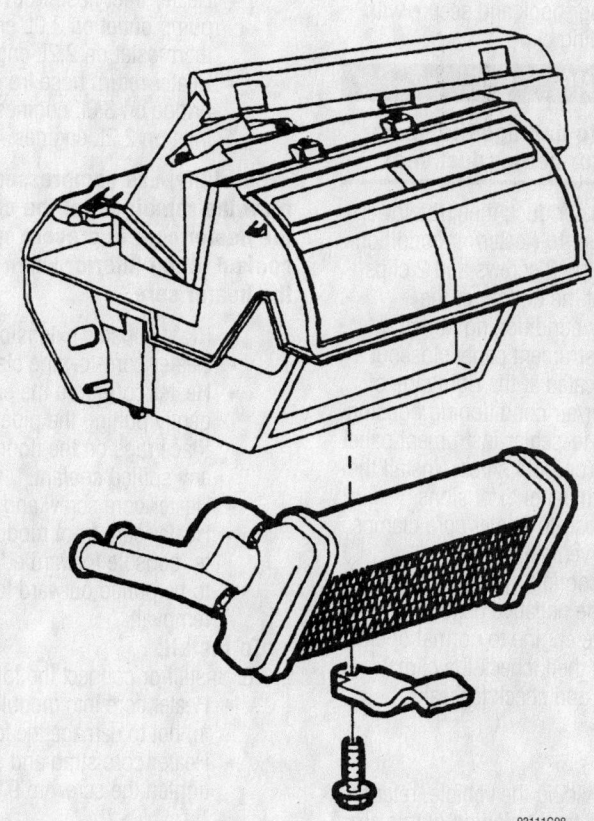

93111G98

Exploded view of the heater core and heater case—S–Series

2 clips (located at the bottom of the heater/air conditioning housing); then, drop the duct down and carefully, slide it out sideways

❋❋ WARNING

Be careful not to damage the heater duct-to-rear floor heater duct seal.

- Release the retaining clip and disconnect the temperature control cable from the heater door hook
- Heater core side cover
- 4 lower heater core cover-to-heater/air conditioning housing assembly screws and the cover
- Heater core clamp and the heater core

To install:

7. Install or connect the following:
- Heater core, the clamp and the heater core-to-heater/air conditioning housing assembly screw
- Lower heater core cover and the 4 cover-to-heater/air conditioning housing assembly screws
- Heater core side cover and the cover screw, located at the left side of the heater/air conditioning housing assembly
- Temperature control cable to the heater door hook and secure with the retaining clip

❋❋ WARNING

Be careful not to damage the heater duct-to-rear floor heater duct seal

- Lower heater duct; then, the lower heater duct-to-heater/air conditioning housing 2 screws and 2 clips, located at the bottom of the heater/air conditioning housing
- Lower instrument panel closeout panel, located at the right side of the heater/air conditioning housing
- Console-to-center instrument panel location (on both sides), install the lower trim panel extensions
- Heater hose-to-heater core clamps

8. Lower the vehicle.
9. Refill the cooling system.
10. Connect the negative battery cable.
11. Operate the engine to normal operating temperatures; then, check the climate control operation and check for leaks.

L–Series

1. Before servicing the vehicle, refer to the precautions in the beginning of this section.
2. Remove or disconnect the following:

- Negative battery cable
- Lower cradle splash shield-to-cradle and wheel house retainers
- Splash shield

➡**Failure to use a drain hose will result in an uncontrolled coolant flow direction. Also a container capable of holding at least 8 quarts of coolant will be required.**

3. Place a pliable piece of hose with an inside diameter of ⅜ inch over the radiator drain tube.
4. Route the hose rearward and downward and position it to drain the coolant into an approved container.
5. Using a flat bladed screwdriver to open the radiator drain valve and drain the coolant.

➡**Approximately 4 quarts of coolant can be drained from the radiator. More coolant will remain in the system. When completely drained, close the drain valve.**

6. Slide the lower cradle splash shield between the wheelhouse linen and the wheelhouse. Install the fasteners and tighten to 44 inch lbs. (5 Nm).
7. Remove or disconnect the following:
- Heater inlet hose from the water pump outlet on 3.0L engines or thermostat on 2.2L engines
- Heater return hose from head bridge on 3.0L engines or block inlet on 2.2L engines

➡**Carefully, use compressed air to blow the remainder of the coolant from the heater core to prevent spillage of coolant in the interior when removing the heater core.**

- Right console extension
- Heater core-to-pipe clamps
- Heater core from the end tank by gently pulling the pipes forward. Place rags on the floor to absorb any spilled coolant.
- Heater core screw and strap
- Heater core from module. The center console forward edge may need to be pulled outward to allow core removal.

To install:

8. Install or connect the following:
- Heater core into module, being careful not to damage the foam seals
- Heater core strap and screw and tighten the screw to 9 inch lbs. (1 Nm)
- New bushings and O–rings onto the core pipes

- Heater pipes to the core and install the plastic clamps. The clamps can only be installed one way, the lower clamp hinge fits into the tab of the core retaining strap
- Right console extension
- Heater inlet hose to the water pump outlet on 3.0L engines or thermostat on 2.2L engines
- Heater return hose to head bridge on 3.0L engines or block inlet on 2.2L engines

9. Rinse off any residual coolant.
10. Refill the cooling system.
11. Connect the negative battery cable.
12. Operate the engine to normal operating temperatures; then, check the climate control operation and check for leaks.

ION

1. Before servicing the vehicle, refer to the precautions in the beginning of this section.
2. Remove or disconnect the following:
- Surge tank cap
- Water pump drain bolt and drain the coolant from the pump. Once drained, close the bolt and torque to 88 inch lbs. (10 Nm).
- Heater outlet and inlet hose clamps
- Heater outlet and inlet hoses

3. Remove the Body Control Module (BCM) as follows:
   a. Disconnect the negative battery cable.
   b. Apply the parking brake and position the transmission into Neutral.
   c. Remove the console shift lever bezel by lifting up around the edge carefully, if equipped with an automatic transmission.
   d. Unsnap the shift boot from the console cup holder, if equipped with a manual transmission.
   e. Lift the front console cup holder to disengage the retainers
   f. Disconnect the cigar lighter connection.
   g. Remove the cup holder, if equipped with a manual transmission, slide the boot through the cup holder.
   h. Remove the console extension retainers from both sides by turning them counterclockwise.
   i. Remove the console extensions by pulling the extension rearwards.
   j. Remove the center extension screws, pull out the center extension to access any remaining fasteners and remove the extension.
   k. Unsnap the parking brake boot.

l. If equipped with an armrest, remove the console compartment screws.

m. Lift up on the rear on the console to release the retainers.

n. Lift the console compartment and push the parking brake through the opening.

o. Slide the compartment over the parking brake lever.

p. Disconnect the rear power supply connector.

q. Remove the console screws and the console.

r. Remove the wiring harness rosebud from the right center support bracket.

s. Pull back the carpet from the right of the left Instrument Panel (IP) support bracket and remove the lower nuts.

t. Remove the center support bracket.

u. Disconnect the small then large harness connectors from the BCM.

v. Disconnect the small then large IP wiring harness connectors from the BCM.

w. Disconnect the Onstar connector, if equipped.

x. Remove the BCM retainers and BCM.

4. Remove or disconnect the following:
- Pull back the carpet at the bottom of the left IP support bracket
- Left IP support bracket nuts
- Accelerator pedal from the front of the dash and position aside
- Center floor outlet ducts by raising the duct and then pushing them to disengage them
- Rotate the center floor duct and pull down to disconnect it from the HVAC module
- Heater core cover screws
- Heater core cover down enough to clear the locating pins from the HVAC module, slide the cover rearwards until the drain tube clears the front of the dash. Slide the cover down, rearwards and to the right and remove it.
- Heater core

To install:

5. Inspect the heater core foam and if damaged use Kent Industries adhesive black foam 46480 to replace.

6. Install or connect the following:
- Heater core
- New drain tube seal on the drain tube

7. Spray the heater core seal and dashmat with soapy water to ease installation.
- Heater core cover from the right side. Slide the cover up, forward

and into position. Align the drain tube with the hole in the dash, raise the cover into position while aligning the holes with the locating pins on the HVAC module. Tighten the cover screws to 15 inch lbs. (1.8 Nm).
- Center floor ducts.
- Accelerator pedal
- Carpet at the bottom of the left IP support bracket
- Left IP center bracket nuts and tighten to 88 inch lbs. (10 Nm)

8. Install the BCM as follows:

a. Install the BCM and tighten the retainers to 88 inch lbs. (10 Nm).

b. Connect the small then large IP wiring harness connectors to the BCM.

c. Connect the small then large harness connectors to the BCM.

d. Connect the Onstar connector, if equipped.

e. Pull back the carpet from the right of the left IP support bracket and the center support bracket. Tighten the nuts to 88 inch lbs. (10 Nm).

f. Connect the wiring harness rosebud to the right center support bracket.
- Install the front floor console over parking lever and shift lever. Push the console down on the carpet and tighten the screws to 22 inch lbs. (2.5 Nm).

g. Connect the rear power supply connector.

h. Slide the compartment over the parking brake lever.

i. Pull the parking brake boot through the opening.

j. Align the retainers with the console and push to engage.

k. If equipped with an armrest, install the console compartment screws and tighten the screws to 22 inch lbs. (2.5 Nm).

l. Snap the parking brake boot to the console.

m. Align the center extension to the IP and push the engage the retainers. Tighten the screws to 22 inch lbs. (2.5 Nm).

n. Install the cup holder, if equipped with a manual transmission, slide the boot through the cup holder.

o. Connect the cigar lighter connection.

p. Align the front console cup holder retainers and push to engage.

q. Snap the shift boot to the console cup holder, if equipped with a manual transmission.

r. Install the console shift lever bezel

by aligning the fingers on the lever bezel with the slots on the lever base and push to engage, if equipped with an automatic transmission.

s. Apply the parking brake and position the transmission into park.

t. Install the console extensions. Turn the retainers clockwise to secure.

9. Install or connect the following:
- Heater outlet and inlet hose clamps
- Heater outlet and inlet hoses
- Surge tank cap

10. Refill the cooling system.

11. Connect the negative battery cable.

12. Operate the engine to normal operating temperatures; then, check the climate control operation and check for leaks.

Water Pump

REMOVAL & INSTALLATION

1.9L Engines

1. Before servicing the vehicle, refer to the precautions in the beginning of this section.

2. Drain the cooling system.

3. Remove or disconnect the following:
- Negative battery cable
- Accessory drive belt
- Right front wheel
- Inner wheel well splash shield
- Water pump pulley bolts and allow the pulley to hang freely on the water pump hub
- Water pump flange bolts
- Water pump

To install:

4. Install or connect the following:
- Water pump with a new gasket and torque the bolts in a crisscross sequence to 22 ft. lbs. (30 Nm)
- Water pump pulley to the hub and torque the bolts to 19 ft. lbs. (25 Nm)

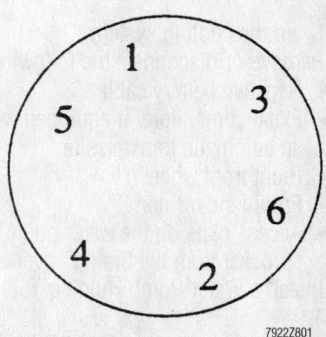

79222Z801

Water pump bolt torque sequence—1.9L engines

- Accessory drive belt
- Right front wheel well splash shield
- Right front wheel
- Negative battery cable

5. Fill the cooling system.

6. Start the vehicle and check for leaks, repair if necessary.

L–Series With 2.2L Engine

1. Before servicing the vehicle, refer to the precautions in the beginning of this section.

2. Drain the cooling system.

3. Remove or disconnect the following:
- Negative battery cable
- Air cleaner assembly
- Exhaust manifold heat shield
- Thermostat housing
- Water pump feed pipe
- Water pump sprocket access plate from the front cover

4. Install a Water Pump Holding Tool J43651.
- Water pump retaining bolts
- Water pump

To install:

5. Install or connect the following:
- Water pump and torque the bolts to 18 ft. lbs. (25 Nm)
- Water pump sprocket and torque the bolts to 89 inch lbs. (10 Nm)
- Water pump sprocket access plate
- Water feed tube after lubricating the O-ring
- Thermostat housing and torque the bolts to 89 inch lbs. (10 Nm)
- Exhaust manifold heat shield
- Air cleaner assembly
- Negative battery cable

6. Fill the cooling system.

7. Start the vehicle and check for leaks, repair if necessary.

ION With 2.2L Engine

1. Before servicing the vehicle, refer to the precautions in the beginning of this section.

2. Drain the cooling system.

3. Remove or disconnect the following:
- Negative battery cable
- Exhaust manifold, if equipped with an automatic transmission
- Right front wheel
- Front fender liner
- Access plate on the water pump sprocket from the timing cover

4. Install a Water Pump Holding Tool J43651.
- Water pump-to-block bolt
- Water pump-to-engine front cover bolts

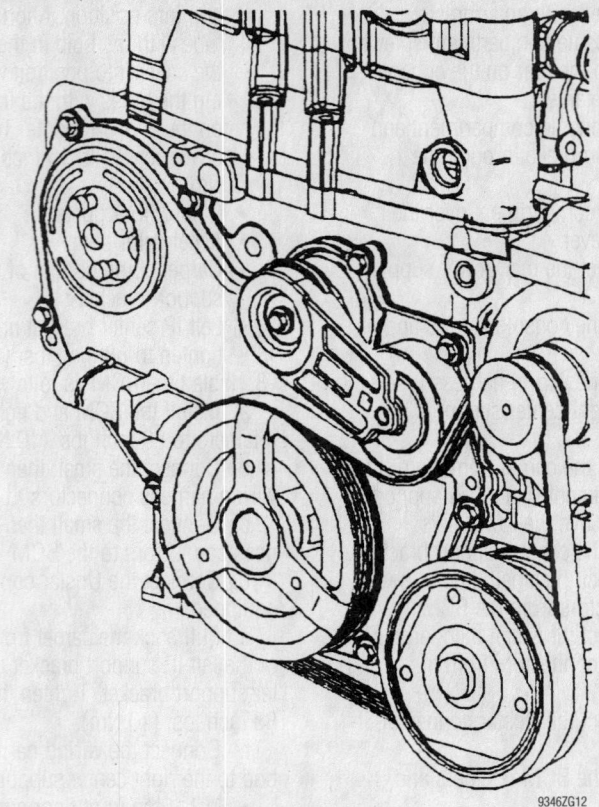

Water pump holding tool J43651—all 2.2L engines

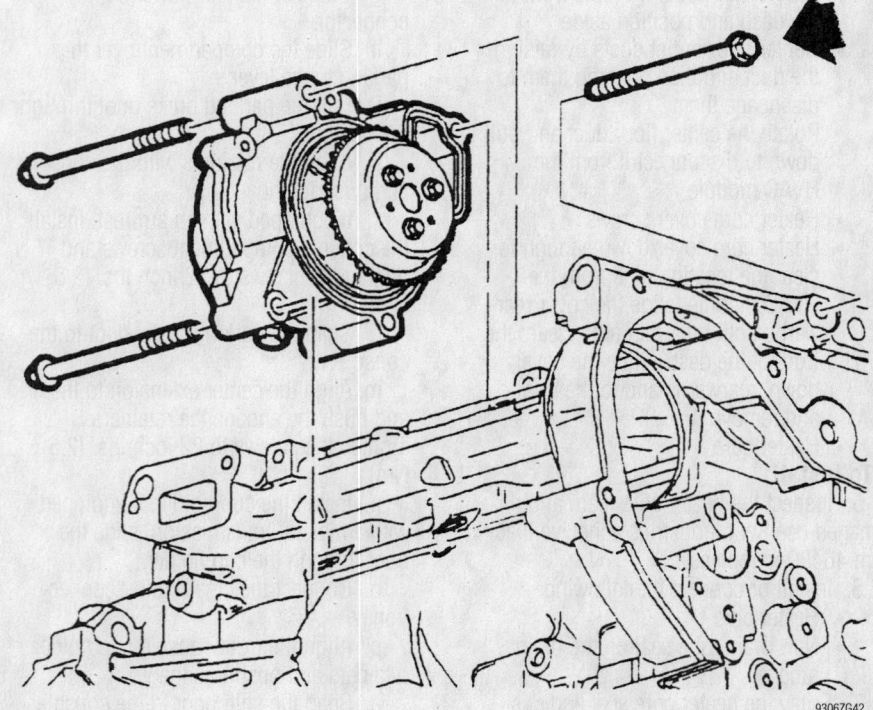

Exploded view of the water pump—all 2.2L engines

- Feed pipe from the thermostat to the water pump
- Water pump-to-block bolts
- Water pump

To install:

5. Install or connect the following:
- Water pump
- Water pump-to-block bolts and tighten to 15 ft. lbs. (20 Nm)
- Feed pipe from the thermostat to the water pump
- Water pump-to-engine front cover bolts and tighten to 15 ft. lbs. (20 Nm)
- Water pump-to-block bolt and tighten to 15 ft. lbs. (20 Nm)
- Water pump sprocket and torque the bolts to 89 inch lbs. (10 Nm)
- Water pump sprocket access plate
- Front fender liner
- Right front wheel
- Exhaust manifold, if equipped with an automatic transmission
- Negative battery cable
6. Fill the cooling system.
7. Start the vehicle and check for leaks, repair if necessary.

3.0L Engine

1. Before servicing the vehicle, refer to the precautions in the beginning of this section.
2. Drain the cooling system.
3. Remove or disconnect the following:
- Negative battery cable
- Air cleaner assembly
- Left front wheel
- Wheel well splash shield
4. Loosen the water pump pulley and power steering pulley bolts, but do not remove them.
5. Install an engine support fixture.
- Right front engine mount
- Accessory drive belt
6. Release the retaining tabs on the wiring harness channel and remove the front cover.
- Wiring harness from the channel
- Water pump pulley
- Power steering pump pulley
- Drive belt tensioner
- Front timing belt cover
- Water pump

To install:

7. Install or connect the following:
- Water pump with a new O-ring and torque the bolts to 18 ft. lbs. (25 Nm)
- Front timing belt cover and torque the bolts to 71 inch lbs. (8 Nm)

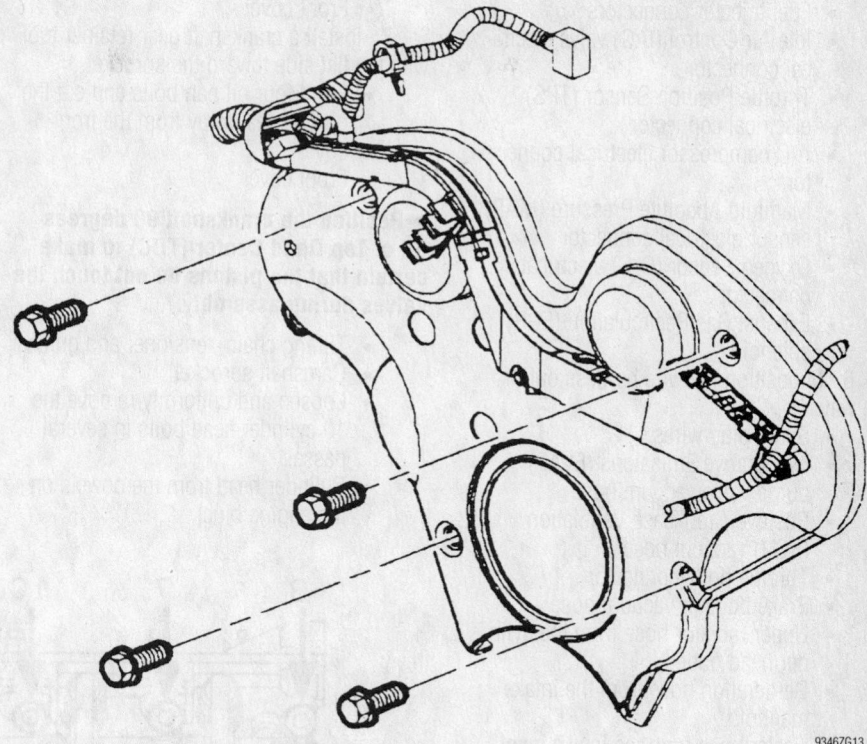

Timing belt cover bolt locations—3.0L engine

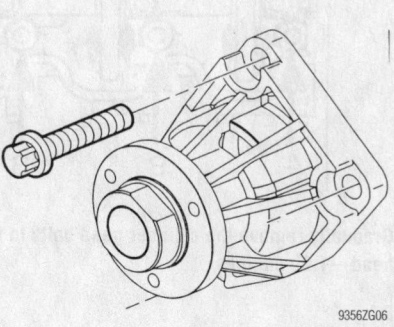

Water pump mounting—3.0L engine

- Drive belt tensioner and torque the bolts to 30 ft. lbs. (40 Nm)
- Water pump pulley's and torque the bolts to 71 inch lbs. (8 Nm)
- Power steering pump and torque the bolts to 15 ft. lbs. (20 Nm)
- Wiring harness into the channel
- Wiring harness channel front cover
- Accessory drive belt
8. Remove the engine support fixture.
- Wheel well splash shield and torque the bolts to 44 inch lbs. (5 Nm)
- Front wheel
- Air cleaner and intake duct
- Negative battery cable

9. Fill the cooling system.
10. Start the vehicle and check for leaks, repair if necessary.

Cylinder Head

REMOVAL & INSTALLATION

1.9L Engines

❋❋ WARNING

Only remove the cylinder head when the engine is cold. Warpage may result if the cylinder head is removed while the engine is hot.

1. Before servicing the vehicle, refer to the precautions in the beginning of this section.
2. Drain the cooling system.
3. Drain the engine oil.
4. Properly relieve the fuel system pressure.
5. Remove or disconnect the following:
- Negative battery cable
- Air cleaner assembly
- Rocker cover fresh air hose
- Accelerator cable from the throttle body and intake manifold bracket
- Coolant temperature gauge electrical connector

- Fuel injector connectors
- Idle Air Control (IAC) valve electrical connector
- Throttle Position Sensor (TPS) electrical connector
- A/C compressor electrical connector
- Manifold Absolute Pressure (MAP) sensor electrical connector
- Oxygen Sensor (O2S) electrical connector
- Exhaust Gas Recirculation (EGR) solenoid

6. Reposition the wire harness out of the way.

- Spark plug wires
- Evaporative Emissions (EVAP) purge valve vacuum hose
- Positive Crankcase Ventilation (PCV) vacuum hose
- Throttle body connector
- Brake booster vacuum hose
- Upper radiator hose from the cylinder head outlet
- De-aeration hose from the intake manifold
- Heater hose from the intake manifold
- Fuel supply line
- Engine torque axis mount
- Accessory drive belt
- Drive belt tensioner
- Drive belt idler pulley
- Rocker arm cover
- De-aeration line from the cylinder head water outlet
- A/C compressor and front bracket without removing the hoses
- Power steering pump without removing the lines
- Right front wheel
- Wheel well splash shield
- Crankshaft pulley
- Front exhaust pipe from the exhaust manifold

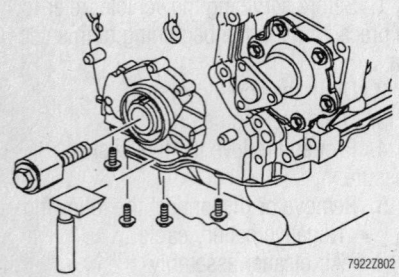

Crankshaft gear retaining and oil pan removal tool shown—be sure to install the crankshaft tool with the flat side toward the gear—1.9L engines

- Front cover

7. Install a crankshaft gear retainer tool with the flat side toward the sprocket.

- Front four oil pan bolts and cut the oil pan seal away from the front cover
- Front cover

➡ **Position the crankshaft 90 degrees off of Top Dead Center (TDC) to make certain that the pistons do not touch the valves during assembly.**

- Timing chain, tensioner and guides
- Camshaft sprocket
- Loosen and uniformly remove the 10 cylinder head bolts in several passes
- Cylinder head from the dowels on the engine block

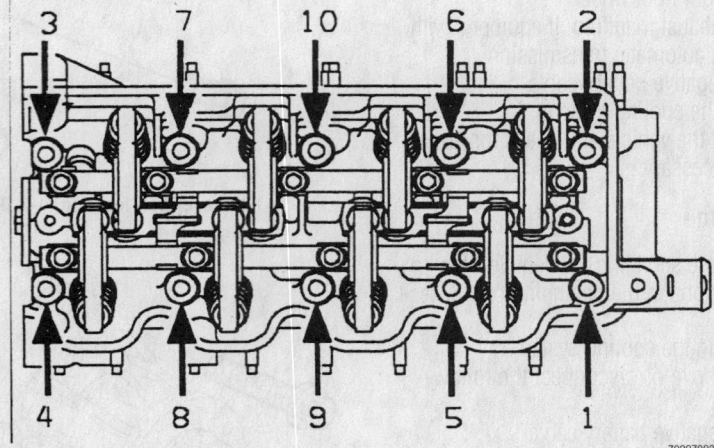

Gradually remove the cylinder head bolts in the sequence shown to prevent warping the head—1.9L engines

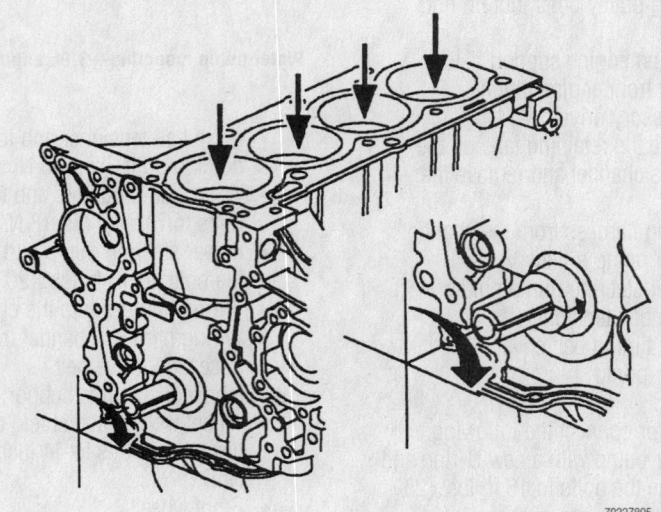

Rotate the crankshaft clockwise to 90 degrees past TDC (the crankshaft sprocket timing mark will be at 3 o'clock) to prevent valve damage during assembly—1.9L engines

To install:

➡ **Before installing the cylinder head, it should be cleaned and inspected for excessive wear or damage.**

8. Clean the gasket mating surfaces. Be careful not to damage the aluminum components and be sure the block bolt holes are clean of any residual sealer, oil or foreign matter.

9. Using a dial gauge, check that the cylinder liners are flush or do not deviate more than 0.0005 in. (0.013mm).

10. Be sure the crankshaft is still 90 degrees past TDC and that the camshaft(s) are properly positioned with the dowel pin(s) at the 12 o'clock position to prevent valve damage.

11. Install the cylinder head gasket and

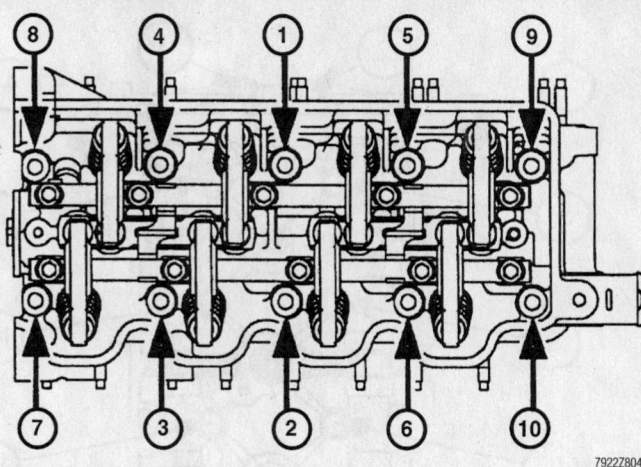

Cylinder head bolt tightening sequence—1.9L engine

carefully guide the head into place over the dowels.

12. If the head bolts and/or the block were replaced, install the bolts and tighten in sequence to 48 ft. lbs. (65 Nm) to insure proper clamp load; then, remove the bolts.

13. Coat the cylinder head bolts with clean engine oil and thread the bolts by hand until finger-tight.

14. Install the cylinder head bolts, in sequence, as follows:

    a. Tighten the bolts to 22 ft. lbs. (30 Nm).

    b. Tighten the bolts to 33 ft. lbs. (45 Nm).

    c. Tighten each bolt an additional 90 degree turn.

15. Install the timing chain, sprockets, guides and tensioners, using the following procedure :

    a. Verify that the crankshaft is positioned 90 degrees clockwise past TDC. The crankshaft key way sprocket timing mark must be aligned with the cylinder block main bearing cap split line to prevent piston and valve damage.

    b. If required, rotate the camshaft up to No. 1 TDC position.

    c. Rotate the crankshaft 90 degrees counterclockwise to No. 1 TDC position. The crankshaft sprocket timing mark must align with the cylinder block timing mark.

16. Install or connect the following:

- Timing chain over the camshaft sprocket and under the crankshaft sprocket. Slide the camshaft sprocket onto the camshaft with the letters **FRT** facing away from the cylinder head
- Camshaft sprocket timing pin
- Camshaft washer and bolt and torque the bolt to 74 ft. lbs. (100 Nm)

- Fixed chain guide and torque the fastener to 19 ft. lbs. (26 Nm). The timing chain should be snug against the fixed guide.
- Pivoting chain guide making certain there is clearance between the block and head and torque the bolts to 19 ft. lbs. (26 Nm)
- Timing chain tensioner and torque the bolts to 168 inch lbs. (19 Nm)
- Crankshaft timing gear retainer tool to align the gerotor oil pump to the front cover
- Front cover and torque the perimeter bolts to 22 ft. lbs. (30 Nm) and the lower center bolt to 89 inch lbs. (10 Nm)
- Oil pan and torque the bolts to 80 inch lbs. (9 Nm)
- Water pump pulley and torque the bolts to 19 ft. lbs. (25 Nm), if removed

17. Remove the crankshaft timing gear retaining tool.

18. Apply RTV across the cylinder head and front cover T-joints and install a new rocker cover gasket

- Crankshaft pulley and torque the bolt to 159 ft. lbs. (215 Nm)
- Rocker cover and torque the bolts to 22 ft. lbs. (30 Nm)
- Drive belt tensioner and torque the fastener to 22 ft. lbs. (30 Nm)
- Drive belt idler pulley and torque the fastener to 20 ft. lbs. (27 Nm)
- Accelerator linkage bracket and the throttle cable. Make certain that the cable is routed properly and is not binding and torque the bolts to 18 ft. lbs. (25 Nm).
- Alternator and torque the bolts to 24 ft. lbs. (32 Nm), if removed

- Power steering pump and torque the bolts to 22 ft. lbs. (30 Nm)
- A/C compressor and torque the rear bracket bolts to 18 ft. lbs. (25 Nm) and the front bracket bolts to 35 ft. lbs. (47 Nm)
- Accessory drive belt
- Upper mount and torque the bracket nuts to 37 ft. lbs. (50 Nm) and then torque the mount to cover bolts to 37 ft. lbs. (50 Nm)
- Wheel well splash shield
- Wheel
- EVAP canister purge hose
- Throttle body vacuum harness hose
- PCV hose
- Brake booster vacuum hose
- Upper radiator hose
- Heater hose
- Ground wire electrical connectors
- O_2S electrical connector
- Fuel injector electrical connectors
- IAC valve electrical connector
- EGR valve electrical connector
- TPS electrical connector
- EVAP canister purge valve connectors
- MAP sensor connectors
- Alternator electrical connectors
- Starter motor electrical connectors
- A/C compressor electrical connectors
- Intake manifold support bracket and torque the bolts to 22 ft. lbs. (30 Nm)
- De-aeration line and torque the cylinder head nut to 80 inch lbs. (9 Nm) and the rear lift bracket nut to 97 inch lbs. (11 Nm)
- Fuel feed and return lines with new retainers and torque the fastener to 53 inch lbs. (6 Nm)
- Air cleaner assembly
- Fresh air hose
- Negative battery cable

19. Fill the cooling system.

20. Fill the engine with new oil.

21. Prime the fuel system by cycling the ignition **ON** for 5 seconds and **OFF** for 10 seconds a few times without cranking the engine.

22. Start the engine, check for leaks, and repair if necessary.

L–Series With 2.2L Engine

※※ **WARNING**

Only remove the cylinder head when the engine is cold. Warpage may result if the cylinder head is removed while the engine is hot.

1. Before servicing the vehicle, refer to the precautions in the beginning of this section.

2. Drain the cooling system.

3. Drain the engine oil.

4. Properly relieve the fuel system pressure.

5. Remove or disconnect the following:
- Negative battery cable
- Intake manifold
- Exhaust manifold flange bolts
- Exhaust manifold down pipe to the catalytic converter
- Fuel lines
- A/C compressor switch electrical connector
- Crankcase vent hose from the camshaft cover
- Coolant air bleed hose
- Upper radiator hose from the cylinder head
- Ignition coil electrical connectors
- Knock Sensor (KS) electrical connector
- Engine Coolant Temperature (ECT) sensor electrical connector
- Fuel injector jumper harness
- Front Oxygen Sensor (O$_2$S) electrical connector
- Ignition coil
- Power steering pump and secure it to the engine without removing the lines
- Camshaft cover ground strap
- Camshaft cover
- Upper timing chain guide
- Upper timing chain tensioner
- Camshaft sprocket bolts
- Camshaft sprockets
- Plug to gain access to the fixed timing chain guide
- Fixed timing chain guide upper bolt only
- Cylinder head bolts using the proper sequence
- Cylinder head

➡ **The manufacturer recommends to perform the remaining steps to complete this procedure. It may be not be necessary to perform the remaining steps if the timing chain can be supported without slipping or moving on the crankshaft pulley.**

- Right front wheel and splash shield
- Drive belt and tensioner
- A/C suction line retaining clips and move the line away to allow clearance to the crankshaft pulley bolt

6. Position the crankshaft 60 degrees Before Top Dead Center (BTDC) and install a crankshaft pulley holder tool.

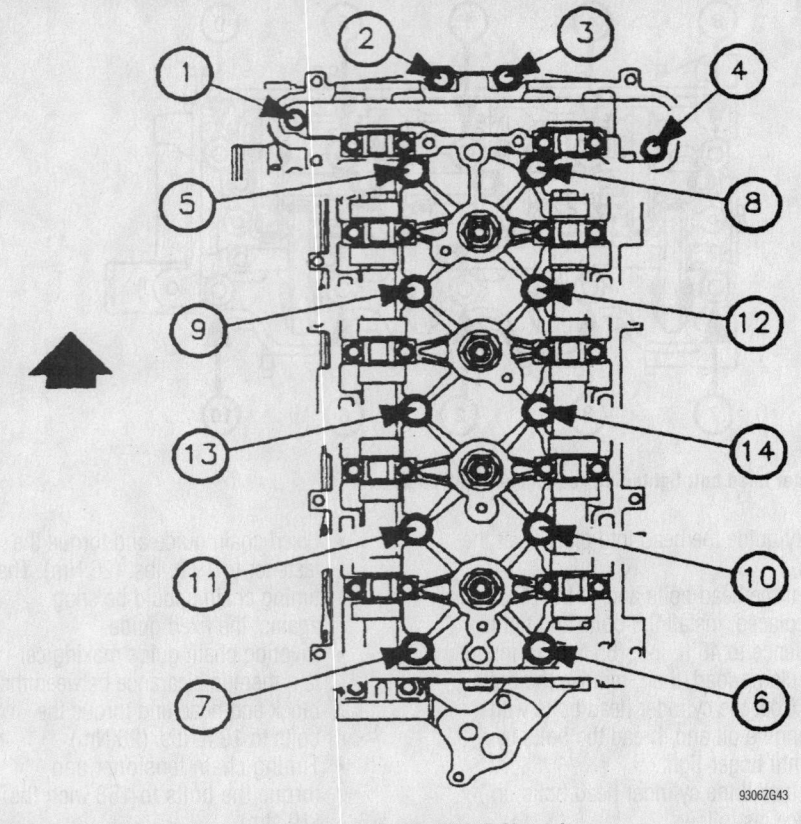

Cylinder head bolt loosening sequence—L–Series With 2.2L engine

- Crankshaft pulley
- Front cover bolts except for the one blocked by the engine mount bracket
- Loosen the water pump bolt but do not remove it

7. Support the engine and remove the right side engine mount.
- Engine mount bracket
- Upper front cover bolt which was blocked by the mount bracket
- Water pump bolt
- Front cover
- Adjustable timing chain guide
- Fixed timing chain guide lower bolt

To install:

➡ **Set the crankshaft to 60 degrees Before Top Dead Center (BTDC) or after Top Dead Center (TDC) to prevent contact between the pistons and valves.**

8. Install or connect the following:
- New cylinder head gasket with the side imprinted **OBEN** facing up
- Cylinder head and align it on the dowels
- New cylinder head bolts and torque them in sequence to 22 ft. lbs. (30 Nm) plus 155 degrees
- Front 4 cylinder head bolts coated

with Loctite® and torque them to 15 ft. lbs. (20 Nm)

9. Position the exhaust camshaft with the offset slot in the 2 o'clock position and the intake camshaft with the offset slot in the 11 o'clock position.
- Timing chain around the intake camshaft sprocket with the copper link aligned with the **INT** diamond timing mark
- Sprocket to the camshaft and align it with the offset slot
- New camshaft sprocket bolt, but do not tighten

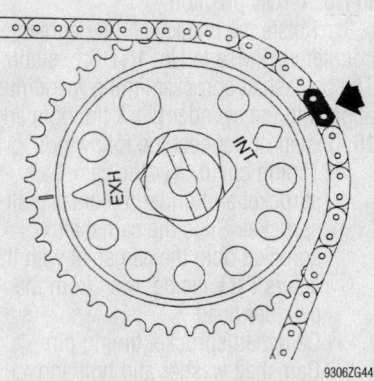

Align the copper link on the timing chain with the INT diamond timing mark

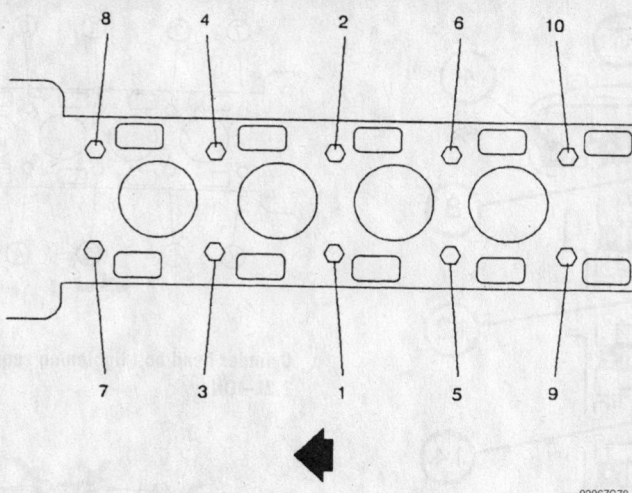

9306ZG78

Cylinder head bolt tightening sequence—L–Series With 2.2L engine

- Timing chain around the crankshaft sprocket and align the silver link to the timing mark
- Adjustable timing chain guide through the opening on top of the cylinder head and torque the bolt to 89 inch lbs. (10 Nm)
- Timing chain around the exhaust camshaft sprocket with the silver link aligned with the offset slot. Install but do not tighten a new sprocket bolt

✳✳ WARNING

Make certain that all timing marks and colored links are aligned properly before proceeding to the next step. If the timing chain is not aligned properly, severe engine damage may occur.

10. Torque the intake and exhaust camshaft bolts to 63 ft. lbs. (85 Nm) plus a 30 degree turn.

11. Install or connect the following:
- Fixed timing guide and torque the bolt to 89 inch lbs. (10 Nm)
- Fixed timing guide bolt access plug after applying Loctite® to the threads and torque it to 30 ft. lbs. (40 Nm)
- Timing chain tensioner and torque the bolts 44 ft. lbs. (60 Nm)

12. Tap the top of the timing chain between the camshaft sprockets to engage the tensioner.
- Upper timing chain guide and torque the bolts to 89 inch lbs. (10 Nm)
- Front cover with a new gasket and torque the bolts to 18 ft. lbs. (25 Nm)

- Water pump bolt and torque it to 18 ft. lbs. (25 Nm)
- Drive belt tensioner and torque the bolts to 30 ft. lbs. (40 Nm)
- Engine mount bracket and torque the bolts to 66 ft. lbs. (90 Nm)
- Engine mount to the side rail and torque the fasteners to 41 ft. lbs. (55 Nm)
- Engine mount to the bracket and torque the bolts to 41 ft. lbs. (55 Nm)
- Crankshaft damper and torque the bolt to 74 ft. lbs. (100 Nm)
- Accessory drive belt
- A/C line clamps
- Right wheel splash shield
- Right front wheel
- Camshaft cover and torque the bolts to 89 inch lbs. (10 Nm)
- Camshaft cover ground strap and torque the bolt to 89 inch lbs. (10 Nm)
- Power steering pump and torque the bolts to 18 ft. lbs. (25 Nm)
- Ignition coil/module assembly and torque the bolts to 71 inch lbs. (8 Nm)
- Intake manifold with a new gasket and torque the bolts to 89 inch lbs. (10 Nm)
- Wiring harness to the intake manifold bracket
- Engine wiring harness and torque the fasteners to 89 inch lbs. (10 Nm)
- All component electrical connections
- Upper radiator hose
- Crankcase ventilation hose
- Fuel pressure regulator hose

- Brake booster vacuum hose
- EVAP purge hose
- Fuel lines
- Fuel lines to the bracket and torque the bracket bolt to 89 inch lbs. (10 Nm)
- Exhaust down pipe to the exhaust manifold and torque the bolts to 25 ft. lbs. (30 Nm)
- Exhaust pipe to the resonator and torque the bolts to 15 ft. lbs. (20 Nm)
- Exhaust manifold heat shield and torque the bolts to 18 ft. lbs. (25 Nm)
- Air cleaner assembly
- Negative battery cable

13. Fill the engine with clean oil.

14. Fill the cooling system.

15. Prime the fuel system by cycling the ignition **ON** for 5 seconds and **OFF** for 10 seconds a few times without cranking the engine.

16. Start the engine, check for leaks, and repair if necessary.

ION With 2.2L Engine

✳✳ WARNING

Only remove the cylinder head when the engine is cold. Warpage may result if the cylinder head is removed while the engine is hot.

1. Before servicing the vehicle, refer to the precautions in the beginning of this section.

2. Drain the cooling system.

3. Drain the engine oil.

4. Properly relieve the fuel system pressure.

5. Remove or disconnect the following:
- Negative battery cable
- Intake manifold
- Exhaust manifold
- Timing chain
- Cylinder head bolts using the proper sequence
- Cylinder head

To install:

➡ **Set the crankshaft to 60 degrees Before Top Dead Center (BTDC) or after Top Dead Center (TDC) to prevent contact between the pistons and valves.**

6. Install or connect the following:
- New cylinder head gasket
- Cylinder head and align it on the dowels
- New cylinder head bolts and torque them in sequence to 22 ft. lbs. (30 Nm) plus 155 degrees

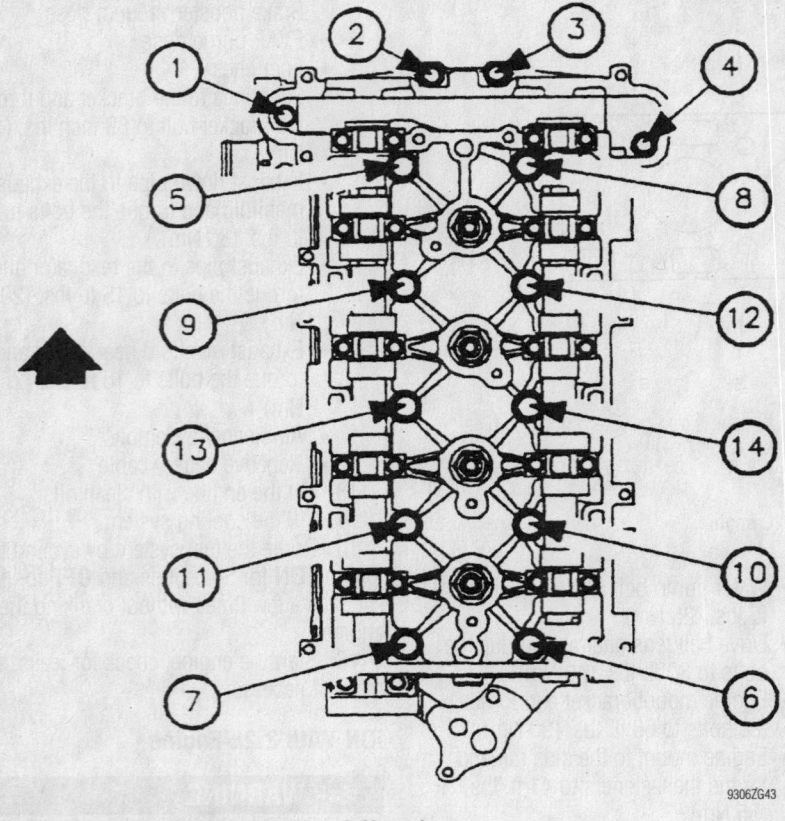

Cylinder head bolt loosening sequence—2.2L engine

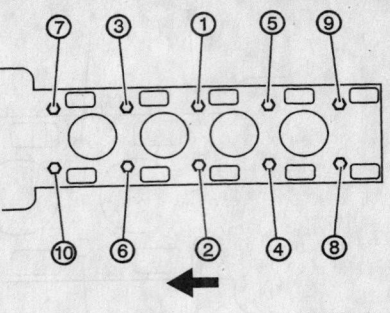

Cylinder head bolt tightening sequence—
2.2L–ION

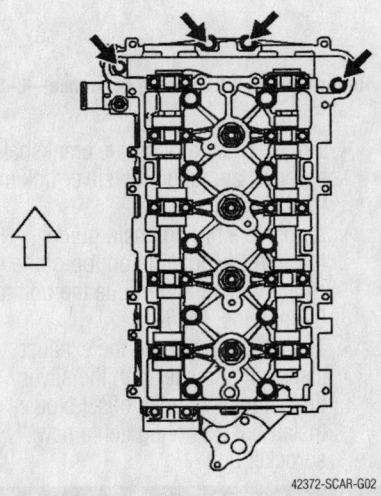

Location of the 4 front cylinder head
bolts—ION

- Front 4 cylinder head bolts coated with Loctite® and torque them to 26 ft. lbs. (35 Nm)
- Timing chain
- Exhaust manifold
- Intake manifold
- Negative battery cable
7. Fill the engine with clean oil.
8. Fill the cooling system.
9. Prime the fuel system by cycling the ignition **ON** for 5 seconds and **OFF** for 10 seconds a few times without cranking the engine.
10. Start the engine, check for leaks, and repair if necessary.

3.0L Engine

FRONT

❋❋ WARNING

Only remove the cylinder head when the engine is cold. Warpage may result if the cylinder head is removed while the engine is hot.

1. Before servicing the vehicle, refer to the precautions in the beginning of this section.
2. Drain the cooling system.
3. Drain the engine oil.

4. Properly relieve the fuel system pressure.
5. Remove or disconnect the following:
- Negative battery cable
- Air cleaner assembly
- Intake plenum
- Intake manifold
- Intake manifold spacer
- Coolant bridge
- Upper radiator hose from the coolant extension housing
6. Properly support the powertrain assembly.
- Front transaxle mount through bolt
- Extension housing over the coolant module
- Oil level indicator tube
- Coolant extension housing by twisting it off
- Front camshaft cover
- Ground wires from the lift bracket
- Oxygen Sensor (O$_2$S) electrical connector
- Down pipe from the exhaust manifold
- Front timing belt cover
- Timing belt
- Timing belt tensioner bracket
- Rear timing belt cover
- Camshaft Position (CMP) sensor electrical connector

- Exhaust camshaft
- Loosen the cylinder head bolts in stages as shown
- Cylinder head
- Exhaust manifold from the cylinder head (if necessary)

To install:
7. Install or connect the following:
- Exhaust manifold with a new gasket and torque the bolts to 15 ft. lbs. (20 Nm)
- New cylinder head gasket with the part number imprint facing the top of the engine
- Cylinder head
8. Torque the new cylinder head bolts, in sequence, as follows:
  a. 18 ft. lbs. (25 Nm)
  b. 90 degree turn
  c. 90 degree turn
  d. 90 degree turn
  e. 15 degree turn
9. Install or connect the following:
- Coolant pipe with new sealing rings

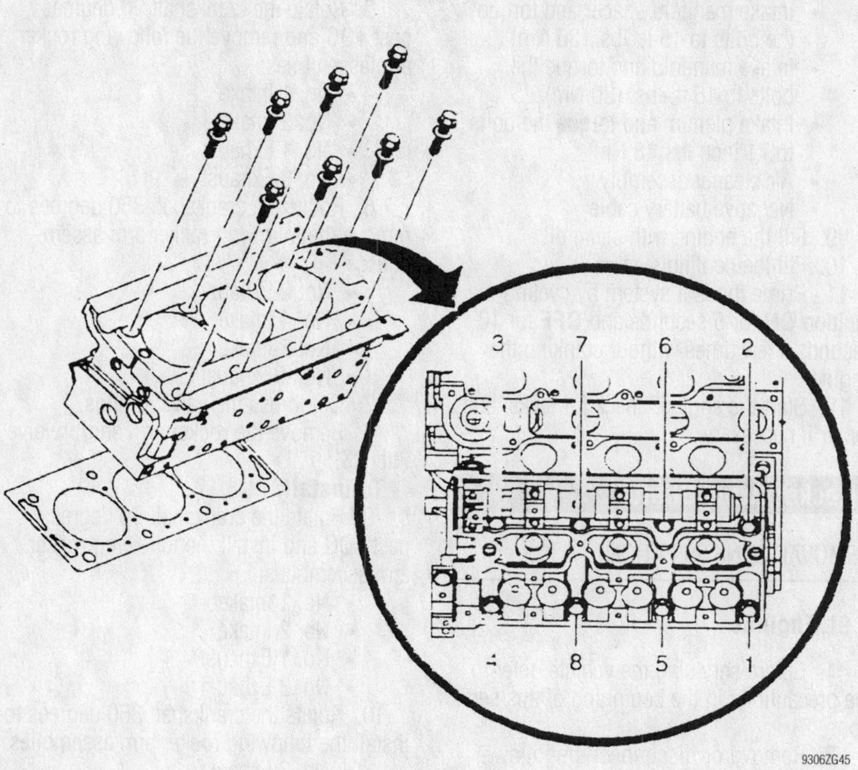

Cylinder head bolt removal for the front and rear cylinder heads—3.0L engine

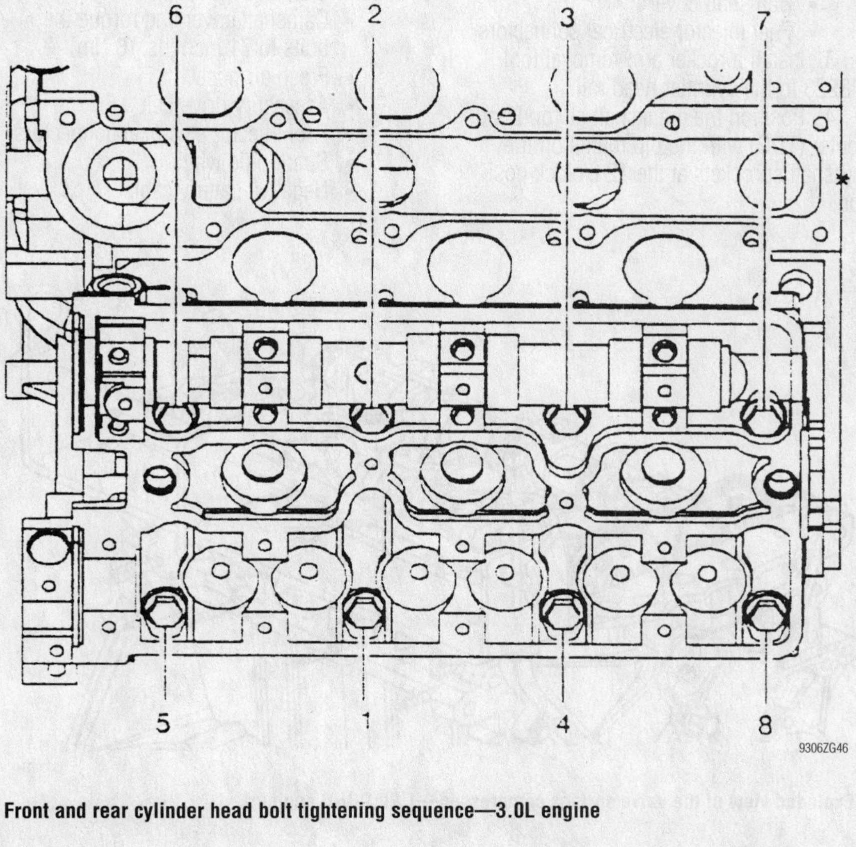

Front and rear cylinder head bolt tightening sequence—3.0L engine

- Engine lift bracket bolt and torque it to 15 ft. lbs. (20 Nm)
- Upper radiator hose to the coolant pipe
- Front transaxle mount through bolt
- Exhaust camshaft
- CMP sensor electrical connector
- Rear timing belt cover and torque the bolts to 71 inch lbs. (8 Nm)
- Camshaft gears
- Timing belt tensioner bracket
- Timing belt
- Front timing belt cover and torque the bolts to 71 inch lbs. (8 Nm)
- Down pipe to the exhaust manifold
- O_2S electrical connector
- Front camshaft cover and torque the bolts to 71 inch lbs. (8 Nm)
- Coolant bridge and torque the bolt to 22 ft. lbs. (33 Nm)
- Intake manifold spacer and torque the bolts in a spiral direction from the inside and working out to 15 ft. lbs. (20 Nm)
- Intake manifold and torque the bolts to 15 ft. lbs. (20 Nm)
- Intake plenum and torque the bolts to 71 inch lbs. (8 Nm)
- Air cleaner assembly
- Negative battery cable

10. Fill the engine with clean oil.
11. Fill the cooling system.
12. Prime the fuel system by cycling the ignition **ON** for 5 seconds and **OFF** for 10 seconds a few times without cranking the engine.
13. Start the engine, check for leaks, and repair if necessary.

REAR

WARNING

Only remove the cylinder head when the engine is cold. Warpage may result if the cylinder head is removed while the engine is hot.

1. Before servicing the vehicle, refer to the precautions in the beginning of this section.
2. Drain the cooling system.
3. Drain the engine oil.
4. Properly relieve the fuel system pressure.
5. Remove or disconnect the following:
   - Negative battery cable
   - Air cleaner assembly
   - Intake plenum
   - Intake manifold
   - Intake manifold spacer
   - Coolant bridge

- Engine ventilation chamber
- Rear camshaft cover
- Front timing belt cover
- Timing belt
- Timing belt tensioner bracket
- Camshaft gears
- Rear timing belt cover
- Exhaust manifold pipe heat shield
- Front exhaust manifold pipe-to-rear exhaust manifold pipe fasteners
- Rear exhaust manifold pipe nuts, pull the manifold pipe down and discard the gasket
- Exhaust Gas Recirculation (EGR)-to-exhaust manifold pipe
- Exhaust camshaft
- Cylinder head bolts
- Cylinder head
- Exhaust manifold from the cylinder head

To install:

6. Install or connect the following:

- Exhaust manifold with a new gasket and torque the bolts to 15 ft. lbs. (20 Nm), if removed
- New cylinder head gasket with the part number imprint facing the top of the engine
- Cylinder head

7. Torque the new cylinder head bolts, in sequence, as follows:

   a. 18 ft. lbs. (25 Nm)
   b. 90 degree turn
   c. 90 degree turn
   d. 90 degree turn
   e. 15 degree turn

8. Install or connect the following:

- Exhaust camshaft
- Rear timing belt cover and torque the bolts to 71 inch lbs. (8 Nm)
- Camshaft gears and torque the bolts to 37 ft. lbs. (50 Nm) plus a 60 degree turn plus another 15 degree turn
- Timing belt tensioner bracket and torque the bolts to 30 ft. lbs. 940 Nm)
- Timing belt
- Front timing belt cover and torque the bolts to 71 inch lbs. (8 Nm)
- Rear camshaft cover and torque the bolts to 71 inch lbs. (8 Nm)
- Exhaust manifold pipe to the manifold
- Exhaust manifold pipe heat shield
- EGR-to-exhaust manifold pipe and torque the nut to 18 ft. lbs. (25 Nm)
- Coolant bridge and torque the bolt to 22 ft. lbs. (33 Nm)
- Engine ventilation chamber and torque the bolts to 71 inch lbs. (8 Nm)

- Intake manifold spacer and torque the bolts to 15 ft. lbs. (20 Nm)
- Intake manifold and torque the bolts to 15 ft. lbs. (20 Nm)
- Intake plenum and torque the bolts to 71 inch lbs. (8 Nm)
- Air cleaner assembly
- Negative battery cable

9. Fill the engine with clean oil.

10. Fill the cooling system.

11. Prime the fuel system by cycling the ignition **ON** for 5 seconds and **OFF** for 10 seconds a few times without cranking the engine.

12. Start the engine, check for leaks, and repair if necessary.

Rocker Arms/Shafts

REMOVAL & INSTALLATION

1.9L Engines

1. Before servicing the vehicle, refer to the precautions in the beginning of this section.

2. Remove or disconnect the following:

- Negative battery cable
- Spark plug wires
- Accessory drive belt
- Fresh air hose
- Camshaft cover
- Fuel injector electrical connectors

3. Install a rocker arm removal tool J43223 to the cylinder head rail.

4. Position the crankshaft at Top Dead Center (TDC) with the pip marks on the camshaft sprockets at the 12 o'clock position.

5. Rotate the crankshaft 90 degrees past TDC and remove the following rocker arm assemblies:

- No. 1 Intake
- No. 2 Intake
- No. 1 Exhaust
- No. 3 Exhaust

6. Rotate the crankshaft 360 degrees to remove the following rocker arm assemblies:

- No. 2 Exhaust
- No. 4 Intake
- No. 3 Intake
- No. 4 Exhaust

7. Compress the valve springs.

8. Remove the rocker arm and/or valve lifters.

To install:

9. Rotate the crankshaft 90 degrees past TDC and install the following rocker arm assemblies:

- No. 1 Intake
- No. 2 Intake
- No. 1 Exhaust
- No. 3 Exhaust

10. Rotate the crankshaft 360 degrees to install the following rocker arm assemblies:

- No. 2 Exhaust
- No. 4 Intake
- No. 3 Intake
- No. 4 Exhaust

11. Remove the rocker arm tool.

12. Install or connect the following:

- Camshaft cover and torque the bolts to 71 inch lbs. (8 Nm)
- Fresh air hose
- Accessory drive belt
- Fuel injector electrical connectors
- Spark plug wires
- Negative battery cable

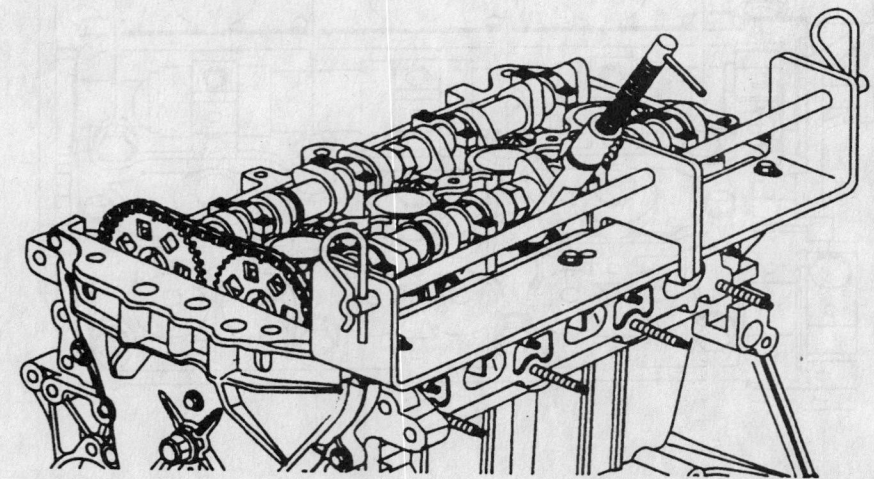

Exploded view of the valve springs compressed—1.9L DOHC engines

9306ZG47

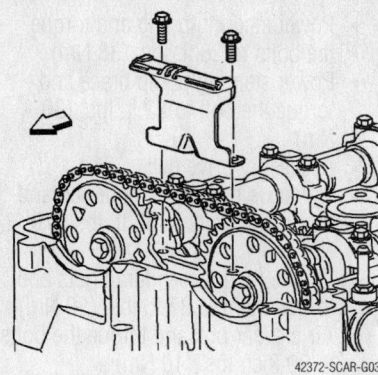

Remove the upper timing chain guide—ION

Install camshaft sprocket holding tool J43655

2.2L Engines

1. Before servicing the vehicle, refer to the precautions in the beginning of this section.

2. Remove or disconnect the following:
- Negative battery cable
- Camshaft cover
- Upper timing chain guide

3. Install camshaft sprocket holding tool J43655 as illustrated.
- Intake and exhaust camshaft sprocket bolts and discard the bolts

4. Slide the sprockets forward and mark the camshaft bearing caps to ensure they are installed in their correct positions.

5. Remove each bearing cap bolt one turn at a time until there is no tension.

6. Remove the bearing cap and the camshaft
- Roller followers
- Lash adjusters

To install:

7. Lubricate the valve tips.

8. Install or connect the following:
- Lash adjusters
- Roller followers

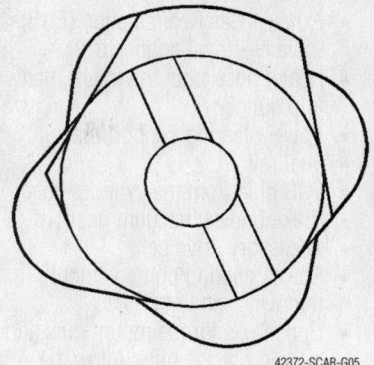

Make sure the alignment notches are aligned with the camshaft sprocket

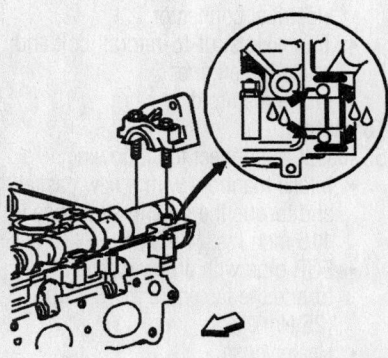

Apply 0.197 inch (5mm) of anaerobic Gasket maker to the rear camshaft bearing cap

9. Make sure the alignment notches are aligned with the camshaft sprocket as illustrated.
- Camshaft
- Bearing caps except the rear camshaft cap

10. Tighten the caps in increments of 3 turns until they are seated, then tighten to 89 inch lbs. (10 Nm).

11. Apply 0.197 inch (5mm) of Permatex Anaerobic Gasket maker® 51813 to the rear camshaft bearing cap. Install the cap and tighten the bolts to 18 ft. lbs. (25 Nm).
- Camshaft sprockets and hand tighten the bolts

12. Remove camshaft sprocket holding tool J43655
- Camshaft sprocket bolts to 63 ft. lbs. (85 Nm) plus and additional 30 degrees

13. Install or connect the following:
- Upper timing chain guide and tighten to 89 inch lbs. (10 Nm)
- Camshaft cover
- Negative battery cable

Intake Manifold

REMOVAL & INSTALLATION

1.9L SOHC Engines

1. Before servicing the vehicle, refer to the precautions in the beginning of this section.

2. Properly relieve the fuel system pressure.

3. Drain the cooling system.

4. Remove or disconnect the following:
- Negative battery cable
- Air cleaner assembly
- Positive Crankcase Ventilation (PCV) valve hose
- Fuel line from the rail
- Throttle cable from the throttle body
- Throttle cable bracket nuts
- Fuel injector electrical connectors
- Throttle Position Sensor (TPS) electrical connectors
- Idle Air Control (IAC) valve electrical connectors
- Manifold Absolute Pressure (MAP) sensor electrical connectors
- Exhaust Gas Recirculation (EGR) valve electrical connectors
- Heater hose from the intake manifold outlet

5. Position the wiring harness over the brake master cylinder.
- Intake manifold support bracket bolt
- Accessory drive belt
- Power steering pump without removing the lines
- Upper intake manifold nuts
- Evaporative Emissions (EVAP) canister purge valve solenoid vacuum hose
- Brake booster vacuum hose
- Lower intake manifold nuts
- Intake manifold

Upper Side				
8	4	1	5	
7	3	2	6	9
Lower Side				

Intake manifold bolt torque sequence—1.9L SOHC engine

To install:

6. Install or connect the following:
- Intake manifold with a new gasket and torque the nuts in sequence to 22 ft. lbs. (30 Nm)
- Power steering pump and torque the fasteners to 27 ft. lbs. (35 Nm)
- Accessory drive belt
- Heater hose
- Intake manifold support bracket bolt and torque it to 22 ft. lbs. (30 Nm)
- Fuel supply and return lines
- Fuel line(s) in the retaining bracket and torque the mounting screw to 36 inch lbs. (4 Nm)
- Throttle cable to the throttle body and accelerator cable bracket and torque the bolts to 22 ft. lbs. (30 Nm)
- IAC valve electrical connector
- TPS electrical connector
- MAP sensor electrical connector
- Fuel injector electrical connectors
- EGR valve electrical connector
- EVAP purge solenoid vacuum hose
- PCV valve hose
- Air cleaner assembly
- Negative battery cable

7. Fill the cooling system.
8. Prime the fuel system by cycling the ignition **ON** for 5 seconds and **OFF** for 10 seconds a few times without cranking the engine.
9. Start the engine, check for leaks, and repair if necessary.

1.9L DOHC Engines

1. Before servicing the vehicle, refer to the precautions in the beginning of this section.
2. Properly relieve the fuel system pressure.
3. Drain the cooling system.
4. Remove or disconnect the following:
- Negative battery cable
- Air cleaner fresh air hose
- Positive Crankcase Ventilation (PCV) valve hose
- Fuel line from the rail
- Throttle cable from the throttle body
- Throttle cable bracket nuts
- Fuel injector electrical connectors
- Throttle Position Sensor (TPS) electrical connector
- Idle Air Control (IAC) valve electrical connector
- Manifold Absolute Pressure (MAP) sensor electrical connector

- Exhaust Gas Recirculation (EGR) valve electrical connector
- Heater hose from the intake manifold outlet
- Power steering support brace
- Fuel rail
- EGR pipe from the cylinder head
- Brake booster vacuum hose
- Accessory drive belt
- Power steering pump without removing the lines
- Upper axis torque mount nuts and midrail bracket nuts. Allow the powertrain assembly to rest on a support fixture.
- Resonator and air cleaner box
- Intake Air Temperature (IAT) sensor electrical connector
- Transaxle strut-to-midrail bolt and rotate the engine
- Intake manifold

To install:

5. Install or connect the following:
- Intake manifold with a new gasket and torque the nuts in sequence to 115 inch lbs. (13 Nm)
- EGR pipe with a new gasket and torque the fasteners to 18 ft. lbs. (25 Nm)
- Heater hose
- Fuel rail and torque the bolts to 106 inch lbs. (12 Nm)
- IAC valve electrical connector
- TPS electrical connector
- MAP sensor electrical connector
- Fuel injector electrical connector
- EGR valve electrical connector
- EVAP canister purge solenoid vacuum hose
- PCV valve hose
- Brake booster vacuum hose
- Fuel rail to the manifold and torque the nut to 36 inch lbs. (4.5 Nm)
- Fuel supply line to the rail and torque the bolt to 36 inch lbs. (4 Nm)

Upper Side

5	1	3
4	2	6

Lower Side

9306ZG49

Intake manifold bolt torque sequence— 1.9L DOHC engine

- Power steering pump and torque the bolts to 28 ft. lbs. (38 Nm)
- Power steering pump brace and torque the bolt to 22 ft. lbs. (30 Nm)
- Accessory drive belt
- Transaxle mount-to-frame rail and torque the bolts to 52 ft. lbs. (70 Nm)
- Engine mount-to-midrail nuts and torque them to 37 ft. lbs. (50 Nm)
- Air cleaner box and torque the bolts to 89 inch lbs. (10 Nm)
- Resonator
- IAT sensor electrical connector
- Throttle cable to the throttle body and support bracket and torque the bolts to 19 ft. lbs. (25 Nm)
- Intake manifold brace and torque the bolts to 58 inch lbs. (6.5 Nm)
- Negative battery cable

6. Fill the cooling system.
7. Prime the fuel system by cycling the ignition **ON** for 5 seconds and **OFF** for 10 seconds a few times without cranking the engine.
8. Start the engine, check for leaks, and repair if necessary.

L–Series With 2.2L Engine

1. Before servicing the vehicle, refer to the precautions in the beginning of this section.
2. Remove or disconnect the following:
- Negative battery cable
- Air cleaner assembly
- Intake Air Temperature (IAT) sensor electrical connector
- Throttle Position Sensor (TPS) electrical connectors
- Idle Air Control (IAC) valve electrical connectors
- Manifold Absolute Pressure (MAP) sensor electrical connectors
- Fuel pressure regulator vacuum pipe
- Evaporative Emissions (EVAP) purge solenoid hose from the throttle body
- Throttle cable and automatic transaxle downshift cable from the throttle body
- Throttle cable bracket nuts
- Brake booster vacuum hose
- Oil level indicator tube
- Throttle body
- Intake manifold

To install:

3. Install or connect the following:
- Intake manifold with a new gasket and torque the nuts to 89 inch lbs.

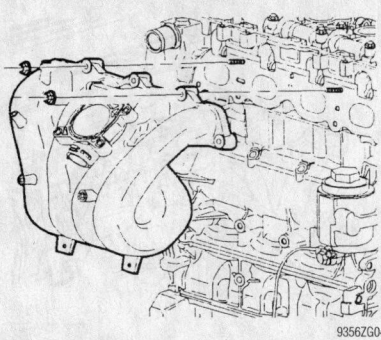

9356ZG04

Intake manifold mounting—all 2.2L

(10 Nm) starting from the center and working outward
- Throttle body to the intake manifold and torque the bolts to 19 ft. lbs. (25 Nm)
- Throttle cable bracket to the throttle body studs
- Oil level indicator tube and torque the fastener to 89 inch lbs. (10 Nm)
- Brake booster vacuum pipe
- Throttle cable to the throttle body and install the support bracket. Torque the bolts to 19 ft. lbs. (25 Nm)
- EVAP canister purge solenoid vacuum hose
- Fuel pressure regulator vacuum pipe to the throttle body
- IAC valve electrical connector
- TPS electrical connector
- MAP sensor electrical connector
- Air inlet hose to the throttle body
- Crankcase vent hose to the camshaft cover
- IAT sensor electrical connector
- Air cleaner assembly
- Negative battery cable

4. Start the engine, check for leaks, and repair if necessary.

ION With 2.2L Engine

1. Before servicing the vehicle, refer to the precautions in the beginning of this section.
2. Remove or disconnect the following:
- Negative battery cable
- Air cleaner assembly
- Idle Air Control (IAC) valve electrical connectors
- Throttle Position Sensor (TPS) electrical connectors
- Vacuum hoses from the throttle body
- Throttle control cable and bracket
- Throttle body
- Positive Crankcase Valve (PCV) hose

- Purge solenoid tube
- Brake booster hose
- Oil dipstick tube bolt
- Fuel rail
- Knock Sensor (KS) connector
- Intake manifold bolts and nuts
- Intake manifold

To install:
3. Install or connect the following:
- Intake manifold with a new gasket and torque the nuts to 89 inch lbs. (10 Nm) starting from the center and working outward
- Knock Sensor (KS) connector
- Fuel rail
- Oil dipstick tube bolt
- Brake booster hose
- Purge solenoid tube
- PCV hose
- Throttle body to the intake manifold and torque the bolts to 89 inch lbs. (10 Nm)
- Throttle cable bracket and torque the bolts to 89 inch lbs. (10 Nm)
- Throttle cable
- IAC sensor electrical connector
- TPS electrical connectors
- Vacuum hoses from the throttle body
- Air cleaner assembly
- Negative battery cable

4. Start the engine, check for leaks, and repair if necessary.

3.0L Engine

1. Before servicing the vehicle, refer to the precautions in the beginning of this section.
2. Properly relieve the fuel system pressure.
3. Remove or disconnect the following:
- Negative battery cable
- Air cleaner assembly
- Throttle body electrical connector
- Fuel pressure regulator vacuum hose
- Both intake runners
- Intake plenum
- Fuel supply and return lines
- Fuel injector electrical connectors
- Fuel rail
- Intake manifold
- Intake manifold spacer and O-ring gaskets

To install:
4. Install or connect the following:
- Intake manifold spacer with new seals. Apply Loctite® 242 to the bolts and torque the bolts to 15 ft. lbs. (20 Nm) in sequence.
- Intake manifold with a new gasket

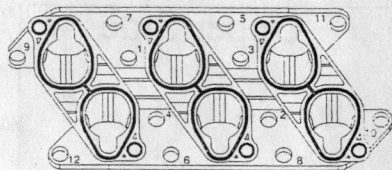

9356ZG07

Tighten the intake manifold spacer bolts as shown—3.0L engines

and torque the bolts to 15 ft. lbs. (20 Nm)
- Fuel rail
- Fuel injector electrical connectors
- Fuel supply and return hoses and torque the fastener to 11 ft. lbs. (15 Nm)
- Intake plenum and torque the bolts to 71 inch lbs. (8 Nm)
- Intake manifold runners and torque the bolts to 71 inch lbs. (8 Nm)
- Negative battery cable

5. Prime the fuel system by cycling the ignition **ON** for 5 seconds and **OFF** for 10 seconds a few times without cranking the engine.
6. Start the engine, check for leaks, and repair if necessary.

Exhaust Manifold

REMOVAL & INSTALLATION

1.9L Engines

1. Before servicing the vehicle, refer to the precautions in the beginning of this section.
2. Remove or disconnect the following:
- Negative battery cable
- Front exhaust pipe-to-engine support bracket mounting fasteners
- Pipe-to-manifold nuts and lower the pipe
- A/C compressor and bracket from the engine without removing the lines
- Oxygen Sensor (O_2S) electrical connector
- Exhaust manifold

To install:
3. Install or connect the following:
- Exhaust manifold with a new gasket and torque the nuts in sequence to 16 ft. lbs. (22 Nm) for the SOHC engine or to 150 inch lbs. (17 Nm) for the DOHC engine
- O_2S electrical connector
- A/C compressor and bracket and torque the front bracket-to-com-

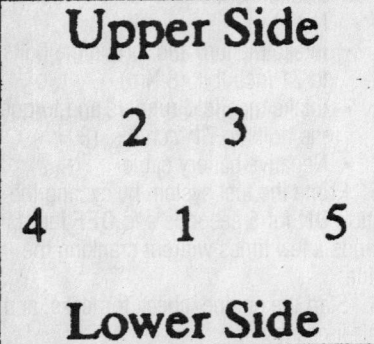

Exhaust manifold bolt torque sequence—
1.9L SOHC engines

Exhaust manifold bolt torque sequence—
1.9L DOHC engines

pressor to 40 ft. lbs. (54 Nm) and all remaining fasteners to 19 ft. lbs. (25 Nm)
- Pipe-to-manifold with a new gasket and torque the fasteners in a cross-wise pattern to 23 ft. lbs. (31 Nm)
- Exhaust pipe-to-engine support bracket and torque the mounting fasteners to 23 ft. lbs. (31 Nm)
- Negative battery cable

4. Start the vehicle and check for leaks, repair if necessary.

L-Series With 2.2L Engine

1. Before servicing the vehicle, refer to the precautions in the beginning of this section.
2. Remove or disconnect the following:
- Negative battery cable
- Exhaust pipe from the manifold
- Exhaust manifold heat shield
- Oxygen Sensor (O₂S) from the manifold
- Exhaust manifold

To install:
3. Install or connect the following:
- Exhaust manifold with a new gasket

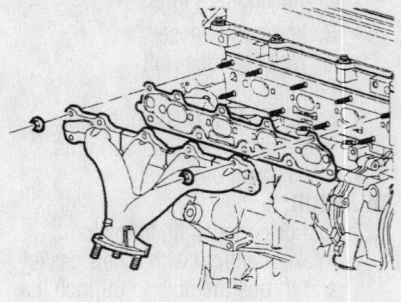

Remove the exhaust manifold and gasket.—L-Series With 2.2L engine

and torque the bolts, starting from the center and working outward, to 13 ft. lbs. (18 Nm)
- O₂S and torque it to 33 ft. lbs. (45 Nm)
- Exhaust manifold heat shield and torque the bolts to 18 ft. lbs. (25 Nm)
- Exhaust pipe to the manifold with a new gasket and torque the nuts to 22 ft. lbs. (30 Nm)
- Negative battery cable

4. Start the vehicle and check for leaks, repair if necessary.

ION With 2.2L Engine

1. Before servicing the vehicle, refer to the precautions in the beginning of this section.
2. Remove or disconnect the following:
- Negative battery cable
- Exhaust manifold heat shield
- Oxygen Sensor (O₂S) from the manifold

➡ **Do not bend the flex coupling more than 3 degrees in any direction to avoid damage.**

- Pipe-to-manifold nuts, pull down and back on the pipe in order to separate the pipe from the manifold
- Exhaust pipe from the manifold
- Exhaust manifold-to-head nuts
- Exhaust manifold

To install:
3. Install or connect the following:
- Exhaust manifold with a new gasket and torque the bolts in the sequence illustrated, to 115 inch lbs. (13 Nm)
- New manifold-to-flex pipe gasket, and place the pipe in position. Tighten the nuts to 32 ft. lbs. (43 Nm).
- O₂S
- Exhaust manifold heat shield and

Remove the exhaust manifold and gasket—ION models

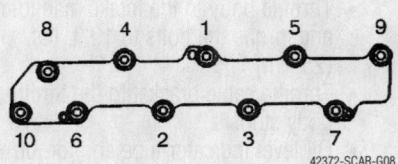

Tighten the exhaust manifold nuts in the sequence shown—ION models

torque the bolts to 18 ft. lbs. (25 Nm)
- Negative battery cable

4. Start the vehicle and check for leaks, repair if necessary.

3.0L Engine

FRONT MANIFOLD

1. Before servicing the vehicle, refer to the precautions in the beginning of this section.
2. Drain the cooling system.
3. Remove or disconnect the following:
- Negative battery cable
- Exhaust manifold Oxygen Sensor (O₂S)
- Upper radiator hose from the coolant extension housing
4. Install and engine support fixture.
- Front transaxle through bolt and raise the powertrain assembly
- Oil level indicator tube and coolant extension tube bolt
- Coolant extension housing
- Power steering pipe bracket bolt
- Upper exhaust manifold nuts
- Front exhaust manifold pipe from the manifold
- Font exhaust manifold pipe from the rear exhaust manifold pipe
- Oil filter housing
- Lower exhaust manifold nuts
- Exhaust manifold

To install:
5. Install or connect the following:
- Exhaust manifold with a new gasket

and torque the bolts to 15 ft. lbs. (20 Nm)
- Oil filter housing and torque the filter cartridge to 33 ft. lbs. (45 Nm)
- Front exhaust manifold pipe and gaskets, do not tighten the bolts
- Front exhaust pipe to the rear exhaust pipe and torque the bolts to 15 ft. lbs. (20 Nm)
- Front manifold pipe and torque the bolts to 15 ft. lbs. (20 Nm)
- Front exhaust manifold O$_2$S and torque it to 73 ft. lbs. (45 Nm)
- Power steering pipe bracket and torque the bolt to 71 inch lbs. (8 Nm)
- New O-rings to the coolant extension housing
- Coolant extension housing and torque the bolt to 15 ft. lbs. (20 Nm)

6. Lower the powertrain assembly.
- Front transaxle mount through bolt and torque it to 41 ft. lbs. (55 Nm)
- Upper radiator hose
- Negative battery cable

7. Fill the cooling system.

8. Start the vehicle and check for leaks, repair if necessary.

REAR MANIFOLD

1. Before servicing the vehicle, refer to the precautions in the beginning of this section.

2. Drain the cooling system.

3. Remove or disconnect the following:
- Negative battery cable
- Oxygen Sensor (O$_2$S)
- Exhaust Gas Recirculation (EGR) pipe
- Exhaust manifold pipe heat shield
- Rear exhaust manifold pipe
- Exhaust manifold

To install:

4. Install or connect the following:
- Exhaust manifold with a new gasket and torque the bolts to 15 ft. lbs. (20 Nm)
- Rear exhaust manifold pipe and torque the nuts to 25 ft. lbs. (30 Nm)
- Front exhaust manifold pipe to the rear exhaust manifold pipe and torque the bolts to 15 ft. lbs. (20 Nm)
- Rear exhaust manifold pipe to the resonator and torque the bolts to 15 ft. lbs. (20 Nm)
- EGR pipe and torque the fasteners to 19 ft. lbs. (25 Nm)
- O$_2$S and torque it to 37 ft. lbs. (50 Nm)

- Exhaust manifold heat shield and torque the bolts to 71 inch lbs. (8 Nm)
- Negative battery cable

5. Start the vehicle and check for leaks, repair if necessary.

Front Crankshaft Seal

REPLACEMENT

On the 1.9L and 2.2L engines (Except ION) the front crankshaft seal is located in the timing chain front cover. Refer to the timing chain procedure for information about removing the front cover and replacing the seal or if replacing the seal on the ION model refer to the procedure below.

ION

1. Before servicing the vehicle, refer to the precautions in the beginning of this section.

2. Remove or disconnect the following:
- Negative battery cable
- Drive belt
- Using harmonic balancer holder J 38122-A to prevent the crankshaft from rotating and remove the balancer bolt and discard
- Crankshaft balancer
- Front seal using a suitable prytool

To install:

3. Install or connect the following:
- Front seal using driver J 35268-A to drive the seal into the position on the front cover
- Crankshaft balancer
- Balancer bolt and using tool J 38122-A to prevent the crankshaft from rotating, install a new balancer bolt and tighten to 74 ft. lbs. (100 Nm) plus an additional 75 degrees

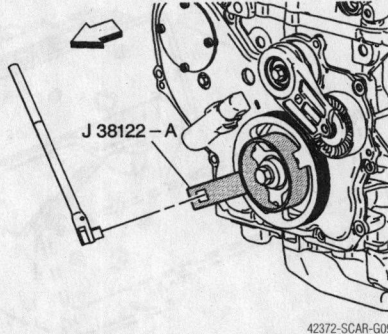

J 38122 — A

42372-SCAR-G09

Using harmonic balancer holder J 38122-A to prevent the crankshaft from rotating— ION models engine

- Drive belt
- Negative battery cable

3.0L Engine

1. Before servicing the vehicle, refer to the precautions in the beginning of this section.

2. Remove or disconnect the following:
- Negative battery cable
- Timing belt
- Crankshaft gear

3. Drill a small pilot hole into the steel ring of the seal.

4. Screw in a self-tapping screw.

5. Use pliers to pull out the oil seal.

To install:

6. Coat the lip of the new oil seal with engine oil.

7. Install the oil seal using a suitable seal installer.

8. Install the crankshaft gear and torque the bolt to 184 ft. lbs. (250 Nm) plus an additional 45 degrees then another 15 degrees.

9. Install the timing belt.

Camshaft and Lifters

REMOVAL & INSTALLATION

1.9L (SOHC) Engines

1. Before servicing the vehicle, refer to the precautions in the beginning of this section.

2. Remove or disconnect the following:
- Both battery cables
- Battery and tray
- Timing chain front cover
- Timing chain and camshaft sprocket
- Rocker arm/shaft assemblies
- Lifters and label or position them for assembly in their original locations

3. Drive the camshaft plug inward, then remove it from the cylinder head with a magnet.

4. Carefully pull the camshaft from the rear of the cylinder head through the oversized camshaft plug hole. Turn the camshaft back and forth slowly while withdrawing to help prevent journal or bearing damage.

To install:

5. Clean and inspect all parts prior to installation. Lubricate the camshaft and carefully insert it through the hole at the rear of the cylinder head.

6. Install or connect the following:
- New rear cylinder head plug coated with Loctite® 242

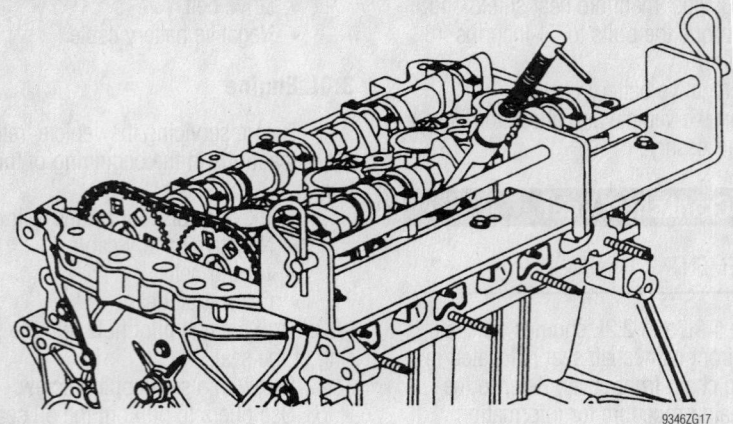

9346ZG17

Remove the rocker arm/shaft assemblies—1.9L (SOHC) engines

- Valve lifters into their original bores, or if the camshaft has been replaced, install new lifters
- Rocker arm/shaft assemblies
- Timing chain and camshaft sprocket
- Timing chain front cover
- Battery and tray and torque the battery hold-down nut and screw to 80 inch lbs. (9 Nm)
- Both battery cables

7. Start the engine and check for leaks, repair if necessary.

1.9L (DOHC) Engines

➡ Be careful when working around the camshaft sprockets and timing chain cover during this procedure. If a bolt or washer is accidentally dropped between the front cover and engine assembly, the cover will have to be removed for retrieval.

1. Before servicing the vehicle, refer to the precautions in the beginning of this section.

2. Remove or disconnect the following:
- Negative battery cable
- Accessory drive belt
- Spark plug wires
- Positive Crankcase Ventilation (PCV) fresh air hose
- Cam cover

3. Turn the crankshaft clockwise until the mark on the crankshaft pulley is in alignment with the pointer on the front cover and the No. 1 cylinder is at Top Dead Center (TDC) of the compression stroke. Both camshaft dowel pins will be at the 12 o'clock position and the timing pin holes will be aligned when the No. 1 cylinder is at TDC. If necessary, the right wheel and splash shield can be removed to help observe the timing marks.

4. Remove the camshaft sprocket retaining bolts. Use a ⅞ in. (21mm) open-end wrench to hold the camshaft from turning while removing the bolts.

5. Position a front angled support fixture in front of the camshaft sprockets.

6. Attach the camshaft sprocket adapters to the end of each camshaft using the pilot bolts, but do not tighten the bolts. The front angled support should come between the sprocket adapters and camshaft sprockets.

7. Remove or disconnect the following:
- Upper timing chain guide

- Both front camshaft bearing caps

8. Secure the support fixture using ⅞ in. bolts/blocks and align the 2 holes in each camshaft sprocket, adapter and the front support fixture. Install the 4 nuts, but do not tighten.

➡ The steel blocks should be installed against the rearward side of the camshaft sprocket.

9. Tighten the sprocket pilot bolts to 19 ft. lbs. (25 Nm) while holding the camshafts from turning with an open end wrench.

10. Move each camshaft sprocket off the end of the camshaft by rocking the sprocket forward or by carefully prying between the end of the camshaft and the sprocket. Then, tighten the 4 nuts and bolts with blocks from the side of the support fixture to 19 ft. lbs. (25 Nm).

11. Install the 2 bolts retaining the support fixture to the engine front cover and tighten the bolts to 89 inch lbs. (10 Nm). Then, remove each camshaft sprocket pilot bolt while holding the camshafts with a wrench.

12. Carefully pry between the sprocket and the end of the camshaft to move the camshaft rearward. Pry only enough to remove its end from inside the sprocket pilot otherwise camshaft or lifter damage may occur.

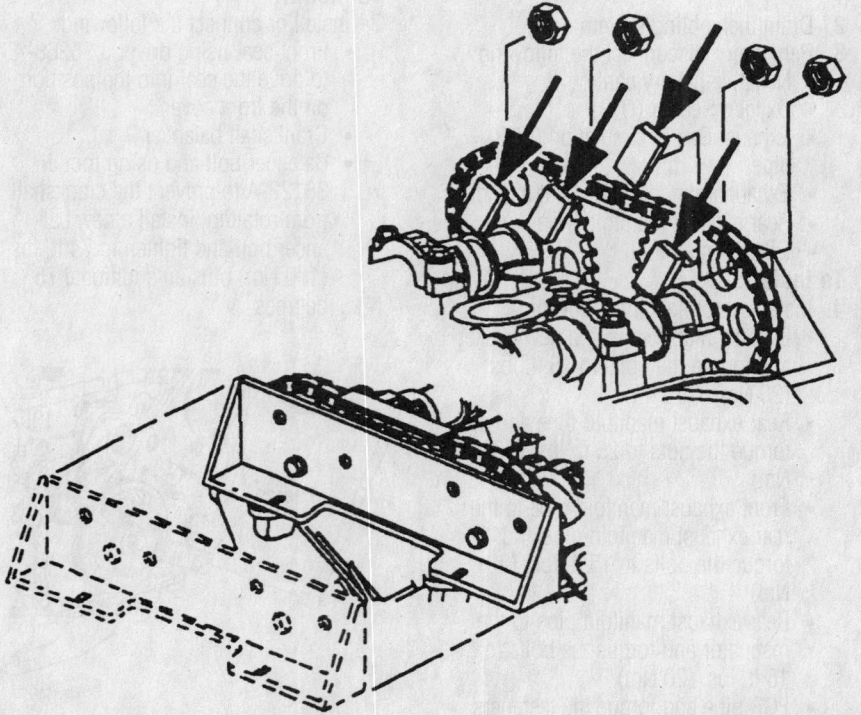

7922Z812

Install the camshaft support fixture to securely hold the camshafts in position—1.9L (DOHC) engines

13. Remove or disconnect the following:
  - Uniformly loosen and remove the remaining camshaft bearing cap bolts. To prevent bolt/cap damage, do not use power tools and make several passes.
  - Camshafts
  - Pull the lifters out to remove them, always place them in the order in which they were removed and with the camshaft side facing down. Oil will drain out of the lifter if it is placed valve side down.

To install:

14. Clean and inspect all parts prior to installation. Oil the camshaft and install with the **IN** camshaft on the intake side and **EX** camshaft on the exhaust side.

➡ **The dowel pin in each camshaft must be located at the 12 o'clock position during installation to prevent valve and piston damage.**

15. Install or connect the following:
  - Lifters in their original locations
  - Bearing caps, except for the forward pair, in their original positions, making sure the arrows on the caps are pointing forward toward the camshaft sprockets. Lightly oil each of the cap bolts, then install and uniformly torque the bolts to 124 inch lbs. (14 Nm).
  - Camshaft sprocket pilot bolt in each camshaft and torque the bolts to 124 inch lbs. (14 Nm) in order to pull the camshaft fully forward and align the sprocket support for installation of the sprocket onto the camshaft

16. Remove the 4 sprocket support bolt/blocks and nuts.

17. Install the forward bearing caps and the upper chain guide. Torque the cap bolts to 124 inch lbs. (14 Nm).

18. Be sure the camshaft dowel pin aligns with the slot in each camshaft sprocket. If necessary, rotate the camshaft slightly (1–2 degrees) and move each sprocket from the adapter onto the end of the camshaft. Fully seat each sprocket on the end of each camshaft.

19. Remove the 2 sprocket pilot bolts and adapters while using a wrench on the camshaft flats to assure the camshaft cannot move.

20. Remove the support angle fixture.

21. Install the camshaft sprocket retaining bolts and washers. Hold the camshafts and torque the bolts to 76 ft. lbs. (103 Nm).

22. Verify all visible timing marks and holes are in alignment. Turn the crankshaft

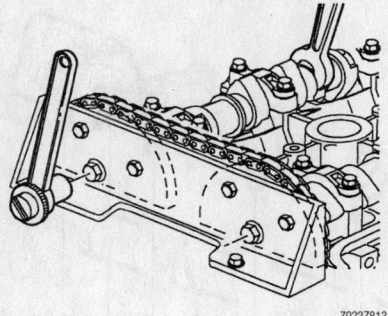

7922ZB13

Hold the camshaft with a wrench while tightening the sprocket bolts—1.9L (DOHC) engine shown

clockwise until the mark on the crankshaft pulley aligns with the mark on the front cover. Check timing by inserting ³⁄₁₆ in. drill bits through the camshaft sprocket alignment holes, into the cylinder head. If the alignment pins cannot be inserted, turn the crankshaft 360 degrees clockwise and repeat. If the pins cannot be inserted within 1–2 degrees of either TDC position, the camshafts are not properly timed. Do not start the engine until the camshafts are timed.

23. Apply a small drop of RTV across the cylinder head and front cover T-joints. Inspect the old camshaft cover gasket and replace if damaged. Install the gasket and the camshaft cover. Torque the fasteners in proper sequence to 89 inch lbs. (10 Nm).

24. Install or connect the following:
  - Right splash shield and wheel, if removed to observe the timing marks
  - PCV and fresh air hoses and the spark plug wires
  - Accessory drive belt
  - Negative battery cable

25. Start the engine and check for leaks, repair if necessary.

L–Series With 2.2L Engine

➡ **Be very careful when working around the camshaft sprockets and timing chain cover during this procedure. If a bolt or washer is accidentally dropped between the front cover and engine assembly, the cover will have to be removed for retrieval.**

1. Before servicing the vehicle, refer to the precautions in the beginning of this section.

2. Remove or disconnect the following:
  - Negative battery cable
  - Ignition coil
  - Ground strap
  - Positive Crankcase Ventilation (PCV) fresh air hose

Intake Side			
9	5	6	10
	3	1	2
8	4	7	11
Exhaust Side			

9356ZG01

Camshaft cover bolt tightening sequence—1.9L DOHC engines

  - Cam cover
  - Fuel line
  - Wire harness bracket
  - Power steering pump without removing the lines

➡ **To avoid valve piston contact, the No. 1 cylinder piston must be positioned at 60 degrees Before Top Dead Center (BTDC). The pistons are properly aligned when the diamond shaped hole on the intake camshaft sprocket is located at 12 o'clock.**

3. Remove the upper timing chain guide and front camshaft caps.

4. Install a camshaft sprocket holding tool through the sprocket holes from the timing chain side. Align the guide pins into the slot on the support head. Torque the pins to 89 inch lbs. (10 Nm).

5. Hold each camshaft in place with a 24mm open end wrench and remove the camshaft timing sprocket retaining bolts and washers. Discard the bolts.

6. Uniformly loosen and remove the remaining camshaft bearing caps.

7. Slide the camshaft sprockets away from the camshafts and remove the camshaft.

To install:

8. Lubricate the camshaft bearing journals with clean engine oil.

9. Install both camshafts and all bearing caps except the front cap on each camshaft.

10. Torque the bearing caps uniformly, except for the front caps and the rear intake cap, to 89 inch lbs. (10 Nm).

➡ **Make certain that the alignment notches are properly positioned with the notches in the camshaft sprockets before final torque is applied. Also, be sure that the timing chain is properly aligned on the fixed guide.**

11. Slide the camshaft sprockets and timing chain on the guide pins toward the camshafts. Rotate the camshafts with a

Remove the camshaft and bearing caps—L–Series With 2.2L engine

9346ZG18

- Air cleaner assembly
- Accelerator cable from the throttle body and bracket
- Accelerator bracket bolts and the bracket
- Positive Crankcase Ventilation (PCV) hose
- Fuel line bracket
- Ignition Coil Module (ICM) connector
- ICM screws and the ICM
- Ignition coil housing bolts and the housing
- Ground strap stud from the camshaft cover
- Ground strap
- Camshaft cover bolts
- Camshaft cover
- Upper timing chain guide

3. Install camshaft sprocket holding tool J43655 as illustrated.
- Intake and exhaust camshaft sprocket bolts and discard the bolts

4. Slide the sprockets forward and mark the camshaft bearing caps to ensure they are installed in their correct positions.

24mm open end wrench to align the camshaft and sprocket.

12. Install new camshaft sprocket bolts and torque them to 63 ft. lbs. (85 Nm) plus 30 degrees.

13. Remove the camshaft sprocket holding tool.

14. Install or connect the following:
- Front camshaft bearing caps and torque the bolts to 89 inch lbs. (10 Nm)
- Upper timing chain guide and apply Loctite® to the bolts
- Rear intake camshaft bearing cap and torque the bolts 19 ft. lbs. (25 Nm)
- Power steering pump and torque the bolts to 19 ft. lbs. (25 Nm)
- Ignition coil
- PCV fresh air hose
- Ground strap
- Fuel line
- Wire harness bracket
- Cam cover
- Negative battery cable

15. Start the vehicle and check for leaks, repair if necessary.

ION With 2.2L Engine

1. Before servicing the vehicle, refer to the precautions in the beginning of this section.

2. Remove or disconnect the following:
- Negative battery cable

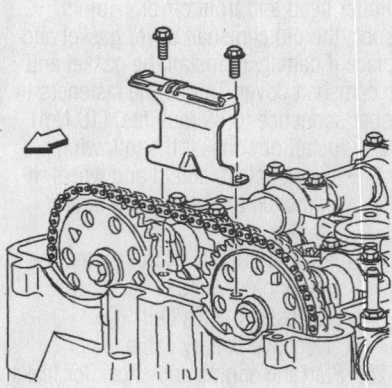

42372-SCAR-G03

Remove the upper timing chain guide— ION

42372-SCAR-G04

Install camshaft sprocket holding tool J43655–ION

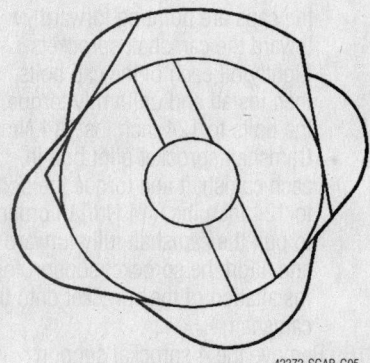

42372-SCAR-G05

Make sure the alignment notches are aligned with the camshaft sprocket—ION

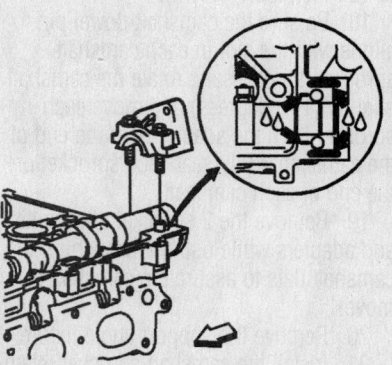

42372-SCAR-G06

Apply 0.197 inch (5mm) of anaerobic Gasket maker to the rear camshaft bearing cap—ION

5. Remove each bearing cap bolt one turn at a time until there is no tension.

6. Remove the bearing cap and the camshaft
- Roller followers
- Lash adjusters

To install:

7. Lubricate the valve tips.

8. Install or connect the following:
- Lash adjusters
- Roller followers

9. Make sure the alignment notches are aligned with the camshaft sprocket as illustrated.
- Camshaft
- Bearing caps except the rear camshaft cap

10. Tighten the caps in increments of 3 turns until they are seated, then tighten to 89 inch lbs. (10 Nm).

11. Apply 0.197 inch (5mm) of Permatex Anaerobic Gasket maker® 51813 to the rear camshaft bearing cap. Install the cap and tighten the bolts to 18 ft. lbs. (25 Nm).
- Camshaft sprockets and hand tighten the bolts

12. Remove camshaft sprocket holding tool J43655
- Camshaft sprocket bolts to 63 ft. lbs. (85 Nm) plus and additional 30 degrees

13. Install or connect the following:
- Upper timing chain guide and tighten to 89 inch lbs. (10 Nm)
- Camshaft cover
- Camshaft cover bolts and tighten to 89 inch lbs. (10 Nm)
- Ground strap
- Ground strap stud to the camshaft cover and tighten to 89 inch lbs. (10 Nm)
- Ignition coil housing and tighten the bolts to 89 inch lbs. (10 Nm)
- ICM and tighten the screws to 13 inch lbs. (1.5 Nm)
- ICM connector
- Fuel line bracket
- PCV hose
- Accelerator bracket and tighten the bolts to 89 inch lbs. (10 Nm)
- Accelerator cable to the throttle body and bracket
- Air cleaner assembly
- Negative battery cable

3.0L Engine

This engine is equipped with front and rear camshafts.

The front camshaft bearing caps for the cylinder head are marked R1–R8 and the rear cylinder head bearing caps are marked L1–L8.

→Be very careful when working around the camshaft sprockets and timing chain cover during this procedure. If a bolt or washer is accidentally dropped between the front cover and engine assembly, the cover will have to be removed for retrieval.

1. Before servicing the vehicle, refer to the precautions in the beginning of this section.

2. Remove or disconnect the following:
- Negative battery cable
- Intake plenum
- Air cleaner assembly
- Camshaft cover
- Front timing belt cover
- Timing belt

→Rotate the crankshaft counterclockwise to 60 degrees Before Top Dead Center (BTDC) to prevent valve to piston contact.

3. Install a Camshaft Gear Locking Tool into the camshaft gears.

4. Remove or disconnect the following:
- Loosen the camshaft gear bolt, remove the holding tool
- Camshaft gear bolt and discard it
- Camshaft gear
- Loosen the camshaft bearing caps sequentially starting in the center and working outward in a spiral direction
- Camshaft bearing caps
- Camshaft

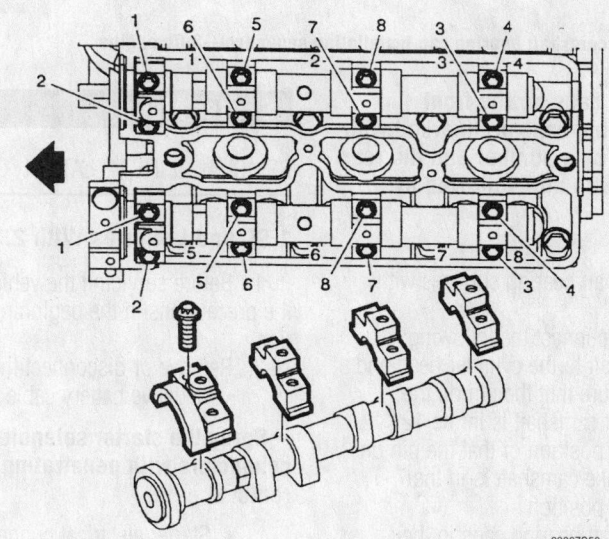

9306ZG50

Front camshaft bearing cap removal sequence—3.0L engine

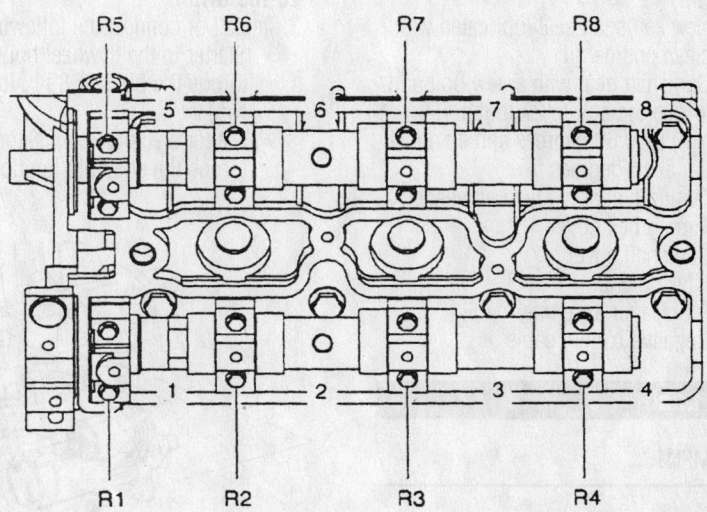

9306ZG51

The rear camshaft bearing caps are marked to ensure proper installation—3.0L engine

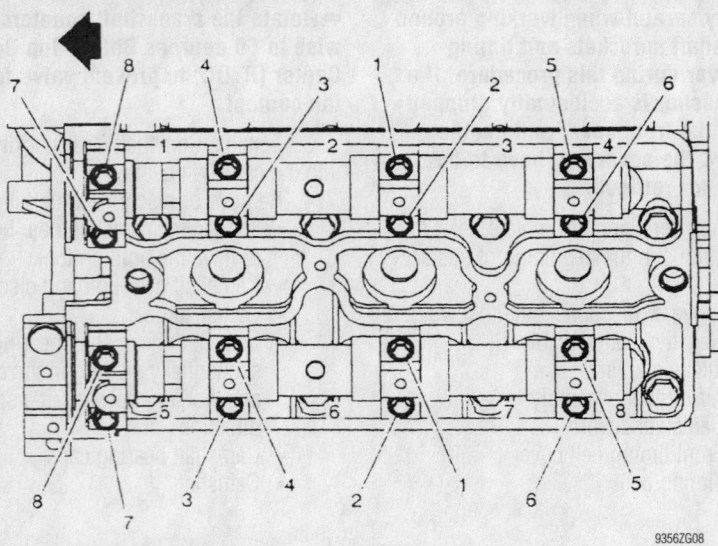

Front and rear camshaft bearing cap installation sequence—3.0L engine

➡**The bearing caps for the front camshaft bearing caps are marked with an R followed by a number and the rear caps marked with an L followed by a number.**

To install:

5. Lubricate all bearing surfaces with clean engine oil.
6. Install or connect the following:
   - Camshaft to the cylinder head and make sure that the pin on the exhaust camshaft is in the 12 o'clock position or that the pin on the intake camshaft is in the 7 o'clock position
   - Camshaft bearing caps in their proper position and torque the bolts, starting in the center and working outward, to 71 inch lbs. (8 Nm)
   - New camshaft seal lubricated with clean engine oil
   - Camshaft gear with a new bolt and torque the bolt to 27 ft. lbs. (50 Nm) plus 60 degrees and an additional 15 degrees
   - Timing belt and adjust as needed
   - Timing belt cover
   - Camshaft cover
   - Intake plenum
   - Air cleaner assembly
   - Negative battery cable

Valve Lash

ADJUSTMENT

All engines utilize hydraulic lash adjusters; no adjustment is necessary.

Starter Motor

REMOVAL & INSTALLATION

1.9L and L–Series With 2.2L Engines

1. Before servicing the vehicle, refer to the precautions in the beginning of this section.
2. Remove or disconnect the following:
   - Negative battery cable

➡**Spray the starter solenoid electrical connectors with penetrating oil before removal.**

   - Starter electrical connections
   - Starter bolts
   - Starter assembly by pulling it toward the left side of the vehicle

To install:

3. Install or connect the following:
   - Starter to the flywheel housing and torque the bolts to 30 ft. lbs. (40 Nm)
   - Starter electrical connectors and torque the solenoid ignition wire to

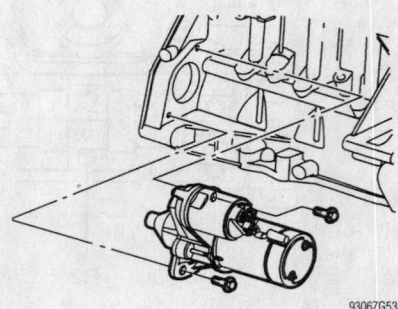

Starter assembly removal—1.9L engines

44 inch lbs. (5 Nm) and the positive battery cable to 89 inch lbs. (10 Nm)
   - Negative battery cable

ION

1. Before servicing the vehicle, refer to the precautions in the beginning of this section.
2. Remove or disconnect the following:
   - Negative battery cable

➡**Spray the starter solenoid electrical connectors with penetrating oil before removal.**

   - Starter electrical connections
   - Starter bolts
   - Starter assembly

To install:

3. Install or connect the following:
   - Starter to the flywheel housing and torque the bolts to 30 ft. lbs. (40 Nm)
   - Starter electrical connectors and torque the solenoid ignition wire to 44 inch lbs. (5 Nm) and the positive battery cable to 89 inch lbs. (10 Nm), on all models except ION. On ION models, attach the electrical connectors to the starter and tighten the battery terminal nut to 13 ft. lbs. (17 Nm) and the S terminal connector to 27 inch lbs. (3 Nm).
   - Negative battery cable

3.0L Engine

1. Before servicing the vehicle, refer to the precautions in the beginning of this section.
2. Remove or disconnect the following:
   - Negative battery cable
   - Right front wheel
   - Starter electrical connections
   - Loosen the fastener securing the electrical harness bracket to the engine

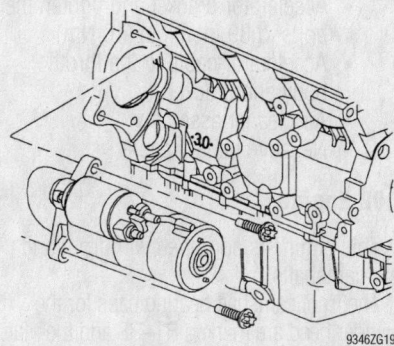

Starter motor mounting—3.0L engine

- Starter bolts
- Starter assembly by pulling it toward the left side of the vehicle

To install:

3. Install or connect the following:
   - Starter to the flywheel housing and torque the bolts to 30 ft. lbs. (40 Nm)
   - Starter electrical connections and tighten the electrical harness bracket bolt
   - Right front wheel
   - Negative battery cable

Oil Pan

REMOVAL & INSTALLATION

1.9L Engines

1. Before servicing the vehicle, refer to the precautions in the beginning of this section.
2. Drain the oil from the engine.
3. Remove or disconnect the following:
   - Negative battery cable
   - Front exhaust pipe
   - Front stiffening bracket
   - Flywheel cover
   - Right front wheel
   - Wheel well splash shield
4. Loosen the 4 front motor mount bolts approximately ½ in. (12mm).
   - Oil pan bolts

➡**If equipped with a manual transaxle, an 8mm flex socket may be used to access the rear oil pan bolts located next to the flywheel.**

5. Using an RTV cutter tool, separate the oil pan from the engine. Drive the tool around the pan to shear the RTV seam, then tap the pan sideways with a rubber mallet to loosen.
6. Pry the engine mount away from the engine as necessary and remove the oil pan. Be careful not to damage or score component surfaces when prying.

To install:

7. Apply a 0.16 in. (4mm) bead of RTV sealer to the pan flange. Be sure the RTV is applied to the inner side of the flange from the bolt holes.
8. Install or connect the following:
   - Oil pan and torque the bolts to 80 inch lbs. (9 Nm)
   - Front engine mount bolts and torque them to 37 ft. lbs. (50 Nm)
   - Right splash shield and wheel
   - Engine stiffening bracket
   - Flywheel cover

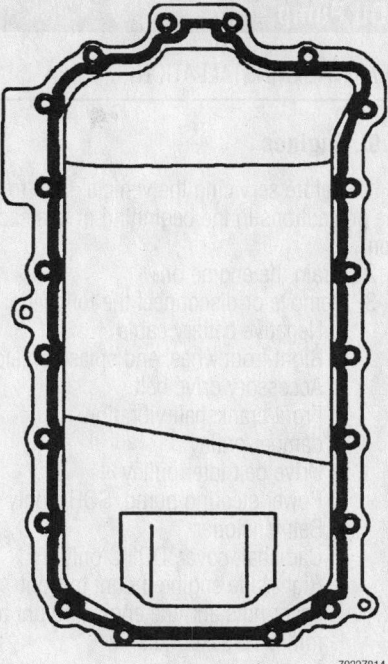

79222Z814

Apply a 0.16 in (4mm) bead of RTV to the oil pan flange to the inner side of the bolt holes—1.9L engines

- Exhaust pipe and torque the pipe-to-manifold nuts in a crosswise pattern to 23 ft. lbs. (31 Nm) and the pipe to converter bolts to 33 ft. lbs. (45 Nm)
- Negative battery cable

9. Fill the engine with clean oil.
10. Start the vehicle and check for leaks, repair if necessary.

L–Series With 2.2L Engine

1. Before servicing the vehicle, refer to the precautions in the beginning of this section.

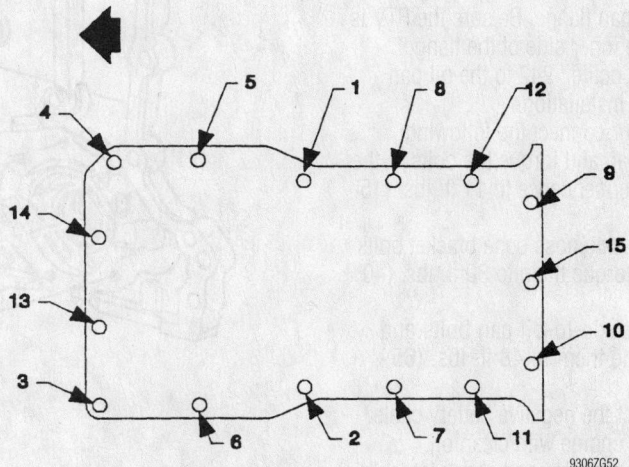

Oil pan bolts torque sequence—2.2L engines

9306ZG52

2. Drain the engine oil.
3. Remove or disconnect the following:
   - Negative battery cable
   - Oil pan bolts
4. Using a flat-bladed tool, pry the oil pan from the engine block.

To install:

5. Apply a 0.08 in. (2mm) bead of RTV sealer to the pan flange. Be sure the RTV is applied to the inner side of the flange.
6. Install or connect the following:
   - Oil pan and torque the bolts in the proper sequence to 18 ft. lbs. (25 Nm)
   - Negative battery cable
7. Fill the engine with clean oil.
8. Start the vehicle and check for leaks, repair if necessary.

ION

1. Before servicing the vehicle, refer to the precautions in the beginning of this section.
2. Drain the engine oil.
3. Remove or disconnect the following:
   - Negative battery cable
   - Drive belt
   - Lower A/C compressor bolt
   - Oil pan bolts
4. Using a flat-bladed tool, pry the oil pan from the engine block.

To install:

5. Apply a 0.08 in. (2mm) bead of RTV sealer to the pan flange. Be sure the RTV is applied to the inner side of the flange.
6. Install or connect the following:
   - Oil pan and torque the bolts in the proper sequence to 18 ft. lbs. (25 Nm)
   - Lower A/C compressor bolt
   - Drive belt
   - Negative battery cable

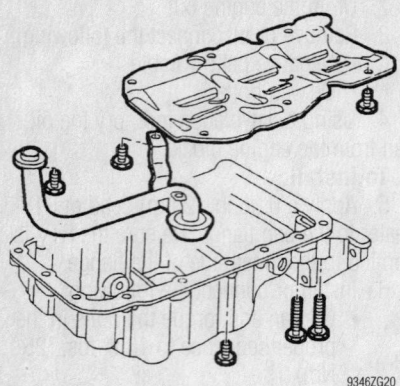

9346ZG20

Oil pan and related components—3.0L engine

7. Fill the engine with clean oil.

8. Start the vehicle and check for leaks, repair if necessary.

3.0L Engine

1. Before servicing the vehicle, refer to the precautions in the beginning of this section.

2. Drain the engine oil.

3. Remove or disconnect the following:

- Negative battery cable
- Nose cone bracket bolts from the oil pan
- Lower transaxle flange-to-oil pan bolts
- Oil pan bolts

➡**Separate the oil pan from the engine with an RTV cutter tool. Drive the tool around the pan to shear the RTV seam, then tap the pan sideways with a rubber mallet to loosen.**

- Oil pan

To install:

4. Apply a 0.10 in. (2mm) bead of RTV sealer to the pan flange. Be sure the RTV is applied to the inner side of the flange.

5. Apply Loctite® 242 to the oil pan bolts prior to installation.

6. Install or connect the following:

- Oil pan and torque the bolts in the proper sequence to 11 ft. lbs. (15 Nm)
- Transaxle nose cone bracket bolts and torque them to 30 ft. lbs. (40 Nm)
- Transaxle-to-oil pan bolts and torque them to 48 ft. lbs. (65 Nm)

7. Connect the negative battery cable.

8. Fill the engine with clean oil.

9. Start the vehicle and check for leaks, repair if necessary.

Oil Pump

REMOVAL & INSTALLATION

1.9L Engines

1. Before servicing the vehicle, refer to the precautions in the beginning of this section.

2. Drain the engine oil.

3. Remove or disconnect the following:

- Negative battery cable
- Right front wheel and splash shield
- Accessory drive belt
- Front crankshaft vibration damper/pulley
- Drive belt idler pulley
- Power steering pump, SOHC only
- Belt tensioner
- Camshaft cover, DOHC only
- Right side engine mount to front cover nuts and the engine mount to midrail bracket nuts
- Front 4 oil pan bolts
- Front cover bolts
- Drive rotor and driven rotor

4. If necessary, remove the relief valve. Because the puller jaws will damage the relief valve sealing seat, the valve cannot be used again when removed.

To install:

5. Install or connect the following:

- New relief valve into the cover bore, if removed. Coat the valve with clean engine oil and tap it into the bore.

➡**Whenever the oil pump is installed, the assembly must be packed with petroleum jelly in order to prime the pump.**

- Drive and driven rotors into the pump with the chamfer toward the front oil seal
- Oil pump body cover using new bolts and torque them to 97 inch lbs. (11 Nm)
- New oil pressure and suction seals in the cylinder block
- Front oil pan bolts and torque them to 80 inch lbs. (9 Nm)
- Front cover and torque the perimeter and center bolts to 19 ft. lbs. (25 Nm) and the lower center bolt to 89 inch lbs. (10 Nm)
- Front oil pan bolts and torque them to 80 inch lbs. (9 Nm)
- Camshaft cover
- Drive belt tensioner and torque the fasteners to 26 ft. lbs. (35 Nm)
- Power steering pump, if removed
- Drive belt and torque the pulley bolts to 33 ft. lbs. (45 Nm)
- Engine mount to midrail bracket nuts first and the engine mount to front cover nuts last and torque all nuts to 37 ft. lbs. (50 Nm)
- Right front splash shield and wheel
- Negative battery cable

6. Fill the engine with clean oil and replace the oil filter.

7. Start the vehicle and check for leaks, repair if necessary.

L–Series With 2.2L Engine

1. Before servicing the vehicle, refer to the precautions in the beginning of this section.

2. Drain the engine oil.

3. Remove or disconnect the following:

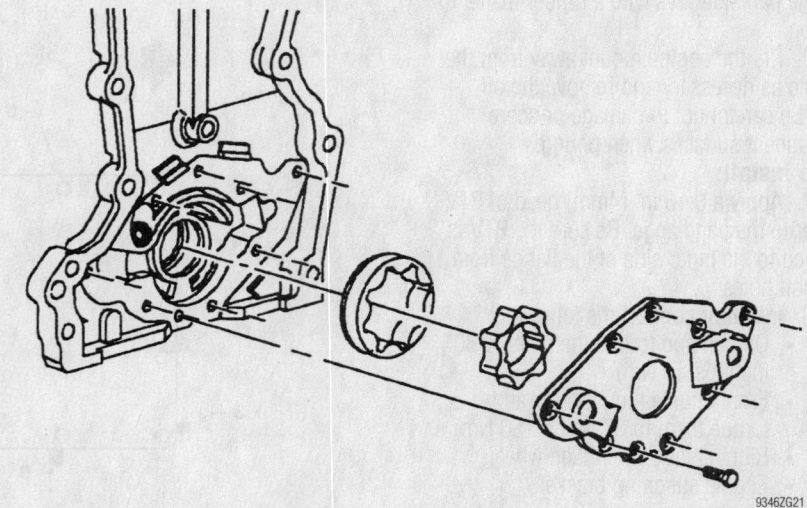

9346ZG21

Oil pump drive rotor and driven rotor—1.9L engines

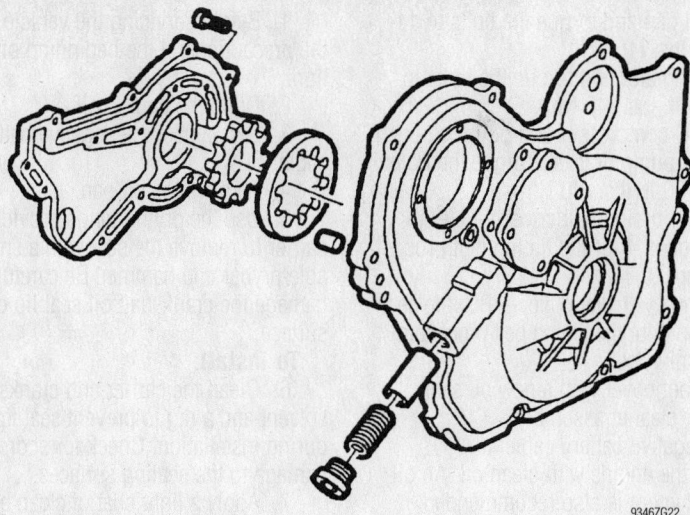

9346ZG22

Front cover and oil pump assembly—2.2L engines

- Negative battery cable
- Air cleaner assembly
- Right front wheel and splash shield
- Accessory drive belt
- Crankshaft damper pulley
- Belt tensioner
4. Install an engine support fixture.
  - Right front engine mount
  - Front cover bolts and the 13mm bolt under the water pump
  - Front cover
  - Oil pump cover plate
  - Drive rotor and driven rotor
  - Pressure relief valve

To install:

5. Install or connect the following:
  - New relief valve into the cover bore, if removed. Coat the valve with clean engine oil and tap it into the bore. Torque the plug to 30 ft. lbs. (40 Nm).

➡**Whenever the oil pump is installed, the assembly must be packed with petroleum jelly in order to prime the pump.**

- Drive and driven rotors into the pump with the chamfer toward the front oil seal
- Oil pump body cover using new bolts and torque the bolts to 53 inch lbs. (6 Nm)
- Front cover with a new oil seal and torque the perimeter and center bolts to 19 ft. lbs. (25 Nm) and the lower center bolt to 89 inch lbs. (10 Nm)
- Right side engine mount and torque the bolts to 41 ft. lbs. (55 Nm)
6. Remove the engine support fixture.

- Drive belt tensioner and torque the bolts 37 ft. lbs. (50 Nm)
- Crankshaft damper pulley and torque the bolt to 74 ft. lbs. (100 Nm) plus 75 degrees
- Accessory drive belt
- Right front splash shield and wheel
- Air cleaner assembly
- Negative battery cable
7. Fill the engine with clean oil and replace the oil filter.
8. Start the vehicle and check for leaks, repair if necessary.

ION

1. Before servicing the vehicle, refer to the precautions in the beginning of this section.
2. Drain the engine oil.
3. Remove or disconnect the following:
  - Negative battery cable
  - Air cleaner assembly
  - Right front wheel and splash shield
  - Accessory drive belt
  - Crankshaft damper pulley
  - Belt tensioner
4. Install an engine support fixture.
  - Right front engine mount
  - Front cover bolts and the 13mm bolt under the water pump
  - Front cover
  - Oil pump cover plate
  - Drive rotor and driven rotor
  - Pressure relief valve

To install:

5. Install or connect the following:
  - New relief valve into the cover bore, if removed. Coat the valve with clean engine oil and tap it into the bore. Torque the plug to 30 ft. lbs. (40 Nm).

➡**Whenever the oil pump is installed, the assembly must be packed with petroleum jelly in order to prime the pump.**

- Drive and driven rotors into the pump with the chamfer toward the front oil seal
- Oil pump body cover using new bolts and torque the bolts to 53 inch lbs. (6 Nm)
- Front cover with a new oil seal and torque the perimeter and center bolts to 19 ft. lbs. (25 Nm) and the lower center bolt to 89 inch lbs. (10 Nm)
- Right side engine mount and torque the bolts to 41 ft. lbs. (55 Nm)
6. Remove the engine support fixture.
  - Drive belt tensioner and torque the bolts 37 ft. lbs. (50 Nm)
  - Crankshaft damper pulley and torque the bolt to 74 ft. lbs. (100 Nm) plus 75 degrees
  - Accessory drive belt
  - Right front splash shield and wheel
  - Air cleaner assembly
  - Negative battery cable
7. Fill the engine with clean oil and replace the oil filter.
8. Start the vehicle and check for leaks, repair if necessary.

3.0L Engine

1. Before servicing the vehicle, refer to the precautions in the beginning of this section.
2. Drain the engine oil.
3. Drain the cooling system.
4. Remove or disconnect the following:
  - Negative battery cable
  - Air cleaner assembly
  - Front timing belt cover
  - Timing belt
  - Rear timing belt cover
  - A/C compressor and power steering pump bracket and move them away from the oil pump housing
  - Alternator bolts and move the alternator out of the way
  - Oil pan
5. Mount a crank hub holding tool to the crankshaft drive gear and remove the drive gear.
  - Oil pump bolts
  - Oil pan housing bolts that thread into the oil pump
  - Oil pump
  - Front main oil seal and collar
  - Pressure relief valve
  - Oil pump cover
  - Drive rotor and driven rotor

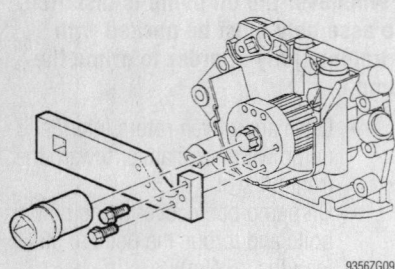

9356ZG09

Mount a crank hub holding tool to the crankshaft drive gear—3.0L engine

To install:

6. Install the new relief valve into the cover bore (if removed) and torque the plug to 30 ft. lbs. (40 Nm).

➡️ **Whenever the oil pump is installed, the new gasket must be coated with a thin bead of sealing Loctite® 518.**

7. Install or connect the following:
- Drive and driven rotors into the pump with the chamfer toward the front oil seal
- Oil pump body cover using new bolts and torque them to 89 inch lbs. (10 Nm)
- Drive gear and torque the new bolt to 184 ft. lbs. (250 Nm) plus 45 degrees then an additional 15 degrees

- Oil pan and torque the bolts to 11 ft. lbs. (15 Nm)
- Alternator and torque the bolts to 30 ft. lbs. (40 Nm)
- A/C compressor and power steering pump bracket. Torque the bolts to 30 ft. lbs. (40 Nm).
- Rear timing belt cover—Refer to section 4 for the timing belt procedure.
- Drive belt idler pulley—Refer to section 4 for the timing belt procedure.
- Timing belt
- Front cover with a new oil seal
- Air cleaner assembly
- Negative battery cable

8. Fill the engine with clean oil. An oil filter replacement is also recommended.

9. Fill the cooling system.

10. Start the vehicle and check for leaks, repair if necessary.

Rear Main Seal

REMOVAL & INSTALLATION

1.9L Engines

The Single Over Head Camshaft (SOHC) and Dual Over Head Camshaft (DOHC) engines use a 1-piece round seal mounted in a seal carrier.

1. Before servicing the vehicle, refer to the precautions in the beginning of this section.

2. Drain the engine oil.

3. Disconnect the negative battery cable.

4. Remove the oil pan.

5. Use the prying tangs provided in the carrier to remove the seal with a small suitable prybar and hammer. Be careful not to damage the crankshaft oil seal lip contact surface.

To install:

6. Clean the carrier and crankshaft with solvent and a rag to prevent seal lip damage during installation. Check for scores or damage to the sealing surfaces.

7. Apply a light coat of clean engine oil to the seal lip and the carrier inner diameter.

8. Install a new rear main seal, using a Seal Installer tool.

9. Install the oil pan.

10. Connect the negative battery cable.

11. Fill the engine with clean oil.

12. Start the vehicle and check for leaks, repair if necessary.

L–Series With 2.2L Engine

1. Before servicing the vehicle, refer to the precautions in the beginning of this section.

2. Remove or disconnect the following:
- Negative battery cable
- Transaxle
- Clutch/pressure plate assembly, if equipped with a manual transaxle
- Flywheel
- Rear main bearing seal by prying it from the engine

➡️ **Be careful not to damage or scratch the seal mounting surfaces.**

To install:

3. Lubricate the new rear main bearing seal with engine oil.

4. Install or connect the following:
- New rear main seal using a Rear

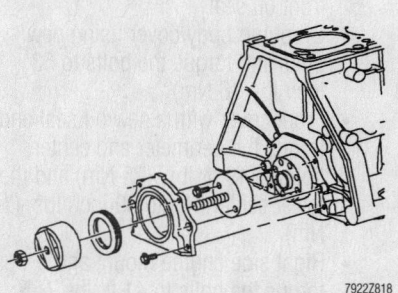

7922Z818

Exploded view of the rear main seal installation—1.9L engines

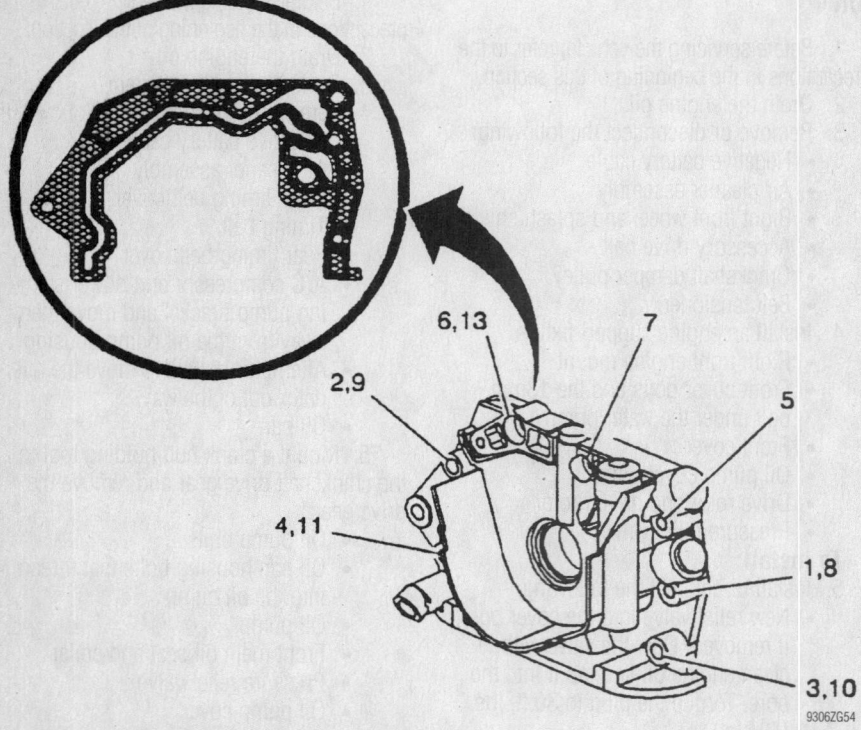

6,13   7

2,9

5

4,11

1,8

3,10

9306ZG54

Oil pump bolt tightening sequence—3.0L engine

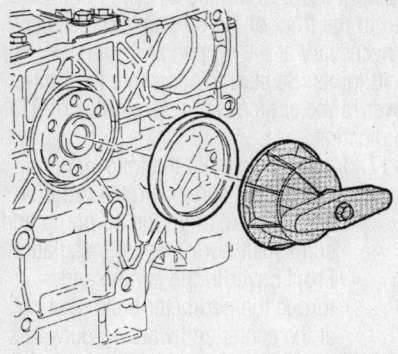

**Rear oil seal and installation tool
J42067—2.2L engine**

Main Bearing Oil Seal Installer Tool
J42067 until it is flush with the block
- Flywheel
- Clutch/pressure plate assembly, if
equipped with a manual transaxle
- Transaxle
- Negative battery cable
5. Start the engine and check for leaks,
repair if necessary.

ION With 2.2L Engine

1. Before servicing the vehicle, refer to the
precautions in the beginning of this section.
2. Remove or disconnect the following:
- Negative battery cable
- Transaxle
- Clutch/pressure plate assembly, if
equipped with a manual transaxle
- Flywheel
- Rear main bearing seal by prying it
from the engine

➡**Be careful not to damage or scratch
the seal mounting surfaces.**

To install:
3. Lubricate the new rear main bearing
seal with engine oil.
4. Install or connect the following:
- New rear main seal using a Rear
Main Bearing Oil Seal Installer Tool
J42067 until it is flush with the
block
- Flywheel
- Clutch/pressure plate assembly, if
equipped with a manual transaxle
- Transaxle
- Negative battery cable
5. Start the engine and check for leaks,
repair if necessary.

3.0L Engine

1. Before servicing the vehicle, refer to
the precautions in the beginning of this section.

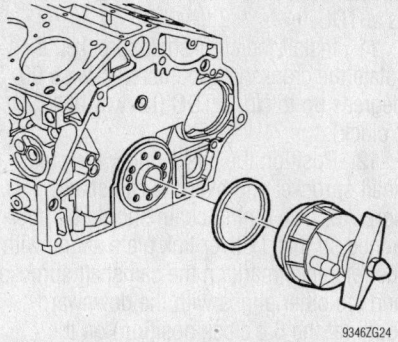

**Rear main oil seal and installation tool
J42067—3.0L engine**

2. Remove or disconnect the following:
- Negative battery cable
- Transaxle
- Clutch/pressure plate assembly, if
equipped with a manual transaxle
- Flywheel
3. Center punch the steel ring of the oil
seal.
4. Drill a small hole into the steel ring.
5. Install a self-tapping screw and using
pliers, pull out the rear main oil seal.

➡**Be careful not to damage or scratch
the seal mounting surfaces.**

To install:
6. Lubricate the new rear main oil seal
with engine oil.
7. Install or connect the following:
- New rear main seal using a Rear
Main Bearing Oil Seal Installer Tool
SA9121E until it is flush with the
block
- Flywheel
- Clutch/pressure plate assembly, if
equipped with a manual transaxle
- Transaxle
- Negative battery cable
8. Start the engine and check for leaks,
repair if necessary.

Timing Chain, Sprockets, Front Cover and Seal

REMOVAL & INSTALLATION

1.9L SOHC Engines

1. Before servicing the vehicle, refer to the
precautions in the beginning of this section.
2. Drain the engine oil.
3. Remove or disconnect the following:
- Negative battery cable
- Right front wheel and splash shield

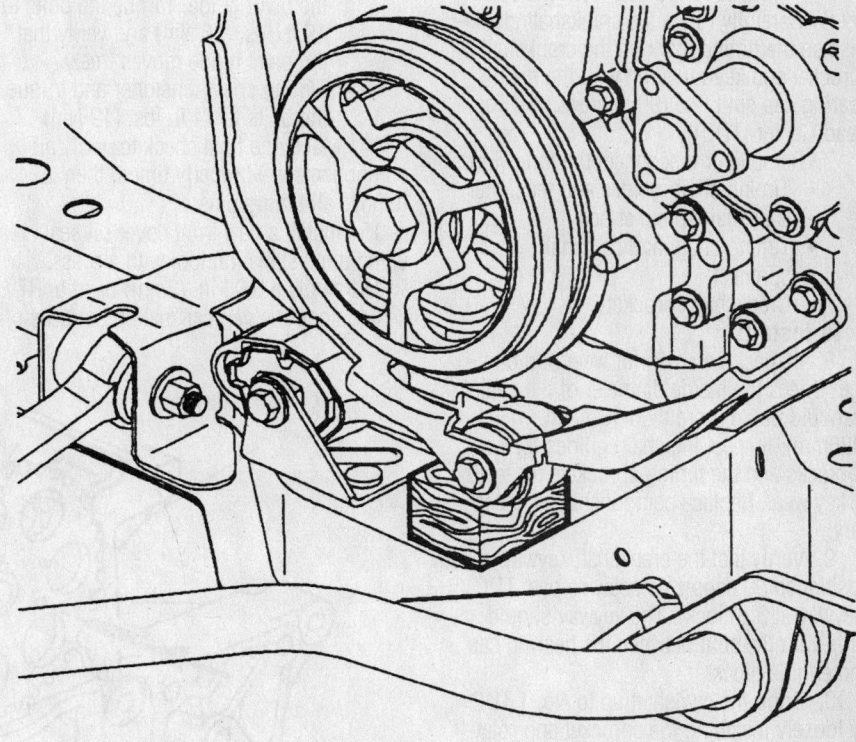

**Place a 1 x 1 x 2 in. (25 x 25 x 51mm) piece of wood between the torque strut and cradle
before removal of the torque engine mount—1.9L (SOHC) engines**

➡️ **Place a 1 x 1 x 2 in. long block of wood between the torque strut and cradle to ease removal and installation of the torque engine mount.**

- 3 right side upper engine torque axis-to-front cover nuts and the 2 mount to midrail bracket nuts, allowing the powertrain to rest on the block of wood
- Drive belt, tensioner and pulley
- Power steering pump attaching bolts and set the pump to the side with the lines still attached
- A/C compressor from the bracket and set aside with the lines attached
- Camshaft cover
- Crankshaft damper/pulley assembly from the crankshaft

4. Install the special oil seal replacement tool SA9104E, to be sure the front crankshaft timing sprocket is held firmly in place and prevent guide damage. Install with the flat side towards the crankshaft sprocket.

5. Remove or disconnect the following:
- Front 4 oil pan bolts and cut the seal away from the front cover
- Front cover bolts and carefully pry the cover away from the cylinder block
- Front cover oil seal from the cover

6. Carefully rotate the crankshaft clockwise so the timing mark on the crankshaft sprocket and keyway align with the main bearing cap split line (90 degrees past Top Dead Center [TDC]).

7. Remove or disconnect the following:
- Timing chain guides and tensioner
- Camshaft sprocket bolt
- Timing chain and camshaft sprocket
- Crankshaft sprocket

To install:

8. Inspect the chain for wear and damage. Check the inside diameter of the chain, it should be no more than 16.77 in. (426mm). Inspect the chain guides for wear or cracks and the timing sprockets for teeth or key wear. Replace components as necessary.

9. Verify that the crankshaft keyway is positioned 90 degrees clockwise past TDC (keyway at 3 o'clock). The keyway should align with the split between the bearing cap and engine block.

10. Bring the camshaft up to No. 1 TDC by loosely installing the sprocket and rotating the sprocket until the timing pin can be inserted. The camshaft contains wrench flats to assist in turning the shaft. The dowel pin

should be at 12 o'clock when the camshaft is at TDC.

11. Install the crankshaft sprocket, then rotate the crankshaft counterclockwise 90 degrees up to No. 1 TDC (keyway at 12 o'clock).

12. Position the chain under the crankshaft sprocket and over the camshaft sprocket. The timing chain should be positioned so that 1 silver link plate aligns with the reference mark on the camshaft sprocket and the other aligns with the downward tooth (at the 6 o'clock position) on the crankshaft sprocket. The letters FRT on the camshaft sprocket must face forward, away from the cylinder head and excess chain slack should be located on the tensioner side of the block.

13. Install or connect the following:
- Timing pin to verify proper alignment of the camshaft and sprocket
- Camshaft sprocket and torque the sprocket bolt to 75 ft. lbs. (102 Nm)

➡️ **Do not allow the camshaft retaining bolt to torque against the timing pin or cylinder head damage will result.**

- Timing chain guides with the words FRONT facing out. Install the fixed guide first and verify the chain is snug against the guide, then install the pivot guide. Torque the bolts to 19 ft. lbs. (26 Nm) and verify that the pivot guide moves freely.
- Timing chain tensioner and torque the bolts to 14 ft. lbs. (19 Nm)

14. Make one final check to verify all components are properly timed, then remove all timing pins.

15. Install a new front cover oil seal using the installation tool with a press.

16. Apply a 0.08 in. (2mm) bead of RTV sealer along the vertical sealing surfaces of

the front cover to the inside of the bolt holes and to the front of the oil pan. Extra sealer is necessary at the oil pan and cylinder head joints. Be sure to assemble the front cover to the engine within 3 minutes of RTV application.

17. Install or connect the following:
- Crankshaft sprocket retaining tool SA9104E, to align the oil pump and crankshaft during cover installation
- Front cover to the engine and torque the perimeter bolts starting at the center and working outwards on both sides to 19 ft. lbs. (25 Nm)
- Front cover center or inner bolts and torque them to 89 inch lbs. (10 Nm), except for the upper inside bolt, which should be tightened to 22 ft. lbs. (30 Nm)
- Oil pan front bolts and torque them to 80 inch lbs. (9 Nm)

18. After front cover installation, spray 6–12 squirts of oil through the front oil seal drain back hole to verify that it is not plugged.

19. Apply a thin film of RTV between the damper/pulley assembly flange and washer

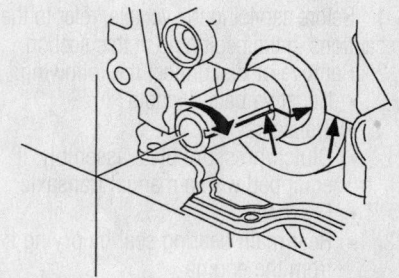

At 90 degrees past TDC, the crankshaft sprocket keyway will align with the main bearing cap split line—1.9L (SOHC) engines

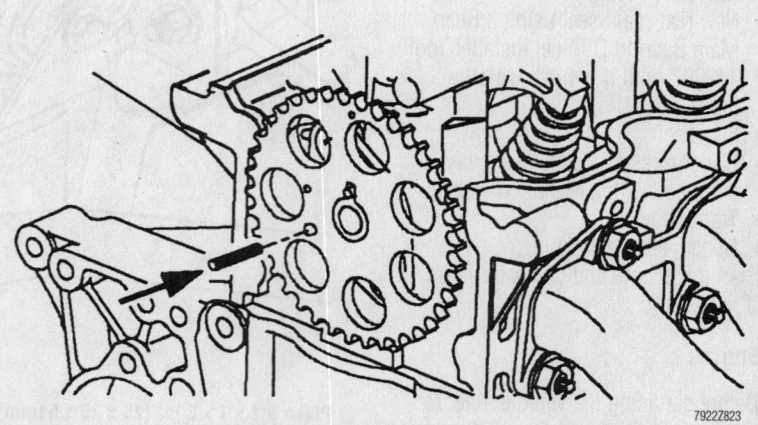

Insert a timing pin to ensure that the camshaft is at No. 1 TDC—1.9L (SOHC) engines

only; the washer and bolt head flange are designed to prevent oil leakage.

20. Remove the crankshaft retaining tool SA9104E.

21. Install the crankshaft damper/pulley assembly and torque the bolt to 159 ft. lbs. (215 Nm).

22. Apply a small drop of RTV across the cylinder head and front cover T-joints. Inspect the old camshaft cover gasket and replace if damaged. Install the gasket and the camshaft cover and torque the fasteners uniformly to 22 ft. lbs. (30 Nm).

23. Install or connect the following:
- A/C compressor assembly and/or the power steering pump assembly
- Drive belt idler pulley
- Drive belt tensioner
- Drive belt
- Engine mount-to-midrail bracket nuts and torque them to 37 ft. lbs. (50 Nm)
- 3 mount-to-front cover nuts and torque them uniformly to 37 ft. lbs. (50 Nm) in order to prevent front cover damage
- Splash shield and the wheel assembly
- Negative battery cable

24. Fill the engine with clean oil.

25. Start the engine and check for leaks, repair if necessary.

1.9L (DOHC) Engines

1. Before servicing the vehicle, refer to the precautions in the beginning of this section.

2. drain the engine oil.

3. Remove or disconnect the following:
- Negative battery cable
- Right front wheel and splash shield

➡**Place a 1 x 1 x 2 in. (25 x 25 x 51mm) block of wood between the torque strut and cradle to ease removal and installation of the torque engine mount.**

- 3 right side upper engine torque axis-to-front cover nuts and the 2 mount-to-midrail bracket nuts, allowing the powertrain to rest on the block of wood
- Accessory drive belt
- Drive belt tensioner and pulley
- Power steering pump attaching bolts and set the pump to the side with the lines still attached
- A/C compressor from the bracket and set it to the side with the lines attached
- Camshaft cover
- Crankshaft damper/pulley

4. Install an Oil Seal Replacement Tool

SA9104E to be sure the front crankshaft timing sprocket is held firmly in place and prevent guide damage. Install with the flat side towards the crankshaft sprocket.
- Front 4 oil pan bolts and cut the seal away from the front cover
- Front cover bolts and carefully pry the cover away from the cylinder block
- Front cover oil seal from the cover

5. Carefully rotate the crankshaft clockwise so the timing mark on the crankshaft sprocket and keyway align with the main bearing cap split line (90 degrees past Top Dead Center [TDC]).
- Timing guides and tensioner
- Camshaft sprocket bolt
- Timing chain and camshaft sprocket
- Crankshaft sprocket

When the camshaft is at TDC, rotate the crankshaft counterclockwise 90 degrees to achieve TDC—1.9L (SOHC) engines

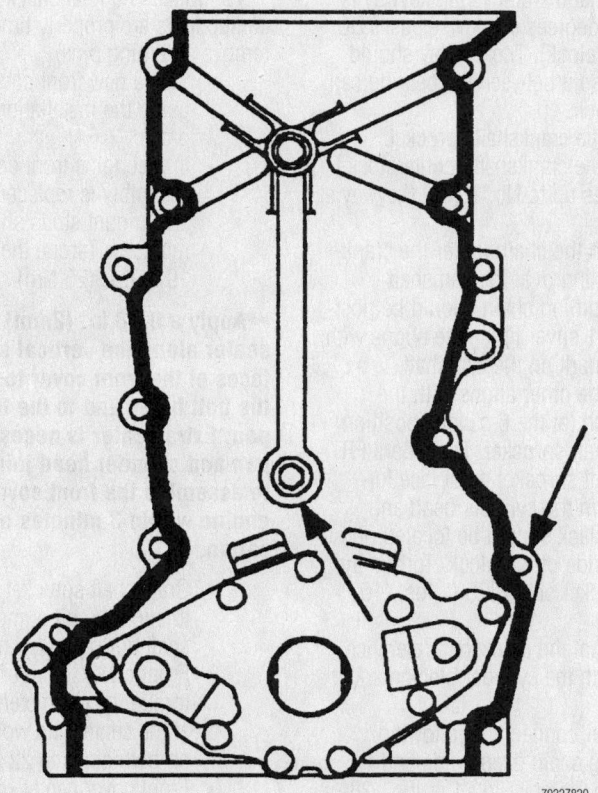

Apply a 0.08 in. (2mm) bead of RTV sealer along the vertical sealing surfaces of the front cover—1.9L (SOHC) engines

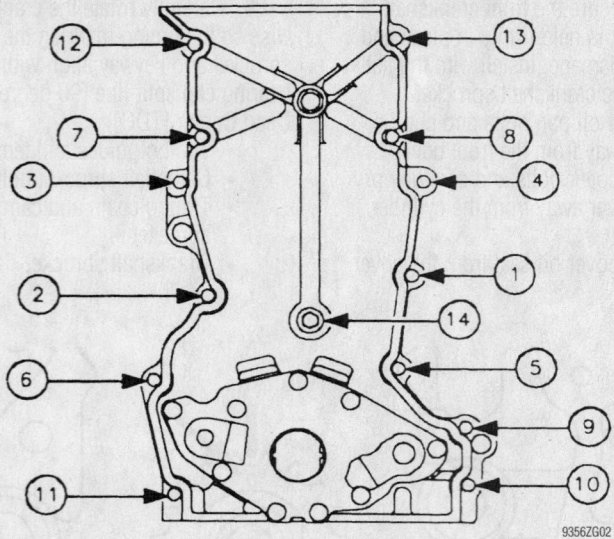

Front cover torque sequence—1.9L (SOHC) engines

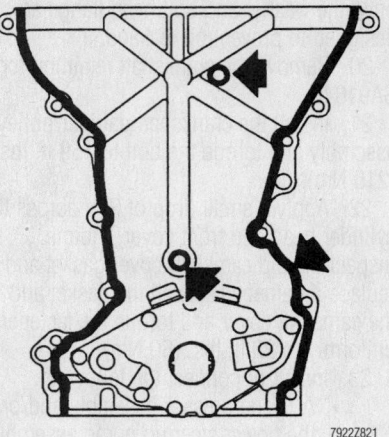

Apply a 0.08 in. (2mm) bead of RTV sealer along the vertical sealing surfaces of the front cover and the front of the oil pan—1.9L (DOHC) engines

To install:

6. Inspect the chain for wear and damage. Check the inside diameter of the chain, it should be no more than 23.15 in. (588mm). Inspect the chain guides for wear or cracks and the timing sprockets for teeth or key wear. Replace components as necessary.

7. Verify that the crankshaft keyway is positioned 90 degrees clockwise past TDC (keyway at 3 o'clock). The keyway should align with the split between the bearing cap and engine block.

8. Install the crankshaft sprocket.

9. Rotate the crankshaft counterclockwise 90 degrees up to No. 1 TDC (keyway at 12 o'clock).

10. Position the chain under the crankshaft sprocket and over the camshaft sprocket. The timing chain should be positioned so that 1 silver link plate aligns with the reference mark on the camshaft sprocket and the other aligns with the downward tooth (at the 6 o'clock position) on the crankshaft sprocket. The letters FRT on the camshaft sprocket must face forward, away from the cylinder head and excess chain slack should be located on the tensioner side of the block. Torque the camshaft sprocket bolt to 75 ft. lbs. (102 Nm).

11. Verify that the crankshaft reference mark aligns with the cylinder block mark at 12 o'clock.

12. Install or connect the following:
- Timing chain fixed guide and torque the bolts to 21 ft. lbs. (28 Nm)
- Pivoting chain guide and torque the bolt to 19 ft. lbs. (26 Nm) and

make certain the guide moves freely
- Two forward bearing caps and the upper timing chain guide and torque the retaining bolts to 124 inch lbs. (14 Nm)
- Timing chain tensioner and torque the bolts to 14 ft. lbs. (19 Nm)

13. Make one final check to verify all components are properly timed, then remove all timing pins.
- Seat a new front cover oil seal using the installation tool with a press
- If the engine front cover casting or assembly is replaced, the 3 torque axis mount studs should also be replaced. Torque the new studs to 19 ft. lbs. (25 Nm).

➡Apply a 0.08 in. (2mm) bead of RTV sealer along the vertical sealing surfaces of the front cover to the inside of the bolt holes and to the front of the oil pan. Extra sealer is necessary at the oil pan and cylinder head joints. Be sure to assemble the front cover to the engine within 3 minutes of RTV application.

- Crankshaft sprocket retaining tool to align the oil pump and crankshaft during cover installation
- Front cover to the engine and torque the perimeter bolts starting at the center and working outwards on both sides to 22 ft. lbs. (30 Nm)
- Front cover center or inner bolts to 89 inch lbs. (10 Nm) except for the upper inside bolt which should be tightened to 22 ft. lbs. (30 Nm)

- 4 oil pan front bolts and torque them to 80 inch lbs. (9 Nm)

14. After front cover installation, spray 6–12 squirts of oil through the front oil seal drain back hole to verify that it is not plugged.

15. Apply a thin film of RTV between the damper/pulley assembly flange and washer only; the washer and bolt head flange are designed to prevent oil leakage.

16. Position the crankshaft damper/pulley assembly, then secure using the wood or strap wrench (as accomplished during removal). Torque the bolt to 159 ft. lbs. (215 Nm).

17. Apply a small drop of RTV across the cylinder head and front cover T-joints. Inspect the old camshaft cover gasket and replace if damaged. Install the gasket and the camshaft cover. Torque the fasteners uniformly to 89 inch lbs. (10 Nm).

18. Remove the crankshaft retaining tool.

19. Install or connect the following:
- A/C compressor assembly and/or the power steering pump assembly
- Drive belt idler pulley
- Drive belt tensioner
- Accessory drive belt
- 2 engine mount-to-midrail bracket nuts and torque them to 37 ft. lbs. (50 Nm)
- 3 mount-to-front cover nuts and torque them uniformly to 37 ft. lbs. (50 Nm) in order to prevent front cover damage
- Splash shield and the wheel
- Negative battery cable

20. Fill the engine with clean oil.

21. Start the engine and check for leaks, repair if necessary.

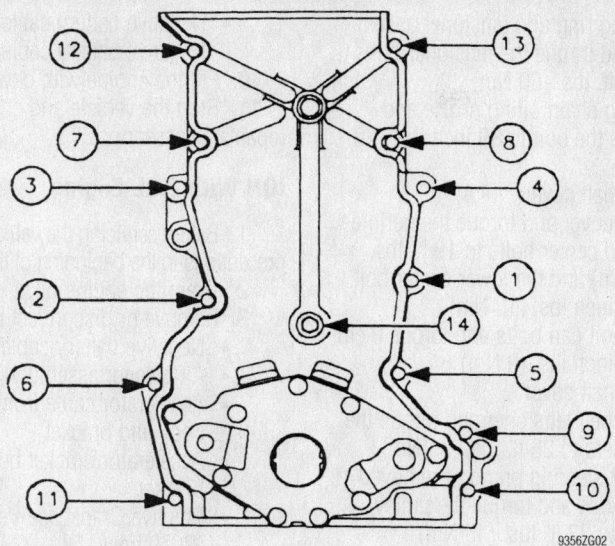

Front cover torque sequence—1.9L (DOHC) engines

L–Series With 2.2L Engine

1. Before servicing the vehicle, refer to the precautions in the beginning of this section.

2. Drain the engine oil.

3. Remove or disconnect the following:
- Negative battery cable
- Air cleaner assembly
- Right front wheel and splash shield
- Accessory drive belt
- Crankshaft damper pulley
- Belt tensioner

4. Install an engine support fixture.
- Right front engine mount
- Front cover bolts and the 13mm bolt under the water pump
- Front cover
- Camshaft cover
- Timing chain tensioner
- Upper timing chain guide
- Exhaust camshaft sprocket
- Adjustable timing chain guide
- Fixed timing chain guide access plug and guide
- Intake camshaft sprocket
- Timing chain through the top of the cylinder head
- Timing chain sprocket from the crankshaft
- Timing chain oiling nozzle

To install:

5. Inspect the chain guides for wear and damage. Replace the guides if wear exceeds 0.045 inch (1.12mm). Inspect the timing chain shoe. Replace the shoe if wear exceeds 0.045 inch (1.12mm).

6. Install or connect the following:
- Timing chain sprocket to the crankshaft. Rotate the crankshaft so that

the mark on the sprocket is at the 5 o'clock position.
- Timing chain to the intake camshaft sprocket alignment copper link to the "INT" diamond timing mark on the camshaft sprocket

➡ **When lowering the timing chain, rotate the assembly 90 degrees to allow the chain to fall between the cylinder block bosses. Rotate the chain back so that the camshaft sprocket is facing forward.**

- Chain through the housing opening on top of the cylinder head. Make certain that the chain goes around both sides of the bosses
- Timing chain around the crankshaft sprocket and align the silver link to the timing mark
- Intake camshaft sprocket loosely on

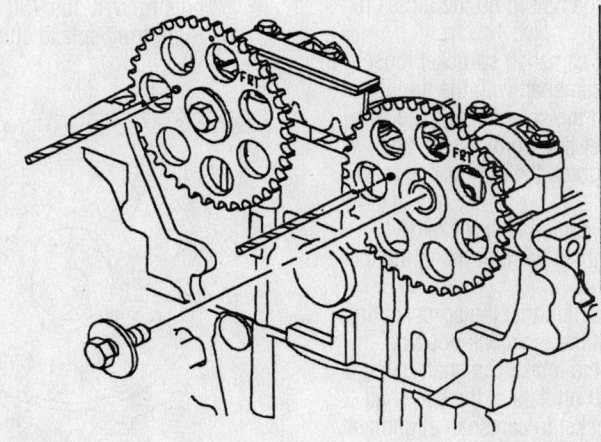

Use drill bits as timing pins to verify that the camshafts are at TDC—1.9L (DOHC) engines

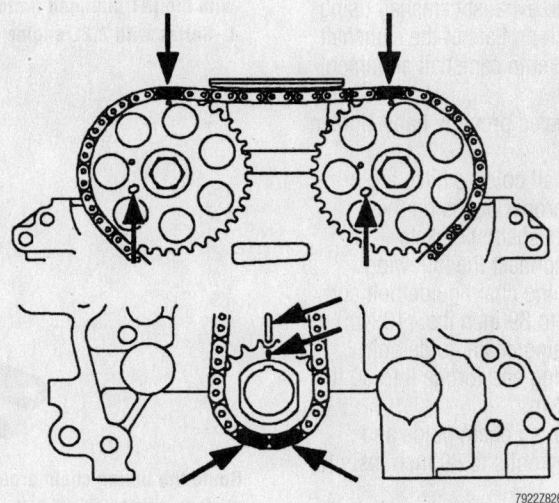

Be sure that the silver link plates and reference marks are all in alignment as shown—1.9L (DOHC) engines

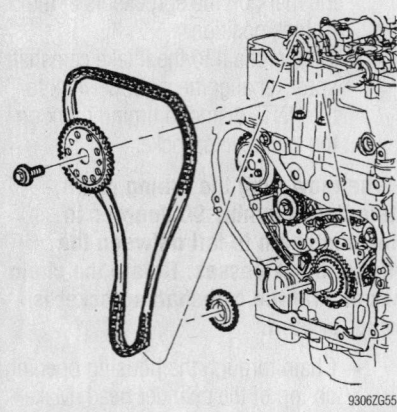

Remove the timing chain through the top of the cylinder head—L–Series with 2.2L engine

the camshaft and hand-tighten the new bolt
- Adjustable timing chain guide and torque the bolt to 89 inch lbs. (10 Nm)
- Exhaust camshaft sprocket loosely on the camshaft with the timing mark on the sprocket aligned with the silver link on the chain and hand-tighten the new bolt at this time

7. Align the camshaft sprocket to camshaft and tighten the bolt using the following procedure:

   a. Make certain that the sprocket timing mark is at the 5 o'clock position.

   b. Rotate the intake camshaft using a 24mm wrench on flats of the camshaft until the sprocket to camshaft alignment notch seats.

   c. When seated properly hand-tighten the bolt.

   d. Rotate the exhaust camshaft using a 24mm wrench on flats of the camshaft until the sprocket to camshaft alignment notch seats.

   e. When seated properly hand-tighten the bolt.

8. Verify that all colored links are aligned with the proper marks on the camshaft and crankshaft sprockets.

9. Install or connect the following:
- Fixed timing chain guide bolt and torque it to 89 inch lbs. (10 Nm)
- Fixed timing chain guide bolt access plug and torque it to 30 ft. lbs. (40 Nm)
- Upper timing chain guide and torque the bolts to 89 inch lbs. (10 Nm)
- Intake and exhaust camshaft sprocket bolts and torque them to 63 ft. lbs. (85 Nm) plus 30 degrees

- Sealing ring and tensioner assembly and torque the tensioner bolts to 44 ft. lbs. (60 Nm)
- Timing chain oiling nozzle and torque the bolt to 89 inch lbs. (10 Nm)
- Camshaft cover
- Front cover and torque the perimeter and center bolts to 19 ft. lbs. (25 Nm) and the lower center bolt to 89 inch lbs. (10 Nm)
- Front oil pan bolts and torque them to 80 inch lbs. (9 Nm)
- Camshaft cover
- Drive belt tensioner and torque the fasteners to 26 ft. lbs. (35 Nm)
- Power steering pump, if removed
- Drive belt and torque the pulley bolts to 33 ft. lbs. (45 Nm)
- Engine mount to midrail bracket nuts first and the engine mount to front cover nuts last and torque all nuts to 37 ft. lbs. (50 Nm)
- Right front splash shield and wheel

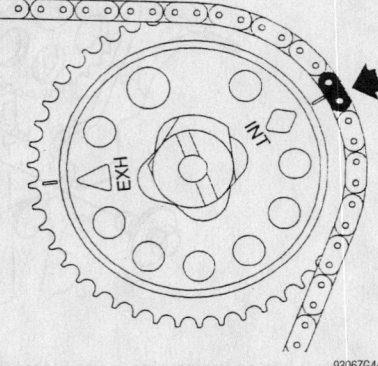

Align the copper link on the timing chain with the INT diamond timing mark—L–Series with 2.2L engine

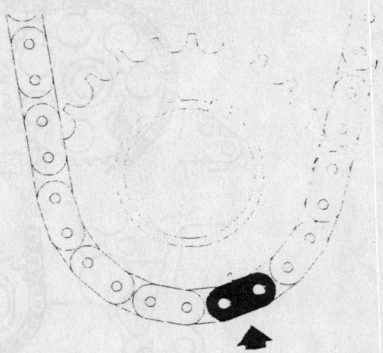

Route the timing chain around the crankshaft sprocket and align the silver link to the timing mark (5 o'clock position)—L–Series with 2.2L engines

- Negative battery cable
- Negative battery cable
10. Fill the engine with clean oil.
11. Start the vehicle and check for leaks, repair if necessary.

ION With 2.2L Engine

1. Before servicing the vehicle, refer to the precautions in the beginning of this section.

2. Drain the engine oil.

3. Remove or disconnect the following:
- Negative battery cable
- Air cleaner assembly
- Accelerator cable from the throttle body and bracket
- Accelerator bracket bolts and the bracket
- Positive Crankcase Ventilation (PCV) hose
- Fuel line bracket
- Ignition Coil Module (ICM) connector
- ICM screws and the ICM
- Ignition coil housing bolts and the housing
- Ground strap stud from the camshaft cover
- Ground strap
- Camshaft cover bolts
- Camshaft cover
- Accessory drive belt
- Using harmonic balancer holder J 38122-A to prevent the crankshaft from rotating and remove the balancer bolt and discard
- Crankshaft balancer
- Belt tensioner

4. Remove the front engine mount as follows:

   a. Support the engine with a jack and a piece of wood.

   b. Remove the mount-to-intermediate bracket bolts

   c. Remove the mount-to-midrail nuts

   d. Remove the mount.
- Front cover bolts and the 13mm bolt under the water pump
- Front cover

➥**The timing chain has 2 matching colored links and 1 unique colored link.**

5. Rotate the engine until the crankshaft sprocket mark aligns with the matching colored link at the 5 o'clock position..

6. Make sure the timing chain to the intake camshaft sprocket alignment copper link to the "INT" diamond timing mark on the camshaft sprocket.

7. Make sure the "EXH" triangle on the exhaust camshaft sprocket is aligned with the colored link.

8. Remove or disconnect the following:
- Timing chain tensioner
- Fixed timing chain guide access plug and the guide
- Upper timing chain guide
- Exhaust camshaft sprocket
- Timing chain tensioner guide
- Intake camshaft sprocket
- Timing chain through the top of the cylinder head
- Timing chain sprocket from the crankshaft
- Balance shaft drive chain tensioner
- Adjustable timing chain guide
- Small balance shaft drive chain guide
- Upper balance shaft drive chain guide

➡**To make it easier to remove the balance shaft chain, get all the slack in the chain between the crankshaft and the water pump sprockets.**

- Balance shaft drive chain

To install:

➡**If the balance shafts are not properly aligned engine noise will occur.**

9. Install or connect the following:
- Upper balance shaft chain guide and tighten the bolts to 89 inch lbs. (10 Nm)
10. Install the balance shaft chain with the colored links lined up on the marks on the balance shaft drive sprockets and crankshaft sprocket. Use the following to align the links with the sprockets while referring to the accompanying illustration for link location:

a. Place the chain so that the copper colored and chrome links can be clearly seen

b. Place link 1 (see illustration) so that it aligns with the timing mark on the intake side balance shaft sprocket.

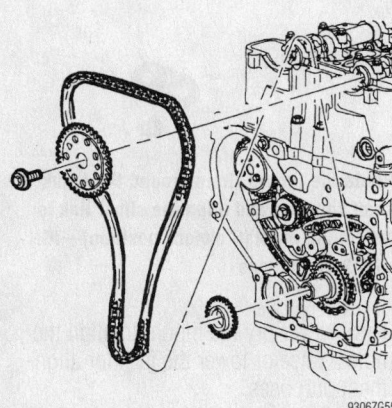

Remove the timing chain through the top of the cylinder head—Ion with 2.2L engine

c. Working in a clockwise direction around the chain, place the link 2 (see illustration) in line with the timing mark on the crankshaft sprocket, this will be approximately at the 5 o'clock position.

d. Place link 3 (see illustration) on the water pump sprocket (alignment is not critical).

e. Align link 4 (see illustration) with the timing mark on the exhaust side balance shaft sprocket.

f. Install the small balance shaft chain guide and tighten the bolts to 89 inch lbs. (10 Nm).

g. Install the adjustable balance shaft drive chain guide and tighten the bolts to 89 inch lbs. (10 Nm).

h. Turn the tensioner plunger 90 degrees in its bore and compress the plunger until a paper clip can be inserted through the hole in the plunger body and the hole in the plunger.

i. Install the timing chain tensioner and tighten the bolts to 89 inch lbs. (10 Nm).

j. Remove the clip from the balance shaft tensioner.

11. Inspect the chain guides for wear and damage. Replace the guides if wear exceeds 0.045 inch (1.12mm). Inspect the timing chain shoe. Replace the shoe if wear exceeds 0.045 inch (1.12mm).

12. Install or connect the following:
- Timing chain sprocket to the crankshaft. Rotate the crankshaft so that the mark on the sprocket is at the 5 o'clock position.

13. Lower the timing chain through the opening in the cylinder head. make sure the chain goes around both sides of the cylinder block bosses (see illustration).

- Intake camshaft sprocket with the "INT" diamond at the 2 o'clock position and hand-tighten a new intake camshaft sprocket bolt
- Timing chain around the crankshaft sprocket and align the silver link to the timing mark
- Timing chain to the intake camshaft sprocket alignment copper link to the "INT" diamond timing mark on the camshaft sprocket
- Timing chain tensioner guide and torque the bolt to 89 inch lbs. (10 Nm)
- Exhaust camshaft sprocket loosely on the camshaft with the timing mark on the sprocket aligned with the silver link (3) on the chain at the 10 o'clock position and hand-tighten the new bolt at this time

14. Align the camshaft sprocket to

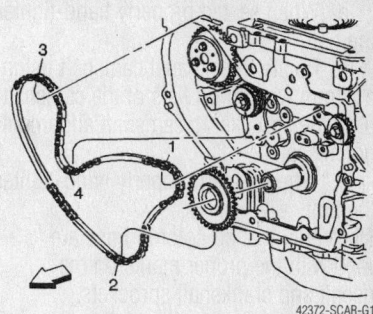

Install the balance shaft chain with the colored links lined up on the marks on the balance shaft drive sprockets and crankshaft sprocket —ION with 2.2L engine

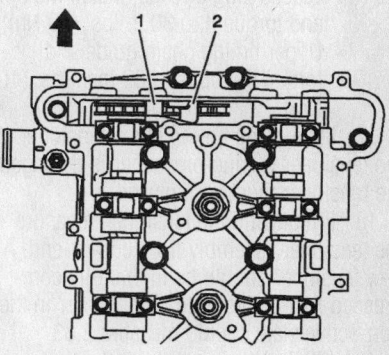

Lower the timing chain through the opening in the cylinder head. make sure the chain goes around both sides of the cylinder block bosses—ION with 2.2L engine

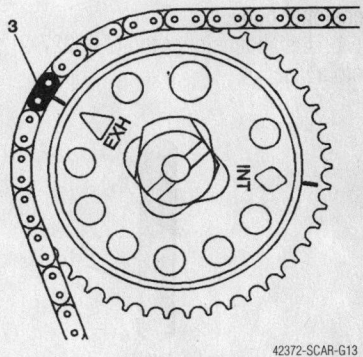

Align the timing mark on the exhaust camshaft sprocket with the silver link (3) on the chain—ION with 2.2L engine

camshaft and tighten the bolt using the following procedure:

a. Make certain that the sprocket timing mark is at the 5 o'clock position.

b. Rotate the intake camshaft using a 24mm wrench on flats of the camshaft until the sprocket to camshaft alignment notch seats.

c. When seated properly hand-tighten the bolt.

d. Rotate the exhaust camshaft using a 24mm wrench on flats of the camshaft until the sprocket to camshaft alignment notch seats.

e. When seated properly hand-tighten the bolt.

15. Verify that all colored links are aligned with the proper marks on the camshaft and crankshaft sprockets.

16. Install or connect the following:
- Fixed timing chain guide and tighten the bolt to 89 inch lbs. (10 Nm)
- Apply Saturn compound 21485277 to the fixed timing chain guide bolt access plug threads, install the bolt and torque it to 30 ft. lbs. (40 Nm)
- Upper timing chain guide and torque the bolts to 89 inch lbs. (10 Nm)

17. Inspect the timing chain tensioner. If the tensioner O–ring or washer is damaged the tensioner must be replaced.

18. If replacing the tensioner, measure the tensioner assembly from end-to-end. A new tensioner should be in the fully compressed non-active state. A tensioner in the non-active state should measure 2.83 inches (72mm) from end-to-end. The tensioner in the active state will measure 3.35 inches (85mm) from end-to-end.

19. If the tensioner is not in the compressed state, perform the following steps:

a. Pull the piston assembly from the tensioner body.

b. Install tensioner tool J 45027-2 into a vise.

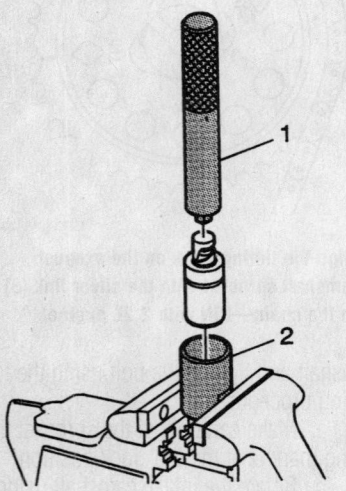

Resetting the timing chain tensioner piston—ION with 2.2L engine

c. Install the notch end of the piston assembly into J 45027-2.

d. Using tool J 45027-1, turn the ratchet cylinder into the pump.

e. Inspect the bore of the tensioner body for damage or dirt. If any damage is present, replace the tensioner.

f. Install the compressed piston assembly back into the tensioner body until it bottoms out in the bore. Do not compress the piston against the bottom of the bore. If the piston is compressed in the bore it will activate the tensioner and the piston will have to be reset.

g. Check that the tensioner measures 2.83 inches (72mm) from end-to-end. If this measurement is incorrect, reset the piston.

h. Install the tensioner assembly and tighten to 55 ft. lbs. (75 Nm)

i. Using a suitable tool with a rubber tip on the end, feed the tool down through the camshaft drive chant to rest on the timing chain. Give a sharp jolt diagonally downwards to release the tensioner.

20. Install or connect the following:
- Intake and exhaust camshaft sprocket bolts and torque them to 63 ft. lbs. (85 Nm) plus 30 degrees. Use a 24mm wrench as a back-up when tightening the bolts.
- Sealing ring and tensioner assembly and torque the tensioner bolts to 44 ft. lbs. (60 Nm)
- Camshaft cover
- Camshaft cover bolts and tighten to 89 inch lbs. (10 Nm)
- Ground strap
- Ground strap stud to the camshaft cover and tighten to 89 inch lbs. (10 Nm)
- Ignition coil housing and tighten the bolts to 89 inch lbs. (10 Nm)
- ICM and tighten the screws to 13 inch lbs. (1.5 Nm)
- ICM connector
- Fuel line bracket
- PCV hose
- Accelerator bracket and tighten the bolts to 89 inch lbs. (10 Nm)
- Accelerator cable to the throttle body and bracket
- Front cover gasket

21. Install the front engine mount as follows:

a. Install the mount.

b. Install the mount-to-midrail nuts and torque them to 74 ft. lbs. (100 Nm).

c. Install the mount-to-intermediate bracket bolts and torque to 37 ft. lbs. (50

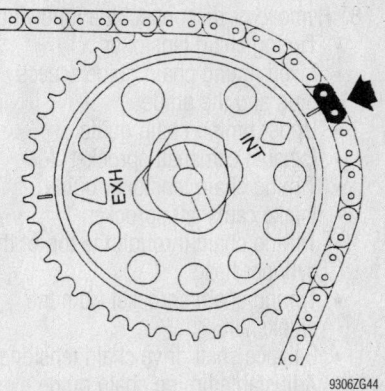

Align the copper link on the timing chain with the INT diamond timing mark—ION with 2.2L engine

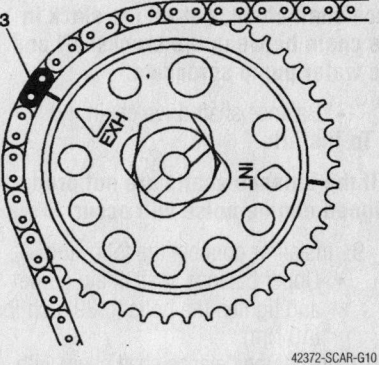

Make sure the "EXH" triangle on the exhaust camshaft sprocket is aligned with the colored link—ION with 2.2L engine

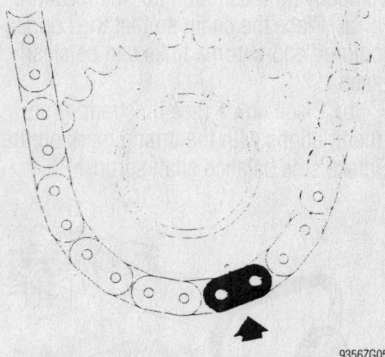

Route the timing chain around the crankshaft sprocket and align the silver link to the timing mark (5 o'clock position)—ION with 2.2L engines

Nm). Never pry the mount to align the holes, raise or lower the jack for alignment purposes.

d. Remove the jack.

22. Install or connect the following:
- Front cover and torque the bolts to

18 ft. lbs. (25 Nm) and the water pump bolt to 15 ft. lbs. (20 Nm)
- Drive belt tensioner and torque the fasteners to 33 ft. lbs. (45 Nm)
- Crankshaft balancer
- Balancer bolt and using tool J 38122-A to prevent the crankshaft from rotating, install a new bal-

ancer bolt and tighten to 74 ft. lbs. (100 Nm) plus an additional 75 degrees
- Drive belt
- Negative battery cable

23. Fill the engine with clean oil.
24. Start the vehicle and check for leaks, repair if necessary.

Timing Belt

REMOVAL & INSTALLATION

3.0L Engine

1. Before servicing the vehicle, refer to the precautions in the beginning of this section.

TIMING BELT INSTALLATION AND ADJUSTMENT TABLE

Step	Action	Value	Yes	No
1	Install the timing belt and align marks on the belt with the marks on the camshaft gears and the crankshaft gear. Check the timing belt deflection between the idler pulley for camshafts 3 & 4 and camshaft number 4. Is the timing belt installed, the marks aligned and the timing belt deflection adjusted?	1 cm (0.4 in) maximum	Go to Step 2	—
2	Set the initial timing belt tension at the timing belt tensioner. Is the initial timing belt tension set?	—	Go to Step 3	—
3	Rotate the engine two complete revolutions and secure the crankshaft at Top Dead Center (TDC) with the J 42069-10. Has the engine been rotated and the crankshaft secured to TDC?	—	Go to Step 4	—
4	Starting with camshafts 3 and 4, check the alignment of the marks on the camshaft gears with the marks on the J 42069-20 checking gauge. Do the marks on the camshaft gears align exactly with the marks on J 42069-20?	—	Go to Step 5	Go to Step 6
5	Check the alignment of the marks on camshafts gears 1 and 2 with the marks on the J 42069-20 checking gauge. Do the marks on the camshaft gears align exactly with the marks on J 42069-20?	—	Go to Step 14	Go to Step 10
6	Do the camshaft gear marks line up to the left (BTDC) of the marks on the J 42069-20 checking gauge?	—	Go to Step 8	Go to Step 7
7	Do the camshaft gear marks line up to the right (ATDC) of the marks on the J 42069-20 checking gauge?	—	Go to Step 9	—
8	Turn the idler pulley eccentric, for camshafts 3 and 4, counterclockwise until the marks on the camshaft gear align exactly with the marks on J 42069-20. Rotate the engine two complete revolutions, lock the crankshaft at TDC with J 42069-10 and recheck the alignment of the camshaft gear marks to the marks on J 42069-20. Do the marks on the camshaft gears align exactly with the marks on J 42069-20?	—	Go to Step 5	Go to Step 6
9	Turn the idler pulley eccentric, for camshafts 3 and 4, clockwise until the marks on the camshaft gear align exactly with the marks on J 42069-20. Rotate the engine two complete revolutions, lock the crankshaft at TDC with J 42069-10 and recheck the alignment of the camshaft gear marks to the marks on J 42069-20. Do the marks on the camshaft gears align exactly with the marks on J 42069-20?	—	Go to Step 5	Go to Step 6

Timing belt installation and adjustment table—3.0L engine

79225G35

Step	Action	Value	Yes	No
10	Do the camshaft gear marks line up to the left (BTDC) of the marks on the J 42069-20 checking gauge?	—	Go to Step 12	Go to Step 11
11	Do the camshaft gear marks line up to the right (ATDC) of the marks on the J 42069-20 checking gauge?	—	Go to Step 13	—
12	Turn the idler pulley eccentric, for camshafts 1 and 2, counterclockwise until the marks on the camshaft gear align exactly with the marks on J 42069-20. Rotate the engine two complete revolutions, lock the crankshaft at TDC with J 42069-10 and recheck the alignment of the camshaft gear marks to the marks on J 42069-20. Do the marks on the camshaft gears align exactly with the marks on J 42069-20?	—	Go to Step 14	Go to Step 10
13	Turn the idler pulley eccentric, for camshafts 1 and 2, clockwise until the marks on the camshaft gear align exactly with the marks on J 42069-20. Rotate the engine two complete revolutions, lock the crankshaft at TDC with J 42069-10 and recheck the alignment of the camshaft gear marks to the marks on J 42069-20. Do the marks on the camshaft gears align exactly with the marks on J 42069-20?	—	Go to Step 14	Go to Step 10
14	Set the final timing belt tension at the timing belt tensioner. Is the final timing belt tension set?	—	Go to Step 15	—
15	Again, rotate the engine two complete revolutions and lock the crankshaft at TDC. Do a final inspection of the camshaft gear marks' relationship to the J 42069-20 marks. The marks must align exactly. Do the marks on the camshaft gears align exactly with the marks on the J 42069-20?	—	Go to Step 16	Go to Step 2
16	Remove all checking tools and ensure all idler pulleys and the tensioner locking nut are tightened to specifications. Continue with re-assembly of the engine.	—	—	—

79225G36

Timing belt installation and adjustment table (continued)—3.0L engine

2. Remove or disconnect the following:
- Negative battery cable
- Intake air resonator
- Intake plenum
- Splash shield
- Secondary air injection (AIR) pipe
- Accessory drive belt
- Water pump and power steering pump pulleys
- Drive belt tensioner
- Front timing belt cover
- Crankshaft balancer

3. Rotate the crankshaft clockwise to 60 degrees before top dead center (BTDC).

4. Loosen the timing belt tensioner and idler pulleys.

5. Remove the timing belt.

To install:

6. Start at the crankshaft sprocket and install the timing belt with the double dash (TDC) aligned with the marks on the oil pump and on the belt drive gear.

7. Set the initial timing belt tension as follows:

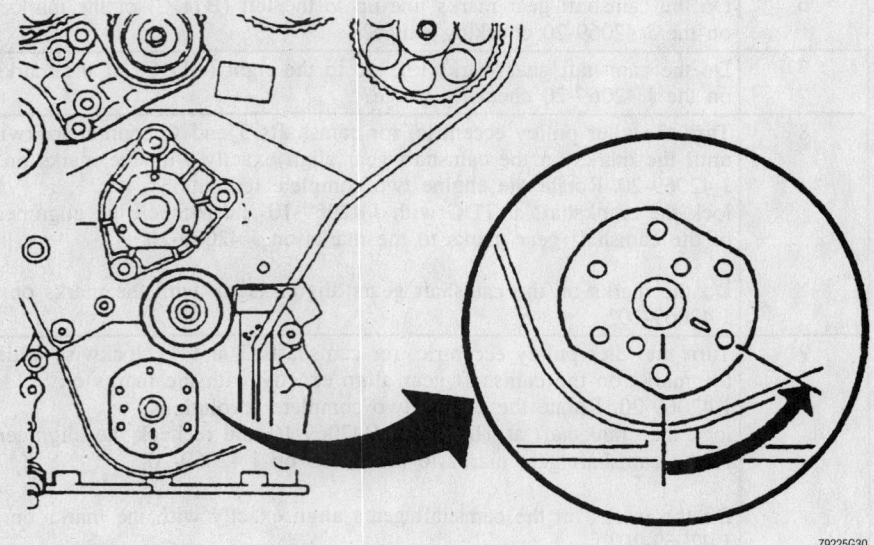

Crankshaft alignment to 60 degrees BTDC—3.0L engine

79225G30

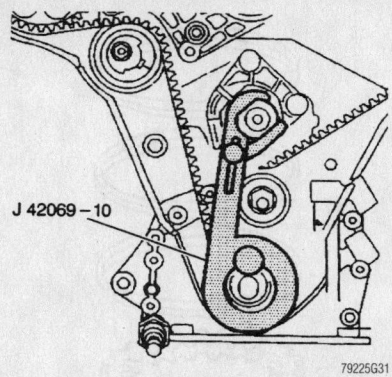

Securing the crankshaft—3.0L engine

a. Turn the tensioner nut COUNTER-
CLOCKWISE to full stop, then, turn the
nut back until the reference mark is 1
mm (0.003 in) over the flange.

b. Tighten the timing belt tensioner
locking nut until snug, the locking nut
will be tightened to specifications after all
final adjustments are made.

8. Rotate the engine in the clockwise
direction two revolutions stopping at 60
degrees BTDC.

9. Inspect the alignment of the reference marks on the camshaft gears with the
notches on the rear timing belt cover, as
well as, the mark on the crankshaft sprocket
and oil pump housing.

10. If timing belt adjustment IS NOT
required, set the final timing belt tension:

a. Loosen the timing belt tensioner
locking nut.

b. Turn the locking nut counterclockwise to full stop, then back until the reference mark is 2-4 mm (0.078-0.157
in) ABOVE the reference mark on the
flange.

11. Tighten the timing belt tensioner
locking nut 15 ft. lbs. (20 Nm).

12. Tighten the idler pulley bolts to 30 ft.
lbs. (40 Nm).

13. Install or connect the following:

- Crankshaft balancer. Torque the
  bolts to 15 ft. lbs. (20 Nm).
- Front timing belt cover. Torque the
  bolts to 71 inch lbs. (8 Nm).
- Drive belt tensioner. Torque the
  bolts to 30 ft. lbs. (40 Nm).
- Water pump pulley. Torque the
  bolts to 71 inch lbs. (8 Nm).
- Power steering pump pulley.
  Torque the bolts to 15 ft. lbs. (20
  Nm).
- Accessory drive belt
- AIR injection pipe
- Splash shield
- Intake plenum
- Intake air resonator

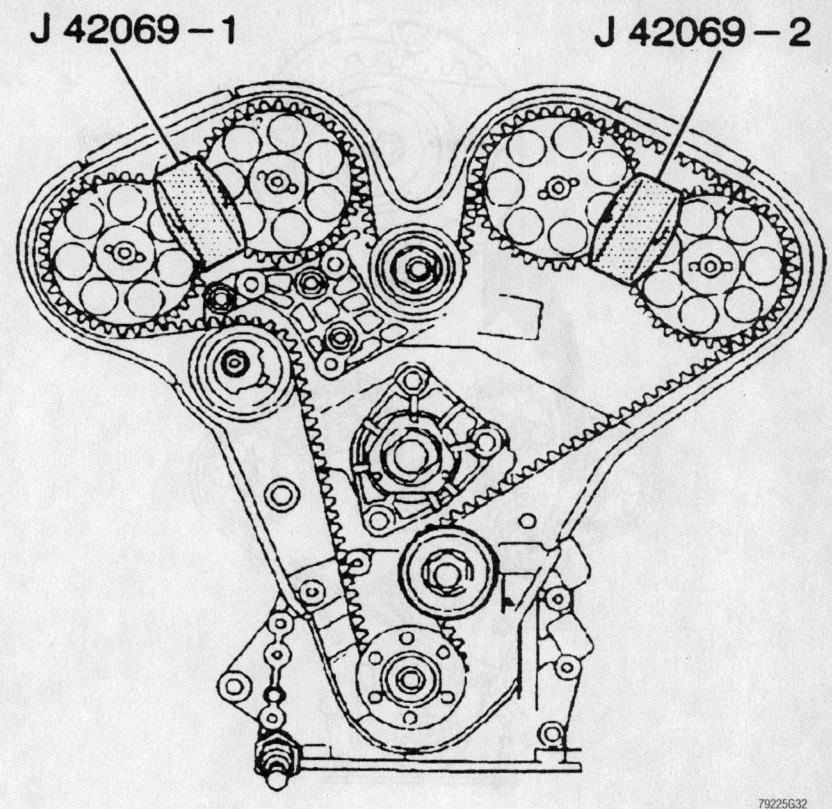

Locking the camshaft—3.0L engine

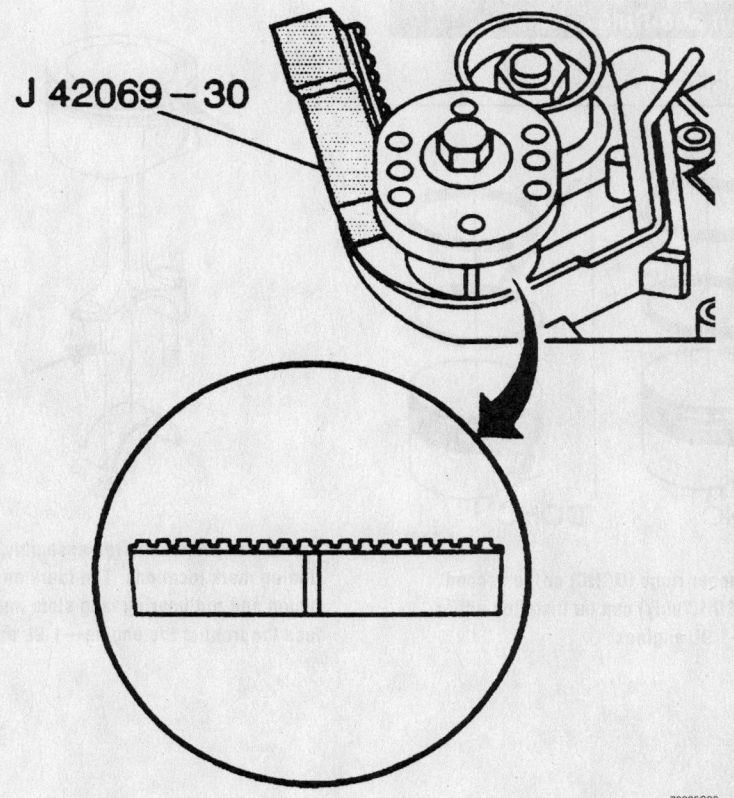

Using the tool to pin the timing belt—3.0L engine

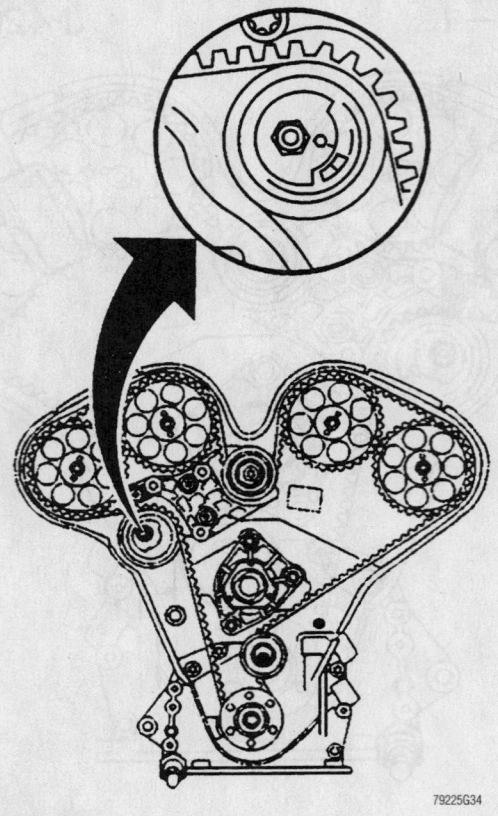

Initial timing belt tension adjustment—3.0L engine

79225G34

Piston ring positioning—2.2L engine

9306ZG77

Piston and Ring

POSITIONING

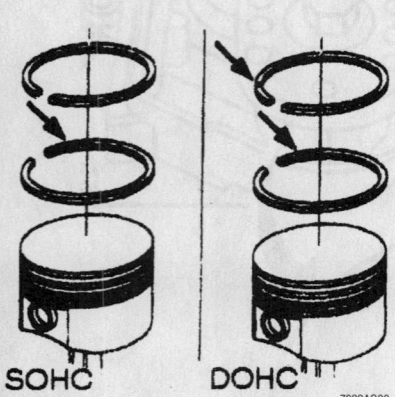

7922AG38

Both upper rings (DOHC) or the second ring (SOHC only) can be installed either way—1.9L engines

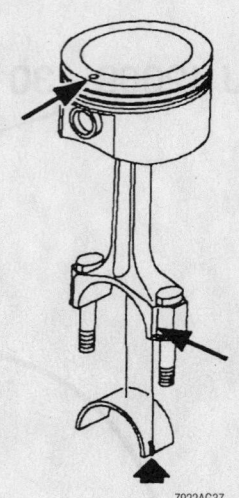

7922AG37

Piston and connecting rod assembly positioning mark locations. The mark on the piston and rod bearing tang slots must face the front of the engine—1.9L engines

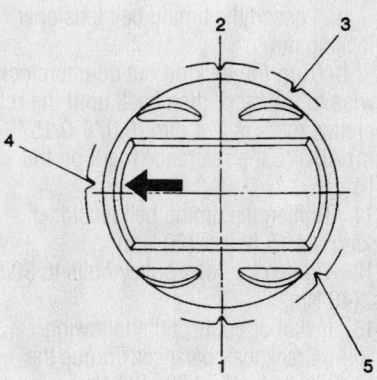

(1) 1st Compression Ring End Gap Location
(2) 2nd Compression Ring End Gap Location
(3) Oil Control Ring Upper Ring End Gap Location
(4) Oil Control Ring Spacer End Gap Location
(5) Oil Control Ring Lower Ring End Gap Location

7922AG55

Piston ring positioning—3.0L engine

FUEL SYSTEM

Fuel System Service Precautions

Safety is the most important factor when performing not only fuel system maintenance but any type of maintenance. Failure to conduct maintenance and repairs in a safe manner may result in serious personal injury or death. Maintenance and testing of the vehicle's fuel system components can be accomplished safely and effectively by adhering to the following rules and guidelines.

• To avoid the possibility of fire and personal injury, always disconnect the negative battery cable unless the repair or test procedure requires that battery voltage be applied

• Always relieve the fuel system pressure prior to disconnecting any fuel system component (injector, fuel rail, pressure regulator, etc.), fitting or fuel line connection. Exercise extreme caution whenever relieving fuel system pressure, to avoid exposing skin, face and eyes to fuel spray. Please be advised that fuel under pressure may penetrate the skin or any part of the body that it contacts

• Always place a shop towel or cloth around the fitting or connection prior to loosening to absorb any excess fuel due to spillage. Ensure that all fuel spillage (should it occur) is quickly removed from engine surfaces. Ensure that all fuel soaked cloths or towels are deposited into a suitable waste container

• Always keep a dry chemical (Class B) fire extinguisher near the work area

• Do not allow fuel spray or fuel vapors to come into contact with a spark or open flame

• Always use a back-up wrench when loosening and tightening fuel line connection fittings. This will prevent unnecessary stress and torsion to fuel line piping

• Always replace worn fuel fitting O-rings with new. Do not substitute fuel hose or equivalent, where fuel pipe is installed

Fuel System Pressure

RELIEVING

1. Before servicing the vehicle, refer to the precautions in the beginning of this section.
2. Unless battery voltage is necessary for testing, disconnect the negative battery cable. This will prevent the fuel pump from running and causing a fuel spill through the disconnected components if the ignition key is accidentally turned **ON**.

3. Remove the air cleaner assembly, for access.
4. Wrap a shop rag around the fuel test port fitting, located at the lower rear of the engine, then remove the cap and connect a fuel pressure gauge.
5. Install the bleed hose from the pressure gauge into an approved container and open the valve to bleed the system pressure.
6. After the pressure is bled, remove the gauge from the test port and recap it.
7. Install the air cleaner assembly.
8. After servicing the vehicle, connect the negative battery cable and prime the fuel system as follows:
    a. Turn the ignition **ON** for 5 seconds, then **OFF** for 10 seconds.
    b. Repeat the **ON/OFF** cycle 2 more times.
    c. Crank the engine until it starts.
    d. If the engine does not readily start, repeat sub-steps A–C.
9. Run the engine and check for leaks.

Fuel Filter

The fuel filter and fuel pressure regulator are one integral component of the new anti-return fuel injection system, and is located underneath the vehicle at the forward edge of the left side of the fuel tank.

REMOVAL & INSTALLATION

1.9L Engines

1. Before servicing the vehicle, refer to the precautions in the beginning of this section.

2. Properly relieve the fuel system pressure.
3. Remove or disconnect the following:
   • Negative battery cable
   • Fuel filter/pressure regulator bracket screws
   • Fuel line bundle retaining clip on the left side of the fuel tank

✷✷ WARNING

Exercise extreme care when opening the retaining clip. The fuel lines must be retained in this clip; if damaged, the fuel tank assembly must be replaced, since the fitting is not serviced separately.

   • Evaporative Emissions (EVAP) purge line at the 90 degree quick connect fitting
   • Outlet of the fuel filter/regulator from the support on the fuel tank bracket and remove the fuel feed line
   • Fuel feed and return line quick-connect fittings on the fuel filter/regulator
   • Fuel filter

➡ **It is not necessary to separate the fuel filter/regulator from the mounting bracket, since both items are serviced as an assembly.**

To install:
4. Install or connect the following:
   • Fuel filter
   • New fuel line retainers (3) into the female portion of the quick-connect fuel line fittings

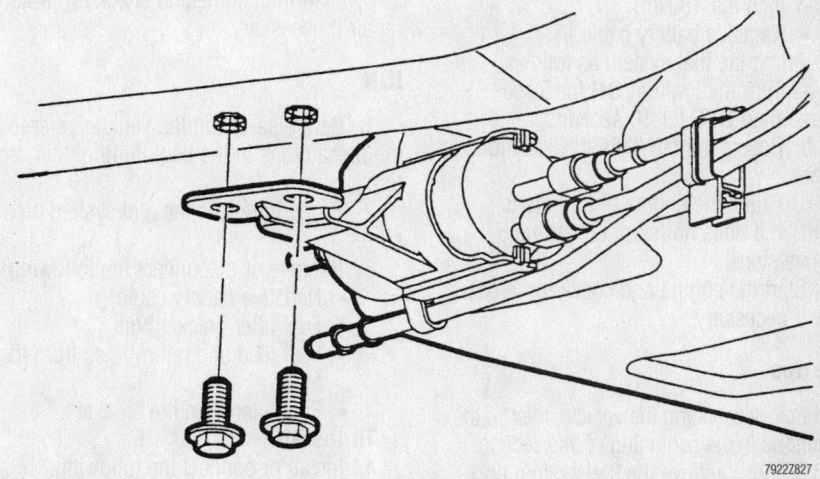

The fuel filter/regulator bracket is held to the frame with 2 bolts—1.9L engines

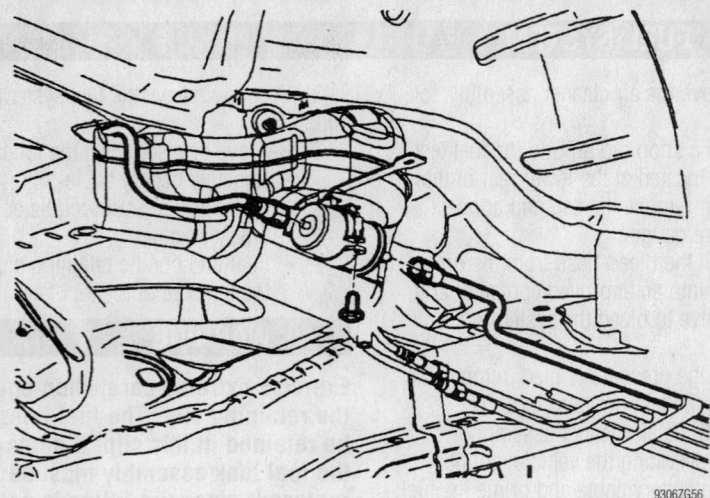

9306ZG56

Remove the fuel filter bracket attaching screw—L–Series shown, others similar

- Fuel feed and return lines onto the fuel filter/regulator and snap them closed
- Fuel feed, return, and EVAP purge lines into the fuel tank's retaining clip

✳✳ WARNING

Be sure to route the chassis fuel feed and purge lines above the parking brake cable. The parking brake cable must be firmly secured to the underbody to support the fuel and purge lines.

- 90 degree EVAP purge line quick-connect fitting to the purge line and snap it closed
- Fuel feed outlet pipe of the fuel filter/regulator into the retaining clip on the fuel tank bracket
- Fuel filter/regulator bracket mounting screws and torque them to 71 inch lbs. (8 Nm)
- Negative battery cable

5. Prime the fuel system as follows:
a. Turn the ignition **ON** for 5 seconds, then **OFF** for 10 seconds.
b. Repeat the **ON/OFF** cycle 2 more times.
c. Crank the engine until it starts.
d. If it does not start, repeat the 3 above steps.
6. Start the engine and check for leaks, repair if necessary.

L–Series

1. Before servicing the vehicle, refer to the precautions in the beginning of this section.
2. Properly relieve the fuel system pressure.

3. Remove or disconnect the following:
- Negative battery cable
- Fuel filter bracket screw
- Fuel feed and return lines from the filter
- Fuel filter from the bracket

To install:
4. Install or connect the following:
- New fuel filter into the bracket
- New fuel line retainers to the female portion of the quick connect fittings
- Fuel feed and return lines
- Fuel filter bracket attaching screw and torque it to 35 inch lbs. (4 Nm)
- Negative battery cable

5. Prime the fuel system as follows:
a. Turn the ignition **ON** for 5 seconds, then **OFF** for 10 seconds.
b. Repeat the **ON/OFF** cycle 2 more times.
c. Crank the engine until it starts.
d. If it does not start, repeat the 3 above steps.
6. Start the engine and check for leaks, repair if necessary.

ION

1. Before servicing the vehicle, refer to the precautions in the beginning of this section.
2. Properly relieve the fuel system pressure.
3. Remove or disconnect the following:
- Negative battery cable
- Fuel filter bracket bolt
- Fuel feed and return lines from the filter
- Fuel filter from the bracket

To install:
4. Install or connect the following:
- New fuel filter into the bracket

- Fuel feed and return lines
- Fuel filter bracket attaching bolt and torque it to 89 inch lbs. (10 Nm)
- Negative battery cable

5. Prime the fuel system as follows:
a. Turn the ignition **ON** for 2 seconds, then **OFF** for 10 seconds.
b. Turn the ignition key to the **ON** position but do not start the engine.
c. Check for leaks.
d. Crank the engine until it starts.
6. Start the engine and check for leaks, repair if necessary.

Fuel Pump

REMOVAL & INSTALLATION

1.9L Engines

To prevent excessive fuel spillage, whenever the tank is removed from the vehicle it should be no more than ½ full. Removal of the fuel pump module assembly requires the removal of the fuel tank.

1. Before servicing the vehicle, refer to the precautions in the beginning of this section.
2. Properly relieve the fuel system pressure.
3. Remove or disconnect the following:
- Negative battery cable
- Fuel pump
- Fuel feed and return lines from the filter/pressure regulator
- Fuel pump vapor line from the fuel tank vent pipe
- Fuel pump module retaining ring with service Tool SA9156E. A ½ inch breaker bar will loosen the lockring

➡ **To prevent bending of the sending unit float arm, lift the pump module up slightly to disengage the orientation tabs in the tank. Rotate the module 90 degrees clockwise until the fuel lines are facing the 1 o'clock position.**

- Fuel pump module from the fuel tank until the bottom of the pump module is close to the fuel tank opening
- Tilt the pump module approximately 45 degrees to the right hand side of the tank and lift the pump from the fuel tank
- Fuel pump to tank seal and discard it

To install:
4. Install or connect the following:
- Fuel pump module into the fuel

tank by orientating the float to face the right side and tilting the module approximately 45 degrees

- Fuel pump into the tank and rotate the assembly 90 degrees counterclockwise to align the module tabs with the slots in the tank

➡ **The fuel pump retaining ring cannot be properly installed if the flange locator tabs are not aligned with the slots in the fuel tank.**

- Retaining lockring with Tool SA9156E
- Vapor line to the fuel tank vent pipe
- Fuel feed and return lines to the filter/pressure regulator
- Fuel feed, return and EVAP canister purge line in the fuel tank retaining clip
- Fuel tank
- Negative battery cable

5. Start the vehicle and check for leaks, repair if necessary.

L–Series

1. Before servicing the vehicle, refer to the precautions in the beginning of this section.

2. Properly relieve the fuel system pressure.

3. Remove or disconnect the following:
- Negative battery cable
- Fuel tank
- Fuel lines from the fuel pump module cover
- Fuel pump module retaining ring with a Sending Unit Wrench, J43827
- Pull the retaining clip toward the float arm and lift up
- Fuel pump straight up from the fuel tank
- Fuel pump tank seal and discard the seal
- Fuel feed line from the bottom of the fuel pump cover with Clamp Pliers J43914
- Fuel pump electrical connector

To install:

4. Install or connect the following:
- Fuel pump feed line to the cover
- Fuel pump electrical connector
- Fuel pump to the new seal
- Fuel pump cover lockring with Tool J-43827
- Fuel lines and wiring harness
- Fuel tank
- Negative battery cable

5. Start the vehicle and check for leaks, repair if necessary.

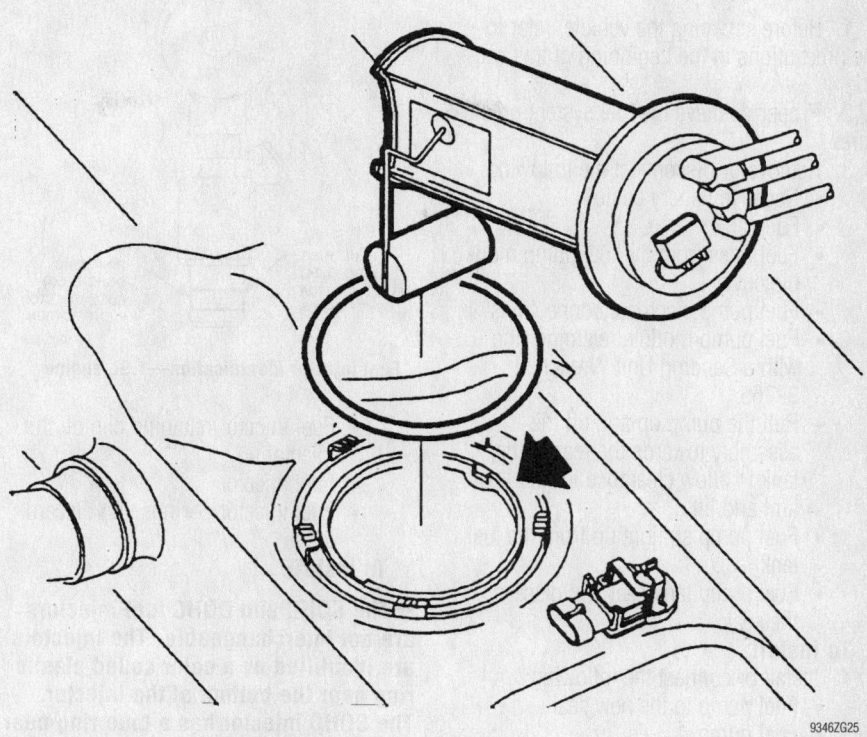

Installing the fuel sending unit and float—1.9L engines

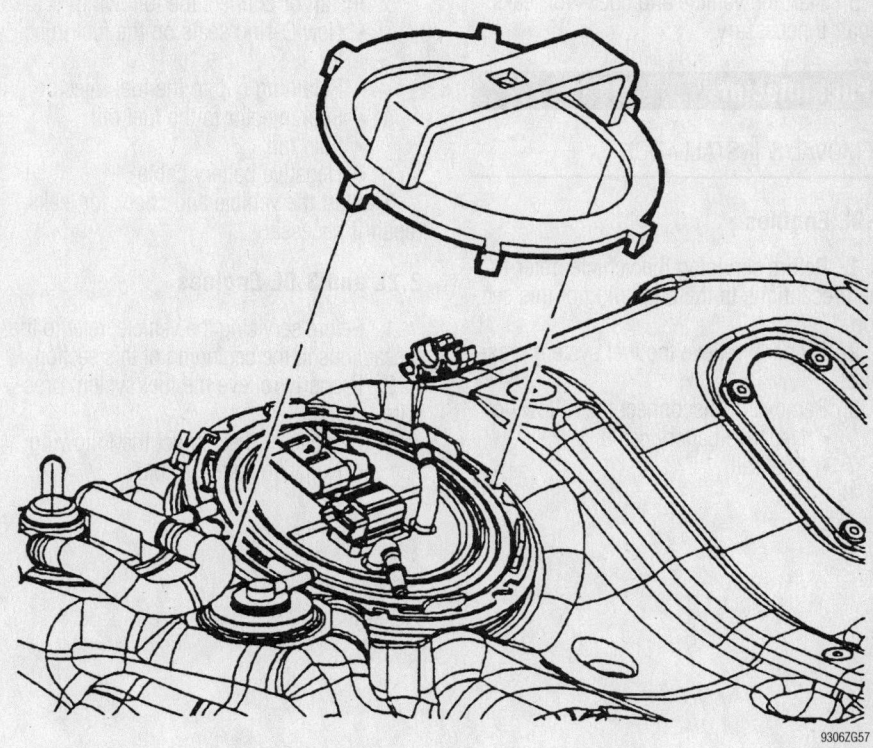

Remove the fuel pump cover lockring—L–Series

ION

1. Before servicing the vehicle, refer to the precautions in the beginning of this section.

2. Properly relieve the fuel system pressure.

3. Remove or disconnect the following:
- Negative battery cable
- Fuel tank
- Fuel lines from the fuel pump module cover
- Fuel pump electrical connections
- Fuel pump module retaining ring with a Sending Unit Wrench, J 39765
- Pull the pump up and tilt the assembly towards the rear of the tank to allow clearance for the float arm and lift up
- Fuel pump straight up from the fuel tank
- Fuel pump tank seal and discard the seal

To install:

4. Install or connect the following:
- Fuel pump to the new seal
- Fuel pump
- Fuel pump cover lockring with Tool J 39765
- Fuel lines and wiring harness
- Fuel tank
- Negative battery cable

5. Start the vehicle and check for leaks, repair if necessary.

Fuel Injector

REMOVAL & INSTALLATION

1.9L Engines

1. Before servicing the vehicle, refer to the precautions in the beginning of this section.

2. Properly relieve the fuel system pressure.

3. Remove or disconnect the following:
- Negative battery cable
- Fuel rail

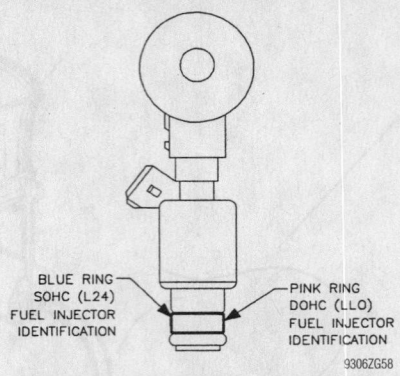

BLUE RING SOHC (L24) FUEL INJECTOR IDENTIFICATION
PINK RING DOHC (LL0) FUEL INJECTOR IDENTIFICATION

9306ZG58

Fuel injector identification—1.9L engine

- Fuel injector retaining clip off the injector
- Fuel injector
- Fuel injector O-rings and discard them

To install:

➥The SOHC and DOHC fuel injectors are not interchangeable. The injectors are identified by a color coded plastic ring near the bottom of the injector. The SOHC injector has a blue ring near the O-ring that mates to the intake manifold and the DOHC has a pink ring.

4. Lubricate the new fuel injector O-ring with clean engine oil.

5. Install or connect the following:
- New O-ring seals on the fuel injector
- Retaining clip to the fuel injector
- Fuel injector to the fuel rail
- Fuel rail
- Negative battery cable

6. Start the vehicle and check for leaks, repair if necessary.

2.2L and 3.0L Engines

1. Before servicing the vehicle, refer to the precautions in the beginning of this section.

2. Properly relieve the fuel system pressure.

3. Remove or disconnect the following:
- Negative battery cable

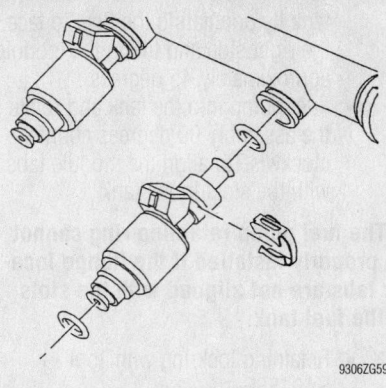

9306ZG59

Remove the retainer clip from the fuel injector—2.2L and 3.0L engines

- Air intake tube
- Fuel pressure regulator hose
- Fuel feed line
- Throttle body
- Throttle control cable bracket
- Fuel injector electrical connectors
- Fuel rail
- Fuel injector retaining clip off the injector
- Fuel injector
- Fuel injector O-rings

To install:

4. Lubricate the new fuel injector O-ring with clean engine oil.

5. Install or connect the following:
- New O-ring seals on the fuel injector
- Retaining clip to the fuel injector
- Fuel injector to the fuel rail
- Fuel rail and torque the bolts to 89 inch lbs. (10 Nm) on 2.2L engines and 71 inch lbs. (8 Nm) on 3.0L engines
- Fuel injector electrical connectors
- Throttle control cable bracket
- Throttle body and torque the bolts to 89 inch lbs. (10 Nm)
- Fuel feed line
- Fuel pressure regulator hose
- Air intake tube
- Negative battery cable

6. Start the vehicle and check for leaks, repair if necessary.

DRIVE TRAIN

Manual Transaxle Assembly

REMOVAL & INSTALLATION

1.9L Engines

1. Before servicing the vehicle, refer to the precautions in the beginning of this section.
2. Drain the transaxle fluid.
3. Remove or disconnect the following:
   - Both battery cables
   - Air cleaner assembly
   - Battery and tray
   - Powertrain Control Module (PCM) **J2** (black 28-way) harness connector. Do not disconnect the **J1** (80-way) connector.
   - PCM attaching bolts and move it aside
   - Transaxle strut to midrail bracket fastener and loosen the strut to transaxle fastener and move the strut aside
   - Back-up lamp electrical connector
   - Vehicle Speed Sensor (VSS) electrical connector, if equipped
   - Ground terminals from the clutch housing bolts
   - Vent tube retaining clip
   - Top 2 clutch housing to the engine bolts
   - Spark plug wires from the coil towers
   - Electronic ignition module electrical connectors
   - Electronic ignition module
   - Shifter cables from the arms and clutch housing
   - Clutch slave cylinder by rotating it ¼-turn counterclockwise and pushing it into the housing
   - 2 clutch hydraulic damper to housing fasteners and wire the actuator cylinder and damper to the upper radiator hose
4. Wire the radiator to the upper support.
5. Install an engine support bar and place the feet on the outer edge of the shock tower.
6. Connect the bar hooks to the engine support bracket.
7. Position the stabilizer foot on the engine to the right of the dipstick tube.
8. Adjust the hooks and stabilizer to remove any looseness.
9. Remove or disconnect the following:
   - Both front wheels and inner splash shields
   - Engine splash shield
   - Engine strut cradle bracket fasteners
   - Transaxle mount-to-cradle fasteners
   - Front exhaust pipe-to-manifold and catalytic converter fasteners
   - Front exhaust pipe
   - Steering gear to cradle fasteners
   - Brake pipe bracket push pin
   - Engine-to-transaxle stiffening brace
   - Dust cover
   - Cotter pin from the lower ball joints
   - Ball joint from the lower control arm with Separator Tool SA9132S
   - Separate the left axle from the transaxle and install an axle seal protector
   - Separate the right axle from the intermediate shaft
   - Intake bracket-to-intake manifold bolt on DOHC engines
   - Intake bracket to intermediate shaft support bolt
   - Support bracket
   - Intermediate shaft from the transaxle
   - 4 cradle-to-body bolts and lower the cradle on a support dolly
   - Bottom clutch housing bolts and install a guide bolt into the rear housing bolt hole
10. Separate the transaxle from the engine.

To install:

11. Place the transaxle assembly securely onto a jack and position under the vehicle. Install axle seal protectors into seals on both sides.
12. Raise the transaxle into the vehicle to align the input shaft to the center of the clutch. Rotate the transaxle back and forth to align the input shaft splines to the clutch disc.
13. Install or connect the following:
    - 2 lower clutch housing-to-engine bolts and torque them to 103 ft. lbs. (140 Nm)
    - Axle Seal Protector tool to the inside seal and remove the transaxle jack
    - Left side axle to the transaxle. When the splines clear the seal protector, remove the tool.
    - Intermediate shaft to the transaxle
    - Intermediate shaft and torque the bolts to 40 ft. lbs. (54 Nm)
    - Starter bracket to the intermediate shaft and torque the bolts to 22 ft. lbs. (30 Nm)
    - Intake manifold support bracket (on DOHC) and torque the bolts to 22 ft. lbs. (30 Nm)

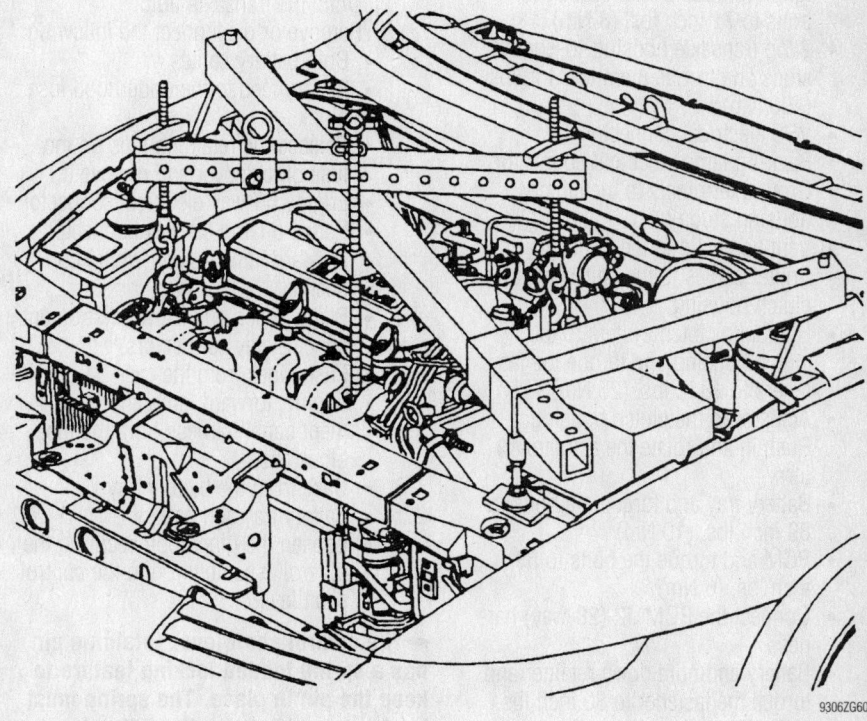

Install an Engine Support tool

9306ZG60

- Intake bracket to the intermediate shaft support and torque the bolt to 40 ft. lbs. (54 Nm)
- Right side axle to the intermediate shaft

14. Raise the cradle up on a support dolly and place the ball joints into the knuckles. Verify the correct positioning of the lower control arm bar studs to the knuckles, the cooling module support bushings, the engine strut bracket and the transaxle mount.

15. Insert 9/16 in. round steel rods into the cradle-to-body alignment holes near the front cradle to body fastener holes. Guide the cradle into position making sure all mount studs are properly guided into their holes.

16. Be sure the washers are in place, then install the 2 rear cradle to body bolts. Verify proper cradle positioning and install the 2 front cradle bolts, then torque the 4 cradle bolts to 155 ft. lbs. (210 Nm).

17. Remove the support dolly and lower the vehicle sufficiently for underhood access. Remove the engine support bar assembly.

18. Install or connect the following:
- Transaxle strut-to-cradle bracket through-bolt and nut and torque them to 40 ft. lbs. (54 Nm)
- Strut cradle bracket-to-cradle bolt and torque the it to 52 ft. lbs. (70 Nm)
- Ignition module and torque the bolts to 71 inch lbs. (8 Nm)
- 2 top transaxle housing-to-engine studs and torque them to 74 ft. lbs. (100 Nm)
- VSS electrical connector
- Back-up lamp electrical connector
- Ground terminals to the clutch housing studs
- Vent hose clip to the housing
- Shifter cables to the shift arms and clutch housing
- Hydraulic clutch system to the clutch housing and torque the fasteners to 18 ft. lbs. (25 Nm)
- Actuator to the clutch housing. Push in and rotate the actuator 1/4 turn
- Battery tray and torque the bolts to 89 inch lbs. (10 Nm)
- PCM and torque the bolts to 53 inch lbs. (6 Nm)
- Connect the PCM **J2** (28-way) harness
- Battery and hold down retainer and torque the fastener to 80 inch lbs. (9 Nm)
- Air cleaner assembly

- Transaxle lower mount-to-cradle fastener and torque it to 37 ft. lbs. (50 Nm)
- 2 engine strut cradle bracket-to-cradle fasteners and torque them to 37 ft. lbs. (50 Nm)
- Steering gear-to-cradle fasteners and torque them to 37 ft. lbs. (50 Nm)
- Brake pipe-to-cradle retainer
- Dust cover and torque the bolts to 8 ft. lbs. (11 Nm)
- Engine-to-transaxle stiffening brace and torque the bolts to 40 ft. lbs. (54 Nm)
- Exhaust manifold pipe and torque the bolts to 23 ft. lbs. (31 Nm)
- Intermediate exhaust pipe to the catalytic converter and torque the bolts to 18 ft. lbs. (25 Nm)
- Front exhaust pipe support to the stiffening bracket and torque the fasteners to 89 inch lbs. (10 Nm)
- Both front wheels
- Both battery cables

19. Fill the transaxle to the proper level.

20. Warm the engine and check the transaxle fluid. Check and adjust vehicle alignment, as necessary.

L–Series

1. Before servicing the vehicle, refer to the precautions in the beginning of this section.

2. Drain the transaxle fluid.

3. Remove or disconnect the following:
- Both battery cables
- Battery feed to the underhood fuse block
- Release the retaining tabs on the fuse block cover and remove it
- Engine 68-way electrical connector
- Forward lamp 68-way connector
- Forward lamp 2-way (white) connector
- Black, green and brown instrument panel 2-way connectors
- Fuse block from the case
- Engine, forward lamp and instrument panel harness from the fuse block case
- Case from the battery tray
- Battery tray and battery
- Loosen the pinch bolt securing the control assembly rod to the control shaft lever

➡The control shaft lever retaining pin has a spring loaded locking feature to keep the pin in place. The spring must be depressed before attempting to remove it.

- Pin retaining the control shaft lever to shift control shaft
- Retaining clips holding the control shaft lever to the transaxle and frame brackets
- Control shaft lever
- Back-up lamp switch electrical connector
- Wire harness from the transaxle
- Spring loaded locking pin from the hole on top of the transaxle, if equipped
- Clutch hydraulic fitting from the transaxle
- Upper transaxle to engine bolts and install and engine support bar, SA9105E, and Adapter Kit J43405
- Matchmark the position of the left transaxle mount and remove the mount
- Frame assembly
- Front transaxle mount
- Rear transaxle mount
- Axle shafts from the transaxle
- A/C line retaining clips and drop the A/C line down from the body

4. Lower the left side of the powertrain assembly to allow the transaxle to clear the engine compartment rail.

5. Remove or disconnect the following:
- Left side transaxle mount bracket
- Control shaft lever assembly pivot pin bracket

6. Properly support the powertrain assembly
- Remaining engine-to-transaxle bolts
- Transaxle from the engine

To install:

7. Raise the transaxle into the vehicle and align the input shaft to the center of the clutch disc.

8. Install or connect the following:
- Lower transaxle to engine bolts and torque them to 48 ft. lbs. (65 Nm)
- Control shaft lever assembly pivot pin bracket and torque the fastener to 18 ft. lbs. (24 Nm)
- Axle shafts to the transaxle. After the splines clear the tool, remove the seal protector and snap the axle into place.

9. Remove the transaxle fill plug, fill the transaxle with 2 quarts of manual fluid and install the plug. Torque the plug to 22 ft. lbs. (30 Nm).

10. Install or connect the following:
- Left side transaxle mount bracket and torque the bolts to 41 ft. lbs. (55 Nm)
- A/C line against the body and install the retaining clips

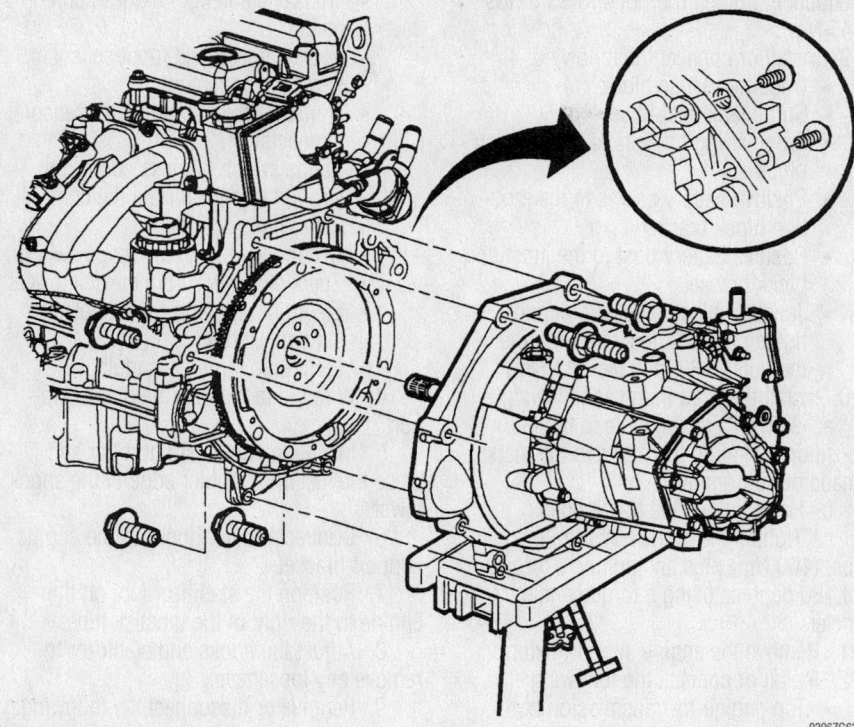

Remove the remaining engine to transaxle bolts—L–Series

9306ZG62

- Rear transaxle mount and torque the bolts to 41 ft. lbs. (55 Nm)
- Front transaxle mount and torque the bolts to 41 ft. lbs. (55 Nm)
- Frame assembly
- Upper transaxle to engine bolts and torque them to 48 ft. lbs. (65 Nm)
- Left side transaxle mount bolts using the matchmarks made during removal and torque them to 41 ft. lbs. (55 Nm)
- Back-up lamp switch electrical connector
- Control shaft lever assembly onto the pivot pin brackets
- Retaining clips holding the control shaft lever to the transaxle and frame brackets
- Pin to retain the control shaft lever to the shift control shaft
- Rotate the shift control shaft clockwise and install the shift linkage lock pin
- Move the shift lever to the 5th gear/reverse gate and install a ⅜ inch punch to hold it in place
- Pinch bolt securing the control assembly rod to the control shaft lever and torque it to 9 ft. lbs. (12 Nm) plus 180 degrees
- Shift lever hole plug
- Clutch hydraulic fitting to the

transaxle and bleed the hydraulic system
- Battery tray and torque the fasteners to 11 ft. lbs. (15 Nm)
- Underhood fuse block to the battery tray and torque the fasteners to 80 inch lbs. (9 Nm)
- Wire harness to the fuse case
- Snap the fuse block onto the hinges
- Engine 68-way connectors
- Forward lamp 68-way connectors
- Instrument panel 68-way connectors
- Forward 2-way (white) connector
- Black, green and brown 2-way connectors
- Snap the fuse block to the case
- Coolant hose to the fuse block
- Battery feed to the fuse block and torque the fastener to 12 inch lbs. (16 Nm)
- Battery tray and hold-down bracket and torque the fastener to 15 ft. lbs. (20 Nm)
- Battery
- Fan control module to the battery hold down bracket
- Both battery cables

11. Fill the transaxle to the proper level.

12. Warm the engine and check the transaxle fluid. Check and adjust vehicle alignment, as necessary.

ION

1. Before servicing the vehicle, refer to the precautions in the beginning of this section.

2. Drain the transaxle fluid.

3. Remove or disconnect the following:
- Negative battery cable
- Positive battery post from the underhood junction block
- Positive cables from the underhood junction block
- Surge tank inlet hose from the surge tank
- Underhood junction block bracket nuts and bolt
- Front wiring harness from the junction block bracket
- Move the junction block aside
- Clutch hose from the actuator cylinder

4. Install an engine support fixture.

5. Secure the cooling module to the upper body structure.
- Upper transmission-to-mount bolts
- Upper transmission-to-engine bolt

6. Remove the front frame as follows:
   a. Mark the frame-to-body position.
   b. Lower the vehicle until it is 3 feet off the ground so that a lift table may be placed in position. As necessary, use blocks of wood between the frame and table to support the assembly.
   c. Slowly remove the front frame bolts, the partially unscrew the bolts until 1 ½ inch of the bolt shank is exposed.
   d. Slowly lower the table to the floor

7. Remove or disconnect the following:
- Drive shafts from the transmission and support with wire
- Starter
- Shift cable from the transmission as follows:
   a. Apply the parking brake.
   b. Shift the transmission to **Neutral**.
   c. Remove the console.
   d. Lift up the cable retainers.
   e. Disconnect the cable ends from the shifter assembly.
   f. Disconnect the cable ends from the transmission shift levers, make sure to mark the cable locations prior to removal, this will aid during assembly.

8. Remove or disconnect the following:
- Back-up lamp switch and Vehicle Speed Sensor connections

9. Use the engine support fixture to lower the powertrain assembly to allow clearance between the side rail and powertrain.

10. Attach a transmission jack and remove the transmission-to-engine bolts.

11. Remove the transmission.

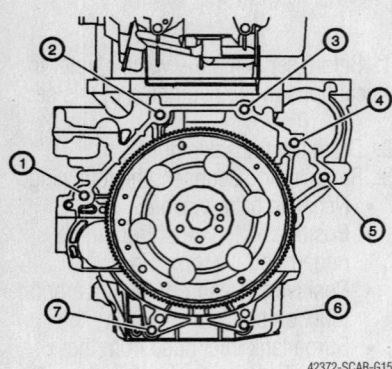

42372-SCAR-G15

The number 3 bolt position does not require a bolt—ION

To install:

➡**The number 3 bolt position does not require a bolt. Refer to the accompanying illustration for bolt locations.**

12. Use the transmission jack to align the transmission, install the transmission-to-engine bolts and torque to 55 ft. lbs. (75 Nm).

13. Install or connect the following:
- Back-up lamp switch and Vehicle Speed Sensor connections

14. Connect the shift cable to the transmission as follows:

a. Connect the cable ends to the transmission shift levers as noted during removal.

b. Connect the cable ends to the shifter assembly.

c. Push the shifter neutral lock clip. Move the shifter slightly to center the clip and make sure the transmission is in **Neutral**.

d. Press down on the locking tabs.

e. Pull the shifter neutral lock into its original position.

f. Install the console.

15. Install or connect the following:
- Starter
- Drive shafts

16. Use the engine support fixture to raise the powertrain assembly into position.

17. Install the side transmission mount as follows:

a. Place the mount in position and tighten the mount-to-rail bolts to 25 ft. lbs. (34 Nm).

18. Use a jack to align the mount when it is in position.

a. Install and hand tighten the mount-to-transmission bolts. Do not pry the transmission or mount to align the holes. Hand start the bolts in the following sequence; rear bolt, middle and then front bolt. Using the same

sequence, tighten the bolts to 33 ft. lbs. (45 Nm).

19. Install or connect the following:
- Underhood fuse block
- Surge tank inlet hose
- Front wiring harness to the junction block
- Positive battery cables to the junction block bracket
- Positive battery post to the junction block bracket
- Junction block bracket bolt and tighten to 18 ft. lbs. (25 Nm) and the nuts to 89 inch lbs. (10 Nm)

20. Install the front frame as follows:

a. Slowly raise the table to the into position and align the frame to the marks made during removal.

b. Hand tighten the frame bolts

c. Tighten the frame bolts to 74 ft. lbs. (100 Nm) plus an additional torque of 180 degrees using a torque angle meter.

21. Remove the engine support fixture.

22. Install or connect the following:
- Top engine-to-transmission bolt and tighten to 55 ft. lbs. (75 Nm)
- Clutch hose to the actuator cylinder and bleed the hydraulic clutch system.
- Cooling module
- Negative battery cable.

23. Fill the transmission to the proper level.

24. Warm the engine and check the transmission fluid. Check and adjust vehicle alignment, as necessary.

Automatic Transaxle Assembly

REMOVAL & INSTALLATION

1.9L Engines

1. Before servicing the vehicle, refer to the precautions in the beginning of this section.

2. Drain the transaxle fluid.

3. Remove or disconnect the following:
- Both battery cables
- Air cleaner assembly
- Battery and tray
- Powertrain Control Module (PCM) **J2** (black 28-way) harness connector. Do not disconnect the **J1** (80-way) connector.
- PCM attaching bolts and move aside
- Transaxle strut to midrail bracket fastener and loosen the strut to transaxle fastener and move the strut aside

- Transaxle solenoid harness connector
- Vehicle speed and input sensor harness connectors
- Transaxle fluid temperature sensor connector
- Range switch harness connector
- Ground terminals from the converter housing
- Ground wire from the range switch
- Spark plug wires from the coil towers
- Electronic ignition module electrical connectors
- Electronic ignition module

4. Wire the radiator to the upper support.

5. Install an engine support bar and place the feet on the outer edge of the shock tower.

6. Connect the bar hooks to the engine support bracket.

7. Position the stabilizer foot on the engine to the right of the dipstick tube.

8. Adjust the hooks and stabilizer to remove any looseness.

9. Remove or disconnect the following:
- Both front wheels and inner splash shields
- Engine splash shield
- Engine strut cradle bracket fasteners
- Transaxle mount to cradle fasteners
- Front exhaust pipe to manifold and catalytic converter fasteners
- Front exhaust pipe
- Steering gear to cradle fasteners
- Brake pipe bracket push pin
- Engine-to-transaxle stiffening brace
- Dust cover
- Torque converter-to-flywheel bolts
- Ball joint from the lower control arm
- Separate the left axle from the transaxle and install an axle seal protector
- Separate the right axle from the intermediate shaft
- Intake bracket-to-intake manifold bolt (on DOHC engines)
- Intake bracket-to-intermediate shaft support bolt
- Support bracket
- Intermediate shaft from the transaxle
- Transaxle cooler lines
- 4 cradle-to-body bolts and lower the cradle on a support dolly
- Lower torque converter bolts
- Separate the transaxle from the engine
- Shifter cable from the converter housing
- Transaxle from the vehicle

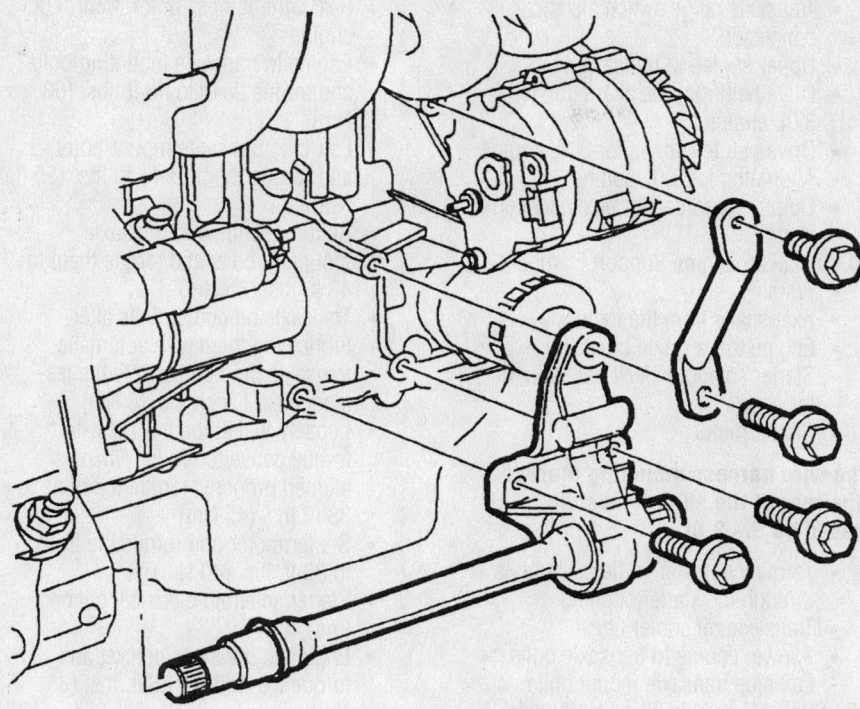

Remove the intake-to-intermediate shaft support bracket and the intermediate shaft—1.9L engines

To install:

10. Flush the transaxle oil cooler lines.

11. Place the transaxle assembly securely onto a jack and position under the vehicle. Install axle seal protectors into seals on both sides.

12. Raise the transaxle high enough to connect the shifter cable.

13. Install or connect the following:
- Shifter cable to the transaxle gear selector lever and insert it in the converter housing
- 2 lower converter housing-to-engine bolts and torque them to 96 ft. lbs. (130 Nm)
- Axle Seal Protector tool to the inside seal and remove the transaxle jack
- Left side axle to the transaxle. When the splines clear the seal protector, remove the tool.
- Intermediate shaft to the transaxle
- Intermediate shaft and torque the bolts to 40 ft. lbs. (54 Nm)
- Intake manifold support bracket (on DOHC) and torque the bolts to 22 ft. lbs. (30 Nm)
- Intake bracket to the intermediate shaft support and torque the bolt to 40 ft. lbs. (54 Nm)
- Right side axle to the intermediate shaft

- Cooler lines to the transaxle

14. Raise the cradle up on a support dolly and place the ball joints into the knuckles. Verify the correct positioning of the lower control arm bar studs to the knuckles, the cooling module support bushings, the engine strut bracket and the transaxle mount.

15. Insert ⁹⁄₁₆ in. round steel rods into the cradle-to-body alignment holes near the front cradle to body fastener holes. Guide the cradle into position making sure all mount studs are properly guided into their holes.

16. Be sure the washers are in place, then install the 2 rear cradle to body bolts. Verify proper cradle positioning and install the 2 front cradle bolts, then torque the 4 cradle bolts to 155 ft. lbs. (210 Nm).

17. Remove the support dolly and lower the vehicle sufficiently for underhood access. Remove the engine support bar assembly.

18. Install or connect the following:
- Transaxle strut-to-cradle bracket through-bolt and nut and torque them to 40 ft. lbs. (54 Nm)
- Strut midrail bracket and torque the bolt to 52 ft. lbs. (70 Nm)
- Ignition module and torque the bolts to 71 inch lbs. (8 Nm)
- Ignition module wire connector
- Spark plug wires to the coil towers

➡ Proper orientation of the ignition module is critical to the operation of the module and the on board diagnostic system. The proper sequence is 4-1-2-3 from left to right.

- 2 top converter housing-to-engine studs and torque them to 74 ft. lbs. (100 Nm)
- Transaxle solenoid harness connector and torque the fastener to 22 inch lbs. (2.5 Nm)
- Vehicle speed and input sensor connectors
- Transaxle fluid temperature sensor connectors
- Transaxle range switch harness connectors
- Ground terminals to the top 2 converter housing bolts and torque the bolts to 19 ft. lbs. (25 Nm)
- Ground wire to the neutral selector switch
- Adjust the control cable
- Air cleaner assembly
- Battery tray and torque the bolts to 89 inch lbs. (10 Nm)
- PCM and torque the bolts to 53 inch lbs. (6 Nm)
- Connect the PCM **J2** (28-way) harness
- Battery and hold down retainer and torque the fastener to 80 inch lbs. (9 Nm)
- Transaxle lower mount to cradle fastener and torque it to 37 ft. lbs. (50 Nm)
- 2 engine strut cradle bracket-to-cradle fasteners and torque them to 37 ft. lbs. (50 Nm)
- Steering gear to cradle fasteners and torque them to 37 ft. lbs. (50 Nm)
- Brake pipe to cradle retainer

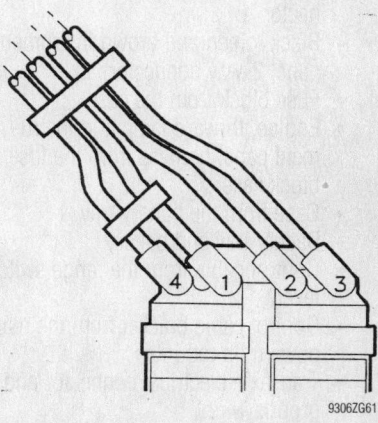

Electronic ignition module wire sequence 4-1-2-3—1.9L engines

- Torque converter to flexplate and torque the bolts to 52 ft. lbs. (70 Nm)
- Dust cover and torque the bolts to 8 ft. lbs. (11 Nm)
- Engine to transaxle stiffening brace and torque the bolts to 40 ft. lbs. (54 Nm)
- Exhaust manifold pipe and torque the bolts to 23 ft. lbs. (31 Nm)
- Intermediate exhaust pipe to the catalytic converter and torque the bolts to 18 ft. lbs. (25 Nm)
- Front exhaust pipe support to the stiffening bracket and torque the fasteners to 23 ft. lbs. (31 Nm)
- Lower ball joint and torque the nut to 55 ft. lbs. (75 Nm)
- New cotter pins
- Splash shields
- Both front wheels
- Both battery cables

19. Fill the transaxle to the proper level.

20. Warm the engine and check the transaxle fluid. Check and adjust vehicle alignment, as necessary.

L–Series

1. Before servicing the vehicle, refer to the precautions in the beginning of this section.
2. Drain the transaxle fluid.
3. Remove or disconnect the following:
   - Both battery cables
   - Move the fan control module up and out of the way
   - Battery feed to the underhood fuse block
   - Coolant hose from the underhood fuse block
   - Release the retaining tabs on the fuse block cover and remove it
   - Engine 68-way electrical connector
   - Forward lamp 68-way connector
   - Forward lamp 2-way (white) connector
   - Black, green and brown instrument panel 2-way connectors
   - Fuse block from the case
   - Engine, forward lamp and instrument panel harness from the fuse block case
   - Case from the battery tray
   - Battery tray and battery
   - Control cable from the range switch lever
   - Control cable bracket from the rear powertrain mount
   - Transaxle electrical connector and ground wire
   - Remaining wire harness connectors from the transaxle

- Transaxle range switch electrical connector
- Upper engine to transaxle bolts
- Drive belt from the alternator, for 3.0L engine
- Drive belt tensioner, for 3.0L engine
- Alternator, for 3.0L engine
- Output speed sensor electrical connector

4. Install an Engine Support Fixture.
   - Frame
   - Axle shafts from the transaxle
   - Engine to transaxle bracket
   - Starter solenoid electrical connector
   - Starter motor

➡ **The wire harness mounting bracket on the rear of the engine must be removed on the 3.0L.**

   - Torque converter-to-flexplate bolts through the starter opening
   - Transaxle oil cooler lines
   - 2 lower engine to transaxle bolts
   - Left side transaxle mount bolts

5. Slide a hydraulic lifting table under the transaxle and remove the remaining mounting bolts.

6. Separate the transaxle from the engine.

To install:

7. With the transaxle on a hydraulic lift, raise the transaxle to the engine.

8. Install or connect the following:

- Rear wire harness bracket, for 3.0L engine
- Engine to transaxle mounting bolts and torque them to 48 ft. lbs. (60 Nm)
- Left side transaxle mount bolts and torque them to 41 ft. lbs. (55 Nm)
- Bottom 2 engine to transaxle mounting bolts and torque them to 48 ft. lbs. (65 Nm)
- Transaxle oil cooler lines after lubricating them with automatic transaxle fluid and torque the fasteners to 71 inch lbs. (8 Nm)
- Loosely install the flex plate-to-torque converter bolts. When aligned properly, torque the bolts to 33 ft. lbs. (45 Nm).
- Starter motor and torque the bolts to 30 ft. lbs. (40 Nm)
- Starter solenoid electrical connectors
- Engine to transaxle bracket and torque the bolts to 26 ft. lbs. (35 Nm)
- New axle shaft retaining rings on the end of the output shafts and install the axle shafts to the transaxle
- Frame assembly
- Output speed sensor electrical connector and secure it to the mounting stud

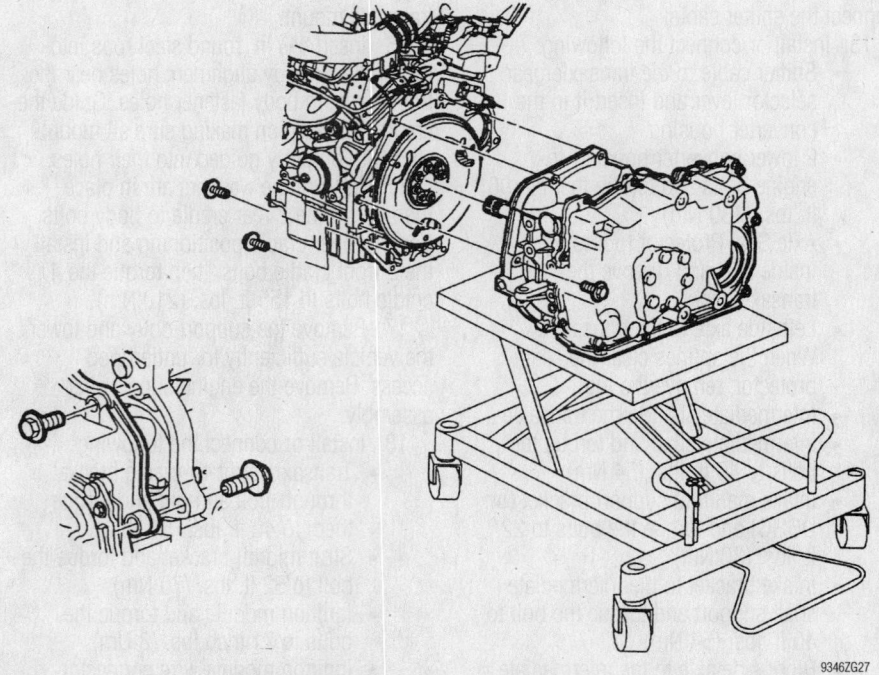

Lifting table mounting points—L–Series

9346ZG27

- Alternator and torque the bolts to 30 ft. lbs. (40 Nm), for 3.0L engine
- Drive belt tensioner and torque the bolts to 30 ft. lbs. (40 Nm), for 3.0L engine
- Drive belt, for 3.0L engine
- Remaining transaxle to engine mounting bolts and torque them to 48 ft. lbs. (65 Nm)
- Transaxle electrical connector and ground wire and torque the fasteners to 18 ft. lbs. (25 Nm)
- Wire harness attachments to the transaxle side cover and torque the fasteners to 15 ft. lbs. (20 Nm)
- Control cable mounting bracket and torque the bolts to 15 ft. lbs. (20 Nm)
- Control cable to the range switch lever
- Battery tray and torque the fasteners to 11 ft. lbs. (15 Nm)
- Underhood fuse block case to the battery tray. Torque the bolt to 80 inch lbs. (9 Nm) and snap the block onto the hinges.
- Engine 68-way connectors
- Forward lamp 68-way connectors
- Instrument panel 68-way connectors
- Forward 2-way (white) connector
- Black, green and brown 2-way connectors
- Snap the fuse block to the case
- Coolant hose to the fuse block
- Battery feed to the fuse block and torque the fastener to 12 inch lbs. (16 Nm)
- Battery tray and hold-down bracket and torque the fastener to 15 ft. lbs. (20 Nm)
- Battery
- Fan control module to the battery hold down bracket
- Both battery cables

9. Fill the transaxle to the proper level.
10. Warm the engine and check the transaxle fluid. Check and adjust vehicle alignment, as necessary.

ION

AUTOMATIC 5-SPEED TRANSMISSION

1. Before servicing the vehicle, refer to the precautions in the beginning of this section.
2. Drain the transaxle fluid.
3. Remove or disconnect the following:
- Negative battery cable
- Transmission Control Module (TCM) connections noting their locations to aid during installation

- TCM by releasing the tab on the junction block bracket
- Positive battery post from the underhood junction block
- Positive cables from the underhood junction block
- Surge tank inlet hose from the surge tank
- Underhood junction block bracket nuts and bolt
- Front wiring harness from the junction block bracket
- Move the junction block aside
- Park/Neutral switch connector
- Main transmission harness connector

4. Install an engine support fixture.
5. Remove or disconnect the following:

- Upper transmission-to-mount bolts
- Upper transmission-to-engine bolt

6. Remove the front frame as follows:
   a. Mark the frame-to-body position.
   b. Lower the vehicle until it is 3 feet off the ground so that a lift table may be placed in position. As necessary, use blocks of wood between the frame and table to support the assembly.
   c. Slowly remove the front frame bolts, the partially unscrew the bolts until 1 ½ inch of the bolt shank is exposed.
   d. Slowly lower the table to the floor.
7. Remove or disconnect the following:
- Cooler lines from the transmission
- Drive shafts from the transmission and support with wire
- Starter
- Torque converter-to-flywheel bolts
- Shift cable from the transmission as follows:
   a. Remove the shift cable retaining nut near the exhaust heat shield.
   b. Remove the shift cable from the transmission cable bracket.
   c. Disconnect the shift cable from the park/neutral switch.

8. Use the engine support fixture to lower the powertrain assembly to allow clearance between the side rail and powertrain.
9. Attach a transmission jack and remove the transmission-to-engine bolts.
10. Remove the transmission.

To install:

➡**The number 3 bolt position does not require a bolt. refer to the accompanying illustration for bolt locations.**

11. Use the transmission jack to align the transmission, install the transmission-to-engine bolts and torque to 55 ft. lbs. (75 Nm).

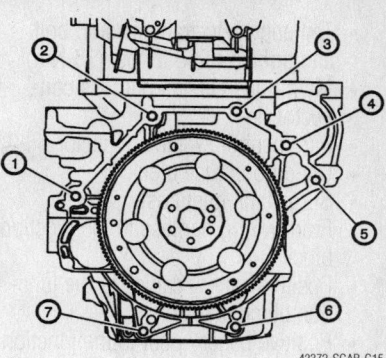

42372-SCAR-G15

The number 3 bolt position does not require a bolt—ION

- Shift cable to the transmission as follows:
   a. Connect the shift cable from the park/neutral switch.
   b. Connect the shift cable to the transmission cable bracket.
   c. Install the shift cable retaining nut near the exhaust heat shield and tighten to 89 inch lbs. (10 Nm).
12. Install or connect the following:
- Torque converter-to-flywheel bolts and tighten to 48 ft. lbs. (65 Nm)
- Starter
- Drive shafts
- Transmission cooler lines to the transmission. Lubricate the cooler pipe seals with transmission fluid prior to connecting the lines.
13. Use the engine support fixture to raise the powertrain assembly into position.
14. Install the side transmission mount as follows:
   a. Place the mount in position and tighten the mount-to-rail bolts to 25 ft. lbs. (34 Nm).
15. Use a jack to align the mount when it is in position.
   a. Install and hand tighten the mount-to-transmission bolts. Do not pry the transmission or mount to align the holes. Hand start the bolts in the following sequence; rear bolt, middle and then front bolt. Using the same sequence, tighten the bolts to 33 ft. lbs. (45 Nm).
16. Install the front frame as follows:
   a. Slowly raise the table to the into position and align the frame to the marks made during removal.
   b. Hand tighten the frame bolts
   c. Tighten the frame bolts to 74 ft. lbs. (100 Nm) plus an additional torque of 180 degrees using a torque angle meter.
17. Remove the engine support fixture.
18. Install or connect the following:

- Top engine-to-transmission bolt and tighten to 55 ft. lbs. (75 Nm)
- Main transmission harness connector
- Park/neutral switch connector
- Underhood fuse block
- Surge tank inlet hose
- Front wiring harness to the junction block
- Positive battery cables to the junction block bracket
- Positive battery post to the junction block bracket
- Junction block bracket bolt and tighten to 18 ft. lbs. (25 Nm) and the nuts to 89 inch lbs. (10 Nm)
- TCM
- Negative battery cable

19. Fill the transaxle to the proper level.
20. Adjust the shift cable as follows:

a. Place the transmission in **Neutral**.

b. Use fascia retainer remover tool J 36346 to disconnect the shift control cable from the transaxle range switch lever.

c. Disconnect the cable from the cable bracket by depressing the cable clip tabs and pulling up.

d. Release the cable assemble adjustment lock by sliding the black tab.

e. Insert a screw driver into the slot under the white tab and pull the tab out.

f. Attach the cable to the cable bracket. The clamp will be secure when an audible click is heard.

g. Attach the cable end fitting into the ball stud of the transaxle range switch selector. The clamp will be secure when an audible click is heard.

h. Attach the cable to the cable bracket. The clamp will be secure when an audible click is heard.

i. Lock the adjustment in place by pushing down on the white tab, slide the black tab up to lock the white tab in place and verify cable operation.

21. Warm the engine and check the transaxle fluid. Check and adjust vehicle alignment, as necessary.

AUTOMATIC 5-SPEED CONTINUOUSLY VARIABLE TRANSMISSION

1. Before servicing the vehicle, refer to the precautions in the beginning of this section.
2. Drain the transaxle fluid.
3. Remove or disconnect the following:
- Negative battery cable
- Electronic Control Module (ECM)
- Transmission Control Module (TCM) connections noting their locations to aid during installation

- TCM by releasing the tab on the junction block bracket, lift it half way, then turn it clockwise to remove
- Positive battery post from the underhood junction block
- Positive cables from the underhood junction block
- Surge tank inlet hose from the surge tank
- Underhood junction block bracket nuts and bolt
- Front wiring harness from the junction block bracket
- Move the junction block aside
- Park/Neutral switch connector

4. Install an engine support fixture.
5. Secure the cooling module to the upper body
6. Remove or disconnect the following:
- Upper transmission-to-mount bolts
- Upper transmission-to-engine bolt

7. Remove the front frame as follows:

a. Mark the frame-to-body position.

b. Lower the vehicle until it is 3 feet off the ground so that a lift table may be placed in position. As necessary, use blocks of wood between the frame and table to support the assembly.

c. Slowly remove the front frame bolts, the partially unscrew the bolts until 1 ½ inch of the bolt shank is exposed.

d. Slowly lower the table to the floor

8. Remove or disconnect the following:
- Cooler lines from the transmission
- Drive shafts from the transmission and support with wire
- Starter
- Torque converter-to-flywheel bolts
- Shift cable from the transmission as follows:

a. Remove the shift cable retaining nut near the exhaust heat shield.

b. Remove the shift cable from the transmission cable bracket.

c. Disconnect the shift cable from the park/neutral switch.

9. Use the engine support fixture to lower the powertrain assembly 2-to-3 inches to allow clearance between the side rail and powertrain.
10. Attach a transmission jack and remove the transmission-to-engine bolts.
11. Remove the transmission.

To install:

➡ **The number 3 bolt position does not require a bolt. refer to the accompanying illustration for bolt locations.**

12. Use the transmission jack to align the transmission, install the transmission-to-engine bolts and torque to 55 ft. lbs. (75 Nm).

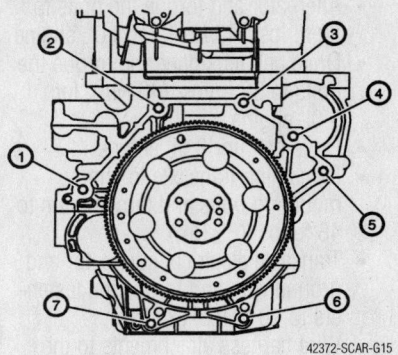

42372-SCAR-G15

The number 3 bolt position does not require a bolt—ION

- Shift cable to the transmission as follows:

a. Connect the shift cable from the park/neutral switch.

b. Connect the shift cable to the transmission cable bracket.

c. Install the shift cable retaining nut near the exhaust heat shield and tighten to 89 inch lbs. (10 Nm).

13. Install or connect the following:
- Torque converter-to-flywheel bolts and tighten to 48 ft. lbs. (65 Nm)
- Starter
- Drive shafts
- Transmission cooler lines to the transmission. Lubricate the cooler pipe seals with transmission fluid prior to connecting the lines.

14. Use the engine support fixture to raise the powertrain assembly into position.

a. Place the mount in position and tighten the mount-to-rail bolts to 25 ft. lbs. (34 Nm).

b. Use a jack raise the engine back into position and to align the mount when it is in position.

c. Install and hand tighten the mount-to-transmission bolts. Do not pry the transmission or mount to align the holes. Hand start the bolts in the following sequence; rear bolt, middle and then front bolt. Using the same sequence, tighten the bolts to 33 ft. lbs. (45 Nm).

15. Install the front frame as follows:

a. Slowly raise the table to the into position and align the frame to the marks made during removal.

b. Hand tighten the frame bolts

c. Tighten the frame bolts to 74 ft. lbs. (100 Nm) plus an additional torque of 180 degrees using a torque angle meter.

16. Install or connect the following:
- Top engine-to-transmission bolt and tighten to 55 ft. lbs. (75 Nm)

- Park/neutral switch connector
- Cooling module

17. Remove the engine support fixture.

- Underhood fuse block
- Surge tank inlet hose
- Front wiring harness to the junction block
- Positive battery cables to the junction block bracket
- Positive battery post to the junction block bracket
- Junction block bracket bolt and tighten to 18 ft. lbs. (25 Nm) and the nuts to 89 inch lbs. (10 Nm)
- TCM and ECM
- Negative battery cable

18. Fill the transaxle to the proper level.

19. Adjust the shift cable as follows:

a. Place the transmission in **Neutral**.

b. Use fascia retainer remover tool J 36346 to disconnect the shift control cable from the transaxle range switch lever.

c. Disconnect the cable from the cable bracket by depressing the cable clip tabs and pulling up.

d. Release the cable assemble adjustment lock by sliding the black tab.

e. Insert a screw driver into the slot under the white tab and pull the tab out.

f. Attach the cable to the cable bracket. The clamp will be secure when an audible click is heard.

g. Attach the cable end fitting into the ball stud of the transaxle range switch selector. The clamp will be secure when an audible click is heard.

h. Attach the cable to the cable bracket. The clamp will be secure when an audible click is heard.

i. Lock the adjustment in place by pushing down on the white tab, slide the black tab up to lock the white tab in place and verify cable operation.

20. Warm the engine and check the transaxle fluid. Check and adjust vehicle alignment, as necessary.

Clutch

ADJUSTMENT

The hydraulic clutch system is self-adjusting.

REMOVAL & INSTALLATION

1.9L Engines

1. Before servicing the vehicle, refer to the precautions in the beginning of this section.

2. Remove the transaxle from the vehicle.

3. Unsnap the release fork from the ball stud, then remove the fork and bearing from the vehicle. Slide the bearing from the fork. The bearing should be checked for excessive play and for minimal bearing drag. It should be replaced if no/little drag or excessive play is found.

➡**The release bearing is packed with grease and should not be washed with solvent.**

4. Using a feeler gauge, measure the distance between the pressure plate and flywheel surfaces in order to determine clutch face thickness. Replace the clutch disc if it is not within 0.205–0.287 in. (5.2–7.3mm) for 2000 or to 0.18 –0.28 in. (4.65mm–7.10mm) for 2001–04.

5. Remove the pressure plate-to-flywheel bolts in a progressive crisscross pattern to prevent warping the cover, then remove the pressure plate and clutch disc.

6. Inspect the pressure plate, as follows:

a. Check for excessive wear, chatter marks, cracks or overheating (indicated by a blue discoloration). Black random spots on the friction surface of the pressure plate is normal.

b. Check the plate for warpage using a straightedge and a feeler gauge; the maximum allowable warpage is 0.006 in. (0.15mm).

c. Replace the plate, if necessary.

7. Inspect the clutch disc, as follows:

a. Check the disc face for oil or burnt spots.

b. Check the disc for loose damper springs, hub or rivets.

c. Replace the disc, if necessary.

8. Check the flywheel, as follows:

a. Check the ring gear for wear or damage.

b. Check the friction surface for excessive wear, chatter marks, cracks or overheating.

c. Check flywheel thickness; the minimum allowable is 1.102 in. (28mm).

d. Measure flywheel run-out using a dial indicator, positioned for at least 2 flywheel revolutions. Push the crankshaft forward to take up thrust bearing clearance. Maximum flywheel run-out is 0.006 in. (0.15mm).

e. Check the flywheel for warpage using a straight-edge and a feeler gauge; the maximum allowable warpage is 0.006 in. (0.15mm).

f. Replace the flywheel, if necessary.

9. If necessary, remove the flywheel

retaining bolts and remove the flywheel from the crankshaft.

To install:

10. If removed, install the flywheel and torque the bolts in a crisscross sequence to 59 ft. lbs. (80 Nm).

11. Install or connect the following:

- Clutch disc and pressure plate with the yellow dot on the pressure plate aligned as close as possible to the mark on the flywheel. The clutch disc is labeled FLYWHEEL SIDE in order to help correctly position the disc. Loosely install the pressure plate bolts.
- Clutch alignment tool in the clutch disc and install the release bearing to the fork and torque the pressure plate bolts to 18 ft. lbs. (25 Nm)
- Snap the release bearing and fork onto the ball stud
- Lubricate the splines of the input shaft lightly with a high temperature grease
- Transaxle assembly
- Negative battery cable

L–Series

1. Before servicing the vehicle, refer to the precautions in the beginning of this section.

2. Remove the transaxle from the vehicle.

3. Check the slave cylinder release bearing minimal bearing drag. Replace the slave cylinder if little or drag is found.

4. Remove the pressure plate and clutch disc.

5. Inspect the pressure plate, as follows:

a. Check for excessive wear, chatter marks, cracks or overheating (indicated by a blue discoloration). Black random spots on the friction surface of the pressure plate is normal.

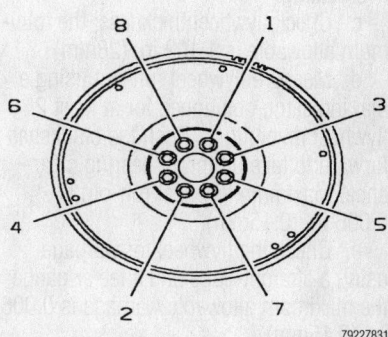

Flywheel mounting bolt tightening sequence—1.9L engines

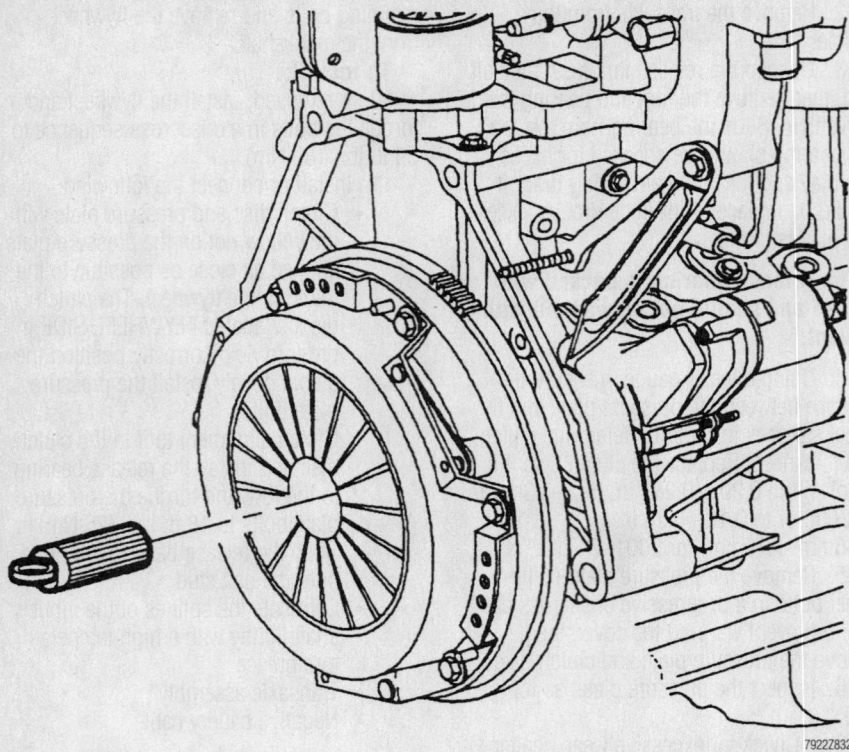

Use a proper clutch alignment tool before tightening the pressure plate retaining bolts—1.9L engines

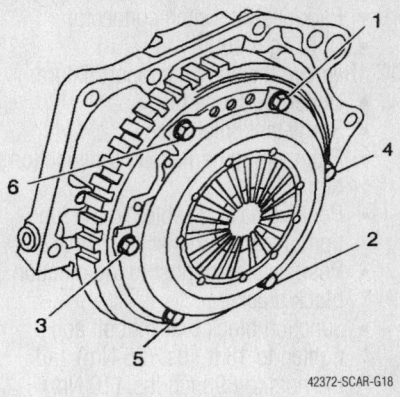

Tighten the pressure plate bolts in sequence —ION

b. Check the plate for warpage using a straightedge and a feeler gauge; the maximum allowable warpage is 0.006 in. (0.15mm).

c. Replace the plate, if necessary.

6. Inspect the clutch disc, as follows:

a. Check the disc face for oil or burnt spots.

b. Check the disc for loose damper springs, hub or rivets.

c. Replace the disc, if necessary.

7. Check the flywheel, as follows:

a. Check the ring gear for wear or damage.

b. Check the friction surface for excessive wear, chatter marks, cracks or overheating.

c. Check flywheel thickness; the minimum allowable is 1.102 in. (28mm).

d. Measure flywheel run-out using a dial indicator, positioned for at least 2 flywheel revolutions. Push the crankshaft forward to take up thrust bearing clearance. Maximum flywheel run-out is 0.006 in. (0.15mm).

e. Check the flywheel for warpage using a straight-edge and a feeler gauge; the maximum allowable warpage is 0.006 in. (0.15mm).

f. Replace the flywheel, if necessary.

8. If necessary, remove the flywheel retaining bolts and remove the flywheel from the crankshaft.

To install:

9. Install or connect the following:

- Flywheel (if removed) and torque the bolts in a crisscross pattern to 39 ft. lbs. (53 Nm) plus 25 degrees
- Clutch disc and pressure plate and loosely install the pressure plate bolts
- Clutch alignment tool in the clutch disc, and push in until it bottoms out in the crankshaft

10. Tighten the pressure plate bolts using multiple passes of a crisscross sequence to 11 ft. lbs. (15 Nm) and remove the alignment tool.

11. Lubricate the splines of the input shaft lightly with a high temperature grease.

12. Install the transaxle assembly.

13. Connect the negative battery cable.

ION

1. Before servicing the vehicle, refer to the precautions in the beginning of this section.

2. Remove the transaxle from the vehicle.

3. Check the slave cylinder release bear-ing minimal bearing drag. Replace the slave cylinder if little or drag is found.

4. Remove the pressure plate and clutch disc.

a. Replace the plate, if necessary.

b. Replace the disc, if necessary.

c. Replace the flywheel, if necessary.

To install:

5. Install or connect the following:

- Clutch disc and pressure plate and loosely install the pressure plate bolts
- Clutch alignment tool in the clutch disc, and push in until it bottoms out in the crankshaft

6. Tighten the pressure plate bolts in sequence to 18 ft. lbs. (24 Nm) and remove the alignment tool.

7. Lubricate the splines of the input shaft lightly with a high temperature grease.

8. Install the transaxle assembly.

9. Connect the negative battery cable.

Hydraulic Clutch System

BLEEDING

1.9L Engines

The clutch hydraulic assembly has been filled with fluid and bled of air at the factory. Do not attempt to bleed the hydraulic system. While the unit does not require periodic checking, it must be serviced, when necessary, as a complete assembly. The system is full when the reservoir is half full.

Only DOT 3 brake fluid should be added to the system. If the fluid level drops, inspect the system, including the slave cylinder, for leakage. A slight wetting of the slave cylinder surface is normal. Fill the clutch master cylinder reservoir with brake fluid. Be careful not to spill brake fluid on the painted surface of the vehicle.

L–Series

TRANSAXLE IN VEHICLE

This procedure outlines how to bleed the hydraulic clutch with the transaxle in the vehicle. Only **DOT 3** brake fluid should be added to the system.

1. Before servicing the vehicle, refer to the precautions in the beginning of this section.

2. Remove or disconnect the following:

3. Remove the reservoir cap and fill the reservoir with new brake fluid.

4. Install a Bleeder Adapter Tool J43915, to the reservoir and connect a pressure bleeder to the adapter.

5. Charge the pressure bleeder to 20–25 psi (138–172 kPa).

6. Attach a transparent hose over the clutch bleeder screw nipple and submerge the opposite end of the hose in a container of brake fluid.

7. Loosen the bleeder screw on the transaxle hydraulic fitting.

8. Bleed the system until no air bubbles are seen in the hose.

9. Tighten the bleeder screw.

10. Check the clutch pedal for a spongy feel. If the pedal feels soft, repeat the bleeding procedure.

11. Remove the bleeder tools and top off the fluid level if necessary.

TRANSAXLE OUT OF VEHICLE

This procedure outlines how to bleed the clutch slave cylinder while the transaxle is out of the vehicle. Only **DOT 3** brake fluid should be added to the system.

1. Before servicing the vehicle, refer to the precautions in the beginning of this section.

2. Remove or disconnect the following:

3. Connect a long piece of transparent hose to the slave cylinder fitting.

4. Depress the release bearing on the slave cylinder towards the clutch housing and release it.

5. The release bearing will spring back, but the piston will remain depressed next to the clutch housing.

6. Fill the hose with 13.8 inches (350mm) of DOT 3 brake fluid.

7. Connect a Vacuum Pump SA9180NE with a Pressure Adapter J35555-92, to the top of the hose.

8. Apply pressure to the hose to force the brake fluid into the slave cylinder. Pressure in the gauge will increase when the slave cylinder piston reaches the end of its travel.

9. Remove the vacuum pressure pump

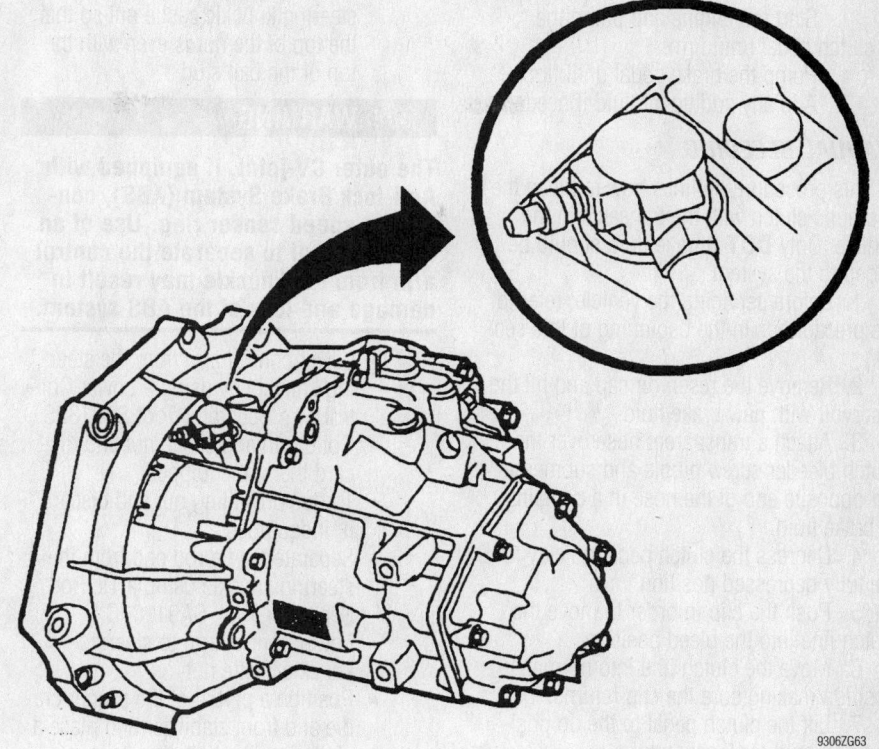

Loosen the clutch bleeder screw on the hydraulic fitting—L–Series

from the hose and depress the release bearing. Air bubbles should appear in the hose.

10. Repeat this procedure until no air bubbles are visible in the hose.

11. Drain and remove the hose.

ION

VACUUM BLEEDING

This procedure outlines how to bleed the hydraulic clutch with the transaxle in the vehicle. Only **DOT 3** brake fluid should be added to the system.

1. Before servicing the vehicle, refer to the precautions in the beginning of this section.

2. Bleed the system as follows:

a. Remove the reservoir cap and fill the reservoir with new brake fluid.

b. Install a metal Mityvac tool J 3555 Tool and adapter J43485.

c. Hold J 43485 in position and apply 15–20 hg (51–68 kPa) of vacuum.

d. Remove the adapter and refill the reservoir.

e. Depress the clutch pedal and cycle it for 30 seconds.

f. Lift the pedal up to the stop position and hold for 30 seconds.

g. Repeat steps A through F until all air is removed.

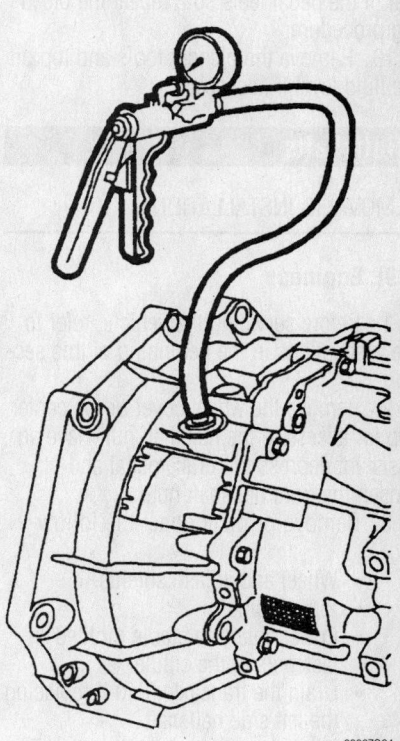

Install a vacuum pump and pressure adapter to bleed the slave cylinder— L–Series

h. Place the transmission in Neutral.

i. Start the engine and pump the clutch pedal until firm.

j. Pump the brake pedal until firm.

k. Add any additional fluid if needed.

MANUAL BLEEDING

This procedure outlines how to bleed the hydraulic clutch with the transaxle in the vehicle. Only **DOT 3** brake fluid should be added to the system.

1. Before servicing the vehicle, refer to the precautions in the beginning of this section.

2. Remove the reservoir cap and fill the reservoir with new brake fluid.

3. Attach a transparent hose over the clutch bleeder screw nipple and submerge the opposite end of the hose in a container of brake fluid.

4. Depress the clutch pedal quickly to the fully depressed position.

5. Push the clip in order to move the clutch line into the bleed position.

6. Move the clutch line into normal position making sure the clip returns.

7. Lift the clutch pedal to the up position and hold for 5 seconds.

8. Bleed the system until no air bubbles are seen in the hose.

9. Check the clutch pedal for a spongy feel. If the pedal feels soft, repeat the bleeding procedure.

10. Remove the bleeder tools and top off the fluid level if necessary.

Halfshafts

REMOVAL & INSTALLATION

1.9L Engines

1. Before servicing the vehicle, refer to the precautions in the beginning of this section.

2. Remove the wheel cover or the center cap for access to the halfshaft nut. Have an assistant depress the brake pedal and loosen the front halfshaft nut.

3. Remove or disconnect the following:

- Wheel and splash shields
- 2 push pins
- Lower shield to cradle molded-in fasteners at the cradle
- Drain the transaxle fluid if replacing the left side halfshaft
- Drive axle nut and washer and discard them
- Lower control arm to steering knuckle cotter pin and discard it

- Loosen the lower control arm to steering knuckle castle nut so that the top of the nut is even with the top of the ball stud

⁕⁕ WARNING

The outer CV-joint, if equipped with Anti-lock Brake System (ABS), contains a speed sensor ring. Use of an incorrect tool to separate the control arm from the knuckle may result in damage and loss of the ABS system.

- Lower control arm from the steering knuckle by using a Lower Control Arm Separator Tool SA9132S
- Cotter pin and castle nut and discard them, if equipped
- Torque prevailing nut and discard it, if equipped
- Separate the tie rod end from the steering knuckle using a Tie Rod Separator Tool, SA91100C
- Lower control arm to steering knuckle castle nut
- Position a prybar at the proper cradle and front stabilizer and place a cloth over the ball stud boot
- Separate the lower control arm from the steering knuckle and pull the knuckle away from the ball stud
- While pulling the knuckle/strut away, pull the outer end of the halfshaft out of the wheel hub and properly support the halfshaft

4. For right side halfshafts, remove the halfshaft from the intermediate shaft by tapping the axle with a block of wood and hammer. Separate the halfshaft from the intermediate shaft.

5. For left side halfshafts, remove the halfshaft by installing a prybar and prying the halfshaft from the transaxle.

To install:

➡**Be careful not to damage the halfshaft oil seal when installing the halfshaft into the transaxle.**

6. Install or connect the following:

- Transaxle Seal Protector Tool SA91112T, when installing the left side halfshaft
- Halfshaft into the transaxle. After the splines have safely passed the oil seal, remove the protector tool
- Insert the inner end of the halfshaft onto the outer end of the intermediate shaft for the right side halfshaft. Push on the axle firmly to engage the retaining ring
- Outer end of the halfshaft into the

wheel hub. Do not install any hardware at this time

- Lower control arm ball stud into the steering knuckle. Hand-tighten the castle nut at this time
- Attach the tie rod end to the steering knuckle, if equipped. Torque the nut to 33 ft. lbs. (45 Nm) and install a new cotter pin
- If equipped with a torque prevailing nut, attach the tie rod to steering knuckle with a new nut. Do not allow the ball stud to turn while tightening the prevailing nut. Torque the nut to 41 ft. lbs. (55 Nm).
- Axle to hub washer and new nut. Torque the nut to 148 ft. lbs. (200 Nm).

➡**When tightening the axle nut, have an assistant depress the brake pedal to prevent axle rotation.**

- Splash shields
- Align the molded in shield fasteners with the holes in the cradle and install push pins
- Wheel

7. Top off the transaxle with the proper fluid.

8. Check and adjust the front end alignment as necessary.

L–Series

1. Before servicing the vehicle, refer to the precautions in the beginning of this section.

2. Remove the wheel cover or the center cap for access to the halfshaft nut. Have an assistant depress the brake pedal and loosen the front halfshaft nut.

3. Remove or disconnect the following:

- Wheel
- Tie rod end torque prevailing nut and discard it

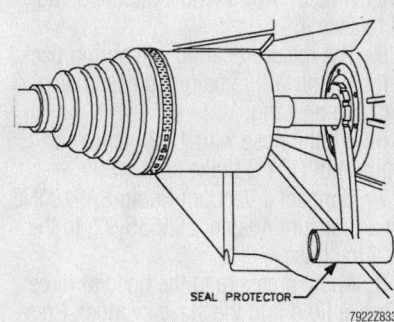

SEAL PROTECTOR

79227833

Failure to use a seal protector may allow the halfshaft splines to damage the transaxle seal

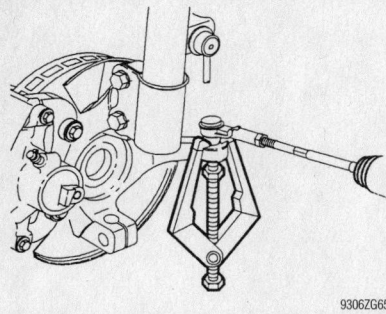

9306ZG65

Separate the tie rod end from the steering knuckle with a Tie Rod Separator tool— L–Series

- Tie rod end from the steering knuckle, using a Tie Rod Separator Tool SA91100C
- Lower control arm to steering knuckle

➡**Do not allow the steering knuckle to contact the ball stud seal. Contact may cause the seal to rip and the ball stud will need replacement.**

- Pull the steering knuckle/strut assembly away from the vehicle and pull the halfshaft out of the hub
- Properly support the halfshaft and remove the halfshaft from the transaxle
- Shaft retaining ring and discard it

To install:

4. Apply Output Shaft Lubricant 7847638, to the splines of the output shaft, if equipped with an automatic transaxle.

5. Install or connect the following:
- New stub shaft retaining ring
- Halfshaft to the transaxle after installing a Seal Protector Tool SA91112T

6. Remove the seal protector tool after the splines have passed the oil seal.

7. Install or connect the following:
- Fully seat the halfshaft into the transaxle
- Outer end of the halfshaft to the wheel hub with a new washer and nut
- Lower control arm ball stud to the steering knuckle. Torque the fastener to 75 ft. lbs. (100 Nm).
- Tie rod end to the steering knuckle. When seated properly, torque the fastener to 45 ft. lbs. (60 Nm).
- Wheel
- Halfshaft to wheel nut. Torque the nut to 74–118 ft. lbs. (100–160 Nm), release the nut until it turns freely and torque the nut to 15 ft. lbs. (20 Nm) plus 90 degrees.

- Cotter pin

8. Top off the transaxle with the proper fluid.

9. Check and adjust the front end alignment as necessary.

ION

1. Before servicing the vehicle, refer to the precautions in the beginning of this section.

2. Remove or disconnect the following:
- Wheel
- Stabilizer link

3. Insert a drift into the caliper to prevent the rotor from turning and remove the half shaft nut.
- Tie rod end from the steering knuckle
- Wheel speed sensor connection. Make sure to position the harness away from the ball joint after disconnecting.
- Lower ball joint from the steering knuckle
- Drive shaft from the knuckle using a hammer and a block of wood. Partially install the nut before striking the wood with the hammer to prevent thread damage.

4. Using slide hammer SA 9173G and drive shaft removal tool J 45341, remove the shaft from the transmission.

To install:

5. Apply Output Shaft Lubricant 7847638, to the splines of the output shaft, if equipped with an automatic transaxle.

6. Install or connect the following:
- New stub shaft retaining ring
- Halfshaft to the transaxle after installing a Seal Protector Tool SA91112T

7. Remove the seal protector tool after the splines have passed the oil seal.

8. Install or connect the following:
- Fully seat the halfshaft into the transaxle
- Outer end of the halfshaft to the wheel hub with a new washer and nut
- Ball joint to the steering knuckle. Torque the fastener to 44 ft. lbs. (60 Nm), then an additional 30 degree turn.
- Tie rod end to the steering knuckle. When seated properly, torque the fastener to 15 ft. lbs. (20 Nm) plus an additional 180 degree turn.
- Halfshaft to wheel nut. Torque the nut to 81 ft. lbs. (110 Nm
- Cotter pin

- Stabilizer link and tighten the nut to 63 ft. lbs. (85 Nm)
- Wheel

9. Top off the transaxle with the proper fluid.

10. Check and adjust the front end alignment as necessary.

CV-Joints

OVERHAUL

L and S–Series

INNER

1. Before servicing the vehicle, refer to the precautions in the beginning of this section.

2. Remove or disconnect the following:
- Front wheel
- Halfshaft
- Swage ring using a hand grinder
- Large CV-joint boot clamp
- CV-joint boot by sliding it away from the tri-pod joint
- Tri-pod housing from the tri-pod spider
- Inboard spacer ring slide it rearward on the shaft
- Outboard retaining ring
- Tri-pod joint spider assembly
- Inboard spacer ring and CV-joint boot

To install:

3. Install or connect the following:
- Swage ring clamp on the CV-joint boot
- CV-joint boot

4. Position the CV-joint boot seal into the axle shaft's joint seal groove and align the swage ring clamp on the boot.

✳✳ WARNING

Make sure that there are no pinch points on the inboard seal.

5. Crimp the swage ring
6. Install or connect the following:
- Inboard spacer ring, slide it rearward on the shaft
- Tri-pod joint spider assembly onto the shaft
- Outboard retaining ring into the axle shaft groove
- Tri-pod joint spider assembly, slide it against the outboard retaining ring
- Inboard spacer ring, seat it in the groove
- ½ kit grease into the boot
- ½ kit grease into the tri-pod housing

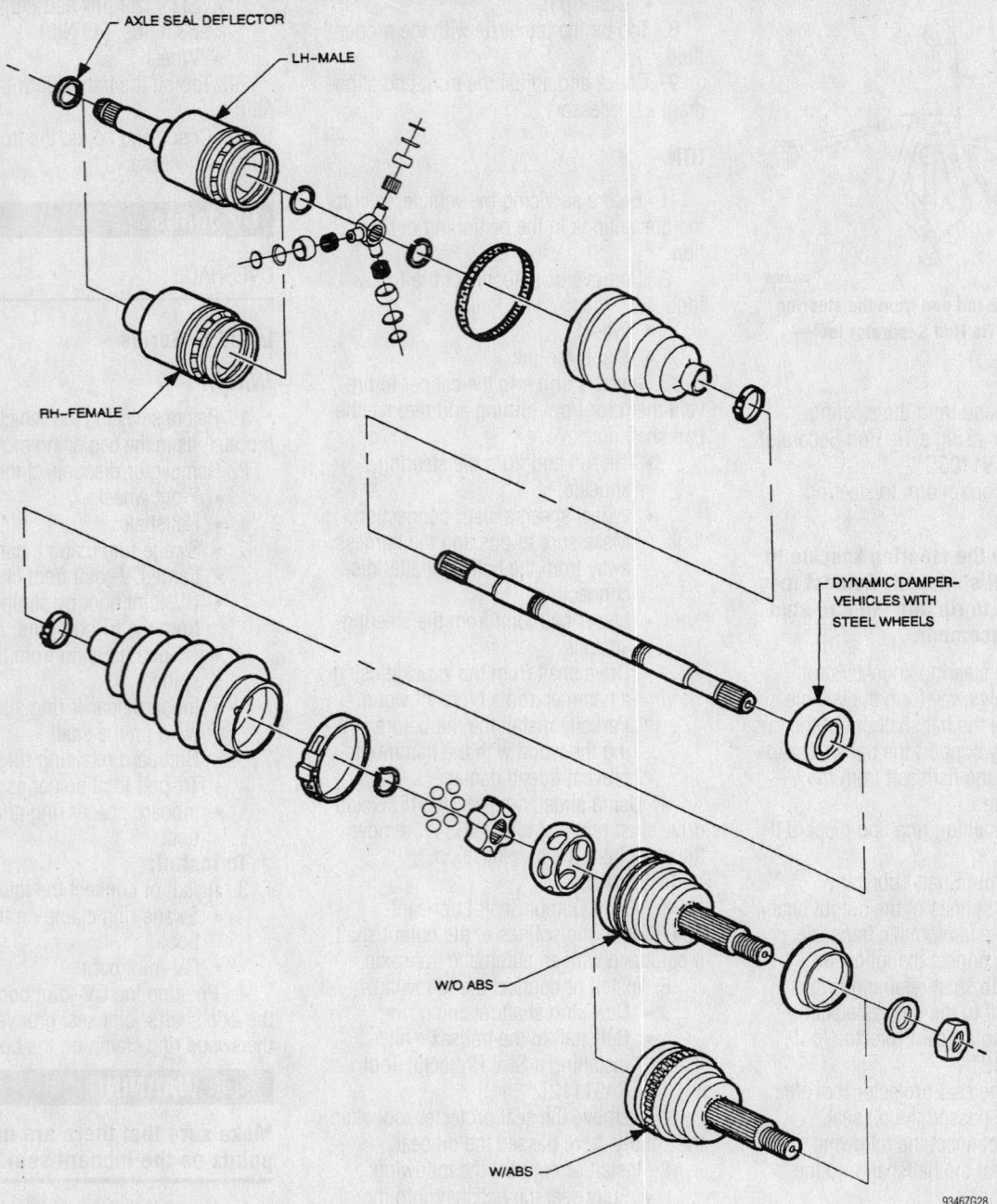

AXLE SEAL DEFLECTOR

LH—MALE

RH—FEMALE

DYNAMIC DAMPER—
VEHICLES WITH
STEEL WHEELS

W/O ABS

W/ABS

9346ZG28

Exploded view of axle shaft—S series shown, L series and ION are similar

- New large seal clamp onto the CV-joint boot
- Tri-pod housing, slide it over the tri-pod joint spider assembly
- CV-joint boot/clamp, slide it into place, over the trilobal tri-pod bushing with the seal lip in the groove

➡**Make sure the boot lies flat against the trilobal bushing.**

7. Position the CV-joint boot so it measures 4.9 in. (125mm).

8. Using a Crimp tool, a torque wrench and a breaker bar, crimp the large CV-joint boot clamp to 130 ft. lbs. (176 Nm).

9. Install the halfshaft and the front wheel.

OUTER

1. Before servicing the vehicle, refer to the precautions in the beginning of this section.

2. Remove or disconnect the following:
- Axle shaft from the vehicle
- Large CV boot retaining clamp
- Small CV boot retaining clamp

- CV boot from the joint
- Axle shaft retaining ring
- Outer joint from the axle shaft
- CV boot

3. Disassemble the chrome alloy balls from the CV-joint cage as follows:

a. Position a brass drift against the CV-joint cage and tap it with a hammer to tilt the cage.

b. Remove the 1st chrome alloy ball from the cage.

c. Tilt the cage in the opposite direction.

d. Remove the opposite chrome alloy ball.

e. Repeat the procedure until all 6 balls are removed.

4. Disassemble the CV-joint cage and inner race as follows:

a. Pivot the cage and race 90 degrees to the center line of the outer race.

b. Align the cage windows with outer race lands.

c. Remove the cage from the outer race.

d. Rotate the inner race upward and remove it from the cage.

To install:

5. Lubricate the parts with a light coat of grease.

6. Assemble the CV-joint cage and inner race, as follows:

a. Rotate the inner race 90 degrees to the cage centerline.

b. Align the cage windows with inner race lands.

c. Insert the inner race into the cage by rotating the inner race downward.

d. Insert the cage/inner race into the outer race.

7. Assemble the chrome alloy balls into the CV-joint cage, as follows:

a. Position a brass drift against the CV-joint cage and tap it with a hammer to tilt the cage.

b. Insert the 1st chrome alloy ball into the cage.

c. Tilt the cage in the opposite direction.

d. Insert the opposite chrome alloy ball.

e. Repeat the procedure until all 6 balls are inserted.

8. Install ½ of the grease provided, into the CV-joint.

9. Install or connect the following:
- Small CV boot retaining ring
- CV boot on the halfshaft
- New retaining ring on the halfshaft
- Large ring clamp on the CV boot
- Outer joint onto the axle shaft
- Retaining ring into the outer race

10. Install the remaining grease into the CV boot.

11. Position the CV boot and the small boot clamp.

12. Crimp the small boot clamp.

13. Position and crimp in place the large boot clamp.

14. Install the Halfshaft in the vehicle.

ION

INNER

1. Before servicing the vehicle, refer to the precautions in the beginning of this section.

2. Remove or disconnect the following:
- Front wheel
- Halfshaft
- Small boot clamp ring using a side cutter and discard
- Earless clamp and discard
- Large CV-joint boot clamp using a side cutter and discard
- CV-joint boot by sliding it away from the tri-pod joint
- Tri-pod housing from the tri-pod spider
- Inboard spacer ring slide it rearward on the shaft
- Outboard retaining ring
- Tri-pod joint spider assembly
- Inboard spacer ring and CV-joint boot

To install:

3. Install or connect the following:
- Small boot clamp onto the neck of the inboard boot but do not tighten
- Inboard boot onto the shaft

➡**The clamp must be positioned correctly during crimping to ensure a correct seal.**

4. Use boot clamp installer SA9203C to tighten the clamp making sure the end gap on the clamp does not exceed 0.118 inch (3mm).
- Spider assembly past the retaining ring groove until it is seated against the shoulder
- Spider retaining ring
- ½ kit grease into the boot
- ½ kit grease into the tri-pod housing

➡**The joint must be assembled with the convoluted retainer in position to avoid boot damage.**

- Tri-lobal housing bushing making sure it is flush with the housing
- New large seal clamp onto the CV-joint boot and slide the housing over the tri-pod joint spider assembly
- CV-joint boot/clamp, slide it into place, over the trilobal tri-pod bushing with the seal lip in the groove

➡**Make sure the boot lies flat against the trilobal bushing.**

5. Position the CV-joint boot so it measures 4.9 in. (125mm).

6. Using a boot clamp installer SA9161C tighten the large clamp.

7. Install the halfshaft and the front wheel.

OUTER

1. Before servicing the vehicle, refer to the precautions in the beginning of this section.

2. Remove or disconnect the following:
- Axle shaft from the vehicle
- Large CV boot retaining clamp
- Small CV boot retaining clamp
- CV boot from the joint
- Axle shaft retaining ring
- Outer joint from the axle shaft
- CV boot

3. Disassemble the chrome alloy balls from the CV-joint cage as follows:

a. Position a brass drift against the CV-joint cage and tap it with a hammer to tilt the cage.

b. Remove the 1st chrome alloy ball from the cage.

c. Tilt the cage in the opposite direction.

d. Remove the opposite chrome alloy ball.

e. Repeat the procedure until all 6 balls are removed.

4. Disassemble the CV-joint cage and inner race as follows:

a. Pivot the cage and race 90 degrees to the center line of the outer race.

b. Align the cage windows with outer race lands.

c. Remove the cage from the outer race.

d. Rotate the inner race upward and remove it from the cage.

To install:

5. Lubricate the parts with a light coat of grease.

6. Assemble the CV-joint cage and inner race, as follows:

a. Rotate the inner race 90 degrees to the cage centerline.

b. Align the cage windows with inner race lands.

c. Insert the inner race into the cage by rotating the inner race downward.

d. Insert the cage/inner race into the outer race.

7. Assemble the chrome alloy balls into the CV-joint cage, as follows:

a. Position a brass drift against the CV-joint cage and tap it with a hammer to tilt the cage.

b. Insert the 1st chrome alloy ball into the cage.

c. Tilt the cage in the opposite direction.

d. Insert the opposite chrome alloy ball.

e. Repeat the procedure until all 6 balls are inserted.

8. Install ½of the grease provided, into the CV-joint.

9. Install or connect the following:
- Small CV boot retaining ring
- CV boot on the halfshaft
- New retaining ring on the half-shaft

- Large ring clamp on the CV boot
- Outer joint onto the axle shaft
- Retaining ring into the outer race

10. Install the remaining grease into the CV boot.

11. Position the CV boot and the small boot clamp.

12. Crimp the small boot clamp making sure not exceed an end gap of 0.118 inch (3mm).

13. Position and crimp in place the large boot clamp making sure not exceed an end gap of 0.118 inch (3mm)

14. Install the halfshaft in the vehicle.

STEERING AND SUSPENSION

Air Bag

☀☀ CAUTION

All vehicles are equipped with an air bag system. The system must be disabled before performing service on or around system components, steering column, instrument panel components, wiring and sensors. Failure to follow safety and disabling procedures could result in accidental air bag deployment, possible personal injury and unnecessary system repairs.

PRECAUTIONS

Several precautions must be observed when handling the inflator module to avoid accidental deployment and possible personal injury.

1. Never carry the inflator module by the wires or connector on the underside of the module.

2. When carrying a live inflator module, hold securely with both hands, and ensure that the bag and trim cover are pointed away.

3. Place the inflator module on a bench or other surface with the bag and trim cover facing up.

4. With the inflator module on the bench, never place anything on or close to the module which may be thrown in the event of an accidental deployment.

DISARMING

Except ION

1. Before servicing the vehicle, refer to the precautions in the beginning of this section.

2. Align the steering wheel so the vehicle wheels are pointing in the straight-ahead position.

3. Turn the ignition switch to the **LOCK** position.

4. Remove the SIR or AIR BAG fuse from the fuse block.

5. Remove the Connector Position Assurance (CPA) device, then disengage the yellow 2-way SIR wiring harness connector at the base of the steering column.

ION

1. Before servicing the vehicle, refer to the precautions in the beginning of this section.

2. Align the steering wheel so the vehicle wheels are pointing in the straight-ahead position.

3. Turn the ignition switch to the **LOCK** position.

4. Remove the SIR or AIR BAG fuse from the fuse block.

5. To disable the roof rail module left and seat belt pre-tensioner (left front), remove the garnish molding from the upper lock pillar. Remove the Connector Position Assurance (CPA) device from the roof rail module. Disconnect the roof rail module from the vehicle harness. Lower the headliner and remove the CPA from the seat belt pre-tensioner connector and disconnect the pre-tensioner connector.

ARMING

Except ION

After the repairs, enable the system as follows:

1. Turn the ignition switch to the **LOCK** position.

2. Engage the yellow 2-way connector at the base of the steering column, then install the CPA device.

3. Reinstall the Supplemental Inflatable Restraint (SIR) or AIR BAG fuse.

4. Turn the ignition switch to the **RUN** position.

5. Verify the SIR indicator light flashes 7–9 times, if not, inspect the system for malfunction.

ION

After the repairs, enable the system as follows:

1. Turn the ignition switch to the **LOCK** position and remove the key.

2. Connect the pre-tensioner and install the CPA.

3. Install the headliner.

4. Connect the roof rail module and install the CPA.

5. Install the garnish molding.

6. Reinstall the Supplemental Inflatable Restraint (SIR) or AIR BAG fuse.

7. Turn the ignition switch to the **RUN** position.

8. Verify the SIR indicator light flashes 7 times, if not, inspect the system for malfunction.

Rack and Pinion Steering Gear

REMOVAL & INSTALLATION

1.9L Engines

1. Before servicing the vehicle, refer to the precautions in the beginning of this section.

2. Remove or disconnect the following:
- Negative battery cable
- Both front wheels
- Outer tie rod from the steering knuckle
- Left side splash shield
- Loosen the intermediate shaft cover from the steering gear and move it aside
- Intermediate shaft pinch bolt

3. On vehicles with power steering, place a suitable container under the steering assembly. Disconnect the pressure and return lines at the steering gear and allow the system to drain.

4. Remove the steering gear to cradle fasteners.

5. Remove the steering gear through the left fender well.

To install:

6. Install or connect the following:
- Steering gear to cradle with new nuts and torque the nuts to 37 ft. lbs. (50 Nm)
- Intermediate shaft to the steering gear and torque the pinch bolt to 35 ft. lbs. (47 Nm)

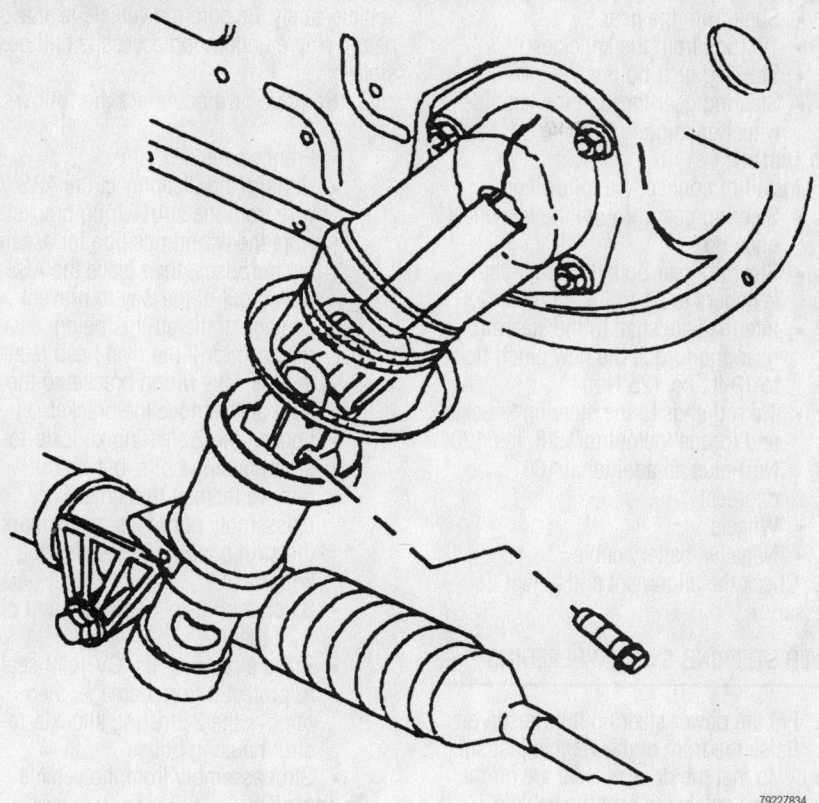

The pinch bolt is located under the intermediate shaft cover—1.9L engines

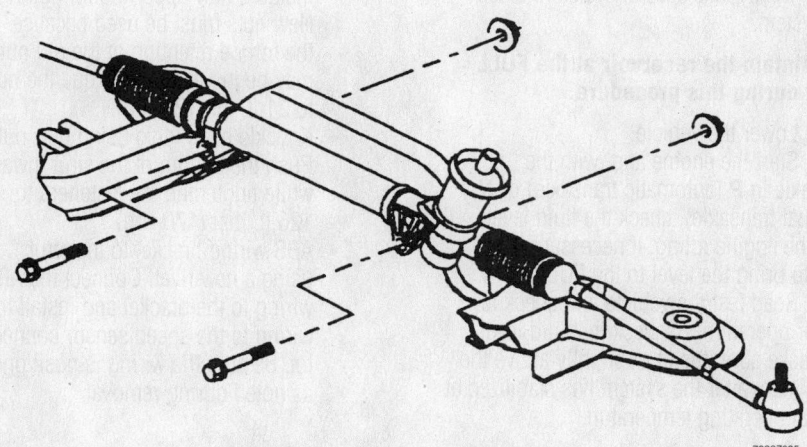

Remove the steering gear-to-cradle fasteners, then remove the gear through the left fender well—1.9L engines

- Power steering pressure and return lines and torque the fittings to 20 ft. lbs. (28 Nm)
- Left side splash shield
- Tie rod ends to the steering knuckle and torque the castle nuts to 41 ft. lbs. (55 Nm)
- New cotter pins to the castle nuts
- Front wheels
- Negative battery cable

7. Check the alignment and adjust as necessary.

8. On vehicles with power steering, bleed the power steering system.

L-Series

1. Before servicing the vehicle, refer to the precautions in the beginning of this section.

2. Remove or disconnect the following:

- Negative battery cable
- Both front wheels
- Intermediate shaft from the steering gear pinch bolt from inside the vehicle,
- Shaft from the gear
- Rear exhaust manifold pipe heat shield (on 3.0L engine)
- Rear transaxle mount through bolt
- Transaxle mount-to-frame bolt
- Power steering pressure and return hoses
- Front wheels
- Right front lower splash shield
- Exhaust manifold pipe (on 2.2L engine with manual transaxle and all 3.0L engines)
- Remaining rear transaxle mount-to-frame bolts
- Tie rods from the steering knuckles
- Steering gear-to-frame bolts
- Steering gear heat shield (on 2.2L engine with automatic transaxle)
- Stabilizer bar links from the struts
- Front suspension support assemblies
- Loosen the remaining frame to body fasteners until there is clearance to remove the steering gear
- Steering gear through the left side wheel opening

To install:

3. One at a time remove and replace the frame bolts and nuts. The torque retention of the old fasteners may not be sufficient.

4. Install or connect the following:

- Steering gear through the left wheel opening
- Raise the frame assembly to the undercarriage and torque the fasteners to 66 ft. lbs. (90 Nm) plus a 45–60 degree turn
- Suspension supports with new bolts and torque the bolts to 66 ft. lbs. (90 Nm) plus a 45–60 degree turn

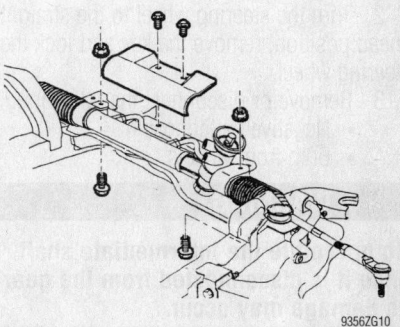

Steering gear-to-frame assembly bolts— L-series models

- Steering gear to the frame with new bolts and torque the bolts to 35 ft. lbs. (45 Nm) plus 45–60 degrees
- Steering gear heat shield (on 2.2L engine with an automatic transaxle) and torque the bolts to 35 inch lbs. (4 Nm)
- Tie rod ends to the steering knuckle and torque the nut to 44 ft. lbs. (60 Nm)
- Stabilizer bar links to the strut and torque the nuts to 48 ft. lbs. (65 Nm)
- Rear transaxle mount-to-frame bolts and torque them to 44 ft. lbs. (60 Nm)
- Power steering pressure and return lines and torque the fittings to 20 ft. lbs. (27 Nm)
- Exhaust manifold pipe (2.2L engines with a manual transaxle and all 3.0L engines) to the manifold and torque the nuts to 22 ft. lbs. (30 Nm)
- Exhaust manifold pipe to the exhaust pipe and torque the bolts to 15 ft. lbs. (20 Nm)
- Front wheels
- Rear transaxle mount-to-frame bolt and torque it to 45 ft. lbs. (60 Nm)
- Rear transaxle mount through bolt and torque it to 66 ft. lbs. (90 Nm)
- Exhaust manifold pipe heat shield (on 3.0L engine) and torque the bolts to 71 inch lbs. (8 Nm)
- Intermediate shaft to the steering gear and torque the new pinch bolt to 21 ft. lbs. (28 Nm)
- Negative battery cable

5. Check the alignment and adjust if necessary.

6. Bleed the power steering system.

ION

1. Before servicing the vehicle, refer to the precautions in the beginning of this section.

2. Turn the steering wheel to the straight ahead position, remove the key and lock the steering wheel.

3. Remove or disconnect the following:
- Negative battery cable
- Both front wheels

✳✳ WARNING

Do not rotate the intermediate shaft once it is disconnected from the gear as damage may occur.

- Intermediate shaft from the steering gear pinch bolt

- Shaft from the gear
- Tie rods from the knuckles
- Steering gear bolts
- Steering gear through the left side wheel opening

To install:

4. Install or connect the following:
- Steering gear through the left wheel opening
- Steering gear bolts to torque the fasteners to 81 ft. lbs. (110 Nm)
- Intermediate shaft to the steering gear and torque the new pinch bolt to 18 ft. lbs. (25 Nm)
- Tie rod ends to the steering knuckle and torque the nut to 15 ft. lbs. (20 Nm) plus an additional 180 degrees
- Wheels
- Negative battery cable

5. Check the alignment and adjust if necessary.

POWER STEERING SYSTEM BLEEDING

1. Fill the power steering fluid reservoir.

2. Raise the front of the vehicle just sufficiently so that the drive wheels are off the ground, then safely support the vehicle.

3. Bleed the system by turning the wheels from side-to-side without hitting the stops. It may take several cycles to bleed the system.

➡**Maintain the reservoir at the FULL mark during this procedure.**

4. Lower the vehicle.

5. Start the engine and, with the transaxle in **P** (automatic transaxle) or **N** (manual transaxle), check the fluid level with the engine idling. If necessary, add fluid to bring the level to the FULL mark.

6. Road test the vehicle and check for proper operation. Recheck the fluid level and make sure it is at or slightly above the FULL mark after the system has stabilized at normal operating temperature.

Strut

REMOVAL & INSTALLATION

Front

1.9L ENGINES

1. Before servicing the vehicle, refer to the precautions in the beginning of this section.

2. If equipped with an Anti-lock Brake System (ABS), disconnect the negative battery cable, then raise and support the

vehicle safely. Be sure the vehicle is at a height where underhood access is still possible.

3. Remove or disconnect the following:
- Front wheel
- Unplug and disconnect the ABS wire from the strut wiring bracket. Note the wiring position for assembly purposes, then place the ABS wiring out of the way to prevent damage. If the strut is being replaced, drill the rivet head retaining the ABS wiring bracket to the strut and remove the bracket
- Loosen the 2 steering knuckle-to-strut housing bolts, but do not remove them at this time. For reassembly purposes, matchmark the strut position to the steering knuckle
- 3 upper strut-to-body nuts and discard them
- Place a rag over the CV-joint seal to protect it from damage, then remove the 2 steering knuckle-to-strut housing bolts
- Strut assembly from the vehicle

To install:

4. Install or connect the following:
- Position the strut in the vehicle and install 3 new upper mount nuts. New nuts must be used because the torque retention of the old nuts may be insufficient. Torque the nuts to 21 ft. lbs. (29 Nm).
- Knuckle bolts, also using new nuts. Push the bottom of the strut inward while tightening the fasteners to 126 ft. lbs. (170 Nm)
- ABS wiring bracket to the strut using a new rivet. Connect the ABS wiring to the bracket and install the wiring to the speed sensor connector. Be sure the wiring is positioned as noted during removal

Remove the 3 nuts to detach the top of the strut from the body—1.9L engine

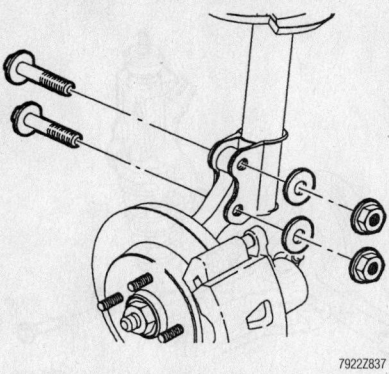

Matchmark, then remove the 2 bolts that secure the strut to the knuckle assembly— 1.9L engine

- Wheel assembly, remove the supports and lower the vehicle.
- Negative battery cable

5. Check and adjust the alignment if necessary.

L–SERIES

1. Before servicing the vehicle, refer to the precautions in the beginning of this section.

2. If equipped with an Anti-lock Brake System (ABS), disconnect the negative battery cable, then raise and support the vehicle safely. Be sure the vehicle is at a height where underhood access is still possible.

3. Remove or disconnect the following:
- Front wheel
- Brake hose and ABS harness from the strut assembly
- Loosen the steering knuckle to strut fasteners, but do not remove them
- Stabilizer bar link to the strut assembly attaching nut and move it toward the rear of the vehicle

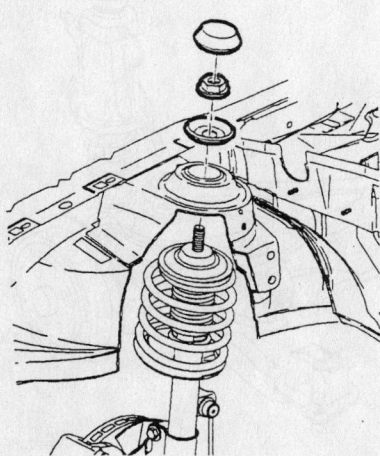

Remove the strut to body attaching nut— L–Series

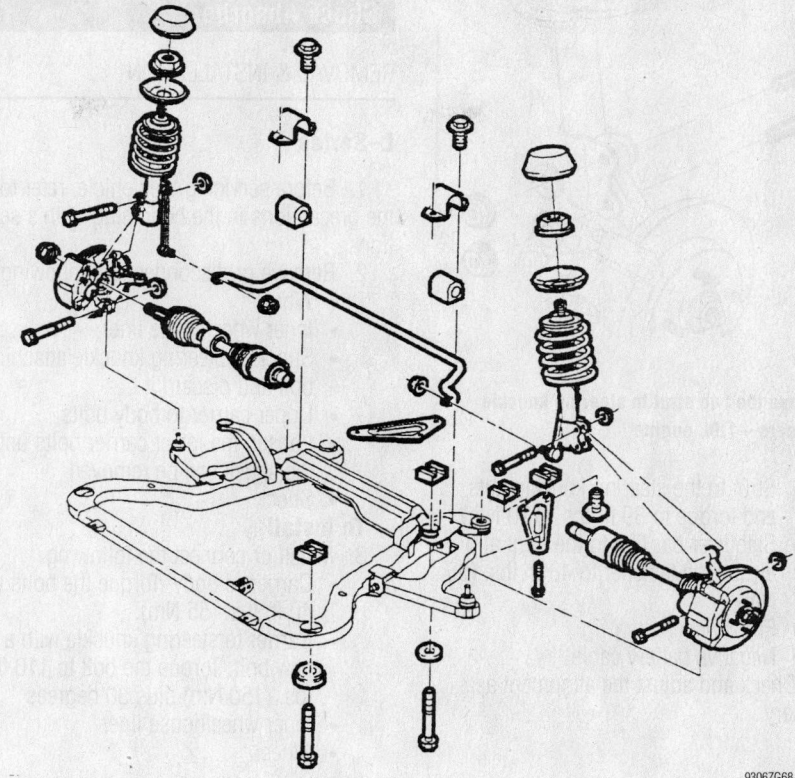

Exploded view of the front suspension—L–Series

- Strut to body attaching nuts
- Place a rag over the CV-joint seal to protect it from damage, then remove the 2 steering knuckle-to-strut housing bolts
- Steering knuckle to strut fasteners
- Strut assembly from the vehicle

To install:

4. Install or connect the following:
- Strut to the body and torque the new attaching nut to 40 ft. lbs. (55 Nm)
- Strut to the steering knuckle and torque the new fasteners to 37 ft. lbs. (50 Nm) then to 66 ft. lbs. (90 Nm) plus a 45 degree turn
- Stabilizer bar link to the strut and torque the fastener to 50 ft. lbs. (65 Nm)
- Brake hose and ABS wiring to the strut
- Front wheel
- Negative battery cable

5. Check and adjust the alignment as necessary.

ION

1. Before servicing the vehicle, refer to the precautions in the beginning of this section.

2. If equipped with an Anti-lock Brake System (ABS), disconnect the negative battery cable, then raise and support the vehicle safely. Be sure the vehicle is at a height where underhood access is still possible.

3. Remove or disconnect the following:
- Front wheel
- Stabilizer bar link to the strut assembly attaching nut and separate it from the strut
- Loosen the steering knuckle to strut nuts
- Wheel Speed Sensor (WSS) harness and bracket, if equipped
- Strut cap and body nut
- Place a rag over the CV-joint seal to protect it from damage, then remove the 2 steering knuckle-to-strut housing bolts
- Steering knuckle to strut fasteners
- Strut assembly from the vehicle

To install:

4. Install or connect the following:
- Strut to the body and torque the new attaching nut to 81 ft. lbs. (110 Nm)
- Strut to the steering knuckle bolts but leave the nuts off
- WSS bracket and harness, if equipped

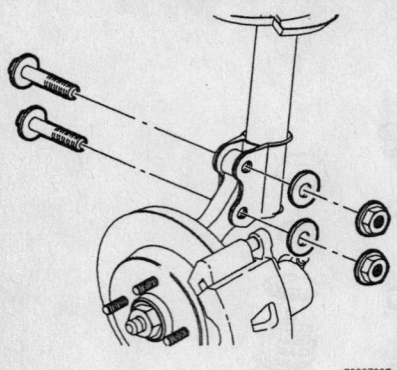

Remove the two strut to steering knuckle fasteners—1.9L engine

- Strut to the steering knuckle nuts and torque to 89 ft. lbs. (120 Nm)
- Stabilizer bar link to the strut and torque the fastener to 48 ft. lbs. (65 Nm)
- Front wheel
- Negative battery cable

5. Check and adjust the alignment as necessary.

Rear

1.9L ENGINES

1. Before servicing the vehicle, refer to the precautions in the beginning of this section.

2. Remove the rear window trim finish panel.

3. Remove or disconnect the following:
- Wheel
- Loosen the 2 strut to steering knuckle fasteners, but do not remove them
- 3 strut to body fasteners
- Slowly raise a hoist to support the steering knuckle and lower the strut assembly
- 2 strut to steering knuckle fasteners
- Strut

To install:

4. Install the strut to the vehicle and hold it in place with strut to knuckle bolts.

5. Install or connect the following:
- Strut to the body with new upper support bolts and torque the bolts to 21 ft. lbs. (29 Nm)
- Strut to steering knuckle fasteners and push the bottom of the strut inward while tightening the bolts. Torque the new bolts to 126 ft. lbs. (170 Nm).
- Front wheel
- Rear window trim panel

6. Check and adjust the alignment if necessary.

Shock Absorber

REMOVAL & INSTALLATION

L–Series

1. Before servicing the vehicle, refer to the precautions in the beginning of this section.

2. Remove or disconnect the following:
- Wheel
- Inner wheelhouse liner
- Shock to steering knuckle attaching bolt and discard it
- Upper carrier to body bolts
- Loosen the lower carrier bolts until the shock can be removed
- Shock

To install:

3. Install or connect the following:
- Carrier to body. Torque the bolts to 40 ft. lbs. (55 Nm).
- Carrier to steering knuckle with a new bolt. Torque the bolt to 110 ft. lbs. (150 Nm) plus 30 degrees
- Inner wheelhouse liner
- Wheel

ION

1. Before servicing the vehicle, refer to the precautions in the beginning of this section.

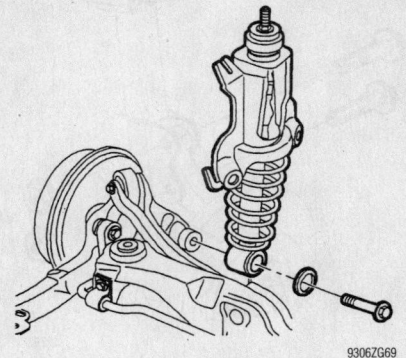

Remove the shock to steering knuckle attaching bolt—L–Series

2. Remove or disconnect the following:
- Wheel

3. Support the rear axle with a jackstand positioned near the shock absorber being removed
- Lower and upper shock absorber bolts
- Shock

To install:

4. Install or connect the following:
- Strut. Using new bolts, torque the upper bolt to 66 ft. lbs. (90 Nm) and the lower bolt to 81 ft. lbs. (110 Nm). Remove the jackstand.
- Wheel

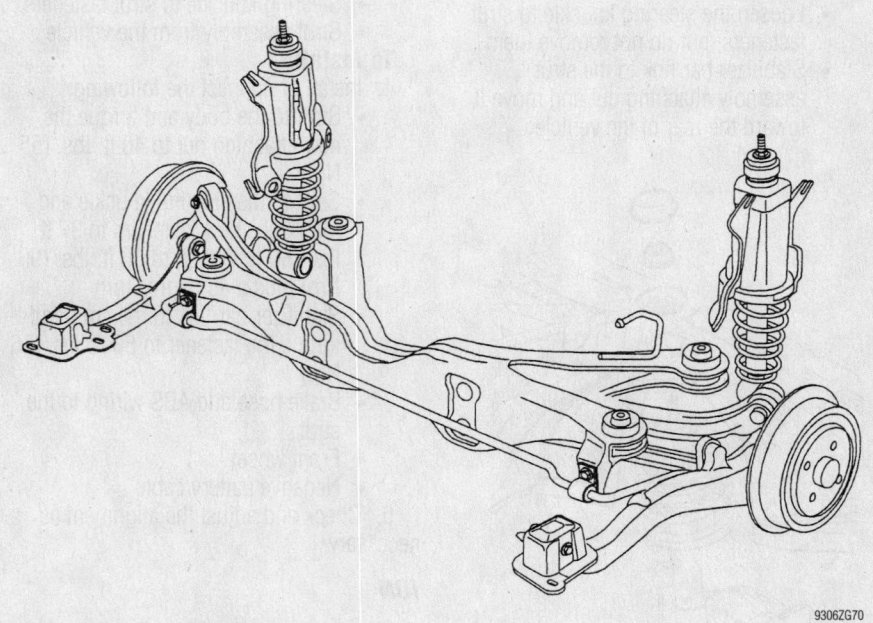

Exploded view of the rear suspension—L–Series

Coil Spring

REMOVAL & INSTALLATION

Front

1.9L ENGINES

1. Before servicing the vehicle, refer to the precautions in the beginning of this section.
2. Remove or disconnect the following:

- Strut from the vehicle
- Mount the strut in a bench vise, then attach a spring compressor/holding fixture; be sure that the strut component is firmly secured
- Compress the spring sufficiently to completely unload the upper strut mount
- Shaft nut while holding the strut stationary with a Torx® head socket wrench
- Upper spring support and inspect the rubber for cracks or deterioration
- Spring from the strut and inspect the spring for damage
- Dust shield assembly and inspect for cracks or deterioration
- Strut from the vise or applicable holding fixture and retract the strut shaft, checking for smooth, even resistance

3. If replacing the coil spring, carefully release the spring compressor.

To install:

4. Secure the strut in the bench vise or applicable holding fixture.
5. Extend the strut shaft to the limit of its travel.
6. Install or connect the following:

- Dust shield assembly onto the strut, then install the spring with the compressor tool installed
- Spring isolator and the strut mount to the top of the assembly
- Guide the strut shaft through the upper strut mount assembly. Compress the coil until the washer and shaft nut can be installed to the end of the shaft, but do not over-compress and damage the spring
- Tighten the shaft to the nut using a Torx® head socket wrench and a torque wrench, while holding the nut steady with an open end wrench. Tighten the fastener to 37 ft. lbs. (50 Nm).
- Release the spring compressor tool

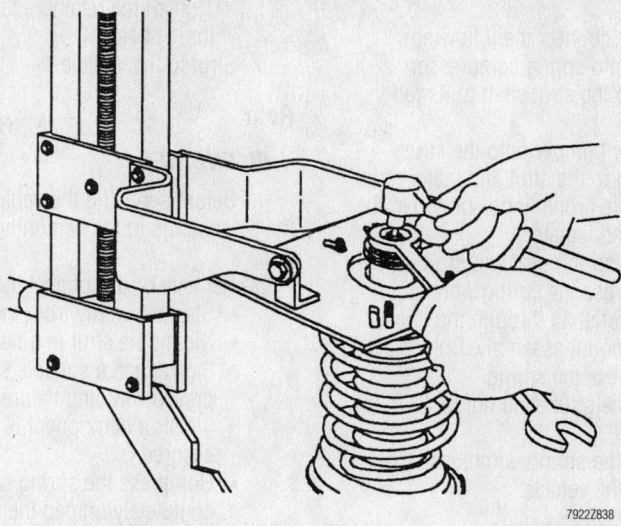

Remove the strut shaft nut while the coil spring is compressed to unload the upper strut mount—1.9L engines

79222Z838

and remove the strut from the fixture
- Strut assembly in the vehicle.

L–SERIES

1. Before servicing the vehicle, refer to the precautions in the beginning of this section.
2. Remove or disconnect the following:

- Strut from the vehicle
- Place the strut into a spring compressor. Fasten the assembly with a strut to steering knuckle bolt through the lower mounting hole
- Compress the spring enough to completely unload the upper strut mount
- Strut shaft nut while holding the shaft stationary with a Torx® socket
- Release the spring compressor and tilt the strut outward
- Upper strut mount assembly
- Spring

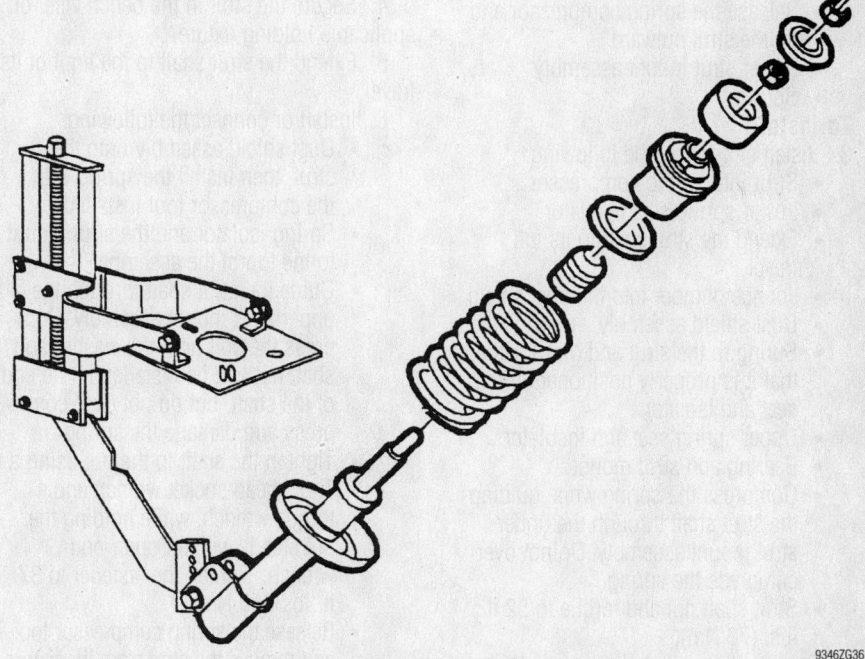

Exploded view of the coil spring and related components—L-Series

9346ZG36

To install:

3. Install or connect the following:
- Strut into spring compressor
- Extend the strut shaft to its full travel
- Hollow bumper onto the strut
- Spring to the strut and make certain that it is properly positioned in the seat and isolator
- Upper spring seat and strut mount
- Compress the spring while guiding the strut shaft through the upper strut mount assembly. Do not over compress the spring

4. Torque the strut shaft nut to 40 ft. lbs. (55 Nm).

5. Release the spring compressor tool.

6. Strut to the vehicle.

ION

1. Before servicing the vehicle, refer to the precautions in the beginning of this section.

2. Remove or disconnect the following:
- Strut from the vehicle
- Upper strut cap retainers and the cap
- Place the strut into a spring compressor. Fasten the assembly with a strut to steering knuckle bolt through the lower mounting hole
- Compress the spring enough to completely unload the upper strut mount
- Strut shaft nut while holding the shaft stationary with a Torx® socket
- Release the spring compressor and tilt the strut outward
- Upper strut mount assembly
- Spring

To install:

3. Install or connect the following:
- Strut into spring compressor
- Lower spring seat insulator
- Extend the strut shaft to its full travel
- Jounce bumper into the dust shield
- Dust shield assembly
- Spring to the strut and make certain that it is properly positioned in the seat and isolator
- Upper spring seat and insulator
- Bearing and strut mount
- Compress the spring while guiding the strut shaft through the upper strut mount assembly. Do not over compress the spring
- Strut shaft nut and torque to 52 ft. lbs. (70 Nm).

4. Release the spring compressor tool.
- Upper strut cap and retainers.

Tighten the retainers to 124 inch lbs. (14 Nm).

5. Strut to the vehicle.

Rear

1.9L ENGINES

1. Before servicing the vehicle, refer to the precautions in the beginning of this section.

2. Remove or disconnect the following:
- Strut assembly from the vehicle
- Mount the strut in a bench vise, then attach a suitable spring compressor/holding fixture; be sure that the strut component is firmly secured
- Compress the spring sufficiently to completely unload the upper strut mount
- Strut shaft nut while holding the strut stationary with a Torx® head socket wrench
- Upper spring support and inspect the rubber for cracks or deterioration
- Spring from the strut and inspect the spring for damage
- Dust shield assembly and inspect for cracks or deterioration
- Strut from the vise or applicable holding fixture and retract the strut shaft, checking for smooth, even resistance

3. If replacing the coil spring, carefully release the spring compressor.

To assemble:

4. Secure the strut in the bench vise, or applicable holding fixture.

5. Extend the strut shaft to the limit of its travel.

6. Install or connect the following:
- Dust shield assembly onto the strut, then install the spring with the compressor tool installed
- Spring isolator and the strut mount to the top of the assembly
- Guide the strut shaft through the upper strut mount assembly. Compress the coil until the washer and shaft nut can be installed to the end of the shaft, but do not over-compress and damage the spring
- Tighten the shaft to the nut using a Torx® head socket wrench and a torque wrench, while holding the nut steady with an open-end wrench. Tighten the fastener to 37 ft. lbs. (50 Nm).
- Release the spring compressor tool and remove the strut from the fixture
- Strut to the vehicle

L–SERIES

1. Before servicing the vehicle, refer to the precautions in the beginning of this section.

2. Remove or disconnect the following:
- Shock absorber assembly from the vehicle
- Mount the carrier assembly Spring Compressor SA9155S, in a holding fixture
- Mount the shock assembly to the spring compressor using an adapter, SA9155–3
- Compress the spring enough to unload the upper spring supports
- Shock absorber shaft nut while holding the shaft securely with a Torx® socket
- Release the spring compressor and remove the carrier assembly
- Rear suspension support and inspect the inner and outer bumpers for cracks or deterioration, replace if necessary
- Rear spring upper insulator
- Extend and retract the shock absorber and check for smooth resistance

To install:

3. Install or connect the following:
- Shock into a spring compressor
- Extend the shock to its full limit of travel
- Inner and outer bumpers and make certain that the springs are properly positioned in their support seats
- Compress the spring while guiding the shock absorber shaft through the upper mount until the washer and nut can be installed. Torque the shaft nut to 15 ft. lbs. (20 Nm) while holding the shaft nut with a Torx® socket.
- Release the compressor tool and remove the carrier from the compressor
- Shock absorber to the vehicle

ION

1. Before servicing the vehicle, refer to the precautions in the beginning of this section.

2. Support the rear axle with jackstands positioned near each shock absorber.

3. Remove or disconnect the following:
- Wheel
- U–clips from the rear brake hose brackets at the axle
- Lower shock bolts

4. Lower the jackstands slowly to remove rear spring tension and remove the spring.

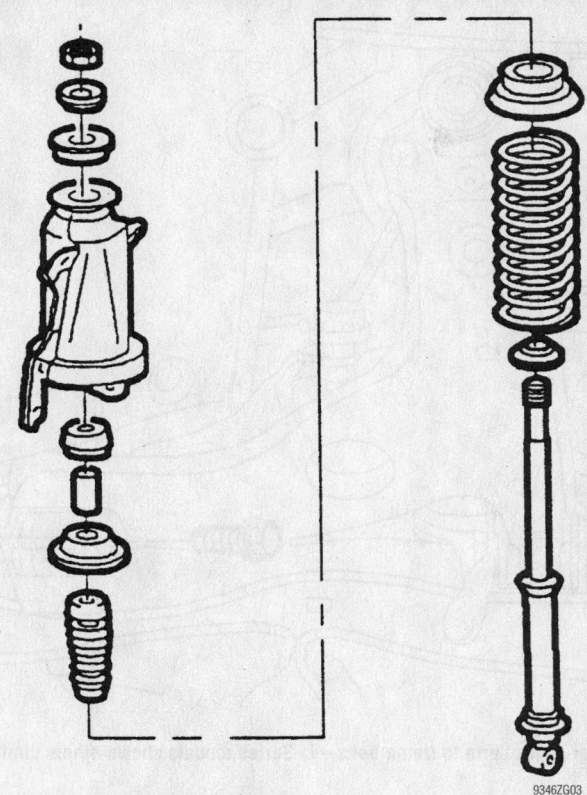

Exploded view of rear shock assembly—L–Series

- Upper spring seat/jounce bumper from the spring while leaving the lower spring seat on the axle

To install:

➡**The rear springs are indexed with the colored tag towards the rear of the vehicle. No orientation for up or down is required.**

5. Install or connect the following:
- Upper spring seat/jounce bumper to the spring
- Spring with the spring tag towards the rear of the vehicle and the lower coil is seated into the lower spring seat. Use the jackstands to raise the rear axle into position.
- Lower shock absorber bolt and tighten to 81 ft. lbs. (110 Nm)
- U–clips to the rear brake hose brackets at the axle
- Wheel

Lower Ball Joint

REMOVAL & INSTALLATION

1.9L Engines

The ball joint is part of the lower control arm and must be replaced as a unit.

L–Series

1. Before servicing the vehicle, refer to the precautions in the beginning of this section.
2. Install or connect the following:
- Lower control arm from the vehicle
- Rivets retaining the ball joint to the control arm using a ½ in. (13mm) drill bit
- Ball joint from the control arm

To install:
3. Install or connect the following:
- Ball joint into the control arm
- Nuts and bolts (included with new

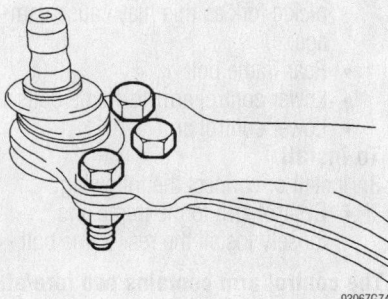

The new ball stud is bolted into the control arm—L–series shown, other similar

ball joint kit) as shown and torque them to 25 ft. lbs. (35 Nm)
- Control arm to the vehicle

ION

1. Before servicing the vehicle, refer to the precautions in the beginning of this section.
2. Install or connect the following:
- Lower control arm from the vehicle
- Rivets retaining the ball joint to the control arm using a ½ in. (13mm) drill bit
- Ball joint from the control arm

To install:
3. Install or connect the following:
- Ball joint into the control arm
- Nuts and bolts (included with new ball joint kit) and torque them to 50 ft. lbs. (68 Nm)
- Control arm to the vehicle

Lower Control Arm

REMOVAL & INSTALLATION

Front

1.9L ENGINES

1. Before servicing the vehicle, refer to the precautions in the beginning of this section.
2. Remove or disconnect the following:
- Wheel
- Lower control arm ball stud cotter pin and discard the pin
- Loosen the castle nut until it is level with the top of the ball stud

➡**Use caution not do damage the Anti-lock Braking System (ABS) speed sensor ring. If the ring is damaged a malfunction of the ABS system is possible.**

3. Separate the lower control arm from the steering knuckle, using a Lower Control Arm Ball Stud Separator Tool SA9132S
4. Remove or disconnect the following:
- Lower control arm castle nut
- Front inner fender splash shield and remove the push pins
- Front section of the shield first on the left hand side and the rear section first on the right side of the vehicle
- Lower control arm to cradle fasteners
- Front stabilizer bar nut at the lower control arm
- Lower control arm

To install:

5. Install or connect the following:
- Control arm to the front stabilizer bar

6. Position the end of the control arm into the cradle. Torque the bolt to 92 ft. lbs. (125 Nm) and the nut to 74 ft. lbs. (100 Nm).
- Control arm to front stabilizer bar nut and torque the nut to 106 ft. lbs. (144 Nm)
- Ball stud to the steering knuckle and torque the castle nut to 55 ft. lbs. (75 Nm) and install a new cotter pin
- Rear section of the shield first on the left side and the front section first on the right side of the vehicle
- Inner fender splash shield
- Align the molded in fasteners with the cradle holes and push them straight in
- Wheel

7. Check and adjust the alignment if necessary.

L–SERIES

1. Before servicing the vehicle, refer to the precautions in the beginning of this section.
2. Remove or disconnect the following:
- Wheel
- Ball stud bolt

➡**Use caution not do damage the Antilock Braking System (ABS) speed sensor ring. If the ring is damaged a malfunction of the ABS system is possible.**

- Separate the ball stud from the steering knuckle by using a prybar
- Lower control arm-to-frame bolts
- Lower control arm

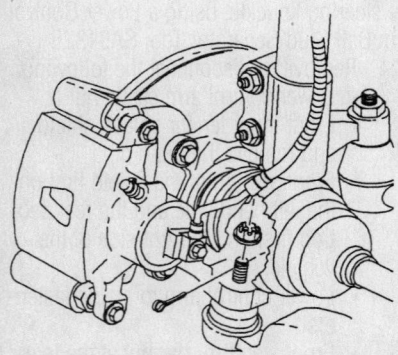

Connect the ball stud to the steering knuckle—1.9L engines

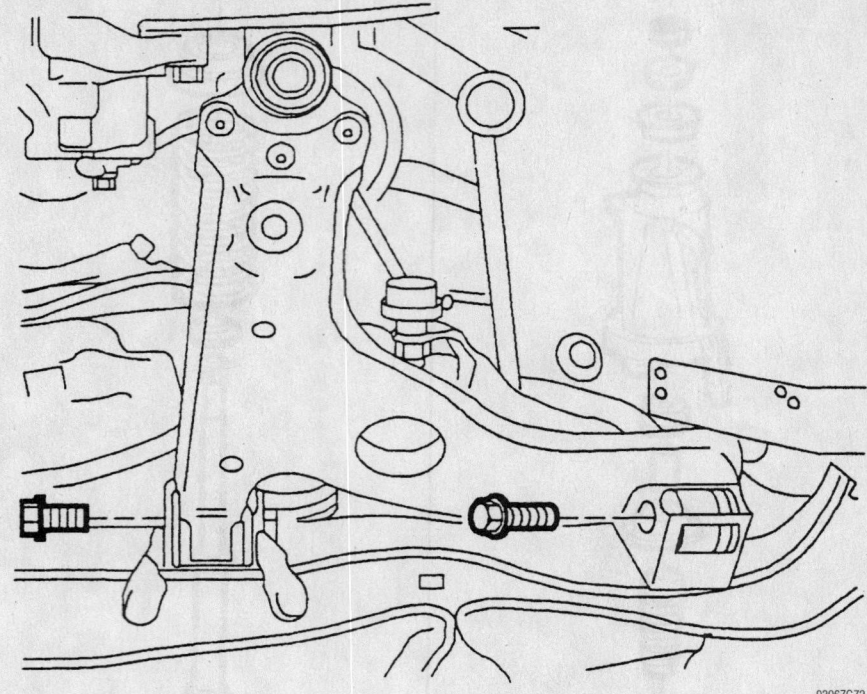

Remove the lower control arm to frame bolts—L–Series models shown others similar

To install:

3. Install or connect the following:
- Control arm to the frame and torque the new bolts to 65 ft. lbs. (90 Nm) plus 75 degrees
- Ball stud to steering knuckle and torque the bolt to 75 ft. lbs. (100 Nm)
- Wheel
- Check and adjust the alignment, if necessary.

ION

1. Before servicing the vehicle, refer to the precautions in the beginning of this section.
2. Remove or disconnect the following:
- Wheel
- Ball stud bolt
- Separate the ball stud from the steering knuckle. Do not use a pickle fork as this may cause damage.
- Rear frame bolt
- Lower control arm-to-frame bolts
- Lower control arm

To install:

3. Install or connect the following:
- Control arm to the frame and loosely install the rear frame bolt

➡**The control arm contains two fore/aft movement limiting brackets. Failure to install these brackets will affect handling.**

- Ball joint stud to the steering knuckle
- Fore/aft movement limiting brackets onto the control arm forward bushing
- Control arm-to-frame-bolts and torque the bolts to 41 ft. lbs. (55 Nm)
- Tighten the rear frame bolt to 74 ft. lbs. (100 Nm) plus an additional 180 degree turn
- Ball joint pinch bolt and nut to 44 ft. lbs. (60 Nm) plus an additional 30 degree turn
- Wheel
- Check and adjust the alignment, if necessary.

Rear

L–SERIES

1. Before servicing the vehicle, refer to the precautions in the beginning of this section.
2. Remove or disconnect the following:
- Rear wheel
- Rear brake hose and bracket from the control arm
- Caliper, if equipped
- Brake drum/rotor
- Antilock Braking System (ABS) sensor harness, if equipped
- Hub
- Parking brake cable and support

- Separate the backing plate from the control arm
- Shock absorber to control arm bolt
- Rear stabilizer bar link to rear control arm bolt
- Loosen the rear suspension upper and lower control arm to rear axle fasteners
- Rear axle to control arm and discard the bolts
- Rear upper and lower suspensions control arm bolts and discard them
- Rear control arm

To install:

3. Install or connect the following:
- Control arm and install new rear axle control arm bolts. Do not tighten them at this time
- Upper and lower rear suspension control arm bolts
- Torque the rear axle control arm bolts and the upper and lower rear suspension bolts to 65 ft. lbs. (90 Nm) plus 30 degrees
- Shock absorber to rear axle control arm and torque the new bolt to 110 ft. lbs. (150 Nm) plus 30 degrees
- Stabilizer bar link to the control arm and torque the bolt 41 ft. lbs. (55 Nm)
- Backing plate and hub assembly and torque the bolts to 37 ft. lbs. (50 Nm) plus 30 degrees
- Rear drum/disc and torque the bolts to 35 inch lbs. (4 Nm)
- Rear brake caliper and torque the bolts to 59 ft. lbs. (80 Nm), if equipped
- Parking brake cable and support bracket and torque the bolts to 71 inch lbs. (8 Nm)
- Brake line bracket to the control arm and torque the bolts to 71 inch lbs. (8 Nm)
- Rear wheel

4. Check and adjust the alignment, as needed.

LOWER CONTROL ARM BUSHING REPLACEMENT

L–Series

FRONT

1. Before servicing the vehicle, refer to the precautions in the beginning of this section.
2. Remove or disconnect the following:
- Control arm
3. Press out the front bushing by using Tools KM-508–3 and KM508–1.

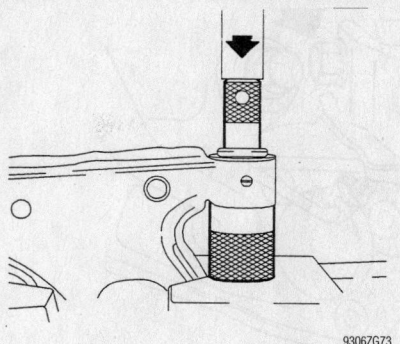

Press out the front control arm front bushing—L–series

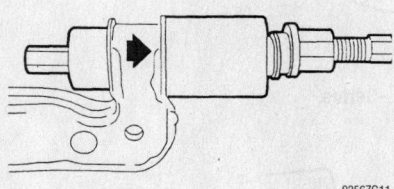

Press out the front control arm rear bushing—L–series

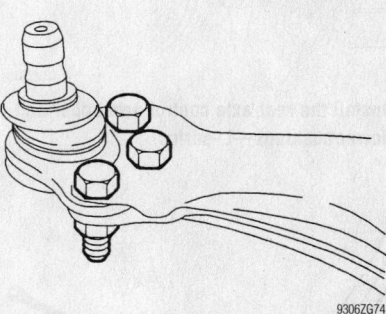

The new ball stud is bolted into the control arm—L–Series

4. Drill out the rivets retaining the ball stud to the control arm.
5. Remove the ball stud from the control.

To install:

➡ **The new ball stud is bolted to the control arm.**

6. Install or connect the following:
- Ball stud to the control arm and torque the new bolts to 25 ft. lbs. (35 Nm)

7. Press in the new control arm bushing using Tools KM508–1, KM508–2 and KM508–3.
- Control arm to the frame and torque the new bolts to 65 ft. lbs. (90 Nm) plus 75 degrees
- Ball stud to the steering knuckle

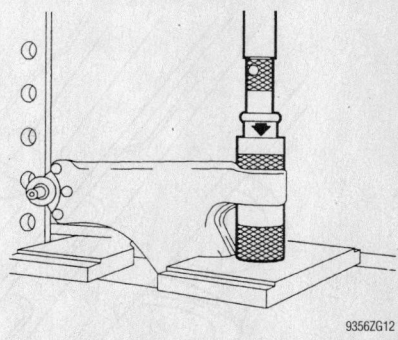

Press in the front control arm front bushing—L–series

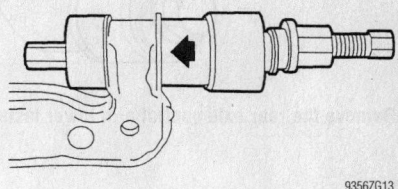

Press in the front control arm rear bushing—L–series

and torque the bolts to 75 ft. lbs. (100 Nm)

REAR

1. Before servicing the vehicle, refer to the precautions in the beginning of this section.
2. Remove or disconnect the following:
- Control arm and place it in a vise
- Rear axle control arm bolt and discard it
3. Press out the rear axle upper and lower control arm bushings with Bushing Removal Tools KM-671—KM-906-62—and KM-906-61.
4. Press out the rear axle control arm front bushing with Bushing Removal Tools KM-671—KM-906-41 and KM-906-44.

To install:

5. Install or connect the following:
6. Press in the new rear axle front control arm bushing using tools KM-671, KM-906-42 and KM-906-43.
7. Press in the rear axle upper and lower control arm bushings with Bushing Removal Tools KM-671, KM-906-64 and KM-906-631.
- Rear axle control arm bracket. Torque the new bolt to 65 ft. lbs. (90 Nm) plus 60 degrees

ION

FRONT

1. Before servicing the vehicle, refer to the precautions in the beginning of this section.

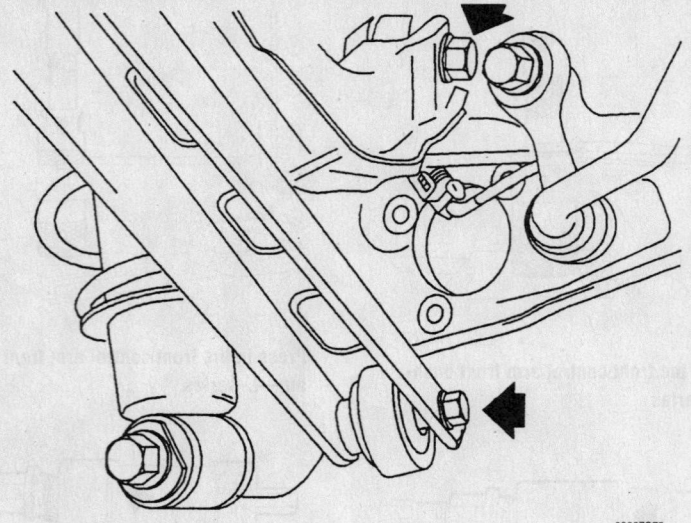

9306ZG75

Remove the rear axle control arm lower fasteners—L–Series

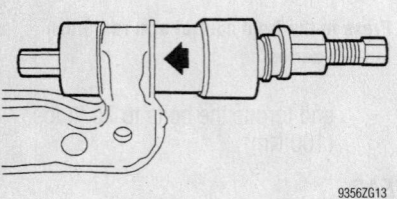

9356ZG13

Remove the rear axle control arm upper and lower bushings—L–series

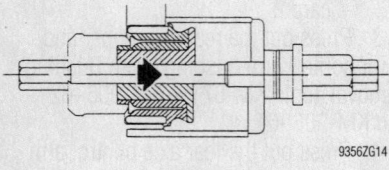

9356ZG14

Remove the rear axle control arm front bushings—L–series

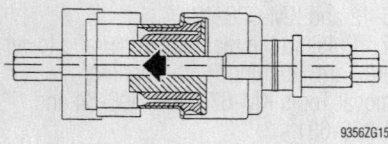

9356ZG15

Install the rear axle control arm front bushings—L–series

2. Remove or disconnect the following:
• Control arm

➡ **Make sure to note the depth and positioning of the old bushing prior to removal.**

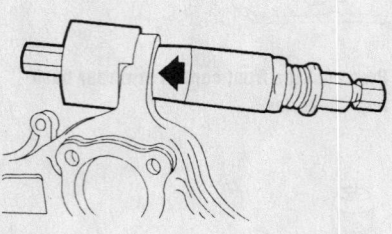

9356ZG16

Install the rear axle control arm upper and lower bushings—L–series

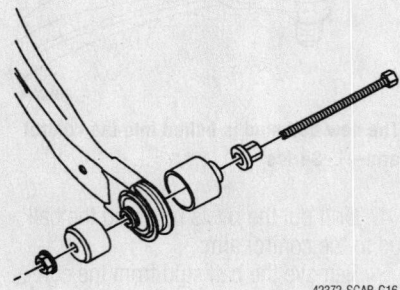

42372-SCAR-G16

Press out the front control arm front bushing—ION

3. Using bushing remover/installer KM-906B install KM-906-70, 906–42, 906-41 and 906-62 onto the bushing.
4. Hold the hex end of the threaded shaft while turning the large nut to pull the bushing through the arm.
5. Remove the tools and the bushing.
To install:
6. Install the new bushing into the tapered side of the arm.
7. Using bushing remover/installer

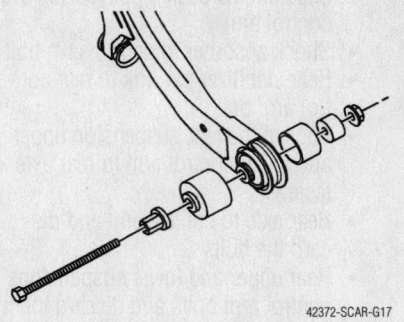

42372-SCAR-G17

Press in the front control arm rear bushing—ION

KM-906B install KM-906-70, 906–42, 906-41 and 906-62 onto the bushing
8. Hold the hex end of the threaded shaft while turning the large nut to pull the bushing into the arm. Install the bushing to the same depth and positioning as the old one.
9. Remove the removal/installer tool.
10. Install the control arm.

Wheel Bearings

ADJUSTMENT

The wheel bearing are sealed at the factory and do not require any adjustment or maintenance.

REMOVAL & INSTALLATION

Front

1.9L ENGINES

1. Before servicing the vehicle, refer to the precautions in the beginning of this section.
2. If equipped with an Antilock Braking System (ABS), disconnect the negative battery cable.
3. Loosen the front halfshaft nut, while an assistant depresses the brake pedal, then raise and support the vehicle safely.
4. Remove or disconnect the following:
• Wheel
• Brake caliper mounting bracket bolts and suspend the assembly from the strut spring with wire
• Loosen the strut-to-knuckle bolts, but do not remove at this time
• Rotor, axle nut and washer
• Cotter pin from the lower control arm ball joint. Back the ball joint nut until the top of the nut is even with the top of the threads
• Separate the lower control arm from the steering knuckle, then remove

the nut. Do not use a wedge tool or seal damage may occur

➡**The outer CV-joint for vehicles equipped with ABS contains a speed sensor ring. Use of an incorrect tool to separate the control arm from the knuckle may result in damage and loss of the ABS system.**

- Tie rod cotter pin and castle nut, then separate the tie rod end from the knuckle
- ABS wheel speed sensor electrical connector
- Suspend the halfshaft from the body with wire, then remove the 2 knuckle-to-strut fasteners and remove the knuckle/hub assembly from the vehicle. If difficulty is encountered, position a block of wood on the end of the halfshaft and tap on the wood with a hammer to free the hub assembly

5. Disassemble the knuckle hub assembly as follows:

a. If equipped, remove the ABS wheel speed sensor from the knuckle.

➡**Any time the hub or bearing is separated from the steering knuckle, a new bearing must be used upon assembly.**

6. Install Wheel Bearing Removing Tools SA9159S, to the knuckle and secure the assembly in a vise.

7. Hold the hub driver with a wrench and tighten the hub driver screw to remove the hub. If the inner bearing race is pulled out with the hub, remove the race with a bearing race remover.

8. Remove the assembly from the vise and separate the wheel hub removal tool from the knuckle.

9. Remove the bearing retainer snapring.

10. Position the knuckle in a shop press on a knuckle support tube and press the bearing from the knuckle with a small driver.

To install:

11. If necessary, assemble the knuckle hub assembly as follows:

a. Use a large driver and press in the new bearing until seated.

b. Use the small driver and the knuckle support tube to press in the hub assembly. The small driver must be used to support the bearing inner race with its small (pilot) side facing towards the press and away from the bearing.

c. Install the bearing retainer snapring.

12. Install the ABS wheel speed sensor into the knuckle. Torque the fastener to 72 inch lbs. (8 Nm), if equipped.

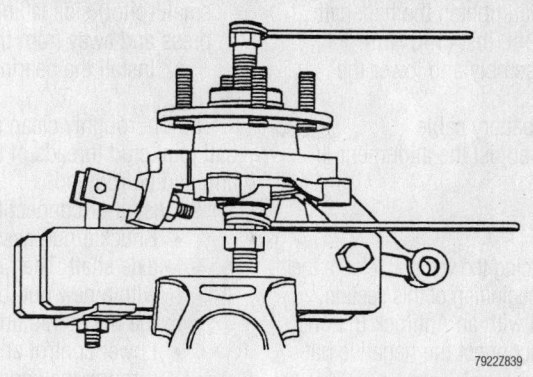

Tighten the hub driver screw to remove the hub while the assembly is held firmly in a vise

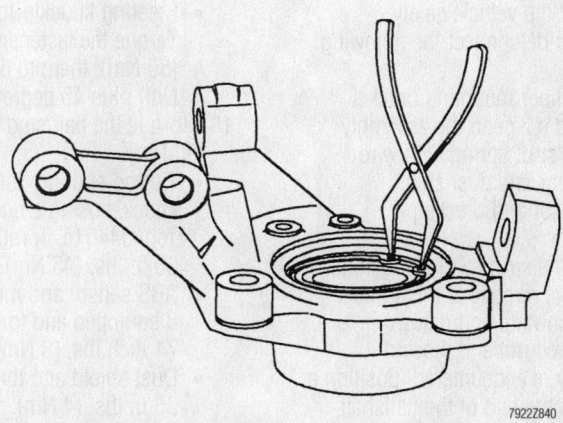

Remove the bearing retainer snapring before pressing out the bearing

➡**Service knuckles may not have holes for brake dust shield mounting. The dust shield is no longer required and does not have to be reinstalled. Also, should the shield become damaged, it may be removed; there is no need to repair or replace it. But, should a shield be removed and discarded, the shield should also be removed from the opposite side to maintain balance/symmetry.**

13. Thoroughly clean and lubricate the ball joint stud threads of the lower control arm and tie rod end. Install the knuckle/hub assembly onto the axle shaft. Then, install the washer with a new nut, but do not tighten the nut at this time.

14. Install or connect the following:

- Lower control arm ball stud through the knuckle bore and install the nut, but do not tighten at this time
- Steering knuckle-to-strut fasteners, but do not tighten at this time
- Tie rod end and nut, then torque the nut to 33 ft. lbs. (45 Nm) and install a new cotter pin. If necessary, tighten the nut additionally, do not back off to insert the cotter pin

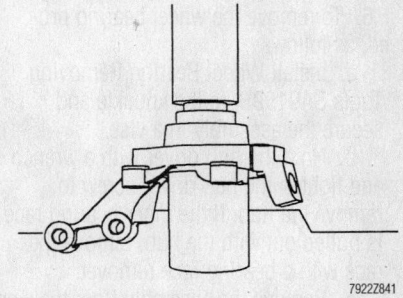

Carefully press the bearing out of the knuckle after removing the snapring

- Push inward on the bottom of the strut and torque the knuckle fasteners to 126 ft. lbs. (170 Nm).
- Torque the lower control arm ball stud nut to 55 ft. lbs. (75 Nm), tighten additionally, if necessary, and install a new cotter pin
- Rotor onto the hub, then install the caliper mount bracket onto the knuckle. Tighten the mount bracket assembly bolts to 81 ft. lbs. (110 Nm).
- While an assistant depresses the

brake pedal, tighten the halfshaft nut to 103 ft. lbs. (140 Nm).
- Wheel assembly and lower the vehicle
- Negative battery cable

15. Check and adjust the alignment, if necessary.

L—SERIES

1. Before servicing the vehicle, refer to the precautions in the beginning of this section.

2. If equipped with an Antilock Braking System (ABS), disconnect the negative battery cable.

3. Loosen the front halfshaft nut, while an assistant depresses the brake pedal, then raise and support the vehicle safely.

4. Remove or disconnect the following:
- Wheel
- Brake caliper mounting bracket bolts and suspend the assembly from the strut spring with wire
- Brake rotor and dust shield
- ABS sensor and bracket, if equipped
- Loosen the strut-to-knuckle bolts, but do not remove at this time
- Axle to hub nut and discard it
- Tie rod end nut and discard it

5. If difficulty is encountered, position a block of wood on the end of the halfshaft and tap on the wood with a hammer to free the hub assembly.
- Steering knuckle/hub assembly

6. To remove the wheel bearing proceed, as follows:

a. Install Wheel Bearing Removing Tools SA9159S, to the knuckle and secure the assembly in a vise.

b. Hold the hub driver with a wrench and tighten the hub driver screw to remove the hub. If the inner bearing race is pulled out with the hub, remove the race with a bearing race remover.

c. Remove the assembly from the vise and separate the wheel hub removal tool from the knuckle.

d. Remove the bearing retainer snapring.

e. Position the knuckle in a shop press on a knuckle support tube and press the bearing from the knuckle with a small driver.

To install:

7. If necessary, assemble the knuckle hub assembly, as follows:

a. Use a suitable large driver and press in the new bearing until seated.

b. Use the small driver and the knuckle support tube to press in the hub assembly. The small driver must be used to support the bearing inner race with its small (pilot) side facing towards the press and away from the bearing.

c. Install the bearing retainer snap ring.

8. Thoroughly clean and lubricate the ball joint stud threads of the lower control arm and tie rod end.

9. Install or connect the following:
- Knuckle/hub assembly onto the axle shaft. Then, install the washer with a new nut, but do not tighten the nut at this time
- Lower control arm ball stud through the knuckle bore and install the nut, but do not tighten at this time
- Steering knuckle-to-strut fasteners. Torque the fasteners to 40 ft. lbs. (50 Nm); then, to 65 ft. lbs. (90 Nm) plus 45 degrees

10. Torque the ball stud fastener to 75 ft. lbs. (100 Nm).
- Tie rod end into the steering knuckle using a Linkage Installer Tool J44015. Torque the fastener to 35 ft. lbs. (45 Nm).
- ABS sensor and mounting bracket, if equipped and torque the bolts to 71 inch lbs. (8 Nm).
- Dust shield and torque the bolts to 35 in lbs. (4 Nm)
- Rotor and screw and torque the screw to 27 inch lbs. (3.5 Nm)
- Caliper mount bracket onto the knuckle and torque the bolts to 70 ft. lbs. (90 Nm)
- Stabilizer bar link to the strut and torque the bolts to 50 ft. lbs. (65 Nm)

- Wheel
- Negative battery cable

11. Depress the brake pedal and torque the axle to hub nut, in the following sequence:
- Step 1: 85 ft. lbs. (115 Nm) to seat the bearing.
- Step 2: Back the nut off .
- Step 3: 15 ft. lbs. (20 Nm).
- Step 4: Turn the nut an additional 90 degrees, using a torque angle gauge.

12. Install the cotter pin.

13. Check and adjust the alignment

ION

1. Before servicing the vehicle, refer to the precautions in the beginning of this section.

2. If equipped with an Antilock Braking System (ABS), disconnect the negative battery cable.

3. Loosen the front halfshaft nut, while an assistant depresses the brake pedal, then raise and support the vehicle safely.

4. Remove or disconnect the following:
- Wheel
- Axle to hub nut and discard it
- Brake caliper mounting bracket bolts and suspend the assembly from the strut spring with wire
- Brake rotor
- ABS sensor connector, if equipped
- ABS sensor jumper connector from the bracket on the strut, if equipped
- Bearing mounting bolts from the rear of the knuckle
- Bearing assembly and spacer. Make sure to note the positioning of the spacer before removal.

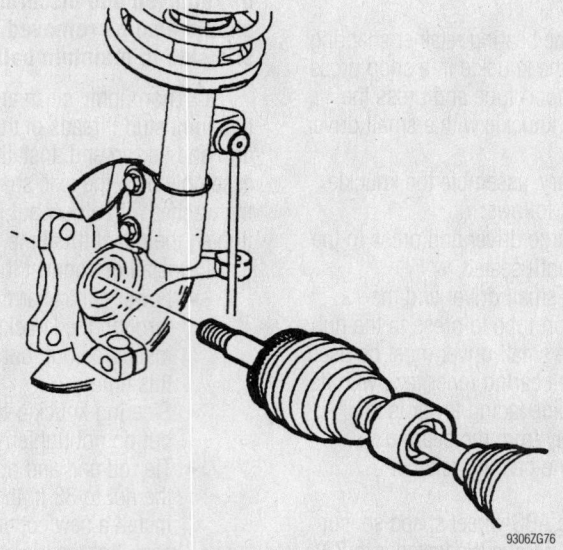

9306ZG76

Separate the axle from the wheel hub—L-Series shown, others similar

To install:

5. Install or connect the following:
- Bearing/hub assembly and spacer to the knuckle making sure to position the spacer as it was before removal. Tighten the bearing assembly-to-knuckle bolts evenly to draw the assembly into the knuckle and final tighten bolts to 85 ft. lbs. (115 Nm).
- ABS sensor jumper connector to the bracket on the strut, if equipped
- ABS sensor connector, if equipped
- Rotor
- Caliper mount bracket onto the knuckle and torque the bolts to 85 ft. lbs. (115 Nm)
- Caliper and tighten the bolts to 25 ft. lbs. (34 Nm)
- Negative battery cable

6. Depress the brake pedal and torque the axle to hub nut to 81 ft. lbs. (110 Nm)
- Wheel

7. Check and adjust the alignment

Rear

1.9L ENGINES

Unlike the front wheel bearings, which may be removed from the hub for replacement, the rear wheel hub and bearing assembly is not serviceable. If damaged or worn, the hub and bearing assembly must be replaced as a unit.

1. Before servicing the vehicle, refer to the precautions in the beginning of this section.
2. Remove or disconnect the following:
- Negative battery cable, if equipped with an Antilock Braking System (ABS)
- Rear wheel
- ABS speed sensor connector, if equipped
- Caliper assembly-to-knuckle mounting bolts and support it with a wire from the strut, then remove the rotor, if equipped with disc brakes
- Brake drum, if equipped with drum brakes
- 4 hub/bearing-to-knuckle bolts, then remove the assembly from the vehicle

To install:

3. Install or connect the following:
- Brake backing plate, hub/bearing assembly and retaining bolts and torque the bolts to 63 ft. lbs. (85 Nm)
- Brake drum or rotor and caliper assembly, if equipped
- ABS speed sensor connector, if equipped

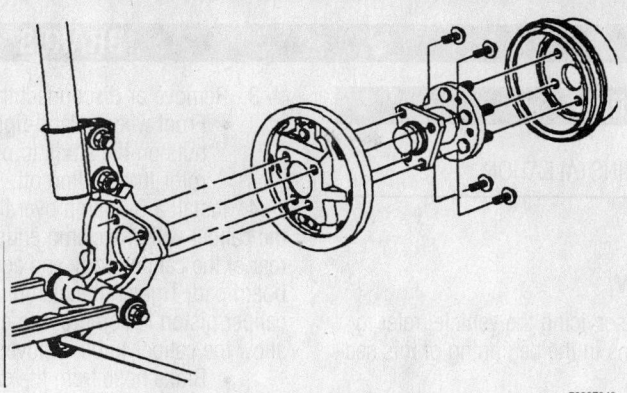

Exploded view of the rear hub/bearing assembly—drum brake set-up shown—1.9L engine

- Brake drum, if equipped
- Brake rotor, if equipped
- Caliper to the steering knuckle, if equipped. Torque the bolts to 63 ft. lbs. (85 Nm).
- Rear wheel
- Negative battery cable

L–SERIES

1. Before servicing the vehicle, refer to the precautions in the beginning of this section.
2. Remove or disconnect the following:
- Negative battery cable, if equipped with an Antilock Braking System (ABS)
- Rear wheel
- Brake line to rear axle control arm attaching clip, if equipped with rear disc brakes
- Caliper to steering knuckle bolts and move the caliper aside, if equipped with rear disc brakes
- Brake drum/rotor
- ABS electrical connector, if equipped
- Hub to control arm nuts and discard them
- Hub

To install:

3. Install or connect the following:
- Hub to control arm and torque the new nuts to 35 ft. lbs. (50 Nm) plus 30 degrees
- ABS electrical connector, if equipped
- Brake drum/rotor attaching screw and torque the screw to 35 inch lbs. (4 Nm)
- Brake caliper, if equipped and torque the bolts to 59 ft. lbs. (80 Nm)
- Brake line to the rear axle control arm bracket, if equipped. Torque the bolt to 71 inch lbs. (8 Nm)

- Rear wheel
- Negative battery cable

ION

1. Before servicing the vehicle, refer to the precautions in the beginning of this section.
2. Remove or disconnect the following:
- Brake drum
3. Remove the access hole plug from the brake backing plate and install a support for the backing plate
- ABS electrical connector, if equipped
- Hub retaining arm nuts and discard them
- Hub

To install:

4. Install or connect the following:
- Hub and torque the new nuts to 37 ft. lbs. (50 Nm) plus 30 degrees
- ABS electrical connector, if equipped

5. Remove the support from the backing plate.

- Access hole plug on the backing plate
- Brake drum
- Rear wheel
- Negative battery cable

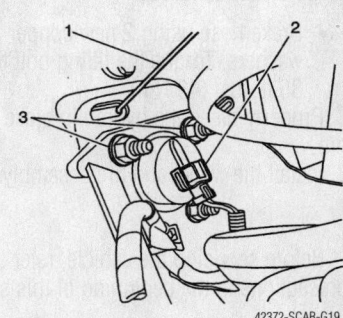

Install a support (1) for the backing plate, disconnect the electrical connector (2) and remove the hub nuts (3)—ION

Brake Caliper

REMOVAL & INSTALLATION

Front

EXCEPT ION

1. Before servicing the vehicle, refer to the precautions in the beginning of this section.

- Front wheel and tire assembly
- Brake hose from the caliper and discard the 2 copper washers. Plug the openings to prevent system contamination or excessive fluid loss.
- Lock pin and guide pin from the caliper
- Caliper from the support, being careful not to damage the pin boots
- Pin boots from the caliper support and inspect for damage

To install:

2. If necessary, bottom the caliper piston by hand, or by using a C-clamp.

3. If removed, install the brake pads and clips to the caliper support.

4. Lubricate the pin boots and guide pins with silicone grease.

5. Install or connect the following:

- Pin boots into the caliper support, using the pin to assure that the boot passes all the way through the support
- Cliper onto the support and over the brake pads

6. Lubricate the non-threaded portion of the guide and lock pins with silicone grease.

- Pins through the caliper and torque to 27 ft. lbs. (36 Nm)

➡**Make sure the brake line is properly routed with loop to the rear and that the hose is not twisted.**

- Brake hose using 2 new copper washers. Torque the fitting bolt to 36 ft. lbs. (49 Nm).

7. Properly bleed the hydraulic brake system.

8. Install the wheel and tire assembly.

ION

1. Before servicing the vehicle, refer to the precautions in the beginning of this section.

2. Use a turkey baster or similar device to remove fluid from the master cylinder until it is about ½ full.

3. Remove or disconnect the following:

- Front wheel. Hand-tighten the lug nuts on the studs to prevent the rotor from falling off.

4. Install a C-clamp over the body of the caliper with the clamp ends against the rear of the caliper body and against the outboard pad. Tighten the clamp until the caliper piston is compressed enough to allow the caliper to be removed.

- Brake hose from the caliper and discard the 2 copper washers. Plug the openings to prevent system contamination or excessive fluid loss.
- Caliper bolts
- Caliper from the bracket
- Pin boots from the caliper support and inspect for damage

To install:

5. Install or connect the following:

- Brake pads and clips to the caliper bracket, if removed

6. Lubricate the pin boots and guide pins with silicone grease.

- Pin boots, if removed
- Caliper onto the bracket and over the brake pads
- Caliper bolts through the caliper and torque to 27 ft. lbs. (34 Nm)

➡**Make sure the brake line is properly routed with loop to the rear and that the hose is not twisted.**

- Brake hose using 2 new copper washers. Torque the fitting bolt to 35 ft. lbs. (50 Nm).

7. Properly bleed the hydraulic brake system.

- Wheel

Rear

EXCEPT ION

1. Before servicing the vehicle, refer to the precautions in the beginning of this section.

2. Remove or disconnect the following:

- Rear wheel and tire assembly
- Brake hose from the caliper and discard the 2 copper washers. Plug the openings to prevent system contamination or excessive fluid loss.

3. Slip the end of the parking cable off the parking brake lever.

- Cable outer housing from the cable bracket with SA9151BR cable release tool

- Lock pin and guide pin
- Caliper from the support, being careful not to damage the pin boots.
- Pin boots from the caliper support for inspection and lubrication

To install:

4. Make sure the piston is bottomed in the bore. Do not compress the piston using a C-clamp; instead the piston must be rotated into the caliper on its threads using a piston driver tool.

5. If removed, install the brake pads and clips to the caliper support.

6. Lubricate the pin boots and guide pins with silicone grease.

7. Install or connect the following:

- Pin boots into the caliper support, using the pin to assure that the boot passes all the way through the support
- Caliper onto the caliper support

8. Lubricate the non-threaded portion of the guide and lock pins.

- Pins and torque to 27 ft. lbs. (36 Nm)
- Brake hose using new copper washers and torque the fitting bolt to 36 ft. lbs. (49 Nm)
- Parking brake cable

9. Properly bleed the hydraulic brake system.

10. Install the wheel and tire assembly.

Disc Brake Pads

REMOVAL & INSTALLATION

Front

EXCEPT ION

1. Before servicing the vehicle, refer to the precautions in the beginning of this section.

2. Remove or disconnect the following:

- Front wheels
- Caliper lower lock pins

3. Either pivot the caliper up on the guide pin or remove the upper guide pin and support the caliper from the strut using a coat hanger or length of wire.

- 2 brake pads and the pad clips from the caliper support. Discard the old pad clips.

4. Check the caliper pins, pin boots and the piston boot for deterioration or damage.

To install:

5. By hand or using a C-clamp, bottom the piston all the way into the caliper bore.

6. Carefully lift the inner edge of the piston boot by hand to release any trapped air.

7. Install or connect the following:
- New pad clips into the caliper support
- Inner and outer brake pads into the support. If installed, remove the temporary support wire from the caliper.
- Caliper body on the support and upper guide pin into position. Compress the boots by hand as the caliper is positioned onto the support.

8. Lubricate the smooth ends of the removed pin(s) with silicone grease.
- Pin(s) and torque to 27 ft. lbs. (36 Nm). Do not get grease on the pin threads.
- Wheels

9. Prior to operating the vehicle, depress the brake pedal a few times until the brake pads are seated against the rotor.

ION

1. Before servicing the vehicle, refer to the precautions in the beginning of this section.

2. Use a turkey baster or similar device to remove fluid from the master cylinder until it is about ½ full.

3. Remove or disconnect the following:
- Front wheel. Hand-tighten the lug nuts on the studs to prevent the rotor from falling off.

4. Install a C–clamp over the body of the caliper with the clamp ends against the rear of the caliper body and against the outboard pad. Tighten the clamp until the caliper piston is compressed enough to allow the caliper to be removed.
- Caliper lower bolt
- 2 brake pads and the pad clips from the caliper support. Discard the old pad clips.

5. Check the caliper pins, pin boots and the piston boot for deterioration or damage.

To install:

6. Install or connect the following:
- Brake pads and clips to the caliper bracket

7. Lubricate the pin boots and guide pins with silicone grease.
- Pin boots, if removed
- Caliper over the brake pads
- Caliper lower bolt through the caliper and torque to 27 ft. lbs. (34 Nm)

➡**Make sure the brake line is properly routed with loop to the rear and that the hose is not twisted.**

- Wheel

8. Prior to operating the vehicle, depress the brake pedal a few times until the brake pads are seated against the rotor.

Rear

EXCEPT ION

1. Before servicing the vehicle, refer to the precautions in the beginning of this section.

2. Remove or disconnect the following:
- Rear wheels
- Caliper lock and guide pins
- Caliper from the support, being careful not to damage the pin boots and suspend the caliper from a wire
- Brake pads from the support

To install:

3. Using SA91110NE piston driver tool, bottom the piston by rotating it clockwise into the caliper bore; do not use a C-clamp to press the piston into the bore.

4. Align the piston slots so they are perpendicular to the brake pads.

5. Carefully lift the inner edge of the piston boot to release any trapped air. The boot must lie flat below the level of the piston face.

6. Install or connect the following:
- New pad clips into the caliper support
- Inner and outer brake pads into the clips on the support. The pad with the wear sensor should be located outboard. The piston indentation slots should be positioned to correctly accept the brake pads.
- Caliper body onto the support. Lubricate the non-threaded portion of the guide and lock pins, then install the pins and torque to 27 ft. lbs. (36 Nm).

7. Check the position of the pad clips. If necessary, use a small suitable tool to reseat or center the pad clips on the support. Repeat the procedure for the opposite side brake pads.

8. Install the rear wheel assemblies.

9. Prior to operating the vehicle, depress the brake pedal a few times until the brake pads are seated against the rotor.

Brake Drums

REMOVAL & INSTALLATION

1. Before servicing the vehicle, refer to the precautions in the beginning of this section.

2. Remove or disconnect the following:
- Rear wheel and tire assembly
- Brake drum. If necessary, turn the starwheel of the brake adjuster

assembly to loosen the brake shoes and allow for drum removal.

To install:

3. Install or connect the following:
- Brake drum over brake shoes and onto hub
- Tire and wheel assembly. Torque to the proper specification.

4. Adjust brakes following the proper procedure.

5. Road test for braking operation.

Brake Shoes

REMOVAL & INSTALLATION

EXCEPT ION

1. Before servicing the vehicle, refer to the precautions in the beginning of this section.

2. Remove or disconnect the following:
- Wheels and brake drums
- Lower return and adjuster springs using a universal brake spring remover. Do not over extend the springs or they will damaged and will need to be replaced.

3. Compress the leading brake shoe hold-down cup and spring while removing the pin from the rear of the backing plate. Release spring compression, then remove the hold-down cup and spring.

4. Pull the leading shoe towards the front of the vehicle and remove the adjuster assembly and lever. It may be necessary to turn the adjuster starwheel to shorten the adjuster's length.
- Leading shoe by twisting the shoe out of engagement with the upper return spring
- Upper return spring from the park brake shoe, then the park brake shoe hold-down cup, spring and pin assembly

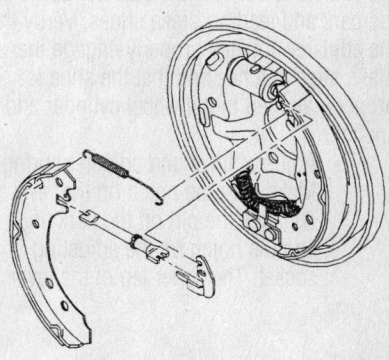

93006G81

Rear brake shoe and adjuster installation—L–Series shown

5. Push the park brake shoe lever into the cable spring while disengaging the cable from the end lever and remove the parking brake shoe, lever and cable spring from the vehicle.

- Retainer and wave washer, then the park brake lever from the shoe

6. Disassemble the brake adjuster socket, screw and nut, then clean the components in denatured alcohol. Inspect the assembly, making sure the screw threads smoothly into the adjusting nut over the full threaded length.

7. Inspect the wheel cylinder for signs of leakage and for cut or damaged boots. Do not attempt to repair a damaged cylinder, the assembly must be replaced.

To install:

8. Lubricate the adjuster assembly, the 6 backing plate raised shoe contact pads, the brake lever pin and surfaces which contact brake shoe webs with brake lubricant.

9. Install or connect the following:

- Park brake lever onto the pin on the brake shoe and secure with the wave washer and retainer clip. Crimp the ends of the retainer to secure the brake lever.
- Cable spring into the cage on the park brake lever, then install the cable through the spring and onto the lever
- Park brake shoe using the hold-down cup assembly; use a universal spring cup remover/installer tool. Make sure the shoe is correctly engaged into the wheel cylinder (top) and the anchor (bottom).
- Long straight end of the upper return spring into the back hole in the park brake shoe, position the other brake shoe and install the other end of the spring into the back of the leading shoe

10. Pull the lead shoe toward the front of the vehicle and install the adjuster between the park and leading brake shoes. Verify that the adjuster notches properly engage the brake shoe notches and that the shoe is properly aligned in the wheel cylinder and anchor.

- Adjuster lever and adjuster spring. Make sure the notch on the lever engages the pin on the park shoe and the notch on the adjusting socket. The lower leg of the lever

should engage the teeth of the star-wheel adjuster assembly.

- Leading brake shoe using the hold-down cup assembly
- Adjuster spring to the upper side of the brake shoes with the short end to the lead shoe and the long end to the adjuster lever
- Lower return spring into the lower holes of the shoes

11. Verify the correct location of all brake components, if necessary, use the other side brake assembly for comparison.

12. Using a suitable drum clearance gauge, measure the inner diameter of the brake drum and adjust the outside diameter of the brake shoes to 0.02 inch (0.50mm) less than the inner diameter of the drum.

13. Repeat the procedure for the opposite brake shoes and install the brake drums.

14. If the wheel cylinders have been replaced, bleed the hydraulic brake system.

15. Install the rear wheels.

16. Apply and release the brake pedal 20 times to allow the adjuster to properly position the brake shoes.

17. Check and adjust the parking brake cable, as necessary.

ION

1. Before servicing the vehicle, refer to the precautions in the beginning of this section.

2. Remove or disconnect the following:

- Wheels and brake drums
- Adjuster spring using a universal brake spring remover. Do not over extend the springs or they will damaged and will need to be replaced.
- Brake adjuster lever from the pivot
- Spread the top of the shoes apart using brake shoe spanner and spring removal tool J 38400
- Adjuster assembly
- Lightly pull the universal spring end out of the shoe web hole using the hook end of J 38400. Hold the universal spring while removing the trailing shoe.
- Park brake cable from the lever
- Lightly pull the universal spring end out of the shoe web hole using the hook end of J 38400. Hold the universal spring while removing the leading shoe.

3. Inspect the wheel cylinder for signs of leakage and for cut or damaged boots. Do not attempt to repair a damaged cylinder, the assembly must be replaced.

To install:

4. Lubricate the adjuster assembly, the 6 backing plate raised shoe contact pads, the brake lever pin and surfaces which contact brake shoe webs with brake lubricant.

5. Install or connect the following:

- Position the hook end of tool J 38400 under the universal spring and lightly pull the spring end out while installing the leading shoe. Make sure the spring is properly engaged in the shoe web hole.
- Park brake cable to the lever
- Position the hook end of tool J 38400 under the universal spring and lightly pull the spring end out while installing the trailing shoe. Make sure the spring is properly engaged in the shoe web hole.
- Spread the top of the shoes apart using tool J 38400
- Adjuster assembly
- Adjuster actuator lever to the shoe and adjuster assembly. Make sure the lever is engaged properly between the adjuster and the shoe.
- Adjuster spring. Make sure the loop end of the spring engages properly to the tab on the actuator lever

6. Release the parking brake.

7. Pull the parking lever boot away from the console after applying light pressure inwards on the boot retainer, this will allow access to the cable adjusting nut.

8. Release the tension on the cable. Make sure to use hand tools only when adjusting the nut.

9. Using a suitable drum clearance gauge, measure the inner diameter of the brake drum and adjust the outside diameter of the brake shoes using the adjuster screw to 0.025 inch (0.50mm) less than the inner diameter of the drum.

10. Repeat the procedure for the opposite brake shoes and install the brake drums.

11. Adjust the parking park and install the lever boot.

12. If the wheel cylinders have been replaced, bleed the hydraulic brake system.

13. Install the rear wheels.

SATURN

Vue

SPECIFICATION CHARTS

ENGINE AND VEHICLE IDENTIFICATION

Engine							Model Year	
Code ①	Liters (cc)	Cu. In.	Cyl.	Fuel Sys.	Engine Type	Eng. Mfg.	Code ②	Year
D	2.2 (2199)	134	4	SFI	DOHC	Saturn	2	2002
B	3.0 (3000)	183	6	SFI	DOHC	Saturn	3	2003
							4	2004

SFI: Sequential Fuel Injection

DOHC: Double Overhead Camshafts

① 8th digit of VIN

② 10th digit of VIN

42372-SVUE-C01

GENERAL ENGINE SPECIFICATIONS

Year	Model	Engine Displacement Liters (cc)	Engine ID/VIN	Fuel System Type	Net Horsepower @ rpm	Net Torque @ rpm (ft. lbs.)	Bore x Stroke (in.)	Com-pression Ratio	Oil Pressure @ rpm
2002	VUE	2.2 (1901)	D	SFI	137@5800	147@4400	3.38x3.50	9.5:1	50-80@1000
		3.0 (3000)	B	SFI	182@6000	184@3400	3.38x3.50	10.0:1	50-80@1000
2003-04	VUE	2.2 (1901)	D	SFI	137@5800	147@4400	3.38x3.50	9.5:1	50-80@1000
		3.0 (3000)	B	SFI	182@6000	184@3400	3.38x3.50	10.0:1	50-80@1000

SFI: Sequential Fuel Injection

42372-SVUE-C02

ENGINE TUNE-UP SPECIFICATIONS

Year	Engine Displacement Liters (cc)	Engine ID/VIN	Spark Plug Gap (in.)	Ignition Timing (deg.)		Fuel Pump (psi) ①	Idle Speed (rpm)		Valve Clearance	
				MT	AT		MT ②	AT ②	In.	Ex.
2002	2.2 (2199)	D	0.045	③	③	50-60	④	④	HYD	HYD
	3.0 (3000)	B	0.043	③	③	50-60	④	④	HYD	HYD
2003-04	2.2 (2199)	D	0.045	③	③	50-60	④	④	HYD	HYD
	3.0 (3000)	B	0.043	③	③	50-60	④	④	HYD	HYD

NOTE: The Vehicle Emission Control Information label often reflects specification changes made during production. The label figures must be used if they differ from those in this chart.

HYD: Hydraulic

① Pressure measured at idle

② Idle speed measured with manual transmission in Neutral; automatic transmission in D (drive)

③ Engines equipped with Distributorless Ignition System (DIS). Ignition timing is not adjustable

④ Refer to the Vehicle Emission Control Information label

42372-SVUE-C03

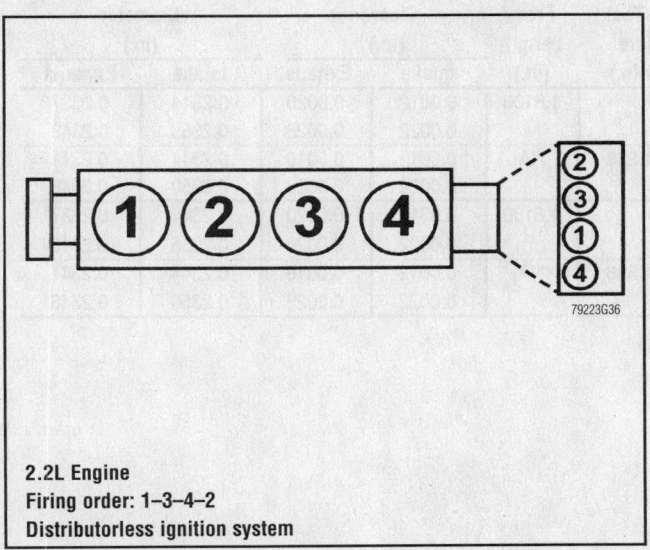

2.2L Engine
Firing order: 1–3–4–2
Distributorless ignition system

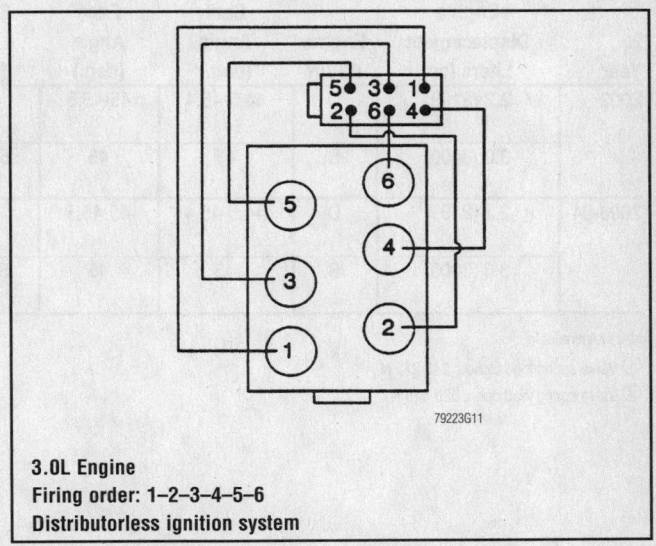

3.0L Engine
Firing order: 1–2–3–4–5–6
Distributorless ignition system

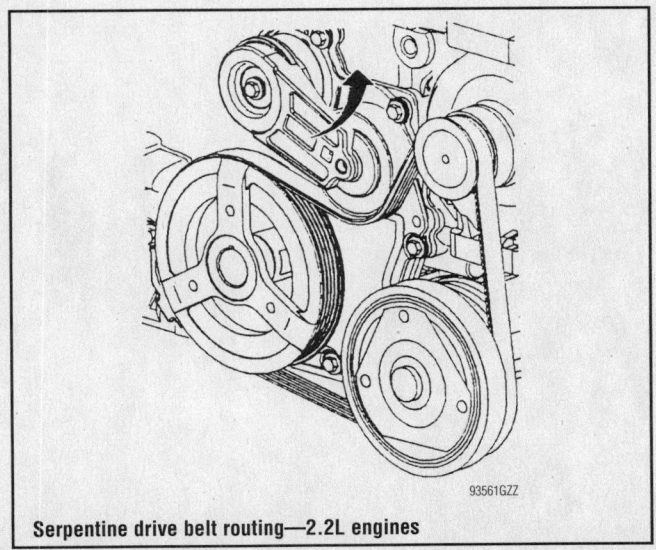

Serpentine drive belt routing—2.2L engines

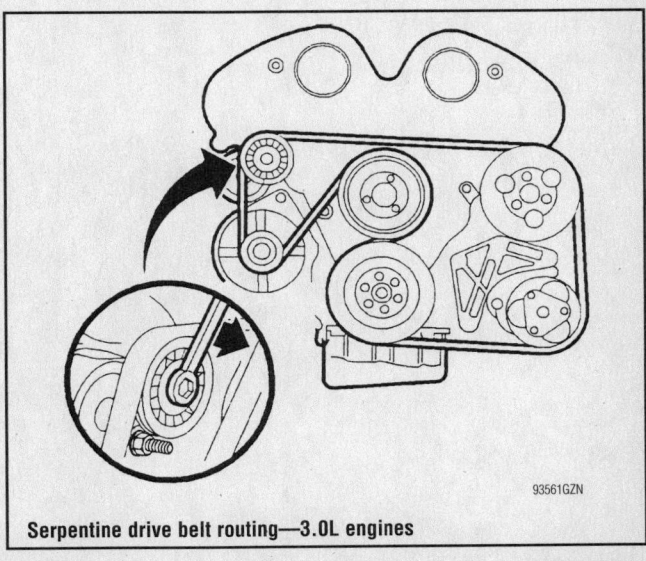

Serpentine drive belt routing—3.0L engines

CAPACITIES

Year	Model	Engine Displacement Liters (cc)	Engine ID/VIN	Engine Oil with Filter (qts.)	Transaxle (qts.)		Fuel Tank (gal.)	Cooling System (qts.)
					Manual	Auto. ①		
2002	VUE	2.2 (2199)	D	5.0	2.0	6.9	15.7	②
		3.0 (3000)	B	5.0	2.0	6.9	15.7	②
2003-04	VUE	2.2 (2199)	D	5.0	2.0	6.9	15.7	②
		3.0 (3000)	B	5.0	2.0	6.9	15.7	②

NOTE: All capacities are approximate. Add fluid gradually and ensure a proper fluid level is obtained.

① Specification is for overhaul. 8.4 pts. with fluid and filter change

② 2.2L with manual transaxle: 7.4 qts.
2.2L with automatic transaxle: 7.3 qts.
3.0L: 7.8 qts.

42372-SVUE-C04

VALVE SPECIFICATIONS

Year	Engine Displacement Liters (cc)	Engine ID/VIN	Seat Angle (deg.)	Face Angle (deg.)	Spring Test Pressure (lbs. @ in.)	Spring Free-Length (in.)	Stem-to-Guide Clearance (in.)		Stem Diameter (in.)	
							Intake	Exhaust	Intake	Exhaust
2002	2.2 (2199)	D	44.5-45.4	45-45.5	①	1.6100	0.0012	0.0020	0.2344	0.2337
					②		0.0022	0.0026	0.2355	0.2343
	3.0 (3000)	B	45	45	56.6 @ 1.338	NA	0.0012	0.0016	0.2344	0.2341
							0.0022	0.0026	0.2350	0.2346
2003-04	2.2 (2199)	D	44.5-45.4	45-45.5	①	1.6100	0.0012	0.0020	0.2344	0.2337
					②		0.0022	0.0026	0.2355	0.2343
	3.0 (3000)	B	45	45	56.6 @ 1.338	NA	0.0012	0.0016	0.2344	0.2341
							0.0022	0.0026	0.2350	0.2346

NA: Not available

① Valve spring load closed: 245-271 N

② Valve spring load open: 525-575 N

42372-SVUE-C05

CRANKSHAFT AND CONNECTING ROD SPECIFICATIONS
All measurements are given in inches.

Year	Engine Displacement Liters (cc)	Engine ID/VIN	Crankshaft				Connecting Rod		
			Main Brg. Journal Dia.	Main Brg. Oil Clearance	Shaft End-play	Thrust on No.	Journal Diameter	Oil Clearance	Side Clearance
2002	2.2 (2199)	D	2.2045-2.2050	0.0012 0.0026	0.0012-0.0150	3	1.9291-1.9297	0.0001-0.0021	0.0028-0.0146
	3.0 (3000)	B	2.6763-2.6766	0.0060 0.0017	0.0004-0.0300	3	1.927-1.9280	0.0001-0.0021	0.0027-0.0110
2003-04	2.2 (2199)	D	2.2045-2.2050	0.0012 0.0026	0.0012-0.0150	3	1.9291-1.9297	0.0001-0.0021	0.0028-0.0146
	3.0 (3000)	B	2.6763-2.6766	0.0060 0.0017	0.0004-0.0300	3	1.927-1.9280	0.0001-0.0021	0.0027-0.0110

42372-SVUE-C06

PISTON AND RING SPECIFICATIONS
All measurements are given in inches.

Year	Engine Displacement Liters (cc)	Engine ID/VIN	Piston Clearance	Ring Gap			Ring Side Clearance		
				Top Compression	Bottom Compression	Oil Control	Top Compression	Bottom Compression	Oil Control
2002	2.2 (2199)	D	0.0004-0.0016	0.008-0.016	0.0014 0.0022	0.0010 0.0030	0.0028-0.0146	0.0005-0.0024	SNUG
	3.0 (3000)	B	0.0010-0.0018	0.0008-0.0015	0.0118 0.0196	0.0157 0.0551	0.0027-0.0110	0.0005-0.0024	SNUG
2003-04	2.2 (2199)	D	0.0004-0.0016	0.008-0.016	0.0014 0.0022	0.0010 0.0030	0.0028-0.0146	0.0005-0.0024	SNUG
	3.0 (3000)	B	0.0010-0.0018	0.0008-0.0015	0.0118 0.0196	0.0157 0.0551	0.0027-0.0110	0.0005-0.0024	SNUG

NA: Not available

① Piston No. 2 and 3: 0.0002-0.0017
Piston No. 1, 4: 0.0003-0.0021

42372-SVUE-C07

TORQUE SPECIFICATIONS
All readings in ft. lbs.

Year	Engine Displacement Liters (cc)	Engine ID/VIN	Cylinder Head Bolts	Main Bearing Bolts	Rod Bearing Bolts	Crankshaft Damper Bolts	Flywheel Bolts	Manifold		Spark Plugs	Lug Nuts
								Intake	Exhaust		
2002	2.2 (2199)	D	①	②	③	④	②	⑤	13	15	92
	3.0 (3000)	B	⑥	⑦	26	15	②	15	15	18	92
2003-04	2.2 (2199)	D	①	②	③	④	②	⑤	13	15	92
	3.0 (3000)	B	⑥	⑦	26	15	②	15	15	18	92

① Step 1: 22 ft. lbs.
  Step 2: 155 degrees
② 39 ft. lbs. Plus 25 degrees
③ 18 ft. lbs. Plus 100 degrees
④ 74 ft. lbs. Plus 75 degrees
⑤ 89 inch lbs.

⑥ Step 1: 18 ft. lbs.
  Step 2: plus 90 degrees
  Step 3: plus 90 degrees
  Step 4: plus 90 degrees
  Step 5: plus 15 degrees

⑦ Step 1: 37 ft. lbs.
  Step 2: plus 60 degrees
  Step 3: plus 15 degrees

42372-SVUE-C08

WHEEL ALIGNMENT

Year	Model		Caster		Camber		Toe-in (in.)	Steering Axis Inclination (Deg.)
			Range (+/-Deg.)	Preferred Setting (Deg.)	Range (+/-Deg.)	Preferred Setting (Deg.)		
2002	VUE	F	2.60-3.40	3.00	-1.00	0.60	0.20 +/- 0.15	—
		R	—	—	—	-0.05	0.10 +/- 0.10	—
2003-04	VUE	F	2.60-3.40	3.00	-1.00	0.60	0.20 +/- 0.15	—
		R	—	—	—	-0.05	0.10 +/- 0.10	—

42372-SVUE-C09

TIRE, WHEEL AND BALL JOINT SPECIFICATIONS

Year	Model		OEM Tires		Tire Pressures (psi)		Wheel Size	Ball Joint Inspection
			Standard	Optional	Front	Rear		
2002	VUE	F	P235/65R16	P215/70R16	①	①	NS	NS
		R	P235/65R16	P215/70R16	①	①	NS	NS
2003-04	VUE	F	P235/65R16	P215/70R16	①	①	NS	NS
		R	P235/65R16	P215/70R16	①	①	NS	NS

OEM: Original Equipment Manufacturer

PSI: Pounds Per Square Inch

① Check the placard on the drivers side sill

NS: Not specified by manufacturer

42372-SVUE-C10

BRAKE SPECIFICATIONS
All measurements in inches unless noted

Year	Model		Brake Disc			Brake Drum Diameter			Minimum Lining Thickness	Brake Caliper	
			Original Thickness	Minimum Thickness	Maximum Runout	Original Inside Diameter	Max. Wear Limit	Maximum Machine Diameter		Bracket Bolt (ft. lbs.)	Mounting Bolt (ft. lbs.)
2002	VUE	F	1.020	0.960	0.001	—	—	—	0.080	118	24
		R	—	—	—	9.84	9.90	9.90	0.040	63	27
2003-04	VUE	F	1.020	0.960	0.001	—	—	—	0.080	118	24
		R	—	—	—	9.84	9.90	9.90	0.040	63	27

NA: Not Available

F: Front

R: Rear

42372-SVUE-C11

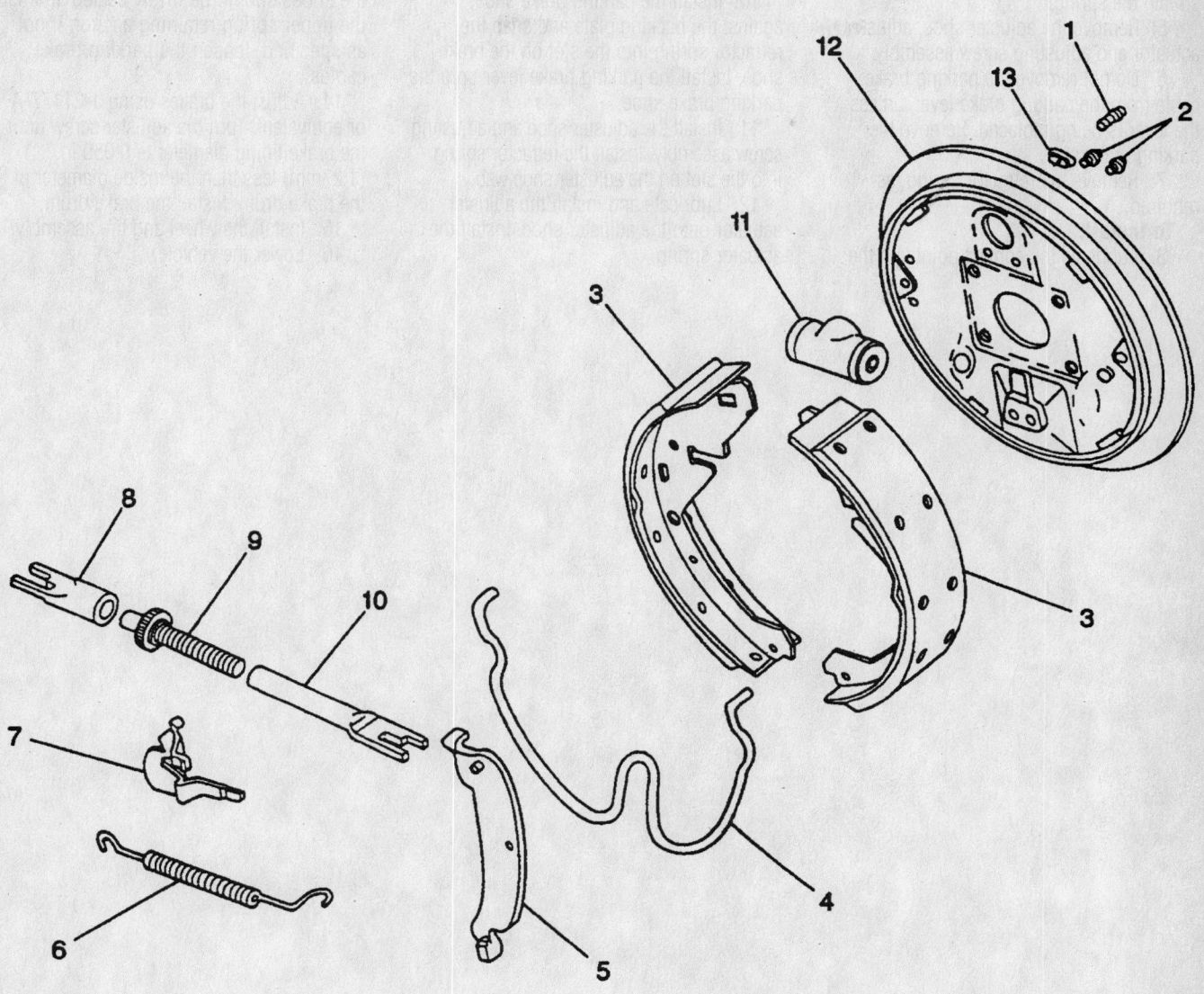

(1) Bleeder Valve
(2) Wheel Cylinder Mounting Bolts
(3) Shoe and Lining
(4) Return Spring
(5) Parking Brake Lever
(6) Adjuster Spring
(7) Adjuster Actuator

(8) Socket Adjuster
(9) Adjuster Screw
(10) Pivot Nut
(11) Wheel Cylinder
(12) Backing Plate
(13) Access Hole Plug

93026G96

Exploded view of the rear drum brake components

4. Remove the actuator spring from the brake shoes. Remove the retractor spring from the shoe web, being careful not to over stretch the spring.

5. Remove the adjuster shoe, adjuster actuator and adjusting screw assembly.

6. Do not remove the parking brake cable from the parking brake lever, unless the lever is being replaced. Remove the parking brake shoe.

7. Remove the retractor spring, as required.

To install:

8. Lubricate the contact points on the backing plate with lithium grease. Clean and lubricate the adjuster with lithium grease.

9. Install the retractor spring, if removed.

10. Install the parking brake shoe against the backing plate and snap the retractor spring into the slot on the brake shoe. Install the parking brake lever onto the parking brake shoe.

11. Install the adjuster shoe and adjusting screw assembly. Install the retractor spring into the slot on the adjuster shoe web.

12. Lubricate and install the adjuster actuator onto the adjuster shoe. Install the actuator spring.

13. Ensure the parking brake system is adjusted properly with no tension on the cables or parking brake lever. The tops of the shoes should be firmly seated against the upper spring retaining anchor, if not as specified, loosen the parking brake cables.

14. Adjust the brakes using J-21177-A or equivalent. Turn the adjuster screw until the brake lining diameter is 0.050 in. (1.27mm) less than the inside diameter of the brake drum. Install the brake drum.

15. Install the wheel and tire assembly.

16. Lower the vehicle.

BUICK AND PONTIAC

14

Le Sabre • Park Ave. • Bonneville

SPECIFICATION CHARTS

ENGINE AND VEHICLE IDENTIFICATION

Engine							Model Year	
Code ①	Liters (cc)	Cu. In.	Cyl.	Fuel Sys.	Engine Type	Eng. Mfg.	Code ②	Year
1 ③	3.8 (3785)	231	6	MFI	OHV	BOC	Y	2000
K	3.8 (3785)	231	6	MFI	OHV	BOC	1	2001
							2	2002
							3	2003
							4	2004

MFI: Multi-point Fuel Injection

BOC: Buick/Oldsmobile/Cadillac

OHV: Overhead Valves

① 8th position of VIN

② 10th position of VIN

③ Supercharged engine

42372-HBOD-C01

GENERAL ENGINE SPECIFICATIONS

Year	Model	Engine Displacement Liters (cc)	Engine Series (ID/VIN)	Fuel System	Net Horsepower @ rpm	Net Torque @ rpm (ft. lbs.)	Bore x Stroke (in.)	Com- pression Ratio	Oil Pressure @ rpm
2000	Bonneville	3.8 (3785)	1 ①	MFI	240@5000	275@3200	3.80x3.40	9.0:1	60@1850
	Bonneville	3.8 (3785)	K	MFI	205@5200	230@4000	3.80x3.40	9.4:1	60@1850
	LeSabre	3.8 (3785)	K	MFI	200@5200	230@4000	3.80x3.40	9.4:1	60@1850
	Park Avenue	3.8 (3785)	1 ①	MFI	240@5200	280@3600	3.80x3.40	9.0:1	60@1850
	Park Avenue	3.8 (3785)	K	MFI	200@5200	230@4000	3.80x3.40	9.4:1	60@1850
2001	Bonneville	3.8 (3785)	1 ①	MFI	240@5000	275@3200	3.80x3.40	8.5:1	60@1850
	Bonneville	3.8 (3785)	K	MFI	205@5200	230@4000	3.80x3.40	9.4:1	60@1850
	LeSabre	3.8 (3785)	K	MFI	200@5200	230@4000	3.80x3.40	9.4:1	60@1850
	Park Avenue	3.8 (3785)	1 ①	MFI	240@5200	280@3600	3.80x3.40	8.5:1	60@1850
	Park Avenue	3.8 (3785)	K	MFI	200@5200	230@4000	3.80x3.40	9.4:1	60@1850
2002	Bonneville	3.8 (3785)	1 ①	MFI	240@5000	275@3200	3.80x3.40	8.5:1	60@1850
	Bonneville	3.8 (3785)	K	MFI	205@5200	230@4000	3.80x3.40	9.4:1	60@1850
	LeSabre	3.8 (3785)	K	MFI	200@5200	230@4000	3.80x3.40	9.4:1	60@1850
	Park Avenue	3.8 (3785)	1 ①	MFI	240@5200	280@3600	3.80x3.40	8.5:1	60@1850
	Park Avenue	3.8 (3785)	K	MFI	200@5200	230@4000	3.80x3.40	9.4:1	60@1850
2003	Bonneville	3.8 (3785)	1 ①	MFI	240@5000	275@3200	3.80x3.40	8.5:1	60@1850
	Bonneville	3.8 (3785)	K	MFI	205@5200	230@4000	3.80x3.40	9.4:1	60@1850
	LeSabre	3.8 (3785)	K	MFI	200@5200	230@4000	3.80x3.40	9.4:1	60@1850
	Park Avenue	3.8 (3785)	1 ①	MFI	240@5200	280@3600	3.80x3.40	8.5:1	60@1850
	Park Avenue	3.8 (3785)	K	MFI	200@5200	230@4000	3.80x3.40	9.4:1	60@1850

MFI: Multi-point Fuel Injection

① Supercharged engine

42372-HBOD-C02

ENGINE TUNE-UP SPECIFICATIONS

Year	Engine Displacement Liters (cc)	Engine ID/VIN	Spark Plug Gap (in.)	Ignition Timing (deg.)	Fuel Pump (psi)	Idle Speed (rpm)	Valve Clearance	
							Intake	Exhaust
2000	3.8 (3786)	1	0.060	①	41-47 ②	③	HYD	HYD
	3.8 (3785)	K	0.060	①	41-47 ②	③	HYD	HYD
2001	3.8 (3786)	1	0.060	①	41-47 ②	③	HYD	HYD
	3.8 (3785)	K	0.060	①	41-47 ②	③	HYD	HYD
2002	3.8 (3786)	1	0.060	①	41-47 ②	③	HYD	HYD
	3.8 (3785)	K	0.060	①	41-47 ②	③	HYD	HYD
2003	3.8 (3786)	1	0.060	①	41-47 ②	③	HYD	HYD
	3.8 (3785)	K	0.060	①	41-47 ②	③	HYD	HYD

NOTE: The Vehicle Emission Control Information label often reflects specification changes made during production.

The label figures must be used if they differ from those in this chart.

HYD: Hydraulic

① DIS Ignition System timing not adjustable

② Pressure at fuel pump

③ Idle speed maintained by ECM. There is no recommended adjustment procedure

42372-HBOD-C03

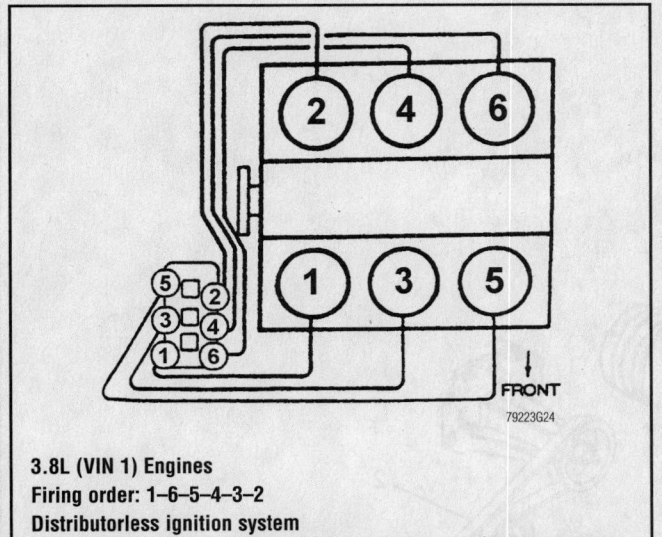

3.8L (VIN 1) Engines
Firing order: 1–6–5–4–3–2
Distributorless ignition system

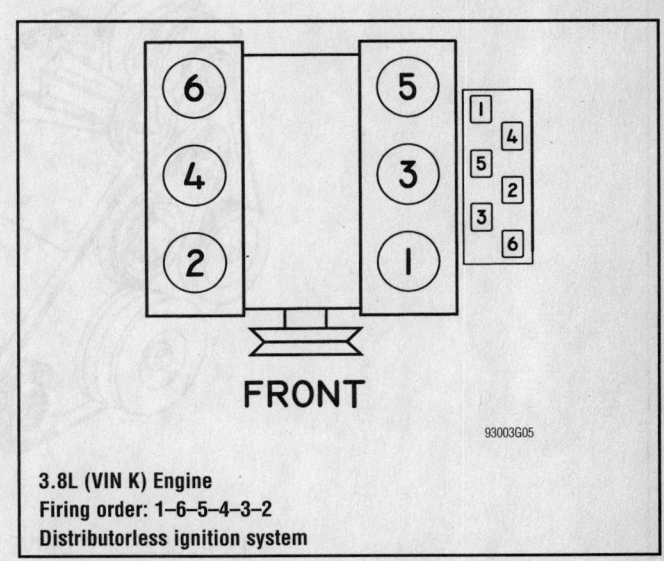

3.8L (VIN K) Engine
Firing order: 1–6–5–4–3–2
Distributorless ignition system

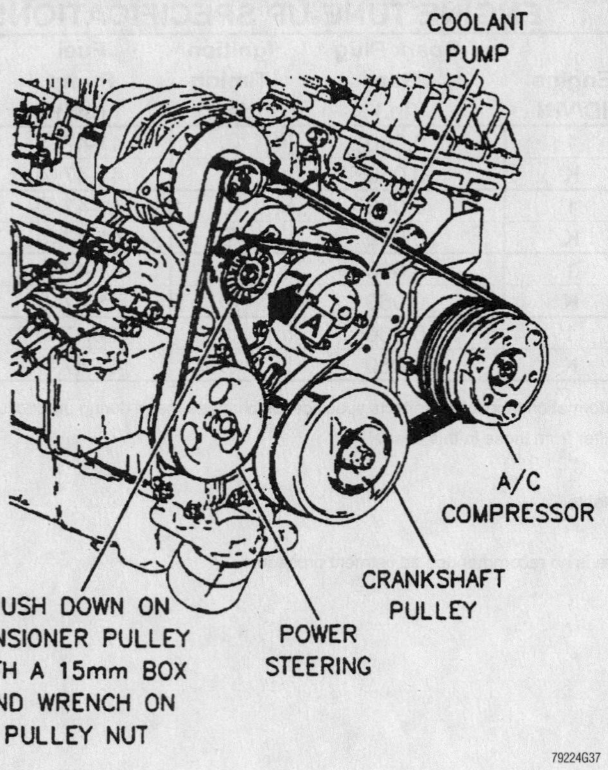

COOLANT
PUMP

PUSH DOWN ON
TENSIONER PULLEY
WITH A 15mm BOX
END WRENCH ON
PULLEY NUT

POWER
STEERING

A/C
COMPRESSOR

CRANKSHAFT
PULLEY

79224G37

Serpentine drive belt routing—3.8L (VIN K) engine

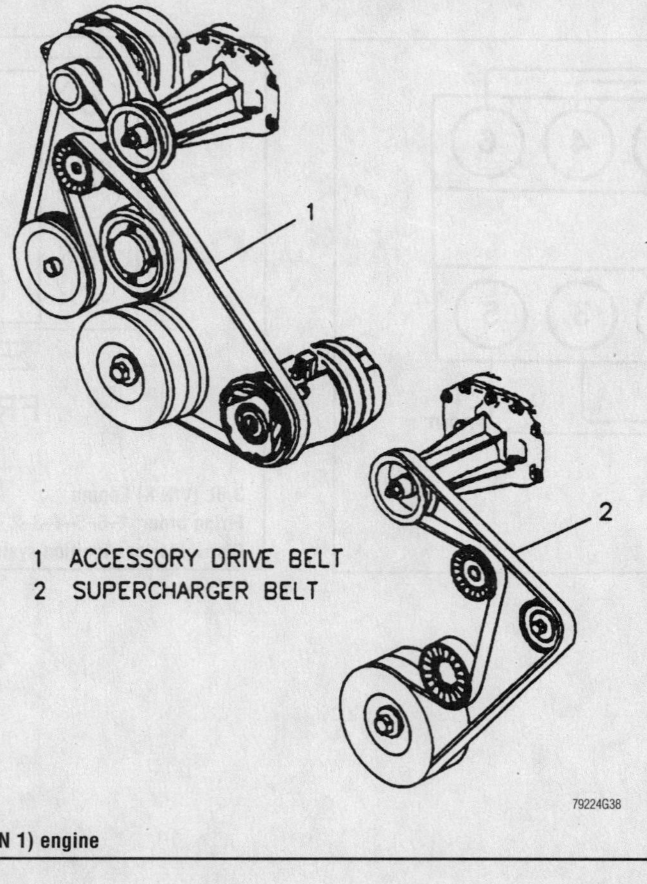

1 ACCESSORY DRIVE BELT
2 SUPERCHARGER BELT

79224G38

Serpentine drive belt routing—3.8L (VIN 1) engine

CAPACITIES

Year	Model	Engine Displacement Liters (cc)	Engine ID/VIN	Engine Oil with Filter (qts.)	Transmission (pts.)	Fuel Tank (gal.)	Cooling System (qts.)
2001	Bonneville	3.8 (3785)	1	4.5	14.8	18.0	10.0
	Bonneville	3.8 (3785)	K	4.5	14.8	18.0	10.0
	LeSabre	3.8 (3785)	K	4.5	14.8	18.0	10.0
	Park Avenue	3.8 (3785)	1	4.5	12.0	18.0	13.0
	Park Avenue	3.8 (3785)	K	4.5	12.0	18.0	13.0
2001	Bonneville	3.8 (3785)	1	4.5	14.8	18.0	10.0
	Bonneville	3.8 (3785)	K	4.5	14.8	18.0	10.0
	LeSabre	3.8 (3785)	K	4.5	14.8	18.0	10.0
	Park Avenue	3.8 (3785)	1	4.5	12.0	18.0	13.0
	Park Avenue	3.8 (3785)	K	4.5	12.0	18.0	13.0
2002	Bonneville	3.8 (3785)	1	4.5	14.8	18.0	10.0
	Bonneville	3.8 (3785)	K	4.5	14.8	18.0	10.0
	LeSabre	3.8 (3785)	K	4.5	14.8	18.0	10.0
	Park Avenue	3.8 (3785)	1	4.5	12.0	18.0	13.0
	Park Avenue	3.8 (3785)	K	4.5	12.0	18.0	13.0
2003	Bonneville	3.8 (3785)	1	4.5	14.8	18.0	10.0
	Bonneville	3.8 (3785)	K	4.5	14.8	18.0	10.0
	LeSabre	3.8 (3785)	K	4.5	14.8	18.0	10.0
	Park Avenue	3.8 (3785)	1	4.5	12.0	18.0	13.0
	Park Avenue	3.8 (3785)	K	4.5	12.0	18.0	13.0

NOTE: All capacities are approximate. Add fluid gradually and ensure a proper fluid level is obtained.

42372-HBOD-C04

CRANKSHAFT AND CONNECTING ROD SPECIFICATIONS

All measurements are given in inches.

Year	Engine Displacement Liters (cc)	Engine ID/VIN	Crankshaft				Connecting Rod		
			Main Brg. Journal Dia.	Main Brg. Oil Clearance	Shaft End-play	Thrust on No.	Journal Diameter	Oil Clearance	Side Clearance
2000	3.8 (3786)	1	2.4988-2.4998	①	0.0030-0.0110	2	2.2487-2.2499	0.0005-0.0026	0.0040-0.0200
	3.8 (3786)	K	2.4988-2.4998	①	0.0030-0.0110	2	2.2487-2.2499	0.0005-0.0026	0.0040-0.0200
2001	3.8 (3786)	1	2.4988-2.4998	①	0.0030-0.0110	2	2.2487-2.2499	0.0005-0.0026	0.0040-0.0200
	3.8 (3786)	K	2.4988-2.4998	①	0.0030-0.0110	2	2.2487-2.2499	0.0005-0.0026	0.0040-0.0200
2002	3.8 (3786)	1	2.4988-2.4998	①	0.0030-0.0110	2	2.2487-2.2499	0.0005-0.0026	0.0040-0.0200
	3.8 (3786)	K	2.4988-2.4998	①	0.0030-0.0110	2	2.2487-2.2499	0.0005-0.0026	0.0040-0.0200
2003	3.8 (3786)	1	2.4988-2.4998	①	0.0030-0.0110	2	2.2487-2.2499	0.0005-0.0026	0.0040-0.0200
	3.8 (3786)	K	2.4988-2.4998	①	0.0030-0.0110	2	2.2487-2.2499	0.0005-0.0026	0.0040-0.0200

① Journal 1: 0.0007 - 0.0016
  Journals 2 and 3: 0.0010 - 0.0020
  Journal 4: 0.0009 - 0.0018

42372-HBOD-C05

VALVE SPECIFICATIONS

Year	Engine Displacement Liters (cc)	Engine ID/VIN	Seat Angle (deg.)	Face Angle (deg.)	Spring Test Pressure (lbs. @ in.)	Spring Installed Height (in.)	Stem-to-Guide Clearance (in.)		Stem Diameter (in.)	
							Intake	Exhaust	Intake	Exhaust
2000	3.8 (3785)	1	45	45	80@1.750	1.690-1.720	0.0015-0.0032	0.0015-0.0032	NA	NA
	3.8 (3785)	K	45	45	80@1.750	1.690-1.720	0.0015-0.0032	0.0015-0.0032	NA	NA
2001	3.8 (3785)	1	45	45	80@1.750	1.690-1.720	0.0015-0.0032	0.0015-0.0032	NA	NA
	3.8 (3785)	K	45	45	80@1.750	1.690-1.720	0.0015-0.0032	0.0015-0.0032	NA	NA
2002	3.8 (3785)	1	45	45	80@1.750	1.690-1.720	0.0015-0.0032	0.0015-0.0032	NA	NA
	3.8 (3785)	K	45	45	80@1.750	1.690-1.720	0.0015-0.0032	0.0015-0.0032	NA	NA
2003	3.8 (3785)	1	45	45	80@1.750	1.690-1.720	0.0015-0.0032	0.0015-0.0032	NA	NA
	3.8 (3785)	K	45	45	80@1.750	1.690-1.720	0.0015-0.0032	0.0015-0.0032	NA	NA

NA: Not Available

42372-HBOD-C06

PISTON AND RING SPECIFICATIONS
All measurements are given in inches.

Year	Engine Displacement Liters (cc)	Engine ID/VIN	Piston Clearance	Ring Gap Top Compression	Bottom Compression	Oil Control	Ring Side Clearance Top Compression	Bottom Compression	Oil Control
2000	3.8 (3786)	1	0.0004-0.0020	0.012-0.022	0.030-0.040	0.010-0.030	0.0013-0.0031	0.0013-0.0031	0.0009-0.0079
	3.8 (3786)	K	0.0004-0.0020	0.012-0.022	0.030-0.040	0.010-0.030	0.0013-0.0031	0.0013-0.0031	0.0009-0.0079
2001	3.8 (3786)	1	0.0004-0.0020	0.012-0.022	0.030-0.040	0.010-0.030	0.0013-0.0031	0.0013-0.0031	0.0009-0.0079
	3.8 (3786)	K	0.0004-0.0020	0.012-0.022	0.030-0.040	0.010-0.030	0.0013-0.0031	0.0013-0.0031	0.0009-0.0079
2002	3.8 (3786)	1	0.0004-0.0020	0.012-0.022	0.030-0.040	0.010-0.030	0.0013-0.0031	0.0013-0.0031	0.0009-0.0079
	3.8 (3786)	K	0.0004-0.0020	0.012-0.022	0.030-0.040	0.010-0.030	0.0013-0.0031	0.0013-0.0031	0.0009-0.0079
2003	3.8 (3786)	1	0.0004-0.0020	0.012-0.022	0.030-0.040	0.010-0.030	0.0013-0.0031	0.0013-0.0031	0.0009-0.0079
	3.8 (3786)	K	0.0004-0.0020	0.012-0.022	0.030-0.040	0.010-0.030	0.0013-0.0031	0.0013-0.0031	0.0009-0.0079

42372-HBOD-C07

TORQUE SPECIFICATIONS
All readings in ft. lbs.

Year	Engine Displacement Liters (cc)	Engine ID/VIN	Cylinder Head Bolts	Main Bearing Bolts	Rod Bearing Bolts	Crankshaft Damper Bolts	Flywheel Bolts	Manifold Intake	Exhaust	Spark Plugs	Lug Nuts
2000	3.8 (3786)	1	①	②	③	④	⑤	⑥	22	11	100
	3.8 (3786)	K	①	⑦	③	④	⑤	11	38	11	100
2001	3.8 (3786)	1	①	②	③	④	⑤	⑥	22	11	100
	3.8 (3786)	K	①	⑦	③	④	⑤	11	38	11	100
2002	3.8 (3786)	1	①	②	③	④	⑤	⑥	22	11	100
	3.8 (3786)	K	①	⑦	③	④	⑤	11	38	11	100
2003	3.8 (3786)	1	①	②	③	④	⑤	⑥	22	11	100
	3.8 (3786)	K	①	⑦	③	④	⑤	11	38	11	100

① Step 1: Tighten all bolts to 37 ft. lbs.
  Step 2: Turn all bolts 120 degrees

② Step 1: Tighten caps in equal increments to 52 ft. lbs.
  Step 2: Loosen 360 degrees
  Step 3: 15 ft. lbs.
  Step 4: 54 ft. lbs.
  Step 5: Plus three turns of 35 degrees for a total of 105 degrees

③ 20 ft. lbs. plus 50 degrees
④ 111 ft. lbs. plus 76 degrees
⑤ 11 ft. lbs. plus 50 degrees
⑥ Upper manifold: 8 ft. lbs.
  Lower manifold: 11 ft. lbs.
⑦ 26 ft. lbs. plus 50 degrees

42372-HBOD-C08

WHEEL ALIGNMENT
Buick Park Avenue

Year	Model		Caster Range (Deg.)	Caster Preferred Setting (Deg.)	Camber Range (Deg.)	Camber Preferred Setting (Deg.)	Toe-in (in.)	Steering Axis Inclination (Deg.)
2000	Park Avenue	F	+0.50	+6.00	+0.50	-0.20	0.20 +/- 0.20	—
		R	—	—	+0.50	-0.30	0.20 +/- 0.20	—
2001	Park Avenue	F	+0.50	+6.00	+0.50	-0.20	0.20 +/- 0.20	—
		R	—	—	+0.50	-0.30	0.20 +/- 0.20	—
2002	Park Avenue	F	+0.50	+6.00	+0.50	-0.20	0.20 +/- 0.20	—
		R	—	—	+0.50	-0.30	0.20 +/- 0.20	—
2003	Park Avenue	F	+0.50	+6.00	+0.50	-0.20	0.20 +/- 0.20	—
		R	—	—	+0.50	-0.30	0.20 +/- 0.20	—

42372-HBOD-C09

WHEEL ALIGNMENT
Pontiac Bonneville

Year	Model		Caster Range (Deg.)	Caster Preferred Setting (Deg.)	Camber Range (Deg.)	Camber Preferred Setting (Deg.)	Toe-in (in.)	Steering Axis Inclination (Deg.)
2000	Bonneville	F	0.50	6.00	0.50	-0.20	0.10 +/- 0.10	—
		R	—	—	0.50	-0.30	0.10 +/- 0.10	—
2001	Bonneville	F	0.50	6.00	0.50	-0.20	0.10 +/- 0.10	—
		R	—	—	0.50	-0.30	0.10 +/- 0.10	—
2002	Bonneville	F	0.50	6.00	0.50	-0.20	0.10 +/- 0.10	—
		R	—	—	0.50	-0.30	0.10 +/- 0.10	—
2003	Bonneville	F	0.50	6.00	0.50	-0.20	0.10 +/- 0.10	—
		R	—	—	0.50	-0.30	0.10 +/- 0.10	—

42372-HBOD-C10

TIRE, WHEEL AND BALL JOINT SPECIFICATIONS
Pontiac
Bonneville SE and SSE

| Year | Model | OEM Tires | | Tire Pressures (psi) | | Wheel Size | Ball Joint Inspection |
		Standard	Optional	Front	Rear		
2000	Bonneville SE	P215/65R15	P225/60R16	30	30	6-JJ	①
	Bonneville SSE	P225/60R16	None	30	30	7-JJ	①
2001	Bonneville SE	P215/65R15	P225/60R16	30	30	6-JJ	①
	Bonneville SSE	P225/60R16	None	30	30	7-JJ	①
2002	Bonneville SE	P215/65R15	P225/60R16	30	30	6-JJ	①
	Bonneville SSE	P225/60R16	None	30	30	7-JJ	①
2003	Bonneville SE	P215/65R15	P225/60R16	30	30	6-JJ	①
	Bonneville SSE	P225/60R16	None	30	30	7-JJ	①

OEM: Original Equipment Manufacturer

PSI: Pounds Per Square Inch

① Do not lift car. Inspect the boss into which the grease fitting is threaded. Replace if the boss is flush or receded below the surface of the ball joint.

42372-HBOD-C11

TIRE, WHEEL AND BALL JOINT SPECIFICATIONS
Buick
LeSabre and Park Avenvue

| Year | Model | OEM Tires | | Tire Pressures (psi) | | Wheel Size | Ball Joint Inspection |
		Standard	Optional	Front	Rear		
2000	LeSabre/Park Avenue	P205/70R15	P215/60R16	30	30	6-JJ	①
2001	LeSabre/Park Avenue	P205/70R15	P215/60R16	30	30	6-JJ	①
2002	LeSabre/Park Avenue	P205/70R15	P215/60R16	30	30	6-JJ	①
2003	LeSabre/Park Avenue	P205/70R15	P215/60R16	30	30	6-JJ	①

OEM: Original Equipment Manufacturer

PSI: Pounds Per Square Inch

① Do not lift car. Inspect the boss into which the grease fitting is threaded. Replace if the boss is flush or receded below the surface of the ball joint.

42372-HBOD-C12

BRAKE SPECIFICATIONS
All measurements in inches unless noted

| Year | Model | | Brake Disc | | | Brake Drum Diameter | | | Minimum Lining Thickness | | Caliper Mounting Bolts |
			Original Thickness	Minimum Thickness	Maximum Runout	Original Inside Diameter	Max. Wear Limit	Maximum Machine Diameter	Front	Rear	(ft. lbs.)
2000	Bonneville	F	1.267	1.209	0.002	—	—	—	0.030	—	63
		R	0.433	0.374	0.002	—	—	—	—	0.030	20
	LeSabre	F	1.267	1.200	0.002	—	—	—	0.030	—	63
		R	0.433	0.374	0.002	—	—	—	—	0.030	20
	Park Avenue	F	1.267	1.200	0.002	—	—	—	0.030	—	63
		R	0.433	0.374	0.002	—	—	—	—	0.030	20
2001	Bonneville	F	1.267	1.200	0.002	—	—	—	0.030	—	63
		R	0.433	0.374	0.002	—	—	—	—	0.030	20
	LeSabre	F	1.267	1.200	0.002	—	—	—	0.030	—	63
		R	0.433	0.374	0.002	—	—	—	—	0.030	20
	Park Avenue	F	1.267	1.200	0.002	—	—	—	0.030	—	63
		R	0.433	0.374	0.002	—	—	—	—	0.030	20
2002	Bonneville	F	1.267	1.200	0.002	—	—	—	0.030	—	63
		R	0.433	0.374	0.002	—	—	—	—	0.030	20
	LeSabre	F	1.267	1.200	0.002	—	—	—	0.030	—	63
		R	0.433	0.374	0.002	—	—	—	—	0.030	20
	Park Avenue	F	1.267	1.200	0.002	—	—	—	0.030	—	63
		R	0.433	0.374	0.002	—	—	—	—	0.030	20
2003	Bonneville	F	1.267	1.200	0.002	—	—	—	0.030	—	63
		R	0.433	0.374	0.002	—	—	—	—	0.030	20
	LeSabre	F	1.267	1.200	0.002	—	—	—	0.030	—	63
		R	0.433	0.374	0.002	—	—	—	—	0.030	20
	Park Avenue	F	1.267	1.200	0.002	—	—	—	0.030	—	63
		R	0.433	0.374	0.002	—	—	—	—	0.030	20

① 0.030 over rivet head; If bonded lining, use 0.062 from shoe

42372-HB0D-C13

SCHEDULED MAINTENANCE INTERVALS
GM H BODIES—BUICK LESABRE, PARK AVENUE
PONTIAC BONNEVILLE

TO BE SERVICED	TYPE OF SERVICE	VEHICLE MILEAGE INTERVAL (x1000)												
		6	12	15	18	21	24	27	30	33	36	39	42	45
Engine oil & filter	R	Change oil and filter every 3,000 miles or 3 months whichever occurs first												
Rotate tires	S/I	✓	✓	✓	✓	✓	✓	✓	✓	✓	✓	✓	✓	✓
Exhaust system & brake hoses	S/I	✓	✓	✓	✓	✓	✓	✓	✓	✓	✓	✓	✓	✓
Driveshaft boots & front suspension components	S/I	✓	✓	✓	✓	✓	✓	✓	✓	✓	✓	✓	✓	✓
Lubricate chassis, suspension, steering linkage, transaxle shift linkage, parking brake cable guides, underbody contact points & linkage	S/I	✓	✓	✓	✓	✓	✓	✓	✓	✓	✓	✓	✓	✓
Coolant level, hoses & clamps	S/I	✓	✓	✓	✓	✓	✓	✓	✓	✓	✓	✓	✓	✓
Throttle linkage	S/I	✓	✓	✓	✓	✓	✓	✓	✓	✓	✓	✓	✓	✓
Brake linings	S/I	✓	✓	✓	✓	✓	✓	✓	✓	✓	✓	✓	✓	✓
Supercharger oil	S/I								✓					
Accessory drive belts	S/I	Every 60,000 miles												
Engine coolant ①	R													
Spark plugs ②	R				✓				✓				✓	
Passenger compartment air filter	R		✓		✓		✓		✓		✓		✓	
Air filter element	S/I			✓					✓					✓
PCV filter	R				✓				✓				✓	
Ignition cables	S/I				✓				✓				✓	
EGR & fuel systems	S/I				✓				✓				✓	
Automatic transaxle fluid & filter	R													✓
Throttle body mount bolt torque	S/I													

R: Replace   S/I: Service or Inspect

① Engine coolant: replace every 100,000 miles. Use O.E. specified (DEX-COOL™) coolant only. If any silicate coolant is used, the service interval is every 30,00

② Platinum tip spark plugs: replace every 100,000 miles.

FREQUENT OPERATION MAINTENANCE (SEVERE SERVICE)

If a vehicle is operated under any of the following conditions it is considered severe service:

- Extremely dusty areas.

- 50% or more of the vehicle operation is in 32°C (90°F) or higher temperatures, or constant operation in temperatures below 0°C (32°F).

- Prolonged idling (vehicle operation in stop and go traffic).

- Frequent short running periods (engine does not warm to normal operating temperatures).

- Police, taxi, delivery usage or trailer towing usage.

CV joints & front suspension components: service or inspect every 3000 miles.

Engine oil & filter change: change every 3000 miles.

Brake linings: check every 6000 miles.

Chassis lubrication: lubricate every 6000 miles.

Suspension, steering linkage, transaxle shift linkage, parking cable guides, underbody contact points: lubricate every 6000 miles.

Throttle body mount bolt torque: tighten at 6000 miles.

Air filter element: service or inspect every 15,000 miles.

Automatic transaxle fluid: change every 50,000 miles.

Inspect throttle body bore & throttle plate for deposits: clean as required every 15,000 miles.

Rotate tires at 6000 miles, then every 15,000 miles.

42372-HBOD-C14

PRECAUTIONS

Before servicing any vehicle, please be sure to read all of the following precautions, which deal with personal safety, prevention of component damage, and important points to take into consideration when servicing a motor vehicle:

• Never open, service or drain the radiator or cooling system when the engine is hot; serious burns can occur from the steam and hot coolant.

• Observe all applicable safety precautions when working around fuel. Whenever servicing the fuel system, always work in a well-ventilated area. Do not allow fuel spray or vapors to come in contact with a spark, open flame, or excessive heat (a hot drop light, for example). Keep a dry chemical fire extinguisher near the work area. Always keep fuel in a container specifically designed for fuel storage; also, always properly seal fuel containers to avoid the possibility of fire or explosion. Refer to the additional fuel system precautions later in this section.

• Fuel injection systems often remain pressurized, even after the engine has been turned **OFF**. The fuel system pressure must be relieved before disconnecting any fuel lines. Failure to do so may result in fire and/or personal injury.

• Brake fluid often contains polyglycol ethers and polyglycols. Avoid contact with the eyes and wash your hands thoroughly after handling brake fluid. If you do get brake fluid in your eyes, flush your eyes with clean, running water for 15 minutes. If eye irritation persists, or if you have taken brake fluid internally, IMMEDIATELY seek medical assistance.

• The EPA warns that prolonged contact with used engine oil may cause a number of skin disorders, including cancer! You should make every effort to minimize your exposure to used engine oil. Protective gloves should be worn when changing oil. Wash your hands and any other exposed skin areas as soon as possible after exposure to used engine oil. Soap and water, or waterless hand cleaner should be used.

• All new vehicles are now equipped with an air bag system, often referred to as a Supplemental Restraint System (SRS) or Supplemental Inflatable Restraint (SIR) system. The system must be disabled before performing service on or around system components, steering column, instrument panel components, wiring and sensors. Failure to follow safety and disabling procedures could result in accidental air bag deployment, possible personal injury and unnecessary system repairs.

• Always wear safety goggles when working with, or around, the air bag system. When carrying a non-deployed air bag, be sure the bag and trim cover are pointed away from your body. When placing a non-deployed air bag on a work surface, always face the bag and trim cover upward, away from the surface. This will reduce the motion of the module if it is accidentally deployed. Refer to the additional air bag system precautions later in this section.

• Clean, high quality brake fluid from a sealed container is essential to the safe and proper operation of the brake system. You should always buy the correct type of brake fluid for your vehicle. If the brake fluid becomes contaminated, completely flush the system with new fluid. Never reuse any brake fluid. Any brake fluid that is removed from the system should be discarded. Also, do not allow any brake fluid to come in contact with a painted surface; it will damage the paint.

• Never operate the engine without the proper amount and type of engine oil; doing so WILL result in severe engine damage.

• Timing belt maintenance is extremely important! Many models utilize an interference-type, non-freewheeling engine. If the timing belt breaks, the valves in the cylinder head may strike the pistons, causing potentially serious (also time-consuming and expensive) engine damage. Refer to the maintenance interval charts in the front of this manual for the recommended replacement interval for the timing belt, and to the timing belt section for belt replacement and inspection.

• Disconnecting the negative battery cable on some vehicles may interfere with the functions of the on-board computer system(s) and may require the computer to undergo a relearning process once the negative battery cable is reconnected.

• When servicing drum brakes, only disassemble and assemble one side at a time, leaving the remaining side intact for reference.

• Only an MVAC-trained, EPA-certified automotive technician should service the air conditioning system or its components.

ENGINE REPAIR

➡Disconnecting the negative battery cable on some vehicles may interfere with the operation of the onboard computer system. The computer may undergo a relearning process once the negative battery cable is reconnected.

Alternator

REMOVAL

1. Before servicing the vehicle, refer to the precautions in the beginning of this section.
2. Remove or disconnect the following:
   • Negative battery cable
   • Supercharger belt, if equipped
   • Accessory drive belt
   • Fuel injector sight shield

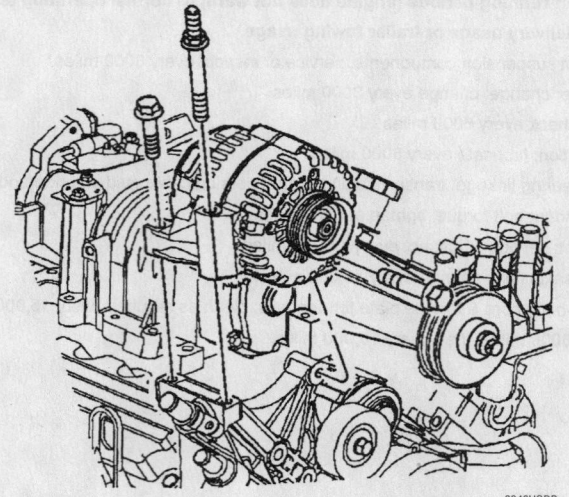

9346UGDB

Exploded view of the alternator mounting—supercharged models

- Alternator brace
- Electrical connections
- Alternator bolts and the alternator

INSTALLATION

1. Install or connect the following:
   - Alternator and torque the bolts to 37 ft. lbs. (50 Nm)
   - Electrical connections and torque the nut to 111 inch lbs. (12.5 Nm)
   - Alternator brace. Torque the nut to 37 ft. lbs. (50 Nm) and the bolt to 22 ft. lbs. (30 Nm)
   - Fuel injector sight shield
   - Accessory drive belt
   - Supercharger belt, if equipped
   - Negative battery cable

Ignition Timing

The 3.8L (VIN K and 1) engines utilize a Distributorless Ignition System (DIS). This system uses 3 twin tower coils which fire 2 spark plugs simultaneously.

ADJUSTMENT

The ignition timing is not adjustable, and is set according to engine demand electronically. The Powertrain Control Module (PCM) controls the ignition timing for all driving conditions.

Engine Assembly

REMOVAL & INSTALLATION

Except Park Avenue And All 2000–04 models

1. Before servicing the vehicle, refer to the precautions in the beginning of this section.
2. Disconnect the negative battery cable.
3. Remove the hood.
4. Relieve the fuel system pressure.
5. Drain the coolant system and crankcase.
6. Remove or disconnect the following:
   - Radiator and heater supply hoses
   - Negative battery cable from the engine block
   - Engine harness connector at the bulkhead
   - Accessory drive belt
   - Supercharger belt, if equipped
   - Power steering pump
   - Air flow duct
   - Throttle cable
   - Cruise control cable, if equipped
   - Mounting bracket

- All other cables from the engine
- Manifold Air Temperature (MAT) sensor connector
- Throttle Position (TPS) sensor connector
- Idle Air Control (IAC) connector
- Oxygen (O$_2$S) sensor connector
- Oil pressure switch connector
- Power Steering cutoff switch connector
- Vehicle Speed Sensor (VSS) connector
- Low oil level sensor connector
- Ignition assembly ground strap from the fender inner panel
- Fuel lines from the fuel rail and the pressure regulator
- Emission control canister hoses from the throttle body
- Brake booster and heater control hoses from the engine vacuum connections
- Front exhaust pipe from the right manifold
- Air conditioning compressor and move aside without disconnecting the lines
- Engine-to-transaxle bracket
- Flexplate cover
- Starter
- Torque converter bolts

➡**Matchmark the flexplate-to-torque converter relationship for reassembly.**

7. Attach a suitable lifting hook and chain to the engine lifting brackets. Raise the engine slightly to take the weight off the engine mounts.
8. Support the transaxle.
9. Remove or disconnect the following:
   - Drive belt
   - Engine torque axis engine mount
   - Transaxle-to-engine bolts
   - Engine from the transaxle
   - Engine from the vehicle

To install:

10. Install or connect the following:
    - Engine. Torque the engine-to-transaxle bolts to 55 ft. lbs. (75 Nm).
    - Torque axis engine mount bolts. Torque them to 52 ft. lbs. (87 Nm).
11. Align the torque converter-to-flexplate matchmark.
    - Torque converter bolts and torque them to 46 ft. lbs. (62 Nm)
    - Starter
    - Flexplate cover
    - Engine-to-transaxle bracket
    - Air conditioning compressor
    - Exhaust pipe to the manifold
    - Brake booster and heater control

hoses to the engine vacuum connections
- Emission control canister hoses to the throttle body
- Fuel lines
- Ignition assembly ground strap to the fender inner panel
- MAT sensor connector
- TPS sensor connector
- IAC valve connector
- O$_2$S sensor connector
- Oil pressure switch connector
- Power steering cutoff switch connector
- VSS sensor connector
- Low oil level sensor connector
- Throttle and cruise control cables
- Mounting bracket
- All other applicable cables
- Air flow duct
- Power steering pump
- Drive belt
- Supercharger belt, if equipped
- Engine harness at the bulkhead
- Negative battery cable to the engine
- Radiator and heater supply hoses
- Hood
- Negative battery cable

12. Refill the crankcase.
13. Refill and bleed the engine cooling system.
14. Start the engine and check for leaks.
15. Road test the vehicle and check operation.
16. Before servicing the vehicle, refer to the precautions in the beginning of this section.
17. Disconnect the negative battery cable.
18. Remove the hood.
19. Relieve the fuel system pressure.
20. Drain the coolant system and crankcase.
21. Remove or disconnect the following:
    - Negative cable
    - Fuel injector sight shield
    - Vacuum brake booster hose from the vacuum connections
    - Fuel feed and return lines from the fuel rail
    - Evaporative emission canister purge valve
    - Cruise control cable from the throttle body bracket and lever
    - Electrical connector from the cruise control module
    - Cruise control module from the mounting studs

➡**Always replace the accelerator control cable with a NEW cable whenever you remove the engine from the vehicle.**

- Accelerator control cable
- Drive belt
- Bolt securing both the battery negative cable and the engine harness ground lead to the engine block

22. Disconnect the wiring harness connectors from the following components:
- A/C compressor clutch
- A/C pressure sensor
- Knock (KS) sensor
- Engine coolant block heater
- Oil level sensor

23. Remove or disconnect the following:
- Wiring harness from the harness clip at the rear of the A/C compressor
- Torque converter cover
- Starter motor
- Bolts securing the flywheel to the torque converter

24. Disconnect then secure the following wiring harness electrical connectors to the cowl panel:
- Knock (KS) sensor number 2 which can be found behind the right exhaust manifold
- Oil pressure sensor
- Vehicle Speed Sensor (VSS)

25. Remove or disconnect the following:
- Bolts securing the transaxle brace to the transaxle
- Nuts attaching the exhaust manifold pipe to the right exhaust manifold
- Exhaust manifold pipe from the right exhaust manifold studs, allowing it to rest on top of the power steering gear heat shield
- Exhaust manifold pipe gasket and discard the gasket
- Right wheelhouse extension
- Font A/C compressor mounting nuts
- Rear A/C compressor mounting bolt
- Compressor off of the mounting studs and rest on top of the engine frame
- Bolt securing the Powertrain Control Module (PCM) ground located at the left front cylinder head

26. Disconnect the wiring harness electrical connectors from the following components on the left side of the engine:
- Fuel injectors
- Ignition harness
- Boost control solenoid (VIN I only)
- Engine Coolant Temperature (ECT) sensor
- Throttle Position (TP) sensor

- Idle Air Control (IAC) valve
- Mass Air Flow (MAF) sensor

27. Disconnect the wiring harness connectors from the following components on the right side of the engine:
- Fuel injectors
- Exhaust Gas Recirculation (EGR) valve
- Manifold Absolute Pressure (MAP) sensor
- Heated Oxygen (O_2S) sensor
- AIR solenoid, if equipped
- Alternator

28. Remove or disconnect the following:
- Alternator
- Air cleaner intake duct

29. Attach an engine support fixture.
- Front power steering pump mounting bolts
- Side power steering pump mounting bolt and piston the power steering pump against the cowl, allowing it to rest on top of the transaxle housing
- Right engine mount bracket
- Right lower engine-to-transaxle mounting bolt
- Coolant and heater hoses

30. Use a block of wood between a floor jack and the transaxle, support the transaxle at the pan.

31. Remove the engine support fixture.

32. Attach a engine lift chain to the engine lift brackets and attach to an engine lift device.

33. Remove all remaining engine-to-transaxle bolts

34. Remove the engine from the vehicle.

To install:

35. Installation is the reverse of removal, please note the following torques:
- 5 upper engine-to-transaxle mounting bolts to 55 ft. lbs. (75 Nm)
- Right lower engine-to-transaxle mounting bolt to 55 ft. lbs. (75 Nm)
- Power steering pump bolts to 20 ft. lbs. (27 Nm)
- Bolt attaching the PCM ground to the left front cylinder head and tighten to 37 ft. lbs. (50 Nm)
- A/C compressor bolts 37 ft. lbs. (50 Nm)
- Transaxle brace bolts to 48 ft. lbs.(65 Nm)
- Flywheel-to-torque converter bolts to 46 ft. lbs. (63 Nm)

36. Refill the crankcase.

37. Refill and bleed the engine cooling system.

38. Start the engine and check for leaks.

39. Road test the vehicle and check operation.

Water Pump

REMOVAL & INSTALLATION

3.8L (VIN K) Engine

1. Before servicing the vehicle, refer to the precautions in the beginning of this section.

2. Drain the cooling system.

3. Remove or disconnect the following:
- Negative battery cable
- Accessory drive belt
- Coolant hoses from the water pump
- Water pump pulley bolts

➡**The long bolt can be removed by aligning the bolt head up with the hole in the frame rail.**

- Pulley
- Water pump bolts
- Water pump

To install:

4. Apply a thin bead of sealer around the outside edge of the water pump.

5. Install or connect the following:
- Water pump with new gasket. Torque the water pump short bolts to 11 ft. lbs. (15 Nm) and the long bolts to 22 ft. lbs. (30 Nm).
- Water pump pulley. Torque the bolts to 115 inch lbs. (13 Nm).
- Coolant hoses to the water pump
- Drive belt
- Supercharger belt, if equipped

6. Refill and bleed the cooling system.

7. Run the engine and check for leaks.

8. Recheck the coolant level when the engine has cooled.

3.8L (VIN 1) Engine

1. Before servicing the vehicle, refer to the precautions in the beginning of this section.

2. Drain the cooling system.

3. Support the engine using an engine support fixture.

4. Remove or disconnect the following:
- Negative battery cable
- Air conditioning compressor splash shield
- Supercharger and accessory drive belts
- Coil pack, if equipped
- Supercharger belt tensioner
- Front engine mount
- Power steering pump
- Engine mount bracket
- Idler pulley
- Water pump pulley
- Water pump

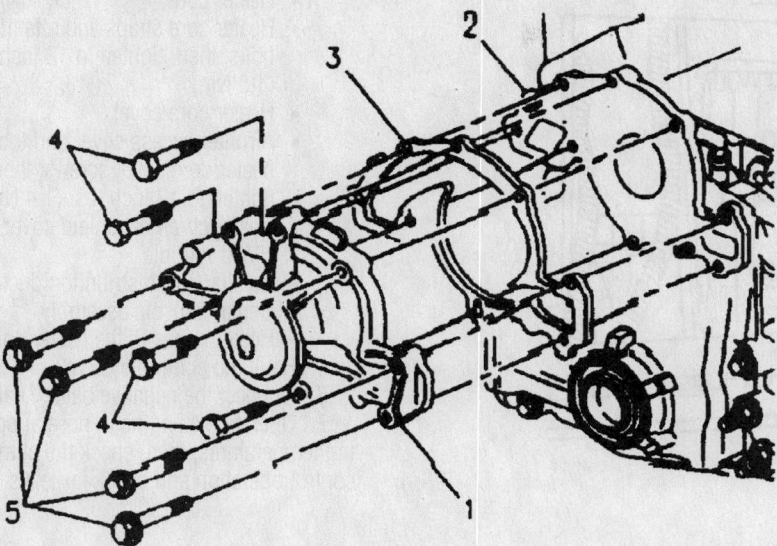

1. Coolant pump
2. Engine front cover
3. Gasket
4. 11 ft. lb.(15 Nm)
5. 22 ft. lb.(30 Nm)

7922UG01

Exploded view of the water pump—3.8L (VIN K and 1) engines

To install:

5. Apply a thin bead of sealer around the outside edge of the water pump.

6. Install or connect the following:

- Water pump with a new gasket. Torque the short bolts to 11 ft. lbs. (15 Nm) and the long bolts to 22 ft. lbs. (30 Nm).
- Water pump pulley and torque the bolts to 115 inch lbs. (13 Nm)
- Engine mount bracket. Torque the nut closest to the a/c compressor to 33 ft. lbs. (45 Nm), and the others to 58 ft. lbs. (78 Nm).
- Idler pulley and torque the bolts to 37 ft. lbs. (50 Nm)
- Power steering pump and torque the bolts to 20 ft. lbs. (27 Nm)
- Front engine mount and torque the nuts to 58 ft. lbs. (78 Nm).
- Supercharger belt tensioner and torque the bolts to 37 ft. lbs. (50 Nm)
- Ignition coil pack assembly, if equipped and torque the nuts to 70 in. lbs. (8 Nm)
- Supercharger drive belt
- Accessory drive belt
- Air conditioning compressor splash shield
- Negative battery cable

7. Refill and bleed the cooling system.
8. Run the engine and check for leaks.
9. Recheck the coolant level when the engine has cooled.

Heater Core

REMOVAL & INSTALLATION

Bonneville and LeSabre

1. Disconnect the negative battery cable.
2. Drain the cooling system into a clean container for reuse.
3. Remove or disconnect the following:

- Heater hoses from the heater core
- Right sound insulator
- All necessary electrical connectors and remove the instrument panel compartment
- Temperature valve actuator
- Electrical connector and remove the HVAC programmer
- Heater core cover-to-heater assembly screws and the cover
- Heater core-to-heater assembly screws and the heater core

To install:

4. Install or connect the following:

- Heater core and the heater core-to-

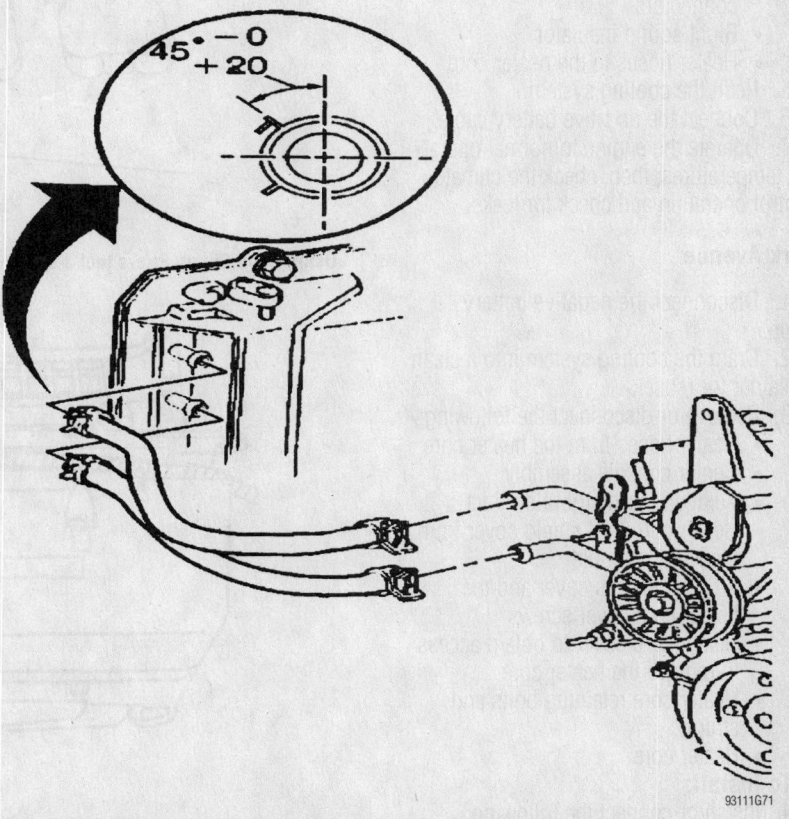

93111G71

View of the heater hoses and clamp positions—Bonneville and LeSabre

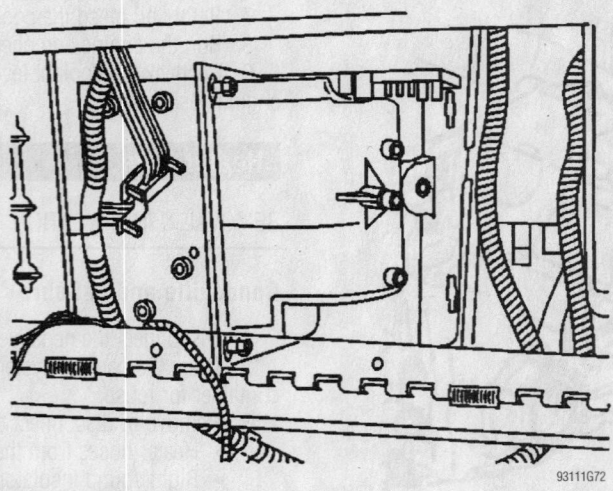

View of the heater core cover—Bonneville and LeSabre

heater assembly screws, then, tighten the screws to 12 inch lbs. (1.4 Nm)
- Heater core cover and the cover-to-heater assembly screws, then, tighten the screws to 12 inch lbs. (1.4 Nm)
- HVAC programmer and connect the electrical connector
- Temperature valve actuator
- Instrument panel compartment and connect any necessary electrical connectors
- Right sound insulator
- Heater hoses to the heater core

5. Refill the cooling system.
6. Connect the negative battery cable.
7. Operate the engine to normal operating temperatures; then, check the climate control operation and check for leaks.

Park Avenue

1. Disconnect the negative battery cable.
2. Drain the cooling system into a clean container for reuse.
3. Remove or disconnect the following:
- Heater hoses from the heater core
- Center console assembly
- Auxiliary air distribution duct
- Heater core heat shield cover from the HVAC module
- Air filter access cover and the heater core cover screws
- Heater core cover to obtain access to remove the heater core
- Heater core retaining bolts and straps
- Heater core

To install:
4. Install or connect the following:

- Heater core
- Heater core straps and retaining bolts, then, tighten to 18 inch lbs. (1.5 Nm)
- Heater core cover
- Air filter access cover and the heater core cover screws, then tighten to 12 inch lbs. (1.4 Nm)
- Heater core heat shield cover to the HVAC module
- Auxiliary air distribution duct
- Center console assembly
- Heater hoses to the heater core

5. Refill the cooling system.
6. Connect the negative battery cable.
7. Operate the engine to normal operating temperatures; then, check the climate control operation and check for leaks.

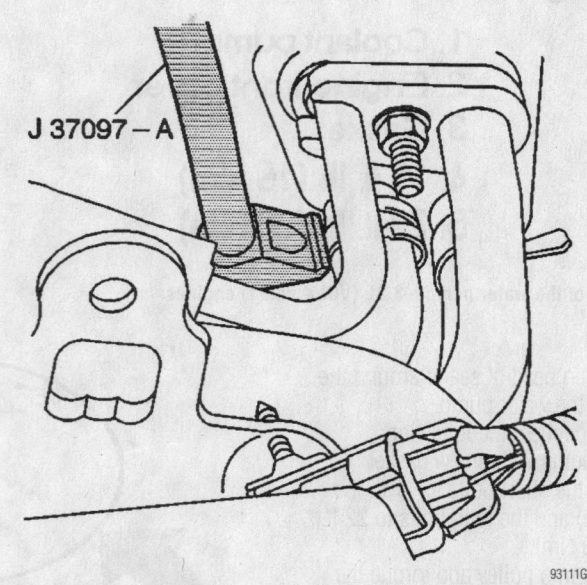

Using Hose Clamp Pliers tool J-37097-A remove the heater hoses clamp—Park Avenue

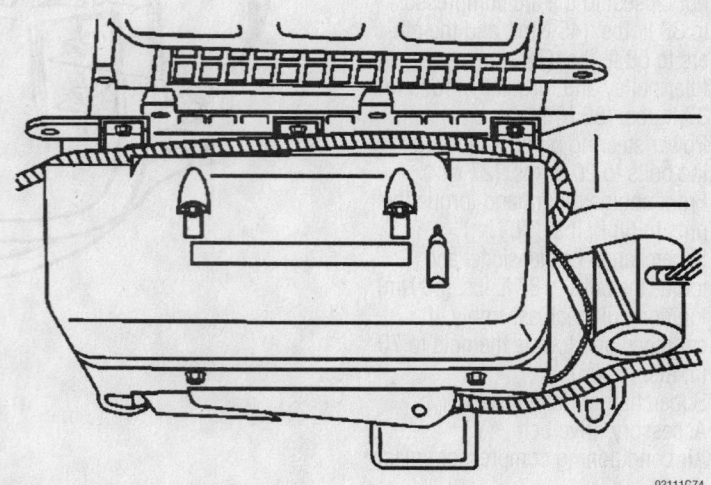

View of the heater core cover—Park Avenue

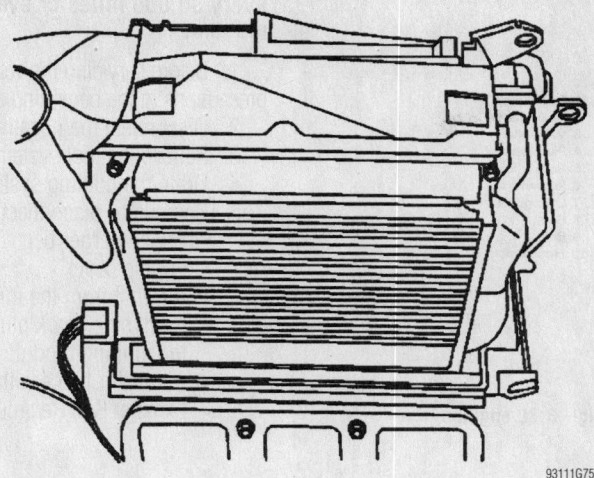

View of the heater core—Park Avenue

LSS and Regency

1. Before servicing the vehicle, refer to the precautions in the beginning of this section.
2. Drain the cooling system.
3. Remove or disconnect the following:
   - Fuel injector sight shield
   - Heater hoses from the heater core
   - Left and right sound insulators
   - Air distributor duct
   - Heater core heat shield
   - Heater core cover
   - Heater core

To install:
4. Install or connect the following:
   - Heater core. Torque the screw to 12 inch lbs. (1.5 Nm).
   - Heater core cover. Torque the screws to 12 inch lbs. (1.5 Nm).
   - Heater core heat shield. Torque the 12 inch lbs. (1.5 Nm).
   - Air distributor duct. Torque the screws to 12 inch lbs. (1.5 Nm).
   - Left and right sound insulators. Torque the fasteners to 18 inch lbs. (2 Nm).
   - Heater hoses
   - Fuel injector sight shield
5. Refill the cooling system.
6. Start the engine and check for leaks.

Cylinder Head

REMOVAL & INSTALLATION

1. Before servicing the vehicle, refer to the precautions in the beginning of this section.
2. Disconnect the negative battery cable.
3. Relieve the fuel system pressure.

4. Drain the cooling system.
5. Remove or disconnect the following:
   - Intake manifold
   - Exhaust manifold
   - Valve covers
   - Ignition wires and ignition coil/module assembly
   - Alternator front mounting bracket and alternator
   - Air conditioning bracket-to-cylinder head bolt
   - Power steering pump
   - Accessory drive belt tensioner
   - Supercharger belt tensioner, if equipped
   - Fuel pipe heat shield
   - Rocker arm assemblies, note their original position
   - Pushrods and guide plate
   - Cylinder head bolts
   - Cylinder head

To install:
6. Place the new cylinder head gasket on the engine block dowels with the note **THIS SIDE UP** facing the cylinder head and the arrow facing the front of the engine.

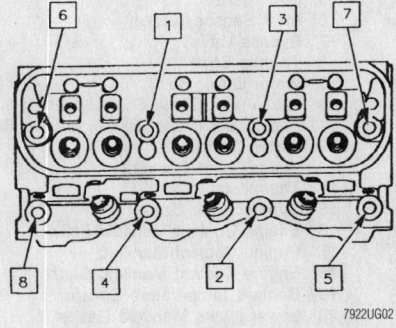

Cylinder head bolt torque sequence—3.8L engine

Position the cylinder head on the engine block.

➡ **The head gasket is identified by either a L or a R stamped on it next to the arrow.**

➡ **This engine uses special torque-to-yield head bolts. The procedure must be followed carefully and new bolts must be used whenever the head is removed. Total bolt torque should not exceed 60 ft. lbs. (81 Nm).**

7. Install new cylinder head bolts and torque them in sequence as follows:
   a. Step 1: 37 ft. lbs. (50 Nm).
   b. Step 2: Plus 120 degrees.
8. Install or connect the following:
   - Pushrods and guide plate
   - Rocker arm assemblies into their original location

➡ **Apply a thread lock compound to the rocker arm pedestal bolts before assembly.**

   - Valve covers
   - Fuel pipe heat shield
   - Accessory drive belt tensioner
   - Supercharger belt tensioner (VIN 1 engine)
   - Power steering pump
   - Air conditioning compressor bracket bolt. Torque it to 52 ft. lbs. (70 Nm).
   - Alternator front mounting bracket, and alternator
   - Ignition coil/module assembly and spark plug wires
   - Exhaust manifold. Torque the bolts to 22 ft. lbs. (30 Nm).
   - Intake manifold
   - Negative battery cable
9. Refill and bleed the cooling system.
10. Start the engine and check for leaks and proper operation.

Rocker Arms

REMOVAL & INSTALLATION

➡ **When removing valvetrain components, it is very important that they are marked for installation reference, so that they can be reinstalled in their original location.**

1. Before servicing the vehicle, refer to the precautions in the beginning of this section.
2. Remove or disconnect the following:
   - Negative battery cable
   - Spark plug wires
   - Rocker arm cover(s)

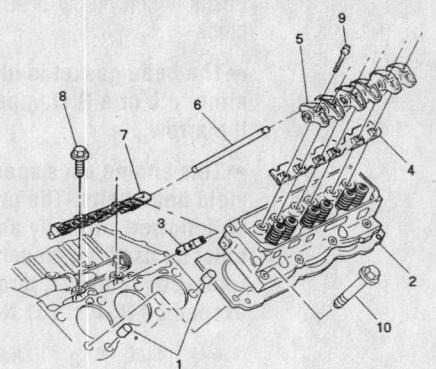

1. Dowel pin
2. Head gasket
3. Valve lifter
4. Pivot retainer
5. Rocker arm
6. Pushrod
7. Lifter guide
8. Bolt
9. Bolt
10. Head bolt

7922UG03

Exploded view of the rocker arms and related components mounting—3.8L engines

- Rocker arm pedestal bolts and assemblies
- Pushrods

To install:

3. Lubricate the pushrod tips and put them in their proper locations.

4. Lubricate the rocker arms and pedestals and install them. Be sure the pushrod tips are properly seated in the rocker arms.

5. Apply thread locking compound to the rocker arm pedestal bolt threads. Torque the bolts to 11 ft. lbs. (15 Nm) plus an additional 90 degree turn.

6. Apply suitable thread locking compound to the rocker arm cover bolts.

7. Install or connect the following:
- Rocker arm cover using a new gasket. Torque the bolts to 89 inch lbs. (10 Nm).
- Spark plug wires
- Accessory drive belt
- Supercharger belt, if equipped
- Negative battery cable

8. Run the engine and check for leaks and proper engine operation.

Supercharger

REMOVAL & INSTALLATION

➡A small amount of oil seepage through the front seal is normal. This seepage is caused by minute traces of oil escaping around the seal due to normal pressure build up in the oil cavity within the supercharger. A build up of dust can stick to the thin oil film, which causes the oil seepage to appear worse than it really is. The supercharger should not be replaced for this seepage. However, if supercharger oil is visually dripping from the supercharger front seal, the supercharger will need to be replaced. The supercharger oil level should be checked

every 30,000 miles or every 36 months.

1. Before servicing the vehicle, refer to the precautions in the beginning of this section.

2. Disconnect the negative battery cable.

3. Relieve the fuel system pressure.

4. Drain the cooling system.

5. Remove or disconnect the following:
- Supercharger belt
- Engine cover
- Air duct from the throttle body
- Right side spark plug wires from the ignition module
- Alternator brace with purge solenoid
- Exhaust Gas Recirculation (EGR)

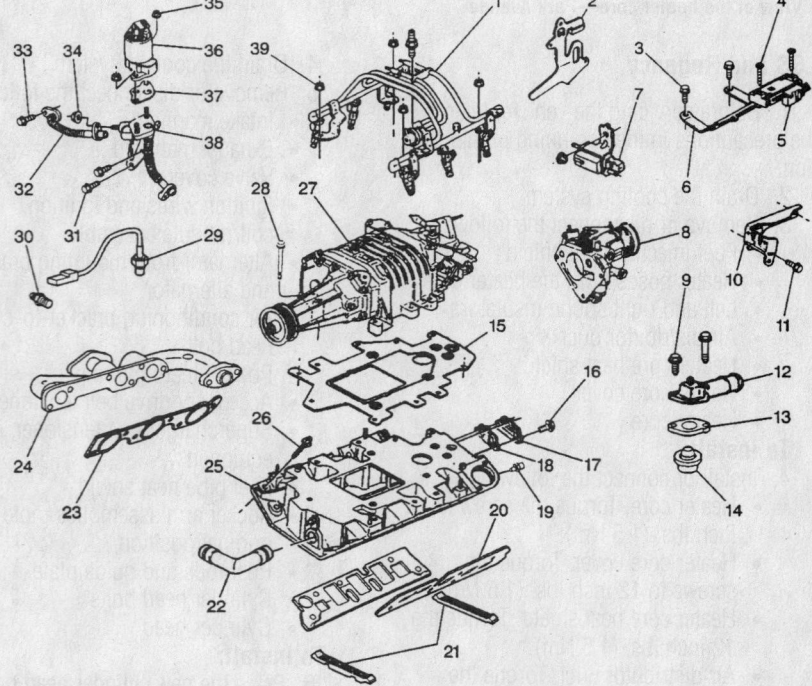

1	Fuel Injection Rail	
2	Fuel Injector Sight Shield Bracket	
3	MAP Sensor Bracket Bolt	
4	MAP Sensor Bolt	
5	MAP Sensor	
6	MAP Sensor Bracket	
7	Bypass Valve	
8	Throttle Body	
9	Water Outlet Bolt	
10	Accelerator Cable Control Bracket	
11	Accelerator Cable Control Bracket Bolt	
12	Water Outlet	
13	Water Outlet Gasket	
14	Thermostat	
15	Supercharger Gasket	
16	Engine Coolant Manifold Bolt	
17	Engine Coolant Manifold	
18	Engine Coolant Manifold Gasket	
19	Coolant Temperature Sensor	
20	Lower Intake Manifold Gasket	
21	Lower Intake Manifold Seal	
22	Heater Inlet Pipe With Seal	
23	Exhaust Manifold Gasket	
24	Exhaust Manifold (Right)	
25	Lower Intake Manifold	
26	Lower Intake Manifold Bolt	
27	Supercharger	
28	Supercharger Bolt	
29	Heated Oxygen Sensor	
30	Exhaust Manifold Bolt/Stud (Right)	
31	EGR Valve Adapter Bolt	
32	EGR Valve Outlet Pipe	
33	EGR Valve Outlet Pipe Bolt	
34	EGR Valve Outlet Pipe Nut	
35	EGR Valve Nut	
36	EGR Valve	
37	EGR Valve Gasket	
38	EGR Valve Adapter	
39	Fuel Injection Rail Nut	

9300UG01

Exploded view of the supercharger and related components—3.8L (VIN 1) engine

wiring harness and shield, if necessary
- Manifold Absolute Pressure (MAP) sensor bracket
- Fuel lines
- Fuel injector connectors
- Fuel rail with injectors
- Booster bypass solenoid
- Regulator valve and harness bolt
- Boost control solenoid
- Throttle body bolts and the assembly
- Supercharger bolts and the assembly

To install:

6. Install or connect the following:
- Supercharger with new gasket and torque the bolts, gradually and evenly, to 17 ft. lbs. (23 Nm).
- MAP sensor bracket
- Throttle body and torque the nuts to 89 inch lbs. (10 Nm)
- Boost control solenoid and torque the nut to 72 inch lbs. (8 Nm)
- Regulator valve and harness bolt
- Booster bypass solenoid
- Fuel rail with injectors
- Fuel lines
- Electrical connectors to the fuel injectors
- Alternator brace with purge solenoid
- EGR wiring harness and shield
- Right side spark plug wires
- Supercharger drive belt
- Air duct
- Engine cover
- Negative battery cable

7. Refill and bleed the cooling system.

8. Run the engine and check for leaks and proper engine operation.

Intake Manifold

REMOVAL & INSTALLATION

3.8L (VIN K) Engine

1. Before servicing the vehicle, refer to the precautions in the beginning of this section.

2. Disconnect the negative battery cable.

3. Drain the cooling system.

4. Relieve the fuel system pressure.

5. Remove or disconnect the following:
- Fuel injector sight shield
- Air inlet duct
- Spark plug wires from the right side
- Manifold Absolute Pressure (MAP) sensor
- Vacuum lines from the intake manifold

- Fuel lines
- Fuel injector electrical connectors
- Fuel regulator vacuum line
- Fuel rail from the intake manifold
- Exhaust Gas Recirculation (EGR) heat shield
- Throttle cable bracket from the cylinder head mounting bracket and the throttle body cables
- Throttle body support bracket
- Upper intake plenum and gasket
- Thermostat housing
- Electrical connector from the Engine Coolant Temperature (ECT) sensor

- Drive belt tensioner assembly
- EGR valve outlet pipe
- Lower intake manifold

To install:

6. Install or connect the following:
- Intake manifold using new manifold gaskets. Torque the bolts in sequence to 11 ft. lbs. (15 Nm); then, re-torque to 11 ft. lbs. (15 Nm).
- EGR valve outlet pipe
- Drive belt tensioner assembly. Torque the tensioner bolts to 37 ft. lbs. (50 Nm).
- Electrical connector to the ECT sensor

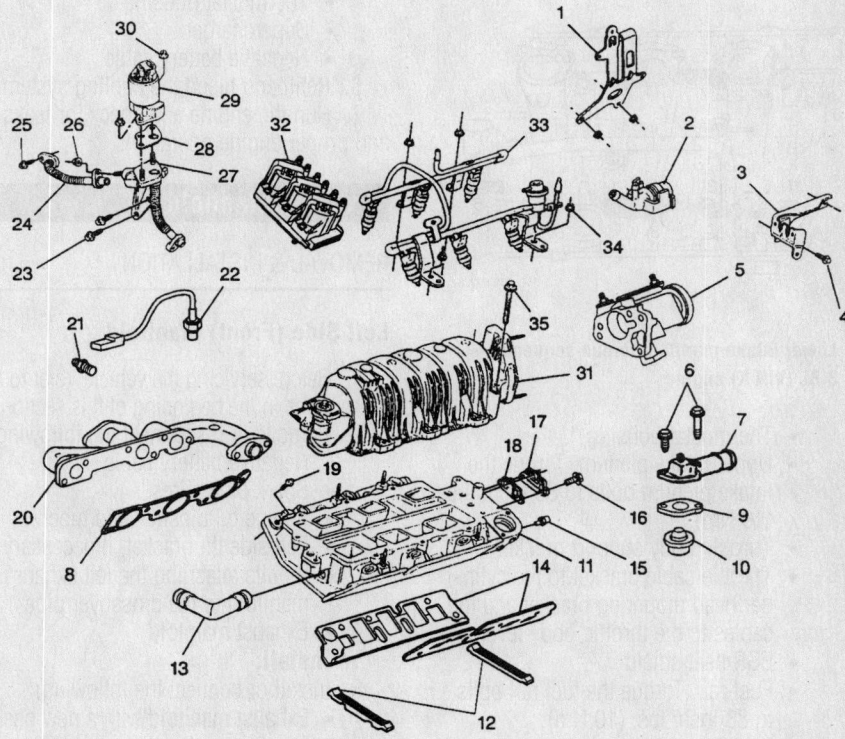

#		#	
1	Fuel Injector Sight Shield Bracket	19	Lower Intake Manifold Bolt
2	Vacuum Source Manifold	20	Exhaust Manifold (Right)
3	Accelerator Cable Control Bracket	21	Exhaust Manifold Bolt/Stud
4	Throttle Body Support Bolt	22	Exhaust Oxygen Sensor
5	Throttle Body	23	EGR Valve Adapter Bolt
6	Water Outlet Bolt	24	EGR Valve Outlet Pipe
7	Water Outlet	25	EGR Valve Outlet Pipe Bolt
8	Exhaust Manifold Gasket	26	EGR Valve Outlet Pipe Nut
9	Water Outlet Gasket	27	EGR Valve Adapter
10	Thermostat	28	EGR Valve Gasket
11	Lower Intake Manifold	29	EGR Valve
12	Intake Manifold Seal	30	EGR Valve Nut
13	Heater Water Inlet Pipe	31	Upper Intake Manifold
14	Lower Intake Manifold Gasket	32	ICM
15	Coolant Temperature Sensor	33	Fuel Injection Rail
16	Engine Coolant Manifold Bolt	34	Fuel Injector Rail Nut
17	Engine Coolant Manifold	35	Upper Intake Manifold Bolt
18	Engine Coolant Manifold Gasket		

9300UG02

Exploded view of the intake manifold and related components—3.8L (VIN K) engine

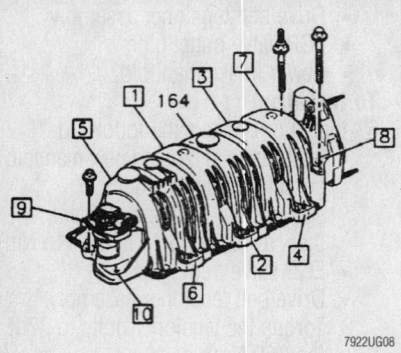

Upper intake manifold torque sequence—3.8L (VIN K) engine

7922UG08

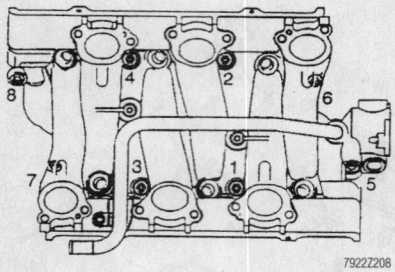

Lower intake manifold torque sequence—3.8L (VIN K) engine

7922Z208

- Thermostat housing
- Upper intake plenum. Torque the intake plenum bolts to 88 inch. lbs. (10 Nm).
- Throttle body support bracket
- Throttle cable bracket to the cylinder head mounting bracket and the cables to the throttle body lever
- EGR heat shield
- Fuel rail. Torque the fuel rail bolts to 88 inch. lbs. (10 Nm).
- Fuel lines
- Fuel regulator vacuum line
- Fuel injector electrical connectors
- Vacuum lines to the intake manifold
- MAP sensor
- Spark plug wires
- Fuel injector sight shield and air inlet duct
- Negative battery cable

7. Refill and bleed the cooling system.
8. Run the engine and check for leaks and proper engine operation.

3.8L (VIN 1) Engine

1. Before servicing the vehicle, refer to the precautions in the beginning of this section.
2. Relieve the fuel system pressure.
3. Drain the cooling system.

4. Remove or disconnect the following:
- Negative battery cable
- Supercharger
- Thermostat housing
- Exhaust Gas Recirculation (EGR) tube at the intake manifold
- Engine Control Temperature (ECT) sensor
- Intake manifold

To install:

5. Install or connect the following:
- Intake manifold with new gaskets. Torque the bolts, working from the center out, to 11 ft. lbs. (15 Nm).
- ECT sensor connector
- EGR tube to the intake manifold
- Thermostat housing
- Supercharger
- Negative battery cable

6. Refill and bleed the cooling system.
7. Run the engine and check for leaks and proper engine operation.

Exhaust Manifold

REMOVAL & INSTALLATION

Left Side (Front) Manifold

1. Before servicing the vehicle, refer to the precautions in the beginning of this section.
2. Remove or disconnect the following:
- Negative battery cable
- Spark plug wires
- Engine oil dipstick and tube
- Left side lift bracket, if necessary
- 2 bolts attaching the left exhaust manifold to the crossover pipe
- Exhaust manifold

To install:

3. Install or connect the following:
- Exhaust manifold with a new gasket. Torque the studs and bolts gradually and evenly to 22 ft. lbs. (30 Nm).

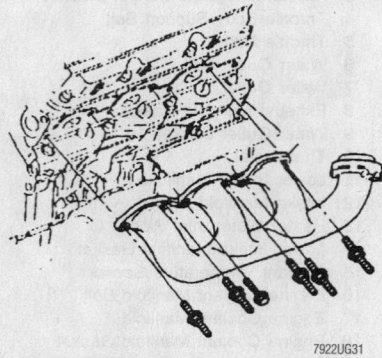

Exploded view of the left exhaust manifold mounting

7922UG31

- 2 bolts attaching the left exhaust manifold to the crossover pipe. Torque the bolts to 15 ft. lbs. (20 Nm).
- Left side lift bracket, if removed
- Engine oil dipstick and tube. Torque the bolts to 15 ft. lbs. (20 Nm).
- Spark plug wires
- Negative battery cable

4. Run the engine and check for exhaust leaks.

Right Side (Rear) Manifold

MODELS WITHOUT NC8

1. Before servicing the vehicle, refer to the precautions in the beginning of this section.
2. Remove or disconnect the following:
- Negative battery cable
- Fuel injector sight shield
- Air cleaner assembly
- Spark plug wires
- Brake booster heat shield
- Crossover pipe
- Engine harness from the right hand engine lift hook bracket
- Transaxle fluid dipstick and tube
- Oxygen (O_2S) sensor
- Exhaust Gas Recirculation (EGR) feed pipe bolt from the manifold
- Transmission oil level tube and seal
- Exhaust manifold flange nuts
- Front exhaust pipe
- Engine lift bracket
- Exhaust manifold

To install:

3. Install or connect the following:
- Manifold to the cylinder head and crossover pipe using new gaskets
- Manifold mounting studs. Torque the studs and bolts to 22 ft. lbs. (30 Nm), beginning at the center and working outwards.
- Engine lift bracket
- Front exhaust pipe
- Front exhaust pipe-to-manifold nuts. Torque the nuts to 22 ft. lbs. (30 Nm).
- Transmission dipstick tube seal and the tube
- EGR feed pipe to the manifold
- O_2S sensor
- Spark plug wires to the spark plugs
- Engine harness to the right hand engine lift hook bracket
- Crossover pipe
- Brake booster heat shield
- Air cleaner assembly
- Fuel injector sight shield
- Negative battery cable

4. Run the engine and check for exhaust leaks.

MODELS WITH NC8

1. Before servicing the vehicle, refer to the precautions in the beginning of this section.
2. Remove or disconnect the following:
   - Negative battery cable
   - Air cleaner assembly
   - Fuel injector sight shield
   - Heated Oxygen (HO2S) sensor electrical connection
   - Spark plug wires
   - Connector from the secondary AIR valve
   - Hose from the secondary AIR valve
   - Bolts attaching the secondary AIR tube to the exhaust manifold
   - Secondary AIR tube and the gasket from the exhaust manifold and discard the gasket
   - Bolt attaching the secondary AIR valve to the fuel injector sight shield bracket
   - Secondary AIR valve from the fuel injector sight shield bracket
   - Brake booster heat shield
   - Crossover pipe
   - Transaxle fluid dipstick and tube
   - Exhaust manifold pipe
   - Right exhaust manifold studs
   - Fuel injector sight shield bracket retaining nuts and the bracket
   - Right engine lift bracket retaining fasteners and the bracket
   - Bolt attaching the Exhaust Gas Recirculation (EGR) inlet pipe to the right exhaust manifold
   - Right exhaust manifold retaining fasteners
   - Right exhaust manifold and the gasket from the engine and discard the gasket
   - Exhaust crossover seal from the exhaust crossover and discard the seal

To install:
3. Install or connect the following:
   - New exhaust crossover seal
   - New manifold gasket
   - Manifold to the cylinder head and crossover pipe
   - Manifold mounting studs. Torque the studs and bolts to 22 ft. lbs. (30 Nm), beginning at the center and working outwards.
   - Bolt attaching the Exhaust Gas Recirculation (EGR) inlet pipe to the right exhaust manifold and tighten to 22 ft. lbs. (30 Nm)
   - Engine lift bracket

- Fuel injector sight shield bracket and tighten the retainers to 22 ft. lbs. (30 Nm)
- Right exhaust manifold studs and tighten to 80 inch lbs. (9 Nm)
- HO2S sensor, if removed
- Exhaust manifold pipe
- Transmission dipstick tube
- Exhaust crossover to the right exhaust manifold bolts and tighten to 15 ft. lbs. (20 Nm)
- Brake booster heat shield
- Air cleaner assembly
- Secondary AIR valve to the fuel injector sight shield bracket and tighten the bolt to 18 ft. lbs. (25 Nm)
- New gasket on the secondary AIR tube
- Bolts attaching the secondary AIR tube to the exhaust manifold and tighten to 89 inch lbs. (10 Nm)
- Hose to the secondary AIR valve
- Connector to the secondary AIR valve
- Spark plug wires to the spark plugs
- HO2S sensor connection
- Fuel injector sight shield
- Negative battery cable

4. Run the engine and check for exhaust leaks.

Camshaft and Valve Lifters

REMOVAL & INSTALLATION

1. Before servicing the vehicle, refer to the precautions in the beginning of this section.
2. Relieve the fuel system pressure.
3. Remove the engine and mount it on an engine stand.
4. Remove or disconnect the following:
   - Negative battery cable
   - Supercharger, if equipped
   - Intake manifold

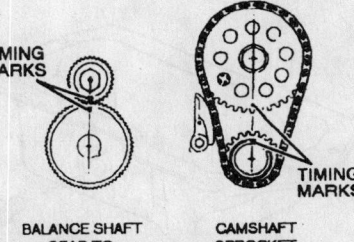

TIMING
MARKS

BALANCE SHAFT
GEAR TO
BALANCE SHAFT
DRIVE GEAR

CAMSHAFT
SPROCKET
TO CRANKSHAFT
SPROCKET

TIMING
MARKS

7922UG09

The timing marks should face each other when the chain and gears are installed properly

- Rocker arm covers
- Rocker arm assemblies
- Pushrods
- Lifters and guides

➡ **A magnet may be helpful when pulling the lifters out of their bores. Identify all parts as they are removed, so they can be reinstalled in their original locations.**

- Crankshaft balancer
- Timing chain front cover

5. Set the engine to Top Dead Center (TDC) No. 1 cylinder (firing position) to align the timing marks, before disassembling the timing chain and sprockets.

✳✳ WARNING

Align the timing marks of the camshaft and crankshaft sprockets to avoid burring the camshaft journals by the crankshaft.

6. Remove or disconnect the following:
   - Camshaft sprocket and timing chain
   - Camshaft thrust plate
   - Camshaft

✳✳ WARNING

If the camshaft was replaced the lifters must also be replaced. The old lifters have developed a wear pattern and will cause the new camshaft to wear prematurely.

To install:
7. Coat the camshaft lobes and bearings with camshaft break-in prelube prior to installation.
8. Install or connect the following:
   - Camshaft
   - Camshaft thrust plate. Torque the bolts to 10 ft. lbs. (14 Nm).
   - Camshaft sprocket and timing chain with timing marks aligned. Torque the camshaft sprocket bolt to 74 ft. lbs. (100 Nm) plus an additional 90 degree (¼) turn.
   - Timing chain front cover
   - Crankshaft balancer. Torque the mounting bolt to 111 ft. lbs. (150 Nm). plus an additional 76 degree turn.
9. Coat the valve lifters with camshaft break-in prelube.
10. Install or connect the following:
   - Valve lifters
   - Lifter guides and lifter guide retainer. Torque the retainer mounting bolts to 22 ft. lbs. (30 Nm).

- Pushrods and rocker arms. Torque the rocker arm bolts to 11 ft. lbs. (15 Nm) plus an additional 90 degree turn.
- Rocker arm covers
- Intake manifold
- Supercharger, if equipped
- Engine
- Negative battery cable

11. Verify that all fluid levels are full and correct.

12. Start the engine and check for leaks. Check engine operation.

Valve Lash

ADJUSTMENT

The valve clearance cannot be adjusted on these engines. The engine is equipped with hydraulic lifters, and adjustment is not necessary.

Starter Motor

REMOVAL & INSTALLATION

1. Before servicing the vehicle, refer to the precautions in the beginning of this section.

2. Remove or disconnect the following:

- Negative battery cable
- Flexplate inspection cover
- Splash shield, if equipped
- Electrical connectors
- Transmission cooler line clip from the transmission, if necessary
- Starter motor wiring
- Starter motor bolts
- Starter

To install:

3. Install or connect the following:

- Starter and torque the bolts to 32 ft. lbs. (43 Nm)

- Wiring and torque the "B" terminal nut to 89 inch lbs. (10 Nm) and the "S" terminal nut to 22 inch lbs. (3 Nm).
- Flexplate inspection cover and torque the bolts to 62 inch lbs. (7 Nm)
- Splash shield
- Negative battery cable

Oil Pan

REMOVAL & INSTALLATION

☀ WARNING

The oil level sensor, located in the oil pan, must be removed prior to removal of the oil pan. If the oil pan is removed first, damage to the oil level sensor may occur.

1. Before servicing the vehicle, refer to the precautions in the beginning of this section.

2. Drain oil into an approved container.

3. Remove or disconnect the following:

- Negative battery cable
- Right engine mount bracket, if necessary
- Flexplate cover
- Oil level sensor
- Oil filter
- Torque axis mount bracket bolts, if necessary
- Oil pan bolts
- Oil pan
- Oil pan gasket

4. Clean the oil pan and cylinder block mating surfaces.

To install:

5. Install or connect the following:

- Oil pan with a new gasket and torque the bolts to 125 inch lbs. (14 Nm)
- Torque axis mount bracket bolts, if removed

- Oil filter
- Flexplate cover
- Oil level sensor
- Oil drain plug and torque the plug to 30 ft. lbs. (40 Nm)
- Right engine mount bracket, if necessary
- Negative battery cable

6. Refill the crankcase.

7. Run the engine and check for leaks.

Oil Pump

REMOVAL & INSTALLATION

1. Before servicing the vehicle, refer to the precautions in the beginning of this section.

2. Support the engine using an engine support fixture.

3. Remove or disconnect the following:

- Negative battery cable
- Engine drive belts and tensioner assembly
- Drive belt idler pulley and bracket

4. Remove or disconnect the following:

- Torque axis mount bracket, if necessary
- Engine front cover assembly
- Oil filter adapter with pressure regulator valve and spring
- Oil pump cover
- Inner and outer pump gears

To install:

5. Lubricate the oil pump gears with petroleum jelly.

6. Install the gears into the oil pump housing.

7. Pack the gear cavity with petroleum jelly after the gears have been installed in the housing.

8. Install or connect the following:

- Oil pump cover. Torque the screws to 97 inch lbs. (11 Nm).

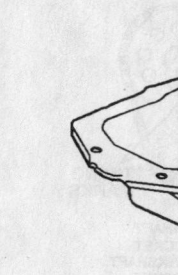

Starter in place with wiring

If equipped, be sure to remove the oil level sensor before removing the pan

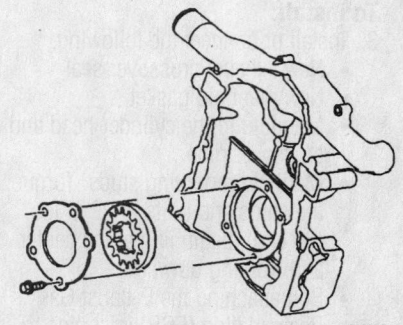

The oil pump is located inside the front engine cover—3.8L (VIN K and 1) engines

- Oil filter adapter with new gasket, pressure regulator valve and spring. Torque the bolts to 11 ft. lbs. (15 Nm).
- Front cover assembly
- Tensioner assembly
- Drive belt idler pulley and bracket, if removed
- Drive belts
- Torque axis mount bracket
- Negative battery cable

9. Remove the engine support fixture.

10. Verify the correct engine oil level.

11. Start the vehicle and verify no leaks and proper oil pressure.

Rear Main Seal

REMOVAL & INSTALLATION

1. Before servicing the vehicle, refer to the precautions in the beginning of this section.

2. Remove or disconnect the following:

- Transaxle assembly
- Flexplate from the crankshaft
- Rear main seal from engine block by inserting a small flat-bladed prytool through the dust lip at an angle, then pry out the crankshaft rear oil seal. Repeat as necessary around the seal until it is removed.

✳✳ WARNING

Do not damage or scratch the sealing surface of the crankshaft or the seal bore.

To install:

3. Lubricate new rear main with clean engine oil prior to installation.

4. Slide the oil seal on the mandrel of seal installer tool J-38196 until the back of the seal is seated squarely against the collar of the tool.

5. Attach the seal installer to the rear of the crankshaft with the 2 mounting bolts, then turn the T-handle until the oil seal is fully seated into the rear of the engine.

6. Loosen the T-handle of the tool completely.

7. Remove both bolts and the tool.

8. Install or connect the following:

- Flexplate. Torque the bolts to 11 ft. lbs. (15 Nm), plus an additional 50 degrees.
- Transaxle

Timing Chain, Sprockets, Front Cover and Seal

REMOVAL & INSTALLATION

1. Before servicing the vehicle, refer to the precautions in the beginning of this section.

2. Drain the cooling system.

3. Support the engine.

4. Remove or disconnect the following:

- Negative battery cable
- Torque axis mount and bracket
- Drive belt
- Supercharger belt, if equipped

- Drive belt idler pulley and bracket (VIN 1 engine)
- Drive belt tensioner (for VIN K engine)
- Crankshaft balancer
- Crankshaft Position (CKP) sensor shield and the CKP sensor
- Oil pan-to-front cover bolts
- Timing chain front cover

5. Align the timing marks on the camshaft and crankshaft sprockets so they are as close together as possible.

- Timing chain damper
- Camshaft sprocket bolt, the camshaft sprocket and timing chain
- Crankshaft sprocket

✳✳ WARNING

Do not rotate the camshaft or crankshaft while the timing chain and sprockets are removed.

To install:

6. Install or connect the following:

- Timing chain and sprockets with the timing marks aligned
- Camshaft sprocket bolt. Torque the bolt to 74 ft. lbs. (100 Nm) plus an additional 90 degree turn.
- Timing chain damper. Torque the bolts to 16 ft. lbs. (22 Nm).

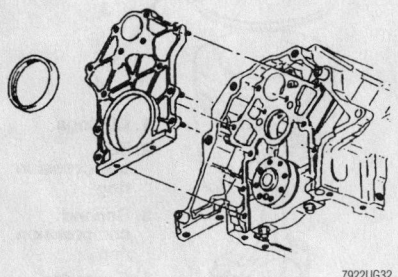

7922UG32

Rear main oil seal and rear cover

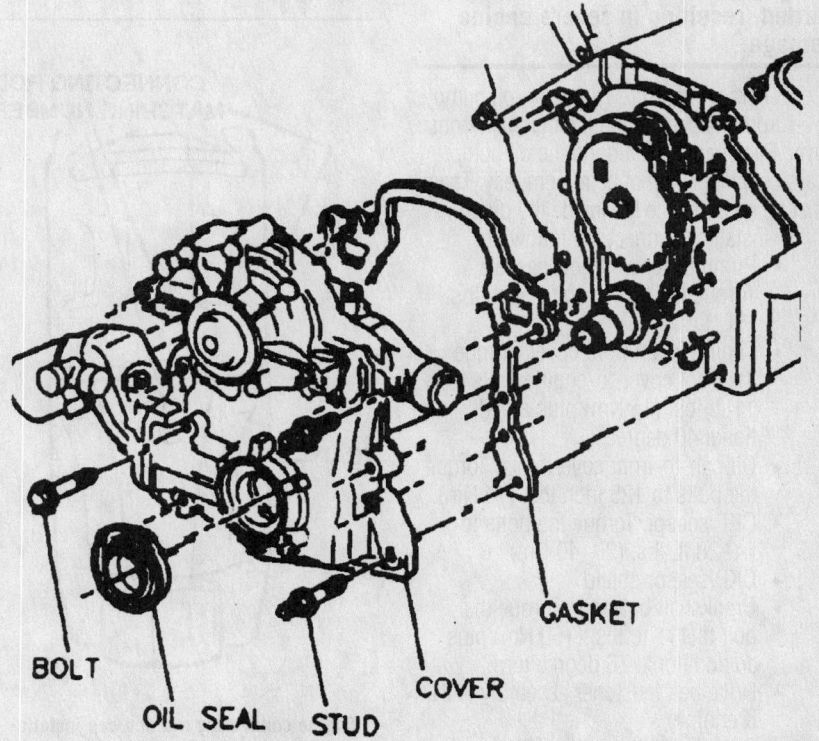

BOLT

OIL SEAL   STUD

COVER

GASKET

7922UG12

Timing chain front cover—3.8L (VIN K and 1) engines

TIMING MARKS

BALANCE SHAFT
GEAR TO
BALANCE SHAFT
DRIVE GEAR

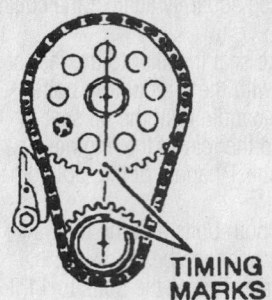

CAMSHAFT
SPROCKET
TO CRANKSHAFT
SPROCKET

TIMING MARKS

7922UG09

Timing chain sprocket and balance shaft gear alignment—3.8L (VIN K and 1) engines

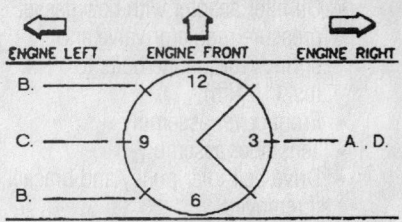

ENGINE LEFT   ENGINE FRONT   ENGINE RIGHT

A. OIL RING SPACER GAP
   (TANG IN HOLE OR SLOT WITH ARC)
B. OIL RING RAIL GAPS
C. 2ND COMPRESSION RING GAP
D. TOP COMPRESSION RING GAP

7922AG46

Piston ring end–gap spacing—3.8L engines

✳✳ WARNING

The oil pump is built into the front cover. When the cover is removed, oil drains from the pump. Since the pump "loses its prime" it may not establish oil pressure as soon as the engine starts. Therefore, it is important to remove the oil pump cover from the back of the timing chain front cover and pack the space around the oil pump gears completely full of petroleum jelly. If this is not done, the oil pump may not pump engine oil when the engine is started, resulting in severe engine damage.

7. Remove the screws and the oil pump cover from the back of the timing chain front cover. Pack the space around the oil pump gears completely full of petroleum jelly. There must be no air space left inside the pump.

8. Install or connect the following:
- Pump cover with new gaskets. Torque the screws to 97 inch lbs. (11 Nm).
- Timing chain front cover. Torque the front cover-to-engine bolts to 11 ft. lbs. (15 Nm) plus an additional 40 degrees.
- Oil pan-to-front cover bolts. Torque the bolts to 125 inch lbs. (14 Nm).
- CKP sensor. Torque the bolts to 14–28 ft. lbs. (20–40 Nm).
- CKP sensor shield
- Crankshaft balancer. Torque the bolt to 111 ft. lbs. (150 Nm) plus an additional 76 degree turn.
- Drive belt tensioner assembly (VIN K engine)
- Drive belt idler pulley (VIN 1 engine)

- Right inner fender access panel and the right front wheel
- Drive belt(s)
- Engine mount
- Coolant hoses
- Negative battery cable

9. Remove the engine support fixture.
10. Refill and bleed the cooling system.
11. Start the vehicle and check for leaks and proper engine operation.

Piston and Ring

POSITIONING

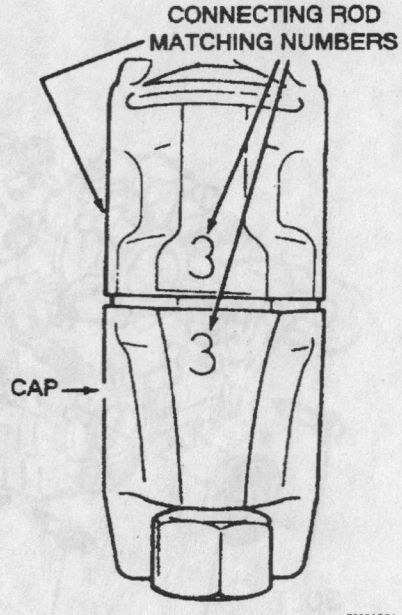

CONNECTING ROD
MATCHING NUMBERS

CAP →

7922AG51

Engine connecting rod and cap installation. Be sure to matchmark the cap and rod prior to disassembly, as shown.

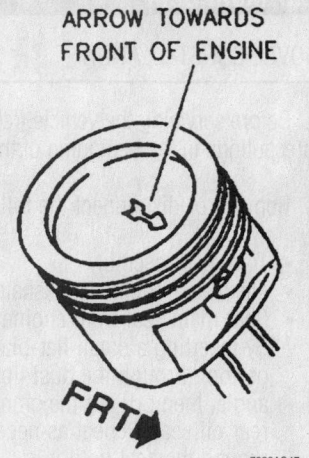

ARROW TOWARDS
FRONT OF ENGINE

FRT

7922AG47

Piston positioning. Often the arrow is replaced by a notch, which also must face toward the front of the engine—3.8L engines

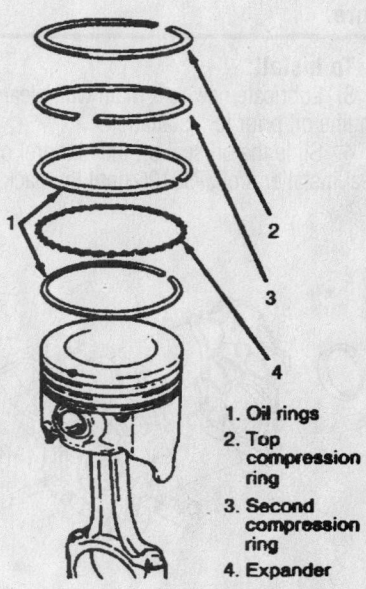

1. Oil rings
2. Top compression ring
3. Second compression ring
4. Expander

7922AG48

Piston ring positioning—3.8L engines

FUEL SYSTEM

Fuel System Service Precautions

Safety is the most important factor when performing not only fuel system maintenance but any type of maintenance. Failure to conduct maintenance and repairs in a safe manner may result in serious personal injury or death. Maintenance and testing of the vehicle's fuel system components can be accomplished safely and effectively by adhering to the following rules and guidelines.

• To avoid the possibility of fire and personal injury, always disconnect the negative battery cable unless the repair or test procedure requires that battery voltage be applied.

• Always relieve the fuel system pressure prior to disconnecting any fuel system component (injector, fuel rail, pressure regulator, etc.), fitting or fuel line connection. Exercise extreme caution whenever relieving fuel system pressure, to avoid exposing skin, face and eyes to fuel spray. Please be advised that fuel under pressure may penetrate the skin or any part of the body that it contacts.

• Always place a shop towel or cloth around the fitting or connection prior to loosening to absorb any excess fuel due to spillage. Ensure that all fuel spillage (should it occur) is quickly removed from engine surfaces. Ensure that all fuel soaked cloths or towels are deposited into a suitable waste container.

• Always keep a dry chemical (Class B) fire extinguisher near the work area.

• Do not allow fuel spray or fuel vapors to come into contact with a spark or open flame.

• Always use a back-up wrench when loosening and tightening the fuel line connection fittings. This will prevent unnecessary stress and torsion to fuel line piping.

• Always replace worn fuel fitting O-rings with new. Do not substitute fuel hose or equivalent, where fuel pipe is installed.

Fuel System Pressure

RELIEVING

1. Disconnect the negative battery cable to avoid possible fuel discharge if an accidental attempt is made to start the engine.
2. Remove the fuel tank cap to relieve tank pressure. Do not tighten until the service procedure has been completed.
3. Connect a fuel pressure gauge with bleed valve to the fuel pressure test port. Wrap a shop towel around the fitting while connecting the gauge to catch any spilled fuel.
4. Install the bleed hose into an approved container and open the valve to bleed off the fuel system pressure.
5. Drain any fuel remaining in the gauge into an approved container.

✳✳ CAUTION

There may still be residual fuel in the system, and a small amount of fuel may be released when servicing fuel lines or connections. In order to reduce the chance of personal injury, cover the fuel line fittings with a shop towel before disconnecting to catch any fuel that may leak out.

Fuel Filter

REMOVAL & INSTALLATION

1. Before servicing the vehicle, refer to the precautions in the beginning of this section.
2. Disconnect the negative battery cable.
3. Relieve the fuel system pressure.
4. Twist the quick connector ¼ turn in each direction to loosen any dirt that may have accumulated in the connector.
5. Use compressed air to remove any dirt in the connector.
6. Squeeze the plastic tabs of the male connector and pull apart.
7. Remove threaded connection from the filter inlet.
8. Remove the filter.

To install:

9. Position the fuel filter, making sure it is facing in the proper direction.
10. Attach the outlet quick connect line to the fuel filter as follows:

    a. Step 1: Be sure the connector is clean and that a new plastic retainer is used on the filter.

    b. Step 2: Apply a couple of drops of engine oil to the male pipe end of filter.

    c. Step 3: Push the fuel line onto the fuel filter until the plastic retainer snaps into place.

    d. Step 4: Check that the connector is locked into place by trying to pull the connector from the filter.

11. Install the threaded connection to the inlet side of filter and tighten to 22 ft. lbs. (30 Nm).
12. Connect the negative battery cable.
13. Pressurize the fuel system by turning the ignition switch to the **ON** position for 2 seconds. Turn OFF the ignition switch for 10 seconds. Turn ON the ignition switch. Check for fuel leaks.

Fuel Pump

REMOVAL & INSTALLATION

1. Before servicing the vehicle, refer to the precautions in the beginning of this section.
2. Relieve the fuel system pressure.
3. Drain the fuel tank.
4. Remove or disconnect the following:

• Negative battery cable
• Spare tire and jack
• Trunk lining
• Fuel sender access panel
• Sender and quick connect fittings from the sender
• Electrical connector from the sender and position harness and hoses aside

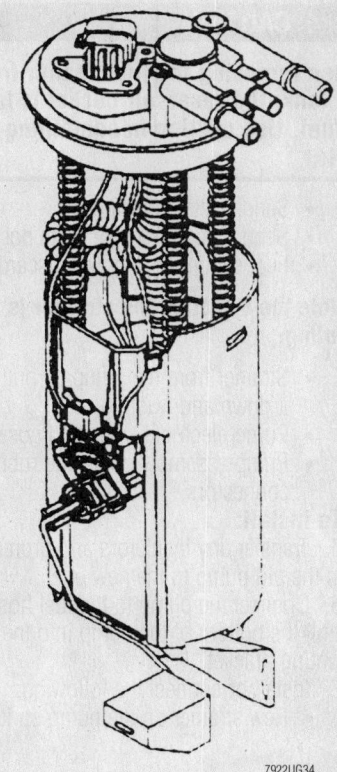

Fuel pump and level sender assembly

7922UG34

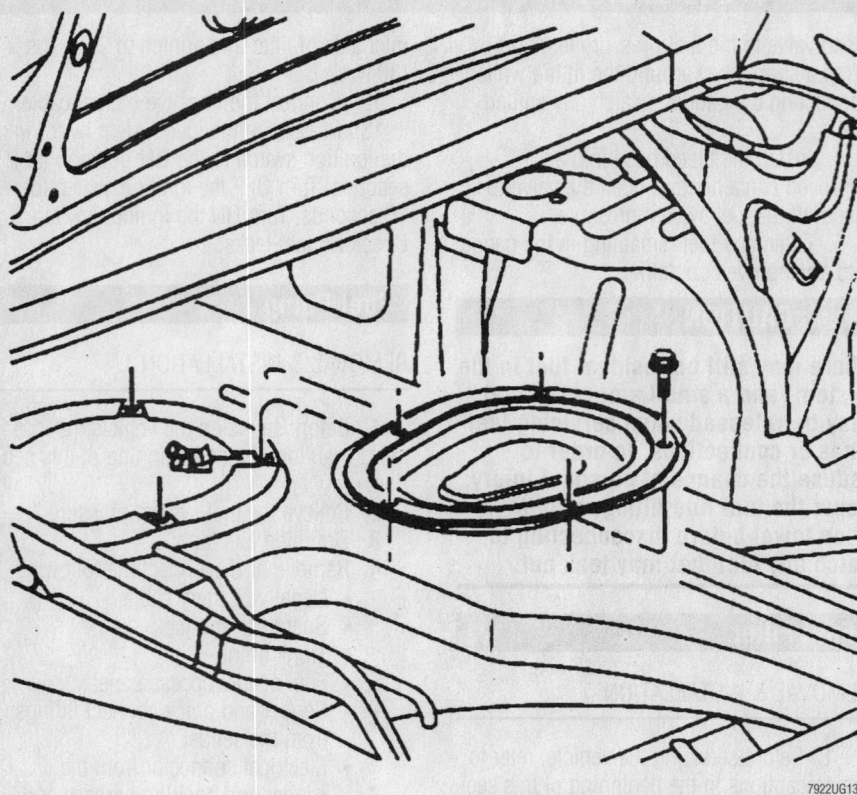

The fuel pump service cover is located in the luggage compartment under the spare tire—Park Avenue shown others similar

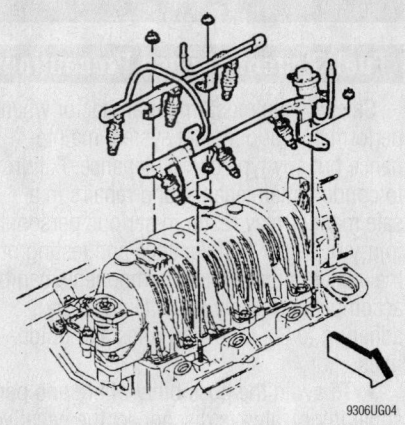

Exploded view of the fuel rail assembly—Vin 1 models

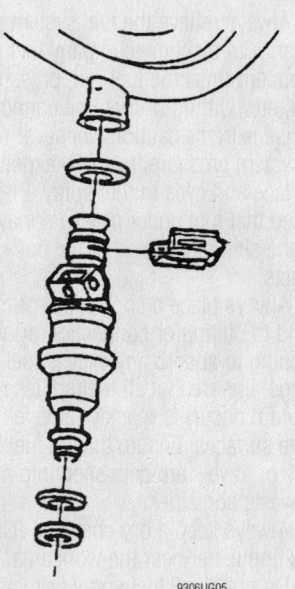

Exploded view of the fuel injector assembly

✳✳ CAUTION

When removing the fuel sender from the tank, the reservoir bucket is full of fuel. Use caution in containing the fuel.

- Sender retaining ring
- Sender and take note of its position
- Fuel sender O-ring and discard it

➡**Note the direction the strainer is pointing.**

- Strainer from the pump by pulling it down and twisting
- Pump electrical wires and hoses
- Pump assembly out of the rubber connectors

To install:

5. Transfer any insulators and grommets from the old pump to the new one.
6. Connect the pump to the fuel hose and tilt the bottom of the pump into the mounting bracket.
7. Install or connect the following:
- New strainer on the pump so it

points in the same direction as noted during removal
- Electrical connectors and fuel lines to the pump
- New O-ring on top of the fuel tank
- Fuel sender assembly into the tank
- Lockring
- Fuel line quick connectors
- Sender electrical connector
- Fuel sender access cover
- Trunk liner
- Spare tire and jack
8. Refill with fuel and check for leaks.

Fuel Injectors

REMOVAL & INSTALLATION

1. Before servicing the vehicle, refer to the precautions in the beginning of this section.
2. Relieve the fuel system pressure.
3. Remove or disconnect the following:
- Negative battery cable
- Fuel lines
- Fuel rail

- Injector retaining clips
- Fuel injector

To install:

4. Install or connect the following:
- Fuel injector with new O-rings, coat the o-rings with clean engine oil prior to installation
- Injector retaining clips
- Fuel rail and torque the bolts to 89 inch lbs. (10 Nm). On Vin 1 models, tighten the fuel rail hold-down stud to 18 ft. lbs. (25 Nm), if equipped.
- Fuel lines
- Negative battery cable

DRIVE TRAIN

Transaxle Assembly

REMOVAL & INSTALLATION

2000 Park Avenue And All 2000 Models

1. Before servicing the vehicle, refer to the precautions in the beginning of this section.
2. Remove or disconnect the following:
   - Negative battery cable
   - Air cleaner cover
   - Transaxle electrical connector
   - Shift control cable bracket with cable attached
   - Top transaxle case bolts
   - Top bolt from rear transaxle mount
   - Left side steering rack mount bolts
   - Oxygen (O$_2$S) sensor connector
3. Install an engine support fixture.
   - Both front tires
   - Both splash shields
   - Power steering line bracket
   - Both transaxle mounts
   - Right steering rack mount bolts
   - Transaxle cooler lines
   - Sway bar mounts
   - Lower ball joints
   - Front subframe bolts and lower subframe
   - Both drive axles
   - Torque converter cover
   - Flexplate bolts
   - Input Speed (ISS) sensor connector
   - Engine to transaxle bracket
   - Transaxle fluid filler tube
   - Rear engine mount
   - Remaining transaxle to engine bolts
   - Transaxle from vehicle

To install

4. Install or connect the following:
   - Transaxle to vehicle
   - Transaxle to engine bolts. Torque the bolts to 55 ft. lbs. (75 Nm).
   - Rear engine mount
   - Transaxle fluid filler tube. Torque the filler tube bolt to 15 ft. lbs. (20 Nm).
   - Engine to transaxle bracket. Torque the bolts to 44 ft. lbs. (60 Nm).
   - ISS electrical connectors
   - Flexplate bolts. Torque the bolts to 46 ft. lbs. (62 Nm).
   - Torque converter cover

➡**Use care when installing the right side drive axle into the transaxle case. The splined shaft of the axle can easily damage the seal.**

   - Left and right drive axles to transaxle
   - Engine frame bolts. Torque the bolts to 83 ft. lbs. (112 Nm).
   - Lower ball joints to steering knuckles. Torque the nuts to 88 in. lbs. (10 Nm).
   - Sway bar mounts to the lower control arms
   - Transaxle cooler lines to the frame
   - Right steering rack mount bolts to engine frame. Torque the bolts to 45 ft. lbs. (65 Nm).
   - Right and left transaxle mount to frame
   - Power steering line bracket to frame
   - Right and left splash shields
   - Both front tires. Torque the wheel nuts to 100 ft. lbs. (140 Nm).
   - O$_2$S sensor electrical connector
   - Left side steering rack mount bolts. Torque the bolts to 48 ft. lbs. (65 Nm).
   - Top bolt of the rear transaxle mount. Torque the bolt to 42 ft. lbs. (58 Nm).
   - Shift control cable and bracket
   - Transaxle electrical connector
   - Air cleaner cover
   - Negative battery cable
5. Check and adjust transaxle fluid level.
6. Road test vehicle and check for transaxle leaks.

All 2001–04 Models

1. Before servicing the vehicle, refer to the precautions in the beginning of this section.

➡**Make sure the wheels of the vehicle are in the straight ahead position and the steering column in the LOCK position before disconnecting the steering column or intermediate shaft from the steering gear. Failure to do so will cause the SIR coil assembly to become uncentered, which may cause damage to the coil assembly.**

2. Lock the steering column by installing tool J-42640 into the underside of the steering column.
3. Remove or disconnect the following:
   - Negative battery cable
   - Air cleaner
   - Range selector cable from the range selector lever
   - Range selector cable with the bracket from the transmission case

   - Transaxle electrical connector
   - Wiring harness from the wiring harness retainer
   - Ground cable bolt from the transaxle
4. Install an engine support fixture.
   - Upper engine-to-transaxle case bolts
   - Front tire and wheel assembly
   - Front fascia extensions from each side
   - Front air deflector

✳✳ CAUTION

Failure to disconnect the intermediate shaft from the rack and pinion steering gear stub shaft can result in damage to the steering gear or to the intermediate shaft. This damage may cause loss of steering control, which could result in an accident and possible personal injury.

   - Intermediate shaft lower pinch bolt
   - Intermediate shaft from the power steering gear
   - Power steering gear heat shield
   - Power steering gear mounting bolts
   - Power steering line retainers from the frame and attach the gear to the exhaust manifold
5. Loosen the two mounting nuts to allow removal of the brake pressure modulator valve from the bracket.
   - Brake line retainers from the frame
   - Left transaxle mount
   - Frame
   - Front transaxle mount bracket with mount attached
   - Right and left drive axles from the transaxle
   - Transmission oil cooler hoses from the transaxle
   - Transaxle fluid filler tube
   - Torque converter cover
   - Flywheel-to-torque converter bolts
6. Support transaxle using an appropriate transaxle jack.
   - Vehicle Speed Sensor (VSS) electrical connector
   - Torque strut bracket-to-transaxle bolts
   - Torque strut bracket-to-engine bolts
   - Torque strut bracket
   - Engine-to-transaxle case bolt
   - Remaining transaxle-to-engine bolt
   - Transaxle

To install:

7. Installation is the reverse of removal, please note the following specifications:
- Left transaxle bracket bolts to 81 ft. lbs. (110 Nm)
- Rear transaxle bracket bolts to 46 ft. lbs. (63 Nm)
- Lower transaxle bolts to 55 ft. lbs. (75 Nm)
- Torque strut bracket-to-engine bolts to 48 ft. lbs. (65 Nm)
- Torque strut bracket-to-transaxle bolts to 26 ft. lbs. (36 Nm)
- Flywheel-to-torque converter bolts to 46 ft. lbs. (63 Nm)
- Brake pressure modulator valve nuts 89 inch lbs. (10 Nm)
- Power steering gear mounting bolts to 95 ft. lbs. (70 Nm)
- Intermediate shaft lower pinch bolt to 33 ft. lbs. (45 Nm)
- Upper transaxle case-to-engine bolts to 55 ft. lbs. (75 Nm)
- Range selector cable bracket nuts to 18 ft. lbs. (25 Nm)

8. Check and adjust transaxle fluid level.
9. Road test vehicle and check for transaxle leaks.

Halfshaft

REMOVAL & INSTALLATION

※※ WARNING

Use care when removing the halfshaft to prevent the inner CV-joint from becoming over-extended. Over-extension of the joint could result in separation of internal components and possible joint failure.

1. Before servicing the vehicle, refer to the precautions in the beginning of this section.

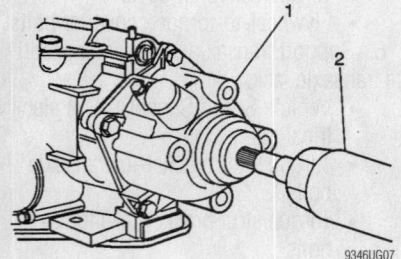

Remove the halfshaft from the transaxle using tool J 42129 Axle Shaft Remover and tool J 2619-01 Slide Hammer

2. Install a boot protector on the outer CV-joint boot.
3. Remove or disconnect the following:
- Front wheel
- Speed sensor connector

4. Loosen the stabilizer shaft link assembly bolt (to accommodate ball joint separation).
5. Remove the ball joint cotter pin and nut. Loosen the joint.

➡**The grease fitting may have to be removed from the ball joint for tool access.**

6. Separate the lower control arm from the joint.
7. Remove or disconnect the following:
- Hub nut
- Halfshaft from the hub using tool J 28733

8. Move the strut and knuckle rearward.
9. Remove the halfshaft from the transaxle using tool J 42129 Axle Shaft Remover and tool J 2619-01 Slide Hammer.

※※ WARNING

If equipped with anti-lock brakes, care must be used to prevent damage to the toothed sensor ring on the halfshaft and the wheel speed sensor on the steering knuckle.

To install:

➡**If installing the right halfshaft, install a seal protector, so that it can be pulled out after the halfshaft is installed.**

10. Install the halfshaft into the transaxle by placing a drift pin or punch into the groove on the joint housing and tapping lightly until seated. Verify that the halfshaft is seated by grasping the inner joint housing and pulling. DO NOT pull on the halfshaft.
11. Install or connect the following:
- Halfshaft into the hub/bearing assembly with new hub nut and tighten to 118 ft. lbs. (160 Nm)
- Ball joint into the steering knuckle. Torque the nut to 88 inch lbs. (10 Nm), plus an additional 120 degree turn during which a torque of 41 ft. lbs. (55 Nm) must be obtained.

➡**Tighten the nut up to one more flat in order to align the slot with the hole in the stud.**

12. Install or connect the following:
- Stabilizer shaft link assembly. Torque the nut to 14 ft. lbs. (17 Nm).

- Speed sensor connector
- Front wheel

13. If a seal protector was installed, remove it by pulling in line with the handle.
14. Road test for proper operation.

CV-Joints

OVERHAUL

Inner (Tripod) Joint

1. Before servicing the vehicle, refer to the precautions in the beginning of this section.
2. Raise and safely support the vehicle.
3. Remove or disconnect the following:
- Front wheel
- Halfshaft and place it in a vise
- Small CV-joint boot clamp, cut and discard it
- Large CV-joint boot clamp, cut and discard it
- CV-joint boot by sliding it away from the tripod joint
- Tripod housing from the tripod spider
- Inboard spacer ring and slide it rearward on the shaft
- Outboard retaining ring
- Tripod joint spider assembly
- Inboard spacer ring and discard it
- Tripod joint spider assembly by tapping it from the halfshaft with a brass drift
- Tripod spider retaining ring and discard it
- Trilobal tripod bushing from the housing
- CV-joint boot

4. Thoroughly clean and inspect all parts.

To install:

5. Install or connect the following:
- Small boot clamp
- CV-joint boot
- New inboard spacer ring. Slide it rearward on the shaft past the 2nd groove
- Tripod joint spider assembly onto the shaft until it passes the 2nd groove

6. Assemble the tripod spider assembly onto the halfshaft as follows:
   a. Position the tripod spider assembly onto the shop press plate.
   b. Position the halfshaft onto the tripod spider assembly, in the shop press.
   c. Press the halfshaft into the tripod spider assembly until the spider assembly passes the 2nd groove.

❋❋ WARNING

When assembling the tripod assembly onto the halfshaft, do not exceed 4,000 lbs. pressure.

7. Remove the halfshaft from the shop press and place it in vise.

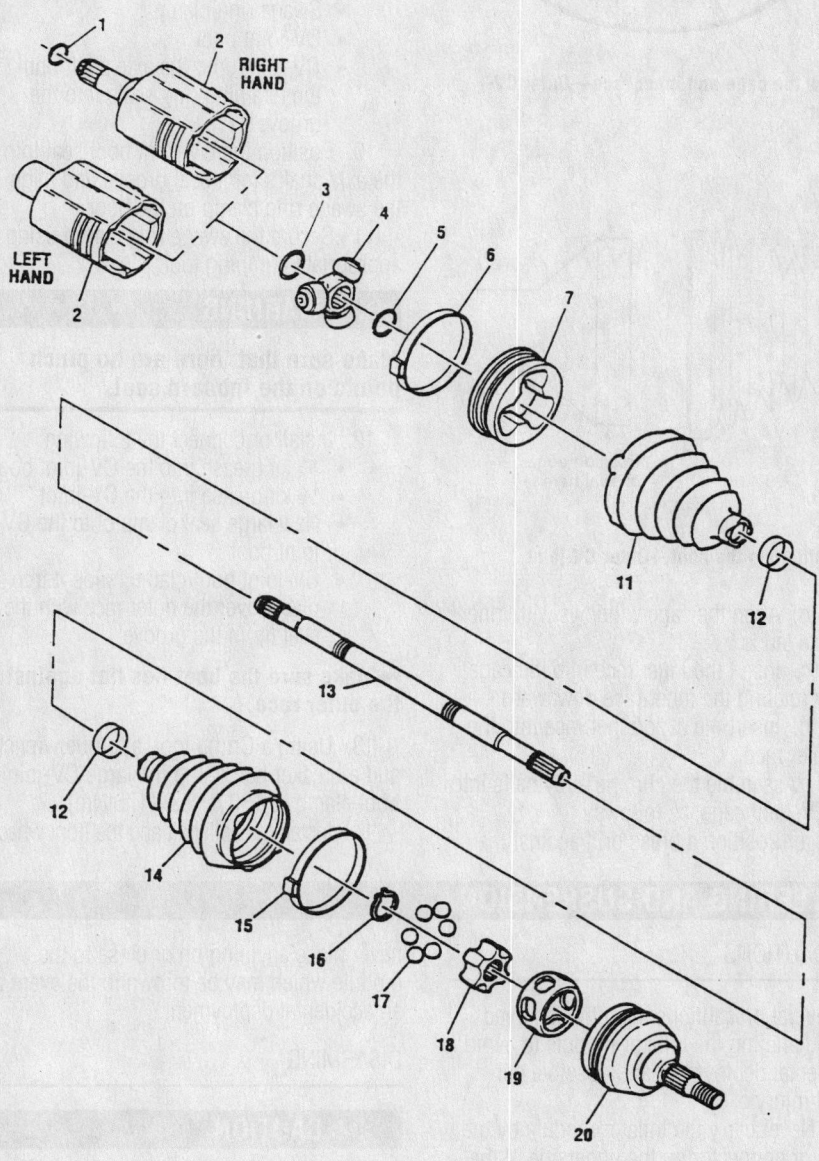

1 - RING, RETAINING
2 - HOUSING ASM, RETAINER &
3 - RING, SHAFT RETAINING
4 - SPIDER, TRIPOT JOINT
5 - RING, SPACER
6 - CLAMP, SEAL RETAINING
7 - BUSHING, TRILOBAL TRIPOT
11 - SEAL, DRIVE AXLE INBOARD
12 - RING, SWAGE

13 - SHAFT, AXLE (RH SHOWN, LH SIMILAR)
14 - SEAL, DRIVE AXLE OUTBOARD
15 - CLAMP, SEAL RETAINING
16 - RING, RACE RETAINING
17 - BALL, CHROME ALLOY
18 - RACE, C/V JOINT INNER
19 - CAGE, C/V JOINT
20 - RACE, C/V JOINT OUTER

9306UG06

Exploded view of the halfshaft assembly

8. Install or connect the following:
- New outboard retaining ring into the axle shaft groove
- Tripod joint spider assembly, slide it against the outboard retaining ring using a brass drift
- Inboard spacer ring, seat it in the groove

9. Use ½ of the grease supplied in the kit into the boot and the other ½ into the tripod housing.
- Trilobal tripod bushing flush with the tripod housing face
- New large seal clamp onto the CV-joint boot
- Tripod housing, slide it over the tripod joint spider assembly
- CV-joint boot/clamp, slide it into place, over the trilobal tripod bushing with the seal lip in the groove

➡**Make sure the boot lies flat against the trilobal bushing.**

10. Using the crimp tool, a torque wrench and a breaker bar, crimp the small CV-joint boot clamp to 100 ft. lbs. (136 Nm).

11. Using the crimp tool, latch the large CV-joint boot clamp.

12. Install the halfshaft and the front wheel.

Outer CV-Joint

1. Before servicing the vehicle, refer to the precautions in the beginning of this section.

2. Remove or disconnect the following:
- Front wheel
- Halfshaft
- Swage ring using a hand grinder
- Large boot clamp
- CV-joint boot, slide it away from the CV-joint
- CV-joint assembly by spreading the inner race-to-axle shaft retaining ring ears using Snapring Pliers
- CV-joint boot from the axle shaft

3. Disassemble the chrome alloy balls from the CV-joint cage as follows:
a. Position a brass drift against the CV-joint cage and tap it with a hammer to tilt the cage.
b. Chrome alloy ball from the cage.
c. Tilt the cage in the opposite direction.
d. Remove the opposite chrome alloy ball.
e. Repeat the procedure until all 6 balls are removed.

4. Disassemble the CV-joint cage and inner race as follows:
a. Pivot the cage and race 90 degrees to the center line of the outer race.
b. Align the cage windows with outer race lands.
c. Remove the cage from the outer race.
d. Rotate the inner race upward and remove it from the cage.

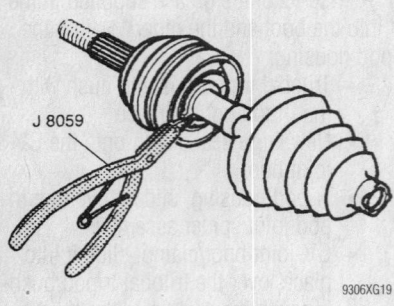

9306XG19

Disconnecting the outer CV-joint from the axle shaft

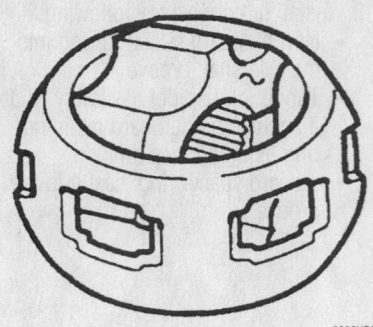

9306XG21

View the cage and inner race—Outer CV-joint

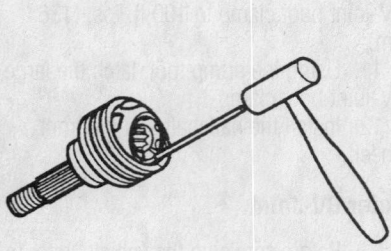

9306XG20

Tilting the cage—Outer CV-joint

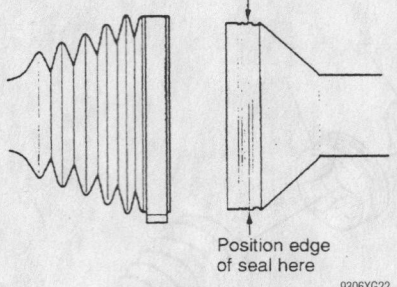

Position edge of seal here

9306XG22

Positioning the boot—Outer CV-joint

5. Thoroughly clean and inspect all parts.

To install:

6. Lubricate the parts with a light coat of grease.

7. Assemble the CV-joint cage and inner race, as follows:

  a. Rotate the inner race 90 degrees to the cage centerline.

  b. Align the cage windows with inner race lands.

  c. Insert the inner race into the cage by rotating the inner race downward.

  d. Insert the cage/inner race into the outer race.

8. Assemble the chrome alloy balls into the CV-joint cage, as follows:

  a. Position a brass drift against the CV-joint cage and tap it with a hammer to tilt the cage.

  b. Insert the 1st chrome alloy ball into the cage.

  c. Tilt the cage in the opposite direction.

  d. Insert the opposite chrome alloy ball.

  e. Repeat the procedure until all 6 balls are inserted.

9. Install or connect the following:
- Swage ring clamp
- CV-joint boot
- CV-joint onto the axle shaft until the retaining ring seats into the groove

10. Position the CV-joint boot seal into the axle shaft's joint seal groove and align the swage ring clamp on the boot.

11. Secure the swage ring clamp using appropriate crimping tool.

✳✳ WARNING

Make sure that there are no pinch points on the inboard seal.

12. Install or connect the following:
- ½ kit grease into the CV-joint boot
- ½ kit grease into the CV-joint
- New large seal clamp onto the CV-joint boot
- CV-joint boot/clamp, slide it into place, over the outer race with the seal lip in the groove

➡ **Make sure the boot lies flat against the outer race.**

13. Using a Crimp tool, a torque wrench and a breaker bar, crimp the large CV-joint boot clamp to 130 ft. lbs. (176 Nm).

14. Install the halfshaft and the front wheel.

STEERING AND SUSPENSION

Air Bag

✳✳ CAUTION

Some vehicles are equipped with an air bag system. The system must be disabled before performing service on or around system components, the steering column, instrument panel components, wiring and sensors. Failure to follow safety precautions and the disarming procedures could result in accidental air bag deployment, possible personal injury and unnecessary system repairs.

PRECAUTIONS

Several precautions must be observed when handling the inflator module to avoid accidental deployment and possible personal injury.

- Never carry the inflator module by the wires or connector on the underside of the module.
- When carrying a live inflator module, hold it securely with both hands, and ensure that the bag and trim cover are pointed away.
- Place the inflator module on a bench or other surface with the bag and trim cover facing up.
- With the inflator module on the bench, never place anything on or close to the module which may be thrown in the event of an accidental deployment.

DISARMING

✳✳ CAUTION

The Supplemental Inflatable Restraint system (SIR) must be disarmed before performing service around the air bag or SIR wiring. Failure to do so may cause accidental deployment of the air bag, resulting in unnecessary SIR repairs and/or personal injury.

1. Turn the steering wheel so the front wheels are in the straight-ahead position.
2. Turn the ignition switch to the **LOCK** position.
3. Remove or disconnect the following:
   - Negative battery cable
   - Air bag fuse from the fuse panel

➡**The position of the fuse on the panel varies according to model and year. Consult the vehicle owner's manual for fuse location.**

   - Left-hand sound insulator trim panel under the instrument panel
   - Connector retainer clip and the yellow 2-way connector at the base of the steering column
   - Connector retainer and detach the passenger side yellow 2-way connector. Located behind the right side sound insulator.

ARMING

After the necessary repairs have been made, re-enable the air bag system as follows:
1. Turn the steering wheel so the front wheels are in the straight-ahead position.
2. Turn the ignition switch to the **LOCK** position.
3. Disconnect the negative battery cable.
4. Install or connect the following:
   - Yellow 2-way connector at the base of the steering column and the connector retainer

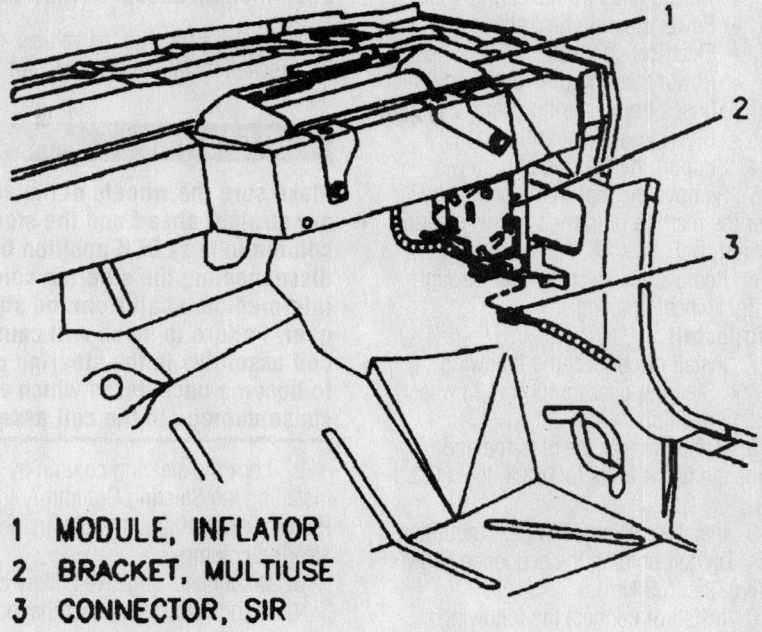

1 MODULE, INFLATOR
2 BRACKET, MULTIUSE
3 CONNECTOR, SIR

7922UG36

Passenger's side air bag connector

   - Left-hand sound insulator
   - Yellow 2-way connector on the right side and the connector retainer
   - Sound insulator and/or glove box
   - Air bag fuse
   - Negative battery cable
5. Turn the ignition switch to the **RUN** position. Verify that the INFLATABLE

RESTRAINT indicator lamp flashes 7–9 times, then remains OFF. If the lamp does not function as specified, there is a malfunction in the air bag system.

Power Rack and Pinion Steering Gear

REMOVAL & INSTALLATION

2000 Park Avenue

✳✳ WARNING

The wheels of the vehicle must be straight-ahead and the steering column in the LOCK position before disconnecting the steering column or intermediate shaft from the steering gear. Failure to do so will cause the Supplemental Inflatable Restraint (SIR) coil assembly in the steering column to become off-center, which will cause damage to the coil assembly.

1. Before servicing the vehicle, refer to the precautions at the beginning of this section.
2. Disconnect negative battery cable.
3. Remove or disconnect the following:
   - Both front tires
   - Exhaust system
   - Wheel house splash shield
   - Intermediate shaft lower connection

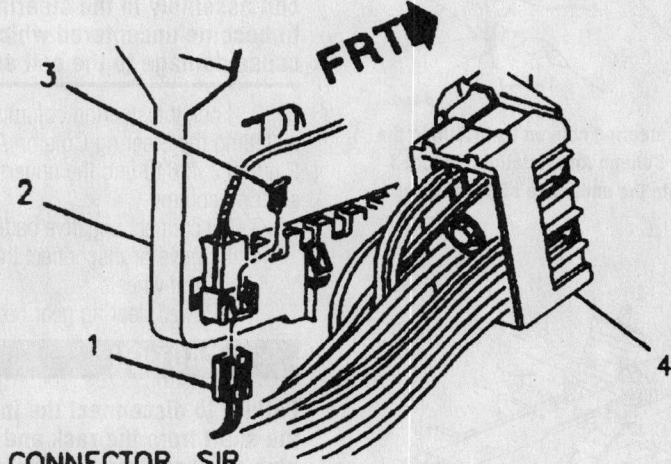

1 CONNECTOR, SIR
2 BRACKET, MULTIUSE MODULE
3 CONNECTOR POSITION ASSURANCE (CPA)
4 CONNECTOR, STEERING COLUMN
  WIRING HARNESS

7922UG35

Drivers side air bag connector

- Tie rod ends from steering knuckles
- Power steering heat shield
- Electrical connector
- Power steering gear outlet and inlet hoses from steering gear
- Steering gear mounting bolts

4. Support the rear the of frame.

5. Remove the rear frame bolts, and lower the rear the of frame to allow steering gear removal.

6. Remove the steering gear through the right wheel opening.

To install

7. Install or connect the following:
- Steering gear through right wheel opening

8. Raise the rear the of frame and torque the frame bolts to 142 ft. lbs. (192 Nm).

9. Install three steering gear mounting bolts. Tighten bolts in the sequence shown to 48 ft. lbs. (65 Nm).

10. Install or connect the following:
- Power steering gear outlet and inlet hoses. Torque the hose connections to 20 ft. lbs. (27 Nm).
- Electrical connector
- Steering gear heat shield
- Tie rod ends to steering knuckles. Torque the tie rod end nuts to 35 ft. lbs. (47 Nm).
- Cotter pin. Tighten the nut up to 52 ft. lbs. (70 Nm) max to align cotter pin slot.
- Intermediate shaft lower connection. Torque the bolt to 35 ft. lbs. (47 Nm).
- Exhaust system. Torque the bolts to 18 ft. lbs. (25 Nm).
- Wheelhouse splash shield
- Tires. Torque the lug nuts to 100 ft. lbs. (140 Nm).

11. Fill and bleed power steering system.

12. Inspect the system for leaks.

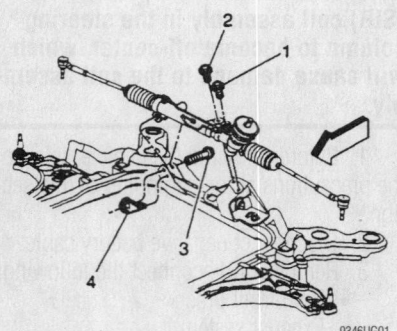

9346UG01

Torque sequence for steering gear mounting bolts—Park Avenue models

2000 Models Except Park Avenue

1. Before servicing the vehicle, refer to the precautions at the beginning of this section.

> ❄❄ **CAUTION**
>
> **Make sure the wheels of the vehicle are straight ahead and the steering column in the LOCK position before disconnecting the steering column or intermediate shaft from the steering gear. Failure to do so will cause the coil assembly in the steering column to become uncentered which will cause damage to the coil assembly.**

2. Lock the steering column by installing the Steering Column Anti Rotation Pin tool J 42640 into the underside of the steering column.

3. Disconnect negative battery cable.

4. Remove or disconnect the following:
- Intermediate shaft lower coupling
- Pressure and return pipes from the rack and pinion gear
- Electrical connector, if equipped
- Stabilizer shaft links at lower control arms

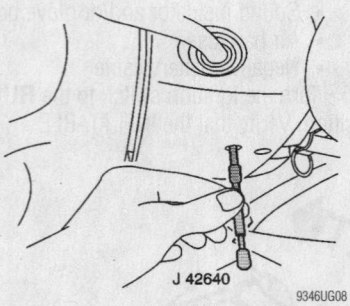

J 42640

9346UG08

Lock the steering column by installing the Steering Column Anti Rotation Pin tool J 42640 into the underside of the steering column

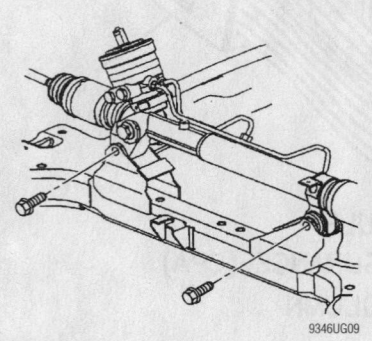

9346UG09

Steering gear mounting bolts—2000 models

5. Rotate the stabilizer shaft to access steering gear bolts.
- Rack and pinion attaching bolts
- Rack and pinion assembly

To install:

6. Install or connect the following:
- Rack and pinion assembly and tighten the bolts to 70 ft. lbs. (95 Nm)
- Electrical connector, if equipped
- Pressure and return lines to the steering gear and tighten the fittings to 22 ft. lbs. (30 Nm)
- Heat shield and bolts. Tighten the bolts to 7 ft. lbs. (10 Nm).

7. Align stabilizer shaft.
- Links to lower control arms
- Intermediate shaft to the steering gear and tighten the pinch bolt to 33 ft. lbs. (45 Nm)

8. Remove the steering column anti rotation pin from the steering column.

9. Fill and bleed the power steering system.

All 2001–04 Models

1. Before servicing the vehicle, refer to the precautions at the beginning of this section.

> ❄❄ **CAUTION**
>
> **Make sure the wheels of the vehicle are straight ahead and the steering column in the LOCK position before disconnecting the steering column or intermediate shaft from the steering gear. Failure to do so will cause the coil assembly in the steering column to become uncentered which will cause damage to the coil assembly.**

2. Lock the steering column by installing the Steering Column Anti Rotation Pin tool J 42640 into the underside of the steering column.

3. Disconnect negative battery cable.

4. Remove or disconnect the following:
- Front wheels
- Power steering gear heat shield

> ❄❄ **CAUTION**
>
> **Failure to disconnect the intermediate shaft from the rack and pinion stub shaft can result in damage to the steering gear and/or intermediate shaft. This damage can cause loss of steering control which could result in personal injury.**

- Intermediate shaft lower pinch bolt
- Intermediate shaft from the power steering gear

- Outer tie rod retaining nuts
- Outer tie rod from the steering knuckles
- Variable effort steering electrical connector from the power steering gear, if equipped
- Power steering gear pressure and return hoses from the gear
- Left stabilizer shaft insulator
- Power steering gear mounting bolts (lift from the mounting holes)
- Power steering gear through the left wheel opening

5. Transfer the outer tie rods if replacing the power steering gear.

To install:

6. Install or connect the following:
- Power steering gear through the left wheel opening
- Power steering gear mounting bolts and tighten to 48 ft. lbs. (65 Nm)
- Left stabilizer shaft insulator
- Power steering gear pressure and return hoses to the gear and tighten to 20–22 ft. lbs. (27–30 Nm)
- Variable effort steering electrical connector, if equipped
- Outer tie rod end to the steering knuckles and tighten the nuts to 52–55 ft. lbs. (70–75 Nm)
- Intermediate shaft to the power steering gear
- Intermediate shaft lower pinch bolt and tighten to 33–35 ft. lbs. (45–47 Nm)
- Power steering gear heat shield
- Wheels

7. Remove tool J-42640 from the steering column.
8. Bleed the power steering system.
9. Inspect the power steering system for leaks.
10. Adjust the front toe.

Strut

REMOVAL & INSTALLATION

Front

❋❋ WARNING

The steering knuckle must be retained after the strut-to-steering knuckle bolts have been removed. Failure to observe this may cause ball joint and/or halfshaft damage.

1. Before servicing the vehicle, refer to the precautions in the beginning of this section.

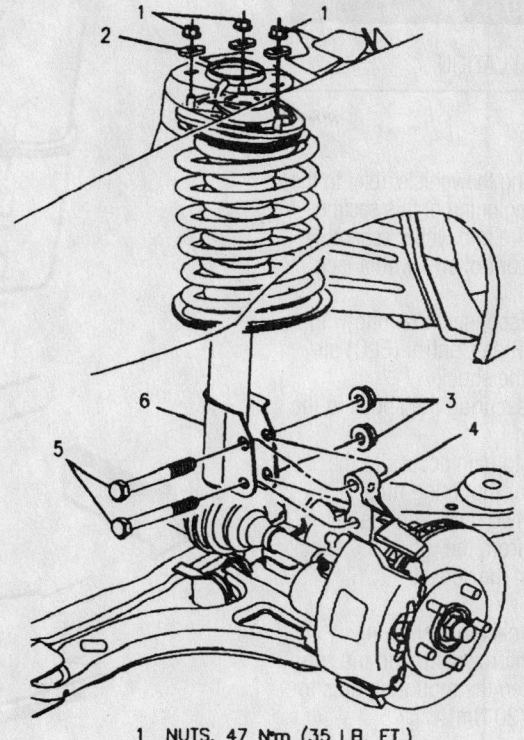

1	NUTS, 47 N·m (35 LB. FT.)	
2	WASHER	
3	NUTS, 185 N·m (136 LB. FT.)	
4	KNUCKLE	
5	BOLT	
6	STRUT	

7922UG17

The strut assembly is mounted between the steering knuckle and the body—front strut shown

2. Matchmark the strut-to-steering knuckle location.
3. If equipped with electronic ride control, detach the electrical connection.
4. Remove or disconnect the following:
- 3 upper strut mount nuts and allow the control arms to hang free
- Anti-lock Brakes System (ABS) front wheel speed sensor
- Wheel speed sensor bracket from the strut
- Brake line bracket from the strut
- Strut-to-steering knuckle bolts and the strut

To install:
5. Install or connect the following:
- Strut. Torque the 3 nuts to 30–35 ft. lbs. (40–47 Nm).
- Electronic ride control electrical connector, if removed
- Strut-to-knuckle bolts. Torque the bolts to 135 ft. lbs. (185 Nm).
- Brake line bracket to the strut
- Wheel speed sensor bracket to the strut
- ABS front wheel speed sensor connector

- Front wheel. Torque the lug nuts to 100 ft. lbs. (140 Nm).
6. Check and adjust the wheel alignment.

Rear

EXCEPT PARK AVENUE AND ALL 2000—04 MODELS

1. Remove rear seat cushion and seatback to gain access to the strut tower mounting nuts.
2. Remove tire.
3. Support the control arm.
4. Remove or disconnect the following:
- Strut to knuckle bolts
- Upper strut nuts
- Strut from vehicle

To install:
5. Install or connect the following:
- Strut
- Upper strut nuts. Torque the nuts to 35 ft. lbs. (47 Nm).
- Strut to knuckle bolts. Torque the bolts to 140 ft. lbs. (190 Nm).
- Rear seatback and cushion
- Tire. Torque the lug nuts to 100 ft. lbs. (140 Nm).

Shock Absorber

REMOVAL & INSTALLATION

Rear

1. Before servicing the vehicle, refer to the precautions in the beginning of this section.
2. Remove the tire and wheel assembly.
3. Support the control arm with a jack stand.
4. Remove or disconnect the following:
   • Electronic Ride Control (ELC) air tube from the shock
   • Two bolts securing the shock to the control arm
   • Trunk trim to gain access to the shock upper mounting nuts.
   • Cover, the two nuts, and the reinforcement from the top of the shock
   • Shock from the vehicle

To install:
5. Install or connect the following:
   • Shock, reinforcement, and the two nuts. Tighten the mounting nuts to 15 ft. lbs. (20 Nm).
   • Shock cover
   • Trunk trim
   • Shock-to-control arm bolts and tighten the bolts to 18 ft. lbs. (24 Nm)
   • ELC air tube to the shock.
   • Tire and wheel assembly

Coil Spring

REMOVAL & INSTALLATION

✶✶ CAUTION

The coil springs are under a considerable amount of tension. Be very

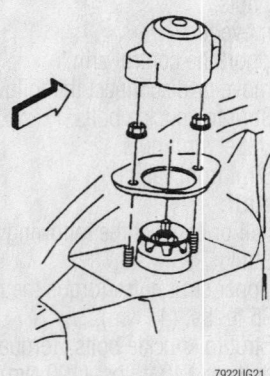

Remove the cover to gain access to the upper shock absorber mounting components—rear suspension

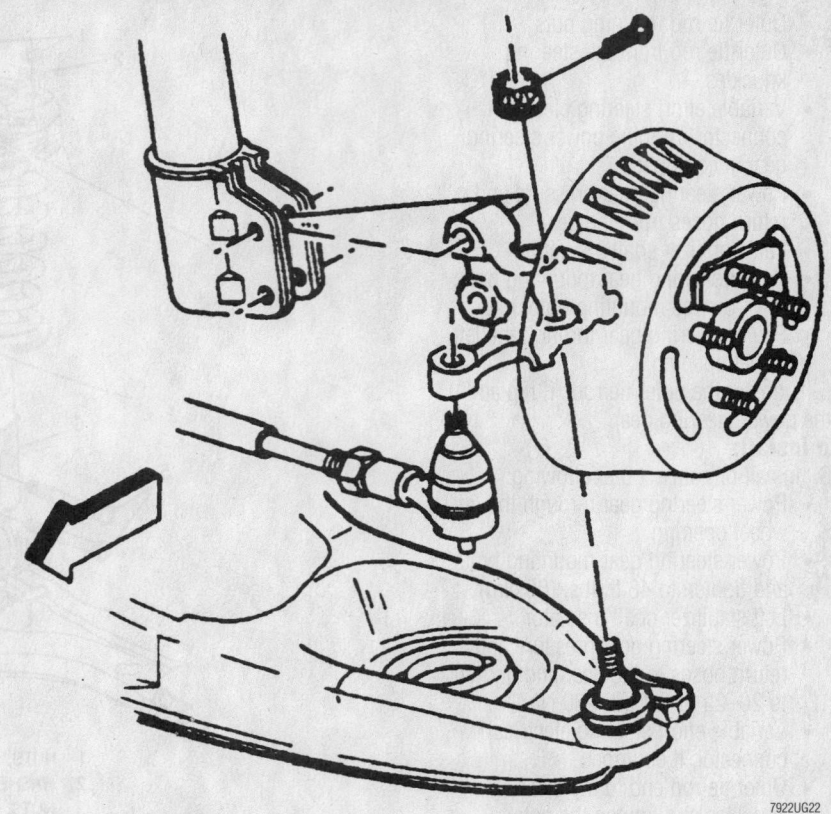

Exploded view of the lower shock absorber mounting to the knuckle assembly—rear suspension

careful when removing or installing them; they can exert enough force to cause very serious injury.

Front

1. Remove the strut from the vehicle.
2. Disassemble the strut as follows:
   a. Step 1: Place the strut assembly into compressor tool, to compress the coil spring.
   b. Step 2: Compress the spring slightly.
   c. Step 3: Hold the strut shaft from

turning using a No. 50 Torx® socket and remove the 24mm nut on the top end of the strut.
   d. Step 4: Install Rod tool J 34013-38 to help guide the strut shaft from the upper mount assembly.
   e. Step 5: Loosen the spring compressor tool until the coil spring and mount can be removed as an assembly. Remove the lower spring insulator, if equipped.

To install:
3. Assemble the strut as follows:

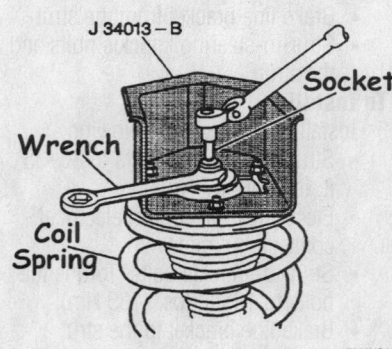

Use a Torx® socket to keep the piston rod from turning while removing the upper nut—front strut shown

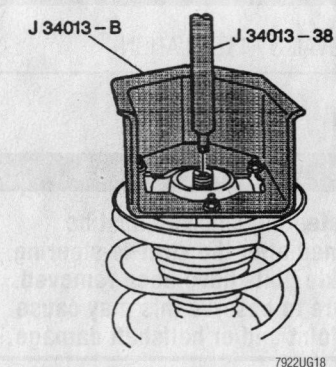

Install Rod J 34013-38 to help guide the strut shaft from the upper mount assembly—front strut shown

a. Step 1: Place the strut in compressor tool.

b. Step 2: Install the coil spring over the strut.

c. Step 3: Compress the coil spring while guiding strut shaft through the top of the strut assembly.

d. Step 4: Install the top strut nut. Torque the nut to 55 ft. lbs. (75 Nm).

e. Step 5: Remove the strut from the compressor.

Rear

1. Before servicing the vehicle, refer to the precautions in the beginning of this section.
2. Remove or disconnect the following:
   • Tire
   • Electronic Leveling Control (ELC) air tube
3. Support the control arm.
   • Two lower shock bolts
   • Trunk trim
   • Cover, 2 nuts and the reinforcement from the top of the shock
   • Cotter pin and hex nut on control arm
4. Separate the adjustment link from the knuckle.

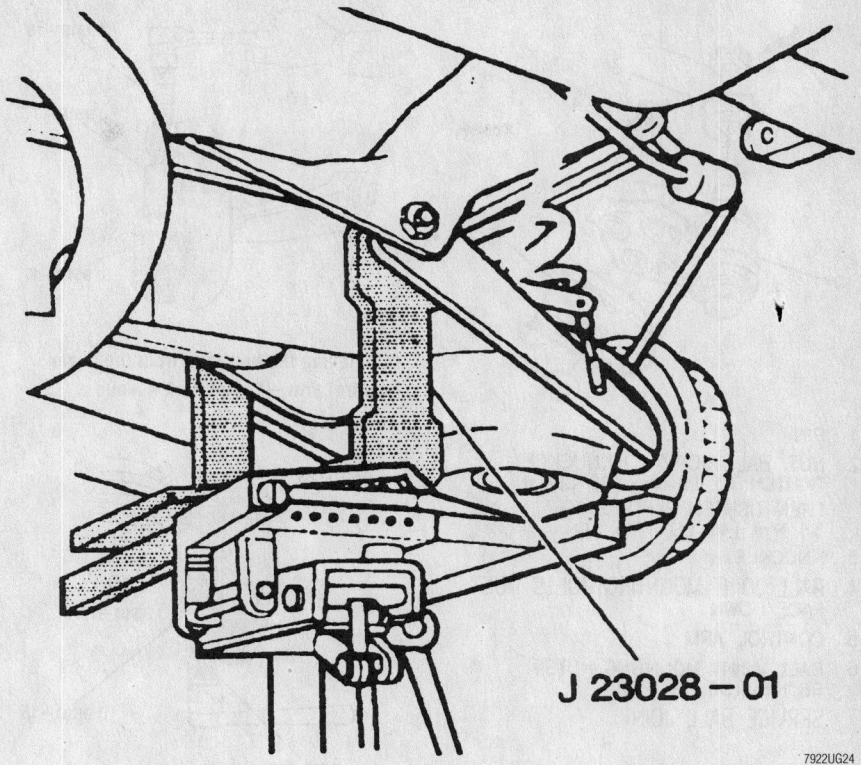

J 23028 – 01

7922UG24

Use support bracket tool J-23028-01 mounted on a jack to support the rear lower control arm—H body

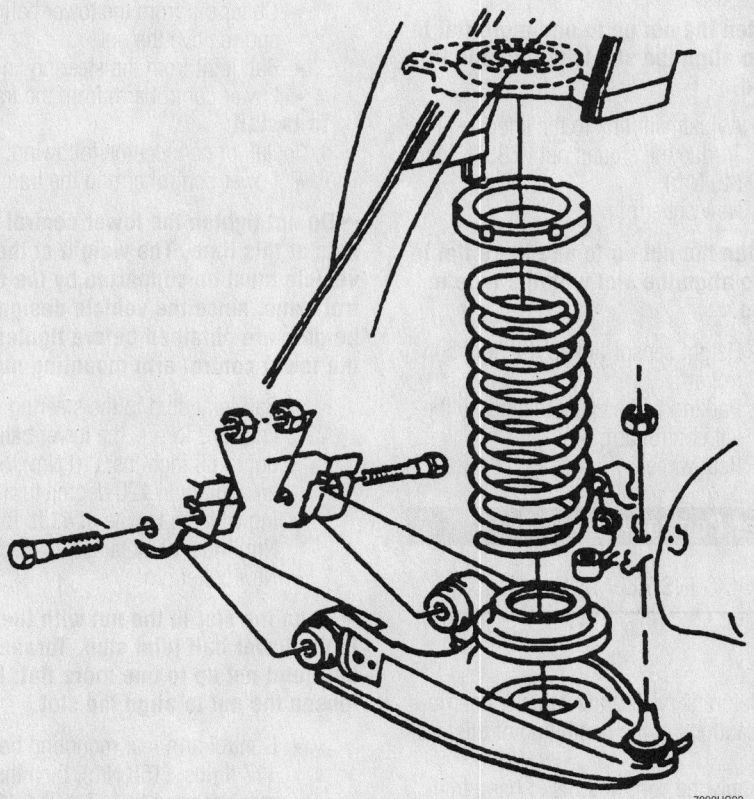

7922UG23

Exploded view of the rear coil spring mounting—H body

5. Slowly lower the control arm.
6. Remove the spring with the lower insulator.

To install

7. Install or connect the following:
   • Coil spring with insulator
   • Lower shock bolts. Torque the bolts to 18 ft. lbs. (24 Nm).
   • Adjustment link. Torque the nut to 88 inch lbs. (10 Nm) 36 ft. lbs. (50 Nm). Tighten an additional « turn to align the cotter pin.
   • ELC air tube
   • Tire. Torque the lug nuts to 100 ft. lbs. (140 Nm).

Lower Ball Joint

REMOVAL & INSTALLATION

Front

The ball joint is an integral part of the lower control arm and if found to be defective the control arm should be replaced.

Rear

1. Before servicing the vehicle, refer to the precautions in the beginning of this section.

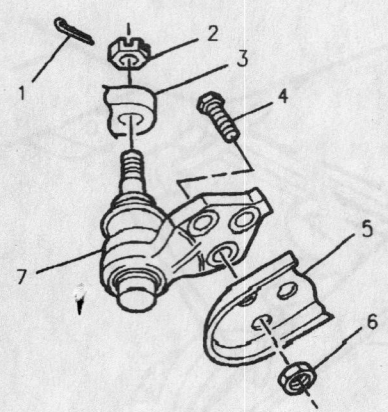

1 PIN
2 NUT, BALL JOINT TO KNUCKLE;
   TIGHTEN TO 10 N•m (88 LB. IN.)
   THEN TIGHTEN 2 FLATS TO
   55 N•m (41 LB. FT.), MIN.
3 KNUCKLE
4 BALL JOINT MOUNTING BOLTS MUST
   FACE DOWN
5 CONTROL ARM
6 BALL JOINT MOUNTING NUTS
   68 N•m (50 LB. FT.)
7 SERVICE BALL JOINT

7922UG25

The replacement ball joint should be attached to the control arm using 3 bolts and nuts—except Park Avenue

2. Remove or disconnect the following:
   - Rear wheel
   - Height sensor link from the right control arm
   - Parking brake cable retaining clip from the left control arm
   - Adjustment link from the knuckle
3. Support the control arm with a jack.
4. Remove the ball joint stud cotter pin and nut.
5. Reinstall the nut on the stud with the flat side of the nut facing up; do not torque the nut.
6. Remove or disconnect the following:
   - Stud from the knuckle
   - Slotted hex nut from the ball joint stud
7. Press the ball joint out of the control arm

To install:
8. Install or connect the following:
   - New ball joint, press it into the control arm
   - Ball joint stud into the knuckle. Torque the nut to 88 inch lbs. (10 Nm), plus an additional 4 flats. The minimum torque on the nut should be 40 ft. lbs. (55 Nm).
   - New cotter pin

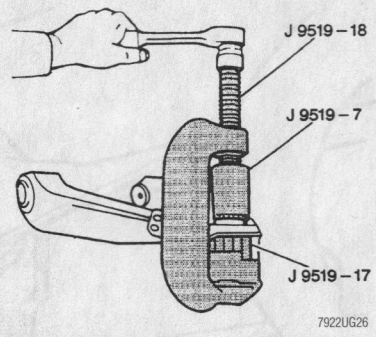

Removing the ball joint from the lower control arm—except Park Avenue

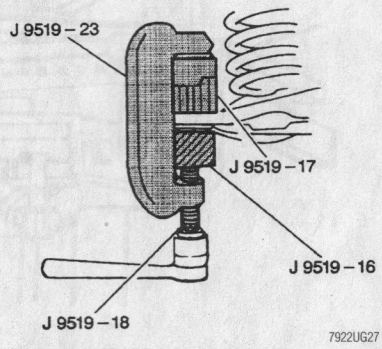

Installing the ball joint into the lower control arm—except Park Avenue

➡ **Tighten the nut up to one more flat in order to align the slot with the hole in the stud.**

   - Adjustment link to the knuckle. Torque the slotted nut to 33 ft. lbs. (45 Nm).
   - New cotter pin

➡ **Tighten the nut up to one more flat in order to align the slot with the hole in the stud.**

   - Height sensor link to the right control arm
   - Parking brake cable retainer to the left control arm
   - Rear wheel

Lower Control Arm

REMOVAL & INSTALLATION

Front

1. Before servicing the vehicle, refer to the precautions in the beginning of this section.
2. Allow the control arms to hang free.
3. Remove or disconnect the following:
   - Front wheel

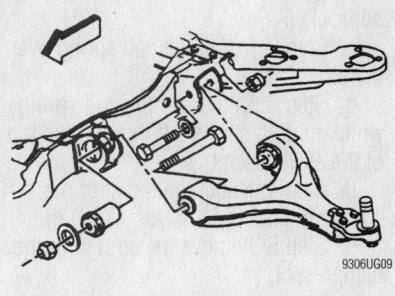

Lower control arm assembly—except Park Avenue

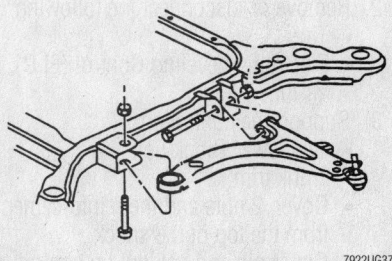

Lower control arm mounting—Park Avenue

   - Stabilizer shaft link assembly from the control arm
   - Cotter pin from the lower ball joint and remove the nut
   - Ball joint from the steering knuckle
   - Lower control arm from the frame

To install:
4. Install or connect the following:
   - Lower control arm to the frame

➡ **Do not tighten the lower control arm nuts at this time. The weight of the vehicle must be supported by the control arms, since the vehicle design trim heights are obtained before tightening the lower control arm mounting nuts.**

   - Ball joint stud to the steering knuckle. Torque the lower ball joint nut to 88 inch lbs. (10 Nm), then an additional 120 degree turn, during which a torque of 41 ft. lbs. (55 Nm) must be obtained.
   - New cotter pin

➡ **Align the slot in the nut with the hole in the lower ball joint stud. Torque the ball joint nut up to one more flat. Never loosen the nut to align the slot.**

   - Control arm rear mounting bolts to 117 ft. lbs. (158 Nm), then the front mounting nut to 117 ft. lbs. (158 Nm)

- Stabilizer shaft link assembly. Torque the nuts to 13 ft. lbs. (17 Nm).
- Front wheel. Torque the lug nuts to 100 ft. lbs. (140 Nm).

5. On models where applicable, torque the front lower control arm mounting nut to 140 ft. lbs. (190 Nm) and the rear lower control arm nut to 93 ft. lbs. (126 Nm).

6. Check and adjust the wheel alignment.

Rear

1. Before servicing the vehicle, refer to the precautions in the beginning of this section.

2. Remove or disconnect the following:
- Rear suspension support assembly
- Electronic Ride Control height sensor
- Tie rod
- Stabilizer link assembly from the control arm
- Antilock Brake System (ABS) electrical connector
- Hub and bearing
- Lower control arm
- Adjustment link from the control arm, if necessary
- Nuts and bolts and the control arm

To install:
3. Install or connect the following:
- Lower control arm to the frame.

➡**Tighten the control arm nuts with the vehicle unsupported and resting on the wheels at the normal trim height.**

- Bolts and the nuts
- Hub and bearing
- Adjustment link to the control arm, if necessary. tighten the nut to 36 ft. lbs. (50 Nm).
- ABS electrical connector
- Tie rod
- Stabilizer link assembly to the control arm
- Height sensor
- Rear wheel

4. Lower the vehicle and tighten the control arm nuts to 78 ft. lbs. (106 Nm).

CONTROL ARM BUSHING REPLACEMENT

Except Park Avenue

FRONT

1. Before servicing the vehicle, refer to the precautions in the beginning of this section.

2. Remove or disconnect the following:
- Front wheel
- Lower control arm

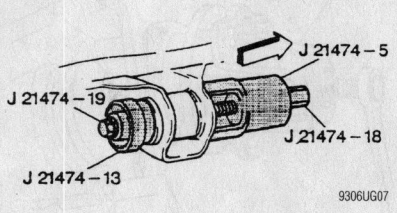

Removing the bushing from the control arm

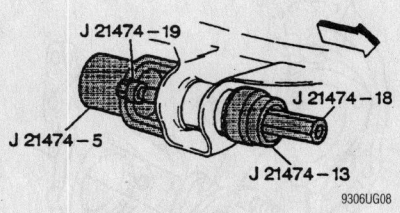

Installing the bushing into the control arm

3. Press the control arm bushing out of the control arm.

To install:
4. Insert the new bushing into control arm.

5. Press the new bushing into the control arm.

6. Install the lower control arm.

7. Install the front wheel. Torque the lug nuts to 100 ft. lbs. (140 Nm).

REAR

1. Before servicing the vehicle, refer to the precautions in the beginning of this section.

2. Remove the control arm.

3. Place tool J 21474-27 with the washer through tool J 41014-2 over the bushing against the control arm.

4. Lubricate the bolt threads with high pressure lubricant.

5. Install tool J 41014-1 (with small end facing the bushing), the thrust bearing and tool J 21474-4 onto tool J 21474-27.

6. Tighten the nut until the bushing is driven out of the control arm.

7. Remove the bushing tools.

To install:
8. Start the bushing into the control arm.

9. Make sure that the flat on the bushing is vertical and facing rearward. Place

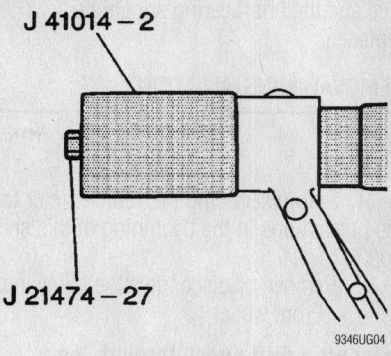

Removing the bushing from the rear control arm

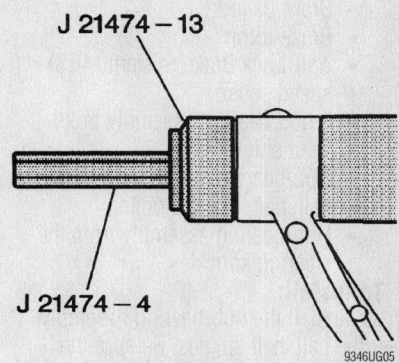

Installing the bushing into the rear control arm

tool J 21474-27 with the washer through tool J41014-1 over the bushing against the control arm.

10. Lubricate the bolt threads with high pressure lubricant.

11. Install tool J 21474-13 (making sure the large end is facing the bushing), the thrust bearing and tool J 21474-4 onto tool J 21474-27.

12. Tighten the bolt until the bushing is fully seated into the control arm.

13. Remove the bushing tools.

14. Install the control arm.

Park Avenue

The lower control arm is replaced as a unit. The bushing can not be removed.

Wheel Bearings

ADJUSTMENT

The wheel bearings are not adjustable. If a wheel bearing is out of specifications, it must be replaced. Using a dial indicator, check for looseness. If play exceeds 0.005 inch (0.127mm), the bearing wear is excessive and the hub/bearing should be replaced.

REMOVAL & INSTALLATION

Front

1. Before servicing the vehicle, refer to the precautions in the beginning of this section.
2. Remove or disconnect the following:
   • Front wheel

➡ **Insert a drift punch through the caliper and into the rotor cooling fins to prevent the rotor from turning.**

   • Halfshaft nut and washer
   • Brake caliper
   • Brake rotor
   • Anti-Lock Brake System (ABS) speed sensor
   • 3 hub/bearing assembly bolts
   • Dust shield
   • Hub/bearing assembly from the halfshaft, using a puller
   • Hub/bearing assembly from the steering knuckle

To install:

3. Install the hub/bearing assembly over the halfshaft splines. Be sure the splines engage smoothly.
4. Apply a light coating of grease to the steering knuckle bore.
5. Slide the hub assembly onto the halfshaft as far as possible. If the hub will not bottom out on the halfshaft, install the hub mounting bolts and use the halfshaft nut to draw the hub onto the halfshaft.
6. Once the hub is flush with the steering knuckle, remove the mounting bolts and install the dust shield.
7. Install or connect the following:
8. Install the mounting bolts. Torque the bolts to 70 ft. lbs. (95 Nm).
9. Place the transaxle in **N**.
10. Install or connect the following:

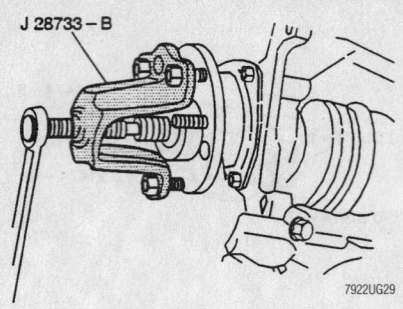

J 28733 – B

7922UG29

Use a puller such as J 28733-B to press the halfshaft from the hub assembly

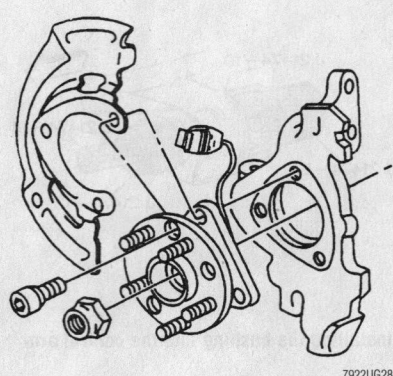

7922UG28

Exploded view of the front hub and wheel bearing assembly

   • ABS front wheel speed sensor connector, and clip to the dust shield
   • Brake rotor
   • Caliper. Torque the bolts to 38 ft. lbs. (51 Nm).
11. Torque the halfshaft nut to 107 ft. lbs. (145 Nm) on all Park Avenue models, or 118 ft. lbs. (160 Nm) on 2000–04 models except Park Avenue.
12. Install the front wheels. Torque the lug nuts to 100 ft. lbs. (140 Nm).
13. Road test the vehicle.

Rear

1. Before servicing the vehicle, refer to the precautions in the beginning of this section.
2. Remove or disconnect the following:
   • Wheel and the tire
   • Brake caliper
   • Brake rotor
   • ABS sensor wire connector.
   • Four bolts from the control arm
   • Hub and bearing from the control arm
   • Brake shield from the control arm

To install:

3. Clean the control arm face and the bore before installing the hub and the bearing.
4. Install or connect the following:
   • Brake shield and the hub and the bearing to the control arm with the four bolts. Reconnect the ABS sensor. Tighten the hub and bearing bolts to 52 ft. lbs. (70 Nm).
   • Brake rotor
   • Brake caliper
   • Wheel and the tire

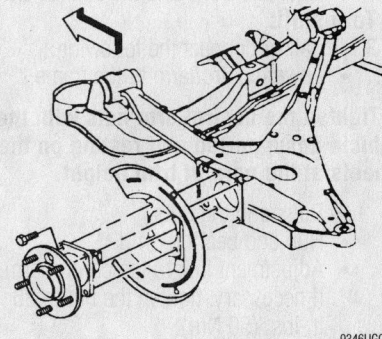

9346UG06

Exploded view of the rear hub/wheel bearing assembly

BRAKES

Brake Caliper

REMOVAL & INSTALLATION

Front & Rear

➥**The Bosch 2U ABS system cannot increase brake pressure above master cylinder pressure applied by during braking. There is no need to depressurize the system prior to service.**

1. Remove brake fluid from the master cylinder reservoir until the reservoir is approximately ⅓ full.

2. Remove the front wheel. Mark the position of the wheel to the wheel studs, prior to removal, for installation reference.

3. Install 2 lug nuts to retain the rotor once the caliper is removed.

4. Using a large C-clamp, bottom the piston in the caliper bore by positioning the C-clamp on the outboard pad and on the round portion of the brake caliper where the piston is housed.

5. Remove the banjo bolt that fastens the brake hose to the brake caliper. Discard the gaskets.

6. Cap the brake line to avoid excessive fluid loss or fluid contamination.

7. Remove the rubber dust boots from the caliper mounting bolt heads (if equipped).

8. Remove the caliper mounting bolts.

9. Remove the park brake cable from the caliper (if working on the rear).

10. Remove the caliper from the vehicle.

11. Remove the brake pads.

To install:

12. Install the brake pads. Lubricate the slides where the caliper mounts on the steering knuckle with silicone grease.

13. Install the caliper over the rotor.

14. Install the caliper mounting bolts.

Torque the mounting bolts to 38 ft. lbs. (51 Nm).

15. Install the park brake cable (if working on the rear).

16. Install the rubber dust boots over the caliper mounting bolt heads (if equipped).

17. Check the clearance between the brake caliper and caliper bracket stops. If the clearance is too tight, check the caliper leading and trailing edges for build up. File down as necessary.

18. Connect the brake hose to the caliper. Install the brake hose banjo bolt, using new gaskets, and torque to 33 ft. lbs. (45 Nm).

19. Remove the lug nuts used to secure the rotor.

20. Install the wheel, aligning the reference marks made during removal, and torque the lug nuts to 100 ft. lbs. (140 Nm).

21. Refill the master cylinder and bleed the brake system using the recommended procedure.

22. Road test the vehicle and check for proper braking performance

Disc Brake Pads

REMOVAL & INSTALLATION

1. Remove brake fluid from the master cylinder reservoir until the reservoir is approximately ⅓ full. Discard the removed fluid.

2. Remove the front wheel. Mark the position of the wheel to the wheel studs, prior to removal, for installation reference.

3. Install 2 lug nuts to retain the rotor once the caliper is removed.

4. Remove the caliper mounting sleeve bolts. Support the caliper out of the way using wire. DO NOT disconnect the brake

hose or allow the caliper to hang from the brake hose.

5. Remove the outboard pad by pushing it in toward the piston until the mounting tabs clear the holes in the caliper body. With the tabs clear of the holes, push the pad out the bottom of the caliper.

6. Remove the inboard pad from the piston by pulling the top of the pad out to disengage the retainer spring.

To install:

7. Before installing the pads in the caliper, the piston must be fully seated in the bore. A large C-clamp can be used to compress the piston.

8. Install the inboard pad in the caliper by inserting the top pad ears in first, then sliding the bottom of the pad into place until the spring clip snaps into place. Make sure the inboard pad seats flush against the caliper piston.

9. Install the outboard pad by lining up the tabs on the rear of the pad with the mounting holes in the caliper body. Press the pad firmly down into the caliper until the tabs snap into the mounting holes.

10. Clean and lubricate the caliper bolt and sleeve assemblies and install them into the caliper. Lubricate the caliper slides and mountings.

11. Install the caliper over the rotor and torque the mounting bolts to 38 ft. lbs. (51 Nm).

12. Remove the 2 lug nuts used to secure the rotor in place.

13. Install the wheel, aligning the marks made during removal, and torque the lug nuts to 100 ft. lbs. (140 Nm).

14. Pump the brake pedal several times to seat the pads against the rotor.

15. Refill the master cylinder using DOT 3 brake fluid only and road test to verify proper brake operation.

Brake Drums

REMOVAL & INSTALLATION

1. Remove the rear wheel. Mark the position of the wheel on the wheel studs, prior to removal, for installation reference.

2. Remove the brake drum from the hub. Mark the position of the drum on the wheel studs, prior to removal, for installation reference.

3. If the brake drum is difficult to remove, check the following:

a. Make sure that the parking brake is released.

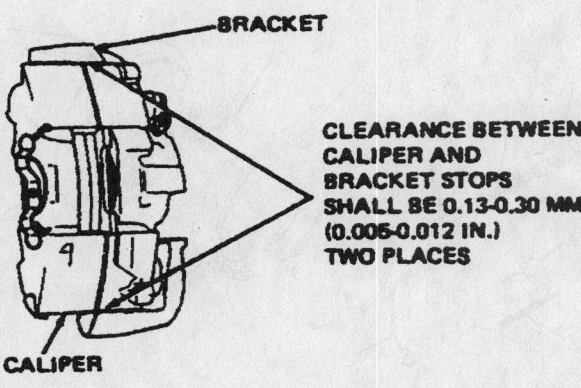

CLEARANCE BETWEEN CALIPER AND BRACKET STOPS SHALL BE 0.13-0.30 MM (0.005-0.012 IN.) TWO PLACES

93006G60

Measuring the caliper clearance

b. Back off on the parking brake cable adjustment.

c. Use a hammer and a small metal punch to bend in the backing plate knockout to provide access to the park brake lever.

d. Insert a screwdriver through the hole and press in to push the park brake lever off its stop. This lets the brake shoes retract slightly.

e. Apply a small amount of penetrating oil around the drum pilot hole.

4. Remove the brake drum from the vehicle.

5. Inspect the wheel cylinder for leakage and inspect the brake shoes for wear; replace as necessary.

6. Inspect the brake drum for scoring, cracks or other wear; machine or replace as necessary. If machining, observe the maximum diameter specification.

To install:

7. Measure the inside diameter of the brake drum using clearance gauge J 21177-A.

8. Turn the starwheel on the adjusting screw assembly until the brake shoe diameter is 0.050 in. (1.27mm) less than the drum inside diameter.

9. Install the brake drum onto the wheel hub, aligning the marks made during removal.

10. Install the wheel, aligning the marks made during removal. Torque the lug nuts to 100 ft. lbs. (140 Nm).

11. Road test the vehicle and check brake operation.

Brake Shoes

REMOVAL & INSTALLATION

1. Remove the rear wheel. Mark the position of the wheel to the wheel studs, prior to removal, for installation reference.

2. Remove the brake drum. Mark the position of the brake drum to the wheel studs, prior to removal, for installation reference.

3. Using tool J 38400 brake spanner tool and remover, remove the actuator spring from the adjuster lever. Use care not to distort the spring when removing it.

4. Lift the end of the retractor spring from the adjuster shoe assembly. Insert the hook end of the J-38400 between the retractor spring and the shoe. Pry slightly to remove the spring end from the hole in the shoe.

5. Pry the end of the retractor spring toward the axle with the flat end of the tool

until the spring snaps down off the shoe web onto the backing plate.

6. Remove the one brake shoe and remove the adjuster assembly.

7. Disconnect the parking brake lever from the shoe. DO NOT remove the parking brake lever from the cable end unless it is being replaced.

8. Using J 38400, lift the end of the retractor spring from the adjuster shoe assembly. Insert the hook end of the tool between the retractor spring and the shoe. Pry slightly to remove the spring end from the hole in the shoe. Pry the end of the retractor spring toward the axle with the flat end of the tool, until the spring snaps down off the shoe web onto the backing plate. Leave the spring there. Do not remove it.

9. Remove the brake shoe.

To install:

10. Check the backing plate attaching bolts to make sure that they are tight. Use fine emery cloth to clean all rust and dirt from the shoe contact surfaces on the plate and lubricate with high temperature grease. Check the wheel cylinder for signs of leakage.

11. Clean all parts completely in brake solvent and air dry.

12. Clean the backing plate shoe contact points.

13. Inspect the inside of the brake drum. If worn, heavily grooved or if the opening is distorted, the drum should be machined or replaced. If machining, observe the maximum drum diameter specification.

14. Disassemble, clean and lubricate the adjuster screw.

15. Position the brake shoe that con-

nects to the parking brake lever, on the backing plate. Using J-38400, pull the end of the retractor spring up to rest on the web of the shoe. Pull the end of the retractor spring up until it snaps into the slot in the brake shoe.

16. Connect the parking brake lever.

17. Install the remaining shoe and the adjuster screw assembly.

18. Position the brake shoe using J-38400 and pull the end of the retractor spring up to rest on the web of the shoe. Pull the end of the retractor spring up until it snaps into the slot in the brake shoe.

19. Using J-38400, spread the brake shoes and work the adjuster screw into position.

20. Install the actuator spring with the U-shaped end going through the web.

21. Measure the inside diameter of the brake drum using clearance gauge J 21177-A.

22. Turn the starwheel on the adjusting screw assembly until the brake shoe diameter is 0.050 in. (1.27mm) less than the drum inside diameter.

23. Install the brake drum onto the wheel hub, aligning the marks made during removal.

24. Install the wheel, aligning the marks made during removal. Torque the lug nuts to 100 ft. lbs. (140 Nm).

25. Repeat the procedure for the brake shoe assembly on the opposite side of the vehicle.

26. Apply and release the brake pedal 30–35 times with normal pedal force. Pause about 1 second between applications.

27. Road test the vehicle and check brake operation.

A	ACCESS HOLE PLUG.	4	RETRACTOR SPRING	9	BACKING PLATE
		5	ADJUSTER SHOE AND LINING	10	PARK BRAKE SHOE AND LINING
1	ADJUSTER SOCKET	6	WHEEL CYLINDER	11	PARK BRAKE LEVER
2	ADJUSTER SCREW	7	BLEEDER VALVE	12	ACTUATOR SPRING
3	PIVOT NUT	8	BOLT	13	ADJUSTER ACTUATOR

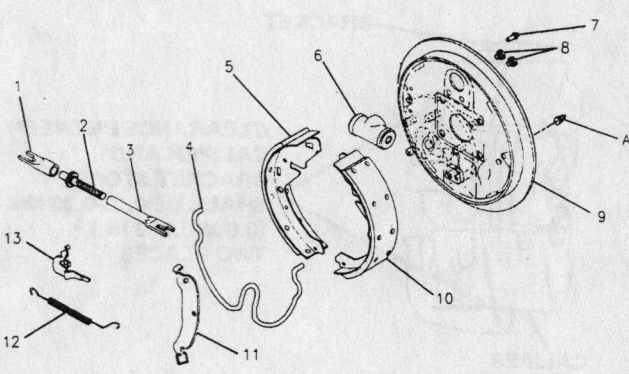

Exploded view of the drum brake components

93006G77

CADILLAC

Deville • Eldorado • Seville

15

SPECIFICATION CHARTS

ENGINE AND VEHICLE IDENTIFICATION

		Engine						Model Year	
Code ①	Liters (cc)	Cu. In.	Cyl.	Fuel Sys.	Engine Type	Eng. Mfg.		Code ②	Year
9	4.6 (4565)	279	8	MFI	DOHC	Cadillac		Y	2000
Y	4.6 (4565)	279	8	MFI	DOHC	Cadillac		1	2001
								2	2002
								3	2003
								4	2004

DOHC: Double Overhead Camshafts

MFI: Multi-point Fuel Injection

① 8th position of VIN

② 10th position of VIN

42372-EKBO-C01

GENERAL ENGINE SPECIFICATIONS

Year	Model	Engine Displacement Liters (cc)	Engine Series (ID/VIN)	Fuel System	Net Horsepower @ rpm	Net Torque @ rpm (ft. lbs.)	Bore x Stroke (in.)	Com-pression Ratio	Oil Pressure @ rpm
2000	DeVille	4.6 (4565)	Y	MFI	275@5600	300@4000	3.66x3.31	10.0:1	35@2000
	DeVille DTS	4.6 (4565)	9	MFI	300@6000	295@4400	3.66x3.31	10.0:1	35@2000
	DeVille DHS	4.6 (4565)	Y	MFI	275@5600	300@4000	3.66x3.31	10.0:1	35@2000
	Eldorado	4.6 (4565)	Y	MFI	275@5600	300@4000	3.66x3.31	10.0:1	35@2000
	Eldorado ETC	4.6 (4565)	9	MFI	300@6000	295@4400	3.66x3.31	10.0:1	35@2000
	Seville SLS	4.6 (4565)	Y	MFI	275@5600	300@4000	3.66x3.31	10.0:1	35@2000
	Seville STS	4.6 (4565)	9	MFI	300@6000	295@4400	3.66x3.31	10.0:1	35@2000
2001	DeVille	4.6 (4565)	Y	MFI	275@5600	300@4000	3.66x3.31	10.0:1	35@2000
	DeVille DTS	4.6 (4565)	9	MFI	300@6000	295@4400	3.66x3.31	10.0:1	35@2000
	DeVille DHS	4.6 (4565)	Y	MFI	275@5600	300@4000	3.66x3.31	10.0:1	35@2000
	Eldorado	4.6 (4565)	Y	MFI	275@5600	300@4000	3.66x3.31	10.0:1	35@2000
	Eldorado ETC	4.6 (4565)	9	MFI	300@6000	295@4400	3.66x3.31	10.0:1	35@2000
	Seville SLS	4.6 (4565)	Y	MFI	275@5600	300@4000	3.66x3.31	10.0:1	35@2000
	Seville STS	4.6 (4565)	9	MFI	300@6000	295@4400	3.66x3.31	10.0:1	35@2000
2002	DeVille	4.6 (4565)	Y	MFI	275@5600	300@4000	3.66x3.31	10.0:1	35@2000
	DeVille DTS	4.6 (4565)	9	MFI	300@6000	295@4400	3.66x3.31	10.0:1	35@2000
	DeVille DHS	4.6 (4565)	Y	MFI	275@5600	300@4000	3.66x3.31	10.0:1	35@2000
	Eldorado	4.6 (4565)	Y	MFI	275@5600	300@4000	3.66x3.31	10.0:1	35@2000
	Eldorado ETC	4.6 (4565)	9	MFI	300@6000	295@4400	3.66x3.31	10.0:1	35@2000
	Seville SLS	4.6 (4565)	Y	MFI	275@5600	300@4000	3.66x3.31	10.0:1	35@2000
	Seville STS	4.6 (4565)	9	MFI	300@6000	295@4400	3.66x3.31	10.0:1	35@2000
2003	DeVille	4.6 (4565)	Y	MFI	275@5600	300@4000	3.66x3.31	10.0:1	35@2000
	DeVille DTS	4.6 (4565)	9	MFI	300@6000	295@4400	3.66x3.31	10.0:1	35@2000
	DeVille DHS	4.6 (4565)	Y	MFI	275@5600	300@4000	3.66x3.31	10.0:1	35@2000
	Seville SLS	4.6 (4565)	Y	MFI	275@5600	300@4000	3.66x3.31	10.0:1	35@2000
	Seville STS	4.6 (4565)	9	MFI	300@6000	295@4400	3.66x3.31	10.0:1	35@2000

MFI: Multi-point Fuel Injection

42372-EKBO-C02

ENGINE TUNE-UP SPECIFICATIONS

Year	Engine Displacement Liters (cc)	Engine ID/VIN	Spark Plug Gap (in.)	Ignition Timing (deg.)	Fuel Pump (psi)	Idle Speed (rpm)	Valve Clearance Intake	Valve Clearance Exhaust
2000	4.6 (4565)	9	0.050	①	40-50	①	HYD	HYD
	4.6 (4565)	Y	0.050	①	40-50	①	HYD	HYD
2001	4.6 (4565)	9	0.050	①	40-50	①	HYD	HYD
	4.6 (4565)	Y	0.050	①	40-50	①	HYD	HYD
2002	4.6 (4565)	9	0.050	①	40-50	①	HYD	HYD
	4.6 (4565)	Y	0.050	①	40-50	①	HYD	HYD
2003	4.6 (4565)	9	0.050	①	40-50	①	HYD	HYD
	4.6 (4565)	Y	0.050	①	40-50	①	HYD	HYD

NOTE: The Vehicle Emission Control Information label often reflects specification changes made during production. The label figures must be used if they differ from those in this chart.

HYD: Hydraulic

① Refer to Vehicle Emission Control Information label

42372-EKBO-C03

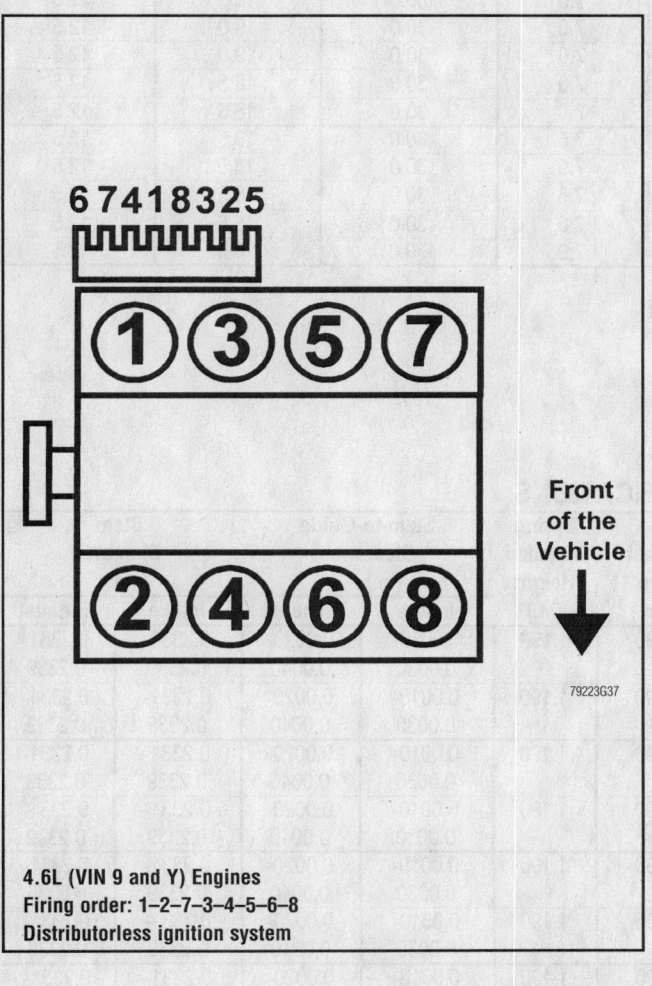

6 7 4 1 8 3 2 5

Front of the Vehicle

79223G37

4.6L (VIN 9 and Y) Engines
Firing order: 1–2–7–3–4–5–6–8
Distributorless ignition system

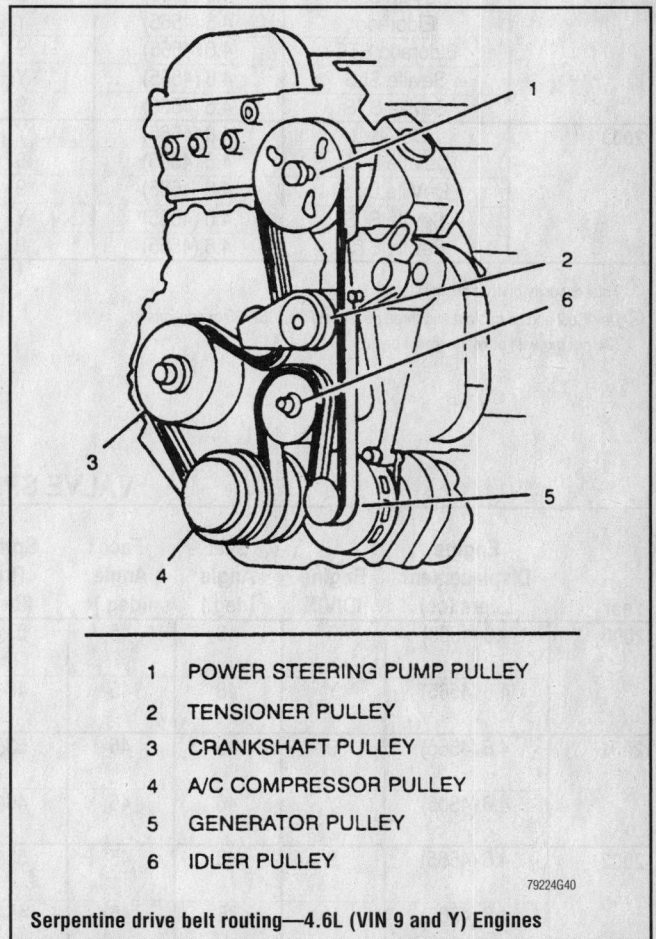

1   POWER STEERING PUMP PULLEY

2   TENSIONER PULLEY

3   CRANKSHAFT PULLEY

4   A/C COMPRESSOR PULLEY

5   GENERATOR PULLEY

6   IDLER PULLEY

79224G40

Serpentine drive belt routing—4.6L (VIN 9 and Y) Engines

CAPACITIES

Year	Model	Engine Displacement Liters (cc)	Engine ID/VIN	Engine Oil with Filter (qts.)	Transmission (pts.) ①	Fuel Tank (gal.)	Cooling System (qts.) ②
2000	DeVille	4.6 (4565)	Y	7.5	30.0	18.5	12.5
	DeVille DHS	4.6 (4565)	9	7.5	30.0	18.5	12.5
	DeVille DTS	4.6 (4565)	9	7.5	30.0	18.5	12.5
	Eldorado	4.6 (4565)	Y	7.0	30.0	19.0	12.5
	Eldorado ETC	4.6 (4565)	9	7.0	30.0	19.0	12.5
	Seville SLS	4.6 (4565)	Y	7.0	30.0	18.5	12.5
	Seville STS	4.6 (4565)	9	7.0	30.0	18.5	12.5
2001	DeVille	4.6 (4565)	Y	7.5	30.0	18.5	12.5
	DeVille DHS	4.6 (4565)	9	7.5	30.0	18.5	12.5
	DeVille DTS	4.6 (4565)	9	7.5	30.0	18.5	12.5
	Eldorado	4.6 (4565)	Y	7.0	30.0	19.0	12.5
	Eldorado ETC	4.6 (4565)	9	7.0	30.0	19.0	12.5
	Seville SLS	4.6 (4565)	Y	7.0	30.0	18.5	12.5
	Seville STS	4.6 (4565)	9	7.0	30.0	18.5	12.5
2002	DeVille	4.6 (4565)	Y	7.5	30.0	18.5	12.5
	DeVille DHS	4.6 (4565)	9	7.5	30.0	18.5	12.5
	DeVille DTS	4.6 (4565)	9	7.5	30.0	18.5	12.5
	Eldorado	4.6 (4565)	Y	7.0	30.0	19.0	12.5
	Eldorado ETC	4.6 (4565)	9	7.0	30.0	19.0	12.5
	Seville SLS	4.6 (4565)	Y	7.0	30.0	18.5	12.5
	Seville STS	4.6 (4565)	9	7.0	30.0	18.5	12.5
2003	DeVille	4.6 (4565)	Y	7.5	30.0	18.5	12.5
	DeVille DHS	4.6 (4565)	9	7.5	30.0	18.5	12.5
	DeVille DTS	4.6 (4565)	9	7.5	30.0	18.5	12.5
	Seville SLS	4.6 (4565)	Y	7.0	30.0	18.5	12.5
	Seville STS	4.6 (4565)	9	7.0	30.0	18.5	12.5

① Total capacity of dry transmission
② Dex-Cool engine coolant and three pellets of P/N 1052753 or equivalent
(Do not mix with ethylene glycol base)

42372-EKBO-C04

VALVE SPECIFICATIONS

Year	Engine Displacement Liters (cc)	Engine ID/VIN	Seat Angle (deg.)	Face Angle (deg.)	Spring Test Pressure (lbs. @ in.)	Spring Installed Height (in.)	Stem-to-Guide Clearance (in.) Intake	Exhaust	Stem Diameter (in.) Intake	Exhaust
2000	4.6 (4565)	9	46	45	53@1.190	1.190	0.0010-0.0030	0.0020-0.0040	0.2331-0.2339	0.2331-0.2339
	4.6 (4565)	Y	46	45	46@1.190	1.190	0.0010-0.0030	0.0020-0.0040	0.2331-0.2339	0.2331-0.2339
2001	4.6 (4565)	9	46	45	53@1.190	1.190	0.0010-0.0030	0.0020-0.0040	0.2331-0.2339	0.2331-0.2339
	4.6 (4565)	Y	46	45	46@1.190	1.190	0.0010-0.0030	0.0020-0.0040	0.2331-0.2339	0.2331-0.2339
2002	4.6 (4565)	9	46	45	53@1.190	1.190	0.0010-0.0030	0.0020-0.0040	0.2331-0.2339	0.2331-0.2339
	4.6 (4565)	Y	46	45	46@1.190	1.190	0.0010-0.0030	0.0020-0.0040	0.2331-0.2339	0.2331-0.2339
2003	4.6 (4565)	9	46	45	53@1.190	1.190	0.0010-0.0030	0.0020-0.0040	0.2331-0.2339	0.2331-0.2339
	4.6 (4565)	Y	46	45	46@1.190	1.190	0.0010-0.0030	0.0020-0.0040	0.2331-0.2339	0.2331-0.2339

42372-EKBO-C05

PISTON AND RING SPECIFICATIONS
All measurements are given in inches.

Year	Engine Displacement Liters (cc)	Engine ID/VIN	Piston Clearance	Ring Gap			Ring Side Clearance		
				Top Compression	Bottom Compression	Oil Control	Top Compression	Bottom Compression	Oil Control
2000	4.6 (4565)	9	0.0008-0.0020	0.010-0.016	0.014-0.020	0.010-0.030	0.0016-0.0037	0.0016-0.0037	①
	4.6 (4565)	Y	0.0008-0.0020	0.010-0.016	0.014-0.020	0.010-0.030	0.0016-0.0037	0.0016-0.0037	①
2001	4.6 (4565)	9	0.0008-0.0020	0.010-0.016	0.014-0.020	0.010-0.030	0.0016-0.0037	0.0016-0.0037	①
	4.6 (4565)	Y	0.0008-0.0020	0.010-0.016	0.014-0.020	0.010-0.030	0.0016-0.0037	0.0016-0.0037	①
2002	4.6 (4565)	9	0.0008-0.0020	0.010-0.016	0.014-0.020	0.010-0.030	0.0016-0.0037	0.0016-0.0037	①
	4.6 (4565)	Y	0.0008-0.0020	0.010-0.016	0.014-0.020	0.010-0.030	0.0016-0.0037	0.0016-0.0037	①
2003	4.6 (4565)	9	0.0008-0.0020	0.010-0.016	0.014-0.020	0.010-0.030	0.0016-0.0037	0.0016-0.0037	①
	4.6 (4565)	Y	0.0008-0.0020	0.010-0.016	0.014-0.020	0.010-0.030	0.0016-0.0037	0.0016-0.0037	①

① Side sealing

42372-EKBO-C06

CRANKSHAFT AND CONNECTING ROD SPECIFICATIONS
All measurements are given in inches.

Year	Engine Displacement Liters (cc)	Engine ID/VIN	Crankshaft				Connecting Rod		
			Main Brg. Journal Dia.	Main Brg. Oil Clearance	Shaft End-play	Thrust on No.	Journal Diameter	Oil Clearance	Side Clearance
2000	9	4.6 (4565)	2.5335-2.5337	0.0006-0.0025	0.0020-0.0200	3	2.1239-2.1235	0.001-0.003	0.0080-0.0200
	Y	4.6 (4565)	2.5335-2.5337	0.0006-0.0025	0.0020-0.0200	3	2.1239-2.1235	0.001-0.003	0.0080-0.0200
2001	9	4.6 (4565)	2.5335-2.5337	0.0006-0.0025	0.0020-0.0200	3	2.1239-2.1235	0.001-0.003	0.0080-0.0200
	Y	4.6 (4565)	2.5335-2.5337	0.0006-0.0025	0.0020-0.0200	3	2.1239-2.1235	0.001-0.003	0.0080-0.0200
2002	9	4.6 (4565)	2.5335-2.5337	0.0006-0.0025	0.0020-0.0200	3	2.1239-2.1235	0.001-0.003	0.0080-0.0200
	Y	4.6 (4565)	2.5335-2.5337	0.0006-0.0025	0.0020-0.0200	3	2.1239-2.1235	0.001-0.003	0.0080-0.0200
2003	9	4.6 (4565)	2.5335-2.5337	0.0006-0.0025	0.0020-0.0200	3	2.1239-2.1235	0.001-0.003	0.0080-0.0200
	Y	4.6 (4565)	2.5335-2.5337	0.0006-0.0025	0.0020-0.0200	3	2.1239-2.1235	0.001-0.003	0.0080-0.0200

42372-EKBO-C07

TORQUE SPECIFICATIONS
All readings in ft. lbs.

Year	Engine Displacement Liters (cc)	Engine ID/VIN	Cylinder Head Bolts	Main Bearing Bolts	Rod Bearing Bolts	Crankshaft Damper Bolts	Flywheel Bolts	Manifold Intake	Manifold Exhaust	Spark Plugs	Lug Nuts
2000	4.6 (4565)	9	①	②	③	④	⑤	⑥	18	11	100
	4.6 (4565)	Y	①	②	③	④	⑤	⑥	18	11	100
2001	4.6 (4565)	9	①	②	③	④	⑤	⑥	18	11	100
	4.6 (4565)	Y	①	②	③	④	⑤	⑥	18	11	100
2002	4.6 (4565)	9	①	②	③	④	⑤	⑥	18	11	100
	4.6 (4565)	Y	①	②	③	④	⑤	⑥	18	11	100
2003	4.6 (4565)	9	①	②	③	④	⑤	⑥	18	11	100
	4.6 (4565)	Y	①	②	③	④	⑤	⑥	18	11	100

① M11 bolts:
Step 1: 30 ft. lbs. plus an additional 70 degrees
Step 2: Plus 70 degrees
Step 3: Plus 60 degrees
Step 4: Plus 60 degrees for a total of 190 degrees

② M10 bolts:
Step 1: 15 ft. lbs.
Step 2: Plus 65 degrees
M8 bolts:
22 ft. lbs.

③ Step 1: 22 ft. lbs.
Step 2: Loosen completely
Step 3: 18 ft. lbs.
Step 4: Plus 110 degrees

④ Step 1: 37 ft. lbs.
Step 2: Plus 120 degrees

⑤ Step 1: 11 ft. lbs.
Step 2: Plus 50 degrees

⑥ 89 inch lbs.

42372-EKBO-C08

TIRE, WHEEL AND BALL JOINT SPECIFICATIONS
Cadillac

Year	Model	OEM Tires Standard	OEM Tires Optional	Tire Pressures (psi) Front	Tire Pressures (psi) Rear	Wheel Size	Ball Joint Inspection
2000	DeVille	P225/60R16	None	30	30	7-JJ	①
	Eldorado	P225/60R16	P225/60ZR16	Std: 28 Opt: 29	Std: 26 Opt: 29	7-JJ	①
	Seville	P225/60R16	P225/60ZR16	Std: 28 Opt: 29	Std: 26 Opt: 29	7-JJ	①
2001	DeVille	P225/60R16	None	30	30	7-JJ	①
	Eldorado	P225/60R16	P225/60ZR16	Std: 28 Opt: 29	Std: 26 Opt: 29	7-JJ	①
	Seville	P225/60R16	P225/60ZR16	Std: 28 Opt: 29	Std: 26 Opt: 29	7-JJ	①
2002	DeVille	P225/60R16	None	30	30	7-JJ	①
	Eldorado	P225/60R16	P225/60ZR16	Std: 28 Opt: 29	Std: 26 Opt: 29	7-JJ	①
	Seville	P225/60R16	P225/60ZR16	Std: 28 Opt: 29	Std: 26 Opt: 29	7-JJ	①
2003	DeVille	P225/60R16	None	30	30	7-JJ	①
	Seville	P225/60R16	P225/60ZR16	Std: 28 Opt: 29	Std: 26 Opt: 29	7-JJ	①

OEM: Original Equipment Manufacturer

PSI: Pounds Per Square Inch

STD: Standard

OPT: Optional

L: Lower

U: Upper

① Replace if any measurable movement is found.

42372-EKBO-C09

WHEEL ALIGNMENT
CADILLAC SEVILLE

Year	Model		Caster Range (Deg.)	Caster Preferred Setting (Deg.)	Camber Range (Deg.)	Camber Preferred Setting (Deg.)	Toe-in (in.)	Steering Axis Inclination (Deg.)
2000	Seville	F	+0.50	+6.00	+0.50	-0.20	0.09 +/- 0.09	—
		R	—	—	+0.50	-0.30	0.09 +/- 0.09	—
2001	Seville	F	+0.50	+6.00	+0.50	-0.20	0.09 +/- 0.09	—
		R	—	—	+0.50	-0.30	0.09 +/- 0.09	—
2002	Seville	F	+0.50	+6.00	+0.50	-0.20	0.09 +/- 0.09	—
		R	—	—	+0.50	-0.30	0.09 +/- 0.09	—
2003	Seville	F	+0.50	+6.00	+0.50	-0.20	0.09 +/- 0.09	—
		R	—	—	+0.50	-0.30	0.09 +/- 0.09	—

42372-EKBO-C10

WHEEL ALIGNMENT
CADILLAC DeVille

Year	Model		Caster Range (Deg.)	Caster Preferred Setting (Deg.)	Camber Range (Deg.)	Camber Preferred Setting (Deg.)	Toe-in (in.)
2000	DeVille	F	0.50	+6.00	+0.50	-0.20	0.09 +/- 0.09
		R	—	—	+0.50	-0.30	0.09 +/- 0.09
2001	DeVille	F	+0.50	+6.00	+0.50	-0.20	0.09 +/- 0.09
		R	—	—	+0.50	-0.30	0.09 +/- 0.09
2002	DeVille	F	+0.50	+6.00	+0.50	-0.20	0.09 +/- 0.09
		R	—	—	+0.50	-0.30	0.09 +/- 0.09
2003	DeVille	F	+0.50	+6.00	+0.50	-0.20	0.09 +/- 0.09
		R	—	—	+0.50	-0.30	0.09 +/- 0.09

42372-EKBO-C11

WHEEL ALIGNMENT
CADILLAC ELDORADO

Year	Model		Caster Range (Deg.)	Caster Preferred Setting (Deg.)	Camber Range (Deg.)	Camber Preferred Setting (Deg.)	Toe-in (in.)	Steering Axis Inclination (Deg.)
2000	Eldorado	F	+1.00	+2.30	+0.50	0	0.20 +/- 0.20	—
		R	—	—	+0.50	0	0.20 +/- 0.20	—
2001	Eldorado	F	+1.00	+2.30	+0.50	0	0.20 +/- 0.20	—
		R	—	—	+0.50	0	0.20 +/- 0.20	—
2002	Eldorado	F	+1.00	+2.30	+0.50	0	0.20 +/- 0.20	—
		R	—	—	+0.50	0	0.20 +/- 0.20	—

42372-EKBO-C12

BRAKE SPECIFICATIONS
All measurements in inches unless noted

Year	Model		Original Thickness	Minimum Thickness	Maximum Runout	Minimum Lining Thickness	Brake Caliper Mounting Bolts (ft. lbs.)
				Brake Disc			
2000	DeVille	F	1.268	1.209	0.002	0.030	38
		R	0.433	0.374	0.002	0.030	20
	DeVille DHS	F	1.268	1.209	0.002	0.030	38
		R	0.433	0.374	0.002	0.030	20
	DeVille DTS	F	1.268	1.209	0.002	0.030	38
		R	0.433	0.374	0.002	0.030	20
	Eldorado	F	1.268	1.209	0.002	0.030	38
		R	0.433	0.374	0.002	0.030	20
	Eldorado ETC	F	1.268	1.209	0.002	0.030	38
		R	0.433	0.374	0.002	0.030	20
	Seville SLS	F	1.268	1.209	0.002	0.030	38
		R	0.433	0.374	0.002	0.030	20
	Seville STS	F	1.268	1.209	0.002	0.030	38
		R	0.433	0.374	0.002	0.030	20
2001	DeVille	F	1.268	1.209	0.002	0.030	38
		R	0.433	0.374	0.002	0.030	20
	DeVille DHS	F	1.268	1.209	0.002	0.030	38
		R	0.433	0.374	0.002	0.030	20
	DeVille DTS	F	1.268	1.209	0.002	0.030	38
		R	0.433	0.374	0.002	0.030	20
	Eldorado	F	1.268	1.209	0.002	0.030	38
		R	0.433	0.374	0.002	0.030	20
	Eldorado ETC	F	1.268	1.209	0.002	0.030	38
		R	0.433	0.374	0.002	0.030	20
	Seville SLS	F	1.268	1.209	0.002	0.030	38
		R	0.433	0.374	0.002	0.030	20
	Seville STS	F	1.268	1.209	0.002	0.030	38
		R	0.433	0.374	0.002	0.030	20
2002	DeVille	F	1.268	1.209	0.002	0.030	38
		R	0.433	0.374	0.002	0.030	20
	DeVille DHS	F	1.268	1.209	0.002	0.030	38
		R	0.433	0.374	0.002	0.030	20
	DeVille DTS	F	1.268	1.209	0.002	0.030	38
		R	0.433	0.374	0.002	0.030	20
	Eldorado	F	1.268	1.209	0.002	0.030	38
		R	0.433	0.374	0.002	0.030	20
	Eldorado ETC	F	1.268	1.209	0.002	0.030	38
		R	0.433	0.374	0.002	0.030	20
	Seville SLS	F	1.268	1.209	0.002	0.030	38
		R	0.433	0.374	0.002	0.030	20
	Seville STS	F	1.268	1.209	0.002	0.030	38
		R	0.433	0.374	0.002	0.030	20
2003	DeVille	F	1.268	1.209	0.002	0.030	38
		R	0.433	0.374	0.002	0.030	20
	DeVille DHS	F	1.268	1.209	0.002	0.030	38
		R	0.433	0.374	0.002	0.030	20
	DeVille DTS	F	1.268	1.209	0.002	0.030	38
		R	0.433	0.374	0.002	0.030	20
	Seville SLS	F	1.268	1.209	0.002	0.030	38
		R	0.433	0.374	0.002	0.030	20
	Seville STS	F	1.268	1.209	0.002	0.030	38
		R	0.433	0.374	0.002	0.030	20

42372-EKBO-C13

SCHEDULED MAINTENANCE INTERVALS
GM E & K BODIES—CADILLAC DEVILLE, ELDORADO & SEVILLE

TO BE SERVICED	TYPE OF SERVICE	VEHICLE MILEAGE INTERVAL (x1000)												
		7.5	15	22.5	30	37.5	45	52.5	60	67.5	75	82.5	90	97.5
Engine oil & filter	R	✓	✓	✓	✓	✓	✓	✓	✓	✓	✓	✓	✓	✓
Coolant level, hoses & clamps	S/I	✓	✓	✓	✓	✓	✓	✓	✓	✓	✓	✓	✓	✓
Drive shaft boots & front suspension components	S/I	✓	✓	✓	✓	✓	✓	✓	✓	✓	✓	✓	✓	✓
Exhaust system, brake hoses & throttle linkage	S/I	✓	✓	✓	✓	✓	✓	✓	✓	✓	✓	✓	✓	✓
Lubricate chassis, suspension, steering linkage, transaxle shift linkage, parking brake cable guides, underbody contact points & linkage	S/I	✓	✓	✓	✓	✓	✓	✓	✓	✓	✓	✓	✓	✓
Brake linings	S/I	✓		✓		✓		✓		✓		✓		✓
Rotate tires	S/I	✓		✓		✓		✓		✓		✓		✓
Inspect throttle body bore & throttle plate for deposits	S/I		✓			✓					✓		✓	
Air filter element	R				✓				✓				✓	
Engine coolant ①	R													
PCV valve	R				✓				✓				✓	
Spark plugs ②	R				✓				✓				✓	
Accessory drive belt(s)	S/I				✓				✓				✓	
Automatic transaxle fluid & filter	S/I				✓				✓				✓	
EGR & fuel systems	S/I				✓				✓				✓	
Ignition cables	S/I				✓				✓				✓	

R: Replace   S/I: Service or Inspect

① Engine coolant: replace every 100,000 miles. Use O.E. specified (DEX-COOL™) coolant only. If any silicate coolant is used, the service interval is every 30,000 miles.

② Platinum tip spark plugs: replace every 100,000 miles.

FREQUENT OPERATION MAINTENANCE (SEVERE SERVICE)

If a vehicle is operated under any of the following conditions it is considered severe service:

- Extremely dusty areas.
- 50% or more of the vehicle operation is in 32°C (90°F) or higher temperatures, or constant operation in temperatures below 0°C (32°F).
- Prolonged idling (vehicle operation in stop and go traffic).
- Frequent short running periods (engine does not warm to normal operating temperatures).
- Police, taxi, delivery usage or trailer towing usage.

CV joints & front suspension components: service or inspect every 3000 miles.

Engine oil & filter change: change every 3000 miles.

Brake linings: check every 6000 miles.

Chassis lubrication: lubricate every 6000 miles.

Suspension, steering linkage, transaxle shift linkage, parking cable guides, underbody contact points: lubricate every 6000 miles.

Air filter element: service or inspect every 15,000 miles.

Automatic transaxle fluid: change every 50,000 miles (1997 only).

Inspect throttle body bore & throttle plate for deposits: clean as required every 15,000 miles.

Rotate tires at 6000 miles, then every 15,000 miles.

42372-EKBO-C14

PRECAUTIONS

Before servicing any vehicle, please be sure to read all of the following precautions, which deal with personal safety, prevention of component damage, and important points to take into consideration when servicing a motor vehicle:

• Never open, service or drain the radiator or cooling system when the engine is hot; serious burns can occur from the steam and hot coolant.

• Observe all applicable safety precautions when working around fuel. Whenever servicing the fuel system, always work in a well-ventilated area. Do not allow fuel spray or vapors to come in contact with a spark, open flame or excessive heat (a hot drop light, for example). Keep a dry chemical fire extinguisher near the work area. Always keep fuel in a container specifically designed for fuel storage; also, always properly seal fuel containers to avoid the possibility of fire or explosion. Refer to the additional fuel system precautions later in this section.

• Fuel injection systems often remain pressurized, even after the engine has been turned **OFF**. The fuel system pressure must be relieved before disconnecting any fuel lines. Failure to do so may result in fire and/or personal injury.

• Brake fluid often contains polyglycol ethers and polyglycols. Avoid contact with the eyes and wash your hands thoroughly after handling brake fluid. If you do get brake fluid in your eyes, flush your eyes with clean, running water for 15 minutes. If eye irritation persists, or if you have taken brake fluid internally, IMMEDIATELY seek medical assistance.

• The EPA warns that prolonged contact with used engine oil may cause a number of skin disorders, including cancer! You should make every effort to minimize your exposure to used engine oil. Protective gloves should be worn when changing oil. Wash your hands and any other exposed skin areas as soon as possible after exposure to used engine oil. Soap and water, or waterless hand cleaner should be used.

• All new vehicles are now equipped with an air bag system, often referred to as a Supplemental Restraint System (SRS) or Supplemental Inflatable Restraint (SIR) system. The system must be disabled before performing service on or around system components, steering column, instrument panel components, wiring and sensors. Failure to follow safety and disabling procedures could result in accidental air bag deployment, possible personal injury and unnecessary system repairs.

• Always wear safety goggles when working with, or around, the air bag system. When carrying a non-deployed air bag, be sure the bag and trim cover are pointed away from your body. When placing a non-deployed air bag on a work surface, always face the bag and trim cover upward, away from the surface. This will reduce the motion of the module if it is accidentally deployed. Refer to the additional air bag system precautions later in this section.

• Clean, high quality brake fluid from a sealed container is essential to the safe and proper operation of the brake system. You should always buy the correct type of brake fluid for your vehicle. If the brake fluid becomes contaminated, completely flush the system with new fluid. Never reuse any brake fluid. Any brake fluid that is removed from the system should be discarded. Also, do not allow any brake fluid to come in contact with a painted surface; it will damage the paint.

• Never operate the engine without the proper amount and type of engine oil; doing so WILL result in severe engine damage.

• Timing belt maintenance is extremely important! Many models utilize an interference-type, non-freewheeling engine. If the timing belt breaks, the valves in the cylinder head may strike the pistons, causing potentially serious (also time-consuming and expensive) engine damage. Refer to the maintenance interval charts in the front of this section for the recommended replacement interval for the timing belt.

• Disconnecting the negative battery cable on some vehicles may interfere with the functions of the on-board computer system(s) and may require the computer to undergo a relearning process once the negative battery cable is reconnected.

• When servicing drum brakes, only disassemble and assemble one side at a time, leaving the remaining side intact for reference.

• Only an MVAC-trained, EPA-certified automotive technician should service the air conditioning system or its components.

ENGINE REPAIR

Alternator

REMOVAL

Eldorado

1. Before servicing the vehicle, refer to the precautions in the beginning of this section.
2. Drain the cooling system.
3. Remove or disconnect the following:
   • Negative battery cable
   • Accessory drive belt
   • Alternator upper mounting bolt
   • Engine splash shield
   • Radiator support access panel
   • Rear alternator bracket from the engine
   • Alternator mounting bolts
   • Duct on the back of the alternator
   • Electrical connectors
   • Front alternator-to-A/C bracket
   • Alternator

2000 DeVille

1. Before servicing the vehicle, refer to the precautions in the beginning of this section.
2. Drain the cooling system.
3. Remove or disconnect the following:
   • Negative battery cable
   • Accessory drive belt
   • Radiator
   • Air conditioning condenser
   • Alternator cooler outlet hose
   • Alternator cooler inlet hose
   • Electrical connectors
   • Alternator mounting bolts
   • Alternator

2001–03 DeVille

1. Before servicing the vehicle, refer to the precautions in the beginning of this section.
2. Drain the cooling system.
3. Remove or disconnect the following:
   • Negative battery cable
   • Accessory drive belt
   • Radiator
   • Electrical connectors
   • Alternator mounting bolts
   • Alternator

Seville

1. Before servicing the vehicle, refer to the precautions in the beginning of this section.

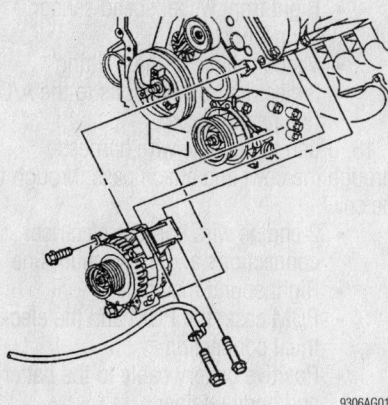

Alternator mounting and wires

9306AG01

2. Drain the cooling system.
3. Remove or disconnect the following:
   - Negative battery cable
   - Accessory drive belt
   - Cooling fans
   - Alternator cooler outlet hose, if equipped
   - Alternator cooler inlet hose, if equipped
   - Electrical connectors
   - Alternator mounting bolts
   - Alternator

INSTALLATION

Eldorado

1. Install or connect the following:
   - Alternator
   - Front alternator-to-A/C bracket. Torque the bolt to 37 ft. lbs. (50 Nm).
   - Electrical connectors. Torque the battery connection nut to 115 inch lbs. (13 Nm).
   - Duct on the back of the alternator
   - Alternator mounting bolts. Torque the bolts to 37 ft. lbs. (50 Nm).
   - Upper front bolt. Torque the bolt to 35 ft. lbs. (47 Nm).
   - Rear bracket to the engine. Torque the bolt to 37 ft. lbs. (50 Nm).
   - Radiator support access panel
   - Engine splash shield
   - Accessory drive belt
   - Negative battery cable
2. Refill the cooling system.

2000 DeVille

1. Install or connect the following:
   - Alternator
   - Alternator mounting bolts. Torque the bolts to 37 ft. lbs. (50 Nm).
   - Electrical connectors. Torque the

battery connection nut to 115 inch lbs. (13 Nm).
   - Alternator cooler inlet hose
   - Alternator cooler outlet hose. Torque the bolt to 80 inch lbs. (9 Nm).
   - Radiator
   - Air conditioning condenser
   - Accessory drive belt
   - Negative battery cable
2. Refill the cooling system.

2001–03 DeVille

1. Install or connect the following:
   - Alternator
   - Alternator mounting bolts. Torque the bolts to 37 ft. lbs. (50 Nm).
   - Electrical connectors. Torque the battery connection nut to 111 inch lbs. (12 Nm).
   - Radiator
   - Accessory drive belt
   - Negative battery cable
2. Refill the cooling system.

Seville

1. Install or connect the following:
   - Alternator. Torque the bolt to 37 ft. lbs. (50 Nm).
   - Electrical connectors. Torque the battery connection nut to 115 inch lbs. (13 Nm).
   - Alternator cooler inlet hose, if equipped
   - Alternator cooler outlet hose, if equipped. Torque the bolt to 80 inch lbs. (9 Nm).
   - Cooling fans
   - Accessory drive belt
   - Negative battery cable
2. Refill the cooling system.

Ignition Timing

ADJUSTMENT

The 4.6L Northstar engine is equipped with a Distributorless Ignition System (DIS). The ignition timing is controlled by the Powertrain Control Module (PCM). Therefore no adjustments are necessary.

Engine Assembly

REMOVAL & INSTALLATION

2000 Eldorado

1. Before servicing the vehicle, refer to the precautions in the beginning of this section.

2. Disable the air bag system.
3. Recover the air-conditioning system.
4. Drain the cooling system.
5. Recover the refrigerant from the air conditioning system.
6. Relieve the fuel pressure.
7. Drain the crankcase.
8. Remove or disconnect the following:
   - Tower-to-tower brace
   - Air intake duct
   - Air cleaner assembly
   - Radiator hose at the crossover
   - Engine cover
   - Radiator upper cover
   - Upper radiator hoses from the engine
   - Forward discriminating sensor from the radiator support and set it aside
   - Cooling fans
   - Transmission cooler lines from the radiator
   - Cruise control servo connections
   - Throttle and cruise control cables from the throttle body
   - Shift cable from the manual shift lever and the bracket
   - Vacuum hose from the brake booster
   - Fuel lines
   - Coolant reservoir
   - Coolant hoses from the pipes at the front of the engine
   - Negative battery cable from the left cylinder head
   - Positive battery cable from the battery and body retainer
   - Powertrain Control Module (PCM) case and the PCM
   - Right sound insulator
   - Engine wire harness electrical connector and the vacuum line under the instrument panel
   - Wire harness from the pass through at the cowl and pull the harness through the cowl
   - 3 electrical connectors from the A/C line near the engine cowl
   - Vacuum hose from the **T** fitting near the engine harness pass through
   - Right and left front wheel speed sensor electrical connectors
   - Manifold Absolute Pressure (MAP) sensor and the control motor near the master cylinder
   - Engine harness retainer from the brake booster
   - 2 hoses from the power steering cooler
   - Engine harness from the hood fuse panel
   - 3 relays from the lower radiator and

the electrical connector from the Antilock Brake System (ABS) modulator
- Tires and wheels
- Drive axles from the steering knuckle
- Post Oxygen Sensor (O_2S) heat shield and wire harness connector from the body
- Intermediate pipe hangers from the intermediate pipe
- Catalytic converter from the pipe
- Steering coupling from the steering gear
- Fuel line bundle from the transmission
- Battery wire from the generator
- A/C manifold from the A/C compressor
- ABS modulator from the frame and support

9. Position the powertrain support table J39580 under the frame and lower the vehicle onto the table.

10. Remove the 6 frame-to-body bolts.

11. Raise the body off the powertrain.

12. Remove or disconnect the following:
- Coolant crossover pipe from the engine
- Park/Neutral switch from the transmission
- Engine wire harness
- Exhaust **Y** pipe
- Accessory drive belt
- Power steering pump and hoses
- Alternator and the bracket
- A/C compressor and bracket
- Idler pulley
- Right engine-to-transmission brace
- Lower engine-to-the transmission brace
- Left engine-to-transmission brace

13. Attach the engine lift to the engine.
- Left engine mount and bracket
- Left exhaust manifold
- Flywheel inspection cover
- Torque converter from the flywheel
- Right engine mount bracket from the engine
- Transmission-to-engine bolts.
- Engine from the transmission and frame
- Exhaust Gas Recirculation (EGR) cross under pipe
- Right exhaust manifold
- Engine

To install:

14. Install or connect the following:
- Engine-to-transmission and frame assembly
- 4 transmission-to-engine bolts and tighten to 55 ft. lbs. (75 Nm)

- Right engine mount bracket to the engine
- Torque converter to the flywheel and tighten the bolts to 35 ft. lbs. (47 Nm)
- Flywheel inspection cover
- Left exhaust manifold
- Left engine mount and bracket

15. Remove the engine lift from the engine.
- Left engine-to-transmission brace
- Lower engine-to-transmission brace
- Right side engine-to-transmission brace
- Right side coolant pipes to the engine
- Power steering lines to the pump and the pump to the engine
- Idler pulley and tighten the bolts to 37 ft. lbs. (50 Nm)
- A/C compressor and the brackets
- Alternator and bracket
- Power steering pump pulley
- Accessory drive belt
- Exhaust manifold rear pipe
- Engine wire harness
- Park/Neutral switch and bracket
- Coolant crossover pipe

16. Position the powertrain assembly under the body and lower the body onto the powertrain assembly.
- Frame-to-body bolts and tighten to 74 ft. lbs. (100 Nm)
- ABS modulator
- A/C manifold to the A/C compressor
- Battery wire to the alternator
- Fuel line bundle to the transmission
- Steering coupling to the steering gear
- Catalytic converter
- Intermediate pipe hangers
- Post converter O_2 sensor electrical connection and heat shield
- Drive axle to the hubs
- Tie rods
- Height sensor links
- Stabilizer shaft links
- Tire and the wheels

17. Lower the vehicle.
- 3 relays to the radiator support
- 2 hoses to the power steering cooler
- Engine harness to under the hood fuse panel
- Engine harness retainer to the brake booster
- Engine harness left front wheel speed sensor, cruise control motor and MAP sensor electrical connections

- Right front wheel speed sensor connection
- Vacuum hose to the **T** fitting
- 3 electrical connections to the A/C line near the cowl

18. Push the engine wire harness through the cowl and attach pass through to the cowl.
- 2 engine wire harness electrical connections and the vacuum line
- Right sound insulator
- PCM case, the PCM and the electrical connection
- Positive battery cable to the battery and body retainer
- Negative battery cable to the left cylinder head
- Coolant hoses to the pipes at the front of the engine
- Coolant reservoir, hoses and electrical connector
- Fuel lines
- Vacuum hose to the brake booster
- Shift cable to the manual shift lever and bracket
- Throttle and cruise cables to the throttle body
- Cooling fans
- Transmission cooler lines to the radiator
- Forward discriminating sensor to the radiator support
- Upper and lower radiator hoses
- Engine mount struts and adjust to zero preload
- Radiator upper cover
- Engine cover
- Air cleaner assembly
- Air intake duct

19. Fill cooling system.

20. Recharge the A/C system.

21. Enable the air bag system

22. Fill the cooling system.

23. Fill the crankcase.

24. Bleed the power steering system.

2000 Models Except Eldorado

1. Before servicing the vehicle, refer to the precautions in the beginning of this section.

2. Disable the air bag system.

3. Recover the air-conditioning system.

4. Drain the cooling system.

5. Recover the refrigerant from the air conditioning system.

6. Relieve the fuel pressure.

7. Drain the crankcase.

8. Remove or disconnect the following:
- Upper filler panel
- Connectors from the Powertrain Control Module (PCM)

- Air cleaner assembly
- 2 nuts from the intake manifold sight shield
- Sight shield from the engine
- Inlet radiator hose from the engine and the outlet radiator hose from the thermostat housing
- Upper transaxle oil cooler line retaining bolt from the fan shroud
- Plastic cap off the upper transaxle oil cooler line fitting
- Internal spring clip from the transaxle oil cooler fitting
- Upper transaxle oil cooler line from the radiator
- Lower transaxle oil cooler line from the radiator
- Surge tank inlet pipe from the engine

❊❊ CAUTION

In order to avoid possible injury or vehicle damage, always replace the accelerator cable with a NEW cable whenever you remove the engine from the vehicle. Also to avoid cruise control cable damage, position the cable out of the way when removing or installing the engine. Never pry or lean against the cruise control cable or kink the cable. If the cable is damaged it must be replaced.

- Accelerator cable from the throttle body
- Cruise control cable from the throttle body
- Heater hoses from the engine
- 2 brake pipes from the master cylinder and plug the lines to prevent system contamination
- Bracket and shift cable from the manual shift lever and set aside
- Vacuum hose from the brake booster
- Evaporative Emission (EVAP) hose from the EVAP purge valve
- Fuel inlet and return fittings from the fuel rail
- Engine ground cable from the body frame rail
- Main engine harness
- Electrical harness from the under-hood fuse block
- Right and left strut tower bolts
- Wheels
- Real time dampening sensor links from the lower control arms
- Both wheel speed sensor connectors
- Both road sensing suspension electrical connectors

- 2 brake pipes from both front frame brackets
- 2 rear brake pipes at the rear of the engine frame
- Air deflector

➡The engine oil cooler quick connect fittings must be replaced whenever they are removed from the engine oil filter adapter.

- Engine oil cooler fittings from the engine oil filter adapter, with the oil lines still attached, and position aside
- Dust cover from the quick connect joint
- Internal spring clip from the engine oil cooler fittings
- Engine oil cooler lines from the cooler fittings
- A/C discharge and suction hoses from the compressor

➡Make sure the wheels of the vehicle are in the straight ahead position and the steering column in the Lock position before disconnecting the steering column or intermediate shaft from the steering gear. Failure to do so will cause the coil assembly in the steering column to become uncentered which will cause damage to the coil assembly.

9. Lock the steering column by removing the ignition key from the lock cylinder.
10. Remove or disconnect the following:
    - Intermediate shaft pinch bolt
    - Intermediate shaft from the steering gear
    - Wheel speed and brake wear sensors at the right and left strut towers
    - Post Heated Oxygen Sensor (HO$_2$S) at the sensor pigtail
    - Exhaust front pipe
    - Brace between the engine oil pan and the transaxle case
    - Torque converter cover
    - Torque converter-to-flywheel bolts
11. Position a powertrain support dolly under the engine assembly. If a powertrain support dolly is unavailable, support the powertrain with four suitable jackstands.
    - Nut attaching the right engine mount to the engine mount bracket
    - Nut attaching the left transaxle mount to the transaxle mount bracket

❊❊ CAUTION

To avoid any vehicle damage, serious personal injury or death when major components are removed from the

vehicle and the vehicle is supported by a hoist, support the vehicle with jack stands at the opposite end from which the components are being removed.

12. Secure the front hoist pads to the vehicle.
13. Remove the 6 frame mount bolts raise the vehicle slowly and attach an engine lift to the engine.
    - Heater pipes
    - Nut attaching the coil cassette ground wire to the cylinder head
    - Engine wiring harness from the engine
    - Bolt attaching the crossover pipe to the cylinder head
    - Power steering pump
    - Bolt attaching the rear transaxle brace to the transaxle
    - Nuts attaching the rear transaxle brace to the cylinder head
    - Rear transaxle brace
    - Bolt attaching the front transaxle brace to the cylinder head
    - Bolts attaching the transaxle brace to the transaxle
    - Bolts securing the transaxle brace to the engine
    - Nut attaching the front engine mount to the engine frame
    - Bolts attaching the engine to the transaxle
    - Engine from the frame and transaxle assembly

To install:
14. Attach the engine to the frame and transaxle assembly..
15. Install or connect the following:
    - Bolts attaching the engine to the transaxle and tighten to 55 ft. lbs. (75 Nm)
    - Nut attaching the front engine mount to the engine frame and tighten to 52 ft. lbs. (70 Nm)
    - 4 bolts attaching the transaxle brace to the engine and tighten to 37 ft. lbs. (50 Nm)
    - Bolt attaching the front transaxle brace to the cylinder head and tighten the bolts to 37 ft. lbs. (50 Nm)
    - Rear transaxle brace and tighten the bolt to 37 ft. lbs. (50 Nm) and the nuts to 37 ft. lbs. (50 Nm)
    - Bolt attaching the crossover pipe to the cylinder head and tighten the bolt to 18 ft. lbs. (25 Nm)
    - Engine wiring harness to the engine
    - Nut attaching the coil cassette

ground wire to the cylinder head and tighten to 13 ft. lbs. (17 Nm)
- Heater pipes
- Power steering pump

16. Remove the engine lift from the engine. Position the engine/transaxle assembly under the vehicle. Carefully lower the vehicle over the engine/transaxle assembly. Place dowel pins in the alignment holes, align the engine frame with the vehicle.
- 6 frame mounting bolts and tighten to 141 ft. lbs. (191 Nm)
- Nut attaching the left transaxle mount to the transaxle mount bracket and tighten to 59 ft. lbs. (80 Nm)
- Nut securing the right engine mount to the engine mount bracket and tighten to 59 ft. lbs. (80 Nm)

17. Remove the powertrain support dolly from under the engine assembly.
- Torque converter to flywheel bolts and tighten to 44 ft. lbs. (60 Nm)
- Torque converter cover
- Oil pan to transaxle brace bolts and tighten and tighten to 37 ft. lbs. (50 Nm)
- Exhaust front pipe
- Post HO$_2$S pigtail
- Wheel speed and brake wear sensors at the strut towers
- Intermediate shaft to the steering gear
- Pinch bolt and tighten to and tighten to 33 ft. lbs. (45 Nm)
- A/C suction and discharge hoses to the compressor and tighten the nuts to 15 ft. lbs. (20 Nm)

➡ The engine oil cooler quick connect fittings must be replaced whenever they are removed from the engine oil filter adapter.

- Engine oil cooler quick connect fittings to the engine oil filter adapter and tighten to 13 ft. lbs. (18 Nm)

18. Push the engine oil cooler lines fully into the oil cooler quick connect fittings, until a click is heard. Slide the dust cover over the quick connect joint.
- Air deflector
- Dampening sensor links
- Road sensing suspension connectors
- Wheel speed sensor connectors
- Brake pipes to both front frame brackets and tighten to 11 ft. lbs. (15 Nm)
- Rear brake pipes at the rear of the engine frame and tighten to 11 ft. lbs. (15 Nm)

- Struts to the strut towers
- Wheels
- Strut tower bolts and washers and tighten to 30 ft. lbs. (40 Nm)
- Engine ground cable
- Main engine harness
- Electrical harness to the underhood fuse block
- Shift cable to the manual shift lever and bracket
- Vacuum hose to the brake booster
- EVAP hose to the EVAP purge valve
- Fuel lines
- Heater hoses
- Brake pipes to the master cylinder and tighten the brake pipes to 11 ft. lbs. (15 Nm)
- Surge tank inlet pipe
- NEW accelerator control cable
- Cruise control cable to the throttle body
- Lower transaxle oil cooler line to the radiator and tighten to 26 ft. lbs. (35 Nm)
- Spring clip to the fitting. Push the upper transaxle oil cooler line into the radiator quick connect fitting, until a "click" is heard. Tug gently on the cooler line to ensure proper retention. Slide the plastic cap (1) over the quick connect joint.
- Upper transaxle oil cooler line retaining bolt to the fan shroud and tighten to 53 inch lbs. (6 Nm)
- Inlet and outlet radiator hoses
- Intake manifold sight shield and tighten the nuts to 27 inch lbs. (3 Nm)
- Air cleaner assembly
- PCM connections
- Upper filler panel

19. Fill the cooling system
20. Recharge the A/C system
21. Connect the negative battery cable.
22. Fill the crankcase.
23. Perform the Crankcase Position (CKP) system variation learn procedure.
24. Bleed the brake system
25. Align the vehicle.

2001–03 DeVille And Seville Models

1. Before servicing the vehicle, refer to the precautions in the beginning of this section.
2. Disable the air bag system.
3. Recover the air-conditioning system.
4. Drain the cooling system.
5. Recover the refrigerant from the air conditioning system.
6. Relieve the fuel pressure.
7. Drain the crankcase.

8. Remove or disconnect the following:
- Negative battery cable
- Vacuum brake booster hose from the vacuum connection
- Fuel inlet and return quick-connect fittings at the fuel
- Hose from the evaporative emission canister purge valve
- Upper filler panel
- Air cleaner assembly
- 2 nuts from the intake manifold sight shield
- Sight shield from the engine
- Nut securing the positive battery cable to the remote positive terminal
- Secondary AIR relay from the relay bracket

9. Disconnect the following electrical connections
- Powertrain Control Module (PCM)
- Engine electrical harness

10. Remove or disconnect the following:
- Bolt attaching the engine ground cable to the right side body frame rail

❋❋ CAUTION

Always replace the accelerator control cable with a NEW cable whenever you remove the engine from the vehicle.

❋❋ CAUTION

Position the cruise control cable out of the way while you remove or install the engine. Do not pry or lean against the cable and do not kink the cable. A damaged cable must be replaced.

- Cruise control cable from the throttle body bracket and lever
- Throttle cable from the throttle body
- Shift cable from the bracket and manual shift lever
- Radiator inlet hose from the water housing crossover
- Radiator outlet hose from the thermostat housing
- Surge tank inlet hose from the tank
- Heater hoses from the heater pipes

➡ Mark the location of the brake pipes to the Brake Pressure Modulator Valve (BPMV) to aid during installation.

- 2 master cylinder brake pipes from the BPMV and plug the outlet ports to prevent fluid loss and contamination

- Upper transaxle oil cooler pipe retaining bolt from the fan shroud
- Plastic cap off the upper transaxle oil cooler pipe quick connect fitting
- Upper transaxle oil cooler pipe from the radiator
- Lower transaxle oil cooler pipe fitting from the radiator

➡**Make sure the vehicle wheels are in the straight position and the steering column in the LOCK position before disconnecting the steering column or intermediate shaft from the steering gear. Failure to do so will cause the SIR coil assembly to become uncentered, which may cause damage to the coil assembly.**

11. Lock the steering column by installing the J 42640 into the underside of the steering column.

- Right and left side strut tower bolts
- Rear exhaust manifold pipe
- Front wheels
- Front Wheel Speed Sensor (WSS) electrical leads from the frame rail
- Road sensing suspension electrical connector at the frame rail, if equipped
- Electronic suspension position sensor links from the lower control arms, if equipped
- Front air deflector
- Front fascia extensions
- Secondary AIR inlet hose from the secondary AIR pump
- 2 nuts attaching the front brake pipe frame brackets to the frame rails
- Front brake pipes from the retainers at the frame rails
- Front brake pipes away from the frame rails by gently pulling
- 2 rear brake pipes at the rear of the engine frame and plug the open outlet ports
- A/C pressure sensor
- A/C discharge hose from the compressor and it secure to the cooling fan assembly
- A/C suction hose from the compressor and secure it to the cooling fan assembly

✶✶ CAUTION

Failure to disconnect the intermediate shaft from the rack and pinion stub shaft could result in damage to the steering gear and/or damage to the intermediate shaft. This damage may cause loss of steering control which could result in personal injury.

- Intermediate shaft pinch bolt
- Steering gear from the intermediate shaft
- Post Heated Oxygen (HO$_2$S) sensor connection
- Engine oil cooler quick connect fittings from the engine oil filter adapter, with the oil pipes still attached, and position aside, if equipped
- Brace between the engine oil pan and the transaxle case
- Torque converter cover
- Torque converter-to-flywheel bolts

12. Position J 39580 powertrain support dolly under the engine frame and lower the vehicle onto the dolly.

13. If a support dolly is unavailable, support the powertrain with four suitable jackstands.

14. Place a 2 in x 4 in block of wood between the front of the engine oil pan and the engine frame.

15. Remove or disconnect the following:

- Nut securing the right engine mount to the right engine mount bracket
- Nut securing the left transaxle mount to the left transaxle mount bracket

✶✶ CAUTION

To avoid any vehicle damage, serious personal injury or death when major components are removed from the vehicle and the vehicle is supported by a hoist, support the vehicle with jack stands at the opposite end from which the components are being removed.

16. Secure the front hoist pads to the vehicle

- 6 frame-to-body mounting bolts

17. Make sure clearance is maintained between the powertrain assembly and the following components:

- A/C accumulator hose
- A/C compressor hose
- Brake pipes
- Heater hoses
- Radiator hoses
- Wheel Speed Sensor (WSS) cables
- Wiring harnesses

18. Carefully raise the vehicle in order to clear the supported engine/transaxle assembly.

19. Drain the engine oil

20. Remove or disconnect the following:

- Heater pipes
- Intermediate hose from the secondary AIR valve at bank 2

- Nut securing the intermediate hose to the secondary AIR valve at bank 1
- Nut securing the coil cassette ground wire to the right cylinder head
- Engine wiring harness from the engine
- Power steering hose from the power steering pump reservoir
- Power steering return hose retaining bolt from the cylinder head
- Power steering pressure hose from the power steering pump
- Nut securing the power steering pressure hose to the right engine mount bracket
- 4 bolts securing the right engine mount bracket to the engine
- Right engine mount bracket
- Bolt securing the rear transaxle brace to the transaxle
- Nuts securing the rear transaxle brace to the stud located on the right cylinder head
- Rear transaxle brace
- Bolts securing the front transaxle brace to the transaxle and right cylinder head
- Nuts securing the Vehicle Speed Sensor (VSS) heat shield to the transaxle
- Bolts securing the transaxle brace to the engine and transaxle

21. Attach an engine lift bracket to the cylinder head

- Engine lift chain to the engine lift brackets and attach to an engine lift devise
- Nut securing the front engine mount to the engine frame
- Bolts attaching the engine to the transaxle

22. Raise the engine from the supported frame and transaxle assembly.

- Front engine mount bracket

To install:

23. Installation is the reverse of removal, please note the following torque specifications:

- Bolts attaching the engine to the transaxle to 55 ft. lbs. (75 Nm)
- Nut securing the front engine mount to the front engine frame to 52 ft. lbs. (70 Nm)
- 4 bolts securing the center transaxle brace to the engine and transaxle to 37 ft. lbs. (50 Nm)
- Retaining nuts attaching the VSS heat shield to the transaxle to 37 ft. lbs. (50 Nm)
- Bolts attaching the front transaxle

brace to the transaxle and right cylinder head to 37 ft. lbs. (50 Nm)
- Rear transaxle brace bolt and nuts to 37 ft. lbs. (50 Nm)
- 4 bolts securing the right engine mount bracket to the engine to 37 ft. lbs. (50 Nm)
- Nut securing the power steering pressure hose to the right engine mount bracket to 80 inch lbs. (9 Nm)
- Power steering pressure hose to the power steering pump to 20 ft. lbs. (27 Nm)
- Power steering return hose retaining bolt to the cylinder head to 37 ft. lbs. (50 Nm)
- Nut securing the coil cassette ground wire to the cylinder head to 13 ft. lbs. (17 Nm)
- Nut securing the intermediate hose to the secondary AIR valve at bank 1 to 80 inch lbs. (9 Nm)
- 6 frame mounting bolts to 141 ft. lbs. (191 Nm)
- Nut securing the right engine mount to the right engine mount bracket to 59 ft. lbs. (80 Nm)
- Nut securing the left transaxle mount to the left transaxle mount bracket to 59 ft. lbs. (80 Nm)
- Flywheel-to-torque converter bolts to 44 ft. lbs. (60 Nm)
- Oil pan-to-transaxle brace bolts to 37 ft. lbs. (50 Nm)
- Pinch bolt to 33 ft. lbs. (45 Nm)
- A/C suction hose nut to 15 ft. lbs. (20 Nm)
- A/C discharge hose nut to 15 ft. lbs. (20 Nm)
- 2 rear brake pipes at the rear of the engine frame to 11 ft. lbs. (15 Nm)
- Brake pipe frame bracket nuts to 11 ft. lbs. (15 Nm)
- Transaxle oil cooler pipe fitting to 26 ft. lbs. (35 Nm)
- 2 master cylinder brake pipes to the BPMV to 11 ft. lbs. (15 Nm)
- Brake pipes to 11 ft. lbs. (15 Nm)

24. Fill the engine with oil.
25. Fill the cooling system.
26. Bleed the brake system.
27. Recharge the A/C refrigerant system.
28. Bleed the power steering system.
29. Check the wheel alignment.
30. Complete the following steps after the engine is installed in the vehicle:

   a. With the ignition **OFF** or disconnected, crank the engine several times. Listen for any unusual noises or evidence that any parts are binding.
   b. Start the engine and listen for abnormal conditions.

   c. Check the vehicle oil pressure gauge or light and confirm that the engine has acceptable oil pressure.
   d. Run the engine at approximately 1000 RPM until the engine reaches normal operating temperature.
   e. While the engine continues to idle, raise and support the vehicle
   f. Inspect for oil, coolant and exhaust leaks while the engine is idling.
   g. Lower the vehicle.
   h. Perform the Crankshaft Position (CKP) system variation learn procedure.
   i. Perform a final inspection for the proper engine oil and coolant levels.
31. Road test the vehicle.

2001–02 Eldorado Models

1. Before servicing the vehicle, refer to the precautions in the beginning of this section.
2. Disable the air bag system.
3. Recover the air-conditioning system.
4. Drain the cooling system.
5. Recover the refrigerant from the air conditioning system.
6. Relieve the fuel pressure.
7. Drain the crankcase.
8. Remove or disconnect the following:
   - Negative battery cable
   - Vacuum brake booster hose from the vacuum connection
   - Air cleaner assembly
   - Fuel inlet and return quick-connect fittings at the fuel
   - Hose from the evaporative emission canister purge valve
   - Radiator support sight shield
   - Sight shield from the engine
9. Disconnect the following electrical connections:
   - Powertrain Control Module (PCM)
   - Engine electrical harness
   - Nut securing the positive battery cable to engine

✳✳ CAUTION

Always replace the accelerator control cable with a NEW cable whenever you remove the engine from the vehicle.

✳✳ CAUTION

Position the cruise control cable out of the way while you remove or install the engine. Do not pry or lean against the cable and do not kink the cable. A damaged cable must be replaced.

- Cruise control cable from the throttle body bracket and lever
- Throttle cable from the throttle body
- Shift cable from the bracket and manual shift lever
- Radiator inlet hose from the water housing crossover
- Radiator outlet hose from the thermostat housing
- Cooling fans
- Surge tank inlet hose from the tank
- Accumulator tube, and position the accumulator aside
- Heater hoses from the heater pipes
- Upper transaxle oil cooler pipe from the radiator
- Lower transaxle oil cooler pipe fitting from the radiator
- Right and left side strut tower nuts
- Engine mount struts
- Rear exhaust manifold pipe
- Front wheels
- Front Wheel Speed Sensor (WSS) electrical leads from the frame rail
- Road sensing suspension electrical connector at the frame rail, if equipped
- Electronic suspension position sensor links from the lower control arms, if equipped
- Splash shield
- Front fascia extensions
- Secondary AIR inlet hose from the secondary AIR pump
- Front left brake pipe fitting from the brake hose
- Front left brake hose retainer from the brake hose retaining bracket
- Bolt attaching the front brake pipes to the frame rails
- Front brake pipes away from the frame rails by gently pulling

➡**Mark the location of the brake pipes to the brake pressure modulator valve (BPMV) to aid during installation.**

- Master cylinder, left and rear brake pipes from the BPMV and plug the outlet ports to prevent fluid loss and contamination
- A/C pressure sensor
- A/C discharge hose from the compressor and it secure to the cooling fan assembly
- A/C suction hose from the compressor and secure it to the cooling fan assembly
- Intermediate shaft pinch bolt
- Steering gear from the intermediate shaft
- Post Heated Oxygen (HO_2S) sensor connection

- Engine oil cooler quick connect fittings from the engine oil filter adapter, with the oil pipes still attached, and position aside, if equipped
- Wiring from the alternator
- Positive battery cable from the front of the frame
- Brace between the engine oil pan and the transaxle case
- Torque converter cover
- Torque converter-to-flywheel bolts

10. Position J 39580 powertrain support dolly under the engine frame and lower the vehicle onto the dolly.

11. If a support dolly is unavailable, support the powertrain with four suitable jackstands.

12. Secure the front hoist pads to the vehicle
- 6 frame-to-body mounting bolts

13. Make sure clearance is maintained between the powertrain assembly and the following components:
- A/C accumulator hose
- A/C compressor hose
- Brake pipes
- Heater hoses
- Radiator hoses
- Wheel Speed Sensor (WSS) cables
- Wiring harnesses

14. Carefully raise the vehicle in order to clear the supported engine/transaxle assembly.

15. Drain the engine oil

16. Remove or disconnect the following:
- Intermediate hose from the AIR valve at bank 2
- Nut securing the intermediate hose to the AIR valve at bank 1
- Bolt securing the AIR pipe to the left engine mount strut bracket
- Exhaust Gas Recirculation (EGR) pipes
- Heater pipes
- Nut securing the coil cassette ground wire to the right cylinder head
- Engine wiring harness from the engine
- Power steering hose from the power steering pump reservoir
- Power steering return hose bolt to the right engine mount bracket
- Power steering pressure hose from the power steering pump
- Bolts securing the right engine mount bracket to the engine
- Bolt securing the top of the generator to the right engine mount strut bracket
- Right engine strut bracket

17. Attach an engine lift bracket to the cylinder head
- Power steering pump
- Engine lift chain to the engine lift brackets and attach to an engine lift devise
- Nuts securing the right transaxle mount to the engine frame
- 3 bolts securing the right transaxle mount to the engine and transaxle
- Right transaxle mount
- Bolt securing the rear transaxle brace to the transaxle
- Nuts securing the rear transaxle brace to the stud located on the right cylinder head
- Rear transaxle brace
- Nuts securing the Vehicle Speed Sensor (VSS) heat shield to the transaxle.
- Bolts securing the front transaxle brace to the engine and transaxle
- 2 nuts securing the front engine mount to the engine frame
- Bolts attaching the engine to the transaxle
- Engine

To install:

18. Installation is the reverse of removal, please note the following torque specifications:
- Bolts attaching the engine to the transaxle to 55 ft. lbs. (75 Nm)
- Nuts securing the front engine mount to the engine frame to 37 ft. lbs. (55 Nm)
- Bolts securing the front transaxle brace to the engine and transaxle to 37 ft. lbs. (50 Nm)
- Nuts securing the VSS heat shield to the transaxle to 37 ft. lbs. (50 Nm)
- Bolts securing the rear transaxle brace to the transaxle to 37 ft. lbs. (50 Nm)
- Bolts securing the right transaxle mount to the engine and transaxle to 37 ft. lbs. (50 Nm)
- Nuts securing the right transaxle mount to the engine frame to 37 ft. lbs. (50 Nm)
- Bolts securing the right engine mount strut bracket to the cylinder head to 37 ft. lbs. (50 Nm)
- Bolt securing the top of the generator to the right engine mount strut bracket to 37 ft. lbs. (50 Nm)
- Power steering pressure hose to the power steering pump to 22 ft. lbs. (30 Nm)
- 6 frame mounting bolts retaining the frame to the vehicle to 72 ft. lbs. (98 Nm)

- Bolts securing the flywheel to the torque converter to 44 ft. lbs. (60 Nm)
- Oil pan-to-transaxle brace to 37 ft. lbs. (50 Nm)
- A/C discharge/suction hose to the compressor to 20 ft. lbs. (30 Nm)
- Lower transaxle oil cooler pipe fitting to the radiator to 26 ft. lbs. (35 Nm)
- Upper transaxle oil cooler pipe fitting to the radiator to 26 ft. lbs. (35 Nm)

19. Fill the engine with oil.
20. Fill the cooling system.
21. Bleed the brake system.
22. Recharge the A/C refrigerant system.
23. Bleed the power steering system.
24. Check the wheel alignment.
25. Complete the following steps after the engine is installed in the vehicle:

a. With the ignition **OFF** or disconnected, crank the engine several times. Listen for any unusual noises or evidence that any parts are binding.

b. Start the engine and listen for abnormal conditions.

c. Check the vehicle oil pressure gauge or light and confirm that the engine has acceptable oil pressure.

d. Run the engine at approximately 1000 RPM until the engine reaches normal operating temperature.

e. While the engine continues to idle raise and support the vehicle

f. Inspect for oil, coolant and exhaust leaks while the engine is idling.

g. Lower the vehicle.

h. Perform the Crankshaft Position (CKP) system variation learn procedure.

i. Perform a final inspection for the proper engine oil and coolant levels.
26. Road test the vehicle.

Water Pump

REMOVAL & INSTALLATION

1. Before servicing the vehicle, refer to the precautions in the beginning of this section.
2. Drain the coolant.
3. Remove or disconnect the following:
- Negative battery cable
- Upper fill panel, if equipped
- AIR injection control valve, if equipped
- Air cleaner
- Oil level indicator tube nut, if necessary

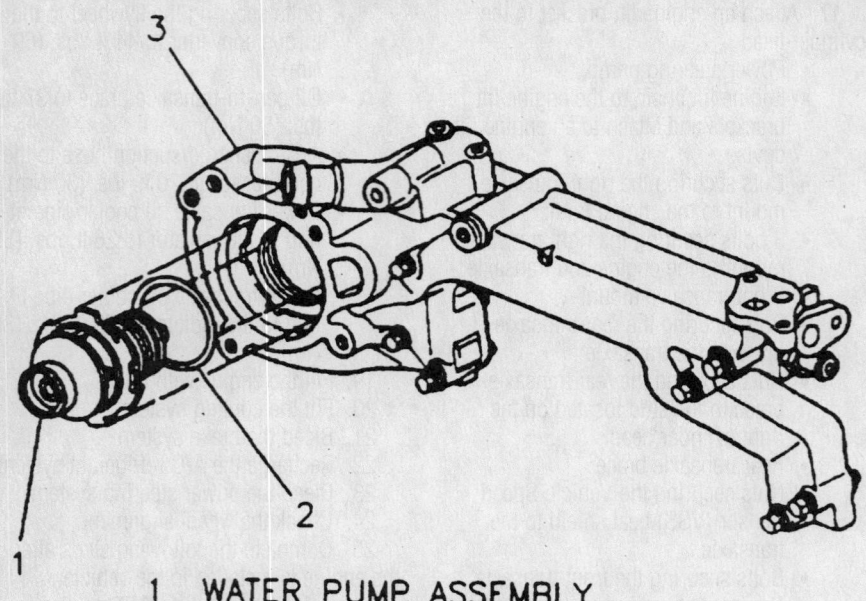

1 WATER PUMP ASSEMBLY
2 O-RING SEAL
3 WATER PUMP HOUSING ASSEMBLY

7922VG01

To ensure proper operation, be sure to install a new O-ring

- Water pump belt cover
- Water pump drive belt
- Lower radiator hose
- Thermostat bypass hose from the coolant inlet housing
- Water pump cover

4. Remove the water pump by rotating it clockwise using special Tool J 38816-1A (water pump remover/installer).

To install:

5. Install the water pump with a new O-ring, by turning it counterclockwise using the special tool, until it stops. Torque the water pump to 73 ft. lbs. (100 Nm).

6. Install or connect the following:
- Water pump cover. Torque the bolts to 89 inch lbs. (10 Nm).
- Lower radiator hose
- Thermostat bypass hose to the coolant inlet housing
- Water pump drive belt
- Water pump belt cover
- Oil level indicator tube nut, if necessary
- Air cleaner
- AIR injection control valve, if equipped
- Upper fill panel, if equipped
- Negative battery cable

7. Refill and bleed the cooling system.

8. Run the engine and check for leaks.

Heater Core

REMOVAL & INSTALLATION

Deville and Eldorado

1. Disconnect the negative battery cable.
2. Drain the cooling system into a clean container for reuse.
3. Remove or disconnect the following:
- Instrument panel compartment
- Sound insulator from the right side
- Electrical connector and remove the heater/air conditioning programmer
- Driver's air mix actuator from the bottom of the heater assembly
- Passenger's air mix actuator from the bottom of the heater assembly
- Heater core cover-to-heater assembly screws and the cover
- Heater hoses from the heater core, located in the engine compartment
- Heater core-to-heater assembly screws
- Heater core

To install:

4. Install or connect the following:
- Heater core

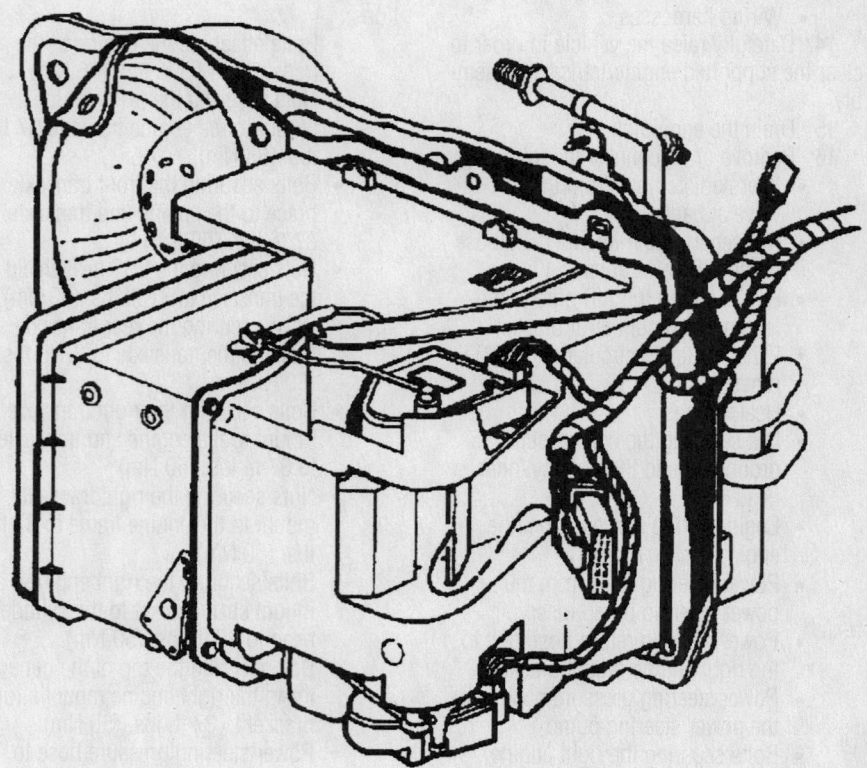

93111G76

View of the heater assembly with the driver's and passenger's air mix actuators—DeVille and Eldorado

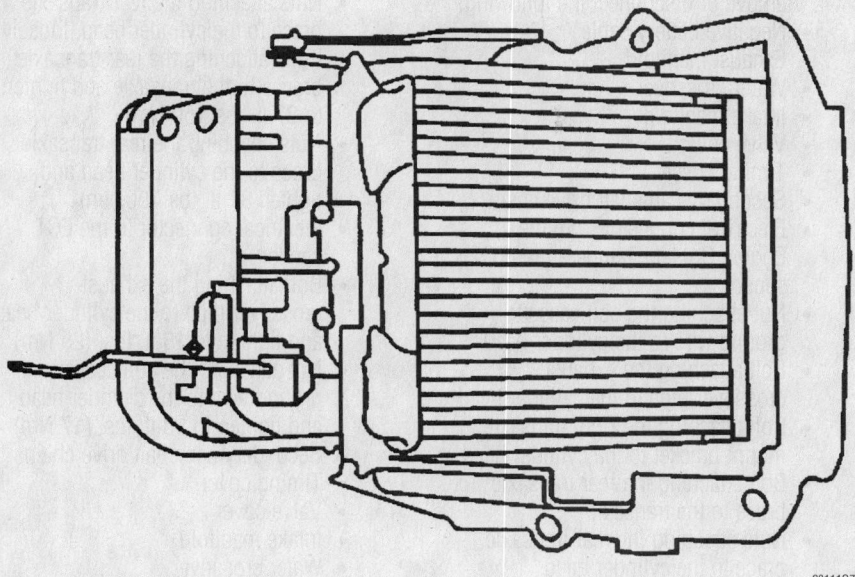

View of the heater core—DeVille and Eldorado

- Heater core-to-heater assembly screws, then, tighten to 18 inch lbs. (2 Nm)
- Heater hoses to the heater core
- Heater core cover and the cover-to-heater assembly screws, then, tighten to 18 inch lbs. (2 Nm)
- Passenger's air mix actuator at the top of the heater assembly
- Driver's air mix actuator at the bottom of the heater assembly
- Heater/air conditioning programmer and connect the electrical connector
- Sound insulator on the right side
- Instrument panel compartment

5. Refill the cooling system.
6. Connect the negative battery cable.

Seville

1. Disconnect the negative battery cable.
2. Drain the cooling system into a clean container for reuse.
3. Disconnect the heater hoses from the heater core.
4. Disable the SIR system.
5. Remove the instrument panel by removing or disconnecting the following:
- Console
- Right insulator panel
- Instrument panel end caps from both sides
- Side windows air outlets
- Upper instrument panel trim pad
- Passenger's side SIR module
- Instrument panel trim plate from the right side

- Instrument panel cluster
- Center instrument panel trim plate
- Radio
- Lap cooler duct
- Knee bolster.
- Steering column
- Headlight switch.
- Fuel door/rear compartment release switch.
- Instrument panel storage compartment
- 10 instrument panel retainer-to-instrument panel carrier fasteners

➡ **One fastener is located in the fuel/rear compartment release switch opening.**

- Instrument panel retainer from the carrier and feed the wiring through the openings, as necessary
- Inside air temperature sensor electrical connector
- Instrument panel
- Heater core-to-heater housing cover screws
- Heater core-to-heater housing screws
- Heater core retaining straps
- Heater core
- Heater core seals

To install:
6. Install or connect the following:
- Heater core seals
- Heater core.
- Heater core retaining straps
- Heater core-to-heater housing screws.
- Heater core-to-heater housing cover screws.

7. Install the instrument panel by installing or connecting the following:
- Instrument panel
- Inside air temperature sensor electrical connector
- Instrument panel retainer to the carrier and feed the wiring through the openings, as necessary
- 10 instrument panel retainer-to-instrument panel carrier fasteners

➡ **One fastener is located in the fuel/rear compartment release switch opening.**

- Instrument panel storage compartment
- Fuel door/rear compartment release switch
- Headlight switch

8. Install or connect the following:
- Steering column
- Knee bolster
- Lap cooler duct
- Radio
- Center instrument panel trim plate
- Instrument panel cluster
- Install the instrument panel trim plate on the right side
- Passenger's side SIR module
- Upper instrument panel trim pad
- Side windows air outlets
- Instrument panel end caps on both sides
- Right insulator panel
- Console

9. Enable the SIR system.
10. Connect the heater hoses to the heater core.
11. Refill the cooling system.
12. Connect the negative battery cable.
13. Operate the engine to normal operating temperatures; then, check the climate control operation and check for leaks.

Cylinder Head

REMOVAL & INSTALLATION

Left Side

1. Before servicing the vehicle, refer to the precautions in the beginning of this section.
2. Drain the cooling system.
3. Properly relieve the fuel system pressure.
4. Remove or disconnect the following:
- Negative battery cable
- Exhaust manifold
- Alternator
- Water crossover

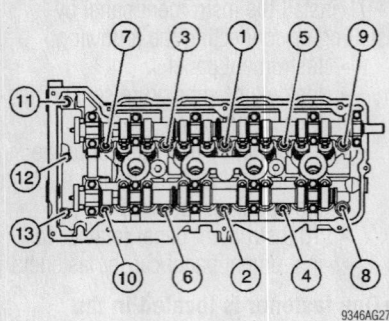

Left cylinder head bolt torque sequence—4.6L engine

- Intake manifold
- Valve cover
- Timing cover
- Secondary camshaft drive chain
- Power steering return hose retaining bolt from the cylinder head
- 3 M6 external drive bolts from the front portion of the cylinder head
- 10 M11 internal drive cylinder head bolts
- Cylinder head and gasket.

5. Clean the head mating surfaces.

To install:

6. Install the cylinder head with a new gasket. Lubricate the bolts with engine oil prior to installation.

7. Torque the M11 bolts in sequence as follows:

   a. Step 1: 30 ft. lbs. (40 Nm), plus an additional 70 degrees

   b. Step 2: Turn an additional 60 degrees

   c. Step 3: Turn an additional 60 degrees (total 190 degrees)

8. Torque the M6 bolts to 106 inch lbs. (12 Nm)

9. Install or connect the following:

- Power steering return hose retaining bolt to the cylinder head and tighten to 37 ft. lbs. (50 Nm)
- Secondary camshaft drive chain
- Timing cover
- Valve cover
- Intake manifold
- Water crossover
- Exhaust manifold
- Negative battery cable

10. Change the oil and filter.

11. Fill the cooling system.

Right Side

1. Before servicing the vehicle, refer to the precautions in the beginning of this section.

2. Drain the cooling system.

3. Properly relieve the fuel system pressure.

4. Remove or disconnect the following:

- Negative battery cable
- Exhaust manifold
- Water crossover
- Intake manifold
- Valve cover
- Timing cover
- Secondary camshaft drive chain
- Electrical connector from the Engine Coolant Temperature (ECT) sensor
- Nut attaching the coil cassette ground wire to the cylinder head
- Bolt attaching the exhaust crossover pipe to the cylinder head
- Bolt attaching the right transaxle mount bracket to the cylinder head
- Bolt attaching the rear transaxle brace to the transaxle
- Nuts attaching the rear transaxle brace to the cylinder head
- Rear transaxle brace
- 3 M6 external drive bolts from the front portion of the cylinder head
- 10 M11 internal drive cylinder head bolts
- Cylinder head and gasket

5. Clean the head mating surfaces.

To install:

6. Install the cylinder head with a new gasket. Lubricate the bolts with engine oil prior to installation.

7. Torque the M11 bolts in sequence as follows:

   a. Step 1: 30 ft. lbs. (40 Nm), plus an additional 70 degrees.

   b. Step 2: Turn an additional 60 degrees.

   c. Step 3: Turn an additional 60 degrees (total 190 degrees).

8. Torque the M6 bolts to 106 inch lbs. (12 Nm).

9. Install or connect the following:

- Rear transaxle brace over the studs located at the rear of the right cylinder head

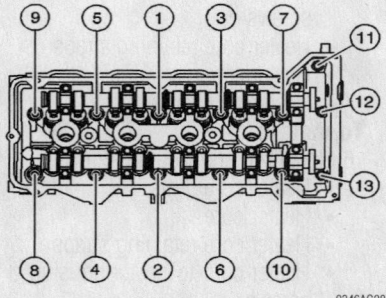

Right cylinder head bolt torque sequence—4.6L engine

- Nuts attaching the rear transaxle brace to the cylinder head, loosely
- Bolts attaching the rear transaxle braces to the transaxle and tighten to 37 ft. lbs. (50 Nm)
- Nuts attaching the rear transaxle brace to the cylinder head and tighten 37 ft. lbs. (50 Nm)
- Electrical connector to the ECT sensor
- Bolt attaching the exhaust crossover pipe to the cylinder head and tighten to 18 ft. lbs. (25 Nm)
- Nut attaching the coil cassette ground wire to the cylinder head and tighten to 13 ft. lbs. (17 Nm)
- Secondary camshaft drive chain
- Timing cover
- Valve cover
- Intake manifold
- Water crossover
- Exhaust manifold
- Negative battery cable

10. Change the oil and filter.

11. Fill the cooling system.

Intake Manifold

REMOVAL & INSTALLATION

1. Before servicing the vehicle, refer to the precautions in the beginning of this section.

2. Relieve the fuel system pressure.

3. Drain the cooling system.

4. Remove or disconnect the following:

- Negative battery cable
- Intake manifold heat shield
- Sight shield
- Coil module connectors from the coil modules located on the valve covers
- Positive Crankcase Ventilation (PCV) hose and valve from the valve cover
- Fuel regulator vacuum tube
- Vacuum tubes from the AIR solenoid
- Fuel inlet and return lines
- Fuel rail bracket retaining nut at the rear lift bracket
- 2 pushnuts attaching the engine coolant heater wire and set it aside, if equipped
- Fuel injector electrical connections
- Fuel rail and injectors
- Plenum duct clamp at the rear of the intake manifold, loosen
- Intake manifold bolts and the manifold

5. Clean the manifold mating surfaces.

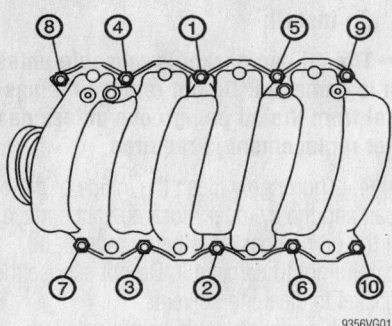

Intake manifold torque sequence—4.6L engines

9356VG01

To install:

6. Grease the inside edge of the rubber plenum duct.

7. Position the intake manifold by performing the following sub-steps:

a. Place the rear of the intake manifold into the plenum duct.

b. Place the front of the intake manifold downward on to the cylinder heads.

8. Install the intake manifold bolts and torque to 89 inch lbs. (10 Nm) in the sequence illustrated. DO NOT torque the intake manifold bolts when the engine is HOT or at operating temperature.

9. Make sure the plenum duct is fully attached to the rear of the intake manifold and tighten the plenum duct clamp to 20 inch lbs. (2.25 Nm).

10. Install or connect the following:
- Fuel rail and injectors and tighten the rail retainers to 89 inch lbs. (10 Nm)
- Fuel injector electrical connections
- Alternator coolant pipe/surge tank pipe and position onto the fuel rail studs
- Engine coolant heater wire onto the studs and install the pushnuts, if equipped
- Fuel rail bracket retaining nut at the rear lift bracket and tighten to 89 inch lbs. (10 Nm)
- Fuel inlet and return lines
- Vacuum tubes to the AIR solenoid
- Fuel regulator vacuum tube
- Positive Crankcase Ventilation (PCV) valve and hose to the valve cover
- Coil module connectors to the coil modules located on the valve covers
- Sight shield
- Intake manifold heat shield
- Negative battery cable

11. Fill and bleed the cooling system.

Exhaust Manifold

REMOVAL & INSTALLATION

Left Side

1. Before servicing the vehicle, refer to the precautions in the beginning of this section.

2. Remove or disconnect the following:
- Negative battery cable
- 2 nuts attaching the AIR tube to the exhaust manifold
- Engine mount and bracket
- 2 bolts at the manifold outlet flange
- Oxygen Sensor (O_2S), if necessary
- Exhaust manifold bolts
- Manifold and the gasket. Discard the gasket.

To install:

3. Install or connect the following:
- Exhaust manifold by inserting the outlet pipe partially into the exhaust crossover pipe
- Exhaust manifold with a new gasket. Torque the nuts to 18 ft. lbs. (25 Nm).
- O_2S. Torque the sensor to 30 ft. lbs. (40 Nm).
- Engine mount and bracket
- 2 nuts attaching the AIR tube to the exhaust manifold and tighten to 106 inch lbs. (12 Nm)
- Negative battery cable

Right Side

1. Before servicing the vehicle, refer to the precautions in the beginning of this section.

2. Remove or disconnect the following:
- Negative battery cable
- Oxygen Sensor (O_2S)
- 2 nuts attaching the AIR tube to the exhaust manifold
- Exhaust front pipe
- Real time dampening sensor links from the lower control arms
- Intermediate shaft from the steering gear

3. Support the rear cross member of the engine cradle with a tall screw jack.
- 4 rearward cradle-to-body bolts, then lower the rear of the engine cradle
- Right side cylinder head to the transaxle brace
- Exhaust manifold nuts
- Manifold and the gasket

To install:

4. Install or connect the following:
- Exhaust manifold with a new gasket. Torque the nuts to 18 ft. lbs. (25 Nm).
- Right side cylinder head to the transaxle brace and tighten the bolt and nut to 35 ft. lbs. (47 Nm)

5. Raise the engine cradle into position.
- 4 rearward cradle-to-body bolts and tighten to 141 ft. lbs. (191 Nm)

6. Remove the screw jack.
- Intermediate shaft to the steering gear and tighten the pinch bolt to 35 ft. lbs. (47 Nm)
- Real time dampening sensor links to the lower control arms
- Exhaust front pipe
- New gasket to the AIR tube
- 2 nuts attaching the AIR tube to the exhaust manifold and tighten to 106 inch lbs. (12 Nm)
- O_2S sensor
- Negative battery cable

Camshaft and Valve Lifters

REMOVAL & INSTALLATION

1. Before servicing the vehicle, refer to the precautions in the beginning of this section.

2. Remove both camshaft covers.

3. Secure the camshaft sprockets to the timing chain by installing tie wraps through the camshaft sprocket holes.

4. Working behind the sprockets, install camshaft Chain Holder J38822 so that it is positioned between the chain tensioner and guide. Apply tension to the tool by tightening the tension adjusting screw.

5. Remove or disconnect the following:
- Both camshaft sprocket bolts
- Both camshaft sprockets

6. Alternately loosen the camshaft bearing cap screws a few turns at a time until all valve spring pressure has been released.

7. Remove the bolts and bearing caps.

8. Remove the camshaft.

To install:

9. Lubricate the camshaft lobes and camshaft journals with clean engine oil.

10. Position the camshaft bearing caps in the cylinder head and loosely install the bolts.

11. Tighten the camshaft bearing cap bolts in the sequence shown, to 44 inch lbs. (5 Nm), plus 30 degrees.

12. Using the hex cast into the camshaft, position the camshafts to accept the camshaft sprockets.

13. Install the camshaft sprockets onto the camshafts. Torque the bolts to 89 ft. lbs. (120 Nm).

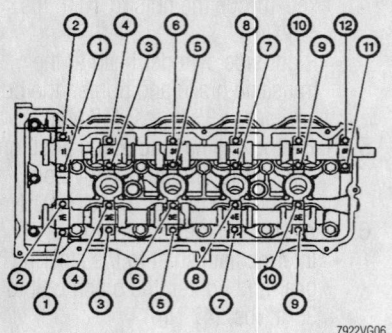

7922VG06

Left cylinder head camshaft bearing cap torque sequence—4.6L engine

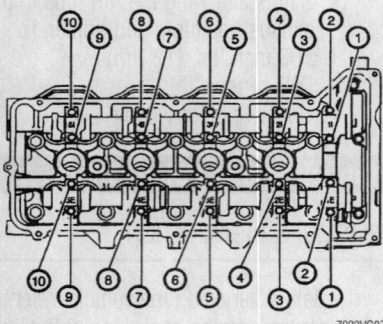

7922VG07

Right cylinder head camshaft bearing cap torque sequence—4.6L engine

14. Remove the Chain Holder J38822 and the tie wraps.

15. Install the left and right camshaft covers.

Valve Lash

ADJUSTMENT

The valve clearance cannot be adjusted. The engines in this section are equipped with hydraulic lifters and no adjustment is possible.

Starter Motor

REMOVAL & INSTALLATION

1. Before servicing the vehicle, refer to the precautions in the beginning of this section.
2. Remove or disconnect the following:
   - Negative battery cable
   - Intake manifold
   - Starter electrical connectors
   - Starter motor

To install:

3. Install or connect the following:
   - "S" terminal wire. Torque the nut to 35 inch lbs. (4 Nm).

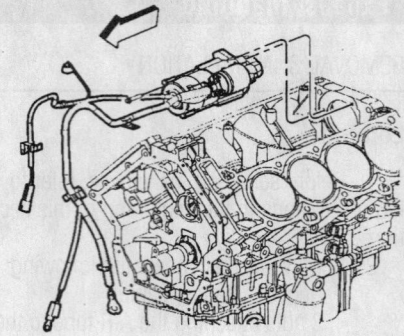

9306AG02

View of starter motor removal and wires

- Starter electrical connectors. Torque the nut to 89 inch lbs. (10 Nm).
- Starter motor
- Mounting bolts. Torque the bolts to 22 ft. lbs. (30 Nm).
- Intake manifold
- Negative battery cable

Oil Pan

REMOVAL & INSTALLATION

1. Before servicing the vehicle, refer to the precautions in the beginning of this section.
2. Drain the crankcase.
3. Remove or disconnect the following:

- Negative battery cable
- Oil level indicator harness connector, if equipped
- Exhaust crossover pipe
- Transaxle assembly from the vehicle, if necessary
- Oil pan bolts and the oil pan

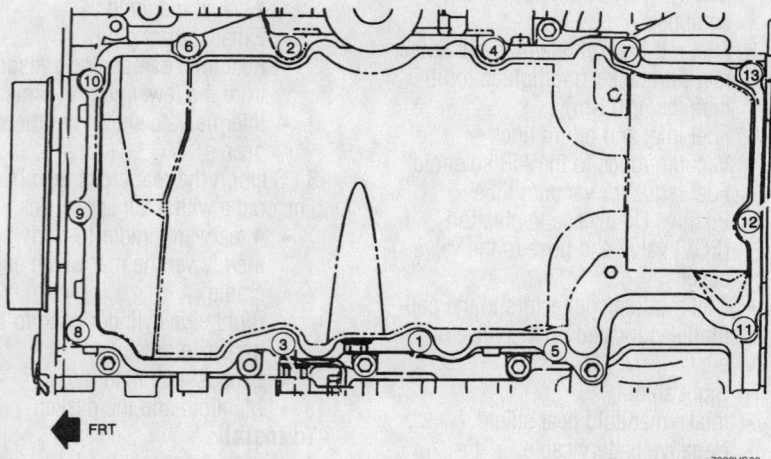

FRT

7922VG08

Oil pan bolt torque sequence

To install:

➡ **The oil pan gasket is reusable unless it is damaged. Do not remove the gasket from the oil pan groove unless gasket replacement is required.**

4. Thoroughly clean the inside of the oil pan and the cylinder block contact surface. If the oil pan gasket is being reused, be careful not to damage it. Do not expose the gasket to cleaning solvents.

5. If a new gasket is being installed, start the gasket into the oil pan groove and work the gasket into the groove in both directions. Once the gasket is exposed to oil, it will expand and no longer stay in the groove without wrinkles. If this condition exists, replace the gasket.

6. Install or connect the following:
   - Oil pan. Torque the bolts in sequence to 89 inch lbs. (10 Nm).
   - Oil level indicator connector, if equipped
   - Exhaust crossover pipe
   - Transaxle assembly, if necessary
   - Oil pan drain plug. Torque it to 15 ft. lbs. (20 Nm).
   - Negative battery cable

7. Refill the crankcase.

8. Run the engine and check for leaks.

Oil Pump

REMOVAL & INSTALLATION

1. Before servicing the vehicle, refer to the precautions in the beginning of this section.

2. Remove or disconnect the following:
   - Negative battery cable
   - Drive belt tensioner
   - Drive belt idler pulley
   - Crankshaft balancer

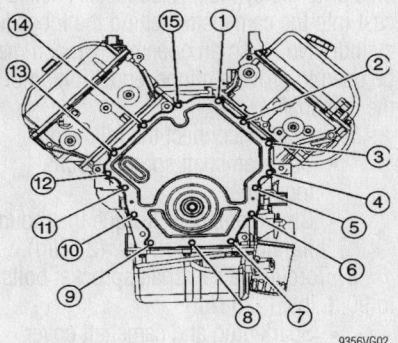

9356VG02

Front cover torque sequence

- Right engine mount bracket
- Engine front cover and gasket. The gasket is reusable. Do not discard unless it is damaged.
- 3 oil pump assembly retaining bolts identified the bolts by the larger head size
- Oil pump assembly off the nose of the crankshaft with the drive collar in place

3. Clean and inspect the oil pump.

To install:

4. Install or connect the following:
- Oil pump drive spacer into the oil pump so that the drive flat engages the pump rotor
- Oil pump on the crankshaft
- Pump retaining bolts and tighten to 89 inch lbs. (10 Nm) plus an additional 35 degrees

5. Place a small amount of sealant at the split line of the upper and lower crankcases.

6. Install or connect the following:
- Engine front cover gasket over the crankcase dowel pins
- Engine front cover in position on the crankcase
- Engine front cover retaining bolts in the sequence illustrated to 89 inch lbs. (10 Nm)
- Crankshaft balancer
- Engine mount bracket
- Drive belt idler pulley and tighten the bolt to 37 ft. lbs. (50 Nm)
- Drive belt tensioner
- Negative battery cable

Rear Main Seal

REMOVAL & INSTALLATION

1. Before servicing the vehicle, refer to the precautions in the beginning of this section.

2. Remove or disconnect the following:
- Transaxle

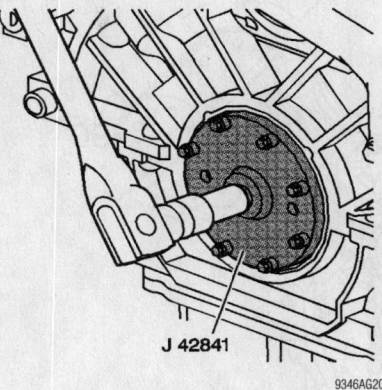

J 42841

9346AG20

Removing the rear main seal—4.6L engines

- Flexplate

3. Place oil seal removal Tool J 42841 on to the crankshaft.

4. Install the tool retaining bolts.

5. Using a drill with a socket adapter, install eight one-inch self-drilling crews into the seal using the guide holes in the removal tool. When drilling, make sure you reduce the drill speed when the screw begins threading into the seal.

6. With all eight removal screws installed, remove the removal tool retaining bolts.

7. Install the removal tool center screw.

8. Tighten the screw until the seal is removed.

To install:

9. Clean any debris from the crankshaft rear oil seal drain using wire or an unbound plastic tie-wrap.

10. Coat the outer diameter of the cylinder block crankshaft rear oil seal area with clean engine oil

➡ **DO NOT allow any engine oil on the area where the crankshaft rear oil seal is to be pressed onto the crankshaft. The green coating pre-applied to the inner diameter of the crankshaft rear oil seal must not be contaminated.**

11. Wipe the outer diameter of the flywheel flange clean with a lint-free cloth.

12. Lubricate the outer rubber surface of the crankshaft rear oil seal with clean engine oil.

13. Loosen the center bolt of the removal tool until the center hub protrudes approximately one-half inch beyond the outer plate. It is not necessary to completely unthread the center bolt and separate the two pieces of the installation Tool J 42842. Place the installation tool on to the crankshaft.

14. Thread the three mounting bolts into the crankshaft flange.

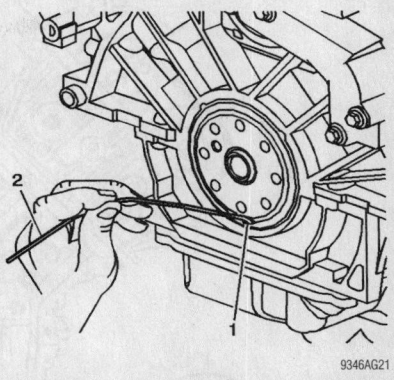

9346AG21

Clean any debris from the crankshaft rear oil seal drain using wire or an unbound plastic tie-wrap

15. Tighten the bolts until the installation tool is firmly mounted on the crankshaft.

16. Install the crankshaft rear oil seal by tightening the center bolt until the installation tool bottoms against the crankcase.

17. Loosen the center bolt to release pressure on the crankcase.

18. Loosen the three mounting bolts.

19. Remove the installation tool from the crankshaft flange.

20. Inspect to ensure the installation depth is equal around the crankshaft rear oil seal's circumference. If the depth is not equal reinstall the tool and repeat the installation procedures.

21. Install the flexplate. Torque the bolts to 11 ft. lbs. (15 Nm) plus an additional 50 degree turn.

22. Install the transaxle.

23. Start the engine and check for leaks.

Timing Chain, Sprockets, Front Cover and Seal

REMOVAL & INSTALLATION

The left and right-side secondary timing chains can be removed with the engine in the vehicle. If the primary timing chain or intermediate shaft sprocket need to be replaced, the engine must be removed from the vehicle and supported on an engine stand.

➡ **Setting the camshaft timing is necessary whenever the camshaft drive system has been disturbed, meaning the relationship between any chain and sprocket has been lost. Correct timing exists when the crankshaft and intermediate shaft sprocket timing marks are in alignment and all 4 camshaft drive pins are perpendicular (90 degrees) to the cylinder head surface.**

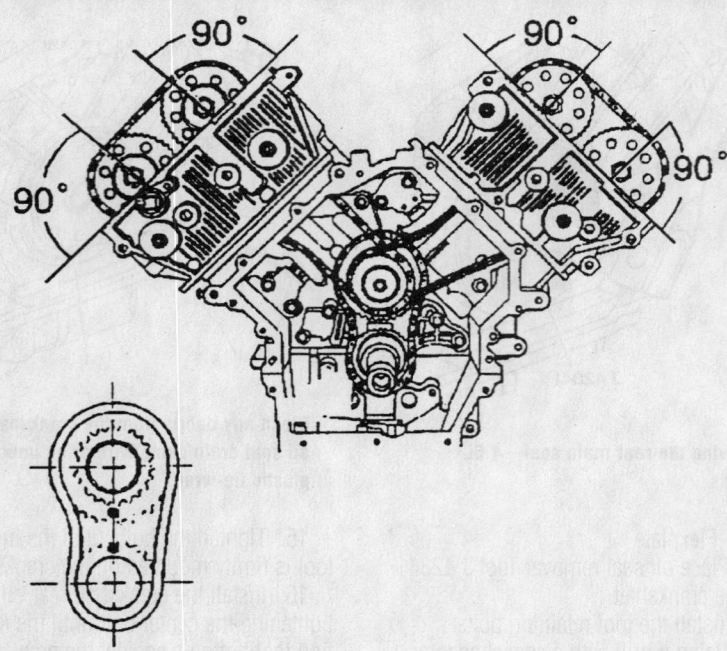

Correct timing chain alignment

Right Side Secondary Chain

1. Before servicing the vehicle, refer to the precautions in the beginning of this section.

2. Remove or disconnect the following:
- Negative battery cable
- Exhaust Y-pipe at the converter
- Tower-to-tower brace
- Ignition Control Module (ICM) wiring connectors and mounting bolts
- ICM
- Spark plug wires on the right bank
- Positive Crankcase Ventilation (PCV) valve
- Canister purge solenoid from the rear of the cover
- Front cover
- Camshaft cover screws
- Right and left torque struts

3. Safely support the front of the engine cradle and remove the 2 mounting bolts at the front of the cradle.

4. Lower the engine cradle or raise the vehicle to provide clearance at the rear of the engine compartment.
- Camshaft cover

➡**The camshaft cover gasket is reusable as long as it is not damaged.**

- Right side secondary chain tensioner
- Right side chain guide. Access the upper chain guide mounting bolt through the hole in the cylinder head capped with the plastic plug.

- Right side camshaft sprocket bolts and camshaft sprockets
- Secondary drive chain

To install:

5. Install the secondary timing chain over the outer row of teeth on the intermediate shaft sprocket. Route the chain over the chain guide and install the exhaust camshaft sprocket so the **RE** (Right Head Exhaust) pin engages the sprocket notch. There should be no slack in the lower section of the timing chain and the camshaft drive pin **must** be perpendicular to the cylinder head face.

➡**The Right Exhaust (RE) camshaft sprocket must contain the Camshaft Position (CMP) sensor pick-up.**

6. Install the intake camshaft sprocket into the chain so the sprocket notch **RI** (Right Head Intake) engages the camshaft and the camshaft drive pin remains perpen-

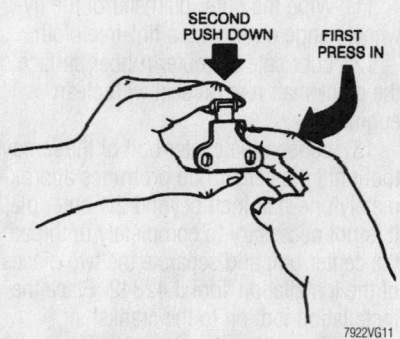

SECOND PUSH DOWN — FIRST PRESS IN

Rotating tensioner release lever

dicular to the cylinder head face. A hex is cast into the camshafts behind the lobes for cylinder No. 1, so an open end wrench may be used to provide minor repositioning of the camshafts.

7. Install or connect the following:
- Both camshaft sprocket bolts loosely
- Chain tensioner. Torque the mounting bolts to 20 ft. lbs. (27 Nm).

8. Torque the camshaft sprocket bolts to 90 ft. lbs. (120 Nm).
- Spark plug and camshaft cover seals
- Camshaft cover. Torque the screws to 84 inch lbs. (10 Nm).
- Engine cradle. Torque both bolts to 75 ft. lbs. (100 Nm).

9. Install the right and left torque struts. Torque the retaining bolts to 45 ft. lbs. (60 Nm).

➡**It is important during installation that the engine torque struts are not preloaded in their installed position. Adjustment is provided at the point the strut fastens to the core support bracket. Be sure this bolt is loose during assembly.**

- Front cover. Torque the bolts to 89 inch lbs. (10 Nm).
- Strut-to-core support bracket bolt. Torque it to 45 ft. lbs. (60 Nm).
- Canister purge solenoid to the rear of the cover
- PCV valve
- ICM and the wiring connectors
- Spark plug wires on the right-bank
- Tower-to-tower brace
- Exhaust Y-pipe to the converter. Torque the bolts to 20 ft. lbs. (27 Nm).
- Negative battery cable

10. Start the engine and check for leaks.

Left Side Secondary Chain

1. Before servicing the vehicle, refer to the precautions in the beginning of this section.

2. Drain the cooling system.

3. Remove or disconnect the following:
- Negative battery cable
- Accessory drive belt
- Front cover bolts
- Front cover and gasket
- Power steering hose
- Right front wheel
- Splash shields from the wheel well
- Flexplate cover
- Crankshaft balancer bolt

4. Support the engine cradle.

- 3 right-side engine cradle bolts
- Road Sensing Suspension (RSS) sensor from the right control arm

5. Lower the cradle to gain access for the crankshaft balancer puller.

- Crankshaft balancer
- Drive belt tensioner
- Drive belt idler pulley
- Upper radiator hose at the water crossover
- Spark plug wires
- Right side cooling fan
- Battery cable at the alternator
- Battery cable harness at the camshaft cover
- Positive Crankcase Ventilation (PCV) fresh air tube from the camshaft cover
- Water pump pulley
- Camshaft seal retainer screws and the seal
- Battery cable retainer at the front of the camshaft cover
- Camshaft cover by pivoting the entire cover around the water pump driveshaft. Continue moving the cover upward and pivoting so that the edge of the cover closely follows the left edge of the intake manifold cover.

➡**The camshaft cover gasket is reusable as long as it is not damaged.**

- Right side secondary chain
- Left side secondary chain tensioner
- Left side chain guide. Access the upper chain guide mounting bolt through the hole in the cylinder head capped with the plastic plug.
- Left side camshaft sprocket bolts and sprockets
- Secondary drive chain

To install:

6. Route the secondary timing chain for the left side over the inner row of intermediate sprocket teeth.

7. Install or connect the following:

- Secondary timing chain over the inner row of teeth on the intermediate shaft sprocket. Route the chain over the chain guide and install the exhaust camshaft sprocket so the **LE** (Left Head Exhaust) pin engages the sprocket notch. There should be no slack in the lower section of the timing chain and the camshaft drive pin **must** be perpendicular to the cylinder head face.
- Intake camshaft sprocket into the chain so the sprocket notch **LI** (Left Head Intake) engages the camshaft

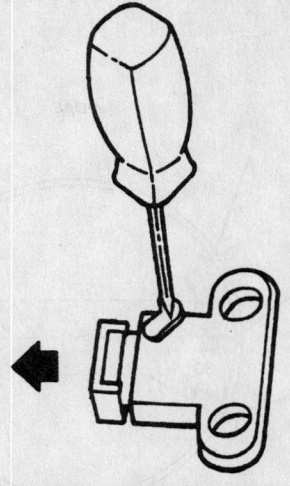

1 RELEASE TO FIRST CLICK
2 INSTALL LOCK PIN

7922VG12

Locking the tensioner in the collapsed position

and the camshaft drive pin remains perpendicular to the cylinder head face. A hex is cast into the camshafts behind the lobes for cylinder No. 2, so an open-end wrench may be used to provide minor repositioning of the camshafts.

- Both camshaft sprocket bolts loosely
- Chain tensioner. Torque the mounting bolts to 20 ft. lbs. (27 Nm).

8. Torque the camshaft sprocket bolts to 90 ft. lbs. (120 Nm).

- Front cover gasket on the dowel pins on the block
- Front cover. Torque the bolts to 89 inch lbs. (10 Nm). Apply a dab of RTV to the split line between the upper and lower crankcase assemblies.
- Drive belt idler pulley. Torque the bolt to 35 ft. lbs. (47 Nm).
- Drive belt tensioner. Torque the nut to 35 ft. lbs. (47 Nm).
- Crankshaft balancer. Torque the bolt to 44 ft. lbs. (60 Nm) plus an additional 120 degree turn.
- Engine cradle. Torque the bolts to 75 ft. lbs. (102 Nm).
- RSS sensor
- Wheel well splash shields
- Flexplate cover
- Spark plug and camshaft cover seals
- Intake camshaft through the hole in the camshaft cover

✳✴ WARNING

Use care to prevent the exposed section of the camshaft cover seal from being damaged by the edge of the cylinder head casting.

- Camshaft cover
- Camshaft cover screws. Torque the screws to 84 inch lbs. (10 Nm).
- Battery cable retainer to the front of the camshaft cover
- Battery cable at the alternator
- Camshaft seal to the end of the intake camshaft. Seal the screw threads with sealer.
- Water pump pulley
- PCV fresh air tube to the camshaft cover
- Right side fan
- Spark plug wires
- Upper radiator hose to the water crossover

9. Refill the cooling system.

Primary Chain/Intermediate Sprocket

1. Before servicing vehicle, refer to the precautions in the beginning of this section.

2. Remove or disconnect the following:

- Engine
- Accessory drive belt
- Drive belt pulley
- Dive belt tensioner
- Front cover
- Right and left camshaft covers

➡**Align all timing marks prior to removal of the timing chains.**

- Timing chain tensioners
- Camshaft sprocket bolts
- Camshaft sprockets (4)
- Right and left secondary chains
- Intermediate shaft sprocket bolt and the sprocket
- Primary timing sprockets and primary chain off the engine

To install:

➡**The following procedure must be followed to set the camshaft timing on the vehicle.**

3. Install the primary and secondary chain guides.

4. Rotate the crankshaft until the sprocket drive key is at the 1 o'clock position.

5. Install or connect the following:

- Crankshaft sprocket and intermediate shaft sprocket in the primary timing chain so the timing marks are aligned
- Primary timing chain assembly

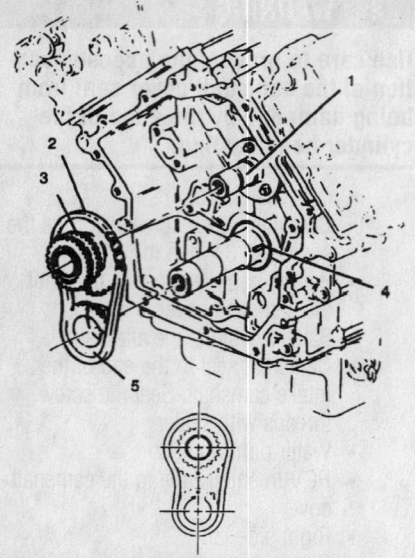

1 INTERMEDIATE SHAFT
2 PRIMARY CHAIN
3 INTERMEDIATE SHAFT SPROCKET
4 CRANKSHAFT SPROCKET KEY
5 SPROCKET

7922VG10

Primary drive chain components

➡The crankshaft sprocket keyway will have to slide over the key on the crankshaft. If it is necessary to turn the crankshaft sprocket, the intermediate shaft sprocket will also have to be turned so the timing mark remains aligned with the crankshaft sprocket.

- Intermediate shaft sprocket bolt. Torque the bolt to 45 ft. lbs. (61 Nm).
- Primary timing chain tensioner. Torque the tensioner mounting bolts to 20 ft. lbs. (27 Nm).
- Right and left secondary chains with the timing marks aligned
- Camshaft sprockets (4). Torque the bolts to 90 ft. lbs. (120 Nm).
- Right and left camshaft covers. Torque the bolts to 89 inch lbs. (10 Nm).
- Front engine cover. Torque the bolts to 89 inch lbs. (10 Nm).
- Accessory drive belt
- Drive belt pulley
- Drive belt tensioner
- Engine

6. Refill the cooling system.
7. Start the engine and check for leaks.

Piston and Ring

POSITIONING

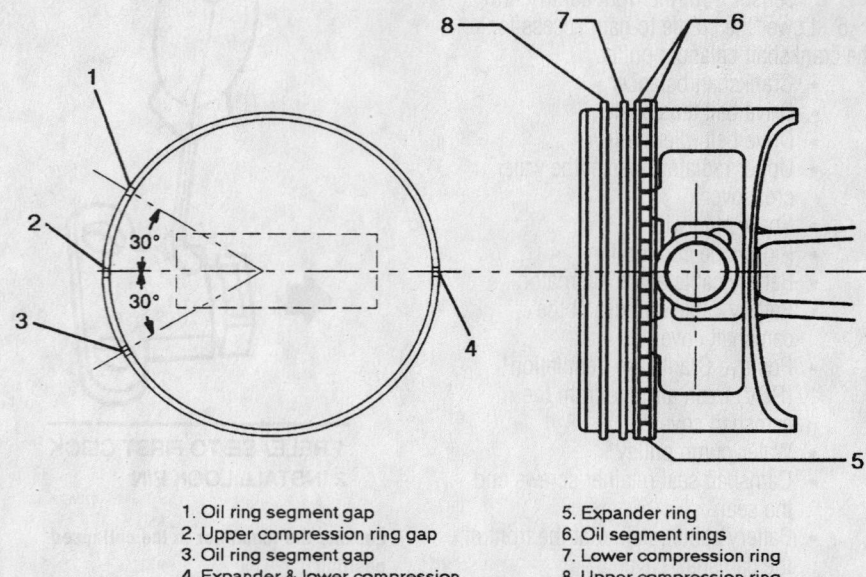

1. Oil ring segment gap
2. Upper compression ring gap
3. Oil ring segment gap
4. Expander & lower compression ring gaps
5. Expander ring
6. Oil segment rings
7. Lower compression ring
8. Upper compression ring

7922AG52

Piston ring and ring end-gap positioning—4.6L engine

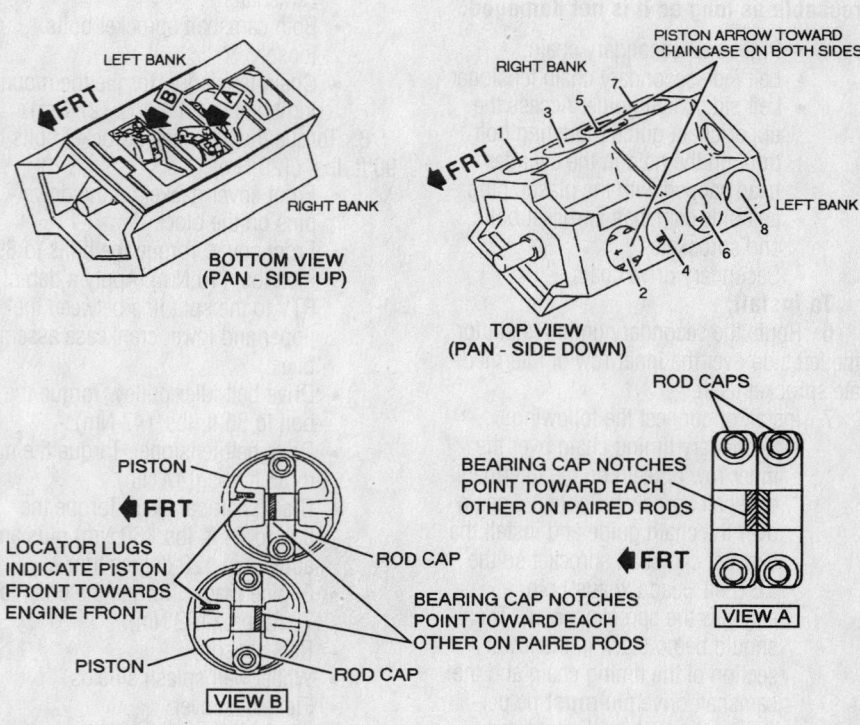

7922AG53

Piston and connecting rod assembly positioning—4.6L engine

FUEL SYSTEM

Fuel System Service Precautions

Safety is the most important factor when performing not only fuel system maintenance but also any type of maintenance. Failure to conduct maintenance and repairs in a safe manner may result in serious personal injury or death. Maintenance and testing of the vehicle's fuel system components can be accomplished safely and effectively by adhering to the following rules and guidelines.

• To avoid the possibility of fire and personal injury, always disconnect the negative battery cable unless the repair or test procedure requires that battery voltage be applied.

• Always relieve the fuel system pressure prior to disconnecting any fuel system component (injector, fuel rail, pressure regulator, etc.), fitting or fuel line connection. Exercise extreme caution whenever relieving fuel system pressure, to avoid exposing skin, face and eyes to fuel spray. Please be advised that fuel under pressure may penetrate the skin or any part of the body that it contacts.

• Always place a shop towel or cloth around the fitting or connection prior to loosening to absorb any excess fuel due to spillage. Ensure that all fuel spillage

(should it occur) is quickly removed from engine surfaces. Ensure that all fuel soaked cloths or towels are deposited into a suitable waste container.

• Always keep a dry chemical (Class B) fire extinguisher near the work area.

• Do not allow fuel spray or fuel vapors to come into contact with a spark or open flame.

• Always use a back-up wrench when loosening. Torque the fuel line connection fittings. This will prevent unnecessary stress and torsion to fuel line piping.

• Always replace worn fuel fitting O-rings with new. Do not substitute fuel hose or equivalent, where fuel pipe is installed.

Fuel System Pressure

RELIEVING

✻✻ CAUTION

The fuel injection system remains under pressure, even when the engine has been turned OFF. The fuel system pressure must be relieved before disconnecting any fuel lines.

Failure to do so may result in fire and/or personal injury.

1. Before servicing vehicle, refer to the precautions at the beginning of this section.
2. Loosen the fuel filler cap to relieve tank vapor pressure.

✻✻ CAUTION

Observe all applicable safety precautions when working around fuel. Whenever servicing the fuel system, always work in a well-ventilated area. Do not allow fuel spray or vapors to come in contact with a spark or open flame. Keep a dry chemical fire extinguisher near the work area. Always keep fuel in a container specifically designed for fuel storage; also, always properly seal fuel containers to avoid the possibility of fire or explosion.

3. Be sure the ignition switch is in the **OFF** position.
4. Disconnect the negative battery cable.
5. Remove the engine cover.

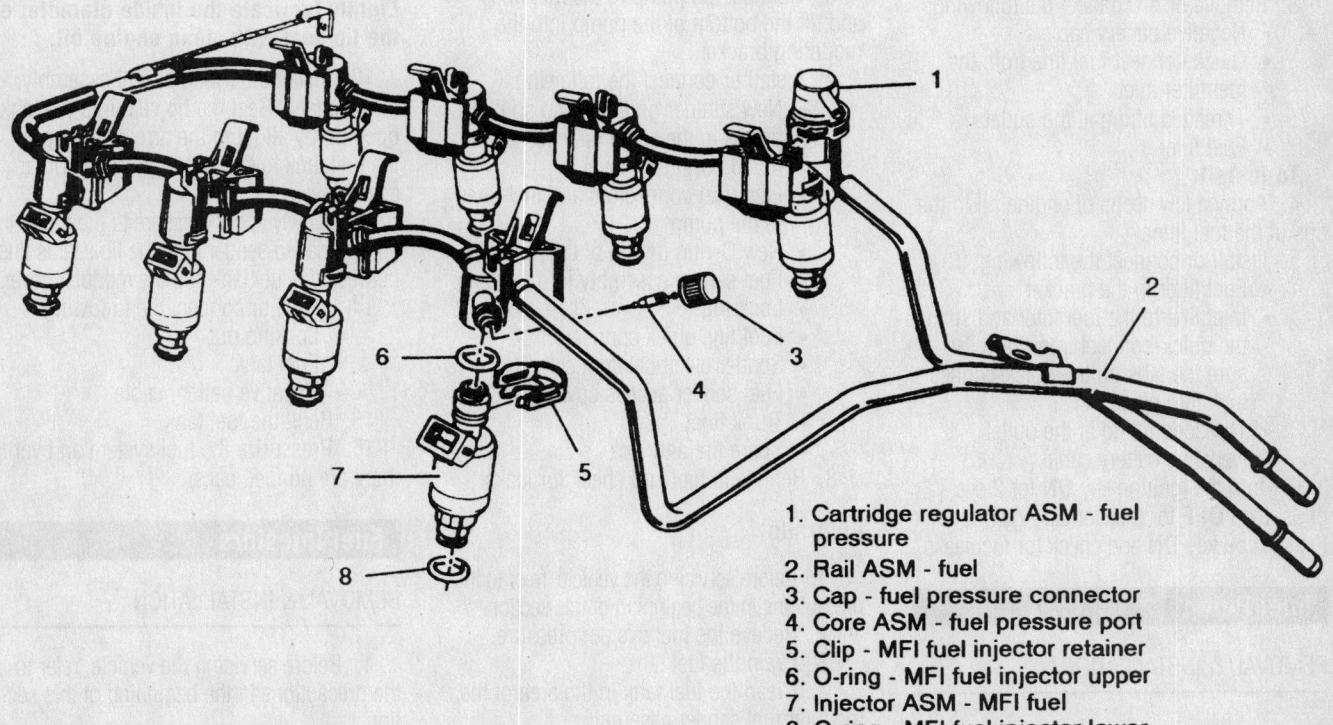

1. Cartridge regulator ASM - fuel pressure
2. Rail ASM - fuel
3. Cap - fuel pressure connector
4. Core ASM - fuel pressure port
5. Clip - MFI fuel injector retainer
6. O-ring - MFI fuel injector upper
7. Injector ASM - MFI fuel
8. O-ring - MFI fuel injector lower

7922VG13

View of the fuel rail assembly, showing the fuel system service port location

✳✳ CAUTION

There may still be residual fuel in the system, and a small amount of fuel may be released when servicing fuel lines or connections. In order to reduce the chance of personal injury, cover the fuel line fittings with a shop towel before disconnecting to catch any fuel that may leak out.

6. Install a fuel pressure gauge with a drain hose attached, J-34730–1, or equivalent. Wrap a shop towel around the fitting while connecting the gauge to avoid spillage.

7. Install the drain hose into an approved container and open the valve to drain the system pressure. Fuel connections are now safe for servicing.

8. Drain any remaining fuel from inside the gauge into the approved container.

Fuel Filter

REMOVAL & INSTALLATION

1. Before servicing the vehicle, refer to the precautions in the beginning of this section.

2. Properly relieve the fuel system pressure.

3. Remove or disconnect the following:
   - Negative battery cable
   - Quick connect fuel line from the fuel filter inlet
   - Threaded fitting at the outlet
   - Fuel filter

To install:

4. Apply a few drops of engine oil to the tips of the fuel filter.

5. Install or connect the following:
   - Fuel filter in the bracket
   - Inlet line to the fuel filter and snap the quick-connect into place. Be sure the tabs on the quick-connects lock into place.
   - Threaded fitting at the outlet
   - Negative battery cable

6. Turn the ignition key **ON** for 2 seconds, then **OFF** for 5 seconds. Again turn the ignition key **ON** and check for fuel leaks.

Fuel Pump

REMOVAL & INSTALLATION

Except Eldorado

1. Before servicing the vehicle, refer to the precautions in the beginning of this section.

2. Relieve the fuel system pressure.
3. Drain the fuel tank.
4. Remove or disconnect the following:
   - Negative battery cable
   - Spare tire and jack
   - Trunk lining
   - Fuel sender access panel
   - Sender and quick connect fittings from the sender
   - Electrical connector from the sender and position harness and hoses aside

✳✳ CAUTION

When removing the fuel sender from the tank, the reservoir bucket is full of fuel. Use caution in containing the fuel.

   - Sender retaining ring
   - Sender and take note of its position
   - Fuel sender O-ring and discard it

➡Note the direction the strainer is pointing.

   - Strainer from the pump by pulling it down and twisting
   - Pump electrical wires and hoses
   - Pump assembly out of the rubber connectors

To install:

5. Transfer any insulators and grommets from the old pump to the new one.

6. Connect the pump to the fuel hose and tilt the bottom of the pump into the mounting bracket.

7. Install or connect the following:
   - New strainer on the pump so it points in the same direction as noted during removal
   - Electrical connectors and fuel lines to the pump
   - New O-ring on top of the fuel tank
   - Fuel sender assembly into the tank
   - Lockring
   - Fuel line quick connectors
   - Sender electrical connector
   - Fuel sender access cover
   - Trunk liner
   - Spare tire and jack

8. Refill with fuel and check for leaks.

Eldorado

1. Before servicing the vehicle, refer to the precautions in the beginning of this section.

2. Relieve the fuel system pressure.
3. Drain the fuel tank.
4. Clean the fuel tank in the area of the modular fuel sender assembly.
5. Remove or disconnect the following:
   - Fuel tank from the vehicle
   - Locking nut
   - Modular fuel sender assembly from the fuel tank

✳✳ CAUTION

The modular fuel sender assembly may spring up from its position. When removing the assembly, be aware that the reservoir bucket is full of fuel. Tip the assembly slightly during removal to avoid damaging the float. Have a shop towel ready to absorb any leakage.

6. Remove modular fuel sender assembly from the fuel tank.

7. Remove fuel sender seal.

8. Slide the fuel sender lip seal downward, past the reservoir and carefully over the float arm assembly.

9. Discard the fuel sender lip seal.

10. Carefully discard the reservoir fuel into an approved container.

To install:

➡The lip seal should be carefully positioned over the float arm assembly, moved up over the reservoir and half way up the guide posts.

11. Install a new lip seal on the modular fuel sender assembly.

➡Always use a new seal when servicing the modular fuel sender assembly. Lightly lubricate the inside diameter of the lip seal with clean engine oil.

12. Install the modular fuel assembly into the tank. Seat the lip seal into the tank opening by aligning the arrows on top of the fuel tank to the arrow on the modular assembly.

13. Slowly apply pressure to the top of the spring-loaded sender until the lip seal is flush between the fuel tank and the modular cover.

14. Install or connect the following:
   - Locking nut
   - Fuel tank
   - Negative battery cable

15. Refill the fuel tank.

16. Pressurize the fuel system and verify there are no fuel leaks.

Fuel Injectors

REMOVAL & INSTALLATION

1. Before servicing the vehicle, refer to the precautions in the beginning of this section.

2. Relieve the fuel system pressure.

3. Remove or disconnect the following:
   - Intake manifold cover

- Intake Air Temperature (IAT) sensor electrical connector
- Mass Airflow (MAF) sensor electrical connector
- Air cleaner intake duct
- Quick-connect fittings from the fuel rail
- Fuel pressure regulator vacuum hose
- Positive Crankcase Ventilation (PCV) air tube
- Fuel rail end-point bracket retainer nut
- Fuel injector electrical connectors
- Fuel rail attaching bolts
- Fuel injectors from the intake manifold with injector removal tool J 43013
- Fuel rail

- Injector from the rail and discard the O-rings

To install:

➡ **Each fuel injector is calibrated for a specific flow rate. Be sure to use the correct part number when ordering replacement fuel injectors.**

4. Install or connect the following:
   - New O-rings lubricated with clean engine oil
   - Fuel injector into the fuel rail
   - Fuel rail into the intake manifold and tighten the attaching bolts to 89 inch lbs. (10 Nm)
   - Fuel injector electrical connectors
   - Fuel rail end-point bracket retainer nut and tighten to 89 inch lbs. (10 Nm)

- PCV air tube
- Fuel pressure regulator vacuum hose
- Quick-connect fittings from the fuel rail
- Air cleaner intake duct
- MAF sensor electrical connector
- IAT sensor electrical connector
- Negative battery cable

5. Inspect for fuel leaks as follows:
   a. Step 1: Turn ignition switch to the **ON** position for 2 seconds.
   b. Step 2: Turn ignition switch **OFF** for 10 seconds.
   c. Step 3: Turn ignition switch **ON**.
   d. Step 4: Check for leaks.
6. Install the intake manifold cover. Torque the nuts to 27 inch lbs. (3 Nm).

DRIVE TRAIN

Transaxle Assembly

REMOVAL & INSTALLATION

2000–01 Models Except Eldorado

1. Before servicing the vehicle, refer to the precautions in the beginning of this section.
2. Drain the cooling system.
3. Remove or disconnect the following:
   - Negative battery cable
   - Upper filler panel
   - Air cleaner upper plenum and intake tube
   - Surge tank outlet pipe
   - Oil cooler lines at the cooler and oil sending line at the transmission and plug the lines to system contamination
   - Shift cable from the shift linkage
   - Shift linkage assembly from the transmission
   - Upper transmission to engine bolts

➡ **Load the support fixture by tightening the wing nuts several turns in order to take the weight of the powertrain off of the mounts and the frame.**

4. Install an engine support fixture. Raise the left side of the assembly 1 inch (25.4mm) above the resting position with the adjusting screws.
   - Tires and wheels
   - Lower splash shield
   - Front suspension position sensors links from the lower control arms and position aside
   - Stabilizer links from the lower control arms and let the stabilizer shaft sit loosely on the engine frame

   - Stabilizer shaft insulator brackets from the engine frame
   - Drive axle nuts
   - Drive axles from the hubs
   - Tie rod cotter pins and nuts
   - Tie rods from the steering knuckles
   - Lower ball joint cotter pins and nuts
   - Ball joints from the steering knuckles
   - Drive axles from the transmission
   - Brake Pressure Modulator Valve (BPMV) and bracket from the engine frame and support the valve with a bungee cord
   - Rear transmission mount bracket nut and bolts
   - Rear transmission mount bolt and nuts from the left frame rail
   - Left engine mount nut from the engine frame
   - Right engine mount nuts from the engine frame
   - Steering rack bolts from the engine frame
   - Brake line and power steering line fasteners from the engine frame
   - Post Oxygen Sensor (O_2S) heat shield and sensor connector
   - One center exhaust hanger to allow easy movement of the exhaust in the rearward direction
5. Support the engine frame.
   - 6 engine frame mount bolts and separate the tie rod ends and lower ball joints from the steering knuckles
6. Raise the vehicle slowly away from the engine frame.
   - Engine frame and support table from under the vehicle

   - Steering rack heat shield from the rack
   - Power steering lines from the steering rack
   - Right rear engine to transmission bracket upper bolt, nuts, and bracket
   - Right transmission mount bracket lower bolts and bracket from the transmission
7. It is necessary to support the steering rack in a position that allows access to the bracket bolts.
   - Steering rack from the vehicle
   - Right front engine-to-transmission brace bolt, nut, brace and heat shield
   - Right front engine-to-transmission bracket
   - Engine oil pan-to-transmission bracket
   - Torque converter cover brace and the cover
   - Flywheel-to-converter bolts
   - Transmission electrical connectors
   - Engine-to-transmission heat shield and bracket
8. Support the transmission with a transmission jack.
   - Engine-to-transmission bolts
   - Transmission

To install:

9. Install or connect the following:
   - Rear transmission mount bracket onto the stud before the transmission is all the way up into the vehicle
   - Rear transmission mount (loosely) to the left frame rail with the nuts and bolt

10. Raise the transmission into position. Guide the rear mount bracket and the rear mount together as the transmission is raised into the vehicle.

11. Align the transmission and position the transmission onto the dowels on the engine.

- Engine-to-transmission bolts and tighten to 55 ft. lbs. (75 Nm), then remove the transmission jack
- Right front engine-to-transmission heat shield and brace, then tighten the bolts to 55 ft. lbs. (75 Nm)
- Right front engine-to-transmission bracket and tighten the bolts to 37 ft. lbs. (50 Nm)
- Right transmission mount bracket and the lower bolts. Tighten the bolts to 81 inch lbs. (110 Nm)
- Steering rack into position (loosely) onto the vehicle and insert the tie rod ends into the steering knuckles
- Right rear engine-to-transmission bracket and tighten the bolt 81 inch lbs. (110 Nm) and the nuts to 37 ft. lbs. (50 Nm)
- Power steering lines to the power steering rack and tighten to 20 ft. lbs. (27 Nm)
- Power steering rack heat shield
- Flywheel to converter bolts and tighten to 44 ft. lbs. (60 Nm)
- Torque converter cover
- All electrical connectors
- Drive axles into the hubs and both drive axle nuts and tighten to 110 ft. lbs. (145 Nm)
- Engine frame on to the engine support table
- Stabilizer bar onto the engine frame in its approximate position

12. Lower the vehicle onto the engine frame while locating all the mounting points.

- Engine frame mounting bolts, finger-tight

13. Raise the vehicle away from the engine support table and tighten the engine frame bolts to 142 ft. lbs. (192 Nm).

- Steering intermediate shaft to the steering gear and install the clamp bolt. Tighten the bolt to 37 ft. lbs. (50 Nm)
- Steering gear bolts and tighten to 77 ft. lbs. (105 Nm)
- Transmission mount nuts and the right engine mount nuts to the frame and tighten to 37 ft. lbs. (50 Nm)

➡When tightening the ball joint nut, a minimum torque of 37 ft. lbs. (50 Nm)

must be reached. If the proper torque is not obtained, inspect for stripped threads. If threads are satisfactory, replace the ball joint and knuckle. If required, turn the nut up to an additional 60 degrees to allow for installation of the cotter pin.

- Lower ball joints into the steering knuckles
- Ball joint nuts and cotter pins. Tighten the ball joint nut to 84 inch lbs. (10 Nm), then tighten the nut an additional 120 degrees.

➡When tightening the tie rod end nut, a minimum torque of 33 ft. lbs. (45 Nm) must be reached. If the proper torque is not obtained, inspect for stripped threads. If the threads are satisfactory, replace the tie rod end and knuckle.

- Tie rod end nut to 74 ft. lbs. (100 Nm), then an additional 1/3 turn (or 2 flats)
- Brake and power steering line retainers to the engine frame
- Brake pressure modulator valve and bracket to the engine frame
- Sway bar insulators onto the sway bar
- Stabilizer links to the control arms and tighten the nuts to 49 inch lbs. (65 Nm)
- Front suspension position sensors to the lower control arms
- Lower close out panel
- Center exhaust hanger
- Post O_2S and heat shield
- Front tire and wheel assemblies

14. Remove the Engine Support Fixture.

- Upper bell housing bolts
- Transmission cooler pipe fittings at the transmission and tighten to 16 ft. lbs. (22 Nm)
- Transmission oil sending pipe to the transmission
- Shift linkage assembly to the transmission
- Shift cable to the shift linkage
- Surge tank outlet hose
- Surge tank inlet hose
- Air cleaner assembly
- Negative battery cable

15. Bleed the power steering system.
16. Fill the transmission with fluid.
17. Fill the cooling system.
18. Check the front suspension alignment and as necessary.

➡The PCM maintains 3 types of transaxle adapt parameters which are used to modify transaxle line pressure.

The line pressure is modified to maintain shift quality regardless of wear or tolerance variations within the transaxle. Whenever the transaxle is replaced, the transaxle adapts must be reset as follows:

a. Step 1: Turn the ignition key **ON**. Enter the self-diagnostic system.

b. Step 2: Select PCM override PS13 (TPS SENSOR LEARN).

c. Step 3: Press the WARMER button. The Driver Information Center (DIC) should display 09, indicating that the Garage Shift Adapt value has been reset.

d. Step 4: Select PCM override PS14 (TRAN ADAPT).

e. Step 5: Press the COOLER button. The DIC should display 90, indicating the Upshift Adapt (UA) value has been reset.

f. Step 6: Press the WARMER button. The DIC should display 09, indicating the Steady State Adapt (SSA) value has been reset.

➡The PCM maintains a value for transaxle oil life. This value indicates the percentage of oil life remaining and is calculated based on transaxle temperature and speed. When the vehicle is new, the transaxle oil life value is 100. As the vehicle operates, the percentage will decrease. Whenever the transaxle is replaced, the transaxle oil life indicator should be reset to 100 as follows:

g. Step 1: Turn the ignition key **ON**, but leave the engine OFF.

h. Step 2: Press and hold the OFF and REAR DEFOG buttons on the DIC until the message TRANSAXLE OIL LIFE RESET is displayed on the DIC.

2002–03 Models Except Eldorado

1. Before servicing the vehicle, refer to the precautions in the beginning of this section.

2. Drain the cooling system.

3. Remove or disconnect the following:

- Negative battery cable
- Front air deflector
- Starter motor
- Upper filler panel
- Upper and lower air cleaner assemblies
- Electrical connector from the Engine Coolant temperature (ECT) sensor.
- Nut securing the coil cassette ground wire to the right cylinder head

- Range selector cable from the shift lever
- Nut that retains the heater inlet pipe bracket
- Locking tabs and pull the heater inlet pipe out of the adapter
- Nut that retains the heater outlet pipe
- Heater outlet pipe from the thermostat housing
- Range selector cable bracket from the studs
- Nuts securing the front transaxle brace to the right cylinder head.

4. Attach an engine lift bracket to the left cylinder head.
- Air inlet grille panel

5. Install an engine support fixture and preload the engine support fixture with the weight of the engine and transaxle.

6. Reposition the main wiring harness to gain access to the upper transaxle to engine bolts.
- Transaxle-to-engine bolts. The remaining bolts will hold the transaxle until it is removed and will be removed from the underside of the vehicle.
- Upper transaxle oil cooler pipe retaining bolt from the radiator fan shroud
- Upper transaxle oil cooler pipe from the radiator plug the lines
- Lower transaxle oil cooler pipe from the transaxle and plug the lines

7. The following are exceptions while following the frame removal procedure:
a. Do not remove the front stabilizer shaft from the frame. Remove the stabilizer shaft links only.
b. Do not remove the A.I.R pump assembly from the vehicle. Retain the pump to the body using mechanics wire.
c. Do not remove the insulators from the frame.
d. Do not remove the control arms from the frame.

8. Remove or disconnect the following:
- Front frame assembly from the vehicle and position frame aside
- Right and left drive shafts from the transaxle end only. Support the shafts using wire
- Engine-to-transaxle bracket lower bolt and bracket
- Torque converter cover assembly
- Flywheel-to-converter bolts
- Main electrical connector and the ground wires from the front of the transaxle
- Vehicle Speed Sensor (VSS) wiring harness

9. Support the transaxle with a suitable transmission jack.
- Nut securing the transaxle mount to the transaxle mount bracket

10. Slowly lower the transmission jack and allow the engine support fixture to hold the weight of the engine and transaxle assembly.

11. Position the transmission jack aside and lower the vehicle.

12. Using the engine support fixture, lower the engine and transaxle assembly until the left mount bracket clears the mount stud.

13. Raise the vehicle.

14. Support the transaxle with a suitable transmission jack.

15. Remove the engine-to-transaxle brace and heat shield.

16. Remove the remaining transaxle to engine bolts.

17. Separate the transaxle from the engine.

18. Carefully lower the transaxle from the vehicle.

To install:

19. Installation is the reverse of removal, please note the following torques:
- Engine-to-transaxle bolts to 55 ft. lbs. (75 Nm)
- Nut securing the transaxle mount to the transaxle mount bracket to 55 ft. lbs. (75 Nm)
- Flywheel-to-converter bolts to 44 ft. lbs. (60 Nm)
- Engine-to-transaxle bracket lower bolt to 37 ft. lbs. (50 Nm)
- Front transaxle brace nuts to 37 ft. lbs. (50 Nm)

20. Fill the transmission with fluid.

21. Fill the cooling system.

22. Check the front suspension alignment and as necessary.

➡ The PCM maintains 3 types of transaxle adapt parameters which are used to modify transaxle line pressure. The line pressure is modified to maintain shift quality regardless of wear or tolerance variations within the transaxle. Whenever the transaxle is replaced, the transaxle adapts must be reset as follows:

a. Step 1: Turn the ignition key **ON**. Enter the self-diagnostic system.
b. Step 2: Select PCM override PS13 (TPS SENSOR LEARN).
c. Step 3: Press the WARMER button. The Driver Information Center (DIC) should display 09, indicating that the Garage Shift Adapt value has been reset.
d. Step 4: Select PCM override PS14 (TRAN ADAPT).

e. Step 5: Press the COOLER button. The DIC should display 90, indicating the Upshift Adapt (UA) value has been reset.
f. Step 6: Press the WARMER button. The DIC should display 09, indicating the Steady State Adapt (SSA) value has been reset.

➡ The PCM maintains a value for transaxle oil life. This value indicates the percentage of oil life remaining and is calculated based on transaxle temperature and speed. When the vehicle is new, the transaxle oil life value is 100. As the vehicle operates, the percentage will decrease. Whenever the transaxle is replaced, the transaxle oil life indicator should be reset to 100 as follows:

g. Step 1: Turn the ignition key **ON**, but leave the engine OFF.
h. Step 2: Press and hold the OFF and REAR DEFOG buttons on the DIC until the message TRANSAXLE OIL LIFE RESET is displayed on the DIC.

Eldorado

1. Before servicing the vehicle, refer to the precautions in the beginning of this section.

2. Drain the cooling system.

3. Remove or disconnect the following:
- Negative battery cable
- Headlamp housing upper filler panel
- Cross vehicle brace
- Air cleaner assembly
- Shift control cable from the shift lever
- Shift control cable from the bracket
- Engine mount struts
- Oil cooler lines at the cooler and the oil sending line at the transmission
- Heater bypass pipe
- Power steering return hose at the auxiliary cooler and plug all lines to avoid contamination
- Both heater tube retainers from the upper case side cover studs
- Electrical connector from the Coolant Temperature Sensor (CTS)
- Ground wire from the stud near the rear transmission-to-engine brace
- Upper nuts from the rear engine-to-transmission brace
- Transmission vent hose at transmission

4. Carefully pull the wiring harness up from beneath the vehicle and set aside.

5. Move the engine harness at the top

of the transmission to access the upper transmission bolts.

- Upper transmission bolts

6. Install an engine support fixture.

7. Raise the left side of the powertrain (the transmission side) 1 inch (25.4 mm) above the powertrain resting position with the adjusting screws.

8. Remove or disconnect the following:
- Both drive axles
- Both front wheel opening splash shields
- Engine splash shield
- Vehicle Speed Sensor (VSS) connector
- Both front suspension sensor rods from lower control arm
- ABS modulator from the bracket and the support
- Torque converter cover
- Engine the transmission brace
- Front engine-to-transmission pencil brace
- Three ground connections at the front of the transmission
- Transmission main harness
- Flywheel to converter bolts
- Powertrain mount nuts from the frame
- Clamp bolt (Rotate the steering intermediate shaft so that the steering gear stub shaft clamp bolt is accessible from the left wheel opening)

✳✳ CAUTION

Failure to disconnect the intermediate shaft from the rack and pinion steering gear stub shaft can result in damage to the steering gear or to the intermediate shaft. This damage may cause loss of steering control, which could result in an accident and possible personal injury.

➡**Never rotate the steering wheel or move the position of the steering gear once the intermediate shaft has been disconnected. This will uncenter the Inflatable Restraint coil in the steering column. If the Inflatable Restraint coil becomes uncentered, it may become damaged during vehicle operation.**

- Steering intermediate shaft from the steering gear
- Electrical harness and connector from the front of the frame

9. Support the rear of frame with a jack stand.

- Four rear frame bolts

10. Lower the jack stand a few inches to

gain access to the power steering gear heat shield and the return line fitting.

- Power steering gear heat shield and the power steering return line at the gear and plug the lines to avoid contamination
- Power steering electrical connector

11. Raise the jack stand and reinstall one rear frame bolt on each side finger-tight in order to support the frame.

- Jack stand
- Support frame
- Six frame mount bolts.
- Lower
- Left transmission mount and bracket from the transmission
- Engine-to-transmission bolts
- Engine from the transmission

12. Lower the transmission down and to the left on a slight angle so the transmission case can clear the end of the starter.

13. Lower the transmission.

To install:

14. Installation is the reverse of removal, please note the following torques:
- Engine-to-transaxle bolts to 55 ft. lbs. (75 Nm)
- Left transmission bracket and mount to the transmission bolts to 35 ft. lbs. (47 Nm)
- Power steering pressure hose fitting to 20 ft. lbs. (27 Nm)
- Steering intermediate shaft to the steering gear clamp bolt to 35 ft. lbs. (47 Nm)
- Left and right transmission mount nuts and the right engine mount nuts to 35 ft. lbs. (47 Nm)
- Manual shaft nut to 15 ft. lbs. (20 Nm)
- Upper transaxle case-to-engine bolts to 55 ft. lbs. (75 Nm)

15. Fill the transmission with fluid.

16. Fill the cooling system.

17. Check the front suspension alignment and as necessary.

➡**The PCM maintains 3 types of transaxle adapt parameters which are used to modify transaxle line pressure. The line pressure is modified to maintain shift quality regardless of wear or tolerance variations within the transaxle. Whenever the transaxle is replaced, the transaxle adapts must be reset as follows:**

a. Step 1: Turn the ignition key **ON**. Enter the self-diagnostic system.

b. Step 2: Select PCM override PS13 (TPS SENSOR LEARN).

c. Step 3: Press the WARMER button. The Driver Information Center (DIC)

should display 09, indicating that the Garage Shift Adapt value has been reset.

d. Step 4: Select PCM override PS14 (TRAN ADAPT).

e. Step 5: Press the COOLER button. The DIC should display 90, indicating the Upshift Adapt (UA) value has been reset.

f. Step 6: Press the WARMER button. The DIC should display 09, indicating the Steady State Adapt (SSA) value has been reset.

➡**The PCM maintains a value for transaxle oil life. This value indicates the percentage of oil life remaining and is calculated based on transaxle temperature and speed. When the vehicle is new, the transaxle oil life value is 100. As the vehicle operates, the percentage will decrease. Whenever the transaxle is replaced, the transaxle oil life indicator should be reset to 100 as follows:**

g. Step 1: Turn the ignition key **ON**, but leave the engine OFF.

h. Step 2: Press and hold the OFF and REAR DEFOG buttons on the DIC until the message TRANSAXLE OIL LIFE RESET is displayed on the DIC.

Halfshaft

REMOVAL & INSTALLATION

✳✳ WARNING

Use care when removing the halfshaft to prevent the inner CV-joint from becoming over-extended. Overextension of the joint could result in separation of internal components and possible joint failure.

1. Before servicing the vehicle, refer to the precautions in the beginning of this section.

2. Remove the front wheel.

3. Install a boot protector on the outer CV-joint boot to protect it from damage.

4. Remove or disconnect the following:
- Hub nut
- Stabilizer link
- Ball joint cotter pin and nut
- Ball joint from the steering knuckle

➡**Partially install the hub nut to protect the threads when removing the halfshaft from the hub.**

5. Remove the halfshaft from the transaxle.

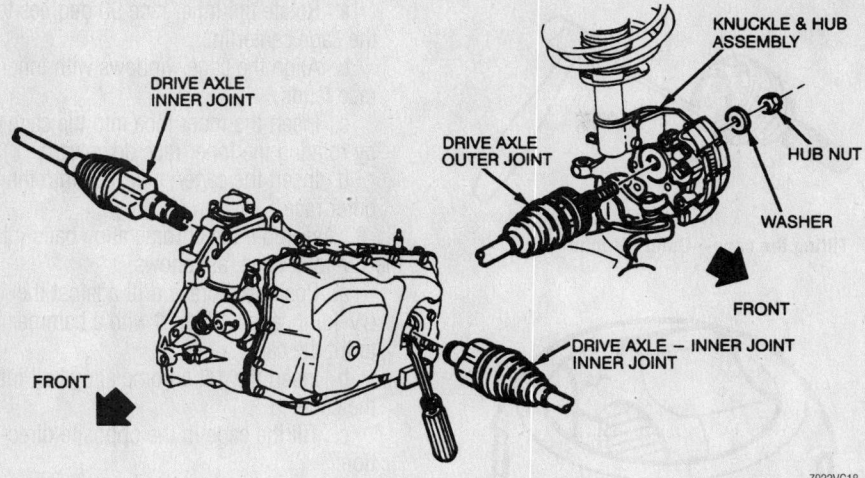

Removing the halfshaft from the transaxle

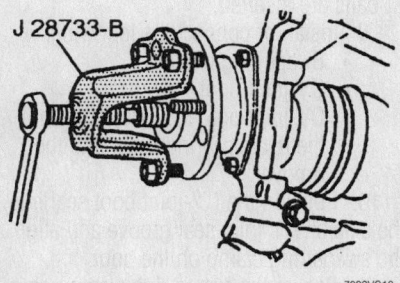

J 28733-B

7922VG19

Removing the halfshaft from the hub

➡️**If equipped with anti-lock brakes, care must be used to prevent damage to the toothed sensor ring on the halfshaft and the wheel speed sensor on the steering knuckle.**

To install:
6. Install or connect the following:
  • Halfshaft into the transaxle

➡️**To verify the halfshaft is properly seated, grasp the inner CV-joint housing and pull it outward. DO NOT pull on the halfshaft. If the CV-joint is properly seated, the halfshaft will not pull back out.**

  • Halfshaft into the hub/bearing assembly
  • New hub nut loosely
  • Ball joint into the steering knuckle
  • Ball joint castle nut. Torque the nut to 88 inch lbs. (10 Nm), plus an additional 150 degrees. A minimum of 41 ft. lbs. (51 Nm) of torque must be attained.

➡️**If necessary, the nut can be tightened up to 20 degrees additional. NEVER loosen the castle nut to install the cotter pin.**

7. Torque the hub nut to 118 ft. lbs. (160 Nm).

8. Install the stabilizer link. Torque the nut to 13 ft. lbs. (17 Nm).

9. If a halfshaft protector was used, remove it now.

10. Install the wheel. Torque the lug nuts to 100 ft. lbs. (140 Nm).

11. Road test and check vehicle operation.

CV-Joints

OVERHAUL

Inner (Tripod) Joint

1. Before servicing the vehicle, refer to the precautions in the beginning of this section.

2. Remove or disconnect the following:
  • Front wheel
  • Halfshaft and place it in a vise
  • Small CV-joint boot clamp

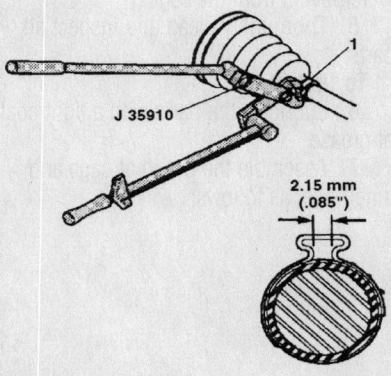

9346AG25

Crimp the small CV-boot clamp as shown

  • Large CV-joint boot clamp
  • CV-joint boot by sliding it away from the tripod joint
  • Tripod housing from the tripod spider
  • Inboard spacer ring, slide it rearward on the shaft
  • Outboard retaining ring
  • Tripod joint spider assembly
  • Inboard spacer ring
  • CV-joint boot
  • Trilobal tripod bushing from the housing

To install:
3. Install or connect the following:
  • New snapring onto the stub shaft
  • Small boot clamp
  • CV-joint boot

4. Using a crimp tool, a torque wrench and a breaker bar, crimp the small CV-joint boot clamp to 100 ft. lbs. (136 Nm) until the crimped gap measures .085 inch (2.15mm).
  • Inboard spacer ring, slide it rearward on the shaft past the 2nd groove
  • Tripod joint spider assembly onto the shaft until it passes the 2nd groove
  • Outboard retaining ring into the axle shaft groove
  • Tripod joint spider assembly, slide it against the outboard retaining ring
  • Inboard spacer ring, seat it in the groove

5. Place ½ of the grease provided in the service kit into the boot and the other ½ of the grease into the tripod housing.
  • Trilobal tripod bushing flush with the tripod housing face
  • New large seal clamp onto the CV-joint boot
  • Tripod housing, slide it over the tripod joint spider assembly
  • CV-joint-boot clamp, slide it into place, over the trilobal tripod bushing with the seal lip in the groove

➡️**Make sure the boot lies flat against the trilobal bushing.**

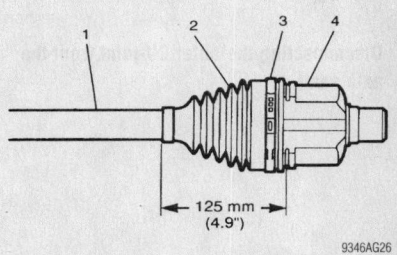

9346AG26

Position the boot as illustrated before clamping the large clamp

6. Make sure the boot is positioned as illustrated, then using a crimp tool, latch the large CV-joint boot clamp.

7. Install the halfshaft and the front wheel. Torque the lug nuts to 100 ft. lbs. (140 Nm).

Outer CV-Joint

1. Before servicing the vehicle, refer to the precautions in the beginning of this section.

2. Remove or disconnect the following:
- Front wheel
- Halfshaft
- Swage ring using a hand grinder
- Large boot clamp
- CV-joint boot, slide it away from the CV-joint
- CV-joint assembly by spreading the inner race-to-axle shaft retaining ring ears using Snapring Pliers
- CV-joint boot from the axle shaft

3. Disassemble the chrome alloy balls from the CV-joint cage as follows:

a. Position a brass drift against the CV-joint cage and tap it with a hammer to tilt the cage.

b. Chrome alloy ball from the cage.

c. Tilt the cage in the opposite direction.

d. Remove the opposite chrome alloy ball.

e. Repeat the procedure until all 6 balls are removed.

4. Disassemble the CV-joint cage and inner race as follows:

a. Pivot the cage and race 90 degrees to the center line of the outer race.

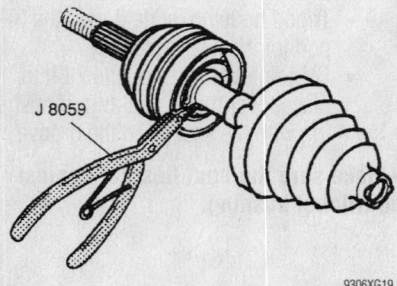

Disconnecting the outer CV-joint from the axle shaft

J 8059

9306XG19

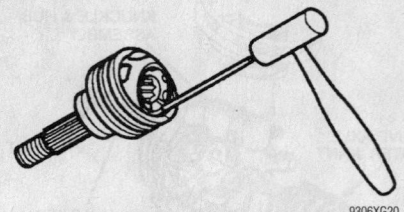

Tilting the cage—Outer CV-joint

9306XG20

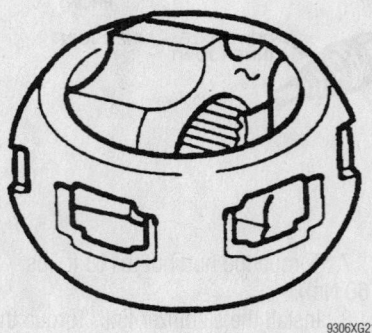

View the cage and inner race—Outer CV-joint

9306XG21

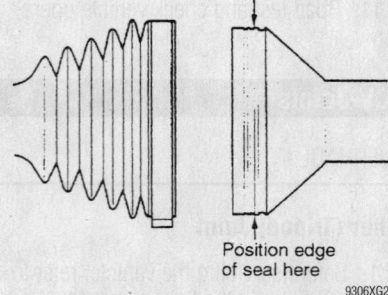

Positioning the boot—Outer CV-joint

Position edge of seal here

9306XG22

b. Align the cage windows with outer race lands.

c. Remove the cage from the outer race.

d. Rotate the inner race upward and remove it from the cage.

5. Thoroughly clean and inspect all parts.

To install:

6. Lubricate the parts with a light coat of grease.

7. Assemble the CV-joint cage and inner race, as follows:

a. Rotate the inner race 90 degrees to the cage centerline.

b. Align the cage windows with inner race lands.

c. Insert the inner race into the cage by rotating the inner race downward.

d. Insert the cage/inner race into the outer race.

8. Assemble the chrome alloy balls into the CV-joint cage, as follows:

a. Position a brass drift against the CV-joint cage and tap it with a hammer to tilt the cage.

b. Insert the 1st chrome alloy ball into the cage.

c. Tilt the cage in the opposite direction.

d. Insert the opposite chrome alloy ball.

e. Repeat the procedure until all 6 balls are inserted.

9. Install or connect the following:
- Swage ring clamp
- CV-joint boot
- CV-joint onto the axle shaft until the retaining ring seats into the groove

10. Position the CV-joint boot seal into the axle shaft's joint seal groove and align the swage ring clamp on the boot.

11. Secure the swage ring clamp using appropriate crimping tool.

✳✳ WARNING

Make sure that there are no pinch points on the inboard seal.

12. Install or connect the following:
- ½ kit grease into the CV-joint boot
- ½ kit grease into the CV-joint
- New large seal clamp onto the CV-joint boot
- CV-joint boot/clamp, slide it into place, over the outer race with the seal lip in the groove

➡ **Make sure the boot lies flat against the outer race.**

13. Using a Crimp tool, a torque wrench and a breaker bar, crimp the large CV-joint boot clamp to 130 ft. lbs. (176 Nm).

14. Install the halfshaft and the front wheel.

STEERING AND SUSPENSION

Air Bag

⁂ CAUTION

Some vehicles are equipped with an air bag system. The system must be disabled before performing service on or around system components, steering column, instrument panel components, wiring and sensors. Failure to follow safety and disabling procedures could result in accidental air bag deployment, possible personal injury and unnecessary system repairs.

PRECAUTIONS

Several precautions must be observed when handling the inflator module to avoid accidental deployment and possible personal injury.

- Never carry the inflator module by the wires or connector on the underside of the module.
- When carrying a live inflator module, hold securely with both hands, and ensure that the bag and trim cover are pointed away.
- Place the inflator module on a bench or other surface with the bag and trim cover facing up.
- With the inflator module on the bench, never place anything on or close to the module, which may be thrown in the event of an accidental deployment.

DISARMING

Eldorado

⁂ CAUTION

The air bag system must be disarmed before performing service procedures around the air bag or wiring. Failure to do so may cause accidental deployment, resulting in unnecessary repairs and/or personal injury.

1. Before servicing the vehicle, refer to the precautions in the beginning of this section.
2. Turn the steering wheel so that the vehicle's wheels are pointing straight-ahead.
3. Turn the ignition switch to the **LOCK** position and remove the key.
4. Remove or disconnect the following:
- Negative battery cable

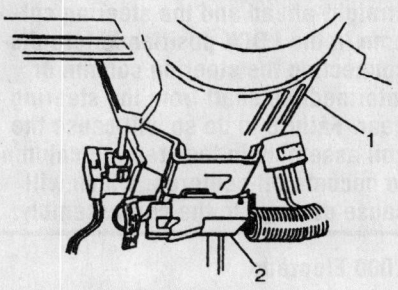

1. Steering column
2. Connector, SRS (yellow)

7922VG20

Detach the SRS yellow 2-way connector

- **AIR BAG** fuse from the fuse block. On some models the fuse compartment is in the trunk, on other models the fuse panel is located under the rear seat.
- Left sound insulator

5. Disconnect the connector retainer clip from the yellow 2-way connector at the base of the steering column, and detach the harness.
6. Remove the connector retainer clip and detach the yellow 2-way connector from the passenger air bag lead located behind the instrument panel.

Except Eldorado

1. Turn the steering wheel so that the vehicle's wheels are pointing straight ahead.
2. Turn the ignition switch to the **OFF** position.
3. Remove the key from the ignition switch.

➡ **With the SIR fuse removed and the ignition switch in the ON position, The AIR BAG warning lamp illuminates. This is normal operation, and does not indicate an SIR system malfunction.**

4. Remove the rear seat.
5. Remove the SIR fuse from the rear fuse block located under the rear seat.
6. Remove the driver sound insulator.
7. Remove the Connector Position Assurance (CPA) from the driver yellow connector located next to steering column.
8. Disconnect the driver frontal air bag yellow connector from the vehicle harness yellow connector.
9. Remove the passenger sound insulator.
10. Remove the CPA from the passenger yellow connector located above the passenger sound insulator.
11. Disconnect the passenger (IP) frontal

air bag yellow connector from the vehicle harness yellow connector.
12. Remove both CPA locks from the driver side (seat) air bag and pretensioner yellow connector located under the driver seat.
13. Disconnect the driver side air bag and pretensioner yellow connector from the vehicle harness yellow connector.
14. Remove both CPA locks from the passenger side (seat) air bag and pretensioner yellow connector located under the passenger seat.
15. Disconnect the passenger side air bag and pretensioner yellow connector from the vehicle harness yellow connector.
16. The above procedures will disable the SIR system. If vehicle is equipped with optional Rear Air Bags the following steps must be done, to completely disable the SIR system:
   a. Remove the rear seat back.
   b. Remove the CPA from the passenger rear side air bag yellow connector.
17. Disconnect the passenger rear side air bag yellow connector from the vehicle harness yellow connector.
18. Remove the CPA from the driver rear side air bag yellow connector.
19. Disconnect the driver rear side air bag yellow connector from the vehicle harness yellow connector.

ARMING

Eldorado

After the applicable service is concluded, enable the air bag system as follows:
1. Turn the ignition switch to the **LOCK** position and remove the key.
2. Install or connect the following:
- Yellow 2-way connector at the base of steering column and secure it with the connector retainer clip. Attach the yellow 2-way connector at the passenger air bag lead and secure it with the connector retainer clip.
- Left sound insulator
- **AIR BAG** fuse in the fuse block
3. Connect the negative battery cable.
4. Turn the ignition switch to the **RUN** position and verify that the **AIR BAG** warning lamp flashes 7 times, then turns **OFF**.

Except Eldorado

➡ **If vehicle is equipped with optional Rear Air Bags (AW9) the following steps must be done, to completely enable the SIR system.**

1. Remove the key from the ignition switch.

2. Connect the passenger rear side air bag yellow connector to the vehicle harness yellow connector.

3. Install the (CPA) to the passenger rear side air bag yellow connector.

4. Connect the driver rear side air bag yellow connector to the vehicle harness yellow connector.

5. Install the CPA to the driver rear side air bag yellow connector.

6. Install the rear seat back.

7. Connect the driver side (seat) air bag and pretensioner yellow connector to the vehicle harness yellow connector.

8. Install both CPA locks to the driver side (seat) air bag and pretensioner yellow connector located under the driver seat.

9. Connect the passenger side (seat) air bag and pretensioner yellow connector to the vehicle harness yellow connector.

10. Install both CPA locks to the passenger side (seat) air bag and pretensioner yellow connector located under the passenger seat.

11. Connect the passenger (IP) frontal air bag yellow connector to the vehicle harness yellow connector located above the passenger sound insulator.

12. Install the CPA to the passenger yellow connector.

13. Install the passenger sound insulator.

14. Connect the driver frontal air bag yellow connector to the vehicle harness yellow connector located next to steering column.

15. Install the CPA to the driver yellow connector.

16. Install the driver sound insulator.

17. Install the SIR fuse to the rear fuse block.

18. Install the rear seat.

19. Turn the ignition switch to the **RUN** position and verify that the **AIR BAG** warning lamp flashes 7 times, then turns **OFF**.

Power Rack and Pinion Steering Gear

REMOVAL & INSTALLATION

✳ CAUTION

Failure to disconnect the intermediate shaft from the rack and pinion stub shaft can result in damage to the steering gear and/or intermediate shaft. This damage can cause loss of steering control, which could result in personal injury.

✳ WARNING

The wheels of the vehicle must be straight-ahead and the steering column in the LOCK position before disconnecting the steering column or intermediate shaft from the steering gear. Failure to do so will cause the coil assembly in the steering column to become off-centered, which will cause damage to the coil assembly.

2000 Eldorado

1. Before servicing the vehicle, refer to the precautions in the beginning of this section.

2. Remove or disconnect the following:
- Negative battery cable
- Front wheels
- Bolt holding the intermediate shaft to the steering shaft
- Road Sensing Suspension (RSS) sensor from the lower control arm
- Outer tie rod ends from the steering knuckles
- Y-pipe from the catalytic converter

3. Support the rear of the subframe with a jack.

4. Remove the rear subframe bolts.

5. Slowly lower the subframe to gain access.

6. Remove or disconnect the following:
- Heat shield
- Power steering line retainer
- Power steering pressure and return lines
- Speed Sensitive Steering (SSS) solenoid valve connector
- 5 power steering rack-to-subframe bolts
- Rack and pinion unit out the left wheel well

To install:

7. Install or connect the following:
- Rack and pinion through the left wheel well. Torque the bolts to 50 ft. lbs. (68 Nm).
- SSS solenoid valve connector
- Power steering pressure and return lines. Torque the fittings to 20 ft. lbs. (27 Nm).
- Power steering line retainer
- Heat shield
- Subframe. Torque the bolts to 76 ft. lbs. (103 Nm).
- Y-pipe to the catalytic converter. Torque the bolts to 20 ft. lbs. (27 Nm).
- Outer tie rod ends to the steering knuckle using new cotter pins. Torque the nuts to 7.5 ft. lbs. (10 Nm).
- RSS sensor to the lower control arm
- Intermediate shaft to the steering shaft. Torque the bolt to 35 ft. lbs. (47 Nm).
- Front wheels. Torque the lug nuts to 100 ft. lbs. (140 Nm).
- Negative battery cable

8. Refill and bleed the power steering system.

9. Check the wheel alignment.

10. Road test the vehicle.

2001–02 Eldorado

1. Before servicing the vehicle, refer to the precautions in the beginning of this section.

2. Remove or disconnect the following:
- Wheel
- Road sensing suspension position sensor
- Outer tie rod ends from the steering knuckles

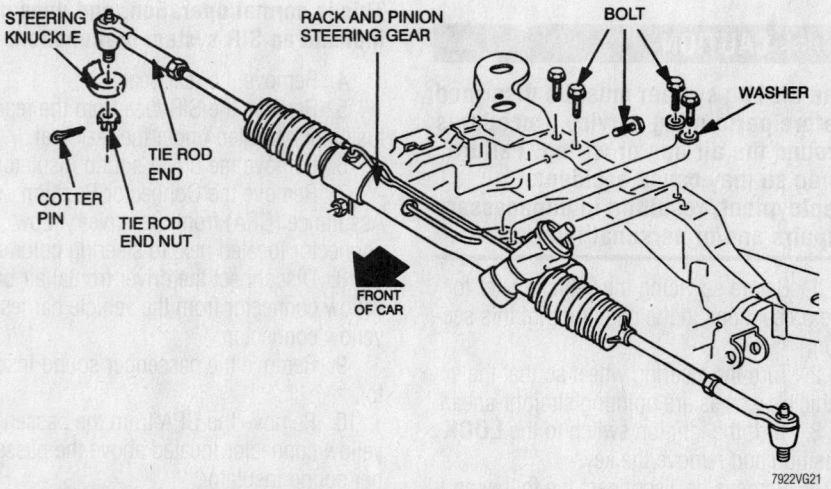

STEERING KNUCKLE

RACK AND PINION STEERING GEAR

BOLT

WASHER

TIE ROD END

COTTER PIN

TIE ROD END NUT

FRONT OF CAR

7922VG21

Exploded view of the power steering rack components

- Intermediate shaft lower pinch bolt
- Intermediate shaft from the power steering gear
- Power steering gear heat shield
- Variable effort steering electrical connector
- Electrical harness clips from the power steering gear
- Power steering pressure hose retaining clip at the power steering gear

3. Support the rear of the frame using a screw jack.
  - Power steering pressure and return hoses from the power steering gear
  - Frame rear body mount bolts

4. Slowly lower the rear of the frame.
  - Power steering gear mounting bolts
  - Power steering gear out the left side wheel opening
  - Outer tie rod ends when replacing the power steering gear

To install:

5. Install or connect the following:
  - Outer tie rod ends when replacing the power steering gear
  - Power steering gear through the left wheel opening
  - Power steering gear mounting bolts and tighten in the sequence illustrated to 50 ft. lbs. (68 Nm)
  - Power steering pressure and return hoses to the power steering gear and tighten to 20 ft. lbs. (27 Nm)

6. Raise the rear of the frame.
  - Frame rear body mounting bolts and tighten to 141 ft. lbs. (191 Nm)
  - Power steering pressure hose retaining clip to the power steering gear
  - Electrical harness clips to the power steering gear
  - Variable effort steering electrical connector
  - Plastic line retainer.
  - Power steering gear heat shield
  - Outer tie rod ends to the steering knuckles
  - Intermediate shaft to the power steering gear
  - Intermediate shaft lower pinch bolt and tighten to 35 ft. lbs. (47 Nm)
  - Road sensing suspension position sensor
  - Wheels

7. Bleed the power steering system.
8. Adjust the front toe.

DeVille

1. Before servicing the vehicle, refer to the precautions in the beginning of this section.

2. Lock the steering column by placing the wheels in the straight ahead position and remove the keys.

3. Remove or disconnect the following:
  - Intermediate shaft pinch bolt and
  - Intermediate shaft lower coupling
  - Heat shield
  - Rear transaxle mount
  - Pressure and return lines from steering gear
  - Electrical connector
  - Steering gear attaching bolts

4. Install a jack under the rear portion of the frame.
  - Rear mounting bolts from the frame

✳✳ CAUTION

The frame must be properly supported before partially lowering. The frame should not be lowered any further than needed to gain access to the steering gear.

5. Lower the rear portion of the frame, if necessary.
  - Steering gear assembly

To install

6. Install or connect the following:
  - Steering gear into vehicle.

7. Raise the rear portion of the frame and tighten the bolts to 142 ft. lbs. (192 Nm), if removed.
  - Rear transaxle mount

8. On 2000 models tighten the gear assembly bolts to 89 ft. lbs. (120 Nm), on 2001–03 models tighten the bolts to 75 ft. lbs. (90 Nm).
  - Electrical connector
  - Pressure and return lines. Torque the lines to 22 ft. lbs. (30 Nm).
  - Heat shield
  - Intermediate shaft to steering gear. Torque the pinch bolt to 33 ft. lbs. (45 Nm).

9. Fill and bleed the power steering system.

Seville

1. Before servicing the vehicle, refer to the precautions in the beginning of this section.

2. Lock the steering column by installing the Steering Column Anti Rotation Pin tool J 42640 into the underside of the steering column.

3. Remove or disconnect the following:
  - Outer tie rod ends from the steering knuckles
  - Intermediate shaft lower pinch bolt
  - Heat shield

- Rear transaxle mount
- Electrical connector
- Pressure and return lines from steering gear
- Steering gear attaching bolts

4. Install a jack under the rear portion of the engine frame.
  - Rear mounting bolts from the engine frame

✳✳ CAUTION

The frame must be properly supported before partially lowering. The frame should not be lowered any further than needed to gain access to the steering gear.

5. Lower the rear portion of the engine frame.
  - Steering gear assembly

To install

6. Install or connect the following:
  - Steering gear into vehicle

7. Raise the rear portion of the engine frame and tighten the bolts to mounting bolts to 142 ft. lbs. (192 Nm)
  - Transaxle mount. Torque the nuts to 55 ft. lbs. (75 Nm).
  - Steering gear bolts to 89 ft. lbs. (120 Nm).
  - Pressure and return lines. Torque the lines to 22 ft. lbs. (30 Nm).
  - Electrical connector
  - Heat shield
  - Outer tie rod ends to the steering knuckles and tighten the nuts to 35 ft. lbs. (47 Nm)
  - Intermediate shaft to steering gear. Torque the pinch bolt to 33 ft. lbs. (45 Nm).

8. Remove the steering column anti rotation pin from the steering column.

9. Fill and bleed the power steering system.

Strut

REMOVAL & INSTALLATION

Front

ELDORADO

✳✳ WARNING

When working near the halfshafts, use care to prevent the inner Tripod CV-joint from being overextended. If the joint is overextended, the internal joint components could separate, resulting in CV-joint failure.

1. Before servicing the vehicle, refer to the precautions in the beginning of this section.

2. Remove or disconnect the following:
- Negative battery cable
- Front wheel
- Electrical connector from the top of the strut, if equipped
- Upper strut mounting nuts
- Road Sensing Suspension (RSS) position sensor from the lower control arm, if equipped
- Anti-lock Brake System (ABS) wheel speed sensor, if equipped
- Wheel speed sensor from the strut bracket, if equipped
- Brake line bracket from the strut
- Stabilizer link from the strut

3. Scribe a mark on the strut referencing the lower strut bracket to the steering knuckle.

4. Remove the strut.

To install:

5. Install or connect the following:
- Strut assembly into the vehicle
- Strut-to-knuckle bolts and nuts, but do not torque yet
- Stabilizer link to the strut assembly, but do not torque the nuts yet
- Brake line bracket to the strut

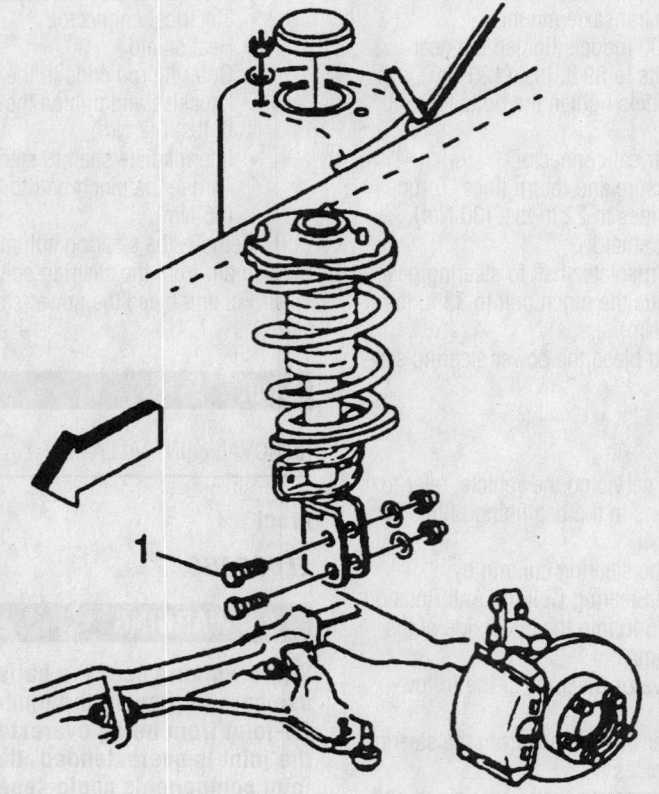

Exploded view of the strut mounting

- Speed sensor bracket on the strut, if equipped
- ABS sensor, if equipped
- RSS sensor to the lower control arm, if equipped
- Electrical connector to the top of the strut, if equipped
- Upper strut mounting nuts. Torque the nuts to 15 ft. lbs. (21 Nm).

6. Torque the stabilizer link nuts to 41 ft. lbs. (55 Nm).

7. Torque the strut-to-knuckle bolt to 140 ft. lbs. (190 Nm).

8. Install the front wheel. Torque the lug nuts to 100 ft. lbs. (140 Nm).

DEVILLE

✳✳ WARNING

When working near the halfshafts, use care to prevent the inner Tripod CV-joint from being overextended. If the joint is overextended, the internal joint components could separate, resulting in CV-joint failure.

1. Before servicing the vehicle, refer to the precautions in the beginning of this section.

2. Remove or disconnect the following:

- Negative battery cable
- 3 strut mount bolts

3. Raise the vehicle and suitably support by the frame allowing the control arms to hang free.
- Front wheel
- Wheel speed sensor connector
- Wheel speed sensor bracket from the strut
- Brake line bracket from the strut

4. Scribe a mark on the strut referencing the lower strut bracket to the steering knuckle.
- Strut-to-knuckle bolts
- Strut from the vehicle

To install:
- Strut
- 3 upper strut mount bolts and washers. Tighten the bolts to 49 ft. lbs. (66 Nm) on models with the FE7 system or 30 ft. lbs. (40 Nm) on FE1 & 3 systems.

5. Align the scribe marks made on the steering knuckle with the strut.
- Strut-to-knuckle bolts and nuts and hand tighten
- Brake line bracket to strut
- Wheel speed sensor bracket to strut
- Wheel speed sensor connector

6. Tighten the strut-to-knuckle bolts to 131 ft. lbs. (177 Nm) on all models with the FE7 system or 108 ft. lbs. (147 Nm) on models with the FE1 & 3 system

7. Tighten the brake line and wheel speed sensor bracket to 17 ft. lbs. (23 Nm).

8. Install the front wheel.

SEVILLE

1. Before servicing the vehicle, refer to the precautions in the beginning of this section.

2. Remove or disconnect the following:
- Tire
- Wheel speed sensor connector
- Wheel speed sensor bracket from the strut
- Brake line bracket from the strut

➡ **Care should be taken to avoid chipping or scratching the coating when handling the suspension coil spring. Damage to the coating can cause premature failure.**

- Strut-to-knuckle bolts
- Three strut mount bolts
- Strut

To install:

3. Install or connect the following:
- Strut
- Three upper strut mount bolts and washers. Tighten the bolts to 33 ft. lbs. (45 Nm)

- New strut-to-knuckle bolts and nuts
- Brake line bracket to strut
- Wheel speed sensor bracket to strut
- Wheel speed sensor connector
- Tighten the strut-to-knuckle bolts to 108 ft. lbs. (147 Nm)
- Tighten the brake line and wheel speed sensor bracket to 13 ft. lbs. (17 Nm)
- Tire

Shock Absorber

REMOVAL & INSTALLATION

Rear

ELDORADO

1. Before servicing the vehicle, refer to the precautions in the beginning of this section.
2. Remove or disconnect the following:
   - Rear wheel
   - Negative battery cable
   - Air line from the top of the strut, if equipped
   - Shock absorber electrical connector or air line from the rear suspension support, if equipped

3. Support the lower control arm with a jack to relieve the tension on the shock absorber.
   - Lower shock absorber mounting bolt and nut
   - Upper mounting nut, retainer and insulator
   - Shock absorber, compress by hand and remove through the upper control arm

To install:
4. Position the top of the shock absorber with the insulator attached into the suspension support.
5. Install or connect the following:
   - Upper shock insulator, retainer and nut. Torque the nut to 55 ft. lbs. (74 Nm).
   - Shock absorber lower mounting nut and bolt. Torque the nut to 75 ft. lbs. (102 Nm).
   - Shock absorber electrical connector or air line to the rear suspension support, if equipped
   - Air line from the top of the strut, if equipped
   - Rear wheel. Torque the lug nuts to 100 ft. lbs. (140 Nm).
   - Negative battery cable

SEVILLE AND DEVILLE

1. Before servicing the vehicle, refer to the precautions in the beginning of this section.
2. Remove the tire and wheel assembly.
3. Support the control arm with a jack stand.
4. Remove or disconnect the following:
   - Electronic Level Control (ELC) air tube from the shock
   - Two bolts securing the shock to the control arm
   - Trunk trim to gain access to the shock upper mounting nuts.
   - Cover, the two nuts, and the reinforcement from the top of the shock
   - Shock from the vehicle

To install:
5. Install or connect the following:
   - Shock, reinforcement, and the two nuts. Tighten the mounting nuts to 18 ft. lbs. (25 Nm) on models with the FE1 & 3 system or 21 ft. lbs. (29 Nm) on models with the FE7 system.
   - Shock cover
   - Trunk trim
   - Shock-to-control arm bolts and tighten the bolts to 18 ft. lbs. (25 Nm) on models with the FE1 & 3 system or 27 ft. lbs. (36 Nm) on models with the FE7 system
   - ELC air tube to the shock.
   - Tire and wheel assembly

Coil Spring

REMOVAL & INSTALLATION

Front

1. Before servicing the vehicle, refer to the precautions in the beginning of this section.
2. Remove the strut from the vehicle.
3. Disassemble the strut as follows:
   a. Step 1: Place the strut assembly into compressor tool, to compress the coil spring.
   b. Step 2: Compress the spring slightly.
   c. Step 3: Hold the strut shaft from turning using a No. 50 Torx® socket and remove the 24mm nut on the top end of the strut.
   d. Step 4: Install Rod Tool J 34013-38 to help guide the strut shaft from the upper mount assembly.
   e. Step 5: Loosen the spring compressor tool until the coil spring and mount can be removed as an assembly.

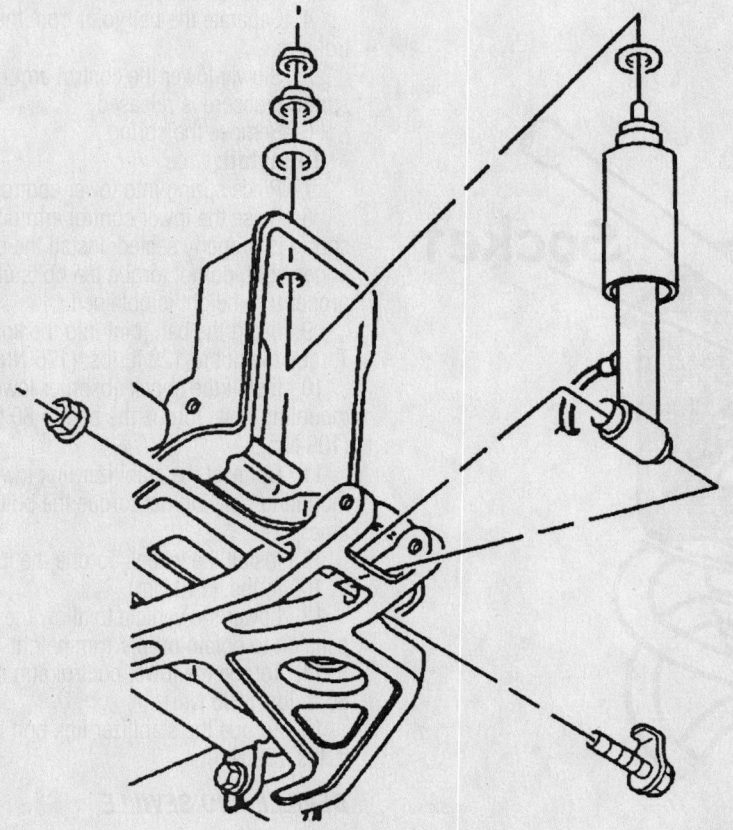

Exploded view of the rear shock absorber mounting

9300VG06

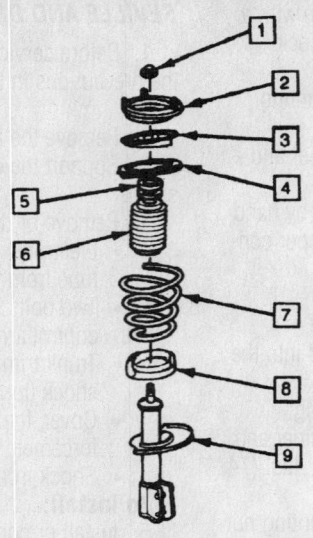

1	NUT, STRUT TO MOUNT
2	STRUT MOUNT
3	FRONT SPRING SEAT
4	FRONT SPRING UPPER INSULATOR
5	JOUNCE BUMPER
6	DUST SHIELD
7	SPRING
8	FRONT SPRING LOWER INSULATOR
9	FRONT STRUT

7922VG23

Disassembled view of strut

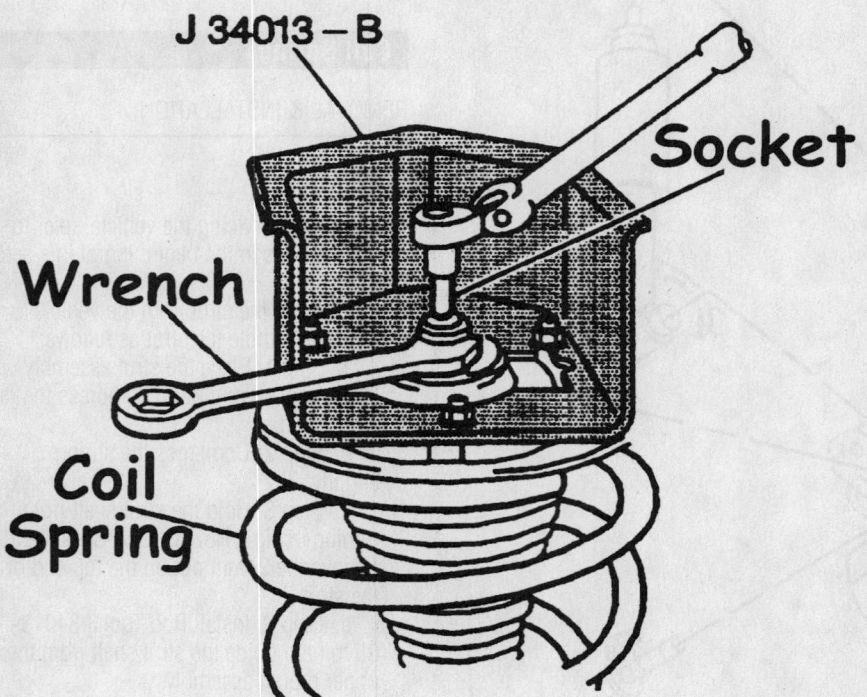

7922UG19

Use a Torx® socket to keep the piston rod from turning while removing the upper nut—front strut shown

Remove the lower spring insulator, if equipped.

To install:

4. Assemble the strut as follows:

   a. Step 1: Place the strut in compressor tool.

   b. Step 2: Install the coil spring over the strut.

   c. Step 3: Compress the coil spring while guiding strut shaft through the top of the strut assembly.

   d. Step 4: Install the top strut nut. Torque the nut to 55 ft. lbs. (75 Nm).

   e. Step 5: Remove the strut from the compressor.

Rear

ELDORADO

1. Before servicing the vehicle, refer to the precautions in the beginning of this section.

2. Support the outboard end of the lower control arm.

3. Remove or disconnect the following:
- Wheel
- Stabilizer link lower mounting bolt
- Shock absorber lower mounting bolt
- Height sensor link attachment from the lower control arm

4. Separate the ball joint from the control arm.

5. Slowly lower the control arm until spring pressure is released.

6. Remove the spring.

To install:

7. Place spring into lower control arm.

8. Raise the lower control arm until the spring is properly seated. Install the bolts finger-tight, do not torque the bolts until proper trim height is obtained.

9. Insert the ball joint into the knuckle. Torque the nut to 129 ft. lbs. (175 Nm).

10. Install the shock absorber lower mounting bolt. Torque the bolt to 80 ft. lbs. (108 Nm).

11. Connect the stabilizer link lower mounting bolt. Do not torque the bolt at this time.

12. Install the wheel. Torque the lug nuts to 100 ft. lbs. (140 Nm).

13. Lower the vehicle to allow the suspension to obtain proper trim height.

14. Torque the lower control arm nuts to 80 ft. lbs. (108 Nm).

15. Torque the stabilizer link bolt to 38 ft. lbs. (52 Nm).

DEVILLE AND SEVILLE

1. Before servicing the vehicle, refer to the precautions in the beginning of this section.

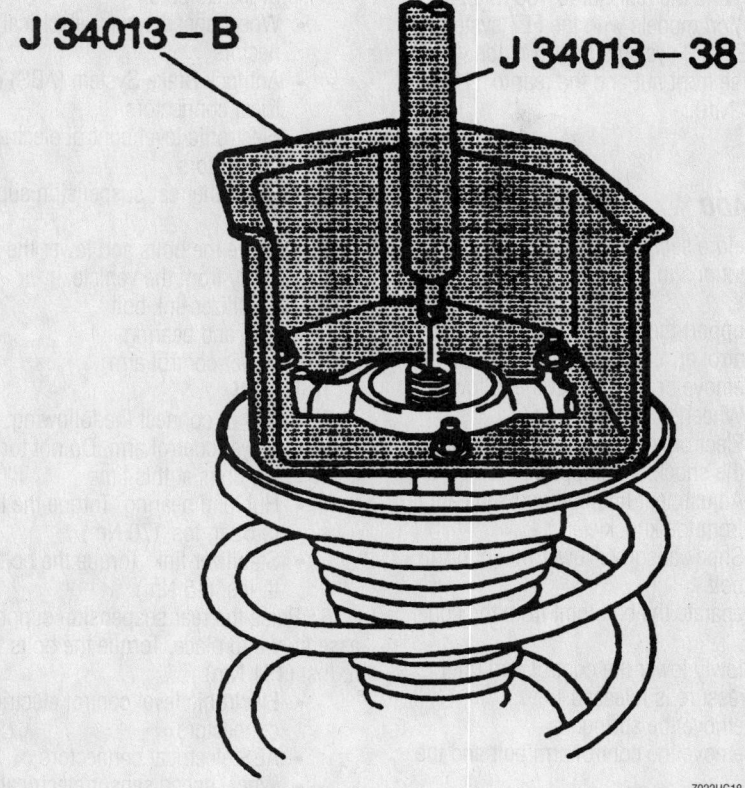

Install Rod J 34013-38 to help guide the strut shaft from the upper mount assembly—front strut shown

2. Support the outboard end of the lower control arm.
3. Remove or disconnect the following:
   - Wheel
   - Electronic level control tube from the shock, if equipped
   - Adjustment link bolt and separate it from the knuckle
   - Shock absorber lower mounting bolt
4. Separate the ball joint from the control arm.
5. Slowly lower the control arm until spring pressure is released.
6. Remove the spring.

To install:
7. Place spring into lower control arm making sure the insulator is properly seated.
8. Raise the lower control arm until the spring is properly seated.
9. Tighten the shock lower bolts.
10. Connect the adjustment link bolt and tighten to 22 ft. lbs. (30 Nm) plus an additional 190 degrees
11. Connect the electronic level control tube to the shock, if equipped.
12. Install the wheel.
13. Check the alignment.

Lower Ball Joint

REMOVAL & INSTALLATION

Commercial Chassis
FRONT

1. Before servicing the vehicle, refer to the precautions in the beginning of this section.

✳✳ CAUTION

Be careful when working in the area of the CV-boot. Damage to the boot could result in eventual joint failure.

2. Remove or disconnect the following:
   - Wheel
   - Ball joint cotter pin and nut
   - Ball joint from steering knuckle
3. Press the ball joint from the lower control arm.

To install:
4. Press the ball joint into the control arm.
5. Install or connect the following:
   - Hex nut. Torque the nut to 129 ft. lbs. (175 Nm).
   - New cotter pin

➡ **If the cotter pin cannot be installed because the hole in the stud does not align with a nut slot, torque the nut an additional 60 degrees to allow for installation. NEVER loosen the nut to provide for cotter pin installation.**

6. Install the wheel and torque the lug nuts to 100 ft. lbs. (140 Nm).

REAR

1. Before servicing the vehicle, refer to the precautions in the beginning of this section.
2. Remove the wheel.
3. Support the lower control arm.

➡ **The tension of the coil spring must be supported with a suitable jack stand.**

4. Remove the cotter pin and ball stud nut.
5. Separate the ball joint from the knuckle.
6. Press the ball joint out of the control arm.

To install:
7. Press the ball joint into the control arm.
8. Install the ball joint into the knuckle.
9. Install the ball stud nut. Torque the nut to 129 ft. lbs. (175 Nm) and install a new cotter pin.
10. Remove the support from the lower control arm.
11. Install the wheel. Torque the lug nuts to 100 ft. lbs. (140 Nm).

All Others

The ball joint is an integral part of the lower control arm and if found to be defective the control arm should be replaced.

Lower Control Arm

REMOVAL & INSTALLATION

Front

ELDORADO

1. Before servicing the vehicle, refer to the precautions in the beginning of this section.
2. Remove or disconnect the following:
   - Wheel
   - Road Sensing Suspension (RSS) sensor
   - Lower ball joint from the steering knuckle
3. On the left control arm, support the transaxle and remove two transaxle mount

nuts. Then raise the transaxle to access the control arm bolt.

4. Remove or disconnect the following:
- Control arm nuts and bolts
- Control arm

To install:

5. Install or connect the following:
- Control arm
- Front and rear bolts, do not tighten yet
- Transaxle mount and lower the transaxle
- Lower ball joint. Torque nut to 84 inch lbs. (10 Nm) then an additional 120 degrees, reaching a minimum of 37 ft. lbs. (50 Nm).
- RSS position sensor
- Wheel. Torque the lug nuts to 100 ft. lbs. (140 Nm).

6. Lower vehicle and allow the vehicle to assume proper trim heights.

7. Torque the control arm front bolt to 93 ft. lbs. (126 Nm).

8. Torque the control arm rear bolt to 116 ft. lbs. (157 Nm).

DEVILLE AND SEVILLE

1. Before servicing the vehicle, refer to the precautions in the beginning of this section.

2. Remove or disconnect the following:
- Wheel
- Stabilizer link bolt from the control arm
- Ball joint from the lower control arm
- Control arm mounting nuts and bolts
- Control arm

To install:

3. Install or connect the following:
- Control arm
- Front and rear bolts, do not tighten yet
- Stabilizer link to control arm
- Ball joint stud in the lower control arm

4. Tighten the stabilizer shaft link assembly nut to 17 ft. lbs. (23 Nm).

5. On models with the FE7 system, tighten the ball joint stud nut to 22 ft. lbs. (30 Nm), then tighten the nut an additional 190 degrees.

6. On models with the FE1 & 3 system, tighten the ball joint stud nut to 88 inch lbs. (10 Nm), then tighten the nut an additional 150 degrees.
- Cotter pin
- Wheel

7. Lower the vehicle.

8. Bounce the vehicle and tighten the front lower control arm nut to 120 ft. lbs.

(162 Nm) and the rear nut to 108 ft. lbs. (146 Nm) on models with the FE7 system or on the FE1 & 3 systems to 116 ft. lbs. (157 Nm) on the front nut and the rear to 116 ft. lbs. (157 Nm).

Rear

ELDORADO

1. Before servicing the vehicle, refer to the precautions in the beginning of this section.

2. Support the outboard end of the lower control arm.

3. Remove or disconnect the following:
- Wheel
- Electronic level control tube from the shock, if equipped
- Adjustment link bolt and separate it from the knuckle
- Shock absorber lower mounting bolt

4. Separate the ball joint from the control arm.

5. Slowly lower the control arm until spring pressure is released.

6. Remove the spring.

7. Remove the control arm bolt and the arm

To install:

8. Lower control arm and inboard control arm bolt and nut and tighten to 75 ft. lbs. (102 Nm).

9. Install the spring and insulators.

10. Raise the lower control arm until the spring is properly seated.

11. Install the stabilizer link lower attachment.

12. Remove the transmission jack. Place the jack under the outboard end of the lower control arm in order to bring the suspension to the suspension's design position.

13. The inner control arm nuts must be tightened in the design position in order to reduce wind up in the bushings as follows:

   a. Stabilizer link lower nut to 44 ft. lbs. (60 Nm).

   b. Tighten the shock absorber lower nut to 75 ft. lbs. (102 Nm).

   c. Tighten the lower control arm inner nuts to 75 ft. lbs. (102 Nm).

14. Install the wheel.

DEVILLE AND SEVILLE

1. Before servicing the vehicle, refer to the precautions in the beginning of this section.

2. Remove or disconnect the following:
- Wheels
- Exhaust system
- Coil springs

- Brake calipers
- Wheel speed sensor electrical connectors
- Antilock Brake System (ABS) electrical connectors
- Electronic level control electrical connectors

3. Support the rear suspension support assembly.

4. Remove the bolts and lower the support assembly from the vehicle.
- Stabilizer link bolt
- Hub and bearing
- Lower control arm

To install

5. Install or connect the following:
- Lower control arm. Do not torque the bolts at this time.
- Hub and bearing. Torque the bolts to 52 ft. lbs. (70 Nm).
- Stabilizer link. Torque the bolt to 11 ft. lbs. (15 Nm).

6. Raise the rear suspension support assembly into place. Torque the bolts to 141 ft. lbs. (191 Nm).
- Electronic level control electrical connectors
- ABS electrical connectors
- Wheel speed sensor electrical connectors
- Brake calipers. Torque the bolts to 63 ft. lbs. (85 Nm).
- Coil springs
- Exhaust system
- Wheels

7. To obtain proper trim height, raise the control arm until the spindle face is parallel to the ground and torque the nuts to 78 ft. lbs. (106 Nm).

CONTROL ARM BUSHING REPLACEMENT

Front

1. Before servicing the vehicle, refer to the precautions in the beginning of this section.

2. Remove or disconnect the following:
- Wheel
- Lower control arm

3. Press the bushing out of the lower control arm.

To install

4. Lubricate the outer case of the new bushing.

5. Press the new bushing into the control arm.

6. Install or connect the following:
- Lower control arm
- Wheel. Torque the lug nuts to 100 ft. lbs. (140 Nm).

Wheel Bearings

ADJUSTMENT

The wheel bearings are not adjustable. If a wheel bearing is out of specifications, it must be replaced. Using a dial indicator, check for looseness. If play exceeds 0.005 inches (0.127mm), the bearing wear is excessive and the hub and bearing should be replaced.

REMOVAL & INSTALLATION

Front

1. Before servicing the vehicle, refer to the precautions in the beginning of this section.
2. Remove or disconnect the following:
   - Front wheel
   - Halfshaft nut and washer
   - Caliper from the steering knuckle and support it on a wire
   - Brake rotor
   - Anti-lock Brake System (ABS) speed sensor electrical connector
   - Hub/bearing assembly to steering knuckle bolts
   - Dust shield
   - Hub and bearing assembly from the halfshaft
   - Hub and bearing assembly from the steering knuckle

To install:

3. Install the hub and bearing assembly over the halfshaft splines.

➡ **Be sure the splines engage smoothly.**

4. Apply a light coating of grease to the steering knuckle bore.
5. Install the hub/bearing assembly onto the halfshaft as far as possible. If the hub will not bottom out on the halfshaft, install the hub bolts and use the hub nut to draw the hub onto the halfshaft.
6. Once the hub is flush with the steering knuckle, remove the mounting bolts and install the dust shield. Reinstall the mounting bolts and torque to 70 ft. lbs. (95 Nm) on Eldorado. Tighten to 95 ft. lbs. (130 Nm) on all Seville and all models with the FE1 & 3 system or on FE7 systems to 112 ft. lbs. (152 Nm).
7. Install or connect the following:
   - ABS speed sensor electrical connector
   - Brake rotor
   - Caliper. Torque the bolts to 38 ft. lbs. (51 Nm).
   - Halfshaft nut. Torque it to 107 ft. lbs. (145 Nm).
   - Wheel. Torque the lug nuts to 100 ft. lbs. (140 Nm).

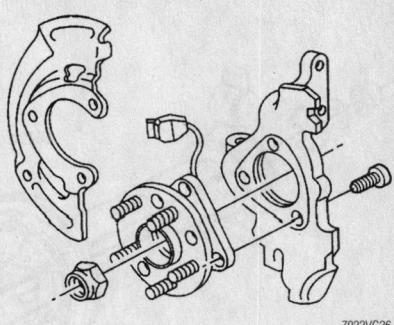

7922VG26

Exploded view of the front wheel bearing mounting

8. Road test the vehicle for proper operation.

Rear

EXCEPT COMMERCIAL CHASSIS

➡ **The wheel bearing and hub are serviced as an assembly. The individual components are not serviceable separately.**

1. Before servicing the vehicle, refer to the precautions in the beginning of this section.
2. Remove or disconnect the following:
   - Rear wheel
   - Wheel speed sensor, if equipped
   - Caliper bracket from the knuckle
   - Brake rotor
   - Hub and bearing assembly bolts
   - Hub and bearing assembly

To install:

3. Install or connect the following:
   - Hub and bearing assembly. Torque the bolts to 52 ft. lbs. (70 Nm).
   - Brake rotor
   - Caliper
   - Wheel speed sensor, if equipped
   - Wheel. Torque the lug nuts to 100 ft. lbs. (140 Nm).
4. Road test the vehicle.

COMMERCIAL CHASSIS

1. Before servicing the vehicle, refer to the precautions in the beginning of this section.
2. Remove or disconnect the following:
   - Wheel
   - Brake drum
   - Brake shoes and cable
   - Brake line
   - Wheel speed sensor
3. Support the lower control arm.
4. Separate the ball joint from the knuckle.
   - Upper knuckle bolt
   - Bearing retainer

WHEEL BEARING LOOSENESS DIAGNOSIS

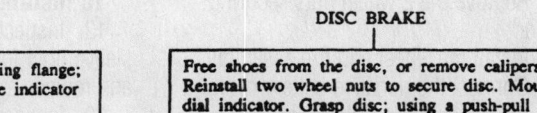

DRUM BRAKE

Mount dial indicator. Grasp bearing flange; using a push-pull movement, note indicator readings.

If looseness exceeds 0.1270 mm (0.005 inch), replace hub and bearing assembly.

DISC BRAKE

Free shoes from the disc, or remove calipers. Reinstall two wheel nuts to secure disc. Mount dial indicator. Grasp disc; using a push-pull movement, note indicator readings.

If looseness exceeds 0.1270 mm (0.005 inch), replace hub and bearing assembly.

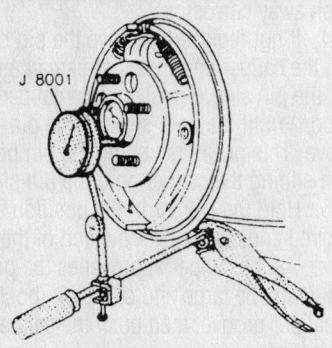

J 8001

MOUNTING DIAL INDICATOR WITH DRUM BRAKES

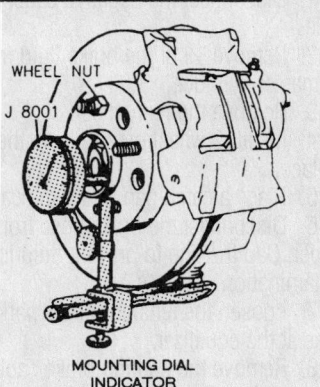

WHEEL NUT

J 8001

MOUNTING DIAL INDICATOR WITH DISK BRAKES

7922VG33

Inspect the wheel bearings for play with a dial indicator

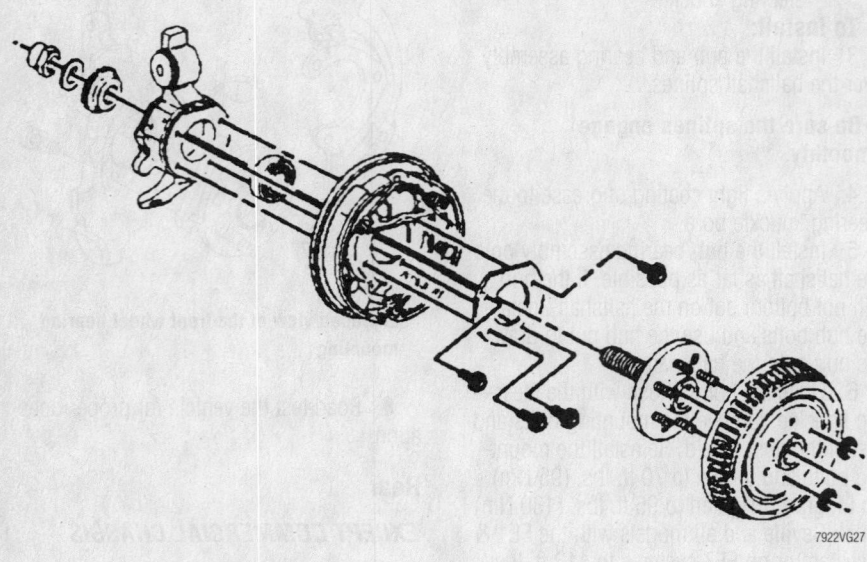

Exploded view of the rear wheel bearing hub mounting

7922VG27

- Nut from backside of knuckle
5. Press the hub from the knuckle.
6. Remove the brake backing plate from the knuckle.
7. Press the bearing from the knuckle.

To install

8. Press the bearing into the knuckle.
9. Install the backing plate and bolts. Torque the bolts to 37 ft. lbs. (50 Nm).
10. Press the hub into the bearing.
11. Install the speed sensor ring and nut. Torque the nut to 145 ft. lbs. (200 Nm).
12. Install or connect the following:
- Upper knuckle bolt and nut. Torque the nut to 80 ft. lbs. (108 Nm).
- Ball joint to knuckle
- Brake line
- Wheel speed sensor
- Brake shoes and cable
- Brake drum
13. Bleed the brake system.
14. Install the wheel. Torque the lug nuts to 100 ft. lbs. (140 Nm).
15. Check the brake system for leaks.

BRAKES

Brake Caliper

REMOVAL & INSTALLATION

Front

1. Disconnect the negative battery cable.
2. Remove ⅔ of the brake fluid from the master cylinder.
3. Remove the front wheel. Mark the relationship between the wheel and the wheel stud for re-installation purposes.
4. Install 2 wheel nuts to keep the rotor in place.
5. Using a large C-clamp against the inboard pad, compress the caliper piston into the caliper to provide clearance during removal.
6. Place a catch pan under the caliper.
7. Disconnect the brake hose from the caliper. Cap the line to prevent excessive fluid loss or contamination.
8. Remove the caliper mounting bolts and remove the caliper from the vehicle.
9. Inspect the mounting bolts; sleeves and boots for wear and/or damage. Replace parts as necessary.

To install:

10. Before installing the caliper, make sure the piston is fully seated in the bore and the brake pads are properly seated.
11. Lubricate the mounting bolt shafts and inner diameter of the sleeves with silicone grease.

12. Install the caliper in the caliper mounting bracket and install the mounting bolts. Torque the mounting bolts to 38 ft. lbs. (51 Nm).
13. Connect the brake hose with the bolt and new gaskets. Torque the brake hose bolt to 33 ft. lbs. (45 Nm).
14. Refill the master cylinder and bleed the brake system.
15. Remove the 2 wheel nuts securing the rotor.
16. Install the wheel and tire assembly.
17. Connect the negative battery cable.
18. Road test the vehicle for proper brake system operation.

Rear

1. Disconnect the negative battery cable.
2. Remove ⅔ of the brake fluid from the master cylinder.
3. Remove the rear wheel.
4. Install 2 wheel nuts to keep the rotor in place.
5. Place a catch pan under the caliper.
6. Disconnect the brake hose from the caliper. Cap the line to prevent fluid loss or contamination.
7. Loosen the tension on the parking brake at the equalizer.
8. Remove the parking brake cable mounting lever, and remove the cable end by lifting up and disengaging the end.
9. Remove the caliper sleeve bolt.

10. Lift the caliper up and slide the caliper inboard off of the pin sleeve to remove the caliper from the vehicle.
11. Use a suitable tool in the caliper piston slots to turn the piston and thread it into the caliper. After bottoming the piston, lift the inner edge of the boot next to the piston and press out any trapped air the boot must lay flat.

To install:

12. Inspect the pin boot, bolt boot and sleeve boot for cuts, tears or deterioration and replace as necessary.
13. Inspect the bolt sleeve and pin sleeve for corrosion or damage. Pull the boots to gain access to the sleeves for inspection or replacement. Replace corroded or damaged sleeves; do not try to polish away corrosion.
14. If not replaced, remove the pin boot from the caliper and install the small end over the pin sleeve (installed on caliper support) until the boot seats in the pin groove. This prevents cutting the pin boot when sliding the caliper onto the pin sleeve.
15. Hold the caliper in the position as removed and start it over the end of the pin sleeve. As the caliper approaches the pin boot, work the large end of the pin boot in the caliper groove, then push the caliper fully onto the pin.
16. Pivot the caliper down, being careful not to damage the piston boot on the inboard disc brake pad. Compress the

sleeve boot by hand as the caliper moves into position to prevent boot damage.

17. After the caliper is in position, recheck the position of the pad clips. If necessary, use a small prybar to reseat or enter the pad clips on the bracket abutments.

18. Install the sleeve bolt and torque to 20 ft. lbs. (27 Nm).

19. Install the parking brake cable bracket, with the cable attached, and torque the bolt to 32 ft. lbs. (43 Nm).

20. Install the parking brake cable onto the parking brake lever and the retaining clip onto the parking brake cable.

21. Connect the brake hose with the bolt and new gaskets and torque the bolt to 32 ft. lbs. (43 Nm).

22. Adjust the parking brake cable.

23. Refill the master cylinder and bleed the brake system.

24. Remove the wheel nuts retaining the rotor and install the wheel.

25. Connect the negative battery cable.

26. Road test the vehicle for proper brake system operation.

Disc Brake Pads

REMOVAL & INSTALLATION

Front

1. Remove ⅔ of the brake fluid from the master cylinder reservoir.

2. Remove the front wheel.

3. Remove the caliper mounting bolts.

4. Remove the caliper from the steering knuckle without disconnecting the brake hose.

5. Suspend the caliper from the coil spring with wire. Do not let the caliper hang from the brake hose.

6. Remove brake pads from anchor bracket.

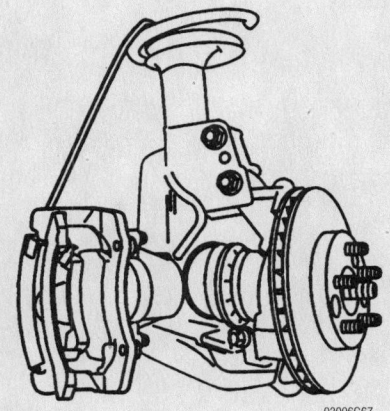

Supporting the front caliper

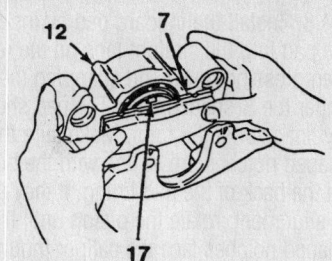

7. Inboard Shoe & Lining
12. Caliper Housing
17. Shoe Retainer Spring

93006G68

Installing the inboard disc brake pad into the front caliper

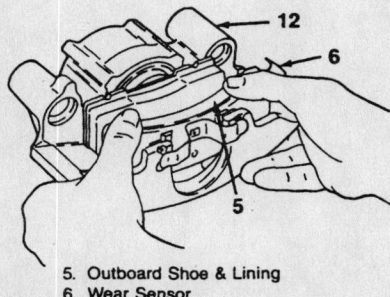

5. Outboard Shoe & Lining
6. Wear Sensor
12. Caliper Housing

93006G69

Installing the outboard disc brake pad in the front caliper

To install:

7. Fully seat the caliper piston into the bore using a large C-clamp.

8. Lubricate the brake caliper mounting surfaces.

9. Install the brake pads onto the anchor bracket using new shims and clips.

10. Place the caliper onto the anchor bracket and install the mounting bolts.

11. Torque the mounting bolts to 63 ft. lbs. (85 Nm).

12. Install the wheel and tire assembly and torque to specification.

13. Lower the vehicle.

14. Pump the brake pedal several times to seat the brake pads.

15. Check the fluid level in the master cylinder and fill as necessary.

16. Road test the vehicle for proper brake operation.

Rear

1. Remove ⅔ of the brake fluid from the master cylinder reservoir.

2. Remove the rear wheel.

3. Remove the park brake cable from the caliper.

➡Some models use a brake pad wear sensor. Disconnect the sensor connector from the vehicle harness.

4. Remove the caliper mounting bolts and position and suspend the caliper from

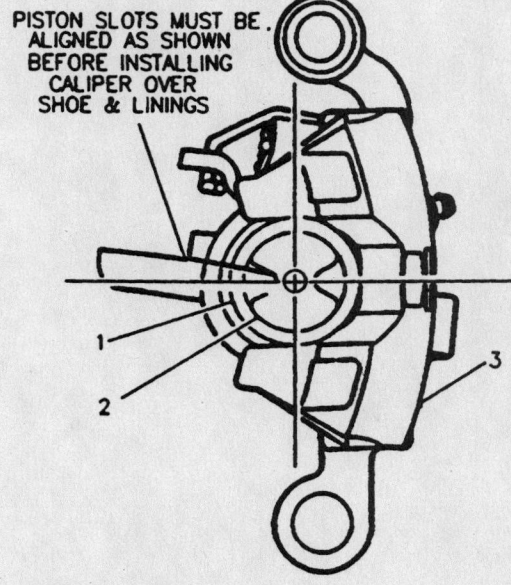

PISTON SLOTS MUST BE ALIGNED AS SHOWN BEFORE INSTALLING CALIPER OVER SHOE & LININGS

1 PISTON BOOT
2 PISTON ASSEMBLY
3 CALIPER BODY ASSEMBLY

93006G70

Aligning the slots on the rear caliper

the strut with a wire. Do not let the caliper hang from the brake hose.

5. Remove the inboard and the outboard brake pads from the caliper mounting bracket.

To install:

6. Lubricate mounting surfaces and install new brake pad clips.

7. The caliper piston must be fully seated in the bore. A spanner type tool can be used to bottom the piston into its bore by turning it in. Make sure the slots in the piston are straight across from each other so the notches on the brake pad will seat.

8. Install the inboard pad by inserting the pad into the straight tabs on the retainer, then pressing down and snapping the pad under the S-shaped tabs. The pad should lay flat against the rotor. Make sure the D-shaped notches are in line with the buttons on the back of the pad lining. If they are not in alignment, rotate the piston until the D-shaped notches face the caliper mounting bolt holes.

9. Install the outer pad into the brake pad clips. Make sure the wear indicator is at the leading edge of the pad during forward wheel rotation.

10. Install the caliper on the mounting bracket and install the bolts. Torque the mounting bolts to 63 ft. lbs. (85 Nm).

11. Install the parking brake cable onto the caliper.

12. Install the wheel and tire assembly and torque to specification.

13. Fill the master cylinder reservoir to the FULL mark and pump the brake pedal several times to seat the brake pads.

14. Check the fluid level in the master cylinder again and fill as necessary.

15. Road test the vehicle for proper brake operation.

SPECIFICATION CHARTS

ENGINE AND VEHICLE IDENTIFICATION

Code ①	Liters (cc)	Cu. In.	Cyl.	Fuel Sys.	Engine Type	Eng. Mfg.
G	5.7 (5665)	350	8	SFI	OHV	CPC
K	3.8 (3785)	231	6	SFI	OHV	CPC

Code ②	Year
Y	2000
1	2001
2	2002

SFI: Sequential Fuel Injection

CPC: Chevrolet/Pontiac/Canada

OHV: Overhead Valve

① 8th position of VIN

② 10th position of VIN

42372-FBOD-C01

GENERAL ENGINE SPECIFICATIONS

Year	Model	Engine Displacement Liters (cc)	Engine Series (ID/VIN)	Fuel System	Net Horsepower @ rpm	Net Torque @ rpm (ft. lbs.)	Bore x Stroke (in.)	Compression Ratio	Oil Pressure @ rpm
2000	Camaro	3.8 (3785)	K	SFI	200@5200	225@4000	3.80x3.40	9.4:1	60@1850
		5.7 (5665)	G	SFI	305@5200	335@4000	3.89x3.62	10.0:1	18@2000
	Firebird	3.8 (3785)	K	SFI	200@5200	225@4000	3.80x3.40	9.4:1	60@1850
		5.7 (5665)	G	SFI	305@5200	335@4000	3.89x3.62	10.0:1	18@2000
2001	Camaro	3.8 (3785)	K	SFI	200@5200	225@4000	3.80x3.40	9.4:1	60@1850
		5.7 (5665)	G	SFI	305@5200	335@4000	3.89x3.62	10.0:1	18@2000
	Firebird	3.8 (3785)	K	SFI	200@5200	225@4000	3.80x3.40	9.4:1	60@1850
		5.7 (5665)	G	SFI	305@5200	335@4000	3.89x3.62	10.0:1	18@2000
2002	Camaro	3.8 (3785)	K	SFI	200@5200	225@4000	3.80x3.40	9.4:1	60@1850
		5.7 (5665)	G	SFI	305@5200	335@4000	3.89x3.62	10.0:1	18@2000
	Firebird	3.8 (3785)	K	SFI	200@5200	225@4000	3.80x3.40	9.4:1	60@1850
		5.7 (5665)	G	SFI	305@5200	335@4000	3.89x3.62	10.0:1	18@2000

SFI: Sequential Fuel Injection

42372-FBOD-C02

ENGINE TUNE-UP SPECIFICATIONS

Year	Engine Displacement Liters (cc)	Engine ID/VIN	Spark Plug Gap (in.)	Ignition Timing (deg.)		Fuel Pump (psi)	Idle Speed (rpm)		Valve Clearance	
				MT	AT		MT	AT	Intake	Exhaust
2000	3.8 (3785)	K	0.045	①	①	41-47	①	①	HYD	HYD
	5.7 (5665)	G	0.060	①	①	48-55	①	①	HYD	HYD
2001	3.8 (3785)	K	0.045	①	①	41-47	①	①	HYD	HYD
	5.7 (5665)	G	0.060	①	①	48-55	①	①	HYD	HYD
2002	3.8 (3785)	K	0.045	①	①	41-47	①	①	HYD	HYD
	5.7 (5665)	G	0.060	①	①	48-55	①	①	HYD	HYD

NOTE: The Vehicle Emission Control Information label often reflects specification changes made during production. The label figures must be used if they differ from those in this chart.

HYD: Hydraulic

① Refer to Vehicle Emission Control Information label

42372-FBOD-C03

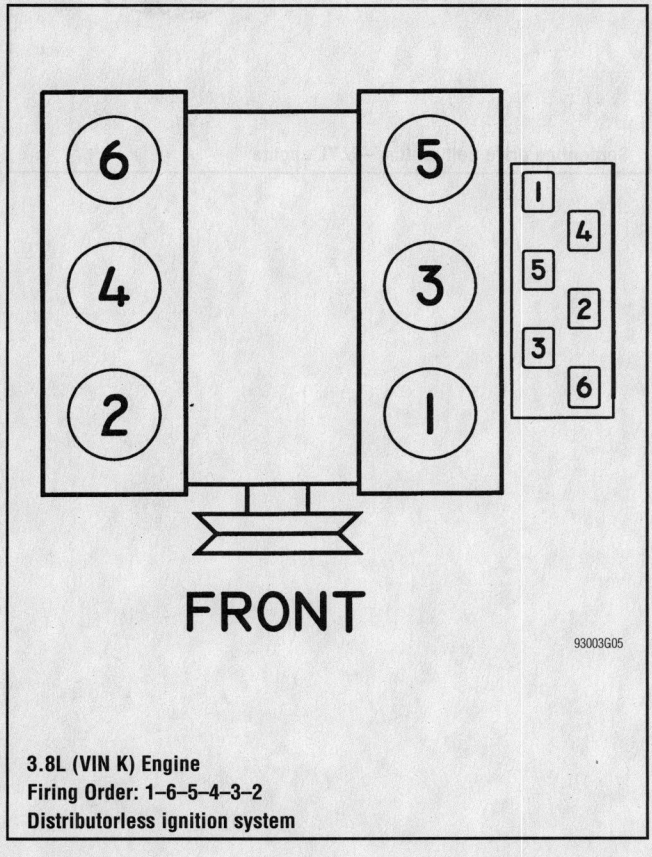

FRONT

93003G05

3.8L (VIN K) Engine
Firing Order: 1–6–5–4–3–2
Distributorless ignition system

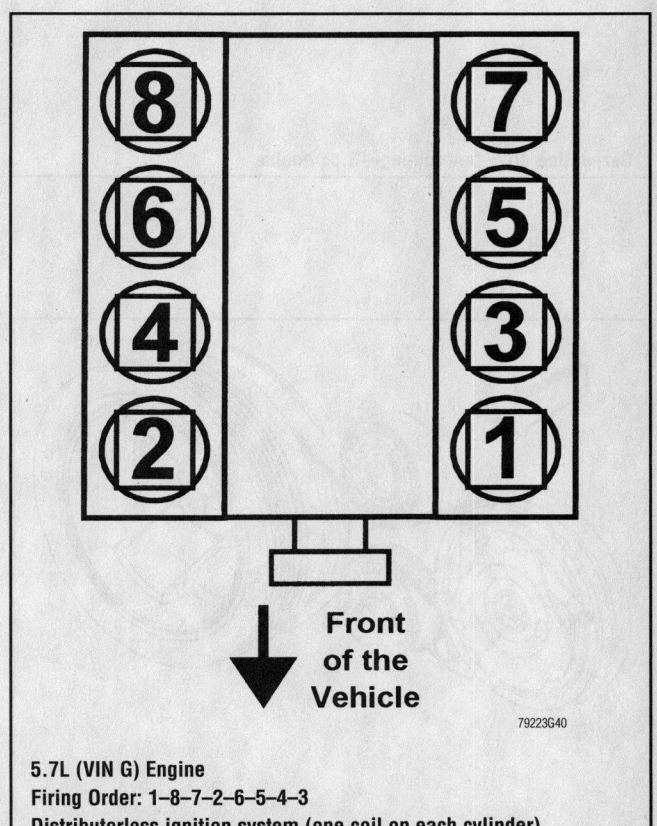

Front of the Vehicle

79223G40

5.7L (VIN G) Engine
Firing Order: 1–8–7–2–6–5–4–3
Distributorless ignition system (one coil on each cylinder)

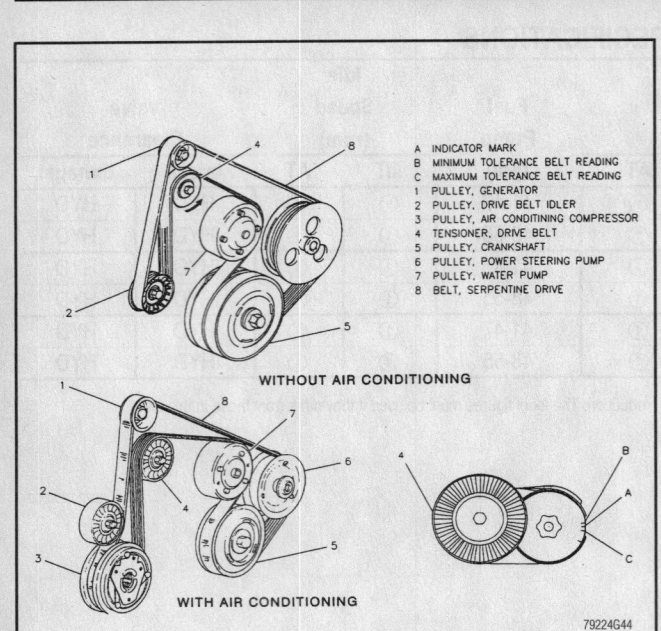

A   INDICATOR MARK
B   MINIMUM TOLERANCE BELT READING
C   MAXIMUM TOLERANCE BELT READING
1   PULLEY, GENERATOR
2   PULLEY, DRIVE BELT IDLER
3   PULLEY, AIR CONDITIONING COMPRESSOR
4   TENSIONER, DRIVE BELT
5   PULLEY, CRANKSHAFT
6   PULLEY, POWER STEERING PUMP
7   PULLEY, WATER PUMP
8   BELT, SERPENTINE DRIVE

WITHOUT AIR CONDITIONING

WITH AIR CONDITIONING

79224G44

Serpentine drive belt routing—3.8L engine

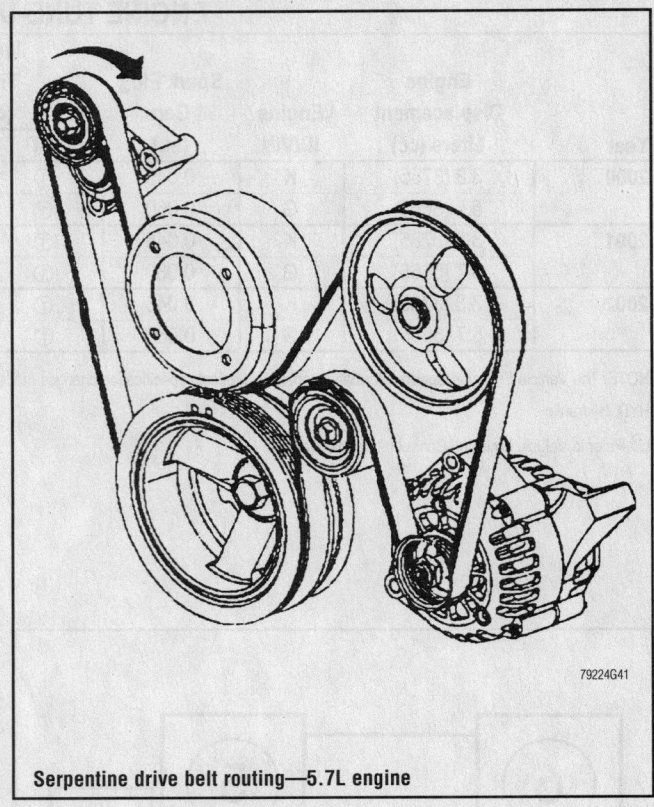

79224G41

Serpentine drive belt routing—5.7L engine

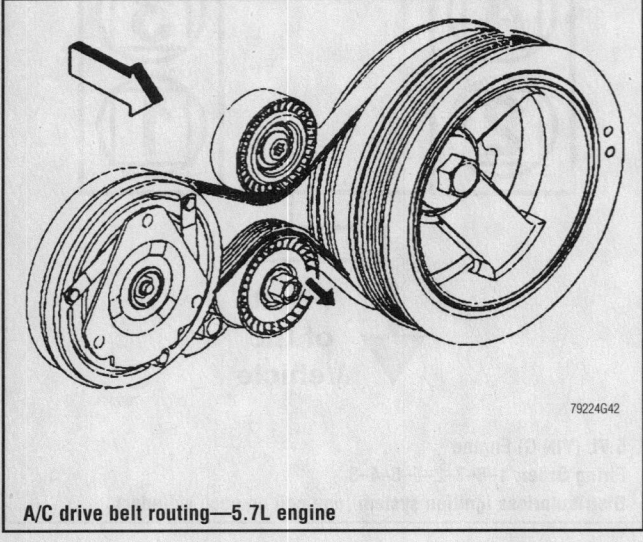

79224G42

A/C drive belt routing—5.7L engine

CAPACITIES

Year	Model	Engine Displacement Liters (cc)	Engine ID/VIN	Engine Oil with Filter (qts.) ①	Transmission (pts.) 5-Spd	Transmission (pts.) Auto.	Drive Axle (pts.)	Fuel Tank (gal.)	Cooling System (qts.) ②
2000	Camaro	3.8 (3785)	K	4.5	6.8	10.0	3.5	16.8	②
		5.7 (5665)	G	5.5	③	10.0	3.5	16.8	④
	Firebird	3.8 (3785)	K	4.5	6.8	10.0	3.5	16.8	②
		5.7 (5665)	G	5.5	③	10.0	3.5	16.8	④
2001	Camaro	3.8 (3785)	K	4.5	6.8	10.0	3.5	16.8	②
		5.7 (5665)	G	5.5	③	10.0	3.5	16.8	④
	Firebird	3.8 (3785)	K	4.5	6.8	10.0	3.5	16.8	②
		5.7 (5665)	G	5.5	③	10.0	3.5	16.8	④
2002	Camaro	3.8 (3785)	K	4.5	6.8	10.0	3.5	16.8	②
		5.7 (5665)	G	5.5	③	10.0	3.5	16.8	④
	Firebird	3.8 (3785)	K	4.5	6.8	10.0	3.5	16.8	②
		5.7 (5665)	G	5.5	③	10.0	3.5	16.8	④

NOTE: All capacities are approximate. Add fluid gradually and ensure a proper fluid level is obtained.

① With vehicle on level surface, check oil level. Add as required to fill.

② With manual transmission: 11.6 qts.
   With automatic transmission: 11.4 qts.

③ ZF 6 speed transmission: 4.4 pts.

④ With manual transmission: 11.9 qts.
   With automatic transmission: 11.8 qts.

42372-FBOD-C04

VALVE SPECIFICATIONS

Year	Engine Displacement Liters (cc)	Engine ID/VIN	Seat Angle (deg.)	Face Angle (deg.)	Spring Test Pressure (lbs. @ in.) ①	Spring Installed Height (in.)	Stem-to-Guide Clearance (in.) Intake	Stem-to-Guide Clearance (in.) Exhaust	Stem Diameter (in.) Intake	Stem Diameter (in.) Exhaust
2000	3.8 (3785)	K	45	45	228@1.277	1.69-1.72	0.0015-0.0035	0.0015-0.0032	NA	NA
	5.7 (5665)	G	46	45	220@1.320	1.80	0.0010-0.0026	0.0010-0.0026	0.0313-0.0314	0.0313-0.0314
2001	3.8 (3785)	K	45	45	228@1.277	1.69-1.72	0.0015-0.0035	0.0015-0.0032	NA	NA
	5.7 (5665)	G	46	45	220@1.320	1.80	0.0010-0.0026	0.0010-0.0026	0.0313-0.0314	0.0313-0.0314
2002	3.8 (3785)	K	45	45	228@1.277	1.69-1.72	0.0015-0.0035	0.0015-0.0032	NA	NA
	5.7 (5665)	G	46	45	220@1.320	1.80	0.0010-0.0026	0.0010-0.0026	0.0313-0.0314	0.0313-0.0314

NA: Not Available

① With valve open

42372-FBOD-C05

TORQUE SPECIFICATIONS
All readings in ft. lbs.

Year	Engine Displacement Liters (cc)	Engine ID/VIN	Cylinder Head Bolts	Main Bearing Bolts	Rod Bearing Bolts	Crankshaft Damper Bolts	Flywheel Bolts	Manifold Intake	Manifold Exhaust	Spark Plugs	Lug Nuts
2000	3.8 (3785)	K	①	②	20	③	④	⑤	17	23	100
	5.7 (5665)	G	⑥	⑦	⑧	⑨	⑩	⑪	⑫	12	100
2001	3.8 (3785)	K	①	②	20	③	④	⑤	17	23	100
	5.7 (5665)	G	⑥	⑦	⑧	⑨	⑩	⑪	⑫	12	100
2002	3.8 (3785)	K	①	②	20	③	④	⑤	17	23	100
	5.7 (5665)	G	⑥	⑦	⑧	⑨	⑩	⑪	⑫	12	100

NA: Not Available

① Step 1: 37 ft. lbs. plus 120 degrees

② Step 1: Tighten to 52 ft. lbs.
Step 2: Loosen bearing cap 360 degrees counter-clockwise
Step 3: Tighten caps to 15 ft. lbs.
Step 4: Tighten caps to 30 ft. lbs.
Step 5: Turn bolts an additional 35 degrees.
Step 6: Turn bolts an additional 35 degrees
Step 7: Turn bolts an additional 40 degrees -- for a total of 110 degrees

③ 111 ft. lbs. plus 76 degrees

④ 11 ft. lbs. plus 50 degrees

⑤ Upper manifold:
Bolts 1-10, in sequence: 11 ft. lbs. (15 Nm)
Coolant outlet bolts: 20 ft. lbs. (27 Nm)
Side bolts: 22 ft. lbs. (30 Nm)
Lower manifold bolt/nut: 11 ft. lbs.

⑥ M11 bolts:
Step 1: Bolts 1-10 22 ft. lbs.
Step 2: Rotate bolts 1-10 90 degrees
Step 3: Rotate bolts 1-8 90 degrees
Step 4: Rotate bolts 9-10 50 degrees
Step 5: Tighten inner (M8) bolts to 22 ft. lbs.

⑦ Step 1: Inner bolts: 15 ft. lbs.
Step 2: Inner bolts: Rotate 80 degrees
Step 3: Outer side bolts: 18 ft. lbs.
Step 4: Outer side studs: 15 ft. lbs. plus 53 degrees

⑧ Step 1: 15 ft. lbs.
Step 2: Rotate 60 degrees

⑨ Step 1: Tighten bolt to 240 ft. lbs. to seat balancer
Step 2: Tighten new bolt to 37 ft. lbs.
Step 3: Rotate an additional 140 degrees

⑩ Step 1: 15 ft. lbs.
Step 2: 37 ft. lbs.
Step 3: 74 ft. lbs.

⑪ Step 1: 44 inch lbs.
Step 2: 89 inch lbs.

⑫ Step 1: 11 ft. lbs.
Step 2: 18 ft. lbs.

42372-FBOD-C06

CRANKSHAFT AND CONNECTING ROD SPECIFICATIONS

All measurements are given in inches.

Year	Engine Displacement Liters (cc)	Engine ID/VIN	Crankshaft Main Brg. Journal Dia.	Crankshaft Main Brg. Oil Clearance	Crankshaft Shaft End-play	Crankshaft Thrust on No.	Connecting Rod Journal Diameter	Connecting Rod Oil Clearance	Connecting Rod Side Clearance
2000	3.8 (3786)	K	2.4988-2.4998	①	0.0030-0.0110	2	2.2487-2.2499	0.0005-0.0026	0.0040-0.0200
	5.7 (5665)	G	2.5580-2.5590	0.0007-0.0021	0.0015-0.0078	3	2.0991-2.0999	0.0006-0.0030	0.0043-0.0200
2001	3.8 (3786)	K	2.4988-2.4998	①	0.0030-0.0110	2	2.2487-2.2499	0.0005-0.0026	0.0040-0.0200
	5.7 (5665)	G	2.5580-2.5590	0.0007-0.0021	0.0015-0.0078	3	2.0991-2.0999	0.0006-0.0030	0.0043-0.0200
2002	3.8 (3786)	K	2.4988-2.4998	①	0.0030-0.0110	2	2.2487-2.2499	0.0005-0.0026	0.0040-0.0200
	5.7 (5665)	G	2.5580-2.5590	0.0007-0.0021	0.0015-0.0078	3	2.0991-2.0999	0.0006-0.0030	0.0043-0.0200

① Journal 1: 0.0007 - 0.0016
Journals 2, 3 and 4: 0.0009 - 0.0018

42372-FBOD-C07

PISTON AND RING SPECIFICATIONS

All measurements are given in inches.

Year	Engine Displacement Liters (cc)	Engine ID/VIN	Piston Clearance	Ring Gap Top Compression	Ring Gap Bottom Compression	Ring Gap Oil Control	Ring Side Clearance Top Compression	Ring Side Clearance Bottom Compression	Ring Side Clearance Oil Control
2000	3.8 (3786)	K	0.0004-0.0020	0.010-0.018	0.023-0.033	0.010-0.030-	0.0013-0.0031	0.0013-0.0031	0.0009-0.0079
	5.7 (5665)	G	0.0007-0.0021	0.009-0.015	0.017-0.025	0.007-0.027	0.0016-0.0034	0.0016-0.0032	0.0004-0.0087
2001	3.8 (3786)	K	0.0004-0.0020	0.010-0.018	0.023-0.033	0.010-0.030-	0.0013-0.0031	0.0013-0.0031	0.0009-0.0079
	5.7 (5665)	G	0.0007-0.0021	0.009-0.015	0.017-0.025	0.007-0.027	0.0016-0.0034	0.0016-0.0032	0.0004-0.0087
2002	3.8 (3786)	K	0.0004-0.0020	0.010-0.018	0.023-0.033	0.010-0.030-	0.0013-0.0031	0.0013-0.0031	0.0009-0.0079
	5.7 (5665)	G	0.0007-0.0021	0.009-0.015	0.017-0.025	0.007-0.027	0.0016-0.0034	0.0016-0.0032	0.0004-0.0087

42372-FBOD-C08

WHEEL ALIGNMENT

Year	Model		Caster Range (+/-Deg.)	Caster Preferred Setting (Deg.)	Camber Range (+/-Deg.)	Camber Preferred Setting (Deg.)	Toe-in (in.)	Steering Axis Inclination (Deg.)
2000	Camaro	F	1.00	+5.00	1.00	+0.80	-0.20 +/- 0.20	—
		R	—	—	0.60	0	0 +/- 0.30	—
	Firebird	F	0.50	+5.00	0.50	+0.41	0 +/- 0.09	—
		R	—	—	0.56	0	0 +/- 0.16	—
2001	Camaro	F	1.00	+5.00	1.00	+0.80	-0.20 +/- 0.20	—
		R	—	—	0.60	0	0 +/- 0.30	—
	Firebird	F	0.50	+5.00	0.50	+0.41	0 +/- 0.09	—
		R	—	—	0.56	0	0 +/- 0.16	—
2002	Camaro	F	1.00	+5.00	1.00	+0.80	-0.20 +/- 0.20	—
		R	—	—	0.60	0	0 +/- 0.30	—
	Firebird	F	0.50	+5.00	0.50	+0.41	0 +/- 0.09	—
		R	—	—	0.56	0	0 +/- 0.16	—

42372-FBOD-C09

TIRE, WHEEL AND BALL JOINT SPECIFICATIONS

Year	Model	OEM Tires Standard	OEM Tires Optional	Tire Pressures (psi) Front	Tire Pressures (psi) Rear	Wheel Size	Ball Joint Inspection
2000	Camaro RS	P215/60TR16	P215/60HR16	30	30	8-J	U: 0.125 in.
			P235/55R16				L: 0.047 in.
	Camaro Z28	P235/55R16	P245/50ZR16	30	30	8-J	U: 0.125 in.
			P275/40ZR17	30	30	9-J	L: 0.047 in.
	Firebird, base	P215/60R16	P235/55R16	30	30	7.5-JJ	0.125 in.
	Firebird Formula	P245/50R16	P275/40ZR17	30	30	8-J	0.125 in.
	Trans Am Coupe	P245/50ZR16	P275/40ZR17	30	30	8-JJ	0.125 in.
2001	Camaro RS	P215/60TR16	P215/60HR16	30	30	8-J	U: 0.125 in.
			P235/55R16				L: 0.047 in.
	Camaro Z28	P235/55R16	P245/50ZR16	30	30	8-J	U: 0.125 in.
			P275/40ZR17	30	30	9-J	L: 0.047 in.
	Firebird, base	P215/60R16	P235/55R16	30	30	7.5-JJ	0.125 in.
	Firebird Formula	P245/50R16	P275/40ZR17	30	30	8-J	0.125 in.
	Trans Am Coupe	P245/50ZR16	P275/40ZR17	30	30	8-JJ	0.125 in.
2002	Camaro RS	P215/60TR16	P215/60HR16	30	30	8-J	U: 0.125 in.
			P235/55R16				L: 0.047 in.
	Camaro Z28	P235/55R16	P245/50ZR16	30	30	8-J	U: 0.125 in.
			P275/40ZR17	30	30	9-J	L: 0.047 in.
	Firebird, base	P215/60R16	P235/55R16	30	30	7.5-JJ	0.125 in.
	Firebird Formula	P245/50R16	P275/40ZR17	30	30	8-J	0.125 in.
	Trans Am Coupe	P245/50ZR16	P275/40ZR17	30	30	8-JJ	0.125 in.

OEM: Original Equipment Manufacturer

PSI: Pounds Per Square Inch

L: Lower

U: Upper

42372-FBOD-C10

BRAKE SPECIFICATIONS

All measurements in inches unless noted

Year	Model		Brake Disc Original Thickness	Brake Disc Minimum Thickness	Brake Disc Maximum Runout	Brake Drum Diameter Original Inside Diameter	Brake Drum Diameter Max. Wear Limit	Brake Drum Diameter Maximum Machine Diameter	Minimum Lining Thickness	Brake Caliper Bracket Bolt (ft. lbs.)	Brake Caliper Mounting Bolt (ft. lbs.)
2000	Camaro	F	1.260	1.223	0.005	—	—	—	0.030	160	23
		R	NA	0.985	0.005	—	—	—	0.030	160	27
	Firebird	F	1.260	1.223	0.005	—	—	—	0.030	160	23
		R	NA	0.985	0.005	—	—	—	0.030	160	27
2001	Camaro	F	1.260	1.223	0.005	—	—	—	0.030	160	23
		R	NA	0.985	0.005	—	—	—	0.030	160	27
	Firebird	F	1.260	1.223	0.005	—	—	—	0.030	160	23
		R	NA	0.985	0.005	—	—	—	0.030	160	27
2002	Camaro	F	1.260	1.223	0.005	—	—	—	0.030	160	23
		R	NA	0.985	0.005	—	—	—	0.030	160	27
	Firebird	F	1.260	1.223	0.005	—	—	—	0.030	160	23
		R	NA	0.985	0.005	—	—	—	0.030	160	27

NA: Not Available

F: Front

R: Rear

42372-FBOD-C11

SCHEDULED MAINTENANCE INTERVALS
GM F BODY—CAMARO & FIREBIRD

TO BE SERVICED	TYPE OF SERVICE	VEHICLE MILEAGE INTERVAL (x1000)												
		7.5	15	22.5	30	37.5	45	52.5	60	67.5	75	82.5	90	97.5
Engine oil & filter	R	✓	✓	✓	✓	✓	✓	✓	✓	✓	✓	✓	✓	✓
Coolant level, hoses & clamps	S/I	✓	✓	✓	✓	✓	✓	✓	✓	✓	✓	✓	✓	✓
Exhaust system, brake hoses & throttle linkage	S/I	✓	✓	✓	✓	✓	✓	✓	✓	✓	✓	✓	✓	✓
Lubricate chassis and suspension	S/I	✓	✓	✓	✓	✓	✓	✓	✓	✓	✓	✓	✓	✓
Lubricate steering linkage and transaxle shift linkage	S/I	✓	✓	✓	✓	✓	✓	✓	✓	✓	✓	✓	✓	✓
Lubricate parking brake cable guides, underbody contact points & linkage	S/I	✓	✓	✓	✓	✓	✓	✓	✓	✓	✓	✓	✓	✓
Brake hoses & brake linings	S/I	✓		✓		✓		✓		✓		✓		✓
Rotate tires ①	S/I	✓		✓		✓		✓		✓		✓		✓
Automatic transmission fluid & filter ②	S/I													
Air filter element & PCV filter	R				✓				✓				✓	
Engine coolant ③	R													
Spark plugs ④	R				✓				✓				✓	
Ignition cables, EGR & fuel systems	S/I				✓				✓				✓	
Serpentine drive belt	S/I				✓				✓				✓	
Rear axle oil (Limited slip) ⑤	R	✓												

R: Replace

S/I: Service or Inspect

① For models with P245/50ZR16 tires, rotate front-to-rear only, & be sure that the tires roll in the direction indicated by the arrows on the side walls

② Automatic transmission fluid & filter: replace every 100,000 miles

③ Engine coolant: replace every 100,000 miles. Use O.E. specified (DEX-COOL™) coolant only. If any silicate coolant is used, the service interval is every 30,000 miles

④ Platinum tip spark plugs: replace every 100,000 miles

⑤ If the vehicle is used to tow a trailer, change the rear axle fluid every 7500 miles in either type of differential

FREQUENT OPERATION MAINTENANCE (SEVERE SERVICE)

If a vehicle is operated under any of the following conditions it is considered severe service:

- Extremely dusty areas.

- 50% or more of the vehicle operation is in 32°C (90°F) or higher temperatures, or constant operation in temperatures below 0°C (32°F).

- Prolonged idling (vehicle operation in stop and go traffic).

- Frequent short running periods (engine does not warm to normal operating temperatures).

- Police, taxi, delivery usage or trailer towing usage.

Oil & oil filter: change every 3000 miles

Chassis lubrication: lubricate every 6000 miles

Automatic transmission fluid & filter: change every 15,000 miles

Air filter element: service or inspect every 15,000 miles

Rotate tires at 6000 miles, then every 15,000 miles

42372-FBOD-C12

PRECAUTIONS

Before servicing any vehicle, please be sure to read all of the following precautions, which deal with personal safety, prevention of component damage, and important points to take into consideration when servicing a motor vehicle:

• Never open, service or drain the radiator or cooling system when the engine is hot; serious burns can occur from the steam and hot coolant.

• Observe all applicable safety precautions when working around fuel. Whenever servicing the fuel system, always work in a well-ventilated area. Do not allow fuel spray or vapors to come in contact with a spark, open flame or excessive heat (a hot drop light, for example). Keep a dry chemical fire extinguisher near the work area. Always keep fuel in a container specifically designed for fuel storage; also, always properly seal fuel containers to avoid the possibility of fire or explosion. Refer to the additional fuel system precautions later in this section.

• Fuel injection systems often remain pressurized, even after the engine has been turned **OFF**. The fuel system pressure must be relieved before disconnecting any fuel lines. Failure to do so may result in fire and/or personal injury.

• Brake fluid often contains polyglycol ethers and polyglycols. Avoid contact with the eyes and wash your hands thoroughly after handling brake fluid. If you do get brake fluid in your eyes, flush your eyes with clean, running water for 15 minutes. If eye irritation persists, or if you have taken brake fluid internally, IMMEDIATELY seek medical assistance.

• The EPA warns that prolonged contact with used engine oil may cause a number of skin disorders, including cancer! You should make every effort to minimize your exposure to used engine oil. Protective gloves should be worn when changing oil. Wash your hands and any other exposed skin areas as soon as possible after exposure to used engine oil. Soap and water, or waterless hand cleaner should be used.

• All new vehicles are now equipped with an air bag system. The system must be disabled before performing service on or around system components, steering column, instrument panel components, wiring and sensors. Failure to follow safety and disabling procedures could result in accidental air bag deployment, possible personal injury and unnecessary system repairs.

• Always wear safety goggles when working with, or around, the air bag system. When carrying a non-deployed air bag, be sure the bag and trim cover are pointed away from your body. When placing a non-deployed air bag on a work surface, always face the bag and trim cover upward, away from the surface. This will reduce the motion of the module if it is accidentally deployed. Refer to the additional air bag system precautions later in this section.

• Clean, high quality brake fluid from a sealed container is essential to the safe and proper operation of the brake system. You should always buy the correct type of brake fluid for your vehicle. If the brake fluid becomes contaminated, completely flush the system with new fluid. Never reuse any brake fluid. Any brake fluid that is removed from the system should be discarded. Also, do not allow any brake fluid to come in contact with a painted surface; it will damage the paint.

• Never operate the engine without the proper amount and type of engine oil; doing so WILL result in severe engine damage.

• Timing belt maintenance is extremely important! Many models utilize an interference-type, non-freewheeling engine. If the timing belt breaks, the valves in the cylinder head may strike the pistons, causing potentially serious (also time-consuming and expensive) engine damage. Refer to the maintenance interval charts in the front of this section for the recommended replacement interval for the timing belt, and to the timing belt procedure in this section for belt replacement and inspection.

• Disconnecting the negative battery cable on some vehicles may interfere with the functions of the on-board computer system(s) and may require the computer to undergo a relearning process once the negative battery cable is reconnected.

• When servicing drum brakes, only disassemble and assemble one side at a time, leaving the remaining side intact for reference.

• Only an MVAC-trained, EPA-certified automotive technician should service the air conditioning system or its components.

ENGINE REPAIR

Alternator

REMOVAL

3.8L Engine

1. Before servicing the vehicle, refer to the precautions in the beginning of this section.
2. Remove or disconnect the following:
   • Negative battery cable
   • Accessory drive belt
   • Alternator electrical connectors
   • Evaporative Emissions (EVAP) canister purge solenoid
   • Rear alternator brace-to-alternator bolt
   • Both front alternator bolts
   • Alternator

5.7L Engine

1. Before servicing the vehicle, refer to the precautions in the beginning of this section.
2. Remove or disconnect the following:
   • Negative battery cable
   • Accessory drive belt
   • Alternator rear inner brace
   • Transmission oil cooler line clips
   • Alternator bolts
   • Alternator electrical connectors
   • Alternator

INSTALLATION

3.8L Engine

Install or connect the following:
   • Alternator. Torque the upper (drive belt tensioner) bolt to 22 ft. lbs. (30 Nm) and the lower bolt to 37 ft. lbs. (50 Nm).
   • Rear alternator brace-to-alternator bolt. Torque the bolt to 22 ft. lbs. (30 Nm).
   • EVAP canister purge solenoid
   • Alternator electrical connectors. Torque the battery cable nut to 16 ft. lbs. (22 Nm).
   • Accessory drive belt
   • Negative battery cable

5.7L Engine

Install or connect the following:
   • Alternator
   • Alternator bolts. Torque the bolts to 37 ft. lbs. (50 Nm).
   • Transmission oil cooler line clips

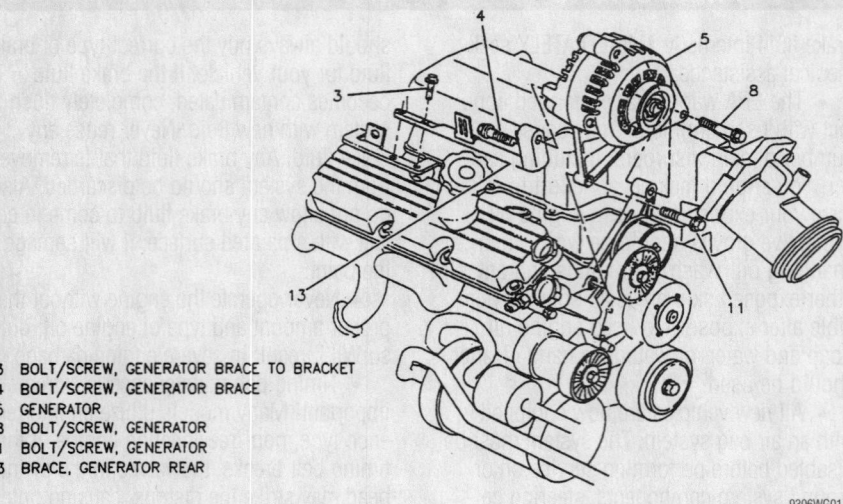

3 BOLT/SCREW, GENERATOR BRACE TO BRACKET
4 BOLT/SCREW, GENERATOR BRACE
5 GENERATOR
8 BOLT/SCREW, GENERATOR
11 BOLT/SCREW, GENERATOR
13 BRACE, GENERATOR REAR

9306WG01

Exploded view of the alternator and related components—3.8L engine

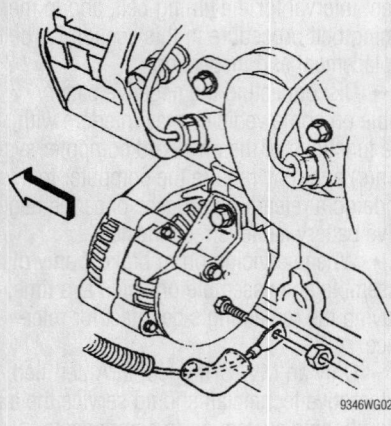

9346WG02

Exploded view of the alternator and related components—5.7L engine

- Alternator rear inner brace. Torque the nut to 18 ft. lbs. (25 Nm).
- Alternator electrical connectors. Torque the battery cable nut to 16 ft. lbs. (22 Nm).
- Accessory drive belt
- Negative battery cable

Ignition Timing

ADJUSTMENT

On these vehicles, base timing is preset when the engine is manufactured. All timing changes are, then controlled directly by the Powertrain Control Module (PCM) based on information from the ignition and knock sensor systems. No adjustments are necessary or possible.

Engine Assembly

REMOVAL & INSTALLATION

All 3.8L Engines And 2000 5.7L Engines

➡The engine, transmission and suspension assembly is removed from the bottom of the vehicle. After the assembly is removed, separate the engine from the transmission and frame.

1. Before servicing the vehicle, refer to the precautions in the beginning of this section.
2. Remove both battery cables.
3. Relieve the fuel system pressure.
4. Recover the refrigerant from the A/C system using approved equipment.
5. Drain the cooling system, crankcase and transmission fluid.
6. Remove or disconnect the following:
   - Intake air temperature (IAT) sensor electrical connector
   - Air intake duct
   - Front wheels
   - 3-way catalytic converter
   - Propeller shaft
   - Torque arm
   - Starter motor
   - Transmission range selector lever cable from the range selector lever at the transmission by unsnapping it, on models equipped with a automatic transmission
   - Retainer from the range selector lever cable
   - Range selector lever cable from the cable bracket
   - Clutch actuator cylinder line from the actuator cylinder using Tool J36221, on models equipped with a manual transmission
   - Left side front air deflector
   - Stabilizer bar bracket bolts and brackets
   - Steering gear coupling shield
   - Intermediate steering shaft bolt and shaft from the rack
   - Wiring harness ground bolt and radio frequency ground strap and cruise control ground lead from the front rail
   - Knock sensors electrical connectors
   - Wheel speed sensors electrical connectors
   - A/C compressor and condenser hose bolt. Discard the O-ring.
   - A/C compressor and condenser hose nut at the condenser. Discard the O-ring.
   - Front fuel pipe heat shield nuts
   - Front fuel pipe heat shield
   - Brake lines from the brake pipe clip
   - Engine wiring harness ground bolt
   - Engine ground
   - A/C compressor and condenser hose bolt at the accumulator. Discard the O-ring.
   - Heater hoses from the drive belt tensioner
   - Fuel feed pipe from the fuel rail
   - Fuel return pipe from the fuel rail
   - Evaporative Emissions (EVAP) pipe at the engine
   - Cruise control and accelerator control cables from the throttle body and bracket
   - Accelerator control cable bracket bolts and bracket
   - Inlet hose from the water outlet
   - Outlet hose from the water pump
   - Brake booster vacuum hose from the intake manifold fitting
   - Front two brake pipes from the brake modulator
   - Secondary captured locks
   - Forward lamp harness from the engine harness
   - Engine harness vacuum tube from the bottom of the vacuum check valve
   - Powertrain Control Module (PCM) connectors
   - PCM
   - Right side insulator panel
   - Hinge pillar trim panel
   - Engine wiring harness from the instrument panel wiring harness
   - Engine wire harness through the front of the dash

7. Place the harness on top of the engine.
- Floor shift control, if equipped
- Transmission control, if equipped
- Right and left lower shock bolts from the lower control arms
- Cotter pins and nuts from the left and right upper ball joints
- Separate the upper control arms from the steering knuckles and support both steering knuckles

8. Position Engine Support Table J 39580 below the vehicle. Lower the vehicle and lay the engine wire harness on top of the engine.
- Transmission support bolts
- Engine and transmission from the vehicle

9. Secure the crossmember to the engine support table.

10. Position a hex-head socket on the belt tensioner pulley bolt.

11. Rotate the drive belt tensioner clockwise to relieve belt tension.

12. Remove or disconnect the following:
- Drive belt
- A/C compressor rear bolts
- A/C compressor rear bracket stud
- A/C compressor rear bracket
- A/C compressor bolts
- A/C compressor
- Drive belt tensioner bolts and the tensioner
- Power steering pump nuts and reposition the power steering pump to the crossmember
- Alternator electrical connector
- Positive cable nut from the alternator
- Alternator

13. Install a suitable lifting device to the engine.
- Right engine mount bracket bolts and bracket
- Left engine mount bracket bolts, stud and the bracket

14. Raise the engine and transmission from the crossmember.

15. If equipped with the automatic transmission remove or disconnect the following:
- Right side transmission support brace bolts, nut and the brace
- Left side transmission support brace bolts, nut and the brace
- Torque converter cover
- Flywheel-to-torque converter bolts

16. If equipped with the manual transmission remove or disconnect the following:
- Right side transmission support bolts and the support

- Left side transmission support bolts and the support
- Electrical connectors from the transmission
- Transmission-to-engine bolts and nut, then separate the transmission from the engine, if equipped with an automatic transmission
- Flywheel housing to engine bolts and nuts, then separate the flywheel housing from the engine, if equipped with a manual transmission
- Pressure plate and clutch, if equipped with a manual transmission
- Flywheel bolts and flywheel, if equipped with a manual transmission

To install:

17. Install or connect the following:
- Flywheel bolts and flywheel, if equipped with a manual transmission. Tighten the flywheel bolts to 11 ft. lbs. (15 Nm), then tighten an additional 50 degrees.
- Clutch and pressure plate, if equipped with a manual transmission. Tighten the pressure plate bolts to 15 ft. lbs. (20 Nm) and tighten an additional 45 degrees.

18. Align the flywheel housing to the engine.
- Flywheel housing-to-engine bolts and nuts, if equipped with a manual transmission and tighten to 70 ft. lbs. (95 Nm).
- Transmission-to-engine bolts and nut and tighten to 70 ft. lbs. (95 Nm), if equipped with an automatic transmission
- Electrical connectors to the transmission

19. If equipped with the manual transmission install or connect the following:
- Left side transmission support
- Left side transmission support brace bolts and tighten to 37 ft. lbs. (50 Nm)
- Left side transmission support brace bolts and tighten to 21 ft. lbs. (28 Nm)
- Right side transmission support bolts and tighten to 37 ft. lbs. (50 Nm)

20. If equipped with the automatic transmission install or connect the following:
- Torque converter-to-flywheel bolts and tighten to 47 ft. lbs. (63 Nm)
- Torque converter cover
- Left side transmission support, bolts and nut

- Right side transmission support

21. Lower the engine and transmission to the crossmember.
- Left engine mount bracket, stud and bolts. Tighten the bolts to 74 ft. lbs. (100 Nm) and the stud 64 ft. lbs. (87 Nm).
- Right engine mount bracket and the bolts

22. Remove the engine lifting device.
- Alternator and wiring
- Power steering pump and tighten the nuts to 22 ft. lbs. (30 Nm)
- Drive belt tensioner and bolts, and tighten to 37 ft. lbs. (50 Nm)
- A/C compressor and tighten to 37 ft. lbs. (50 Nm)
- A/C compressor rear bracket. and tighten the stud to 37 ft. lbs. (50 Nm) and the bolts to 22 ft. lbs. (30 Nm)
- Drive belt

23. Remove the crossmember from the table security straps.

24. Lower the vehicle and install the engine and transmission to the vehicle.

25. Align the crossmember to the vehicle body.

26. Install or connect the following:
- Front crossmember bolts. Tighten the lower 2 to 107 ft. lbs. (145 Nm) and the upper 4 cradle bolts to 92 ft. lbs. (125 Nm).
- Transmission support bolts and tighten to 42 ft. lbs. (55 Nm)
- Ball stud to the steering knuckle
- Nuts and cotter pins to the left and right upper ball joints. Tighten to 39 ft. lbs. (53 Nm).
- Right and left lower shock bolts to the lower control arms
- Transmission control, if equipped
- Floor shift control, if equipped
- Engine wire harness through the front of dash
- Engine wiring harness to the instrument panel wiring harness
- Hinge pillar trim panel
- Right side insulator panel
- PCM and electrical connectors
- Engine harness vacuum tube
- Forward lamp harness
- Engine harness
- Secondary captured locks
- Front two brake pipes
- Brake booster vacuum hose
- Outlet hose to the water pump
- Inlet hose to the water outlet
- Accelerator control cable bracket and bolts. Tighten to 12 ft. lbs. (16 Nm).
- Accelerator control and cruise con-

trol cables to the throttle body and bracket
- EVAP pipe
- Fuel return pipe
- Fuel feed pipe
- Heater hoses
- New O-ring
- A/C compressor and condenser hose bolt at the accumulator and tighten to 36 ft. lbs. (48 Nm)
- Engine ground and bolt and tighten to 18 ft. lbs. (25 Nm)
- Brake lines
- Front fuel pipe heat shield and bolts
- A/C compressor and condenser hose nut at the condenser. Tighten to 12 ft. lbs. (16 Nm).
- Wheel speed sensors electrical connectors
- Knock sensors electrical connectors
- Radio frequency ground strap, cruise control ground lead and wiring harness ground bolt to the front rail
- Intermediate steering shaft and bolt to the rack. Tighten to 35 ft. lbs. (47 Nm).
- Steering gear coupling shield
- Stabilizer bar brackets and bolts Tighten to 41 ft. lbs. (55 Nm).
- Left side front air deflector
- Clutch actuator cylinder line to the actuator cylinder, if equipped
- Range selector lever cable, if equipped
- Starter motor
- Torque arm
- Propeller shaft
- 3-way catalytic converter
- Front wheels
- Air intake duct
- IAT sensor electrical connector

27. Refill the engine oil, transmission fluid and coolant
28. Recharge the A/C system.
29. Bleed the brakes.
30. Bleed the clutch hydraulic system.
31. Connect the battery cables.

2001–02 5.7L Engines

➡The engine, transmission and suspension assembly is removed from the bottom of the vehicle. After the assembly is removed, separate the engine from the transmission and frame.

1. Before servicing the vehicle, refer to the precautions in the beginning of this section.
2. Remove both battery cables.
3. Relieve the fuel system pressure.

4. Recover the refrigerant from the A/C system using approved equipment.
5. Drain the cooling system, crankcase and transmission fluid.
6. Remove or disconnect the following:
- Fan shroud
- Upper and lower radiator hoses
- Shift control closeout boot, if equipped with a manual transmission
- A/C compressor and condenser hose bolt at the accumulator and discard the O-ring
- Upper oil cooler line from the radiator
- A/C condenser tube nut
- Condenser tube from the condenser
- Inlet and outlet heater hoses from the water pump
- Fuel line from the fuel rail
- Vapor line from the fuel vapor purge valve
- Cruise control cable from the throttle lever

✳✳ CAUTION

Always replace the accelerator and cruise control cables with NEW cables whenever you remove the engine from the vehicle.

- Accelerator control cable from the throttle lever
- Accelerator and cruise control cable from the throttle body bracket
- Secondary air injection (AIR) hose from the AIR shut-off valve
- Brake booster vacuum hose
- Front brake lines from the brake pressure modulator valve
- Secondary captured locks
- Forward lamp harness from the engine harness
- Engine harness vacuum tube from the bottom of the vacuum check valve
- Powertrain Control Module (PCM) connectors
- PCM from the bracket
- Right side insulator panel
- Hinge pillar trim panel
- Engine wiring harness from the instrument panel wiring harness

7. Pull the engine wiring harness through the front of dash. Lay the engine wire harness on top of the engine.
- Lower oil cooler line from the radiator
- Front wheels
- All Oxygen Sensor (O$_2$S) connectors

- Exhaust crossover pipe
- Driver side front fascia air deflector
- Stabilizer bar bracket bolts/stud
- Stabilizer bar brackets
- Wiring harness ground bolt

8. Reposition the radio frequency ground strap and cruise control ground lead away from the frame rail.
- Automatic transmission range selector lever cable from the range selector lever, if equipped
- Automatic transmission range selector lever cable retainer, if equipped
- Automatic transmission range selector lever cable from the bracket, if equipped

9. If equipped with a manual transmission, depress the white circular release ring on the actuator hose using tool J36221 and simultaneously pull lightly on the master cylinder hose to disconnect.
- Propeller shaft
- Torque arm
- Intermediate steering shaft bolt and shaft from the power steering rack
- Left and right side wheel speed sensor connectors
- Battery positive cable from the alternator
- Starter motor
- Right and left side shock absorber lower bolts
- Cotter pins and nuts from the left and right upper ball joints
- Upper control arms from the steering knuckles and secure both steering knuckles
- Front fuel pipe heat shield
- Brake lines from the brake pipe clip

10. Position engine support table J 38580 below the vehicle. Lower the vehicle until the crossmember is resting on the table and secure the vehicle to the hoist.
- Front crossmember bolts
- Transmission crossmember bolts

11. Carefully raise the vehicle off of the engine and transmission assembly. Secure the crossmember to the support table.
- Ignition coil(s) (as required) and install engine lift brackets before lifting the engine
- Drive belt from the tensioner and pulleys
- A/C drive belt from the pulleys
- Power steering pump pulley
- Power steering pump bolts and brace from the pump, then reposition the power steering pump to the crossmember
- Power steering pump bracket bolts and bracket

- A/C compressor electrical connector
- Rosebud clip from the engine bracket
- A/C compressor
- A/C compressor bracket
- Left transmission cover bolt and cover, if equipped.
- Right transmission cover bolt and cover, if equipped
- Torque converter bolts.
- Engine wiring harness clip bolt from the automatic transmission, if equipped
- 20-way connector and Vehicle Speed Sensor (VSS) electrical connectors from the automatic transmission, if equipped
- Automatic transmission fill tube bolt
- Rosebud clip from the right side of the manual transmission, if equipped
- Back-up lamp switch connector
- All electrical connections from the manual transmission, if equipped
- Push-in retainer from the manual transmission tab
- Alternator rear bracket bolt
- Alternator

12. Install an engine hoist
- Right side engine mount bolts and mount
- Engine mount heat shield nuts and shield
- Left side engine mount studs and mount

13. Raise the engine using the lifting brackets and hoist.

14. With the aide of an assistant, lift the engine and transmission from the crossmember make sure to support the transmission on the table.

15. Remove or disconnect the following:
- Automatic transmission bolts, if equipped
- Manual transmission bolts, if equipped

16. Separate the transmission from the engine
- Pressure plate bolts, if equipped
- Pressure plate and clutch disc from the flywheel, if equipped
- Engine flywheel bolts
- Flywheel

17. Install the engine on a engine stand. Remove the engine hoist from the engine block.

To install:

18. Installation is the reverse of removal but please note the following torques:

- Automatic transmission flywheel bolts first pass in a star pattern to 26 ft. lbs. (20 Nm), second pass in sequence to 37 ft. lbs. (50 Nm) and finally to 74 ft. lbs. (100 Nm)
- Clutch pressure plate bolts evenly over 3 increments with the fourth increment to 52 ft. lbs. (70 Nm)
- Manual transmission bolts , if equipped to 37 ft. lbs. (50 Nm)
- Automatic transmission bolts, if equipped to 70 ft. lbs. (90 Nm)
- Left and right side engine mount studs/bolts to 37 ft. lbs. (50 Nm)
- Automatic transmission fill tube bolt to 37 ft. lbs. (50 Nm)
- Torque converter bolts to 47 ft. lbs. (63 Nm)
- A/C compressor bracket bolts to 37 ft. lbs. (50 Nm)
- A/C compressor bolts to 37 ft. lbs. (50 Nm)
- Power steering pump bolts to 37 ft. lbs. (50 Nm)
- Power steering pump brace bolts to 18 ft. lbs. (25 Nm)
- Transmission crossmember bolts to 42 ft. lbs. (57 Nm)
- Front crossmember bolts, four upper crossmember bolts to 92 ft. lbs. (125 Nm) and the two lower crossmember bolts to 107 ft. lbs. (145 Nm)
- Stabilizer bar bracket bolts/stud to 41 ft. lbs. (55 Nm)
- A/C condenser tube nut to 12 ft. lbs. (16 Nm)
- Upper oil cooler line fitting to 20 ft. lbs. (27 Nm)

19. Refill the engine oil, transmission fluid and coolant.
20. Recharge the A/C system.
21. Bleed the brakes.
22. Bleed the clutch hydraulic system.
23. Connect the negative battery cable.

Water Pump

REMOVAL & INSTALLATION

3.8L Engine

1. Before servicing the vehicle, refer to the precautions in the beginning of this section.
2. Drain the cooling system.
3. Remove or disconnect the following:
- Negative battery cable
- Accessory drive belt
- Radiator inlet hose from the water pump
- Water pump pulley
- Water pump

To install:

4. Install or connect the following:
- Water pump using a new gasket. Torque the water pump bolts to 11 ft. lbs. (15 Nm), plus an additional 80 degree rotation.
- Water pump pulley. Torque the bolts to 115 inch lbs. (13 Nm).
- Accessory drive belt
- Radiator inlet hose to the water pump
- Negative battery cable

5. Refill the cooling system.
6. Start the engine and check for leaks.

5.7L Engines

1. Before servicing the vehicle, refer to the precautions in the beginning of this section.
2. Drain the cooling system by removing the block coolant drain plug and the knock sensor.
3. Remove or disconnect the following:
- Negative battery cable
- Electrical connector from the cooling fan
- Intake Air Temperature (IAT) sensor electrical connector

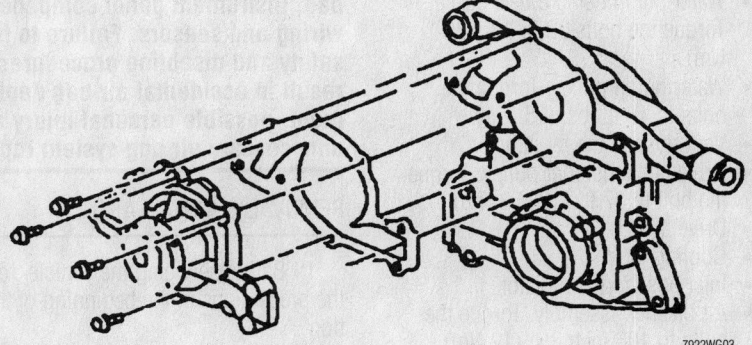

Exploded view of the water pump mounting—3.8L engine

7922WG03

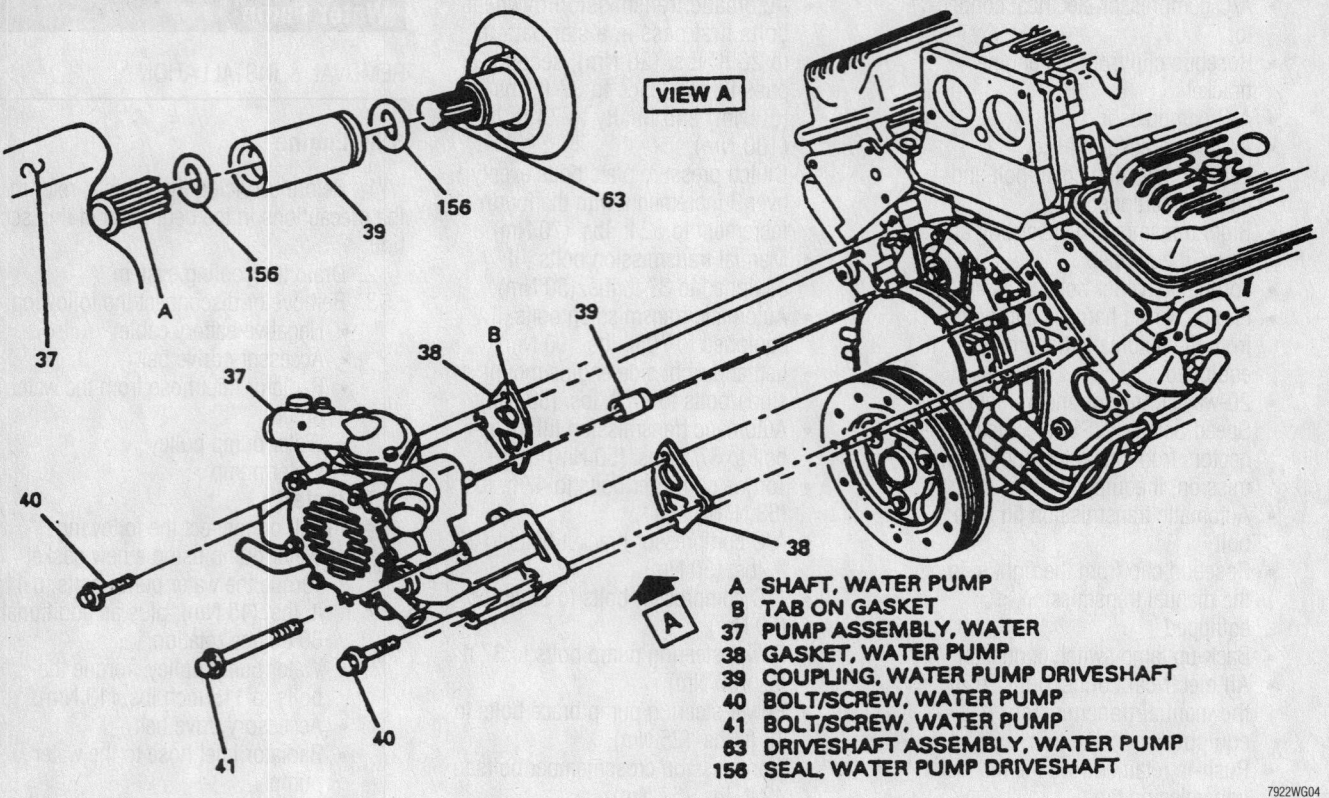

VIEW A

A SHAFT, WATER PUMP
B TAB ON GASKET
37 PUMP ASSEMBLY, WATER
38 GASKET, WATER PUMP
39 COUPLING, WATER PUMP DRIVESHAFT
40 BOLT/SCREW, WATER PUMP
41 BOLT/SCREW, WATER PUMP
63 DRIVESHAFT ASSEMBLY, WATER PUMP
156 SEAL, WATER PUMP DRIVESHAFT

7922WG04

Exploded view of the water pump mounting and related components—5.7L engines

- Mass Airflow (MAF) sensor electrical connector
- Air intake duct
- Outlet hose from the water outlet
- Inlet hose from the water pump
- Air cleaner assembly
- Inlet hose from the radiator
- Both cooling fan assemblies
- Drive belt
- Drive belt tensioner pulley
- Vent hose from the radiator
- Water pump pulley
- Water pump

To install:
4. Install or connect the following:
- Drain plug
- Water pump using a new gasket. Torque the bolts 22 ft. lbs. (30 Nm).
- Water pump pulley. Torque the bolts to 18 ft. lbs. (25 Nm).
- Vent hose to the radiator
- Drive belt tensioner pulley. Torque the bolt to 37 ft. lbs. (50 Nm).
- Drive belts
- Cooling fans
- Inlet hose to the radiator
- Air cleaner assembly. Torque the bolts to 106 inch lbs. (12 Nm).
- Inlet hose to the water pump
- Outlet hose to the water outlet

- Air intake duct
- IAT and MAF sensor electrical connectors
- Coolant fan electrical connectors
- Negative battery cable
5. Refill the cooling system.
6. Start the engine and check for leaks.

Heater Core

✳✳ CAUTION

Some vehicles are equipped with an SIR or air bag system. The air bag system must be disabled before performing service on or around the air bag, instrument panel components, wiring and sensors. Failure to follow safety and disabling procedures could result in accidental air bag deployment, possible personal injury and unnecessary air bag system repairs.

REMOVAL & INSTALLATION

1. Before servicing the vehicle, refer to the precautions in the beginning of this section.
2. With the ignition key removed, disconnect the negative battery cable.

3. Disable the SIR system by removing or disconnecting the following:
- SIR fuse from the fuse panel
- Left side sound insulator
- Connector Positive Assurance (CPA) from the yellow 2-way SIR harness connector at the base of the steering column and separate the connector
- Squeeze the sides of the glove box and release it to access the heater core assembly
4. Drain the cooling system.
5. Remove or disconnect the following:
- 2 heater module cover retaining screws and the cover
- Clamp at the left side of the heater core

➡Do not apply excessive pressure on the tubes or the heater core will be damaged.

- Clamp from the heater core tubes on the engine compartment side. Carefully, remove the heater hoses from the core tubes. Plug the hoses to prevent leakage.
- Pull heater core toward the rear of the vehicle to remove it

To install:
6. Install or connect the following:
- Position the heater core into place

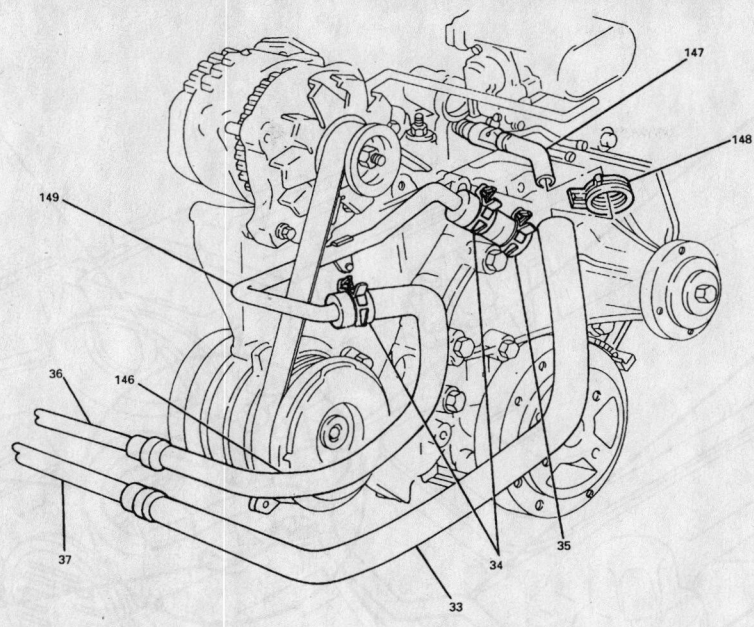

33 HOSE ASSEMBLY, HEATER INLET
34 CLAMP, HEATER OUTLET FRONT HOSE
35 CLAMP, HEATER INLET FRONT HOSE
36 PIPE, HEATER HOSE OUTLET
37 PIPE, HEATER HOSE INLET
146 HOSE ASSEMBLY, HEATER OUTLET
147 HOSE, THROTTLE BODY HEATER RETURN
148 CLAMP, THROTTLE BODY HEATER RETURN HOSE
149 PIPE, HEATER OUTLET

88146G08

Heater hose and pipe routing—Camaro/Firebird with the 3.4L engine

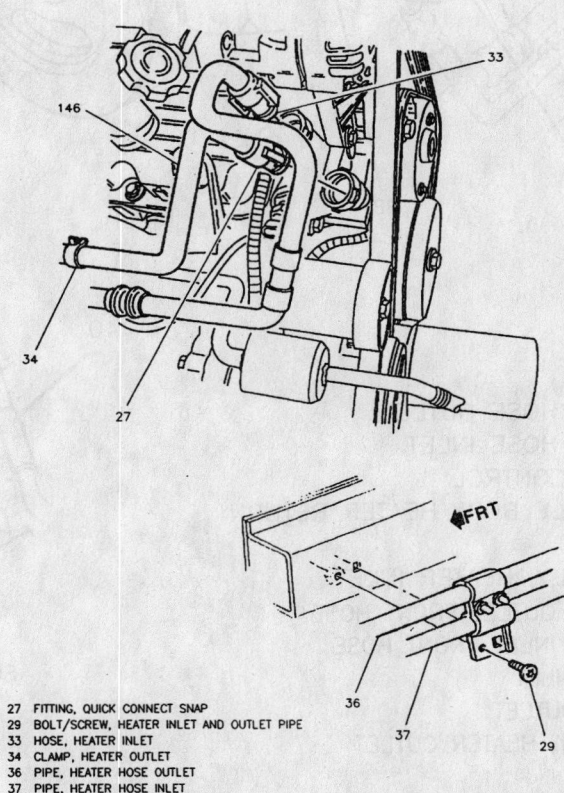

27 FITTING, QUICK CONNECT SNAP
29 BOLT/SCREW, HEATER INLET AND OUTLET PIPE
33 HOSE, HEATER INLET
34 CLAMP, HEATER OUTLET
36 PIPE, HEATER HOSE OUTLET
37 PIPE, HEATER HOSE INLET
146 HOSE, HEATER OUTLET

88146G09

Heater hose and pipe routing—Camaro/Firebird with the 3.8L engine

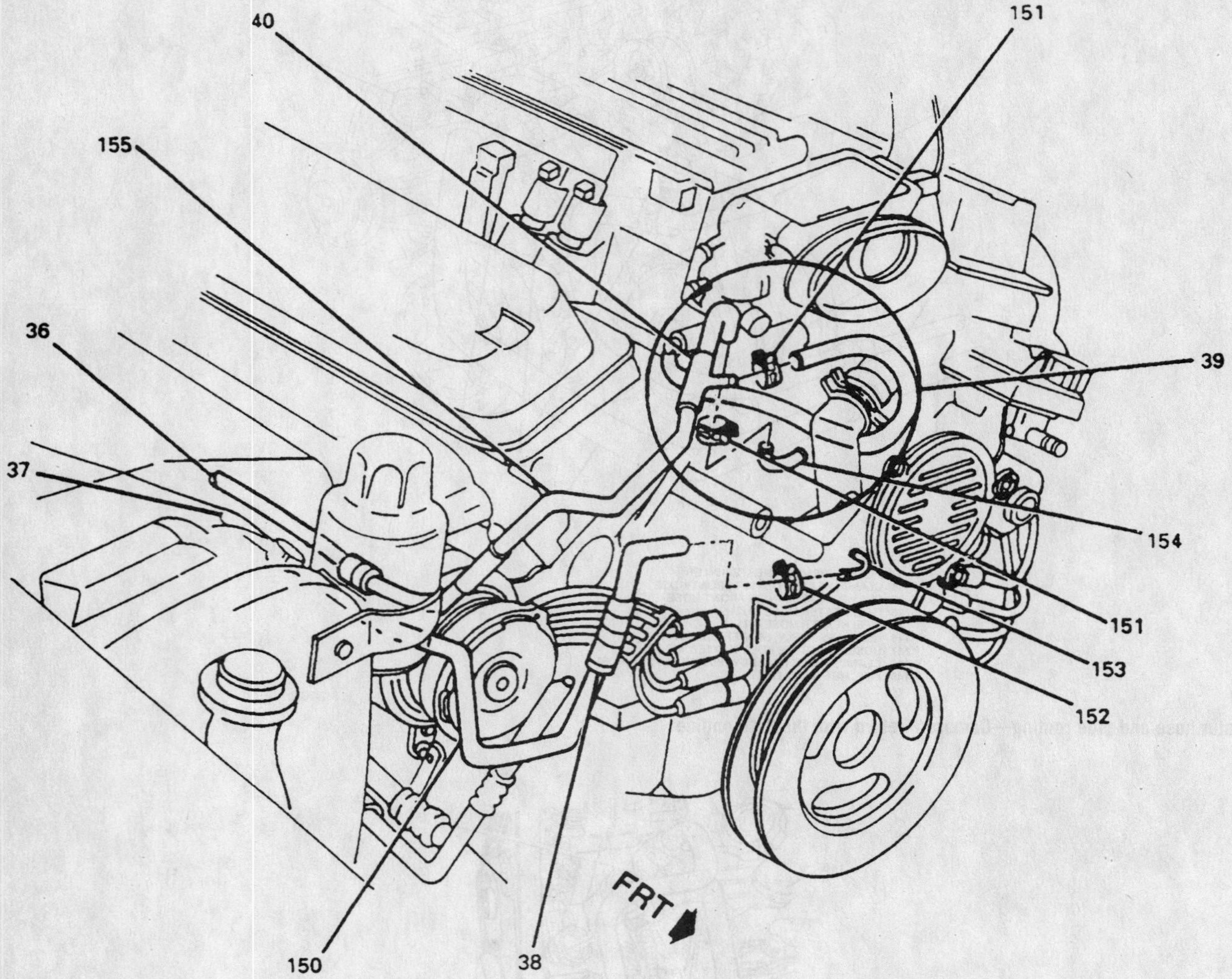

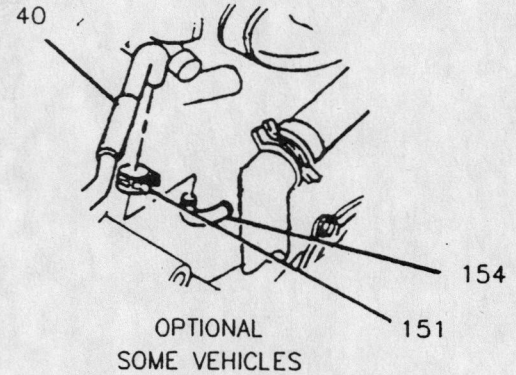

OPTIONAL
SOME VEHICLES

36 PIPE, HEATER HOSE OUTLET
37 PIPE, HEATER HOSE INLET
38 VALVE, FLOW CONTROL
39 HOSE, THROTTLE BODY HEATER RETURN
40 VALVE, BLEED
150 HOSE ASSEMBLY, HEATER INLET
151 CLAMP, HEATER OUTLET FRONT HOSE
152 CLAMP, HEATER INLET FRONT HOSE
153 PIPE, HEATER INLET
154 PIPE, HEATER OUTLET
155 HOSE ASSEMBLY, HEATER OUTLET

88146G10

Heater hose and pipe routing—Camaro/Firebird with the 5.7L engine

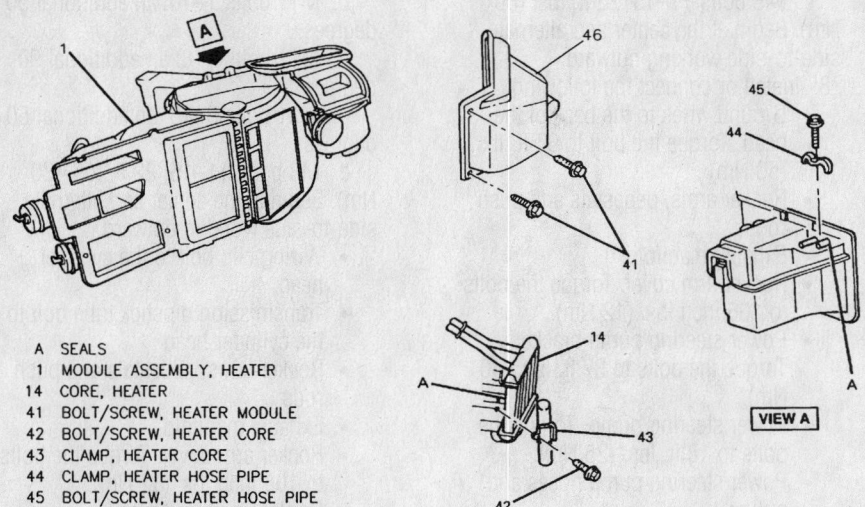

A SEALS
1 MODULE ASSEMBLY, HEATER
14 CORE, HEATER
41 BOLT/SCREW, HEATER MODULE
42 BOLT/SCREW, HEATER CORE
43 CLAMP, HEATER CORE
44 CLAMP, HEATER HOSE PIPE
45 BOLT/SCREW, HEATER HOSE PIPE
46 COVER, HEATER MODULE ASSEMBLY

88146G07

Exploded view of the heater core removed from the heater module assembly—Camaro/Firebird

and then attach the heater hoses to the heater core tubes. Install the hose clamp

➡**Lubricate the heater tubes with petroleum jelly for best sealing. Be sure the seals around the heater pipes remain in place.**

- Heater core clamp, then the heater core module cover

7. Properly, refill the cooling system, then operate the system and check for leaks.

8. If equipped, enable the SIR system by installing connecting the following:
- Yellow 2-way SIR connector and insert the Connector Positive Assurance (CPA) at the base of the steering column
- Left side sound insulator
- SIR fuse in the fuse panel
- Negative battery cable

Cylinder Head

REMOVAL & INSTALLATION

3.8L Engine

LEFT SIDE

1. Before servicing the vehicle, refer to the precautions in the beginning of this section.

2. Relieve the fuel system pressure.
3. Drain the cooling system.
4. Remove or disconnect the following:
- Negative battery cable
- Fuel lines from the fuel rail
- Upper and lower intake manifolds
- Spark plugs
- Left exhaust manifold
- Rocker arm cover, rocker arms and push rods
- Cylinder head bolts and cylinder head

To install:

5. Install the cylinder head with a new gasket and secure with NEW head bolts.

6. Torque the head bolts in sequence as follows:
a. Step 1: 37 ft. lbs. (50 Nm).
b. Step 2: Plus 120 degree turn.

7. Install or connect the following:
- Rocker arms and push rods
- Rocker arm cover. Torque the bolts to 89 inch lbs. (10 Nm).

- Left exhaust manifold
- Spark plugs and torque them to 20 ft. lbs. (27 Nm).
- Lower and upper intake manifolds
- Negative battery cable

8. Refill the cooling system.
9. Start the engine and check for leaks.

RIGHT SIDE

1. Before servicing the vehicle, refer to the precautions in the beginning of this section.

2. Relieve the fuel system pressure.
3. Drain the cooling system.
4. Remove or disconnect the following:
- Negative battery cable
- Fuel lines from the fuel rail
- Upper and lower intake manifolds
- Spark plugs
- Right exhaust manifold
- Rocker arm cover, rocker arms and push rods
- Cylinder head bolts and cylinder head

To install:

5. Install the cylinder head with a new gasket and secure with NEW head bolts. Torque the bolts in sequence as follows:
a. Step 1: 37 ft. lbs. (50 Nm).
b. Step 2: Plus a 120 degree turn

6. Install or connect the following:
- Rear alternator brace. Torque the bolts to 18 ft. lbs. (25 Nm).
- Drive belt tensioner. Torque the bolts to 37 ft. lbs. (50 Nm).
- Rocker arms and push rods
- Rocker arm cover. Torque the bolts to 89 inch lbs. (10 Nm).

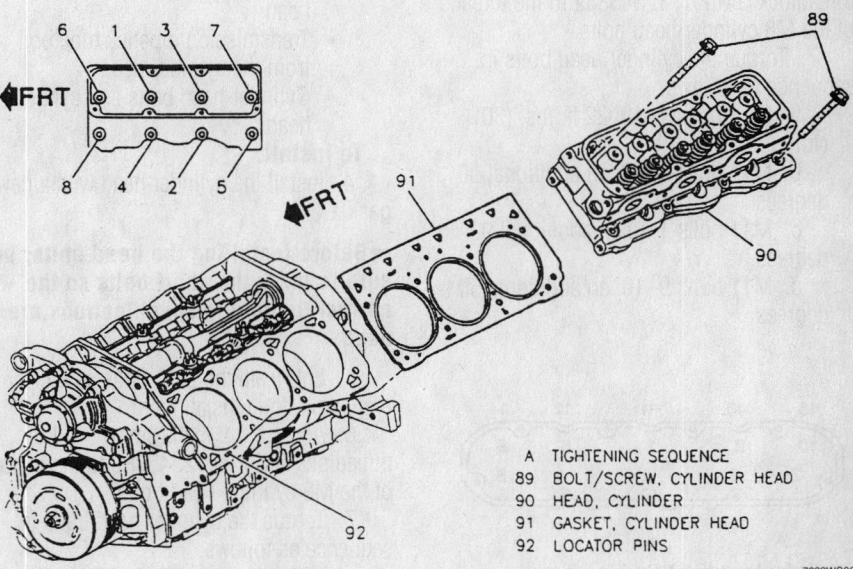

A TIGHTENING SEQUENCE
89 BOLT/SCREW, CYLINDER HEAD
90 HEAD, CYLINDER
91 GASKET, CYLINDER HEAD
92 LOCATOR PINS

7922WG06

Cylinder head bolt torque sequence—3.8L (VIN K) engine

- Right exhaust manifold. Torque the bolts to 106 inch lbs. (12 Nm).
- Spark plugs. Torque the plugs to 20 ft. lbs. (27 Nm).
- Lower and upper intake manifolds
- Negative battery cable

7. Refill the cooling system.
8. Start the engine and check for leaks.

5.7L Engines

LEFT SIDE

1. Before servicing the vehicle, refer to the precautions in the beginning of this section.
2. Drain the cooling system.
3. Remove or disconnect the following:
- Negative battery cable
- Coolant air bleed pipe
- Power steering pump pulley and hoses
- Power steering pump and bracket
- Exhaust manifold
- Intake manifold
- Rocker arm cover
- Rocker arms, pedestals and push rods
- Ground wires from the back of the head
- Cylinder head bolts and cylinder head

To install:

4. Install the cylinder head with a new gasket.

➥Before installing the head bolts, be sure to mark the short bolts so the correct tightening specifications are used.

5. Lubricate the M11 bolts with clean engine oil, then install them.
6. Apply a 0.2 inch (5mm) bead of threadlock GM P/N 12345382 to the threads of the M8 cylinder head bolts.
7. Torque the cylinder head bolts in sequence as follows:
   a. M11 bolts 1–10: 22 ft. lbs. (30 Nm).
   b. M11 bolts 1–10: an additional 90 degrees.
   c. M11 bolts 1–8: an additional 90 degrees.
   d. M11 bolts 9–10: an additional 50 degrees.

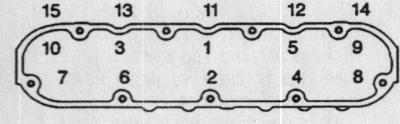

Cylinder head bolt tightening sequence—5.7L (VIN G) engine

   e. M8 bolts 11–15: 22 ft. lbs. (30 Nm). Begin at the center and alternate side-to-side working outward.
8. Install or connect the following:
- Ground wires to the back of the head. Torque the bolt to 37 ft. lbs. (50 Nm).
- Rocker arms, pedestals and push rods
- Exhaust manifold
- Rocker arm cover. Torque the bolts to 106 inch lbs. (12 Nm).
- Power steering pump bracket. Torque the bolts to 37 ft. lbs. (50 Nm).
- Power steering pump. Torque the bolts to 18 ft. lbs. (25 Nm).
- Power steering pump hoses and pulley
- Coolant air bleed pipe
- Intake manifold
- Negative battery cable

9. Refill the cooling system.
10. Change the oil and filter.
11. Start the engine and check for leaks.

RIGHT SIDE

1. Before servicing the vehicle, refer to the precautions in the beginning of this section.
2. Drain the cooling system.
3. Remove or disconnect the following:
- Negative battery cable
- Coolant air bleed pipe
- Intake manifold
- Exhaust manifold
- Rocker arm cover
- Rocker arms, pedestals and push rods
- Wiring clip bolt from the cylinder head
- Transmission dipstick tube bolt from the cylinder head
- Cylinder head bolts and cylinder head

To install:

4. Install the cylinder head with a new gasket.

➥Before installing the head bolts, be sure to mark the short bolts so the correct tightening specifications are used.

5. Lubricate the M11 bolts with clean engine oil, then install them.
6. Apply a 0.2 inch (5mm) bead of threadlock GM P/N 12345382 to the threads of the M8 cylinder head bolts.
7. Torque the cylinder head bolts in sequence as follows:
   a. M11 bolts 1–10: 22 ft. lbs. (30 Nm).

   b. M11 bolts 1–10: an additional 90 degrees.
   c. M11 bolts 1–8: an additional 90 degrees.
   d. M11 bolts 9–10: an additional 50 degrees.
   e. M8 bolts 11–15: 22 ft. lbs. (30 Nm). Begin at the center and alternate side-to-side working outward.
- Wiring clip bolt to the cylinder head
- Transmission dipstick tube bolt to the cylinder head
- Rocker arms, pedestals and push rods
- Exhaust manifold
- Rocker arm cover. Torque the bolts to 106 inch lbs. (12 Nm).
- Coolant air bleed pipe
- Intake manifold
- Negative battery cable

8. Refill the cooling system.
9. Change the oil and filter.
10. Start the engine and check for leaks.

Rocker Arms

REMOVAL & INSTALLATION

3.8L Engines

➥When removing valvetrain components, it is very important that they are marked for installation reference, so that they can be reinstalled in their original location.

1. Before servicing the vehicle, refer to the precautions in the beginning of this section.
2. Remove or disconnect the following:
- Negative battery cable
- Spark plug wires
- Rocker arm cover(s)
- Rocker arm pedestal bolts and assemblies
- Pushrods

To install:

3. Lubricate the pushrod tips and put them in their proper locations.
4. Lubricate the rocker arms and pedestals and install them. Be sure the pushrod tips are properly seated in the rocker arms.
5. Apply thread locking compound to the rocker arm pedestal bolt threads. Torque the bolts to 11 ft. lbs. (15 Nm) plus an additional 90 degree turn.
6. Apply suitable thread locking compound to the rocker arm cover bolts.
7. Install or connect the following:
- Rocker arm cover using a new gas-

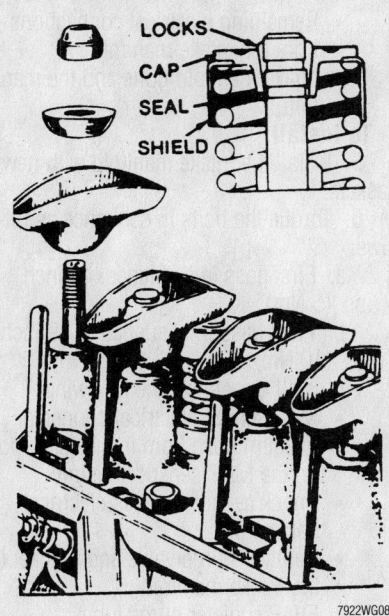

Exploded view of common rocker arm components and valve spring retainer

7922WG08

ket. Torque the bolts to 89 inch lbs. (10 Nm).
- Spark plug wires
- Negative battery cable

8. Run the engine and check for leaks and proper engine operation.

5.7L Engines

➡ Be sure to keep all the components in the exact order of removal so they may be installed in their original locations. Coat the replacement rocker arm and ball with engine oil before installation.

1. Before servicing the vehicle, refer to the precautions in the beginning of this section.
2. Remove or disconnect the following:
   - Negative battery cable
   - Rocker arm cover
   - Rocker arm bolts and rocker arms
   - Retainer
   - Push rods

To install:

3. Lubricate the bearing surfaces with a thin coating of Molykote® or equivalent.
4. Install or connect the following:
5. The push rods, retainer, rocker arms and bolts.
6. On 3.8L engines, tighten the bolts to 11 ft. lbs. (15 Nm) plus an additional 90 degrees.
7. Adjust the rocker arm bolts for the 5.7L engine as follows:
   a. Step 1: Rotate the crankshaft until number 1 piston is at Top Dead Center (TDC).

b. Step 2: Tighten exhaust rocker arms 1,2,7 and 8 to 22 ft. lbs. (30 Nm).
   c. Step 3: Tighten intake rocker arms 1,3,4 and 5 to 22 ft. lbs. (30 Nm).
   d. Step 4: Rotate the crankshaft 360 degrees.
   e. Step 5: Tighten exhaust rocker arms 3,4,5 and 6 to 22 ft. lbs. (30 Nm).
   f. Step 6: Tighten intake rocker arms 2,6,7 and 8 to 22 ft. lbs. (30 Nm).

8. Install the rocker arm cover and torque the bolts to 106 inch lbs. (12 Nm).
9. Install the negative battery cable.

Intake Manifold

REMOVAL & INSTALLATION

3.8L Engine

UPPER MANIFOLD

1. Before servicing the vehicle, refer to the precautions in the beginning of this section.
2. Relieve the fuel system pressure.
3. Drain the cooling system.
4. Remove or disconnect the following:
   - Negative battery cable
   - Intake Air Temperature (IAT) sensor electrical connector
   - Drive belt tensioner
   - Fuel pressure regulator vacuum and Evaporative Emission (EVAP) 1canister purge harness from the manifold vacuum source, purge solenoid valve and the fuel pressure regulator valve
   - Manifold vacuum source electrical connector
   - Manifold vacuum source screws
   - Manifold vacuum source from the upper intake manifold
   - Canister purge solenoid valve electrical connector and hose
   - Evaporative emissions canister purge solenoid valve
   - Idle Air Control (IAC) valve
   - Fuel lines from the fuel rail
   - EVAP pipe at the engine
   - Accelerator control cable bracket and control cables from the throttle body
   - Ignition control module and coil pack assembly
   - Brake booster hose from the upper manifold
   - Alternator brace
   - Throttle Position Sensor (TPS) electrical connector
   - Mass Air Flow (MAF) sensor electrical connector

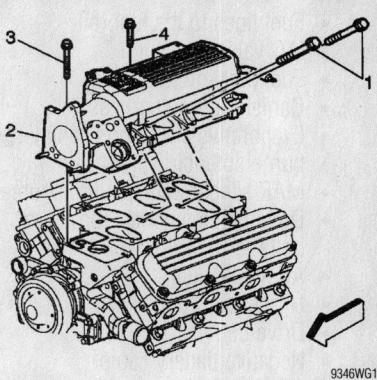

9346WG10

Upper intake manifold torque sequence—3.8L engine

   - Wiring harness from the fuel rail clips
   - Thermostat housing
   - Electrical connectors from the fuel injectors
   - Fuel rail
   - Upper intake manifold

To install:

5. Install the upper intake manifold using a new gasket. Torque the bolts as follows:
   a. Tighten the upper intake manifold bolts in sequence: 11 ft. lbs. (15 Nm).
   b. Tighten the nine upper intake manifold bolts to 89 inch lbs. (10 Nm).
   c. Tighten the two upper intake manifold bolts to 22 ft. lbs. (30 N).
6. Install the fuel rail as follows:
   a. Install new fuel injector O-rings lubricated with engine oil.
   b. Install the fuel injectors into the manifold bores.
   c. Insert the fuel rail until the injectors are properly seated.
   d. Torque the nuts to 75 inch lbs. (8.5 Nm).
7. Install or connect the following:
   - Fuel injector electrical connectors
   - Thermostat and thermostat housing. Torque the housing bolts to 20 ft. lbs. (27 Nm).
   - Throttle body and tighten the retainers to 89 inch lbs. (10 Nm)
   - Wiring harness to the fuel rail clips
   - Electrical connectors to the IAC, TPS, MAF and IAT sensors
   - Alternator brace
   - Brake booster hose
   - Ignition control module and coil pack assembly
   - Accelerator cable bracket and cruise control cables to the throttle body. Torque the bracket bolts to 12 ft. lbs. (16 Nm).
   - Accelerator and cruise control cables

- Fuel lines to the fuel rail
- IAC valve
- Fuel pressure regulator
- Canister purge harness
- Evaporative emissions canister purge solenoid valve
- MAP sensor and vacuum source
- Drive belt tensioner. Torque the bolts to 37 ft. lbs. (50 Nm).
- Air intake duct
- IAT sensor connector
- Drive belt
- Negative battery cable

8. Refill the coolant system.
9. Start the engine and check for leaks.

LOWER MANIFOLD

1. Before servicing the vehicle, refer to the precautions in the beginning of this section.

➡ **There are 2 bolts that are hidden under the upper manifold. They are located in the front and left rear corners of the lower manifold. It is necessary to remove the upper manifold to service the lower manifold.**

2. Remove or disconnect the following:
- Negative battery cable
- Upper intake manifold
- Engine Coolant Temperature (ECT) sensor
- Lower intake manifold

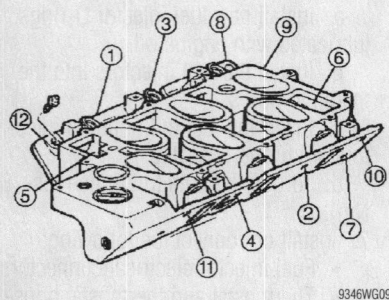

Lower intake manifold torque sequence—3.8L engine

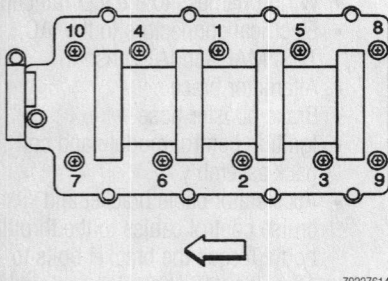

Intake manifold torque sequence—5.7L (VIN G) engine

To install:
3. Install or connect the following:
- New gaskets and seals
- Lower manifold. Torque the manifold bolts in sequence to 11 ft. lbs. (15 Nm).
- ECT sensor
- Upper intake manifold
- Negative battery cable

5.7L Engines

2000–01 MODELS

1. Before servicing the vehicle, refer to the precautions in the beginning of this section.
2. Relieve the fuel system pressure.
3. Drain the coolant system.
4. Remove or disconnect the following:
- Negative battery cable
- Transmission fluid level indicator tube bolt
- Fuel lines from the fuel rail
- Vapor line from the fuel vapor purge valve
- Intake Air Temperature (IAT) sensor electrical connector
- Air inlet duct
- Idle Air Control (IAC) valve and Throttle Position Sensor (TPS) electrical connections from the throttle body
- Drive belt
- Exhaust Gas Recirculation (EGR) solenoid electrical connector
- EGR valve pipe
- O-ring seal from the EGR valve pipe. Discard the exhaust manifold gasket and O-ring seal.
- Cruise control and throttle cables from the throttle body
- Cruise control servo bracket
- Fuel injector electrical connectors
- Secondary Air Injection (AIR) solenoid electrical connector
- Fuel rail
- Fresh air tube from the throttle body and rocker cover
- Vapor vent tube from the throttle body
- Throttle body heater outlet hose from the throttle body
- Evaporative emission (EVAP) canister purge tube
- Canister purge valve and bracket from the intake manifold
- Knock sensor harness electrical connector
- Manifold Absolute Pressure (MAP) sensor electrical connector
- Vacuum hose from the vacuum port on the MAP sensor housing

- Remaining electrical connections from the intake manifold
- Intake manifold bolts and the manifold

To install:
5. Install the intake manifold with new gaskets.
6. Torque the bolts in sequence as follows:
   a. First pass in sequence: 44 inch lbs. (5 Nm)
   b. Final pass in sequence to 89 inch lbs. (10 Nm).
7. Install or connect the following:
- MAP sensor electrical connector
- Vacuum hose from the vacuum port on the MAP sensor housing
- Knock sensor harness electrical connector
- Canister purge valve and bracket to the intake manifold
- EVAP canister purge tube
- Throttle body heater outlet hose to the throttle body
- Vapor vent tube to the throttle body
- Fresh air tube to the throttle body and rocker cover
- Fuel rail
- AIR solenoid electrical connector
- Fuel injector electrical connectors
- All remaining electrical connections
- Cruise control servo bracket
- Cruise control and throttle cables to the throttle body

8. Apply a light coating of clean engine oil to a NEW O-ring seal and install the seal onto the EGR valve pipe.
- EGR valve pipe into the intake manifold. Finger start the EGR valve pipe to the intake manifold bolt and the EGR valve pipe to the cylinder head bolts.

9. Finger start a NEW EGR valve pipe exhaust manifold gasket and bolts and tighten as follows:
   a. Tighten the EGR valve pipe-to-intake manifold bolt to 89 inch lbs. (10 Nm).
   b. Tighten the EGR valve pipe-to-cylinder head bolts to 37 ft. lbs. (50 Nm).
   c. Tighten the EGR valve pipe to the exhaust manifold bolts to 22 ft. lbs. (30 Nm).
- EGR solenoid electrical connector
- Drive belt
- IAC valve and TPS electrical connections to the throttle body
- Air inlet duct
- IAT and MAF sensor electrical connector
- Fuel lines to the fuel rail

- Vapor line to the fuel vapor purge valve
- Transmission fluid level indicator tube bolt
10. Refill the cooling system.
- Negative battery cable

2002 MODELS

1. Before servicing the vehicle, refer to the precautions in the beginning of this section.
2. Relieve the fuel system pressure.
3. Drain the coolant system.
4. Remove or disconnect the following:
- Negative battery cable
- Throttle body
- Fuel rail
- Transmission fluid level indicator tube bolt
- Accelerator control cable bracket nut
- Accelerator control cable and cruise control servo bracket
- Positive Crankcase Ventilation (PCV) tube from the right rocker arm cover and throttle body
- PCV valve pipe from the left rocker arm cover
- PCV valve pipe strap nut
- PCV valve pipe from the right rocker arm cover and intake manifold
- Vapor line from the Evaporative Emission (EVAP) canister purge solenoid valve
- EVAP canister purge solenoid valve connection
- EVAP canister purge tube from the intake manifold and EVAP canister purge solenoid valve
- EVAP canister purge solenoid valve (with bracket) from the intake manifold
- Knock Sensor (KS) harness connector
- KS wire harness connector from the fuel rail stop bracket
- Secondary Air Injection (AIR) vacuum control solenoid valve connection
- AIR shut off valve vacuum harness from the shut off valve and vacuum tee
- AIR vacuum control solenoid valve (with bracket) from the intake manifold
- Power brake booster vacuum hose from the vacuum port
- Manifold Absolute Pressure (MAP) sensor connector
- Vacuum hose from the tee on the rear vacuum port
- MAP sensor from the rear of the intake manifold
- Intake manifold bolts and fuel rail stop bracket
- Intake manifold
- Intake manifold-to-cylinder head gaskets and discard

To install:
- New intake manifold-to-cylinder head gaskets
- Intake manifold

5. Apply threadlock GM U.S. P/N 12345382 to the threads of the intake manifold bolts.

> ☀ CAUTION
>
> **The fuel rail stop bracket must be installed onto the engine assembly. The stop bracket serves as a protective shield for the fuel rail in the event of a vehicle frontal crash. If the fuel rail stop bracket is not installed and the vehicle is involved in a frontal crash, fuel could be sprayed possibly causing a fire and personal injury from burns.**

6. Install or connect the following:
- Fuel rail stop bracket
- Intake manifold bolts and tighten in a first pass in sequence to 44 inch lbs. (5 Nm) and a final pass in sequence to 89 inch lbs. (10 Nm)
- MAP sensor
- All vacuum hoses
- AIR vacuum control solenoid valve (with bracket) to the intake manifold
- AIR shut off valve vacuum harness to the AIR shut off valve and vacuum tee
- AIR vacuum control solenoid valve connection
- KS sensor wire harness connector to the fuel rail stop bracket
- KS sensor harness connection
- EVAP canister purge solenoid valve (with bracket) to the intake manifold
- Vapor line to the EVAP canister purge solenoid valve
- EVAP canister purge tube to the intake manifold and EVAP canister purge solenoid valve
- EVAP canister purge solenoid valve connection
- Vapor line to the EVAP canister purge solenoid valve
- PCV valve pipe to the right rocker arm cover and intake manifold
- PCV valve pipe strap nut
- PCV valve pipe to the left rocker arm cover and throttle body
- PCV tube to the right rocker arm cover
- Accelerator control cable and cruise control servo bracket
- Accelerator control cable bracket nut
- Transmission fluid level indicator tube bolt
- Fuel rail
- Throttle body
- Negative battery cable

Exhaust Manifold

REMOVAL & INSTALLATION

3.8L Engine

LEFT SIDE

1. Before servicing the vehicle, refer to the precautions in the beginning of this section.
2. Remove or disconnect the following:
- Negative battery cable
- Catalytic converter pipe
- Connector Position Assurance (CPA) retainer, if equipped
- Oxygen Sensor (O_2S) electrical connector
- O_2 sensor
- Spark plugs
- Exhaust Gas Recirculation (EGR) valve adapter from the exhaust manifold, if equipped
- 2 rear exhaust manifold studs and nut from the exhaust manifold
- Exhaust manifold heat shields
- Exhaust manifold and gasket

To install:
3. Install or connect the following:
- Exhaust manifold with a new gasket. Tighten the 2 front exhaust manifold studs and nuts to the exhaust manifold to 11 ft lbs. (15 Nm) and the nuts to 13 ft. lbs. (18 Nm).
- Exhaust manifold heat shield and nuts. Tighten the to 89 inch lbs. (10 Nm).
- Spark plugs and wires
- 2 rear exhaust manifold studs and nut to the exhaust manifold. Tighten the 2 rear exhaust manifold studs to 11 ft lbs. (15 Nm) and the nuts to 13 ft. lbs. (18 Nm).
- O_2 sensor
- Catalytic converter pipe
- CPA retainer, if equipped
- EGR valve adapter, if equipped and

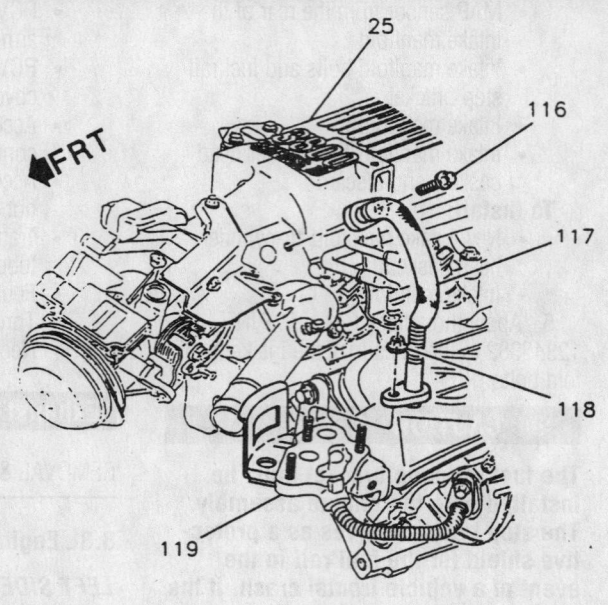

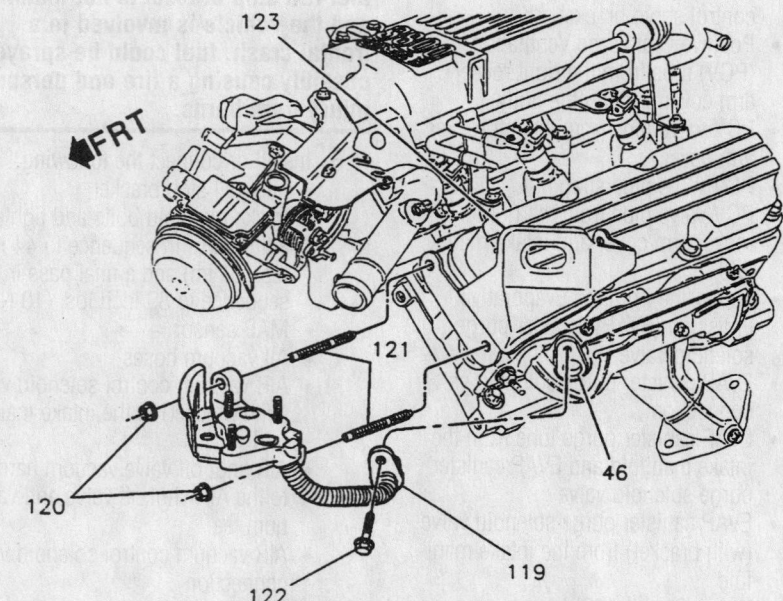

25   MANIFOLD, UPPER INTAKE
46   MANIFOLD, LEFT EXHAUST
116 BOLT/SCREW, EGR VALVE OUTLET PIPE
117 PIPE, EGR VALVE OUTLET
118 NUT, EGR VALVE OUTLET PIPE
119 ADAPTER, EGR VALVE
120 NUT, EGR VALVE ADAPTER
121 STUD, EGR VALVE ADAPTER
122 BOLT/SCREW, EGR VALVE ADAPTER
123 BRACKET, ENGINE LIFT

9300WG01

Exploded view of the EGR valve adapter and outlet pipe—3.8L engine

torque the bolt to 18 ft. lbs. (25 Nm)
- Negative battery cable

RIGHT SIDE

1. Before servicing the vehicle, refer to the precautions in the beginning of this section.
2. Remove or disconnect the following:
   - Negative battery cable
   - Connector Position Assurance (CPA) retainer, if equipped
   - Exhaust catalytic converter pipe
   - Oxygen Sensor (O2S) electrical connector
   - Spark plugs and wires

- 2 rear exhaust manifold studs and nut from the exhaust manifold
- Exhaust manifold heat shields
- Exhaust manifold and gasket

To install:

3. Install or connect the following:
   - Exhaust manifold with a new gasket. Tighten the 2 front exhaust manifold studs and nuts to the exhaust manifold to 11 ft lbs. (15 Nm) and the nuts to 13 ft. lbs. (18 Nm).
   - Exhaust manifold heat shield and nuts. Tighten the to 89 inch lbs. (10 Nm).
   - Spark plugs and wires

- 2 rear exhaust manifold studs and nut to the exhaust manifold. Tighten the 2 rear exhaust manifold studs to 11 ft lbs. (15 Nm) and the nuts to 13 ft. lbs. (18 Nm).
- O2 sensor
- Catalytic converter pipe
- CPA retainer, if equipped
- Negative battery cable

5.7L Engines

LEFT SIDE

1. Before servicing the vehicle, refer to the precautions in the beginning of this section.

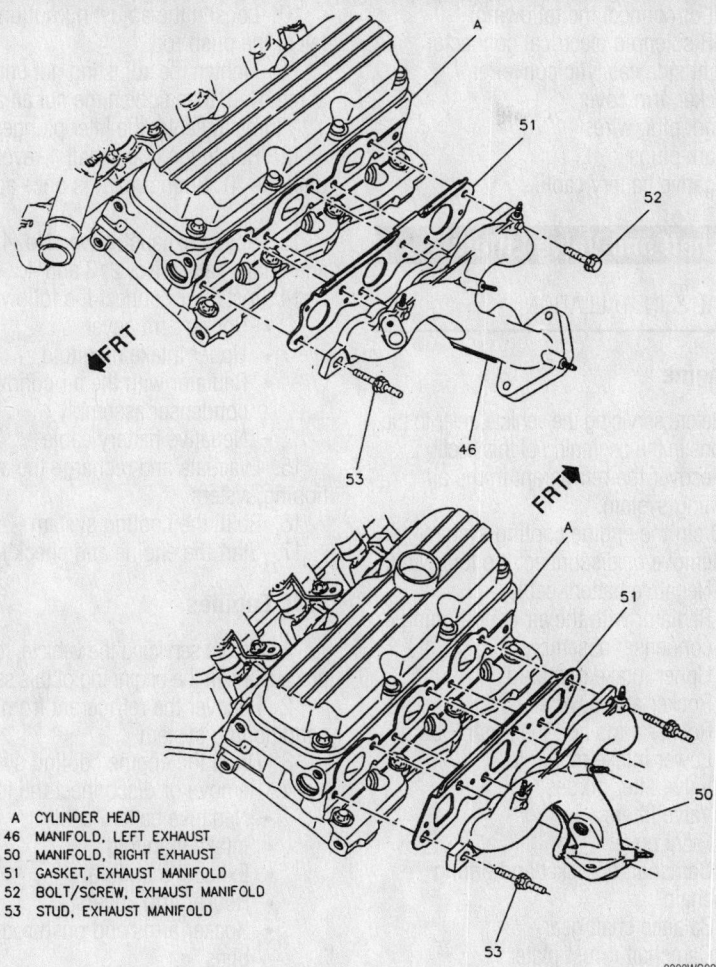

A CYLINDER HEAD
46 MANIFOLD, LEFT EXHAUST
50 MANIFOLD, RIGHT EXHAUST
51 GASKET, EXHAUST MANIFOLD
52 BOLT/SCREW, EXHAUST MANIFOLD
53 STUD, EXHAUST MANIFOLD

9300WG02

Exploded view of the right and left exhaust manifold mounting—3.8L engine

2. Remove or disconnect the following:
- Negative battery cable
- Rocker arm cover
- Spark plug wires
- Spark plugs
- Left side catalytic converter
- Exhaust manifold bolts
- Exhaust manifold and gasket

To install:

3. Apply a 0.2 inch (5mm) wide band of threadlock GM P/N 12345493 to the threads of the exhaust manifold bolts.

4. Install a NEW exhaust manifold gasket and the exhaust manifold.

5. Install the exhaust manifold bolts and tighten as follows:

a. Tighten the exhaust manifold bolts on first pass to 11 ft. lbs. (15 Nm). Tighten the manifold bolts beginning with the center 2 bolts. Alternate from side-to-side, and work toward the outside bolts.

b. Tighten the exhaust manifold bolts on final pass to 18 ft. lbs. (25 Nm). Tighten the exhaust manifold bolts beginning with the center 2 bolts. Alternate from side-to-side, and work toward the outside bolts.

6. Using a flat punch, bend over the exposed edge of the exhaust manifold gasket at the rear of the cylinder head.

Install or connect the following:
- Left side catalytic converter
- Rocker arm cover
- Spark plug wires
- Spark plugs
- Negative battery cable

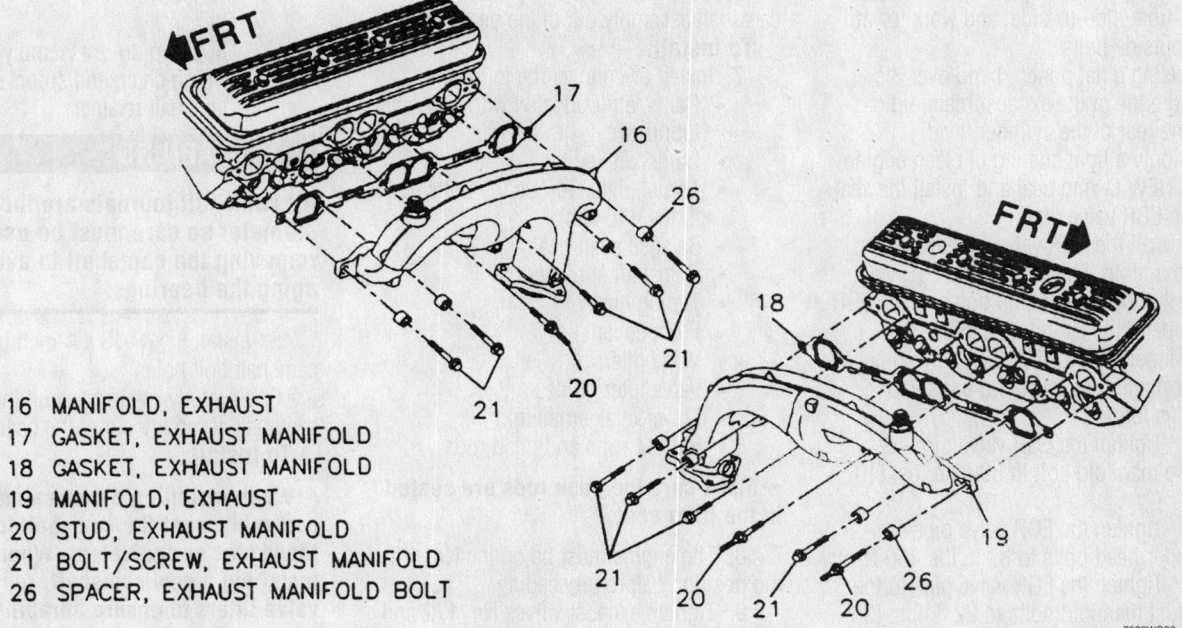

16 MANIFOLD, EXHAUST
17 GASKET, EXHAUST MANIFOLD
18 GASKET, EXHAUST MANIFOLD
19 MANIFOLD, EXHAUST
20 STUD, EXHAUST MANIFOLD
21 BOLT/SCREW, EXHAUST MANIFOLD
26 SPACER, EXHAUST MANIFOLD BOLT

7922WG36

Exploded view of right and left exhaust manifold mounting—5.7L engines

RIGHT SIDE

1. Before servicing the vehicle, refer to the precautions in the beginning of this section.

2. Remove or disconnect the following:
- Negative battery cable
- Rocker arm cover
- Spark plug wires
- Spark plugs
- Right side catalytic converter
- Exhaust Gas Recirculation (EGR) valve electrical connector
- EGR valve pipe bolts
- EGR valve pipe up out of the intake manifold, using mild force
- O-ring seal from the EGR valve pipe. Discard the exhaust manifold gasket and O-ring seal.
- Exhaust manifold bolts
- Exhaust manifold and gasket

To install:

3. Apply a 0.2 inch (5mm) wide band of threadlock GM P/N 12345493 to the threads of the exhaust manifold bolts.

4. Install a NEW exhaust manifold gasket and the exhaust manifold.

5. Install the exhaust manifold bolts and tighten as follows:

    a. Tighten the exhaust manifold bolts on first pass to 11 ft. lbs. (15 Nm). Tighten the manifold bolts beginning with the center 2 bolts. Alternate from side-to-side, and work toward the outside bolts.

    b. Tighten the exhaust manifold bolts on final pass to 18 ft. lbs. (25 Nm). Tighten the exhaust manifold bolts beginning with the center 2 bolts. Alternate from side-to-side, and work toward the outside bolts.

6. Using a flat punch, bend over the exposed edge of the exhaust manifold gasket at the rear of the cylinder head.

7. Apply a light coating of clean engine oil to a NEW O-ring seal and install the seal onto the EGR valve pipe.

8. Install the EGR valve pipe into the intake manifold. Finger start the EGR valve pipe to the intake manifold bolt and the EGR valve pipe to the cylinder head bolts.

9. Finger start a NEW EGR valve pipe exhaust manifold gasket and bolts and tighten as follows:

    a. Tighten the EGR valve pipe-to-intake manifold bolt to 89 inch lbs. (10 Nm).

    b. Tighten the EGR valve pipe-to-cylinder head bolts to 37 ft. lbs. (50 Nm).

    c. Tighten the EGR valve pipe to the exhaust manifold bolts to 22 ft. lbs. (30 Nm).

Install or connect the following:
- EGR solenoid electrical connector
- Right side catalytic converter
- Rocker arm cover
- Spark plug wires
- Spark plugs
- Negative battery cable

Camshaft and Valve Lifters

REMOVAL & INSTALLATION

3.8L Engine

1. Before servicing the vehicle, refer to the precautions in the beginning of this section.

2. Recover the refrigerant in the air conditioning system.

3. Drain the engine cooling system.

4. Remove or disconnect the following:
- Negative battery cable
- Radiator with the air conditioning condenser assembly
- Upper intake manifold
- Rocker arm covers
- Rocker arms and the push rods
- Lower intake manifold
- Valve lifter guides
- Valve lifters
- Front cover
- Camshaft sprocket and timing chain
- Balance shaft gear
- Camshaft thrust plate
- Camshaft key

5. Install 3 $\frac{5}{16}$–18 x 4-inch bolts in the camshaft bolt holes.

6. Carefully rotate and pull the camshaft assembly out of the bearings.

To install:

7. Install or connect the following:
- Camshaft lubricated with assembly lubricant
- Camshaft key
- Thrust plate. Torque the bolts to 11 ft. lbs. (15 Nm).
- Balance shaft gear
- Camshaft sprocket
- Timing chain
- Front cover
- Valve lifters
- Valve lifter guides
- Lower intake manifold
- Rocker arms and push rods

➡ **Make sure the push rods are seated in the lifter seat.**

8. The engine must be on the No. 1 firing position before proceeding.

9. Tighten exhaust valves No. 1, 2 and 3; then intake valves No. 1, 5 and 6.

10. Loosen the adjusting nut until lash is felt at the push rod.

11. Tighten the adjusting nut until all lash is removed, then tighten the nut an additional 1 ½ turns to center the lifter plunger.

12. Rotate the crankshaft 1 revolution until the "0" timing mark is once again aligned.

13. Adjust exhaust valves No. 4, 5 and 6 and intake valves No. 2, 3 and 4.

14. Install or connect the following:
- Rocker arm cover
- Upper intake manifold
- Radiator with the air conditioning condenser assembly
- Negative battery cable

15. Evacuate and recharge the air conditioning system.

16. Refill the cooling system.

17. Start the engine and check for leaks.

5.7L Engines

1. Before servicing the vehicle, refer to the precautions in the beginning of this section.

2. Recover the refrigerant from the air conditioning system.

3. Drain the engine cooling system.

4. Remove or disconnect the following:
- Negative battery cable
- Intake manifold
- Exhaust manifold
- Rocker arm covers
- Rocker arms and push rod assemblies
- Cylinder head
- Valve lifters
- Radiator with the condenser
- Front cover
- Oil pan
- Oil pump drive assembly
- Timing chain and sprocket
- Camshaft retainer

✳✳ WARNING

All camshaft journals are the same diameter so care must be used when removing the camshaft to avoid damaging the bearings.

5. Install 3 $\frac{5}{16}$–18 x 4-inch bolts in the camshaft bolt holes.

6. Carefully rotate and pull the camshaft assembly out of the bearings.

To install:

➡ **When installing a new camshaft assembly, coat the camshaft lobes with Molykote® or equivalent. When installing a new camshaft, replace all valve lifters to ensure durability of the camshaft lobes and lifters.**

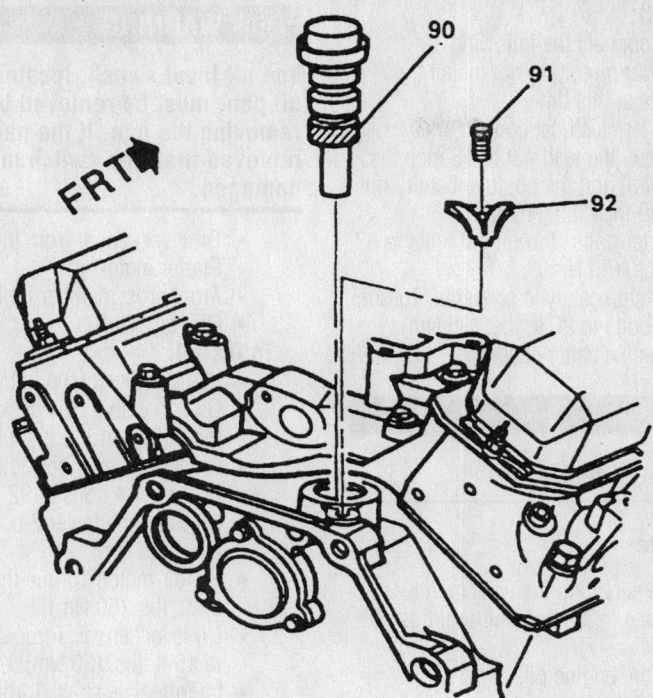

90 DRIVE ASSEMBLY, OIL PUMP
91 BOLT/SCREW, OIL PUMP DRIVE CLAMP
92 CLAMP, OIL PUMP DRIVE

7922WG12

Unfasten the retaining bolt and clamp, then remove the oil pump drive—5.7L engines

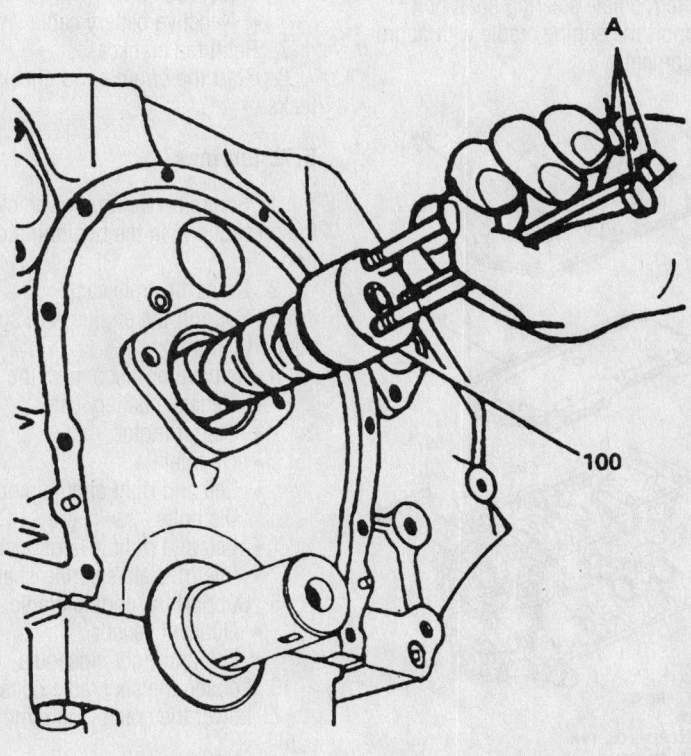

A BOLTS
100 CAMSHAFT ASSEMBLY

7922WG11

Thread 3 bolts into the camshaft to facilitate removal and installation—5.7L engines

7. Install or connect the following:
   - Camshaft
   - Camshaft retainer. Torque the bolts to 18 ft. lbs. (25 Nm).
   - Timing chain and camshaft sprocket
   - Oil pump drive assembly. Torque the bolt to 25 ft. lbs. (34 Nm).
   - Oil pan
   - Front cover
   - Radiator with the air conditioning condenser
   - Air conditioning receiver and dehydrator hose to the condenser
   - Air conditioning compressor and condenser hose to the condenser
   - Valve lifters
   - Cylinder head with a new gasket
   - Push rods and rocker arms

➡**Make sure the push rods are seated in the lifter seat.**

8. Place the engine on Top Dead Center (TDC) of the No. 1 firing position.
9. Tighten the exhaust valves No. 1, 2, 7 and 8; then intake valves No. 1, 3, 4 and 5 to 22 ft. lbs. (30 Nm).
10. Rotate the crankshaft 1 revolution until the "0" timing mark is once again aligned.
11. Adjust exhaust valves No. 3, 4, 5 and 6; then intake valves 2, 6, 7 and 8 to 22 ft. lbs. (30 Nm).
12. Install or connect the following:
    - Exhaust manifold
    - Rocker arm covers
    - Intake manifold with a new gasket
    - Negative battery cable
13. Recharge the air conditioning system.
14. Fill the engine cooling system.

Valve Lash

ADJUSTMENT

➡**The engines in this section utilize hydraulic valve lifters, therefore no adjustment is necessary.**

Starter Motor

REMOVAL & INSTALLATION

3.8L Engine

1. Before servicing the vehicle, refer to the precautions in the beginning of this section.
2. Remove or disconnect the following:
   - Negative battery cable

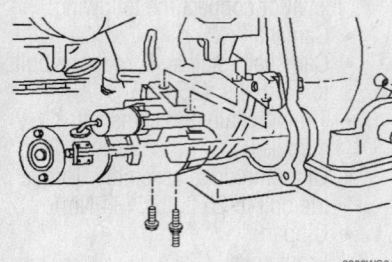

View of the starter—3.8L engine

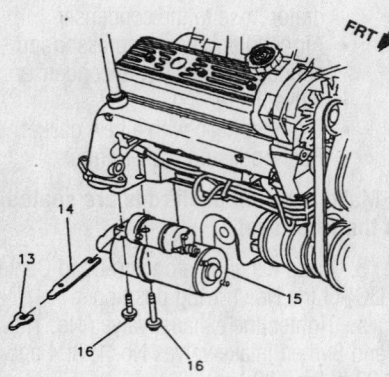

13 SHIM, STARTER MOTOR SINGLE
14 SHIM, STARTER MOTOR DOUBLE
15 MOTOR, STARTER
16 BOLT/SCREW, STARTER MOTOR
21 STUD, STARTER MOTOR

View of the starter—5.7L engine

- Starter shield
- Starter bolt and nut
- Starter electrical connectors
- Starter motor

To install:

Install or connect the following:

- Starter motor, do not install the bolts at this time
- Starter electrical connectors. Torque the lead nut to 17 inch lbs. (2 Nm) and the positive battery nut to 89 inch lbs. (10 Nm).
- Starter bolt and nut. Torque the starter motor bolt to 35 ft. lbs. (47 Nm) and the nut to 33 ft. lbs. (45 Nm).
- Starter shield
- Negative battery cable

5.7L Engine

1. Before servicing the vehicle, refer to the precautions in the beginning of this section.
2. Remove or disconnect the following:
   - Negative battery cable
   - Left side catalytic converter
   - Starter bolts
   - Starter electrical connectors
   - Starter motor

To install:
Install or connect the following:

- Starter motor, do not install the bolts at this time
- Starter electrical connectors. Torque the lead nut to 18 inch lbs. (2 Nm) and the positive battery nut to 89 inch lbs. (10 Nm).
- Starter bolts. Torque the bolts to 37 ft. lbs. (50 Nm).
- Left side catalytic converter. Torque the bolts to 26 ft. lbs. (35 Nm).
- Negative battery cable

Oil Pan

REMOVAL & INSTALLATION

3.8L Engine

1. Before servicing the vehicle, refer to the precautions in the beginning of this section.
2. Drain the engine oil.
3. Support the engine with suitable support fixture.
4. Remove or disconnect the following:
   - Negative battery cable
   - Right and left engine mount-to-cradle bolts
   - Right and left lower shock bolts
   - Intermediate steering shaft bolt
5. Support the engine cradle with appropriate equipment.

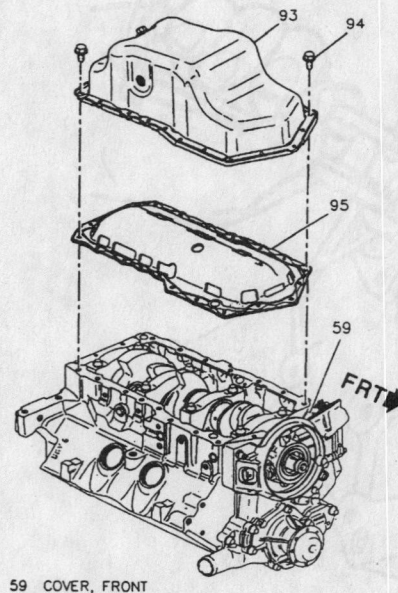

59 COVER, FRONT
93 PAN, OIL
94 BOLT/SCREW, OIL PAN
95 GASKET, OIL PAN

Exploded view of the oil pan and gasket—3.8L engine

※※ **WARNING**

The oil level switch, located in the oil pan, must be removed before removing the pan. If the pan is removed first, the switch may be damaged.

- Oil level sensor from the oil pan
- Starter motor
- Front crossmember bolts
- Oil pan

To install:
6. Install or connect the following:
   - Oil pan with a new gasket. Torque the bolts to 10 ft. lbs. (14 Nm).
   - Front crossmember bolts. Torque the upper 4 bolts to 92 ft. lbs. (125 Nm) and the lower 2 bolts to 107 ft. lbs. (145 Nm).
   - Starter motor. Torque the bolts to 37 ft. lbs. (50 Nm).
   - Oil level sensor. Torque the sensor to 15 ft. lbs. (20 Nm).
   - Intermediate shaft. Torque the bolt to 35 ft. lbs. (47 Nm).
   - Right and left lower shock bolts. Torque the bolts to 48 ft. lbs. (65 Nm).
   - Right and left engine mount to cradle bolts. Torque the bolts to 43 ft. lbs. (58 Nm).
   - Negative battery cable
7. Refill the crankcase.
8. Start the engine and check for leaks.

5.7L Engines

1. Before servicing the vehicle, refer to the precautions in the beginning of this section.
2. Drain the crankcase.
3. Support the engine with suitable support fixture.
4. Remove or disconnect the following:
   - Negative battery cable
   - Starter motor
   - Oil filter
   - Left and right engine mount-to-cradle bolts
   - Left and right lower shock bolts
   - Intermediate steering shaft bolt
5. Support the engine cradle.
   - Oil level sensor
   - Left and right closeouts
6. Loosen the six cradle bolts.
7. Lower the cradle and remove the oil pan.
8. Drill out the oil pan gasket rivets and remove the gasket from the oil pan.
9. Discard the old gasket and rivets.

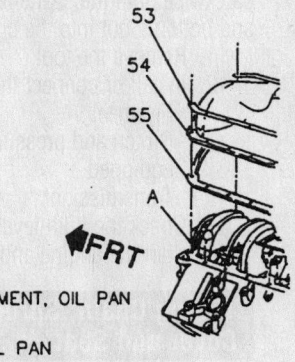

A   SEALER
53   REINFORCEMENT, OIL PAN
54   PAN, OIL
55   GASKET, OIL PAN

Oil pan sealer and gasket location—5.7L engines

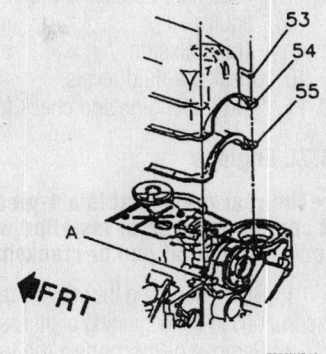

7922WG13

To install:

10. Apply a 0.2 inch (5mm) bead of sealant RTV sealer 0.8 inch (20mm) long to the engine block. Apply the sealant directly onto the tabs of the front cover gasket that protrudes into the oil pan surface

11. Pre-assemble the oil pan gasket to the pan.

12. Install or connect the following:
- Gasket onto the pan
- Oil pan bolts to the pan through the gasket
- Oil pan, gasket and bolts to the engine block. Snug the oil pan bolts finger-tight. Do not over-tighten.
- 2 lower bellhousing bolts to position the oil pan correctly. Snug the lower bellhousing bolt finger-tight. Do not overtighten.

13. Tighten the bolts as follows:
 a. Tighten the oil pan-to-block and oil pan-to-oil pan front cover bolts to 18 ft. lbs. (25 Nm).
 b. Tighten the oil pan-to-rear cover bolts to 106 inch lbs. (12 Nm).
 c. Tighten the bellhousing bolts to 37 ft. lbs. (50 Nm) for models with a manual transmission or 70 ft. lbs. (95 Nm) for models with an automatic transmission.

14. Install or connect the following:
- Engine cradle. Torque the upper four bolts to 92 ft. lbs. (125 Nm) and the lower two bolts to 107 ft. lbs. (145 Nm).
- Side closeouts and bolts. Tighten the bolts to 106 inch lbs. (12 Nm).
- Starter motor. Torque the bolts to 37 ft. lbs. (50 Nm).
- Oil level sensor. Torque it to 26 ft. lbs. (35 Nm).
- Intermediate steering shaft bolt. Torque the bolt to 35 ft. lbs. (47 Nm).
- Left and right lower shock bolts.

Torque the bolts to 48 ft. lbs. (65 Nm).
- Left and right engine mount-to-cradle bolts. Torque the bolts to 43 ft. lbs. (58 Nm).
- Oil filter
- Negative battery cable

15. Refill the crankcase.

16. Start the engine and check for leaks.

Oil Pump

REMOVAL & INSTALLATION

3.8L Engine

1. Before servicing the vehicle, refer to the precautions in the beginning of this section.

2. Remove or disconnect the following:
- Negative battery cable
- Front cover
- Oil filter adapter, pressure valve and spring

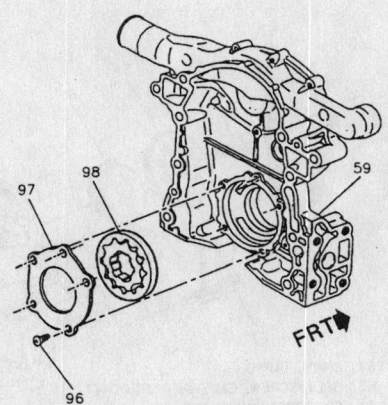

59   COVER, FRONT
96   BOLT/SCREW, OIL PUMP COVER
97   COVER, OIL PUMP
98   GEAR SET, OIL PUMP

7922WG16

Exploded view of the oil pump gear set—3.8L engine

- Oil pump cover
- Oil pump gear set

To install:

3. Lubricate the gear set with petroleum jelly.

4. Install the oil pump gear set into the front cover.

5. Pack the gear cavity with petroleum jelly for proper priming.

6. Install or connect the following:
- Oil pump cover. Torque the screws to 98 inch lbs. (11 Nm).
- Oil filter adapter with a new gasket, pressure valve and spring. Torque the bolts to 11 ft. lbs. (15 Nm).
- Front cover. Torque the bolts to 11 ft. lbs. (15 Nm).
- Negative battery cable

5.7L Engines

1. Before servicing the vehicle, refer to the precautions in the beginning of this section.

2. Drain the crankcase.

3. Remove or disconnect the following:
- Front cover
- Oil pan
- Oil pump screen
- Oil pan deflector
- Oil pump and propeller shaft

To install:

4. Install or connect the following:
- Oil pump and propeller shaft.

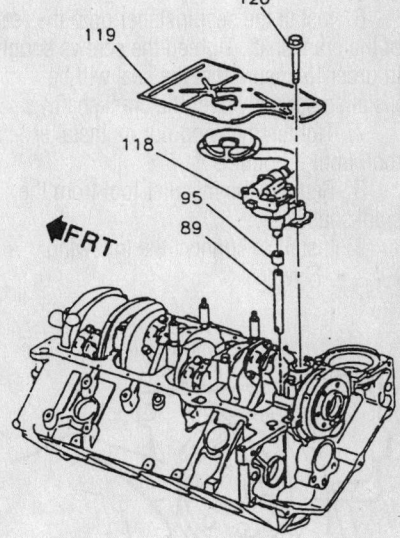

89   DRIVESHAFT, OIL PUMP
95   RETAINER, OIL PUMP DRIVESHAFT
118   PUMP, OIL
119   DEFLECTOR, CRANKSHAFT OIL
120   BOLT/SCREW, OIL PUMP

7922WG17

Exploded view of the oil pump and pro-peller shaft mounting—5.7L engine

Torque the bolts to 18 ft. lbs. (25 Nm).
- Oil pan deflector. Torque the nuts to 18 ft. lbs. (25 Nm).
- Oil pump screen. Torque the bolt to 106 inch lbs. (12 Nm).
- Oil pan
- Front cover. Torque the bolts to 18 ft. lbs. (25 Nm).
5. Refill the crankcase.

Rear Main Seal

REMOVAL & INSTALLATION

3.8L Engines

➡The rear main seal is a 1-piece unit. It can be removed or installed without removing the oil pan or crankshaft.

1. Before servicing the vehicle, refer to the precautions in the beginning of this section.
2. Remove or disconnect the following:
   - Transmission
   - Clutch and pressure plate, if equipped
   - Flywheel
   - Rear main seal
3. Inspect the crankshaft for nicks or burrs and correct as required.

To install:
4. Coat the new seal entirely with engine oil.
5. Install the seal onto Crankshaft Rear Oil Seal Installer J 38196.
6. Install the seal installer onto the rear of the crankshaft. Tighten the screws snugly in order to ensure that the seal will be installed squarely over the crankshaft.
7. Tighten the wing nut on installer tool, until it bottoms.
8. Remove the installer tool from the crankshaft.
9. Install or connect the following:
   - Flywheel

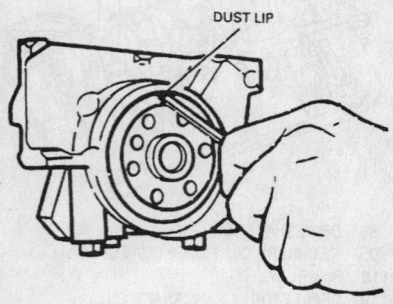

DUST LIP

7922WG18

When prying the rear main seal out, be careful not to damage the crankshaft sealing surface or the seal bore

- Clutch and pressure plate, if equipped
- Transmission
10. Check the fluid levels.
11. Start the engine and check for leaks.

5.7L Engines

➡The rear main seal is a 1-piece unit. It can be removed or installed without removing the oil pan or crankshaft.

1. Before servicing the vehicle, refer to the precautions in the beginning of this section.
2. Remove or disconnect the following:
   - Transmission
   - Clutch and pressure plate, if equipped
   - Flywheel
   - Rear main seal
3. Inspect the crankshaft for nicks or burrs and correct as required.

To install:
4. Lubricate the outside diameter of the oil seal with clean engine oil. Do not allow oil or other lubricants to contact the seal surface.
5. Lubricate the rear cover oil seal bore with clean engine oil. DO NOT allow oil or other lubricants to contact the crankshaft surface.
6. Install Crankshaft Rear Oil Seal Installer J 41479 cone and bolts onto the rear of the crankshaft. Tighten the bolts until snug. Do not overtighten.
7. Install the rear oil seal onto the tapered cone and push the seal to the rear cover bore.
8. Thread the installer tool threaded rod into the tapered cone until the tool contacts the oil seal.
9. Align the oil seal onto the tool.
10. Rotate the tool handle of the tool

clockwise until the seal enters the rear cover and bottoms out into the cover bore.
11. Remove the tool
12. Install or connect the following:
   - Flywheel
   - Clutch and pressure plate, if equipped
   - Transmission
13. Check the fluid levels.
14. Start the engine and check for leaks.

Timing Chain, Sprockets, Front Cover and Seal

REMOVAL & INSTALLATION

3.8L Engine

1. Before servicing the vehicle, refer to the precautions in the beginning of this section.
2. Drain the crankcase and cooling system.
3. Remove or disconnect the following:
   - Negative battery cable
   - Air cleaner and intake air duct
   - Coolant pump pulley bolts, loosen
   - Drive belt tensioner
   - Power steering pump pulley, if necessary
   - Inlet hose from the power steering pump, if necessary
4. Reposition the hose clamp on the return line, if necessary.
   - Reservoir hose return line from the power steering pump, if necessary
   - Power steering pump, if necessary
   - Crankshaft balancer
   - Crankshaft Position (CKP) sensor shield and the sensor

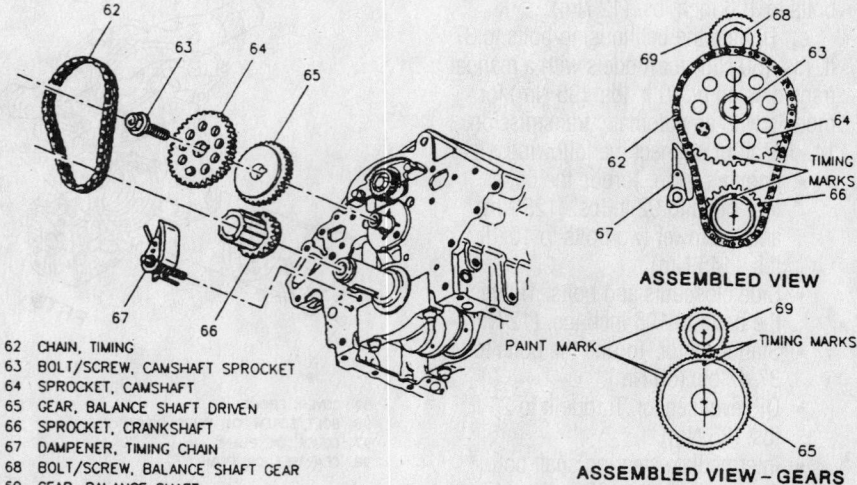

62 CHAIN, TIMING
63 BOLT/SCREW, CAMSHAFT SPROCKET
64 SPROCKET, CAMSHAFT
65 GEAR, BALANCE SHAFT DRIVEN
66 SPROCKET, CRANKSHAFT
67 DAMPENER, TIMING CHAIN
68 BOLT/SCREW, BALANCE SHAFT GEAR
69 GEAR, BALANCE SHAFT

ASSEMBLED VIEW

TIMING MARKS

PAINT MARK

TIMING MARKS

ASSEMBLED VIEW – GEARS

7922WG20

Exploded view of the timing chain and sprockets and timing mark alignment—3.8L engine

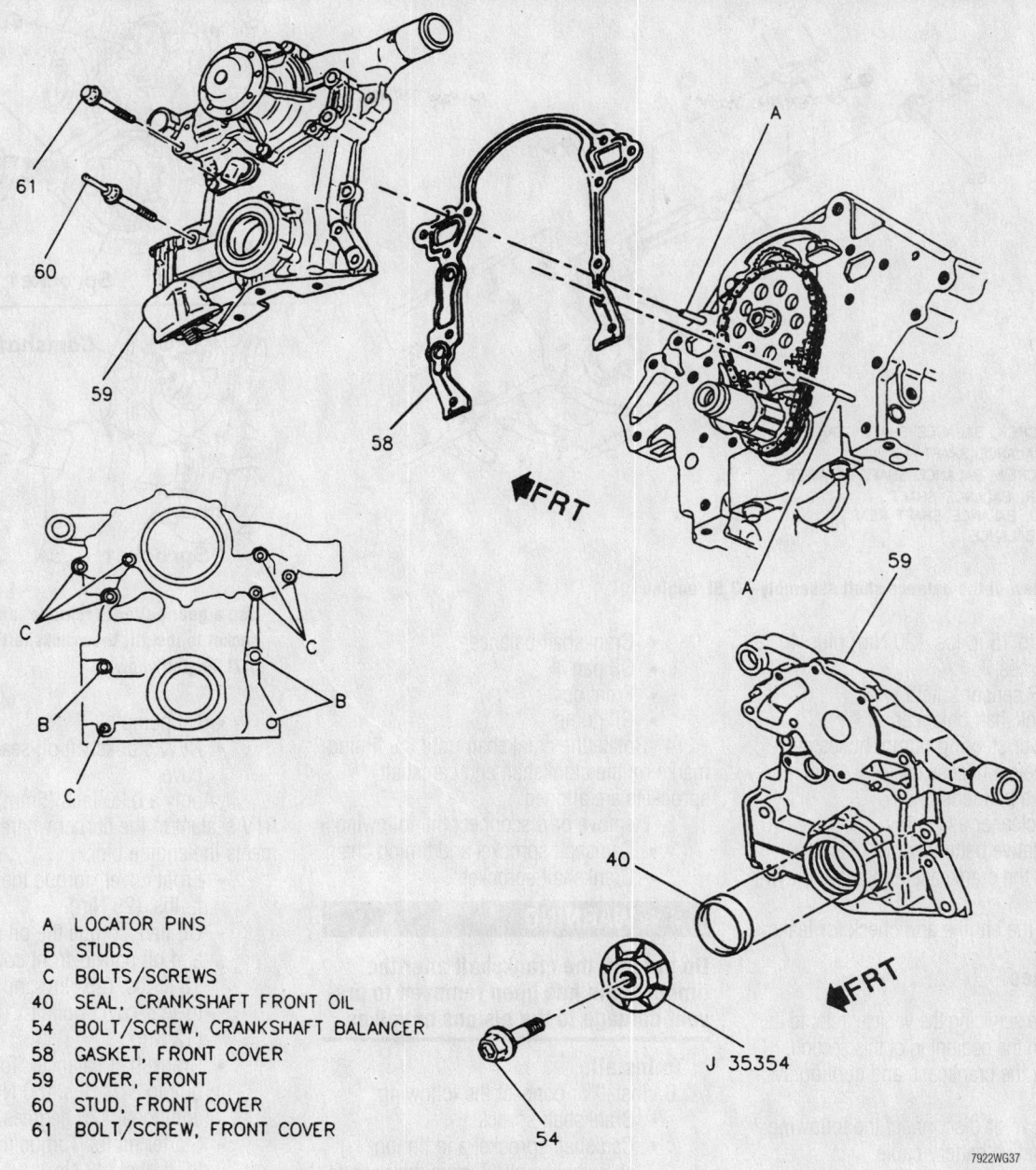

A   LOCATOR PINS
B   STUDS
C   BOLTS/SCREWS
40   SEAL, CRANKSHAFT FRONT OIL
54   BOLT/SCREW, CRANKSHAFT BALANCER
58   GASKET, FRONT COVER
59   COVER, FRONT
60   STUD, FRONT COVER
61   BOLT/SCREW, FRONT COVER

7922WG37

Exploded view of the front cover mounting and bolt locations—3.8L engine

- Oil pan-to-front cover bolts
- Oil pan bolts, loosen and lower the pan slightly
- Water pump
- Radiator outlet hose from the front cover
- Front cover bolts and studs
- Cover and gasket
- Crankshaft seal from the cover

5. Align the timing marks on the sprockets so they are as close together as possible.

- Timing chain damper
- Camshaft sprocket and timing chain
- Crankshaft sprocket

To install:

6. If the crankshaft has been turned in the engine, perform the following:

    a. Turn the crankshaft so the No. 1 piston is at top Dead Center (TDC) of its compression stroke.

    b. Turn the camshaft so that, with the sprocket temporarily installed, the timing mark is straight down.

7. Install or connect the following:

- Timing chain on the sprockets with the timing marks aligned
- Timing chain and sprockets. Torque the bolt to 74 ft. lbs. (100 Nm) plus an additional 90 degrees.

- Timing chain damper. Torque the bolt to 16 ft. lbs. (22 Nm).

8. Rotate the engine 2 revolutions, then check to be sure the timing marks are aligned.

- Front cover with a new crankshaft oil seal. Torque the fasteners to 15 ft. lbs. (20 Nm) plus 40 degrees.
- Radiator outlet hose to the front cover
- Water pump
- Oil pan-to-front cover bolts. Torque the bolts to 125 inch lbs. (14 Nm).
- Oil pan bolts
- CKP sensor and torque the fasten-

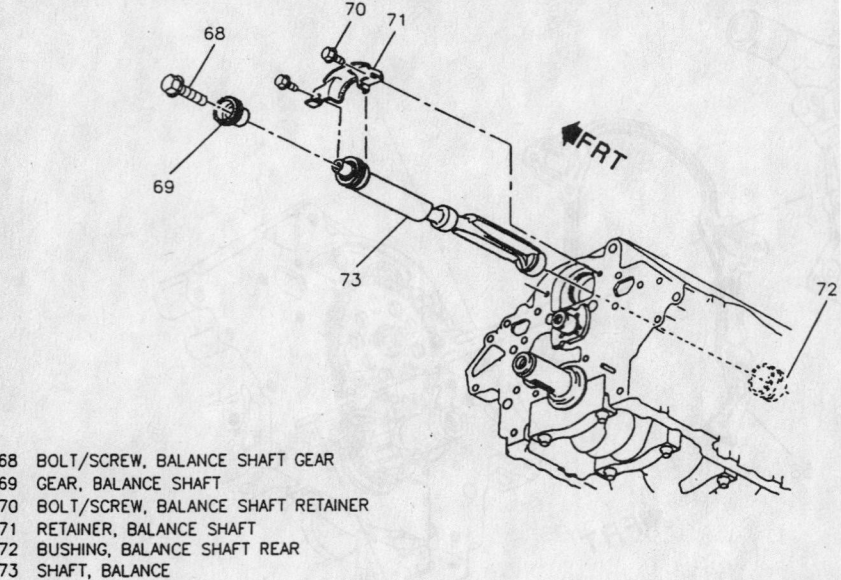

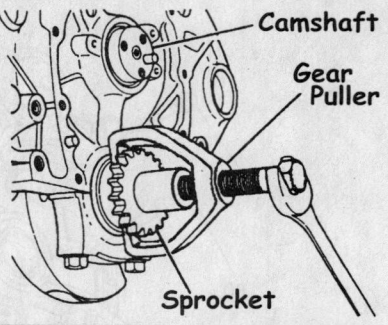

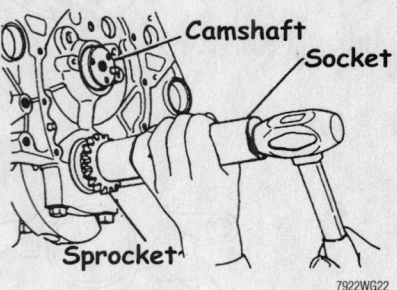

68 BOLT/SCREW, BALANCE SHAFT GEAR
69 GEAR, BALANCE SHAFT
70 BOLT/SCREW, BALANCE SHAFT RETAINER
71 RETAINER, BALANCE SHAFT
72 BUSHING, BALANCE SHAFT REAR
73 SHAFT, BALANCE

Exploded view of the balance shaft assembly—3.8L engine

Use a gear puller to remove, and a large socket to install, the crankshaft sprocket—5.7L engine shown

ers to 15 ft. lbs. (20 Nm) plus 40 degrees
• CKP sensor shield
• Crankshaft balancer
• Power steering pump, hoses and pulley, if necessary
• Drive belt tensioner
• Air cleaner assembly
• Negative battery cable

9. Refill the crankcase and cooling system.
10. Start the engine and check for leaks.

5.7L Engines

1. Before servicing the vehicle, refer to the precautions in the beginning of this section.
2. Drain the crankcase and cooling system.
3. Remove or disconnect the following:
• Negative battery cable
• Mass Air flow (MAF) sensor electrical connector
• Intake Air Temperature (IAT) sensor electrical connector
• Air intake duct
• Drive belt
• Radiator hoses from the water pump and water outlet
• Upper radiator support
• Fan shroud
• Drive belt tensioner
• Overflow hose from the radiator
• Throttle body heater hose from the radiator
• Drive belt idler pulley
• Water pump
• Starter motor

• Crankshaft balancer
• Oil pan
• Front cover
• Oil pump

4. Rotate the crankshaft until the timing marks on the crankshaft and camshaft sprockets are aligned.
5. Remove or disconnect the following:
• Camshaft sprocket and timing chain
• Crankshaft sprocket

✷✷ WARNING

Do not turn the crankshaft after the timing chain has been removed to prevent damage to the pistons or valves.

To install:
6. Install or connect the following:
• Crankshaft sprocket
• Camshaft sprocket and timing chain assembly. Torque the bolts to 26 ft. lbs. (35 Nm).

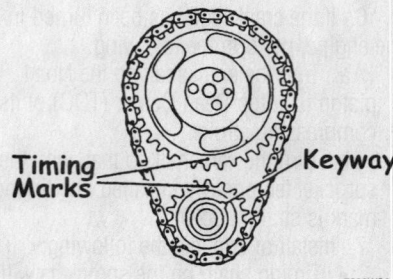

When assembled, be sure the marks on the gears are facing each other—5.7L engine

• Oil pump
• NEW crankshaft oil seal to the front cover

7. Apply a 0.20 inch (5mm) bead of RTV sealant to the corner where the oil pan meets the engine block.
• Front cover. Torque the bolts to 18 ft. lbs. (25 Nm).
• Oil pan. Torque the oil pan-to-block and oil pan-to-front cover bolts to 18 ft. lbs. (25 Nm) and the oil pan-to-rear cover bolts to 106 inch lbs. (12 Nm).
• Crankshaft balancer. Torque the bolt to 37 ft. lbs. (50 Nm), then an additional 120 degrees.
• Starter motor. Torque the bolts to 35 ft. lbs. (47 Nm).
• Water pump. Torque the bolts to 105 inch lbs. (12 Nm).

✷✷ WARNING

The camshaft sprocket and water pump propeller shaft gears must mesh or damage to the camshaft retainer may occur.

• Drive belt idler pulley. Torque the bolt to 37 ft. lbs. (50 Nm).
• Throttle body heater hose to the radiator
• Overflow hose to the radiator
• Drive belt tensioner. Torque the bolts to 37 ft. lbs. (50 Nm).

- Fan shroud
- Upper radiator support
- Radiator hoses to the water pump and water outlet
- Drive belt
- Air intake duct
- MAF sensor electrical connector
- IAT sensor electrical connector
- Negative battery cable

8. Refill the engine crankcase.
9. Refill the cooling system.
10. Operate the engine and check for leaks.

Piston and Ring

POSITIONING

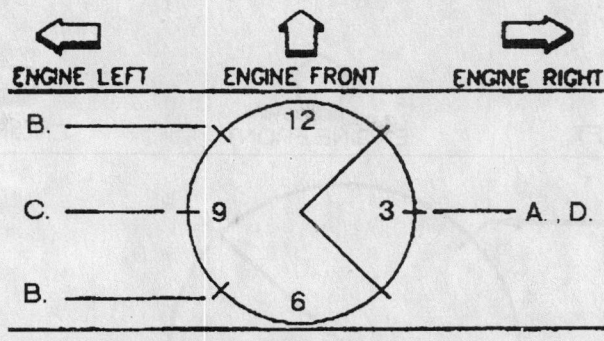

A. OIL RING SPACER GAP
(TANG IN HOLE OR SLOT WITH ARC)
B. OIL RING RAIL GAPS
C. 2ND COMPRESSION RING GAP
D. TOP COMPRESSION RING GAP

Piston ring end-gap spacing—3.8L engine

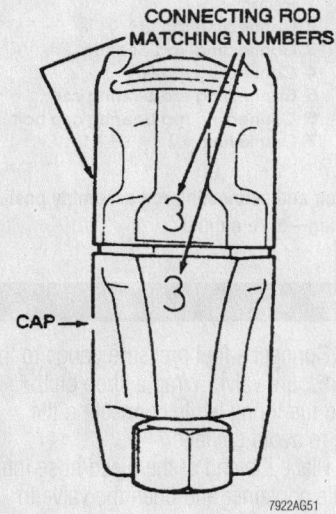

Engine connecting rod and cap installation. Be sure to matchmark the cap and rod prior to disassembly, as shown

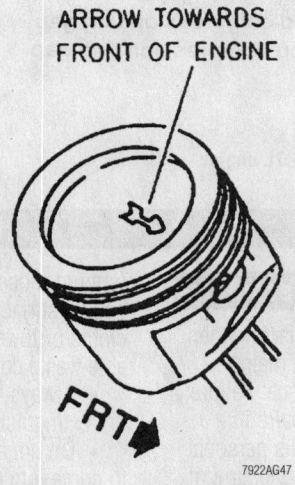

ARROW TOWARDS FRONT OF ENGINE

Piston positioning. Often the arrow is replaced by a notch, which also must face the front of the engine—3.8L engine

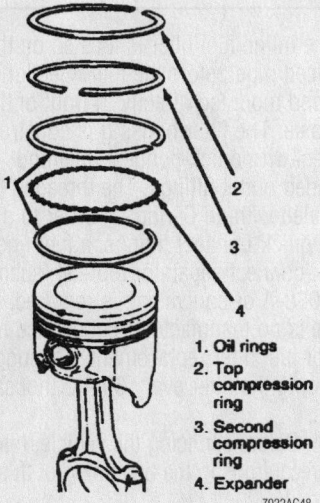

1. Oil rings
2. Top compression ring
3. Second compression ring
4. Expander

Piston ring positioning—3.8L engine

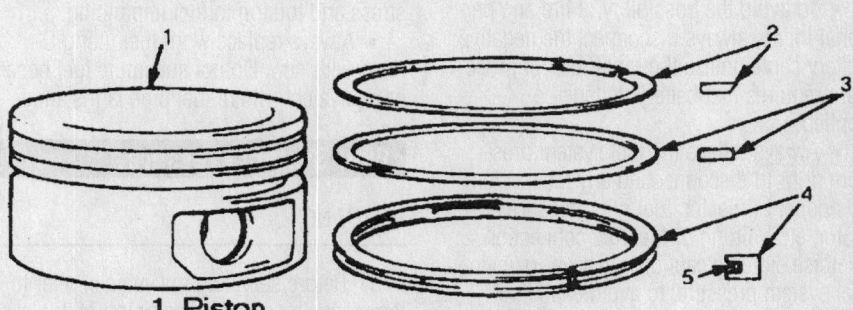

1. Piston
2. Upper compression piston ring
3. Lower compression piston ring
4. Oil control piston ring
5. Oil control ring spring w/spacer

Piston ring positioning—5.7L engine

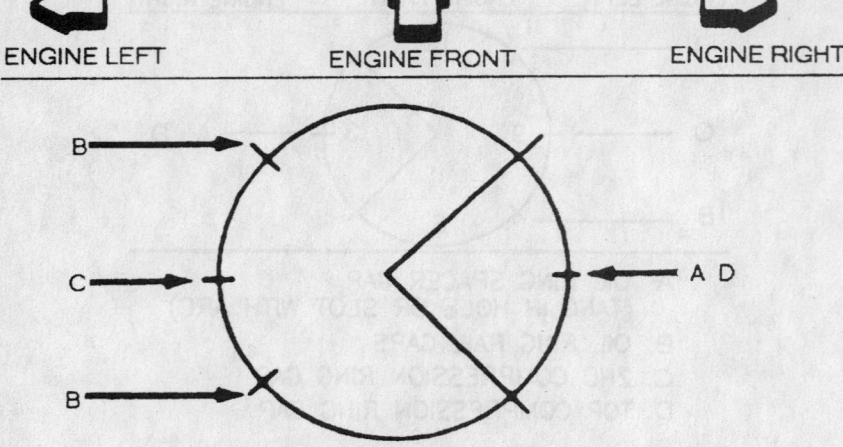

A. Oil ring spacer gap
B. Oil ring rail gaps
C. 2nd compression ring gap
D. Top compression ring gap

7922AG42

Piston ring end-gap spacing—5.7L engine

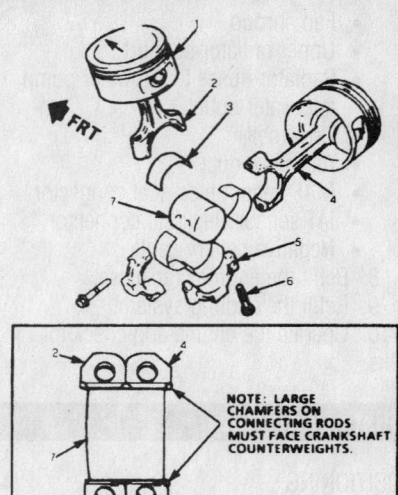

NOTE: LARGE CHAMFERS ON CONNECTING RODS MUST FACE CRANKSHAFT COUNTERWEIGHTS.

1. Piston
2. Connecting rod LH
3. Connecting rod bearing
4. Connecting rod RH
5. Connecting rod bearing cap
6. Connecting rod bearing cap bolt
7. Crankshaft

7922AG44

Piston and connecting rod assembly positioning—5.7L engine

FUEL SYSTEM

Fuel System Service Precautions

Safety is the most important factor when performing not only fuel system maintenance but any type of maintenance. Failure to conduct maintenance and repairs in a safe manner may result in serious personal injury or death. Maintenance and testing of the vehicle's fuel system components can be accomplished safely and effectively by adhering to the following rules and guidelines.

• To avoid the possibility of fire and personal injury, always disconnect the negative battery cable unless the repair or test procedure requires that battery voltage be applied.

• Always relieve the fuel system pressure prior to disconnecting any fuel system component (injector, fuel rail, pressure regulator, etc.), fitting or fuel line connection. Exercise extreme caution whenever relieving fuel system pressure, to avoid exposing skin, face and eyes to fuel spray. Please be advised that fuel under pressure may penetrate the skin or any part of the body that it contacts.

• Always place a shop towel or cloth around the fitting or connection prior to loosening to absorb any excess fuel due to spillage. Ensure that all fuel spillage (should it occur) is quickly removed from engine surfaces. Ensure that all fuel soaked cloths or towels are deposited into a suitable waste container.

• Always keep a dry chemical (Class B) fire extinguisher near the work area.

• Do not allow fuel spray or fuel vapors to come into contact with a spark or open flame.

• Always use a back-up wrench when loosening and tightening fuel line connection fittings. This will prevent unnecessary stress and torsion to fuel line piping.

• Always replace worn fuel fitting O-rings with new. Do not substitute fuel hose or equivalent, where fuel pipe is installed.

Fuel System Pressure

RELIEVING

1. Before servicing the vehicle, refer to the precautions in the beginning of this section.

2. Disconnect the negative battery cable to prevent fuel discharge if the key is accidentally turned to the **RUN** position.

3. Loosen the fuel filler cap to relieve the tank pressure and do not tighten until service has been completed.

4. Connect a fuel pressure gauge to the fuel pressure valve. Wrap a shop cloth around the fitting while connecting the gauge to avoid spillage.

5. Place the end of the bleed hose into a suitable container and open the valve to relieve the fuel system pressure.

Fuel Filter

REMOVAL & INSTALLATION

The inline fuel filter is located on the fuel feed pipe before the fuel injection system and mounted directly in front of the rear axle. The filter housing is constructed of steel with quick-connect inlet and threaded outlet fittings. The threaded fitting is sealed with an O-ring. In order to disengage quick-connect fittings, a fuel line quick-connect separator tool set, such as J-37088-A or equivalent, is required. There is no manufacturer set service interval for fuel filter replacement. We suggest replacing the filter every 30–40 thousand miles.

1. Before servicing the vehicle, refer to the precautions in the beginning of this section.

2. Relieve the fuel system pressure.

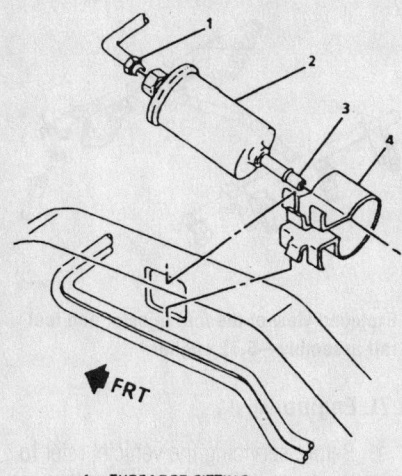

1 THREADED FITTING
2 IN-LINE FUEL FILTER
3 QUICK-CONNECT FITTING
4 IN-LINE FUEL FILTER BRACKET

7922WG23

The inline fuel filter is mounted in a bracket located under the vehicle, directly in front of the rear axle

3. Clean both the inlet and the outlet fittings on the fuel filter.

4. Disengage the quick-connect fittings at the fuel filter inlet as follows:

a. Slide the dust covers from the quick-connect fittings.

b. Grasp both sides of the fitting. Twist the female connector ¼ turn in each direction to loosen any dirt within the fitting. Using compressed air and safety glasses, blow any accumulated dirt out of the fitting.

c. If equipped with the plastic hand releasable fitting, squeeze the plastic retainer release tabs and pull the connection apart.

d. If equipped with metal fittings, choose the correct size quick release tool and insert the tool into the female connector, then push inward to release the locking tabs. Pull the connector apart.

e. Use a clean lint free rag to clean male pipe ends. Inspect both ends of the fitting for dirt and burrs. Clean or replace components as required. If it is necessary to remove rust or burrs from the fuel pipe, use emery cloth in a radial motion with the pipe end to prevent damage to the O-ring sealing surface.

5. Remove the threaded outlet fitting from the chassis fuel pipe. Slide the fuel filter from the bracket.

6. Inspect the fuel pipe O-ring for cuts, nicks, swelling or distortion. Replace if necessary.

To install:

7. Slide the fuel filter into the bracket.

8. Tighten the outlet fitting to the chas-

sis fuel pipe. Torque the fitting to 22 ft. lbs. (30 Nm).

9. Engage the quick-connect inlet fitting as follows:

a. Apply a few drops of clean engine oil to the male pipe end. This will ensure proper reconnection and prevent a possible fuel leak.

b. Push both sides of the fitting together to cause the retainer tabs to snap in place. Once installed, pull on both sides of the fitting to ensure connection is secure.

c. Reposition dust cover over the quick-connect fitting.

10. Tighten the fuel filler cap.

11. Confirm that the ignition is in the **OFF** position. Connect the negative battery cable.

12. Pressurize the fuel system by cycling the ignition without attempting to start the engine. Turn the ignition switch to the **ON** position for 2 seconds, then turn to the **OFF** position for 10 seconds. Again, turn to the **ON** position and check for fuel leaks.

Fuel Pump

REMOVAL & INSTALLATION

1. Before servicing the vehicle, refer to the precautions in the beginning of this section.

2. Relieve the fuel system pressure.

3. Drain the fuel tank.

4. Clean the area surrounding the sender assembly to prevent contamination of the fuel system.

5. Remove or disconnect the following:
- Fuel tank
- Fuel sender-to-fuel tank retaining ring
- Fuel sender from the fuel tank
- Fuel pump from the sending unit

To install:

6. Install or connect the following:
- Fuel pump with new strainer onto the sending unit
- New O-ring in the tank opening groove
- New O-ring on the fuel sender feed tube

➡**The fuel pump strainer must be in a horizontal position and must not block the float arm travel.**

7. Fold the strainer over itself and slowly position the sending assembly in the tank so the strainer is not damaged or trapped by the sump walls.
- Retaining ring
- Fuel tank. Torque the bolts to 33 ft. lbs. (45 Nm).
- Fuel filler cap
- Negative battery cable

8. Turn the ignition switch to the **ON** position for 2 seconds, **OFF** for 10 seconds, then back to the **ON** position. Check for fuel leaks.

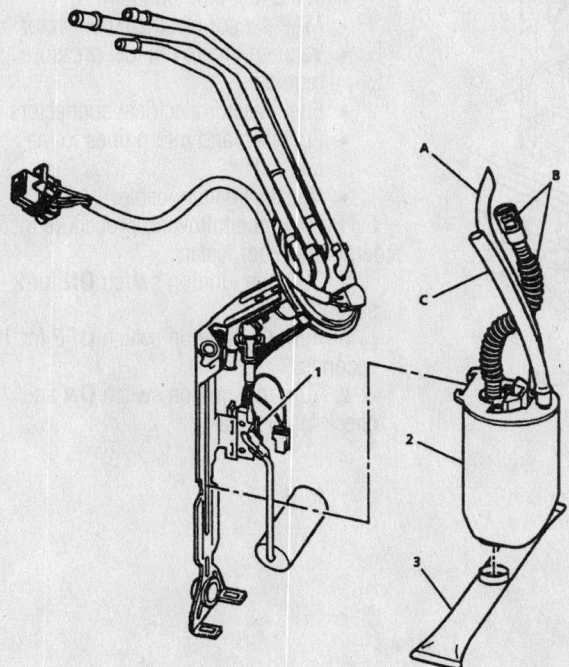

1 SENDER – FUEL

2 FUEL PUMP AND RESERVOIR ASSEMBLY
A VAPOR VENT HOSE
B FUEL PUMP FLEX PIPE AND QUICK – CONNECT FITTING
C FUEL RETURN HOSE

3 STRAINER – FUEL PUMP

7922WG24

Fuel sender/pump assembly component identification

Fuel Injector

REMOVAL & INSTALLATION

3.8L Engine

1. Before servicing the vehicle, refer to the precautions in the beginning of this section.
2. Relieve the fuel system pressure.
3. Remove or disconnect the following:
   - Fuel feed and return lines from the fuel rail
   - Fuel injector electrical connectors
   - Vacuum line from the fuel pressure regulator
   - Manifold Absolute Pressure (MAP) sensor electrical connector
   - Vacuum line from the vacuum switch at the fuel pipe bundle
   - Fuel injector harness fasteners
   - Fuel rail hold-down bolts
   - Fuel rail with fuel injectors
   - Fuel injectors by removing the retainer clips

To install:

✳✳ WARNING

If the new O-rings are different colors (black and brown), install the black

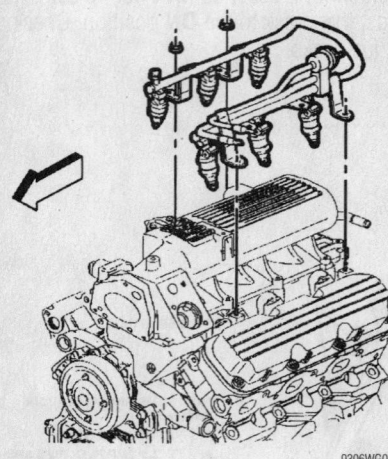

View of the fuel rail assembly—3.8L engine

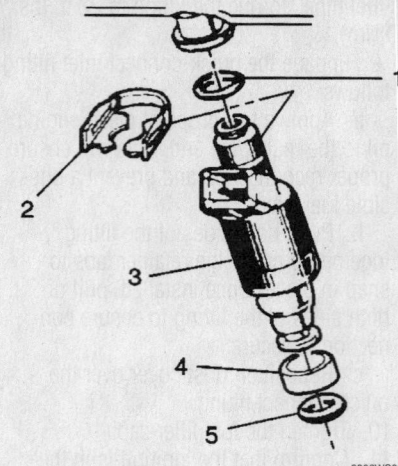

Exploded view of the fuel injector—3.8L engine

one in the upper portion and the brown one in the lower portion of the fuel injector.

4. Coat the injector O-rings with clean engine oil prior to installation.
5. Install or connect the following:
   - Fuel injectors with new O-rings to the fuel rail by installing the retainer clips
   - Fuel rail with fuel injectors. Torque the bolts to 89 inch lbs. (10 Nm).
   - Fuel injector harness fasteners
   - Vacuum line to the vacuum switch at the fuel pipe bundle
   - MAP sensor electrical connector
   - Vacuum line to the fuel pressure regulator
   - Fuel injector electrical connectors
   - Fuel feed and return lines to the fuel rail
   - Negative battery cable
6. Perform the following procedure in order to check for leaks:
   a. Turn the ignition switch **ON** for 2 seconds.
   b. Turn the ignition switch **OFF** for 10 seconds.
   c. Turn the ignition switch **ON** and check for fuel leaks.

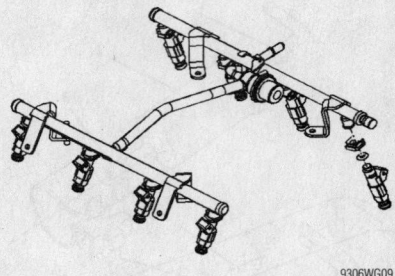

Exploded view of the fuel injector and fuel rail assembly—5.7L engine

5.7L Engine

1. Before servicing the vehicle, refer to the precautions in the beginning of this section.
2. Relieve the fuel system pressure.
3. Remove or disconnect the following:
   - Fuel feed hose from the fuel rail
   - Accelerator cable and cable bracket
   - Fuel injector electrical connectors
   - Electrical harness from the fuel rail
   - Fuel rail assembly
   - Fuel injectors from the fuel rail by spreading the injector retainer clips

To install:

4. Lubricate the new injector O-ring seals with clean engine oil prior to installation.
5. Install or connect the following:
   - New O-rings on the fuel injectors
   - Fuel injectors to the fuel rail using new retaining clips

➡**Position the fuel injector electrical connector facing outward.**

   - Fuel rail assembly. Torque the bolts to 89 inch lbs. (10 Nm).
   - Electrical harness to the fuel rail
   - Fuel injector electrical connectors
   - Accelerator cable and cable bracket
   - Fuel feed hose to the fuel rail
   - Negative battery cable
6. Perform the following procedure:
   a. Turn the ignition switch **ON** for 2 seconds.
   b. Turn the ignition switch **OFF** for 10 seconds.
   c. Turn the ignition switch **ON** and check for fuel leaks.

DRIVE TRAIN

Transmission

REMOVAL & INSTALLATION

Manual

1. Before servicing the vehicle, refer to the precautions in the beginning of this section.
2. Drain the oil from the transmission.
3. Remove or disconnect the following:
   - Negative battery cable
   - Shift control lever boot assembly
   - Shift control lever
   - Propeller shaft
   - Rear axle torque arm
   - Right side catalytic converter
   - Electrical connectors from the transmission
   - Starter motor
   - Wiring harness and bracket from the transmission
   - Transmission brace bolts and the braces, on M49 transmissions
   - Flywheel housing cover
4. Depress the white circular release ring on the actuator hose and simultaneously pull on the master cylinder hose to disconnect.
5. Support the transmission.
6. Remove or disconnect the following:
   - Transmission support
   - Transmission mounting bolts
   - Transmission from the vehicle

To install:

7. Install or connect the following:
   - Transmission to the vehicle. Torque the bolts to 55 ft. lbs. (73 Nm) on M49 transmissions and 37 ft. lbs. (50 Nm) on MM6 transmissions.
   - Transmission support. Torque the bolts to 43 ft. lbs. (57 Nm).
   - Clutch actuator hose to the clutch master cylinder hose.
   - Flywheel housing cover and tighten the bolts to 80 inch lbs. (9 Nm)
   - On M49 transmissions Transmission brace bolts and the braces. Tighten the right transmission brace bolts to 37 ft. lbs. (50 Nm). Tighten the left transmission brace bolts (to transmission) to 37 ft. lbs. (50 Nm) and the bolts to the engine to 21 ft. lbs. (28 Nm).
   - Wiring harness and bracket
   - Electrical connectors
   - Starter motor

   - Right side catalytic converter. Torque the bolts to 18 ft. lbs. (25 Nm).
   - Rear axle torque arm. Torque the bolts to 37 ft. lbs. (50 Nm) and the nuts to 30 ft. lbs. (41 Nm).
   - Propeller shaft. Torque the bolts to 16 ft. lbs. (22 Nm).
   - Transmission oil drain plug. Torque the plug to 20 ft. lbs. (27 Nm).
8. Refill the transmission with oil.
9. Bleed the clutch hydraulic system.
10. Install or connect the following:
   - Shift control lever. Torque the bolts to 13 ft. lbs. (18 Nm).
   - Shift control lever boot assembly
   - Negative battery cable

Automatic

1. Before servicing the vehicle, refer to the precautions in the beginning of this section.
2. Remove or disconnect the following:
   - Negative battery cable
   - Intake Air Temperature (IAT) sensor electrical connector
   - Air intake duct
   - Catalytic converter
   - Range selector cable from the transmission
   - Propeller shaft
   - Rear axle torque arm
   - Torque converter cover
   - Torque converter bolts
   - Transmission oil cooler lines
   - Wiring harness from the transmission
   - Transmission 20-way connector
   - Vehicle Speed sensor (VSS) electrical connector
   - Transmission support
   - Transmission fill tube
   - Transmission nuts and bolts
   - Transmission from the vehicle

To install:

3. Install or connect the following:
   - Transmission to the vehicle. Torque the bolts and nuts to 70 ft. lbs. (95 Nm) on 3.8L engines and 37 ft. lbs. (50 Nm) on 5.7L engines.
   - Transmission fill tube
   - Transmission support. Torque the bolts to 66 ft. lbs. (90 Nm) on 3.8L engines and 77 ft. lbs. (105 Nm) on 5.7L engines.
   - VSS electrical connector
   - 20-way connector
   - Wiring harness clamp to the trans-

---

   mission. Torque the bolt to 22 inch lbs. (2.5 Nm).
   - Transmission oil cooler lines
   - Torque converter bolts. Torque the bolts to 47 ft. lbs. (63 Nm) on 3.8L engines and 44 ft. lbs. (60 Nm) on 5.7L engines.
   - Torque converter cover
   - Rear axle torque arm. Torque the bolts to 37 ft. lbs. (50 Nm) and the nuts to 30 ft. lbs. (41 Nm).
   - Propeller shaft. Torque the bolts to 16 ft. lbs. (22 Nm).
   - Range selector cable
   - Catalytic converter. Torque the bolts to 18 ft. lbs. (25 Nm).
   - Air intake duct
   - IAT sensor electrical connector
   - Negative battery cable

Clutch

REMOVAL & INSTALLATION

1. Before servicing the vehicle, refer to the precautions in the beginning of this section.
2. Remove or disconnect the following:
   - Negative battery cable
   - Transmission assembly
   - Flywheel housing cover
   - Flywheel housing
3. Install a clutch disc alignment tool through the center of the disc and into the pilot bearing to prevent the disc from falling when the pressure plate is removed.
4. Remove or disconnect the following:
   - Pressure plate retaining bolts
   - Pressure plate with clutch disc

To install:

5. Install or connect the following:
   - Clutch plate to the flywheel
   - Pressure plate and cover. Do not torque the bolts at this time.
6. Align the clutch plate with the pilot bearing and clutch pressure plate.
7. On 3.8L models, torque the clutch pressure plate and cover bolts in a star pattern to 15 ft. lbs. (20 Nm), plus an additional 45 degree turn.
8. On 5.7L models, torque the bolts using a star pattern torque sequence to 52 ft. lbs. (70 Nm).
9. Install or connect the following:
   - Flywheel housing
   - Flywheel housing cover
   - Transmission
   - Negative battery cable

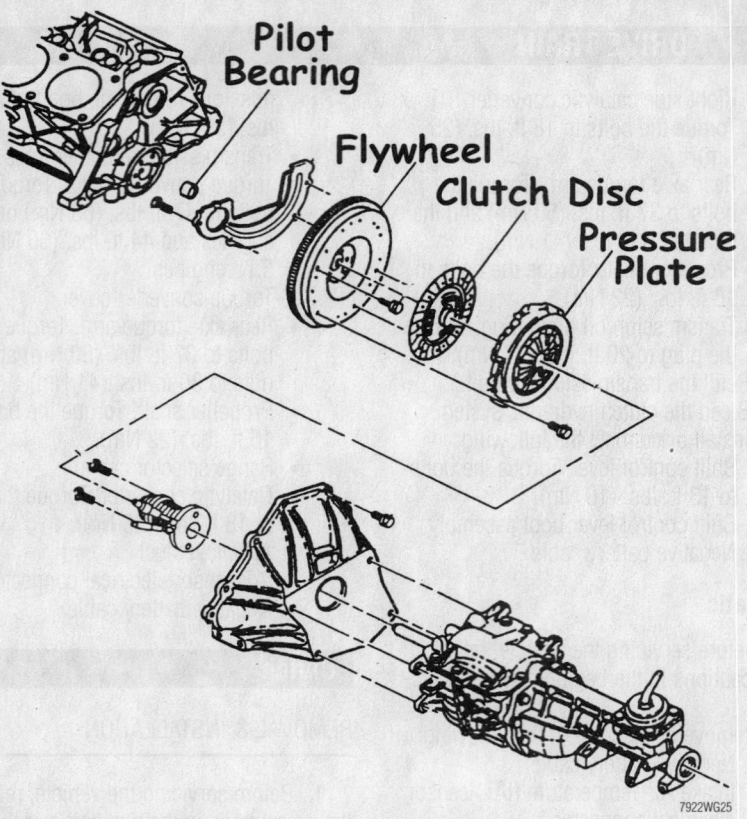

Exploded view of the clutch disc, pressure plate and related components—5-speed transmission

Hydraulic Clutch System

BLEEDING

Bleeding air from the hydraulic clutch system is necessary whenever any part of the system has been disconnected or the fluid level (in the reservoir) has been allowed to fall so low that air has been drawn into the master cylinder.

⁕⁕ WARNING

NEVER use fluid that has been bled from a clutch system to fill the master cylinder reservoir, as it may be aerated, contain excessive moisture and/or be contaminated in some other way.

1. Before servicing the vehicle, refer to the precautions in the beginning of this section.
2. Fill the clutch master cylinder reservoir with new hydraulic clutch fluid.
3. Attach a hose to the bleeder on the clutch actuator and submerge the other end of the hose in a container of hydraulic clutch fluid.
4. Have an assistant slowly depress and hold the clutch pedal.

5. Loosen the bleeder to purge air.
6. Tighten the bleeder.
7. Repeat the above 3 steps until all air is completely purged from the sys-tem.
8. Refill the clutch master cylinder reservoir.

Axle Shaft, Bearing and Seal

REMOVAL & INSTALLATION

1. Before servicing the vehicle, refer to the precautions in the beginning of this section.
2. Remove or disconnect the following:
   - Wheel
   - Brake rotor
   - Rear axle housing cover
   - Pinion gear shaft lock bolt
   - Pinion gear shaft
3. Push the axle shaft into the housing in order to gain access to the rear axle shaft C-clip.
4. Remove the C-clip and remove the axle from the housing.

⁕⁕ WARNING

Do not damage the rear wheel speed sensor reluctor wheel during axle shaft removal.

5. Remove or disconnect the following:
   - Brake backing plate

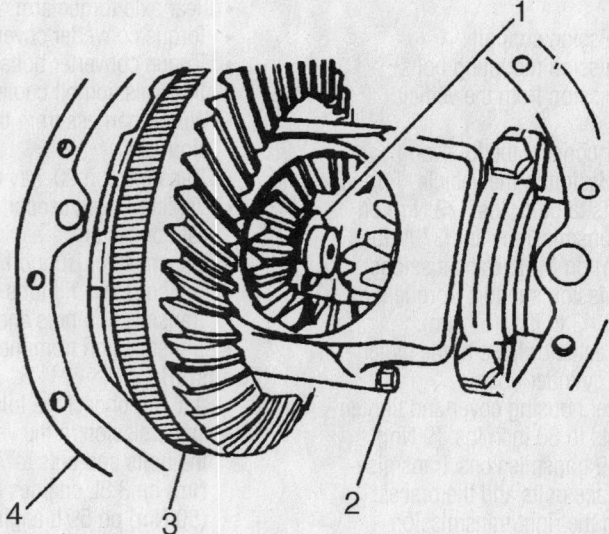

1. Axle shaft C-clip
2. Pinion gear shaft lock bolt
3. Speed sensor reluctor wheel
4. Axle housing

View of the rear axle housing with cover removed

- Brake caliper mounting bracket
- Oil seal, by prying it out
- Rear wheel bearing with suitable puller

To install:

6. Lubricate the new wheel bearing with gear oil.

7. Install or connect the following:
- Wheel bearing, using suitable installation tool
- Oil seal, using suitable installation tool
- Rear brake caliper mounting bracket
- Brake backing plate
- Axle shaft and C-clip

➡**Be sure the splines on the end of the axle engage with the splines of the differential side gear.**

- Pinion gear shaft
- Pinion gear shaft lock bolt. Torque the bolt to 27 ft. lbs. (36 Nm).
- Rear axle housing cover with new gasket. Torque the bolts in a star pattern to 22 ft. lbs. (30 Nm).
- Rear brake rotor
- Wheel

Pinion Seal

REMOVAL & INSTALLATION

1. Before servicing the vehicle, refer to the precautions in the beginning of this section.

2. Matchmark the propeller shaft-to-drive pinion gear yoke, drive pinion gear and drive pinion yoke nut.

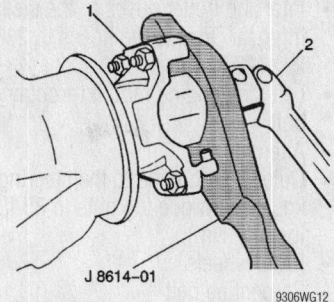

J 8614-01

9306WG12

View of the drive pinion gear yoke nut removal tools

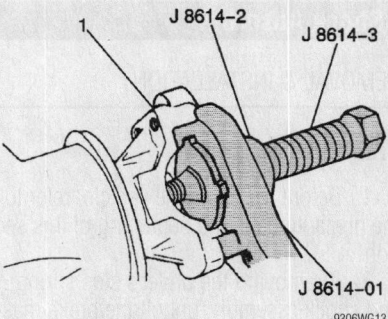

J 8614-2   J 8614-3

J 8614-01

9306WG13

View of the drive pinion gear yoke removal tools

3. Remove or disconnect the following:
- Propeller shaft
- Drive pinion gear yoke nut using Pinion Flange Remover/Installer Tool J-8614-01 to hold the yoke and a socket wrench
- Drive pinion gear yoke using Tools J-8614-01, J-8614-2 and J-8614-3

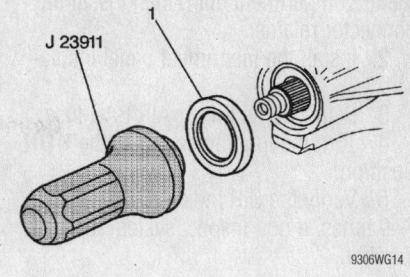

J 23911

9306WG14

View of the pinion seal nut installer tool

➡**Use a container to catch the oil that may flow from the axle housing once the yoke is removed.**

- Pinion seal

➡**It may be necessary to use a blunt chisel to drive the seal from the housing.**

To install:

4. Install or connect the following:
- New pinion seal, lubricated with chassis grease using pinion oil seal installer Tool J 8614-01 until it is flush with the housing
- Drive pinion gear yoke using Tool J 8614-01 and a socket wrench
- Drive pinion gear nut, tighten to 1/16 in. (1.59mm) beyond the alignment mark made earlier
- Propeller shaft. Torque the propeller shaft-to-pinion yoke bolts to 16 ft. lbs. (22 Nm) and the propeller shaft-to-center support to 37 ft. lbs. (50 Nm) for 2-piece shaft.

5. Refill the axle housing.

STEERING AND SUSPENSION

Air Bag

※※ CAUTION

The vehicles covered in this section are equipped with an air bag system. The system must be disabled before performing service on or around system components, steering column, instrument panel components, wiring and sensors. Failure to follow safety and disabling procedures could result in accidental air bag deployment, possible personal injury and unnecessary system repairs.

PRECAUTIONS

Several precautions must be observed when handling the inflator module to avoid accidental deployment and possible personal injury.

- Never carry the inflator module by the wires or connector on the underside of the module.
- When carrying a live inflator module, hold securely with both hands, and ensure that the bag and trim cover are pointed away.
- Place the inflator module on a bench or other surface with the bag and trim cover facing up.
- With the inflator module on the bench, never place anything on or close to the module, which may be thrown in the event of an accidental deployment.

DISARMING

1. Align the steering wheel so the vehicle wheels are pointing in the straight-ahead position.

2. Turn the ignition switch to the **LOCK** position and remove the key.

3. Remove the SIR or AIR BAG fuse from the fuse block.

4. Remove the left instrument panel insulator.

5. Remove the connector retainer, then disengage the yellow 2-way SIR wiring harness connector at the base of the steering column.

6. Remove the right instrument panel insulator.

7. Remove the connector retainer, then disengage the yellow 2-way SIR wiring harness connector located behind the IP compartment door.

ARMING

1. Engage the yellow 2-way connector at the base of the steering column and behind

the IP compartment door, then install the connector retainer.

2. Install the instrument panel insulators.

3. Reinstall the SIR or AIR BAG fuse.

4. Turn the ignition switch to the **RUN** position.

5. Verify the SIR indicator light flashes 7–9 times, if not, inspect system for malfunction.

Power Rack and Pinion Steering Gear

REMOVAL & INSTALLATION

1. Before servicing the vehicle, refer to the precautions in the beginning of this section.

2. Remove or disconnect the following:
- Serpentine belt
- Air intake resonator
- Front wheels
- Outer tie rod ends from the knuckles
- Alternator
- Left side engine mount through bolt
- Inlet and outlet hoses from the steering gear
- Steering gear coupling shaft from the steering gear
- Steering gear retainers
- Steering gear from the vehicle

To install:

3. Install or connect the following:
- Steering gear and its retainers
- Steering gear to the crossmember. Torque the bolts to 63 ft. lbs. (85 Nm).
- Steering gear to the coupling shaft. Torque the bolt to 35 ft. lbs. (47 Nm).

- Inlet and outlet hoses to the steering gear. Torque the hoses to 21 ft. lbs. (28 Nm).
- Left side engine mount through bolt
- Alternator
- Outer tie rod ends to the steering knuckle. Torque the nuts to 35 ft. lbs. (47 Nm).
- Front wheels
- Serpentine belt
- Air intake resonator

4. Refill and bleed the power steering system.

Shock Absorber

REMOVAL & INSTALLATION

Front

1. Before servicing the vehicle, refer to the precautions in the beginning of this section.

2. If removing the driver's side spring and shock assembly, unbolt the brake master cylinder from the booster and position it aside, but do not disconnect the brake lines.

3. Remove or disconnect the following:
- Upper spring/shock mounting bolts and nuts
- Front wheels
- Stabilizer bar (shaft link) from the control arm

4. Matchmark the lower coil spring mount location to the upper coil spring mount location before removing the shock absorber.

5. Remove or disconnect the following:
- Lower shock absorber nuts and bolts
- Lower ball joint stud from the steering knuckle

- Spring/shock absorber unit from the vehicle

6. Remove the spring from the shock absorber assembly by performing the following steps:

➡ **If using other than a GM spring compressor, follow the manufacturers instructions regarding the use of the specific tool you are using.**

   a. Assemble J-34013-B and J-34013-114 on the spring unit.

   b. Use the wing nuts to secure the tool to mounting holes **C-H** (lower left corner) and **P** (upper right corner) for the driver's side shock and to mounting holes **A-X-P** (upper left) and **C-H** (lower right) for the passenger's side shock.

   c. Install J-34013-114 and J 34013-88.

➡ **Make certain that J-34013-114 and J-34013-88 are aligned so that they can open and close together. If not properly aligned, they will not function.**

   d. Attach the shock unit to the tools.

➡ **Make certain that the top of the shock is flat against J-34013-114.**

   e. Close the tools and install the locking pin.

➡ **Make certain that the mounting ears of the shock are facing downward toward the rear of J-34013-B, otherwise the shock will not align properly.**

   f. Turn the screw of J-34013-B counterclockwise to raise the shock up to J-34013-114. Be sure that the studs go through the guide holes in J-34013-114 and the top of the shock is flat against the tool.

✳✳ CAUTION

Do not over-compress the spring! Over-compression can cause tool failure, resulting in severe bodily injury!

   g. Compress the spring approximately ½ in. (13mm) or 3–4 complete turns of the screw on J-34013-114.

   h. Insert J-39642-1 on the shock nut, then insert J-39642-2 through J-39642-1 to hold the shock absorber rod in place.

   i. Remove the shock absorber nut with J-39642-1, while holding the shock rod from rotating with J-39642-2.

   j. Discard the shock absorber nut.

   k. Turn J-34013-B clockwise to fully relieve spring pressure and remove the spring from the shock.

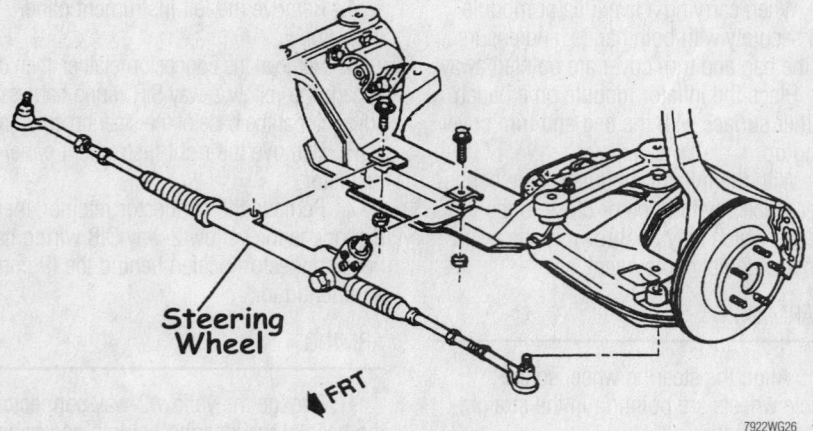

Steering Wheel

FRT

7922WG26

Exploded view of the power steering gear mounting

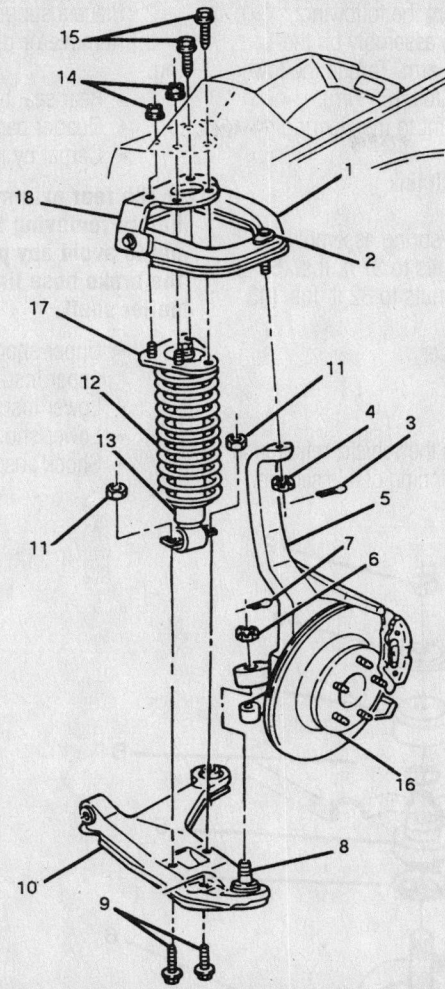

1 ARM ASSEMBLY, FRONT UPPER CONTROL
2 STUD ASSEMBLY, FRONT UPPER CONTROL ARM BALL
3 PIN, FRONT UPPER CONTROL ARM COTTER
4 NUT, FRONT UPPER CONTROL ARM, 53 N•m (39 LB. FT.)
5 KNUCKLE ASSEMBLY, STEERING
6 NUT, FRONT LOWER CONTROL ARM, 110 N•m (81 LB. FT.)
7 PIN, FRONT LOWER CONTROL ARM COTTER
8 STUD ASSEMBLY, FRONT LOWER CONTROL ARM BALL
9 BOLT/SCREW, FRONT SHOCK ABSORBER, 65 N•m (48 LB. FT.)
10 ARM ASSEMBLY, FRONT LOWER CONTROL
11 NUT, FRONT SHOCK ABSORBER LOWER BRACKET, 65 N•m (48 LB. FT.)
12 SPRING ASSEMBLY, FRONT
13 ABSORBER ASSEMBLY, FRONT SHOCK
14 NUT, FRONT SHOCK ABSORBER UPPER MOUNT, 43 N•m (32 LB. FT.)
15 BOLT/SCREW, FRONT SHOCK ABSORBER UPPER MOUNT, 50 N•m (37 LB. FT.)
16 HUB ASSEMBLY, FRONT WHEEL
17 MOUNT ASSEMBLY, FRONT UPPER SHOCK ABSORBER
18 SUPPORT, FRONT UPPER CONTROL ARM

7922WG27

Exploded view of the shock absorber unit and related suspension components

To install:

7. Assemble the shock absorber/spring assembly by performing the following procedure:

a. Assemble J-34013-B spring compressor and J-34013-114 adapter on the spring unit. Use wing nuts to secure the tool to mounting holes **C-H** (lower left corner and **P** (upper right corner) for the driver's side shock and to mounting holes **A-X-P** (upper left) and **C-H** (lower right) for the passenger's side shock.

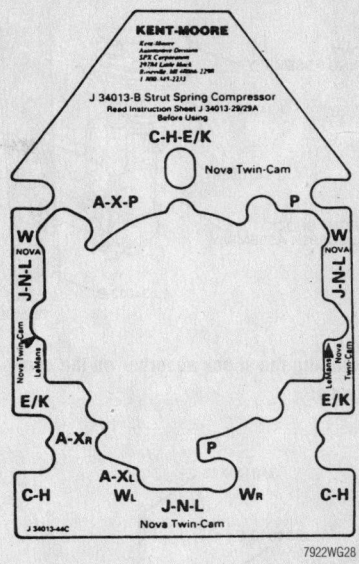

Shock absorber compressor mounting hole locations

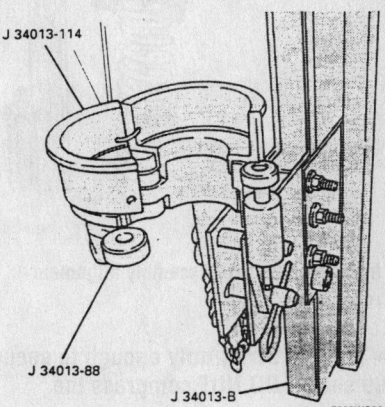

Install the strut compressor adapter on the spring compressor

b. Install Tools J-34013-114 and J-34013-88.

c. Attach the shock unit to the tools.

➡**Make certain that the mounting ears of the shock are facing downward towards the rear of J 34013-B!**

d. Close the tools and install the locking pin.

e. Be sure that the spring seats are positioned properly.

➡**Make certain that the top of the shock is flat against J-34013-114.**

f. Turn the screws of J-34013-B counterclockwise to raise the shock up to J-34013-114. Be sure that the studs go through the guide holes in J-34013-114 and the top of the shock is flat against the tool.

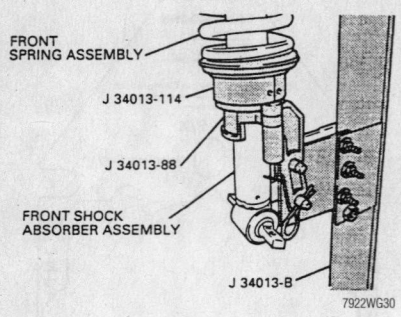

FRONT
SPRING ASSEMBLY

J 34013-114

J 34013-88

FRONT SHOCK
ABSORBER ASSEMBLY

J 34013-B

7922WG30

Installing the shock absorber on the compressor

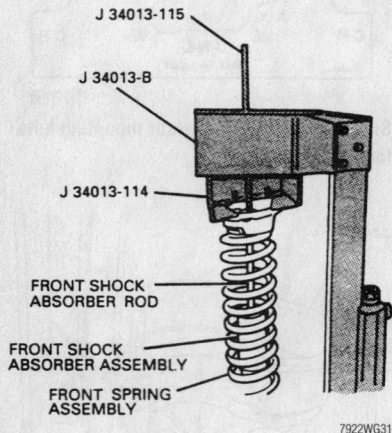

J 34013-115

J 34013-B

J 34013-114

FRONT SHOCK
ABSORBER ROD

FRONT SHOCK
ABSORBER ASSEMBLY

FRONT SPRING
ASSEMBLY

7922WG31

Inserting the shock assembly alignment rod

➡ **Turn the screw only enough to secure the shock. DO NOT compress the spring.**

g. Place J-34013-115 down through the top of J-34013-B, through the top of the shock absorber and onto the rod.

➡ **Make certain that J-34013-115 is straight with the shock.**

✷✷ CAUTION

Do not over-compress the spring! Over-compression can cause tool failure, resulting in severe bodily injury!

h. Turn the operating screw clockwise to compress the spring until the threaded portion of the rod is through the top of the shock. Remove J-34013-115.

i. Install a new shock absorber nut.

j. Install J-39642-1 on the nut, insert J-39642-2 through J-39642-1 and tighten the nut while holding J-39642-2.

8. Remove the shock/spring assembly from the tool.

9. Install or connect the following:
- Shock/spring assembly on the lower control arm. Torque the lower nuts to 48 ft. lbs. (65 Nm).
- Lower ball joint to the steering knuckle
- Stabilizer shaft link
- Front wheel
- Upper shock/spring assembly. Torque the bolts to 37 ft. lbs. (50 Nm) and the nuts to 32 ft. lbs. (43 Nm).
- Master cylinder

Rear

1. Before servicing the vehicle, refer to the precautions in the beginning of this section.

2. Place a support under the rear axle.

3. Remove or disconnect the following:
- Rear seat back by folding it down
- Quarter panel trim assembly
- Carpet by folding it back

➡ **The rear axle must be supported before removing the upper mounting nut to avoid any possible damage to the brake hose lines, tie rods and propeller shaft.**

- Upper shock nut, retainer and upper insulator
- Lower insulator and retainer
- Lower shock-to-rear axle nut
- Shock absorber from the vehicle

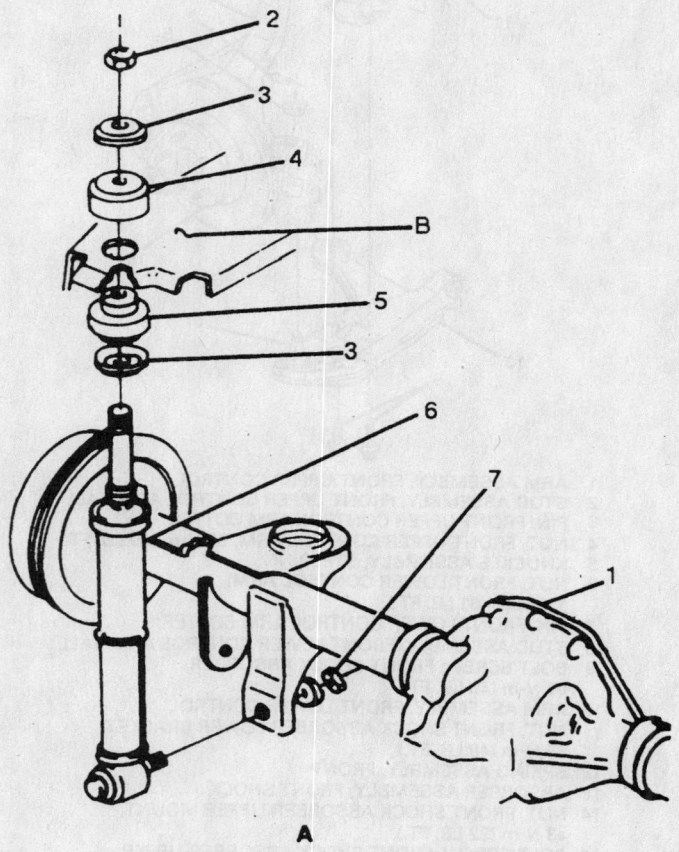

A

A Typical rear shock absorber assembly installation (right-hand shown)
B Underbody pan assembly
1 Rear axle assembly
2 Rear shock absorber nut 17 Nm (13 lb. ft.)
3 Rear shock absorber upper insulator retainer
4 Rear shock absorber upper insulator
5 Rear shock absorber lower insulator
6 Rear shock absorber assembly
7 Rear shock absorber nut 90 Nm (66 lb. ft.)

7922WG32

Exploded view of the rear shock absorber mounting

To install:

4. Install or connect the following:
   - Rear shock absorber. Torque the lower shock-to-axle nut to 66 ft. lbs. (90 Nm).
   - Lower insulator and retainer
   - Upper insulator and retainer
   - Upper shock mounting nut. Torque the nut to 13 ft. lbs. (17 Nm).

※※ WARNING

Turning the shock absorber while tightening the nut could damage the shock. To prevent damage, keep the shock absorber stationary when tightening the nut.

   - Carpet
   - Quarter panel trim assembly

Coil Spring

REMOVAL & INSTALLATION

Front

Refer to the from shock absorber removal and installation procedure in this manual for coil spring Removal & Installation.

Rear

1. Before servicing the vehicle, refer to the precautions in the beginning of this section.
2. Raise and safely support the vehicle by the frame so that the rear axle can be independently raised and lowered.
3. Support the rear axle with an adjustable support.
4. Remove or disconnect the following:
   - Brake hose brackets, if equipped, allowing the hoses to hang free

➡**Perform the previous step only if the hoses would be stretched and damaged when the axle is lowered.**

   - Lower shock absorber bolts
5. Lower the rear axle.
6. Remove or disconnect the following:
   - Upper insulator
   - Spring

➡**The springs are painted with a protective coating. Take care to avoid damaging this coating. If the coating is chipped or damaged, paint the exposed spring to prevent rust.**

To install:

7. Install or connect the following:
   - Spring on the axle with the open lower end facing forward
   - Upper insulator

8. Raise and safely support the rear axle.
   - Lower shock absorber nuts
   - Brake hose brackets, if removed

Upper Ball Joint

REMOVAL & INSTALLATION

1. Before servicing the vehicle, refer to the precautions in the beginning of this section.
2. Remove the front wheel.
3. Position a floor jack under the shock mount for support.

※※ WARNING

The jack must remain in place for the entire duration of the procedure to hold the spring and lower control arm in proper position.

4. Remove or disconnect the following:
   - Cotter pin
   - Ball stud nut

   - 4 ball joint rivets by drilling them 0.25 in. (6mm) deep using a ⅛ in. (3.175mm) bit
   - Rivet heads by drilling them out using a ½ in. (12.7mm) bit
   - Rivets by using a small drift to punch them out
   - Upper ball joint from the control arm

To install:

5. Install or connect the following:
   - New upper ball joint to the control arm. Torque the nuts and bolts according to the specifications given in the replacement kit.
   - Ball joint to the steering knuckle. Torque the nut to 39 ft. lbs. (53 Nm).

➡**Advance the nut to align the nearest cotter pin hole; NEVER back off the nut to align a hole!**

   - New cotter pin
   - Front wheel

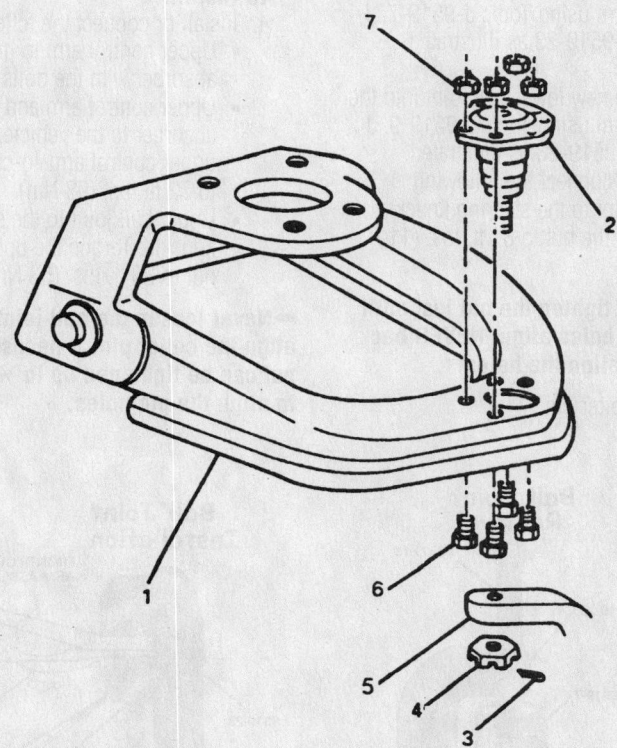

1 Front upper control arm assembly
2 Front upper control ball stud assembly
3 Front upper control arm cotter pin
4 Front upper control arm nut 53 Nm (39 lb. ft.)
5 Steering knuckle assembly
6 Service kit bolt/screw
7 Service kit nut

7922WG33

Position the new upper ball joint stud in the control arm and secure with the replacement nuts and bolts

Lower Ball Joint

REMOVAL & INSTALLATION

→**To prevent component damage, an on-car ball joint press should be used.**

1. Before servicing the vehicle, refer to the precautions in the beginning of this section.
2. Remove the front wheel.
3. Position a support under the shock mount for support.

❋❋ WARNING

The support must remain in place for the entire duration of the procedure to hold the spring and lower control arm in proper position.

4. Remove or disconnect the following:
   • Lower ball joint cotter pin
   • Lower ball joint stud nut
   • Lower ball joint from the steering knuckle
5. Press the lower ball joint from the lower control arm using Tools J-9519-7, J-9519-18 and J-9519-23 as illustrated.

To install:

6. Press the new lower ball joint into the lower control arm using Tools J 9519-9, J 9519-18 and J 9519-23 as illustrated.
7. Install or connect the following:
   • Ball joint to the steering knuckle. Torque the nut to 81 ft. lbs. (110 Nm).

→**Continue to tighten the nut just until the cotter pin holes align; NEVER back off the nut to align the holes!**

   • New cotter pin

   • Front wheel
8. Check and/or adjust the alignment.

Upper Control Arm

REMOVAL & INSTALLATION

1. Before servicing the vehicle, refer to the precautions in the beginning of this section.
2. Remove or disconnect the following:
   • Brake master cylinder, if removing the drivers side control arm
   • Upper shock absorber mount bolts
   • Front wheel
   • Stabilizer shaft link
   • Lower shock absorber bolts
   • Upper ball joint nut
3. Support the steering knuckle.
   • Upper ball joint from the steering knuckle
   • Upper control arm with the shock absorber
   • Upper control arm from the shock absorber

To install:

4. Install or connect the following:
   • Upper control arm to the shock absorber with the bolts finger-tight
   • Upper control arm and shock absorber to the vehicle. Torque the upper control arm-to-chassis bolts to 72 ft. lbs. (98 Nm).
   • Upper ball joint to the steering knuckle. Torque the upper ball joint nut to 39 ft. lbs. (53 Nm).

→**Never loosen the ball joint nut to align the cotter pin. If necessary, the nut can be tightened up to ⅙ of a turn to align the pin holes.**

   • New cotter pin
5. Remove the steering knuckle support.
   • Lower shock absorber to the steering knuckle. Torque the bolts to 48 ft. lbs. (65 Nm).
   • Stabilizer shaft link. Torque the nuts to 17 ft. lbs. (23 Nm).
   • Front wheel
   • Upper shock absorber. Torque the nuts to 30 ft. lbs. (41 Nm) and the bolts to 37 ft. lbs. (50 Nm).
   • Brake master cylinder. Torque the nuts to 21 ft. lbs. (29 Nm).

CONTROL ARM BUSHING REPLACEMENT

1. Before servicing the vehicle, refer to the precautions in the beginning of this section.
2. Remove or disconnect the following:
   • Upper control arm
   • Upper control arm from the upper control arm support
3. Remove the bushing from the upper control arm by performing the following procedure:
   a. Thread the Upper Control Arm Screw Tool J-21474-19 through the Upper Control Arm Bushing Receiver/Installer Tool J-39930; the 3 tangs must be against the screw head.
   b. Thread the Upper Control Arm Screw Tool J-21474-19 through the upper control arm bushing.
   c. Run the smaller end of the Control Arm Bushing Receiver Tool J-21474-5 onto the Upper Control Arm Screw Tool J-21474-19; then, place the thrust washer onto Tool J-21474-19 with the seams facing Tool J-21474-5.
   d. Position the Half Moon Spacer Tool J-39872 around the outside of the bushing to prevent metal distortion during removal.
   e. Ensure that the tools are aligned; then, install the Upper Control Arm Nut Tool J-21474-18 on the Upper Control Arm Screw Tool J-21474-19.
   f. Tighten the Upper Control Arm Nut Tool J-21474-18 and Upper Control Arm Screw Tool J-21474-19 until the busing is pressed from the upper control arm.

To install:

4. Install the bushing to the upper control arm by performing the following procedure:
   a. Thread the Upper Control Arm Screw Tool J-21474-19 through the Upper Control Arm Bushing Receiver/Installer Tool J-39930; the 3 tangs must NOT be against the screw head.

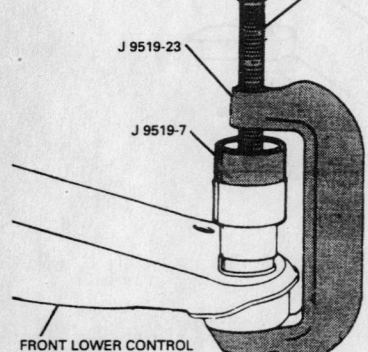

**Ball Joint
Removal**

J 9519-18
J 9519-23
J 9519-7
FRONT LOWER CONTROL
ARM ASSEMBLY

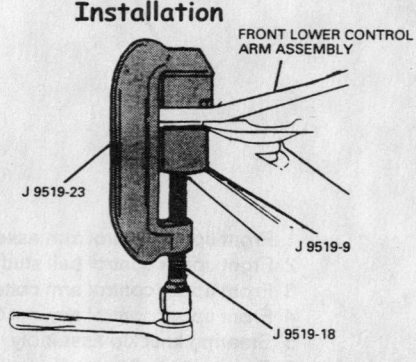

**Ball Joint
Installation**

FRONT LOWER CONTROL
ARM ASSEMBLY
J 9519-23
J 9519-9
J 9519-18

7922WG34

Lower ball joint replacement requires the use of special tools, such as the ones shown

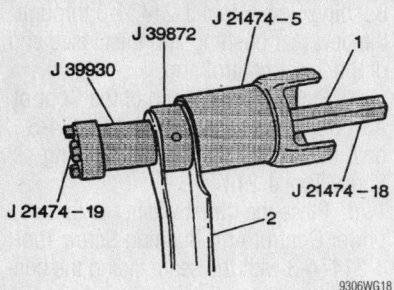

Removing the upper control arm bushing

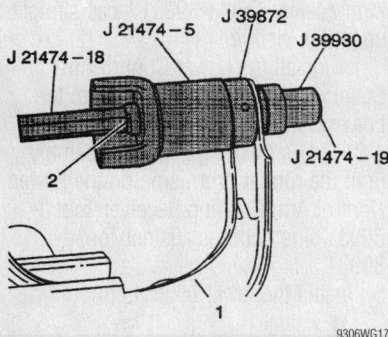

Installing the upper control arm bushing

b. Install a new bushing onto the Upper Control Arm Screw Tool J-21474-19 with the 3 indentations facing the 3 tangs on the Upper Control Arm Bushing Receiver/Installer Tool J-39930.

c. Install the threaded end of the Upper Control Arm Screw Tool J-21474-19 to the upper control arm from the outside.

d. Install the Control Arm Bushing Receiver Tool J-21474-5 onto the Upper Control Arm Screw Tool J-21474-19 from the inside.

e. Place the thrust washer onto Tool J-21474-19 with the seam facing the bushing.

f. Position the Half Moon Spacer Tool J-39872 around the outside of the bushing to prevent metal distortion during installation.

g. Install the Upper Control Arm Nut Tool J-21474-18 on the Upper Control Arm Screw Tool J-21474-19; ensure that the 3 tangs on the Upper Control Arm Bushing Receiver/Installer Tool J-39930 fit into the bushing indentations.

h. Tighten the assembly until the bushing is pressed into the upper control arm.

5. Install or connect the following:
- Upper control arm to the upper control arm support. Torque the nuts/bolts to 72 ft. lbs. (98 Nm).
- Upper control arm

Lower Control Arm

REMOVAL & INSTALLATION

1. Before servicing the vehicle, refer to the precautions in the beginning of this section.
2. Remove or disconnect the following:
- Front wheel
- Stabilizer shaft link
- Outer tie rod end from the steering knuckle
- Lower shock absorber bolts
- Lower ball joint from the steering knuckle
- Lower control arm nuts
- Lower control arm

To install:
3. Install or connect the following:
- Lower control arm
- Lower control arm-to-crossmember mounting nuts and tighten to 74 ft. lbs. (100 Nm)
- Lower ball joint to the steering knuckle. Torque the ball joint nut to 81 ft. lbs. (110 Nm).

➡**Never loosen the ball joint nut to align the cotter pin. If necessary, the nut can be tightened up to ⅙ of a turn to align the pin holes.**

- New lower ball joint cotter pin
- Lower shock absorber. Torque the bolts to 48 ft. lbs. (65 Nm).
- Outer tie rod end to the steering knuckle. Torque the nut to 35 ft. lbs. (47 Nm).
- New outer tie rod end cotter pin
- Stabilizer shaft link. Torque the nut to 17 ft. lbs. (23 Nm).
- Front wheel

CONTROL ARM BUSHING REPLACEMENT

Front Bushing

1. Before servicing the vehicle, refer to the precautions in the beginning of this section.
2. Remove the lower control arm.
3. Remove the front bushing from the upper control arm by performing the following procedure:

a. Install the Lower Control Arm Bushing Screw Tool J-21474-3 through the large end of the Control Arm Bushing Receiver Tool J-21474-6.

b. Install the Lower Control Arm Bushing Screw Tool J-21474-3 through the front bushing and the open end of the Lower Control Arm Bushing Receiver Tool J-39876.

c. Place the thrust washer onto the Lower Control Arm Bushing Screw Tool J-21474-3 so the seam faces the front bushing.

d. Install the Half Moon Spacer Tool J-39875 around the outside of the bushing to avoid metal distortion during removal.

e. Ensure that the tools align; then, install the Lower Control Arm Bushing Nut Tool J-21474-4 and tighten the assembly until the bushing is pressed from the control arm.

To install:
4. Install the front bushing to the lower control arm by performing the following procedure:

a. Thread the Lower Control Arm Bushing Screw Tool J-21474-3 through the Control Arm Bushing Receiver Tool J-21474-5.

b. Install the Lower Control Arm Bushing Screw Tool J-21474-3 through the new front bushing and the lower control arm.

c. Thread the Lower Control Arm Bushing Screw Tool J-21474-3 and

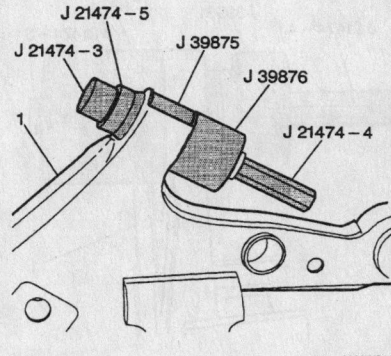

Removing the lower control arm front bushing

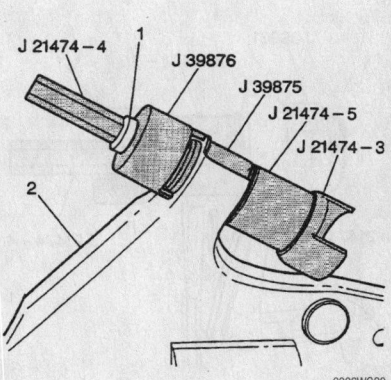

Installing the lower control arm front bushing

accessories into the front lower control arm.

d. Install the Lower Control Arm Bushing Receiver Tool J-39876 onto the Lower Control Arm Bushing screw Tool J-21474-3 with the open end facing the control arm.

e. Install the Half Moon Spacer Tool J-39875 around the outside of the bushing to avoid metal distortion during installation.

f. Place the thrust washer onto the Lower Control Arm Bushing Screw Tool J-21474-3 threaded end with the seam facing the control arm.

g. Ensure that the tools align; then, install the Lower Control Arm Bushing Nut and the Lower Control Arm Bushing Screw Tool J-21474-3.

h. Tighten the assembly until the front bushing is flush with the lower control arm.

5. Install the lower control arm.

Rear Bushing

1. Before servicing the vehicle, refer to the precautions in the beginning of this section.

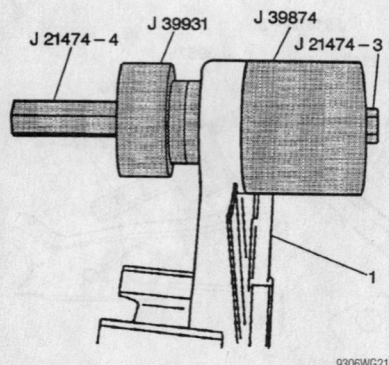

Removing the lower control arm rear bushing

9306WG21

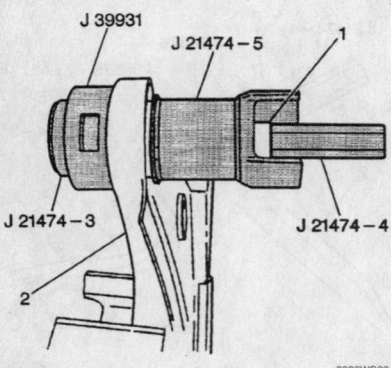

Installing the lower control arm rear bushing

9306WG22

2. Remove the lower control arm.

3. Remove the rear bushing from the upper control arm by performing the following procedure:

a. Install the Lower Control Arm Bushing Receiver Tool J-39874 onto the Lower Control Arm Bushing Screw Tool J-21474-3.

b. Install the Lower Control Arm Bushing Screw Tool J-21474-3 through the rear bushing, the control arm and into the small end of the Lower Control Arm Bushing Receiver Tool J-39931.

c. Place the thrust washer onto the Lower Control Arm Bushing Screw Tool J-21474-3 so the seam faces the control arm.

d. Install the Lower Control Arm Bushing Nut Tool J-21474-4 onto the Lower Control Arm Bushing Screw Tool J-21474-3; then, ensure that the assembly aligns with the bushing.

e. Tighten the assembly until the bushing is pressed from the control arm.

To install:

4. Install the rear bushing to the lower control arm by performing the following procedure:

a. Thread the Lower Control Arm Bushing Screw Tool J-21474-3 through the small end of the Lower Control Arm Bushing Receiver Tool J-39931.

b. Install the Lower Control Arm

Bushing Screw Tool J-21474-3 through the new rear bushing and the closed end of the lower control arm.

c. Place the small end of the Control Arm Bushing Receiver Tool J-21474-5 onto the Lower Control Arm Bushing Screw Tool J-21474-3.

d. Place the thrust washer onto the Lower Control Arm Bushing Screw Tool J-21474-3 with the seam facing the control arm.

e. Ensure that the tools align and the window on the Lower Control Arm Bushing Receiver Tool J-39931 faces straight up and is visible.

f. Install the Lower Control Arm Bushing Nut Tool J-21474-4 onto the Lower Control Arm Bushing Screw Tool J-21474-3; then, tighten the assembly until the rubber protrusion on the Lower Control Arm Bushing Receiver Tool J-39931 side bottoms against Tool J-39931.

5. Install the lower control arm.

Wheel Bearings

ADJUSTMENT

The front wheel bearing assembly used on these vehicles is a sealed, non-serviceable unit. No wheel bearing adjustments (front or rear) are necessary or possible.

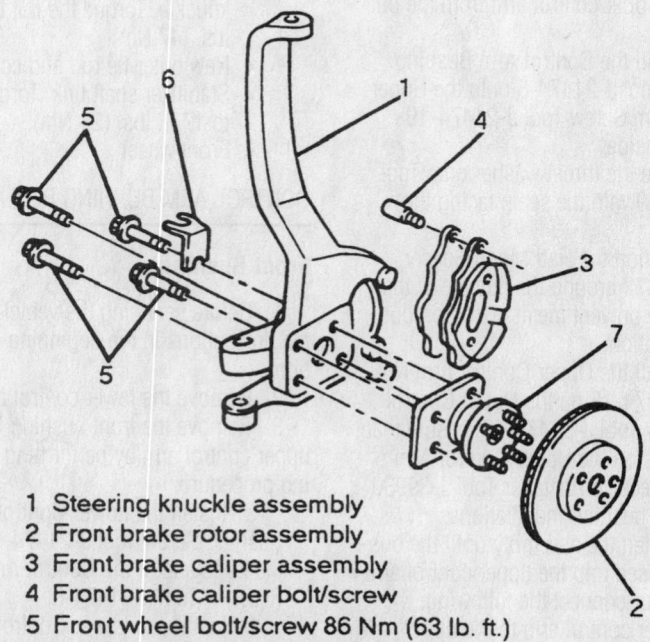

1 Steering knuckle assembly
2 Front brake rotor assembly
3 Front brake caliper assembly
4 Front brake caliper bolt/screw
5 Front wheel bolt/screw 86 Nm (63 lb. ft.)
6 Wheel speed sensor wire bracket
7 Front wheel hub assembly

7922WG35

Exploded view of the wheel hub/bearing unit mounting

REMOVAL & INSTALLATION

Front

1. Before servicing the vehicle, refer to the precautions in the beginning of this section.
2. Remove or disconnect the following:
   - Front wheel
   - Brake caliper
   - Brake rotor
   - Wheel speed sensor electrical connector
   - Hub bolts
   - Hub/bearing assembly by pulling it from the spindle

To install:

3. Install or connect the following:
   - Hub/bearing assembly on the spindle. Torque the bolts to 63 ft. lbs. (86 Nm).
   - Wheel speed sensor electrical connector

✳✳ WARNING

Be sure the wheel speed sensor electrical connector is reattached to the sensor wire bracket and sensor or the wires could be damaged.

   - Brake rotor
   - Brake caliper
   - Front wheel

Rear

1. Before servicing the vehicle, refer to the precautions in the beginning of this section.
2. Remove or disconnect the following:
   - Rear wheel
   - Brake rotor or brake drum and components, as equipped
3. Clean the axle housing cover and surrounding area to prevent dirt or contamination from entering the housing.
4. Remove the axle housing cover.
5. Install an Anti-lock Brake System (ABS) tone ring protector kit.
6. Remove or disconnect the following:
   - Rear axle pinion shaft lock screw
   - Pinion gear shaft
7. Push the flanged end of the axle shaft into the axle housing in order to access the rear axle shaft lock.
8. Remove or disconnect the following:
   - Shaft lock C-clip from the axle shaft
   - Axle shaft from the axle housing
9. Use a small suitable pry tool to remove the oil seal from the axle housing.

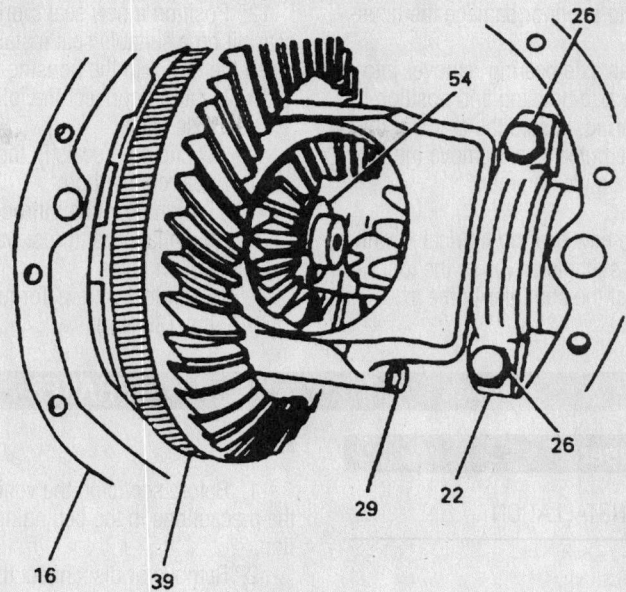

16 HOUSING, REAR AXLE
22 CAP, DIFFERENTIAL CARRIER BEARING
26 BOLT/SCREW, DIFFERENTIAL BEARING CAP
29 BOLT/SCREW, DIFFERENTIAL PINION GEAR SHAFT LOCK
39 WHEEL, REAR WHEEL SPEED SENSOR RELUCTOR
54 LOCK, REAR AXLE SHAFT

7922WG38

Differential component identification

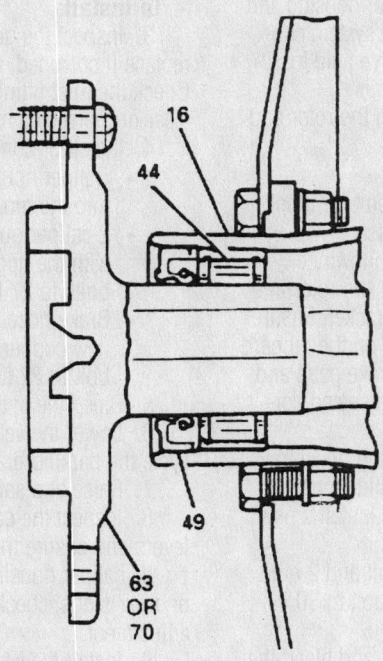

16 HOUSING, REAR AXLE
44 BEARING, REAR AXLE SHAFT
49 SEAL, REAR AXLE SHAFT BEARING
63 SHAFT, REAR AXLE (DRUM BRAKE ASSEMBLY)
70 SHAFT, REAR AXLE (DISC BRAKE ASSEMBLY)

7922WG39

Cut-away view of the rear axle bearing and seal

Be careful not to score or damage the housing.

10. Install an axle bearing remover into the bore of the axle housing and position it behind the bearing, ensure the tangs of the tool engage the outer race. Remove the bearing using a slide hammer.

To install:

11. Install a new bearing lubricated with gear oil with a suitable driver so the tool bottoms against the shoulder in the axle housing.

12. Position a new seal lubricated with gear oil on a suitable seal installer, then insert the seal into the housing bore.

13. Install or connect the following:
- Axle shaft
- C-clip so it seats in the axle side gear counterbore
- Pinion gear shaft through the differential case, thrust washer and pinion gears
- Shaft lock screw. Torque it to 27 ft. lbs. (36 Nm).

- New axle housing cover with gasket. Torque the bolts in a crosswise pattern to 22 ft. lbs. (30 Nm).

➡ **When refilling a limited slip differential rear axle with gear oil, 4 oz. (118ml) of limited slip additive should be added.**

14. Refill the rear axle with SAE 80-90W GL-5 gear lubricant, then install the plug.
- Rear brake assemblies
- Rear wheels

BRAKES

Brake Caliper

REMOVAL & INSTALLATION

Front

1. Before servicing the vehicle, refer to the precautions in the beginning of this section.

2. Remove ⅔ of the brake fluid from the master cylinder.

3. Remove or disconnect the following:
- Wheel
- Banjo bolt, inlet fitting and 2 gaskets from the caliper housing. Plug the openings in the caliper housing and inlet fitting to prevent system contamination or excessive fluid loss.
- Circlip and retainer pin
- Caliper housing from the rotor and mounting bracket

To install:

4. Check the inlet fitting bolt for blockage, clear or replace as necessary.

5. Install or connect the following:
- Caliper housing over the rotor and onto the mounting bracket. Ensure the guiding surfaces on the inboard and outboard disc brake pads and mounting bracket are seated correctly.

6. Press the caliper housing down to compress the bias springs. Slide a new retainer pin into position and install a new circlip.
- Inlet fitting, banjo bolt and 2 new gaskets. Torque the bolt to 30 ft. lbs. (40 Nm).

7. Fill the master cylinder and bleed the brake system.

8. Install the wheel and tire assembly.

9. With the engine running pump the brake pedal slowly and firmly 3 times to seat then brake pads.

Rear

1. Before servicing the vehicle, refer to the precautions in the beginning of this section.

2. Remove or disconnect the following:
- Wheel and install 2 wheel nuts to retain the rotor
- Banjo bolt inlet fitting and 2 gaskets from the caliper housing. Plug the openings in the caliper housing and inlet fitting to prevent system contamination or excessive fluid loss.
- 2 caliper guide pin bolts
- Caliper housing from the rotor and mounting bracket

To install:

3. Inspect the guide pins and boots and replace if corroded, worn or damaged. Check the inlet fitting bolt for blockage, clear or replace as necessary.

4. Install or connect the following:
- Caliper housing over the rotor and into the mounting bracket
- 2 caliper guide pin bolts starting with the upper pin and torque the bolts to 27 ft. lbs. (37 Nm)
- Brake hose to the caliper using 2 new copper gaskets and torque the bolt to 22 ft. lbs. (30 Nm)

5. Bleed the entire brake system.

6. Lower the vehicle sufficiently and cycle the parking brake.

7. Raise and safely support the vehicle.

8. Inspect the caliper parking brake levers and ensure they are against the stops on the caliper housing. If the levers are not on their stops, check the parking brake adjustment.

9. Install the wheel assembly.

10. With the engine running pump the brake pedal slowly and firmly 3 times to seat then brake pads.

11. Check the hydraulic system for leaks.

Disc Brake Pads

REMOVAL & INSTALLATION

Front

1. Before servicing the vehicle, refer to the precautions in the beginning of this section.

2. Remove or disconnect the following:
- Caliper assembly from the rotor and mounting bracket without disconnecting the brake line, then position aside. Do not allow the caliper to hang by the brake hose, suspend it with a length of wire or a fabricated hook.
- Outer pad assembly using a small prytool, if necessary, to disengage the show buttons from the caliper assembly
- Inner pad

To install:

3. If not done already, bottom the piston into the caliper using a large C-clamp positioned on the housing and inside the piston well.

4. With the piston bottomed in the caliper bore, lift the inner edge of the boot next to the piston and press out any trapped air. Make sure the boot is flat.

5. Install or connect the following:
- Inner pad assembly by positioning the retainer spring into the piston. The pad must lay flat against the piston and the boot must not touch the pad. If the boot and pad are in contact, remove the pad and reposition the boot.
- Outer pad assembly, making sure the back of the pad is flat against the caliper. The wear sensor should be at the trailing lower edge of the outer pad during forward wheel

rotation or the outer pad has been installed on the wrong side.

- Caliper assembly to the rotor and mounting bracket
- Wheel

6. With the engine running pump the brake pedal slowly and firmly 3 times to seat then brake pads.

Rear

1. Before servicing the vehicle, refer to the precautions in the beginning of this section.

2. Remove ⅔ of the brake fluid from the master cylinder reservoir.

3. Remove or disconnect the following:
- Wheel and install 2 wheel nuts to retain the rotor

4. Position a C-clamp and tighten until the piston bottoms in the base of the caliper housing. Make sure 1 end of the C-clamp rests on the inlet fitting bolt and the other against the outboard disc brake pad.

➡It is not necessary to remove the parking brake caliper lever return spring to replace the disc brake pads.

- Upper caliper guide pin bolt and discard

5. Rotate the caliper housing on the lower caliper mounting bolt. Be careful not to strain the hose or cable conduit. It may be necessary to loosen the lower caliper guide pin slightly.
- Disc brake pads and discard the shim

To install:

6. Clean all residue from the pad guide surfaces on the mounting bracket and caliper housing. Inspect the guide pins for free movement in the mounting bracket. Replace the guide pins or boots, if they are corroded or damaged.

7. Install or connect the following:
- New shim, then the disc brake pads. The outboard pad with insulator is installed toward the caliper housing. The inboard pad with the wear sensor is installed nearest the caliper piston. The wear sensor must be on the leading edge with forward wheel rotation or the pad has been installed on the wrong side.

8. Rotate the caliper housing into its operating position. The springs on the outboard brake pad must not stick through the inspection hole in the caliper housing. If the springs are sticking through the inspection hole in the caliper housing, lift the caliper housing and make the necessary corrections to the outboard brake pad positions.
- New upper caliper guide pin bolt and torque to 27 ft. lbs. (37 Nm). Ensure that the lower caliper guide bolt is torqued to 27 ft. lbs. (37 Nm).

9. With the engine running, pump the brake pedal slowly and firmly to seat the brake pads.

10. Check the caliper parking brake levers to make sure they are against the stops on the caliper housing. If the levers are not on their stops, check the parking brake adjustment.

11. Remove the 2 wheel nuts from the rotor, then align the marks and install the wheel assembly.

12. Check the master cylinder fluid level and road test the vehicle.

Brake Drums

REMOVAL & INSTALLATION

1. Before servicing the vehicle, refer to the precautions in the beginning of this section.

A WEAR SENSOR
8 BRACKET, PARKING BRAKE CABLE
9 BOLT/SCREW, LOWER CALIPER GUIDE PIN
10 SHIM
13 BOLT/SCREW, UPPER CALIPER GUIDE PIN
14 HOUSING, REAR BRAKE CALIPER
15 PIN, REAR BRAKE CALIPER GUIDE
16 BOOT, REAR BRAKE CALIPER GUIDE PIN
17 BRACKET, REAR BRAKE CALIPER ANCHOR
29 PAD, REAR DISC BRAKE INNER
30 PAD, REAR DISC BRAKE OUTER
40 DAMPENER, REAR BRAKE VIBRATION
41 NUT, REAR BRAKE VIBRATION DAMPENER

93006G92

Exploded view of the rear caliper assembly—Camaro, Firebird

2. Remove the wheel and tire assembly.

3. Matchmark the relationship of the drum to the axle flange and remove the brake drum. If the brake drum is difficult to remove, try the following:

    a. Ensure the parking brake is released.

    b. Back off the parking brake cable adjustment.

    c. Remove the adjusting hole cover or knockout plate from the backing plate. Back off the adjustment screw, using suitable brake adjusting tools.

    d. Use a rubber mallet to tap gently on the outer rim of the drum and/or around the inner drum diameter by the spindle. Be careful not to deform the drum by excessive use of force.

To install:

4. Adjust the brake shoes. The outside diameter of the shoe and linings should be 0.050 inch (1.27mm) less than the inside diameter of the brake drum on each wheel.

5. Install the drum, aligning the marks on the drum and the axle flange.

6. Install the wheel and tire assembly, aligning the marks on the wheel and axle flange.

Brake Shoes

REMOVAL & INSTALLATION

1. Before servicing the vehicle, refer to the precautions in the beginning of this section.

2. Remove or disconnect the following:
- Wheel
- Brake drum
- Return springs using a suitable tool
- Hold-down springs and pins
- Actuator pivot
- Actuator link while lifting up on the actuator
- Actuator lever and lever return spring
- Shoe guide, parking brake strut and strut spring
- Brake shoes and disconnect the parking brake lever from the appropriate shoe
- Adjusting screw assembly and spring

To install:

3. Install or connect the following:
- Parking brake lever on the appropriate shoe by hooking the lever tab into the slot
- Adjusting screw and spring. Lubricate the adjusting screw with suitable brake grease.

4. Clean and lubricate the contact points of the backing plate.
- Brake shoe assemblies and the parking brake cable to the backing plate
- Parking brake strut and strut spring by spreading the shoes apart. The strut end with the spring engages the parking brake lever and shoe. The other end engages the opposite shoe.
- Shoe guide, actuator lever and lever return spring
- Hold-down pins, actuator pivot and springs
- Actuator link on the anchor pin
- Actuator link into the actuator lever while holding up on the lever
- Shoe return springs and adjust the brakes. When properly adjusted, the brake shoe linings will be approximately 0.050 inch (1.27mm) less than the inner diameter of the brake drum.
- Brake drum
- Wheel

5. Pump the brake pedal to ensure proper operation.

SPECIFICATION CHARTS

ENGINE AND VEHICLE IDENTIFICATION

		Engine						Model Year	
Code ①	Liters (cc)	Cu. In.	Cyl.	Fuel Sys.	Engine Type	Eng. Mfg.		Code ②	Year
H	3.5 (3475)	212	6	SFI	DOHC	BOC		1	2001
1 ③	3.8 (3785)	231	6	MFI	OHV	BOC		2	2002
C	4.0 (3995)	244	8	MFI	DOHC	BOC		3	2003
								4	2004

NOTE: Oldsmobile did not market an Aurora for 2000.

DOHC: Double Overhead Camshafts

OHV: Overhead Valves

MFI: Multi-point Fuel Injection

SFI: Sequential Fuel Injection

BOC: Buick/Oldsmobile/Cadillac

① 8th position of VIN

② 10th position of VIN

③ Supercharged Engine

42372-AUR-C01

GENERAL ENGINE SPECIFICATIONS

Year	Model	Engine Displacement Liters (cc)	Engine Series (ID/VIN)	Fuel System	Net Horsepower @ rpm	Net Torque @ rpm (ft. lbs.)	Bore x Stroke (in.)	Com-pression Ratio	Oil Pressure @ rpm
2001	Aurora	3.5 (3475)	H	SFI	215@5600	230@4400	3.52x3.62	9.3:1	30@2000
		4.0 (3995)	C	SFI	250@5600	260@4000	3.43x3.31	10.2:1	35@2000
2002	Aurora	3.5 (3475)	H	SFI	215@5600	230@4400	3.52x3.62	9.3:1	30@2000
		4.0 (3995)	C	SFI	250@5600	260@4000	3.43x3.31	10.2:1	35@2000
2003	Aurora	4.0 (3995)	C	SFI	250@5600	260@4400	3.43x3.31	10.2:1	35@2000

MFI: Multi-point Fuel Injection

SFI: Sequential Fuel Injection

① Supercharged engine

42372-AUR-C02

ENGINE TUNE-UP SPECIFICATIONS

Year	Engine Displacement Liters (cc)	Engine ID/VIN	Spark Plug Gap (in.)	Ignition Timing (deg.)	Fuel Pump (psi)	Idle Speed (rpm)	Valve Clearance	
							Intake	Exhaust
2001	3.5 (3475)	H	0.050	①	41-47	②	HYD	HYD
	4.0 (3995)	C	0.050	①	41-47	②	HYD	HYD
2002	3.5 (3475)	H	0.050	①	41-47	②	HYD	HYD
	4.0 (3995)	C	0.050	①	41-47	②	HYD	HYD
2003	4.0 (3995)	C	0.050	①	41-47	②	HYD	HYD

NOTE: The Vehicle Emission Control Information label often reflects specification changes made during production.

The label figures must be used if they differ from those in the chart.

HYD: Hydraulic

① DIS Ignition System timing is not adjustable

② Idle speed is maintained by the ECM. There is no recommended adjustment procedure

42372-AUR-C03

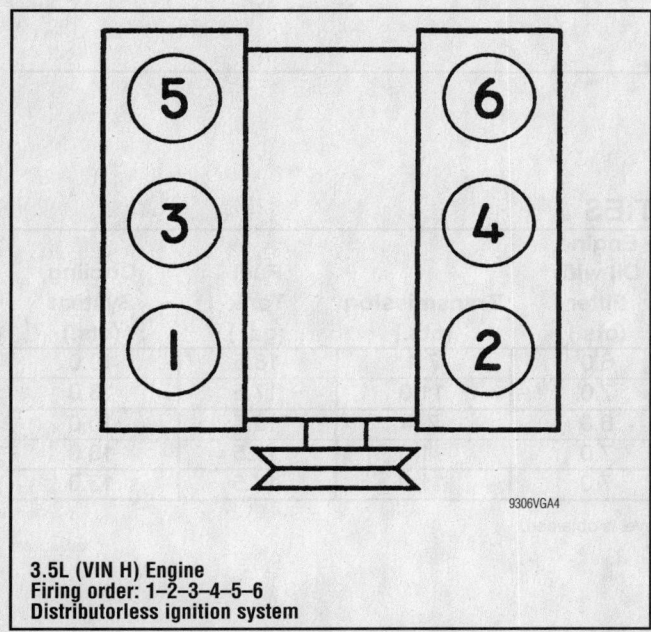

9306VGA4

3.5L (VIN H) Engine
Firing order: 1–2–3–4–5–6
Distributorless ignition system

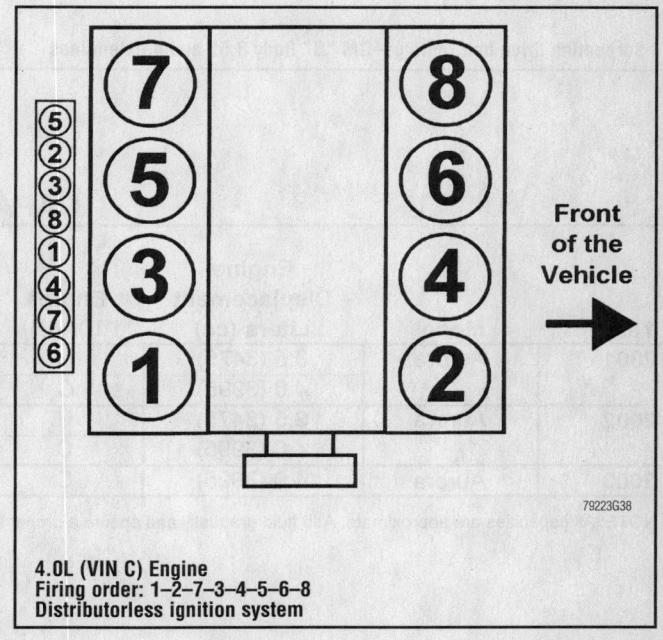

Front
of the
Vehicle
→

79223G38

4.0L (VIN C) Engine
Firing order: 1–2–7–3–4–5–6–8
Distributorless ignition system

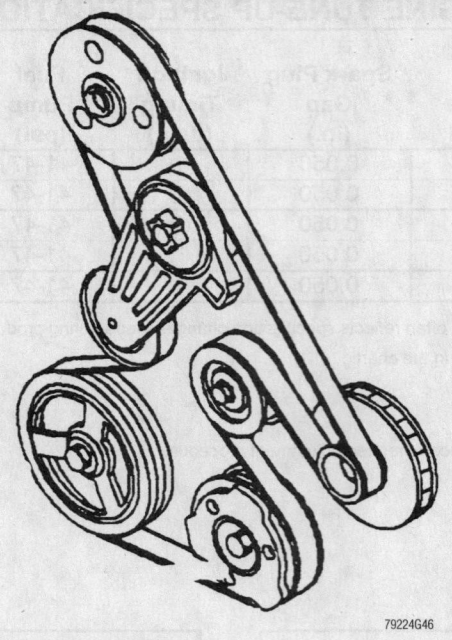

79224G46

Serpentine drive belt routing—GM "G" Body 3.5L and 4.0L engines

CAPACITIES

Year	Model	Engine Displacement Liters (cc)	Engine ID/VIN	Engine Oil with Filter (qts.)	Transmission (qts.)	Fuel Tank (gal.)	Cooling System (qts.)
2001	Aurora	3.5 (3475)	H	6.0	7.4	18.5	10.0
		4.0 (3995)	C	7.0	11.0	17.5	13.0
2002	Aurora	3.5 (3475)	H	6.0	7.4	18.5	10.0
		4.0 (3995)	C	7.0	11.0	17.5	13.0
2003	Aurora	4.0 (3995)	C	7.0	11.0	17.5	13.0

NOTE: All capacities are approximate. Add fluid gradually and ensure a proper fluid level is obtained.

42372-AUR-C04

VALVE SPECIFICATIONS

Year	Engine Displacement Liters (cc)	Engine ID/VIN	Seat Angle (deg.)	Face Angle (deg.)	Spring Test Pressure (lbs. @ in.)	Spring Installed Height (in.)	Stem-to-Guide Clearance (in.)		Stem Diameter (in.)	
							Intake	Exhaust	Intake	Exhaust
2001	3.5 (3475)	H	45.75	45	136@0.964	1.377	0.0010-0.0030	0.0020-0.0040	0.233-0.234	0.233-0.234
	4.0 (3995)	C	45.75	45	136@0.964	1.190	0.0010-0.0030	0.0020-0.0040	0.233-0.234	0.233-0.234
2002	3.5 (3475)	H	45.75	45	136@0.964	1.377	0.0010-0.0030	0.0020-0.0040	0.233-0.234	0.233-0.234
	4.0 (3995)	C	45.75	45	136@0.964	1.190	0.0010-0.0030	0.0020-0.0040	0.233-0.234	0.233-0.234
2003	4.0 (3995)	C	45.75	45	136@0.964	1.138	0.0010-0.0030	0.0020-0.0040	0.233-0.234	0.233-0.234

NA: Not Available

42372-AUR-C05

CRANKSHAFT AND CONNECTING ROD SPECIFICATIONS
All measurements are given in inches.

Year	Engine Displacement Liters (cc)	Engine ID/VIN	Crankshaft				Connecting Rod		
			Main Brg. Journal Dia.	Main Brg. Oil Clearance	Shaft End-play	Thrust on No.	Journal Diameter ①	Oil Clearance	Side Clearance
2001	3.5 (3475)	H	2.7550-2.7560	0.0006-0.0021	0.0050-0.0200	3	2.1829-2.1835	0.0009-0.0025	0.0040-0.0130
	4.0 (3995)	C	2.5335-2.5337	0.0006-0.0025	0.0020-0.0200	2	2.2490	0.0010-0.0030	0.0080-0.0200
2002	3.5 (3475)	H	2.7550-2.7560	0.0006-0.0021	0.0050-0.0200	3	2.1829-2.1835	0.0009-0.0025	0.0040-0.0130
	4.0 (3995)	C	2.5335-2.5337	0.0006-0.0025	0.0020-0.0200	2	2.2490	0.0010-0.0030	0.0080-0.0200
2003	4.0 (3995)	C	2.5335-2.5341	0.0006-0.0025	0.0020-0.0197	2	2.1239-2.1245	0.0010-0.0030	0.0079-0.0197

① Large end of connecting rod ID

42372-AUR-C06

PISTON AND RING SPECIFICATIONS

All measurements are given in inches.

Year	Engine ID/VIN	Engine Displacement Liters (cc)	Piston Clearance	Ring Gap Top Compression	Ring Gap Bottom Compression	Ring Gap Oil Control	Ring Side Clearance Top Compression	Ring Side Clearance Bottom Compression	Ring Side Clearance Oil Control
2001	3.5 (3475)	H	0.0010-0.0025	0.008-0.018	0.014-0.020	0.010-0.030	0.0016-0.0037	0.0016-0.0037	Side-sealing
	4.0 (3995)	C	0.0008-0.0020	0.0098 0.0157	0.0138 0.0020	0.0098 0.0299	0.0016-0.0037	0.0016-0.0037	Side-sealing
2002	3.5 (3475)	H	0.0010-0.0025	0.008-0.018	0.014-0.020	0.010-0.030	0.0016-0.0037	0.0016-0.0037	Side-sealing
	4.0 (3995)	C	0.0008-0.0020	0.0098 0.0157	0.0138 0.0020	0.0098 0.0299	0.0016-0.0037	0.0016-0.0037	Side-sealing
2003	4.0 (3995)	C	0.0008-0.0020	0.0098 0.0157	0.0138 0.0020	0.0098 0.0299	0.0016-0.0037	0.0016-0.0037	Side-sealing

42372-AUR-C07

TORQUE SPECIFICATIONS

All readings in ft. lbs.

Year	Engine Displacement Liters (cc)	Engine ID/VIN	Cylinder Head Bolts	Main Bearing Bolts	Rod Bearing Bolts	Crankshaft Damper Bolts	Flywheel Bolts	Manifold Intake	Manifold Exhaust	Spark Plugs	Lug Nuts
2001	3.5 (3475)	H	①	②	③	④	⑤	5	18	11	100
	4.0 (3995)	C	⑥	⑦	⑧	⑨	⑤	7.5	18	11	100
2002	3.5 (3475)	H	①	②	③	④	⑤	5	18	11	100
	4.0 (3995)	C	⑥	⑦	⑧	⑨	⑤	7.5	18	11	100
2003	4.0 (3995)	C	⑥	⑦	⑧	⑨	⑤	7.5	18	11	100

① M11 bolts:
  Step 1: 30 ft. lbs.
  Step 2: 100 degrees
  Step 3: 100 degrees
  M6 bolts: 22 ft. lbs.

② Cap Bolts
  Step 1: 15 ft. lbs.
  Step 2: 70 degrees
  Perimeter bolts: 22 ft. lbs.

③ Step 1: 22 ft. lbs.
  Step 2: Loosen completely
  Step 3: 18 ft. lbs.
  Step 4: 110 degrees

④ Step 1: 37 ft. lbs.
  Step 2: 120 degrees

⑤ Step 1: 11 ft. lbs.
  Step 2: 50 degrees

⑥ M11 bolts:
  Step 1: 30 ft. lbs.
  Step 2: 70 degrees
  Step 3: 60 degrees
  Step 4: 60 degrees
  M6 bolts: 106 inch lbs.

⑦ Step 1: 15 ft. lbs.
  Step 2: 65 degrees

⑧ Step 1: 22 ft. lbs.
  Step 2: Loosen completely
  Step 3: 18 ft. lbs.
  Step 4: 110 degrees

⑨ Step 1: 37 ft. lbs.
  Step 2: 120 degrees

⑪ New cylinder head 1st time installation: 20 ft. lbs.
  All others: 11 ft. lbs.

42372-AUR-C08

WHEEL ALIGNMENT

Year	Model		Caster Range (+/-Deg.)	Caster Preferred Setting (Deg.)	Camber Range (+/-Deg.)	Camber Preferred Setting (Deg.)	Toe-in (in.)	Steering Axis Inclination (Deg.)
2001	Aurora	F	+0.50	+6.00	+0.50	+0.20	0 +/- 0.09	12.50
		R	—	—	+0.50	-0.31	0.09 +/- 0.09	—
2002	Aurora	F	+0.50	+6.00	+0.50	+0.20	0 +/- 0.09	12.50
		R	—	—	+0.50	-0.31	0.09 +/- 0.09	—
2003	Aurora	F	+0.50	+5.00	+0.50	-0.20	0.20 +/- 0.20	NA
		R	—	—	+0.50	-0.30	0.20 +/- 0.20	—

F: Front

R: Rear

42372-AUR-C09

TIRE, WHEEL AND BALL JOINT SPECIFICATIONS

Year	Model	OEM Tires Standard	OEM Tires Optional	Tire Pressures (psi) Front	Tire Pressures (psi) Rear	Wheel Size	Ball Joint Inspection
2001	Aurora	P235/60R16	P235/60VR16	30	30	7-J	①
2002	Aurora	P235/60R16	P235/60VR16	30	30	7-J	①
2003	Aurora	P235/55HR17	NA	②	②	NA	①

OEM: Original Equipment Manufacturer

PSI: Pounds Per Square Inch

① Replace if any measurable movement is found.

② See vehicle placard

42372-AUR-C10

BRAKE SPECIFICATIONS

All measurements in inches unless noted

Year	Model		Brake Disc Original Thickness	Brake Disc Minimum Thickness	Brake Disc Maximum Run-out	Minimum Lining Thickness	Brake Caliper Mounting Bolt (ft. lbs.)
2001	Aurora	F	1.267	1.224	0.002	0.030	63
		R	0.433	0.423	0.002	0.030	20
2002	Aurora	F	1.267	1.224	0.002	0.030	63
		R	0.433	0.423	0.002	0.030	20
2003	Aurora	F	1.267	1.224	0.002	0.030	63
		R	0.433	0.423	0.002	0.030	20

F: Front

R: Rear

42372-AUR-C11

SCHEDULED MAINTENANCE INTERVALS
GM G BODY—OLDSMOBILE AURORA

TO BE SERVICED	TYPE OF SERVICE	7.5	15	22.5	30	37.5	45	52.5	60	67.5	75	82.5	90	97.5
							VEHICLE MILEAGE INTERVAL (x1000)						
Engine oil & filter	R	✓	✓	✓	✓	✓	✓	✓	✓	✓	✓	✓	✓	✓
Coolant level, hoses & clamps	S/I	✓	✓	✓	✓	✓	✓	✓	✓	✓	✓	✓	✓	✓
Driveshaft boots & front suspension components	S/I	✓	✓	✓	✓	✓	✓	✓	✓	✓	✓	✓	✓	✓
Exhaust system & brake hoses	S/I	✓	✓	✓	✓	✓	✓	✓	✓	✓	✓	✓	✓	✓
Lubricate chassis and suspension	S/I	✓	✓	✓	✓	✓	✓	✓	✓	✓	✓	✓	✓	✓
Lubricate steering linkage and transaxle linkage	S/I	✓	✓	✓	✓	✓	✓	✓	✓	✓	✓	✓	✓	✓
Lubricate parking brake cable guides, underbody contact points & linkage	S/I	✓	✓	✓	✓	✓	✓	✓	✓	✓	✓	✓	✓	✓
Throttle linkage	S/I	✓	✓	✓	✓	✓	✓	✓	✓	✓	✓	✓	✓	✓
Brake linings	S/I	✓		✓		✓		✓		✓		✓		✓
Rotate tires	S/I	✓		✓		✓		✓		✓		✓		✓
Air filter element	R				✓				✓				✓	
Engine coolant ①	R													
Spark plugs ②	R				✓				✓				✓	
Accessory drive belt(s)	S/I				✓				✓				✓	
Automatic transaxle fluid A8& filter ③	S/I													
Fuel system	S/I				✓				✓				✓	
Ignition cables	R				✓				✓				✓	
Inspect throttle body bore & throttle plate for deposits	S/I				✓				✓				✓	
Supercharger oil	S/I				✓				✓				✓	

R: Replace

S/I: Service or Inspect

① Engine coolant: replace every 100,000 miles. Use O.E. specified (DEX-COOL™) coolant only. If any silicate coolant is used, the service interval is every 30,000

② Platinum tip spark plugs: replace every 100,000 miles.

③ Replace fuid every 50, 000 miles if driven in etxreme traffic or in places where the temperature exceeds 90 degrees F

FREQUENT OPERATION MAINTENANCE (SEVERE SERVICE)

If a vehicle is operated under any of the following conditions it is considered severe service:

- Extremely dusty areas.

- 50% or more of the vehicle operation is in 32°C (90°F) or higher temperatures, or constant operation in temperatures below 0°C (32°F).

- Prolonged idling (vehicle operation in stop and go traffic).

- Frequent short running periods (engine does not warm to normal operating temperatures).

- Police, taxi, delivery usage or trailer towing usage.

CV joints & front suspension components: service or inspect every 3000 miles.

Engine oil & filter change: change every 3000 miles.

Brake linings: check every 6000 miles.

Chassis lubrication: lubricate every 6000 miles.

Suspension, steering linkage, transaxle shift linkage, parking cable guides, underbody contact points: lubricate every 6000 miles.

Throttle body mount bolt torque: tighten at 6000 miles.

Air filter element: service or inspect every 15,000 miles.

Inspect throttle body bore & throttle plate for deposits: clean as required every 15,000 miles.

Rotate tires at 6000 miles, then every 15,000 miles.

42372-AUR-C12

PRECAUTIONS

Before servicing any vehicle, please be sure to read all of the following precautions, which deal with personal safety, prevention of component damage, and important points to take into consideration when servicing a motor vehicle:

• Never open, service or drain the radiator or cooling system when the engine is hot; serious burns can occur from the steam and hot coolant.

• Observe all applicable safety precautions when working around fuel. Whenever servicing the fuel system, always work in a well-ventilated area. Do not allow fuel spray or vapors to come in contact with a spark, open flame or excessive heat (a hot drop light, for example). Keep a dry chemical fire extinguisher near the work area. Always keep fuel in a container specifically designed for fuel storage; also, always properly seal fuel containers to avoid the possibility of fire or explosion. Refer to the additional fuel system precautions later in this section.

• Fuel injection systems often remain pressurized, even after the engine has been turned **OFF**. The fuel system pressure must be relieved before disconnecting any fuel lines. Failure to do so may result in fire and/or personal injury.

• Brake fluid often contains polyglycol ethers and polyglycols. Avoid contact with the eyes and wash your hands thoroughly after handling brake fluid. If you do get brake fluid in your eyes, flush your eyes with clean, running water for 15 minutes. If eye irritation persists, or if you have taken brake fluid internally, IMMEDIATELY seek medical assistance.

• The EPA warns that prolonged contact with used engine oil may cause a number of skin disorders, including cancer! You should make every effort to minimize your exposure to used engine oil. Protective gloves should be worn when changing oil. Wash your hands and any other exposed skin areas as soon as possible after exposure to used engine oil. Soap and water, or waterless hand cleaner should be used.

• All new vehicles are now equipped with an air bag system. The system must be disabled before performing service on or around system components, steering column, instrument panel components, wiring and sensors. Failure to follow safety and disabling procedures could result in accidental air bag deployment, possible personal injury and unnecessary system repairs.

• Always wear safety goggles when working with, or around, the air bag system. When carrying a non-deployed air bag, be sure the bag and trim cover are pointed away from your body. When placing a non-deployed air bag on a work surface, always face the bag and trim cover upward, away from the surface. This will reduce the motion of the module if it is accidentally deployed. Refer to the additional air bag system precautions later in this section.

• Clean, high quality brake fluid from a sealed container is essential to the safe and proper operation of the brake system. You should always buy the correct type of brake fluid for your vehicle. If the brake fluid becomes contaminated, completely flush the system with new fluid. Never reuse any brake fluid. Any brake fluid that is removed from the system should be discarded. Also, do not allow any brake fluid to come in contact with a painted surface; it will damage the paint.

• Never operate the engine without the proper amount and type of engine oil; doing so WILL result in severe engine damage.

• Timing belt maintenance is extremely important! Many models utilize an interference-type, non-freewheeling engine. If the timing belt breaks, the valves in the cylinder head may strike the pistons, causing potentially serious (also time-consuming and expensive) engine damage. Refer to the maintenance interval charts in the front of this section for the recommended replacement interval for the timing belt, and to the timing belt procedure in this section for belt replacement and inspection.

• Disconnecting the negative battery cable on some vehicles may interfere with the functions of the on-board computer system(s) and may require the computer to undergo a relearning process once the negative battery cable is reconnected.

• When servicing drum brakes, only disassemble and assemble one side at a time, leaving the remaining side intact for reference.

• Only an MVAC-trained, EPA-certified automotive technician should service the air conditioning system or its components.

ENGINE REPAIR

Alternator

REMOVAL

3.5L Engine

1. Before servicing the vehicle, refer to the precautions in the beginning of this section.
2. Remove or disconnect the following:
   • Negative battery cable
   • Drive belt
   • Engine cooling fan assembly
   • Alternator electrical connectors
   • Idler pulley
   • 2 upper alternator mounting bolts
   • A/C compressor
   • Alternator

4.0L Engine

1. Before servicing the vehicle, refer to the precautions in the beginning of this section.
2. Remove or disconnect the following:
   • Rear seat cushion
   • Negative battery cable
   • Radiator
   • Alternator electrical connectors
   • Alternator bolts
   • Alternator

INSTALLATION

3.5L Engine

1. Install or connect the following:
   • Alternator. Do not torque the bolts at this time.

• A/C compressor. Torque the bolts and the front nut to 37 ft. lbs. (50 Nm) and the rear nut to 18 ft. lbs. (25 Nm).
• Idler pulley. Torque the bolt to 37 ft. lbs. (50 Nm).
2. Torque the alternator bolts to 37 ft. lbs. (50 Nm).
3. Install or connect the following:
   • Alternator electrical connectors. Torque the positive battery terminal to 111 inch lbs. (12.5 Nm).
   • Engine cooling fan assembly. Torque the bolts to 53 inch lbs. (6 Nm).
   • Drive belt
   • Negative battery cable

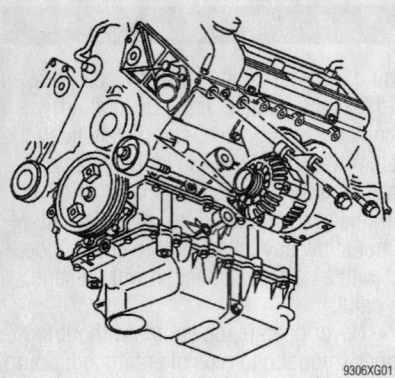

9306XG01

View of the alternator mounting—3.5L engine

4.0L Engine

1. Install or connect the following:
   - Alternator. Torque the bolts to 37 ft. lbs. (50 Nm).
   - Alternator electrical connectors
   - Radiator
   - Negative battery cable
   - Rear seat cushion

Ignition Timing

ADJUSTMENT

The engines are equipped with a Distributorless Ignition System (DIS). The system consists of 2 Crankshaft Position (CKP) sensors, crankshaft reluctor ring, Camshaft Position (CMP) sensor, Ignition Control Module (ICM), 4 ignition coils, spark plug wires and spark plugs, Knock Sensor (KS) and the Powertrain Control Module (PCM).

The PCM controls spark advance under all driving conditions. The PCM incorporates a permanent spark control override, which electronically lowers the base timing if spark knock (detonation) is encountered during normal operation due to the use of low octane fuel.

Engine Assembly

REMOVAL & INSTALLATION

3.5L Engine

1. Before servicing the vehicle, refer to the precautions in the beginning of this section.

2. Drain the crankcase and cooling system.

3. Recover the air conditioning system refrigerant.

4. Remove or disconnect the following:
   - Negative battery cable
   - Vacuum brake booster hose from the vacuum brake booster and secure to the top of the engine
   - Fuel inlet and return quick-connect fittings at the fuel rail and secure to the air inlet grille
   - Hose from the evaporative emission canister purge valve and secure to the air inlet grille
   - Air cleaner assembly
   - 2 fuel injector sight shield nuts at the front of the fuel injector sight shield
   - Fuel injector sight shield (lift up) at the front and slide out of the engine bracket
   - Battery positive cable from the remote positive terminal and secure to the top of the engine
   - Secondary AIR relay from the relay bracket and secure to the top of the engine

5. Disconnect and secure the following wiring harness electrical connectors to the top of the engine:
   - Powertrain Control Module (PCM)
   - Connection C101
   - Engine electrical harness

6. Remove or disconnect the following:
   - Engine ground cable from the right side body frame rail

✳✳ CAUTION

To avoid possible injury or vehicle damage, always replace the accelerator control cable with a NEW cable whenever you remove the engine from the vehicle. Also to avoid cruise control cable damage, position the cable out of the way while you remove or install the engine. Do not pry or lean against the cruise control cable and do not kink the cable. You must replace a damaged cable.

   - Cruise control cable from the throttle body bracket and lever
   - Accelerator control cable from the throttle body
   - Transaxle range selector cable terminal from the transaxle manual shift lever pin
   - Transaxle range selector cable from the bracket and position the cable aside
   - Radiator inlet hose from the water housing crossover
   - Radiator outlet hose from the thermostat housing
   - Surge tank inlet hose from the surge tank
   - Heater outlet hose from the thermostat housing
   - Heater inlet hose from the water crossover
   - Two master cylinder brake pipes from the Brake Pressure Modulator Valve (BPMV) and plug the lines
   - Transaxle oil cooler pipe bracket bolt from the fan shroud
   - Plastic cap off the transaxle oil cooler pipe quick connect fittings
   - Upper and lower transaxle oil cooler pipes from the radiator

➡ **Make sure the wheels of the vehicle are in the straight position ahead and the steering column in the LOCK position before disconnecting the steering column or intermediate shaft from the steering gear. Failure to do so will cause the SIR coil assembly to become uncentered, which may cause damage to the coil assembly.**

7. Lock the steering column by installing tool J42460 into the underside of the steering column.
   - Right and left side strut tower bolts
   - Transaxle oil cooler pipe bracket bolt from the lower tie bar
   - Exhaust manifold pipe
   - Front wheels
   - Front wheel speed sensor electrical leads from the frame rail
   - Front air deflector
   - Front fascia extensions
   - Secondary AIR inlet hose from the secondary AIR pump
   - 2 nuts securing the front brake pipe frame brackets to the frame rails
   - Front brake pipes from the retainers at the frame rails
   - Front brake pipes away from the body frame rails by carefully pulling them
   - 2 rear brake pipes at the rear of the engine frame
   - A/C pressure sensor
   - A/C discharge hose from the compressor and secure to the cooling fan assembly
   - A/C suction hose from the compressor and secure to the cooling fan assembly

Failure to disconnect the intermediate shaft from the rack and pinion stub shaft can result in damage to the steering gear and/or damage to the intermediate shaft. This damage may cause loss of steering control which could result in personal injury.

- Intermediate shaft pinch bolt
- Steering gear from the intermediate shaft
- Post Heated Oxygen Sensor (HO2S) at the pigtail
- Torque converter cover
- Torque converter-to-flywheel bolts

8. Position a suitable powertrain support dolly under the engine frame and lower the vehicle onto the dolly,

9. If the powertrain support dolly is unavailable. Support the powertrain with 4 suitable jackstands. Place a 2 in x 4 in block of wood between the front of the engine oil pan and the engine frame.

10. Remove or disconnect the following:
- Nut securing the right engine mount to the engine mount bracket
- Nut securing the left transaxle mount to the transaxle mount bracket

To avoid any vehicle damage, serious personal injury or death when major components are removed from the vehicle and the vehicle is supported by a hoist, support the vehicle with jack stands at the opposite end from which the components are being removed.

11. Secure the front hoist pads to the vehicle.
- 6 frame to body mounting bolts (3).

12. Make clearance is maintained between the powertrain assembly and the following components:
- A/C accumulator hose
- A/C compressor hose
- Brake pipes
- Heater hoses
- Radiator hoses
- Wheel speed sensor leads
- Wiring harnesses

13. Carefully raise the vehicle in order to clear the supported engine/transaxle assembly.

14. Remove or disconnect the following:
- Secondary air injection crossover pipe/hose
- Engine wiring harness from the engine
- Power steering return hose from the power steering pump reservoir
- Power steering return hose retaining bolt from the cylinder head
- Power steering pressure hose from the power steering pump
- Bolt securing the power steering pressure hose to the secondary AIR valve bracket
- 3 bolts securing the right engine mount bracket to the engine
- Right engine mount bracket
- Bolts securing the transaxle brace to the engine and transaxle

15. Attach an engine lift chain to the engine lift brackets, then attach the chain to an engine lift device.
- Nut securing the front engine mount to the front engine mount bracket
- Bolts attaching the engine to the transaxle
- Engine

To install:

16. Installation is the reverse of removal, please note the following torques:
- Engine-to-transaxle bolts to 55 ft. lbs. (75 Nm)
- Front engine mount-to-bracket nut to 41 ft. lbs. (55 Nm)
- Transaxle brace bolts to 37 ft. lbs. (50 Nm)
- Right engine mount bracket bolts to 44 ft. lbs. (60 Nm)
- Power steering pressure hose to power steering pump to 22 ft. lbs. (30 Nm)
- Power steering pressure hose retaining bolt to 80 inch lbs. (9 Nm)
- Power steering return hose retaining bolt to 13 ft. lbs. (17 Nm)
- Frame mounting bolts to 141 ft. lbs. (191 Nm)
- Engine mount nuts to 59 ft. lbs. (80 Nm)
- Flywheel-to-torque converter bolts to 48 ft. lbs. (65 Nm)
- Intermediate shaft to steering gear pinch bolt to 33 ft. lbs. (45 Nm)
- A/C hoses to 15 ft. lbs. (20 Nm)

17. Connect the battery negative cable.
18. Fill the engine with oil.
19. Fill the cooling system.
20. Bleed the brake system.
21. Recharge the A/C system.
22. Bleed the power steering system.
23. Check the wheel alignment.
24. Complete the following procedure after the engine is installed in the vehicle:

a. With the ignition **OFF** or disconnected, crank the engine several times. Listen for any unusual noises or evidence that any parts are binding.

b. Start the engine and listen for abnormal conditions.

c. Check the vehicle oil pressure gauge or light and confirm that the engine has acceptable oil pressure.

d. Run the engine at approximately 1000 RPM until the engine reaches normal operating temperature.

e. While the engine continues to idle raise and support the vehicle.

f. Inspect for oil, coolant and exhaust leaks while the engine is idling.

g. Lower the vehicle.

h. Perform the Crankshaft Position (CKP) system variation learn procedure.

i. Perform a final inspection for the proper engine oil and coolant levels.

j. Road test the vehicle.

4.0L Engine

1. Before servicing the vehicle, refer to the precautions in the beginning of this section.

2. Drain the crankcase and cooling system.

3. Recover the air conditioning system refrigerant.

4. Remove or disconnect the following:
- Negative battery cable
- Vacuum brake booster hose from the vacuum brake booster
- Fuel inlet and return quick-connect fittings at the fuel rail and secure to the air inlet grille
- Hose from the evaporative emission canister purge valve and secure to the air inlet grille
- Air cleaner assembly
- 2 fuel injector sight shield nuts at the front of the fuel injector sight shield
- Fuel injector sight shield (lift up) at the front and slide out of the engine bracket
- Battery positive cable from the remote positive terminal and secure to the top of the engine
- Secondary AIR relay from the relay bracket and secure to the top of the engine

5. Disconnect and secure the following wiring harness electrical connectors to the top of the engine:
- Powertrain control Module (PCM)
- Connection C101
- Engine electrical harness

6. Remove or disconnect the following:

- Engine ground cable from the right side body frame rail

⁂ CAUTION

To avoid possible injury or vehicle damage, always replace the accelerator control cable with a NEW cable whenever you remove the engine from the vehicle.

Also to avoid cruise control cable damage, position the cable out of the way while you remove or install the engine. Do not pry or lean against the cruise control cable and do not kink the cable. You must replace a damaged cable.

- Cruise control cable from the throttle body bracket and lever
- Accelerator control cable from the throttle body
- Transaxle range selector cable terminal from the transaxle manual shift lever pin
- Transaxle range selector cable from the bracket and position the cable aside
- Radiator inlet hose from the water housing crossover
- Radiator outlet hose from the thermostat housing
- Surge tank inlet hose from the surge tank
- Heater hoses
- Two master cylinder brake pipes from the Brake Pressure Modulator Valve (BPMV) and plug the lines
- Transaxle oil cooler pipe bracket bolt from the fan shroud
- Plastic cap off the transaxle oil cooler pipe quick connect fittings
- Upper and lower transaxle oil cooler pipes from the radiator

➡ **Make sure the wheels of the vehicle are in the straight position ahead and the steering column in the LOCK position before disconnecting the steering column or intermediate shaft from the steering gear. Failure to do so will cause the SIR coil assembly to become uncentered, which may cause damage to the coil assembly.**

7. Lock the steering column by installing tool J42460 into the underside of the steering column.
- Right and left side strut tower bolts
- Exhaust manifold pipe
- Front wheels

- Front wheel speed sensor electrical leads from the frame rail
- Front air deflector
- Front fascia extensions
- Secondary AIR inlet hose from the secondary AIR pump
- 2 nuts securing the front brake pipe frame brackets to the frame rails
- Front brake pipes from the retainers at the frame rails
- Front brake pipes away from the body frame rails by carefully pulling them
- 2 rear brake pipes at the rear of the engine frame
- A/C pressure sensor
- A/C discharge hose from the compressor and secure to the cooling fan assembly
- A/C suction hose from the compressor and secure to the cooling fan assembly

⁂ CAUTION

Failure to disconnect the intermediate shaft from the rack and pinion stub shaft can result in damage to the steering gear and/or damage to the intermediate shaft. This damage may cause loss of steering control which could result in personal injury.

- Intermediate shaft pinch bolt
- Steering gear from the intermediate shaft
- Post Heated Oxygen Sensor (HO_2S) at the pigtail
- Brace between the engine oil pan and the transaxle case
- Torque converter cover
- Torque converter-to-flywheel bolts

8. Position a suitable powertrain support dolly under the engine frame and lower the vehicle onto the dolly,

9. If the powertrain support dolly is unavailable. Support the powertrain with 4 suitable jackstands. Place a 2 in x 4 in block of wood between the front of the engine oil pan and the engine frame.

10. Remove or disconnect the following:
- Nut securing the right engine mount to the engine mount bracket
- Nut securing the left transaxle mount to the transaxle mount bracket

⁂ CAUTION

To avoid any vehicle damage, serious personal injury or death when major components are removed from the vehicle and the vehicle is supported

by a hoist, support the vehicle with jack stands at the opposite end from which the components are being removed.

11. Secure the front hoist pads to the vehicle.
- 6 frame to body mounting bolts (3).

12. Make clearance is maintained between the powertrain assembly and the following components:
- A/C accumulator hose
- A/C compressor hose
- Brake pipes
- Heater hoses
- Radiator hoses
- Wheel speed sensor leads
- Wiring harnesses

13. Carefully raise the vehicle in order to clear the supported engine/transaxle assembly.

14. Remove or disconnect the following:
- Heater pipes
- Intermediate hose from the secondary AIR valve at bank 2
- Nut securing the intermediate hose to the secondary AIR valve at bank 1
- Nut securing the coil cassette ground wire to the right cylinder head
- Engine wiring harness from the engine
- Power steering return hose from the power steering pump reservoir
- Power steering return hose retaining bolt from the cylinder head
- Power steering pressure hose from the power steering pump
- Bolt securing the power steering pressure hose to the to the right engine mount bracket
- Bolts securing the right engine mount bracket to the engine
- Right engine mount bracket
- Bolts/nuts securing the transaxle brace's to the engine and transaxle

15. Attach an engine lift chain to the engine lift brackets, then attach the chain to an engine lift devise.
- Nut securing the front engine mount to the front engine mount bracket
- Bolts attaching the engine to the transaxle
- Engine

To install:
16. Installation is the reverse of removal, please note the following torques:
- Engine-to-transaxle bolts to 55 ft. lbs. (75 Nm)
- Front engine mount-to-bracket nut to 52 ft. lbs. (70 Nm)

- Transaxle brace bolts to 37 ft. lbs. (50 Nm)
- Right engine mount bracket bolts to 37 ft. lbs. (50 Nm)
- Power steering pressure hose to power steering pump to 22 ft. lbs. (30 Nm)
- Power steering pressure hose retaining bolt to 80 inch lbs. (9 Nm)
- Power steering return hose retaining bolt to 37 ft. lbs. (50 Nm)
- Frame mounting bolts to 141 ft. lbs. (191 Nm)
- Engine mount nuts to 59 ft. lbs. (80 Nm)
- Flywheel-to-torque converter bolts to 44 ft. lbs. (60 Nm)
- Intermediate shaft to steering gear pinch bolt to 33 ft. lbs. (45 Nm)
- A/C hoses to 15 ft. lbs. (20 Nm)

17. Connect the battery negative cable.
18. Fill the engine with oil.
19. Fill the cooling system.
20. Bleed the brake system.
21. Recharge the A/C system.
22. Bleed the power steering system.
23. Check the wheel alignment.
24. Complete the following procedure after the engine is installed in the vehicle:

a. With the ignition **OFF** or disconnected, crank the engine several times. Listen for any unusual noises or evidence that any parts are binding.

b. Start the engine and listen for abnormal conditions.

c. Check the vehicle oil pressure gauge or light and confirm that the engine has acceptable oil pressure.

d. Run the engine at approximately 1000 RPM until the engine reaches normal operating temperature.

e. While the engine continues to idle raise and support the vehicle.

f. Inspect for oil, coolant and exhaust leaks while the engine is idling.

g. Lower the vehicle.

h. Perform the Crankshaft Position (CKP) system variation learn procedure.

i. Perform a final inspection for the proper engine oil and coolant levels.

j. Road test the vehicle.

Water Pump

REMOVAL & INSTALLATION

3.5L Engine

1. Before servicing the vehicle, refer to the precautions in the beginning of this section.

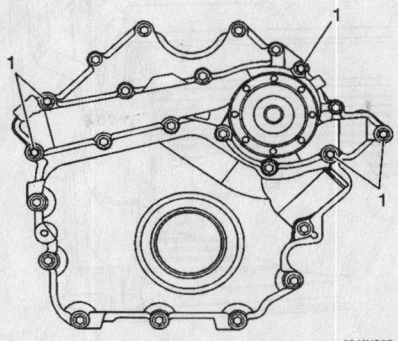

Be sure to install the 5 long water pump bolts (1) in the correct locations—3.5L engine

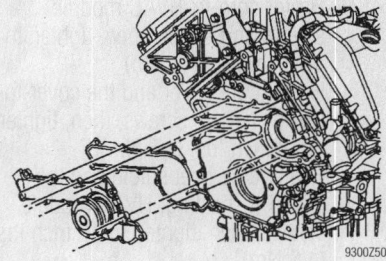

Water pump mounting—3.5L engine

2. Drain the cooling system.
3. Remove or disconnect the following:
- Water pump pulley bolts, loosen but do not remove at this time
- Drive belt
- Idler pulley
- Water pump pulley

➡The water pump is attached to the engine with both long and short bolts, be sure to note their locations.

- Water pump

To install:

4. Install or connect the following:
- Water pump using a new gasket. Torque the bolts to 124 inch lbs. (14 Nm).
- Water pump pulley. Torque the bolts to 106 inch lbs. (12 Nm).
- Idler pulley. Torque the bolt to 37 ft. lbs. (50 Nm).
- Drive belt
5. Refill the cooling system.
6. Start the engine and check for leaks.

4.0L Engine

1. Before servicing the vehicle, refer to the precautions in the beginning of this section.
2. Drain the cooling system.
3. Remove or disconnect the following:
- Negative battery cable
- Air cleaner assembly
- Secondary AIR injection control valve, if necessary
- Oil level indicator tube nut, if necessary
- Water pump drive belt cover
- Water pump drive belt
- Lower radiator hose
- Bypass hose
- Water pump cover
- Water pump from the water pump housing, by turning the locking ring with a Water Pump Remover/Installer tool J-38816

To install:

4. Install or connect the following:
- Water pump with new o-ring. Torque the pump to 73 ft. lbs. (100 Nm).
- Water pump cover. Torque the bolts to 89 inch lbs. (10 Nm).
- Lower radiator hose

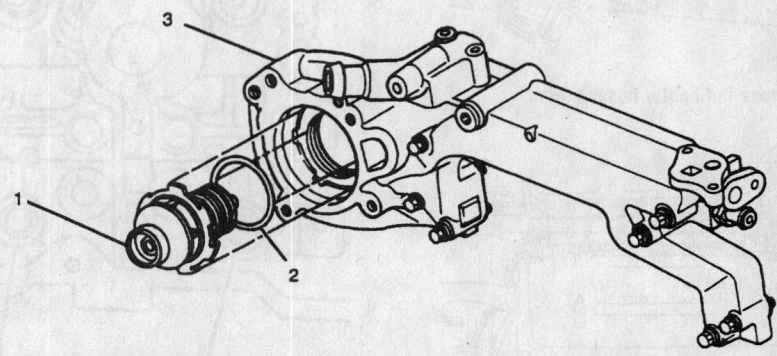

| 1 | WATER PUMP ASM. | 3 | WATER PUMP HOUSING ASM. |
| 2 | O-RING SEAL | | |

Exploded view of the water pump housing and water pump—4.0L engine

- Coolant bypass hose
- Water pump drive belt
- Water pump drive belt cover
- Oil level indicator tube nut, if necessary
- Secondary AIR injection control valve, if necessary
- Air cleaner assembly
- Negative battery cable
5. Refill and bleed the cooling system.

Heater Core

REMOVAL & INSTALLATION

1. Disconnect the negative battery cable.
2. Drain the cooling system into a clean container for reuse.
3. Remove or disconnect the following:

- Heater hoses from the heater core
- Center console assembly
- Auxiliary air duct connector
- Right sound insulator
- Heater core heat shield-to-heating, HVAC module screws and the shield
- Heater core cover-to-HVAC module screws and the cover
- Heater core-to-HVAC module screws and straps
- Heater core

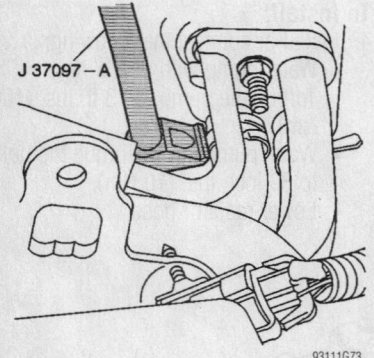

Remove the heater hoses clamp

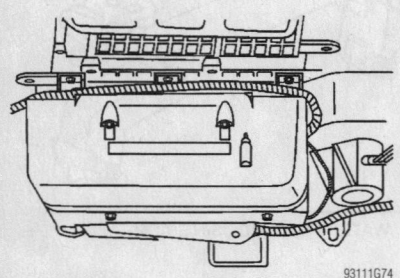

View of the heater core cover

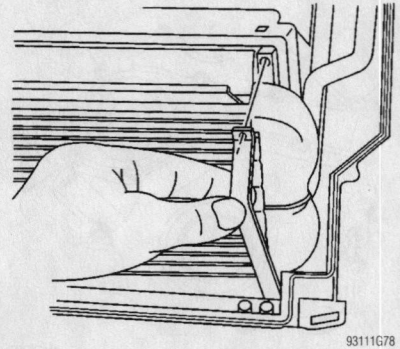

View of the heater core retaining straps

To install:
4. Install or connect the following:

- Heater core
- Heater core-to-HVAC module screws and straps, then, tighten to 18 inch lbs. (1.5 Nm)
- Heater core cover and the cover-to-HVAC module screws, then, tighten to 18 inch lbs. (1.5 Nm)
- Heater core heat shield and the shield-to-heating, HVAC module screws, then, tighten to 18 inch lbs. (1.5 Nm)
- Right sound insulator
- Auxiliary air duct connector
- Center console assembly
- Heater hoses to the heater core

5. Refill the cooling system.
6. Connect the negative battery cable.

Cylinder Head

REMOVAL & INSTALLATION

3.5L Engine

LEFT SIDE (FRONT)

1. Before servicing the vehicle, refer to the precautions in the beginning of this section.
2. Drain the crankcase and the cooling system.
3. Remove or disconnect the following:

- Negative battery cable
- Intake manifold
- Coolant crossover pipe
- Left exhaust manifold
- Camshaft cover

4. Install a holding tool on the camshafts to hold them in position.
5. Remove or disconnect the following:

- Camshaft primary chain
- Camshafts from the left cylinder head
- Rocker arms and valve lifters

➡Be sure to keep the rocker arms and lifters in order so they can be installed in their original locations.

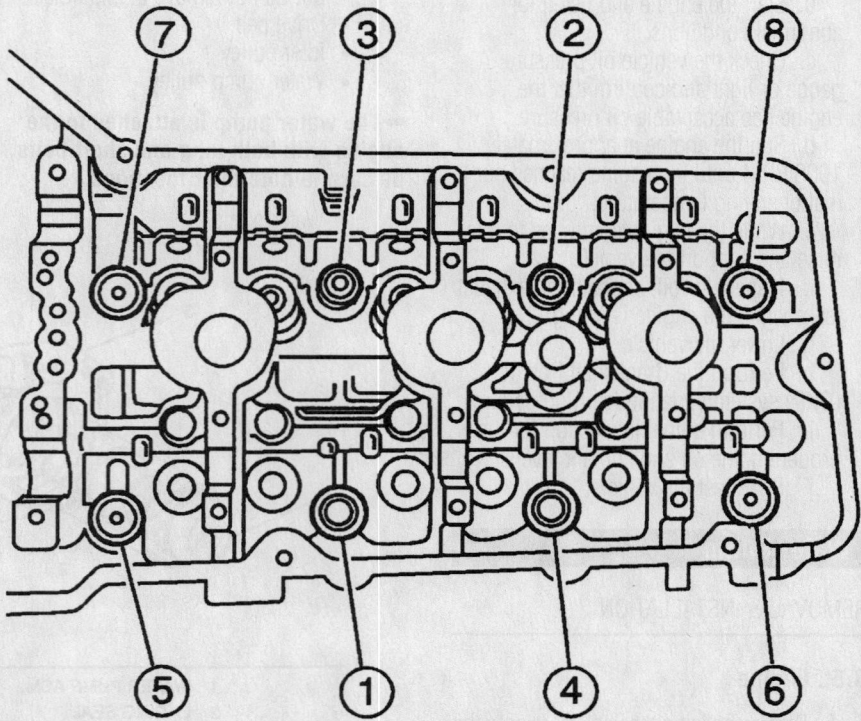

Front cylinder head bolt torque sequence—3.5L (VIN H) engine

Camshaft holding fixture J-42038—3.5L engine

- M6 bolts from the front of the cylinder head, note the location of the longer bolt
- M11 cylinder head bolts
- Cylinder head

To install:

6. Install the cylinder head with a new gasket.

7. Install new M11 bolts and M6 bolts.

8. Torque the M11 bolts in sequence to:

   a. Step 1: 30 ft. lbs. (40 Nm).
   b. Step 2: 100 degree turn.
   c. Step 3: 100 degree turn.

9. Torque the long M6 bolt to 22 ft. lbs. (30 Nm).

10. Torque both shorter M6 bolts to 106 inch lbs. (12 Nm).

11. Install or connect the following:
- Lifters
- Rocker arms
- Camshafts
- Primary chain

12. Remove the timing chain holding tool.

13. Install or connect the following:
- Camshaft covers. Torque the bolts to 80 inch lbs. (9 Nm).
- Exhaust manifold. Torque the bolts to 22 ft. lbs. (30 Nm).
- Coolant crossover pipe. Torque the bolts to 18 ft. lbs. (25 Nm).
- Intake manifold. Torque the bolts to 62 inch lbs. (7 Nm).
- New oil filter
- Negative battery cable

14. Refill the crankcase.

15. Refill the cooling system.

16. Start the engine and check for leaks.

RIGHT SIDE (REAR)

1. Before servicing the vehicle, refer to the precautions in the beginning of this section.

2. Drain the crankcase and the cooling system.

3. Remove or disconnect the following:
- Negative battery cable
- Intake manifold
- Coolant crossover pipe

➡ **Do not remove the rear exhaust manifold. Detach it from the cylinder head and the connection from the front manifold; then, move it aside.**

- Rear exhaust manifold from the cylinder head and front manifold
- Camshaft cover

4. Install a holding tool on the camshafts to hold them in position.

5. Remove or disconnect the following:
- Primary camshaft chain
- Right side camshafts
- Rocker arms and lifters

➡ **Keep the rocker arms and lifters in order so they can be installed in their original positions.**

- Engine Coolant Temperature (ECT) sensor from the cylinder head
- M6 bolts from the front of the cylinder head, note the location of the longer bolt
- M11 cylinder head bolts
- Cylinder head

To install:

6. Install the cylinder head with a new gasket.

7. Install new M11 bolts.

8. Install M6 bolts in the front of the cylinder head.

9. Tighten the M11 bolts in sequence using the following steps:

   a. Step 1: 30 ft. lbs. (40 Nm).
   b. Step 2: 100 degree turn.
   c. Step 3: 100 degree turn.

10. Torque the long M6 bolt to 22 ft. lbs. (30 Nm).

11. Torque both short M6 bolts to 106 inch lbs. (12 Nm).

12. Install or connect the following:
- ECT sensor. Torque the sensor to 15 ft. lbs. (20 Nm).
- Lifters and rocker arms in their original positions
- Camshafts
- Primary chain
- Camshaft covers. Torque the bolts to 80 inch lbs. (9 Nm).
- Exhaust manifold on the cylinder head. Torque the bolts to 22 ft. lbs. (30 Nm).
- Coolant crossover pipe. Torque the bolts to 18 ft. lbs. (25 Nm).
- Intake manifold. Torque the bolts to 62 inch lbs. (7 Nm).
- New oil filter
- Negative battery cable

13. Refill the crankcase.

14. Refill the cooling system.

15. Start the engine and check for leaks.

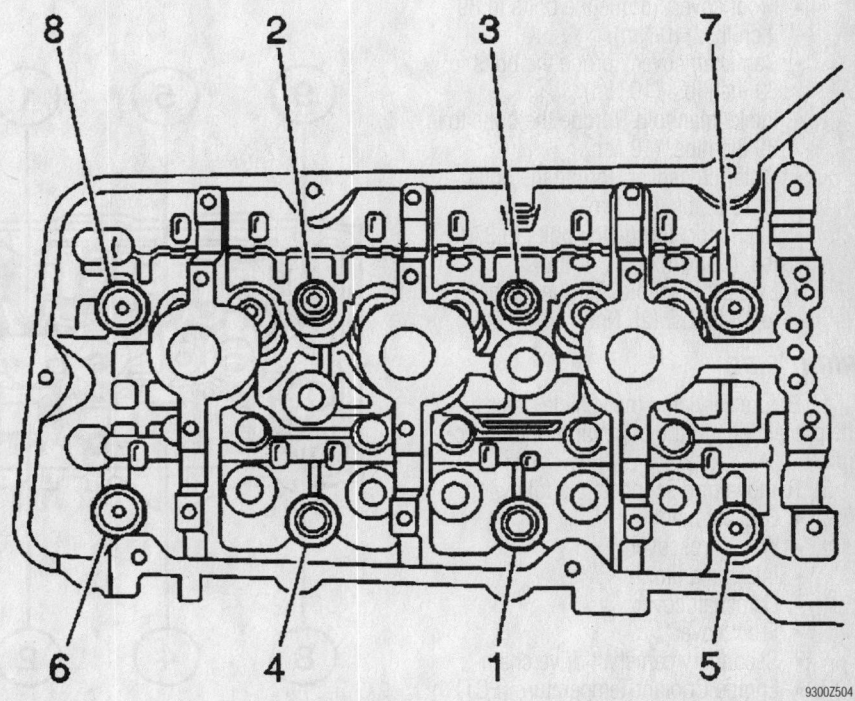

Rear cylinder head bolt torque sequence—3.5L (VIN H) engine

4.0L Engine

LEFT SIDE

1. Before servicing the vehicle, refer to the precautions in the beginning of this section.

2. Remove or disconnect the following:
   - Exhaust manifold
   - Alternator
   - Water crossover
   - Intake manifold
   - Camshaft cover
   - Front cover
   - Secondary camshaft drive chain
   - Power steering hose retaining bolt from the head
   - Cylinder head

To install:

3. Install the cylinder head with a new gasket.

4. Install new M11 head bolts and M6 head bolts.

5. Torque the M11 bolts in sequence as follows:
   a. Step 1: 30 ft. lbs. (40 Nm).
   b. Step 2: 70 degree turn.
   c. Step 3: 60 degree turn.
   d. Step 4: 60 degree turn.

6. Torque the M6 bolts to 106 inch lbs. (12 Nm).

7. Install or connect the following:
   - Power steering hose retaining bolt. Torque the bolt to 37 ft. lbs. (50 Nm).
   - Secondary camshaft drive chain
   - Front cover. Torque the bolts to 89 inch lbs. (10 Nm).
   - Camshaft cover. Torque the bolts to 89 inch lbs. (10 Nm).
   - Intake manifold. Torque the bolts to 89 inch lbs. (10 Nm).
   - Water crossover. Torque the bolts to 18 ft. lbs. (25 Nm).
   - Alternator. Torque the bolts to 37 ft. lbs. (50 Nm).
   - Exhaust manifold. Torque the bolts to 18 ft. lbs. (25 Nm).

RIGHT SIDE

1. Before servicing the vehicle, refer to the precautions in the beginning of this section.

2. Remove or disconnect the following:
   - Exhaust manifold
   - Water crossover
   - Intake manifold
   - Camshaft cover
   - Front cover
   - Secondary camshaft drive chain
   - Engine Coolant Temperature (ECT) sensor electrical connector
   - Ground wire from the head

   - Bolt securing the exhaust crossover pipe to the cylinder head
   - Front and rear transmission braces from the head
   - Cylinder head

To install:

3. Install the cylinder head with a new gasket.

4. Install new M11 head bolts and M6 head bolts.

5. Torque the M11 head bolts in sequence as follows:
   a. Step 1: 30 ft. lbs. (40 Nm).
   b. Step 2: 70 degree turn.
   c. Step 3: 60 degree turn.
   d. Step 4: 60 degree turn.

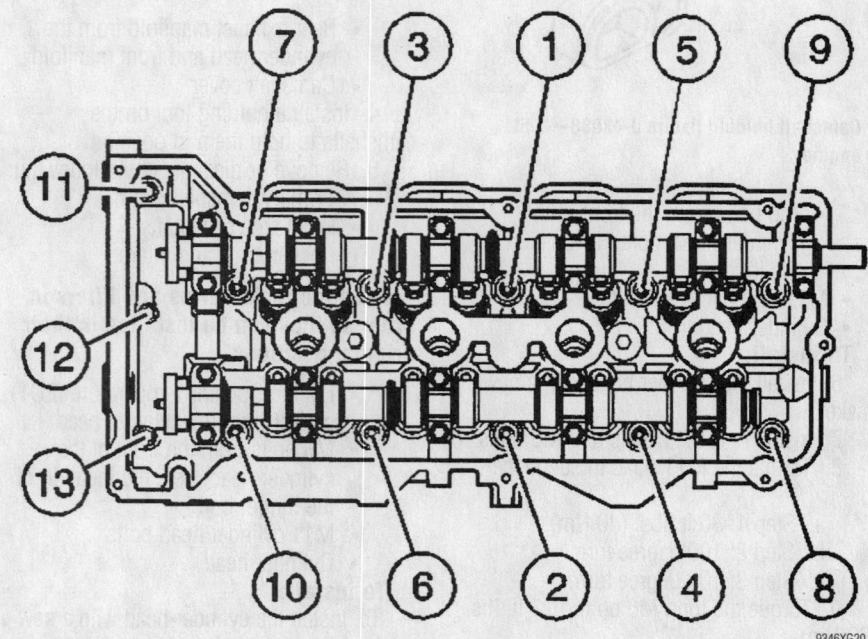

Left cylinder head bolt torque sequence—4.0L models

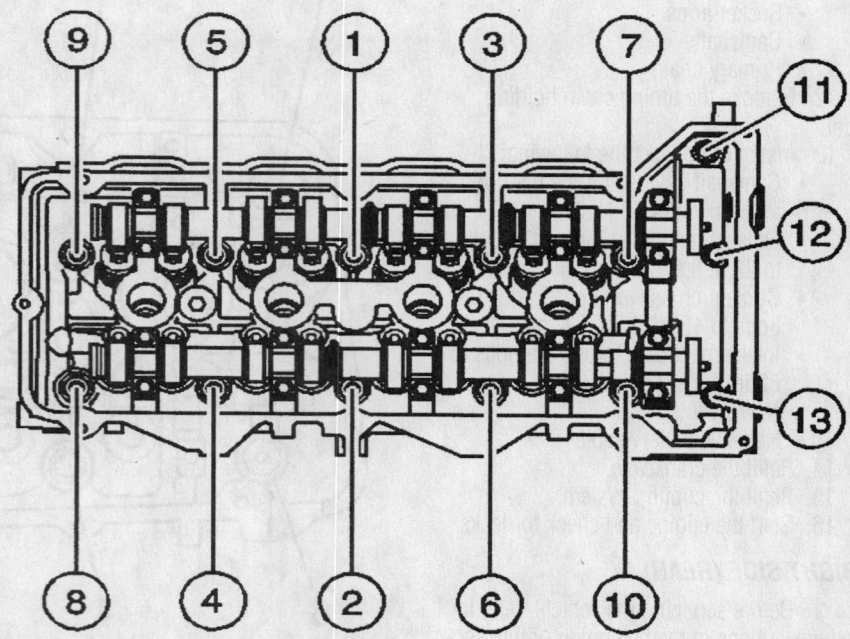

Right cylinder head bolt torque sequence—4.0L models

6. Torque the M6 head bolts to 106 inch lbs. (12 Nm).
7. Install or connect the following:
   - Front and rear transmission braces to the head. Torque the bolts to 37 ft. lbs. (50 Nm).
   - ECT electrical connector
   - Ground wire to the head. Torque the nut to 13 ft. lbs. (17 Nm).
   - Bolt securing the exhaust crossover pipe to the cylinder head. Torque the bolt to 18 ft. lbs. (25 Nm).
   - Secondary camshaft drive chain
   - Front cover. Torque the bolts to 18 ft. lbs. (25 Nm).
   - Camshaft cover. Torque the bolts to 89 inch lbs. (10 Nm).
   - Intake manifold. Torque the bolts to 89 inch lbs. (10 Nm).
   - Water crossover. Torque the bolts to 18 ft. lbs. (25 Nm).
   - Exhaust manifold. Torque the bolts to 22 ft. lbs. (30 Nm).

Rocker Arms

REMOVAL & INSTALLATION

➡**All valve train components should be kept in the order that they were removed, so that they can be reinstalled in their original position.**

3.5L Engine

Refer to the camshaft removal and installation procedure for rocker arm service.

4.0L Engine

The 4.0L engine is not equipped with rocker arms. The camshaft directly actuates the valves.

Intake Manifold

REMOVAL & INSTALLATION

3.5L Engine

1. Before servicing the vehicle, refer to the precautions in the beginning of this section.
2. Partially drain the engine coolant.
3. Relieve the fuel system pressure.
4. Remove or disconnect the following:
   - Air duct from the throttle body
   - Fuel injector sight shield
   - Accelerator cable and bracket
   - Cruise control cable and bracket
   - Coolant hoses from the throttle body

- Fuel lines from the fuel supply rail
- Fuel vapor line from the Evaporative Emission (EVAP) canister purge solenoid
- Brake booster vacuum hose
- Air conditioning vacuum hose from the engine
- Surge tank inlet pipe retainer from the fuel supply rail
- Fuel injector electrical connectors
- Throttle Position Sensor (TPS) electrical connector
- Idle Air Control (IAC) valve electrical connector
- Evaporative Emission (EVAP) canister purge solenoid connector
- Manifold Absolute Pressure (MAP) sensor connector
- Wiring harness from the camshaft covers
- Vacuum hose from the fuel pressure regulator and throttle body
- Positive Crankcase Ventilation (PCV) tubes from both camshaft covers and the intake manifold
- Exhaust Gas Recirculation (EGR) valve outlet pipe
- Fuel supply rail with injectors

➡**Disengage the snap-lock retainers by pushing toward the camshaft covers and lifting.**

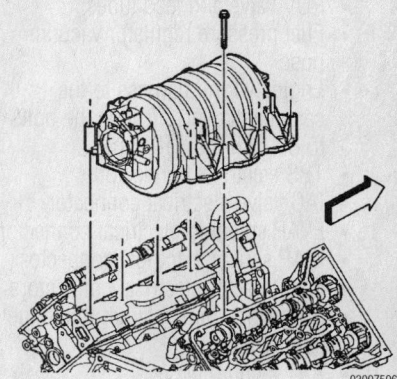

Intake manifold assembly—3.5L engine

- Intake manifold

➡**The manifold-to-cylinder head seals are reusable unless cut or damaged.**

To install:
5. Install or connect the following:
   - Intake manifold with new gaskets (if necessary). Torque the bolts in sequence to 89 inch lbs. (10 Nm).
   - New O-rings on the fuel injectors
   - Fuel supply rail with the injectors
   - EGR pipe. Torque the intake manifold bolts to 89 inch lbs. (10 Nm) and the coolant crossover bolt to 18 ft. lbs. (24 Nm).

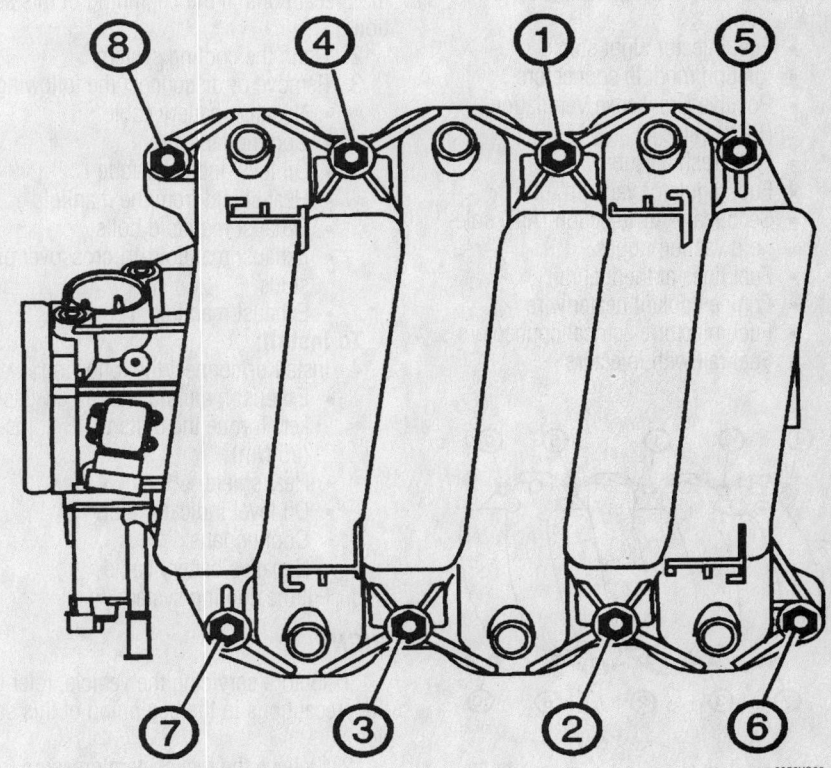

Intake manifold torque sequence—3.5L engine

- PCV valve and feed tubes
- Fuel pressure regulator vacuum hose
- Engine wiring harness to the camshaft covers. Torque the bolts to 89 inch lbs. (10 Nm).
- TPS electrical connector
- IAC valve electrical connector
- EVAP solenoid electrical connector
- MAP sensor electrical connector
- Fuel injector electrical connectors
- Surge tank pipe retainer to the fuel supply rail
- A/C vacuum hose
- Brake booster vacuum hose
- Vapor line to the EVAP canister purge solenoid
- Fuel lines to the fuel supply rail
- Coolant hoses to the throttle body
- Cruise control cable and bracket
- Accelerator cable and bracket
- Fuel injector sight shield
- Air duct to the throttle body
- Negative battery cable

6. Refill the cooling system.
7. Start the engine and check for leaks.

4.0L Engine

1. Before servicing the vehicle, refer to the precautions in the beginning of this section.
2. Relieve the fuel system pressure.
3. Remove or disconnect the following:

- Fuel injector sight shield
- Ignition module connectors
- Positive Crankcase Ventilation (PCV) valve
- PCV fresh air tube
- Fuel regulator vacuum tube
- Secondary Air Injection (AIR) solenoid vacuum tubes
- Fuel lines at the fuel rail
- Engine coolant heater wire
- Fuel injector electrical connectors
- Fuel rail with injectors

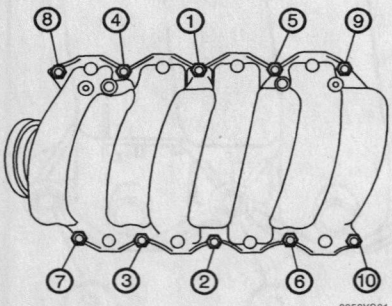

Intake manifold torque sequence—4.0L engine

- Plenum duct clamp
- Intake manifold

To install:
4. Install or connect the following:

- Intake manifold with a new gasket. Torque the bolts in sequence to 89 inch lbs. (10 Nm).
- Torque the plenum duct clamp to 20 inch lbs. (2.25 Nm)
- Fuel rail with the injectors. Torque the bolts to 89 inch lbs. (10 Nm).
- Fuel injector electrical connectors
- Engine coolant heater wire
- Fuel lines to the fuel rail
- AIR solenoid vacuum tubes
- Fuel regulator vacuum tube
- PCV fresh air tube
- PCV valve
- Ignition module connectors

5. Pressurize the fuel system and check for leaks.
6. Install the fuel injector sight shield. Torque the bolts to 27 inch lbs. (3 Nm).

Exhaust Manifold

REMOVAL & INSTALLATION

3.5L Engine

LEFT

1. Before servicing the vehicle, refer to the precautions in the beginning of this section.
2. Drain the cooling system.
3. Remove or disconnect the following:

- Negative battery cable
- Cooling fans
- Oil level indicator tube
- Heat shield from the manifold
- Exhaust manifold bolts
- Exhaust manifold-to-crossover pipe studs
- Exhaust manifold

To install:
4. Install or connect the following:

- Exhaust manifold with a new gasket. Torque the bolts to 18 ft. lbs. (25 Nm).
- Heat shield
- Oil level indicator tube
- Cooling fans
- Negative battery cable

5. Fill the cooling system.

RIGHT

1. Before servicing the vehicle, refer to the precautions in the beginning of this section.
2. Relieve the fuel system pressure.
3. Remove or disconnect the following:

- Cruise control cable from the throttle body lever and bracket and position aside
- Accelerator cable from the throttle body lever and bracket and position aside
- Air cleaner intake duct
- Air cleaner assembly
- Fasteners attaching the accelerator control cable bracket to the throttle body
- Accelerator control cable bracket
- Vacuum brake booster
- Transaxle filler tube
- Transaxle range selector cable
- Bolt connecting the ground lead to the top of the transaxle
- Connector C101 at the left strut tower
- Left wheel speed sensor connector
- Left wheel speed sensor lead from the retaining brackets
- Connector C100 at the top of the transaxle
- Harness from the harness clip and position aside
- Bolts attaching the exhaust manifold heat shield to the left exhaust manifold and position aside
- Right exhaust manifold crossover bolts
- Two bolts attaching the left exhaust manifold to the right exhaust manifold crossover
- Exhaust manifold pipe
- Right exhaust manifold studs
- Heated Oxygen Sensor (HO2S)
- Exhaust Gas Recirculation (EGR) inlet pipe from the right exhaust manifold
- Right exhaust manifold retaining bolts
- Right exhaust manifold and the gasket from the engine and discard the gasket
- Exhaust manifold crossover seal from the left exhaust manifold and discard the seal

To install:
4. Install a new exhaust manifold crossover seal over the left exhaust manifold
5. Align the exhaust manifold to the cylinder head.
6. Place the crossover portion of the exhaust manifold around the left exhaust manifold.
7. Hold the exhaust manifold to the right cylinder head using a retaining bolt.
8. Grasp the exhaust manifold and remove the retaining bolt.
9. Place a new exhaust manifold gasket

between the exhaust manifold and the cylinder head.

10. Install or connect the following:
- Right exhaust manifold bolts and tighten to 18 ft. lbs. (25 Nm)
- EGR inlet pipe to the exhaust manifold nut to 44 ft. lbs. (60 Nm)
- Right exhaust manifold studs and tighten to 80 inch. lbs. (Nm)
- HO$_2$S
- Exhaust manifold pipe
- Two bolts attaching the left exhaust manifold to the right exhaust manifold crossover and tighten to 18 ft. lbs. (25 Nm)
- Heat shield to the left exhaust manifold and tighten the bolts to 80 inch. lbs. (9 Nm)
- Harness to the harness clip
- C100 to the transaxle
- Left wheel speed sensor connector
- Left wheel speed sensor lead to the retaining brackets
- C101 at the left strut tower
- Bolt connecting the ground lead to the top of the transaxle and tighten to 13 ft. lbs. (17 Nm)
- Transaxle range selector cable
- Transaxle filler tube
- Accelerator control cable bracket to the throttle body
- Fasteners attaching the accelerator control cable bracket to the throttle body. Tighten the accelerator control cable bracket bolt and to 115 inch. lbs. (13 Nm) and the accelerator control cable bracket nuts to 89 inch lbs. (10 Nm).
- Air cleaner assembly
- Air cleaner intake duct

➡**Make sure the throttle should operate freely without binding between full closed and wide open throttle.**

- Accelerator cable to the throttle body lever and bracket

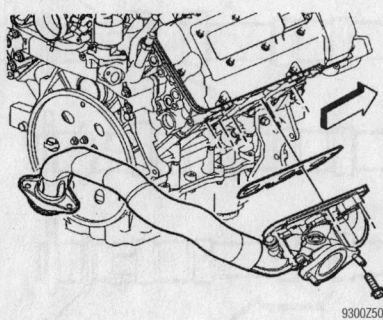

93002507

Exploded view of the right exhaust manifold—3.5L engine

- Cruise control cable to the throttle body lever and bracket

4.0L Engine
LEFT SIDE (FRONT)

1. Before servicing the vehicle, refer to the precautions in the beginning of this section.
2. Remove or disconnect the following:
- Negative battery cable
- Two nuts attaching the secondary AIR tube to the exhaust manifold
- Secondary AIR tube and the gasket from the exhaust manifold and discard the gasket
- Front engine mount bracket
- Oxygen Sensor (O$_2$S), if necessary
- Two bolts attaching the left exhaust manifold to the front exhaust manifold pipe
- Exhaust manifold retaining bolts
- Exhaust manifold and the gasket from the engine and discard the gasket
- Front exhaust manifold pipe seal from the left exhaust manifold and discard the seal
- Exhaust manifold flange seal retainer

To install:

3. Install or connect the following:
Retainer and a new front exhaust manifold pipe seal over the exhaust manifold
- New exhaust manifold gasket between the exhaust manifold and the cylinder head
- Exhaust manifold into the front exhaust manifold pipe and up to the cylinder head
- Left exhaust manifold bolts and tighten to 18 ft. lbs. (25 Nm)
- Two bolts attaching the left exhaust manifold to the front exhaust manifold pipe and tighten to 18 ft. lbs. (25 Nm)
- O$_2$S
- New gasket to the secondary AIR tube
- Two nuts attaching the secondary AIR tube to the exhaust manifold and tighten to 106 inch lbs. (12 Nm)

RIGHT SIDE (REAR)

1. Before servicing the vehicle, refer to the precautions in the beginning of this section.
2. Relieve the fuel system pressure.
3. Remove or disconnect the following:
- Cruise control cable from the throttle body lever and bracket and position aside

- Accelerator cable from the throttle body lever and bracket and position aside
- Air cleaner intake duct
- Air cleaner assembly
- Fasteners attaching the accelerator control cable bracket to the throttle body
- Accelerator control cable bracket
- Vacuum brake booster
- Transaxle filler tube
- Transaxle range selector cable
- Bolt connecting the ground lead to the top of the transaxle
- Connector C101 at the left strut tower
- Left wheel speed sensor connector
- Left wheel speed sensor lead from the retaining brackets
- Connector C100 at the top of the transaxle
- Harness from the harness clip and position aside
- Bolts attaching the exhaust manifold heat shield to the left exhaust manifold and position aside
- Right exhaust manifold crossover bolts
- Two bolts attaching the left exhaust manifold to the right exhaust manifold crossover
- Exhaust manifold pipe
- Right exhaust manifold studs
- Heated Oxygen Sensor (HO$_2$S)
- Exhaust Gas Recirculation (EGR) inlet pipe from the right exhaust manifold
- Right exhaust manifold retaining bolts
- Right exhaust manifold and the gasket from the engine and discard the gasket
- Exhaust manifold crossover seal from the left exhaust manifold and discard the seal

To install:

4. Install a new exhaust manifold crossover seal over the left exhaust manifold
5. Align the exhaust manifold to the cylinder head.
6. Place the crossover portion of the exhaust manifold around the left exhaust manifold.
7. Hold the exhaust manifold to the right cylinder head using a retaining bolt.
8. Grasp the exhaust manifold and remove the retaining bolt.
9. Place a new exhaust manifold gasket between the exhaust manifold and the cylinder head.
10. Install or connect the following:

- Right exhaust manifold bolts and tighten to 18 ft. lbs. (25 Nm)
- EGR inlet pipe to the exhaust manifold nut to 44 ft. lbs. (60 Nm)
- Right exhaust manifold studs and tighten to 80 inch. lbs. (Nm)
- HO$_2$S
- Exhaust manifold pipe
- Two bolts attaching the left exhaust manifold to the right exhaust manifold crossover and tighten to 18 ft. lbs. (25 Nm)
- Heat shield to the left exhaust manifold and tighten the bolts to 80 inch. lbs. (9 Nm)
- Harness to the harness clip
- C100 to the transaxle
- Left wheel speed sensor connector
- Left wheel speed sensor lead to the retaining brackets
- C101 at the left strut tower
- Bolt connecting the ground lead to the top of the transaxle and tighten to 13 ft. lbs. (17 Nm)
- Transaxle range selector cable
- Transaxle filler tube
- Accelerator control cable bracket to the throttle body
- Fasteners attaching the accelerator control cable bracket to the throttle body. Tighten the accelerator control cable bracket bolt and to 115 inch. lbs. (13 Nm) and the accelerator control cable bracket nuts to 89 inch lbs. (10 Nm).
- Air cleaner assembly
- Air cleaner intake duct

➡ **Make sure the throttle should operate freely without binding between full closed and wide open throttle.**

- Accelerator cable to the throttle body lever and bracket
- Cruise control cable to the throttle body lever and bracket

Camshaft and Valve Lifters

REMOVAL & INSTALLATION

➡ **All valve train components should be kept in the order that they were removed, so that they can be reinstalled in their original position.**

3.5L Engine

LEFT SIDE (FRONT)

1. Before servicing the vehicle, refer to the precautions in the beginning of this section.
2. Disconnect the negative battery cable.

3. Remove or disconnect the following:
- Fuel injector sight shield
- Oil level indicator tube
- Positive Crankcase Ventilation (PCV) valve
- Spark plug wires
- Ignition coil assembly
- Fuel injector wiring harness
- Engine wiring harness
- Camshaft cover
4. Install the camshaft holding fixture (J 42038) on the camshafts.
5. Remove or disconnect the following:
- Camshaft sprockets with the secondary drive chains
- Camshaft bearing caps
- Camshaft holding fixture
- Camshafts
- Rocker arms
- Lifters

To install:

6. Coat the lifters with clean engine oil and place them into position in the engine.
7. Coat the rocker arms with clean engine oil and set them in place in the head.
8. Coat the camshafts with clean engine oil and place them in their proper position in the head.
9. Install or connect the following:
- Camshaft holding fixture
- Camshaft bearing caps. Torque the bolts to 71 inch lbs. (8 Nm) plus an additional 22 degree turn.
- Camshaft sprockets with the secondary drive chains. Torque the bolts to 18 ft. lbs. (25 Nm) plus an additional 45 degree turn.
10. Remove the camshaft holding fixture from the head.
11. Install or connect the following:

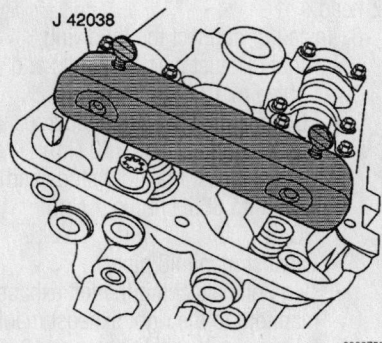

Camshaft holding fixture J-42038 installed on the camshafts—3.5L engine

- Camshaft cover (seals are reusable if they are not damaged). Torque the bolts to 80 inch lbs. (9 Nm).
- Engine wiring harness
- Fuel injector wiring harness
- Ignition coil assembly
- Spark plug wires
- PCV valve and feed tube
- Oil level indicator tube
- Fuel injector sight shield. Torque the nuts to 27 inch lbs. (3 Nm).
- Negative battery cable

RIGHT SIDE (REAR)

1. Before servicing the vehicle, refer to the precautions in the beginning of this section.

2. Remove or disconnect the following:
- Negative battery cable
- Transmission filler tube
- Fuel injector sight shield
- Positive Crankcase Ventilation (PCV) feed tube

1. Left intake
2. Left exhaust
3. Right intake
4. Right exhaust

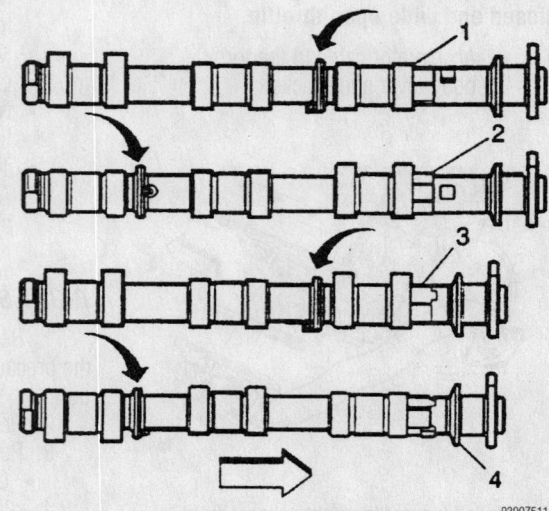

Camshaft identification—3.5L engine

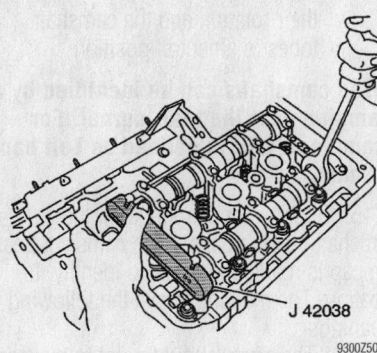

Use the flats on the camshaft if rotation is necessary for installation of the holding tool—3.5L engine

9300Z509

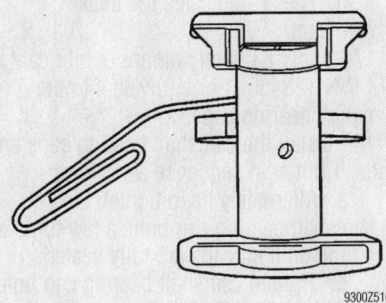

Before installation, compress the tensioner and lock it in place with a piece of wire—3.5L engine

9300Z510

- Engine wiring harness clips
- Oxygen Sensor (O2S) electrical connector
- Spark plug wires
- Ignition coil assembly
- Camshaft cover

3. Install the camshaft holding fixture (tool J-42038).

4. Remove or disconnect the following:
- Camshaft position sensor
- Camshaft sprockets with the secondary drive chains
- Camshaft bearing caps
- Camshaft holding fixture
- Camshafts
- Rocker arms
- Lifters

To install:

5. Coat the lifters with clean engine oil and place them into position in the engine.

6. Coat the rocker arms with clean engine oil and set them in place in the head.

7. Coat the camshafts with clean engine oil and place them in position in the head.

8. Install or connect the following:
- Camshaft holding fixture
- Camshaft bearing caps. Torque the

bolts to 71 inch lbs. (8 Nm). plus an additional 22 degree turn.
- Camshaft sprockets with secondary drive chains. Torque the bolts to 18 ft. lbs. (25 Nm) plus an additional 45 degree turn.
- Camshaft position sensor

9. Remove the camshaft holding fixture.

10. Install or connect the following:
- Camshaft cover (seals are reusable if undamaged). Torque the bolts to 80 inch lbs. (9 Nm).
- Ignition coil assembly
- Spark plug wires
- O2S electrical connector
- Engine wiring harness clips
- PCV feed tube
- Fuel injector sight shield. Torque the nuts to 27 inch lbs. (3 Nm).
- Transmission filler tube
- Negative battery cable

4.0L Engine

LEFT SIDE (FRONT)

1. Before servicing the vehicle, refer to the precautions in the beginning of this section.

2. Drain the cooling system.

3. Remove or disconnect the following:
- Negative battery cable
- Fuel injector sight shield
- Inlet radiator hose from the water housing crossover
- Positive Crankcase Ventilation (PCV) fresh air tube from the left side camshaft cover
- Ignition coil cassette and spark plug boots
- 2 pushnuts securing the engine coolant heater wire, if equipped and position aside
- Surge tank pipe from the fuel rail studs and carefully position aside
- Cable harness clips at the front of the camshaft cover and position the cable harness aside
- AIR valve bracket nut closest to the center of the engine. Pry outward slightly on the AIR valve bracket in order to gain clearance to remove the water pump drive belt shield nut.
- Water pump drive belt shield and belt
- Water pump belt tensioner
- 3 camshaft seal retainer bolts

➡**DO NOT reuse the camshaft seal.**

- Camshaft seal
- Camshaft cover

4. Rotate the crankshaft to Top Dead

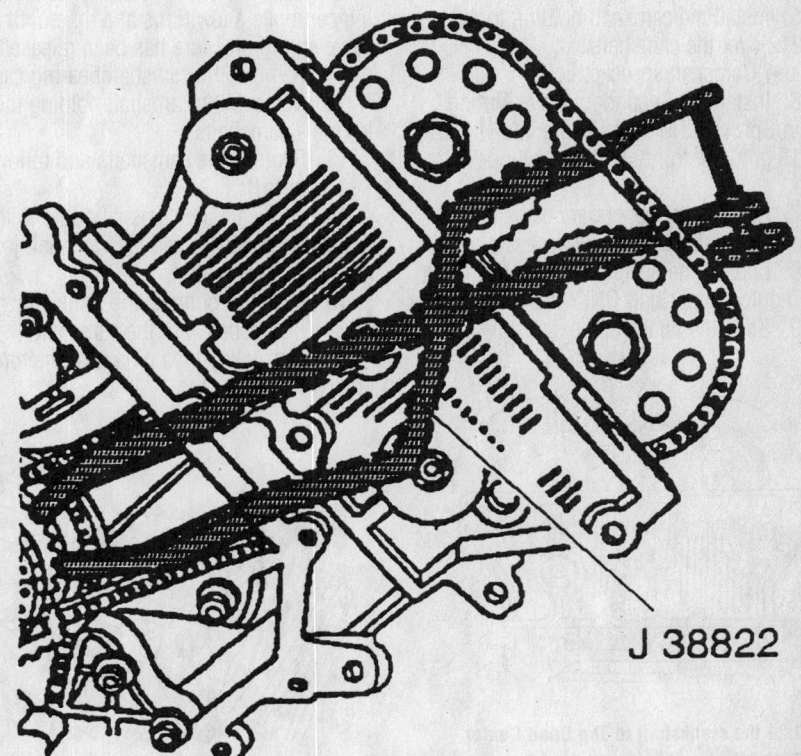

Camshaft chain holding tool J-38822—4.0L engine

J 38822

9300XG05

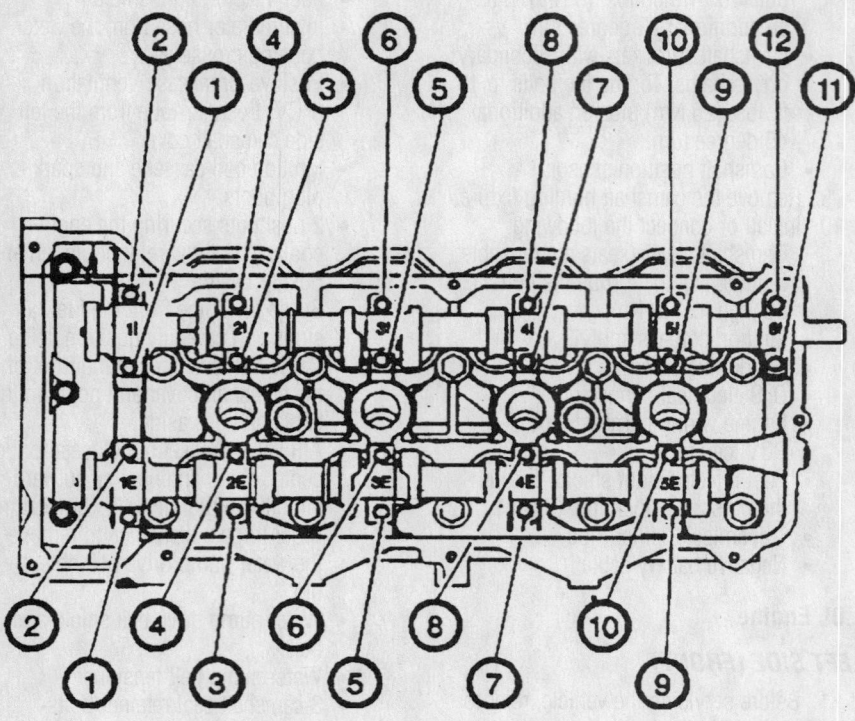

Left cylinder head camshaft bearing cap tightening sequence—4.0L (VIN C) engine

7922XG06

Center (TDC) of the number 1 cylinders compression stroke, both camshaft sprocket drive pins should be at the top of their rotation.

5. Install the camshaft holding tool J44212 over the camshafts.
 • Camshaft sprocket bolts

6. Install the Secondary Drive Timing Chain/Sprocket Holding Fixture Tool J 44213 onto the front rail of the cylinder head.

7. Remove the secondary camshaft drive chain guide upper bolt access plug.

8. Loosen the secondary camshaft drive chain guide upper bolt ONLY two turns.

9. Slide the intake and exhaust

camshaft sprockets onto the pins of the Secondary Drive Timing chain/Sprocket Holding Fixture Tool J 44213.

10. Alternately loosen the camshaft bearing cap bolts a few turns at a time until all valve spring pressure has been released.

11. Remove the camshaft bearing caps.

12. Remove the camshaft holding tool from the camshafts.

13. Remove the camshafts and followers.

To install:

14. Lubricate the camshaft lobes with assembly lubricant, and the camshaft journals with engine oil.

15. Install or connect the following:
 • Camshaft with the camshaft sprocket drive pins near the top of

their rotation and the camshaft lobes in a neutral position.

➡The camshafts can be identified by a stamping near the rear journal. For example: L-EXH is defined as Left bank Exhaust.

16. Observe the markings on the camshaft bearing caps. Each camshaft bearing cap is marked in order to identify its location. The markings have the following meanings:
 a. The arrow should point to the front of the engine.
 b. The number indicates the position from the front of the engine.
 c. The "E" indicates the exhaust camshaft.
 d. The "I" indicates the intake camshaft.

17. Apply a liberal amount of lubricant GM P/N 12345001 or equivalent to the camshaft bearing caps.

18. Install the camshaft bearing caps and bolts. Tighten in sequence as follows:
 a. Alternately hand tighten the camshaft bearing cap bolts a few turns at a time until all caps are fully seated.
 b. Tighten camshaft bearing cap bolts to 44 inch lbs. (5 Nm).
 c. Tighten camshaft bearing cap bolts an additional 30 degrees.

19. Align the camshafts.

20. Install the camshaft holding fixture.

➡Ensure the camshaft sprockets properly engage the camshaft sprocket drive pins and camshafts.

21. Slide the intake and exhaust camshaft sprockets off the pins of the Secondary Drive Timing chain/Sprocket Holding Fixture Tool J 44213 and onto the pins of the camshafts.

22. Tighten the secondary camshaft drive

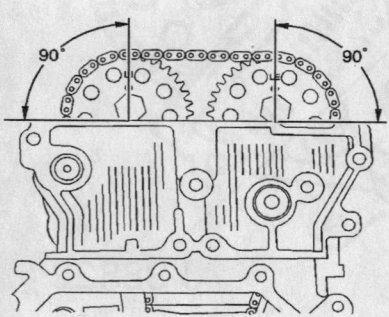

Rotate the crankshaft to Top Dead Center (TDC) of the number 1 cylinders compression stroke, both camshaft sprocket drive pins should be at the top of their rotation

9346XG23

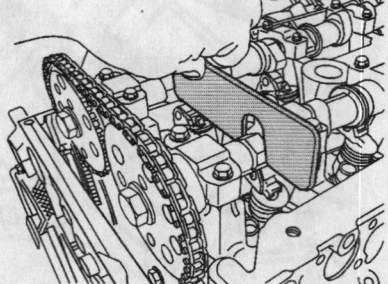

Install the camshaft holding tool J44212 over the camshafts

9346XG24

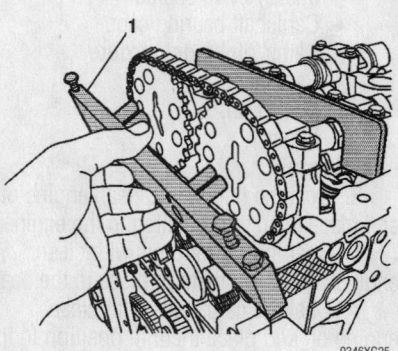

Install Secondary Drive Timing Chain/Sprocket Holding Fixture Tool J 44213 onto the front rail of the cylinder head

9346XG25

chain guide upper bolt to 18 ft. lbs. (25 Nm).

23. Install the secondary camshaft drive chain guide upper bolt access plug and tighten to 39 inch lbs. (4.5Nm).

24. Remove the Secondary Drive Timing chain/Sprocket Holding Fixture Tool J 44213 while carefully holding the camshaft sprockets against the camshafts.

25. Install the camshaft sprocket bolts and tighten to 89 ft. lbs. (120 Nm).

26. Verify the camshaft sprocket alignment.

27. Remove the camshaft holding tool.

28. Install or connect the following:
- Camshaft cover and tighten the bolts to 89 inch lbs. (10 Nm)
- Camshaft seal lips lubricated with clean oil
- Camshaft seal retainers coated with sealer GM P/N 1052080 and tighten to 27 inch lbs. (3 Nm)
- Water pump belt tensioner
- Water pump drive belt and shield
- AIR valve bracket nut to 80 inch lbs. (9Nm)
- Cable harness clips at the front of the camshaft cover
- 2 pushnuts securing the engine coolant heater wire, if equipped
- Inlet radiator hose to the water housing crossover
- Fuel injector sight shield
- Surge tank pipe to the fuel rail studs
- Ignition coil cassette and spark plug boots
- PCV fresh air tube to the left side camshaft cover
- Fuel injector sight shield
- Negative battery cable

29. Refill the cooling system.

RIGHT SIDE (REAR)

1. Before servicing the vehicle, refer to the precautions in the beginning of this section.

2. Drain the cooling system.

3. Remove or disconnect the following:
- Negative battery cable
- Fuel injector sight shield
- Positive Crankcase Ventilation (PCV) valve from the camshaft cover
- Vacuum tubes from the AIR vent solenoid
- AIR vent solenoid electrical connector
- AIR control valve bracket
- Nut securing the AIR tube
- Ignition coil cassette and spark plug boots

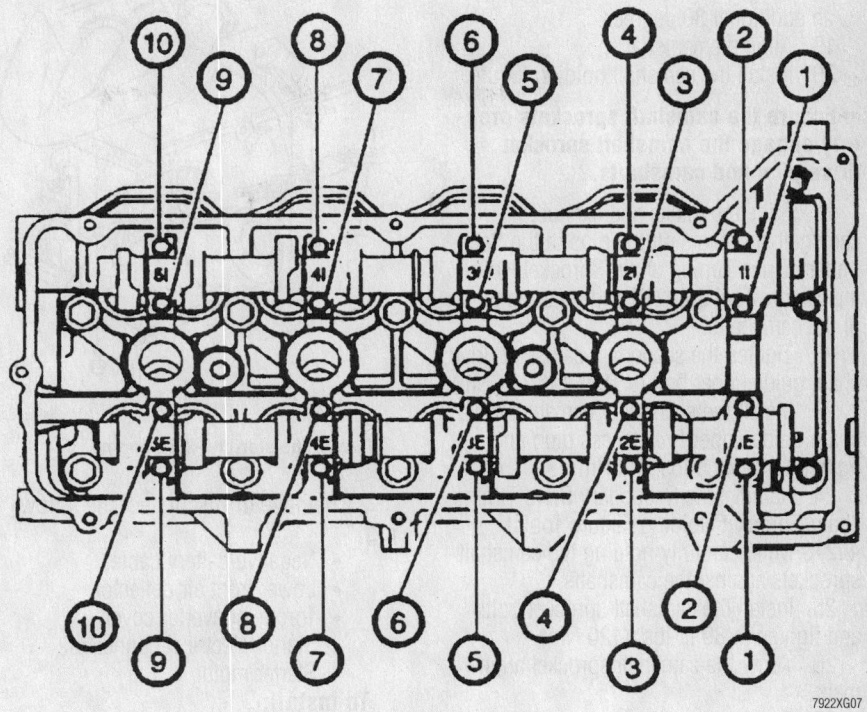

Right cylinder head camshaft bearing cap tightening sequence—4.0L engine

- Cable harness clips at the front of the camshaft cover and position the cable harness aside
- Camshaft cover

4. Rotate the crankshaft to Top Dead Center (TDC) of the number 1 cylinders compression stroke, both camshaft sprocket drive pins should be at the top of their rotation.

5. Install the camshaft holding tool J44212 over the camshafts.
- Camshaft sprocket bolts

6. Install the Secondary Drive Timing Chain/Sprocket Holding Fixture Tool J 44213 onto the front rail of the cylinder head.

7. Remove the secondary camshaft drive chain guide upper bolt access plug.

8. Loosen the secondary camshaft drive chain guide upper bolt ONLY two turns.

9. Slide the intake and exhaust camshaft sprockets onto the pins of the Secondary Drive Timing chain/Sprocket Holding Fixture Tool J 44213.

10. Alternately loosen the camshaft bearing cap bolts a few turns at a time until all valve spring pressure has been released.

11. Remove the camshaft bearing caps.

12. Remove the camshaft holding tool from the camshafts.

13. Remove the camshafts and followers.

To install:

14. Lubricate the camshaft lobes with assembly lubricant, and the camshaft journals with engine oil.

15. Install or connect the following:
- Camshaft with the camshaft sprocket drive pins near the top of their rotation and the camshaft lobes in a neutral position.

➡ **The camshafts can be identified by a stamping near the rear journal. For example: L-EXH is defined as Left bank Exhaust.**

16. Observe the markings on the camshaft bearing caps. Each camshaft bearing cap is marked in order to identify its location. The markings have the following meanings:

a. The arrow should point to the front of the engine.

b. The number indicates the position from the front of the engine.

c. The "E" indicates the exhaust camshaft.

d. The "I" indicates the intake camshaft.

17. Apply a liberal amount of lubricant GM P/N 12345001 or equivalent to the camshaft bearing caps.

18. Install the camshaft bearing caps and bolts. Tighten in sequence as follows:

a. Alternately hand tighten the camshaft bearing cap bolts a few turns at a time until all caps are fully seated.

b. Tighten camshaft bearing cap bolts to 44 inch lbs. (5 Nm).

c. Tighten camshaft bearing cap bolts an additional 30 degrees.
19. Align the camshafts.
20. Install the camshaft holding fixture.

➡ **Ensure the camshaft sprockets properly engage the camshaft sprocket drive pins and camshafts.**

21. Slide the intake and exhaust camshaft sprockets off the pins of the Secondary Drive Timing chain/Sprocket Holding Fixture Tool J 44213 and onto the pins of the camshafts.
22. Tighten the secondary camshaft drive chain guide upper bolt to 18 ft. lbs. (25 Nm).
23. Install the secondary camshaft drive chain guide upper bolt access plug and tighten to 39 inch lbs. (4.5Nm).
24. Remove the Secondary Drive Timing chain/Sprocket Holding Fixture Tool J 44213 while carefully holding the camshaft sprockets against the camshafts.
25. Install the camshaft sprocket bolts and tighten to 89 ft. lbs. (120 Nm).
26. Verify the camshaft sprocket alignment.
27. Remove the camshaft holding tool.
28. Install or connect the following:
- Camshaft cover and tighten the bolts to 89 inch lbs. (10 Nm)
- Cable harness clips at the front of the camshaft cover
- Ignition coil cassette and spark plug boots
- Nut securing the AIR tube
- AIR control valve bracket
- AIR vent solenoid electrical connector
- Vacuum tubes from the AIR vent solenoid
- Positive Crankcase Ventilation (PCV) valve to the camshaft cover
- Fuel injector sight shield
- Negative battery cable
29. Refill the cooling system.

Valve Lash

ADJUSTMENT

The valve lash in these models cannot be adjusted.

Starter Motor

REMOVAL & INSTALLATION

3.5L Engine

1. Before servicing the vehicle, refer to the precautions in the beginning of this section.

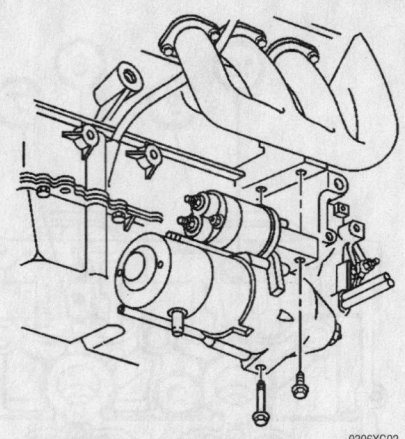

9306XG02

View of the starter—3.5L engine

2. Remove or disconnect the following:
- Negative battery cable
- Lower front air deflector
- Torque converter cover
- Starter electrical connectors
- Starter motor

To install:
3. Install or connect the following:
- Starter motor. Torque the bolts to 37 ft. lbs. (50 Nm).
- Starter electrical connectors. Torque the positive battery terminal nut to 89 inch lbs. (10 Nm) and the "S" terminal nut to 30 inch lbs. (3.5 Nm).
- Torque converter cover
- Lower front air deflector
- Negative battery cable

4.0L Engine

1. Before servicing the vehicle, refer to the precautions in the beginning of this section.
2. Remove or disconnect the following:
- Negative battery cable
- Intake manifold
- Starter electrical connectors
- Starter motor

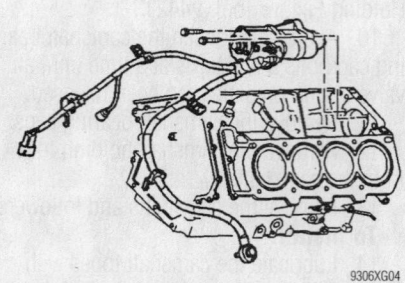

9306XG04

Exploded view of the starter—4.0L engine

To install:

Before installing the starter, torque the inner solenoid and battery terminal nuts to 70 inch lbs. (8 Nm). If not properly tightened, the starter may fail, due to terminal or cap damage.

3. Install or connect the following:
- Starter motor. Torque the bolts to 22 ft. lbs. (30 Nm).
- Starter electrical connectors. Torque the positive battery terminal nut to 89 inch lbs. (10 Nm) and the "S" terminal nut to 35 inch lbs. (4 Nm).
- Intake manifold
- Negative battery cable

Oil Pan

REMOVAL & INSTALLATION

3.5L Engine

1. Before servicing the vehicle, refer to the precautions in the beginning of this section.
2. Drain the crankcase.
3. Remove or disconnect the following:
- Oil filter
- Oil level sensor
- Transmission brace
- Oil pan

To install:
4. Install or connect the following:
- Oil pan with a new gasket. Do not tighten the bolts at this time.
- Transmission brace on the engine block only. Torque the bolts to 37 ft. lbs. (50 Nm).
- Oil level sensor. Torque the bolt to 80 inch lbs. (9 Nm).

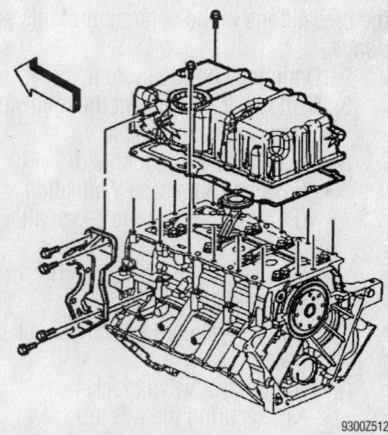

9300Z512

Exploded view of the oil pan—3.5L engine

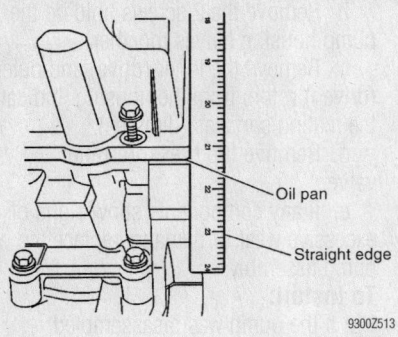

Use a straight edge to align the rear of the oil pan to the rear of the engine—3.5L engine

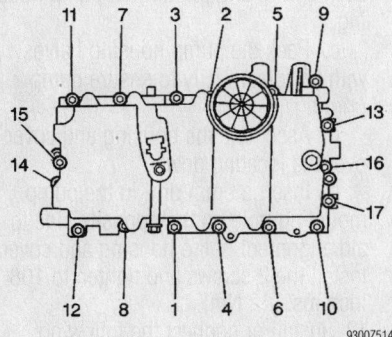

Oil pan mounting bolt tightening sequence—3.5L engine

- Brace-to-oil pan bolts, loosely install them

5. Align the rear of the oil pan flush with the rear of the engine block. Use a straight edge for reference.

6. Press the front of the oil pan against the transmission brace; then torque the brace-to-oil pan bolts to 18 ft. lbs. (25 Nm). Be sure to keep the rear of the pan flush with the rear of the engine.

7. Torque the oil pan bolts to 18 ft. lbs. (25 Nm).

8. Install or connect the following:
- Brace-to-transmission bolts. Torque them to 37 ft. lbs. (50 Nm).
- Oil level sensor electrical connector
- Drain plug. Torque it to 15 ft. lbs. (20 m).
- New oil filter. Torque the cap to 18 ft. lbs. (25 Nm).

9. Refill the crankcase.

10. Start the engine and inspect for leaks.

4.0L Engine

1. Before servicing the vehicle, refer to the precautions in the beginning of this section.

2. Raise the rear seat cushion to access the battery.

3. Disconnect the negative battery cable.

4. Drain the crankcase.

5. Remove or disconnect the following:

- Transmission
- Exhaust crossover pipe
- Oil pan

➡ **The oil pan gasket is reusable unless it is damaged. Do not remove the gasket from the oil pan groove unless gasket replacement is required.**

To install:

6. Install or connect the following:
- Oil pan. Torque the bolts in sequence to 89 inch. lbs. (10 Nm).
- Exhaust crossover pipe. Torque the bolts to 18 ft. lbs. (25 Nm).
- Transmission
- Oil pan drain plug. Torque it to 15 ft. lbs. (20 Nm).
- Negative battery cable

7. Refill the crankcase.

Oil Pump

REMOVAL & INSTALLATION

3.5L Engine

1. Before servicing the vehicle, refer to the precautions in the beginning of this section.

2. Remove or disconnect the following:
- Front cover
- Rocker arm covers

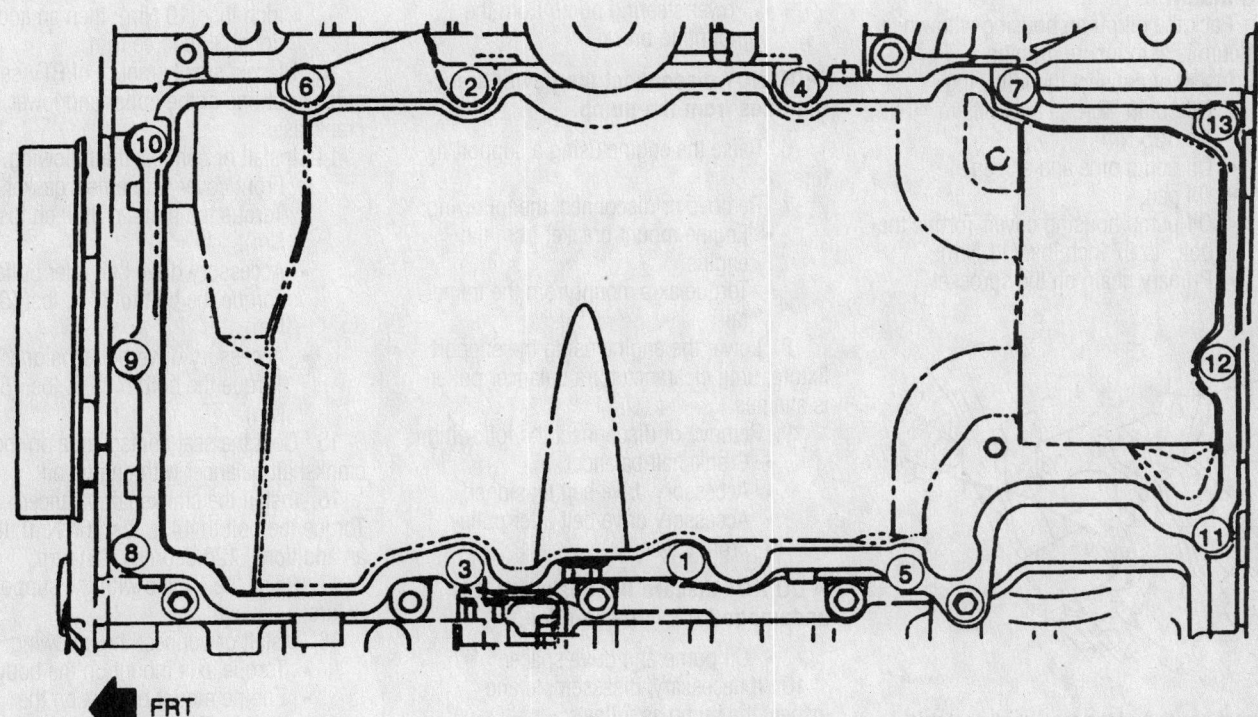

Oil pan bolt tightening sequence—4.0L engine

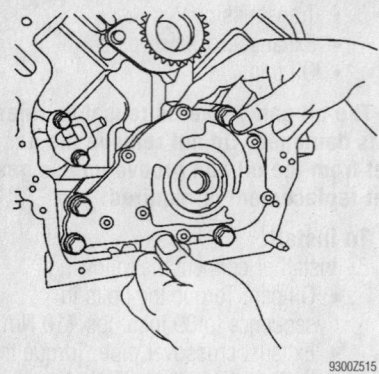

9300Z515

The oil pump is mounted on the front of the engine and driven by the crankshaft— 3.5L engine

3. Install camshaft holding fixtures on both sets of camshafts.

4. Remove or disconnect the following:
- Primary chain tensioner
- Primary chain from the drive sprocket
- Oil pan
- Oil pump pipe and screen
- Oil pump by sliding it off the crankshaft

➡**The internal parts of the oil pump are not serviced separately. The oil pump may be opened for inspection. If damage or wear is noted, replace the entire pump as an assembly.**

To install:

5. Pack the oil pump housing with white petroleum jelly to insure priming.

6. Install or connect the following:
- Oil pump. Torque the bolts to 18 ft. lbs. (25 Nm).
- Oil pump pipe and screen
- Oil pan
- Oil pump housing cover. Torque the bolts to 97 inch lbs. (11 Nm).
- Primary chain on the sprocket

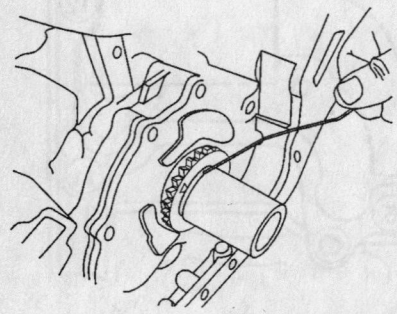

9300Z516

Correct position of the crankshaft sprocket when the oil pump is installed correctly— 3.5L engine

➡**Be sure to maintain correct timing.**
- Chain tensioner

7. Remove the camshaft holding tools.

8. Install or connect the following:
- Rocker arm covers. Torque the bolts to 80 inch lbs. (9 Nm).
- Front cover. Torque the bolts to 124 inch lbs. (14 Nm).

4.0L Engine

1. Before servicing the vehicle, refer to the precautions in the beginning of this section.

2. Disconnect the negative battery cable.

3. Drain the engine oil.

4. Install an engine support fixture.

5. Remove or disconnect the following:
- Engine mount-to-frame through-bolt and nut
- Engine mount-to-engine through-bolt and nut
- Front wheels
- Right inner fender well splash shield
- Lower center air deflector
- Left transmission mount-to-frame through-bolt
- Power steering line retainer from the bracket
- Accessory drive belt from the power steering pump pulley
- Fuel injector sight shield from the intake manifold
- Power steering pump from the mounting bracket

➡**DO NOT disconnect the power steering lines from the pump.**

6. Raise the engine using a support fixture.

7. Remove or disconnect the following:
- Engine mount bracket from the engine
- Torque axis mount from the frame rail

8. Lower the engine using the support fixture, until clearance for a balancer puller is attained.

9. Remove or disconnect the following:
- Crankshaft balancer
- Accessory drive belt tensioner
- Accessory drive belt idler pulley
- Front cover

➡**DO NOT discard the gasket, if it is undamaged it can be re-used.**
- Oil pump and drive spacer

10. If necessary, disassemble and inspect the pump as follows:
   a. Remove the drive spacer from the pump housing.

   b. Remove the 2 screws holding the pump housing halves together.

   c. Remove the inner (drive) and outer (driven) rotors from the housing. Indicate the mating surfaces (dimples).

   d. Remove the pressure relief valve.

   e. If any components show signs of excessive wear or damage, replace the pump assembly.

To install:

11. If the pump was disassembled, reassemble as follows:
   a. Install the inner and outer rotors to the pump cover in the same orientation as removed.

   b. Install the pressure relief valve seat, spring and pilot in the pump housing.

   c. Pack the pump housing halves with petroleum jelly to ensure pump priming.

   d. Assemble the housing and cover over the locating dowel.

   e. Insert a 9mm drill in the pump mounting hole on the opposite side to aid alignment of the housing and cover. Install the 2 screws and tighten to 108 inch lbs. (12 Nm).

12. Install or connect the following:
- Oil pump drive spacer into the oil pump from the rear so the drive flat engages the pump rotor
- Oil pump. Torque the bolts to 89 inch lbs. (10 Nm); then an additional 35 degree turn.

13. Place a small amount of RTV sealant at the split line of the upper and lower crankcases.

14. Install or connect the following:
- Front cover with a new gasket. Torque the bolts to 89 inch lbs. (10 Nm).
- Accessory drive belt idler pulley. Torque the bolt to 37 ft. lbs. (50 Nm).
- Accessory drive belt tensioner. Torque the bolt to 37 ft. lbs. (50 Nm).

15. Coat the seal contact area on the crankshaft balancer with engine oil.

16. Install the crankshaft balancer. Torque the bolt to 44 ft. lbs. (60 Nm); then an additional 120 degrees (⅔) turn.

17. Raise the engine with the support fixture.

18. Install or connect the following:
- Torque axis mount on the body
- Engine mount bracket on the engine. Torque the nuts to 30 ft. lbs. (40 Nm) and the bolts to 41 ft. lbs. (55 Nm).

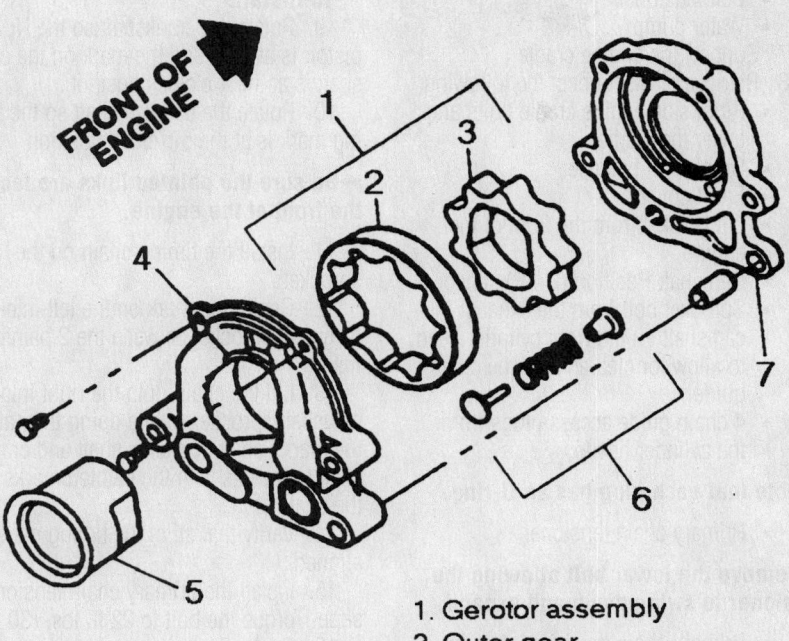

1. Gerotor assembly
2. Outer gear
3. Inner gear
4. Housing
5. Drive spacer
6. Relief valve
7. Cover

7922XG14

Exploded view of the oil pump—4.0L engine

19. Lower the engine support fixture until the engine is at its normal height.
20. Install or connect the following:
   - Power steering pump to the bracket
   - Fuel injector sight shield
   - Accessory drive belt
   - Power steering line retainer to the engine mount bracket
   - Left transmission mount-to-frame through-bolt. Torque the bolt to 63 ft. lbs. (85 Nm).
   - Lower center air deflector
   - Right inner fender well splash shield
   - Front wheels. Torque the nuts to 100 ft. lbs. (140 Nm).
   - Engine mount-to-engine bracket. Torque the through-bolt to 70 ft. lbs. (95 Nm).
   - Engine mount-to-frame. Torque the through-bolt to 37 ft. lbs. (50 Nm).
   - Negative battery cable
21. Remove the engine support fixture.
22. Replace the oil filter and refill the crankcase.
23. Run the engine and check for leaks and proper engine operation.

Rear Main Seal

REMOVAL & INSTALLATION

3.5L and 4.0L Engines With the Cartridge Type Seal

1. Before servicing the vehicle, refer to the precautions in the beginning of this section.
2. Remove or disconnect the following:
   - Engine and transmission assembly
   - Transmission from the engine
   - Flexplate
3. Remove the seal as follows:
   a. Place Seal Removal tool J-42841 onto the crankshaft.
   b. Install 8, 1-inch self starting screws through the guide holes of the tool and into the seal. A variable speed drill will be helpful for installing the screws.
   c. Tighten the center screw of the tool to remove the seal.
4. Clean out the drain at the bottom of the seal bore with a piece of wire or pipe cleaner.

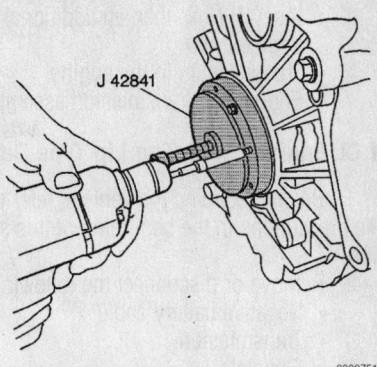

93002517

Use the guide holes in tool J-42841 to install the screws in the seal—3.5L and 4.0L engines

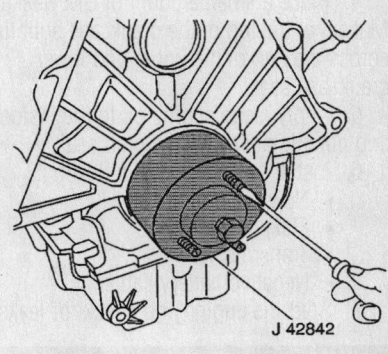

J 42842

93002518

Use rear main seal installer J-42842 to install the seal—3.5L engine

To install:

5. Apply a small amount of RTV gasket maker at the crankcase split line across the end of the upper and lower crankcase seal.
6. Coat the outer diameter of the rear crankshaft seal area with engine oil.
7. Wipe the outer diameter of the crankshaft flexplate flange with a lint-free cloth.
8. Lubricate the outer rubber surface of the seal with clean engine oil. Do not apply any oil to the green coating, pre-applied to the inner diameter of the seal.
9. Loosen the center bolt of the seal installer until the center hub protrudes about ½ inch past the outer plate.
10. Thread the 3 mounting bolts into the crankshaft flange until the tool is firmly mounted on the crankshaft.
11. Install the seal by tightening the center bolt until the tool bottoms against the crankshaft.
12. Remove the tool and verify that the seal is installed evenly.
13. Install or connect the following:
   - Flexplate. Torque the bolts to 11 ft.

lbs. (15 Nm); then an additional 50 degree turn.

- Transmission to the engine
- Engine and transmission assembly

4.0L Engines With The Lip Type Seal

1. Before servicing the vehicle, refer to the precautions in the beginning of this section.

2. Remove or disconnect the following:
- Negative battery cable
- Transmission
- Flexplate

3. Carefully pry the oil seal from the housing using a suitable prying tool, taking care not to damage the housing or the crankshaft sealing surface.

To install:

4. Place a small amount of GM Gasket Maker® at the top of the crankcase split line across the end of the upper and lower crankcase seal.

5. Apply clean engine oil to the inside and outside diameter of the oil seal.

6. Install or connect the following:
- New seal
- Flexplate
- Transmission
- Negative battery cable

7. Start the engine and check for leaks.

Timing Chain, Sprockets, Front Cover and Seal

REMOVAL & INSTALLATION

3.5L Engine

PRIMARY CHAIN

1. Before servicing the vehicle, refer to the precautions in the beginning of this section.

2. Remove or disconnect the following:
- Negative battery cable
- Rocker arm covers

3. Rotate the crankshaft so the No. 1 piston is at Top Dead Center (TDC) and the flats on the rear of the camshafts are parallel with the camshaft cover sealing surface.

4. Install camshaft holding fixtures on both sets of camshafts.

5. Drain the cooling system.

6. Remove or disconnect the following:
- Front diagonal brace
- Battery and tray
- Washer and coolant reservoirs
- Underhood accessory wiring junction block and move it aside
- Drive belt
- Power steering pump pulley
- Idler pulley

- Belt tensioner
- Water pump

7. Support the engine cradle

8. Remove or disconnect the following:
- Right side engine cradle bolts and lower the cradle
- Crankshaft balancer
- Front cover
- Lift bracket from the front of the engine
- Camshaft Position (CMP) sensor
- Sprocket bolt from the exhaust camshaft on the right cylinder head to allow for clearance of the chain guide
- 4 chain guide access plugs from the cylinder heads

➥Note that each plug has an O-ring.

- Primary chain tensioner

➥Remove the lower bolt allowing the tensioner to swing down and expand.

- Primary chain tensioner shoe by removing the bolt, pushing the guide downward slightly and pulling it up through the cylinder head
- Primary chain from the right camshaft, allowing it to fall into the oil pump area
- Primary chain

To install:

9. Rotate the crankshaft so the No. 1 piston is at TDC and the mark on the crankshaft is at the 4 o'clock position.

10. Rotate the balance shaft so the timing mark is at the 5 o'clock position.

➥Be sure the painted links are facing the front of the engine.

11. Install the timing chain on the sprockets.

12. Center the mark on the left intake camshaft sprocket between the 2 painted links.

13. Lift the chain onto the right intake camshaft sprocket. While doing this, align the marks on the balance shaft and crankshaft sprockets with the painted marks on the chain.

14. Verify that all of the timing marks are aligned.

15. Install the primary chain tensioner shoe. Torque the bolt to 22 ft. lbs. (30 Nm).

16. Compress the primary chain tensioner using the following steps:

   a. Rotate the ratchet release lever counterclockwise and hold it.

   b. Press the tensioner shoe in and hold it.

   c. Release the ratchet lever and slowly release the pressure on the shoe.

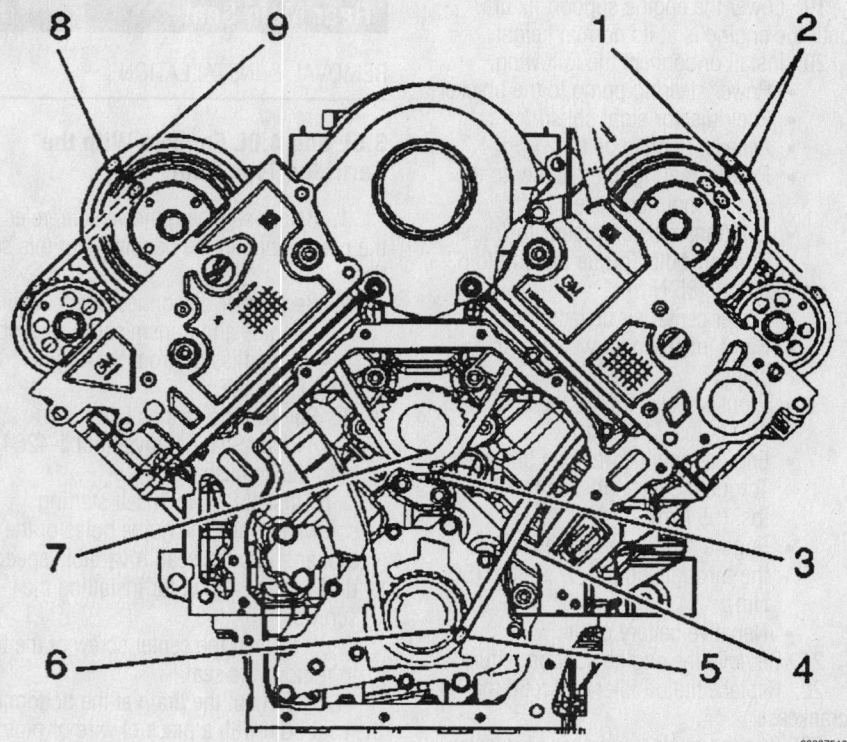

Primary timing chain alignment marks—3.5L engine

9300Z519

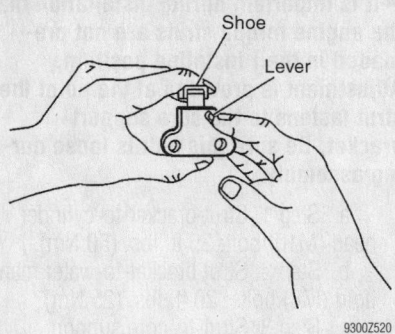

Compressing the primary chain tensioner—3.5L engine

d. Insert a pin through the hole in the lever as the lever moves to the first click. The ratchet should hold the shoe in the compressed position.

➡ **Be sure the lever on the tensioner is facing you when installed.**

17. Install or connect the following:
- Primary chain tensioner. Torque the bolts to 18 ft. lbs. (25 Nm); then remove the chain tensioner pin.
- 4 chain guide access plugs. Torque the plugs to 44 inch lbs. (5 Nm).
- Front engine lift bracket. Torque the hex head bolt to 37 ft. lbs. (50 Nm)

and the internal drive bolt to 18 ft. lbs. (25 Nm).
- CMP sensor. Torque the bolts to 80 inch lbs. (9 Nm).

18. Remove the camshaft holding tools.

19. Place a small bead of RTV sealant on the 3 areas indicated in the diagram.

20. Install or connect the following:
- Front cover with a new gasket. Torque the bolts to 124 inch lbs. (14 Nm) and the coolant drain plug to 89 inch lbs. (10 Nm).
- Crankshaft balancer. Torque the bolt to 37 ft. lbs. (50 Nm); then an additional 120 degree turn.

21. Raise the engine cradle and install new bolts. Torque the bolts to 133 ft. lbs. (180 Nm).

22. Coat the sub-frame bushings with rubber lubricant.

23. Install or connect the following:
- Water pump with a new gasket. Torque the bolts to 124 inch lbs. (14 Nm).
- Belt tensioner. Torque the bolts to 37 ft. lbs. (50 Nm).
- Idler pulley. Torque the bolt to 37 ft. lbs. (50 Nm).
- Power steering pump pulley
- Drive belt

- Underhood accessory wiring junction block
- Washer and coolant reservoirs
- Battery and tray
- Front diagonal brace
- Rocker arm covers. Torque the bolts to 80 inch lbs. (9 Nm).
- Negative battery cable

24. Refill the cooling system.
25. Refill the crankcase.
26. Start the engine and verify no leaks.

SECONDARY TIMING CHAIN

1. Before servicing the vehicle, refer to the precautions in the beginning of this section.

2. Disconnect the negative battery cable.

3. Remove the rocker arm cover and install camshaft holding fixture J-42038.

4. Remove the camshaft sprocket bolts and install the timing chain holding fixture J-42042 on the cylinder head.

5. Remove or disconnect the following:
- Sprockets and chain
- Secondary timing sprocket and chain

To install:

6. Install or connect the following:
- Secondary timing chain on the sprockets, with the drive pins at the 12 o'clock positions

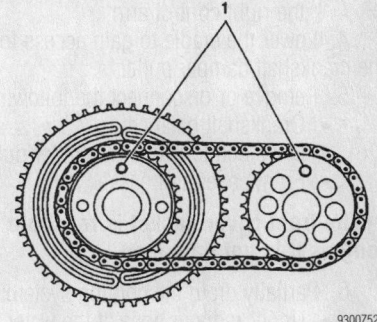

Correct sprocket alignment for the left secondary timing chain—3.5L engine

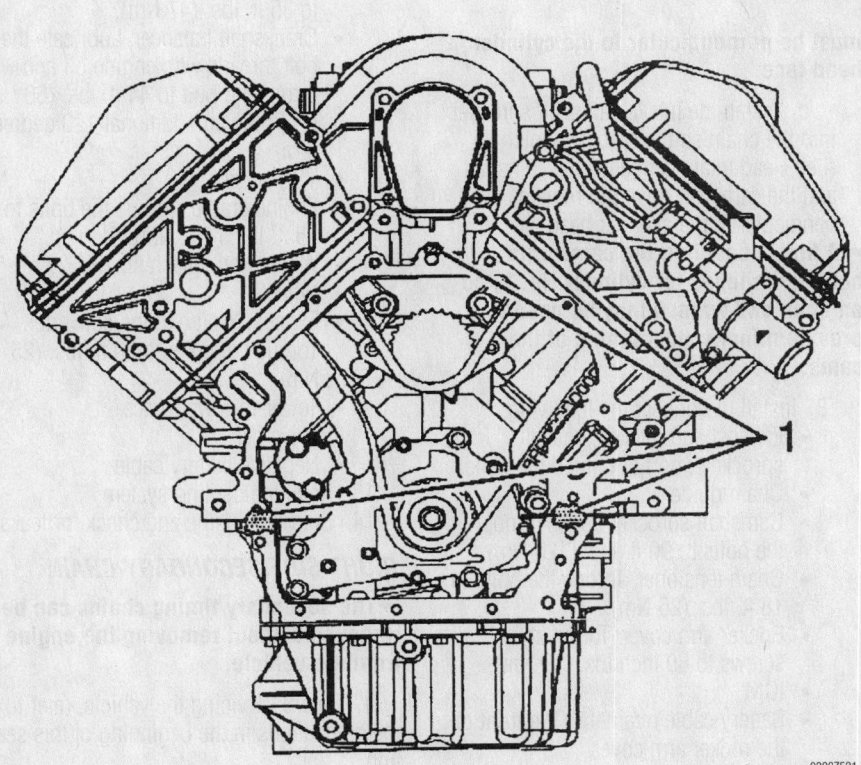

Apply RTV sealant to the 3 areas indicated before installing the front cover and gasket—3.5L engine

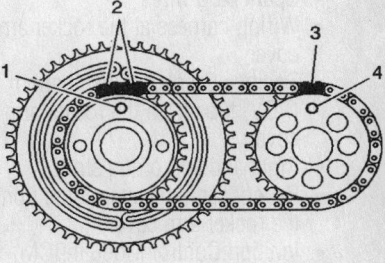

Correct sprocket alignment for the right secondary timing chain—3.5L engine

- Sprockets and chain assembly onto the camshafts, with the chain properly aligned on the tensioner

7. Remove the timing chain holding fixture

8. Install the sprocket bolts. Torque the bolts to 18 ft. lbs. (25 Nm); then an additional 45 degree turn.

9. Remove the camshaft holding fixture.

10. Install or connect the following:
- Rocker arm cover
- Negative battery cable

4.0L Engine

LEFT SIDE SECONDARY CHAIN

➡The secondary timing chains can be removed without removing the engine from the vehicle.

1. Before servicing the vehicle, refer to the precautions in the beginning of this section.

2. Remove or disconnect the following:
- Negative battery cable
- Drive belt
- Tower-to-tower brace
- Exhaust Y-pipe at the converter
- Front wheels
- Wheel well splash shields
- Crankshaft balancer bolt

3. Support the engine cradle.
- 3 right-side engine cradle bolts
- Vehicle Speed Sensor (VSS) from the right control arm

4. Lower the cradle to gain access for the crankshaft damper puller.

5. Remove or disconnect the following:
- Crankshaft balancer
- Drive belt tensioner and idler pulley
- Front cover and gasket

➡The front cover gasket is reusable as long as it is not damaged.

6. Partially drain the cooling system.
- Upper radiator hose at the water crossover
- Spark plug wires
- Right side fan
- Battery cable at the alternator
- Spark plug wires
- Wiring harness at the rocker arm cover
- Positive Crankcase Ventilation (PCV) tube from the rocker arm cover
- Right and left torque struts
- Battery cable retainer at the front of the rocker arm cover
- Ignition Control Module (ICM)
- Rocker arm cover by pivoting it around the water pump driveshaft

➡Continue moving the cover upward and pivoting so that the edge of the cover closely follows the left edge of the intake manifold cover. The gasket is reusable as long as it is not damaged.

- Left side secondary chain tensioner
- Left side chain guide

➡Access the upper chain guide mounting bolt through the hole in the cylinder head capped with the plastic plug.

- Secondary drive chain and sprockets

To install:

➡Correct timing exists when the crankshaft and intermediate shaft sprocket timing marks are in alignment and all 4 camshaft drive pins are perpendicular (90 degrees) to the cylinder head surface.

7. Assemble the left side secondary timing chain as follows:

a. Route the timing chain over the intermediate sprocket teeth outer row.

b. Route the timing chain over the chain guide and install the exhaust camshaft sprocket so the **LE** (Left Head Exhaust) pin engages the sprocket notch.

➡There should be no slack in the lower section of the timing chain and the camshaft drive pin

must be perpendicular to the cylinder head face.

c. Install the intake camshaft sprocket into the chain so the sprocket notch **LI** (Left Head Intake) engages the camshaft and the camshaft drive pin remains perpendicular to the cylinder head face.

➡A hex is cast into the camshafts behind the lobes for cylinder No. 2, so an open end wrench may be used to provide minor repositioning of the cams.

8. Install or connect the following:
- Exhaust and intake camshaft sprockets, do not tighten the bolts
- Chain guide
- Camshaft sprocket bolts. Torque the bolts to 90 ft. lbs. (120 Nm).
- Chain tensioner. Torque the bolts to 18 ft. lbs. (25 Nm).
- Rocker arm cover. Torque the screws to 89 inch lbs. (10 Nm).
- ICM
- Battery cable retainer to the front of the rocker arm cover

9. Install the right and left torque struts and torque the bolts as follows:

➡It is important during installation that the engine torque struts are not pre-loaded in their installed position. Adjustment is provided at the point the strut fastens to the core support bracket. Be sure this bolt is loose during assembly.

a. Step 1: Strut bracket-to-cylinder head (M10) bolt: 35 ft. lbs. (50 Nm).

b. Step 2: Strut bracket-to-water manifold (M8) bolts: 20 ft. lbs. (25 Nm).

c. Step 3: Strut-to-core support bracket bolt: 45 ft. lbs. (60 Nm).

10. Install or connect the following:
- PCV fresh air tube to the rocker arm cover
- Wiring harness to the cover
- Spark plug wires
- Battery cable at the alternator
- Right side fan
- Upper radiator hose to the water crossover
- Front cover with a new gasket. Torque the bolts to 89 inch lbs. (10 Nm).

11. Apply a dab of RTV to the split line between the upper and lower crankcase assemblies.

12. Install or connect the following:
- Drive belt idler pulley. Torque the bolt to 35 ft. lbs. (47 Nm).
- Drive belt tensioner. Torque the nut to 35 ft. lbs. (47 Nm).
- Crankshaft balancer. Lubricate the bolt threads with engine oil and torque the bolt to 44 ft. lbs. (60 Nm) then an additional 120 degree turn.
- VSS sensor
- Engine cradle. Torque the bolts to 75 ft. lbs. (102 Nm).
- Wheel well splash shields
- Wheels
- Exhaust Y-pipe to the converter. Torque the bolts to 20 ft. lbs. (25 Nm).
- Tower-to-tower brace
- Drive belt
- Negative battery cable

13. Refill the cooling system.

14. Start the engine and check for leaks.

RIGHT SIDE SECONDARY CHAIN

➡The secondary timing chains can be removed without removing the engine from the vehicle.

1. Before servicing the vehicle, refer to the precautions in the beginning of this section.

2. Remove or disconnect the following:
- Exhaust Y-pipe from the converter

90°
90°
90°
90°
B
A
3
90°
1
2
LI
LE
VIEW A
VIEW B
1 INTAKE POSITION
2 EXHAUST POSITION
3 TIMING MARKS
7922XG18

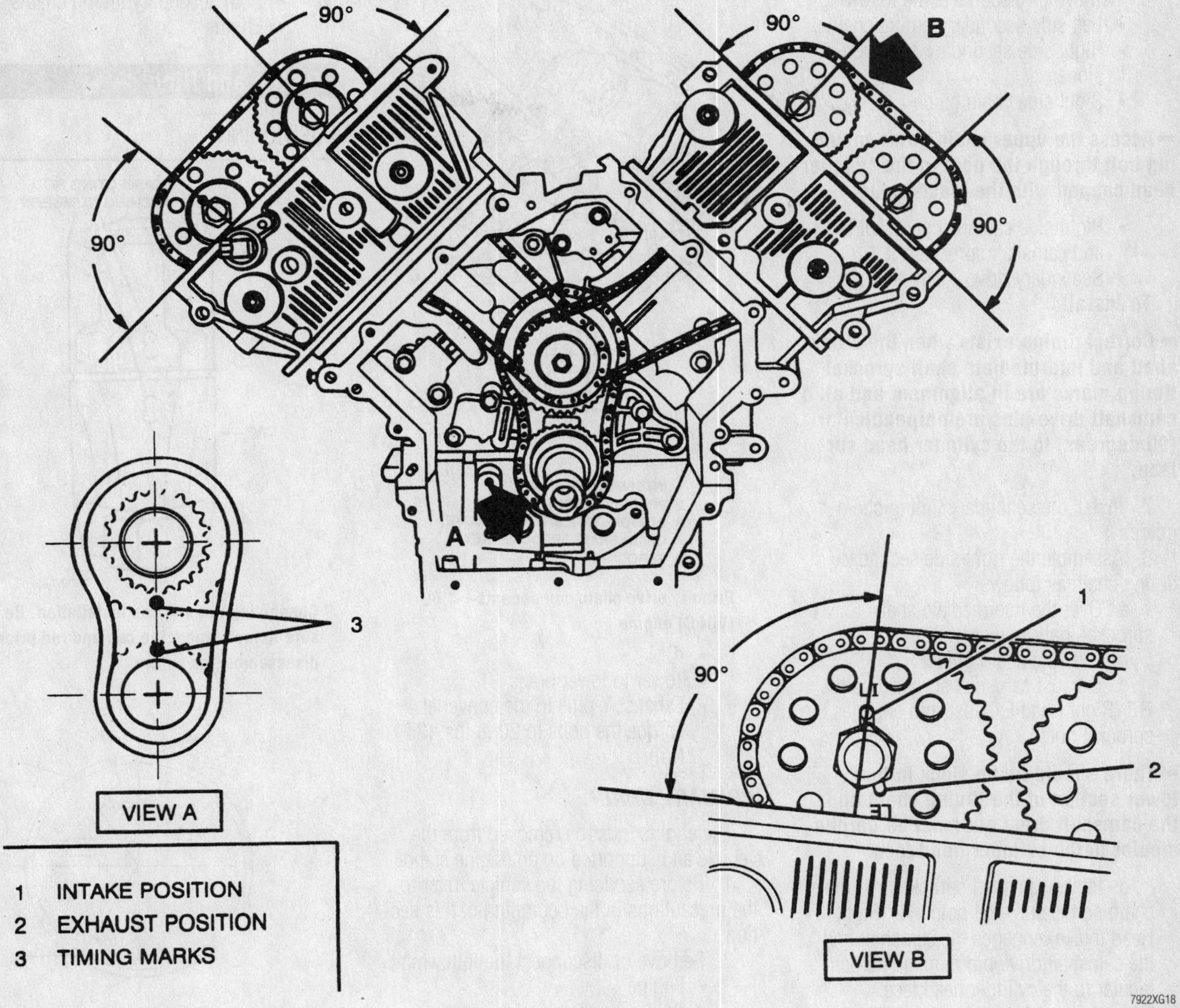

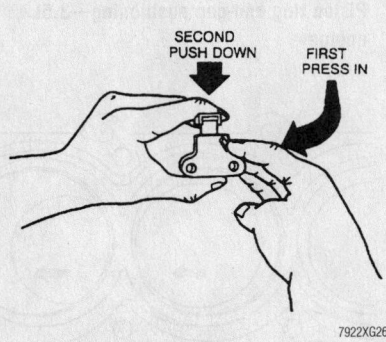

Primary and secondary timing mark alignment—4.0L (VIN C) engine

SECOND
PUSH DOWN
FIRST
PRESS IN
7922XG26

Rotating tensioner release lever—4.0L (VIN C) engine

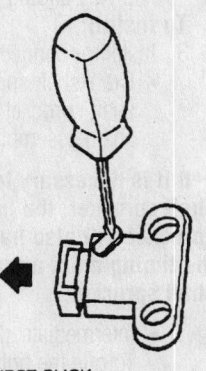

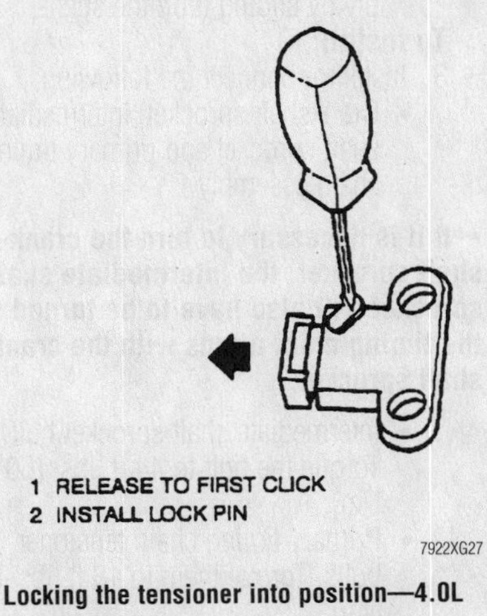

1 RELEASE TO FIRST CLICK
2 INSTALL LOCK PIN
7922XG27

Locking the tensioner into position—4.0L (VIN C) engine

- Tower-to-tower brace
- Ignition Control Module (ICM)
- Spark plug wires from the right bank
- Positive Crankcase Ventilation (PCV) valve
- Purge canister solenoid from the rear of the cover
- Rocker arm cover

3. Safely support the front of the engine cradle.
4. Remove or disconnect the following:
   - 2 mounting bolts at the front of the cradle
   - Right and left torque struts
5. Lower the engine cradle (or raise the vehicle) to provide clearance at the rear of the engine compartment.

6. Remove or disconnect the following:
- Left side secondary timing chain
- Right side secondary chain tensioner
- Right side chain guide

➡️**Access the upper chain guide mounting bolt through the hole in the cylinder head capped with the plastic plug.**

- Right side camshaft sprocket bolts and camshaft sprockets
- Secondary drive chain

To install:

➡️**Correct timing exists when the crankshaft and intermediate shaft sprocket timing marks are in alignment and all 4 camshaft drive pins are perpendicular (90 degrees) to the cylinder head surface.**

7. Install the secondary timing chain guide.

8. Assemble the right side secondary timing chain as follows:

   a. Over the intermediate shaft sprocket inner row of teeth.

   b. Over the chain guide.

   c. Exhaust camshaft sprocket so the **RE** (Right Head Exhaust) pin engages the sprocket notch.

➡️**There should be no slack in the lower section of the timing chain and the camshaft drive pin must be perpendicular to the cylinder head face.**

   d. Intake camshaft sprocket into the chain so the sprocket notch **RI** (Right Head Intake) engages the camshaft and the camshaft drive pin remains perpendicular to the cylinder head face.

➡️**A hex is cast into the camshafts behind the lobes for cylinder No. 1, so an open end wrench may be used to provide minor repositioning of the cams.**

9. Install or connect the following:
- Exhaust and intake camshaft sprockets, do not tighten
- Timing chain tensioner. Torque the bolts to 18 ft. lbs. (25 Nm).
- Camshaft sprocket bolts. Torque the bolts to 90 ft. lbs. (120 Nm).
- Left side secondary timing chain
- Torque struts
- Engine cradle. Torque the bolts to 75 ft. lbs. (102 Nm).
- Rocker arm cover. Torque the screws to 84 inch lbs. (10 Nm).
- Purge canister solenoid
- PCV valve
- Spark plug wires

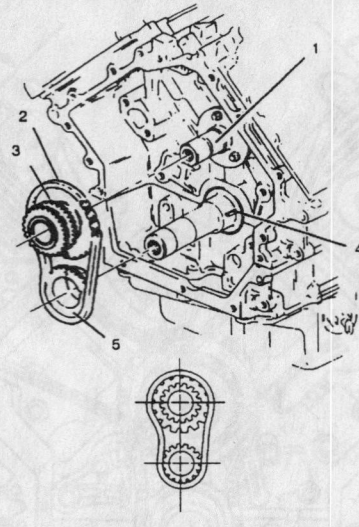

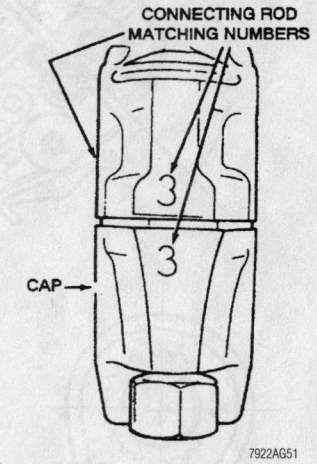

1 INTERMEDIATE SHAFT
2 PRIMARY CHAIN
3 INTERMEDIATE SHAFT SPROCKET
4 CRANKSHAFT SPROCKET KEY
5 SPROCKET

7922XG25

Primary drive chain components—4.0L (VIN C) engine

- ICM
- Tower to tower brace
- Exhaust Y-pipe to the converter. Torque the bolts to 20 ft. lbs. (25 Nm).

PRIMARY CHAIN

The engine must be removed from the vehicle and supported on an engine stand.

1. Before servicing the vehicle, refer to the precautions in the beginning of this section.

2. Remove or disconnect the following:
- Engine
- Both secondary timing chains
- Primary timing chain tensioner
- Intermediate shaft sprocket
- Primary chain and sprocket assembly by sliding it off the shafts

To install:

3. Install or connect the following:
- Crankshaft sprocket, intermediate shaft sprocket and primary timing chain assembly

➡️**If it is necessary to turn the crankshaft sprocket, the intermediate shaft sprocket will also have to be turned so the timing mark aligns with the crankshaft sprocket.**

- Intermediate shaft sprocket bolt. Torque the bolt to 44 ft. lbs. (60 Nm).
- Primary timing chain tensioner bolts. Torque them to 18 ft. lbs. (25 Nm).

- Both secondary timing chains
- Engine

Piston and Ring

POSITIONING

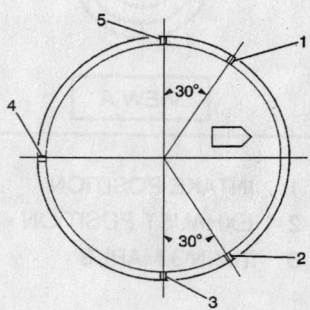

CONNECTING ROD MATCHING NUMBERS

CAP

7922AG51

Connecting rod and cap installation. Be sure to matchmark the cap and rod prior to disassembly, as shown

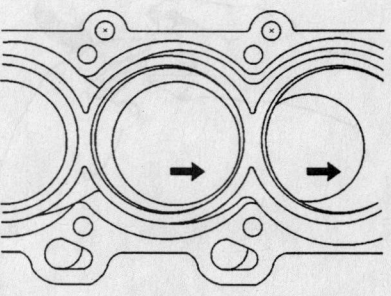

1. Lower oil control ring
2. Upper oil control ring
3. Top Ring
4. Oil control ring expander
5. Second ring

9306XG05

Piston ring end-gap positioning—3.5L engine

Piston positioning—3.5L engine

9306XG06

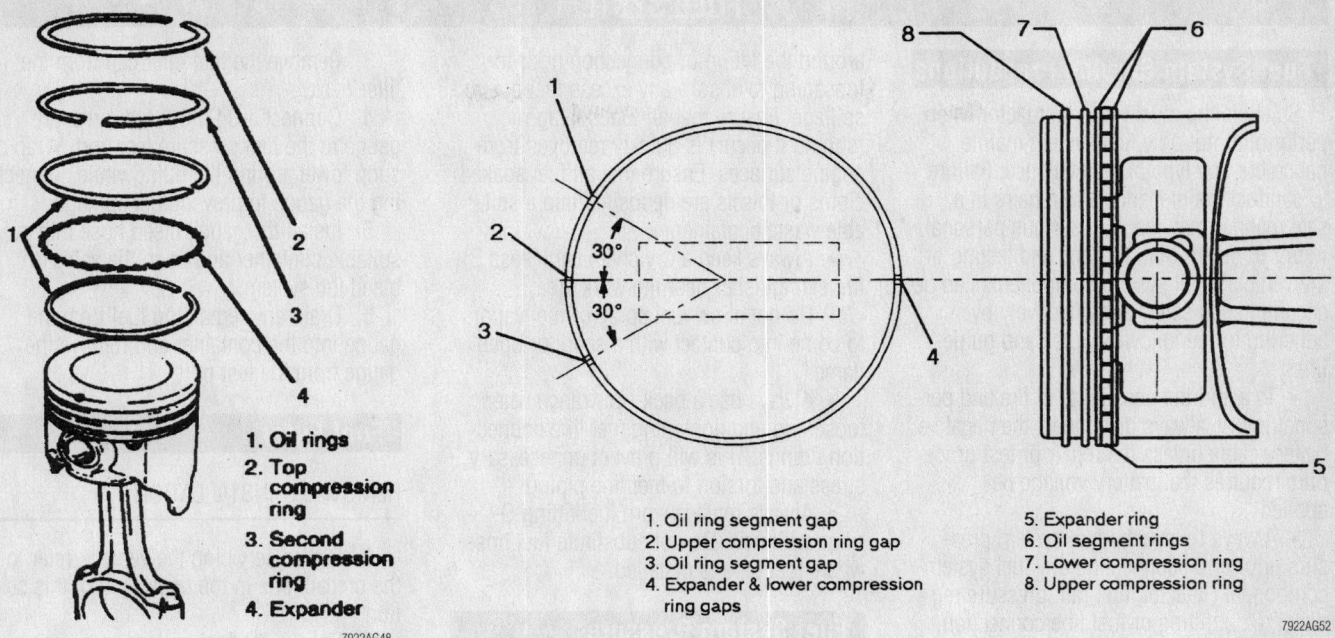

1. Oil rings
2. Top compression ring
3. Second compression ring
4. Expander

7922AG48

Piston ring positioning—3.5L engines

1. Oil ring segment gap
2. Upper compression ring gap
3. Oil ring segment gap
4. Expander & lower compression ring gaps

5. Expander ring
6. Oil segment rings
7. Lower compression ring
8. Upper compression ring

7922AG52

Piston ring and end-gap positioning—4.0L engine

LEFT BANK

FRT

RIGHT BANK

BOTTOM VIEW
(PAN - SIDE UP)

RIGHT BANK

PISTON ARROW TOWARD
CHAINCASE ON BOTH SIDES

FRT

LEFT BANK

TOP VIEW
(PAN - SIDE DOWN)

ROD CAPS

PISTON

FRT

BEARING CAP NOTCHES
POINT TOWARD EACH
OTHER ON PAIRED RODS

LOCATOR LUGS
INDICATE PISTON
FRONT TOWARDS
ENGINE FRONT

ROD CAP

FRT

PISTON

ROD CAP

BEARING CAP NOTCHES
POINT TOWARD EACH
OTHER ON PAIRED RODS

VIEW A

VIEW B

7922AG53

Piston and connecting rod assembly positioning—4.0L engine

FUEL SYSTEM

Fuel System Service Precautions

Safety is the most important factor when performing not only fuel system maintenance but any type of maintenance. Failure to conduct maintenance and repairs in a safe manner may result in serious personal injury or death. Maintenance and testing of the vehicle's fuel system components can be accomplished safely and effectively by adhering to the following rules and guidelines.

• To avoid the possibility of fire and personal injury, always disconnect the negative battery cable unless the repair or test procedure requires that battery voltage be applied.

• Always relieve the fuel system pressure prior to disconnecting any fuel system component (injector, fuel rail, pressure regulator, etc.), fitting or fuel line connection. Exercise extreme caution whenever relieving fuel system pressure, to avoid exposing skin, face and eyes to fuel spray. Please be advised that fuel under pressure may penetrate the skin or any part of the body that it contacts.

• Always place a shop towel or cloth around the fitting or connection prior to loosening to absorb any excess fuel due to spillage. Ensure that all fuel spillage (should it occur) is quickly removed from engine surfaces. Ensure that all fuel soaked cloths or towels are deposited into a suitable waste container.

• Always keep a dry chemical (Class B) fire extinguisher near the work area.

• Do not allow fuel spray or fuel vapors to come into contact with a spark or open flame.

• Always use a back-up wrench when loosening and tightening fuel line connection fittings. This will prevent unnecessary stress and torsion to fuel line piping.

• Always replace worn fuel fitting O-rings with new. Do not substitute fuel hose where fuel pipe is installed.

Fuel System Pressure

RELIEVING

1. Before servicing the vehicle, refer to the precautions in the beginning of this section.
2. Disconnect the negative battery cable.

3. Remove the fuel filler cap from the filler neck.
4. Connect J-34730-1 fuel pressure gauge to the fuel pressure test port. Wrap a shop towel around the fitting while connecting the gauge to prevent fuel spillage.
5. Install the gauge bleed hose into a suitable container and open the valve to bleed the system.
6. Drain any remaining fuel from the gauge into the container and remove the gauge from the test port.

Fuel Filter

REMOVAL & INSTALLATION

1. Before servicing the vehicle, refer to the precautions in the beginning of this section.
2. Relieve the fuel system pressure.
3. Remove or disconnect the following:
   • Quick connect fitting at the fuel filter inlet
   • Fuel filter outlet fitting from the fuel filter while holding the filter fitting with a back-up wrench
   • Fuel filter from the vehicle

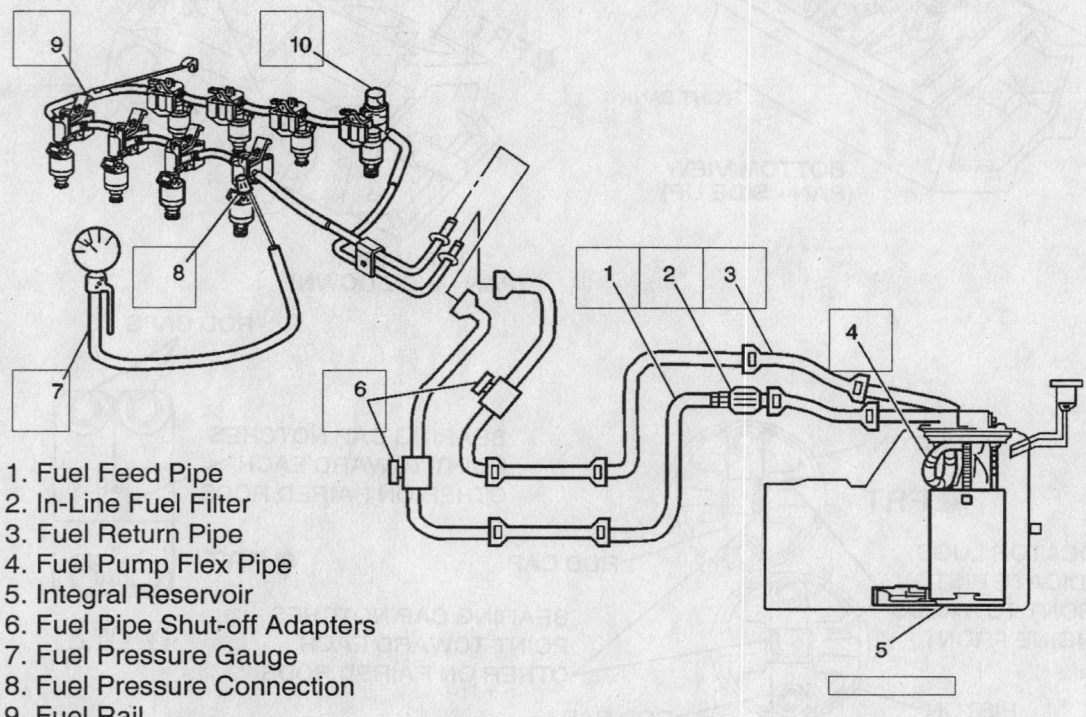

1. Fuel Feed Pipe
2. In-Line Fuel Filter
3. Fuel Return Pipe
4. Fuel Pump Flex Pipe
5. Integral Reservoir
6. Fuel Pipe Shut-off Adapters
7. Fuel Pressure Gauge
8. Fuel Pressure Connection
9. Fuel Rail
10. Fuel Pressure Regulator

9300XG07

Fuel system circuit showing the pressure test port on the supply rail—4.0L engine

To install:

4. Position the new fuel filter in the bracket.

5. Install a new plastic retainer on the fuel inlet line.

6. Apply a drop of oil on the fuel filter inlet fitting and snap the fitting onto the fuel filter.

7. Reconnect the fuel outlet line to the filter by holding the filter with a back-up wrench and tightening the line fitting to 22 ft. lbs. (30 Nm).

8. Pressurize the fuel system and verify no leaks.

Fuel Pump

REMOVAL & INSTALLATION

1. Before servicing the vehicle, refer to the precautions in the beginning of this section.

2. Relieve the fuel system pressure.

3. Drain the fuel tank.

4. Remove or disconnect the following:
- Spare tire and jack
- Floor trunk liner by pulling it back
- Fuel sender access panel
- Fuel sender assembly quick connect fittings
- Fuel sender assembly electrical connector

✳✳ WARNING

When the lock-ring is removed from the fuel sender, the sender assembly will spring up. Downward pressure should be kept on the assembly and slowly released to ensure the sender assembly does not get damaged.

- Lock-ring from the fuel sender

✳✳ CAUTION

The reservoir bucket on the fuel sender assembly will be full of fuel when it is removed from the tank. Be sure to have a catch pan nearby to drain the sender into.

- Sender assembly from the tank

To install:

5. Install or connect the following:
- New O-ring on top of the tank
- Sender assembly into the tank
- Retainer on top of the fuel tank, compress the sender until the retainer can be engaged; then lock it in place
- Quick connect fittings to the fuel sender assembly

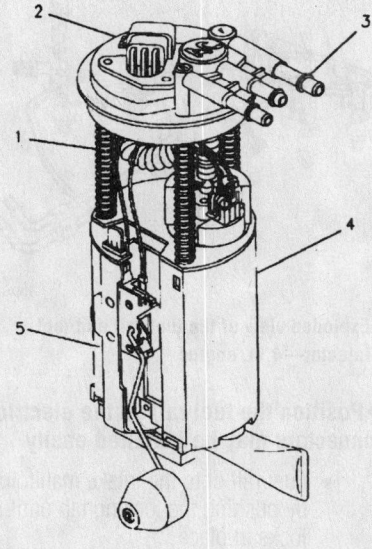

1 SUPPORT ASSEMBLY – FUEL SENDER
2 COVER ASSEMBLY – FUEL SENDER
3 FUEL PIPES (ABOVE COVER)
4 RESERVOIR – FUEL PUMP FUEL
5 SENSOR ASSEMBLY – FUEL LEVEL

7922XG19

Fuel pump and sending unit module assembly

- Sender assembly electrical connector
- Negative battery cable

6. Pressurize the fuel system and verify no leaks.

7. Install or connect the following:
- Fuel sender access panel
- Trunk liner
- Spare tire and jack

Fuel Injector

REMOVAL & INSTALLATION

3.5L Engine

1. Before servicing the vehicle, refer to the precautions in the beginning of this section.

2. Relieve the fuel system pressure.

3. Remove or disconnect the following:
- Fuel injector sight shield
- Fuel line fittings
- Fuel injector electrical connectors
- All necessary vacuum hoses
- Wiring harness from the fuel rail, if necessary
- Fuel rail from the intake manifold by pushing the locking tab away

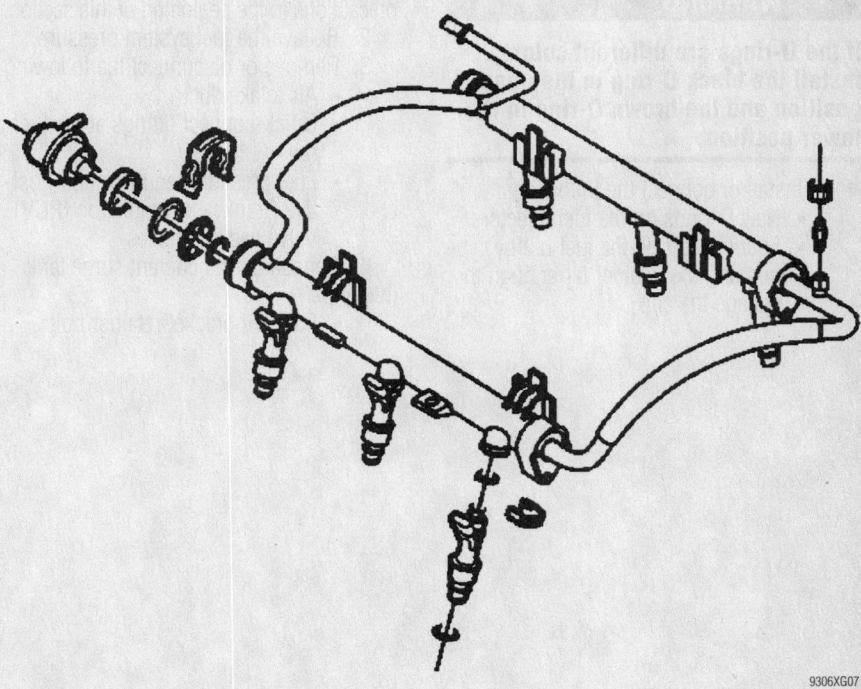

9306XG07

Exploded view of the fuel rail assembly–3.5L engine

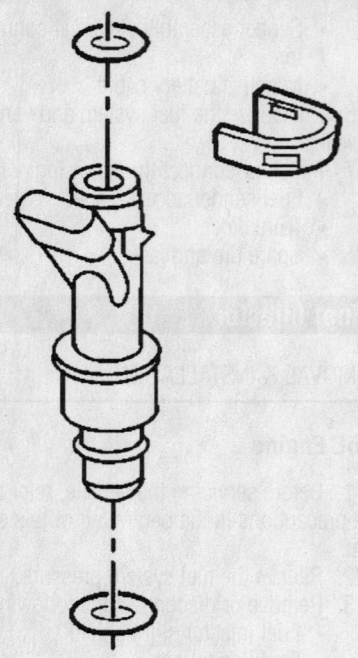

Exploded view of the fuel injector—3.5L engine

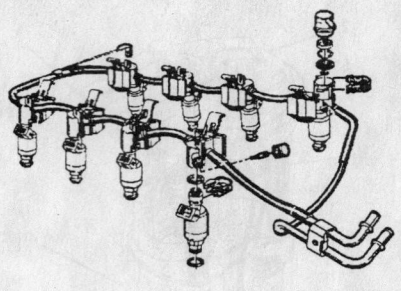

Exploded view of the fuel rail and fuel injector—4.0L engine

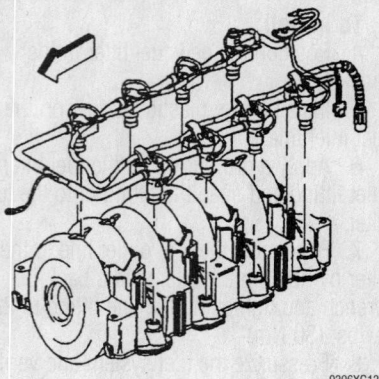

Exploded view of the fuel rail and intake manifold—4.0L engine

from the center of the intake manifold
- Fuel injector from the fuel rail and by spreading the retainer clip to release it

To install:

✳✳ WARNING

If the O-rings are different colors, install the black O-ring in the upper position and the brown O-ring in the lower position.

4. Install or connect the following:
- New O-rings on the fuel injectors
- Fuel injector on the fuel rail by pushing the retainer far enough to engage the clip

➡ **Position the fuel rail so the electrical connectors may be installed easily**
- Fuel rail onto the intake manifold by pushing the locking tab until it locks in place
- Wiring harness to the fuel rail, if necessary
- All vacuum hoses
- Fuel injector electrical connectors
- Fuel line fittings
- Negative battery cable

5. Pressurize the fuel system and check for leaks.

6. Install the fuel injector sight shield. Torque the bolts to 27 inch lbs. (3 Nm).

4.0L Engine

1. Before servicing the vehicle, refer to the precautions in the beginning of this section.
2. Relieve the fuel system pressure.
3. Remove or disconnect the following:
- Air intake duct
- Quick-connect fittings at the fuel rail
- Fuel pressure regulator and Positive Crankcase Ventilation (PCV) Valve vacuum hoses
4. Reposition the coolant surge tank inlet pipe.
- Fuel rail bracket retainer bolt

- Fuel injector electrical connectors
- Fuel rail bolts
- Rail and injectors as an assembly from the manifold
- Fuel injector from the fuel rail by spreading the retainer
- Fuel injector

To install:
5. Lubricate the new O-rings with engine oil.

6. Install or connect the following:
- Fuel injector using a new retainer clip and new O-rings
- Rail and injectors as an assembly into the manifold until fully seated
- Fuel rail bolts and tighten to 89 inch lbs. (10 Nm)
- Fuel injector electrical connectors
- Fuel rail bracket retainer bolt and tighten to 89 inch lbs. (10 Nm)
- Fuel pressure regulator and PCV vacuum hoses
- Coolant surge tank inlet pipe
- Quick-connect fittings at the fuel rail
- Air intake duct
- Negative battery cable

7. Pressurize the system and check for leaks.

DRIVE TRAIN

Transmission Assembly

REMOVAL & INSTALLATION

4T80–E Transmission

1. Before servicing the vehicle, refer to the precautions in the beginning of this section.
2. Remove or disconnect the following:
   - Negative battery cable
   - Air intake duct
   - Range selector cable from the transmission
   - Upper transmission oil cooler line from the radiator
   - Lower transmission oil cooler line from the transmission
   - Both heater tube retainers from the upper case side cover
   - Coolant temperature sensor electrical connector
   - Ground wire at the rear of the transmission
   - Left front Wheel Speed Sensor (WSS) electrical connector
   - Vehicle Speed Sensor (VSS) from the transmission
   - Rack and pinion electrical connector
   - Transmission vent hose
   - Upper transmission bolts
3. Install an engine support fixture.
4. Remove or disconnect the following:
   - Vacuum reservoir
   - Ball joints from the steering knuckles
   - Antilock Brake System (ABS) module
   - Secondary Air Injection (AIR) pumps
   - Left and right splash shields
   - Air deflector
   - Power steering line brackets at the frame
   - Front and rear transmission mount nuts
   - Rack and pinion from the frame
5. Support the engine frame.
6. Remove the 6 frame bolts and lower the frame from the body.
7. Remove or disconnect the following:
   - Left and right halfshafts
   - Exhaust heat shield
   - Transmission-to-engine brace
   - Front engine-to-transmission pencil brace
   - Three ground connections at the front of the transmission

- Transmission main wiring harness
- Engine to transmission bracket
- Flexplate cover
- Torque converter bolts
- Right transmission bracket
8. Support the transmission.
9. Remove the lower transmission bolts and lower the transmission towards the left so it can clear the starter motor.

To install:
10. Install or connect the following:
   - Transmission. Torque the transmission-to-engine bolts to 55 ft. lbs. (75 Nm).
   - Right transmission bracket. Torque the bolt to 55 ft. lbs. (75 Nm).
   - Torque converter bolts. Torque the bolts to 47 ft. lbs. (63 Nm).
   - Flexplate cover. Torque the bolts to 20 ft. lbs. (27 Nm).
   - Transmission brace to the engine. Torque the bolts to 44 ft. lbs. (60 Nm).
   - Transmission main wiring harness
   - Three ground connections to the transmission
   - Front engine-to-transmission pencil brace
   - Engine-to-transmission bracket
   - Exhaust heat shield
   - Left and right halfshafts
   - Engine frame. Torque the bolts to 142 ft. lbs. (192 Nm).
11. Remove the frame support fixture.
12. Install or connect the following:
   - Rack and pinion. Torque the bolts to 48 ft. lbs. (65 Nm).
   - Front and rear transmission mounts. Torque the nuts to 48 ft. lbs. (65 Nm).
   - Power steering line brackets. Torque the bolts to 53 inch lbs. (6 Nm).
   - Air deflector
   - Left and right splash shields
   - AIR pumps
   - ABS module
   - Lower ball joints
13. Remove the engine support fixture.
14. Install or connect the following:
   - Vacuum reservoir
   - Upper transmission bolts. Torque the bolts to 55 ft. lbs. (75 Nm).
   - Transmission vent hose
   - Rack and pinion electrical connector
   - VSS to the transmission
   - WSS electrical connector
   - Ground wire at the rear of the transmission

- Coolant temperature sensor electrical connector
- Both heater tube retainers from the upper case side cover
- Lower transmission oil cooler line
- Upper transmission oil cooler line
- Range selector cable and bracket
- Air intake duct
- Negative battery cable
15. Check and adjust the transmission fluid level.
16. Check the front end alignment.

4T65-E Transmission

1. Before servicing the vehicle, refer to the precautions in the beginning of this section.

➡️**Make sure the wheels of the vehicle must are in the straight ahead position and the steering column in the LOCK position before disconnecting the steering column or intermediate shaft from the steering gear. Failure to do so will cause the SIR coil assembly to become uncentered, which may cause damage to the coil assembly.**

2. Lock the steering column by installing tool J 42640 into the underside of the steering column.
3. Remove or disconnect the following:
   - Negative battery cable
   - Air cleaner assembly
   - Range selector cable from the range selector lever
   - Range selector cable with bracket from the transmission case and set aside
   - Nut and bolt from the AIR pipe
   - Bolt from the AIR pipe
   - Ground cable bolt from the transaxle
   - Transaxle electrical connector
   - Wiring harness from the wiring harness retainer on the transaxle
4. Install the engine support fixture.
   - Upper engine-to-transaxle case bolts
   - Front wheels
   - Both of the left facia extensions
   - Front air deflector
   - Intermediate shaft lower pinch bolt
   - Intermediate shaft from the power steering gear
   - Power steering gear heat shied
   - Power steering gear mounting bolts
   - Power steering line retainers from the frame

5. Secure the power steering gear to the exhaust manifold.

6. Loosen the two mounting nuts in order to allow removal of the Brake Pressure Modulator Valve (BPMV) from the bracket.

7. Remove or disconnect the following:
- Brake line retainers from the frame
- Left transaxle mount
- Frame
- Right and left drive axles from the transaxle
- Transmission oil cooler hoses from the transaxle
- Transaxle fluid filler tube
- Torque converter cover
- Flywheel-to-torque converter bolts

8. Support transaxle using an appropriate transaxle jack.
- Vehicle Speed Sensor (VSS) electrical connector
- Transaxle brace-to-transaxle bolts
- Transaxle brace-to-engine bolts
- Transaxle brace
- Engine-to-transaxle case bolt which can be accessed through the right wheel opening
- Remaining transaxle-to-engine bolt
- Transaxle from vehicle using an appropriate transaxle jack
- Rear transaxle bracket from the transaxle
- Left transaxle bracket from the transaxle

To install:

9. Installation is the reverse of removal, please note the following specifications:
- Left transaxle bracket bolts to 81 ft. lbs. (110 Nm)
- Rear transaxle bracket bolts to 46 ft. lbs. (63 Nm)
- Lower transaxle bolts to 55 ft. lbs. (75 Nm)
- Transaxle brace-to-engine bolts to 48 ft. lbs. (65 Nm)
- Transaxle brace-to-transaxle bolts to 26 ft. lbs. (36 Nm)
- Flywheel-to-torque converter bolts to 47 ft. lbs. (63 Nm)
- Power steering gear mounting bolts to 70 ft. lbs. (95 Nm)
- Intermediate shaft lower pinch bolt to 33 ft. lbs. (45 Nm)
- Upper transaxle case-to-engine bolts to 55 ft. lbs. (75 Nm)
- Bolt to AIR pipe to 17 ft. lbs. (23 Nm)
- Negative battery cable

10. Check and adjust the transmission fluid level.

11. Check the front end alignment.

Halfshaft

REMOVAL & INSTALLATION

1. Before servicing the vehicle, refer to the precautions in the beginning of this section.

2. Remove or disconnect the following:
- Wheel
- Outer tie rod end from the steering knuckle

3. Insert a drift or punch into the brake rotor and against the brake caliper in order to prevent the wheel hub and bearing from turning.
- Wheel driveshaft spindle nut and discard
- Stabilizer shaft link
- Electrical connector from the Wheel Speed Sensor (WSS) and position the wiring harness away from the ball joint
- Lower ball joint from the steering knuckle

4. Install wheel hub removal tool J 42129 onto the wheel hub and secure with wheel nuts.
- Wheel driveshaft from the wheel hub and bearing using the removal tool and support the wheel driveshaft.

5. Assemble tool J 2619-01 slide hammer with adapter, J 29794 extension, and J 33008-A wheel driveshaft removal

6. Using the assembled tools, separate the shaft from the transaxle.

7. Remove the wheel driveshaft from the vehicle.

To install:

8. Install or connect the following:
- Driveshaft to the transaxle

9. Verify that the driveshaft is properly engaged to the transaxle by grasping the inner tripod housing and pulling outward. Do not pull on the driveshaft bar. The driveshaft will remain firmly in place when properly engaged

10. Remove or disconnect the following:
- Driveshaft to the hub and bearing
- Ball joint to the steering knuckle
- Wheel speed sensor electrical connector
- Stabilizer shaft link

11. Insert a drift or punch into the rotor and against the caliper in order to prevent the hub and bearing from turning.

12. Remove or disconnect the following:
- A new driveshaft spindle nut and tighten to 118 ft. lbs. (160 Nm)
- Outer tie rod end to the steering knuckle

- Wheel

13. Check the alignment.

CV-Joint

REMOVAL & INSTALLATION

Inner (Tri-Pod) Joint

1. Before servicing the vehicle, refer to the precautions in the beginning of this section.

2. Remove or disconnect the following:
- Front wheel
- Halfshaft
- Swage ring
- Large CV-joint boot clamp
- CV-joint boot by sliding it away from the tri-pod joint
- Tri-pod housing from the tri-pod spider
- Trilobal tri-pod bushing from the housing
- Inboard spacer ring slide, it rearward on the shaft
- Outboard retaining ring
- Tri-pod joint spider assembly
- Inboard spacer ring and CV-joint boot

To install:

3. Install or connect the following:
- Swage ring clamp
- CV-joint boot

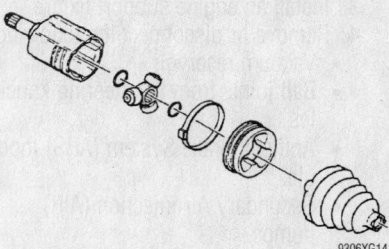

9306XG14

Exploded view of the inner (tri-pod) joint

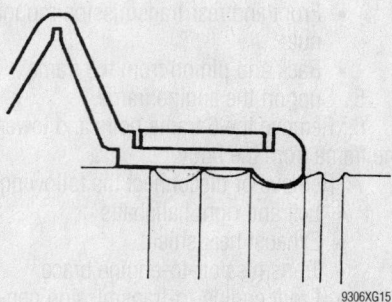

9306XG15

Positioning the inner CV-joint boot seal and swage ring—Inner (tri-pod) joint

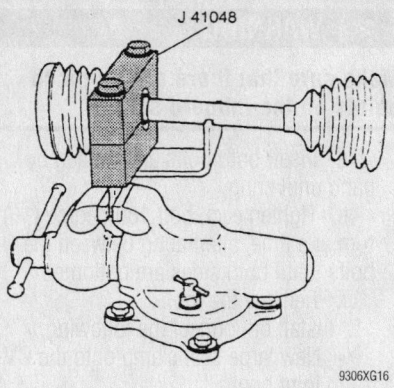

View of the swage ring crimping tool—Inner (tri-pod) joint

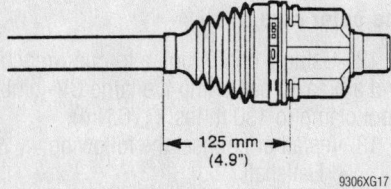

← 125 mm
(4.9")

9306XG17

Boot measurement—Inner (tri-pod) joint

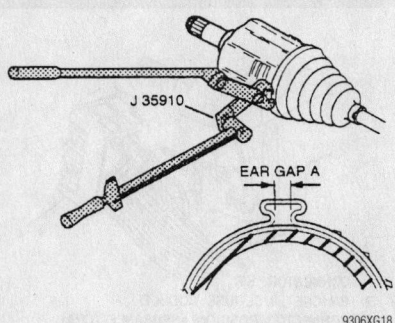

J 35910

EAR GAP A

9306XG18

Crimping the large CV-joint boot ring—Inner (tri-pod) joint

4. Position the CV-joint boot seal into the axle shaft's joint seal groove and align the swage ring clamp on the boot.

5. Secure the swage ring clamp as follows:

   a. Mount the lower half of tool J-41048 in a vise.

   b. Position the outboard of the halfshaft in the tool.

   c. Position the upper end of tool J-41048 onto the lower half.

☀☀ WARNING

Make sure that there are no pinch points on the inboard seal.

   d. Insert both bolts and tighten by hand until snug.

   e. Tighten each bolt 180 degree (½) turn at a time, alternating between the bolts, until both sides are bottomed.

   f. Remove the tool.

6. Install or connect the following:

   - Inboard spacer ring slide it rearward on the shaft
   - Tri-pod joint spider assembly onto the shaft
   - Outboard retaining ring into the axle shaft groove
   - Tri-pod joint spider assembly, slide it against the outboard retaining ring
   - Inboard spacer ring
   - ½ kit grease into the boot
   - ½ kit grease into the tri-pod housing
   - Trilobal tri-pod bushing flush with the tri-pod housing face
   - New large seal clamp onto the CV-joint boot
   - Tri-pod housing, slide it over the tri-pod joint spider assembly
   - CV-joint boot/clamp, slide it into place over the trilobal tri-pod bushing with the seal lip in the groove

➡ **Make sure the boot lies flat against the trilobal bushing.**

7. Position the CV-joint boot so it measures 4.9 in. (125mm).

8. Using a crimp tool, a torque wrench and a breaker bar, crimp the large CV-joint boot clamp to 130 ft. lbs. (176 Nm).

9. Install or connect the following:

   - Halfshaft
   - Front wheel

Outer Joint

1. Before servicing the vehicle, refer to the precautions in the beginning of this section.

2. Remove or disconnect the following:

   - Front wheel
   - Halfshaft
   - Swage ring
   - Large boot clamp
   - CV-joint boot, slide it away from the CV-joint
   - CV-joint assembly by spreading the inner race-to-axle shaft retaining ring ears
   - CV-joint boot from the axle shaft

3. Disassemble the chrome alloy balls from the CV-joint cage as follows:

   a. Position a brass drift against the CV-joint cage and tap it with a hammer to tilt the cage.

   b. Remove the 1st chrome alloy ball from the cage.

   c. Tilt the cage in the opposite direction.

   d. Remove the opposite chrome alloy ball.

   e. Repeat the procedure until all 6 balls are removed.

4. Disassemble the CV-joint cage and inner race as follows:

   a. Pivot the cage and race 90 degrees to the center line of the outer race.

   b. Align the cage windows with outer race lands.

   c. Remove the cage from the outer race.

   d. Rotate the inner race upward and remove it from the cage.

To install:

5. Lubricate the parts with a light coat of grease.

6. Assemble the CV-joint cage and inner race, as follows:

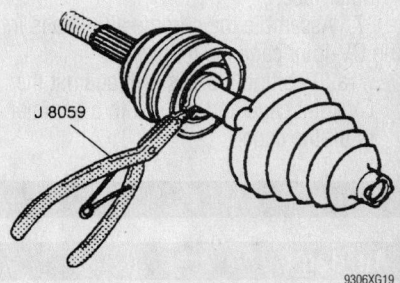

J 8059

9306XG19

Disconnecting the outer CV-joint from the axle shaft

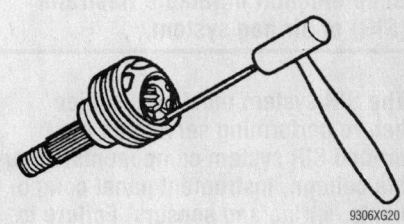

9306XG20

Tilting the cage—Outer CV-joint

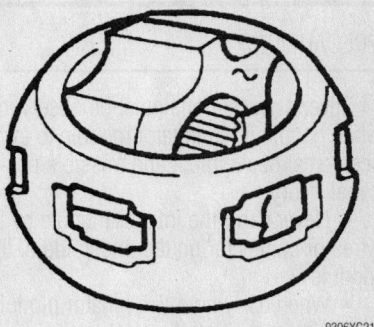

9306XG21

View the cage and inner race—Outer CV-joint

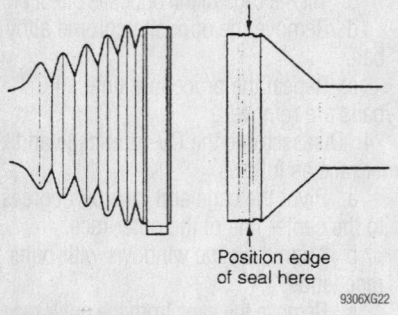

Positioning the boot—Outer CV-joint

9306XG22

a. Rotate the inner race 90 degrees to the cage centerline.

b. Align the cage windows with inner race lands.

c. Insert the inner race into the cage by rotating the inner race downward.

d. Insert the cage/inner race into the outer race.

7. Assemble the chrome alloy balls into the CV-joint cage, as follows:

a. Position a brass drift against the CV-joint cage and tap it with a hammer to tilt the cage.

b. Insert the 1st chrome alloy ball into the cage.

c. Tilt the cage in the opposite direction.

d. Insert the opposite chrome alloy ball.

e. Repeat the procedure until all 6 balls are inserted.

8. Install or connect the following:
- ½ kit grease into the CV-joint boot
- ½ kit grease into the CV-joint
- Swage ring clamp
- CV-joint boot
- CV-joint onto the axle shaft until the retaining ring seats into the groove

9. Position the CV-joint boot seal into the axle shaft's joint seal groove and align the swage ring clamp on the boot.

10. Secure the swage ring clamp as follows:

a. Mount the lower half of tool J-41048 in a vise.

b. Position the outboard of the half-shaft in the tool.

c. Position the upper end of tool J-41048 onto the lower half.

✳✳ WARNING

Make sure that there are no pinch points on the inboard seal.

d. Insert both bolts and tighten by hand until snug.

e. Tighten each bolt 180 degree (½) turn at a time, alternating between the bolts, until both sides are bottomed.

f. Remove the tool.

11. Install or connect the following:
- New large seal clamp onto the CV-joint boot
- CV-joint boot/clamp, slide it into place over the outer race with the seal lip in the groove

➡ **Make sure the boot lies flat against the outer race.**

12. Using a crimp tool, a torque wrench and a breaker bar, crimp the large CV-joint boot clamp to 130 ft. lbs. (176 Nm).

13. Install or connect the following:
- Halfshaft
- Front wheel

STEERING AND SUSPENSION

Air Bag

✳✳ CAUTION

The vehicles are equipped with the Supplemental Inflatable Restraint (SIR) or air bag system.

The SIR system must be disabled before performing service on or around SIR system components, steering column, instrument panel components, wiring and sensors. Failure to follow safety and disabling procedures could result in accidental air bag deployment, possible personal injury and unnecessary SIR system repairs.

PRECAUTIONS

Several precautions must be observed when handling the inflator module to avoid accidental deployment and possible personal injury.

- Never carry the inflator module by the wires or connector on the underside of the module.
- When carrying a live inflator module, hold securely with both hands and ensure that the bag and trim cover are pointed away.

- Place the inflator module on a bench or other surface with the bag and trim cover facing up.
- With the inflator module on the bench, never place anything on or close to the module, which may be thrown in the event of an accidental deployment.

DISARMING

1. Turn the steering wheel so that the vehicle's wheels are pointing straight ahead.

2. Turn the ignition switch to the **OFF** position.

3. Remove the key from the ignition switch.

➡ **With the SIR fuse removed and the ignition switch in the ON position, The AIR BAG warning lamp illuminates. This is normal operation, and does not indicate an SIR system malfunction.**

4. Remove the rear seat.

5. Remove the SIR fuse from the rear fuse block located under the rear seat.

6. Remove the driver sound insulator.

7. Remove the Connector Position Assurance (CPA) from the driver yellow connector located next to steering column.

8. Disconnect the driver frontal air bag yellow connector from the vehicle harness yellow connector.

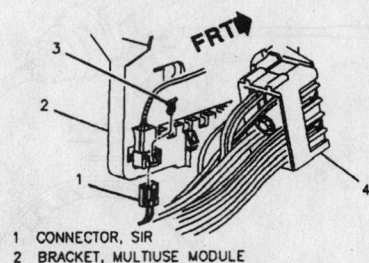

1 CONNECTOR, SIR
2 BRACKET, MULTIUSE MODULE
3 CONNECTOR POSITION ASSURANCE (CPA)
4 CONNECTOR, STEERING COLUMN WIRING HARNESS

7922XG21

SIR 2-way connector location—driver's side

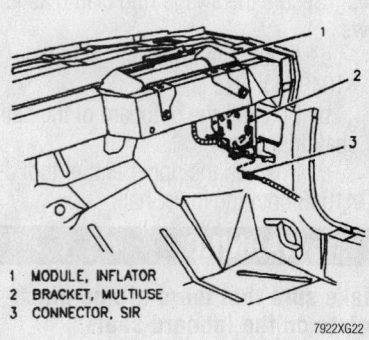

1 MODULE, INFLATOR
2 BRACKET, MULTIUSE
3 CONNECTOR, SIR

7922XG22

SIR 2-way connector location—passenger's side

9. Remove the passenger sound insulator.

10. Remove the CPA from the passenger yellow connector located above the passenger sound insulator.

11. Disconnect the passenger (IP) frontal air bag yellow connector from the vehicle harness yellow connector.

12. Remove both CPA locks from the driver side (seat) air bag and pretensioner yellow connector located under the driver seat.

13. Disconnect the driver side air bag and pretensioner yellow connector from the vehicle harness yellow connector.

14. Remove both CPA locks from the passenger side (seat) air bag and pretensioner yellow connector located under the passenger seat.

15. Disconnect the passenger side air bag and pretensioner yellow connector from the vehicle harness yellow connector.

16. The above procedures will disable the SIR system. If vehicle is equipped with optional Rear Air Bags the following steps must be done, to completely disable the SIR system:

a. Remove the rear seat back.

b. Remove the CPA from the passenger rear side air bag yellow connector.

17. Disconnect the passenger rear side air bag yellow connector from the vehicle harness yellow connector.

18. Remove the CPA from the driver rear side air bag yellow connector.

19. Disconnect the driver rear side air bag yellow connector from the vehicle harness yellow connector.

ARMING

1. Remove the key from the ignition switch.

2. Connect the passenger rear side air bag yellow connector to the vehicle harness yellow connector.

3. Install the (CPA) to the passenger rear side air bag yellow connector.

4. Connect the driver rear side air bag yellow connector to the vehicle harness yellow connector.

5. Install the CPA to the driver rear side air bag yellow connector.

6. Install the rear seat back.

7. Connect the driver side (seat) air bag and pretensioner yellow connector to the vehicle harness yellow connector.

8. Install both CPA locks to the driver side (seat) air bag and pretensioner yellow connector located under the driver seat.

9. Connect the passenger side (seat) air

bag and pretensioner yellow connector to the vehicle harness yellow connector.

10. Install both CPA locks to the passenger side (seat) air bag and pretensioner yellow connector located under the passenger seat.

11. Connect the passenger (IP) frontal air bag yellow connector to the vehicle harness yellow connector located above the passenger sound insulator.

12. Install the CPA to the passenger yellow connector.

13. Install the passenger sound insulator.

14. Connect the driver frontal air bag yellow connector to the vehicle harness yellow connector located next to steering column.

15. Install the CPA to the driver yellow connector.

16. Install the driver sound insulator.

17. Install the SIR fuse to the rear fuse block.

18. Install the rear seat.

19. Turn the ignition switch to the **RUN** position and verify that the **AIR BAG** warning lamp flashes 7 times, then turns **OFF**.

Power Rack and Pinion Steering Gear

REMOVAL & INSTALLATION

1. Before servicing the vehicle, refer to the precautions at the beginning of this section.

✳✳ CAUTION

Make sure the wheels of the vehicle are straight ahead and the steering column in the LOCK position before disconnecting the steering column or intermediate shaft from the steering gear. Failure to do so will cause the coil assembly in the steering column to become uncentered which will cause damage to the coil assembly.

2. Lock the steering column by installing the Steering Column Anti Rotation Pin tool J 42640 into the underside of the steering column.

3. Disconnect negative battery cable.

4. Remove or disconnect the following:
   - Intermediate shaft lower coupling
   - Outer tie rod end from the steering knuckle
   - Pressure and return pipes from the rack and pinion gear
   - Electrical connector, if equipped

   - Stabilizer shaft links at lower control arms

5. Rotate the stabilizer shaft to access steering gear bolts.
   - Rack and pinion attaching bolts
   - Rack and pinion assembly

To install:

6. Install or connect the following:
   - Rack and pinion assembly and tighten the bolts to 70 ft. lbs. (95 Nm)
   - Electrical connector, if equipped
   - Pressure and return lines to the steering gear and tighten the fittings to 20 ft. lbs. (27 Nm)
   - Heat shield and bolts. Tighten the bolts to 7 ft. lbs. (10 N)

7. Align stabilizer shaft.
   - Links to lower control arms
   - Tie rod ends to the steering knuckles. Torque the castle nuts to 52 ft. lbs. (70 Nm).
   - Intermediate shaft to the steering gear and tighten the pinch bolt to 35 ft. lbs. (47 Nm)

8. Remove the steering column anti rotation pin from the steering column.

9. Fill and bleed the power steering system.

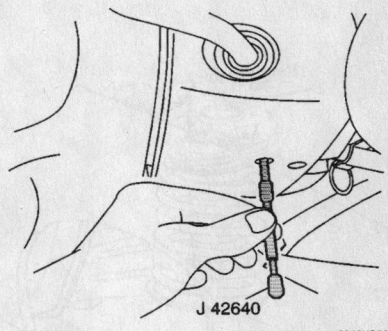

J 42640

9346UG08

Lock the steering column by installing the Steering Column Anti Rotation Pin tool J 42640 into the underside of the steering column

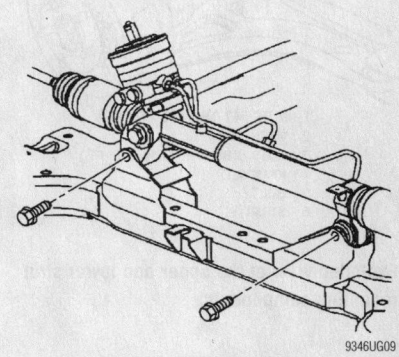

9346UG09

Steering gear mounting bolts

Strut

REMOVAL & INSTALLATION

Front

1. Before servicing the vehicle, refer to the precautions in the beginning of this section.
2. Remove or disconnect the following:

   - Negative battery cable
   - Front wheel
   - Anti-lock Brake System (ABS) wheel speed sensor electrical connector
   - ABS speed sensor bracket from the strut
   - Brake line bracket, if removing the left strut

3. Matchmark the lower strut bracket to the steering knuckle.
4. Support the steering knuckle.
5. Remove or disconnect the following:

   - Lower strut bracket nut and through-bolts
   - 3 upper strut plate mounting nuts and washers
   - Strut from the vehicle

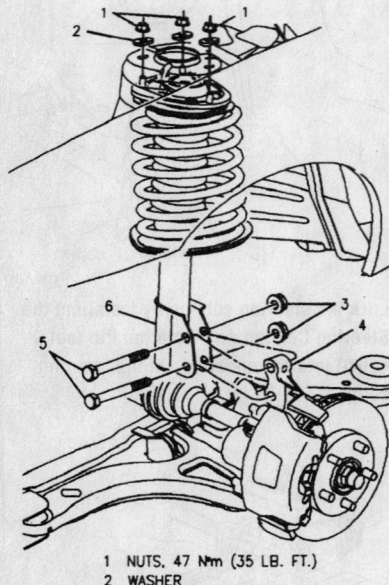

1 NUTS, 47 N·m (35 LB. FT.)
2 WASHER
3 NUTS, 185 N·m (136 LB. FT.)
4 KNUCKLE
5 BOLT
6 STRUT

7922XG23

Exploded view of the upper and lower strut mounting components

To install:

6. Install or connect the following:

   - Strut, but do not tighten the upper strut plate-to-body nuts yet
   - Lower strut through-bolts by aligning the matchmarks. Torque the nuts to 136 ft. lbs. (185 Nm).

7. Remove the steering knuckle support.
8. Install or connect the following:

   - Brake line bracket to the strut, if removed
   - ABS wheel speed sensor bracket on the strut
   - ABS sensor electrical connector
   - Front wheel. Torque the nuts to 100 ft. lbs. (140 Nm).
   - Upper strut plate nuts. Torque the nuts to 35 ft. lbs. (47 Nm).
   - Negative battery cable

9. Check the front end alignment and adjust as necessary.

Shock Absorber

REMOVAL & INSTALLATION

Rear

1. Before servicing the vehicle, refer to the precautions in the beginning of this section.
2. Support the lower control arm at such a height that the upper shock bolts will still be accessible.
3. Remove or disconnect the following:

   - Rear wheel
   - Electronic Level Control (ELC) air tube from the shock
   - Both lower shock mount bolts
   - Trunk trim panel to access the upper shock mount bolts
   - Upper shock cap
   - Both upper shock mount nuts and reinforcement
   - Shock

To install:

4. Install or connect the following:

   - Shock
   - Reinforcement and upper shock mounting nuts. Torque the nuts to 15 ft. lbs. (20 Nm).
   - Upper shock cap and inner trunk trim
   - Lower shock bolts. Torque the bolts to 18 ft. lbs. (24 Nm).
   - ELC air tube to the shock
   - Rear wheel. Torque the nuts to 100 ft. lbs. (140 Nm).

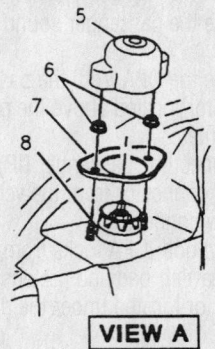

VIEW A

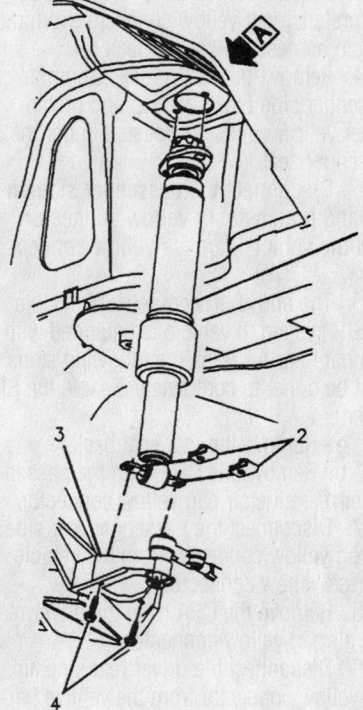

1 SHOCK
2 U–NUTS
3 CONTROL ARM
4 BOLTS 24 N·m (18 LB. FT.)
5 COVER
6 NUTS 20 N·m (15 LB. FT.)
7 REINFORCEMENT
8 MOUNT, UPPER

7922XG34

Exploded view of the rear shock mounting

Coil Spring

REMOVAL & INSTALLATION

Front

1. Remove the strut from the vehicle.
2. Disassemble the strut as follows:
   a. Step 1: Place the strut assembly

into compressor tool, to compress the coil spring.

b. Step 2: Compress the spring slightly.

c. Step 3: Hold the strut shaft from turning using a No. 50 Torx® socket and remove the 24mm nut on the top end of the strut.

d. Step 4: Install Rod tool J 34013-38 to help guide the strut shaft from the upper mount assembly.

e. Step 5: Loosen the spring compressor tool until the coil spring and mount can be removed as an assembly. Remove the lower spring insulator, if equipped.

To install:

3. Assemble the strut as follows:

a. Step 1: Place the strut in compressor tool.

b. Step 2: Install the coil spring over the strut.

c. Step 3: Compress the coil spring while guiding strut shaft through the top of the strut assembly.

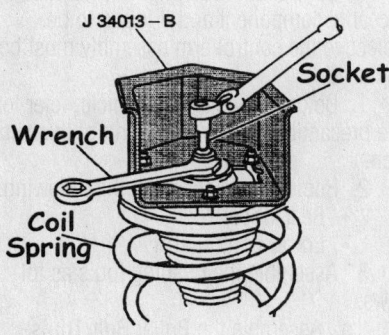

Use a Torx® socket to keep the piston rod from turning while removing the upper nut—front strut shown

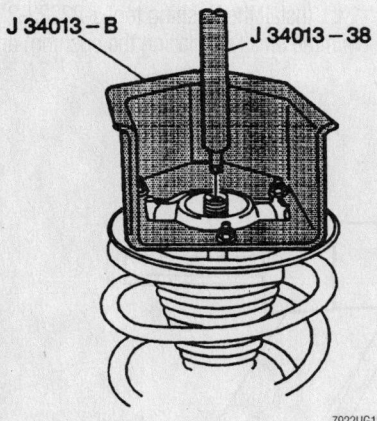

Install Rod J 34013-38 to help guide the strut shaft from the upper mount assembly—front strut shown

d. Step 4: Install the top strut nut. Torque the nut to 55 ft. lbs. (75 Nm).

e. Step 5: Remove the strut from the compressor.

Rear

1. Before servicing the vehicle, refer to the precautions in the beginning of this section.

2. Support the lower control arm.

3. Remove or disconnect the following:
- Rear wheel
- Electronic Level Control (ELC) air tube from the shock absorber
- Both lower shock absorber-to-control arm bolts
- Adjustment link outer ball stud nut
- Ball stud from the knuckle

4. Lower the control arm until the arm bottoms out on the rear suspension support.

5. Using a suitable prying tool, pry under the lower spring insulator to unseat it from the control arm.

6. Remove or disconnect the following:
- Insulator and coil spring
- Upper spring insulator, if necessary

To install:

7. Install or connect the following:
- Upper spring insulator, if removed

➡️**Engage the retainer on the back of the insulator in the upper mount hole.**

- Coil spring and lower insulator

➡️**Be sure to seat the lower insulator in the control arm hole.**

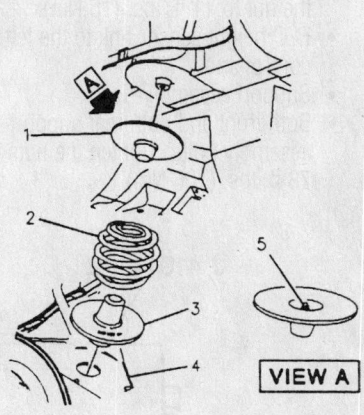

1 JOUNCE BUMPER
2 SPRING
3 LOWER SPRING INSULATOR
4 CONTROL ARM
5 RETAINER

Exploded view of the rear coil spring and related components

8. Raise the control arm into position.

9. Install or connect the following:
- Lower shock absorber-to-control arm bolts. Torque both bolts to 18 ft. lbs. (24 Nm).
- Adjustment link ball stud to the knuckle. Torque the castle nut to 55 ft. lbs. (70 Nm).

➡️**If necessary to align the cotter pin holes, tighten the nut up to an additional 60 degree (⅙) turn. NEVER loosen the castle nut to align the cotter pin holes.**

- ELC air tube to the shock absorber
- Rear wheel. Torque the nuts to 100 ft. lbs. (140 Nm).

Lower Ball Joint

REMOVAL & INSTALLATION

The ball joint is an integral part of the control arm. If defective the control arm must be replaced.

Lower Control Arm

REMOVAL & INSTALLATION

Front

1. Before servicing the vehicle, refer to the precautions in the beginning of this section.

2. Remove or disconnect the following:
- Front wheel
- Sway bar link kit

➡️**Take note of the positions of the washers and insulators for installation purposes.**

- Lower ball joint from the steering knuckle

✳️ WARNING

Be careful not to overextend the half-shaft tri-pod joint.

- Lower control arm-to-engine frame nuts and bolts
- Lower control arm

To install:

3. Install or connect the following:
- Lower control arm, do not tighten the nuts and bolts
- Lower ball joint to the steering knuckle. Torque the castle nut to 41 ft. lbs. (55 Nm).

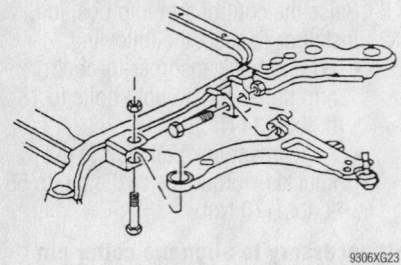

Exploded view of the lower control arm—front suspension

➤If necessary, tighten the castle nut up to an additional 60 degree (⅙) turn to align the cotter pin holes. NEVER loosen the nut to make the alignment.

- Sway bar link kit. Torque the nut to 13 ft. lbs. (17 Nm).
- Front wheel. Torque the nuts to 100 ft. lbs. (140 Nm).

4. Lower the vehicle in order to achieve proper trim height.

5. On all except Aurora, torque the lower control arm-to-frame rear bolt to 117 ft. lbs. (158 Nm) and the lower control arm-to-frame front nut to 93 ft. lbs. (126 Nm).

6. On Aurora, torque the lower control arm-to-frame rear bolt to 117 ft. lbs. (158 Nm) and the lower control arm-to-frame front nut to 117 ft. lbs. (158 Nm).

Rear

1. Before servicing the vehicle, refer to the precautions in the beginning of this section.

2. Remove or disconnect the following:
- Rear wheels
- Exhaust system
- Coil springs
- Parking brake cable from the brake calipers
- Brake calipers from the control arms
- Parking brake cable from the rear suspension support assembly
- Support assembly electrical connectors
- Electronic Level Control (ELC) electrical connector and vent hose
- ELC air tube from the air compressor

3. Support the rear suspension support assembly.

4. Remove or disconnect the following:
- 3 support bracket-to-chassis bolts at each side
- Both front and both rear support assembly bolts
- Support assembly

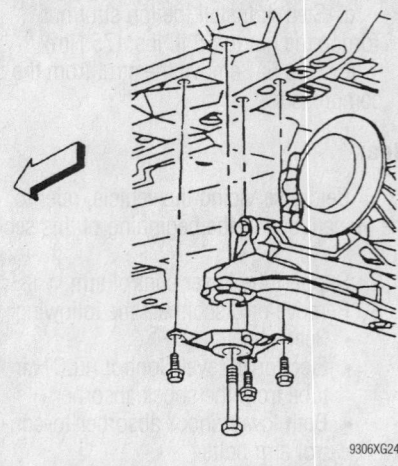

Exploded view of the support assembly—rear suspension

- ELC height sensor link from the left control arm
- Stabilizer link bolt and nut
- Anti-lock Brake System (ABS) electrical connector
- Wheel/hub assembly
- Lower control arm-to-rear suspension support assembly nuts and bolts
- Lower control arm

To install:

5. Install or connect the following:
- Lower control arm
- Lower control arm-to-rear suspension support assembly nuts and bolts, do not tighten
- Wheel/hub assembly
- ABS electrical connector
- Stabilizer link nut and bolt. Torque the nut to 11 ft. lbs. (15 Nm).
- ELC height sensor link to the left control arm
- Support assembly
- Both front and both rear support assembly bolts. Tighten the nuts to 78 ft. lbs. (106 Nm).

- 3 support bracket-to-chassis bolts at each side. Torque the bolts to 63 ft. lbs. (86 Nm).
- ELC air tube to the air compressor
- ELC electrical connector and vent hose
- Support assembly electrical connectors
- Parking brake cable to the rear suspension support assembly
- Brake calipers to the control arms
- Parking brake cable to the brake calipers
- Coil springs
- Exhaust system
- Rear wheels. Torque the nuts to 100 ft. lbs. (140 Nm).

6. Lower the vehicle in order to achieve proper trim height.

7. Torque the lower control arm nuts to 78 ft. lbs. (106 Nm).

CONTROL ARM BUSHING REPLACEMENT

Front

The control arm bushings are not a serviceable component and if found to be defective the control arm assembly must be replaced.

1. Before servicing the vehicle, refer to the precautions in the beginning of this section.

2. Remove or disconnect the following:
- Rear wheel
- Lower control arm

3. Assemble the bushing tools as follows:

a. Assemble the Puller Bolt/Thrust Bearing tool J-21474-19 through the Bushing Receiver tool J-14014-2 over the bushing against the control arm.

b. Lubricate the bolt threads with high pressure lubricant.

c. Install the Bushing tool J-22222-2 (with the small end facing the bushing) and

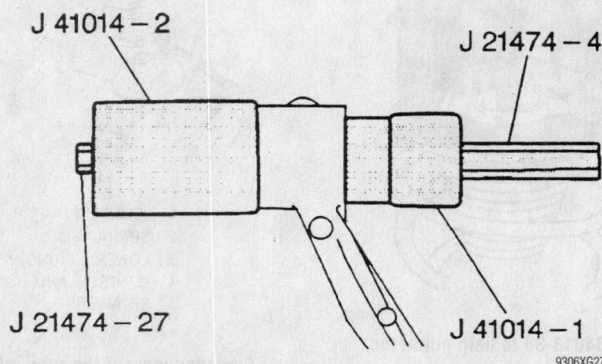

Removing the bushing from the lower control arm—rear suspension

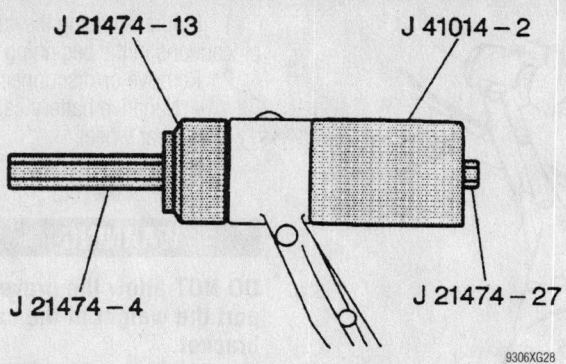

J 21474 – 13   J 41014 – 2

J 21474 – 4   J 21474 – 27

9306XG28

Installing the bushing to the lower control arm—rear suspension

the Long Nut J-21474-18 onto the Puller Bolt/Thrust Bearing tool J-21474-19.

4. Tighten the Long Nut J-21474-18 until the bushing is pressed from the control arm.

5. Remove the bushing tools.

To install:

6. Start the bushing onto the control arm.

➡**Position the bushing flat vertical and rearward.**

7. Assemble the bushing tools as folows:

a. Position the Puller Bolt/Thrust Bearing tool J-21474-19 through the Bushing Receiver tool J-14014-2 over the bushing against the control arm.

b. Lubricate the bolt threads with high pressure lubricant.

c. Install the Bushing Installer tool J-28685 (large end facing the bushing) and Long Nut J-21474-18 onto the Puller Bolt/Thrust Bearing tool J-21474-19.

8. Tighten the Puller Bolt/Thrust Bearing tool J-21474-19 until the bushing is fully seated in the control arm.

9. Remove the tools.

10. Install the lower control arm.

11. Install the front wheel.

Wheel Bearings

ADJUSTMENT

The wheel bearings are not adjustable. If the wheel bearings are defective, the hub and bearing assembly must be replaced.

REMOVAL & INSTALLATION

Front

The front wheel bearings are not serviced separately. If the front wheel bearings are defective, the hub and bearing assembly must be replaced.

1. Before servicing the vehicle, refer to the precautions in the beginning of this section.

2. Remove or disconnect the following:
- Negative battery cable
- Front wheel

3. Lubricate the threads on the halfshaft with clean engine oil.

4. Remove or disconnect the following:
- Halfshaft hub nut
- Caliper from the steering knuckle

✳✳ WARNING

DO NOT allow the brake hose to support the weight of the caliper.

- Brake rotor
- Anti-lock Brake System (ABS) sensor from the backing plate
- Hub and bearing assembly from the backing plate
- Hub and bearing assembly from the halfshaft
- Hub and bearing assembly

To install:

5. Install or connect the following:
- Hub and bearing assembly onto the halfshaft
- Hub and bearing assembly to the backing plate. Torque the 3 bolts alternately and evenly to 70 ft. lbs. (95 Nm).
- ABS sensor
- Brake rotor
- Caliper on the steering knuckle. Torque the bolts to 38 ft. lbs. (51 Nm).
- Halfshaft nut. Torque it to 118 ft. lbs. (160 Nm).
- Front wheel. Torque the nuts to 100 ft. lbs. (140 Nm).
- Negative battery cable

Rear

The rear wheel bearings are not serviced separately. If the rear wheel bearings are defective, the hub and bearing assembly must be replaced.

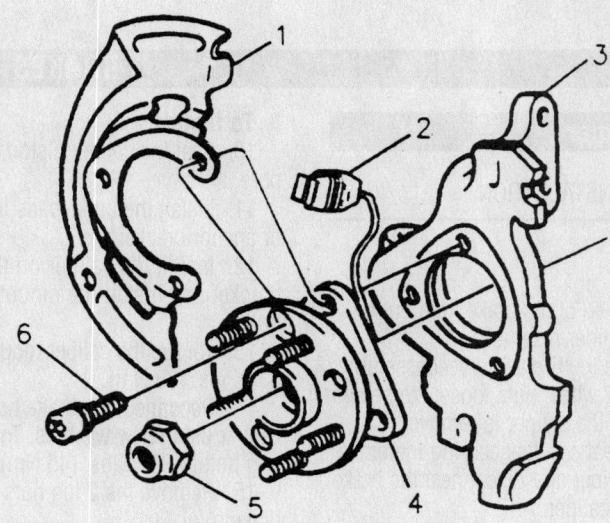

1 DUST SHIELD

2 WHEEL SPEED
   SENSOR CONNECTOR

3 STEERING KNUCKLE

4 HUB AND BEARING

5 NUT, DRIVE AXLE,
   145 N•m (107 LB. FT.)

6 RETAINING BOLT,
   95 N•m (75 LB. FT.)

7922XG36

Exploded view of the front hub mounting and related components

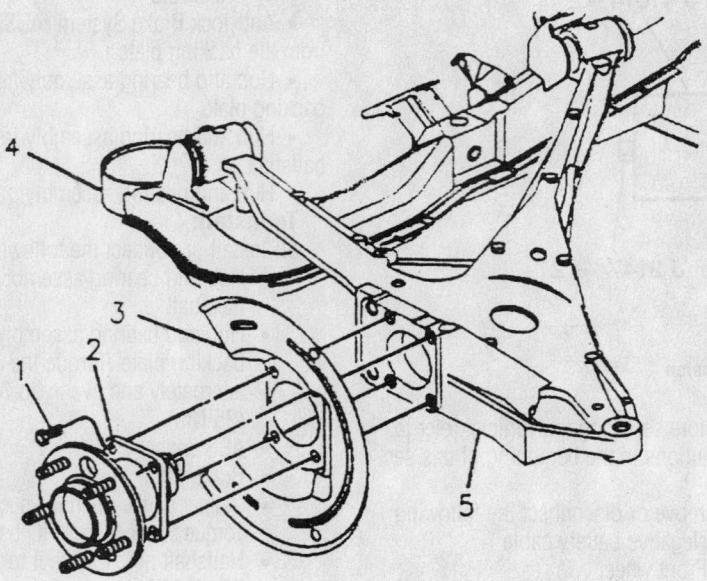

1 BOLT
2 HUB & BEARING
3 BRAKE SHIELD
4 REAR SUSPENSION SUPPORT ASSEMBLY
5 CONTROL ARM

7922XG37

Exploded view of the rear hub mounting and related components

1. Before servicing the vehicle, refer to the precautions in the beginning of this section.
2. Remove or disconnect the following:
   - Negative battery cable
   - Rear wheel
   - Caliper and bracket assembly from the brake rotor

※※ WARNING

DO NOT allow the brake hose to support the weight of the caliper and bracket.

   - Brake rotor
   - Anti-lock Brake System (ABS) sensor electrical connector
   - Hub and bearing assembly from the rear control arm

To install:

3. Install or connect the following:
   - Hub and bearing assembly onto the control arm, loosely install the bolts
   - ABS sensor electrical connector
   - Hub and bearing bolts. Torque the bolts alternately and evenly to 52 ft. lbs. (70 Nm).
   - Brake rotor
   - Caliper and bracket assembly. Torque the bolts to 35 ft. lbs. (48 Nm).
   - Rear wheel. Torque the nuts to 100 ft. lbs. (140 Nm).
   - Negative battery cable

BRAKES

Brake Caliper

REMOVAL & INSTALLATION

Front

1. Siphon ⅔ of the brake fluid out of the master cylinder reservoir.
2. Remove the tire and wheel assembly.
3. Install 2 wheel nuts loosely to secure the rotor when the caliper is removed.
4. Remove the bolts securing the brake hose to the caliper and disconnect the brake hose from the caliper.
5. Plug the hose to prevent excessive fluid loss and possible fluid contamination.
6. Compress the piston into the caliper bore to provide clearance for removal.
7. Remove the caliper mounting bolts.
8. Remove the caliper from the anchor bracket.
9. If the caliper is being replaced remove the brake pads from the caliper or anchor bracket.

To install:
10. Seat the caliper piston fully in its bore.
11. Install the brake pads in the caliper or anchor bracket.
12. Install the caliper on the anchor bracket and install the mounting bolts.
13. Torque the caliper mounting bolts to 63 ft. lbs. (85 Nm).
14. Reconnect the brake hose to the caliper using new washers. Torque the fitting bolt to 33 ft. lbs. (45 Nm).
15. Remove the 2 lug nuts securing the brake rotor.
16. Install the wheel and tire assembly and torque the wheel nuts to 100 ft. lbs. (140 Nm).
17. Refill the master cylinder with fluid and bleed the brake system.
18. Pump the brake pedal several times to seat the brake pads against the rotor. Verify no hydraulic leaks and a good firm brake pedal.

Rear

1. Siphon ⅔ of the brake fluid out of the master cylinder reservoir.
2. Remove the tire and wheel assembly.
3. Install 2 wheel nuts loosely to secure the rotor when the caliper is removed.
4. Remove the bolt securing the brake hose to the caliper and disconnect the brake hose from the caliper.
5. Plug the hose to prevent excessive fluid loss and possible fluid contamination.
6. Remove the park brake cable.
7. Remove upper and lower caliper bolts and lift caliper from mounting bracket.

To install:
8. Seat the caliper piston fully in its bore.
9. Make sure the notches in the caliper piston are at 6 and 12 o'clock.
10. Mount caliper onto anchor bracket. Install the 2 mounting bolts and torque bolts to 20 ft. lbs.
11. Reconnect the brake hose to the

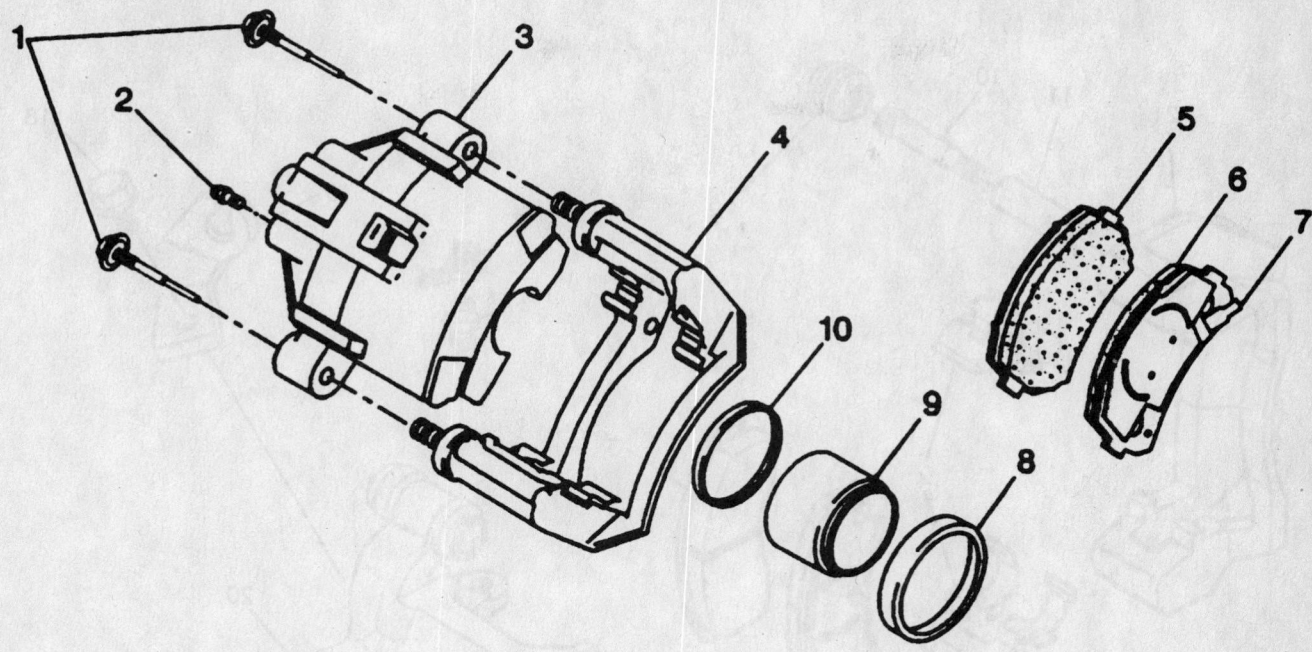

(1) Pin Bolts
(2) Outboard Pad
(3) Caliper Boot
(4) Inboard Pad
(5) Piston Seal

(6) Piston
(7) Anchor bracket
(8) Bleeder Valve
(9) Caliper Housing
(10) Caliper Anchor Bracket

93006G87

Front caliper and brake pad components

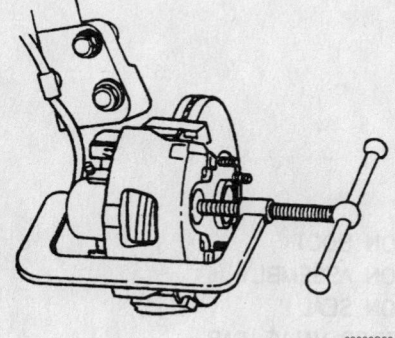

93006G88

Compressing the front caliper piston

caliper using new copper washers. Torque the mounting bolt to 33 ft. lbs. (45 Nm).

12. Install the park brake cable.

13. Remove the wheel nuts securing the rotor.

14. Reinstall the tire and wheel assembly and torque the wheel nuts to 100 ft. lbs. (140 Nm).

15. Refill the master cylinder with fluid and bleed the brake system.

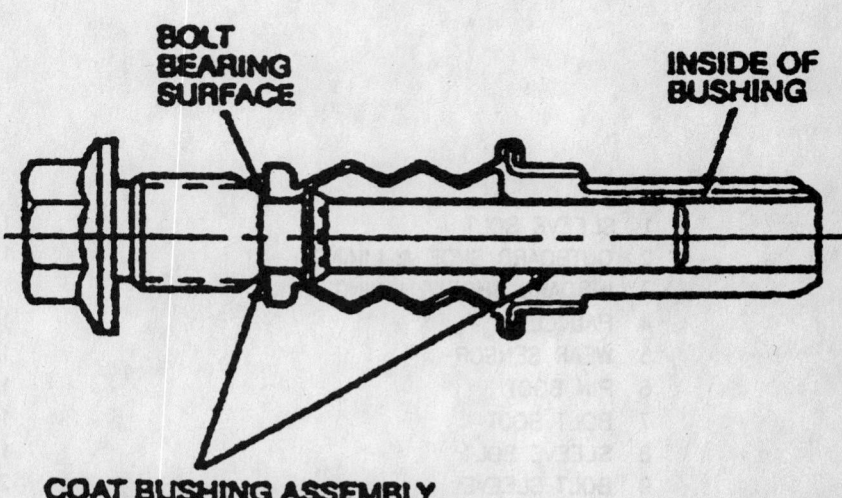

BOLT BEARING SURFACE

INSIDE OF BUSHING

COAT BUSHING ASSEMBLY WITH SILICONE GREASE 2 PLACES

93006G89

Caliper mounting bolt and sleeve lubrication points

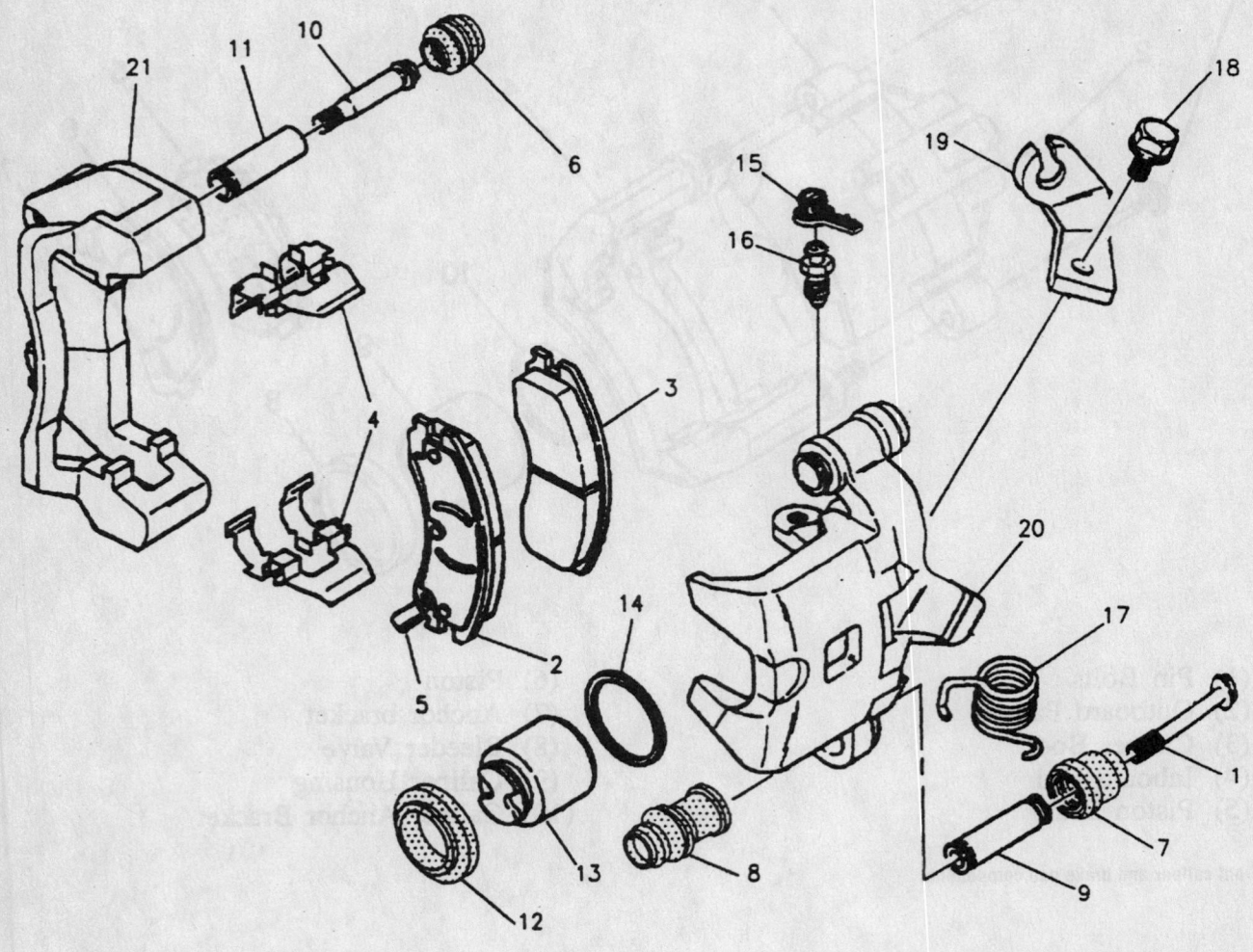

1 SLEEVE BOLT
2 OUTBOARD SHOE & LINING
3 INBOARD SHOE & LINING
4 PAD CLIP
5 WEAR SENSOR
6 PIN BOOT
7 BOLT BOOT
8 SLEEVE BOLT
9 BOLT SLEEVE
10 PIN BOLT
11 PIN SLEEVE

12 PISTON BOOT
13 PISTON ASSEMBLY
14 PISTON SEAL
15 BLEEDER VALVE CAP
16 BLEEDER VALVE
17 LEVER RETURN SPRING
18 BOLT AND WASHER
19 CABLE SUPPORT BRACKET
20 CALIPER BODY ASSEMBLY
21 CALIPER SUPPORT

93006G90

Rear caliper and brake pad components

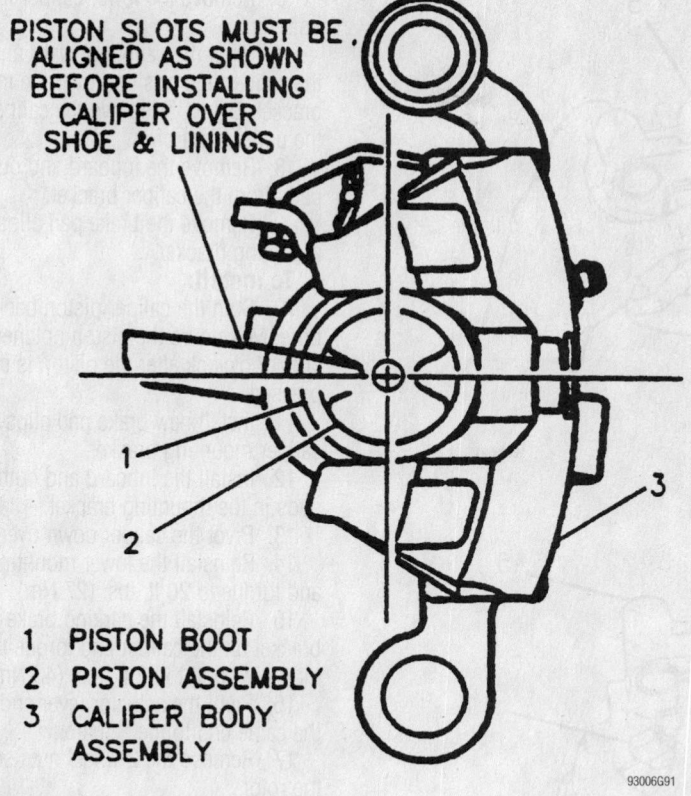

PISTON SLOTS MUST BE ALIGNED AS SHOWN BEFORE INSTALLING CALIPER OVER SHOE & LININGS

1
2
3

1 PISTON BOOT
2 PISTON ASSEMBLY
3 CALIPER BODY ASSEMBLY

93006G91

Aligning the notches in the piston

16. Pump the brake pedal several times to seat the brake pads against the rotor. Verify no hydraulic leaks and a good firm brake pedal.

Disc Brake Pads

REMOVAL & INSTALLATION

Front

1. Siphon ⅔ of the brake fluid from the master cylinder.
2. Remove the tire and wheel assembly.
3. Install 2 wheel nuts to secure the rotor on the hub.
4. Remove the caliper mounting bolts and sleeves.
5. Remove the caliper from the mounting bracket and support the caliper. DO NOT allow the caliper to hang unsupported from the brake hose.
6. Remove the outboard brake pad from the caliper by pushing the pad inward toward the piston to unseat the buttons on the back of pad from the holes in the caliper. Once the buttons are unseated push the pad out of the caliper.
7. Remove the inboard pad from the caliper by pulling the top of the pad away

from the piston and disengaging the spring clip from the caliper.
8. Compress the piston back into the bore using a C-Clamp.
To install:
9. Install the inboard pad into the caliper so the spring clip seats in the pis-

ton. Make sure lower edge of the spring clip is engaged in the piston and the pad is even against the base of the piston. Push the pad flat against caliper piston.
10. Install the outboard pad into the caliper so the wear indicator is at the trailing edge of the pad during forward wheel rotation. Push the pad the straight down into the caliper so the spring clips rode along the outside of the caliper. The buttons on the pad will snap into the mounting holes when the pad is correctly installed.
11. Reinstall the caliper onto the steering knuckle.
12. Remove the 2 wheel nuts securing the rotor.
13. Reinstall the tire and wheel assembly and torque the wheel nuts to 100 ft. lbs. (140 Nm).
14. Refill the master cylinder. Pump the brake pedal several times to seat the pads against the rotor.
15. Check the master cylinder level and add fluid as necessary.

REAR

1. Siphon ⅔ of the brake fluid from the master cylinder.
2. Remove the tire and wheel assembly.
3. Install 2 wheel nuts to secure the rotor on the hub.
4. Pivot the parking brake actuator level and disconnect the parking brake cable from the lever.
5. Remove the mounting bolt securing the parking brake cable bracket to the caliper and position the cable and bracket out of the way.

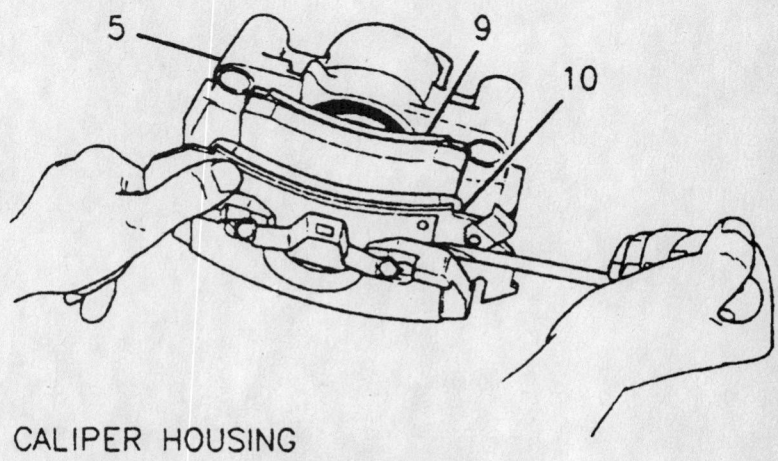

5
9
10

5 CALIPER HOUSING
9 INBOARD SHOE AND LINING
10 OUTBOARD SHOE AND LINING

93006G93

Removing the outboard front pad

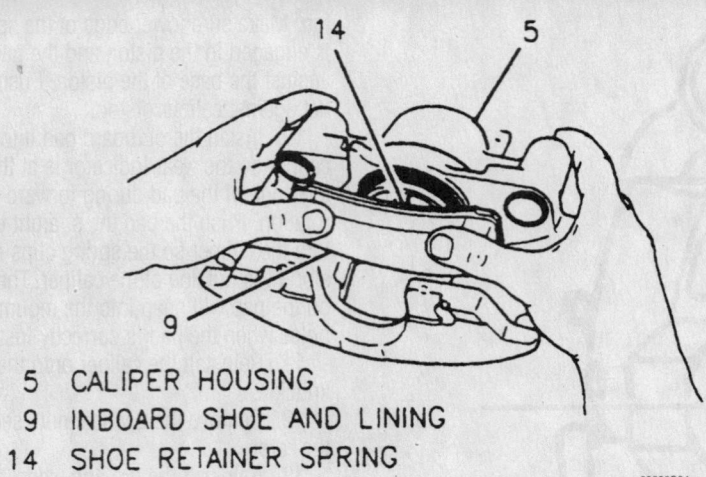

5 CALIPER HOUSING
9 INBOARD SHOE AND LINING
14 SHOE RETAINER SPRING

93006G94

Installing the inboard front pad

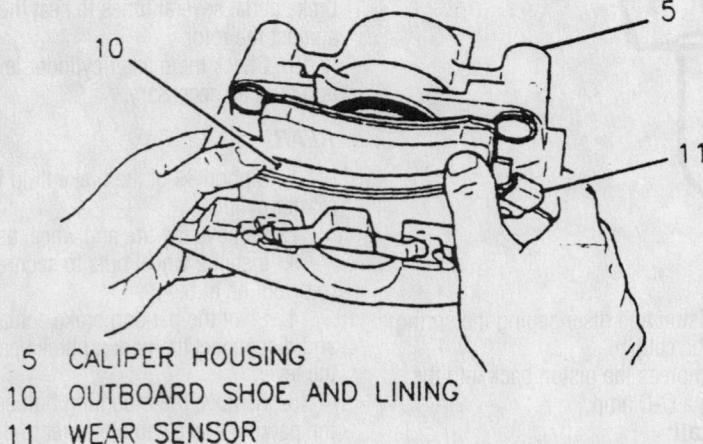

5 CALIPER HOUSING
10 OUTBOARD SHOE AND LINING
11 WEAR SENSOR

93006G95

Installing the outboard front pad

6. Remove the lower caliper mounting bolt.

7. Pivot the caliper upward and secure the caliper in a position over the mounting bracket. DO NOT remove the caliper from the upper pivot pin.

8. Remove the inboard and outboard pads from the caliper bracket.

9. Remove the brake pad clips from the mounting bracket.

To install:

10. Spin the caliper piston back into the bore. Make sure the piston notches are at 6 and 12 o'clock after the piston is compressed.

11. Install new brake pad clips in the caliper mounting bracket.

12. Install the inboard and outboard pads in the mounting bracket.

13. Pivot the caliper down over the pads.

14. Reinstall the lower mounting bolt and torque to 20 ft. lbs. (27 Nm).

15. Reinstall the parking brake cable bracket on the caliper and torque the mounting bolt to 32 ft. lbs. (43 Nm).

16. Pivot the actuator lever and connect the cable end to the actuator.

17. Remove the 2 wheel nuts securing the rotor.

18. Reinstall the tire and wheel assembly and torque the wheel nuts to 100 ft. lbs. (140 Nm).

19. Refill the master cylinder. Pump the brake pedal several times to seat the pads against the rotor.

20. Check the master cylinder level and add fluid as necessary.

CHEVROLET AND PONTIAC

18

Cavalier • Sunfire

SPECIFICATION CHARTS

ENGINE AND VEHICLE IDENTIFICATION CHART

Engine							Model Year	
Code ①	Liters (cc)	Cu. In.	Cyl.	Fuel Sys.	Engine Type	Eng. Mfg.	Code ②	Year
4	2.2 (2180)	133	4	MFI	OHV	CUS	Y	2000
T	2.4 (2392)	146	4	MFI	DOHC	CUS	1	2001
							2	2002
							3	2003
							4	2004

CUS: Chevrolet/United States

MFI: Multi-point Fuel Injection

OHV: Overhead Valves

DOHC: Double Overhead Camshafts

① 8th position of VIN

② 10th position of VIN

42372-JB0D-C01

GENERAL ENGINE SPECIFICATIONS

Year	Model	Engine Displacement Liters (cc)	Engine Series (ID/VIN)	Fuel System	Net Horsepower @ rpm	Net Torque @ rpm (ft. lbs.)	Bore x Stroke (in.)	Com- pression Ratio	Oil Pressure @ rpm
2000	Cavalier	2.2 (2180)	4	MFI	115@5000	136@3600	3.50x3.46	9.0:1	56@3000
		2.4 (2392)	T	MFI	150@5600	155@4400	3.54x3.70	9.5:1	30@3000
	Sunfire	2.2 (2180)	4	MFI	115@5000	136@3600	3.50x3.46	9.0:1	56@3000
		2.4 (2392)	T	MFI	150@5600	155@4400	3.54x3.70	9.5:1	30@3000
2001	Cavalier	2.2 (2180)	4	MFI	115@5000	136@3600	3.50x3.46	9.0:1	56@3000
		2.4 (2392)	T	MFI.	150@5600	155@4400	3.54x3.70	9.5:1	30@3000
	Sunfire	2.2 (2180)	4	MFI	115@5000	136@3600	3.50x3.46	9.0:1	56@3000
		2.4 (2392)	T	MFI	150@5600	155@4400	3.54x3.70	9.5:1	30@3000
2002	Cavalier	2.2 (2180)	4	MFI	115@5000	136@3600	3.50x3.46	9.0:1	56@3000
		2.4 (2392)	T	MFI	150@5600	155@4400	3.54x3.70	9.5:1	30@3000
	Sunfire	2.2 (2180)	4	MFI	115@5000	136@3600	3.50x3.46	9.0:1	56@3000
		2.4 (2392)	T	MFI	150@5600	155@4400	3.54x3.70	9.5:1	30@3000
2003	Cavalier	2.2 (2180)	4	MFI	115@5000	136@3600	3.50x3.46	9.0:1	56@3000
		2.4 (2392)	T	MFI	150@5600	155@4400	3.54x3.70	9.5:1	30@3000
	Sunfire	2.2 (2180)	4	MFI	115@5000	136@3600	3.50x3.46	9.0:1	56@3000
		2.4 (2392)	T	MFI	150@5600	155@4400	3.54x3.70	9.5:1	30@3000

MFI: Multi-point Fuel Injection

42372-JB0D-C02

ENGINE TUNE-UP SPECIFICATIONS

Year	Engine Displacement Liters (cc)	Engine ID/VIN	Spark Plug Gap (in.)	Ignition Timing (deg.)	Fuel Pump (psi)	Idle Speed (rpm)	Valve Clearance	
							Intake	Exhaust
2000	2.2 (2180)	4	0.060	①	41-47	①	HYD	HYD
	2.4 (2392)	T	0.060	①	41-47	①	HYD	HYD
2001	2.2 (2180)	4	0.060	①	41-47	①	HYD	HYD
	2.4 (2392)	T	0.060	①	41-47	①	HYD	HYD
2002	2.2 (2180)	4	0.060	①	41-47	①	HYD	HYD
	2.4 (2392)	T	0.060	①	41-47	①	HYD	HYD
2003	2.2 (2180)	4	0.060	①	41-47	①	HYD	HYD
	2.4 (2392)	T	0.060	①	41-47	①	HYD	HYD

NOTE: The Vehicle Emission Control Information label often reflects specification changes made during production.

The label figures must be used if they differ from those in this chart.

HYD: Hydraulic

① Refer to Vehicle Emission Control Information label

42372-JB0D-C03

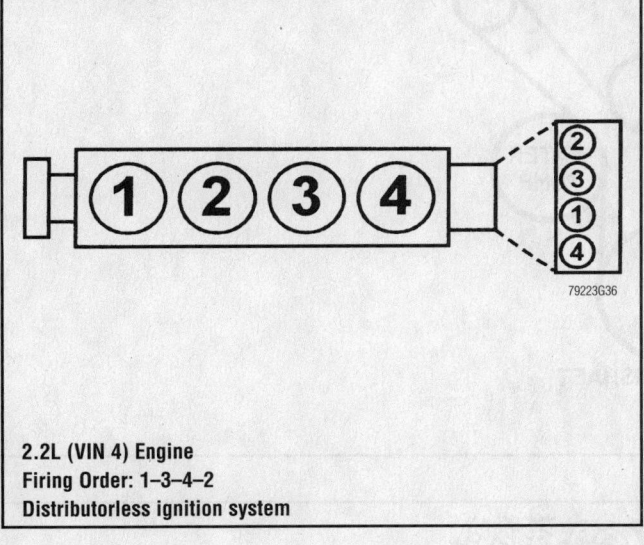

2.2L (VIN 4) Engine
Firing Order: 1-3-4-2
Distributorless ignition system

79223G36

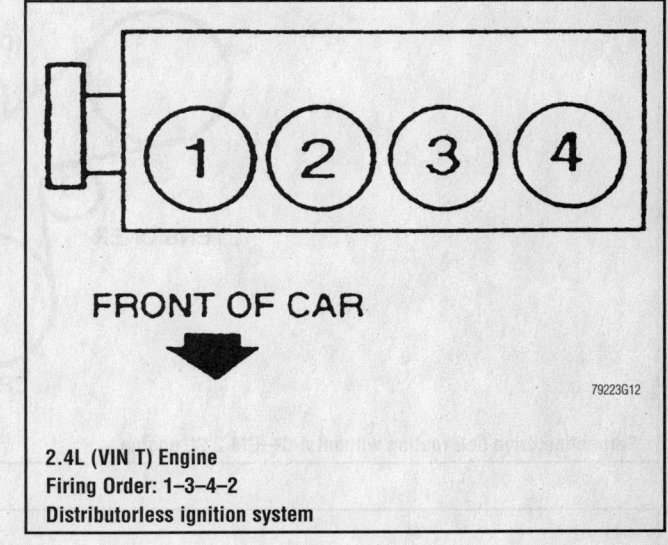

FRONT OF CAR

79223G12

2.4L (VIN T) Engine
Firing Order: 1-3-4-2
Distributorless ignition system

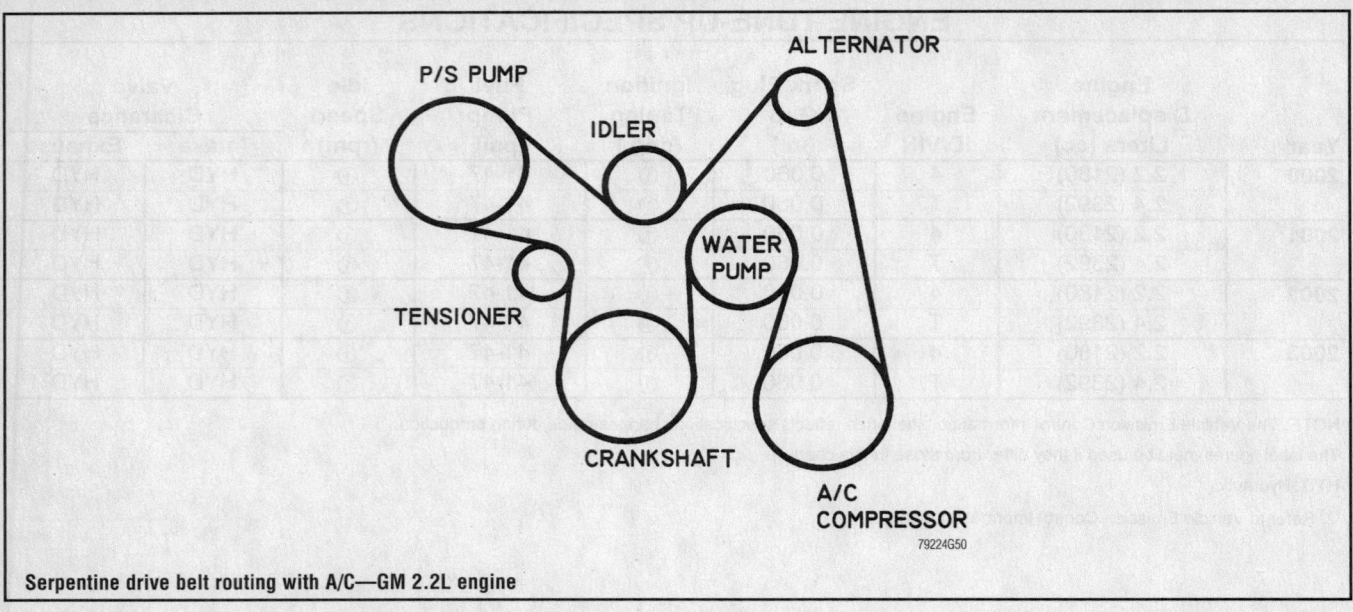

Serpentine drive belt routing with A/C—GM 2.2L engine

79224G50

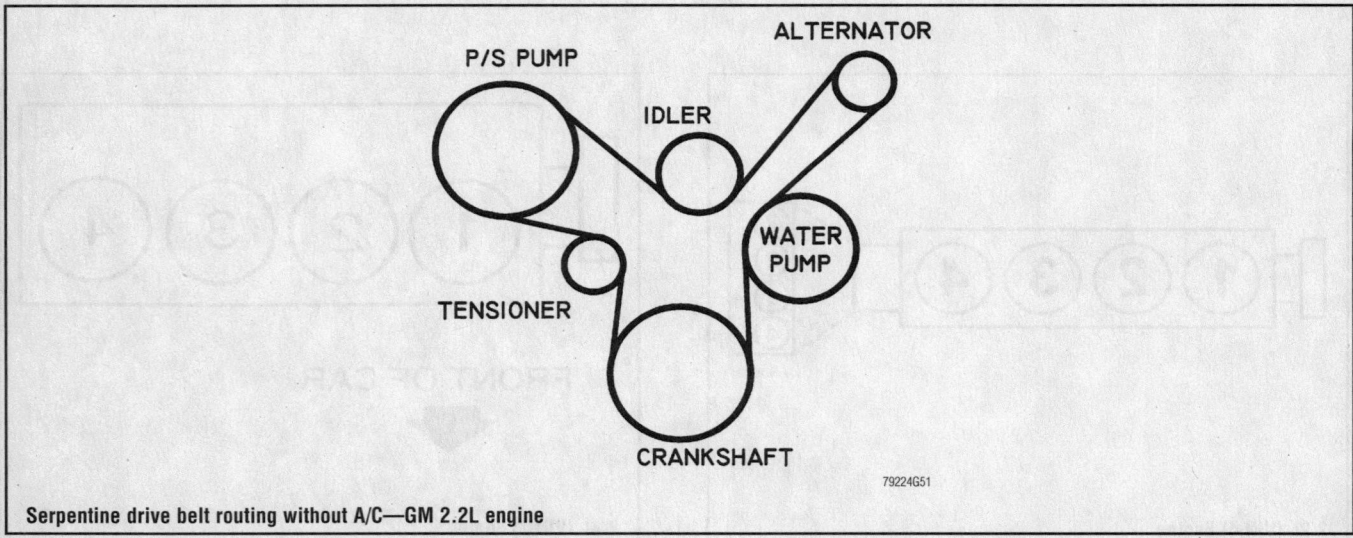

Serpentine drive belt routing without A/C—GM 2.2L engine

79224G51

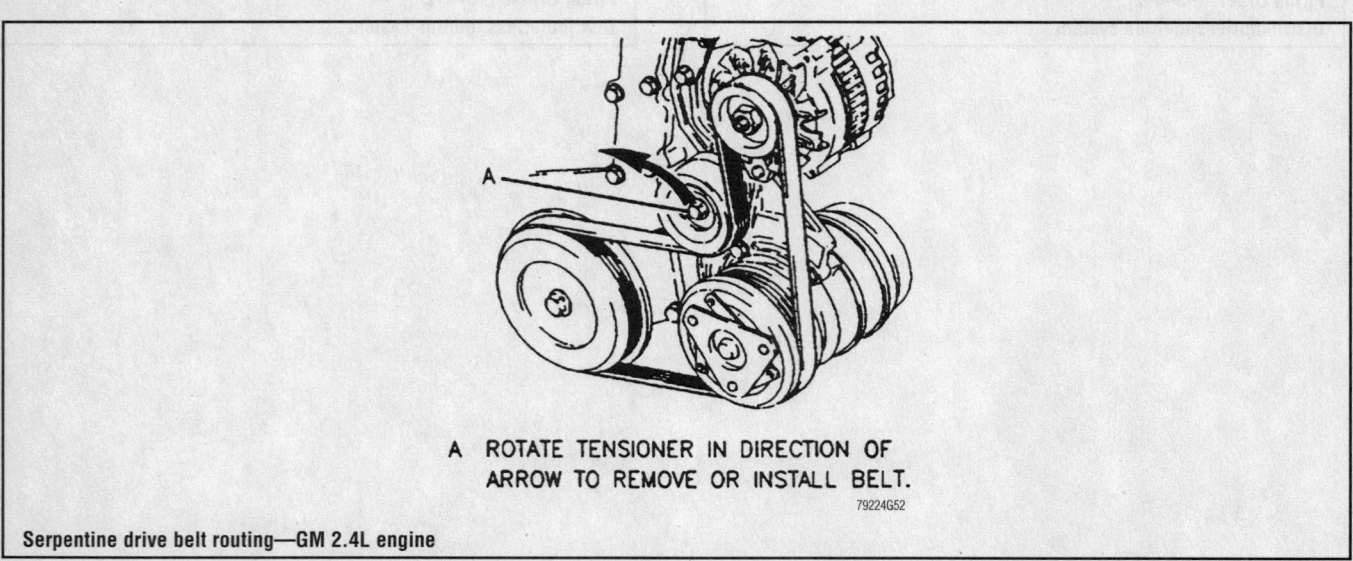

A ROTATE TENSIONER IN DIRECTION OF ARROW TO REMOVE OR INSTALL BELT.

79224G52

Serpentine drive belt routing—GM 2.4L engine

CAPACITIES

Year	Model	Engine Displacement Liters (cc)	Engine ID/VIN	Engine Oil with Filter (qts.)	Transmission (pts.) Manual	Auto.	Fuel Tank (gal.)	Cooling System (qts.)
2000	Cavalier	2.2 (2180)	4	4.0	3.6	①	15.0	10.2
		2.4 (2392)	T	4.0	3.6	①	15.0	10.7
	Sunfire	2.2 (2180)	4	4.0	3.6	①	15.0	10.2
		2.4 (2392)	T	4.0	3.6	①	15.0	10.7
2001	Cavalier	2.2 (2180)	4	4.0	3.6	①	14.3	10.2
		2.4 (2392)	T	4.0	3.6	①	14.3	10.2
	Sunfire	2.2 (2180)	4	4.0	3.6	①	14.3	10.2
		2.4 (2392)	T	4.0	3.6	①	14.3	10.2
2002	Cavalier	2.2 (2180)	4	4.0	3.6	①	14.3	10.2
		2.4 (2392)	T	4.0	3.6	①	14.3	10.2
	Sunfire	2.2 (2180)	4	4.0	3.6	①	14.3	10.2
		2.4 (2392)	T	4.0	3.6	①	14.3	10.2
2003	Cavalier	2.2 (2180)	4	4.0	3.6	①	14.3	10.2
		2.4 (2392)	T	4.0	3.6	①	14.3	10.2
	Sunfire	2.2 (2180)	4	4.0	3.6	①	14.3	10.2
		2.4 (2392)	T	4.0	3.6	①	14.3	10.2

NOTE: All capacities are approximate. Add fluid gradually and ensure a proper fluid level is obtained.

① 3 Speed: 8.0 pts.
   4 Speed: 14.8 pts.

42372-JBOD-C04

VALVE SPECIFICATIONS

Year	Engine Displacement Liters (cc)	Engine ID/VIN	Seat Angle (deg.)	Face Angle (deg.)	Spring Test Pressure (lbs. @ in.)	Spring Installed Height (in.)	Stem-to-Guide Clearance (in.) Intake	Exhaust	Stem Diameter (in.) Intake	Exhaust
2000	2.2 (2180)	4	46	45	72-81@1.60	1.600	0.0007-0.0020	0.0014-0.0029	0.2740-0.2743	0.2731-0.2736
	2.4 (2392)	T	45	46	50-55@1.44	1.437	0.0009-0.0025	0.0016-0.0032	0.2331-0.2339	0.2326-0.2334
2001	2.2 (2180)	4	46	45	72-81@1.60	1.600	0.0007-0.0020	0.0014-0.0029	0.2740-0.2743	0.2731-0.2736
	2.4 (2392)	T	45	46	50-55@1.44	1.437	0.0009-0.0025	0.0016-0.0032	0.2331-0.2339	0.2326-0.2334
2002	2.2 (2180)	4	46	45	72-81@1.60	1.600	0.0007-0.0020	0.0014-0.0029	0.2740-0.2743	0.2731-0.2736
	2.4 (2392)	T	45	46	50-55@1.44	1.437	0.0009-0.0025	0.0016-0.0032	0.2331-0.2339	0.2326-0.2334
2003	2.2 (2180)	4	46	45	72-81@1.60	1.600	0.0007-0.0020	0.0014-0.0029	0.2740-0.2743	0.2731-0.2736
	2.4 (2392)	T	45	46	50-55@1.44	1.437	0.0009-0.0025	0.0016-0.0032	0.2331-0.2339	0.2326-0.2334

42372-JBOD-C05

CRANKSHAFT AND CONNECTING ROD SPECIFICATIONS

All measurements are given in inches.

Year	Engine Displacement Liters (cc)	Engine ID/VIN	Crankshaft				Connecting Rod		
			Main Brg. Journal Dia.	Main Brg. Oil Clearance	Shaft End-play	Thrust on No.	Journal Diameter	Oil Clearance	Side Clearance
2000	2.2 (2180)	4	2.4945-2.4954	0.0006-0.0019	0.0020-0.0070	4	1.9983-1.9994	0.0010-0.0030	0.0039-0.0149
	2.4 (2392)	T	2.3612-2.3631	0.0004-0.0023	0.0034-0.0095	3	1.8887-1.8897	0.0004-0.0026	0.0059-0.0177
2001	2.2 (2180)	4	2.4945-2.4954	0.0006-0.0019	0.0020-0.0070	4	1.9983-1.9994	0.0010-0.0030	0.0039-0.0149
	2.4 (2392)	T	2.3612-2.3631	0.0004-0.0023	0.0034-0.0095	3	1.8887-1.8897	0.0004-0.0026	0.0059-0.0177
2002	2.2 (2180)	4	2.4945-2.4954	0.0006-0.0019	0.0020-0.0070	4	1.9983-1.9994	0.0010-0.0030	0.0039-0.0149
	2.4 (2392)	T	2.3612-2.3631	0.0004-0.0023	0.0034-0.0095	3	1.8887-1.8897	0.0004-0.0026	0.0059-0.0177
2003	2.2 (2180)	4	2.4945-2.4954	0.0006-0.0019	0.0020-0.0070	4	1.9983-1.9994	0.0010-0.0030	0.0039-0.0149
	2.4 (2392)	T	2.3612-2.3631	0.0004-0.0023	0.0034-0.0095	3	1.8887-1.8897	0.0004-0.0026	0.0059-0.0177

42372-JBOD-C06

PISTON AND RING SPECIFICATIONS

All measurements are given in inches.

Year	Engine Displacement Liters (cc)	Engine ID/VIN	Piston Clearance	Ring Gap			Ring Side Clearance		
				Top Compression	Bottom Compression	Oil Control	Top Compression	Bottom Compression	Oil Control
2000	2.2 (2180)	4	0.0006-0.0018 ①	0.0100-0.0200	0.0120-0.0177	0.0100-0.0300	0.0020-0.0035	0.0016-0.0031	0.0005-0.0087
	2.4 (2392)	T	0.0006-0.0009 ②	0.0060-0.0120	0.0098-0.0157	0.0098-0.0299	0.0016-0.0031	0.0012-0.0028	NA
2001	2.2 (2180)	4	0.0006-0.0018 ①	0.0100-0.0200	0.0120-0.0177	0.0100-0.0300	0.0020-0.0035	0.0016-0.0031	0.0005-0.0087
	2.4 (2392)	T	0.0006-0.0009 ②	0.0060-0.0120	0.0098-0.0157	0.0098-0.0299	0.0016-0.0031	0.0012-0.0028	NA
2002	2.2 (2180)	4	0.0006-0.0018 ①	0.0100-0.0200	0.0120-0.0177	0.0100-0.0300	0.0020-0.0035	0.0016-0.0031	0.0005-0.0087
	2.4 (2392)	T	0.0006-0.0009 ②	0.0060-0.0120	0.0098-0.0157	0.0098-0.0299	0.0016-0.0031	0.0012-0.0028	NA
2003	2.2 (2180)	4	0.0006-0.0018 ①	0.0100-0.0200	0.0120-0.0177	0.0100-0.0300	0.0020-0.0035	0.0016-0.0031	0.0005-0.0087
	2.4 (2392)	T	0.0006-0.0009 ②	0.0060-0.0120	0.0098-0.0157	0.0098-0.0299	0.0016-0.0031	0.0012-0.0028	NA

① 42.7mm from top of piston

② 42.7mm from top of piston

42372-JBOD-C07

TORQUE SPECIFICATIONS
All readings in ft. lbs.

Year	Engine Displacement Liters (cc)	Engine ID/VIN	Cylinder Head Bolts	Main Bearing Bolts	Rod Bearing Bolts	Crankshaft Damper Bolts	Flywheel Bolts	Manifold Intake	Manifold Exhaust	Spark Plugs	Lug Nuts
2000	2.2 (2180)	4	①	70	38	77 ②	55	③	10	13	100
	2.4 (2392)	T	④	⑤	⑥	⑦	⑧	⑨	⑩	13	100
2001	2.2 (2180)	4	①	70	38	77 ②	55	③	10	13	100
	2.4 (2392)	T	④	⑤	⑥	⑦	⑧	⑨	⑩	13	100
2002	2.2 (2180)	4	①	70	38	77 ②	55	③	10	13	100
	2.4 (2392)	T	④	⑤	⑥	⑦	⑧	⑨	⑩	13	100
2003	2.2 (2180)	4	①	70	38	77 ②	55	③	10	13	100
	2.4 (2392)	T	④	⑤	⑥	⑦	⑧	⑨	⑩	13	100

NA: Not Available

① Step 1: Long bolts: 46 ft. lbs.

   Step 2: Short bolts: 43 ft. lbs.

   Step 3: Long bolts an additional 90 degree turn

   Step 4: Short bolts an additional 90 degree turn

② Center bolt spec shown; Pulley-to-hub bolts: 37 ft. lbs.

③ Bolts: 17 ft. lbs.

   Nuts: 17 ft. lbs.

   Studs: 9 ft. lbs.

④ Step 1: bolts 1-8; 40 ft. lbs.

   Step 2: bolts 9-10: 30 ft. lbs.

   Step 3: An additional 90 degree turn

⑤ 15 ft. lbs. plus 90 degrees

⑥ 18 ft. lbs. plus 80 degrees

⑦ 129 ft. lbs. plus 90 degrees

⑧ 15 ft. lbs. plus 45 degrees

⑨ Nuts: 18 ft. lbs.

   Studs: 97 inch lbs.

⑩ Nuts: 31 ft. lbs.

   Studs: 97 inch lbs.

42372-JBOD-C08

WHEEL ALIGNMENT

Year	Model		Caster Range (+/-Deg.)	Caster Preferred Setting (Deg.)	Camber Range (+/-Deg.)	Camber Preferred Setting (Deg.)	Toe-in (in.)	Steering Axis Inclination (Deg.)
2000	All	F	+1.00	+4.30	+1.00	0	0 +/- 0.25	—
		R	—	—	+0.75	-0.40	0.20 +/- 0.30	—
2001	All	F	+1.00	+4.30	+1.00	0	0 +/- 0.25	—
		R	—	—	+0.75	-0.40	0.20 +/- 0.30	—
2002	All	F	+1.00	+4.30	+1.00	0	0 +/- 0.25	—
		R	—	—	+0.75	-0.40	0.20 +/- 0.30	—
2003	All	F	+1.00	+4.30	+1.00	0	0 +/- 0.25	—
		R	—	—	+0.75	-0.40	0.20 +/- 0.30	—

42372-JBOD-C09

TIRE, WHEEL AND BALL JOINT SPECIFICATIONS

Year	Model	OEM Tires Standard	OEM Tires Optional	Tire Pressures (psi) Front	Tire Pressures (psi) Rear	Wheel Size	Ball Joint Inspection
2001	Cavalier base	P195/70R14	None	30	30	6-JJ	①
	Cavalier LS, RS	P195/65R15	None	30	30	6-JJ	①
	Cavalier Z24	P205/55R16	None	30	30	6-JJ	①
	Sunfire SE	P195/70R14	P195/65R15	30	30	6-JJ	0.125 in.
	Sunfire GT	P205/55R16	None	30	30	6-JJ	0.125 in.
2002	Cavalier base	P195/70R14	None	30	30	6-JJ	①
	Cavalier LS, RS	P195/65R15	None	30	30	6-JJ	①
	Cavalier Z24	P205/55R16	None	30	30	6-JJ	①
	Sunfire SE	P195/70R14	P195/65R15	30	30	6-JJ	0.125 in.
	Sunfire GT	P205/55R16	None	30	30	6-JJ	0.125 in.
2003	Cavalier base	P195/70R14	None	30	30	6-JJ	①
	Cavalier LS, RS	P195/65R15	None	30	30	6-JJ	①
	Cavalier Z24	P205/55R16	None	30	30	6-JJ	①
	Sunfire SE	P195/70R14	P195/65R15	30	30	6-JJ	0.125 in.
	Sunfire GT	P205/55R16	None	30	30	6-JJ	0.125 in.

OEM: Original Equipment Manufacturer

PSI: Pounds Per Square Inch

① Replace if any measurable movement is found

42372-JBOD-C10

BRAKE SPECIFICATIONS
All measurements in inches unless noted

| Year | Model | Brake Disc | | | Brake Drum | | | Minimum Lining Thickness | | Caliper Mounting Bolts (ft. lbs.) |
		Original Thickness	Minimum Thickness	Maximum Run-out	Original Inside Diameter	Max. Wear Limit	Maximum Machine Diameter	Front	Rear	
2000	Cavalier	0.786	0.751	0.002	NA	8.909	8.879	0.030	0.030	38
	Sunfire	0.786	0.751	0.002	NA	8.909	8.879	0.030	0.030	38
2001	Cavalier	0.786	0.751	0.002	NA	8.909	8.879	0.030	0.030	38
	Sunfire	0.786	0.751	0.002	NA	8.909	8.879	0.030	0.030	38
2002	Cavalier	0.786	0.751	0.002	NA	8.909	8.879	0.030	0.030	38
	Sunfire	0.786	0.751	0.002	NA	8.909	8.879	0.030	0.030	38
2003	Cavalier	0.786	0.751	0.002	NA	8.909	8.879	0.030	0.030	38
	Sunfire	0.786	0.751	0.002	NA	8.909	8.879	0.030	0.030	38

NA: Not Available

42372-JB0D-C11

SCHEDULED MAINTENANCE INTERVALS
GM J BODY—CHEVROLET CAVALIER & PONTIAC SUNFIRE

TO BE SERVICED	TYPE OF SERVICE	7.5	15	22.5	30	37.5	45	52.5	60	67.5	75	82.5	90	97.5
Engine oil & filter	R	✓	✓	✓	✓	✓	✓	✓	✓	✓	✓	✓	✓	✓
Coolant level, hoses & clamps	S/I	✓	✓	✓	✓	✓	✓	✓	✓	✓	✓	✓	✓	✓
Exhaust system & brake hoses	S/I	✓	✓	✓	✓	✓	✓	✓	✓	✓	✓	✓	✓	✓
Lubricate chassis and suspension	S/I	✓	✓	✓	✓	✓	✓	✓	✓	✓	✓	✓	✓	✓
Lubricate steering linkage and transaxle linkage	S/I	✓	✓	✓	✓	✓	✓	✓	✓	✓	✓	✓	✓	✓
Throttle linkage	S/I	✓	✓	✓	✓	✓	✓	✓	✓	✓	✓	✓	✓	✓
Brake linings	S/I	✓		✓		✓		✓		✓		✓		✓
Rotate tires	S/I	✓		✓		✓		✓		✓		✓		✓
Air filter element	R				✓				✓				✓	
Engine coolant ①	R													
Spark plugs ②	R				✓				✓				✓	
Accessory drive belt(s)	S/I				✓				✓				✓	
Automatic transaxle fluid & filter ③	S/I													
Fuel system	S/I				✓				✓				✓	
Ignition cables	R				✓				✓				✓	
Inspect throttle body bore & throttle plate for deposits	S/I				✓				✓				✓	
Supercharger oil	S/I				✓				✓				✓	

R: Replace

S/I: Service or Inspect

① Engine coolant: replace every 100,000 miles. Use O.E. specified (DEX-COOL™) coolant only. If any silicate coolant is used, the service interval is every 30,000

② Platinum tip spark plugs: replace every 100,000 miles.

③ Replace fuid every 50, 000 miles if driven in etxreme traffic or in places where the temperature exceeds 90 degrees F

FREQUENT OPERATION MAINTENANCE (SEVERE SERVICE)

If a vehicle is operated under any of the following conditions it is considered severe service:

- Extremely dusty areas.

- 50% or more of the vehicle operation is in 32°C (90°F) or higher temperatures, or constant operation in temperatures below 0°C (32°F).

- Prolonged idling (vehicle operation in stop and go traffic).

- Frequent short running periods (engine does not warm to normal operating temperatures).

- Police, taxi, delivery usage or trailer towing usage.

CV joints & front suspension components: service or inspect every 3000 miles.

Engine oil & filter change: change every 3000 miles.

42372-JBOD-C12

PRECAUTIONS

Before servicing any vehicle, please be sure to read all of the following precautions, which deal with personal safety, prevention of component damage, and important points to take into consideration when servicing a motor vehicle:

• Never open, service or drain the radiator or cooling system when the engine is hot; serious burns can occur from the steam and hot coolant.

• Observe all applicable safety precautions when working around fuel. Whenever servicing the fuel system, always work in a well-ventilated area. Do not allow fuel spray or vapors to come in contact with a spark, open flame, or excessive heat (a hot drop light, for example). Keep a dry chemical fire extinguisher near the work area. Always keep fuel in a container specifically designed for fuel storage; also, always properly seal fuel containers to avoid the possibility of fire or explosion. Refer to the additional fuel system precautions later in this section.

• Fuel injection systems often remain pressurized, even after the engine has been turned **OFF**. The fuel system pressure must be relieved before disconnecting any fuel lines. Failure to do so may result in fire and/or personal injury.

• Brake fluid often contains polyglycol ethers and polyglycols. Avoid contact with the eyes and wash your hands thoroughly after handling brake fluid. If you do get brake fluid in your eyes, flush your eyes with clean, running water for 15 minutes. If eye irritation persists, or if you have taken brake fluid internally, IMMEDIATELY seek medical assistance.

• The EPA warns that prolonged contact with used engine oil may cause a number of skin disorders, including cancer! You should make every effort to minimize your exposure to used engine oil. Protective gloves should be worn when changing oil. Wash your hands and any other exposed skin areas as soon as possible after exposure to used engine oil. Soap and water, or waterless hand cleaner should be used.

• All new vehicles are now equipped with an air bag system, often referred to as a Supplemental Restraint System (SRS) or Supplemental Inflatable Restraint (SIR) system. The system must be disabled before performing service on or around system components, steering column, instrument panel components, wiring and sensors. Failure to follow safety and disabling procedures could result in accidental air bag deployment, possible personal injury and unnecessary system repairs.

• Always wear safety goggles when working with, or around, the air bag system. When carrying a non-deployed air bag, be sure the bag and trim cover are pointed away from your body. When placing a non-deployed air bag on a work surface, always face the bag and trim cover upward, away from the surface. This will reduce the motion of the module if it is accidentally deployed. Refer to the additional air bag system precautions later in this section.

• Clean, high quality brake fluid from a sealed container is essential to the safe and proper operation of the brake system. You should always buy the correct type of brake fluid for your vehicle. If the brake fluid becomes contaminated, completely flush the system with new fluid. Never reuse any brake fluid. Any brake fluid that is removed from the system should be discarded. Also, do not allow any brake fluid to come in contact with a painted surface; it will damage the paint.

• Never operate the engine without the proper amount and type of engine oil; doing so WILL result in severe engine damage.

• Timing belt maintenance is extremely important! Many models utilize an interference-type, non-freewheeling engine. If the timing belt breaks, the valves in the cylinder head may strike the pistons, causing potentially serious (also time-consuming and expensive) engine damage. Refer to the maintenance interval charts in the front of this manual for the recommended replacement interval for the timing belt, and to the timing belt section for belt replacement and inspection.

• Disconnecting the negative battery cable on some vehicles may interfere with the functions of the on-board computer system(s) and may require the computer to undergo a relearning process once the negative battery cable is reconnected.

• When servicing drum brakes, only disassemble and assemble one side at a time, leaving the remaining side intact for reference.

• Only an MVAC-trained, EPA-certified automotive technician should service the air conditioning system or its components.

ENGINE REPAIR

Alternator

REMOVAL

2.2L Engines

1. Before servicing the vehicle, refer to the precautions in the beginning of this section.
2. Remove or disconnect the following:
   • Negative battery cable
   • Accessory drive belt
   • Alternator mounting bolts
   • Alternator electrical connectors
   • Alternator

2.4L Engines

1. Before servicing the vehicle, refer to the precautions in the beginning of this section.

2. Remove or disconnect the following:
   • Negative battery cable
   • Accessory drive belt
   • Alternator electrical connectors
   • Power steering line clip
   • Rear alternator brace
   • Alternator mounting bolts
   • Alternator

INSTALLATION

2.2L Engines

1. Install or connect the following:
   • Alternator. Torque the bolts to 37 ft. lbs. (50 Nm).
   • Alternator electrical connectors
   • Accessory drive belt
   • Negative battery cable

2.4L Engines

   • Alternator
   • Alternator mounting bolts. Torque the bolts to 37 ft. lbs. (50 Nm).
   • Rear alternator brace
   • Power steering line clip
   • Alternator electrical connectors
   • Accessory drive belt
   • Negative battery cable

Ignition Timing

ADJUSTMENT

Ignition timing is controlled by the Powertrain Control Module (PCM). No adjustment is necessary or possible.

Engine Assembly

REMOVAL & INSTALLATION

2.2L Engine

WITH MANUAL TRANSMISSION

1. Before servicing the vehicle, refer to the precautions in the beginning of this section.
2. Relieve the fuel system pressure.
3. Drain the cooling system.
4. Remove or disconnect the following:
   - Both battery cables
   - Throttle body air inlet duct
   - Upper radiator hose
   - Vacuum hose from the power brake booster
   - Idle Air Control (IAC) valve electrical connector
   - Alternator electrical connectors
   - Throttle Position (TPS) sensor electrical connector
   - Manifold Absolute Pressure (MAP) sensor electrical connector
   - EVAP emission solenoid electrical connection
   - Fuel injector electrical connectors
   - Exhaust Gas Recirculation (EGR) valve electrical connector
   - Engine Coolant Temperature (ECT) sender electrical connector
   - Oxygen (O2S) sensor electrical connector
   - Engine grounds electrical connection
   - Accessory drive belt
   - Shift control cable
   - Coolant surge tank

- Vacuum line near the master cylinder
5. Install an engine support fixture.
   - Lower radiator hose
   - Wheels
   - Wheel well splash shields
   - Exhaust pipe from the manifold
   - Engine mount strut
   - Wheel Speed Sensor (WSS) electrical connection
   - Ball joints from the steering knuckles
   - Tie rod ends from the steering knuckles
   - Brake lines from the suspension support
   - Air conditioning compressor. DO NOT disconnect the lines.
   - Ignition Control Module (ICM) electrical connector
   - Vehicle Speed Sensor (VSS) electrical connector
   - Cooling fan electrical connector
   - Starter electrical connector
   - Oil pressure sensor electrical connector
   - Power steering lines from the rack and pinion
   - Intermediate shaft
   - Accelerator and cruise control cables
   - Rack and pinion bolts
   - Suspension support extension brace
   - Stabilizer bar
   - Suspension support
   - Heater hoses
   - Halfshafts from the transmission
   - Fuel lines
   - Transmission mount
   - Engine mount
6. Remove the engine support fixture.
7. Remove or disconnect the following:
   - Engine and transmission assembly from the vehicle
   - Engine from the transmission

To install:
8. Install or connect the following:

- Transaxle to the engine. Torque the bolts to 55 ft. lbs. (75 Nm).
- Engine and transmission assembly to the vehicle
- Engine support fixture
- Engine mount. Torque the bolts to 49 ft. lbs. (66 Nm).
- Transmission mount. Torque the bolts to 49 ft. lbs. (66 Nm).
- Halfshafts
- Heater hoses
- Fuel lines
- Suspension support. Torque the bolts to 71 ft. lbs. (110 Nm) plus an additional 90 degree turn.
- Stabilizer bar. Torque the bolts to 49 ft. lbs. (66 Nm).
- Suspension support extension brace. Torque the bolts to 53 ft. lbs. (72 Nm).
- Rack and pinion bolts. Torque the bolts to 89 ft. lbs. (120 Nm).
- Accelerator and cruise control cables
- Intermediate shaft. Torque the pinch bolt to 30 ft. lbs. (41 Nm).
- Power steering lines
- Ignition module electrical connector
- VSS electrical connector
- Cooling fan electrical connector
- Starter electrical connector
- Oil pressure sensor electrical connector
- A/C compressor. Torque the front bolts to 33 ft. lbs. (45 Nm) and the rear bolt to 48 ft. lbs. (65 Nm).
- Brake lines to the suspension support
- Tie rod ends to the steering knuckles. Torque the nuts to 7 ft. lbs. (10 Nm) plus an additional 120 degree turn.
- Ball joints to the steering knuckles. Torque the nuts to 46 ft. lbs. (62 Nm).
- WSS electrical connector
- Engine mount strut. Torque the bolts to 74 ft. lbs. (100 Nm) plus an additional turn of 90 degrees.
- Exhaust pipe to the manifold. Torque the bolts to 26 ft. lbs. (35 Nm).
- Wheel well splash shields
- Wheels
- Lower radiator hose
9. Remove the engine support fixture.
10. Install or connect the following:
    - Coolant surge tank. Torque the bolts to 89 inch lbs. (10 Nm).
    - Shift control cable

Exploded view of the upper engine mount—2.2L engine

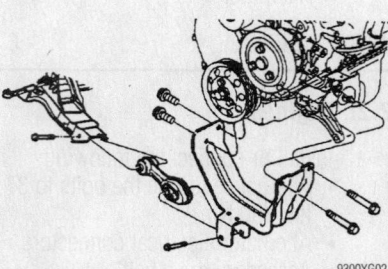

Exploded view of the lower engine mount and strut—2.2L engine

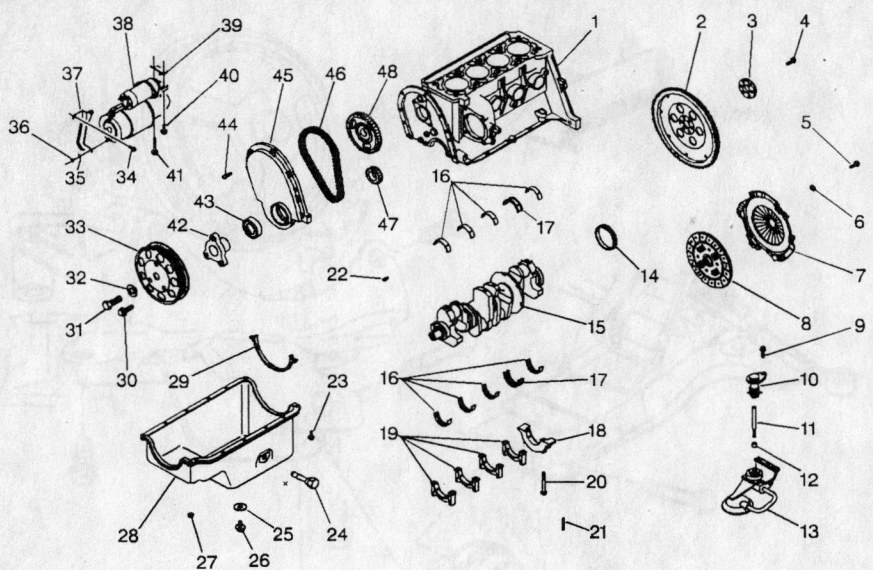

1 Cylinder Block
2 Flywheel
3 Flywheel Retainer
4 Flywheel Bolt
5 Clutch Pressure Plate Bolt
6 Clutch Pressure Plate Washer
7 Clutch Pressure Plate
8 Clutch Disc
9 Oil Pump Drive Bolt
10 Oil Pump Drive
11 Oil Pump Drive Shaft
12 Oil Pump Drive Shaft Retainer
13 Oil Pump
14 Crankshaft Rear Oil Seal
15 Crankshaft
16 Crankshaft Bearing
17 Crankshaft Thrust Bearing
18 Main Bearing Cap
19 Main Bearing Cap
20 Crankshaft Bearing Cap Bolt
21 Oil Pan Stud
22 Connecting Rod Nut
23 Oil Pan Bolt
24 Connecting Rod Nut

25 Oil Pan Drain Plug Gasket
26 Oil Pan Drain Plug
27 Oil Pan Nut
28 Oil Pan
29 Oil Pan Rear Seal
30 Crankshaft Pulley Bolt
31 Crankshaft Pulley Hub Bolt
32 Crankshaft Pulley Hub Bolt Washer
33 Crankshaft Pulley
34 Bracket Bolt
35 Starter Motor Bracket Nut
36 Starter Motor Bracket Nut
37 Starter Motor Bracket
38 Starter Motor
39 Starter Motor Shim
40 Starter Motor Bolt
41 Starter Motor Bolt
42 Crankshaft Pulley Hub
43 Crankshaft Front Oil Seal
44 Crankcase Front Cover Bolt
45 Crankcase Front Cover
46 Timing Chain
47 Crankshaft Sprocket
48 Camshaft Sprocket

9300YG06

Exploded view of the crankshaft and related components—2.2L engine

- Accessory drive belt
- O_2S electrical connector
- ECT sender electrical connector
- EGR valve electrical connector
- Fuel injector electrical connectors
- MAP sensor electrical connector
- TPS sensor electrical connector
- Alternator electrical connector
- IAC electrical connector
- Brake booster vacuum hose
- Upper radiator hose
- Throttle body air inlet duct
- Both battery cables

11. Refill the cooling system.

12. Start the engine and check for proper operation.

WITH AUTOMATIC TRANSMISSION

1. Before servicing the vehicle, refer to the precautions in the beginning of this section.

2. Drain the engine oil and coolant.

3. Recover the refrigerant from the A/C system.

4. Matchmark the bolts and remove the hood.

5. Remove or disconnect the following:

- Both battery cables
- Air cleaner assembly
- Brake booster vacuum line
- Fuel injector sight shield
- Accelerator and cruise control cables

- Engine wiring harness
- Cruise control module
- A/C line from the accumulator
- Accessory drive belt
- Coolant surge tank
- Upper and lower radiator hoses
- Fuel feed and return lines from the engine
- Coolant pipe assembly
- Wheels
- Wheel well splash shields
- Engine mount strut
- Torque converter dust cover
- Exhaust pipe from the manifold
- Starter motor
- A/C compressor and bracket
- Torque converter bolts
- Transmission-to-engine support brace
- Upper front engine mount
- Transmission-to-engine bolts
- Transmission from the engine

6. Lift the engine from the vehicle.

To install:

7. Position the engine in the vehicle, and hand tighten the upper transmission bolts.

8. Install or connect the following:

- Transmission-to-engine bolts. Torque the bolts to 43 ft. lbs. (58 Nm).
- Upper front engine mount. Torque the bolts to 55 ft. lbs. (75 Nm).
- Transmission-to-engine brace. Torque the bolts to 71 ft. lbs. (96 Nm).
- Torque converter bolts. Torque the bolts to 46 ft. lbs. (62 Nm).
- A/C compressor bracket
- A/C compressor. Torque the bolts to 37 ft. lbs. (50 Nm).
- Starter motor. Torque the bolts to 37 ft. lbs. (50 Nm).
- Exhaust pipe to the manifold. Torque the bolts to 26 ft. lbs. (35 Nm).
- Torque converter dust cover
- Engine mount strut. Torque the bolts to 74 ft. lbs. (100 Nm) plus an additional turn of 90 degrees.
- Wheel well splash shields
- Wheels
- Coolant pipe assembly
- Fuel feed and return lines
- Upper and lower radiator hoses
- Coolant surge tank. Torque the bolts to 89 inch lbs. (10 Nm).
- Accessory drive belt
- A/C line to the accumulator
- Cruise control module
- Engine wiring harness
- Accelerator and cruise control cables

- Brake booster vacuum line
- Air cleaner assembly
- Hood. Torque the bolts to 15 ft. lbs. (20 Nm).
- Battery cables

9. Evacuate and recharge the A/C system.

10. Refill the engine oil and coolant.

11. Start the engine and check for leaks.

12. Install the fuel injector sight shield.

2.4L Engine

WITH MANUAL TRANSMISSION

1. Before servicing the vehicle, refer to the precautions in the beginning of this section.

2. Discharge and recover the air conditioning refrigerant.

3. Properly drain the cooling system and engine oil.

4. Relieve the fuel system pressure.

5. Remove or disconnect the following:

- Both battery cables
- Clutch pushrod from the pedal assembly
- Heater hose at the thermostat housing
- Upper radiator hose
- Air cleaner assembly
- Coolant fan
- Refrigerant hose assembly at the compressor
- Both vacuum hoses from the front of the engine
- Alternator electrical connector
- Air conditioning compressor electrical connector
- Fuel injector harness
- Idle Air Control (IAC) electrical connector
- Throttle Position (TPS) sensor electrical connector
- Manifold Absolute Pressure (MAP) sensor electrical connector

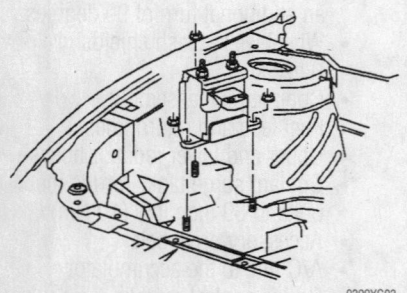

Exploded view of the upper engine mount—2.4L engine

9300YG03

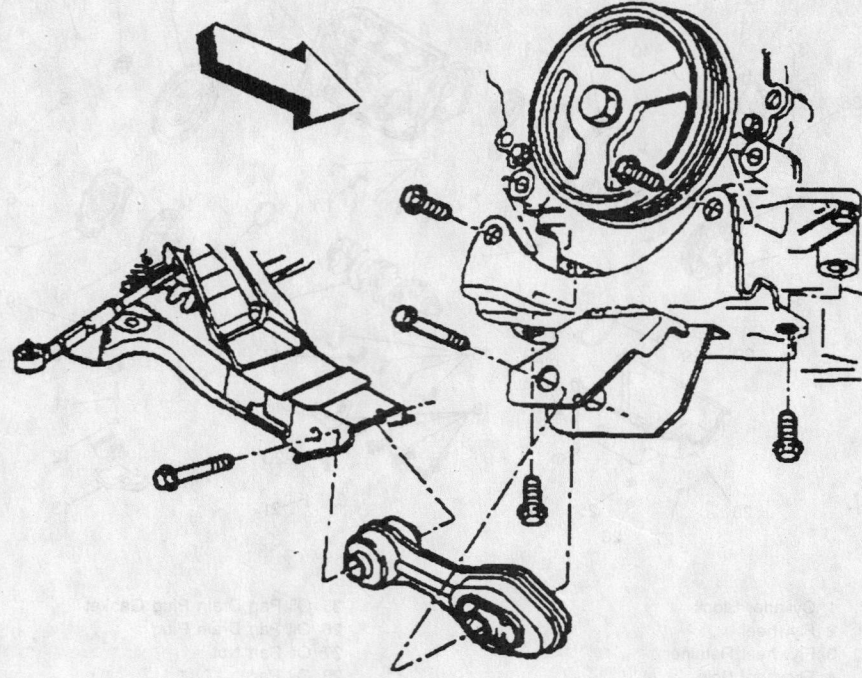

Exploded view of the lower engine mount and strut—2.4L engine

9300YG04

- Intake Air Temperature (IAT) sensor electrical connector
- Evaporative Emissions (EVAP) canister purge solenoid electrical connector
- Negative battery cable from the transaxle
- Electronic ignition module electrical connector
- Engine Coolant Temperature (ECT) sensors electrical connectors
- Oil pressure sensor/switch electrical connector
- Oxygen (O_2S) sensor electrical connector
- Crankshaft Position (CKP) sensor electrical connector
- Back-up lamp switch electrical connector
- Power brake vacuum hose from the throttle body
- Throttle cable and bracket
- Power steering pump and move it aside with the lines attached
- Fuel lines
- Shift cables
- Clutch actuator line
- Exhaust manifold and heat shield
- Lower radiator hose

6. Install an engine support fixture.

7. Remove or disconnect the following:

- Coolant recovery/surge tank, move it aside with the hoses attached
- Engine strut mount

- Front wheels
- Right splash shield
- Vehicle Speed Sensor (VSS) electrical connector
- Knock Sensor (KS) electrical connector
- Starter solenoid electrical connector
- Both front Anti-lock Brake System (ABS) wheel speed sensor electrical connectors
- Engine mount strut
- Transmission mount
- Ball joints from the steering knuckles
- Suspension supports, crossmember and stabilizer shaft as an assembly
- Heater outlet hose from the radiator outlet pipe
- Halfshaft from the transmission and intermediate shaft
- Flywheel housing cover

8. Position a suitable support below the engine, then carefully lower the vehicle onto the support.

9. Raise the vehicle slowly off the engine/transaxle assembly.

➡️ It may be necessary to move the engine/transaxle assembly rearward to clear the intake manifold.

10. Remove the engine from the transaxle.

To install:

✷✷ WARNING

Be sure the retaining bolts are in their correct locations. If not, engine damage may occur.

11. Install or connect the following:
- Engine to the transaxle. Torque the mounting bolts to 55 ft. lbs. (75 Nm).
- Engine/transaxle assembly under the vehicle, then lower the vehicle over the assembly
- Engine support fixture, making sure to adjust it to the previous setting
- Halfshafts to the transaxle
- Heater outlet hose to the radiator outlet pipe
- Suspension supports, crossmember and stabilizer shaft assembly. Torque the bolts to 53 ft. lbs. (72 Nm).
- Ball joints to the steering knuckles. Torque the nuts to 44 ft. lbs. (60 Nm) plus an additional turn of 180 degrees.
- Engine mount assembly. Torque the bolts to 46 ft. lbs. (62 Nm).
- Transmission mount. Torque the bolts to 66 ft. lbs. (90 Nm).
- Both front ABS wheel speed sensor electrical connectors
- Starter solenoid electrical connector
- KS electrical connector
- VSS electrical connector
- Right splash shield
- Front wheels

12. Carefully raise the vehicle off the support.

13. Install or connect the following:
- Engine strut mount. Torque the bolts to 74 ft. lbs. (100 Nm).
- Coolant recovery/surge tank
- Lower radiator hose
- Exhaust manifold. Torque the nuts to 31 ft. lbs. (42 Nm).
- Exhaust manifold heat shield. Torque the bolts to 19 ft. lbs. (26 Nm).
- Clutch actuator line
- Shift cables
- Fuel lines
- Power steering pump. Torque the bolts to 22 ft. lbs. (30 Nm).
- Throttle cable and bracket
- Vacuum hoses to the intake manifold and the brake booster
- Back-up lamp switch electrical connector
- CKP sensor electrical connector

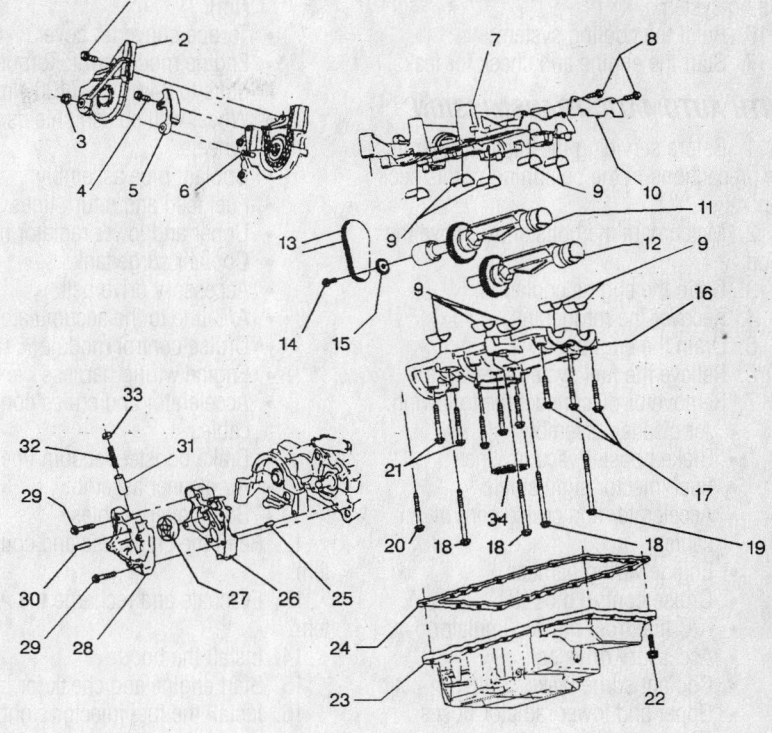

1 Balance Shaft Chain Cover Bolt
2 Balance Shaft Chain Cover
3 Balance Shaft Guide Adjuster Nut
4 Balance Shaft Chain Guide Adjuster Bolt
5 Balance Shaft Chain Adjuster Guide
6 Stud
7 Balance Shaft Upper Housing
8 Balance Shaft Thrust Plate Retainer Bolt
9 Balance Shaft Bearings
10 Balance Shaft Thrust Plate Retainer
11 Balance Shaft Drive
12 Balance Shaft Driven
13 Balance Shaft Drive Chain
14 Driven Sprocket Bolt-Note: Left Hand Thread
15 Balance Shaft Driven Sprocket
16 Balance Shaft Lower Housing

17 Balance Shaft Housing Bolt
18 Balance Shaft Housing to Block Bolt
19 Balance Shaft Housing to Block Bolt
20 Balance Shaft Housing to Block Bolt
21 Balance Shaft Housing Bolt
22 Oil Pan Bolt
23 Oil Pan
24 Oil Pan Gasket
25 Dowel Pin
26 Oil Pump Body
27 Gerotor
28 Oil Pump Cover
29 Oil Pump Cover to Body Bolt
30 Relief Valve Pin
31 Relief Valve Spring Guide
32 Relief Valve Spring
33 Relief Valve
34 Pickup Screen

9300YG09

Exploded view of the balance shafts and related components—2.4L engine

- O₂S electrical connector
- Oil pressure sensor/switch electrical connector
- ECT sensors electrical connectors
- Electronic ignition module electrical connector
- Negative battery cable to the transaxle
- EVAP canister purge solenoid electrical connector
- IAT sensor electrical connector
- MAP sensor electrical connector
- TPS sensor electrical connector
- IAC sensor electrical connector
- Fuel injector harness

- Air conditioning compressor electrical connector
- Alternator electrical connector
- Refrigerant hose assembly to the compressor
- Coolant fan
- Air cleaner assembly
- Upper radiator hose
- Heater hose at the thermostat housing
- Clutch pushrod to the pedal assembly
- Both battery cables

14. Refill the transaxle and the crankcase.

15. Evacuate and recharge the air conditioning system.
16. Refill the cooling system.
17. Start the engine and check for leaks.

WITH AUTOMATIC TRANSMISSION

1. Before servicing the vehicle, refer to the precautions in the beginning of this section.
2. Matchmark the bolts and remove the hood.
3. Drain the engine coolant.
4. Recover the refrigerant.
5. Drain the engine oil.
6. Relieve the fuel system pressure.
7. Remove or disconnect the following:
   - Air cleaner assembly
   - Brake booster vacuum line
   - Fuel injector sight shield
   - Accelerator and cruise control cables
   - Engine wiring harness
   - Cruise control module
   - A/C line from the accumulator
   - Accessory drive belt
   - Coolant surge tank
   - Upper and lower radiator hoses
   - Fuel feed and return lines
   - Coolant pipe assembly
   - Wheels
   - Wheel well splash shields
   - Engine mount strut
   - Torque converter cover
   - Exhaust pipe from the manifold
   - Starter motor
   - A/C compressor
   - Torque converter bolts
   - Transmission-to-engine brace
   - Upper front engine mount
   - Transmission-to-engine bolts
8. Separate the engine from the transmission and lift the engine from the vehicle.

To install:
9. Position the engine in the vehicle.
10. Hand tighten the upper transmission bolts.
11. Install or connect the following:
   - Upper front engine mount. Torque the bolts to 46 ft. lbs. (62 Nm).
   - Lower transmission-to-engine bolts. Torque all bolts to 55 ft. lbs. (75 Nm).
   - Transmission-to-engine brace. Torque the bolts to 71 ft. lbs. (96 Nm).
   - Torque converter bolts. Torque the bolts to 46 ft. lbs. (62 Nm).
   - A/C compressor. Torque the bolts to 37 ft. lbs. (50 Nm).
   - Starter motor. Torque the bolts to 37 ft. lbs. (50 Nm).
   - Exhaust pipe to the manifold.

Torque the bolts to 26 ft. lbs. (35 Nm).
   - Torque converter cover
   - Engine mount strut. Torque the bolts to 74 ft. lbs. (100 Nm).
   - Wheel well splash shields
   - Wheels
   - Coolant pipe assembly
   - Fuel feed and return lines
   - Upper and lower radiator hoses
   - Coolant surge tank
   - Accessory drive belt
   - A/C line to the accumulator
   - Cruise control module
   - Engine wiring harness
   - Accelerator and cruise control cables
   - Brake booster vacuum line
   - Air cleaner assembly
   - Both battery cables
12. Refill the crankcase and cooling system.
13. Evacuate and recharge the A/C system.
14. Install the hood.
15. Start engine and check for leaks.
16. Install the fuel injector sight shield.

Water Pump

REMOVAL & INSTALLATION

2.2L Engine

1. Before servicing the vehicle, refer to the precautions in the beginning of this section.

2. Drain the cooling system.
3. Remove or disconnect the following:
   - Negative battery cable
   - Water pump pulley bolts, loosen
   - Accessory drive belt
   - Water pump pulley
   - Surge tank hose
   - Water pump bolts, pump and gasket

To install:
4. Apply a thin bead of sealer around the outer edge of the water pump gasket seating area and place the gasket on the pump.
5. Install or connect the following:
   - Water pump. Torque the bolts to 18 ft. lbs. (25 Nm).
   - Water pump pulley and hand-tighten the bolts
   - Surge tank hose
   - Accessory drive belt
   - Water pump pulley bolts. Torque the bolts to 22 ft. lbs. (30 Nm).
6. Connect the negative battery cable.
7. Refill and bleed the cooling system.

2.4L Engine

1. Before servicing the vehicle, refer to the precautions in the beginning of this section.
2. Drain the cooling system.
3. Remove or disconnect the following:
   - Negative battery cable
   - Oxygen (O_2S) sensor electrical connector
   - Upper exhaust manifold heat shield
   - Exhaust manifold brace-to-manifold bolt

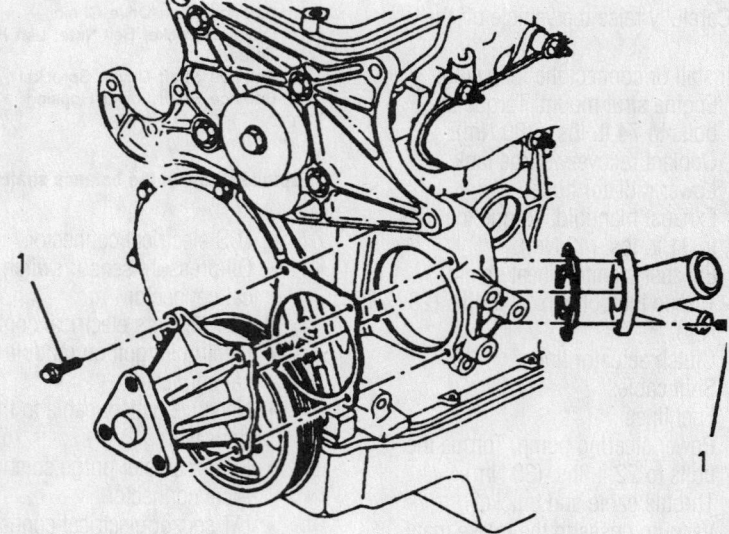

1 BOLT – 25 N·m (18 LBS. FT.)

7922YG01

Exploded view of the water pump mounting—2.2L engine

1 TIMING CHAIN HOUSING
2 GASKET, TIMING CHAIN
   HOUSING TO WATER PUMP COVER
3 NUT (3)
4 WATER PUMP BODY ASM.
5 GASKET, WATER PUMP BODY
   TO WATER PUMP COVER
6 WATER PUMP COVER
7 BOLT (M6 X 1 X 65) – 3
   LOWER POSITIONS
8 BOLT (M6 X 1 X 25)
9 BOLT (M6 X 1 X 90)
10 GASKET, WATER PUMP
   COVER TO BLOCK
11 BOLTS, WATER PUMP COVER
   TO BLOCK (2)

7922YG02

Exploded view of the water pump mounting and related components—2.4L engine

- Manifold-to-exhaust pipe spring loaded bolts
- Both radiator outlet pipe-to-water pump cover bolts
- Exhaust manifold pipe

☀ WARNING

DO NOT rotate the flex coupling more than 4 degrees or damage may occur.

- Radiator outlet pipe from the oil pan
- Brake vacuum pipe from the camshaft housing
- Exhaust manifold from the cylinder head
- Front cover
- Timing chain tensioner
- Water pump cover and water pump as an assembly

4. Separate the water pump from the water pump cover.

To install:

5. Install or connect the following:
- Water pump with a new gasket to the cover and tighten the bolts finger-tight

➡**Lubricate the splines with clean grease**

- Water pump and cover assembly using new gaskets and tighten the fasteners finger-tight

6. Torque the bolts in the following sequence;
   a. Pump assembly-to-chain housing nuts: 19 ft. lbs. (26 Nm).
   b. Pump cover-to-pump assembly: 124 inch lbs. (14 Nm).
   c. Cover-to-block, bottom bolt first: 19 ft. lbs. (26 Nm).

7. Install or connect the following:
- Timing chain tensioner
- Front cover. Torque the bolts to 97 inch lbs. (11 Nm).
- Exhaust manifold with a new gas-

ket. Torque the bolts to 31 ft. lbs. (42 Nm).
- Brake vacuum pipe to the camshaft housing
- Radiator outlet pipe to the water pump cover. Torque the bolts to 125 inch lbs. (14 Nm).
- Radiator outlet pipe to the oil pan. Torque the bolts to 18 ft. lbs. (25 Nm).
- Exhaust pipe to the manifold. Tighten the exhaust pipe flange bolts evenly and gradually to avoid binding, until fully seated. Torque the bolts to 26 ft. lbs. (35 Nm).
- Exhaust manifold brace to the manifold. Torque the bolts to 41 ft. lbs. (56 Nm).
- Upper heat shield. Torque the bolts to 125 inch lbs. (14 Nm).
- O$_2$S electrical connector
- Negative battery cable

8. Refill the cooling system until it comes out of the heater hose outlet; then connect the heater hose.

9. Start the engine. Run the vehicle until the thermostat opens, refill the radiator and recovery tank, then turn the engine **OFF**.

10. Once the vehicle has cooled, recheck the coolant level.

Heater Core

REMOVAL & INSTALLATION

1. Disable the SIR system by performing the following procedure:
   a. Point the wheel in the straight-ahead position.
   b. Turn the ignition switch to the LOCK position.
   c. Remove the AIR BAG fuse from the fuse block.
   d. At the base of the steering column, remove the left sound insulator.
   e. At the base of the steering column, disconnect the Connector Position Assurance (CPA), the yellow 2-way electrical connectors and the passenger's side module electrical connector.

2. Disconnect the negative battery cable.

3. Drain the cooling system into a clean container for reuse.

4. Disconnect the heater hoses from the heater core.

5. If equipped, remove the Diagnostic Energy Reserve Module (DERM) with attaching brackets.

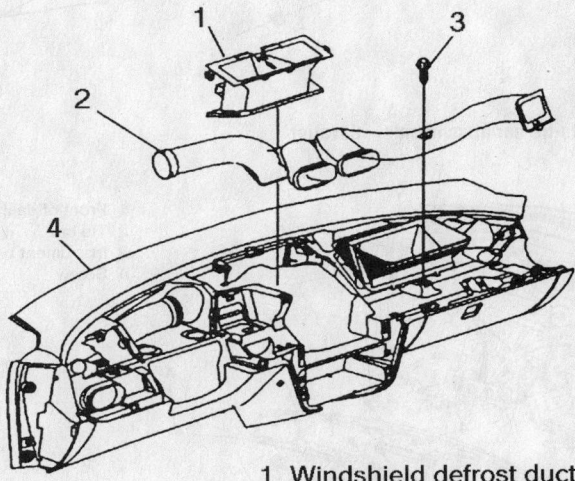

1 Windshield defrost duct
2 Air distribution duct
3 Screw
4 Lower I/P

87950083

Air distribution duct mounting—Cavalier shown

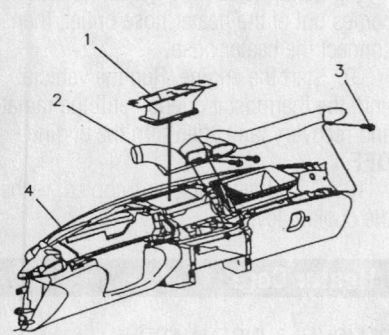

1 DUCT, WINDSHIELD DEFROST
2 DUCT, AIR DISTRIBUTION
3 SCREW
4 LOWER I/P

87950084

Location of the air distribution duct mounting—Sunfire shown

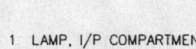

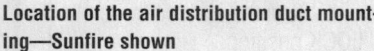

1 LAMP, I/P COMPARTMENT
2 RETAINER

87950085

Detach the instrument panel lamp connector—Cavalier and Sunfire

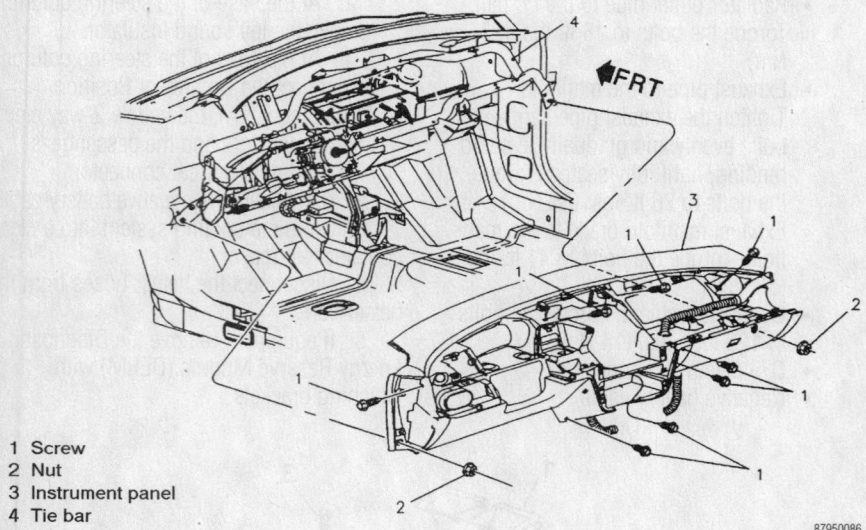

1 Screw
2 Nut
3 Instrument panel
4 Tie bar

87950086

Instrument panel-to-tie bar attachments—Cavalier

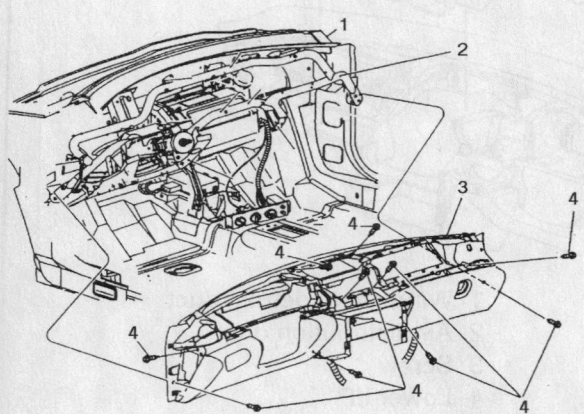

1 Front of dash
2 Tie bar
3 Instrument panel
4 Screw

87950087

Instrument panel-to-tie bar mounting—Sunfire

6. Remove the steering wheel by removing or disconnecting the following:
- SIR module-to-steering wheel screws
- SIR module and disconnect the electrical connector

✳✳ CAUTION

Place the SIR module in a safe place with the front facing upward.

- Steering wheel-to-steering column nut
- Steering wheel from the steering column

7. Remove the instrument panel by removing or disconnecting the following:
- Defroster grille
- Instrument panel end caps from both sides
- Instrument panel trim pad, (Cavalier models only)
- Accessory trim plate, (Sunfire models only)
- Instrument panel trim plate
- Heater/air conditioning control assembly
- Radio
- Air ventilation ducts from the heater housing
- Instrument panel light electrical connector
- Tie bar screws
- Instrument panel from the tie bar
- Heater core outlet screws and the outlet
- Heater core cover-to-heater case screws and the cover.

➡ **The heater core mounting screw is located in the recess in the center of the heater core cover.**

- Heater core clamp from the heater assembly
- Heater core from the heater assembly

To install:

8. Install or connect the following:
- Heater core to the heater assembly
- Heater core clamp to the heater assembly and torque the clamp screws to 9 inch lbs. (1 Nm)
- Heater core cover and the cover-to-heater case screw and torque the screws to 9 inch lbs. (1 Nm)
- Heater core outlet and torque the screws to 9 inch lbs. (1 Nm)
- Diagnostic Energy Reserve Module (DERM) with attaching bracket, if equipped

9. Install the instrument panel by installing or connecting the following:

- Instrument panel to the tie bar
- Tie bar screws
- Instrument panel light electrical connector
- Air ventilation ducts to the heater housing
- Radio
- Heater/air conditioning control assembly
- Instrument panel trim plate
- Instrument panel trim pad, (Cavalier models only)
- Accessory trim plate, (Sunfire models only)
- Instrument panel end caps on both sides
- Defroster grill
- Heater hoses to the heater core

10. Install the steering wheel by installing or connecting the following:
- Steering wheel to the steering column
- Steering wheel-to-steering column nut and torque the nut to 30 ft. lbs. (41 Nm)
- Electrical connector and install the SIR module
- SIR module-to-steering wheel screws and torque the screws to 89 inch lbs. (10 Nm)

11. Refill the cooling system.
12. Connect the negative battery cable.
13. Enable the SIR system by performing the following procedure:

a. Turn the ignition switch to the LOCK position.

b. At the base of the steering column, connect the Connector Position Assurance (CPA), the yellow 2-way electrical connectors and the passenger's side module electrical connector.

c. At the base of the steering column, install the left sound insulator.

d. Install the AIR BAG fuse to the fuse block.

e. Turn the ignition switch to the RUN position; the INFL REST warning light should flash 7–9 times then turn OFF.

14. Operate the engine to normal operating temperatures; then, check the climate control operation and check for leaks.

Cylinder Head

REMOVAL & INSTALLATION

2.2L Engine

1. Before servicing the vehicle, refer to the precautions in the beginning of this section.

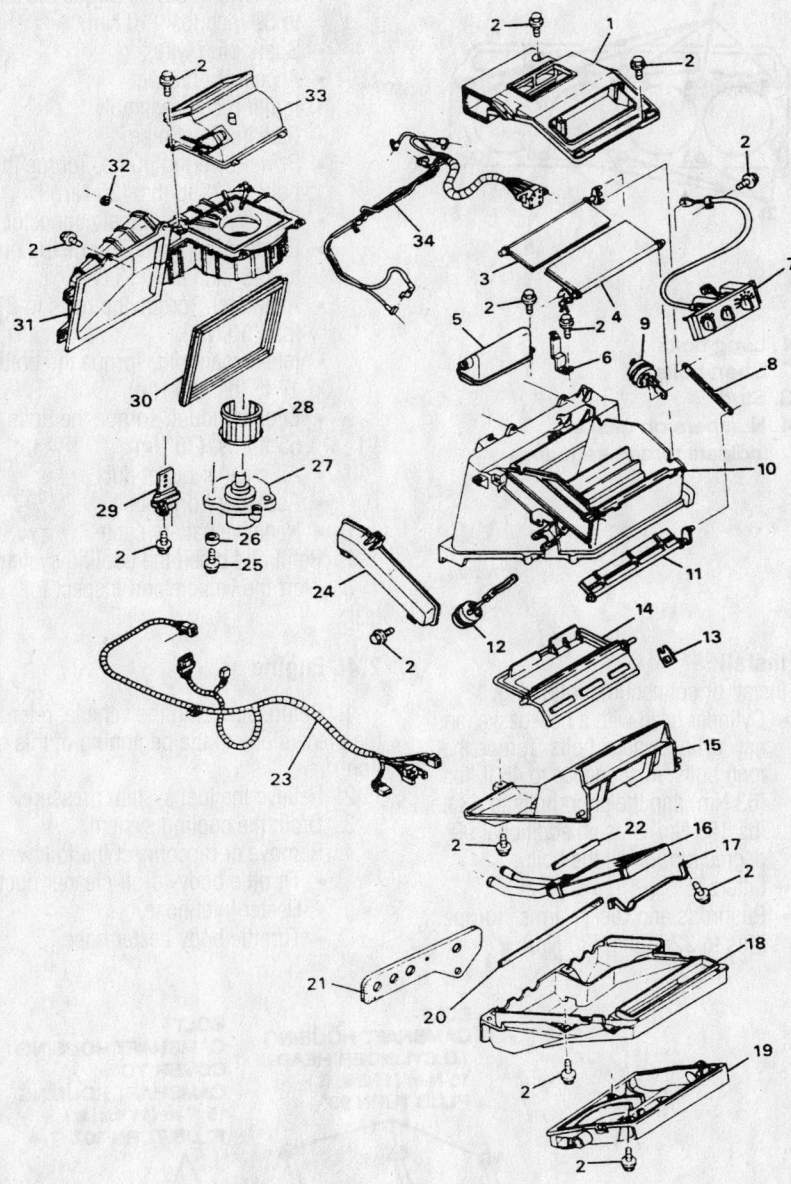

1 VALVE HOUSING COVER	18 HEATER COVER
2 HEATER/BLOWER MODULE BOLT	19 HEATER OUTLET
3 DEFROSTER VALVE	20 HEATER CORE SHROUD SEAL
4 MODE VALVE	21 HEATER CORE TUBE AND MOUNT SEAL
5 HEATER-VACUUM TANK	22 HEATER CORE SEAL
6 HEATER MODULE MOUNTING BRACKET	23 HEATER AND A/C CONTROL SWITCH HARNESS
7 HEATER-CONTROL	24 DEFROSTER DUCT
8 HEATER VALVE LEVER LINK	25 BLOWER MOTOR BOLTS
9 DEFROSTER VALVE ACTUATOR	26 BLOWER MOTOR ISOLATOR
10 HEATER CASE	27 BLOWER MOTOR
11 HEATER VALVE	28 BLOWER FAN
12 MODE VALVE ACTUATOR	29 BLOWER RESISTOR
13 TEMPERATURE VALVE CLIP	30 HEATER CASE SEAL
14 TEMPERATURE VALVE	31 BLOWER AND AIR INLET CASE
15 HEATER CORE SHROUD	32 MOUNTING SEAL
16 HEATER CORE	33 AIR INLET HOUSING
17 HEATER CORE STRAP	34 VACUUM HARNESS

93111GB4

Exploded view of the heater/evaporator housing assembly—Cavalier and Sunfire

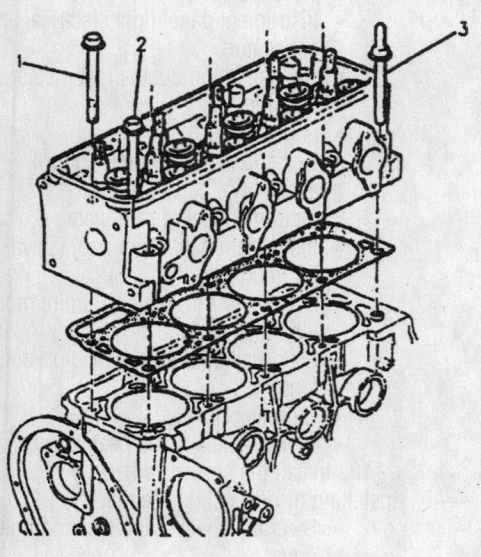

Cylinder head torquing sequence—2.2L engine

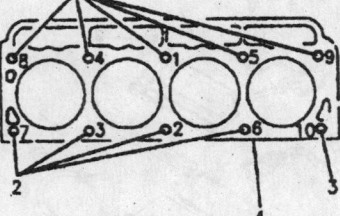

1. Long bolts
2. Short bolts
3. Stud
4. Numbers on gasket indicate torque sequence

7922YG03

2. Relieve the fuel system pressure.
3. Drain the cooling system.
4. Remove or disconnect the following:

- Accessory drive belt
- Air cleaner outlet duct
- Engine mount
- Intake manifold
- Exhaust manifold
- Engine Coolant Temperature (ECT) sensor electrical connector
- Power steering pump
- Radiator inlet pipe
- Spark plug wires
- Rocker arm cover

➡ Whenever valve train components are removed, keep them in order for installation purposes.

- Rocker arms and pushrods
- Lifters
- Alternator rear brace and the alternator
- Power steering pump
- Radiator inlet pipe
- Ignition coil assembly
- Accessory bracket
- Spark plug wires
- Cylinder head bolts

❊❊ WARNING

Two sizes of bolts are used; note the location of each.

- Cylinder head
5. Inspect the cylinder head and block surface for cracks, nicks, heavy scratches and flatness.

To install:
6. Install or connect the following:

- Cylinder head with a new gasket and new cylinder head bolts. Torque the long bolts, in sequence to 46 ft. lbs. (63 Nm) and the short bolts to 43 ft. lbs. (58 Nm) plus an additional 90 degree turn on all the bolts.
- Lifters
- Pushrods and rocker arms. Torque nuts to 22 ft. lbs. (30 Nm).

- Rocker arm cover. Torque the bolts to 89 inch lbs. (10 Nm).
- Spark plug wires
- Accessory bracket
- Ignition coil assembly
- Radiator inlet pipe
- Power steering pump. Torque the bolts to 22 ft. lbs. (30 Nm).
- ECT sensor electrical connector
- Exhaust manifold. Torque the nuts to 118 inch lbs. (13 Nm).
- Alternator. Torque the bolts to 37 ft. lbs. (50 Nm).
- Intake manifold. Torque the bolts to 17 ft. lbs. (24 Nm).
- Engine mount. Torque the bolts to 55 ft. lbs. (75 Nm).
- Air cleaner outlet duct
- Accessory drive belt
- Negative battery cable
7. Refill and bleed the cooling system.
8. Start the vehicle and inspect for leaks.

2.4L Engine

1. Before servicing the vehicle, refer to the precautions in the beginning of this section.
2. Relieve the fuel system pressure.
3. Drain the cooling system.
4. Remove or disconnect the following:
- Throttle body-to-air cleaner duct
- Heater inlet hose
- Throttle body heater hose

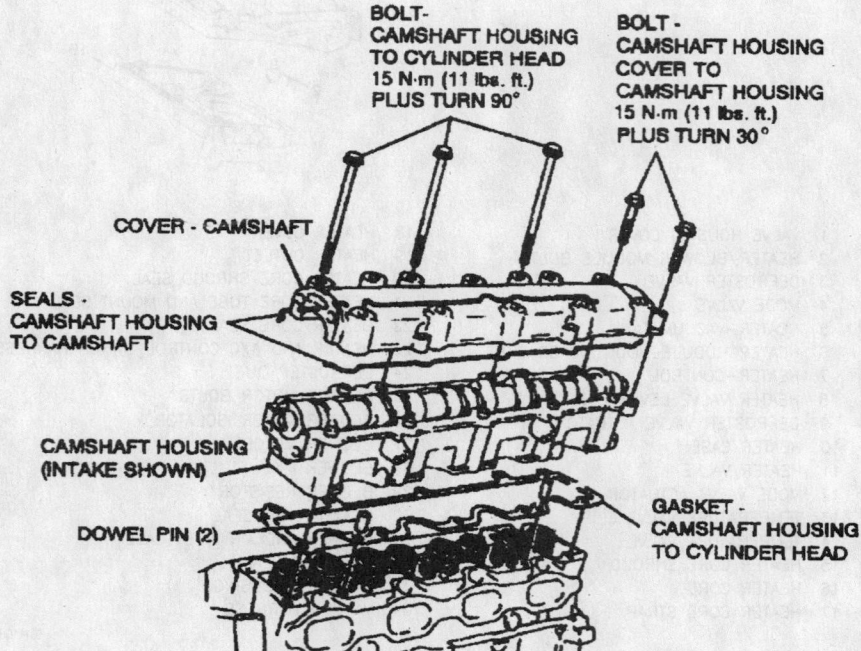

Exploded view of the camshaft housing cover mounting—2.4L engine

7922YG05

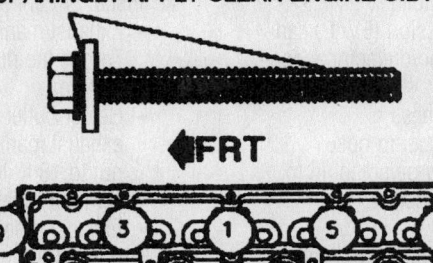

SPARINGLY APPLY CLEAN ENGINE OIL HERE

◄FRT

A. Tighten the bolts to the following N.m (lb. ft.) specification in sequence:
bolts 1 through 8: 65 N·m (40 ft. lb.)
bolts 9 and 10: 40 N·m (30 ft. lb.)
B. Then turn all 10 bolts an additional 90 degrees in sequence

7922YG04

Head bolt torque sequence—2.4L engine

- Power brake vacuum hose from throttle body
- Manifold Absolute Pressure (MAP) sensor electrical connector
- Intake Air Temperature (IAT) sensor electrical connector
- Evaporative Emission (EVAP) canister purge solenoid
- Camshaft Position (CMP) sensor electrical connector
- Intake manifold
- Exhaust manifold
- Ignition coil assembly
- CMP sensor
- Power steering pump
- Vacuum line from the fuel pressure regulator
- Fuel injector harness connector
- Fuel rail
- Timing chain housing from the intake camshaft housing. DO NOT remove it from the vehicle.
- Oil pressure switch electrical connector
- Transmission fluid level indicator tube
- Intake and exhaust camshaft housing
- Upper radiator hose
- Coolant temperature sensor connector
- Cylinder head bolts
- Cylinder head and gasket

To install:

➡Do not use abrasive pads to clean the cylinder head or block surfaces. An abrasive pad may damage the cylinder head and/or block.

5. Install or connect the following:
- Cylinder head with a new gasket. Torque the new bolts 1-8 to 40 ft. lbs. (54 Nm) and bolts 9-10 to 30 ft. lbs. (40 Nm); then, turn all bolts an additional 90 degrees (¼ turn) in sequence.
- Upper radiator hose to coolant outlet
- Intake and exhaust camshaft housings
- Timing chain housing. Torque the bolts to 19 ft. lbs. (26 Nm).
- Fuel rail with new o-rings. Torque the bolts to 19 ft. lbs. (26 Nm).
- Vacuum line to the fuel pressure regulator
- Fuel injector electrical connectors
- CMP
- Power steering pump. Torque the bolts to 19 ft. lbs. (26 Nm).
- Oil pressure switch electrical connector
- Transmission fluid level indicator tube
- Ignition coil assembly. Torque the bolts to 11 ft. lbs. (15 Nm) plus an additional 30 degree turn.
- Exhaust manifold. Torque the bolts to 31 ft. lbs. (42 Nm).
- Intake manifold. Torque the bolts to 18 ft. lbs. (24 Nm).
- Upper radiator hose
- MAP sensor electrical connector
- IAT sensor electrical connector
- Canister purge solenoid
- Power brake vacuum hose
- Throttle body heater hose

- Heater inlet hose
- Throttle body-to-air cleaner duct
6. Refill the coolant system.
7. Connect the negative battery cable.
8. Start the engine and check for leaks.

Rocker Arms

REMOVAL & INSTALLATION

➡**Place the components in a rack in order to be sure they are installed in the same location.**

1. Before servicing the vehicle, refer to the precautions in the beginning of this section.
2. Remove or disconnect the following:
- Negative battery cable
- Rocker arm cover(s)
- Rocker arm nuts
- Rocker arm pivot ball(s)
- Rocker arm(s)
- Pushrods

To install:
3. Install the pushrods.
4. Coat the bearing surfaces of the rocker arms and pivot balls with camshaft lubricant.
5. Install or connect the following:
- Rocker arms. Torque the nuts to 22 ft. lbs. (30 Nm).
- Rocker arm covers. Torque the bolts to 89 inch lbs. (10 Nm).
- Negative battery cable

Intake Manifold

REMOVAL & INSTALLATION

2.2L Engine

1. Before servicing the vehicle, refer to the precautions in the beginning of this section.
2. Properly relieve the fuel system pressure.
3. Remove or disconnect the following:
- Air cleaner inlet duct
- Air inlet resonator and bracket
- Throttle and cruise control cable from the throttle body

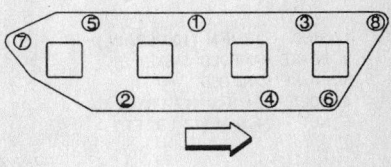

7922YG09

Intake manifold tightening sequence— 2.2L engine

- Manifold Absolute Pressure (MAP) sensor
- Throttle Position (TPS) sensor
- Idle Air Control (IAC) valve
- Fuel supply line
- Throttle body
- Fuel inlet pipe
- Intake manifold

To install:

4. Install or connect the following:
- Intake manifold with a new gasket. Torque the bolts/nuts, in sequence, to 18 ft. lbs. (24 Nm).
- Fuel inlet pipe. Torque the bolts to 18 ft. lbs. (24 Nm).
- Throttle body. Torque the bolts to 89 inch lbs. (10 Nm).
- Fuel supply line
- MAP sensor
- TPS sensor
- IAC valve
- Cruise control and throttle cables to the throttle body
- Air inlet resonator bracket and resonator
- Air cleaner inlet duct
- Negative battery cable

2.4L Engine

1. Before servicing the vehicle, refer to the precautions in the beginning of this section.
2. Properly relieve the fuel system pressure.
3. Drain the cooling system.
4. Remove or disconnect the following:
- Manifold Absolute Pressure (MAP) sensor electrical connector

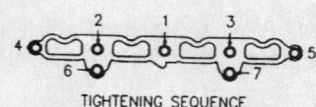

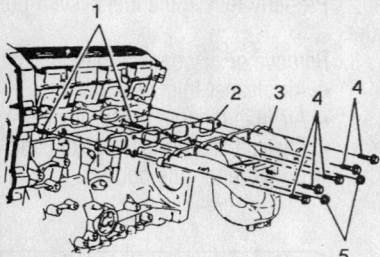

1. STUD – 12 N•M (100 LB. IN.)
2. INTAKE MANIFOLD GASKET
3. INTAKE MANIFOLD
4. BOLT – 24 N•M (17 LB. FT.)
5. NUT – 24 N•M (17 LB. FT.)

7922YG08

Intake manifold torque sequence—2.4L engine

- Intake Air Temperature (IAT) sensor electrical connector
- Evaporative Emission (EVAP) canister purge solenoid electrical connector
- Fuel injector harness
- Fuel regulator vacuum hose
- EVAP canister purge solenoid to canister vacuum hose
- Air cleaner duct
- Accelerator control cable bracket
- Stud-ended alternator mount bolt
- Exhaust Gas Recirculation (EGR) pipe from the EGR adapter
- Intake manifold support brace
- Intake manifold

To install:

5. Install or connect the following:
- Intake manifold with a new gasket. Torque the bolts/nuts, in sequence, to 18 ft. lbs. (24 Nm).
- EGR pipe to the adapter. Torque the fasteners to 19 ft. lbs. (26 Nm).
- Stud-ended alternator bolt
- Accelerator control cable bracket
- Fuel injector electrical connectors
- Vacuum hoses to the fuel regulator and EVAP canister purge solenoid
- MAP sensor electrical connector
- IAT sensor electrical connector
- EVAP canister purge solenoid electrical connector
- Air cleaner duct
- Negative battery cable
6. Refill the cooling system.
7. Start the engine and inspect for leaks.

Exhaust Manifold

REMOVAL & INSTALLATION

2.2L Engine

1. Before servicing the vehicle, refer to the precautions in the beginning of this section.
2. Drain the cooling system.
3. Remove or disconnect the following:
- Negative battery cable
- Oxygen (O2S) sensor electrical connector
- Accessory drive belt
- Alternator
- Radiator inlet pipe
- Exhaust pipe-to-exhaust manifold bolts
- Oil filler tube
- Heater outlet hose assembly-to-exhaust manifold nut
- Exhaust manifold

To install:

4. Install or connect the following:
- Exhaust manifold with new gaskets. Torque the nuts to 115 inch lbs. (13 Nm).
- Heater outlet hose assembly-to-exhaust manifold nut. Torque the nut to 18 ft. lbs. (25 Nm).
- Oil filler tube
- Exhaust pipe-to-exhaust manifold bolts. Torque the bolts to 26 ft. lbs. (35 Nm).
- Radiator inlet pipe
- Alternator. Torque the bolts to 37 ft. lbs. (50 Nm).
- Accessory drive belt
- O2S sensor electrical connector
- Negative battery cable
5. Refill the cooling system.
6. Start the engine and check for exhaust leaks.

2.4L Engine

1. Before servicing the vehicle, refer to the precautions in the beginning of this section.
2. Remove or disconnect the following:
- Negative battery cable
- Oxygen (O2S) sensor electrical connector
- Upper heat shield
- Exhaust manifold brace-to-manifold bolt
- Exhaust manifold-to-oil pan nuts

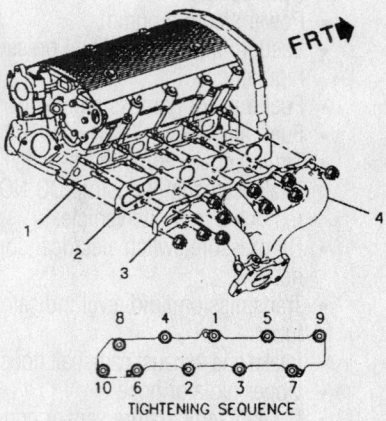

1. STUD, EXHAUST MANIFOLD
2. GASKET, EXHAUST MANIFOLD
3. MANIFOLD, EXHAUST
4. NUT, EXHAUST MANIFOLD, MUST BE TIGHTENED IN SEQUENCE SHOWN TO 12.5 N•m (110 LB. IN.)

7922YG11

Exploded view of the exhaust manifold, showing the torque sequence—2.4L engine

➡**Do not bend the exhaust flex coupler more than necessary to remove it. Excessive movement will damage the flex coupler.**

- Exhaust pipe from the exhaust manifold
- Exhaust manifold

To install:

3. Install or connect the following:
- Exhaust manifold with new gaskets. Torque the nuts to 110 inch lbs. (12 Nm), in sequence.
- Heat shield. Torque the bolts to 124 inch lbs. (14 Nm).
- Exhaust manifold brace-to-manifold bolt and oil pan nuts. Torque the bolts to 41 ft. lbs. (56 Nm) and nuts to 19 ft. lbs. (26 Nm).
- Manifold-to-flex coupler fasteners. Torque the bolts to 26 ft. lbs. (35 Nm).
- O$_2$S sensor electrical connector
- Negative battery cable

4. Start the engine and check for exhaust leaks.

Camshaft and Valve Lifters

REMOVAL & INSTALLATION

2.2L Engine

1. Before servicing the vehicle, refer to the precautions in the beginning of this section.
2. Drain the engine oil and coolant.
3. Remove or disconnect the following:

- Engine
- Accessory drive belt
- Alternator
- Power steering pump
- Drive belt tensioner
- Water pump pulley
- Crankshaft pulley
- Rocker arm cover
- Pushrods, by pivoting the rocker arms to the sides, keeping them in order
- Cylinder head
- Valve lifters, keeping them in order
- Front cover
- Camshaft timing sprocket
- Timing chain
- Timing chain tensioner
- Camshaft thrust plate
- Camshaft Position (CMP) sensor
- Camshaft from the block, being sure the lobes do not contact the bearings

To install:

✳✳ WARNING

The camshaft lobes and journals must be adequately lubricated or engine damage could occur upon start up.

4. Install or connect the following:
- Camshaft, being careful not to contact the bearings with the cam lobes
- CMP sensor
- Camshaft thrust plate. Torque the bolts to 106 inch lbs. (12 Nm).
- Timing chain and sprocket. Torque the bolts to 96 ft. lbs. (130 Nm).
- Timing chain tensioner. Torque the bolts to 18 ft. lbs. (24 Nm).
- Front cover. Torque the bolts to 97 inch lbs. (11 Nm).
- Valve lifters
- Cylinder head. Torque the long bolts to 46 ft. lbs. (63 Nm) and the short bolts to 43 ft. lbs. (58 Nm), plus an additional 90 degree turn on all bolts.
- Pushrods
- Rocker arms

5. Adjust the valve lash after installing the engine.

6. Install or connect the following:
- Rocker arm cover. Torque the bolts to 89 inch lbs. (10 Nm).
- Crankshaft pulley. Torque the bolts to 37 ft. lbs. (50 Nm).
- Water pump pulley. Torque the bolts to 22 ft. lbs. (30 Nm).
- Drive belt tensioner. Torque the bolts to 37 ft. lbs. (50 Nm).
- Power steering pump. Torque the bolts to 22 ft. lbs. (30 Nm).
- Alternator. Torque the bolts to 37 ft. lbs. (50 Nm).
- Accessory drive belt
- Engine

7. Replace the oil filter and refill engine with new oil.

8. Refill the cooling system.

2.4L Engine

INTAKE CAMSHAFT

1. Before servicing the vehicle, refer to the precautions in the beginning of this section.
2. Relieve the fuel system pressure.
3. Remove or disconnect the following:
- Ignition coil and module assembly
- Camshaft Position (CMP) sensor electrical connector

- Power steering pump. DO NOT disconnect the lines.
- Fuel pressure regulator vacuum line
- Fuel rail
- Timing chain housing, DO NOT remove from the engine
- Intake camshaft housing cover
- Intake camshaft housing, by reversing the torque sequence

➡**Leave 2 of the bolts loosely in place to hold the camshaft housing while separating the camshaft cover from the housing.**

- Intake camshaft oil seal from the camshaft
- Camshaft

➡**The seal must be replaced any time the housing and cover are separated.**

- Camshaft housing from the cylinder head

To install:

4. Install or connect the following:
- Lifters into their bores

✳✳ WARNING

The camshaft lobes and journals must be adequately lubricated or engine damage could occur upon start up.

- Camshaft
- Camshaft housing with a new gasket
- New intake camshaft seal lubricated with engine oil

➡**The timing chain sprocket dowel pin should be straight up and align with the centerline of the lifter bores.**

- Camshaft housing cover with a new gasket. Torque the bolts, in sequence, to 11 ft. lbs. (15 Nm) plus an additional 90 degree turn (except for the 2 rear fuel pipe-to-camshaft housing bolts). Torque the 2 rear bolts to 11 ft. lbs. (15 Nm), plus an additional 30 degree turn.
- Timing chain housing and the timing chain
- Fuel rail with new injector o-rings
- Vacuum line to the fuel pressure regulator
- Fuel injector harness connectors
- Power steering pump. Torque the bolts to 22 ft. lbs. (30 Nm).
- Camshaft position sensor
- Ignition coil/module assembly.

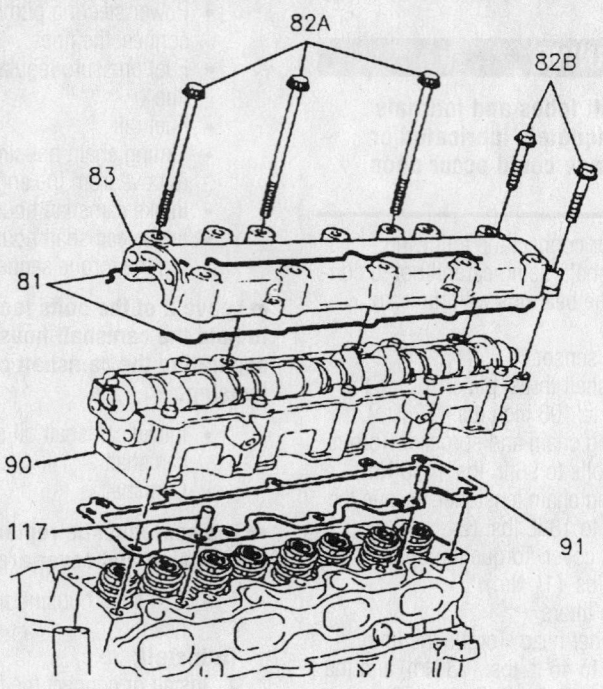

81   SEALS – CAMSHAFT HOUSING TO CAMSHAFT

82A BOLT – CAMSHAFT HOUSING TO CYLINDER
      HEAD – 15 N·m (11 LBS. FT.) PLUS TURN 90°

82B BOLT – CAMSHAFT HOUSING COVER TO
      CAMSHAFT HOUSING – 15 N·m (11 LBS. FT.)
      PLUS TURN 30°

83   COVER – CAMSHAFT

90   CAMSHAFT HOUSING (INTAKE SHOWN)

91   GASKET – CAMSHAFT HOUSING TO CYLINDER
      HEAD

117 DOWEL PIN (2)

7922YG31

Exploded view of the camshaft housing, cover and gaskets—2.4L engine

Torque the bolts to 11 ft. lbs. (15 Nm) plus 30 degrees.
- Negative battery cable

5. Start the engine and check for leaks.

EXHAUST CAMSHAFT

1. Before servicing the vehicle, refer to the precautions in the beginning of this section.
2. Relieve the fuel system pressure.
3. Remove or disconnect the following:
- Ignition coil/module assembly
- Oil pressure switch electrical connector
- Transaxle fluid level indicator tube assembly, (if equipped) from exhaust camshaft cover
- Timing chain housing; do not remove from the engine
- Exhaust camshaft cover
- Exhaust camshaft housing by reversing of the torque sequence

4. Press the cover off the housing by threading 4 housing-to-head bolts into the camshaft housing cover tapped holes. Tighten the bolts evenly so the cover does not bind on the dowel pins.
5. Remove the camshaft housing cover.
6. Loosely install a camshaft housing-to-cylinder head bolt to retain the housing during camshaft and lifter removal.

➡ **Note the position of the chain sprocket dowel pin for reassembly.**

7. Remove or disconnect the following:
- Lifters
- Camshaft
- Camshaft housing from the cylinder head

To install:

8. Install or connect the following:
- Camshaft housing with a new gas-

ket on the cylinder head with 1 bolt loosely to hold it in place
- Lifters

- Camshaft

➡ The timing chain sprocket dowel pin should be straight up and align with the centerline of the lifter bores.

➡ Apply thread locking compound to the camshaft housing cover bolt threads.

- Camshaft housing cover with a new gasket. Torque the bolts, in sequence, to 11 ft. lbs. (15 Nm) plus an additional 90 degree turn (except for the 2 rear fuel pipe-to-camshaft housing bolts). Torque the 2 rear bolts to 11 ft. lbs. (15 Nm), plus an additional 30 degree turn.
- Timing chain housing
- Transaxle fluid level indicator tube assembly to the exhaust camshaft cover
- Oil pressure switch electrical connector
- Ignition coil/module assembly. Torque the bolts to 11 ft. lbs. (15 Nm) plus an additional 30 degrees.
- Negative battery cable

9. Start the engine and check for leaks.

Valve Lash

ADJUSTMENT

All of the engines are equipped with hydraulic valve lifters. There is no adjustment.

Starter Motor

REMOVAL & INSTALLATION

1. Before servicing the vehicle, refer to the precautions in the beginning of this section.
2. Remove or disconnect the following:
- Negative battery cable
- Air inlet duct
- Top starter bolt
- Lower starter bolt
- Starter electrical connectors
- Starter motor

Exploded view of the starter motor—2.4L engine

To install:

3. Install or connect the following:
- Starter motor. Torque the bolts to 66 ft. lbs. (90 Nm) (for 2.4L) or 37 ft. lbs. (50 Nm) (for the 2.2 L)
- Starter electrical connectors
- Air inlet duct to the throttle body
- Negative battery cable

Oil Pan

REMOVAL & INSTALLATION

2.2L Engine

1. Before servicing the vehicle, refer to the precautions in the beginning of this section.
2. Drain the engine oil.
3. Remove or disconnect the following:
- Negative battery cable

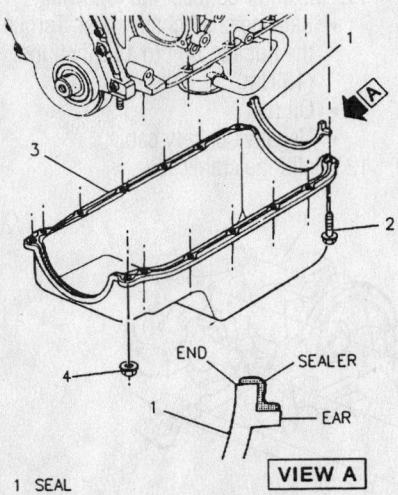

1 SEAL
2 BOLT, 10 N•m (89 LB. IN.)
3 OIL PAN
4 NUT, OIL PAN 10 N•m (89 LB. IN.)

Exploded view of the oil pan mounting and related components—2.2L engine

- Right front wheel
- Right inner fender splash shield
- Starter motor and bracket
- Engine mount strut bracket
- Oil pan

To install:

4. Place a 2mm bead of RTV sealer to the oil pan sealing surface except at the rear seal mounting surface. Using a new oil pan rear seal, apply a thin coat of RTV sealer on the end down to the ears.
5. Install or connect the following:
- Oil pan with a new gasket. Torque the nuts and bolts to 89 inch lbs. (10 Nm).
- Engine mount strut bracket. Torque the bolts to 49 ft. lbs. (66 Nm).
- Starter motor and bracket. Torque the bolts to 37 ft. lbs. (50 Nm).
- Right fender splash shield
- Right front wheel. Torque the nuts to 100 ft. lbs. (140 Nm).
- Negative battery cable
6. Refill the crankcase.
7. Start the vehicle and verify no leaks.

2.4L Engine

1. Before servicing the vehicle, refer to the precautions in the beginning of this section.
2. Drain the engine oil.
3. Drain the cooling system.
4. Remove or disconnect the following:
- Negative battery cable
- Flywheel/converter cover
- Right wheel
- Right wheel well splash shield
- Accessory drive belt
- Air conditioning compressor lower bolts

- Transmission-to-engine brace
- Engine mount strut bracket
- Radiator outlet pipe bolts
- Radiator outlet pipe from the oil pan
- Oil pan to the flywheel cover bolt and nut
- Flywheel cover stud for clearance
- Radiator outlet pipe from the lower radiator hose and oil pan
- Oil level sensor connector
- Oil pan

To install:

5. Inspect the oil pan gasket; it is reusable if not damaged.
6. Install or connect the following:
- Oil pan with the gasket. Torque the M8 bolts to 18 ft. lbs. (24 Nm) and the M6 bolts to 106 inch lbs. (12 Nm).
- Oil pan to the transmission nut
- Oil level sensor connector
- Radiator outlet pipe to the lower radiator hose and oil pan
- Exhaust manifold brace
- Radiator outlet pipe. Torque the bolts to 124 inch lbs. (14 Nm).
- Engine mount strut bracket. Torque the bolts to 55 ft. lbs. (75 Nm).
- Transmission to the engine brace
- Air conditioning compressor lower bolts. Torque the bolts to 37 ft. lbs. (50 Nm).
- Accessory drive belt
- Right splash shield
- Right front wheel
- Flywheel/converter cover
- Negative battery cable
7. Refill the crankcase.
8. Refill the cooling system.
9. Start the vehicle and verify no leaks.

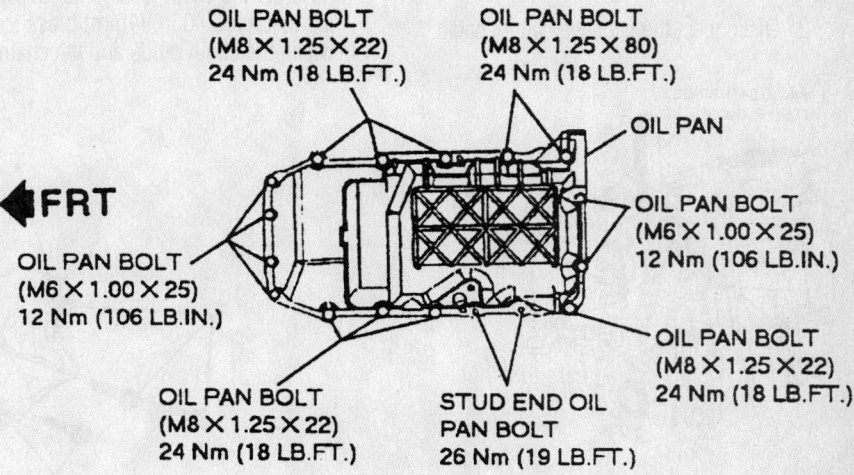

Oil pan fastener torque specifications—2.4L engine

Oil Pump

REMOVAL & INSTALLATION

2.2L Engine

1. Before servicing the vehicle, refer to the precautions in the beginning of this section.
2. Drain the crankcase.
3. Remove or disconnect the following:
   - Negative battery cable
   - Oil pan
   - Oil pump and extension shaft

To install:

✳✳ WARNING

A plastic extension shaft retainer connects the oil pump driveshaft to the oil pump. Heat the extension shaft retainer in hot water prior to assembly. Be sure the retainer does not crack upon installation.

4. Fill the oil pump cavities with petroleum jelly before installing the gears into the pump body.
5. Install or connect the following:
   - Extension shaft and oil pump. Torque the oil pump-to-bearing cap bolt to 32 ft. lbs. (43 Nm) and the upper oil pump drive bolt to 18 ft. lbs. (25 Nm).
   - Oil pan
   - Negative battery cable
6. Refill the crankcase.
7. Start the engine and check oil pressure and check for leaks.

2.4L Engine

1. Before servicing the vehicle, refer to the precautions in the beginning of this section.
2. Disconnect the negative battery cable.

3. Install an engine support fixture.
4. Drain the crankcase.
5. Remove or disconnect the following:
   - Oil pan
   - Balance shaft chain cover and tensioner
   - Oil pump
6. Disassemble the oil pump as follows:
   a. Remove the oil pump cover.
   b. Remove the pump gear.
   c. Remove the sub-assembly from the balance shaft assembly
   d. Remove the gerotor from the oil pump housing.
   e. Remove the oil pump from the balance shaft housing.
   f. Remove the pressure relief valve.

To install:

7. Lubricate the gears with clean engine oil.
8. Assemble the oil pump as follows:
   a. Place the gerotor gear into the housing.
   b. Fill the oil pump cavities with petroleum jelly.
   c. Install the pressure relief valve using a ⁹⁄₁₆ in. deep well socket to seat the valve.
   d. Connect the pump housing to the balance shaft assembly.
   e. Install the pump cover.
9. Install or connect the following:
   - Oil pump. Torque the bolts to 40 ft. lbs. (54 Nm).
   - Balance shaft chain tensioner and chain

➡**A brass feeler gauge must be used to ensure that correct measurements are obtained. If a steel gauge is used, it will not bend to conform to the guide and will allow for incorrect measurements.**

10. Adjust the chain tension as follows:
    a. Insert a 0.40 in. (1mm) brass feeler between the chain guide and the chain.

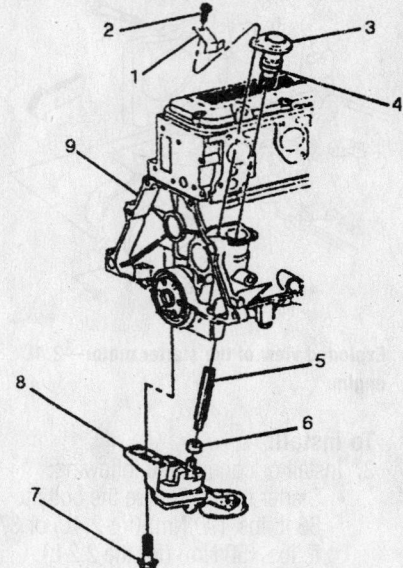

1 Bracket
2 Bolt
3 Oil pump drive assembly
4 O-ring
5 Shaft
6 Retainer: Heat and water soak prior to installation
7 Bolt
8 Oil pump
9 Cylinder block

7922YG13

Exploded view of the oil pump to engine mounting block—2.2L engine

b. Press the guide against the chain using about 3 pounds of force.
c. Torque the chain tensioner fastener to 115 inch lbs. (13 Nm).
11. Install or connect the following:
   - Balance shaft chain cover. Torque the nut and bolt to 115 inch lbs. (13 Nm).
   - Oil pan
   - Negative battery cable.
12. Refill the crankcase.

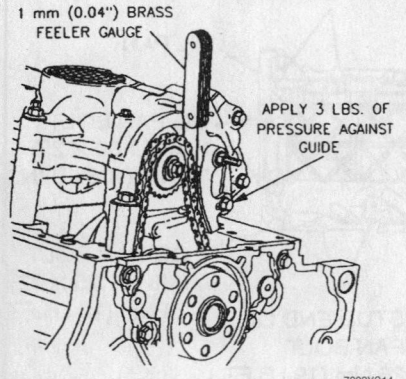

1 mm (0.04") BRASS FEELER GAUGE

APPLY 3 LBS. OF PRESSURE AGAINST GUIDE

7922YG14

Using a feeler gauge to check the chain tension—2.4L engine

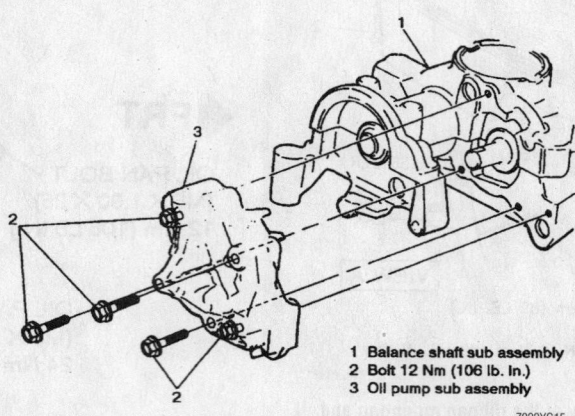

1 Balance shaft sub assembly
2 Bolt 12 Nm (106 lb. ln.)
3 Oil pump sub assembly

7922YG15

Exploded view of the oil pump assembly mounting—2.4L engine

13. Start the vehicle and verify oil pressure and no leaks.

Rear Main Seal

REMOVAL & INSTALLATION

2.2L Engine

1. Before servicing the vehicle, refer to the precautions in the beginning of this section.
2. Remove or disconnect the following:
   - Negative battery cable
   - Transaxle
   - Clutch/pressure plate assembly, if equipped with a manual transmission
   - Flywheel
   - Rear main bearing seal by prying it from the engine

➡ **Be careful not to damage or scratch the seal mounting surfaces.**

To install:

3. Lubricate the new rear main bearing seal with engine oil.
4. Install or connect the following:
   - New rear main bearing seal using seal installer J-34686 until its flush with the block
   - Flywheel
   - Clutch/pressure plate assembly, if equipped with a manual transmission
   - Transaxle
   - Negative battery cable
5. Start the engine and check for leaks.

2.4L Engine

1. Before servicing the vehicle, refer to the precautions in the beginning of this section.
2. Remove or disconnect the following:
   - Negative battery cable
   - Transaxle
   - Clutch/pressure plate assembly, if equipped with a manual transmission
   - Flywheel
   - Seal housing and discard the gasket
   - Oil seal from the transaxle side of the seal housing

To install:

3. Clean and inspect the gasket mounting surfaces.
4. If necessary, add silicone sealer along the oil pan-to-cylinder block mating surface.
5. Lubricate the new rear main bearing seal with engine oil.
6. Install or connect the following:
   - New rear main bearing seal into the seal housing using installer tool J-

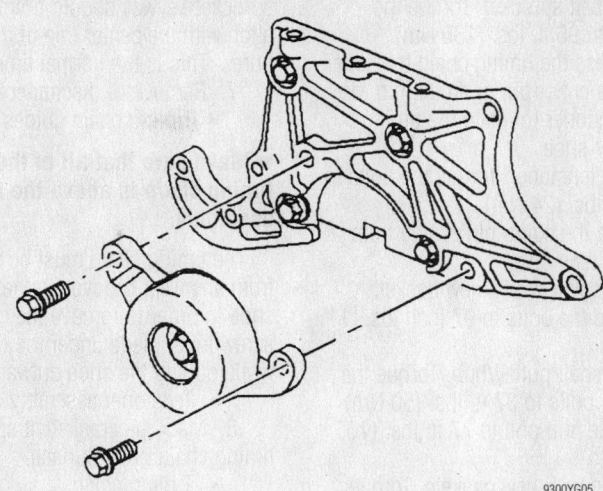

Exploded view of the accessory drive belt tensioner mounting—2.2L engine

9300YG05

36005 until its flush with the housing.
   - Oil seal housing with a new gasket. Torque the bolts to 106 inch lbs. (12 Nm).
   - Flywheel, using new bolts. Torque the bolts to 22 ft. lbs. (30 Nm) plus an additional 45 degree turn.
   - Clutch/pressure plate assembly, if equipped with a manual transmission
   - Transaxle
   - Negative battery cable
7. Start the engine and check for leaks.

Timing Chain, Sprockets, Front Cover and Seal

REMOVAL & INSTALLATION

2.2L Engine

1. Before servicing the vehicle, refer to the precautions in the beginning of this section.
2. Remove or disconnect the following:
   - Negative battery cable
   - Accessory drive belt
   - Drive belt tensioner
3. Install an engine support fixture.
4. Remove or disconnect the following:
   - Engine mount assembly
   - Alternator
   - Power steering pump and move it aside with the lines attached
   - Oil pan
   - Crankshaft pulley/hub
   - Front cover
5. Position the No. 1 piston at Top Dead Center (TDC) of the compression

stroke so the camshaft and crankshaft sprockets timing marks are aligned.
6. Remove or disconnect the following:
   - Timing chain tensioner
   - Camshaft sprocket and chain as an assembly
   - Crankshaft sprocket, using a puller

To install:

7. Lubricate the timing chain with clean engine oil.
8. Install the crankshaft sprocket, press it onto the crankshaft.
9. Install the timing chain over the camshaft sprocket, then around the crankshaft sprocket. Be sure the marks on the 2 sprockets are in alignment. Lubricate the thrust surface with Molykote® or equivalent.
10. Install or connect the following:

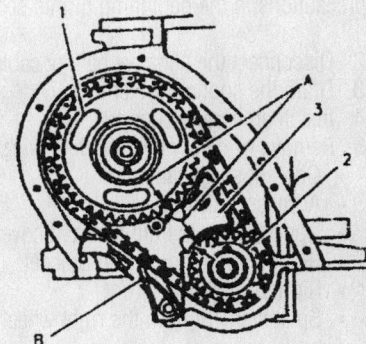

1. Camshaft sprocket
2. Crankshaft sprocket
3. Timing chain tensioner
A. Line up timing marks on sprockets with tabs on timing chain tensioner
B. Remove pin after timing chain is installed

7922YG16

Align the sprocket timing marks with the alignment tabs on the tensioner during timing chain installation—2.2L engine

- Camshaft sprocket. Torque the bolts to 96 ft. lbs. (130 Nm).

11. Compress the timing chain tensioner spring. Insert a cotter pin or a nail into the hole in the tensioner to retain the timing chain tensioner shoe.

- Chain tensioner. Torque the bolts to 18 ft. lbs. (24 Nm).

12. Remove the cotter pin or nail from the hole in the tensioner.

- Front cover with a new gasket. Torque the bolts to 97 inch lbs. (11 Nm).
- Crankshaft pulley/hub. Torque the pulley bolts to 37 ft. lbs. (50 Nm) and the hub bolt to 77 ft. lbs. (105 Nm).
- Oil pan with new gaskets. Torque the nuts and bolts to 89 inch lbs. (10 Nm).
- Power steering pump. Torque the bolts to 22 ft. lbs. (30 Nm).
- Alternator. Torque the bolts to 37 ft. lbs. (50 Nm).
- Engine mount assembly. Torque the bolts to 55 ft. lbs. (75 Nm).

13. Remove the engine support fixture.

14. Install or connect the following:

- Accessory drive belt tensioner. Torque the bolts to 37 ft. lbs. (50 Nm).
- Accessory drive belt
- Negative battery cable

2.4L Engine

➡ **It is recommended that the entire procedure be reviewed before attempting to service the timing chain.**

1. Before servicing the vehicle, refer to the precautions in the beginning of this section.

2. Disconnect the negative battery cable.

3. Drain the cooling system.

4. Install an engine support.

5. Remove or disconnect the following:

- Coolant surge tank
- Accessory drive belt
- Upper fasteners from the front cover
- Right engine mount and bracket
- Right front wheel
- Splash shield from the right wheel well
- Crankshaft balancer
- Front cover

6. Rotate the crankshaft clockwise, as viewed from front of engine (normal rotation), until the camshaft sprocket's timing dowel pin holes align with the holes in the timing chain housing. The mark on the crankshaft sprocket should align with the mark on the cylinder block. The crankshaft sprocket keyway should point upwards and align with the center line of the cylinder bores. This is the normal timed position.

7. Remove or disconnect the following:

- Timing chain guides

➡ **Make sure that all of the slack in the timing chain is above the tensioner assembly.**

The timing chain must be disengaged from any wear grooves in the tensioner shoe in order to remove the shoe. Slide a screwdriver blade under the timing chain while pulling the shoe outward.

- Tensioner assembly

8. Mark the crankshaft sprocket and the timing chain outer surface.

- Timing chain

To install:

9. Install the intake camshaft sprocket. Torque the bolt to 52 ft. lbs. (70 Nm).

➡ **Install the Special tool J 36008-A through the holes in the camshaft sprockets and into the holes in the timing chain housing. This positions the camshafts for correct timing.**

10. If the camshafts are out of position and must be rotated more than ⅛ turn in order to install the alignment dowel pins, perform the following:

a. Rotate the crankshaft 90 degrees clockwise off Top Dead Center (TDC) in order to give the valves adequate clearance to open.

b. Once the camshafts are in position and the dowels installed, rotate the crankshaft counterclockwise back to TDC.

✳✳ WARNING

Do not rotate the crankshaft clockwise to TDC or valve and piston damage may occur.

11. Install the timing chain over the exhaust camshaft sprocket, around the coolant pump sprocket and around the crankshaft sprocket.

12. Remove the alignment dowel pin from the intake camshaft. Using tool J 39579, rotate the intake camshaft sprocket counterclockwise enough to slide the timing chain over the intake camshaft sprocket. Release the camshaft sprocket wrench. The length of chain between the 2 camshaft sprockets will tighten. If properly timed, the intake camshaft alignment dowel pin should slide in easily. If the dowel pin does not fully index, the camshafts are not timed correctly and the procedure must be repeated.

13. Leave the alignment dowel pins installed.

14. With slack removed from the chain between the intake camshaft sprocket and the crankshaft sprocket, the crankshaft keyway and the cylinder block mark should be aligned. If not aligned, move the chain 1 tooth forward or rearward. Remove the slack and recheck marks.

15. Reload timing chain tensioner assembly to its **0** position as follows:

a. Insert the tensioner plunger assembly into the tensioner housing.

b. With the tensioner plunger fully extended, turn the complete assembly upside down on a flat surface.

c. Press the bottom of the tensioner housing to compress the plunger into the housing until it is seated.

d. Make sure that the plunger does not extend out of the tensioner housing more than 0.07 in. (1.7mm).

e. Loosely install the tensioner assembly to the timing chain housing.

16. Install or connect the following:

- Tensioner shoe on the stud.
- Tensioner assembly by applying hand pressure on the timing chain tensioner shoe until the locking tab seats in the stud groove. Tighten the bolts to 89 inch lbs. (10 Nm).

✳✳ WARNING

If the timing chain tensioner plunger is not released from the installation position, engine damage will occur upon start up.

17. Release the tensioner plunger by firmly pressing a flat blade tool against the plunger face.

18. Remove tool J 36008-A from the camshaft sprockets.

a. Rotate crankshaft clockwise 2 full rotations. Align crankshaft keyway with mark on cylinder block and reinstall alignment dowel pins. Alignment dowel pins will slide in easily if engine is timed correctly.

19. Install or connect the following:

- Timing chain guides
- New seal lubricated with engine oil
- Front cover using new gaskets. Tighten the nuts and bolts to 115 inch lbs. (13 Nm).
- Crankshaft balancer. Torque the bolt to 129 ft. lbs. (175 Nm) plus a 90 degree turn.
- Right front wheel well splash shield
- Wheel
- Right engine mount bracket. Torque the bolts to 44 ft. lbs. (60 Nm) plus an additional 60 degree turn.
- Right engine mount. Torque the bolts to 55 ft. lbs. (75 Nm).

20. Remove the engine support.
21. Install or connect the following:
    • Accessory drive belt
    • Coolant surge tank
    • Negative battery cable
22. Refill the cooling system and check for leaks.

Piston and Ring

POSITIONING

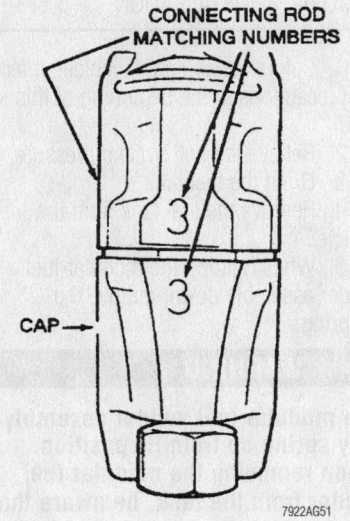

Connecting rod and cap installation. Be sure to matchmark the cap and rod prior to disassembly—2.2L and 2.4L engines

7922AG51

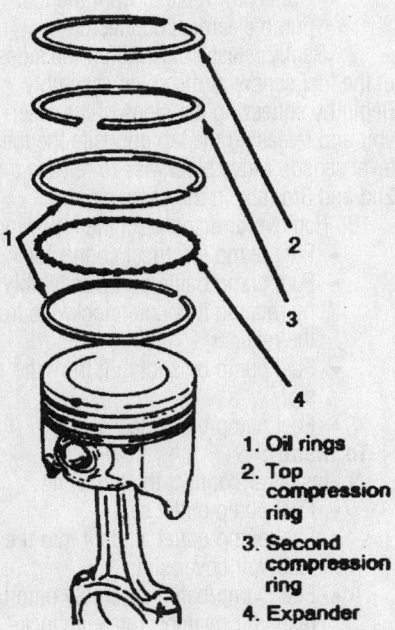

1. Oil rings
2. Top compression ring
3. Second compression ring
4. Expander

7922AG48

Piston ring positioning—2.2L engine

ENGINE LEFT ENGINE FRONT ENGINE RIGHT

A. OIL RING SPACER GAP (TANG IN HOLE OR SLOT WITH ARC)
B. OIL RING RAIL GAPS
C. 2ND COMPRESSION RING GAP
D. TOP COMPRESSION RING GAP

7922AG46

Piston ring end-gap spacing—2.2L engine

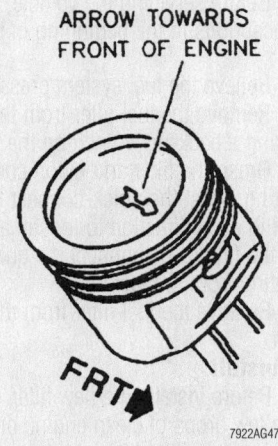

ARROW TOWARDS FRONT OF ENGINE

FRT

7922AG47

Piston positioning. Often the arrow is replaced by a notch, which also must face toward the front of the engine—2.2L engine

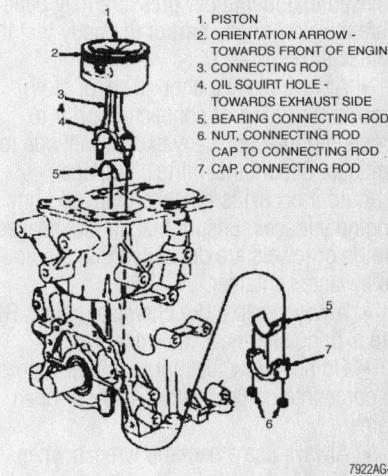

1. PISTON
2. ORIENTATION ARROW - TOWARDS FRONT OF ENGINE
3. CONNECTING ROD
4. OIL SQUIRT HOLE - TOWARDS EXHAUST SIDE
5. BEARING CONNECTING ROD
6. NUT, CONNECTING ROD CAP TO CONNECTING ROD
7. CAP, CONNECTING ROD

7922AG49

Piston and connecting rod assembly positioning—2.4L engine

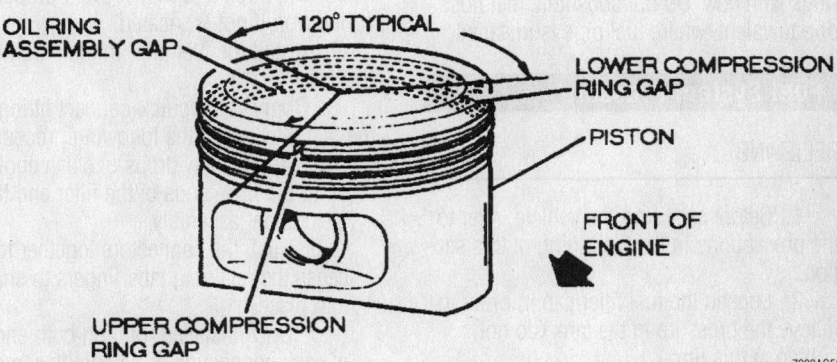

OIL RING ASSEMBLY GAP
120° TYPICAL
LOWER COMPRESSION RING GAP
PISTON
FRONT OF ENGINE
UPPER COMPRESSION RING GAP

7922AG50

Piston ring end-gap spacing—2.4L engine

FUEL SYSTEM

Fuel System Service Precautions

Safety is the most important factor when performing not only fuel system maintenance but any type of maintenance. Failure to conduct maintenance and repairs in a safe manner may result in serious personal injury or death. Maintenance and testing of the vehicle's fuel system components can be accomplished safely and effectively by adhering to the following rules and guidelines.

• To avoid the possibility of fire and personal injury, always disconnect the negative battery cable unless the repair or test procedure requires that battery voltage be applied.

• Always relieve the fuel system pressure prior to disconnecting any fuel system component (injector, fuel rail, pressure regulator, etc.), fitting or fuel line connection. Exercise extreme caution whenever relieving fuel system pressure, to avoid exposing skin, face and eyes to fuel spray. Please be advised that fuel under pressure may penetrate the skin or any part of the body that it contacts.

• Always place a shop towel or cloth around the fitting or connection prior to loosening to absorb any excess fuel due to spillage. Ensure that all fuel spillage (should it occur) is quickly removed from engine surfaces. Ensure that all fuel soaked cloths or towels are deposited into a suitable waste container.

• Always keep a dry chemical (Class B) fire extinguisher near the work area.

• Do not allow fuel spray or fuel vapors to come into contact with a spark or open flame.

• Always use a back-up wrench when loosening and tightening fuel line connection fittings. This will prevent unnecessary stress and torsion to fuel line piping.

• Always replace worn fuel fitting O-rings with new. Do not substitute fuel hose or equivalent, where fuel pipe is installed.

Fuel System Pressure

RELIEVING

1. Before servicing the vehicle, refer to the precautions in the beginning of this section.
2. Loosen the fuel filler cap in order to relieve the pressure in the tank (do not tighten at this time).
3. Raise and safely support the vehicle.

4. Detach the fuel pump electrical connector.
5. Start and run the vehicle until it stalls, then engage the starter for an additional 3 seconds to ensure the relief of any remaining pressure.
6. Disconnect the negative battery cable.
7. Once the tests or repairs are completed, reattach the fuel pump electrical connector.
8. Connect the negative battery cable.
9. Lower the vehicle.
10. Tighten the fuel filler cap.
11. prime the fuel system by cycling the ignition switch **ON** for 2 seconds, **OFF** for 10 seconds, then **ON** again. Repeat, if necessary to build system pressure.

Fuel Filter

REMOVAL & INSTALLATION

1. Before servicing the vehicle, refer to the precautions in the beginning of this section.
2. Relieve the fuel system pressure.
3. Remove the fuel filter from the fuel line using a back-up wrench on the filter.
4. Grasp the filter and nylon connection line fitting. Twist the quick-connect fitting ¼ turn in each direction to loosen any dirt within the fitting. Disconnect the quick-connect fitting from the fuel filter.
5. Remove the fuel filter from the bracket.

To install:

6. Before installing a new filter, always apply a few drops of clean engine oil to the male tube end of the filter and to the fuel sending unit assembly connection. This will help ensure proper connection and prevent possible fuel leaks. During normal operation, the O-rings located in the female connector will swell and may prevent proper connection if not lubricated.
7. Install the fuel filter in the mounting bracket.
8. Connect the quick-connect fitting to the fuel filter using the following procedure:
   a. Apply a few drops of clean engine oil to the male ends of the filter and the fuel sender assembly.
   b. Push the connectors together to cause the retaining tabs/fingers to snap into place.
   c. Once installed, pull on both ends of each connection to be sure they are secure.

9. Install or connect the following:
   • New O-ring
   • Fuel filter. Torque the fitting to 20 ft. lbs. (27 Nm).
   • Negative battery cable
10. Pressurize the fuel system and verify no leaks.

Fuel Pump

REMOVAL & INSTALLATION

1. Before servicing the vehicle, refer to the precautions in the beginning of this section.
2. Relieve the fuel system pressure.
3. Drain the fuel tank.
4. Remove the fuel tank from the vehicle.
5. While holding the modular fuel sender assembly down, remove the snapring.

✳✳ WARNING

The modular fuel sender assembly may spring up from its position. When removing the modular fuel sender from the tank, be aware that the reservoir bucket is full of fuel. It must be tipped slightly during removal to avoid damage to the float.

6. Remove or disconnect the following:
   • External fuel strainer
   • Connector retainer from the fuel pump electrical connector
7. Gently release the tabs on the sides of the fuel sender at the cover assembly. Begin by squeezing the sides of the reservoir and releasing the tab opposite the fuel level sensor. Move clockwise to release the 2nd and 3rd tabs in the same manner.
8. Remove or disconnect the following:
   • Fuel pump electrical connector
   • Fuel pump baffle/pump assembly by rotating it counterclockwise from the retainer
   • Fuel pump by sliding it from the slot
   • Fuel pump outlet seal

To install:

9. Install or connect the following:
   • Fuel pump outlet seal
   • Fuel pump outlet, slide it into the reservoir cover slots
   • Fuel pump/baffle assembly onto the reservoir retainer; rotate it clockwise until seated
   • Lower retainer assembly partially

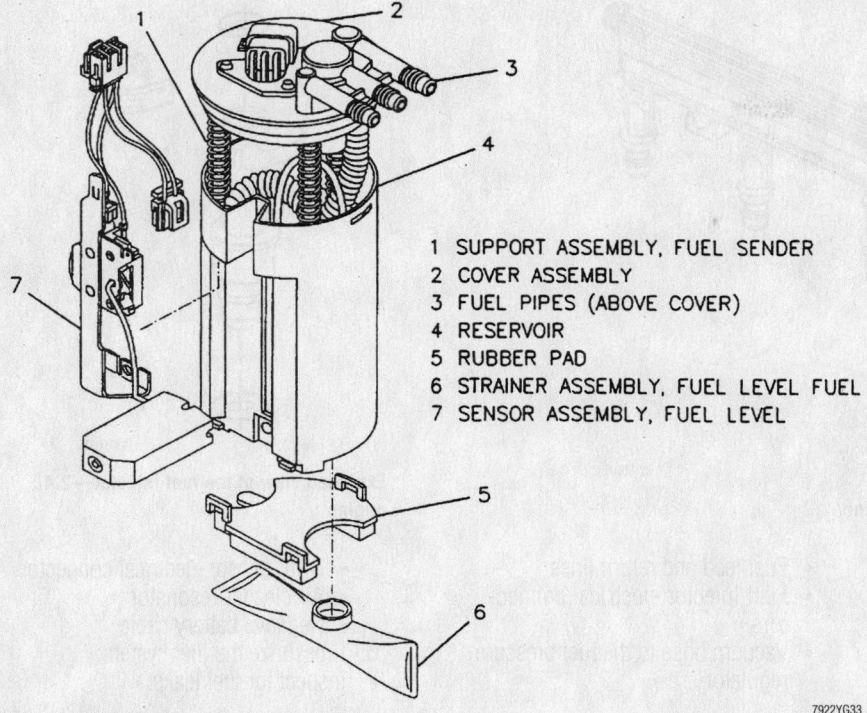

1 SUPPORT ASSEMBLY, FUEL SENDER
2 COVER ASSEMBLY
3 FUEL PIPES (ABOVE COVER)
4 RESERVOIR
5 RUBBER PAD
6 STRAINER ASSEMBLY, FUEL LEVEL FUEL
7 SENSOR ASSEMBLY, FUEL LEVEL

7922YG33

Modular fuel pump component identification

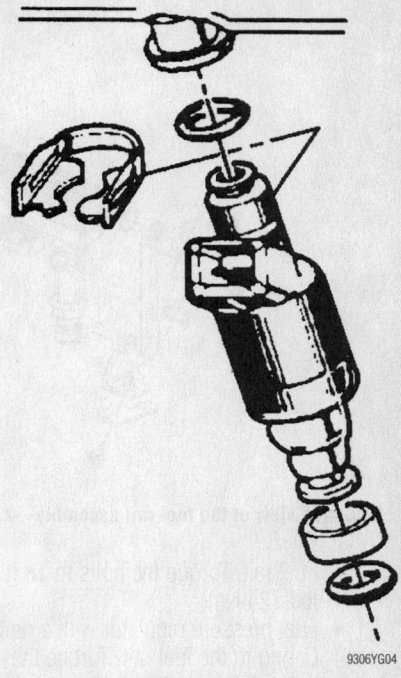

9306YG04

Exploded view of the fuel injector—2.2L engine

into the reservoir; align the 3 tabs and press the retainer onto the reservoir making sure all 3 tabs are firmly seated

➡**Gently, pull on the fuel pump reservoir to assure it is secure to the retainer. If not secure, replace the entire fuel sender.**

- Fuel pump electrical connector
- Connector retainer to the fuel sender cover
- New external fuel strainer
- Modular fuel sender
- Fuel tank in the vehicle
- Negative battery cable

10. Pressurize the fuel system and verify no leaks.

Fuel Injector

REMOVAL & INSTALLATION

2.2L Engine

1. Before servicing the vehicle, refer to the precautions in the beginning of this section.
2. Relieve the fuel system pressure.
3. Remove or disconnect the following:
- Air cleaner resonator and bracket
- Fuel injector electrical connectors
- Fuel feed inlet pipe

- Fuel return pipe from the pressure regulator
- Fuel Rail
- Fuel injector-to-fuel rail retaining clip
- Fuel injector

To install:

4. Lubricate the O-rings with engine oil prior to installation.
5. Install or connect the following:
- Fuel injectors with new O-rings
- Fuel Rail. Torque the bolts to 18 ft. lbs. (24 Nm).
- Fuel return pipe to the pressure regulator. Torque the nut to 22 ft. lbs. (30 Nm).

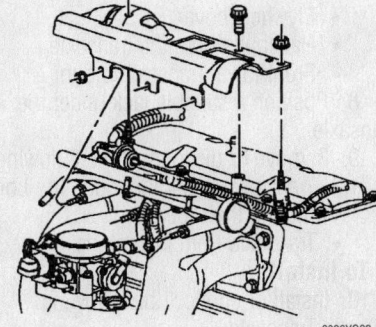

9306YG03

View of the fuel rail assembly—2.2L engine

- Fuel feed inlet pipe
- Fuel injector electrical connectors
- Air cleaner resonator and bracket
- Negative battery cable

6. Pressurize the fuel system.
7. Inspect for fuel leaks.

2.4L Engine

1. Before servicing the vehicle, refer to the precautions in the beginning of this section.
2. Relieve the fuel system pressure.
3. Remove or disconnect the following:
- Air cleaner resonator
- Fuel feed and return lines
- Camshaft Position (CMP) sensor electrical connector
- Fuel injector electrical connectors
- Vacuum hose from the fuel pressure regulator
- Fuel rail
- Fuel pressure regulator from the fuel rail, twist it back and forth
- Fuel injector-to-fuel rail retaining clip
- Fuel injector

To install:

4. Lubricate the O-rings with engine oil prior to installation.
5. Install or connect the following:
- Fuel injector(s) with new O-rings

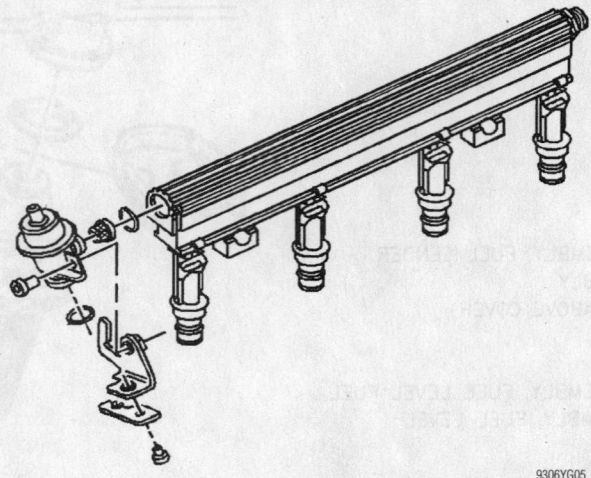

Exploded view of the fuel rail assembly—2.4L engine

9306YG05

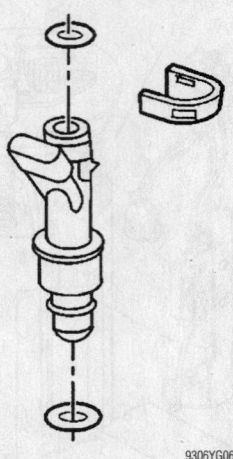

Exploded view of the fuel injector—2.4L engine

9306YG06

- Fuel rail. Torque the bolts to 18 ft. lbs. (24 Nm).
- Fuel pressure regulator with a new O-ring to the fuel rail. Torque the bolt to 53 inch lbs. (6 Nm).

- Fuel feed and return lines
- Fuel injector electrical connectors
- Vacuum hose to the fuel pressure regulator

- CMP sensor electrical connector
- Air cleaner resonator
- Negative battery cable
6. Pressurize the fuel system.
7. Inspect for fuel leaks.

DRIVE TRAIN

Transmission Assembly

REMOVAL & INSTALLATION

Manual

1. Before servicing the vehicle, refer to the precautions in the beginning of this section.
2. Install and engine support fixture.
3. Raise the engine enough to take pressure off of the transaxle mounts.
4. Install an engine support fixture and raise the engine enough to take the pressure off the transaxle mounts.
5. Remove or disconnect the following:
   - Battery
   - Air cleaner and duct assembly
   - Pushrod from the clutch pedal
   - Wiring harness from the upper transmission mount bracket
   - Upper transaxle mount-to-transaxle bolts
   - Negative battery cable from the transmission
   - Starter motor
   - Pressure line from the clutch actuator cylinder
   - Back-up light switch connector
   - Rear transmission mount bolts
   - Vehicle Speed Sensor (VSS) electrical connector
   - Upper transmission mounting bolts

- Shift cables
- Cable bracket
- Rack and pinion mounting bolts
- Front wheels
- Splash shields
- Engine strut
- Front exhaust pipe
- Both front Anti-lock Brake System (ABS) wheel speed sensor harness
- Intermediate steering shaft
- Ball joints from the steering knuckles
- Tie rod ends from the steering knuckles
- Brake lines from the support frame
- Power steering hoses from the rack and pinion
6. Support the suspension crossmember.
7. Remove or disconnect the following:
   - Left side suspension support bolts
   - Flywheel cover
   - Halfshafts from the transaxle
   - Front lower transaxle mount
8. Position a suitable jack under the transaxle.
9. Remove or disconnect the following:
   - Transaxle-to-engine mounting bolts (noting their location)
   - Transaxle from the engine

To install:
10. Install or connect the following:
    - Transaxle
    - Transaxle-to-engine mounting bolts. Torque the bolts to 55 ft. lbs. (75 Nm).

- Front transaxle mount. Torque the bolt to 44 ft. lbs. (60 Nm).
- Flywheel cover
- Halfshafts into the transaxle
- Suspension support. Torque the bolts to 81 ft. lbs. (110 Nm).
- Power steering hoses to the rack and pinion
- Brake lines to the support frame
- Ball joints to the steering knuckle. Torque the nuts to 48 ft. lbs. (65 Nm).
- Tie rod ends to the steering knuckles. Torque the nuts to 33 ft. lbs. (45 Nm).
- Intermediate steering shaft. Torque the pinch bolt to 12 ft. lbs. (17 Nm).
- ABS wheel speed sensor wiring harness connectors
- Front exhaust pipe. Torque the bolts to 26 ft. lbs. (35 Nm).
- Engine strut. Torque the bolts to 74 ft. lbs. (100 Nm) plus an additional turn of 90 degrees.
- Inner splash shield
- Front wheels
- Rack and pinion mounting bolts. Torque the bolts to 89 ft. lbs. (120 Nm).
- Starter motor. Torque the bolts to 37 ft. lbs. (50 Nm).
- Upper transaxle bolts. Torque the bolts to 71 ft. lbs. (96 Nm).

- Pressure line to clutch actuator cylinder
- VSS electrical connector
- Back-up light switch connector
- Rear transaxle mount. Torque the bolts to 55 ft. lbs. (75 Nm).
- Wiring harness to the mount bracket

11. Remove the engine support fixture.
12. Install or connect the following:
- Shift cable clamp. Torque the nut to 89 inch lbs. (10 Nm).
- Shift cables
- Negative battery cable to the transmission
- Air cleaner and duct assembly to the throttle body
- Pushrod to the clutch pedal
- Battery

13. Refill the transaxle.
14. Refill and bleed the power steering system.
15. Road test the vehicle and verify proper operation.

Automatic

1. Before servicing the vehicle, refer to the precautions in the beginning of this section.
2. Remove or disconnect the following:
- Negative battery cable
- Air intake duct
- Throttle Valve (TV) cable
- Shift cable and bracket
- Vacuum lines
- Electrical connectors
- Filler tube

3. Attach an engine support fixture.
4. Remove or disconnect the following:
- Upper engine-to-transaxle bolts
- Front wheels
- Left side splash shield
- Both front Anti-lock Brake System (ABS) wheel speed sensors and harness from left suspension support
- Both lower ball joints
- Stabilizer shaft links
- Front air deflector
- Both halfshafts
- Engine-to-transaxle brace
- Torque converter cover
- Starter motor
- Torque converter bolts
- Transaxle oil cooler lines and brace
- Ground wires from the transaxle
- Engine-to-transaxle mount bolts

5. Support the transaxle with a jack.
6. Remove or disconnect the following:
- Transaxle mount-to-body bolts
- Heater core hose brace from transaxle

- Engine-to-transmission bolts
- Transmission

To install:

7. Install or connect the following:
- Transaxle while installing right half-shaft
- Lower engine-to-transaxle bolts. Torque the bolts to 71 ft. lbs. (96 Nm).
- Transmission mount-to-body bolts. Torque the bolts to 49 ft. lbs. (66 Nm).
- Heater core hose brace
- Ground wires to the transaxle
- Oil cooler lines
- Torque converter. Torque the bolts to 46 ft. lbs. (62 Nm).
- Torque converter cover
- Starter motor. Torque the bolts to 37 ft. lbs. (50 Nm).
- Engine-to-transaxle brace. Torque the bolts to 32 ft. lbs. (43 Nm).
- Halfshafts
- Front air deflector
- Stabilizer links. Torque the bolts to 49 ft. lbs. (66 Nm).
- Lower ball joints. Torque the nuts to 45 ft. lbs. (61 Nm).
- Both ABS wheel speed sensors
- Left splash shield
- Front wheels
- Upper engine-to-transaxle bolts. Torque the bolts to 71 ft. lbs. (96 Nm).

8. Remove the engine support fixture.
9. Install or connect the following:
- Filler tube
- Electrical connectors
- Vacuum lines
- Shift cable and bracket
- TV cable
- Intake air duct
- Negative battery cable

10. Refill the transaxle, start the vehicle and verify that there are no leaks.
11. Road test the vehicle.

Clutch

REMOVAL & INSTALLATION

1. Before servicing the vehicle, refer to the precautions in the beginning of this section.
2. Remove or disconnect the following:

- Negative battery cable
- Clutch master cylinder pushrod from the clutch pedal
- Transaxle

3. If any of the parts are to be reused,

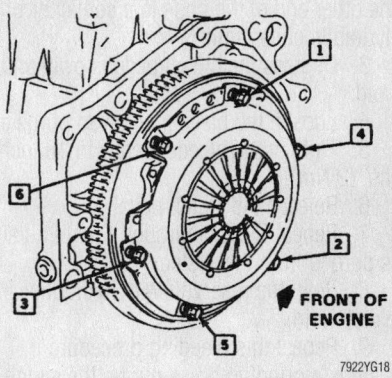

FRONT OF ENGINE

7922YG18

Clutch cover bolt tightening sequence

mark the pressure plate assembly and the flywheel so they can be assembled in the same position.

4. Loosen the attaching bolts 1 turn at a time until spring tension is relieved.
5. Support the pressure plate and remove the bolts.
6. Remove the pressure plate and clutch disc as an assembly.

➡**Do not disassemble the pressure plate assembly. Replace it if defective.**

To install:

7. Install the pressure plate and clutch disc assembly, supporting it with a clutch aligning tool.
8. Torque the pressure plate-to-flywheel bolts, gradually, in a cross pattern, as follows:
   a. Step 1: Lightly seat all bolts.
   b. Step 2: Tighten the bolts to 16 ft. lbs. (20 Nm).
   c. Step 3: Tighten the bolts an additional 45 degrees in sequence

9. Lubricate the outside groove and the inside recess of the release bearing with high temperature grease.
10. Install or connect the following:
- Release bearing
- Transaxle
- Clutch master cylinder pushrod to the clutch pedal and secure with the retaining clip
- Negative battery cable

11. Bleed clutch system as necessary and road test vehicle.

Hydraulic Clutch System

BLEEDING

1. Before servicing the vehicle, refer to the precautions in the beginning of this section.
2. Attach a hose to the bleeder screw on the clutch actuator assembly and submerge

the other end of the hose in a container of hydraulic clutch fluid.

3. Depress the clutch pedal slowly and hold.

4. Loosen the bleeder screw to purge air.

5. Tighten the bleeder screw to 18 inch lbs. (2 Nm).

6. Release the clutch pedal.

7. Repeat Steps 3 through 6 until all air is purged from the system.

8. Refill the reservoir with hydraulic clutch fluid.

9. Repeat this bleeding procedure if there is a grinding noise during the clutch spin down procedure.

Halfshaft

REMOVAL & INSTALLATION

1. Before servicing the vehicle, refer to the precautions in the beginning of this section.

2. Remove or disconnect the following:
- Negative battery cable
- Front wheel
- Hub nut and washer
- Lower ball joint
- Wheel Speed Sensor (WSS) electrical connector
- Stabilizer bar link
- Halfshaft, press it from wheel bearing/hub assembly
- Halfshaft from the transaxle

✶✶ WARNING

Do not pull the halfshaft by the CV-joint boot or on the joint itself.

To install:

3. Install the halfshaft into the transaxle (or intermediate shaft, if equipped) by plac-

ing a brass drift pin into the groove on the joint housing and tapping until seated. Be careful not to damage the axle seal or dislodge the seal garter spring when installing the axle.

➡**Be sure the halfshaft is fully engaged in the transaxle. Verify that the halfshaft is seated by grasping the inner joint housing and pulling outward. Do not pull on the shaft or the boot, but on the inner joint housing only.**

4. Install or connect the following:
- Halfshaft into the hub/bearing assembly
- Stabilizer link. Torque the bolts to 13 ft. lbs. (17 Nm).
- WSS electrical connector
- Lower ball joint to the steering knuckle. Torque the nut to 41–48 ft. lbs. (55–65 Nm).
- Washer and a new hub nut. Torque the nut to 144 ft. lbs. (200 Nm).
- Front wheel
- Negative battery cable

CV-Joints

OVERHAUL

Outer CV-Joint

1. Before servicing the vehicle, refer to the precautions in the beginning of this section.

2. Remove or disconnect the following:
- Front wheel
- Halfshaft and position it in a vise
- Large CV-joint boot clamp
- Small CV-joint boot clamp
- CV-joint boot and slide it back on the shaft

3. Perform the following:

a. Choose a reference mark on the halfshaft.

b. Measure the distance between the reference mark and the CV-joint inner race face; retain this measurement.

4. Attach CV Puller tool J-41398 to the outer race threaded area.

5. Remove or disconnect the following:
- CV-joint outer race from the halfshaft using a slide hammer puller
- CV Puller tool J-41398 and slide hammer puller from the outer race
- Retaining ring from the halfshaft
- CV-joint boot from the halfshaft

6. Disassemble the chrome alloy balls from the CV-joint cage as follows:

a. Position a brass drift against the CV-joint cage and tap it with a hammer to tilt the cage.

b. Remove the 1st chrome alloy ball from the cage.

c. Tilt the cage in the opposite direction.

d. Remove the opposite chrome alloy ball.

e. Repeat the procedure until all 6 balls are removed.

7. Disassemble the CV-joint cage and inner race as follows:

a. Pivot the cage and race 90 degrees to the center line of the outer race.

b. Align the cage windows with outer race lands.

c. Remove the cage from the outer race.

d. Rotate the inner race upward and remove it from the cage.

To install:

8. Lubricate the parts with a light coat of grease.

9. Assemble the CV-joint cage and inner race, as follows:

a. Rotate the inner race 90 degrees to the cage centerline.

b. Align the cage windows with inner race lands.

c. Insert the inner race into the cage by rotating the inner race downward.

d. Insert the cage/inner race into the outer race.

10. Assemble the chrome alloy balls into the CV-joint cage, as follows:

a. Position a brass drift against the CV-joint cage and tap it with a hammer to tilt the cage.

b. Insert the 1st chrome alloy ball into the cage.

c. Tilt the cage in the opposite direction.

d. Insert the opposite chrome alloy ball.

1 RIGHT DRIVE AXLE
2 LEFT DRIVE AXLE
3 J 28468 OR J 33008
4 J 29794
5 J2619-01

7922YG34

To prevent damaging the transaxle or halfshaft, use the tools as shown to remove the halfshafts

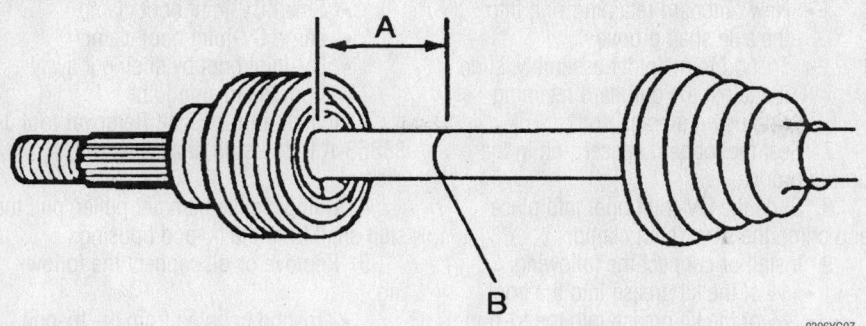

Measuring the halfshaft-to-inner race reference distance—Outer CV-joint

9306YG07

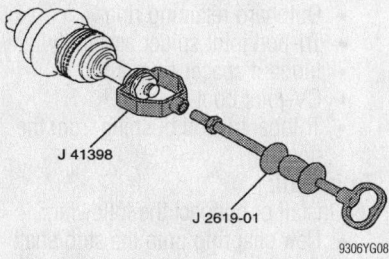

J 41398

J 2619-01

9306YG08

Removing the outer race from the half-shaft—Outer CV-joint

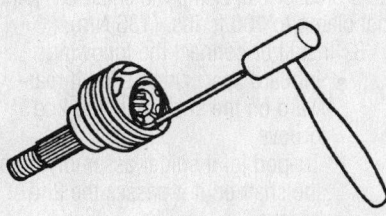

9306XG20

Tilting the cage—Outer CV-joint

e. Repeat the procedure until all 6 balls are inserted.

11. Install ½ of the kit grease into the CV-joint.

12. Install or connect the following:
- Small ring clamp on the CV boot, do not crimp at this time
- CV-joint boot and slide it up the shaft to expose the reference mark
- New retaining ring on the half-shaft
- Large ring clamp on the CV boot
- Outer race assembly by pressing it onto the halfshaft until the ring engages

❋❋ WARNING

When installing the outer race assembly onto the halfshaft, do not exceed 4,000 lbs. pressure.

13. Measure the distance between the reference mark and the CV-joint inner race

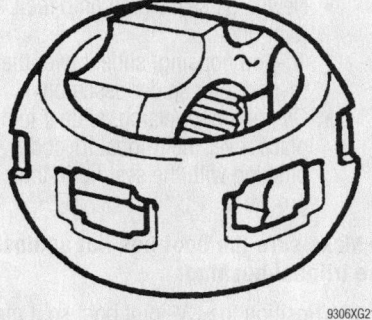

9306XG21

View the cage and inner race—Outer CV-joint

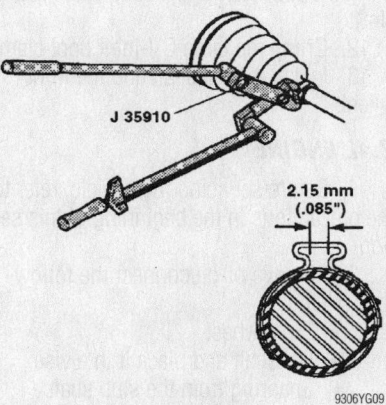

J 35910

2.15 mm (.085")

9306YG09

Crimping the small boot clamp—Outer CV-joint

face; the distance must be +/- 0.039 in. (1mm) of the original measurement. If the measurement is not correct, repress the outer race assembly and recheck it.

14. Slide the small end of the CV-joint boot/clamp into place, with the seal lip in the halfshaft groove.

➡️**Make sure the boot lies flat against the halfshaft.**

15. Using a Crimp tool, a torque wrench and a breaker bar, crimp the small CV-joint boot clamp to 100 ft. lbs. (136 Nm).

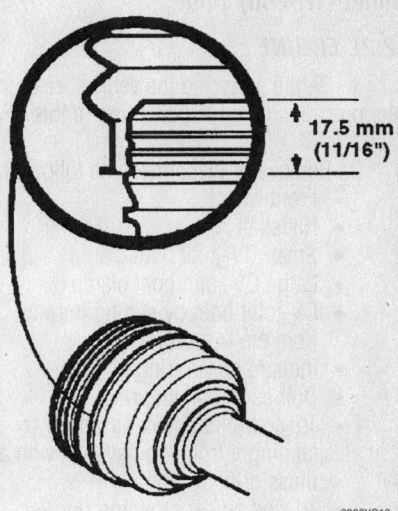

17.5 mm (11/16")

9306YG10

Positioning the CV boot onto the outer race—Outer CV-joint

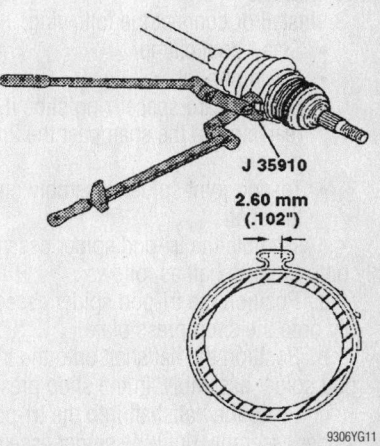

J 35910

2.60 mm (.102")

9306YG11

Crimping the large boot clamp—Outer CV-joint

16. Check the clamp gap dimension; if it is not 0.085 in. (2.15mm), continue tightening the clamp until it is.

17. Install ½ of the kit grease into the CV-joint boot.

18. Slide the large end of the CV boot/clamp into place, with the seal lip in place over the outer race.

➡️**Make sure the boot lies flat against the outer race.**

19. Using a Crimp tool, a torque wrench and a breaker bar, crimp the large CV-joint boot clamp to 130 ft. lbs. (176 Nm).

20. Check the clamp gap dimension; if it is not 0.102 in. (2.60mm), continue tightening the clamp until it is.

21. Install the halfshaft and the front wheel.

Inner (Tri-Pod) Joint

2.2L ENGINE

1. Before servicing the vehicle, refer to the precautions in the beginning of this section.

2. Remove or disconnect the following:
- Front wheel
- Halfshaft
- Small CV-joint boot clamp
- Large CV-joint boot clamp
- CV-joint boot by sliding it away from the tri-pod joint
- Inboard spacer ring
- Outboard retaining ring
- Tri-pod joint spider assembly by tapping it from the halfshaft with a brass drift
- Tri-pod spider retaining ring
- Trilobal tri-pod bushing from the housing
- CV-joint boot

To install:

3. Install or connect the following:
- Small boot clamp
- CV-joint boot
- New inboard spacer ring slide it rearward on the shaft past the 2nd groove
- Tri-pod joint spider assembly onto the shaft

4. Assemble the tri-pod spider assembly onto the halfshaft as follows:

a. Position the tri-pod spider assembly onto the shop press plate.

b. Position the halfshaft onto the tri-pod spider assembly, in the shop press.

c. Press the halfshaft into the tri-pod spider assembly until the spider assembly passes the 2nd groove.

✳✳ WARNING

When assembling the tri-pod assembly onto the halfshaft, do not exceed 4,000 lbs. pressure.

5. Remove the halfshaft from the shop press and place it in vise.

6. Install or connect the following:

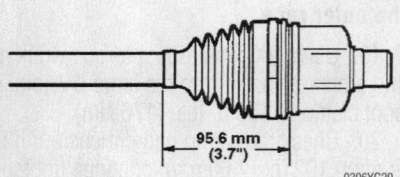

CV-joint boot measurement—Inner (tri-pod) joint—2.2L engine

95.6 mm
(3.7")

9306YG20

- New outboard retaining ring into the axle shaft groove
- Tri-pod joint spider assembly, slide it against the outboard retaining ring using a brass drift

7. Seat the inboard spacer ring in the proper groove.

8. Slide the CV-joint boot into place and crimp the small boot clamp.

9. Install or connect the following:
- ½ of the kit grease into the boot
- ½ of the kit grease into the tri-pod housing
- Trilobal tri-pod bushing flush with the tri-pod housing face
- New large seal clamp onto the CV-joint boot
- Tri-pod housing, slide it over the tri-pod joint spider assembly
- CV-joint boot/clamp, slide it into place, over the trilobal tri-pod bushing with the seal lip in the groove

➡ **Make sure the boot lies flat against the trilobal bushing.**

10. Position the CV-joint boot so it measures or 3.7 in. (95.7).

11. Using a Crimp tool, a torque wrench and a breaker bar, crimp the small CV-joint boot clamp to 100 ft. lbs. (136 Nm).

12. Crimp the large CV-joint boot clamp.

13. Install the halfshaft and the front wheel.

2.4L ENGINE

1. Before servicing the vehicle, refer to the precautions in the beginning of this section.

2. Remove or disconnect the following:
- Front wheel
- Halfshaft and place it in a vise
- Snapring from the stub shaft

- Small CV-joint boot clamp
- Large CV-joint boot clamp
- CV-joint boot by sliding it away from the tri-pod joint

3. Install a Stub Shaft Removal tool J-38868-A to the stub shaft snapring groove.

4. Using a slide hammer puller, pull the stub shaft from the tri-pod housing.

5. Remove or disconnect the following:
- Tri-pod housing from the tri-pod spider
- Inboard spacer ring slide it rearward on the shaft
- Outboard retaining ring
- Tri-pod joint spider assembly
- Inboard spacer ring
- CV-joint boot
- Trilobal tri-pod bushing from the housing

To install:

6. Install or connect the following:
- New snapring onto the stub shaft
- Small boot clamp
- CV-joint boot

7. Using a Crimp tool, a torque wrench and a breaker bar, crimp the small CV-joint boot clamp to 100 ft. lbs. (136 Nm).

8. Install or connect the following:
- Inboard spacer ring slide it rearward on the shaft, past the 2nd groove
- Tri-pod joint spider assembly onto the shaft until it passes the 2nd groove
- Outboard retaining ring into the axle shaft groove
- Tri-pod joint spider assembly, slide it against the outboard retaining ring
- Inboard spacer ring, seat it in the groove
- ½ kit grease into the boot
- ½ kit grease into the tri-pod housing

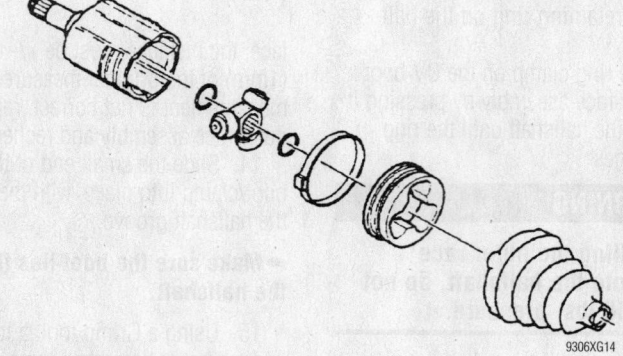

Exploded view of the inner (tri-pod) joint

9306XG14

- Trilobal tri-pod bushing flush with the tri-pod housing face
- New large seal clamp onto the CV-joint boot
- Tri-pod housing, slide it over the tri-pod joint spider assembly

- CV-joint boot/clamp, slide it into place, over the trilobal tri-pod bushing with the seal lip in the groove

➡**Make sure the boot lies flat against the trilobal bushing.**

9. Position the CV-joint boot so it measures 4.9 in. (125mm).
10. Crimp the large CV-joint boot clamp.
11. Install the halfshaft and the front wheel.

STEERING AND SUSPENSION

Air Bag

※ CAUTION

Some vehicles are equipped with an air bag system. The system must be disabled before performing service on or around system components, steering column, instrument panel components, wiring and sensors. Failure to follow safety and disabling procedures could result in accidental air bag deployment, possible personal injury and unnecessary system repairs.

PRECAUTIONS

Several precautions must be observed when handling the inflator module to avoid accidental deployment and possible personal injury.

- Never carry the inflator module by the wires or connector on the underside of the module.
- When carrying a live inflator module, hold securely with both hands, and ensure that the bag and trim cover are pointed away.
- Place the inflator module on a bench or other surface with the bag and trim cover facing up.
- With the inflator module on the bench, never place anything on or close to the module which may be thrown in the event of an accidental deployment.

DISARMING

※ CAUTION

The Supplemental Inflatable Restraint (SIR) system must be disarmed before performing many in-vehicle service procedures. Failure to do so may cause accidental deployment of the air bag, resulting in unnecessary SIR system repairs and/or personal injury.

1. Turn the steering wheel so the vehicle's wheels are pointing straight-ahead.

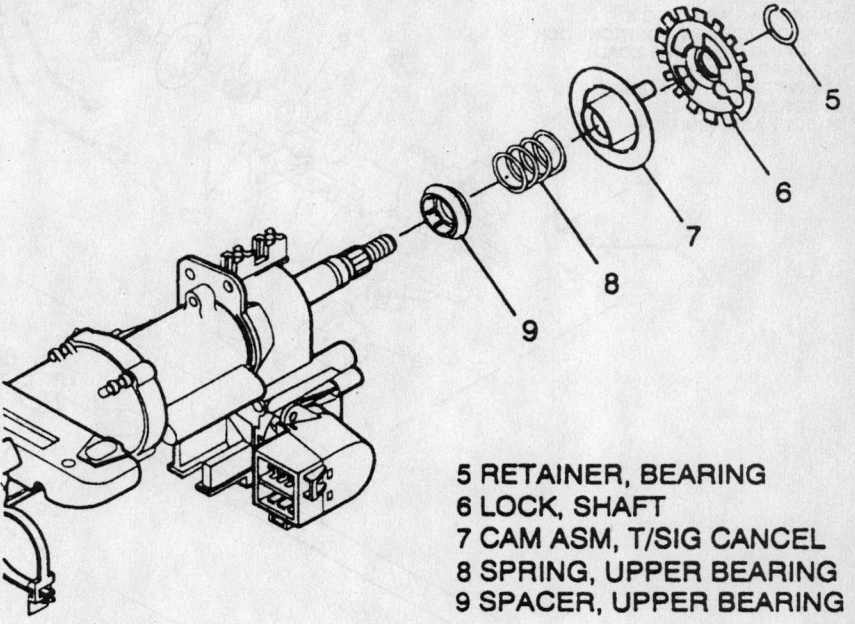

5 RETAINER, BEARING
6 LOCK, SHAFT
7 CAM ASM, T/SIG CANCEL
8 SPRING, UPPER BEARING
9 SPACER, UPPER BEARING

9300YG12

Exploded view of the upper steering column components

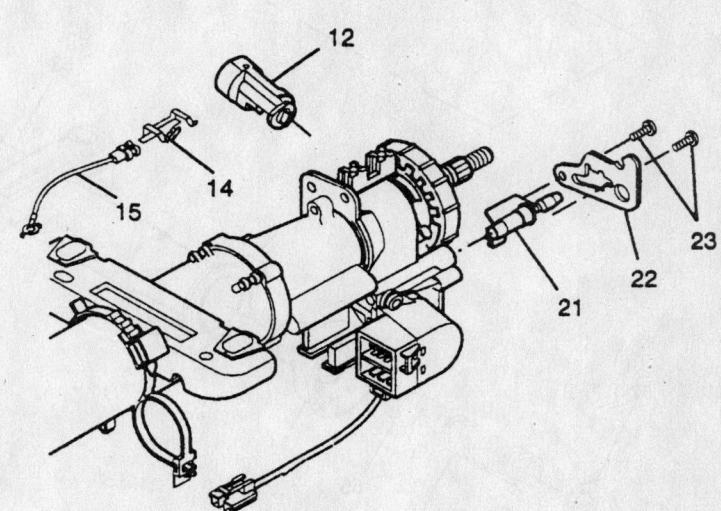

12 ACTUATOR ASM, IGNITION LOCK
14 SPRING, LOCK PRE-LOAD
15 STRAP, GROUND
21 BOLT ASM, LOCK
22 BRACKET, LOCK BOLT SUPPORT
23 SCREW, TAPPING

9300YG11

Exploded view of the ignition lock cylinder and related components

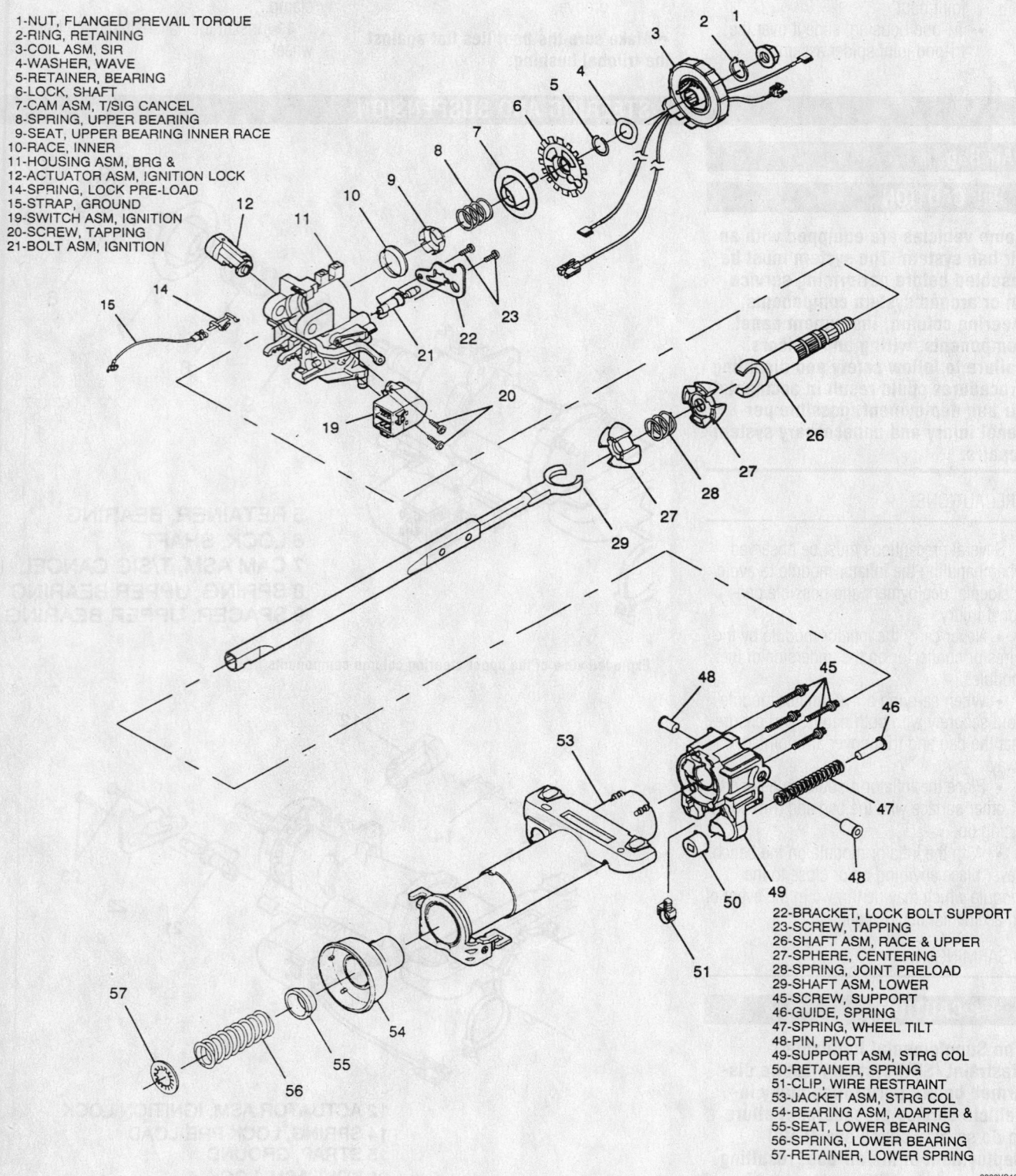

1-NUT, FLANGED PREVAIL TORQUE
2-RING, RETAINING
3-COIL ASM, SIR
4-WASHER, WAVE
5-RETAINER, BEARING
6-LOCK, SHAFT
7-CAM ASM, T/SIG CANCEL
8-SPRING, UPPER BEARING
9-SEAT, UPPER BEARING INNER RACE
10-RACE, INNER
11-HOUSING ASM, BRG &
12-ACTUATOR ASM, IGNITION LOCK
14-SPRING, LOCK PRE-LOAD
15-STRAP, GROUND
19-SWITCH ASM, IGNITION
20-SCREW, TAPPING
21-BOLT ASM, IGNITION

22-BRACKET, LOCK BOLT SUPPORT
23-SCREW, TAPPING
26-SHAFT ASM, RACE & UPPER
27-SPHERE, CENTERING
28-SPRING, JOINT PRELOAD
29-SHAFT ASM, LOWER
45-SCREW, SUPPORT
46-GUIDE, SPRING
47-SPRING, WHEEL TILT
48-PIN, PIVOT
49-SUPPORT ASM, STRG COL
50-RETAINER, SPRING
51-CLIP, WIRE RESTRAINT
53-JACKET ASM, STRG COL
54-BEARING ASM, ADAPTER &
55-SEAT, LOWER BEARING
56-SPRING, LOWER BEARING
57-RETAINER, LOWER SPRING

9300YG10

Exploded view of the steering column with tilt wheel

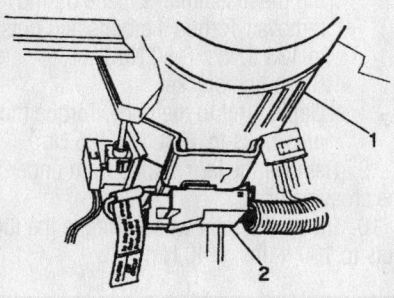

1. Steering column
2. Connector,sir(yellow)

7922YG19

Driver's side Yellow 2-way air bag connector retainer location

2. Turn the ignition switch to the **LOCK** position.
3. Remove or disconnect the following:
 • Ignition key
 • AIR BAG fuse from the instrument panel fuse block
 • Left-hand sound insulator
 • Connector retainer and yellow 2-way connector at the base of the steering column.
 • Right-hand sound insulator, if equipped with passenger side air bags
 • Connector retainer and yellow 2-way connector from the passenger inflator module pigtail, if equipped with passenger side air bags

ARMING

1. Turn the ignition switch to the **LOCK** position and remove the key.

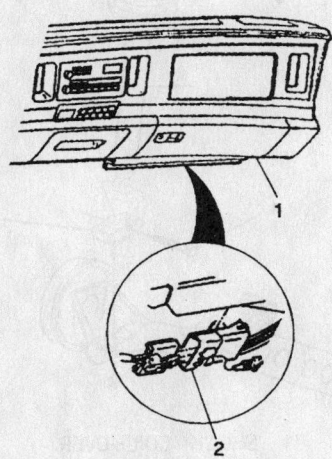

1. I/P compartment
2. Connector,sir(yellow)

7922YG20

Passenger's side Yellow 2-way air bag connector location

2. Install or connect the following:
 • Connector retainer and yellow 2-way connector to the passenger inflator module pigtail, if equipped with passenger side air bags
 • Right-hand sound insulator, if equipped with passenger side air bags
 • Yellow 2-way connector and connector retainer at the base of the steering column
 • Left-hand insulator
 • AIR BAG fuse into the instrument panel fuse block

3. Turn the ignition switch to the **RUN** position and verify that the AIR BAG warning lamp flashes 7–9 times, then the light turns **OFF**.

Rack and Pinion Steering Gear

REMOVAL & INSTALLATION

1. Before servicing the vehicle, refer to the precautions in the beginning of this section.
2. Remove or disconnect the following:
 • Negative battery cable
 • Front wheels
 • Inlet and outlet hose assemblies from the rack and pinion
 • Tie rod ends from the steering knuckles
 • Intermediate shaft
 • Brake pipes from the retainers on the front suspension support

 • Crossmember bolts, loosen them to gain additional clearance
 • Rack and pinion mounting bolts
 • Rack and pinion

To install:
3. Install the rack and pinion. Torque the bolts to 89 ft. lbs. (120 Nm).
4. Install the crossmember bolts in the following order:
 a. Step 1: Left rear outboard to 71 ft. lbs. (110 Nm).
 b. Step 2: Right rear outboard 2nd to 71 ft. lbs. (110 Nm).
 c. Step 3: Front upper bolts to 71 ft. lbs. (110 Nm).
 d. Step 4: Rear inboard bolts to 71 ft. lbs. (110 Nm).
5. Install or connect the following:
 • Intermediate shaft. Torque the pinch bolt to 30 ft. lbs. (41 Nm).
 • Brake pipes to the retainers on the front suspension crossmember
 • Tie rod ends. Torque the nuts to 44 ft. lbs. (60 Nm), use new cotter pins.
 • Inlet and outlet hoses to the rack and pinion. Torque the fittings to 20 ft. lbs. (27 Nm).
 • Wheels
 • Negative battery cable
6. Refill and bleed the power steering system.
7. Check and/or adjust the toe setting.
8. Road test vehicle and verify no leaks.

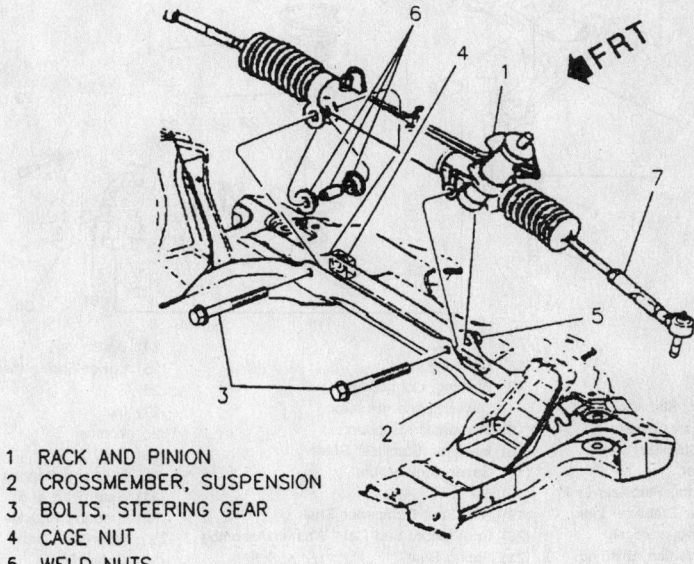

1 RACK AND PINION
2 CROSSMEMBER, SUSPENSION
3 BOLTS, STEERING GEAR
4 CAGE NUT
5 WELD NUTS
6 BUSHING AND SLEEVE
7 TIE ROD

7922YG21

Exploded view of the rack and pinion steering gear mounting

Strut

REMOVAL & INSTALLATION

Front

1. Before servicing the vehicle, refer to the precautions in the beginning of this section.

2. Remove the upper strut-to-body nuts and/or bolts.

3. Support the front crossmember with jack stands.

4. Lower the vehicle so the weight of the vehicle rests on the jack stands and NOT the control arms.

5. Before removing front suspension components, their positions should be marked so they may be assembled correctly.

✳✳ WARNING

Whenever working near the halfshaft, use care to prevent damage from

over-extension of the halfshaft joints. When either end of the shaft is disconnected, over-extension of the joint could result in separation of the internal components and possible joint failure.

6. Install a drive axle joint protective cover.

7. Remove or disconnect the following:
- Front wheel
- Brake line bracket, if necessary
- Strut from the steering knuckle. Matchmark the assembly before removal.

✳✳ WARNING

The steering knuckle MUST be supported to prevent axle joint over-extension.

To install:

8. Install or connect the following:
- Strut to the steering knuckle, align-

ing the matchmarks made during removal. Torque the bolts and nuts to 133 ft. lbs. (180 Nm).
- Brake line bracket
- Upper strut to the body, Torque the nuts/bolts to 18 ft. lbs. (25 Nm).

9. Remove the jack stands from under the crossmember.

10. Install the front wheel. Torque the lug nuts to 100 ft. lbs. (140 Nm).

Shock Absorber

REMOVAL & INSTALLATION

Rear

1. Before servicing the vehicle, refer to the precautions in the beginning of this section.

2. Open the deck lid and position the carpet out of the way.

3. Raise and safely support the rear axle with jack stands.

4. Remove or disconnect the following:
- Wheel
- Upper mounting bolts
- Lower mounting bolt
- Shock absorber

To install:

5. Install or connect the following:
- Shock
- Lower mounting bolt. Torque it to 51 ft. lbs. (70 Nm).

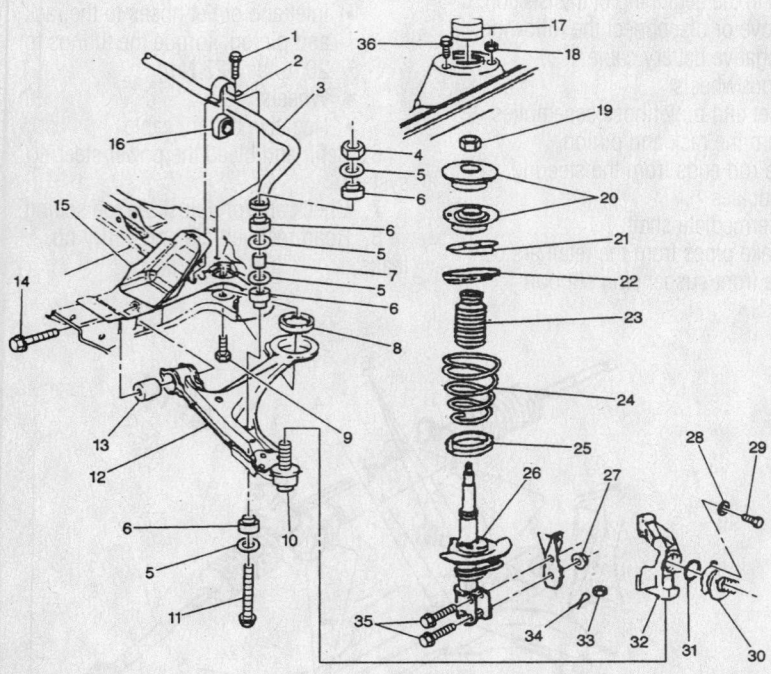

Legend

(1) Bolt	(12) Control Arm
(2) Clamp, Stabilizer Shaft	(13) Bushing, Control Arm
(3) Stabilizer Shaft	(14) Lower Spring Insulator
(4) Nut, Stabilizer Link	(15) Suspension Support
(5) Washer	(16) Insulator, Stabilizer Shaft
(6) Insulator, Stabilizer Link	(17) Cover, Strut Mount
(7) Spacer Stabilizer Link	(18) Nut
(8) Bushing, Vertical	(19) Nut, Strut Dampener Shaft
(9) Bolt, Vertical Bushing	(20) Strut Mount and Rate Washer Assembly
(10) Ball Joint	(21) Spring Seat
(11) Bolt, Stabilizer Link	(22) Upper Spring Insulator
	(23) Strut Bumper and Shield

(24) Spring	
(25) Lower Spring Insulator	
(26) Strut	
(27) Nut	
(28) Washer	
(29) Bolt	
(30) Hub and Bearing Assembly	
(31) Seal (Part of 5)	
(32) Steering Knuckle	
(33) Nut, Ball Joint	
(34) Cotter Pin	
(35) Bolt	
(36) Bolt	

7922YG22

Exploded view of the front suspension

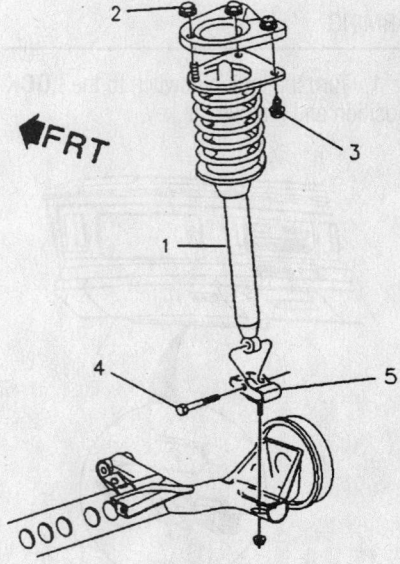

1	SHOCK, COIL-OVER
2	NUT
3	BOLT, UPPER STRUT (2)
4	BOLT, AXLE

7922YG23

Exploded view of the rear shock mounting

- Upper mounting bolts. Torque them to 18 ft. lbs. (25 Nm).
- Wheel
- Rear compartment carpet
6. Road test the vehicle.

Coil Spring

REMOVAL & INSTALLATION

Front

1. Before servicing the vehicle, refer to the precautions in the beginning of this section.
2. Remove the strut from the vehicle.
3. Mount the strut assembly into a strut compressor. Note that the strut compressor has strut mounting holes drilled for specific vehicle lines.
4. Compress the strut approximately ½ its height after initial contact with the top cap.

✳✳ WARNING

Never bottom the spring or damper rod!

5. Remove the nut from the strut damper shaft and place an alignment/guiding rod on top of the damper shaft. Use the rod to guide the damper shaft straight

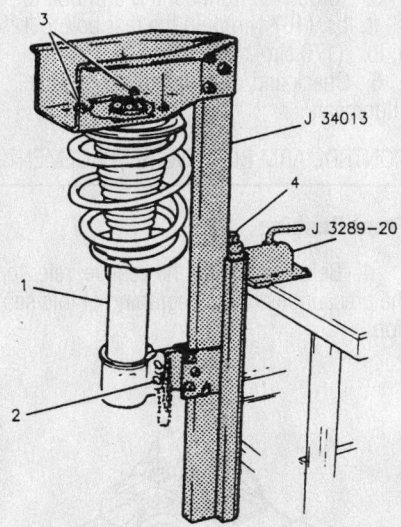

1 STRUT ASSEMBLY
2 INSTALL LOCKING PINS THROUGH STRUT ASSEMBLY
3 TIGHTEN NUTS UNTIL FLUSH WITH STRUT COMPRESSOR
4 COMPRESSOR FORCING SCREW

7922YG24

View of a typical strut assembly mounted in a compressor

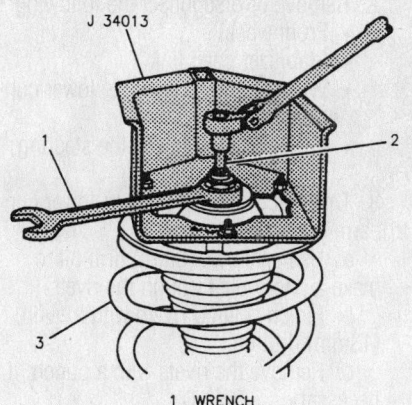

1 WRENCH
2 SOCKET
3 STRUT ASSEMBLY

7922YG25

Use a socket and a wrench to remove the damper shaft nut spring cap while compressing the spring

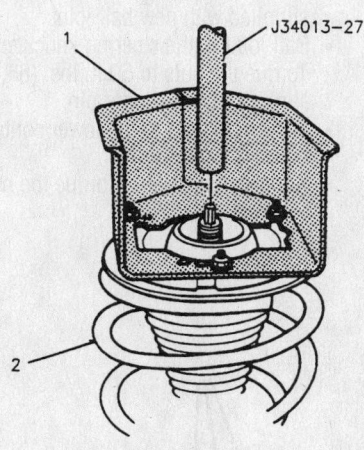

1 STRUT COMPRESSOR
2 STRUT ASSEMBLY

7922YG26

Install the rod to guide the damper shaft straight down through the spring cap while compressing the spring

down through the spring cap while compressing the spring. Remove the components.

To install:
6. Mount the strut assembly in the strut compressor, using the bottom locking pin only.
7. Install the spring over the damper and swing the assembly up so the upper locking pin can be installed.
8. Install all shields, bumpers and insulators on the spring seat.
9. Install the spring seat on top of the spring. The spring seat flat should be facing the same direction as the centerline of the strut assembly spindle.

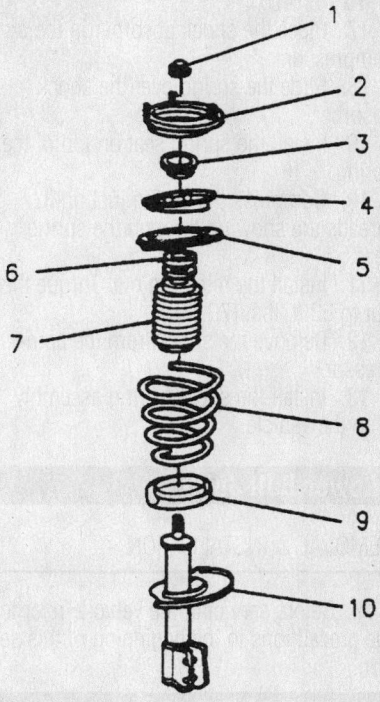

1 STRUT MOUNT NUT
2 STRUT MOUNT
3 RATE WASHER
4 SPRING SEAT
5 SPRING UPPER INSULATOR
6 JOUNCE BUMPER
7 STRUT DUST SHIELD
8 SPRING
9 SPRING LOWER INSULATOR
10 STRUT

7922YG27

Exploded view of the front strut assembly

10. Install the guiding rod and turn the forcing screw while the guiding rod centers the assembly. When the threads on the damper shaft are visible, remove the guiding rod and install the nut. tighten the nut to 34 ft. lbs. (47 Nm.
11. Remove the strut from the compressor.
12. Install the strut to the vehicle.

Rear

1. Remove the shock absorber from the vehicle.
2. Place the shock absorber in a strut compressor tool using a shock adapter.
3. Compress the spring.
4. Remove the nut.
5. Slowly release the tension on the spring.
6. Remove the spring and shock absorber from the compressor.

To install:

7. Place the shock absorber in the strut compressor.

8. Slide the spring over the shock absorber.

9. Install the spring seat on top of the spring.

10. Compress the spring just until threads are showing through the spring seat.

11. Install the retaining nut. Torque the nut to 52 ft. lbs. (70 Nm).

12. Remove the shock from the compressor.

13. Install the shock/spring assembly onto the vehicle.

Lower Ball Joint

REMOVAL & INSTALLATION

1. Before servicing the vehicle, refer to the precautions in the beginning of this section.

✳✳ WARNING

Care must be exercised to prevent the axle shaft joints from being over-extended.

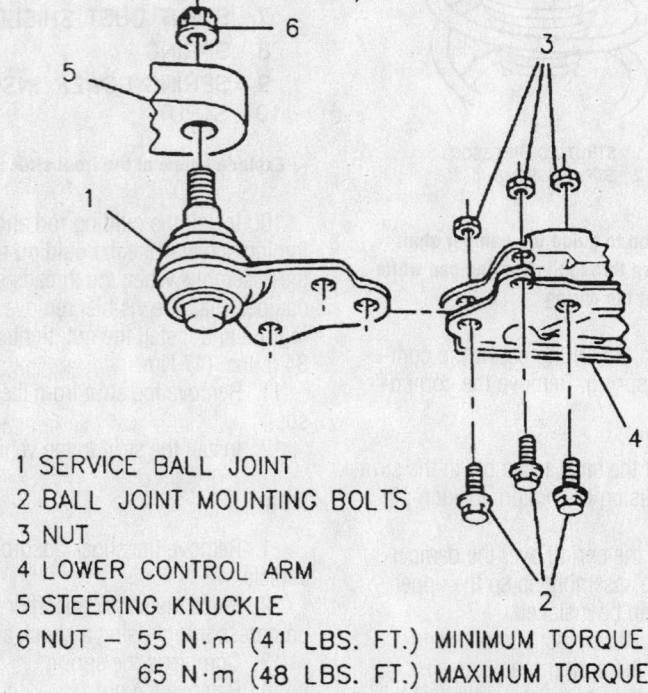

1 SERVICE BALL JOINT
2 BALL JOINT MOUNTING BOLTS
3 NUT
4 LOWER CONTROL ARM
5 STEERING KNUCKLE
6 NUT — 55 N·m (41 LBS. FT.) MINIMUM TORQUE
        65 N·m (48 LBS. FT.) MAXIMUM TORQUE
        TO INSTALL PIN
7 PIN

7922YG28

Exploded view of the ball joint mounting

2. Remove or disconnect the following:
- Front wheel
- Stabilizer shaft link
- Wiring harness from the lower control arm
- Lower ball joint from the steering knuckle

3. Drill out the 3 ball joint-to-lower control arm rivets as follows:

   a. Use an ⅛ in. (3mm) drill bit to make a pilot hole through the rivets.

   b. Finish drilling rivets with a ½ in. (13mm) drill bit.

   c. Remove the rivets with a punch, if necessary.

4. Remove the ball joint from the steering knuckle.

To install:

5. Install or connect the following:
- Ball joint into the control arm
- 3 new nuts/bolts. Tighten the nuts/bolts to the specifications supplied with new ball joint
- Ball joint to the steering knuckle. Torque the nuts to 50 ft. lbs. (65 Nm), use a new cotter pin.
- Wiring harness to the lower control arm
- Stabilizer shaft link. Torque the nuts to 13 ft. lbs. (17 Nm).

- Front wheel
6. Check and/or adjust the front wheel alignment.

Lower Control Arm

REMOVAL & INSTALLATION

1. Before servicing the vehicle, refer to the precautions in the beginning of this section.

2. Remove or disconnect the following:
- Front wheel
- Stabilizer shaft link
- Ball joint from the steering knuckle
- Wiring harness from the lower control arm
- Lower control arm

To install:

3. Install or connect the following:
- Lower control arm to the crossmember. DO NOT torque the bolts at this time.
- Wiring harness to the lower control arm
- Ball joint to the steering knuckle. Torque the nut to 50 ft. lbs. (65 Nm), use a new cotter pin.
- Stabilizer shaft link. Torque the nut to 13 ft. lbs. (17 Nm).
- Front wheel

4. Lower the vehicle and allow it to assume proper trim height.

5. Torque the front control arm bolt to 79 ft. lbs. (107 Nm) and the rear bolt to 125 ft. lbs. (170 Nm).

6. Check and/or adjust the front wheel alignment.

CONTROL ARM BUSHING REPLACEMENT

Front Bushing

1. Before servicing the vehicle, refer to the precautions in the beginning of this section.

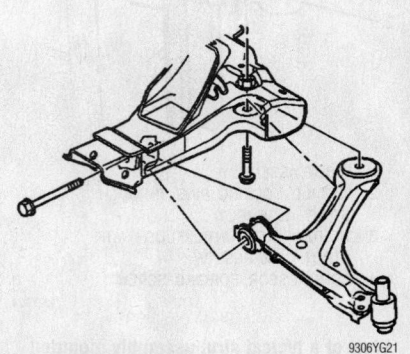

9306YG21

View of the lower control arm

2. Remove or disconnect the following:
- Front wheel
- Lower control arm and place it in a vise

3. Lubricate the threads of tool J-21474-19 with High Pressure Lubricant J-23444-A.

4. Assemble ⅜ in. Bolt J-21474-19, Remover/Installer tool J-41397-1A, Remover/Installer tool J-41397-2A and ⅜ in. Nut J-21474-18 onto the lower control arm bushing.

5. Tighten the ⅜ in. Nut J-21474-18 to press the front bushing from the lower control arm.

6. Disassemble the tools.

To install:

7. Lubricate the outer case of the new front bushing.

8. Insert the new front bushing into the lower control arm.

9. Assemble ⅜ in. Bolt J-21474-19, Remover/Installer tool J-41397-1A, Remover/Installer tool J-41397-2A and ⅜ in. Nut J-21474-18 onto the lower control arm.

10. Tighten the ⅜ in. Nut J-21474-18 to press the bushing into the lower control arm.

11. Disassemble the tools.

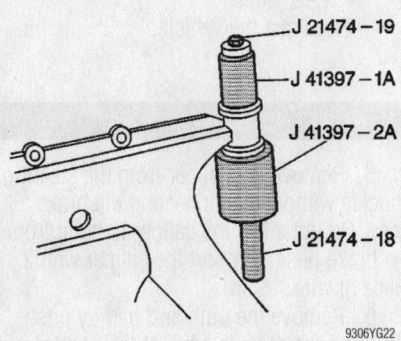

Removing the lower control arm front bushing

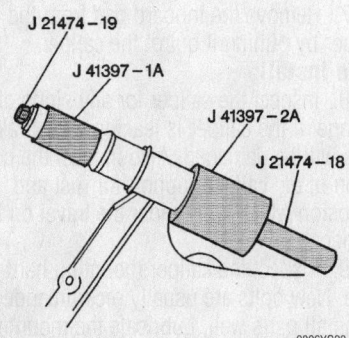

Installing the lower control arm front bushing

Rear Bushing

1. Before servicing the vehicle, refer to the precautions in the beginning of this section.

2. Remove or disconnect the following:
- Front wheel
- Lower control arm and place it in a vise

3. Lubricate the threads of tool J-21474-27 with High Pressure Lubricant J-23444-A.

4. Assemble bolt J-21474-27, Remover/Installer tool J-41211-1, Remover/Installer tool J-41211-3 and ½ in. Nut J-21474-4.

5. Tighten the Bolt J-21474-27 to press the rear bushing from the lower control arm.

6. Disassemble the tools.

To install:

7. Insert the rear lower control arm bushing.

8. Assemble bolt J-21474-27, Remover/Installer tool J-41211-1, Remover/Installer tool J-41211-3 and ½ in. Nut J-21474-4.

9. Tighten the Nut J-21474-4 to press the rear bushing into the lower control arm.

10. Disassemble the tools.

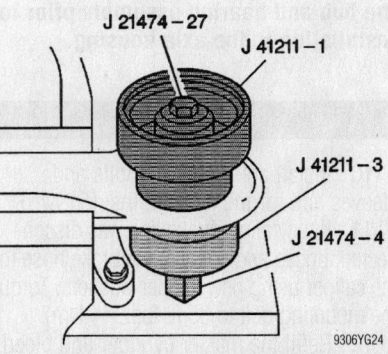

Removing the lower control arm rear bushing

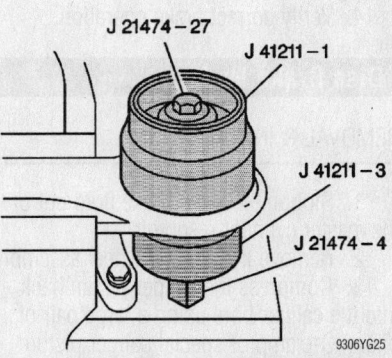

Installing the lower control arm rear bushing

Wheel Bearings

ADJUSTMENT

These vehicles are equipped with sealed hub and bearing assemblies. The hub and bearing assemblies are non-serviceable. If the assembly is damaged, the complete unit must be replaced.

REMOVAL & INSTALLATION

Front

1. Before servicing the vehicle, refer to the precautions in the beginning of this section.

2. Remove or disconnect the following:
- Front wheel
- Halfshaft nut
- Brake caliper without disconnecting the fluid line
- Brake rotor
- 3 hub/bearing assembly-to-steering knuckle bolts
- Hub/bearing assembly from the steering knuckle

To install:

3. Install or connect the following:
- Hub/bearing assembly to the steering knuckle. Torque the bolts to 70 ft. lbs. (95 Nm).
- Brake rotor
- Brake caliper
- Halfshaft through hub/knuckle assembly. Torque the nut to 144 ft. lbs. (200 Nm).
- Front wheel

4. Road test the vehicle and verify proper operation.

Rear

A single-unit hub/bearing assembly is bolted to both ends of the rear axle assembly or rear knuckle assembly. The hub/bearing assembly is a sealed unit.

1. Before servicing the vehicle, refer to the precautions in the beginning of this section.

2. Remove or disconnect the following:
- Rear wheel
- Brake drum
- Anti-lock Brake System (ABS) wheel speed sensor electrical connector
- Hub/bearing assembly-to-rear axle assembly nuts/bolts

➡ **The top rear mounting bolt will not clear the brake shoe when removing the hub and bearing assembly. Par-**

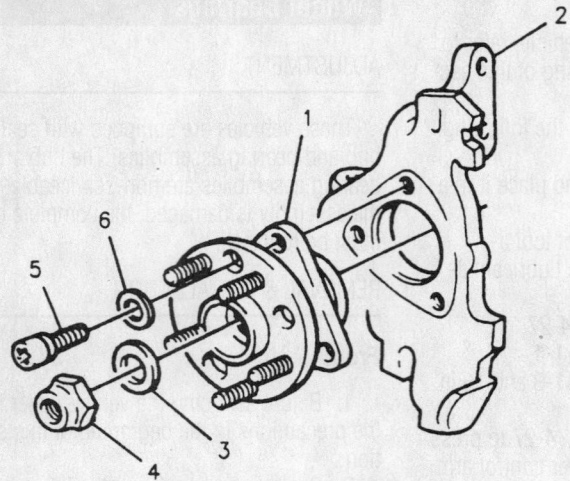

1 HUB AND BEARING ASSEMBLY
2 STEERING KNUCKLE
3 WASHER
4 DRIVE AXLE NUT – 260 N·m (192 LBS. FT.)
5 HUB AND BEARING RETAINING BOLT
6 WASHER

7922YG29

1 NUT
2 REAR AXLE ASSEMBLY
3 BACKING PLATE
4 HUB AND BEARING ASSEMBLY
5 BOLT

7922YG30

Exploded view of the front hub and bearing assembly

Exploded view of the rear hub and bearing assembly

tially remove the hub and bearing
assembly prior to removing this bolt.

- Hub/bearing assembly from the
  rear axle

To install:

3. Install or connect the following:

- Hub/bearing assembly. Torque the
  nuts/bolts to 44 ft. lbs. (60 Nm).

➡ **Position the top rear mounting bolt in
the hub and bearing assembly prior to
installation to the axle housing.**

- Rear ABS wheel speed sensor wire
  connector
- Brake drum
- Rear wheel

4. Road test the vehicle.

BRAKES

Brake Caliper

REMOVAL & INSTALLATION

1. Siphon ⅔ of the brake fluid out of
the master cylinder.
2. Remove the tire and wheel assembly.
3. Compress the caliper piston back
into the caliper bore using a large pair of
pliers, C-clamp or special piston retracting
tool.
4. If the caliper is to be completely
removed from the vehicle for bench service,
disconnect and cap the brake line from the
caliper. Discard the old washers.
5. Remove the caliper mounting bolts
and sleeves.
6. Remove the caliper from the knuckle.
7. Remove the brake pads from the
caliper, if caliper is being replaced.

To install:

8. Install the brake pads in the caliper.
9. Install the caliper on the steering
knuckle.

10. Install the mounting bolts and
sleeves and torque to 40 ft. lbs. (51 Nm).
11. If the caliper brake line was disconnected, uncap and connect the brake hose to
the caliper using new copper washers. Torque
the mounting bolt to 35 ft. lbs. (44 Nm).
12. Refill the master cylinder and bleed
the brake system.
13. Install the tire and wheel assembly
and tighten to specification.
14. Verify correct brake operation.

Disc Brake Pads

REMOVAL & INSTALLATION

1. Siphon ⅔ of the brake fluid out of
the master cylinder reservoir.
2. Remove the tire and wheel assembly.
3. Compress the caliper piston back
into the caliper bore using a large pair of
pliers, C-clamp or special caliper piston
retracting tool.
4. Remove the caliper mounting bolts
and sleeves.

5. Remove the caliper from the steering
knuckle without disconnecting the brake
hose. Do not allow the caliper to hang from
the brake hose. Support the caliper with a
piece of wire.
6. Remove the outboard pad by pushing in on the outside edge of the pad to
release the mounting dowel from the hole in
the caliper. When both dowels are unseated,
push the pad out the bottom of the caliper.
7. Remove the inboard pad from the
caliper by pulling it out of the caliper.

To install:

8. Inspect the caliper for any signs of
leakage. If the caliper is leaking, new brake
pads will be damaged. Also inspect the condition of the caliper support for rust and
corrosion which will hinder the travel on the
caliper.
9. Inspect the caliper mounting hardware. New bolts are usually recommended.
Clean all parts well. Lubricate the mounting
bushings and sleeves with silicone grease
as required.
10. Install the inboard pad in the caliper

so the spring clip on the pad back engages in the caliper piston.

11. Install the outboard pad over the caliper end until the mounting dowels snap into the mounting holes in the caliper.

12. Install the caliper over the rotor onto the steering knuckle.

13. Install the mounting bolts and sleeves and torque to 40 ft. lbs. (51 Nm).

14. Install the tire and wheel assembly.

15. Pump the brake pedal several times

to seat the pads against the rotor before attempting to move the vehicle.

16. Check the master cylinder level and add fluid as necessary.

Brake Drums

REMOVAL & INSTALLATION

1. Remove the wheel and tire assembly.
2. Pull the brake drum off. It may be nec-

essary to gently tap the rear edge of the drum to start it off the studs. If extreme resistance to removal is encountered, it will be necessary to retract the brake shoe self-adjuster screw, sometimes called the star-wheel. Knock out the access hole in the brake drum and turn the adjuster to retract the linings from the drum. Install a replacement hole cover before reinstalling the drum.

➡**Do not hammer on the brake drum to remove it.**

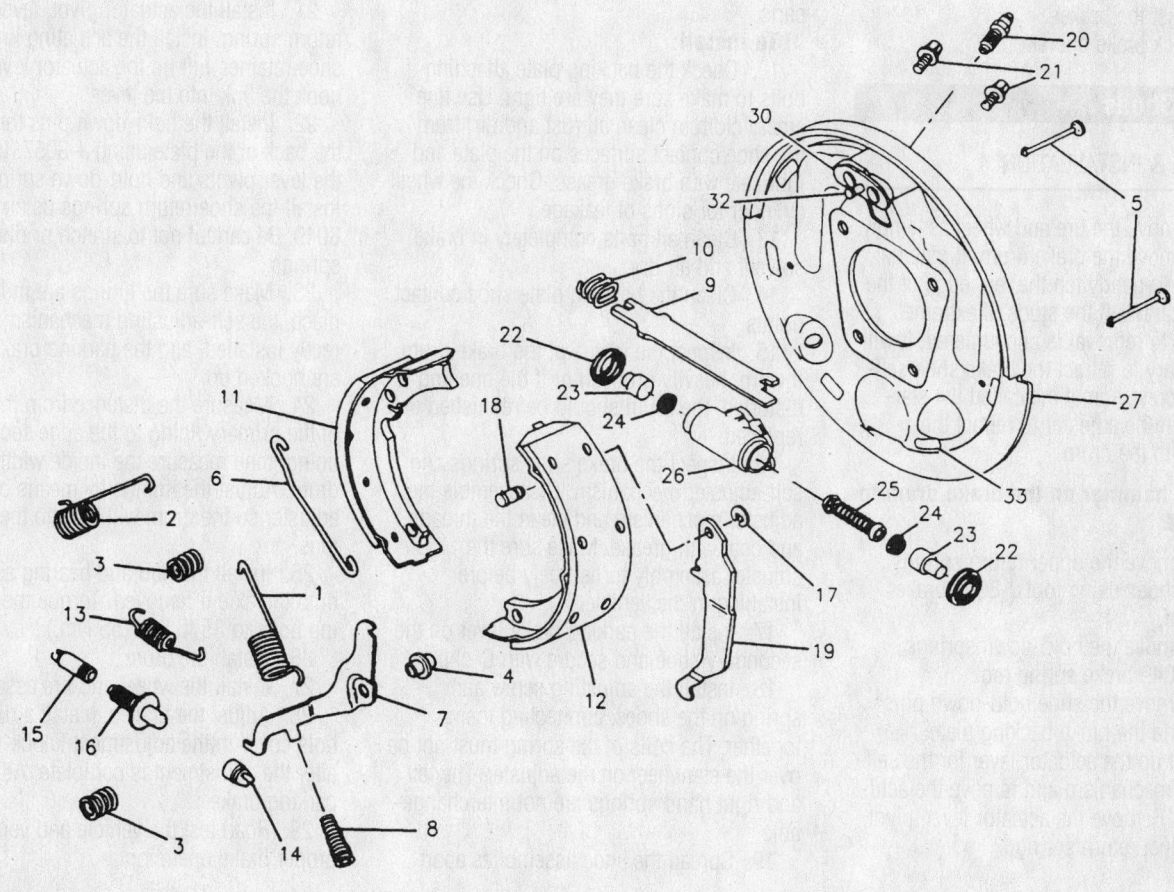

1	RETURN SPRING	11	PRIMARY SHOE AND LINING	21	BOLT
2	RETURN SPRING	12	SECONDARY SHOE AND LINING	22	BOOT
3	HOLD DOWN SPRING	13	ADJUSTING SCREW SPRING	23	PISTON
4	BEARING SLEEVE	14	SOCKET	24	SEAL
5	HOLD–DOWN PIN	15	PIVOT NUT	25	SPRING ASSEMBLY
6	ACTUATOR LINK	16	ADJUSTING SCREW	26	WHEEL CYLINDER
7	ACTUATOR LEVER	17	RETAINING RING	27	BACKING PLATE
8	LEVER RETURN SPRING	18	PIN	30	SHOE RETAINER
9	PARKING BRAKE STRUT	19	PARKING BRAKE LEVER	32	ANCHOR PIN
10	STRUT SPRING	20	BLEEDER VALVE	33	SHOE PADS

93006G97

Exploded view of drum brake components—Cavalier, Sunfire

To install:

3. Inspect the inside of the brake drum. If worn, heavily grooved or if the opening is distorted, the drum should be refinished or replaced. If refinishing, observe the maximum drum diameter specification.

4. Inspect the wheel cylinder for signs of brake fluid leakage. Inspect the brake shoe springs and self-adjuster mechanism. The adjuster should usually be disassembled, cleaned and lubricated when the drum is removed for brake service.

5. Install the drum over the brake shoes.
6. Install the wheel and tire assembly.
7. Adjust the brakes.
8. Check brake operation.

Brake Shoes

REMOVAL & INSTALLATION

1. Remove the tire and wheel assembly.
2. Remove the brake drum. It may be necessary to gently tap the rear edge of the drum to start it off the studs. If extreme resistance to removal is encountered, it will be necessary to retract the brake shoe self-adjuster screw, sometime called the starwheel. Turn the adjuster to retract the linings from the drum.

➡**Do not hammer on the brake drum to remove it.**

3. Remove the upper return springs from the shoes using tool J-8057 brake spring tool.

4. Remove the hold-down springs using J-8049 brake spring tool.

5. Remove the shoe hold-down pins from behind the brake backing plate.

6. Lift up the actuator lever for the self-adjusting mechanism and remove the actuating link. Remove the actuator lever, pivot, and the pivot return spring.

7. Spread the shoes apart to clear the wheel cylinder pistons, then remove the parking brake strut and spring.

8. Disconnect the parking brake cable from the lever. Remove the shoes, still connected by their adjusting screw spring.

9. With the shoes removed, note the position of the adjusting spring and remove the spring and adjusting screw.

10. Remove the C-clip from the parking brake lever and remove the lever from the secondary shoe.

11. Use a damp cloth to remove all dirt and dust from the backing plate and brake parts.

To install:

12. Check the backing plate attaching bolts to make sure they are tight. Use fine emery cloth to clean all rust and dirt from the shoe contact surfaces on the plate and lubricate with brake grease. Check the wheel cylinder for signs of leakage.

13. Clean all parts completely in brake solvent and air dry.

14. Clean the backing plate shoe contact points.

15. Inspect the inside of the brake drum. If worn, heavily grooved or if the opening is distorted, the drum should be refinished or replaced.

16. Inspect the brake shoe springs and self-adjuster mechanism. Disassemble the adjuster mechanism and clean the threads and coat with grease. Make sure the adjuster assembly turns freely before installing in the vehicle.

17. Install the parking brake lever on the secondary shoe and secure with C-clip.

18. Install the adjusting screw and spring on the shoes, connecting them together. The coils of the spring must not be over the starwheel on the adjuster. The left and right hand springs are not interchangeable.

19. Spread the shoe assemblies apart

and connect the parking brake cable. Install the shoes on the backing plate, engaging the shoes at the top temporarily with the wheel cylinder pistons. Make sure the starwheel on the adjuster is lined up with the adjusting hole in the backing plate, if equipped.

20. Spread the shoes apart slightly and install the parking brake strut and spring. Make sure the end of the strut without the spring engages the parking brake lever. The end with the spring engages the primary shoe (the one with the shorter lining).

21. Install the actuator pivot, lever and return spring. Install the actuating link in the shoe retainer. Lift up the actuator lever and hook the link into the lever.

22. Install the hold-down pins through the back of the plate using J-8057. Install the lever pivots and hold-down springs. Install the shoe return springs using J-8049. Be careful not to stretch or distort the springs.

23. Make sure the linings are in the right place, the self-adjusting mechanism is correctly installed, and the parking brake parts are hooked up.

24. Measure the distance from the edge of the primary lining to the edge secondary lining, then measure the inside width of the drum. Adjust the linings by means of the adjuster so the drum will fit onto the linings.

25. Install the hub and bearing assembly onto the axle if removed. Torque the retaining bolts to 35 ft. lbs. (55 Nm.).

26. Install the drum.
27. Install the wheel and tire assembly.
28. Adjust the brakes. Install a rubber hole cover in the adjustment knock-out hole after the adjustment is complete. Adjust the parking brake.

29. Road test the vehicle and verify proper brake operation.

SPECIFICATION CHARTS

ENGINE AND VEHICLE IDENTIFICATION

Engine							Model Year	
Code ①	Liters (cc)	Cu. In.	Cyl.	Fuel Sys.	Engine Type	Eng. Mfg.	Code ②	Year
M/J	3.1 (3130)	191	6	SFI	OHV	BOC	Y	2000
							1	2001
							2	2002
							3	2003
							4	2004

BOC: Buick/Oldsmobile/Cadillac

CUS: Chevrolet/United States

SFI: Sequential Fuel Injection

① 8th position of VIN

② 10th position of VIN

42372-MALI-C01

GENERAL ENGINE SPECIFICATIONS

Year	Model	Engine Displacement Liters (cc)	Engine Series (ID/VIN)	Fuel System	Net Horsepower @ rpm	Net Torque @ rpm (ft. lbs.)	Bore x Stroke (in.)	Com-pression Ratio	Oil Pressure @ rpm
2000	Malibu	3.1 (3130)	M/J	SFI	170@5200	190@4000	3.50x3.31	9.6:1	15@1100
2001	Malibu	3.1 (3130)	J	SFI	170@5200	190@4000	3.50x3.31	9.6:1	15@1100
2002	Malibu	3.1 (3130)	J	SFI	170@5200	190@4000	3.50x3.31	9.6:1	15@1100
2003	Malibu	3.1 (3130)	J	SFI	170@5200	190@4000	3.50x3.31	9.6:1	15@1100

SFI: Multi-point Fuel Injection

42372-MALI-C02

ENGINE TUNE-UP SPECIFICATIONS

Year	Engine Displacement Liters (cc)	Engine ID/VIN	Spark Plug Gap (in.)	Ignition Timing (deg.) MT	AT	Fuel Pump (psi)	Idle Speed (rpm) MT	AT	Valve Clearance In.	Ex.
2000	3.1 (3130)	M/J	0.060	①	①	41-47	①	①	HYD	HYD
2001	3.1 (3130)	J	0.060	①	①	41-47	①	①	HYD	HYD
2002	3.1 (3130)	J	0.060	①	①	41-47	①	①	HYD	HYD
2003	3.1 (3130)	J	0.060	①	①	41-47	①	①	HYD	HYD

NOTE: The Vehicle Emission Control Information label often reflects specification changes made during production. The label figures must be used if they differ from those in this chart.

HYD: Hydraulic

① Refer to Vehicle Emission Control Information label

42372-MALI-C03

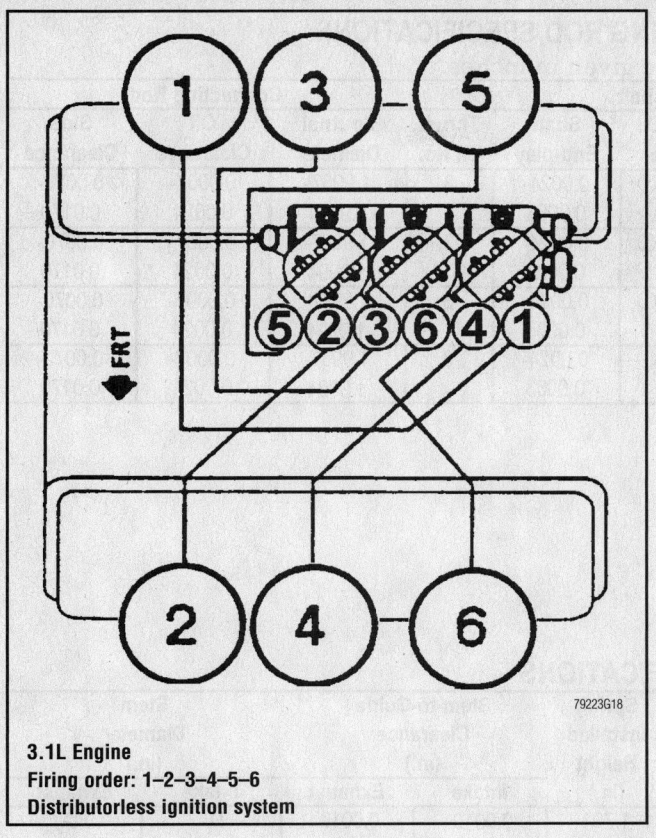

3.1L Engine
Firing order: 1–2–3–4–5–6
Distributorless ignition system

79223G18

Serpentine drive belt routing—3.1L Engine

79224G54

CAPACITIES

Year	Model	Engine Displacement Liters (cc)	Engine ID/VIN	Engine Oil with Filter (qts.)	Transmission (pts.)	Fuel Tank (gal.)	Cooling System (qts.)
2000	Malibu	3.1 (3130)	M/J	4.5	13.8	15.0	13.6
2001	Malibu	3.1 (3130)	J	4.5	13.8	14.3	13.6
2002	Malibu	3.1 (3130)	J	4.5	13.8	14.3	13.6
2003	Malibu	3.1 (3130)	J	4.5	13.8	14.3	13.6

NOTE: All capacities are approximate. Add fluid gradually and ensure a proper fluid level is obtained.

42372-MALI-C04

CRANKSHAFT AND CONNECTING ROD SPECIFICATIONS

All measurements are given in inches.

Year	Engine Displacement Liters (cc)	Engine ID/VIN	Crankshaft Main Brg. Journal Dia.	Crankshaft Main Brg. Oil Clearance	Crankshaft Shaft End-play	Crankshaft Thrust on No.	Connecting Rod Journal Diameter	Connecting Rod Oil Clearance	Connecting Rod Side Clearance
2000	3.1 (3130)	M/J	2.6473-2.6383	0.0008-0.0025 ①	0.0024-0.0083	3	1.9987-1.9994	0.0007-0.0024	0.0070-0.0170
2001	3.1 (3130)	J	2.6473-2.6383	0.0008-0.0025 ①	0.0024-0.0083	3	1.9987-1.9994	0.0007-0.0024	0.0070-0.0170
2002	3.1 (3130)	J	2.6473-2.6383	0.0008-0.0025 ①	0.0024-0.0083	3	1.9987-1.9994	0.0007-0.0024	0.0070-0.0170
2003	3.1 (3130)	J	2.6473-2.6383	0.0008-0.0025 ①	0.0024-0.0083	3	1.9987-1.9994	0.0007-0.0024	0.0070-0.0170

① Thrust bearing: 0.0012 - 0.0030

42372-MALI-C05

VALVE SPECIFICATIONS

Year	Engine Displacement Liters (cc)	Engine ID/VIN	Seat Angle (deg.)	Face Angle (deg.)	Spring Test Pressure (lbs. @ in.)	Spring Installed Height (in.)	Stem-to-Guide Clearance (in.) Intake	Stem-to-Guide Clearance (in.) Exhaust	Stem Diameter (in.) Intake	Stem Diameter (in.) Exhaust
2000	3.1 (3130)	M/J	46	45	230@1.260	1.701	0.0010-0.0027	0.0010-0.0027	NA	NA
2001	3.1 (3130)	J	46	45	230@1.260	1.701	0.0010-0.0027	0.0010-0.0027	NA	NA
2002	3.1 (3130)	J	46	45	230@1.260	1.701	0.0010-0.0027	0.0010-0.0027	NA	NA
2003	3.1 (3130)	J	46	45	230@1.260	1.701	0.0010-0.0027	0.0010-0.0027	NA	NA

NA: Not Available

42372-MALI-C06

PISTON AND RING SPECIFICATIONS

All measurements are given in inches.

Year	Engine Displacement Liters (cc)	Engine ID/VIN	Piston Clearance	Ring Gap Top Compression	Ring Gap Bottom Compression	Ring Gap Oil Control	Ring Side Clearance Top Compression	Ring Side Clearance Bottom Compression	Ring Side Clearance Oil Control
2000	3.1 (3130)	M/J	0.0013-0.0027	0.006-0.014	0.020-0.028	0.010-0.050	0.0020-0.0033	0.0020-0.0035	0.0080
2001	3.1 (3130)	J	0.0013-0.0027	0.006-0.014	0.020-0.028	0.010-0.050	0.0020-0.0033	0.0020-0.0035	0.0080
2002	3.1 (3130)	J	0.0013-0.0027	0.006-0.014	0.020-0.028	0.010-0.050	0.0020-0.0033	0.0020-0.0035	0.0080
2003	3.1 (3130)	J	0.0013-0.0027	0.006-0.014	0.020-0.028	0.010-0.050	0.0020-0.0033	0.0020-0.0035	0.0080

42372-MALI-C07

TORQUE SPECIFICATIONS
All readings in ft. lbs.

Year	Engine Displacement Liters (cc)	Engine ID/VIN	Cylinder Head Bolts	Main Bearing Bolts	Rod Bearing Bolts	Crankshaft Damper Bolts	Flywheel Bolts	Manifold Intake	Manifold Exhaust	Spark Plug	Lug Nut
2000	3.1 (3130)	M/J	①	②	③	76	52	④	12	⑤	100
2001	3.1 (3130)	J	①	②	③	76	52	④	12	⑤	100
2002	3.1 (3130)	J	①	②	③	76	52	④	12	⑤	100
2003	3.1 (3130)	J	①	②	③	76	52	④	12	⑤	100

NA: Not Available

① Coat threads with sealer and torque to 37 ft. lbs., then turn an additional 95 degrees

② 37 ft. lbs. plus 75 degrees

③ 15 ft. lbs. plus 75 degrees

④ Lower intake manifold bolts: 10 ft. lbs.
Upper intake manifold bolts: 18 ft. lbs.

⑤ New cylinder first-time installation: 20 ft. lbs.
All others: 11 ft. lbs.

42372-MALI-C08

WHEEL ALIGNMENT

Year	Model		Caster Range (+/-Deg.)	Caster Preferred Setting (Deg.)	Camber Range (+/-Deg.)	Camber Preferred Setting (Deg.)	Toe-in (in.)	Steering Axis Inclination (Deg.)
2000	Malibu	F	+1.00	+4.10	+0.80	-0.20	0.06 +/- 0.12	—
		R	—	—	+0.50	-0.20	0.03 +/- 0.05	—
2001	Malibu	F	+1.00	+4.10	+0.80	-0.20	0.06 +/- 0.12	—
		R	—	—	+0.50	-0.20	0.03 +/- 0.05	—
2002	Malibu	F	+1.00	+4.10	+0.80	-0.20	0.06 +/- 0.12	—
		R	—	—	+0.50	-0.20	0.03 +/- 0.05	—
2003	Malibu	F	+1.00	+4.10	+0.80	-0.20	0.06 +/- 0.12	—
		R	—	—	+0.50	-0.20	0.03 +/- 0.05	—

F: Front

R: Rear

42372-MALI-C09

TIRE, WHEEL AND BALL JOINT SPECIFICATIONS

Year	Model	OEM Tires Standard	OEM Tires Optional	Tire Pressures (psi) Front	Tire Pressures (psi) Rear	Wheel Size	Ball Joint Inspection
2000	Malibu	P215/60TR15	P215/60HR15	30	26	6-JJ	①
2001	Malibu	P215/60TR15	P215/60HR15	30	26	6-JJ	①
2002	Malibu	P215/60TR15	P215/60HR15	30	26	6-JJ	①
2003	Malibu	P215/60TR15	P215/60HR15	30	26	6-JJ	①

OEM: Original Equipment Manufacturer

PSI: Pounds Per Square Inch

① Replace if any measurable movement is found.

42372-MALI-C10

BRAKE SPECIFICATIONS
All measurements in inches unless noted

| Year | Model | Brake Disc | | | Brake Drum Diameter | | | Minimum Lining Thickness | | Brake Caliper | |
		Original Thickness	Minimum Thickness	Maximum Runout	Original Inside Diameter	Max. Wear Limit	Maximum Machine Diameter	Front	Rear	Bracket Bolts (ft. lbs.)	Mounting Bolts (ft. lbs.)
2000	Malibu	1.031	0.980	0.003	8.900-8.909	8.920	8.909	0.030	①	85	23
2001	Malibu	1.031	0.980	0.003	8.900-8.909	8.920	8.909	0.030	①	85	23
2002	Malibu	1.031	0.980	0.003	8.900-8.909	8.920	8.909	0.030	①	85	23
2003	Malibu	1.031	0.980	0.003	8.900-8.909	8.920	8.909	0.030	①	85	23

① 0.030 over rivet head; If bonded lining, use 0.062 from shoe

42372-MALI-C11

SCHEDULED MAINTENANCE INTERVALS
GM L/N BODY—CHEVROLET MALIBU

| TO BE SERVICED | TYPE OF SERVICE | VEHICLE MILEAGE INTERVAL (x1000) | | | | | | | | | | | | | | | | | |
		7.5	15	22.5	30	37.5	45	52.5	60	67.5	75	82.5	90	97.5	100	105	113	120	150
Accessory drive belt(s)	S/I								✓									✓	
Air cleaner element	R				✓				✓				✓					✓	
Cooling system ①	S/I																		✓
Cooling system hoses	S/I																		✓
Engine coolant	R																		✓
Engine oil and filter	R	✓	✓	✓	✓	✓	✓	✓	✓	✓	✓	✓	✓	✓		✓	✓	✓	
Fuel tank, cap & lines	S/I				✓				✓				✓					✓	
Radiator, condenser, pressure cap and neck	C																		✓
Spark plug wires	S/I														✓				
Spark plugs	R														✓				
Tires and wheels ②	S/I	✓	✓	✓	✓	✓	✓	✓	✓	✓	✓	✓	✓	✓		✓	✓	✓	
Automatic transaxle fluid & filter ③	S/I																		

R: Replace

S/I: Service or Inspect

C: Clean

① Pressure test the cooling system and pressure cap.

② This includes rotating the tires.

③ Replace fuid every 50, 000 miles if driven in etxreme traffic or in places where the temperature exceeds 90 degrees F

FREQUENT OPERATION MAINTENANCE (SEVERE SERVICE) ADDITIONS

If a vehicle is operated under any of the following conditions it is considered severe service:

- Towing a trailer or using a camper or car-top carrier.
- Extensive idling or low-speed driving for long distances as in heavy commercial use, such as delivery, taxi or police cars.
- Operating on rough, muddy or salt-covered roads.
- Operating on unpaved or dusty roads.
- 50% or more of the vehicle operation is in 32°C (90°F) or higher temperatures, or constant operation in temperatures below 0°C (32°F).

Engine oil and filter: change every 3000 miles or 3 months, whichever occurs first.

Wheels and tires: inspect and rotate every 6000 miles.

Air cleaner element: inspect every 15,000 miles and replace or clean as needed. Replace it at least every 30,000 miles.

Automatic transaxle fluid & filter: replace every 50,000 miles.

42372-MALI-C12

PRECAUTIONS

Before servicing any vehicle, please be sure to read all of the following precautions, which deal with personal safety, prevention of component damage, and important points to take into consideration when servicing a motor vehicle:

• Never open, service or drain the radiator or cooling system when the engine is hot; serious burns can occur from the steam and hot coolant.

• Observe all applicable safety precautions when working around fuel. Whenever servicing the fuel system, always work in a well-ventilated area. Do not allow fuel spray or vapors to come in contact with a spark, open flame or excessive heat (a hot drop light, for example). Keep a dry chemical fire extinguisher near the work area. Always keep fuel in a container specifically designed for fuel storage; also, always properly seal fuel containers to avoid the possibility of fire or explosion. Refer to the additional fuel system precautions later in this section.

• Fuel injection systems often remain pressurized, even after the engine has been turned **OFF**. The fuel system pressure must be relieved before disconnecting any fuel lines. Failure to do so may result in fire and/or personal injury.

• Brake fluid often contains polyglycol ethers and polyglycols. Avoid contact with the eyes and wash your hands thoroughly after handling brake fluid. If you do get brake fluid in your eyes, flush your eyes

with clean, running water for 15 minutes. If eye irritation persists, or if you have taken brake fluid internally, IMMEDIATELY seek medical assistance.

• The EPA warns that prolonged contact with used engine oil may cause a number of skin disorders, including cancer! You should make every effort to minimize your exposure to used engine oil. Protective gloves should be worn when changing oil. Wash your hands and any other exposed skin areas as soon as possible after exposure to used engine oil. Soap and water, or waterless hand cleaner should be used.

• All new vehicles are now equipped with an air bag system, often referred to as a Supplemental Restraint System (SRS) or Supplemental Inflatable Restraint (SIR) system. The system must be disabled before performing service on or around system components, steering column, instrument panel components, wiring and sensors. Failure to follow safety and disabling procedures could result in accidental air bag deployment, possible personal injury and unnecessary system repairs.

• Always wear safety goggles when working with, or around, the air bag system. When carrying a non-deployed air bag, be sure the bag and trim cover are pointed away from your body. When placing a non-deployed air bag on a work surface, always face the bag and trim cover upward, away from the surface. This will reduce the motion of the module if it is accidentally

deployed. Refer to the additional air bag system precautions later in this section.

• Clean, high quality brake fluid from a sealed container is essential to the safe and proper operation of the brake system. You should always buy the correct type of brake fluid for your vehicle. If the brake fluid becomes contaminated, completely flush the system with new fluid. Never reuse any brake fluid. Any brake fluid that is removed from the system should be discarded. Also, do not allow any brake fluid to come in contact with a painted surface; it will damage the paint.

• Never operate the engine without the proper amount and type of engine oil; doing so WILL result in severe engine damage.

• Timing belt maintenance is extremely important! Many models utilize an interference-type, non-freewheeling engine. If the timing belt breaks, the valves in the cylinder head may strike the pistons, causing potentially serious (also time-consuming and expensive) engine damage. Refer to the maintenance interval for the recommended replacement interval for the timing belt.

• Disconnecting the negative battery cable on some vehicles may interfere with the functions of the on-board computer system(s) and may require the computer to undergo a relearning process once the negative battery cable is reconnected.

• When servicing drum brakes, only disassemble and assemble one side at a time, leaving the remaining side intact for reference.

ENGINE REPAIR

Alternator

REMOVAL

1. Before servicing the vehicle, refer to the precautions in the beginning of this section.
2. Remove or disconnect the following:
   • Negative battery cable
   • Accessory drive belt
   • Alternator electrical connectors
   • Power steering line clip
   • Alternator

INSTALLATION

1. Install or connect the following:
   • Alternator. Torque the nuts to 22 ft. lbs. (30 Nm) and the bolts to 37 ft. lbs. (50 Nm).
   • Alternator electrical connector

• Power steering line clip
• Accessory drive belt
• Negative battery cable

Ignition Timing

ADJUSTMENT

Ignition timing is controlled by the Powertrain Control Module (PCM). No adjustment is necessary or possible.

Engine Assembly

REMOVAL & INSTALLATION

1. Before servicing the vehicle, refer to the precautions in the beginning of this section.
2. Relieve the fuel system pressure.
3. Drain the engine coolant.
4. Remove or disconnect the following:

• Air cleaner assembly
• Hood
• Accessory drive belt
• Coolant inlet line from the coolant surge tank
• Cruise control module
• Upper engine wiring harness
• Throttle and cruise control cables
• Starter motor
• A/C compressor. DO NOT remove the lines.
• Lower engine wiring harness and position aside
• Catalytic converter
• Torque converter cover
• Torque converter bolts
• Engine splash shields
• Transaxle-to-engine brace
• 2 outer transmission mounting bolts
• Upper and lower radiator hoses
• Fuel lines

- Brake booster vacuum line
- Heater hoses

5. Install an engine support fixture.
6. Remove or disconnect the following:
- Engine mount
- Power steering pump
- Transmission-to-engine bolts
- Engine from the vehicle

To install:

7. Install or connect the following:
- Transaxle to the engine. Torque the bolts to 55 ft. lbs. (75 Nm).
- Power steering pump. Torque the bolts to 25 ft. lbs. (34 Nm).
- Engine mount. Torque the bolts to 35 ft. lbs. (47 Nm).

8. Remove the engine support fixture.
9. Install or connect the following:
- Heater hoses
- Brake booster vacuum line
- Fuel lines
- Radiator hoses
- Engine splash shields
- Torque converter bolts. Torque the bolts to 46 ft. lbs. (62 Nm).
- Torque converter cover. Torque the bolts to 89 inch lbs. (10 Nm).
- Catalytic converter
- Lower engine wiring harness
- A/C compressor. Torque the bolts to 37 ft. lbs. (50 Nm).
- Starter motor
- Throttle and cruise control cables
- Upper engine wiring harness
- Cruise control module
- Coolant surge tank hose
- Accessory drive belt
- Hood
- Air cleaner assembly

10. Refill the coolant.
11. Start the engine and check for proper operation.

Water Pump

REMOVAL & INSTALLATION

1. Before servicing the vehicle, refer to the precautions in the beginning of this section.
2. Drain the cooling system.
3. Remove or disconnect the following:
- Negative battery cable
- Water pump pulley bolts, loosen
- Accessory drive belt
- Water pump pulley
- Water pump bolts, pump and gasket

To install:

4. Apply a thin bead of sealer around the outside edge of the water pump along the gasket sealing area.

1   COOLANT PUMP
2   GASKET
3   BOLT — 10 N·m (89 LBS. IN.)
4   LOCATOR (MUST BE VERTICAL)

79222202

Exploded view of the water pump mounting

5. Install or connect the following:
- Water pump with a new gasket. Torque the bolts to 89 inch lbs. (10 Nm).
- Water pump pulley and finger-tighten the bolts
- Accessory drive belt
- Water pump pulley bolts and torque to 18 ft. lbs. (24 Nm)
- Negative battery cable

6. Refill and bleed the cooling system.
7. Operate the engine and check for leaks.

Heater Core

REMOVAL & INSTALLATION

1. Disable the air bag by performing the following procedure:
a. Place the front wheel in the straight-ahead position.
b. Turn the ignition switch to the LOCK position.
c. Remove the air bag fuse from the fuse block.

d. At the left side, remove the sound insulator.
e. At the driver's side, disconnect the Connector Position Assurance (CPA) and the yellow 2-way electrical connectors and the lead to the passenger's side SIR module.

2. Disconnect the negative battery cable.
3. Drain the cooling system into a clean container for reuse.
4. Disconnect the heater hoses from the heater core and remove the drain tube from the heater/air conditioning housing.
5. Remove the steering wheel by removing or disconnecting the following:
- 2 SIR module-to-steering wheel screws (located at the rear of the steering wheel), and the module
- SIR module and disconnect the electrical connector and remove the SIR module

✳✳ CAUTION

Place the SIR module in a safe place with the front facing upward.

- Horn and cruise control electrical connectors
- Steering wheel-to-steering column nut
- Steering wheel from the steering column

6. Remove the console by removing or disconnecting the following:
- Front seats
- Fold the console compartment upward
- Console trim plate
- Rear cup holder
- Console-to-chassis screws and the console

7. Remove the instrument panel by removing or disconnecting the following:
- Defroster grille
- Instrument panel valance screws and the valance
- Instrument panel end caps from both sides
- Screws located under the instrument panel end caps
- Instrument panel compartment
- Steering column covers
- Steering column stalks
- Instrument cluster and accessory trim plates
- Instrument cluster fasteners, disconnect the electrical connectors and remove the cluster
- Heater/air conditioning control assembly
- Stereo/tape deck assembly
- Ignition switch
- Upper windshield side garnish molding
- Loosen the center console, pull it rearward to disengage it from the instrument panel
- Instrument panel-to-tie bar screws
- Instrument panel
- Outlet from the heater/air conditioning housing
- Heater core cover-to-heater/air conditioning housing

➡There is a screw located in the recess in the center of the cover.

- Heater core-to-heater/air conditioning housing clamps
- Heater core

To install:

8. Install or connect the following:
- Heater core
- Heater core-to-heater/air conditioning housing clamps
- Heater core cover-to-heater/air conditioning housing

➡There is a screw located in the recess in the center of the cover.

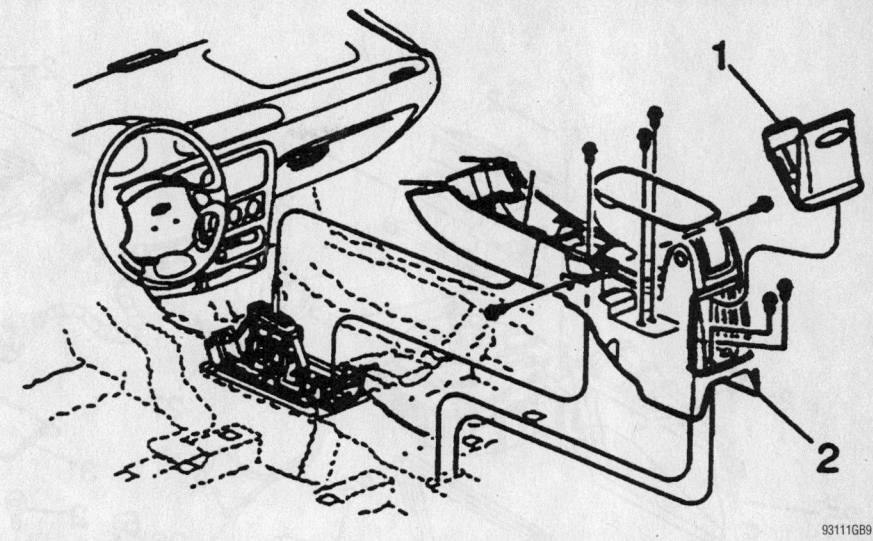

Exploded view of the console and related components—Malibu

93111GB9

- Outlet to the heater/air conditioning housing

9. Install the instrument panel by installing or connecting the following:
- Instrument panel
- Instrument panel-to-tie bar screws
- Move the center console forward and engage it to the instrument panel
- Upper windshield side garnish molding

- Ignition switch
- Stereo/tape deck assembly
- Heater/air conditioning control assembly
- Instrument cluster, connect the electrical connectors and install the cluster fasteners
- Instrument cluster accessory trim plates
- Steering column stalks
- Steering column covers

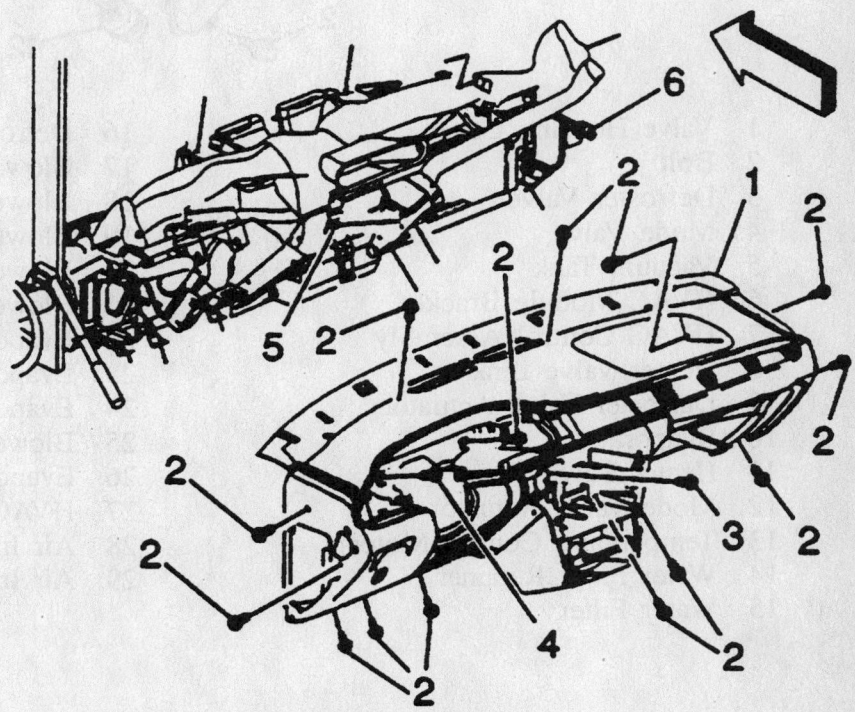

View of instrument panel—Malibu

93111GB0

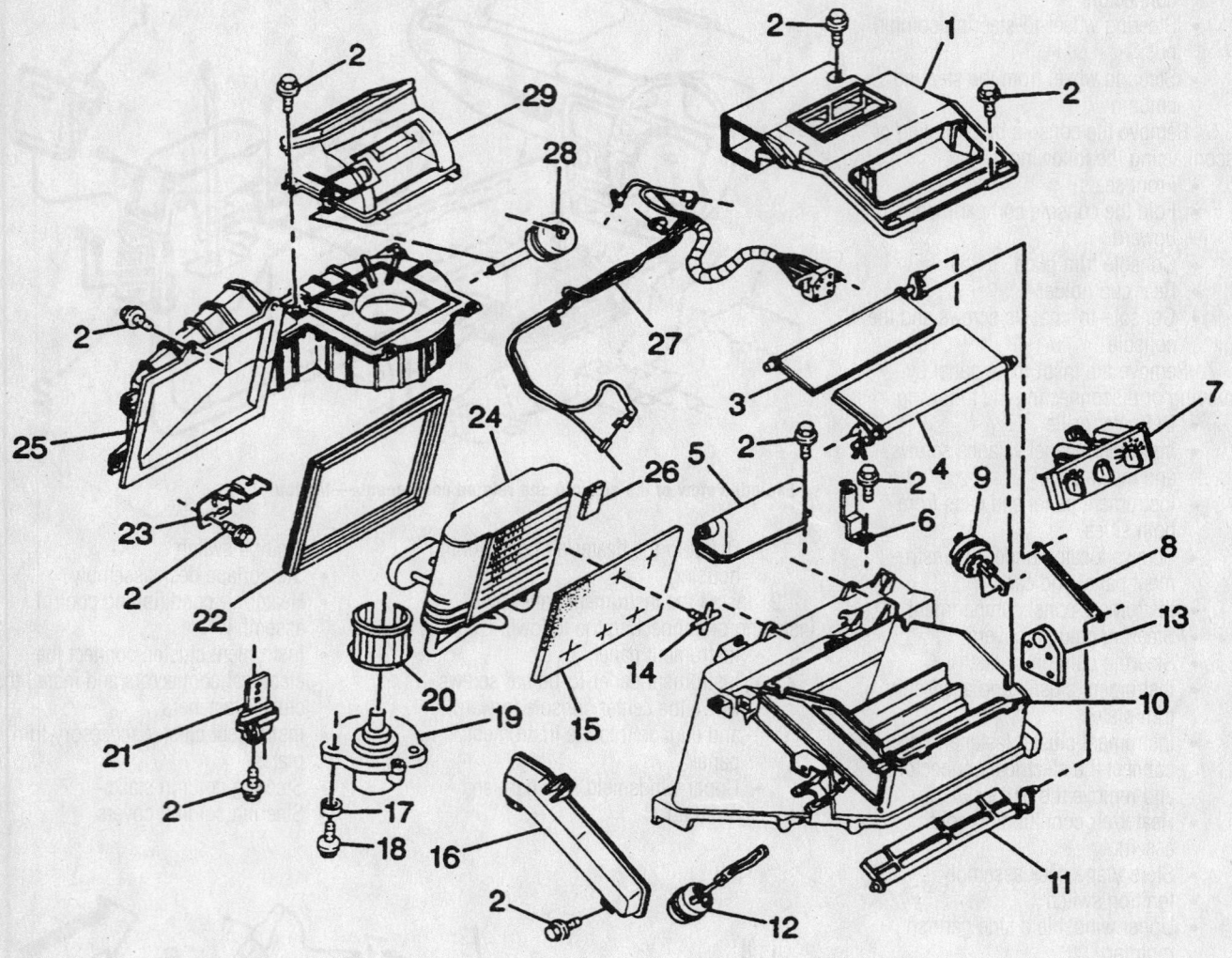

1 Valve Housing Cover
2 Bolt
3 Defroster Valve
4 Mode Valve
5 Vacuum Tank
6 HVAC Module Bracket
7 HVAC Control Assembly
8 Heater Valve Link
9 Defroster Valve Actuator
10 Evaporator Case
11 Heater Valve
12 Mode Valve Actuator
13 Temperature Control Motor
14 Water Filter Retainer
15 Water Filter

16 Defroster Case
17 Blower Motor Bolt Insulator
18 Blower Motor Bolt
19 Blower Motor
20 Blower Motor Fan
21 Blower Motor Resistor
22 Evaporator Core Seal
23 Evaporator Core Bracket
24 Evaporator Core
25 Blower and Air Inlet Case
26 Evaporator Core Spacer
27 HVAC Vacuum Harness
28 Air Inlet Valve Actuator
29 Air Inlet Case

93111GC1

View of the evaporator core, upper heater/air conditioning housing and related components—Malibu

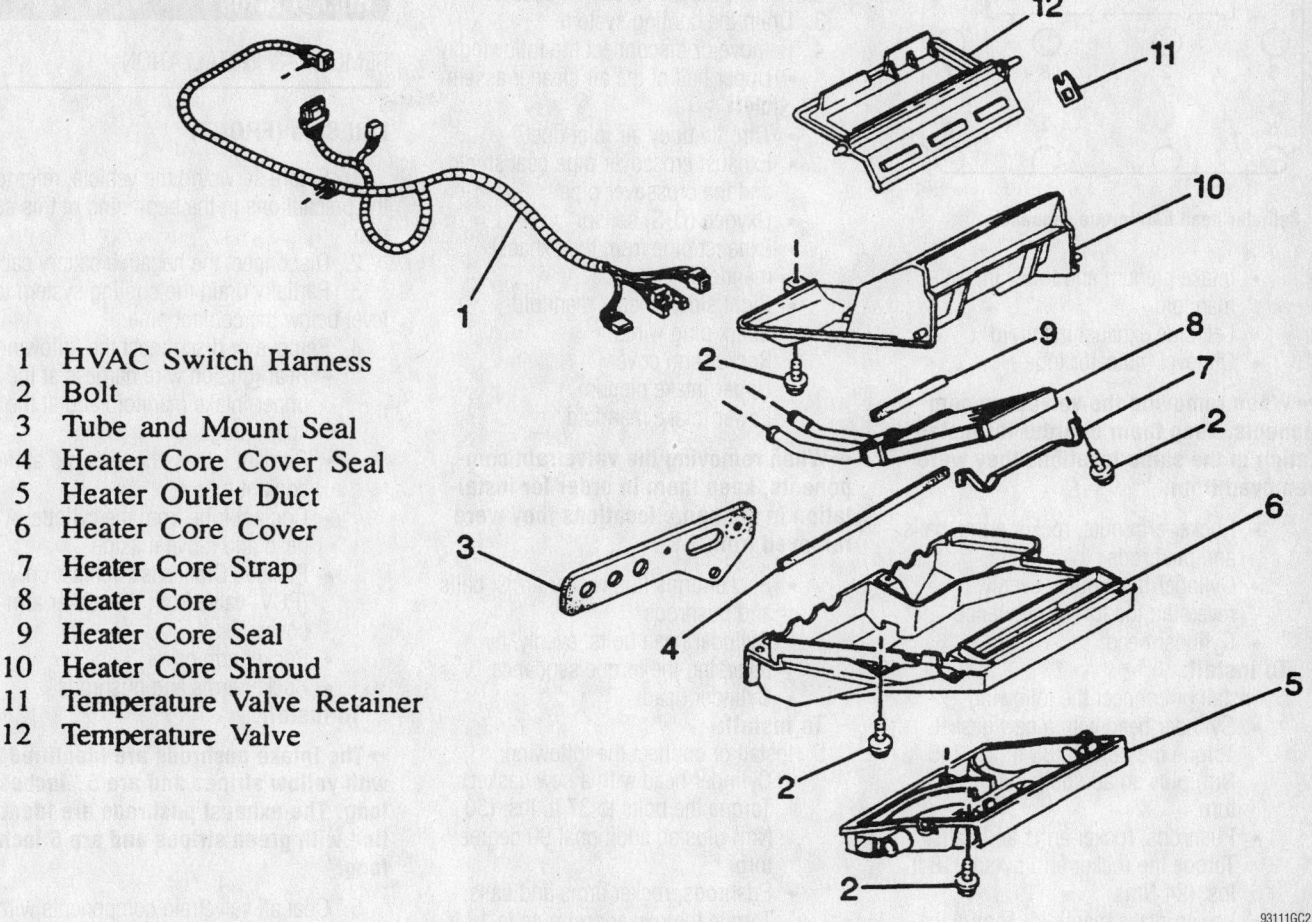

1 HVAC Switch Harness
2 Bolt
3 Tube and Mount Seal
4 Heater Core Cover Seal
5 Heater Outlet Duct
6 Heater Core Cover
7 Heater Core Strap
8 Heater Core
9 Heater Core Seal
10 Heater Core Shroud
11 Temperature Valve Retainer
12 Temperature Valve

93111GC2

View of the heater core, lower heater/air conditioning housing and related components—Malibu

- Instrument panel compartment
- Instrument panel end caps, install the screws
- Instrument panel end caps on both sides
- Instrument panel valance and the valance screws
- Defroster grille

10. Install the console by installing or connecting the following:

- Console and the console-to-chassis screws
- Rear cup holder
- Console trim plate
- Front seats

11. Install the steering wheel by installing or connecting the following:

- Steering wheel to the steering column
- Steering wheel-to-steering column nut and torque to 30 ft. lbs. (41 Nm)
- Horn and cruise control electrical connectors
- SIR module and connect the electrical connector

- SIR module and torque the 2 module-to-steering wheel screws to 89 inch lbs. (10 Nm)
- Heater hoses to the heater core and install the drain tube to the heater/air conditioning housing

12. Refill the cooling system.
13. Connect the negative battery cable.
14. Enable the SIR or air bag by installing or connecting the following:

- Connector Position Assurance (CPA), the yellow 2-way electrical connectors and the lead to the passenger's side SIR module (located on the drivers' side)
- Sound insulator, located on the left side
- Air bag fuse to the fuse block

a. Turn the ignition switch to RUN and verify that the Air Bag Warning light flashes 7–9 times and turns OFF.

➡If the SIR system does not operate as described, perform the SIR diagnostic system check.

15. Operate the engine to normal operating temperatures; then, check the climate control operation and check for leaks.

Cylinder Head

REMOVAL & INSTALLATION

Left (FRONT)

1. Before servicing the vehicle, refer to the precautions in the beginning of this section.
2. Relieve the fuel system pressure.
3. Drain the cooling system.
4. Remove or disconnect the following:

- Upper half of the air cleaner assembly
- Air inlet duct
- Exhaust crossover pipe heat shield and the crossover pipe
- Spark plug wires
- Rocker arm covers

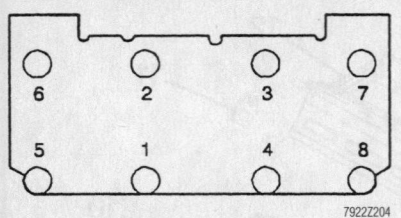

Cylinder head bolt torque sequence

- Intake plenum and lower intake manifold
- Left side exhaust manifold
- Oil level indicator tube

➡**When removing the valvetrain components, keep them in order for installation in the same locations they were removed from.**

- Rocker arm nuts, rocker arms, balls and pushrods
- Cylinder head bolts, evenly, by reversing the torque sequence
- Cylinder head

To install:

5. Install or connect the following:
- Cylinder head with a new gasket. Torque the bolts to 33 ft. lbs. (45 Nm) plus an additional 90 degree turn.
- Pushrods, rocker arms and balls. Torque the rocker arm nuts to 18 ft. lbs. (24 Nm).
- Lower intake manifold. Torque the bolts to 115 inch lbs. (13 Nm).
- Upper intake plenum. Torque the bolts to 18 ft. lbs. (24 Nm).
- Rocker arm covers. Torque the bolts to 89 inch lbs. (10 Nm).
- Oil level indicator tube
- Spark plug wires
- Left side exhaust manifold. Torque the bolts to 12 ft. lbs. (16 Nm).
- Exhaust crossover pipe. Torque the bolts to 18 ft. lbs. (24 Nm).
- Crossover pipe heat shield. Torque the bolts to 89 inch lbs. (10 Nm).
- Upper half of the air cleaner assembly
- Air inlet duct
- Negative battery cable

6. Refill the cooling system.

➡**An oil change is recommended.**

7. Start the vehicle and check for leaks.

Right (REAR)

1. Before servicing the vehicle, refer to the precautions in the beginning of this section.

2. Relieve the fuel system pressure.
3. Drain the cooling system.
4. Remove or disconnect the following:
- Upper half of the air cleaner assembly
- Throttle body air inlet duct
- Exhaust crossover pipe heat shield and the crossover pipe
- Oxygen (O_2S) sensor
- Exhaust pipe from the exhaust manifold
- Right side exhaust manifold
- Spark plug wires
- Rocker arm cover
- Upper intake plenum
- Lower intake manifold

➡**When removing the valvetrain components, keep them in order for installation in the same locations they were removed from.**

- Rocker arms nut, rocker arms, balls and pushrods
- Cylinder head bolts, evenly, by reversing the torque sequence
- Cylinder head

To install:

5. Install or connect the following:
- Cylinder head with a new gasket. Torque the bolts to 37 ft. lbs. (50 Nm) plus an additional 90 degree turn.
- Pushrods, rocker arms and balls. Torque the rocker arm nuts to 18 ft. lbs. (24 Nm).
- Lower intake manifold. Torque the bolts to 115 inch lbs. (13 Nm).
- Upper intake plenum. Torque the bolts to 18 ft. lbs. (24 Nm).
- Rocker arm covers. Torque the bolts to 89 inch lbs. (10 Nm).
- Spark plug wires
- Exhaust manifold. Torque the bolts to 12 ft. lbs. (16 Nm).
- Exhaust pipe to the exhaust manifold. Torque the bolts to 26 ft. lbs. (35 Nm).
- O_2S sensor
- Exhaust crossover pipe. Torque the bolts to 18 ft. lbs. (24 Nm).
- Heat shield. Torque the bolts to 89 inch lbs. (10 Nm).
- Upper half of the air cleaner assembly
- Throttle body air inlet duct
- Negative battery cable

6. Refill the cooling system.
7. An oil and filter change is recommended.
8. Start the vehicle and check for leaks.

Rocker Arms

REMOVAL & INSTALLATION

Left Side (FRONT)

1. Before servicing the vehicle, refer to the precautions in the beginning of this section.
2. Disconnect the negative battery cable.
3. Partially drain the cooling system to a level below the coolant pipe.
4. Remove or disconnect the following:
- Rear ignition wire harness at the upper intake manifold and at the spark plugs
- Coolant bypass hose clamp at the coolant tube
- Coolant tube from the cylinder head and move it aside
- Positive Crankcase Ventilation (PCV) valve from the rocker arm cover
- Rocker arm cover
- Rocker arms and pushrods

To install:

➡**The intake pushrods are identified with yellow stripes and are 5 ' inches long. The exhaust pushrods are identified with green stripes and are 6 inches long.**

5. Coat all valvetrain components with engine oil.
6. Install or connect the following:
- Pushrods and rocker arms. Torque the rocker arm bolts to 89 inch lbs. (10 Nm) plus an additional 30 degree turn.
- Rocker arm cover with a new gasket. Torque the bolts to 89 inch lbs. (10 Nm).
- PCV valve
- Coolant tube and thermostat bypass hose. Torque the screw at the water pump to 106 inch lbs. (12 Nm) and the nut/bolt at the cylinder head to 18 ft. lbs. (24 Nm).
- Rear ignition wire harness to the upper intake manifold and the spark plugs
- Negative battery cable

7. Refill the cooling system.
8. Start the vehicle and verify no leaks.

Right Side (REAR)

1. Before servicing the vehicle, refer to the precautions in the beginning of this section.
2. Remove or disconnect the following:

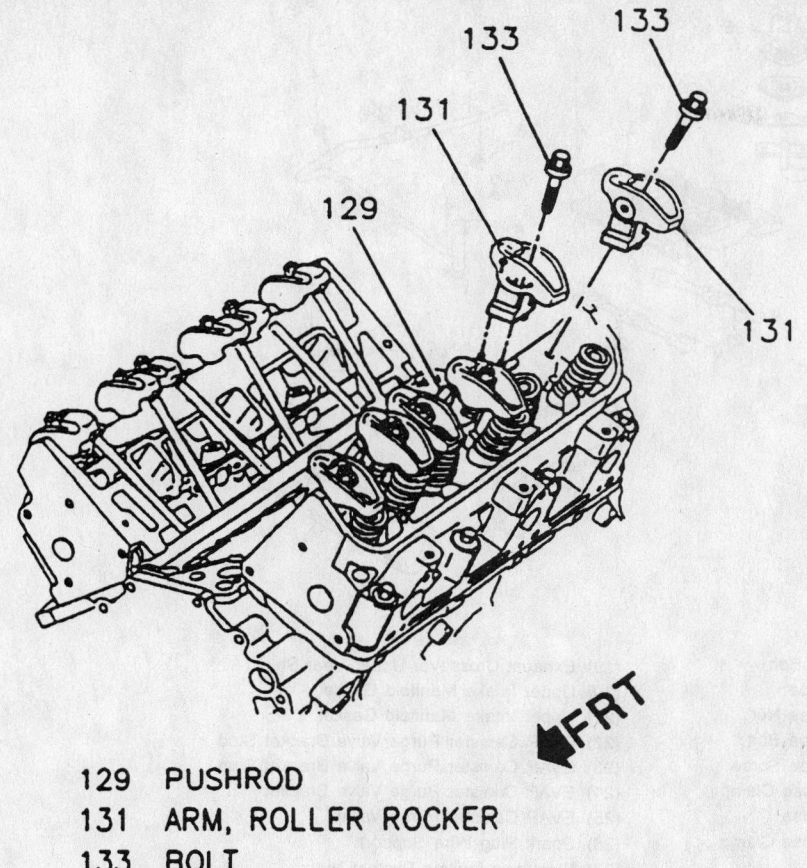

133
133
131
129
131

FRT

129   PUSHROD
131   ARM, ROLLER ROCKER
133   BOLT

7922Z205

Rocker arm components

- Negative battery cable
- Spark plug wires
- Power brake booster vacuum pipe from the intake plenum
- Accessory drive belt
- Alternator
- Ignition coil assembly
- Evaporative Emission (EVAP) canister purge solenoid
- Rocker arm cover
- Rocker arms and pushrods

To install:

➡**The intake pushrods are identified with yellow stripes and are 5 ' inches long. The exhaust pushrods are identified with green stripes and are 6 inches long.**

3. Coat all valvetrain components with engine oil.
4. Install or connect the following:
    - Pushrods and rocker arms. Torque the bolts to 89 inch lbs. (10 Nm) plus an additional 30 degree turn.
    - Rocker arm cover with a new gasket. Torque the bolts to 89 inch lbs. (10 Nm).

- Ignition coil assembly
- EVAP canister purge solenoid
- Alternator. Torque the bolts to 18 ft. lbs. (24 Nm).
- Accessory drive belt
- Power brake booster vacuum pipe to the plenum
- Spark plug wires
- Negative battery cable

Intake Manifold

REMOVAL & INSTALLATION

1. Before servicing the vehicle, refer to the precautions in the beginning of this section.
2. Drain the cooling system.
3. Relieve the fuel system pressure.
4. Remove or disconnect the following:
    - Upper half of the air cleaner assembly
    - Brake vacuum pipe at the intake plenum
    - Fuel pressure regulator vacuum line
    - Spark plug wires

- Ignition coil assembly
- Evaporative Emission (EVAP) canister purge solenoid
- Throttle Position (TPS) sensor electrical connector
- Idle Air Control (IAC) valve electrical connector
- Fuel Injector electrical connectors
- Engine Coolant Temperature (ECT) sensor electrical connector
- Manifold Absolute Pressure (MAP) sensor electrical connector and vacuum line
- Camshaft Position (CMP) sensor electrical connector
- Vacuum modulator supply hose
- Control cables from the throttle body and intake plenum bracket
- Upper intake manifold
- Left and right rocker arm covers
- Fuel injector electrical connectors
- Fuel feed and return line
- Fuel rail with the injectors
- Power steering pump and move it aside without disconnecting the lines
- Heater inlet pipe
- Thermostat bypass hose
- Upper radiator hose at the thermostat housing
- Lower intake manifold bolts

➡**When removing the valvetrain components, keep them in order for installation in their original locations.**

- Rocker arm nuts, rocker arms and pushrods
- Intake manifold

To install:

5. Place a 0.08–0.11 inch (8–12mm) bead of RTV on each ridge, where the front and rear of the intake manifold contact the block.
6. Install or connect the following:
    - Lower intake manifold using a new gasket
    - Pushrods and rocker arms

➡**Be sure the pushrods are properly seated in the valve lifters and rocker arms.**

- Rocker arm nuts. Torque them to 18 ft. lbs. (24 Nm).
- Lower intake manifold bolts, lubricated with sealant. Torque the bolts to 115 inch lbs. (13 Nm).
- Upper radiator hose at the thermostat housing
- Thermostat bypass hose
- Heater inlet pipe
- Power steering pump. Torque the bolts to 25 ft. lbs. (34 Nm).

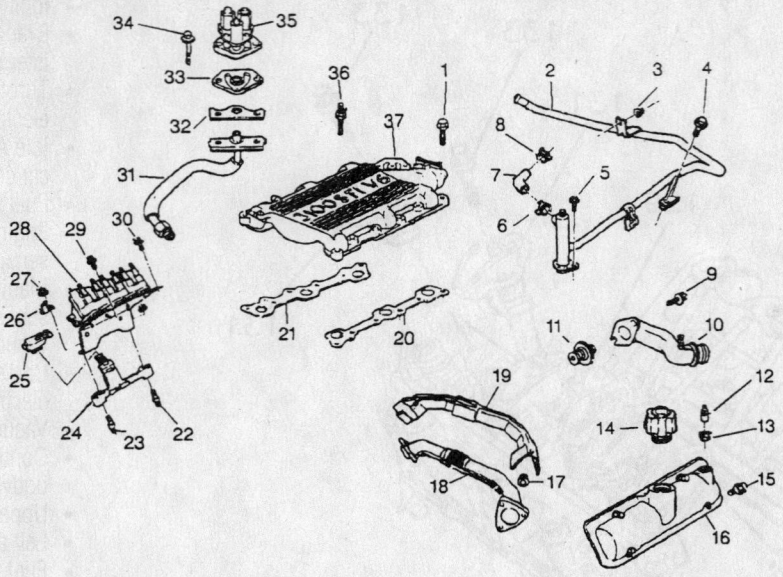

(1) Upper Intake Manifold Bolt
(2) Thermostat Bypass Pipe
(3) Thermostat Bypass Pipe Nut
(4) Thermostat Bypass Pipe Bolt
(5) Thermostat Bypass Pipe Screw
(6) Thermostat Bypass Hose Clamp
(7) Thermostat Bypass Hose
(8) Thermostat Bypass Hose Clamp
(9) Coolant Outlet Bolt
(10) Coolant Outlet Assembly
(11) Thermostat
(12) PCV Valve
(13) PCV Valve Grommet
(14) Oil Fill Cap
(15) Rocker Arm Cover Bolt
(16) Rocker Arm Cover
(17) Exhaust Crossover Nut
(18) Exhaust Crossover Pipe

(19) Exhaust Crossover Upper Heat Shield
(20) Upper Intake Manifold Gasket
(21) Upper Intake Manifold Gasket
(22) EVAP Canister Purge Valve Bracket Stud
(23) EVAP Canister Purge Valve Bracket Stud
(24) EVAP Canister Purge Valve Bracket
(25) EVAP Canister Purge Valve
(26) Spark Plug Wire Support
(27) Electronic Ignition System Nut
(28) Electronic Ignition System
(29) Electronic Ignition System Bolt
(30) Electronic Ignition System Bolt
(31) EGR Valve Pipe Assembly
(32) EGR Valve Pipe Gasket
(33) EGR Valve Gasket
(34) EGR Valve Assembly Bolt
(35) EGR Valve Assembly
(36) Upper Intake Manifold Stud
(37) Upper Intake Manifold

79222207

Exploded view of the upper intake and related components

- Fuel rail with the injectors
- Fuel feed and return lines
- Fuel injector electrical connectors
- Rocker arm covers. Torque the bolts to 89 inch lbs. (10 Nm).
- Upper intake manifold with a new

gasket. Torque the mounting bolts to 18 ft. lbs. (24 Nm).
- Control cables to the throttle body
- MAP sensor vacuum line and electrical connector
- Fuel pressure regulator vacuum line
- Vacuum modulator supply line
- CMP sensor electrical connector
- ECT sensor electrical connector
- IAC valve electrical connector
- TPS sensor electrical connector
- EVAP canister purge solenoid
- Ignition coil assembly
- Spark plug wires
- Brake vacuum pipe at the intake plenum
- Accessory drive belt
- Brake booster vacuum line

- Upper half of the air cleaner assembly
- Negative battery cable
7. Fill the cooling system.

➡**An engine oil and filter change is recommended.**

8. Start the vehicle and verify no leaks.

Exhaust Manifold

REMOVAL & INSTALLATION

Left Side (FRONT)

1. Before servicing the vehicle, refer to the precautions in the beginning of this section.

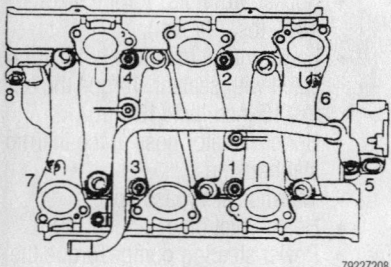

79222208

Lower intake manifold torque sequence

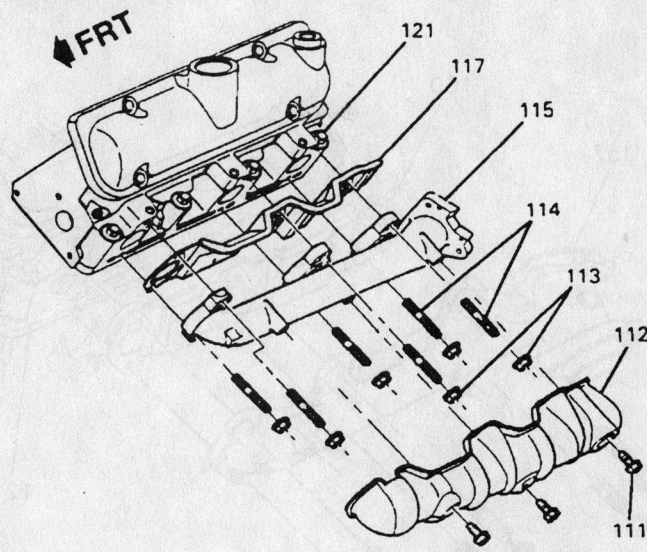

111 SCREW, LH EXHAUST MANIFOLD HEAT SHIELD
112 SHIELD LH EXHAUST MANIFOLD
113 NUT, LH EXHAUST MANIFOLD
114 STUD, LH EXHAUST MANIFOLD
115 MANIFOLD, LH EXHAUST
117 GASKET, LH EXHAUST MANIFOLD
121 HEAD, LH CYLINDER

79222211

Exploded view of the left-hand exhaust manifold

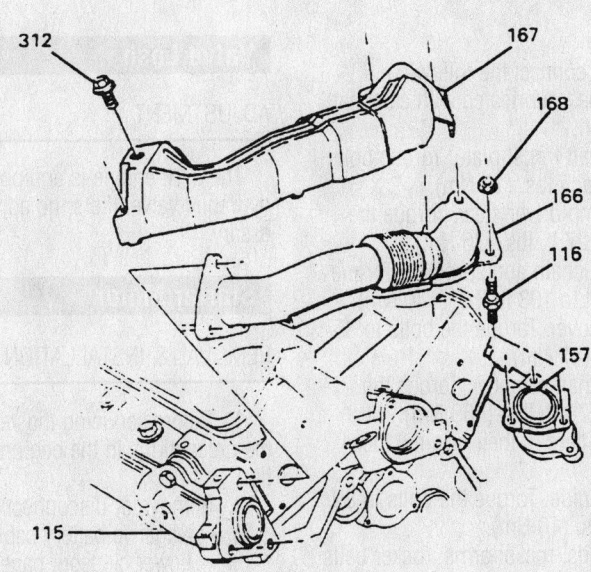

115 MANIFOLD, LEFT HAND EXHAUST
116 STUD, EXHAUST CROSSOVER
157 MANIFOLD, RIGHT HAND EXHAUST
166 CROSSOVER PIPE, EXHAUST
167 SHIELD, EXHAUST CROSSOVER UPPER HEAT
168 NUT, EXHAUST CROSSOVER
312 BOLT/SCREW, EXHAUST CROSSOVER
       UPPER HEAT SHIELD

79222212

Exploded view of the exhaust crossover pipe

2. Partially drain the cooling system.
3. Remove or disconnect the following:
   - Negative battery cable
   - Air cleaner assembly
   - Throttle body duct
   - Exhaust crossover heat shield
   - Exhaust crossover pipe from the manifold
   - Upper radiator hose from the thermostat housing
   - Spark plug wires
   - Exhaust manifold heat shield
   - Exhaust manifold

To install:
4. Install or connect the following:
   - Exhaust manifold using a new gasket. Torque the nuts to 12 ft. lbs. (16 Nm).
   - Exhaust manifold heat shield. Torque the bolts to 89 inch lbs. (10 Nm).
   - Spark plug wires
   - Upper radiator hose
   - Exhaust crossover pipe to the manifold. Torque the bolts to 18 ft. lbs. (24 Nm).
   - Exhaust crossover pipe heat shield. Torque the bolts to 89 inch lbs. (10 Nm).
   - Throttle body duct
   - Air cleaner assembly
   - Negative battery cable
5. Fill the cooling system.

Right Side (REAR)

1. Before servicing the vehicle, refer to the precautions in the beginning of this section.
2. Remove or disconnect the following:
   - Negative battery cable
   - Air cleaner assembly
   - Heated Oxygen (HO$_2$S) sensor
   - Crossover pipe heat shield
   - Crossover pipe
   - Spark plug wires
   - Exhaust manifold heat shield
   - Exhaust manifold

To install:
3. Install or connect the following:
   - Exhaust manifold with a new gasket. Torque the bolts to 12 ft. lbs. (16 Nm).
   - Exhaust manifold heat shield. Torque the bolts to 89 inch lbs. (10 Nm).
   - Spark plug wires
   - Crossover pipe. Torque the bolts to 18 ft. lbs. (24 Nm).
   - Crossover pipe heat shield. Torque the bolts to 89 inch lbs. (10 Nm).
   - HO$_2$S sensor

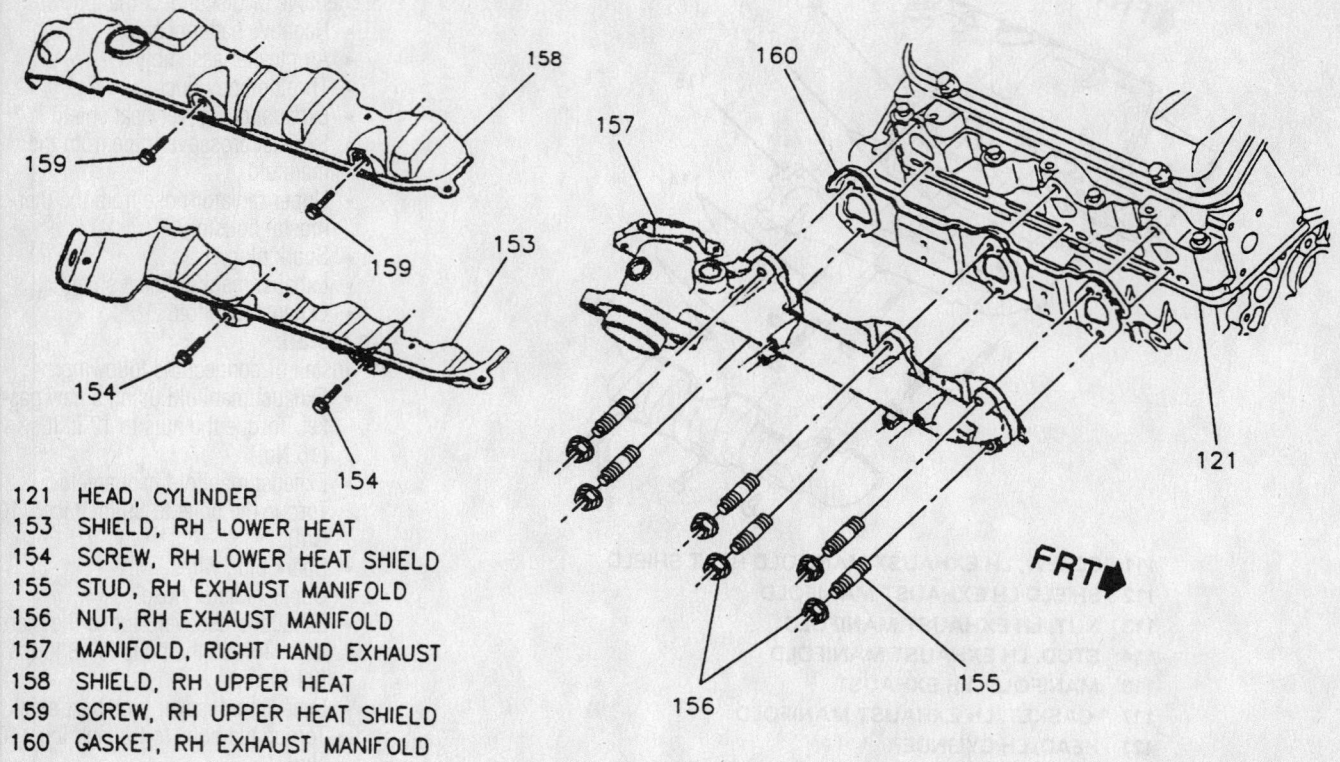

121 HEAD, CYLINDER
153 SHIELD, RH LOWER HEAT
154 SCREW, RH LOWER HEAT SHIELD
155 STUD, RH EXHAUST MANIFOLD
156 NUT, RH EXHAUST MANIFOLD
157 MANIFOLD, RIGHT HAND EXHAUST
158 SHIELD, RH UPPER HEAT
159 SCREW, RH UPPER HEAT SHIELD
160 GASKET, RH EXHAUST MANIFOLD

Exploded view of the right-hand exhaust manifold

- Air cleaner assembly
- Negative battery cable

Camshaft and Valve Lifters

REMOVAL & INSTALLATION

1. Before servicing the vehicle, refer to the precautions in the beginning of this section.
2. Relieve the fuel system pressure.

➡ **When removing valvetrain components, marked them for installation in the same location they are removed from.**

3. Remove or disconnect the following:

- Rocker arm covers
- Intake manifold
- Rocker arms and push rods
- Lifter guide bolts and the guide
- Valve lifter(s) from the lifter bores
- Crankshaft balancer
- Front cover
- Timing chain and sprockets
- Oil pump driven gear bolt and gear
- Camshaft thrust plate
- Camshaft

To install:

4. Install or connect the following:

- Camshaft, lubricated with camshaft lubricant
- Camshaft thrust plate. Torque bolts to 89 inch lbs. (10 Nm).
- Oil pump driven gear. Torque the bolt to 27 ft. lbs. (36 Nm).
- Timing chain and sprocket. Torque the bolt to 103 ft. lbs. (140 Nm).
- Front cover. Torque the bolts to 35 ft. lbs. (47 Nm).
- Crankshaft balancer. Torque the bolt to 76 ft. lbs. (103 Nm).
- Valve lifters in their original locations
- Lifter guide. Torque the bolts to 89 inch lbs. (10 Nm).
- Pushrods, rocker arms, rocker balls and rocker arm nuts. Torque the rocker arm nuts to 89 inch lbs. (10 Nm) plus an additional 30 degrees.
- Intake manifold. Torque the upper and lower intake bolts to 115 inch lbs. (13 Nm).
- Rocker arm covers. Torque the bolts to 89 inch lbs. (10 Nm).
- Negative battery cable

5. Start the engine and verify no oil leaks.

Valve Lash

ADJUSTMENT

The 3.1L engine is equipped with hydraulic valve lifters; no adjustment is necessary.

Starter Motor

REMOVAL & INSTALLATION

1. Before servicing the vehicle, refer to the precautions in the beginning of this section.
2. Remove or disconnect the following:

- Negative battery cable
- Lower closeout panel
- Starter bolts
- Starter electrical connectors
- Starter motor

To install:

3. Install or connect the following:

- Starter electrical connectors. Torque the solenoid cable to 106 inch lbs. (12 Nm).
- Starter motor. Torque the bolts to 37 ft. lbs. (50 Nm).

- Lower closeout panel
- Negative battery cable

Oil Pan

REMOVAL & INSTALLATION

1. Before servicing the vehicle, refer to the precautions in the beginning of this section.
2. Evacuate and recover the A/C system.
3. Drain the engine oil.
4. Remove or disconnect the following:
   - Negative battery cable
   - Accessory drive belt
   - Upper and lower radiator hoses
   - Right front wheel
   - Inner fender splash shield
   - Antilock Brake System (ABS) Wheel Speed Sensor (WSS) electrical connector
   - Right front ball joint from the steering knuckle
   - Right side outer tie rod end
   - A/C compressor bolts and move the compressor aside

✳✳ WARNING

DO NOT disconnect the refrigerant lines or allow the compressor to hang unsupported.

- Evaporator-to-accumulator A/C line
- Flywheel cover
- Engine cradle bolts
- Crankshaft balancer
- Starter motor
- Oil pan

To install:
5. Install or connect the following:
   - Oil pan using a new gasket

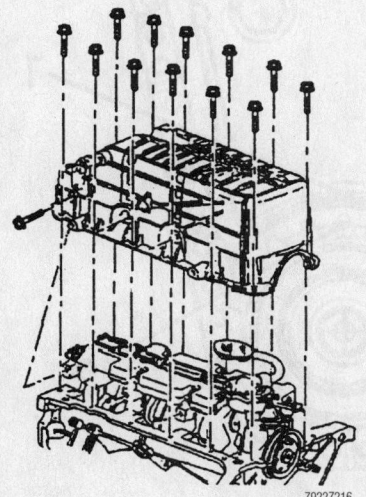
Exploded view of the oil pan mounting

➡**Apply silicone sealer to the portion of the pan that contacts the rear of the block.**

- Oil pan bolts. Torque the bolts to 18 ft. lbs. (24 Nm) and the side bolts to 37 ft. lbs. (50 Nm).
- Starter motor. Torque the bolts to 37 ft. lbs. (50 Nm).
- Flywheel cover
- Crankshaft balancer. Torque the bolt to 76 ft. lbs. (103 Nm).
- A/C compressor. Torque the bolts to 37 ft. lbs. (50 Nm).
- Evaporator-to-accumulator line
- Engine cradle bolts. Torque the bolts to 89 ft. lbs. (120 Nm).
- Ball joint to the steering knuckle using a new cotter pin. Torque the nut 48 ft. lbs. (60 Nm).
- Outer tie rod end. Torque the nut to 33 ft. lbs. (45 Nm).
- ABS wheel speed sensor electrical connector
- Inner fender splash shield
- Right front wheel
- Accessory drive belt
- Upper and lower radiator hoses
- Negative battery cable

6. Refill the crankcase.
7. Evacuate and recharge the A/C system.
8. Start the engine and check for leaks.

➡**Whenever the vehicle subframe is removed or lowered, the wheel alignment should be checked.**

Oil Pump

REMOVAL & INSTALLATION

1. Before servicing the vehicle, refer to the precautions in the beginning of this section.
2. Disconnect the negative battery cable.
3. Drain the engine oil.
4. Remove or disconnect the following:
   - Oil pan
   - Crankshaft oil deflector
   - Oil pump and pump driveshaft

To install:
5. Install or connect the following:
   - Oil pump and pump driveshaft. Torque the bolts to 30 ft. lbs. (41 Nm).
   - Crankshaft oil deflector. Torque the nuts to 18 ft. lbs. (24 Nm).
   - Oil pan. Torque the bolts to 18 ft. lbs. (24 Nm) and the side bolts to 37 ft. lbs. (50 Nm).
   - Negative battery cable

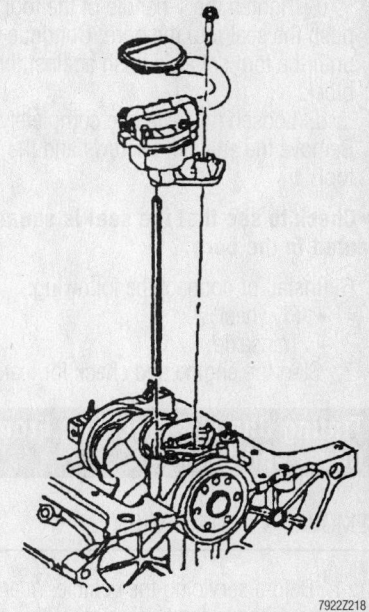
Exploded view of the oil pump mounting

6. Refill the crankcase.

➡**A filter change is recommended.**

7. Start the engine, check the oil pressure and check for leaks.

Rear Main Seal

REMOVAL & INSTALLATION

1. Before servicing the vehicle, refer to the precautions in the beginning of this section.
2. Support the engine.
3. Remove or disconnect the following:
   - Transaxle
   - Flywheel
   - Rear main seal by prying it from the housing

✳✳ WARNING

Use care not to damage the crankshaft seal surface with a prytool.

To install:
4. Lubricate the seal bore and new seal with engine oil.
5. Install the new seal by performing the following procedure:
   a. Slide the new seal over the mandrel until the dust lip bottoms squarely against the tool collar.
   b. Align the dowel pin of the tool with the dowel pin hole in the crankshaft and attach the tool to the crankshaft. Tighten the attaching screws to 24–60 inch lbs. (2.7–6.8 Nm).

c. Tighten the T-handle of the tool to push the seal into the bore. Continue until the tool collar is flush against the block.

d. Loosen the T-handle completely. Remove the attaching screws and the tool.

➡**Check to see that the seal is squarely seated in the bore.**

6. Install or connect the following:
   • Flywheel
   • Transaxle
7. Start the engine and check for leaks.

Timing Chain, Sprockets, Front Cover and Seal

REMOVAL & INSTALLATION

1. Before servicing the vehicle, refer to the precautions in the beginning of this section.
2. Disconnect the negative battery cable.
3. Drain the cooling system.
4. Drain the engine oil.
5. Recover the A/C system refrigerant.
6. Install an engine support fixture.
7. Remove or disconnect the following:
   • Right engine mount assembly
   • Accessory drive belt
   • Air cleaner assembly
   • Throttle body tube
   • Power steering line at the pump
   • Alternator and bracket
   • Right front wheel
   • Right inner fender well splash shield
   • Right engine mount bracket
   • Crankshaft balancer
   • Drive belt tensioner
   • Oil pan
   • Crankshaft Position (CKP) sensor
   • Coolant bypass pipe from the water pump and the intake manifold
   • Lower radiator hose from the front cover outlet
   • Front cover
8. Rotate the crankshaft until the timing marks on the camshaft and crankshaft sprockets are in alignment.
9. Remove or disconnect the following:
   • Camshaft sprocket and timing chain
   • Crankshaft sprocket
   • Timing chain damper

To install:
10. Install or connect the following:
   • Timing chain damper. Torque the bolts to 15 ft. lbs. (21 Nm).

   • Crankshaft sprocket
11. Be sure the crankshaft sprocket timing mark is pointing straight up.
12. Install the timing chain over the camshaft sprocket and hold the sprocket in such a way, that the timing mark is pointing down, and the timing chain is hanging down off the sprocket.
13. Loop the timing chain under the crankshaft sprocket and install the camshaft sprocket on the camshaft. The sprocket will only fit on the camshaft if the dowel on the camshaft aligns with the hole in the sprocket.
14. Verify that the marks are aligned (the camshaft sprocket will be at the 6 o'clock position and the crankshaft sprocket will be in the 12 o'clock position).
15. Torque the camshaft sprocket mounting bolt to 103 ft. lbs. (140 Nm).
16. Lubricate the timing chain components with engine oil.

17. Apply a thin bead of sealer around the gasket sealing area of the front cover.
18. Install or connect the following:
   • Front cover with a new seal and gasket. Torque the small bolts to 15 ft. lbs. (21 Nm) and the large bolts to 35 ft. lbs. (47 Nm).
   • Radiator hose to the coolant outlet
   • Coolant bypass pipe to the water pump and intake manifold
   • CKP sensor
   • Oil pan. Torque the bolts to 18 ft. lbs. (24 Nm) and the side bolts to 37 ft. lbs. (50 Nm).
   • Crankshaft balancer. Torque the bolt to 76 ft. lbs. (103 Nm).
   • Accessory drive belt tensioner. Torque the bolt to 40 ft. lbs. (54 Nm).
   • Right engine mount bracket. Torque the bolts to 96 ft. lbs. (130 Nm).
   • Right inner fender well splash shield

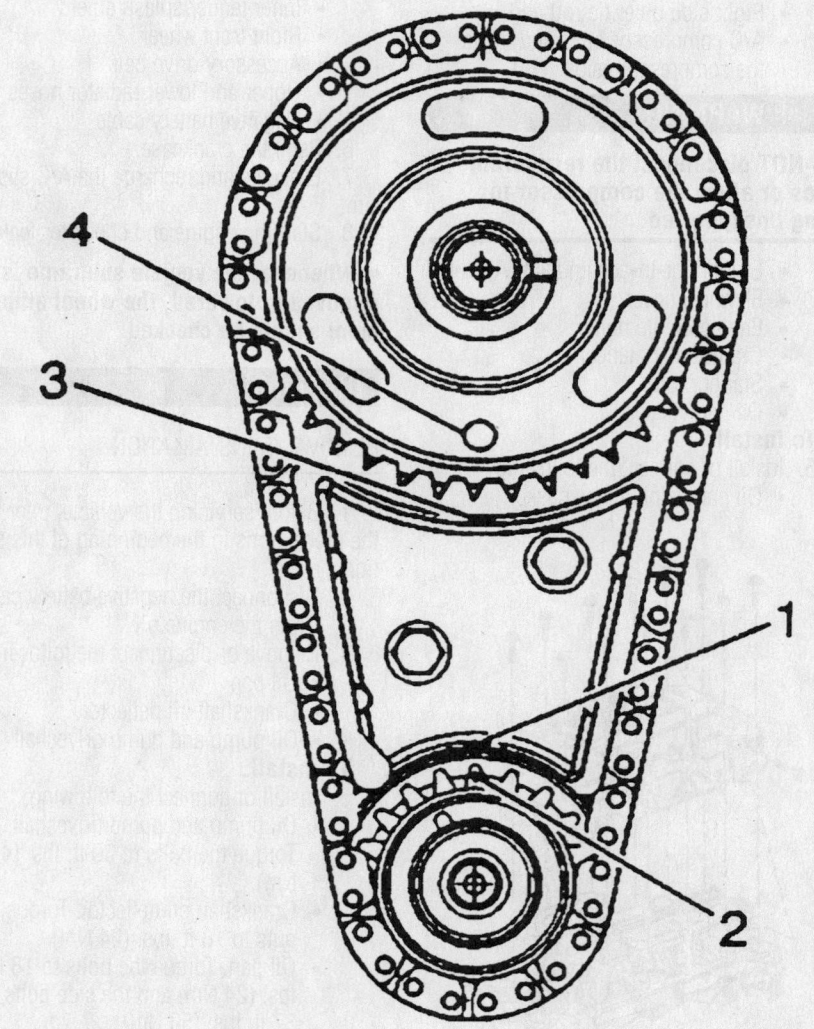

Timing chain and sprocket timing mark alignment

79222Z220

- Wheel
19. Remove the engine support fixture.
20. Install or connect the following:
    - Alternator. Torque the front bolt to 37 ft. lbs. (50 Nm) and the rear bolt to 18 ft. lbs. (24 Nm).
    - Power steering line
    - Throttle body tube
    - Air cleaner assembly
    - Accessory drive belt
    - Negative battery cable
21. Refill the cooling system.
22. Check the engine oil level.

➡**An oil and filter change is recommended.**

23. Start the engine and verify that there are no leaks.

Piston and Ring

POSITIONING

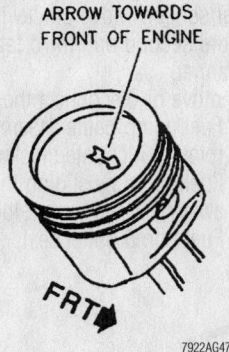

Piston positioning. Often the arrow is replaced by a notch, which also must face toward the front of the engine

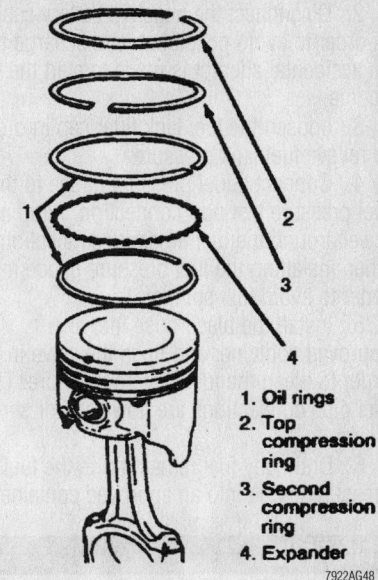

Connecting rod and cap installation. Be sure to matchmark the cap and rod prior to disassembly, as shown

1. Oil rings
2. Top compression ring
3. Second compression ring
4. Expander

Piston ring positioning

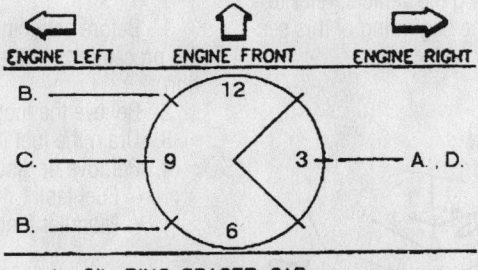

A. OIL RING SPACER GAP (TANG IN HOLE OR SLOT WITH ARC)
B. OIL RING RAIL GAPS
C. 2ND COMPRESSION RING GAP
D. TOP COMPRESSION RING GAP

Piston ring end-gap spacing

FUEL SYSTEM

Fuel System Service Precautions

Safety is the most important factor when performing not only fuel system maintenance but any type of maintenance. Failure to conduct maintenance and repairs in a safe manner may result in serious personal injury or death. Maintenance and testing of the vehicle's fuel system components can be accomplished safely and effectively by adhering to the following rules and guidelines.

- To avoid the possibility of fire and personal injury, always disconnect the negative battery cable unless the repair or test procedure requires that battery voltage be applied.
- Always relieve the fuel system pressure prior to disconnecting any fuel system component (injector, fuel rail, pressure regulator, etc.), fitting or fuel line connection. Exercise extreme caution whenever relieving fuel system pressure, to avoid exposing skin, face and eyes to fuel spray. Please be advised that fuel under pressure may penetrate the skin or any part of the body that it contacts.
- Always place a shop towel or cloth around the fitting or connection prior to loosening to absorb any excess fuel due to spillage. Ensure that all fuel spillage (should it occur) is quickly removed from engine surfaces. Ensure that all fuel soaked cloths or towels are deposited into a suitable waste container.
- Always keep a dry chemical (Class B) fire extinguisher near the work area.
- Do not allow fuel spray or fuel vapors to come into contact with a spark or open flame.
- Always use a back-up wrench when loosening and tightening fuel line connection fittings. This will prevent unnecessary stress and torsion to fuel line piping. Always follow the proper torque specifications.
- Always replace worn fuel fitting O-rings with new. Do not substitute fuel hose, where fuel pipe is installed.

Fuel System Pressure

RELIEVING

1. Before servicing the vehicle, refer to the precautions in the beginning of this section.

2. Disconnect the negative battery cable in order to avoid possible fuel discharge if an accidental attempt is made to start the engine.

3. Loosen the fuel tank filler cap in order to relieve fuel tank pressure.

4. Connect a fuel pressure gauge to the fuel pressure test port connection. Wrap a towel around the fuel pressure connection when installing the fuel pressure gauge in order to avoid fuel spillage.

5. Install the bleed hose into an approved container and open the valve in order to bleed the fuel system pressure. The fuel pipe connections are now safe for servicing.

6. Drain any fuel remaining in the fuel pressure gauge into an approved container.

Fuel Filter

REMOVAL & INSTALLATION

1. Before servicing the vehicle, refer to the precautions in the beginning of this section.

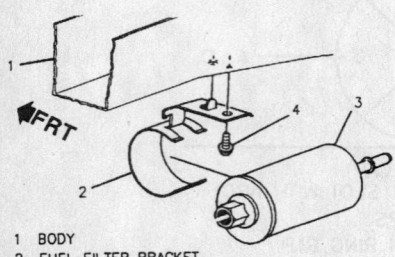

1 BODY
2 FUEL FILTER BRACKET
3 FUEL FILTER
4 SCREW – FULLY DRIVEN, SEATED AND NOT STRIPPED

79222Z221

Exploded view of the fuel filter mounting

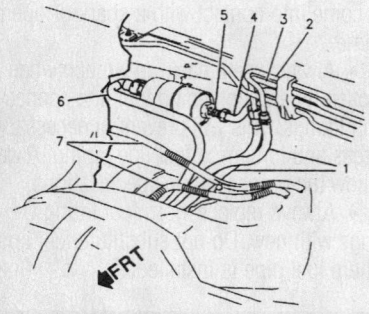

1 HOSE, PART OF FUEL SENDER
2 FUEL VAPOR PIPE
3 FUEL RETURN PIPE
4 FUEL FEED PIPE
5 FUEL FEED PIPE NUT
   27 N•m (20 LBS. FT.)
6 HOSE, PART OF FUEL SENDER
7 ABS AND FUEL SENDER HARNESS

79222Z222

Fuel filter mounting location and component identification

2. Relieve the fuel system pressure.

3. Remove or disconnect the following:
- Fuel line from the filter using a back-up wrench
- Quick-connect fitting from the fuel filter
- Fuel filter from the mounting bracket

To install:

4. Install or connect the following:
- Fuel filter to the mounting bracket
- Fuel line. Torque the fuel line fitting to 20 ft. lbs. (27 Nm).
- Quick-connect fitting to the fuel filter
- Negative battery cable

5. Pressurize the fuel system and verify no leaks.

Fuel Pump

REMOVAL & INSTALLATION

1. Before servicing the vehicle, refer to the precautions in the beginning of this section.

2. Relieve the fuel system pressure.

3. Drain the fuel tank.

4. Remove or disconnect the following:
- Fuel tank
- Modular fuel sender assembly-to-fuel tank snapring by using fuel sender lock nut Tool J-39765, press down and rotate the cam lock ring until free of the fuel sender retaining tabs.

➡ **When removing the modular fuel sender from the tank, be aware that it may spring upward.**

- Modular fuel sender assembly

✳✳ CAUTION

When removing the modular fuel sender from the tank, be aware that the reservoir bucket is full of fuel.

- External fuel strainer
- Connector retainer from the wiring harness
- Fuel pump electrical connector

5. Gently release the tabs on the sides of the fuel sender at the cover assembly. Begin by squeezing the sides of the reservoir and releasing the tab opposite the fuel level sensor. Move clockwise to release the second and third tab in the same manner.

6. Remove or disconnect the following:
- Fuel pump/baffle assembly by rotating it counterclockwise
- Fuel pump by sliding the outlet away from the cover slots
- Fuel pump outlet seal

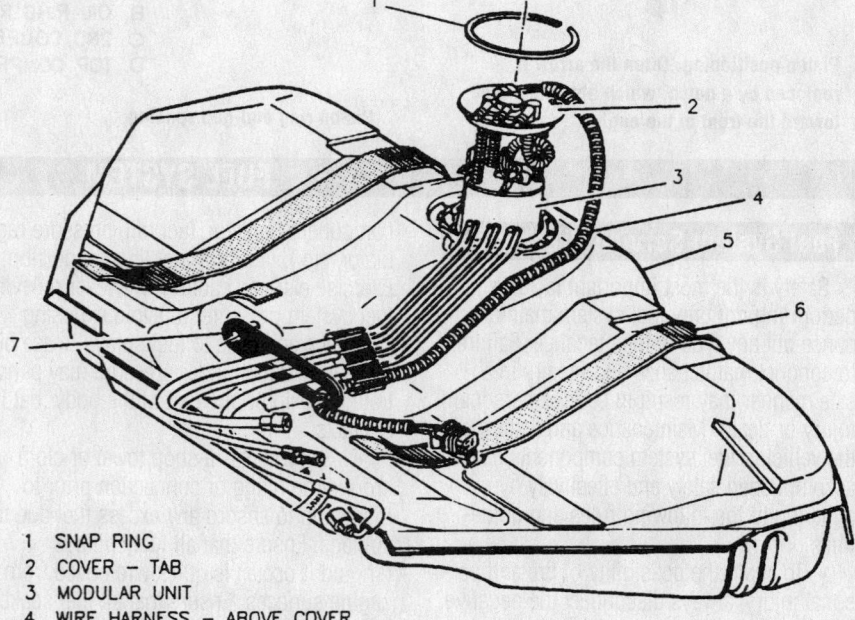

1 SNAP RING
2 COVER – TAB
3 MODULAR UNIT
4 WIRE HARNESS – ABOVE COVER
5 FUEL TANK
6 TANK ISOLATION STRIPS (3)
7 RUBBER ISOLATOR

79222Z223

Exploded view of the fuel sender assembly mounting to the tank

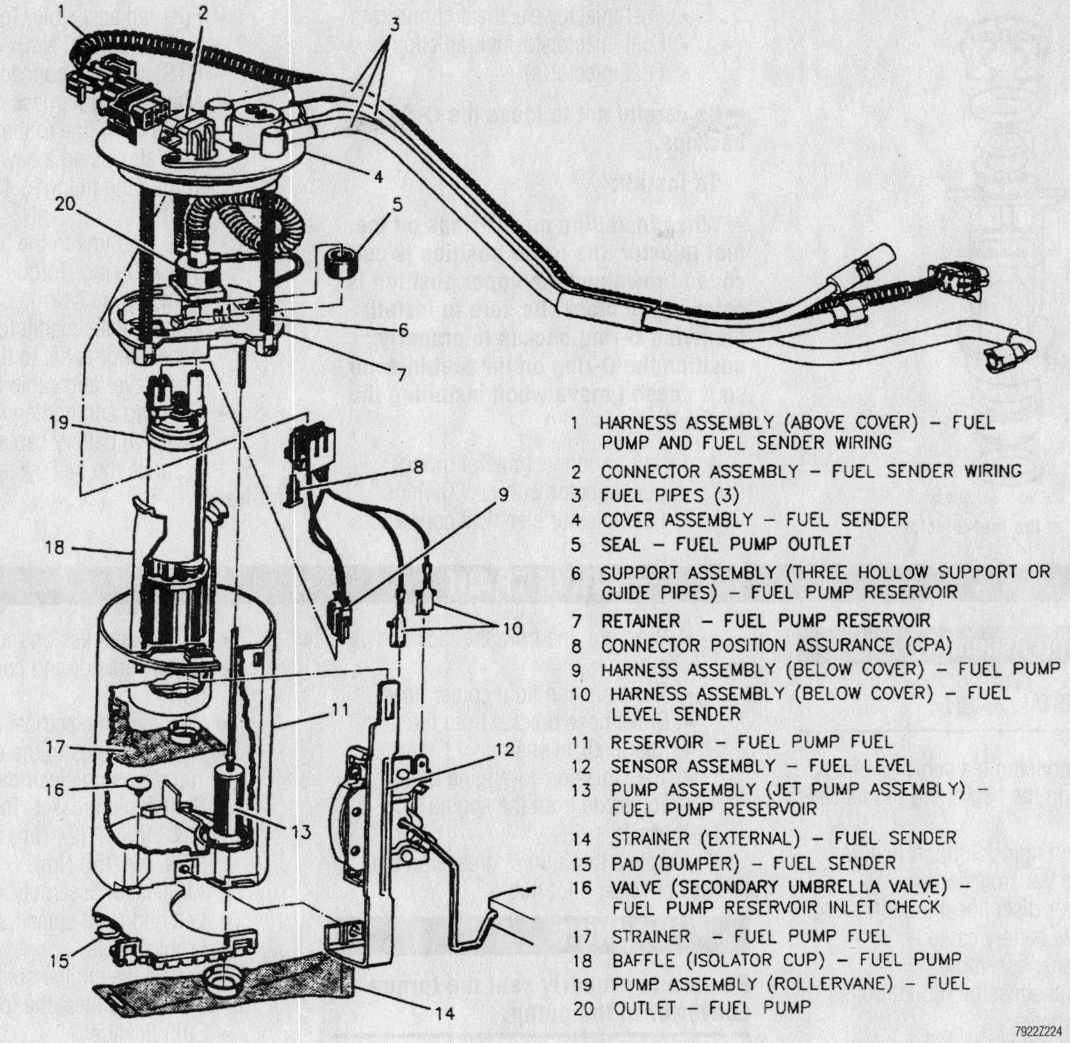

Exploded view of the fuel pump assembly

1 HARNESS ASSEMBLY (ABOVE COVER) – FUEL PUMP AND FUEL SENDER WIRING
2 CONNECTOR ASSEMBLY – FUEL SENDER WIRING
3 FUEL PIPES (3)
4 COVER ASSEMBLY – FUEL SENDER
5 SEAL – FUEL PUMP OUTLET
6 SUPPORT ASSEMBLY (THREE HOLLOW SUPPORT OR GUIDE PIPES) – FUEL PUMP RESERVOIR
7 RETAINER – FUEL PUMP RESERVOIR
8 CONNECTOR POSITION ASSURANCE (CPA)
9 HARNESS ASSEMBLY (BELOW COVER) – FUEL PUMP
10 HARNESS ASSEMBLY (BELOW COVER) – FUEL LEVEL SENDER
11 RESERVOIR – FUEL PUMP FUEL
12 SENSOR ASSEMBLY – FUEL LEVEL
13 PUMP ASSEMBLY (JET PUMP ASSEMBLY) – FUEL PUMP RESERVOIR
14 STRAINER (EXTERNAL) – FUEL SENDER
15 PAD (BUMPER) – FUEL SENDER
16 VALVE (SECONDARY UMBRELLA VALVE) – FUEL PUMP RESERVOIR INLET CHECK
17 STRAINER – FUEL PUMP FUEL
18 BAFFLE (ISOLATOR CUP) – FUEL PUMP
19 PUMP ASSEMBLY (ROLLERVANE) – FUEL
20 OUTLET – FUEL PUMP

79222224

To install:
7. Install or connect the following:
- Fuel pump outlet seal
- Fuel pump outlet by sliding it into the reservoir cover slots
- Fuel pump/baffle assembly onto the reservoir retainer by rotating it clockwise until seated
- Lower retainer assembly by aligning the 3 sleeve tabs and pressing the retainer onto the reservoir until tabs are firmly seated
- Fuel pump electrical connector
- Connector retainer to the wiring harness
- External fuel strainer
- Modular fuel sender assembly
- Modular fuel sender assembly-to-fuel tank snapring using Tool J-39765
- Fuel tank
- Negative battery cable

8. Refill the fuel tank.
9. Pressurize the fuel system and verify no leaks.

Fuel Injector

REMOVAL & INSTALLATION

1. Before servicing the vehicle, refer to the precautions in the beginning of this section.
2. Relieve the fuel system pressure.
3. Remove or disconnect the following:

- Accelerator cable from the throttle body lever and cable bracket
- Upper intake manifold
- Fuel feed line from the fuel rail and discard the O-ring
- Fuel return line from the fuel pressure regulator

- Main wiring harness connectors located near the alternator
- Coolant Temperature Sensor (CTS) electrical connector
- Fuel rail assembly

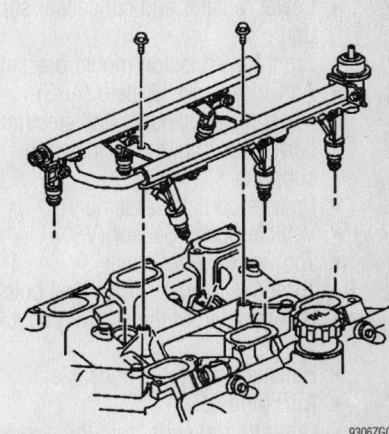

9306ZG02

Exploded view of the fuel rail assembly

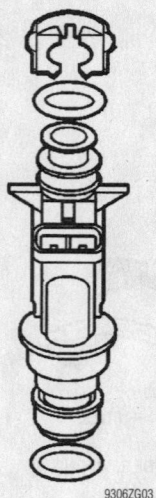

Exploded view of the fuel injector

- Fuel injector electrical connectors
- Fuel injector-to-fuel rail clips
- Fuel injector(s)

➡ **Be careful not to loose the O-ring backups.**

To install:

➡ **When installing new O-rings on the fuel injector, the lower position is color coded brown and the upper position is color coded black. Be sure to install the nylon O-ring backup to properly position the O-ring on the fuel injector so it doesn't move when installing the fuel rail.**

4. Install or connect the following:
- Fuel injector with new O-rings
- Fuel injector electrical connectors
- Fuel rail assembly. Torque the bolts to 7 ft. lbs. (10 Nm).
- CTS electrical connector
- Main wiring harness connectors
- Fuel return line to the fuel pressure regulator using a new O-ring. Torque the fitting to 13 ft. lbs. (17 Nm).
- Fuel feed line to the fuel rail using a new O-ring. Torque the fitting to 13 ft. lbs. (17 Nm).
- Upper intake manifold
- Accelerator cable to the throttle body lever and cable bracket
- Fuel cap and tighten it
- Negative battery cable

5. Pressurize the fuel system and check for leaks.

DRIVE TRAIN

Transmission Assembly

REMOVAL & INSTALLATION

1. Before servicing the vehicle, refer to the precautions in the beginning of this section.
2. Install an Engine Support Fixture.
3. Drain the transmission.
4. Remove or disconnect the following:
- Negative battery cable
- Air cleaner assembly
- Front transmission mount bolts
- Front wheels
- Left and right splash shields
- Shift linkage from the transaxle
- Wiring harness connection from the transaxle
- Ground cables from the engine block
- Park Neutral Position (PNP) switch electrical connector
- Lower radiator and condenser support
- Front transmission mount bracket
- Anti-lock Brake System (ABS) wheel speed sensors and electrical harnesses from the suspension supports
- Brake modulator assembly
- Vehicle Speed Sensor (VSS)
- Torque converter cover
- Torque converter-to-flywheel bolts
- Ball joints from the steering knuckles
- Halfshafts from the transaxle
- ABS module
- Outer tie rod ends from the steering knuckles

- Pressure line from the rack and pinion
- Transmission fluid cooler lines
- Brake hose bracket from the body
- Intermediate shaft
- Transmission-to-engine bolts
- Transaxle from the engine

To install:

5. Apply a thin film of grease on the torque converter pilot hub.

✳✳ WARNING

Be sure to properly seat the torque converter in the pump.

6. Install or connect the following:
- Transaxle. Torque the transmission-to-engine bolts to 66 ft. lbs. (90 Nm).
- Intermediate shaft. Torque the pinch bolt to 15 ft. lbs. (20 Nm).
- Brake hose bracket to the body
- Transaxle fluid cooler lines
- Pressure line to the rack and pinion
- Outer tie rod ends. Torque the nuts to 14 ft. lbs. (20 Nm) plus an additional 180 degree turn.
- ABS module
- Halfshafts to the transaxle
- Ball joints. Torque the nuts to 48 ft. lbs. (65 Nm).
- Torque converter. Torque the bolts to 46 ft. lbs. (62 Nm).
- Torque converter cover
- VSS electrical connector
- Brake modulator assembly
- Front ABS wheel speed sensors and harnesses to the suspension support

- Front transmission mount bracket
- Lower radiator and condenser support
- PNP switch electrical connector
- Ground cables to the engine block
- Transmission electrical connections
- Shift cable bracket. Torque the bolt to 18 ft. lbs. (24 Nm) and the nut to 37 ft. lbs. (50 Nm).
- Air cleaner assembly
- Left and right splash shields
- Front wheels

7. Remove the engine support fixture.
8. Install or connect the following:
- Shift linkage
- Negative battery cable
9. Refill the transmission.
10. Apply the brakes and start the engine.
11. Shift the transaxle from **R** to **D** and back to **P**.
12. Recheck the fluid level.

Halfshaft

REMOVAL & INSTALLATION

1. Before servicing the vehicle, refer to the precautions in the beginning of this section.
2. Remove or disconnect the following:
- Wheel
- Tie rod from the steering knuckle
- Halfshaft hub nut and washer
- Stabilizer link
- Lower ball joint from the steering knuckle
- Halfshaft from the hub/bearing assembly using a puller

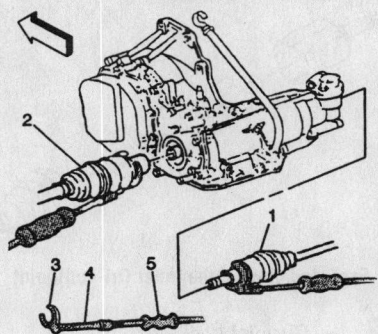

(1) Drive Axle, Right Side   (4) J 29794
(2) Drive Axle, Left Side   (5) J 2619-01
(3) J 28468 or J 33008

79222225

Removing the left and right halfshafts

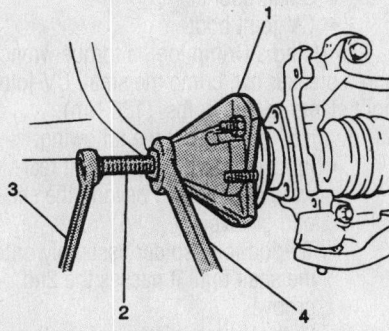

(1) J 28733A   (3) Turn Box Wrench
(2) Forcing Screw   (4) Hold Wrench

79222226

Removing the halfshaft from the hub utilizing the appropriate tools

- Halfshaft from the transaxle

To install:

3. Install the halfshaft into the transaxle by placing a tool in the joint housing groove and tapping until seated.

❋❋ WARNING

Be careful not to damage the axle seal or dislodge the transaxle seal garter spring when installing the halfshaft.

4. Verify that the halfshaft is seated in the transaxle by grasping on the housing and pulling outward.

5. Install or connect the following:
- Halfshaft into the hub/bearing assembly
- Tie rod to the steering knuckle. Torque the nut to 33 ft. lbs. (45 Nm).
- Lower ball joint to the steering knuckle. Torque the nut to 48 ft. lbs. (65 Nm).
- New cotter pin to the lower ball joint

- Stabilizer link. Torque the nut to 13 ft. lbs. (17 Nm).
- New halfshaft nut. Torque it to 284 ft. lbs. (385 Nm).
- Wheel

CV-Joint

OVERHAUL

Outer CV-Joint

1. Before servicing the vehicle, refer to the precautions in the beginning of this section.

2. Remove or disconnect the following:
- Front wheel
- Halfshaft and position it in a vise
- Large CV-joint boot clamp
- Small CV-joint boot clamp
- CV-joint boot and slide it back on the shaft

- Outer race from the halfshaft by spreading the outer race-to-half-shaft retaining ring
- Retaining ring from the halfshaft
- CV-joint boot from the halfshaft

3. Disassemble the chrome alloy balls from the CV-joint cage as follows:
  a. Position a brass drift against the CV-joint cage and tap it with a hammer to tilt the cage.
  b. Remove the 1st chrome alloy ball from the cage.
  c. Tilt the cage in the opposite direction.
  d. Remove the opposite chrome alloy ball.
  e. Repeat the procedure until all 6 balls are removed.

4. Disassemble the CV-joint cage and inner race as follows:
  a. Pivot the cage and race 90 degrees to the center line of the outer race.

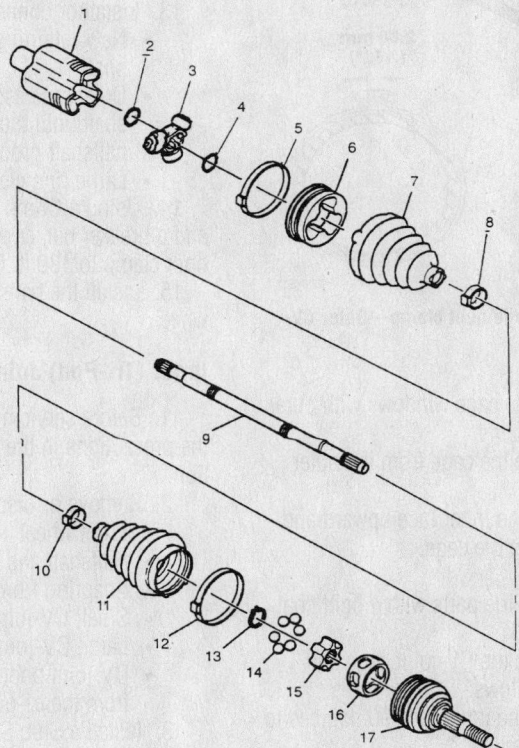

1 Retainer and Housing Assembly	10	Seal Retaining Clamp
2 Shaft Retaining Ring	11	Drive Axle Outboard Seal
3 Tripot Joint Spider Assembly	12	Seal Retaining Clamp
4 Spacer Ring	13	Race Retaining Ring
5 Seal Retaining Clamp	14	Chrome Alloy Ball
6 Tripot Trilobal Bushing	15	CV Joint Inner Race
7 Drive Axle Inboard Seal	16	CV Joint Cage
8 Seal Retaining Clamp	17	CV Joint Outer Race
9 Axle Shaft		

9306ZG01

Exploded view of the halfshaft assembly

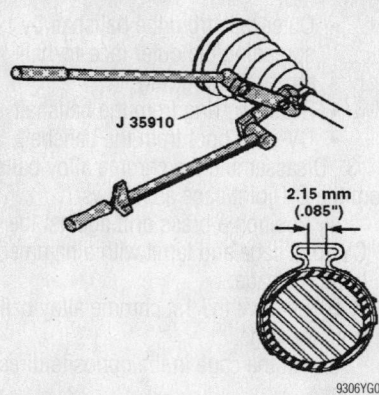

Crimping the small boot clamp—Outer CV-joint

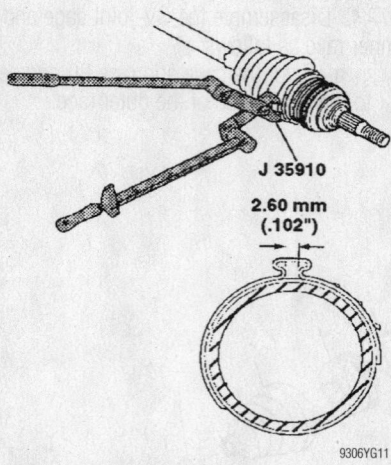

Crimping the large boot clamp—Outer CV-joint

b. Align the cage windows with outer race lands.

c. Remove the cage from the outer race.

d. Rotate the inner race upward and remove it from the cage.

To install:

5. Lubricate the parts with a light coat of grease.

6. Assemble the CV-joint cage and inner race, as follows:

a. Rotate the inner race 90 degrees to the cage centerline.

b. Align the cage windows with inner race lands.

c. Insert the inner race into the cage by rotating the inner race downward.

d. Insert the cage/inner race into the outer race.

7. Assemble the chrome alloy balls into the CV-joint cage, as follows:

a. Position a brass drift against the CV-joint cage and tap it with a hammer to tilt the cage.

b. Insert the 1st chrome alloy ball into the cage.

c. Tilt the cage in the opposite direction.

d. Insert the opposite chrome alloy ball.

e. Repeat the procedure until all 6 balls are inserted.

8. Install ½ of the grease provided into the CV-joint.

9. Install or connect the following:
- Small ring clamp on the CV boot
- CV boot onto the halfshaft

10. Slide the small end of the CV-joint boot/clamp into place, with the seal lip in the halfshaft groove.

➡ **Make sure the boot lies flat against the halfshaft.**

11. Using a Crimp tool, a torque wrench and a breaker bar, crimp the small CV-joint boot clamp to 100 ft. lbs. (136 Nm).

12. Install the remainder of the grease into the CV-joint boot.

13. Install or connect the following:
- New retaining ring on the halfshaft
- Outer race assembly onto the halfshaft until the ring engages the halfshaft groove
- Large ring clamp on the CV boot

14. Using a Crimp tool, a torque wrench and a breaker bar, crimp the large CV-joint boot clamp to 130 ft. lbs. (176 Nm).

15. Install the halfshaft and the front wheel.

Inner (Tri-Pod) Joint

1. Before servicing the vehicle, refer to the precautions in the beginning of this section.

2. Remove or disconnect the following:
- Front wheel
- Halfshaft and place it in a vise
- Snapring from the stub shaft
- Small CV-joint boot clamp
- Large CV-joint boot clamp
- CV-joint boot by sliding it away from the tri-pod joint

3. Install a Stub Shaft Removal Tool J-38868-A to the stub shaft snapring groove.

4. Using a slide hammer puller, pull the stub shaft from the tri-pod housing.

5. Remove or disconnect the following:
- Tri-pod housing from the tri-pod spider
- Inboard spacer ring slide it rearward on the shaft
- Outboard retaining ring
- Tri-pod joint spider assembly
- Inboard spacer ring and discard it

Exploded view of the inner (tri-pod) joint

- CV-joint boot
- Trilobal tri-pod bushing from the housing

To install:

6. Install or connect the following:
- New snapring onto the stub shaft
- Small boot clamp
- CV-joint boot

7. Using a Crimp tool, a torque wrench and a breaker bar, crimp the small CV-joint boot clamp to 100 ft. lbs. (136 Nm).

8. Install or connect the following:
- Inboard spacer ring slide it rearward on the shaft beyond the second groove
- Tri-pod joint spider assembly onto the shaft until it passes the 2nd groove
- Outboard retaining ring into the axle shaft groove
- Tri-pod joint spider assembly, slide it against the outboard retaining ring
- Inboard spacer ring, seat it in the groove
- ½ of the grease provided into the boot
- The remaining grease into the tri-pod housing
- Trilobal tri-pod bushing flush with the tri-pod housing face
- New large seal clamp onto the CV-joint boot
- Tri-pod housing, slide it over the tri-pod joint spider assembly
- CV-joint boot/clamp, slide it into place, over the trilobal tri-pod bushing with the seal lip in the groove

➡ **Make sure the boot lies flat against the trilobal bushing.**

9. Using a Crimp tool, a torque wrench and a breaker bar, crimp the large CV-joint boot clamp to 130 ft. lbs. (176 Nm).

10. Check the clamp gap dimension; if it is not 0.085 in. (2.16mm), continue tightening the clamp until it is.

11. Install the halfshaft and the front wheel.

STEERING AND SUSPENSION

Air Bag

❋❋ CAUTION

Some vehicles are equipped with an air bag system, also known as the Supplemental Inflatable Restraint (SIR) or Supplemental Restraint System (SRS). The system must be disabled before performing service on or around system components, steering column, instrument panel components, wiring and sensors. Failure to follow safety and disabling procedures could result in accidental air bag deployment, possible personal injury and unnecessary system repairs.

PRECAUTIONS

Several precautions must be observed when handling the inflator module to avoid accidental deployment and possible personal injury.

• Never carry the inflator module by the wires or connector on the underside of the module.

• When carrying a live inflator module, hold securely with both hands, and ensure that the bag and trim cover are pointed away.

• Place the inflator module on a bench or other surface with the bag and trim cover facing up.

• With the inflator module on the bench, never place anything on or close to the module, which may be thrown in the event of an accidental deployment.

DISARMING

❋❋ CAUTION

The Supplemental Restraint System (SRS) must be disarmed before performing service procedures around the air bag or SRS wiring. Failure to do so may cause accidental deployment of the air bag, resulting in unnecessary SRS repairs and/or personal injury.

1. Disconnect the negative battery cable.
2. Turn the steering wheel so the vehicle's wheels are pointing straight-ahead.
3. Turn the ignition switch to the **LOCK** position.
4. Remove or disconnect the following:

• Key
• **AIR BAG** fuse from the fuse block
• Left sound insulator
• Connector retainer clip from the yellow 2-way connector at the base of the steering column
• Connector retainer and yellow 2-way connector from the passenger air bag lead, if equipped with a passenger's side air bag

ARMING

1. Turn the ignition switch to the **LOCK** position and remove the key.
2. Install or connect the following:

• Yellow 2-way connector at the base of steering column and the connector retainer clip
• Yellow 2-way connector at the passenger air bag lead and secure it with the connector retainer clip, if equipped with a passenger's side air bag
• Left sound insulator
• **AIR BAG** fuse in the fuse block

3. Turn the ignition switch to the **RUN** position and verify that the **AIR BAG** warning lamp flashes 7 times, then turns **OFF**.

4. Connect the negative battery cable.

Power Rack and Pinion Steering Gear

REMOVAL & INSTALLATION

1. Before servicing the vehicle, refer to the precautions in the beginning of this section.
2. Disconnect the negative battery cable.
3. Siphon the power steering fluid from the reservoir.
4. Remove or disconnect the following:

• Left front wheel
• Intermediate shaft assembly
• Tie rod ends from the steering knuckles
• Stabilizer shaft links
• Steering gear mounting bolts

5. Support the rear of the support frame.
6. Remove or disconnect the following:

• Transaxle-to-crossmember mounting bolt

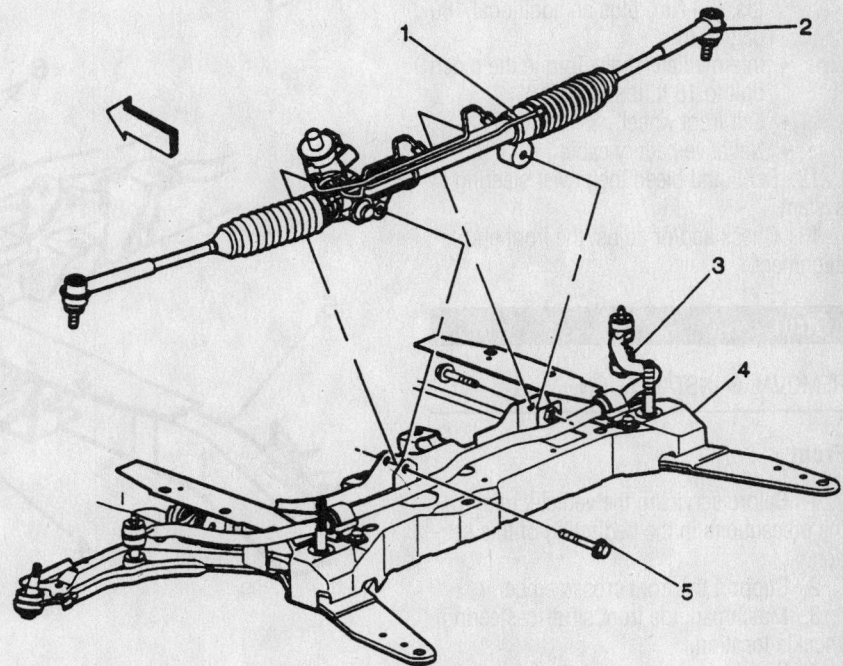

(1) Gear Assembly, Power Steering
(2) Tie Rod
(3) Shaft, Front Stabilizer
(4) Frame Assembly, Drivetrain and Front Crossmember
(5) Bolt, Power Steering Gear Assembly

79222227

Exploded view of the power rack and pinion steering gear mounting

- Rear crossmember-to-body bolts
- Front crossmember bolts, loosen them
7. Lower the support frame for access.
8. Remove or disconnect the following:
   - Power steering hoses from the steering gear
   - Steering gear through the left wheel opening

To install:
9. Install or connect the following:
   - Steering gear through the left wheel opening
   - Power steering hoses to the steering gear
10. Raise the support frame.
11. Install or connect the following:
    - Front crossmember bolts. Torque them to 71 ft. lbs. (110 Nm) plus an additional 90 degree turn.
    - Rear crossmember bolts. Torque them to 71 ft. lbs. (110 Nm) plus an additional 90 degree turn.
    - Transaxle mount-to-crossmember bolt. Torque it to 49 ft. lbs. (66 Nm).
    - Steering gear bolts. Torque them to 89 ft. lbs. (120 Nm).
    - Stabilizer shaft links. Torque the nuts to 13 ft. lbs. (17 Nm).
    - Tie rod ends to the steering knuckle. Torque the nut to 15 ft. lbs. (20 Nm) plus an additional 180 degree turn.
    - Intermediate shaft. Torque the pinch bolt to 16 ft. lbs. (22 Nm).
    - Left front wheel
    - Negative battery cable
12. Refill and bleed the power steering system.
13. Check and/or adjust the front end alignment.

Strut

REMOVAL & INSTALLATION

Front

1. Before servicing the vehicle, refer to the precautions in the beginning of this section.
2. Support the front crossmember.
3. Matchmark the front strut-to-steering knuckle location.
4. Remove or disconnect the following:
   - Front wheel
   - Brake line bracket from the strut
   - Strut lower mounting bracket nut and through-bolts
   - Upper strut-to-chassis nuts

- Strut

To install:
5. Install or connect the following:
   - Strut and finger-tighten the upper strut-to-chassis nuts
   - Strut to the steering knuckle. Torque the through-bolts/nuts to 133 ft. lbs. (180 Nm) with the reference marks aligned.
   - Brake line bracket to the strut. Torque the bolt to 10 ft. lbs. (14 Nm).
   - Wheel
6. Torque the upper strut fasteners to 18 ft. lbs. (24 Nm).
7. Check and/or adjust the front end alignment.

Rear

1. Before servicing the vehicle, refer to the precautions in the beginning of this section.
2. Matchmark the strut-to-knuckle position.
3. Remove or disconnect the following:
   - Rear wheels
   - Upper strut-to-chassis nuts
   - Strut-to-knuckle bolts
   - Strut

To install:
4. Install or connect the following:
   - Strut and finger-tighten the strut-to-knuckle bolts
   - Upper strut-to-chassis nuts. Torque the nuts to 18 ft. lbs. (24 Nm)

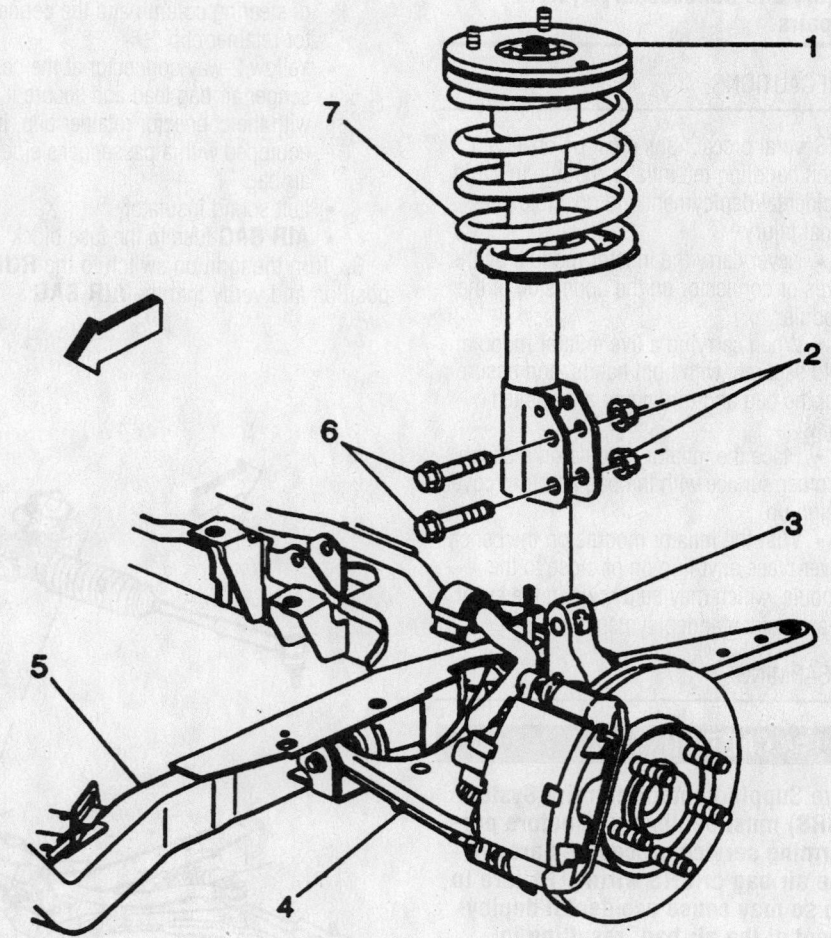

(1) **Bearing, Strut Mount**	(5) **Crossmember**
(2) **Nuts, Strut Mount**	(6) **Bolts, Strut Mount**
(3) **Knuckle**	(7) **Spring, Strut Coil**
(4) **Arm, Control**	

Exploded view and component identification of the front strut

79222228

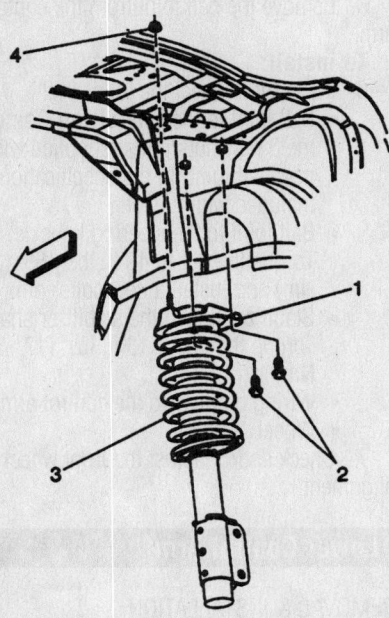

(1) Mount, Strut Upper   (3) Spring, Coil
(2) Bolt, Strut Mount   (4) Nut

79222229

Exploded view of the upper strut mounting

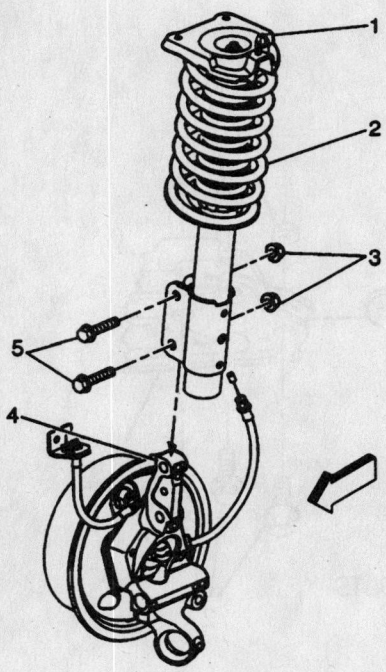

(1) Mount, Strut Upper
(2) Spring, Coil
(3) Nuts, Strut Bolt
(4) Knuckle
(5) Bolt, Strut Mount

79222230

Exploded view of the lower strut mounting

→**Align the matchmarks to ensure proper alignment.**

- Strut-to-knuckle bolts. Torque them to 89 ft. lbs. (120 Nm).
- Rear wheels

5. Check and/or adjust the alignment.

Coil Spring

REMOVAL & INSTALLATION

Front and Rear

1. Before servicing the vehicle, refer to the precautions in the beginning of this section.

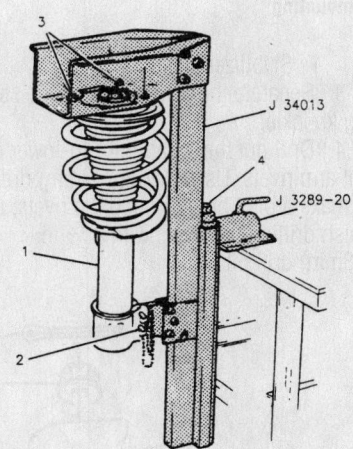

1 STRUT ASSEMBLY
2 INSTALL LOCKING PINS THROUGH STRUT ASSEMBLY
3 TIGHTEN NUTS UNTIL FLUSH WITH STRUT COMPRESSOR
4 COMPRESSOR FORCING SCREW

79222231

View of the strut mounted in the compressor

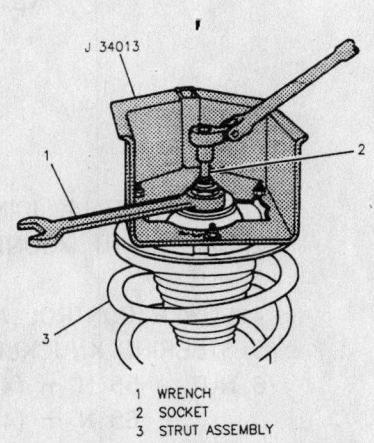

1 WRENCH
2 SOCKET
3 STRUT ASSEMBLY

79222232

Use a socket and a wrench to remove the damper shaft nut spring cap while compressing the spring

2. Remove the strut.
3. Mount the strut in a strut compressor.

→**The strut compressor has strut mounting holes drilled for specific vehicle lines.**

4. Compress the strut approximately ½ of its height after initial contact with the top cap.

❊❊ **WARNING**

Never bottom the spring or damper rod.

5. Remove the strut damper shaft nut and place a guiding rod on top of the damper shaft. Use the rod to guide the damper shaft straight down through the spring cap while compressing the spring. Remove the components.

To install:

6. Install the bearing cap onto the strut compressor, if removed.

7. Mount the strut in the strut compressor using the bottom locking pin only.

8. Install the spring over the damper and swing the assembly up so the upper locking pin can be installed.

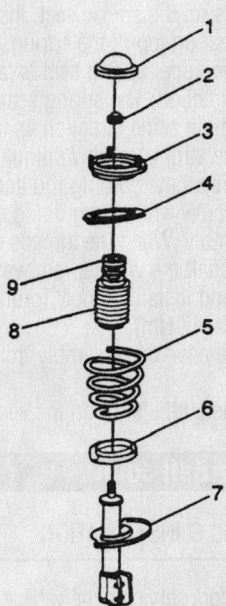

(1) Cap
(2) Nut, Strut Mount
(3) Bearing, Strut Mount
(4) Insulator, Spring (Upper)
(5) Spring, Strut Coil
(6) Insulator, Spring (Lower)
(7) Strut
(8) Bumper, Jounce
(9) Shield, Strut Dust

79222233

Exploded view of the front strut

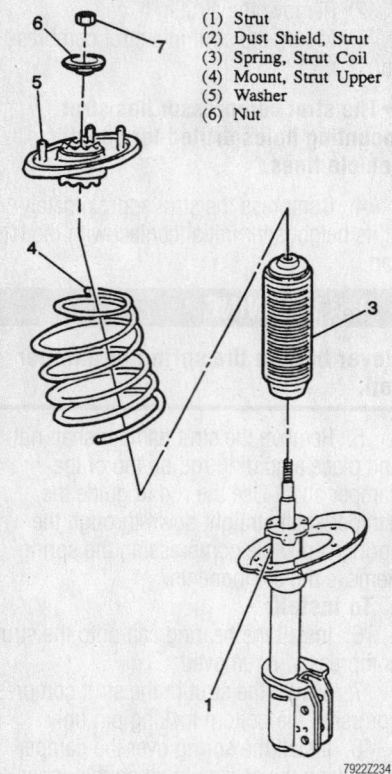

(1) Strut
(2) Dust Shield, Strut
(3) Spring, Strut Coil
(4) Mount, Strut Upper
(5) Washer
(6) Nut

Exploded view of the rear strut

9. Install all shields, bumpers and insulators on the spring seat. Install the spring seat on top of the spring. Be sure the flat on the upper spring seat is facing in the proper direction. The spring seat flat should be facing the same direction as the centerline of the strut assembly spindle.

10. Install the guiding rod and turn the forcing screw while the guiding rod centers the assembly. When the threads on the damper shaft are visible, remove the guiding rod and install the nut. Torque the nut to 34 ft. lbs. (47 Nm).

11. Remove the assembly from the compressor.

12. Install the strut on the vehicle.

Lower Ball Joint

REMOVAL & INSTALLATION

1. Before servicing the vehicle, refer to the precautions in the beginning of this section.

2. Remove or disconnect the following:
   • Wheel

➡Care must be exercised to prevent the halfshaft joints from being over-extended.

   • Wiring harness from the control arm

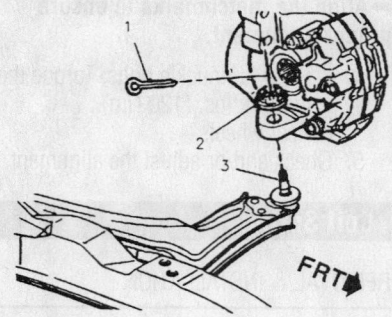

1 PIN
2 NUT – 55 N·m (41 LBS. FT.) MINIMUM TORQUE
  65 N·m (48 LBS. FT.) MAXIMUM TORQUE TO
  INSTALL PIN
3 LOWER BALL JOINT

Exploded view of the ball joint-to-knuckle mounting

   • Stabilizer shaft link

3. Separate the ball joint from the steering knuckle.

4. Drill out the 3 ball joint-to-lower control arm rivets. Use an ⅛ in. (3mm) drill bit to make a pilot hole through the rivets; then, finish drilling the rivets with a ½ in. (13mm) drill bit.

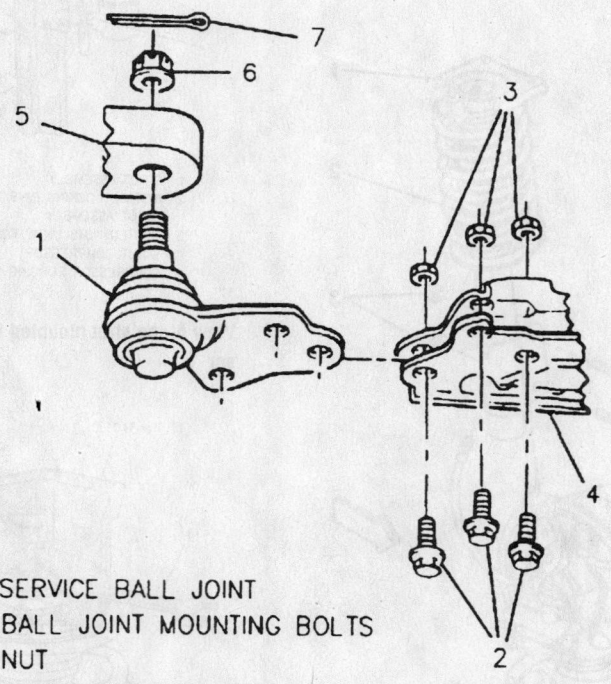

1 SERVICE BALL JOINT
2 BALL JOINT MOUNTING BOLTS
3 NUT
4 LOWER CONTROL ARM
5 STEERING KNUCKLE
6 NUT – 55 N·m (41 LBS. FT.) MINIMUM TORQUE
       65 N·m (48 LBS. FT.) MAXIMUM TORQUE
       TO INSTALL PIN
7 PIN

Exploded view of the ball joint mounting

5. Remove the ball joint from the control arm.

To install:

6. Install or connect the following:
   • Ball joint to the control arm. Torque the 3 new nuts/bolts (supplied with new ball joint) to the specifications included with the kit.
   • Ball joint to the steering knuckle. Torque the nut to 48 ft. lbs. (65 Nm) and install a new cotter pin.
   • Stabilizer link to the stabilizer shaft. Torque the nut to 13 ft. lbs. (17 Nm).
   • Wiring harness to the control arm
   • Wheel

7. Check and/or adjust the front wheel alignment.

Lower Control Arm

REMOVAL & INSTALLATION

1. Before servicing the vehicle, refer to the precautions in the beginning of this section.

2. Remove or disconnect the following:

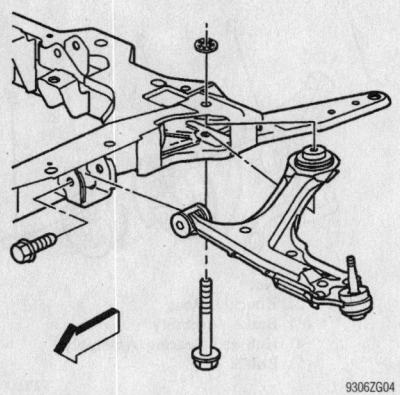

Exploded view of the lower control arm and related components

- Front wheel
- Stabilizer shaft link
- Lower ball joint from the steering knuckle
- Lower control arm-to-suspension crossmember bolts
- Lower control arm

To install:

3. Install or connect the following:

- Lower control arm-to-suspension crossmember and hand-tighten the bolts
- Lower ball joint to the steering knuckle. Torque the nut to 48 ft. lbs. (65 Nm).
- Stabilizer shaft link. Torque the nut to 13 ft. lbs. (17 Nm).
- Front wheel

4. Position the vehicle at curb height.

5. Torque the bolts as follows:

  a. Front lower control arm-to-suspension crossmember bolt to 84 ft. lbs. (115 Nm) plus an additional 120 degrees.

  b. Rear lower control arm-to-suspension crossmember bolt to 180 ft. lbs. (245 Nm) plus an additional 180 degrees.

6. Check and/or adjust the front alignment.

CONTROL ARM BUSHING REPLACEMENT

Front Bushing

1. Before servicing the vehicle, refer to the precautions in the beginning of this section.

2. Remove the lower control arm and place it in a vise.

3. Lubricate the threads of Screw J-21474-19 with high pressure lubricant.

4. Assemble Tools Screw J-21474-19, Remover/Installer J-41397-1A, Receiver J-

41397-2A and J-21474-18 onto the front control arm bushing.

5. Tighten Tool J-21474-18 until the front bushing is pressed from the control arm.

6. Disassemble the tools.

To install:

7. Lubricate the new front bushing outer casing.

8. Insert the new bushing into the control arm.

9. Assemble Tools Screw J-21474-19, Remover/Installer J-41397-1A, Receiver J-41397-2A and J-21474-18 onto the front control arm bushing.

10. Tighten Screw J-21474-19 until the front bushing is pressed into the control arm.

11. Disassemble the tools.

12. Install the lower control arm.

Rear Bushing

1. Before servicing the vehicle, refer to the precautions in the beginning of this section.

2. Remove the lower control arm and place it in a vise.

3. Assemble Tools Screw J-21474-27,

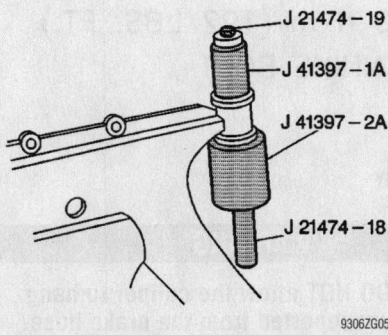

Removing the front bushing from the lower control arm

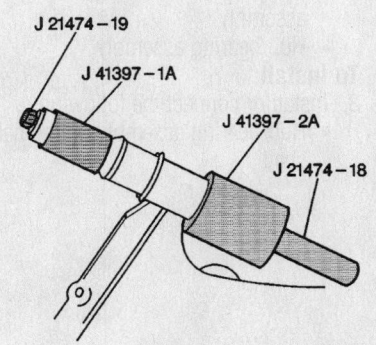

Installing the front bushing to the lower control arm

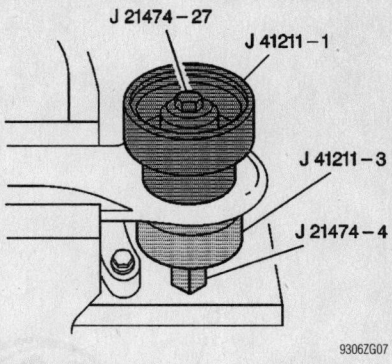

Removing the rear bushing from the lower control arm

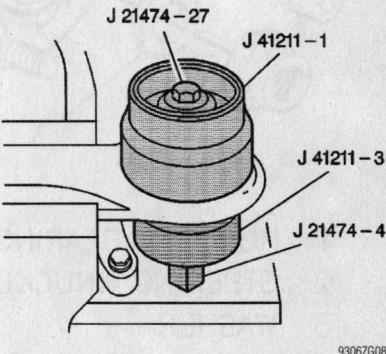

Installing the rear bushing to the lower control arm

Remover/Installer J-41211-1, Receiver J-41211-3 and J-21474-4 onto the rear control arm bushing.

4. Tighten Tool J-21474-27 until the rear bushing is pressed from the control arm.

5. Disassemble the tools.

To install:

6. Insert the new bushing into the control arm.

7. Assemble Tools Screw J-21474-27, Remover/Installer J-41211-1, Receiver J-41211-3 and J-21474-4 onto the rear control arm bushing.

8. Tighten Screw J-21474-4 until the rear bushing is pressed into the control arm.

9. Disassemble the tools.

10. Install the lower control arm.

Wheel Bearings

ADJUSTMENT

These vehicles are equipped with sealed hub and bearing assemblies. The hub and bearing assemblies are non-serviceable. If

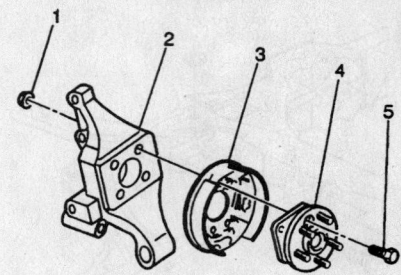

(1) Nut
(2) Knuckle, Rear
(3) Brake Assembly
(4) Hub and Bearing Assembly
(5) Bolt

79222Z238

Hub and bearing components

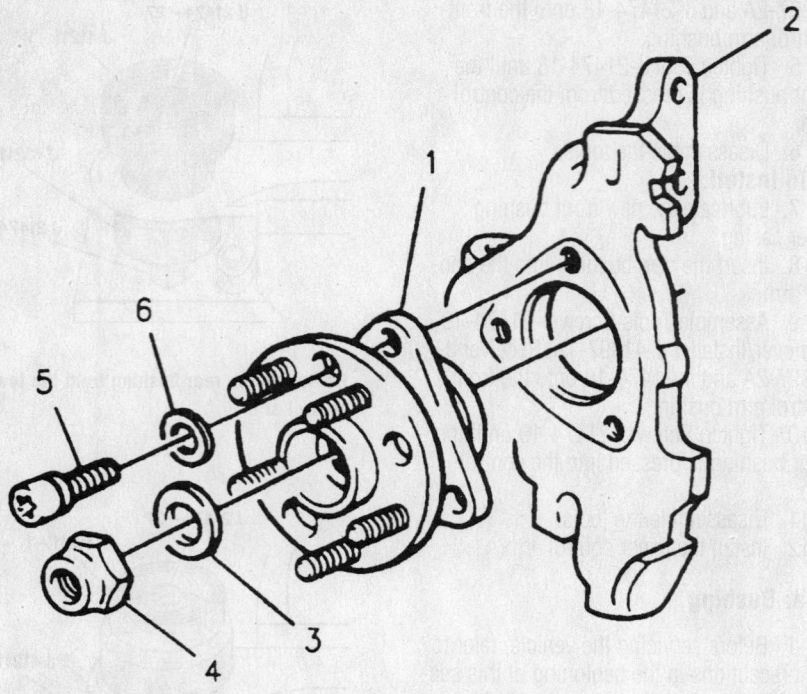

1 HUB AND BEARING ASSEMBLY
2 STEERING KNUCKLE
3 WASHER
4 DRIVE AXLE NUT — 260 N·m (192 LBS. FT.)
5 HUB AND BEARING RETAINING BOLT
6 WASHER

79222Z237

Exploded view of the front hub and bearing assembly

the assembly is damaged, the complete unit must be replaced.

REMOVAL & INSTALLATION

Front

1. Before servicing the vehicle, refer to the precautions in the beginning of this section.
2. Remove or disconnect the following:

- Front wheel
- Halfshaft nut
- Caliper from the steering knuckle and support it aside

※※ WARNING

DO NOT allow the caliper to hang unsupported from the brake hose.

- Brake rotor
- 3 hub bearing-to-steering knuckle bolts
- Halfshaft from the hub/bearing assembly
- Hub/bearing assembly

To install:
3. Install or connect the following:
- Hub/bearing assembly on the half-shaft

- 3 hub/bearing-to-steering knuckle bolts. Torque the bolts to 70 ft. lbs. (95 Nm).
- Brake rotor
- Caliper onto the steering knuckle. Torque the bolts to 38 ft. lbs. (51 Nm).
- Halfshaft nut. Torque it to 284 ft. lbs. (385 Nm).
- Front wheel. Torque the nuts to 100 ft. lbs. (140 Nm).

Rear

1. Before servicing the vehicle, refer to the precautions in the beginning of this section.
2. Remove or disconnect the following:

- Wheel
- Brake drum
- 4 hub/bearing assembly-to-knuckle nuts
- Anti-lock Brake System (ABS) speed sensor wire from the hub/bearing assembly
- Hub/bearing assembly

To install:
3. Install or connect the following:
- Hub/bearing assembly
- ABS wheel speed sensor
- Hub/bearing-to-knuckle nuts. Torque the nuts 62 ft. lbs. (85 Nm).
- Brake drum
- Wheel. Torque the bolts to 100 ft. lbs. (140 Nm).

BRAKES

Brake Caliper

REMOVAL & INSTALLATION

1. Before servicing the vehicle, refer to the precautions in the beginning of this section.

2. Siphon ⅔ of the brake fluid out of the master cylinder.

3. Remove the tire and wheel assembly.

4. Compress the caliper piston back into the caliper bore using a large pair of pliers, C-clamp or special piston retracting tool.

5. Remove the brake hose from the caliper and discard the copper washers.

6. Plug the hose to prevent excessive fluid loss and possible fluid contamination.

7. Remove the caliper mounting bolts and remove the caliper from the knuckle.

8. Remove the brake pads from the caliper, if the caliper is being replaced.

To install:

9. Inspect the condition of the caliper support for rust and corrosion that will hinder the travel of the caliper.

10. Inspect the caliper mounting hardware. New bolts are usually recommended.

11. Lubricate the mounting bushings and sleeves with silicone grease as required.

12. Install the brake pads in the caliper.

13. Install the caliper on the steering knuckle.

14. Install the mounting bolts and torque to 40 ft. lbs. (51 Nm).

15. Connect the brake hose to the caliper using new copper washers and torque the mounting bolt to 35 ft. lbs. (44 Nm).

16. Refill the master cylinder and bleed the brake system.

17. Install the tire and wheel assembly.

18. Verify correct brake operation.

Disc Brake Pads

REMOVAL & INSTALLATION

1. Before servicing the vehicle, refer to the precautions in the beginning of this section.

2. Siphon ⅔ of the brake fluid out of the master cylinder reservoir.

3. Remove the tire and wheel assembly.

4. Remove the caliper from the steering knuckle without disconnecting the brake hose. DO NOT allow the caliper to hang

from the brake hose. Support the caliper with a piece of wire.

5. Remove the outboard pad by pushing in on the outside edge of the pad to release the mounting dowel from the hole in the caliper. When both dowels are unseated push the pad out the bottom of the caliper.

6. Remove the inboard pad from the caliper by pulling it out of the caliper.

To install:

7. Install the inboard pad in the caliper so the spring clip on the pad back engages in the caliper piston.

8. Install the outboard pad over the caliper end until the mounting dowels snap into the mounting holes in the caliper.

9. Install the caliper over the rotor onto the steering knuckle. Install the mounting bolts and sleeves and torque to 40 ft. lbs. (51 Nm).

10. Install the tire and wheel assembly.

11. Pump the brake pedal several times to seat the pads against the rotor before attempting to move the vehicle.

12. Check the master cylinder level and add fluid as necessary.

Brake Drums

REMOVAL & INSTALLATION

1. Before servicing the vehicle, refer to the precautions in the beginning of this section.

2. Remove the wheel.

3. Pull the brake drum off. It may be necessary to gently tap the rear edge of the drum to start it off the studs. If extreme resistance to removal is encountered, it will be necessary to retract the brake shoe self-adjuster screw, sometimes called the star wheel. Knock out the access hole in the brake drum and turn the adjuster to retract the linings from the drum. Install a replacement hole cover before reinstalling the drum.

➡ **DO NOT hammer on the brake drum to remove it.**

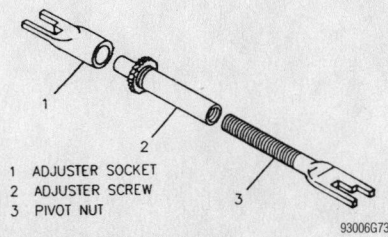

1 ADJUSTER SOCKET
2 ADJUSTER SCREW
3 PIVOT NUT

93006G73

Brake adjuster—Malibu

To install:

4. Inspect the inside of the brake drum. If worn, heavily grooved or if the opening is distorted, the drum should be refinished or replaced. If refinishing, observe the maximum drum diameter specification.

5. Inspect the wheel cylinder for signs of brake fluid leakage. Inspect the brake shoe springs and self-adjuster mechanism. The adjuster should usually be disassembled, cleaned and lubricated when the drum is removed for brake service.

6. Install the drum over the brake shoes.

7. Install the wheel. Adjust the brakes.

8. Check brake operation.

Brake Shoes

REMOVAL & INSTALLATION

1. Before servicing the vehicle, refer to the precautions in the beginning of this section.

2. Remove the tire and wheel assembly.

3. Remove the brake drum.

4. Remove the upper return springs from the shoes using tool J-8057 brake spring tool.

5. Remove the hold-down springs using J-8049 brake spring tool.

6. Remove the shoe hold-down pins from behind the brake backing plate.

7. Lift up the actuator lever for the self-adjusting mechanism and remove the actuating link. Remove the actuator lever, pivot, and the pivot return spring.

8. Spread the shoes apart to clear the wheel cylinder pistons and remove the parking brake strut and spring.

9. Disconnect the parking brake cable from the lever. Remove the shoes, still connected by their adjusting screw spring.

10. With the shoes removed, note the position of the adjusting spring and remove the spring and adjusting screw.

11. Remove the C-clip from the parking brake lever and the lever from the secondary shoe.

12. Use a damp cloth to remove all dirt and dust from the backing plate and brake parts.

To install:

13. Check the backing plate attaching bolts to make sure they are tight. Use fine emery cloth to clean all rust and dirt from the shoe contact surfaces on the plate and lubricate with brake grease. Check the wheel cylinder for signs of leakage.

14. Clean all parts completely in brake

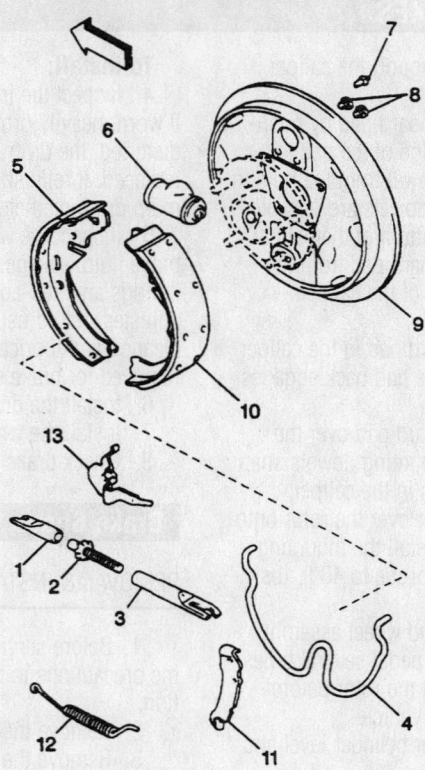

(1) Socket, Brake Adjuster
(2) Screw, Brake Adjuster
(3) Nut, Brake Pivot
(4) Spring, Retractor
(5) Brake Shoe and Lining
(6) Cylinder, Wheel Brake
(7) Valve, Bleeder
(8) Bolts, Wheel Cylinder
(9) Plate, Brake Backing
(10) Brake Shoe and Lining
(11) Lever, Park Brake
(12) Spring
(13) Adjuster Actuator

93006G78

Drum brake components, exploded view—Malibu

solvent and air dry. Clean the backing plate shoe contact points.

15. Inspect the inside of the brake drum. If worn, heavily grooved or if the opening is distorted, the drum should be refinished or replaced.

16. Inspect the brake shoe springs and self-adjuster mechanism. Disassemble the adjuster mechanism and clean the threads and coat with grease. Make sure the adjuster assembly turns freely before installing in the vehicle.

17. Install the parking brake lever on the secondary shoe and secure with C-clip.

18. Install the adjusting screw and spring on the shoes, connecting them together. The coils of the spring must not be over the star wheel on the adjuster. The left and right hand springs are not interchangeable.

19. Spread the shoe assemblies and connect the parking brake cable. Install the shoes on the backing plate, engaging the shoes at the top temporarily with the wheel cylinder pistons. Make sure the star wheel on the adjuster is lined up with the adjusting hole in the backing plate, if equipped.

20. Spread the shoes slightly and install the parking brake strut and spring. Make sure the end of the strut without the spring engages the parking brake lever. The end with the spring engages the primary shoe (the one with the shorter lining).

21. Install the actuator pivot, lever and return spring. Install the actuating link in the shoe retainer. Lift up the actuator lever and hook the link into the lever.

22. Install the hold-down pins through the back of the plate using J-8057. Install the lever pivots and hold-down springs. Install the shoe return springs using J-8049. Be careful not to stretch or distort the springs.

23. Make sure the linings are in the right place, the self-adjusting mechanism is correctly installed, and the parking brake parts are hooked up.

24. Measure the distance from the edge of the primary lining to the edge secondary lining, then measure the inside width of the drum. Adjust the linings by means of the adjuster so the drum will fit onto the linings.

25. Install the hub and bearing assembly onto the axle if removed. Torque the retaining bolts to 35 ft. lbs. (55 Nm).

26. Install the drum and the wheels.

27. Adjust the brakes. Install a rubber hole cover in the adjustment knock-out hole after the adjustment is complete. Adjust the parking brake.

28. Road test the vehicle and verify proper brake operation.

SPECIFICATION CHARTS

ENGINE AND VEHICLE IDENTIFICATION

Code ①	Liters (cc)	Cu. In. (cc)	Cyl.	Fuel Sys.	Engine Type	Eng. Mfg.
E	3.4 (3350)	207	6	MFI	OHV	BOC
M	3.1 (3130)	191	6	MFI	OHV	BOC
F	2.2 (2180)	134	4	MFI	OHV	BOC
T	2.4 (2392)	146	4	MFI	DOHC	CUS

Code ②	Year
Y	2000
1	2001
2	2002
3	2003
4	2004

BOC: Buick/Oldsmobile/Cadillac

CUS: Chevrolet/United States

MFI: Multi-point Fuel Injection

① 8th position of VIN

② 10th position of VIN

42372-NBOD-C01

GENERAL ENGINE IDENTIFICATION

Year	Model	Engine Displacement Liters (cc)	Engine Series (ID/VIN)	Fuel System	Net Horsepower @ rpm	Net Torque @ rpm (ft. lbs.)	Bore x Stroke (in.)	Compression Ratio	Oil Pressure @ rpm
2000	Alero	2.4 (2392)	T	MFI	150@5600	155@4400	3.54x3.70	9.5:1	30@3000
	Alero	3.1 (3130)	M	MFI	160@5200	185@4000	3.50x3.31	9.5:1	15@1100
	Grand Am	2.4 (2392)	T	MFI	150@5600	155@4400	3.54x3.70	9.5:1	30@3000
	Grand Am	3.1 (3130)	M	MFI	160@5200	185@4000	3.50x3.31	9.5:1	15@1100
2001	Alero	2.4 (2392)	T	MFI	150@5600	155@4400	3.54x3.70	9.5:1	30@3000
	Alero	3.4 (3934)	E	MFI	170@4800	200@4000	3.62x3.31	9.5:1	15@1100
	Grand Am	2.4 (2392)	T	MFI	150@5600	155@4400	3.54x3.70	9.5:1	30@3000
	Grand Am	3.4 (3934)	E	MFI	170@4800	200@4000	3.62x3.31	9.5:1	15@1100
2002	Alero	2.4 (2392)	T	MFI	150@5600	155@4400	3.54x3.70	9.5:1	30@3000
	Alero	3.4 (3934)	E	MFI	170@4800	200@4000	3.62x3.31	9.5:1	15@1100
	Alero	2.2 (2180)	F	MFI	148@5600	150@4000	3.50x3.46	10.0:1	50-80@1000
	Grand Am	2.4 (2392)	T	MFI	150@5600	155@4400	3.54x3.70	9.5:1	30@3000
	Grand Am	2.2 (2180)	F	MFI	148@5600	150@4000	3.50x3.46	10.0:1	50-80@1000
	Grand Am	3.4 (3934)	E	MFI	170@4800	200@4000	3.62x3.31	9.5:1	15@1100
2003-04	Alero	2.4 (2392)	T	MFI	150@5600	155@4400	3.54x3.70	9.5:1	30@3000
	Alero	3.4 (3934)	E	MFI	170@4800	200@4000	3.62x3.31	9.5:1	15@1100
	Alero	2.2 (2180)	F	MFI	148@5600	150@4000	3.50x3.46	10.0:1	50-80@1000
	Grand Am	2.4 (2392)	T	MFI	150@5600	155@4400	3.54x3.70	9.5:1	30@3000
	Grand Am	2.2 (2180)	F	MFI	148@5600	150@4000	3.50x3.46	10.0:1	50-80@1000
	Grand Am	3.4 (3934)	E	MFI	170@4800	200@4000	3.62x3.31	9.5:1	15@1100

MFI: Multi-point Fuel Injection

42372-NBOD-C02

ENGINE TUNE-UP SPECIFICATIONS

Year	Engine Displacement Liters (cc)	Engine ID/VIN	Spark Plugs Gap (in.)	Ignition Timing (deg.) MT	AT	Fuel Pump (psi)	Idle Speed (rpm) MT	AT	Valve Clearance In.	Ex.
2000	2.4 (2392)	T	0.050	①	①	41-47	①	①	HYD	HYD
	3.1 (3130)	M	0.060	①	①	41-47	①	①	HYD	HYD
2001	2.4 (2392)	T	0.050	①	①	41-47	①	①	HYD	HYD
	3.4 (3934)	E	0.060	①	①	41-47	①	①	HYD	HYD
2002	2.4 (2392)	T	0.050	①	①	41-47	①	①	HYD	HYD
	2.2 (2180)	F	0.060	①	①	50-60	①	①	HYD	HYD
	3.4 (3934)	E	0.060	①	①	41-47	①	①	HYD	HYD
2003-04	2.4 (2392)	T	0.050	①	①	41-47	①	①	HYD	HYD
	2.2 (2180)	F	0.060	①	①	50-60	①	①	HYD	HYD
	3.4 (3934)	E	0.060	①	①	41-47	①	①	HYD	HYD

NOTE: The Vehicle Emission Control Information label often reflects specification changes made during production. The label figures must be used if they differ from those in this chart.

HYD: Hydraulic

① Refer to Vehicle Emission Control Information label

42372-NB0D-C03

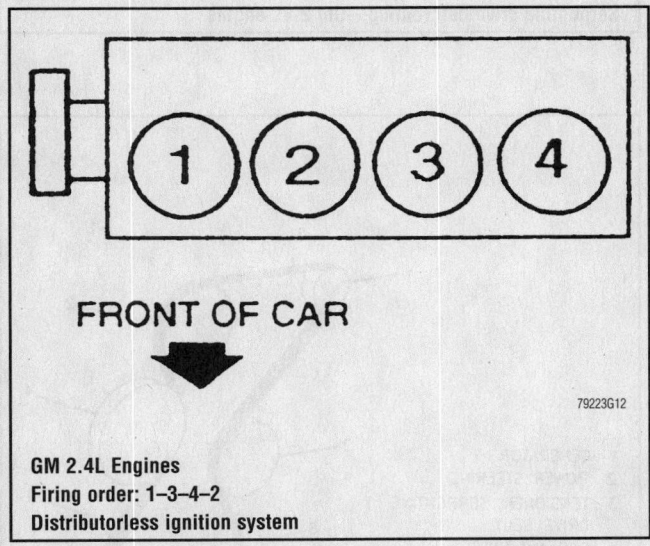

FRONT OF CAR

GM 2.4L Engines
Firing order: 1–3–4–2
Distributorless ignition system

79223G12

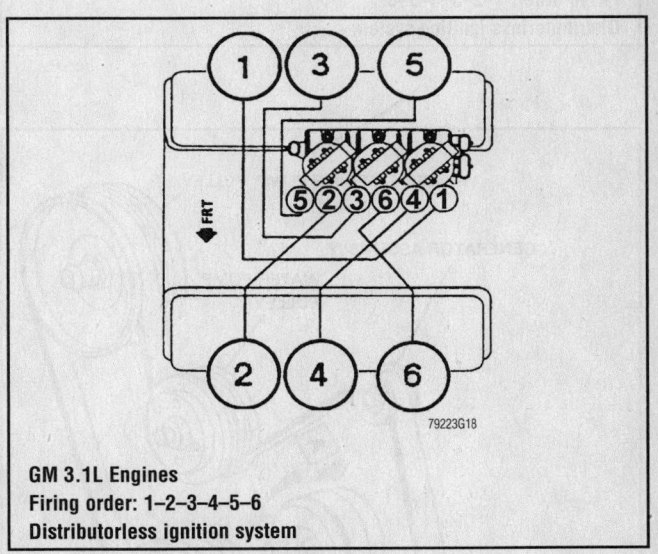

GM 3.1L Engines
Firing order: 1–2–3–4–5–6
Distributorless ignition system

79223G18

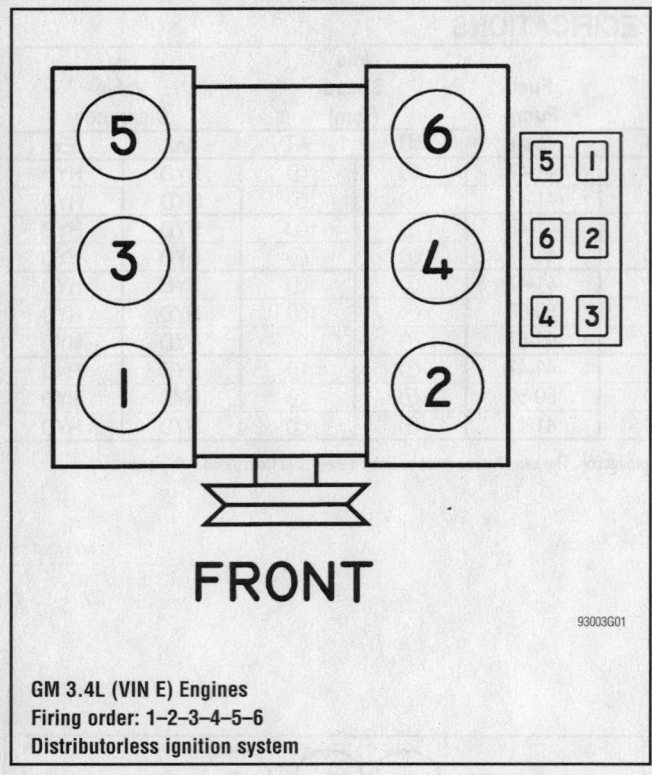

FRONT

93003G01

GM 3.4L (VIN E) Engines
Firing order: 1–2–3–4–5–6
Distributorless ignition system

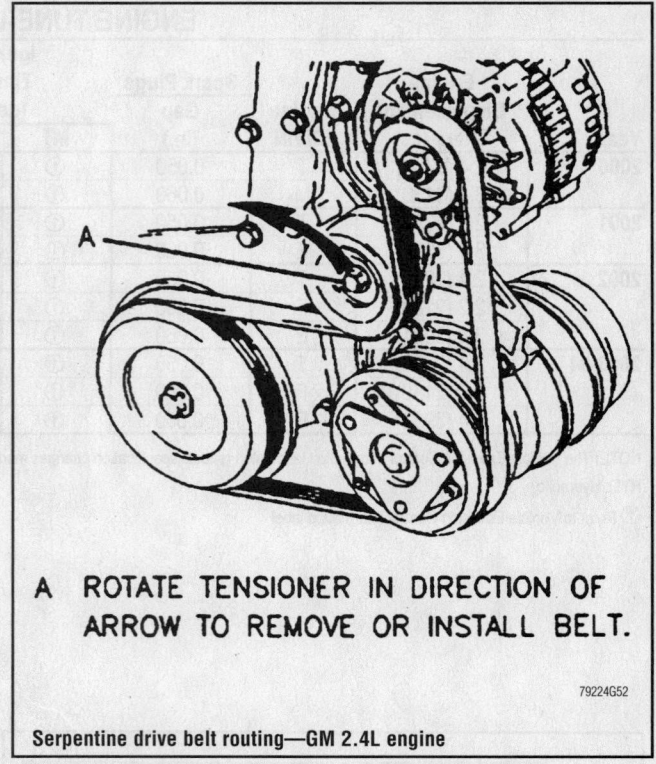

A ROTATE TENSIONER IN DIRECTION OF ARROW TO REMOVE OR INSTALL BELT.

79224G52

Serpentine drive belt routing—GM 2.4L engine

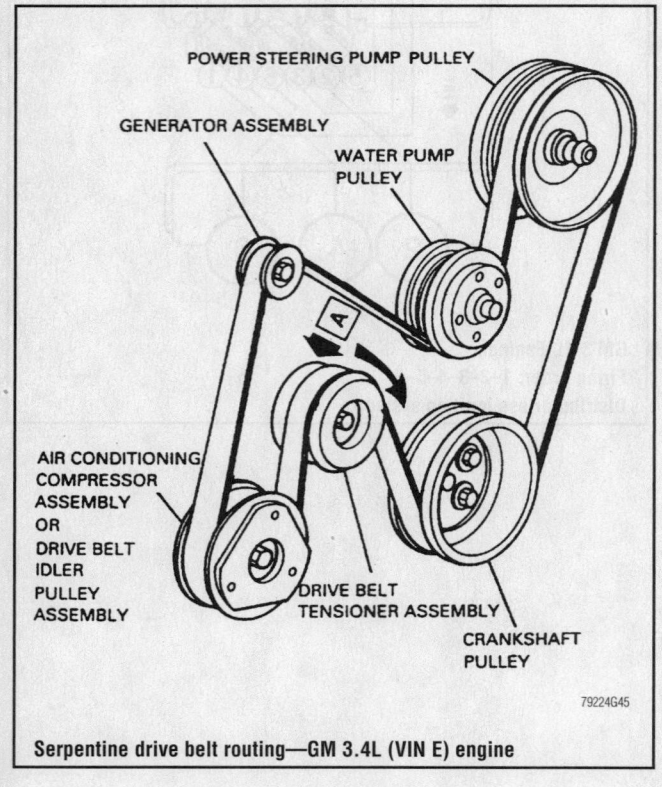

POWER STEERING PUMP PULLEY

GENERATOR ASSEMBLY

WATER PUMP PULLEY

AIR CONDITIONING COMPRESSOR ASSEMBLY OR DRIVE BELT IDLER PULLEY ASSEMBLY

DRIVE BELT TENSIONER ASSEMBLY

CRANKSHAFT PULLEY

79224G45

Serpentine drive belt routing—GM 3.4L (VIN E) engine

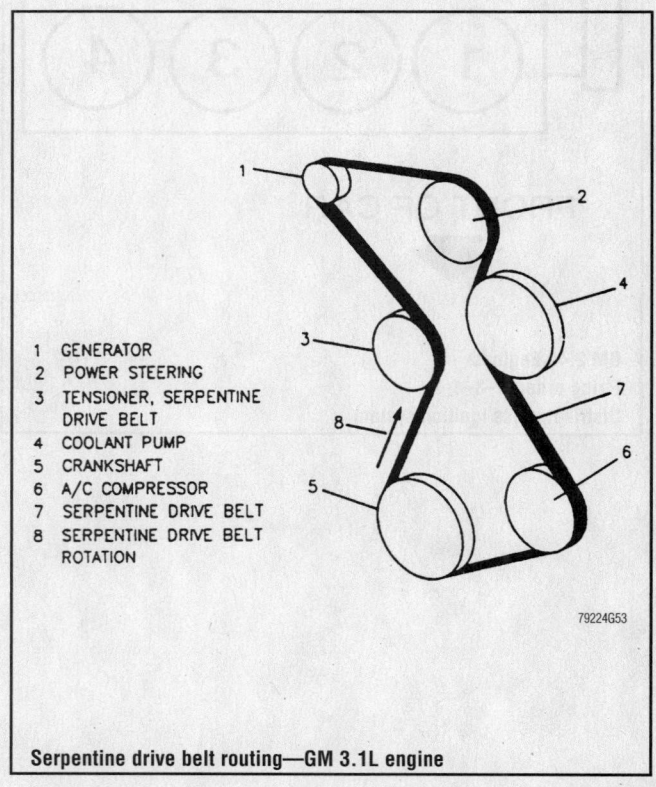

1   GENERATOR
2   POWER STEERING
3   TENSIONER, SERPENTINE DRIVE BELT
4   COOLANT PUMP
5   CRANKSHAFT
6   A/C COMPRESSOR
7   SERPENTINE DRIVE BELT
8   SERPENTINE DRIVE BELT ROTATION

79224G53

Serpentine drive belt routing—GM 3.1L engine

CAPACITIES

Year	Model	Engine Displacement Liters (cc)	Engine ID/VIN	Engine Oil with Filter (qts.)	Transmission (pts.) Manual	Transmission (pts.) Auto.	Fuel Tank (gal.)	Cooling System (qts.)
2000	Alero	2.4 (2392)	T	4.0	3.6	13.8	15.0	11.3
	Alero	3.1 (3130)	M	4.5	3.6	13.8	15.0	11.3
	Grand Am	2.4 (2392)	T	4.0	3.6	13.8	15.0	10.4
	Grand Am	3.1 (3130)	M	4.5	3.6	13.8	15.0	11.3
2001	Alero	2.4 (2392)	T	4.0	3.6	13.8	14.3	11.3
	Alero	3.4 (3934)	E	4.5	3.6	13.8	14.3	13.6
	Grand Am	2.4 (2392)	T	4.0	3.6	13.8	14.3	11.3
	Grand Am	3.4 (3934)	E	4.5	3.6	13.8	14.3	13.6
2002	Alero	2.4 (2392)	T	4.0	3.6	13.8	14.3	11.3
	Alero	3.4 (3934)	E	4.5	3.6	13.8	14.3	13.6
	Alero	2.2 (2180)	F	4.0	3.6	13.8	14.3	11.3
	Grand Am	2.2 (2180)	F	4.0	3.6	13.8	14.3	11.3
	Grand Am	2.4 (2392)	T	4.0	3.6	13.8	14.3	11.3
	Grand Am	3.4 (3934)	E	4.5	3.6	13.8	14.3	13.6
2003-04	Alero	2.4 (2392)	T	4.0	3.6	13.8	14.3	11.3
	Alero	3.4 (3934)	E	4.5	3.6	13.8	14.3	13.6
	Alero	2.2 (2180)	F	4.0	3.6	13.8	14.3	11.3
	Grand Am	2.2 (2180)	F	4.0	3.6	13.8	14.3	11.3
	Grand Am	2.4 (2392)	T	4.0	3.6	13.8	14.3	11.3
	Grand Am	3.4 (3934)	E	4.5	3.6	13.8	14.3	13.6

NOTE: All capacities are approximate. Add fluid gradually and ensure a proper fluid level is obtained.

42372-NBOD-C04

VALVE SPECIFICATIONS

Year	Engine Displacement Liters (cc)	Engine ID/VIN	Seat Angle (deg.)	Face Angle (deg.)	Spring Test Pressure (lbs. @ in.)	Spring Installed Height (in.)	Stem-to-Guide Clearance (in.) Intake	Stem-to-Guide Clearance (in.) Exhaust	Stem Diameter (in.) Intake	Stem Diameter (in.) Exhaust
2000	2.4 (2392)	T	45	46	50-55@ 1.437	1.437	0.0009- 0.0025	0.0016- 0.0032	0.2331- 0.2339	0.2326- 0.2334
	3.1 (3130)	M	45	45	250@ 1.239	1.710	0.0010- 0.0027	0.0010- 0.0027	NA	NA
2001	2.4 (2392)	T	45	46	50-55@ 1.437	1.437	0.0009- 0.0025	0.0016- 0.0032	0.2331- 0.2339	0.2326- 0.2334
	3.4 (3934)	E	45	45	75@ 1.701	1.701	0.0010- 0.0027	0.0010- 0.0027	NA	NA
2002	2.4 (2392)	T	45	46	50-55@ 1.437	1.437	0.0009- 0.0025	0.0016- 0.0032	0.2331- 0.2339	0.2326- 0.2334
	2.2 (2180)	F	45	46	NA	NA	0.2362- 0.2367	0.2362- 0.2367	0.2344- 0.2355	0.2337- 0.2343
	3.4 (3934)	E	45	45	75@ 1.701	1.701	0.0010- 0.0027	0.0010- 0.0027	NA	NA
2003-04	2.4 (2392)	T	45	46	50-55@ 1.437	1.437	0.0009- 0.0025	0.0016- 0.0032	0.2331- 0.2339	0.2326- 0.2334
	2.2 (2180)	F	45	46	NA	NA	0.2362- 0.2367	0.2362- 0.2367	0.2344- 0.2355	0.2337- 0.2343
	3.4 (3934)	E	45	45	75@ 1.701	1.701	0.0010- 0.0027	0.0010- 0.0027	NA	NA

NA: Not available

42372-NBOD-C05

CRANKSHAFT AND CONNECTING ROD SPECIFICATIONS

All measurements are given in inches.

| Year | Engine Displacement Liters (cc) | Engine ID/VIN | Crankshaft | | | | Connecting Rod | | |
			Main Brg. Journal Dia.	Main Brg. Oil Clearance	Shaft End-play	Thrust on No.	Journal Diameter	Oil Clearance	Side Clearance
2000	2.4 (2392)	T	2.3622-2.3631	0.0004-0.0023	0.0034-0.0095	3	1.8887-1.8897	0.0004-0.0026	0.0059-0.0177
	3.1 (3130)	M	2.6473-2.6483	0.0008-① 0.0025	0.0024-0.0083	3	1.9987-1.9994	0.0007-0.0024	0.0070-0.0170
2001	2.4 (2392)	T	2.3622-2.3631	0.0004-0.0023	0.0034-0.0095	3	1.8887-1.8897	0.0004-0.0026	0.0059-0.0177
	3.4 (3934)	E	2.6473-2.6483	0.0008-① 0.0025	0.0024-0.0083	3	1.9987-1.9994	0.0007-0.0024	0.0070-0.0170
2002	2.4 (2392)	T	2.3622-2.3631	0.0004-0.0023	0.0034-0.0095	3	1.8887-1.8897	0.0004-0.0026	0.0059-0.0177
	2.2 (2180)	F	2.2045-2.2050	0.0012-0.0026	0.0012-0.0150	3	2.0519-2.0525	0.0004-0.0026	0.0028-0.0146
	3.4 (3934)	E	2.6473-2.6483	0.0008-① 0.0025	0.0024-0.0083	3	1.9987-1.9994	0.0007-0.0024	0.0070-0.0170
2003-04	2.4 (2392)	T	2.3622-2.3631	0.0004-0.0023	0.0034-0.0095	3	1.8887-1.8897	0.0004-0.0026	0.0059-0.0177
	2.2 (2180)	F	2.2045-2.2050	0.0012-0.0026	0.0012-0.0150	3	2.0519-2.0525	0.0004-0.0026	0.0028-0.0146
	3.4 (3934)	E	2.6473-2.6483	0.0008-① 0.0025	0.0024-0.0083	3	1.9987-1.9994	0.0007-0.0024	0.0070-0.0170

① Thrust bearing: 0.0012 - 0.0030

42372-NBOD-C06

PISTON AND RING SPECIFICATIONS

All measurements are given in inches.

| Year | Engine Displacement Liters (cc) | Engine ID/VIN | Piston Clearance | Ring Gap | | | Ring Side Clearance | | |
				Top Compression	Bottom Compression	Oil Control	Top Compression	Bottom Compression	Oil Control
2000	2.4 (2392)	T	0.0006-0.0015	0.006-0.012	0.010-0.016	0.010-0.030	0.0016-0.0031	0.0012-0.0028	0.0005-0.0089
	3.1 (3130)	M	0.0013-0.0027	0.006-0.014	0.0197-0.0280	0.0098-0.0500	0.0020-0.0033	0.0020-0.0035	0.0080
2001	2.4 (2392)	T	0.0006-0.0015	0.006-0.012	0.010-0.016	0.010-0.030	0.0016-0.0031	0.0012-0.0028	0.0005-0.0089
	3.4 (3934)	E	0.0013-0.0027	0.006-0.014	0.0197-0.0280	0.0098-0.0500	0.0020-0.0033	0.0020-0.0035	0.0080
2002	2.4 (2392)	T	0.0006-0.0015	0.006-0.012	0.010-0.016	0.010-0.030	0.0016-0.0031	0.0012-0.0028	0.0005-0.0089
	2.2 (2180)	F	0.0004-0.0016	0.008-0.016	0.014-0.022	0.010-0.0300	0.0015-0.0031	0.0012-0.0028	0.0035-0.0052
	3.4 (3934)	E	0.0013-0.0027	0.006-0.014	0.0197-0.0280	0.0098-0.0500	0.0020-0.0033	0.0020-0.0035	0.0080
2003-04	2.4 (2392)	T	0.0006-0.0015	0.006-0.012	0.010-0.016	0.010-0.030	0.0016-0.0031	0.0012-0.0028	0.0005-0.0089
	2.2 (2180)	F	0.0004-0.0016	0.008-0.016	0.014-0.022	0.010-0.0300	0.0015-0.0031	0.0012-0.0028	0.0035-0.0052
	3.4 (3934)	E	0.0013-0.0027	0.006-0.014	0.0197-0.0280	0.0098-0.0500	0.0020-0.0033	0.0020-0.0035	0.0080

42372-NBOD-C07

TORQUE SPECIFICATIONS
All readings in ft. lbs.

Year	Engine Displacement Liters (cc)	Engine ID/VIN	Cylinder Head Bolts	Main Bearing Bolts	Rod Bearing Bolts	Crankshaft Damper Bolts	Flywheel Bolts	Manifold Intake	Manifold Exhaust	Spark Plug	Lug Nut
2000	2.4 (2392)	T	①	②	③	④	⑤	⑥	⑦	13	100
	3.1 (3130)	M	⑧	⑨	⑩	76	52	⑪	12	20	100
2001	2.4 (2392)	T	①	②	③	④	⑤	⑥	⑦	13	100
	3.4 (3934)	E	⑫	⑬	⑩	76	52	⑪	12	20	100
2002	2.4 (2392)	T	①	②	③	④	⑤	⑥	⑦	13	100
	2.2 (2180)	F	⑭	⑮	⑯	⑰	⑱	⑲	⑳	15	100
	3.4 (3934)	E	⑫	⑬	⑩	76	52	⑪	12	20	100
2003-04	2.4 (2392)	T	①	②	③	④	⑤	⑥	⑦	13	100
	2.2 (2180)	F	⑭	⑮	⑯	⑰	⑱	⑲	⑳	15	100
	3.4 (3934)	E	⑫	⑬	⑩	76	52	⑪	12	20	100

① Nos. 1-8: 40 ft. lbs.
Nos. 9-10: 30 ft. lbs.
Tighten all bolts an additional 90 degrees
② 15 ft. lbs. plus 90 degrees
③ 18 ft. lbs. plus 80 degrees
④ 129 ft. lbs. plus 90 degrees
⑤ 22 ft. lbs. plus 45 degrees
⑥ Nuts: 19 ft. lbs.
Studs: 97 inch lbs.
⑦ Nuts: 11 ft. lbs.
Studs: 97 inch lbs.
⑧ Coat threads with sealer torque to 33 ft. lbs., then turn 1/4 turn (90 degrees)
⑨ 37 ft. lbs. plus 77 degrees
⑩ 15 ft. lbs. plus 75 degrees

⑪ Lower: 115 inch lbs.
Upper: 18 ft. lbs.
⑫ 37 ft. lbs. plus 90 degrees
⑬ 37 ft. lbs. plus 77 degrees
⑭ First pass: 15 ft. lbs.
Second pass: 155 degrees
⑮ 15 ft. lbs. plus 70 degrees
⑯ 18 ft. lbs. plus 100 degrees
⑰ First pass: 74 ft. lbs.
Second pass: 75 degrees
⑱ 39 ft. lbs. plus 25 degrees
⑲ Manifold nut/bolt: 89 inch lbs.
Manifold stud: 53 inch lbs.
⑳ Manifold stud: 89 inch lbs.
Manifold 9 ft. lbs.

42372-NBOD-C08

WHEEL ALIGNMENT

Year	Model		Caster Range (+/-Deg.)	Caster Preferred Setting (Deg.)	Camber Range (+/-Deg.)	Camber Preferred Setting (Deg.)	Toe-in (in.)	Steering Axis Inclination (Deg.)
2000	Alero	F	1.00	+4.10	1.00	-0.20	0.05 +/- 0.12	—
		R	—	—	0.50	-0.20	0.03 +/- 0.04	—
	Grand Am	F	1.00	+4.10	1.00	-0.20	0.05 +/- 0.13	—
		R	—	—	0.50	-0.20	0.03 +/- 0.10	—
2001	Alero	F	1.00	+4.10	1.00	-0.20	0.05 +/- 0.12	—
		R	—	—	0.50	-0.20	0.03 +/- 0.04	—
	Grand Am	F	1.00	+4.10	1.00	-0.20	0.05 +/- 0.13	—
		R	—	—	0.50	-0.20	0.03 +/- 0.10	—
2002	Alero	F	1.00	+4.10	1.00	-0.20	0.05 +/- 0.12	—
		R	—	—	0.50	-0.20	0.03 +/- 0.04	—
	Grand Am	F	1.00	+4.10	1.00	-0.20	0.05 +/- 0.13	—
		R	—	—	0.50	-0.20	0.03 +/- 0.10	—
2003-04	Alero	F	1.00	+4.10	1.00	-0.20	0.05 +/- 0.12	—
		R	—	—	0.50	-0.20	0.03 +/- 0.04	—
	Grand Am	F	1.00	+4.10	1.00	-0.20	0.05 +/- 0.13	—
		R	—	—	0.50	-0.20	0.03 +/- 0.10	—

42372-NBOD-C09

TIRE, WHEEL AND BALL JOINT SPECIFICATIONS

Year	Model	OEM Tires Standard	OEM Tires Optional	Tire Pressures (psi) Front	Tire Pressures (psi) Rear	Wheel Size	Ball Joint Inspection
2000	Alero GX, GL	P215/60R15	P225/50HR15	30	30	6-JJ	①
	Alero GL1, GL2, GL3	P225/50VR16	None	30	30	6-JJ	①
	Alero GLS	P225/50R16	None	30	30	6-JJ	①
	Grand AM SE/SE1	P215/60R15	None	30	30	6-JJ	0.125 in.
	Grand Am SE2	P225/50R16	None	30	30	6-JJ	0.125 in.
	Grand Am GT, GT1	P225/50R16	None	30	30	6-JJ	0.125 in.
2001	Alero GX, GL	P215/60R15	P225/50HR15	30	30	6-JJ	①
	Alero GL1, GL2, GL3	P225/50VR16	None	30	30	6-JJ	①
	Alero GLS	P225/50R16	None	30	30	6-JJ	①
	Grand AM SE/SE1	P215/60R15	None	30	30	6-JJ	0.125 in.
	Grand Am SE2	P225/50R16	None	30	30	6-JJ	0.125 in.
	Grand Am GT, GT1	P225/50R16	None	30	30	6-JJ	0.125 in.
2002	Alero GX, GL	P215/60R15	P225/50HR15	30	30	6-JJ	①
	Alero GL1, GL2, GL3	P225/50VR16	None	30	30	6-JJ	①
	Alero GLS	P225/50R16	None	30	30	6-JJ	①
	Grand AM SE/SE1	P215/60R15	None	30	30	6-JJ	0.125 in.
	Grand Am SE2	P225/50R16	None	30	30	6-JJ	0.125 in.
	Grand Am GT, GT1	P225/50R16	None	30	30	6-JJ	0.125 in.
2003-04	Alero GX, GL	P215/60R15	P225/50HR15	30	30	6-JJ	①
	Alero GL1, GL2, GL3	P225/50VR16	None	30	30	6-JJ	①
	Alero GLS	P225/50R16	None	30	30	6-JJ	①
	Grand AM SE/SE1	P215/60R15	None	30	30	6-JJ	0.125 in.
	Grand Am SE2	P225/50R16	None	30	30	6-JJ	0.125 in.
	Grand Am GT, GT1	P225/50R16	None	30	30	6-JJ	0.125 in.

OEM: Original Equipment Manufacturer

PSI: Pounds Per Square Inch

STD: Standard

OPT: Optional

42372-NBOD-C10

BRAKE SPECIFICATIONS
All measurements in inches unless noted

Year	Model		Brake Disc Original Thickness	Brake Disc Minimum Thickness	Brake Disc Maximum Runout	Brake Drum Diameter Original Inside Diameter	Max. Wear Limit	Maximum Machine Diameter	Minimum Lining Thickness Front	Minimum Lining Thickness Rear	Brake Caliper Bracket Bolts (ft. lbs.)	Brake Caliper Mounting Bolts (ft. lbs.)
2000	Alero	F	1.031	0.972	0.003	—	—	—	0.030	—	85	23
		R	0.430	0.410	—	8.863	8.920	8.909	—	①	85	81
	Grand Am	F	1.031	0.972	0.003	—	—	—	0.030	—	85	23
		R	0.430	0.410	—	8.863	8.920	8.909	—	①	85	81
2001	Alero	F	1.031	0.972	0.003	—	—	—	0.030	—	85	23
		R	0.430	0.410	—	8.863	8.920	8.909	—	①	85	81
	Grand Am	F	1.031	0.972	0.003	—	—	—	0.030	—	85	23
		R	0.430	0.410	—	8.863	8.920	8.909	—	①	85	81
2002	Alero	F	1.031	0.972	0.003	—	—	—	0.030	—	85	23
		R	0.430	0.410	—	8.863	8.920	8.909	—	①	85	81
	Grand Am	F	1.031	0.972	0.003	—	—	—	0.030	—	85	23
		R	0.430	0.410	—	8.863	8.920	8.909	—	①	85	81
2003-04	Alero	F	1.031	0.972	0.003	—	—	—	0.030	—	85	23
		R	0.430	0.410	—	8.863	8.920	8.909	—	①	85	81
	Grand Am	F	1.031	0.972	0.003	—	—	—	0.030	—	85	23
		R	0.430	0.410	—	8.863	8.920	8.909	—	①	85	81

NA: Not available

① 0.030 over rivet head; If bonded lining, use 0.030 from shoe

42372-NBOD-C11

SCHEDULED MAINTENANCE INTERVALS
GM N BODY—OLDSMOBILE ALERO & PONTIAC GRAND AM

TO BE SERVICED	TYPE OF SERVICE	VEHICLE MILEAGE INTERVAL (x1000)												
		7.5	15	22.5	30	37.5	45	52.5	60	67.5	75	82.5	90	97.5
Engine oil & filter	R	✓	✓	✓	✓	✓	✓	✓	✓	✓	✓	✓	✓	✓
Automatic transaxle fluid & filter ①	S/I	✓	✓	✓	✓	✓	✓	✓	✓	✓	✓	✓	✓	✓
Brake hoses	S/I	✓	✓	✓	✓	✓	✓	✓	✓	✓	✓	✓	✓	✓
Chassis lubrication	S/I	✓	✓	✓	✓	✓	✓	✓	✓	✓	✓	✓	✓	✓
Coolant level, hoses & clamps	S/I	✓	✓	✓	✓	✓	✓	✓	✓	✓	✓	✓	✓	✓
Driveshaft boots & front suspension components	S/I	✓	✓	✓	✓	✓	✓	✓	✓	✓	✓	✓	✓	✓
Exhaust system	S/I	✓	✓	✓	✓	✓	✓	✓	✓	✓	✓	✓	✓	✓
Lubricate chassis & suspension	S/I	✓	✓	✓	✓	✓	✓	✓	✓	✓	✓	✓	✓	✓
Lubricate steering linkage & transaxle shift linkage	S/I	✓	✓	✓	✓	✓	✓	✓	✓	✓	✓	✓	✓	✓
Lubricate parking brake cable guides, underbody contact points & linkage	S/I	✓	✓	✓	✓	✓	✓	✓	✓	✓	✓	✓	✓	✓
Manual transaxle oil	S/I	✓	✓	✓	✓	✓	✓	✓	✓	✓	✓	✓	✓	✓
Throttle linkage	S/I	✓	✓	✓	✓	✓	✓	✓	✓	✓	✓	✓	✓	✓
Brake linings	S/I	✓		✓				✓		✓		✓		✓
Rotate tires	S/I	✓		✓				✓		✓		✓		✓
Air filter element & PCV filter	R				✓				✓				✓	
Engine coolant ②	R													
Spark plugs ③	R				✓				✓				✓	
Accessory drive belt(s)	S/I				✓				✓				✓	
EGR & fuel systems	S/I				✓				✓				✓	
Ignition cables	S/I				✓				✓				✓	

R: Replace

S/I: Service or Inspect

① Automatic transaxle fluid & filter: replace at 100,000 miles (if not changed previously).

② Engine coolant: replace every 100,000 miles. Use O.E. specified (DEX-COOL™) coolant only. If any silicate coolant is used, the service interval is every 30,000 miles.

③ Platinum tip spark plugs: replace every 100,000 miles.

FREQUENT OPERATION MAINTENANCE (SEVERE SERVICE) ADDITIONS

If a vehicle is operated under any of the following conditions it is considered severe service:

- Towing a trailer or using a camper or car-top carrier.

- Extensive idling or low-speed driving for long distances as in heavy commercial use, such as delivery, taxi or police cars.

- Operating on rough, muddy or salt-covered roads.

- Operating on unpaved or dusty roads.

- 50% or more of the vehicle operation is in 32°C (90°F) or higher temperatures, or constant operation in temperatures below 0°C (32°F).

Engine oil and filter: change every 3000 miles or 3 months, whichever occurs first.

Wheels and tires: inspect and rotate every 6000 miles.

Air cleaner element: inspect every 15,000 miles and replace or clean as needed. Replace it at least every 30,000 miles.

Automatic transaxle fluid & filter: replace every 50,000 miles.

42372-NBOD-C12

PRECAUTIONS

Before servicing any vehicle, please be sure to read all of the following precautions, which deal with personal safety, prevention of component damage, and important points to take into consideration when servicing a motor vehicle:

• Never open, service or drain the radiator or cooling system when the engine is hot; serious burns can occur from the steam and hot coolant.

• Observe all applicable safety precautions when working around fuel. Whenever servicing the fuel system, always work in a well-ventilated area. Do not allow fuel spray or vapors to come in contact with a spark, open flame or excessive heat (a hot drop light, for example). Keep a dry chemical fire extinguisher near the work area. Always keep fuel in a container specifically designed for fuel storage; also, always properly seal fuel containers to avoid the possibility of fire or explosion. Refer to the additional fuel system precautions later in this section.

• Fuel injection systems often remain pressurized, even after the engine has been turned **OFF**. The fuel system pressure must be relieved before disconnecting any fuel lines. Failure to do so may result in fire and/or personal injury.

• Brake fluid often contains polyglycol ethers and polyglycols. Avoid contact with the eyes and wash your hands thoroughly after handling brake fluid. If you do get brake fluid in your eyes, flush your eyes with clean, running water for 15 minutes. If eye irritation persists, or if you have taken brake fluid internally, IMMEDIATELY seek medical assistance.

• The EPA warns that prolonged contact with used engine oil may cause a number of skin disorders, including cancer! You should make every effort to minimize your exposure to used engine oil. Protective gloves should be worn when changing oil. Wash your hands and any other exposed skin areas as soon as possible after exposure to used engine oil. Soap and water, or waterless hand cleaner should be used.

• All new vehicles are now equipped with an air bag system. The system must be disabled before performing service on or around system components, steering column, instrument panel components, wiring and sensors. Failure to follow safety and disabling procedures could result in accidental air bag deployment, possible personal injury and unnecessary system repairs.

• Always wear safety goggles when working with, or around, the air bag system. When carrying a non-deployed air bag, be sure the bag and trim cover are pointed away from your body. When placing a non-deployed air bag on a work surface, always face the bag and trim cover upward, away from the surface. This will reduce the motion of the module if it is accidentally deployed. Refer to the additional air bag system precautions later in this section.

• Clean, high quality brake fluid from a sealed container is essential to the safe and proper operation of the brake system. You should always buy the correct type of brake fluid for your vehicle. If the brake fluid becomes contaminated, completely flush the system with new fluid. Never reuse any brake fluid. Any brake fluid that is removed from the system should be discarded. Also, do not allow any brake fluid to come in contact with a painted surface; it will damage the paint.

• Never operate the engine without the proper amount and type of engine oil; doing so WILL result in severe engine damage.

• Timing belt maintenance is extremely important! Many models utilize an interference-type, non-freewheeling engine. If the timing belt breaks, the valves in the cylinder head may strike the pistons, causing potentially serious (also time-consuming and expensive) engine damage.

• Disconnecting the negative battery cable on some vehicles may interfere with the functions of the on-board computer system(s) and may require the computer to undergo a relearning process once the negative battery cable is reconnected.

• When servicing drum brakes, only disassemble and assemble one side at a time, leaving the remaining side intact for reference.

• Only an MVAC-trained, EPA-certified automotive technician should service the air conditioning system or its components.

ENGINE REPAIR

Alternator

REMOVAL

2.2L Engine

1. Before servicing the vehicle, refer to the precautions in the beginning of this section.
2. Remove or disconnect the following:
   • Negative battery cable
   • Drive belt
   • Electrical connectors from the alternator
   • Alternator bolts and tighten to 15 ft. lbs. (20 Nm)
   • Alternator

2.4L Engine

1. Before servicing the vehicle, refer to the precautions in the beginning of this section.

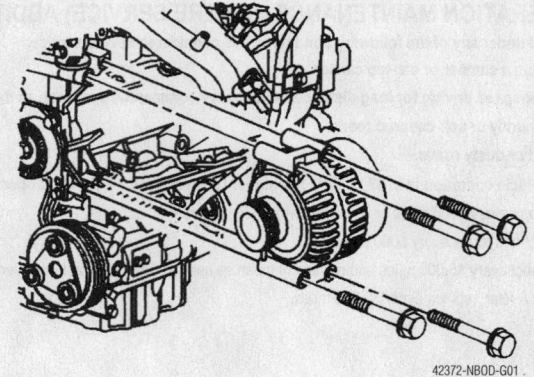

Alternator mounting—2.2L engine

2. Remove or disconnect the following:
   • Negative battery cable
   • Accessory drive belt
   • Alternator mounting bolts
   • Alternator electrical connectors
   • Alternator

3.1L Engine

1. Before servicing the vehicle, refer to the precautions in the beginning of this section.
2. Remove or disconnect the following:
   • Negative battery cable

42372-NBOD-G01

- Accessory drive belt
- Alternator electrical connectors
- Power steering-to-alternator line clip
- Rear alternator brace
- Alternator air intake connector
- Alternator-to-bracket nut/bolt
- Alternator

3.4L Engine

1. Before servicing the vehicle, refer to the precautions in the beginning of this section.
2. Remove or disconnect the following:
   - Negative battery cable
   - Accessory drive belt
   - Alternator electrical connectors
   - Power steering-to-alternator line clip
   - Alternator mounting nuts and bolts
   - Alternator

INSTALLATION

2.2L Engine

1. Install or connect the following:
   - Alternator
   - Alternator bolts
   - Electrical connectors to the alternator
   - Drive belt
   - Negative battery cable

2.4L Engine

1. Install or connect the following:
   - Alternator
   - Alternator electrical connectors
   - Alternator mounting bolts. Torque the bolts to 37 ft. lbs. (50 Nm).
   - Accessory drive belt
   - Negative battery cable

3.1L Engine

1. Install or connect the following:
   - Alternator. Torque the nut to 37 ft. lbs. (50 Nm) and the bolt to 18 ft. lbs. (25 Nm).
   - Alternator air intake connector
   - Rear alternator brace. Torque the nut/bolt to 18 ft. lbs. (25 Nm).
   - Power steering-to-alternator line clip
   - Alternator electrical connectors
   - Accessory drive belt
   - Negative battery cable

3.4L Engine

1. Install or connect the following:
   - Alternator. Torque the nuts to 22 ft.

lbs. (30 Nm) and the bolts to 37 ft. lbs. (50 Nm).
   - Power steering-to-alternator line clip
   - Alternator electrical connectors
   - Accessory drive belt
   - Negative battery cable

Ignition Timing

ADJUSTMENT

The ignition timing is not adjustable, and is set electronically according to engine demand.

Engine Assembly

REMOVAL & INSTALLATION

2.2L Engine

MANUAL TRANSAXLE

1. Before servicing the vehicle, refer to the precautions in the beginning of this section.
2. Remove or disconnect the following:
   - Hood
   - Negative battery cable
   - Air inlet duct and resonator
   - Accelerator and cruise control cable
   - Hose from the brake booster
   - Power steering pump bolts and set the pump aside
   - Fuel lines
   - Transmission shift control cables
   - Transmission shift control cables from the bracket
   - Clutch actuator cylinder from the transmission
3. Drain the cooling system.
   - Radiator inlet hose
   - Hose from the surge tank to the cylinder head
   - Outlet hose from the surge tank to the radiator
   - Bolt retaining the surge tank outlet hose to the intake manifold
   - Radiator outlet hose
   - Heater hoses
4. Disconnect the following electrical connectors:
   - Idle Air Control (IAC) motor
   - Throttle Position (TPS) sensor
   - Manifold Absolute Pressure (MAP) sensor
   - Crankshaft Position (CKP) sensor
   - Camshaft Position (CMP) sensor
   - Oil pressure sensor
   - Purge solenoid

- Ignition coil and module assembly
- Oxygen (O₂S) sensor
- Vehicle speed sensor
- Engine Coolant Temperature (ECT) sensor
- Back-up lamp switch
- Electrical harness from the engine and set the harness aside
5. Raise the vehicle.
6. Remove or disconnect the following:
   - Front suspension crossmember
   - Drive axles
   - Engine drive belt
   - AC compressor bolts and set the compressor aside.
   - Alternator and starter electrical connectors
7. Drain the engine oil.
   - Front exhaust pipe from the exhaust manifold
8. Use a block of wood to support the front of the engine at the front of the oil pan.
9. Lower the vehicle onto an engine support table.
   - Front engine mount
   - Bolts which secure the transmission mounts to the frame
10. Raise the vehicle away from the engine and transmission assembly.
11. Install an engine hoist to the engine.
   - Transmission bellhousing bolts
   - Engine and the transmission

To install:
12. Installation is the reverse of removal. Please note the following torques:
   - Engine to the transaxle. Torque the bolts to 66 ft. lbs. (90 Nm).
   - Transaxle mount. Torque the bolt to 81 ft. lbs. (110 Nm).
   - Engine mount nuts and bolts to 49 ft. lbs. (66 Nm) and the engine mount bolts to 81 ft. lbs. (110 Nm)
   - A/C compressor. Torque the bolts to 16 ft. lbs. (22 Nm).
   - Power steering pump. Torque the bolts to 25 ft. lbs. (34 Nm).
13. Refill the cooling system and engine oil.
14. Start the engine and check for proper operation.

AUTOMATIC TRANSAXLE

1. Before servicing the vehicle, refer to the precautions in the beginning of this section.
2. Remove or disconnect the following:
   - Hood
   - Negative battery cable
   - Air inlet duct and resonator
   - Accelerator and cruise control cable
   - Hose from the brake booster

- Power steering pump bolts and set the pump aside
- Fuel lines
- Transmission shift control cables
- Transmission shift control cables from the bracket
- Clutch actuator cylinder from the transmission

3. Drain the cooling system.
- Radiator inlet hose
- Hose from the surge tank to the cylinder head
- Outlet hose from the surge tank to the radiator
- Bolt retaining the surge tank outlet hose to the intake manifold
- Radiator outlet hose
- Heater hoses

4. Disconnect the following electrical connectors:
- Idle Air Control (IAC) motor
- Throttle Position (TPS) sensor
- Manifold Absolute Pressure (MAP) sensor
- Crankshaft Position (CKP) sensor
- Camshaft Position (CMP) sensor
- Oil pressure sensor
- Purge solenoid
- Ignition coil and module assembly
- Oxygen (O2S) sensor
- Vehicle speed sensor
- Engine Coolant Temperature (ECT) sensor
- Back-up lamp switch
- Electrical harness from the engine and set the harness aside

5. Remove or disconnect the following:
- Upper transmission bellhousing bolts

6. Raise the vehicle.
- Engine drive belt
- AC compressor bolts and set the compressor aside
- Crankshaft balancer
- Alternator and starter electrical connectors

7. Drain the engine oil.
- Front exhaust pipe from the exhaust manifold
- Starter
- Flywheel-to-torque convertor bolts
- Lower transmission bellhousing bolts
- Transmission to engine brace

8. Lower the vehicle.
9. Install an engine hoist to the engine.
- Front engine mount
- Upper transmission bellhousing bolts
- Engine and the transmission
- Engine

To install:

10. Installation is the reverse of removal. Please note the following torques:
- Upper bell housing bolts and tighten to 66 ft. lbs. (90 Nm)
- Front engine mount and tighten the bolts to 81 ft. lbs. (110 Nm)
- Lower bell housing bolts and tighten to 66 ft. lbs. (90 Nm)
- Torque convertor bolts and tighten to 46 ft. lbs. (62 Nm)
- Brace from the transmission to the engine bolts to 53 ft. lbs. (72 Nm)
- A/C compressor. Torque the bolts to 16 ft. lbs. (22 Nm).
- Power steering pump bolts to 19 ft. lbs. (25 Nm)

11. Refill the cooling system and engine oil.
12. Start the engine and check for proper operation.

2.4L Engine

AUTOMATIC TRANSAXLE

1. Before servicing the vehicle, refer to the precautions in the beginning of this section.
2. Properly drain the cooling system.
3. Relieve the fuel system pressure.
4. Drain the engine oil.
5. Remove or disconnect the following:
- Fuel rail assembly
- Air intake duct and bracket
- Ignition coil assembly
- Camshaft Position (CMP) sensor
- Power steering pump. DO NOT remove the lines.
- Oil pressure sending switch
- Cruise control assembly

6. Install an engine support fixture.
- Engine mount assembly
- Fuel pressure regulator vacuum line

7. Raise the engine with the engine support fixture J-hook.
- Engine mount bracket
- Accessory drive belt
- Front wheels
- Right front splash shield
- Crankshaft balancer
- Manifold Absolute Pressure (MAP) sensor electrical connector
- Intake Air Temperature (IAT) sensor electrical connector
- Exhaust Gas Recirculation (EGR) sensor electrical connector
- Evaporator canister electrical connector
- Alternator
- Accelerator cable and bracket
- Starter motor

- Exhaust manifold heat shield
- Exhaust pipe from the manifold
- Engine intake coolant pipe
- Torque converter cover
- Torque converter bolts
- Oxygen Sensor (O2S)
- A/C compressor. DO NOT remove the hoses.
- Oil pan-to-bell housing bolts
- Transaxle mount
- Transaxle bolts

8. Install an engine lifting device and a transaxle support.
9. Raise the engine slightly and separate it from the transaxle to remove it from the vehicle.

To install:

10. Install or connect the following:
- Engine to the transaxle. Torque the bolts to 75 ft. lbs. (100 Nm).
- Transaxle mount. Torque the through bolt to 55 ft. lbs. (75 Nm).
- Oil pan-to-bell housing bolts. Torque the bolts to 17 ft. lbs. (23 Nm).
- A/C compressor. Torque the bolts to 37 ft. lbs. (50 Nm).
- O2S
- Torque converter bolts. Torque the bolts to 46 ft. lbs. (62 Nm).
- Torque converter cover. Torque the bolts to 115 inch lbs. (13 Nm).
- Intake coolant pipe
- Exhaust pipe to the manifold. Torque the bolts to 26 ft. lbs. (35 Nm).
- Exhaust manifold heat shield. Torque the bolts to 124 inch lbs. (14 Nm).
- Starter motor. Torque the bolts to 66 ft. lbs. (90 Nm).
- Accelerator cable and bracket
- Alternator
- Fuel pressure regulator vacuum line
- Evaporator canister electrical connector
- EGR sensor electrical connector
- IAT sensor electrical connector
- MAP sensor electrical connector
- Crankshaft balancer. Torque the bolt to 129 ft. lbs. (175 Nm).
- Right front splash shield
- Front wheels
- Accessory drive belt

- Engine mount bracket. Torque the bolts to 96 ft. lbs. (130 Nm).
- Engine mount assembly. Torque the bolts to 49 ft. lbs. (66 Nm).
- Cruise control assembly
- Oil pressure sending switch
- Power steering pump. Torque the bolts to 19 ft. lbs. (26 Nm).
- CMP sensor
- Ignition coil assembly
- Air intake assembly
- Fuel rail assembly

11. Refill the cooling system and engine oil.

12. Start the engine and check for proper operation.

MANUAL TRANSAXLE

1. Before servicing the vehicle, refer to the precautions in the beginning of this section.

2. Properly drain the cooling system.

3. Relieve the fuel system pressure.

4. Drain the engine oil.

5. Discharge and recover the refrigerant.

6. Remove or disconnect the following:

- Air cleaner duct
- Air cleaner
- Underhood fuse block
- Air cleaner bracket
- Shift cables at the shift control
- Shift cables from the bracket and remove the bracket
- Back up lamp switch electrical connection
- Vehicle Speed Sensor (VSS) electrical connection
- Engine controls harness
- Vacuum lines
- Fuel line quick disconnect fittings
- Radiator hoses
- Heater hoses from the core
- Slave cylinder hydraulic line
- Evaporative (EVAP) solenoid
- Ground cables at the rear of the engine block
- Power steering pump with lines attached and position the pump aside
- Electrical connector from the A/C compressor and the Crankshaft Position (CKP) sensor and position the harness aside
- Starter with the wires attached and position it aside
- Surge tank bypass hose from the engine
- Cruise and accelerator cables from the throttle body
- Cruise control module

7. Tie the radiator to the hood latch panel with mechanics wire

8. Raise and support the vehicle. Safety strap the front of the vehicle to the hoist.

- Oil filter
- Front splash shields
- Front closeout panel
- Lower radiator support
- Right front brake hose
- Retaining nut from the Brake Pressure Modulator Valve (BPMV) at the mounting bracket
- Wheel Speed Sensors (WSS)
- WSS harnesses from the control arm retainers and the frame retainers and position them aside
- Driveshafts
- Ball joints from the control arms
- Outer tie rod ends from the control arms
- Catalytic converter from the exhaust manifold
- A/C compressor hose from the A/C compressor
- Bolt from the power steering pressure line retainer

9. Lower the vehicle until the front suspension crossmember rests on the support table. Position a three inch block of wood between the front of the oil pan and the crossmember.

- Front engine mount-to-bracket bolts
- Front suspension crossmember retaining bolts

10. Carefully raise the vehicle off of the engine/transaxle assembly. Install the engine hoist to the engine/transaxle assembly.

- Front and rear transaxle mount through-bolts
- 2 side transaxle mount lower nuts
- Engine/transaxle assembly off of the front suspension crossmember
- Transaxle from the engine
- Clutch drive plate and clutch driven plate
- Flywheel

11. Mount the engine on a suitable engine stand.

To install:

12. Remove the engine from the engine stand

13. Install or connect the following:

- Flywheel
- Clutch driven plate and clutch drive plate
- Transaxle to the engine

14. Lower the engine/transaxle assembly on to the front suspension crossmember.

- 2 side transaxle mount lower nuts and tighten the nuts to 60 ft. lbs. (44 Nm)

- Front and rear transaxle mount through-bolts. Tighten the through-bolts to 75 ft. lbs. (55 Nm).

15. Remove the engine hoist from the engine/transaxle assembly.

16. Carefully lower the vehicle on to the engine/transaxle assembly.

- Front suspension crossmember retaining bolts
- Front engine mount-to-bracket bolts
- Power steering pressure line retainer bolt.
- A/C compressor hose to the A/C compressor
- Catalytic converter to the exhaust manifold
- Outer tie rod ends to the control arms
- Ball joints to the control arms
- Driveshafts
- WSS harnesses to the frame retainers and the control arm retainers
- Wheel speed sensors
- Retaining nut to the BPMV at the mounting bracket
- Right front brake hose to the vehicle
- Lower radiator support
- Front closeout panel and the panel fasteners
- Front splash shields
- New oil filter

17. Remove the safety straps from the front of the vehicle and the hoist. Lower the vehicle. Untie the radiator from the hood latch panel.

- Cruise control module
- Cruise and accelerator cables to the throttle body
- Surge tank bypass hose to the engine
- Starter
- Connectors to the A/C compressor and the CKP sensor
- Power steering pump
- Ground cables at the rear of the engine block
- EVAP solenoid
- Slave cylinder hydraulic line
- Heater hoses to the heater core
- Radiator hoses
- Fuel line fittings
- Vacuum lines
- Engine controls harness
- VSS
- Back up lamp switch
- Shift cable bracket and the shift cables to the bracket
- Shift cables at the shift control
- Air cleaner bracket
- Underhood fuse block

- Air cleaner
- Air cleaner duct
- Negative battery cable

18. Refill the cooling system.
19. Refill the crankcase.
20. Recharge the A/C system.
21. Start the vehicle and verify no leaks.
22. Check and/or adjust the wheel alignment.

3.1L Engines

Please note that the engine and transaxle are removed as an assembly from under the vehicle.

1. Before servicing the vehicle, refer to the precautions in the beginning of this section.
2. Relieve the fuel system pressure.
3. Drain the cooling system.
4. Drain the engine oil.
5. Remove or disconnect the following:

- Air cleaner assembly
- Upper and lower radiator hoses
- Coolant inlet line from the coolant surge tank
- Vacuum hoses from the Evaporative Emissions (EVAP) canister purge valve, vacuum modulator and power brake booster
- Heater outlet hose from the water pump
- Accessory drive belt
- Control cables from the throttle body lever and intake manifold bracket
- Ignition coil assembly electrical connectors
- Heated Oxygen Sensor (HO2S) electrical connector
- Fuel injector electrical connectors
- Idle Air Control (IAC) valve electrical connector
- Throttle Position Sensor (TPS) electrical connector
- Engine Coolant Temperature (ECT) sensor electrical connector
- Park/neutral position switch electrical connector
- Exhaust Gas Recirculation (EGR) valve electrical connector
- Ground wire from the transaxle
- Alternator
- Power steering lines from the power steering pump
- Fuel lines from the fuel rail
- Cooling fan assembly
- Shift control cable from the transaxle shift lever and cable bracket
- Transaxle vent tube

- Vacuum hose from the vacuum reservoir

6. Install an engine support fixture.
7. Loosen, but do not remove the top 2 A/C compressor mounting bolts.
8. Remove or disconnect the following:

- Front wheels
- Right and left inner fender splash shields
- Engine mount strut
- Anti-lock Brake System (ABS) sensor wires from the Wheel Speed Sensors (WSS)
- Lower ball joints from the steering knuckles
- Suspension support assembly
- Halfshafts from the transaxle
- Oil filter and oil filter adapter
- Flywheel cover
- Starter motor
- Heater hoses from the heater core
- Knock sensor electrical connector
- 2 Crankshaft Position (CKP) sensor electrical connectors
- Oil level sensor electrical connector
- Vehicle Speed Sensor (VSS) electrical connector
- Ground wires from the transaxle
- A/C compressor and move it aside with the refrigerant lines connected
- Vacuum reservoir tank
- Exhaust pipe from the exhaust manifold
- Engine mount strut bracket from the engine
- Transaxle cooler lines from the radiator
- Transaxle oil fill tube

9. Lower the vehicle until the powertrain assembly is resting on an engine table.
10. Remove or disconnect the following:

- Transaxle mount-to-body bolts
- Intermediate bracket from the right engine mount
- Engine support fixture

11. Raise the vehicle leaving the powertrain assembly on the engine table.
12. Separate the engine from the transaxle.

To install:

13. Install the transaxle to the engine. Torque the bolts to 55 ft. lbs. (75 Nm).
14. Lower the vehicle over the powertrain assembly.
15. Install or connect the following:

- Engine support fixture
- Intermediate bracket to the right engine mount
- Transaxle mount-to-body bolts. Torque the bolts to 49 ft. lbs. (66 Nm).

- Transaxle oil fill tube
- Transaxle cooler lines to the radiator
- Engine mount strut bracket to the engine. Torque the bolts to 44 ft. lbs. (60 Nm).
- Exhaust pipe to the exhaust manifold. Torque the bolts to 18 ft. lbs. (25 Nm).
- Vacuum reservoir tank
- A/C compressor. Torque the bolts to 37 ft. lbs. (50 Nm).
- Ground wires to the transaxle
- VSS electrical connector
- Oil level sensor electrical connector
- CKP electrical connectors
- Knock sensor electrical connector
- Heater hoses to the heater core
- Starter motor. Torque the bolts to 32 ft. lbs. (43 Nm).
- Flywheel cover. Torque the bolts to 115 inch lbs. (13 Nm).
- Oil filter and oil filter adapter
- Halfshafts to the transaxle
- Suspension support assembly. Torque the bolts to 89 ft. lbs. (120 Nm).
- Lower ball joints to the steering knuckles, using new cotter pins. Torque the nuts to 41 ft. lbs. (55 Nm).
- WSS electrical connectors
- Engine mount strut. Torque the bolts to 44 ft. lbs. (60 Nm) plus an additional 90 degree turn.
- Right and left inner fender splash shields
- Front wheel
- Top 2 A/C compressor mounting bolts. Torque the bolts to 37 ft. lbs. (50 Nm).

16. Remove the engine support fixture.
17. Install or connect the following:

- Vacuum hose to the vacuum reservoir
- Transaxle vent tube to the transaxle
- Shift control cable to the transaxle shift lever and cable bracket
- Cooling fan assembly
- Fuel lines to the fuel rail
- Power steering lines to the power steering pump
- Alternator. Torque the nut to 37 ft. lbs. (50 Nm).
- Ground wire to the transaxle
- EGR valve electrical connector
- Park/neutral position switch electrical connector
- ECT electrical connector
- TPS electrical connector

- IAC electrical connector
- Fuel injector electrical connectors
- HO$_2$S electrical connector
- Ignition coil assembly electrical connector
- Control cables to the throttle body lever and intake manifold bracket
- Accessory drive belt
- Heater outlet hose to the water pump
- Vacuum hoses to the EVAP canister purge valve, vacuum modulator and power brake booster
- Coolant inlet line to the coolant surge tank
- Upper and lower radiator hoses
- Air cleaner assembly
- Negative battery cable

18. Check and fill all the engine fluids as necessary.

19. Start the vehicle and bleed the power steering system.

3.4L Engine

1. Before servicing the vehicle, refer to the precautions in the beginning of this section.

2. Disconnect the negative battery cable.

3. Drain the engine coolant.

4. Remove or disconnect the following:
- Air cleaner assembly
- Hood
- Accessory drive belt
- Hoses from the surge tank
- Cruise control module
- Upper wiring harness from the engine
- Throttle and cruise control cables
- Starter motor
- A/C compressor and move it aside with the lines attached
- Lower wiring harness from the engine components
- Catalytic converter from the rear exhaust manifold
- Torque converter cover
- Torque converter-to-flywheel bolts
- Engine splash shields
- Transaxle-to-engine brace and the 2 outer transaxle mounting bolts
- Upper and lower radiator hoses
- Fuel lines from the engine
- Vacuum hose from the power brake booster
- Heater hoses

5. Install an engine hoist and raise the engine slightly.

6. Remove or disconnect the following:
- Engine mount and bracket
- Power steering pump

- Transaxle-to-engine bolts
- Engine from the vehicle

To install:

7. Install or connect the following:
- Engine. Torque the upper transaxle-to-engine bolts to 66 ft. lbs. (90 Nm).
- Engine support fixture and remove the engine hoist
- Power steering pump. Torque the bolts to 25 ft. lbs. (34 Nm).
- Engine mount bracket and mount. Torque the bolts to 43 ft. lbs. (58 Nm) and the nuts to 35 ft. lbs. (47 Nm).
- Heater hoses
- Fuel lines
- Vacuum hose to the power brake booster
- Upper and lower radiator hoses
- Both outer transaxle to engine bolts. Torque the bolts to 66 ft. lbs. (90 Nm).
- Transaxle to the engine brace. Torque the bolts to 32 ft. lbs. (43 Nm).
- Engine splash shields
- Flexplate to the torque converter. Torque the bolts to 46 ft. lbs. (62 Nm).
- Torque converter cover. Torque the bolts to 89 inch lbs. (10 Nm).
- Catalytic converter to the exhaust manifold. Torque the bolts to 25 ft. lbs. (34 Nm).
- A/C compressor. Torque the bolts to 37 ft. lbs. (50 Nm).
- Starter motor. Torque the bolts to 32 ft. lbs. (43 Nm).
- Lower and upper wiring harnesses

❊❊ CAUTION

To avoid personal injury and/or damage to the vehicle, replace the throttle cable with a new one any time the engine has been removed from the vehicle.

- Cruise control and throttle cables
- Cruise control module
- Accessory drive belt
- Hoses to the surge tank
- Hood. Torque the bolts to 13 ft. lbs. (17 Nm).
- Air cleaner assembly

8. Refill the cooling system.

9. Refill the crankcase.

10. Start the engine and inspect for coolant and/or oil leaks.

11. Stop the engine and recheck the fluid levels after the engine has cooled.

Water Pump

REMOVAL & INSTALLATION

2.2L Engine

1. Before servicing the vehicle, refer to the precautions in the beginning of this section.

2. Disconnect the negative battery cable.

3. Drain the engine coolant.

4. Remove or disconnect the following:
- Exhaust manifold, if equipped with n automatic transaxle
- Right front wheel
- Splash shield
- Water pump sprocket access plate from the timing cover

5. Attach Water Pump Sprocket Holding Tool J 43651to the sprocket using the access plate bolts to secure the tool to the engine front cover.
- Sprocket to water pump bolts
- Engine block to water pump bolt
- Engine front cover to water pump bolt
- Thermostat housing to water pump feed pipe
- 2 water pump to engine block bolts
- Water pump

To install:

6. Use a threaded stud in the water pump hub to align the hub to the water pump sprocket.
- Water pump and bolts
- Thermostat housing to water pump feed pipe
- Engine front cover bolt and engine block to water pump bolt. Tighten the bolts to 15 ft. lbs. (20 Nm)
- 2 of the water pump sprocket to water pump bolts

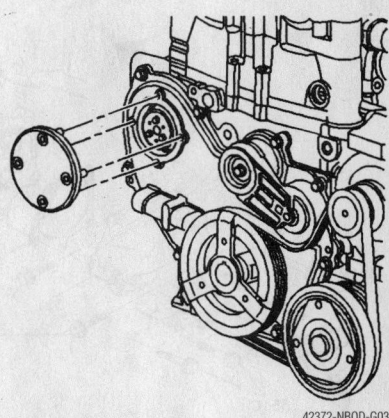

42372-NBOD-G03

Remove the water pump sprocket access plate from the timing cover —2.2L engine

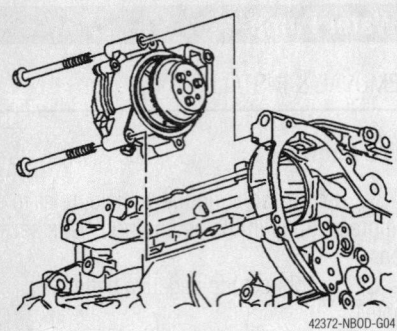

Water pump mounting—2.2L engine

7. Remove the threaded stud.
- Last water pump sprocket to water pump bolt and tighten the sprocket bolts to 89 inch lbs. (10 Nm).
8. Remove the tool.
- Water pump sprocket access plate to the timing cover and tighten the bolts to 89 inch lbs. (10 Nm)
- Splash shield
- Right front wheel
- Exhaust manifold, if removed
9. Fill the cooling system.
10. Connect the negative battery cable.

2.4L Engine

1. Before servicing the vehicle, refer to the precautions in the beginning of this section.
2. Disconnect the negative battery cable.
3. Drain the cooling system.
4. Remove or disconnect the following:
- Oxygen Sensor (O₂S) electrical connector
- Exhaust manifold heat shield
- Coolant inlet housing bolt through the exhaust manifold

- Exhaust manifold brace-to-manifold bolt
- Manifold-to-exhaust pipe studs
- Coolant inlet housing-to-water pump cover bolt
- Exhaust pipe from the exhaust manifold by pulling it downward

❋❋ WARNING

Do not rotate the flex coupling more than 4 degrees or damage may occur.

- Coolant inlet pipe from the oil pan
- Brake vacuum pipe from the camshaft housing
- Exhaust manifold from the cylinder head
- Heater hose from the heater outlet pipe
- Timing chain cover and tensioner
- Water pump cover-to-engine bolts
- 3 water pump-to-timing chain housing nuts
- Water pump and cover assembly
- Water pump cover from the pump

To install:
5. Install or connect the following:
- Water pump with a new gasket to the cover and tighten the bolts finger-tight
- Water pump cover-to-engine bolts and tighten finger-tight
- Water pump-to-timing chain housing nuts and tighten finger-tight
- Coolant inlet pipe into the water pump cover and tighten the bolts finger-tight
6. With all gaps closed, torque the bolts, in the following sequence, to the proper values:

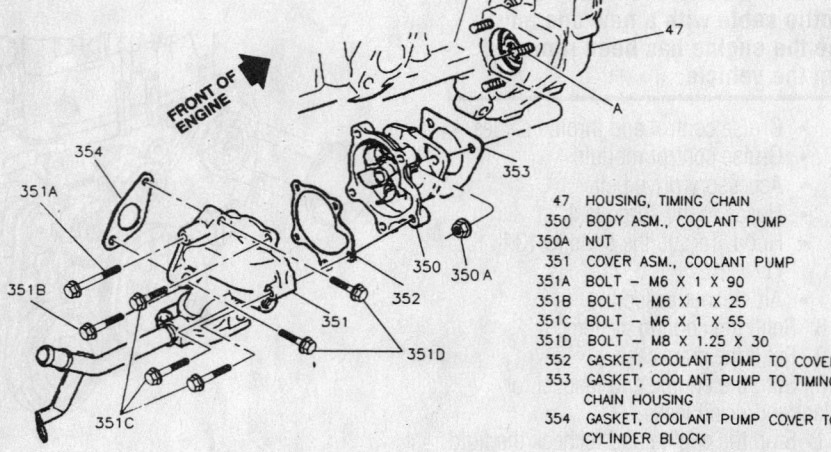

47	HOUSING, TIMING CHAIN
350	BODY ASM., COOLANT PUMP
350A	NUT
351	COVER ASM., COOLANT PUMP
351A	BOLT — M6 X 1 X 90
351B	BOLT — M6 X 1 X 25
351C	BOLT — M6 X 1 X 55
351D	BOLT — M8 X 1.25 X 30
352	GASKET, COOLANT PUMP TO COVER
353	GASKET, COOLANT PUMP TO TIMING CHAIN HOUSING
354	GASKET, COOLANT PUMP COVER TO CYLINDER BLOCK

Exploded view of the water pump mounting—2.4L engine

a. Pump assembly-to-chain housing nuts to 19 ft. lbs. (26 Nm).
b. Pump cover-to-pump assembly to 124 inch lbs. (14 Nm).
c. Water pump cover-to-engine bolts to 19 ft. lbs. (26 Nm).
d. Coolant inlet pipe assembly-to-water pump cover bolts to 124 inch lbs. (14 Nm).
7. Install or connect the following:
- Exhaust manifold to the cylinder head. Torque the nuts to 11 ft. lbs. (15 Nm).
- Brake vacuum pipe to the camshaft housing
- Coolant inlet pipe to the oil pan. Torque the nuts to 19 ft. lbs. (25 Nm).
- Timing chain tensioner. Torque the bolts to 89 inch lbs. (10 Nm.).
- Front cover. Torque the bolts to 106 inch lbs. (12 Nm).
- Exhaust pipe to the manifold. Torque the bolts to 26 ft. lbs. (35 Nm).
- Exhaust manifold brace to the manifold. Torque the bolt to 41 ft. lbs. (56 Nm) and the nuts to 19 ft. lbs. (26 Nm).
- Exhaust manifold to the exhaust pipe. Torque the nuts to 26 ft. lbs. (35 Nm).
- Heater hose to the heater pipe
- Exhaust manifold heat shield. Torque the bolts to 124 inch lbs. (14 Nm).
- O₂S electrical connector
- Negative battery cable
8. Refill the cooling system.
9. Start the engine and check for leaks.

3.1L and 3.4L Engines

1. Before servicing the vehicle, refer to the precautions in the beginning of this section.
2. Disconnect the negative battery cable.
3. Drain the cooling system.
4. Remove or disconnect the following:
- Accessory drive belt
- Water pump pulley
- Water pump

To install:
5. Install or connect the following:
- Water pump with a new gasket. Torque the bolts to 89 inch lbs. (10 Nm).
- Water pump pulley. Torque the bolts to 18 ft. lbs. (25 Nm).
- Accessory drive belt
- Negative battery cable
6. Refill and bleed the cooling system.
7. Test run the engine and check for leaks.

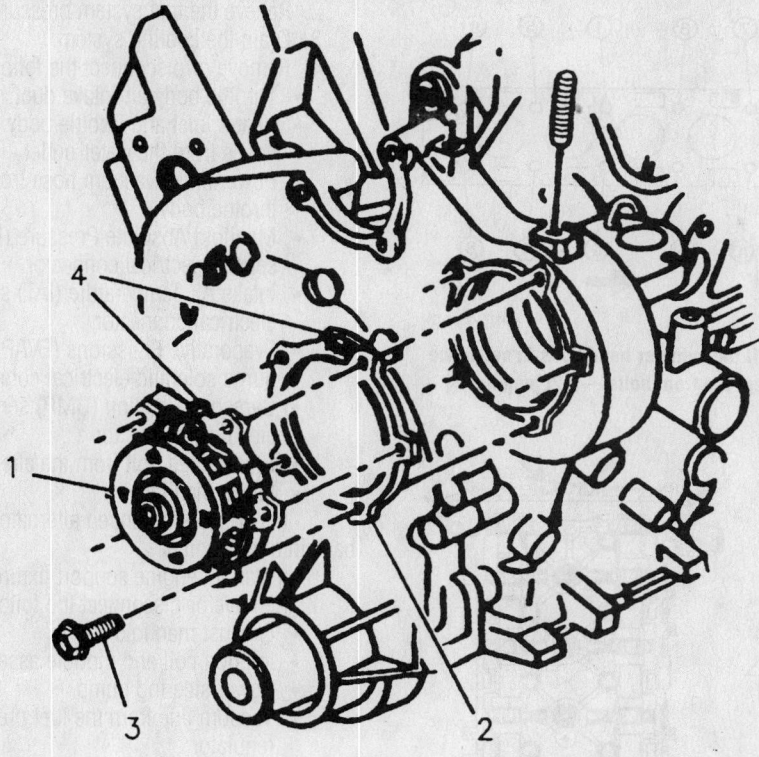

1 COOLANT PUMP
2 GASKET
3 BOLT — 10 N·m (89 LBS. IN.)
4 LOCATOR (MUST BE VERTICAL)

7922Z303

Exploded view of the water pump mounting—3.1L and 3.4L engines

Heater Core

REMOVAL & INSTALLATION

1. Before servicing the vehicle, refer to the precautions in the beginning of this section.
2. Drain the engine cooling system.
3. Disable the Supplemental Inflatable Restraint (SIR) system.
4. Recover the A/C refrigerant.
5. Remove or disconnect the following:
   - Evaporator hose from the evaporator
   - Heater hoses
   - Drain tube elbow
   - Heater case plate for the heater pipes
   - Heater case plate for the evaporator block
   - End caps
   - Sound insulators
   - Instrument panel compartment
   - Passenger inflator module
   - Tilt steering lever
   - Upper and the lower steering column covers
   - Hazard warning switch
   - Instrument panel cluster trim plate
   - Instrument panel cluster
   - Driver inflator module
   - Steering wheel
   - Multifunction lever
   - Accessory trim plate
   - Radio
   - Cup holder
   - Gearshift handle
   - Ignition switch bolts
   - Instrument panel upper bolt covers
   - Electrical junction box connections
   - Fog lamp switch
   - Windshield side upper garnish molding
   - Console from the instrument panel
   - Instrument panel from the tie bar
   - Air distribution duct
   - Floor air duct
   - Tie bar
   - Daytime Running Lights (DRL) sensor wiring harness from the Heating, Ventilation, Air Conditioning (HVAC) module
   - Blower motor electrical connectors
   - Temperature actuator electrical connector
   - HVAC module from the vehicle
   - Heater core case cover
   - Heater core bracket
   - Heater core

To install:

6. Install or connect the following:
   - Heater core
   - Heater core bracket. Torque the screw to 9 inch lbs. (1 Nm).
   - Heater core case cover. Torque the screws to 9 inch lbs. (1 Nm).
   - HVAC module
   - Temperature actuator electrical connector
   - Blower motor electrical connectors
   - DRL wiring harness
   - Tie bar. Torque the bolts to 15 ft. lbs. (20 Nm).
   - Floor air duct. Torque the bolts to 18 inch lbs. (2 Nm).
   - Air distribution duct. Torque the bolts to 27 inch lbs. (3 Nm).
   - Instrument panel
   - Console to the instrument panel
   - Windshield side upper garnish molding
   - Fog lamp switch
   - Electrical junction box connections
   - Instrument panel upper bolt covers
   - Ignition switch bolts
   - Gearshift handle
   - Cup holder
   - Radio. Torque the bolts to 27 inch lbs. (3 Nm).
   - Accessory trim plate
   - Multifunction lever. Torque the screw to 35 inch lbs. (4 Nm).
   - Steering wheel. Torque the nut to 27 ft. lbs. (37 Nm).
   - Driver inflator module. Torque the screw to 89 inch lbs. (10 Nm).
   - Instrument panel cluster
   - Instrument panel cluster trim plate
   - Hazard warning switch
   - Upper and the lower steering column covers. Torque the screw to 18 inch lbs. (2 Nm).
   - Tilt steering lever
   - Passenger inflator module. Torque the fasteners to 89 inch lbs. (10 Nm).
   - Instrument panel compartment
   - Sound insulators
   - End caps
   - Heater case plate for the evaporator

block. Torque the nuts to 89 inch lbs. (10 Nm).

- Heater case plate for the heater pipes
- Drain tube elbow
- Heater hoses
- Evaporator hose. Torque the fitting to 18 ft. lbs. (25 Nm).

7. Fill the engine cooling system.

8. Evacuate and recharge the A/C system.

Cylinder Head

REMOVAL & INSTALLATION

2.2L Engine

1. Before servicing the vehicle, refer to the precautions in the beginning of this section.

2. Remove or disconnect the following:
- Negative battery cable
- Intake manifold
- Power steering pump bolts and set pump aside
- Exhaust manifold
- Timing chain

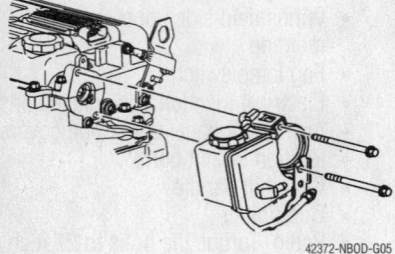

Remove the power steering pump bolts and set pump aside—2.2L engine

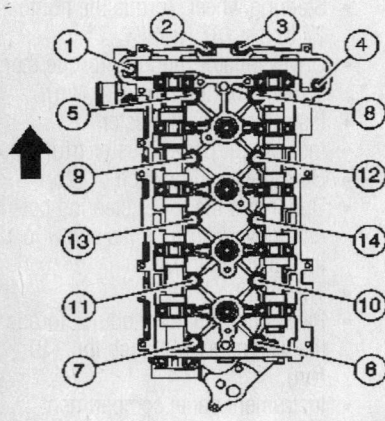

Remove the cylinder head bolts in sequence and discard —2.2L engine

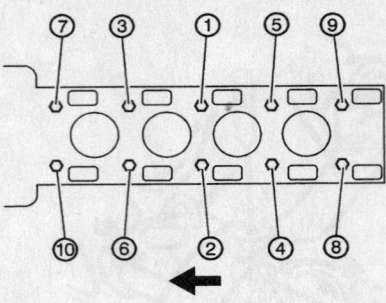

Install the cylinder head bolts in sequence (except the front bolts)—2.2L engine

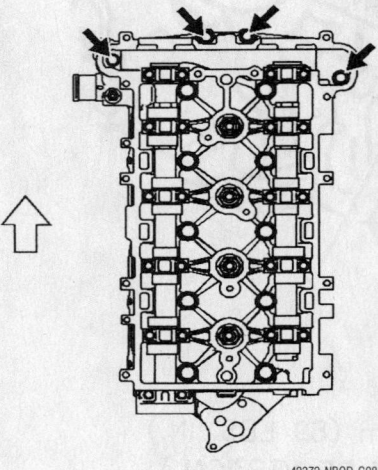

Install the front cylinder head bolts—2.2L engine

3. Drain the cooling system.
- Cylinder head bolts in sequence and discard
- Cylinder head and gasket

To install:

4. Install or connect the following:
- Cylinder head and gasket
- NEW cylinder head bolts (except the front bolts) and tighten in sequence to 22 ft. lbs. (30 Nm) and then an additional 155 degrees
- NEW front cylinder head bolts and tighten to 26 ft. lbs. (35 Nm)
- Timing chain
- Exhaust manifold
- Intake manifold
- Power steering pump and tighten the bolts to 19 ft. lbs. (25 Nm).

5. Fill the cooling system.

6. Connect the negative battery cable.

2.4L Engine

1. Before servicing the vehicle, refer to the precautions in the beginning of this section.

2. Relieve the fuel system pressure.

3. Drain the cooling system.

4. Remove or disconnect the following:
- Throttle body air intake duct
- Heater inlet and throttle body heater hoses from the water outlet
- Power brake vacuum hose from the throttle body
- Manifold Absolute Pressure (MAP) sensor electrical connector
- Intake Air Temperature (IAT) sensor electrical connector
- Evaporative Emissions (EVAP) purge solenoid electrical connector
- Camshaft Position (CMP) sensor electrical connector
- Stud-ended bolt from the alternator
- Intake manifold

5. Install the stud-ended alternator bolt back into the engine.

6. Install an engine support fixture.

7. Remove or disconnect the following:
- Exhaust manifold
- Ignition coil and module assembly
- Power steering pump
- Vacuum line from the fuel pressure regulator
- Fuel injector electrical connectors
- Fuel line clamp from the intake camshaft housing bracket
- Fuel rail without disconnecting the fuel lines
- Timing chain and sprocket
- Water pump
- Electrical connection from the oil switch
- Transaxle fluid level indicator tube assembly from the exhaust camshaft cover and move it aside, if equipped with an automatic transaxle

➡️**Any time the camshaft housing-to-cylinder head bolts are loosened or removed, the camshaft housing-to-cylinder head gasket must be replaced.**

- Intake camshaft housing

➡️**Turn the camshaft housing upside down as soon as it is removed from the cylinder head; otherwise, the lifters may fall out.**

- Exhaust camshaft housing
- Upper radiator hose
- Engine Coolant Temperature (ECT) sensor electrical connector
- Cylinder head bolts by reversing of the tightening sequence
- Cylinder head

To install:

8. Install or connect the following:
- Cylinder head with a new gasket

Cylinder head bolt tightening sequence— 2.4L engine

- New cylinder head bolts lubricated with clean engine oil

9. Torque the cylinder head bolts, in sequence, as follows:

    a. Step 1: Bolts 1 through 8 to 40 ft. lbs. (65 Nm).

    b. Step 2: Bolts 9 and 10 to 30 ft. lbs. (40 Nm).

    c. Step 3: All bolts an additional 90 degree (¼) turn.

10. Install the intake and exhaust camshaft housings and torque the bolts as follows:

    a. Long bolts: 11 ft. lbs. (15 Nm) plus an additional 90 degree turn.

    b. Short bolts: 11 ft. lbs. (15 Nm) plus an additional 30 degree turn.

11. Install or connect the following:

- Water pump. Torque the pump-to-cover bolts to 124 inch lbs. (14 Nm) and the water pump cover-to-block bolts to 19 ft. lbs. (26 Nm).
- Timing chain and sprocket. Torque the sprocket bolt to 52 ft. lbs. (70 Nm).
- Upper radiator hose to the water outlet
- ECT sensor connector
- Fuel rail
- Fuel injector electrical connectors
- Fuel pressure regulator vacuum line
- MAP sensor electrical connector
- IAT sensor electrical connector
- EVAP purge solenoid electrical connector
- CMP sensor electrical connector
- Oil switch electrical connector
- Exhaust manifold. Torque the nuts to 110 inch lbs. (13 Nm).
- Intake manifold. Torque the bolts to 18 ft. lbs. (24 Nm).
- Power steering pump. Torque the bolts to 19 ft. lbs. (26 Nm).
- Throttle body air intake duct
- Oil fill tube into the engine
- Heater inlet and throttle body heater hoses to the water outlet
- Power brake vacuum hose
- Ignition coil and module assembly

12. Remove the engine support fixture.

13. Fill all fluids to their proper levels.

➡ **An oil and filter change is recommended.**

14. Connect the negative battery cable. Start the vehicle and verify no leaks.

3.1L and 3.4L Engines

LEFT (FRONT) CYLINDER HEAD

1. Before servicing the vehicle, refer to the precautions in the beginning of this section.

2. Relieve the fuel system pressure.

3. Drain the crankcase.

4. Drain the cooling system.

5. Remove or disconnect the following:

- Upper half of the air cleaner assembly
- Throttle body air inlet duct
- Exhaust crossover pipe heat shield
- Crossover pipe
- Spark plug wires from the spark plugs
- Rocker arm covers
- Upper intake plenum and lower intake manifold
- Left side exhaust manifold
- Oil level indicator tube

➡ **When removing the valvetrain components, keep them in order for installation purposes.**

- Rocker arms and pushrods
- Cylinder head

To install:

6. Install the cylinder head with a new gasket.

7. Torque the cylinder head bolts to 33 ft. lbs. (45 Nm) plus an additional 90 degree turn.

8. Install or connect the following:

- New intake manifold gasket
- Pushrods and rocker arms. Torque the bolts to 89 inch lbs. (10 Nm) plus an additional 30 degree turn.
- Lower intake manifold. Torque the bolts to 115 inch lbs. (13 Nm).
- Upper intake plenum. Torque the bolts to 18 ft. lbs. (25 Nm).
- Rocker arm covers. Torque the bolts to 89 inch lbs. (10 Nm).

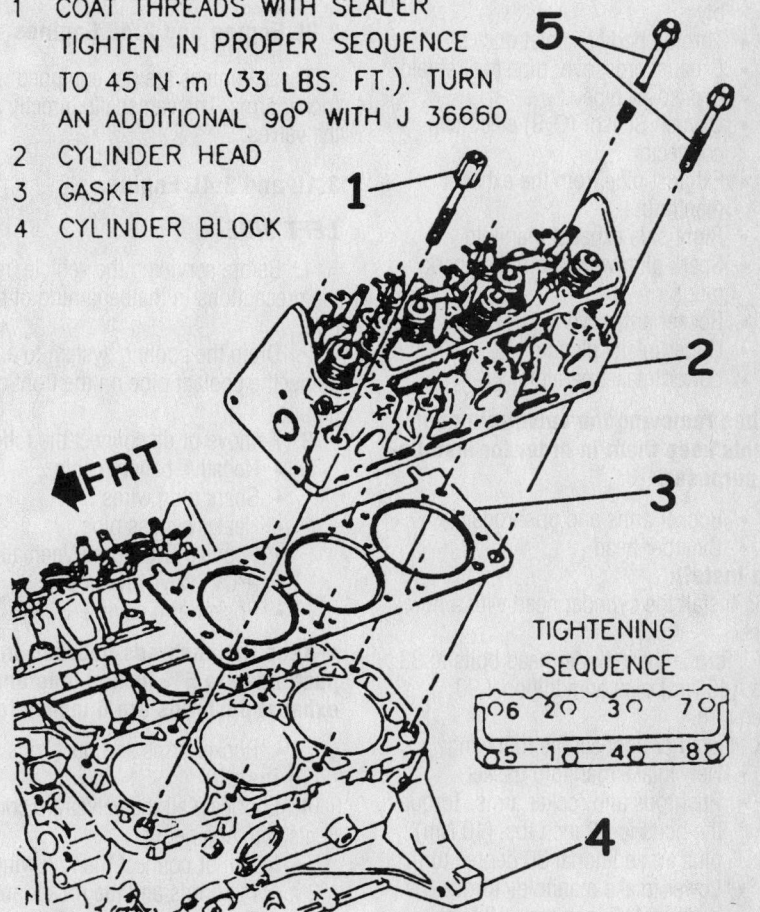

1 COAT THREADS WITH SEALER
TIGHTEN IN PROPER SEQUENCE
TO 45 N·m (33 LBS. FT.). TURN
AN ADDITIONAL 90° WITH J 36660

2 CYLINDER HEAD

3 GASKET

4 CYLINDER BLOCK

Cylinder head bolt tightening sequence—3.1L and 3.4L engines

- Oil level indicator tube
- Left side exhaust manifold. Torque the nuts to 12 ft. lbs. (16 Nm).
- Spark plug wires
- Exhaust crossover pipe. Torque the bolts to 18 ft. lbs. (25 Nm).
- Crossover pipe heat shield. Torque the bolts to 89 inch lbs. (10 Nm).
- Upper half of the air cleaner assembly
- Throttle body air inlet duct
- Negative battery cable

9. Refill the cooling system.
10. Refill the crankcase.

➡**A filter change is recommended.**

11. Start the engine and verify no leaks.

RIGHT (REAR) CYLINDER HEAD

1. Before servicing the vehicle, refer to the precautions in the beginning of this section.
2. Relieve the fuel system pressure.
3. Drain the crankcase.
4. Drain the cooling system.
5. Remove or disconnect the following:
- Upper half of the air cleaner assembly
- Throttle body air inlet duct
- Exhaust crossover pipe heat shield
- Crossover pipe
- Oxygen Sensor (O2S) electrical connector
- Exhaust pipe from the exhaust manifold
- Right side exhaust manifold
- Spark plug wires from the spark plugs
- Rocker arm covers
- Upper intake plenum
- Lower intake manifold

➡**When removing the valvetrain components keep them in order for installation purposes.**

- Rocker arms and pushrods
- Cylinder head

To install:
6. Install the cylinder head with a new gasket.
7. Torque the cylinder head bolts to 33 ft. lbs. (45 Nm) plus an additional 90 degree turn.
8. Install or connect the following:
- New intake manifold gasket
- Pushrods and rocker arms. Torque the bolts to 89 inch lbs. (10 Nm) plus an additional 30 degree turn.
- Lower intake manifold. Torque the bolts to 115 inch lbs. (13 Nm).
- Upper intake plenum. Torque the bolts to 18 ft. lbs. (25 Nm).

- Rocker arm covers. Torque the bolts to 89 inch lbs. (10 Nm).
- Spark plug wires
- Exhaust manifold. Torque the nuts to 12 ft. lbs. (16 Nm).
- Exhaust pipe to the exhaust manifold. Torque the nuts to 33 ft. lbs. (45 Nm).
- O2S electrical connector
- Exhaust crossover pipe. Torque the bolts to 18 ft. lbs. (25 Nm).
- Crossover pipe heat shield. Torque the nuts to 89 inch lbs. (10 Nm).
- Upper half of the air cleaner assembly
- Throttle body air inlet duct
- Negative battery cable

9. Refill the cooling system.
10. Refill the crankcase.

➡**An oil filter change is recommended.**

11. Start the engine and verify no leaks.

Rocker Arms

REMOVAL & INSTALLATION

2.2L Engine and 2.4L Engines

These engines are not equipped with rocker arms. The camshafts directly actuate the valves.

3.1L and 3.4L Engines

LEFT SIDE

1. Before servicing the vehicle, refer to the precautions in the beginning of this section.
2. Drain the cooling system to a level below the coolant pipe on the front of the engine.
3. Remove or disconnect the following:
- Negative battery cable
- Spark plug wires
- Heater bypass pipe
- Positive Crankcase Ventilation (PCV) valve and hose
- Rocker arm cover

➡**Keep the pushrods in order. Intake pushrods are 5¾ inches long and exhaust pushrods are 6 inches long.**

- Rocker arms and pushrods

To install:
4. Lubricate all the valvetrain components with engine oil.
5. Install or connect the following:
- Pushrods and the rocker arms. Torque the bolts to 89 inch lbs. (10 Nm) plus an additional 30 degree turn.

- Rocker arm cover using a new gasket. Torque the rocker cover bolts to 89 inch lbs. (10 Nm).
- PCV valve and hose
- Heater bypass pipe. Torque the screw at the water pump to 106 inch lbs. (12 Nm), the bolt at the cylinder head corner to 18 ft. lbs. (25 Nm) and the nut to 18 ft. lbs. (25 Nm).
- Spark plug wires
- Negative battery cable

6. Refill the cooling system.
7. Start the vehicle and verify no leaks.

RIGHT SIDE

1. Before servicing the vehicle, refer to the precautions in the beginning of this section.
2. Remove or disconnect the following:
- Negative battery cable
- Spark plug wires from the spark plugs and the upper intake plenum wire retainer
- Power brake booster vacuum pipe from the intake plenum
- Accessory drive belt
- Alternator, if necessary
- Ignition coil assembly and Evaporative Emissions (EVAP) canister purge solenoid as an assembly
- Rocker arm cover

➡**Keep the pushrods in order. Intake pushrods are 5¾ inches long and exhaust pushrods are 6 inches long.**

- Rocker arms and pushrods

To install:
3. Lubricate all the valvetrain components with engine oil.
4. Install or connect the following:
- Pushrods and the rocker arms. Torque the bolts to 89 inch lbs. (10 Nm) plus an additional 30 degree turn.

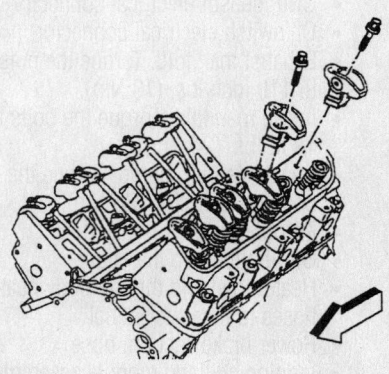

9306ZG09

Rocker arm components—3.1L and 3.4L engines

- Rocker arm cover using a new gasket. Torque the rocker cover bolts to 89 inch lbs. (10 Nm).
- Ignition coil and EVAP solenoid assembly
- Alternator, if removed. Torque the bolts to 37 ft. lbs. (50 Nm).
- Accessory drive belt
- Power brake booster vacuum pipe to the plenum
- Spark plug wires
- Negative battery cable
5. Start the vehicle and verify no leaks.

Intake Manifold

REMOVAL & INSTALLATION

These vehicles were filled at the factory with an antifreeze/coolant called GM Goodwrench DEX-COOL®. When adding coolant to vehicles, it is important that you use GM Goodwrench DEX-COOL (orange-colored, silicate-free) coolant. **Propylene glycol is not recommended for use in GM vehicles**. A 50/50 mixture of DEX-COOL and clean water will provide all the recommended protection. **DO NOT mix DEX-COOL with any other type of antifreeze**.

2.2L Engine

1. Before servicing the vehicle, refer to the precautions in the beginning of this section.
2. Remove or disconnect the following:
   - Air inlet duct and resonator
   - Idle Air Control (IAC) electrical connector
   - Throttle Position (TP) sensor electrical connector
   - Manifold Absolute Pressure (MAP) sensor electrical connector
   - Evaporative Emissions (EVAP) hose
   - Positive Crankcase Ventilation (PCV) hose

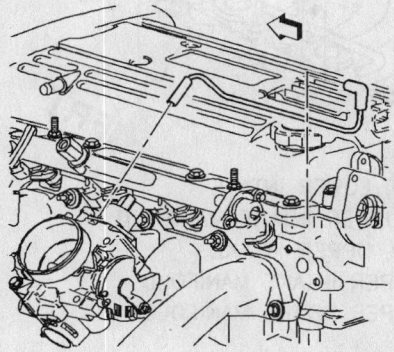

Remove the throttle body—2.2L engine

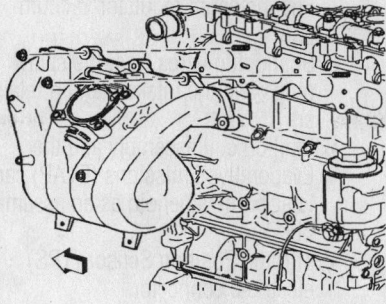

42372-NBOD-G10

Remove the intake manifold—2.2L engine

   - Purge solenoid tube
   - Brake booster hose
   - Oil level indicator tube bolt
   - Accelerator and cruise control cables
   - Throttle body
   - Fuel rail
   - Knock Sensor (KS) connector from the intake manifold
   - Intake manifold nuts and bolts
   - Intake manifold and gasket. The gasket is reusable if the gasket is not damaged.

To install:
3. Install or connect the following:
   - Intake manifold gasket
   - Intake manifold
   - Intake manifold nuts and bolts and tighten to 89 inch lbs. (10 Nm)
   - KS connector
   - Fuel rail
   - Throttle body
   - IAC, TP, and MAP electrical connectors
   - EVAP hose
   - PCV hose
   - Purge solenoid tube
   - Brake booster hose
   - Oil level indicator tube bolt
   - Accelerator and the cruise control cables
   - Air inlet duct and resonator

2.4L Engine

1. Before servicing the vehicle, refer to the precautions in the beginning of this section.
2. Relieve the fuel system pressure.
3. Drain the cooling system to a level below the intake manifold.
4. Remove or disconnect the following:
   - Air cleaner duct
   - Manifold Absolute Pressure (MAP) sensor electrical connector
   - Intake Air Temperature (IAT) sensor electrical connector

   - Evaporative Emissions (EVAP) canister purge solenoid electrical connector
   - Fuel injector electrical connectors
   - Vacuum hoses from the intake manifold, fuel pressure regulator and EVAP canister purge solenoid
   - Accelerator control cable bracket
   - Coolant lines from the throttle body
   - Stud-end alternator mount bolt
   - Exhaust Gas Recirculation (EGR) pipe from the EGR adapter
   - Intake manifold support brace, if equipped
   - Intake manifold

To install:
5. Install or connect the following:
   - Intake manifold with a new gasket. Torque the fasteners in sequence to 18 ft. lbs. (25 Nm).
   - Intake manifold brace, if equipped. Torque the bolts to 19 ft. lbs. (26 Nm).

➡ **The brace-to-block bolts must be tightened first, then the brace-to-manifold bolt.**

   - EGR pipe to the EGR adapter
   - Stud-end alternator mount bolt
   - Coolant lines to the throttle body
   - Accelerator control cable bracket
   - Vacuum hoses to the intake manifold, fuel pressure regulator and EVAP canister purge solenoid
   - Electrical connectors to the fuel injectors
   - MAP sensor electrical connector
   - IAT sensor electrical connector

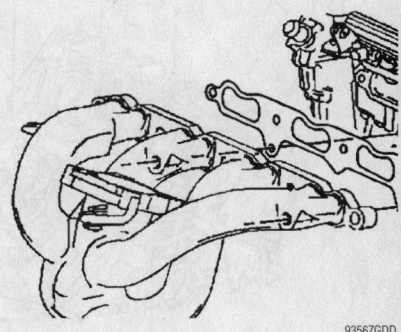

9356ZGDD

Exploded view of the intake manifold mounting—2.4L engine

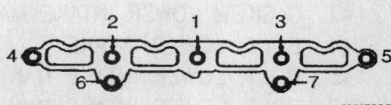

9356ZGDE

Intake manifold tightening sequence—2.4L engine

- EVAP canister purge solenoid electrical connector
- Air cleaner duct

6. Refill the cooling system.
7. Connect the negative battery cable.

3.1L and 3.4L Engines

1. Before servicing the vehicle, refer to the precautions in the beginning of this section.
2. Relieve the fuel system pressure.
3. Drain the cooling system.
4. Remove or disconnect the following:
   - Air cleaner assembly
   - Accessory drive belt
   - Exhaust Gas Recirculation (EGR) valve
   - Brake vacuum pipe at the intake plenum
   - Fuel pressure regulator vacuum line
   - Spark plug wires from the spark plugs and the intake plenum retainers
   - Ignition coil assembly and the Evaporative Emissions (EVAP) canister purge solenoid as an assembly
   - Throttle Position Sensor (TPS) electrical connector
   - Idle Air Control (IAC) sensor electrical connector
   - Fuel injector electrical connectors
   - Engine Coolant Temperature (ECT) sensor electrical connector
   - Camshaft Position (CMP) sensor electrical connector
   - Vacuum modulator
   - Manifold Absolute Pressure (MAP) sensor
   - Coolant hoses from the throttle body
   - Control cables from the throttle body and intake plenum bracket
   - Upper intake plenum
   - Both rocker arm covers
   - Fuel lines from the fuel rail and fuel line bracket
   - Fuel rail with the injectors
   - Inlet cooling pipe from the outlet housing
   - Heater bypass hose from the water pump and the cylinder head
   - Upper radiator hose at the thermostat housing
   - Thermostat housing
   - Lower intake manifold

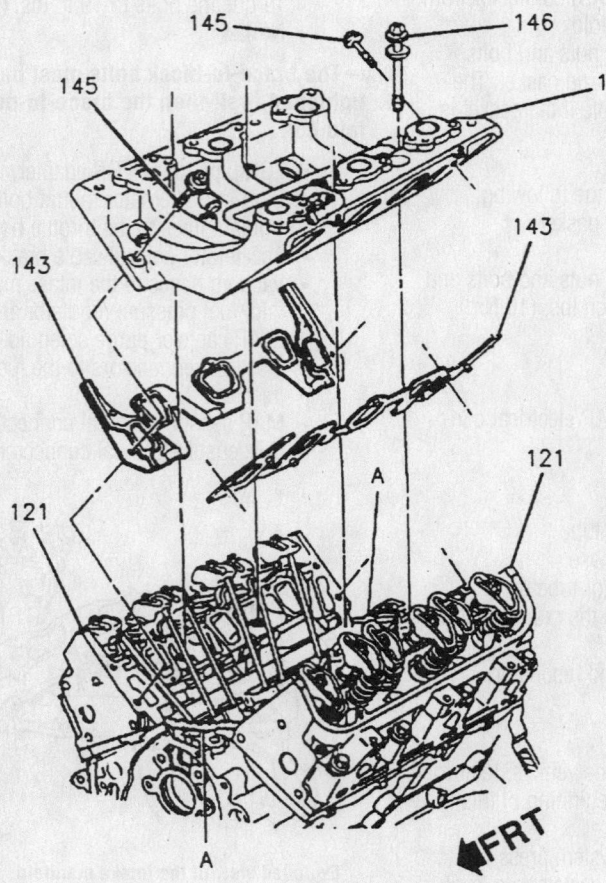

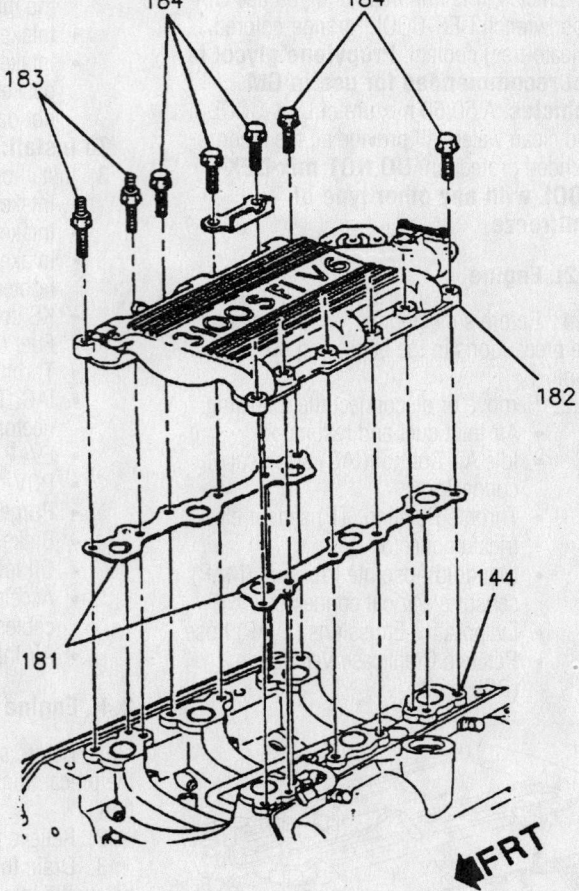

A   APPLY SEALANT
121 HEAD ASSEMBLY, CYLINDER
143 GASKET, LOWER INTAKE MANIFOLD
144 BOLT, LOWER INTAKE
145 BOLT, LOWER INTAKE MANIFOLD
146 BOLT, LOWER INTAKE MANIFOLD

144 MANIFOLD, LOWER INTAKE
181 GASKET, UPPER INTAKE MANIFOLD
182 MANIFOLD, UPPER INTAKE
183 STUD, UPPER INTAKE MANIFOLD
184 BOLT, UPPER INTAKE MANIFOLD

79227310

Exploded view of the upper and lower intake—3.1L and 3.4L engines

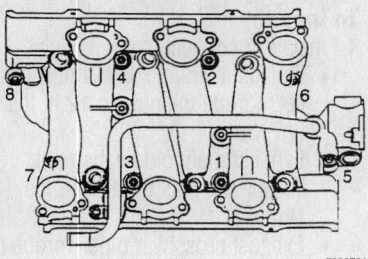

Lower intake manifold tightening sequence—3.1L and 3.4L engines

➡ **When removing the valvetrain components, keep them in order for installation purposes.**

- Rocker arms and pushrods

To install:

5. Place a 3mm bead of RTV, on each ridge, where the front and rear of the intake manifold contact the block.

6. Install or connect the following:
- New intake manifold gasket
- Pushrods and rocker arms. Torque the bolts to 89 inch lbs. (10 Nm) plus an additional 30 degree turn.
- Lower intake manifold. Apply sealant to the threads of the bolts and torque the bolts to 115 inch lbs. (13 Nm).

✳✳ WARNING

In order to prevent oil leaks at the intake manifold, tighten the vertical bolts before the diagonal bolts.

- Thermostat housing with a new gasket. Torque the bolts to 19 ft. lbs. (26 Nm).
- Upper radiator hose at thermostat housing
- Coolant bypass hose to the water pump and the cylinder head
- Coolant inlet pipe to coolant outlet housing
- Fuel rail with the injectors
- Fuel lines to the fuel rail and fuel line bracket
- Both rocker arm covers. Torque the bolts to 89 inch lbs. (10 Nm).
- Upper intake manifold with a new gasket. Torque the bolts to 18 ft. lbs. (25 Nm).
- Coolant hoses to the throttle body
- MAP sensor
- Vacuum modulator
- CMP sensor electrical connector
- ECT sensor electrical connector
- Fuel injector electrical connectors
- IAC sensor electrical connector

- TPS electrical connector
- Ignition coil assembly and the EVAP canister purge solenoid as an assembly
- Spark plug wires
- Fuel pressure regulator vacuum line
- EGR valve. Torque the bolts to 22 ft. lbs. (30 Nm)
- Control cables to the throttle body and intake plenum bracket
- Brake vacuum pipe at the intake plenum
- Accessory drive belt
- Air cleaner assembly
- Negative battery cable

7. Refill the cooling system.

➡ **An engine oil and filter change is recommended.**

8. Start the vehicle and verify no leaks.

Exhaust Manifold

REMOVAL & INSTALLATION

2.2L Engine

1. Before servicing the vehicle, refer to the precautions in the beginning of this section.

2. Remove or disconnect the following:
- Negative battery cable
- Exhaust manifold heat shield
- Oxygen (O$_2$S) sensor
- Manifold to the exhaust flex decoupler retainers

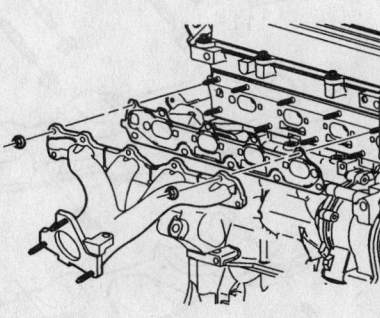

Remove the exhaust manifold—2.2L engine

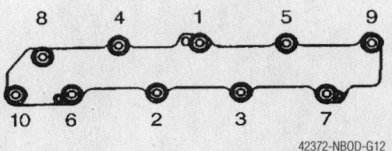

Tighten the exhaust manifold retainers in this sequence—2.2L engine

3. Pull down and back on the exhaust pipe in order to disconnect the pipe from the exhaust manifold.
- Exhaust manifold to cylinder head retaining nuts
- Exhaust manifold and gasket

To install:

4. Install or connect the following:
- New exhaust manifold gasket
- Exhaust manifold and tighten the retainers in sequence to 106 inch lbs. (12 Nm)
- New exhaust manifold to flex coupler gasket

5. Push the flex coupler into position on the exhaust manifold.
- Retaining nuts and tighten to 26 ft. lbs. (35 Nm)
- O$_2$S sensor
- Exhaust manifold heat shield and tighten the bolts to 18 ft. lbs. (25 Nm)
- Negative battery cable

2.4L Engine

1. Before servicing the vehicle, refer to the precautions in the beginning of this section.

2. Remove or disconnect the following:
- Negative battery cable
- Oxygen Sensor (O$_2$S) electrical connector
- Exhaust manifold brace
- Exhaust manifold-to-exhaust pipe spring loaded bolts
- Exhaust pipe by pulling it down and back from the exhaust manifold

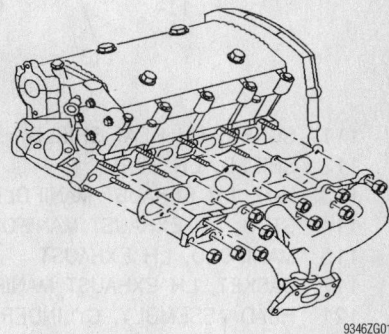

Exploded view of the exhaust manifold assembly mounting— 2.4L engine

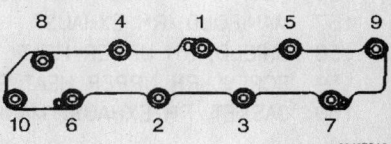

Exhaust manifold assembly torque sequence— 2.4L engine

➡**Do not bend the exhaust flex coupler more than 3 degrees in any direction, for it may damage the flex coupler.**

- Exhaust manifold

To install:

3. Install or connect the following:
- Exhaust manifold with new gaskets. Torque the nuts, in sequence, to 110 inch lbs. (13 Nm).
- Exhaust manifold brace. Torque the bolt to 41 ft. lbs. (56 Nm) and the nuts to 19 ft. lbs. (26 Nm).
- Exhaust pipe. Torque the bolts evenly to 26 ft. lbs. (35 Nm).
- O_2S electrical connector
- Negative battery cable

4. Start the vehicle and verify no exhaust leaks.

3.1L and 3.4L Engines

LEFT SIDE

1. Before servicing the vehicle, refer to the precautions in the beginning of this section.
2. Partially drain the cooling system.
3. Remove or disconnect the following:
- Negative battery cable
- Air cleaner assembly
- Throttle body duct
- Exhaust crossover heat shield
- Exhaust crossover pipe from the manifold
- Radiator hose from the thermostat housing
- Spark plug wires
- Exhaust manifold heat shield
- Exhaust manifold

To install:

4. Install or connect the following:
- Exhaust manifold with a new gasket. Torque the nuts to 12 ft. lbs. (16 Nm).
- Exhaust manifold heat shield. Torque the nuts to 89 inch lbs. (10 Nm).
- Exhaust crossover pipe. Torque the bolts to 18 ft. lbs. (25 Nm).
- Crossover pipe heat shield. Torque the bolts to 89 inch lbs. (10 Nm).
- Spark plug wires
- Radiator hose to the coolant outlet housing
- Air cleaner assembly
- Throttle body duct
- Negative battery cable

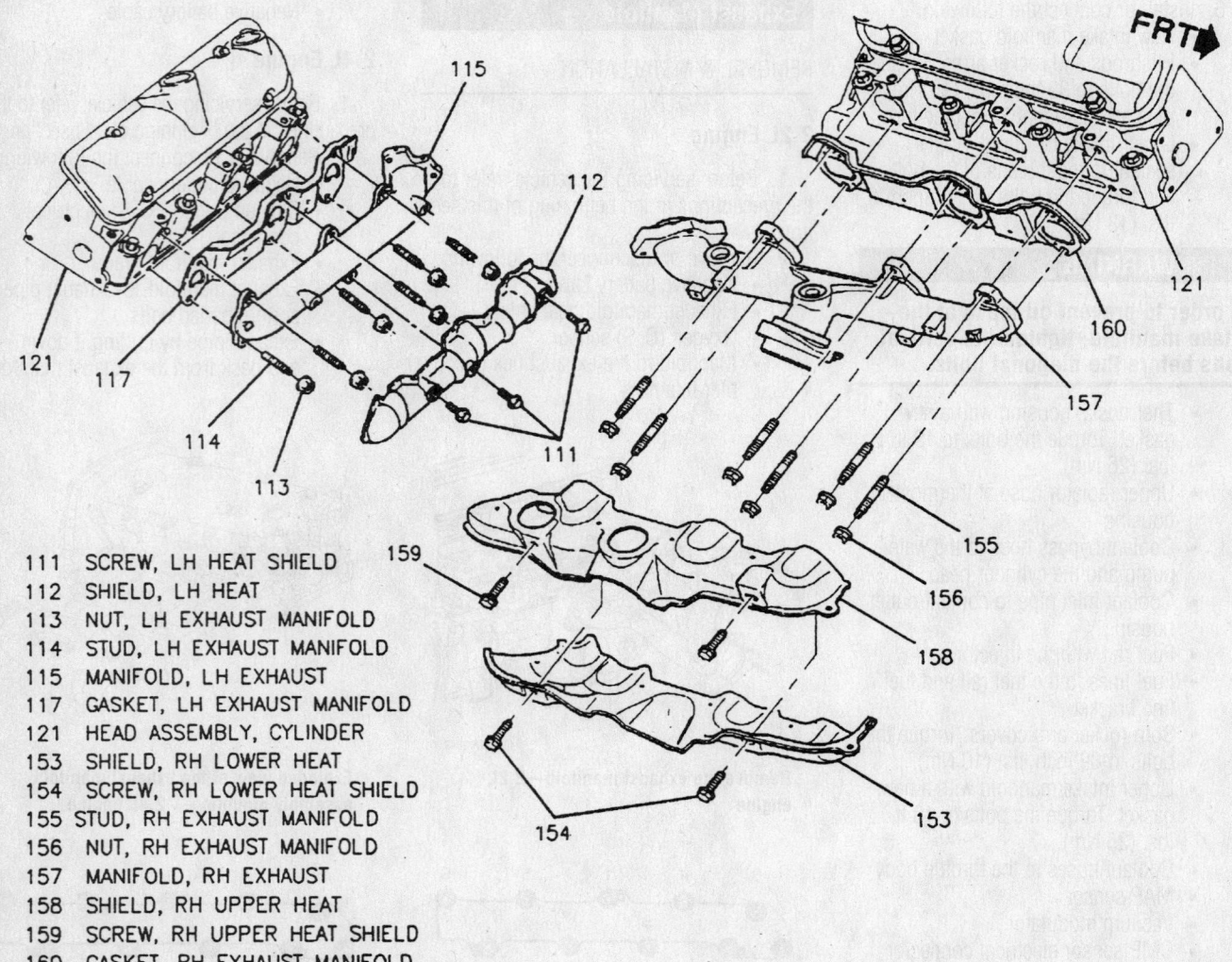

111	SCREW, LH HEAT SHIELD
112	SHIELD, LH HEAT
113	NUT, LH EXHAUST MANIFOLD
114	STUD, LH EXHAUST MANIFOLD
115	MANIFOLD, LH EXHAUST
117	GASKET, LH EXHAUST MANIFOLD
121	HEAD ASSEMBLY, CYLINDER
153	SHIELD, RH LOWER HEAT
154	SCREW, RH LOWER HEAT SHIELD
155	STUD, RH EXHAUST MANIFOLD
156	NUT, RH EXHAUST MANIFOLD
157	MANIFOLD, RH EXHAUST
158	SHIELD, RH UPPER HEAT
159	SCREW, RH UPPER HEAT SHIELD
160	GASKET, RH EXHAUST MANIFOLD

79222314

Exploded view of the exhaust manifold mounting—3.1L and 3.4L engines

RIGHT SIDE

1. Before servicing the vehicle, refer to the precautions in the beginning of this section.

2. Remove or disconnect the following:

- Negative battery cable
- Air cleaner assembly
- Throttle body duct
- Heated Oxygen Sensor (HO$_2$S)
- Exhaust crossover heat shield
- Exhaust Gas Recirculation (EGR) pipe from the exhaust manifold
- Exhaust crossover pipe
- Exhaust manifold heat shield
- Transaxle oil fill tube
- Front exhaust pipe from the exhaust manifold
- Wires from the spark plugs
- Exhaust manifold

To install:

3. Install or connect the following:

- Exhaust manifold with a new gasket. Torque the nuts to 12 ft. lbs. (16 Nm).
- Exhaust manifold heat shield. Torque the nuts to 89 inch lbs. (10 Nm).
- Exhaust crossover pipe to the manifold. Torque the bolts to 18 ft. lbs. (25 Nm).
- Exhaust crossover pipe heat shield. Torque the bolts to 89 inch lbs. (10 Nm).
- Exhaust pipe to the exhaust manifold. Torque the bolts to 33 ft. lbs. (45 Nm).
- Transaxle oil level indicator and fill tube assembly
- HO$_2$S
- EGR pipe to the exhaust manifold
- Exhaust crossover heat shield
- Throttle body duct
- Air cleaner assembly
- Negative battery cable

Camshaft and Valve Lifters

REMOVAL & INSTALLATION

2.2L Engine

1. Before servicing the vehicle, refer to the precautions in the beginning of this section.

2. Remove or disconnect the following:

- Camshaft cover
- Upper timing chain guide

3. Install Camshaft Sprocket Holding Tool J 43665.

Install Camshaft Sprocket Holding Tool J 43665—2.2L engine

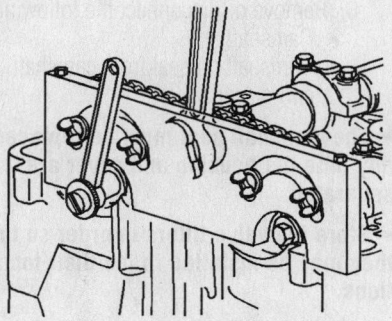

Remove both the intake and exhaust camshaft sprocket bolts and discard—2.2L engine

- Both the intake and exhaust camshaft sprocket bolts and discard

4. Slide the camshaft sprockets forward.

- Power steering pump bolts and set pump aside

5. Mark bearing caps to ensure they are installed in the original position.

➡**Remove each bolt on each cap one turn at a time until there is no spring tension on the camshaft.**

- Bearing caps
- Camshaft
- Camshaft roller followers
- Hydraulic lash adjusters

To install:

6. Lubricate the valve tips.

7. Install or connect the following:

- Hydraulic lash adjusters
- Camshaft roller followers

8. Make sure that the alignment notches are aligned with the camshaft sprocket.

- Camshaft
- Camshaft bearing caps

9. Tighten the camshaft bearing cap bolts in 3 steps to 89 inch lbs. (10 Nm).

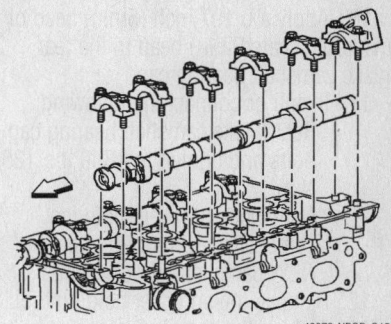

Remove the bearing caps and camshaft—2.2L engine

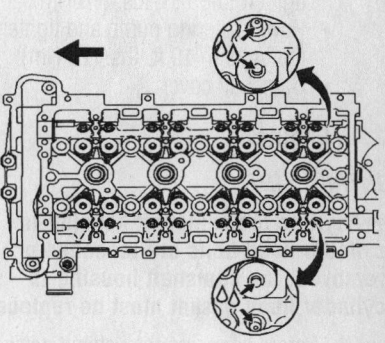

Lubricate the valve tips—2.2L engine

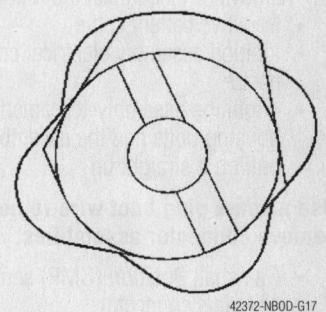

Make sure that the alignment notches are aligned with the camshaft sprocket—2.2L engine

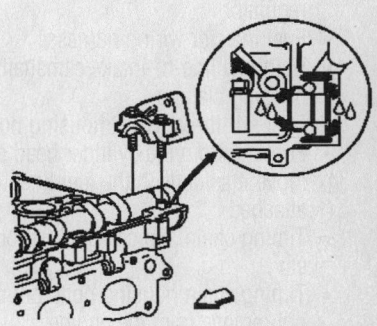

Apply a 0.197 inch (5mm) bead of anaerobic sealer97 in) bead to the rear intake camshaft bearing cap—2.2L engine

10. Apply a 0.197 inch (5mm) bead of anaerobic sealer97 in) bead to the rear intake camshaft bearing cap.

11. Install or connect the following:
- Rear intake camshaft bearing cap bolts and tighten to 18 ft. lbs. (25 Nm)
- Camshaft sprockets onto the camshafts and hand-tighten the NEW camshaft sprocket bolts

12. Remove the sprocket holding tool.

13. Tighten the camshaft sprocket bolts to 63 ft. lbs. (85 Nm), then an additional 30 degrees.
- Upper timing chain guide and tighten to 89 in. lbs. (10 Nm).
- Power steering pump and tighten the bolts to 19 ft. lbs. (25 Nm).
- Camshaft cover

2.4L Engine

INTAKE SIDE

➡**Anytime the camshaft housing-to-cylinder head bolts are loosened or removed, the camshaft housing-to-cylinder head gasket must be replaced.**

1. Before servicing the vehicle, refer to the precautions in the beginning of this section.

2. Remove or disconnect the following:
- Negative battery cable
- Ignition assembly electrical connector
- 4 ignition assembly-to-camshaft housing bolts and the assembly by pulling it straight up

➡**Use a spark plug boot wire remover to remove connector assemblies.**

- Camshaft Position (CMP) sensor electrical connector
- Power steering pump and move it aside without disconnecting the lines
- Vacuum line from fuel pressure regulator
- Fuel injector wiring harness
- Both fuel line-to-intake camshaft housing clamps
- Fuel rail-to-camshaft housing bolts
- Fuel rail from the cylinder head and move it aside with the fuel lines attached
- Timing chain and camshaft sprockets
- Timing chain housing bolts but do not remove from the engine
- Camshaft housing cover-to-camshaft housing bolts
- Camshaft housing-to-cylinder head

bolts by reversing of the tightening sequence

➡**Leave 2 bolts loosely in place to hold the camshaft housing while separating the camshaft cover from housing.**

3. Press the cover off the housing by threading 4 of the housing-to-cylinder head bolts into the tapped camshaft housing cover holes. Tighten the bolts in evenly so the cover does not bind on the dowel pins.

4. Remove the camshaft housing cover and discard the gaskets.

5. Note the position of the chain sprocket dowel pin for reassembly.

6. Remove or disconnect the following:
- Camshaft
- Camshaft oil seal from camshaft and discard it

➡**The camshaft seal must be replaced any time the housing and cover are separated.**

➡**Store the valve lifters in order so that they may be installed in the their locations.**

- Valve lifters from the camshaft housing
- Camshaft carrier from the cylinder head and discard the gasket

To install:

7. Clean all the gasket surfaces completely.

8. Install or connect the following:
- New camshaft housing gasket
- Camshaft housing

➡**Install 1 bolt loosely to hold the housing in place.**

➡**If the camshaft was replaced the valve lifters must also be replaced.**

9. Install the lifters into their original bores.

10. Lubricate the camshaft lobes, journals and lifters with camshaft and lifter pre-lube. The camshaft lobes and journals must be adequately lubricated or engine damage could occur upon start up.

11. Install or connect the following:
- Camshaft in its original position with the timing chain sprocket dowel pin straight up and aligned with the centerline of the lifter bores.
- New Green camshaft housing cover seal

➡**The seals for the intake and exhaust covers are different and the correct seals must be used.**

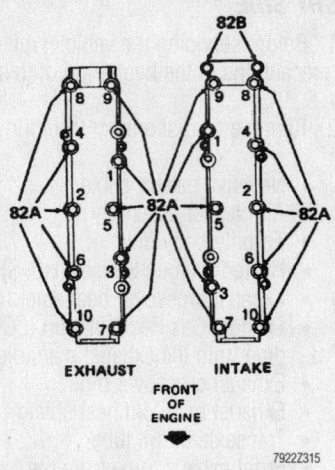

Camshaft housing bolt tightening sequence—2.4L engine

12. Remove the bolt holding the housing in place.

13. Apply thread locking compound to the camshaft housing and cover bolt threads.

14. Install or connect the following:
- Camshaft housing cover. Tighten the bolts, in sequence, to 11 ft. lbs. (15 Nm) plus an additional 90 degree turn (long bolts) and 11 ft. lbs. (15 Nm) plus an additional 30 degree turn (short bolts).
- Timing chain housing bolts
- Timing chain and sprockets
- New fuel injector O-ring seals lubricated with engine oil
- Fuel rail. Tighten the bolts to 19 ft. lbs. (26 Nm).
- Ignition assembly on the camshaft housing. Tighten the bolts to 11 ft. lbs. (15 Nm), plus an additional 30 degree turn.
- Ignition assembly electrical connector
- Negative battery cable

15. Start the vehicle and verify proper operation and no leaks.

EXHAUST SIDE

➡**Anytime the camshaft housing-to-cylinder head bolts are loosened or removed, the camshaft housing-to-cylinder head gasket must be replaced.**

1. Before servicing the vehicle, refer to the precautions in the beginning of this section.

2. Remove or disconnect the following:
- Negative battery cable
- Ignition assembly electrical connector
- 4 ignition assembly-to-camshaft

housing bolts and the assembly by pulling it straight up

➡ **Use a spark plug boot wire remover to remove connector assemblies.**

- Oil pressure switch electrical connector
- Transaxle fill tube
- Timing chain and camshaft sprockets
- Timing chain housing bolts but do not remove from the engine
- Camshaft housing cover-to-camshaft housing bolts
- Camshaft housing-to-cylinder head bolts by reversing of the tightening sequence

➡ **Leave 2 bolts loosely in place to hold the camshaft housing while separating the camshaft cover from housing.**

3. Press the cover off the housing by threading 4 of the housing-to-cylinder head bolts into the tapped camshaft housing cover holes. Tighten the bolts in evenly so the cover does not bind on the dowel pins.

4. Remove the camshaft housing cover and discard the gaskets.

5. Note the position of the chain sprocket dowel pin for reassembly.

6. Remove or disconnect the following:
- Camshaft
- Camshaft oil seal from camshaft and discard it

➡ **The camshaft seal must be replaced any time the housing and cover are separated.**

➡ **Store the valve lifters in order so that they may be installed in the their locations.**

- Valve lifters from the camshaft housing
- Camshaft carrier from the cylinder head and discard the gasket

To install:

7. Clean all the gasket surfaces completely.

8. Install or connect the following:
- New camshaft housing gasket
- Camshaft housing

➡ **Install 1 bolt loosely to hold the housing in place.**

➡ **If the camshaft was replaced the valve lifters must also be replaced.**

9. Install the lifters into their original bores.

10. Lubricate the camshaft lobes, jour-

nals and lifters with camshaft and lifter pre-lube. The camshaft lobes and journals must be adequately lubricated or engine damage could occur upon start up.

11. Install or connect the following:
- Camshaft in it original position with the timing chain sprocket dowel pin straight up and aligned with the centerline of the lifter bores.
- New Orange camshaft housing cover seal

➡ **The seals for the intake and exhaust covers are different and the correct seals must be used.**

12. Remove the bolt holding the housing in place.

13. Apply thread locking compound to the camshaft housing and cover bolt threads.

14. Install or connect the following:
- Camshaft housing cover. Tighten the bolts, in sequence, to 11 ft. lbs. (15 Nm) plus an additional 90 degree turn (long bolts) and 11 ft. lbs. (15 Nm) plus an additional 30 degree turn (short bolts).
- Timing chain housing mounting bolts
- Timing chain and sprockets
- New fuel injector O-ring seals lubricated with engine oil
- Fuel rail. Tighten the bolts to 19 ft. lbs. (26 Nm).
- Transaxle fill tube
- Oil pressure switch electrical connector
- Ignition assembly on the camshaft housing. Tighten the bolts to 11 ft. lbs. (15 Nm), plus an additional 30 degree turn.
- Ignition assembly electrical connector
- Negative battery cable

15. Start the vehicle and verify proper operation and no leaks.

3.1L and 3.4L Engines

1. Before servicing the vehicle, refer to the precautions in the beginning of this section.

2. Relieve the fuel system pressure.

3. Remove or disconnect the following:
- Engine assembly

✳✳ **WARNING**

When removing valvetrain components they must be marked for

installation in their original location.

- Rocker arm covers
- Intake manifold
- Rocker arms and pushrods
- Lifter guides
- Valve lifter(s) from the bores
- Crankshaft balancer and front cover
- Timing chain and sprockets
- Oil pump driven gear
- Camshaft thrust plate
- Camshaft

✳✳ **WARNING**

Avoid damaging the camshaft bearing surfaces.

To install:

4. Coat the camshaft with camshaft lubricant.

5. Install or connect the following:
- Camshaft
- Camshaft thrust plate. Torque the bolts to 89 inch lbs. (10 Nm).
- Oil pump driven gear. Torque the bolt to 27 ft. lbs. (36 Nm).
- Timing chain and sprocket. Torque the sprocket bolt to 103 ft. lbs. (140 Nm).
- Front cover. Torque the bolts to 15 ft. lbs. (21 Nm).
- Crankshaft balancer. Torque the bolt to 76 ft. lbs. (103 Nm).

6. Lubricate the bearing surfaces with Molykote®.

7. Install or connect the following:
- Lifters in their original locations
- Lifter guides. Torque the guide bolts to 89 inch lbs. (10 Nm).
- Pushrods and rocker arms. Torque the nuts to 89 inch lbs. (10 Nm) plus an additional 30 degree turn.
- Intake manifold
- Rocker arm covers. Torque the bolts to 89 inch lbs. (10 Nm).
- Engine assembly
- Negative battery cable

8. Adjust the valves, as required. Start the engine and verify no oil leaks.

Valve Lash

ADJUSTMENT

The engines are equipped with hydraulic valve lifters that do not require periodic valve lash adjustment. Adjustment to zero lash is maintained automatically by hydraulic pressure in the lifters.

Starter Motor

REMOVAL & INSTALLATION

2.2L Engine

1. Before servicing the vehicle, refer to the precautions in the beginning of this section.
2. Remove or disconnect the following:
   • Negative battery cable
   • Starter electrical connectors
   • Starter motor

To install:

3. Install or connect the following:
   • Starter motor. Torque the bolts to 30 ft. lbs. (40 Nm).
   • Starter electrical connectors
   • Negative battery cable

2.4L Engine

1. Before servicing the vehicle, refer to the precautions in the beginning of this section.
2. Remove or disconnect the following:
   • Negative battery cable
   • Air inlet duct from the throttle body
   • Upper starter bolt
   • Lower closeout panel
   • Lower starter bolt
   • Starter electrical connectors
   • Starter motor

To install:

3. Install or connect the following:
   • Starter electrical connectors
   • Starter motor. Torque the bolts to 66 ft. lbs. (90 Nm).
   • Lower closeout panel
   • Air inlet duct to the throttle body
   • Negative battery cable

3.1L Engine

1. Before servicing the vehicle, refer to the precautions in the beginning of this section.

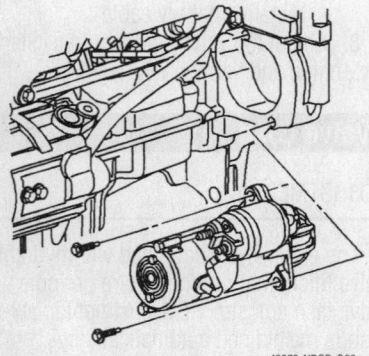

Starter mounting—2.2L engine

2. Remove or disconnect the following:
   • Negative battery cable
   • Starter electrical connectors
   • Starter motor

To install:

3. Install or connect the following:
   • Starter motor. Torque the bolts to 32 ft. lbs. (43 Nm).
   • Starter electrical connectors
   • Negative battery cable

3.4L Engine

1. Before servicing the vehicle, refer to the precautions in the beginning of this section.
2. Remove or disconnect the following:
   • Negative battery cable
   • Flywheel inspection cover
   • Starter electrical connectors
   • Starter motor

To install:

3. Install or connect the following:
   • Starter motor. Torque the bolts to 32 ft. lbs. (43 Nm).
   • Starter electrical connectors
   • Flywheel inspection cover. Torque the bolts to 89 inch lbs. (10 Nm).
   • Negative battery cable

Oil Pan

REMOVAL & INSTALLATION

2.2L Engine

1. Before servicing the vehicle, refer to the precautions in the beginning of this section.
2. Drain the engine oil.
3. Remove or disconnect the following:
   • Engine mount strut bracket
   • Drive belt
   • Lower then upper AC compressor bolts
   • Oil pan bolts
   • Oil pan

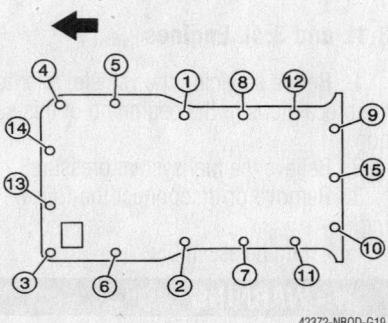

Remove the oil pan bolts in this sequence—2.2L engine

42372-NBOD-G19

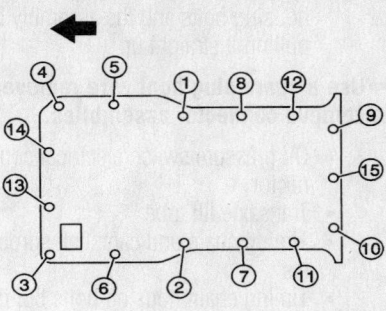

42372-NBOD-G20

Install the oil pan bolts using this sequence—2.2L engine

To install:

4. Install or connect the following:
   • Oil pan and tighten the bolts to 18 ft. lbs. (25 Nm)
   • AC compressor bolts
   • Engine mount bracket
   • Drive belt
5. Refill the crankcase.
6. Start the vehicle and verify no leaks.

2.4L Engine

1. Before servicing the vehicle, refer to the precautions in the beginning of this section.
2. Drain the engine oil.
3. Drain the cooling system.
4. Remove or disconnect the following:
   • Negative battery cable
   • Flywheel/converter cover
   • Right wheel
   • Right wheel well splash shield
   • Accessory drive belt
   • Air conditioning compressor lower bolts
   • Transaxle-to-engine brace
   • Engine mount strut bracket
   • Radiator outlet pipe bolts
   • Radiator outlet pipe from the oil pan
   • Oil pan to the flywheel cover bolt and nut
   • Flywheel cover stud for clearance
   • Radiator outlet pipe from the lower radiator hose and oil pan
   • Oil level sensor connector
   • Oil pan

To install:

5. Inspect the oil pan gasket; it is reusable if not damaged.
6. Install or connect the following:
   • Oil pan with the gasket. Torque the M8 bolts to 18 ft. lbs. (24 Nm) and the M6 bolts to 106 inch lbs. (12 Nm).
   • Oil pan to the transaxle nut
   • Oil level sensor connector

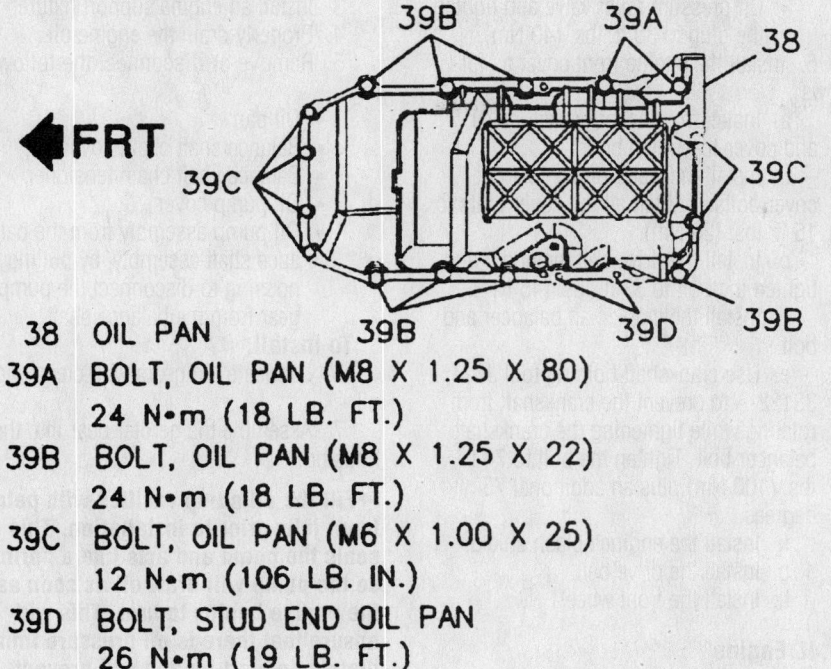

◄ **FRT**

38 OIL PAN

39A BOLT, OIL PAN (M8 X 1.25 X 80)
24 N•m (18 LB. FT.)

39B BOLT, OIL PAN (M8 X 1.25 X 22)
24 N•m (18 LB. FT.)

39C BOLT, OIL PAN (M6 X 1.00 X 25)
12 N•m (106 LB. IN.)

39D BOLT, STUD END OIL PAN
26 N•m (19 LB. FT.)

Oil pan mounting bolt locations—2.4L engine

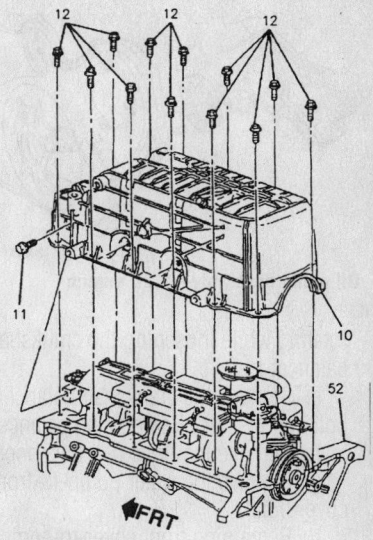

10 PAN, OIL
11 BOLT, OIL PAN SIDE
12 BOLT, OIL PAN RETAINING
52 BLOCK, ENGINE

*Exploded view of the oil pan mounting—
3.1L & 3.4L engines*

- Radiator outlet pipe to the lower radiator hose and oil pan
- Exhaust manifold brace
- Radiator outlet pipe. Torque the bolts to 124 inch lbs. (14 Nm).
- Engine mount strut bracket. Torque the bolts to 55 ft. lbs. (75 Nm).
- Transaxle to the engine brace
- Air conditioning compressor lower bolts. Torque the bolts to 37 ft. lbs. (50 Nm).
- Accessory drive belt
- Right splash shield
- Right front wheel
- Flywheel/converter cover
- Negative battery cable

7. Refill the crankcase.
8. Refill the cooling system.
9. Start the vehicle and verify no leaks.

3.1L and 3.4L Engine

1. Before servicing the vehicle, refer to the precautions in the beginning of this section.
2. Drain the engine oil.
3. Evacuate the A/C system.
4. Remove or disconnect the following:
   - Negative battery cable
   - Accessory drive belt
   - Right front wheel
   - Right splash shield
   - Anti-lock Brake System (ABS) Wheel Speed (WSS) sensor from the right subframe

- Right lower ball joint from the steering knuckle
- Right outer tie rod end
- A/C compressor without disconnecting the lines
- Evaporator-to-accumulator A/C line
- Flywheel cover
- Right side engine cradle bolts
- Crankshaft balancer
- Starter motor
- Oil pan

To install:

5. Apply silicone sealer to the portion of the pan that contacts the rear of the block.
6. Install or connect the following:
   - Oil pan with a new gasket. Torque the flange bolts to 18 ft. lbs. (25 Nm) and the side bolts to 37 ft. lbs. (50 Nm).
   - Starter motor. Torque the bolts to 32 ft. lbs. (43 Nm).
   - Flywheel cover. Torque the bolts to 89 inch lbs. (10 Nm).
   - Crankshaft balancer. Torque the bolt to 76 ft. lbs. (103 Nm).
   - Engine cradle bolts. Torque the bolts to 84 ft. lbs. (115 Nm) plus an additional 120 degree turn.
   - A/C compressor. Torque the bolts to 37 ft. lbs. (50 Nm).
   - Evaporator-to-accumulator line
   - Ball joint. Torque the nut to 48 ft. lbs. (60 Nm).

- Outer tie rod end. Torque the nut to 44 ft. lbs. (60 Nm).
- WSS electrical connector
- Wheel well splash shield
- Right front wheel
- Accessory drive belt
- Negative battery cable

7. Fill the crankcase.
8. Evacuate and recharge the A/C system.
9. Start the engine and check for leaks.

➡ **Whenever the vehicle subframe is removed or lowered, the wheel alignment should be checked.**

10. Check and/or adjust the front end alignment.

Oil Pump

REMOVAL & INSTALLATION

2.2L Engine

1. Before servicing the vehicle, refer to the precautions in the beginning of this section.
2. Remove the engine front cover as follows:
   a. Remove the front wheel.
   b. Remove the engine splash shield.
   c. Remove the drive belt.
   d. Use crankshaft holding tool J 38122-A to prevent the crankshaft from

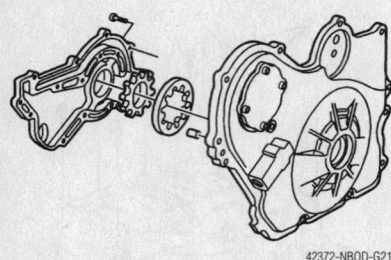

Oil pump assembly—2.2L engine

rotating while loosening the crankshaft balancer bolt.

e. Remove the crankshaft balancer bolt and discard. Remove the balancer.

f. Remove the drive belt tensioner.

g. Remove the water pump-to-front cover bolts.

h. Remaining front cover-to-engine bolts, the cover and gasket.

3. Remove or disconnect the following:
- Oil pressure relief valve
- Oil pump cover
- Oil pump gears

To install:

4. Install or connect the following:
- Oil pump gears
- Oil pump cover and tighten the bolts to 53 inch lbs. (6 Nm)

- Oil pressure relief valve and tighten the plug to 30 ft. lbs. (40 Nm).

5. Install the engine front cover as follows:

a. Install front cover gasket, cover and cover-to-engine bolts.

b. Install the water pump-to-front cover bolts. Tighten all the cover bolts to 15 ft. lbs. (20 Nm).

c. Install the drive belt tensioner and tighten the bolt to 33 ft. lbs. (45 Nm).

d. Install the crankshaft balancer and bolt.

e. Use crankshaft holding tool J 38122-A to prevent the crankshaft from rotating while tightening the crankshaft balancer bolt. Tighten the bolt to 74 ft. lbs. (100 Nm) plus an additional 75 degrees.

f. Install the engine splash shield.

g. Install the drive belt.

h. Install the front wheel.

2.4L Engine

1. Before servicing the vehicle, refer to the precautions in the beginning of this section.

2. Disconnect the negative battery cable.

3. Install an engine support fixture.

4. Properly drain the engine oil.

5. Remove or disconnect the following:
- Oil pan
- Balance shaft chain cover
- Balance shaft chain tensioner
- Oil pump cover
- Oil pump assembly from the balance shaft assembly, by pulling the housing to disconnect the pump gear from the balance shaft

To install:

6. Lubricate the gears with clean engine oil.

7. Assemble the geroter gear into the housing.

➡ **Fill the oil pump cavities with petroleum jelly prior to installation. This seals the pump and acts like a "prime" so the pump will draw oil as soon as the engine begins to turn. This will ensure that there is oil pressure immediately on start-up and will prevent engine damage.**

8. Install or connect the following:
- Oil pump to the balance shaft assembly. Torque the bolts to 40 ft. lbs. (54 Nm).

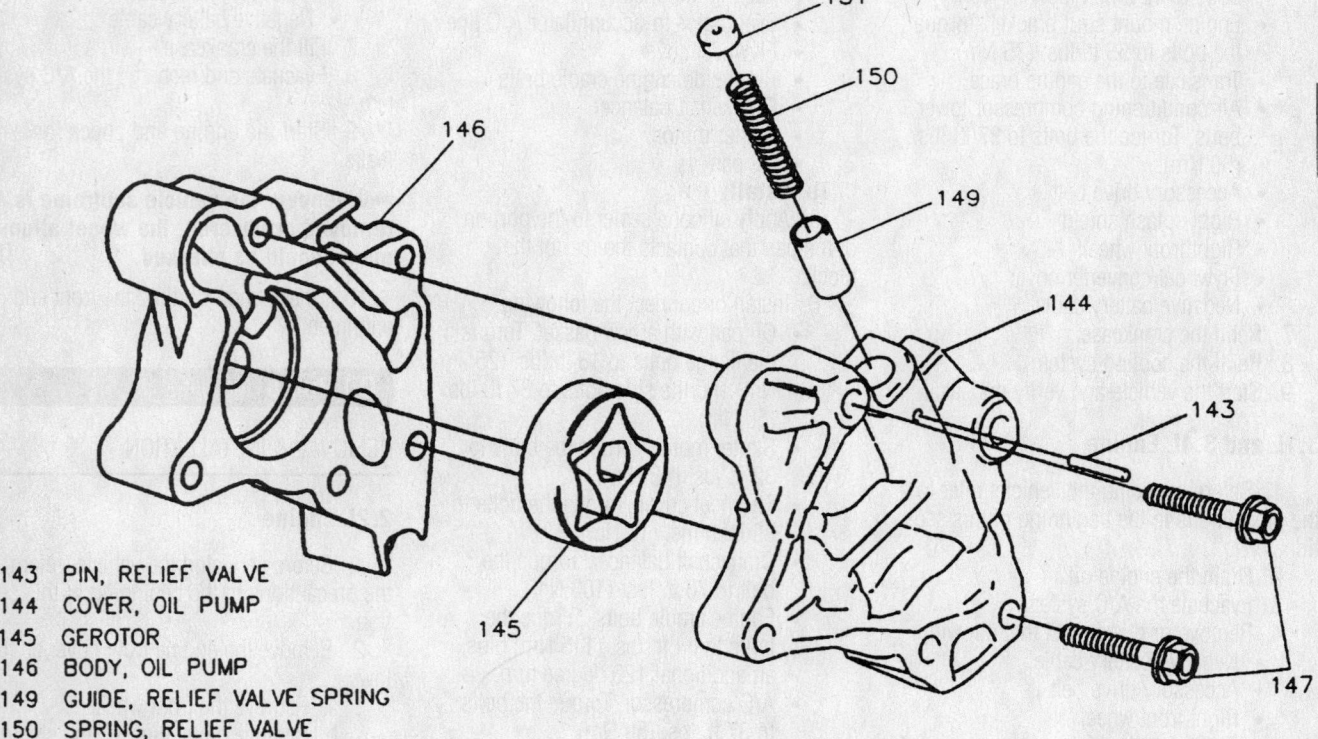

143	PIN, RELIEF VALVE
144	COVER, OIL PUMP
145	GEROTOR
146	BODY, OIL PUMP
149	GUIDE, RELIEF VALVE SPRING
150	SPRING, RELIEF VALVE
151	VALVE, RELIEF VALVE

Exploded view of the oil pump components—2.4L engine

- Oil pump cover. Torque the bolts to 40 ft. lbs. (54 Nm).
- Balance shaft chain tensioner with the bolts finger tight

9. Adjust the chain tension inserting a 0.40 in. (1mm) brass feeler, between the chain guide and the chain.

➡A brass feeler gauge must be used to ensure that correct measurements are obtained. If a steel gauge is used, it will not bend to conform to the guide and will allow for incorrect measurements.

10. Press the guide against the chain using about 10 lbs. of force. Torque the chain tensioner fastener to 89 inch lbs. (10 Nm).

11. Install or connect the following:
- Balance shaft chain cover. Torque the nut/bolt to 10 ft. lbs. (13 Nm).
- Oil pan. Torque the bolts to 18 ft. lbs. (24 Nm).
- Negative battery cable

12. Fill the crankcase.

➡An oil filter change is recommended.

13. Start the engine and verify oil pressure and no leaks.

3.1L and 3.4L Engines

1. Before servicing the vehicle, refer to the precautions in the beginning of this section.

2. Disconnect the negative battery cable.
3. Drain the engine oil.
4. Remove or disconnect the following:
- Oil pan
- Oil pump and pump driveshaft

To install:
5. Install or connect the following:
- Oil pump by engaging the oil pump driveshaft. Torque the oil pump bolts to 30 ft. lbs. (41 Nm).
- Oil pan. Torque the bolts to 18 ft. lbs. (25 Nm).
- Negative battery cable

6. Fill the crankcase.

➡An oil filter change is recommended.

7. Start the engine, check the oil pressure and check for leaks.

Rear Main Seal

REMOVAL & INSTALLATION

2.2L Engine

1. Before servicing the vehicle, refer to the precautions in the beginning of this section.

2. Support the engine.
3. Remove or disconnect the following:
- Transaxle
- Flywheel
- Rear main seal using a prytool

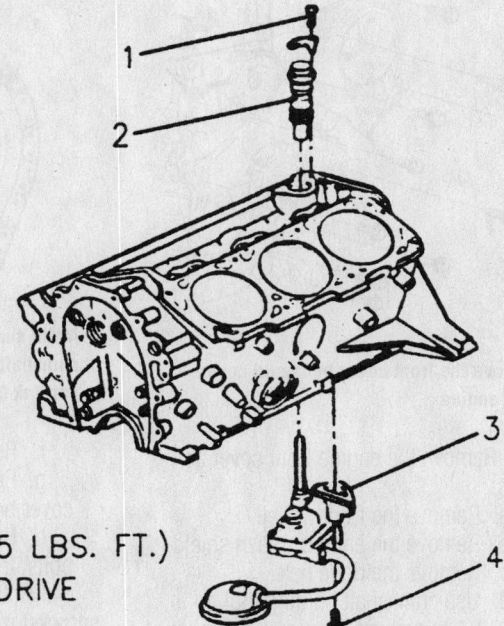

1	34 N·m (25 LBS. FT.)
2	OIL PUMP DRIVE
3	OIL PUMP
4	41 N·m (30 LBS. FT.)

79222319

Exploded view of the oil pump mounting—3.1L and 3.4L engines

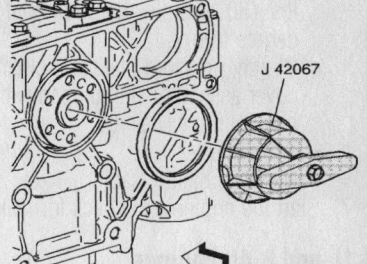

J 42067

42372-NBOD-G22

Install rear main seal using tool J 42067—2.2L engine

※※ **WARNING**

Be careful not to damage the crankshaft sealing surface.

To install:
4. Install or connect the following:
- Seal using tool J 42067
- Flywheel. Tighten the flywheel-to-crankshaft bolts to 39 ft. lbs. (53 Nm) plus an additional 25 degrees.
- Transaxle

5. Start the engine and check for leaks.

2.4L Engine

1. Before servicing the vehicle, refer to the precautions in the beginning of this section.

2. Remove or disconnect the following:
- Negative battery cable
- Transaxle
- Pressure plate and clutch disc, if equipped
- Flywheel
- Oil pan-to-seal housing bolts
- Seal housing
- Rear main seal from the housing

※※ **WARNING**

Be careful not to damage the seal housing sealing surface; damage may result in an oil leak.

To install:
3. Install the new rear main seal into the housing.

4. Inspect the oil pan gasket inner silicone bead for damage and repair using a silicone sealant, if necessary.

5. Lubricate the lip of the seal with clean engine oil.

6. Install or connect the following:
- Seal housing with a new gasket. Torque the bolts to 106 inch lbs. (12 Nm).
- Flywheel. Torque the bolts to 22 ft.

lbs. (30 Nm) plus an additional 45 degree turn.

- Clutch, pressure plate and clutch cover assembly, if equipped with a manual transaxle
- Transaxle
- Negative battery cable

7. Start the engine and check for leaks.

3.1L and 3.4L Engines

1. Before servicing the vehicle, refer to the precautions in the beginning of this section.

2. Support the engine.

3. Remove or disconnect the following:

- Transaxle
- Flywheel
- Rear main seal using a prytool

☀☀ WARNING

Be careful not to damage the crankshaft sealing surface.

To install:

4. Lubricate the seal and bore with engine oil.

5. Install the new seal by sliding it over the mandrel of rear main seal installer tool J 34686 until the dust lip bottoms squarely against the tool collar.

6. Align the dowel pin of the tool with the dowel pin hole in the crankshaft and attach the tool to the crankshaft. Tighten the attaching screws to 24–60 inch lbs. (2.7–6.8 Nm).

7. Tighten the tool T-handle to press the seal into the bore until the tool collar is flush against the engine.

8. Remove the installation tool.

➡**Make sure that the seal is squarely seated in the bore.**

9. Install or connect the following:

- Flywheel. Tighten the flywheel-to-crankshaft bolts to 61 ft. lbs. (83 Nm).
- Transaxle

10. Start the engine and check for leaks.

Timing Chain, Sprockets, Front Cover and Seal

REMOVAL & INSTALLATION

1. Before servicing the vehicle, refer to the precautions in the beginning of this section.

2. Remove the valve cover.

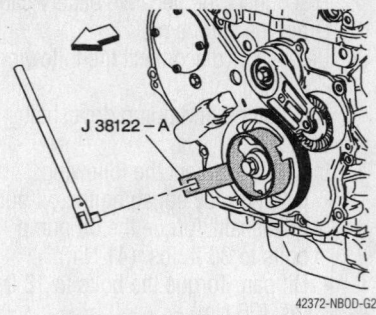

Use crankshaft holding tool J 38122-A to prevent the crankshaft from rotating while loosening the crankshaft balancer bolt—2.2L engine

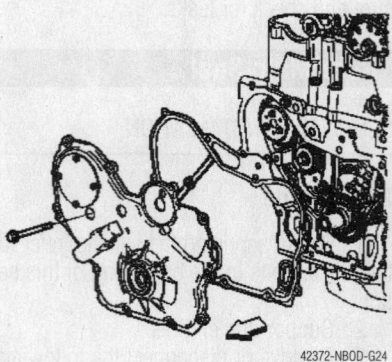

Remove the water pump-to-front cover bolt—2.2L engine

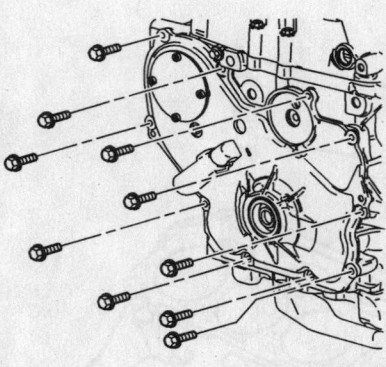

Remove the front cover bolts and cover—2.2L engine

3. Remove the engine front cover as follows:

a. Remove the front wheel.

b. Remove the engine splash shield.

c. Remove the drive belt.

d. Use crankshaft holding tool J 38122-A to prevent the crankshaft from rotating while loosening the crankshaft balancer bolt.

e. Remove the crankshaft balancer bolt and discard. Remove the balancer.

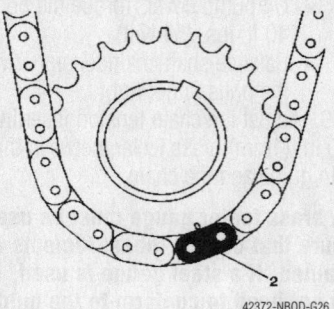

Rotate the engine until the crankshaft sprocket mark aligns with the second silver link (2) at the 5 o'clock position—2.2L engine

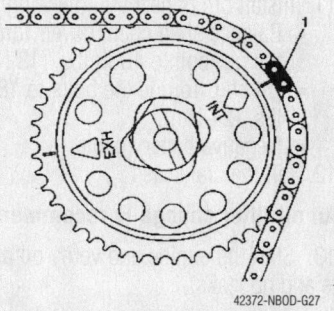

Make sure the INT diamond on the intake camshaft sprocket is aligned with the copper link at (1) at the 2 o'clock position—2.2L engine

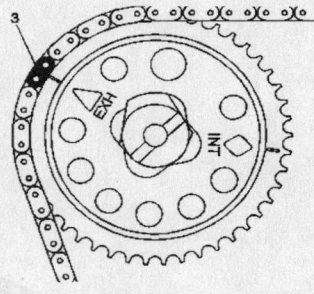
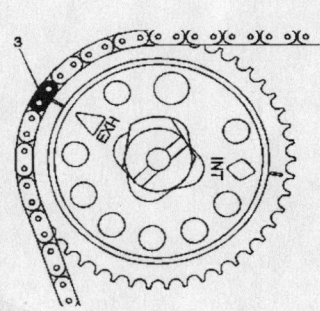

Make sure the EXH triangle on the exhaust camshaft sprocket is aligned with the silver link (3)—2.2L engine

f. Remove the drive belt tensioner.

g. Remove the water pump-to-front cover bolt.

h. Remaining front cover-to-engine bolts, the cover and gasket.

4. Rotate the engine until the crankshaft sprocket mark aligns with the second silver link (2) at the 5 o'clock position. Refer to the illustration.

5. Make sure the INT diamond on the intake camshaft sprocket is aligned with the

copper link at (1) at the 2 o'clock position. Refer to the illustration.

6. Make sure the EXH triangle on the exhaust camshaft sprocket is aligned with the silver link (3). Refer to the illustration.

7. Remove or disconnect the following:
- Timing chain tensioner
- Timing chain tensioner guide
- Fixed timing chain guide access plug
- Fxed timing chain guide
- Upper timing chain guide

8. Use a 24 mm wrench to hold the camshafts from turning.
- Exhaust camshaft sprocket bolt and discard
- Exhaust camshaft sprocket
- Intake camshaft sprocket bolt and discard
- Intake camshaft sprocket
- Timing chain through the top of the cylinder head
- Crankshaft sprocket
- Balance shaft drive chain tensioner
- Adjustable balance shaft chain guide
- Small balance shaft drive chain guide
- Upper balance shaft drive chain guide
- Balance shaft drive chain.

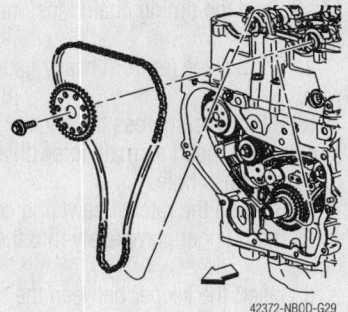

Remove the intake camshaft sprocket, then the chain through the top of the head—2.2L engine

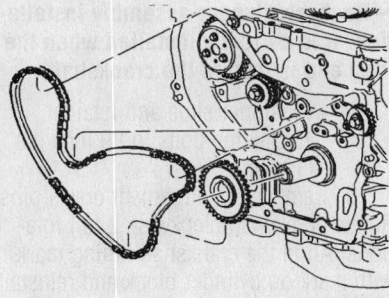

Remove the upper balance shaft chain guide, then the chain—2.2L engine

To install:

9. Install or connect the following:
- Upper balance shaft chain guide and tighten to 89 inch lbs. (10 Nm)
- Small balance shaft chain guide and bolts and tighten to 89 inch lbs. (10 Nm)
- Adjustable balance shaft drive chain guide and bolts and tighten to 89 inch lbs. (10 Nm)

10. Turn the tensioner plunger 90 degrees in its bore and compress the plunger until a paper clip can be inserted through the hole in the plunger body and into hole in the tensioner plunger.
- Timing chain tensioner and bolts and tighten to 89 inch lbs. (10 Nm)

11. Remove the paper clip from the balance shaft drive chain tensioner.

12. Install the crankshaft sprocket with timing mark at the 5 o'clock position.

13. Lower the timing chain through the opening in the top of the cylinder head. Carefully ensure that the chain goes around both sides of the cylinder block bosses.

14. Install or connect the following:
- Intake camshaft sprocket with the INT diamond at the 2 o'clock position

15. Hand tighten a NEW intake camshaft sprocket bolt.

16. Route the timing chain around the crankshaft sprocket with the second silver link aligning with the timing mark.

17. Route the timing chain around the intake camshaft sprocket with the copper colored link aligning with the INT diamond.
- Timing chain tensioner guide through the opening in the top of the cylinder head and tighten the bolts to 89 inch lbs. (10 Nm).
- Exhaust camshaft sprocket with the timing chain silver link at EXH triangle aligned at the 10 o'clock position.

18. Use a 24 mm wrench to rotate the

Install the crankshaft sprocket with timing mark at the 5 o'clock position—2.2L engine

camshaft slightly, until exhaust sprocket aligns with the camshaft.

19. Hand tighten the NEW exhaust camshaft sprocket bolt.

20. Install the fixed timing chain guide and tighten the bolts to 89 inch lbs. (10 Nm).

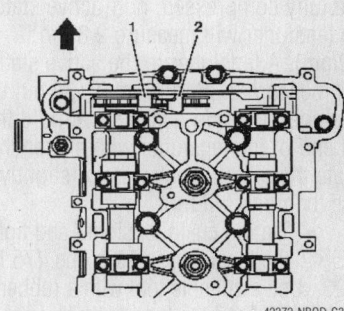

Lower the timing chain through the opening in the top of the cylinder head. Carefully ensure that the chain goes around both sides of the cylinder block bosses (1 & 2)—2.2L engine

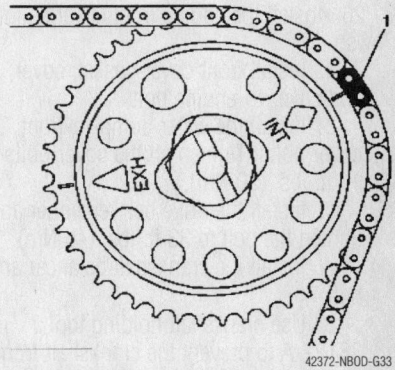

Install the intake camshaft sprocket with the INT diamond at the 2 o'clock position—2.2L engine

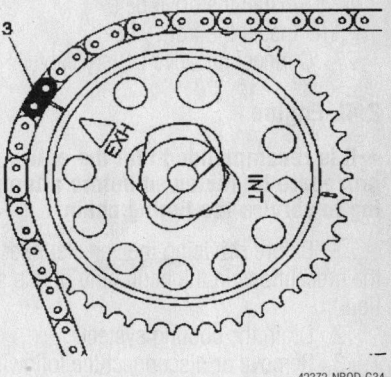

Install the exhaust camshaft sprocket with the timing chain silver link at EXH triangle aligned at the 10 o'clock position—2.2L engine

21. Apply sealant, GM P/N 12345382 compound to thread and install the timing chain guide bolt access hole plug. Tighten the access hole plug to 30 ft. lbs. (40 Nm).
   • Timing chain upper guide and bolts and tighten to 89 inch lbs. (10 Nm)
22. Measure the timing chain tensioner. In a fully compressed, non-active state, the tensioner will measure 2.83 in (72mm). A tensioner in the active state will measure 3.35 in (85mm). To put the tensioner in a non-active state, hold the flat end of the tensioner with a wrench and rotate the piston clockwise for slightly less than one full turn.
   • Timing chain tensioner and tighten the tensioner to 55 ft. lbs. (75 Nm)
23. Use a suitable tool with a rubber tip on the end. Feed the tool down through the camshaft drive chant to rest on the timing chain. Then give a sharp jolt diagonally downwards to release the tensioner.
24. Use a 24 mm wrench to hold the camshaft and tighten the new bolts to 63 ft. lbs. (85 Nm) plus an additional 30 degrees.
25. Install the valve cover.
26. Install the engine front cover as follows:
   a. Install front cover gasket, cover and cover-to-engine bolts.
   b. Install the water pump-to-front cover bolts. Tighten all the cover bolts to 15 ft. lbs. (20 Nm).
   c. Install the drive belt tensioner and tighten the bolt to 33 ft. lbs. (45 Nm).
   d. Install the crankshaft balancer and bolt.
   e. Use crankshaft holding tool J 38122-A to prevent the crankshaft from rotating while tightening the crankshaft balancer bolt. Tighten the bolt to 74 ft. lbs. (100 Nm) plus an additional 75 degrees.
   f. Install the engine splash shield.
   g. Install the drive belt.
   h. Install the front wheel.
27. Connect negative battery cable.

2.4L Engine

➡ **It is recommended that the entire procedure be reviewed before attempting to service the timing chain.**

1. Before servicing the vehicle, refer to the precautions in the beginning of this section.
2. Drain the cooling system.
3. Remove or disconnect the following:
   • Negative battery cable
   • Coolant surge tank
   • Accessory drive belt
   • Alternator

4. Install an engine support.
5. Remove or disconnect the following:
   • Upper cover fasteners
   • Front cover vent hose
   • Right engine mount and bracket
   • Right front wheel
   • Right lower splash shield
   • Crankshaft balancer
   • Lower cover fasteners
   • Front cover
6. Rotate the crankshaft clockwise, as viewed from front of engine (normal rotation) until the camshaft sprocket's timing dowel pin holes align with the timing chain housing holes. The crankshaft sprocket mark should align with the engine mark. The crankshaft sprocket keyway should point upward and align with the cylinder bores centerline. This is the normal timed position.
7. Remove the timing chain guides.
8. Remove the timing chain tensioner.

➡ **Be sure all the slack in the timing chain is above the tensioner assembly when removing it.**

✳✳ CAUTION

The tensioner plunger is spring loaded and could fly out causing personal injury.

9. Remove or disconnect the following:
   • Timing chain
   • Camshaft sprockets
To install:
10. Install or connect the following:
   • Camshaft sprockets. Torque the bolts to 52 ft. lbs. (70 Nm).
   • Camshaft sprocket alignment pin through the camshaft sprockets holes into the timing chain housing holes to position the camshafts for timing.
11. If the camshafts are out of position and must be rotated more than ⅛ turn in order to install the alignment dowel pins, perform the following:
   a. Rotated the crankshaft 90 degrees clockwise off Top Dead Center (TDC) in order to give the valves adequate clearance to open.
   b. Once the camshafts are positioned and the dowels installed, rotate the crankshaft counterclockwise back to TDC.

✳✳ WARNING

Do not rotate the crankshaft clockwise to TDC or valve and piston damage may occur.

12. Install the timing chain over the exhaust camshaft sprocket, around the idler sprocket and around the crankshaft sprocket.
13. Remove the alignment dowel pin from the intake camshaft. Using a dowel pin remover tool, rotate the intake camshaft sprocket counterclockwise enough to slide the timing chain over the intake camshaft sprocket. Release the camshaft sprocket wrench. The length of chain between the 2 camshaft sprockets will tighten.

➡ **If properly timed, the intake camshaft alignment dowel pin should slide in easily. If the dowel pin does not fully index, the camshafts are not timed correctly and the procedure must be repeated.**

14. Leave the alignment dowel pins installed.
15. With slack removed from chain between intake camshaft sprocket and crankshaft sprocket, the timing marks on the crankshaft and the cylinder block should be aligned. If marks are not aligned, move the chain 1 tooth forward or rearward, remove slack and recheck the marks.
16. Tighten the chain housing to engine stud. The stud is installed under the timing chain. Torque it to 19 ft. lbs. (26 Nm).
17. Reload the timing chain tensioner as follows:
   a. Form a keeper from heavy gauge wire.
   b. Slightly, compress the shoe plunger and insert a small screwdriver into the access hole.
   c. Release the ratchet pawl and compress the plunger completely into the hole.
   d. Insert the keeper between the access hole and the blade.
18. Install or connect the following:
   • Tensioner assembly to the chain housing. Torque the bolts to 89 inch lbs. (10 Nm).

➡ **Recheck plunger assembly installation. It is correctly installed when the long end is toward the crankshaft.**

   • Tensioner shoe and retainer. Torque the bolts to 89 inch lbs. (10 Nm).
19. Remove the alignment dowel pins. Rotate crankshaft clockwise 2 full rotations. Align the crankshaft timing mark with mark on cylinder block and reinstall alignment dowel pins. Alignment dowel pins will slide in easily if engine is timed correctly.

※※ **WARNING**

If the engine is not correctly timed, severe engine damage could occur.

20. Install or connect the following:
- Timing chain guides
- New seal into the front cover by lubricating the seal lip and tapping it into place
- Front cover and gaskets. Torque the nuts and bolts to 106 inch lbs. (12 Nm).
- Crankshaft balancer. Torque the bolt to 129 ft. lbs. (175 Nm).
- Right front lower splash shield
- Front wheel. Torque the nuts to 100 ft. lbs. (140 Nm).
- Right engine mount bracket. Torque the bolts to 81 ft. lbs. (110 Nm) plus an additional 90 degree turn.
- Right engine mount. Torque the bolt to 49 ft. lbs. (60 Nm).
- Upper cover vent hose

21. Remove the engine support.

22. Install or connect the following:
- Alternator. Torque the bolts to 37 ft. lbs. (50 Nm).
- Accessory drive belt
- Coolant surge tank
- Negative battery cable

23. Refill the cooling system.

24. Start the engine and check for leaks.

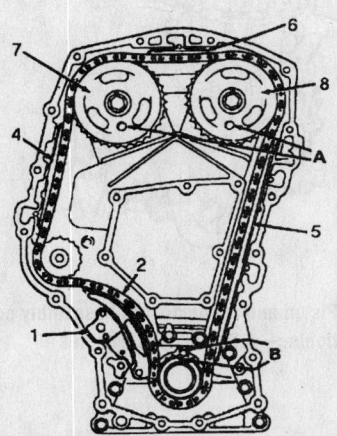

A. Camshaft timing alignment pin locations
B. Crankshaft gear timing marks
1. Shoe asm. timing chain tensioner
2. Timing chain
3. Timing chain tensioner
4. R.H. timing chain guide
5. L.H. timing chain guide
6. Upper timing chain guide
7. Exhaust camshaft sprocket
8. Intake camshaft sprocket

7922Z321

Timing chain and sprocket alignment positions—2.4L DOHC engine

3.1L and 3.4L Engines

1. Before servicing the vehicle, refer to the precautions in the beginning of this section.
2. Disconnect the negative battery cable.
3. Drain the cooling system.
4. Drain the engine oil.
5. Discharge and recover the A/C refrigerant.
6. Install an engine support fixture.
7. Remove or disconnect the following:
- Front engine mount and bracket
- Accessory drive belt
- Air cleaner assembly
- Air intake duct
- 2 upper air conditioning compressor mounting bolts
- Power steering pump
- Alternator and bracket
- Right front wheel
- Right wheel well splash shield
- Crankshaft balancer
- Drive belt tensioner
- Right Wheel Speed Sensor (WSS) harness at the suspension support
- Lower ball joint
- Stabilizer bar from the control arm and suspension support
- Suspension support
- A/C compressor-to-oil pan bolts
- Oil filter and adapter
- Starter motor
- Oil pan
- Crankshaft Position (CKsP) sensor
- Lower front cover bolts
- Coolant bypass hose
- Upper radiator hose
- Engine front cover

8. Place the number one piston at Top Dead Center (TDC).
9. Remove or disconnect the following:
- Camshaft sprocket
- Timing chain
- Crankshaft sprocket

To install:

10. Install the crankshaft sprocket.

➡ **Be sure the timing mark on the crankshaft sprocket is pointing toward the mark on the chain damper.**

11. Place the timing chain over the camshaft sprocket and hold the sprocket so the timing mark is pointing down and the timing chain is hanging off the sprocket.

12. Loop the timing chain under the crankshaft sprocket and install the camshaft sprocket on the camshaft.

13. Verify that the marks are aligned; the camshaft sprocket will be at the 6 o'clock position and the crankshaft sprocket at the 12 o'clock position.

➡ **The No. 1 piston will be at TDC and the No. 4 piston will also be at TDC but on the compression stroke.**

14. Torque the camshaft sprocket bolt to 103 ft. lbs. (140 Nm).

15. Install or connect the following:
- Front cover, using a new gasket and seal. Torque the small bolts to 15 ft. lbs. (21 Nm) and the long bolts to 35 ft. lbs. (47 Nm).
- Upper radiator hose
- Coolant bypass hose
- CKP sensor
- Oil pan. Torque the bolts to 18 ft. lbs. (25 Nm).
- Starter motor. Torque the bolts to 32 ft. lbs. (43 Nm).
- Oil filter adapter and filter
- A/C compressor-to-oil pan bolts. Torque the bolts to 37 ft. lbs. (50 Nm).
- Suspension support. Torque the bolts to 61 ft. lbs. (82 Nm).
- Stabilizer bar. Torque the bolts to 13 ft. lbs. (17 Nm).
- Ball joint. Torque the nuts to 41 ft. lbs. (55 Nm).
- WSS electrical connector
- Drive belt tensioner. Torque the bolts to 33 ft. lbs. (45 Nm).
- Crankshaft balancer. Torque the bolt to 76 ft. lbs. (103 Nm).
- Right wheel well splash shield
- Right front wheel
- Alternator. Torque the front bolt to 37 ft. lbs. (50 Nm) and the rear bolt to 18 ft. lbs. (25 Nm).

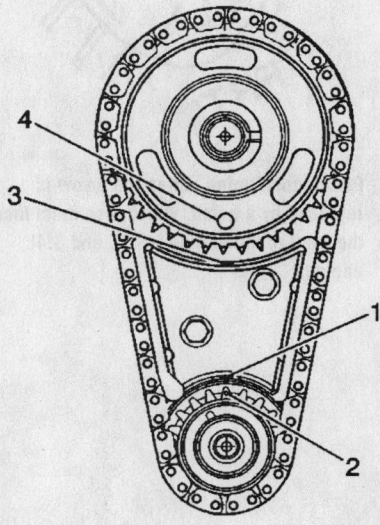

9300Z301

Be sure to align the damper mark (1) with the crankshaft mark (2) and the damper mark (3) with the camshaft sprocket mark (4)—3.1L and 3.4L engines

- 2 upper air conditioning compressor mounting bolts. Torque the bolts to 37 ft. lbs. (50 Nm).
- Power steering pump. Torque the bolts to 25 ft. lbs. (34 Nm).
- Air intake duct
- Air cleaner assembly
- Accessory drive belt
- Front engine mount and bracket. Torque the 8mm bolts to 15 ft. lbs. (20 Nm) and the 12mm bolts to 30 ft. lbs. (40 Nm).
- Negative battery cable

16. Evacuate and recharge the A/C system.
17. Refill the engine oil and coolant.
18. Start the engine and check for leaks.

Piston and Ring

POSITIONING

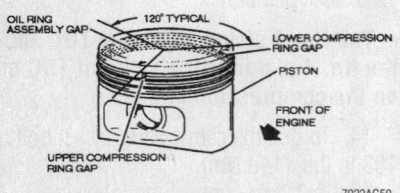

Piston ring end-gap spacing—2.2L and 2.4L engines

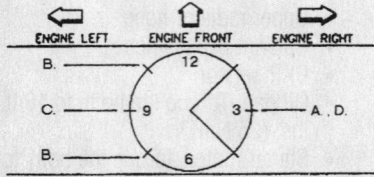

A. OIL RING SPACER GAP (TANG IN HOLE OR SLOT WITH ARC)
B. OIL RING RAIL GAPS
C. 2ND COMPRESSION RING GAP
D. TOP COMPRESSION RING GAP

Piston ring end-gap spacing—3.1L and 3.4L engines

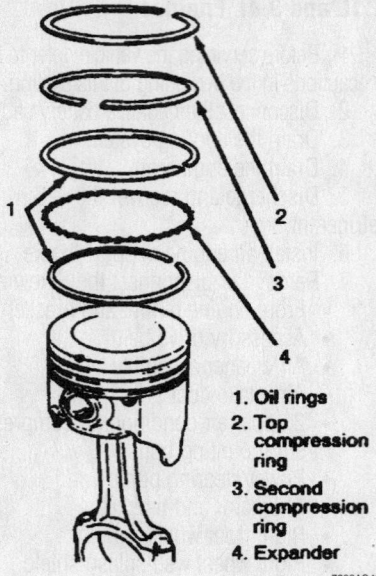

1. Oil rings
2. Top compression ring
3. Second compression ring
4. Expander

Piston ring positioning—3.1L and 3.4L engines

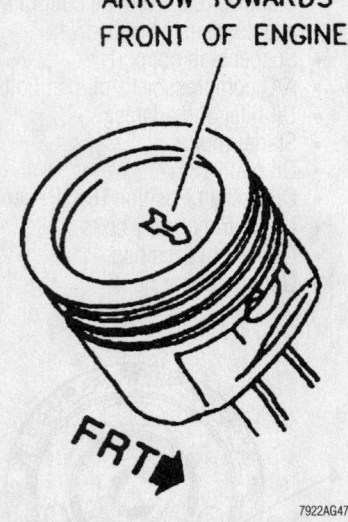

ARROW TOWARDS FRONT OF ENGINE

Piston positioning. Often the arrow is replaced by a notch, which also must face the front of the engine—3.1L and 3.4L engines

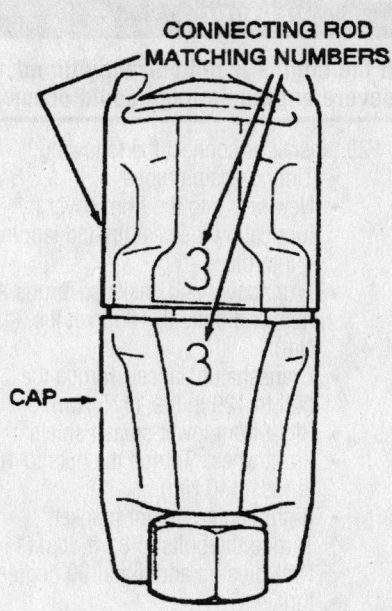

CONNECTING ROD MATCHING NUMBERS

Connecting rod and cap installation. Be sure to matchmark the cap and rod prior to disassembly, as shown—All engines

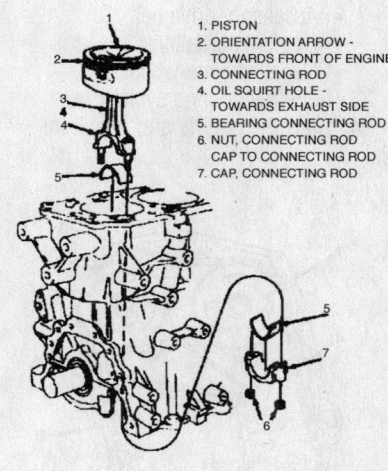

1. PISTON
2. ORIENTATION ARROW - TOWARDS FRONT OF ENGINE
3. CONNECTING ROD
4. OIL SQUIRT HOLE - TOWARDS EXHAUST SIDE
5. BEARING CONNECTING ROD
6. NUT, CONNECTING ROD CAP TO CONNECTING ROD
7. CAP, CONNECTING ROD

Piston and connecting rod assembly positioning—2.2L and 2.4L engines

FUEL SYSTEM

Fuel System Service Precautions

Safety is the most important factor when performing not only fuel system maintenance but any type of maintenance. Failure to conduct maintenance and repairs in a safe manner may result in serious personal injury or death. Maintenance and testing of the vehicle's fuel system components can be accomplished safely and effectively by adhering to the following rules and guidelines.

• To avoid the possibility of fire and personal injury, always disconnect the negative battery cable unless the repair or test procedure requires that battery voltage be applied.

• Always relieve the fuel system pressure prior to disconnecting any fuel system component (injector, fuel rail, pressure regulator, etc.), fitting or fuel line connection. Exercise extreme caution whenever relieving fuel system pressure, to avoid exposing skin, face and eyes to fuel spray. Please be advised that fuel under pressure may penetrate the skin or any part of the body that it contacts.

• Always place a shop towel or cloth around the fitting or connection prior to loosening to absorb any excess fuel due to spillage. Ensure that all fuel spillage (should it occur) is quickly removed from engine surfaces. Ensure that all fuel soaked cloths or towels are deposited into a waste container.

• Always keep a dry chemical (Class B) fire extinguisher near the work area.

• Do not allow fuel spray or fuel vapors to come into contact with a spark or open flame.

• Always use a backup wrench when loosening and tightening fuel line connection fittings. This will prevent unnecessary stress and torsion to fuel line piping. Always follow the proper torque specifications.

• Always replace worn fuel fitting O-rings with new. Do not substitute fuel hose or equivalent, where fuel pipe is installed.

Fuel System Pressure

RELIEVING

2.4L Engine

1. Before servicing the vehicle, refer to the precautions in the beginning of this section.

2. Loosen the fuel filler cap in order to relieve the pressure in the tank (do not tighten at this time).

3. Detach the fuel pump electrical connector.

4. Start and run the vehicle until it stalls, then engage the starter for an additional 3 seconds to ensure the relief of any remaining pressure.

5. Disconnect the negative battery cable.

6. Once the tests or repairs are completed, reattach the fuel pump electrical connector.

7. Connect the negative battery cable.

8. Tighten the fuel filler cap.

9. Prime the fuel system by cycling the ignition switch **ON** for 2 seconds, **OFF** for 10 seconds, then **ON** again. Repeat, if necessary to build system pressure.

2.2L, 3.1L and 3.4L Engines

1. Before servicing the vehicle, refer to the precautions in the beginning of this section.

2. Disconnect the negative battery cable in order to avoid possible fuel discharge if an accidental attempt is made to start the engine.

3. Loosen the fuel tank filler cap in order to relieve fuel tank pressure.

4. Connect a fuel pressure gauge (with bleed hose) to the fuel pressure test port connection. Wrap a towel around the fuel pressure connection when installing the fuel pressure gauge in order to avoid fuel spillage.

5. Install the bleed hose into an approved container and open the valve in order to bleed the fuel system pressure. The fuel pipe connections are now safe for servicing.

6. Drain any fuel remaining in the fuel pressure gauge into an approved container.

Fuel Filter

REMOVAL & INSTALLATION

1. Before servicing the vehicle, refer to the precautions in the beginning of this section.

2. Relieve the fuel system pressure.

3. Remove or disconnect the following:
   • Fuel line from the filter, using a backup wrench
   • Quick-connect fitting from the fuel filter by compressing the tabs while pulling outward on the line

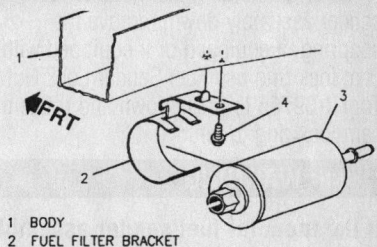

1 BODY
2 FUEL FILTER BRACKET
3 FUEL FILTER
4 SCREW – FULLY DRIVEN, SEATED AND NOT STRIPPED

79222323

Exploded view of the fuel filter mounting

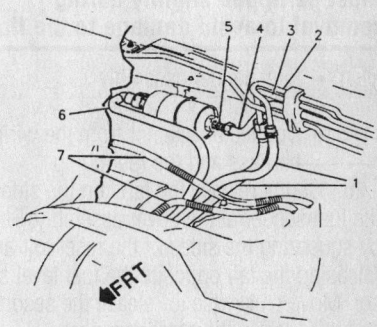

1 HOSE, PART OF FUEL SENDER
2 FUEL VAPOR PIPE
3 FUEL RETURN PIPE
4 FUEL FEED PIPE
5 FUEL FEED PIPE NUT 27 N·m (20 LBS. FT.)
6 HOSE, PART OF FUEL SENDER
7 ABS AND FUEL SENDER HARNESS

79222324

Fuel filter mounting location and component identification

• Fuel filter from the mounting bracket

To install:

4. Install or connect the following:
   • Fuel filter to the mounting bracket
   • Fuel line using a backup wrench. Torque the fitting to 20 ft. lbs. (27 Nm).
   • Quick-connect fitting to the fuel filter
   • Negative battery cable

5. Pressurize the fuel system and verify no leaks.

Fuel Pump

REMOVAL & INSTALLATION

1. Before servicing the vehicle, refer to the precautions in the beginning of this section.

2. Relieve the fuel system pressure.

3. Drain the fuel tank.

4. Remove or disconnect the following:

5. While holding the modular fuel sender assembly down, remove the snapring, if equipped or if equipped with a cam lock ring use Fuel Sender Lock Nut Tool J-39765 to press down and rotate the cam lock ring to remove it.

☀ WARNING

If the modular fuel sender assembly is retained by a snapring, it may spring up from its position. When removing the modular fuel sender from the tank, be aware that the reservoir bucket is full of fuel. It must be tipped slightly during removal to avoid damage to the float.

- Fuel sender assembly
- External fuel strainer
- Connector retainer from the wiring harness and the fuel pump

6. Gently release the tabs on the sides of the fuel sender at the cover assembly. Begin by squeezing the sides of the reservoir and releasing the tab opposite the fuel level sensor. Move clockwise to release the second and third tab in the same manner.

7. Remove or disconnect the following:
- Fuel pump electrical connection by lifting the cover assembly
- Baffle and pump assembly from the retainer by rotating the fuel pump baffle counterclockwise
- Fuel pump outlet by sliding it out of slot
- Fuel pump outlet seal

To install:

8. Install or connect the following:
- Fuel pump outlet with a new seal by sliding it in the reservoir cover slots
- Fuel pump and baffle assembly onto the reservoir retainer by rotating it clockwise until seated
- Lower retainer assembly partially into the reservoir by aligning all 3 sleeve tabs and pressing the retainer onto the reservoir making sure all 3 tabs are firmly seated

➡**Gently pull on the fuel pump reservoir to assure it is secure. If not secure, replace the entire fuel sender.**

- Connector retainer to the wiring harness and the fuel pump
- External fuel strainer
- Snapring, if equipped to the retainer slots while holding the modular fuel sender assembly down or use Fuel Sender Lock Nut

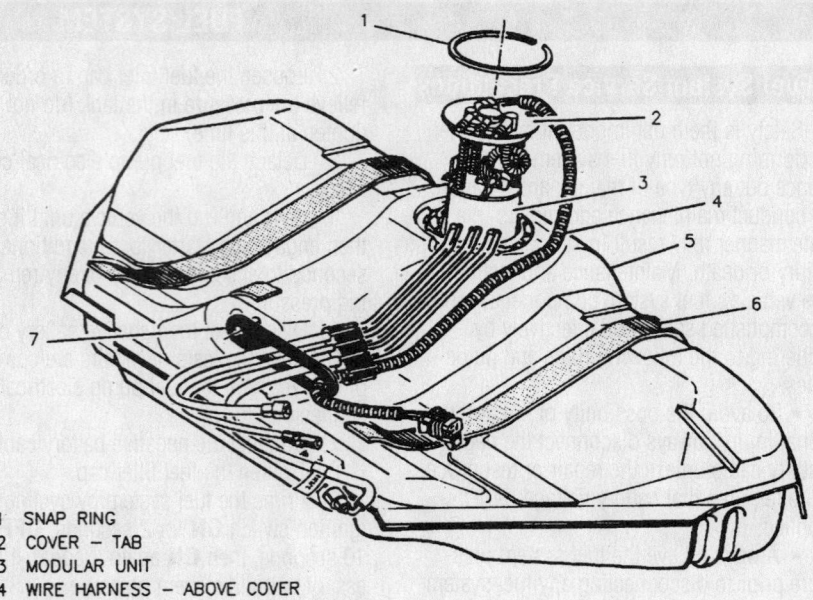

1 SNAP RING
2 COVER – TAB
3 MODULAR UNIT
4 WIRE HARNESS – ABOVE COVER
5 FUEL TANK
6 TANK ISOLATION STRIPS (3)
7 RUBBER ISOLATOR

79222Z325

Exploded view of the fuel sender assembly mounting to the tank

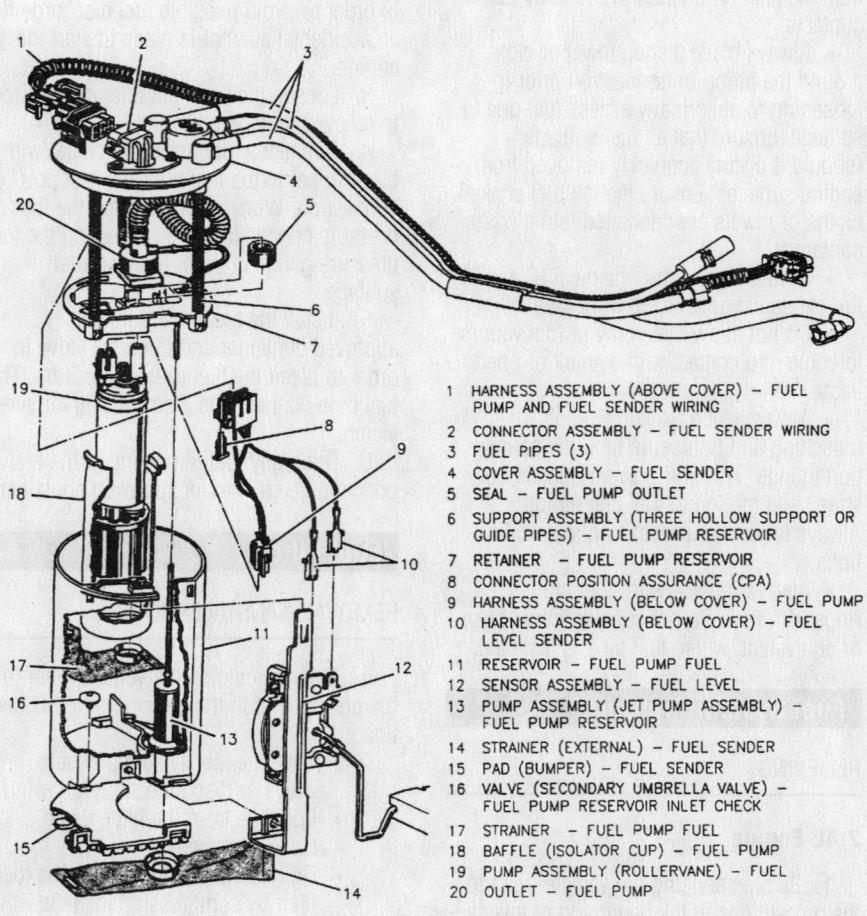

1 HARNESS ASSEMBLY (ABOVE COVER) – FUEL PUMP AND FUEL SENDER WIRING
2 CONNECTOR ASSEMBLY – FUEL SENDER WIRING
3 FUEL PIPES (3)
4 COVER ASSEMBLY – FUEL SENDER
5 SEAL – FUEL PUMP OUTLET
6 SUPPORT ASSEMBLY (THREE HOLLOW SUPPORT OR GUIDE PIPES) – FUEL PUMP RESERVOIR
7 RETAINER – FUEL PUMP RESERVOIR
8 CONNECTOR POSITION ASSURANCE (CPA)
9 HARNESS ASSEMBLY (BELOW COVER) – FUEL PUMP
10 HARNESS ASSEMBLY (BELOW COVER) – FUEL LEVEL SENDER
11 RESERVOIR – FUEL PUMP FUEL
12 SENSOR ASSEMBLY – FUEL LEVEL
13 PUMP ASSEMBLY (JET PUMP ASSEMBLY) – FUEL PUMP RESERVOIR
14 STRAINER (EXTERNAL) – FUEL SENDER
15 PAD (BUMPER) – FUEL SENDER
16 VALVE (SECONDARY UMBRELLA VALVE) – FUEL PUMP RESERVOIR INLET CHECK
17 STRAINER – FUEL PUMP FUEL
18 BAFFLE (ISOLATOR CUP) – FUEL PUMP
19 PUMP ASSEMBLY (ROLLERVANE) – FUEL
20 OUTLET – FUEL PUMP

79222Z326

Exploded view of the fuel pump assembly

Tool J-39765 in order to install the cam lock ring
- Fuel tank
- Negative battery cable

9. Pressurize the fuel system and verify no leaks.

Fuel Injector

REMOVAL & INSTALLATION

2.2L Engine

1. Before servicing the vehicle, refer to the precautions in the beginning of this section.
2. Relieve the fuel system pressure.
3. Remove or disconnect the following:
   - Air cleaner outlet resonator
   - Vacuum pipe from the fuel pressure regulator
   - Engine fuel supply and return pipes
   - Fuel injector harness connectors
   - Fuel rail attaching studs.

➡ **Be careful when removing the fuel rail so that you do not damage the injector tips or electrical connections.**

4. Remove the fuel rail as follows:
   a. Pull the fuel rail back and upward to remove the fuel injectors from the ports.
   b. Rotate the fuel rail to position the injectors downward.
   c. Remove the fuel rail.
5. Remove or disconnect the following:
   - Fuel injector retainer clip
   - Fuel injector from the fuel rail
6. Inspect the fuel injector in order to determine if the upper O-ring was also removed. If the upper O-ring in not removed, remove the O-ring from the fuel rail assembly.
   - Fuel injector O-rings and discard

To install:

7. Install or connect the following:
   - O-rings on the fuel injector
   - Fuel injector clip on the fuel injector

➡ **The fuel injector will click when the injector is installed correctly.**

   - Fuel injector to the fuel rail with the connector facing upward
8. Install the fuel rail as follows:
   a. With the fuel injectors positioned downward, lower the fuel injectors into the ports.
   b. Align the injectors by rotating the fuel rail forward.
   c. Carefully push the fuel injectors into the cylinder head ports.

d. Install and tighten the fuel rail retainers to 89 inch lbs. (10 Nm).
   - Fuel injector harness connectors. Gently pull on the connectors to make sure they are fully engaged.
   - Fuel supply and return pipes
   - Vacuum pipe to the fuel pressure regulator
   - Air cleaner outlet resonator
   - Negative battery cable
9. Inspect for fuel leaks using the following procedure:
   a. Turn ON the ignition, with the engine OFF for 2 seconds.
   b. Turn OFF the ignition for 10 seconds.
   c. Turn ON the ignition.
   d. Inspect for fuel leaks.

2.4L Engine

1. Before servicing the vehicle, refer to the precautions in the beginning of this section.
2. Relieve the fuel system pressure.
3. Remove or disconnect the following:
   - Air cleaner outlet resonator
   - Fuel pressure regulator vacuum hose
   - Camshaft Position (CMP) sensor electrical connector
   - Fuel injector electrical connectors
   - Fuel inlet pipe at the fuel rail
   - Fuel return pipe bracket screw and separate the fuel pipe from the pressure regulator
   - Fuel rail assembly
   - Fuel injector retaining clip
   - Fuel injector

To install:

4. Lubricate the O-rings with engine oil prior to installation.
5. Install or connect the following:
   - Fuel injector with new O-rings
   - Fuel injector retaining clip
   - Fuel rail assembly. Torque the bolts to 19 ft. lbs. (26 Nm).
   - New fuel pipe O-rings lubricated with engine oil
   - Fuel inlet pipe. Torque the fitting to 22 ft. lbs. (30 Nm).
   - Fuel return pipe to the pressure regulator. Torque the screw to 53 inch lbs. (6 Nm).
   - Fuel injector electrical connectors
   - CMP sensor electrical connector
   - Fuel pressure regulator vacuum hose

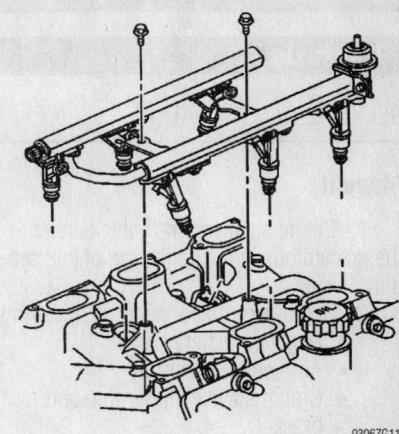

9306ZG11

Exploded view of the fuel rail assembly—3.4L engine shown, 3.1L engine is similar

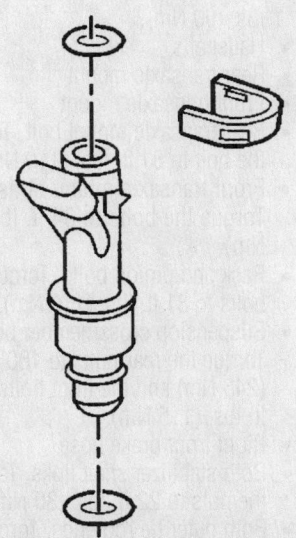

9306ZG10

Exploded view of the fuel injector—2.4L engine

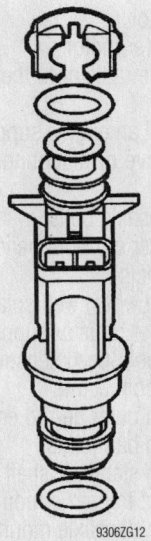

9306ZG12

Exploded view of the fuel injector—3.4L engine shown, 3.1L engine is similar

- Air cleaner outlet resonator
- Negative battery cable

6. Re-pressurize the fuel system and check for leaks.

3.1L and 3.4L Engines

1. Before servicing the vehicle, refer to the precautions in the beginning of this section.
2. Relieve the fuel system pressure.
3. Remove or disconnect the following:
- Upper intake manifold
- Fuel feed pipe
- Fuel return pipe from the pressure regulator
- Main fuel injector wiring harness electrical connector
- Engine Coolant Temperature (ECT) sensor electrical connector

- Fuel rail assembly
- Fuel injector electrical connectors
- Fuel injector-to-fuel rail clip
- Fuel injector

To install:

✳✳ WARNING

If the fuel injector O-rings are color coded, install the black O-ring in the upper position and the brown O-ring in the lower position.

4. Lubricate the O-rings with engine oil prior to installation.
5. Install or connect the following:
- New fuel injector O-rings
- Fuel injector

- Fuel injector retaining clip
- Fuel injector electrical connectors
- Fuel rail assembly. Torque the bolts to 89 inch lbs. (10 Nm).
- ECT sensor electrical connector
- Main fuel injector wiring harness electrical connector
- New fuel pipe O-rings lubricated with engine oil
- Fuel feed pipe at the fuel rail. Torque the pipe nut to 13 ft. lbs. (17 Nm).
- Fuel return pipe to the pressure regulator. Torque the pipe nut to 13 ft. lbs. (17 Nm).
- Upper intake manifold

6. Re-pressurize the fuel system and check for leaks.

DRIVE TRAIN

Transaxle

REMOVAL & INSTALLATION

Manual

1. Before servicing the vehicle, refer to the precautions in the beginning of this section.
2. Remove or disconnect the following:
- Negative battery cable
- Air cleaner assembly
- Clutch slave cylinder line and bracket
- Shifter cables from the transaxle
- Cable bracket
- Vehicle Speed Sensor (VSS) electrical connector
- Backup lamp switch electrical connector
- Starter motor

3. Tie the radiator to the upper hood latch panel.
4. Install an engine support fixture.
5. Remove or disconnect the following:
- Transaxle bolts
- Lower closeout panel
- Wheels
- Both wheel well splash shields
- Lower radiator support
- Wheel Speed sensor (WSS) electrical connectors
- Both outer tie rod ends
- Both ball joints
- Both stabilizer shaft links
- Front transaxle mount bolts
- Side transaxle mount nuts
- Rear transaxle mount stud
- Right brake hose

6. Support the suspension crossmember.
7. Remove or disconnect the following:
- Suspension crossmember bolts
- Rack and pinion bolts

8. Raise the vehicle off of the front suspension crossmember.
9. Remove or disconnect the following:
- Front transaxle mount
- Rear transaxle mount
- Halfshafts from the vehicle

10. Lower the engine and transaxle assembly enough to clear the left side inner body panel.
11. Support and remove the transaxle from the vehicle.

To install:

12. Install or connect the following:
- Transaxle. Torque the bolts to 66 ft. lbs. (90 Nm).
- Halfshafts
- Rear transaxle mount
- Front transaxle mount
- Rear transaxle mount bolt. Torque the bolt to 81 ft. lbs. (110 Nm).
- Front transaxle mount bolts. Torque the bolts to 84 ft. lbs. (115 Nm).
- Rack and pinion bolts. Torque the bolts to 81 ft. lbs. (110 Nm).
- Suspension crossmember bolts. Torque the rear bolts to 180 ft. lbs. (245 Nm) and the front bolts to 84 ft. lbs. (115 Nm).
- Right front brake hose
- Both stabilizer shaft links. Torque the nuts to 22 ft. lbs. (30 Nm).
- Both outer tie rod ends. Torque the nuts to 15 ft. lbs. (20 Nm) plus an additional 180 degree turn.

- Both ball joints. Torque the nuts to 45 ft. lbs. (61 Nm).
- WSS electrical connectors
- Lower radiator support
- Both wheel well splash shields
- Front wheels
- Lower closeout panel

13. Remove the engine support fixture.
14. Untie the radiator support.
15. Install or connect the following:
- Starter motor. Torque the bolts to 66 ft. lbs. (90 Nm) on 2.4L engines and 32 ft. lbs. (43 Nm) on 3.4L engines.
- Backup lamp switch electrical connector
- VSS electrical connector
- Shifter cables
- Clutch slave cylinder bracket and line
- Air cleaner assembly
- Negative battery cable

Automatic

2000–01 MODELS

1. Before servicing the vehicle, refer to the precautions in the beginning of this section.
2. Disconnect the negative battery cable.
3. Drain the transaxle.
4. Install an engine support fixture.
5. Remove or disconnect the following:
- Front transaxle mount bolts
- Splash shield
- Air cleaner and duct assembly
- Upper transaxle mount
- Shifter cable

6. Secure the radiator and condenser to the upper radiator support.

7. Remove or disconnect the following:

- Ground cables from the engine
- Park/Neutral Position (PNP) switch electrical connectors
- Front suspension crossmember
- Front wheels and splash shields
- Lower radiator and condenser support
- Front transaxle mount and bracket
- Both front Antilock Brake System (ABS) Wheel Speed sensor (WSS) electrical connectors
- Brake modulator assembly
- Vehicle Speed Sensor (VSS)
- Torque converter bolts
- Ball joints from the control arms
- Halfshafts
- Front transaxle mount
- ABS brake control module
- Tie rod ends from the steering knuckles
- Power steering pressure line from the steering rack
- Brake lines from the retainer under the steering rack
- Steering column intermediate shaft from the steering gear

8. Place a jack under the transaxle.

⁕⁕ CAUTION

The torque converter may fall out of the transaxle if it is tilted downward.

9. Remove or disconnect the following:
- Transaxle-to-engine bolts
- Transaxle

To install:

10. Install the transaxle onto the engine and torque the bolts to 66 ft. lbs. (90 Nm).

11. Remove the jack from the transaxle.

12. Install or connect the following:

- Intermediate shaft to the steering gear. Torque the pinch bolt to 30 ft. lbs. (41 Nm).
- Brake lines in the retainer under the steering gear
- Power steering pressure line to the steering gear
- Tie rod ends to the steering knuckles. Torque the nuts to 15 ft. lbs. (20 Nm) plus an additional 180 degree turn.
- ABS control module
- Front transaxle mount bracket. Torque the bolts to 96 ft. lbs. (130 Nm).
- Front transaxle mount. Torque the bolt to 55 ft. lbs. (75 Nm).
- Halfshafts
- Ball joints to the control arms.

Torque the nuts to 41 ft. lbs. (55 Nm).
- Torque converter bolts. Torque them to 46 ft. lbs. (62 Nm).
- VSS
- Brake modulator assembly. Torque the nut to 22 ft. lbs. (30 Nm) and the fluid lines to 18 ft. lbs. (24 Nm).
- WSS electrical connectors
- Lower radiator and condenser support
- Splash shields and the front wheels
- Front suspension crossmember. Torque the bolts to 81 ft. lbs. (110 Nm).
- PNP switch electrical connector
- Ground wiring to the engine
- Shifter cable and the upper transaxle mount
- Upper transaxle mount. Torque the bolts to 89 ft. lbs. (120 Nm).
- Air cleaner and duct assembly
- Splash shield
- Front transaxle mount. Torque the bolts to 89 ft. lbs. (120 Nm).

13. Remove the engine support fixture.

14. Connect the negative battery cable.

15. Refill the transaxle with fluid.

16. Refill and bleed the power steering system.

2002–04 MODELS

1. Before servicing the vehicle, refer to the precautions in the beginning of this section.

2. Disconnect the negative battery cable.

3. Drain the transaxle.

4. Remove or disconnect the following:

- Air inlet duct hose from the intake plenum
- Wiring harness from the transaxle and the Park Neutral Position (PNP) switch
- Upper transaxle-to-engine bolts and stud

5. Install an engine support fixture.
- Front wheels
- Both front fender liners
- Steering gear mounting bolts and secure the steering gear with mechanics wire
- Wheel Speed Sensor (WSS) wires from the front wheels and unclip them from the frame
- Ball joints from the steering knuckles
- Antilock Brake System (ABS) sensor from the WSS and frame
- Brake modulator assembly from the support

- Tie rod ends from the steering knuckles
- Front and rear transaxle mount bracket bolts
- Brake lines from the retainers on the crossmember

6. Lower the vehicle until the suspension crossmember rests on the jack stands.

7. Remove or disconnect the following:
- Front suspension crossmember support bolts
- Rear suspension crossmember support bolts
- Suspension crossmember-to-body bolts

8. Raise the vehicle off of the suspension
- Frame
- Bolts from the transaxle brace
- Shift cable from the shift linkage
- Cable from the bracket
- Flywheel inspection cover
- Starter
- Torque converter-to-flywheel bolts
- Transaxle cooler lines by removing the nut holding the bracket to the transaxle case
- Vehicle Speed Sensor (VSS) wiring harness from the sensor
- Drive shafts from the transaxle
- Body-to-transaxle mount bolts

9. Lower the transaxle with the engine support fixture enough to remove the transaxle.

10. Support the transaxle with a suitable jack.

11. Remove or disconnect the following:
- Transaxle-to-engine nut
- Engine from the transaxle
- Transaxle

To install:

12. Install or connect the following:
- Transaxle
- Lower transaxle-to-engine bolts and nuts and tighten to 66 ft. lbs. (90 Nm)
- Transaxle cooler pipes to the transaxle and tighten to 71 inch lbs. (8 Nm)
- Torque converter-to-flywheel bolts tighten to 46 ft. lbs. (62 Nm)
- Drive shafts
- VSS wiring
- Starter
- Flywheel inspection cover bolts tighten to 97 inch lbs. (11 Nm)

13. Use the engine support fixture to raise the engine and transaxle assembly.

14. Remove or disconnect the following:
- Transaxle mount-to-body bolts tighten to 66 ft. lbs. (90 Nm)
- Frame and hand-tighten the bolts

15. Once all the bolts have been installed tighten them as follows:
- Crossmember rear bolts to 180 ft. lbs. (245 Nm) plus an additional 180 degrees
- Front crossmember bolts to 84 ft. lbs. (115 Nm) plus an additional 120 degrees
- Front crossmember-to-body bolts to 81 ft. lbs. (110 Nm)

16. Raise the vehicle and support with jack stands
- Brake lines to the retainers on the crossmember
- Front and rear transaxle mount bracket bolts
- Power steering gear mounting bolts to 81 ft. lbs. (110 Nm)
- Tie rod ends to the steering knuckles
- Brake modulator assembly to the support bracket
- Lower ball joints to the steering knuckles
- ABS sensor to the WSS and frame
- Transaxle-to-engine brace bolts and tighten to 53 ft. lbs. (72 Nm)
- Both front fender liners
- Front wheels

- Upper transaxle-to-engine bolts and stud and tighten to 66 ft. lbs. (90 Nm)

17. Remove the engine support fixture.
- Shift linkage
- PNP switch and transaxle connections
- Air duct hose to the intake plenum
- Negative battery cable

18. Refill the transaxle with fluid.
19. Refill and bleed the power steering system.

Clutch

REMOVAL & INSTALLATION

1. Before servicing the vehicle, refer to the precautions in the beginning of this section.
2. Remove or disconnect the following:
- Negative battery cable
- Hydraulic line from the clutch actuator (slave cylinder)
- Transaxle from the vehicle
3. If any clutch components are to be reused, use the following procedure:
a. Matchmark the clutch pressure

plate to the flywheel. This is to retain the balance of the original parts. If all parts are to be replaced with new, this step is not necessary.
b. If the pressure plate is to be reused, loosen the pressure plate mounting bolts by turning each bolt 1 full turn until all the spring pressure is removed. This helps avoid warping the pressure plate.
c. Remove the clutch disc and pressure plate.

To install:

4. Apply a small amount of high-temperature grease to the pilot bearing as well as the tip of the transaxle input shaft and the clutch splines.
5. Install or connect the following:
- Clutch alignment tool into the flywheel
- Clutch disc onto the tool
- Pressure plate

➡**New pressure plate-to-flywheel replacement bolts are recommended.**

6. Torque the pressure plate-to-flywheel bolts, in sequence to 12 ft. lbs. (16 Nm).
7. Remove the clutch alignment tool.
8. Lubricate the inside diameter of the

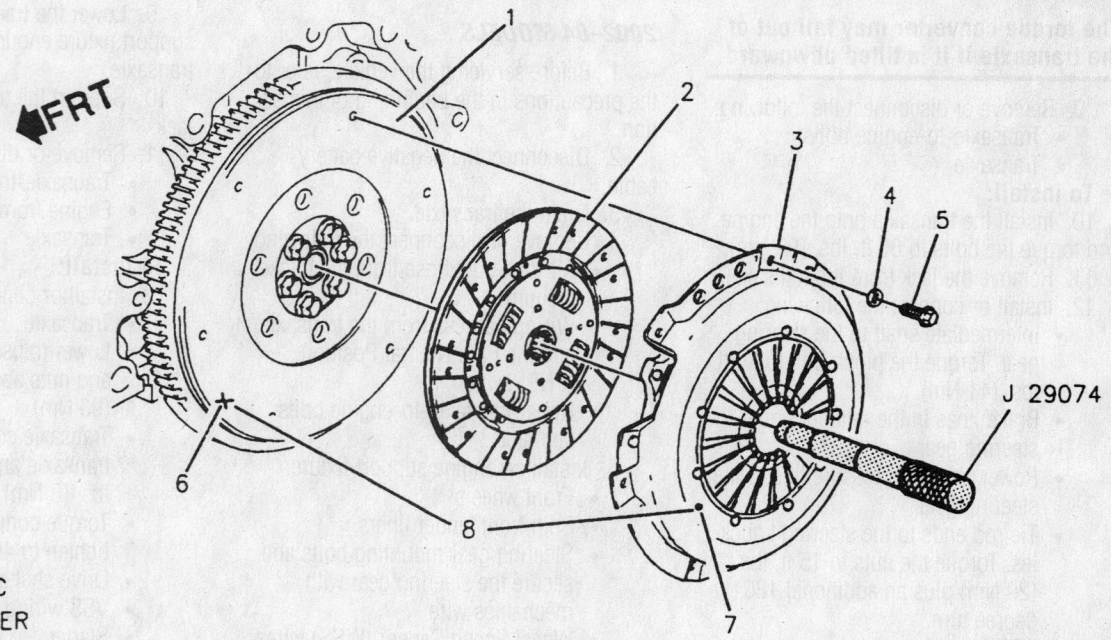

1 FLYWHEEL
2 CLUTCH DISC
3 CLUTCH COVER
4 WASHER
5 BOLT
6 FLYWHEEL "HEAVY SIDE" IDENTIFICATION
7 CLUTCH COVER "LIGHT SIDE" IDENTIFICATION
8 ALIGN IDENTIFICATION MARKS ON ASSEMBLY

J 29074

Exploded view of the clutch components—showing the clutch disk alignment tool

79222328

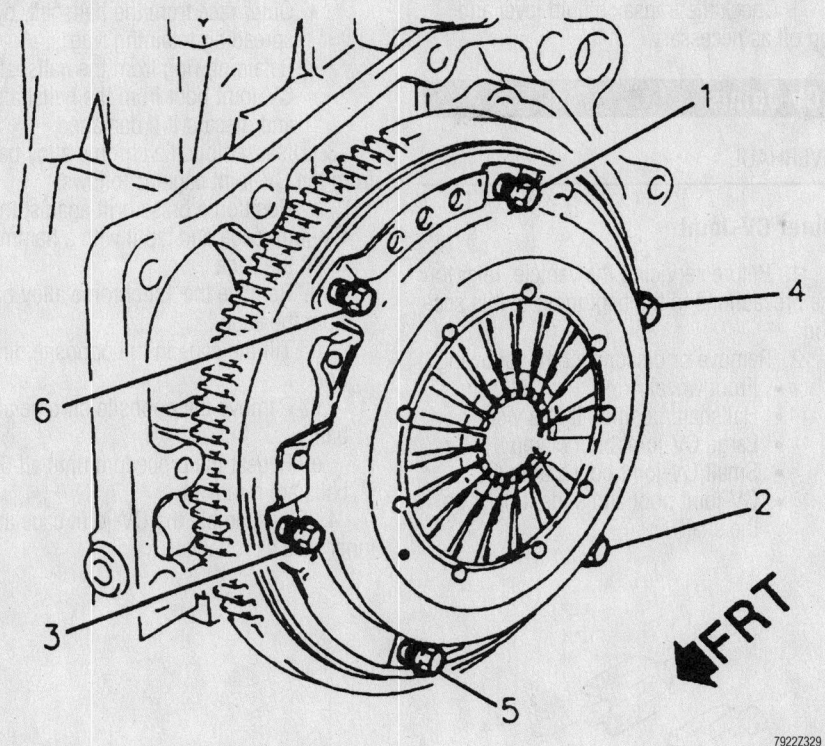

79222Z329

Clutch cover tightening sequence

actuator (slave cylinder/throw out bearing) with clutch bearing lubricant.

9. Install or connect the following:
- Transaxle
- Hydraulic line to the actuator (slave cylinder)
- Negative battery cable

10. Bleed the clutch hydraulic system.

11. Road test the vehicle to verify correct operation and easy shifting.

Hydraulic Clutch System

BLEEDING

With Bleeder Screw

1. Before servicing the vehicle, refer to the precautions in the beginning of this section.

2. Be sure the reservoir is kept full throughout this procedure.

3. Depress the clutch pedal.

4. Loosen the bleed screw, located on the actuator cylinder body next to the inlet connection.

5. Torque the bleeder screw to 17 inch lbs. (2 Nm).

6. Repeat previous 3 steps until all air is removed from the system.

7. Refill the fluid reservoir.

8. To check the system, start the engine and wait 10 seconds.

9. Depress the clutch pedal and shift into **R**. If there is any gear clash, air may still be present.

Without Bleeder Screw

1. Before servicing the vehicle, refer to the precautions in the beginning of this section.

2. Remove or disconnect the following:
- Actuator cylinder from the transaxle
- Loosen the master cylinder attaching nuts to the ends of the studs
- Reservoir cap and diaphragm

3. Depress the actuator cylinder pushrod about ¾ in. into its bore and hold the position.

4. Install the reservoir diaphragm and cap while holding the actuator pushrod.

5. Release the pushrod when the diaphragm and cap are properly installed.

6. With the actuator lower than the master cylinder, hold the actuator vertically with the pushrod end facing the ground.

7. Press the actuator pushrod into its bore with ½ in. strokes. Check the reservoir for bubbles. Continue until no bubbles enter the reservoir.

8. Install the master cylinder and actuator. Refill the fluid reservoir.

9. To check the system, start the engine and wait 10 seconds.

10. Depress the clutch pedal and shift

into reverse. If there is any gear clash, air may still be present.

Halfshaft

REMOVAL & INSTALLATION

Left and Right Halfshafts

1. Before servicing the vehicle, refer to the precautions in the beginning of this section.

2. Remove or disconnect the following:
- Negative battery cable
- Wheel
- Hub nut and washer
- Lower ball joint from the steering knuckle
- Anti-lock Brake System (ABS) electrical connector
- Sway bar link kit

3. Press the halfshaft from the hub/bearing assembly.

4. Pull the halfshaft from the transaxle using a slide hammer and axle removal tool.

To install:

5. Install or connect the following:
- Halfshaft into the transaxle or intermediate shaft using a brass drift positioned in the inboard joint groove and tap the joint in until it is seated

➡ **Verify the joint is seated properly by grasping the inboard joint and pulling on it firmly.**

✳✳ WARNING

DO NOT pull on the halfshaft or damage to the inner joint may result.

- Halfshaft into the hub assembly
- Lower ball joint to the steering knuckle. Torque the nut to 44 ft. lbs. (60 Nm).

➡ **If necessary, tighten the nut up to 60 degree (⅙) turn additional rotation to align the cotter pins. NEVER loosen the nut to make the holes align.**

- New cotter pin
- Washer and hub nut. Torque the hub nut to 284 ft. lbs. (385 Nm).
- Sway bar link kit. Torque the nut to 13 ft. lbs. (17 Nm).
- Wheel. Torque the nuts to 100 ft. lbs. (140 Nm).
- Negative battery cable

6. Check the transaxle fluid level and top off as necessary.

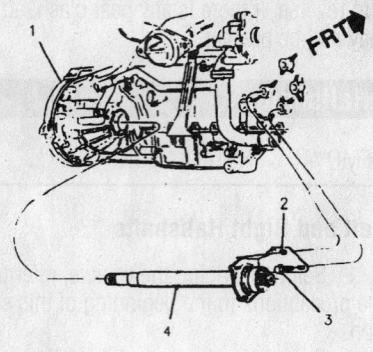

1	TRANSAXLE
2	INTERMEDIATE SHAFT SUPPORT BRACKET
3	BOLT
4	INTERMEDIATE SHAFT

7922Z330

Intermediate shaft components

Intermediate Shaft

1. Before servicing the vehicle, refer to the precautions in the beginning of this section.

2. Install an engine support fixture.

3. Remove or disconnect the following:

- Right side wheel
- Sway bar link kit
- Lower ball joint from the steering knuckle
- Halfshaft from the intermediate shaft
- Intermediate shaft support bracket-to-engine bolts
- Intermediate shaft from the transaxle

To install:

4. Install or connect the following:

- Intermediate shaft into the transaxle. Torque the intermediate shaft support bracket-to-engine bolts to 49 ft. lbs. (66 Nm). Coat the splines of the intermediate shaft with chassis grease.
- Halfshaft to the intermediate shaft
- Lower ball joint to the steering knuckle. Torque the castle nut to 44 ft. lbs. (60 Nm).

➡️**If necessary, tighten the nut up to 60 degrees (⅙) turn additional rotation to align the cotter pins. NEVER loosen the nut to make the holes align.**

- New cotter pin
- Sway bar link kit. Torque the nut to 13 ft. lbs. (17 Nm).
- Wheel. Torque the nuts to 100 ft. lbs. (140 Nm).

5. Remove the engine support fixture.

6. Connect the negative battery cable.

7. Check the transaxle fluid level and top off as necessary.

CV-Joints

OVERHAUL

Outer CV-Joint

1. Before servicing the vehicle, refer to the precautions in the beginning of this section.

2. Remove or disconnect the following:

- Front wheel
- Halfshaft, position it in a vise
- Large CV-joint boot clamp
- Small CV-joint boot clamp
- CV-joint boot and slide it back on the shaft
- Outer race from the halfshaft, by spreading retaining ring
- Retaining ring from the halfshaft
- CV-joint boot from the halfshaft and discard it if damaged

3. Disassemble the chrome alloy balls from the CV-joint cage as follows:

a. Position a brass drift against the CV-joint cage and tap it with a hammer to tilt the cage.

b. Remove the 1st chrome alloy ball from the cage.

c. Tilt the cage in the opposite direction.

d. Remove the opposite chrome alloy ball.

e. Repeat the procedure until all 6 balls are removed.

4. Disassemble the CV-joint cage and inner race as follows:

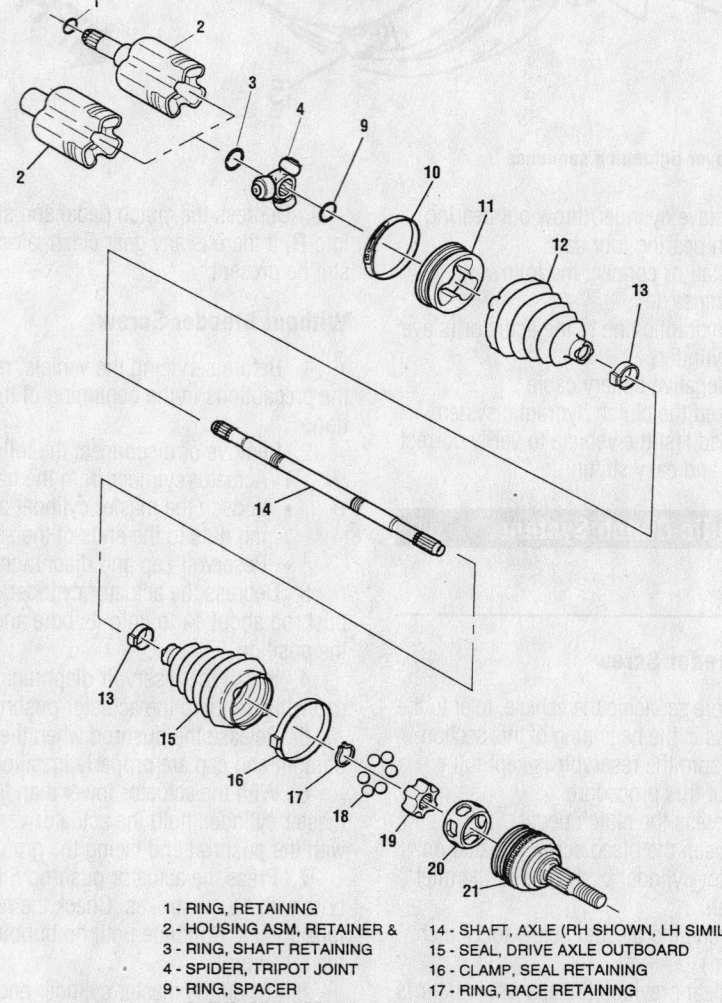

1 - RING, RETAINING	
2 - HOUSING ASM, RETAINER &	
3 - RING, SHAFT RETAINING	14 - SHAFT, AXLE (RH SHOWN, LH SIMILAR)
4 - SPIDER, TRIPOT JOINT	15 - SEAL, DRIVE AXLE OUTBOARD
9 - RING, SPACER	16 - CLAMP, SEAL RETAINING
10 - CLAMP, SEAL RETAINING	17 - RING, RACE RETAINING
11 - BUSHING, TRILOBAL TRIPOT	18 - BALL, CHROME ALLOY
12 - SEAL, DRIVE AXLE INBOARD	19 - RACE, C/V JOINT INNER
13 - CLAMP, SEAL RETAINING	20 - CAGE, C/V JOINT
	21 - RACE, C/V JOINT OUTER

9306ZG13

Exploded view of the halfshaft assembly

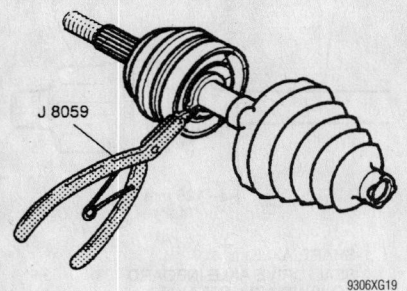

J 8059

9306XG19

Disconnecting the outer CV-joint from the axle shaft

   a. Pivot the cage and race 90 degrees to the center line of the outer race.
   b. Align the cage windows with outer race lands.
   c. Remove the cage from the outer race.
   d. Rotate the inner race upward and remove it from the cage.

To install:

   5. Lubricate the parts with a light coat of grease.
   6. Assemble the CV-joint cage and inner race, as follows:
   a. Rotate the inner race 90 degrees to the cage centerline.
   b. Align the cage windows with inner race lands.
   c. Insert the inner race into the cage by rotating the inner race downward.
   d. Insert the cage/inner race into the outer race.
   7. Assemble the chrome alloy balls into the CV-joint cage, as follows:
   a. Position a brass drift against the CV-joint cage and tap it with a hammer to tilt the cage.
   b. Insert the 1st chrome alloy ball into the cage.
   c. Tilt the cage in the opposite direction.
   d. Insert the opposite chrome alloy ball.
   e. Repeat the procedure until all 6 balls are inserted.
   8. Install ½ of the grease provided, into the CV-joint.
   9. Install or connect the following:
- Small ring clamp on the CV boot
- New retaining ring on the halfshaft
- Large ring clamp on the CV boot
- Outer race assembly onto the half-shaft until the ring engages the halfshaft groove

   10. Slide the small end of the CV-joint boot/clamp into place, with the seal lip in the halfshaft groove.

➡ **Make sure the boot lies flat against the halfshaft.**

   11. Using a crimp tool, a torque wrench and a breaker bar, crimp the small CV-joint boot clamp to 100 ft. lbs. (136 Nm).
   12. Check the clamp gap dimension; if it is not 0.085 in. (2.15mm), continue tightening the clamp until it is.
   13. Install ½ kit grease into the CV-joint boot.
   14. Measure approximately 0.687 in. (17.5mm) up from the bottom edge of the outer CV-joint assembly.
   15. Slide the large end of the CV boot/clamp into place, with the seal lip in place over the outer race.

➡ **Make sure the boot lies flat against the outer race.**

   16. Using a crimp tool, a torque wrench and a breaker bar, crimp the large CV-joint boot clamp to 130 ft. lbs. (176 Nm).
   17. Check the clamp gap dimension; if it is not 0.102 in. (2.60mm), continue tightening the clamp until it is.

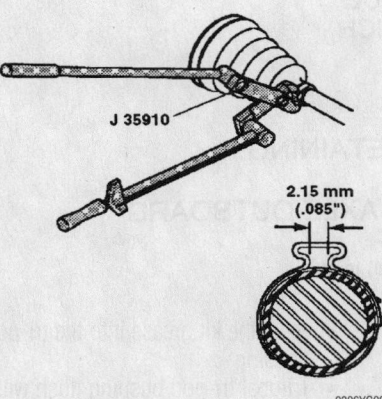

J 35910

2.15 mm (.085")

9306YG09

Crimping the small boot clamp—Outer CV-joint

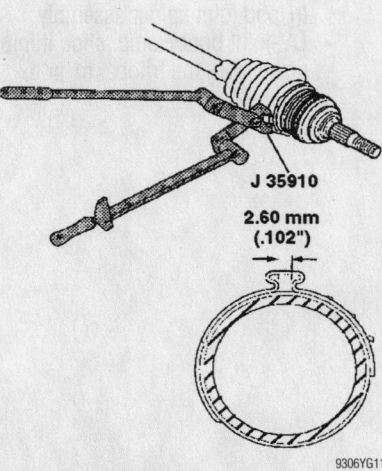

J 35910

2.60 mm (.102")

9306YG11

Crimping the large boot clamp—Outer CV-joint

   18. Install the halfshaft and the front wheel.

Inner (Tri-Pod) Joint

   1. Before servicing the vehicle, refer to the precautions in the beginning of this section.
   2. Remove or disconnect the following:
- Front wheel
- Halfshaft and place it in a vise
- Small CV-joint boot clamp
- Large CV-joint boot clamp
- CV-joint boot by sliding it away from the tri-pod joint
- Tri-pod housing from the tri-pod spider
- Inboard spacer ring and slide it rearward on the shaft, using snapring pliers
- Outboard retaining ring
- Tri-pod joint spider assembly by tapping it from the halfshaft with a brass drift
- Inboard tri-pod spider retaining ring
- Trilobal tri-pod bushing from the housing
- CV-joint boot

   3. Thoroughly clean and inspect all parts.

To install:

   4. Install or connect the following:
- Small boot clamp
- CV-joint boot
- New tri-pod spider retaining ring onto the halfshaft, slide it past the 2nd ring groove

   5. Assemble the tri-pod spider assembly onto the halfshaft as follows:
   a. Position the tri-pod spider assembly onto the shop press plate.
   b. Position the halfshaft onto the tri-pod spider assembly, in the shop press.
   c. Press the halfshaft into the tri-pod spider assembly until it passes the 2nd ring groove.
   6. Remove the halfshaft from the shop press.

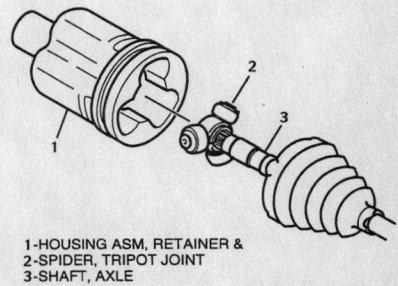

1-HOUSING ASM, RETAINER &
2-SPIDER, TRIPOT JOINT
3-SHAFT, AXLE

9306YG14

Exploded view of the inner (tri-pod) joint

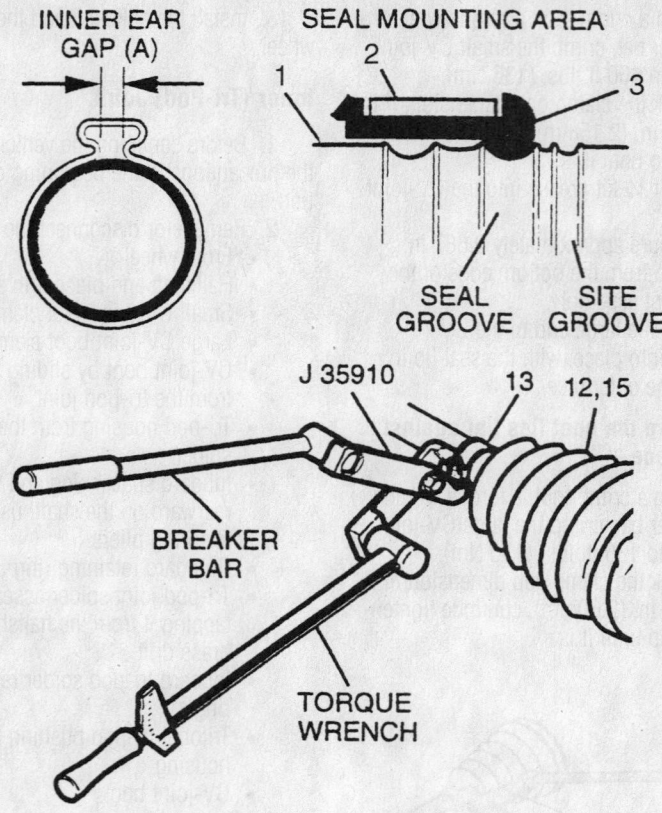

13-CLAMP, SEAL RETAINING
14-SHAFT, AXLE
12,15-SEAL, DRIVE AXLE OUTBOARD

9306YG16

Crimping the small CV-joint boot ring—Inner (tri-pod) joint

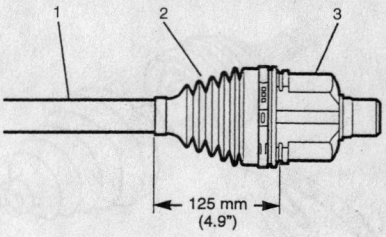

1-SHAFT, AXLE
2-SEAL, DRIVE AXLE INBOARD
3-HOUSING ASM, RETAINER

9306YG17

CV-joint boot measurement—Inner (tri-pod) joint

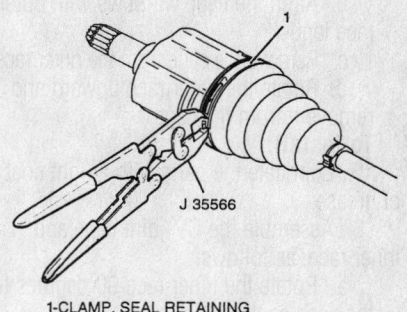

1-CLAMP, SEAL RETAINING

9306YG18

Latching the large CV-joint boot ring—Inner (tri-pod) joint

7. Install or connect the following:
- Outboard retaining ring into the axle shaft groove using snapring pliers
- Tri-pod joint spider assembly, slide it against the outboard retaining ring
- Inboard spacer ring, seat it in the groove
- ½ of the kit grease into the boot
- ½ of the kit grease into the tri-pod housing
- Trilobal tri-pod bushing flush with the tri-pod housing face
- New large seal clamp onto the CV-joint boot
- Tri-pod housing, slide it over the tri-pod joint spider assembly
- CV-joint boot/clamp, slide it into place, over the trilobal tri-pod

bushing with the seal lip in the groove

➡Make sure the boot lies flat against the trilobal bushing.

8. Position the CV-joint boot so it measures 4.9 in. (125mm).

9. Using a crimp tool, a torque wrench and a breaker bar, crimp the small CV-joint boot clamp to 100 ft. lbs. (136 Nm).

10. Using a crimp tool, latch the large CV-joint boot clamp.

11. Install the halfshaft and the front wheel.

STEERING AND SUSPENSION

Air Bag

✳✳ CAUTION

Some vehicles are equipped with an air bag system. The system must be disabled before performing service on or around system components, steering column, instrument panel components, wiring and sensors. Failure to follow safety and disabling procedures could result in accidental air bag deployment, possible personal injury and unnecessary system repairs.

PRECAUTIONS

Several precautions must be observed when handling the inflator module to avoid accidental deployment and possible personal injury.

- Never carry the inflator module by the wires or connector on the underside of the module.
- When carrying a live inflator module, hold securely with both hands, and ensure that the bag and trim cover are pointed away.
- Place the inflator module on a bench or other surface with the bag and trim cover facing up.

- With the inflator module on the bench, never place anything on or close to the module which may be thrown in the event of an accidental deployment.

DISARMING

✳✳ CAUTION

The Supplemental Restraint System (SRS) must be disarmed before performing service procedures around the air bag or SRS wiring. Failure to do so may cause accidental deployment of the air bag, resulting in unnecessary SRS repairs and/or personal injury.

1. Disconnect the negative battery cable.
2. Turn the steering wheel so the vehicle's wheels are pointing straight ahead.
3. Turn the ignition switch to the **LOCK** position and remove the key.
4. Remove the **AIR BAG** fuse from the fuse block.
5. Remove the left sound insulator.
6. Remove the Connector Position Assurance (CPA) clip from the yellow 2-way connector at the base of the steering column, and detach the connector. If equipped with a passenger's side air bag, remove the CPA and detach the yellow 2-way connector from the passenger air bag lead.

REARMING

1. Turn the ignition switch to the **LOCK** position and remove the key.
2. Attach the yellow 2-way connector at the base of steering column and secure it with the Connector Position Assurance (CPA) clip. If equipped with a passenger's side air bag, attach the yellow 2-way connector at the passenger air bag lead and secure it with the CPA clip.
3. Install the left sound insulator.
4. Install the **AIR BAG** fuse in the fuse block.
5. Turn the ignition switch to the **RUN** position and verify that the **AIR BAG** warning lamp flashes 7 times, then turns **OFF**.
6. Connect the negative battery cable.

Power Rack and Pinion Steering Gear

REMOVAL & INSTALLATION

1. Before servicing the vehicle, refer to the precautions in the beginning of this section.
2. Remove or disconnect the following:
   - Negative battery cable
   - Both front wheels
   - Stabilizer shaft links from the control arms

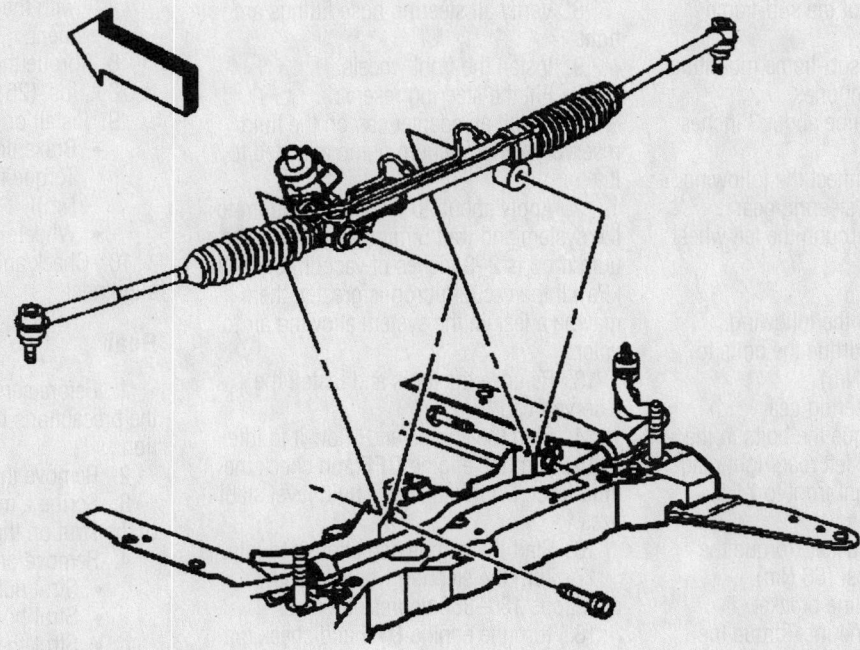

9300Z303

Power steering gear mounting

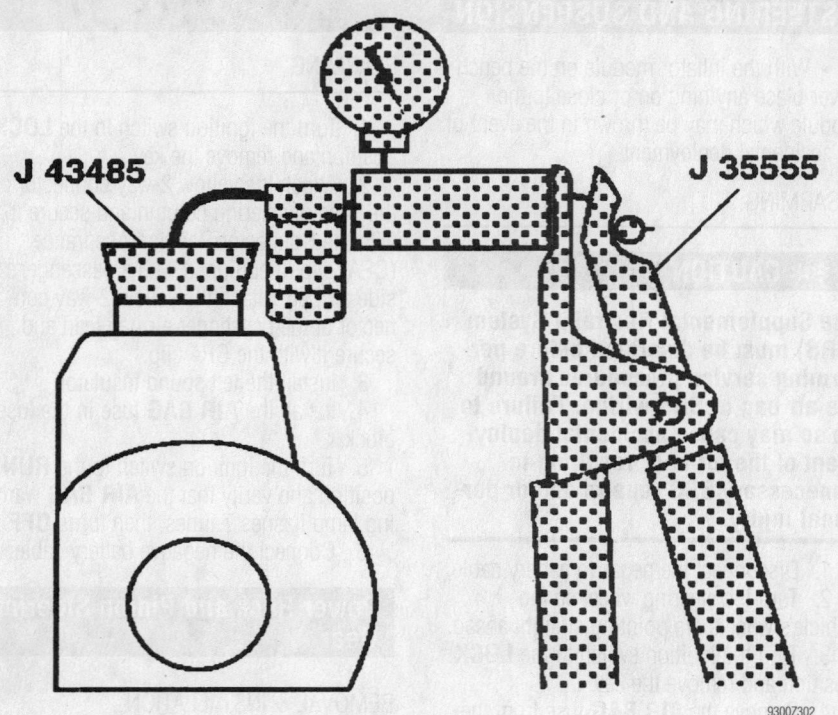

J 43485

J 35555

9300Z302

Install the special tools as shown on the steering fluid reservoir when bleeding the system

- Tie rod ends from the steering knuckles
- Intermediate shaft lower pinch bolt
- Through-bolt from the rear transaxle mount
- Power steering line bracket from the crossmember
- Stabilizer bar

3. Support the rear of the sub-frame with jacks.

4. Remove the rear sub-frame mounting bolts and loosen the front ones.

5. Lower the sub-frame about 3 inches using the jacks.

6. Remove or disconnect the following:
- Hoses from the steering gear
- Steering gear through the left wheel opening

To install:

7. Install or connect the following:
- Steering gear. Torque the bolts to 81 ft. lbs. (110 Nm).
- Hoses to the steering gear
- Sub-frame. Torque the bolts in the following order: left rear, right rear, left front and right front to 71 ft. lbs. (110 Nm).
- Stabilizer bar bracket. Torque the bolts to 49 ft. lbs. (66 Nm).
- Power steering line bracket
- Rear transaxle mount. Torque the bolt to 89 ft. lbs. (120 Nm).

- Intermediate shaft. Torque the pinch bolt to 15 ft. lbs. (20 Nm).
- Tie rod ends. Torque the nuts to 15 ft. lbs. (20 Nm) plus an additional 180 degree turn.
- Stabilizer bar links. Torque the nuts to 13 ft. lbs. (17 Nm).
- Negative battery cable

8. Verify all steering hose fittings are tight.

9. Install the front wheels.

10. Fill the steering reservoir.

11. Install an adapter cap on the fluid reservoir with a vacuum pump attached to it.

12. Apply about 20 inches of vacuum to the system and wait 5 minutes. Typical vacuum drop is 2–3 inches of vacuum or)7–10 kPa). If the vacuum drop is greater, there may be a leak in the system allowing air to enter.

13. Remove the tools and install the reservoir cap.

14. Start the engine and allow it to idle.

15. Turn the engine **OFF** and check the fluid level. Do this until the fluid level stabilizes.

16. Start the engine and allow it to idle.

17. Turn the steering wheel in both directions 180–360 degrees 5 times.

18. Turn the engine **OFF** and check the fluid level.

19. Install an adapter cap on the fluid reservoir with a vacuum pump attached to it once again.

20. Apply about 20 inches of vacuum to the system and wait 5 minutes.

21. Remove the tools and check the fluid level. Install the cap.

Strut

REMOVAL & INSTALLATION

Front

1. Before servicing the vehicle, refer to the precautions in the beginning of this section.

2. Support the front crossmember.

3. Remove or disconnect the following:
- Upper strut mounting fasteners
- Front wheel
- Brake line bracket from the strut

4. Scribe reference marks on the front strut and steering knuckle for installation purposes.

5. Remove the strut lower mounting bracket nut and through-bolts.

6. Remove the strut from the vehicle.

To install:

7. Install or connect the following:
- Strut assembly
- Upper strut plate mounting nuts finger tight
- Lower strut bracket to the steering knuckle. Torque the through-bolt/nuts to 133 ft. lbs. (180 Nm) with the reference marks in alignment.

8. Torque the upper mounting fasteners to 18 ft. lbs. (25 Nm).

9. Install or connect the following:
- Brake line bracket to the strut. Torque the bolt to 10 ft. lbs. (14 Nm).
- Wheel

10. Check and/or adjust the front end alignment.

Rear

1. Before servicing the vehicle, refer to the precautions in the beginning of this section.

2. Remove the rear wheel.

3. Scribe a mark indicating the position of the strut on the knuckle.

4. Remove or disconnect the following:
- Strut nuts from inside the trunk
- Strut bolts from the wheel well area
- Strut-to-knuckle bolts

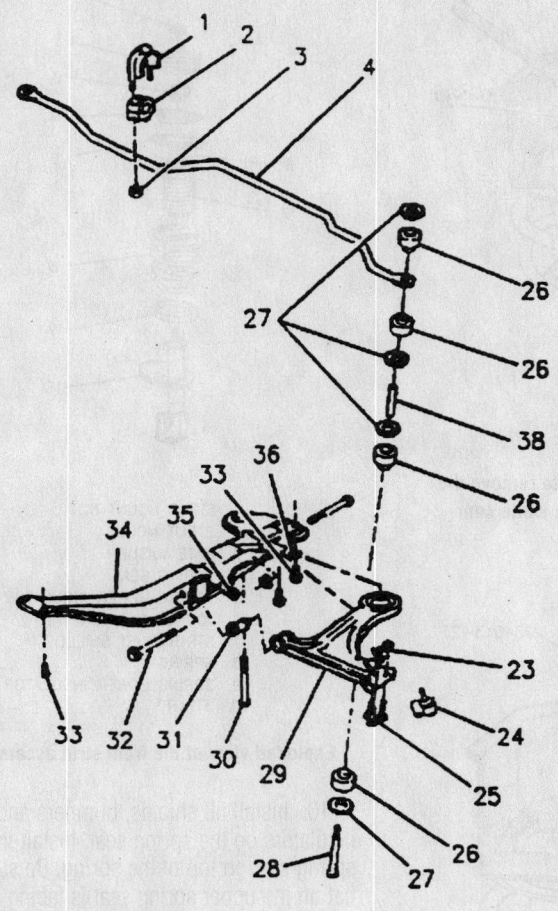

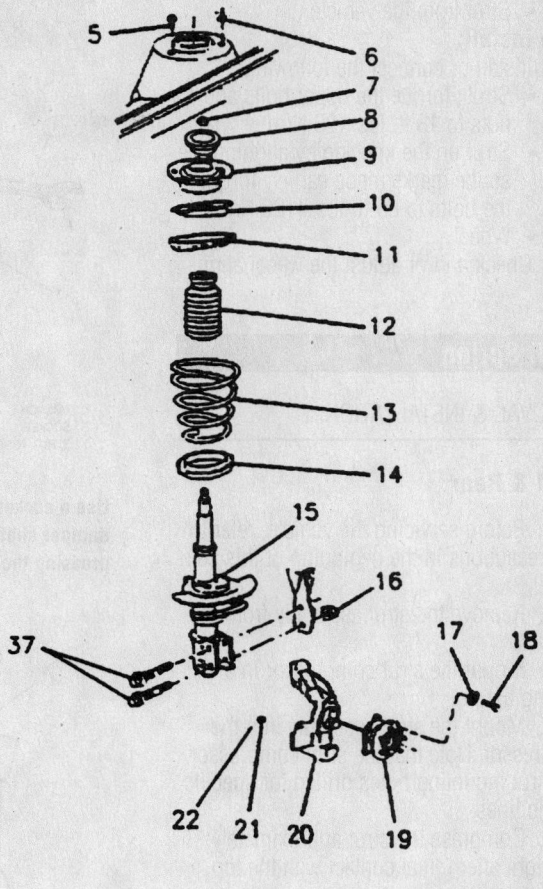

1 CLAMP, STABILIZER SHAFT
2 INSULATOR, STABILIZER SHAFT
3 NUT
4 STABILIZER SHAFT
5 BOLT
6 NUT
7 NUT, STRUT DAMPENER SHAFT
8 RATE WASHER
9 STRUT MOUNT
10 UPPER SPRING SEAT
11 UPPER SPRING INSULATOR
12 DUST TUBE ASSEMBLY
13 SPRING
14 LOWER SPRING INSULATOR
15 STRUT
16 NUT
17 WASHER
18 BOLT
19 HUB AND BEARING ASSEMBLY

20 STEERING KNUCKLE
21 NUT, BALL JOINT
22 COTTER PIN
23 NUT
24 BALL JOINT
25 BOLT
26 INSULATOR, STABILIZER LINK
27 WASHER, STABILIZER LINK
28 BOLT, STABILIZER LINK
29 CONTROL ARM
30 BOLT
31 BUSHING, CONTROL ARM
32 BOLT
33 BOLT
34 SUSPENSION SUPPORT
35 NUT
36 WASHER
37 BOLT
38 SPACER, STABILIZER LINK

Exploded view of the front suspension

79222332

- Strut from the vehicle

To install:

5. Install or connect the following:
  - Strut. Torque the upper bolts and nuts to 18 ft. lbs. (25 Nm).
  - Strut on the knuckle by aligning the scribe marks made earlier. Torque the bolts to 89 ft. lbs. (120 Nm).
  - Wheel

6. Check and/or adjust the wheel alignment.

Coil Spring

REMOVAL & INSTALLATION

Front & Rear

1. Before servicing the vehicle, refer to the precautions in the beginning of this section.

2. Remove the strut assembly from the vehicle.

3. Mount the strut compressor in a holding fixture.

4. Mount the strut assembly into the compressor. Note that the strut compressor has strut mounting holes drilled for specific vehicle lines.

5. Compress the strut approximately ½ its height after initial contact with the top cap.

✳✳ WARNING

Never bottom the spring or damper rod.

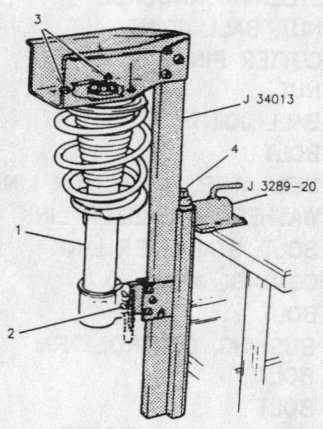

1 STRUT ASSEMBLY
2 INSTALL LOCKING PINS THROUGH STRUT ASSEMBLY
3 TIGHTEN NUTS UNTIL FLUSH WITH STRUT COMPRESSOR
4 COMPRESSOR FORCING SCREW

79222334

View of the strut assembly mounted in a compressor

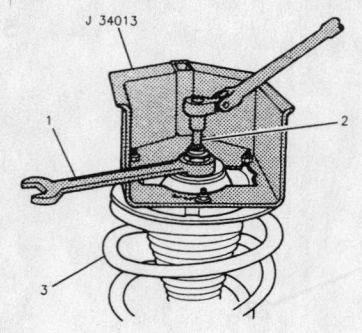

1 WRENCH
2 SOCKET
3 STRUT ASSEMBLY

79222335

Use a socket and a wrench to remove the damper shaft nut spring cap while compressing the spring

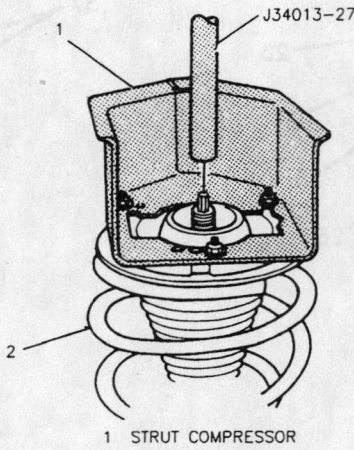

1 STRUT COMPRESSOR
2 STRUT ASSEMBLY

79222336

Install the rod to guide the damper shaft straight down through the spring cap while compressing the spring

6. Remove the nut from the strut damper shaft and place alignment/guiding rod J-34013-27 on top of the damper shaft. Use the rod to guide the damper shaft straight down through the spring cap while compressing the spring. Remove the components.

To install:

7. Install the bearing cap into the strut compressor, if removed.

8. Mount the strut assembly in strut compressor, using bottom locking pin only. Extend the damper shaft and install clamp J-34013-20 on the damper shaft.

9. Install the spring over the damper and swing the assembly up so the upper locking pin can be installed.

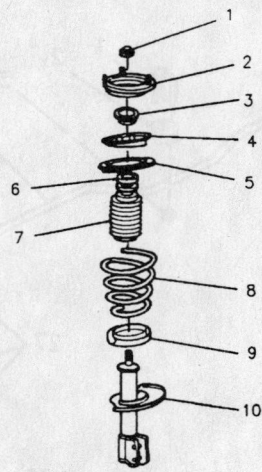

1 STRUT MOUNT NUT
2 STRUT MOUNT
3 RATE WASHER
4 SPRING SEAT
5 SPRING UPPER INSULATOR
6 JOUNCE BUMPER
7 STRUT DUST SHIELD
8 SPRING
9 SPRING LOWER INSULATOR
10 STRUT

79222337

Exploded view of the front strut assembly

10. Install all shields, bumpers and insulators on the spring seat. Install the spring seat on top of the spring. Be sure the flat on the upper spring seat is facing in the proper direction. The spring seat flat should be facing the same direction as the centerline of the strut assembly spindle.

11. Install the guiding rod and turn the forcing screw while the guiding rod centers the assembly. When the threads on the damper shaft are visible, remove the guiding rod and install the nut. tighten the nut to 52 ft. lbs. (70 Nm) on rear assemblies or 55 ft. lbs. (75 Nm) on front assemblies. Use a crow's foot line wrench while holding the damper shaft with a socket.

12. Remove the clamp.

Torsion Bars

REMOVAL & INSTALLATION

1. Before servicing the vehicle, refer to the precautions in the beginning of this section.

2. Remove or disconnect the following:
  - Torsion bar-to-knuckle bolt, washer and bushing
  - Torsion bar-to-chassis bolt
  - Torsion bar from the chassis

To install:

3. Install or connect the following:

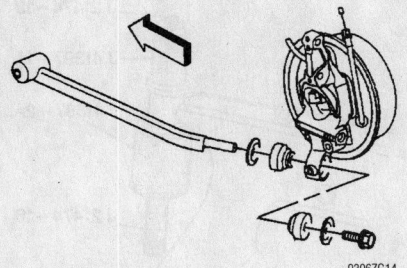

Exploded view of the torsion bar

- Torsion bar to the chassis. Torque the nut/bolt to 48 ft. lbs. (65 Nm) plus an additional 120 degree turn.
- Torsion bar to the knuckle. Torque the bolt to 51 ft. lbs. (69 Nm).

Lower Ball Joint

REMOVAL & INSTALLATION

1. Before servicing the vehicle, refer to the precautions in the beginning of this section.
2. Remove or disconnect the following:
   - Wheel
   - Wiring harness from the control arm
   - Stabilizer shaft link
   - Lower ball joint from the steering knuckle
3. Drill a ⅛ in. pilot hole in the center of each of the 3 ball joint mounting rivets.
4. Using a ½ in. drill bit, drill the heads off the rivets.
5. With a hammer and punch, knock the rivets out of the control arm. Remove the sway bar link kit.
6. Pull the control arm down so the ball stud clears the steering knuckle. Slide the ball joint out of the control arm.

To install:

7. Position the new ball joint into the lower control arm and install the bolts. The nuts must be on top of the control arm. Tighten the mounting bolts to the specification provided with the ball joint service kit.
8. Connect the ball joint to the steering knuckle and torque the castle nut to 41 ft. lbs. (55 Nm).

➡**If necessary to align the cotter pin holes tighten the nut up to 60 degree (⅙) turn additional rotation. NEVER loosen the nut to make the holes align.**

9. Install or connect the following:
   - New cotter pin
   - Stabilizer shaft link. Torque the nut to 22 ft. lbs. (30 Nm).
   - Wiring harness to the control arm

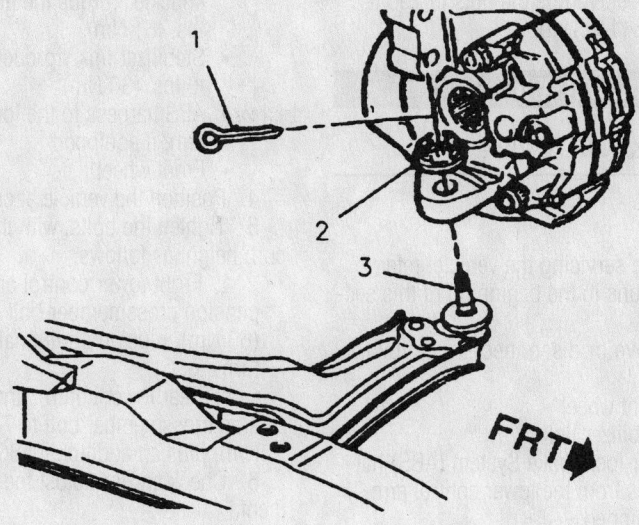

1 PIN
2 NUT – 55 N·m (41 LBS. FT.) MINIMUM TORQUE
   65 N·m (48 LBS. FT.) MAXIMUM TORQUE TO
   INSTALL PIN
3 LOWER BALL JOINT

Exploded view of the ball joint-to-knuckle mounting

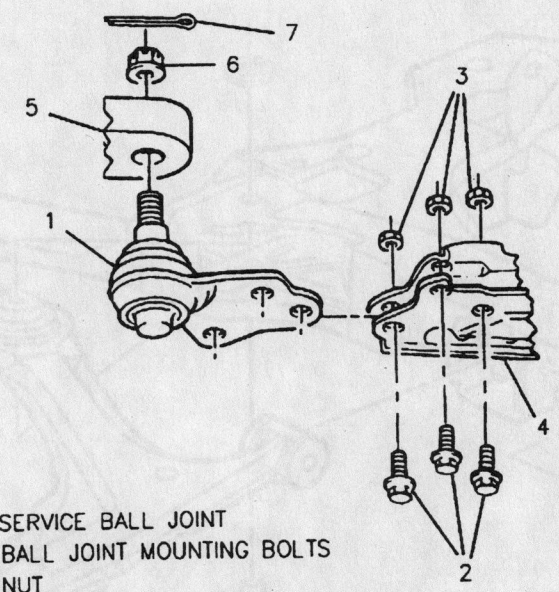

1 SERVICE BALL JOINT
2 BALL JOINT MOUNTING BOLTS
3 NUT
4 LOWER CONTROL ARM
5 STEERING KNUCKLE
6 NUT – 55 N·m (41 LBS. FT.) MINIMUM TORQUE
   65 N·m (48 LBS. FT.) MAXIMUM TORQUE
   TO INSTALL PIN
7 PIN

Exploded view of the replacement ball joint mounting

• Wheel. Torque the nuts to 100 ft. lbs. (140 Nm).

Lower Control Arm

REMOVAL & INSTALLATION

Front

1. Before servicing the vehicle, refer to the precautions in the beginning of this section.

2. Remove or disconnect the following:

- Front wheel
- Stabilizer link
- Anti-lock Brake System (ABS) harness from the lower control arm, if equipped
- Lower ball joint from the steering knuckle
- Lower control arm-to-suspension crossmember bolts
- Lower control arm

To install:

3. Install or connect the following:

- Lower control arm-to-suspension crossmember and hand-tighten the bolts
- Lower ball joint to the steering

knuckle. Torque the nut to 45 ft. lbs. (61 Nm).

- Stabilizer link. Torque the nut to 22 ft. lbs. (30 Nm).
- ABS harness to the lower control arm, if equipped
- Front wheel

4. Position the vehicle at curb height.

5. Tighten the bolts, with the vehicle at curb height as follows:

   a. Front lower control arm-to-suspension crossmember bolt to 45 ft. lbs. (60 Nm), plus an additional 120 degree turn.

   b. Rear lower control arm-to-suspension crossmember bolt to 74 ft. lbs. (100 Nm), plus an additional 180 degree turn.

6. Check and/or adjust the front alignment.

CONTROL ARM BUSHING REPLACEMENT

Front Bushing

1. Before servicing the vehicle, refer to the precautions in the beginning of this section.

2. Remove the lower control arm and place it in a vise.

3. Lubricate the threads of Screw J-21474-19 with high pressure lubricant.

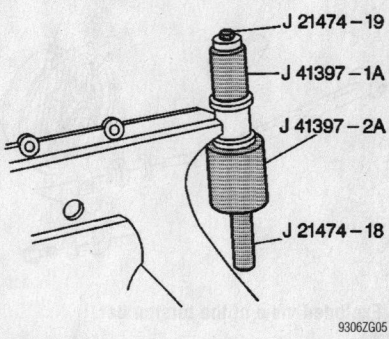

Removing the front bushing from the lower control arm

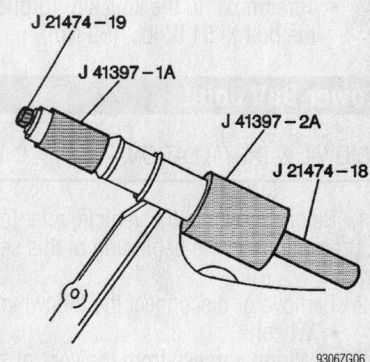

Installing the front bushing to the lower control arm

4. Assemble Tools Screw J-21474-19, Remover/Installer J-41397-1A, Receiver J-41397-2A and J-21474-18 onto the front control arm bushing.

5. Tighten Tool J-21474-18 until the front bushing is pressed from the control arm.

6. Disassemble the tools.

To install:

7. Lubricate the new front bushing outer casing.

8. Insert the new bushing into the control arm.

9. Assemble Tools Screw J-21474-19, Remover/Installer J-41397-1A, Receiver J-41397-2A and J-21474-18 onto the front control arm bushing.

10. Tighten Screw J-21474-19 until the front bushing is pressed into the control arm.

11. Disassemble the tools.

12. Install the lower control arm.

Rear Bushing

1. Before servicing the vehicle, refer to the precautions in the beginning of this section.

2. Remove the lower control arm and place it in a vise.

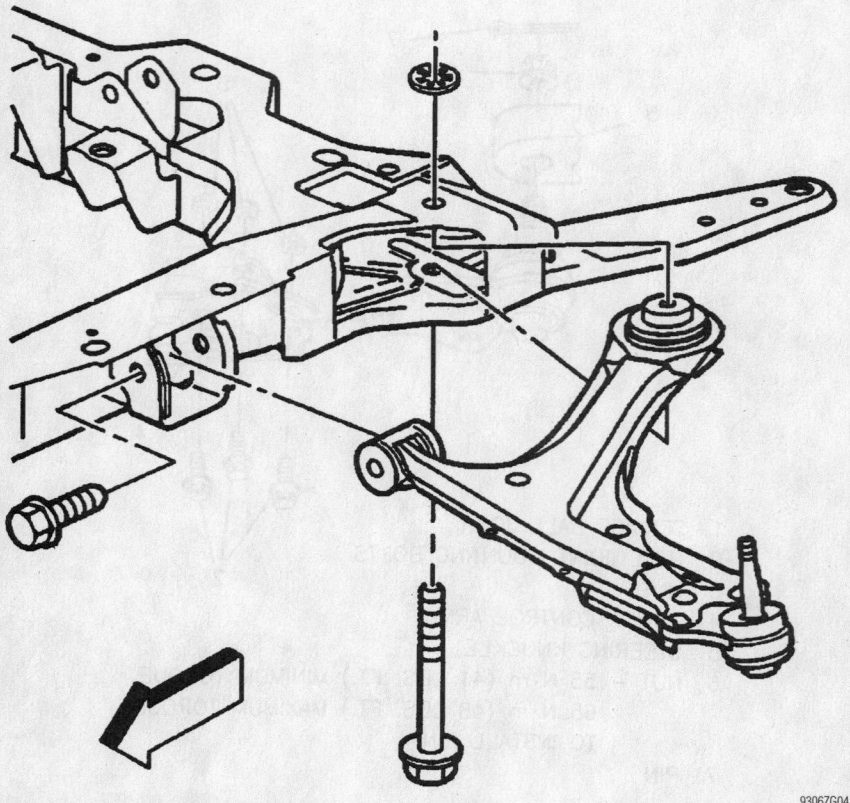

Exploded view of the lower control arm and related components

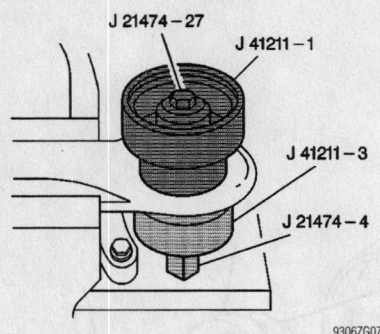

Removing the rear bushing from the lower control arm

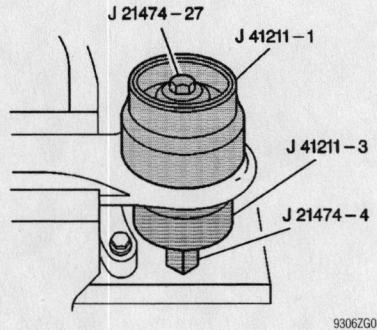

Installing the rear bushing to the lower control arm

3. Assemble Tools Screw J-21474-27, Remover/Installer J-41211-1, Receiver J-41211-3 and J-21474-4 onto the rear control arm bushing.

4. Tighten Tool J-21474-27 until the rear bushing is pressed from the control arm.

5. Disassemble the tools.

To install:

6. Insert the new bushing into the control arm.

7. Assemble Tools Screw J-21474-27, Remover/Installer J-41211-1, Receiver J-41211-3 and J-21474-4 onto the rear control arm bushing.

8. Tighten Screw J-21474-4 until the rear bushing is pressed into the control arm.

9. Disassemble the tools.

10. Install the lower control arm.

Wheel Bearings

ADJUSTMENT

These vehicles are equipped with sealed hub and bearing assemblies. The hub and bearing assemblies are non-serviceable. If the assembly is damaged, the complete unit must be replaced.

REMOVAL & INSTALLATION

Front

1. Before servicing the vehicle, refer to the precautions in the beginning of this section.

2. Remove or disconnect the following:
   - Front wheel
   - Halfshaft nut and washer
   - Caliper from the steering knuckle and support it aside

✴✴ WARNING

DO NOT allow the caliper to hang unsupported from the brake hose.

   - Brake rotor
   - Anti-lock Brake System (ABS) connector, if equipped
   - 3 hub/bearing assembly bolts
   - Halfshaft from the hub/bearing assembly
   - Hub/bearing assembly

To install:

3. Install or connect the following:
   - Hub/bearing assembly onto the halfshaft, making sure the splines engage smoothly
   - Hub/bearing assembly to the steering knuckle. Torque the bolts to 70 ft. lbs. (95 Nm).
   - Anti-lock Brake System (ABS) connector, if equipped
   - Brake rotor
   - Caliper onto the steering knuckle. Torque the bolts to 38 ft. lbs. (51 Nm).
   - Halfshaft nut. Torque the hub nut to 284 ft. lbs. (385 Nm).
   - Front wheel. Torque the nuts to 100 ft. lbs. (140 Nm).

Rear

DRUM BRAKE

1. Before servicing the vehicle, refer to the precautions in the beginning of this section.

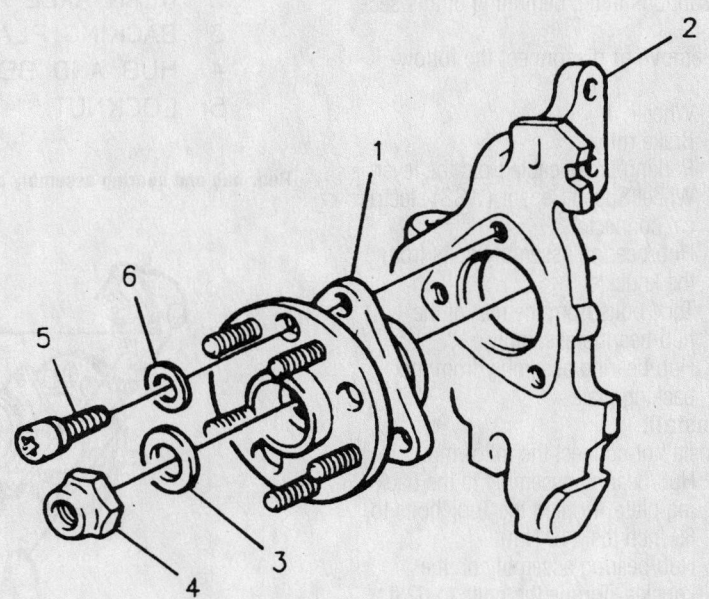

1 HUB AND BEARING ASSEMBLY
2 STEERING KNUCKLE
3 WASHER
4 DRIVE AXLE NUT — 260 N·m (192 LBS. FT.)
5 HUB AND BEARING RETAINING BOLT
6 WASHER

Exploded view of the front hub/bearing assembly

2. Remove or disconnect the following:
- Wheel
- Brake drum
- 4 hub/bearing assembly to knuckle nuts

➡ **The top rear bolt will not clear the brake shoes and must be removed with the bearing assembly.**

- Anti-lock Brake System (ABS) speed sensor wire from the hub/bearing assembly
- Hub/bearing assembly

To install:

3. Install or connect the following:
- Hub/bearing assembly on the knuckle
- ABS wheel speed sensor
- Hub/bearing nuts. Torque the nuts to 44 ft. lbs. (60 Nm).
- Brake drum
- Wheel. Torque the nuts to 100 ft. lbs. (140 Nm).

DISC BRAKE

1. Before servicing the vehicle, refer to the precautions in the beginning of this section.

2. Remove or disconnect the following:
- Wheel
- Brake rotor
- Parking brake cable from the lever
- Wheel Speed Sensor (WSS) electrical connector
- Hub/bearing assembly bolts from the knuckle
- Torx®bolts from the rear of the hub/bearing assembly
- Hub/bearing assembly from the backing plate

To install:

3. Install or connect the following:
- Hub/bearing assembly to the backing plate. Tighten the Torx®bolts to 89 inch lbs. (10 Nm).
- Hub/bearing assembly on the knuckle. Torque the bolts to 62 ft. lbs. (85 Nm).
- WSS electrical connector
- Parking brake cable to the lever
- Brake rotor
- Wheel assembly

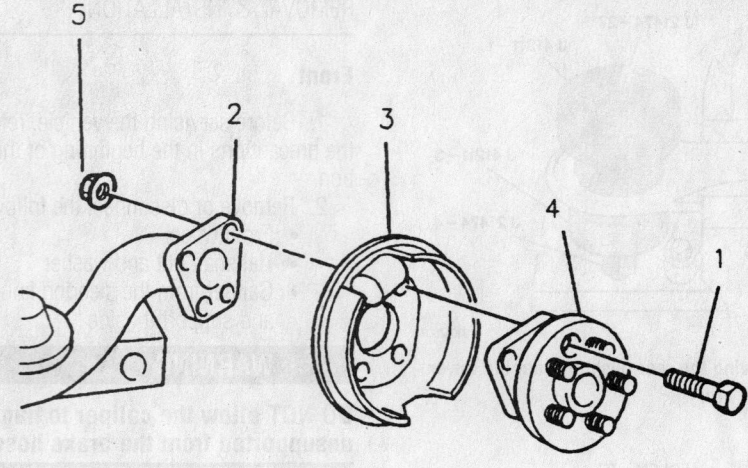

1   BOLT
2   REAR AXLE ASSEMBLY
3   BACKING PLATE
4   HUB AND BEARING ASSEMBLY
5   LOCKNUT

79222341

Rear hub and bearing assembly on models equipped with drum brakes

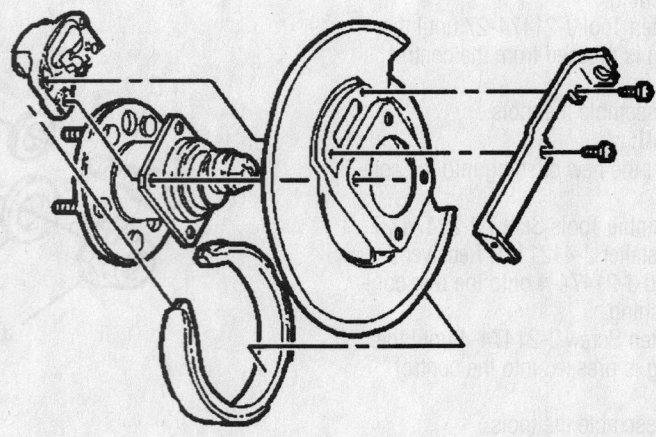

9300Z304

Rear wheel bearing assembly with disc brakes

BRAKES

Brake Caliper

REMOVAL & INSTALLATION

1. Before servicing the vehicle, refer to the precautions in the beginning of this section.

2. Siphon ⅔ of the brake fluid out of the master cylinder.

3. Remove or disconnect the following:
   • Wheel assembly

4. Compress the caliper piston back into the caliper bore using a large pair of pliers, C-clamp or special piston retracting tool.

   • Brake hose from the caliper and discard the copper washers. Plug the hose to prevent excessive fluid loss and possible fluid contamination.
   • Caliper mounting bolts and the caliper from the knuckle
   • Brake pads from the caliper, if the caliper is being replaced

To install:

5. Inspect the condition of the caliper support for rust and corrosion that will hinder the travel of the caliper.

6. Inspect the caliper mounting hardware. New bolts are usually recommended.

7. Lubricate the mounting bushings and sleeves with silicone grease as required.

8. Install or connect the following:
   • Brake pads in the caliper
   • Caliper on the steering knuckle
   • Mounting bolts and torque to 40 ft. lbs. (51 Nm)

   • Brake hose to the caliper using new copper washers and torque the mounting bolt to 35 ft. lbs. (44 Nm)

9. Refill the master cylinder and bleed the brake system.

10. Install the wheel assembly.

11. Verify correct brake operation.

Disc Brake Pads

REMOVAL & INSTALLATION

1. Before servicing the vehicle, refer to the precautions in the beginning of this section.

2. Siphon ⅔ of the brake fluid out of the master cylinder reservoir.

3. Remove or disconnect the following:
   • Wheel assembly
   • Caliper from the steering knuckle without disconnecting the brake hose. DO NOT allow the caliper to hang from the brake hose. Support the caliper with a piece of wire.
   • Outboard pad by pushing in on the outside edge of the pad to release the mounting dowel from the hole in the caliper. When both dowels are unseated push the pad out the bottom of the caliper.
   • Inboard pad from the caliper by pulling it out of the caliper

To install:

4. Install or connect the following:
   • Inboard pad in the caliper so the spring clip on the pad back engages in the caliper piston

   • Outboard pad over the caliper end until the mounting dowels snap into the mounting holes in the caliper
   • Caliper over the rotor onto the steering knuckle
   • Caliper mounting bolts and sleeves and torque to 40 ft. lbs. (51 Nm)
   • Wheel assembly

5. Pump the brake pedal several times to seat the pads against the rotor before attempting to move the vehicle.

6. Check the master cylinder level and add fluid as necessary.

Brake Drums

REMOVAL & INSTALLATION

1. Before servicing the vehicle, refer to the precautions in the beginning of this section.

2. Remove the wheel.

3. Pull the brake drum off. It may be necessary to gently tap the rear edge of the drum to start it off the studs. If extreme resistance to removal is encountered, it will be necessary to retract the brake shoe self-adjuster screw, sometimes called the star-wheel. Knock out the access hole in the brake drum and turn the adjuster to retract the linings from the drum. Install a replacement hole cover before reinstalling the drum.

➡**DO NOT hammer on the brake drum to remove it.**

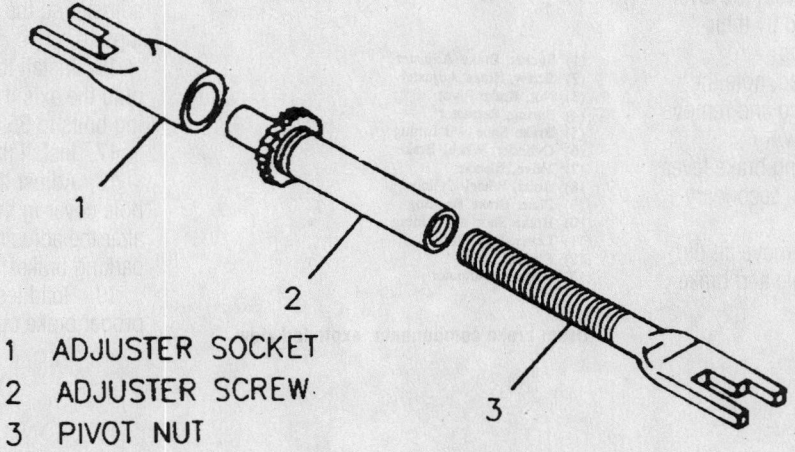

1 ADJUSTER SOCKET
2 ADJUSTER SCREW
3 PIVOT NUT

93006G73

Brake adjuster

To install:

4. Inspect the inside of the brake drum. If worn, heavily grooved or if the opening is distorted, the drum should be refinished or replaced. If refinishing, observe the maximum drum diameter specification.

5. Inspect the wheel cylinder for signs of brake fluid leakage. Inspect the brake shoe springs and self-adjuster mechanism. The adjuster should usually be disassembled, cleaned and lubricated when the drum is removed for brake service.

6. Install the drum over the brake shoes.

7. Install the wheel. Adjust the brakes.

8. Check brake operation.

Brake Shoes

REMOVAL & INSTALLATION

1. Before servicing the vehicle, refer to the precautions in the beginning of this section.

2. Remove or disconnect the following:

- Wheel assembly
- Brake drum
- Upper return springs from the shoes using tool J-8057 brake spring tool
- Hold-down springs using J-8049 brake spring tool
- Shoe hold-down pins from behind the brake backing plate

3. Lift up the actuator lever for the self-adjusting mechanism and remove the actuating link. Remove the actuator lever, pivot, and the pivot return spring.

4. Spread the shoes apart to clear the wheel cylinder pistons and remove the parking brake strut and spring.

- Parking brake cable from the lever
- Shoes, still connected by their adjusting screw spring

5. With the shoes removed, note the position of the adjusting spring and remove the spring and adjusting screw.

- C-clip from the parking brake lever and the lever from the secondary shoe

6. Use a damp cloth to remove all dirt and dust from the backing plate and brake parts.

To install:

7. Check the backing plate attaching bolts to make sure they are tight. Use fine emery cloth to clean all rust and dirt from the shoe contact surfaces on the plate and lubricate with brake grease. Check the wheel cylinder for signs of leakage.

8. Clean all parts completely in brake solvent and air dry. Clean the backing plate shoe contact points.

9. Inspect the inside of the brake drum. If worn, heavily grooved or if the opening is distorted, the drum should be refinished or replaced.

10. Inspect the brake shoe springs and self-adjuster mechanism. Disassemble the adjuster mechanism and clean the threads and coat with grease. Make sure the adjuster assembly turns freely before installing in the vehicle.

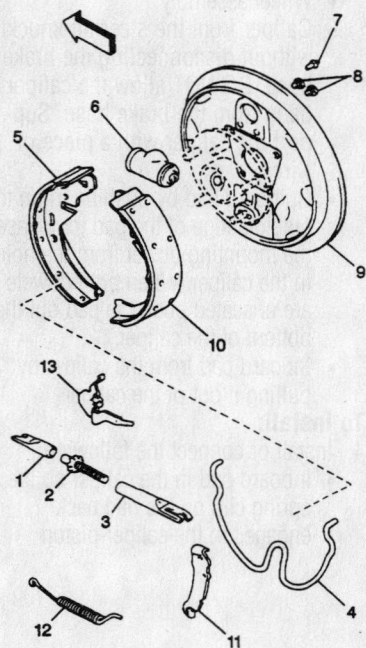

(1) Socket, Brake Adjuster
(2) Screw, Brake Adjuster
(3) Nut, Brake Pivot
(4) Spring, Retractor
(5) Brake Shoe and Lining
(6) Cylinder, Wheel Brake
(7) Valve, Bleeder
(8) Bolts, Wheel Cylinder
(9) Plate, Brake Backing
(10) Brake Shoe and Lining
(11) Lever, Park Brake
(12) Spring
(13) Adjuster Actuator

93006G78

Drum brake components, exploded view

11. Install or connect the following:

- Parking brake lever on the secondary shoe and secure with C-clip
- Adjusting screw and spring on the shoes, connecting them together. The coils of the spring must not be over the starwheel on the adjuster. The left and right hand springs are not interchangeable.

12. Spread the shoe assemblies and connect the parking brake cable.

- Shoes on the backing plate, engaging the shoes at the top temporarily with the wheel cylinder pistons. Make sure the starwheel on the adjuster is lined up with the adjusting hole in the backing plate, if equipped.

13. Spread the shoes slightly and install the parking brake strut and spring. Make sure the end of the strut without the spring engages the parking brake lever. The end with the spring engages the primary shoe (the one with the shorter lining).

- Actuator pivot, lever and return spring
- Actuating link in the shoe retainer. Lift up the actuator lever and hook the link into the lever.
- Hold-down pins through the back of the plate using J-8057
- Lever pivots and hold-down springs. Install the shoe return springs using J-8049. Be careful not to stretch or distort the springs.

14. Make sure the linings are in the right place, the self-adjusting mechanism is correctly installed, and the parking brake parts are hooked up.

15. Measure the distance from the edge of the primary lining to the edge secondary lining, then measure the inside width of the drum. Adjust the linings by means of the adjuster so the drum will fit onto the linings.

16. Install the hub and bearing assembly onto the axle if removed. Torque the retaining bolts to 35 ft. lbs. (55 Nm).

17. Install the drum and the wheels.

18. Adjust the brakes. Install a rubber hole cover in the adjustment knock-out hole after the adjustment is complete. Adjust the parking brake.

19. Road test the vehicle and verify proper brake operation.

CADILLAC

V-Body • Catera

21

SPECIFICATION CHARTS

ENGINE AND VEHICLE IDENTIFICATION

		Engine					Model Year	
Code ①	Liters (cc)	Cu. In.	Cyl.	Fuel Sys.	Engine Type	Eng. Mfg.	Code ②	Year
R	3.0 (2972)	181	V6	SMFI	DOHC	Opel	Y	2000
							1	2001

SMFI: Sequential Multi-port Fuel Injection

DOHC: Double Overhead Camshaft

① 8th position of VIN

② 10th position of VIN

42372-VBOD-C01

GENERAL ENGINE SPECIFICATIONS

Year	Model	Engine Displacement Liters (cc)	Engine Series (ID/VIN)	Fuel System	Net Horsepower @ rpm	Net Torque @ rpm (ft. lbs.)	Bore x Stroke (in.)	Com-pression Ratio	Oil Pressure @ rpm
2000	Catera	3.0 (2972)	R	MFI	200@6000	192@3600	3.38x3.34	10.0:1	22@900
2001	Catera	3.0 (2972)	R	MFI	200@6000	192@3600	3.38x3.34	10.0:1	22@900

MFI: Multi-point Fuel Injection

42372-VBOD-C02

ENGINE TUNE-UP SPECIFICATIONS

Year	Engine Displacement Liters (cc)	Engine ID/VIN	Spark Plug Gap (in.)	Ignition Timing (deg.) AT	Fuel Pump (psi)	Idle Speed (rpm) AT	Valve Clearance	
							In.	Ex.
2000	3.0 (2972)	R	0.034 - 0.043	①	41-47	①	HYD	HYD
2001	3.0 (2972)	R	0.034 - 0.043	①	41-47	①	HYD	HYD

NOTE: The Vehicle Emission Control Information label often reflects specification changes made during production. The label figures must be used if they differ from those in this chart.

HYD: Hydraulic

① Refer to Vehicle Emission Control Information label

42372-VBOD-C03

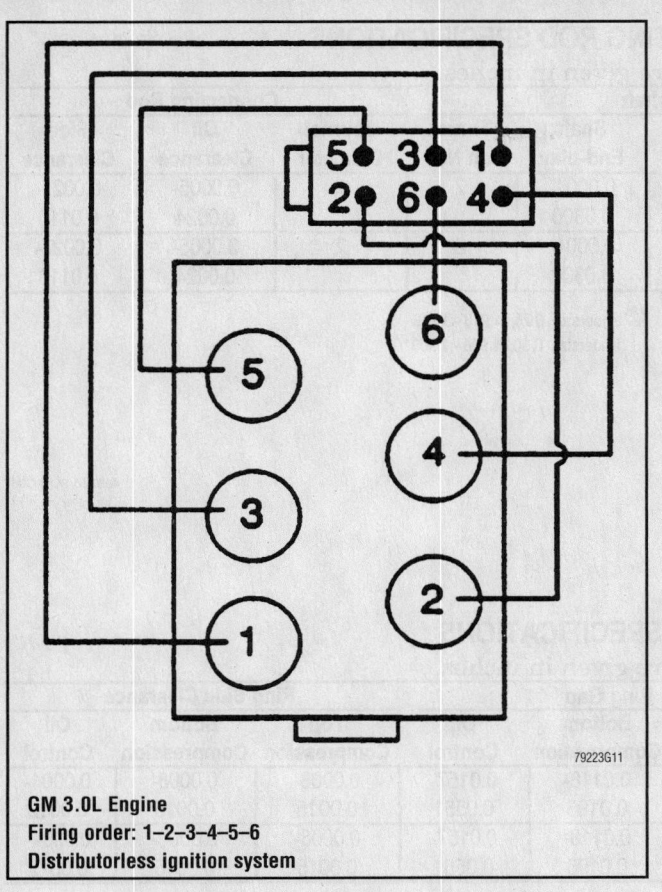

GM 3.0L Engine
Firing order: 1–2–3–4–5–6
Distributorless ignition system

79223G11

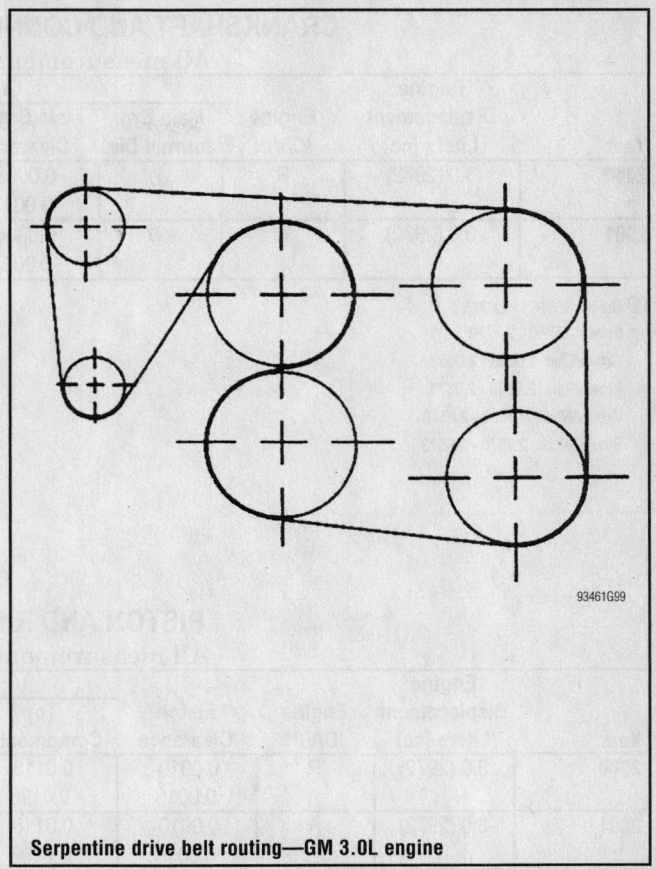

93461G99

Serpentine drive belt routing—GM 3.0L engine

42372-VBOD-C04

CAPACITIES

Year	Model	Engine Displacement Liters (cc)	Engine ID/VIN	Engine Oil with Filter (qts.)	Transmission (pts.)	Drive Axle Rear (pts.)	Fuel Tank (gal.)	Cooling System (qts.)
2000	Catera	3.0 (2972)	R	6.0	14	3.5	16	10.5
2001	Catera	3.0 (2972)	R	6.0	14	3.5	16	10.5

NOTE: All capacities are approximate. Add fluid gradually and ensure a proper fluid level is obtained.

VALVE SPECIFICATIONS

Year	Engine Displacement Liters (cc)	Engine ID/VIN	Seat Angle (deg.)	Face Angle (deg.)	Spring Test Pressure (lbs. @ in.)	Spring Installed Height (in.)	Stem-to-Guide Clearance (in.)		Stem Diameter (in.)	
							Intake	Exhaust	Intake	Exhaust
2000	3.0 (2972)	R	①	45	56.6@1.338	1.543	0.0012-0.0022	0.0016-0.0026	0.2344-0.2350	0.2341-0.2346
2001	3.0 (2972)	R	①	45	56.6@1.338	1.543	0.0012-0.0022	0.0016-0.0026	0.2344-0.2350	0.2341-0.2346

① 45 degrees 20 minutes

42372-VBOD-C05

CRANKSHAFT AND CONNECTING ROD SPECIFICATIONS

All measurements are given in inches.

Year	Engine Displacement Liters (cc)	Engine ID/VIN	Crankshaft: Main Brg. Journal Dia.	Crankshaft: Main Brg. Oil Clearance	Crankshaft: Shaft End-play	Crankshaft: Thrust on No.	Connecting Rod: Journal Diameter	Connecting Rod: Oil Clearance	Connecting Rod: Side Clearance
2000	3.0 (2972)	R	①	0.0006-0.0017	0.0004-0.0300	2	②	0.0005-0.0024	0.0027-0.0110
2001	3.0 (2972)	R	①	0.0006-0.0017	0.0004-0.0300	2	②	0.0005-0.0024	0.0027-0.0110

① Green 2.6763 - 2.6766
Brown 26766 - 2.6770
Green/Blue 2.6665 - 2.6668
Brown/Blue 2.6668 - 2.6671
Green/White 2.6566 - 2.6570
Brown/White 2.6570 - 2.6573

② Undersize 0.25; 1.918 - 1.919
Undersize 0.50; 1.908 - 1.909

42372-VBOD-C06

PISTON AND RING SPECIFICATIONS

All measurements are given in inches.

Year	Engine Displacement Liters (cc)	Engine ID/VIN	Piston Clearance	Ring Gap: Top Compression	Ring Gap: Bottom Compression	Ring Gap: Oil Control	Ring Side Clearance: Top Compression	Ring Side Clearance: Bottom Compression	Ring Side Clearance: Oil Control
2000	3.0 (2972)	R	0.0010-0.0018	0.0118-0.0196	0.0118-0.0196	0.0157-0.0551	0.0008-0.0015	0.0008-0.0015	0.0004-0.0012
2001	3.0 (2972)	R	0.0010-0.0018	0.0118-0.0196	0.0118-0.0196	0.0157-0.0551	0.0008-0.0015	0.0008-0.0015	0.0004-0.0012

42372-VBOD-C07

TORQUE SPECIFICATIONS

All readings in ft. lbs.

Year	Engine Displacement Liters (cc)	Engine ID/VIN	Cylinder Head Bolts	Main Bearing Bolts	Rod Bearing Bolts	Crankshaft Damper Bolts	Flywheel Bolts	Manifold: Intake	Manifold: Exhaust	Spark Plugs	Lug Nut
2000	3.0 (2972)	R	①	②	③	15	④	15	15	18	100
2001	3.0 (2972)	R	①	②	③	15	④	15	15	18	100

① Step 1: 18 ft. lbs.
Step 2: Rotate 90 degrees
Step 3: Rotate 90 degrees
Step 4: Rotate 90 degrees
Step 5: Rotate 15 degrees

② Step 1: 37 ft. lbs.
Step 2: Rotate 60 degrees
Step 3: Rotate 15 degrees

③ Step 1: 26 ft. lbs.
Step 2: Rotate 45 degrees
Step 3: Rotate 15 degrees

④ Step 1: 48 ft. lbs.
Step 2: Rotate 30 degrees
Step 3: Rotate 15 degrees

42372-VBOD-C08

WHEEL ALIGNMENT

Year	Model		Caster Range (+/-Deg.)	Caster Preferred Setting (Deg.)	Camber Range (+/-Deg.)	Camber Preferred Setting (Deg.)	Toe-in (in.)	Steering Axis Inclination (Deg.)
2000	Catera	F	1.00	+5.00	0.55	-0.60	0.06 +/- 0.05	—
		R	—	—	0.55	-1.75	0 +/- 0.05	—
2001	Catera	F	1.00	+5.00	0.55	-0.60	0.06 +/- 0.05	—
		R	—	—	0.55	-1.75	0 +/- 0.05	—

42372-VBOD-C09

TIRE, WHEEL AND BALL JOINT SPECIFICATIONS

Year	Model	OEM Tires Standard	OEM Tires Optional	Tire Pressures (psi) Front	Tire Pressures (psi) Rear	Wheel Size	Ball Joint Inspection
2000	Catera	P225/55HR16	None	32	32	7-JJ	0.125 in. ①
2001	Catera	P225/55HR16	None	32	32	7-JJ	0.125 in. ①

OEM: Original Equipment Manufacturer

PSI: Pounds Per Square Inch

① Replace if any measurable movement is found.

42372-VBOD-C10

BRAKE SPECIFICATIONS
All measurements in inches unless noted

Year	Model		Brake Disc Original Thickness	Brake Disc Minimum Thickness	Brake Disc Maximum Runout	Minimum Lining Thickness	Brake Caliper Bracket Bolts (ft. lbs.)	Brake Caliper Mounting Bolts (ft. lbs.)
2000	Catera	F	1.102	0.984	0.001	0.315 ①	②	22
		R	0.472	0.393	0.004	NA	NA	59
2001	Catera	F	1.102	0.984	0.001	0.315 ①	②	22
		R	0.472	0.393	0.004	NA	NA	59

F: Front

R: Rear

NA: Not available

① Total thickness of lining and backing plate

② Tighten to 70 ft. lbs. (plus an additional 37 degrees

42372-VBOD-C11

SCHEDULED MAINTENANCE INTERVALS
GM V BODY—CADILLAC CATERA

TO BE SERVICED	TYPE OF SERVICE	VEHICLE MILEAGE INTERVAL (x1000)																			
		5	10	15	20	25	30	35	40	45	50	55	60	65	70	75	80	85	90	95	100
Engine oil & filter	R	✓	✓	✓	✓	✓	✓	✓	✓	✓	✓	✓	✓	✓	✓	✓	✓	✓	✓	✓	✓
Rotate tires	S/I	✓		✓		✓		✓		✓		✓		✓		✓		✓		✓	
Brake hoses	S/I	✓		✓		✓		✓		✓		✓		✓		✓		✓		✓	
Passenger Compartment Air Filter (Pollen Filter)	R			✓			✓			✓			✓			✓			✓		
Air filter element	S/I			✓			✓			✓			✓			✓			✓		
	R						✓						✓						✓		
Fuel tank, cap and lines	S/I						✓						✓								
Rear axle fluid level	S/I						✓						✓								
Automatic transaxle fluid & filter ①	R										✓										✓
Accessory drive belt(s)	S/I												✓								
Spark plugs ②	R																				✓
Ignition cables	S/I																				✓
Camshaft timing belt ③	R												✓								✓
Fuel filter	R																				✓
Engine coolant ④	R																				

R: Replace

S/I: Service or Inspect

① Automatic transaxle fluid & filter: replace at 50,000 miles (83,000 km) if the vehicle has experienced severe service usage.

② Platinum tip spark plugs: replace every 100,000 miles.

③ Replace at 60,000 miles (96,000 km) if the engine was driven without an engine coolant heater being used and where temperatures fall below -20°F (-28°C).
Otherwise, replace the belt at 100,000 miles (160,000 km).

④ Engine coolant: replace every 150,000 miles. Use O.E. specified (DEX-COOL™) coolant only. If any silicate coolant is used, the service interval is every 30,000 miles.

FREQUENT OPERATION MAINTENANCE (SEVERE SERVICE)

If a vehicle is operated under any of the following conditions it is considered severe service:

- Extremely dusty areas.

- 50% or more of the vehicle operation is in 32°C (90°F) or higher temperatures, or constant operation in temperatures below 0°C (32°F).

- Prolonged idling (vehicle operation in stop and go traffic).

- Frequent short running periods (engine does not warm to normal operating temperatures).

- Police, taxi, delivery usage or trailer towing usage.

Oil & oil filter: change every 5000 miles

Rotate tires at 5000 miles, then every 10,000 miles.

Air filter element: service or inspect every 15,000 miles.

Camshaft timing belt: change every 60,000 miles for severe service.

42372-VBOD-C12

PRECAUTIONS

Before servicing any vehicle, please be sure to read all of the following precautions, which deal with personal safety, prevention of component damage, and important points to take into consideration when servicing a motor vehicle:

• Never open, service or drain the radiator or cooling system when the engine is hot; serious burns can occur from the steam and hot coolant.

• Observe all applicable safety precautions when working around fuel. Whenever servicing the fuel system, always work in a well-ventilated area. Do not allow fuel spray or vapors to come in contact with a spark, open flame or excessive heat (a hot drop light, for example). Keep a dry chemical fire extinguisher near the work area. Always keep fuel in a container specifically designed for fuel storage; also, always properly seal fuel containers to avoid the possibility of fire or explosion. Refer to the additional fuel system precautions later in this section.

• Fuel injection systems often remain pressurized, even after the engine has been turned **OFF**. The fuel system pressure must be relieved before disconnecting any fuel lines. Failure to do so may result in fire and/or personal injury.

• Brake fluid often contains polyglycol ethers and polyglycols. Avoid contact with the eyes and wash your hands thoroughly after handling brake fluid. If you do get brake fluid in your eyes, flush your eyes with clean, running water for 15 minutes. If eye irritation persists, or if you have taken brake fluid internally, IMMEDIATELY seek medical assistance.

• The EPA warns that prolonged contact with used engine oil may cause a number of skin disorders, including cancer! You should make every effort to minimize your exposure to used engine oil. Protective gloves should be worn when changing oil. Wash your hands and any other exposed skin areas as soon as possible after exposure to used engine oil. Soap and water, or waterless hand cleaner should be used.

• All new vehicles are now equipped with an air bag system. The system must be disabled before performing service on or around system components, steering column, instrument panel components, wiring and sensors. Failure to follow safety and disabling procedures could result in accidental air bag deployment, possible personal injury and unnecessary system repairs.

• Always wear safety goggles when working with, or around, the air bag system. When carrying a non-deployed air bag, be sure the bag and trim cover are pointed away from your body. When placing a non-deployed air bag on a work surface, always face the bag and trim cover upward, away from the surface. This will reduce the motion of the module if it is accidentally deployed. Refer to the additional air bag system precautions later in this section.

• Clean, high quality brake fluid from a sealed container is essential to the safe and proper operation of the brake system. You should always buy the correct type of brake fluid for your vehicle. If the brake fluid becomes contaminated, completely flush the system with new fluid. Never reuse any brake fluid. Any brake fluid that is removed from the system should be discarded. Also, do not allow any brake fluid to come in contact with a painted surface; it will damage the paint.

• Never operate the engine without the proper amount and type of engine oil; doing so WILL result in severe engine damage.

• Timing belt maintenance is extremely important! Many models utilize an interference-type, non-freewheeling engine. If the timing belt breaks, the valves in the cylinder head may strike the pistons, causing potentially serious (also time-consuming and expensive) engine damage. Refer to the maintenance interval charts in the front of this section for the recommended replacement interval for the timing belt, and to the timing belt procedure in this section for belt replacement and inspection.

• Disconnecting the negative battery cable on some vehicles may interfere with the functions of the on-board computer system(s) and may require the computer to undergo a relearning process once the negative battery cable is reconnected.

• When servicing drum brakes, only disassemble and assemble one side at a time, leaving the remaining side intact for reference.

• Only an MVAC-trained, EPA-certified automotive technician should service the air conditioning system or its components.

ENGINE REPAIR

➡**Disconnecting the negative battery cable on some vehicles may interfere with the operation of the on board computer system. The computer may undergo a relearning process once the negative battery cable is reconnected.**

Alternator

REMOVAL

1. Before servicing the vehicle, refer to the precautions in the beginning of this section.

2. Remove or disconnect the following:
   • Negative battery cable
   • Intake air resonator
   • Drive belt
   • Coolant heater, if equipped

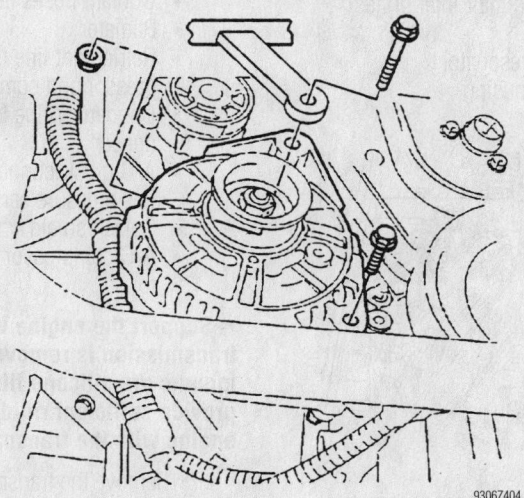

Alternator mounting hardware

93062404

- Alternator cooling duct
- Alternator electrical connectors
- Alternator

INSTALLATION

1. Install or connect the following:
   - Alternator. Torque the lower nut to 30 ft. lbs. (40 Nm) and the upper nut to 30 ft. lbs. (40 Nm).
   - Cooling duct
   - Alternator electrical connectors
   - Coolant heater, if equipped
   - Drive belt
   - Intake air resonator
   - Negative battery cable
2. Perform a charging system test and verify that the system is operating properly.

Ignition Timing

ADJUSTMENT

➡ **The 3.0L DOHC engine used in the Catera utilizes a Distributorless Ignition System (DIS). No ignition timing adjustment is possible.**

Engine Assembly

REMOVAL & INSTALLATION

1. Before servicing the vehicle, refer to the precautions in the beginning of this section.
2. Drain the cooling system.
3. Drain the engine oil.
4. Recover the A/C refrigerant.
5. Relieve the fuel system pressure.
6. Remove or disconnect the following:
   - Battery
   - Wiper arms
   - Left and right air inlet grilles
   - Hood
   - Intake air resonator
   - Air filter housing

93062401

Red, white and black harness connectors (1)

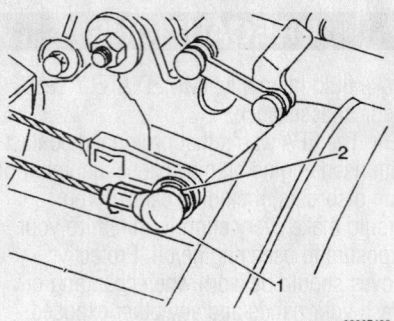

93062402

Accelerator (1) and cruise control (2) cables

- Red, white and black wiring harness connectors
- Body ground cable
- Power supply wires at the battery cable end
- Power steering hoses from the power steering pump
- Brake booster vacuum lines
- Electronic Control Module (ECM) and harness from the electrical center
- Relays and wiring harness from electrical center box
- Cruise control and accelerator cables
- Fuel lines from the fuel rail. Use a backup wrench to prevent damage to the fuel rail.
- Coolant hose from the throttle body
- Vacuum hose from the purge valve on the engine ventilation chamber
- Vacuum hose from the heater control valve
- Coolant reservoir hose from the coolant inlet pipe
- Electric water pump with hoses attached
- Coolant hoses from the heater core
- Radiator
- Refrigerant line from the A/C compressor and compressor bracket
- Condenser line from the A/C condenser
- A/C quick-connect hose near the low pressure service valve
- Splash shield
- A/C compressor electrical connector

➡ **Support the engine whenever the transmission is removed. The motor mounts are silicone filled and do not provide sufficient rigidity to support the engine with the transmission removed.**

7. Remove the transmission from the vehicle.

8. Attach a hoist to the engine lifting eyes. Raise the engine slightly and remove the engine support fixture.
9. Remove the engine mount nuts and lift the engine out of the vehicle. Raise the engine slowly after being certain all wiring, cables and hoses have been removed.

To install:

10. Install the engine.
11. Install the engine mount. Torque the upper motor mount nuts to 30 ft. lbs. (40 Nm). Torque the lower nuts to 41 ft. lbs. (55 Nm).
12. Install an engine support fixture and remove the chain hoist.
13. Install or connect the following:
   - Transmission. Torque the mounting bolts to 44 ft. lbs. (60 Nm).
   - A/C compressor electrical connector
   - Splash shield and tighten securely
   - Quick connects for the high and low pressure fittings
   - A/C condenser line to the condenser
   - A/C compressor line
   - Radiator
   - Coolant hoses to the heater core
   - Electric water pump
   - Coolant reservoir hose to the coolant inlet pipe
   - Vacuum hose to the heater control valve
   - Vacuum hose for the purge valve on the engine ventilation chamber
   - Coolant hose to the throttle body
   - Fuel return and supply lines. Torque the lines to 11 ft. lbs. (15 Nm).
   - Cruise control and accelerator cables to the throttle body
   - Relays and wiring harness to the electrical center box
   - ECM
   - Brake booster vacuum lines
   - Power steering hoses to the steering pump. Torque the discharge hose fasteners to 21 ft. lbs. (28 Nm).
   - Power supply wires at the battery cable ends
   - Body ground cable
   - Red, white and black wiring harness connectors
   - Air filter housing
   - Intake air resonator
   - Hood
   - Left and right air inlet grilles
   - Wiper arms
   - Battery
   - Negative battery cable

14. Fill and bleed the power steering system.

15. Refill the engine oil.

➡**An oil filter change is recommended.**

16. Fill and bleed the coolant system.

➡**When refilling the coolant system add 2 crushed engine coolant supplement sealant pellets (PN 3634621) into the reservoir.**

17. Recharge the A/C system and check for leaks.

18. Start the vehicle and inspect for leaks.

Water Pump

REMOVAL & INSTALLATION

1. Before servicing the vehicle, refer to the precautions in the beginning of this section.

2. Drain the coolant.

3. Remove or disconnect the following:
   - Negative battery cable
   - Intake air resonator
   - Water pump pulley bolts, loosen only
   - Front timing belt cover
   - Water pump pulley
   - Water pump

To install:

4. Install or connect the following:
   - Water pump with a new O-ring. Torque the bolts to 18 ft. lbs. (25 Nm).
   - Water pump pulley
   - Front timing belt cover
   - Intake air resonator
   - Negative battery cable

5. Fill the cooling system through the reservoir tank.

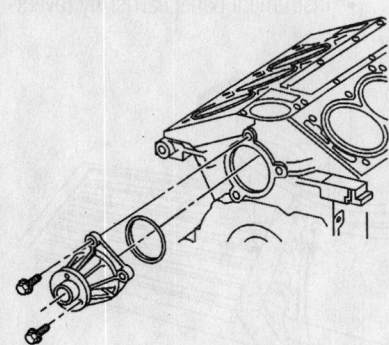

Exploded view of the water pump mounting

93002401

➡**When refilling the cooling system, add 2 crushed engine coolant supplement sealant pellets (PN 3634621) into the coolant reservoir.**

6. Start the vehicle and inspect the coolant systems for leaks.

Heater Core

REMOVAL & INSTALLATION

1. Before servicing the vehicle, refer to the precautions in the beginning of this section.

2. Disable the SIR system.

3. Disconnect the negative battery cable.

4. Drain the cooling system.

5. Remove the heater hose quick connects from the heater core pipes by performing the following procedure:

   a. At the passenger side, raise the air inlet screen and open the access door near the pollen filter.

   b. Unlock the quick connect collars by squeezing the tabs and carefully pulling back on the tabs to disconnect the sleeve.

   c. If green assembly marks are attached, discard them.

❋❋ WARNING

The front wheels must be maintained in the straight-ahead position and the steering column must be in the LOCK position. Failure to do so will cause improper alignment of some components during installation and may result to damage to the SIR coil assembly.

6. Remove the steering column by removing or disconnecting the following:
   - Instrument panel driver knee bolster energy absorber and sound insulator
   - Steering column electrical connector(s)
   - Coupler bolt from the lower steering column connection and slightly separate the coupler to aid in the shaft removal
   - Using a chisel and a hammer, rotate the forward support strap shear nut and bolt (located under the steering column) counterclockwise in order to remove them
   - Steering column-to-rear support bracket bolt
   - Pull the steering column straight

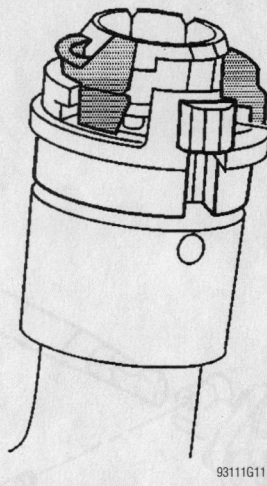

93111G11

View of a heater hose connector—Catera

back and through the dash panel and remove it from the vehicle.

❋❋ WARNING

Handle the steering column with care for it is very susceptible to damage. Dropping, leaning or hammering on it could cause damage to its collapsible design.

7. Remove the instrument panel carrier by removing or disconnecting the following:
   - Windshield pillar moldings
   - Access panel, the air deflector outlet screw, the air deflector outlet and the air outlet duct, located at the right-side of the instrument panel
   - Instrument panel SIR (air bag) module cover, the instrument panel compartment and the SIR module
   - Upper center console and the lower center console
   - Center console air duct screw and the center console air duct
   - Radio tape player bezel and the radio
   - Climate control head
   - Center air outlet deflector, the outlet screws and the outlet housing
   - Driver's side access panel
   - Driver's side air outlet deflector, the outlet screw and the outlet housing
   - Driver's side lower outlet duct
   - Instrument cluster
   - Fuse and relay panel screws and the panels
   - Instrument panel carrier bolts
   - Electrical connectors from the instrument panel and/or the wiring harness clips from the instrument panel carrier, if necessary
   - Instrument panel carrier

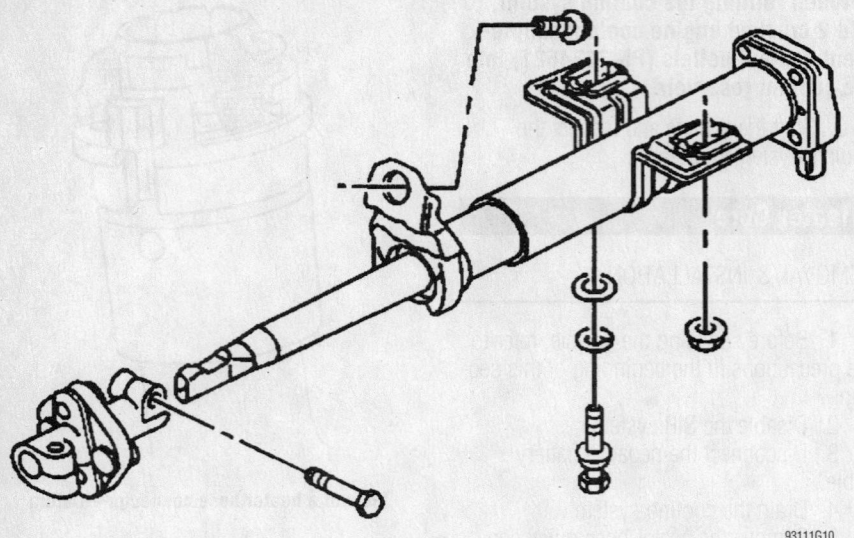

93111G10

View of the steering column assembly—Catera

→It is not necessary to physically remove the instrument panel but it is necessary to pull the carrier rearward to enable the heater core to be removed from its housing.

- Blower motor housing screws and the housing with the motor
- Heater core pipe bracket-to-chassis screw and the bracket
- Heater core inlet/outlet pipe bracket-to-heater core screw and the inlet/outlet pipe bracket
- Instrument panel support brace bolts from the instrument panel and the transmission well, then remove the brace
- Heater core-to-housing retaining screw. Plug the heater core pipes and protect the interior from coolant spills
- Heater core and the rubber seal from the heater housing

To install:
8. Install or connect the following:
   - Rubber seal and heater core into the heater housing; be careful not to damage the fins
   - Heater core-to-housing retainer screw
   - Instrument panel support brace to the transmission well and instrument panel, then torque the bolts to 16 ft. lbs. (22 Nm)
   - Heater core inlet/outlet pipe bracket to the heater core (using a new O-ring lightly coated with coolant), and torque the screw to 44 inch lbs. (5 Nm)
   - Heater core pipe bracket with the screw to the chassis, be careful not to strip the screw
   - Blower motor/housing assembly and torque the screws to 35 inch lbs. (4 Nm)
   - Instrument panel carrier
   - Instrument panel carrier by revers-

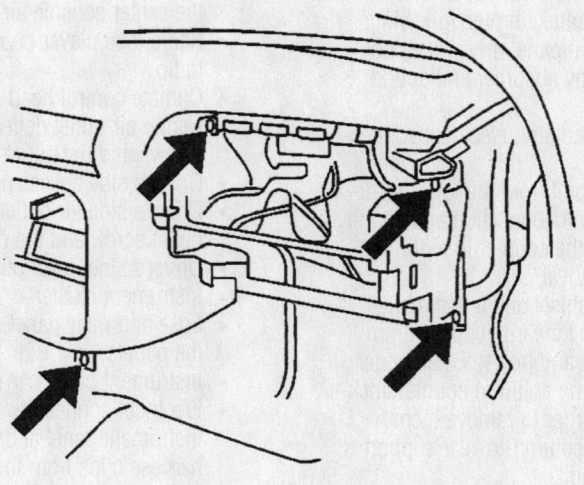

93111G09

View of the instrument panel carrier bolt locations—Catera

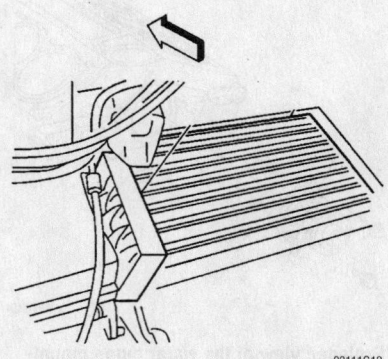

93111G12

View of a heater core and seal—Catera

ing the removal procedures and torque the instrument panel carrier bolts to 16 ft. lbs. (22 Nm)

9. Remove the steering column by installing or connecting the following:
- Steering column into the vehicle, through the dash panel and into the lower steering coupling
- Hand start the rear support bracket bolt, the forward support strap nut and shear bolt
- Rear support bracket bolt. Torque to 16 ft. lbs. (22 Nm)
- Forward support strap nut. Torque to 16 ft. lbs. (22 Nm)
- New forward support shear bolt. Torque to 15 ft. lbs. (21 Nm)
- Lower steering column shaft bolt and torque to 16 ft. lbs. (22 Nm)
- Steering column electrical connector(s)
- Sound insulator and the instrument panel driver knee bolster energy absorber

10. Connect the heater hoses to the heater core pipes by performing the following procedure:
a. If not attached to the quick connect, discard the green assembly marker(s).
b. Push the quick connects into the pipes until they are fully seated.
c. Squeeze the locking tabs and press the retaining sleeve into the locked position.
d. At the passenger side, close the access door near the pollen filter and lower the air inlet screen.

11. Refill the cooling system by performing the following procedure:
a. Add a 50/50 mixture of water and DEX-COOL® antifreeze to the KALT/COLD mark (seam) on the surge tank.
b. Start the engine and allow it to idle for 1 min.
c. Add more coolant to the surge tank as necessary.
d. Install the radiator sure tank cap.
e. Cycle the engine, from idle to 3000 rpm, in 30 second intervals, until the engine reaches normal operating temperatures.

➡**The cooling system will bleed itself automatically during warm-up.**

f. Turn the engine OFF and recheck the coolant level when the engine is cool

12. Connect the negative battery cable.
13. Enable the SIR system.
14. Reprogram the necessary accessories.

REMOVAL & INSTALLATION

1. Before servicing the vehicle, refer to the precautions in the beginning of this section.
2. Drain the engine coolant.
3. Relieve the fuel system pressure.
4. Remove or disconnect the following:
- Intake plenum
- Intake air resonator
- Intake manifold and spacer
- Both Engine Coolant Temperature (ECT) sensor electrical connectors
- Coolant crossover
- Ignition coil assembly
- Camshaft cover
- Front timing belt cover
5. Position the crankshaft 60 degrees

Before Top Dead Center (BTDC) to avoid contact between the valves and the pistons.
6. Remove or disconnect the following:
- Timing belt
- Timing belt tensioner bracket
- Four camshaft gears
- Water pump
- Rear timing belt cover
- Camshaft Position (CMP) sensor electrical connector
- Exhaust camshaft
- Coolant pipe/engine lift bracket from the cylinder head
- Oil level indicator tube
- Upper radiator hose from the coolant pipe
- Coolant pipe
- Exhaust manifold from the cylinder head
- Cylinder head

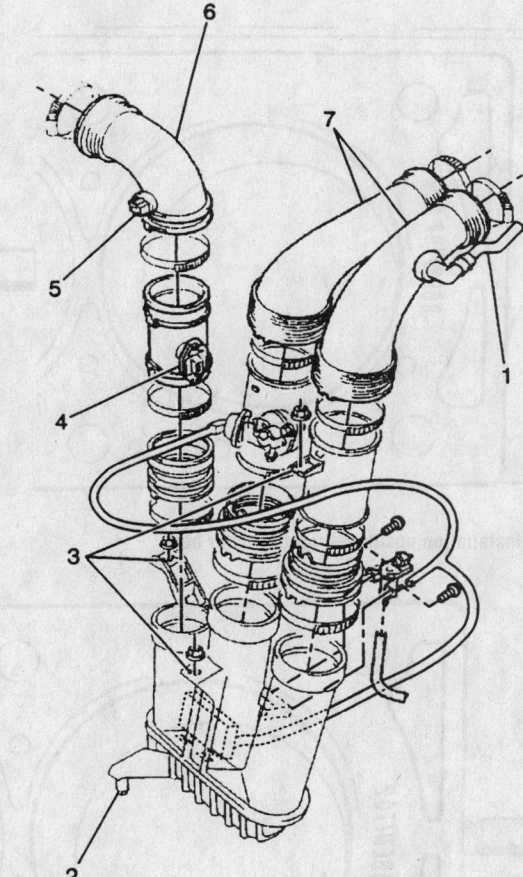

(1) Idle Air Control (IAC) Inlet Hose
(2) Resonance Chamber Guide Pin
(3) Resonance Chamber Nut
(4) Mass Air Flow (MAF) Sensor
(5) Intake Air Temperature (IAT) Sensor
(6) Resonance Chamber Air Intake Hose
(7) Intake Plenum Air Inlet Hose

79227406

Intake air resonator and related components

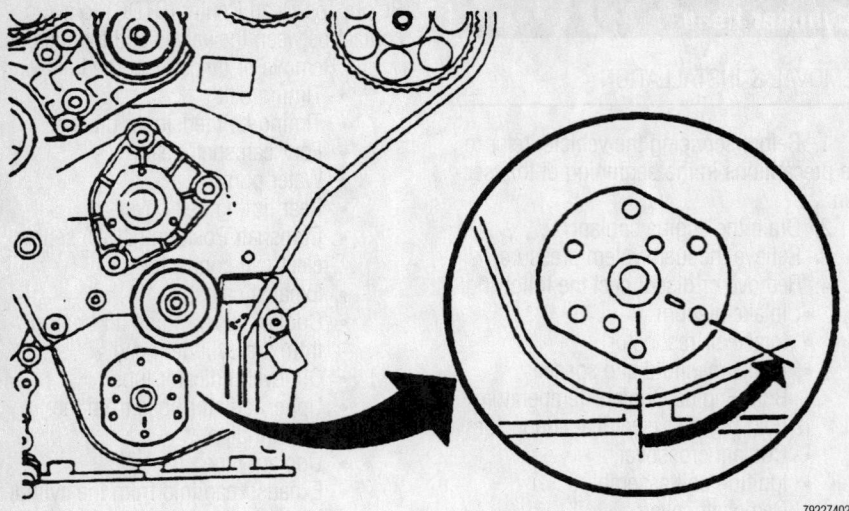

Before removing the timing belt, be sure to turn the crankshaft 60 degrees BTDC to avoid valve-to-piston contact and subsequent engine damage.

Proper head gasket installation position—right cylinder head

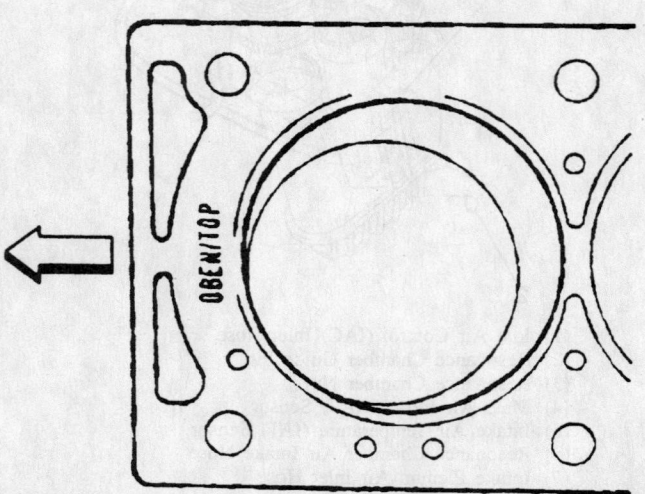

Proper head gasket installation position—left cylinder head

To install:

→ Be sure to use new cylinder head bolts when installing the cylinder head. The old bolts have been stretched and are not reusable.

7. Install a new cylinder head gasket.
8. Install the cylinder head.
9. Install new cylinder head bolts and tighten the bolts as follows:
   a. Step 1: Torque the bolts to 18 ft. lbs. (25 Nm).
   b. Step 2: An additional 90 degrees.
   c. Step 3: An additional 90 degrees.
   d. Step 4: An additional 90 degrees.
   e. Step 5: An additional 15 degrees.
   f. Step 6: An additional 15 degrees.
10. Install or connect the following:
    - Exhaust manifold using a new gasket. Torque the nuts to 15 ft. lbs. (20 Nm).
    - Coolant pipe with new O-rings
    - Oil level indicator tube
    - Coolant pipe/engine lift bracket. Torque the bolt to 15 ft. lbs. (20 Nm).
    - Upper radiator hose to the coolant pipe
    - Exhaust camshaft. Torque the bearing cap bolts to 71 inch lbs. (8 Nm).
    - CMP electrical connector
    - Rear timing belt cover. Torque the bolts to 71 inch lbs. (8 Nm).
    - Water pump. Torque the bolts to 18 ft. lbs. (25 Nm).
    - Camshaft gears. Torque the bolts to 37 ft. lbs. (50 Nm) plus a 60 degree turn, plus another 15 degree turn.
    - Timing belt tensioner bracket. Torque the bolts to 30 ft. lbs. (40 Nm).
    - Timing belt
    - Front timing belt cover. Torque the bolts to 71 inch lbs. (8 Nm).
    - Left camshaft cover with new O-rings and gaskets. Torque the fasteners to 71 inch lbs. (8 Nm).
    - Ignition coil assembly
    - Coolant crossover. Torque the bolts to 22 ft. lbs. (30 Nm).
    - Both ECT sensor electrical connectors
    - Intake manifold spacer. Torque the bolts in a spiral direction, starting from the inside and working outward to 15 ft. lbs. (20 Nm).
    - Intake manifold. Torque the bolts to 15 ft. lbs. (20 Nm).
    - Intake air resonator. Torque the nuts to 27 inch lbs. (3 Nm).

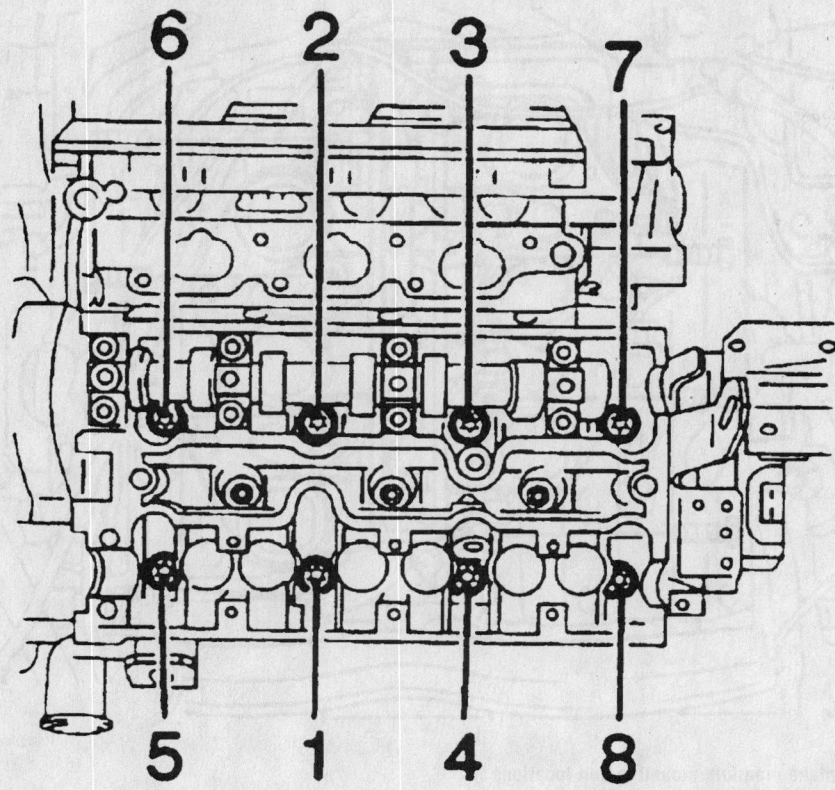

Cylinder head torque sequence—left cylinder head

• Intake plenum. Torque the bolts to 71 inch lbs. (8 Nm).
• Negative battery cable

➡ **An oil and filter change is recommended.**

11. Refill the cooling system.
12. Start the engine and inspect for leakage.
13. Inspect all fluid levels and top off, if necessary.

Intake Manifold Assembly

REMOVAL & INSTALLATION

Intake Plenum

1. Before servicing the vehicle, refer to the precautions in the beginning of this section.
2. Remove or disconnect the following:
   • Negative battery cable
   • Intake plenum air inlet hoses from the throttle body
   • Brake booster vacuum hose from the intake plenum
   • Wiring harness channel from the plenum
   • Switch over valve electrical connector and vacuum line
   • Accelerator, cruise control cables and the bracket from the throttle body
   • Throttle body control electrical connector
   • Idle Air Control (IAC) inlet hose and electrical connector
   • Throttle Position Sensor (TPS) electrical connection
   • Throttle body from the intake plenum
   • Crankcase vent tube adapter and cover from the intake plenum
   • Intake plenum and O-rings

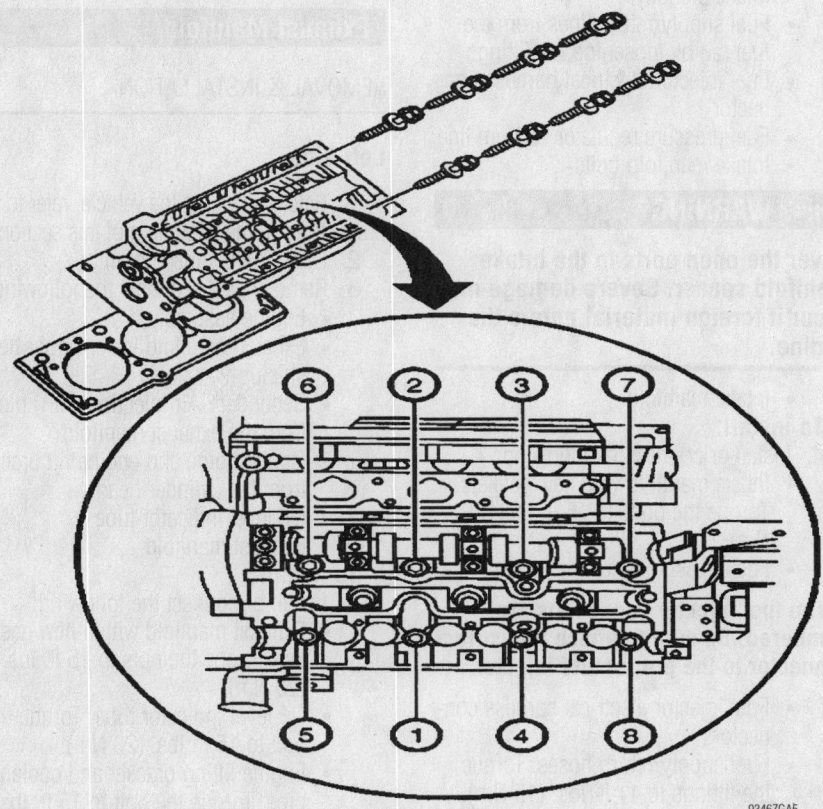

Cylinder head torque sequence—right cylinder head

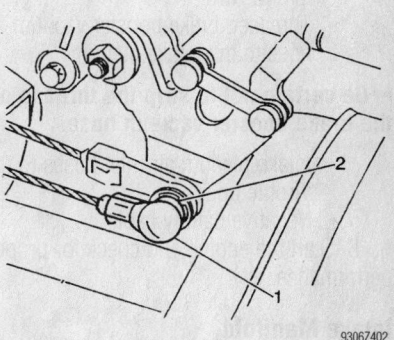

Disconnect the accelerator and cruise control cables

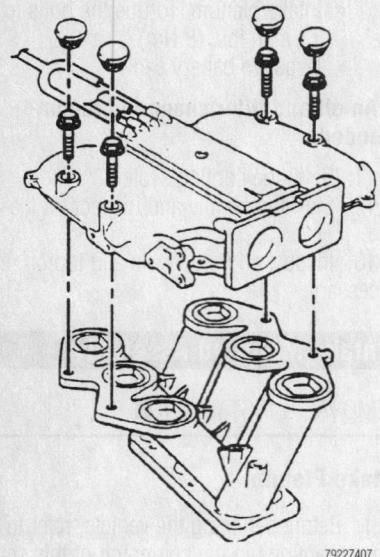

Intake plenum mounting bolt locations

To install:

3. Install or connect the following:
- Intake plenum with new O-rings. Torque the bolts to 71 inch lbs. (8 Nm).
- Crankcase vent tube adapter to the intake plenum. Torque the fastener to 71 inch lbs. (8 Nm).
- Throttle body to the intake plenum. Torque the fasteners to 106 inch lbs. (12 Nm).
- TPS electrical connector
- IAC electrical connector and vacuum line
- Throttle body control electrical connector
- Accelerator and cruise control bracket and cables. Torque the bolts to 71 inch lbs. (8 Nm).
- Electrical connector and vacuum hose to the intake plenum switchover valve solenoid
- Wiring channel to the intake plenum. Torque the bolts to 71 inch lbs. (8 Nm).
- Threaded brake booster vacuum hose to the intake plenum

➥**Be certain not to strip the threads on the brake booster vacuum hose.**

- Intake plenum air inlet hoses to the throttle body
- Negative battery cable

4. Start the engine and check for proper performance.

Intake Manifold

1. Before servicing the vehicle, refer to the precautions in the beginning of this section.

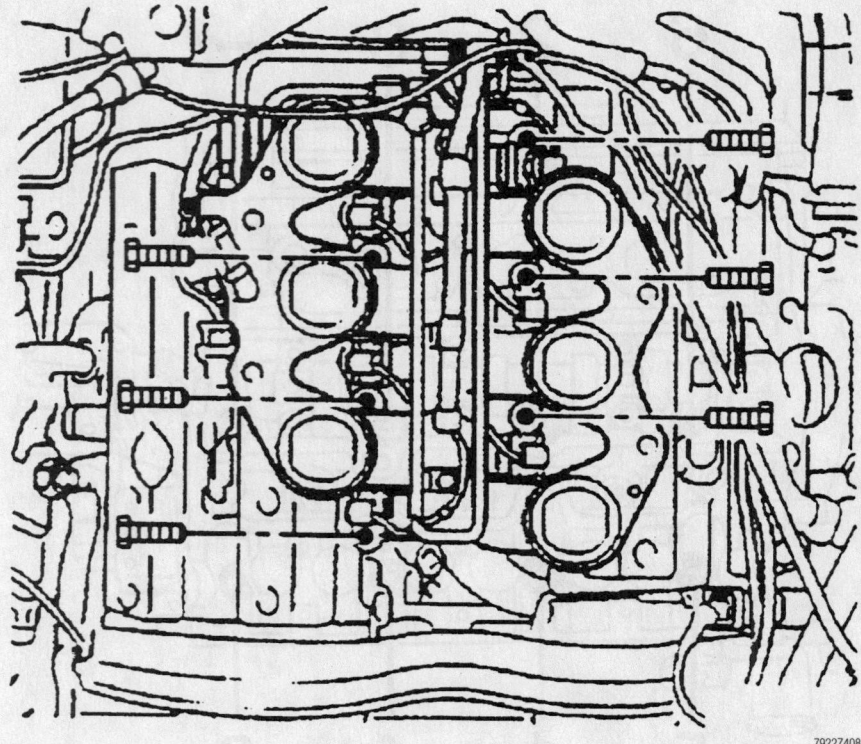

Intake manifold mounting bolt locations

2. Relieve the fuel system pressure.
3. Remove or disconnect the following:
- Intake plenum
- Fuel supply/return lines from the fuel rail by loosening the fittings
- Fuel injector electrical harness connector
- Fuel pressure regulator vacuum line
- Intake manifold bolts

✳✳ WARNING

Cover the open ports to the intake manifold spacer. Severe damage may occur if foreign material enters the engine.

- Intake manifold

To install:

4. Install or connect the following:
- Intake manifold with new gaskets. Torque the bolts to 15 ft. lbs. (20 Nm).
- Fuel pressure regulator vacuum line

➥**The fuel injector connectors are numbered. Be sure to attach the correct connector to the proper fuel injector.**

- Fuel injector electrical harness connectors
- Fuel supply/return hoses. Torque the fittings to 11 ft. lbs. (15 Nm).
- Intake plenum

- Negative battery cable

5. Start the vehicle and inspect for leaks.

Exhaust Manifold

REMOVAL & INSTALLATION

Left Side

1. Before servicing the vehicle, refer to the precautions in the beginning of this section.
2. Drain the engine coolant.
3. Remove or disconnect the following:
- Engine assembly
- Exhaust manifold lower/upper heat shields
- Secondary Air Injection (AIR) pipe from the exhaust manifold
- Coolant pipe and engine lift bracket from the cylinder head
- Oil level indicator tube
- Exhaust manifold

To install:

4. Install or connect the following:
- Exhaust manifold with a new gasket. Torque the nuts to 15 ft. lbs. (20 Nm).
- Oil level indicator tube. Torque the bolt to 15 ft. lbs. (20 Nm).
- Engine lifting bracket and coolant pipe. Torque the bolt to 15 ft. lbs. (20 Nm).

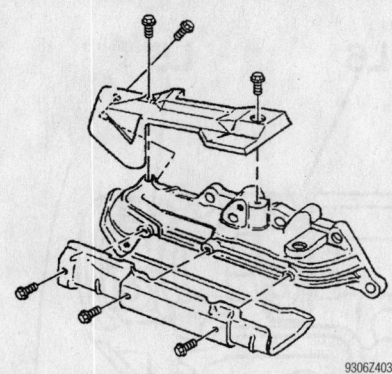

Exhaust manifold heat shield bolts

- Air injection pipe to the exhaust manifold. Torque the bolts to 15 ft. lbs. (20 Nm).
- Exhaust manifold upper/lower heat shields. Torque the bolts to 71 inch lbs. (8 Nm).
- Engine assembly

5. Refill the engine coolant.

6. Start the vehicle and inspect for any leaks.

Right Side

1. Before servicing the vehicle, refer to the precautions in the beginning of this section.

2. Drain the engine coolant.

3. Remove or disconnect the following:
- Transmission assembly
- Coolant intake pipe

4. Remove or disconnect the following:
- Exhaust manifold lower heat shield
- Exhaust manifold lower rear nuts
- Exhaust manifold upper rear heat shield bolt
- Catalytic converter nuts from the exhaust manifold
- Drive belt tensioner
- Exhaust manifold front two nuts
- Exhaust manifold upper heat shield bolts

➡**It is not necessary to remove the upper heat shield.**

- Secondary Air Injection (AIR) injection pipe from the exhaust manifold
- Exhaust manifold

To install:

5. Install or connect the following:
- Exhaust manifold with a new gasket. Torque the 2 upper nuts to 15 ft. lbs. (20 Nm).
- AIR injection pipe to the exhaust manifold. Torque the bolts to 15 ft. lbs. (20 Nm).
- Upper heat shield bolts. Torque the bolts to 71 inch lbs. (8 Nm).

- Exhaust manifold upper heat shield bolts. Torque the bolts to 71 inch lbs. (8 Nm).
- Exhaust manifold lower front nuts. Torque the nuts to 15 ft. lbs. (20 Nm).
- Drive belt tensioner. Torque the bolts to 30 ft. lbs. (40 Nm).
- Catalytic converter nuts. Torque the nuts to 15 ft. lbs. (20 Nm).
- Exhaust manifold lower rear nuts. Torque the nuts to 15 ft. lbs. (20 Nm).
- Exhaust manifold lower heat shield. Torque the bolts to 71 inch lbs. (8 Nm).
- Coolant intake pipe. Torque the bolts to 15 ft. lbs. (20 Nm).
- Transmission assembly

6. Refill the engine coolant.

7. Start the vehicle and inspect for leaks.

Front Crankshaft Seal

REMOVAL AND INSTALLATION

1. Before servicing the vehicle, refer to the precautions in the beginning of this section.

2. Remove the timing belt and crankshaft gear.

3. Drill a small shallow hole into the steel ring of the seal.

➡**Use caution so as not to damage the area around and behind the seal.**

4. Insert a self-tapping screw.

5. Using pliers, pull the front crankshaft seal out.

To install:

6. Install or connect the following:
- New seal coated with grease using Seal Installer Tool (such as J35268-A)
- Crankshaft gear
- Timing belt

Camshaft

REMOVAL & INSTALLATION

1. Before servicing the vehicle, refer to the precautions in the beginning of this section.

2. Remove or disconnect the following:
- Negative battery cable
- Intake plenum
- Intake air resonator

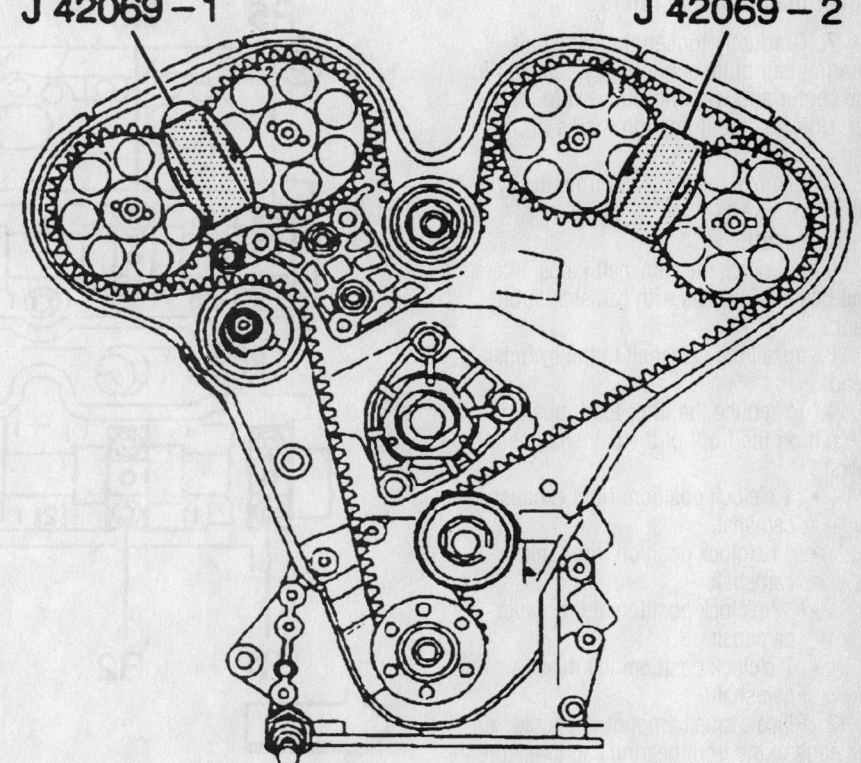

Camshaft gear holding tools must be installed before attempting to remove the gears from the camshafts

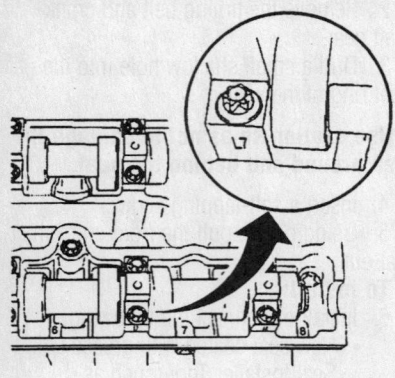

The camshaft bearing caps have identification marks stamped on the side

- Camshaft cover
- Front timing belt cover
- Timing belt

3. Rotate the crankshaft 60 degrees counterclockwise Before Top Dead Center (BTDC) to prevent valve/piston contact.

4. Use a Camshaft Locking tool to hold the camshaft gears in place and loosen the camshaft gear bolt.

5. Remove the locking tool from the gears.

6. Remove the camshaft gear(s).

➡Be certain that the camshaft is not under load from the lifters.

7. Gradually loosen the camshaft bearing cap bolts sequentially, starting in the center and working outward in a spiral. Note the identification marks on the caps.

8. Remove the camshaft from the cylinder head.

To install:

9. Lubricate the camshaft lobes, lifters and bearing journals with camshaft lubricant.

10. Install the camshaft to the cylinder head.

11. To reduce the lifter load, position the pin on the front of the camshaft as follows:

- 1 o'clock position: right exhaust camshaft
- 11 o'clock position: right intake camshaft
- 12 o'clock position: left exhaust camshaft
- 7 o'clock position: left intake camshaft

12. Place a small amount of Loctite® on the edge of the front bearing cap to ensure a good seal between the cap and surface of the cylinder head. Do not allow the sealer to get into the oil journal of the cap.

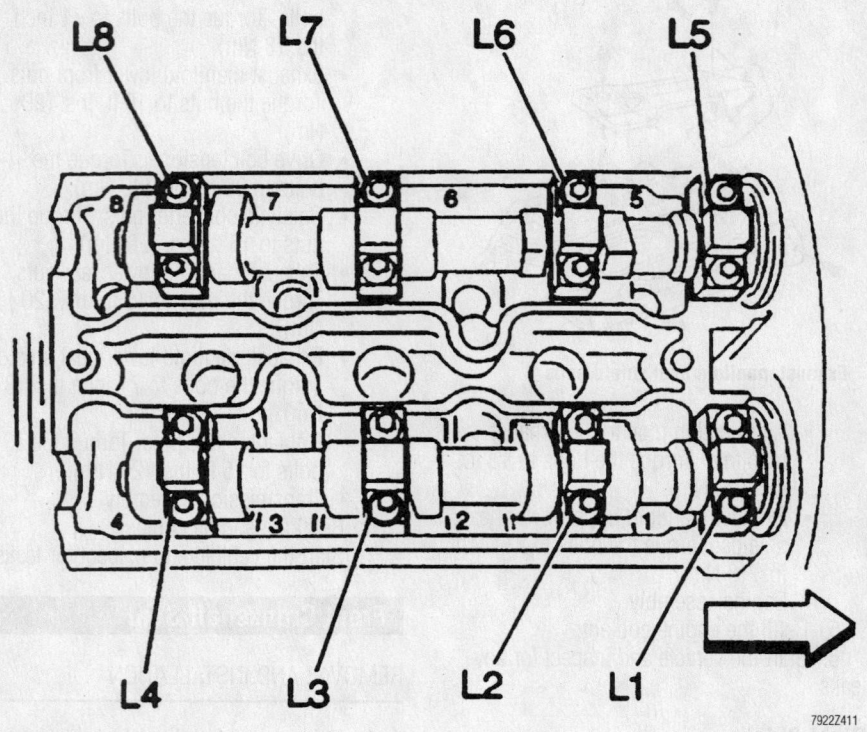

Camshaft bearing cap locations-right cylinder head

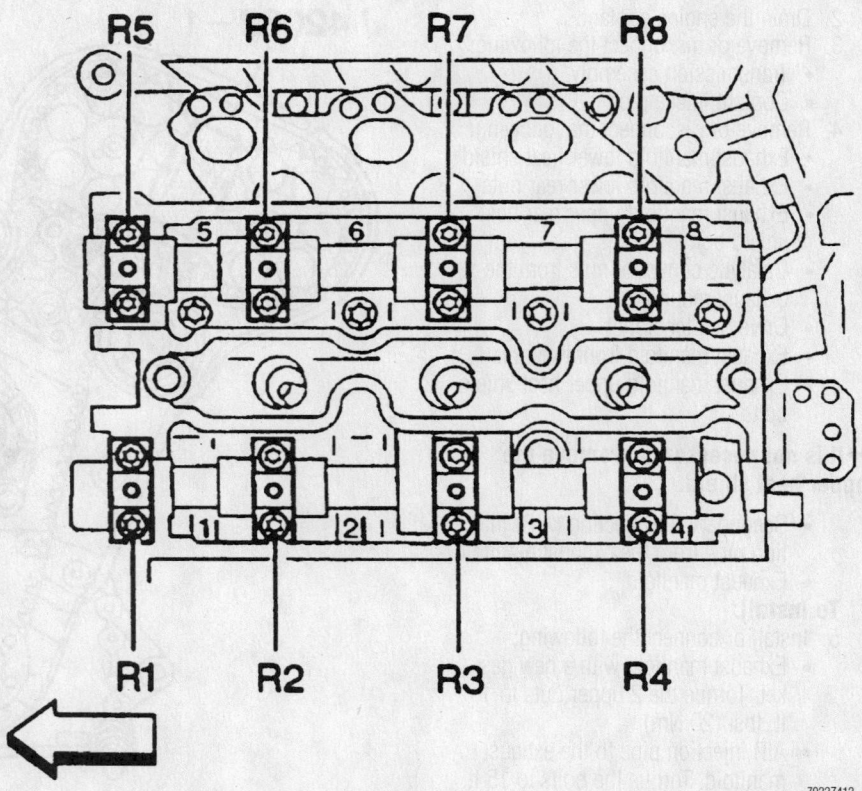

Camshaft bearing cap locations—left cylinder head

✻✻ WARNING

The bearing caps must be installed in their original positions.

13. Install or connect the following:
   - Camshaft bearing caps. Torque the bolts in sequence starting in the center and working outwards to 71 inch lbs. (8 Nm).
   - New camshaft seal lubricated with engine oil

➡**Make certain that the camshaft seal is fully seated.**

   - Camshaft gear. Tighten the new bolt to 37 ft. lbs. (50 Nm), plus a 60 degree turn, plus a 15 degree turn.

✻✻ WARNING

A new camshaft gear bolt must be installed. The required tightening method will stretch the bolt making the original bolt unusable.

   - Timing belt and adjust as needed
   - Front timing belt cover. Torque the bolts to 71 inch lbs. (8 Nm).
   - Camshaft cover. Torque the bolts to 71 inch lbs. (8 Nm).
   - Intake plenum. Torque the bolts to 71 inch lbs. (8 Nm).
   - Intake air resonator
   - Negative battery cable

14. Start the vehicle and verify the engine is running properly.

Valve Lash

ADJUSTMENT

➡**The valve lash is non-adjustable.**

Starter Motor

REMOVAL & INSTALLATION

1. Before servicing the vehicle, refer to the precautions in the beginning of this section.
2. Remove or disconnect the following:
   - Negative battery cable
   - Starter electrical connectors
   - Right hand catalytic converter
   - Right side engine mount nuts
   - Intake air resonator to the throttle body ducts
3. Install an Engine Support Fixture to the right side of the engine.
4. Raise the right side of the engine approximately 1.5 inches (38mm) to gain access to the starter motor bolts.

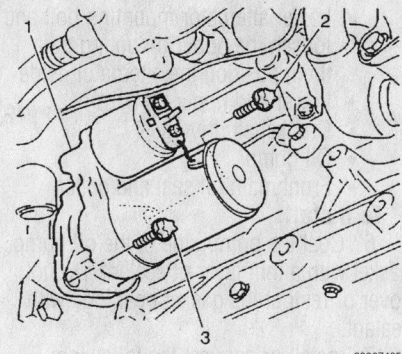

Remove the starter motor and bolts

5. Remove or disconnect the following:
   - Engine mount from the bracket and cradle
   - Engine mount bracket bolts and reposition the bracket
   - Starter motor

To install:

6. Install or connect the following:
   - Starter motor. Torque the bolts to 44 ft. lbs. (60 Nm).
   - Engine mount to the bracket and front crossmember. Torque the bolts to 30 ft. lbs. (40 Nm).

7. Lower the engine into position and make certain that the mount locator tab is fully seated in the front crossmember slot.
8. Remove the special tools from the engine.
9. Install or connect the following:
   - Engine mount nuts. Torque the upper nut to 30 ft. lbs. (40 Nm) and the lower nut to 41 ft. lbs. (55 Nm).
   - Intake air resonator to the throttle body ducts
   - Catalytic converter. Torque the bolt to 25 ft. lbs. (34 Nm) and the nuts to 18 ft. lbs. (25 Nm).
   - Starter electrical connectors. Torque the battery cable nut to 115 inch lbs. (13 Nm) and the starter solenoid nut to 35 inch lbs. (4 Nm).
   - Negative battery cable

Oil Pan

REMOVAL & INSTALLATION

Lower

1. Before servicing the vehicle, refer to the precautions in the beginning of this section.
2. Drain the engine oil.
3. Remove or disconnect the following:
   - Negative battery cable

   - Splash shield
   - Oil level sensor wiring C-clip from the oil pan housing
   - Oil level sensor electrical connector
   - Oil pan from the oil pan housing

To install:

4. Clean the oil pan and housing sealing surfaces with a non-abrasive cleaner.
5. Install or connect the following:
   - Oil level sensor wire connector to the oil pan housing C-clip
   - Oil pan with a new gasket. Torque the bolts to 71 inch lbs. (8 Nm).
   - Oil level sensor electrical connector
   - Splash shield. Torque the bolts until they are fully seated.
   - Negative battery cable
6. Refill the engine oil.
7. Start the vehicle and check for leaks.

Upper

1. Before servicing the vehicle, refer to the precautions in the beginning of this section.
2. Remove or disconnect the following:
   - Negative battery cable
   - Lower oil pan
   - Engine mount lower nuts
   - A/C compressor hose strap from the oil pan
   - Transmission bolts from the oil pan
   - All but the 4 corner bolts from the oil pan
   - Propeller shaft
3. Support the propeller shaft out of the way.
4. Remove or disconnect the following:
   - Catalytic converter
   - Idler arm bolts and lower the relay rod out of the way
5. Install an engine support fixture and raise the engine slightly to allow removal of the oil pan.
6. Remove or disconnect the following:
   - Remaining oil pan bolts
   - Oil intake pipe
   - Oil pan from the vehicle

To install:

7. Apply a bead of silicone sealer in the bottom of the upper oil pan groove.
8. Install a new rubber seal into the groove in the pan.
9. Apply a 3mm (0.12 in) bead of silicone sealant on the OUTSIDE edge of the seal, at the front of the upper oil pan and apply an identical bead on the INSIDE edge of the seal at the rear of the upper oil pan. The beads will overlap at the middle of the upper oil pan.
10. Install or connect the following:
   - Upper oil pan

- Oil intake pipe. Torque the bolt to 71 inch lbs. (8 Nm).
- Four oil pan corner bolts finger tight

11. Lower the engine into place.

12. Install or connect the following:
- Remaining oil pan bolts. Torque the bolts to 11 ft. lbs. (15 Nm).
- Transmission-to-oil pan bolts. Torque the bolts to 30 ft. lbs. (40 Nm).
- Idler arm bolts. Torque the bolts to 44 ft. lbs. (60 Nm).
- Catalytic converter. Torque the nuts to 18 ft. lbs. (25 Nm).
- Propeller shaft. Torque the bolts to 70 ft. lbs. (95 Nm).
- A/C compressor hose strap. Torque the bolt to 71 inch lbs. (8 Nm).
- Engine mount lower nuts. Torque the nuts to 41 ft. lbs. (55 Nm).
- Lower oil pan

13. Refill the engine oil.

Oil Pump

REMOVAL & INSTALLATION

1. Before servicing the vehicle, refer to the precautions in the beginning of this section.
2. Drain the engine oil.
3. Drain the engine coolant.
4. Discharge and recover the A/C system.
5. Remove or disconnect the following:
- Negative battery cable
- Intake air resonator
- Front timing belt cover
- Intake plenum
- Timing belt
- Rear timing belt cover
- A/C compressor
- A/C compressor and power steering pump bracket and position it out of the way of the oil pump housing

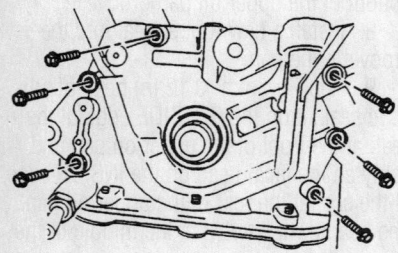

Oil pump mounting bolt locations

- Lower alternator mounting bolt and loosen the upper mounting bolt then, position the alternator aside
- Lower and upper oil pans
- Crankshaft drive gear
- Oil pump
- Front main oil seal and collar

To install:

6. Coat the pump side of the oil pump gasket with a thin layer of sealant. Do not cover or restrict the gasket openings with sealant.

7. Install or connect the following:
- Oil pump collar
- New front main oil seal
- Oil pump with a new gasket by aligning the guide pins. Torque the bolts to 9 ft. lbs. (12 Nm).
- Crankshaft gear. Torque the bolt to 184 ft. lbs. (250 Nm) plus an additional 60 degree turn.
- Upper oil pan. Torque the bolts to 11 ft. lbs. (15 Nm).
- Lower oil pan. Torque the bolts to 71 inch lbs. (8 Nm).
- Alternator. Torque the upper and lower bolts to 30 ft. lbs. (40 Nm).

8. Re-tighten the oil pump bolts to 15 ft. lbs. (20 Nm).

9. Install or connect the following:
- A/C compressor and power steering pump bracket. Torque the fasteners to 30 ft. lbs. (40 Nm).
- A/C compressor. Torque the bolts to 30 ft. lbs. (40 Nm).
- Rear timing belt cover. Torque the bolts to 71 inch lbs. (8 Nm).
- Timing belt
- Front timing belt cover. Torque the bolts to 71 inch lbs. (8 Nm).
- Intake plenum. Torque the bolts to 71 inch lbs. (8 Nm).
- Intake air resonator
- Negative battery cable

10. Fill the engine oil.
11. Fill the coolant.
12. Recharge the A/C system.
13. Start the vehicle and checks for leaks.

Rear Main Seal

REMOVAL & INSTALLATION

1. Before servicing the vehicle, refer to the precautions in the beginning of this section.
2. Remove or disconnect the following:
- Negative battery cable.
- Flywheel

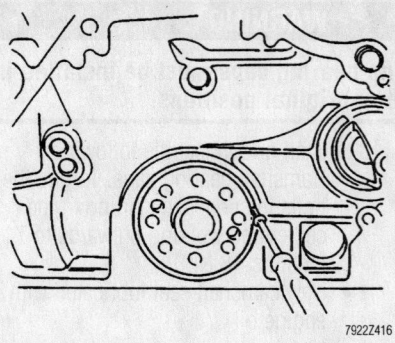

Thread a self-tapping screw into the metal part of the seal and remove the seal

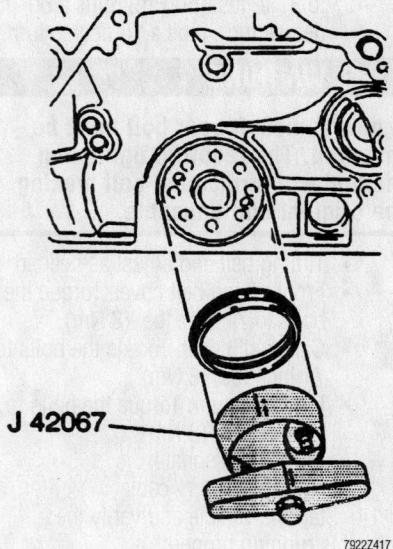

J 42067

Press the rear main seal into position using a threaded seal installer like J 42067

3. Center punch the steel ring of the rear main seal.

4. Drill a small hole into the steel ring.

➡**Make certain not to damage any engine components on the opposite side of the seal.**

5. Install a small self tapping screw.
6. Remove the rear main oil seal.

To install:

7. Clean all sealing surfaces with a non-abrasive cleaner.

8. Install or connect the following:
- New seal lubricated with chassis grease
- Rear main oil seal using an Oil Seal Installer tool J 42067
- Flywheel. Torque the new bolts to 48 ft. lbs. (65 Nm), plus a 30 degree turn, plus a 15 degree turn.
- Negative battery cable

TIMING BELT

Timing Belt

REMOVAL & INSTALLATION

1. Before servicing the vehicle, refer to the precautions in the beginning of this section.

2. Remove or disconnect the following:
- Negative battery cable
- Intake air resonator
- Intake plenum
- Splash shield
- Secondary air injection (AIR) pipe
- Accessory drive belt
- Water pump and power steering pump pulleys
- Drive belt tensioner
- Front timing belt cover
- Crankshaft balancer

3. Rotate the crankshaft clockwise to 60 degrees before top dead center (BTDC).

TIMING BELT INSTALLATION AND ADJUSTMENT TABLE

Step	Action	Value	Yes	No
1	Install the timing belt and align marks on the belt with the marks on the camshaft gears and the crankshaft gear. Check the timing belt deflection between the idler pulley for camshafts 3 & 4 and camshaft number 4. Is the timing belt installed, the marks aligned and the timing belt deflection adjusted?	1 cm (0.4 in) maximum	Go to Step 2	—
2	Set the initial timing belt tension at the timing belt tensioner. Is the initial timing belt tension set?	—	Go to Step 3	—
3	Rotate the engine two complete revolutions and secure the crankshaft at Top Dead Center (TDC) with the J 42069-10. Has the engine been rotated and the crankshaft secured to TDC?	—	Go to Step 4	—
4	Starting with camshafts 3 and 4, check the alignment of the marks on the camshaft gears with the marks on the J 42069-20 checking gauge. Do the marks on the camshaft gears align exactly with the marks on J 42069-20?	—	Go to Step 5	Go to Step 6
5	Check the alignment of the marks on camshafts gears 1 and 2 with the marks on the J 42069-20 checking gauge. Do the marks on the camshaft gears align exactly with the marks on J 42069-20?	—	Go to Step 14	Go to Step 10
6	Do the camshaft gear marks line up to the left (BTDC) of the marks on the J 42069-20 checking gauge?	—	Go to Step 8	Go to Step 7
7	Do the camshaft gear marks line up to the right (ATDC) of the marks on the J 42069-20 checking gauge?	—	Go to Step 9	—
8	Turn the idler pulley eccentric, for camshafts 3 and 4, counterclockwise until the marks on the camshaft gear align exactly with the marks on J 42069-20. Rotate the engine two complete revolutions, lock the crankshaft at TDC with J 42069-10 and recheck the alignment of the camshaft gear marks to the marks on J 42069-20. Do the marks on the camshaft gears align exactly with the marks on J 42069-20?	—	Go to Step 5	Go to Step 6
9	Turn the idler pulley eccentric, for camshafts 3 and 4, clockwise until the marks on the camshaft gear align exactly with the marks on J 42069-20. Rotate the engine two complete revolutions, lock the crankshaft at TDC with J 42069-10 and recheck the alignment of the camshaft gear marks to the marks on J 42069-20. Do the marks on the camshaft gears align exactly with the marks on J 42069-20?	—	Go to Step 5	Go to Step 6

79225G35

Timing belt installation and adjustment table—GM 3.0L (VIN R) engine

Step	Action	Value	Yes	No
10	Do the camshaft gear marks line up to the left (BTDC) of the marks on the J 42069-20 checking gauge?	—	Go to Step 12	Go to Step 11
11	Do the camshaft gear marks line up to the right (ATDC) of the marks on the J 42069-20 checking gauge?	—	Go to Step 13	—
12	Turn the idler pulley eccentric, for camshafts 1 and 2, counterclockwise until the marks on the camshaft gear align exactly with the marks on J 42069-20. Rotate the engine two complete revolutions, lock the crankshaft at TDC with J 42069-10 and recheck the alignment of the camshaft gear marks to the marks on J 42069-20. Do the marks on the camshaft gears align exactly with the marks on J 42069-20?	—	Go to Step 14	Go to Step 10
13	Turn the idler pulley eccentric, for camshafts 1 and 2, clockwise until the marks on the camshaft gear align exactly with the marks on J 42069-20. Rotate the engine two complete revolutions, lock the crankshaft at TDC with J 42069-10 and recheck the alignment of the camshaft gear marks to the marks on J 42069-20. Do the marks on the camshaft gears align exactly with the marks on J 42069-20?	—	Go to Step 14	Go to Step 10
14	Set the final timing belt tension at the timing belt tensioner. Is the final timing belt tension set?	—	Go to Step 15	—
15	Again, rotate the engine two complete revolutions and lock the crankshaft at TDC. Do a final inspection of the camshaft gear marks' relationship to the J 42069-20 marks. The marks must align exactly. Do the marks on the camshaft gears align exactly with the marks on the J 42069-20?	—	Go to Step 16	Go to Step 2
16	Remove all checking tools and ensure all idler pulleys and the tensioner locking nut are tightened to specifications. Continue with re-assembly of the engine.	—	—	—

79225G36

Timing belt installation and adjustment table (continued)—GM 3.0L (VIN R) engine

4. Loosen the timing belt tensioner and idler pulleys.

5. Remove the timing belt.

To install:

6. Start at the crankshaft sprocket and install the timing belt with the double dash (TDC) aligned with the marks on the oil pump and on the belt drive gear.

7. Set the initial timing belt tension as follows:

a. Turn the tensioner nut COUNTER-CLOCKWISE to full stop, then, turn the nut back until the reference mark is 1 mm (0.003 in) over the flange.

b. Tighten the timing belt tensioner locking nut until snug, the locking nut will be tightened to specifications after all final adjustments are made.

8. Rotate the engine in the clockwise direction two revolutions stopping at 60 degrees BTDC.

9. Inspect the alignment of the reference marks on the camshaft gears with the

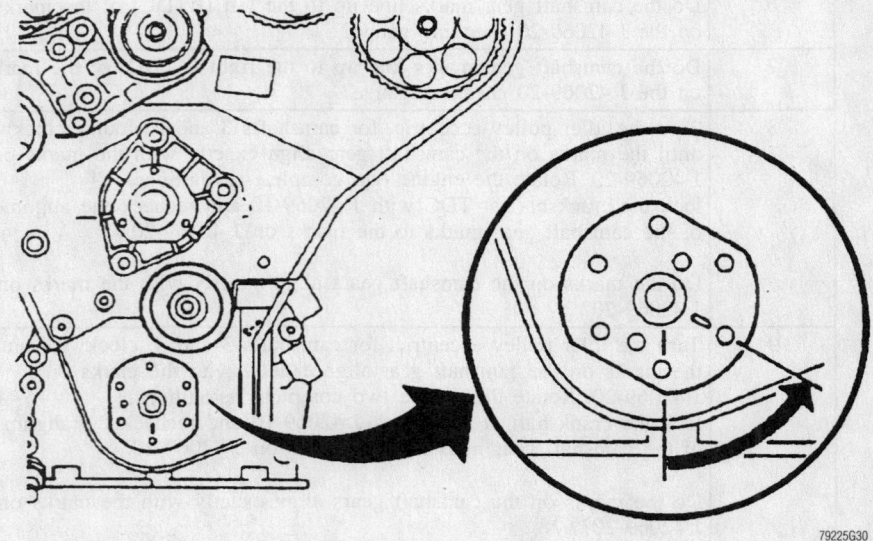

79225G30

Crankshaft alignment to 60 degrees BTDC—GM 3.0L (VIN R) engine

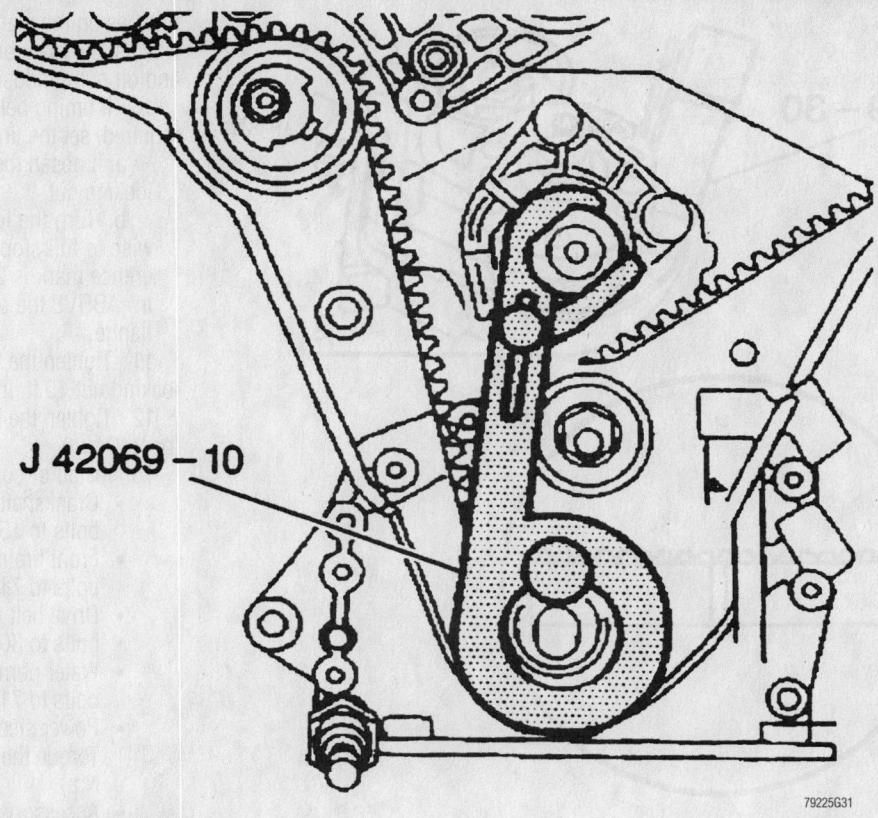

J 42069 – 10

79225G31

Securing the crankshaft—GM 3.0L (VIN R) engine

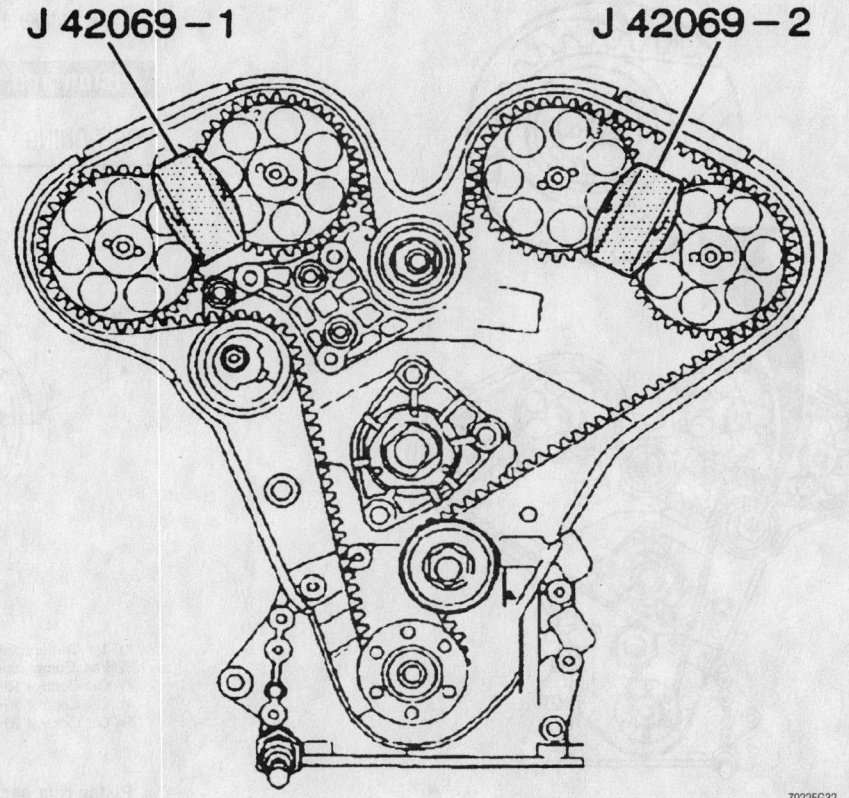

J 42069 – 1   J 42069 – 2

79225G32

Locking the camshaft—GM 3.0L (VIN R) engine

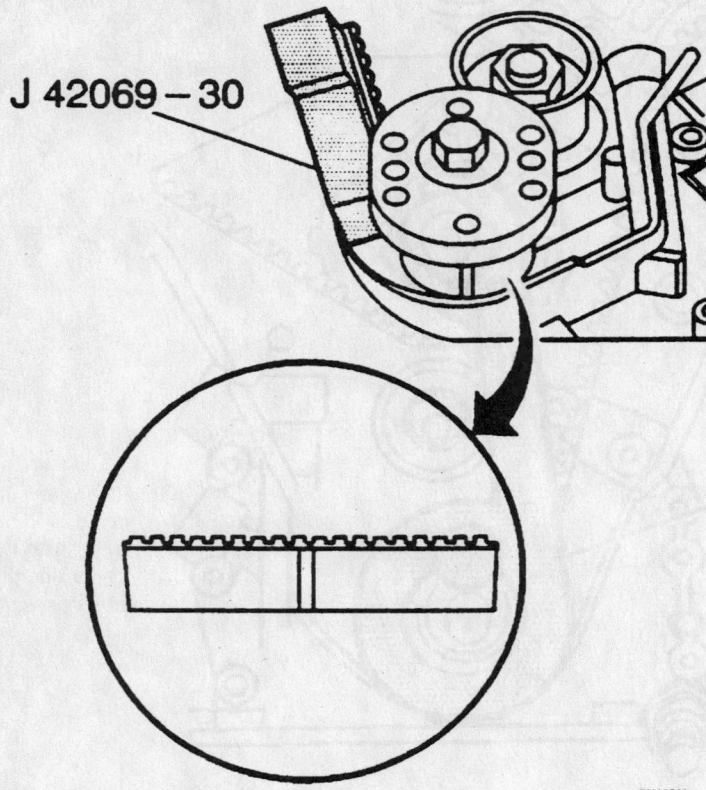

J 42069 – 30

79225G33

Using the tool to pin the timing belt—GM 3.0L (VIN R) engine

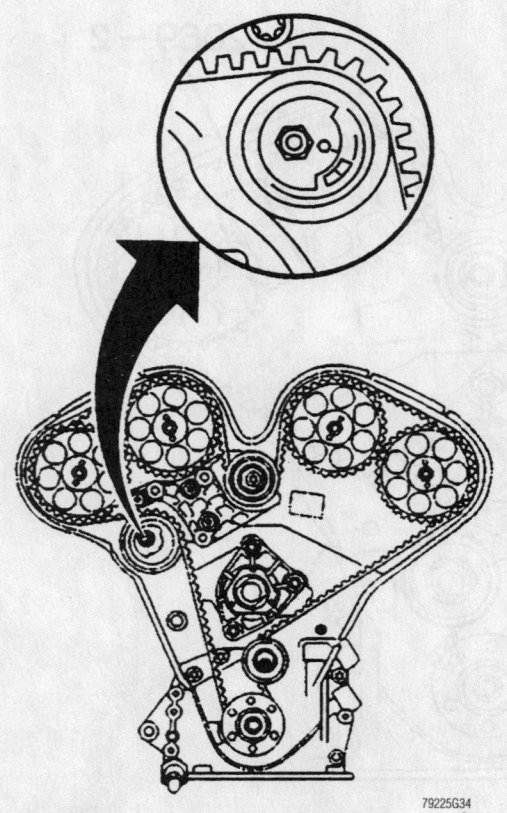

79225G34

Initial timing belt tension adjustment—GM 3.0L (VIN R) engine

notches on the rear timing belt cover, as well as, the mark on the crankshaft sprocket and oil pump housing.

10. If timing belt adjustment IS NOT required, set the final timing belt tension:

a. Loosen the timing belt tensioner locking nut.

b. Turn the locking nut counterclockwise to full stop, then back until the reference mark is 2-4 mm (0.078-0.157 in) ABOVE the reference mark on the flange.

11. Tighten the timing belt tensioner locking nut 15 ft. lbs. (20 Nm).

12. Tighten the idler pulley bolts to 30 ft. lbs. (40 Nm).

13. Install or connect the following:

- Crankshaft balancer. Torque the bolts to 15 ft. lbs. (20 Nm).
- Front timing belt cover. Torque the bolts to 71 inch lbs. (8 Nm).
- Drive belt tensioner. Torque the bolts to 30 ft. lbs. (40 Nm).
- Water pump pulley. Torque the bolts to 71 inch lbs. (8 Nm).
- Power steering pump pulley. Torque the bolts to 15 ft. lbs. (20 Nm).
- Accessory drive belt
- AIR injection pipe
- Splash shield
- Intake plenum
- Intake air resonator

Piston and Ring

POSITIONING

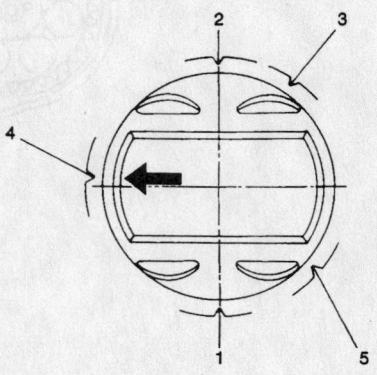

(1) 1st Compression Ring End Gap Location
(2) 2nd Compression Ring End Gap Location
(3) Oil Control Ring Upper Ring End Gap Location
(4) Oil Control Ring Spacer End Gap Location
(5) Oil Control Ring Lower Ring End Gap Location

7922AG55

Piston ring end-gap spacing—Catera 3.0L engine

FUEL SYSTEM

Fuel System Service Precautions

Safety is the most important factor when performing not only fuel system maintenance but any type of maintenance. Failure to conduct maintenance and repairs in a safe manner may result in serious personal injury or death. Maintenance and testing of the vehicle's fuel system components can be accomplished safely and effectively by adhering to the following rules and guidelines.

• To avoid the possibility of fire and personal injury, always disconnect the negative battery cable unless the repair or test procedure requires that battery voltage be applied.

• Always relieve the fuel system pressure prior to disconnecting any fuel system component (injector, fuel rail, pressure regulator, etc.), fitting or fuel line connection. Exercise extreme caution whenever relieving fuel system pressure, to avoid exposing skin, face and eyes to fuel spray. Please be advised that fuel under pressure may penetrate the skin or any part of the body that it contacts.

• Always place a shop towel or cloth around the fitting or connection prior to loosening to absorb any excess fuel due to spillage. Ensure that all fuel spillage (should it occur) is quickly removed from engine surfaces. Ensure that all fuel soaked cloths or towels are deposited into a suitable waste container.

• Always keep a dry chemical (Class B) fire extinguisher near the work area.

• Do not allow fuel spray or fuel vapors to come into contact with a spark or open flame.

• Always use a backup wrench when loosening and tightening fuel line connection fittings. This will prevent unnecessary stress and torsion to fuel line piping. Always follow the proper torque specifications.

• Always replace worn fuel fitting O-rings with new. Do not substitute fuel hose or equivalent, where fuel pipe is installed.

Fuel System Pressure

RELIEVING

1. Before servicing the vehicle, refer to the precautions in the beginning of this section.

2. Loosen the fuel filler cap to relieve the tank pressure.

3. Remove or disconnect the following:
   • Negative battery cable
   • Intake manifold top cover

4. Install a Fuel Pressure Gauge to the fuel pressure fitting. Wrap a shop towel around the fitting while installing the gauge.

5. Connect a bleed hose into an approved container and open the valve to bleed the system.

6. Close the valve and disconnect the gauge.

7. Drain any remaining fuel from the gauge into the approved container.

Fuel Filter

REMOVAL & INSTALLATION

1. Before servicing the vehicle, refer to the precautions in the beginning of this section.

2. Loosen the fuel filler cap to relieve pressure in the tank.

3. Relieve the fuel system pressure.

4. Remove or disconnect the following:
   • Fuel feed and return lines
   • Fuel filter from the bracket

To install:

5. Clean the bolts and fittings on both fuel lines.

6. Install or connect the following:
   • Filter in the retaining strap making certain that the flow is in the proper direction
   • Both fuel lines. Torque the fittings to 18 ft. lbs. (25 Nm).
   • Fuel filter bracket mounting bolt. Torque the bolt to 13 ft. lbs. (18 Nm).
   • Fuel filler cap

7. Crank the engine for a few seconds and check for leakage.

Fuel Pump

REMOVAL & INSTALLATION

1. Before servicing the vehicle, refer to the precautions in the beginning of this section.

2. Relieve the fuel system pressure.

3. Drain the fuel tank.

4. Remove or disconnect the following:
   • Fuel tank
   • Fuel tank pressure sensor
   • Fuel tank connection cover electrical connector
   • Spring loaded clamp around the tank boot
   • Tank boot from the housing

➡ The fuel pump assembly may spring up from its position. The reservoir bucket on the assembly is full of fuel. Tip the assembly slightly so the float is not damaged during removal.

   • Fuel sender locking nut from the tank using a Fuel Tank Sender Wrench (J 42219)
   • Fuel pump electrical connectors
   • Fuel tank connection cover
   • Hose from the pump
   • Fuel pump from the tank by pushing the tabs inward
   • Locking ring by pushing the tabs in and pushing the locking ring from the housing simultaneously
   • Fuel pump from the housing
   • Lip seal from the cover by sliding it downward, past the reservoir and over the float arm and discard it

5. Drain the remaining fuel from the reservoir into an approved container.

To install:

6. Install or connect the following:
   • New lip seal lubricated with engine oil
   • Fuel pump into the housing
   • Locking ring into the housing
   • New fuel inlet screen
   • Fuel pump assembly into the tank
   • Fuel line to the pump using a new clamp
   • Fuel pump electrical connectors
   • Fuel tank connection cover
   • Fuel sender locking nut. Torque the nut to 37 ft. lbs. (50 Nm).

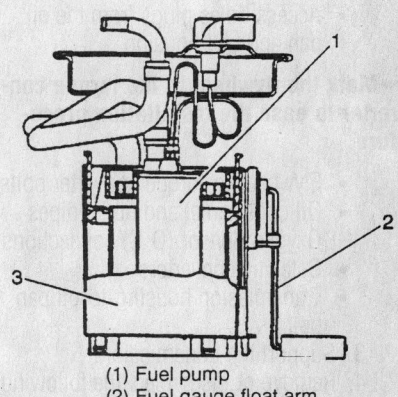

(1) Fuel pump
(2) Fuel gauge float arm
(3) Fuel reservoir

79222419

Fuel pump module components

- Fuel tank boot to the housing
- Fuel tank pressure sensor. Torque the fastener to 18 inch lbs. (4 Nm).
- Air reference hose to the pressure sensor
- Fuel feed and return lines using new clamps
- Fuel tank pressure sensor and connection cover electrical connectors
- Fuel tank to the vehicle. Torque the bolts to 22 ft. lbs. (30 Nm).
- Negative battery cable

7. Refill the fuel tank.
8. Start the vehicle and inspect for leaks.

Fuel Injector

REMOVAL & INSTALLATION

※※ CAUTION

Remove the fuel rail assembly carefully to prevent damage to the injector electrical connector terminals and spray tips. Support the fuel rail after it is removed in order to avoid damaging the fuel rail components. Cap the fittings and plug the holes when servicing the fuel system to prevent debris from entering open ports.

Fuel injection systems often remain pressurized, even after the engine has been turned **OFF**. The fuel system pressure must be relieved before disconnecting any fuel lines. Failure to do so may result in fire and/or personal injury.

1. Before servicing the vehicle, refer to the precautions in the beginning of this section.
2. Relieve fuel system pressure.
3. Remove or disconnect the following:

- Intake plenum
- Fuel rail supply and return lines from the fuel rail
- Fuel injector electrical connectors
- Harness from the fuel rail
- Fuel rail attaching bolts from the intake manifold
- Fuel pressure regulator vacuum lines
- Fuel rail assembly
- Injectors from the fuel rail and discard the O-rings

To install:

4. Install or connect the following:

- New O-rings lubricated with clean engine oil onto the fuel injectors
- Fuel injectors with new retaining clips
- Fuel rail to the intake manifold. Torque the bolt to 71 inch lbs. (8 Nm).
- Fuel pressure regulator vacuum line
- Fuel injector electrical connectors
- Harness to the fuel rail
- Fuel supply/return lines to the fuel rail. Torque the fasteners to 11 ft. lbs. (15 Nm).
- Fuel line bracket. Torque the bolts to 37 ft. lbs. (50 Nm).
- Intake plenum. Torque the bolts to 71 inch lbs. (8 Nm).
- Negative battery cable

5. Crank the engine several times to pressurize the system.
6. Check for leaks and repair, if necessary.

DRIVE TRAIN

Transmission Assembly

REMOVAL & INSTALLATION

1. Before servicing the vehicle, refer to the precautions in the beginning of this section.
2. Remove or disconnect the following:
- Negative battery cable
- Shift linkage from the transmission
- Propeller shaft coupling bolts
- Propeller shaft coupling from the drive flange
- Access holes plugs from the oil pan and bell housing

➡**Mark the flywheel to the torque converter to ease the installation procedure.**

- Flywheel-to-torque converter bolts
- Oil cooler inlet and outlet pipes
- Oxygen Sensor (O2S) connections
- Catalytic converters
- Transmission housing-to-oil pan bolts

3. Support the transmission.
4. Remove or disconnect the following:
- Transmission crossmember to mount nuts
- Transmission crossmember
- Vent hose

- Electrical connector from the transmission control selector switch
- Electrical connector from the adapter case
- Electrical connector from the main case
- Electrical connector from the transmission speed sensor

5. Support the front of the engine.
6. Lower the transmission to gain access to the upper housing bolts.
7. Remove the upper transmission housing bolts.
8. Carefully remove the transmission from the vehicle.

To install:

※※ WARNING

The transmission fluid cooler and lines must be flushed out prior to installing a new or reconditioned transmission.

9. Apply thread locking compound to the transmission-to-engine mounting bolts.
10. Install or connect the following:
- Transmission. Torque the transmission-to-engine bolts to 44 ft. lbs. (60 Nm).
- Electrical connector to the transmission control selector switch

- Electrical connector to the adapter case
- Electrical connector to the main case
- Electrical connector to the speed sensor
- Vent hose
- Crossmember. Torque the bolts to 33 ft. lbs. (45 Nm).
- Crossmember-to-mount and torque the nuts to 15 ft. lbs. (20 Nm)

11. Remove the transmission support.
12. Install or connect the following:
- Transmission housing-to-oil pan bolts and torque the bolts to 15 ft. lbs. (20 Nm)
- Catalytic converters. Torque the bolts to 18 ft. lbs. (25 Nm).
- O2S connectors
- Oil cooler inlet and outlet pipes to the center pipes
- Torque converter to the flywheel by aligning the matchmarks. Torque the new bolts to 22 ft. lbs. (30 Nm).

➡**Be sure the torque converter weld nuts are flush with the flywheel.**

- Oil pan access hole plugs
- Bell housing access hole plugs
- Propeller shaft coupling-to-drive flange an torque the bolts to 70 ft. lbs. (95 Nm)

- Shift linkage to the transmission
- Negative battery cable

13. Adjust the shift lever rod as follows:

a. Place the shift control lever in the **P** position.

b. Loosen the adjusting bolt on the rod.

c. Hold the selector lever on the transmission toward the rear stop to eliminate any free-play.

d. Tighten the adjusting bolt to 71 inch lbs. (8 Nm).

14. Refill the transmission. Tighten the plug to 33 ft. lbs. (45 Nm).

15. Start the vehicle and check for leaks. Be sure the vehicle will only start in the **P** and **N** positions.

Axle Shaft, Bearing and Seal

REMOVAL & INSTALLATION

1. Before servicing the vehicle, refer to the precautions in the beginning of this section.

2. Place the gear selector in **N**.

3. Remove or disconnect the following:
- Rear wheel
- Drive axle flange bolts
- Outer end of the drive axle from the wheel bearing/hub inner flange

➡**Make certain that the drive axle separator tool is properly aligned to the differential and drive axle. The tool is clearly marked "Differential Side". If installed incorrectly, damage to the Anti-lock Brake System (ABS) sensor reluctor ring is possible.**

4. Remove the axle shaft.

5. Use a suitable prytool to remove the axle seal from the differential.

To install:

6. Use a Seal Installation Tool (such as J 26234) to install the new seal.

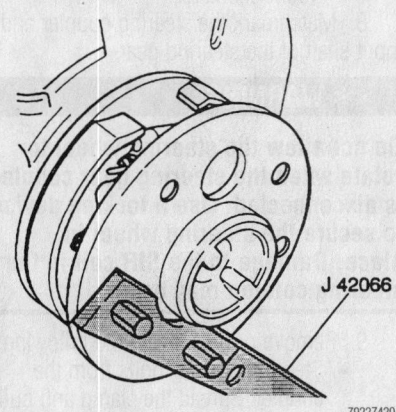

J 42066

7922Z420

Prevent the outer hub from turning by installing a tool such as Hub Holding Tool J 42066

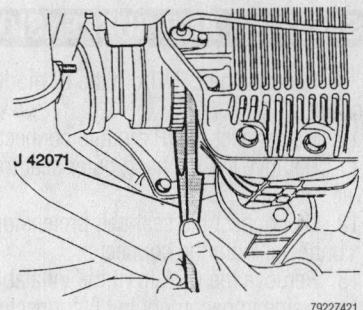

J 42071

7922Z421

Pry the halfshaft out of the differential

7. Lubricate the splines of the halfshaft and the sealing surfaces with differential oil.

8. Install or connect the following:
- Axle shaft into the differential bore by aligning the splines

➡**If necessary use a soft faced tool, such as a rubber mallet, to seat the halfshaft fully in the differential bore.**

- Outer end of the axle shaft into the wheel bearing/hub inner flange. Torque the bolts to 37 ft. lbs. (50 Nm) plus an additional 67 degree turn.
- Rear wheel

9. Road test the vehicle.

CV-Joints

REMOVAL & INSTALLATION

Inner & outer Joint

1. Before servicing the vehicle, refer to the precautions in the beginning of this section.

2. Disconnect the negative battery cable.

3. Remove the axle shaft and place it in a vise.

4. Remove the large boot retaining clamp.

5. Remove the small boot retaining clamp.

6. Slide the boot up the shaft away from the joint.

7. With a brass drift and a hammer tap on the inner race of the cross grove.

8. Remove the CV joint retaining ring.

9. Remove the CV boot.

➡**The joint is not serviceable, if damaged it must be replaced.**

To install:

10. Pack the CV joint with ½ of the grease provided in the service kit.

11. Install new small and large boot retaining clamps on the ends of the boot.

12. Install a new boot on the halfshaft.

13. Install a new joint retaining ring onto the halfshaft.

14. Place halfshaft into a shop press and press the joint into place.

15. Slide the CV boot over the joint .

16. Using a crimping tool, a breaker bar and a torque wrench, torque the small retaining clamp to 100 ft. lbs. (140 Nm).

17. Place the remaining grease into the CV boot.

18. Using a crimping tool a breaker bar and a torque wrench, torque the large retaining clamp to 130 ft. lbs. (176 Nm).

19. Install the halfshaft in the vehicle.

Pinion Seal

REMOVAL & INSTALLATION

1. Before servicing the vehicle, refer to the precautions in the beginning of this section.

2. Place the transmission in **N**.

3. Remove or disconnect the following:
- Underbody heat shields
- Bolts from the propeller shaft center bearing bracket
- Propeller shaft from the front and rear propeller shaft couplings
- Rear propeller shaft coupling

4. Mark the position of the gear yoke, gear and nut for correct bearing pre-load.

5. Remove or disconnect the following:
- Drive pinion flange nut
- Drive pinion flange using a gear puller
- Pinion seal from the rear axle housing by using a blunt chisel to drive the seal out

To install:

6. Apply lubricant to the outer diameter of the gear yoke and the outer lip of the new seal.

7. Install or connect the following:
- New seal lubricated with chassis grease
- Drive pinion flange
- Drive pinion gear nut and hand-tighten

8. Tighten the nut to the position marked in the removal procedure. Tighten the nut an additional 0.062 inch (1.59mm).

9. Install or connect the following:
- Rear propeller shaft coupling
- Propeller shaft center bearing bracket. Torque the bolts to 15 ft. lbs. (20 Nm).
- Propeller shaft to the rear coupling. Torque the bolts to 70 ft. lbs. (95 Nm).
- Propeller shaft to the front coupling. Torque the bolts to 70 ft. lbs. (95 Nm).
- Underbody heat shields. Torque the bolts to 18 inch lbs. (2 Nm).

10. Road test the vehicle.

STEERING AND SUSPENSION

Air Bag

✳✳ CAUTION

These vehicles are equipped with an air bag system, known as a Supplemental Inflatable Restraint (SIR) system. The system must be disabled before performing service on or around system components, steering column, instrument panel components, wiring and sensors. Failure to follow safety and disabling procedures could result in accidental air bag deployment, possible personal injury and unnecessary system repairs.

PRECAUTIONS

Several precautions must be observed when handling the inflator module to avoid accidental deployment and possible personal injury.

- Never carry the inflator module by the wires or connector on the underside of the module.
- When carrying a live inflator module, hold securely with both hands, and ensure that the bag and trim cover are pointed away.
- Place the inflator module on a bench or other surface with the bag and trim cover facing up.
- With the inflator module on the bench, never place anything on or close to the module which may be thrown in the event of an accidental deployment.

DISARMING

1. Disconnect the negative battery cable.
2. Turn the steering wheel to align the wheels in the straight-ahead position.
3. Turn the ignition switch to the **LOCK** position and remove the key.
4. Remove the two TORX® screws from the back of the steering wheel.
5. Remove the inflatable restraint steering wheel module from the steering wheel.
6. Remove the Connector Position Assurance (CPA) from the inflatable restraint steering wheel module 2-way connector.
7. Disconnect the inflatable restraint steering wheel module 2-way connector
8. Remove the inflatable restraint Instrument Panel (IP) module cover.
9. Remove the CPA from the IP module connector.
10. Disconnect the IP module connector.
11. Remove the left hand outer seat track cover.
12. Disconnect the seat belt pretensioner LF connector from the connector.
13. Remove the CPA from the inflatable restraint side impact module LF connector.
14. Disconnect the side impact module LF connector.
15. Remove the right hand outer seat track cover.
16. Disconnect the seat belt pretensioner RF connector from the connector.
17. Remove the CPA from the inflatable restraint side impact module RF connector.
18. Disconnect the side impact module RF connector.

REARMING

1. Turn the steering wheel so that the wheels are pointing straight ahead.
2. Turn the ignition switch to the **LOCK** position and remove the key.
3. Connect the side impact module RF connector.
4. Install the Connector Position Assurance (CPA) from the inflatable restraint side impact module RF connector.
5. Connect the seat belt pretensioner RF connector to the connector.
6. Install the right hand outer seat track cover.
7. Connect the side impact module LF connector.
8. Install the CPA from the inflatable restraint side impact module LF connector.
9. Connect the seat belt pretensioner LF connector to the connector.
10. Install the left hand outer seat track cover.
11. Connect the IP module connector.
12. Install the CPA to the IP module connector.
13. Install the inflatable restraint Instrument Panel (IP) module cover.
14. Connect the inflatable restraint steering wheel module 2-way connector
15. Connect the CPA to the inflatable restraint steering wheel module 2-way connector.
16. Install the inflatable restraint steering wheel module to the steering wheel.
17. Install the two TORX® screws to the back of the steering wheel and tighten to 72 inch lbs. (8 Nm).
18. Reconnect the negative battery cable.

19. Wait at least 1 minute for the capacitors to recharge.
20. While staying away from the inflator modules, turn the key to the **RUN** position. The air bag warning lamp should turn ON for 3–4 seconds, then turn OFF.

Recirculating Ball Power Steering Gear

REMOVAL & INSTALLATION

1. Before servicing the vehicle, refer to the precautions in the beginning of this section.
2. Center the front wheels, then turn the ignition key to the **LOCK** position.
3. Drain the coolant.
4. Evacuate the A/C system.
5. Siphon the power steering fluid from the reservoir.
6. Siphon the brake fluid from the reservoir.
7. Remove or disconnect the following:
- Negative battery cable
- Windshield wiper assembly
- 3 body harness electrical connectors
- Electronic Control Module (ECM) from the electrical box
- Upper radiator hose
- Evaporative line extension bolt
- Power steering fluid reservoir and bracket
- Brake booster vacuum connection from the intake plenum
- Brake pipes from the master cylinder
- Electrical connector from the master cylinder reservoir cap
- Sound insulator
8. Matchmark the steering coupler and input shaft of the steering gear.

✳✳ WARNING

Do not allow the steering wheel to rotate when the steering gear coupler is disconnected. Use a locking devise to secure the steering wheel in place. Damage to the SIR coil in the steering column may occur.

9. Remove or disconnect the following:
- Steering coupler bolts from the coupler. Spread the clamp and pull the steering shaft upward and away from the steering gear connection.
- Brake pedal from the power brake

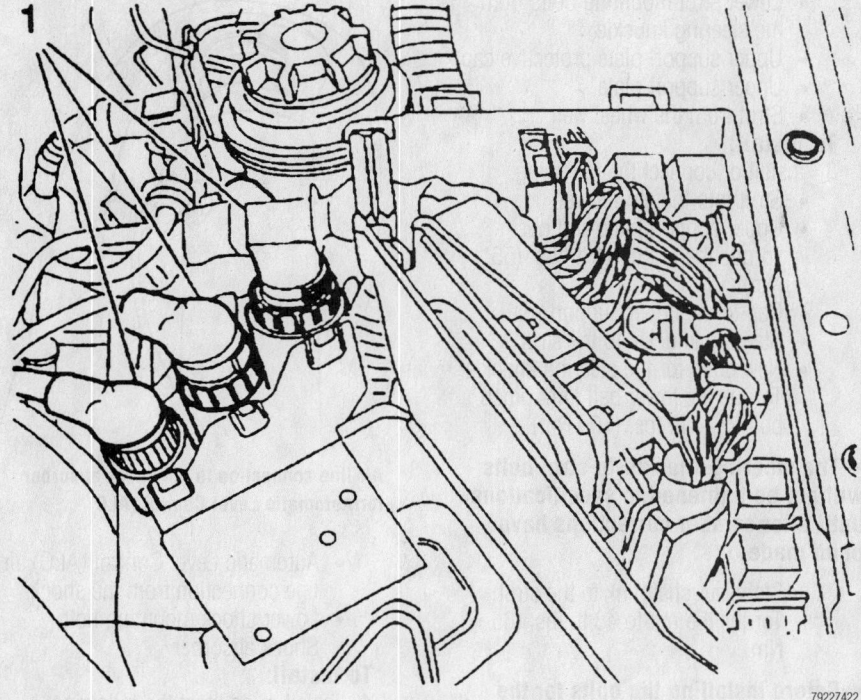

1

7922Z422

Body harness electrical connector locations (1)

booster link rod by removing the retaining clip from the pin and driving the pin out of the linkage rod
- Instrument panel drivers side knee bolster energy absorber
- Relay panel. Move the fuse and relay panel away from the brake booster.
- Brake booster nuts from the inward side of the cowling
- Brake booster and master cylinder as one unit
- A/C evaporator line quick connect fitting
- Power steering hoses from the steering gear
- Electronic Brake Traction Control Module (EBTCM)

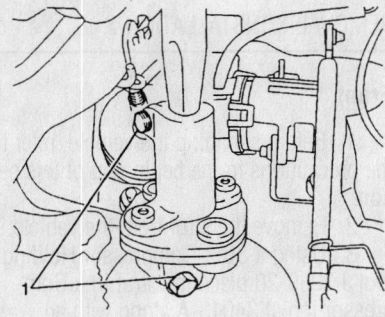

1

7922Z423

Matchmark the coupler to the steering gear before removing the coupler bolts (1)

- Heat shield upper fastening bolt
- Pitman arm washer and nut. Mark the alignment of the pitman arm to the steering gear for ease of installation.
- Pitman arm using a suitable puller
- Lower heat shield mounting nuts
- Steering gear lower washers, bolts and nuts
- Heat Shield
- Electrical connector from the power steering fluid flow control valve actuator
- Upper steering gear shims and bolt
- Steering gear

To install:

10. Find the center of travel of the steering gear by turning the stubshaft and counting the number of turns from lock-to-lock. Divide the total number in half and turn the stub shaft from either lock position by this amount. Finally align the mark on the stub shaft to the "V" mark on the steering gear.

11. Install or connect the following:
- Steering gear
- Steering gear shims and bolt. Torque the bolt snugly at this time. Do not tighten the bolt to proper specification at this time.
- Electrical connector to the power steering flow control valve actuator
- Heat shield with the upper mounting bolt. Torque the bolt to 71 inch lbs. (8 Nm).

- Lower steering gear washers, bolts and nuts. Torque all fasteners to 30 ft. lbs. (40 Nm).
- Heat shield lower mounting bolt. Torque the bolt to 11 ft. lbs. (15 Nm).
- Pitman arm to the steering gear. Make certain that the steering gear shaft protector is in place.
- Pitman arm washer and nut. Torque the fastener to 118 ft. lbs. (160 Nm).
- EBTCM assembly
- Inlet and outlet hoses to the steering gear. Torque the hoses to 21 ft. lbs. (28 Nm).
- A/C evaporator line quick connect with a new O-ring lubricated with mineral oil
- Vacuum booster/master cylinder assembly. Make certain the vacuum booster nuts are facing the inward side of the cowl. Torque the fasteners to 15 ft. lbs. (20 Nm).
- Fuse and relay panels and tighten the fasteners securely
- Drivers side instrument panel knee bolster energy absorber

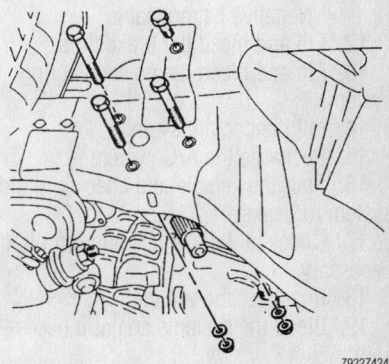

7922Z424

Steering gear mounting bolt locations

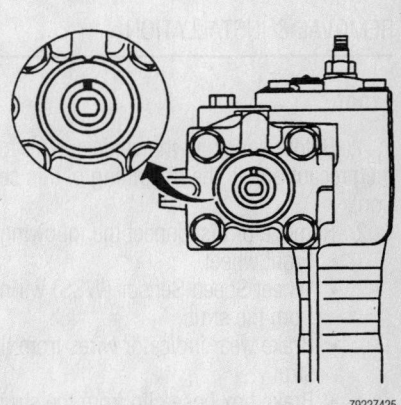

7922Z425

Align the mark on the stub shaft with the "V" mark on the steering gear mounting bolt locations

- Brake pedal to the power booster by driving the pin into the linkage rod
- Retaining clip to the pin
- Steering coupler to the gear by aligning the matchmarks
- Coupler bolts. Torque the bolts to 16 ft. lbs. (22 Nm).
- Sound insulator
- Brake pipes to the master cylinder. Torque the fasteners to 12 ft. lbs. (16 Nm).
- Electrical connector to the master cylinder reservoir cap
- Brake booster vacuum connection to the intake plenum
- Power steering fluid reservoir and clamp. Torque the clamp to 62 inch lbs. (7 Nm).
- New O-ring lubricated with mineral oil to the A/C evaporator line extension
- Evaporator line to the cowl. Torque the bolt to 15 ft. lbs. (20 Nm).
- Upper radiator hose
- ECM to the electrical box
- Body harness electrical connectors
- Wiper assembly
- Negative battery cable

12. Fill and bleed the brake fluid.
13. Fill and bleed the power steering system.
14. Fill the coolant system.
15. Recharge the A/C system.
16. Start the vehicle and check all fluid system for leaks.
17. Check all fluid levels and top off, if necessary.
18. Road test the vehicle.
19. Bleed the systems again, if necessary.

Strut

REMOVAL & INSTALLATION

Front

1. Before servicing the vehicle, refer to the precautions in the beginning of this section.
2. Remove or disconnect the following:
   - Front wheel
   - Wheel Speed Sensor (WSS) wiring from the strut
   - Brake wear indicator wires from the strut
   - Brake flex hose clip from the strut
   - Brake caliper and bracket from the steering knuckle
   - Stabilizer shaft link

- Lower strut mounting bolts from the steering knuckle
- Upper support plate protective cap
- Upper support plate
- Strut from the wheel well

To install:
3. Install or connect the following:
   - Strut into the strut tower
   - Upper support plate and nut. Torque the nut to 41 ft. lbs. (55 Nm).
   - Cap to the upper support nut
   - Steering knuckle to the strut
   - New bolts for the steering knuckle. Torque the lower ball joint pinch bolt to 74 ft. lbs. (100 Nm).

➡**The steering knuckle-to-strut bolts will not be tightened to specifications until after camber corrections have been made.**

   - Stabilizer shaft link to the strut. Torque the nut to 48 ft. lbs. (65 Nm).

➡**Before installing the bolts for the caliper, run an M12 x 1.5 tap through the holes and coat the new bolts with a thread locking compound.**

   - Brake caliper and bracket to the steering knuckle. Torque the bracket bolts to 70 ft. lbs. (95 Nm) plus a 37 degree turn.
   - Brake flex hose to the strut and retain it with the clip
   - WSS wiring to the strut
   - Brake wear indicator wires to the strut
   - Front wheel

4. Adjust the wheel alignment.
5. Torque the steering knuckle-to-strut bolts to 52 ft. lbs. (70 Nm).
6. Road test the vehicle.

Shock Absorber

REMOVAL & INSTALLATION

Rear

1. Before servicing the vehicle, refer to the precautions in the beginning of this section.
2. Position the rear seat backs forward to allow access to the upper shock absorber mounting.
3. Remove or disconnect the following:
   - Protective cap and nut from shock tower
   - Upper mounting nut, washer and grommet

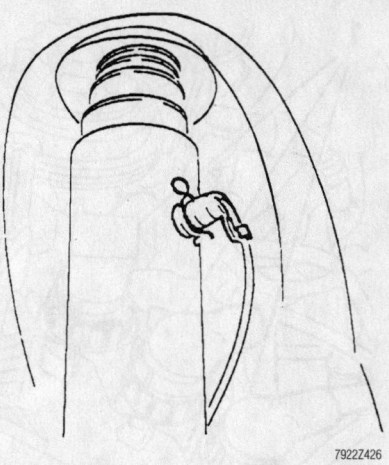

Air line connection to the shock absorber for Automatic Level Control (ALC)

   - Automatic Level Control (ALC) air line connection from the shock
   - Lower shock mounting bolt
   - Shock absorber

To install:
4. Install or connect the following:
   - Lower rubber mounting grommet and washer
   - Shock into the tower
   - Upper shock mounting grommet, washer and nut. Torque the nut 15 ft. lbs. (20 Nm).
   - Protective cap to the shock tower
   - Lower shock mounting bolt. Torque the bolt 81 ft. lbs. (110 Nm).

➡**It may be necessary to support the lower control arm for ease of installation of the lower nut.**

   - ALC air line connection to the shock
5. Position the rear seats in their proper position.
6. Road test the vehicle and verify a smooth ride.

Coil Spring

REMOVAL & INSTALLATION

Front

1. Before servicing the vehicle, refer to the precautions in the beginning of this section.
2. Remove the strut from the vehicle.
3. Using a Strut Compressor Holding tool J 3289-20 place the strut in a Compressor tool J 34013-A along with an Adapter J 3413-88
4. Remove or disconnect the following:

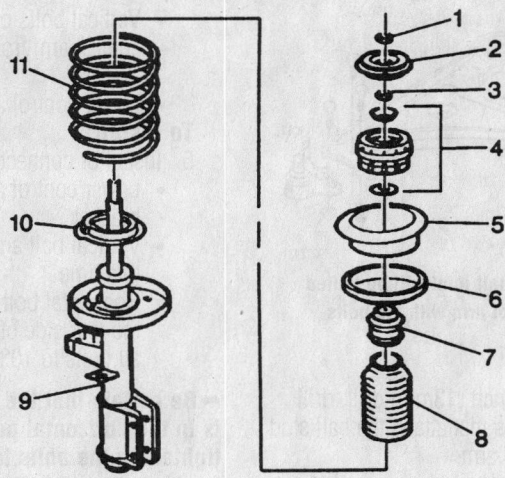

(1) Upper Support Plate Nut
(2) Upper Support Plate
(3) Upper Bearing Support Nut
(4) Bearing and Bearing Plate Assembly
(5) Upper Spring Support Plate
(6) Upper Insulator
(7) Strut Bumper
(8) Strut Cover
(9) Strut
(10) Lower Insulator
(11) Spring

79222427

Exploded view of the strut assembly

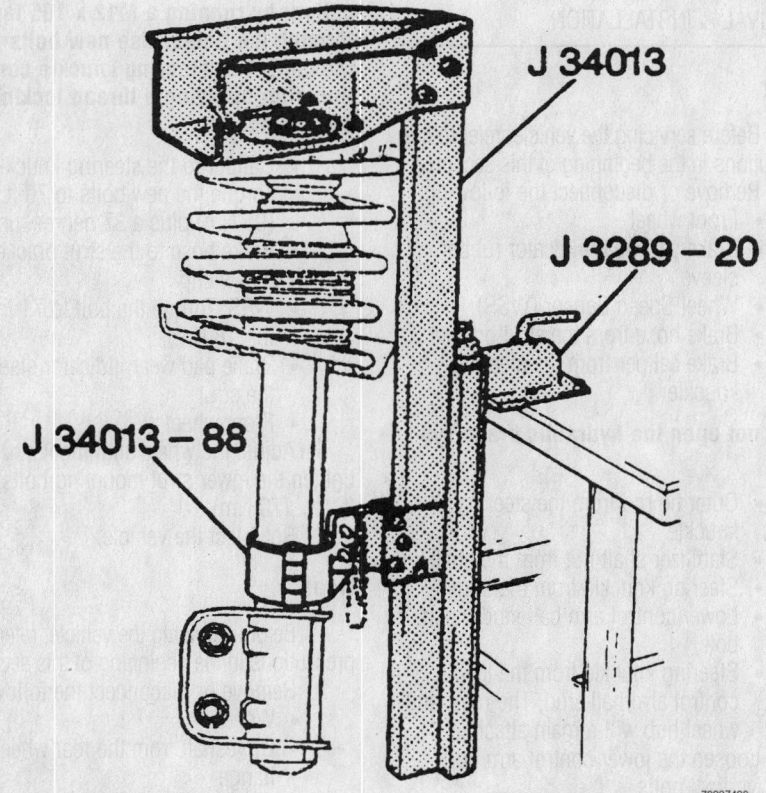

79222428

Be sure to compress the coil spring before removing the upper bearing support nut

- Compress the spring
- Upper bearing support nut
- Bearing and plate assembly

5. Decompress the spring.

6. Remove or disconnect the following:
   - Upper spring support plate
   - Upper insulator
   - Strut bumper
   - Cover from the strut
   - Spring from the strut
   - Lower insulator from the strut

To install:

7. Before servicing the vehicle, refer to the precautions in the beginning of this section.

- Lower insulator to the strut
- Spring on the strut
- Upper spring support plate
- Upper insulator
- Strut bumper
- Cover

8. Align the spring between the upper and lower rubber isolation rings.

9. Compress the spring.

10. Install or connect the following:
    - Bearing and plate assembly
    - Upper bearing support nut. Torque the nut to 52 ft. lbs. (70 Nm).

11. Release the tension from the spring.

12. Remove the strut compressor tools from the strut.

13. Install the strut to the vehicle.

14. Road test the vehicle and verify a smooth ride and no abnormal noises from the strut and spring.

Rear

1. Before servicing the vehicle, refer to the precautions in the beginning of this section.

2. Remove or disconnect the following:
   - Retainers from the lower control arms
   - Brake pipes from the lower control arms without disconnect the brake fitting
   - Stabilizer shaft link bolts
   - Stabilizer shaft link from the lower control arms
   - Rubber exhaust insulators from the hangers. Support the exhaust system so no damage occurs to the pipes or gaskets.
   - Rear wheel speed sensor electrical connections
   - Shock lower mounting bolt after properly supporting the lower control arms
   - Support from the lower control arms and place on the rear differential

- Rear axle cradle mounts to the body bolts

3. Lower the rear differential so that coil spring can be removed.

4. Remove or disconnect the following:
- Rear springs with the seats on the spring
- Seats from the spring

To install:

5. Install or connect the following:
- Seats to the spring
- Rear springs to the lower control arm and body. Make certain that the spring leg is aligned properly.
- Protective shields
- Rear differential with the cradle mounting bolts. Torque the bolts to 48 ft. lbs. (65 Nm).
- Shock absorber. Torque the bolt to 81 ft. lbs. (110 Nm).
- Rear wheel speed sensor electrical connections
- Rubber exhaust insulators to the hangers
- Stabilizer shaft link to the lower control arm. Torque the bolts to 15 ft. lbs. (20 Nm).
- Brake pipes to the lower control arm and secure with the retainers

6. Check and/or adjust the rear toe.

Lower Ball Joint

REMOVAL & INSTALLATION

1. Before servicing the vehicle, refer to the precautions in the beginning of this section.

2. Remove or disconnect the following:
- Front wheel
- Lower control arm

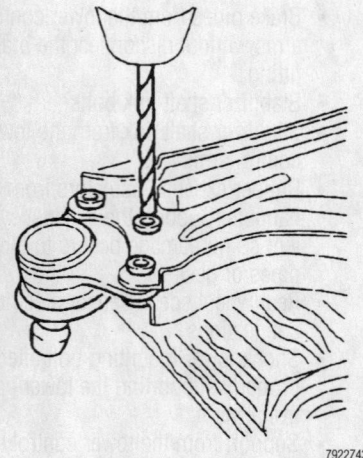

Drill out the rivets to remove the ball joint from the lower control arm

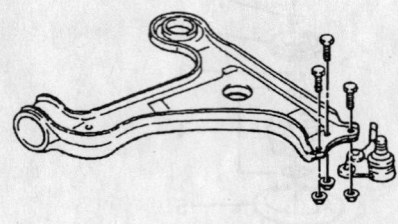

The replacement ball joint will be bolted to the lower control arm with the bolts supplied in the kit

3. Using a 0.5 inch (13mm) drill, drill out the 3 rivet heads that attach the ball stud to the lower control arm.

4. Remove the ball stud from the lower control arm.

To install:

5. Install or connect the following:
- Ball stud to the lower control arm
- Bolts through the upper side of the lower control arm. Torque the new bolts to 26 ft. lbs. (35 Nm).
- Lower control arm
- Front wheel

6. Check and/or adjust the front end alignment.

Lower Control Arm

REMOVAL & INSTALLATION

Front

1. Before servicing the vehicle, refer to the precautions in the beginning of this section.

2. Remove or disconnect the following:
- Front wheel
- Brake pad wear indicator rubber sleeve
- Wheel Speed Sensor (WSS)
- Brake hose from the strut bracket
- Brake caliper from the steering knuckle

➡ **Do not open the hydraulic brake system.**

- Outer tie rod from the steering knuckle
- Stabilizer shaft nut from the shaft
- Steering knuckle from the strut
- Lower control arm ball stud pinch bolt
- Steering knuckle from the lower control arm ball stud. The rotor and wheel hub will remain attached.

3. Loosen the lower control arm horizontal/vertical bolts.

4. Remove or disconnect the following:
- Horizontal bolts

- Vertical bolts by rotating the lower control arm from the support bracket
- Lower control arm

To install:

5. Install or connect the following:
- Lower control arm into the support bracket
- Vertical bolt and hand-tighten at this time
- Horizontal bolt by inserting it from the rear side of the vehicle. Torque all bolts to 103 ft. lbs. (140 Nm).

➡ **Be certain that the lower control arm is in the horizontal position before tightening the bolts to prevent the bushings from binding.**

- Steering knuckle to the lower control arm ball stud. Torque the bolt to 74 ft. lbs. (100 Nm).
- Steering knuckle to the strut with new bolts. Hand tighten the bolts at this time.
- Stabilizer shaft link to the shaft. Torque the nut to 48 ft. lbs. (65 Nm).
- Outer tie rod to the steering knuckle
- New outer tie rod nut. Torque the nut 44 ft. lbs. (60 Nm).

➡ **Clean the bolt holes for the brake caliper by running a M12 x 105 tap through the holes. Use new bolts for the caliper to steering knuckle connection after applying a thread locking compound.**

- Caliper to the steering knuckle. Torque the new bolts to 70 ft. lbs. (95 Nm) plus a 37 degree turn.
- Brake hose to the strut bracket with a new clip
- WSS. Torque the bolt to 71 inch lbs. (8 Nm).
- Brake pad wear indicator sleeve to the strut
- Front wheel

6. Adjust the wheel alignment and tighten the lower strut mounting bolts to 52 ft. lbs. (70 Nm).

7. Road test the vehicle.

Rear

1. Before servicing the vehicle, refer to the precautions in the beginning of this section.

2. Remove or disconnect the following:
- Wheel
- Driveshaft from the rear wheel hub flange
- Brake pipe from the lower control arm

- Brake caliper
- Brake rotor
- Parking brake cable from the actuator bracket
- Rear hub and flange
- Wheel bearing
- Brake backing plate
- Exhaust system from the rubber mounts
- Rear outer tie rod from the rear control arm
- Stabilizer shaft link connection at the lower control arm
- Rear axle cradle mounting bolts

➡ **Make certain that the rear differential is properly supported.**

- Lower shock mounting bolts
- Rear coil spring
- Underbody heat shield
- Propeller shaft center bearing bracket bolts
- Rear support flange bolts
- Rear support bushing bolt
- Lower control arm

To install:

3. Install or connect the following:
- Lower control arm. Torque the bolts to 74 ft. lbs. (100 Nm).
- Rear support bushing bolt. Torque the bolt to 92 ft. lbs. (125 Nm).
- Rear support flange bolts. Torque the bolts to 48 ft. lbs. (65 Nm).
- Propeller shaft center bearing bracket. Torque the bolts to 15 ft. lbs. (20 Nm).
- Underbody heat shield. Torque the bolts to 18 inch lbs. (2 Nm).
- Rear coil spring
- Rear axle cradle mount. Torque the body bolts to 48 ft. lbs. (65 Nm).
- Shock absorber. Torque the lower mount bolt to 81 ft. lbs. (110 Nm).
- Stabilizer shaft link to the lower control arm. Torque the bolt to 15 ft. lbs. (20 Nm).
- Outer tie rod. Torque the nut to 44 ft. lbs. (60 Nm).
- Exhaust system to the rubber mounts
- Brake backing plate
- Wheel bearing
- Hub and flange
- Parking brake cable and route it through the bracket in the rear lower control arm
- Brake rotor and set screw. Torque the screw to 35 inch lbs. (4 Nm).
- Brake caliper. Torque the bolts to 59 ft. lbs. (80 Nm).
- Brake pipe and clip to the lower control arm

- Driveshaft. Torque the bolts to 37 ft. lbs. (50 Nm) plus an additional 70 degree turn.
- Wheel

4. Check and/or adjust the rear toe.

CONTROL ARM BUSHING REPLACEMENT

Front

1. Before servicing the vehicle, refer to the precautions in the beginning of this section.
2. Remove the control arm from the vehicle.
3. Press the bushings from the control arm.

To install:

4. Press the new bushings into the control arm.
5. Install the control arm in the vehicle.

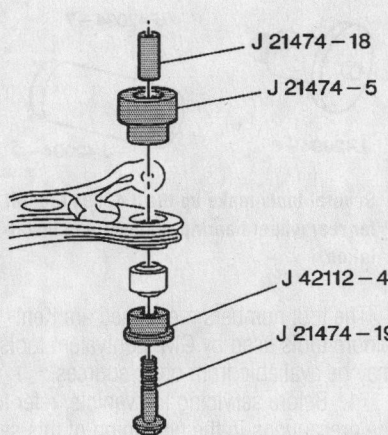

9356ZGAB

Use the following tools to install the front control arm vertical bushing

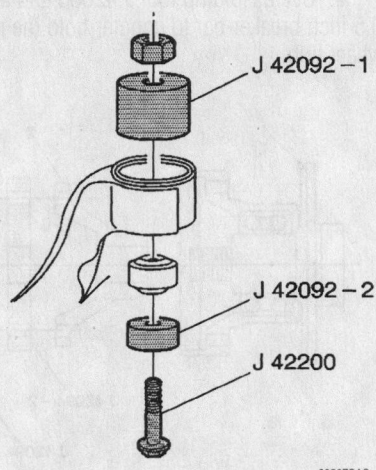

9356ZGAC

Use the following tools to install the front control arm horizontal bushing

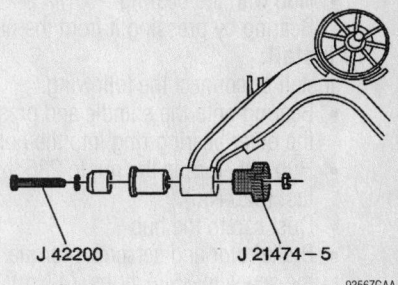

9356ZGAA

Use the following tools to remove and install the rear control arm bushing

Rear

1. Before servicing the vehicle, refer to the precautions in the beginning of this section.
2. Remove or disconnect the following:
- Wheel
- Rear axle lower control arm
- Collar for the inboard and outboard lower control arm bushings by cutting it off
3. Press the bushings out of the control arm.

To install:

4. Press the inboard side bushing into the control arm with the collar toward the rear differential.
5. Press the outboard side bushing into the control arm.
6. Install or connect the following:
- Lower control arm
- Wheel

Wheel Bearings

ADJUSTMENT

The wheel bearings are not adjustable.

REMOVAL & INSTALLATION

Front

1. Before servicing the vehicle, refer to the precautions in the beginning of this section.
2. Remove or disconnect the following:
- Front wheel
- Brake pad wear indicator wire from the strut bracket
- Brake hose from the strut bracket
- Caliper bolts from the steering knuckle
- Brake caliper. The brake system should remain closed.
- Brake rotor and setscrew
- Dust cap from the hub

- Hub with the bearing
- Bearing by pressing it from the hub

To install:

3. Install or connect the following:
- Bearing onto the spindle and press the outer bearing ring into the hub
- Hub nut. Torque the nut to 236 ft. lbs. (320 Nm).
- Dust cap to the hub
- Brake rotor and setscrew. Torque the screw to 35 inch lbs. (4 Nm).
- Brake caliper to the steering knuckle after cleaning the bolt holes with an M12 x 1.5 tap and coating the new bolts with a thread locking compound. Torque the bolts to 70 ft. lbs. (95 Nm), plus a 37 degree turn.
- Brake hose to the strut bracket with a retaining clip
- Brake pad wear indicator wire to the strut
- Front wheel

4. Road test the vehicle.
5. Check and/or adjust the front end alignment.

Rear

➡ **Several special tools are necessary to remove the rear wheel bearing from the knuckle. The tools required are as follows:**

- J 36660 Torque Angle Meter
- J 42066 Holding tool
- J 42094-1 Holding Fixture
- J 42094-2 Spacer
- J 42094-3 Threaded Driver
- J 42094-4 Threaded Arbor
- J 42094-5 Ball Head
- J 42094-6 Bearing Remover
- J 42094-7 Threaded Spacer Pin
- J 42094-8 Bearing Installer
- J 42094-9 Hub Installer
- J 42094-10 Thrust Bearing
- J 42072 Triple Hex Head (10mm) deep socket

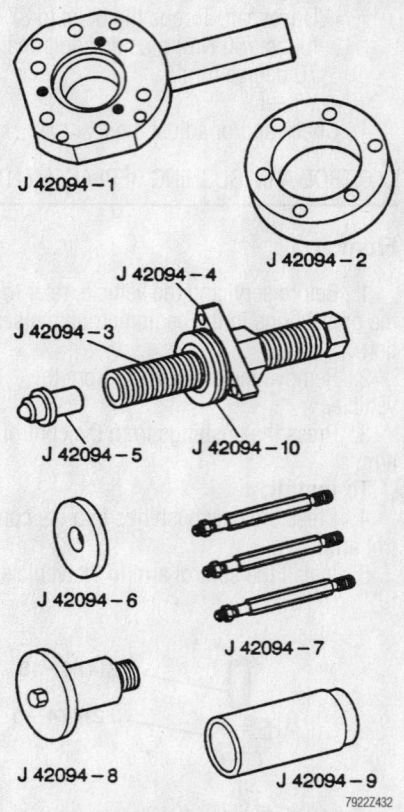

J 42094 – 1
J 42094 – 2
J 42094 – 4
J 42094 – 3
J 42094 – 5   J 42094 – 10
J 42094 – 6
J 42094 – 7
J 42094 – 8   J 42094 – 9

Several tools make up the J 42094 tool kit for rear wheel bearing removal and installation

The tool numbers mentioned are Kent-Moore tools used by GM. Equivalent tools may be available from other sources.

1. Before servicing the vehicle, refer to the precautions in the beginning of this section.

2. Remove the rear wheel.

3. Remove the outer tie rod from the knuckle.

4. Use a Holding tool J 42066 and a 0.5 inch breaker bar to counter hold the rear wheel hub.

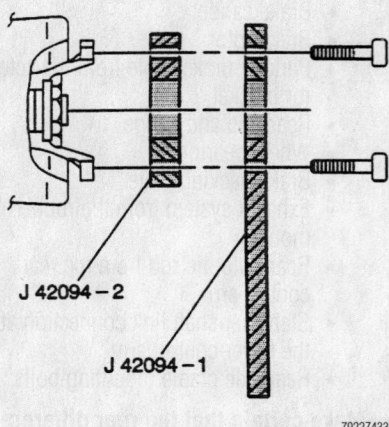

J 42094 – 2
J 42094 – 1

Attach the Spacer and Holding Fixture to the flange before removing the hub retaining nut

5. Remove or disconnect the following:
- Driveshaft from the rear hub
- Brake pipe and retaining clip from the lower control arm
- Rear brake caliper and support it aside
- Set screw and brake rotor
- Three of the four brake backing plate bolts approximately 9 revolutions by installing a deep well socket
- Rear wheel hub nut by installing a Spacer J 42094-2 and Holding Fixture J 42094-1 to the rear wheel hub

6. Use a Thrust Bearing J 42094-10 as a spacer to attach a Threaded Arbor J42094-4 to the holding Fixture J 42094-7 with 3 bolts.

7. Attach a Threaded Driver into the threaded arbor.

8. Screw the bolts into the backing plate and attach the holding fixture with the stem upward.

9. Press out the rear wheel hub.

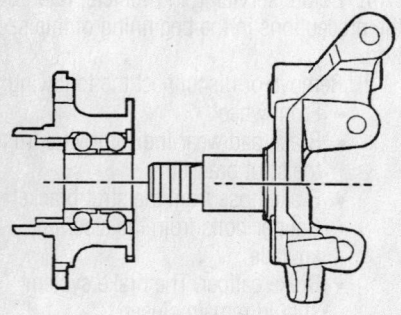

Cut-away view of the hub/wheel bearing assembly and steering knuckle

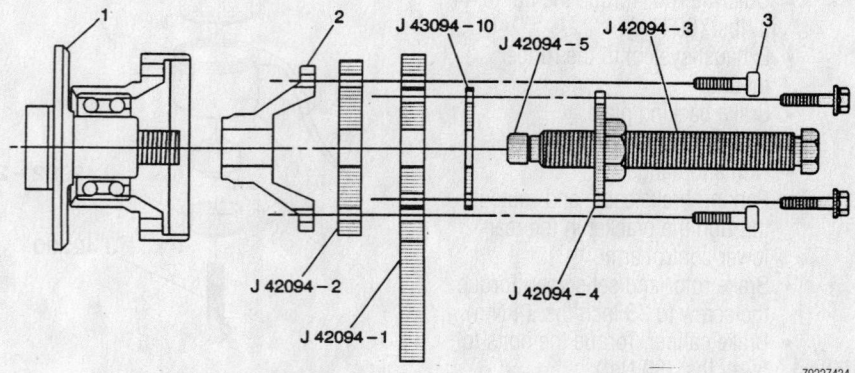

1   2   J 43094 – 10   J 42094 – 3   3
J 42094 – 5

J 42094 – 2
J 42094 – 1   J 42094 – 4

Set up the special tools as shown to remove the flange

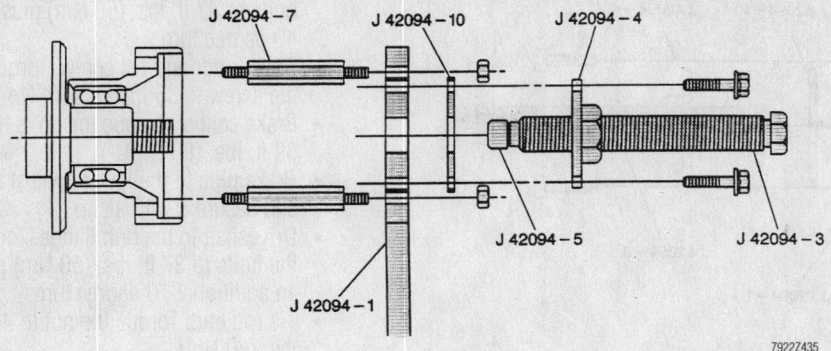

Set up the special tools as shown to remove the hub

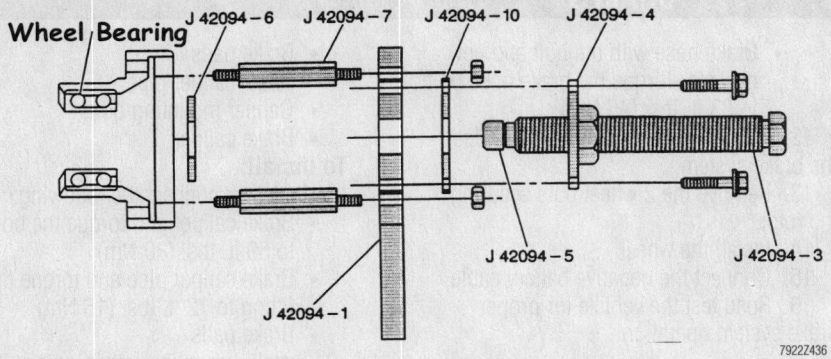

Set up the special tools as shown to remove the bearing assembly

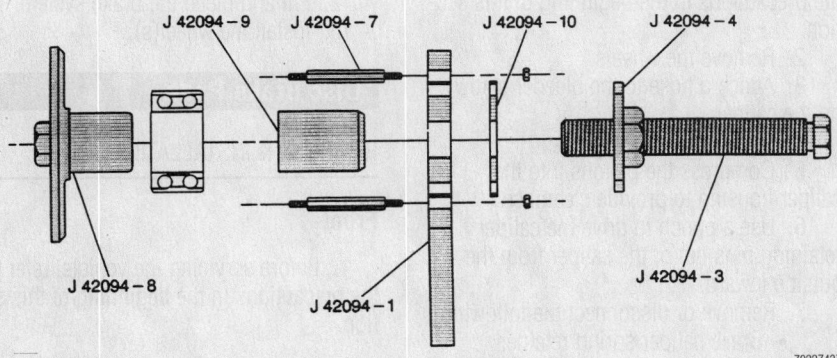

Set up the special tools as shown to install the bearing assembly

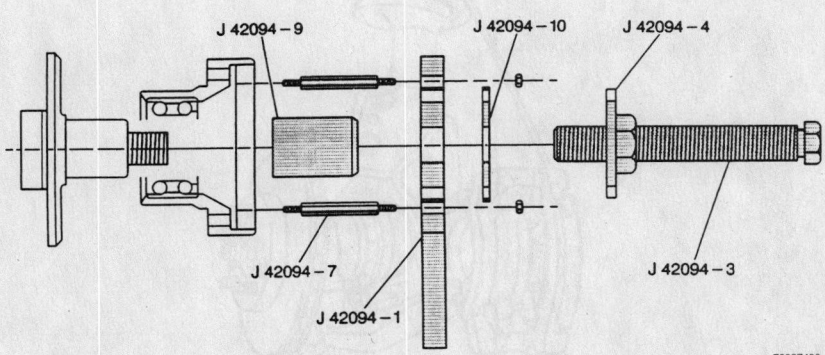

Set up the special tools as shown to pull the hub into the bearing assembly

➥Damage to the wheel bearing seal is possible while pressing out the rear wheel hub. Inspect the seal and replace if needed.

10. Remove the wheel bearing retaining ring.

11. Attach a Bearing Remover J 42094-6 to the end of the threaded driver.

12. Turn the driver in a clockwise manner to press out the wheel bearing.

To install:

13. Install or connect the following:
   - Wheel Bearing Installer Tool J 42094-8 through the wheel bearing and on to the threaded driver
   - Wheel bearing until fully seated by turning the threaded driver
   - Wheel bearing retaining ring
   - Hub Installer Tool J 42094-9 on the shaft of the threaded driver. Make certain that the rear wheel hub installer is on the inner ring of the wheel bearing

14. Attach the Holding Fixture to a Threaded Spacer Pin J 42094-7 and remove the anchoring bolts from the threaded arbor.

15. Install or connect the following:
   - Rear wheel hub into the driver and make certain the driver is seated properly on the wheel bearing. If not centered properly, it may cause the hub to bind
   - Wheel hub into the wheel bearing by holding the driver and turning the arbor clockwise. When fully seated, remove the tools

16. Connect the Threaded Arbor J 42094-4, Threaded Driver J 42094-3, Holding Fixture J 42094-1, Spacer J 42094-2, and Thrust Bearing J 42094-10 to the wheel flange with the halfshaft mounting bolts.

17. Install or connect the following:
   - Hub to the flange and make certain that the splines are aligned properly
   - Flange fully onto the hub by turning the arbor clockwise while counter holding it with the driver

18. Remove the thrust bearing, arbor and driver while leaving the holding fixture and spacer attached to the hub.

19. Install the rear wheel hub nut. Torque the nut 221 ft. lbs. (300 Nm).

20. Remove the spacer and holding fixture from the hub.

21. Install or connect the following:
   - Retaining washer to the hub
   - Backing plate bolts. Torque the

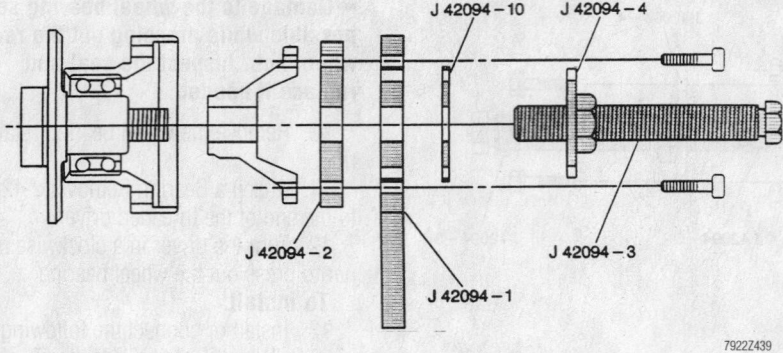

J 42094 – 10   J 42094 – 4

J 42094 – 2

J 42094 – 1

J 42094 – 3

7922Z439

Set up the special tools as shown to install the flange on the hub

bolts to 37 ft. lbs. (50 Nm) plus a 40 degree turn.
- Brake rotor and set screw. Torque the screw to 35 inch lbs. (4 Nm).
- Brake caliper. Torque the bolts to 59 ft. lbs. (80 Nm).
- Brake pipe to the lower control arm and secure it with a clip
- Driveshaft to the hub flange. Torque the bolts to 37 ft. lbs. (50 Nm) plus an additional 70 degree turn.
- Tie rod end. Torque the nut to 44 ft. lbs. (60 Nm).
- Rear wheel

BRAKES

Brake Caliper

REMOVAL & INSTALLATION

FRONT

1. Before servicing the vehicle, refer to the precautions in the beginning of this section.
2. Remove or disconnect the following:
3. Remove ⅔ of the brake fluid from the master cylinder.
   - Front wheel. Mark the relationship between the wheel and the wheel stud for re-installation purposes.
4. Install 2 wheel nuts to keep the rotor in place.
5. Using a large C-clamp against the inboard pad, compress the caliper piston into the caliper to provide clearance during removal.
6. Place a catch pan under the caliper.
7. Remove or disconnect the following:
   - Brake hose from the caliper. Cap the line to prevent excessive fluid loss or contamination.
   - Caliper mounting bolts and the caliper from the vehicle
8. Inspect the mounting bolts; sleeves and boots for wear and/or damage. Replace parts as necessary.

   To install:
9. Before installing the caliper, make sure the piston is fully seated in the bore and the brake pads are properly seated.
10. Lubricate the mounting bolt shafts and inner diameter of the sleeves with silicone grease.
11. Install or connect the following:
    - Caliper in the caliper mounting bracket
    - Caliper mounting bolts and torque the mounting bolts to 38 ft. lbs. (51 Nm)

- Brake hose with the bolt and new gaskets. Torque the brake hose bolt to 33 ft. lbs. (45 Nm).
12. Refill the master cylinder and bleed the brake system.
13. Remove the 2 wheel nuts securing the rotor.
14. Install the wheel.
15. Connect the negative battery cable.
16. Road test the vehicle for proper brake system operation.

REAR

1. Before servicing the vehicle, refer to the precautions in the beginning of this section.
2. Remove the wheels.
3. Attach a hose to the bleeder screw on the caliper.
4. Open the bleeder screw.
5. Compress the pistons into the caliper housing to provide clearance.
6. Use a punch to drive the caliper retaining pins out of the caliper from the outside inward.
7. Remove or disconnect the following:
   - Brake caliper spring retainer

- Brake pads
- Brake caliper pipe
- Caliper mounting bolts
- Brake caliper

To install:
8. Install or connect the following:
   - Brake caliper and torque the bolts to 59 ft. lbs. (80 Nm)
   - Brake caliper pipe and torque the fitting to 12 ft. lbs. (16 Nm)
   - Brake pads
9. Install one caliper retaining pin.
10. Install the caliper spring retainer.
11. Install the second retaining pin.
12. Fill and bleed the brake system.
13. Install the wheel(s).

Disc Brake Pads

REMOVAL & INSTALLATION

Front

1. Before servicing the vehicle, refer to the precautions in the beginning of this section.

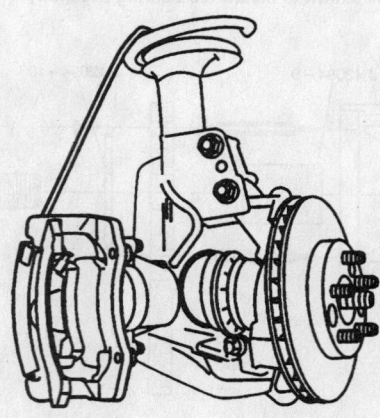

93006G67

Supporting the front caliper—Catera

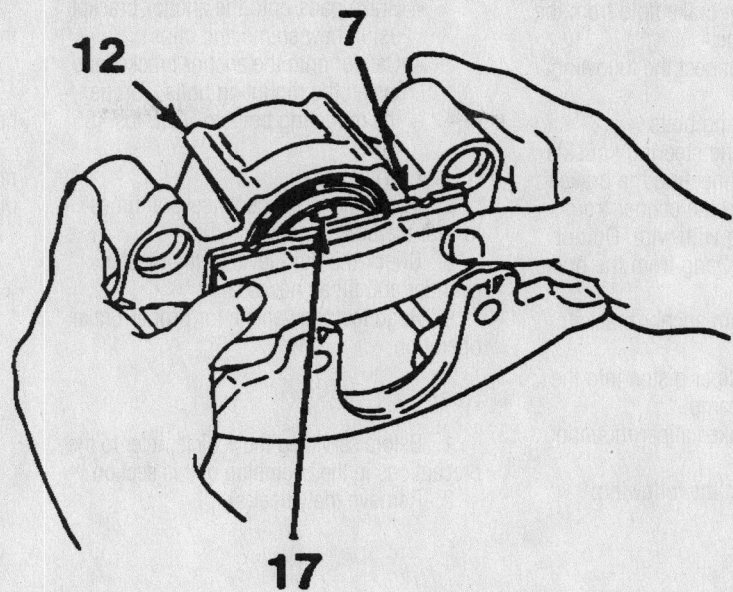

7. **Inboard Shoe & Lining**
12. **Caliper Housing**
17. **Shoe Retainer Spring**

93006G68

Installing the inboard disc brake pad into the front caliper—Catera

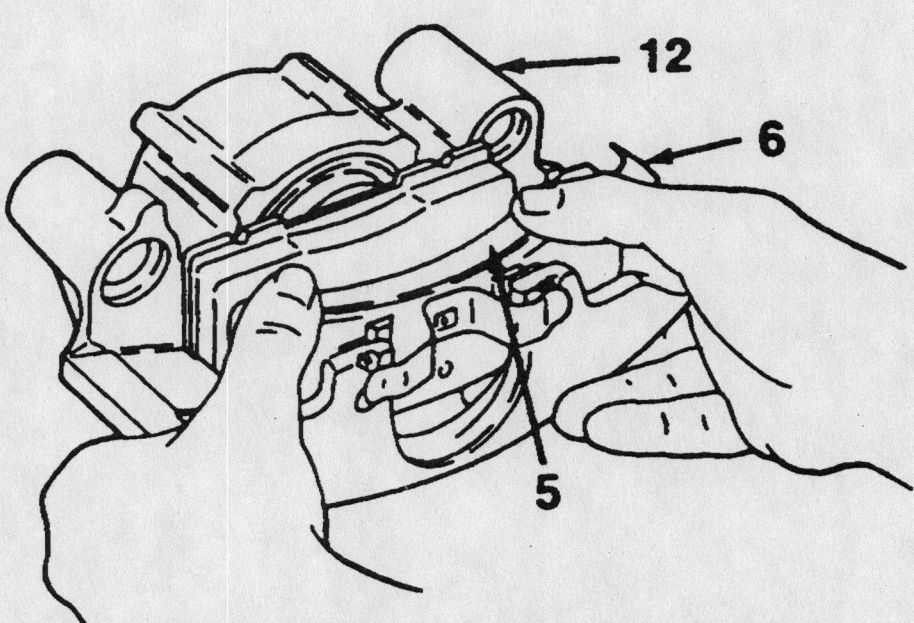

5. **Outboard Shoe & Lining**
6. **Wear Sensor**
12. **Caliper Housing**

93006G69

Installing the outboard disc brake pad in the front caliper—Catera

2. Remove ⅔ of the brake fluid from the master cylinder reservoir.

3. Remove or disconnect the following:
- Front wheel
- Caliper mounting bolts
- Caliper from the steering knuckle without disconnecting the brake hose. Suspend the caliper from the coil spring with wire. Do not let the caliper hang from the brake hose.
- Brake pads from anchor bracket

To install:

4. Fully seat the caliper piston into the bore using a large C-clamp.

5. Lubricate the brake caliper mounting surfaces.

6. Install or connect the following:

- Brake pads onto the anchor bracket using new shims and clips
- Caliper onto the anchor bracket and install the mounting bolts. Torque the mounting bolts to 63 ft. lbs. (85 Nm).
- Wheel

7. Pump the brake pedal several times to seat the brake pads.

8. Check the fluid level in the master cylinder and fill as necessary.

9. Road test the vehicle for proper brake operation.

Rear

1. Before servicing the vehicle, refer to the precautions in the beginning of this section.

2. Remove the wheel(s).

3. Attach a hose to the bleeder screw on the caliper.

4. Open the bleeder screw.

5. Compress the pistons into the caliper housing to provide clearance.

6. Use a punch to drive the caliper retaining pins out of the caliper from the outside inward.

7. Remove or disconnect the follow-ing:
- Brake caliper spring retainer
- Brake pads

To install:

8. Install or connect the following:
- Brake pads
- One caliper retaining pin
- Caliper spring retainer
- Second retaining pin

9. Fill and bleed the brake system.
- Wheel(s)

BUICK, CHEVROLET, OLDSMOBILE AND PONTIAC

22

Century • Regal • Impala • Lumina • Monte Carlo • Intrigue • Grand Prix

SPECIFICATION CHARTS

ENGINE AND VEHICLE IDENTIFICATION

Code ①	Liters (cc)	Cu. In.	Cyl.	Fuel Sys.	Engine Type	Eng. Mfg.	Code ②	Year
1	3.8 (3785)	231	6	MFI	OHV	BOC	Y	2000
K	3.8 (3785)	231	6	MFI	OHV	CPC	1	2001
H	3.5 (3475)	212	6	MFI	DOHC	BOC	2	2002
M	3.1 (3130)	191	6	MFI	OHV	BOC	3	2003
J	3.1 (3130)	191	6	SFI	OHV	BOC	4	2004
E	3.4 (3393)	207	6	MFI	OHV	CPC		

Header groupings: *Engine Code* spans Code ① through Eng. Mfg.; *Model Year* spans Code ② and Year.

MFI: Multi-point Fuel Injection

BOC: Buick/Oldsmobile/Cadillac

CPC: Chevrolet/Pontiac/Canada

DOHC: Dual Overhead Camshafts

OHV: Overhead Valves

① 8th position of VIN

② 10th position of VIN

42372-WB0D-C01

GENERAL ENGINE SPECIFICATIONS

Year	Model	Engine Displacement Liters (cc)	Engine Series (ID/VIN)	Fuel System	Net Horsepower @ rpm	Net Torque @ rpm (ft. lbs.)	Bore x Stroke (in.)	Compression Ratio	Oil Pressure @ rpm
2000	Century	3.1 (3130)	M	MFI	170@5200	185@4000	3.50x3.31	9.5:1	15@1100
	Century	3.8 (3785)	1	MFI	240@5200	280@3200	3.80x3.40	9.0:1	60@1850
	Century	3.8 (3785)	K	MFI	205@5200	230@4000	3.80x3.40	9.4:1	60@1850
	Grand Prix	3.8 (3785)	1	MFI	240@5200	280@3200	3.80x3.40	9.0:1	60@1850
	Grand Prix	3.8 (3785)	K	MFI	205@5200	230@4000	3.80x3.40	9.5:1	60@1850
	Grand Prix	3.1 (3130)	J	SFI	170@5200	185@4000	3.50x3.31	9.5:1	15@1100
	Impala	3.4 (3393)	E	MFI	210@5200	215@4000	3.62x3.31	9.5:1	15@1100
	Impala	3.8 (3785)	K	MFI	205@5200	230@4000	3.80x3.40	9.4:1	60@1850
	Intrigue	3.5 (3475)	H	MFI	215@5600	230@4400	3.52x3.62	9.3:1	29@2000
	Lumina	3.1 (3130)	J	SFI	170@5200	185@4000	3.50x3.31	9.5:1	15@1100
	Monte Carlo	3.8 (3785)	K	MFI	205@5200	230@4000	3.80x3.40	9.4:1	60@1850
	Monte Carlo	3.4 (3393)	E	MFI	210@5200	215@4000	3.62x3.31	9.5:1	15@1100
	Regal	3.8 (3785)	1	MFI	240@5200	280@3200	3.80x3.40	9.0:1	60@1850
	Regal	3.8 (3785)	K	MFI	205@5200	230@4000	3.80x3.40	9.4:1	60@1850
	Regal	3.1 (3130)	M	MFI	170@5200	185@4000	3.50x3.31	9.5:1	15@1100
2001	Century	3.1 (3130)	M	MFI	170@5200	185@4000	3.50x3.31	9.5:1	15@1100
	Century	3.8 (3785)	1	MFI	240@5200	280@3200	3.80x3.40	9.0:1	60@1850
	Century	3.8 (3785)	K	MFI	205@5200	230@4000	3.80x3.40	9.4:1	60@1850
	Grand Prix	3.8 (3785)	1	MFI	240@5200	280@3200	3.80x3.40	9.0:1	60@1850
	Grand Prix	3.8 (3785)	K	MFI	205@5200	230@4000	3.80x3.40	9.5:1	60@1850
	Grand Prix	3.1 (3130)	J	SFI	170@5200	185@4000	3.50x3.31	9.5:1	15@1100
	Impala	3.4 (3393)	E	MFI	210@5200	215@4000	3.62x3.31	9.5:1	15@1100
	Impala	3.8 (3785)	K	MFI	205@5200	230@4000	3.80x3.40	9.4:1	60@1850
	Intrigue	3.5 (3475)	H	MFI	215@5600	230@4400	3.52x3.62	9.3:1	29@2000
	Lumina	3.1 (3130)	J	SFI	170@5200	185@4000	3.50x3.31	9.5:1	15@1100
	Monte Carlo	3.8 (3785)	K	MFI	205@5200	230@4000	3.80x3.40	9.4:1	60@1850
	Monte Carlo	3.4 (3393)	E	MFI	210@5200	215@4000	3.62x3.31	9.5:1	15@1100
	Regal	3.8 (3785)	1	MFI	240@5200	280@3200	3.80x3.40	9.0:1	60@1850
	Regal	3.8 (3785)	K	MFI	205@5200	230@4000	3.80x3.40	9.4:1	60@1850
	Regal	3.1 (3130)	M	MFI	170@5200	185@4000	3.50x3.31	9.5:1	15@1100
2002	Century	3.1 (3130)	M	MFI	170@5200	185@4000	3.50x3.31	9.5:1	15@1100
	Century	3.8 (3785)	1	MFI	240@5200	280@3200	3.80x3.40	9.0:1	60@1850
	Century	3.8 (3785)	K	MFI	205@5200	230@4000	3.80x3.40	9.4:1	60@1850
	Grand Prix	3.8 (3785)	1	MFI	240@5200	280@3200	3.80x3.40	9.0:1	60@1850
	Grand Prix	3.8 (3785)	K	MFI	205@5200	230@4000	3.80x3.40	9.5:1	60@1850
	Grand Prix	3.1 (3130)	J	SFI	170@5200	185@4000	3.50x3.31	9.5:1	15@1100
	Impala	3.4 (3393)	E	MFI	210@5200	215@4000	3.62x3.31	9.5:1	15@1100
	Impala	3.8 (3785)	K	MFI	205@5200	230@4000	3.80x3.40	9.4:1	60@1850
	Intrigue	3.5 (3475)	H	MFI	215@5600	230@4400	3.52x3.62	9.3:1	29@2000
	Lumina	3.1 (3130)	J	SFI	170@5200	185@4000	3.50x3.31	9.5:1	15@1100
	Monte Carlo	3.8 (3785)	K	MFI	205@5200	230@4000	3.80x3.40	9.4:1	60@1850
	Monte Carlo	3.4 (3393)	E	MFI	210@5200	215@4000	3.62x3.31	9.5:1	15@1100
	Regal	3.8 (3785)	1	MFI	240@5200	280@3200	3.80x3.40	9.0:1	60@1850
	Regal	3.8 (3785)	K	MFI	205@5200	230@4000	3.80x3.40	9.4:1	60@1850
	Regal	3.1 (3130)	M	MFI	170@5200	185@4000	3.50x3.31	9.5:1	15@1100

42372-WBOD-C02

GENERAL ENGINE SPECIFICATIONS

Year	Model	Engine Displacement Liters (cc)	Engine Series (ID/VIN)	Fuel System	Net Horsepower @ rpm	Net Torque @ rpm (ft. lbs.)	Bore x Stroke (in.)	Compression Ratio	Oil Pressure @ rpm
2003	Century	3.1 (3130)	M	MFI	170@5200	185@4000	3.50x3.31	9.5:1	15@1100
	Century	3.8 (3785)	1	MFI	240@5200	280@3200	3.80x3.40	9.0:1	60@1850
	Century	3.8 (3785)	K	MFI	205@5200	230@4000	3.80x3.40	9.4:1	60@1850
	Grand Prix	3.8 (3785)	1	MFI	240@5200	280@3200	3.80x3.40	9.0:1	60@1850
	Grand Prix	3.8 (3785)	K	MFI	205@5200	230@4000	3.80x3.40	9.5:1	60@1850
	Grand Prix	3.1 (3130)	J	SFI	170@5200	185@4000	3.50x3.31	9.5:1	15@1100
	Impala	3.4 (3393)	E	MFI	210@5200	215@4000	3.62x3.31	9.5:1	15@1100
	Impala	3.8 (3785)	K	MFI	205@5200	230@4000	3.80x3.40	9.4:1	60@1850
	Lumina	3.1 (3130)	J	SFI	170@5200	185@4000	3.50x3.31	9.5:1	15@1100
	Monte Carlo	3.8 (3785)	K	MFI	205@5200	230@4000	3.80x3.40	9.4:1	60@1850
	Monte Carlo	3.4 (3393)	E	MFI	210@5200	215@4000	3.62x3.31	9.5:1	15@1100
	Regal	3.8 (3785)	1	MFI	240@5200	280@3200	3.80x3.40	9.0:1	60@1850
	Regal	3.8 (3785)	K	MFI	205@5200	230@4000	3.80x3.40	9.4:1	60@1850
	Regal	3.1 (3130)	M	MFI	170@5200	185@4000	3.50x3.31	9.5:1	15@1100

MFI: Multi-point Fuel Injection

42372-WBOD-C03

ENGINE TUNE-UP SPECIFICATIONS

Year	Engine Displacement Liters (cc)	Engine ID/VIN	Spark Plug Gap (in.)	Ignition Timing (deg.)	Fuel Pump (psi)	Idle Speed (rpm)	Valve Clearance In.	Valve Clearance Ex.
2000	3.1 (3130)	M	0.060	①	41-47	②	HYD	HYD
	3.1 (3130)	J	0.060	①	41-47	②	HYD	HYD
	3.4 (3393)	E	0.045	①	41-47	②	HYD	HYD
	3.5 (3475)	H	0.050	①	41-47	②	HYD	HYD
	3.8 (3785)	1	0.060	①	41-47	②	HYD	HYD
	3.8 (3785)	K	0.060	①	41-47	②	HYD	HYD
2001	3.1 (3130)	M	0.060	①	41-47	②	HYD	HYD
	3.1 (3130)	J	0.060	①	41-47	②	HYD	HYD
	3.4 (3393)	E	0.045	①	41-47	②	HYD	HYD
	3.5 (3475)	H	0.050	①	41-47	②	HYD	HYD
	3.8 (3785)	1	0.060	①	41-47	②	HYD	HYD
	3.8 (3785)	K	0.060	①	41-47	②	HYD	HYD
2002	3.1 (3130)	M	0.060	①	41-47	②	HYD	HYD
	3.1 (3130)	J	0.060	①	41-47	②	HYD	HYD
	3.4 (3393)	E	0.045	①	41-47	②	HYD	HYD
	3.5 (3475)	H	0.050	①	41-47	②	HYD	HYD
	3.8 (3785)	1	0.060	①	41-47	②	HYD	HYD
	3.8 (3785)	K	0.060	①	41-47	②	HYD	HYD
2003	3.1 (3130)	M	0.060	①	41-47	②	HYD	HYD
	3.1 (3130)	J	0.060	①	41-47	②	HYD	HYD
	3.4 (3393)	E	0.045	①	41-47	②	HYD	HYD
	3.8 (3785)	1	0.060	①	41-47	②	HYD	HYD
	3.8 (3785)	K	0.060	①	41-47	②	HYD	HYD

NOTE: The Vehicle Emission Control Information label often reflects specification changes made during production. The label figures must be used if they differ from those in this chart.

HYD: Hydraulic

① Distributorless Ignition System (DIS) timing is not adjustable
② Idle speed is maintained by the Engine Control Module (ECM). There is no recommended adjustment procedure.

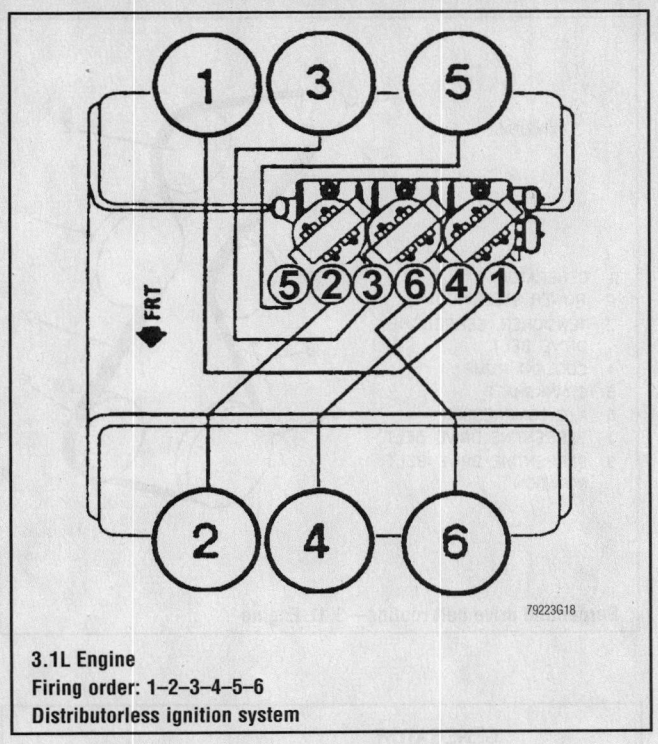

3.1L Engine
Firing order: 1–2–3–4–5–6
Distributorless ignition system

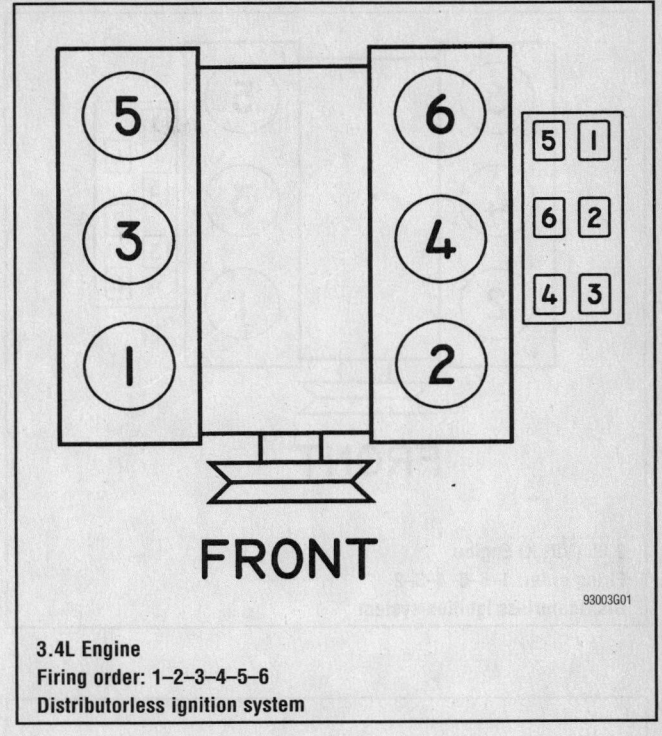

3.4L Engine
Firing order: 1–2–3–4–5–6
Distributorless ignition system

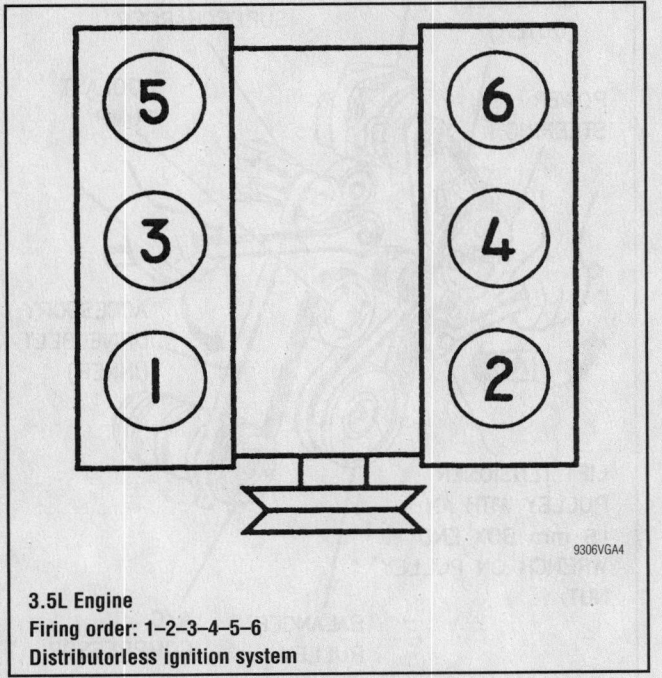

3.5L Engine
Firing order: 1–2–3–4–5–6
Distributorless ignition system

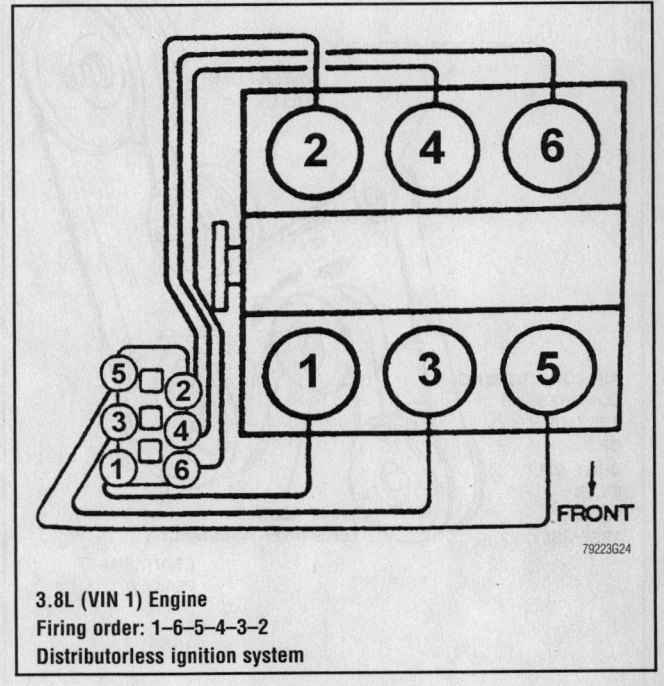

3.8L (VIN 1) Engine
Firing order: 1–6–5–4–3–2
Distributorless ignition system

FRONT

93003G05

3.8L (VIN K) Engine
Firing order: 1–6–5–4–3–2
Distributorless ignition system

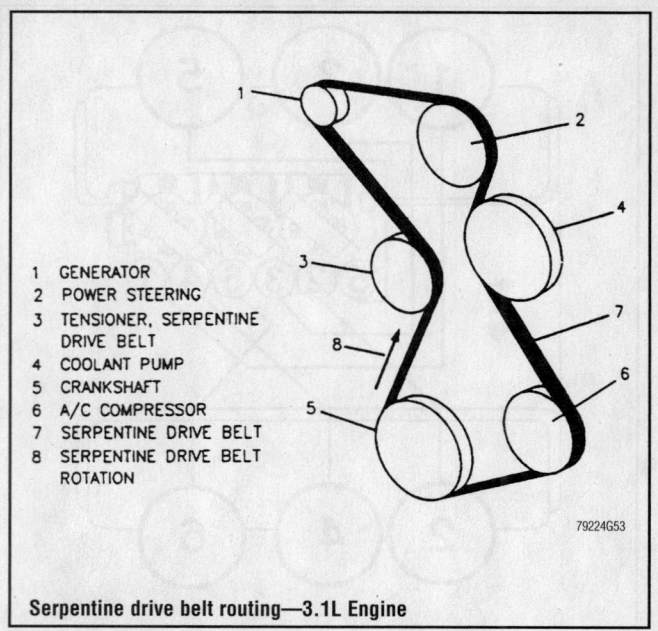

1 GENERATOR
2 POWER STEERING
3 TENSIONER, SERPENTINE
  DRIVE BELT
4 COOLANT PUMP
5 CRANKSHAFT
6 A/C COMPRESSOR
7 SERPENTINE DRIVE BELT
8 SERPENTINE DRIVE BELT
  ROTATION

79224G53

Serpentine drive belt routing—3.1L Engine

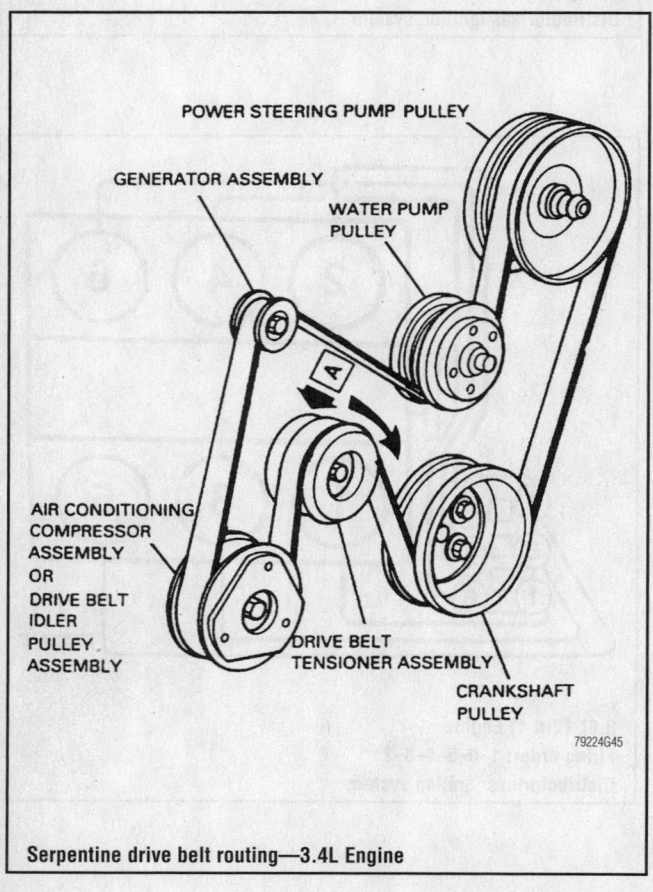

POWER STEERING PUMP PULLEY

GENERATOR ASSEMBLY

WATER PUMP PULLEY

AIR CONDITIONING
COMPRESSOR
ASSEMBLY
OR
DRIVE BELT
IDLER
PULLEY
ASSEMBLY

DRIVE BELT
TENSIONER ASSEMBLY

CRANKSHAFT
PULLEY

79224G45

Serpentine drive belt routing—3.4L Engine

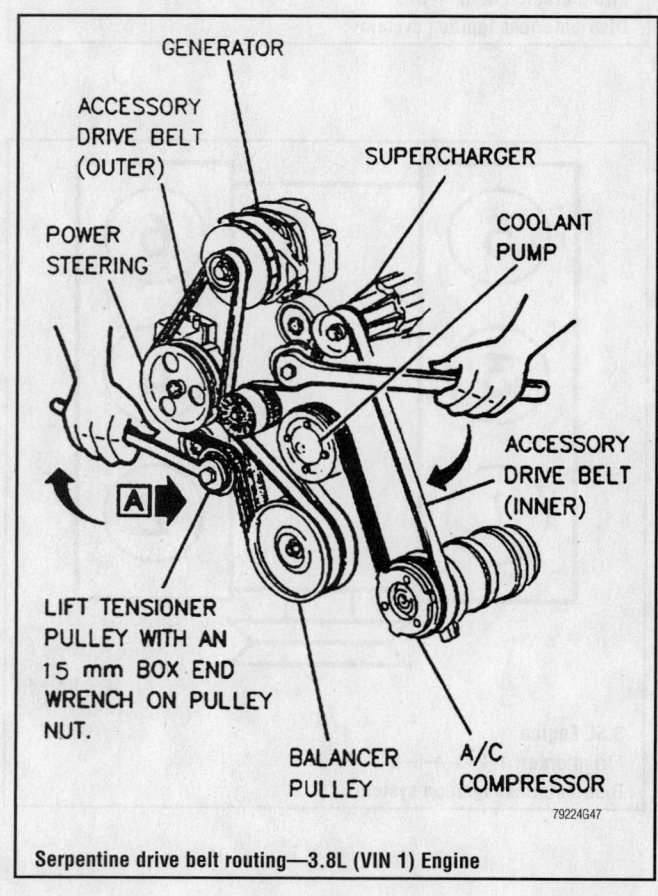

GENERATOR

ACCESSORY
DRIVE BELT
(OUTER)

SUPERCHARGER

POWER
STEERING

COOLANT
PUMP

ACCESSORY
DRIVE BELT
(INNER)

LIFT TENSIONER
PULLEY WITH AN
15 mm BOX END
WRENCH ON PULLEY
NUT.

BALANCER
PULLEY

A/C
COMPRESSOR

79224G47

Serpentine drive belt routing—3.8L (VIN 1) Engine

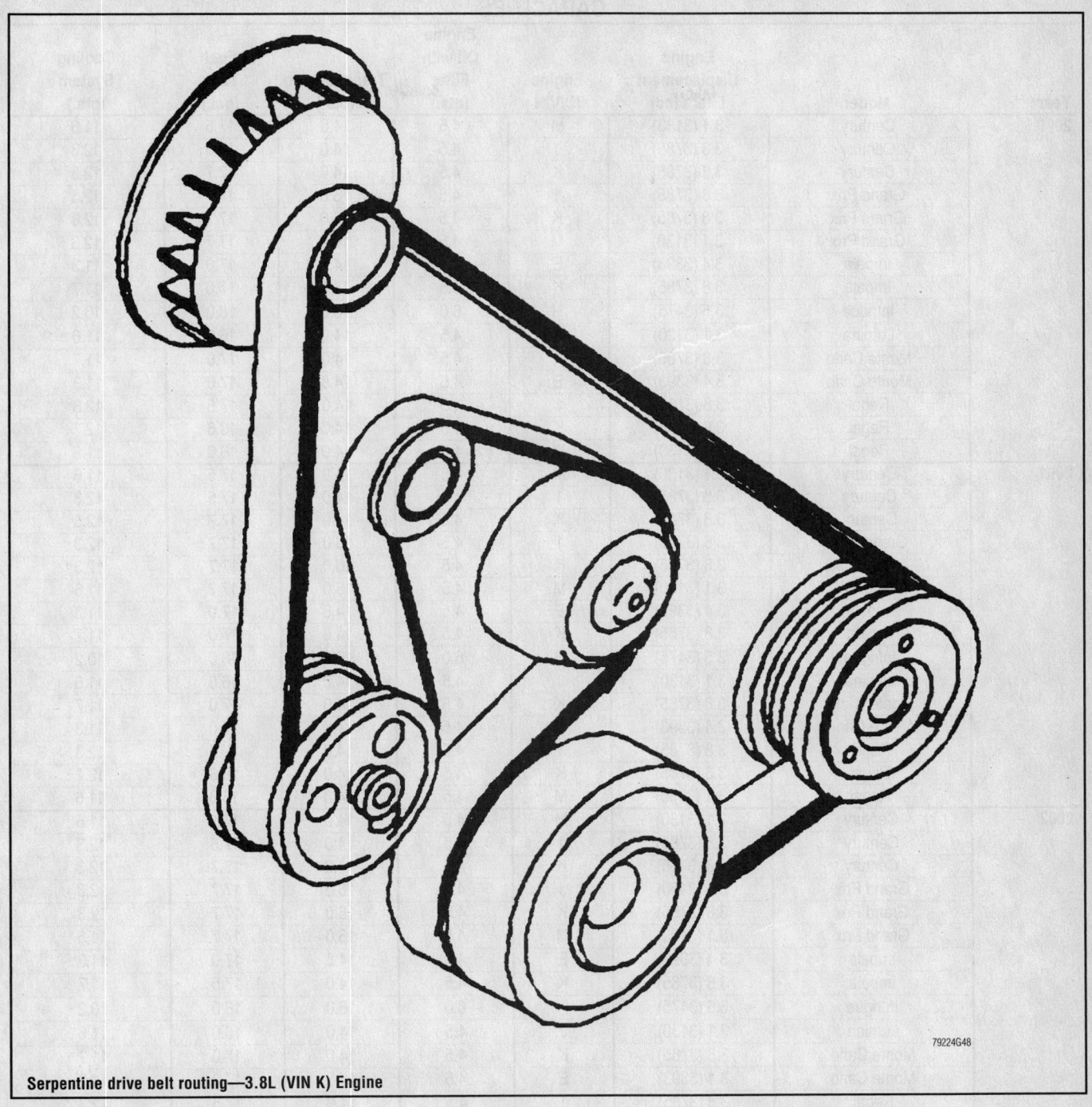

79224G48

Serpentine drive belt routing—3.8L (VIN K) Engine

CAPACITIES

Year	Model	Engine Displacement Liters (cc)	Engine ID/VIN	Engine Oil with Filter (qts.)	Transmission (pts.)	Fuel Tank (gal.)	Cooling System (qts.)
2000	Century	3.1 (3130)	M	4.5	14.0	17.5	11.6
	Century	3.8 (3785)	1	4.5	14.0	17.5	12.3
	Century	3.8 (3785)	K	4.5	14.0	17.7	12.3
	Grand Prix	3.8 (3785)	1	4.5	16.0	17.7	12.3
	Grand Prix	3.8 (3785)	K	4.5	16.0	17.7	12.3
	Grand Prix	3.1 (3130)	M	4.5	16.0	17.7	12.3
	Impala	3.4 (3393)	E	4.5	14.8	17.0	11.3
	Impala	3.8 (3785)	K	4.5	14.0	17.0	11.7
	Intrigue	3.5 (3475)	H	6.0	16.0	18.0	10.2
	Lumina	3.1 (3130)	J	4.5	14.0	16.0	11.6
	Monte Carlo	3.8 (3785)	K	4.5	14.0	17.0	11.7
	Monte Carlo	3.4 (3393)	E	4.5	14.8	17.0	11.3
	Regal	3.8 (3785)	1	4.5	14.0	17.5	12.3
	Regal	3.8 (3785)	K	4.5	14.0	16.6	12.7
	Regal	3.1 (3130)	M	4.5	14.0	16.0	11.6
2001	Century	3.1 (3130)	M	4.5	14.0	17.5	11.6
	Century	3.8 (3785)	1	4.5	14.0	17.5	12.3
	Century	3.8 (3785)	K	4.5	14.0	17.7	12.3
	Grand Prix	3.8 (3785)	1	4.5	16.0	17.7	12.3
	Grand Prix	3.8 (3785)	K	4.5	16.0	17.7	12.3
	Grand Prix	3.1 (3130)	M	4.5	16.0	17.7	11.6
	Impala	3.4 (3393)	E	4.5	14.8	17.0	11.3
	Impala	3.8 (3785)	K	4.5	14.0	17.0	11.7
	Intrigue	3.5 (3475)	H	6.0	16.0	18.0	10.2
	Lumina	3.1 (3130)	J	4.5	14.0	16.0	11.6
	Monte Carlo	3.8 (3785)	K	4.5	14.0	17.0	11.7
	Monte Carlo	3.4 (3393)	E	4.5	14.8	17.0	11.3
	Regal	3.8 (3785)	1	4.5	14.0	17.5	12.3
	Regal	3.8 (3785)	K	4.5	14.0	16.6	12.7
	Regal	3.1 (3130)	M	4.5	14.0	16.0	11.6
2002	Century	3.1 (3130)	M	4.5	14.0	17.5	11.6
	Century	3.8 (3785)	1	4.5	14.0	17.5	12.3
	Century	3.8 (3785)	K	4.5	14.0	17.7	12.3
	Grand Prix	3.1 (3130)	J	4.5	16.0	17.7	12.3
	Grand Prix	3.8 (3785)	K	4.5	16.0	17.7	12.3
	Grand Prix	3.1 (3130)	M	4.5	16.0	17.7	11.6
	Impala	3.4 (3393)	E	4.5	14.8	17.0	11.3
	Impala	3.8 (3785)	K	4.5	14.0	17.0	11.7
	Intrigue	3.5 (3475)	H	6.0	16.0	18.0	10.2
	Lumina	3.1 (3130)	J	4.5	14.0	16.0	11.6
	Monte Carlo	3.8 (3785)	K	4.5	14.0	17.0	11.7
	Monte Carlo	3.4 (3393)	E	4.5	14.8	17.0	11.3
	Regal	3.8 (3785)	1	4.5	14.0	17.5	12.3
	Regal	3.8 (3785)	K	4.5	14.0	16.6	12.7
	Regal	3.1 (3130)	M	4.5	14.0	16.0	11.6

42372-WBOD-C05

CAPACITIES

Year	Model	Engine Displacement Liters (cc)	Engine ID/VIN	Engine Oil with Filter (qts.)	Transmission (pts.)	Fuel Tank (gal.)	Cooling System (qts.)
2003	Century	3.1 (3130)	M	4.5	14.0	17.5	11.6
	Century	3.8 (3785)	1	4.5	14.0	17.5	12.3
	Century	3.8 (3785)	K	4.5	14.0	17.7	12.3
	Grand Prix	3.1 (3130)	J	4.5	16.0	17.7	12.3
	Grand Prix	3.8 (3785)	K	4.5	16.0	17.7	12.3
	Grand Prix	3.1 (3130)	M	4.5	16.0	17.7	11.6
	Impala	3.4 (3393)	E	4.5	14.8	17.0	11.3
	Impala	3.8 (3785)	K	4.5	14.0	17.0	11.7
	Lumina	3.1 (3130)	J	4.5	14.0	16.0	11.6
	Monte Carlo	3.8 (3785)	K	4.5	14.0	17.0	11.7
	Monte Carlo	3.4 (3393)	E	4.5	14.8	17.0	11.3
	Regal	3.8 (3785)	1	4.5	14.0	17.5	12.3
	Regal	3.8 (3785)	K	4.5	14.0	16.6	12.7
	Regal	3.1 (3130)	M	4.5	14.0	16.0	11.6

NOTE: All capacities are approximate. Add fluid gradually and ensure a proper fluid is obtained.

42372-WB0D-C06

CRANKSHAFT AND CONNECTING ROD SPECIFICATIONS
All measurements are given in inches.

Year	Engine Displacement Liters (cc)	Engine ID/VIN	Crankshaft				Connecting Rod		
			Main Brg. Journal Dia.	Main Brg. Oil Clearance	Shaft End-play	Thrust on No.	Journal Diameter	Oil Clearance	Side Clearance
2000	3.1 (3130)	M	2.6473-2.6383	0.0008- ① 0.0025	0.0024-0.0083	3	1.9987-1.9994	0.0007-0.0024	0.0070-0.0170
	3.4 (3393)	E	2.6472-2.6479	0.0008-0.0025	0.0024-0.0083	3	1.9987-1.9994	0.0007-0.0024	0.0070-0.0170
	3.5 (3475)	H	2.7550-2.7560	0.0006-0.0021	0.0050-0.0200	3	2.1829-2.1835	0.0009-0.0025	0.0040-0.0130
	3.8 (3785)	J	2.4988-2.4998	0.0008-0.0022	0.0030-0.0110	2	2.3738-2.3745	0.0005-0.0026	0.0030-0.0150
	3.8 (3785)	1	2.4988-2.4998	0.0008-0.0022	0.0030-0.0110	2	2.3738-2.3745	0.0005-0.0026	0.0030-0.0150
	3.8 (3785)	K	2.4988-2.4998	0.0008-0.0022	0.0030-0.0110	2	2.3738-2.3745	0.0005-0.0026	0.0030-0.0150
2001	3.1 (3130)	M	2.6473-2.6383	0.0008- ① 0.0025	0.0024-0.0083	3	1.9987-1.9994	0.0007-0.0024	0.0070-0.0170
	3.4 (3393)	E	2.6472-2.6479	0.0008-0.0025	0.0024-0.0083	3	1.9987-1.9994	0.0007-0.0024	0.0070-0.0170
	3.5 (3475)	H	2.7550-2.7560	0.0006-0.0021	0.0050-0.0200	3	2.1829-2.1835	0.0009-0.0025	0.0040-0.0130
	3.8 (3785)	J	2.4988-2.4998	0.0008-0.0022	0.0030-0.0110	2	2.3738-2.3745	0.0005-0.0026	0.0030-0.0150
	3.8 (3785)	1	2.4988-2.4998	0.0008-0.0022	0.0030-0.0110	2	2.3738-2.3745	0.0005-0.0026	0.0030-0.0150
	3.8 (3785)	K	2.4988-2.4998	0.0008-0.0022	0.0030-0.0110	2	2.3738-2.3745	0.0005-0.0026	0.0030-0.0150
2002	3.1 (3130)	M	2.6473-2.6383	0.0008- ① 0.0025	0.0024-0.0083	3	1.9987-1.9994	0.0007-0.0024	0.0070-0.0170
	3.4 (3393)	E	2.6472-2.6479	0.0008-0.0025	0.0024-0.0083	3	1.9987-1.9994	0.0007-0.0024	0.0070-0.0170
	3.5 (3475)	H	2.7550-2.7560	0.0006-0.0021	0.0050-0.0200	3	2.1829-2.1835	0.0009-0.0025	0.0040-0.0130
	3.8 (3785)	J	2.4988-2.4998	0.0008-0.0022	0.0030-0.0110	2	2.3738-2.3745	0.0005-0.0026	0.0030-0.0150
	3.8 (3785)	1	2.4988-2.4998	0.0008-0.0022	0.0030-0.0110	2	2.3738-2.3745	0.0005-0.0026	0.0030-0.0150
	3.8 (3785)	K	2.4988-2.4998	0.0008-0.0022	0.0030-0.0110	2	2.3738-2.3745	0.0005-0.0026	0.0030-0.0150
2003	3.1 (3130)	M	2.6473-2.6383	0.0008- ① 0.0025	0.0024-0.0083	3	1.9987-1.9994	0.0007-0.0024	0.0070-0.0170
	3.4 (3393)	E	2.6472-2.6479	0.0008-0.0025	0.0024-0.0083	3	1.9987-1.9994	0.0007-0.0024	0.0070-0.0170
	3.8 (3785)	J	2.4988-2.4998	0.0008-0.0022	0.0030-0.0110	2	2.3738-2.3745	0.0005-0.0026	0.0030-0.0150
	3.8 (3785)	1	2.4988-2.4998	0.0008-0.0022	0.0030-0.0110	2	2.3738-2.3745	0.0005-0.0026	0.0030-0.0150
	3.8 (3785)	K	2.4988-2.4998	0.0008-0.0022	0.0030-0.0110	2	2.3738-2.3745	0.0005-0.0026	0.0030-0.0150

① Thrust bearing: 0.0012 - 0.0030

42372-WB0D-C07

VALVE SPECIFICATIONS

Year	Engine Displacement Liters (cc)	Engine ID/VIN	Seat Angle (deg.)	Face Angle (deg.)	Spring Test Pressure (lbs. @ in.)	Spring Installed Height (in.)	Stem-to-Guide Clearance (in.)		Stem Diameter (in.)	
							Intake	Exhaust	Intake	Exhaust
2000	3.1 (3130)	M	45	45	250@1.239	1.710	0.0001-0.0027	0.0010-0.0027	NA	NA
	3.4 (3393)	E	46	45	75@1.40	1.400	0.0011-0.0026	0.0014-0.0031	NA	NA
	3.5 (3475)	H	45.75	45	130-142@0.964	1.377	0.0010-0.0030	0.0020-0.0040	0.233-0.234	0.233-0.234
	3.8 (3785)	J	45	45	80@1.750	1.690-1.720	0.0015-0.0032	0.0015-0.0032	NA	NA
	3.8 (3785)	1	45	45	80@1.750	1.690-1.720	0.0015-0.0032	0.0015-0.0032	NA	NA
	3.8 (3785)	K	45	45	210@1.32	1.690-1.720	0.0015-0.0035	0.0015-0.0032	NA	NA
2001	3.1 (3130)	M	45	45	250@1.239	1.710	0.0001-0.0027	0.0010-0.0027	NA	NA
	3.4 (3393)	E	46	45	75@1.40	1.400	0.0011-0.0026	0.0014-0.0031	NA	NA
	3.5 (3475)	H	45.75	45	130-142@0.964	1.377	0.0010-0.0030	0.0020-0.0040	0.233-0.234	0.233-0.234
	3.8 (3785)	J	45	45	80@1.750	1.690-1.720	0.0015-0.0032	0.0015-0.0032	NA	NA
	3.8 (3785)	1	45	45	80@1.750	1.690-1.720	0.0015-0.0032	0.0015-0.0032	NA	NA
	3.8 (3785)	K	45	45	210@1.32	1.690-1.720	0.0015-0.0035	0.0015-0.0032	NA	NA
2002	3.1 (3130)	M	45	45	250@1.239	1.710	0.0001-0.0027	0.0010-0.0027	NA	NA
	3.4 (3393)	E	46	45	75@1.40	1.400	0.0011-0.0026	0.0014-0.0031	NA	NA
	3.5 (3475)	H	45.75	45	130-142@0.964	1.377	0.0010-0.0030	0.0020-0.0040	0.233-0.234	0.233-0.234
	3.8 (3785)	J	45	45	80@1.750	1.690-1.720	0.0015-0.0032	0.0015-0.0032	NA	NA
	3.8 (3785)	1	45	45	80@1.750	1.690-1.720	0.0015-0.0032	0.0015-0.0032	NA	NA
	3.8 (3785)	K	45	45	210@1.32	1.690-1.720	0.0015-0.0035	0.0015-0.0032	NA	NA
2003	3.1 (3130)	M	45	45	250@1.239	1.710	0.0001-0.0027	0.0010-0.0027	NA	NA
	3.4 (3393)	E	46	45	75@1.40	1.400	0.0011-0.0026	0.0014-0.0031	NA	NA
	3.8 (3785)	J	45	45	80@1.750	1.690-1.720	0.0015-0.0032	0.0015-0.0032	NA	NA
	3.8 (3785)	1	45	45	80@1.750	1.690-1.720	0.0015-0.0032	0.0015-0.0032	NA	NA
	3.8 (3785)	K	45	45	210@1.32	1.690-1.720	0.0015-0.0035	0.0015-0.0032	NA	NA

NA: Not Available

42372-WBOD-C08

TORQUE SPECIFICATIONS
All readings in ft. lbs.

Year	Engine Displacement Liters (cc)	Engine ID/VIN	Cylinder Head Bolts	Main Bearing Bolts	Rod Bearing Bolts	Crankshaft Damper Bolts	Flywheel Bolts	Manifold Intake	Manifold Exhaust	Spark Plug	Lug Nut
2000	3.1 (3130)	M	①	②	③	76	61	④	10	⑤	100
	3.1 (3130)	J	①	②	③	76	61	④	10	⑤	100
	3.4 (3393)	E	⑥	②	39	78	61	18	⑦	11	100
	3.5 (3475)	H	⑧	⑨	⑩	⑪	⑫	5	18	15	100
	3.8 (3785)	1	⑬	⑭	⑮	⑯	⑰	⑱	22	11	100
	3.8 (3785)	K	⑬	⑭	⑮	⑯	⑰	⑲	38	11	100
2001	3.1 (3130)	M	①	②	③	76	61	④	10	⑤	100
	3.1 (3130)	J	①	②	③	76	61	④	10	⑤	100
	3.4 (3393)	X	⑥	②	39	78	61	18	⑦	11	100
	3.5 (3475)	H	⑧	⑨	⑩	⑪	⑫	5	18	15	100
	3.8 (3785)	1	⑬	⑭	⑮	⑯	⑰	⑱	22	11	100
	3.8 (3785)	K	⑬	⑭	⑮	⑯	⑰	⑲	38	11	100
2002	3.1 (3130)	M	①	②	③	76	61	④	10	⑤	100
	3.1 (3130)	J	①	②	③	76	61	④	10	⑤	100
	3.4 (3393)	X	⑥	②	39	78	61	18	⑦	11	100
	3.5 (3475)	H	⑧	⑨	⑩	⑪	⑫	5	18	15	100
	3.8 (3785)	1	⑬	⑭	⑮	⑯	⑰	⑱	22	11	100
	3.8 (3785)	K	⑬	⑭	⑮	⑯	⑰	⑲	38	11	100
2003	3.1 (3130)	M	①	②	③	76	61	④	10	⑤	100
	3.1 (3130)	J	①	②	③	76	61	④	10	⑤	100
	3.4 (3393)	X	⑥	②	39	78	61	18	⑦	11	100
	3.8 (3785)	1	⑬	⑭	⑮	⑯	⑰	⑱	22	11	100
	3.8 (3785)	K	⑬	⑭	⑮	⑯	⑰	⑲	38	11	100

① Coat threads with sealer torque to 33 ft. lbs., then turn 1/4 turn (90 degrees)

② 37 ft. lbs. plus 77 degrees

③ 15 ft. lbs. plus 75 degrees

④ Torque all bolts to 15 ft. lbs. Retorque to 24 ft. lbs.

⑤ 20 ft. lbs. plus 50 degrees

⑥ 37 ft. lbs. plus 90 degrees

⑦ 115 inch lbs.

⑧ Torque the M11 bolts in sequence to:
Step 1: 22 ft. lbs.
Step 2: 100 degree turn.
Step 3: 100 degree turn.
Torque the long M6 bolt to 106 inch lbs.
Torque both shorter M6 bolts to 106 inch lbs.

⑨ Cap bolts
Step 1: 15 ft. lbs.
Step 2: plus 70 degrees
Perimeter bolts: 22 ft. lbs.

⑩ Step 1: 22 ft. lbs.
Step 2: Loosen completely
Step 3: 18 ft. lbs.
Step 4: plus 110 degrees

⑪ Step 1: 37 ft. lbs.
Step 2: plus 150 degrees

⑫ Step 1: 11 ft. lbs.
Step 2: plus 50 degrees

⑬ Step 1: 35 ft. lbs.
Step 2: plus 130 degrees

⑭ 30 ft. lbs. plus 110 degrees
Side bolts: 11 ft. lbs. plus 45 degrees

⑮ 20 ft. lbs. plus 50 degrees

⑯ 110 ft. lbs. plus 76 degrees

⑰ 11 ft. lbs. plus 50 degrees

⑱ Upper manifold: 8 ft. lbs.
Lower manifold: 11 ft. lbs.

⑲ Upper manifold: 18 ft. lbs.
Lower manifold bolt/nut: 22 ft. lbs.
Upper manifold studs: 89 inch lbs.

42372-WBOD-C09

PISTON AND RING SPECIFICATIONS

All measurements are given in inches.

Year	Engine Displacement Liters (cc)	Engine ID/VIN	Piston Clearance	Ring Gap			Ring Side Clearance		
				Top Compression	Bottom Compression	Oil Control	Top Compression	Bottom Compression	Oil Control
2000	3.1 (3130)	M	0.0010-0.0018	0.0118-0.0196	0.0118-0.0196	0.0157-0.0551	0.0008-0.0015	0.0008-0.0015	0.0004-0.0012
	3.4 (3393)	E	0.0008-0.0020	0.0080-0.0180	0.0220-0.0320	0.0098-0.0229	0.0013-0.0031	0.0013-0.0031	0.0011-0.0081
	3.5 (3475)	H	0.0010-0.0025	0.0080-0.0180	0.0140-0.0200	0.0100-0.0300	0.0016-0.0037	0.0016-0.0037	side-sealing
	3.8 (3785)	J	0.0004-0.0020	0.0100-0.0160	0.0300-0.0400	0.0100-0.0300	0.0013-0.0031	0.0013-0.0031	0.0009-0.0079
	3.8 (3785)	1	0.0004-0.0020	0.0100-0.0160	0.0300-0.0400	0.0100-0.0300	0.0013-0.0031	0.0013-0.0031	0.0009-0.0079
	3.8 (3785)	K	0.0004-0.0020	0.0120-0.0220	0.0300-0.0400	0.0100-0.0300	0.0013-0.0031	0.0013-0.0031	0.0009-0.0079
2001	3.1 (3130)	M	0.0010-0.0018	0.0118-0.0196	0.0118-0.0196	0.0157-0.0551	0.0008-0.0015	0.0008-0.0015	0.0004-0.0012
	3.4 (3393)	E	0.0008-0.0020	0.0080-0.0180	0.0220-0.0320	0.0098-0.0229	0.0013-0.0031	0.0013-0.0031	0.0011-0.0081
	3.5 (3475)	H	0.0010-0.0025	0.0080-0.0180	0.0140-0.0200	0.0100-0.0300	0.0016-0.0037	0.0016-0.0037	side-sealing
	3.8 (3785)	J	0.0004-0.0020	0.0100-0.0160	0.0300-0.0400	0.0100-0.0300	0.0013-0.0031	0.0013-0.0031	0.0009-0.0079
	3.8 (3785)	1	0.0004-0.0020	0.0100-0.0160	0.0300-0.0400	0.0100-0.0300	0.0013-0.0031	0.0013-0.0031	0.0009-0.0079
	3.8 (3785)	K	0.0004-0.0020	0.0120-0.0220	0.0300-0.0400	0.0100-0.0300	0.0013-0.0031	0.0013-0.0031	0.0009-0.0079
2002	3.1 (3130)	M	0.0010-0.0018	0.0118-0.0196	0.0118-0.0196	0.0157-0.0551	0.0008-0.0015	0.0008-0.0015	0.0004-0.0012
	3.4 (3393)	E	0.0008-0.0020	0.0080-0.0180	0.0220-0.0320	0.0098-0.0229	0.0013-0.0031	0.0013-0.0031	0.0011-0.0081
	3.5 (3475)	H	0.0010-0.0025	0.0080-0.0180	0.0140-0.0200	0.0100-0.0300	0.0016-0.0037	0.0016-0.0037	side-sealing
	3.8 (3785)	J	0.0004-0.0020	0.0100-0.0160	0.0300-0.0400	0.0100-0.0300	0.0013-0.0031	0.0013-0.0031	0.0009-0.0079
	3.8 (3785)	1	0.0004-0.0020	0.0100-0.0160	0.0300-0.0400	0.0100-0.0300	0.0013-0.0031	0.0013-0.0031	0.0009-0.0079
	3.8 (3785)	K	0.0004-0.0020	0.0120-0.0220	0.0300-0.0400	0.0100-0.0300	0.0013-0.0031	0.0013-0.0031	0.0009-0.0079
2003	3.1 (3130)	M	0.0010-0.0018	0.0118-0.0196	0.0118-0.0196	0.0157-0.0551	0.0008-0.0015	0.0008-0.0015	0.0004-0.0012
	3.4 (3393)	E	0.0008-0.0020	0.0080-0.0180	0.0220-0.0320	0.0098-0.0229	0.0013-0.0031	0.0013-0.0031	0.0011-0.0081
	3.8 (3785)	J	0.0004-0.0020	0.0100-0.0160	0.0300-0.0400	0.0100-0.0300	0.0013-0.0031	0.0013-0.0031	0.0009-0.0079
	3.8 (3785)	1	0.0004-0.0020	0.0100-0.0160	0.0300-0.0400	0.0100-0.0300	0.0013-0.0031	0.0013-0.0031	0.0009-0.0079
	3.8 (3785)	K	0.0004-0.0020	0.0120-0.0220	0.0300-0.0400	0.0100-0.0300	0.0013-0.0031	0.0013-0.0031	0.0009-0.0079

42372-WBOD-C10

WHEEL ALIGNMENT
BUICK REGAL

Year	Model		Caster Range (Deg.)	Caster Preferred Setting (Deg.)	Camber Range (Deg.)	Camber Preferred Setting (Deg.)	Toe-in (in.)	Steering Axis Inclination (Deg.)
2000	Regal	F	+0.50	+3.00	+0.50	-0.90	0.10 +/- 0.20	—
		R	—	—	+0.50	-0.90	0 +/- 0.20	
2001	Regal	F	+0.50	+3.00	+0.50	-0.90	0.10 +/- 0.20	
		R	—	—	+0.50	-0.90	0 +/- 0.20	
2002	Regal	F	+0.50	+3.00	+0.50	-0.90	0.10 +/- 0.20	
		R	—	—	+0.50	-0.90	0 +/- 0.20	
2003	Regal	F	+0.50	+3.00	+0.50	-0.90	0.10 +/- 0.20	
		R	—	—	+0.50	-0.90	0 +/- 0.20	

42372-WBOD-C11

WHEEL ALIGNMENT
CHEVROLET IMPALA

Year	Model		Caster Range (Deg.)	Caster Preferred Setting (Deg.)	Camber Range (Deg.)	Camber Preferred Setting (Deg.)	Toe-in (in.)	Steering Axis Inclination (Deg.)
2000	Impala	F	+0.50	+3.20 ①	+0.50	+0.85	0.10 +/- 0.20	0
		R	—	—	+0.50	0	0.10 +/- 0.20	—
2001	Impala	F	+0.50	+3.20 ①	+0.50	+0.85	0.10 +/- 0.20	0
		R	—	—	+0.50	0	0.10 +/- 0.20	—
2002	Impala	F	+0.50	+3.20 ①	+0.50	+0.85	0.10 +/- 0.20	0
		R	—	—	+0.50	0	0.10 +/- 0.20	—
2003	Impala	F	+0.50	+3.20 ①	+0.50	+0.85	0.10 +/- 0.20	0
		R	—	—	+0.50	0	0.10 +/- 0.20	—

① Not adjustable, for reference only

42372-WBOD-C12

WHEEL ALIGNMENT
BUICK CENTURY

Year	Model		Caster Range (Deg.)	Caster Preferred Setting (Deg.)	Camber Range (Deg.)	Camber Preferred Setting (Deg.)	Toe-in (in.)	Steering Axis Inclination (Deg.)
2000	Century	F	+0.50	+3.00	+0.50	-0.90	0.10 +/- 0.20	—
		R	—	—	+0.50	-0.90	0 +/- 0.20	—
2001	Century	F	+0.50	+3.00	+0.50	-0.90	0.10 +/- 0.20	—
		R	—	—	+0.50	-0.90	0 +/- 0.20	—
2002	Century	F	+0.50	+3.00	+0.50	-0.90	0.10 +/- 0.20	—
		R	—	—	+0.50	-0.90	0 +/- 0.20	—
2003	Century	F	+0.50	+3.00	+0.50	-0.90	0.10 +/- 0.20	—
		R	—	—	+0.50	-0.90	0 +/- 0.20	—

42372-WBOD-C13

WHEEL ALIGNMENT
CHEVROLET LUMINA

Year	Model		Caster Range (Deg.)	Caster Preferred Setting (Deg.)	Camber Range (Deg.)	Camber Preferred Setting (Deg.)	Toe-in (in.)	Steering Axis Inclination (Deg.)
2000	Lumina	F	+0.50	+1.80	+0.50	+0.70	0 +/- 0.09	—
		R	—	—	+0.50	-0.15	0.06 +/- 0.06	—
2001	Lumina	F	+0.50	+1.80	+0.50	+0.70	0 +/- 0.09	—
		R	—	—	+0.50	-0.15	0.06 +/- 0.06	—
2002	Lumina	F	+0.50	+1.80	+0.50	+0.70	0 +/- 0.09	—
		R	—	—	+0.50	-0.15	0.06 +/- 0.06	—
2003	Lumina	F	+0.50	+1.80	+0.50	+0.70	0 +/- 0.09	—
		R	—	—	+0.50	-0.15	0.06 +/- 0.06	—

42372-WB0D-C14

WHEEL ALIGNMENT
CHEVROLET MONTE CARLO

Year	Model		Caster Range (Deg.)	Caster Preferred Setting (Deg.)	Camber Range (Deg.)	Camber Preferred Setting (Deg.)	Toe-in (in.)	Steering Axis Inclination (Deg.)
2000	Monte Carlo	F	0.50	+1.80	+0.50	+0.70	0 +/- 0.20	—
		R	—	—	+0.50	-0.15	0.10 +/- 0.10	—
2001	Monte Carlo	F	0.50	+1.80	+0.50	+0.70	0 +/- 0.20	—
		R	—	—	+0.50	-0.15	0.10 +/- 0.10	—
2002	Monte Carlo	F	0.50	+1.80	+0.50	+0.70	0 +/- 0.20	—
		R	—	—	+0.50	-0.15	0.10 +/- 0.10	—
2003	Monte Carlo	F	0.50	+1.80	+0.50	+0.70	0 +/- 0.20	—
		R	—	—	+0.50	-0.15	0.10 +/- 0.10	—

42372-WB0D-C15

WHEEL ALIGNMENT
OLDSMOBILE INTRIGUE

Year	Model		Caster Range (Deg.)	Caster Preferred Setting (Deg.)	Camber Range (Deg.)	Camber Preferred Setting (Deg.)	Toe-in (in.)	Steering Axis Inclination (Deg.)
2000	Intrigue	F	0.50	3.00	0.50	-0.90	0.05 +/- 0.10	—
		R	—	—	0.50	-0.90	0 +/- 0.10	—
2001	Intrigue	F	0.50	3.00	0.50	-0.90	0.05 +/- 0.10	—
		R	—	—	0.50	-0.90	0 +/- 0.10	—
2002	Intrigue	F	0.50	3.00	0.50	-0.90	0.05 +/- 0.10	—
		R	—	—	0.50	-0.90	0 +/- 0.10	—

42372-WB0D-C16

WHEEL ALIGNMENT
PONTIAC GRAND PRIX

Year	Model		Caster Range (Deg.)	Caster Preferred Setting (Deg.)	Camber Range (Deg.)	Camber Preferred Setting (Deg.)	Toe-in (in.)	Steering Axis Inclination (Deg.)
2000	Grand Prix	F	0.50	3.00	0.50	-0.90	0.05 +/- 0.10	—
		R	—	—	0.50	-0.90	0 +/- 0.10	
2001	Grand Prix	F	0.50	3.00	0.50	-0.90	0.05 +/- 0.10	—
		R	—	—	0.50	-0.90	0 +/- 0.10	
2002	Grand Prix	F	0.50	3.00	0.50	-0.90	0.05 +/- 0.10	—
		R	—	—	0.50	-0.90	0 +/- 0.10	
2003	Grand Prix	F	0.50	3.00	0.50	-0.90	0.05 +/- 0.10	—
		R	—	—	0.50	-0.90	0 +/- 0.10	

42372-WBOD-C17

TIRE, WHEEL AND BALL JOINT SPECIFICATIONS
Buick Century and Regal

Year	Model	OEM Tires Standard	OEM Tires Optional	Tire Pressures (psi) Front	Tire Pressures (psi) Rear	Wheel Size	Ball Joint Inspection
2000	Regal, exc GS	P205/70R15	P215/70R15 P225/60R16	30	30	6-JJ	①
	Regal GS	P225/60R16	None	30	30	6.5-JJ	①
	Century 2dr	P225/60R16	None	30	30	6.5-JJ	①
	Century 4dr	P205/70R15	P225/60R16	30	30	Std: 6-JJ Opt: 6.5-JJ	①
2001	Regal, exc GS	P205/70R15	P215/70R15 P225/60R16	30	30	6-JJ	①
	Regal GS	P225/60R16	None	30	30	6.5-JJ	①
	Century 2dr	P225/60R16	None	30	30	6.5-JJ	①
	Century 4dr	P205/70R15	P225/60R16	30	30	Std: 6-JJ Opt: 6.5-JJ	①
2002	Regal, exc GS	P205/70R15	P215/70R15 P225/60R16	30	30	6-JJ	①
	Regal GS	P225/60R16	None	30	30	6.5-JJ	①
	Century 2dr	P225/60R16	None	30	30	6.5-JJ	①
	Century 4dr	P205/70R15	P225/60R16	30	30	Std: 6-JJ Opt: 6.5-JJ	①
2003	Regal, exc GS	P205/70R15	P215/70R15 P225/60R16	30	30	6-JJ	①
	Regal GS	P225/60R16	None	30	30	6.5-JJ	①
	Century 2dr	P225/60R16	None	30	30	6.5-JJ	①
	Century 4dr	P205/70R15	P225/60R16	30	30	Std: 6-JJ Opt: 6.5-JJ	①

OEM: Original Equipment Manufacturer

PSI: Pounds Per Square Inch

STD: Standard

OPT: Optional

L: Lower

U: Upper

① Replace if any measurable movement is found.

42372-WBOD-C18

TIRE, WHEEL AND BALL JOINT SPECIFICATIONS
Pontiac Grand Prix

| Year | Model | OEM Tires | | Tire Pressures (psi) | | Wheel Size | Ball Joint Inspection |
		Standard	Optional	Front	Rear		
2000	Grand Prix	P205/70R15	P225/60R16	30	30	7-JJ	0.125 in.
2001	Grand Prix	P205/70R15	P225/60R16	30	30	7-JJ	0.125 in.
2002	Grand Prix	P205/70R15	P225/60R16	30	30	7-JJ	0.125 in.
2003	Grand Prix	P205/70R15	P225/60R16	30	30	7-JJ	0.125 in.

OEM: Original Equipment Manufacturer

PSI: Pounds Per Square Inch

42372-WBOD-C19

TIRE, WHEEL AND BALL JOINT SPECIFICATIONS
Chevrolet Impala, Lumina and Monte Carlo

| Year | Model | OEM Tires | | Tire Pressures (psi) | | Wheel Size | Ball Joint Inspection |
		Standard	Optional	Front	Rear		
2000	Monte Carlo	P195/70R14	None	30	30	6-JJ	①
			P225/60R16N	30	30	6-JJ	
	Impala	P225/60R16	P225/60R16	30	30	6-JJ	①
			P225/60R16N	30	30	6-JJ	
	Lumina 3.1L	P205/70R15	P215/65R16	30	30	6-JJ	①
			P225/60R16	30	30	6.5-JJ	
	Lumina 3.4L	P225/60R16	None	30	30	6.5-JJ	①
2001	Monte Carlo	P195/70R14	None	30	30	6-JJ	①
			P225/60R16N	30	30	6-JJ	
	Impala	P225/60R16	P225/60R16	30	30	6-JJ	①
			P225/60R16N	30	30	6-JJ	
	Lumina 3.1L	P205/70R15	P215/65R16	30	30	6-JJ	①
			P225/60R16	30	30	6.5-JJ	
	Lumina 3.4L	P225/60R16	None	30	30	6.5-JJ	①
2002	Monte Carlo	P195/70R14	None	30	30	6-JJ	①
			P225/60R16N	30	30	6-JJ	
	Impala	P225/60R16	P225/60R16	30	30	6-JJ	①
			P225/60R16N	30	30	6-JJ	
	Lumina 3.1L	P205/70R15	P215/65R16	30	30	6-JJ	①
			P225/60R16	30	30	6.5-JJ	
	Lumina 3.4L	P225/60R16	None	30	30	6.5-JJ	①
2003	Monte Carlo	P195/70R14	None	30	30	6-JJ	①
			P225/60R16N	30	30	6-JJ	
	Impala	P225/60R16	P225/60R16	30	30	6-JJ	①
			P225/60R16N	30	30	6-JJ	
	Lumina 3.1L	P205/70R15	P215/65R16	30	30	6-JJ	①
			P225/60R16	30	30	6.5-JJ	
	Lumina 3.4L	P225/60R16	None	30	30	6.5-JJ	①

OEM: Original Equipment Manufacturer

PSI: Pounds Per Square Inch

① Replace if any measurable movement is found

42372-WBOD-C20

TIRE, WHEEL AND BALL JOINT SPECIFICATIONS
Oldsmobile Intrigue

| Year | Model | OEM Tires | | Tire Pressures (psi) | | Wheel Size | Ball Joint Inspection |
		Standard	Optional	Front	Rear		
2000	Intrigue	P225/60R16	None	30	30	6.5-JJ	①
2001	Intrigue	P225/60R16	None	30	30	6.5-JJ	①
2002	Intrigue	P225/60R16	None	30	30	6-JJ	①

OEM: Original Equipment Manufacturer

PSI: Pounds Per Square Inch

STD: Standard

OPT: Optional

① Replace if any measurable movement is found.

42372-WBOD-C21

BRAKE SPECIFICATIONS
GM W BODY
All measurements in inches unless noted

Year	Model		Brake Disc Original Thickness	Brake Disc Minimum Thickness	Brake Disc Maximum Runout	Brake Drum Original Inside Diameter	Brake Drum Max. Wear Limit	Brake Drum Maximum Machine Diameter	Min Lining Front	Min Lining Rear	Bracket Bolts (ft. lbs.)	Mounting Bolts (ft. lbs.)
2000	Century	F	1.039	0.987	0.003	—	—	—	0.030	0.030	137	63
		R	0.492	0.429	0.003	8.863	8.920	8.909	0.030	0.030	—	—
	Grand Prix	F	1.270	1.210	0.003	—	—	—	0.030	0.030	137	63
		R	0.430	0.350	0.003	—	—	—	0.030	0.030	92	33
	Impala	F	1.270	1.210	0.003	—	—	—	0.030	0.030	137	63
		R	0.430	0.350	0.003	—	—	—	0.030	0.030	92	32
	Intrigue	F	1.270	1.210	0.003	—	—	—	0.030	0.030	137	63
		R	0.430	0.350	0.003	—	—	—	0.030	0.030	92	32
	Lumina	F	1.390	0.987	0.003	—	—	—	0.030	0.030	137	63
		R	0.433	0.385	0.003	8.863	8.920	8.909	0.030	0.030	—	—
	Monte Carlo	F	1.390	0.987	0.003	—	—	—	0.030	0.030	137	63
		R	0.433	0.385	0.003	8.863	8.920	8.909	0.030	0.030	—	—
	Regal	F	1.390	0.987	0.003	—	—	—	0.030	0.030	137	63
		R	0.433	0.385	0.003	8.863	8.920	8.909	0.030	0.030	—	—
2001	Century	F	1.270	1.210	0.002	—	—	—	0.030	0.030	133	70
		R	0.430	0.350	0.002	8.863	8.920	8.909	0.030	0.030	85	32
	Grand Prix	F	1.270	1.210	0.002	—	—	—	0.030	0.030	133	70
		R	0.430	0.350	0.002	8.863	8.920	8.909	0.030	0.030	85	32
	Impala	F	1.270	1.210	0.002	—	—	—	0.030	0.030	133	70
		R	0.430	0.350	0.002	8.863	8.920	8.909	0.030	0.030	85	32
	Intrigue	F	1.270	1.210	0.002	—	—	—	0.030	0.030	133	70
		R	0.430	0.350	0.002	—	—	—	0.030	0.030	85	32
	Lumina	F	1.270	1.210	0.002	—	—	—	0.030	0.030	133	70
		R	0.430	0.350	0.002	8.863	8.920	8.909	0.030	0.030	85	32
	Monte Carlo	F	1.270	1.210	0.002	—	—	—	0.030	0.030	133	70
		R	0.430	0.350	0.002	8.863	8.920	8.909	0.030	0.030	85	32
	Regal	F	1.270	1.210	0.002	—	—	—	0.030	0.030	133	70
		R	0.430	0.350	0.002	8.863	8.920	8.909	0.030	0.030	85	32
2002	Century	F	1.270	1.210	0.002	—	—	—	0.030	0.030	133	70
		R	0.430	0.350	0.002	8.863	8.920	8.909	0.030	0.030	85	32
	Grand Prix	F	1.270	1.210	0.002	—	—	—	0.030	0.030	133	70
		R	0.430	0.350	0.002	8.863	8.920	8.909	0.030	0.030	85	32
	Impala	F	1.270	1.210	0.002	—	—	—	0.030	0.030	133	70
		R	0.430	0.350	0.002	8.863	8.920	8.909	0.030	0.030	85	32
	Intrigue	F	1.270	1.210	0.002	—	—	—	0.030	0.030	133	70
		R	0.430	0.350	0.002	—	—	—	0.030	0.030	85	32
	Lumina	F	1.270	1.210	0.002	—	—	—	0.030	0.030	133	70
		R	0.430	0.350	0.002	8.863	8.920	8.909	0.030	0.030	85	32
	Monte Carlo	F	1.270	1.210	0.002	—	—	—	0.030	0.030	133	70
		R	0.430	0.350	0.002	8.863	8.920	8.909	0.030	0.030	85	32
	Regal	F	1.270	1.210	0.002	—	—	—	0.030	0.030	133	70
		R	0.430	0.350	0.002	8.863	8.920	8.909	0.030	0.030	85	32

42372-WBOD-C22

BRAKE SPECIFICATIONS
GM W BODY

All measurements in inches unless noted

Year	Model		Brake Disc Original Thickness	Brake Disc Minimum Thickness	Brake Disc Maximum Runout	Brake Drum Diameter Original Inside Diameter	Brake Drum Diameter Max. Wear Limit	Brake Drum Diameter Maximum Machine Diameter	Minimum Lining Thickness Front	Minimum Lining Thickness Rear	Brake Caliper Bracket Bolts (ft. lbs.)	Brake Caliper Mounting Bolts (ft. lbs.)
2003	Century	F	1.270	1.210	0.002	—	—	—	0.030	0.030	133	70
		R	0.430	0.350	0.002	8.863	8.920	8.909	0.030	0.030	85	32
	Grand Prix	F	1.270	1.210	0.002	—	—	—	0.030	0.030	133	70
		R	0.430	0.350	0.002	8.863	8.920	8.909	0.030	0.030	85	32
	Impala	F	1.270	1.210	0.002	—	—	—	0.030	0.030	133	70
		R	0.430	0.350	0.002	8.863	8.920	8.909	0.030	0.030	85	32
	Lumina	F	1.270	1.210	0.002	—	—	—	0.030	0.030	133	70
		R	0.430	0.350	0.002	8.863	8.920	8.909	0.030	0.030	85	32
	Monte Carlo	F	1.270	1.210	0.002	—	—	—	0.030	0.030	133	70
		R	0.430	0.350	0.002	8.863	8.920	8.909	0.030	0.030	85	32
	Regal	F	1.270	1.210	0.002	—	—	—	0.030	0.030	133	70
		R	0.430	0.350	0.002	8.863	8.920	8.909	0.030	0.030	85	32

F: Front

R: Rear

42372-WBOD-C23

SCHEDULED MAINTENANCE INTERVALS
GM W BODY—BUICK CENTURY, REGAL, CHEVROLET IMPALA, LUMINA, MONTE CARLO, OLDSMOBILE INTRIGUE & PONTIAC GRAND PRIX

TO BE SERVICED	TYPE OF SERVICE	VEHICLE MILEAGE INTERVAL (x1000)												
		7.5	15	22.5	30	37.5	45	52.5	60	67.5	75	82.5	90	97.5
Engine oil & filter	R	✓	✓	✓	✓	✓	✓	✓	✓	✓	✓	✓	✓	✓
Automatic transaxle fluid & filter ①	S/I	✓	✓	✓	✓	✓	✓	✓	✓	✓	✓	✓	✓	✓
Brake hoses	S/I	✓	✓	✓	✓	✓	✓	✓	✓	✓	✓	✓	✓	✓
Coolant level, hoses & clamps	S/I	✓	✓	✓	✓	✓	✓	✓	✓	✓	✓	✓	✓	✓
Drive shaft boots & front suspension components	S/I	✓	✓	✓	✓	✓	✓	✓	✓	✓	✓	✓	✓	✓
Exhaust system & throttle linkage	S/I	✓	✓	✓	✓	✓	✓	✓	✓	✓	✓	✓	✓	✓
Lubricate chassis, suspension, steering linkage, transaxle shift linkage, parking brake cable guides, underbody contact points & linkage	S/I	✓	✓	✓	✓	✓	✓	✓	✓	✓	✓	✓	✓	✓
Rotate tires	S/I	✓		✓		✓		✓		✓		✓		✓
Air filter element	R					✓			✓				✓	
Engine coolant ②	R					✓			✓				✓	
PCV filter	R					✓			✓				✓	
Spark plugs ③	R					✓			✓				✓	
Accessory drive belt(s)	S/I					✓			✓				✓	
Ignition cables, EGR & fuel systems	S/I					✓			✓				✓	
Camshaft timing belt	R								✓					

R: Replace   S/I: Service or Inspect

① Automatic transaxle fluid & filter: replace at 100,000 miles (if not changed previously).

② Engine coolant: replace every 100,000 miles. Use O.E. specified (DEX-COOL™) coolant only. If any silicate coolant is used, the service interval is every 30,000 miles

③ Platinum tip spark plugs: replace every 100,000 miles.

FREQUENT OPERATION MAINTENANCE (SEVERE SERVICE)
If a vehicle is operated under any of the following conditions it is considered severe service:
- Extremely dusty areas.
- 50% or more of the vehicle operation is in 32°C (90°F) or higher temperatures, or constant operation in temperatures below 0°C (32°F).
- Prolonged idling (vehicle operation in stop and go traffic).
- Frequent short running periods (engine does not warm to normal operating temperatures).
- Police, taxi, delivery usage or trailer towing usage.

Oil & oil filter: change every 3000 miles
Chassis lubrication: lubricate every 6000 miles.
Rotate tires at 6000 miles, then every 12,000 miles.
Air filter element: service or inspect every 15,000 miles.
Camshaft timing belt: change every 60,000 miles.

42372-WBOD-C24

PRECAUTIONS

Before servicing any vehicle, please be sure to read all of the following precautions, which deal with personal safety, prevention of component damage, and important points to take into consideration when servicing a motor vehicle:

• Never open, service or drain the radiator or cooling system when the engine is hot; serious burns can occur from the steam and hot coolant.

• Observe all applicable safety precautions when working around fuel. Whenever servicing the fuel system, always work in a well-ventilated area. Do not allow fuel spray or vapors to come in contact with a spark, open flame or excessive heat (a hot drop light, for example). Keep a dry chemical fire extinguisher near the work area. Always keep fuel in a container specifically designed for fuel storage; also, always properly seal fuel containers to avoid the possibility of fire or explosion. Refer to the additional fuel system precautions later in this section.

• Fuel injection systems often remain pressurized, even after the engine has been turned **OFF**. The fuel system pressure must be relieved before disconnecting any fuel lines. Failure to do so may result in fire and/or personal injury.

• Brake fluid often contains polyglycol ethers and polyglycols. Avoid contact with the eyes and wash your hands thoroughly after handling brake fluid. If you do get brake fluid in your eyes, flush your eyes with clean, running water for 15 minutes. If eye irritation persists, or if you have taken brake fluid internally, seek medical assistance IMMEDIATELY.

• The EPA warns that prolonged contact with used engine oil may cause a number of skin disorders, including cancer! You should make every effort to minimize your exposure to used engine oil. Protective gloves should be worn when changing oil. Wash your hands and any other exposed skin areas as soon as possible after exposure to used engine oil. Soap and water, or waterless hand cleaner should be used.

• All new vehicles are now equipped with an air bag system. The system must be disabled before performing service on or around system components, steering column, instrument panel components, wiring and sensors. Failure to follow safety and disabling procedures could result in accidental air bag deployment, possible personal injury and unnecessary system repairs.

• Always wear safety goggles when working with, or around, the air bag system. When carrying a non-deployed air bag, be sure the bag and trim cover are pointed away from your body. When placing a non-deployed air bag on a work surface, always face the bag and trim cover upward, away from the surface. This will reduce the motion of the module if it is accidentally deployed. Refer to the additional air bag system precautions later in this section.

• Clean, high quality brake fluid from a sealed container is essential to the safe and proper operation of the brake system. You should always buy the correct type of brake fluid for your vehicle. If the brake fluid becomes contaminated, completely flush the system with new fluid. Never reuse any brake fluid. Any brake fluid that is removed from the system should be discarded. Also, do not allow any brake fluid to come in contact with a painted surface; it will damage the paint.

• Never operate the engine without the proper amount and type of engine oil; doing so WILL result in severe engine damage.

• Timing belt maintenance is extremely important! Many models utilize an interference-type, non-freewheeling engine. If the timing belt breaks, the valves in the cylinder head may strike the pistons, causing potentially serious (also time-consuming and expensive) engine damage. Refer to the maintenance interval charts for the recommended replacement interval for the timing belt.

• Disconnecting the negative battery cable on some vehicles may interfere with the functions of the on-board computer system(s) and may require the computer to undergo a relearning process once the negative battery cable is reconnected.

• When servicing drum brakes, only disassemble and assemble one side at a time, leaving the remaining side intact for reference.

ENGINE REPAIR

Distributor

The engines in this section all utilize a Distributorless Ignition System (DIS).

Alternator

REMOVAL

3.1L Engine

1. Before servicing the vehicle, refer to the precautions in the beginning of this section.
2. Remove or disconnect the following:
   • Negative battery cable
   • Engine compartment cross vehicle brace, if equipped
   • Drive belt from the alternator
   • Coolant recovery reservoir and place it away from the alternator

• Alternator electrical connector
• Alternator bolts and the alternator

3.4L Engine

1. Before servicing the vehicle, refer to the precautions in the beginning of this section.
2. Remove or disconnect the following:
   • Negative battery cable
   • Engine compartment cross brace
   • Accessory drive belt
   • Coolant recovery reservoir and place it aside
   • Alternator bolts
   • Alternator electrical connector
   • Alternator

3.5L Engine

1. Before servicing the vehicle, refer to the precautions in the beginning of this section.
2. Drain the cooling system.

3. Remove or disconnect the following:
   • Battery and tray
   • Drive belt
   • Engine cooling fan assembly
   • Thermostat housing and radiator hose
   • Outboard/inboard alternator bolts
   • Idler pulley bolt and pulley
   • Alternator electrical connectors
   • Alternator bolts and the alternator

3.8L Engine

1. Before servicing the vehicle, refer to the precautions in the beginning of this section.
2. Remove or disconnect the following:
   • Negative battery cable
   • Drive belt
   • Alternator electrical connectors
   • Rear alternator brace
   • Alternator bolts and the alternator

INSTALLATION

3.1L Engine

1. Install or connect the following:
   - Alternator
   - Alternator output BAT terminal and torque the nut to 15 ft. lbs. (20 Nm)
   - Protective boot from the output terminal
   - Electrical connector
   - Alternator bolts and torque them to 37 ft. lbs. (50 Nm)
   - Coolant recovery reservoir
   - Drive belt
   - Engine compartment cross vehicle brace, if removed
   - Negative battery cable

3.4L Engine

1. Install or connect the following:
   - Alternator
   - Alternator output BAT wire and torque the nut to 15 ft. lbs. (20 Nm)
   - Alternator electrical connector
   - Alternator bolts and torque them to 37 ft. lbs. (50 Nm)
   - Coolant recovery reservoir
   - Drive belt
   - Negative battery cable

3.5L Engine

1. Install or connect the following:
   - Alternator
   - Alternator electrical connectors. Torque the positive battery terminal to 15 ft. lbs. (20 Nm).
   - Idler pulley and torque the bolt to 37 ft. lbs. (50 Nm)
   - Alternator bolts and torque them to 37 ft. lbs. (50 Nm)
   - Thermostat housing with radiator hose and torque the bolts to 80 inch lbs. (9 Nm)
   - Engine cooling fan assembly

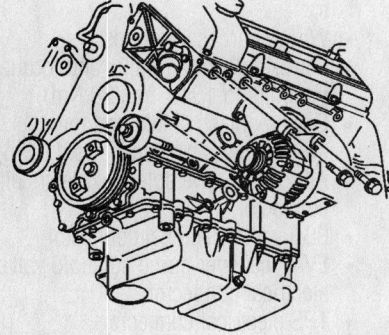

9306VG01

View of the alternator—3.5L engine

- Drive belt
- Battery tray and torque the bolts to 44 inch lbs. (5 Nm)
- Battery
2. Refill the cooling system.

3.8L Engine

1. Install or connect the following:
   - Alternator
   - Alternator bolts and torque them to 37 ft. lbs. (50 Nm)
   - Alternator electrical connectors and torque the positive battery terminal to 15 ft. lbs. (20 Nm)
   - Rear alternator brace and torque the bolts to 37 ft. lbs. (50 Nm)
   - Accessory drive belt
   - Negative battery cable

Ignition Timing

ADJUSTMENT

The ignition timing is not adjustable.

Engine Assembly

REMOVAL & INSTALLATION

3.1L Engine

1. Before servicing the vehicle, refer to the precautions in the beginning of this section.
2. Drain the engine coolant.
3. Drain the engine oil.
4. Relieve the fuel system pressure.
5. Remove or disconnect the following:
   - Negative battery cable
   - Hood
   - Throttle body air inlet duct
   - Air cleaner
   - Engine mount struts
   - Drive belt
   - Knock Sensor (KS) electrical connector
   - Heated Oxygen Sensor (HO2S) electrical connector
   - Camshaft Position (CMP) sensor electrical connector
   - Crankshaft Position (CKP) sensor electrical connector
   - Manifold Air Pressure (MAP) sensor electrical connector
   - Exhaust Gas Recirculation (EGR) valve electrical connector
   - Evaporative (EVAP) emissions canister purge solenoid valve
   - Throttle Position Sensor (TPS) electrical connector

   - Idle Air Control (IAC) valve electrical connector
   - Starter motor electrical connector
   - Alternator electrical connector
   - Ignition coil electrical connector
   - Wiring harness ground connections
   - Two body wiring harness-to-engine harness connectors
   - Oil filter
   - Catalytic converter pipe from the rear exhaust manifold
   - Engine mount lower nuts
   - Torque converter cover
   - A/C compressor
   - Right side splash shield
   - Torque converter bolts
   - Transaxle brace
   - Radiator outlet hose from the engine
   - Accelerator and cruise control cables
   - Heater inlet and outlet hoses
   - Vacuum hoses from the upper intake manifold
   - Fuel feed and return hoses
   - Power steering pump and reposition
   - Radiator inlet and outlet hoses
6. Attach an engine lifting device.
7. Remove the transmission-to-engine bolts.
8. Remove the engine from the vehicle.

To install:

9. Install or connect the following:
   - Engine to the vehicle
   - Transmission-to-engine bolts and torque them to 55 ft. lbs. (75 Nm)
10. Remove the engine lifting device.
    - Power steering pump and torque the bolts to 25 ft. lbs. (34 Nm)
    - Power steering lines to the pump and torque the fasteners to 20 ft. lbs. (27 Nm)
    - Radiator inlet hose
    - Heater inlet and outlet hoses
    - Fuel feed and return hoses
    - Vacuum hoses to the upper intake manifold

➡ **In order to avoid possible injury or vehicle damage, always replace the accelerator control cable with a NEW cable whenever you remove the engine from the vehicle.**

11. Remove the trim panel under the left instrument panel and detach the throttle cable from the top of the pedal then squeeze the retainer and push the cable through the bulkhead to remove it.
12. Install or connect the following:
    - New accelerator cable
    - Cruise control cable

- Radiator outlet hose
- Transmission brace and torque the transmission bolts to 32 ft. lbs. (43 Nm) and the engine bolts to 46 ft. lbs. (63 Nm)
- Right side splash shield and torque the fasteners to 44 inch lbs. (5 Nm)
- Torque converter bolts and torque them to 47 ft. lbs. (63 Nm)
- A/C compressor and torque the bolts to 37 ft. lbs. (50 Nm)
- Starter motor and torque the bolts to 32 ft. lbs. (43 Nm)
- Torque converter cover and torque the bolts to 89 inch lbs. (10 Nm)
- Engine mount lower nuts and torque them to 32 ft. lbs. (43 Nm)
- Catalytic converter pipe to the rear exhaust manifold and torque the fasteners to 26 ft. lbs. (35 Nm)
- Oil Filter
- Two body wiring to engine harness
- Wiring harness grounds
- Ignition coil electrical connector
- Alternator electrical connector
- A/C compressor electrical connector
- Starter motor electrical connector
- IAC valve electrical connector
- TPS electrical connector
- EVAP canister purge solenoid valve electrical connector
- EGR valve electrical connector
- MAP sensor electrical connector
- CKP sensor electrical connector
- CMP sensor electrical connector
- HO2sensor electrical connector
- KS electrical connector
- Drive belt
- Engine mount struts and torque the fasteners to 35 ft. lbs. (48 Nm)
- Hood and torque the bolts to 18 ft. lbs. (25 Nm)
- Throttle body air inlet duct
- Air cleaner
- Negative battery cable

13. Fill the engine with clean oil.
14. Fill the coolant system.
15. Start the vehicle and check for leaks, repair if necessary.

3.4L Engine

1. Before servicing the vehicle, refer to the precautions in the beginning of this section.
2. Drain the cooling system.
3. Drain the engine oil.
4. Relieve the fuel system pressure.
5. Remove or disconnect the following:
   - Hood
   - Cross vehicle brace

- Engine mount struts
- Drive belt
- Brake booster vacuum hose
- Heated Oxygen Sensor (HO2S) electrical connector
- Secondary Air Injection (AIR) check valve electrical connector
- Exhaust Gas Recirculation (EGR) valve electrical connector
- Evaporative Emissions (EVAP) canister purge solenoid valve electrical connector
- Throttle Position Sensor (TPS) electrical connector
- Idle Air Control (IAC) valve electrical connector
- Alternator electrical connector
- Ignition coil electrical connector
- Wiring harness grounds
- Two engine wiring harness connectors
- Lower radiator air baffle
- Engine splash shields
- Oil filter
- Vehicle Speed Sensor (VSS) electrical connector
- Oil level sensor electrical connector
- Oil pressure switch electrical connector
- Engine block heater (if equipped)
- Knock Sensor (KS) electrical connector
- Crankshaft Position (CKP) sensor
- A/C compressor electrical connector
- Catalytic converter
- Engine mount lower nuts
- Torque converter cover
- Starter motor
- A/C compressor without disconnecting the lines
- Torque converter bolts
- Transaxle brace
- Radiator outlet hose
- Accelerator and cruise control cables
- Vacuum hoses from the upper intake manifold
- Fuel feed and return hoses
- Power steering pump without disconnecting the hoses
- Heater inlet and outlet hoses
- Radiator inlet hose

6. Attach an engine lifting device.
7. Remove the transmission bolts.
8. Remove the engine from the vehicle.

To install:

9. Install or connect the following:
   - Engine to the transaxle and torque the bolts to 55 ft. lbs. (75 Nm)
   - Radiator inlet hose
   - Heater inlet and outlet hoses

- Power steering pump and torque the bolts to 25 ft. lbs. (34 Nm)
- Fuel feed and return hoses
- Vacuum hoses to the upper intake manifold

➡In order to avoid possible injury or vehicle damage, always replace the accelerator control cable with a NEW cable whenever you remove the engine from the vehicle.

10. Remove the trim panel under the left instrument panel and detach the throttle cable from the top of the pedal then squeeze the retainer and push the cable through the bulkhead to remove it.
11. Install or connect the following:
    - New accelerator cable
    - Cruise control cable
    - Radiator outlet hose
    - Transmission brace and torque the transmission bolts to 32 ft. lbs. (43 Nm) and the engine bolts to 46 ft. lbs. (63 Nm)
    - Torque converter-to-flywheel bolts and torque them to 47 ft. lbs. (63 Nm)
    - A/C compressor and torque the bolts to 37 ft. lbs. (50 Nm)
    - Starter motor and torque the bolts to 32 ft. lbs. (43 Nm)
    - Torque converter cover and torque the bolts to 89 inch lbs. (10 Nm)
    - Engine mount lower nuts and torque them to 32 ft. lbs. (43 Nm)
    - Catalytic converter and torque the nuts to 24 ft. lbs. (32 Nm)
    - New oil filter
    - VSS electrical connector
    - Oil level sensor electrical connector
    - Oil pressure switch electrical connector
    - Engine block heater electrical connector
    - KS electrical connector
    - HO2S electrical connector
    - CKP sensor electrical connector
    - A/C compressor electrical connector
    - Wiring harness grounds
    - Lower radiator air baffle and torque the bolts to 15 ft. lbs. (20 Nm)
    - Engine splash shield and torque the fasteners to 18 inch lbs. (2 Nm)
    - AIR check valve solenoid electrical connector
    - EGR valve electrical connector
    - EVAP canister purge solenoid valve electrical connector
    - TPS electrical connector
    - IAC valve electrical connector
    - Alternator electrical connector

- Ignition coil electrical connector
- Wiring harness grounds
- Engine wiring harness connectors
- Drive belt
- Engine mount strut and torque the bolt to 35 ft. lbs. (48 Nm)
- Throttle body air inlet duct
- Cross vehicle brace and torque the nuts to 29 ft. lbs. (40 Nm)
- Hood and torque the bolts to 18 ft. lbs. (25 Nm)
  - Negative battery cable
12. Fill the engine oil.
13. Fill the engine coolant.
14. Inspect the transmission fluid level and top off if necessary.
15. Turn the ignition to the **ON** position several times to pressurize the fuel system.
16. Start the engine and inspect for any leaks, repair if necessary.
17. Check and top off the fluid levels if required.

3.5L Engine

2000 MODELS

1. Before servicing the vehicle, refer to the precautions in the beginning of this section.
2. Relieve the fuel system pressure.
3. Drain the engine oil.
4. Drain the engine cooling system.
5. Remove or disconnect the following:
- Drive belt
- Fuel injector sight shield
- Power steering pump and move it aside
- Air intake duct from the throttle body
- Engine mount strut from the bracket
- Fuel lines from the fuel supply rail
- Fuel vapor line
- Throttle and cruise control cables with the bracket from the throttle body
- Range selector cable from the Park/Neutral Position (PNP) switch
- Brake booster vacuum hose from the engine
- Air conditioning vacuum hose from the engine
- Wiring harness from the engine and transaxle
- Radiator inlet hose from the engine
- Transaxle fluid cooler lines from the radiator
- Thermostat housing
- Surge tank inlet hose
- Heater hoses from the engine
- Lower radiator air deflector

- Secondary Air Injection (AIR) pipe from the AIR inlet valve
- Battery cables from the retainers
- Radiator outlet hose from the engine
- Alternator
- A/C compressor, move it aside without disconnecting the lines
- Torque converter cover
- Starter motor
- Flexplate-to-torque converter bolts
- Catalytic converter from the rear exhaust manifold
- Lower transaxle-to-engine bolts
- Front wheels and splash shields
- Fog lamp electrical connectors
- Wheel Speed Sensor (WSS) harness from the lower control arms
- Tie rod ends from the steering knuckles
- Lower ball joints from the steering knuckles
- Halfshafts
- Intermediate shaft from the steering rack

> ❈❈ **WARNING**
>
> **Secure the front of the vehicle to the lift. The vehicle may become unstable as the engine/transaxle assembly is removed from the vehicle.**

6. Position a frame table under the vehicle.
7. Lower the vehicle so the frame is resting on the frame table.
8. Remove the frame-to-chassis bolts.

> ❈❈ **WARNING**
>
> **Do not damage the air conditioning compressor or lines when removing the frame from the vehicle.**

9. Lift the vehicle from the engine/transaxle assembly.
10. Remove or disconnect the following:
- Remove the engine mount bracket bolts
- Remove the transmission brace
- Remove the engine from the transaxle

To install:
11. Install or connect the following:
- Transaxle to the engine and torque the bolts to 55 ft. lbs. (75 Nm)
- Transaxle brace and torque the bolts to 32 ft. lbs. (43 Nm)
- Engine mount bracket to the front of the engine and torque the bolts to 43 ft. lbs. (58Nm)

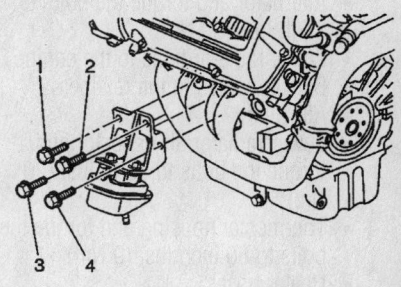

9300Z501

Engine mount bracket bolt tightening sequence—3.5L engine

- Engine/transaxle assembly under the vehicle
12. Coat the sub-frame bushings with rubber lubricant.
13. Lower the vehicle onto the assembly. Align the sub-frame on the vehicle using 2 bolts or drill bits, ¾ inches thick by 8 inches long through the alignment holes on the right side of the frame.
14. Install new frame-to-body bolts and torque them to 133 ft. lbs. (180 Nm) starting with the rear bolts and then the front bolts
15. Raise the vehicle and remove the frame table.
16. Install or connect the following:
- Intermediate shaft to the steering rack and torque the pinch bolt to 35 ft. lbs. (48 Nm)

➡ **Be sure the shaft is fully seated on the stub before installing the pinch bolt.**

- Halfshafts
- Lower ball joints and torque the nuts to 40 ft. lbs. (50 Nm)
- Tie rod ends and torque the nuts to 22 ft. lbs. (30 Nm) plus an additional 120 degree turn
- WSS wiring harness to the lower control arm
- Splash shields
- Front wheels
- Fog lamp electrical connectors
- Catalytic converter to the rear exhaust manifold and torque the nuts to 53 inch lbs. (6 Nm)
- Flexplate-to-torque converter by aligning the matchmarks and torque the bolts to 47 ft. lbs. (63 Nm)
- Starter motor and torque the bolts to 37 ft. lbs. (50 Nm)
- Torque converter cover and torque the bolts to 89 inch lbs. (10 Nm)
- A/C compressor and torque the bolts to 37 ft. lbs. (50 Nm)

- Alternator and torque the bolts to 37 ft. lbs. (50 Nm)
- Lower radiator hose to the engine
- Battery cables in the retainers
- AIR pipe
- Lower radiator air deflector and torque the bolts to 15 ft. lbs. (20 Nm)
- Thermostat housing and torque the bolts to 80 inch lbs. (9 Nm)
- Heater hoses
- Surge tank hose
- Transaxle fluid cooler lines to the radiator
- Upper radiator hose to the engine
- Wiring harness connectors to the engine/transaxle
- A/C vacuum hose to the engine
- Brake booster hose to the engine
- Shifter cable to the transaxle PNP switch

➡ In order to avoid possible injury or vehicle damage, always use a new accelerator cable when replacing the engine assembly.

17. Remove the trim panel under the left instrument panel and detach the throttle cable from the top of the pedal then squeeze the retainer and push the cable through the bulkhead to remove it.

18. Install or connect the following:
- New throttle cable
- Cruise control cable to the throttle body
- Fuel vapor line
- Fuel lines to the supply rail
- Engine mount strut and torque the bolt to 35 ft. lbs. (48 Nm)
- Air duct to the throttle body
- Power steering pump and torque the bolts to 25 ft. lbs. (34 Nm)
- Fuel injector shield and torque the bolts to 27 inch lbs. (3 Nm)
- Drive belt
- Negative battery cable

19. Refill the cooling system.
20. Install a new oil filter and refill the engine with new oil.
21. Start the engine and check for leaks, repair if necessary.

3.5L Engine

2001–02 MODELS

1. Before servicing the vehicle, refer to the precautions in the beginning of this section.
2. Relieve the fuel system pressure.
3. Drain the engine oil.
4. Drain the engine cooling system.
5. Remove or disconnect the following:

- Front wheels
- Lower air deflector
- Fuel injector sight shield
- Left diagonal brace
- Air cleaner assembly
- Radiator inlet hose
- Alternator
- Fuel lines from the rail and position
- Fuel vapor line
- Throttle and cruise cables with the mounting bracket from the throttle body
- Automatic range selector cable from the Park/Neutral Position (PNP) switch
- Vacuum brake booster hose from the booster
- Transaxle oil cooler lines from the radiator
- Heater hoses from the engine
- Right side AIR control valve, if equipped
- Upper engine electrical connectors, including grounds
- Lower AIR hose from the elbow, if equipped and discard the clamp
- Lower electrical connectors, including grounds and the harness retainer
- Fog lamp electrical connectors, if equipped
- Torque converter cover
- Starter
- Bolts attaching the engine flywheel to the torque converter
- Engine splash shields from the inner fender
- A/C compressor and position aside without disconnecting the lines
- Catalytic converter pipe from the rear exhaust manifold
- Bolts attaching the lower transaxle to the engine
- Connectors from the Wheel Speed Sensors, and unclip the harnesses from the lower control arms
- Drive axles from the steering knuckles

※ CAUTION

Failure to disconnect the intermediate shaft from the rack and pinion stub shaft can result in damage to the steering gear and/or damage to the intermediate shaft. This damage may cause loss of steering control which could result in personal injury.

- Pinch bolt from the intermediate steering shaft

- Intermediate shaft from the steering gear
- Upper and lower engine wiring harnesses position them aside

6. Place a frame table under the engine/transaxle/frame. With an assistant, lower the vehicle to the frame table in order to support the engine/transaxle/frame.
- Bolts attaching the frame to the body
- Frame table with the engine/transaxle/frame from under the vehicle
- Power steering pump from the engine and lay the pump aside
- Secondary air injection check valve/pipe, if equipped
- Dipstick tube
- Exhaust manifolds and crossover pipe

7. Install an engine lifting device to the engine.
- Bolts attaching the engine mount bracket to the engine
- Transaxle brace
- Bolts attaching the upper transmission to the engine
- Engine from the transaxle/frame

To install:

8. Installation is the reverse of removal, please note the following torques:
9. Install or connect the following:
- Bolts attaching the transmission to the engine to 55 ft. lbs. (75 Nm)
- Frame bolts to 133 ft. lbs. (180 Nm)

10. Center the steering angle sensor.
11. Fill the engine with new oil.

➡ A filter change is recommended.

12. Fill the cooling system.
13. Perform the Crankshaft Position (CKP) system variation learn procedure.
14. Start the engine and check for leaks, repair if necessary.

3.8L Engine

2000 MODELS

1. Before servicing the vehicle, refer to the precautions in the beginning of this section.
2. Relieve the fuel system pressure.
3. Drain the cooling system.
4. Drain the engine oil.
5. Remove or disconnect the following:

- Hood
- Fuel injector sight shield
- Throttle body air inlet duct
- Air cleaner
- Radiator inlet/outlet hoses

- A/C compressor without disconnecting the lines
- Starter motor wiring harness
- Accelerator and cruise control cables
- Wiring harness from the top of the engine
- Fuel lines from the fuel rail
- Fuel vapor line
- Power brake booster vacuum hose
- A/C vacuum hose
- Heater hoses
- Engine mount struts from the brackets
- Power steering pump and move it aside
- Torque converter cover
- Starter motor
- Flywheel-to-torque converter bolts
- Engine mount lower nuts
- Transaxle brace
- Catalytic converter pipe from the rear exhaust manifold
- Lower transaxle to engine bolts

6. Lower the vehicle and install a suitable engine lifting device.

7. Remove the upper transaxle to engine bolts.

8. Remove the engine from the vehicle.

To install:

9. Install the engine to the vehicle.

10. Torque the transmission-to-engine bolts to 55 ft. lbs. (75 Nm).

11. Remove the engine lifting device.

12. Install or connect the following:
- Catalytic converter to the rear exhaust manifold. Torque the nuts to 26 ft. lbs. (35 Nm).
- Transaxle brace and torque the transaxle bolts to 32 ft. lbs. (43 Nm) and the engine bolts to 46 ft. lbs. (63 Nm)
- Engine mount lower nuts and torque them to 35 ft. lbs. (47 Nm)
- Flywheel-to-torque converter bolts and torque them to 47 ft. lbs. (63 Nm)
- Starter motor and torque the bolts to 32 ft. lbs. (43 Nm)
- Torque converter cover and torque the bolts to 89 inch lbs. (10 Nm)
- Power steering pump and torque the bolt to 25 ft. lbs. (34 Nm)
- Engine mount struts and torque the bolts to 35 ft. lbs. (48 Nm)
- Heater hoses
- A/C vacuum hose
- Power brake booster vacuum hose
- Fuel vapor line
- Fuel lines
- Wiring harness to the top of the engine

➡**In order to avoid possible injury or vehicle damage, always replace the accelerator control cable with a NEW cable whenever you remove the engine from the vehicle.**

13. Remove the trim panel under the left instrument panel and detach the throttle cable from the top of the pedal then squeeze the retainer and push the cable through the bulkhead to remove it.

14. Install or connect the following:
- New throttle cable
- Cruise control cable to the throttle body
- Starter motor wiring harness
- A/C compressor and torque the bolts to 23 ft. lbs. (31 Nm) and the nuts to 74 ft. lbs. (100 Nm)
- Inlet and outlet radiator hoses
- Throttle body air inlet duct
- Air cleaner
- Hood and torque the bolts to 20 ft. lbs. (27 Nm)
- Fuel injector sight shield
- Negative battery cable

15. Fill the engine with new oil.

➡**A filter change is recommended.**

16. Fill the cooling system.

17. Start the engine and check for leaks, repair if necessary.

2001–03 MODELS

1. Before servicing the vehicle, refer to the precautions in the beginning of this section.

2. Relieve the fuel system pressure.

3. Matchmark the hood hinges and remove the hood.

4. Drain the cooling system.

5. Drain the engine oil.

6. Remove or disconnect the following:
- Negative battery cable
- Hood
- Fuel injector sight shield
- Torque converter cover
- Starter motor
- Engine flywheel-to-torque converter bolts
- Transaxle bracket
- Engine mount lower nuts
- Catalytic converter pipe from the right exhaust manifold
- Oil level harness retainer from the engine

7. Unplug and reposition the following electrical connections:
- Vehicle Speed Sensor (VSS)
- Oil pressure sensor
- Oil level sensor
- Knock sensors
- Heated Oxygen Sensor (HO2S)

8. Remove or disconnect the following:
- Oil filter adapter housing
- Engine ground nut and engine ground from the transaxle stud
- Bolt and the stud which attach the lower transaxle to the engine

9. Lower the vehicle while supporting the transaxle.
- Air cleaner assembly
- Fuel lines from the fuel rail
- Fuel vapor line
- Throttle and cruise control cables and bracket from the throttle body
- Water pump pulley bolts, loosen only
- Drive belt
- Water pump pulley bolts and the pulley
- Right and left engine mount struts
- Engine mount brackets from the radiator support
- Positive battery cable
- Engine cooling fans
- Vacuum booster hose from the engine
- A/C vacuum hose from the engine
- Power steering pump and lay aside with the lines attached

10. Unplug the upper engine electrical connectors from the following components:
- Fuel injectors
- Throttle Position (TP) sensor
- Idle Air Control (IAC) valve
- Evaporative Emissions (EVAP) purge solenoid
- Exhaust Gas Recirculation (EGR) valve
- Manifold Absolute Pressure (MAP) sensor
- Upper engine wiring harness from retaining clips and reposition the engine wiring harness for engine removal
- Radiator inlet and outlet hoses from the engine
- Heater hoses from the drive belt tensioner
- A/C compressor from the engine and set aside without disconnecting the lines

11. Install an engine lifting device.
- Bolts that attach the upper transaxle to the engine
- Engine from the vehicle

To install:

12. Installation is the reverse of removal, please note the following torques:
- Bolts attaching the upper transaxle to the engine to 55 ft. lbs. (75 Nm)
- A/C compressor-to-bracket nuts to 22 ft. lbs. (30 Nm)

- Power steering bolts to 25 ft. lbs. (34 Nm)
- Water pump pulley bolts to 116 inch lbs. (13 Nm)
- Bolt and the stud attaching the lower transaxle to the engine to 55 ft. lbs. (75 Nm)
- Engine ground nut to 26 ft. lbs. (35 Nm)
- Engine mount lower nuts to 58 ft. lbs. (78 Nm)

13. Fill the engine with new oil.

➡ **A filter change is recommended.**

14. Fill the cooling system.
15. Perform the Crankshaft Position (CKP) system variation learn procedure.
16. Start the engine and check for leaks, repair if necessary.

Water Pump

REMOVAL & INSTALLATION

3.1L and 3.4L Engines

1. Before servicing the vehicle, refer to the precautions in the beginning of this section.
2. Drain the cooling system.
3. Remove or disconnect the following:
   - Negative battery cable
   - Accessory drive belt guard
   - Water pump pulley bolts, loosen
   - Accessory drive belt
   - Water pump pulley
   - Water pump

To install:

4. Install or connect the following:
   - Water pump with a new gasket and torque the bolts to 89 inch lbs. (10 Nm)
   - Water pump pulley and torque the bolts to 18 ft. lbs. (25 Nm)
   - Accessory drive belt
   - Drive belt guard
5. Fill the cooling system.
6. Start the engine and check for leaks, repair if necessary.

3.5L Engine

1. Before servicing the vehicle, refer to the precautions in the beginning of this section.
2. Partially drain the cooling system.
3. Remove or disconnect the following:
   - Surge tank outlet hose
   - Drive belt
   - Idler pulley
   - Water pump pulley

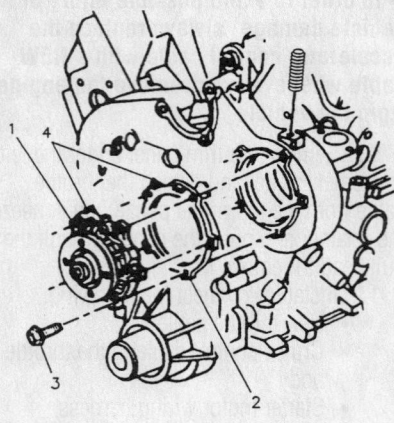

1 WATER PUMP
2 GASKET
3 10 N·m (89 LB. IN.)
4 LOCATOR — MUST BE VERTICAL

7924LG01

Water pump assembly mounting—3.1L and 3.4L engines

➡ **The water pump is attached to the engine with both long and short bolts, be sure to note their locations.**

   - Water pump

To install:

4. Install or connect the following:
   - Water pump using a new gasket and torque the bolts to 124 inch lbs. (14 Nm)

- Water pump pulley and torque the bolts to 106 inch lbs. (12 Nm)
- Idler pulley and torque the bolt to 37 ft. lbs. (50 Nm)
- Drive belt
- Surge tank hose
5. Refill the cooling system.
6. Start the engine and check for leaks.

3.8L Engines

1. Before servicing the vehicle, refer to the precautions in the beginning of this section.
2. Drain the engine coolant.
3. Remove or disconnect the following:
   - Negative battery cable.
   - Drive belt
   - Water pump pulley
   - Power steering pump and move it aside
   - Water pump

To install:

4. Install or connect the following:
   - Water pump using a new gasket and torque the long bolts to 15 ft. lbs. (20 Nm) plus a 40 degree turn on 2000 models or 25 ft. lbs. (34 Nm) on 2001–04 models and the short bolts to 11 ft. lbs. (15 Nm)

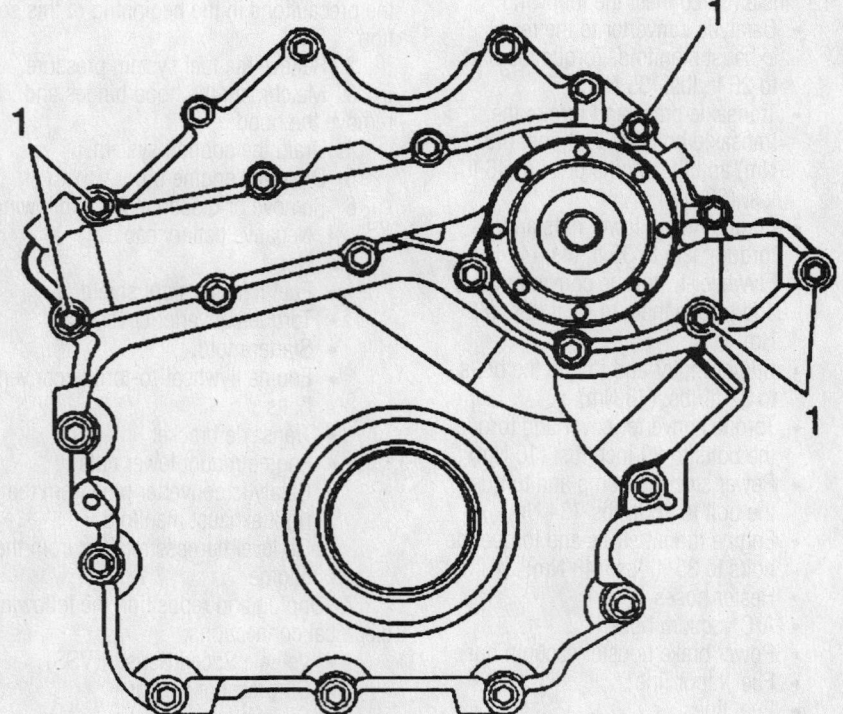

Be sure to install the 5 long water pump bolts in the correct locations—3.5L engine

9300Z502

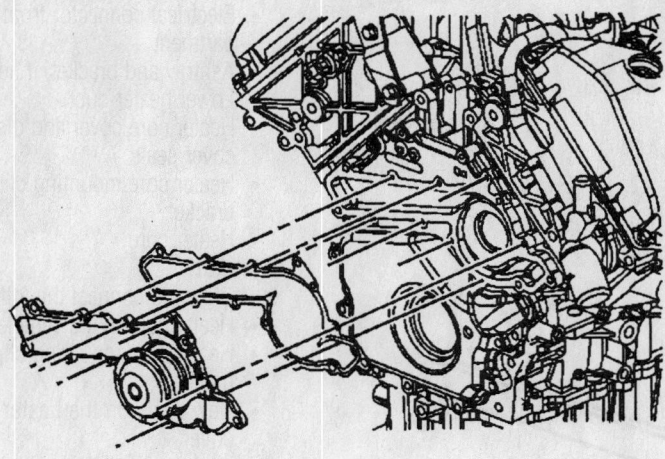

Water pump mounting—3.5L engine

9300Z503

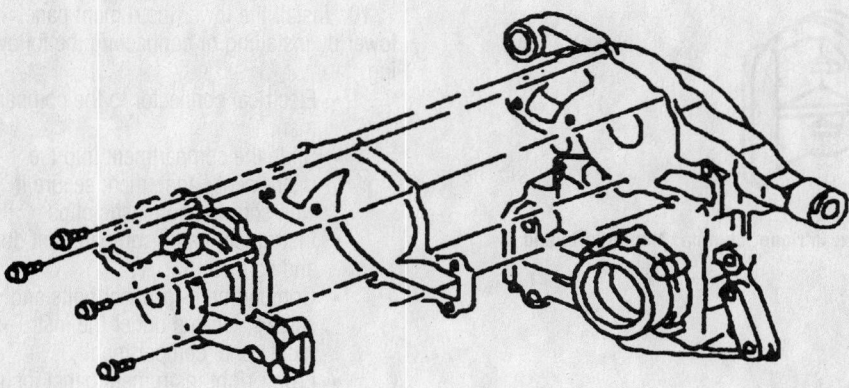

Exploded view of the water pump assembly mounting—3.8L engine

79222504

plus an additional 80 degree turn on 2000 models or 16 ft. lbs. (22 Nm) on 2001–04 models
• Water pump pulley and torque the bolts to 115 inch lbs. (13 Nm)
• Power steering pump and torque the bolts to 25 ft. lbs. (34 Nm)
• Drive belt
• Negative battery cable
5. Fill the cooling system.
6. Start the vehicle and check for leaks, repair if necessary.

Heater Core

REMOVAL & INSTALLATION

Century, Lumina and Regal

1. Before servicing the vehicle, refer to the precautions in the beginning of this section.
2. Disconnect the negative battery cable.

3. If equipped with a 3.1L engine, remove the air cleaner and duct assembly.
4. If equipped with a 3.8L engine, remove the fuel injector sight shield by performing the following procedures:
   a. Clean the area around the oil filler cap/tube assembly location.
   b. Rotate the oil filler cap/tube assembly counterclockwise from the valve cover.
   c. Lift the fuel injector sight shield up at the front and slide it from the rear engine bracket.
   d. Reinstall the oil filler cap/tube assembly in the valve cover.

✳✳ CAUTION

Before draining the cooling system, allow the engine to cool to relieve the systems internal pressure and to avoid scalding coolant.

5. Drain the cooling system by performing the following procedure:

a. Raise and safely support the front of the vehicle.
b. Remove and clean the coolant recovery tank.
c. Place a 2 gallon pan under the radiator to catch the coolant.
d. At the bottom of the radiator, open the drain valve and drain the coolant to a level lower than the heater core.
e. Remove the radiator cap and open the air bleed screw (2–3 turns) located on top of the thermostat housing.
f. After sufficient coolant has been drained from the system, close the drain valve.

✳✳ CAUTION

Engine coolant is a hazardous waste; it should be stored for reuse or submitted for recycling. NEVER dispose of it by dumping it into the environment.

6. Disconnect and plug the heater hoses at the heater core.
7. If equipped, remove the lower center console by removing or disconnecting the following:
   • Cigarette lighter
   • Automatic transaxle shift handle
   • Upper console trim plate from the front floor console by unsnapping it
   • Electrical connectors from the upper console trim plate and remove the trim plate
   • CD storage compartment from the front floor console by unsnapping it
   • Raise the front floor console armrest and remove the compartment mat
   • Console-to-chassis bolts and screws
   • Electrical connectors from the console and remove the console from the vehicle
8. Remove the lower instrument panel lower compartment by removing or disconnecting the following:
   • Lower right instrument panel insulator
   • Compartment-to-panel bolts and screws, from under the instrument panel compartment
   • Instrument panel compartment door screws and the door
   • Instrument panel compartment screws and plastic clips; then slide the compartment from the instrument panel

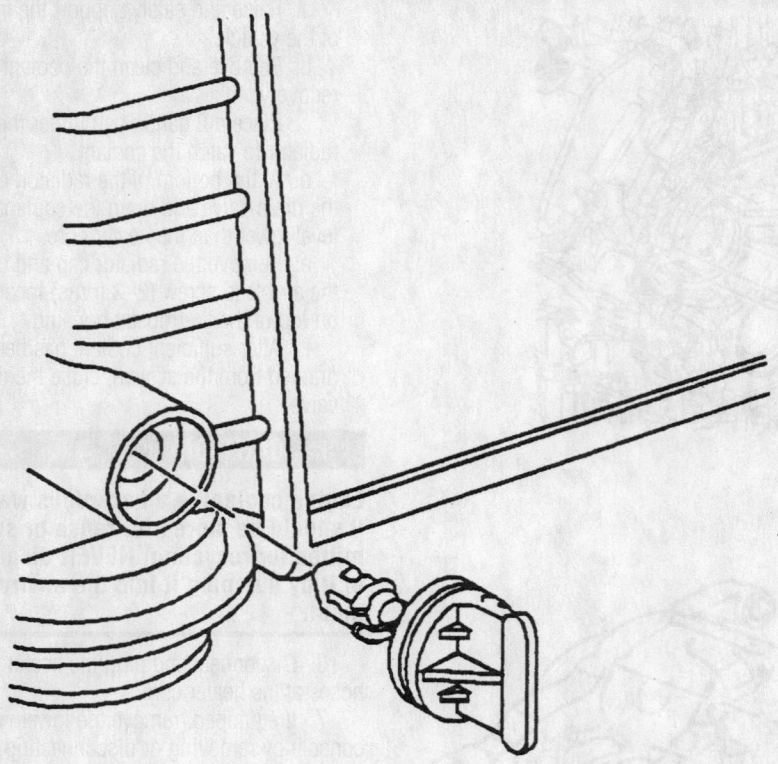

View of the radiator drain valve—Century, Grand Prix, Intrigue, Lumina, Monte Carlo and Regal

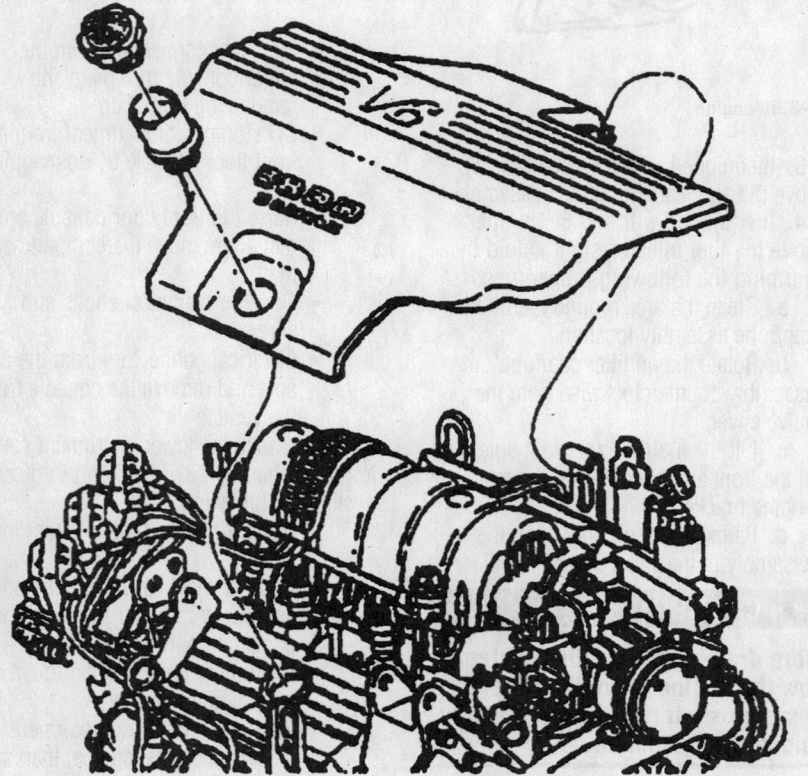

View of the fuel injector sight shield—Century, Grand Prix, Intrigue, Lumina, Monte Carlo and Regal

- Electrical connector from the compartment
- Ashtray and bracket, if necessary
- Lower heater duct
- Heater core cover and discard the cover seals
- Heater core mounting clip and bracket
- Heater core

To install:

9. Install and connect the following:
- Heater core in the vehicle
- Heater core mounting clip and bracket
- New seals on the heater core cover
- Heater core cover and torque the bolts to 13 inch lbs. (1.5 Nm)
- Lower heater duct, if removed
- Ashtray and bracket, if removed

10. Install the lower instrument panel lower by installing or connecting the following:

- Electrical connector to the compartment
- Slide the compartment into the instrument panel; then, secure it with screws and plastic clips
- Instrument panel compartment door and screws
- Compartment-to-panel bolts and screws, located under the instrument panel compartment
- Lower right instrument panel insulator

11. If equipped, install the lower center console install or connect the following:
- Console in the vehicle and connect the electrical connectors
- Console-to-chassis bolts and screws, tighten in sequence beginning at the front right and continue in a clockwise order to 106 inch lbs. (12 Nm)
- Armrest compartment mat and lower the front floor console armrest
- C/D storage compartment into the front floor console
- Electrical connectors to the upper console trim plate and snap the trim plate onto the console
- Automatic transaxle shift handle and the cigarette lighter
- Heater hoses to the heater core and secure with the clamps

12. Refill the cooling system by performing the following procedure:
a. Close the radiator drain valve.
b. If the air bleed screws (located at the top of the thermostat housing) is closed, open it by turning it 2–3 turns.

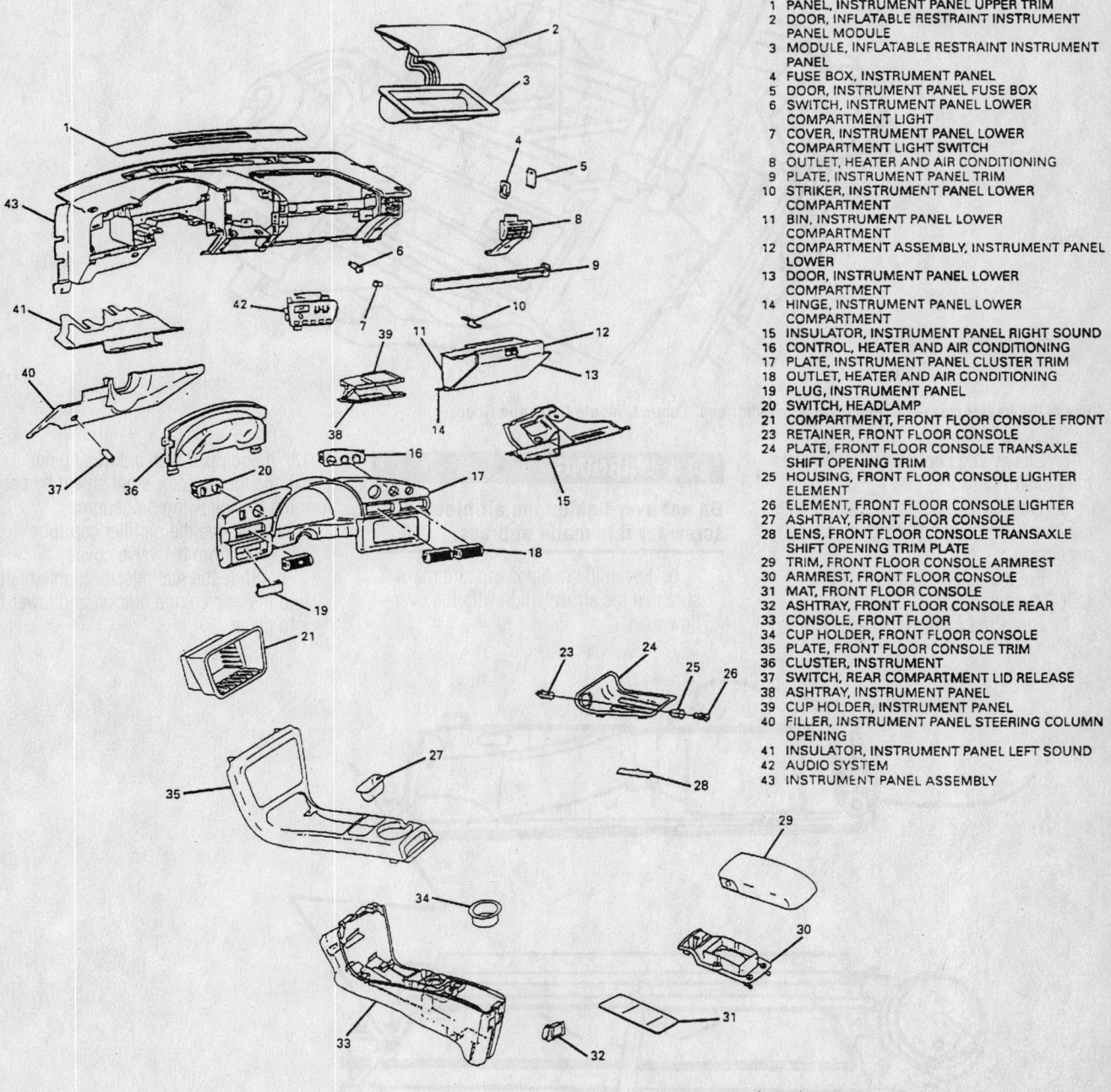

1 PANEL, INSTRUMENT PANEL UPPER TRIM
2 DOOR, INFLATABLE RESTRAINT INSTRUMENT PANEL MODULE
3 MODULE, INFLATABLE RESTRAINT INSTRUMENT PANEL
4 FUSE BOX, INSTRUMENT PANEL
5 DOOR, INSTRUMENT PANEL FUSE BOX
6 SWITCH, INSTRUMENT PANEL LOWER COMPARTMENT LIGHT
7 COVER, INSTRUMENT PANEL LOWER COMPARTMENT LIGHT SWITCH
8 OUTLET, HEATER AND AIR CONDITIONING
9 PLATE, INSTRUMENT PANEL TRIM
10 STRIKER, INSTRUMENT PANEL LOWER COMPARTMENT
11 BIN, INSTRUMENT PANEL LOWER COMPARTMENT
12 COMPARTMENT ASSEMBLY, INSTRUMENT PANEL LOWER
13 DOOR, INSTRUMENT PANEL LOWER COMPARTMENT
14 HINGE, INSTRUMENT PANEL LOWER COMPARTMENT
15 INSULATOR, INSTRUMENT PANEL RIGHT SOUND
16 CONTROL, HEATER AND AIR CONDITIONING
17 PLATE, INSTRUMENT PANEL CLUSTER TRIM
18 OUTLET, HEATER AND AIR CONDITIONING
19 PLUG, INSTRUMENT PANEL
20 SWITCH, HEADLAMP
21 COMPARTMENT, FRONT FLOOR CONSOLE FRONT
23 RETAINER, FRONT FLOOR CONSOLE
24 PLATE, FRONT FLOOR CONSOLE TRANSAXLE SHIFT OPENING TRIM
25 HOUSING, FRONT FLOOR CONSOLE LIGHTER ELEMENT
26 ELEMENT, FRONT FLOOR CONSOLE LIGHTER
27 ASHTRAY, FRONT FLOOR CONSOLE
28 LENS, FRONT FLOOR CONSOLE TRANSAXLE SHIFT OPENING TRIM PLATE
29 TRIM, FRONT FLOOR CONSOLE ARMREST
30 ARMREST, FRONT FLOOR CONSOLE
31 MAT, FRONT FLOOR CONSOLE
32 ASHTRAY, FRONT FLOOR CONSOLE REAR
33 CONSOLE, FRONT FLOOR
34 CUP HOLDER, FRONT FLOOR CONSOLE
35 PLATE, FRONT FLOOR CONSOLE TRIM
36 CLUSTER, INSTRUMENT
37 SWITCH, REAR COMPARTMENT LID RELEASE
38 ASHTRAY, INSTRUMENT PANEL
39 CUP HOLDER, INSTRUMENT PANEL
40 FILLER, INSTRUMENT PANEL STEERING COLUMN OPENING
41 INSULATOR, INSTRUMENT PANEL LEFT SOUND
42 AUDIO SYSTEM
43 INSTRUMENT PANEL ASSEMBLY

93111G03

Exploded view of the instrument panel—Century, Grand Prix, Intrigue, Lumina, Monte Carlo and Regal

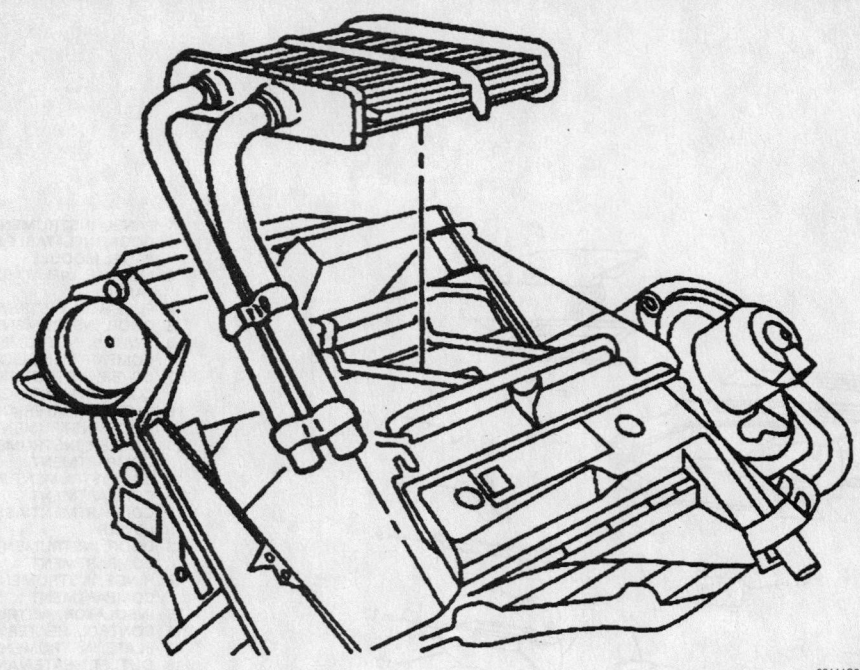

93111G04

View of the heater core—Century, Grand Prix, Intrigue, Lumina, Monte Carlo and Regal

c. Slowly add coolant until it reaches the radiator neck.

d. Wait for 2 minutes and recheck the coolant level; then, add more coolant if necessary.

e. Fill the coolant reservoir to the COLD mark.

f. Close the air bleed screw.

✶✶ WARNING

Do not over-tighten the air bleed screw for it is made of brass.

g. Install the radiator cap and make sure that the arrows align with the overflow tube.

13. If equipped with a 3.8L engine, install the fuel injector sight shield by performing the following procedures:

a. Remove the oil filler cap/tube assembly from the valve cover.

b. Slide the fuel injector sight shield into the rear engine bracket and lower it into place.

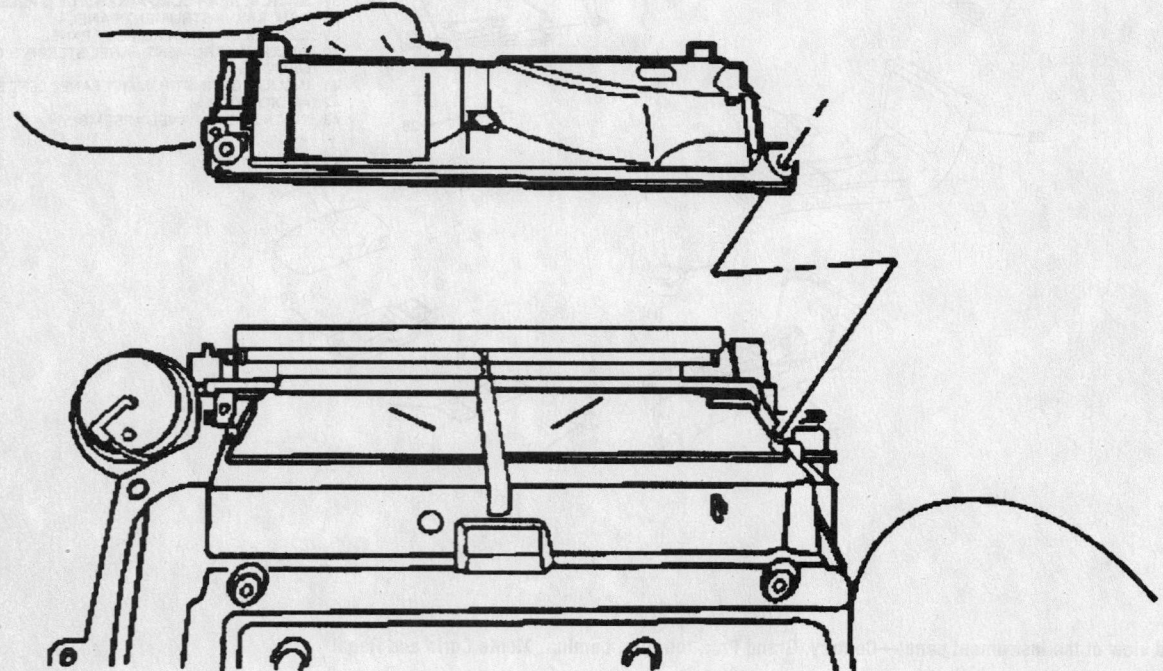

93111G65

View of the heater core outlet cover—Grand Prix, Intrigue and Regal

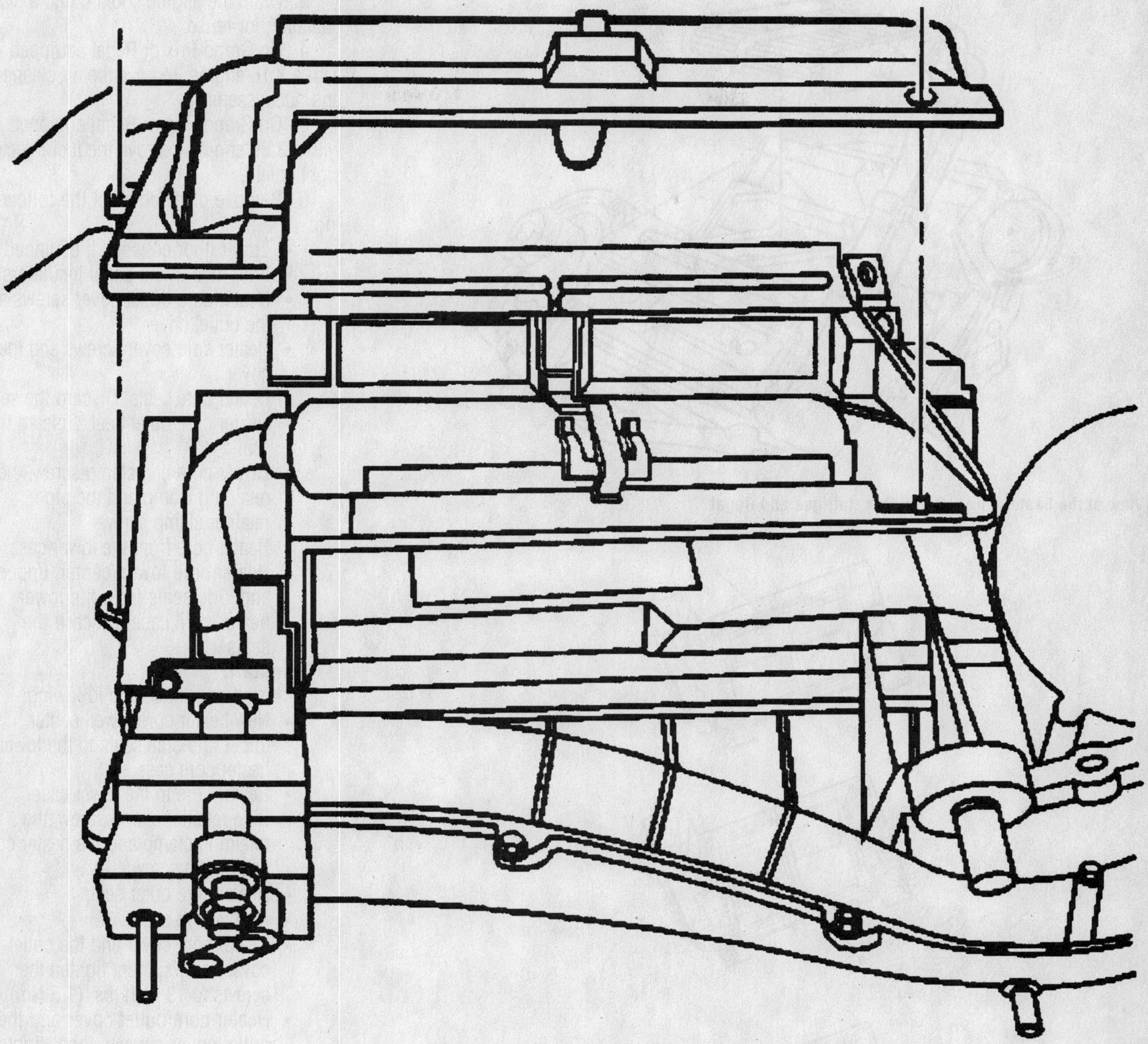

93111G66

View of the heater core cover—Grand Prix, Intrigue and Regal

c. Reinstall the oil filler cap/tube assembly in the valve cover. Twist it clockwise to lock the detent on the tube into the notch in the valve cover.

14. If equipped with a 3.1L engine, install the air cleaner and duct assembly.

15. Connect the negative battery cable.

Grand Prix, Intrigue and Regal

1. Before servicing the vehicle, refer to the precautions in the beginning of this section.

2. Disconnect the negative battery cable.

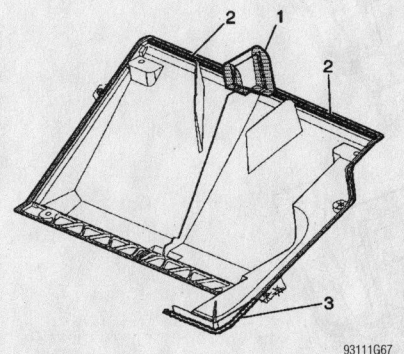

93111G67

Location of the heater core cover seals— Grand Prix, Intrigue and Regal

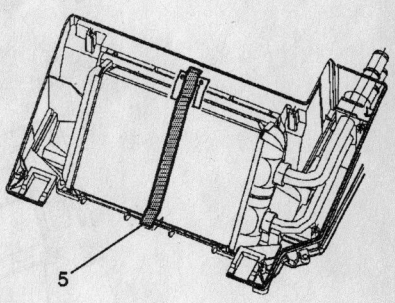

93111G68

View of the heater core outer seal—Grand Prix, Intrigue and Regal

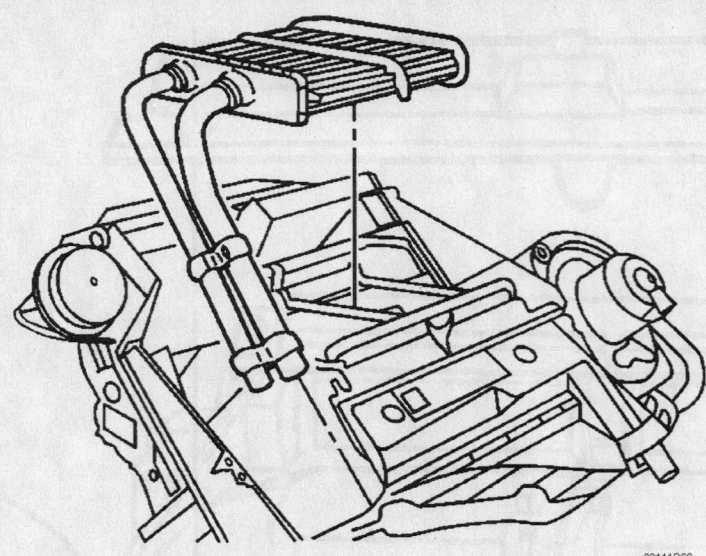

93111G69

View of the heater core—Grand Prix, Intrigue and Regal

3. Drain the engine coolant into a clean container for reuse.

4. On Grand Prix or Regal equipped with a 3.1L engine, remove the air cleaner and duct assembly.

5. On Grand Prix or Regal equipped with a 3.8L engine, remove the fuel injector sight shield.

6. Remove or disconnect the following:

- Lower floor console, if equipped
- Both instrument panel insulators
- Heater core outlet cover screws and the outlet cover
- Heater core cover screws and the cover
- Heater core seals. Discard the seals
- Heater core outer seal. Discard the seal
- Heater core line clamp screw, the retaining clamp and the pipe retainer clamp screw
- Heater core from the lower case
- Heater core lower, center, upper and side seals from the lower heater core case. Discard the seals

To install:

7. Install or connect the following:

- New heater core lower, center, upper and side seals to the lower heater core case
- Heater core to the lower case
- Pipe retainer clamp screw, the retaining clamp and the heater core line clamp screw
- Heater core outer seal
- Heater core seals
- Heater core cover and the outlet cover screws, then, tighten the screws to 13 inch lbs. (1.5 Nm)
- Heater core outlet cover and the outlet cover screws, then, tighten the screws to 13 inch lbs. (1.5 Nm)
- Both instrument panel insulators
- Lower floor console, if equipped

8. On Grand Prix or Regal equipped with a 3.8L engine, install the fuel injector sight shield.

9. On Grand Prix or Regal equipped with a 3.1L engine, install the air cleaner and duct assembly.

10. Refill the engine cooling system.

11. Connect the negative battery cable.

12. Operate the engine to normal operating temperatures; then, check the climate control operation and check for leaks.

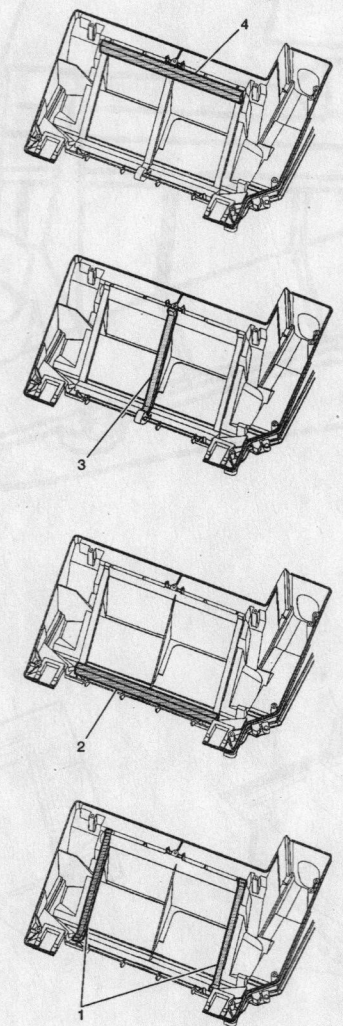

93111G70

View of the heater core lower, center, upper and side seals of the lower heater core case—Grand Prix, Intrigue and Regal

Cylinder Head

REMOVAL & INSTALLATION

3.1L and 3.4L Engines

LEFT (FRONT) CYLINDER HEAD

1. Before servicing the vehicle, refer to the precautions in the beginning of this section.
2. Relieve the fuel system pressure.
3. Drain the engine oil.
4. Drain the cooling system.
5. Remove or disconnect the following:
   - Upper half of the air cleaner
   - Throttle body air inlet duct
   - Upper intake manifold
   - Spark plug wires from the spark plugs
   - Rocker arm covers
   - Lower intake manifold
   - Rocker arm bolt, rocker arms, balls and pushrods
   - Exhaust crossover pipe
   - Engine mount strut bracket from the cylinder head
   - Left side exhaust manifold
   - Oil level indicator tube
   - Cylinder head bolts evenly
   - Cylinder head and discard the gasket

To install:

6. Install the cylinder head with a new gasket.
7. Apply thread sealer to the head bolts.
8. Tighten the cylinder head bolts to 37 ft. lbs. (50 Nm) plus an additional 90 degree turn.
9. Install or connect the following:
   - Left side exhaust manifold and torque the nuts to 12 ft. lbs. (16 Nm)
   - Oil level indicator tube
   - Engine mount strut bracket and torque the bolts to 52 ft. lbs. (70 Nm)
   - Exhaust crossover pipe and torque the nuts to 18 ft. lbs. (25 Nm)
   - Exhaust crossover pipe heat shield

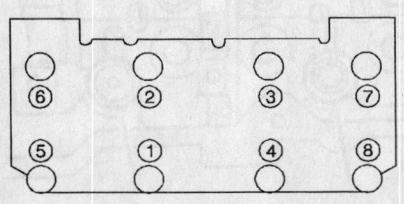

Tighten the cylinder head bolts using the following sequence—31L and 3.4L

and torque the bolts to 89 inch lbs. (10 Nm)
   - Pushrods, rocker arms, balls and bolts and tighten the bolts to 89 inch lbs. (10 Nm) plus an additional 30 degree turn
   - Lower intake manifold and torque the bolts to 115 inch lbs. (13 Nm)
   - Upper intake plenum and torque the bolts to 18 ft. lbs. (25 Nm)
   - Rocker arm cover and torque the bolts to 89 inch lbs. (10 Nm)
   - Spark the plug wires
   - Upper half of the air cleaner assembly
   - Throttle body air inlet duct
   - Negative battery cable
10. Refill the cooling system.
11. Refill the engine with clean oil.

➡ **A filter change is also recommended.**

12. Start the engine and check for leaks, repair if necessary.

RIGHT (REAR) CYLINDER HEAD

1. Before servicing the vehicle, refer to the precautions in the beginning of this section.
2. Relieve the fuel system pressure.
3. Drain the engine oil.
4. Drain the cooling system.
5. Remove or disconnect the following:
   - Upper half of the air cleaner
   - Throttle body air inlet duct
   - Upper intake plenum
   - Lower intake manifold
   - Spark plug wires from the spark plugs
   - Rocker arm covers
   - Exhaust crossover pipe heat shield
   - Crossover pipe
   - Right side exhaust manifold
   - Fuel line bracket
   - Alternator and bracket
   - Rocker arms bolt, rocker arms, balls and pushrods
   - Cylinder head bolts evenly
   - Cylinder head

To install:

6. Install the cylinder head with a new gasket.
7. Apply thread sealer to the head bolts.
8. Tighten the cylinder head bolts to 37 ft. lbs. (50 Nm) plus an additional 90 degree turn.
9. Install or connect the following:
   - Pushrods, rocker arms, balls and rocker arm bolts and tighten the bolts to 89 inch lbs. (10 Nm) plus an additional 30 degrees

   - Exhaust manifold and torque the nuts to 12 ft. lbs. (16 Nm)
   - Alternator bracket and torque the bolts to 37 ft. lbs. (50 Nm)
   - Alternator and torque the bolts to 37 ft. lbs. (50 Nm)
   - Rocker arm cover and torque the bolts to 89 inch lbs. (10 Nm)
   - Spark plug wires
   - Exhaust crossover pipe and torque the nuts to 18 ft. lbs. (25 Nm)
   - Exhaust crossover pipe heat shield and torque the bolts to 89 inch lbs. (10 Nm)
   - Pushrods, rocker arms, balls and bolts and tighten the bolts to 89 inch lbs. (10 Nm) plus an additional 30 degree turn
   - Lower intake manifold and torque the bolts to 115 inch lbs. (13 Nm)
   - Upper intake plenum and torque the bolts to 18 ft. lbs. (25 Nm)
   - Upper half of the air cleaner assembly
   - Throttle body air inlet duct
   - Negative battery cable
10. Refill the cooling system.
11. Refill the engine with clean oil.

➡ **An oil filter change is recommended.**

12. Start the engine and check for leaks, repair if necessary.

3.5L Engine

FRONT

1. Before servicing the vehicle, refer to the precautions in the beginning of this section.
2. Drain the engine oil.
3. Drain the cooling system.
4. Remove or disconnect the following:
   - Negative battery cable
   - Intake manifold
   - Water outlet housing
   - Engine mount strut bracket
   - Coolant crossover pipe
   - Front exhaust manifold
   - Camshaft cover
5. Install a holding tool on the camshafts to hold them in position.
6. Remove or disconnect the following:
   - Camshaft primary chain
   - Camshafts from the front cylinder head
   - Rocker arms and valve lifters

➡ **Be sure to keep the arms and lifters in order so they can be installed the their original locations.**

   - M6 bolts from the front of the cylinder head

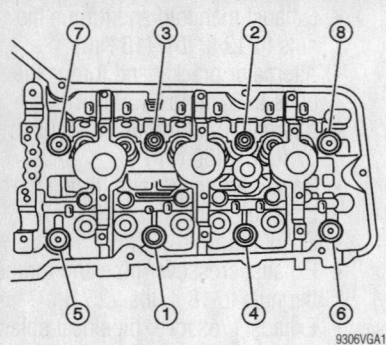

Front cylinder head bolt torque sequence—3.5L engine

- M11 cylinder head bolts and discard
- Cylinder head

To install:

7. Be sure the dowels are securely mounted in the engine block.

8. Install or connect the following:
- New gasket
- Cylinder head
- New M11 bolts
- M6 bolts in the front of the cylinder head

9. Torque the M11 bolts in sequence to:
   a. Step 1: 22 ft. lbs. (30 Nm).
   b. Step 2: 100 degree turn.
   c. Step 3: 100 degree turn.

10. Torque the long M6 bolt to 106 inch lbs. (12 Nm).

11. Torque both shorter M6 bolts to 106 inch lbs. (12 Nm).

12. Install or connect the following:
- Lifters and rocker arms in their original positions
- Camshafts and torque the bearing cap bolts to 71 inch lbs. (8 Nm) plus an additional 22 degree turn
- Primary camshaft drive chain
- Camshaft cover and torque the bolts to 80 inch lbs. (9 Nm)
- Exhaust manifold and torque the bolts to 18 ft. lbs. (25 Nm)
- Coolant crossover pipe and torque the bolts to 18 ft. lbs. (25 Nm)
- Engine mount strut bracket and torque the bolts to 37 ft. lbs. (50 Nm)
- Water outlet housing and torque the bolts to 80 inch lbs. (9 Nm)
- Intake manifold and torque the bolts to 62 inch lbs. (7 Nm)
- New oil filter
- Negative battery cable

13. Refill the engine with clean oil.

14. Refill the cooling system.

15. Start the engine and check for leaks, repair if necessary.

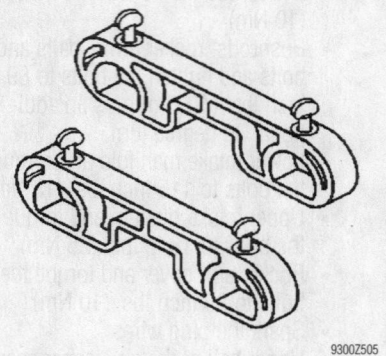

Camshaft holding fixture J-42038—3.5L engine

REAR

1. Before servicing the vehicle, refer to the precautions in the beginning of this section.

2. Drain the engine oil.

3. Drain the cooling system.

4. Remove or disconnect the following:
- Negative battery cable
- Intake manifold

➡**Do not remove the rear exhaust manifold. Detach it from the cylinder head and the connection from the front manifold; then, move it aside.**

- Exhaust manifold
- Coolant crossover pipe
- Camshaft cover

5. Install a holding tool on the camshafts to hold them in position.

6. Remove or disconnect the following:
- Primary camshaft drive chain
- Camshafts
- Rocker arms and valve lifters

➡**Keep the valve train parts in order so they can be installed in their original positions.**

- Engine Coolant Temperature (ECT) sensor from the cylinder head
- M6 bolts from the front of the cylinder head, note the location of the longer bolt
- M11 cylinder head bolts and discard the bolts
- Cylinder head

To install:

7. Be sure the dowels are securely mounted in the engine block.

8. Install or connect the following:
- New gasket
- Cylinder head
- New M11 bolts
- M6 bolts in the front of the cylinder head

9. Torque the M11 bolts in sequence to:
   a. Step 1: 22 ft. lbs. (30 Nm).
   b. Step 2: 100 degree turn.
   c. Step 3: 100 degree turn.

10. Torque the long M6 bolt to 106 inch lbs. (12 Nm).

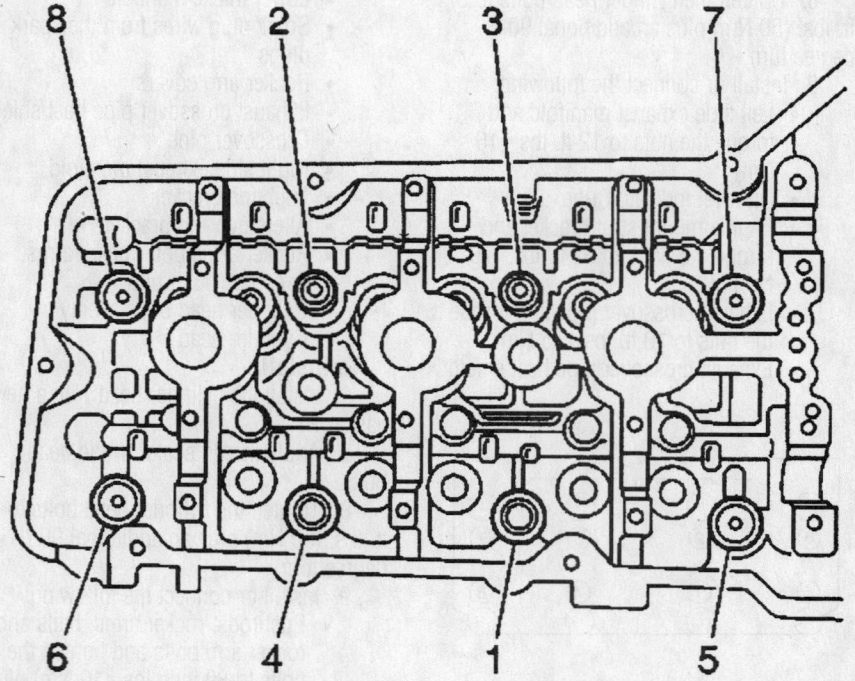

Rear cylinder head bolt torque sequence—3.5L engine

11. Torque both shorter M6 bolts to 106 inch lbs. (12 Nm).

12. Install or connect the following:
- ECT sensor and torque the fastener to 15 ft. lbs. (20 Nm)
- Lifters and rocker arms in their original positions
- Camshafts and torque the bearing cap bolts to 71 inch lbs. (8 Nm) plus an additional 22 degree turn
- Primary camshaft drive chain
- Camshaft cover and torque the bolts to 80 inch lbs. (9 Nm)
- Coolant crossover pipe and torque the bolts to 18 ft. lbs. (25 Nm)
- Exhaust manifold and torque the bolts to18 ft. lbs. (25 Nm)
- Intake manifold and torque the bolts to 62 inch lbs. (7 Nm)
- New oil filter
- Negative battery cable

13. Refill the engine with clean oil.

14. Refill the cooling system.

15. Start the engine and check for leaks, repair if necessary.

3.8L Engine

LEFT SIDE

1. Before servicing the vehicle, refer to the precautions in the beginning of this section.

2. Relieve the fuel system pressure.

3. Drain the cooling system.

4. Remove or disconnect the following:
- Fuel injector sight shield
- Throttle body air inlet duct
- Right and left engine mount strut brackets
- Fuel lines from the fuel rail
- Upper intake manifold
- Lower intake manifold
- Left exhaust manifold
- Rocker arm cover
- Rocker arms and pushrods
- Cylinder head bolts
- Cylinder head

To install:

5. Install the new cylinder head gasket with the arrow pointing to the front of the engine.

6. Install the cylinder head.

7. Install NEW cylinder head bolts. Torque the bolts in sequence, as follows:
  a. Step 1: 37 ft. lbs. (50 Nm).
  b. Step 2: Plus 120 degree turn.

8. Install or connect the following:
- Push rods and rocker arms
- Rocker arm cover and torque the bolts to 89 inch lbs. (10 Nm)
- Left exhaust manifold and torque the bolts to 22 ft. lbs. (30 Nm)

- Lower intake manifold and torque the bolts to 11 ft. lbs. (15 Nm)
- Upper intake manifold and torque the bolts to 89 inch lbs. (10 Nm)
- Left and right engine mount strut brackets and torque the bolts to 37 ft. lbs. (50 Nm)
- Throttle body air inlet duct
- Fuel injector sight shield and torque the bolts to 27 inch lbs. (3 Nm)
- Negative battery cable

9. Refill the cooling system.

10. Start the engine and check for leaks.

RIGHT SIDE

1. Before servicing the vehicle, refer to the precautions in the beginning of this section.

2. Relieve the fuel system pressure.

3. Drain the cooling system.

4. Remove or disconnect the following:
- Throttle body air inlet duct
- Fuel injector sight shield
- Accessory drive belt
- Alternator
- Catalytic converter from the exhaust manifold
- Engine mount struts

5. Place the transmission in neutral and rotate the engine forward for access.

6. Remove or disconnect the following:
- Drive belt tensioner
- Power steering pump without disconnecting the lines
- Heater hoses from the engine

- Spark plug wires
- Oxygen Sensor (O2S) electrical connector
- Throttle and cruise control cables
- Fuel lines from the fuel rail
- Fuel rail
- Exhaust Gas Recirculation (EGR) valve heat shield
- EGR valve
- EGR valve outlet pipe
- EGR valve adapter
- Upper intake manifold
- Lower intake manifold
- Exhaust crossover pipe
- Right exhaust manifold
- Rocker arm cover
- Rocker arms and pushrods
- Cylinder head

To install:

7. Install a new cylinder head gasket with the arrow pointing to the front of the engine.

8. Install the cylinder head and secure with NEW head bolts. Torque the bolts, in sequence, as follows:
  a. Step 1: 37 ft. lbs. (50 Nm).
  b. Step 2: Plus a 120 degree turn.

9. Install or connect the following:
- Push rods and rocker arms
- Rocker arm cover and torque the bolts to 89 inch lbs. (10 Nm)
- Right exhaust manifold and torque the bolts to 22 ft. lbs. (30 Nm)
- Exhaust crossover pipe and torque the nuts to 18 ft. lbs. (25 Nm)
- Lower intake manifold and torque the bolts to 11 ft. lbs. (15 Nm)

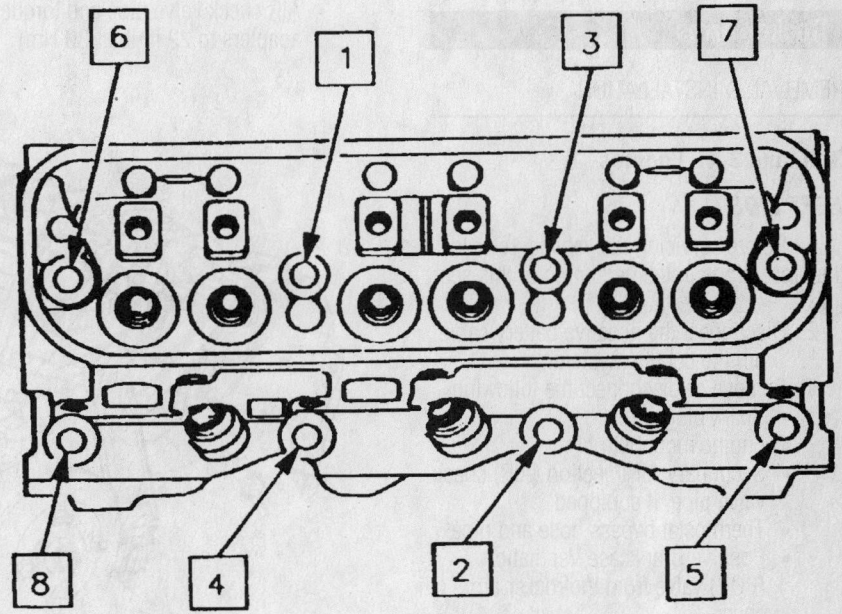

Cylinder head bolt torque sequence—3.8L engine

79222507

- Upper intake manifold and torque the bolts to 89 inch lbs. (10 Nm)
- EGR valve adapter and torque the bolts to 37 ft. lbs. (50 Nm)
- EGR valve outlet pipe and torque the bolts to 21 ft. lbs. (29 Nm)
- EGR valve and torque the nuts to 21 ft. lbs. (29 Nm)
- EGR valve heat shield and torque the bolts to 89 inch lbs. (10 Nm)
- Fuel rail
- Fuel lines to the fuel rail
- Throttle and cruise control cables
- O2S electrical connector
- Spark plugs and torque them to 20 ft. lbs. (27 Nm)
- Spark plug wires
- Heater hoses
- Power steering pump and torque the bolts to 25 ft. lbs. (34 Nm)
- Drive belt tensioner and torque the bolts to 37 ft. lbs. (50 Nm)

10. Carefully return the engine to its proper position.
11. Place the transmission in park.
12. Install or connect the following:
- Throttle body air inlet duct
- Engine mount struts and torque the bolts to 35 ft. lbs. (48 Nm)
- Catalytic converter and torque the bolts to 26 ft. lbs. (35 Nm)
- Alternator and torque the bolts to 37 ft. lbs. (50 Nm)
- Accessory drive belt
- Fuel injector sight shield
- Negative battery cable

13. Refill the cooling system.
14. Start the engine and check for leaks.

Rocker Arms

REMOVAL & INSTALLATION

3.1L and 3.4L Engines

LEFT SIDE

1. Before servicing the vehicle, refer to the precautions in the beginning of this section.
2. Disconnect the negative battery cable.
3. Drain the cooling system.
4. Remove or disconnect the following:
- Spark plug wires
- Engine mount strut
- Secondary Air Injection (AIR) check valve pipe, if equipped
- Thermostat bypass hose and pipe
- Positive Crankcase Ventilation (PCV) valve from the rocker arm cover
- Rocker arm cover

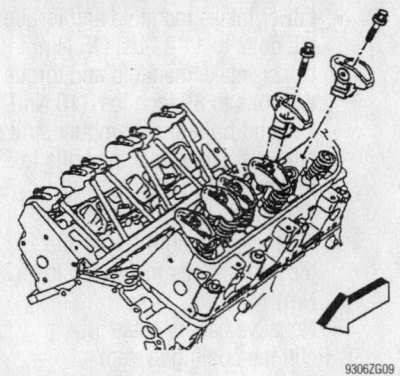

9306ZG09

Rocker arm components—3.1L and 3.4L engines

➡**Keep the pushrods in order. Intake pushrods are 5¾ inches long and exhaust pushrods are 6 inches long.**

- Rocker arm bolts
- Rocker arms
- Pushrods

To install:

5. Lubricate all the valvetrain components with engine oil.
6. Install or connect the following:
- Pushrods and the rocker arms and tighten the bolts to 14 ft. lbs. (19 Nm) plus 30 degrees
- Rocker arm cover using a new gasket and tighten the rocker cover bolts to 89 inch lbs. (10 Nm)
- PCV valve to the rocker arm cover
- Coolant tube with the thermostat bypass hose and tighten the bolt to 98 inch lbs. (11 Nm) and the nut to 18 ft. lbs. (25 Nm)
- AIR check valve pipe and torque the adapters to 22 ft. lbs. (30 Nm)

- Engine mount strut and torque the bolt to 35 ft. lbs. (48 Nm)
- Spark plug wires
- Negative battery cable

7. Refill the cooling system.
8. Start the vehicle and verify no leaks.

RIGHT SIDE

1. Before servicing the vehicle, refer to the precautions in the beginning of this section.
2. Remove or disconnect the following:
- Negative battery cable
- Accessory drive belt
- Alternator
- Alternator Bracket
- Spark plug wires
- Ignition coil and Evaporative (EVAP) emissions canister purge solenoid as an assembly
- Power brake booster vacuum pipe from the intake plenum
- Rocker arm cover
- Rocker arm bolts, balls, rocker arms and pushrods

To install:

3. Lubricate all the valve train components with engine oil.
4. Install or connect the following:
- Pushrods and the rocker arms and tighten the bolts to 14 ft. lbs. (19 Nm) plus 30 degrees
- Rocker arm cover using a new gasket and tighten the rocker cover bolts to 89 inch lbs. (10 Nm)
- Power brake booster vacuum pipe to the plenum
- EVAP solenoid and ignition coil assembly
- Spark plug wires

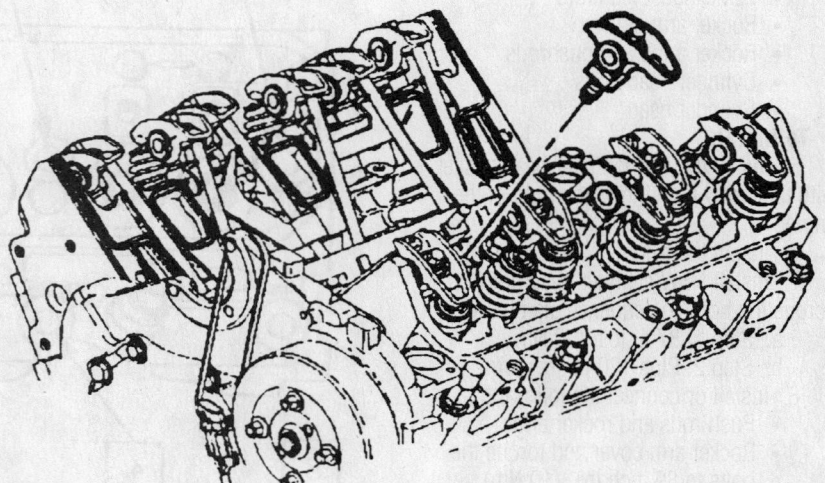

79222Z508

Rocker arm mounting—3.1L engine

- Alternator bracket and torque the bolts to 37 ft. lbs. (50 Nm)
- Alternator and torque the bolts to 37 ft. lbs. (50 Nm)
- Accessory drive belt
- Negative battery cable

5. Start the vehicle and verify no leaks.

3.5L Engine

Refer to the camshaft removal and installation procedure for rocker arm service.

3.8L Engine

❊❊ WARNING

The rocker arm bolts have been permanently stretched during installation. New bolts must be used each time the rocker arm assemblies are removed.

LEFT SIDE (FRONT) ROCKER ARMS

1. Before servicing the vehicle, refer to the precautions in the beginning of this section.
2. Remove or disconnect the following:
   - Negative battery cable
   - Engine lift bracket from the exhaust manifold studs
   - Fuel injector sight shield
   - Engine mount strut bracket
   - Spark plug wires
   - Spark plug wire cover from the valve rocker arm cover
   - Rocker arm cover

➡️The rocker arms, pushrods, pedestals and bolts must be kept in order for installation in the same locations they were removed from.

   - Rocker arm bolts, pedestals, rocker arms and pushrods

To install:

3. Apply thread locking compound to the rocker arm bolts.
4. Install or connect the following:
   - Pushrod in the lifter
   - Rocker arm, pedestal and mounting bolt and tighten the mounting bolt to 11 ft. lbs. (15 Nm) plus an additional 90 degrees
   - Rocker arm cover with a new gasket and torque the mounting bolts to 89 inch lbs. (10 Nm)
   - Spark plug wire cover to the rocker arm cover
   - Spark plug wires
   - Engine mount strut bracket and torque the bolt to 75 ft. lbs. (102 Nm)
   - Fuel injector sight shield

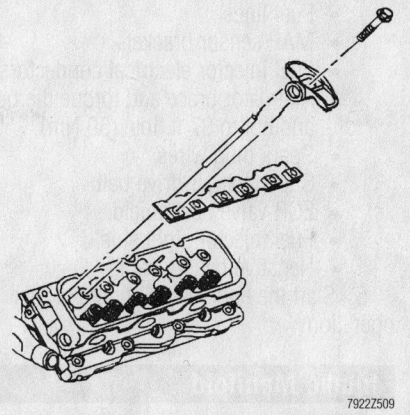

Exploded view of the rocker arm, pushrod and pushrod guide plate assembly—3.8L engine

   - Engine lift bracket to the exhaust manifold and torque the bolt to 22 ft. lbs. (30 Nm)
   - Negative battery cable

5. Start the vehicle and check for leaks, repair if necessary.

RIGHT SIDE (REAR) ROCKER ARMS

1. Before servicing the vehicle, refer to the precautions in the beginning of this section.
2. Remove or disconnect the following:
   - Negative battery cable
   - Fuel injector sight shield
   - Accessory drive belt
   - Alternator
   - Drive belt tensioner
   - Engine mount struts
   - Throttle body air inlet duct

3. Rotate the engine forward for proper access.

   - Evaporative (EVAP) emissions purge solenoid and brace
   - Right side spark plug wires
   - Fuel injector sight shield bracket
   - Rocker arm cover

➡️The rocker arms, pushrods, pedestals and bolts must be kept in order for installation in the same locations they were removed from.

   - Rocker arm bolts, pedestals, rocker arms and pushrods

To install:

4. Seat the pushrod in the lifter and install the rocker arm, pedestal and mounting bolt. Tighten the mounting bolt to 11 ft. lbs. (15 Nm) plus an additional 90 degrees.
5. Install or connect the following:
   - Rocker arm cover using a new gasket and torque the bolts to 89 inch lbs. (10 Nm)
   - Fuel injector sight shield bracket

and torque the bolts to 22 ft. lbs. (30 Nm)
   - EVAP purge solenoid and brace
   - Right side spark plug wires

6. Return the engine back to its original position.
   - Throttle body air inlet duct
   - Engine mount struts and torque the bolt to 35 ft. lbs. (48 Nm)
   - Drive belt tensioner and torque the bolts to 37 ft. lbs. (50 Nm)
   - Alternator and torque the bolt to 37 ft. lbs. (50 Nm)
   - Fuel injector sight shield
   - Negative battery cable

7. Start the vehicle and check for leaks, repair if necessary.

Supercharger

REMOVAL & INSTALLATION

3.8L (VIN 1) Engine

1. Before servicing the vehicle, refer to the precautions in the beginning of this section.
2. Relieve the fuel system pressure.
3. Remove or disconnect the following:

   - Fuel injector sight shield
   - Exhaust Gas Recirculation (EGR) valve heat shield
   - Supercharger drive belt
   - Right side spark plug wires from the ignition module
   - Alternator brace
   - Fuel injector electrical connectors
   - Manifold Absolute Pressure (MAP) sensor bracket
   - Fuel lines
   - Fuel rail with the injectors
   - Boost control solenoid

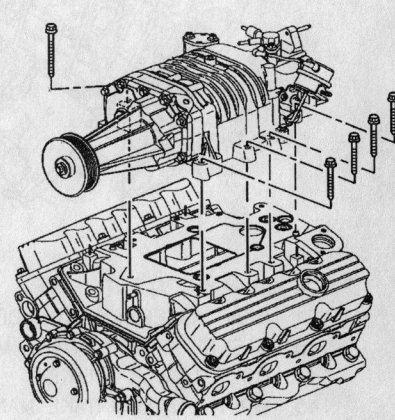

Supercharger bolt locations–3.8L (VIN 1) engine

- Electrical and vacuum connections as necessary
- Throttle and cruise control cables with the bracket
- Throttle body air inlet duct
- Supercharger

To install:

4. Install or connect the following:
   - Supercharger with new a new gasket and O-rings and torque the bolts to 17 ft. lbs. (23 Nm)
   - Throttle body air inlet duct
   - Throttle and cruise control cables with the bracket and torque the bolts to 142 inch lbs. (16 Nm)
   - Electrical and vacuum connectors
   - Boost control solenoid and torque the nut to 72 inch lbs. (8 Nm)
   - Fuel rail with fuel injectors and torque the hold-down bolts to 7 ft. lbs. (10 Nm) and the stud to 18 ft. lbs. (25 Nm)

- Fuel lines
- MAP sensor bracket
- Fuel injector electrical connectors
- Alternator brace and torque the bolt and nut to 37 ft. lbs. (50 Nm)
- Spark plug wires
- Supercharger drive belt
- EGR valve heat shield
- Fuel injector sight shield
- Negative battery cable

5. Start the engine and ensure proper operation.

Intake Manifold

REMOVAL & INSTALLATION

3.1L and 3.4L Engines

1. Before servicing the vehicle, refer to the precautions in the beginning of this section.

2. Relieve the fuel system pressure.
3. Drain the cooling system.
4. Remove or disconnect the following:
   - Intake Air Temperature (IAT) sensor electrical connector
   - Throttle body air inlet duct
   - Throttle and cruise control cables with bracket
   - Throttle Position Sensor (TPS) electrical connector
   - Idle Air Control (IAC) valve electrical connector
   - Camshaft Position (CMP) sensor wiring harness from the intake
   - Spark plug wires from the intake
   - Thermostat bypass hoses from the throttle body
   - Ignition coil bracket with the coils, the purge solenoid and the vacuum canister solenoid
   - Vacuum lines from the upper intake manifold

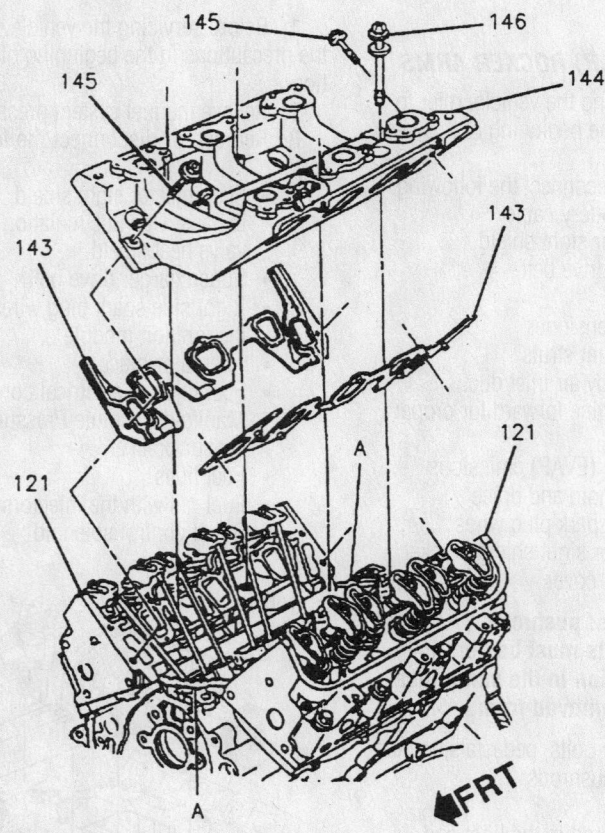
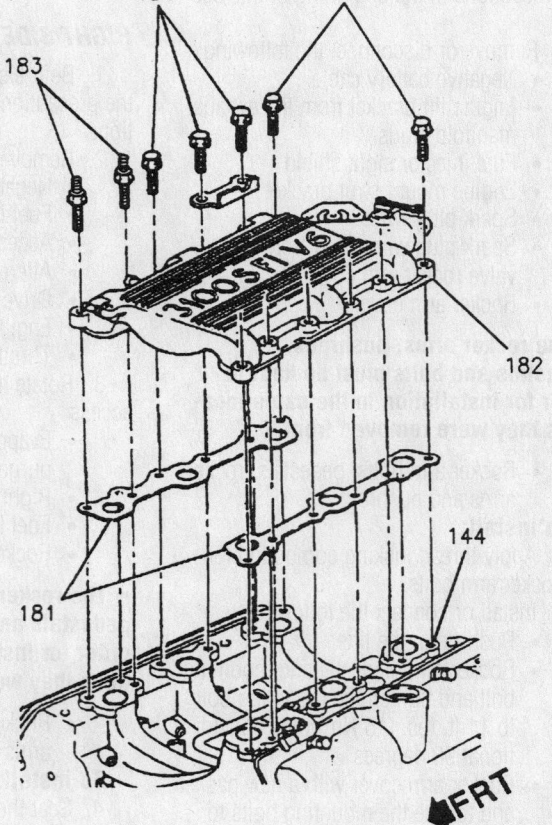

A   APPLY SEALANT
121   HEAD ASSEMBLY, CYLINDER
143   GASKET, LOWER INTAKE MANIFOLD
144   BOLT, LOWER INTAKE
145   BOLT, LOWER INTAKE MANIFOLD
146   BOLT, LOWER INTAKE MANIFOLD

144   MANIFOLD, LOWER INTAKE
181   GASKET, UPPER INTAKE MANIFOLD
182   MANIFOLD, UPPER INTAKE
183   STUD, UPPER INTAKE MANIFOLD
184   BOLT, UPPER INTAKE MANIFOLD

79222310

Lower intake manifold torque sequence—3.1L and 3.4L engines

- Manifold Absolute Pressure (MAP) sensor
- Alternator brace
- Exhaust Gas Recirculation (EGR) valve
- Upper intake manifold
- Both rocker arm covers
- Engine Coolant Temperature (ECT) sensor electrical connector
- Fuel injector and MAP wiring harness
- Fuel injector electrical connectors
- Fuel lines from the fuel rail and fuel line bracket
- Fuel rail with the injectors
- Power steering pump and move it aside without disconnecting the lines
- Coolant inlet pipe from the coolant outlet housing
- Coolant bypass hose
- Upper radiator hose at thermostat housing
- Lower intake manifold bolts

➡ **When removing the valvetrain components, keep them in order for installation purposes.**

- Rocker arm bolts, rocker arms and pushrods
- Intake manifold

To install:

5. Place a 8–12mm bead of RTV, on each ridge, where the front and rear of the intake manifold contact the block.

6. Install or connect the following:
- Intake manifold gasket
- Pushrods and rocker arms
- Rocker arm bolts and torque the bolts to 89 inch lbs. (10 Nm) plus an additional 30 degree turn

✳ WARNING

In order to prevent oil leaks, tighten the vertical bolts before the diagonal bolts.

- Lower intake manifold and torque the bolts to 115 inch lbs. (13 Nm) and if equipped, tighten the vertical and diagonal lower intake manifold bolts to 62 inch lbs. (7 Nm)
- Both rocker arm covers and torque the bolts to 89 inch lbs. (10 Nm)
- Thermostat bypass hose
- Upper radiator hose
- Heater inlet pipe and heater hose to the lower intake
- Power steering pump and torque the bolts to 25 ft. lbs. (34 Nm)
- Fuel fail with the injectors

- Fuel feed and return lines
- Fuel injector electrical connectors
- ECT sensor electrical connector
- Upper intake manifold using a new gasket and torque the bolts to 18 ft. lbs. (25 Nm)
- EGR valve
- Alternator brace
- Vacuum lines
- MAP sensor
- Ignition coil bracket with the coils, the purge solenoid and the vacuum canister solenoid
- Thermostat bypass pipe coolant hoses to the throttle body
- CMP sensor wiring harness to the intake
- Spark plug wires
- TPS electrical connector
- IAC valve electrical connector
- Throttle and cruise control cables with the bracket and torque the bolts to 115 inch lbs. (13 Nm) and the nuts to 89 inch lbs. (10 Nm)
- Throttle body air inlet duct
- IAT sensor electrical connector
- Negative battery cable

7. Refill the cooling system.

➡ **An engine oil and filter change is recommended.**

8. Start the vehicle and check for leaks, repair if necessary.

3.5L Engine

1. Before servicing the vehicle, refer to the precautions in the beginning of this section.
2. Partially drain the engine coolant.
3. Remove or disconnect the following:
- Negative battery cable
- Throttle body air inlet duct
- Fuel injector sight shield
- Throttle and cruise control cables from the throttle body with the bracket
- Coolant hoses from the throttle body
- Fuel lines from the fuel supply rail
- Fuel vapor line from the Evaporative Emission (EVAP) canister purge solenoid
- Brake booster vacuum hose
- Air conditioning vacuum hose from the engine
- Surge tank inlet pipe retainer from the fuel supply rail
- Fuel injector electrical connectors
- Throttle Position Sensor (TPS) electrical connectors
- Idle Air Control (IAC) valve electrical connector

- EVAP canister purge solenoid connector
- Manifold Absolute Pressure (MAP) sensor connector
- Wiring harness channels from the camshaft covers
- Vacuum tube from the fuel pressure regulator
- Positive Crankcase Ventilation (PCV) valve and both feed tubes
- Exhaust Gas Recirculation (EGR) valve outlet pipe
- Fuel supply rail with injectors

➡ **Disengage the snap-lock retainers by pushing toward the camshaft covers and lifting.**

- Throttle body coolant hose
- Intake manifold

➡ **The manifold-to-cylinder head seals are reusable unless cut or damaged.**

To install:

4. Install or connect the following:
- New intake manifold-to-cylinder head seals
- Intake manifold and torque the bolts, in a circular pattern, starting from the center to 62 inch lbs. (7 Nm)
- Throttle body heater hose
- New O-rings on the fuel injectors
- Fuel supply rail with injectors
- EGR pipe and torque the intake manifold bolt to 89 inch lbs. (10 Nm) and the coolant crossover bolt to 18 ft. lbs. (24 Nm)
- PCV valve and related tubing
- Brake booster vacuum hose
- Fuel pressure regulator vacuum hose
- Air conditioning vacuum hose
- Engine wiring harness with channel to the camshaft covers and torque the bolts to 89 inch lbs. (10 Nm)

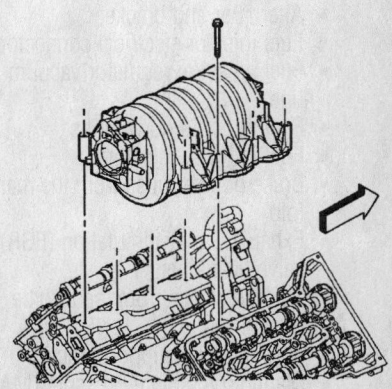

93002506

Intake manifold assembly—3.5L engine

- Surge tank pipe retainer to the fuel supply rail
- TPS electrical connector
- IAC electrical connector
- EVAP solenoid electrical connector
- MAP sensor electrical connector
- Fuel injector electrical connectors
- Coolant hoses to the throttle body
- Throttle and cruise control cables to the throttle body
- Fuel lines to the fuel supply rail
- Fuel vapor line to the purge solenoid
- Fuel injector sight shield
- Throttle body air inlet duct
- Negative battery cable

5. Refill the cooling system.
6. Start the engine and check for leaks, repair if necessary.

3.8L Engine

WITHOUT SUPERCHARGER

➡There are 2 bolts hidden beneath the upper intake manifold. These bolts are located in the right front and left rear corners of the lower intake manifold. It is necessary to remove the upper intake manifold to service the lower intake manifold.

1. Before servicing the vehicle, refer to the precautions in the beginning of this section.
2. Relieve the fuel system pressure.
3. Drain the cooling system.
4. Remove or disconnect the following:
   - Negative battery cable
   - Fuel injector sight shield and air inlet duct
   - Spark plug wires from the right side of the engine
   - Evaporative emissions (EVAP) canister purge solenoid electrical connector
   - Fuel feed and return lines
   - Vacuum lines from the throttle body
   - Accessory drive belt
   - Alternator and bracket
   - Fuel injector electrical connectors
   - Fuel pressure regulator vacuum line
   - Fuel rail hold down bolts
   - Fuel rail with injectors
   - Brake booster hose from the manifold
   - Exhaust Gas Recirculation (EGR) valve heat shield
   - Throttle cables from the throttle body lever
   - Throttle body support bracket
   - Manifold Absolute Pressure (MAP) sensor electrical connector

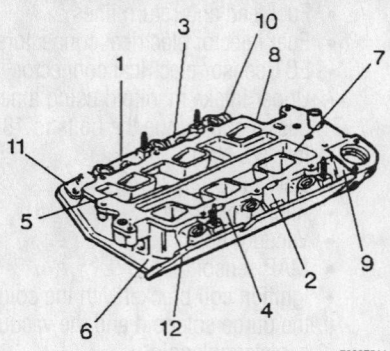

Always tighten the lower intake manifold bolts in sequence—3.8L (VIN 1 and K) engines

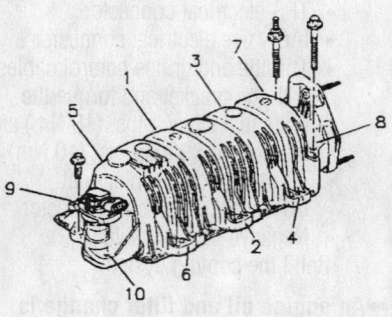

Upper intake manifold bolt tightening sequence—3.8L (VIN 1 and K) engines

- Upper intake plenum
- Drive belt tensioner
- EGR valve outlet pipe
- Upper radiator hose
- Engine Coolant Temperature (ECT) sensor electrical connector
- Evaporative emissions canister purge solenoid valve
- Alternator brace
- Lower intake manifold

To install:
5. Install or connect the following:
   - New intake manifold gaskets
   - Intake manifold and torque the bolts, in sequence, to 11 ft. lbs. (15 Nm)
   - Evaporative emissions canister purge solenoid valve
   - Alternator brace
   - ECT sensor electrical connector
   - Upper radiator hose
   - EGR valve outlet pipe
   - Drive belt tensioner and torque the bolts to 37 ft. lbs. (50 Nm)
   - Intake plenum using a new gasket and torque the bolts, in sequence, to 89 inch lbs. (10 Nm)
   - MAP sensor electrical connector

- Throttle body support bracket
- Throttle cables to the throttle body lever
- EGR heat shield
- Brake booster vacuum hose
- Fuel rail assembly and torque the bolts to 84 inch lbs. (10 Nm)
- Fuel pressure regulator vacuum line
- Fuel injector electrical connectors
- Alternator bracket and torque the nut and bolt to 37 ft. lbs. (50 Nm)
- Alternator and torque the bolts to 37 ft. lbs. (50 Nm)
- Accessory drive belt
- Vacuum lines to the throttle body
- Fuel feed and return lines
- EVAP canister purge solenoid electrical connector
- Spark plug wires to the rear bank spark plugs
- Air inlet duct
- Fuel injector sight shield
- Negative battery cable

6. Pressurize the fuel system.
7. Refill and bleed the cooling system.
8. Start the vehicle and check for leaks, repair if necessary.

WITH SUPERCHARGER

1. Before servicing the vehicle, refer to the precautions in the beginning of this section.
2. Relieve the fuel system pressure.
3. Drain the cooling system.
4. Remove or disconnect the following:
   - Negative battery cable
   - Exhaust Gas Recirculation (EGR) valve heat shield
   - Supercharger drive belt
   - Right side spark plug wires
   - Alternator rear brace
   - Fuel injector electrical connectors
   - Manifold Absolute Pressure (MAP) sensor and bracket
   - Fuel feed and return lines
   - Fuel rail with the injectors
   - Boost control solenoid
   - Cruise control and throttle cables
   - Air inlet duct
   - Supercharger
   - Upper radiator hose
   - Thermostat housing
   - EGR tube
   - Temperature sensor electrical connector
   - Intake manifold

To install:
5. Install or connect the following:
   - Intake manifold with new gaskets and torque the bolts to 11 ft. lbs. (15 Nm)

- Temperature sensor electrical connector
- EGR tube
- Thermostat housing with a new gasket and torque the bolts to 20 ft. lbs. (27 Nm)
- Upper radiator hose
- Supercharger with new gaskets and torque the bolts to 17 ft. lbs. (23 Nm)
- Air inlet duct
- Cruise control and throttle cables
- Boost control solenoid and torque the bolts to 72 inch lbs. (8 Nm)
- Fuel rail and torque the hold down bolts to 7 ft. lbs. (10 Nm)
- Fuel feed and return lines
- MAP sensor and bracket
- Fuel injector electrical connectors
- Alternator rear brace and torque the bolts to 37 ft. lbs. (50 Nm)
- Spark plug wires
- Supercharger drive belt
- EGR valve heat shield
- Fuel injector sight shield
- Negative battery cable

6. Fill the cooling system.
7. Start the engine and check for leaks.

Exhaust Manifold

REMOVAL & INSTALLATION

3.1L Engine

LEFT SIDE

1. Before servicing the vehicle, refer to the precautions in the beginning of this section.
2. Remove or disconnect the following:

- Negative battery cable
- Throttle body air inlet duct
- Right engine mount strut bracket
- Crossover pipe heat shield
- Crossover pipe nuts to the left exhaust manifold
- Exhaust manifold heat shield
- Exhaust manifold

To install:
3. Install or connect the following:

- Exhaust manifold with a new gasket and torque the nuts to 12 ft. lbs. (16 Nm)
- Exhaust manifold heat shield and torque the nuts to 89 inch lbs. (10 Nm)
- Exhaust crossover pipe to the manifold and torque the bolt to 18 ft. lbs. (25 Nm)
- Crossover pipe heat shield and

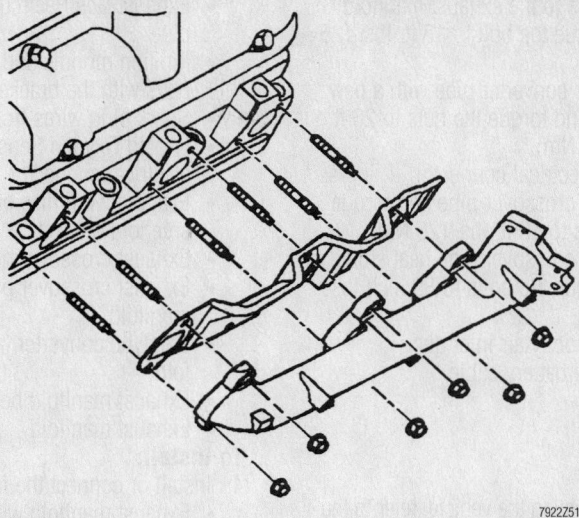

Exploded view of the left exhaust manifold mounting—3.1L engine

torque the bolt to 89 inch lbs. (10 Nm)
- Left side engine mount strut bracket and torque the bolt to 37 ft. lbs. (50 Nm)
- Throttle body air inlet duct
- Negative battery cable

RIGHT SIDE

1. Before servicing the vehicle, refer to the precautions in the beginning of this section.
2. Remove or disconnect the following:

- Negative battery cable
- Throttle body air inlet duct
- Exhaust crossover pipe heat shield

- Exhaust crossover pipe-to-right exhaust manifold nuts
- Heated Oxygen Sensor (HO2S) electrical connector
- Catalytic converter pipe from the exhaust manifold
- Exhaust Gas Recirculation (EGR) pipe from the exhaust manifold
- Exhaust manifold heat shields
- Exhaust manifold

To install:
3. Install or connect the following:

- Exhaust manifold with a new gasket and torque the nuts to 12 ft. lbs. (16 Nm)
- Exhaust manifold heat shields and torque the bolts to 89 inch lbs. (10 Nm)

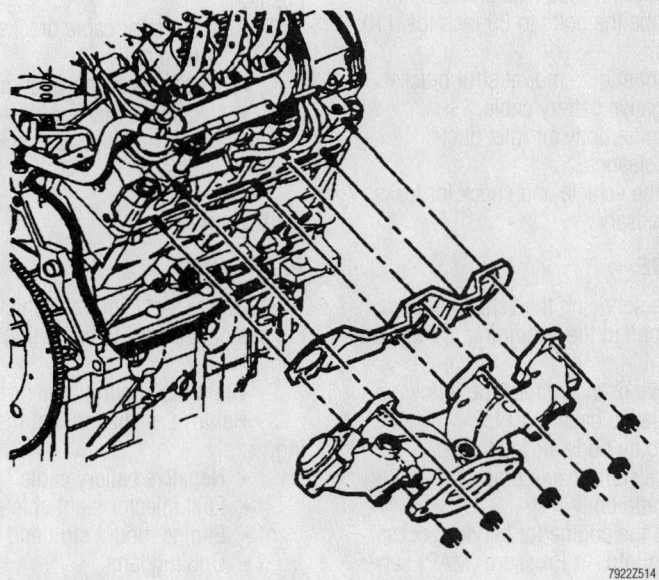

Exploded view of the right exhaust manifold mounting—3.1L engine

- EGR pipe to the exhaust manifold and torque the bolt to 18 ft. lbs. (25 Nm)
- Catalytic converter pipe with a new gasket and torque the nuts to 26 ft. lbs. (35 Nm).
- HO2S electrical connector
- Exhaust crossover pipe and torque the bolts to 18 ft. lbs. (25 Nm)
- Exhaust crossover pipe heat shield and torque the bolts to 89 inch lbs. (10 Nm)
- Throttle body air inlet duct
- Negative battery cable

3.4L Engine

LEFT SIDE

1. Before servicing the vehicle, refer to the precautions in the beginning of this section.
2. Remove or disconnect the following:
   - Negative battery cable
   - Air cleaner
   - Throttle body air inlet duct
   - Right engine mount strut bracket
   - Exhaust crossover pipe heat shield
   - Exhaust crossover pipe nuts to the left exhaust manifold
   - Exhaust manifold heat shield
   - Exhaust manifold

To install:

3. Install or connect the following:
   - Exhaust manifold with a new gasket and torque the bolts to 12 ft. lbs. (16 Nm)
   - Exhaust manifold heat shield and torque the bolts to 89 inch lbs. (10 Nm)
   - Crossover pipe and torque the bolts to 18 ft. lbs. (25 Nm)
   - Crossover pipe heat shield and torque the bolts to 89 inch lbs. (10 Nm)
   - Right engine mount strut bracket
   - Negative battery cable
   - Throttle body air inlet duct
   - Air cleaner
4. Start the vehicle and check for leaks, repair if necessary.

RIGHT SIDE

1. Before servicing the vehicle, refer to the precautions in the beginning of this section.
2. Remove or disconnect the following:
   - Negative battery cable
   - Throttle body air inlet duct
   - Accelerator cable bracket from the throttle body
3. Rotate the engine for service access.
   - Manifold Air Pressure (MAP) sensor

- Exhaust Gas Recirculation (EGR) pipe
- Ignition module and the ignition coils with the bracket
- Spark plug wires from the plugs
- Heated Oxygen Sensor (HO2S) electrical connector
- Evaporative Emissions (EVAP) solenoid bracket
- Exhaust crossover heat shield
- Exhaust crossover pipe from the manifold
- Catalytic converter from the manifold
- Exhaust manifold heat shields
- Exhaust manifold

To install:

4. Install or connect the following:
   - Exhaust manifold with a new gasket and torque the nuts to 12 ft. lbs. (16 Nm)
   - Exhaust manifold heat shields and torque the bolts to 89 inch lbs. (10 Nm)
   - Catalytic converter and torque the nuts to 24 ft. lbs. (32 Nm)
   - Crossover pipe to the manifold and torque the bolts to 18 ft. lbs. (25 Nm)
   - Crossover pipe heat shield and torque the bolts to 89 inch lbs. (10 Nm)
   - EVAP solenoid bracket
   - HO2S electrical connector
   - Spark plug wires from the plugs
   - Ignition module, ignition coils and bracket
   - EGR pipe
   - MAP sensor
5. Rotate the engine back to the original position.
   - Accelerator cable bracket to the throttle body
   - Throttle body air inlet duct
   - Negative battery cable
6. Start the vehicle and check for leaks, repair if necessary.

3.5L Engine

LEFT SIDE

1. Before servicing the vehicle, refer to the precautions in the beginning of this section.
2. Drain the cooling system.
3. Remove or disconnect the following:
   - Negative battery cable
   - Fuel injector sight shield
   - Engine mount strut and bracket
   - Cooling fans
   - Upper and lower radiator hoses

- Transaxle oil cooler lines from the radiator
- Radiator
- Alternator
- Exhaust manifold heat shield
- Oil level indicator tube
- Secondary Air Injection (AIR) control valve assembly from the engine mount strut bracket
- Exhaust manifold-to-crossover pipe studs
- Exhaust manifold

To install:

4. Install or connect the following:
   - Manifold to the crossover pipe using a new gasket and torque the studs to 18 ft. lbs. (25 Nm)
   - Exhaust manifold with a new gasket and torque the bolts to 18 ft. lbs. (25 Nm)
   - AIR valve and torque the pipe nut to 44 ft. lbs. (60 Nm) and the bolt to 80 inch lbs. (9 Nm)
   - Oil level indicator tube and torque the bolt to 80 inch lbs. (9 Nm)
   - Exhaust manifold heat shield and torque the bolts to 80 inch lbs. (9 Nm)
   - Alternator and torque the bolts to 37 ft. lbs. (50 Nm)
   - Radiator and torque the bolts to 89 inch lbs. (10 Nm)
   - Transaxle oil cooler lines and torque the fitting to 17 ft. lbs. (23 Nm)
   - Upper and lower radiator hoses
   - Cooling fans and torque the bolts to 89 inch lbs. (10 Nm)
   - Engine mount strut bracket and torque the bolts to 37 ft. lbs. (50 Nm)
   - Engine mount strut and torque the bolts to 35 ft. lbs. (48 Nm)
   - Fuel injector sight shield and torque the nuts to 27 inch lbs. (3 Nm)
   - Negative battery cable
5. Fill the cooling system.
6. Start the vehicle and check for leaks, repair if necessary.

RIGHT SIDE

1. Before servicing the vehicle, refer to the precautions in the beginning of this section.
2. Relieve the fuel system pressure.
3. Drain the cooling system.
4. Remove or disconnect the following:
   - Negative battery cable
   - Fuel injector cover
   - Air intake duct
   - Engine mount strut

- Fuel lines from the supply rail
- Cruise control and accelerator cables from the throttle body
- Transaxle selector range cable and cable brackets
- Transaxle shift cable from the shift module
- Brake booster vacuum hose from the engine
- Wiring harness connectors from the engine and transaxle
- Upper radiator hose
- Transaxle oil cooler lines from the radiator
- Surge tank inlet hose
- Heater hoses from the engine
- Lower radiator air deflector
- Battery cables from the retainers
- Lower radiator hose
- A/C compressor without disconnecting the lines and move it aside
- Starter wiring
- Catalytic converter from the manifold
- Front wheels and splash shields
- Wheel Speed Sensor (WSS) wiring from the lower control arms
- Tie rod ends from the steering knuckles
- Lower ball joints from the knuckles
- Halfshafts
- Intermediate shaft from the steering rack

5. Secure the vehicle to the lift in preparation for engine removal.

6. Position an engine/frame support table under the vehicle and lower the vehicle to meet the table.

7. Remove or disconnect the following:
- Frame-to-body bolts
- Crossover pipe from the front manifold
- Exhaust Gas Recirculation (EGR) pipe from the crossover pipe
- Right exhaust manifold from the engine

To install:

8. Install or connect the following:
- Right exhaust manifold with a new gasket and torque the bolts to 18 ft. lbs. (25 Nm)
- Crossover pipe to the front exhaust manifold and torque the bolts to 18 ft. lbs. (25 Nm)
- EGR pipe to the crossover pipe and torque the pipe nut to 44 ft. lbs. (60 Nm)

9. Position the engine/transaxle assembly under the vehicle.

10. Coat the sub-frame bushings with rubber lubricant.

11. Lower the vehicle onto the assembly. Align the sub-frame on the vehicle using 2 bolts or drill bits, ¾ inches thick by 8 inches long through the alignment holes on the right side of the frame.

12. Install new frame-to-body bolts. Torque the bolts to 133 ft. lbs. (180 Nm) starting with the rear bolts and then the front bolts.

13. Raise the vehicle and remove the frame table.

14. Install or connect the following:
- Intermediate shaft to the steering rack and torque the bolts to 35 ft. lbs. (48 Nm)

➡**Be sure the shaft is fully seated on the stub before installing the pinch bolt.**

- Halfshafts
- Ball joints and torque the nuts to 40 ft. lbs. (55 Nm)
- Tie rod ends and torque the nuts to 22 ft. lbs. (30 Nm) plus an additional 120 degree turn
- WSS wiring harness
- Splash shields and front wheels

- Catalytic converter and torque the nuts to 53 inch lbs. (6 Nm)
- Starter wiring
- A/C compressor and torque the bolts to 37 ft. lbs. (50 Nm)
- Lower radiator hose
- Battery cables in their retainers
- Lower radiator air deflector and torque the bolts to 15 ft. lbs. (20 Nm)

15. Remove the straps securing the vehicle to the lift.
- Heater hoses
- Surge tank hose
- Transaxle oil cooler lines to the radiator and torque the fittings to 17 ft. lbs. (23 Nm)
- Upper radiator hose
- Wiring harness connectors to the engine and transaxle
- Transaxle shift cable to the shift module and bracket
- Transaxle range selector cable
- Cruise control and accelerator cables to the throttle body
- Fuel lines to the supply rail
- Engine mount strut and torque the bolts to 35 ft. lbs. (48 Nm)
- Air inlet duct
- Fuel injector sight shield
- Negative battery cable

16. Refill the cooling system.

17. Start the engine and check for leaks, repair if necessary.

3.8L Engines

LEFT SIDE

1. Before servicing the vehicle, refer to the precautions in the beginning of this section.

2. Remove or disconnect the following:

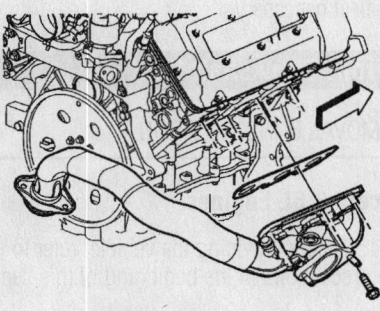

93002507

Exploded view of the right exhaust manifold—3.5L engine

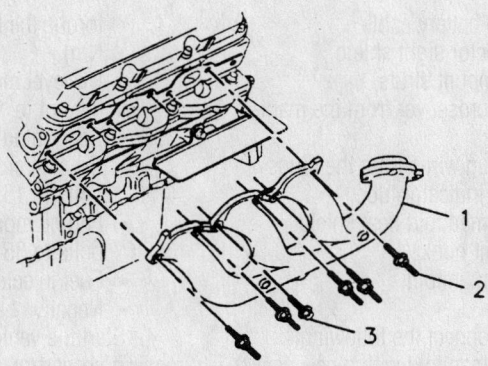

1 LEFT (FRONT) EXHAUST MANIFOLD
2 STUD 30 N•m (22 LB. FT.)
3 BOLT 30 N•m (22 LB. FT.)

7922XG30

Exploded view of the left exhaust manifold mounting—3.8L engine

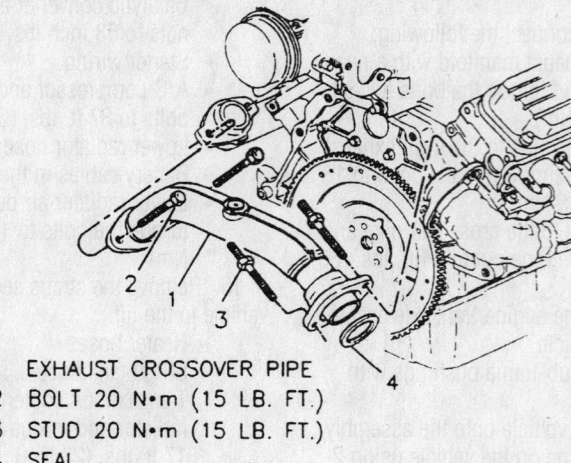

1 EXHAUST CROSSOVER PIPE
2 BOLT 20 N•m (15 LB. FT.)
3 STUD 20 N•m (15 LB. FT.)
4 SEAL

7922XG31

Exploded view of the crossover pipe mounting—3.8L engine

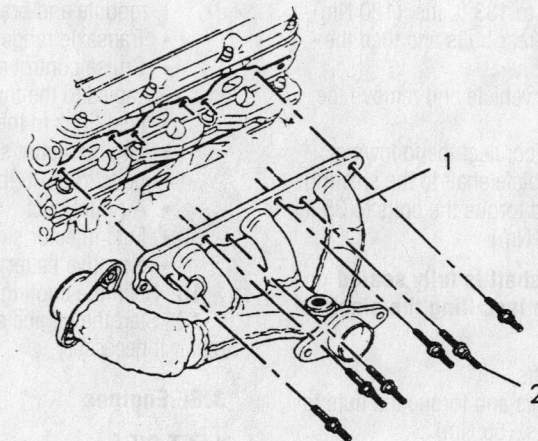

1 RIGHT (REAR) EXHAUST MANIFOLD
2 STUD 30 N•m (22 LB. FT.)

7922XG29

Exploded view of the right exhaust manifold mounting—3.8L engine

- Negative battery cable
- Fuel injector sight shield
- Engine mount struts
- Exhaust crossover from the manifold
- Spark plug wires from the plugs
- Oil level indicator tube
- Exhaust manifold heat shield
- Engine lift hook
- Exhaust manifold

To install:

3. Install or connect the following:
- Exhaust manifold with a new gasket and torque the bolts to 22 ft. lbs. (30 Nm)
- Engine lift hook and torque the bolts to 22 ft. lbs. (30 Nm)
- Exhaust manifold heat shield and torque the bolts to 15 ft. lbs. (20 Nm).
- Oil level indicator tube and torque the nut to 14 ft. lbs. (19 Nm)
- Spark plug wires
- Exhaust crossover and torque the bolts to 15 ft. lbs. (20 Nm)
- Engine mount struts and torque the bolts to 35 ft. lbs. (48 Nm)
- Fuel injector sight shield
- Negative battery cable

4. Start the vehicle and check for leaks, repair if necessary.

RIGHT SIDE

1. Before servicing the vehicle, refer to the precautions in the beginning of this section.

2. Remove or disconnect the following:
- Negative battery cable
- Fuel injector sight shield
- Cross vehicle brace
- Exhaust crossover pipe from the right exhaust manifold
- AIR pipe, if equipped
- Exhaust manifold pipe stud nuts and the catalytic converter pipe from the right exhaust manifold
- Exhaust Gas Recirculation (EGR) adapter pipe from the manifold

3. Rotate the engine to access the right side of the engine.
- Heated Oxygen Sensor (HO$_2$S) electrical connector
- Spark plug wires from the plugs
- Right engine lift bracket
- Fuel injector sight shield bracket
- Transaxle fluid filler tube
- Exhaust manifold

To install:

4. Install or connect the following:
- Exhaust manifold with a new gasket and torque the bolts to 22 ft. lbs. (30 Nm)
- Engine lift bracket and torque the bolts to 22 ft. lbs. (30 Nm)
- Transaxle fluid filler tube
- Fuel injector sight shield bracket and torque the bolts to 22 ft. lbs. (30 Nm)
- Spark plug wires
- Heated Oxygen Sensor (HO$_2$S) electrical connector

5. Rotate the engine back to the original position.
- EGR adapter pipe and torque the bolt to 21 ft. lbs. (29 Nm)
- Catalytic converter and torque the bolts to 26 ft. lbs. (35 Nm)
- Exhaust crossover pipe and torque the bolts to 13 ft. lbs. (18 Nm)
- AIR pipe, if equipped
- Cross vehicle brace
- Fuel injector sight shield
- Negative battery cable

6. Start the vehicle and check for leaks, repair if necessary.

Front Crankshaft Seal

REMOVAL & INSTALLATION

Except 3.5L Engines

1. Before servicing the vehicle, refer to the precautions in the beginning of this section.

2. Remove the crankshaft balancer using an appropriate puller.

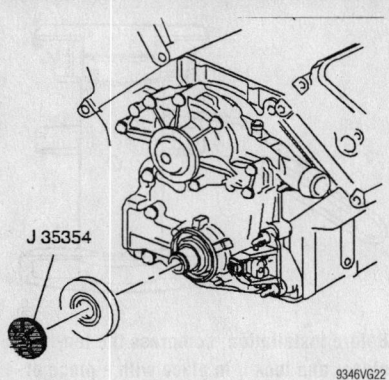

Installing the front crankshaft seal—All except 3.5L engines

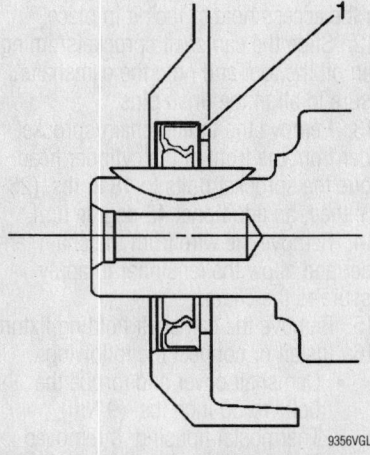

Make sure the crankshaft front oil seal is installed flush to the engine front cover (1)—All except 3.5L engines

3. Pry the front crankshaft seal out.

To install:

4. Coat the new seal with engine oil to ease installation.

5. Install the front crankshaft seal using seal installer J-35468 on 3.1L and 3.4L engines. On 3.8L engines use tool J 35354. Make sure that the crankshaft front oil seal lip faces the engine.

6. Make sure the seal is flush with the engine when installed.

• Crankshaft balancer and torque the bolt to 76 ft. lbs. (103 Nm)

3.5L Engines

1. Before servicing the vehicle, refer to the precautions in the beginning of this section.

2. Remove the crankshaft balancer using an appropriate puller.

3. Pry the front crankshaft seal out.

To install:

4. Coat the new seal with engine oil to ease installation.

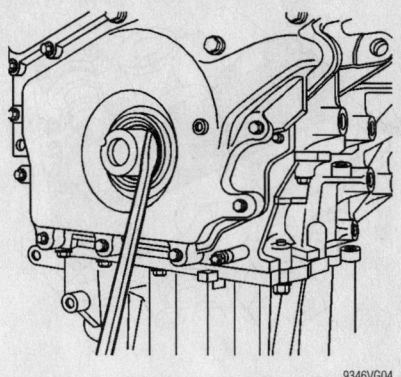

Remove the front oil seal, using a suitable prying tool

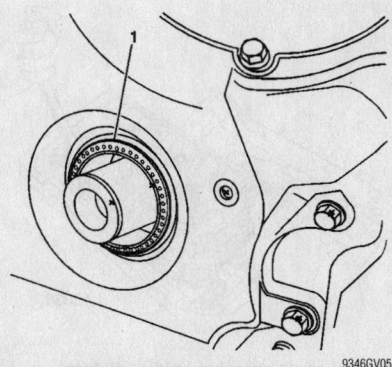

The installed depth of the seal (1) should protrude 1-2 mm (0.039-0.078 in.) from the front cover—3.5L engine

5. Place the front seal onto the mandrel of the installation tool J-42041.

6. Slide the mandrel onto the screw of the installation tool.

7. Thread the installation tool in the crankshaft. Ensure you engage at least ten threads of the tool before pressing the seal in place.

8. Push the seal into position by tightening the nut on the installation tool until the mandrel bottoms out on the crankshaft sprocket.

9. Check the installed depth of the seal, which should protrude 1–2 mm (0.039–0.078 in.), from the front cover.

10. Install the crankshaft balancer to 37 ft. lbs. (50 Nm) plus an additional 125 degrees.

Camshaft and Valve Lifters

REMOVAL & INSTALLATION

3.1L and 3.4L Engines

1. Before servicing the vehicle, refer to the precautions in the beginning of this section.

2. Relieve the fuel system pressure.

3. Remove or disconnect the following:
• Engine assembly
• Rocker arm covers
• Intake manifold
• Rocker arm bolts, balls, rocker arms and pushrods
• Lifter guide bolts and the guide
• Valve lifters from the bores
• Crankshaft balancer
• Front cover
• Timing chain and sprockets
• Oil pump driven gear
• Camshaft thrust plate
• Camshaft

✳✳ WARNING

Avoid damaging the camshaft bearing surfaces.

To install:

4. Coat the camshaft with assembly lubricant.

5. Install or connect the following:
• Camshaft
• Camshaft thrust plate and torque the bolts to 89 inch lbs. (10 Nm)
• Oil pump driven gear and torque the bolt to 27 ft. lbs. (36 Nm)
• Timing chain and sprocket and torque the bolt to 103 ft. lbs. (140 Nm)
• Front cover and torque the small bolts to 15 ft. lbs. (21 Nm), the medium bolts to 35 ft. lbs. (47 Nm) and the large bolts to 41 ft. lbs. (55 Nm)
• Crankshaft balancer and torque the bolt to 76 ft. lbs. (103 Nm)

6. Lubricate the bearing surfaces with Molykote®.

➡**Installation of a new camshaft or a wear pattern on the old valve lifter will require the replacement of the camshaft and lifters together. If camshaft replacement is not necessary, be sure to install the used valve lifters in their original position.**

7. Install or connect the following:
• Lifters in their original locations
• Lifter guide and torque guide bolts to 89 inch lbs. (10 Nm)
• Pushrods, rocker arms, balls and bolts and torque the nuts to 89 inch lbs. (10 Nm) plus an additional 30 degree turn
• Intake manifold and torque the lower manifold bolts to 115 inch lbs. (13 Nm) and the upper manifold bolts to 18 ft. lbs. (25 Nm)

- Rocker arm covers and torque the bolts to 89 inch lbs. (10 Nm)
- Engine assembly
- Negative battery cable

➡ The only time valve adjustment in needed is if there was a valve job performed or the rocker studs have been replaced with an adjustable rocker arm stud. The rocker arm stud installed from the factory should be shouldered and not need any adjustment.

8. Start the engine and check for leaks, repair if necessary.

3.5L Engine

1. Before servicing the vehicle, refer to the precautions in the beginning of this section.
2. Drain the cooling system.
3. Remove or disconnect the following:
   - Negative battery cable
   - Fuel injector sight shield
   - Thermostat housing for clearance when installing the camshaft Holding Fixture
   - Camshaft cover
4. Rotate the crankshaft so the camshaft flats are parallel to the camshaft's sealing surface, then install a camshaft holding fixture.
5. Remove the camshaft sprocket bolts.
6. Install a timing chain/sprocket holding fixture J 42042.
7. Evenly slide the camshaft sprocket and chain from the camshafts onto the holding tool.

➡ The camshaft bearing caps are marked. Be sure the raised portion of the cap faces the outside of the engine. They must always be installed in their original positions.

8. Remove or disconnect the following:
   - Camshaft bearing caps
   - Camshaft holding fixture
   - Camshafts

To install:

9. Coat the rocker arms with engine oil and place them in their original positions.

➡ Be sure to install the rounded end on the lash adjuster and the flat end on the tip of the valve.

10. Install or connect the following:
    - Camshafts, lubricated with engine oil, with the sprocket drive pin notch located at the top
    - Bearing caps and torque the bolts to 71 inch lbs. (8 Nm) plus an additional 22 degree turn

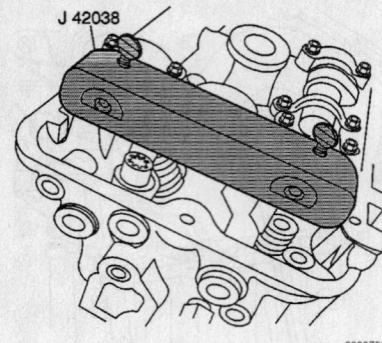

Camshaft holding fixture installed on the camshafts—3.5L engine

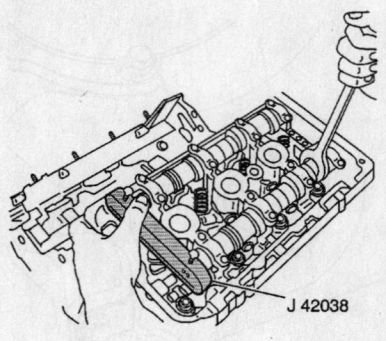

Timing chain/sprocket holding fixture installed on the cylinder head; use the flats on the camshaft if rotation is necessary for installation of the holding tool—3.5L engine

- camshaft Holding Fixture tool onto the camshaft(s) at the rear of the cylinder head

➡ Use the camshaft flats to turn the camshaft.

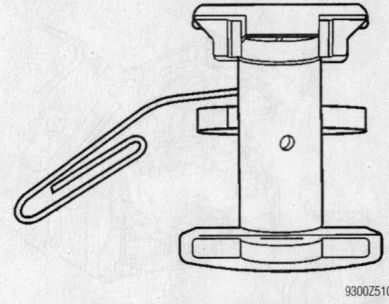

Before installation, compress the tensioner and lock it in place with a piece of wire—3.5L engine

11. Compress the secondary timing chain tensioner by hand and insert a wire into the access hole to lock it in place.
12. Slide the camshaft sprockets/timing chain off the tool and onto the camshafts. Be sure to align the drive pins.
13. Remove the timing chain/sprocket holder from the front of the cylinder head. Torque the sprocket bolts to 18 ft. lbs. (25 Nm); then, an additional 45 degree turn.
14. Remove the wire from the chain tensioner and allow the tensioner to apply pressure to the chain.
15. Remove the camshaft holding fixture.
16. Install or connect the following:
    - Camshaft cover and torque the bolts to 80 inch lbs. (9 Nm)
    - Thermostat housing, if removed
    - Fuel injector sight shield and torque the nuts to 27 inch lbs. (3 Nm)
    - Negative battery cable
17. Refill the cooling system.
18. Start the vehicle and check for leaks, repair if necessary.

1. Left intake
2. Left exhaust
3. Right intake
4. Right exhaust

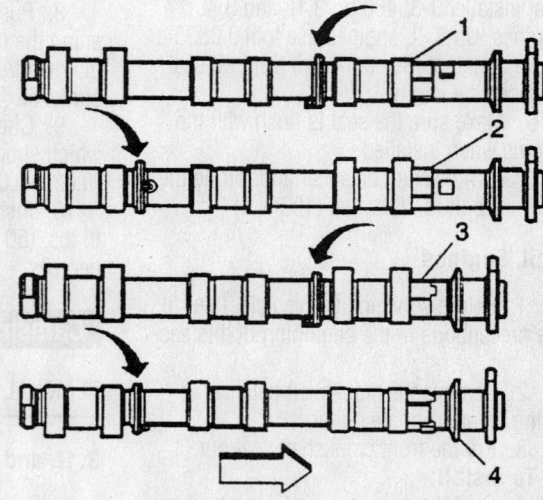

Camshaft identification—3.5L engine

3.8L Engine

1. Before servicing the vehicle, refer to the precautions in the beginning of this section.
2. Discharge the A/C system.
3. Drain the cooling system.
4. Remove or disconnect the following:
   - Negative battery cable
   - Radiator with the air conditioning condenser assembly
   - Rocker arm cover
   - Valve lifters
   - Front cover
   - Camshaft sprocket and timing chain
   - Camshaft thrust plate
5. Remove the camshaft assembly, as follows:
   a. Step 1: Install 1₂–20 x 6 inch bolt in the camshaft front bolt hole
   b. Step 2: Carefully rotate and pull the camshaft assembly out of the bearings.
6. Inspect the camshaft for damage and replace if necessary.

To install:

7. Install or connect the following:
   - Camshaft lubricated with assembly lubricant
   - Thrust plate and torque the bolts to 11 ft. lbs. (15 Nm)
   - Front cover and torque the bolts to 15 ft. lbs. (20 Nm) plus an additional 40 degree turn
   - Valve lifters
   - Rocker arm cover and torque the bolts to 89 inch lbs. (10 Nm)

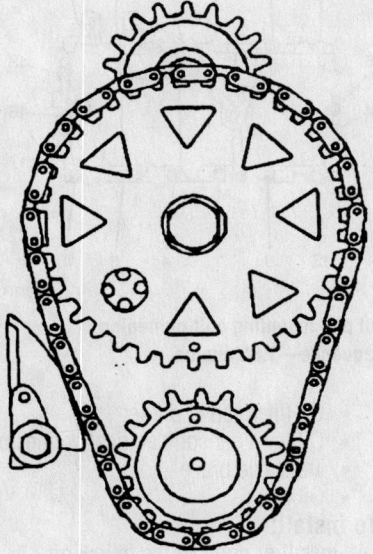

The timing marks should face each other if the chain and gears are installed properly—3.8L engine

   - Radiator with the air conditioning condenser and torque the bolts to 18 ft. lbs. (25 Nm)
   - Negative battery cable
8. Evacuate and recharge the A/C system.
9. Refill the cooling system.
10. Start the engine and check for leaks, repair if necessary.

Valve Lash

ADJUSTMENT

The valve clearance cannot be adjusted on these engines.

Starter Motor

REMOVAL & INSTALLATION

1. Before servicing the vehicle, refer to the precautions in the beginning of this section.

➡**The starter motors used in these vehicles are of varying lengths, make sure to use the correct size starter motor.**

2. Remove or disconnect the following:
   - Negative battery cable
   - Radiator lower air deflector
   - Torque converter cover
   - Electrical connections from the starter
   - Starter motor

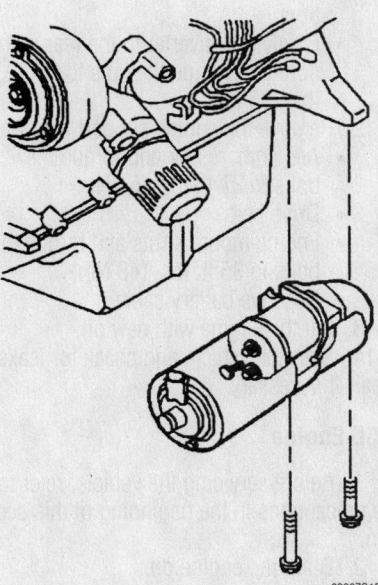

Exploded view of the starter—3.1L and 3.4L engines

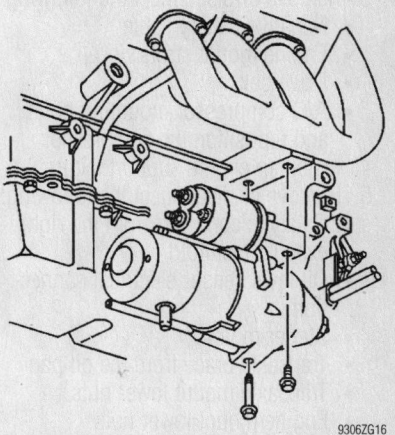

Exploded view of the starter—3.5L engine

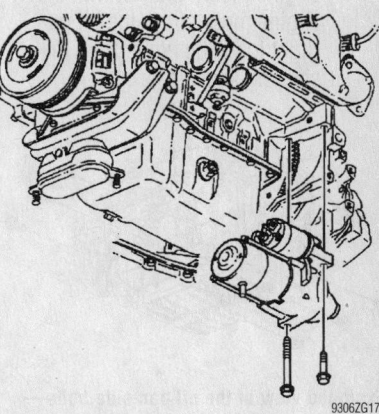

Exploded view of the starter—3.8L engine

To install:

3. Install or connect the following:
   - Starter motor and torque the bolts to 32 ft. lbs. (43 Nm) on all models except Intrigue. On Intrigue models, tighten the bolts to 37 ft. lbs. (50 Nm).
   - Starter solenoid BAT terminal
   - Starter solenoid S terminal
   - Torque converter cover and torque the bolts to 89 inch lbs. (10 Nm)
   - Radiator lower air deflector and torque the bolts to 15 ft. lbs. (20 Nm)
   - Negative battery cable

Oil Pan

REMOVAL & INSTALLATION

3.1L and 3.4L Engines

1. Before servicing the vehicle, refer to the precautions in the beginning of this section.
2. Drain the engine oil.

3. Remove or disconnect the following:
- Negative battery cable
- Engine mount struts
- Drive belt
- A/C compressor mounting bolts and reposition the compressor

4. Install an engine support fixture.

5. Remove or disconnect the following:
- Catalytic converter from the right exhaust manifold
- Oil level sensor electrical connection
- Starter motor
- Transaxle brace from the oil pan
- Transaxle mount lower nuts
- Engine mount lower nuts

6. Raise the engine to gain access to the oil pan.
- Engine mount bracket with the engine mount from the oil pan
- Right lower ball joint from the steering knuckle

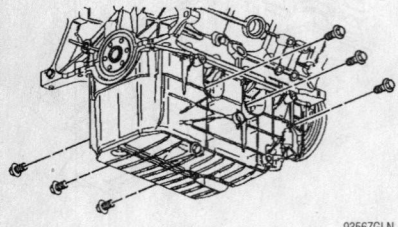

Exploded view of the oil pan side bolts— 3.1L and 3.4L engines

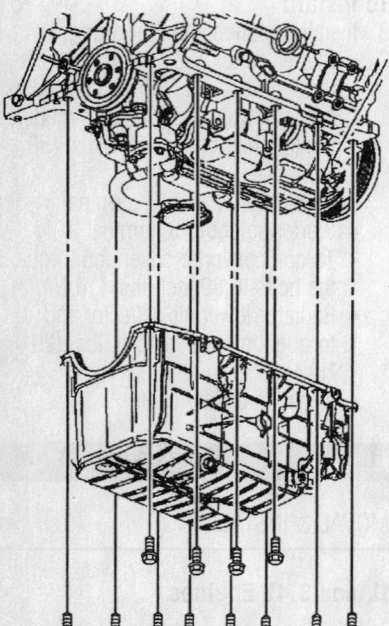

Exploded view of the oil pan retaining bolts—3.1L and 3.4L engines

- Right outer tie rod from the steering knuckle
- Right side frame bolts, loosen only
- Left side frame bolts, loosen only

7. Remove the oil pan.

To install:

8. Apply a small amount of sealer on both sides of the rear main bearing cap between the gasket and the engine.

9. Install or connect the following:
- Oil pan with a new gasket and torque the retaining bolts to 18 ft. lbs. (25 Nm)
- Front and rear oil pan bolts and torque the bolts to 37 ft. lbs. (50 Nm)
- Frame bolts to 133 ft. lbs. (180 Nm)
- Right outer tie rod to the steering knuckle
- Right lower ball joint to the steering knuckle
- Engine mount and bracket to the oil pan and torque the bolts to 43 ft. lbs. (58 Nm)

10. Lower the engine into position.

11. Install or connect the following:
- Engine mount lower nuts and torque the nuts to 32 ft. lbs. (43 Nm)
- Transaxle mount lower nuts and torque the nuts to 35 ft. lbs. (47 Nm)
- Transaxle brace to the oil pan and torque the bolts to 46 ft. lbs. (63 Nm)
- Starter motor and torque the bolts to 32 ft. lbs. (43 Nm)
- Oil level sensor electrical connection
- Catalytic converter to the rear manifold and torque the nuts to 26 ft. lbs. (35 Nm)

12. Remove the engine support fixture.
- A/C compressor and torque the bolts to 37 ft. lbs. (50 Nm)
- Drive belt
- Engine mount struts and torque the bolts to 35 ft. lbs. (48 Nm)
- Negative battery cable

13. Fill the engine with new oil.

14. Start the vehicle and check for leaks, repair if necessary.

3.5L Engine

1. Before servicing the vehicle, refer to the precautions in the beginning of this section.

2. Drain the engine oil.

3. Remove or disconnect the following:
- Negative battery cable

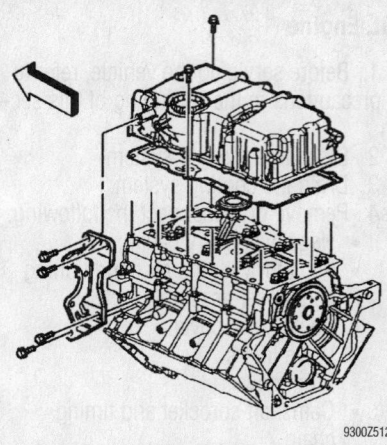

Exploded view of the oil pan—3.5L engine

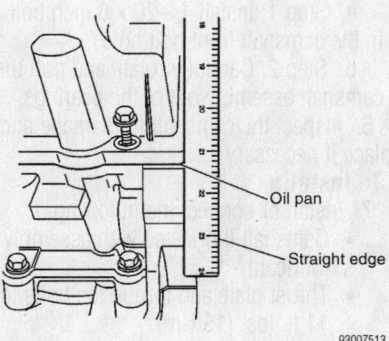

Use a straight-edge to align the rear of the oil pan to the rear of the engine—3.5L engine

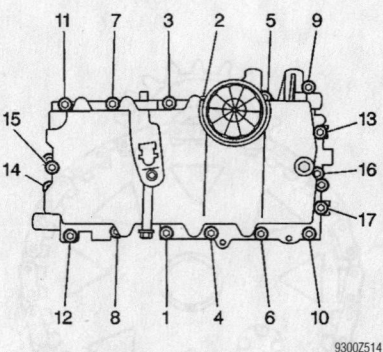

Oil pan mounting bolt tightening sequence—3.5L engine

- Oil filter cap and filter
- Oil level sensor electrical connector
- Transaxle brace
- Oil pan

To install:

4. Install or connect the following:
- Oil pan with a new gasket, Do not tighten the bolts
- Transaxle brace on the engine block

only and torque the bolts to 18 ft. lbs. (25 Nm)
- Brace-to-oil pan bolts, loosely install them

5. Align the rear of the oil pan flush with the rear of the engine block. Use a straight edge for reference.

6. Press the front of the oil pan against the transaxle brace; then, torque the brace-to-oil pan bolts to 18 ft. lbs. (25 Nm). Be sure to keep the rear of the pan flush with the rear of the engine.

7. Torque the oil pan bolts in sequence to 18 ft. lbs. (25 Nm).

8. Torque the brace-to-transaxle bolts to 32 ft. lbs. (43 Nm).

9. Install or connect the following:
- Oil level sensor electrical connection
- Drain plug and torque it to 15 ft. lbs. (20 m)
- New oil filter and torque the cap to 18 ft. lbs. (25 Nm)
- Negative battery cable

10. Refill the engine with clean oil.

11. Start the vehicle and check for leaks, repair if necessary.

3.8L Engine

1. Before servicing the vehicle, refer to the precautions in the beginning of this section.

2. Drain the engine oil.

3. Remove or disconnect the following:
- Negative battery cable
- Throttle body air inlet duct
- Engine mount struts
- Accessory drive belt

4. Install an engine support fixture.
- Catalytic converter from the right exhaust manifold
- Right front wheel
- Right side splash shield

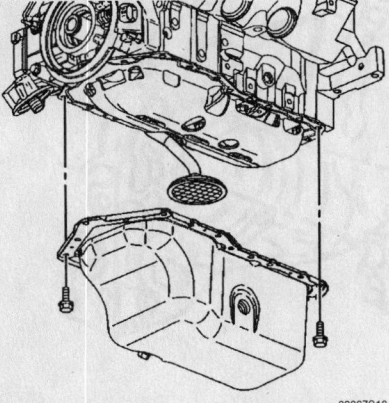

Exploded view of the oil pan mounting— 3.8L engine

- Oil filter and discard
- A/C compressor bracket and reposition the compressor
- Power steering oil cooler pipe brackets from the frame
- Engine mount bracket bolts
- Lower engine mount nuts
- Torque converter cover
- Oil level sensor electrical connector
- Oil level sensor

5. Raise the vehicle and place a support under the frame.
- Right side frame bolts and loosen the left side bolts
- Engine mount and bracket
- Oil pan
- Oil pump pipe and screen

To install:

6. Install or connect the following:
- New oil pan gasket and the oil pump pipe and screen assembly and torque the bolts to 11 ft. lbs. (15 Nm)

➡ **Do not over-tighten the oil pan bolts or damage to the pan may occur.**

- Oil pan and torque the bolts to 125 inch lbs. (14 Nm)
- Engine mount and bracket and torque the bolts to 50 ft. lbs. (68 Nm)
- New right side frame bolts and torque all the bolts to 133 ft. lbs. (180 Nm)
- Lower engine mount and torque the nuts to 35 ft. lbs. (47 Nm)
- Oil level sensor and torque the fastener to 15 ft. lbs. (20 Nm)
- Oil level wire connector
- Torque converter cover and torque the bolts to 89 inch lbs. (10 Nm)
- Power steering oil cooler pipe brackets to the frame
- A/C compressor and torque the upper bolt to 37 ft. lbs. (50 Nm) and the lower bolt to 59 ft. lbs. (80 Nm)
- New oil filter
- Drain plug and torque the plug to 22 ft. lbs. (30 Nm)
- Splash shield
- Right front wheel
- Catalytic converter pipe to the right exhaust manifold and torque the bolts to 26 ft. lbs. (35 Nm)

7. Remove the engine support fixture.
- Drive belt
- Engine mount struts to the engine and torque the bolts to 41 ft. lbs. (56 Nm)
- Throttle body air inlet duct
- Negative battery cable

8. Fill the engine with new oil.

9. Start the vehicle and check for leaks, repair if necessary.

10. Road test the vehicle, check the front end alignment and adjust if necessary.

Oil Pump

REMOVAL & INSTALLATION

3.1L and 3.4L Engines

1. Before servicing the vehicle, refer to the precautions in the beginning of this section.

2. Drain the engine oil.

3. Remove or disconnect the following:
- Negative battery cable
- Oil pan
- Bolt attaching the oil pump to the rear crankshaft bearing cap
- Oil pump and driveshaft

To install:

➡ **Rotate the driveshaft as required to obtain the proper engagement with the oil pump drive unit.**

4. Install or connect the following:
- Oil pump and driveshaft and torque the bolt to 30 ft. lbs. (41 Nm)
- Oil pan
- Negative battery cable

5. Fill the engine with new oil.

6. Start the vehicle and check for leaks, repair if necessary.

3.5L Engine

1. Before servicing the vehicle, refer to the precautions in the beginning of this section.

2. Drain the engine oil.

3. Remove or disconnect the following:
- Negative battery cable
- Front cover
- Rocker arm covers

4. Install camshaft holding fixtures J 42038 on both sets of camshafts. Turn the hex portion of the camshaft to align them for tool installation. When installed, the flats on the rear of the camshafts will be parallel with the camshaft cover sealing surface.

5. Remove or disconnect the following:
- Oil pan
- Oil pump pipe and screen
- Primary chain tensioner
- Primary chain from the drive sprocket
- Oil pump by sliding it off the crankshaft

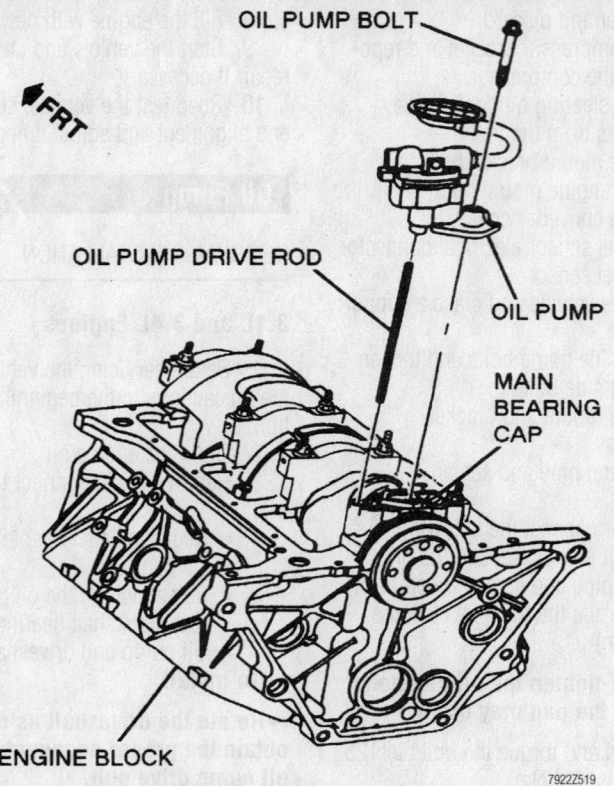

OIL PUMP BOLT

OIL PUMP DRIVE ROD

OIL PUMP

MAIN BEARING CAP

ENGINE BLOCK

FRT

Oil pump and driveshaft components—3.1L and 3.4L engines

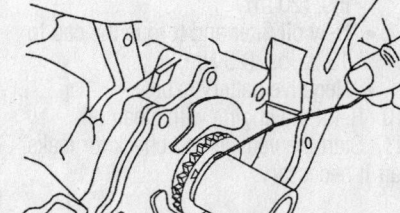

Align the crankshaft sprocket splines with the oil pump gear and install the sprocket in the oil pump—3.5L engine

Correct position of the crankshaft sprocket when the oil pump is installed correctly—3.5L engine

To install:

6. Pack the oil pump housing with white petroleum jelly to insure priming.

7. Align the oil pump sprocket with the crankshaft and install the pump on the engine until a positive stop is felt. When installed properly, the sprocket will protrude slightly from the oil pump and the face of the sprocket will be behind the machined step in the crankshaft.

8. Install or connect the following:
- Oil pump and torque the bolts to 18 ft. lbs. (25 Nm)
- Primary chain on the sprocket

➡ Be sure to maintain correct timing.

- Chain tensioner
- Oil pump pipe with the screen and torque the nut and bolt to 89 inch lbs. (10 Nm)

1. 97 inch lbs. (11 Nm)
2. Oil pump cover
3. Pump outer gear
4. Pump inner gear
5. Front cover

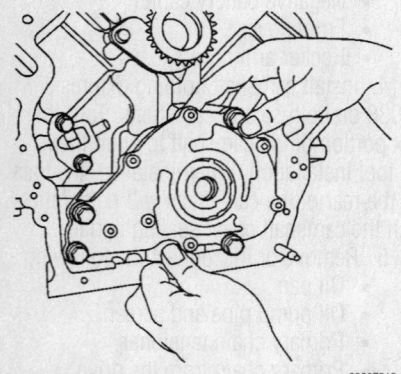

The oil pump is mounted on the front of the engine and driven by the crankshaft—3.5L engine

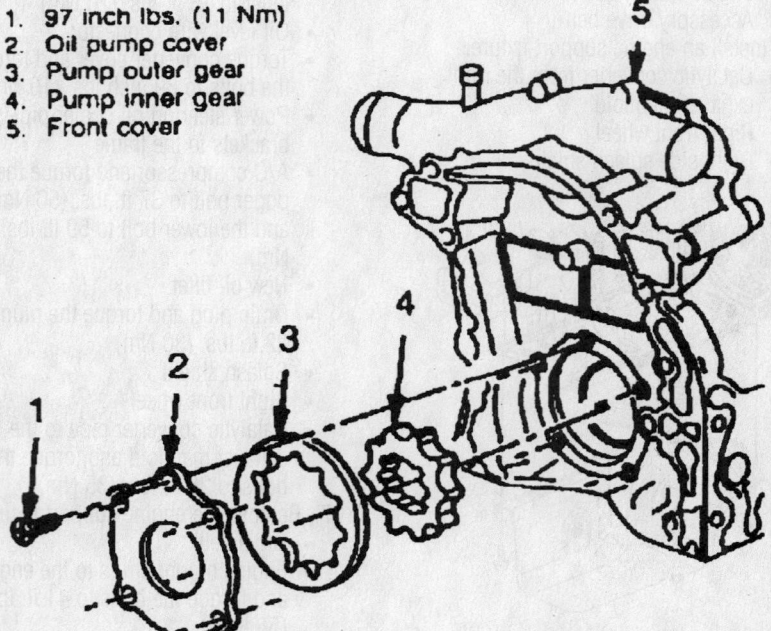

Oil pump assembly—3.8L engine

- Oil pan and torque the bolts to 18 ft. lbs. (25 Nm)
9. Remove the camshaft holding tools.
10. Install or connect the following:
   - Camshaft covers and torque the bolts to 80 inch lbs. (99 Nm)
   - Engine front cover and torque the bolts to 124 inch lbs. (14 Nm)
   - Negative battery cable
11. Fill the engine with new oil.
12. Start the vehicle and check for leaks, repair if necessary.

3.8L Engines

1. Before servicing the vehicle, refer to the precautions in the beginning of this section.
2. Drain the engine oil.
3. Remove or disconnect the following:
   - Negative battery cable
   - Front cover. Refer to the timing chain removal and installation procedure for front cover removal.
   - Oil pump cover
   - Oil pump gear set

To install:

4. Lubricate the oil pump gears with petroleum jelly and install the gears into the housing.
5. Pack the gear cavity with petroleum jelly after the gears have been installed.
6. Install or connect the following:
   - Oil pump cover and torque the screws to 98 inch lbs. (11 Nm)
   - Front cover. Refer to the timing chain removal and installation procedure for front cover installation.
   - Negative battery cable
7. Fill the engine oil. A new oil filter is recommended.
8. Start the vehicle and check for leaks, repair if necessary.

Rear Main Seal

REMOVAL & INSTALLATION

Except 3.5L Engine

1. Before servicing the vehicle, refer to the precautions in the beginning of this section.
2. Remove or disconnect the following:
   - Negative battery cable
   - Transmission
   - Flexplate
   - Rear main seal by prying it out

To install:

3. Lubricate the lip and the outer edge of the new seal with clean engine oil.
4. Install or connect the following:

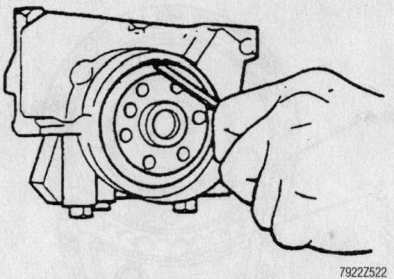

Carefully pry the seal from the bore without damaging the crankshaft seal surface

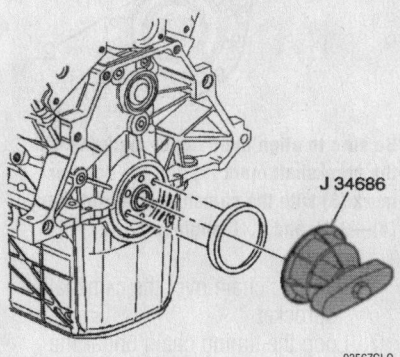

Installing the rear main seal—except 3.5L engines

- Position the new seal on the mandrel of Rear Main Seal Installer tool J 34686 until the back of the seal is flush against the collar of the tool
- Seal installer tool to the rear of the crankshaft with the 2 mounting bolts. Turn the handle until the seal is seated in the rear of the engine. Remove the installer tool
- Flexplate
- Transmission
- Negative battery cable
5. Start the vehicle and check for leaks, repair if necessary.

3.5L Engine

1. Before servicing the vehicle, refer to the precautions in the beginning of this section.
2. Remove or disconnect the following:
   - Negative battery cable
   - Transaxle
   - Engine flywheel
3. Place Rear Seal Remover Tool J 42841 on the crankshaft with retaining bolts.
4. Install eight one-inch self starting screws through the guide holes of the tool. Tighten the screws.

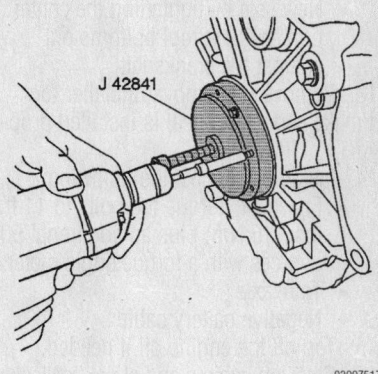

Use the guide holes in Tool J 42841 to install the screws in the seal—3.5L

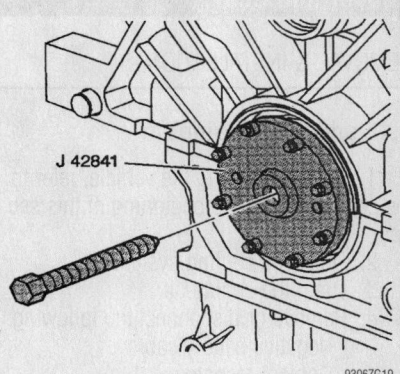

Install a center forcing screw into the removal tool—3.5L engine

5. Install the two retaining bolts.
6. Install a center forcing screw into the removal tool and pull the seal off the end of the crankshaft.

To install:

7. Clean debris from the crankshaft rear seal drain. The seal may leak if the drain is not properly cleaned.
8. Place a small amount of gasket maker to the crankcase split line across the end of the upper and lower crankcase seal.
9. Coat the outer diameter of the block with clean engine oil.
10. Clean the outer diameter of the flywheel flange with a lint-free cloth.

✳✳ CAUTION

Do not apply any oil on the green coating of the new seal.

11. Loosen the center bolt of the seal installer tool until the hub protrudes past the outer plate (approximately ½ inch).
12. Install or connect the following:
   - Three mounting bolts into the crankshaft flange until the tool is fully seated on the crankshaft

- New seal by tightening the center bolt until the tool bottoms out against the crankshaft

13. Remove the removal/installer tool and make certain the seal is installed properly

14. Install or connect the following:
- Flywheel. Torque the bolts to 11 ft. lbs. (15 Nm) plus an additional 50 degrees with a torque angle meter.
- Transaxle
- Negative battery cable

15. Top off the engine oil if needed.

16. Start the vehicle and check for leaks, repair if necessary.

Timing Chain, Sprockets, Front Cover and Seal

REMOVAL & INSTALLATION

3.1L and 3.4L Engines

1. Before servicing the vehicle, refer to the precautions in the beginning of this section.

2. Drain the cooling system.

3. Drain the engine oil.

4. Remove or disconnect the following:
- Negative battery cable
- Coolant reservoir
- Accessory drive belt
- Crankshaft balancer
- Drive belt tensioner
- Power steering pump, DO NOT disconnect the lines
- Thermostat bypass pipe from the front cover
- Upper radiator hose
- Water pump pulley
- Lower Crankshaft Position (CKP) sensor
- Front cover

5. Rotate the crankshaft until the timing marks on the camshaft and crankshaft sprockets are in alignment (facing each other).
- Camshaft sprocket bolt, sprocket and timing chain
- Crankshaft sprocket
- Timing chain damper bolts and damper, if necessary

To install:

6. Install or connect the following:
- Timing chain damper (if removed) and torque the bolts to 15 ft. lbs. (21 Nm)
- Crankshaft sprocket

➡**Be sure the timing mark on the crankshaft sprocket is pointing toward the mark on the chain damper.**

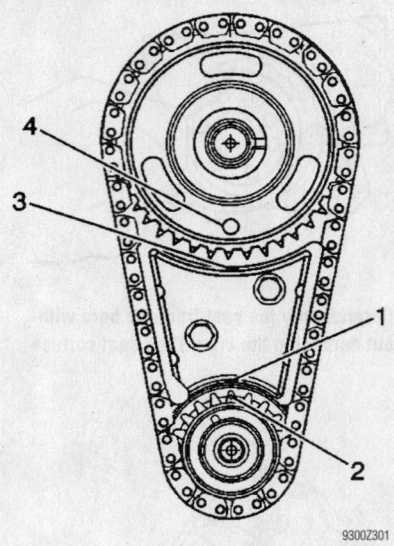

93002301

Be sure to align the damper mark (1) with the crankshaft mark (2) and the damper mark (3) with the camshaft sprocket mark (4)—3.1L and 3.4L engines

- Timing chain over the camshaft sprocket

7. Loop the timing chain under the crankshaft sprocket and install the camshaft sprocket on the camshaft.

8. Verify that the marks are aligned; the camshaft sprocket will be at the 6 o'clock position and the crankshaft sprocket at the 12 o'clock position.

➡**The No. 1 piston will be at Top Dead Center (TDC) and the No. 4 piston will also be at TDC but on the compression stroke.**

9. Tighten the camshaft sprocket bolt to 103 ft. lbs. (140 Nm).

10. Lubricate the timing chain components with engine oil.

11. Install or connect the following:
- New front cover seal
- Front cover using a new gasket and torque the small bolts to 20 ft. lbs. (27 Nm) and the large bolts to 41 ft. lbs. (55 Nm)
- Water pump pulley and torque the bolts to 18 ft. lbs. (25 Nm)
- Upper radiator hose
- Thermostat bypass pipe
- Power steering pump and torque the bolts to 25 ft. lbs. (34 Nm)
- Drive belt tensioner and torque the bolt to 37 ft. lbs. (50 Nm)
- CKP sensor
- Crankshaft balancer and torque the bolt to 76 ft. lbs. (103 Nm)
- Accessory drive belt
- Coolant reservoir

- Negative battery cable

12. Refill the fluids.

13. Start the engine and check for leaks, repair if necessary.

3.5L Engine

PRIMARY CHAIN

1. Before servicing the vehicle, refer to the precautions in the beginning of this section.

2. Drain the engine oil.

3. Drain the cooling system.

4. Disconnect the negative battery cable.

5. Remove the camshaft covers.

6. Rotate the crankshaft so the No. 1 piston is at Top Dead Center (TDC) and the flats on the rear of the camshafts are parallel with the camshaft cover sealing surface.

7. Install camshaft holding fixtures on both sets of camshafts. Turn the hex portion of the camshaft to align them for tool installation.

➡**When installed, the flats on the rear of the camshafts will be parallel with the camshaft cover sealing surface.**

8. Remove or disconnect the following:
- Right diagonal brace
- Battery and tray
- Coolant reservoir
- Underhood accessory wiring junction block, move it aside
- Drive belt
- Power steering pump pulley
- Idler pulley and belt tensioner
- Water pump pulley
- Water pump drive belt shield
- Water pump

9. Support the engine cradle.

10. Remove the right side engine cradle bolts.

11. Lower the cradle.

12. Remove or disconnect the following:
- Crankshaft balancer
- Front cover
- Engine lift bracket from the front of the engine
- Camshaft Position (CMP) sensor
- Sprocket bolt from the exhaust camshaft on the right cylinder head to allow for clearance of the chain guide
- Four chain guide access plugs from the cylinder heads

➡**Note that each plug has an O-ring.**

- Primary chain tensioner

➡**Remove the lower bolt allowing the tensioner to swing down and expand.**

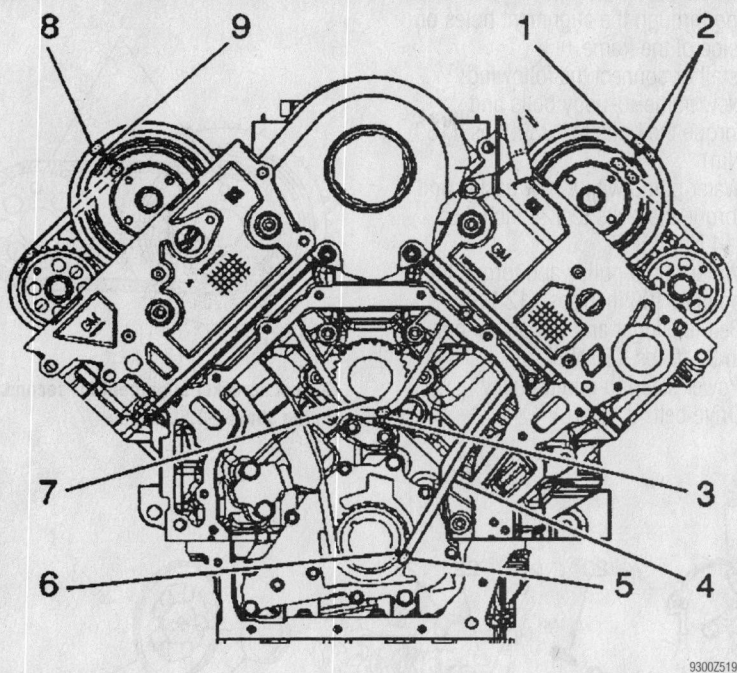

Primary timing chain alignment marks—3.5L engine

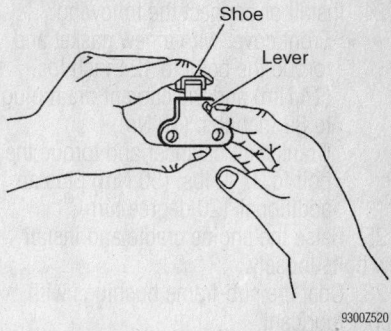

Compressing the primary chain tensioner—3.5L engine

- Primary chain tensioner shoe, by removing the bolt, pushing the guide downward slightly and pulling it up through the cylinder head
- Primary chain from the right camshaft, allowing it to fall into the oil pump area
- Primary chain

To install:

13. Rotate the crankshaft so the No. 1 piston is at Top Dead Center (TDC) and the mark on the crankshaft is at the 4 o'clock position.

14. Rotate the balance shaft so the timing mark is at the 5 o'clock position.

➡️**Be sure the painted links are facing the front of the engine.**

15. Install the timing chain on the sprockets.

16. Center the mark on the left intake camshaft sprocket between the 2 painted links.

17. Verify that all of the timing marks are aligned.

18. Install the primary chain tensioner shoe. Torque the bolt to 22 ft. lbs. (30 Nm).

19. Compress the primary chain tensioner using the following sub-steps:

   a. Step 1: Rotate the ratchet release lever counterclockwise and hold it.

   b. Step 2: Press the tensioner shoe in and hold it.

   c. Step 3: Release the ratchet lever and slowly release the pressure on the shoe.

   d. Step 4: Insert a pin through the hole in the lever as the lever moves to the first click. The ratchet should hold the shoe in the compressed position.

➡️**Be sure the lever on the tensioner is facing you when installed.**

20. Install or connect the following:
   - Primary chain tensioner and torque the bolts to 18 ft. lbs. (25 Nm); then, remove the chain tensioner pin
   - Four chain guide access plugs and torque the plugs to 44 inch lbs. (5 Nm)
   - Front engine lift bracket and torque the hex head bolt to 37 ft. lbs. (50 Nm) and the internal drive bolt to 18 ft. lbs. (25 Nm)
   - CMP sensor and torque the bolts to 80 inch lbs. (9 Nm)

21. Remove the camshaft holding tools.

22. Install the rocker arm covers. Torque the bolts to 80 inch lbs. (9 Nm).

23. Place a small bead of RTV sealant on the 3 areas indicated in the diagram.

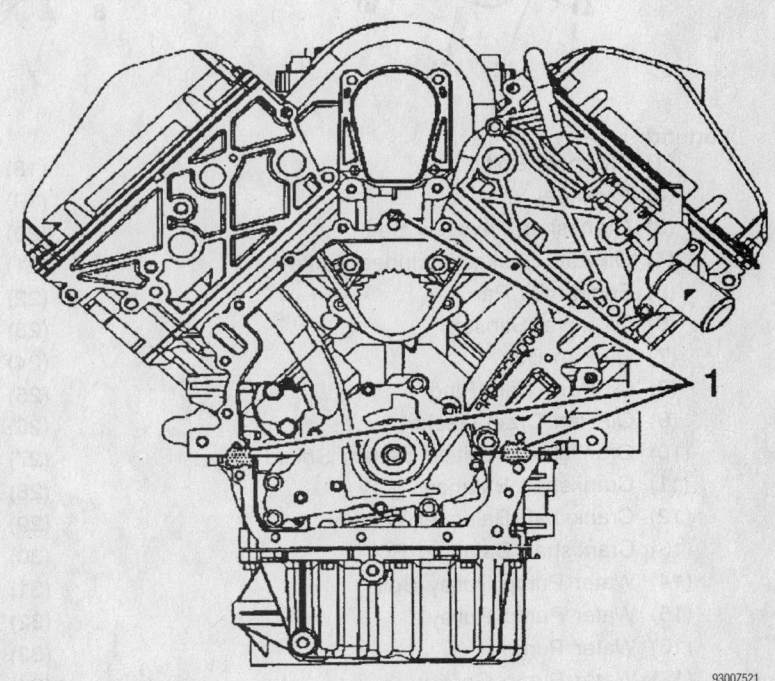

Apply RTV sealant to the 3 areas indicated before installing the front cover and gasket—3.5L engine

24. Install or connect the following:
    - Front cover with a new gasket and torque the bolts to 124 inch lbs. (14 Nm) and the coolant drain plug to 89 inch lbs. (10 Nm)
    - Crankshaft balancer and torque the bolt to 37 ft. lbs. (50 Nm) plus an additional 120 degree turn

25. Raise the engine cradle and install new bolts loosely.

26. Coat the sub-frame bushings with rubber lubricant.

27. Lower the vehicle onto the assembly. Align the sub-frame on the vehicle using 2 bolts or drill bits, ¾ inches thick by 8 inches long through the alignment holes on the right side of the frame.

28. Install or connect the following:
    - New frame-to-body bolts and torque the bolts to 133 ft. lbs. (180 Nm)
    - Water pump with a new gasket and torque the bolts to 124 inch lbs. (14 Nm)
    - Water pump pulley and torque the bolts to 106 inch lbs. (12 Nm)
    - Belt tensioner and torque the bolt to 37 ft. lbs. (50 Nm)
    - Power steering pump pulley
    - Drive belt

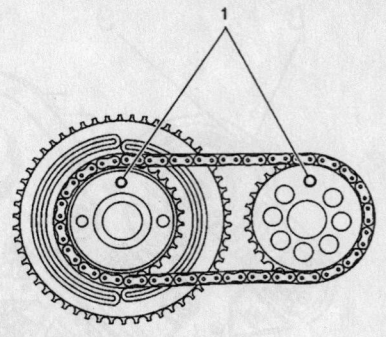

Correct sprocket alignment for secondary timing chain—3.5L engine

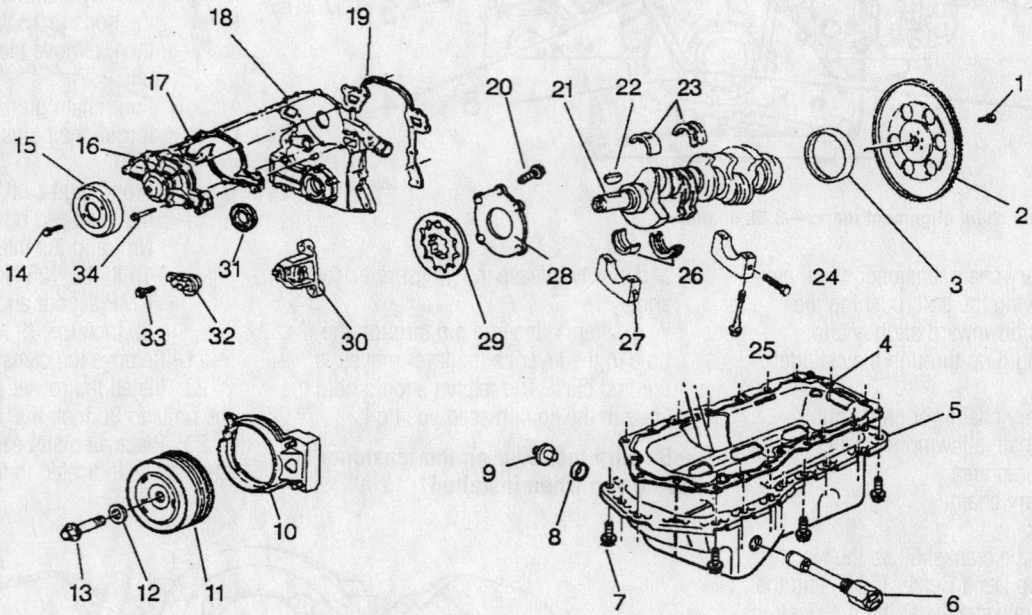

Legend

(1) Flywheel Bolt
(2) Flywheel
(3) Crankshaft Rear Oil Seal
(4) Oil Pan Gasket (Includes Baffle)
(5) Engine Oil Pan
(6) Oil Level Sensor
(7) Oil Pan Bolt
(8) Oil Pan Drain Plug
(9) Oil Pan Drain Gasket
(10) Crankshaft Position Sensor Shield
(11) Crankshaft Balancer
(12) Crankshaft Balancer Washer
(13) Crankshaft Balancer Bolt
(14) Water Pump Pulley Bolt
(15) Water Pump Pulley
(16) Water Pump
(17) Water Pump Gasket
(18) Engine Front Cover
(19) Engine Front Cover Gasket
(20) Oil Pump Cover Bolt
(21) Engine Crankshaft
(22) Crankshaft Balancer Key
(23) Crankshaft Upper Bearing
(24) Side Main Bolt
(25) Crankshaft Main Bearing Cap Bolt
(26) Crankshaft Lower Bearing
(27) Crankshaft Main Bearing Cap
(28) Oil Pump Cover
(29) Oil Pump Gear Set
(30) Crankshaft Position Sensor
(31) Crankshaft Front Oil Seal
(32) Camshaft Position Sensor
(33) Camshaft Position Sensor Bolt
(34) Water Pump Bolt

Exploded view of lower engine components—3.8L engine

- Underhood accessory wiring junction
- Coolant reservoir and torque the nuts to 30 inch lbs. (3 Nm)
- Battery and tray
- Right diagonal brace and torque the bolts to 35 ft. lbs. (47 Nm)
- Camshaft covers and torque the bolts to 80 inch lbs. (9 Nm)
- Negative battery cable

29. Refill the engine cooling system.
30. Refill the engine with new oil.
31. Start the vehicle and check for leaks, repair if necessary.

SECONDARY TIMING CHAIN

1. Before servicing the vehicle, refer to the precautions in the beginning of this section.
2. Drain the cooling system.
3. Remove or disconnect the following:

- Negative battery cable
- Thermostat housing for clearance (when working on the front cylinder head)
- Rocker arm cover and install a Camshaft Holding Fixture
- Camshaft Position (CMP) sensor
- Camshaft sprocket bolts

4. Install the timing chain/sprocket holding fixture on the cylinder head.
5. Evenly slide the secondary drive chain and sprockets off the camshafts.

To install:

6. Install the secondary timing chain on the sprockets, with the drive pins at the 12 o'clock positions.
7. Install the sprockets/chain assembly onto the camshafts, with the chain properly aligned on the tensioner.
8. Remove the sprocket holding fixture from the cylinder head.
9. Install the sprocket bolts and torque the bolts to 18 ft. lbs. (25 Nm) plus an additional 45 degree turn.
10. Remove the camshaft holding fixture.
11. Install or connect the following:

- Rocker arm cover and torque the bolts to 80 inch lbs. (9 Nm)
- Thermostat housing (if removed) and torque the bolts to 80 inch lbs. (9 Nm)
- CMP sensor and torque the bolts to 80 inch lbs. (9 Nm)
- Negative battery cable

12. Refill the cooling system.
13. Start the vehicle and check for leaks, repair if necessary.

3.8L Engines

1. Before servicing the vehicle, refer to the precautions in the beginning of this section.
2. Drain the coolant system.
3. Remove or disconnect the following:

3. Drain the engine oil.
4. Install an Engine Support Fixture.
5. Raise the engine so that the weight is removed from the engine mount.
6. Remove or disconnect the following:

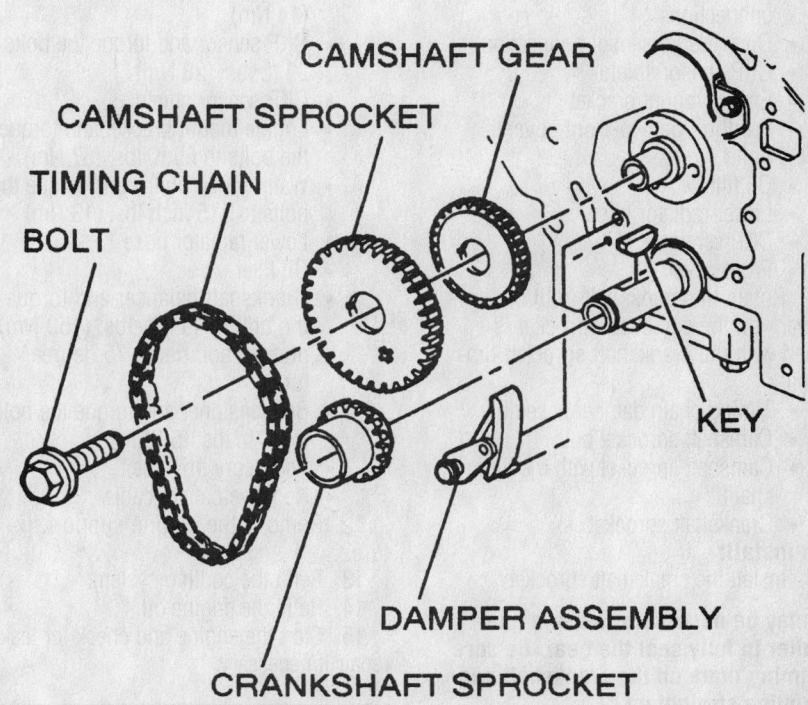

Exploded view of the timing chain and sprockets—3.8L engines

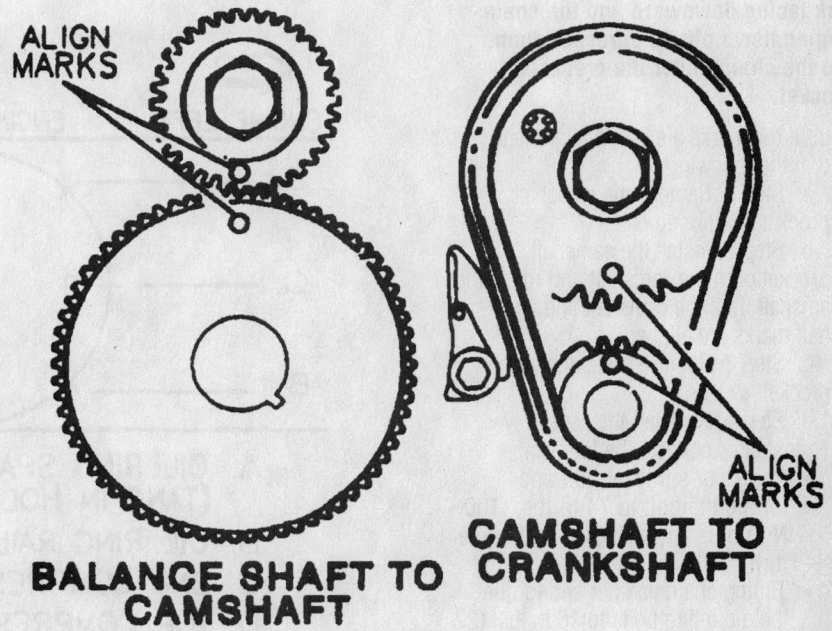

Balance shaft-to-camshaft and camshaft-to-crankshaft timing mark alignment—3.8L engine

- Negative battery cable
- Water pump pulley bolts
- Accessory drive belt(s)
- Drive belt tensioner
- Water pump pulley
- Crankshaft balancer
- Crankshaft Position (CKP) and Camshaft Position (CMP) sensor connections
- Oil pressure sensor connection
- CKP sensor shield
- Engine mount bracket
- Front oil pan-to-front cover bolts
- Oil filter
- Lower radiator hose
- CKP sensor
- Front cover

7. Rotate the crankshaft until the timing mark on the camshaft sprocket is aligned with the crankshaft sprocket timing mark.

- Timing chain damper assembly
- Camshaft sprocket bolt
- Camshaft sprocket with the timing chain
- Crankshaft sprocket

To install:

8. Install the crankshaft sprocket.

➡ **It may be necessary to use a gear installer to fully seat the gear. Be sure the timing mark on the crankshaft gear is pointing straight up.**

9. Install the camshaft gear with the timing chain.

➡ **Hold the sprocket with the timing mark facing downward and the chain hanging down off the sprocket; then, loop the chain under the crankshaft sprocket.**

10. If the marks are not in alignment perform the following:

a. Step 1: Remove the camshaft sprocket and timing chain.

b. Step 2: Install the camshaft sprocket onto the camshaft and rotate the camshaft until the camshaft and crankshaft marks are aligned.

c. Step 3: Remove the camshaft sprocket.

d. Step 4: Reinstall the assembly.

11. Install or connect the following:

- Camshaft sprocket bolt and torque the bolt to 74 ft. lbs. (100 Nm) plus an additional 90 degree turn
- Timing chain damper and torque the mounting bolts to 16 ft. lbs. (22 Nm)
- Front cover seal lubricated with

engine oil, using the appropriate seal driver

- New front cover gasket
- Front cover and torque the bolts to 15 ft. lbs. (20 Nm) plus an additional 40 degree turn
- Oil pan-to-front cover bolts and torque the bolts to 125 inch lbs. (14 Nm)
- CKP sensor and torque the bolts to 21 ft. lbs. (28 Nm)
- CKP sensor shield
- Engine mount bracket and torque the bolts to 65 ft. lbs. (87 Nm)
- Water pump pulley and torque the bolts to 115 inch lbs. (13 Nm)
- Lower radiator hose
- Oil filter
- Crankshaft balancer and torque the bolt to 111 ft. lbs. (150 Nm) plus an additional 75 degree turn
- Belt tensioner and torque the bolts to 37 ft. lbs. (50 Nm)
- Accessory drive belt
- Negative battery cable

12. Remove the engine support fixture.

13. Refill the cooling system.

14. Refill the engine oil.

15. Start the engine and check for leaks, repair if necessary.

Piston and Ring

POSITIONING

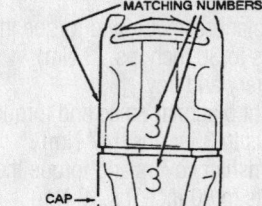

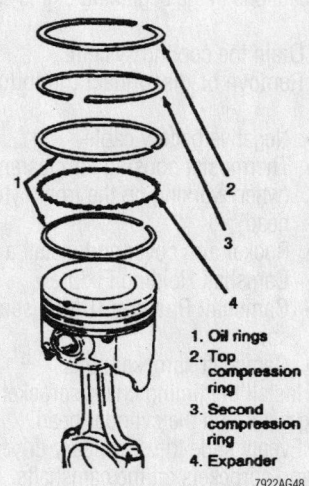

Connecting rod and cap installation. Be sure to matchmark the cap and rod prior to disassembly, as shown

Piston ring positioning—3.1L, 3.4L, 3.5L and 3.8L engines

1. Oil rings
2. Top compression ring
3. Second compression ring
4. Expander

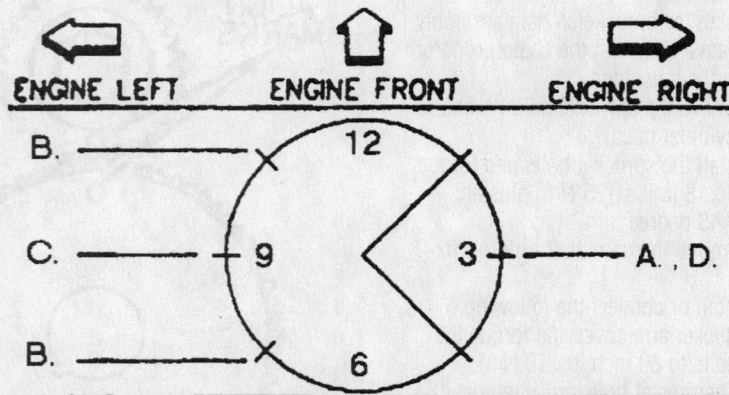

ENGINE LEFT   ENGINE FRONT   ENGINE RIGHT

A. OIL RING SPACER GAP (TANG IN HOLE OR SLOT WITH ARC)
B. OIL RING RAIL GAPS
C. 2ND COMPRESSION RING GAP
D. TOP COMPRESSION RING GAP

Piston ring end-gap spacing—3.1L, 3.4L, 3.8L engines

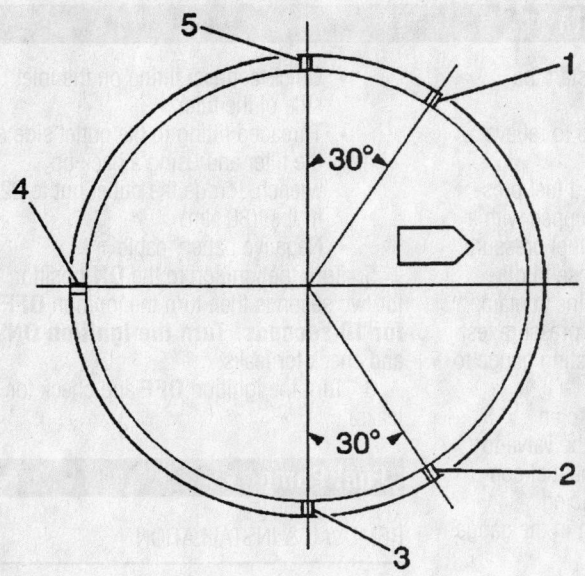

1. Lower oil control ring
2. Upper oil control ring
3. Top Ring
4. Oil control ring expander
5. Second ring

Piston ring end-gap positioning—3.5L engine

9306XG05

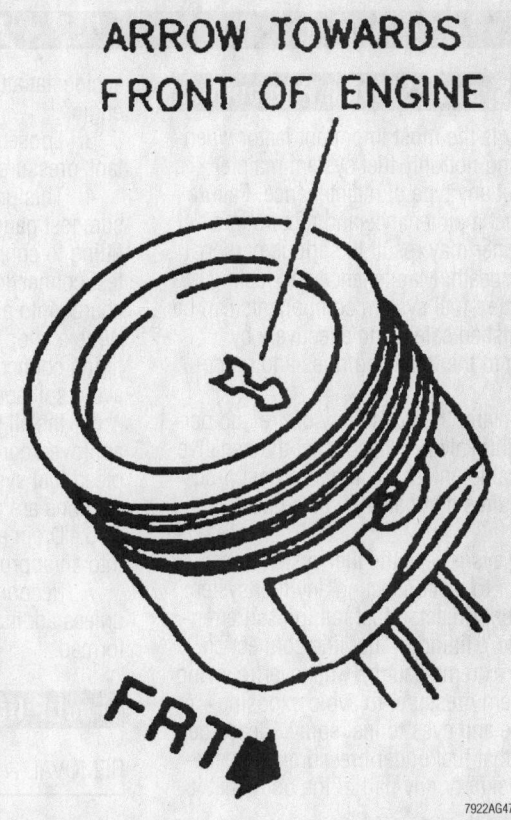

ARROW TOWARDS FRONT OF ENGINE

FRT

7922AG47

Piston positioning. Often the arrow is replaced by a notch, which also must face toward the front of the engine—3.1L, 3.4L and 3.8L

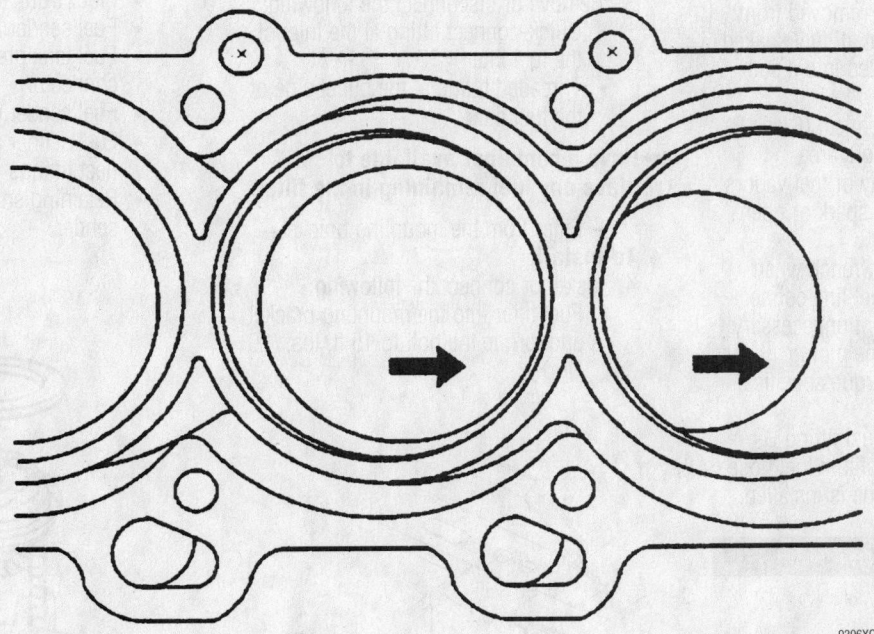

9306XG06

Piston positioning—3.5L engine

FUEL SYSTEM

Fuel System Service Precautions

Safety is the most important factor when performing not only fuel system maintenance but any type of maintenance. Failure to conduct maintenance and repairs in a safe manner may result in serious personal injury or death. Maintenance and testing of the vehicle's fuel system components can be accomplished safely and effectively by adhering to the following rules and guidelines.

• To avoid the possibility of fire and personal injury, always disconnect the negative battery cable unless the repair or test procedure requires that battery voltage be applied.

• Always relieve the fuel system pressure prior to disconnecting any fuel system component (injector, fuel rail, pressure regulator, etc.), fitting or fuel line connection. Exercise extreme caution whenever relieving fuel system pressure, to avoid exposing skin, face and eyes to fuel spray. Please be advised that fuel under pressure may penetrate the skin or any part of the body that it contacts.

• Always place a shop towel or cloth around the fitting or connection prior to loosening to absorb any excess fuel due to spillage. Ensure that all fuel spillage (should it occur) is quickly removed from engine surfaces. Ensure that all fuel soaked cloths or towels are deposited into a suitable waste container.

• Always keep a dry chemical (Class B) fire extinguisher near the work area.

• Do not allow fuel spray or fuel vapors to come into contact with a spark or open flame.

• Always use a backup wrench when loosening and tightening fuel line connection fittings. This will prevent unnecessary stress and torsion to fuel line piping. Always follow the proper torque specifications.

• Always replace worn fuel fitting O-rings with new. Do not substitute fuel hose or equivalent, where fuel pipe is installed.

Fuel System Pressure

RELIEVING

1. Before servicing the vehicle, refer to the precautions in the beginning of this section.

2. Disconnect the negative battery cable to prevent possible discharge of fuel if an accidental attempt is made to start the engine.

3. Loosen the fuel filler cap to relieve tank pressure.

4. This procedure calls for a fuel pressure test gauge with a line equipped with a fitting to connect to the to the fuel pressure test connection and another hose to discharge into an approved gasoline container. Wrap a shop towel around the pressure test fitting connection while connecting gauge to avoid spillage.

5. Install the bleed hose into an approved container and open the valve to bleed fuel system pressure. The fuel connections are now safe for servicing.

6. Drain any fuel remaining in the gauge into an approved container.

7. Reconnect the negative battery cable unless additional service work is being performed.

Fuel Filter

REMOVAL & INSTALLATION

1. Before servicing the vehicle, refer to the precautions in the beginning of this section.

2. Relieve fuel system pressure.

3. Remove or disconnect the following:
   • Quick-connect fitting at the inlet of the fuel filter
   • Threaded fitting at the outlet side of the fuel filter.

➡**Have a container available to retrieve any fuel remaining in the filter.**

   • Filter from the mounting bracket

To install:

4. Install or connect the following:
   • Fuel filter into the mounting bracket and torque the bolt to 15 ft. lbs. (20 Nm)

   • Quick-connect fitting on the inlet side of the filter
   • Threaded fitting to the outlet side of the filter and using a back up wrench, torque the outlet nut to 22 ft. lbs. (30 Nm)
   • Negative battery cable

5. Turn the ignition to the **ON** position for two seconds then turn the ignition **OFF for 10 seconds. Turn the ignition ON** and check for leaks.

6. Turn the ignition **OFF** and check for leaks.

Fuel Pump

REMOVAL & INSTALLATION

Except Lumina

1. Before servicing the vehicle, refer to the precautions in the beginning of this section.

2. Relieve the fuel system pressure.

3. Drain the fuel tank with a hand held siphon until the level is less than ¼ full.

4. Remove or disconnect the following:

   • Negative battery cable
   • Spare tire and jack
   • Floor trunk liner by pulling it back
   • Fuel sender access panel
   • Fuel tank pressure sensor electrical connector
   • Fuel sender electrical connector
   • Fuel sender assembly quick connect fittings
   • Retaining snapring from the fuel sender

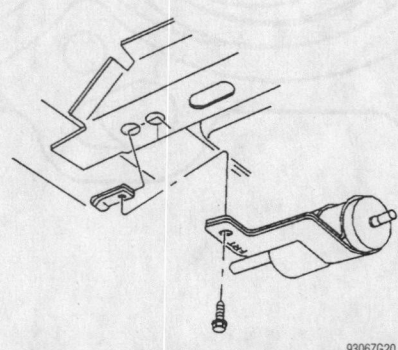

Common fuel filter mounting

9306ZG20

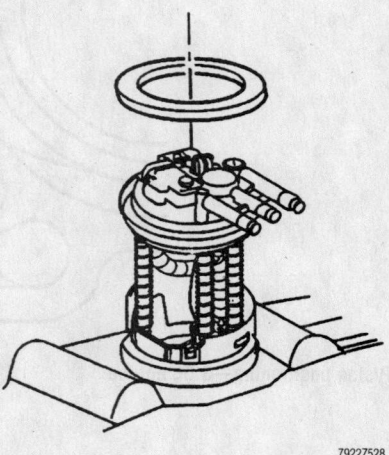

7922Z528

Remove the fuel pump from the tank after removing the locking ring

✶✶ WARNING

When the lockring is removed from the fuel sender, the sender assembly will spring up. Downward pressure should be kept on the assembly and slowly released to ensure the sender assembly does not get damaged.

- Modular fuel sender assembly

To install:

5. Install or connect the following:
- New O-ring on the fuel tank
- Fuel sender
- Snapring on the fuel sender
- Fuel sender electrical connector
- Fuel tank pressure sensor electrical connector
- Quick connect fittings at the fuel sender
- Negative battery cable

6. Add a small amount of fuel to the fuel tank.

7. Turn the ignition to the **ON** position for 2 seconds then turn the ignition **OFF** for 10 seconds. Turn the ignition **ON** and check for leaks.

8. Turn the ignition **OFF** and check for leaks.

9. Install or connect the following:
- Fuel sender access panel and torque the nuts to 89 inch lbs. (10 Nm)
- Trunk liner
- Spare tire, jack and spare tire cover

10. Refill the fuel tank.

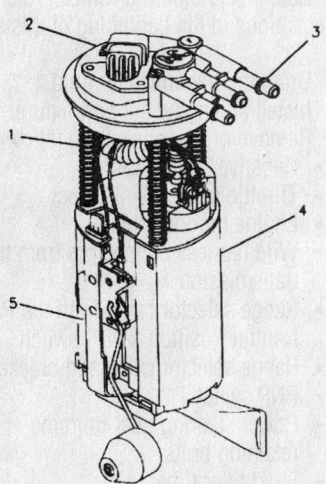

1 SUPPORT ASSEMBLY – FUEL SENDER
2 COVER ASSEMBLY – FUEL SENDER
3 FUEL PIPES (ABOVE COVER)
4 RESERVOIR – FUEL PUMP FUEL
5 SENSOR ASSEMBLY – FUEL LEVEL

7922XG19

Fuel pump and sending unit module assembly

Lumina

1. Before servicing the vehicle, refer to the precautions in the beginning of this section.

2. Relieve the fuel system pressure.

3. Drain the fuel tank.

4. Disconnect the negative battery cable.

5. Remove the fuel tank from the vehicle.

6. Clean all of pipe and hose connections. Also clean all areas around the connections.

7. Disconnect the quick-connect fittings at the fuel sender assembly.

8. Remove the fuel sender assembly retaining snap ring, the fuel sender assembly and the O-ring from the fuel tank. Discard the fuel sender assembly O-ring.

To install:

9. Position the new fuel sender assembly O-ring on the fuel tank.

➡**Make sure not to fold over or twist the fuel pump strainer when installing the fuel sender assembly, as this will restrict fuel flow. Also, ensure that the fuel pump strainer does not block full travel of float arm.**

10. Install the fuel sender assembly and the fuel sender assembly retaining snap ring.

11. Connect the quick-connect fittings at the fuel sender assembly.

12. Install the fuel tank and add fuel.

13. Pressurize the system as follows to check for leaks:
    a. Turn the ignition switch **ON** for 2 seconds.
    b. Turn the ignition switch **OFF** for 10 seconds.
    c. Turn the ignition switch **ON**.
    d. Check for fuel leaks.

Fuel Injector

REMOVAL & INSTALLATION

3.1L and 3.4L Engines

1. Before servicing the vehicle, refer to the precautions in the beginning of this section.

2. Relieve the fuel system pressure.

3. Remove or disconnect the following:
- Upper intake manifold
- Fuel inlet and return lines
- Fuel injector electrical connectors
- Coolant temperature sensor electrical connector
- Fuel rail

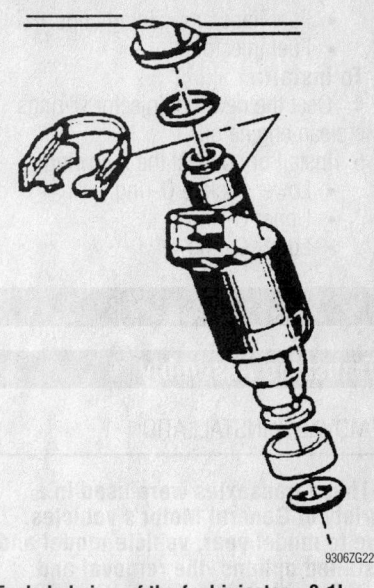

9306ZG22

Exploded view of the fuel injector—3.1L and 3.4L engines

- Fuel injector retaining clips
- Fuel injectors

To install:

4. Coat the new fuel injector O-rings with clean engine oil.

5. Install or connect the following:
- Lower backup O-ring
- Upper O-ring
- Lower O-ring
- Fuel injector
- Fuel injector retaining clips
- Fuel rail and torque the bolts to 7 ft. lbs. (10 Nm)
- Coolant temperature sensor electrical connector
- Fuel injector electrical connectors
- Fuel inlet and return lines with new O-rings and torque the fittings to 13 ft. lbs. (17 Nm)
- Upper intake manifold and torque the bolts to 18 ft. lbs. (25 Nm)
- Negative battery cable

6. Start the vehicle and check for leaks, repair if necessary.

3.5L and 3.8L Engines

1. Before servicing the vehicle, refer to the precautions in the beginning of this section.

2. Relieve the fuel system pressure.

3. Remove or disconnect the following:
- Fuel injector sight shield
- Fuel feed and return lines
- Fuel pressure regulator vacuum connection
- Ignition coil wires from the coil
- Fuel injector electrical connectors
- Fuel rail

- Fuel injector retaining clips
- Fuel injectors

To install:

4. Coat the new fuel injector O-rings with clean engine oil.

5. Install or connect the following:
  - Lower backup O-ring
  - Upper O-ring
  - Lower O-ring

- Fuel injector
- Fuel injector retaining clips
- Fuel rail and torque the bolts to 7 ft. lbs. (10 Nm) if equipped with bolt retainers. If equipped with snap-lock tab retainers, push down until the retainers snap into place.
- Fuel injector electrical connectors

- Ignition coil wires to the coil
- Fuel pressure regulator vacuum connection
- Fuel feed and return lines
- Negative battery cable

6. Start the vehicle and check for leaks, repair if necessary.

7. Install the fuel injector sight shield and torque the nuts to 27 inch lbs. (3 Nm).

DRIVE TRAIN

Transaxle Assembly

REMOVAL & INSTALLATION

➡These transaxles were used in a variety of General Motor's vehicles. Due to model year, vehicle model and installed options, the removal and installation procedures may vary slightly. The procedures given here should suffice for most all vehicles using these transaxles.

4T60-E Transmission

1. Before servicing the vehicle, refer to the precautions in the beginning of this section.

2. Drain the transmission fluid.

3. Install an engine support fixture.

4. Remove or disconnect the following:

- Negative battery cable
- Throttle body air inlet duct
- Engine mount struts
- Wire harness connectors from the transmission
- Vacuum hose and pipe from the modulator
- Range selector cable from the Park/Neutral Position (PNP) switch
- PNP switch
- Fluid filler tube
- Upper transmission bolts
- Wire harness grounds
- Both front wheels
- Engine splash shields
- Both tie rod ends from the steering knuckles
- Power steering gear from the frame and secure it to the body of the vehicle
- Power steering cooler line clamps
- Engine mount lower nuts
- Lower ball joints from the steering knuckles
- Torque converter cover
- Starter motor
- Torque converter bolts
- Oil cooler hoses

- Drive axles and secure them to the steering knuckles and struts
- Wheel Speed Sensor (WSS) connectors
- Vehicle Speed Sensor (VSS) connectors

5. Use a transmission table to support the transmission.

- Transmission brace
- Lower transmission bolts
- Frame-to-body bolts

6. Lower the frame from the vehicle.

7. Remove the transmission from the frame.

To install:

8. Install or connect the following:

- Transmission
- Lower transmission bolts and torque them to 55 ft. lbs. (75 Nm)
- New frame to body bolts and torque them to 125 ft. lbs. (170 Nm)
- Transmission brace and torque the bolts to 35 ft. lbs. (47 Nm)
- VSS electrical connector
- WSS electrical connectors
- Drive axles to the transmission
- Oil cooler hoses
- Torque converter bolts and torque them to 46 ft. lbs. (63 Nm)
- Starter motor and torque the bolts to 32 ft. lbs. (43 Nm)
- Torque converter cover and torque the bolts to 89 inch lbs. (10 Nm)
- Lower ball joints to the steering knuckle and torque the nuts to 40 ft. lbs. (55 Nm)
- Engine mount lower nuts and torque them to 32 ft. lbs. (43 Nm)
- Power steering cooler line clamps to the frame
- Power steering gear to the frame and torque the bolts to 59 ft. lbs. (80 Nm)
- Tie rod ends to the steering knuckles and torque the nuts to 63 ft. lbs. (85 Nm)
- Engine splash shields
- Front wheels
- Fluid filler tube

- Upper transaxle bolts and torque them to 55 ft. lbs. (75 Nm)
- PNP switch and torque the bolts to 18 ft. lbs. (25 Nm)
- Range selector cable with the bracket and torque the nut to 15 ft. lbs. (20 Nm)
- Range selector cable to the PNP switch
- Wire harness connectors to the transmission
- Vacuum hose and pipe to the modulator
- Engine mount struts and torque the bolts to 37 ft. lbs. (50 Nm)

9. Remove the engine support fixture.

- Throttle body air inlet duct
- Negative battery cable

10. Fill the transmission with fluid.

11. Start the engine and check the engine and transaxle oil level. Add oil if necessary.

4T65-E Transmission

1. Before servicing the vehicle, refer to the precautions in the beginning of this section.

2. Drain the transmission fluid.

3. Install an engine support fixture.

4. Remove or disconnect the following:

- Negative battery cable
- Throttle body air inlet duct
- Engine mount struts
- Wire harness connectors from the transmission
- Range selector cable from the Park Neutral Position (PNP) switch
- Range selector cable and bracket
- PNP switch
- Power steering gear to frame retaining bolts
- Fluid filler tube
- Upper transmission bolts
- Wire harness grounds
- Both front wheels
- Engine splash shields
- Both tie rod ends from the steering knuckles
- Power steering gear from the frame

and secure it to the body of the vehicle
- Power steering cooler line clamps
- Engine mount lower nuts
- Lower ball joints from the steering knuckles
- Torque converter cover
- Starter motor
- Torque converter bolts
- Oil cooler hoses
- Drive axles and secure them to the steering knuckles
- Wheel Speed Sensor (WSS) electrical connectors
- Vehicle Speed Sensor (VSS) electrical connectors

5. Use a transmission table to support the transmission.
- Engine mount lower nuts
- Transmission brace
- Lower transmission bolts
- Frame-to-body bolts

6. Separate the transmission from the engine.

7. Lower the transmission and frame from the vehicle.

8. Remove the transmission.

To install:

9. Install or connect the following:
- Transmission to the frame and raise the assembly into position
- New frame-to-body bolts and torque them to 133 ft. lbs. (180 Nm)
- Lower transmission-to-engine bolts and torque them to 55 ft. lbs. (75 Nm)
- Transmission brace and torque the transmission bolts to 32 ft. lbs. (43 Nm) and the engine bolts to 46 ft. lbs. (63 Nm)
- VSS electrical connector
- WSS electrical connectors
- Drive axles to the transmission
- Oil cooler hoses
- Torque converter bolts and torque them to 46 ft. lbs. (63 Nm)
- Starter motor and torque the bolts to 32 ft. lbs. (43 Nm)
- Torque converter cover and torque the bolts to 89 inch lbs. (10 Nm).
- Lower ball joints to the steering knuckle and torque the nuts to 40 ft. lbs. (55 Nm)
- Engine mount lower nuts and torque them to 35 ft. lbs. (47 Nm)
- Power steering cooler line clamps to the frame
- Power steering gear to the frame and torque the bolts to 59 ft. lbs. (80 Nm)
- Tie rod ends to the steering knuck-

les and torque the nuts to 63 ft. lbs. (85 Nm)
- Engine splash shields
- Front wheels
- Fluid filler tube
- Upper transmission bolts and torque them to 55 ft. lbs. (75 Nm)
- PNP switch and torque the bolts to 18 ft. lbs. (25 Nm)
- Range selector cable with the bracket and torque the nut to 15 ft. lbs. (20 Nm)
- Range selector cable to the PNP switch
- Wire harness connectors to the transmission
- Engine mount strut and torque the bolts to 37 ft. lbs. (50 Nm)

10. Remove the engine support fixture.
- Throttle body air inlet duct
- Negative battery cable

11. Fill the transmission with fluid.

12. Adjust the wheel alignment.

13. Start engine and check the engine and transaxle oil levels. Add oil if necessary.

Halfshaft

REMOVAL & INSTALLATION

Lumina

➡**If equipped with Antilock Brake System (ABS) brakes, use care to avoid damage to the ABS toothed ring. Damage to the ring may cause the self-diagnostic feature of the ABS system to set a system fault code.**

1. Before servicing the vehicle, refer to the precautions in the beginning of this section.

2. Remove or disconnect the following:
- Front wheel
- Drive shaft nut
- Brake caliper and bracket assembly
- Brake rotor
- Wheel Speed Sensor (WSS) electrical connector
- Wheel hub/bearing
- Axle shaft using an appropriate puller

3. Remove the halfshaft/bearing assembly through the steering knuckle.

To install:

4. Install or connect the following:
- Halfshaft/bearing assembly through the knuckle and into the transaxle
- Bearing-to-knuckle bolts, loosely
- Halfshaft into the transaxle

5. Verify that the snapring is seated

properly by tapping on the inner groove with a prying tool. Grasp the inner housing of the axle shaft and pull outward. Do not pull on the axle shaft. If the snapring is properly seated, the axle will remain in place.

6. Install or connect the following:
- WSS electrical connector
- Front wheel driveshaft bearing and torque the bolts to 52 ft. lbs. (70 Nm)
- New driveshaft nut and torque it to 159 ft. lbs. (215 Nm)
- Brake rotor
- Brake caliper and attaching bracket and torque the bracket bolts to 148 ft. lbs. (200 Nm) and the caliper bolts to 80 ft. lbs. (108 Nm)
- Front wheel
- Negative battery cable

7. Check the transaxle fluid level.

8. Check the front alignment and adjust if necessary.

Century, Regal, Impala, Intrigue, Grand Prix and Monte Carlo

1. Before servicing the vehicle, refer to the precautions in the beginning of this section.

2. Remove or disconnect the following:
- Front wheel
- Stabilizer shaft link
- Drive shaft nut
- Outer tie rod end from the steering knuckle
- Ball joint from the steering knuckle

3. Press the axle shaft through the hub.

❋❋ WARNING

To prevent damage to the inner CV-joint, do not pull on the axle shaft to remove it from the transaxle.

4. Place a drain pan under the transaxle to catch any transaxle fluid that leaks out when the axle shaft is removed.

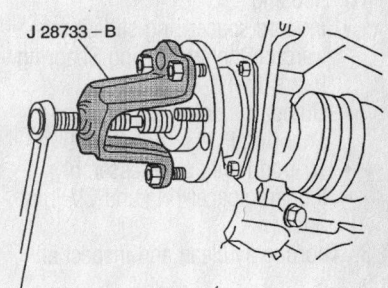

J 28733-B

9300Z523

Use a puller to press the axle shaft through the hub/bearing assembly

5. Remove the axle shaft from the transaxle by prying between the transaxle and the inner CV-joint housing.

To install:

6. Install or connect the following:
- Axle shaft in the transaxle. Verify that it is seated by pulling on the housing
- Axle shaft through the hub/bearing assembly
- Ball joint and torque the nut to 40 ft. lbs. (55 Nm)
- Tie rod end and torque the nut to 22 ft. lbs. (30 Nm) plus an additional 120 degree turn
- New drive shaft nut and torque it to 118 ft. lbs. (160 Nm) on all 2000 models except Impala, or to 159 ft. lbs. (215 Nm) on all 2001–04 models and all Impala models
- Stabilizer shaft link and torque the nut to 17 ft. lbs. (23 Nm)
- Front wheel

7. Check the transaxle fluid level.

8. Check the front alignment and adjust, if necessary.

CV-Joint

OVERHAUL

Inner (Tri-Pod) Joint

1. Before servicing the vehicle, refer to the precautions in the beginning of this section.

2. Remove or disconnect the following:
- Front wheel
- Halfshaft
- Swage ring using a hand grinder
- Large CV-joint boot clamp, cut and discard it
- CV-joint boot by sliding it away from the tri-pod joint
- Tri-pod housing from the tri-pod spider
- Trilobal tri-pod bushing from the housing
- Inboard spacer ring slide it rearward on the shaft using Snapring Pliers Tool J-8059
- Outboard retaining ring using Snapring Pliers Tool J-8059
- Tri-pod joint spider assembly
- Inboard spacer ring and CV-joint boot

3. Thoroughly clean and inspect all parts.

To install:

4. Install or connect the following:
- Swage ring clamp

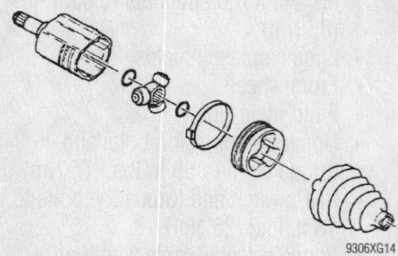

9306XG14

Exploded view of the inner (tri-pod) joint

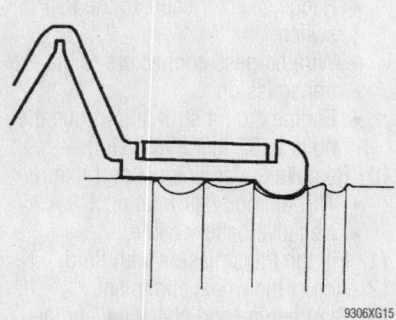

9306XG15

Positioning the inner CV-joint boot seal and swage ring—Inner (tri-pod) joint

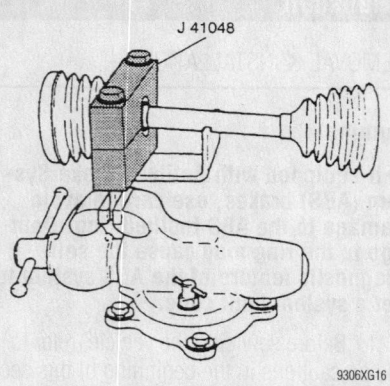

J 41048

9306XG16

View of the swage ring crimping tool—Inner (tri-pod) joint

- CV-joint boot

5. Position the CV-joint boot seal into the axle shaft's joint seal groove and align the swage ring clamp on the boot.

6. Secure the swage ring clamp as follows:

a. Mount the lower half of Tool J-41048 in a vise.

b. Position the outboard of the halfshaft in the tool.

c. Position the upper end of Tool J-41048 onto the lower half.

✳✳ WARNING

Make sure that there are no pinch points on the inboard seal.

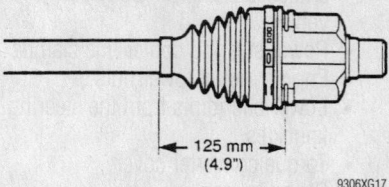

125 mm (4.9")

9306XG17

Boot measurement—Inner (tri-pod) joint

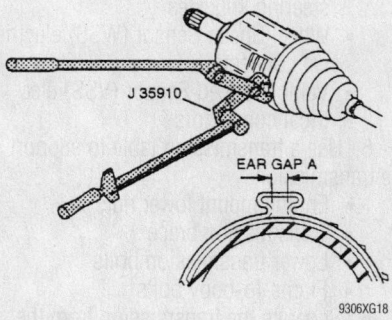

J 35910

EAR GAP A

9306XG18

Crimping the large CV-joint boot ring—Inner (tri-pod) joint

d. Insert both bolts and tighten by hand until snug.

e. Tighten each bolt 180 degree (½ turn) at a time, alternating between the bolts, until both sides are bottomed.

f. Remove the tool.

7. Install or connect the following:
- Inboard spacer ring, slide it rearward on the shaft using Snapring Pliers Tool J-8059
- Tri-pod joint spider assembly onto the shaft
- Outboard retaining ring into the axle shaft groove using Snapring Pliers Tool J-8059
- Tri-pod joint spider assembly, slide it against the outboard retaining ring
- Inboard spacer ring, seat it in the groove
- ½ kit grease into the boot
- ½ kit grease into the tri-pod housing
- Trilobal tip-pot bushing flush with the tri-pod housing face
- New large seal clamp onto the CV-joint boot
- Tri-pod housing, slide it over the tri-pod joint spider assembly
- CV-joint boot/clamp, slide it into place, over the trilobal tri-pod bushing with the seal lip in the groove

➡Make sure the boot lies flat against the trilobal bushing.

8. Position the CV-joint boot so it measures 4.9 in. (125mm).

9. Using the Crimp Tool J-35910, a torque wrench and a breaker bar, crimp the large CV-joint boot clamp to 130 ft. lbs. (176 Nm).

10. Install or connect the following:
- Halfshaft
- Front wheel

Outer Joint

1. Before servicing the vehicle, refer to the precautions in the beginning of this section.

2. Remove or disconnect the following:
- Front wheel
- Halfshaft
- Swage ring using a hand grinder
- Large boot clamp, cut and discard it
- CV-joint boot, slide it away from the CV-joint
- CV-joint assembly by spreading the inner race-to-axle shaft retaining ring ears using Snapring Pliers Tool J-8059
- CV-joint boot from the axle shaft and discard it

3. Disassemble the chrome alloy balls from the CV-joint cage as follows:

a. Position a brass drift against the CV-joint cage and tap it with a hammer to tilt the cage.

b. Chrome alloy ball from the cage.

c. Tilt the cage in the opposite direction.

d. Remove the opposite chrome alloy ball.

e. Repeat the procedure until all 6 balls are removed.

4. Disassemble the CV-joint cage and inner race as follows:

a. Pivot the cage and race 90 degrees to the center line of the outer race.

b. Align the cage windows with outer race lands.

c. Remove the cage from the outer race.

d. Rotate the inner race upward and remove it from the cage.

5. Thoroughly clean and inspect all parts.

To install:

6. Lubricate the parts with a light coat of grease.

7. Assemble the CV-joint cage and inner race, as follows:

a. Rotate the inner race 90 degrees to the cage centerline.

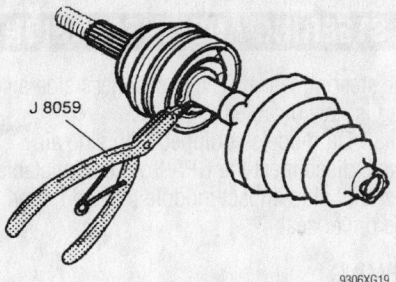

Disconnecting the outer CV-joint from the axle shaft

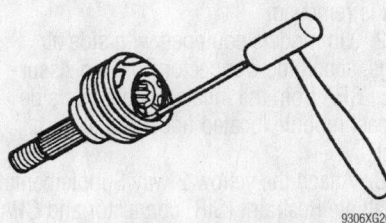

Tilting the cage—Outer CV-joint

b. Align the cage windows with inner race lands.

c. Insert the inner race into the cage by rotating the inner race downward.

d. Insert the cage/inner race into the outer race.

8. Assemble the chrome alloy balls into the CV-joint cage, as follows:

a. Position a brass drift against the CV-joint cage and tap it with a hammer to tilt the cage.

b. Insert the 1st chrome alloy ball into the cage.

c. Tilt the cage in the opposite direction.

d. Insert the opposite chrome alloy ball.

e. Repeat the procedure until all 6 balls are inserted.

9. Install or connect the following:
- ½ kit grease into the CV-joint boot
- ½ kit grease into the CV-joint
- Swage ring clamp
- CV-joint boot
- CV-joint onto the axle shaft until the retaining ring seats into the groove

10. Position the CV-joint boot seal into the axle shaft's joint seal groove and align the swage ring clamp on the boot.

11. Secure the swage ring clamp as follows:

a. Mount the lower half of tool J-41048 in a vise.

b. Position the outboard of the half-

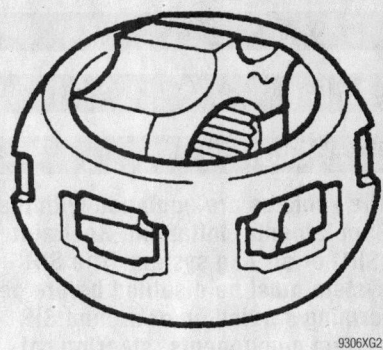

View the cage and inner race—Outer CV-joint

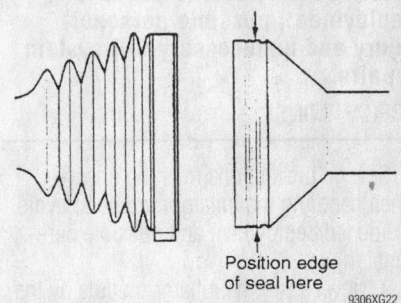

Positioning the boot—Outer CV-joint

shaft in the tool.

c. Position the upper end of tool J-41048 onto the lower half.

✳✳ WARNING

Make sure that there are no pinch points on the inboard seal.

d. Insert both bolts and tighten by hand until snug.

e. Tighten each bolt 180 degree (½) turn at a time, alternating between the bolts, until both sides are bottomed.

f. Remove the tool.

12. Install or connect the following:
- New large seal clamp onto the CV-joint boot
- CV-joint boot/clamp, slide it into place, over the outer race with the seal lip in the groove

➡ **Make sure the boot lies flat against the outer race.**

13. Using the Crimp Tool J-35910, a torque wrench and a breaker bar, crimp the large CV-joint boot clamp to 130 ft. lbs. (176 Nm).

14. Install or connect the following:
- Halfshaft
- Front wheel

STEERING AND SUSPENSION

Air Bag

✳✳ CAUTION

The vehicles are equipped with the Supplemental Inflatable Restraint (SIR) or air bag system. The SIR system must be disabled before performing service on or around SIR system components, steering column, instrument panel components, wiring and sensors. Failure to follow safety and disabling procedures could result in accidental air bag deployment, possible personal injury and unnecessary SIR system repairs.

PRECAUTIONS

Several precautions must be observed when handling the inflator module to avoid accidental deployment and possible personal injury.

• Never carry the inflator module by the wires or connector on the underside of the module.

• When carrying a live inflator module, hold securely with both hands, and ensure that the bag and trim cover are pointed away.

• Place the inflator module on a bench or other surface with the bag and trim cover facing up.

• With the inflator module on the bench, never place anything on or close to the module which may be thrown in the event of an accidental deployment.

DISARMING

1. Turn the steering wheel so the vehicle wheels are pointing straight-ahead.

2. Turn the ignition key to the **LOCK** position and remove the key.

3. Remove the AIR BAG fuse from the fuse block.

4. Remove the left side sound insulator.

5. Detach the Connector Position Assurance (CPA) and yellow 2-way Supplemental Inflatable Restraint (SIR) connector at the multi-use bracket near the base of the steering column. The driver's side air bag is now disabled.

6. Remove the right side sound insulator.

7. Detach the CPA and yellow 2-way SIR connector which is located at the base of

the steering wheel. The passenger's side air bag is now disabled.

8. On models equipped with side air bags, disconnect the CPA from the inflatable restraint side impact module located under the driver seat.

ARMING

After necessary repairs are made, re-enable the air bag system as follows:

1. Be sure the ignition is locked and the key is removed.

2. On models equipped with side air bags, attach the Connector Position Assurance CPA from the inflatable restraint side impact module located under the driver seat.

3. Attach the yellow 2-way Supplemental Inflatable Restraint (SIR) connector and CPA at the base of the steering wheel.

4. Install the right side sound insulator.

5. Attach the yellow 2-way SIR connector and CPA at the multi-use bracket at the base of the column.

6. Install the left side sound insulator.

7. Install the AIR BAG fuse.

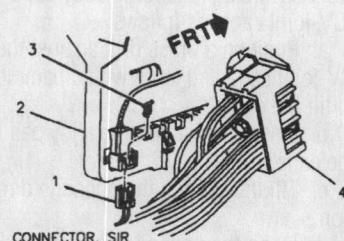

1 CONNECTOR, SIR
2 BRACKET, MULTIUSE MODULE
3 CONNECTOR POSITION ASSURANCE (CPA)
4 CONNECTOR, STEERING COLUMN
  WIRING HARNESS

7922XG21

SIR 2-way connector location—driver's side

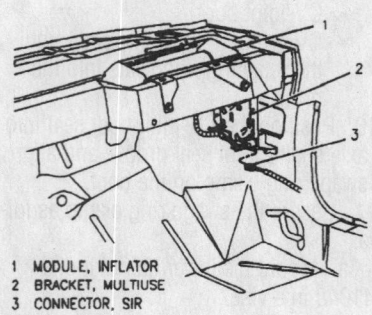

1 MODULE, INFLATOR
2 BRACKET, MULTIUSE
3 CONNECTOR, SIR

7922XG22

SIR 2-way connector location—passenger's side

8. Turn the ignition switch to the **RUN** position and verify the AIR BAG light flashes 7 times, then shuts off.

Power Rack and Pinion Steering Gear

REMOVAL & INSTALLATION

1. Before servicing the vehicle, refer to the precautions in the beginning of this section.

2. Remove or disconnect the following:
   • Negative battery cable
   • Front wheels

✳✳ CAUTION

Failure to disconnect the intermediate shaft from the rack and pinion stub shaft may result in damage to the steering gear. This damage may cause a loss of steering control and may cause personal injury.

➡**Set the steering shaft so that the block tooth on the upper steering shaft is at the 12 o'clock position. The wheels should be straight ahead. Set the ignition key lock to the LOCK position. Failure to follow these procedures could result in damage to the SIR coil assembly.**

3. Remove or disconnect the following:
   • Intermediate steering shaft lower pinch bolt from the steering gear stub shaft
   • Intermediate steering shaft
   • Both tie rod ends from the steering knuckles

4. Support the frame at the center rear.

➡**DO NOT lower the frame too far. Engine components near the firewall may be damaged.**

5. Remove or disconnect the following:
   • Frame bolts from the rear of the frame. Lower the frame slightly
   • Power steering pressure line from the steering gear
   • Power steering return hose
   • Magnasteer Variable Assist electrical connector from the power steering gear assembly, if equipped
   • Steering gear mounting bolts
   • Power steering gear through the left wheel opening

To install:

6. Install or connect the following:

- Power steering gear through the left wheel opening
- Mounting bolts and torque them to 59 ft. lbs. (80 Nm)
- Magnasteer Variable Assist elec-

trical connector to the power steering gear assembly, if equipped
- Power steering lines with new O-rings to the steering gear. Torque

the fasteners to 20 ft. lbs. (27 Nm).
- Power steering return hose
7. Raise the rear frame to its original position.

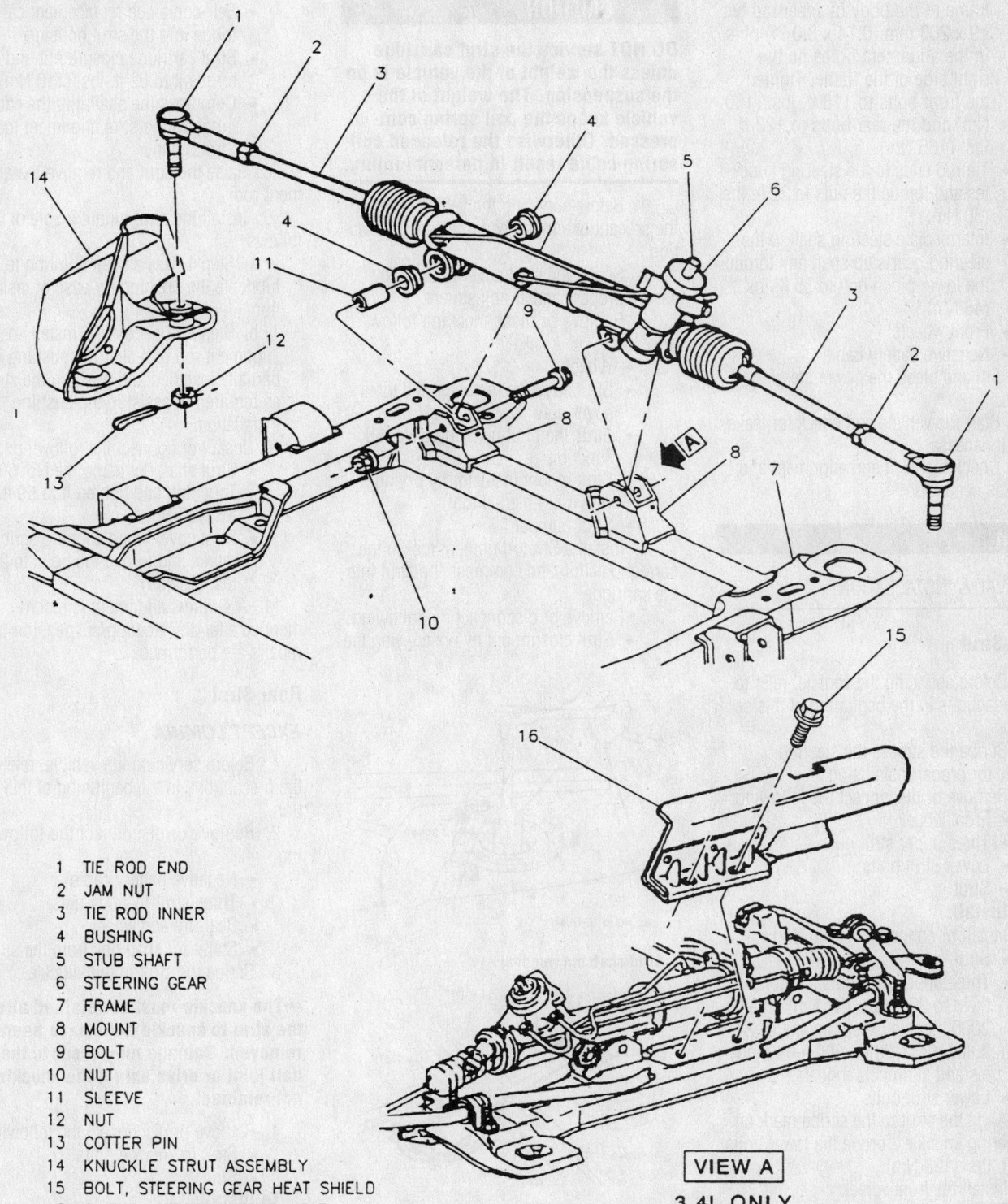

1 TIE ROD END
2 JAM NUT
3 TIE ROD INNER
4 BUSHING
5 STUB SHAFT
6 STEERING GEAR
7 FRAME
8 MOUNT
9 BOLT
10 NUT
11 SLEEVE
12 NUT
13 COTTER PIN
14 KNUCKLE STRUT ASSEMBLY
15 BOLT, STEERING GEAR HEAT SHIELD
16 SHIELD, STEERING GEAR HEAT

VIEW A
3.4L ONLY

79222529

Common rack and pinion steering gear mounting

8. Install or connect the following:
- New rear frame bolts and torque them to 133 ft. lbs. (180 Nm) on all models except 2001–04 Lumina, Impala and Monte Carlo models. On Lumina, Impala and Monte Carlo models, align the frame to the body by inserting two 19 x 203 mm (0.74 x 8.0 in) pins in the alignment holes on the right side of the frame. Tighten the front bolts to 118 ft. lbs. (160 Nm) and the rear bolts to 122 ft. lbs. (165 Nm).
- Tie rod ends to the steering knuckles and torque the nuts to 22 ft. lbs. (30 Nm)
- Intermediate steering shaft to the steering gear stub shaft and torque the lower pinch bolt to 35 ft. lbs. (48 Nm)
- Front wheels
- Negative battery cable

9. Fill and bleed the power steering system.

10. Start the vehicle and check for leaks, repair if necessary.

11. Check the front end alignment and adjust as needed.

Strut

REMOVAL & INSTALLATION

Front Strut

1. Before servicing the vehicle, refer to the precautions in the beginning of this section.

2. Scribe the strut to the steering knuckle for proper installation.

3. Remove or disconnect the following:
- Front wheel
- Three upper strut nuts
- Lower strut bolts
- Strut

To install:

4. Install or connect the following:
- Strut
- Three upper strut nuts and torque them to 30 ft. lbs. (41 Nm) on all 2000 models except Impala or 24 ft. lbs. (33 Nm) on 2001–04 models and all Impala models.
- Lower strut bolts

5. Align the strut to the scribe mark on the steering knuckle. Torque the lower bolts to 90 ft. lbs. (123 Nm).

6. Install the front wheel.

7. Road test the vehicle and check the front end alignment, adjust if necessary.

Front Strut Cartridge

➡The Chevrolet Lumina utilizes a replaceable cartridge housed within the strut assembly. The cartridge can be replaced without removing the strut assembly from the vehicle.

❋❋ CAUTION

DO NOT service the strut cartridge unless the weight of the vehicle is on the suspension. The weight of the vehicle keeps the coil spring compressed. Otherwise the released coil spring could result in personal injury.

1. Before servicing the vehicle, refer to the precautions in the beginning of this section.

2. Scribe the strut cover to body to assure proper camber adjustment.

3. Remove or disconnect the following:
- Wheel
- Strut cover by removing the three cover nuts
- Strut shaft nut by using a No. 50 Torx® bit
- Strut mount insulator by prying with a flat-bladed tool
- Strut bumper

4. Install a Strut Alignment tool in the correct position and compress the strut into the cartridge.

5. Remove or disconnect the following:
- Strut closure nut by unscrewing the

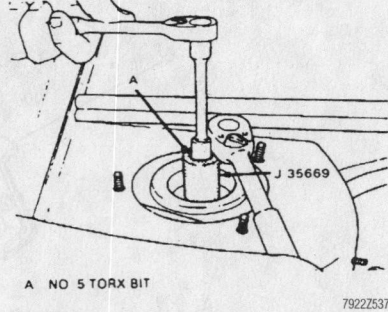

A   NO 5 TORX BIT

Strut shaft nut removal

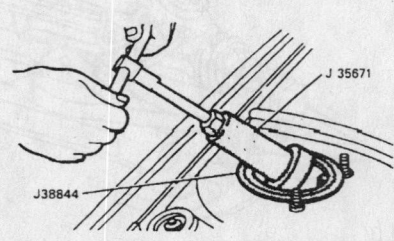

Strut closure nut removal

Strut closure nut removal

closure nut using a Strut Cap Nut wrench
- Strut cartridge

6. Remove any oil in the strut housing using a suction pump.

To install:

7. Install or connect the following:
- Self-contained replacement cartridge into the strut housing
- Strut cartridge closure nut and torque it to 82 ft. lbs. (110 Nm)
- Compress the shaft into the cartridge with a strut alignment tool
- Strut bumper

8. Raise the strut and remove the alignment rod.

9. Install the strut mount insulator as follows:
a. Step 1: Use a soap solution to lubricate the bushing for ease of installation.
b. Step 2: If necessary, install an alignment rod tool after the bushing is partially installed and position the strut as required to assist in the bushing installation.

10. Install or connect the following:
- Strut shaft nut using the No. 50 Torx® bit, and tighten it to 59 ft. lbs. (80 Nm)
- Strut cover while aligning scribe marks and torque the bolts to 24 ft. lbs. (33 Nm)

11. A 4-wheel alignment is recommended after any steering/suspension repairs are performed.

Rear Strut

EXCEPT LUMINA

1. Before servicing the vehicle, refer to the precautions in the beginning of this section.

2. Remove or disconnect the following:
- Negative battery cable
- Three strut-to-body nuts
- Rear tire and wheel
- Stabilizer shaft link from the strut

3. Scribe the strut to the knuckle.

➡The knuckle must be retained after the strut to knuckle bolts have been removed. Damage may occur to the ball joint or drive axle if the knuckle is not retained.

4. Remove or disconnect the following:
- Strut to knuckle bolts
- Strut

To install:

5. Install or connect the following:
- Strut

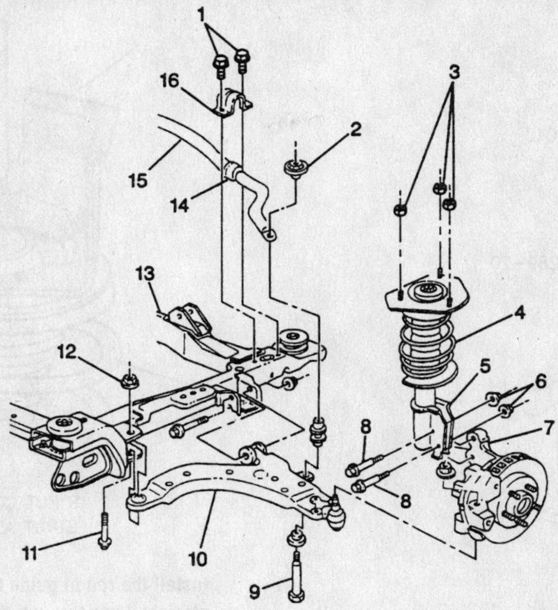

(1) Front Stabilizer Shaft Insulator Clamp Bolt/screw
(2) Front Stabilizer Shaft Link Nut
(3) Front Suspension Strut Mount Nut
(4) Front Suspension Spring
(5) Front Suspension Strut
(6) Strut To Knuckle Nut
(7) Front Steering Knuckle
(8) Strut To Knuckle Bolt/screw
(9) Front Stabilizer Shaft Link]
(10) Front Lower Control Arm
(11) Front Lower Control Arm Bolt/screw
(12) Front Lower Cotrol Arm Nut
(13) Frame
(14) Front Stabilizer Shaft Insulator
(15) Front Stabilizer Shaft
(16) Front Stabilizer Shaft Clamp

7922Z530

Exploded view of the front suspension without replaceable strut cartridge

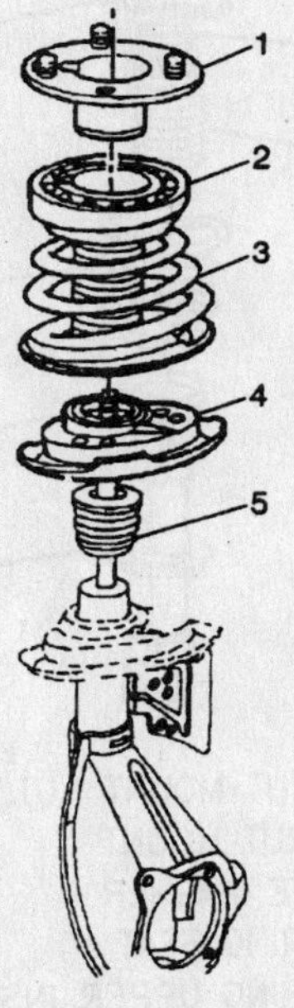

(1) Front Suspension Strut Mount Retainer
(2) Front Spring Upper Insulator
(3) Front Spring
(4) Front Spring Seat
(5) Front Suspension Strut Bumper

7922Z531

Knuckle and strut assembly with replaceable strut cartridge

- Strut to knuckle bolts and torque the bolts to 90 ft. lbs. (122 Nm)
- Stabilizer shaft link to the strut
- Rear tire and wheel
- Three strut to body mount nuts and torque them to 30 ft. lbs. (41 Nm)

6. Road test the vehicle and adjust the rear wheel alignment if needed.

LUMINA

1. Before servicing the vehicle, refer to the precautions in the beginning of this section.
2. Remove or disconnect the following:
   - Negative battery cable
   - Wheel
   - Brake hose bracket at the strut
   - Strut mount to body nuts
   - Strut/stabilizer shaft bracket from the knuckle
   - Strut

To install:

3. Install or connect the following:
   - Strut mount-to-body nuts and tighten to 37 ft. lbs. (50 Nm)
   - Strut/stabilizer shaft bracket to the knuckle
   - Strut-to-knuckle bolts and nuts. Tighten the nuts to 82 ft. lbs. (112 Nm).
   - Brake hose bracket
   - Wheel

Coil Spring

REMOVAL & INSTALLATION

Except Models Equipped With A Front Strut Cartridge

1. Before servicing the vehicle, refer to the precautions in the beginning of this section.
2. Remove the strut from the vehicle.
3. Mount the strut assembly into a strut compressor. Note that the strut compressor has strut mounting holes drilled for specific vehicle lines.
4. Compress the strut approximately ½ its height after initial contact with the top cap.

⁂ WARNING

Never bottom the spring or damper rod!

5. Remove the nut from the strut damper shaft and place an alignment/guiding rod on top of the damper shaft. Use the rod to guide the damper shaft straight down through the spring cap while compressing the spring. Remove the components.

To install:

6. Mount the strut assembly in the strut compressor, using the bottom locking pin only.
7. Install the spring over the damper and swing the assembly up so the upper locking pin can be installed.
8. Install all shields, bumpers and insulators on the spring seat.
9. Install the spring seat on top of the spring. The spring seat flat should be facing the same direction as the centerline of the strut assembly spindle.
10. Install the guiding rod and turn the forcing screw while the guiding rod centers the assembly. When the threads on the damper shaft are visible, remove the guiding rod and install the nut. tighten the front strut nut to 63 ft. lbs. (85 Nm) on all models except 2000 Impala models. On 2000 Impala models tighten the front strut nut to 52 ft. lbs. (70 Nm).Tighten the rear strut nut to 55 ft. lbs. (75 Nm).
11. Remove the strut from the compressor.
12. Install the strut to the vehicle.

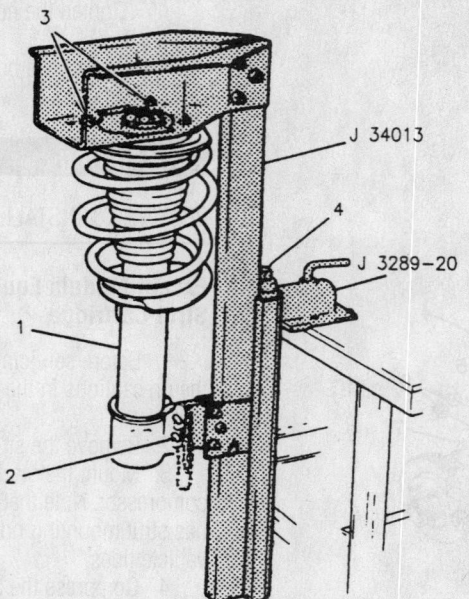

J 34013

J 3289-20

1. STRUT ASSEMBLY
2. INSTALL LOCKING PINS THROUGH STRUT ASSEMBLY
3. TIGHTEN NUTS UNTIL FLUSH WITH STRUT COMPRESSOR
4. COMPRESSOR FORCING SCREW

7922YG24

View of a typical strut assembly mounted in a compressor

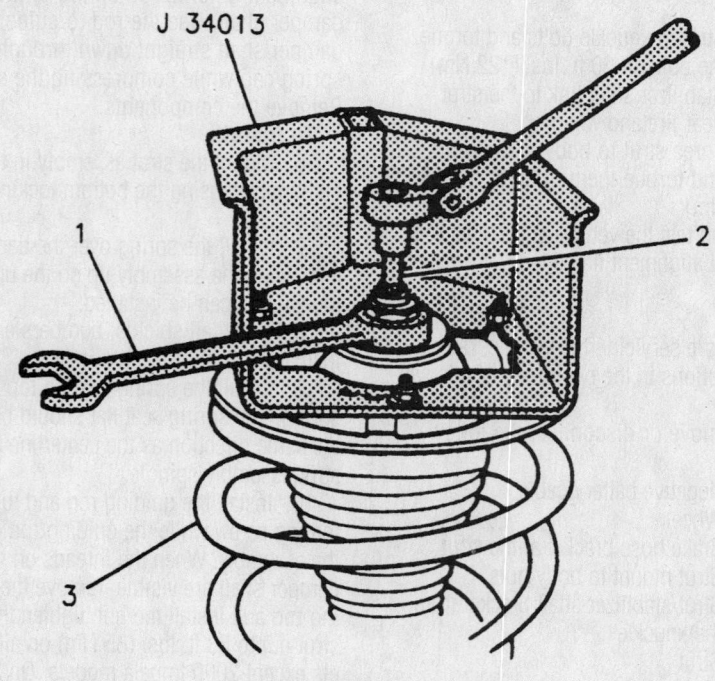

J 34013

1. WRENCH
2. SOCKET
3. STRUT ASSEMBLY

7922YG25

Use a socket and a wrench to remove the damper shaft nut spring cap while compressing the spring

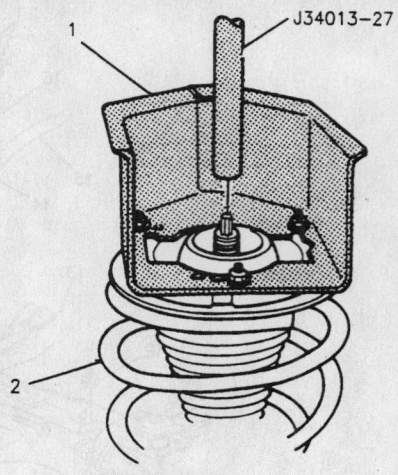

J34013-27

1. STRUT COMPRESSOR
2. STRUT ASSEMBLY

7922YG26

Install the rod to guide the damper shaft straight down through the spring cap while compressing the spring

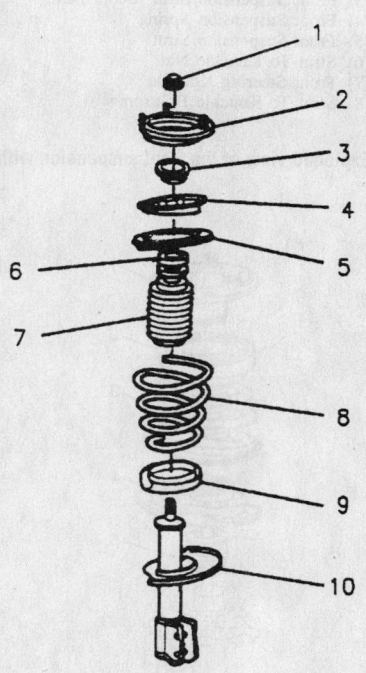

1. STRUT MOUNT NUT
2. STRUT MOUNT
3. RATE WASHER
4. SPRING SEAT
5. SPRING UPPER INSULATOR
6. JOUNCE BUMPER
7. STRUT DUST SHIELD
8. SPRING
9. SPRING LOWER INSULATOR
10. STRUT

7922YG27

Exploded view of the front strut assembly

Models Equipped With A Front Strut Cartridge

1. Before servicing the vehicle, refer to the precautions in the beginning of this section.
2. Remove the strut assembly from the vehicle.

✳✳ CAUTION

Do not over compress the coil spring. Only compress the spring until it comes away from the seat.

3. Compress the coil spring approximately ½inch (13mm) using a spring compressor.
4. Remove or disconnect the following:
   - Strut rod nut with the proper socket while not allowing the rod to rod to rotate. Discard the nut
5. Relieve the spring tension and remove the spring.

To install:
   - Spring to the strut and make certain that the upper and lower seats are positioned correctly
   - Strut rod nut and torque to 59–63 ft. lbs. (80–85 Nm)
   - Strut

Lower Ball Joint

REMOVAL & INSTALLATION

1. Before servicing the vehicle, refer to the precautions in the beginning of this section.
2. Remove the wheel.
3. Remove the lower control arm from the vehicle.
4. Remove the ball joint from the lower control arm by drilling out the 4 rivets retaining the ball joint to the control arm. Use an ⅛ in. drill bit to make a pilot hole through the rivets. Finish drilling the rivets using a ½ in. drill bit.
5. Remove the ball joint.

To install:
6. Install or connect the following:
   - Ball joint to the control arm
   - Bolts with the heads facing down and torque them to 50 ft. lbs. (68 Nm)
   - Lower control arm to the vehicle
   - Wheel

➡**A 4-wheel alignment is recommended after any steering/suspension repairs are performed.**

Lower Control Arm

REMOVAL & INSTALLATION

1. Before servicing the vehicle, refer to the precautions in the beginning of this section.
2. Remove or disconnect the following:
   - Front wheel
   - Antilock Brake System (ABS) wheel speed sensor connector and jumper harness from the retainer
   - Stabilizer shaft link
   - Cotter pin from the ball stud and loosen the nut
3. Install a ball joint removal tool over the ball joint and lower control arm. Rotate the ball stud nut counterclockwise to separate the ball stud from the steering knuckle.
4. Remove the lower control arm.

To install:
5. Install the lower control arm and bolts. Do not tighten the nuts at this time.

➡**Align the ball stud cotter pin hole parallel to the knuckle to ease the cotter pin installation.**

6. Install or connect the following:
   - Ball stud to the knuckle and torque the nut to 40 ft. lbs. (55 Nm) on 2000 Monte Carlo, Grand Prix, Century and Regal models. On Impala and 2001– 03 models, tighten to 15 ft. lbs. (20 Nm) plus an additional 120 degrees on Impala models. On Lumina models tighten to 63 ft. lbs. (85 Nm).
   - New cotter pin and bend the ends. Make certain the ends do not make contact with the ABS wheel speed sensor
   - Stabilizer shaft link
   - ABS wheel speed sensor wire harness to the retainer clips
   - ABS wheel speed sensor connector
   - Lower control arm nuts and torque them to 83 ft. lbs. (113 Nm) on all models except Lumina. On Lumina models tighten the control arm-to-frame nuts to 52 ft. lbs. (70 Nm).
   - Front wheel

CONTROL ARM BUSHING REPLACEMENT

1. Before servicing the vehicle, refer to the precautions in the beginning of this section.
2. Remove or disconnect the following:
   - Front wheel
   - Lower control arm

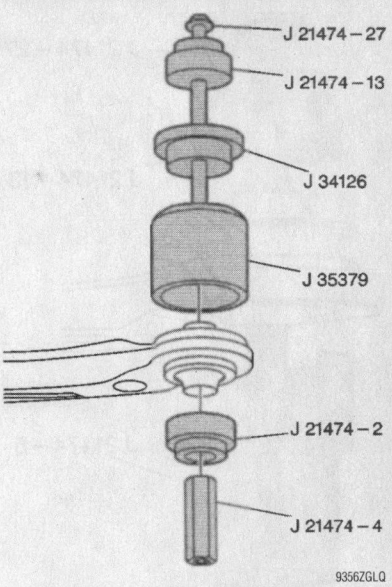

Removing the control arm bushing

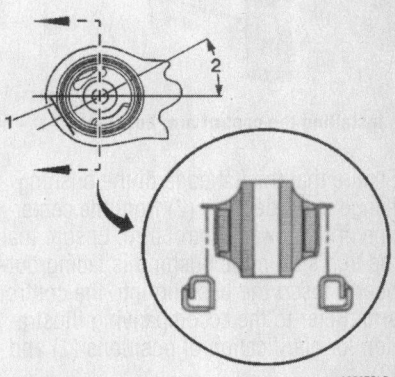

Positioning the control arm bushing

3. Mark the lower control arm along the flat edge of the bushing flange.
4. Coat the threads of tool J 21474-27 with a high pressure lubricant.
5. Assemble the following bushing removal tools as illustrated:
   - J 21474-27
   - J 21474-13
   - J 34126
   - J 35379
   - J 21474-2
   - J 21474-4
6. Tighten J 21474-4 to remove the bushing.

To install:

➡**You MUST install the lower control arm vertical bushing in the same position in order to maintain the original vehicle ride, handling, and road feel.**

7. Align the flat edge of the bushing flange to the mark in the control arm (1).

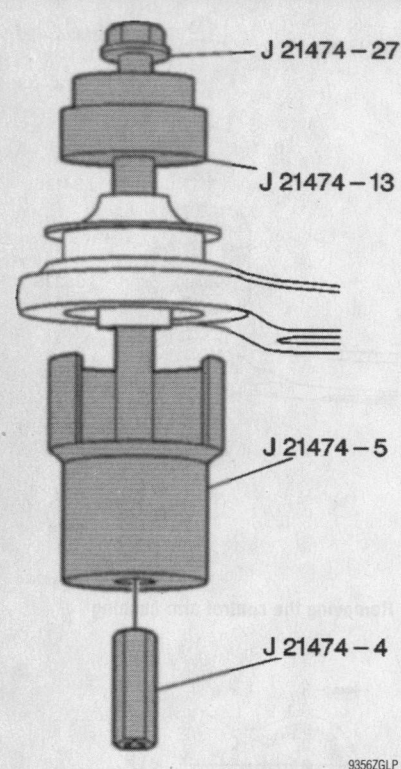

Installing the control arm bushing

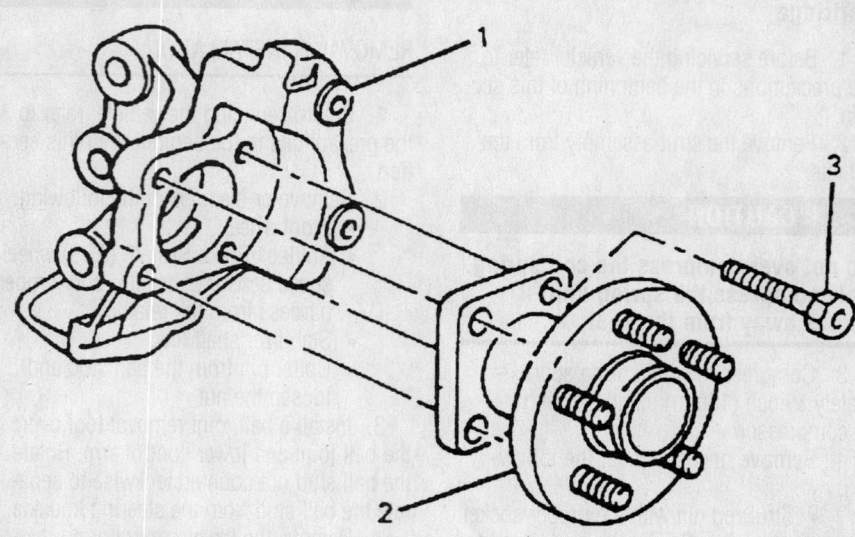

1 KNUCKLE ASSEMBLY, REAR SUSPENSION
2 HUB AND BEARING ASSEMBLY
3 BOLT/SCREW, WHEEL

The rear hub/bearing assembly is bolted to the knuckle

Ensure that the flat edge of the bushing flange is 30 degrees (2) from the centerline of the lower control arm. Ensure that the thin slot in the bushing is facing outboard. Insert the bushing into the control arm. Refer to the accompanying illustration for clarification of positions (1) and (2).

8. Coat the threads of tool J 21474-27 with a high pressure lubricant.

9. Assemble the following bushing installation tools as illustrated:
- J 21474-27
- J 21474-13
- J 21474-5
- J 21474-4

10. Tighten J 21474-4 to remove the bushing.

11. Install the lower control arm.
- Front wheel

Wheel Bearings

ADJUSTMENT

The wheel bearings are not adjustable. If a wheel bearing is out of specification, it must be replaced. Using a dial indicator, check for looseness. If it exceeds 0.005 in. (0.127mm) on drum or disc brakes the bearing wear is excessive and the hub and bearing should be replaced.

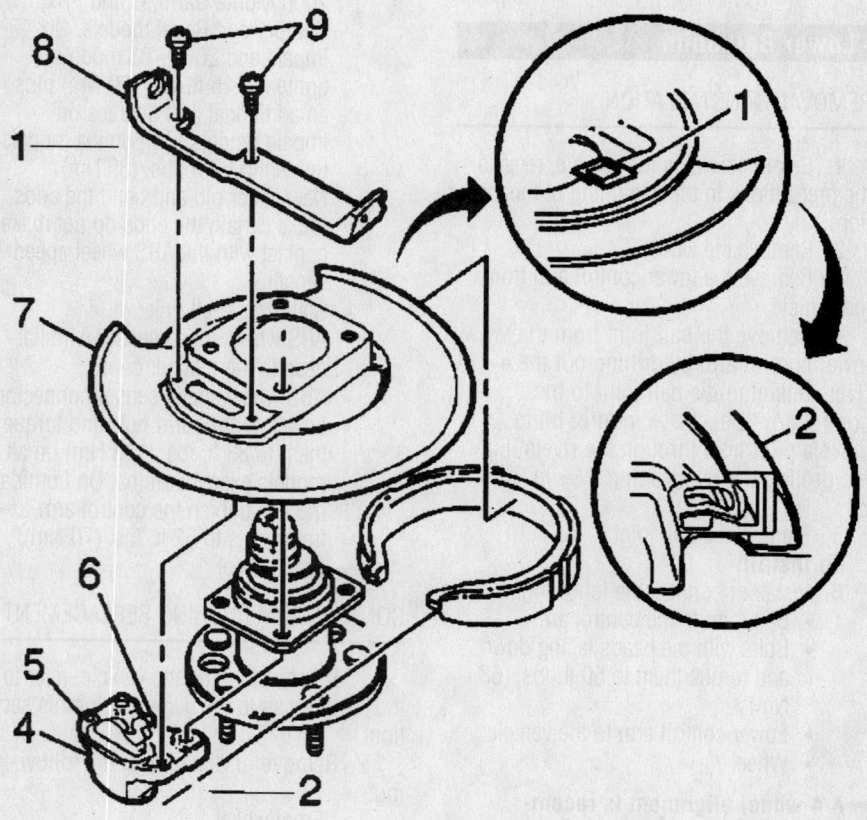

Parking brake lever bracket (1) and parking brake actuator (2)

REMOVAL & INSTALLATION

Front

1. Before servicing the vehicle, refer to the precautions in the beginning of this section.

2. Remove or disconnect the following:
   - Front wheel
   - Wheel Speed Sensor (WSS) electrical connector
   - Brake caliper and bracket
   - Rotor
   - Driveshaft nut

3. Install a front hub removal tool to the wheel bearing/hub assembly with three wheel nuts. Use the tool to push the driveshaft out of the wheel bearing/hub.

4. Remove the wheel bearing/hub assembly and discard the bolts.

To install:

5. Install or connect the following:
   - Wheel bearing/hub assembly using new bolts and torque them to 96 ft. lbs. (130 Nm)
   - New drive shaft nut and torque it to 118 ft. lbs. (160 Nm) on all 2000 models except Impala models or 159 ft. lbs. (215 Nm) on all 2001–04 models and all Impala models
   - Brake rotor
   - Front caliper with the bracket and torque the bracket bolts to 133 ft. lbs. (180 Nm) and the caliper mounting bolts to 63 ft. lbs. (85 Nm) on 2000 models and 70 ft. lbs. (97 Nm) on 2001–04 models
   - WSS electrical connector
   - Front wheel

Rear

The rear wheel bearing/hub is integrated into one unit. The unit is non-serviceable. If the hub or bearing is damaged, the complete hub and bearing unit must be replaced.

1. Before servicing the vehicle, refer to the precautions in the beginning of this section.

2. Remove or disconnect the following:
   - Rear wheel
   - Brake drum, if equipped
   - Rear caliper and bracket, if equipped
   - Brake rotor, if equipped
   - Antilock Brake System (ABS) Wheel Speed Sensor (WSS) electrical connector
   - Rear wheel hub to knuckle bolts
   - Parking brake lever bracket and parking brake actuator
   - Rear wheel hub from the knuckle

To install:

3. Install or connect the following:
   - Parking brake lever bracket and actuator
   - Hub and bearing assembly and torque the bolts to 55 ft. lbs. (75 Nm)
   - WSS electrical connector
   - Brake rotor, if equipped
   - Brake caliper with the bracket and torque the bracket bolts to 85 ft. lbs. (115 Nm) and the caliper bolts to 32 ft. lbs. (44 Nm), if equipped
   - Brake drum, if equipped
   - Rear wheel

4. A 4-wheel alignment is recommended after any steering/suspension repairs have been performed.

BRAKES

Brake Caliper

REMOVAL & INSTALLATION

Front

1. Remove ⅔ of the brake fluid from the master cylinder assembly.

2. Remove the tire and wheel assembly. Mark a relationship between the wheel and the wheel stud for re-installation purposes.

3. Install 2 wheel nuts to retain the rotor on the vehicle.

4. Install a large C-clamp over top of the caliper housing and against the back of the outboard shoe. Slowly tighten the C-clamp until the piston(s) are pushed into the caliper bore.

5. Disconnect the bolt attaching the brake hose fitting to the brake caliper.

6. Plug the opening in the caliper housing and brake line to prevent brake fluid loss and contamination.

7. Remove the caliper mounting bolts.

8. Remove the brake caliper housing from the rotor and mounting bracket.

9. If the caliper is to be replaced or repaired remove the brake pads from the caliper or mounting bracket.

To install:

10. Fully inspect the brake caliper bushing assemblies for cuts, tears or deterioration and replace parts as needed.

11. Inspect the slide bolts for corrosion. If corrosion is found, replace the slide bolts and bushings before installing the brake caliper assembly.

12. Before installing the caliper, make sure the piston(s) are seated in the bore, and that the brake pads are correctly installed.

13. Lubricate the caliper slide bolts with silicone grease. Do not lubricate the bolt threads.

14. Install brake pads onto caliper or mounting bracket.

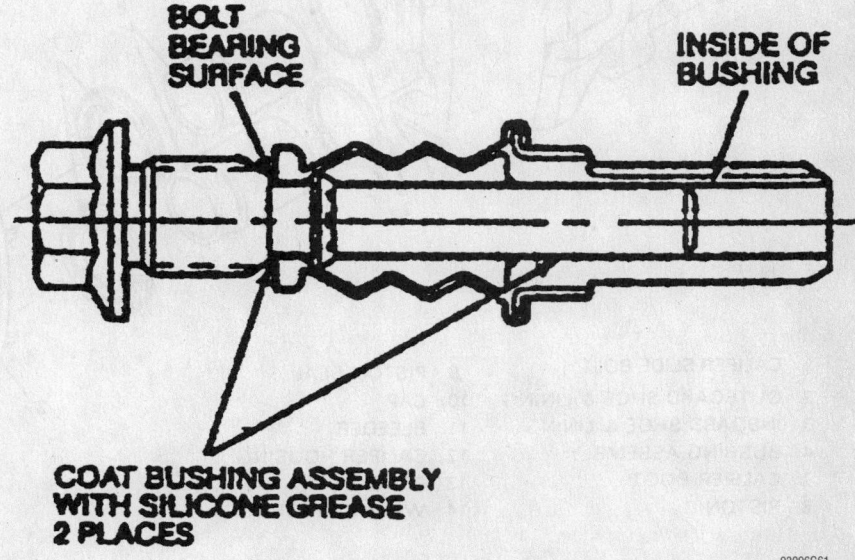

BOLT BEARING SURFACE

INSIDE OF BUSHING

COAT BUSHING ASSEMBLY WITH SILICONE GREASE 2 PLACES

93006G61

Slide bolts lubrication points

15. Install the caliper bolts and torque the bolts to 63 ft. lbs. (85 Nm).

16. Install the brake hose inlet fitting to the caliper and torque the bolt to 24 ft. lbs. (32 Nm).

17. Remove wheel nuts securing the brake rotor to the hub.

18. Install the tire and wheel assembly.

19. Fill the master cylinder to the proper level.

20. Bleed the brake system.

21. Verify correct brake operation.

Rear

1. Remove ⅔ of the brake fluid from the master cylinder assembly.

2. Remove the tire and wheel assembly. Mark a relationship between the wheel and the wheel stud for re-installation purposes.

3. Install 2 wheel nuts to retain the rotor.

4. Remove the brake hose from the caliper and discard the copper washers.

5. Plug the openings in the caliper and the brake hose to prevent brake fluid loss and contamination.

6. The following steps are for models with the park brake cable attached to the caliper:

   a. Disconnect the parking brake cable from the parking brake lever on the caliper. Lift up one end of the cable spring clip free end of the cable from the lever.

   b. Remove the bolt and washer attaching the cable support bracket to the caliper body assembly.

   c. Remove the caliper sleeve bolts.

7. For all others, remove the two caliper mounting bolts.

8. Remove the caliper body assembly from the vehicle. Pivot the caliper assembly up to clear the rotor and then slide it inboard off the pin sleeve.

To install:

9. Inspect the caliper bolt boots, pins and sleeve bolt for cuts, tears or deterioration. Replace as necessary.

10. Lubricate the mounting surfaces and the mounting sleeves.

11. Hold the caliper body assembly in the position from which it was removed, and start it over the end of the pin sleeve.

12. As the caliper body assembly approaches the pin boot, work the large end of the pin boot in the caliper body groove. Push the caliper body fully onto the pin.

13. Pivot the caliper body assembly down, using care not to damage the piston boot on the inboard shoe. Compress the sleeve boot by hand as the caliper body moves into position to prevent boot damage.

14. After installing the caliper assembly into position, recheck the installation of the pad clips. If necessary, use a small prying tool to reset or center the pad clips.

15. Install the brake caliper sleeve bolts and torque to 20 ft. lbs. (27 Nm).

16. On models without caliper mounted park brake cable, torque caliper mounting bolts to 32 ft. lbs. (43 Nm)

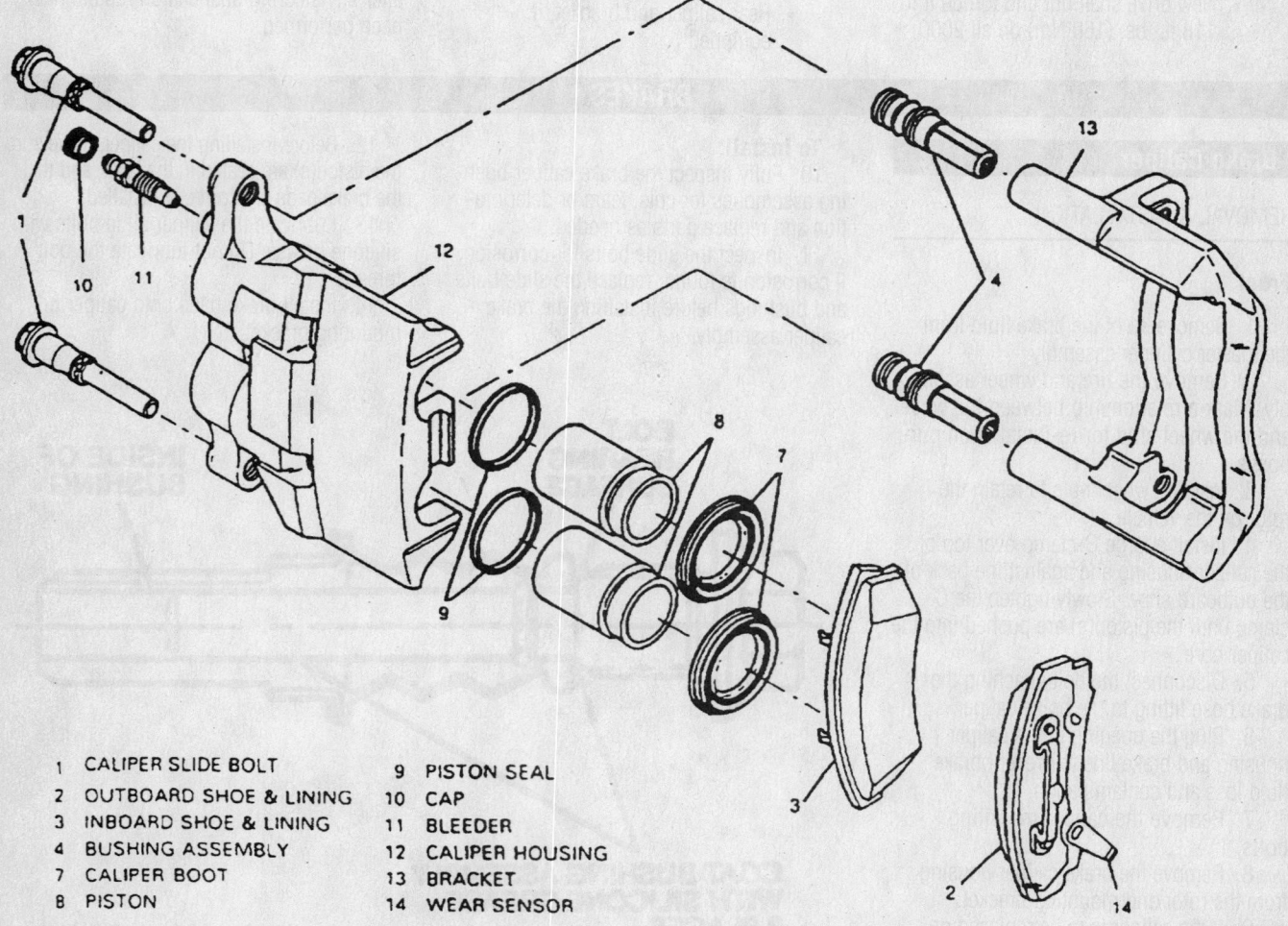

1	CALIPER SLIDE BOLT	
2	OUTBOARD SHOE & LINING	
3	INBOARD SHOE & LINING	
4	BUSHING ASSEMBLY	
7	CALIPER BOOT	
8	PISTON	
9	PISTON SEAL	
10	CAP	
11	BLEEDER	
12	CALIPER HOUSING	
13	BRACKET	
14	WEAR SENSOR	

Caliper attachments

93006G62

17. To install the park brake cable, the following steps apply (some models).

a. Install the cable support bracket with the cable attached. Torque bolt to 32 ft. lbs. (43 Nm).

b. Lift up on the end of the cable spring clip and work the end of the parking brake cable into the notch of the parking brake lever.

18. Connect the brake hose to the brake caliper and torque the bolt to 32 ft. lbs. (44 Nm).

19. Remove the wheel nuts securing the rotor to the hub and bearing assembly.

20. Install the tire and wheel assembly.

21. Fill the master cylinder to the proper level with clean brake fluid.

22. Bleed the brake system using the recommended procedure.

23. Apply approximately 175 lbs. (79 kg) of force, 3 times, to properly seat the brake shoe and linings against the rotor.

24. Adjust the parking brake cable as necessary (some models).

Disc Brake Pads

REMOVAL & INSTALLATION

Front

1. Siphon ⅔ of the brake fluid out of the master cylinder.

2. Mark the relationship of the wheel to the wheel stud for re-installation purposes. Remove the tire and wheel assembly.

3. Install 2 lug nuts to secure the rotor in place when the caliper is removed.

4. Install a large C-clamp over the top of the caliper housing and against the back of the outboard shoe. Slowly tighten the C-clamp until the caliper pistons are pushed into the caliper bore enough to slide the caliper assembly off the rotor. Use care not to tighten the C-clamp too far or the outboard shoe retaining spring will be deformed and require replacement.

5. Remove the caliper mounting bolts and remove the brake caliper from the mounting bracket.

6. DO NOT disconnect the brake hose from the caliper or allow the brake hose to support the weight of the caliper. Support the caliper on a piece of wire out of the way.

7. Remove the outer brake pad from the caliper using a suitable prying tool to lift the outboard shoe retaining spring so that it will clear the caliper center lug and pull the brake pad out of the caliper.

8. Remove the inner brake pad by

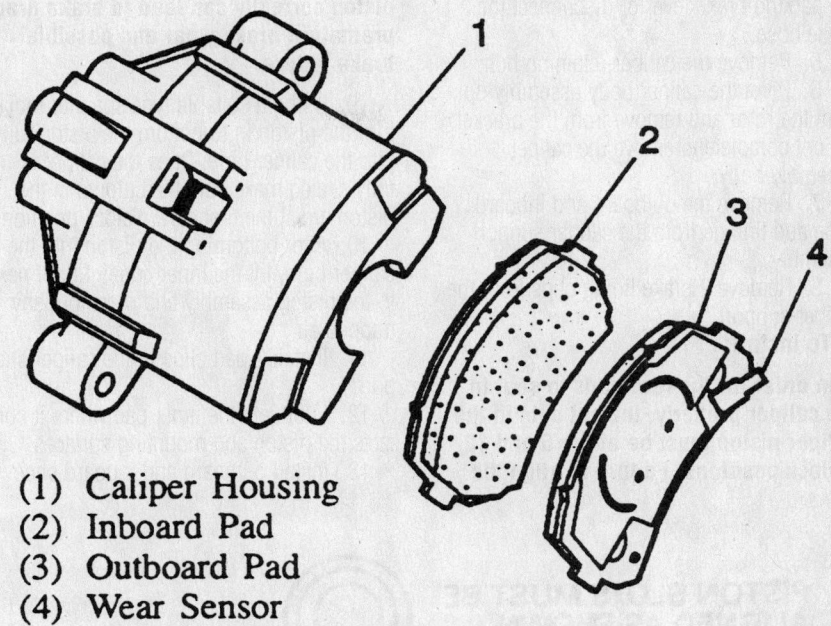

(1) Caliper Housing
(2) Inboard Pad
(3) Outboard Pad
(4) Wear Sensor

93006G71

Front brake pads and caliper—Century, Regal, Lumina, Monte Carlo Intrigue, Grand Prix

unsnapping the shoe springs from the piston.

To install:

9. Clean all parts well. If the brake pads were worn so badly that the brake rotor is damaged, it must be replaced. Light scoring of the rotor surfaces not exceeding 0.060 inch (1.5mm) in depth is not harmful to brake operation and may result from normal use. Brake rotors may be refinished. Do not use a rotor that, after refinishing, will not meet the thickness specification cast in the rotor. Always replace with a new rotor.

10. If not done at removal, now use a C-clamp and clamp both pistons at the same time with a metal plate or wooden block across the face of both pistons. Take care not to damage the pistons or caliper boots.

➡ **After bottoming the pistons into the caliper bore, lift the inner edge of each caliper boot next to the piston and press out any trapped air. Make sure each boot convolution is tucked back into place. Boots must lay flat.**

11. Inspect the caliper bushings for wear. Replace as necessary. Carefully inspect the slide bolts for corrosion. If corrosion if found, use new parts including the bushing assemblies when installing the caliper. Do not attempt to polish away corrosion. Lubricate caliper slide bolts with silicone grease.

12. Install the new inner disc brake pad in the caliper by snapping the shoe retainer springs into the piston making sure both

sets of locking tabs are seated in the caliper pistons. The pad must seat flat against the pistons.

13. Install the outer pad into the caliper by snapping the outboard shoe retaining spring over the caliper center lug and into the housing slot. The pad will slide up onto the caliper and the retaining ring will lock into place on the groove in the caliper.

14. The outer pad wear sensor should be at the trailing edge of the shoe during forward wheel rotation.

15. Install the caliper mounting bolts and torque to 80 ft. lbs. (108 Nm).

16. Remove the 2 nuts temporarily securing the rotor.

17. Install the tire and wheel assembly and tighten to specification.

18. Pump the brake pedal several times to seat the pads against the rotor.

19. Check the brake fluid level and top off as necessary.

20. Road test the vehicle to ensure the proper brake performance.

Rear

1. Siphon ⅔ of the brake fluid out of the master cylinder.

2. Remove the tire and wheel assembly.

3. Install 2 lug nuts to secure the rotor in place when the caliper is removed.

4. Remove bolt and washer attaching cable support bracket to caliper body assembly. It is not necessary to disconnect

the parking brake lever or disconnect the brake hose.

5. Remove the caliper retaining bolt.

6. Pivot the caliper body assembly up from the rotor and remove from the bracket. Do not completely remove the caliper assembly body.

7. Remove the outboard and inboard shoe and linings from the caliper support assembly.

8. Remove 2 brake lining clips from the caliper support.

To install:

➡**In order for the rear pads to seat in the caliper properly, the cut outs in the caliper piston must be at the 6 and 12 o'clock positions. Failure to align the**

piston correctly can lead to brake drag, premature brake wear and possible brake failure.

9. Using a suitable type spanner tool turn the piston in to bottom the piston fully into the caliper bore. Once the caliper is fully seated make sure the cutouts in the piston are at the 6 and 12 o'clock positions.

10. After bottoming the piston into the caliper bore, lift the inner edge of boot next to the piston assembly and press out any trapped air.

11. Install 2 pad clips in the caliper support.

12. Lubricate the inner pad where it contacts the piston and mounting surfaces.

13. Install outboard and inboard shoe

and linings in caliper support. Position the wear sensors downward at the leading edge of the rotor during forward wheel rotation.

14. Hold the metal shoe edge against the spring end of clips in the caliper support. Push brake pad in towards the hub, bending spring ends slightly and engage shoe notches with support abutments.

15. Pivot the caliper body assembly down over the brake pad. Compress the sleeve boot by hand as the caliper body moves into position to prevent boot damage.

➡**After the caliper body assembly is in position, recheck installation of the brake pad clips. If necessary, use a small prying tool to reseat or center the pad clip on the support abutments.**

16. Install the sleeve bolts and torque bolt to 20 ft. lbs. (27 Nm).

17. Install the cable support bracket with the cable attached and bolt washer. Torque bolt to 32 ft. lbs. (43 Nm).

18. Remove the 2 lug nuts securing the rotor.

19. Install the tire and wheel assembly and tighten to specification.

20. Pump the brake pedal several times to seat the pads against the rotor.

21. Check the brake fluid level and top off as necessary.

22. Adjust parking brake as necessary.

23. Road test the vehicle for proper brake performance.

Brake Drums

REMOVAL & INSTALLATION

1. Mark the relationship of the wheel to the axle flange to help maintain wheel balance after assembly.

2. Remove the tire and wheel assembly.

3. Mark the relationship of the brake drum to the axle flange.

4. If difficulty is encountered in removing the brake drum, the following steps may be of assistance.

   a. Make sure the parking brake is released.

   b. Back off the parking brake cable adjustment.

   c. Remove the access hole plug from the backing plate.

   d. Using a screwdriver, back off the adjusting screw.

   e. Install the access hole plug to prevent dirt or contamination from entering the drum brake assembly.

   f. Use a small amount of penetrating

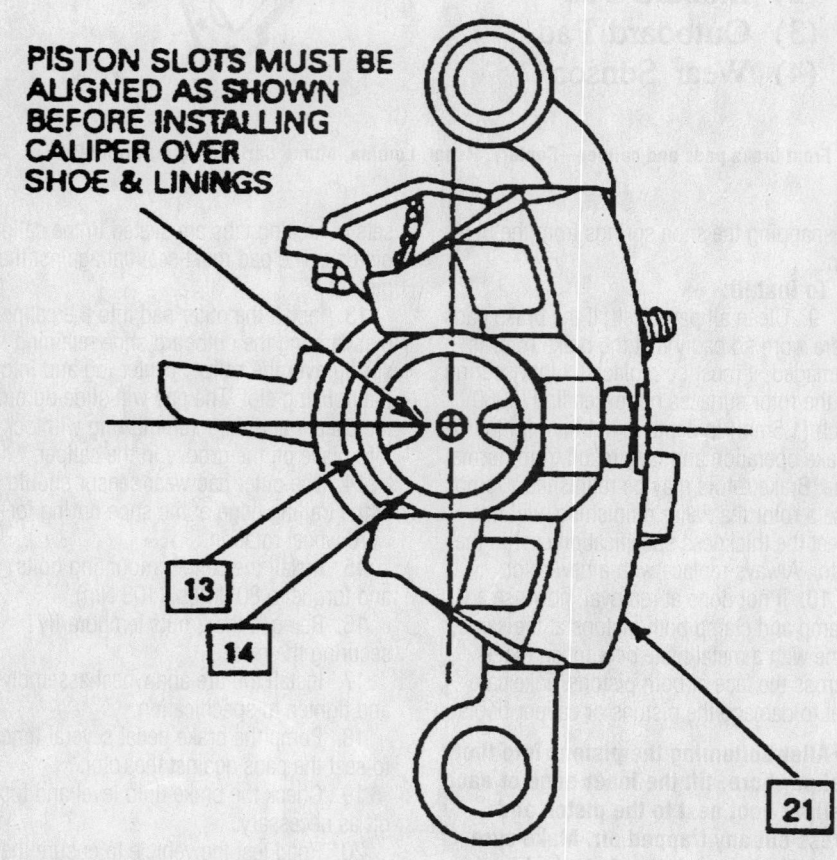

PISTON SLOTS MUST BE ALIGNED AS SHOWN BEFORE INSTALLING CALIPER OVER SHOE & LININGS

13	PISTON BOOT
14	PISTON ASSEMBLY
21	CALIPER BODY ASSEMBLY

93006G72

Positioning rear caliper piston slots—Century, Regal, Lumina, Monte Carlo, Intrigue, Grand Prix

oil applied around the brake drum pilot hole.

g. Carefully remove the brake drum from the vehicle.

5. After removing the brake drum it should be checked for the following:

a. Inspecting for cracks and deep grooves.

b. Inspect for out of round and taper.

c. Inspecting for hot spots (black in color).

To install:

6. Install the brake drum onto the vehicle aligning the reference marks on the axle flange.

7. Install the tire and wheel assembly, torquing it to specifications.

8. Road test for proper brake operation.

Brake Shoes

REMOVAL & INSTALLATION

1. Remove the tire and wheel assembly.
2. Remove the brake drum.
3. Using tool J-38400 brake spanner and remover, remove the actuator spring from the adjuster lever. Use care not to distort the spring when removing it.

✳✳ CAUTION

During the following steps when removing the retractor spring from either shoe and lining assembly, do not over stretch the spring. This will reduce its effectiveness. Keep fingers away from retractor spring to prevent fingers from being pinched between the spring and shoe web or spring and the backing plate.

4. Lift the end of the retractor spring from the adjuster shoe assembly. Insert the hook end of the J-38400, between the retractor spring and the shoe. Pry slightly to remove the spring end from the hole in the shoe.

5. Pry the end of the retractor spring toward the axle with the flat end of the tool until the spring snaps down off the shoe web onto the backing plate.

6. Remove the one brake shoe and remove the adjuster assembly.

7. Disconnect the parking brake lever from the shoe. DO NOT remove the parking brake lever from the cable end unless it is being replaced.

8. Using J-38400, lift the end of the retractor spring from the adjuster shoe assembly. Insert the hook end of the J-38400, between the retractor spring and the shoe. Pry slightly to remove the spring end from the hole in the shoe. Pry the end of the retractor spring toward the axle with the flat end of the tool until the spring snaps down off the shoe web onto the backing plate.

9. Remove the brake shoe.

To install:

10. Clean all the brake spring completely with brake solvent and allow to air dry.

11. Disassemble, clean and lubricate the adjuster screw. Once lubricated, reassemble.

12. Clean the backing plate and after it is dry apply a thin coat of brake grease to the brake shoe contact points on the backing plate.

13. Position the brake shoe that connects to the parking brake lever, on the backing plate. Using J-38400, pull the end of the retractor spring up to rest on the web of the shoe. Pull the end of the retractor spring up until it snaps into the slot in the brake shoe.

14. Connect the parking brake lever.

15. Install the remaining shoe and the adjuster screw assembly.

1   SPRING, ACTUATOR
2   ACTUATOR, ADJUSTER
6   SHOE AND LINING, ADJUSTER
7   LEVER, PARKING BRAKE
8   SHOE AND LINING, PARKING BRAKE
9   SPRING, RETRACTOR
13 PLATE ASSEMBLY, BACKING
14 PLUG, ACCESS HOLE
15 BOLT/SCREW ASSEMBLY, ADJUSTING

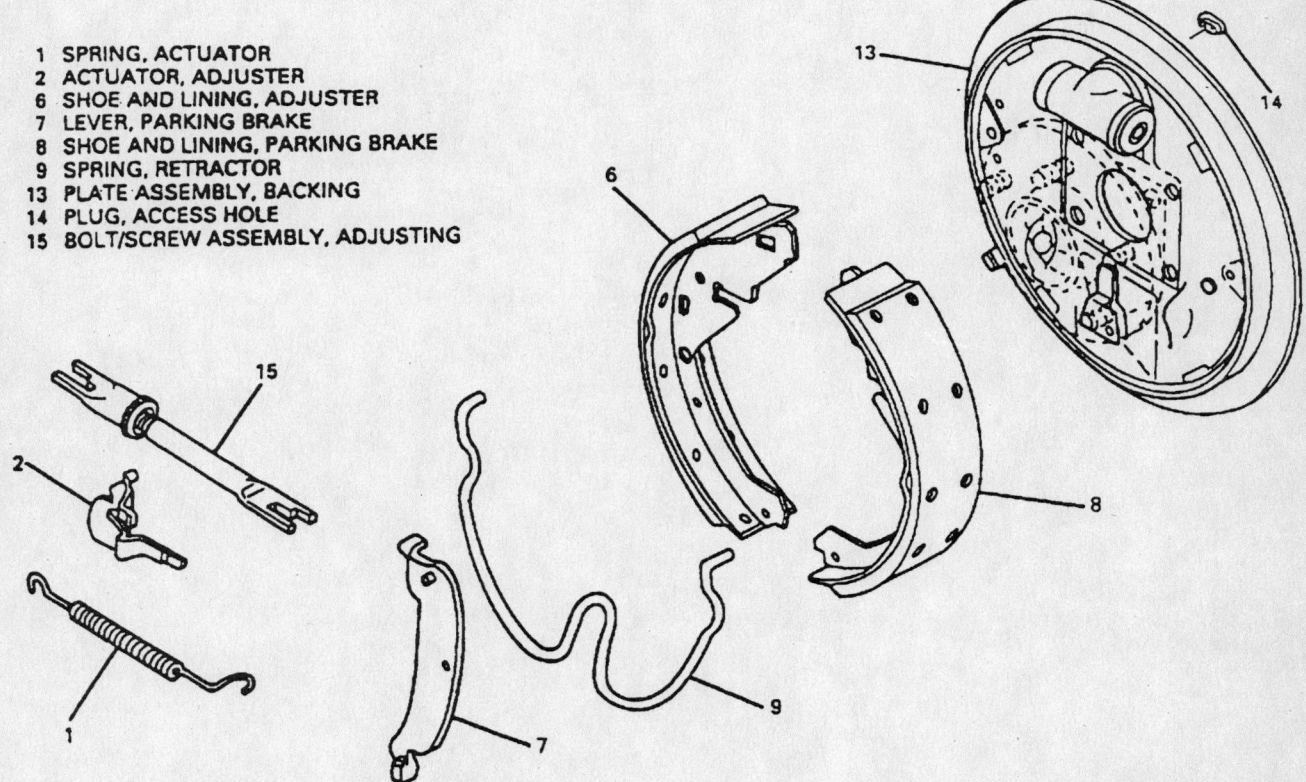

93006G79

Rear brake assembly component alignment

16. Position the brake shoe, using J-38400, pull the end of the retractor spring up to rest on the web of the shoe. Pull the end of the retractor spring up until it snaps into the slot in the brake shoe.

17. Using J-38400, spread the brake shoes and work the adjuster screw into position.

18. Install the actuator spring with the U-shaped end going through the web.

19. Install the brake drum.
20. Install the tire and wheel assembly.
21. Adjust the brakes.
22. Road test for proper brake operation.

SPECIFICATION CHARTS

ENGINE AND VEHICLE IDENTIFICATION

Code ①	Liters (cc)	Cu. In.	Cyl.	Fuel Sys.	Engine Type	Eng. Mfg.
G	5.7 (5665)	350	8	SFI	OHV	CPC
S ③	5.7 (5665)	350	8	SFI	OHV	CPC

CPC: Chevrolet/Pontiac/Canada
MFI: Multi-point Fuel Injection
OHV: Over Head Valves

① 8th position of VIN
② 10th position of VIN
③ High output engine

Code ②	Year
Y	2000
1	2001
2	2002
3	2003
4	2004

Model Year

42372-YBOD-C01

GENERAL ENGINE SPECIFICATIONS

Year	Model	Engine Displacement Liters (cc)	Engine Series (ID/VIN)	Fuel System	Net Horsepower @ rpm	Net Torque @ rpm (ft. lbs.)	Bore x Stroke (in.)	Compression Ratio	Oil Pressure @ rpm
2000	Corvette	5.7 (5665)	G	SFI	345@5600	350@4400	3.89x3.62	10.1:1	18@2000
2001	Corvette	5.7 (5665)	G	SFI	345@5600	350@4400	3.89x3.62	10.1:1	18@2000
2002	Corvette	5.7 (5665)	G	SFI	345@5600	350@4400	3.89x3.62	10.1:1	18@2000
		5.7 (5665)	S	SFI	NA	NA	3.89x3.62	10.5:1	18@2000
2003-04	Corvette	5.7 (5665)	G	SFI	345@5600	350@4400	3.89x3.62	10.1:1	18@2000
		5.7 (5665)	S	SFI	NA	NA	3.89x3.62	10.5:1	18@2000

SFI: Sequential Fuel Injection
NA: Not Available

42372-YBOD-C02

ENGINE TUNE-UP SPECIFICATIONS

Year	Engine Displacement Liters (cc)	Engine ID/VIN	Spark Plug Gap (in.)	Ignition Timing (deg.) MT	Ignition Timing (deg.) AT	Fuel Pump (psi)	Idle Speed (rpm) MT	Idle Speed (rpm) AT	Valve Clearance In.	Valve Clearance Ex.
2000	5.7 (5665)	G	0.060	①	①	48-55	①	①	HYD	HYD
2001	5.7 (5665)	G	0.060	①	①	48-55	①	①	HYD	HYD
2002	5.7 (5665)	G	0.060	①	①	48-55	①	①	HYD	HYD
	5.7 (5665)	S	0.060	①	①	48-55	①	①	HYD	HYD
2003-04	5.7 (5665)	G	0.060	①	①	48-55	①	①	HYD	HYD
	5.7 (5665)	S	0.060	①	①	48-55	①	①	HYD	HYD

NOTE: The Vehicle Emission Control Information label often reflects specification changes made during production. The label figures must be used if they differ from those in this chart.
HYD: Hydraulic

① Refer to Vehicle Emission Control Information label

42372-YBOD-C03

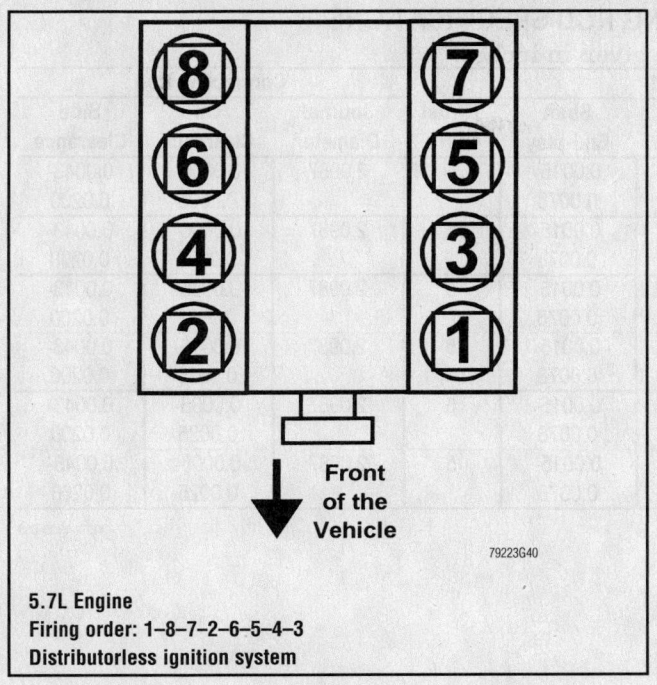

5.7L Engine
Firing order: 1–8–7–2–6–5–4–3
Distributorless ignition system

79223G40

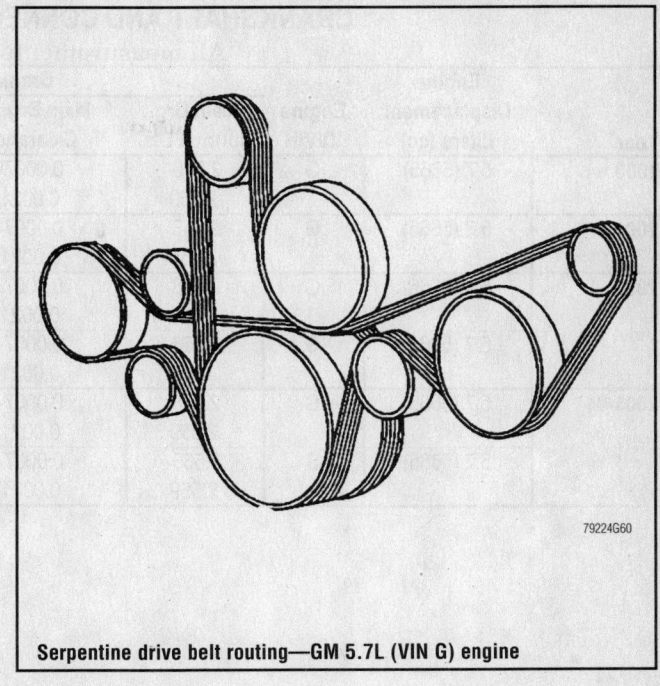

79224G60

Serpentine drive belt routing—GM 5.7L (VIN G) engine

CAPACITIES

Year	Model	Engine Displacement Liters (cc)	Engine ID/VIN	Engine Oil with Filter (qts.)	Transmission (pts.) 6-Spd	Auto.	Drive Axle (pts.)	Fuel Tank (gal.)	Cooling System (qts.)
2000	Corvette	5.7 (5737)	G	6.5	①	②	3.8	19.1	14.7
2001	Corvette	5.7 (5737)	G	6.5	①	②	3.8	18.5	12.6
2002	Corvette	5.7 (5737)	G	6.5	①	②	3.8	18.5	12.6
		5.7 (5737)	S	6.5	①	②	3.8	18.5	12.6
2003-04	Corvette	5.7 (5737)	G	6.5	①	②	3.8	18.5	12.6
		5.7 (5737)	S	6.5	①	②	3.8	18.5	12.6

NOTE: All capacities are approximate. Add fluid gradually and ensure a proper fluid level is obtained.

① MM 6 speed trans: 8.2 pts.

② 4L60E trans: 10.0 pts.

42372-YBOD-C04

CRANKSHAFT AND CONNECTING ROD SPECIFICATIONS
All measurements are given in inches.

| Year | Engine Displacement Liters (cc) | Engine ID/VIN | Crankshaft | | | | Connecting Rod | | |
			Main Brg. Journal Dia.	Main Brg. Oil Clearance	Shaft End-play	Thrust on No.	Journal Diameter	Oil Clearance	Side Clearance
2000	5.7 (5665)	G	2.558-2.559	0.0007-0.0021	0.0015-0.0078	5	2.0987	0.0006-0.0025	0.0043-0.0200
2001	5.7 (5665)	G	2.558-2.559	0.0007-0.0021	0.0015-0.0078	5	2.0987	0.0006-0.0025	0.0043-0.0200
2002	5.7 (5665)	G	2.558-2.559	0.0007-0.0021	0.0015-0.0078	5	2.0987	0.0006-0.0025	0.0043-0.0200
	5.7 (5665)	S	2.558-2.559	0.0007-0.0021	0.0015-0.0078	5	2.0987	0.0006-0.0025	0.0043-0.0200
2003-04	5.7 (5665)	G	2.558-2.559	0.0007-0.0021	0.0015-0.0078	5	2.0987	0.0006-0.0025	0.0043-0.0200
	5.7 (5665)	S	2.558-2.559	0.0007-0.0021	0.0015-0.0078	5	2.0987	0.0006-0.0025	0.0043-0.0200

42372-YBOD-C05

VALVE SPECIFICATIONS

| Year | Engine Displacement Liters (cc) | Engine ID/VIN | Seat Angle (deg.) | Face Angle (deg.) | Spring Test Pressure (lbs. @ in.) | Spring Installed Height (in.) | Stem-to-Guide Clearance (in.) | | Stem Diameter (in.) | |
							Intake	Exhaust	Intake	Exhaust
2000	5.7 (5665)	G	46	45	76@1.80	1.80	0.0010-0.0026	0.0010-0.0026	NA	NA
2001	5.7 (5665)	G	46	45	76@1.80	1.80	0.0010-0.0026	0.0010-0.0026	NA	NA
2002	5.7 (5665)	G	46	45	76@1.80	1.80	0.0010-0.0026	0.0010-0.0026	NA	NA
	5.7 (5665)	S	46	45	76@1.80	1.80	0.0010-0.0026	0.0010-0.0026	NA	NA
2003-04	5.7 (5665)	G	46	45	76@1.80	1.80	0.0010-0.0026	0.0010-0.0026	NA	NA
	5.7 (5665)	S	46	45	76@1.80	1.80	0.0010-0.0026	0.0010-0.0026	NA	NA

NA: Not Available

42372-YBOD-C06

PISTON AND RING SPECIFICATIONS

All measurements are given in inches.

Year	Engine Displacement Liters (cc)	Engine ID/VIN	Piston Clearance	Ring Gap			Ring Side Clearance		
				Top Compression	Bottom Compression	Oil Control	Top Compression	Bottom Compression	Oil Control
2000	5.7 (5665)	G	0.0007-0.0021	0.009-0.015	0.017-0.025	0.007-0.027	0.0016-0.0033	0.0016-0.0031	0.0004-0.0087
2001	5.7 (5665)	G	0.0007-0.0021	0.009-0.015	0.017-0.025	0.007-0.027	0.0016-0.0033	0.0016-0.0031	0.0004-0.0087
2002	5.7 (5665)	G	0.0007-0.0021	0.009-0.015	0.017-0.025	0.007-0.027	0.0016-0.0033	0.0016-0.0031	0.0004-0.0087
	5.7 (5665)	S	0.0007-0.0021	0.009-0.015	0.017-0.025	0.007-0.027	0.0016-0.0033	0.0016-0.0031	0.0004-0.0087
2003-04	5.7 (5665)	G	0.0007-0.0021	0.009-0.015	0.017-0.025	0.007-0.027	0.0016-0.0033	0.0016-0.0031	0.0004-0.0087
	5.7 (5665)	S	0.0007-0.0021	0.009-0.015	0.017-0.025	0.007-0.027	0.0016-0.0033	0.0016-0.0031	0.0004-0.0087

42372-YBOD-C07

TORQUE SPECIFICATIONS

All readings in ft. lbs.

Year	Engine Displacement Liters (cc)	Engine ID/VIN	Cylinder Head Bolts	Main Bearing Bolts	Rod Bearing Bolts	Crankshaft Damper Bolts	Flywheel Bolts	Manifold		Spark Plugs	Lug Nut
								Intake	Exhaust		
2000	5.7 (5665)	G	①	②	③	④	⑤	⑥	⑦	11	100
2001	5.7 (5665)	G	①	②	③	④	⑤	⑥	⑦	11	100
2002	5.7 (5665)	G	①	②	③	④	⑤	⑥	⑦	11	100
	5.7 (5665)	S	①	②	③	④	⑤	⑥	⑦	11	100
2003-04	5.7 (5665)	G	①	②	③	④	⑤	⑥	⑦	11	100
	5.7 (5665)	S	①	②	③	④	⑤	⑥	⑦	11	100

① M11 bolts: 22 ft. lbs.
M11 bolts: plus 90 degrees
M11 bolts 1–8: plus 90 degrees
M11 bolts 9–10: plus 50 degrees
M8 bolts (11–15): 22 ft. lbs.

② Inner Bolts:
Step 1: 15 ft. lbs.
Step 2: 80 degrees
Side bolts: 18 ft. lbs.
Outer studs:
Step 1: 15 ft. lbs.
Step 2: 53 degrees

③ Step 1: 15 ft. lbs.
Step 2: Rotate 75 degrees

④ Step 1: 240 ft. lbs.
Step 2: Install a new bolt and tighten to 37 ft. lbs.
Step 3: Rotate 140 degrees

⑤ Step 1: 15 ft. lbs.
Step 2: 37 ft. lbs.
Step 3: 74 ft. lbs.

⑥ Step 1: 44 inch lbs.
Step 2: 89 inch lbs.

⑦ Step 1: 11 ft. lbs.
Step 2: 18 ft. lbs.

42372-YBOD-C08

WHEEL ALIGNMENT

Year	Model		Caster Range (+/-Deg.)	Caster Preferred Setting (Deg.)	Camber Range (+/-Deg.)	Camber Preferred Setting (Deg.)	Toe-in (in.)	Steering Axis Inclination (Deg.)
2000	Corvette	F	0.50	+6.90	0.50	-0.20	0.05 +/- 0.10	—
		R	—	—	0.50	-0.18	0 +/- 0.10	—
2001	Corvette	F	0.50	+6.90	0.50	-0.20	0.05 +/- 0.10	—
		R	—	—	0.50	-0.18	0 +/- 0.10	—
2002	Corvette	F	0.50	+6.90	0.50	-0.20	0.05 +/- 0.10	—
		R	—	—	0.50	-0.18	0 +/- 0.10	—
2003-04	Corvette	F	0.50	+6.90	0.50	-0.20	0.05 +/- 0.10	—
		R	—	—	0.50	-0.18	0 +/- 0.10	—

42372-YBOD-C09

TIRE, WHEEL AND BALL JOINT SPECIFICATIONS

Year	Model	OEM Tires Standard	OEM Tires Optional	Tire Pressures (psi) Front	Tire Pressures (psi) Rear	Wheel Size	Ball Joint Inspection
2000	Corvette	Fr: P245/45ZR17 Rr: P275/40ZR18	None	30	30	8.5	U: 0.125 in. L: 0.047 in.
2001	Corvette	Fr: P245/45ZR17 Rr: P275/40ZR18	None	30	30	8.5	U: 0.125 in. L: 0.047 in.
2002	Corvette	Fr: P245/45ZR17 Rr: P275/40ZR18	None	30	30	8.5	U: 0.125 in. L: 0.047 in.
2003-04	Corvette	Fr: P245/45ZR17 Rr: P275/40ZR18	None	30	30	8.5	U: 0.125 in. L: 0.047 in.

OEM: Original Equipment Manufacturer
PSI: Pounds Per Square Inch
L: Lower
U: Upper
Fr: Front
Rr: Rear

42372-YBOD-C10

BRAKE SPECIFICATIONS

All measurements in inches unless noted

Year	Model	Brake Disc Original Thickness	Brake Disc Minimum Thickness	Brake Disc Maximum Runout	Minimum Lining Thickness Front	Minimum Lining Thickness Rear	Brake Caliper Bracket Bolts (ft. lbs.)	Brake Caliper Mounting Bolts (ft. lbs.)
2000	Corvette	1.26	1.19	0.002	0.030	0.030	125	23
2001	Corvette	1.26	1.19	0.002	0.030	0.030	125	23
2002	Corvette	1.26	1.19	0.002	0.030	0.030	125	23
2003-04	Corvette	1.26	1.19	0.002	0.030	0.030	125	23

① Heavy duty: 1.110; Std.: 0.795
② Heavy duty: 1.059; Std.: 0.744
③ Front: not available, Rear: upper 26 ft. lbs., lower 16 ft. lbs.

42372-YBOD-C11

SCHEDULED MAINTENANCE INTERVALS
GM Y BODY—CHEVROLET CORVETTE

TO BE SERVICED	TYPE OF SERVICE	VEHICLE MILEAGE INTERVAL (x1000)												
		7.5	15	22.5	30	37.5	45	52.5	60	67.5	75	82.5	90	97.5
Engine oil & filter ①	R	✓	✓	✓	✓	✓	✓	✓	✓	✓	✓	✓	✓	✓
Brake hoses & brake lining	S/I	✓	✓	✓	✓	✓	✓	✓	✓	✓	✓	✓	✓	✓
Coolant level, hoses & clamps	S/I	✓	✓	✓	✓	✓	✓	✓	✓	✓	✓	✓	✓	✓
Exhaust system & throttle linkage	S/I	✓	✓	✓	✓	✓	✓	✓	✓	✓	✓	✓	✓	✓
Lubricate chassis, suspension, steering linkage, transaxle shift linkage, parking brake cable guides, underbody contact points & linkage	S/I	✓	✓	✓	✓	✓	✓	✓	✓	✓	✓	✓	✓	✓
Rear axle fluid level	S/I	✓	✓	✓	✓	✓	✓	✓	✓	✓	✓	✓	✓	✓
Automatic transaxle fluid & filter ②	S/I		✓		✓		✓		✓		✓		✓	
Air filter element	R				✓				✓				✓	
Engine coolant ③	R				✓				✓				✓	
Ignition cables, EGR & fuel systems	S/I				✓				✓				✓	
Serpentine drive belt	S/I					✓			✓				✓	
Spark plugs ④	R													

R: Replace   S/I: Service or Inspect

① Some engines use an engine oil life monitor to inforn you to change the oil. On vehicles not equipped with the monitor: every 3000 miles or 3 months

Use engine oil meeting the GM Standard GM 4718M.

② Replace fuid every 50, 000 miles if driven in etxreme traffic or in places where the temperature exceeds 90 degrees F

③ Engine coolant: replace every 100,000 miles. Use O.E. specified (DEX-COOL™) coolant only. If any silicate coolant is used, the service interval is every 30,000 miles.

④ Platinum tip spark plugs: replace every 100,000 miles.

FREQUENT OPERATION MAINTENANCE (SEVERE SERVICE)

If a vehicle is operated under any of the following conditions it is considered severe service:

- Extremely dusty areas.

- 50% or more of the vehicle operation is in 32°C (90°F) or higher temperatures, or constant operation in temperatures below 0°C (32°F).

- Prolonged idling (vehicle operation in stop and go traffic).

- Frequent short running periods (engine does not warm to normal operating temperatures).

- Police, taxi, delivery usage or trailer towing usage.

Engine oil & oil filter: change every 3000 miles

Chassis lubrication: lubricate every 6000 miles.

Lubricate suspension, parking brake cable guides, underbody contact points & linkage: lubricate every 6000 miles.

Air filter element: service or inspect every 15,000 miles.

Automatic transmission fluid & filter: change every 15,000 miles.

42372-YBOD-C12

PRECAUTIONS

Before servicing any vehicle, please be sure to read all of the following precautions, which deal with personal safety, prevention of component damage, and important points to take into consideration when servicing a motor vehicle:

• Never open, service or drain the radiator or cooling system when the engine is hot; serious burns can occur from the steam and hot coolant.

• Observe all applicable safety precautions when working around fuel. Whenever servicing the fuel system, always work in a well-ventilated area. Do not allow fuel spray or vapors to come in contact with a spark, open flame or excessive heat (a hot drop light, for example). Keep a dry chemical fire extinguisher near the work area. Always keep fuel in a container specifically designed for fuel storage; also, always properly seal fuel containers to avoid the possibility of fire or explosion. Refer to the additional fuel system precautions later in this section.

• Fuel injection systems often remain pressurized, even after the engine has been turned **OFF**. The fuel system pressure must be relieved before disconnecting any fuel lines. Failure to do so may result in fire and/or personal injury.

• Brake fluid often contains polyglycol ethers and polyglycols. Avoid contact with the eyes and wash your hands thoroughly after handling brake fluid. If you do get brake fluid in your eyes, flush your eyes with clean, running water for 15 minutes. If eye irritation persists, or if you have taken brake fluid internally, IMMEDIATELY seek medical assistance.

• The EPA warns that prolonged contact with used engine oil may cause a number of skin disorders, including cancer! You should make every effort to minimize your exposure to used engine oil. Protective gloves should be worn when changing oil. Wash your hands and any other exposed skin areas as soon as possible after exposure to used engine oil. Soap and water, or waterless hand cleaner should be used.

• All new vehicles are now equipped with an air bag system. The system must be disabled before performing service on or around system components, steering column, instrument panel components, wiring and sensors. Failure to follow safety and disabling procedures could result in accidental air bag deployment, possible personal injury and unnecessary system repairs.

• Always wear safety goggles when working with, or around, the air bag system. When carrying a non-deployed air bag, be sure the bag and trim cover are pointed away from your body. When placing a non-deployed air bag on a work surface, always face the bag and trim cover upward, away from the surface. This will reduce the motion of the module if it is accidentally deployed. Refer to the additional air bag system precautions later in this section.

• Clean, high quality brake fluid from a sealed container is essential to the safe and proper operation of the brake system. You should always buy the correct type of brake fluid for your vehicle. If the brake fluid becomes contaminated, completely flush the system with new fluid. Never reuse any brake fluid. Any brake fluid that is removed from the system should be discarded. Also, do not allow any brake fluid to come in contact with a painted surface; it will damage the paint.

• Never operate the engine without the proper amount and type of engine oil; doing so WILL result in severe engine damage.

• Timing belt maintenance is extremely important! Many models utilize an interference-type, non-freewheeling engine. If the timing belt breaks, the valves in the cylinder head may strike the pistons, causing potentially serious (also time-consuming and expensive) engine damage.

• Disconnecting the negative battery cable on some vehicles may interfere with the functions of the on-board computer system(s) and may require the computer to undergo a relearning process once the negative battery cable is reconnected.

• When servicing drum brakes, only disassemble and assemble one side at a time, leaving the remaining side intact for reference.

• Only an MVAC-trained, EPA-certified automotive technician should service the air conditioning system or its components.

ENGINE REPAIR

Distributor

All 5.7L engines use a Direct Ignition System.

Alternator

REMOVAL

1. Before servicing the vehicle, refer to the precautions in the beginning of this section.
2. Remove or disconnect the following:
   • Negative battery cable
   • Regulator connector and battery-to-alternator terminal from the rear of the alternator
   • Accessory drive belt
   • Alternator mounting bolts
   • Alternator

To install:
3. Install or connect the following:
   • Alternator
   • Alternator mounting bolts and torque them to 37 ft. lbs. (50 Nm)

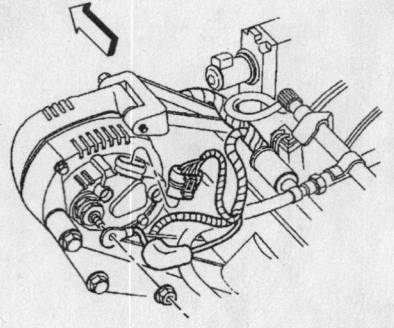

9306ZG24

Exploded view of the alternator—5.7L engine

   • Regulator connector and battery-to-alternator terminals and torque the alternator terminal nut to 10 inch lbs. (13 Nm)
   • Drive belt
   • Negative battery cable

Engine Assembly

REMOVAL & INSTALLATION

1. Before servicing the vehicle, refer to the precautions in the beginning of this section.

The following tools will be required in addition to the basic hand tools:
   • Transverse spring compressor and adapters
   • Ball joint separator
   • Engine support table
   • Driveshaft support strap

- Fuel pressure gauge
2. Relieve the fuel system pressure.
3. Drain the cooling system.
4. Drain the engine oil.
5. Evacuate the A/C system.
6. Remove or disconnect the following:
- Intake Air Temperature (IAT) sensor electrical connector
- Mass Air Flow (MAF) sensor electrical connector
- Air intake duct and air cleaner
- Upper radiator support
- Radiator
- Electronic Brake Control Traction Module/Brake Pressure Modulator Valve (EBTCM/BPMV) and bracket
- Brake pipes
- Accessory drive belt
- Fuel line from the connector at the front of the dash

➡ **Cap and plug all openings for the fuel system to prevent contaminants from entering the system.**

- Fuel rail covers
- Fuel line at the fuel rail
- Radiator hoses from the water pump
- Heater hoses from the water pump
- Fuel injector electrical connectors
- Ignition coil main harness connector
- Evaporative Emission (EVAP) solenoid electrical connector
- Electric throttle motor electrical connector
- Throttle Position Sensor (TPS) electrical connector
- Engine Coolant Temperature (ECT) sensor electrical connector
- A/C compressor electrical connector
- Alternator electrical connectors
- Alternator
- Brake booster vacuum hose
- Intermediate shaft to steering gear bolt
- Intermediate shaft from the steering gear
- Secondary Air Injection (AIR) hose from the left exhaust manifold
- Both front wheels
- Heated Oxygen Sensor (HO2S) connectors from the intermediate exhaust pipes
- Intermediate exhaust pipes
- Close-out panel
- Starter motor electrical connectors
- Starter motor
- Oil level sensor connector
- Crankshaft Position (CKP) sensor electrical connector
- A/C compressor hose assembly

- Engine oil temperature sensor electrical connector
- Left Heated Oxygen Sensor (HO2S)
- Ground straps
- Front stabilizer shaft
- Connectors from the cooling fans
- Cooling fan assembly
- Tie rod ends from the steering knuckles
- Antilock Brake System (ABS) electrical connectors from the crossmember, if equipped
- Electronic Variable Orifice (EVO) connector clips from the crossmember, if equipped
- Real Time Damping (RTD) connector clips from the crossmember, if equipped
- Shock absorber lower mounting bolts
- Front transverse leaf spring
- Automatic transmission cooler pipes at the flywheel housing junction
- Automatic transmission cooler pipe clamps from the engine oil pan
- Automatic transmission cooler pipe from the radiator

Failure to use the minimum fastener length specified will prevent the proper retention of the propeller shaft during assembly.

7. For vehicles with an automatic transmission, perform the following steps:
  a. Remove the two plug bolts in the driveline support assembly.
  b. Install an M10 x 55 bolt, or longer, in each plug location. Torque the bearing support bolts to 26 ft. lbs. (35 Nm).
8. On vehicles with automatic transmission, remove the bell housing inspection cover, then turn the flywheel hub collar to access the bolt and loosen it.
9. Remove the engine flywheel housing access plug.
10. Rotate the automatic transmission collar so that the bolt is facing down.
11. Loosen the bolt and unclip the wire harness from the engine and reposition it to the driveline.
12. Install a driveline support tool to the close out panel flange.

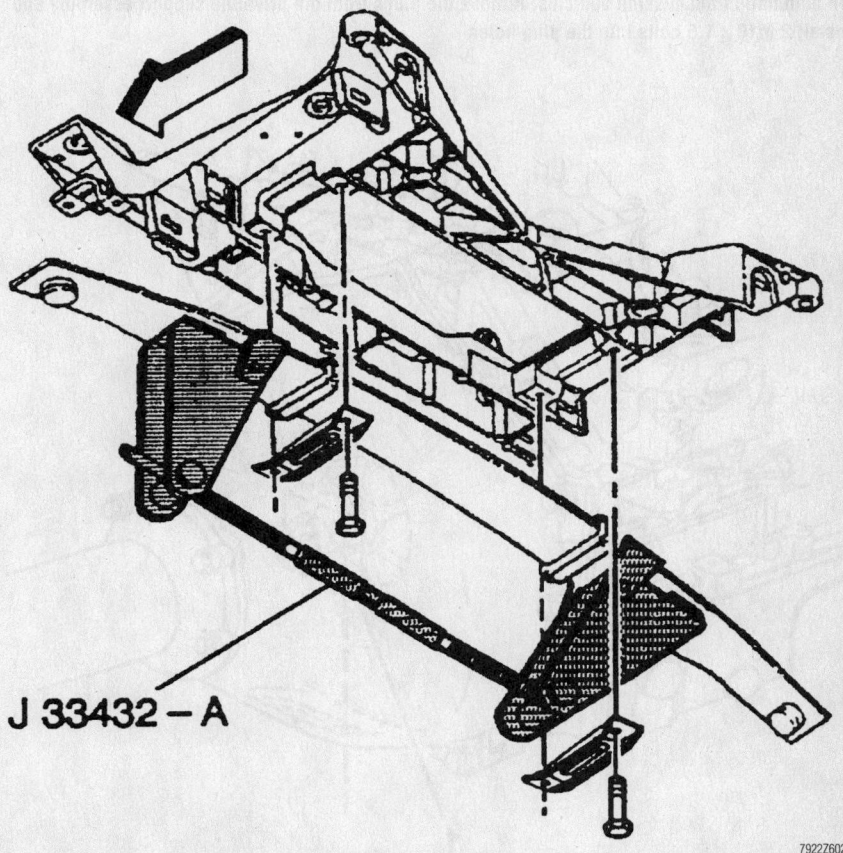

J 33432 - A

79222Z602

Use a compressor such as J 33432-A to compress the front transverse spring—5.7L (VIN G) engine

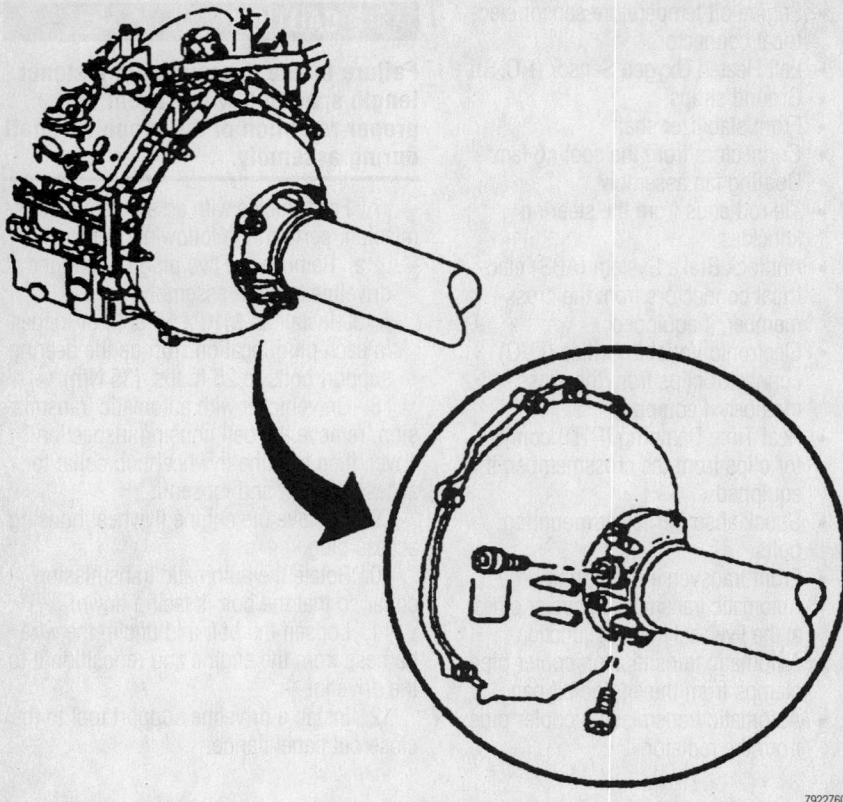

On automatic transmission vehicles, remove the plugs from the driveline support assembly and install 2 M10 x 1.5 bolts into the plug holes

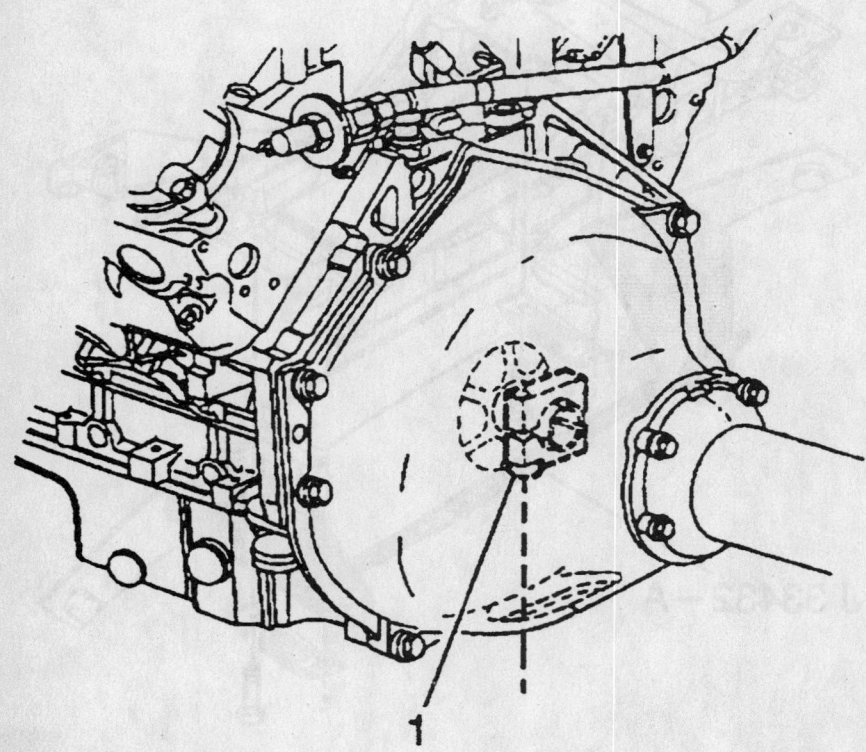

Loosen the bolt on the flywheel hub collar after turning it for access

Never use a driveline support tool to support the weight of the engine assembly.

13. If equipped with a manual transmission, perform the following steps:

a. Unclip the clutch actuator hose from the engine flywheel housing clip.

b. Using a hydraulic clutch line separator tool, depress the white release ring on the actuator hose and pull lightly on the master cylinder hose.

c. Remove the flywheel housing bolts from the driveline support.

14. Support the engine and crossmember assembly.

15. Remove or disconnect the following:

- Front crossmember nuts BY HAND and lower the engine assembly slightly
- Secondary Air Injection (AIR) tube bracket bolt and reposition the bracket to gain access to the ground strap
- Ground strap from the rear of the left cylinder head
- Engine oil pressure sensor electrical connector
- Camshaft Position (CMP) sensor electrical connector
- Manifold Absolute Pressure (MAP) sensor electrical connector
- Knock Sensor (KS) electrical connector
- Front driveline support assembly bolts
- Engine from the driveline, by placing a flat blade tool between edge of the driveline and the flywheel housing

16. Pull the engine away from the propeller shaft.

17. Raise the vehicle off of the engine and crossmember.

- Power steering pump (with reservoir) from the engine and reposition them to the crossmember
- A/C, alternator and power steering pump brackets
- AIR pipe and gasket
- Spark plugs

18. Remove the engine mount-to-crossmember nuts.

19. Remove the engine from the crossmember.

To install:

20. Install the engine on the crossmember and torque the nuts to 48 ft. lbs. (65 Nm).

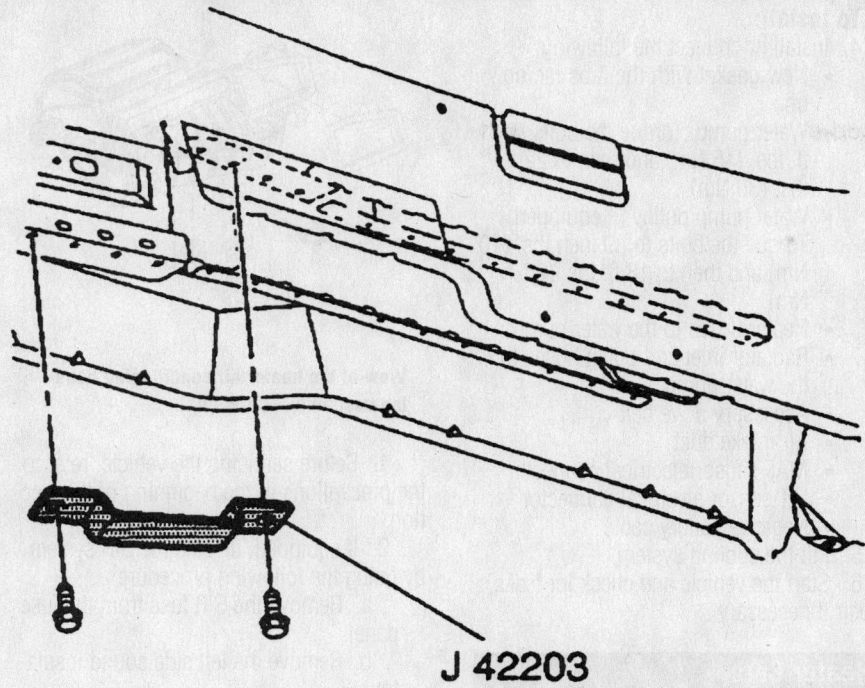

J 42203

79222605

Install a Driveline Support tool J 42203 to the under cover flange

21. Lower the vehicle onto the engine/crossmember assembly.

22. Install or connect the following:
- Spark plugs
- AIR pipe and gasket
- A/C, alternator and power steering pump brackets
- Power steering pump and reservoir on the engine and torque the bolts to 18 ft. lbs. (25 Nm)
- Ground strap to the rear of the left cylinder head and torque the bolt to 24 ft. lbs. (32 Nm)
- Engine oil pressure sensor electrical connector
- CMP sensor electrical connector
- MAP sensor electrical connector
- KS electrical connector
- Install or connect the following:
- Front driveline support bolts and torque them to 37 ft. lbs. (50 Nm)
- Air tube and bracket and torque the bolt to 15 ft. lbs. (20 Nm)
- New crossmember nuts and torque them to 81 ft. lbs. (110 Nm)
- Flywheel housing bolts and torque them to 37 ft. lbs. (50 Nm)
- Clutch actuator hose to the flywheel housing (if equipped)
- Master cylinder hose to the clutch actuator hose (if equipped)
- Wire harness to the rear of the engine

23. Remove the driveline support tool.

24. Remove the M10 x 55mm bolts from the driveline support assembly.
- Two plugs in the driveline support assembly after removing the M10 x 55 bolts and torque them to 37 ft. lbs. (50 Nm)
- Automatic transmission oil cooler pipes at the flywheel housing junction and torque them to 20 ft. lbs. (27 Nm), if equipped
- Automatic transmission cooler pipes to the front and rear of the oil pan
- Front transverse spring and torque the bolts to 46 ft. lbs. (62 Nm)
- Shock absorber lower mounting nuts and torque them to 21 ft. lbs. (28 Nm)
- Front stabilizer shaft and torque the bolts to 53 ft. lbs. (72 Nm)
- ABS electrical connector
- EVO electrical connector
- RTD electrical connector
- Tie rods to the steering knuckle and torque the nuts to 33 ft. lbs. (45 Nm).
- Radiator
- Cooling fan assembly to the radiator and attach the electrical connectors to the fans
- Ground wires to the left side of the engine
- HO2S connectors

- Engine oil temperature sensor connector
- A/C compressor hoses to the compressor
- A/C compressor line retaining bolt and torque it to 26 ft. lbs. (35 Nm)
- CKP sensor electrical connector
- Oil level sensor electrical connector
- Wire harness ground wires to the right side of the engine
- Starter motor and torque the bolts to 37 ft. lbs. (50 Nm)
- Driveline close out panel and torque the bolts to 106 inch lbs. (12 Nm)
- Intermediate exhaust pipe and hangers and torque the bolts to 37 ft. lbs. (50 Nm) and the nuts to 15 ft. lbs. (20 Nm)
- Front wheels
- Automatic transmission cooler pipes to the radiator (if equipped) and torque the fittings to 26 ft. lbs. (35 Nm)
- AIR hose to the left exhaust manifold
- Intermediate steering shaft to the steering gear and torque the bolt to 35 ft. lbs. (48 Nm)
- Brake booster vacuum hose
- Alternator bracket and torque the bolts to 37 ft. lbs. (50 Nm)
- Alternator and torque the bolts to 37 ft. lbs. (50 Nm)
- Alternator connectors and torque the nuts to 10 ft. lbs. (13 Nm)
- Fuel injector electrical connectors
- Ignition coil main electrical connectors
- EVAP solenoid electrical connectors
- Throttle motor electrical connector
- ECT sensor electrical connector
- TPS electrical connector
- A/C compressor electrical connector
- Heater hoses to the water pump
- Radiator hoses to the water pump
- Fuel line to the fuel rail
- Fuel rail covers
- Drive belt
- EBTCM/BPMV bracket and torque the bolts to 20 ft. lbs. (27 Nm)
- EBTCM/BPMV to the bracket and torque the nuts to 89 inch lbs. (10 Nm)
- Upper radiator hose
- Upper radiator support
- Air cleaner and air intake duct
- IAT sensor electrical connector
- MAF sensor electrical connector
- Negative battery cable

25. Fill the engine with new oil.

26. Fill the cooling system.

27. Evacuate and recharge the A/C system.

28. On vehicles with an automatic transmission, start the engine and allow it to idle for 10 minutes. Torque the hub collar bolt to 96 ft. lbs. (130 Nm) and install the flywheel inspection plug.

29. A wheel alignment is recommended when the engine and crossmember have been removed from the vehicle.

30. Check the vehicle for leaks and repair if necessary.

Water Pump

REMOVAL & INSTALLATION

1. Before servicing the vehicle, refer to the precautions in the beginning of this section.

2. Drain the cooling system.

3. Remove or disconnect the following:

- Negative battery cable
- Intake Air Temperature (IAT) sensor electrical connector
- Mass Air Flow (MAF) sensor electrical connector
- Air intake duct
- Accessory drive belt
- Inlet and outlet hoses from the water pump
- Heater hoses from the water pump
- Water pump pulley bolts and pulley (if equipped)
- Six water pump retaining bolts
- Water pump

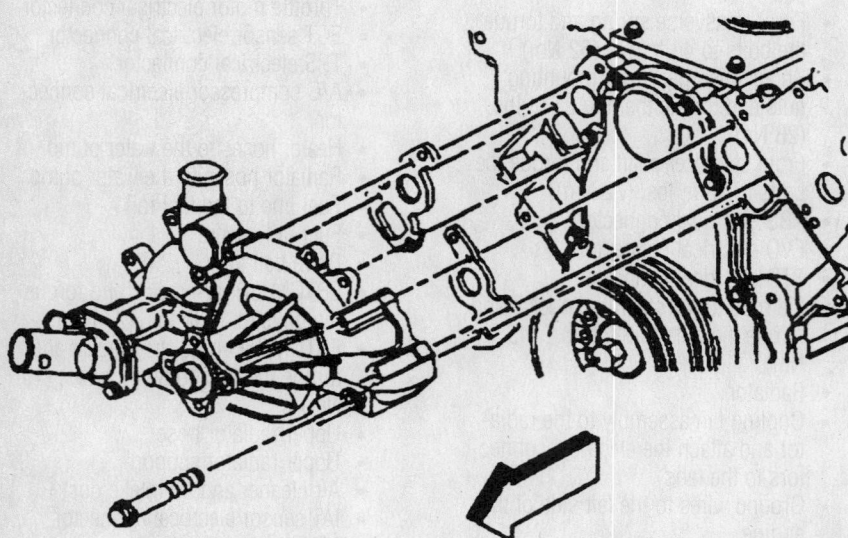

Exploded view of the water pump mounting assembly—5.7L engine

To install:

4. Install or connect the following:

- New gasket with the tabs facing up
- Water pump. Torque the bolts to 11 ft. lbs. (15 Nm) and then to 22 ft. lbs. (30 Nm).
- Water pump pulley (if equipped). Torque the bolts to 89 inch lbs. (10 Nm) and then to 18 ft. lbs. (25 Nm).
- Heater hoses to the water pump
- Radiator inlet and outlet hoses to the water pump
- Accessory drive belt
- Air intake duct
- MAF sensor electrical connector
- IAT sensor electrical connector
- Negative battery cable

5. Fill the cooling system.

6. Start the vehicle and check for leaks, repair if necessary.

Heater Core

REMOVAL & INSTALLATION

✳✳ CAUTION

The air bag system must be disabled before performing service on or around the air bag, instrument panel components, wiring and sensors. Failure to follow safety and disabling procedures could result in accidental air bag deployment, possible personal injury and unnecessary air bag system repairs.

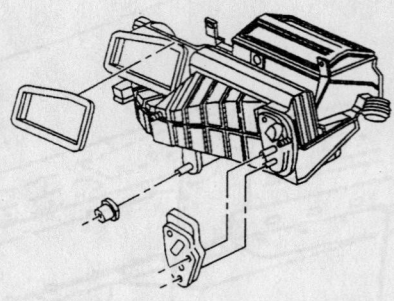

View of the heater/air conditioning housing assembly—Corvette

1. Before servicing the vehicle, refer to the precautions in the beginning of this section.

2. If equipped, disable the SIR system by using the following procedure:

a. Remove the SIR fuse from the fuse panel.

b. Remove the left side sound insulator.

c. Disconnect the Connector Positive Assurance (CPA) from the yellow 2-way SIR harness connector at the base of the steering column and separate the connector.

d. Disconnect the steering wheel module coil connector located at the base of the steering column.

e. Disconnect the CPA from the inflatable restraint instrument panel module connector located at the near the base of the steering column.

f. Disconnect the instrument panel module connector located near the base of the steering column.

g. Disconnect the negative battery cable.

3. Discharge and recover the air conditioning system refrigerant.

4. Drain the cooling system into a clean container for reuse.

5. Remove or disconnect the following:

- Battery heat shield
- Intake manifold
- Hose from the right hand air injection check valve
- Bolt attaching the right hand air injection check valve to the cylinder head(s), if equipped
- Nut attaching the heater pipe bracket to the cowl and set the bracket aside
- Heater pipe-to-heater core bolt
- Heater pipe assembly from the core, allow any remaining coolant to drain and cap the lines. Discard the sealing washers.

- Accumulator hose-to-evaporator bolt
- Accumlator hose and evaporator tube from the evaporator. Discard the O–ring seals and cap hose and tube ends and also the evaporator.
- HVAC module drain tube from the module

6. Remove the following components to access the heater core assembly as follows:

a. Remove the folding top stowage compartment lid extension by opening the compartment, removing the screws from the lower sides and the top of the panel. Lift the panel up off the bracket.

b. Open the console door.

c. Pull up on the rear of the traction control ride switch to release it from the retaining clips. If the switch will not release, carefully insert a suitable prytool into the recess at the rear of the switch and carefully pull up at the rear of the switch.

d. Disconnect the connector from the traction switch.

e. Disconnect the LED connector from the harness connector and remove the switch.

f. Remove the console retaining nut covers using a suitable prytool.

g. Remove the console retaining nuts.

h. Lift the rear of the console slightly and pull back gently to release the console from under the accessory trim plate.

i. Disconnect the accessory power plug connector, unscrew the plug housing retainer and remove the plug housing.

j. Disconnect the fuel door release electrical connector and if equipped, the rear lift window switch. Remove the switch.

k. Remove the console.

l. Apply the parking brake for additional clearance.

m. Place the transmission in second gear on models with an Automatic Transmission (AT) or fourth on models with a Manual Transmission (MT).

n. On models equipped with MT, grasp the shift control boot and apply light pressure in towards the shift lever to release the tabs from the IP accessory trim plate.

o. Release the remaining tabs using light pressure and remove the boot from the trim plate.

p. Remove the ashtray.

q. Remove the IP accessory trim plate grille.

r. Remove the accessory trim plate screws next to the cigar lighter and behind the ashtray.

s. Remove the accessory trim plate retaining screws in the grille opening.

t. Grasp the sides of the trim plate near the curve at the base.

u. Pull the trim plate rearwards to release the tabs, making sure to lift the rear of the trim plate to clear the driveline tunnel studs.

v. Disconnect the cigar lighter electrical connector.

w. Rotate the shift control boot and reposition one end down into the shifter opening in the trim plate, on models with MT.

x. Lift the trim plate over the shifter and boot and remove the trim plate.

y. Remove the fog lamp, rear compartment lid release switch.

z. Remove the driver knee bolster trim panel screw from behind the switch.

aa. Remove the lower driver knee bolster screws,.

bb. Grasp the knee bolster trim panel and pull firmly rearward to disengage the locking tabs. Disconnect the connector from the inside air temperature sensor and remove the panel.

cc. Open the glove box and disconnect the light switch.

dd. Remove the trim plugs from the bottom of the door by reaching behind the door and pushing the plugs out while using a prytool on the front.

ee. Remove the lower bolts and the upper and side screws.

ff. Slide the glove box forward and unplug the wiring harness connector at the passenger side air bag.

gg. Remove the glove box.

hh. Remove the inboard side window glass defroster duct from the windshield defroster duct on both sides

ii. Remove the DRL electrical connector and the sun load temperature sensor from the windshield defroster duct, if equipped

jj. Remove the windshield side garnish moldings.

kk. Remove the screws attaching the upper trim pad to the defroster duct.

ll. Remove the screws attaching the upper trim pad to the left and right hand hinge pillars.

mm. Remove the screws attaching the IP cluster bezel to the upper trim pad.

nn. Remove the screws attaching the upper trim panel to the driver knee bolster outer bracket and center support bracket.

oo. Remove the screw attaching the upper trim panel to the passenger SIR bracket.

pp. Tilt the steering wheel to its lowest position.

qq. Lift the rearward edge of the trim panel approximately 2 inches (5 CM) to provide clearance for the air distribution duct.

rr. Slowly pull the pad away from the windshield while guiding the tabs past the hinge pillars.

ss. Disconnect the remaining electrical connectors to the upper trim panel and remove the panel.

tt. Remove the left hand side window defogger outlet duct.

uu. Remove the inside air temperature sensor duct.

vv. Disconnect the DRL sensor, if equipped.

ww. Disconnect the left hand temperature sensor connector, if equipped.

xx. Turn the key to the ON position.

yy. Using a suitable prytool, depress the park/lock cable retaining tab located on the underside of the ignition switch and pull to release the cable from the ignition switch.

zz. Using a suitable prytool, release the tab retaining the park/lock cable to the slot in the shift control cable.

aaa. Lift the park lock cable from the slot.

bbb. Grasp the cable end and pull rearwards to unlock the cable from the shifter pivot arm stud.

ccc. Remove the park/lock cable.

ddd. Remove the CPA from the electrical connector on the inflatable restraint and Sensing and Diagnostic Module (SDM).

eee. Disconnect the SDM.

fff. Remove the ignition switch bolts and position the switch aside.

ggg. Mark the position of the IP center support bracket.

hhh. Remove the right bolt that attaches the bracket to the driveline tunnel.

iii. Remove the bolts that attach the bracket to the passenger knee bolster bracket.

jjj. Remove the bolt that attaches the bracket to the IP lower support beam.

kkk. Remove the left bolt that attaches the bracket to the driveline tunnel.

lll. Mark the location of the ignition switch bracket.

mmm. Remove the bolt that attaches the ignition switch housing bracket to the steering column bracket.

nnn. Remove the bolt that attaches the c enter support bracket to the IP lower support beam.

ooo. Pull the bracket away from the IP to access the radio control and HVAC connectors.

ppp. Disconnect connectors from the radio.

qqq. Disconnect the electrical connectors and vacuum harness connectors from the HVAC control.

rrr. Disconnect the IP wiring harness and parking brake connector clip from the bracket.

sss. Remove the bolts attaching the ignition switch housing bracket to the center support bracket.

ttt. Note the routing of the IP wiring harness and remove the harness.

uuu. Remove the center bracket.

vvv. Remove the left hand floor outlet duct.

www. Remove the right hand side window defogger outlet duct.

xxx. Remove the SIR bracket and the passenger knee bolster bracket.

yyy. Disconnect the sunload electrical connector.

zzz. Disconnect the right hand temperature actuator electrical connector.

aaaa. Remove the right hand floor air duct.

bbbb. Move the carpet away from the right hand side of the driveline tunnel to remove the carpet air duct (rear).

cccc. Remove the right hand floor air duct (rear).

dddd. Disconnect the blower motor connector.

eeee. Remove the blower motor.

ffff. Disconnect the vacuum supply from the HVAC module.

gggg. Disconnect the vacuum solenoid connector.

hhhh. Remove the windshield defroster duct.

iiii. Remove the HVAC module and discard seals.

jjjj. Remove the heater core outlet cover screws and the cover.

kkkk. Remove the heater core cover screws and the cover. Discard the heater core cover seals.

llll. Remove and discard the heater core outer seal.

mmmm. Remove the core clamp screw and clamp.

nnnn. Remove the core pipe retaining clamp screw and remove the core from the case.

oooo. Remove and discard the core lower, center, upper and side seals.

pppp. Remove the core pipe clamp using a suitable tool.

To install:

a. Install the core pipe clamp.

b. Install new core lower, center, upper and side seals.

c. Install the core in the case.

d. Install the core pipe retaining clamp and screw.

e. Install the core clamp and screw.

f. Install a new heater core outer seal.

g. Install new cavity seals.

h. Install the heater core cover and screws.

i. Install the heater core outlet cover and screws.

j. Install the HVAC module using new seals. Make sure the dashmat is aligned. The opening in the dashmat for the HVAC drain should be aligned so the drain opening in the cowl is approximately centered in the dashmat opening allowing room from the module drain seal to fully align against the cowl. Make sure the heater core joint fitting, condenser fitting module drain and module studs correspond with the openings in the cowl. Insert the left hand stud into the cowl first, then the module drain, then the right hand stud. Install the right hand then the left hand bolts and tighten to 89 inch lbs. (10 Nm).

k. Install the windshield defroster duct.

l. Connect the vacuum solenoid connector.

m. Connect the vacuum supply to the HVAC module.

n. Install the blower motor.

o. Connect the blower motor connector.

p. Install the right hand floor air duct (rear).

q. Install the carpet air duct (rear) and replace carpet.

r. Install the right hand floor air duct.

s. Connect the right hand temperature actuator electrical connector.

t. Connect the sunload electrical connector.

u. Install the SIR bracket and the passenger knee bolster bracket and tighten the bolts to 106 ft. lbs. (12 Nm).

v. Install the right hand side window defogger outlet duct.

w. Install the left hand floor outlet duct.

x. Position the ignition switch housing bracket to the center bracket aligning the marks made during removal, tighten the bolts to 106 inch lbs. (12 Nm).

y. Install the center bracket.

z. Rout the IP wiring harness. Connect the IP wiring harness and parking brake connector clip to the bracket.

aa. Connect the electrical connectors and vacuum harness connectors from the HVAC control.

bb. Connect connectors to the radio.

cc. Install the center support bracket aligning the marks made during removal and tighten the left bolt that attaches the bracket to the IP lower support beam to 106 inch lbs. (12 Nm).

dd. Install the ignition switch housing aligning the marks made during removal and tighten the bolt to 17 inch lbs. (2 Nm).

ee. Install the left bolt that attaches the bracket to the driveline tunnel and tighten to 106 ft. lbs. (12 Nm).

ff. Install the bolt that attaches the bracket to the IP lower support beam and tighten to 106 ft. lbs. (12 Nm).

gg. Install the bolts that attach the bracket to the passenger knee bolster bracket and tighten to 106 ft. lbs. (12 Nm).

hh. Install the right bolt that attaches the bracket to the driveline tunnel and tighten to 106 ft. lbs. (12 Nm).

ii. Install the ignition switch.

jj. Connect the SDM.

kk. Install the CPA electrical.

ll. Route the park/lock cable through the IP center support bracket. Make sure to route the cable under the wiring harness and over the floor rear outlet.

mm. Engage the park/lock cable to the ignition switch.

nn. Turn the ignition switch OFF.

oo. Engage the park/lock cable to the shift control pivot arm stud once aligned pull back to lock.

pp. Insert the cable into the slot on the shift control making sure it locks in place.

qq. Turn the ignition ON.

rr. Place the shift lever in each position but park and try to turn the key to the OFF position.

ss. If the key turns to the OFF position except in park the cable is working properly.

tt. Place the shift lever in PARK. Turn the key OFF and remove the key.

uu. Install the trim panel and attach the hazard warning switch connector.

vv. Install the upper trim panel.

ww. Install the screws attaching the upper trim pad to the defroster duct.

xx. Install the screws attaching the upper trim pad to the left and right hand hinge pillars.

yy. Install the screws attaching the upper trim panel to the driver knee bolster outer bracket and center support bracket.

zz. Install the screw attaching the upper trim panel to the passenger SIR bracket.

aa. Install the screws attaching the IP cluster bezel to the upper trim pad.

bb. Install the windshield side garnish moldings.

cc. Install the DRL electrical connector and the sun load temperature sensor to the windshield defroster duct, if equipped

dd. Install the inboard side window glass defroster duct to the windshield defroster duct on both sides

ee. Connect the passenger side air bag connector and install the glove box.

ff. Install the upper and side screws and tighten to 7 inch lbs. (2 Nm).

gg. Install the lower retaining bolts and tighten to 106 inch lbs. (12 Nm). Install the trim plugs.

hh. Connect the light switch and close the door.

ii. Connect the inside air temperature switch connector and install the drivers knee bolster trim in position. Insert the connector for the fog lamp, rear compartment lid release switch through the opening in the panel.

jj. Install the trim panel and tighten the screws to 16 inch lbs. (1.8 Nm).

kk. Install the fog lamp, rear compartment lid release switch.

ll. Lower the trim plate over the shifter and under the parking brake lever.

mm. Place the shifter boot in position and insert a tool up through the shifter opening in the trim plate.

nn. Connect the cigar lighter electrical connector.

oo. Install the trim plate into position, install the upper locator tabs and upper locking tabs, then work downwards to engage the remaining tabs.

pp. Install the trim plate screws and tighten to 17 inch lbs. (2 Nm).

qq. Install trim plate grille.

rr. Install the ashtray.

ss. Install the shift boot in position and press to engage the tabs.

tt. Place the transmission in Park on models with an Automatic Transmission (AT) or reverse on models with a Manual Transmission (MT).

uu. Release the parking brake.

vv. Install the console.

ww. Install the fuel door release electrical switch and if equipped, the rear lift window switch and attach the connections.

xx. Install the accessory power plug and attach connector.

yy. Slide the console under the accessory trim plate.

zz. Install the console retaining nuts.

aa. Install the console retaining nut covers.

bb. Install the traction switch.

cc. Connect the LED connector to the harness connector.

dd. Connect the traction switch connector.

ee. Engage traction switch to the retaining clips.

ff. Close the console door.

gg. Install the stowage compartment and tighten the screws to 35 inch lbs. (4 Nm).

7. Install or connect the following:
- Battery heat shield
- Intake manifold
- Hose to the right hand air injection check valve
- Bolt attaching the right hand air injection check valve to the cylinder head(s), if equipped
- Heater pipe bracket to the cowl and tighten the bolt
- Heater pipe-to-heater core bolt
- Heater pipe assembly to the core.
- Accumulator hose-to-evaporator bolt
- Accumlator hose and evaporator tube to the evaporator using new O–ring seals.
- HVAC module drain tube to the module

8. Refill the cooling system.

9. Evacuate, charge and leak test the air conditioning system refrigerant.

10. Connect the negative battery cable.

11. If equipped, enable the SIR system by performing the following procedure:

a. Connect the Connector Positive Assurance (CPA) to the yellow 2-way SIR harness connector at the base of the steering column.

b. Connect the steering wheel module coil connector located at the base of the steering column.

c. Connect the CPA from the inflatable restraint instrument panel module connector located at the near the base of the steering column.

d. Connect the instrument panel module connector located near the base of the steering column.

e. Install the left side sound insulator.

f. Install the SIR fuse to the fuse panel.

12. Operate the engine to normal operating temperatures; then, check the climate control operation and check for leaks.

Cylinder Head

REMOVAL & INSTALLATION

1. Before servicing the vehicle, refer to the precautions in the beginning of this section.

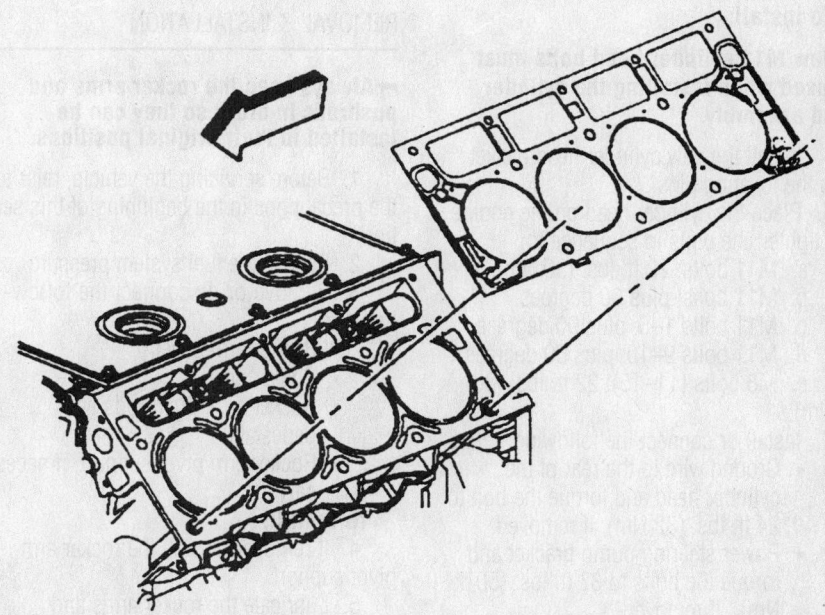

79222611

Be sure the tab on the edge of the gasket is closer to the front of the engine when installed

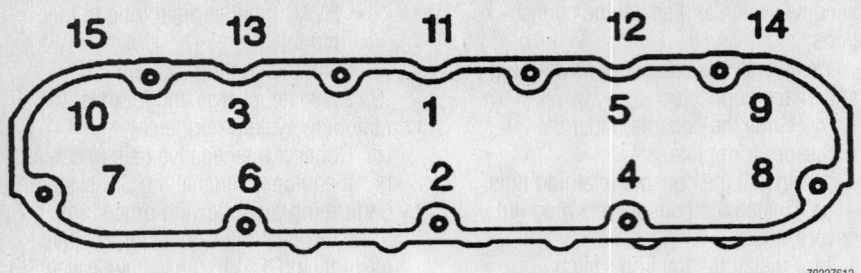

Cylinder head bolt torque sequence—5.7L engine

2. Drain the engine coolant.
3. Relieve the fuel system pressure.
4. Remove or disconnect the following:
- Negative battery cable
- Rocker arm cover
- Alternator bracket, if removing the left side
- Exhaust manifold from the cylinder head
- Oil level indicator tube, if removing the right side
- Wiring harness from the clip at the rear of the cylinder head, if removing the right side
- Intake manifold
- Vapor vent pipe
- Power steering pump pulley, if removing the left side
- Power steering pump and reposition it, if removing the left side
- Power steering pump bracket, if removing the left side
- Ground wire bolt from the rear of the cylinder head, if removing the left side
- Cylinder head

To install:

➡️**New M11 cylinder head bolts must be used when installing the cylinder head assembly.**

5. Install the new cylinder head gasket onto the locating pins.
6. Place the cylinder head on the engine and tighten the bolts in sequence to:
   a. M11 bolts: 22 ft. lbs. (30 Nm).
   b. M11 bolts: plus 90 degrees.
   c. M11 bolts 1–8: plus 90 degrees.
   d. M11 bolts 9–10: plus 50 degrees.
   e. M8 bolts (11–15): 22 ft. lbs. (30 Nm).
7. Install or connect the following:
- Ground wire to the rear of the cylinder head and torque the bolt to 24 ft. lbs. (32 Nm), if removed
- Power steering pump bracket and torque the bolts to 37 ft. lbs. (50 Nm), if removed
- Power steering pump and torque

the bolts to 18 ft. lbs. (25 Nm), if removed
- Power steering pump pulley, if removed
- Exhaust manifold and torque the bolts, in sequence, to 11 ft. lbs. (15 Nm), then to 18 ft. lbs. (25 Nm)
- Oil level indicator tube, if removed
- Wiring harness from the clip to the rear of the cylinder head, if removed
- Valve rocker arm cover and torque the bolts to 106 inch lbs. (12 Nm)
- Intake manifold and torque the bolts in sequence to 44 inch lbs. (5 Nm) then to 89 inch lbs. (10 Nm)
- Vapor vent pipe and torque the bolt to 106 inch lbs. (12 Nm)
- Alternator bracket, if removed
- Negative battery cable
8. Fill the cooling system.
9. Start the vehicle and check for leaks, repair if necessary.

Rocker Arms

REMOVAL & INSTALLATION

➡️**Always keep the rocker arms and pushrods in order so they can be installed in their original positions.**

1. Before servicing the vehicle, refer to the precautions in the beginning of this section.
2. Relieve the fuel system pressure.
3. Remove or disconnect the following:
- Rocker arm cover
- Rocker arm
- Rocker arm, push rods and pedestals
- Rocker arm pivot support, if necessary

To install:

4. If removed, install the rocker arm pivot support.
5. Lubricate the rocker arms and pushrods with clean engine oil.

6. Lubricate the flange and washer surface of the rocker arm mounting bolts.
7. Install the rocker arms but do not tighten the bolts at this time.
8. Rotate the crankshaft so the No. 1 piston is at Top Dead Center (TDC).
9. With the engine in this position, tighten the exhaust valve rocker arm bolts on cylinders No. 1, 2, 7 and 8 to 22 ft. lbs. (30 Nm), then, tighten the intake valve rocker arm bolts on cylinders No. 1, 3, 4 and 5 to 22 ft. lbs. (30 Nm).
10. Rotate the crankshaft 1 revolution (360 degrees).
11. With the engine in this position, tighten the exhaust valve rocker arm bolts on cylinders No. 3, 4, 5 and 6 to 22 ft. lbs. (30 Nm), then, tighten the intake valve rocker arm bolts on cylinders No. 2, 6, 7 and 8 to 22 ft. lbs. (30 Nm).
12. Install the rocker arm cover using a new gasket and torque the bolts to 106 inch lbs. (12 Nm).
13. Negative battery cable.
14. Start the vehicle and check for leaks, repair if necessary.

Intake Manifold

REMOVAL & INSTALLATION

2000 Models

➡️**The intake manifold, throttle body, fuel injectors and rail may be removed from the engine as an assembly.**

1. Before servicing the vehicle, refer to the precautions in the beginning of this section.
2. Drain the cooling system.
3. Relieve the fuel system pressure.
4. Remove or disconnect the following:
- Intake Air Temperature (IAT) sensor electrical connector
- Mass Air Flow (MAF) sensor electrical connector
- Air intake duct
- Fuel line from the fuel rail
- Vacuum ventilation hose
- Fuel injector electrical connectors
- Throttle Position Sensor (TPS) electrical connector
- Knock Sensor (KS) connector
- KS jumper pigtail from the fuel rail stop bracket
- Positive Crankcase Ventilation (PCV) valve pipes from the rocker arm covers
- PCV tube from the throttle body
- Engine cooling air bleed hose from the throttle body
- Throttle body heater outlet hose

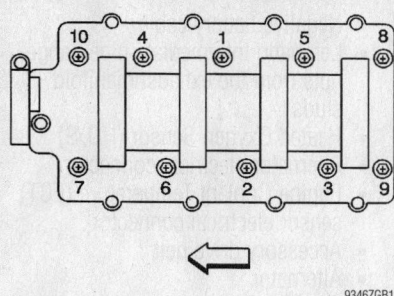

Intake manifold torque sequence—5.7L engine

9346ZGB1

- Intake manifold and fuel stop bracket

To install:

5. Apply a 0.20 in. (5mm) bead of threadlock to the threads of the intake manifold bolts.

❋❋ CAUTION

The fuel rail stop bracket must be installed on the intake manifold in its original position. Failure to reinstall this part may result in fuel spray causing personal injury.

6. Install or connect the following:
- Intake manifold with a new gasket
- Fuel stop bracket

- Intake manifold bolts and torque the bolts first to 44 inch lbs. (5 Nm) in the sequence shown, then make a second pass and tighten the bolts to 89 inch lbs. (10 Nm)
- Throttle body heater outlet hose
- Engine coolant air bleed hose
- PCV tube
- PCV valve pipes
- KS sensor jumper pigtail
- KS electrical connector
- TPS electrical connector
- Fuel injector electrical connectors
- Vacuum vent hose
- Fuel line
- Air intake duct
- IAT sensor electrical connector
- MAF sensor electrical connector
- Negative battery cable

7. Fill the cooling system.
8. Start the vehicle and check for leaks, repair if necessary.

2001–04 Models

➡ **The intake manifold, throttle body, fuel injectors and rail may be removed from the engine as an assembly.**

1. Before servicing the vehicle, refer to the precautions in the beginning of this section.
2. Drain the cooling system.

3. Relieve the fuel system pressure.
4. Remove or disconnect the following:
- Fuel rail covers
- Fuel feed hose
- Evaporative Emission (EVAP) canister purge tube from the intake manifold
- EVAP canister purge tube from the EVAP canister purge solenoid valve
- EVAP canister purge tube
- EVAP canister purge tube from the EVAP canister purge solenoid valve
- Disconnect the EVAP canister purge tube from the fuel feed pipe
- EVAP canister purge tube
- Throttle Position (TP) sensor connector
- Coolant air bleed hose
- Throttle body heater outlet hose
- Fuel injector electrical connectors
- EVAP canister purge solenoid valve connector
- Electronic Throttle Control (ETC) connector
- EVAP canister purge solenoid valve from the bracket
- Harness clips at the fuel rails
- Power brake booster vacuum hose at the booster
- Knock Sensor (KS) wire harness clip from the fuel rail stop bracket
- TP sensor harness clip from the Positive Crankcase Ventilation (PCV) tube
- PCV tube from the right rocker arm cover and throttle body
- PCV valve pipe from the left rocker arm cover, if equipped with the VIN G engine
- PCV valve pipe strap nut, if equipped with the VIN G engine
- PCV valve pipe from the right rocker arm cover and intake manifold
- PCV valve hose from the valley cover and intake manifold, if equipped with the VIN S engine
- Intake manifold bolts and fuel rail stop bracket

5. Position the intake manifold forward.
- Manifold Absolute Pressure (MAP) sensor vacuum hose
- MAP sensor connector
- Intake manifold
- Intake manifold gaskets and discard

To install:

6. Install or connect the following:
- New intake manifold gaskets
- Intake manifold
7. Position the intake manifold forward.
- MAP sensor vacuum hose
- MAP sensor connector

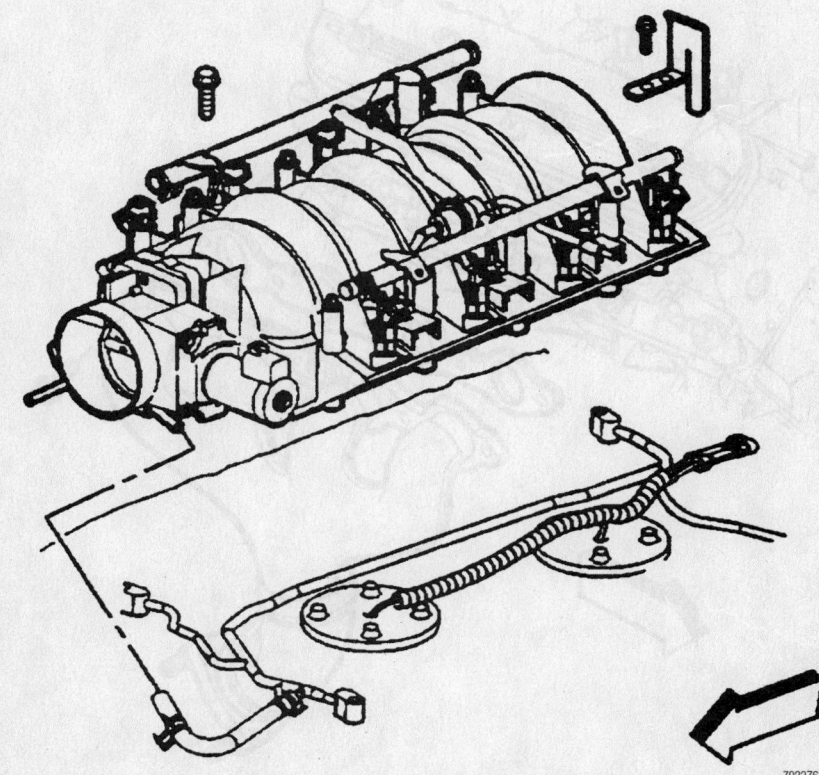

7922Z613

Be sure to reinstall the fuel stop bracket when installing the intake manifold

8. Position the intake manifold into place.

9. Apply threadlock GM P/N 12345382 to the threads of the intake manifold bolts.

✷✷ CAUTION

The fuel rail stop bracket must be installed onto the engine assembly. The stop bracket serves as a protective shield for the fuel rail in the event of a vehicle frontal crash. If the fuel rail stop bracket is not installed and the vehicle is involved in a frontal crash, fuel could be sprayed possibly causing a fire and personal injury from burns.

10. Install or connect the following:
- Fuel rail stop bracket
- Intake manifold bolts

11. Tighten the intake manifold bolts as follows:

a. Step 1: in sequence to 44 inch lbs. (5 Nm).

b. Step 2: in sequence to 89 inch lbs. (10 Nm).

12. Install or connect the following:
- PCV valve hose to the valley cover and intake manifold, if equipped
- PCV valve pipe to the right valve rocker arm cover and intake manifold, if equipped
- PCV pipe strap nut and tighten to 106 inch lbs. (12 Nm)
- PCV valve pipe to the left rocker arm cover, if equipped
- PCV tube to the right rocker arm cover
- TP sensor harness clip to the PCV tube
- KS wire harness to the fuel rail stop bracket
- Power brake booster vacuum hose to the booster
- Intake manifold branches of the wiring harness
- Harness clips at the fuel rails
- EVAP canister purge solenoid valve to the bracket
- ETC connector
- EVAP canister purge solenoid valve connector
- Fuel injector connectors
- Throttle body heater outlet hose
- Throttle body heater outlet hose to the throttle body
- Throttle body heater outlet hose clamp at the throttle body
- Coolant air bleed hose
- TP sensor connector
- EVAP canister purge tube

- EVAP canister purge tube to the fuel feed pipe
- EVAP canister purge tube to the EVAP canister purge solenoid valve
- EVAP canister purge tube
- EVAP canister purge tube to the EVAP canister purge solenoid valve
- EVAP canister purge tube to the intake manifold
- Fuel feed hose
- Fuel rail covers
- Negative battery cable

13. Fill the cooling system.

14. Start the vehicle and check for leaks, repair if necessary.

Exhaust Manifold

REMOVAL & INSTALLATION

Left Side

2000 MODELS

1. Before servicing the vehicle, refer to the precautions in the beginning of this section.

2. Relieve the fuel system pressure.

3. Remove or disconnect the following:

- Negative battery cable
- Left hand intermediate pipe flange nuts from the exhaust manifold studs
- Heated Oxygen Sensor (HO2S)
- Alternator electrical connector
- Engine Coolant Temperature (ECT) sensor electrical connector
- Accessory drive belt
- Alternator
- Secondary Air Injection (AIR) hose from the check valve
- AIR pipe from the exhaust manifold and reposition it
- AIR pipe retainer from the rear of the cylinder head and reposition it
- Spark plug wires
- Spark plugs
- Exhaust manifold

To install:

4. Apply threadlock to the threads of the bolts.

5. Install the exhaust manifold using a new gasket.

6. Beginning with the center 2 bolts, torque the bolts to 11 ft. lbs. (15 Nm) alternating from side-to-side until all the bolts are tight. Torque the bolts again in the same sequence to 18 ft. lbs. (25 Nm).

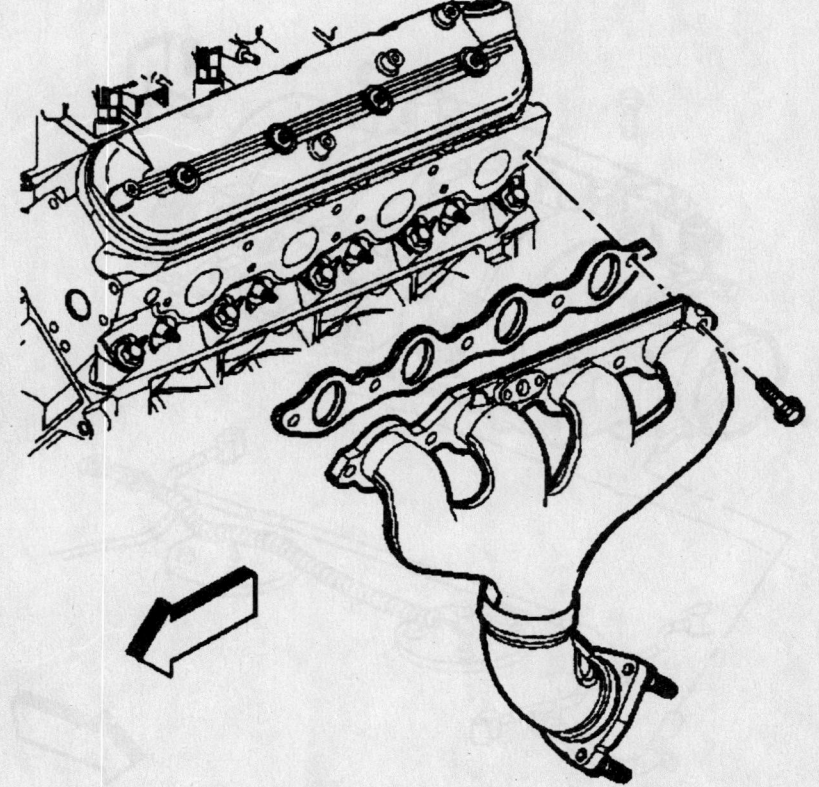

Exploded view of the exhaust manifold—5.7L engine

9306ZG25

7. Bend over the exposed portion of the gasket at the rear of the cylinder head.

8. Install or connect the following:
- Spark plugs and torque them to 11 ft. lbs. (15 Nm)
- Spark plug wires
- AIR pipe retainer bolt to the rear of the cylinder head and torque the bolt to 15 ft. lbs. (20 Nm)
- AIR pipe with a new gasket and torque the bolts to 15 ft. lbs. (20 Nm)
- AIR hose to the check valve and torque it to 15 ft. lbs. (20 Nm)
- Alternator and torque the bolts to 37 ft. lbs. (50 Nm)
- Accessory drive belt
- Alternator electrical connector
- ECT sensor electrical connectors
- HO2S and torque it to 30 ft. lbs. (42 Nm)
- Fuel lines
- Intermediate pipe flange-to-exhaust manifold nuts and torque them to 15 ft. lbs. (20 Nm)
- Negative battery cable

9. Start the vehicle and check for leaks, repair if necessary.

2001–04 MODELS

1. Before servicing the vehicle, refer to the precautions in the beginning of this section.

2. Relieve the fuel system pressure.

3. Remove or disconnect the following:
- Negative battery cable
- Fuel rail cover
- Alternator
- Exhaust manifold nuts
- Connector Position Assurance (CPA) lock
- Oxygen Sensor (O2S)
- O2S connector clip at the body
- Secondary Air Injection (AIR) hose clamp
- AIR hose from the pipe
- Hose clamps at the right and left check valves
- AIR injection pipe hose from the right check valve
- AIR injection pipe hose from the left check valve
- AIR pipe
- AIR pipe and gasket, and discard the old gasket
- Brake booster vacuum hose from the booster
- Ignition coil bracket
- Spark plugs
- Exhaust manifold

To install:

4. Apply threadlock to the threads of the bolts.

5. Install the exhaust manifold using a new gasket.

6. Beginning with the center 2 bolts, torque the bolts to 11 ft. lbs. (15 Nm) alternating from side-to-side until all the bolts are tight. Torque the bolts again in the same sequence to 18 ft. lbs. (25 Nm).

7. Bend over the exposed portion of the gasket at the rear of the cylinder head.

8. Install or connect the following:
- Spark plugs and torque them to 11 ft. lbs. (15 Nm)
- Ignition coil bracket and tighten to 106 inch lbs. (12 Nm)
- Brake booster vacuum hose
- AIR pipe and gasket
- AIR pipe bolts and tighten to 15 ft. lbs. (20 Nm)
- AIR injection pipe hose to the left check valve
- AIR injection pipe hose to the right check valve
- Hose clamps at the right and left check valves
- AIR hose to the pipe
- AIR hose clamp
- O2S and tighten to 30 ft. lbs. (40 Nm)
- O2S connector clip at the body
- CPA lock
- Exhaust manifold nuts and tighten to 15 ft. lbs. (20 Nm)
- Alternator
- Fuel rail cover
- Negative battery cable

Right Side

2000 MODELS

1. Before servicing the vehicle, refer to the precautions in the beginning of this section.

2. Remove or disconnect the following:
- Negative battery cable
- Right side intermediate exhaust pipe nuts from the manifold studs
- Heated Oxygen Sensor (HO2S)
- Secondary Air Injection (AIR) pipe from the exhaust manifold
- Spark plug wires
- Spark plugs
- Oil level indicator and tube
- Exhaust manifold

To install:

3. Apply threadlock to the threads of the bolts.

4. Install the exhaust manifold using a new gasket.

5. Beginning with the center 2 bolts, tighten the bolts to 11 ft. lbs. (15 Nm) alternating from side-to-side until all the bolts are tight. Then, tighten the bolts

again in the same sequence to 18 ft. lbs. (25 Nm).

6. Bend over the exposed portion of the gasket at the rear of the cylinder head.

7. Install or connect the following:
- Oil level indicator tube and torque the bolt to 12 ft. lbs. (16 Nm)
- Spark plugs and torque them to 11 ft. lbs. (15 Nm)
- Spark plug wires
- AIR pipe using a new gasket and torque the bolts to 15 ft. lbs. (20 Nm)
- HO2S in the manifold and torque it to 30 ft. lbs. (42 Nm)
- Intermediate pipe flange-to-exhaust manifold nuts and torque them to 15 ft. lbs. (20 Nm)
- Negative battery cable

8. Start the vehicle and check for leaks, repair if necessary.

2001–04 MODELS

1. Before servicing the vehicle, refer to the precautions in the beginning of this section.

2. Relieve the fuel system pressure.

3. Remove or disconnect the following:
- Negative battery cable
- Fuel rail cover
- Exhaust manifold nuts
- Connector Position Assurance (CPA) lock
- Oxygen (O2S) sensor
- Secondary Air Injection (AIR) hose from the right check valve

4. Loosen, DO NOT remove the AIR pipe bolt at the rear of the left cylinder head.
- AIR pipe and gasket, and discard the old gasket
- Oil level indicator tube
- Ignition coil bracket
- Spark plugs
- Exhaust manifold

To install:

5. Apply threadlock to the threads of the bolts.

6. Install the exhaust manifold using a new gasket.

7. Beginning with the center 2 bolts, torque the bolts to 11 ft. lbs. (15 Nm) alternating from side-to-side until all the bolts are tight. Torque the bolts again in the same sequence to 18 ft. lbs. (25 Nm).

8. Bend over the exposed portion of the gasket at the rear of the cylinder head.

9. Install or connect the following:
- Spark plugs and torque them to 11 ft. lbs. (15 Nm)
- Ignition coil bracket and tighten to 106 inch lbs. (12 Nm)
- Oil level indicator tube

- AIR pipe and gasket
- AIR pipe bolts and tighten to 15 ft. lbs. (20 Nm)
- AIR injection pipe to the right check valve
- Hose clamps at the right check valve
- O$_2$S and tighten to 30 ft. lbs. (40 Nm)
- CPA lock
- Exhaust manifold nuts and tighten to 15 ft. lbs. (20 Nm)
- Alternator
- Fuel rail cover
- Negative battery cable

Camshaft

REMOVAL & INSTALLATION

2000 Models

1. Before servicing the vehicle, refer to the precautions in the beginning of this section.
2. Drain the cooling system.
3. Relieve the fuel system pressure.
4. Remove or disconnect the following:
   - Fuel lines from the fuel rail
   - Rocker arm covers
   - Rocker arms and pushrods
   - Both cylinder heads
   - Valve lifter guides and the lifters
   - Radiator
   - Engine front cover
   - Oil pan
   - Oil pump
   - Camshaft sprocket and timing chain
   - Camshaft Position (CMP) sensor bolt and sensor
   - Camshaft retainer plate

> ❊❊ **WARNING**
>
> **All camshaft bearings are the same diameter, extreme care must be taken during camshaft removal or installation to avoid damaging them.**

5. Install 3 M8 x 100mm long bolts in the front of the camshaft to use as a handle.
6. Carefully remove the camshaft from the engine, be sure not to damage the camshaft bearings.

To install:

7. Lubricate the camshaft with clean engine oil.
8. Install or connect the following:
   - Three M8–1.25 x 100mm long bolts in the front of the camshaft to use as a handle

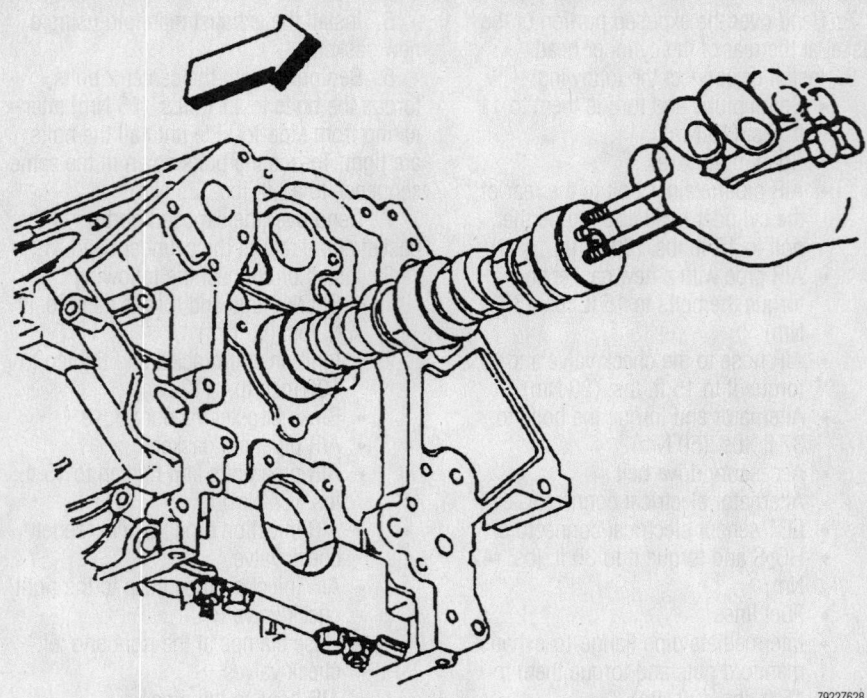

Install three M8–1.25 x 100mm long bolts into the camshaft to use as a handle—2000 models

79222Z628

- Camshaft into the engine and remove the 3 bolts
- Camshaft retainer using a new gasket and torque the bolts to 18 ft. lbs. (25 Nm)
- CMP sensor with a new O-ring and torque the bolt to 18 ft. lbs. (25 Nm)
- Timing chain and sprocket and torque the sprocket bolts to 26 ft. lbs. (35 Nm)
- Oil pump and torque the bolts to 18 ft. lbs. (25 Nm)
- Oil pan with a new gasket and torque the bolts pan-to-rear cover bolts to 106 inch lbs. (12 Nm) and the remaining bolts to 18 ft. lbs. (25 Nm)
- Front cover with a new gasket and torque the bolts to 18 ft. lbs. (25 Nm)
- Radiator
- Valve lifters and lifter guides and torque the bolts to 106 inch lbs. (12 Nm)
- Cylinder heads
- Pushrods and rocker arms
- Rocker arm covers with new gaskets and torque the bolts to 106 inch lbs. (12 Nm)
- Fuel lines to the fuel rail
- Negative battery cable

9. Refill the cooling system.
10. Start the engine and check for leaks, repair if necessary.

2001–04 Models

1. Before servicing the vehicle, refer to the precautions in the beginning of this section.
2. Drain the cooling system.
3. Relieve the fuel system pressure.
4. Remove or disconnect the following:

- Engine
- Crankshaft balancer
- Oil level indicator tube
- Left and right exhaust manifolds
- Water pump
- Intake manifold
- Coolant air bleed pipe
- Left and right valve rocker arm covers
- Valve rocker arms and push rods
- Left and right cylinder heads
- Valve lifters
- Oil pan-to-front cover bolts
- Front cover and gasket. Discard the old gasket.

5. Rotate the engine to Top Dead Center (TDC).

- Camshaft sprocket bolts
- Timing chain from the camshaft sprocket, and allow the timing chain to rest on the crankshaft sprocket
- Camshaft Position (CMP) sensor bolt and sensor
- Camshaft retainer plate

⁂⁂ WARNING

All camshaft bearings are the same diameter, extreme care must be taken during camshaft removal or installation to avoid damaging them.

6. Install 3 M8 x 100mm long bolts in the front of the camshaft to use as a handle.

7. Carefully remove the camshaft from the engine, be sure not to damage the camshaft bearings.

8. Clean and inspect the camshaft and bearings.

To install:

9. Install or connect the following:

10. Lubricate the camshaft with clean engine oil.

- Three M8–1.25 x 100mm long bolts in the front of the camshaft to use as a handle
- Camshaft into the engine and remove the 3 bolts
- Camshaft retainer using a new gasket and torque the bolts to 18 ft. lbs. (25 Nm)
- CMP sensor with a new O-ring and torque the bolt to 18 ft. lbs. (25 Nm)

11. Align the camshaft sprocket alignment mark in the 6 o'clock position.

- Camshaft sprocket and timing chain
- Camshaft sprocket bolts and tighten to 26 ft. lbs. (35 Nm)

12. Apply a 5 mm bead of sealant GM P/N 12378190, 0.8 inch (20mm) long to the oil pan-to-engine block junction.

- Front cover with a new gasket

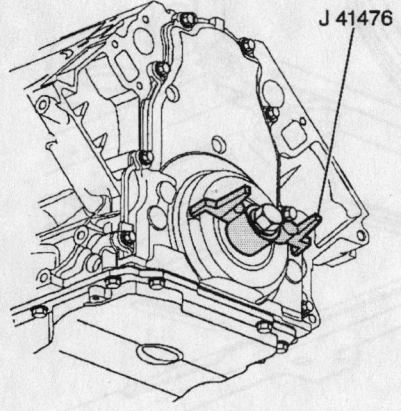

J 41476

9356ZGZZ

Install tool J 41476 and crankshaft balancer bolt to the front cover. Align the tapered legs of the tool with the machined alignment surfaces on the front cover—2001–04 models

- Front cover bolts until snug
- Oil pan-to-front cover bolts until snug

13. Install tool J 41476 and crankshaft balancer bolt to the front cover. Align the tapered legs of the tool with the machined alignment surfaces on the front cover.

14. Install the crankshaft balancer bolt until snug. Tighten the oil pan-to-front cover bolts to 18 ft. lbs. (25 Nm) and the engine front cover bolts to 18 ft. lbs. (25 Nm).

15. Remove the tool.

16. Install or connect the following:

- New crankshaft front oil seal
- Valve lifters
- Right and left cylinder heads
- Valve rocker arms and push rods
- Right and left valve rocker arm covers
- Coolant air bleed pipe
- Intake manifold
- Water pump
- Exhaust manifolds
- Oil level indicator tube
- Crankshaft balancer
- Engine assembly

Valve Lash

ADJUSTMENT

The rocker arms on the 5.7L (VIN G) engine are bolted into position so that no adjustment is possible. If the valves are noisy, suspect low oil pressure or worn valve train components.

Starter Motor

REMOVAL & INSTALLATION

1. Before servicing the vehicle, refer to the precautions in the beginning of this section.

2. Remove or disconnect the following:

- Negative battery cable
- Intermediate exhaust pipe
- Starter electrical connectors
- Starter bolts
- Starter motor

To install:

3. Install or connect the following:

- Starter motor and torque the bolts to 37 ft. lbs. (50 Nm)

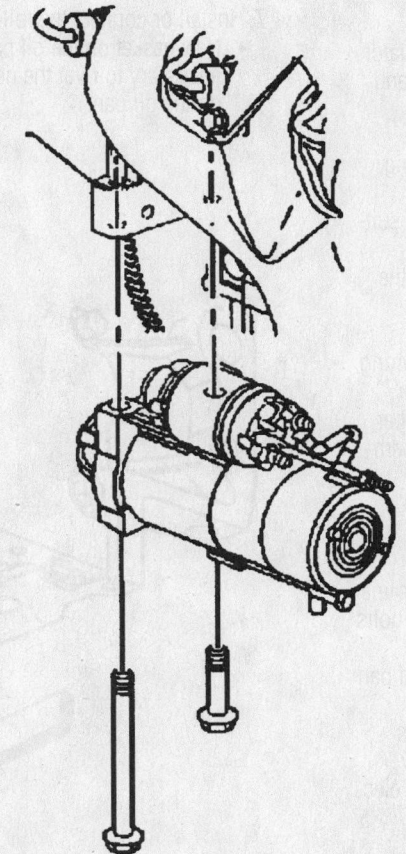

9306ZG26

Exploded view of the starter assembly—5.7L engine

- Starter electrical connectors
- Intermediate exhaust pipe and torque intermediate pipe-to-manifold nuts to 15 ft. lbs. (20 Nm) and all other bolts to 37 ft. lbs. (50 Nm)
- Negative battery cable

Oil Pan

REMOVAL & INSTALLATION

1. Before servicing the vehicle, refer to the precautions in the beginning of this section.
2. Drain the engine oil.
3. Remove or disconnect the following:
- Negative battery cable
- Alternator
- Washer reservoir
- Engine Coolant Temperature (ECT) sensor electrical connector
- Headlamp electrical connector
- Tie rods from the steering linkage
- Real Time Damping (RTD) sensor links, if equipped
- Stabilizer shaft
- Intermediate shaft from the steering gear
- Electronic Brake Control Module/Brake Pressure Modulator Valve (EBCM/BPMV) bracket and reposition
- Power steering gear
- Lower shock absorber mounting bolts
- Transverse leaf spring using a suitable compressor
- Lower control arm bolts from the suspension crossmember
- Engine mount lower nuts
- Wheel Speed Sensor (WSS) wiring harness from the crossmember
- Brake line from the crossmember
4. Support the suspension crossmember.
- Crossmember mounting nuts
- Front crossmember
- Oil filter
- Automatic transmission cooler line front and rear retaining clamp bolts (if equipped)
- Engine flywheel housing-to-oil pan bolts
- Engine flywheel close out bolts
- Engine oil level sensor
- Engine oil temperature sensor electrical connector
- Oil pan assembly
- Lower oil pan bolts and separate the pans (if necessary)

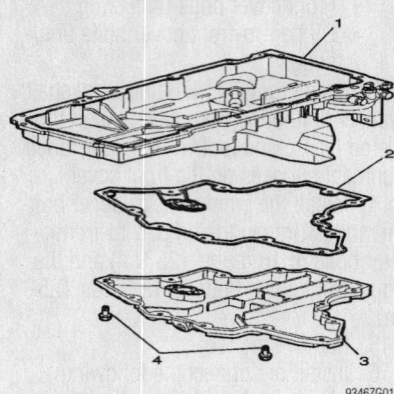

9346ZG01

Separate the upper (1) and lower (3) oil pans

➡ **To remove the oil pan gasket it is necessary to drill out 3 rivets securing it to the pan.**

To install:

5. If separated, connect the upper and lower oil pans with a new gasket and torque the bolts to 106 inch lbs. (12 Nm).
6. Apply a 0.20 in. (5mm) bead of RTV sealant directly on the tabs of the front and rear cover gaskets that extend onto the oil pan mounting surface.
7. Install or connect the following:
- New gasket on the oil pan. It is not necessary to rivet the new gasket on the oil pan.

- Oil pan and hand-tighten the bolts
- Oil pan-to-engine bolts and torque them to 18 ft. lbs. (25 Nm)
- Oil pan-to-front cover bolts and torque them to 18 ft. lbs. (25 Nm)
- Oil pan-to-rear cover bolts and torque them to 106 inch lbs. (12 Nm)
- Flywheel housing bolts and torque them to 37 ft. lbs. (50 Nm)
- Left and right close outs and bolts and torque the bolts to 106 inch lbs. (12 Nm)
- Engine oil temperature sensor electrical connector
- Engine oil level sensor and torque it to 26 ft. lbs. (35 Nm)
- Automatic transmission cooler pipe front retaining clamp bolt and torque it to 106 inch lbs. (12 Nm)
- Automatic transmission cooler pipe rear retaining clamp bolt and torque it to 22 inch lbs. (3 Nm)
- Oil filter and torque it to 22 ft. lbs. (30 Nm)
- Oil drain plug and torque it to 18 ft. lbs. (25 Nm)
8. Raise the suspension crossmember into position.
- Crossmember mounting nuts and torque them to 81 ft. lbs. (110 Nm)

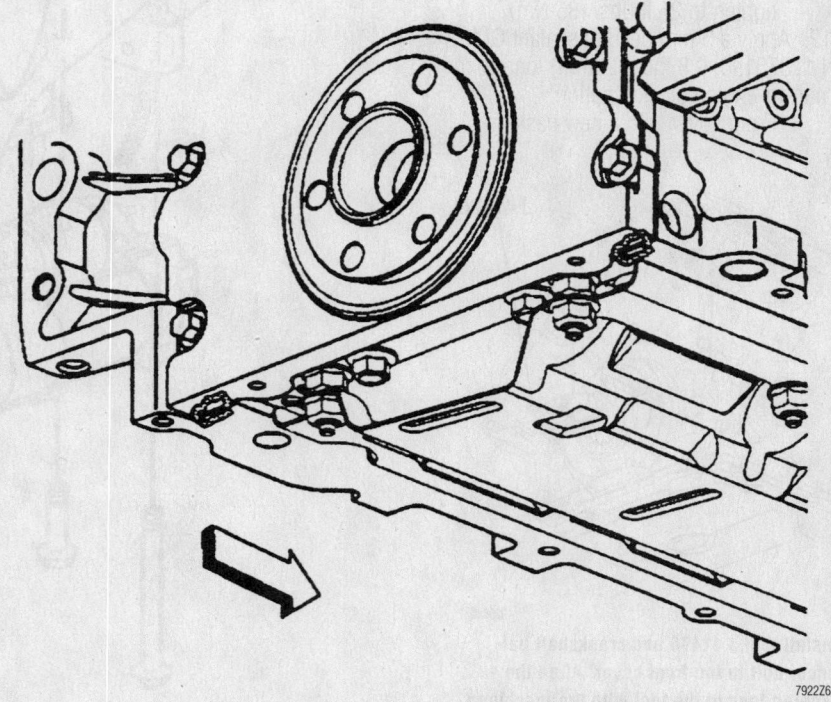

79222629

Apply sealant to the areas where the front and rear covers attach to the engine block—5.7L engine

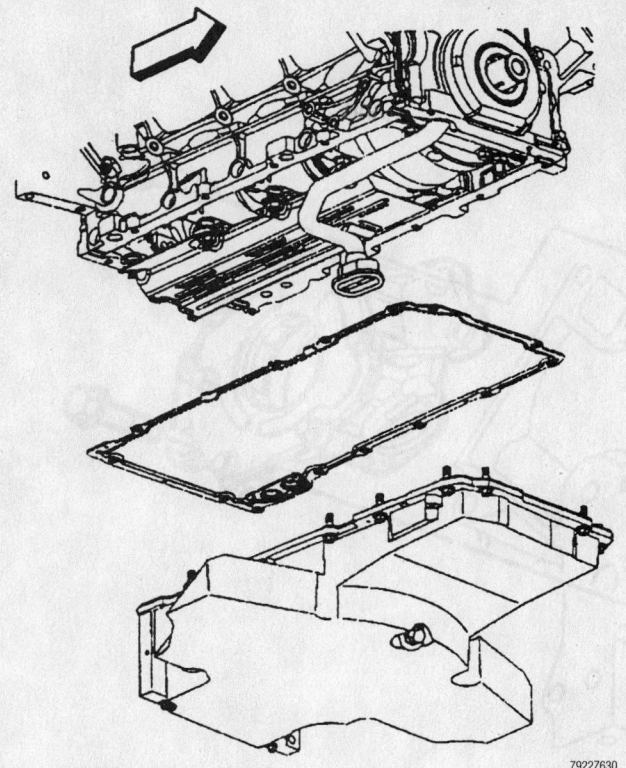

Exploded view of the oil pan mounting—5.7L engine

- Engine mount lower nuts and torque them to 48 ft. lbs. (65 Nm)
- WSS wiring harness to the cross-member
- Brake line to the crossmember
- Transverse leaf spring
- Lower control arm to the cross-member, DO NOT torque the bolts at this time
- Lower shock absorber mounting bolts and torque them to 21 ft. lbs. (28 Nm)
- Power steering gear and torque the bolts to 74 ft. lbs. (100 Nm)
- EBCM/BPMV bracket and torque the bolts to 20 ft. lbs. (27 Nm)
- Intermediate shaft and torque the pinch bolt to 25 ft. lbs. (35 Nm)
- Outer tie rods and torque the nuts to 18 ft. lbs. (25 Nm) plus an additional 180 degree turn
- RTD sensor electrical connector, if equipped
- Stabilizer shaft and torque the link nuts to 53 ft. lbs. (72 Nm) and the clamp bolts to 43 ft. lbs. (58 Nm)
- Alternator and torque the bolts to 37 ft. lbs. (50 Nm)
- Washer reservoir and torque the nuts to 66 inch lbs. (7.5 Nm)
- ECT sensor electrical connector
- Front headlamp electrical connector

- Negative battery cable
9. Fill the engine with new oil.
10. Start the vehicle and check for leaks, repair if necessary.
11. Check and adjust the front end alignment.

12. Torque the lower control arm bolts to 125 ft. lbs. (170 Nm).

Oil Pump

REMOVAL & INSTALLATION

1. Before servicing the vehicle, refer to the precautions in the beginning of this section.
2. Drain the engine oil.
3. Drain the cooling system.
4. Remove or disconnect the following:
   - Negative battery cable
   - Engine front cover
   - Oil pan
   - Oil pump screen
   - Oil pump

To install:

5. Install or connect the following:
   - Oil pump and torque the bolts to 18 ft. lbs. (25 Nm)

➡ **Be sure the crankshaft sprocket and oil pump drive gear are aligned properly.**

   - Oil pump screen with a new O-ring and torque the bolt to 106 inch lbs. (12 Nm) and the nut to 18 ft. lbs. (25 Nm)
   - Oil pan
   - Engine front cover
   - Negative battery cable
6. Fill the engine with new oil.
7. Fill the cooling system.

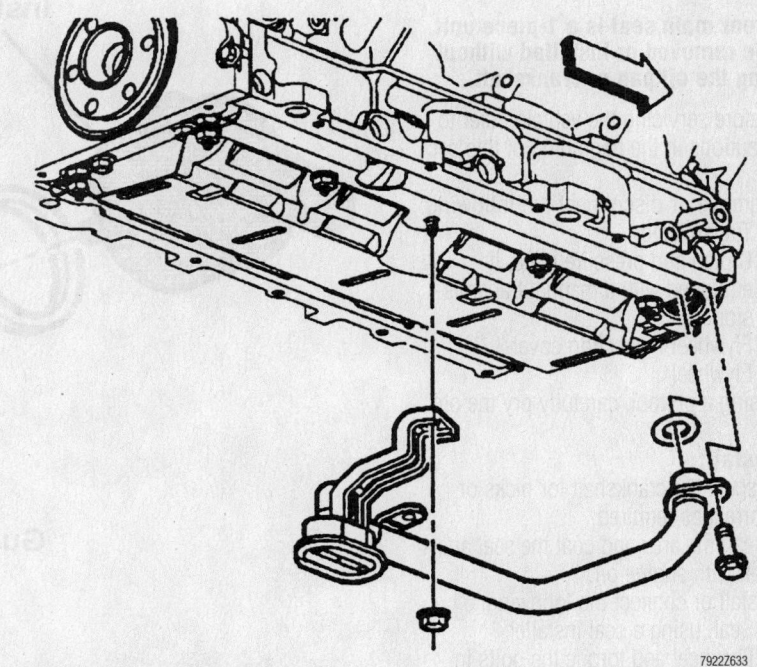

Be sure to seat the tube into the pump before installing the bolt

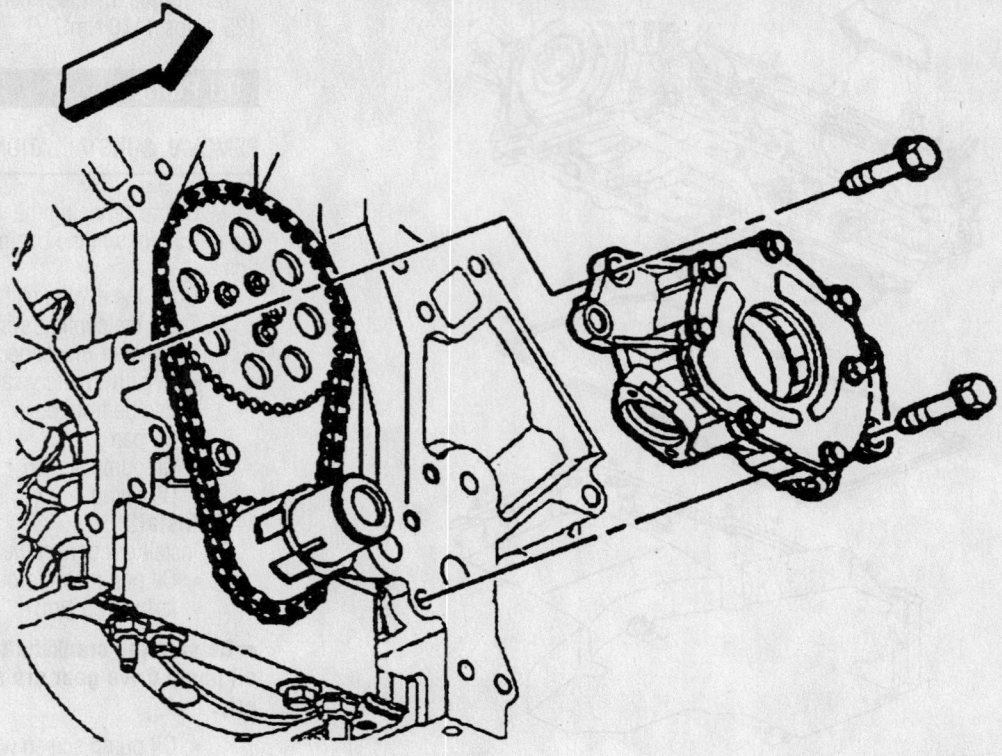

Oil pump assembly mounting—5.7L engine

8. Start the vehicle and check for leaks, repair if necessary.

Rear Main Seal

REMOVAL & INSTALLATION

➡ **The rear main seal is a 1-piece unit. It can be removed or installed without removing the oil pan or crankshaft.**

1. Before servicing the vehicle, refer to the precautions in the beginning of this section.
2. Remove or disconnect the following:
   - Transmission
   - Clutch and pressure plate, if equipped with a manual transmission
   - Flywheel inspection cover
   - Flywheel
3. Using a prytool, carefully pry the old seal out.

To install:

4. Inspect the crankshaft for nicks or burrs, correct as required.
5. Clean the area and coat the seal and crankshaft with engine oil.
6. Install or connect the following:
   - Seal, using a seal installer
   - Flywheel and torque the bolts in sequence to 15 ft. lbs. (20 Nm)

then to 37 ft. lbs. (50 Nm) and finally to 74 ft. lbs. (100 Nm)
   - Flywheel inspection cover and torque the bolts 18 ft. lbs. (25 Nm)
   - Clutch and pressure plate if

equipped with a manual transmission
   - Transmission
7. Start the vehicle and check for leaks, repair if necessary.

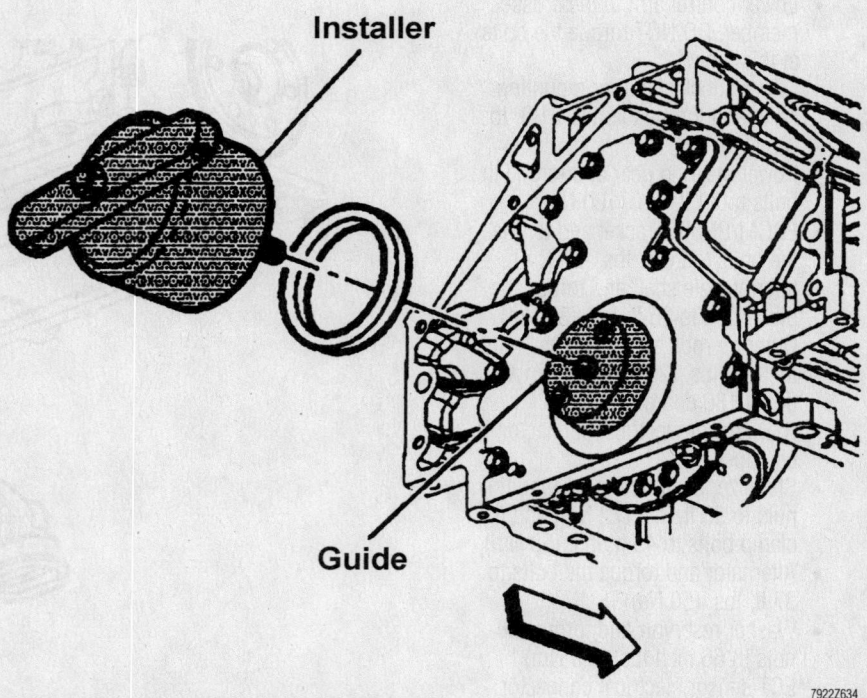

Use a seal installer to press the seal on the crankshaft

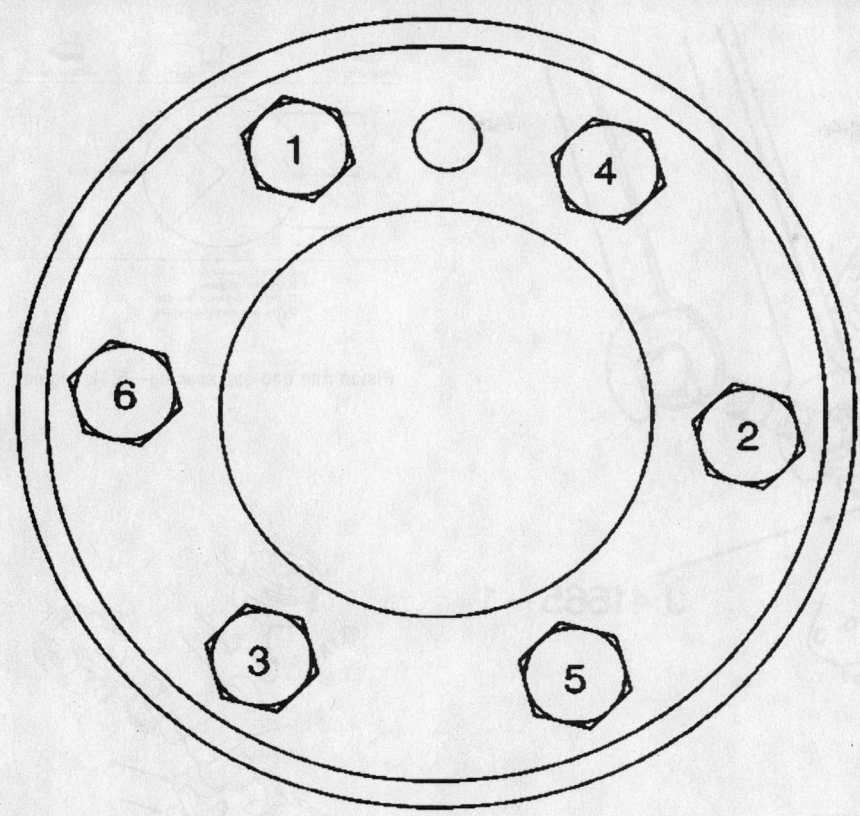

Flywheel torque sequence

9346ZG02

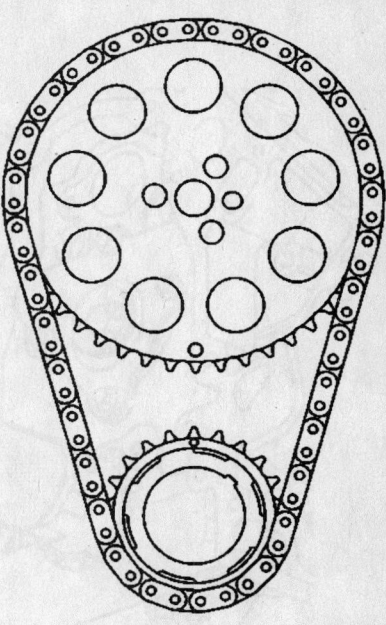

9346ZGB2

Timing chain alignment—5.7L engine

Timing Chain, Sprockets, Front Cover and Seal

REMOVAL & INSTALLATION

1. Before servicing the vehicle, refer to the precautions in the beginning of this section.
2. Drain the cooling system.
3. Drain the engine oil.
4. Remove or disconnect the following:
   - Negative battery cable
   - Engine front cover
   - Oil pan
   - Oil pump
5. Rotate the crankshaft until the timing marks are aligned. The marks should face each other. Remove the camshaft sprocket bolts.

➡**Do not turn the crankshaft after the timing chain has been removed. Damage to the pistons or valves may occur.**

6. Remove or disconnect the following:
   - Camshaft sprocket and timing chain
   - Crankshaft sprocket using a suitable puller

To install:

7. Be sure the timing mark on the crankshaft sprocket is at the 12 o'clock position. If not rotate the crankshaft to the correct position.

8. Install or connect the following:
   - Crankshaft sprocket with a Crankshaft Balancer and Sprocket Installer tool
   - Timing chain on the camshaft sprocket.
   - Timing chain and sprocket assembly so the timing marks are aligned and torque the camshaft sprocket bolts to 26 ft. lbs. (35 Nm)
   - Oil pump and torque the bolts to 18 ft. lbs. (25 Nm)

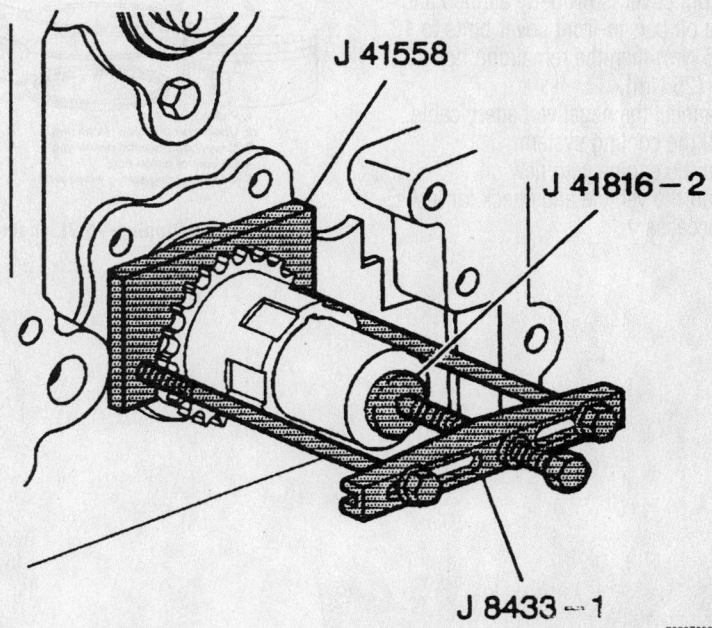

J 41558

J 41816 – 2

J 8433 – 1

7922Z638

A puller is needed to remove the crankshaft sprocket

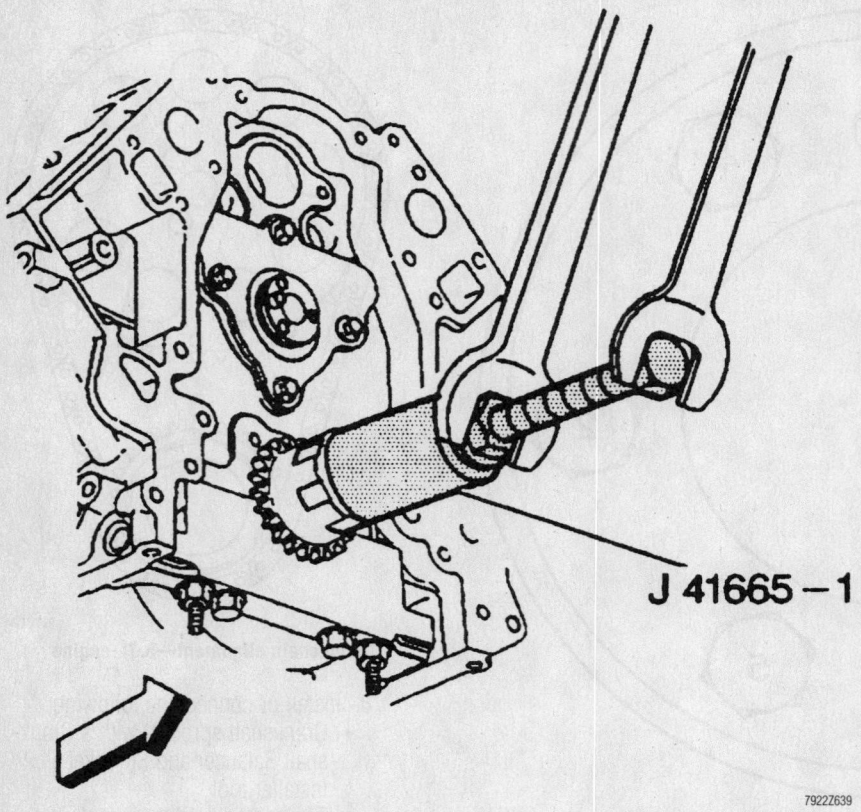

Be sure to install the crankshaft sprocket properly

- Oil pan with a new gasket
- Front cover with a new gasket and seal and hand-tighten the bolts at this time

9. Install the oil pan to cover bolts and hand-tighten them at this time. Make certain that the front cover is properly aligned and torque the oil pan-to-front cover bolts to 18 ft. lbs. (25 Nm) then the remaining bolts to 18 ft. lbs. (25 Nm).

10. Connect the negative battery cable.

11. Fill the cooling system.

12. Fill the engine with new oil.

13. Start the vehicle and check for leaks, repair if necessary.

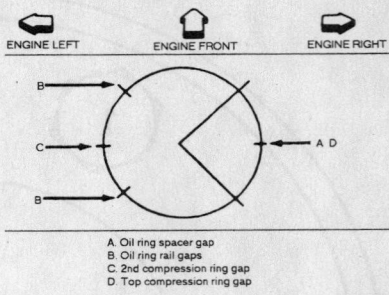

A. Oil ring spacer gap
B. Oil ring rail gaps
C. 2nd compression ring gap
D. Top compression ring gap

Piston ring end-gap spacing—5.7L engine

Piston and Ring

POSITIONING

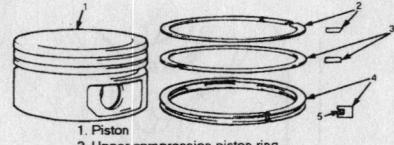

1. Piston
2. Upper compression piston ring
3. Lower compression piston ring
4. Oil control piston ring
5. Oil control ring spring w/spacer

Piston ring positioning—5.7L engine

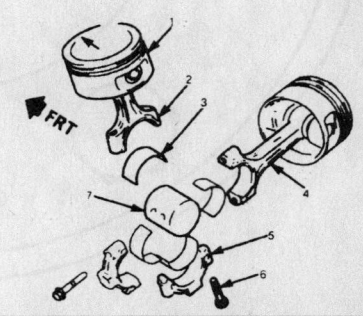

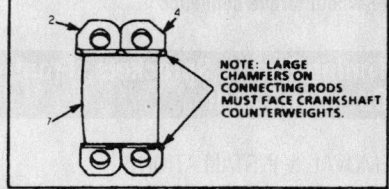

NOTE: LARGE CHAMFERS ON CONNECTING RODS MUST FACE CRANKSHAFT COUNTERWEIGHTS.

1. Piston
2. Connecting rod LH
3. Connecting rod bearing
4. Connecting rod RH
5. Connecting rod bearing cap
6. Connecting rod bearing cap bolt
7. Crankshaft

Piston and connecting rod positioning— 5.7L engine

FUEL SYSTEM

Fuel System Service Precautions

Safety is the most important factor when performing not only fuel system maintenance but any type of maintenance. Failure to conduct maintenance and repairs in a safe manner may result in serious personal injury or death. Maintenance and testing of the vehicle's fuel system components can be accomplished safely and effectively by adhering to the following rules and guidelines.

• To avoid the possibility of fire and personal injury, always disconnect the negative battery cable unless the repair or test procedure requires that battery voltage be applied.

• Always relieve the fuel system pressure prior to disconnecting any fuel system component (injector, fuel rail, pressure regulator, etc.), fitting or fuel line connection. Exercise extreme caution whenever relieving fuel system pressure, to avoid exposing skin, face and eyes to fuel spray. Please be advised that fuel under pressure may penetrate the skin or any part of the body that it contacts.

• Always place a shop towel or cloth around the fitting or connection prior to loosening to absorb any excess fuel due to spillage. Ensure that all fuel spillage (should it occur) is quickly removed from engine surfaces. Ensure that all fuel soaked cloths or towels are deposited into a suitable waste container.

• Always keep a dry chemical (Class B) fire extinguisher near the work area.

• Do not allow fuel spray or fuel vapors to come into contact with a spark or open flame.

• Always use a back-up wrench when loosening and tightening fuel line connection fittings. This will prevent unnecessary stress and torsion to fuel line piping. Always follow the proper torque specifications.

• Always replace worn fuel fitting O-rings with new. Do not substitute fuel hose or equivalent, where fuel pipe is installed.

Fuel System Pressure

RELIEVING

1. Before servicing the vehicle, refer to the precautions in the beginning of this section.
2. Disconnect the negative battery cable.

3. Loosen the fuel filler cap to relieve the tank pressure.
4. Remove the left fuel rail cover.
5. Wrap a shop towel around the fuel pressure valve fitting (located on the side or end of the fuel rail assembly) to catch any fuel spray and connect a fuel pressure gauge.
6. Place the bleed hose into a suitable container, then open the valve to bleed the fuel system pressure.
7. Close the valve and disconnect the fuel gauge. Drain any remaining fuel from the gauge into the bleed container.

Fuel Filter

REMOVAL & INSTALLATION

1. Before servicing the vehicle, refer to the precautions in the beginning of this section.
2. Relieve the fuel system pressure.
3. Clean the filter connections, then depress the locking tabs and detach the quick-connect fittings from the filter.
4. Remove or disconnect the following:
   • Rear stabilizer bar from the rear cradle
   • Intermediate pipe-to-muffler bolts and lower the left muffler
   • Fuel feed and return quick connect fittings from the fuel filter
   • Fuel filter/pressure regulator bracket mount nut
   • Fuel feed pipe from the outlet side of the fuel filter/pressure regulator
   • Fuel filter/pressure regulator and bracket
   • Fuel system ground strap
   • Fuel filter/pressure regulator from the bracket

To install:
5. Install or connect the following:
   • New plastic quick connector retainers on the inlet and outlet tubes
   • Fuel filter/pressure regulator into the bracket
   • Fuel system ground strap
   • Fuel feed pipe to the outlet side of the fuel filter/pressure regulator
   • Fuel filter/pressure regulator and bracket to the mounting stud and torque the nut to 40 inch lbs. (4.5 Nm)
   • Fuel return and feed quick connect fittings
   • Muffler and torque the hanger nuts to 12 ft. lbs. (16 Nm) and the inter-

mediate pipe bolts to 37 ft. lbs. (50 Nm)
   • Stabilizer shaft and torque the nuts to 70 ft. lbs. (95 Nm)
6. Connect the negative battery cable.
7. Turn the ignition switch **ON** for 2 seconds, then **OFF** for 10 seconds. Turn the ignition switch back **ON** and inspect for leaks.

Fuel Pump

REMOVAL & INSTALLATION

These models have 2 fuel tanks (right and left). The fuel pump is part of the fuel sender assembly in the left fuel tank. The right fuel tank contains a fuel sender assembly with a siphon jet pump which supplies fuel to the left tank through the fuel sender feed pipe. Although there are 2 sending units, the removal and installation procedure is the same for both.

1. Before servicing the vehicle, refer to the precautions in the beginning of this section.
2. Properly relieve the fuel system pressure.
3. Drain the fuel tank(s).
4. Remove or disconnect the following:
   • Both rear wheels
   • Fuel tank shield(s)
5. Clean the area around the fuel sender assembly.
6. Mark each fuel line to help identify them during installation.
7. Remove or disconnect the following:
   • Quick connect fittings from the fuel sender
   • Fuel sender electrical connector
   • Fuel tank strap and properly support the fuel tank
   • Fuel sender attaching bolts and discard them

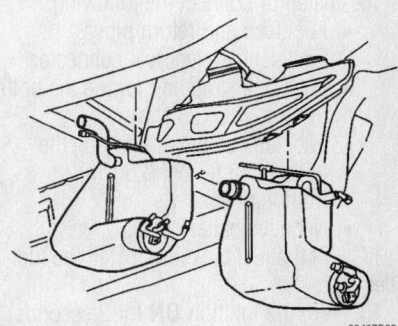

9346ZG03

These models have 2 fuel tanks connected by a fuel sender feed pipe

- Float arm retaining clip and the float arm (left side sender only)
- Fuel sender and gasket

To install:

➡**Always install a new fuel strainer before installing the fuel sender assembly. A strainer that has been exposed to fuel will not unfold completely and may interfere with the full travel of the float arm.**

8. Install a new gasket on the fuel sender.

⁕⁕ WARNING

Do not damage the float arm during installation.

9. Fold the long strainer over itself and hold the strainer in this position.
10. Pinch both strainers upward toward each other.
11. Install the float arm through the fuel tank opening with the folded strainers.

➡**It may be necessary to rotate the sender for proper installation.**

12. Look into the fuel tank opening and make certain that the long strainer is visible. If it is not visible, rotate the fuel sender until the strainer is free
13. Align the fuel sender gasket tab with the fuel sender cover mark and align the cover mark with the fuel tank mark.
14. Install the new attaching bolts to the fuel sender and hand-tighten them.

⁕⁕ CAUTION

The upper hex head portion of the fuel sender attaching bolts is designed to shear off the lower section of the bolt when the proper torque is reached. Do not tighten the bolts after the head is sheared off.

15. Tighten the new break away attaching bolts in sequence to 62 inch lbs. (7 Nm) .
16. Install or connect the following:
- Fuel feed and return pipes
- Fuel sender electrical connector
- Fuel tank strap and torque the bolts to 18 ft. lbs. (25 Nm)
- Fuel tank shield and torque the bolts to 18 ft. lbs. (25 Nm)
- Both rear wheels
- Negative battery cable
17. Fill the fuel tank and install the fuel filler cap.
18. Turn the ignition **ON** for 2 seconds, **OFF** for 10 seconds, then **ON** again and inspect the system for leaks.

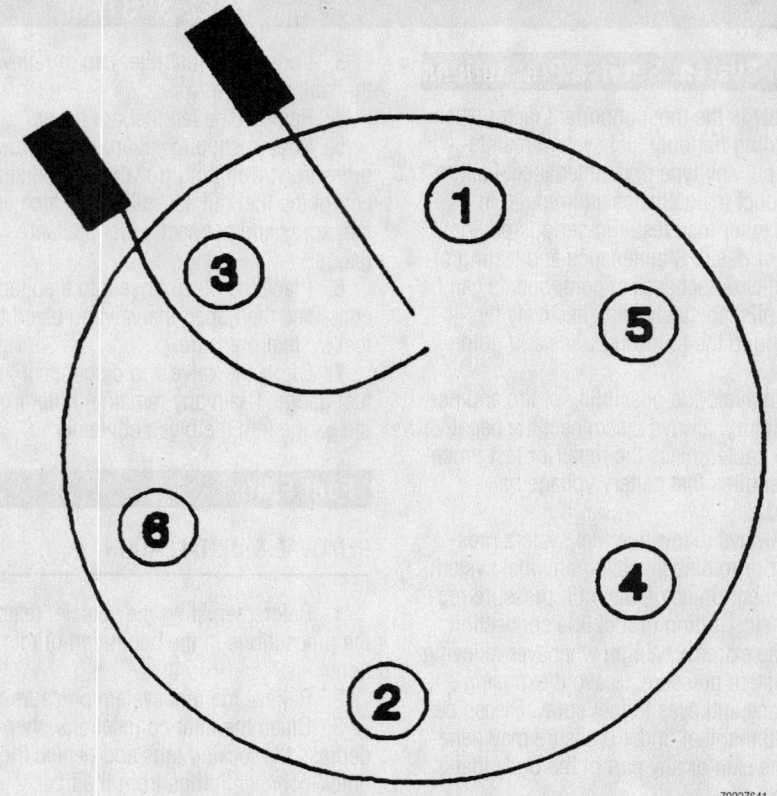

Tighten the fuel pump mounting bolts in the sequence shown—5.7L engine

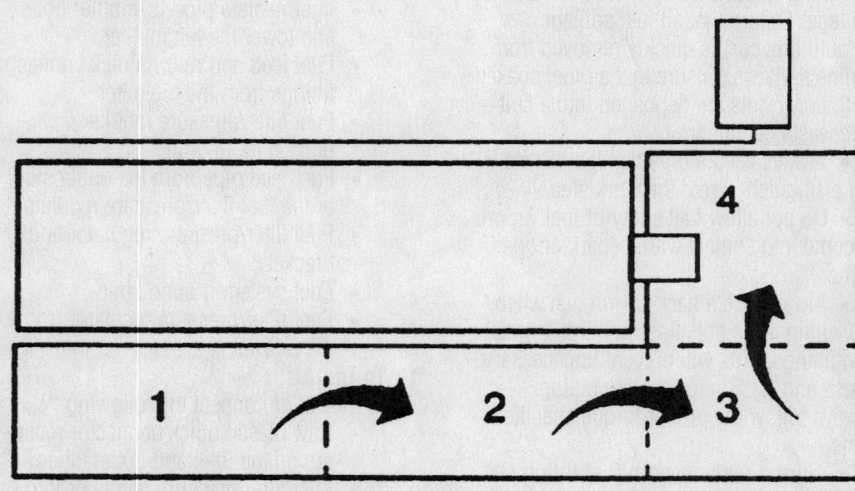

Fold the strainer on itself 3 times so it will fit into the opening in the fuel tank

Fuel Injector

REMOVAL & INSTALLATION

1. Before servicing the vehicle, refer to the precautions in the beginning of this section.
2. Relieve the fuel system pressure.
3. Remove or disconnect the following:
- Both fuel rail covers
- Fuel feed hose from the fuel rail
- Fuel injector electrical connectors
- Fuel rail mounting bolts
- Fuel rail
4. Spread the injector clip to release the injector from the fuel rail.
5. Fuel injector by spreading the retaining clip.

To install:

➡The fuel injector is stamped with a part number identification, manufacturing date, week code and plant number. Make certain the correct injector is ordered when replacing them.

6. Lubricate the new injector seals with clean oil.

7. Install or connect the following:
- New O-ring seals on the injectors
- New retainer clip on the injector
- Fuel injector into the fuel rail socket

✳✳ CAUTION

The fuel rail stop bracket must be installed onto the engine. The bracket serves as protection for the fuel rail in the event of a frontal crash. If the bracket is not installed, fuel could spray possibly causing a fire and personal injury.

- Fuel rail and ground strap to the intake manifold and torque the fuel rail attaching bolts to 89 inch lbs. (10 Nm)
- Electrical connectors to the injectors
- Fuel feed hose to the fuel rail
- Negative battery cable
- Left and right fuel rail covers

8. Turn the ignition **ON** for 2 seconds, **OFF** for 10 seconds, then **ON** again and inspect the system for leaks.

DRIVE TRAIN

Transmission Assembly

REMOVAL & INSTALLATION

Automatic

1. Before servicing the vehicle, refer to the precautions in the beginning of this section.

2. Disconnect the negative battery cable.

3. Shift the transmission into **N**.

4. Remove or disconnect the following:
- Rear wheels
- Intermediate exhaust pipe
- Right side muffler and tie the left side muffler to the underbody
- Driveline tunnel closeout panel
- Rear bell housing access plug and matchmark the flexplate to the torque converter
- Flexplate-to-torque converter bolts
- 2 plug bolts from the front of the driveline support assembly

5. Install two M10 x 1.5 x 55mm or longer bolts into the bolt holes. Torque the bolts to 26 ft. lbs. (35 Nm). These bolts must remain installed until instructed to remove them in order to maintain the position of the input shaft bearing.

6. Remove or disconnect the following:
- Engine flywheel housing access plug
- Propeller shaft hub clamp bolt
- Shift cable bracket nuts
- Shift control cable from the shift lever
- Rear transverse spring and support the lower control arms
- Outer tie rod ends from the knuckle
- Shock absorber lower mounting bolts
- Lower ball joints from the knuckles

7. Install a transmission support fixture to the transmission.

- Wiring harness and brake pipe clip retainers from the rear crossmember
- Lower transmission-to-differential nut
- Transaxle mount-to-rear crossmember nut
- Rear crossmember retaining nuts while supporting the crossmember
- Crossmember
- Transaxle mount bracket

8. Separate the axle shafts from the differential and tie them to the underbody.

9. Release the retainer securing the wiring harness from the "L" shaped brackets along the driveline support and move the harness out of the way.

10. Lower the driveline slightly and tilt it to gain access to the electrical connectors.

11. Remove or disconnect the following:
- Vehicle Speed Sensor (VSS) electrical connector
- Wire harness retainer from the differential rear cover stud
- Wire harness retainer clip from the top of the differential
- Transmission harness 20 way connector
- Park Neutral Position (PNP) switch electrical connectors

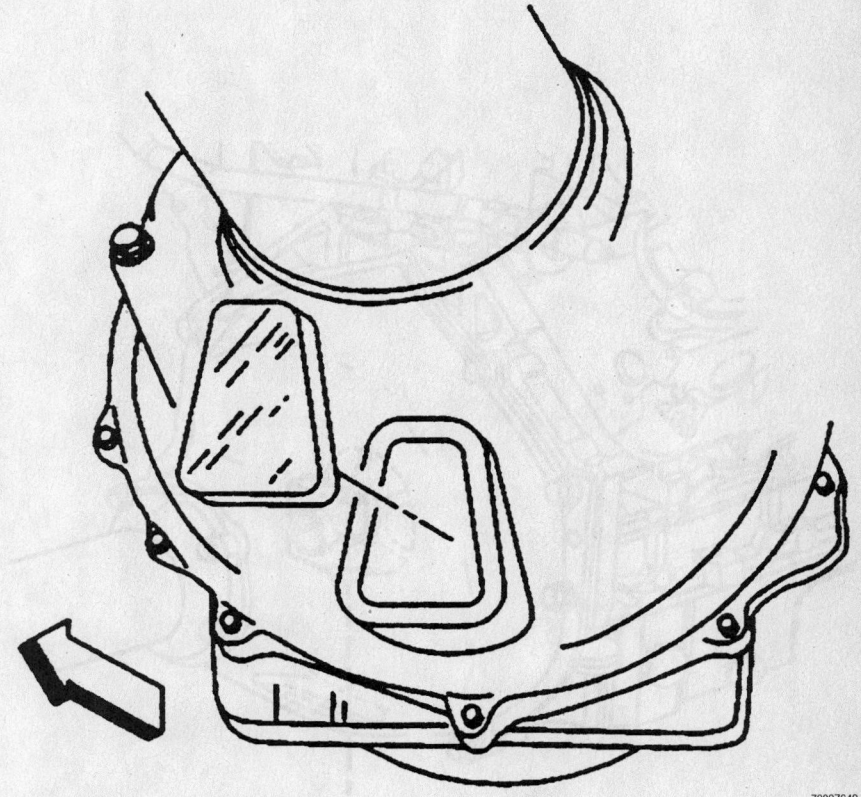

79222642

Remove the inspection plug, then remove the flexplate-to-torque converter bolts

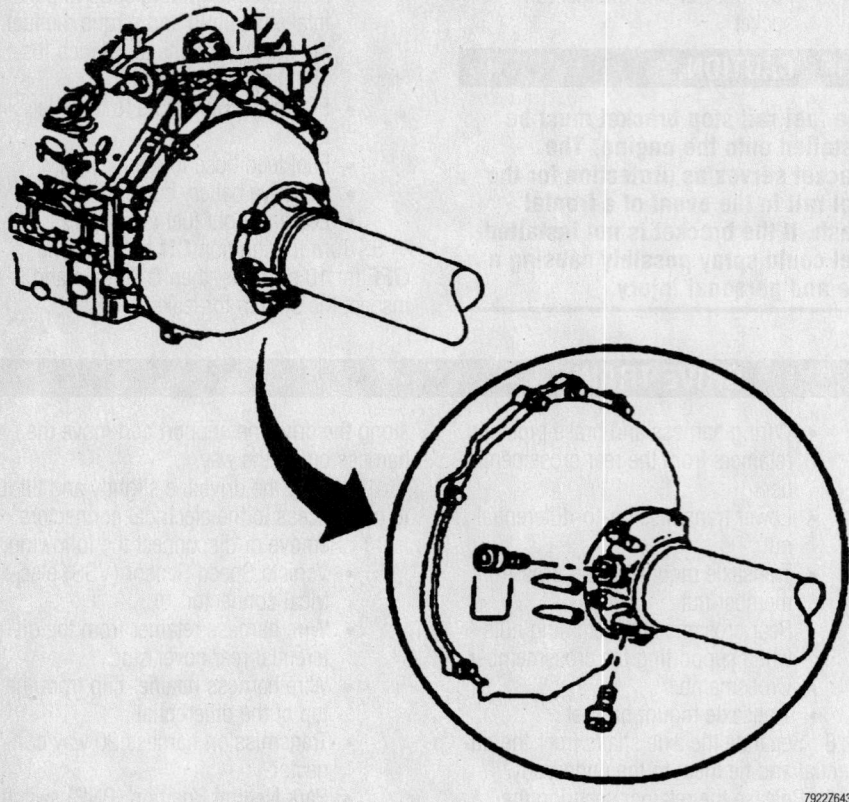

79222Z643

Remove the plugs and install two M10 x 1.5 x 55mm or longer bolts into the bolt holes to secure the bearing—5.7L engine

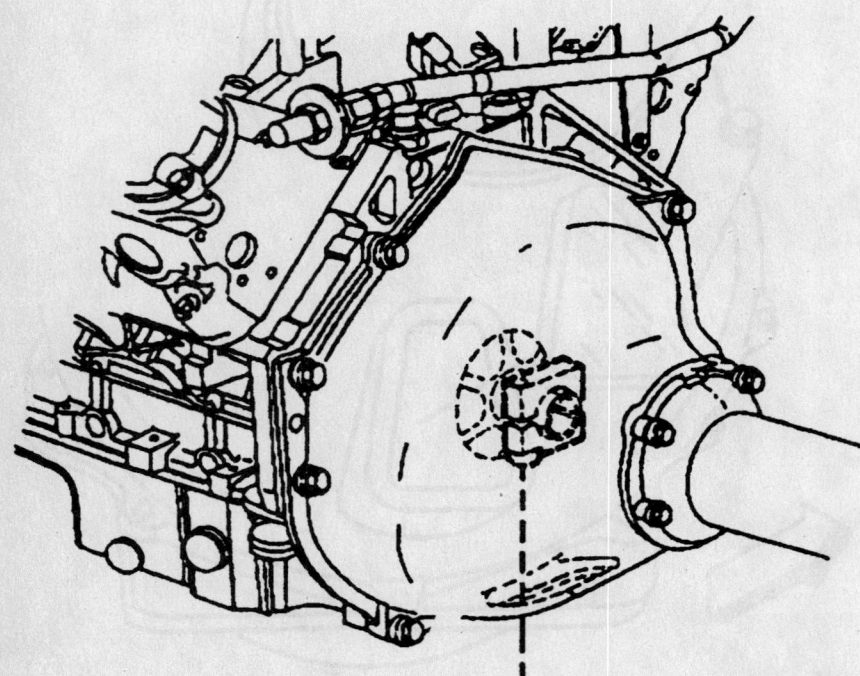

79222Z644

Rotate the flywheel to gain access to the clamp bolt, then loosen it—5.7L engine

- Bolt securing the wire harness to the left side of the case

12. Lower the driveline and angle it while observing the top rear of the differential and the lowest part of the rear compartment floor panel. The engine Positive Crankcase Ventilation (PCV) pipes will most likely contact the dash panel. Make certain not to damage the dash.

13. Remove or disconnect the following:
- Wire harness from the retainer along the top of transmission
- Transmission rear oil cooler pipes from the junction fittings at the flywheel housing
- 5 driveline to flywheel housing bolts after supporting the rear of the engine

14. Bend the wiring harness bracket away from the driveline, toward the tunnel wall to have access to remove the driveline.

15. Separate the driveline from the engine by prying them apart with a flat blade tool.

16. Lower the driveline and tilt it away from the engine until the input shaft clears the flywheel housing.

17. Remove or disconnect the following:
- Driveline
- Transmission oil cooler rear pipes from the fittings
- Transmission to driveline bolts
- Separate the transmission from the driveline with a flat blade tool

18. Using a lifting device, place the assembly on a bench.
- Rear transmission oil cooler fittings from the transmission
- Transmission-to-driveline assembly support bolts
- Driveline from the transmission while supporting the torque converter
- Differential-to-transmission bolts
- Differential from the transmission

To install:

19. Install or connect the following:
- Differential to the transmission and torque the bolts to 37 ft. lbs. (50 Nm)
- Transmission to the driveline support and torque the bolts to 37 ft. lbs. (50 Nm)
- Rear transmission oil cooler fittings to the transmission and torque the fittings to 30 ft. lbs. (40 Nm)

20. Using a chain hoist, place the assembly on a transmission jack.

21. Carefully raise the assembly into the vehicle while placing the wiring harness loosely into the harness retaining slots.

22. Align the assembly for installation

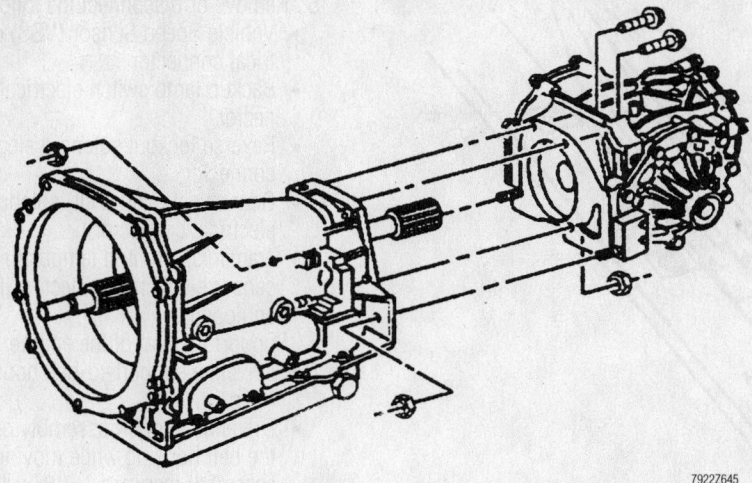

Transmission-to-differential mounting bolt locations—5.7L engine

79222645

into the engine. The driveline will slide into the rear of the engine as long as the angles are the same.

23. Install or connect the following:
- Driveline assembly in the vehicle. Reposition the wiring harness bracket to align with the appropriate hole in the driveline assembly.
- Driveline support-to-flywheel housing bolts and torque them to 37 ft. lbs. (50 Nm)
- Oil cooler lines and torque the fittings to 20 ft. lbs. (27 Nm)
- Wiring harness to the left side of the transmission and torque the bolt to 22 inch lbs. (2.5 Nm)
- PNP switch electrical connector
- Transmission harness 20 way connector
- Wiring harness clip to the top of the differential
- VSS electrical connector

24. Raise the transmission to installation height.
- Axle shafts
- Transmission mount to the crossmember and torque the nuts to 37 ft. lbs. (50 Nm)
- Suspension crossmember and torque the nuts to 81 ft. lbs. (110 Nm)
- Lower transmission-to-differential nut and torque it to 37 ft. lbs. (50 Nm)
- Wiring harness and brake pipe clip retainers to the crossmember
- Lower ball joints to the suspension knuckles and torque the nuts to 52 ft. lbs. (70 Nm)
- Shock absorber lower mounting bolts and torque them to 162 ft. lbs. (220 Nm)

- Outer tie rod ends to the knuckles and torque the nuts to 15 ft. lbs. plus an additional 160 degrees
- Rear transverse spring and torque the bolts to 46 ft. lbs. (62 Nm)
- Wiring harness into the "L" shaped brackets
- Transmission cable and bracket and torque the nuts to 15 ft. lbs. (20 Nm)
- Transmission flexplate to the torque converter using the matchmarks made during removal and torque the bolts to 47 ft. lbs. (63 Nm)
- Rear bell housing access plug
- Propeller shaft hub bolt until it is finger-tight

25. Remove the two M10 x 55mm bolts from the input shaft front bearing.

26. Install or connect the following:
- 2 plug bolts to the driveline support assembly and torque the bolts to 37 ft. lbs. (50 Nm)
- Driveline tunnel closeout panel and torque the bolts to 89 inch lbs. (10 Nm)
- Right hand muffler assembly
- Intermediate pipe and torque the bolts to 37 ft. lbs. (50 Nm)
- Both rear wheels
- Negative battery cable

27. Start the vehicle and allow it to reach normal operating temperature. Turn the engine **OFF**. Allow the engine to cool to ambient temperature, then tighten the flywheel hub collar bolt to 96 ft. lbs. (130 Nm).

28. Install the bell housing inspection plug.

29. Flush the automatic transmission oil cooler.

30. A front end alignment is recommended when the crossmember is removed.

Manual

1. Before servicing the vehicle, refer to the precautions in the beginning of this section.

2. Remove or disconnect the following:
- Negative battery cable
- Both rear wheels
- Folding top stowage compartment lid extension panel (on convertible models)
- Traction control/ride control switch
- Console retaining nut covers
- Console retaining nuts
- Accessory plug electrical connector
- Fuel door release electrical connector
- Console
- Shift control knob button
- Shift control knob retainer
- Shift control knob
- Shift control boot by grabbing both sides and pulling it in toward the lever
- Ashtray
- Trim plate grill
- Retaining screws located behind the grill and behind the ashtray
- Instrument panel accessory trim plate
- Shift control closeout boot retaining nuts and boot
- Shift rod clamp bolt
- Shift control mounting bolts
- Shift control assembly
- Courtesy lamp
- Left side lower closeout panel while guiding the courtesy lamp through the hole
- Clutch master cylinder pushrod retainer and the pushrod from the pedal
- Clutch actuator cylinder hose from the retainer clip
- Clutch actuator hose from the master cylinder hose
- Rear wheels
- Intermediate exhaust pipe and secure the mufflers out of the way
- Driveline closeout panel
- Rear transverse spring
- Outer tie rod ends
- Lower shock absorber bolts
- Lower ball joints

3. Support the transmission
- Wiring harness and brake lines from the suspension crossmember
- Lower transmission-to-differential nut

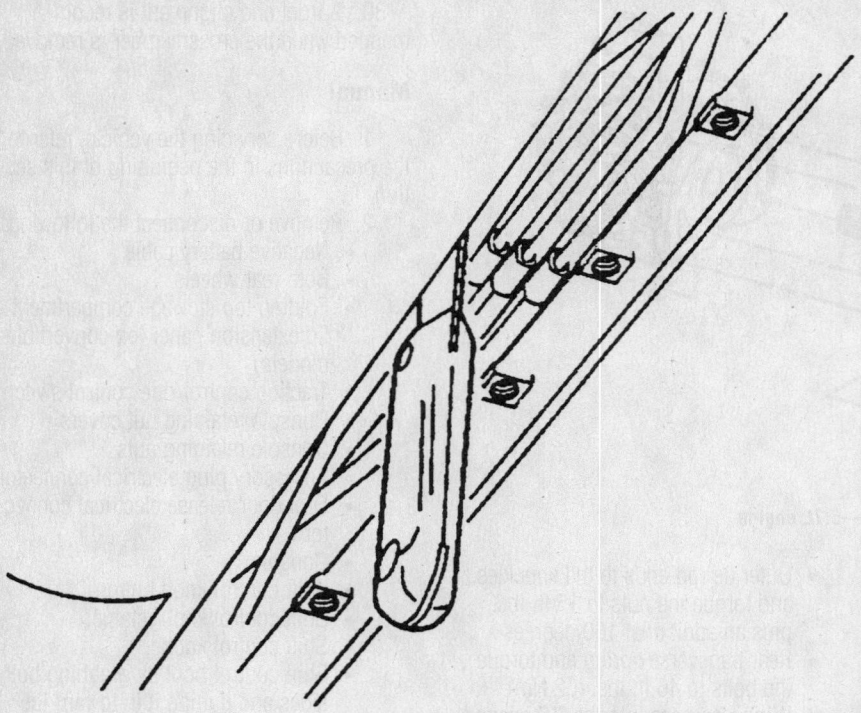

Insert a flat-bladed tool between the shifter bracket and brake liner retainer before lowering the transmission assembly—5.7L engine

79222646

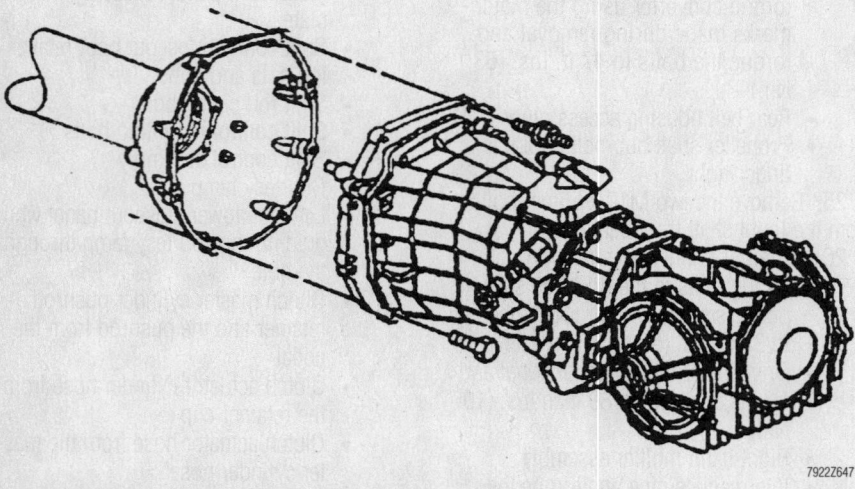

Driveline support-to-transmission mounting—5.7L engine

79222647

- Transmission mount-to-crossmember nuts
- Rear suspension crossmember nuts
- Transmission mount and bracket from the differential
- Axle shafts from the differential and position them out of the way
- Wiring harness from the driveline support assembly
- Harness clip from the top of the differential

⁕⁕ WARNING

Do not lower the top rear portion of the differential past the bottom of the storage compartment or the PCV pipe will hit the dash panel possibly causing damage.

4. Lower the transmission enough to remove the wiring harness from the top of the assembly.

5. Remove or disconnect the following:
- Vehicle Speed Sensor (VSS) electrical connector
- Backup lamp switch electrical connector
- Reverse lockout solenoid electrical connector
- Gear select (skip shift) solenoid electrical connector
- Transmission fluid temperature sensor electrical connector (if equipped)

6. Support the rear of the engine
- Driveline support-to-bell housing bolts
- Driveline support assembly out of the bell housing while moving the assembly rearward

7. Carefully lower the assembly away from the vehicle while simultaneously adjusting the angle.

8. Attach a chain hoist to the assembly, then remove it from the jack and place it on a workbench.

9. Remove or disconnect the following:
- Driveline support-to-transmission mounting bolts. Carefully pry the driveline assembly away from the transmission while guiding the shift rod through the opening in the driveline support
- Transmission shift rod from the transmission
- Transmission-to-differential mounting bolts and separate the differential from the transmission

To install:

10. Install or connect the following:
- Differential to the transmission and torque the bolts to 37 ft. lbs. (50 Nm)
- Transmission shift rod
- Driveline support to the transmission and torque the bolts to 37 ft. lbs. (50 Nm)

11. To aid in later installation place a rubber band around the shift rod then tape the shift rod to the driveline support assembly.

12. Place the assembly on the transmission jack with a chain hoist.

13. Begin to raise the assembly into the vehicle while loosely installing the wiring harness along the driveline assembly retaining slots.

14. Have an assistant guide the front of the driveline support to the bell housing.

15. Be sure the driveline assembly is at the same angle as the engine before trying to install it.

16. Install or connect the following:

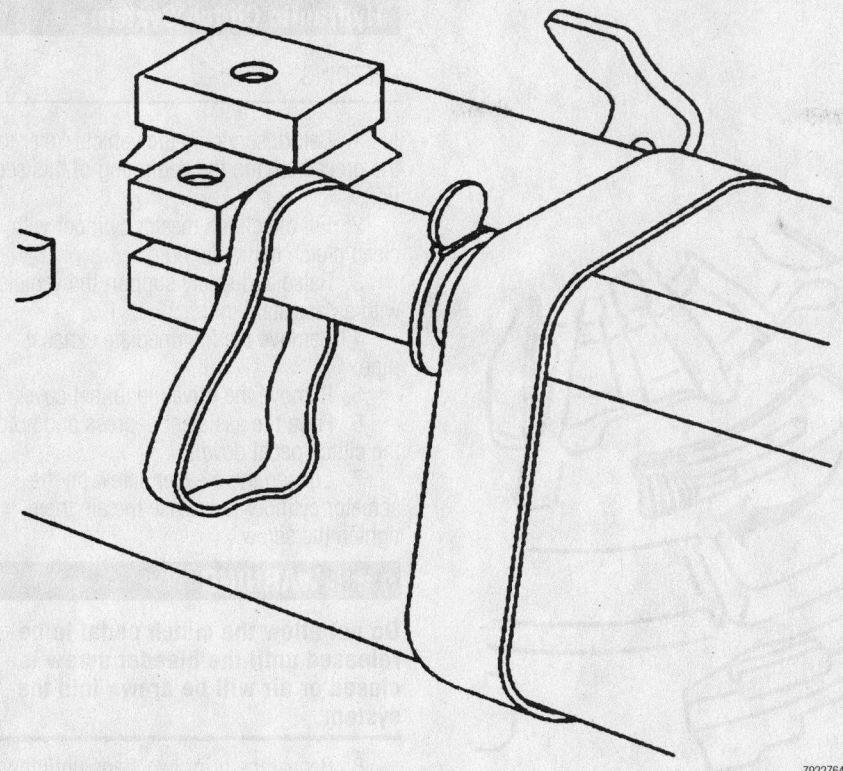

Place a rubber band on the shift rod, then tape the rod to the driveline support assembly—5.7L engine

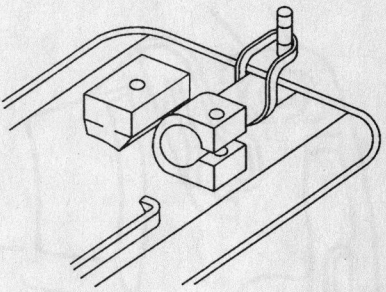

Pull the shift rod up to break the tape and hook the rubber band on the rear stud—5.7L engine

- Input propeller shaft into the clutch disc
- Wiring harness bracket to align with the appropriate hole in the driveline assembly
- Driveline support-to-flywheel housing bolts and torque them to 37 ft. lbs. (50 Nm)
- Wiring harness to the retainer on the top of the transmission and attach the connectors
- Transmission fluid temperature sensor electrical connector (if equipped)
- Gear select/skip shift solenoid electrical connector
- Reverse lockout solenoid electrical connector
- Backup lamp switch electrical connector
- Wiring harness to the top of the differential
- VSS electrical connector
- Axle shafts in the differential
- Transmission mount on the differential and torque the bolts to 37 ft. lbs. (50 Nm)
- Rear suspension crossmember and torque the nuts to 81 ft. lbs. (110 Nm)

- Transmission mount-to-crossmember nuts and torque them to 37 ft. lbs. (50 Nm)
- Lower transmission-to-differential nut and torque it to 37 ft. lbs. (50 Nm)
- Wiring harness and brake line retainers to the crossmember
- Lower ball joints and torque the nuts to 52 ft. lbs. (70 Nm)
- Shock absorber lower mounting bolts and torque them to 162 ft. lbs. (220 Nm)
- Outer tie rod ends and torque the nuts to 15 ft. lbs. (20 Nm) plus an additional 160 degrees
- Transverse spring and torque the mounting bolts to 46 ft. lbs. (62 Nm)
- Clutch actuator hose to the master cylinder hose
- Clutch actuator hose to the retaining clip
- Driveline tunnel closeout panel and torque the bolts to 89 inch lbs. (10 Nm)
- Muffler assemblies
- Intermediate exhaust pipe and torque the bolts to 37 ft. lbs. (50 Nm)

- Rear wheels
- Clutch master cylinder pushrod to the clutch pedal
- Clutch master cylinder pushrod retainer
- Left lower closeout panel
- Courtesy lamp

17. Pull up the shift rod to break the tape and hook the rubber band on the rear stud on the top of the driveline tunnel.
- Shift control assembly and torque the bolts to 22 ft. lbs. (30 Nm)
- Shift control closeout boot and torque the nuts to 106 inch lbs. (12 Nm)
- Instrument panel accessory trim plate
- Shift control boot
- Shift control knob
- Shift control knob retainer and button
- Console
- Fuel door release electrical connector
- Accessory plug electrical connector
- Console retaining nuts and torque them to 89 inch lbs. (10 Nm)
- Console retaining nut covers
- Traction control/ride control switch
- Folding top stowage compartment lid extension panel
- Rear wheels
- Negative battery cable

18. Bleed the clutch system.

Clutch

ADJUSTMENT

1. Before servicing the vehicle, refer to the precautions in the beginning of this section.
2. Remove the flywheel inspection cover.
3. Have an assistant depress the clutch

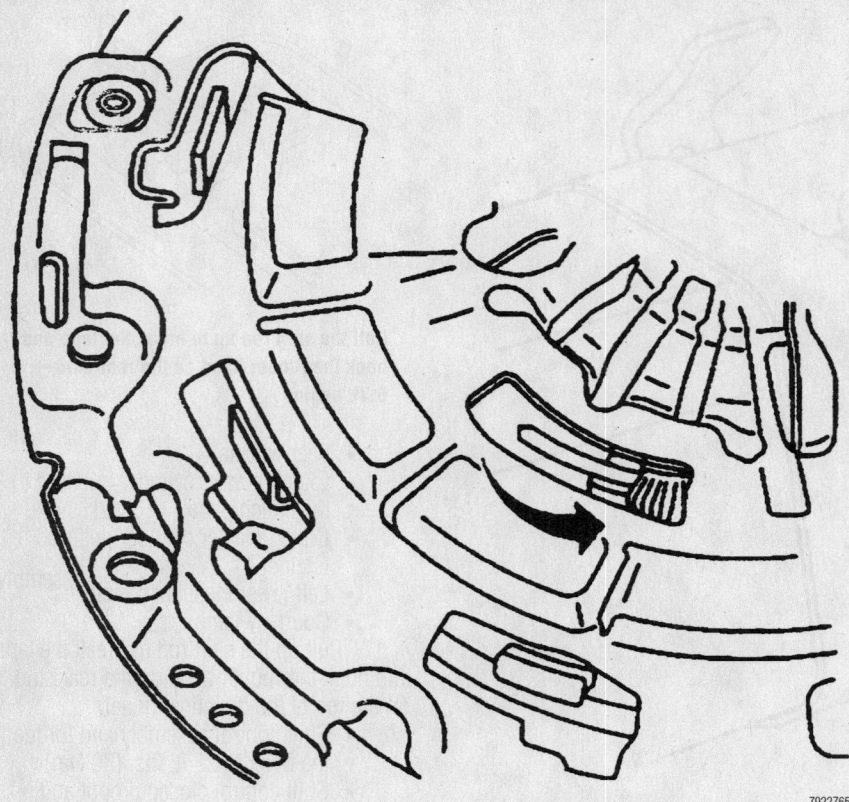

Rotate the stepped adjusting ring counterclockwise to compress the springs

79222652

pedal until the tension is released from the stepped adjusting ring.

4. Using 2 screwdrivers, rotate the stepped adjusting ring counterclockwise until fully adjusted out.

5. Continue to hold them in this position and have the assistant release the clutch pedal.

6. Remove the screwdrivers.

7. Install the inspection cover. Torque the bolts to 18 ft. lbs. (25 Nm).

REMOVAL & INSTALLATION

1. Before servicing the vehicle, refer to the precautions in the beginning of this section.

2. Remove or disconnect the following:
   - Negative battery cable
   - Exhaust system
   - Driveline support and transmission as an assembly
   - Flywheel inspection cover
   - Pressure plate bolts. It will be necessary to rotate the flywheel to access all the bolts
   - Pressure plate and disc

To install:

3. Install or connect the following:
   - Clutch disc and pressure plate on the flywheel

- Clutch disc alignment tool through the disc to keep it in place
- Pressure plate bolts finger-tight. Turn the flywheel to access all the bolt holes

4. Torque the bolts in the sequence shown, using 3 steps to 52 ft. lbs. (70 Nm).
   - Inspection cover and torque the bolts to 18 ft. lbs. (25 Nm)
   - Driveline support and transmission assembly
   - Exhaust system
   - Negative battery cable

5. Bleed the clutch system.

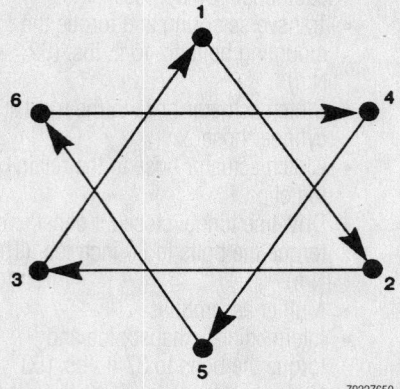

Clutch assembly torque sequence

79222650

Hydraulic Clutch System

BLEEDING

1. Before servicing the vehicle, refer to the precautions in the beginning of this section.

2. Fill the clutch master cylinder with clean clutch hydraulic fluid.

3. Raise and safely support the vehicle with an assistant in it.

4. Remove the intermediate exhaust pipe.

5. Remove the driveline tunnel cover.

6. Have the assistant depress and hold the clutch pedal down.

7. Loosen the bleeder screw on the actuator cylinder to release the air, then tighten the screw.

✳✳ WARNING

Do not allow the clutch pedal to be released until the bleeder screw is closed or air will be drawn into the system.

8. Repeat the prior two steps until the air has been purged from the system. Check the master cylinder fluid level and refill as needed.

9. Install the driveline tunnel cover.

10. Install the intermediate exhaust pipe.

11. Lower the vehicle.

Halfshaft

REMOVAL & INSTALLATION

1. Before servicing the vehicle, refer to the precautions in the beginning of this section.

2. Apply the parking brake.

3. Remove or disconnect the following:
   - Wheel
   - Axle nut
   - Rear transverse spring
   - Outer tie rod end from the knuckle
   - Antilock Brake System (ABS) Wheel Speed Sensor (WSS) electrical connector
   - Park brake cable from the lever and bracket

➡ **Be sure to support the halfshaft until it is removed. Do not let it hang by the CV-joint.**

4. Attach a puller on the wheel studs and start to push the axle shaft into the hub assembly. This will provide clearance for the

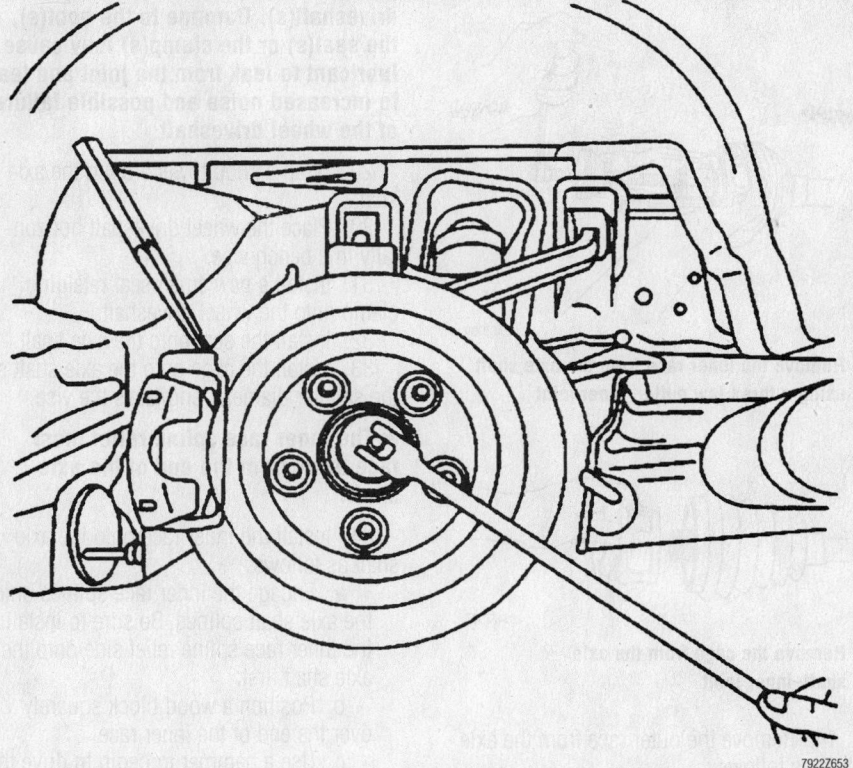

Insert a large drift through the cooling fins to keep the hub assembly from turning while removing the retaining nut

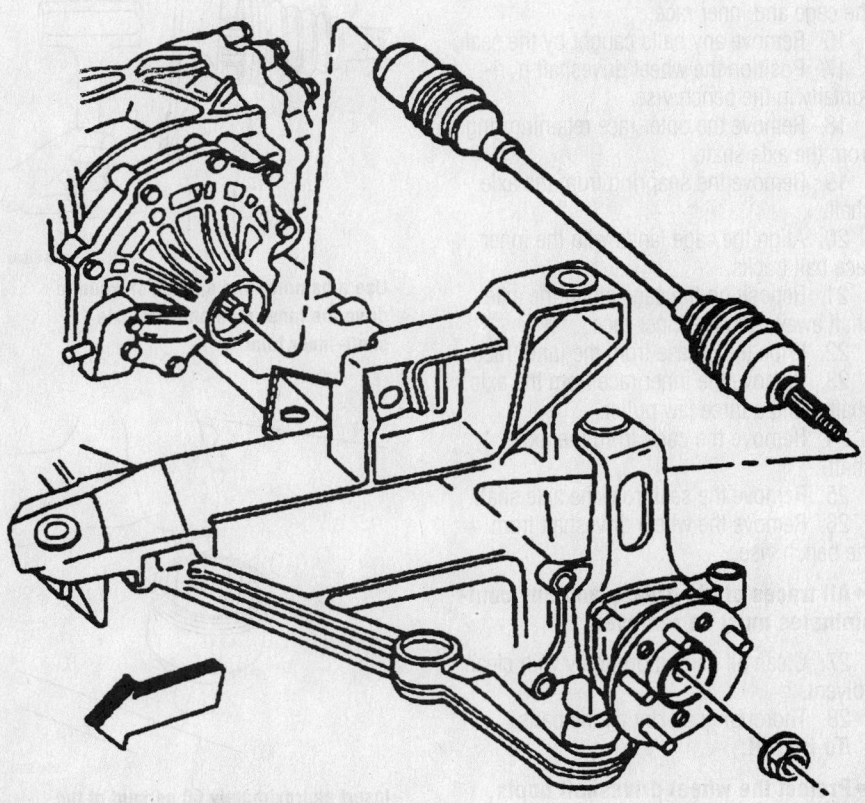

Exploded view of the halfshaft mounting

ball joint to be separated from the knuckle assembly.

5. Remove or disconnect the following:
   - Ball joint from the knuckle
   - Axle shaft from the hub
   - Halfshaft assembly from the differential by inserting a suitable tool between the CV-joint and differential and prying them apart

To install:

6. Install or connect the following:
   - Halfshaft on the differential output shaft. Use light force to be sure it is fully seated
   - Halfshaft through the hub assembly but do not install completely. This will provide clearance for installing the ball joint
   - Ball joint to the knuckle and torque the nut to 41 ft. lbs. (55 Nm)
   - Halfshaft through the hub assembly completely
   - Park brake cable to the bracket and the lever
   - WSS electrical connector
   - Tie rod end to the knuckle and torque the nut to 15 ft. lbs. (20 Nm) plus an additional 160 degrees
   - Rear transverse spring and torque the bolts to 46 ft. lbs. (62 Nm)
   - Halfshaft retaining nut and torque the nut to 118 ft. lbs. (160 Nm)
   - Wheel

CV-Joints

OVERHAUL

Inner Joint

1. Before servicing the vehicle, refer to the precautions in the beginning of this section.
2. Remove axle shaft.

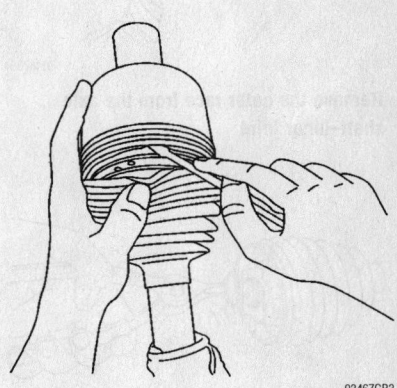

Separate the seal from the joint outer race at the large diameter end—inner joint

3. Wrap a shop towel around the axle shaft.

4. Place the wheel driveshaft horizontally in a bench vise.

5. Remove the large seal retaining clamp from the CV joint seal.

6. Use a side cutter or other suitable tool and discard the clamp.

7. Remove the small seal retaining clamp from the joint seal.

8. Use a side cutter or other suitable tool and discard the clamp.

9. Separate the seal from the joint outer race at the large diameter end.

10. Position the seal behind the joint face.

11. Position the wheel driveshaft vertically in the bench vise so the inner joint is up.

12. Slide the joint outer race down toward the vise.

13. Disengage the outer race retaining ring as follows:

   a. Insert a small flat-bladed screwdriver between the retaining ring and the outer race.

   b. Pry the retaining ring from the outer race.

   c. Position the retaining ring along the axle shaft away from the outer race.

➡️**The balls may fall out of the cage and inner race when the outer race is removed.**

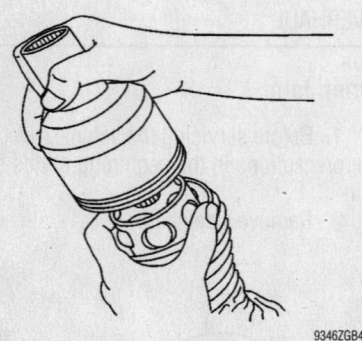

Remove the outer race from the axle shaft–inner joint

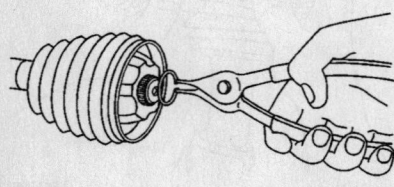

Remove the snapring from the axle shaft–inner joint

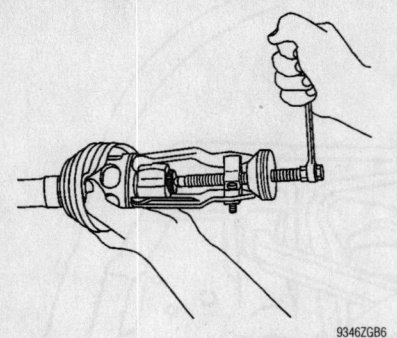

Remove the inner race from the axle shaft using a three jaw puller–inner joint

Remove the cage from the axle shaft–inner joint

14. Remove the outer race from the axle shaft as follows:

   a. Use the seal to catch any balls which are not retained by grease.

   b. Lift the outer race off the axle shaft.

15. Remove any remaining balls from the cage and inner race.

16. Remove any balls caught by the seal.

17. Position the wheel driveshaft horizontally in the bench vise.

18. Remove the outer race retaining ring from the axle shaft.

19. Remove the snapring from the axle shaft.

20. Align the cage lands with the inner race ball tracks.

21. Reposition the cage along the axle shaft away from the inner race.

22. Wipe the grease from the inner race.

23. Remove the inner race from the axle shaft using a three jaw puller.

24. Remove the cage from the axle shaft.

25. Remove the seal from the axle shaft.

26. Remove the wheel driveshaft from the bench vise.

➡️**All traces of old grease and any contaminates must be removed.**

27. Clean all parts thoroughly with clean solvent.

28. Thoroughly air dry all the parts.

To install:

➡️**Protect the wheel driveshaft boots, seals and clamps from sharp objects when servicing on or near the wheel driveshaft(s). Damage to the boot(s), the seal(s) or the clamp(s) may cause lubricant to leak from the joint and lead to increased noise and possible failure of the wheel driveshaft.**

29. Wrap a shop towel around the axle shaft.

30. Place the wheel driveshaft horizontally in a bench vise.

31. Install a new small seal retaining clamp onto the wheel driveshaft.

32. Install the seal onto the axle shaft.

33. Install the cage onto the axle shaft so the smaller diameter end faces the vise.

➡️**The inner race spline relief must face away from the end of the axle shaft.**

34. Install the inner race onto the axle shaft as follows:

   a. Engage the inner race splines onto the axle shaft splines. Be sure to install the inner race spline relief side onto the axle shaft first.

   b. Position a wood block squarely over the end of the inner race.

   c. Use a hammer to begin to drive the inner race onto the axle shaft.

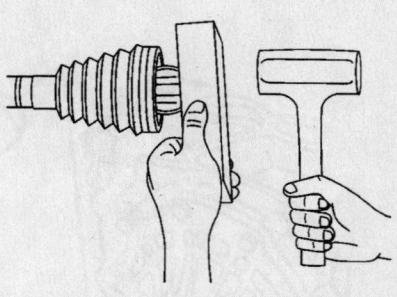

Use a hammer and a block of wood to drive the inner race onto the axle shaft–inner joint

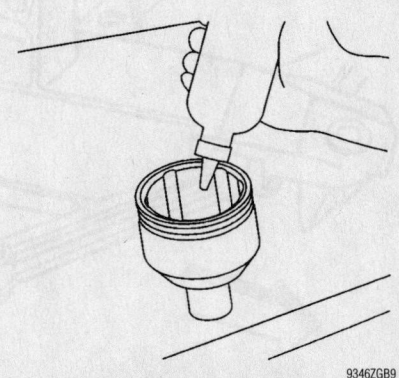

Insert approximately 60 percent of the grease from the service kit into the outer race–inner joint

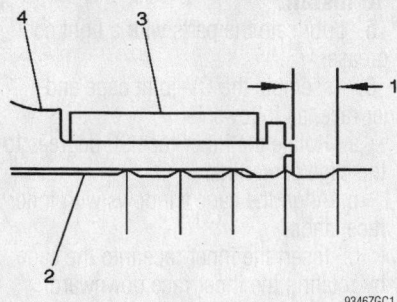

Measure the distance between the edge of the seal and the edge of the last axle shaft groove closing edge and adjust fit to 0.10 inch (2.5mm)–inner joint

d. Reposition the wood block along the face of the inner race to avoid the axle shaft.

e. Work evenly around the inner race and continue to drive the inner race, until you feel the inner race seat fully onto the axle shaft.

35. Inspect to be sure that the axle shaft snapring groove is exposed.

36. Install the snapring to the axle shaft.

37. Position the cage so the cage lands align with the inner race ball tracks.

38. Install the cage onto the inner race.

39. Position the cage windows to align with the inner race ball tracks.

40. Insert approximately 60 percent of the grease from the service kit into the outer race.

41. Position the wheel driveshaft vertically in the bench vise so the inner joint end is up.

42. Apply a small amount of the grease from the service kit to the cage windows and inner race ball tracks.

43. Insert the remaining grease from the service kit into the seal.

44. Install the outer race retaining ring onto the axle shaft.

45. Position the retaining ring below the cage, toward the vise.

46. Install the balls through the cage windows to the inner race ball tracks.

47. Use the seal to keep the balls in position if necessary.

48. Install the outer race onto the axle shaft as follows:

a. Be careful not to allow the grease in the outer race to leak out.

b. Align the outer race ball tracks to the balls.

c. Slide the outer race down over the balls.

49. Position the wheel driveshaft horizontally in the bench vise. Engage the outer race retaining ring as follows:

a. Slide the outer race toward the vise.

b. Insert the outer race retaining ring into the groove along the outer edge of the outer race.

c. Position the outer race retaining ring so the opening in the ring aligns with an outer race land (not a ball track).

50. Position the large diameter end of the seal onto the outer race.

51. Position the small seal retaining clamp onto the neck of the seal.

52. Position the seal and small retaining clamp to the axle shaft.

53. Measure the distance between the edge of the seal and the edge of the last axle shaft groove closing edge and adjust fit to 0.10 inch (2.5mm).

➡**The seal retaining clamp must not be over-tightened or under-tightened.**

54. Crimp the small seal retaining clamp using tool J 42572. Tighten the small seal retaining clamp until the base of the omega shape has a gap width between 0.079–0.118 inch (2–3mm), with a differ-

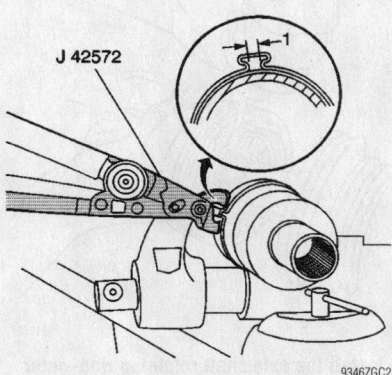

Tighten the small seal retaining clamp until the base of the omega shape has a gap width between 0.079–0.118 inch (2–3mm), with a difference in the gap width from side to side no greater than 0.016 inch (0.4mm)–inner joint

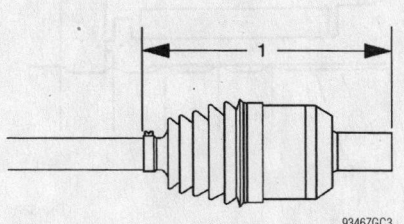

Measure the distance between the end of the seal and the end of the joint outer race and adjust the plunging motion of the joint to 8.90 inch within 0.16 inch (226mm within 4mm–inner joint

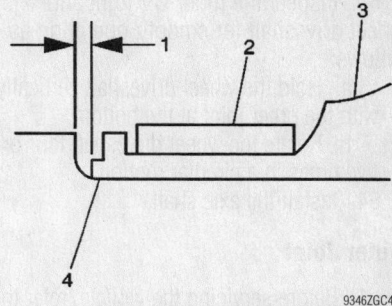

Measure the distance between the edge of the seal and the edge of the joint outer race last groove closing edge and adjust fit to 0.03 inch (0.8mm)–inner joint

ence in the gap width from side to side no greater than 0.016 inch (0.4mm). The clamping hold time must be no less than 2 seconds.

55. Measure the distance between the end of the seal and the end of the joint outer race and adjust the plunging motion of the joint to 8.90 inch within 0.16 inch (226mm within 4mm).

56. Position the large seal retaining clamp onto the seal.

57. Position the seal and large retaining clamp to the joint outer race. Measure the distance between the edge of the seal and the edge of the joint outer race last groove closing edge and adjust fit to 0.03 inch (0.8mm).

➡**The seal must not be dimpled, stretched or out of shape in any way.**

58. Inspect the seal for proper shape. If the seal is NOT shaped correctly, equalize the pressure in the seal and shape the seal properly by hand. Inspect the seal for damage. If the seal has been cut or punctured during assembly, you must discard and replace the seal.

➡**The seal retaining clamp must not be over-tightened or under-tightened.**

59. Crimp the large seal retaining clamp using tool J 42572. Tighten the small seal retaining clamp until the base of the omega shape has a gap width between 0.079–0.118 inch (2–3mm), with a difference in the gap width from side to side no greater than 0.016 inch (0.4mm). The clamping hold time must be no less than 2 seconds.

60. Remove the wheel driveshaft from the bench vise.

61. Distribute the grease within the inner CV joint.

62. Plunge the joint back and forth four or five times.

63. Inspect the inner CV joint and wheel driveshaft for smooth operation as follows:

a. Hold the wheel driveshaft vertically, with the outer joint at the bottom.

b. Rotate the wheel driveshaft four or five times in a circular motion.

64. Install the axle shaft.

Outer Joint

1. Before servicing the vehicle, refer to the precautions in the beginning of this section.

2. Remove or disconnect the following:

- Axle shaft from the vehicle
- Large CV boot retaining clamp
- Small CV boot retaining clamp
- CV boot from the joint
- Outer joint from the axle shaft using a hammer and wood block
- Axle shaft retaining ring
- CV boot

3. Disassemble the chrome alloy balls from the CV-joint cage as follows:

a. Position a brass drift against the CV-joint cage and tap it with a hammer to tilt the cage.

b. Remove the 1st chrome alloy ball from the cage.

c. Tilt the cage in the opposite direction.

d. Remove the opposite chrome alloy ball.

e. Repeat the procedure until all 6 balls are removed.

4. Disassemble the CV-joint cage and inner race as follows:

a. Pivot the cage and race 90 degrees to the center line of the outer race.

b. Align the cage windows with outer race lands.

c. Remove the cage from the outer race.

d. Rotate the inner race upward and remove it from the cage.

To install:

5. Lubricate the parts with a light coat of grease.

6. Assemble the CV-joint cage and inner race, as follows:

a. Rotate the inner race 90 degrees to the cage centerline.

b. Align the cage windows with inner race lands.

c. Insert the inner race into the cage by rotating the inner race downward.

d. Insert the cage/inner race into the outer race.

7. Assemble the chrome alloy balls into the CV-joint cage, as follows:

a. Position a brass drift against the CV-joint cage and tap it with a hammer to tilt the cage.

b. Insert the 1st chrome alloy ball into the cage.

c. Tilt the cage in the opposite direction.

9346ZGC7

Install the axle shaft retaining ring—outer joint

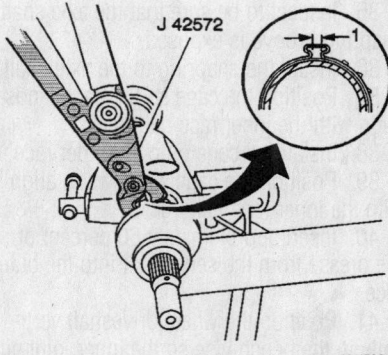

9346ZGC9

Tighten the small seal retaining clamp until the base of the omega shape has a gap width between 0.079–0.118 inch (2–3mm), with a difference in the gap width from side to side no greater than 0.016 inch (0.4mm)—outer joint

9346ZGC5

Separate the outer joint from the axle shaft using a hammer and wood block—outer joint

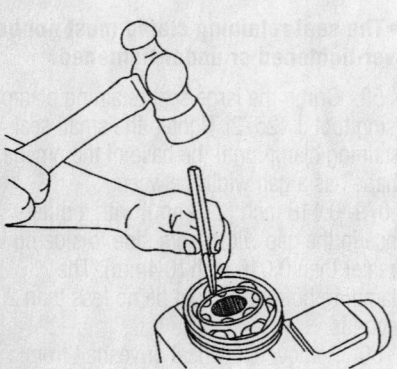

9346ZGC6

Position a brass drift against the CV-joint cage and tap it with a hammer to tilt the cage—outer joint

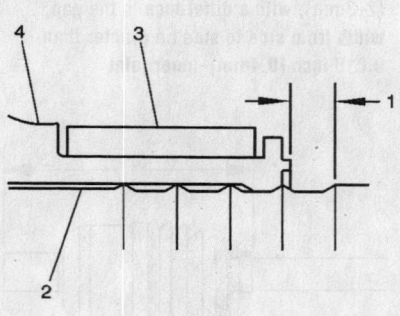

9346ZGC8

Measure the distance between the edge of the seal and the edge of the last axle shaft groove closing edge and adjust fit to 0.10 inch (2.5mm)

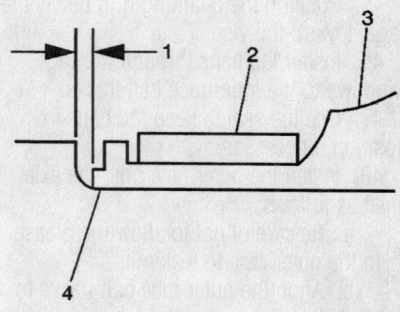

9346ZGD1

Measure the distance between the edge of the seal and the edge of the last axle shaft groove closing edge and adjust fit to 0.3 inch (8mm)—outer joint

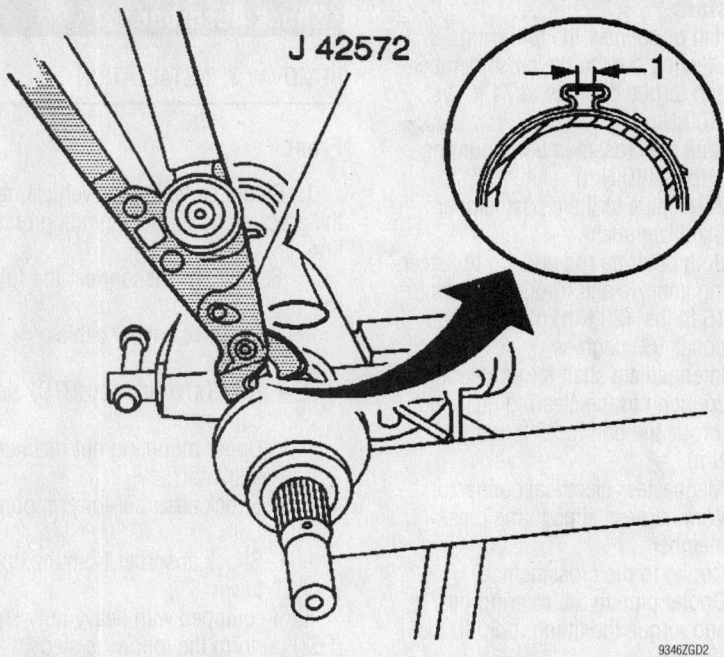

J 42572

Tighten the small seal retaining clamp until the base of the omega shape has a gap width between 0.079–0.118 inch (2–3mm), with a difference in the gap width from side to side no greater than 0.016 inch (0.4mm)–outer joint

9346ZGD2

d. Insert the opposite chrome alloy ball.

e. Repeat the procedure until all 6 balls are inserted.

8. Install ½ of the grease provided, into the CV-joint.

9. Install or connect the following:

- Small CV boot retaining ring
- CV boot on the halfshaft
- New retaining ring on the half-shaft
- Large ring clamp on the CV boot
- Outer joint onto the axle shaft

10. Compress the retaining ring using a small flat-bladed tool while pushing the outer joint onto the axle shaft.

11. Seat the outer joint on the shaft using a hammer and block of wood.

12. Install the remaining grease into the CV boot.

13. Position the CV boot and the small boot clamp.

14. Crimp the small boot clamp.

15. Position and crimp in place the large boot clamp.

16. Install the halfshaft in the vehicle.

STEERING AND SUSPENSION

Air Bag

✳✳ CAUTION

All vehicles are equipped with an air bag system. The system must be disabled before performing service on or around system components, steering column, instrument panel components, wiring and sensors. Failure to follow safety and disabling procedures could result in accidental air bag deployment, possible personal injury and unnecessary system repairs.

PRECAUTIONS

Several precautions must be observed when handling the inflator module to avoid accidental deployment and possible personal injury.

1. Never carry the inflator module by the wires or connector on the underside of the module.

2. When carrying a live inflator module, hold securely with both hands, and ensure that the bag and trim cover are pointed away.

3. Place the inflator module on a bench or other surface with the bag and trim cover facing up.

4. With the inflator module on the bench, never place anything on or close to the module which may be thrown in the event of an accidental deployment.

DISARMING

1. Before servicing the vehicle, refer to the precautions in the beginning of this section.

2. Turn the steering wheel to align the wheels in the straight-ahead position.

3. Turn the ignition switch to the **LOCK** position.

a. Remove the SIR fuse from the fuse panel.

b. Remove the left side sound insulator.

c. Disconnect the Connector Positive Assurance (CPA) from the yellow 2-way SIR harness connector at the base of the steering column and separate the connector.

d. Disconnect the steering wheel module coil connector located at the base of the steering column.

e. Disconnect the CPA from the inflatable restraint instrument panel module connector located at the near the base of the steering column.

ARMING

After the necessary repairs have been made, re-enable the air bag system as follows:

1. Turn the ignition switch to the **LOCK** position.

a. Connect the Connector Positive Assurance (CPA) to the yellow 2-way SIR harness connector at the base of the steering column.

b. Connect the steering wheel module coil connector located at the base of the steering column.

c. Connect the CPA from the inflatable restraint instrument panel module connector located at the near the base of the steering column.

d. Connect the instrument panel module connector located near the base of the steering column.

e. Install the left side sound insulator.

f. Install the SIR fuse to the fuse panel.

Power Rack and Pinion Steering Gear

REMOVAL & INSTALLATION

1. Before servicing the vehicle, refer to the precautions in the beginning of this section.
2. Drain the power steering fluid.
3. Remove or disconnect the following:
   - Negative battery cable
   - Brake Pressure Modulator Valve (BPMV) bracket, if equipped
   - Both front wheels
   - Intermediate shaft shield
   - Intermediate shaft lower coupling from the power steering gear
   - Power steering inlet hose from the steering gear
   - Steering cooler pipe from the steering gear
   - Power steering cooler
   - Both outer tie rod ends from the knuckles
   - Magnasteer electrical connector
   - Stabilizer shaft from the crossmember
   - Wire harness clips from the crossmember
   - Brake pipe from the crossmember
   - Steering gear mounting bolts and loosen the crossmember mounting nuts
   - Power steering gear through the left wheel opening

To install:

4. Install or connect the following:
   - Steering gear to the crossmember and torque the nuts to 74 ft. lbs. (100 Nm)
5. Torque the crossmember mounting nuts 81 ft. lbs. (110 Nm)
   - Brake pipe to the crossmember
   - Stabilizer shaft
   - Both outer tie rod ends to the steering knuckle and torque the nuts to 15 ft. lbs. (20 Nm) plus and additional 160 degrees
   - Intermediate shaft lower steering coupling to the steering gear and torque the bolt to 25 ft. lbs. (34 Nm)
   - Magnasteer electrical connector
   - Wire harness clips to the crossmember
   - Cooler to the crossmember
   - Cooler pipe to the steering gear and torque the fitting to 20 ft. lbs. (27 Nm)
   - Inlet hose to the steering gear and torque the fitting to 20 ft. lbs. (27 Nm)
   - Intermediate shaft shield and torque the clamp to 32 inch lbs. (3.5 Nm)
   - Both front wheels
   - Negative battery cable
6. Refill and bleed the power steering system.
7. Check and adjust the front end alignment.

Shock Absorber

REMOVAL & INSTALLATION

Front

1. Before servicing the vehicle, refer to the precautions in the beginning of this section.
2. Remove or disconnect the following:
   - Negative battery cable
   - Wheel
   - Real Time Damper (RTD) electrical connector
   - Upper mounting nut retainer and insulator
   - Shock absorber lower mounting bolts
   - Shock absorber from the upper tower
3. If equipped with heavy duty shocks (FE3) perform the following steps:
   a. Compress the shock absorber from the bottom upward.
   b. Install a shock support tool to the shock while it is compressed.
   c. Remove the shock from the vehicle.
   d. Remove the tool from the shock.
4. Remove the shock absorber.

To install:

5. Install or connect the following:
   - Retainer and insulator to the shock absorber
   - Shock absorber to the upper shock tower
   - Upper insulator, retainer and nut and torque the nut to 19 ft. lbs. (26 Nm)
   - Lower mounting bolts and torque them to 21 ft. lbs. (28 Nm)
   - RTD electrical connector

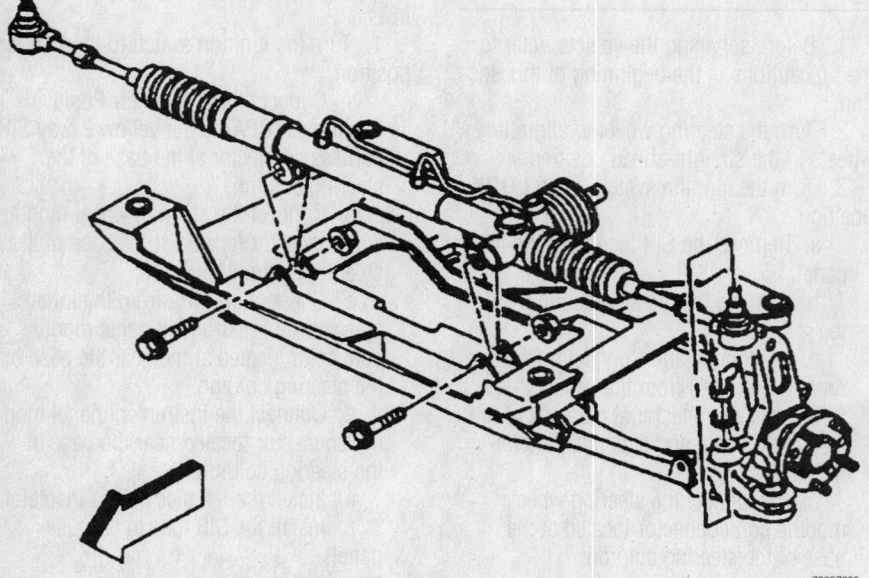

79222655

Exploded view of the power steering gear mounting—5.7L engine

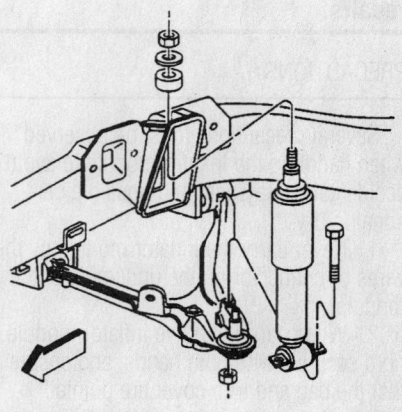

79222656

Exploded view of the front shock absorber mounting—5.7L engine

6. If equipped with heavy duty shocks (FE3) perform the following steps:

a. Install a shock support tool to the shock absorber.

b. Install the shock to the vehicle.

c. Install the upper insulator, retainer and nut and torque the nut to 19 ft. lbs. (26 Nm).

d. Remove the support tool from the shock and install a spring compressor to the spring.

7. Raise the lower control arm and install the shock absorber lower mounting bolts and torque them to 21 ft. lbs. (28 Nm).

8. Remove the spring compressor tool.

9. Install or connect the following:
   - Wheel
   - Negative battery cable

Rear

1. Before servicing the vehicle, refer to the precautions in the beginning of this section.

2. Remove or disconnect the following:
   - Negative battery cable
   - Wheel
   - Rear position sensor electrical connector, if equipped
   - Lower mounting bolt
   - Upper mounting bolts
   - Shock absorber from the lower control arm and shock tower
   - Upper insulator and retainer from the shock absorber

To install:

3. Install or connect the following:
   - Upper insulator and retainer to the shock absorber, if removed
   - Shock absorber to the shock tower and lower control arm
   - Upper mounting bolts and torque them to 22 ft. lbs. (30 Nm)
   - Lower shock absorber mounting

bolt and torque it to 162 ft. lbs. (220 Nm)
   - Rear position sensor electrical connector
   - Wheel
   - Negative battery cable

Transverse Spring

REMOVAL & INSTALLATION

Front

1. Before servicing the vehicle, refer to the precautions in the beginning of this section.

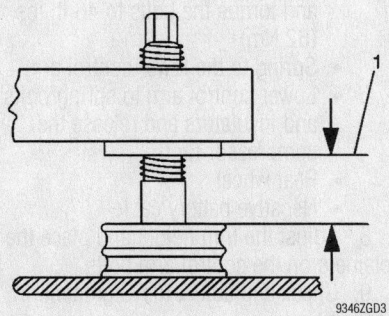

Measure the front spring adjuster bolt gap to ease the installation procedure and setting the proper vehicle trim height—front assembly

2. Remove or disconnect the following:
   - Negative battery cable
   - Front wheels

3. Measure the front spring adjuster bolt gap to ease the installation procedure and setting the proper vehicle trim height.

4. Install a spring compressor tool J 33432-A to the spring and compress it.

5. Remove or disconnect the following:
   - Lower shock absorber mounting bolts from one of the lower control arms
   - Stabilizer shaft link from the lower control arm
   - Lower ball joint from the steering knuckle
   - Cam bolts from the lower control arm after matchmarking them
   - Lower control arm
   - Transverse spring bolts and retainers
   - Transverse spring and the compressor tool

To install:

6. Install or connect the following:
   - Spring to the crossmember
   - New spring retainers and bolts to the crossmember and torque them to 46 ft. lbs. (62 Nm)
   - Lower control arm to the crossmember

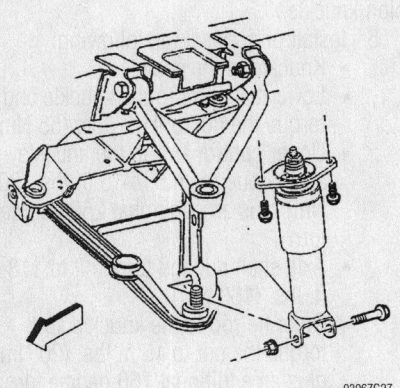

Exploded view of the rear shock absorber—5.7L engine

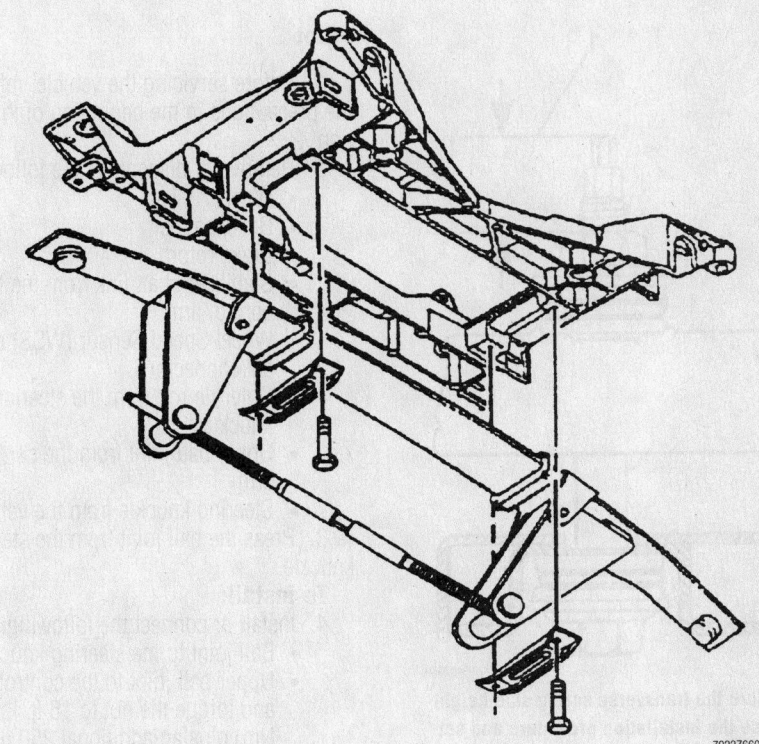

Exploded view of the front transverse spring—5.7L engine

- Cam bolts to the matchmarks made during the removal procedure. Hand-tighten the cam bolts at this time
- Lower control arm ball joint stud to the steering knuckle and torque the nut to 15 ft. lbs. (20 Nm) plus an additional 210 degrees
- Shock absorber
- Shock absorber lower mounting bolts and torque them to 21 ft. lbs. (28 Nm)
- Stabilizer shaft link to the lower control arm and torque the nut to 53 ft. lbs. (72 Nm)

7. Remove the spring compressor tool from the transverse spring and move the lower control arm supports.

8. Install or connect the following:
- Front wheels
- Negative battery cable

9. Adjust the front trim height.

10. Perform a front end alignment.

11. Torque the control arm cam bolts to 125 ft. lbs. (170 Nm)

Rear

1. Before servicing the vehicle, refer to the precautions in the beginning of this section.

2. Remove or disconnect the following:
- Negative battery cable
- Rear wheels

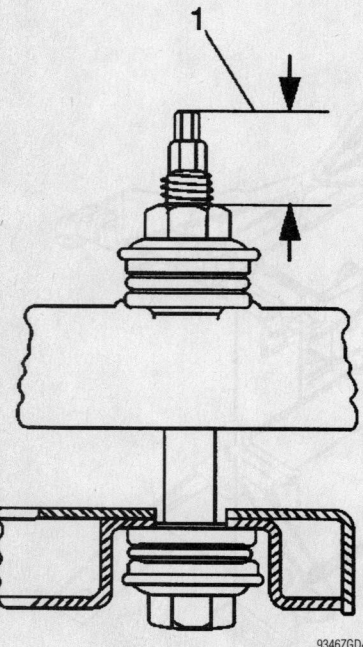

Measure the transverse spring stud height to ease the installation procedure and setting the proper vehicle trim height—rear assembly

3. Measure the transverse spring stud height to ease the installation procedure and setting the proper vehicle trim height.

4. Install a spring compressor tool J 33432-A to the spring and compress it.

5. Remove the spring mounting bolts, spacers and insulators from the crossmember and control arm.

6. Remove the spring from the vehicle and remove the compressor tool.

To install:

7. Install or connect the following:
- Spring compressor to the transverse spring
- Spring to the vehicle
- Spring spacers, insulators and mounting bolts to the crossmember and torque the bolts to 46 ft. lbs. (62 Nm)
- Spring to the lower control arm
- Lower control arm to spring bolts and insulators and release the compressor tool
- Rear wheel
- Negative battery cable

8. Adjust the trim height and place the retainers on the control arm bolts.

9. Check and adjust the alignment, if needed.

Upper Ball Joint

REMOVAL & INSTALLATION

Front

1. Before servicing the vehicle, refer to the precautions in the beginning of this section.

2. Remove or disconnect the following:
- Wheel
- Brake caliper
- Brake rotor
- Stabilizer shaft link from the lower control arm
- Wheel Speed Sensor (WSS) electrical connector
- Outer tie rod from the steering knuckle
- Upper ball joint from the control arm
- Steering knuckle from the vehicle

3. Press the ball joint from the steering knuckle.

To install:

4. Install or connect the following:
- Ball joint to the steering knuckle
- Upper ball joint to the control arm and torque the nut to 15 ft. lbs. (20 Nm) plus an additional 250 degree turn

- Lower ball joint to the steering knuckle and torque the nut to 15 ft. lbs. (20 Nm) plus an additional 210 degree turn
- Outer tie rod to the steering knuckle and torque the nut to 18 ft. lbs. (25 Nm) plus an additional 180 degree turn
- Stabilizer shaft link to the control arm and torque the nut to 53 ft. lbs. (72 Nm)
- WSS electrical connector
- Brake rotor
- Brake caliper and torque the mounting bracket bolts to 125 ft. lbs. (175 Nm)
- Wheel

5. Check and adjust the front alignment.

Rear

1. Before servicing the vehicle, refer to the precautions in the beginning of this section.

2. Remove or disconnect the following:
- Wheel
- Wheel Speed Sensor (WSS) electrical connector
- Real Time Damping (RTD) position sensor, if equipped
- Brake caliper
- Brake rotor
- Shock absorber solenoid electrical connector, if equipped
- Outer tie rod from the suspension knuckle
- Axle shaft nut
- Upper control arm from the knuckle
- Lower ball joint from the knuckle
- Suspension knuckle from the vehicle

3. Press the upper ball joint from the knuckle.

To install:

4. Press the ball joint into the suspension knuckle.

5. Install or connect the following:
- Knuckle to the vehicle
- Lower ball joint to the knuckle and torque the nut to 41 ft. lbs. (55 Nm)
- Upper control arm to the knuckle and torque the nut to 15 ft. lbs. (20 Nm) plus an additional 250 degree turn
- Axle shaft nut and torque it to 118 ft. lbs. (160 Nm)
- Outer tie rod to the knuckle and torque the nut to 15 ft. lbs. (20 Nm) plus an additional 160 degree turn
- Brake rotor
- Brake caliper and torque the

mounting bracket bolts to 125 ft. lbs. (175 Nm)
- WSS electrical connector
- Shock absorber solenoid electrical connector, if equipped
- RTD position sensor, if equipped
- Wheel

6. Check and adjust the rear wheel alignment.

Lower Ball Joint

REMOVAL & INSTALLATION

Front and Rear

1. Before servicing the vehicle, refer to the precautions in the beginning of this section.
2. Remove the lower control arm from the vehicle.
3. Press the ball joint from the control arm.

To install:

4. Press the ball joint into the control arm.
5. Install the control arm in the vehicle.

Upper Control Arm

REMOVAL & INSTALLATION

Front

1. Before servicing the vehicle, refer to the precautions in the beginning of this section.
2. Remove or disconnect the following:
   - Negative battery cable
   - Front wheel
   - Real Time Damping (RTD) sensor link and support the lower control arm
   - Upper ball joint from the upper control arm
   - Upper control arm bolts and shims
   - Upper control arm

To install:

3. Install or connect the following:
   - Upper control arm
   - Upper control arm shims and bolts and torque the bolts to 48 ft. lbs. (65 Nm)
   - Ball joint stud into the upper control arm and torque the nut to 15 ft. lbs. (20 Nm) plus an additional 250 degrees
   - RTD sensor link
   - Front wheel
   - Negative battery cable
4. Check and adjust the front wheel alignment.

Rear

1. Before servicing the vehicle, refer to the precautions in the beginning of this section.
2. Remove or disconnect the following:
   - Negative battery cable
   - Rear wheel
   - Wheel Speed Sensor (WSS) electrical connector
   - Real Time Damping (RTD) position sensor link, if equipped
3. Support the lower control arm with a jack stand.
   - Ball joint stud from the control arm
   - Upper control arm bolts
   - Upper control arm

To install:

4. Install or connect the following:
   - Upper control arm and torque the bolts to 81 ft. lbs. (110 Nm)
   - Suspension knuckle ball joint stud nut into the upper control arm. It may be necessary to use an Allen wrench to keep the ball joint stud from spinning while tightening the ball joint stud nut. Torque the nut to 15 ft. lbs. (20 Nm) plus an additional 250 degrees
   - WSS electrical connector
   - RTD position sensor link, if equipped
   - Rear wheel
   - Negative battery cable
5. Check and adjust the rear wheel alignment.

CONTROL ARM BUSHING REPLACEMENT

The manufacturer does not provide a bushing replacement procedure. The bushing is replaced when a new control arm is installed.

Lower Control Arm

REMOVAL & INSTALLATION

Front

1. Before servicing the vehicle, refer to the precautions in the beginning of this section.
2. Remove or disconnect the following:
   - Front wheel
   - Front transverse spring
   - Wheel Speed Sensor (WSS) electrical connector
   - Real Time Damping (RTD) electrical connector, if equipped
3. Support the lower control arm with a jack stand

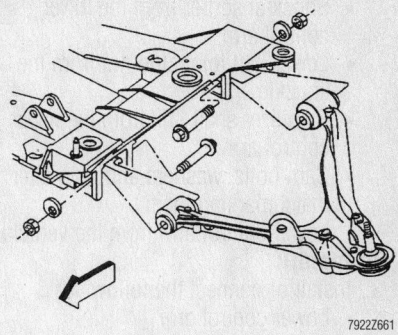

Lower control arm mounting bolts—5.7L engine

- Shock absorber from the lower control arm
- Stabilizer shaft link from the lower control arm
- Ball joint stud from the knuckle
- Cam bolts, washers and nuts after matchmarking them
- Lower control arm from the vehicle

To install:

4. Install or connect the following:
   - Lower control arm
5. Support the lower control arm with a jack stand
   - Cam bolts to the position matchmarked during the removal procedure and hand-tighten the bolts
   - Lower control arm ball joint stud to the steering knuckle and torque the nut to 15 ft. lbs. (20 Nm) plus an additional 210 degrees
   - Stabilizer shaft link to the lower control arm and torque the nut to 53 ft. lbs. (72 Nm)
   - Transverse spring and remove
   - Shock absorber lower mounting bolts and torque them to 21 ft. lbs. (28 Nm)
   - RTD electrical connector, if equipped
   - WSS electrical connector
   - Front wheel
   - Negative battery cable
6. Perform a front wheel alignment.
7. Tighten the lower control arm nuts to 125 ft. lbs. (170 Nm).

Rear

1. Before servicing the vehicle, refer to the precautions in the beginning of this section.
2. Remove or disconnect the following:
   - Negative battery cable
   - Rear wheel
   - Transverse spring from the lower control arm

- Shock absorber from the lower control arm
- Lower ball joint stud nut from the knuckle
- Stabilizer shaft link from the lower control arm
- Cam bolts, washers and nuts after matchmarking them
- Lower control arm from the vehicle

To install:

3. Install or connect the following:
   - Lower control arm
   - Cam bolts to the position matchmarked during the removal procedure and hand-tighten the bolts
   - Lower ball joint stud to the steering knuckle and torque the nut to 15 ft. lbs. (20 Nm) plus an additional 210 degrees
   - Stabilizer shaft link to the lower control arm and torque the link nut to 53 ft. lbs. (72 Nm)
   - Shock absorber lower mounting bolts and torque the bolts to 162 ft. lbs. (220 Nm)
   - Transverse spring to the lower control arm
   - Rear wheel
   - Negative battery cable
4. Perform a rear wheel alignment.
5. Torque the lower control arm front cam bolt to 107 ft. lbs. (145 Nm) and the rear cam bolt to 70 ft. lbs. (95 Nm).

CONTROL ARM BUSHING REPLACEMENT

The manufacturer does not provide a bushing replacement procedure. The bushing is replaced when a new control arm is installed.

Wheel Bearings

ADJUSTMENT

No periodic wheel bearing adjustment is necessary. The wheel bearings are a sealed unit that must be replaced if loose or noisy.

REMOVAL & INSTALLATION

Front

1. Before servicing the vehicle, refer to the precautions in the beginning of this section.
2. Remove or disconnect the following:
   - Negative battery cable
   - Front wheel
   - Wheel Speed Sensor (WSS) electrical connector

- Brake caliper
- Brake rotor
- Stabilizer shaft link from the lower control arm and support the lower control arm
- Separate the outer tie rod ball stud from the steering knuckle
- Lower ball joint stud from the steering knuckle
- Wheel hub mounting bolts
- Wheel hub/bearing assembly from the knuckle

To install:

3. Install or connect the following:
   - Hub/bearing assembly to the steering knuckle. Make certain that the speed sensor cable connection is facing rearward and torque the mounting bolts to 96 ft. lbs. (130 Nm)
   - Lower control arm ball stud to the steering knuckle and torque the nut to 55 ft. lbs. (75 Nm)
   - Outer tie rod to the steering knuckle and torque the nut to 33 ft. lbs. (45 Nm)
   - WSS electrical connector
   - Stabilizer shaft link to the lower control arm and torque the nut to 55 ft. lbs. (75 Nm)

- Brake rotor
- Brake caliper and torque the bracket mounting bolts to 125 ft. lbs. (175 Nm)
- Front wheel
- Negative battery cable

Rear

1. Before servicing the vehicle, refer to the precautions in the beginning of this section.
2. Remove or disconnect the following:
   - Negative battery cable
   - Rear wheel
   - Wheel Speed Sensor (WSS) electrical connector
   - Real Time Damping (RTD) position sensor link, if equipped
   - Brake caliper
   - Brake rotor
   - Shock absorber solenoid electrical connector, if equipped
   - Outer tie rod end from the suspension knuckle
   - Axle nut
   - Separate the upper control arm from the suspension knuckle
   - Steering knuckle from the lower control arm ball joint stud
   - Suspension knuckle

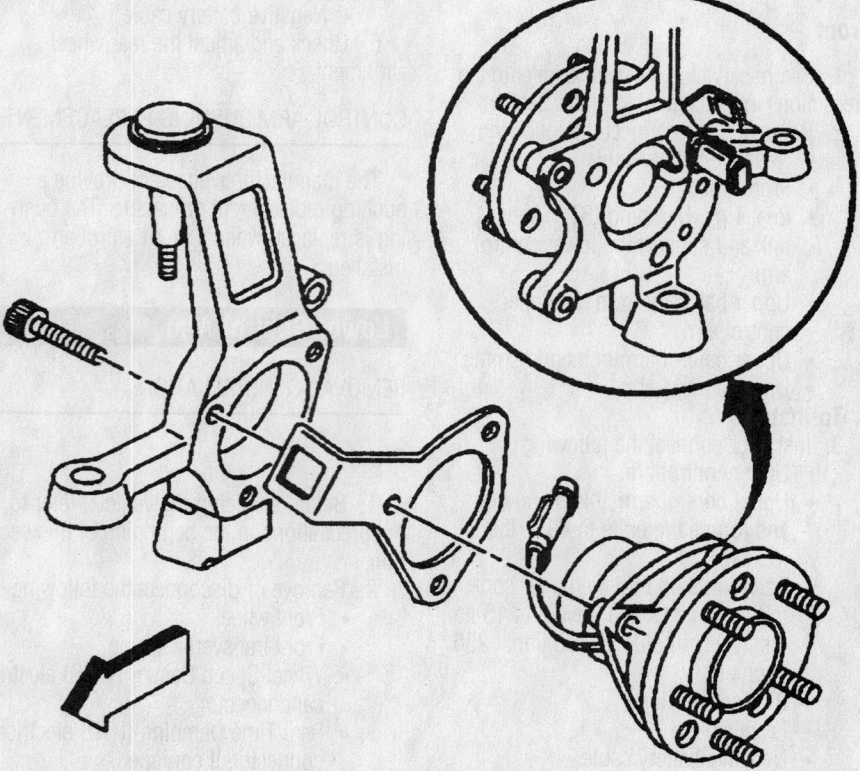

Exploded view of the front hub/wheel bearing and knuckle assembly—5.7L engine

79222657

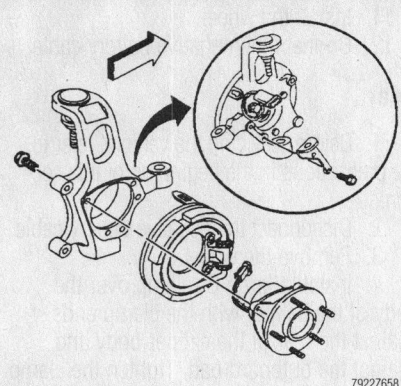

79222658

Exploded view of the rear hub/wheel bearing and knuckle assembly—5.7L engine

- Hub/bearing bolts
- Hub/bearing from the knuckle

To install:

3. Install or connect the following:
- Hub/bearing assembly to the steering knuckle. Make certain that the speed sensor cable connection is facing rearward and torque the hub mounting bolts to 96 ft. lbs. (130 Nm)
- Upper control arm to the steering knuckle and torque the bolt to 49 ft. lbs. (65 Nm)
- Lower control arm ball stud to the suspension knuckle and torque the nut to 55 ft. lbs. (75 Nm)

- Axle nut and torque it to 118 ft. lbs. (160 Nm)
- Outer tie rod ball stud to the steering knuckle and torque the nut to 33 ft. lbs. (45 Nm)
- Brake rotor
- Brake caliper and torque the mounting bracket bolts to 125 ft. lbs. (175 Nm)
- WSS electrical connector
- Shock absorber solenoid electrical connector
- RTD position sensor link, if equipped
- Rear wheel
- Negative battery cable

BRAKES

Brake Caliper

REMOVAL & INSTALLATION

Front

1. Before servicing the vehicle, refer to the precautions in the beginning of this section.

2. Disconnect the negative battery cable and remove ⅔ of the brake fluid from the master cylinder reservoir.

3. Remove or disconnect the following:

- Wheel assembly
- Brake hose fitting at the caliper by removing the bolt. Discard the 2 copper washers and plug the hose to prevent fluid contamination or loss.

➡️ **Do not allow the fluid to come into contact with the front transverse spring, as damage to the spring may occur.**

- 2 caliper mounting bolts
- Caliper housing from the rotor and the caliper mounting bracket.

To install:

4. Compress the pistons using a suitable tool, if necessary.

5. Install or connect the following:

- Caliper over the brake rotor and into the caliper mounting bracket. Make sure the shoe lining guiding surfaces are correctly seated in the bracket. Tighten the bolts to 23 ft. lbs. (31 Nm).
- Brake hose inlet fitting using 2 new copper washers and the inlet fitting

bolt. Torque the bolt to 37 ft. lbs. (50 Nm).

6. Properly bleed the entire brake system.

- Wheel assembly

7. Check the brake fluid and add as necessary.

8. Connect the negative battery cable, start the engine and pump the brake pedal slowly and firmly 3 times to seat the shoe and lining assemblies.

Rear

1. Before servicing the vehicle, refer to the precautions in the beginning of this section.

2. Disconnect the negative battery cable and remove ⅔ of the brake fluid from the master cylinder reservoir.

3. Remove or disconnect the following:

- Wheel assembly
- Brake hose fitting at the caliper by

removing the bolt. Discard the 2 copper washers and plug the hose to prevent fluid contamination or loss.

➡️ **Do not allow the fluid to come into contact with the front transverse spring, as damage to the spring may occur.**

- 2 caliper mounting bolts
- Caliper housing from the rotor and the caliper mounting bracket.

To install:

4. Compress the pistons using a suitable tool, if necessary.

5. Install or connect the following:

- Caliper over the brake rotor and into the caliper mounting bracket. Make sure the shoe lining guiding surfaces are correctly seated in the bracket. Tighten the bolts to 23 ft. lbs. (31 Nm).

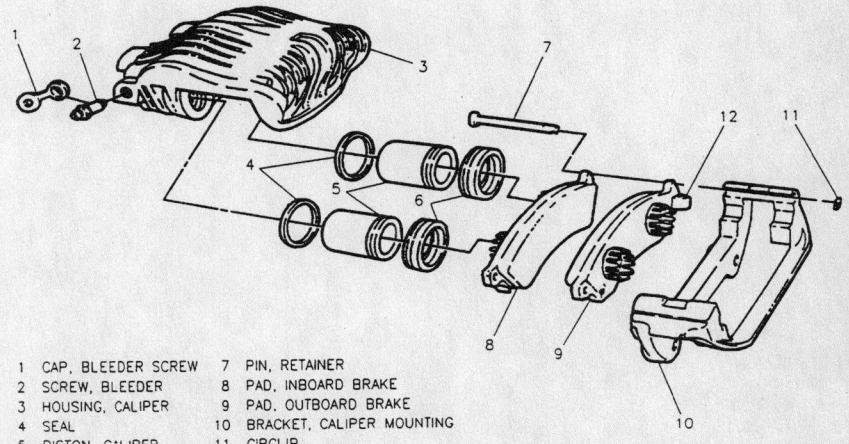

1	CAP, BLEEDER SCREW	7	PIN, RETAINER
2	SCREW, BLEEDER	8	PAD, INBOARD BRAKE
3	HOUSING, CALIPER	9	PAD, OUTBOARD BRAKE
4	SEAL	10	BRACKET, CALIPER MOUNTING
5	PISTON, CALIPER	11	CIRCLIP
6	BOOT	12	SPRING, BIAS

93006G63

Exploded view of the front caliper assembly

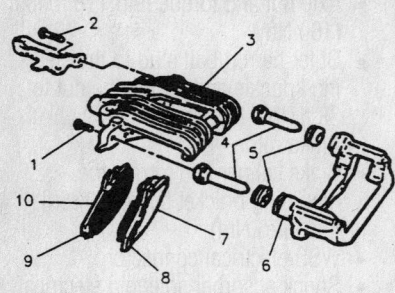

1 BOLT, UPPER GUIDE PIN
2 BOLT, LOWER GUIDE PIN
3 HOUSING, CALIPER
4 PIN, GUIDE
5 BOOT
6 BRACKET, MOUNTING
7 INSULATOR
8 PAD, OUTBOARD BRAKE
9 SENSOR, WEAR
10 PAD, INBOARD BRAKE

93006G64

Exploded view of the rear caliper assembly

- Brake hose inlet fitting using 2 new copper washers and the inlet fitting bolt. Torque the bolt to 37 ft. lbs. (50 Nm).

6. Properly bleed the entire brake system.

- Wheel assembly

7. Check the brake fluid and add as necessary.

8. Connect the negative battery cable, start the engine and pump the brake pedal slowly and firmly 3 times to seat the shoe and lining assemblies.

Disc Brake Pad

REMOVAL & INSTALLATION

Front

1. Before servicing the vehicle, refer to the precautions in the beginning of this section.
2. Disconnect the negative battery cable.
3. Remove the wheel.
4. Install a large C–clamp over the body of the caliper with the clamp ends against the rear of the caliper body and against the outboard pad. Tighten the clamp evenly to compress the pistons.
5. Remove the upper caliper bolt, then pivot the caliper downwards until enough clearance is achieved and support the caliper with wire.
6. Remove the pad and lining assemblies from the caliper.

To install:

7. Clean all residue from the pad and lining assembly guiding surfaces on the caliper housing and the mounting bracket.
8. Ensure the caliper pistons are fully compressed in their bores.
9. Install the outboard pad with the insulator to the caliper housing and the inboard pad with the wear sensor into the caliper pistons. Press the pads firmly until they are they are fully seated.
10. Install the caliper in position, insert and tighten the upper bolt to 23 ft. lbs. (31 Nm).

11. Install the wheel.
12. Connect the negative battery cable.

Rear

1. Before servicing the vehicle, refer to the precautions in the beginning of this section.
2. Disconnect the negative battery cable.
3. Remove the wheel.
4. Install a large C–clamp over the body of the caliper with the clamp ends against the rear of the caliper body and against the outboard pad. Tighten the clamp evenly to compress the pistons.
5. Remove the upper caliper bolt, then pivot the caliper downwards until enough clearance is achieved and support the caliper with wire.
6. Remove the pad and lining assemblies from the caliper.

To install:

7. Clean all residue from the pad and lining assembly guiding surfaces on the caliper housing and the mounting bracket.
8. Ensure the caliper pistons are fully compressed in their bores.
9. Install the outboard pad with the insulator to the caliper housing and the inboard pad with the wear sensor into the caliper pistons. Press the pads firmly until they are they are fully seated.
10. Install the caliper in position, insert and tighten the upper bolt to 23 ft. lbs. (31 Nm).
11. Install the wheel.
12. Connect the negative battery cable.

GLOSSARY

ABS: Anti-lock braking system. An electro-mechanical braking system which is designed to minimize or prevent wheel lock-up during braking.

ABSOLUTE PRESSURE: Atmospheric (barometric) pressure plus the pressure gauge reading.

ACCELERATOR PUMP: A small pump located in the carburetor that feeds fuel into the air/fuel mixture during acceleration.

ACCUMULATOR: A device that controls shift quality by cushioning the shock of hydraulic oil pressure being applied to a clutch or band.

ACTUATING MECHANISM: The mechanical output devices of a hydraulic system, for example, clutch pistons and band servos.

ACTUATOR: The output component of a hydraulic or electronic system.

ADVANCE: Setting the ignition timing so that spark occurs earlier before the piston reaches top dead center (TDC).

ADAPTIVE MEMORY (ADAPTIVE STRATEGY): The learning ability of the TCM or PCM to redefine its decision-making process to provide optimum shift quality.

AFTER TOP DEAD CENTER (ATDC): The point after the piston reaches the top of its travel on the compression stroke.

AIR BAG: Device on the inside of the car designed to inflate on impact of crash, protecting the occupants of the car.

AIR CHARGE TEMPERATURE (ACT) SENSOR: The temperature of the airflow into the engine is measured by an ACT sensor, usually located in the lower intake manifold or air cleaner.

AIR CLEANER: An assembly consisting of a housing, filter and any connecting ductwork. The filter element is made up of a porous paper, sometimes with a wire mesh screening, and is designed to prevent airborne particles from entering the engine through the carburetor or throttle body.

AIR INJECTION: One method of reducing harmful exhaust emissions by injecting air into each of the exhaust ports of an engine. The fresh air entering the hot exhaust manifold causes any remaining fuel to be burned before it can exit the tailpipe.

AIR PUMP: An emission control device that supplies fresh air to the exhaust manifold to aid in more completely burning exhaust gases.

AIR/FUEL RATIO: The ratio of air-to-gasoline by weight in the fuel mixture drawn into the engine.

ALDL (assembly line diagnostic link): Electrical connector for scanning ECM/PCM/TCM input and output devices.

ALIGNMENT RACK: A special drive-on vehicle lift apparatus/measuring device used to adjust a vehicle's toe, caster and camber angles.

ALL WHEEL DRIVE: Term used to describe a full time four wheel drive system or any other vehicle drive system that continuously delivers power to all four wheels. This system is found primarily on station wagon vehicles and SUVs not utilized for significant off road use.

ALTERNATING CURRENT (AC): Electric current that flows first in one direction, then in the opposite direction, continually reversing flow.

ALTERNATOR: A device which produces AC (alternating current) which is converted to DC (direct current) to charge the car battery.

AMMETER: An instrument, calibrated in amperes, used to measure the flow of an electrical current in a circuit. Ammeters are always connected in series with the circuit being tested.

AMPERAGE: The total amount of current (amperes) flowing in a circuit.

AMPLIFIER: A device used in an electrical circuit to increase the voltage of an output signal.

AMP/HR. RATING (BATTERY): Measurement of the ability of a battery to deliver a stated amount of current for a stated period of time. The higher the amp/hr. rating, the better the battery.

AMPERE: The rate of flow of electrical current present when one volt of electrical pressure is applied against one ohm of electrical resistance.

ANALOG COMPUTER: Any microprocessor that uses similar (analogous) electrical signals to make its calculations.

ANODIZED: A special coating applied to the surface of aluminum valves for extended service life.

ANTIFREEZE: A substance (ethylene or propylene glycol) added to the coolant to prevent freezing in cold weather.

ANTI-FOAM AGENTS: Minimize fluid foaming from the whipping action encountered in the converter and planetary action.

ANTI-WEAR AGENTS: Zinc agents that control wear on the gears, bushings, and thrust washers.

ANTI-LOCK BRAKING SYSTEM: A supplementary system to the base hydraulic system that prevents sustained lock-up of the wheels during braking as well as automatically controlling wheel slip.

ANTI-ROLL BAR: See stabilizer bar.

ARC: A flow of electricity through the air between two electrodes or contact points that produces a spark.

ARMATURE: A laminated, soft iron core wrapped by a wire that converts electrical energy to mechanical energy as in a motor or relay. When rotated in a magnetic field, it changes mechanical energy into electrical energy as in a generator.

ATDC: After Top Dead Center.

ATF: Automatic transmission fluid.

ATMOSPHERIC PRESSURE: The pressure on the Earth's surface caused by the weight of the air in the atmosphere. At sea level, this pressure is 14.7 psi at 32°F (101 kPa at 0°C).

ATOMIZATION: The breaking down of a liquid into a fine mist that can be suspended in air.

AUXILIARY ADD-ON COOLER: A supplemental transmission fluid cooling device that is installed in series with the heat exchanger (cooler), located inside the radiator, to provide additional support to cool the hot fluid leaving the torque converter.

AUXILIARY PRESSURE: An added fluid pressure that is introduced into a regulator or balanced valve system to control valve movement. The auxiliary pressure itself can be either a fixed or a variable value. (See balanced valve; regulator valve.)

AWD: All wheel drive.

AXIAL FORCE: A side or end thrust force acting in or along the same plane as the power flow.

AXIAL PLAY: Movement parallel to a shaft or bearing bore.

AXLE CAPACITY: The maximum load-carrying capacity of the axle itself, as specified by the manufacturer. This is usually a higher number than the GAWR.

AXLE RATIO: This is a number (3.07:1, 4.56:1, for example) expressing the ratio between driveshaft revolutions and wheel revolutions. A low numerical ratio allows the engine to work easier because it doesn't have to turn as fast. A high numerical ratio means that the engine has to turn more rpm's to move the wheels through the same number of turns.

BACKFIRE: The sudden combustion of gases in the intake or exhaust system that results in a loud explosion.

BACKLASH: The clearance or play between two parts, such as meshed gears.

BACKPRESSURE: Restrictions in the exhaust system that slow the exit of exhaust gases from the combustion chamber.

BAKELITE®: A heat resistant, plastic insulator material commonly used in printed circuit boards and transistorized components.

BALANCED VALVE: A valve that is positioned by opposing auxiliary hydraulic pressures and/or spring force. Examples include mainline regulator, throttle, and governor valves. (See regulator valve.)

BAND: A flexible ring of steel with an inner lining of friction material. When tightened around the outside of a drum, a planetary member is held stationary to the transmission/transaxle case.

BALL BEARING: A bearing made up of hardened inner and outer races between which hardened steel balls roll.

BALL JOINT: A ball and matching socket connecting suspension components (steering knuckle to lower control arms). It permits rotating movement in any direction between the components that are joined.

BARO (BAROMETRIC PRESSURE SENSOR): Measures the change in the intake manifold pressure caused by changes in altitude.

BAROMETRIC MANIFOLD ABSOLUTE PRESSURE (BMAP) SENSOR: Operates similarly to a conventional MAP sensor; reads intake manifold pressure and is also responsible for determining altitude and barometric pressure prior to engine operation.

BAROMETRIC PRESSURE: (See atmospheric pressure.)

BALLAST RESISTOR: A resistor in the primary ignition circuit that lowers voltage after the engine is started to reduce wear on ignition components.

BATTERY: A direct current electrical storage unit, consisting of the basic active materials of lead and sulfuric acid, which converts chemical energy into electrical energy. Used to provide current for the operation of the starter as well as other equipment, such as the radio, lighting, etc.

BEAD: The portion of a tire that holds it on the rim.

BEARING: A friction reducing, supportive device usually located between a stationary part and a moving part.

BEFORE TOP DEAD CENTER (BTDC): The point just before the piston reaches the top of its travel on the compression stroke.

BELTED TIRE: Tire construction similar to bias-ply tires, but using two or more layers of reinforced belts between body plies and the tread.

BEZEL: Piece of metal surrounding radio, headlights, gauges or similar components; sometimes used to hold the glass face of a gauge in the dash.

BIAS-PLY TIRE: Tire construction, using body ply reinforcing cords which run at alternating angles to the center line of the tread.

BI-METAL TEMPERATURE SENSOR: Any sensor or switch made of two dissimilar types of metal that bend when heated or cooled due to the different expansion rates of the alloys. These types of sensors usually function as an on/off switch.

BLOCK: See Engine Block.

BLOW-BY: Combustion gases, composed of water vapor and unburned fuel, that leak past the piston rings into the crankcase during normal engine operation. These gases are removed by the PCV system to prevent the buildup of harmful acids in the crankcase.

BOOK TIME: See Labor Time.

BOOK VALUE: The average value of a car, widely used to determine trade-in and resale value.

BOOST VALVE: Used at the base of the regulator valve to increase mainline pressure.

BORE: Diameter of a cylinder.

BRAKE CALIPER: The housing that fits over the brake disc. The caliper holds the brake pads, which are pressed against the discs by the caliper pistons when the brake pedal is depressed.

BRAKE HORSEPOWER (BHP): The actual horsepower available at the engine flywheel as measured by a dynamometer.

BRAKE FADE: Loss of braking power, usually caused by excessive heat after repeated brake applications.

BRAKE HORSEPOWER: Usable horsepower of an engine measured at the crankshaft.

BRAKE PAD: A brake shoe and lining assembly used with disc brakes.

BRAKE PROPORTIONING VALVE: A valve on the master cylinder which restricts hydraulic brake pressure to the wheels to a specified amount, preventing wheel lock-up.

BREAKAWAY: Often used by Chrysler to identify first-gear operation in D and 2 ranges. In these ranges, first-gear operation depends on a one-way roller clutch that holds on acceleration and releases (breaks away) on deceleration, resulting in a freewheeling coast-down condition.

BRAKE SHOE: The backing for the brake lining. The term is, however, usually applied to the assembly of the brake backing and lining.

BREAKER POINTS: A set of points inside the distributor, operated by a cam, which make and break the ignition circuit.

BRINNELLING: A wear pattern identified by a series of indentations at regular intervals. This condition is caused by a lack of lube, overload situations, and/or vibrations.

BTDC: Before Top Dead Center.

BUMP: Sudden and forceful apply of a clutch or band.

BUSHING: A liner, usually removable, for a bearing; an anti-friction liner used in place of a bearing.

CALIFORNIA ENGINE: An engine certified by the EPA for use in California only; conforms to more stringent emission regulations than Federal engine.

CALIPER: A hydraulically activated device in a disc brake system, which is mounted straddling the brake rotor (disc). The caliper contains at least one piston and two brake pads. Hydraulic pressure on the piston(s) forces the pads against the rotor.

CAPACITY: The quantity of electricity that can be delivered from a unit, as from a battery in ampere-hours, or output, as from a generator.

CAMBER: One of the factors of wheel alignment. Viewed from the front of the car, it is the inward or outward tilt of the wheel. The top of the tire will lean outward (positive camber) or inward (negative camber).

CAMSHAFT: A shaft in the engine on which are the lobes (cams) which operate the valves. The camshaft is driven by the crankshaft, via a belt, chain or gears, at one half the crankshaft speed.

CAPACITOR: A device which stores an electrical charge.

CARBON MONOXIDE (CO): A colorless, odorless gas given off as a normal byproduct of combustion. It is poisonous and extremely dangerous in confined areas, building up slowly to toxic levels without warning if adequate ventilation is not available.

CARBURETOR: A device, usually mounted on the intake manifold of an engine, which mixes the air and fuel in the proper proportion to allow even combustion.

CASTER: The forward or rearward tilt of an imaginary line drawn through the upper ball joint and the center of the wheel. Viewed from the sides, positive caster (forward tilt) lends directional stability, while negative caster (rearward tilt) produces instability.

CATALYTIC CONVERTER: A device installed in the exhaust system, like a muffler, that converts harmful byproducts of combustion into carbon dioxide and water vapor by means of a heat-producing chemical reaction.

CENTRIFUGAL ADVANCE: A mechanical method of advancing the spark timing by using flyweights in the distributor that react to centrifugal force generated by the distributor shaft rotation.

CENTRIFUGAL FORCE: The outward pull of a revolving object, away from the center of revolution. Centrifugal force increases with the speed of rotation.

CETANE RATING: A measure of the ignition value of diesel fuel. The higher the cetane rating, the better the fuel. Diesel fuel cetane rating is roughly comparable to gasoline octane rating.

CHECK VALVE: Any one-way valve installed to permit the flow of air, fuel or vacuum in one direction only.

CHOKE: The valve/plate that restricts the amount of air entering an engine on the induction stroke, thereby enriching the air/fuel ratio.

CHUGGLE: Bucking or jerking condition that may be engine related and may be most noticeable when converter clutch is engaged; similar to the feel of towing a trailer.

CIRCLIP: A split steel snapring that fits into a groove to hold various parts in place.

CIRCUIT BREAKER: A switch which protects an electrical circuit from overload by opening the circuit when the current flow exceeds a pre-determined level. Some circuit breakers must be reset manually, while most reset automatically.

CIRCUIT: Any unbroken path through which an electrical current can flow. Also used to describe fuel flow in some instances.

CIRCUIT, BYPASS: Another circuit in parallel with the major circuit through which power is diverted.

CIRCUIT, CLOSED: An electrical circuit in which there is no interruption of current flow.

CIRCUIT, GROUND: The non-insulated portion of a complete circuit used as a common potential point. In automotive circuits, the ground is composed of metal parts, such as the engine, body sheet metal, and frame and is usually a negative potential.

CIRCUIT, HOT: That portion of a circuit not at ground potential. The hot circuit is usually insulated and is connected to the positive side of the battery.

CIRCUIT, OPEN: A break or lack of contact in an electrical circuit, either intentional (switch) or unintentional (bad connection or broken wire).

CIRCUIT, PARALLEL: A circuit having two or more paths for current flow with common positive and negative tie points. The same voltage is applied to each load device or parallel branch.

CIRCUIT, SERIES: An electrical system in which separate parts are connected end to end, using one wire, to form a single path for current to flow.

CIRCUIT, SHORT: A circuit that is accidentally completed in an electrical path for which it was not intended.

CLAMPING (ISOLATION) DIODES: Diodes positioned in a circuit to prevent self-induction from damaging electronic components.

CLEARCOAT: A transparent layer which, when sprayed over a vehicle's paint job, adds gloss and depth as well as an additional protective coating to the finish.

CLUTCH: Part of the power train used to connect/disconnect power to the rear wheels.

CLUTCH, FLUID: The same as a fluid coupling. A fluid clutch or coupling performs the same function as a friction clutch by utilizing fluid friction and inertia as opposed to solid friction used by a friction clutch. (See fluid coupling.)

CLUTCH, FRICTION: A coupling device that provides a means of smooth and positive engagement and disengagement of engine torque to the vehicle powertrain. Transmission of power through the clutch is accomplished by bringing one or more rotating drive members into contact with complementing driven members.

COAST: Vehicle deceleration caused by engine braking conditions.

COEFFICIENT OF FRICTION: The amount of surface tension between two contacting surfaces; identified by a scientifically calculated number.

COIL: Part of the ignition system that boosts the relatively low voltage supplied by the car's electrical system to the high voltage required to fire the spark plugs.

COMBINATION MANIFOLD: An assembly which includes both the intake and exhaust manifolds in one casting.

COMBINATION VALVE: A device used in some fuel systems that routes fuel vapors to a charcoal storage canister instead of venting them into the atmosphere. The valve relieves fuel tank pressure and allows fresh air into the tank as the fuel level drops to prevent a vapor lock situation.

COMBUSTION CHAMBER: The part of the engine in the cylinder head where combustion takes place.

COMPOUND GEAR: A gear consisting of two or more simple gears with a common shaft.

COMPOUND PLANETARY: A gearset that has more than the three elements found in a simple gearset and is constructed by combining members of two planetary gearsets to create additional gear ratio possibilities.

COMPRESSION CHECK: A test involving removing each spark plug and inserting a gauge. When the engine is cranked, the gauge will record a pressure reading in the individual cylinder. General operating condition can be determined from a compression check.

COMPRESSION RATIO: The ratio of the volume between the piston and cylinder head when the piston is at the bottom of its stroke (bottom dead center) and when the piston is at the top of its stroke (top dead center).

COMPUTER: An electronic control module that correlates input data according to prearranged engineered instructions; used for the management of an actuator system or systems.

CONDENSER: An electrical device which acts to store an electrical charge, preventing voltage surges.

2. A radiator-like device in the air conditioning system in which refrigerant gas condenses into a liquid, giving off heat.

CONDUCTOR: Any material through which an electrical current can be transmitted easily.

CONNECTING ROD: The connecting link between the crankshaft and piston.

CONSTANT VELOCITY JOINT: Type of universal joint in a halfshaft assembly in which the output shaft turns at a constant angular velocity without variation, provided that the speed of the input shaft is constant.

CONTINUITY: Continuous or complete circuit. Can be checked with an ohmmeter.

CONTROL ARM: The upper or lower suspension components which are mounted on the frame and support the ball joints and steering knuckles.

CONVENTIONAL IGNITION: Ignition system which uses breaker points.

CONVERTER: (See torque converter.)

CONVERTER LOCKUP: The switching from hydrodynamic to direct mechanical drive, usually through the application of a friction element called the converter clutch.

COOLANT: Mixture of water and anti-freeze circulated through the engine to carry off heat produced by the engine.

CORROSION INHIBITOR: An inhibitor in ATF that prevents corrosion of bushings, thrust washers, and oil cooler brazed joints.

COUNTERSHAFT: An intermediate shaft which is rotated by a mainshaft and transmits, in turn, that rotation to a working part.

COUPLING PHASE: Occurs when the torque converter is operating at its greatest hydraulic efficiency. The speed differential between the impeller and the turbine is at its minimum. At this point, the stator freewheels, and there is no torque multiplication.

CRANKCASE: The lower part of an engine in which the crankshaft and related parts operate.

CRANKSHAFT: Engine component (connected to pistons by connecting rods) which converts the reciprocating (up and down) motion of pistons to rotary motion used to turn the driveshaft.

CURB WEIGHT: The weight of a vehicle without passengers or payload, but including all fluids (oil, gas, coolant, etc.) and other equipment specified as standard.

CURRENT: The flow (or rate) of electrons moving through a circuit. Current is measured in amperes (amp).

CURRENT FLOW CONVENTIONAL: Current flows through a circuit from the positive terminal of the source to the negative terminal (plus to minus).

CURRENT FLOW, ELECTRON: Current or electrons flow from the negative terminal of the source, through the circuit, to the positive terminal (minus to plus).

CV-JOINT: Constant velocity joint.

CYCLIC VIBRATIONS: The off-center movement of a rotating object that is affected by its initial balance, speed of rotation, and working angles.

CYLINDER BLOCK: See engine block.

CYLINDER HEAD: The detachable portion of the engine, usually fastened to the top of the cylinder block and containing all or most of the combustion chambers. On overhead valve engines, it contains the valves and their operating parts. On overhead cam engines, it contains the camshaft as well.

CYLINDER: In an engine, the round hole in the engine block in which the piston(s) ride.

DATA LINK CONNECTOR (DLC): Current acronym/term applied to the federally mandated, diagnostic junction connector that is used to monitor ECM/PC/TCM inputs, processing strategies, and outputs including diagnostic trouble codes (DTCs).

DEAD CENTER: The extreme top or bottom of the piston stroke.

DECELERATION BUMP: When referring to a torque converter clutch in the applied position, a sudden release of the accelerator pedal causes a forceful reversal of power through the drivetrain (engine braking), just prior to the apply plate actually being released.

DELAYED (LATE OR EXTENDED): Condition where shift is expected but does not occur for a period of time, for example, where clutch or band engagement does not occur as quickly as expected during part throttle or wide open throttle apply of accelerator or when manually downshifting to a lower range.

DETENT: A spring-loaded plunger, pin, ball, or pawl used as a holding device on a ratchet wheel or shaft. In automatic transmissions, a detent mechanism is used for locking the manual valve in place.

DETENT DOWNSHIFT: (See kickdown.)

DETERGENT: An additive in engine oil to improve its operating characteristics.

DETONATION: An unwanted explosion of the air/fuel mixture in the combustion chamber caused by excess heat and compression, advanced timing, or an overly lean mixture. Also referred to as "ping".

DEXRON®: A brand of automatic transmission fluid.

DIAGNOSTIC TROUBLE CODES (DTCs): A digital display from the control module memory that identifies the input, processor, or output device circuit that is related to the powertrain emission/driveability malfunction detected. Diagnostic trouble codes can be read by the MIL to flash any codes or by using a handheld scanner.

DIAPHRAGM: A thin, flexible wall separating two cavities, such as in a vacuum advance unit.

DIESELING: The engine continues to run after the car is shut off; caused by fuel continuing to be burned in the combustion chamber.

DIFFERENTIAL: A geared assembly which allows the transmission of motion between drive axles, giving one axle the ability to rotate faster than the other, as in cornering.

DIFFERENTIAL AREAS: When opposing faces of a spool valve are acted upon by the same pressure but their areas differ in size, the face with the larger area produces the differential force and valve movement. (See spool valve.)

DIFFERENTIAL FORCE: (See differential areas)

DIGITAL READOUT: A display of numbers or a combination of numbers and letters.

DIGITAL VOLT OHMMETER: An electronic diagnostic tool used to measure voltage, ohms and amps as well as several other functions, with the readings displayed on a digital screen in tenths, hundredths and thousandths.

DIODE: An electrical device that will allow current to flow in one direction only.

DIRECT CURRENT (DC): Electrical current that flows in one direction only.

DIRECT DRIVE: The gear ratio is 1:1, with no change occurring in the torque and speed input/output relationship.

DISC BRAKE: A hydraulic braking assembly consisting of a brake disc, or rotor, mounted on an axle shaft, and a caliper assembly containing, usually two brake pads which are activated by hydraulic pressure. The pads are forced against the sides of the disc, creating friction which slows the vehicle.

DISPERSANTS: Suspend dirt and prevent sludge buildup in a liquid, such as engine oil.

DOUBLE BUMP (DOUBLE FEEL): Two sudden and forceful applies of a clutch or band.

DISPLACEMENT: The total volume of air that is displaced by all pistons as the engine turns through one complete revolution.

DISTRIBUTOR: A mechanically driven device on an engine which is responsible for electrically firing the spark plug at a pre-determined point of the piston stroke.

DOHC: Double overhead camshaft.

DOUBLE OVERHEAD CAMSHAFT: The engine utilizes two camshafts mounted in one cylinder head. One camshaft operates the exhaust valves, while the other operates the intake valves.

DOWEL PIN: A pin, inserted in mating holes in two different parts allowing those parts to maintain a fixed relationship.

DRIVELINE: The drive connection between the transmission and the drive wheels.

DRIVE TRAIN: The components that transmit the flow of power from the engine to the wheels. The components include the clutch, transmission, driveshafts (or axle shafts in front wheel drive), U-joints and differential.

DRUM BRAKE: A braking system which consists of two brake shoes and one or two wheel cylinders, mounted on a fixed backing plate, and a brake drum, mounted on an axle, which revolves around the assembly.

DRY CHARGED BATTERY: Battery to which electrolyte is added when the battery is placed in service.

DVOM: Digital volt ohmmeter

DWELL: The rate, measured in degrees of shaft rotation, at which an electrical circuit cycles on and off.

DYNAMIC: An application in which there is rotating or reciprocating motion between the parts.

EARLY: Condition where shift occurs before vehicle has reached proper speed, which tends to labor engine after upshift.

EBCM: See Electronic Control Unit (ECU).

ECM: See Electronic Control Unit (ECU).

ECU: Electronic control unit.

ELECTRODE: Conductor (positive or negative) of electric current.

ELECTROLYSIS: A surface etching or bonding of current conducting transmission/transaxle components that may occur when grounding straps are missing or in poor condition.

ELECTROLYTE: A solution of water and sulfuric acid used to activate the battery. Electrolyte is extremely corrosive.

ELECTROMAGNET: A coil that produces a magnetic field when current flows through its windings.

ELECTROMAGNETIC INDUCTION: A method to create (generate) current flow through the use of magnetism.

ELECTROMAGNETISM: The effects surrounding the relationship between electricity and magnetism.

ELECTROMOTIVE FORCE (EMF): The force or pressure (voltage) that causes current movement in an electrical circuit.

ELECTRONIC CONTROL UNIT: A digital computer that controls engine (and sometimes transmission, brake or other vehicle system) functions based on data received from various sensors. Examples used by some manufacturers include Electronic Brake Control Module (EBCM), Engine Control Module (ECM), Powertrain Control Module (PCM) or Vehicle Control Module (VCM).

ELECTRONIC IGNITION: A system in which the timing and firing of the spark plugs is controlled by an electronic control unit, usually called a module. These systems have no points or condenser.

ELECTRONIC PRESSURE CONTROL (EPC) SOLENOID: A specially designed solenoid containing a spool valve and spring assembly to control fluid mainline pressure. A variable current flow, controlled by the ECM/PCM, varies the internal force of the solenoid on the spool valve and resulting mainline pressure. (See variable force solenoid.)

ELECTRONICS: Miniaturized electrical circuits utilizing semiconductors, solid-state devices, and printed circuits. Electronic circuits utilize small amounts of power.

ELECTRONIFICATION: The application of electronic circuitry to a mechanical device. Regarding automatic transmissions, electrification is incorporated into converter clutch lockup, shift scheduling, and line pressure control systems.

ELECTROSTATIC DISCHARGE (ESD): An unwanted, high-voltage electrical current released by an individual who has taken on a static charge of electricity. Electronic components can be easily damaged by ESD.

ELEMENT: A device within a hydrodynamic drive unit designed with a set of blades to direct fluid flow.

ENAMEL: Type of paint that dries to a smooth, glossy finish.

END BUMP (END FEEL OR SLIP BUMP): Firmer feel at end of shift when compared with feel at start of shift.

END-PLAY: The clearance/gap between two components that allows for expansion of the parts as they warm up, to prevent binding and to allow space for lubrication.

ENERGY: The ability or capacity to do work.

ENGINE: The primary motor or power apparatus of a vehicle, which converts liquid or gas fuel into mechanical energy.

ENGINE BLOCK: The basic engine casting containing the cylinders, the crankshaft main bearings, as well as machined surfaces for the mounting of other components such as the cylinder head, oil pan, transmission, etc.

ENGINE BRAKING: Use of engine to slow vehicle by manually downshifting during zero-throttle coast down.

ENGINE CONTROL MODULE (ECM): Manages the engine and incorporates output control over the torque converter clutch solenoid. (Note: Current designation for the ECM in late model vehicles is PCM.)

ENGINE COOLANT TEMPERATURE (ECT) SENSOR: Prevents converter clutch engagement with a cold engine; also used for shift timing and shift quality.

EP LUBRICANT: EP (extreme pressure) lubricants are specially formulated for use with gears involving heavy loads (transmissions, differentials, etc.).

ETHYL: A substance added to gasoline to improve its resistance to knock, by slowing down the rate of combustion.

ETHYLENE GLYCOL: The base substance of antifreeze.

EXHAUST MANIFOLD: A set of cast passages or pipes which conduct exhaust gases from the engine.

FAIL-SAFE (BACKUP) CONTROL: A substitute value used by the PCM/TCM to replace a faulty signal from an input sensor. The temporary value allows the vehicle to continue to be operated.

FAST IDLE: The speed of the engine when the choke is on. Fast idle speeds engine warm-up.

FEDERAL ENGINE: An engine certified by the EPA for use in any of the 49 states (except California).

FEEDBACK: A circuit malfunction whereby current can find another path to feed load devices.

FEELER GAUGE: A blade, usually metal, of precisely predetermined thickness, used to measure the clearance between two parts.

FILAMENT: The part of a bulb that glows; the filament creates high resistance to current flow and actually glows from the resulting heat.

FINAL DRIVE: An essential part of the axle drive assembly where final gear reduction takes place in the powertrain. In RWD applications and north-south FWD applications, it must also change the power flow direction to the axle shaft by ninety degrees. (Also see axle ratio).

FIRING ORDER: The order in which combustion occurs in the cylinders of an engine. Also the order in which spark is distributed to the plugs by the distributor.

FIRM: A noticeable quick apply of a clutch or band that is considered normal with medium to heavy throttle shift; should not be confused with harsh or rough.

FLAME FRONT: The term used to describe certain aspects of the fuel explosion in the cylinders. The flame front should move in a controlled pattern across the cylinder, rather than simply exploding immediately.

FLARE (SLIPPING): A quick increase in engine rpm accompanied by momentary loss of torque; generally occurs during shift.

FLAT ENGINE: Engine design in which the pistons are horizontally opposed. Porsche, Subaru and some old VW are common examples of flat engines.

FLAT RATE: A dealership term referring to the amount of money paid to a technician for a repair or diagnostic service based on that particular service versus dealership's labor time (NOT based on the actual time the technician spent on the job).

FLAT SPOT: A point during acceleration when the engine seems to lose power for an instant.

FLOODING: The presence of too much fuel in the intake manifold and combustion chamber which prevents the air/fuel mixture from firing, thereby causing a no-start situation.

FLUID: A fluid can be either liquid or gas. In hydraulics, a liquid is used for transmitting force or motion.

FLUID COUPLING: The simplest form of hydrodynamic drive, the fluid coupling consists of two look-alike members with straight radial varies referred to as the impeller (pump) and the turbine. Input torque is always equal to the output torque.

FLUID DRIVE: Either a fluid coupling or a fluid torque converter. (See hydrodynamic drive units.)

FLUID TORQUE CONVERTER: A hydrodynamic drive that has the ability to act both as a torque multiplier and fluid coupling. (See hydrodynamic drive units; torque converter.)

FLUID VISCOSITY: The resistance of a liquid to flow. A cold fluid (oil) has greater viscosity and flows more slowly than a hot fluid (oil).

FLYWHEEL: A heavy disc of metal attached to the rear of the crankshaft. It smoothes the firing impulses of the engine and keeps the crankshaft turning during periods when no firing takes place. The starter also engages the flywheel to start the engine.

FOOT POUND (ft. lbs., lbs. ft. or sometimes, ft. lb.): The amount of energy or work needed to raise an item weighing one pound, a distance of one foot.

FREEZE PLUG: A plug in the engine block which will be pushed out if the coolant freezes. Sometimes called expansion plugs, they protect the block from cracking should the coolant freeze.

FRICTION: The resistance that occurs between contacting surfaces. This relationship is expressed by a ratio called the coefficient of friction (CL).

FRICTION, COEFFICIENT OF: The amount of surface tension between two contacting surfaces; expressed by a scientifically calculated number.

FRONT END ALIGNMENT: A service to set caster, camber and toe-in to the correct specifications. This will ensure that the car steers and handles properly and that the tires wear properly.

FRICTION MODIFIER: Changes the coefficient of friction of the fluid between the mating steel and composition clutch/band surfaces during the engagement process and allows for a certain amount of intentional slipping for a good "shift-feel".

FRONTAL AREA: The total frontal area of a vehicle exposed to air flow.

FUEL FILTER: A component of the fuel system containing a porous paper element used to prevent any impurities from entering the engine through the fuel system. It usually takes the form of a canister-like housing, mounted in-line with the fuel hose, located anywhere on a vehicle between the fuel tank and engine.

FUEL INJECTION: A system replacing the carburetor that sprays fuel into the cylinder through nozzles. The amount of fuel can be more precisely controlled with fuel injection.

FULL FLOATING AXLE: An axle in which the axle housing extends through the wheel giving bearing support on the outside of the housing. The front axle of a four-wheel drive vehicle is usually a full floating axle, as are the rear axles of many larger (1 ton and over) pick-ups and vans.

FULL-TIME FOUR-WHEEL DRIVE: A four-wheel drive system that continuously delivers power to all four wheels. A differential between the front and rear driveshafts permits variations in axle speeds to control gear wind-up without damage.

FULL THROTTLE DETENT DOWNSHIFT: A quick apply of accelerator pedal to its full travel, forcing a downshift.

FUSE: A protective device in a circuit which prevents circuit overload by breaking the circuit when a specific amperage is present. The device is constructed around a strip or wire of a lower amperage rating than the circuit it is designed to protect. When an amperage higher than that stamped on the fuse is present in the circuit, the strip or wire melts, opening the circuit.

FUSIBLE LINK: A piece of wire in a wiring harness that performs the same job as a fuse. If overloaded, the fusible link will melt and interrupt the circuit.

FWD: Front wheel drive.

GAWR: (Gross axle weight rating) the total maximum weight an axle is designed to carry.

GCW: (Gross combined weight) total combined weight of a tow vehicle and trailer.

GARAGE SHIFT: initial engagement feel of transmission, neutral to reverse or neutral to a forward drive.

GARAGE SHIFT FEEL: A quick check of the engagement quality and responsiveness of reverse and forward gears. This test is done with the vehicle stationary.

GEAR: A toothed mechanical device that acts as a rotating lever to transmit power or turning effort from one shaft to another. (See gear ratio.)

GEAR RATIO: A ratio expressing the number of turns a smaller gear will make to turn a larger gear through one revolution. The ratio is found by dividing the number of teeth on the smaller gear into the number of teeth on the larger gear.

GEARBOX: Transmission

GEAR REDUCTION: Torque is multiplied and speed decreased by the factor of the gear ratio. For example, a 3:1 gear ratio changes an input torque of 180 ft. lbs. and an input speed of 2700 rpm to 540 Ft. lbs. and 900 rpm, respectively. (No account is taken of frictional losses, which are always present.)

GEARTRAIN: A succession of intermeshing gears that form an assembly and provide for one or more torque changes as the power input is transmitted to the power output.

GEL COAT: A thin coat of plastic resin covering fiberglass body panels.

GENERATOR: A device which produces direct current (DC) necessary to charge the battery.

GOVERNOR: A device that senses vehicle speed and generates a hydraulic oil pressure. As vehicle speed increases, governor oil pressure rises.

GROUND CIRCUIT: (See circuit, ground.)

GROUND SIDE SWITCHING: The electrical/electronic circuit control switch is located after the circuit load.

GVWR: (Gross vehicle weight rating) total maximum weight a vehicle is designed to carry including the weight of the vehicle, passengers, equipment, gas, oil, etc.

HALOGEN: A special type of lamp known for its quality of brilliant white light. Originally used for fog lights and driving lights.

HARD CODES: DTCs that are present at the time of testing; also called continuous or current codes.

HARSH(ROUGH): An apply of a clutch or band that is more noticeable than a firm one; considered undesirable at any throttle position.

HEADER TANK: An expansion tank for the radiator coolant. It can be located remotely or built into the radiator.

HEAT RANGE: A term used to describe the ability of a spark plug to carry away heat. Plugs with longer nosed insulators take longer to carry heat off effectively.

HEAT RISER: A flapper in the exhaust manifold that is closed when the engine is cold, causing hot exhaust gases to heat the intake manifold providing better cold engine operation. A thermostatic spring opens the flapper when the engine warms up.

HEAVY THROTTLE: Approximately three-fourths of accelerator pedal travel.

HEMI: A name given an engine using hemispherical combustion chambers.

HERTZ (HZ): The international unit of frequency equal to one cycle per second (10,000 Hertz equals 10,000 cycles per second).

HIGH-IMPEDANCE DVOM (DIGITAL VOLT-OHMMETER): This styled device provides a built-in resistance value and is capable of limiting circuit current flow to safe milliamp levels.

HIGH RESISTANCE: Often refers to a circuit where there is an excessive amount of opposition to normal current flow.

HORSEPOWER: A measurement of the amount of work; one horsepower is the amount of work necessary to lift 33,000 lbs. one foot in one minute. Brake horsepower (bhp) is the horsepower delivered by an engine on a dynamometer. Net horsepower is the power remaining (measured at the flywheel of the engine) that can be used to turn the wheels after power is consumed through friction and running the engine accessories (water pump, alternator, air pump, fan etc.)

HOT CIRCUIT: (See circuit, hot; hot lead.)

HOT LEAD: A wire or conductor in the power side of the circuit. (See circuit, hot.)

HOT SIDE SWITCHING: The electrical/electronic circuit control switch is located before the circuit load.

HUB: The center part of a wheel or gear.

HUNTING (BUSYNESS): Repeating quick series of up-shifts and downshifts that causes noticeable change in engine rpm, for example, as in a 4-3-4 shift pattern.

HYDRAULICS: The use of liquid under pressure to transfer force of motion.

HYDROCARBON (HC): Any chemical compound made up of hydrogen and carbon. A major pollutant formed by the engine as a by-product of combustion.

HYDRODYNAMIC DRIVE UNITS: Devices that transmit power solely by the action of a kinetic fluid flow in a closed recirculating path. An impeller energizes the fluid and discharges the high-speed jet stream into the turbine for power output.

HYDROMETER: An instrument used to measure the specific gravity of a solution.

HYDROPLANING: A phenomenon of driving when water builds up under the tire tread, causing it to lose contact with the road. Slowing down will usually restore normal tire contact with the road.

HYPOID GEARSET: The drive pinion gear may be placed below or above the centerline of the driven gear; often used as a final drive gearset.

IDLE MIXTURE: The mixture of air and fuel (usually about 14:1) being fed to the cylinders. The idle mixture screw(s) are sometimes adjusted as part of a tune-up.

IDLER ARM: Component of the steering linkage which is a geometric duplicate of the steering gear arm. It supports the right side of the center steering link.

IMPELLER: Often called a pump, the impeller is the power input (drive) member of a hydrodynamic drive. As part of the torque converter cover, it acts as a centrifugal pump and puts the fluid in motion.

INCH POUND (inch lbs.; sometimes in. lb or in. lbs.): One twelfth of a foot pound.

INDUCTANCE: The force that produces voltage when a conductor is passed through a magnetic field.

INDUCTION: A means of transferring electrical energy in the form of a magnetic field. Principle used in the ignition coil to increase voltage.

INITIAL FEEL: A distinct firmer feel at start of shift when compared with feel at finish of shift.

INJECTOR: A device which receives metered fuel under relatively low pressure and is activated to inject the fuel into the engine under relatively high pressure at a predetermined time.

INPUT: In an automatic transmission, the source of power from the engine is absorbed by the torque converter, which provides the power input into the transmission. The turbine drives the input(turbine)shaft.

INPUT SHAFT: The shaft to which torque is applied, usually carrying the driving gear or gears.

INTAKE MANIFOLD: A casting of passages or pipes used to conduct air or a fuel/air mixture to the cylinders.

INTERNAL GEAR: The ring-like outer gear of a planetary gearset with the gear teeth cut on the inside of the ring to provide a mesh with the planet pinions.

ISOLATION (CLAMPING) DIODES: Diodes positioned in a circuit to prevent self-induction from damaging electronic components.

IX ROTARY GEAR PUMP: Contains two rotating members, one shaped with internal gear teeth and the other with external gear teeth. As the gears separate, the fluid fills the gaps between gear teeth, is pulled across a crescent-shaped divider, and then is forced to flow through the outlet as the gears mesh.

IX ROTARY LOBE PUMP: Sometimes referred to as a gerotor type pump. Two rotating members, one shaped with internal lobes and the other with external lobes, separate and then mesh to cause fluid to flow.

JOURNAL: The bearing surface within which a shaft operates.

JUMPER CABLES: Two heavy duty wires with large alligator clips used to provide power from a charged battery to a discharged battery mounted in a vehicle.

JUMPSTART: Utilizing the sufficiently charged battery of one vehicle to start the engine of another vehicle with a discharged battery by the use of jumper cables.

KEY: A small block usually fitted in a notch between a shaft and a hub to prevent slippage of the two parts.

KICKDOWN: Detent downshift system; either linkage, cable, or electrically controlled.

KILO: A prefix used in the metric system to indicate one thousand.

KNOCK: Noise which results from the spontaneous ignition of a portion of the air-fuel mixture in the engine cylinder caused by overly advanced ignition timing or use of incorrectly low octane fuel for that engine.

KNOCK SENSOR: An input device that responds to spark knock, caused by over advanced ignition timing.

LABOR TIME: A specific amount of time required to perform a certain repair or diagnostic service as defined by a vehicle or after-market manufacturer .

LACQUER: A quick-drying automotive paint.

LATE: Shift that occurs when engine is at higher than normal rpm for given amount of throttle.

LIGHT-EMITTING DIODE (LED): A semiconductor diode that emits light as electrical current flows through it; used in some electronic display devices to emit a red or other color light.

LIGHT THROTTLE: Approximately one-fourth of accelerator pedal travel.

LIMITED SLIP: A type of differential which transfers driving force to the wheel with the best traction.

LIMP-IN MODE: Electrical shutdown of the transmission/ transaxle output solenoids, allowing only forward and reverse gears that are hydraulically energized by the manual valve. This permits the vehicle to be driven to a service facility for repair.

LIP SEAL: Molded synthetic rubber seal designed with an outer sealing edge (lip) that points into the fluid containing area to be sealed. This type of seal is used where rotational and axial forces are present.

LITHIUM-BASE GREASE: Chassis and wheel bearing grease using lithium as a base. Not compatible with sodium-base grease.

LOAD DEVICE: A circuit's resistance that converts the electrical energy into light, sound, heat, or mechanical movement.

LOAD RANGE: Indicates the number of plies at which a tire is rated. Load range B equals four-ply rating; C equals six-ply rating; and, D equals an eight-ply rating.

LOAD TORQUE: The amount of output torque needed from the transmission/transaxle to overcome the vehicle load.

LOCKING HUBS: Accessories used on part-time four-wheel drive systems that allow the front wheels to be disengaged from the drive train when four-wheel drive is not being used. When four-wheel drive is desired, the hubs are engaged, locking the wheels to the drive train.

LOCKUP CONVERTER: A torque converter that operates hydraulically and mechanically. When an internal apply plate (lockup plate) clamps to the torque converter cover, hydraulic slippage is eliminated.

LOCK RING: See Circlip or Snapring

MAGNET: Any body with the property of attracting iron or steel.

MAGNETIC FIELD: The area surrounding the poles of a magnet that is affected by its attraction or repulsion forces.

MAIN LINE PRESSURE: Often called control pressure or line pressure, it refers to the pressure of the oil leaving the pump and is controlled by the pressure regulator valve.

MALFUNCTION INDICATOR LAMP (MIL): Previously known as a check engine light, the dash-mounted MIL illuminates and signals the driver that an emission or driveability problem with the powertrain has been detected by the ECM/PCM. When this occurs, at least one diagnostic trouble code (DTC) has been stored into the control module memory.

MANIFOLD ABSOLUTE PRESSURE (MAP) SENSOR: Reads the amount of air pressure (vacuum) in the engine's intake manifold system; its signal is used to analyze engine load conditions.

MANIFOLD VACUUM: Low pressure in an engine intake manifold formed just below the throttle plates. Manifold vacuum is highest at idle and drops under acceleration.

MANIFOLD: A casting of passages or set of pipes which connect the cylinders to an inlet or outlet source.

MANUAL LEVER POSITION SWITCH (MLPS): A mechanical switching unit that is typically mounted externally to the transmission/transaxle to inform the PCM/ECM which gear range the driver has selected.

MANUAL VALVE: Located inside the transmission/transaxle, it is directly connected to the driver's shift lever. The position of the manual valve determines which hydraulic circuits will be charged with oil pressure and the operating mode of the transmission.

MANUAL VALVE LEVER POSITION SENSOR (MVLPS): The input from this device tells the TCM what gear range was selected.

MASS AIR FLOW (MAF) SENSOR: Measures the airflow into the engine.

MASTER CYLINDER: The primary fluid pressurizing device in a hydraulic system. In automotive use, it is found in brake and hydraulic clutch systems and is pedal activated, either directly or, in a power brake system, through the power booster.

MacPherson STRUT: A suspension component combining a shock absorber and spring in one unit.

MEDIUM THROTTLE: Approximately one-half of accelerator pedal travel.

MEGA: A metric prefix indicating one million.

MEMBER: An independent component of a hydrodynamic unit such as an impeller, a stator, or a turbine. It may have one or more elements.

MERCON: A fluid developed by Ford Motor Company in 1988. It contains a friction modifier and closely resembles operating characteristics of Dexron.

METAL SEALING RINGS: Made from cast iron or aluminum, their primary application is with dynamic components involving pressure sealing circuits of rotating members. These rings are designed with either butt or hook lock end joints.

METER (ANALOG): A linear-style meter representing data as lengths; a needle-style instrument interfacing with logical numerical increments. This style of electrical meter uses relatively low impedance internal resistance and cannot be used for testing electronic circuitry.

METER (DIGITAL): Uses numbers as a direct readout to show values. Most meters of this style use high impedance internal resistance and must be used for testing low current electronic circuitry.

MICRO: A metric prefix indicating one-millionth (0.000001).

MILLI: A metric prefix indicating one-thousandth (0.001).

MINIMUM THROTTLE: The least amount of throttle opening required for upshift; normally close to zero throttle.

MISFIRE: Condition occurring when the fuel mixture in a cylinder fails to ignite, causing the engine to run roughly.

MODULE: Electronic control unit, amplifier or igniter of solid state or integrated design which controls the current flow in the ignition primary circuit based on input from the pick-up coil. When the module opens the primary circuit, high secondary voltage is induced in the coil.

MODULATED: In an electronic-hydraulic converter clutch system (or shift valve system), the term modulated refers to the pulsing of a solenoid, at a variable rate. This action controls the buildup of oil pressure in the hydraulic circuit to allow a controlled amount of clutch slippage.

MODULATED CONVERTER CLUTCH CONTROL (MCCC): A pulse width duty cycle valve that controls the converter lockup apply pressure and maximizes smoother transitions between lock and unlock conditions.

MODULATOR PRESSURE (THROTTLE PRESSURE): A hydraulic signal oil pressure relating to the amount of engine load, based on either the amount of throttle plate opening or engine vacuum.

MODULATOR VALVE: A regulator valve that is controlled by engine vacuum, providing a hydraulic pressure that varies in relation to engine torque. The hydraulic torque signal functions to delay the shift pattern and provide a line pressure boost. (See throttle valve.)

MOTOR: An electromagnetic device used to convert electrical energy into mechanical energy.

MULTIPLE-DISC CLUTCH: A grouping of steel and friction lined plates that, when compressed together by hydraulic pressure acting upon a piston, lock or unlock a planetary member.

MULTI-WEIGHT: Type of oil that provides adequate lubrication at both high and low temperatures.

needed to move one amp through a resistance of one ohm.

MUSHY: Same as soft; slow and drawn out clutch apply with very little shift feel.

MUTUAL INDUCTION: The generation of current from one wire circuit to another by movement of the magnetic field surrounding a current-carrying circuit as its ampere flow increases or decreases.

NEEDLE BEARING: A bearing which consists of a number (usually a large number) of long, thin rollers.

NITROGEN OXIDE (NOx): One of the three basic pollutants found in the exhaust emission of an internal combustion engine. The amount of NOx usually varies in an inverse proportion to the amount of HC and CO.

NONPOSITIVE SEALING: A sealing method that allows some minor leakage, which normally assists in lubrication.

O2 SENSOR: Located in the engine's exhaust system, it is an input device to the ECM/PCM for managing the fuel delivery and ignition system. A scanner can be used to observe the fluctuating voltage readings produced by an O2 sensor as the oxygen content of the exhaust is analyzed.

O-RING SEAL: Molded synthetic rubber seal designed with a circular cross-section. This type of seal is used primarily in static applications.

OBD II (ON-BOARD DIAGNOSTICS, SECOND GENERATION): Refers to the federal law mandating tighter control of 1996 and newer vehicle emissions, active monitoring of related devices, and standardization of terminology, data link connectors, and other technician concerns.

OCTANE RATING: A number, indicating the quality of gasoline based on its ability to resist knock. The higher the number, the better the quality. Higher compression engines require higher octane gas.

OEM: Original Equipment Manufactured. OEM equipment is that furnished standard by the manufacturer.

OFFSET: The distance between the vertical center of the wheel and the mounting surface at the lugs. Offset is positive if the center is outside the lug circle; negative offset puts the center line inside the lug circle.

OHM'S LAW: A law of electricity that states the relationship between voltage, current, and resistance. Volts = amperes x ohms

OHM: The unit used to measure the resistance of conductor-to-electrical flow. One ohm is the amount of resistance that limits current flow to one ampere in a circuit with one volt of pressure.

OHMMETER: An instrument used for measuring the resistance, in ohms, in an electrical circuit.

ONE-WAY CLUTCH: A mechanical clutch of roller or sprag design that resists torque or transmits power in one direction only. It is used to either hold or drive a planetary member.

ONE-WAY ROLLER CLUTCH: A mechanical device that transmits or holds torque in one direction only.

OPEN CIRCUIT: A break or lack of contact in an electrical circuit, either intentional (switch) or unintentional (bad connection or broken wire).

ORIFICE: Located in hydraulic oil circuits, it acts as a restriction. It slows down fluid flow to either create back pressure or delay pressure buildup downstream.

OSCILLOSCOPE: A piece of test equipment that shows electric impulses as a pattern on a screen. Engine performance can be analyzed by interpreting these patterns.

OUTPUT SHAFT: The shaft which transmits torque from a device, such as a transmission.

OUTPUT SPEED SENSOR (OSS): Identifies transmission/transaxle output shaft speed for shift timing and may be used to calculate TCC slip; often functions as the VSS (vehicle speed sensor).

OVERDRIVE: (1.) A device attached to or incorporated in a transmission/transaxle that allows the engine to turn less than one full revolution for every complete revolution of the wheels. The net effect is to reduce engine rpm, thereby using less fuel. A typical overdrive gear ratio would be .87:1, instead of the normal 1:1 in high gear. (2.) A gear assembly which produces more shaft revolutions than that transmitted to it.

OVERDRIVE PLANETARY GEARSET: A single planetary gearset designed to provide a direct drive and overdrive ratio. When coupled to a three-speed transmission/transaxle configuration, a four-speed/overdrive unit is present.

OVERHEAD CAMSHAFT (OHC): An engine configuration in which the camshaft is mounted on top of the cylinder head and operates the valve either directly or by means of rocker arms.

OVERHEAD VALVE (OHV): An engine configuration in which all of the valves are located in the cylinder head and the camshaft is located in the cylinder block. The camshaft operates the valves via lifters and pushrods.

OVERRUNCLUTCH: Another name for a one-way mechanical clutch. Applies to both roller and sprag designs.

OVERSTEER: The tendency of some vehicles, when steering into a turn, to over-respond or steer more than required, which could result in excessive slip of the rear wheels. Opposite of under-steer.

OXIDATION STABILIZERS: Absorb and dissipate heat. Automatic transmission fluid has high resistance to varnish and sludge buildup that occurs from excessive heat that is generated primarily in the torque converter. Local temperatures as high as 6000F (3150C) can occur at the clutch plates during engagement, and this heat must be absorbed and dissipated. If the fluid cannot withstand the heat, it burns or oxidizes, resulting in an almost immediate destruction of friction materials, clogged filter screen and hydraulic passages, and sticky valves.

OXIDES OF NITROGEN: See nitrogen oxide (NOx).

OXYGEN SENSOR: Used with a feedback system to sense the presence of oxygen in the exhaust gas and signal the computer which can use the voltage signal to determine engine operating efficiency and adjust the air/fuel ratio.

PARALLEL CIRCUIT: (See circuit, parallel.)

PARTS WASHER: A basin or tub, usually with a built-in pump mechanism and hose used for circulating chemical solvent for the purpose of cleaning greasy, oily and dirty components.

PART-TIME FOUR WHEEL DRIVE: A system that is normally in the two wheel drive mode and only runs in four-wheel drive when the system is manually engaged because more traction is desired. Two or four wheel drive is normally selected by a lever to engage the front axle, but if locking hubs are used, these must also be manually engaged in the Lock position. Otherwise, the front axle will not drive the front wheels.

PASSIVE RESTRAINT: Safety systems such as air bags or automatic seat belts which operate with no action required on the part of the driver or passenger. Mandated by Federal regulations on all vehicles sold in the U.S. after 1990.

PAYLOAD: The weight the vehicle is capable of carrying in addition to its own weight. Payload includes weight of the driver, passengers and cargo, but not coolant, fuel, lubricant, spare tire, etc.

PCM: Powertrain control module.

PCV VALVE: A valve usually located in the rocker cover that vents crankcase vapors back into the engine to be reburned.

PERCOLATION: A condition in which the fuel actually "boils," due to excessive heat. Percolation prevents proper atomization of the fuel causing rough running.

PICK-UP COIL: The coil in which voltage is induced in an electronic ignition.

PING: A metallic rattling sound produced by the engine during acceleration. It is usually due to incorrect ignition timing or a poor grade of gasoline.

PINION: The smaller of two gears. The rear axle pinion drives the ring gear which transmits motion to the axle shafts.

PINION GEAR: The smallest gear in a drive gear assembly.

PISTON: A disc or cup that fits in a cylinder bore and is free to move. In hydraulics, it provides the means of converting hydraulic pressure into a usable force. Examples of piston applications are found in servo, clutch, and accumulator units.

PISTON RING: An open-ended ring which fits into a groove on the outer diameter of the piston. Its chief function is to form a seal between the piston and cylinder wall. Most automotive pistons have three rings: two for compression sealing; one for oil sealing.

PITMAN ARM: A lever which transmits steering force from the steering gear to the steering linkage.

PLANET CARRIER: A basic member of a planetary gear assembly that carries the pinion gears.

PLANET PINIONS: Gears housed in a planet carrier that are in constant mesh with the sun gear and internal gear. Because they have their own independent rotating centers, the pinions are capable of rotating around the sun gear or the inside of the internal gear.

PLANETARY GEAR RATIO: The reduction or overdrive ratio developed by a planetary gearset.

PLANETARY GEARSET: In its simplest form, it is made up of a basic assembly group containing a sun gear, internal gear, and planet carrier. The gears are always in constant mesh and offer a wide range of gear ratio possibilities.

PLANETARY GEARSET (COMPOUND): Two planetary gearsets combined together.

PLANETARY GEARSET (SIMPLE): An assembly of gears in constant mesh consisting of a sun gear, several pinion gears mounted in a carrier, and a ring gear. It provides gear ratio and direction changes, in addition to a direct drive and a neutral.

PLY RATING: A. rating given a tire which indicates strength (but not necessarily actual plies). A two-ply/four-ply rating has only two plies, but the strength of a four-ply tire.

POLARITY: Indication (positive or negative) of the two poles of a battery.

PORT: An opening for fluid intake or exhaust.

POSITIVE SEALING: A sealing method that completely prevents leakage.

POTENTIAL: Electrical force measured in volts; sometimes used interchangeably with voltage.

POWER: The ability to do work per unit of time, as expressed in horsepower; one horsepower equals 33,000 ft. lbs. of work per minute, or 550 ft. lbs. of work per second.

POWER FLOW: The systematic flow or transmission of power through the gears, from the input shaft to the output shaft.

POWER-TO-WEIGHT RATIO: Ratio of horsepower to weight of car.

POWERTRAIN: See Drivetrain.

POWERTRAIN CONTROL MODULE (PCM): Current designation for the engine control module (ECM). In many cases, late model vehicle control units manage the engine as well as the transmission. In other settings, the PCM controls the engine and is interfaced with a TCM to control transmission functions.

Ppm: Parts per million; unit used to measure exhaust emissions.

PREIGNITION: Early ignition of fuel in the cylinder, sometimes due to glowing carbon deposits in the combustion chamber. Preignition can be damaging since combustion takes place prematurely.

PRELOAD: A predetermined load placed on a bearing during assembly or by adjustment.

PRESS FIT: The mating of two parts under pressure, due to the inner diameter of one being smaller than the outer diameter of the other, or vice versa; an interference fit.

PRESSURE: The amount of force exerted upon a surface area.

PRESSURE CONTROL SOLENOID (PCS): An output device that provides a boost oil pressure to the mainline regulator valve to control line pressure. Its operation is determined by the amount of current sent from the PCM.

PRESSURE GAUGE: An instrument used for measuring the fluid pressure in a hydraulic circuit.

PRESSURE REGULATOR VALVE: In automatic transmissions, its purpose is to regulate the pressure of the pump output and supply the basic fluid pressure necessary to operate the transmission. The regulated fluid pressure may be referred to as mainline pressure, line pressure, or control pressure.

PRESSURE SWITCH ASSEMBLY (PSA): Mounted inside the transmission, it is a grouping of oil pressure switches that inputs to the PCM when certain hydraulic passages are charged with oil pressure.

PRESSURE PLATE: A spring-loaded plate (part of the clutch) that transmits power to the driven (friction) plate when the clutch is engaged.

PRIMARY CIRCUIT: The low voltage side of the ignition system which consists of the ignition switch, ballast resistor or resistance wire, bypass, coil, electronic control unit and pick-up coil as well as the connecting wires and harnesses.

PROFILE: Term used for tire measurement (tire series), which is the ratio of tire height to tread width.

PROM (PROGRAMMABLE READ-ONLY MEMORY): The heart of the computer that compares input data and makes the engineered program or strategy decisions about when to trigger the appropriate output based on stored computer instructions.

PULSE GENERATOR: A two-wire pickup sensor used to produce a fluctuating electrical signal. This changing signal is read by the controller to determine the speed of the object and can be used to measure transmission/transaxle input speed, output speed, and vehicle speed.

PSI: Pounds per square inch; a measurement of pressure.

PULSE WIDTH DUTY CYCLE SOLENOID (PULSE WIDTH MODULATED SOLENOID): A computer-controlled solenoid that turns on and off at a variable rate producing a modulated oil pressure; often referred to as a pulse width modulated (PWM) solenoid. Employed in many electronic automatic transmissions and transaxles, these solenoids are used to manage shift control and converter clutch hydraulic circuits.

PUSHROD: A steel rod between the hydraulic valve lifter and the valve rocker arm in overhead valve (OHV) engines.

PUMP: A mechanical device designed to create fluid flow and pressure buildup in a hydraulic system.

QUARTER PANEL: General term used to refer to a rear fender. Quarter panel is the area from the rear door opening to the tail light area and from rear wheel well to the base of the trunk and roof-line.

RACE: The surface on the inner or outer ring of a bearing on which the balls, needles or rollers move.

RACK AND PINION: A type of automotive steering system using a pinion gear attached to the end of the steering shaft. The pinion meshes with a long rack attached to the steering linkage.

RADIAL TIRE: Tire design which uses body cords running at right angles to the center line of the tire. Two or more belts are used to give tread strength. Radials can be identified by their characteristic sidewall bulge.

RADIATOR: Part of the cooling system for a water-cooled engine, mounted in the front of the vehicle and connected to the engine with rubber hoses. Through the radiator, excess combustion heat is dissipated into the atmosphere through forced convection using a water and glycol based mixture that circulates through, and cools, the engine.

RANGE REFERENCE AND CLUTCH/BAND APPLY CHART: A guide that shows the application of clutches and bands for each gear, within the selector range positions. These charts are extremely useful for understanding how the unit operates and for diagnosing malfunctions.

RAVIGNEAUX GEARSET: A compound planetary gearset that features matched dual planetary pinions (sets of two) mounted in a single planet carrier. Two sun gears and one ring mesh with the carrier pinions.

REACTION MEMBER: The stationary planetary member, in a planetary gearset, that is grounded to the transmission/transaxle case through the use of friction and wedging devices known as bands, disc clutches, and one-way clutches.

REACTION PRESSURE: The fluid pressure that moves a spool valve against an opposing force or forces; the area on which the opposing force acts. The opposing force can be a spring or a combination of spring force and auxiliary hydraulic force.

REACTOR, TORQUE CONVERTER: The reaction member of a fluid torque converter, more commonly called a stator. (See stator.)

REAR MAIN OIL SEAL: A synthetic or rope-type seal that prevents oil from leaking out of the engine past the rear main crankshaft bearing.

RECIRCULATING BALL: Type of steering system in which recirculating steel balls occupy the area between the nut and worm wheel, causing a reduction in friction.

RECTIFIER: A device (used primarily in alternators) that permits electrical current to flow in one direction only.

REDUCTION: (See gear reduction.)

REGULATOR VALVE: A valve that changes the pressure of the oil in a hydraulic circuit as the oil passes through the valve by bleeding off (or exhausting) some of the volume of oil supplied to the valve.

REFRIGERANT 12 (R-12) or 134 (R-134): The generic name of the refrigerant used in automotive air conditioning systems.

REGULATOR: A device which maintains the amperage and/or voltage levels of a circuit at predetermined values.

RELAY: A switch which automatically opens and/or closes a circuit.

RELAY VALVE: A valve that directs flow and pressure. Relay valves simply connect or disconnect interrelated passages without restricting the fluid flow or changing the pressure.

RELIEF VALVE: A spring-loaded, pressure-operated valve that limits oil pressure buildup in a hydraulic circuit to a predetermined maximum value.

RELUCTOR: A wheel that rotates inside the distributor and triggers the release of voltage in an electronic ignition.

RESERVOIR: The storage area for fluid in a hydraulic system; often called a sump.

RESIN: A liquid plastic used in body work.

RESIDUAL MAGNETISM: The magnetic strength stored in a material after a magnetizing field has been removed.

RESISTANCE: The opposition to the flow of current through a circuit or electrical device, and is measured in ohms. Resistance is equal to the voltage divided by the amperage.

RESISTOR SPARK PLUG: A spark plug using a resistor to shorten the spark duration. This suppresses radio interference and lengthens plug life.

RESISTOR: A device, usually made of wire, which offers a preset amount of resistance in an electrical circuit.

RESULTANT FORCE: The single effective directional thrust of the fluid force on the turbine produced by the vortex and rotary forces acting in different planes.

RETARD: Set the ignition timing so that spark occurs later (fewer degrees before TDC).

RHEOSTAT: A device for regulating a current by means of a variable resistance.

RING GEAR: The name given to a ring-shaped gear attached to a differential case, or affixed to a flywheel or as part of a planetary gear set.

ROADLOAD: grade.

ROCKER ARM: A lever which rotates around a shaft pushing down (opening) the valve with an end when the other end is pushed up by the pushrod. Spring pressure will later close the valve.

ROCKER PANEL: The body panel below the doors between the wheel opening.

ROLLER BEARING: A bearing made up of hardened inner and outer races between which hardened steel rollers move.

ROLLER CLUTCH: A type of one-way clutch design using rollers and springs mounted within an inner and outer cam race assembly.

ROTARY FLOW: The path of the fluid trapped between the blades of the members as they revolve with the rotation of the torque converter cover (rotational inertia).

ROTOR: (1.) The disc-shaped part of a disc brake assembly, upon which the brake pads bear; also called, brake disc. (2.) The device mounted atop the distributor shaft, which passes current to the distributor cap tower contacts.

ROTARY ENGINE: See Wankel engine.

RPM: Revolutions per minute (usually indicates engine speed).

RTV: A gasket making compound that cures as it is exposed to the atmosphere. It is used between surfaces that are not perfectly machined to one another, leaving a slight gap that the RTV fills and in which it hardens. The letters RTV represent room temperature vulcanizing.

RUN-ON: Condition when the engine continues to run, even when the key is turned off. See dieseling.

SEALED BEAM: A automotive headlight. The lens, reflector and filament from a single unit.

SEATBELT INTERLOCK: A system whereby the car cannot be started unless the seatbelt is buckled.

SECONDARY CIRCUIT: The high voltage side of the ignition system, usually above 20,000 volts. The secondary includes the ignition coil, coil wire, distributor cap and rotor, spark plug wires and spark plugs.

SELF-INDUCTION: The generation of voltage in a current-carrying wire by changing the amount of current flowing within that wire.

SEMI-CONDUCTOR: A material (silicon or germanium) that is neither a good conductor nor an insulator; used in diodes and transistors.

SEMI-FLOATING AXLE: In this design, a wheel is attached to the axle shaft, which takes both drive and cornering loads. Almost all solid axle passenger cars and light trucks use this design.

SENDING UNIT: A mechanical, electrical, hydraulic or electromagnetic device which transmits information to a gauge.

SENSOR: Any device designed to measure engine operating conditions or ambient pressures and temperatures. Usually electronic in nature and designed to send a voltage signal to an on-board computer, some sensors may operate as a simple on/off switch or they may provide a variable voltage signal (like a potentiometer) as conditions or measured parameters change.

SERIES CIRCUIT: (See circuit, series).

SERPENTINE BELT: An accessory drive belt, with small multiple v-ribs, routed around most or all of the engine-powered accessories such as the alternator and power steering pump. Usually both the front and the back side of the belt comes into contact with various pulleys.

SERVO: In an automatic transmission, it is a piston in a cylinder assembly that converts hydraulic pressure into mechanical force and movement; used for the application of the bands and clutches.

SHIFT BUSYNESS: When referring to a torque converter clutch, it is the frequent apply and release of the clutch plate due to uncommon driving conditions.

SHIFT VALVE: Classified as a relay valve, it triggers the automatic shift in response to a governor and a throttle signal by directing fluid to the appropriate band and clutch apply combination to cause the shift to occur.

SHIM: Spacers of precise, predetermined thickness used between parts to establish a proper working relationship.

SHIMMY: Vibration (sometimes violent) in the front end caused by misaligned front end, out of balance tires or worn suspension components.

SHORT CIRCUIT: An electrical malfunction where current takes the path of least resistance to ground (usually through damaged insulation). Current flow is excessive from low resistance resulting in a blown fuse.

SHUDDER: Repeated jerking or stick-slip sensation, similar to chuggle but more severe and rapid in nature, that may be most noticeable during certain ranges of vehicle speed; also used to define condition after converter clutch engagement.

SIMPSON GEARSET: A compound planetary gear train that integrates two simple planetary gearsets referred to as the front planetary and the rear planetary.

SINGLE OVERHEAD CAMSHAFT: See overhead camshaft.

SKIDPLATE: A metal plate attached to the underside of the body to protect the fuel tank, transfer case or other vulnerable parts from damage.

SLAVE CYLINDER: In automotive use, a device in the hydraulic clutch system which is activated by hydraulic force, disengaging the clutch.

SLIPPING: Noticeable increase in engine rpm without vehicle speed increase; usually occurs during or after initial clutch or band engagement.

SLUDGE: Thick, black deposits in engine formed from dirt, oil, water, etc. It is usually formed in engines when oil changes are neglected.

SNAP RING: A circular retaining clip used inside or outside a shaft or part to secure a shaft, such as a floating wrist pin.

SOFT: Slow, almost unnoticeable clutch apply with very little shift feel.

SOFTCODES: DTCs that have been set into the PCM memory but are not present at the time of testing; often referred to as history or intermittent codes.

SOHC: Single overhead camshaft.

SOLENOID: An electrically operated, magnetic switching device.

SPALLING: A wear pattern identified by metal chips flaking off the hardened surface. This condition is caused by foreign particles, overloading situations, and/or normal wear.

SPARK PLUG: A device screwed into the combustion chamber of a spark ignition engine. The basic construction is a conductive core inside of a ceramic insulator, mounted in an outer conductive base. An electrical charge from the spark plug wire travels along the conductive core and jumps a preset air gap to a grounding point or points at the end of the conductive base. The resultant spark ignites the fuel/air mixture in the combustion chamber.

SPECIFIC GRAVITY (BATTERY): The relative weight of liquid (battery electrolyte) as compared to the weight of an equal volume of water.

SPLINES: Ridges machined or cast onto the outer diameter of a shaft or inner diameter of a bore to enable parts to mate without rotation.

SPLIT TORQUE DRIVE: In a torque converter, it refers to parallel paths of torque transmission, one of which is mechanical and the other hydraulic.

SPONGY PEDAL: A soft or spongy feeling when the brake pedal is depressed. It is usually due to air in the brake lines.

SPOOLVALVE: A precision-machined, cylindrically shaped valve made up of lands and grooves. Depending on its position in the valve bore, various interconnecting hydraulic circuit passages are either opened or closed.

SPRAG CLUTCH: A type of one-way clutch design using cams or contoured-shaped sprags between inner and outer races. (See one-way clutch.)

SPRUNG WEIGHT: The weight of a car supported by the springs.

SQUARE-CUT SEAL: Molded synthetic rubber seal designed with a square- or rectangular-shaped cross-section. This type of seal is used for both dynamic and static applications.

SRS: Supplemental restraint system

STABILIZER (SWAY) BAR: A bar linking both sides of the suspension. It resists sway on turns by taking some of added load from one wheel and putting it on the other.

STAGE: The number of turbine sets separated by a stator. A turbine set may be made up of one or more turbine members. A three-element converter is classified as a single stage.

STALL: In fluid drive transmission/transaxle applications, stall refers to engine rpm with the transmission/transaxle engaged and the vehicle stationary; throttle valve can be in any position between closed and wide open.

STALL SPEED: In fluid drive transmission/transaxle applications, stall speed refers to the maximum engine rpm with the transmission/transaxle engaged and vehicle stationary, when the throttle valve is wide open. (See stall; stall test.)

STALL TEST: A procedure recommended by many manufacturers to help determine the integrity of an engine, the torque converter stator, and certain clutch and band combinations. With the shift lever in each of the forward and reverse positions and with the brakes firmly applied, the accelerator pedal is momentarily pressed to the wide open throttle (WOT) position. The engine rpm reading at full throttle can provide clues for diagnosing the condition of the items listed above.

STALL TORQUE: The maximum design or engineered torque ratio of a fluid torque converter, produced under stall speed conditions. (See stall speed.)

STARTER: A high-torque electric motor used for the purpose of starting the engine, typically through a high ratio geared drive connected to the flywheel ring gear.

STATIC: A sealing application in which the parts being sealed do not move in relation to each other.

STATOR (REACTOR): The reaction member of a fluid torque converter that changes the direction of the fluid as it leaves the turbine to enter the impeller vanes. During the torque multiplication phase, this action assists the impeller's rotary force and results in an increase in torque.

STEERING GEOMETRY: Combination of various angles of suspension components (caster, camber, toe-in); roughly equivalent to front end alignment.

STRAIGHT WEIGHT: Term designating motor oil as suitable for use within a narrow range of temperatures. Outside the narrow temperature range its flow characteristics will not adequately lubricate.

STROKE: The distance the piston travels from bottom dead center to top dead center.

SUBSTITUTION: Replacing one part suspected of a defect with a like part of known quality.

SUMP: The storage vessel or reservoir that provides a ready source of fluid to the pump. In an automatic transmission, the sump is the oil pan. All fluid eventually returns to the sump for recycling into the hydraulic system.

SUN GEAR: In a planetary gearset, it is the center gear that meshes with a cluster of planet pinions.

SUPERCHARGER: An air pump driven mechanically by the engine through belts, chains, shafts or gears from the crankshaft. Two general types of supercharger are the positive displacement and centrifugal type, which pump air in direct relationship to the speed of the engine.

SUPPLEMENTAL RESTRAINT SYSTEM: See air bag.

SURGE: Repeating engine-related feeling of acceleration and deceleration that is less intense than chuggle.

SWITCH: A device used to open, close, or redirect the current in an electrical circuit.

SYNCHROMESH: A manual transmission/transaxle that is equipped with devices (synchronizers) that match the gear speeds so that the transmission/transaxle can be downshifted without clashing gears.

SYNTHETIC OIL: Non-petroleum based oil.

TACHOMETER: A device used to measure the rotary speed of an engine, shaft, gear, etc., usually in rotations per minute.

TDC: Top dead center. The exact top of the piston's stroke.

TEFLON SEALING RINGS: Teflon is a soft, durable, plastic-like material that is resistant to heat and provides excellent sealing. These rings are designed with either scarf-cut joints or as one-piece rings. Teflon sealing rings have replaced many metal ring applications.

TERMINAL: A device attached to the end of a wire or cable to make an electrical connection.

TEST LIGHT, CIRCUIT-POWERED: Uses available circuit voltage to test circuit continuity.

TEST LIGHT, SELF-POWERED: Uses its own battery source to test circuit continuity.

THERMISTOR: A special resistor used to measure fluid temperature; it decreases its resistance with increases in temperature.

THERMOSTAT: A valve, located in the cooling system of an engine, which is closed when cold and opens gradually in response to engine heating, controlling the temperature of the coolant and rate of coolant flow.

THERMOSTATIC ELEMENT: A heat-sensitive, spring-type device that controls a drain port from the upper sump area to the lower sump. When the transaxle fluid reaches operating temperature, the port is closed and the upper sump fills, thus reducing the fluid level in the lower sump.

THROTTLE POSITION (TP) SENSOR: Reads the degree of throttle opening; its signal is used to analyze engine load conditions. The ECM/PCM decides to apply the TCC, or to disengage it for coast or load conditions that need a converter torque boost.

THROTTLE PRESSURE/MODULATOR PRESSURE: A hydraulic signal oil pressure relating to the amount of engine load, based on either the amount of throttle plate opening or engine vacuum.

THROTTLE VALVE: A regulating or balanced valve that is controlled mechanically by throttle linkage or engine vacuum. It sends a hydraulic signal to the shift valve body to control shift timing and shift quality. (See balanced valve; modulator valve.)

THROW-OUT BEARING: As the clutch pedal is depressed, the throwout bearing moves against the spring fingers of the pressure plate, forcing the pressure plate to disengage from the driven disc.

TIE ROD: A rod connecting the steering arms. Tie rods have threaded ends that are used to adjust toe-in.

TIE-UP: Condition where two opposing clutches are attempting to apply at same time, causing engine to labor with noticeable loss of engine rpm.

TIMING BELT: A square-toothed, reinforced rubber belt that is driven by the crankshaft and operates the camshaft.

TIMING CHAIN: A roller chain that is driven by the crankshaft and operates the camshaft.

TIRE ROTATION: Moving the tires from one position to another to make the tires wear evenly.

TOE-IN (OUT): A term comparing the extreme front and rear of the front tires. Closer together at the front is toe-in; farther apart at the front is toe-out.

TOP DEAD CENTER (TDC): The point at which the piston reaches the top of its travel on the compression stroke.

TORQUE: Measurement of turning or twisting force, expressed as foot-pounds or inch-pounds.

TORQUE CONVERTER: A turbine used to transmit power from a driving member to a driven member via hydraulic action, providing changes in drive ratio and torque. In automotive use, it links the driveplate at the rear of the engine to the automatic transmission.

TORQUE CONVERTER CLUTCH: The apply plate (lockup plate) assembly used for mechanical power flow through the converter.

TORQUE PHASE: Sometimes referred to as slip phase or stall phase, torque multiplication occurs when the turbine is turning at a slower speed than the impeller, and the stator is reactionary (stationary). This sequence generates a boost in output torque.

TORQUE RATING (STALL TORQUE): The maximum torque multiplication that occurs during stall conditions, with the engine at wide open throttle (WOT) and zero turbine speed.

TORQUE RATIO: An expression of the gear ratio factor on torque effect. A 3:1 gear ratio or 3:1 torque ratio increases the torque input by the ratio factor of 3. Input torque (100 ft. lbs.) x 3 = output torque (300 ft. lbs.)

TRACTION: The amount of usable tractive effort before the drive wheels slip on the road contact surface.

TORSION BAR SUSPENSION: Long rods of spring steel which take the place of springs. One end of the bar is anchored and the other arm (attached to the suspension) is free to twist. The bars' resistance to twisting causes springing action.

TRACK: Distance between the centers of the tires where they contact the ground.

TRACTION CONTROL: A control system that prevents the spinning of a vehicle's drive wheels when excess power is applied.

TRACTIVE EFFORT: The amount of force available to the drive wheels, to move the vehicle.

TRANSAXLE: A single housing containing the transmission and differential. Transaxles are usually found on front engine/front wheel drive or rear engine/rear wheel drive cars.

TRANSDUCER: A device that changes energy from one form to another. For example, a transducer in a microphone changes sound energy to electrical energy. In automotive air-conditioning controls used in automatic temperature systems, a transducer changes an electrical signal to a vacuum signal, which operates mechanical doors.

TRANSMISSION: A powertrain component designed to modify torque and speed developed by the engine; also provides direct drive, reverse, and neutral.

TRANSMISSION CONTROL MODULE (TCM): Manages transmission functions. These vary according to the manufacturer's product design but may include converter clutch operation, electronic shift scheduling, and mainline pressure.

TRANSMISSION FLUID TEMPERATURE (TFT) SENSOR: Originally called a transmission oil temperature (TOT) sensor, this input device to the ECM/PCM senses the fluid temperature and provides a resistance value. It operates on the thermistor principle.

TRANSMISSION INPUT SPEED (TIS) SENSOR: Measures turbine shaft (input shaft) rpm's and compares to engine rpm's to determine torque converter slip. When compared to the transmission output speed sensor or VSS, gear ratio and clutch engagement timing can be determined.

TRANSMISSION OIL TEMPERATURE (TOT) SENSOR: (See transmission fluid temperature (TFT) sensor.)

TRANSMISSION RANGE SELECTOR (TRS) SWITCH: Tells the module which gear shift position the driver has chosen.

TRANSFER CASE: A gearbox driven from the transmission that delivers power to both front and rear driveshafts in a four-wheel drive system. Transfer cases usually have a high and low range set of gears, used depending on how much pulling power is needed.

TRANSISTOR: A semi-conductor component which can be actuated by a small voltage to perform an electrical switching function.

TREAD WEAR INDICATOR: Bars molded into the tire at right angles to the tread that appear as horizontal bars when 1/16 in. of tread remains.

TREAD WEAR PATTERN: The pattern of wear on tires which can be "read" to diagnose problems in the front suspension.

TUNE-UP: A regular maintenance function, usually associated with the replacement and adjustment of parts and components in the electrical and fuel systems of a vehicle for the purpose of attaining optimum performance.

TURBINE: The output (driven) member of a fluid coupling or fluid torque converter. It is splined to the input (turbine) shaft of the transmission.

TURBOCHARGER: An exhaust driven pump which compresses intake air and forces it into the combustion chambers at higher than atmospheric pressures. The increased air pressure allows more fuel to be burned and results in increased horsepower being produced.

TURBULENCE: The interference of molecules of a fluid (or vapor) with each other in a fluid flow.

TYPE F: Transmission fluid developed and used by Ford Motor Company up to 1982. This fluid type provides a high coefficient of friction.

TYPE 7176: The preferred choice of transmission fluid for Chrysler automatic transmissions and transaxles. Developed in 1986, it closely resembles Dexron and Mercon. Type 7176 is the recommended service fill fluid for all Chrysler products utilizing a lockup torque converter dating back to 1978.

U-JOINT (UNIVERSAL JOINT): A flexible coupling in the drive train that allows the driveshafts or axle shafts to operate at different angles and still transmit rotary power.

UNDERSTEER: The tendency of a car to continue straight ahead while negotiating a turn.

UNIT BODY: Design in which the car body acts as the frame.

UNLEADED FUEL: Fuel which contains no lead (a common gasoline additive). The presence of lead in fuel will destroy the functioning elements of a catalytic converter, making it useless.

UNSPRUNG WEIGHT: The weight of car components not supported by the springs (wheels, tires, brakes, rear axle, control arms, etc.).

UPSHIFT: A shift that results in a decrease in torque ratio and an increase in speed.

VACUUM: A negative pressure; any pressure less than atmospheric pressure.

VACUUM ADVANCE: A device which advances the ignition timing in response to increased engine vacuum.

VACUUM GAUGE: An instrument used for measuring the existing vacuum in a vacuum circuit or chamber. The unit of measure is inches (of mercury in a barometer).

VACUUM MODULATOR: Generates a hydraulic oil pressure in response to the amount of engine vacuum.

VALVES: Devices that can open or close fluid passages in a hydraulic system and are used for directing fluid flow and controlling pressure.

VALVE BODY ASSEMBLY: The main hydraulic control assembly of the transmission/transaxle that contains numerous valves, check balls, and other components to control the distribution of pressurized oil throughout the transmission.

VALVE CLEARANCE: The measured gap between the end of the valve stem and the rocker arm, cam lobe or follower that activates the valve.

VALVE GUIDES: The guide through which the stem of the valve passes. The guide is designed to keep the valve in proper alignment.

VALVE LASH (clearance): The operating clearance in the valve train.

VALVE TRAIN: The system that operates intake and exhaust valves, consisting of camshaft, valves and springs, lifters, pushrods and rocker arms.

VAPOR LOCK: Boiling of the fuel in the fuel lines due to excess heat. This will interfere with the flow of fuel in the lines and can completely stop the flow. Vapor lock normally only occurs in hot weather.

VARIABLE DISPLACEMENT (VARIABLE CAPACITY) VANE PUMP: Slipper-type vanes, mounted in a revolving rotor and contained within the bore of a movable slide, capture and then force fluid to flow. Movement of the slide to various positions changes the size of the vane chambers and the amount of fluid flow. **Note:** GM refers to this pump design as variable displacement, and Ford terms it variable capacity.

VARIABLE FORCE SOLENOID (VFS): Commonly referred to as the electronic pressure control (EPC) solenoid, it replaces the cable/linkage style of TV system control and is integrated with a spool valve and spring assembly to control pressure. A variable computer-controlled current flow varies the internal force of the solenoid on the spool valve and resulting control pressure.

VARIABLE ORIFICE THERMAL VALVE: Temperature-sensitive hydraulic oil control device that adjusts the size of a circuit path opening. By altering the size of the opening, the oil flow rate is adapted for cold to hot oil viscosity changes.

VARNISH: Term applied to the residue formed when gasoline gets old and stale.

VCM: See Electronic Control Unit (ECU).

VEHICLE SPEED SENSOR (VSS): Provides an electrical signal to the computer module, measuring vehicle speed, and affects the torque converter clutch engagement and release.

VESPEL SEALING RINGS: Hard plastic material that produces excellent sealing in dynamic settings. These rings are found in late versions of the 4T60 and in all 4T60-E and 4T80-E transaxles.

VISCOSITY: The ability of a fluid to flow. The lower the viscosity rating, the easier the fluid will flow. 10 weight motor oil will flow much easier than 40 weight motor oil.

VISCOSITY INDEX IMPROVERS: Keeps the viscosity nearly constant with changes in temperature. This is especially important at low temperatures, when the oil needs to be thin to aid in shifting and for cold-weather starting. Yet it must not be so thin that at high temperatures it will cause excessive hydraulic leakage so that pumps are unable to maintain the proper pressures.

VISCOUS CLUTCH: A specially designed torque converter clutch apply plate that, through the use of a silicon fluid, clamps smoothly and absorbs torsional vibrations.

VOLT: Unit used to measure the force or pressure of electricity. It is defined as the pressure

VOLTAGE: The electrical pressure that causes current to flow. Voltage is measured in volts (V).

VOLTAGE, APPLIED: The actual voltage read at a given point in a circuit. It equals the available voltage of the power supply minus the losses in the circuit up to that point.

VOLTAGE DROP: The voltage lost or used in a circuit by normal loads such as a motor or lamp or by abnormal loads such as a poor (high-resistance) lead or terminal connection.

VOLTAGE REGULATOR: A device that controls the current output of the alternator or generator.

VOLTMETER: An instrument used for measuring electrical force in units called volts. Voltmeters are always connected parallel with the circuit being tested.

VORTEX FLOW: The crosswise or circulatory flow of oil between the blades of the members caused by the centrifugal pumping action of the impeller.

WANKEL ENGINE: An engine which uses no pistons. In place of pistons, triangular-shaped rotors revolve in specially shaped housings.

WATER PUMP: A belt driven component of the cooling system that mounts on the engine, circulating the coolant under pressure.

WATT: The unit for measuring electrical power. One watt is the product of one ampere and one volt (watts equals amps times volts). Wattage is the horsepower of electricity (746 watts equal one horsepower).

WHEEL ALIGNMENT: Inclusive term to describe the front end geometry (caster, camber, toe-in/out).

WHEEL CYLINDER: Found in the automotive drum brake assembly, it is a device, actuated by hydraulic pressure, which, through internal pistons, pushes the brake shoes outward against the drums.

WHEEL WEIGHT: Small weights attached to the wheel to balance the wheel and tire assembly. Out-of-balance tires quickly wear out and also give erratic handling when installed on the front.

WHEELBASE: Distance between the center of front wheels and the center of rear wheels.

WIDE OPEN THROTTLE (WOT): Full travel of accelerator pedal.

WORK: The force exerted to move a mass or object. Work involves motion; if a force is exerted and no motion takes place, no work is done. Work per unit of time is called power. Work = force x distance = ft. lbs. 33,000 ft. lbs. in one minute = 1 horsepower

ZERO-THROTTLE COAST DOWN: A full release of accelerator pedal while vehicle is in motion and in drive range.

Commonly Used Abbreviations

2

2WD	Two Wheel Drive

4

4WD	Four Wheel Drive

A

A/C	Air Conditioning
ABDC	After Bottom Dead Center
ABS	Anti-lock Brakes
AC	Alternating Current
ACL	Air cleaner
ACT	Air Charge Temperature
AIR	Secondary Air Injection
ALCL	Assembly Line Communications Link
ALDL	Assembly Line Diagnostic Link
AT	Automatic Transaxle/Transmission
ATDC	After Top Dead Center
ATF	Automatic Transmission Fluid
ATS	Air Temperature Sensor
AWD	All Wheel Drive

B

BAP	Barometric Absolute Pressure
BARO	Barometric Pressure
BBDC	Before Bottom Dead Center
BCM	Body Control Module
BDC	Bottom Dead Center
BPT	Backpressure Transducer
BTDC	Before Top Dead Center
BVSV	Bimetallic Vacuum Switching Valve

C

CAC	Charge Air Cooler
CARB	California Air Resources Board
CAT	Catalytic Converter
CCC	Computer Command Control
CCCC	Computer Controlled Catalytic Converter
CCCI	Computer Controlled Coil Ignition
CCD	Computer Controlled Dwell
CDI	Capacitor Discharge Ignition
CEC	Computerized Engine Control
CFI	Continuous Fuel Injection
CIS	Continuous Injection System
CIS-E	Continuous Injection System - Electronic
CKP	Crankshaft Position
CL	Closed Loop
CMP	Camshaft Position
CPP	Clutch Pedal Position
CTOX	Continuous Trap Oxidizer System
CTP	Closed Throttle Position
CVC	Constant Vacuum Control
CYL	Cylinder

D

DBC	Dual Bed Catalyst
DC	Direct Current
DFI	Direct Fuel Injection
DIS	Distributorless Ignition System
DLC	Data Link Connector
DMM	Digital Multimeter
DOHC	Double Overhead Camshaft
DRB	Diagnostic Readout Box
DTC	Diagnostic Trouble Code
DTM	Diagnostic Test Mode
DVOM	Digital Volt/Ohmmeter

E

EBCM	Electronic Brake Control Module
ECM	Engine Control Module
ECT	Engine Coolant Temperature
ECU	Engine Control Unit or Electronic Control Unit
EDIS	Electronic Distributorless Ignition System
EEC	Electronic Engine Control
EEPROM	Electrically Erasable Programmable Read Only Memory
EFE	Early Fuel Evaporation
EGR	Exhaust Gas Recirculation
EGRT	Exhaust Gas Recirculation Temperature
EGRVC	EGR Valve Control
EPROM	Erasable Programmable Read Only Memory
EVAP	Evaporative Emissions
EVP	EGR Valve Position

F

FBC	Feedback Carburetor
FEEPROM	Flash Electrically Erasable Programmable Read Only Memory
FF	Flexible Fuel
FI	Fuel Injection
FT	Fuel Trim
FWD	Front Wheel Drive

G

GND	Ground

H

HAC	High Altitude Compensation
HEGO	Heated Exhaust Gas Oxygen sensor
HEI	High Energy Ignition
HO2 Sensor	Heated Oxygen Sensor

I

IAC	Idle Air Control
IAT	Intake Air Temperature
ICM	Ignition Control Module
IFI	Indirect Fuel Injection
IFS	Inertia Fuel Shutoff
ISC	Idle Speed Control
IVSV	Idle Vacuum Switching Valve

Commonly Used Abbreviations

K

KOEO	Key On, Engine Off
KOER	Key ON, Engine Running
KS	Knock Sensor

M

MAF	Mass Air Flow
MAP	Manifold Absolute Pressure
MAT	Manifold Air Temperature
MC	Mixture Control
MDP	Manifold Differential Pressure
MFI	Multiport Fuel Injection
MIL	Malfunction Indicator Lamp or Maintenance
MST	Manifold Surface Temperature
MVZ	Manifold Vacuum Zone

N

NVRAM	Nonvolatile Random Access Memory

O

O2 Sensor	Oxygen Sensor
OBD	On-Board Diagnostic
OC	Oxidation Catalyst
OHC	Overhead Camshaft
OL	Open Loop

P

P/S	Power Steering
PAIR	Pulsed Secondary Air Injection
PCM	Powertrain Control Module
PCS	Purge Control Solenoid
PCV	Positive Crankcase Ventilation
PIP	Profile Ignition Pick-up
PNP	Park/Neutral Position
PROM	Programmable Read Only Memory
PSP	Power Steering Pressure
PTO	Power Take-Off
PTOX	Periodic Trap Oxidizer System

R

RABS	Rear Anti-lock Brake System
RAM	Random Access Memory
ROM	Read Only Memory
RPM	Revolutions Per Minute
RWAL	Rear Wheel Anti-lock Brakes
RWD	Rear Wheel Drive

S

SBC	Single Bed Converter
SBEC	Single Board Engine Controller
SC	Supercharger
SCB	Supercharger Bypass
SFI	Sequential Multiport Fuel Injection
SIR	Supplemental Inflatible Restraint
SOHC	Single Overhead Camshaft
SPL	Smoke Puff Limiter
SPOUT	Spark Output
SRI	Service Reminder Indicator
SRS	Supplemental Restraint System
SRT	System Readiness Test
SSI	Solid State Ignition
ST	Scan Tool
STO	Self-Test Output

T

TAC	Thermostatic Air Clearner
TBI	Throttle Body Fuel Injection
TC	Turbocharger
TCC	Torque Converter Clutch
TCM	Transmission Control Module
TDC	Top Dead Center
TFI	Thick Film Ignition
TP	Throttle Position
TR Sensor	Transaxle/Transmission Range Sensor
TVV	Thermal Vacuum Valve
TWC	Three-way Catalytic Converter

V

VAF	Volume Air Flow, or Vane Air Flow
VAPS	Variable Assist Power Steering
VRV	Vacuum Regulator Valve
VSS	Vehicle Speed Sensor
VSV	Vacuum Switching Valve

W

WOT	Wide Open Throttle
WU-TWC	Warm Up Three-way Catalytic Converter

ENGLISH TO METRIC CONVERSION: TORQUE

To convert foot-pounds (ft. lbs.) to Newton-meters (Nm), multiply the number of ft. lbs. by 1.36

To convert Newton-meters (Nm) to foot-pounds (ft. lbs.), multiply the number of Nm by 0.7376

ft. lbs.	Nm	ft. lbs.	Nm	ft. lbs.	Nm	ft. lbs.	Nm
0.1	0.1	34	46.2	76	103.4	118	160.5
0.2	0.3	35	47.6	77	104.7	119	161.8
0.3	0.4	36	49.0	78	106.1	120	163.2
0.4	0.5	37	50.3	79	107.4	121	164.6
0.5	0.7	38	51.7	80	108.8	122	165.9
0.6	0.8	39	53.0	81	110.2	123	167.3
0.7	1.0	40	54.4	82	111.5	124	168.6
0.8	1.1	41	55.8	83	112.9	125	170.0
0.9	1.2	42	57.1	84	114.2	126	171.4
1	1.4	43	58.5	85	115.6	127	172.7
2	2.7	44	59.8	86	117.0	128	174.1
3	4.1	45	61.2	87	118.3	129	175.4
4	5.4	46	62.6	88	119.7	130	176.8
5	6.8	47	63.9	89	121.0	131	178.2
6	8.2	48	65.3	90	122.4	132	179.5
7	9.5	49	66.6	91	123.8	133	180.9
8	10.9	50	68.0	92	125.1	134	182.2
9	12.2	51	69.4	93	126.5	135	183.6
10	13.6	52	70.7	94	127.8	136	185.0
11	15.0	53	72.1	95	129.2	137	186.3
12	16.3	54	73.4	96	130.6	138	187.7
13	17.7	55	74.8	97	131.9	139	189.0
14	19.0	56	76.2	98	133.3	140	190.4
15	20.4	57	77.5	99	134.6	141	191.8
16	21.8	58	78.9	100	136.0	142	193.1
17	23.1	59	80.2	101	137.4	143	194.5
18	24.5	60	81.6	102	138.7	144	195.8
19	25.8	61	83.0	103	140.1	145	197.2
20	27.2	62	84.3	104	141.4	146	198.6
21	28.6	63	85.7	105	142.8	147	199.9
22	29.9	64	87.0	106	144.2	148	201.3
23	31.3	65	88.4	107	145.5	149	202.6
24	32.6	66	89.8	108	146.9	150	204.0
25	34.0	67	91.1	109	148.2	151	205.4
26	35.4	68	92.5	110	149.6	152	206.7
27	36.7	69	93.8	111	151.0	153	208.1
28	38.1	70	95.2	112	152.3	154	209.4
29	39.4	71	96.6	113	153.7	155	210.8
30	40.8	72	97.9	114	155.0	156	212.2
31	42.2	73	99.3	115	156.4	157	213.5
32	43.5	74	100.6	116	157.8	158	214.9
33	44.9	75	102.0	117	159.1	159	216.2

METRIC TO ENGLISH CONVERSION: TORQUE

To convert foot-pounds (ft. lbs.) to Newton-meters (Nm), multiply the number of ft. lbs. by 1.36
To convert Newton-meters (Nm) to foot-pounds (ft. lbs.), multiply the number of Nm by 0.7376

Nm	ft. lbs.	Nm	ft. lbs.	Nm	ft. lbs.	Nm	ft. lbs.	Nm	ft. lbs.
0.1	0.1	34	25.0	76	55.9	118	86.8	160	117.6
0.2	0.1	35	25.7	77	56.6	119	87.5	161	118.4
0.3	0.2	36	26.5	78	57.4	120	88.2	162	119.1
0.4	0.3	37	27.2	79	58.1	121	89.0	163	119.9
0.5	0.4	38	27.9	80	58.8	122	89.7	164	120.6
0.6	0.4	39	28.7	81	59.6	123	90.4	165	121.3
0.7	0.5	40	29.4	82	60.3	124	91.2	166	122.1
0.8	0.6	41	30.1	83	61.0	125	91.9	167	122.8
0.9	0.7	42	30.9	84	61.8	126	92.6	168	123.5
1	0.7	43	31.6	85	62.5	127	93.4	169	124.3
2	1.5	44	32.4	86	63.2	128	94.1	170	125.0
3	2.2	45	33.1	87	64.0	129	94.9	171	125.7
4	2.9	46	33.8	88	64.7	130	95.6	172	126.5
5	3.7	47	34.6	89	65.4	131	96.3	173	127.2
6	4.4	48	35.3	90	66.2	132	97.1	174	127.9
7	5.1	49	36.0	91	66.9	133	97.8	175	128.7
8	5.9	50	36.8	92	67.6	134	98.5	176	129.4
9	6.6	51	37.5	93	68.4	135	99.3	177	130.1
10	7.4	52	38.2	94	69.1	136	100.0	178	130.9
11	8.1	53	39.0	95	69.9	137	100.7	179	131.6
12	8.8	54	39.7	96	70.6	138	101.5	180	132.4
13	9.6	55	40.4	97	71.3	139	102.2	181	133.1
14	10.3	56	41.2	98	72.1	140	102.9	182	133.8
15	11.0	57	41.9	99	72.8	141	103.7	183	134.6
16	11.8	58	42.6	100	73.5	142	104.4	184	135.3
17	12.5	59	43.4	101	74.3	143	105.1	185	136.0
18	13.2	60	44.1	102	75.0	144	105.9	186	136.8
19	14.0	61	44.9	103	75.7	145	106.6	187	137.5
20	14.7	62	45.6	104	76.5	146	107.4	188	138.2
21	15.4	63	46.3	105	77.2	147	108.1	189	139.0
22	16.2	64	47.1	106	77.9	148	108.8	190	139.7
23	16.9	65	47.8	107	78.7	149	109.6	191	140.4
24	17.6	66	48.5	108	79.4	150	110.3	192	141.2
25	18.4	67	49.3	109	80.1	151	111.0	193	141.9
26	19.1	68	50.0	110	80.9	152	111.8	194	142.6
27	19.9	69	50.7	111	81.6	153	112.5	195	143.4
28	20.6	70	51.5	112	82.4	154	113.2	196	144.1
29	21.3	71	52.2	113	83.1	155	114.0	197	144.9
30	22.1	72	52.9	114	83.8	156	114.7	198	145.6
31	22.8	73	53.7	115	84.6	157	115.4	199	146.3
32	23.5	74	54.4	116	85.3	158	116.2	200	147.1
33	24.3	75	55.1	117	86.0	159	116.9	201	147.8

ENGLISH/METRIC CONVERSION: TEMPERATURE

To convert Fahrenheit (F°) to Celsius (C°), take F° temperature and subtract 32, multiply the result by 5 and divide the result by 9

To convert Celsius (C°) to Fahrenheit (F°), take C° temperature and multiply it by 9, divide the result by 5 and add 32

F°	C°	F°	C°	C°	F°	C°	F°
-40	-40.0	150	65.6	-38	-36.4	46	114.8
-35	-37.2	155	68.3	-36	-32.8	48	118.4
-30	-34.4	160	71.1	-34	-29.2	50	122
-25	-31.7	165	73.9	-32	-25.6	52	125.6
-20	-28.9	170	76.7	-30	-22	54	129.2
-15	-26.1	175	79.4	-28	-18.4	56	132.8
-10	-23.3	180	82.2	-26	-14.8	58	136.4
-5	-20.6	185	85.0	-24	-11.2	60	140
0	-17.8	190	87.8	-22	-7.6	62	143.6
1	-17.2	195	90.6	-20	-4	64	147.2
2	-16.7	200	93.3	-18	-0.4	66	150.8
3	-16.1	205	96.1	-16	3.2	68	154.4
4	-15.6	210	98.9	-14	6.8	70	158
5	-15.0	212	100.0	-12	10.4	72	161.6
10	-12.2	215	101.7	-10	14	74	165.2
15	-9.4	220	104.4	-8	17.6	76	168.8
20	-6.7	225	107.2	-6	21.2	78	172.4
25	-3.9	230	110.0	-4	24.8	80	176
30	-1.1	235	112.8	-2	28.4	82	179.6
35	1.7	240	115.6	0	32	84	183.2
40	4.4	245	118.3	2	35.6	86	186.8
45	7.2	250	121.1	4	39.2	88	190.4
50	10.0	255	123.9	6	42.8	90	194
55	12.8	260	126.7	8	46.4	92	197.6
60	15.6	265	129.4	10	50	94	201.2
65	18.3	270	132.2	12	53.6	96	204.8
70	21.1	275	135.0	14	57.2	98	208.4
75	23.9	280	137.8	16	60.8	100	212
80	26.7	285	140.6	18	64.4	102	215.6
85	29.4	290	143.3	20	68	104	219.2
90	32.2	295	146.1	22	71.6	106	222.8
95	35.0	300	148.9	24	75.2	108	226.4
100	37.8	305	151.7	26	78.8	110	230
105	40.6	310	154.4	28	82.4	112	233.6
110	43.3	315	157.2	30	86	114	237.2
115	46.1	320	160.0	32	89.6	116	240.8
120	48.9	325	162.8	34	93.2	118	244.4
125	51.7	330	165.6	36	96.8	120	248
130	54.4	335	168.3	38	100.4	122	251.6
135	57.2	340	171.1	40	104	124	255.2
140	60.0	345	173.9	42	107.6	126	258.8
145	62.8	350	176.7	44	111.2	128	262.4

LENGTH CONVERSION

To convert inches (in.) to millimeters (mm), multiply the number of inches by 25.4

To convert millimeters (mm) to inches (in.), multiply the number of millimeters by 0.04

Inches	Millimeters	Inches	Millimeters	Inches	Millimeters	Inches	Millimeters
0.0001	0.00254	0.005	0.1270	0.09	2.286	4	101.6
0.0002	0.00508	0.006	0.1524	0.1	2.54	5	127.0
0.0003	0.00762	0.007	0.1778	0.2	5.08	6	152.4
0.0004	0.01016	0.008	0.2032	0.3	7.62	7	177.8
0.0005	0.01270	0.009	0.2286	0.4	10.16	8	203.2
0.0006	0.01524	0.01	0.254	0.5	12.70	9	228.6
0.0007	0.01778	0.02	0.508	0.6	15.24	10	254.0
0.0008	0.02032	0.03	0.762	0.7	17.78	11	279.4
0.0009	0.02286	0.04	1.016	0.8	20.32	12	304.8
0.001	0.0254	0.05	1.270	0.9	22.86	13	330.2
0.002	0.0508	0.06	1.524	1	25.4	14	355.6
0.003	0.0762	0.07	1.778	2	50.8	15	381.0
0.004	0.1016	0.08	2.032	3	76.2	16	406.4

ENGLISH/METRIC CONVERSION: LENGTH

To convert inches (in.) to millimeters (mm), multiply the number of inches by 25.4
To convert millimeters (mm) to inches (in.), multiply the number of millimeters by 0.04

Inches		Millimeters	Inches		Millimeters	Inches		Millimeters
Fraction	Decimal	Decimal	Fraction	Decimal	Decimal	Fraction	Decimal	Decimal
1/64	0.016	0.397	11/32	0.344	8.731	11/16	0.688	17.463
1/32	0.031	0.794	23/64	0.359	9.128	45/64	0.703	17.859
3/64	0.047	1.191	3/8	0.375	9.525	23/32	0.719	18.256
1/16	0.063	1.588	25/64	0.391	9.922	47/64	0.734	18.653
5/64	0.078	1.984	13/32	0.406	10.319	3/4	0.750	19.050
3/32	0.094	2.381	27/64	0.422	10.716	49/64	0.766	19.447
7/64	0.109	2.778	7/16	0.438	11.113	25/32	0.781	19.844
1/8	0.125	3.175	29/64	0.453	11.509	51/64	0.797	20.241
9/64	0.141	3.572	15/32	0.469	11.906	13/16	0.813	20.638
5/32	0.156	3.969	31/64	0.484	12.303	53/64	0.828	21.034
11/64	0.172	4.366	1/2	0.500	12.700	27/32	0.844	21.431
3/16	0.188	4.763	33/64	0.516	13.097	55/64	0.859	21.828
13/64	0.203	5.159	17/32	0.531	13.494	7/8	0.875	22.225
7/32	0.219	5.556	35/64	0.547	13.891	57/64	0.891	22.622
15/64	0.234	5.953	9/16	0.563	14.288	29/32	0.906	23.019
1/4	0.250	6.350	37/64	0.578	14.684	59/64	0.922	23.416
17/64	0.266	6.747	19/32	0.594	15.081	15/16	0.938	23.813
9/32	0.281	7.144	39/64	0.609	15.478	61/64	0.953	24.209
19/64	0.297	7.541	5/8	0.625	15.875	31/32	0.969	24.606
5/16	0.313	7.938	41/64	0.641	16.272	63/64	0.984	25.003
21/64	0.328	8.334	21/32	0.656	16.669	1/1	1.000	25.400
			43/64	0.672	17.066			

ALL LABOR TIMES DEVELOPED BY CHILTON EDITORS

Manual ISBN 1-4018-4356-5/Part No. 24356

Professional technicians have relied on the *Chilton Labor Guide* estimated repair times for decades. The labor times reflect actual vehicle conditions found in the aftermarket for domestic and import vehicles, including rust, wear, and grime. Times also reflect technicians' use of aftermarket tools and training. Coverage includes 1981 through 2004 model years. This year's edition is available as both a hardcover manual and on CD-ROM.

Labor Guide Manual Benefits:

up-to-date edition reflects today's repair industry standards
prior model coverage has been re-evaluated by experts to ensure accuracy
wide acceptance by extended warranty companies

Hardcover manual is 8 7/8" x 11", printed in 1-Color, ©2004

Labor Guide CD-ROM Benefits:

- same great features and low price as the hardcover manual, but with much more functionality
- updated software allows users to create estimates and invoices quickly and easily
- software keeps track of customers and prior estimates, saving valuable time
- helpful "How to Use" drop-down screens make it easy for new users and those accustomed to printed manuals

CD-ROM ISBN 1-4018-4357-3/Part No. 24357

Previous Year Editions

2003 Chilton Labor Guide Manual, ISBN 0-8019-9360-1/Part No. 9360

2003 Chilton Labor Guide CD-ROM, ISBN 0-8019-9361-X/Part No. 9361

The 2004 editions of the *Chilton Service Manuals* have been reconfigured to better meet the needs of automotive professionals like never before! The same, great mechanical service and repair information that technicians have come to rely on is now offered in a series of five manufacturer-based books, with updated coverage that reflects actual vehicle conditions found in the aftermarket, including rust, wear, and grime.

Comprehensive, technically detailed content is supported by exploded-view illustrations, diagrams, and specification charts, all carefully organized by system and model for easy reference. Step-by-step procedures, from drive train to chassis and related components, help yield fast, accurate repairs.

Professional Service and Repair Manual Benefits:

- current coverage includes aftermarket repair information that reflects today's industry standards
- popular mechanical systems are included, such as engines, suspensions, and steering components
- special tools are described and illustrated so that performing the repairs is easier and quicker
- complete engine overhaul procedures are broken down step-by-step for clarification

Chilton Ford Service Manual, 2000-2004
    ISBN 1-4018-4238-0/Part No. 24238
Chilton General Motors Service Manual, 2000-2004
    ISBN 1-4018-4237-2/Part No. 24237
Chilton Chrysler Service Manual, 2000-2004
    ISBN 1-4018-4239-9/Part No. 24239
Chilton Asian Service Manual, 2000-2004
    ISBN 1-4018-4235-6/Part No. 24235
Chilton European Service Manual, 2000-2004
    ISBN 1-4018-4234-8/Part No. 24234

Manuals are 8 1/2" x 11", printed in 1-Color, ©2004

 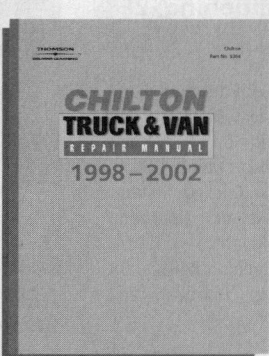  

The *Chilton Perennial Editions* contain repair and maintenance information for popular mechanical systems that may not be available elsewhere. They offer a wide range of repair information on cars, trucks, vans, and SUVs dating back to the early 1960's, and as current as 2002. Information for 1993 and later model years includes scheduled maintenance interval charts.

Benefits:

- covers the most common vehicle models found in the repair aftermarket today
- gain quick understanding of systems using exploded-view illustrations, diagrams, and charts
- simplify tough jobs with easy-to-follow removal and installation instructions for heater core and other components
- obtain complete coverage of repair procedures from drive train to chassis and associated components

Auto Repair Manual, 1998-2002, 1,426 pages **NEW!**
    ISBN 0-8019-9362-8/Part No. 9362

Auto Repair Manual, 1993-1997, 2,064 pages
    ISBN 0-8019-7919-6/Part No. 7919

Auto Repair Manual, 1988-1992, 1,284 pages
    ISBN 0-8019-7906-4/Part No. 7906

Auto Repair Manual, 1980-1987, 1,344 pages
    ISBN 0-8019-7670-7/Part No. 7670

Import Car Repair Manual, 1998-2002, 1,792 pps **NEW!**
    ISBN 0-8019-9363-6/Part No. 9363

Import Car Repair Manual, 1993-1997, 2,080 pps
    ISBN 0-8019-7920-X/Part No. 7920

Import Car Repair Manual, 1988-1992, 1,632 pages
    ISBN 0-8019-7907-2/Part No. 7907

Import Car Repair Manual, 1980-1987, 1,488 pages
    ISBN 0-8019-7672-3/Part No. 7672

Truck & Van Repair Manual, 1998-2002, 1,408 pages **NEW!**
    ISBN 0-8019-9364-4/Part No. 9364

Truck & Van Repair Manual, 1993-1997, 2,096 pages
    ISBN 0-8019-7921-8/Part No. 7921

Truck & Van Repair Manual, 1991-1995, 1,664 pages
    ISBN 0-8019-7911-0/Part No. 7911

Truck & Van Repair Manual, 1986-1990, 1,536 pages
    ISBN 0-8019-7902-1/Part No. 7902

Truck & Van Repair Manual, 1979-1986, 1,440 pages
    ISBN 0-8019-7655-3/Part No. 7655

SUV Repair Manual, 1998-2002, 1,292 pages **NEW!**
    ISBN 0-8019-9365-2/Part No. 9365

Hardcover manuals are 8 1/2" x 11", printed in 1-Color

Chilton Collector's Editions - *Reference Manuals for Vintage Vehicles*
Auto Repair Manual, 1964-1971 ISBN 0-8019-5974-8/Part No. 5974,
Truck & Van Repair Manual, 1961-1971 ISBN 0-8019-6198-X/Part No. 6198
Truck & Van Repair Manual, 1971-1978 ISBN 0-8019-7012-1/Part No. 7012

ASE Test Preparation Series
Delmar Learning
1-4018-5182-7
Part No. 25182
(Complete Set)

Delmar Learning has developed comprehensive ASE Test Preparation Manuals to help automotive technicians increase their success on these certification programs. The material covers the topics one might find during the test process. The booklets include many review questions and answers, as well as detailed descriptions of the repairs involved. Designed to look like the actual test, participants will feel more comfortable with practice, which will translate into greater success in taking the actual tests. The design of the Delmar Learning product also includes helpful test taking hints and student preparation ideas designed to enhance success.

BENEFITS
- The history of the ASE
- Test-taking strategies
- Tasks lists and overview
- Sample test questions
- ASE-style exams
- Explanations to the answers (right and wrong)
- Glossary of terms

(A1) Automotive Engine Repair, 2E
1-4018-2040-9
Part No. 22040
General Engine Diagnosis, Cylinder Head and Valve Train Diagnosis and Repair, Engine Block Diagnosis and Repair, Lubrication and Cooling Systems Diagnosis and Repair, and Fuel, Electrical, Ignition and Exhaust Systems Inspection and Service.

(A2) Automotive Transmissions and Transaxles, 2E
1-4018-2041-7
Part No. 22041
General Transmission/ Transaxle Diagnosis (Mechanical/Hydraulic Systems and Electronic Systems), Transmission/Transaxle Maintenance and Adjustment, In-Vehicle Transmission/Transaxle Repair, Off-Vehicle Transmission/Transaxle Repair.

(A3) Automotive Manual Drive Trains and Axles, 2E
1-4018-2042-5
Part No. 22042
Clutch Diagnosis and Repair, Transmission Diagnosis and Repair, Transaxle Diagnosis and Repair, Drive Shaft/Half Shaft and Universal Joint/Constant Velocity (CV) Joint Diagnosis and Repair (Front and Rear Wheel Drive), Rear Axle Diagnosis and Repair, Four Wheel Drive/All Wheel Drive Component Diagnosis and Repair.

(A4) Automotive Suspension and Steering, 2E
1-4018-2043-3
Part No. 22043
Steering Systems Diagnosis and Repair (Steering Columns and Manual Steering Gears, Power Assisted Steering Units, Steering Linkage), Suspension Systems Diagnosis and Repair (Front Suspensions, Rear Suspensions, Miscellaneous Services), Wheel Alignment Diagnosis, Adjustment and Repair, and Wheel and Tire Diagnosis and Repair.

(A5) Automotive Brakes, 2E
1-4018-2044-1
Part No. 22044
Hydraulic System Diagnosis and Repair, Drum Brake Diagnosis and Repair, Disc Brake Diagnosis and Repair, Power Assist Units Diagnosis and Repair, Miscellaneous Systems Diagnosis and Repair, Antilock Brake Systems (ABS) Diagnosis and Repair.

(A6) Automotive Electrical-Electronic Systems, 2E
1-4018-2045-X
Part No. 22045
General Electrical/Electronic Systems Diagnosis, Battery Diagnosis and Service, Starting Systems Diagnosis and Repair, Charging Systems Diagnosis and Repair, Lighting Systems Diagnosis and Repair, Gauges, Warning Devices and Driver Information Systems Diagnosis and Repair, Horn and Wiper/Washer Diagnosis and Repair.

(A7) Automotive Heating and Air Conditioning, 2E
1-4018-2046-8
Part No. 22046
The manual for A7 includes the following topics: A/C System Diagnosis and Repair, Refrigeration System Component Diagnosis and Repair, Heating and Engine Cooling Systems Diagnosis and Repair, Operating Systems and Related Controls Diagnosis and Repair, Refrigerant Recovery, Recycling, Handling and Retrofit.

(A8) Automotive Engine Performance, 2E
1-4018-2047-6
Part No. 22047
The manual for A8 includes the following topics: General Engine Diagnosis, Ignition System Diagnosis and Repair, Fuel, Air Induction, and Exhaust Systems Diagnosis and Repair, Emissions Control Systems Diagnosis and Repair (Including OBDII), Computerized Engine controls Diagnosis and Repair (Including OBDII), Engine Electrical Systems diagnosis and Repair.

(L1) Automotive Advance Engine Performance, 2E
1-4018-2049-2
Part No. 22049
The manual for L1 includes the following topics: General Powertrain Diagnosis, Computerized Powertrain Controls Diagnosis (Including OBDII), Ignition System Diagnosis, Fuel Systems and Air Induction Systems Diagnosis, Emission Control Systems Diagnosis, I/M Failure Diagnosis.

(P2) Automobile Parts Specialist, 2E
1-4018-2048-4
Part No. 22048
The manual for P2 includes the following topics: General Operations, Customer Relations and Sales Skills, Vehicle Systems Knowledge, Vehicle Identification, Cataloging Skills, Inventory Management, Merchandising.

(X1) Exhaust Systems
1-4018-2050-6
Part No. 22050
Exhaust Systems includes the following topics: Exhaust Systems Inspection and Repair, Emissions Systems Diagnosis, Exhaust System Fabrication, Exhaust System Installation, Exhaust System Repair Regulations.

(C1) Service Consultant
See next page for details

ASE Test Preparation Series in Español!
1-4018-1530-8 *(Complete Set)*

Now available in Español – the first of its kind for Spanish-speaking technicians! This comprehensive package of ASE test preparation booklets are intended for any Spanish-speaking automotive technician who is preparing to take an ASE examination. The series includes questions that relate to each competency required for certification by ASE. In addition to a multitude of questions, the reason why each answer is right or wrong is explained, along with task lists and overview, test-taking strategies, and more.

(A1) Reparación de Motores, 2A Edición
1-4018-1014-4/Part No. 21014

(A2) Transmision Automática/ Eje de Transmision Automática, 2A Edición
1-4018-1015-2/Part No. 21015

(A3) Tren de y Mando Ejes Manuales, 2A Edición
1-4018-1016-0/Part No. 21016

(A4) Suspensión y Dirección, 2A Edición
1-4018-1017-9/Part No. 21017

(A5) Frenos, 2A Edición
1-4018-1018-7/Part No. 21018

(A6) Sistemas Eléctricos/ Electrónicos, 2A Edición
1-4018-1019-5/Part No. 21019

(A7) Calefacción y Aire Acondicionado, 2A Edición
1-4018-1020-9/Part No. 21020

(A8) Funcionamiento de Motores, 2A Edición
1-4018-1021-7/Part No. 21021

(L1) Especialista en el Funciommiato Avansado de Motores, 2A Edición
1-4018-1022-5/Part No. 21022

(P2) Especialista en Partes de Automovil, 2A Edición
1-4018-1023-3/Part No. 21023

(X1) Sistemas de Escape, 2A Edición
1-4018-1024-1/Part No. 21024

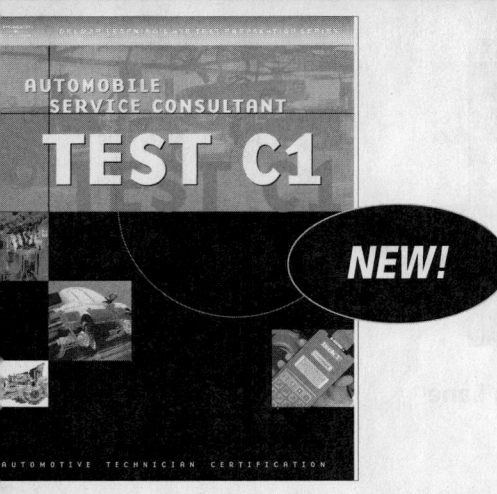

TEST C1
AUTOMOBILE SERVICE CONSULTANT

NEW!

AUTOMOTIVE TECHNICIAN CERTIFICATION

Engine Machinist
Test M1

MACHINIST TECHNICIAN CERTIFICATION

Mitchell/Delmar Collision Test Preparation Series 2nd Edition
Non-Structural Analysis & Damage Repair B3

SE Test Preparation Manual - C1 Service Consultant
BN 1-4018-2029-8/Part No.22029

epare to pass the new Service Consultant ASE Exam
ith help from this new test preparation booklet.
ne new C1 Exan is designed to measure systems
nowledge and people skills of those who come in
ntact with the customer. It will contain questions
n Communications, Product Knowledge, Sales Skills,
d Shop Operations.

is new manual from Delmar Learning features the
and new task list for the ASE C1 Exam, along with
ritten overviews that describe the responsibilities
at a reader will be tested on by ASE. Those seeking
rtification will benefit from the valuable
eparation offered, including ASE test taking
rategies, hundreds of ASE-style exam questions,
d detailed explanations as to why a particular
nswer is correct or incorrect.

rvice Consultant ASE Test Preparation Manual Benefits:

the ASE task list is fully up-to-date, while current
test prep questions reflect the most recent ASE task
changes for the broadest knowledge possible
hundreds of ASE-style exam questions adequately
prepare readers to successfully pass the ASE exam
readers are given multiple opportunities to check
their understanding of critical concepts through
sample problems, refresher materials, and
competency-specific test questions
overviews of each task provide a great reference
point to help answer difficult ASE questions
explanations for each answer help the user under-
stand why the response is correct or incorrect

oftcover manual is 8 1/2" x 11", printed in 1-Color,
2004

ASE Test Preparation Manuals - Engine Machinist Complete Set (M1-M3) 0-7668-6283-6/ Part No. 16283

With an abundance of up-to-date content, Delmar Learning's ASE Test Preparation Series contains the most current ASE test preparation material avail-able. Each manual combines refresher materials with an abundance of sample test questions, as well as a wealth of information regarding test-taking strate-gies and the types of ques-tions found in an ASE exam. In addition to the questions, thorough explanations are provided as to why each answer is correct or incorrect

Benefits:
- The History section explains why the exams are important to the industry
- test-taking strategies help prepare technicians for the environment they will encounter during the actu-al examrience testing first-hand

(M1) Cylinder Head Specialist 0-7668-6280-1/Part No. 16280

(M2) Cylinder Block Specialist 0-7668-6281-X/ Part No. 16281

(M3) Assembly Specialist 0-7668-6282-8/ Part No. 16282

Softcover manuals are 8 1/2" x 11", printed in 1-Color, ©2002

ASE Test Preparation Manuals-Collison Repair

With fully-updated and expanded content, the second edition of Delmar Learning's ASE Test Preparation Series contains the most current ASE test preparation material avail-able. Each manual combines refresher materials with an abundance of sample test questions, as well as a wealth of information regarding test-taking strategies and the types of questions found in an ASE exam. In addition to the ques-tions, thorough explanations are provided as to why each answer is correct or incorrect.

Benefits:
- The History section explains why the exams are impor-tant to the industry
- test-taking strategies help prepare technicians for the environment they will encounter during the actual exam

(B2) Painting and Refinishing, 2E, 0-7668-4885-X/Part No. 14885
(B3) Non-Structural Analysis and Damage Repair, 2E, 0-7668-4886-8/Part No. 14886
(B4) Structural Analysis and Damage Repair, 2E, 0-7668-4887-6/Part No. 14887
(B5) Mechanical and Electrical Components, 2E, 0-7668-4888-4/Part No. 14888
(B6) Damage Analysis and Estimation, 2E, 0-7668-4889-2/Part No. 14889

Softcover manuals are 8 1/2" x 11", printed in 1-Color, ©2001

Automotive ASE Preparation Video Series
Delmar Learning

0-7668-3168-X *(Complete Set of 12 Tapes)*
0-7668-8042-7 *(Complete Set of 3 CD-ROMs)*

Delmar's Automotive ASE Test Prep Videos present test takers with a review of the A1-A8, L1, and P2 tests prior to taking the exam. Each tape summarizes key topics and key task areas through live action and animation. Actual technicians, authentic automotive shops, and late-model vehicles are featured for an up-to-date look and feel. Safety is emphasized throughout each tape. An overview tape introduces test takers to the ASE testing style.

BENEFITS OF THE VIDEO SERIES
- lively, easy to follow videos emphasize safety throughout
- covers major task areas and topics for each of the ASE exams
- accompanying Instructor's Guide helps users comprehend and retain information presented

Complete Set of 12 Tapes (with Instructor's Guide), ©2001

Tape 1: Overview of ASE, 0-7668-2484-5
Tape 2: A1 Engine Repair, 0-7668-2485-3
Tape 3: A2 Automatic Transmission, 0-7668-2498-5
Tape 4: A3 Manual Transmission, 0-7668-2499-3
Tape 5: A4 Steering and Suspension, 0-7668-2500-0
Tape 6: A5 Automotive Brakes, 0-7668-2501-9
Tape 7: A6 Electricity/Electronics, 0-7668-2493-4
Tape 8: A7 Air Conditioning, 0-7668-2486-1
Tape 9: A8 Engine Performance, 0-7668-2494-2
Tape 10: P2 Parts Specialist, 0-7668-2487-X
Tape 11: L1 Advanced Engine Performance (Part 1), 0-7668-2491-8
Tape 12: L1 Advanced Engine Performance (Part 2), 0-7668-2492-6

BUNDLES
Bundle 1: Specialty Topics (Set of 4 Tapes) includes Overview of ASE, A1 Engine Repair, A7 Air Conditioning, and P2 Parts Specialist, 0-7668-2483-7
Bundle 2: Engine Performance/Electronics (Set of 4 Tapes) includes L1 Part 1, L1 Part 2, A6 Electricity/ Electronics, and A8 Engine Performance, 0-7668-2490-X
Bundle 3: Undercar (Set of 4 Tapes) includes A2 Automatic Transmissions, A3 Manual Transmissions, A4 Steering and Suspension, and A5 Automotive Brakes, 0-7668-2497-7

CD-ROM COURSEWARE
Based on the ASE Test Prep Series, the CD-ROMs offer the following in addition to the video content:
- Gradebook
- Video Glossary
- Video File Server compatible
- Pre-test/Post-test
- Variety of question types
- Ability to modify
- Automatic remediation
diation

See Inside Front Cover for System Requirements

CD-ROM 1: Specialty Topics CD-ROM includes Overview of ASE, A1 Engine Repair, A7 Air Conditioning, and P2 Parts Specialist, 0-7668-2489-6
CD-ROM 2: Engine Performance/Electronics CD-ROM includes L1 Part 1, L1 Part 2, A6 Electricity/ Electronics, and A8 Engine Performance, 0-7668-2496-9
CD-ROM 3: Undercar CD-ROM includes A2 Automatic Transmissions, A3 Manual Transmissions, A4 Steering and Suspension, and A5 Automotive Brakes, 0-7668-2503-5

The ASE "Passing Lane" Package
Delmar Learning

0-7668-4338-6
(Complete Set)

The most comprehensive test preparation for Automotive Tests A1-A8, L1, and P2. Combining the most thorough ASE Test Preparation books with the latest in ASE videos, this package provides a program of self-study for the automotive ASE Tests.

EACH BOOK IN THE SERIES BENEFITS:
- test-taking strategies
- tasks lists and overview
- sample test questions
- ASE-style exams
- explanations to the answers
- glossary of terms

EACH VIDEO IN THE SERIES BENEFITS:
- lively, easy to follow videos emphasize safety throughout
- covers major task areas and topics for each of the ASE exams
- accompanying Activity Sheets help comprehend and retain information

(A1) Automotive Engine Repair Book/Video,
0-7668-4181-2
(A2) Automotive Transmissions and Transaxles Book/Video,
0-7668-4182-0
(A3) Automotive Manual Drive Trains and Axles Book/Video,
0-7668-4183-9
(A4) Automotive Suspension and Steering Book/Video,
0-7668-4184-7
(A5) Automotive Brakes Book/Video,
0-7668-4185-5
(A6) Automotive Electrical-Electronics Systems Book/Video,
0-7668-4186-3
(A7) Automotive Heating and Air Conditioning Book/Video,
0-7668-4187-1
(A8) Automotive Engine Performance Book/Video,
0-7668-4188-X
(L1) Automotive Advanced Engine Performance Book/Video,
0-7668-4189-8
(P2) Automobile Parts Specialist Book/Video,
0-7668-4190-1

Automotive Technician Certification Test Preparation Manual, 2E
Don Knowles

**0-7668-1948-5/
Part No. 11948**

The second edition of Certified ASE Master Technician Don Knowles' popular ASE test preparation book adds coverage of the L1 Advanced Engine Performance test to its coverage of automotive tests A1 through A8. All nine tests covered in this book reflect year 2000 task lists, including the updated composite vehicle in the L1 test. This revised edition contains at least one practice question for every ASE task in the tests. Also included is the updated and expanded coverage of electronic automatic transmissions, electronically controlled automatic transmissions, electronically controlled 4 wheel drive and steering, ABS systems, wiring diagrams, and repairing electronic components.

BENEFITS
- a new section has been added on computer-controlled automatic transmissions and transaxles including those used in OBD II vehicles
- new information has been included on electronically-controlled 4WD systems and ABS systems
- the chapter on Electrical/Electronic Systems has been expanded to include information on reading wiring diagrams and inspecting, testing, and repairing electronic components
- a complete chapter has been added to prepare technicians for the Advanced Engine Performance (L1) test

CONTENTS
Engine Repair Automatic Transmission/ Transaxle. Manual Drive Train and Axles. Suspension and Steering. Brakes. Electrical/Electronic Systems. Heating, Ventilation, and Air Conditioning Systems. Engine Performance. Advanced Engine Performance.

788 pp, 8½" x 11", SC, 1-Color, ©2001

ASE CERTIFICATION TEST PREPARATION

THOMSON
DELMAR LEARNING

NEW!

Prepare to Pass the ASE Exam Online

ATCChallenge.com
Delmar Learning

Delmar Learning's online ASE test preparation web site has been carefully reviewed and researched by ASE master technicians to include fully updated content on tests A1-A8, L1, P2, and X1. The site offers two different study options so users can choose their study method each time they sign on.

- practice questions provide helpful hints, insight into right and wrong answers, and links back to further reading for each task area
- sample tests prepare users for test day by using ASE-style questions - reflecting the type of questions and task areas on the actual exam - making this the most up-to-date and realistic ASE test preparation study aid available
- both study options offer detailed reports that provide accurate test results, feedback, and links on areas of further study down to task areas
- one year secure access through any web-enabled computer
- a complete task list including an overview of each task to further enhance study
- includes coverage of the new ASE Exhaust Systems exam (Test X1)
- automotive dictionary with more than 5,000 terms and Spanish translation
- technical support provided by Delmar Learning

ATCChallenge.com Plus
Delmar Learning

New from Delmar Learning, *ATCChallenge.com Plus* is the ideal way to gain the expertise required to pass the A5 (brakes) and A7 (heating and air conditioning) ASE exams. Thoroughly reviewed and researched by master automotive technicians, this total online courseware solution may be used effectively in professional training and education courses as well as by individuals preparing for selected ASE exams.
While this version includes all the features that *ATCChallenge.com* contains, it also brings with it a variety of new tools for use by students and educators. The biggest addition from the original version is the immediate remediation to the ASE-style questions that brings the user to a file or a video clip that explains the answer to each question.

- combines the *Today's Technician Series*, Erjavec's *Automotive Technology*, and Delmar Learning's *Automotive Video Series* to bring technicians the most comprehensive ASE coverage available in one place
- individuals can track their scores by ASE task, by test, or by question; create personalized testbanks; and generate their own progress reports; making *ATCChallenge.com Plus* an ideal self-study guide and test preparation tool
- a single training director or administrator can easily track average and/or raw scores at the shop-level, by region, or system-wide to gain a measure of learning achievement and program effectiveness
- a complete task list, with an overview of each task to enhances the learning experience, ensures that users are 100% prepared to pass each ASE test

Call Your Delmar Sales Rep for Part Numbers & Pricing

Visit ATCChallenge.com to see the latest modules and a free demo!

ATC Challenge 3.0 CD-ROM
Delmar Learning
0-7668-2982-0

These exciting interactive CD-ROMs have been designed to prepare technicians for successful completion of the Automotive ASE task areas (A1-A8, L1, P2, and F1). This multimedia software assesses strengths and weaknesses by identifying topics needing further study while allowing users to review ASE task areas at their own pace. Explanations, hints, notes, and a glossary aid the user in comprehension, critical thinking and retention. These CD-ROMs offer hundreds of ASE-style questions, a test taking strategy section and LAN compatibility. Not only is *ATC Challenge 3.0* the ultimate in test preparation, but it is also an excellent learning tool!

CD-ROM, ©2001

Site License Available for Multiple Unit Purchases or Multiple Workstations for ATC Challenge 3.0:
User 1: Full Price (List or Net)
Users 2-5:
$80/workstation + Full Price
Users 6-10:
$70/workstation + Full Price
Users 11-20:
$60/workstation + Full Price
Users 21+:
$50/workstation + Full Price

ATC Challenge for P2
Delmar Learning
0-7668-1827-6

This interactive CD-ROM contains material that will help prepare technicians for the Automotive Parts Specialist (P2) certification exam.
CD-ROM, ©2000

NATEF Standards Job Sheets
Delmar Learning
0-7668-6375-1
(Complete Set, Tests A1-A8)

Each of our eight *NATEF (National Automotive Technicians Education Foundation) Standards Job Sheets* workbooks has been thoughtfully designed to assist users in gaining valuable job preparedness skills and mastering specific technical competencies required for success as a professional automotive technician. The entire series is based on current NATEF standards.
Central to each manual are well-designed and easy-to-read job sheets, each of which contains specific, performance-based objectives, lists of required tools and materials, safety precautions, plus step-by-step procedures to lead users to completion of shop activities.

KEY FEATURES
- easy to use in any automotive education or training program in which NATEF coverage is desired
- completed Job Sheets may be kept as records, providing tangible evidence that instructors are addressing all NATEF tasks while paving the way for program certification

JOB SHEETS AVAILABLE FOR:
(A1) **Automotive Engine Repair,**
(A2) **Automatic Transmissions and Transaxles, 0-7668-6368-9**
(A3) **Manual Drive Trains and Axles, 0-7668-6369-7**
(A4) **Automotive Suspension and Steering, 0-7668-6370-0**
(A5) **Automotive Brakes, 0-7668-6371-9**
(A6) **Automotive Electrical and 0-7668-6367-0 Electronic Systems, 0-7668-6372-7**
(A7) **Automotive Heating and Air Conditioning, 0-7668-6373-5**
(A8) **Automotive Engine Performance, 0-7668-6374-3**

All share the following information: 8½" x 11", SC, 1-Color, ©2002

This pioneering eight-book series offers automotive repair shop owners and those wanting to be shop owners the necessary business and customer service skills to run a successful automotive service facility.

The series covers three main topical areas: personnel management, business management, and sales and marketing. Each book provides a framework to help technicians make consistent, high-quality, and productive service a part of every day shop operations. According to the author, "Great performance coupled with increased customer loyalty, trust, and operational excellence will almost always reult in increased profits."

Automotive Service Management Series Benefits:

- real-world approach reflects author's experience as a fourth generation technician, a repair & service company owner, and an automotive industry trainer
- all-inclusive coverage spans from designing an automotive repair facility floor plan through financial management techniques, customer/staff relations, and more
- length of each book makes it easy to incorporate this series into workshops, seminars, and training/education courses
- information is available "as is" or for customization

Total Customer Relationship Management
ISBN 1-4018-2657-1/Part No. 22657
From Intent to Implementation
ISBN 1-4018-2658-X/Part No. 22658
Operational Excellence
ISBN 1-4018-2659-8/Part No. 22659
Building a Team
ISBN 1-4018-2660-1/Part No. 22660
The High Performance Shop
ISBN 1-4018-2661-X/Part No. 22661
Safety Communications
ISBN 1-4018-2662-8/Part No. 22662
Managing Dollars with Sense
ISBN 1-4018-2663-6/Part No. 22663
Operations Management
ISBN 1-4018-2665-2/Part No. 22665
Entire Set of 8 Books
ISBN 1-4018-2499-4/Part No. 2499

Softcover manuals are 8 1/2" x 11", printed in 1-Color, ©2003

ABOUT THE AUTHOR

Mitch Schneider is a fourth generation mechanic/technician and is a frequent speaker at major conventions and meet of automotive industry trade organizati Schneider is also an award-winning jou ist and is a regular contributor and sen contributing editor for *Motor Age* m zine. He provides commentary on the e ing relationship between service dealer jobbers, warehouse directors and manu turers.

Schneider has also appeared on the TN cable show "Truckin' USA" where he h the "Tech Tips" segment. In addition to operating the award-winning Schneide Automotive for 22 years in Simi Valley, he is also the president and founder of Schneider's Future-Tech, a service comp specializing in conducting management seminars for automotive service dealers jobbers, warehouse distribution compar and manufacturers.

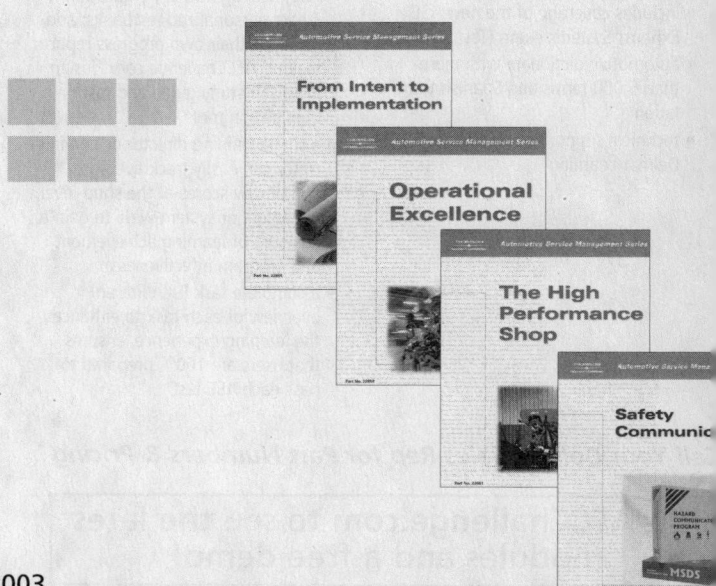

Delmar's Automotive Dictionary
David W. South & Boyce Dwiggins

0-8273-7405-4

This handy, ready-reference dictionary provides the automotive engineer, technician, mechanic, student, enthusiast or layperson with a single source for the most up-to-date definitions available of technical, professional and informal terminology used in today s automotive world. It is descriptive and covers the wide scope of terms pertinent to the automotive field. With multiple definitions and aids, and proper pronunciation of terms, this dictionary is a must for all!

BENEFITS

- over 3000 terms comprehensively covering more than 100 subject areas
- enhanced by a list of acronyms and abbreviations
- up-to-date definitions of today's automotive terminology
- aids for proper pronunciation
- each term has multiple definitions

281 pp, 6" x 9", SC, 1-Color, ©1997

Practical Problems in Mathematics for Automotive Technicians, 5E
George Morre, Todd Sformo & Larry Sformo

0-8273-7944-7

By showing how to apply math solutions to everyday problems, this all-in-one math reference transforms the "remove it and replace it" mechanic into a complete automotive technician. The book builds from math basics to cover more complex topics--not to mention such workplace issues as invoices and scale reading of test meters. Each easy-to-read chapter features step-by-step instructions, diagrams, charts and examples to make the problem-solving process a snap.

256 pp, 7⅞" x 9¼", SC, 1-Color, ©1998
Instructor's Manual   **0-8273-7945-5**

Math for the Automotive Trade, 3E
John C. Peterson & William deKryger

0-8273-6712-0

Math for Automotive Trades, 3E provides excellent examples and problems that reflect technological requirements of workers in automotive technology. The text has three parts: review of basic mathematics skills, math applications to specific automotive situations, and an examination of measurement aspects beginning with angle and linear measurements and ending with an extensive look at measurement tools used in the automotive trade.

345 pp, 8½" x 11", SC, 1-Color, ©1995
Instructor's Manual   **0-8273-6713-9**